Standard Handbook for Aerospace Engineers

Brij N. Agrawal Editor
Max F. Platzer Editor

Second Edition

New York Chicago San Francisco
Athens London Madrid
Mexico City Milan New Delhi
Singapore Sydney Toronto

Library of Congress Control Number: 2017957635

Standard Handbook for Aerospace Engineers, Second Edition

2 3 4 5 6 7 8 9 LKV 29 28 27 26 25 24

ISBN 978-1-259-58517-3
MHID 1-259-58517-4

The pages within this book were printed on acid-free paper.

The first edition of this book was titled *The Standard Handbook for Aeronautical and Astronautical Engineers.*

Sponsoring Editor
Robert Argentieri

Editing Supervisor
Stephen M. Smith

Production Supervisor
Pamela A. Pelton

Acquisitions Coordinator
Lauren Rogers

Project Managers
Anubhav Singh and
Sonam Arora,
Cenveo® Publisher Services

Copy Editor
Surendra Shivam,
Cenveo Publisher Services

Proofreader
Barnali Ojha,
Cenveo Publisher Services

Art Director, Cover
Jeff Weeks

Composition
Cenveo Publisher Services

Contents

Contributors

Brij N. Agrawal *Distinguished Professor, Department of Mechanical and Aerospace Engineering, Naval Postgraduate School, Monterey, California* (Secs. 7, 11, 12)

Sachin Agrawal *Senior Control Engineer, formerly at Space System Loral and Lockheed Martin, Palo Alto, California* (Sec. 11)

D. N. Baker *Laboratory for Atmospheric and Space Physics, University of Colorado, Boulder, Colorado* (Sec. 10)

Eranga Batuwangala *Researcher, Aerospace Engineering and Aviation Discipline, RMIT University, Bundoora, Victoria, Australia* (Sec. 5)

Suraj Bijjahalli *Researcher, Aerospace Engineering and Aviation Discipline, RMIT University, Bundoora, Victoria, Australia* (Sec. 5)

Frederic Boniol *Research Engineer, ONERA, France* (Sec. 5)

Dominique Brière *Head of Flight Control and Automatic Flight Control Systems Department, Airbus, France* (Sec. 5)

Dennis M. Bushnell *Chief Scientist, NASA Langley Research Center, Hampton, Virginia* (Sec. 1)

J. P. Catani *Head of Department, Power Supply and Electromagnetic Compatibility, Centre National d'Etudes Spatiales, France* (Sec. 5)

Muguru S. Chandrasekhara *Research Professor, Department of Mechanical and Aerospace Engineering, Naval Postgraduate School, Monterey, California* (Sec. 3)

Florent Christophe *Deputy Head, Department of Electromagnetism and Radar, ONERA, France* (Sec. 5)

Jonathan Cooper *Professor of Engineering, School of Engineering, University of Manchester, United Kingdom* (Sec. 3)

Alan R. Crocker *Senior System Engineer, NASA Ames Research Center, Moffett Field, California* (Sec. 12)

M. Crokaert *Doctor in Atomic Physics–Engineer, Centre National d'Etudes Spatiales, France* (Sec. 5)

Atri Dutta *Assistant Professor, Aerospace Engineering, Wichita State University, Wichita, Kansas* (Sec. 8)

Peter Eckart *Division of Astronautics, Technical University of Munich, Germany* (Sec. 10)

John A. Ekaterinaris *Distinguished Professor, Department of Aerospace Engineering, Embry-Riddle Aeronautical University, Daytona Beach, Florida* (Sec. 3)

Jack Foisseau *Head of Modelling and Requirement Engineering, ONERA, France* (Sec. 5)

Anthony J. Gannon *Associate Professor, Department of Mechanical and Aerospace Engineering, Naval Postgraduate School, Monterey, California* (Sec. 3)

Alessandro Gardi *Research Officer, Aerospace Engineering and Aviation Discipline, RMIT University, Bundoora, Victoria, Australia* (Sec. 5)

Guru P. Guruswamy *Senior Scientist, NASA Ames Research Center, Moffett Field, California* (Sec. 3)

Kenneth R. Hamm, Jr. *NESC Chief Engineer, NASA Ames Research Center, Moffett Field, California* (Sec. 11)

Thomas M. Hancock III *Private Consultant, Systems Engineering and Flight Software Safety, Huntsville, Alabama* (Sec. 11)

Rüdiger Jehn *European Space Operation Center, Germany* (Sec. 10)

Michael W. Jenkins *Professor Emeritus of Aerospace Design, Georgia Institute of Technology, Atlanta, Georgia* (Sec. 6)

Rohan Kapoor *Researcher, Aerospace Engineering and Aviation Discipline, RMIT University, Bundoora, Victoria, Australia* (Sec. 5)

Trevor Kistan *Research and Technology Manager, THALES Australia, Melbourne, Victoria, Australia* (Sec. 5)

Gary H. Kitmacher *International Space Station Program, National Aeronautics and Space Administration, Johnson Space Center, Houston, Texas* (Sec. 7)

G. Komatsu *International Research School of Planetary Sciences, Università d'Annunzio, Italy* (Sec. 10)

Yixiang Lim *Researcher, Aerospace Engineering and Aviation Discipline, RMIT University, Bundoora, Victoria, Australia* (Sec. 5)

Gerald Lo *Formerly at INTELSAT, Washington, D.C.* (Sec. 11)

Louis L. Maack *Fellow, Lockheed Martin Space Systems, Sunnyvale, California* (Sec. 7)

Jean-Claude Mollier *Head of Department, Systemes Electronics Photoniques, SUPAERO, France* (Sec. 5)

Roy Y. Myose *Professor of Aerospace Engineering, Wichita State University, Wichita, Kansas* (Sec. 8)

Andrew J. Niven *Senior Lecturer, Department of Mechanical and Aeronautical Engineering, University of Limerick, Ireland* (Sec. 3)

Tina L. Panontin *Chief Engineer, NASA Ames Research Center, Moffett Field, California* (Sec. 12)

J. P. Parmantier *Doctor/Engineer in Electromagnetism, ONERA, France* (Sec. 5)

Marc Pélegrin *Docteur ès Sciences Automatics, FEDESPACE, France* (Sec. 5)

Joseph N. Pelton *Former Dean, International Space University, and Executive Board, International Association for the Advancement of Space Safety* (Sec. 7)

Max F. Platzer *Distinguished Professor Emeritus, Department of Mechanical and Aerospace Engineering, Naval Postgraduate School, Monterey, California* (Sec. 3)

Sylvain Prudhomme *Head of Identification and Control Research Group, ONERA, France* (Sec. 5)

Jeffery J. Puschell *Principal Engineering Fellow, Raytheon Company, El Segundo, California* (Sec. 11)

Subramanian Ramasamy *Research Officer, Aerospace Engineering and Aviation Discipline, RMIT University, Bundoora, Victoria, Australia* (Sec. 5)

Michael J. Rycroft *Cambridge Atmospheric, Environmental and Space Activities and Research, United Kingdom* (Sec. 10)

Roberto Sabatini *Professor of Aerospace Engineering and Aviation, RMIT University, Bundoora, Victoria, Australia* (Sec. 5)

Abbas A. Salim *Engineering Fellow/Principal Engineer (Retired), Lockheed Martin Space Systems, Denver, Colorado* (Sec. 11)

Nesrin Sarigul-Klijn *Professor, Department of Mechanical and Aerospace Engineering, University of California, Davis, California* (Sec. 3)

Dieter Scholz *Professor, Aircraft Design and Systems, Hamburg University of Applied Sciences, Germany* (Sec. 2)

Michael J. Sekerak *Mission Systems Engineer, NASA Goddard Space Flight Center, Greenbelt, Maryland* (Sec. 9)

Jerry Jon Sellers *Senior Space Systems Engineer, Teaching Science and Technology, Inc., Manitou Springs, Colorado* (Sec. 9)

Stevan M. Spremo *Senior Systems Engineer, NASA Ames Research Center, Moffett Field, California* (Sec. 12)

Constantinos Stavrinidis *Former Head of Mechanical Engineering Department, ESTEC, European Space Agency, The Netherlands* (Sec. 11)

Robert Stevens *Director of Model-Based Systems Engineering Office, The Aerospace Corporation, El Segundo, California* (Sec. 12)

Subchan Subchan *Vice Rector of Academic Affairs, Kalimantan Institute of Technology, Malaysia* (Sec. 4)

Douglas G. Thomson *Chief Adviser of Studies, School of Engineering, University of Glasgow, United Kingdom* (Sec. 4)

Trevor M. Young *Associate Professor, Department of Mechanical and Aeronautical Engineering, University of Limerick, Ireland* (Sec. 4)

Rafał Żbikowski *Professor of Control Engineering, Cranfield University, Cranfield, United Kingdom* (Sec. 4)

About the Editors

Dr. Brij N. Agrawal is a Distinguished Professor in the Department of Mechanical and Aerospace Engineering and Director of the Spacecraft Research and Design Center at the Naval Postgraduate School (NPS). Prior to joining NPS in 1989, he worked for 20 years in the research, design, and development of communications satellites at COMSAT and INTELSAT. Dr. Agrawal is the author of *Design of Geosynchronous Spacecraft*. He is a Fellow of the American Institute of Aeronautics and Astronautics and a Member of the International Academy of Astronautics.

Dr. Max F. Platzer is a Distinguished Professor Emeritus in the Department of Mechanical and Aerospace Engineering at the Naval Postgraduate School (NPS). Prior to joining NPS in 1970, he was a member of the Saturn space launch vehicle development team at the NASA Marshall Space Flight Center and head of the Aeromechanics Research Group at the Lockheed Georgia Research Center. He is a Fellow of both the American Institute of Aeronautics and Astronautics and the American Society of Mechanical Engineers.

About SAE International

SAE International (sae.org) is a global association committed to being the ultimate knowledge source for the engineering profession. By uniting over 127,000 engineers and technical experts, we drive knowledge and expertise across a broad spectrum of industries. We act on two priorities: encouraging a lifetime of learning for mobility engineering professionals and setting the standards for industry engineering. We strive for a better world through the work of our charitable arm, the SAE Foundation, which helps fund programs like A World in Motion® and the Collegiate Design Series™.

Preface to the Second Edition

In the 15 years since the publication of the first edition of this handbook, many new developments have occurred, especially in the astronautics field. We have included them in this second edition, which is divided into three major areas.

In the first section the chief scientist of the NASA Langley Research Center presents his view of the likely aerospace developments in the coming years. The subsequent five sections provide the reader with an update of the major developments in aeronautics. These include major advances in predicting and measuring very complex flow phenomena due to the rapid increases in computing power in recent years. Therefore, parts on computational fluid dynamics, modern flow measuring techniques, computational aeroelasticity, and computational acoustics have been added to the coverage of classical aerodynamic analysis methods retained from the first edition. Similarly, a part on optimal control theory was added to the coverage of aircraft performance, stability, and control in order to draw attention to the progress achieved in this field. This is followed by a major revision of avionics coverage because, here again, major advances have occurred. Also, in this section new parts on air traffic management have been added. Two sections retained with only minor changes cover aircraft systems and aircraft design.

The subsequent six sections provide the reader with an update of the major developments in astronautics. The sections titled Astrodynamics, Rockets and Launch Vehicles, and Earth's Environment and Space have been retained from the first edition with an updating of the material. Three new sections titled Spacecraft Systems, Spacecraft Subsystems, and Spacecraft Design have been added. The Spacecraft Systems section covers satellite missions, test and product certification of space vehicles, space safety engineering and design, and spacecraft for human operation and habitation. The Spacecraft Subsystems section covers attitude dynamics and control, observation payloads, spacecraft structures, satellite electric power subsystems, systems engineering requirements, independent verification and validation, software safety for aerospace systems, thermal control, and communications. Spacecraft Design covers the spacecraft design process, a design example, concurrent engineering, and small spacecraft.

We would like to recognize the contributions of the editor of the first edition, Mark Davies. We are greatly indebted to the contributors of the new sections in the second edition for their efforts and cooperation and to the authors of the sections retained from the first edition for updating their work. Also, we express our special thanks to the Editorial Director—Engineering at McGraw-Hill, Robert Argentieri, and the Senior Project Manager at Cenveo Publisher Services, Sonam Arora, for their outstanding support during the preparation and production of this book. And then we are especially indebted to two wonderful ladies, our wives Shail Agrawal and Dorothea Platzer, who made it all possible through their love and understanding.

Brij N. Agrawal
Max F. Platzer
Editors

Preface to the First Edition

The *Standard Handbook for Aeronautical and Astronautical Engineers* represents the efforts of many people working toward the common goal of amalgamating aeronautical and astronautical engineering into a single handbook. This is the first publication of such a book. A handbook on only astronautical was published by the same publishers in the early 1960s, which now represents a fascinating insight into the minds of those early pioneers.

The challenge to put the aeronautical and astronautical together was considerable. Although they overlap in so many ways, they also have many differences that needed to be addressed. The publisher's brief was for a book that successfully brought about this combination and that would be of value to professional engineers and engineering students alike. It must, therefore, cover something of every aspect of the vast spectrum of knowledge and methods that is aerospace engineering. Working between the covers of a book that can be carried by an unaided individual, of average strength, has meant that much cannot be included.

At an early stage in the *Handbook*'s development, I decided that there would not be sufficient pages available to do justice to the military aspects of aerospace engineering. Consequently, the reader will not find many references to the military for the aeronautical and, similarly, for astronautical observation. Perhaps 75% of the book's contents would be on most engineers' list of essential engineering; the remaining 25% is there because of the section editors' and my opinions and prejudices.

The *Handbook* opens with a look at what the future may hold for the development of aeronautical and space systems. This sets the scene for what is to follow. Before addressing these issues directly, there are five sections on basic engineering science and mathematics that are the foundation of aerospace operations and design. Applications have been excluded, for the most part, from these sections to emphasize their generality. In the specialist section, wherever possible, aeronautical and space issues have been addressed in the same section, as in Aerospace Structures (Section 9) and Avionics and Astrionics (11); elsewhere, they have been divided, as in Aeronautical Propulsion (7) and Rockets and Launch Vehicles (8). Subsystems for aircraft are covered in a single section (12), whereas for spacecraft, they are part of Section 15. Because aircraft design is more standardized and mature, it occupies its own section (13). Astrodynamics (14) and Spacecraft (15) are unique to space, whereas the discussions on safety (17) and maintenance (18) are unique to aircraft.

Due to its limited size, the book cannot give a definitive account of any specific area. Thus, experienced aerodynamicists may not find everything of interest in the aerodynamics section; nevertheless, they will find much of interest, for example, in the structures sections—the very structures that interact with the aerodynamic forces.

In this, the first edition, I feel that only the first stage in the journey to provide a comprehensive handbook has been made. Lionel Marks' *Standard Handbook for Mechanical Engineers,* in print through many editions for almost a century, is a reference that has been invaluable to that discipline.

It is my hope that one day I will have made a similar contribution to aeronautical and astronautical engineering.

For the present, I thank all of those who have helped in this endeavor, beginning with my commissioning editor, Shelley Carr, with whom at times I have been in daily correspondence; she never wavered in her confidence and support for me, or if she did, I never knew. Then, I thank all of the section editors, the contributors, all of their colleagues and students who have helped, all of the institutes and companies that employ them, and my own institution, the University of Limerick, and my family: Judith, Elisabeth, and Helena.

Mark Davies
Editor

SECTION 1

Futures of Aerospace

Dennis M. Bushnell

"Entering the Age of the Small, the Fast, the Smart, and the Many"
[and the Inexpensive and the Ubiquitous]

—*Arthur Cebrowski, former director of the Department of Defense Office of Force Transformation*

1.1 Potential Impacts of Global Technology and Resultant Economic Context on Aerospace Going Forward

The world is in the throes of a rapid-to-exponential simultaneous set of revolutionary technology developments, including the areas of information technology (IT), biotechnology, nanotechnology, quantum technology, and energetics. Many of these developments are occurring at the frontiers of the small and with considerable evolving synergies. Fundamentally, society has transitioned out of the Industrial Age, is currently in the IT age, and is rapidly entering the virtual age, which is typified by an immersive, "tele"-vice physical presence. Up-to-five-senses virtual reality and advanced holography, along with direct brain communications, are in active development-to-commercialization. Society increasingly employs telecommuting, telework, telemedicine, tele-education, telemanufacturing (on-site printing), teleshopping, teletravel, telepolitics, and telecommerce writ large, along with telesocialization and tele-entertainment. These "tele"/virtual presence activities along with the Internet of Things are changing society in increasingly major ways, including shifts away from retail physical shopping. "Peak car" miles driven per person per year have been dropping year on year, partially due to increased virtual/tele living. In addition, these technologies are greatly increasing the options for navigation and communication going forward, including atom optics/cold atoms proffering orders-of-magnitude improvements in inertial navigation and massive bandwidth increases from optical free space communications.

The impacts of these technological developments and consequent societal shifts on aeronautics are potentially significant. Teletravel, five-senses virtual reality where you can hug family members across the country any time that is convenient (virtual reality haptic touch), and immersive virtual presence have the potential to reduce air travel, although the extent of such reduction is yet to be determined. Coupled with this are the increasing concerns that rapidly improving machine intelligence and autonomous robotics are "taking the jobs" which, in the runout, could possibly reduce economic capacity to travel by air for increasing segments of the population. The rapid developments in printing manufacture suggest a move to more onsite manufacture, which could reduce air cargo. However, these same technological developments should enable inexpensive, quiet, personal air vehicles (PAVs), operable from an individual holding, with a projected market in the $1 trillion per year range. These would possibly subsume much of domestic air travel (the bulk of such travel is shorter range) and enable the population to spread out much more and reduce the infrastructure costs for highways and bridges. For safety reasons these PAVs would be operated autonomously—humans would be passengers. In a similar time frame, over the next one to two decades, creative designs are evolving for very fuel-efficient transport aircraft, which will be required for long haul/over ocean, with up to 80% or greater fuel-burn reduction. These design changes, combined with advanced batteries and renewable electrical generation for electric aircraft, solar generated hydrogen, and carbon neutral or better biofuels, will result in major reductions in aircraft climate emissions.

The potential impacts of these evolving technologies on space travel, space commercialization, and space industry are also very significant going forward. Currently, commercial space is primarily "positional" earth utilities, including telecom, a hundreds of billions of dollars per year industry. A major impediment to increased space commercialization, aside from commercial support for government activities, is the cost of space access. Developing technologies are miniaturizing nearly everything except humans and the equipage that scales with their size. This miniaturization is replacing the usual metric of dollars per pound to orbit with value per pound. Satellites are downsizing, being employed in cooperative arrangements for array gain. Instruments and sensors are downsizing while improving in performance. There are consequent ongoing major increases in space launch/satellite emplacements and scientific and commercial space activity overall with nearly all nations somehow involved in/with space and students manufacturing, instrumenting, and "flying" micro/cube satellites. An additional technology that is also reducing costs of space access is reusable rocket stages. Payload is reduced as a penalty of reusability, but many reuses of the same equipment produce a large overall reduction in launch cost.

In addition to reducing launch costs, the ongoing technical developments are producing ever more capable (verging on autonomous) robotics, on site printing and, via improved sensors writ large, much more knowledge regarding in-space/on-planet, other solid body resources. Mars, for example, is now known to have vast amounts of water which, using advanced technology and carbon from the atmosphere along with other on-planet resources, could enable production on Mars of just about everything needed to colonize the planet. In fact, enough, using Mars produced fuel, to become the Walmart for the inner solar system. There are many extant plans to harvest and process all manner of products from planetary, moon, asteroid, etc., resources, mainly for use in space but with transportation costs reducing, possibly eventually on the home planet.

In summary, the evolving technologies are at this point expected to possibly reduce the demand for conventional air transport (compared to earlier projections) but enable a wholly new, major PAV market and a consequent revolution in personal transportation. For space the evolving technologies are reducing the costs of space access, greatly altering the nature of space payloads and enabling colonization of Mars and other destinations, both safely and affordably.

1.2 Civilian Aeronautical Futures

Civilian aeronautics has more recently pursued a self-fulfilling prophecy, becoming a "mature" commodity industry. Advancements have been largely incremental for decades. This incrementalism is in fact "usual" as an industry matures. Aeronautics was a technological "fast-mover" in the twentieth century with many "players," most of which have merged or gone out of business. New products can be a "bet the company" situation and the industry is currently far more "comfortable" with long technology maturation processes for risk reduction. The industry is based largely on long-haul transport aircraft with an emerging small jet component, legacy general aviation markets, and the newbee—UAS/UAVs (unmanned air systems/unmanned air vehicles).

Going forward, the industry is beset with a large and growing number of problems. These problems include emissions/warming (CO_2, NO_x, and water/contrails), increasing competition from vastly improving telepresence/teletravel alternatives, which save both time and money, air traffic control delays and inefficiencies, expanding noise restrictions, security and safety concerns, and an often overall less-than-robust business case highly dependent on fuel prices.

The ongoing IT, bio, nano, energetics, and quantum technology revolutions are changing both the nature of the industry problem set and solution spectrum options. The foremost solution component is enabled primarily by the IT revolution and associated swarm technologies—a digital airspace wholly autonomous in terms of air traffic control, navigation, and vehicle operations. Autonomous aircraft operation is becoming feasible due to the major improvements in sensors and machine intelligence. Future military and homeland defense functionalities appear to require ever-increasing autonomous aircraft operation(s)—in turn requiring (and limited by the lack of) an autonomous digital airspace.

Such a digital airspace would enable in turn a complete revolution in personal mobility—the PAV easily usable by "everyone" given wholly autonomous operation. The technologies to enable a fly/drive, superSTOL (short takeoff and landing), street-in-front-of-your-house operation, safe, quiet, affordable personal transportation vehicle appear to be within sight given reasonable research support and the digital air space. Missions for such a vehicle include automatic package delivery, flying Humvees, a superb transportation system for areas lacking intercity roads such as island nations and cold regions, and eventual supplementation-to-supplantation of the automobile. The estimated worldwide market for such vehicles is in the trillion dollar range and their use would erode the scheduled domestic airline customer base. The variety of machines currently under study is accessible at www.roadabletimes.com. This vision, increasingly enabled by the ongoing technology revolutions, would provide a true revolution in civilian aeronautics and personal mobility, enable 200-mile-plus "commutes," and provide huge cost avoidance for roads and bridges. Such capability has been on societies' radar screen for nearly a century but the technology was simply not there to do it. This is no longer the case.

Even given eventual development of an affordable, safe, fly/drive, airport-independent personal transportation system there is still a need for reinvention of long-haul transports, especially for transoceanic stage lengths. The current machines of this genre are direct descendants of the Boeing 707 and after these many decades of evolutionary improvement this design approach simply lacks the margins to address the multitudinous issues that need to be addressed. There are several alternatives, such as the blended wing body and strut/truss braced wings, for example, which proffer major potential increases in lift-to-drag ratio and improvements in structural weight fraction. Such improvements in these parameters would provide design margins to address most of the issues except for emissions/warming. The water emissions concerns can be alleviated by designs that cruise below some 27,000 ft, where water is cooling instead of warming or recourse is made to electrical propulsion utilizing batteries. Hydrogen/fuel cells and biofuels combustion would still emit water, requiring flight below 27,000 ft in the tropopause. Such low-altitude cruise designs, if needed (we do not go to electric propulsion using batteries), could be enabled by downsizing the wing for the higher dynamic pressure/air density and utilizing circulation control for high lift and employing it at cruise for load alleviation and ride quality improvements required due to flight in "weather" at those altitudes. The resulting STOL performance could also improve airport productivity (several takeoffs on the same runway). The CO_2 emissions issue is addressable either via electric propulsion using renewable electric energy to charge the batteries, solar hydrogen, and fuel cells for electrics or by utilizing biofuels, whose CO_2 "price" was paid via CO_2 uptake from the atmosphere during plant growth. NO_x reductions are available via clever combustor design.

There is an energetics wild card which at this point is being studied for potential commercial application, termed low-energy nuclear reactions (LENRs). This is the 27-years-after version of the "cold fusion" of the late 1980s. We now have a quarter century of experiments worldwide indicating heat and transmutations without the expected radiation and at levels of input energy not at all in agreement with the usual nuclear theories, which are the *strong force* and *particle physics*. There are theories that suggest that this is the weak force and collective effects, but currently we do not have a validated theory to enable scaling, engineering, and making safe. If this technology is eventually understood and found to be real and scalable then aerospace writ large changes greatly. Thus far we have been in thrall to chemical energetics, LENR is, from theory and some experiments, orders of magnitude times chemical. As an example of its potential impacts, consider SSTs (supersonic transports), which have the following issues—sonic boom, emissions, takeoff noise, weight/costs, ozone, and possibly atmospheric radiation. The much greater energy density of LENR, as currently and inadequately understood, would enable a wholly new "energy rich" design paradigm, have no emissions, greatly reduce sonic boom by projecting energy far forward to extend the apparent length of the vehicle and have very different engine characteristics, along with nearly negligible fuel fraction/much lower weight. In the meantime, there are extant efforts to redesign the airframe for lower boom and there is an extreme arrow wing SST design due to Pfenninger that has far greater lift to drag ratio/aerodynamic efficiency. The emissions issue is still there but again possibly addressable via electrics, using batteries not solar hydrogen or biofuels as the latter two emit much water, which is of more concern than CO_2 at the design altitudes (50,000 to some 62,000 ft).

Some are speculating and working on hypersonic transports for civilian markets. Given the various city pairs of interest, the need to keep the machine in the air for revenue purposes and thus far not very successful many SST campaigns, hypersonic transports appear to have some interesting issues to work. First, when the distances to accelerate and decelerate are included/considered, a Mach number in the lower 4.0s appears to be most efficacious because the planet is apparently not large enough to go much faster efficiently. Also, keeping the machines in the air would stress scheduling and possibly require departures or arrivals at inconvenient times. Costs of such machines are currently to be determined, their higher cruise altitude would mitigate sonic boom somewhat. There are various hypersonic air breathing engine cycles pioneered by the military over the years, which would be applicable and of interest. Like many goals in aerospace, versions could be engineered and produced. But the issue is whether they conform to the various safety, regulatory, environmental writ large, and business strictures, i.e., can they move from the possible to the probable, practical, useful, and what technologies and design approaches would it take to accomplish this. As stated previously, for almost a century, we have been working, trying to do personal aircraft, combinations of drive and fly machines. Now, after such a time frame, we are getting quite close. Need both the technology and a market. The market is dictated by functionality and costs, the technologies are required to enable marketable functionalities *and* costs.

1.3 Military Aeronautics Futures

The ongoing technological revolutions are changing the nature and equipage of warfare, including military aeronautics. In the nearer term, aircraft are becoming increasingly "uninhabited" [UCAVs (uninhabited combat air vehicles), UAVs (unmanned air vehicles), etc.], enabled by the IT revolution, with accompanying benefits, including affordability, survivability, redefinition of "risk," and lethality. The major issue for such aircraft is enhanced persistence/increased range within the context of the overall system metrics. In the longer term the increasingly capable worldwide "sensor web" will place at risk all air vehicles, in or out of theater and whether or not stealthy. This combined with advanced conventional electromagnetic pulse (EMP) and affordable swarms of "brilliant" munitions will probably require yet another redefinition of military aeronautics.

It is the near coincidence of many commercially driven/worldwide technological and affordability revolutions on ever shorter time scales available commercially which is the cause of the tremendous ongoing changes in air warfare. Such emerging technologies and capabilities include

- Printing fabrication
- Beyond silicon computing and machine intelligence
- Optical communication and possibly navigation
- Increasingly nano "everything"—(hypersensitive/hyperspectral) sensors, identification "tags," materials, robots, guidance, navigation, control (GNC), satellites....
- High-energy density materials (HEDM), propellants/explosives, "volumetric" munitions
- Fast lasers/other lasers, high-power microwave
- Antipersonnel/antimaterial bio/microwave (MW)
- Miniaturized, brilliant, lightweight, low-power, inexpensive everything (satellites, weapons, robots, sensors, mines, etc.)
- A ubiquitous, inexpensive hyperspectral, multiphysics, hypersensitive, miniaturized land, sea, air, space commercial, scientific, military Global Sensor Web—potential demise of "stealth"
- Information operations/information warfare as a weapon of mass destruction
- Inexpensive global reach (miniature rockets, "Slingatron," transoceanic UAVs, etc.)
- Swarms

There are many orders of magnitude improvements in the offing for computing, communications, sensors, energetic and other materials, and "machine intelligence." All of this and more are resulting in doctrine-level changes, including depopulation of the battlefield, "dispersal, effective targeting," military equipage increasingly "commercial," warfare increasingly robotic, area denial.

All of this is causing a shift from "Industrial Age" warfare constructs and equipage, away from "megatonnage" of Expensive Industrial Age steel and aluminum artifacts and toward systems, which are increasingly brilliant, robotic, long range, inexpensive, numerous-to-swarms, miniaturized, precise, multifunctional, and networked. The increasing shift toward smart precision guided munitions (PGMs) and UAVs/UCAVs is a clear indication of these changes. As the "sensor web" develops, "stealth" will become highly problematical as will the survival of air vehicles in- or out-of-theater.

An emerging and increasingly important mantra dating from the Gulf War era is "you can see everything and everything you can see you can kill." In the short(er) term, stealth is obviously a prime military metric (hard to be effective if it cannot survive). Stealth per se originated as a set of approaches and technologies to reduce monostatic radar return/signature via a combination of absorption and redirection. Infrared (IR) signature reduction was added into the mix, as most missile seekers were either IR or radar based. Increasingly, the technology revolutions will enable surveillance/reconnaissance for ALL signatures with ever-increasing sensitivity and, where efficacious, in a multistatic manner via both active and passive means. Such emerging hyperspectral, hypersensitive, multiphysics, ubiquitous, miniaturized, inexpensive capability could eventually negate "stealth." There are simply far too many passive and active "signatures" associated with the presence and passage of any air vehicle to contemplate otherwise. Sensor technology is one of THE most rapidly evolving areas, greatly aided and abetted by the nano and now quantum technology "revolutions." Attriting the sensor networks and associated communications, as opposed to spoofing the sensors themselves, along with killing the sensors via high power microwave (HPMW) and lasers would appear to be the next evolution of "stealth." However, this type of stealth would not impose the impacts on vehicle design/operability, which the current approaches require.

Therefore, "conventional" vehicle-based stealth is admittedly "goodness" for the near(er) term but may not be so for the longer term.

During the Industrial Age the military ran on petroleum products, powered by internal combustion and gas turbine engines. Several nascent and converging technologies, aided and abetted in many cases by the technology revolutions, could change the propulsion situation considerably. The nearest term of these propulsion developments is probably fuel cells—producing electricity, which then turns electric motors or heats air for propulsion. Batteries are now also developing rapidly, driven by the needs of personal electronics and storage for wind and solar renewable energy and are increasingly being deployed in aeronautics, including manned aircraft. There are several benefits for adopting electric propulsion for air warfare. Also possibly applicable to propulsion, as well as to explosives are metastable interstitial composite (MIC), cubanes, thermobarics, and strain bond energy release (SBER), ranging from 5 to 100 times the energy of usual chemical explosives.

There is increasing agreement that the "ground is ready," the technologies are "there" for not only UAVs, which have been in increasing use for decades, but also UCAVs. Hellfire missiles fired from Predators, etc., is the "new normal." The drivers for this rapid development of UAVs and UCAVs are truly legion mainly due to the absence of an onboard pilot. They include the increasing vulnerability of the forward bases used for operation of conventional manned/short-legged fighters, the CNN syndrome (leading to a desire for little-to-no casualties/prisoners) and the increasing capabilities of air defense systems/increasing vulnerability of conventional fighters, reduced initial, life cycle and personnel costs, increased maneuverability, improved stealth, redefines "risk," and promotes smaller/lighter vehicles with greater endurance/loiter.

UCAVs can range from UAVs with "add-on" munitions through automation of existing manned aircraft to development of new designs. Such new air vehicle designs, thus far, are driven primarily by stealth considerations. However, many-to-most of the emerging missions and desired capabilities of UCAVs also involve requirements for greatly enhanced range and loiter, one driver for which is to combine intelligence/surveillance/reconnaissance (ISR) and "real-time" attack. Such enhanced capability could be available in the future from one or a combination of advanced energetic fuels, greater propulsion efficiency, enhanced structural efficiency/multifunctional brilliant structures/materials, and drag reduction—both friction and drag due to lift. Perhaps the major future UCAV design challenge is the simultaneous optimization of both stealth and loiter/range metrics. Yet another metric for UCAVs is V/STOL capability, useful to avoid dependence on ultra-vulnerable fixed forward bases.

Estimates indicate that there is sufficient explosive in a current PGM to take out not one bridge but 10, if properly placed. Such munition size/weight reduction efforts, combined with the capability to loiter, waiting for targets to be identified by either onboard or offboard sensors, will constitute a devastating/very cost effective weapons system. UCAVs/UAVs are increasingly replacing "inhabited" aircraft for the entire litany of missions. There are extensive civilian catalogs of commercially available uninhabited air, ground, and surface/undersea vehicles, most at reasonable cost and increasingly capable.

The current concept of operations for military transports assumes sanctuary outside of the theater of operations, i.e., within the continental United States (CONUS), over most international waters, etc. The developing worldwide ISR capabilities (aka, the Global Sensor Web) and "loitering"/global reach precision strike will increasingly place logistic assets at risk "everywhere." Destroying an opponent's capability while it is conveniently packaged together is a potentially efficient method of warfighting. Current military (and co-opted civilian) logistic aircraft are non-LO (low observable) and essentially undefended—altogether quite vulnerable targets. Again, this vulnerability derives from emerging enhancements in the "sensor web" and weaponry.

The outlook for airbreathing hypersonics is still unclear. For some more than 50 years, since Ferri and others first demonstrated the feasibility of supersonic combustion ramjets, hypersonic airbreathing propulsion in the Mach 6 to 12 range has been a military dream for an entire range of missions—from "aerospace planes" through global strike/reconnaissance to time-critical target missiles and boost phase interception for theater ballistic missiles (TBMs). There is a vastly improved hypersonic technology base across the spectrum (computational fluid dynamics, facilities, instrumentation, designs, materials, etc.) compared to 50 years ago. There have been research flight experiments into the Mach 10 range. The all-up deployment costs of airbreathing hypersonic systems has instigated other, more cost effective, alternative approaches.

The basic issues concerning future PGMs/missiles are similar to those for UCAVs—significant range increases within the context of LOs and enhanced lethality. The distinction between UCAVs and PGMs is not at all clear. UCAVs are derived from "crewed" aircraft by deleting the crew. They start usually as a rather sizable winged vehicle which is then miniaturized. The PGM approach starts with various flavors of relatively small "missiles" and adds increasing brilliance and wings/airbreathing for longer range. The two approaches are converging.

What appears to be in the offing is development of a plethora of smart-to-brilliant munitions in several size ranges/special uses with increasing capabilities in terms of their multitudinous system metrics. The HEDM materials discussed previously could conceivably provide significant enhancements in both range/loiter and lethality. The reducing costs of these smart munitions/weapons offer the possibility of "swarm" deployment/operation—potentially overcoming usual defensive systems with sheer numbers.

Currently, global precision strike is/can be executed via intercontinental ballistic missiles (ICBMs), tanking B-2 and B-52 aircraft, and steaming aircraft carriers. Future options include several wholly new systems, enabled by advances in range, precision, and lethality. These future options for global precision strike include swarms of intercontinental, LO, small UCAVs; swarms of microrocket ICBMs; and Tidman's Slingatron, a mechanical ("hula hoop"–like) spiral accelerator which, for some $20 million and an 80-m-diameter cleared space (above or below ground), is apparently capable of accelerating reasonable-sized payloads to ICBM speeds at the rate of up to hundreds per minute.

1.4 Futures of Space Access

Current space access capability/approaches devolved directly from the German missile program of World War II and subsequent ICBM developments in several countries. For many decades there have been serious efforts to greatly improve on this "evolved" ICBM technology/capability, thus far largely unsuccessful with the exception of the recent successes of SpaceX. The current "cost" of access to space is in the range of thousands of dollars per pound of payload. Some of the larger, non–man-rated systems and systems from nations with low labor costs are in the lower portion of that range while man-rated systems and some of the smaller payload systems are in the upper range. These costs are currently considered inhibiting for various nascent "space-related activities," such as space industrialization, space tourism, space solar power, moon/asteroid "mining," and much else (including space "colonization"). Current civilian space utilization involves various flavors of "earth utilities," primarily telecom but including earth resource monitoring, navigation/GPS, environmental monitoring, weather monitoring/prediction, etc. These are obviously deemed to provide sufficient value to merit "operation" at current launch costs.

There are a plethora of existing space access design options, including various classes/types of (conventional) rockets, air breathing (as opposed to "rocket") propulsion, staging, reusability, takeoff/landing options, different (conventional) fuels, and material and controls options. Over the past several decades a large number of design teams in various countries have tried innumerable combinations within this rich parameter/variable set in search of a "winning combination," which would significantly reduce the cost(s) of space access. Thus far these efforts have not been particularly successful, leading to comments such as that from Mark Albert (*Scientific American*)—"If God wanted people to go to space, She would have given them more money." Something different, something(s) not contained in the "usual" parameter set are evidently required to answer the space access (cost reduction) requirements.

Another major "problem" (besides affordability) with current space access approaches is "safety/reliability." This is obviously an absolutely first-order concern for human space flight but is also a major issue with civilian space access in general, where the current reliability situation leads to loss of expensive payloads and concomitant insurance rates. The demonstrated loss-of-vehicle accident rates for the Space Shuttle System were greater than projected values, which themselves are in excess of that probably required to enable a serious space tourism market. Mention is frequently made of the extremely low accident rates of/for scheduled airlines with an expressed desire to emulate such for space access. Space access is actually a *very different* situation/mission from (subsonic) aircraft flight involving as it does imparting to the vehicle the order of 625 times the specific kinetic energy required for subsonic cruise.

Additional considerations include the fundamental differences between evolving "military space" access requirement(s)/desirements and civilian space access needs. In addition to the civilian cost and safety/reliability metrics the military is interested in such features as launch-on-demand, increased launch site options, large cross range, self-ferry/in-atmosphere cruise, enhanced launch windows and possibly orbit/de-orbit/re-orbit, "storable fuels" and perhaps even surreptitious operations. Approaches/systems which would satisfy these military "needs" are not necessarily optimal or even reasonable with regard to the dominant civilian metrics of cost and reliability/safety.

Near(er)-Term Potential Space Access "Solutions"

Payload Size/Mass Reduction

Several of the major ongoing technology revolutions, particularly IT and nano are changing the entire business case and option set for (nonhuman) space access and utilization. These technologies are enabling tremendous functionality and greatly improved performance to be placed in ever-smaller, lighter payloads and packages. Thus far, order(s)-of-magnitude reductions in size/weight are either available or projected for many space mission elements or, in some cases, whole satellites/payloads with even further improvements in performance potentially on the horizon. Such improvements could/should change to a major extent the space access situation. Companies and universities are placing many such payloads on conventional launch vehicles. Aperture or array gain is available via either the burgeoning lightweight inflatable membrane/smart surface(s) technology or cooperative flight management/formation flight. Such changes in the payload essentially convert the space access cost problem from dollars per pound to *value* per pound. Current launch costs per pound are more acceptable if there are not many pounds to loft. The alternative is to use the "microrockets" under development at, for example, the Massachusetts Institute of Technology, to inexpensively launch the micro/nano payloads.

The obvious exception to this "space business revolution" is of course "humans." Thus far the humans are not "shrinking" and therefore human-related space access (humans themselves and as much of their "support/infrastructure" scales with their physical size/weight) is largely not affected by this technology-engendered *major* change in the space business model/requirements for space access. This same (IT/nano) technology set does, however, enable a possible near(er)-term (exploration/terraforming) "replacement/stand-in" for ("onsite"/in-space) humans—deployed robotic arrays of distributed sensors and actuators increasingly autonomous and producing data streams made available to *everyone* via virtual reality/immersive presence, including haptic touch, etc. This method will affordably (robotic missions are an order of a factor of 50 or so less expensive than human ops) enable *everyone* to be a synchronous/asynchronous (virtual) space explorer.

Approaches to Reducing Cost(s) of (Conventional) Space Access

An examination of the cost elements for space access indicates that a major contributor is the cost of human time and labor. The cost per pound does not refer to placing these monies in the combustion chamber; the funds are used to pay people. Several studies of the Space Shuttle cost problems point to the "standing army" issue. The ongoing technology revolutions should enable extremely robotic fabrication and operation of space access systems, thereby greatly reducing the direct human labor costs. Such approaches as integrated vehicle health management (IVHM) are being worked as is "free from fabrication." An ab initio approach to life cycle cost reduction (design, fab, erect, checkout, operate, store, etc.) with an eye to reducing man-hours via the increasingly effective IT/nano-engendered automatics/robotics should be efficacious. Such approaches, for other "consumer goods," have resulted in and continues to result in major cost reductions. Another perhaps essential ingredient in reducing the costs, and along the way increasing reliability in major ways, is to provide "performance margins," possibly via use of more robust, less costly, and less sophisticated approaches and operation "below the limits." Overall, "cost" and "performance" are not necessarily synonymous.

Farther-Term Potential Space Access "Solutions"

There are an amazing number of options, possibilities on the table, horizon for farther-term space access, requiring some 10 years or more of research to sort through, evaluate, and sort out. These possibilities span the spectrum from propulsion cycle to fuels, materials, and launch assist. Launch assist options include Tidman's "Slingatron" for smaller, g-tolerant payloads, MW energy radiated from the ground or from orbiting "beamers" to onboard rectennas with the energy used to power an exit Magneto-hydrodynamic (MHD) accelerator (2000+ seconds of specific impulse at high thrust), space elevator(s), "tethers," and high pressure, polymer stabilized, laser-guided water jets. The foremost emerging material option is of course structural carbon nanotubes (potentially a factor of 5 or so dry weight reduction). Advanced propulsion cycle options include pulse detonation (PDW), rockets (possibly with detonation within a liquid fuel), and MHD adjuncts/variants. Emerging fuel alternatives include cubanes/N_4, metallic H_2, solid H_2 with embedded atomic species, and even some emerging very "clean" fusion approaches such as H/B-11. Obviously, "rockets" are very far from being "mature."

The extent to which these and other emerging/conceptual technologies could improve space access cost/reliability is to be determined. As an example, pulse detonation wave rockets could greatly reduce the pressure in the turbine feed pumps, very significantly improving a *major cost and reliability problem* on conventional pump pressure-fed rockets, the Space Shuttle main engine (SSME) in particular. However, specific impulse per se is *not* always directly translatable to a cost reduction. In the nearer term SpaceX is leading the way to reduced launch costs via highly efficient industrial processes and reusable rocket stages. Their techniques extant and under development proffer cost reductions up to the order of a factor of 4, which would open up more options for commercial space developments.

1.5 Aerospace beyond LEO

Thus far, aerospace beyond low-earth orbit (LEO) has consisted of telecom, commercial, military satellites in HEO and GEO, the Apollo project (humans to the moon), and a plethora of smallish scientific projects both in-space and orbiters/landers. Costs, including those for space access, have inhibited the various "space dreams," including colonizing both in-space and on various bodies/planets which have been published over the years. Humans evolved in a 1 *g* and galactic cosmic rays(GCR) protected environment; hence, there are serious safety issues associated with human presence in space. Both micro g and GCR decimate the immune system and collectively degrade or worse almost all physiological systems. The current bio revolutions are now proffering biological countermeasures (BCMs), which would enable increasing protection, greatly improved safety. Along with health issues, safety concerns in-space for humans and robots writ large include reliability, etc.

Reusable Space Infrastructures

Thus far, space activities have largely to almost exclusively been conducted using expendable, one time vice reusable, launch vehicles, capsules, and on planet equipage, e.g., for transportation, habitability, and operability, etc. There was an attempt on the Space Shuttle to attain partial reusability, especially with respect to the orbiter. The extensive refurbishment required flight-to-flight obviated much of the benefits of such. As has now been demonstrated by SpaceX and some others for launch vehicles, it is conceivable going forward to employ a much more reusable space exploration-to-commercialization mantra, which should reduce cost(s) and increase safety. Benefits of reusability are obviously a function of the number of reuses and any rework, maintenance required. IVHM possibly including self-healing would be required to ensure safety and operability. What is suggested is essentially a panoply of reusable space "utilities." An initial inventory of such might include

- Terrestrial, in-space, on planet/body beamers for propulsion and energetics. This allows the separation of propulsive mass and energy as well as reusability. Such beamers could also power tethers for orbit raising, where magnetic fields are extant.
- GPS, RF, or optical for navigation and location on planets/bodies. Navigation alternatives, which could obviate the need for such a central, reusable system, include "atom optics," aka in USAF parlance—cold atoms proffering orders of magnitude improvements in inertial guidance. Also quantum-enhanced graviometers and magnetometers could utilize the emerging detailed scans, documentation of surface magnetic and gravity fields.
- Space solar satellites around planets, bodies. These would not be impacted by the dust issues that have an effect on planet/body photovoltaics, have higher 24×7 and 365 days output, and be capable of servicing distributed areas, providing both redundancy and reductions in surface infrastructure(s). The nanotechnology is greatly improving the efficiency of photovoltaics as are combinational photovoltaics and thermal designs, which utilize the photovoltaics "waste heat." Obviously the in-space beamers referred to above could also perform this function.
- Establishment of a semiregular "slow boat" cyclic transportation system, utilizing sails of various flavors or low thrust/high efficiency electric propulsion, between earth and other "bodies." This could over time supply the necessary initial ingredients for serious *in situ resource utilization* (ISRU) as well as initial equipage and that required for expansion along with whatever critical supplies that could not be produced on site. Such a transportation system would presumably be inexpensive compared to current practices/approaches.
- Distributed space "service stations," for repairs, fueling, and most importantly, for possibly saving lives, equipped as life boats in the event of serious vehicle or habitat malfunctions.
- Virtual exploration, utilization of in situ inexpensive nano/other instrumentation and other sensors/sensing such as from satellites as input to software, which enables virtual reality exploration 24/7/365 days for everyone on (personal) demand. Optical free space communications would provide greater band width.
- Reusable launch vehicles, the key to serious launch cost reduction for conventional space access approaches, requires reasonable launch rates and durable/low refurbishment designs for viability.

- Momentum tethers, from planet to solar system scales. These have interesting-to-serious engineering issues of various flavors. Another related technology in some aspects/issues is space elevators.

Advanced technologies are improving mightily and rapidly both the cost and safety outlook for humans in space. Machine intelligence and robotics and printing manufacture developments proffer on planet ISRU vice hauling everything from earth, greatly lowering the costs of space access, which the SpaceX approaches are already reducing. Studies indicate it is potentially feasible, given the ongoing tech advances, to manufacture, on Mars, given the extant massive resources there, almost everything humans would need to colonize the planet before the humans leave home. This approach, besides reducing costs greatly, improves safety via the opportunity to, after manufacture and arrangement there, obtain in situ functionality and reliability data, again prehuman presence. This in addition to the ongoing miniaturization and capability improvements of nearly everything enables humans beyond LEO, *both* affordably and safe, opening the vista of a new era for space exploration, pioneering and colonization. Operationally, at some point we will have to deploy some or more versions of the several technical approaches to reducing space debris in various earth orbits and perhaps reuse, repurpose this material, which is mainly high-quality aluminum placed in orbit at significant expense. Printing manufacture in space is an obvious method for practical reuse.

Synopsis of Frontier Aerospace Technologies

The following is a listing of the revolutionary frontier technologies, which, in the aggregate, have the potential to change mightily essentially all aspects of society, including aerospace. These technologies are in progress now, and their potential individual, let-alone combinatorial impacts have not yet been fully projected and documented.

- Revolutionary energetics—LENR, halophytes (salt plants grown on wastelands using direct seawater irrigation), positrons, energy beaming, ultra-efficiency PV and energy conversion
- 3D and 4D printing on the way to molecular manufacturing, potential for order of magnitude improvement in material properties via controlling microstructure
- Structural nanotubes, potential for factors of 3 to 5 dry weight reduction
- Quantum computing, for an increasing number of problems a huge number of orders of magnitude faster
- Atom optics/cold atoms, orders of magnitude improved inertial navigation
- Deep learning/soft computing and biomimetics machine intelligence
- Five-senses virtual reality/immersive presence
- Designer/modified/more robust humans homospacious
- Vector/scalar quantum potential non E-M communications
- Autonomous robotics
- Global sensor grid/global mind
- Synthetic biology for bioproduction and biofunctionalism

Examples of what these technologies and their combinations proffer include climate solutions (also, in the case of halophytes, solutions for land, food, and water), energy rich aerospace, massive cost reductions in many systems/functionalities, tele-everything/the virtual age, mod sim vice physical experiments, increasing replacement of humans by robotics, increased tele vice physical travel, less cargo/more at home manufacture, human space exploration both safe and affordable.

Bibliography

1. Bushnell, Dennis M., "Advanced to Revolutionary Space Technology Options—The Responsibly Imaginable," NASA T M 2013-217981, 2013.
2. Bushnell, Dennis M., "Emerging Options and Opportunities in Civilian Aeronautics," NASA T M 2012-217759, 2012.
3. Bushnell, Dennis M., "Frontier Aerospace Opportunities," NASA T M 2014-218519, 2014.
4. Bushnell, Dennis M., Moses, Robert W, "Fresh Thinking about Mars," *Aerospace America*, March 2016, pp. 34–39.
5. Bekey, Ivan, "Advanced Space Concepts and Technologies, 2010–2030," *Aerospace Press*, 2003.
6. Impey, Chris, *Beyond Our Future in Space*, W.W. Norton and Company, New York, NY, 2015.
7. Krone, Bob (ed.), *Beyond Earth: The Future of Humans in Space*, Apogee Books, Tucson, AZ, 2006.
8. Millis, Marc G. and Davis, Eric W. (eds.), *Frontiers of Propulsion Science*, Volume 227, AIAA Progress in Astronautics and Aeronautics., Washington, D.C.
9. NASA Technology Roadmaps, TA 15, Aeronautics, 2015.
10. Truman, T. and de Graaff, A. (eds.), *Out of the Box Ideas about the Future of Air Transport*, European Commission, 2007.

SECTION 2

Aircraft Systems

Dieter Scholz

2.1 Introduction

Aircraft Systems—General

What Are Aircraft Systems?

Broadly speaking, an aircraft can be subdivided into three categories:

1. The airframe (the aircraft structure)
2. The power plant (the engines)
3. The aircraft systems (the equipment)

This section deals with the last of these categories.

The airframe provides the aircraft with its (relative) rigidity. It also enables the generation of lift through its aerodynamic shape. A glider flies without a power plant, but in order to maintain weather-independent sustained level flight, a power plant is necessary to produce thrust to overcome the drag.

The airframe and power plant might seem to be all that is needed, but this is not so. Even the earliest aircraft needed more. Some means to steer the aircraft (flight controls) and to handle it on the ground (landing gear) were needed. These aircraft systems play a key role today and must be considered in the very early stages of aircraft design. A fuel system was also needed from the beginning of the history of powered flight. With aircraft flying longer distances, navigation and communication systems became important; with aircraft flying higher and taking passengers on board, cabin systems such as air conditioning and oxygen systems were introduced.

Above is given a general idea of what aircraft systems are. A more rigorous definition of the term is given further.

Significance of Aircraft Systems

Aircraft systems account for one-third of the aircraft's empty mass. Aircraft systems have a high economic impact: more than one-third of the development and production costs of a medium-range civil transport craft can be allocated to aircraft systems, and this ratio can be even higher for military aircraft. The price of the aircraft is driven in the same proportion by aircraft systems. Aircraft systems account for roughly one-third of the direct operating costs (DOC) and direct maintenance costs (DMC).

Historical Trends

Aircraft silhouettes and general design concepts have been stable since the 1960s. Nevertheless, remarkable progress has been made since that time. Just as aerodynamics, structures, and power plants have been optimized, aircraft systems have been gradually improved in economics, reliability, and safety. This has been made possible by constant evolution and optimization through inservice experience, research, and development and by employment of new technologies.

Probably the most important factor in the changes has been made by digital data processing. Today computers are part of almost every aircraft system in larger aircraft. Computers also play a key role in the design and manufacturing process of aircraft systems. The evolution of aircraft systems has not come to an end yet. Modern achievements in computer technology will continue to make their way into aircraft.

Striving for improved safety, economics, and passenger comfort will demand even more sophisticated technologies and complexity. The airlines have been reluctant to accept the ever-increasing complexity, since it does not make troubleshooting the aircraft any easier. The aviation industry has taken the approach that technology has to buy its way onto the aircraft—i.e., only if new technologies can prove their overall benefit will they be considered in new aircraft design.

The separate tasks of the structure, the engines, and the systems are being more and more integrated to handle the tasks together. Here are some examples:

- Electronic flight control systems stabilize a fighter aircraft with an unstable layout or stabilize aircraft structural or rigid body modes.
- A gust load alleviation system as part of the flight control systems helps reduce the design loads for the wing structure.
- A highly reliable yaw damper system enables the aircraft to be built with a fin smaller than would otherwise be required.
- Engine parameters are changed in accordance with air conditioning demands.

To achieve an overall optimum in aircraft design, it is no longer possible to look at the structure, the engines, and the aircraft systems separately. Today's challenge lies in optimizing the aircraft as a whole by means of multidisciplinary design optimization (MDO).

The Industry

Aircraft systems are defined by the aircraft manufacturer. This commonly takes place in joint teams with engineers from specialized subcontractors. The subcontractors work on the final design, manufacture the system or component, and deliver their parts to the aircraft manufacturer's final assembly line. The trend is for aircraft manufacturers to select major subcontractors who are made responsible for designing and manufacturing a complete aircraft system. These subcontractors may even become risk-sharing partners in the aircraft program. Aircrafts are maintained by dedicated maintenance organizations. Maintenance is done on and off aircraft. Off-aircraft maintenance is performed on aircraft components in specialized shops.

Scope of This Section

Section 2 provides background information and describes the general principles of transport category aircraft systems. The Airbus A321 (Figure 2.2) from the family of Airbus narrow-body aircraft is used to provide an example of the systems under discussion. *At no time should the information given be used for actual aircraft operation or maintenance. The information given is intended for familiarization and training purposes only.* Space in this handbook is too limited for all aircraft systems to be covered in depth. For some aircraft systems only the definition is given and the reader is referred to other parts of the handbook that also deal with the subject. For other aircraft systems the definition is given together with selected views on the Airbus A321. Emphasis is put on selected major mechanical aircraft systems. The References and Further Reading show the way to actual design work and detailed studies.

Definitions

The term *system* is frequently used in engineering sciences. In thermodynamics, for example, a system is characterized by its defined boundary. The definition of the term with respect to aircraft is more specific.

The World Airlines Technical Operations Glossary (WATOG) defines:

- *System:* A combination of inter-related items arranged to perform a specific function
- *Subsystem:* A major functional portion of a system, which contributes to operational completeness of the system

The WATOG also gives an example together with further subdivisions of the system and subsystem:

- *System:* auxiliary power unit
- *Subsystem:* power generator
- *Component:* fuel control unit
- *Subassembly:* valve
- *Part:* seal

Note that these definitions refer to civil aircraft. With respect to military aircraft, instead of *aircraft systems* the term is *aircraft subsystems*. In the example above, the auxiliary power unit hence would be considered a subsystem.

In dealing with aircraft systems, all categories of aircrafts need to be considered. ICAO defines:

- *Aircraft:* Any machine that can derive support in the atmosphere from the reaction of the air (ICAO Annex 2)
- *Aircraft category:* Classification of aircraft according to specified basic characteristics, e.g., aeroplane, glider, rotorcraft, free balloon (ICAO Annex 1)

Combining the above definitions, a definition for aircraft systems might be:

- *Aircraft system:* A combination of interrelated items arranged to perform a specific function on an aircraft

This section deals with aircraft systems in powered heavier-than-air aircraft. Although aircraft systems in gliders, rotorcrafts, and free balloons have to take into account the specifics of their respective categories, they are not fundamentally different from aircraft systems in aeroplanes.

Breakdown

Aircraft systems are distinguished by function. It is common practice in civil aviation to group aircraft systems according to Specification 100 of the Air Transport Association of America (ATA) (ATA 100), which thoroughly structures aircraft documentation. According to ATA 100,[1] aircraft

[1]Recently ATA 100 became part of the new ATA 2200. ATA 2200 has introduced minor changes and updates to the definitions of aircraft systems. This text uses the well-established ATA 100 and presents differences to ATA 2200 in footnotes.

Identifier	Name of system
21	air conditioning
22	auto flight
23	communications
24	electrical power
25	equipment/furnishings
26	fire protection
27	flight controls
28	fuel
29	hydraulic power
30	ice and rain protection
31	indicating/recording systems
32	landing gear
33	lights
34	navigation
35	oxygen
36	pneumatic
38	water/waste
49	airborne auxiliary power

[a]Not included in this table are Chapters 37, 41, 45, and 46 from ATA 100, which are not of relevance here. Also not included here are new Chapters 44 and 50 from ATA 2200.

Table 2.1 Aircraft Systems[a] (ATA 100)

equipment is identified by an equipment identifier consisting of three elements of two digits each. The identifier 29-31-03 points to system 29, subsystem 31, and unit 03. The aircraft systems— or, in ATA terms, *airframe systems*—are listed in Table 2.1 together with their system identifiers. It is common practice to refer to just the system identifier *ATA 28,* instead of to the "fuel system." Furthermore, *Chapter 28* (from ATA 100) is often referred to, because that is the chapter allocated to the fuel system in any aircraft documentation showing ATA conformity.

Autopilot, communications, navigation, and indicating/recording systems (ATA 22, 23, 34, 31, [44, 45, 46]) are electronic systems, known in aviation as *avionic systems,* and are characterized by processing information (compare with SAE 1998).

Other systems provide fuel, power, and essential comfort to crew and passengers. These non-avionic systems are the *general* or *utility systems.* Today there is an increase in the number of electronic control units within the utility systems; nevertheless, the primary purpose of these systems remains some kind of energy transfer (Moir and Seabridge 2001).

Secondary power systems include the nonpropulsive power generation and transmission. They include electrical power, hydraulic power, pneumatic, and auxiliary power (SAE 1998) (ATA 24, 29, 36, 49). Secondary power systems provide power to other aircraft systems.

The *environmental control system* (ECS) is an engineering system that maintains the immediate environment of an organism within defined limits of temperature, pressure, and gaseous composition suitable for continuance of comfort and efficiency (AGARD 1980). The air conditioning system and oxygen system (ATA 21, 35) are assigned these tasks.

Other aircraft systems are grouped and assigned a specific name often without a formal definition.

Hydraulic systems comprise all systems that apply hydraulic power. In general, these are hydraulic power, flight controls, and landing gear (ATA 29, 27, 32).

Electric systems comprise all systems that apply electric power. In general, these are electric power (ATA 24) and all systems with major electrical consumers. Electrical systems are characterized by electrical power generation, distribution, and consumption and have to be distinguished from avionic systems.

Pneumatic systems comprise all systems that apply pneumatic power. In general, these are pneumatic and other systems with pneumatic components (ATA 36, 21, 30).

Cabin systems[2] comprise all systems with an impact on the cabin of the aircraft and hence with an influence on the passenger (ATA 21, 25, 35, 38, and partially 23, 26, 31, 33).

These groupings depend to a certain extent on the system technologies applied in the aircraft being considered.

Certification

After one or several prototype aircraft are designed and manufactured, they go through a series of *certification tests* in order to show compliance with the *certification requirements*. Compliance with the requirements may be shown by analysis, ground, or flight test, depending on the requirements or negotiations with the *aviation administration*. System tests are a substantial part of the certification program. In Europe, certification of large aeroplanes is based on the Joint Aviation Requirements (CS-25), and in the United States it is based on the Airworthiness Standards: Transport Category Airplanes (FAR Part 25). Large aeroplanes are those aircraft with a maximum takeoff mass of more than 5,700 kg. CS and FAR are very similar; the basic code for CS-25 is FAR Part 25, and further harmonization of the requirements is in progress. The certification of one or several prototype aircraft leads to a *type certificate* being issued. Aircraft in series production have to show *airworthiness* and *conformity with the prototype aircraft*. In service the aircrafts have to be maintained according to an agreed maintenance schedule to prove continuous airworthiness.

CS-25 and FAR Part 25 are grouped into several subparts (the following is based on CS-25).

Subpart F, "Equipment," contains many requirements for aircraft systems.

Subpart E, "Power plant," contains requirements for power plant-related systems.

Also Subpart D, "Design and Construction," contains requirements for aircraft systems.

Subpart J, "Gas Turbine Auxiliary Power Unit Installation," contains requirements for airborne auxiliary power—i.e., the auxiliary power unit (APU).

General information on aircraft systems can be found in Section 1301 "Function and installation" and Section 1309 "Equipment, systems and installations" of CS-25 and FAR Part 25. Section 1309 provides information on safety requirements, loads, and environmental conditions. Table 2.2 provides access to the certification requirements for large airplanes when specific information related to a particular aircraft system is needed.

Interpretative material to most paragraphs is provided:

- FAR: Advisory Circulars (AC) (especially in AC 25-17 and AC 25-22)
- CS: CS-25, Book 2, Acceptable Means of Compliance (AMC-25)

[2]Following the new ATA 2200, "Cabin Systems (ATA 44)" are defined as "Those units and components which furnish means of entertaining the passengers and providing communication within the aircraft and between the aircraft cabin and ground stations. Includes voice, data, music and video transmissions."

Identifier name of system	Applicable sections
21 Air conditioning	831–833: Sections under the heading "Ventilation and heating" 841–843: Sections under the heading "Pressurisation" 1461: Equipment containing high energy rotors
22 Auto flight	1329: Automatic pilot system 1335: Flight director systems
23 Communications	1307: Miscellaneous equipment (radio communication) 1457: Cockpit voice recorders
24 Electrical power	1351: General 1353: Electrical equipment and installations 1355: Distribution system 1357: Circuit protective devices 1359: Electrical system fire and smoke protection 1363: Electrical system tests
25 Equipment/furnishings	771–793: Sections under the heading "Personnel and cargo accommodations" 819: Lower deck service compartments (including galleys) 1411: General (under heading safety equipment) 1413: Safety belts 1415: Ditching equipment 1421: Megaphones
26 Fire protection	851–867: Sections under the heading "Fire protection" 1181–1207: Sections under the heading "Powerplant fire protection" 1307: Miscellaneous equipment (portable fire extinguishers) A1181–A1207: Sections related to APU fire protection
27 Flight controls	671–703: Sections under the heading "Control systems"
28 Fuel	951–981: Sections under the heading "Fuel system" 991–1001: Sections under the heading "Fuel system components" A952–A999: Sections related to the APU fuel system
29 Hydraulic power	1435: Hydraulic systems
30 Ice & rain protection	1307: Miscellaneous equipment (including windshield wiper) 1416: Pneumatic de-icer boot system 1419: Ice protection
31 Indicating/recording systems	1303: Flight and navigation instruments 1305: Powerplant instruments 1321: Arrangement and visibility 1331: Instruments using a power supply 1333: Instrument systems 1337: Powerplant instruments
32 Landing gear	721–X745: Sections under the heading "Landing gear"

Table 2.2 Selected Certification Requirements for Aircraft Systems Based on CS-25

Identifier name of system	Applicable sections
33 Lights	812: Emergency lighting 1322: Warning, caution, and advisory lights 1381: Instrument lights 1383: Landing lights 1385, 1387, 1389, 1391, 1393, 1395, 1397: Position lights 1401: Anti-collision light system 1403: Wing icing detection lights
34 Navigation	1307: Miscellaneous equipment (radio navigation) 1323: Airspeed indicating system 1325: Static pressure system 1326: Pilot heat indication system 1327: Magnetic direction indicator 1459: Flight recorders
35 Oxygen	1439: Protective breathing equipment 1441: Oxygen equipment and supply 1443: Minimum mass flow of supplemental oxygen 1445: Equipment standards for the oxygen distributing system 1447: Equipment standards for oxygen dispensing units 1449: Means for determining use of oxygen 1450: Chemical oxygen generators 1451: Fire protection for oxygen equipment 1453: Protection of oxygen equipment from rupture
36 Pneumatic	X1436: Pneumatic systems—high pressure 1438: Pressurisation and low pressure pneumatic systems
38 Water/waste	1455: Draining of fluids subject to freezing X799: Water systems
49 Airborne auxiliary power	Paragraphs in Subpart J—Gas turbine auxiliary power unit installations

Table 2.2 Selected Certification Requirements for Aircraft Systems Based on CS-25 (*Continued*)

Safety and Reliability

Safety and reliability considerations of aircraft systems are an integral part of the safety and reliability considerations of the whole aircraft. Modern sophisticated aircraft depend very much on the proper functioning of their aircraft systems, so that safety and reliability considerations of aircraft systems have become highly important in their own right. For this reason an aircraft systems-specific approach to the topic is presented here.

Safety is a state in which the risk is lower than a permissible risk. The risk is defined by the probability of a failure and the expected effect.

The *effect* of failure describes the consequences of the failure (damage or injury).

The *probability of failure*, $F(t)$, is equal to the number of failures within a given period of time divided by the total number of parts in a test.

The *safety requirements* for aircraft systems are stated in Section 1309 of the certification requirements CS-25 and FAR Part 25 and are listed in Table 2.3.

Effect on aircraft and occupants	Normal	Nuisance	Operating limitations Emergency procedures	Significant reduction in safety margins Difficult for crew to cope with adverse conditions Passenger injuries	Large reduction in safety margins Crew extended because of workload or environmental conditions Serious injury or death of small number of occupants	Multiple deaths, usually with loss of aircraft
Category of effect Probability of a failure according to CS-25 (per flight hour)	Minor Frequent 10^0–10^{-2}	Minor Frequent 10^{-2}–10^{-3}	Minor Reasonably probable 10^{-3}–10^{-5}	Major Remote 10^{-5}–10^{-7}	Hazardous Extremely remote 10^{-7}–10^{-9}	Catastrophe Extremely improbable $< 10^{-9}$

Source: AMC-25.

Table 2.3 Safety Requirements for Large Airplane's Systems

The probability of a failure in a system increases with the time period of operation and is specified for an operation time of one flight hour (FH). Obviously, the higher the effect of a failure is on aircraft operation, passengers, and the aircraft itself, the lower the permissible probability of such a failure has to be.

The *reliability* is the probability of survival, $R(t)$. It is an item's ability to fulfill defined requirements for a specific period of time under specified conditions. A statement referring to the reliability of a system can only be made if the failure criteria are precisely defined.

The reliability or *probability of survival*, $R(t)$, can also be defined as the number of parts surviving within a given period of time divided by the total number of parts in a test:

$$R(t) + F(t) = 1$$

Although referring to the reliability $R(t)$, mostly the value of the probability of failure $F(t)$ is given (10^{-7}) because the reliability yields values more difficult to handle (0.9999999).

The *hazard rate function*, $z(t)$, is a measure of the probability that a component will fail in the next time interval, given that it has survived up to the beginning of that time interval. If the hazard rate function is constant (which is often assumed), it is called the *failure rate*, λ. Failure rates of mechanical components are listed in Rome (1985), and failure rates for electric and electronic equipment can be estimated using MIL-HDBK-217. The failure rate has units of one per flight hour (1/FH). The inverse of the failure rate, called the mean time between failures (MTBF), is often used in reliability and maintenance circles.

$$\text{MTBF} = 1/\lambda$$

The *failure to removal ratio* (FTRR) is a maintenance quantity. It shows the ratio of faults found in a component during a shop visit, divided by the number of component removals. Unfortunately, the FTRR is especially low in case of electrical components (0.6–0.7) and electronic components (0.3–0.4). Hydraulic components (0.8–0.9) and mechanical components (1.0) show better values. The product of MTBF and FTRR yields the maintenance cost driver, the *mean time between unscheduled removals* (MTBUR).

$$\text{MTBUR} = \text{MTBF} \cdot \text{FTTR}$$

The reliability and the probability of failure can be calculated from the failure rate:

$$R(t) = e^{-\lambda t}, \;\; F(t) = 1 - e^{-\lambda t}$$

For low failure rates, which are common in aviation, the probability of failure calculated for a period of one hour ($F(t)$/FH) equals almost exactly the failure rate, λ.

Systems are a combination of many components either in parallel, in series, or in a combination of both. The reliability of a *series system* is equal to the product of is component values.

$$R_S(t) = R_1(t)R_2(t)R_3(t)\ldots$$

The failure rate of a series system is approximately the sum of the failure rates of its (reliable) components.

$$\lambda_S \approx \lambda_1 + \lambda_2 + \lambda_3 \;\ldots$$

The probability of failure of a *parallel system* is equal to the product of is component values.

$$F_P(t) = F_1(t)F_2(t)F_3(t)\ldots$$

The failure rate of a parallel system is approximately the product of is (reliable) component values.

$$\lambda_P \approx \lambda_1\lambda_2\lambda_3 \;\ldots$$

Systems can be depicted by *reliability block diagrams* (RBDs). The analysis of large systems is carried out in successive stages. At each stage a small number of components connected either in parallel or in series is combined with equations as shown above. In this way the complexity of the system can be reduced step by step. The *fault tree analysis* (FTA) is an alternative method to deal with complex systems. Parallel systems are combined by an OR gate symbol. Series systems are combined by an AND gate symbol. Top events are shown in a rectangle and basic failure causes are shown in circles. Software tools exist that support a FTA or the analysis of a RBD. Systems might show cross-linkages so that some units are in more than one subsystem. One way of dealing with this problem is to use a theorem on conditional probability or to apply a truth table (Davidson 1988).

These approximate equations for series and parallel systems are quite useful in day-to-day business. The last equation also shows the ability of parallel systems to achieve low failure rates and thus high reliability. For example, three components combined in parallel with a failure rate of 10^{-3} 1/FH each, yield an overall failure rate of 10^{-9} 1/FH. This is a failure rate that could not have been achieved by a single component no matter how carefully this component was manufactured and tested. This thought leads us to the concept of redundancy, which is so typical in safety critical aircraft systems.

Redundancy is the existence of more means for accomplishing a given function than would simply be necessary. It is divided into

- Homogeneous redundancy (the multiple means are identical) and
- Inhomogeneous redundancy (the multiple means are of different type)

Inhomogeneous redundancy is divided into:

- Dissimilar redundancy or
- Diversitary redundancy

Safety-critical aircraft systems often show *triplex* subsystems. The system architecture of safety-critical computers may be even of *quadruplex* or *duo duplex* type.

The subsystems of a system with built-in redundancy may all work together. If one subsystem fails, the others will just have to cope with a somewhat higher load. These systems are called *active-active* systems. Other systems may be of the *activestandby* type and need to perform a changeover in case of a failure. If the standby subsystem is constantly waiting to be activated, it is on *hot standby;* otherwise it is on *cold standby.* The changeover should not be dependent on a changeover unit, because this unit with its own limited reliability might fail and prevent the changeover. If an active-standby concept is applied, the subsystems should take turns doing the job. This could be achieved with a planned changeover before every takeoff. If the same subsystem stays in standby all the time, it may show an (undetected) *dormant failure* and hence will not be able to take up the job in case of failure of the first subsystem. Systems with a potential of dormant failures need regular maintenance checks and should be avoided.

An assumption has been made in the calculation of parallel systems that the failures of individual subsystems are independent of each other, that is, that two or more subsystems do not fail simultaneously from precisely the same cause (except purely by chance). However, most systems have the potential of having more than one failure due to a common cause. These failures are called *common cause failures* (CCFs). They tend to arise from errors made during design, manufacture, maintenance, operation, or environmental effects. For example, loss of power supply could cause both a running and a standby pump to fail (design error), or an empty fuel tank could cause all engines to quit (error in operation). Because these failure modes may appear to be outside the system being assessed, they can easily be overlooked, leading to too-optimistic assessments. Methods to avoid common cause failures in the design stage are the application of

- Inhomogeneous redundancy (see above)
- Segregation in the rooting of redundant wires, pipes, and ducts
- Separation of redundant components
- Placement of safety-critical components in safe areas
- Design of redundant components or software programs by independent teams with different (software) tools

An aircraft should not only be safe to fly, it should also show very few errors that need the attention of maintenance personnel. In this respect we face a problem with high safety requirements. High safety requirements lead to the application of redundancy and hence more subsystems. The probability of a failure leading to the loss of the overall function can be reduced by redundancy, but the probability of occurrence of any failure anywhere in the system is increased. Two subsystems with a failure rate of 10^{-3} 1/FH each yield an overall probability of failure of about 10^{-6} and a probability of any failure of $2 \cdot 10^{-3}$ (based on a 1-hour operation). Three subsystems yield an overall probability of failure of 10^{-9} and a probability of any failure of already $3 \cdot 10^{-3}$. The level of safety during flight can only be achieved if all subsystems work properly before takeoff, but, as we have seen, the probability for any failure increases with an increased number of subsystems. These thoughts lead to what is called availability and dispatch reliability.

The *steady state availability* is defined as the probability that a system will be available when required, or as the proportion of total time that the system is available for use. Therefore, the availability of a system is a function of its failure rate λ and of its repair rate $\mu = 1/\text{MTTR}$, where MTTR is the mean time to repair:

$$A_{SS} = \frac{\text{MTBF}}{\text{MTBF} + \text{MTTR}} = \frac{\mu}{\lambda + \mu}$$

The *instantaneous availability*, or probability that the system will be available at time t, is

$$A_I = \frac{\mu}{\lambda + \mu} + \frac{\lambda}{\lambda + \mu} e^{-(\lambda + \mu)t}$$

Often it is more revealing to consider system unavailability, $U = 1 - A$. The instantaneous availability of an aircraft at the moment of dispatch from the gate is called *dispatch reliability.* Dispatch reliability, for technical reasons, primarily depends on the combined dispatch reliability of the aircraft systems. The airlines monitor their fleets' dispatch reliability very carefully because high *dispatch unreliability* leads to delays and cancellations of flights and incurs delay and cancellation costs (see below). Dispatch reliability depends on the maturity of an aircraft program and is on the order of 0.99. A method to increase dispatch reliability is the introduction of *built-in test equipment* (BITE) into electronic systems. Though this adds complexity and might result in spurious failure indications, it can greatly reduce maintenance times by providing an instantaneous indication of failure location. Another method is to provide *extra redundancy* above the level required for safety reasons. This would than allow to dispatch with one subsystem inoperative. Components that are not needed for takeoff may be known as *flying spares.* The pilot gets a clear indication about which subsystems or components need to be available at takeoff from the *minimum equipment list* (MEL), written by the airline on the basis of the master minimum equipment list (MMEL) provided by the manufacturer and approved by the authorities.

Reliability assurance during the aircraft system design applies a couple of different methods, including:

- Drawing a fault tree for a *fault tree analysis* (FTA) (see above) starts from consideration of system failure effects, referred to as top event. The analysis proceeds by determining how these can be caused by lower-level failures. In this way it is a top-down approach.
- The *reliability apportionment* breaks an overall system reliability requirement down into individual subsystem reliabilities. This is common in large systems when different design teams of subcontractors are involved. Clearly it follows a top-down approach.
- In contrast, the *failure mode, effects, and criticality analysis* (FMECA) (MILSTD-1629) follows a bottom-up approach. It considers each mode of failure of every component of a system to ascertain the effects on system operation and defines a *failure mode criticality number.*
- The *zonal safety analysis* (ZSA), rather than looking at an aircraft from a functional point of view, looks at the components' location. The ZSA checks installation rules and checks the effects of events originating within the zone, in other zones, or on the outside.

Software defies the above calculations and methods. However, information can be drawn from RTCA/DO-178B, which deals with *software considerations* in airborne systems and equipment. *Environmental conditions* for airborne equipment are presented in RTCA/DO-160D.

Mass

Mass estimation of aircraft systems is part of the mass (or weight) estimation of the whole aircraft.

The mass of all the aircraft systems m_{SYS} amounts to 23–40% of the aircraft's empty mass m_{OE}, where m_{OE} is the mass related to the operational empty weight (OEW). The figure 23% is true in case of a modern long-range airliner, whereas 40% is about right for a smaller aircraft such as business jet. Hence, for civil jet transport we may write

$$\frac{m_{SYS}}{m_{OE}} \approx 0.23 - 0.4$$

On average this ratio comes to ⅓, as stated above. Taking into account the ratio of the aircraft's empty mass m_{OE} and the maximum takeoff mass m_{MTO}, the mass related to the maximum takeoff weight (MTOW).

$$\frac{m_{SYS}}{m_{MTO}} \approx 0.11 - 0.23$$

Figure 2.1 shows the mass of aircraft systems of selected civil jet aircraft as a function of their maximum takeoff mass. We follow a *top-down approach* and fit a curve to these data to obtain

$$m_{SYS} = 0.92\, m_{MTO}^{0.85} \qquad \text{for } m_{SYS} \text{ and } m_{MTO} \text{ in kg}$$

This function is shown in Figure 2.1. The average relative mass of the individual systems of civil jet aircraft is given in Table 2.4.

Some aircraft systems, like the landing gear system (ATA 32) and the equipment and furnishings (ATA 25), account for a large percentage of the total aircraft system mass. The avionic system relative mass is 6% on average, but this figure depends on aircraft size because the amount of avionics needed in jet aircraft tends to be nearly constant. For this reason, the relative mass of avionic systems of business aircraft may be as high as 14% and as low as 5% in case of a large civil transport. As can be seen in Table 2.4, a number of systems are of minor importance for aircraft system mass predictions.

Alternatively, it is also possible to follow a *bottom-up approach*. This statistical technique uses system parameters to predict the mass of the system. Equations are given in Raymer (1992), Roskam (1989), and Torenbeek (1988). In addition, the knowledge gathered in papers from the Society of Allied Weight Engineers should be tapped (see SAWE 2002).

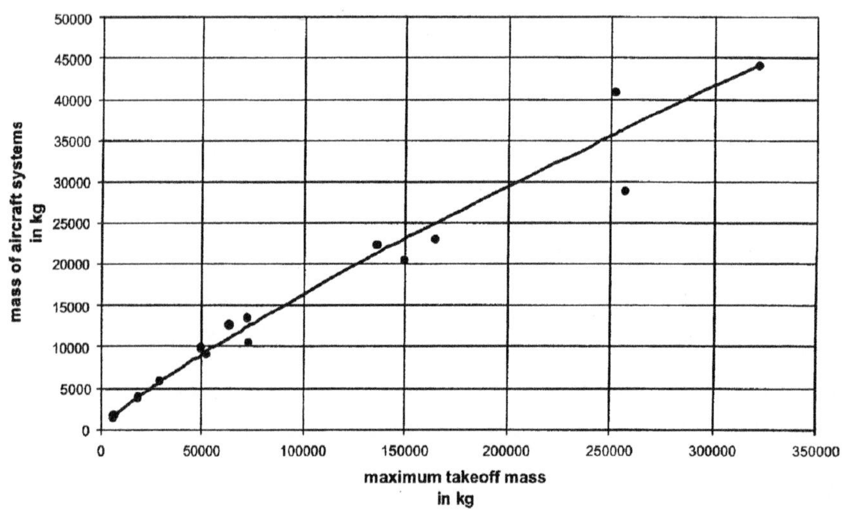

FIGURE 2.1 Mass of aircraft systems of selected civil jet aircraft plotted against their maximum takeoff mass.

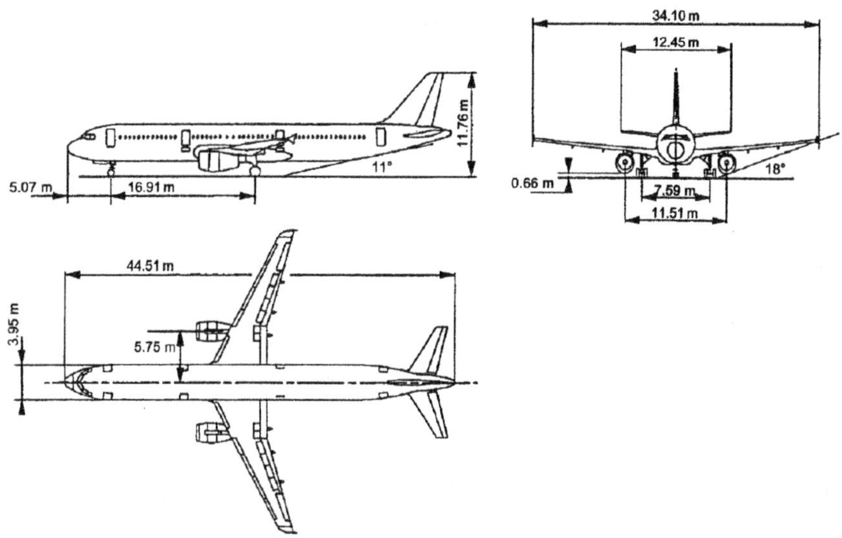

FIGURE 2.2 The Airbus A321 is used throughout this section to provide aircraft system examples. One hundred eighty-six passengers in two-class layout, MTOW: 83,000 kg, M_{MO} = 0.82, maximum FL 390.

Identifier	Name of system	Average relative mass of system
21	Air conditioning	6%
22	Auto flight	1%
23	Communications	2%
24	Electrical power	10%
25	Equipment/furnishings	24%
26	Fire protection	1%
27	Flight controls	8%
28	Fuel	3%
29	Hydraulic power	7%
30	Ice and rain protection	<1%
31	Indicating/recording systems	<1%
32	Landing gear	27%
33	Lights	2%
34	Navigation	3%
35	Oxygen	1%
36	Pneumatic	2%
38	Water/waste	1%
49	Airborne auxiliary power	2%

TABLE 2.4 Average Relative Mass of Aircraft Systems of Civil Jets

Statistics of aircraft system mass have to take as many aircraft into account as possible in order to broaden the statistical base. This, however, is really possible only if mass data are based on comparable and detailed mass breakdowns. Unfortunately, there are many quite different breakdowns in use, and it is found that system boundaries overlap from one method to another or are not well defined in the first place. So in the present situation it is very difficult to use and compare mass data and mass equations based on one of these breakdowns in another setting. This situation adds to the difficulties that exist with statistical methods anyhow and explains why statistical mass equations for systems or subsystems do not provide particularly reliable data.

Boeing has used a *breakdown format* called Weight Research Data 1 (WRD1). In the literature, breakdowns very similar to WRD1 can be found. Airbus uses so-called Weight Chapters. Another approach is given with MIL-STD-1374. Above we have used a mass breakdown according to the ATA 100 chapter numbering. ATA 100 also includes a widely accepted mass breakdown for weight and balance manuals. This breakdown, however, provides only as much detail as needed in aircraft operation but not enough detail for aircraft system design.

Note that aircraft system *mass predictions* deteriorate in *accuracy* when the level of detail is increased. For its old class I weight prediction method, Boeing estimates the prediction of single systems to be off by as much as ±90%. In contrast, the resultant mass of all systems combined is claimed to be off by not more than ±16% (Boeing 1968). This is because many inaccuracies combined fortunately cancel out to a certain extent.

Detailed system mass predictions are also necessary for *center of gravity* (CG) *calculation* for the aircraft. The main landing gear accounts for about 87% and the nose landing gear for the remaining 13% of the complete landing gear mass. With known positions of nose and main landing gear, this information can be fed into the CG calculation of the aircraft. The CG of the other systems can roughly be assumed at a point 40–50% of the fuselage length aft of the aircraft nose.

Practical mass predictions will look like this: In the early design stage, statistical methods are used. The aircraft manufacturer can also use the information contained in the mass database of older aircraft for the new design. In a later design stage a subcontractor will offer a system or an item of equipment. The subcontractor probably has quite a good idea what the item's mass will be from a comparison with similar items already built. If the required size of equipment is different from an older one, a mass estimate may be obtained from scaling. In the final development stage, mass accounting can be based on the actual mass of components that are already delivered to the manufacturer.

There is another virtue in mass predictions: the system mass has been used for rough *cost calculations*. This is possible when, from statistics, costs per unit mass are known and costs are assumed to be proportional with mass. Evidently, the concept of calculating costs from mass fails if expensive mass reduction programs are being applied. The concept also fails if highly sophisticated technologies are applied to reduce mass that are not considered in the established cost per unit mass.

Power

Gliders use the energy of up-currents, while solar-powered vehicles use the energy from the sun. Human-powered flight has also been demonstrated. Propulsive power for any other "down to earth" flying depends on fuel. This fuel is used in the aircraft main engines. *Secondary power* systems (hydraulic power, electrical power, pneumatic power) in turn draw on engine power to supply their client systems with *nonpropulsive power* in all those cases where functions are not directly actuated by the pilot's muscles. This is the simple picture of the aircraft power management. However, there is more to it, due to safety requirements and the need for autonomous operation of the aircraft on the ground with engines shut down.

Various secondary power sources are available in the air and on the ground. Secondary power loads may be grouped into two major categories. Power conversion transforms secondary power from one form into another.

An *auxiliary power unit* (APU) (see above) is used to produce power from fuel independent of the main engines. An APU is a gas turbine engine. Most often it produces electrical power and pneumatic power. A *ram air turbine* (RAT) (see Subsection 2.8) is used to produce hydraulic or electrical power from the kinetic energy of the air passing by the aircraft. This is possible even without fuel and without the main engines running—at least as long as the aircraft soars down consuming its potential energy. Except for the pilot's own energy, the *aircraft batteries* are the last and very limited source of energy on board.

Ground power may be available on the apron or in the hangar. The aircraft may be supplied directly with electricity, high-pressure hydraulic fluid, pressurized air, and/or air conditioned air. Human power could work a hand pump in the hydraulic system. If only electrical ground power is available, the aircraft depends on its secondary power conversion capabilities to activate the hydraulic and pneumatic system. Without ground equipment and with engines shut down, the aircraft may operate autonomously if it is equipped with an auxiliary power unit (APU).

First of all, *secondary power loads* may be grouped into:

- Technical loads consumed by equipment required to operate the aircraft safely
- Commercial loads consumed by equipment required to increase passenger comfort and satisfaction, given the airline's need to provide these services

Power conversion among different[3] secondary power systems is used to increase overall system reliability. If we consider electrical power, hydraulic power, and pneumatics:

- Six different unidirectional conversions are possible. Examples are:
  - Electrical to hydraulic power conversion: electric motor-driven pump
  - Pneumatic to hydraulic power conversion: air turbine motor-driven pump
  - Hydraulic to electrical power conversion: hydraulic motor-driven generator
- Three different bidirectional conversions are possibilities that allow a two-way power conversion among two different secondary power systems within one conversion unit.

For many years hydraulic, pneumatic, and electrical power supply in commercial aircraft had been sufficient to meet the demands from technical and commercial loads. System design emphasized reliable, lightweight solutions. From fuel input to system output, very low overall efficiencies were accepted in exchange.

In recent years it has been observed that aircraft face increasing technical loads. Also, market trends together with increasing flight durations have resulted in higher commercial loads, caused, for example, by today's standards in in-flight entertainment. Possibilities for power off-takes do not increase proportionally with aircraft size. Large modern civil aircraft are therefore likely to face limitations of cost effectiveness, geometry, or weight with present-day technologies in an attempt to meet these new power load levels. The aerospace industry has identified a potential deadlock, where power needs will exceed the maximum available power supply.

In the future a move toward electrical power as a single source to meet secondary power demands is expected to be a solution to the problem. The last aircraft generation brought steering by wire. The next generation of aircraft might bring power by wire.

---

[3]Power conversion is even applied within one type of secondary power system: the hydraulic system. Transport category aircraft apply several independent hydraulic systems. Among pairs of these hydraulic systems unidirectional or bidirectional hydraulic power transfer without the interchange of hydraulic fluid can be desirable. For this purpose, power transfer units (PTU) (ARP 1280) are used. They are built by coupling a hydraulic motor and a hydraulic pump via a connecting shaft.

Costs and Trade-Off Studies

Trade-off studies play an important roll in aircraft system design. Trade-off studies try to find the best among several system design proposals. Safety aspects allow no compromise because certification regulations have to be closely followed. Also, performance aspects leave little room because usually only as much performance as necessary to do the job will be allowed for. More powerful aircraft systems will unnecessarily produce costs that add to the overall costs of the aircraft. Clearly, costs need to be reduced as much as possible to come up with a viable product. Therefore, it is the costs aspect that is usually decisive in trade-off studies of which system design will get on board the aircraft.

At the aircraft system level, evaluations are done in the early design stage by looking separately at various aspects:

- Mass
- Maintainability
- Reliability
- System price
- Other specific criteria depending on the aircraft system in question

Based on these separate evaluations, the simplest way to come up with one single figure of merit for a proposal is to define subjectively a *weighted sum* of the results based on the individual criteria.

In contrast to the above approach, at the aircraft level an evaluation is traditionally based primarily on one single figure: the direct operating costs (DOC). DOCs take account of criteria such as mass, maintainability, and aircraft price, but combine these separate parameters unambiguously by calculating their economical implications. Subjective manipulations of the results are largely avoided in this way.

Unfortunately, aircraft DOC methods cannot be taken as is for applying this advantage to an aircraft system evaluation. In contrast to aircraft DOC methods, a DOC method on the systems level must incorporate many system-specific parameters. Therefore, a *DOC method for aircraft systems* called DOC_{SYS} has been developed (Scholz 1998) which follows the principles of aircraft DOC methods as closely as possible while taking aircraft system peculiarities into account as much as necessary.

$$C_{DOC,SYS} = C_{DEP} + C_F + C_M + C_{DEL} + C_{SH}$$

where C_{DEP} = depreciation of the system (a function of system price)
C_F = fuel costs caused by the system
C_M = direct maintenance costs caused by the system
C_{DEL} = delay and cancellation costs caused by the system
C_{SH} = capital costs caused by necessary system spare parts on stock (spare holding)

The fuel costs, C_F, are due to:

- Transportation of the system's mass (fixed or variable during flight) (taking into account the lift-to-drag ratio of the aircraft and the specific fuel consumption of the engines)
- Power off-takes from the engines (by electrical generators or hydraulic pumps)
- Bleed air off-takes (for the pneumatic system)
- Ram air off-takes (e.g., for the air conditioning system)

- Additional drag caused by the presents of aircraft systems, subsystems, or single parts (e.g., due to drain masts)

In contrast to Scholz (1998), who combines various system aspects to U.S. dollars, Shustrov (1998) combines system mass effects and effects related to the system's energy consumption to a quantity called *starting mass*.

Proprietary methods for the evaluation of aircraft systems are in use at aircraft manufacturers and subcontractors.

2.2 Air Conditioning (ATA 21)

Air conditioning as defined by ATA 100:

> Those units and components which furnish a means of pressurizing, heating, cooling, moisture controlling, filtering and treating the air used to ventilate the areas of the fuselage within the pressure seals. Includes cabin supercharger, equipment cooling, heater, heater fuel system, expansion turbine, valves, scoops, ducts, etc.

Fundamentals

Impact of Atmospheric Parameters

In the troposphere, the air temperature decreases with increasing altitude. In the stratosphere above 11,000 m (36,089 ft), the air temperature is at constant –56.5 °C. The air pressure also decreases with altitude. Although oxygen amounts to approximately 21% independent of altitude, the partial pressure[4] of oxygen drops with increasing altitude. Our body is used to a partial oxygen pressure of about 0.21 times sea level pressure. If we want to survive at high altitudes, either (a) the oxygen fraction has to be increased (using an oxygen system), or (b) the total pressure has to be maintained close to sea level pressure (using a pressurization system). For civil aircraft generally option (b) is applied; flights in nonpressurized cabins[5] without supplemental oxygen are limited to an altitude of 10,000 ft. Military aircraft use a combination of (a) and (b); cabin altitude[6] does not exceed about 20,000 ft.

Purpose of Air Conditioning Systems

The purpose of the air conditioning system is to make the interior environment of the aircraft comfortable for human beings. Depending on the type of aircraft and altitude of operation, this may involve only *ventilation* of the cabin by supplying a flow of fresh air using air vents. If the temperature must be adjusted, some method of *heating* or *cooling* is required. At high altitudes the aircraft can fly above most of the weather conditions that contain turbulence and make flight uncomfortable. Additionally, the fuel efficiency of the aircraft is increased. *Pressurization* is necessary if the aircraft is operated at these high altitudes. In some parts of the world the relative humidity[7] is quite high. Water extractors are therefore used for *dehumidification* of the cabin air. This is necessary to

---

[4]Partial pressure: "The pressure exerted by one gas in a mixture of gases; equal to the fraction … of one gas times the total pressure" (AIR 171).

[5]Nonpressurized cabin: "An airplane cabin that is not designed … for pressurizing and which will, therefore, have a cabin pressure equal to that of the surrounding atmosphere" (SAE 1998).

[6]Cabin altitude: "The standard altitude at which atmospheric pressure is equal to the cabin pressure" (SAE 1998).

[7]Relative humidity: "The ratio, expressed as percentage, of the amount of water vapor … actually present in the air, to the amount of water vapor that would be present if the air were saturated with respect to water at the same temperature and pressure" (SAE 1998).

prevent damage to electrical and electronic equipment, aircraft insulation and structure. Reduced humidity also limits window and windscreen misting. At an altitude of 40,000 ft the relative humidity is quite low (1–2%) compared to the comfort level for crew and passengers (30%). Nevertheless, *humidification* of the cabin air would be impractical for the other reasons named and for the costs involved in carrying that water (AIR 1609).

The air conditioning system is a safety-critical system because passengers and crew depend on its proper function. Transport category aircraft will have two independent subsystems to meet these safety requirements. The certification requirements include minimum standards. The aircraft manufacturer may choose higher standards in order to increase passenger comfort.

Ventilation

- Under normal conditions 4.7 L/s (10 ft^3/min ≈ 0.6 lb/min) are required for each crew member (CS-25 Section 831(a)).
- Manufacturers will typically provide a minimum of about 7.8 L/s (1.0 lb/min) for each person in the aircraft.
- In case of a failure (with a probability of not more than 10^{-5} 1/FH) the supply of fresh air should not be less than 3.1 L/s (0.4 Lb/min) per person excluding supply from the recirculation system (CS-25 Section 831(c)).
- In order to avoid drafts, the air velocity in the cabin should be limited to 0.2 m/s (40 ft/min) in the vicinity of the passengers (AIR 1168/3). Individual air outlets, however, show air velocities of about 1.0 m/s. Conditioned air may enter the cabin through cabin outlets at not more than 2.0 m/s.

Temperature Control

- The temperature control from the cockpit may typically be possible in the range between 18°C and 30°C.
- Heating and cooling requirements have to be met as specified for various steady state and transient scenarios. Here are some lessons learned:
  - During cruise cooling almost always is required (an exception is flights without passengers).
  - Cooling loads on the ground on a hot day with passengers on board are higher then in flight.
  - Transient scenarios will probably determine the heating and cooling performance of the air conditioning system in civil subsonic aircraft:
    - Heating the cabin of a cold-soaked airplane from –32°C to 21°C in 30 minutes (no internal heat loads, doors closed).
    - Cooling the aircraft from 46°C to 27°C in 30 minutes (full passenger load, doors closed) (ARP 85).
    - Cooling requirements for high-speed aircraft are driven by kinetic heating. Kinetic heating occurs when the aircraft skin heats up due to friction with air molecules. In the flight range below Mach 2, the skin temperature is equal to the recovery temperature[8]: $T_{skin} \approx T_{ambient} (1 + 0.18\,M^2)$

[8]Recovery temperature: "The equilibrium temperature of an object placed in a flow … always less than the total temperature" (AGARD 1980).

Pressure Control (for aircraft with a pressurized cabin)[9]

- Under normal conditions the cabin altitude in pressurized cabins must not be more than 2,440 m (8,000 ft) (CS-25 Section 841(a)).
- In case of a failure (with a probability of not more than 10^{-5} 1/FH), cabin altitude must not be more than 4,570 m (15,000 ft) (CS-25 Section 841(a)).
- For passenger comfort, the cabin rate of climb should not be more than 2.5 m/s (500 ft/min) and the cabin rate of descent should not be more than 1.5 m/s (300 ft/min) (ARP 1270).
- The flow rate of air for cabin pressurization shall be enough to account for cabin leakage (allowing for an in-service increase of 10–15%) and cabin repressurization with 1.5 m/s (300 ft/min) (ARP 85).

Heating Systems

The simplest type of heating system, often employed in light aircraft, consist of a *heater muff* around the engine exhaust, an air scoop to draw ram air into the heater muff, ducting to carry the heated air into the cabin, and a valve to control the flow of heated air. Alternatively to the heater muff, a portion of the exhaust gases could also be fed to a *heat exchanger* to heat the ram air or the recirculated air from the cabin.

In larger aircraft *combustion heaters* are often employed. The heater burns fuel in a combustion chamber, and airflow around the chamber is heated and carried through ducts into the cabin.

Turbine engine–powered aircraft with a nonpressurized cabin normally make use of hot pressurized air tapped from the turbine engine compressor. This air is call *bleed air*. Temperature control is achieved by mixing the bleed air with ambient or recirculated air before it enters the cabin.

A pressurized aircraft cabin is usually heated by regulating the temperature of the air used to pressurize the cabin. This again is combined with an effort to cool the cabin. The combined process will be addressed in the following subsections.

Cooling Systems

There are several *heat sources* that cause a need for cooling. *External* heat sources include heat transfer through cabin walls and heat received through solar radiation. *Internal* heat sources include passengers and crew, heat generated by electronic, electric, and mechanical equipment.

Cooling systems require *energy* for their operation. This energy may come from ram air, engine bleed air, an engine-driven compressor, or the auxiliary power unit.

Cooling may apply different *heat sinks* to get rid of the heat: ram air, engine fan air, cabin exhaust air, fuel, or expendable cooling media (water or liquid hydrogen). Note that any ambient air taken aboard is at total temperature.[10]

The *cooling air* (ram air) may be moved by a fan driven by an electric or hydraulic motor, the air cycle machine, or an ejector pump.

The above means may be combined in systems applying two basic cooling principles. These systems are known as:

- The vapor cycle system, in which the heat of vaporization is lost by evaporating a liquid refrigerant.

---

[9]Pressurized cabin: "An airplane cabin that is constructed, sealed, and equipped with an auxiliary system to maintain a pressure within the cabin greater than that of the surrounding atmosphere" (SAE 1998).

[10]Total temperature = stagnation temperature: "The temperature which would arise if the fluid were brought to rest adiabatically" (AGARD 1980).

- The air cycle system, which is based on the reduction of heat by the transformation of heat energy into work.

Combination of both principles is possible.

The *vapor cycle system* (Figure 2.3) is what is used in refrigerators. The cooling process is best explained starting at the *compressor,* where the refrigerant (a special fluid) is in gaseous form. The compressor increases pressure and temperature of the refrigerant and pushes it through the entire system. A heat exchanger called a *condenser* extracts heat from the compressed refrigerant and carries the heat overboard. The refrigerant cools down a little and changes into liquid form. Still under pressure, the refrigerant goes past the *expansion valve,* where it is sprayed into little droplets. Behind the expansion valve, pressure is low. With reduced pressure the temperature is also considerably reduced. The *evaporator* is the second heat exchanger in the system. The refrigerant, in the form of cold droplets, cools the air destined for the cabin that goes past the evaporator. By taking up the energy from the passing air in the evaporator, the refrigerant changes to gaseous form again. It now enters the compressor, where the cycle starts anew. Example: Dassault Falcon 10.

The vapor cycle is a closed cycle that works with a phase change from gas to liquid and vice versa. The latent heat[11] involved in the phase change makes the vapor cycle very efficient.

If we substitute air for the refrigerant and a turbine for the expansion valve, we basically get a *closed air cycle system.* In aircraft air conditioning, however, the cold air leaving the turbine is used directly as cabin air, forming an *open air cycle system.* Various air cycle systems have been conceived. The discussion here is limited to three open air cycle systems: the *basic air cycle systems,* the *bootstrap system,* and the *three-wheel system.*

In the *open basic air cycle system* (Figure 2.4), bleed air is cooled in a *heat exchanger* with ram air. The bleed air drives a *turbine,* using the pressure differential between bleed and cabin pressure. The bleed air is cooled during the expansion in the turbine. The work extracted from the turbine drives a *fan* that augments the airflow through the heat exchanger. In the cold air behind the turbine, water is condensed in form of minute drops (fog). A *low-pressure water separator* extracts this water.

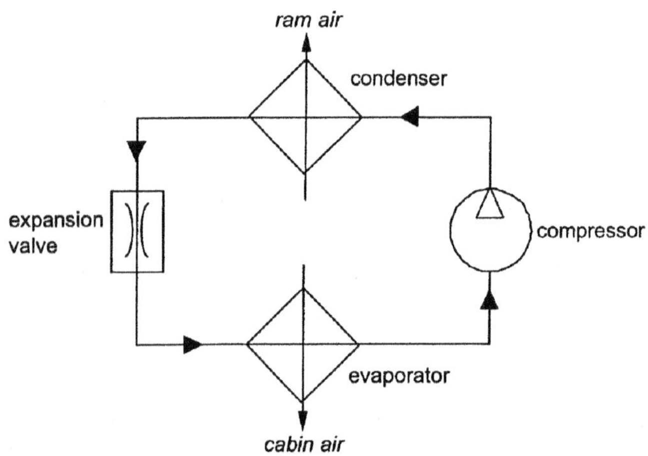

FIGURE 2.3 Vapor cycle system.

[11]Latent heat: "The unit quantity of heat required for isothermal change in a state of a unit mass of matter" (SAE 1998).

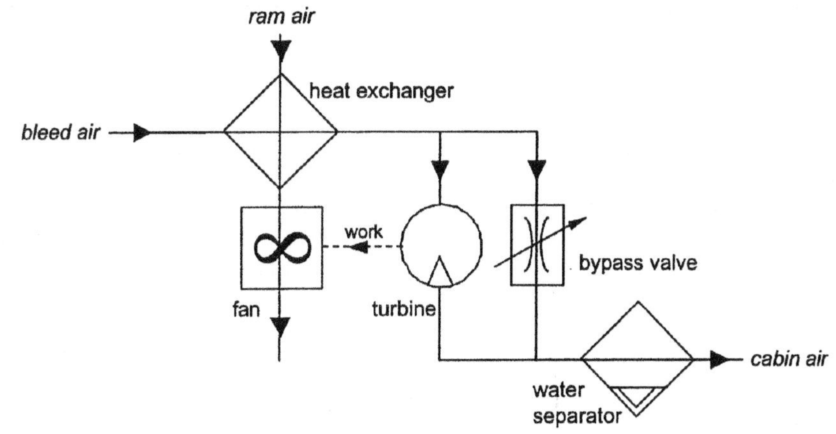

Figure 2.4 Open basic air cycle system.

A *bypass valve* is used for temperature regulation and to prevent ice buildup in the water separator. Example: Lockheed C-130.

The turbine can also be used to drive a compressor that further increases the pressure of the air supplied to the cooling turbine. A higher pressure ratio leads to a higher temperature drop across the turbine and hence an improved performance. An air cycle system with a turbine coupled to a compressor is called a bootstrap system.

The *open bootstrap air cycle system* (Figure 2.5) directs bleed air through a *primary heat exchanger.* The air is compressed and then passed through a secondary heat exchanger (or *main heat exchanger*). The air then enters the *turbine,* where it is expanded to cabin pressure. A *low-pressure water separator* reduces the water content. Heat energy is converted into shaft work and used to drive the bootstrap *compressor.* The primary and main heat exchangers are cooled by ram air. The *fan,* used to augment the airflow through the heat exchangers, may be driven by an electric motor. Bypass lines are integrated for temperature control. Example: Boeing 727.

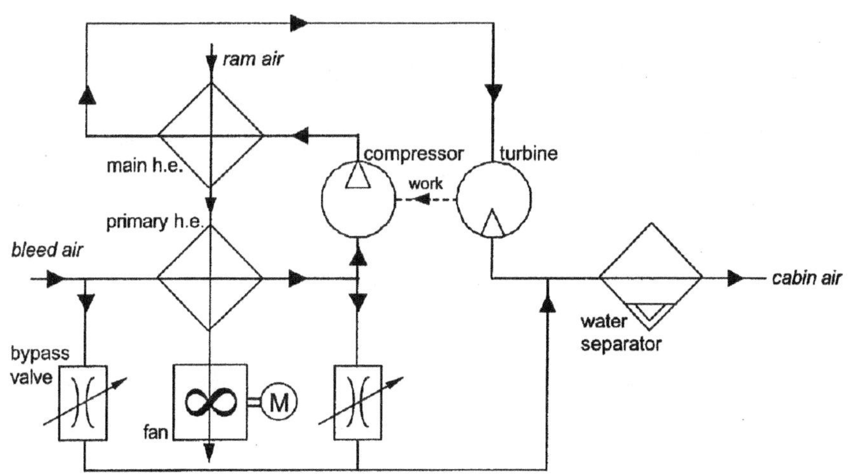

Figure 2.5 Open bootstrap air cycle system.

Two types of *water separators* exist. So far we have seen the application of a low-pressure water separator that is installed behind the turbine and limits cabin air to temperature above 0°C. In contrast, a high-pressure water separator is installed before the turbine. Separating the water before the turbine requires at least one more heat exchanger: a condenser or a condenser and a reheater. The advantage of the high-pressure water separator is that the air may be cooled down to temperatures of –50°C. This results in higher temperature differences at the heat exchangers and higher efficiency of the system.

More recent transport category aircraft use the *open three-wheel air cycle system* with a high-pressure water separator (examples: B757, B767, A320). The three-wheel system is a bootstrap system where the turbine drives not only the compressor but also the fan. Figure 2.6 shows this configuration.

Pressurization Systems

As we saw above, pressurization is necessary to fly at high altitudes (compare with Figure 2.8). The use of pressurization is found in aircraft ranging from light single-engine aircraft up to big turbine-powered transport aircraft. Although the basic controlling mechanisms for each of these types are the same, the *sources of pressure* and details of the system vary. Pressure generation and distribution are the responsibility of the pneumatic system and are discussed in Subsection 2.13 in more detail. Reciprocating engines can supply pressure from a supercharger, a turbocharger, or an engine-driven compressor. Turbine-powered aircraft usually use bleed air as a source for compressed air. Bleed air is air that is tapped from the compressor section of the turbine engine.

Heating and cooling with an open air cycle system provide conditioned air to the cabin that is used at the same time for pressurization. Heating, cooling, and pressurization all have to be integrated in such a way that an optimum overall system solution results.

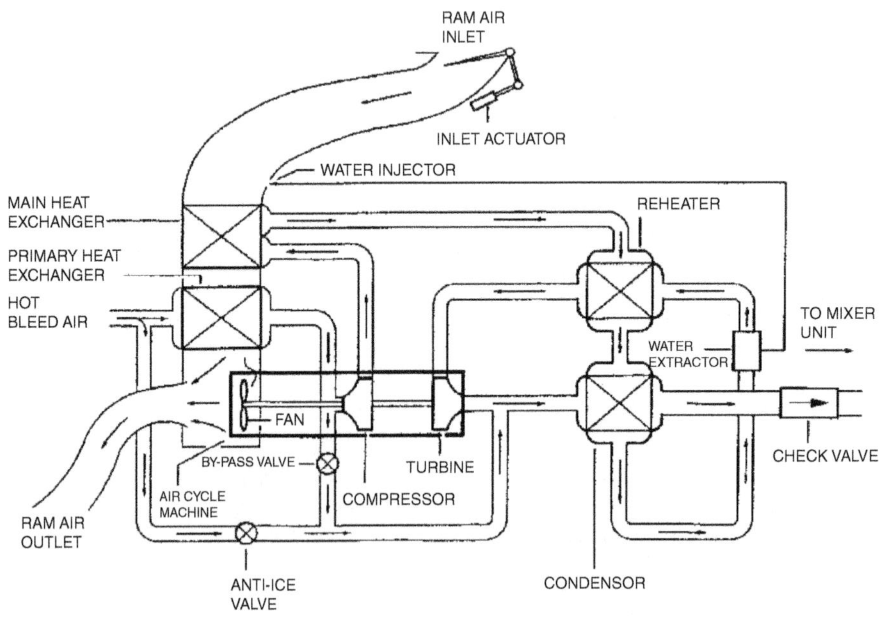

FIGURE 2.6 A321 air cooling in the pack.

The flow of air into the cabin is approximately constant. *Pressure control* is hence achieved by varying the amount of flow out of the cabin. This is done with a regulated outflow valve. The outflow valve may be operated directly, by pneumatic pressure, or by electric motors.

An aircraft must have enough *structural strength* to withstand the stresses caused by a pressurized cabin. The limiting factor in how high an aircraft can operate is the maximum allowed cabin differential pressure, i.e., the difference between the cabin pressure and the pressure at maximum altitude for which certification is sought: $\Delta p = p_{\text{cabin}} - p_{\text{max,alt}}$. Aircraft are not intended to fly with a cabin pressure below ambient pressure.

Safety valves are used to safeguard against unauthorized positive or negative differential pressure. A pressure relief valve opens automatically if the cabin differential pressure gets above permitted limits. An automatic negative pressure relief valve opens automatically if the negative cabin differential pressure gets above permitted limits. A dump valve is used to release remaining cabin differential pressure when the aircraft lands. Note that one pressurization control valve may serve more than one function in a specific aircraft design.

Example: Airbus A321

The Airbus A321 has two air conditioning packs which are open three-wheel air cycle systems. Figure 2.6 shows an air conditioning pack with the air cycle machine, the heat exchangers, and a high-pressure water separator.

The cabin temperature can be adjusted by computer individually in three different cabin zones (Figure 2.7). The air conditioning packs (Figure 2.6) deliver air at a temperature to satisfy the zone with the lowest temperature demand. Air from the packs is delivered to the mixing unit. Also, recirculated air from the cabin enters the mixing unit through filters and cabin fans. The recirculated air amounts to 40% of the total air supplied to the cabin. Recirculated air restores some

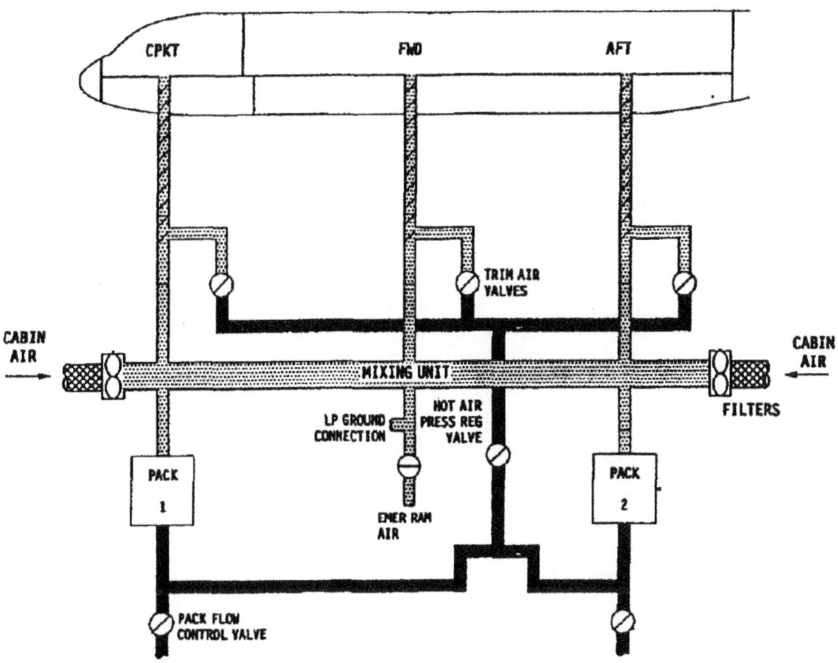

Figure 2.7 A321 air conditioning.

humidity into the cabin. Trim air valves mix hot bleed air with the air from the mixing unit to attain the individually requested zone temperatures.

The pressurization control system includes two cabin pressure controllers. Operation may be fully automatic, semiautomatic, or manual. The outflow valve is equipped with three electrical motors. Two safety valves avoid excessive positive (593 hPa = 8.6 psi) or negative (–17 hPa = –0.25 psi) differential pressure (compare with Figure 2.8).

Figure 2.9 shows the air distribution in the cabin.

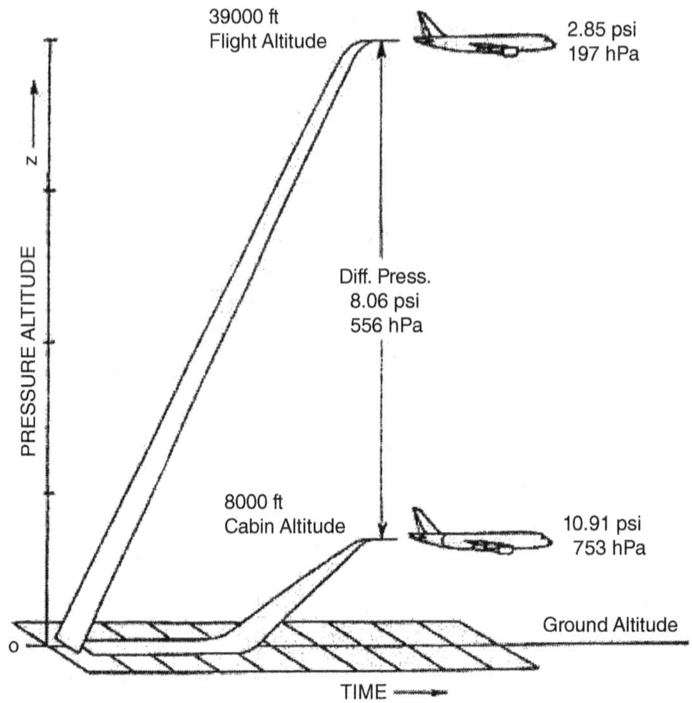

Figure 2.8 A321 pressure control.

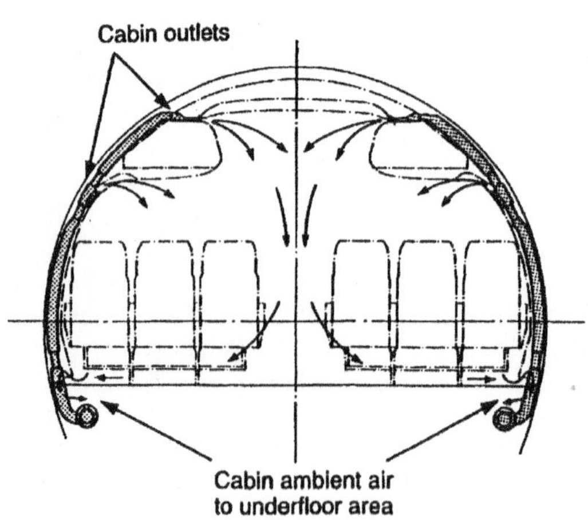

Figure 2.9 A321 cabin air distribution.

2.3 Electrical Power (ATA 24)

Electrical power as defined by ATA 100:

> Those electrical units and components which generate, control and supply AC and/or DC electrical power for other systems, including generators and relays, inverters, batteries, etc., through the secondary busses. Also includes common electrical items such as wiring, switches, connectors, etc.

System Classification

Electrical power includes (ATA 100):

- Power generation:
  - Generator drive systems: constant speed drives (CSD)
  - Alternating current (ac) generation
  - Direct current (dc) generation
  - External power
- Power distribution:
  - Alternating current (ac) electrical load distribution
  - Direct current (dc) electrical load distribution

Power Generation

Power is generated with different electrical components. Light aircraft use 14 V or 28 V direct current (dc) generators or alternators. Large aircraft employ generators that produce an alternating current (ac) of 115 V at 400 Hz. Compared to a 28-V dc system, a higher-voltage ac system will develop several times as much power for the same weight and hence provide a great advantage where heavy electrical loads are imposed.

Aircraft *dc generators* have for the most part been replaced by dc alternators on modern aircraft. Although generators and alternators are technically different, the terms alternator and generator are used interchangeably.

A *starter-generator* is a combination of a dc generator and a dc motor in one housing. Starter-generators are typically employed on small turboprop and turbine-powered aircraft.

There are two major types of *alternators* currently used on aircraft: the dc alternator and the ac alternator. Dc alternators are most often found on light aircraft where the electric load is relatively small. Ac alternators are found on large commercial airliners and many military aircraft.

Both ac and dc alternators for aircraft show a construction with a rotating field (supplied with current from the outside via slip rings) and a stationary armature. The aircraft alternator is a three-phase unit having three separate windings 120° apart.

Light airplanes use an alternator with a three-phase full-wave rectifier to produce dc power. The rectifier is built into the alternator, so that dc current leaves the alternator with a nominal voltage of either 14 V for a 12-V battery system or with 28 V for a 24-V battery system.

Transport category aircraft use three-phase ac alternators with Y-connected stator windings. (Note: High output ac alternators are mostly called ac generators. If they are of a design without slip rings, they are called *brushless generators*.) The output frequency depends on the drive speed of the generator. The required constant frequency of 400 Hz requires the use of a constant speed drive (CSD). The integrated drive generator (IDG) contains both, the CSD and the generator in one unit. Details of this state-of-the-art system are explained using the Airbus example below.

Advantages of ac high-voltage systems include:

- Weight savings
- Voltage transformation possibilities
- Low current, low power losses in the wiring

Electrical power generation systems on large aircraft show a range of typical components:

- A *generator control unit* (GCU) is a solid-state device that carries out voltage regulation, current limiting, and frequency control.
- An *inverter* is a device for converting direct current into alternating current at a demanded frequency (400 Hz). A *static inverter* achieves this with standard electric and electronic components.
- A *transformer rectifier* (TR) unit is a device for converting alternating current into direct current.
- A *variable-speed constant-frequency* (VSCF) system employs a generator driven directly from the engine without a constant-speed drive (CSD). The generator is driven at variable engine speeds, thus producing a variable-frequency output. A generator converter control unit converts the variable frequency into a constant frequency of 400 Hz. A VSCF system is found on the Boeing 737.

Power Distribution

The design of the power distribution system depends on

1. The size of the aircraft and hence upon its system complexity
2. On the type of primary power generation applied (ac or dc)

A simple power distribution system consists of a *bus bar* or *bus*. The bus is a conductor designed to carry the entire electrical load and distribute that load to the individual power users. Each electric power user is connected to the bus through a circuit breaker. Simple distribution systems like this are found on small single-engine aircraft.

More complex power distribution systems consist of bus bars, bus tie breakers, and various solid-state controllers such as *generator control units* (GCUs).

Electrical power distribution systems on large aircraft show a range of typical components:

- *Bus tie contactors* (BTCs) (also known as bus tie breaker) are electric solenoids used to connect two bus bars.
- *Generator line contactors* (GLCs) (also known as generator breakers) are similar to BTCs but connect the generators to the buses.
- *Bus power control units* (BPCUs) are supplied with information from all parts of the distribution system. Taking this information into account, BPCUs will ensure the appropriate distribution system configuration. In some architectures, the GCUs include the BPCU functions. The BPCUs enable *reconfiguration* of the power distribution between individual busses. For example, if a generator fails or a bus shorts to ground, the appropriate BTCs and GLCs must be set to the correct position. In the event of a system overload, the controller must reduce the electrical load to an acceptable level. This is called *load shedding*. The aircraft's galley power is usually the first nonessential load to be disconnected.

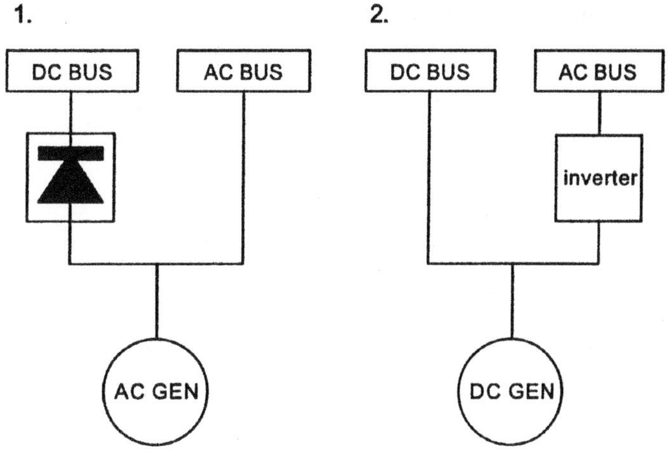

FIGURE 2.10 Distribution systems with 1. primary ac power generation; 2. primary dc power generation.

Figure 2.10 shows the two principal distribution systems with

1. Primary ac generation and dc generation through transformer-rectifiers
2. Primary dc generation and ac generation through inverters

Three different power distribution systems exist for large aircraft, all of which apply primary ac generation:

1. The split-bus system
2. The parallel system
3. The split parallel system

The *split-bus system* (Figure 2.11) contains two completely isolated power-generating systems. Each system contains its own ac generator. The generator 1 (GEN 1) and generator 2 (GEN 2) power

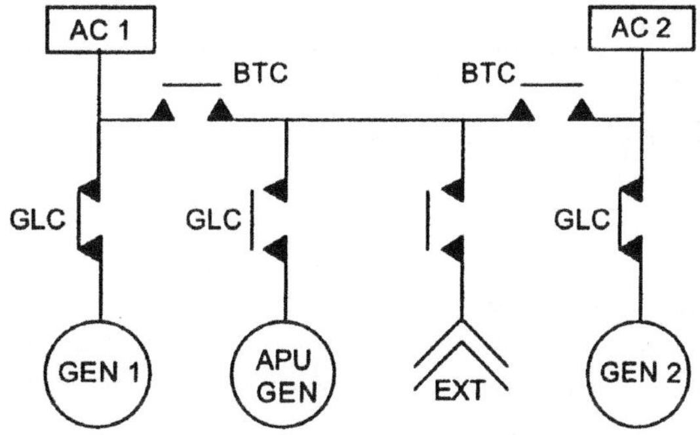

FIGURE 2.11 General layout of a split-bus system.

their respective loads independently of other system operations. In the event of a generator failure, the remaining operating generator is connected to both buses AC 1 and AC 2, or the APU generator (APU GEN) may be employed to carry the electrical load of the inoperative generator. The major advantage of a split-bus system is that the generators operate independently, so that generator output frequencies and phase relationships need not be so closely regulated. A split-bus system is used on the Airbus A321 and most other modern twin-engine transport category aircraft. The A321's electrical system diagram is shown below in more detail.

In a *parallel system* (Figure 2.12), all ac generators are connected to one tie bus. This type of system maintains equal load sharing for three or more ac generators. Since the generators are connected in parallel to a common bus, all generator voltages, frequencies, and their phase sequence must be within very strict limits to ensure proper system operation. If one generator fails, the generator is isolated from its load bus. Nevertheless, that load bus still continues to receive power while connected to the tie bus. A parallel system is used on, for example, the Boeing 727.

A *split parallel system* (Figure 2.13) allows for flexibility in load distribution and yet maintains isolation between systems when needed. The ac buses are paralleled through the bus tie breakers (BTB) and the split system breaker (SSB). When the SSB is open, the right system operates independently of the left. With this system any generator can supply power to any load bus (AC 1, AC 2, . . .), and any combination of the generators (GEN 1, GEN 2, . . .) can operate in parallel. A split parallel system is used on the Boeing 747-400.

Let's look at the *dc distribution systems* on aircraft with primary ac power generation. Transformer rectifiers (TRs) powered by an ac bus, feed their main dc bus bars. In the event of a complete generator system failure, the aircraft's batteries would supply the essential dc power. An inverter would also be powered from the batteries in an emergency situation to operate all essential ac loads.

The aircraft electrical system is designed with a *power distribution hierarchy.* The system is designed so that the most critical components are the least likely to fail. The generators feed their respective bus AC 1, AC 2, . . . The least critical ac loads are powered by these busses. The critical ac loads are powered by the essential ac bus (AC ESS). The same is true for the dc busses: the least

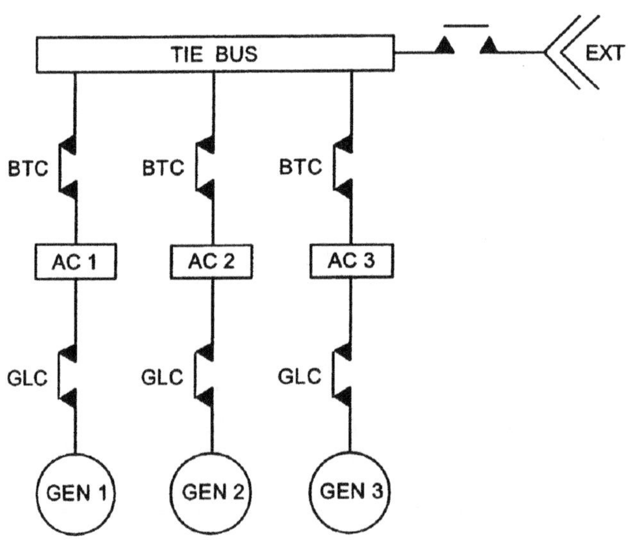

FIGURE 2.12 General layout of a parallel system.

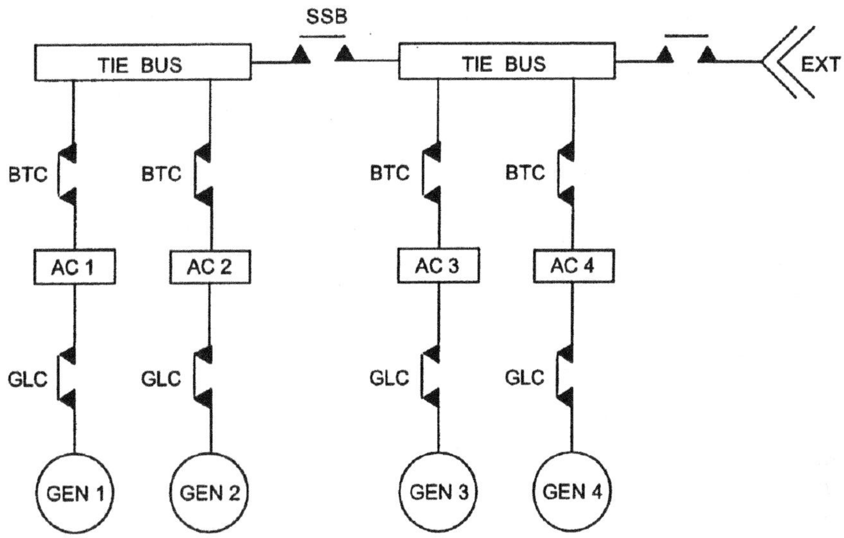

FIGURE 2.13 General layout of a split parallel system.

critical dc loads are powered by the DC 1, DC 2, . . . buses, which are fed by their respective transformer rectifier (TR 1, TR 2, . . .). The next-most critical systems are powered by the essential dc bus (DC ESS), which can be powered by any transformer rectifier. The most critical loads are powered by the battery bus (BAT BUS).

Example: Airbus A321

In the A321, primary ac power generation is applied, where ac is converted to dc by means of transformer rectifiers (TRs). The distribution system is a split-bus system and consists of two separated distribution networks. Normally, one main generator supplies each network. The two distribution networks may be connected when the aircraft is on external power, APU power, or if one main generator fails. Under no circumstances may two generators be connected.

A321 *power generation* encompasses primary ac power generation in flight and on the ground, dc power generation, and ac power generation from dc. The location of related components in the aircraft is shown in Figure 2.14.

In flight, two engine-driven generators (GEN 1 and GEN 2), also known as integrated drive generators (IDGs), supply the aircraft electrical power system. A third APU-driven generator (APU GEN) can replace one engine-driven generator. In the event of a major failure, a unit consisting of a constant-speed hydraulic motor coupled to a generator (constant-speed motor/generator, CSM/G) is able to supply the most essential parts of the electrical systems. The CSM/G is powered by the ram air turbine (RAT) via the Blue hydraulic system.

On the ground, an external electrical ground power unit (GPU) can supply the aircraft. Alternatively, the APU generator can serve as an independent source for electrical power supply on the ground.

All the power sources named above supply the distribution network with ac power. Dc power is supplied by transformer rectifiers (TR). Two batteries are used as a dc emergency power source and for APU start in flight and on the ground.

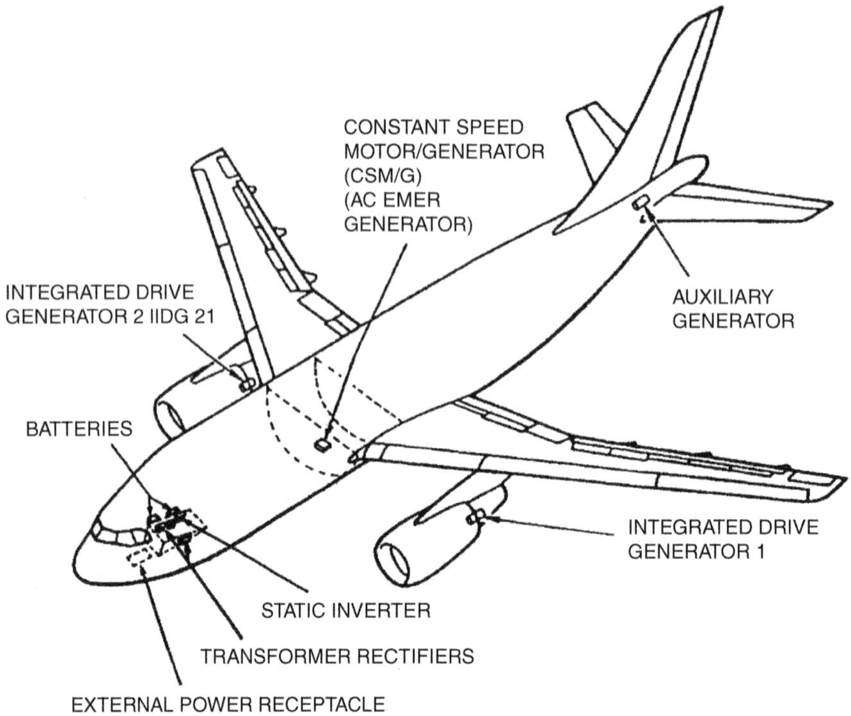

FIGURE 2.14 A321 electrical power sources and their location in the aircraft.

Essential ac power can be obtained in an emergency situation from the batteries through a static inverter.

A321 *power distribution* encompasses (Figure 2.15):

- The distribution network 1, which consists of AC BUS 1, AC ESS BUS, AC ESS SHED. The AC ESS SHED may be shed due to a lack of power in an emergency.
- The distribution network 2, which consists of AC BUS 2.
- The transformer rectifier 1 (TR 1), which is powered from the AC BUS 1 supplies through its contactor: DC BUS 1, DC BAT BUS, DC ESS BUS, DC ESS SHED. The DC ESS SHED may be shed due to a lack of power in an emergency.
- Two batteries, which are associated with the DC BAT BUS.
- The transformer rectifier 2 (TR 2), which is powered from the AC BUS 2 supplies through its contactor the DC BUS 2.
- A third essential transformer rectifier (ESS TR), which can be powered from the AC BUS 1, or the emergency generator (EMER GEN) may supply the DC ESS BUS and the DC ESS SHED through its contactor only in certain failure cases.

In failure cases, various possibilities for reconfiguration exist.

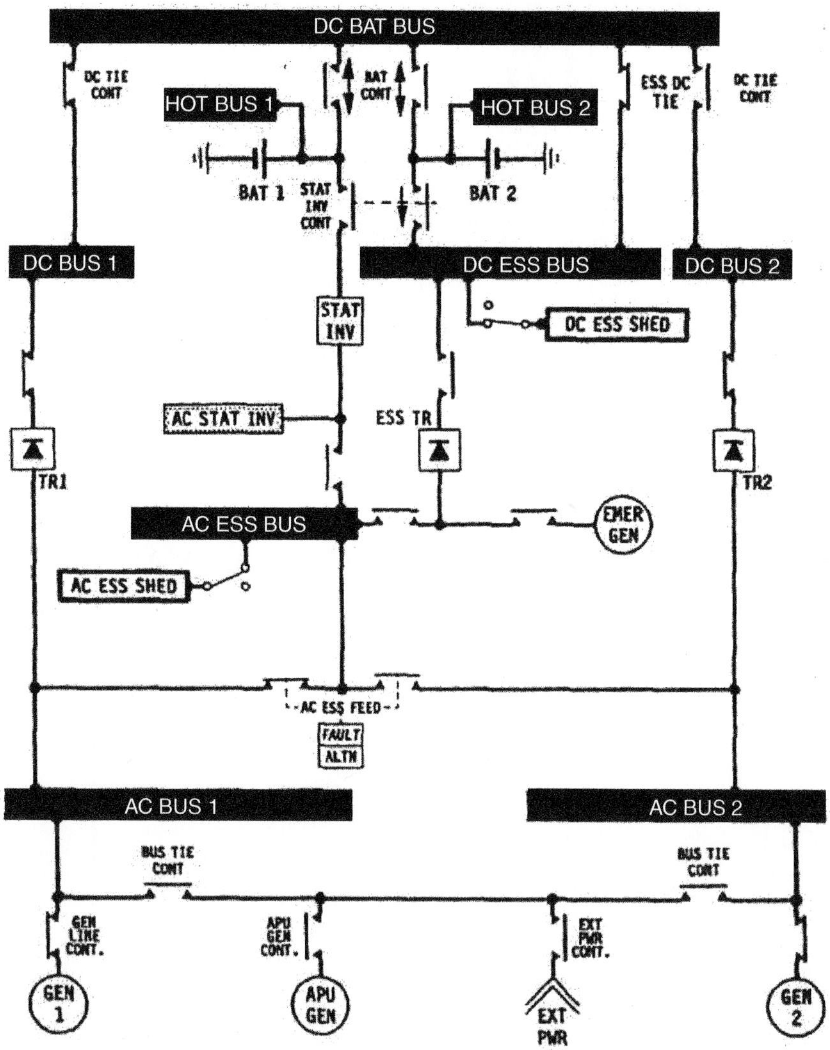

FIGURE 2.15 A321 electrical system diagram.

Each engine's high-pressure stage drives its associated *integrated drive generator* (IDG) through the accessory gearbox (Figure 2.16). The drive speed varies according to the engine rating. The IDG provides a 115/200 V, three-phase, 400 Hz AC supply. The IDG consists of two parts: the constant-speed drive (CSD) and the generator. The hydromechanical CSD drives the ac four-pole generator at a nominal speed of constant 12,000 rpm.

The *constant-speed drive* (CSD) consists of a mechanical differential gear that transmits power to the generator of the IDG. The output speed of the differential gear is modified by two mechanically coupled twin hydraulic subassemblies: a pump and a motor. Each subassembly includes a hydraulic swashplate: the pump is equipped with a variable-angle swashplate, and the motor is equipped

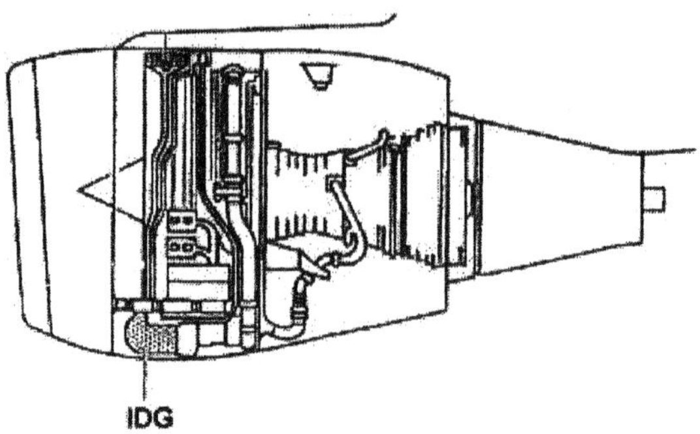

Figure 2.16 A321: location of the integrated drive generator (IDG).

with a fixed swashplate. A governor controls the CSD output speed by the swashplate angle of the pump (Figure 2.17).

The *generator* is a three-stage assembly that includes three machines connected in cascade. The first machine is a 12-pole permanent magnet generator (PMG). The second machine is a 10-pole stator and receives its field excitation from the first machine via the voltage regulator in the generator control unit (GCU). Its dc output feeds the rotating field of the third machine (the main alternator). The main alternator has a three-phase star-connected stator winding. The three phases and star point are taken to the generator output terminal block.

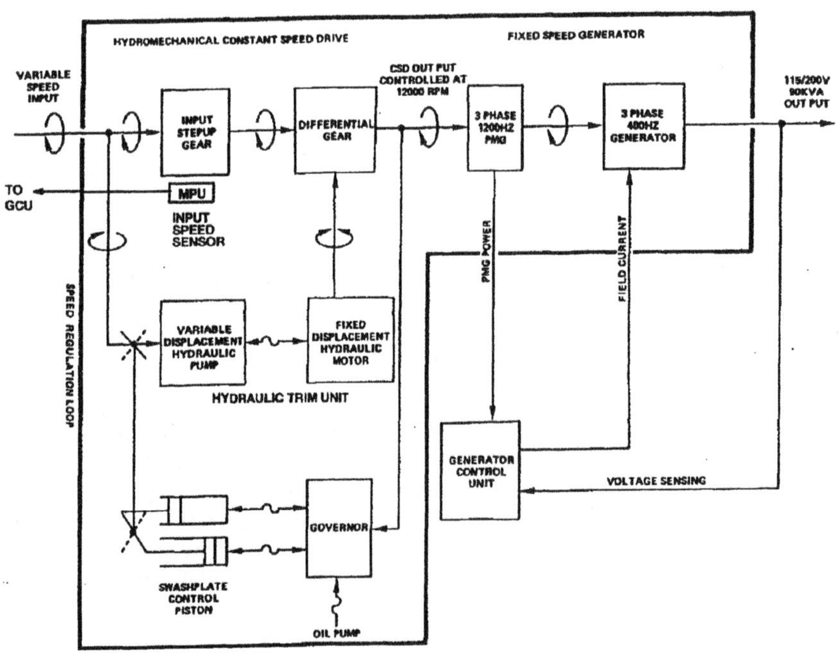

Figure 2.17 A321 integrated drive generator (IDG): speed conversion and power generation.

2.4 Equipment/Furnishings (ATA 25)

Equipment and furnishings as defined by ATA 2200:

> Those removable items of equipment and furnishings contained in the flight and passenger compartments. Includes emergency, galley and lavatory equipment. Does not include structures or equipment assigned specifically to other [systems].

Elements of Equipment

Equipment and furnishings include items in several parts of the aircraft. Examples of such equipment include:

- In the flight compartment: flight crew seats, tables, wardrobes, electronic equipment racks, and stowage facilities for manuals and other equipment.
- In the passenger compartment: seats, overhead storage compartments, wall coverings, carpets, wardrobes, movable partitions.
- In buffets and galleys: cabinets, ovens, refrigerators, coffee maker, electrical outlets and wiring, trolleys, garbage containers.
- In the lavatories: mirrors, seats, cabinets, dispensing equipment, electrical outlets and wiring (the wash basin and the closets are part of the water/waste system).
- In the cargo compartment[12]: equipment used to load and unload the aircraft; includes restrains and latches, rollers, and drive systems.
- In all parts of the aircraft, thermal insulation[13] minimizes the losses of heat from the fuselage, stops the formation of condensation, and reduces the noise level in the fuselage. Thermal insulation is dimensioned in conjunction with the design of the air conditioning system.

Some aircraft, especially very large commercial transports, also offer space for additional equipment in the *under floor area.* The space can be used for crew rest facilities, galleys, a bar, or an exercise room. The need might arise to incorporate an elevator (lift) in multideck aircraft in order to move goods or passenger.

Emergency equipment includes items for use in emergency procedures, such as evacuation equipment, life rafts, jackets, crash ax, flashlights, megaphone, protective gloves, emergency locator transmitters, underwater locator devices, first aid kits, and supplementary medical equipment. Fire extinguishers and oxygen equipment are part of their respective systems. Evacuation equipment facilitates passenger and crew evacuation. These procedures are explained below.

Cabin Design

The cabin is the place where the paying customer has to be satisfied. Much attention is given to its design, starting during aircraft design, where an optimum cabin cross-section has to be found. Designers have to find ways to create an aesthetically pleasing impression and a suggestion of spaciousness within the always limited dimensions of an aircraft (Figure 2.18). These design activities have an influence on the shape of ceiling panels, sidewall panels, stowage compartment doors, and passenger service units (PSUs) located underneath the stowage compartment. Cabin lighting design is also part of this effort. The airlines would like to see their corporate design reflected not only outside but inside the aircraft. They may choose their own material, pattern, and texture for

[12]Under the new ATA 2200, allocated to "Cargo and Accessory Compartment (ATA 50)."

[13]Under the new ATA 2200, allocated to "Cargo and Accessory Compartment—Insulation (ATA 50-60)."

FIGURE 2.18 Boeing 717: the result of a thorough cabin design (Granzeier 2001).

panel coverings, dividers, curtains, and seats and will select a suitable carpet. All cabin materials have to fulfill requirements related to fire, wear, and cleaning.

Passenger Seats

Passenger seats are probably the most important single item of equipment in the cabin. They should provide comfortable seating for many hours during normal flights and the best protection during a crash. Elements of a seat are shown in Figure 2.19. Not visible in the figure are the literature pocket and the folding table on the back of the seat. Seats are installed on seat tracks in the cabin floor structure. This allows flexibility in spacing the seats.

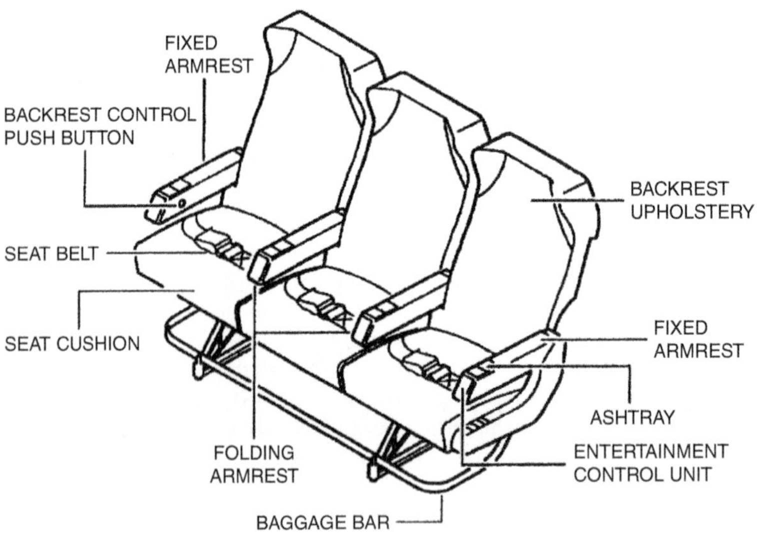

FIGURE 2.19 Economy class passenger seats (A321).

Seat pitch is a comfort measure for seat spacing. It is the distance between corresponding points on two seats installed one in front of the other. The seat pitch is internationally given in inches. Seats in first, business, and economy class feature different levels of comfort, and the seat pitch also varies among these classes. Typical values today are:

- First class: 62 in. (1.57 m)
- Business class: 40 in. (1.02 m)
- Economy class: 32 in. (0.81 m)
- High density: 30 in. (0.76 m)

These numbers are not fixed, but change with product policy of the airlines. During the last decades seat pitch has increased in first class, but decreased in economy class in a fight for low fares.

Seats are bought by the airline from specialized seat manufacturers as buyer-furnished equipment (BFE) and are then installed by the aircraft manufacturer in the new aircraft.

Emergency Evacuation

Rapid evacuation of passengers and cabin crew has to be possible in case of a crash landing. For airplanes with 44 passengers or more it must be shown that passengers and cabin crew can be evacuated to the ground within 90 seconds, with up to 50% of the emergency exits blocked (CS-25, Section 803; AC 25.803). In an emergency, passengers usually leave the aircraft through *emergency exits* (these can also be the normal passenger doors) via inflatable escape slides (Figure 2.20).

Evacuation of flight crew from commercial aircraft designed to be achieved through passenger emergency exits, through a *hatch*, or by using an escape rope to slide down from the flight deck through the opening side *windows*.

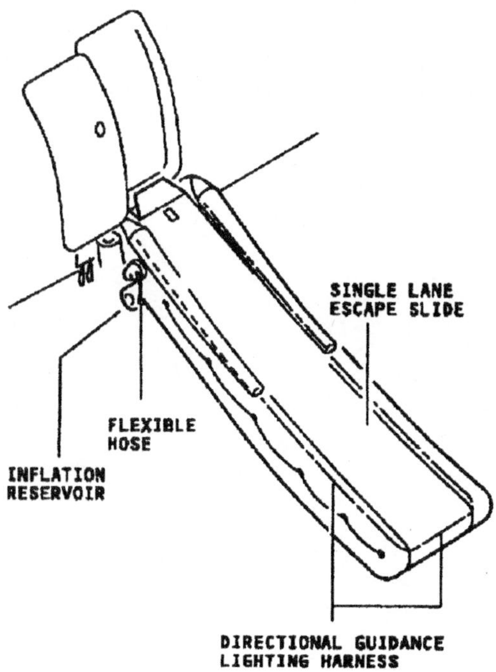

Figure 2.20 Escape slide (Airbus A321).

Evacuation of crew from military combat aircraft is usually achieved with *ejector seats* that allow the crew to abandon their aircraft at all flight conditions, ranging from high speed, high altitude to zero speed and zero height. The ejector seat is mounted in the aircraft on a slide rail and is propelled out of the aircraft by a rocket motor. After a predetermined time, the seat detaches from the person, who is brought to the ground by parachute. In some multicrew combat aircraft the crew are evacuated in an *escape module* that is jettisoned and parachuted to the ground.

Example: Airbus A321

Equipment and furnishings give comfort and safety to passengers in the cabin and to the crew in the cockpit. Equipment is also used for handling of cargo in the cargo compartments.

The cockpit is equipped with adjustable seats for two crew members (Figure 2.21). The A321 has a fly-by-wire flight control system steered with a side stick. The side stick armrest located on the outboard side of the seat can be adjusted in height and tilt angle so that the pilots can rest their respective arm in an optimum position with respect to the side stick controller. A third occupant seat and a folding seat for a fourth occupant are also available.

The cabin also includes the galleys (Figure 2.22) and lavatories (Figure 2.23), in addition to the passenger seats (Figure 2.19).

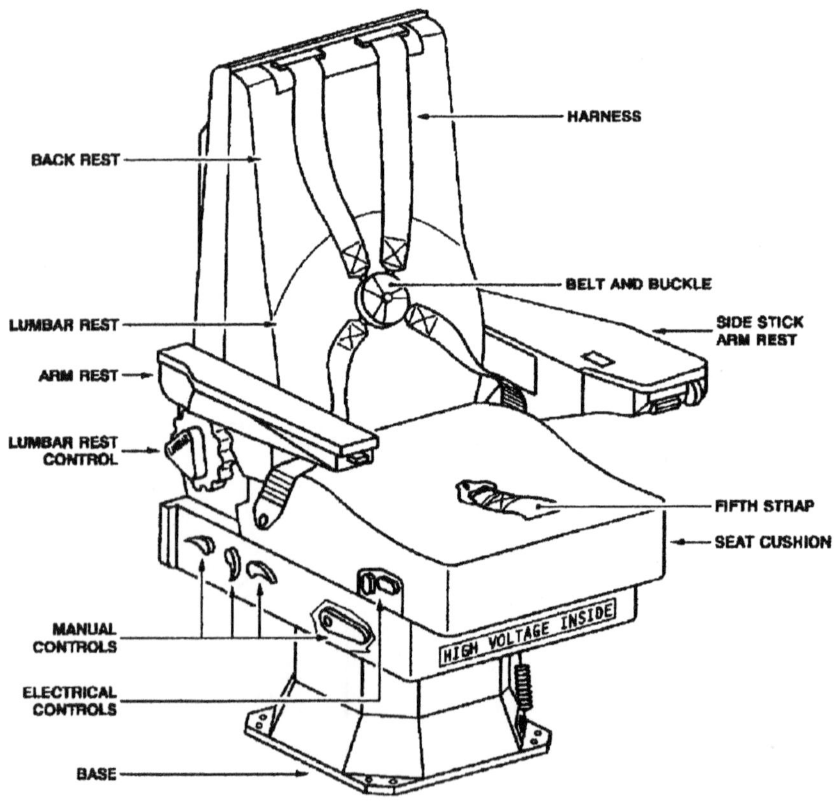

FIGURE 2.21 A321 captain/first officer seat.

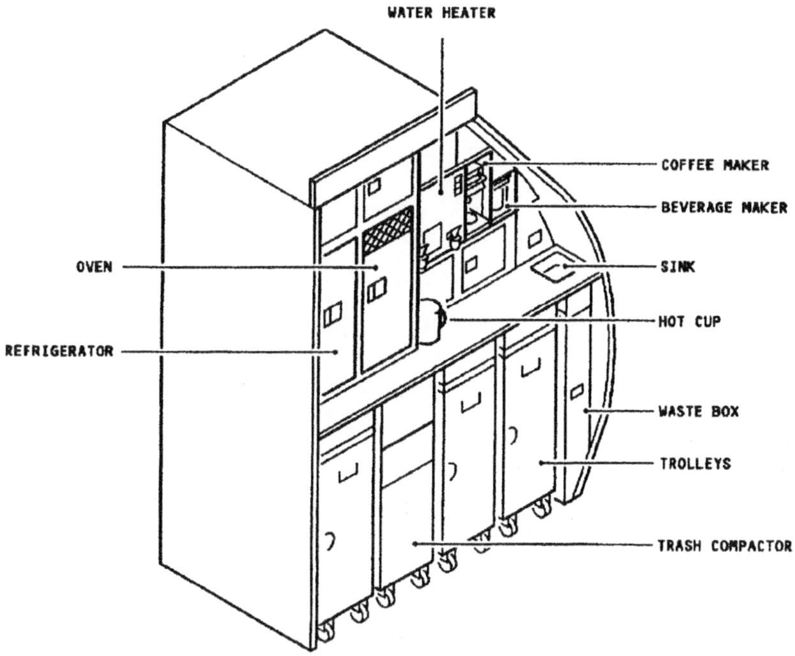

Figure 2.22 Galley equipment (A321).

2.5 Fire Protection (ATA 26)

Fire protection equipment as defined by ATA 100:

> Those fixed and portable units and components which detect and indicate fire or smoke and store and distribute fire extinguishing agent to all protected areas of the aircraft; including bottles, valves, tubing, etc.

Detection Fundamentals

Fire detection includes that part of the fire protection system which is used to sense and indicate the presence of overheat, smoke, or fire (ATA 100).

There are various ways in detecting a fire, including:

- Direct observation by cockpit and cabin crew (optical indication, sensing of heat or smell)
- Overheat detector
- Smoke detector
- Rate-of-temperature-rise detector
- Inspection by video camera
- Fiberoptic detectors
- Thermal imaging devices
- Radiation sensing devices
- Ultraviolet aircraft fire detection system
- Detection of combustion gases like CO or CO_2

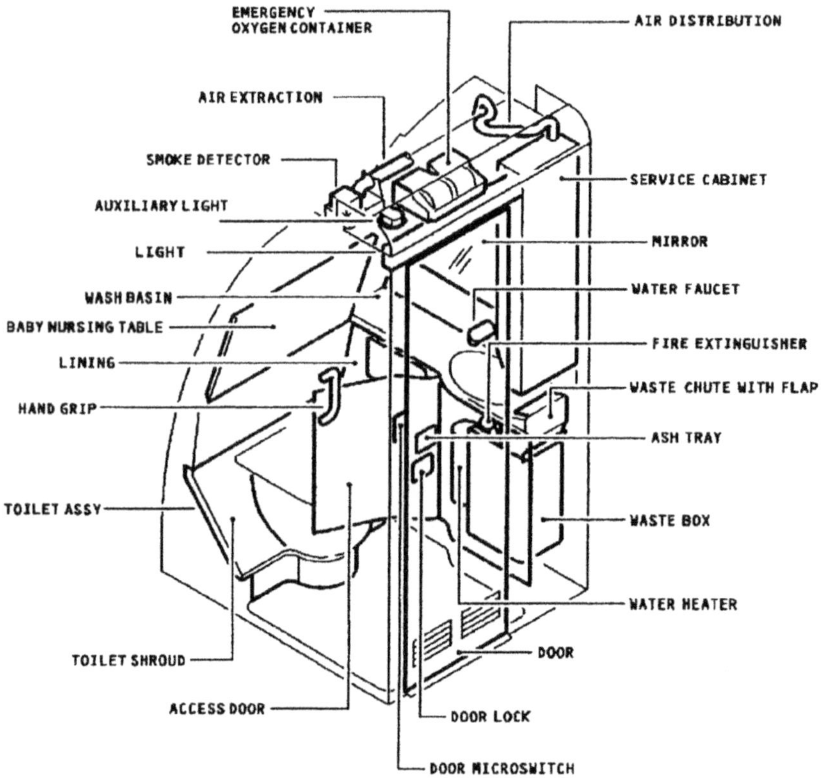

FIGURE 2.23 Lavatory equipment (A321).

Designated fire zones must be equipped with fire detection and extinguishing equipment. Designated fire zones are (CS-25, FAR Part 25):

- Power plant compartment (Section 1181)
- Auxiliary power unit (APU) compartment (Section A1181)
- Combustion heater chamber (Section 859)

Fire detection and extinguishing equipment is required for cargo compartments according to the *cargo compartment classification* (Section 857, CS-25, FAR Part 25):

- Class A compartments are accessible in flight. A fire in the compartment would be easily discovered by a crew member while at his station.
- Class B compartments provide access in flight to enable a crew member to use a hand fire extinguisher. The compartments are equipped with a smoke or fire detector.
- Class C compartments are equipped with a smoke or fire detector and a built-in fire extinguishing system.
- Class D compartments are able to confine a fire completely without the safety of the aircraft being endangered.

```
fire detector
├──>overheat detector
│    ├──>point detector
│    │    ├──>thermal switch
│    │    │    ├──>spot detector
│    │    │    └──>bimetallic thermostat
│    │    └──>thermocouple
│    └──>continuous-loop detector
│         ├──>electric
│         │    ├──>averaging
│         │    └──>discrete
│         └──>pneumatic
├──>smoke detector
│    ├──>ionization type
│    └──>photoelectric type
└──>...
```

FIGURE 2.24 Roadmap to the most widely used detection devices.

Lavatories must be equipped with a smoke detector system, and lavatories must be equipped with a built-in fire extinguisher for each disposal receptacle for towels, paper, or waste located within the lavatory (Section 854, CS-25, FAR Part 25).

Other areas equipped with fire detectors may include the avionic compartment or the landing gear bay.

Fire detectors are generally either overheat detectors or smoke detectors. From the beginning until today, these and other fire-detection devices for aircraft have been developed by only a few U.S. companies: Walter Kidde, Fenwal, and Systron-Donner. Their component designs will be presented here (Hillman et al. 2001). The roadmap to the following discussion of the most widely used detection devices is presented in Figure 2.24.

Overheat Detection

In the 1940s, overheat detection coverage in the engine nacelle was done with thermal switches or thermocouples. Several of these switches were positioned in parallel at different places around the engine. A fire alarm was activated if one of the switches was triggered. However, it was recognized that these *point detectors* were very limited with regard to area of coverage. The placement of the point detector therefore became the most critical factor in how successful the detection system would be.

In the early and mid 1950s, *continuous-loop detectors* were introduced in the aircraft industry. This technology became the most popular detection approach for aircraft engines and has remained so to this day. Continuous-loop detectors are either electric or pneumatic continuous-loop detectors. Electric continuous-loop detectors are of either averaging type or the discrete type (Figure 2.25).

Some versions of electric continuous-loop detector depend on the amount of element heated to reach their alarm threshold level. These have been termed *averaging electrical continuous-loop detectors*. Their alarm threshold averages the temperature over its entire length. These detectors monitor either changing electrical resistance alone or resistance and capacitance in conjunction.

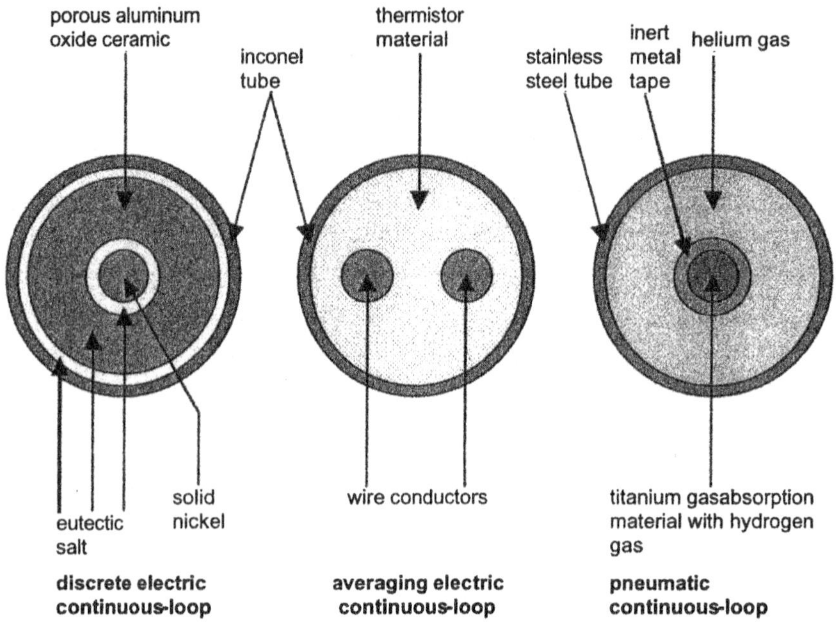

FIGURE 2.25 Cross-section of continuous-loop detectors.

Electrical continuous-sensing elements have one or two internal wire conductors embedded in a ceramic-like thermistor material contained in a metallic outer tube. As the surrounding temperature increases, the resistance between the inner conductor and the outer tube conductor decreases while the capacitance increases. When two internal wire conductors are embedded in the sensing element, the resistance change between these two wires is typically measured. When the resistance between the internal conductor and the external sensing element tube drops to some predetermined level (and/or the capacitance increases) corresponding to the desired alarm temperature, a monitoring control unit issues a hazard signal. When the hazard condition is eliminated and the temperature returns to normal, the resistance increases and the capacitance decreases, thereby canceling the alarm. Multiple trip resistance/capacitance settings can be used when multiple thresholds are pursued to indicate fire versus overheat.

Shortly after the first averaging-type detection systems, *discrete electrical continuous-loop detectors* were introduced (Figure 2.26). To achieve its alarm threshold, the discrete system utilizes sensing elements that are essentially independent of the length of element heated. These systems employ a sensing element which, as in the electrical averaging systems, has either one or two internal wire conductors embedded in a ceramic-like core material surrounded by a metallic outer tube. The ceramic core is impregnated with eutectic salt. The salt melts at its eutectic melt temperature, even when only a very short length of element is heated. When this occurs, the electrical resistance between the inner conductor and the outer tube very rapidly breaks down (also, the capacitance increases), and a monitoring control unit signals a fire or overheat, depending on which is appropriate for the intended application. The characteristics of the discrete type are paramount for reliable early warning of small, discrete overheat events, such as bleed air duct failures. By its nature, the discrete type cannot provide multiple alarm thresholds or any kind of analog temperature trend information.

Pneumatic-based continuous-loop detectors rely on increasing gas pressure to achieve the alarm threshold. These sensing elements have a hydrogen-charged core surrounded by helium gas,

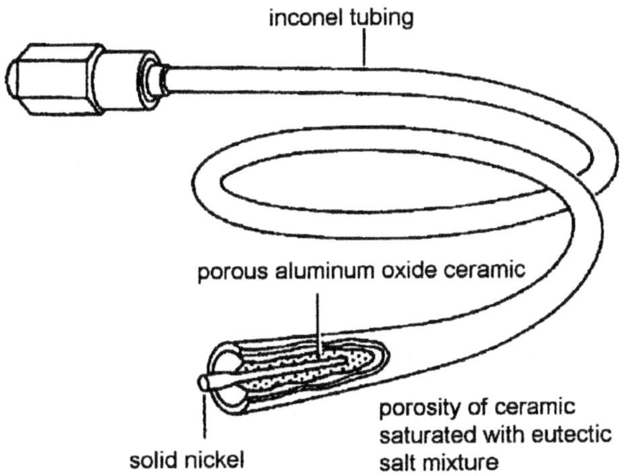

FIGURE 2.26 Discrete electric continuous-loop detector (A321, pneumatic system, leak detection).

contained in a metallic outer tube. As the surrounding temperature increases, the helium gas pressure increases, closing a pressure switch and thereby issuing an alarm. As the temperature returns to normal, the pressure decreases and the alarm is canceled. If a localized high-temperature event is present, the hydrogen core also outgasses its hydrogen gas, increasing the internal pressure and closing the pressure switch. As the sensing element cools, the hydrogen absorbs back into the core so that the internal pressure decreases, removing the alarm output. A leak in the detector can be discovered with an integrity switch opening due to a loss of pressure (Figure 2.27).

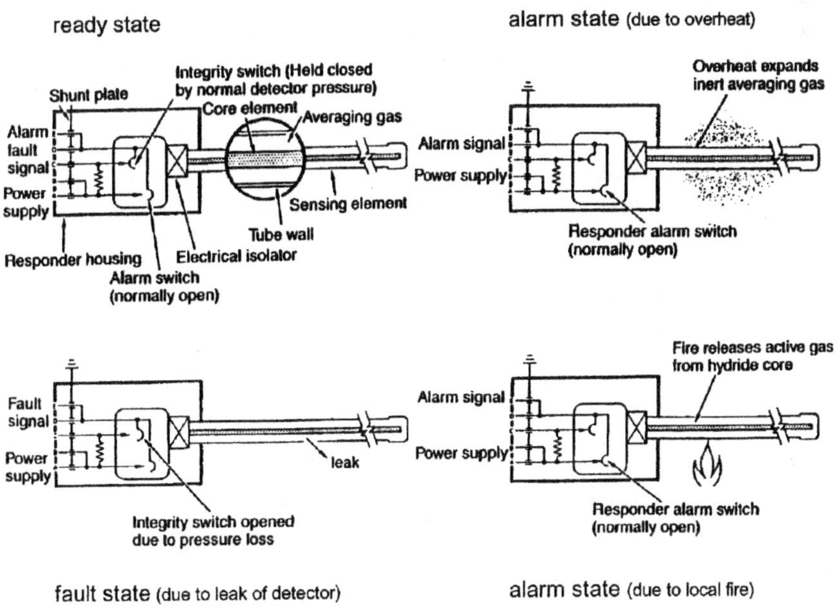

FIGURE 2.27 Principle of pneumatic continuous-loop detector (A321).

Overheat detection may be applied in the areas of the engine, auxiliary power unit (APU), bleed air ducts, and the landing gear bay.

Smoke Detection

Smoke detection systems are the primary means of fire detection used in cargo compartments. This has not changed much over the last 50 years. While solid state electronics and new optics and new processing algorithms have been introduced, the basic mechanism that these detectors operate under has remained the same. There are two basic designs of smoke detectors: ionization and photoelectric.

Ionization-type smoke detectors monitor ionized combustion byproducts as they pass through a charged electrical field. Photoelectric detectors measure light attenuation, reflection, refraction, and/or absorption of certain wavebands. Ionization smoke detectors have been used from the early years. The typical approach was to use a radioactive isotope as the source to charge the combustion products (Figure 2.28). However, this source may also charge everything else, including dust and fine water droplets, and can make ionization-type detectors unreliable. Ionization-type smoke detectors have been used by the commercial aviation community primarily in lavatories and cargo compartments.

Photoelectric-type smoke detectors have become the industry standard. This is not to imply that photoelectric-based detectors have been free from false alarms. These detectors, too, have been quite troublesome over the years. Most cargo compartment applications use aerospace-quality photoelectric-type smoke detectors that rely on scattered or reflected light radiation caused by a particulate matter between a radiation-emitting source and a detector device. Solid state photoelectric smoke detectors use a long-life light-emitting diode (LED) as the source of light.

Smoke detectors still have many limitations. Their operational success depends highly on their placement with respect to where a fire event is. But there are also problems with other detectors. Since one cannot count on visual line of sight of a cargo bay fire, future cargo-detection technologies cannot rely on the use of video camera or thermal imaging devices. Deep-seated fires and/or fires inside LD3 containers will still be hidden. This makes standalone thermal-based systems impractical. While combustion gases, such as CO or CO_2, could be monitored, these gases may be introduced from sources other than fires.

Smoke detection can be applied in the cargo compartment, lavatories, galleys, and avionic compartments.

Extinguishing Fundamentals

Fire extinguishing includes that part of the fire protection system using fixed or portable systems used to extinguish a fire (ATA 100).

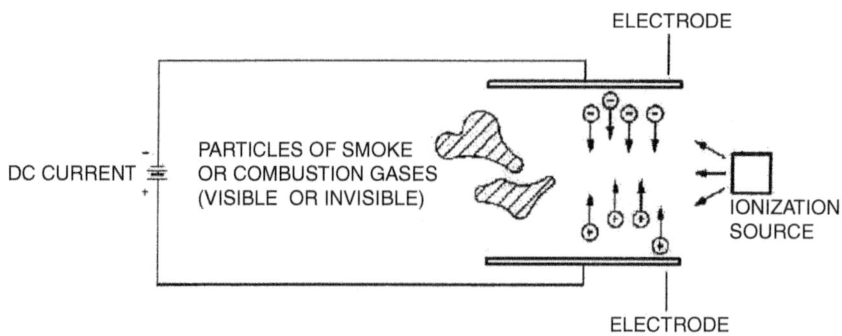

Figure 2.28 Principle of ionization-type smoke detector (A321).

A *fire classification* includes three types of fire relevant to aircraft application:

- Class A: Fires involving ordinary combustible solid materials, such as wood, paper, rubber, and many plastics
- Class B: Fires involving flammable liquids, oils, greases, paints, lacquers, and flammable gases
- Class C: Fires involving energized electrical equipment

Each of these types of fire requires its own suitable *type of extinguisher:*

- Water extinguishers are used on Class A fires only. Water must never be used on Class C fires and can be counterproductive on Class B fires.
- CO_2 extinguishers are specifically used to combat Class C fires. A hand-held CO_2 extinguisher includes a megaphone-shaped nozzle that permits discharge of the CO_2 close to the fire. Be aware that excessive use of CO_2 extinguishers robs a closed area of oxygen. In an aircraft, this could affect passengers.
- Dry chemical fire extinguishers can be used on Class A, B, or C fires. Use of such an extinguisher on the flight deck could lead to temporary severe visibility restrictions. In addition, because the agent is nonconductive, it may interfere with electrical contacts of surrounding equipment.
- Halon has almost exclusively been in use in portable aircraft fire extinguishers.

In the late 1940s, the very effective halogenated hydrocarbon (later termed *halon*) fire extinguishing agents were introduced. The primary agents used for fixed fire extinguishing systems were methylbromide (Halon 1001) and bromochloro-methane (Halon 1011). Halon 1011 eventually displaced Halon 1001 for engine extinguishing systems primarily because of lower toxicity and corrosion.

The halons introduced in the early 1950s were less toxic than Halon 1011. Over the next 30 years, the higher-vapor pressure bromotrifluoromethane (Halon 1301) essentially displaced most of the Halon 1011. Because of the high vapor pressure of Halon 1301, the use of elaborate spray nozzles and spray bars was no longer required. The new Halon 1301 extinguisher systems were designed to discharge at a very high rate. This concept was called the *high rate discharge* (HRD) concept. The high rate discharge systems utilized halon pressurized to 600 psig (40 bar).

Hand-held dibromofluoromethane (Halon 1211) and/or water extinguishers have been the approved approach for accessible firefighting.

In recent years, due to international agreement on banning the production and use of ozone-depleting substances, including all the halons, the need for alternative extinguishing agents to the halons has arisen. However, the use of halons is still permitted for essential applications, such as aircraft, until a suitable replacement agent can be developed, approved, and certified for aircraft use. Until that time comes, existing stocks of halon, recovered from decommissioned fire protection systems, are sufficient to support many years of aircraft production and use. Upon review of alternative agents, it is evident that there is no clear winner with respect to a replacement for Halon 1301 in fire suppression systems that will use similar hardware and architecture. Each candidate has at least one characteristic that makes it inferior to Halon 1301.

Engine and APU Extinguishing

First step: The engine is shut down and combustible fluid entry (jet fuel, hydraulic fluid, and engine oil) into the engine compartment is stopped. This is necessary for the engine extinguisher to be effective. If the engine were not shut off, the fire would probably just relight after the extinguishing agent dissipated. Because of this practice, only multiengine aircraft utilize extinguishing systems.

Second step: The extinguishing agent flows from a pressure vessel through rigid pipes and is sprayed in the engine-protected zones.

Third step: If after some time (30 s) the fire warning still remains on, extinguishing agent from a second pressure vessel (if still available for that engine) may be used for further fire extinguishing.

The extinguishing agent is stored in high-pressure vessels commonly called *bottles*. A spherical-shaped pressure vessel design represents the most weight- and volume-efficient geometrical configuration for containing the greatest amount of agent. It is also the optimum shape with respect to stress levels in the vessel's material. The spherical pressure vessel is the most popular design (Figure 2.29). Other details of the design are stated in Section 1199 of CS-25 and FAR Part 25.

APU fire extinguishing is technically similar to engine fire extinguishing, but the APU may only be equipped with one bottle.

Cargo Extinguishing and Inerting

Cargo compartments have traditionally been protected with hand-held fire extinguishers if the compartment was accessible and with a fixed Halon 1301 fire extinguishing/inerting system if the compartment was not accessible.

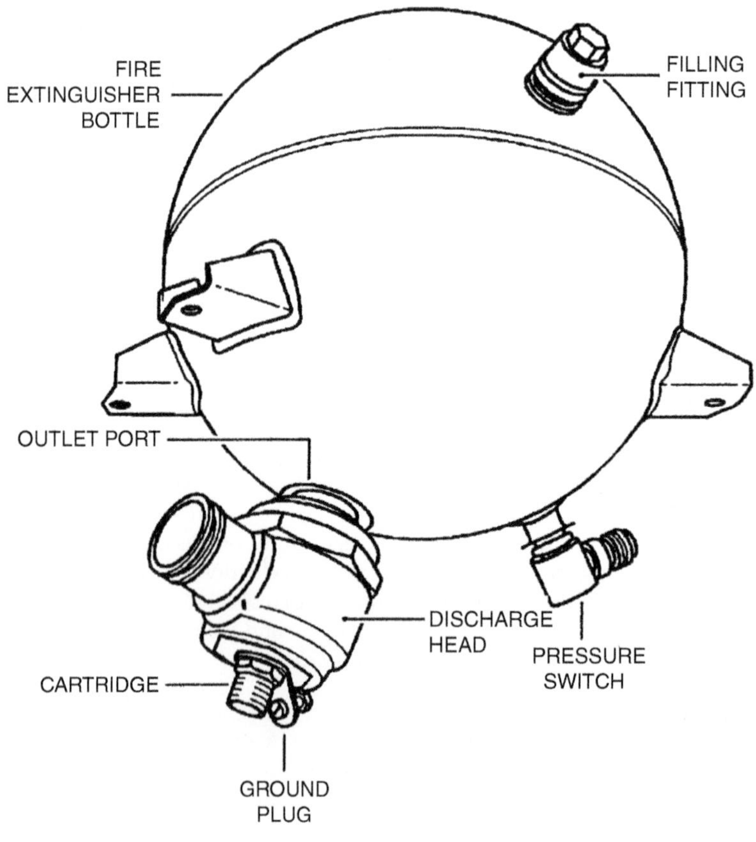

Figure 2.29 Fire extinguishing bottle (A321).

Like engine extinguishing systems, a cargo compartment suppression system is required to provide an initial peak volumetric agent concentration to knock down the fire. Since complete fire extinction cannot be assured, a cargo suppression system is required to maintain a lower concentration for some extended period of time. The compartment is thus inerted to prevent the fire from reigniting or growing. The typical time period for keeping the compartment inert against flaming combustion is 60 minutes. In case of extended range twin-engine operations (ETOPS), inerting periods are much higher.

A typical cargo fire-suppression system will consist of two fire extinguishers connected to single or multiple cargo compartments by distribution plumbing. The knock-down or high rate discharge (HRD) extinguisher provides the initial high volumetric concentration, and the second low rate discharge (LRD) extinguisher provides the metered lower inerting concentration.

Passenger Compartment Extinguishing

Fires that could occur in an aircraft cockpit or cabin are Classes A, B, and C. The number of hand-held fire extinguishers to be carried in an aircraft is determined by Section 851 of the certification regulations (CS-25, FAR Part 25).

For airplanes with a passenger capacity of 20 or more, each lavatory must be equipped with a built-in fire extinguisher for each disposal receptacle for towels, paper, or waste, located within the lavatory. The extinguisher must be designed to discharge automatically into each disposal receptacle upon occurrence of a fire in that receptacle (Section 854, CS-25, FAR Part 25).

Example: Airbus A321

For each engine, two fire extinguisher bottles contain fire extinguishing agent. The fire extinguisher bottles are connected to the extinguishing lines. The lines are routed in the pylon, leading to the outlet nozzles around the engine. The agent from the second bottle can be used if, after application of the first bottle, the fire warning remains on. The fire extinguisher bottles are controlled from the cockpit by pressing the DISCH (discharge) button. This supplies 28 V dc to two filaments in the cartridge on the bottle (see Figure 2.30). The filaments ignite 400 mg of explosive powder, which in turn causes rupture of the frangible disk in the cartridge and frees the agent with a high discharge rate.

2.6 Flight Controls (ATA 27)

Flight controls (the flight control system) are addressed in two other sections of this handbook. Please consult Sections 6 and 7.

Flight controls as defined by ATA 100:

> Those units and components which furnish a means of manually controlling the flight attitude characteristics of the aircraft, including items such as hydraulic boost system, rudder pedals, controls, mounting brackets, etc. Also includes the functioning and maintenance aspects of the flaps, spoilers, and other control surfaces, but does not include the structure.

Flight controls extend from the controls in the cockpit to the control surface actuators. The definition reads "means of *manually* controlling"; this sets the flight control system apart from the auto flight system. Thus, the flight control system is concerned only with direct inputs from the pilot via control column, rudder pedals, or other such control devices and the transformation of these inputs to adequate control surface movements.

Flight controls are subdivided into the mechanical aspects of the system and—in case of fly-by-wire (FBW) aircraft—the electronic (avionic) part.

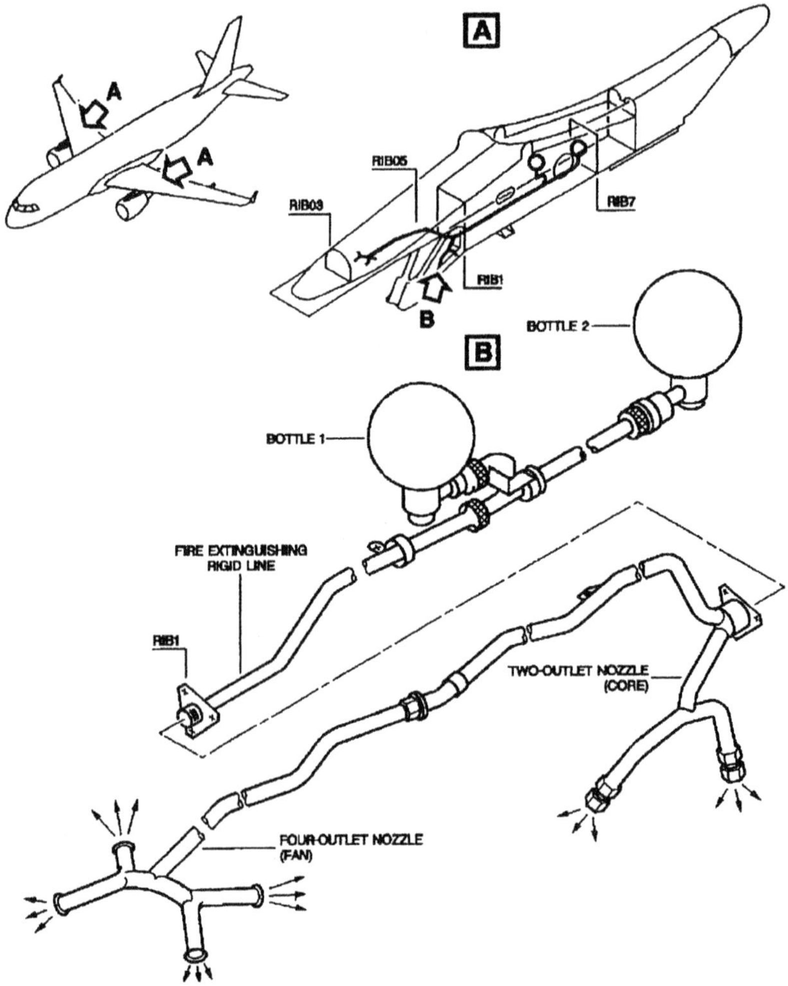

Figure 2.30 A321 engine fire extinguishing distribution system.

Following ATA 100, the *mechanical subsystems* include:

- The ailerons
- The rudder
- The elevator
- The spoilers
- The horizontal stabilizer
- The high-lift system
- Gust locks and dampers

The *electronic* (avionic) *subsystem* is the Electronic Flight Control System (EFCS).

Even in modern FBW aircraft there exist many mechanical parts because in the end control surfaces have to be moved against heavy air loads in limited time. The high-lift systems (flaps and slats) also show a considerable amount of mechanical parts.

See the References and Further Reading for more on the mechanical aspects of modern flight control systems design.

2.7 Fuel (ATA 28)

The fuel system as defined by ATA 100:

> Those units and components which store and deliver fuel to the engine. Includes engine driven fuel pumps for reciprocating engines, includes tanks (bladder), valves, boost pumps, etc., and those components which furnish a means of dumping fuel overboard. Includes integral and tip fuel tank leak detection and sealing. Does not include the structure of integral or tip fuel tanks and the fuel cell backing boards which are [part of the structure], and does not include fuel flow rate sensing, transmitting and/or indicating, which are covered [by the powerplant systems].

Fuel—General

The purpose of the fuel system is to provide reliably the proper amount of clean fuel at the right pressure to the engines during all phases of flight and during all maneuvers.

The fuel system includes (ATA 100) all components necessary to achieve:

- Fuel storage (tanks, components for tank ventilation, over-wing filler necks and caps)
- Fuel distribution (all components from the filler to the tank and from the tank to the engine quick disconnect: plumbing, pumps, valves, and controls)
- Fuel dump (all components used to dump fuel overboard during flight)
- Indicating (all components used to indicate the quantity, temperature, and pressure of the fuel)

Without fuel supply, powered sustained flight would not be possible. For this reason, the fuel system, together with the flight control system and the landing gear, can be considered the most essential systems of an aircraft. This fact is also reflected in the many sections of the certification requirements dedicated to the fuel system: For transport category aircraft these are Sections 951 through 1001 of CS-25 and FAR Part 25.

All aircraft use hydrocarbon fuels. Piston engine aircraft use a high-octane number *gasoline.* Common for these aircraft is AVGAS 100LL. Jet engine aircraft use *kerosene.* Depending upon the application (civil or military), various grades are utilized. Common jet fuel for civil applications is JET A-1. Table 2.5 contains some fuel data relevant to aircraft fuel systems.

Kerosene has a sufficiently high *flashpoint.* At sea-level pressure and normal temperatures, kerosene can be considered a safe fuel. Gasoline, in contrast, could easily ignite and needs to be handled especially careful.

When fuel in the fuel lines is heated enough to cause it to vaporize, a bubble of fuel vapor appears, blocking the fuel from flowing to the engine. Such a situation is called vapor lock and must obviously be avoided. The *vapor pressure* is a measure showing if a fuel is prone to vapor lock.

Fuel contains a certain amount of energy per unit mass known, as specific heat or *heating value* H. The fuel tank offers a limited fuel volume V. Hence, fuel mass m and fuel energy E in the fuel tank vary with fuel density ρ.

$$m = \rho \cdot V \qquad E = m \cdot H \qquad E = \rho \cdot V \cdot H$$

	AVGAS 100LL gasoline for piston engines	JET A-l kerosene for jet engines
Flashpoint (at standard sea level pressure)	−40°C (−40°F)	+38°C (+100°F)
Vapor pressure (Reid standard conditions)	500 hPa (7.25 psi)	10 hPa (0.145 psi)
Density (at 15°C)	720 kg/m³ (6.0 lb/U.S. gal)	810 kg/m³ (6.7 lb/U.S. gal)
Heating value	43.5 MJ/kg (18,700 BTU/lb)	42.5 MJ/kg (18,300 BTU/lb)

Table 2.5 Fuel Characteristics Related to Aircraft Fuel Systems

Since fuel density decreases with increasing temperature, so do storable fuel mass and energy. For aircraft operation, the amount of energy on board is of importance. Accordingly, *indicating fuel mass* to the pilots does make sense (in contrast to indicating fuel volume). The drawback: not only measurements of fuel level and hence fuel volume are required, but additionally, measurements of fuel density.

Water may be contained in the fuel dissolved, entrained, or free. As fuel is taken from the tank, air (at given humidity) enters the space above the fuel in the tank. With decreasing temperature, water condenses from this air and enters into the fuel. During flight at high altitudes and low temperatures, ice crystals can form that clog fuel filters. To prevent clogging, the fuel may be passed through a fuel heater prior to entering the filter. Fuel systems must be capable of sustained operation with a specified amount of free water under critical conditions for icing (Section 951). With the aircraft at rest, water in the fuel collects in the fuel tank sump that is the lowest part of the fuel tank. This happens because density of water (1,000 kg/m³) is greater than fuel density. "Each fuel tank sump must have an accessible drain" (Section 971). Water drain valves are used to extract the water.

Microorganisms, bacteria or fungi, may grow in jet fuel tanks. These organisms live and multiply in the water contained in the fuel and feed on the hydrocarbons. The buildup of microorganisms not only interferes with fuel flow and quantity indication but can start electrolytic corrosion. The organisms form a dark slime on the bottom of the lowest parts of the fuel tank, especially near water drain valves. Regularly draining water from the fuel together with fuel additives may solve the problem of microbial growth.

Unintended ignition of fuel must be prevented. Therefore, Section 954 of CS-25 and FAR Part 25 reads: "Each fuel system must be designed and arranged to prevent the ignition of fuel vapour" by lightning strikes or other effects at outlets of the vent and jettison systems or directly through the structure.

Fuel Storage

Fuel *tank location* can be in the wing, fuselage, horizontal stabilizer, or fin. Tanks can be permanently attached or mounted onto the wing tip (tip tank). In the case of combat aircraft, additional tanks can be under-wing mounted, over-wing mounted, or belly mounted. Transport aircraft often use the center section of the wing for a center tank (Figure 2.33). These aircraft may trade payload versus fuel capacity (i.e., maximum range) by using part or all of the cargo compartment for additional center tanks (ACTs).

"Fuel tanks must have an *expansion space* of not less than 2% of the tank capacity. It must be impossible to fill the expansion space inadvertently with the aeroplane in the normal ground attitude" (Section 969). A 2% expansion is equivalent to an increase in fuel temperature of 20°C.

Fuel initially filled into the empty tanks cannot practically be expected to be taken out again "to the last drop" under all operating conditions. The amount of fuel that remains in the tank is called *unusable fuel*. "The unusable fuel quantity for each tank and its fuel system components [is] the quantity at which the first evidence of engine malfunction occurs under the most adverse fuel feed condition for all intended operations and flight manoeuvres" (Section 959). Aircraft manufacturers try to reduce the unusable fuel volume as much as possible. So-called scavenge pumps are used to collect fuel from different areas of the tank.

The fuel in the fuel tanks can be used for center of gravity (CG) control. Supersonic aircraft may use *CG control* to minimize trim drag that is caused by the rearward shift of lift at supersonic speeds. The Concorde uses trim tanks in the forward part of the wing for CG control. Subsonic aircraft may use a trim tank in the empennage to maintain an optimum rearward CG in cruise. An aft CG reduces trim drag and thus enhances aircraft performance. The Airbus A340 applies a trim tank in the horizontal tail to move the CG in cruise back to approximately 2% mean aerodynamic chord (MAC) forward of the certified aft limit.

The weight of fuel in the wings directly balances lift. This reduces *wing-bending* moments and allows for the design of a lighter structure. In order to make as much use as possible of this phenomenon, fuel is preferably taken from the center tank or an inner wing tank first, whereas the fuel in outboard wing tanks is used only during the last part of the flight. During the last part of the flight, lift is already reduced anyway due to a reduction of aircraft weight as a result of fuel consumption.

Fuel tank construction can be divided into three basic types: rigid removable, bladder, and integral.

A *rigid removable fuel tank* is one that is installed in a compartment designed to hold the tank. The tank must be fuel-tight, but the compartment in which it fits is not fuel-tight. The tank is commonly made of aluminum components welded together or composites. Rigid fuel tanks are used on small aircraft or as additional center tanks (ACT). ACTs inside the fuselage must be double-walled.

A *bladder tank* is a reinforced rubberized bag placed in a non-fuel-tight compartment designed to structurally carry the weight of the fuel. Bladder tanks are found on medium- to high-performance light aircraft or inside a rigid ACT structure to produce a double-walled tank.

An *integral fuel tank* is a tank that is part of the basic structure of the aircraft. Integral fuel tanks, e.g., in the wing, use structural members of the wing and sealing materials where members join to form a fuel-tight tank. Tank access panels seal the oval cutouts in the lower wing surface used for tank inspection. Baffles are frequently installed inside fuel tanks to reduce fuel sloshing. Baffles may have check valves that open in the inboard direction only. These check valves keep fuel in the inboard part of the tank, where the pumps are located.

The *tank vent system* "maintains acceptable differences of pressure between the interior and exterior of the tank" (Section 975) under all operating conditions, including:

- Cruise (fuel burn)
- Maximum rate of climb and descent (change of outside pressure)
- Refueling and defueling

Overpressure and underpressure in the tanks can cause structural damage. Under-pressure can cause engine fuel starvation. The vent system for a light aircraft may be as simple as a hole drilled into the fuel cap. Large aircraft connect each main tank via vent pipes with a vent surge tank for tank venting (Figure 2.34). The vent surge tanks take up any overflow fuel from the main tanks and direct it back to these tanks through vent float valves. The vent surge tanks are each connected to the outside via a NACA air intake, which achieves a pressure in the fuel tank slightly above ambient pressure.

Aircraft fuel is also used as a *heat sink*. The hydraulic system and the air conditioning system, especially of jet fighters, use fuel for cooling purposes. It is obviously important to monitor the fuel temperature carefully in order to avoid overtemperatures.

Fuel Distribution

The fuel distribution system may consist of:

- The engine feed system
- The fuel transfer system
- The crossfeed system
- The refuel/defuel system

Engine feed, i.e., fuel flow to the engines, may be either gravity feed or pressure feed.

In the case of *gravity feed*, the fuel flows by gravity to the engine. This is possible if the tank is located sufficiently above the engine. Gravity feed is used on small high-wing aircraft and on large aircraft in emergency cases with system fuel pumps inoperative (suction provided from engine fuel pumps).

In the case of *pressure feed*, fuel pumps are used to move fuel through the fuel system. For turbine-engine fuel systems, there must be one *main pump* for each engine (Section 953) and one *emergency pump* (Section 991) immediately available to supply fuel to the engine if the main pump fails. Various fuel pump principles exist including: vane pump, centrifugal pump, and ejector pump.

The *centrifugal pump* (Figure 2.31) draws fuel into the center inlet of a centrifugal impeller and expels it at the outer edge. Fuel can flow through the pump when the pump is not in operation. This eliminates the need for a bypass valve.

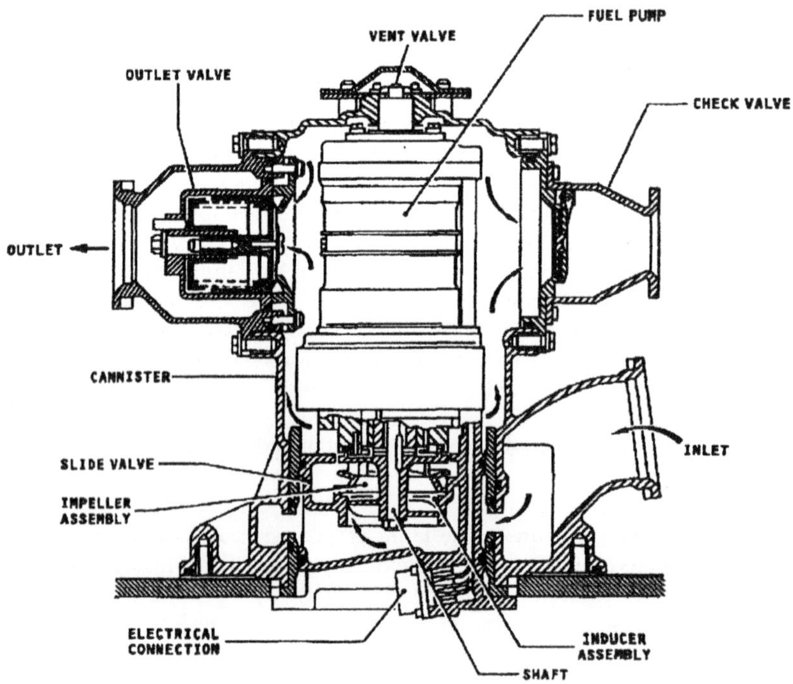

FIGURE 2.31 Centrifugal fuel pump (A321).

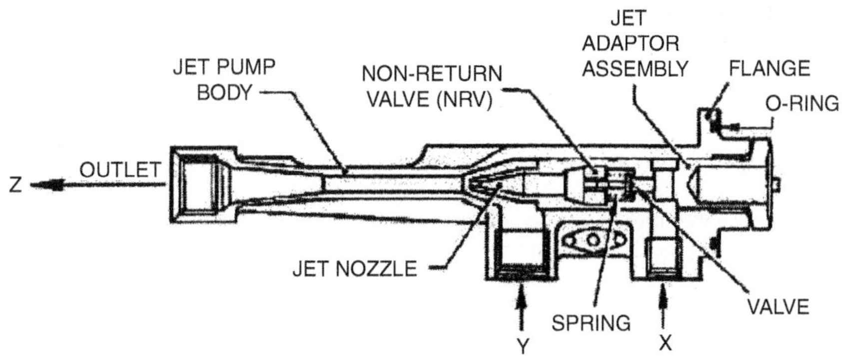

FIGURE 2.32 Jet pump (A321). X = input from main pump; Y = suction input; Z = output.

An *ejector pump* (Figure 2.32) is used to scavenge fuel from other areas of the fuel tank or from adjacent fuel tanks. This type of pump has no moving parts. Instead it relies on the fuel flow from a main pump.

Fuel *selector valves* provide means to select a tank from which to draw fuel in a multiple-tank installation, transferring fuel from one tank to another and directing fuel to one or more engines. A *shutoff valve* (Section 1189) disconnects fuel flow to an engine. The shutoff valve is also closed by the fire handle in case of engine fire. "There must be a fuel strainer or filter" (Section 997).

The *fuel transfer system* allows fuel to be pumped from one tank into another. The main feature of the crossfeed system is its fuel manifold. Fuel is supplied from the tanks to the crossfeed manifold. Crossfeed valves on the crossfeed manifold can be set such that each engine can be fed from all tanks.

There are two basic refuel procedures for aircraft: over-wing refueling and pressure refueling. In addition, some aircraft are able to use in-flight refueling.

The historical form of refueling an aircraft from above simply by gravity is called *over-wing refueling*. Small aircrafts apply this simple method. It is slow, and depending on aircraft size and wing location it may be difficult to reach on top of the wing.

Pressure refueling uses pressure from the fueling station or truck to force fuel into the aircraft tanks. This is usually done through a fueling coupling located under the wing at the right wing leading edge. Pressure refueling is fast and the refuel coupling is in easy reach.

During *in-flight refueling*, a military aircraft is supplied with fuel in the air from a tanker aircraft. Tanker aircraft are converted large civil transports. The connection between the receiving and providing aircraft can be established with a flexible hose or a rigid boom. In-flight refueling was first used for fighter aircraft to extend their limited range capabilities. Later, in-flight refueling was applied to cover large distances in global conflicts or to maintain constant combat air patrol.

Defueling is the opposite of refueling: fuel is pumped out of the aircraft fuel tanks and back into the station or truck. During *fuel ground transfer*, fuel is pumped from one aircraft tank into another tank. Defueling and fuel ground transfer may become necessary prior to tank maintenance.

Fuel Jettison

Fuel weight amounts to a large fraction of aircraft gross weight, especially at the beginning of a long-range flight (with a long-range aircraft). If an *emergency* occurs shortly after takeoff, the aircraft may be forced to return and land as soon as possible. In such a situation, the present aircraft weight will still be considerably above maximum landing weight. An overweight landing might unduly stress and endanger the aircraft, and in the case of a discontinued approach, the heavily laden aircraft will not be able to fly a successful go-around maneuver with sufficient climb rate (Section 1001).

A fuel *jettison system* (fuel dump system) helps to solve the situation. The fuel jettison system allows dumping of all but some reserve fuel overboard in not more than 15 minutes. This now brings the aircraft weight down quickly as a prerequisite for a successful emergency landing.

Two fuel-jettison principles have been used: systems can work with gravity or with pump pressure. A *gravity jettison system* is equipped with long dump chutes that are deployed at both wing tips. The long chutes produce the necessary pressure differential for the flow. A *pump jettison system* is equipped with dump nozzles at both wing tips.

Indicating

Quantity, temperature, and pressure of the fuel can be measured for the fuel system. Other fuel parameters are measured by the engine.

A *fuel quantity indicator* can be a mechanical quantity indicator, a resistance quantity indicator, or a capacitance quantity indicator.

A *capacitance quantity indicator* is a condenser installed in the tank so that the condenser is immersed in the fuel. Fuel respectively air in the tank serve as dielectric material for the condenser. When the probe is dry, its capacitance value is low, but as fuel moves up the probe its capacitance value increases. A controller monitors the capacitance value and converts it into a fuel volume.

In addition to the fuel quantity indicator, which is primarily used in flight, it is desirable to have an alternative provision to determine the fuel quantity visually. On light aircraft this may be accomplished by viewing the fuel surface through the fuel filler cap opening, but on large aircraft this would be extremely difficult. For this reason, calibrated hollow fiberglass *dripsticks* have been used that are unlocked and slowly lowered from under the wing. The position of the stick when it drips marks the fuel level inside the tank.

More sophisticated are *magnetic level indicators* (MLIs). MLIs are also unlocked and lowered from under the wing. A magnetic float on the fuel surface gets hold of the magnetic top of a stick. The position of the stick attached to the float determines the fuel level.

Example: Airbus A321

The Airbus A321 has three *fuel tanks* (Figure 2.33): the left wing tank, the right wing tank, and the center tank. The total usable fuel capacity of these tanks is 23,700 L. The total unusable fuel capacity is 89.7 L. This is less than 0.4%.

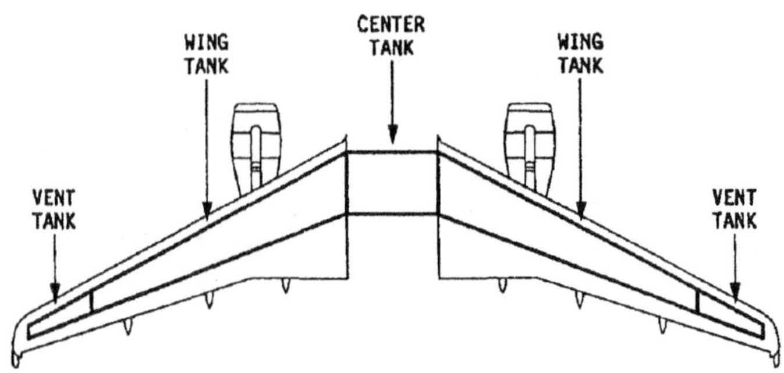

FIGURE 2.33 A321 fuel tanks.

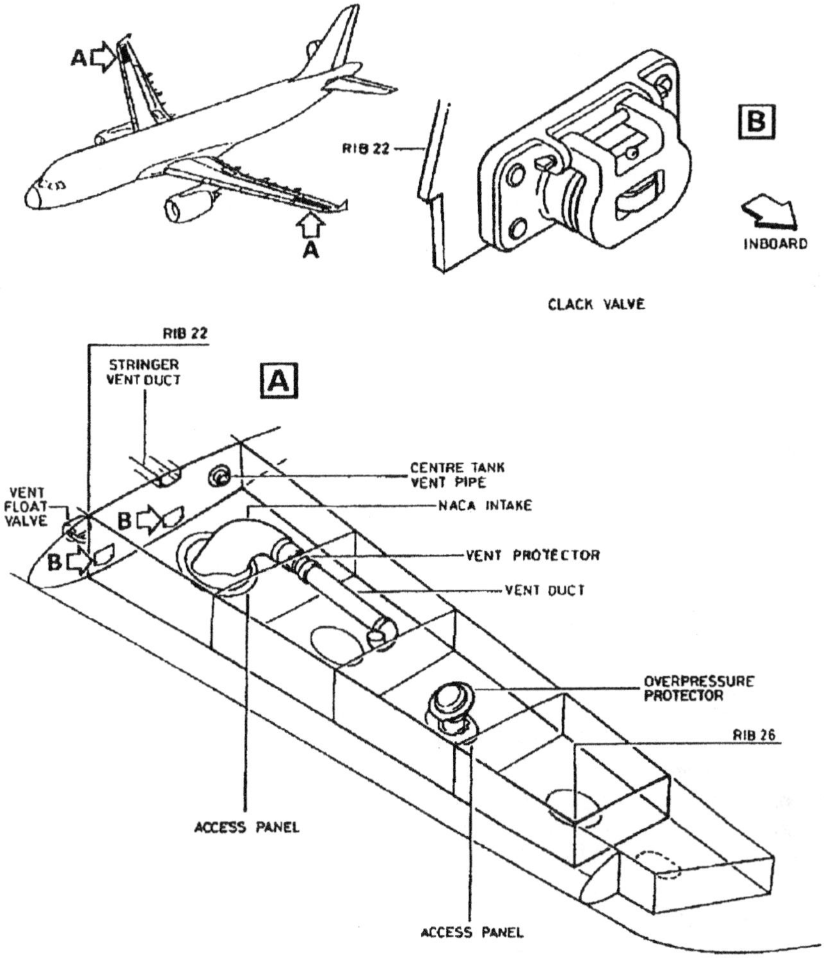

Figure 2.34 A321 vent surge tank.

The *vent surge tanks* (Figure 2.34) do not normally contain fuel. They are connected to the wing tank and center tank through the stringer vent duct and the center tank vent pipe. The vent surge tanks can vent these tanks because they are open to the external air through a vent duct. The vent duct contains a vent protector with a flame arrestor and an ice protector. The vent duct is connected to a NACA intake on the bottom of the tank. The vent surge tanks are also a temporary reservoir for the fuel that could enter through the vent pipes. This fuel is drained back to the wing tanks through vent float valves (clack valves). In case of an obstruction in the vent duct, the overpressure protector ensures that the pressure in the vent surge tank does not exceed specified limits.

The *fuel distribution system* of the A321 is shown in Figure 2.35:

- The engine feed system takes fuel from the wing tanks and supplies it to the engines. Two main pumps (Figure 2.31) are located in each wing tank.
- The main transfer system enables transfer of fuel from the center tank to the left and right wing tank. This fuel transfer is a normal procedure necessary to make use of the fuel in the

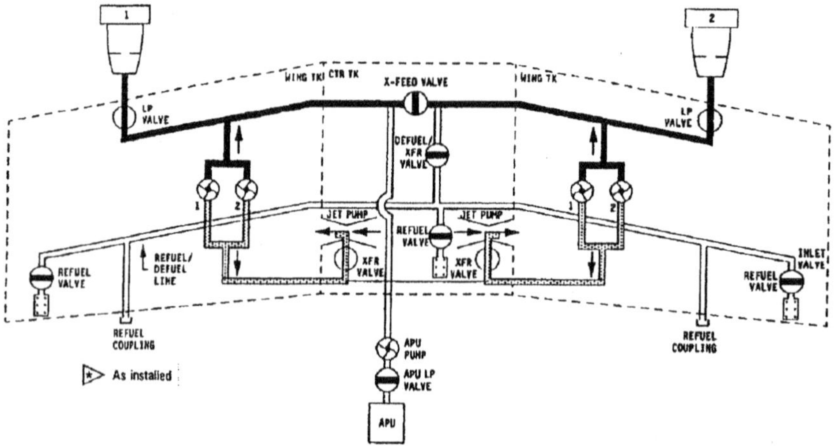

Figure 2.35 A321: Overview of the fuel distribution system. Black lines: engine feed system. Gray lines: main transfer system. White lines: refuel/defuel system and APU feed. X-FEED: crossfeed. XFR: transfer.

center tank. Fuel transfer is achieved with ejector pumps (jet pumps). The jet pumps in the center tank are driven by fuel from the main pumps.

- The crossfeed system connects the left and right fuel feed system. The engine feed line has a crossfeed valve that permits the isolation or interconnection of the left (engine 1) and right (engine 2) fuel supply system. Under normal conditions, the crossfeed valve is closed.
- The refuel/defuel system:
    - Refueling: Fuel is supplied to the fuel tanks via the refuel coupling in the right wing. A second refuel coupling in the left wing is optionally available.
    - Defueling: Fuel is pumped out of the tanks by way of the refuel coupling. The defuel transfer valve is open.
    - Fuel transfer: The system may be used to transfer fuel from one tank into any other tank. The defuel transfer valve is open.
- The APU feed system takes fuel from the engine feed line and supply fuel to the auxiliary power unit (APU).

2.8 Hydraulic Power (ATA 29)

The hydraulic system as defined by ATA 100:

> Those units and components which furnish hydraulic fluid under pressure (includes pumps, regulators, lines, valves, etc.) to a common point (manifold) for redistribution to other defined systems.

Purpose

The purpose of the hydraulic system is to assist the pilot in accomplishing mechanical tasks that would otherwise be impractical or impossible because of the level of force, work, or power required.

On smaller aircraft the flight control surfaces are moved by pilot force. On larger and faster aircraft this becomes impossible and so hydraulic power is applied. A total failure of the flight control system evidently has a catastrophic effect. Consequently, a failure of the hydraulic power supply of large aircraft has to be extremely improbable. This required level of safety is achieved with redundancy through three or even four independent hydraulic sub-systems.

Principle

Figure 2.36 shows the principle of a hydraulic system. Hydraulic fluid is contained in a *reservoir.* Through a suction line the *pump* draws fluid from the reservoir and puts it at a higher pressure. Today aircraft hydraulic systems are typically designed to a nominal pressure of 206 bar (3,000 psi). The trend is toward higher system pressure: 345 bar (5,000 psi). An *accumulator* serves as temporary energy storage and is able to store or redistribute surplus high-pressure fluid. A *pressure-relief valve* is able to shortcut the high-pressure line to the reservoir in case of a system malfunction leading to higher pressure than specified. The pressure differential supplied by the pump is used by hydraulic consumers. The example shows a typical consumer in the flight control system. An *actuator* piston rod has to move in and out in order to deflect a control surface (not shown). The actuator piston is moved through hydraulic fluid that enters the left actuator chamber and fluid that leaves the right actuator chamber (or vice versa). A valve schedules the required fluid flow. Shown is a *servo valve.* The valve has four connections to hydraulic tubes: one connection to each of the two actuator chambers, one connection to the high-pressure line, and one connection to the return line. The valve may be moved into one of three positions that lead to piston rod extension, piston rod retraction, or no piston rod movement. In the case of a flight control consumer, it is necessary that the valve move gradually from one position into the other to allow a proportional control of the surface. In the case of landing gear extension and retraction, a *selector valve* would be used. The selector valve allows three distinct valve positions without any intermediate positions.

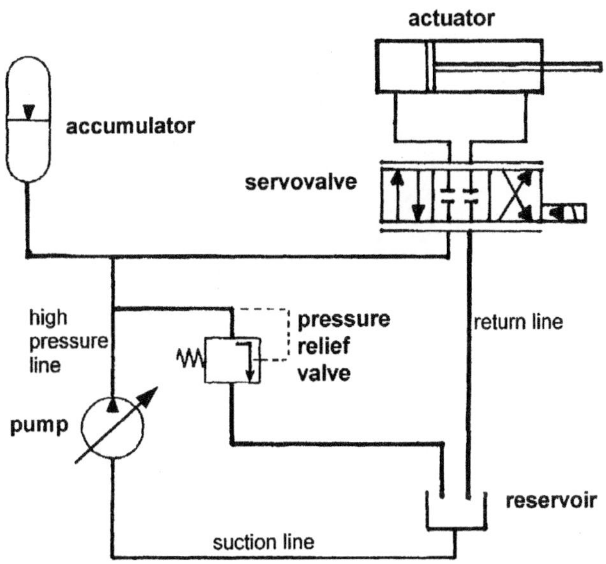

FIGURE 2.36 A basic hydraulic system.

During system design the complete circuit, including hydraulic power generation, distribution, and consumption, has to be analyzed. According to the ATA breakdown, the consumers with their valves are allocated to their respective system. ATA 29 deals only with power generation and distribution.

Three types of *hydraulic fluids* exist: vegetable based, mineral based, and synthetic or phosphate ester-based. Transport category aircraft use the purple-colored phosphate ester-based fluid—most commonly Skydrol® LD. Skydrol® shows good performance even at low temperatures, excellent flammability characteristics, and minimal effects on most common aircraft metals, but does react with certain types of paint and can be an eye and respiratory irritant.

Components

The *reservoir* acts as a storage tank for the system's fluid. Reservoirs can be broken down into two basic types, *in-line* and *integral,* and these can be further classified as *pressurized* and *unpressurized.* Integral reservoirs, found on small aircraft, are combined with the pump. Aircraft that operate at low altitudes could use unpressurized reservoirs that vent the reservoir to the atmosphere. Other aircraft positively pressurize the reservoir with air from the pneumatic system, hydraulic pressure (bootstrap reservoir) (Figure 2.37), or a spring. In a *bootstrap reservoir,* high-pressure (HP) fluid acts on a small plunger that is coupled with a large plunger that in turn acts on the low pressure (LP) fluid in the reservoir. Commonly, air pressure is used for reservoir pressurization. The air pressure usually needs to be reduced by a pressure regulator. It then enters the airspace above the fluid in the reservoir.

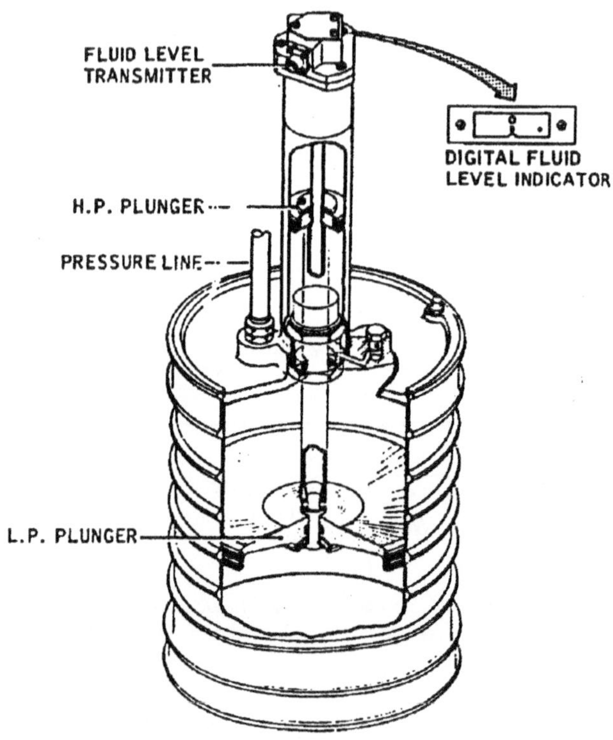

Figure 2.37 Hydraulically pressured reservoir known as bootstrap reservoir (VFW 614).

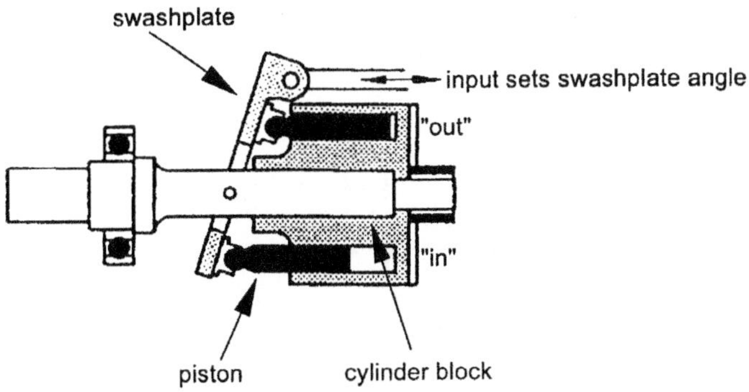

Figure 2.38 Variable displacement axial multiple-piston pump (TUHH).

Commonly used are axial multiple-piston *pumps*. The two principles applied are *constant displacement* and *variable displacement*. The shaft can be driven by the aircraft engine, by an electric motor, or through a device powered by the pneumatic system. The shaft turns the cylinder block with the pistons. Whenever an elevated piston is pushed into the cylinder block, fluid is ejected into an out port. Accordingly, during the other half of the revolution on its way back to the elevated position, the piston draws fluid from an in port into the cylinder block. Constant displacement pumps deliver exactly the same amount of fluid every revolution and must incorporate a pressure regulator. Most widely used, however, are *variable displacement axial multiple-piston pumps* (Figure 2.38). The variable displacement is achieved by a swashplate. The angle of the swashplate is adjusted by a pressure controller. At highest swashplate angle, the pump achieves its maximum flow rate, and at zero angle there is no fluid flow.

For minor tasks, *hand pumps* may be applied. A *ram air turbine* (RAT) (Figure 2.43) may be turned into the free stream of air to power a hydraulic pump. This is done in the event of an engine failure or a major electrical system failure.

Three types of *accumulators* are known: the *diaphragm-type accumulator*, the *bladder type accumulator*, and the *piston type accumulator* (Figure 2.39). The diaphragm, bladder, or piston divides the fluid chamber from the nitrogen chamber of the accumulator. Hydraulic fluid is allowed to flow freely into and out of the fluid chamber of the accumulator. The compressible nitrogen acts like a spring against the hydraulic fluid. The accumulator acts as a high-pressure and fluid storage and eliminates shock waves from the system.

Filters are installed in the high-pressure and the return line. Three filter types are in use: micron, porous metal, and magnetic. *Micron filters* contain a treated paper element to trap particles as the fluid flows through the element. *Porous metal filter* are composed of metal particles joined together by a sintering process. *Magnetic filters* attract metal particles. Filters consist of a head assembly that contains the fluid line connections and a bypass valve to prevent the system from becoming inoperative should the filter become clogged, a bowl assembly, and the filter element.

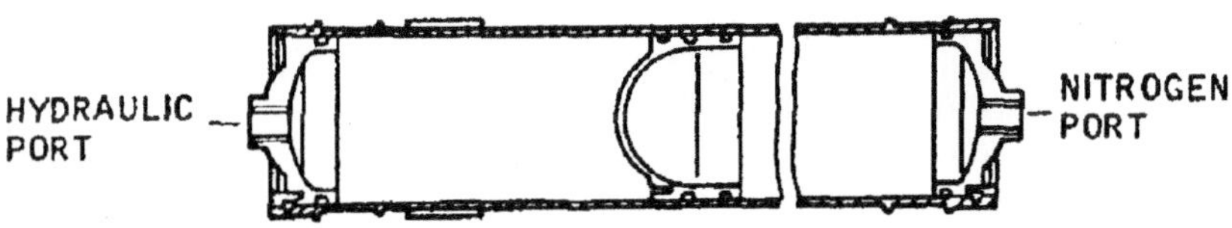

Figure 2.39 Piston-type accumulator (VFW 614).

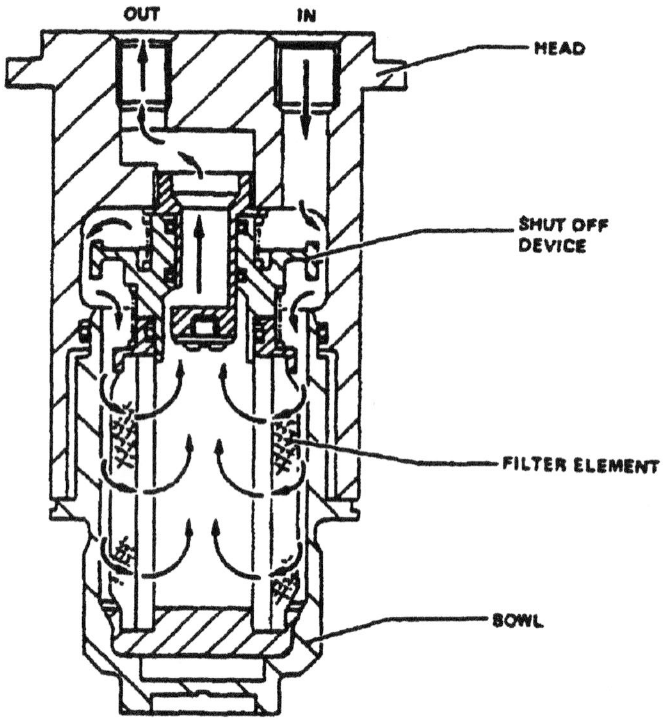

FIGURE 2.40 Low-pressure filter (A321).

Fluid enters through the head into the bowl and leaves through the filter element and out of the head (Figure 2.40).

Two principal types of *valves* are used in the hydraulic system: flow control valves and pressure control valves. *Flow-control valves* route the fluid through the system. Examples are *selector valves,* which permit the user to channel the fluid selectively, and *servo valves,* as explained above. *Check valves* permit flow only in one direction. A *hydraulic fuse* is a safety valve that prevents fluid flow in the event of a serious system leak. Examples of *pressure-control valves* are the *pressure-relief valve* and the *pressure regulator.* A *priority valve* is mechanically identical to a pressure relief valve, set to an opening pressure below nominal pressure. The priority valve is closed at low pressure and allows flow to secondary consumers only if a minimum system pressure has been reached. In this way it gives priority to primary consumers located upstream of the priority valve.

Hydraulic *fluid lines* are classified as rigid or flexible. *Rigid lines* are made of either aluminum for return and suction lines or of stainless steel for high-pressure lines. *Flexible lines* are hoses typically wrapped with stainless steel braid. *Fittings* are used to connect fluid lines with other hydraulic components.

"The *power transfer unit* (PTU) is a device which uses some of the hydraulic power in one hydraulic system to supplement the hydraulic power in a second system without interchange of fluid between the systems" (ARP 1280).

PTUs can be designed either to transfer power from one system to a second system in one direction only (unidirectional PTU) or to transfer power in either direction between two systems (bidirectional PTU) (Figure 2.41). The basic concept consists of a hydraulic motor driving a pump, mounted back-to-back. The displacement of each of these may be the same or different. Accordingly,

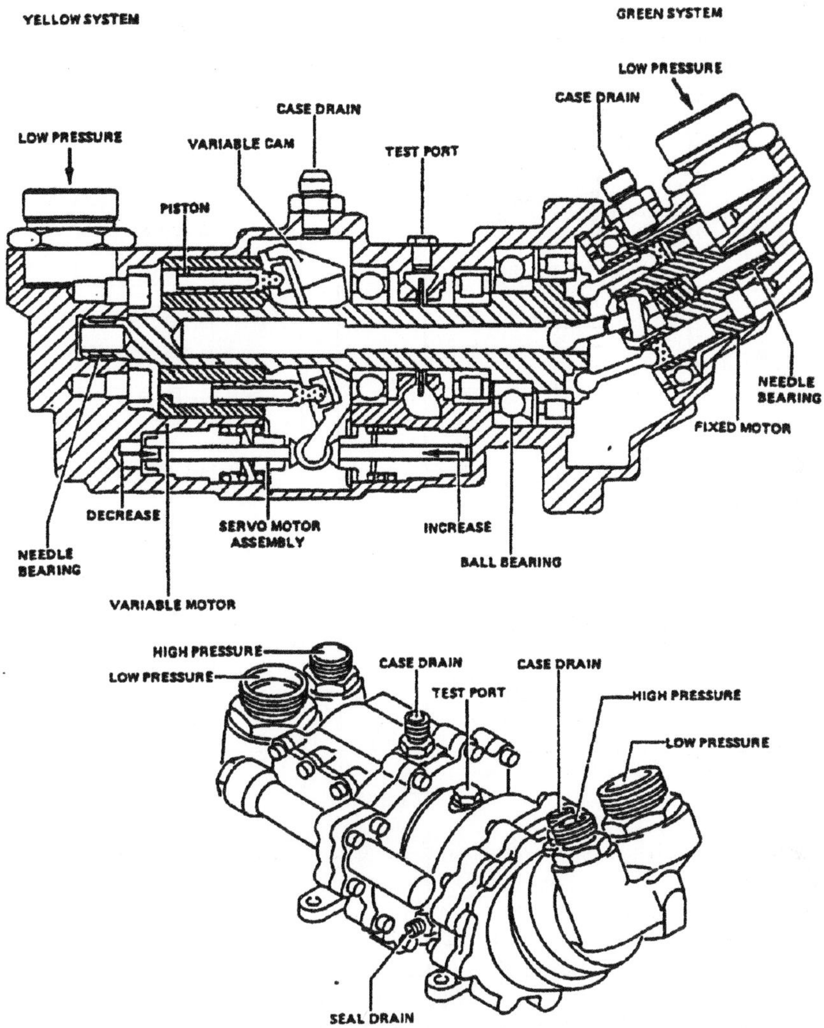

FIGURE 2.41 Bidirectional power transfer unit (PTU) (A321).

PTUs can be used as pressure reducers, as pressure intensifiers, or to maintain the same pressure in both systems. If bidirectional operation is required, both the pump and the motor reverse their functions. That unit which was previously the pump will operate as motor and vice versa. If the pressure relationship between the two systems must remain the same in both directions of operation, at least one of the units must be of a variable displacement design.

Example: Airbus A321

The Airbus A321 has three main hydraulic (sub)systems (Figure 2.42):

- The Green system
- The Blue system
- The Yellow system

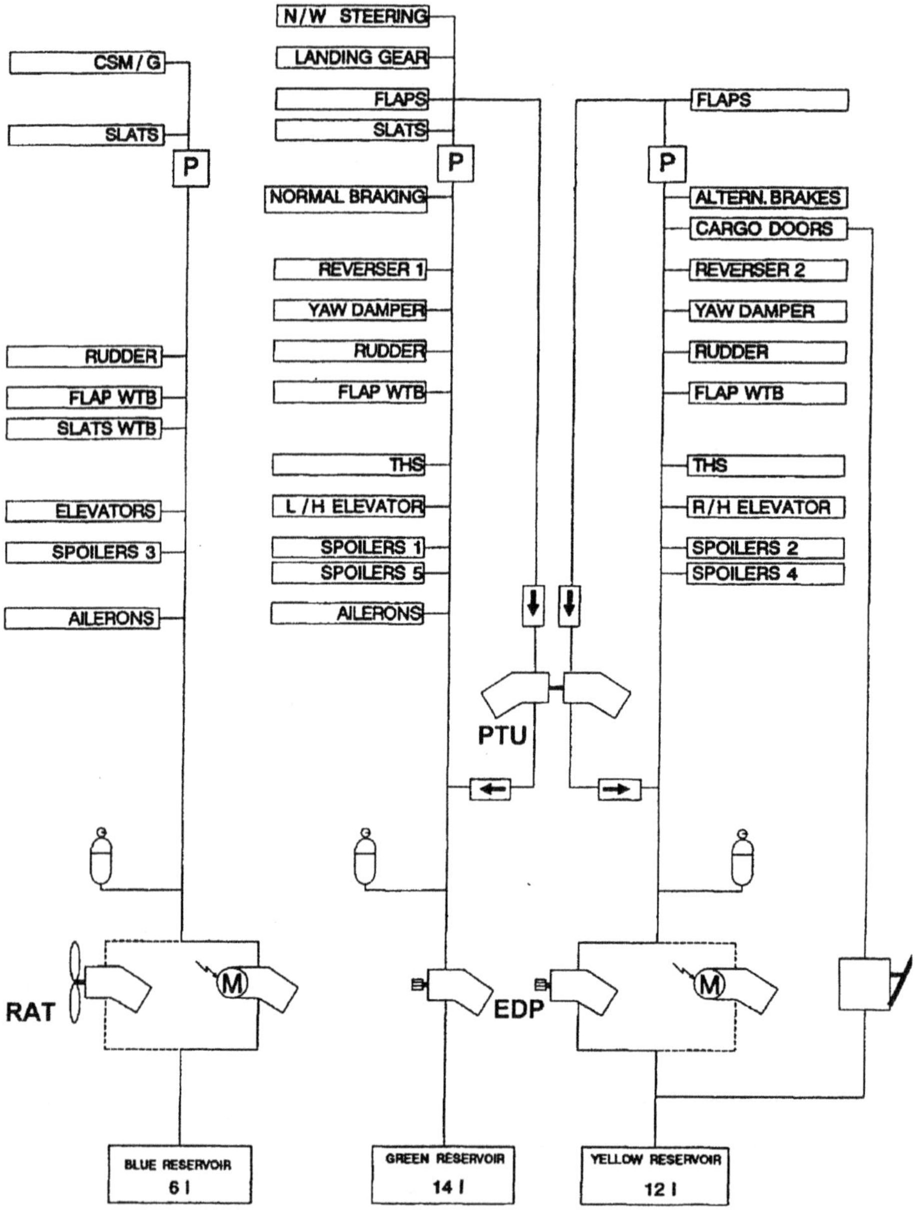

FIGURE 2.42 A321 hydraulic system schematic. EDP: Engine-driven pump. M: Electric pump. RAT: Ram air turbine. PTU: Power transfer unit. P: Priority valve. →: Check valve (indicating flow direction). CSM/G: Constant-speed motor/generator (emergency generator). THS: Trimmable horizontal stabilizer (horizontal tail). WTB: Wing tip brake (in high-lift system).

Together they supply hydraulic power at 20.7 MPa (3,000 psi) to the main power users. These include:

- Flight controls
- Landing gear
- Cargo doors
- Brakes
- Thrust reversers

Main system pumps are the engine-driven pumps (EDPs) in the Green and Yellow systems as well as the electric pump in the Blue system. The EDP of the Green system is connected to the left (No. 1) engine. The EDP of the Yellow system is connected to the right (No. 2) engine. The three main systems automatically supply hydraulic power when the engines operate. The two EDP are connected directly to their related engine (through the accessory gearbox), and the Blue electric pump operates when any one of the two engines starts. The three system main pumps are usually set to operate permanently. If necessary (because of a system fault, or for servicing), the pumps can be set to off from the flight compartment.

If the main pumps cannot be used, it is possible to pressurize each hydraulic system with one or more of the *auxiliary system pumps.*

- The Green system can also be pressurized by the power transfer unit (PTU).
- The Blue system can also be pressurized by the ram air turbine (RAT) (Figure 2.43).
- The Yellow system can also be pressurized by the Yellow electric pump or the power transfer unit (PTU).

Pressurization of the hydraulic systems *on the ground* is possible as follows:

- Yellow system—with the Yellow electric pump
- Green system—with the Yellow electric pump (through the PTU)
- Blue main system—with the Blue electric pump

For maintenance, all of the systems can be pressurized from a ground hydraulic supply. Connectors are installed on the ground service panels of the three systems. The cargo doors can also be operated with a hand pump in the Yellow system.

2.9 Ice and Rain Protection (ATA 30)

Ice and rain protection as defined by ATA 100:

> Those units and components which provide a means of preventing or disposing of formation of ice and rain on various parts of the aircraft. Includes alcohol pump, valves, tanks, propeller/rotor anti-icing system, wing heaters, water line heaters, pitot heaters, scoop heaters, windshield wipers and the electrical and heated air portion of windshield ice control. Does not include the basic windshield panel. For turbine type power plants using air as the anti-icing medium, engine anti-icing is [part of the powerplant].

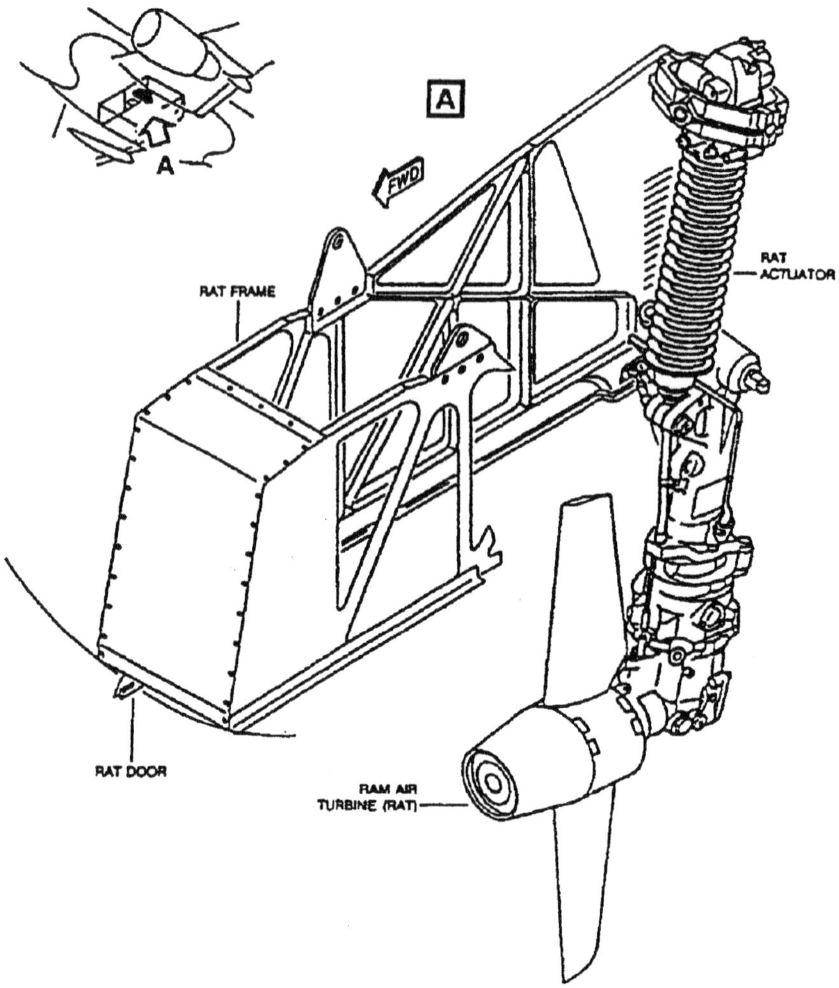

Figure 2.43 A321 ram air turbine (RAT).

System Classification

Ice and rain protection may be classified as follows:

- Nontransparent surfaces: ice protection (leading edges, radome, inlets, etc.):
  - Pneumatic boot systems
  - Thermal ice protection systems:
    - Hot air systems
    - Electrical resistance systems
  - Fluid systems
  - Electroimpulse deicing (EIDI) systems
  - Microwave systems

- External components: ice protection (antennas, sensors, drain masts, etc.)
- Internal components: ice protection (water lines, etc.)
- Windshield: ice and fog protection
- Windshield: rain removal
- Ice detection

External and internal components are generally protected against icing by electrical resistance systems. Some technical solutions for windshield ice protection serve at the same time for windshield rain removal.

The two main ice *protection principles* are deicing and anti-icing. Various technical solutions exist. Some ice protection *technical solutions* can perform both deicing and anti-icing. Other technical solutions only manage deicing (Table 2.6). The terms deicing and anti-icing are defined in AIR 1168/4:

- *Deicing* is the periodic shedding, either by mechanical or thermal means, of small ice buildups by destroying the bond between the ice and the protected surface.
- *Anti-icing* is the prevention of ice buildup on the protected surface, either by evaporating the impinging water or by allowing it to run back and freeze on noncritical areas.

Icing Fundamentals

From our daily experience we know that water freezes to ice below 0°C (32°F) and melts again above 0°C. When it comes to aircraft icing, we learn that this need not be so. Small droplets can still be in the liquid phase below 0°C! Most droplets will have turned to ice below –20°C (–4°F), though very small and pure droplets may reach temperatures as low as –40°C (–40°F) and still remain liquid. Below –40°C finally all water in the air will be frozen. "Liquid" water below 0°C is called *supercooled water*. Supercooled water can exist because the water has been totally undisturbed during cooling—nothing has caused it to turn to ice. When an aircraft hits the droplet, however, the droplet receives the necessary input for the phase change and turns to ice. (The phase change from water to ice *usually* requires some latent heat extraction, but when the droplets are supercooled water, the heat extraction has already taken place.) The ice will be slightly warmer than the supercooled water was just a second earlier. Summing up: Supercooled water turns instantly to ice due to the interaction with the aircraft. The result will be *ice accretion* on the aircraft surface if the surface is below 0°C.

Ice protection technical solutions	Deicing	Anti-icing
Pneumatic boot systems	x	o
Hot air systems	x	x
Electrical resistance systems	x	(x)
Fluid systems	x	(x)

x solution usually applied to protection principle
(x) solution usually not applied to protection principle
o solution can not be applied to protection principle

Table 2.6 Ice Protection Technical Solutions and Protection Principles

Aircraft icing is thus possible if

1. The air contains water (clouds are an indication of water in the air).
2. The air temperature is below 0°C.
3. The air temperature is above –40°C
4. The aircraft surface is below 0°C.

There are *other icing mechanisms* besides the standard one just discussed:

- Icing will occur during descent from high altitudes if the aircraft encounters humid air even above 0°C. The aircraft surface will be below 0°C due to a long flight at high altitudes. The fuel in the wings will also be below freezing. The fuel is in close contact with the skin as a consequence of integral fuel tank design (see Subsection 2.7). The fuel does not warm up quickly and is likely to remain below 0°C until landing.
- Carburetor icing can occur at temperatures between –7°C (20°F) and –21°C (70°F) when there is visible moisture or high humidity. Carburetor icing is caused by cooling from vaporization of fuel, combined with the expansion of air as it flows through the carburetor.
- Water and slush that the aircraft picks up during taxi out can freeze at higher altitudes with detrimental effects to the aircraft.
- Frost, ice, and snow that have settled on an aircraft on the ground have to be removed before takeoff. Ground deicing equipment and procedures have been developed (see Section AC 135–16).

The two *basic forms of ice build-up* on the aircraft surface are clear ice and rime ice (Figure 2.44).

- Clear ice forms between 0°C and –10°C, usually from larger water droplets or freezing rain, and can drastically change the form of the leading edge. It can spread over the surface.
- Mixed ice forms between –10°C and –15°C. A mixture of clear ice and rime ice have the bad characteristics of both types and can form rapidly.
- Rime ice forms between –15°C and –20°C from small droplets that freeze immediately when contacting the aircraft surface. This type of ice is brittle, rough looking, and colored milky white.

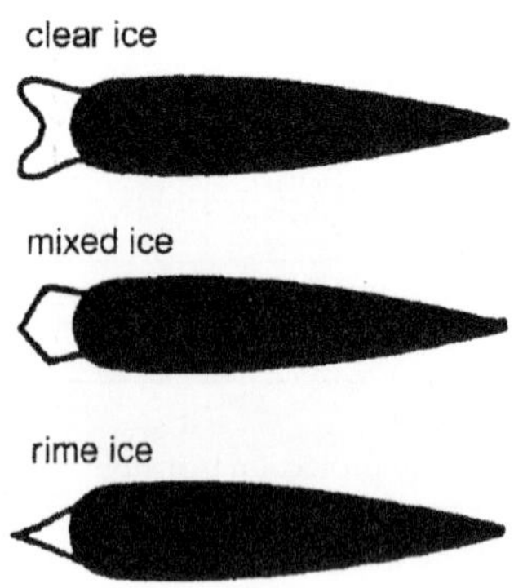

FIGURE 2.44 Ice shapes on the leading edge of airfoils (TÜV 1980).

In order to *calculate the total water catch* of the wing, let us cut off a piece of a wing with a spanwise extension Δy and maximum thickness t. This piece of wing will fly at a speed v through a unit volume of air with a certain mass of supercooled water. The mass of supercooled water per volume is called liquid water content (LWC) and is something like a density we name ρ_{LWC}. We consider $t \cdot \Delta y$ as the area of an imaginary sieve at an angle perpendicular to its flight path. The mass flow rate of supercooled water through the sieve would be $\dot{m} = vt\ \Delta y\ \rho_{LWC}$. The impingement of water on the leading edge of the wing will, however, be different from the flow through the sieve as shown in Figure 2.45. The air and with it very small droplets pass around the wing; only larger droplets hit the surface. This phenomenon is expressed by the water catch efficiency E_m. The imaginary sieve shows an efficiency $E_m = 1$. The total water catch of a piece of wing is calculated by including E_m:

$$\dot{m} = vt\ \Delta y\ \rho_{LWC} E_m$$

E_m is a function of aircraft speed and droplet size, airfoil shape and thickness, viscosity, and density of the air.

- High aircraft speeds and large droplet size cause an increase in water catch efficiency.
- High aircraft velocities, however, lead to aerodynamic heating of the leading edges. This reduces icing.
- Thin wings divert the flow less and increase the water catch efficiency.

AIR 1168/4 presents detailed methods to calculate E_m.

A *simplified method to calculate the water catch efficiency* E_m is presented here based on Figure 3F-3 of AIR 1168/4 as a function of aircraft speed v and wing thickness t (Figure 2.46):

$$E_m = 0.00324\left(\frac{v}{t}\right)^{0.613} \quad \text{for } v \text{ in m/s and } t \text{ in m}$$

This equation is based on typical airfoils with a relative thickness of 6–16% at an angle of attack $\alpha = 4°$. The mean effective drop diameter $d_{med} = 20\ \mu m$, altitude $h = 10{,}000$ ft. Other altitudes from sea level to $h = 20{,}000$ ft will result in an error less than 10%.

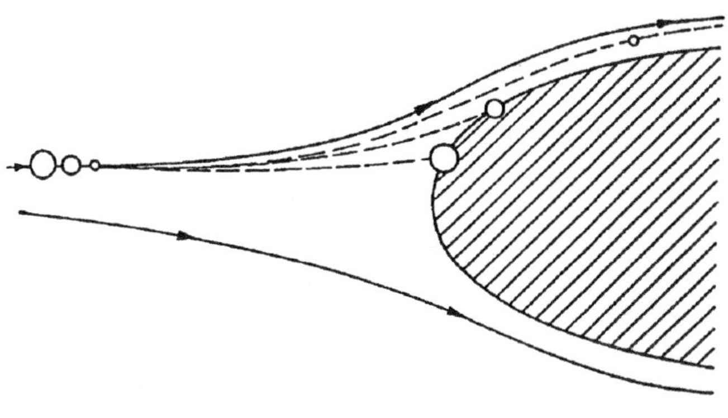

FIGURE 2.45 Flow around a wing leading edge: streamlines of dry airflow; trajectories of differently sized droplets (TUHH).

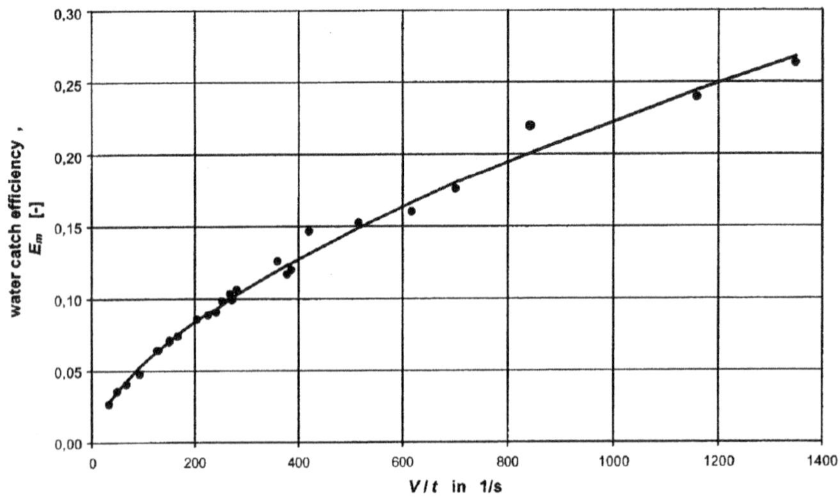

FIGURE 2.46 Water catch efficiency E_m as a function of aircraft speed v and wing thickness t for typical applications. The diagram is calculated from AIR 1168/4, Figure 3F-3.

AMC 25, Section 1419 assumes for certification a typical mean effective drop diameter $d_{\text{med}} = 20\ \mu\text{m}$. The *liquid water content* (LWC) that an aircraft is supposed to meet continuously in flight ranges from $\rho_{\text{LWC}} = 0.2\ \text{g/m}^3$ at $-30°\text{C}$ to $\rho_{\text{LWC}} = 0.8\ \text{g/m}^3$ at $0°\text{C}$.

The *detrimental effects of icing* on the aircraft are manifold. Ice can:

- Alter the shape of an airfoil. This can change the angle of attack at which the aircraft stalls, and cause the aircraft to stall at a significantly higher airspeed. Ice can reduce the amount of lift that an airfoil will produce and increase drag several-fold.
- Partially block control surfaces or limit control surfaces deflection.
- Add weight to the aircraft. The aircraft may not be able to maintain altitude. The stall speed is higher.
- Block the pitot tube and static ports.
- Cause the breakage of antennas on the aircraft.
- Cause a tailplane stall. The airplane will react by pitching down, sometimes in an uncontrollable manner.
- Reduce propeller efficiency. Ice that is hurled away from the propeller is a hazard to everything in its plane of rotation.
- Endanger the internal parts of a jet engine.

In order to protect the aircraft properly against these effects, ice protection may become necessary in areas shown in Figure 2.47.

The design of ice protection systems will always have to be based on the *certification requirements.* For transport category aircraft, the fundamental statement reads: "If certification for flight in icing conditions is desired, the airplane must be able to safely operate in the continuous maximum and intermittent maximum icing conditions" (Section 1419, FAR Part 25, CS-25). Icing conditions are given in Appendix C of these documents.

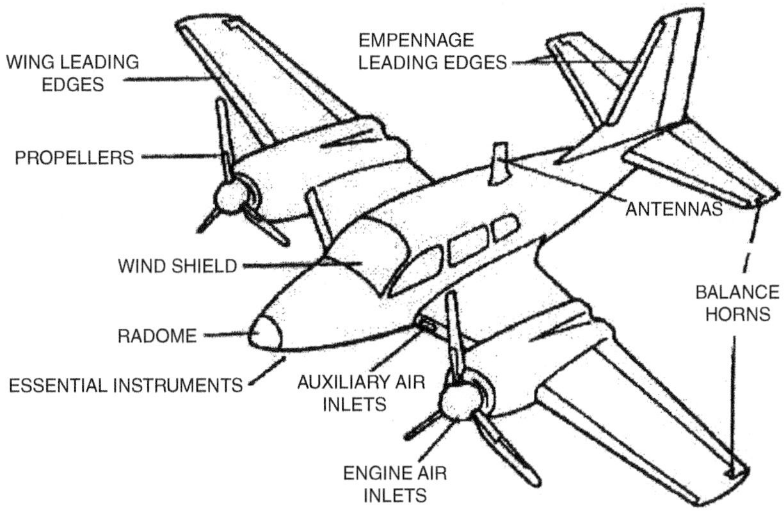

Figure 2.47 Areas of the airframe that may require ice protection (FAA 1993).

Critical parts of the aircraft (like the wing) will probably need some kind of ice protection device. Other parts or the aircraft (like the empennage) may fulfill the requirements without being protected.

Pneumatic Boot Systems

Pneumatic boot systems have been the standard ice protection method for piston engine aircraft since the 1930s. The *boot surfaces* remove ice accumulations mechanically by alternately inflating and deflating tubes within a boot that covers the surface to be protected (Figure 2.48). Inflation of the tubes under the accreted ice breaks the ice into particles and destroys the ice bond to the surface. Aerodynamic forces and centrifugal forces on rotating airfoils then remove the ice.

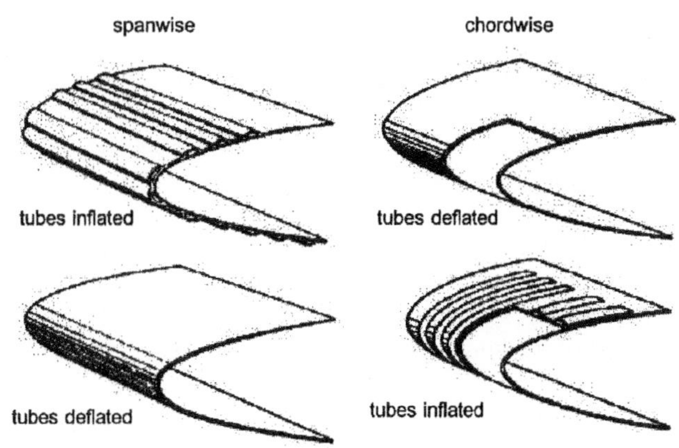

Figure 2.48 Inflatable deicing boots (FAA 1993).

In principle, this method of deicing is designed to remove ice after it has accumulated rather than to prevent its accretion in the first place. Thus, by definition a pneumatic boot system cannot be used as an anti-icing device. Conventional pneumatic boots are constructed of fabric-reinforced synthetic rubber or other flexible material. The material is wrapped around and bonded to the leading-edge surfaces to be deiced on wings or empennage. Total thickness of typical pneumatic boots is usually less than 1.9 mm (0.075 in.). Pneumatic boots require very little power and are a lightweight system of reasonable cost. The tubes in the pneumatic boot are usually oriented spanwise but may be oriented chordwise. The inflatable tubes are mani-folded together in a manner to permit alternate or simultaneous inflation as shown in Figure 2.48. Alternate inflation is less commonly used.

In addition to the boots, the primary *components of a pneumatic system* are a regulated pressure source, a vacuum source, and an air distribution system. Miscellaneous components may include check and relief valves, air filters, control switches and timer, and electrical interfaces, including fuses and circuit breakers. A regulated pressure source is required to ensure expansion of all tubes in the system to design limits and within design rise times. Pneumatic boots should inflate and deflate rapidly to function effectively. The time to reach full pressure should be about 5 to 6 seconds. If tube expansion is too slow, deicing effectiveness is lessened. The vacuum source is essential to ensure positive deflation and keep the tubes collapsed during nonicing flight conditions to minimize the aerodynamic penalty. Air pumps generally multiply the atmospheric pressure by a fixed factor, so the pressure delivered becomes a function of altitude. Therefore, for air pump systems, the pressure produced at service ceiling altitude is a design condition.

Some aerodynamic *drag penalty* is to be expected with pneumatic boot deicing systems on an airfoil, but it can be lessened by recessing the surface leading edge to offset the boot thickness. Pneumatic boot deicing systems have been in use for many years, and their repair, inspection, *maintenance,* and replacement are well understood. Pneumatic boot material deteriorates with time, and periodic inspection is recommended to determine the need for replacement. System *weight and power* requirements are minimal.

Ice bridging is the formation of an arch of ice over the boot, which is not removed by boot inflation. In the lore of flight of early piston-powered air transports, it used to be recommended that some ice should be allowed to accrete before the deicing system was turned on in order to avoid ice bridging. The aircraft flight manual (AFM) for modern aircraft now requires that the system be activated at the first sign of ice formation. Ice bridging for modern, properly functioning deicing boots has not been reported (FAA 1993).

Hot Air Systems

Hot air systems and electrical resistance systems are *thermal ice protection systems.* Thermal ice protection systems are classified into three groups:

1. *Evaporative anti-icing systems* supply sufficient heat to evaporate all water droplets impinging upon the heated surface.
2. *Running-wet anti-icing systems* provide only enough heat to prevent freezing on the heated surface. Beyond the heated surface of a running-wet system, the water can freeze, resulting in runback ice. For this reason, running-wet systems must be used carefully so as not to permit buildup of runback ice in critical locations. For example, a running-wet system may be used for a turbine engine inlet duct where the runback is permitted to enter the engine.
3. *Cyclic deicing systems* periodically shed small ice buildups by melting the surface–ice interface with a high rate of heat input. When the adhesion at the interface becomes zero, aerodynamic or centrifugal forces remove the ice.

An evaporative anti-icing system uses the most energy of the three ice protection principles presented, cyclic deicing uses the least energy.

Hot air systems are used on most of the large jet transports because of the availability of hot air from the engines and the relative efficiency and reliability of these systems. Hot air is used to anti-ice or deice leading-edge wing panels and high-lift devices, empennage surfaces, engine inlet and air scoops, radomes, and selected components. Details of the hot air system are given below, using the Airbus A321 as an example.

Electrical Resistance Systems

Electrical resistance systems are thermal ice protection systems. They may also be classified as *evaporative, running wet,* or *cyclic.*

Electrical resistance systems have a wide range of application:

- *External components* protected with an electrical resistance system will use the evaporative technique.
- *Internal components* like water lines are simply heated above 0°C to prevent freezing.
- *Nontransparent surfaces* will most probably use cyclic deicing because the electrical loads would otherwise become unbearable.

Electrical resistance systems use electrical resistance heaters in the form of foil, film, resistance wire, or mesh embedded in fiberglass, plastic, rubber, or metal to heat the surface or component.

Electrical deicing systems for nontransparent surfaces may use *parting strips* to divide the total protected area into smaller sequentially heated areas. The spanwise and chordwise parting strips must prevent any ice bridging from one shedding zone to another (Figure 2.49). Parting zones reduce the total instantaneous power requirement and maintain a stable load on the electrical system. Aircraft wings with about 30° or more sweepback will normally use only chordwise parting strips.

For efficient deicing protection, the correct *amount of heat* must be supplied. If there is too little heat, the ice may not shed as required, perhaps causing large chunks of ice to shed. If too much heat is supplied, there can be too much melting, resulting in undesirable amounts of runback ice. It has been found desirable to have a high specific heat input applied over a short period.

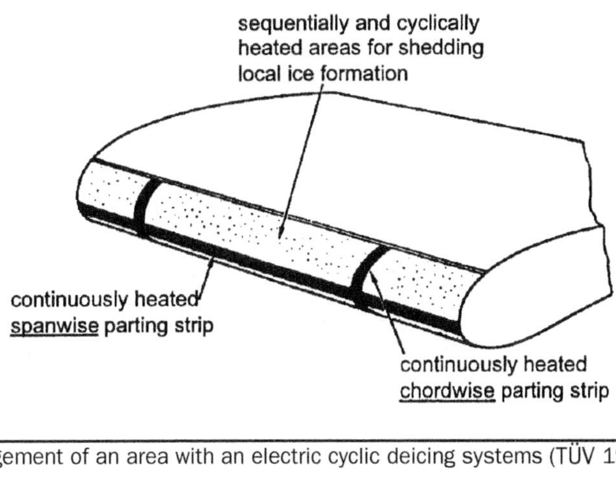

FIGURE 2.49 Arrangement of an area with an electric cyclic deicing systems (TÜV 1980).

The *off-time* of a shedding zone depends upon the rate at which the surface cools to 0°C (32°F). It also depends upon the icing rate. The off-time may be tailored to the maximum ice thickness allowed for the application and can be as long as 3 to 4 minutes for fixed-wing aircraft.

The biggest *disadvantage* of an electrical resistance system for large surfaces is the high power demand. If additional generators are installed just for the purpose of ice detection, the system will get very heavy.

Fluid Systems

Fluid ice protection systems operate on the *principle* that the surface to be protected is coated with a fluid that acts as a freezing point depressant (FPD). Current systems use a glycol-based fluid. When supercooled water droplets impinge on a surface, they combine with the FPD fluid to form a mixture with a freezing temperature below the temperature of the ambient air. The mixture then flows aft and is either evaporated or shed from the trailing edge of the surface. FPD fluid is distributed onto the surface leading edge by pumping it through porous material or spraying the fluid onto the surface.

The use of a freezing point depressant can provide anti-icing or deicing protection. The anti-icing mode is the normal mode of operation in light to moderate icing conditions. The deicing mode is a condition allowing ice to accumulate and bond to the wing surface. When the fluid ice protection system is turned on, a flow is introduced between the ice and the surface to weaken the bond so that the ice is shed by aerodynamic forces.

FPD fluid is stored in a tank. A pump meters the system's fluid flow requirements. Porous panels are constructed typically of sintered stainless steel mesh or laser-drilled titanium for the outer skin, a stainless steel or titanium backplate to form a reservoir, and a porous plastic liner to provide uniform control of panel porosity (Figure 2.50).

The principle *disadvantage* of the fluid protection system is the fluid storage requirement. The stored fluid weight may be significant when compared to other candidate ice protection systems. The system has a finite period of protection, dependent on fluid supply (FAA 1993).

Windshield Ice and Fog Protection

Windshield panels are usually provided with anti-icing protection on those aircraft that are required to operate in all weather conditions. The most widely used system is an *electrical resistance system* for anti-icing, whereby electric current is passed through a transparent conductive film or

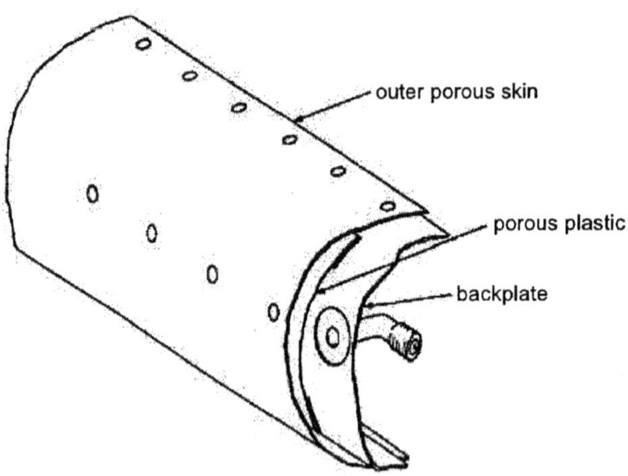

FIGURE 2.50 Construction of a typical porous panel (FAA 1993).

resistance wire that is part of the laminated windshield. The heat from the anti-icing film or resistance wire also achieves internal defogging. Electrical heat may also be used to maintain the windshield layers of glass and plastic near the optimum temperature for resistance against bird strikes.

Where electric power seems not to be the adequate solution, an *external hot air blast system* can be an alternative. This system may also be used for rain removal.

Windshield Rain Protection

Rain-removal systems are designed to allow the pilots to have a clear view out of the cockpit at the airport and during departure and approach. The systems are not commonly used during flight at altitude. Rain may be removed by the use of windshield wipers. Alternatively, an external hot air blast can clear the windshield. In addition to either one of the two systems, a chemical rain repellent may be used.

Windshield wipers perform adequately, although their ability is limited. High oscillation rates are desirable to keep up with high rates of rain impingement during heavy rainfall. Sufficient blade pressure on the windshield must be maintained to produce satisfactory wiping when the aerodynamic forces are high at high aircraft speeds. Unfortunately, wipers also cause considerable aerodynamic drag.

An *external hot air blast* operates on the principle of blanketing the outside surface of the windshield with a protective wall of high-velocity, high-temperature air. The air blast prevents water impingement by deflecting many of the incoming raindrops. Water on the surface that has penetrated the air blast will be evaporated.

Rain repellent may be sprayed on the windshield to form a transparent film that reduces the adhesive force between the water and the glass. The water draws up into beads that cover only a portion of the glass. The high-velocity slipstream continually removes the beads. Depending on the rain intensity, the rain impingement breaks down the repellent film, causing the window to return gradually to a wettable condition. Unless the windshield is wiped off frequently, the effectiveness or repeated repellent application decreases. Windshield wipers spread the repellent and improve its efficiency. Rain repellent used together with an external hot air blast is used in the critical landing phase when engine bleed air pressure is low and the jet blast is reduced.

Ice-Detection Systems

Some method of ice detection is necessary so that the ice protection system is operated only when necessary. Two methods exist: visual detection and electronic detection.

Visual detection is achieved by the flight crew monitoring such things as windshield wipers, wing leading edges, pylons, or landing lights that could serve as an ice datum. Those surfaces of the airplane directly exposed to stagnation flow conditions usually accumulate the largest quantity of ice. Wing and engine scan lights are used to monitor the engine intakes and the wing leading edges at night.

Electronic ice detectors consist of a probe extending into the free stream. The probe vibrates at a known frequency. When ice starts to build on the probe, the frequency will decrease. This will be detected by an attached controller. The controller will energize a heating element in the probe to remove the ice so that the probe can check again for icing conditions.

Example: Airbus A321

The ice and rain protection system lets the aircraft operate normally in ice conditions or heavy rain. Ice protection is given by the use of hot air or electrical power to make the necessary areas of the aircraft hot. The areas supplied by hot air are (Figure 2.51):

- The leading edge of slats 3, 4, and 5 on each wing
- The engine air intakes

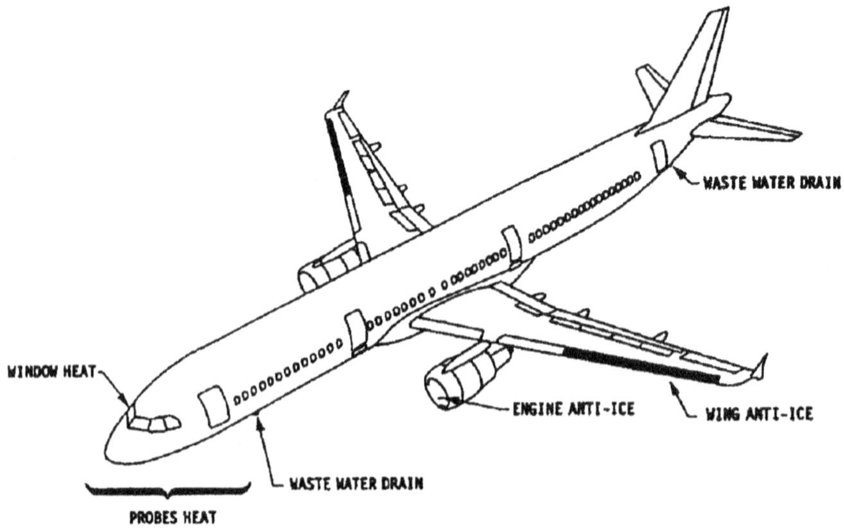

FIGURE 2.51 A321 ice and rain protection component locations.

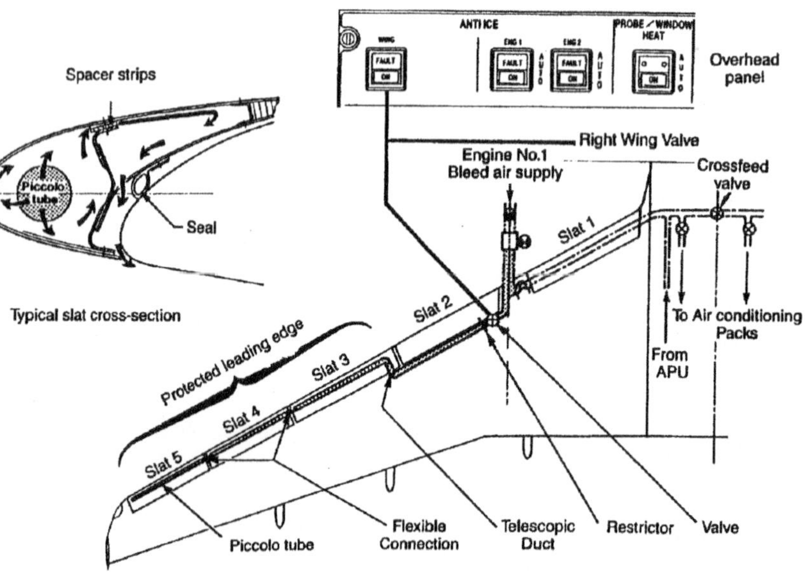

FIGURE 2.52 A321 wing anti-ice.

The engine bleed air system supplies the hot air to the anti-ice system.

The items with electrical heaters are:

- The cockpit windshield and side windows
- The total air temperature (TAT) probes
- The angle of attack (alpha) probes
- The pitot and static probes of the air data system (ADS)
- The wastewater drain masts

Rain is removed from the windshield with windshield wipers.

The A321 wing ice protection system is a hot air evaporative anti-ice system. Only slats 3, 4, and 5 on the outboard wing need to be ice protected. The hot air is bled from the engine. Each engine supplies its related wing. On both wings, an anti-ice valve isolates the anti-ice system from the bleed air supply. When the crossfeed valve is open, it is possible to supply the two wings from only one engine bleed-air system. Lagged ducts connect the anti-ice valve to a telescopic duct at slat 3. A piccolo tube runs along slats 3, 4, and 5 and supplies the hot air to the leading edge. A piccolo tube is a tube with calibrated holes that ensures that hot air is evenly distributed along the leading edge, although bleed pressure decreases toward the wing tip. The bleed air in the slats is released overboard through the holes in the bottom surface of the slat. The operation of the anti-ice valve is controlled by the WING push-button switch on the ANTI-ICE overhead panel in the cockpit.

2.10 Landing Gear (ATA 32)

Landing gear is defined by ATA 100:

> Those units and components which furnish a means of supporting and steering the aircraft on the ground or water, and make it possible to retract and store the landing gear in flight. Includes tail skid assembly, brakes, wheels, floats, skids, skis, doors, shock struts, tires, linkages, position indicating and warning systems. Also includes the functioning and maintenance aspects of the landing gear doors but does not include the structure [of the doors].

Following ATA 100, the landing gear system may be subdivided into:

- Main gear and doors
- Nose gear and doors
- Extension and retraction system
- Wheels and brakes
- Steering system
- Position indicating and warning
- Supplementary gear (devices used to stabilize the aircraft while on the ground and prevent damage by ground contact)

Landing gear design has always been an integral part of aircraft design. The aircraft configuration cannot be laid out without due considerations given to the landing gear. Details of the steering system, the extension and retraction system, as well as the wheels and brakes may be the subject of separate studies.

Further Reading includes literature on landing gear design.

2.11 Lights (ATA 33)

Lights as defined by ATA 100:

> Those units and components (electrically powered) which provide for external and internal illumination such as landing lights, taxi lights, position lights, rotating lights, ice lights, master warning lights, passenger reading and cabin dome lights, etc. Includes light fixtures, switches and wiring. Does not include warning lights for individual systems or self-illuminating signs.

Example: Airbus A321

Detailed requirements for instrument lights, landing lights, position lights, anti-collision lights, ice-detection lights, and emergency lighting are laid down in the certification requirements Sections 1381 to 1403 and 812. Much room for varying system designs is thus not permitted. Innovation has been brought in, however, through new lighting technologies and new circuit designs to control light intensities. The Airbus A321 lighting system provides illumination inside and outside of the aircraft. The system includes different parts.

The *cockpit lighting* consists of the following subsystems:

- General illumination of cockpit panels, instruments, and work surfaces
- Integral lighting of panels and instruments
- Test system for annunciator lights
- Dimming system for annunciator lights

The *cabin lighting* consists of the following subsystems:

- General illumination of cabin, galley areas, and entrances
- Illumination of the lavatories
- Passenger reading lights (customer option)
- Cabin lighted signs
- Work lights for the cabin attendants

The *cargo and service compartment lighting* provides illumination and power outlets for maintenance purposes. The system includes:

- Service area lighting for equipment and APU compartments
- Air conditioning duct and accessory compartment lights
- Cargo compartment lights
- Equipment compartment lights
- Wheel well lighting

The *external lighting* system illuminates the runways and/or taxiway, some aircraft surfaces, and gives an indication of the aircraft's position. The system (see Figure 2.53) consists of different lights:

- Two anticollision beacon lights (1) which flash red, installed one at the top and one at the bottom of the fuselage
- Two wing and engine scan lights (2) installed one at each side of the fuselage to illuminate the wing leading edge and engine air intakes to detect ice accretion

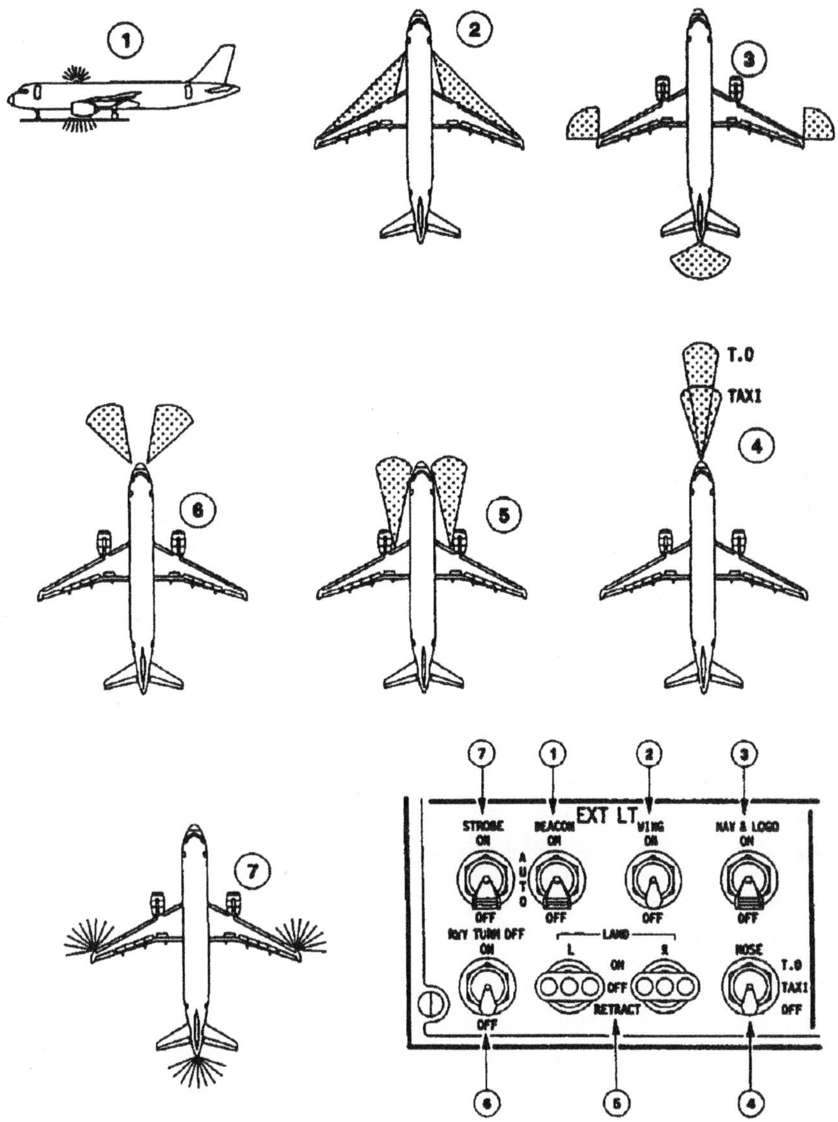

FIGURE 2.53 A321 external lights.

- Three navigation lights (3), colored red (port), green (starboard), and white (tail), installed one at the tip of each wing and one aft of the fuselage
- Two logo lights (not shown) installed in the upper surface of each horizontal stabilizer to illuminate the company logo on the vertical stabilizer, provided the main gear struts are compressed or the flaps are extended
- One fixed-position takeoff light (4) (600 W) and one fixed-position taxi light (4) (400 W) installed on the nose landing gear

- Two retractable landing lights (5) (600 W) installed one under each wing
- Two fixed runway turnoff lights (6) installed on the nose landing gear
- Three synchronized strobe lights (7), one on each wing tip and one below the tail cone

The *emergency lighting* system provides illumination with batteries independently of the aircraft power supplies in the event of a failure of the main lighting system. Illumination is provided for:

- The cabin and the exit areas
- The exit location signs and the exit marking signs at all doors
- The door escape slides
- The marking system of the emergency escape path
- The lavatories

2.12 Oxygen (ATA 35)

The oxygen system as defined by ATA 100:

> Those units and components which store, regulate, and deliver oxygen to the passengers and crew, including bottles, relief valves, shut-off valves, outlets, regulators, masks, walk-around bottles, etc.

Human Oxygen Requirements

The *human reaction to a lack of oxygen* depends on altitude. Normally, individuals living at sea level may become aware of the effects of altitude at about 3,048 m (10,000 ft). Above 10,000 ft, piloting skills are degraded. Up to 4,267 m (14,000 ft), the body is more or less able to compensate for the diminishing partial oxygen pressure by a higher breathing frequency. Above 14,000 ft, compensation is not possible anymore and hypoxia symptoms (headache, etc.) become apparent. Above 6,096 m (20,000 ft) unconsciousness and death are only a function of time. If a person is exposed to an altitude of 9,144 m (30,000 ft), unconsciousness may well set in after 1 minute. At an altitude of 15,240 m (50,000 ft), unconsciousness may set in after 10 s.

In order to compensate these effects, the partial oxygen pressure can be increased by breathing higher oxygen concentrations. The partial oxygen pressure[14] p at sea level (SL) is

$$0.21 p_{SL} = 0.21 \cdot 1013 \text{ hPa} = 212.7 \text{ hPa}$$

If this partial pressure is to be maintained with altitude h, the required oxygen concentration x is

$$x = \frac{212.7 \text{ hPa}}{p(h)}$$

As can be seen from Figure 2.54, 100% (pure) oxygen is required at an altitude of about 37,000 ft. Beyond 37,000 ft, it becomes necessary to increase the pressure of the oxygen delivered to the mask in order to provide a sea-level equivalent environment. The lungs are in effect supercharged by the differential pressure between the mask and the surrounding pressure in the (nonpressurized) cabin.

It is evident that *cabin decompression* at high altitudes requires immediate action by the crew. Passengers and crew have to be provided with oxygen, and an emergency descent has to be initiated.

---

[14]Many special terms relevant to the oxygen system are defined in Subsection 2.2.

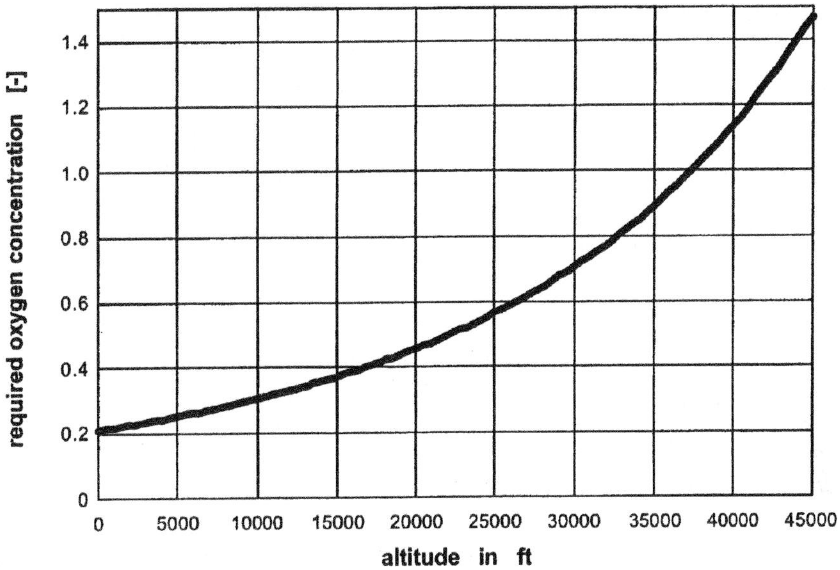

Figure 2.54 Required oxygen concentration with altitude.

The lower the aircraft gets, the longer the survival time. Circumstances are eased by the fact that even a big hole in the structure does not instantly lead to ambient pressure in the cabin.

Certification requirements for transport category aircraft (with pressurized cabins) state, e.g., "If certification for operation above 30,000 ft is requested, the dispensing units providing the required oxygen flow must be automatically presented to the occupants before the cabin pressure altitude exceeds 15,000 ft" (CS-25, Section 1447).

System Classification

A classification of oxygen systems may take various aspects into account. We will look at classifications based on:

- Various reasons for oxygen supply
- Fixed versus portable oxygen equipment
- Oxygen regulator types
- Oxygen mask types
- Different oxygen sources
- The type of person supplied with oxygen:
    - Passenger oxygen system
    - Crew oxygen system

An *oxygen supply* may be necessary *for various reasons.* During high-altitude flights in nonpressurized cabins, normal oxygen supply is part of the normal flight procedures. In case of a failure of the normal supply, emergency oxygen is needed. In pressurized cabins emergency oxygen is supplied to all passengers and crew in case of cabin decompression. Provisions may have to be made for the supply of sustenance oxygen to a limited number of passengers after an emergency descent.

Provisions also have to be made to supply first-aid oxygen to individual passengers for medical reasons. "'Supplemental oxygen' means the additional oxygen required to protect each occupant against the adverse effects of excessive cabin altitude and to maintain acceptable physiological conditions" (CS 1).

Oxygen equipment may be grouped into fixed and portable equipment. Fixed equipment is provided in those aircraft in which oxygen is frequently required or many passengers are involved. Additional portable equipment is used to allow the crew to move in the aircraft cabin under varying conditions. This could include the use of portable equipment when fighting small cabin fires. Portable equipment is also used for first-aid oxygen supplies to individual passengers. Small aircraft with nonpressurized cabins may not have a fixed oxygen system installed, so portable equipment is taken aboard whenever the situation arises due to planned high-altitude flights.

Oxygen taken from a bottle that provides a continuous flow via a supply hose directly into the mouth would technically be the easiest way to inhale. Although this was historically the first method applied, it has several disadvantages. The most apparent are:

1. Oxygen will be wasted during exhalation.
2. There is no need to inhale 100% oxygen at low altitudes.
3. There will be a need to hold the hose.
4. Communication will be hampered.
5. In a toxic environment (smoke) a face protection will be missing.

In order to overcome disadvantages 1 and 2, different *types of oxygen systems* based on the regulator design have evolved: the continuous flow system, the demand system, the pressure-demand system, the diluter-demand system, and the pressure-demand system with dilution at low altitudes. The most common systems in transport aircraft are the continuous flow system for passengers and the diluter-demand system for members of the flight crew.

Problems 3, 4, and 5 are addressed with the specific design of the *oxygen masks.*

Regulators

A *continuous-flow system* provides—as the name indicates—a continuous flow of oxygen to the mask. In order not to waste the volume of oxygen flowing toward the mask during exhalation, a flexible plastic or rubber reservoir is incorporated between the mask and the supply hose. The reservoir that is used to collect the oxygen has typically a volume of 0.5 to 1.0 L. During inspiration the stored oxygen can be used together with the oxygen currently flowing. Three valves are built into a continuous-flow mask: an exhalation valve and a nonreturn valve to the reservoir and a dilution valve. The exhalation valve opens the mask to ambient air during exhalation. At the same time, the nonreturn valve to the reservoir closes to prevent used air to enter the oxygen reservoir. When the reservoir has been emptied during the first part of the inhalation phase, the dilution valve opens and allows ambient air to dilute the already inhaled oxygen from the reservoir during the second part of the inhalation phase. The primary disadvantage of the constant-flow system is its inability to adjust itself automatically to various levels of physical activity. A regulator could, however, be provided for manual adjustment of flow to the reservoir. A constant-flow regulator provides automatic control of the flow depending on altitude. This capability evidently depends on the ability of the oxygen source to allow for varying flows. Varying the flow of oxygen is not always possible; the chemical oxygen generators commonly used in aircraft cabins do not allow flow control.

A *demand system* provides—as the name indicates—a flow of oxygen only on demand, i.e., during the inhalation phase, conserving oxygen during exhalation. A demand system requires a demand oxygen regulator for each user. The regulator may be panel-mounted, man-mounted, or seat-mounted. The regulator includes an outlet control valve that responds to minute changes in pressure. The slight negative pressure (compared to ambient cabin pressure) created within the mask at the onset of inhalation opens the valve and permits a flow of oxygen into the mask. At the end of the inhalation phase, the pressure has become slightly positive and the valve shuts off the flow. Masks for demand systems have to fit tightly. If the breather drew too much ambient air around the mask, the mask could not hold negative pressure and hence the regulator could not function properly.

A *pressure-demand system* is a demand system that has the ability also to supply oxygen under positive pressure (compared to ambient cabin pressure) to the mask. The principal components of the system are a mask that has the ability to hold positive pressure and an oxygen pressure regulator. A pressure-demand system is necessary for operation at altitudes above 10,668 m (35,000 ft) to maintain safe partial pressure for the user (compare with Figure 2.54).

The *diluter-demand system* is a demand system that has the ability to control the air–oxygen ratio automatically depending on altitude. The purpose of air dilution is to conserve the aircraft oxygen supply further and still maintain a safe partial pressure. For safe operating conditions, dilution occurs up to 9,754 m (32,000 ft). At this altitude the dilution port in the diluter-demand oxygen regulator, which is automatically controlled, is shut off and the regulator delivers 100% oxygen. Besides an on-off-type supply lever, these regulators have an oxygen-selection lever to obtain 100% oxygen delivery throughout the whole altitude range. Some models are also provided with an emergency lever which, when actuated, will deliver a limited amount of positive pressure (safety pressure) for emergency toxic atmosphere protection (Figure 2.55).

Masks

Different oxygen masks exist. Apart from the differences resulting from the type of oxygen system for which they are used (see above), we may differentiate various types.

The *nasal mask* fits snugly around the nose and is intended for flights below 4,877 m (16,000 ft), where air intake through the mouth is acceptable. The *oronasal mask* fits completely over the mouth

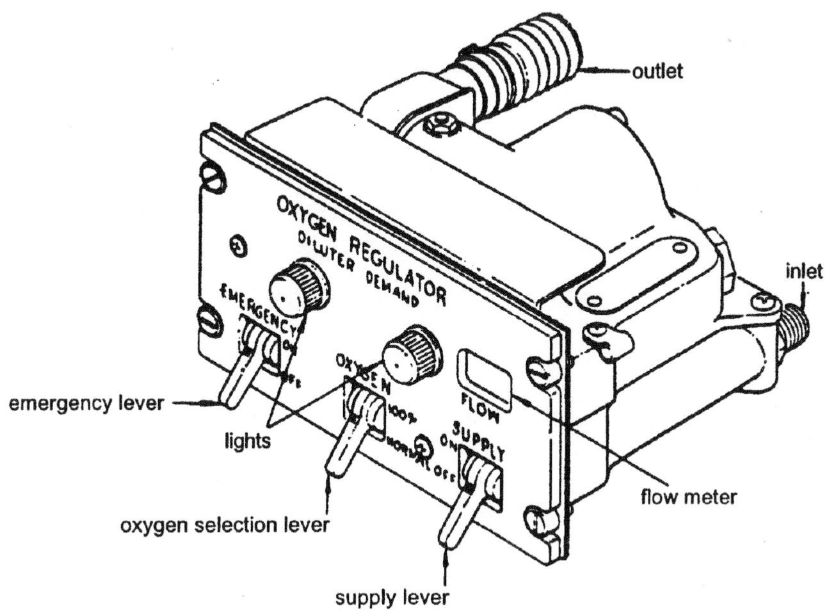

Figure 2.55 Basic panel-mounted diluter-demand oxygen regulator (VFW 614).

and nose. Provisions are made for the inclusion of a microphone for communication purposes. *Full-face masks* cover the mouth, nose, and eyes. These masks can meet protective breathing equipment requirements but cannot be used in a pressure-demand system because the eyes should not be exposed to a positive pressure. *Goggles* combined with an oronasal mask can meet both protective breathing and pressure-demand requirements.

If certification for transport category aircraft is sought for operation above 25,000 ft, each flight crew member must be provided with a *quick-donning mask* (see Figure 2.57) that can be put on within 5 seconds (CS-25, Section 1447). Quick-donning masks are equipped with an inflatable harness. The crew member presses a side lever on the mask when passing the harness over the head. The side lever guides pressurized oxygen into the harness, causing the harness to stretch. When the side lever is released, the oxygen escapes from the harness and integrated straps pull the harness tightly to the head.

A *smoke hood*, a mask used to fight small cabin fires, protects the head and parts of the body and includes some type of oxygen supply.

Sources

Oxygen supply may be in the form of gaseous oxygen supply, liquid oxygen (LOX) supply, chemical oxygen supply, and on-board oxygen generation (OBOG).

Gaseous oxygen is stored in the aircraft in special oxygen cylinders. U.S. oxygen cylinders are colored green. They are properly marked and must only be filled with aviators breathing oxygen. Charge pressure is 12.8 MPa (1,850 psi). Oxygen cylinders are fitted with a combined flow-control and pressure-reducing valve as well as a pressure gauge. Two types of high-pressure cylinders exist: standard weight cylinders and lightweight cylinders. These cylinders are certified to Department of Defense (DoD) standards. They must regularly be checked and are life limited. Safety precautions have to be adhered to because of the general danger associated with such pressure vessels and the risk involved with handling oxygen. Crew oxygen systems on transport aircraft use gaseous oxygen.

Oxygen boils at sea-level pressure at –183°C. The highest boiling point is –118°C at 5.07 MPa. Hence, *liquid oxygen* has to be below that temperature. Liquid oxygen is stored in insulated tanks. Special equipment is required to convert liquid oxygen to gaseous oxygen on-board the aircraft. Liquid oxygen systems show weight and space savings compared to equivalent gaseous oxygen systems. Evaporation losses, however, can amount to 5% per 24 hours and need constant refilling in service. For these reasons, liquid oxygen systems are used on most combat aircraft but seem impractical for civil operation.

Chemical oxygen generation on aircraft is done with sodium chlorate. Sodium chlorate decomposes when heated to 478°C into salt and oxygen:

$$2NaClO_3 + heat \rightarrow 2NaCl + 3O_2$$

The heat is generated with some kind of fuel, commonly iron. The chemical reaction is:

$$NaClO_3 + Fe \rightarrow NaCl + FeO + O_2 + heat$$

The overall mass balance of both equations combined: 100% sodium chlorate yields 45% oxygen by weight, 38% of which is delivered and 7% of which is used in oxidizing of the iron. The chlorate core is located in the center of the generator and is insulated against the outside steel housing. Nevertheless, the outside of the generator reaches temperatures of up to 260°C, so that adjacent aircraft components need to be protected against the generator. The oxygen cools quickly and has reached normal temperatures when it arrives at the mask. The chemical reaction is self-sustained and can be started mechanically (in most aircraft by pulling a lanyard) or electrically (Lockheed L-1011) with an adequate device on the generator. An outlet filter holds back particles and gaseous

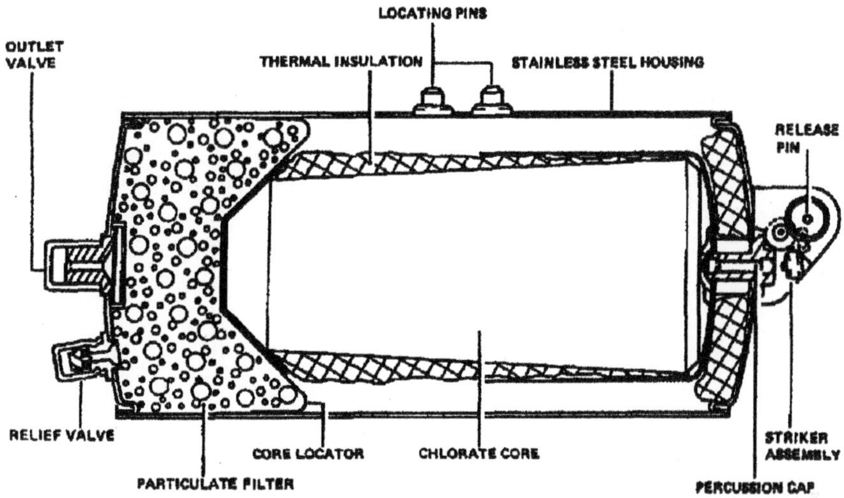

Figure 2.56 Chemical oxygen generator (Airbus A321).

impurities. The reaction cannot be stopped once it is in progress. In case the outlet gets blocked, a pressure-relief valve averts an explosion of the generator. Figure 2.56 shows a cross-section of a chemical oxygen generator. Its diameter determines the flow rate and its length and the duration of the supply. Generators are designed for a flow duration of about 15 minutes. The overall flow rate depends on the number of masks attached to the generator (1, 2, 3, or 4) and on certification requirements. The flow rate decreases over the duration of the supply. Most transport aircraft use chemical oxygen generation for the passenger oxygen system because of weight and maintenance savings compared with gaseous oxygen supply.

On-board oxygen generation systems (OBOGS) apply electrical power and bleed air to produce breathable oxygen from ambient air. Various techniques exist. Air can be processed through molecular sieve beds to provide oxygen-enriched breathing gas.

Example: Airbus A321

The aircraft has three separate oxygen systems: a flight crew oxygen system, a passenger oxygen system, and a portable oxygen system.

The *flight crew oxygen* system (Figure 2.57) supplies oxygen to the flight crew if there is a sudden decrease in cabin pressurization. It also supplies oxygen if there is smoke or dangerous gases in the cockpit. Each crew station has a quick-donning mask with a demand regulator installed. The oxygen is supplied from a high-pressure oxygen cylinder to the masks through a pressure regulator/transmitter assembly and a distribution circuit.

The *passenger oxygen* system provides emergency oxygen for passengers and cabin attendants (Figure 2.58). Emergency oxygen containers are installed:

- Above the passenger seats
- In the lavatories
- At the cabin attendant stations
- In the galley working areas

Each container has a chemical oxygen generator and two or more continuous-flow oxygen masks, each with a flexible supply hose.

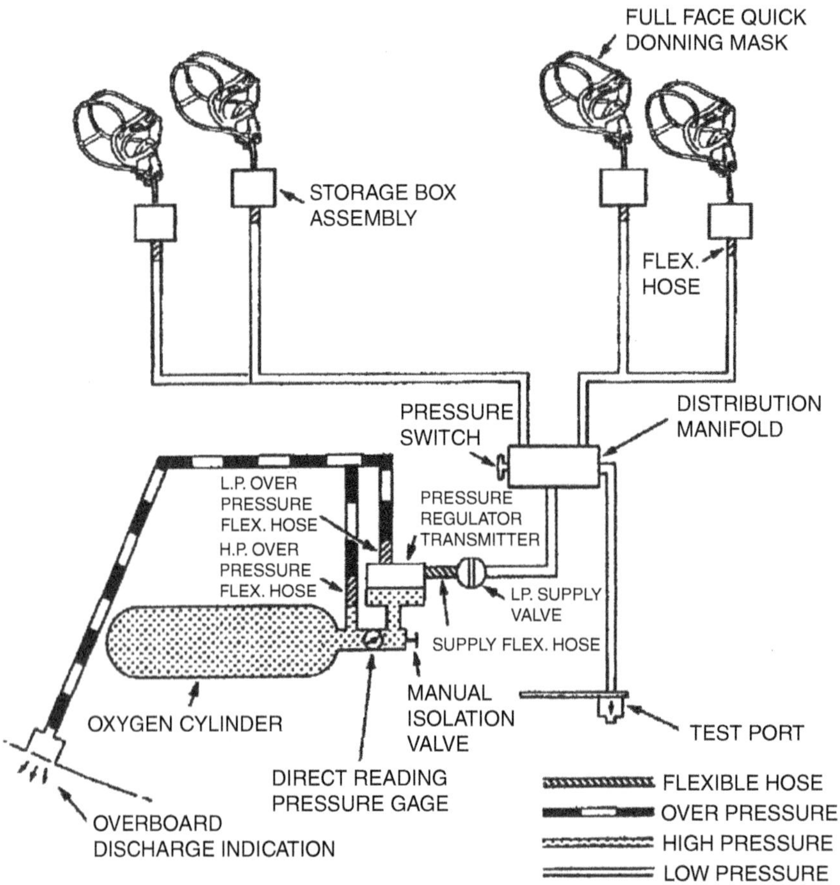

FIGURE 2.57 A321 crew oxygen system.

2.13 Pneumatic (ATA 36)

The pneumatic system as defined by ATA 100:

> Those units and components (ducts and valves) which deliver large volumes of compressed air from a power source to connecting points for such other systems as air conditioning, pressurization, deicing, etc.

High-Pressure Pneumatic Systems

High-pressure pneumatic systems must be differentiated from low-pressure pneumatic systems. High-pressure pneumatic systems, much like hydraulic systems, may apply a nominal system pressure of 20.7 MPa (3,000 psi). In contrast, low-pressure pneumatic systems may operate at only 0.3 MPa (44 psi).

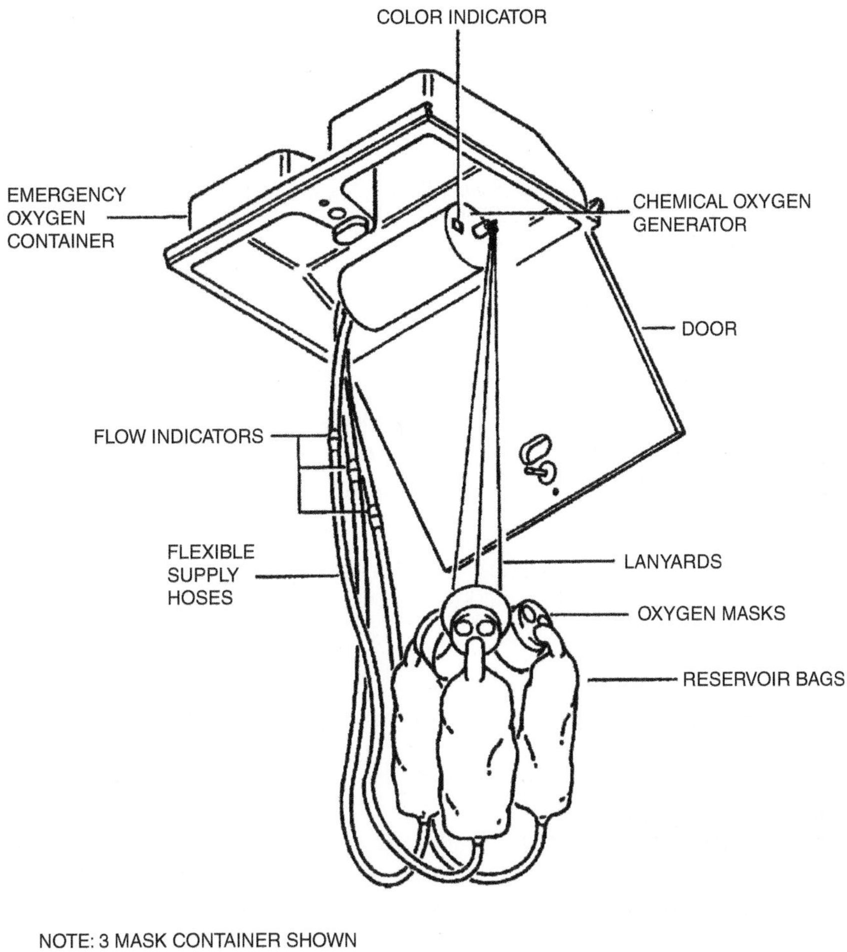

FIGURE 2.58 A321 emergency passenger oxygen container.

High-pressure pneumatic systems work very similarly to hydraulic systems. The difference is that in pneumatic systems compressible air is used instead of incompressible hydraulic fluid. Pneumatic systems do not need a reservoir because air is directly available from the operating environment. The air is put to high pressure in a compressor. The pneumatic pressure is stored in an air storage bottle. The bottle can provide a short-burst reserve flow for heavy operations, or limited emergency flow in case of compressor failure. The compressed air is routed through tubes, filters, moisture separators, and valves to the consumer. After having done its duty at the consumer, the air is simply released. In a high-pressure system it is of the utmost importance that the air in the system be completely dry. Moisture in the system can cause freezing of units and thus interfere with normal operation. High-pressure pneumatics have been applied, e.g., for landing gear extension and retraction, nose wheel steering, as well as to wheel and propeller braking. The Fairchild Hiller FH-227 is equipped with such a high-pressure pneumatic system.

High-pressure pneumatics shows advantages and disadvantages compared to hydraulics in aircraft operation:

- *Advantages:*
  - Air is a readily available, nonaggressive, clean, and lightweight fluid.
  - There is no need for return lines.
- *Disadvantages:*
  - Due to compressibility of the air, pneumatic systems lack the instant response that hydraulic systems provide.
  - The rate of movement of pneumatic actuators is highly load-dependent.
  - An actuator position cannot easily be controlled since even when the flow has stopped, the actuator will move in response to load variations.
  - Pneumatic systems are inefficient in transmitting power because energy is lost in compressing the air.

The many more disadvantages than advantages explain why high-pressure pneumatic systems are rarely used. This is much different than the low-pressure pneumatics used extensively on most aircraft.

Low-Pressure Pneumatic Systems

Low-pressure consumers include:

- Air conditioning (including cabin pressurization)
- Wing and engine anti-icing
- Engine starting
- Hydraulic reservoir pressurization
- Potable water pressurization
- Air-driven hydraulic pumps

One aircraft type will not necessarily use all these pneumatic functions.

Pressurized air is generated and used in aircraft ranging from light single-engine aircraft up to big turbine-powered transport aircraft. The simplest *source of pressurized air* is ram air. Reciprocating engines can supply pressure from a supercharger (driven by the engine primarily used to produce compressed air for the combustion process), a turbocharger (similar to a supercharger but driven by exhaust gases), or an engine-driven compressor. Turbine-powered aircraft usually use bleed air as a source for compressed air. The bleed air system will now be explained in more detail.

The engine *bleed air system* extracts pressurized air from one or more bleed ports at different stages of the engine compressor of each engine on the aircraft. The system controls the pressure and temperature of the air and delivers it to a distribution manifold. The pressure is controlled by a pressure-regulating valve and the temperature is lowered in a precooler with fan air or ram air. Bleed air from alternate sources such as the auxiliary power unit (APU) or a ground cart is also connected to the distribution manifold. The consumers are supplied from the distribution manifold. Additional bleed air from each engine may be taken directly off the engine (independent from the pneumatic system) for engine demands such as engine intake anti-ice. Isolation valves and a

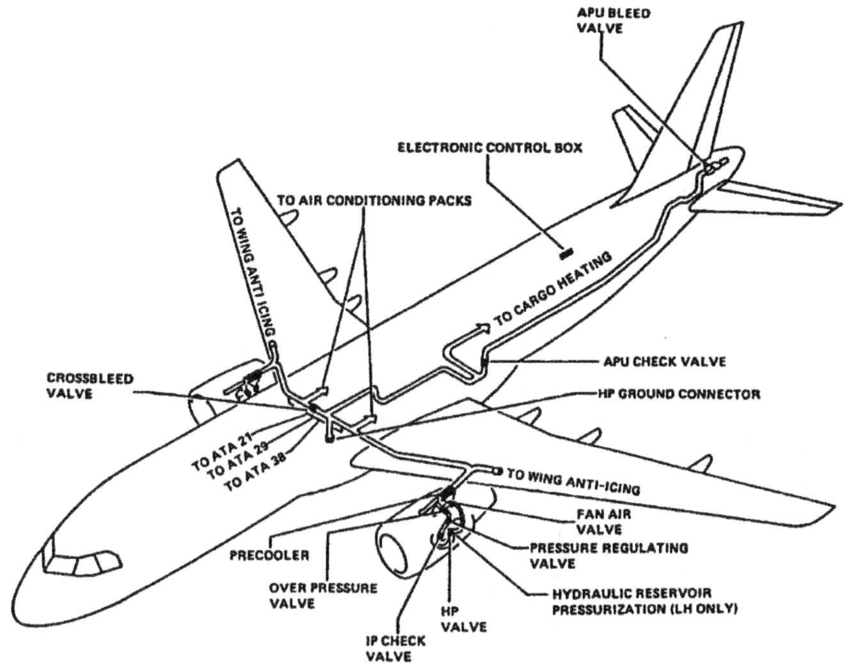

FIGURE 2.59 A321 pneumatic system overview.

crossbleed valve are required in the distribution manifold to maintain essential functions in the event of a failure in the supply or in a consumer. Check valves are required to prevent reverse flow. The Airbus A321 (Figure 2.59) shows all those elements that are typical for a conventional bleed air system.

Pressure control is set to the lowest level acceptable to all consumers. Engine bleed port switching is designed to use intermediate pressure (IP) bleed air during cruise. When intermediate stage bleed pressure is not adequate, the system switches automatically to off-takes from the high-pressure (HP) stage. A check valve prevents air from flowing back to the IP port. Pressure control may be pneumatic or computer controlled electropneumatic.

With modern high-bypass-ratio engines the *fuel burn penalty* of a given amount of engine bleed air has been decreased. However, high-bypass-ratio engines also show decreased total compressor airflow relative to engine thrust. Hence, less bleed air is available from these engines. The economic impact of the bleed air system is by no means negligible. An overall economically optimum solution has to take into account all aspects that were named in Subsection 2.1 under Costs and Tradeoff Studies. These design details could be considered:

- Use of lowest acceptable compressor stage bleed port
- Strict control of leakage from pneumatic systems
- Optimized precooler design with a trade-off among weight, price, and coolant air usage
- Optimum proportioning of bleed flows from multiengine installations
- Use of multiple bleed ports, i.e., tapping at more than the typical two compressor stages
- Consideration of alternate sources of compressed air (APU, mixing ejector, auxiliary compressor: engine driven, pneumatic, hydraulic, or electric driven)

At the airport, an external supply with pressurized air (in contrast to an APU supply) is environmentally more friendly and can also be more economical.

Example: Airbus A321

The A321 pneumatic system supplies high-pressure hot air to these *consumers:*

- Air conditioning
- Engine starting
- Wing anti-icing
- Hydraulic reservoir pressurization
- Potable water pressurization

There are two *engine bleed systems* (Figure 2.59): the left side (engine 1) (Figures 2.60 and 2.61) and the right side (engine 2). A crossbleed duct connects both engine bleed systems. A crossbleed valve mounted on the crossbleed duct allows the left and the right side to be either interconnected or separated. During normal operation, the crossbleed valve is closed and the systems are separated. There are two interconnected bleed monitoring computers. BMC 1 is used primarily for engine 1 bleed system, and BMC 2 is used primarily for engine 2 bleed system.

Air is normally bled from the IP valve. When IP pressure is not sufficient, the HP valve opens. This happens at low engine speeds, especially during descent, with engines at idle. *Pressure regulation* is done downstream of the junction of HP and IP ducting with the pressure-regulating valve (PRV), which acts as pressure regulator and shut-off valve. Delivery pressure is regulated to 0.3 MPa (44 psi). When pressure is excessive in a failure case, an over-pressure valve (OPV) closes.

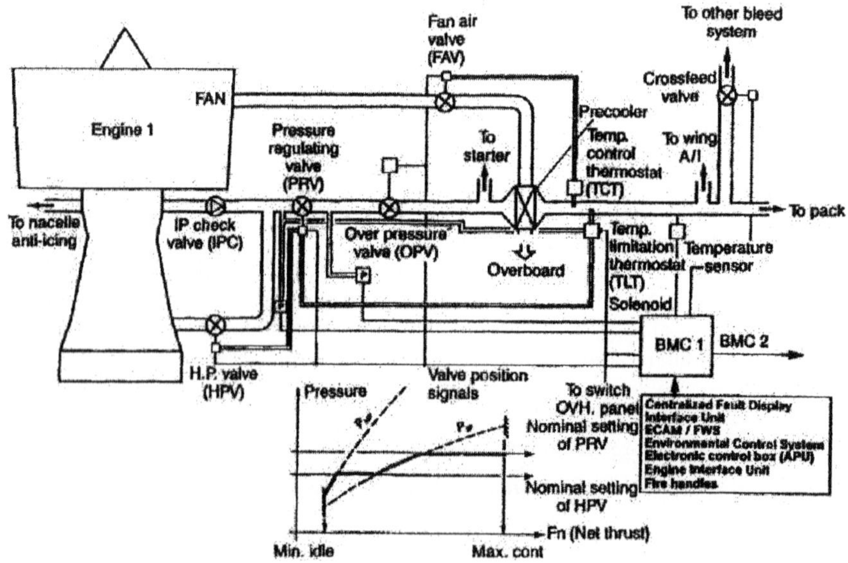

Figure 2.60 A321 schematic diagram of the pneumatic system.

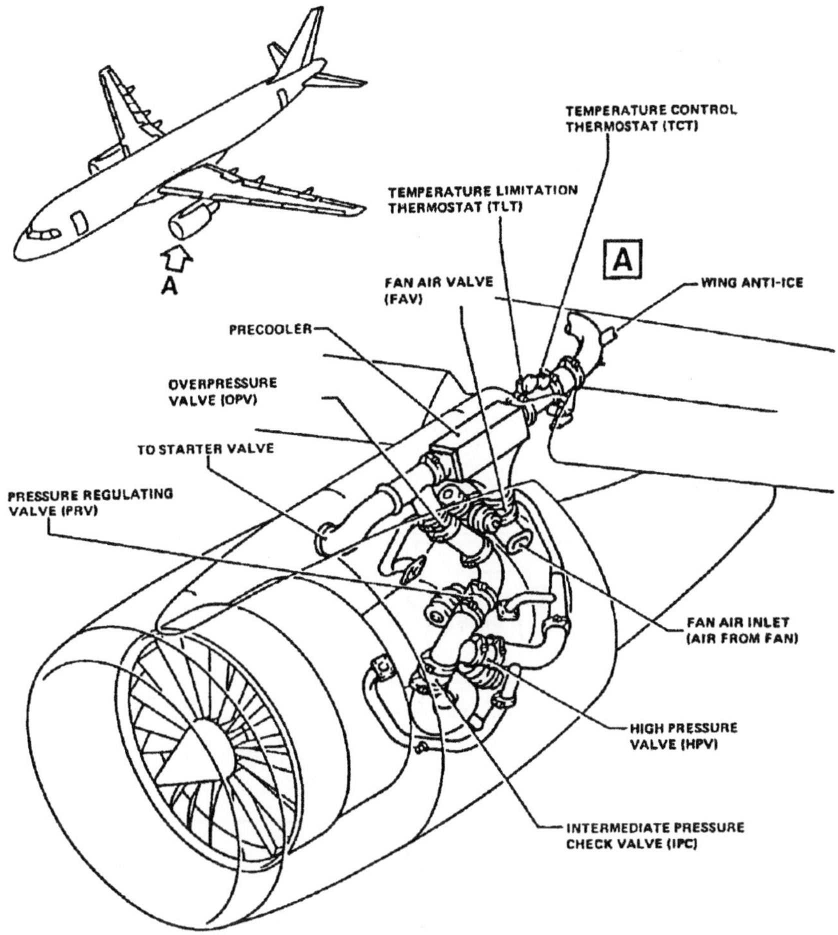

FIGURE 2.61 A321 engine bleed air supply components.

Temperature regulation of the bleed air is achieved with a fan air valve (FAV) and an air-to-air crossflow tubular heat exchanger called a precooler. The precooler uses cooling air bled from the engine fan to regulate the original bleed air with a temperature of up to 400°C down to a delivery temperature of 200°C.

2.14 Water/Waste (ATA 38)

The water/waste system as defined by ATA 100:

> Those fixed units and components which store and deliver for use, fresh water, and those fixed components which store and furnish a means of removal of water and waste. Includes wash basins, toilet assemblies, tanks, valves, etc.

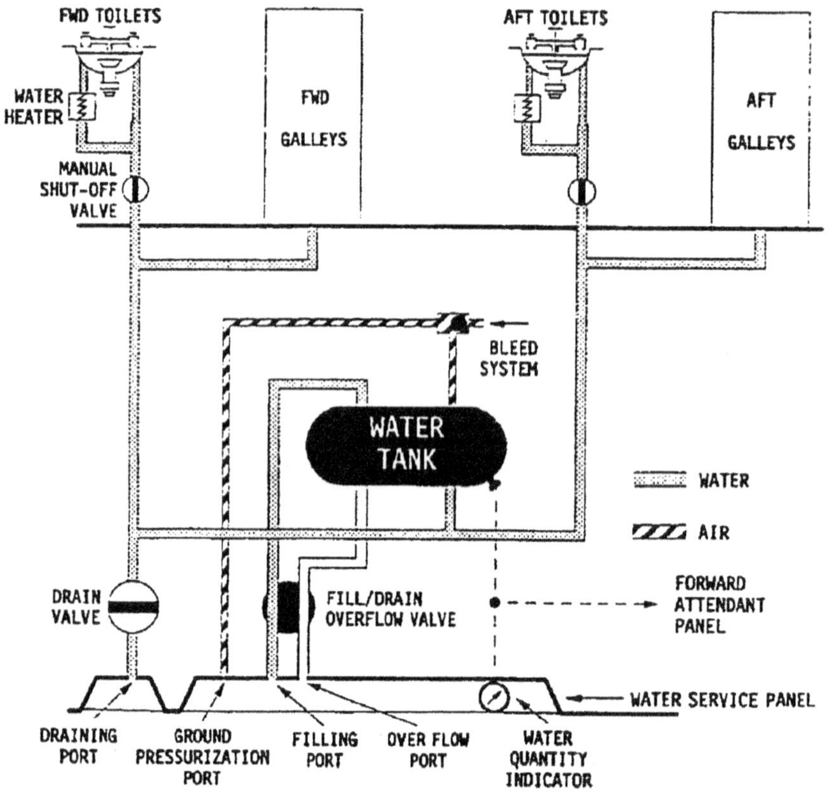

Figure 2.62 A321 potable water system.

System Classification

The water/waste system may be divided into three subsystems:

1. The potable water system is used to store and deliver fresh drinking water.
2. The wastewater drain system disposes the wastewater from lavatory washbasins and galley sinks.
3. The toilet system gives sanitary facilities to passengers and crew.

Potable Water Systems

The potable water system delivers drinking water to faucets and coffee makers in the galleys and to faucets and (in some cases) toilet bowls in the lavatories.

The water is stored in tanks made from composite material. Sensors on the tank measure the water quantity. The distribution system delivers the water through lines to the consumers. In critical areas, lines and valves are protected against freezing by insulation material and electrical heating elements. Nevertheless, water must be drained from the potable water system if the aircraft is parked overnight at temperatures below freezing.

If water left the tank just by gravity, the exit pressure would be very low. For this reason, gravity dispensing is applied only on small aircraft. On most aircraft, potable water tanks located below the cabin floor are pressurized with air. The pressurized air exerts a pressure on the water surface in the tank and thus enables water distribution at a higher pressure. The tanks may be pressurized with bleed air from the engines or the APU. Alternatively, air could be pressurized with a dedicated compressor. On the ground it is also possible to pressurize the tanks from an external pressure source.

In-service measurements have shown an average water consumption of about 0.2 L per passenger (pax) per hour in aircraft with a vacuum toilet system. This amount is made up of:

- 0.11 L/pax/h consumed in the washbasin
- 0.07 L/pax/h used for toilet rinsing
- 0.02 L/pax/h consumed in the galley

Wastewater Systems

The wastewater system disposes the wastewater from lavatory washbasins and galley sinks. Commonly, wastewater is drained overboard through drain valves via drain lines to drain masts on the lower side of the fuselage. The drain masts are electrically heated to prevent water from freezing on exit. The drain valve in the drain line prevents leakage of cabin air through the drain line. *Note:* Toilet waste is never drained overboard.

Principally, the wastewater could also be disposed into the waste tanks together with toilet waste. This technique, however, would increase aircraft weight compared with draining the wastewater. The wastewater could also be reused on board for flushing of vacuum toilets. This would save potable water taken on board and would therefore reduce aircraft weight.

Toilet Systems

Two types of toilet systems are in use: the chemical toilet system and the vacuum toilet system.

Waste tanks of recirculating liquid *chemical toilet systems* are precharged with a dye–deodorant–disinfectant chemical flushing liquid. Sensors on the tank measure the waste quantity. A tank-mounted motor/pump/filter assembly develops pressure to flush the toilets. A flush signal is generated when the flush control lever on a toilet is pressed. This signal is electronically processed and opens the flush valve. Subsequently, pressurized and filtered flushing liquid rinses the toilet bowl. The waste and the flushing liquid enter the waste tank. The waste tanks are vented overboard.

Simpler chemical toilet systems are operated with a toilet-mounted foot pedal that is connected to a mechanical pump.

The *vacuum toilet system* (Figure 2.63) is described in the Airbus example.

Example: Airbus A321

The *potable water system* supplies water from a water tank (200 L) through a distribution system. Potable water is supplied to water faucets in the galleys and lavatories. The system also supplies potable water to the water heaters, which are located below the lavatory washbasins, and to the toilet bowls for rinsing. Water lines in cold areas of the aircraft are insulated and heated to avoid freezing. Air pressure is used to pressurize the potable water system. The air is supplied from the bleed air system or the ground pressure connection.

The A321 is equipped with a *vacuum toilet system.* It removes waste from the toilet bowls through a vacuum drain to an under floor waste tank (170 L). Toilet wastes are flushed to the waste storage tank under the effect of differential pressure between the cabin and the waste tank. On

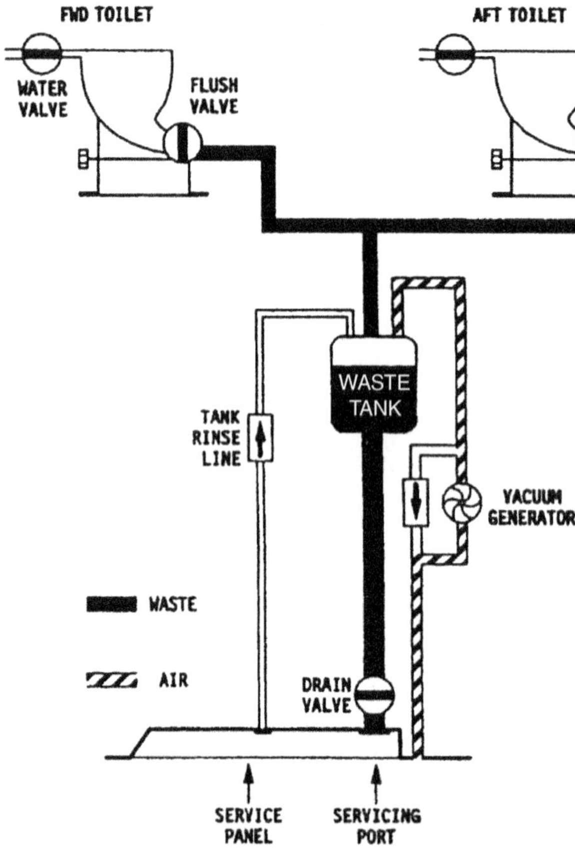

FIGURE 2.63 A321 vacuum toilet system.

ground and at low altitudes (below 16,000 ft) a vacuum generator produces the necessary differential pressure. At high altitudes (above 16,000 ft), ambient pressure alone ensures the differential pressure. A vacuum system controller (VSC) controls the operation of the vacuum generator. The system uses water from the aircraft potable water system to flush the toilet. A flush control unit (FCU) in each toilet controls the flush process. During ground service, the waste holding tank is emptied, cleaned, and filled with a prescribed quantity of sanitary fluid.

2.15 Airborne Auxiliary Power (ATA 49)

Airborne auxiliary power as defined by ATA 100:

> Those airborne power plants (engines) which are installed on the aircraft for the purpose of generating and supplying a single type or combination of auxiliary electric, hydraulic, pneumatic or other power. Includes power and drive section, fuel, ignition and control systems; also wiring, indicators, plumbing, valves, and ducts up to the power unit. Does not include generators, alternators, hydraulic pumps, etc. or their connecting systems which supply and deliver power to their respective aircraft systems.

Fundamentals

An auxiliary power unit (APU) is a compact, self-contained gas turbine-powered unit delivering rotating shaft power, compressed air, or both. Rotating shaft power can be used to drive a generator, a hydraulic pump, and/or a load compressor. An APU includes the air intake and exhaust, the fuel and oil system, engine controls and indications, as well as ignition and starting equipment.

An APU may be used on the ground and in the air, or only on the ground. For the overall aircraft system safety concept it makes a difference if the APU is dependable or not. If overall safety depends on the APU, then the APU is essential; otherwise it is nonessential.

An *essential APU* is "an APU which produces bleed air and/or power to drive accessories necessary for the dispatch of the aircraft to maintain safe aircraft operation" (CS-1).

A *nonessential APU* is "an APU which may be used on the aircraft as a matter of convenience, either on the ground or in flight, and may be shut down without jeopardising safe aircraft operations" (CS-1).

An essential APU is necessary for dispatch. For the pilot this will be indicated on the minimum equipment list (MEL).

The APU is installed in the tail cone of most airplanes, isolated from flight-critical structure and control surfaces by a firewall. The APU is started by battery. When running, the APU is able to start the main engines with its pneumatic power supply.

The significance of APU power within the concept of the secondary power systems is explained in Subsection 2.1 under Power.

Example: Airbus A321

The A321 is equipped with an APU (Figure 2.64) to permit aircraft *ground operation* independent from external power supply, allowing the operator to service airports without adequate ground power facilities. The APU is also available in flight. This is of importance for flights under extended-range twin-engine operations (ETOPS) rules, where the aircraft flies on remote routes with no alternative airfield available within a flight time of up to 180 minutes.

The APU essentially generates shaft power. A load compressor is flanged to the shaft to generate *pneumatic power*. With APU pneumatic power it is possible to start the aircraft main engines and operate the air conditioning system.

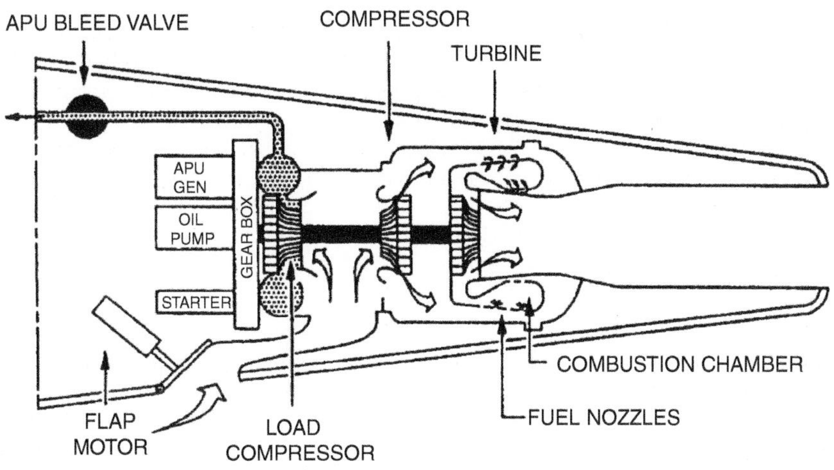

FIGURE 2.64 A321 auxiliary power unit (APU).

The APU shaft also drives a 90-kVA generator via a gearbox to generate *electrical power.* The APU is regulated to a constant speed, so that the generator is able to produce 110 V ac at a constant frequency of 400 Hz. If an increase in demand to the aircraft systems is necessary, the supply of the electrical power has priority over the supply of bleed air.

The APU is fitted with a dc starter motor, which draws its power from the electrical system battery bus. The APU starts in flight up to an altitude of 7,620 m (25,000 ft) with the use of the aircraft batteries alone. The starter motor turns the engine to such speed that self-sustained engine operation becomes possible. The electronic control box (ECB) automatically controls and monitors the APU. Manual control of the APU is possible through the crew interfaces in the cockpit. The APU is supplied with fuel from the aircraft tanks. The APU compartment is equipped with a fire detection and extinguishing system.

2.16 Avionic Systems

Avionic systems are dealt with in Section 7 of this handbook. For the sake of completeness, definitions of the avionic system are given here in the same way as above for the nonavionic systems. Introductory information can also be obtained from the related literature given in Further Reading.

Auto Flight (ATA 22)

Details of the auto flight system are covered in Section 7. The auto flight as defined by ATA 100:

> Those units and components which furnish a means of automatically controlling the flight of the aircraft. Includes those units and components which control direction, heading, attitude, altitude and speed.

The most important parts of the auto flight system are the autopilot and the auto throttle (auto thrust) system.

The autopilot is (ATA 100):

> that portion of the system that uses radio/radar signals, directional and vertical references, air data (pitot-static), computed flight path data, or manually induced inputs to the system to automatically control the flight path of the aircraft through adjustment to the pitch/roll/yaw axis or wing lift characteristics and provide visual cues for flight path guidance, i.e.: Integrated Flight Director. This includes power source devices, interlocking devices and amplifying, computing, integrating, controlling, actuating, indicating and warning devices such as computers, servos, control panels, indicators, warning lights, etc.

and the auto throttle is

> that portion of the system that automatically controls the position of the throttles to properly manage engine power during all phases of flight/attitude. This includes engaging, sensing, computing, amplifying, controlling, actuating and warning devices such as amplifiers, computers, servos, limit switches, clutches, gear boxes, warning lights, etc.

Communication (ATA 23)

Details of the communication system are covered in Section 7. Communication systems as defined by ATA 100:

> Those units and components which furnish a means of communicating from one part of the aircraft to another and between the aircraft or ground stations, includes voice, data, C-W communicating components, PA [Passenger Address] system, intercom and tape reproducer-record player.

The communication system includes (ATA 100):

- Speech communication: Radio communication air-to-air, air to ground. HF, VHF, UHF radio communication, in-flight telephone, and satellite receiver
- Data transmission and automatic calling: Selcal (Selected Call) and ACARS (Aircraft Communicating Addressing and Reporting System)
- Passenger address and entertainment system[15]:
  - Entertainment: Audio, overhead video, in-seat video, interactive video, in-seat telephone, video on demand, Internet systems, and seat power supply system for passenger laptops
  - Passenger address system: The system to address the passengers from the cockpit or the cabin crew station, playback of automatic recordings, boarding music, or acoustic signs
- Audio integrating: Controls the output of the communications and navigation receivers into the flight crew headphones and speakers and the output of the flight crew microphones into the communications transmitters; also includes the interphone, used by flight and ground personnel to communicate between areas on the aircraft
- Integrated automatic tuning of navigation transmitters and receivers
- Cockpit voice recorder

Indicating/Recording Systems (ATA 31)

The indicating/recording system deals primarily with the instrument panels and controls. This aspect is covered in Section 7 of this handbook. Indicating/recording systems as defined by ATA 100:

> Coverage of all instruments, instrument panels and controls... Includes systems/units which integrate indicating instruments into a central display system and instruments not related to any specific system.

The indicating/recording system includes (ATA 100):

- The instrument and control panels (Figure 2.65)
- Independent instruments (not related to any other aircraft system)
- Flight data recorder, recorders for performance or maintenance data
- Central computers, central warning and display systems

Navigation (ATA 34)

Details of the navigation system are covered in Section 7. The navigation system as defined by ATA 100:

> Those units and components which provide aircraft navigational information. Includes VOR, pitot, static, ILS, ... compasses, indicator, etc.

Data handling of the navigation system includes (ATA 100):

- Flight environment data (pitot/static system, rate of climb, airspeed, etc.)
- Magnetic data (magnetic compass)

---

[15]In ATA 2200 the passenger address and entertainment system has become the "cabin systems (ATA 44)" in its own right. Definition: "Those units and components which furnish a means of entertaining the passengers and providing communication within the aircraft and between the aircraft cabin and ground stations. Includes voice, data, music and video transmissions."

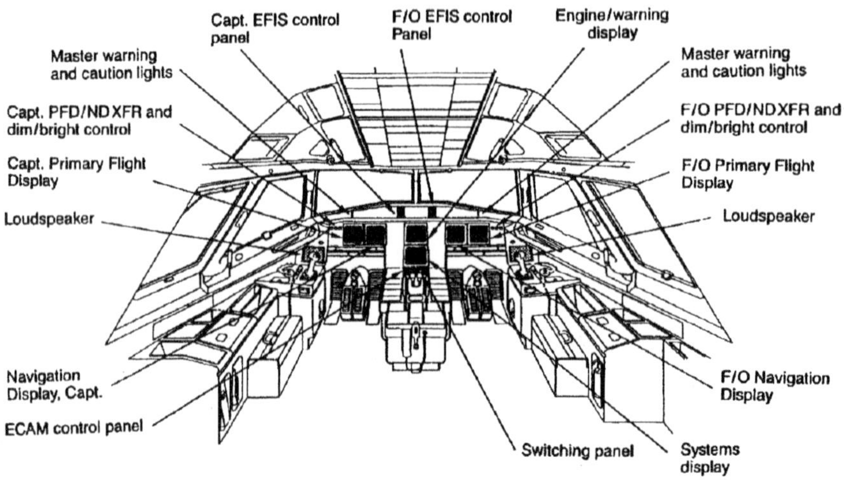

FIGURE 2.65 A321 general cockpit arrangement and instrument layout.

- Independent data (inertia guidance systems, weather radar, Doppler, proximity warning, collision avoidance)
- Dependent data (DME, transponder, radio compass, LORAN, VOR, ADF, OMEGA, GPS)
- Data from landing and taxiing aids (ILS, marker)

Acknowledgment

All figures named "A321" are by courtesy of Airbus. They are taken from the *Aircraft Maintenance Manual* (AMM), the *Flight Crew Operations Manual* (FCOM), or other material prepared or used for flight maintenance training. *At no time should the information given be used for actual aircraft operation or maintenance. The information given is intended for familiarization and training purposes only.*

References

AC 25-17. Federal Aviation Administration, Department of Transportation. 1991. *Transport Airplane Cabin Interiors Crashworthiness Handbook* (AC 25-17), available online from http://www.faa.gov.

AC 25-22. Federal Aviation Administration, Department of Transportation. 2000. *Certification of Transport Airplane Mechanical Systems* (AC 25-22), available online from http://www.faa.gov.

AC 135-16. Federal Aviation Administration, Department of Transportation. 1994. *Ground Deicing and Anti-Icing Training and Checking* (AC 135-16), available online from http://www.faa.gov.

AC 25.803. Federal Aviation Administration, Department of Transportation. 1989. *Emergency Evacuation Demonstration* (AC 25.803), available online from http://www.faa.gov.

AGARD. 1980. *Multilingual Aeronautical Dictionary,* Advisory Group for Aerospace Research and Development, Neuilly sur Seine, available online from NATO's Research and Technology Organisation, http://www.rta.nato.int.

AIR 171. Society of Automotive Engineers (SAE), *Glossary of Technical and Physiological Terms Related to Aerospace Oxygen Systems*, SAE, Warrendale, PA (2000) (AIR 171D).

AIR 1168/3. Society of Automotive Engineers (SAE). 1989. *Aerothermodynamic Systems Engineering and Design*, SAE, Warrendale, PA (AIR 1168/3).

AIR 1168/4. Society of Automotive Engineers. 2014. *Ice, Rain, Fog, and Frost Protection,* SAE (AIR 1168/4A).

AIR 1609. Society of Automotive Engineers (SAE). 2005. *Aircraft Humidification,* SAE, Warrendale, PA (AIR 1609A).

AMC-25. European Aviation Safety Agency. CS-25, Book 2, Acceptable Means of Compliance, Large Aeroplanes, available online from http://www.easa.europa.eu.

ARP 85. Society of Automotive Engineers (SAE). 2012. *Air Conditioning Systems for Subsonic Airplanes,* SAE, Warrendale, PA (ARP 85F).

ARP 1270. Society of Automotive Engineers (SAE). 2010. *Aircraft Pressurization Control Criteria,* SAE, Warrendale, PA (ARP 1270B).

ARP 1280. Society of Automotive Engineers (SAE). 2009. *Application Guide for Hydraulic Power Transfer Units.* SAE, Warrendale, PA (AIR 1280B).

ATA 100. Air Transport Association of America (ATA). 1999. *Manufacturers' Technical Data* (ATA Spec 100), ATA, Washington, DC, available from ATA, http://www.airlines.org.

ATA 2200. Air Transport Association of America (ATA). 2016. *Information Standards for Aviation Maintenance* (ATA iSpec 2200), ATA, Washington, DC, available from ATA, http://www.airlines.org.

Boeing Company, Weight Research Group. 1968. *Weight Prediction Manual—Class I,* The Boeing Company, Commercial Airplane Division, Renton, WA (D6-23201 TN).

CS-25. European Aviation Safety Agency. CS-25, Book 1, Certification Specifications, Large Aeroplanes, available online from http://www.easa.europa.eu.

Davidson, J. 1988. *The Reliability of Mechanical Systems,* Mechanical Engineering Publications, London.

FAR Part 25. Federal Aviation Administration, Department of Transportation. Part 25, Airworthiness Standards: Transport Category Airplanes, available online from http://www.faa.gov.

Federal Aviation Administration, Department of Transportation. 1993. *Aircraft Icing Handbook,* FAA Tech Report DOT/FAA/CT-88/8-2, updated sections available online from http://www.fire.tc.faa.gov.

Granzeier, W. 2001. "Flugzeugkabine Boeing B717-200," in *Flugzeugkabine/Kabinensysteme—Die naechsten Schritte Workshop DGLR S2.1/T8, Hamburg, 2001,* ed. D. Scholz, Deutsche Gesellschaft fuer Luft-und Raumfahrt, Bonn, pp. 79–87, available online from http://l2.dglr.de.

Hillman, T. C., Hill, S. W., and Sturla, M. J. 2001. *Aircraft Fire Detection and Suppression,* Kidde plc, URL http://www.walterkidde.com (2002-02-28).

ICAO Annex 1. International Civil Aviation Organization (ICAO). 2001. *Convention on International Civil Aviation,* Annex 1: Personnel Licensing, 9th ed., ICAO, Montreal, available from http://www.icao.int.

ICAO Annex 2. International Civil Aviation Organization (ICAO). 1990. *Convention on International Civil Aviation,* Annex 1: Rules of the Air. 9th ed., ICAO, Montreal, available from http://www.icao.int.

CS-1. Joint Aviation Authorities. Definitions and Abbreviations (CS-1), available from http://regulations.ProfScholz.de).

MIL-HDBK 217. Rome Air Development Center. 1991. *Reliability Prediction for Electronic Equipment* (MIL-HDBK-217F), available online from the Society of Reliability Engineers, http://www.sre.org.

MIL-STD-1629. Department of Defense. 1980. *Procedures for Performing a Failure Mode, Effects and Criticality Analysis* (MIL-STD-1629A), available online from the Society of Reliability Engineers, http://www.sre.org.

Moir, I. and Seabridge, A. 2001. *Aircraft Systems: Mechanical, Electrical, and Avionics Subsystems Integration,* AIAA Education Series, AIAA, Washington, DC.

O'Connor, P. D. T. 1991. *Practical Reliability Engineering,* John Wiley & Sons, Chichester.

Raymer, D. P. 1992. *Aircraft Design: A Conceptual Approach,* AIAA Education Series, AIAA, Washington DC.

Rome Air Development Center; Hughes Aircraft Company. 1985. *Nonelectronic Reliability Notebook,* Revision B. (ADA 163900), available from the National Technical Information Service, http://www.ntis.gov.

Roskam, J. 1989. *Airplane Design,* vol. 5, *Component Weight Estimation,* Roskam Aviation and Engineering Corporation, Ottawa, KS, available from DARcorporation, http://www.darcorp.com.

RP-8. Society of Allied Weight Engineers. 2015. *Weight and Balance Data Reporting Forms for Aircraft* (RP-8, derived from MIL-STD-1374A), available online from http://www.sawe.org.

RTCA/DO-160. Radio Technical Commission for Aeronautics. 2010. *Environmental Conditions and Test Procedures for Airborne Equipment.* RTCA, Washington, DC (RTCA/DO-160G), available online from http://www.rtca.org.

RTCA/DO-178. Radio Technical Commission for Aeronautics. 1992. *Software Considerations in Airborne Systems and Equipment Certification,* RTCA, Washington, DC (RTCA/DO-178B), available online from http://www.rtca.org.

SAE Dictionary of Aerospace Engineering, ed. J. L. Tomsic. 1998. Society of Automotive Engineers, Warrendale, PA.

SAWE 2002. http://www.sawe.org (2002-02-28).

Scholz, D. 1998. DOCsys—A Method to Evaluate Aircraft Systems, in *Bewertung von Flugzeugen (Workshop: DGLR Fachausschuß S2—Luftfahrtsysteme, Mu¨nchen, 26./27. October 1998),* ed. D. Schmitt, Deutsche Gesellschaft für Luft-und Raumfahrt, Bonn, available online from http://www.ProfScholz.de.

Shustrov, Y. M. 1999. "'Starting Mass'—a Complex Criterion of Quality for Aircraft On-Board Systems," *Aircraft Design,* vol. 1, pp. 193–203.

Torenbeek, E. 1988. *Synthesis of Subsonic Airplane Design,* Delft University Press, Delft.

TUHH. *Flugzeugsysteme* (Lecture notes), Technische Universita¨t Hamburg—Harburg, Germany.

TÜV 1980. Luftfahrt Bundesamt, Bundesminister für Verkehr. *Grundlagen der Luftfahrzeug¨ technik in Theorie und Prasix,* vol. 2, *Flugwerk,* TUV Rheinland, Köln, Germany.

VFW 614. *Schulungsunterlagen VFW614,* Vereinigte Flugtechnische Werke—Fokker GmbH, Germany.

WATOG. Air Transport Association of America. 1992. *Airline Industry Standard, World Airlines Technical Operations Glossary* (WATOG), ATA, Washington, DC, available from ATA, http://www.airlines.org.

Further Reading

Aircraft Systems—General

Cundy, D. R. and Brown, R. S. *Introduction to Avionics,* Prentice Hall, Upper Saddle River, NJ (1997).

Federal Aviation Administration, Department of Transportation, *Airframe and Powerplant Mechanics Airframe Handbook,* AC 65-15A, FAA (1976), available online from http://www.faa.gov.

Kroes, M. J., Watkins, W. A., and Delp, F., *Aircraft Maintenance and Repair,* McGraw-Hill, Singapore (1993).

Lombardo, D., *Advanced Aircraft Systems,* TAB Books, McGraw-Hill, New York (1993).

Middleton, D. H., ed., *Avionic Systems,* Longman, Harlow (1989).

Roskam, J., *Airplane Design,* vol. 4, *Layout Design of Landing Gear and Systems,* Roskam Aviation and Engineering Corporation, Ottawa, KS (1989), available from DARcorporation, http://www.darcorp.com.

Wild, T. W., *Transport Category Aircraft Systems,* IAP, Casper, WY (1990).

Wilkinson, R. *Aircraft Structures and Systems,* Addison Wesley Longman, Harlow (1996).

Definitions and Breakdown

Society of Automotive Engineers (SAE), *Aerospace Landing Gear Systems Terminology,* SAE, Warrendale, PA (2012) (AIR 1489C).

Society of Automotive Engineers (SAE), *Nomenclature, Aircraft Air Conditioning Equipment,* SAE, Warrendale, PA (2001) (ARP 147E).

Society of Automotive Engineers (SAE), *Terminology and Definitions for Aerospace Fluid Power, Actuation, and Control Technologies,* SAE, Warrendale, PA (2010) (ARP 4386C).

Certification

Federal Aviation Administration, Department of Transportation. *Comprehensive List of Advisory Circulars,* FAA (2015), available online from http://www.faa.gov.

Safety and Reliability

Federal Aviation Administration, Department of Transportation. 1998. *System Design and Analysis,* FAA (AC 25.1309-1A), available online from http://www.faa.gov.

Power

Society of Automotive Engineers (SAE), *Aerospace Auxiliary Power Sources,* SAE, Warrendale, PA (2010) (AIR 744C).

Society of Automotive Engineers (SAE), *Power Sources for Fluidic Controls,* SAE, Warrendale, PA (2012) (AIR 1245B).

Air Conditioning

Department of Defense, *Environmental Control System, Aircraft, General Requirements for* (1986) (MIL-E-18927E), available from http://www.everyspec.com.

Society of Automotive Engineers (SAE), *Aerospace Pressurization System Design,* SAE, Warrendale, PA (2011) (AIR 1168/7A).

Society of Automotive Engineers (SAE), *Aircraft Fuel Weight Penalty Due to Air Conditioning,* SAE, Warrendale, PA (2011) (AIR 1168/8A).

Electrical Power

Eismin, T. K., *Aircraft Electricity and Electronics,* Glencoe/Macmillan/McGraw-Hill, New York (1994).

Pallett, E. H. J., *Aircraft Electrical Systems,* GB: Longman, Harlow (1998).

Equipment/Furnishings

Society of Automotive Engineers (SAE), *Crew Rest Facilities,* SAE, Warrendale, PA (1992) (ARP 4101/3).

Society of Automotive Engineers (SAE), *Galley Installations,* SAE, Warrendale, PA (1986) (ARP 695C).

Society of Automotive Engineers (SAE), *Lavatory Installation,* SAE, Warrendale, PA (1998) (ARP 1315C).

Society of Automotive Engineers (SAE), *Passenger Evacuation Devices—Civil Air Transport,* SAE, Warrendale, PA (1989) (ARP 495C).

Society of Automotive Engineers (SAE), *Performance Standard for Seats in Civil Rotorcraft, Transport Aircraft, and General Aviation Aircraft,* SAE, Warrendale, PA (1997) (AS 8049A).

Flight Controls

Raymond, E. T. and Chenoweth, C. C., *Aircraft Flight Control Actuation System Design,* Society of Automotive Engineers, Warrendale, PA (1993).

Schmitt, V. R., Morris, J. W., and Jenny G. D., *Fly-by-Wire: A Historical and Design Perspective,* Society of Automotive Engineers, Warrendale, PA (1998).

Scholz, D. "Development of a CAE-Tool for the Design of Flight Control and Hydraulic Systems," in Institution of Mechanical Engineers, *Avionic Systems, Design and Software,* Mechanical Engineering Publications, London (1996), pp. 1–22. [Introduction to the mechanical design aspects of fly-by-wire aircraft.]

Hydraulic Power

Federal Aviation Administration, Department of Transportation, *Hydraulic System Certification Tests and Analysis,* FAA (2001) (AC 25.1435-1), available online from http://www.faa.gov.

Green, W. L., *Aircraft Hydraulic Systems: An Introduction to the Analysis of Systems and Components,* John Wiley & Sons, Chichester (1985).

Guillon, M., *Hydraulic Servo Systems: Analysis and Design*. Butterworths, London (1968). [Translation of the French edition: *Etude et Détermination des Systèmes Hydrauliques,* Dunod, Paris (1961).]

Scholz, D., "Computer Aided Engineering for the Design of Flight Control and Hydraulic Systems," *SAE 1996 Transactions, Journal of Aerospace,* sec. 1, vol. 105, pp. 203–212, available online from http://paper.ProfScholz.de [SAE Paper: 961327].

Society of Automotive Engineers (SAE), *Aerospace—Design and Installation of Commercial Transport Aircraft Hydraulic Systems,* SAE, Warrendale, PA (2013) (ARP 4752B).

Society of Automotive Engineers (SAE), *Hydraulic Systems, Aircraft, Design and Installation, Requirements for,* SAE, Warrendale, PA (2011) (AS 5440A) [formerly MIL-H-5440].

Ice and Rain Protection

Federal Aviation Administration, Department of Transportation, *Aircraft Ice Protection,* FAA (2006) (AC 20-73A), available online from http://www.faa.gov.

Federal Aviation Administration, Department of Transportation, *Certification of Transport Category Airplanes for Flight in Icing Conditions,* FAA (2004) (AC 25.1419-1A), available online from http://www.faa.gov.

Federal Aviation Administration, Department of Transportation, *Effect of Icing on Aircraft Control and Airplane Deice and Anti-Ice Systems,* FAA (1996) (AC 91-51A), available online from http://www.faa.gov.

Landing Gear

Conway, H. G., *Landing Gear Design,* Chapman & Hall (1958).

Currey, N. S., *Aircraft Landing Gear Design: Principles and Practices,* AIAA Education Series, AIAA, Washington, DC (1988).

Department of Defense, *Landing Gear Systems* (1984) (MIL-L-87139), available from http://www.everyspec.com.

Pazmany, L., *Landing Gear Design for Light Aircraft,* Pazmany Aircraft Corporation, San Diego, CA (1986).

Society of Automotive Engineers (SAE), *Landing Gear System Development Plan,* SAE, Warrendale, PA (1997) (ARP 1598A).

Lights

Society of Automotive Engineers (SAE), *1994 SAE Aircraft Lighting Handbook,* SAE, Warrendale, PA (1994) [collection of *all* aerospace standards prepared by the SAE A-20 Committee].

Oxygen

Society of Automotive Engineers (SAE), *Chemical Oxygen Supplies,* SAE, Warrendale, PA (1991) (AIR 1133A).

Society of Automotive Engineers (SAE), *Introduction to Oxygen Equipment for Aircraft*, SAE, Warrendale, PA (2001) (AIR 825/1).

Society of Automotive Engineers (SAE), *Oxygen Equipment for Aircraft*, SAE, Warrendale, PA (2012) (AIR 825D).

Pneumatics

Department of Defense, *Bleed Air Systems, General Specification for* (1966) (MIL-B-81365), available from http://www.everyspec.com.

Society of Automotive Engineers (SAE), *Engine Bleed Air Systems for Aircraft*, SAE, Warrendale, PA (2015) (ARP 1796B), available from SAE, http://www.sae.org.

Society of Automotive Engineers (SAE), *High Pressure Pneumatic Compressors Users Guide for Aerospace Applications*, SAE, Warrendale, PA (2013) (AIR 4994A).

Airborne Auxiliary Power

Society of Automotive Engineers (SAE), *Commercial Aircraft Auxiliary Power Unit Installations*, SAE, Warrendale, PA (1991) (AIR 4204).

Availability of SAE documents

Aerospace Information Reports (AIR) and Aerospace Recommended Practice (ARP) are listed together with a summary of the document on http://www.sae.org. In most cases, the documents may be ordered online.

SECTION 3

Aerodynamics, Aeroelasticity, and Acoustics

Section Editor: Max F. Platzer

3.1 Introduction

This section covers airplane aerodynamics, aeroelasticity, and acoustics. Part 1 provides a brief overview of the physics of drag and lift generation. Parts 2 and 3 introduce the reader to the standard aerodynamic analysis methods for airfoils and wings at subsonic and supersonic flight speeds, mostly based on linear or linearized equations, which are still of great utility today. However, impressive progress has been achieved in the numerical solution of the nonlinear Navier–Stokes equations for the analysis of viscous compressible flows. Therefore, an overview of computational aerodynamics is presented in Part 4. Parts 5 and 6 provide an overview of the major experimental methods with an emphasis on flow visualization and optical velocity measurement techniques as well as the measurement of fluctuating pressures. Modern computational methods have also become increasingly useful for the analysis of aeroelastic phenomena and for the determination of the aircraft noise characteristics. Therefore, overviews of the current status of computational aeroelasticity and computational acoustics are presented in Parts 8 and 9.

PART 1

The Physics of Drag and Lift Generation

Max F. Platzer

The basic physics of aerodynamic flows is well described by applying Newton's second law to a "fluid particle," which is assumed to be small enough so that specific values of velocity, pressure, density, and temperature can be ascribed to it, yet not so small that its molecular nature needs to be taken into account. This makes it possible to regard the fluid particle as a tiny cube whose surfaces are being acted upon by normal and in-plane forces. Defining a stress as the force per unit area the fluid particle is being deformed ("strained") by normal and tangential stresses. Making the assumption that a linear relationship exists between the stresses and the rates of strain (in generalizing Hooke's linear stress–strain relationship governing certain elastic materials) the equations of motion of the fluid particle were formulated in the 1830s by Claude Navier and George Stokes. These Navier–Stokes equations express the momentum change of the fluid particle in terms of the pressure and friction forces acting on the particle. In other words, the fluid particle experiences a constant interplay between three forces, namely, the inertia, pressure, and friction forces.

Depending on the magnitude of each force relative to the others an amazing variety of flows is being generated. It is therefore useful to estimate the magnitude of each force. The inertia force acting on a fluid particle is equal to the rate of change of the momentum in unit time. If L is denoted as the characteristic length of the fluid particle and U the characteristic velocity the time scale is given by L/U. Denoting ρ as the density and μ as the coefficient of viscosity of air then the mass of the fluid particle will be ρL^3 and the momentum becomes ρUL^3. The rate of change of the momentum then is obtained by multiplication with U/L to yield ρL^2U^2. The friction force acting on a unit area is proportional to $\mu U/L$ because it is equal to the velocity gradient multiplied with the coefficient of friction. The friction force on the fluid particle is found by multiplying this expression with the characteristic area of the fluid particle, hence it becomes μUL. The ratio between the inertia and the friction forces, therefore, is proportional to $\rho UL/\mu$. This is a nondimensional quantity denoted as the Reynolds number Re, hence

$$\mathrm{Re} = \rho UL/\mu$$

The Mach number is the other important similarity parameter. It is obtained by comparing the inertia force with the pressure force. Multiplying the pressure with the characteristic area L^2 and dividing by the inertia force yields $p/\rho U^2$. The speed of sound is given by the square root of the derivative $dp/d\rho$ while keeping the entropy constant. For constant entropy pressure and density are related through the equation

$$p/\rho^\gamma = \text{const}$$

where γ is the ratio of specific heats. Therefore, $dp/d\rho = \gamma p/\rho = a^2$ and

$$p/(\rho U^2) = a^2/(\gamma U^2) = 1/\gamma M^2$$

It is seen that the Mach number can be interpreted as the second important similarity parameter governing aerodynamic flows. Of course, the Mach number is usually introduced as representing the ratio between the flight speed and the speed of sound $M = U/a$.

3.2 Drag Generation

An aircraft whose flight speed and altitude vary from very low speeds at takeoff and landing to supersonic speeds at some portions of its flight envelope will experience large variations of its Reynolds number and Mach number. The aerodynamicist, therefore, has to understand and account for the change of the aerodynamic characteristics as a function of Reynolds and Mach number.

As first recognized by Gotthilf Hagen in Germany as early as 1854 and further investigated in a systematic series of experiments by Osborne Reynolds in England in 1883, the flow may change from a steady and orderly flow at small velocities to a rather irregular flow as soon as the flow speed exceeds a critical value. Reynolds recognized that this change also depends on the flow density and its coefficient of viscosity and the change to irregular flow occurs as soon as the Reynolds number exceeds a certain critical value. The orderly flow is referred to as laminar flow and the irregular flow as turbulent flow. In laminar flow the fluid particles stay in the same "lamina," whereas in turbulent flow a constant intermingling of the particles in neighboring laminas occurs, which can be easily visualized by adding dye particles to the flow. For example, it is found that the laminar flow velocity profile in a tube is parabolic, whereas the turbulent profile is much flatter. On the tube walls the flow velocity must be zero, referred to as the "no-slip" condition. Hence, the velocity gradient adjacent to the surface is much steeper in turbulent flow than in laminar flow. This condition applies to any surface and, consequently, the friction loss on a wing surface is much greater in a turbulent flow. The drag generated by the attached flow over an airfoil or wing, therefore, is called the *skin friction drag* and it is strongly dependent on the Reynolds number.

Often the flow is unable to stay attached to the body surface. For example, consider the flow over a cylinder or sphere. In an ideal frictionless flow the flow speeds change from zero at the forward stagnation points to a maximum value "on top of the hill" and decrease again to zero at the rearward stagnation point. In a viscous flow, the fluid particles are unable to flow into a region of rapidly increasing static pressure and instead start to separate soon after having reached the top of the hill. The violent intermingling of particles in a turbulent flow enables the turbulent flow to stick to the surface longer than in a laminar flow. As a consequence, the region of fully separated flow on the rearward side of the cylinder or sphere is much larger in laminar flow than in turbulent flow. In both cases a large amount of drag is generated because the static pressure in the separated flow region is close to the free-stream pressure, whereas the pressure in the forward region reaches stagnation pressure values near the stagnation point. As a result, the pressure difference between the forward and rearward sides causes a second type of drag, referred to as *pressure drag*. In the case of the cylinder or sphere there occurs a significant drop in pressure drag as the Reynolds number is increased beyond the critical Reynolds number.

The transition to turbulent flow can be triggered by inserting disturbances into the laminar flow. For example, Ludwig Prandtl inserted a thin wire around a sphere a short distance upstream of the separation point of the laminar layer, thus causing the flow to become turbulent. This caused a sudden drop of the drag. Therefore, although the wire ring was an additional obstacle, the total drag of the sphere was reduced because of the reduction of the separated flow region. For this reason, the grooves on the golf ball have the purpose of triggering turbulent flow regardless of ball orientation and thus ensuring drag reduction.

In 1904, Prandtl stimulated a major advance in the understanding and analysis of airfoil flows by observing that the viscous effects occur in a very thin layer close to the airfoil surface, which he termed the boundary layer, provided the Reynolds number is sufficiently large. He suggested simplifying the Navier–Stokes equations by retaining only the lowest-order terms. This made it possible to obtain solutions for simple geometries, such as the laminar flow over flat plates aligned with the free stream. Attempts to predict the transition to turbulent flow remained unsuccessful for a few more decades until Tollmien and Schlichting showed that one possible transition mechanism is the amplification of waves in the boundary layer. There are other transition mechanisms that complicate matters and make it difficult to develop reliable computational methods for all possible flow situations. Similarly, the computations of fully turbulent flows still rely on time-averaging the Navier–Stokes equations necessitating empirical inputs for the resulting turbulence modeling.

A third type of drag is caused in the process of generating lift. As explained further below, the total aerodynamic force generated by a finite-span wing is slightly tilted backward, thus producing a force component in the streamwise direction. This drag is caused by the need to generate a continuing shedding of so-called trailing vortices from the wing trailing edge and is, therefore, often called *vortex drag* or *induced drag*. It is the price that needs to be paid for generating lift.

The fourth type of drag occurs as soon as the airfoil or wing starts to fly at speeds approaching or exceeding the speed of sound. In subsonic flight any disturbance created by a body propagates upstream, thus giving the fluid particle an opportunity to adjust speed, pressure, and density as it approaches the body. This possibility starts to diminish as the flight Mach number starts to approach unity and it completely vanishes at supersonic flight Mach numbers. As a consequence, the flow has to adjust virtually instantaneously as it encounters the body which can only be done by the formation of shock waves. The resulting pressure changes on the body surface always add up to a net force in the streamwise direction. This force therefore is called the *wave drag*.

Because of the huge drag increase caused by the wave drag methods had to be found to delay or at least to minimize the wave drag in transonic and supersonic flight. During World War II, it was recognized in Germany that sweeping the wings was an effective way of delaying the onset of the drag rise as the flight speeds started to approach the speed of sound. Another important discovery was also made during the same time period by Otto Frenzl at the Junkers Aircraft Company in Germany. After the war it was more fully established experimentally by Richard Whitcomb at NASA and theoretically by Klaus Oswatitsch at the Royal Institute of Technology in Stockholm. It is known as the area rule, which states that fuselage-wing combinations need to have a smooth cross-sectional area distribution in the streamwise direction in order to minimize the transonic drag.

3.3 Lift Generation on Airfoils in Two-Dimensional Low-Speed Flow

Holding one's hand into a wind stream at a positive angle of attack yields the immediate impression that the force on the hand is caused by an overpressure on the lower surface. It is, therefore, not surprising that Isaac Newton argued that the air particles flow along straight lines until they hit the lower surface and then are deflected. Therefore, the first analytical prediction of the force acting on a body immersed in a moving fluid can be found in Newton's *Principia*, postulating that the force acting on a body is equal to the change of its momentum due to the deflection of the fluid. When applied to a flat plate inclined at an angle of attack this flow model leads to the famous sine squared law. As human flight attracted increasing interest in the last decades of the 19th century it became apparent that this theory could not explain the experimental information on airfoil lift which had been obtained by that time. However, already a few years earlier the German physicist Hermann Helmholtz had become interested in the study of vortical flow phenomena, which led him to the recognition that vorticity can only be created by friction or by the presence of sharp edges on a body. Toward the end of the 19th century it was gradually recognized that vortex generation had something to do with lift generation.

Consider the flow pictures of Figure 3.1 taken by Ludwig Prandtl. They show the flow generated by an airfoil at small angle of attack after the airfoil is suddenly started from rest. A similar flow is generated by an airfoil flying at a steady speed whose angle is suddenly increased by a small amount. Note the appearance of a counterclockwise vortex at the sharp trailing edge that starts to separate from the trailing edge and to flow downstream with the flow speed. A similar phenomenon occurs when the angle of attack is suddenly reduced by a small amount, causing the

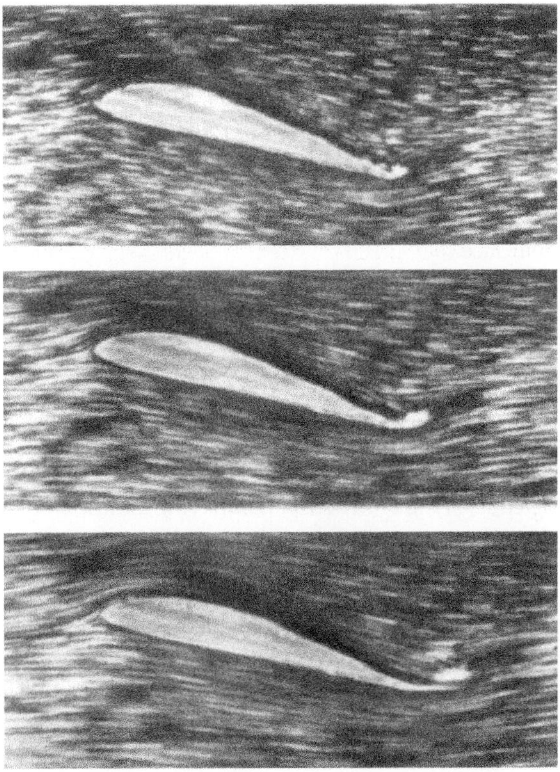

FIGURE 3.1 Generation of the starting vortex (from Prandtl and Tietjens 1934).

shedding of a clockwise vortex. With no further change in angle of attack the flow around the airfoil and the pressure distribution on the airfoil reach a steady state as soon as the shed vortex is some 20 chord lengths downstream from the trailing edge.

It is the function of the sharp trailing edge to prevent the fluid particles from flowing around the trailing edge. Indeed, an airfoil with a rounded trailing edge allows flow around it, resulting in a much diminished lift. Nowadays, the precise viscous flow details can be computed by means of the Navier–Stokes equations. Prior to the development of this capability in the 1970s, the analysis had to be based on a purely inviscid flow analysis using Laplace's equation. Wilhelm Kutta recognized, in 1902, that good agreement with the experiments could be achieved by prescribing the condition of smooth flow off the trailing edge. This assumption made it possible to analyze lifting airfoil flows quite rapidly in the precomputer days. It is, therefore, still used today in situations where the viscous effects are small, i.e., where separation bubbles or regions of flow separation are absent. This condition occurs on airfoils flying at high Reynolds numbers (greater than one million) at angles of attack below the stall angle. In this case, the viscous effects are confined to a layer, called the boundary layer, whose thickness is only a few percent of the chord length. The actual flow, therefore, closely resembles the inviscid flow over the airfoil and the aerodynamic force perpendicular to the free stream (i.e., the lift) can be predicted quite well with inviscid flow theories. On the other hand, the force parallel to the free stream (i.e., the drag) requires a viscous flow analysis. It is quite fortunate that most aircraft fly at Reynolds numbers above one million, where inviscid lift prediction methods are readily applicable. At Reynolds numbers below one million airfoil flows are much more sensitive to laminar to turbulent flow transition and the formation of separation bubbles and separated flow regions, thus requiring more sophisticated and time-consuming analysis methods. The recent development of small unmanned air vehicles and micro air vehicles, therefore, has opened up interesting new challenges for the aerodynamicist.

It is obvious that connecting lift generation to vortex generation changed the whole physical picture from the earlier "natural" view that the air hits the inclined lower surface of the wing and thus generates an overpressure which "holds up" the wing. Actually, the contribution to the lift from the negative pressure on the upper surface is larger than the contribution from the positive pressure at the lower surface.

However, Newton's concept of lift generation retains its validity and applicability for bodies flying at speeds much greater than the speed of sound. In such hypersonic flow conditions the oncoming fluid particles receive no "warning" of the presence of the body and are, therefore, deflected very close to the body surface.

In contrast, for lift generation on low-speed airfoils the following four fundamental observations are important:

1. An airfoil produces a significant amount of lift if it has a well-rounded leading edge and a sharp trailing edge, being either cusped or having a finite wedge angle.
2. The rounded leading edge is needed to prevent flow separation. The purpose of the sharp trailing edge is to prevent flow around the trailing edge. If the trailing edge is cusped, then the velocities of the fluid particles leaving the top and bottom surfaces at the trailing edge are finite and equal in magnitude and direction. If the trailing edge angle is finite, then the trailing edge is a stagnation point.
3. The sharp trailing edge causes the shedding of a starting vortex which, in turn, generates a flow around the airfoil in compliance with the Helmholtz vortex theorem such that a circulation is induced around the airfoil. After completion of the starting process the steady-state flow pattern can be modeled using inviscid flow theory, first accomplished by Wilhelm Kutta in Germany in 1902 and independently by Nikolai Joukowski in Russia in 1906. It yielded the result that the airfoil lift is directly related to the circulation by the equation

   $$\text{Lift} = \text{density} \times \text{flight speed} \times \text{circulation}$$

   It is the famous Kutta–Joukowski law that is the key to the understanding and analysis of low-speed lifting aerodynamics.
4. The generation of the circulation is responsible for the increase in local flow speed and reduction of the static pressure on the airfoil upper surface and the decrease in flow speed and the increase in static pressure on the lower surface according to the Bernoulli equation. However, the lift generation cannot merely be explained by stating that two neighboring fluid particles, one just above and one just below the stagnation streamline near the forward stagnation point, have to travel different distances and, therefore, the upper surface particle has to have a higher local speed because they have to arrive at the same time at the trailing edge. In fact, the upper surface particle arrives at the trailing edge before the lower surface particle.

A fuller appreciation for the critical role of the starting vortices can be gained by looking at the vortex street generated by an airfoil flying at a steady forward speed while executing small amplitude vertical oscillation. Due to the continuously changing angle of attack between positive and negative values, a row of vortices is shed from the trailing edge such that a vortex street is generated, which consists of counterclockwise rotating vortices in the upper row and clockwise rotating vortices in the lower row. It is referred to as the reverse Karman vortex street in distinction from the Karman vortex street shed from a nonoscillating cylinder whose upper row vortices are rotating clockwise and the lower row vortices are counterclockwise. The time-averaged velocity distribution at some station downstream from the cylinder yields the well-known velocity defect distribution indicative of a drag. In contrast, the time-averaged velocity distribution in the reverse Karman vortex street yields a jet profile. The oscillating airfoil captures a certain amount of air per unit time and gives it an additional velocity. In response, the airfoil experiences a forward thrust. Evidently, birds have evolved a very effective means of "jet propulsion" by flapping their wings.

3.4 Lift Generation on Finite-Span Wings in Low-Speed Flow

The critical role of vortex generation in generating lift on a nonoscillating airfoil in steady forward flight becomes apparent only during the starting process. The starting vortex induces a circulation about the airfoil, which can be thought of as being due to a vortex (or more precisely, due to a vortex sheet). Lift generation on a finite-span is also due to vortex generation. For a first approximation, the wing is again replaced by a vortex (or more precisely, by a vortex sheet). However, according to the Helmholtz vortex laws the vortex may not end at the wing tips. Instead, it has to be a closed-loop vortex, which extends downstream from the wing tips as so-called trailing vortices and is closed by the starting vortex. This vortex loop would presuppose that the vortex strength stays constant along the span implying that the local lift will remain constant along the span and then suddenly drop to zero at the wing tips. A more realistic model is one where the lift gradually decreases toward the wing tips. However, again according to the Helmholtz laws any change in vortex strength must be accompanied by the shedding of a trailing vortex. This leads to the lifting line model first suggested by Ludwig Prandtl, which was later complemented by a lifting surface model where the wing surface is replaced by a vortex sheet.

Using this lifting line model, Prandtl obtained several very important results. The vortices induce a downwash velocity field w behind the wing trailing edge, which is constant along the span if the lift varies elliptically along the span. This downwash velocity is given by

$$w/U = C_l/(\pi AR)$$

where U is the free-stream speed, C_l is the lift coefficient and AR is the wing aspect ratio given by $AR = b^2/S$, b = wing span, S = wing area. This equation can easily be rewritten as

$$L = 2w\rho U\pi b^2/4$$

The lifting line vortex model, therefore, captures the mass flow through a circle of diameter b to give it a downward velocity $2w$ far downstream from the trailing edge. Again, as in the case of the flapping airfoil, this is merely a manifestation of Newton's second law. In reaction to this downward rate of flow momentum, a vertical force (i.e., lift) is generated. Another important result was the recognition that the total aerodynamic force that is being generated is slightly tilted backward, producing a drag component. Prandtl showed that this induced drag (or vortex drag) becomes a minimum if the lift varies elliptically along the span. Furthermore, it is inversely proportional to the wing aspect ratio. High-aspect-ratio wings with elliptic spanwise lift distribution, therefore, will minimize the induced drag. Airplane designers are well aware of this requirement in order to maximize the endurance. Another result published by Prandtl is less well known. For minimum friction drag the wing should have a small chord. These two requirements are in conflict with each other. Prandtl, therefore, proposed to impose the averaged bending moment along the span as a constraint because the wing weight largely depends on the bending moment along the span. This consideration leads to a wing with a slightly increased span and a more tapered spanwise loading as the structurally optimal configuration.

The above formula also provides the connection with the lift generation mechanism on airfoils because it remains valid for an infinitely large aspect ratio. For this asymptotic case the downwash velocity w becomes zero, and one might suspect a contradiction with Newton's second law. However, the amount of flow which is being captured is infinitely large yielding a finite value of lift for the product of downwash velocity and mass flow.

3.5 Lift Generation on Slender Wings

Lift generation on airfoils and finite-span wings is caused by vortex generation from sharp trailing edges while making sure that well-rounded leading edges are used to avoid flow separation. It turns out that this wing design doctrine is too restrictive. Any efficiently generated vortex can be exploited for lift generation. Flow separation from the leading edges of highly swept low-aspect wings produces a pair of distinct conical vortices, if the wings are flying at high angles of attack. If properly arranged, these vortices induce low-pressure regions on the wing upper surfaces yielding very useful amounts of lift. This type of lift generation is being applied in the design of high-speed fighter-attack aircraft.

3.6 Lift Generation in Transonic and Supersonic Flight

The critical role of vortex generation from sharp trailing or leading edges on lift generation in low-speed flight remains valid at flight speeds well below the speed of sound. However, as soon as the fluid particles are forced to flow at speeds close to or greater than the speed of sound near the airfoil surface, the sound wave propagation phenomena start to affect the flow features, leading to the formation of locally supersonic flow regions which are terminated by weak shocks and to the formation of shock waves extending from the leading and trailing edges in fully supersonic flows.

3.7 Lift Generation in Hypersonic Flight

At very large flight speeds the shock emanating from the leading edge of the flat plate inclined at a positive angle of attack starts to coincide with the lower surface of the plate. In this case, Newton's original assumption becomes a good approximation of the actual flow pattern. The fluid particles can be assumed to flow along straight lines until they hit the lower surface and are being deflected. The resulting momentum change leads to the generation of a normal force acting on the plate which is proportional to the second power of the sine of the angle of attack.

3.8 Summary

In spite of the great advances in computing power, the understanding and prediction of the amazing variety of flow phenomena encountered on a typical aircraft is still hampered in many cases by the impossibility of solving the underlying equations in sufficiently fine detail due to the very complicated nonlinear interactions between vortices of many sizes in a transitional or fully turbulent flow. For this reason, it is necessary to look for classes of flows that are amenable to simplified flow modeling. A good visualization of the flow, which is to be studied, therefore, is a requirement in order to begin "understanding" the flow. Van Dyke's album of fluid motion provides excellent examples of the laminar, transitional, turbulent, separated, subsonic, transonic, and supersonic flows that can occur on an aircraft. The study of these flow pictures also provides the key to an appreciation of how the pioneers of aerodynamics went about formulating flow models that were sufficiently simple to enable their mathematical analysis yet not so simple as to eliminate the essential underlying physics.

In his book, Theodore von Karman explains the major aerodynamic phenomena and the development of the modern theories of lift and drag with a clarity which makes this book indispensable reading together with the books of the other two pioneers of modern aerodynamics, Ludwig Prandtl and Robert T. Jones. Readers who wish to inform themselves about the development of the swept wing concept and the area rule will want to read the book edited by H. U. Meier and Holt Ashley and Marten Landahl's book on the aerodynamics of wings and bodies. John Anderson's book on the fundamentals of aerodynamics is a widely used textbook that covers the major aspects of steady-state aerodynamics. Readers who want to study unsteady flow physics and analysis, modeling and computation of boundary layer flows, the physics and analysis of flow stability, and transition and computational fluid dynamics may find Tuncer Cebeci's books of value.

The important lesson to be drawn is the recognition that the development and application of simplified flow modeling was the key to the successful prediction of certain classes of flows while recognizing its limitations. This started with the recognition in the early years of powered flight that inviscid flow modeling enabled the prediction of the lift on low-speed airfoils and finite-span wings with the help of one empirical input in the form of the Kutta trailing edge condition. The prediction of skin friction drag became possible by recognizing the possibility of simplifying the Navier–Stokes equations by means of the boundary layer concept. The prediction of lift and wave drag of thin wings in supersonic flight became possible by linearizing the nonlinear inviscid flow equations and finally in the last few decades of the last century it became possible to develop numerical solutions of the complete Navier–Stokes equations for three-dimensional flow. Yet, it is important to keep in mind that the flow over most aircraft configurations involves large regions of transitional and turbulent flow which would require the numerical solution of the unsteady three-dimensional Navier–Stokes equation at a time scale sufficient to resolve the turbulent fluctuations. Such direct numerical solutions are still beyond present computing capacities and therefore time-averaging, originally suggested by Reynolds, needs to be used. The resulting Reynolds-averaged Navier–Stokes (RANS) equations lead to additional terms which account for the turbulent stresses. Unfortunately, this procedure requires relating these additional Reynolds stress terms in some way to the time-averaged flow field by means of turbulence models, which contain experimental information. In spite of extensive research, no universally valid turbulence model has been found, necessitating the use of different turbulence models for different classes of flows. Therefore, current aerodynamic analysis methods still have to rely on flow modeling, albeit much more sophisticated than the modeling used in precomputer days. The aerodynamicist, therefore, still has to be aware of the limitations of currently available prediction methods and he/she has to use good judgment as to the need for additional expensive wind tunnel and flight testing in order to ensure project success.

References

Anderson, Jr., J. D., *Foundations of Aerodynamics*, McGraw-Hill, 2001. Boston, Massachusetts.

Ashley, H. and Landahl, M. 1965. *Aerodynamics of Wings and Bodies*, Addison-Wesley. Reading, Massachusetts.

Cebeci, T. 2004. *Stability and Transition: Theory and Application*, Springer. New York, N.Y.

Cebeci, T. and Cousteix, J. 1999. *Modeling and Computation of Boundary-Layer Flows*, Springer. New York, N.Y.

Cebeci, T., Platzer, M., Chen, H., Chang, K. C., and Shao, J. P. 2005. *Analysis of Low-Speed Unsteady Airfoil Flows*, Springer. New York, N.Y.

Cebeci, T., Shao, J. P., Kafyeke, F., and Laurendeau, E. 2005. *Computational Fluid Dynamics for Engineers*, Springer. New York, N.Y.

Dyke, M., *An Album of Fluid Motion*, The Parabolic Press, Stanford, California, 1982. Jones, R. T., *Wing Theory*, Princeton University Press, 1990.Princeton, N.J.

Prandtl, L. and Tietjens, O. G. 1934. *Fundamentals of Hydro and Aerodynamics*, McGraw-Hill. New York and London von Karman, Th. 2004. *Aerodynamics*, Dover Publications, Mineola, N.Y.

PART 2

Aerodynamic Analysis of Airfoils and Wings

Andrew J. Niven

Notation

a	local speed of sound, m/s
A	area, m^2
A'	axial force per unit span, N/m
A_R	aspect ratio of wing
b	wing span, m
c	airfoil chord, m
$c_{l,i}$	ideal lift coefficient
c_g	mean geometric chord, m
c_p	specific heat at constant pressure, J/K/kg
$c_{l\alpha}$	airfoil lift curve slope, per radian or per degree
$\overline{c}$	mean aerodynamic chord, m
C_L	wing lift coefficient
D'	drag per unit span, N/m
f_x, f_y, f_z	components of body force, N/kg
$\dot{G}$	heat generated per unit volume, J/m^3
h	maximum z-distance between camber line and chord line, m
$\hat{h}$	nondimensional maximum camber, h/c
i	specific internal energy, J/kg
k	fluid thermal conductivity, W/m/K
l	distance along specified path, m
L'	lift per unit span, N/m
m	mass, kg
M'	aerodynamic moment per unit span, N
M_∞	free-stream Mach number
N'	normal force per unit span, N/m
p	pressure, N/m^2
$\vec{q}$	local or instantaneous velocity vector, m/s
$\vec{q}$	turbulence velocity vector, m/s
$\vec{Q}$	free-stream or mean velocity, m/s
R	specific gas constant, J/K/kg
R'	resultant aerodynamic force per unit span, N/m
s	specific entropy, J/K/kg
S	wing area, m^2
t	maximum airfoil thickness, m
$\hat{t}$	nondimensional maximum thickness, t/c
T	air temperature, K
u, v, w	components of local or instantaneous velocity vector, m/s
$\tilde{u}, \tilde{v}, \tilde{w}$	local disturbance velocities, m/s
$u', v'we'$	components of turbulence velocity vector, m/s

$U_\infty, V_\infty, W_\infty$	components of free-stream or mean velocity, m/s
x, y, z	Cartesian coordinate system
z	a complex number, x + iy
$\hat{x}$	nondimensional x-coordinate, x/c
$\hat{y}$	nondimensional y-coordinate, 2y/b
$\hat{z}$	nondimensional z-coordinate, z/c
$\hat{z}_t$	nondimensional thickness, z_t/c

Greeks

α	angle of attack, degrees or radians
α_i	ideal angle of attack, degrees or radians
α_{l0}	airfoil zero lift angle of attack, degrees or radians
α_{L0}	wing zero lift angle of attack, degrees or radians
β_0	defined as $\sqrt{1-M_\infty^2}$
β_1	defined as $\sqrt{M_\infty^2-1}$
γ	vortex sheet strength, m/s
γ	isentropic index
Γ	circulation, m^2/s
δ	flow turning angle, degrees or radians
ε	apex half angle of delta wing, degrees or radians
ε_t	wing tip rake angle, degrees or radians
ε_a	aerodynamic twist, degrees or radians
ε_g	geometric twist, degrees or radians
$\dot{\varepsilon}$	fluid strain rate, m/s
H	doublet strength, m^3/s
θ	an arbitrary angle, degrees or radians
λ	wing taper ratio
Λ	sweep angle, degrees
μ	viscosity, N/s
ξ	source sheet strength, m^2/s
ρ	density, kg/m^3
σ	oblique shock wave angle, degrees or radians
σ_M	Mach wave angle, degrees or radians
τ	viscous shear stress, N/m^2
τ	constant used in wing lift curve slope calculation, radians
ψ	stream function, m^2/s
ϕ	velocity potential function, m^2/s
$\tilde{\phi}$	disturbance velocity potential function, m^2/s
$\vec{\omega}$	vorticity vector $(\omega_x, \omega_y, \omega_z)$

Subscripts

f	skin friction value
inc	incompressible flow value
l	lower surface
le	value at leading edge
n	value normal to leading edge of swept wing

pl	value appropriate to a flat plate supersonic airfoil
r	wing root
t	wing tip
u	upper surface
w	value at wall
$c/4$	value at the quarter chord location
$3c/4$	value at the three-quarter chord location
∞	free-stream value

3.9 Airfoil Geometric and Aerodynamic Definitions

Airfoil Geometry

Figure 3.2 illustrates the terminology and the geometric parameters used to systematically define an airfoil (sometimes referred to as a wing section). In general, most airfoil profiles are generated by combining a mean line (or camber line) and a thickness distribution. The upper and lower surface coordinates are related to the camber line and thickness in the following manner:

$$\hat{x}_{u,l} = \hat{x}_c \mp \hat{z}_t \sin\theta \qquad \text{and} \qquad \hat{z}_{u,l} = \hat{z}_c \pm \hat{z}_t \cos\theta.$$

where the upper sign indicates the upper surface. To increase the accuracy of the leading edge geometry, a circle is centered on a line defined by the tangent of the camber line at $\hat{x}_c = 0.005$ and the radius specified such that the circumference passes through the leading edge.

The camber line is defined by a function $\hat{z}_c = f_c(\hat{x})$ which has a maximum value, $\hat{h}$, at a particular chord position, $\hat{x}_{c,m}$. This value is normally referred to as the camber of the airfoil rather than the maximum camber.

The thickness distribution is function of chord and a specified maximum thickness, i.e., $\hat{z}_t = f_t(\hat{t}, \hat{x})$. The maximum thickness, $\hat{t}$ (simply known as the thickness), occurs at a particular chord position, $\hat{x}_{t,m}$. The thickness function actually defines a symmetrical airfoil in its own right which is often referred to as the basic thickness form.

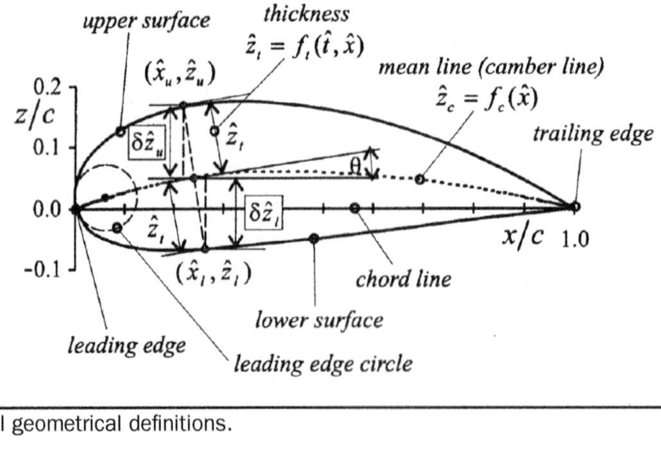

FIGURE 3.2 Airfoil geometrical definitions.

The NACA Series of Airfoils

In 1929 the National Advisory Committee for Aeronautics (NACA) embarked upon the systematic development of various families of airfoils using a combination of theoretical methods and wind tunnel testing. Each airfoil was assigned a number which represented specific geometrical and aerodynamic properties of the airfoil. The equations for the camber line, $\hat{z}_c = f_c(\hat{x})$, and the basic thickness form, $\hat{z}_t = f_t(\hat{t}, \hat{x})$, for most of the NACA airfoils are given in Abbott and von Doenhoff (1959). Once these equations are known, the airfoils may be easily plotted via spreadsheet software.

The first family of airfoils to be developed was designated a four-digit number, e.g., NACA 2412. The meaning of this number is described in Table 3.1 and the airfoil profile is shown in Figure 3.3(b). The basic thickness form is shown in Figure 3.3(a). The effect of varying the camber and thickness is illustrated in Figure 3.3(c), which displays the NACA 6408 airfoil.

The NACA five-digit family of airfoils was designed to have point of maximum camber within the first quarter chord. The example profile shown in Figure 3.3(d) is the NACA 23012, and the meaning of the number is given in Table 3.2.

The NACA 6 series was designed, using the methods described in Subsection 3.14, to possess low drag at the design lift coefficient. Figure 3.3(e) displays the NACA 65_2-215 profile, and Table 3.3 explains the numbering scheme. The mean lines and basic thickness forms for the NACA six series are given in Abbott and von Doenhoff (1959).

There are many more types of NACA airfoils: modified four- and five-digit families; the 1, 2, 3, 4, 5, and 7 series airfoils; and modified 6 series airfoils. Further details on the theoretical development of the NACA airfoils, along with extensive wind tunnel data, are given by Abbott and von Doenhoff (1959); this reference is the definitive work on airfoil design and development and should always be consulted when choosing an airfoil for a particular application.

Airfoil Aerodynamic Forces and Moments

When an airfoil moves through the air, each surface element is subjected to a normal pressure stress, p, and a tangential viscous shear stress, τ. Although the magnitude of these stresses will vary greatly around the airfoil contour, Figure 3.4 shows they may be integrated around the entire surface such that a resultant aerodynamic force and moment is obtained. Since an airfoil is a two-dimensional shape, surface areas are based on unit wing span which gives rise to forces and moments per unit length (denoted by a prime). The magnitude of the moment will depend on the chosen reference point about which the elemental surface moments are calculated. The angle α is known as the angle of attack and is measured between the chord line and free-stream velocity; it is defined as positive when the chord line is rotated anticlockwise onto the free-stream velocity vector (often referred to as *nose up*). The moment is taken as positive in the direction which increases the angle of attack. As indicated in Figure 3.4 the resultant aerodynamic force can be split into various components: normal force, axial (or chord) force, lift, and drag.

NACA four-digit	Meaning	Equation
2	Maximum camber	$100(h/c)$
4	Distance along chord from leading edge to point of maximum camber	$10(x_{c,m}/c)$
12	Maximum thickness at $\hat{x}_{t,m} = 0.3$	$100(t/c)$

TABLE 3.1 The NACA Four-Digit Numbering System

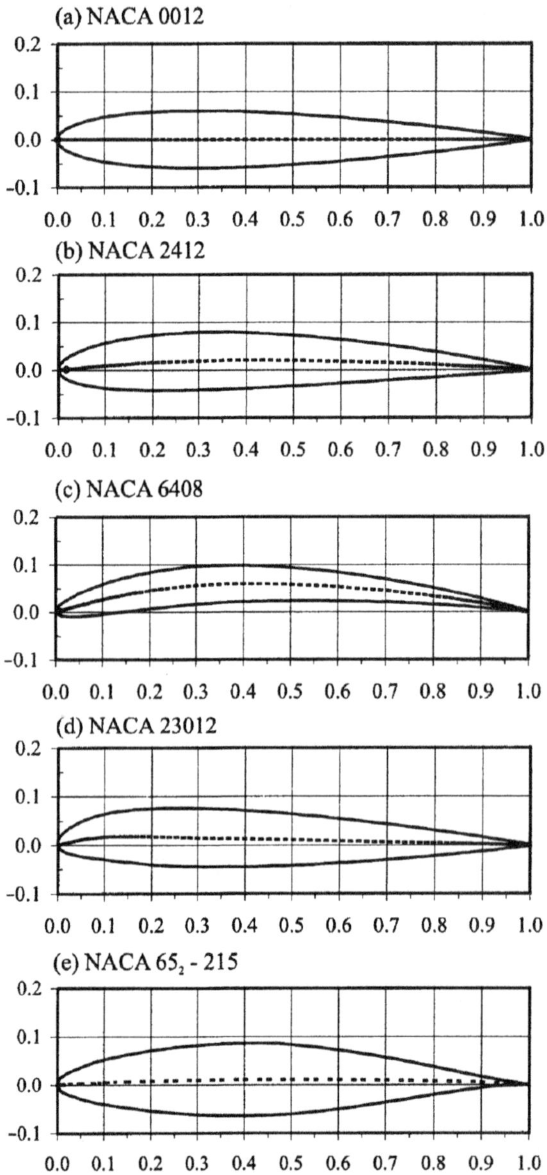

FIGURE 3.3 Examples of NACA airfoil profiles.

NACA five-digit	Meaning	Equation
2	Strictly related to the ideal lift coefficient. Also approximately equal to the maximum camber	$20(c_{l,i}/3)$ $\approx 100(h/c)$
30	Distance along chord from leading edge to point of maximum camber	$200(x_{c,m}/c)$
12	Maximum thickness at $\hat{x}_{t,m} = 0.3$	$100(t/c)$

Table 3.2 The NACA Five-Digit Series Numbering System

NACA six series	Meaning	Equation
6	Series designation	
5	Nondimensional distance along chord line from leading edge to minimum pressure for the basic symmetrical section at zero lift	$10(\hat{x}_{mp})$
2	Defines the width of the low drag region whose limits are given by $(c_{l,i} - \Delta c_l)$ and $(c_{l,i} + \Delta c_l)$ on a drag polar	$10(\Delta c_l)$
2	Defines the ideal lift coefficient	$10(c_{l,i})$
15	Maximum thickness at $\hat{x}_{t,m} = 0.4$	$100(t/c)$

Table 3.3 The NACA Six Series Numbering System

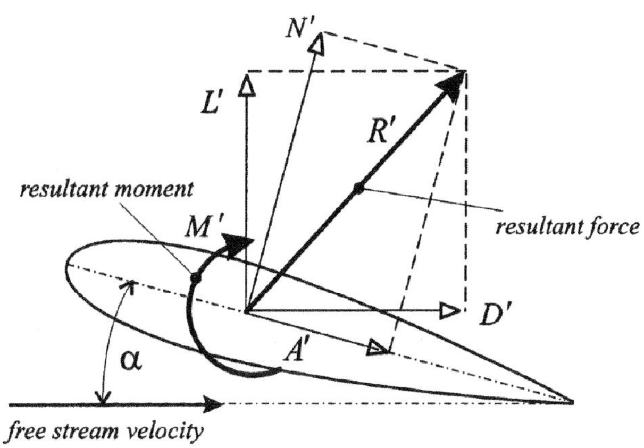

Figure 3.4 Forces and moments on an airfoil.

When discussing the aerodynamic characteristic of an airfoil it is common practice to deal with the following nondimensional groups.

Pressure coefficient: $$C_p = \frac{p - p_\infty}{\frac{1}{2}\rho_\infty Q_\infty^2}$$

Section lift coefficient: $$c_f = \frac{\tau}{\frac{1}{2}\rho_\infty Q_\infty^2}$$

Skin friction coefficient: $$c_l = \frac{L'}{\frac{1}{2}\rho_\infty Q_\infty^2 c}$$

Section drag coefficient: $$c_d = \frac{D'}{\frac{1}{2}\rho_\infty Q_\infty^2 c}$$

Section pitching moment coefficient: $$c_m = \frac{M'}{\frac{1}{2}\rho_\infty Q_\infty^2 c^2}$$

Section normal force coefficient: $$c_n = \frac{N'}{\frac{1}{2}\rho_\infty Q_\infty^2 c}$$

Section axial force coefficient: $$c_a = \frac{A'}{\frac{1}{2}\rho_\infty Q_\infty^2 c}$$

The above integrated force coefficients are related to the local surface pressure and skin friction coefficients in the following manner:

$$c_n = \int_0^1 (C_{p,l} - C_{p,u})\, d\hat{x} + \int_0^1 \left(c_{f,u} \frac{d\hat{z}_u}{d\hat{x}} + c_{f,l} \frac{d\hat{z}_l}{d\hat{x}} \right) d\hat{x}$$

$$c_n = \int_0^1 \left(C_{p,u} \frac{d\hat{z}_u}{d\hat{x}} - C_{p,l} \frac{d\hat{z}_l}{d\hat{x}} \right) d\hat{x} + \int_0^1 (c_{f,u} - d_{f,t})\, d\hat{x}$$

$$c_{m,le} = \int_0^1 (C_{p,u} - C_{p,l})\, \hat{x} d\hat{x} - \int_0^1 \left(c_{f,u} \frac{d\hat{z}_u}{d\hat{x}} + c_{f,l} \frac{d\hat{z}_l}{d\hat{x}} \right) \hat{x}\, d\hat{x}$$
$$+ \int_0^1 \left(C_{p,u} \frac{d\hat{z}_u}{d\hat{x}} - c_{f,u} \right) \hat{z}_u d\hat{x} + \int_0^1 \left(-C_{p,l} \frac{d\hat{z}_l}{d\hat{x}} + c_{f,l} \right) \hat{z}_l\, d\hat{x}$$

The lift and drag coefficients can be obtained from

$$c_l = c_n \cos\alpha - c_a \sin\alpha$$
$$c_d = c_n \sin\alpha + c_a \cos\alpha$$

The Center of Pressure

When integrating the surface pressures and shear stresses, a particular moment reference point can be chosen about which there will be no net moment. This point is known as the center of pressure.

If the resultant force is split into the normal and axial components, the distance from the leading edge to the center of pressure is given by

$$\hat{x}_{cp} = \frac{x_{cp}}{c} = \frac{M'_{le}}{N'c}$$

For small angles of attack, N' can be replaced by L' if required.

The Aerodynamic Center

The aerodynamic center is defined as the moment reference point about which the pitching moment does not significantly change with lift. Essentially, the pitching moment remains at the zero lift value. The aerodynamic center for a thin airfoil is theoretically located at the quarter chord position, $\hat{x}_{ac} = \frac{1}{4}$, for subsonic flow (Subsection 3.14) and the half chord position for supersonic flow (Subsection 3.17), $\hat{x}_{ac} = \frac{1}{2}$.

The Reynolds Number and the Mach Number

The aerodynamic characteristics of an airfoil are governed by its shape and two nondimensional groups known as the Reynolds number and the Mach number, respectively defined as

$$\mathrm{Re}_\infty = \frac{\rho_\infty Q_\infty c}{\mu_\infty}$$

$$M_\infty = \frac{Q_\infty}{a_\infty}$$

It is shown in Subsection 3.11 that if two airfoils are geometrically similar and have equal Reynolds and Mach numbers they will produce identical aerodynamic force and moment coefficients. When this condition is satisfied, the flow fields are said to be dynamically similar. The influences the Reynolds number and Mach number have on the flow field are, respectively, discussed in Subsections 3.13 and 3.16.

Typical Wind Tunnel Aerodynamic Data

Figures 3.5 and 3.6 display typical wind tunnel data obtained for the NACA 23012 airfoil. Figure 3.5 illustrates the terminology used when describing aerodynamic load characteristics.

As indicated in Figure 3.5, every airfoil will have one particular angle of attack at which zero lift will be produced; this is known as the zero lift angle, α_{l0}. In general, symmetrical airfoils will have a zero lift angle of zero and airfoils with a net positive camber will have a small negative value. The lift curve slope is defined as the slope of the linear portion of the lift curve which passes through the zero lift angle. Within this linear region the lift coefficient can be computed from

$$c_l = \frac{dc_l}{d\alpha}(\alpha - \alpha_{l0}) = c_{l\alpha}(\alpha - \alpha_{l0})$$

where $c_{l\alpha}$ is known as the lift curve slope.

When the drag coefficient is plotted against the lift coefficient, as in Figure 3.6, the resulting curve is known as a drag polar. The marked decrease in lift coefficient and accompanying increase in drag coefficient are referred to as airfoil stall, and this, as described in Subsections 3.13 and 3.14, is due to various boundary layer phenomena. Figure 3.7 displays the generally accepted terminology used to describe an airfoil pressure distribution. This figure illustrates the variation in upper surface pressure distribution with angle of attack for a modified NACA 23012 airfoil (Niven 1988); this particular airfoil stalls due to trailing edge separation which reduces the net upper surface suction force and thus the lift.

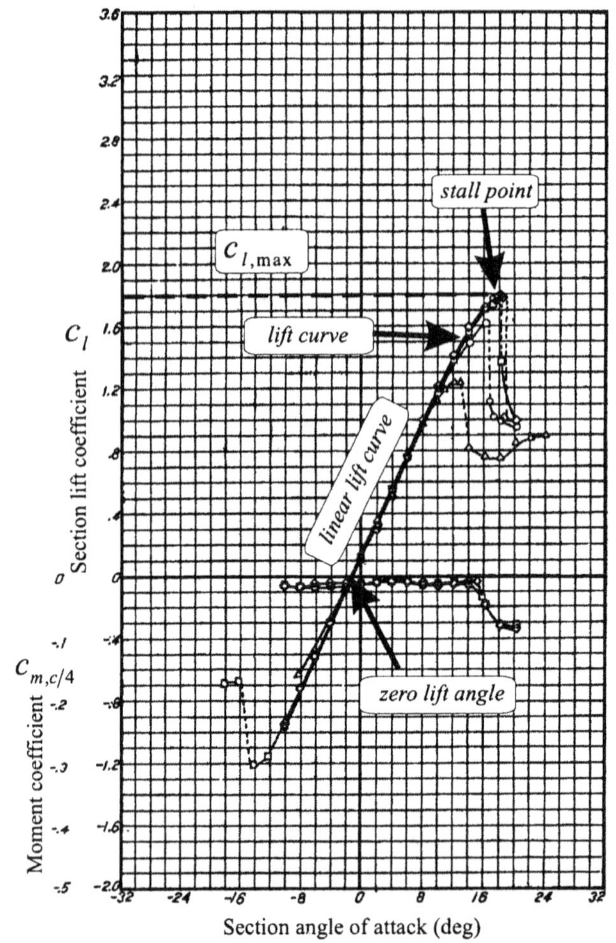

Figure 3.5 Lift and pitching moment characteristics for a NACA 23012 (from Abbott et al. 1945, courtesy of NASA).

3.10 Wing Geometric and Aerodynamic Definitions

Wing Geometry

Figure 3.8 illustrates the main terminology and geometric parameters used to define a wing. The following parameters are also used to help define the wing:

Taper ratio:	$\lambda = c_t/c_r$
Nondimensional spanwise position:	$\hat{y} = 2y/b$
Wing area:	$S = 2\int_0^{b/2} c\,dy = b\int_0^1 c\,d\hat{y}$
Mean geometric chord:	$c_g = S/b$
Aspect ratio:	$A_g = b/c_g = b^2/S$

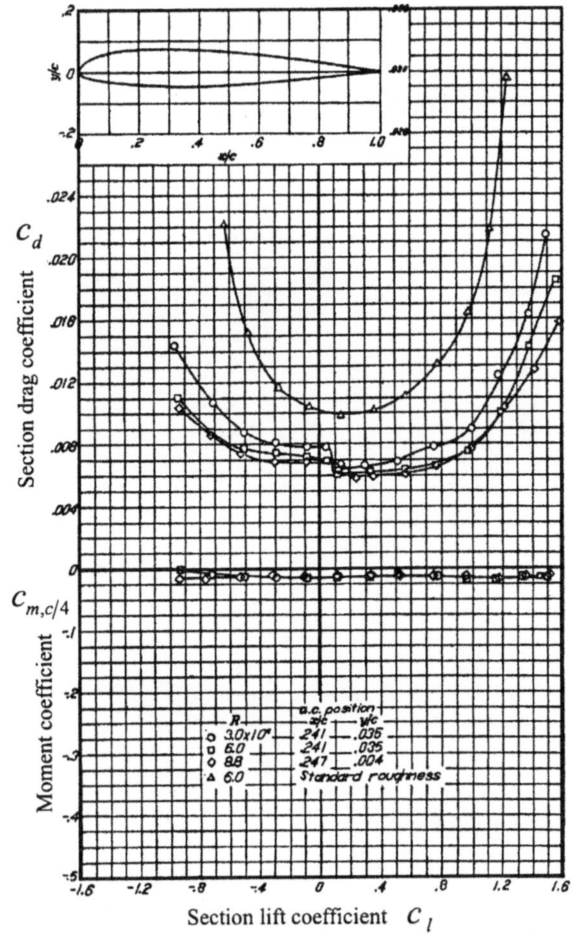

FIGURE 3.6 Drag polar for NACA 23012 (from Abbott et al. 1945, courtesy of NASA).

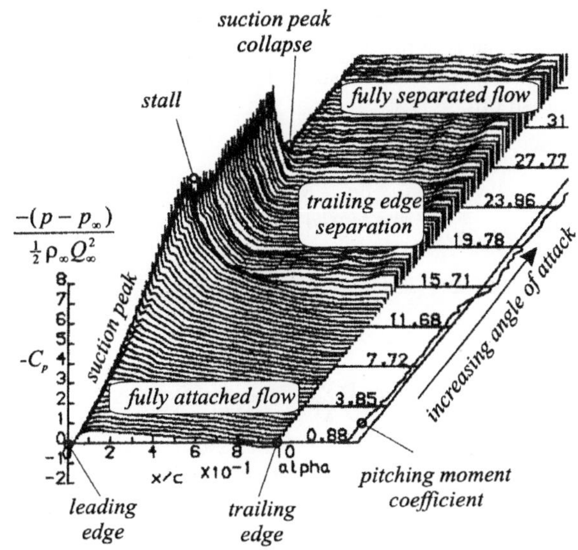

FIGURE 3.7 Typical upper surface pressure distribution (from Niven 1988).

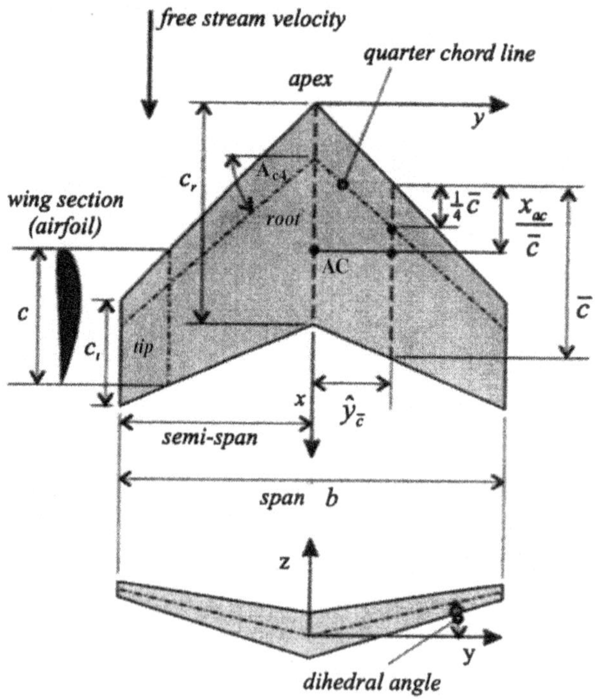

FIGURE 3.8 Wing geometric definitions.

Definitions of Wing Twist

It will be shown in Subsection 3.14 that it is aerodynamically beneficial to twist a tapered wing. Figure 3.9 shows that the geometric twist at an arbitrary spanwise position (which lies between the root and tip), $\varepsilon_g(\hat{y})$ is defined as the angle between the root chord and the chord of the wing section at the location of interest. Twist is defined as positive when the section is rotated nose-up (relative to the root chord), and is referred to as wash-in. If the section is rotated nose-down the twist is negative and is called wash-out.

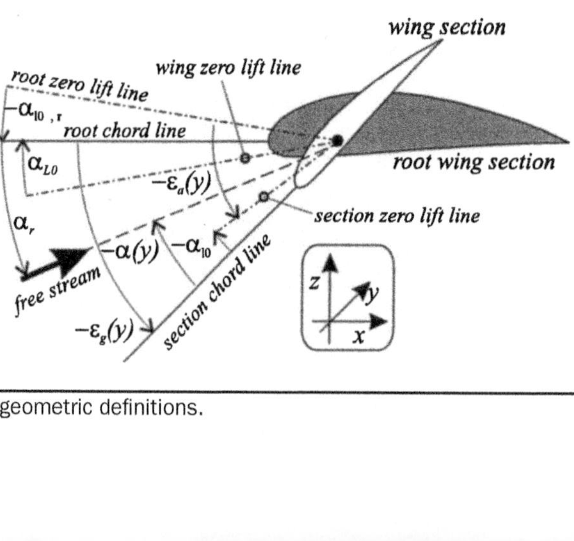

FIGURE 3.9 Wing twist geometric definitions.

The aerodynamic twist, $\varepsilon_a(\hat{y})$, is defined as the angle between the section zero lift line and the root zero lift line. The definition allows for the airfoil profile to change along the span and is thus defined as

$$\varepsilon_a(\hat{y}) = \varepsilon_g(\hat{y}) + \alpha_{l0,r} - \alpha_{l0}(\hat{y})$$

The aerodynamic twist at the wing tip, $\varepsilon_{a,t}$, is often simply referred to as the wing twist and denoted by ε_t. A common spanwise distribution of aerodynamic twist would be a linear variation given by

$$\varepsilon_a(\hat{y}) = \hat{y}(\varepsilon_{a,t})$$

Wing Aerodynamic Forces and Moments

When discussing the aerodynamic characteristic of a wing, the following nondimensional groups are used:

Wing lift coefficient: $$C_L = \frac{L}{\frac{1}{2}\rho_\infty Q_\infty^2 S}$$

Wing drag coefficient: $$C_D = \frac{D}{\frac{1}{2}\rho_\infty Q_\infty^2 S}$$

Figure 3.9 shows that the wing angle of attack is defined as the angle between the root chord line and the free-stream velocity. The section angle of attack is related to the wing angle of attack by

$$\alpha(\hat{y}) = \varepsilon_g(\hat{y}) + \alpha_r$$

Every wing will have a particular angle of attack at which no net lift is produced; this is known as the wing zero lift angle, α_{L0}, and can be obtained using the methods described in Subsection 3.15. In a manner analogous to an airfoil, the wing lift coefficient can be plotted against the wing angle of attack. Over the linear portion of the lift curve the wing lift coefficient is given by

$$C_L = \frac{dC_L}{d\alpha}(\alpha_r - \alpha_{L0}) = C_{L\alpha}(\alpha_r - \alpha_{L0})$$

where $C_{L\alpha}$ is known as the wing lift curve slope and may be computed using the methods described in Subsection 3.15.

The pitching moment coefficient of the wing is defined as

$$C_m = \frac{M}{\frac{1}{2}\rho_\infty Q_\infty^2 S\bar{c}}$$

where the mean aerodynamic chord of a wing is given by

$$\bar{c} = \frac{2}{S}\int_0^{b/2} c^2\,dy = \frac{b}{S}\int_0^1 c^2\,d\hat{y}$$

and is located at a spanwise position equal to

$$\hat{y}_{\bar{c}} = \frac{b^2}{2S}\int_0^1 c\hat{y}\,d\hat{y}$$

The mean aerodynamic chord represents the chord of a untwisted, unswept, rectangular wing which produces the same lift and pitching moment as the actual wing. For a straight tapered wing it is given by

$$\frac{\bar{c}}{c_r} = \frac{2}{3}\left(\frac{1+\lambda+\lambda^2}{1+\lambda^2}\right)$$

The Wing Aerodynamic Center

The wing aerodynamic center is located on the root chord and is defined as the moment reference point about which the wing pitching moment does not significantly change with wing lift. Figure 3.8 shows that the approximate position of the wing aerodynamic center can be found (Schlichting and Truckenbrodt 1979) by first locating the aerodynamic center on the mean aerodynamic chord, $x_{ac}/\bar{c}$, and then projecting this point onto the root chord. When the flow is subsonic $x_{ac}/\bar{c} = \frac{1}{4}$, while for supersonic flow $x_{ac}/\bar{c} = \frac{1}{2}$.

3.11 Fundamentals of Vector Fluid Dynamics

Control Volume Analysis

The laws of mass, momentum, and energy conservation are applied to a fluid using well-defined regions of the flow field known as control volumes. When the chosen volume is very small it is referred to as a differential control volume or a fluid element. The geometry of the fluid element can be defined using either Cartesian (as used here), cylindrical, or spherical coordinate systems. The conservation laws can be applied either to a moving or a stationary fluid element.

Generalized Motion of a Fluid

Consider a point in a steady flow where the fluid velocity in the x-direction is u_1. A small distance away from this point the velocity, u_2, can be related to u_1 via the truncated Taylor series expansion below:

$$u_2 = u_1 + \left(\frac{\partial u}{\partial x}\right)dx + \left(\frac{\partial u}{\partial y}\right)dy + \left(\frac{\partial u}{\partial z}\right)dz$$

The fluid velocity can now be written in the form

$$\begin{aligned} u_2 = u_1 &+ \left[0 - \frac{1}{2}\left(\frac{\partial v}{\partial x} - \frac{\partial u}{\partial y}\right)dy + \frac{1}{2}\left(\frac{\partial u}{\partial z} - \frac{\partial w}{\partial x}\right)dz\right] \\ &\left[\left(\frac{\partial u}{\partial x}\right)dx + \frac{1}{2}\left(\frac{\partial v}{\partial x} + \frac{\partial u}{\partial y}\right)dy + \frac{1}{2}\left(\frac{\partial w}{\partial x} + \frac{\partial u}{\partial z}\right)dz\right] \end{aligned}$$

Similar equations can be written for the v-velocity and the w-velocity to give the following tensor equation (i.e., a matrix of vector quantities):

$$\begin{bmatrix} u_2 \\ v_2 \\ w_2 \end{bmatrix} = \begin{bmatrix} u_1 \\ v_1 \\ w_1 \end{bmatrix} + \begin{bmatrix} 0 & -\Omega_z & \Omega_y \\ \Omega_z & 0 & -\Omega_x \\ -\Omega_y & \Omega_x & 0 \end{bmatrix} \begin{bmatrix} dx \\ dy \\ dz \end{bmatrix} + \begin{bmatrix} \dot{\varepsilon}_{xx} & \dot{\varepsilon}_{xy} & \dot{\varepsilon}_{xz} \\ \dot{\varepsilon}_{yx} & \dot{\varepsilon}_{yy} & \dot{\varepsilon}_{yz} \\ \dot{\varepsilon}_{zx} & \dot{\varepsilon}_{zy} & \dot{\varepsilon}_{zz} \end{bmatrix} \begin{bmatrix} dx \\ dy \\ dz \end{bmatrix}$$

When written in this form, the motion of the fluid can be considered to be composed of a translation, a rotation (the Ω matrix) and a deformation (the $\dot{\varepsilon}$ matrix). The mathematical form of the rotation and deformation matrices is described below.

The Curl of the Velocity Vector

By considering the motion of a small rectangular fluid element it can be shown (Eskinazi 1967) that the local angular velocity of the fluid is related to the velocity field in the following manner:

$$(\Omega_x,\Omega_y,\Omega_z)=\vec{i}\,\frac{1}{2}\left(\frac{\partial w}{\partial y}-\frac{\partial v}{\partial z}\right)+\vec{j}\,\frac{1}{2}\left(\frac{\partial u}{\partial z}-\frac{\partial w}{\partial x}\right)+\vec{k}\,\frac{1}{2}\left(\frac{\partial v}{\partial x}-\frac{\partial u}{\partial y}\right)$$

$$=\frac{1}{2}(\nabla\times\vec{q})=\frac{1}{2}\begin{vmatrix}\vec{i} & \vec{j} & \vec{k}\\ \dfrac{\partial}{\partial x} & \dfrac{\partial}{\partial y} & \dfrac{\partial}{\partial z}\\ u & v & w\end{vmatrix} \tag{3.1}$$

where

$$\nabla=\vec{i}\,\frac{\partial}{\partial x}+\vec{j}\,\frac{\partial}{\partial y}+\vec{k}\,\frac{\partial}{\partial z}$$

is referred to as a vector differential operator and is commonly called del. The vector product $(\nabla\times\vec{q})$ is known as the curl of the velocity vector. In fluid dynamics twice the angular velocity is known as the vorticity vector and is denoted $\vec{\omega}=(\omega_x,\omega_y,\omega_z)$. The right-hand screw rule is used to indicate the positive sense of both the angular velocity and the vorticity vectors. An irrotational flow is one in which $\nabla\times\vec{q}=0$ throughout the flow field.

The Divergence of the Velocity Vector

Utilizing the fluid element again, it can be shown (White 1991) that the deformation matrix consists of normal and shear strain rates which consist of various groups of the local velocity gradients. This can be compactly written in matrix form as

$$\begin{bmatrix}\dot{\varepsilon}_{xx} & \dot{\varepsilon}_{xy} & \dot{\varepsilon}_{xz}\\ \dot{\varepsilon}_{yx} & \dot{\varepsilon}_{yy} & \dot{\varepsilon}_{yz}\\ \dot{\varepsilon}_{zx} & \dot{\varepsilon}_{zy} & \dot{\varepsilon}_{zz}\end{bmatrix}=\begin{bmatrix}\dfrac{\partial u}{\partial x} & \dfrac{1}{2}\left(\dfrac{\partial v}{\partial x}+\dfrac{\partial u}{\partial y}\right) & \dfrac{1}{2}\left(\dfrac{\partial w}{\partial x}+\dfrac{\partial u}{\partial z}\right)\\ \dfrac{1}{2}\left(\dfrac{\partial u}{\partial y}+\dfrac{\partial v}{\partial x}\right) & \dfrac{\partial v}{\partial y} & \dfrac{1}{2}\left(\dfrac{\partial w}{\partial y}+\dfrac{\partial v}{\partial z}\right)\\ \dfrac{1}{2}\left(\dfrac{\partial u}{\partial z}+\dfrac{\partial w}{\partial x}\right) & \dfrac{1}{2}\left(\dfrac{\partial v}{\partial z}+\dfrac{\partial w}{\partial y}\right) & \dfrac{\partial w}{\partial z}\end{bmatrix} \tag{3.2}$$

The volumetric strain rate is given by

$$\dot{\varepsilon}_V=\frac{\partial u}{\partial x}+\frac{\partial v}{\partial y}+\frac{\partial w}{\partial z}=\nabla\cdot\vec{q} \tag{3.3}$$

where $\nabla\cdot\vec{q}$ is known as the divergence of the velocity vector.

The Gradient of a Scalar Field

At any point within a scalar field, there will be a single value and direction for the maximum spatial rate of change of the scalar. The gradient of a scalar field is a vector field given, in Cartesian coordinates, as

$$\nabla\phi=\vec{i}\,\frac{\partial\phi}{\partial x}+\vec{j}\,\frac{\partial\phi}{\partial y}+\vec{k}\,\frac{\partial\phi}{\partial z}$$

As described in Subsection 3.12, an important application of the gradient of a scalar field is the velocity potential function. Another important parameter, used in many aerodynamic theories, is the unit vector normal to an arbitrary surface in the flow field. This surface may be defined by

$$F(x_x, y_s, z_s) = z_s - f_x(x_s, y_s) = 0$$

The unit normal surface vector can be obtained at any location using the expression

$$\vec{n} = \frac{\nabla F}{|\nabla F|} = \frac{1}{|\nabla F|}\left(\frac{\partial F}{\partial x}, \frac{\partial F}{\partial y}, \frac{\partial F}{\partial z}\right)$$

Circulation and Integral Theorems

Circulation is defined as the line integral, around a closed curve, of the component of the local velocity tangential to the path of integration. The right-hand screw rule is used to indicate the positive direction along the integration path. If l and $\vec{l}$ respectively denote the distance around the path and the unit vector along the path, the circulation is given by

$$\Gamma = \oint_c (\vec{q} \cdot \vec{l})\, dl \tag{3.4}$$

Figure 3.10 illustrates a small differential area placed in a flow which lies in the y–z plane. The circulation around $abcd$ is given by

$$\delta\Gamma = v\delta y + w\delta z - \left(v + \frac{\partial v}{\partial z}\delta z\right)\delta y - \left(w + \frac{\partial w}{\partial y}\delta y\right)\delta z$$

$$= \left(\frac{\partial w}{\partial y} - \frac{\partial v}{\partial z}\right)\delta y\, \partial z$$

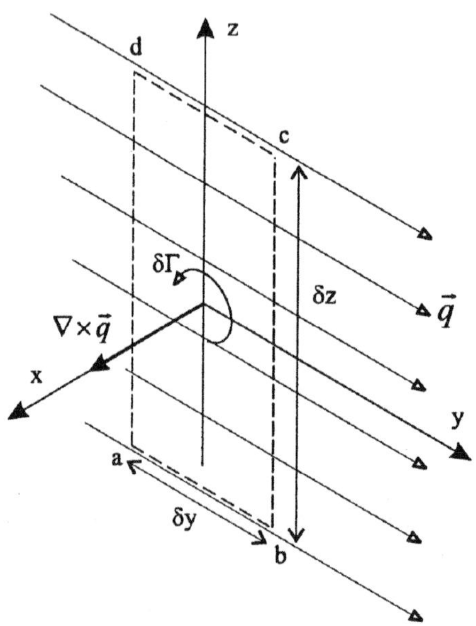

FIGURE 3.10 Relationship between circulation and vorticity.

This result states that circulation is the product of vorticity and area and may be generalized in the form

$$\frac{d\Gamma}{dA} = (\nabla \times \vec{q}) \cdot \vec{n}$$

where $\vec{n}$ is the unit normal of area dA. Stokes's theorem states that the net vorticity over an arbitrary three-dimensional surface can be equated to the circulation around any closed curve which bounds the surface, i.e.,

$$\Gamma = \oint_C \vec{q} \cdot \vec{l}\, dl = \oiint_{CS} (\nabla \times \vec{q}) \cdot \vec{n}\, dA \tag{3.5}$$

There are two other useful relationships, known as the gradient theorem and Gauss's theorem, which are respectively given by

$$\oint_{CS} p\, dA = \oiiint_{CV} (\nabla p)\, dV \text{ and } \oint_{CS} (\vec{q} \cdot \vec{n})\, dA = \oiiint_{CV} (\nabla \cdot \vec{q})\, dV$$

Definitions of Inviscid and Incompressible Flow Fields

Figure 3.11 displays the surface stresses experienced by a fluid element. These stresses are due to the pressure of the surrounding fluid and the action of viscosity. Viscous stresses arise when the fluid element is distorted over a period of time and, as described in Subsection 3.13, are a direct result of the microscopic behavior of the fluid molecules. When these forces are assumed to be zero, the fluid is said to be inviscid and the fluid element only experiences pressure forces. Furthermore, an inviscid fluid will not exhibit the effects of mass diffusion or thermal conduction. An incompressible flow arises when the density of the fluid is assumed to be constant throughout.

The Conservation of Mass

Applying the conservation of mass principle to a stationary fluid element results in the equation

$$\frac{\partial \rho}{\partial t} + \nabla \cdot (p\vec{q}) = 0$$

This equation is often referred to as the continuity equation, which for an incompressible flow reduces to $\nabla \cdot \vec{q} = 0$. This agrees with the earlier definition of volumetric strain rate, which must be zero if the flow is incompressible.

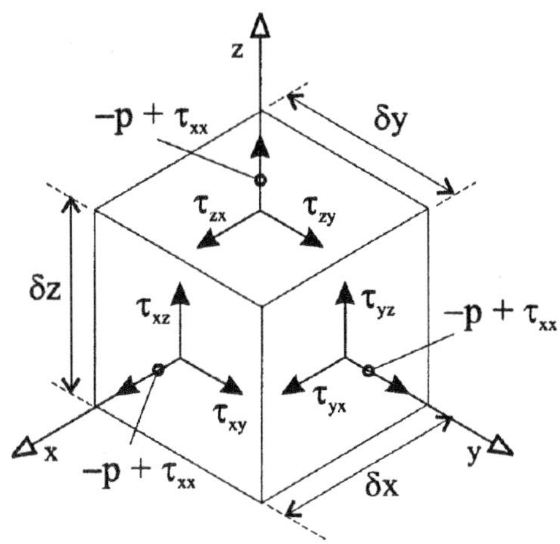

FIGURE 3.11 Pressure and viscous stresses on a fluid element.

The Navier-Stokes Equations

The Navier–Stokes equations are the result of applying the conservation of linear momentum to a differential control volume along each direction of the chosen coordinate system.

In addition to the surface forces which act on a fluid element (see Figure 3.11), there are body forces which act directly on the mass of the fluid element, e.g., gravity, centrifugal, and electromagnetic. For a stationary fluid element, the sum of these forces is set to the rate of change of momentum flow through the element, which is represented by the left-hand side of the equations below.

$$\frac{\partial(\rho u)}{\partial t}+\nabla\cdot(\rho u\vec{q})=-\frac{\partial p}{\partial x}+\frac{\partial \tau_{xx}}{\partial x}+\frac{\partial \tau_{xy}}{\partial y}+\frac{\tau_{xz}}{\partial z}+\rho f_x$$

$$\frac{\partial(\rho v)}{\partial t}+\nabla\cdot(\rho v\vec{q})=-\frac{\partial p}{\partial y}+\frac{\partial \tau_{yx}}{\partial x}+\frac{\partial \tau_{yy}}{\partial y}+\frac{\partial \tau_{yz}}{\partial z}+\rho f_y$$

$$\frac{\partial(\rho w)}{\partial t}+\nabla\cdot(\rho w\vec{q})=-\frac{\partial p}{\partial z}+\frac{\partial \tau_{zx}}{\partial x}+\frac{\partial \tau_{zy}}{\partial y}+\frac{\partial \tau_{zz}}{\partial z}+\rho f_z$$

The viscous stresses are related to the fluid strain rates using Stokes's deformation laws, which consist of the following three postulations:

1. The stress is a linear function of the strain rate (originally proposed by Newton in 1686, hence the name Newtonian fluid).
2. The fluid displays no preferential direction of deformation, i.e., the fluid is isotropic.
3. When the strain rates reduce to zero (stationary fluid) the normal stresses must become equal and opposite to the local fluid static pressure (the hydrostatic condition, where the static pressure is a function of temperature and depth from a reference point).

These three postulations can be used to develop the following relationships between viscous stresses and strain rates (Prandtl and Tietjens 1934; White 2017):

$$\begin{bmatrix} \tau_{xx} & \tau_{xy} & \tau_{xz} \\ \tau_{yx} & \tau_{yy} & \tau_{xz} \\ \tau_{zx} & \tau_{yz} & \tau_{zz} \end{bmatrix} = \begin{bmatrix} 2\mu\frac{\partial u}{\partial x}-\tfrac{2}{3}\mu\nabla\cdot\vec{q} & \mu\left(\frac{\partial u}{\partial y}+\frac{\partial v}{\partial x}\right) & \mu\left(\frac{\partial u}{\partial z}+\frac{\partial w}{\partial x}\right) \\ \mu\left(\frac{\partial v}{\partial x}+\frac{\partial u}{\partial y}\right) & 2\mu\frac{\partial v}{\partial y}-\tfrac{2}{3}\mu\nabla\cdot\vec{q} & \mu\left(\frac{\partial v}{\partial z}+\frac{\partial w}{\partial y}\right) \\ \mu\left(\frac{\partial w}{\partial x}+\frac{\partial u}{\partial z}\right) & \mu\left(\frac{\partial w}{\partial y}+\frac{\partial v}{\partial z}\right) & 2\mu\frac{\partial w}{\partial z}-\tfrac{2}{3}\mu\nabla\cdot\vec{q} \end{bmatrix} \tag{3.6}$$

The Conservation of Energy

Applying the conservation of energy principle to a stationary fluid element results in the following equation, which involves the internal energy of the fluid:

$$\frac{\partial(\rho i)}{\partial t}+\nabla\cdot(\rho i\vec{q})=-p\nabla\cdot\vec{q}+\nabla\cdot(k\nabla T)+\Phi+\rho(\dot{G}+\vec{J}\cdot\vec{q})$$

The term Φ is known as the viscous dissipation term and is given by

$$\Phi = \mu \begin{bmatrix} 2\left(\frac{\partial u}{\partial x}\right)^2 + 2\left(\frac{\partial v}{\partial y}\right)^2 + 2\left(\frac{\partial w}{\partial z}\right)^2 \\ + \left(\frac{\partial u}{\partial y} + \frac{\partial v}{\partial x}\right)^2 + \left(\frac{\partial u}{\partial z} + \frac{\partial w}{\partial x}\right)^2 + \left(\frac{\partial v}{\partial z} + \frac{\partial w}{\partial y}\right)^2 \\ -\frac{2}{3}\left(\frac{\partial u}{\partial x} + \frac{\partial v}{\partial y} + \frac{\partial w}{\partial z}\right)^2 \end{bmatrix}$$

The Euler and Bernoulli Equations

The Euler equations can be obtained from the Navier–Stokes equations by using the assumption that the fluid is incompressible and inviscid, which gives

$$\frac{\partial u}{\partial t} + \vec{q} \cdot \nabla u = \frac{1}{\rho}\frac{\partial p}{\partial x} + f_x$$

$$\frac{\partial v}{\partial t} + \vec{q} \cdot \nabla v = -\frac{1}{\rho}\frac{\partial p}{\partial y} + f_y$$

$$\frac{\partial w}{\partial t} + \vec{q} \cdot \nabla w = -\frac{1}{\rho}\frac{\partial p}{\partial z} + f_z$$

Let h denote the positive vertical direction opposite to that of gravity. Considering only gravitational force on the fluid element, the body forces terms can now be written as

$$f_x = -\frac{\partial(gh)}{\partial x}, \quad f_y = -\frac{\partial(gh)}{\partial y} \quad \text{and} \quad f_z = -\frac{\partial(gh)}{\partial z}$$

Now assume that the flow is rotational (see Subsection 3.12), which gives

$$u = \frac{\partial \phi}{\partial x}, \; v = \frac{\partial \phi}{\partial y} \quad \text{and} \quad w = \frac{\partial \phi}{\partial z}$$

Using this condition along with the three xyz Euler equations and then integrating each resulting equation with respect to the appropriate coordinate direction leads to the conclusion that

$$\frac{\partial \phi}{\partial t} + \frac{p}{\rho} + \frac{1}{2}q^2 + gh = F(t) \tag{3.7}$$

This is known as Bernoulli's equation and in this form is applicable to a fluid which is unsteady, inviscid, irrotational, and incompressible. For steady flow, $(\partial\phi/\partial t) = 0$ and the value of F is constant throughout the entire flow field. It can be shown (White 1991) that Bernoulli's equation is also valid for rotational flow, but the value of F is only constant along a given streamline and varies from streamline to streamline.

The Reynolds Number and Mach Number

The nondimensionalized version of the x-direction Navier–Stokes equation for two-dimensional steady flow is given by

$$\frac{\partial(\hat{\rho}\hat{u}\hat{u})}{\partial \hat{x}} + \frac{\partial(\hat{\rho}\hat{u}\hat{v})}{\partial \hat{y}} = -\frac{1}{\gamma M_\infty^2}\frac{\partial \hat{p}}{\partial \hat{x}} + \frac{1}{\text{Re}_\infty}\frac{\partial}{\partial \hat{y}}\left[\hat{\mu}\left(\frac{\partial \hat{v}}{\partial \hat{x}} + \frac{\partial \hat{u}}{\partial \hat{y}}\right)\right]$$

where

$$\bar{x} = x/c,\ \hat{y} = y/c,\ \hat{u} = u/Q_\infty,\ \hat{v} = v/Q_\infty,\ \hat{p} = p/p_\infty$$
$$\hat{\rho} = \rho/\rho_\infty,\ \hat{\mu} = \mu/\mu_\infty$$

and

$$M_\infty = Q_\infty/a_\infty \quad \text{and} \quad \mathrm{Re}_\infty = (\rho_\infty Q_\infty c)/\mu_\infty$$

If the flow over two geometrically similar bodies has the same Mach number and Reynolds numbers the solution of the nondimensional Navier–Stokes equations will be numerically identical. When this occurs, the two flows are said to be dynamically similar.

3.12 Fundamentals of Potential Flow

The Potential Function

Consider an inviscid irrotational flow defined by $\vec{\omega} = \nabla \times \vec{q} = 0$ (see Subsection 3.11 for expansion of vector product). Using the rules of partial differentiation, this condition is also satisfied if

$$\vec{\omega} = \nabla \times (\nabla \phi) = 0$$

Thus, for an irrotational flow, the velocity vector field is the gradient of a scalar field (see Subsection 3.11), which is known as the velocity potential, ϕ, i.e.,

$$\vec{q} = (u, v, w) = \nabla \phi = \left(\frac{\partial \phi}{\partial x}, \frac{\partial \phi}{\partial y}, \frac{\partial \phi}{\partial z} \right)$$

The velocity potential can also be defined as $\vec{q} = -\nabla \phi$ without any effect on the irrotational flow condition. If the flow is both irrotational and incompressible, we have

$$\nabla \cdot \vec{q} = \nabla \cdot (\nabla \phi) = \frac{\partial^2 \phi}{\partial x^2} + \frac{\partial^2 \phi}{\partial y^2} = \nabla^2 \phi = 0$$

This is a second-order linear partial differential equation for the velocity potential and is known as Laplace's equation. A flow which is both incompressible and irrotational is known as a potential flow. Laplace's equation may be solved analytically or numerically subject to the boundary conditions discussed later on in this section. Many analytical solutions of the Laplace equation can be associated with simple flow field patterns; the most frequently used solutions are described below.

The Disturbance Velocity Potential

A concept used frequently in aerodynamic theories is one which considers the flow field velocity potential function to consist of two parts: one due to the free-stream flow and one, known as the disturbance potential, which accounts for the presence of any body within the flow. Mathematically this is written as $\phi = \phi_\infty + \tilde{\phi}$, where $\phi_\infty = U_\infty x + V_\infty y + W_\infty z$ and $\tilde{\phi}$ is the disturbance velocity potential. The local velocity field is given by $u = U_\infty + \tilde{u}$, $v = V_\infty + \tilde{v}$, and $w = W_\infty + \tilde{w}$, where $\tilde{u} = \partial\tilde{\phi}/\partial x$, $\tilde{v} = \partial\tilde{\phi}/\partial y$, and $\tilde{w} = \partial\tilde{\phi}/\partial z$. The solution to the flow field is often found using the small disturbance approximation as described later in this section.

The Stream Function

A stream line is defined as a line whose spatial variation in gradient, at any given instant in time, corresponds to the variation in flow direction of the local velocity vector. For two-dimensional flow, a streamline is mathematically defined (using Cartesian coordinates) by the relationship

$$\frac{dy}{dx}=\frac{v}{u}$$

The conservation of mass principle (Subsection 3.11) for two-dimensional incompressible flow is given by

$$\frac{\partial u}{\partial x}+\frac{\partial v}{\partial y}=0$$

This equation can also be satisfied, using the rules of partial differentiation, by another function, called the stream function, and defined by

$$u=\frac{\partial \psi}{\partial y} \quad \text{and} \quad v=\frac{\partial \psi}{\partial x}$$

When the flow field is irrotational, the stream function satisfies Laplace's equation, i.e.,

$$\frac{\partial v}{\partial x}-\frac{\partial u}{\partial y}=\frac{\partial^2 \psi}{\partial x^2}+\frac{\partial^2 \psi}{\partial y^2}=\nabla^2\psi=0$$

It should be noted that whereas both streamlines and the potential function exist for all irrotational three-dimensional flow fields, the stream function can only be defined in three dimensions when the flow is axisymmetric (Vallentine 1959).

Two-Dimensional Solutions of the Laplace Equation

Table 3.4 gives the potential and stream functions in polar coordinates for a number of elementary types of flow which form the basis of most aerodynamic mathematical models. In the uniform flow case, α is the angle between the free-stream flow and the horizontal. For the vortex, Γ is taken positive according to the righthand screw rule, which is anticlockwise when the vortex lies in the $x-y$ plane and $\psi=0$ when $r=r_0$. Figure 3.12 illustrates the stream line patterns associated with these flows.

The components of the local velocity vector, $\vec{q}$, in polar coordinates are related to the potential and stream functions as follows:

$$q_r=\frac{\partial \phi}{\partial r}=\frac{1}{r}\frac{\partial \psi}{\partial \theta} \quad \text{and} \quad q_\theta=\frac{1}{r}\frac{\partial \phi}{\partial \theta}=-\frac{\partial \psi}{\partial r}$$

Flow	Potential function	Stream function
Uniform flow	$\phi=Q_\infty r\cos(\theta-\alpha)$	$\psi=Q_\infty r\sin(\theta-\alpha)$
Source	$\phi=\frac{\Lambda}{2\pi}\ln r$	$\psi=\frac{\Lambda}{2\pi}\theta$
Vortex	$\phi=\frac{\Gamma}{2\pi}\theta$	$\psi=-\frac{\Gamma}{2\pi}\ln\frac{r}{r_0}$
Doublet	$\phi=\frac{H}{2\pi}\frac{\cos\theta}{r}$	$\psi=-\frac{H}{2\pi}\frac{\sin\theta}{r}$

TABLE 3.4 Elementary Flow Patterns

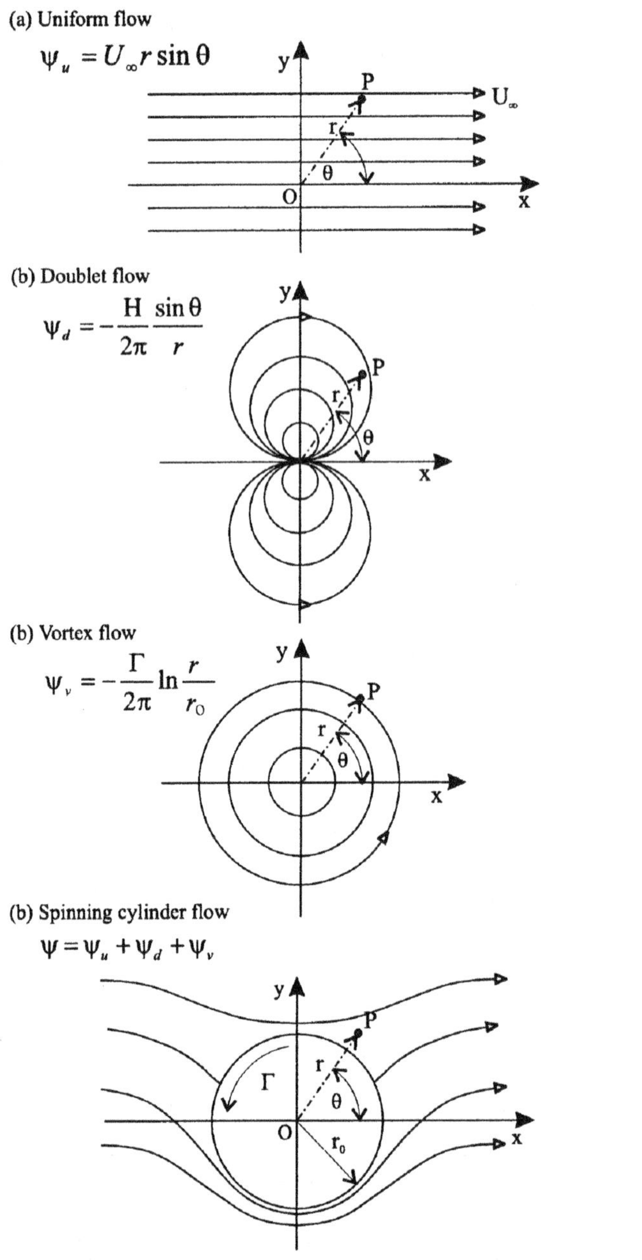

FIGURE 3.12 The superposition principle.

With respect to the vortex flow field, the circulation around any closed curve which encloses the center has the value Γ. However, if the closed curve does not surround the vortex center the circulation will be zero. Thus, the entire flow field is irrotational and the vortex center is said to be a singularity.

The Principle of Superposition

Solutions of the Laplace equation, which are relevant to aerodynamic work, are normally achieved by adding together elementary potential flow solutions which is known as superposition. For example, Figure 3.12 shows that the lifting flow over a cylinder can be solved by adding together the flow fields due to uniform flow, a doublet and a vortex (the circulation can be set to any value). The final stream function for the spinning cylinder would be given by

$$\psi = U_\infty \left(r - \frac{r_0^2}{r} \right) \sin\theta - \frac{\Gamma}{2\pi} \ln\frac{r}{r_0} \tag{3.8}$$

where $r_0^2 = \dfrac{H}{2\pi U_\infty}$

The Kutta–Joukowski Theorem of Lift

Using the stream function, given by equation (3.8), the tangential component of the velocity vector, at any point in the flow field, is given by

$$q_\theta = -\frac{\partial\psi}{\partial r} = -U_\infty \left(1 + \frac{r_0^2}{r^2} \right) \sin\theta + \frac{\Gamma}{2\pi r}$$

The velocity on the cylinder surface is given when $r = r_0$, i.e.,

$$q_0 = -2U_\infty \sin\theta + \frac{\Gamma}{2\pi r_0}$$

The surface pressure coefficient becomes

$$C_p = 1 - \left(\frac{q_\theta}{U_\infty} \right)^2 = 1 - 4\sin^2\theta + \frac{4\Gamma}{2\pi U_\infty r_0} \sin\theta - \left(\frac{\Gamma}{2\pi U_\infty r_0} \right)^2$$

The lift coefficient for the cylinder can be obtained from

$$c_l = -\frac{1}{2} \int_0^{2\pi} C_p \sin\theta \, d\theta = \frac{-\Gamma}{U_\infty r_0}$$

which gives the lift per unit span as

$$L' = \frac{1}{2} \rho_\infty Q_\infty^2 2 r_0 c_l = -\rho_\infty U_\infty \Gamma$$

This result can be shown to apply to apply to the flow around any body which has some value of circulation associated with it and is known as the Kutta–Joukowski theorem. The pressure drag on the cylinder is given by

$$c_d = -\frac{1}{2} \int_0^{2\pi} C_p \cos\theta \, d\theta = 0$$

The fact that inviscid flow theory always predicts zero drag which is never the case in reality is known as d'Alembert's paradox. As discussed in Subsections 3.13 and 3.14, this difference can be explained using the viscous phenomena of boundary layer formation and separation.

Three-Dimensional Vortex Flows

A vortex line is defined as a line whose spatial variation in gradient, at any given instant in time, corresponds to the spatial variation in direction of the local vorticity vector.

When a number of vortex lines bunch together and pass through a common differential cross-sectional area, dA, a vortex filament is said to have formed. The strength of this filament is given by the circulation around the perimeter of dA and is related to the local vorticity via Stokes's theorem (Subsection 3.11), i.e., $d\Gamma = \vec{\omega} \cdot \hat{n} dA$. A vortex tube is defined as a bundle of vortex filaments.

The behavior of a vortex tube is governed by Helmholtz's laws of vorticity, which state that:

1. The strength of a vortex filament is constant along its length.
2. A vortex filament must either form a closed path, extend to infinity, or terminate on a solid boundary. This is a consequence of the first law, which effectively states that the product of vorticity and the filament cross-sectional area must remain constant.
3. A vortex filament always consists of the same fluid elements.
4. The strength of a vortex filament remains constant as it moves throughout the flow field.

The last two laws are consequences of the inviscid flow assumption, which does not allow any diffusion of flow properties. For more information on vortex flows and proofs of Helmholtz's laws, see Lugt (1995) and Eskinazi (1967).

The velocity field induced by a vortex filament is given by the Biot–Savart law. Figure 3.13 shows a straight-line vortex filament lying along the z-axis. The velocity induced by a small element, δl, has the magnitude

$$\delta q_\theta = \frac{\Gamma}{4\pi} \frac{\sin\theta}{r^2} \delta l$$

and points in the direction $\delta\hat{l} \times \vec{r}$, where $\delta\hat{l}$ is the unit vector in the direction l. Integrating between the points A and B results in

$$q_\theta = \frac{\Gamma}{4\pi r_n}(\cos\theta_1 + \cos\theta_2)$$

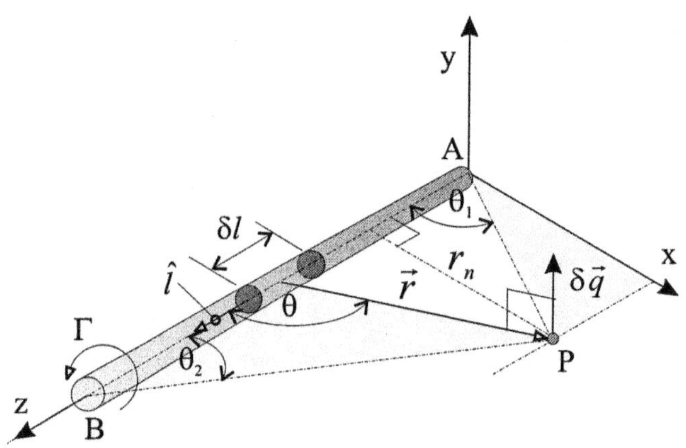

FIGURE 3.13 A straight-line vortex filament.

As will be discussed in Subsection 3.15, the concept of a semi-infinite vortex filament will be utilized to model the flow field around a finite wing. Referring again to Figure 3.13, a semi-infinite vortex is obtained $\theta_1 = \pi/2$ and $\theta_2 \to 0$, which causes q_θ to lie in the x–y plane and have the magnitude

$$q_\theta = \frac{\Gamma}{4\pi r_n}$$

Proofs of the Biot–Savart law can be found in Eskinazi (1967), Karamcheti (1980), and Katz and Plotkin (1991).

An important concept in aerodynamic theory is the vortex sheet, shown in Figure 3.14, which is defined as a large number of vortex filaments whose axes lie parallel to each other within a mutual plane. The circulation (see Subsection 3.11) around the path *abcd* is given by

$$\delta\Gamma = \left(\frac{\partial u}{\partial z} - \frac{\partial w}{\partial x}\right)\delta z \delta x = \delta u \delta x - \delta v \delta z$$

For a thin vortex sheet this can be approximated by

$$\delta\Gamma = \delta u \delta x = (u_u - u_l)\delta x = \gamma \delta x$$

where γ is known as the vortex sheet strength. Referring to Figure 3.14, the velocity potential and induced velocity at point P given by a small segment of the vortex sheet lying at the origin are respectively given by

$$\delta\phi = \frac{\gamma \delta x}{2\pi}\theta \quad \text{and} \quad \delta q_\theta = \frac{\gamma \delta x}{2\pi r}$$

Conformal Transformation

In a similar fashion to the function $y = f(x)$, a function of a complex number $w = f(z)$ can be formulated. When this occurs, z is known as a complex variable and w can be written in the following form:

$$w = f(z) = f(x + iy) = \phi(x, y) + i\psi(x, y)$$

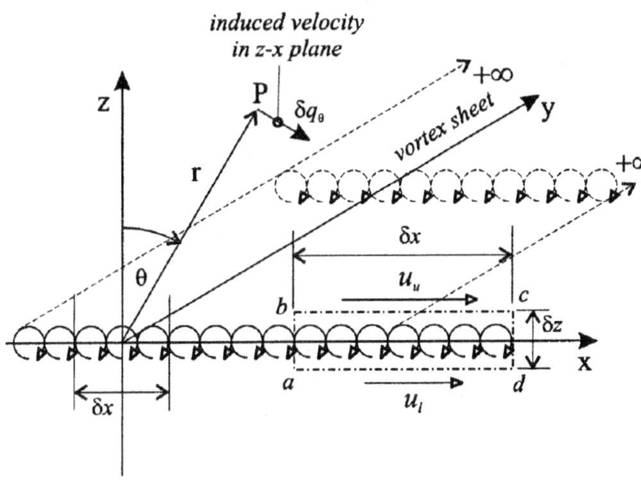

Figure 3.14 A vortex sheet lying in the x–z plane.

Just as z defines a point on the z-plane with x as the abscissa and y as the ordinate, w defines a point on the w-plane which has ϕ as the abscissa and ψ as the ordinate. The phrase function of a complex variable is conventionally restricted to a type of function known as analytic (or holomorphic or regular). A function $w = f(z)$ is classified as analytic when the following two conditions are satisfied:

1. For each value of z there is only one finite value of w.
2. dw/dz is single-valued and neither zero or infinite.

Although there can be exceptions to these conditions, known as singularities, these points can often be mathematically excluded from the transformation process. Condition (1) is normally always satisfied, and it can be shown that condition (2) is satisfied when

$$\frac{\partial \phi}{\partial x} = \frac{\partial \psi}{\partial y} \quad \text{and} \quad \frac{\partial \phi}{\partial y} = -\frac{\partial \psi}{\partial x}$$

These are known as the Cauchy–Riemann equations, which, after differentiation, result in

$$\nabla^2 \phi = 0 \quad \text{and} \quad \nabla^2 \psi = 0$$

This result indicates that ϕ can be regarded as representing the velocity potential function and ψ as the stream function. Thus, the pattern on the w-plane represents uniform inviscid incompressible irrotational flow and the function $w = f(z)$ is often referred to as the complex potential. Since both ϕ and ψ satisfy Laplace's equation and are related by the Cauchy–Rieman equations, they are known as conjugate harmonic functions. Under the inverse function $z = f^{-1}(w)$, Figure 3.15 shows that the pattern on the z-plane can be associated with a particular case of nonuniform potential flow; in this example, the function $w = z^2$ is used and the flow on the z-plane can be taken to represent the internal flow around a right-angle corner.

The derivative dw/dz can be regarded as a complex operator where a small line δz on the z-plane is mapped onto a corresponding line δw on the w-plane, i.e., $\delta w = (dw/dz)\delta z$. Since dw/dz is itself a complex number, this mapping consists of a rotation and a change of scale. Furthermore, it can be shown that the angle of intersection between any two lines is preserved on both the z and w planes. When this characteristic occurs the transformation is known as conformal. Alternatively, dw/dz can be thought of as a complex velocity since

$$\begin{aligned} dw &= \left(\frac{\partial w}{\partial x}\right)dx + \left(\frac{\partial ws}{\partial y}\right)dy = \left(\frac{\partial \phi}{\partial x} + i\frac{\partial \psi}{\partial x}\right)dx + \left(\frac{\partial \phi}{\partial y} + i\frac{\partial \psi}{\partial y}\right)dy \\ &= (u - iv)dz \end{aligned}$$

Thus,

$$\frac{dw}{dz} = V^{-i\theta} \quad \text{where } V = \sqrt{u^2 + v^2} \quad \text{and } \theta = \tan^{-1}(v/u)$$

A typical solution procedure for a flow field using conformal transformation could be as follows:

1. Find $w = f(z)$, which represents the flow field of interest, by adding together standard transformation functions (see Table 3.5).
2. Split w into ϕ and ψ components.
3. Use both lines of constant ϕ and ψ to draw flow pattern on z-plane.
4. Differentiate ϕ or ψ with respect to either x or y to obtain local velocity components u and v, respectively, velocity magnitude Q_∞, and flow angle θ at any point of interest.

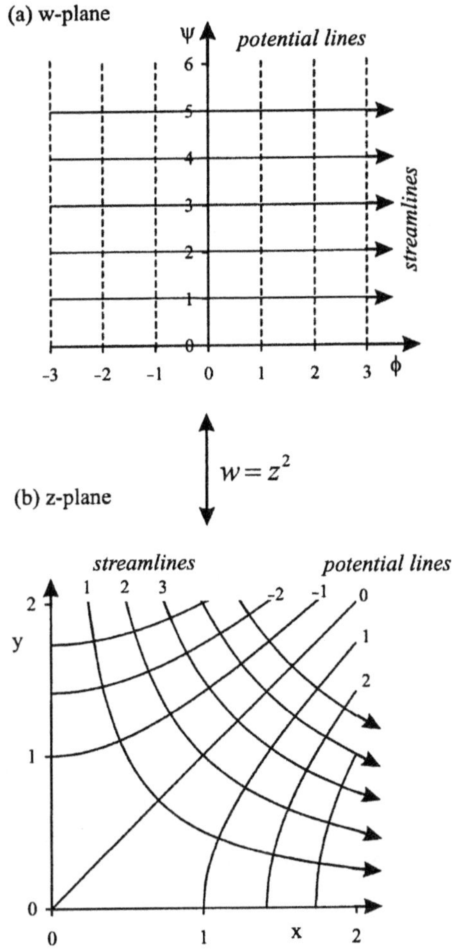

FIGURE 3.15 An example of conformal transformation.

Flow in z-plane	Transformation
Uniform flow	$w = Az$
Source, centered at z_0	$w = m\ln(z - z_0)$
Vortex, centered at z_0 (positive Γ anticlockwise)	$w = -\frac{i\Gamma}{2\pi}\ln(z - z_0)$
Doublet, centered as z_0	$w = \frac{\mu}{z - z_0}$
Cylinder, radius r_0^2, with circulation	$w = Q_\infty\left(z + \frac{r_0^2}{z}\right) w = -\frac{i\Gamma}{2\pi}\ln\frac{z}{r_0}$

TABLE 3.5 Elementary Complex Potential Functions

Flow Field Boundary Conditions

Any potential flow field can be obtained by solving the Laplace equation either analytically or numerically using the following boundary conditions:

1. There is zero normal fluid velocity relative to the body surface, which, for steady flow, is given by $\vec{q}\,.\,\hat{n}=(\nabla\phi)\cdot\hat{n}=0$. This criterion is sometimes referred to as the flow tangency condition or the Neumann problem.
2. The velocity components away from the body should equal those of the free stream.

In many aerodynamic theories the first boundary condition is applied using a technique known as the small disturbance approximation. This approximation can be described by first considering the flow around a finite wing, shown in Figure 3.16, whose surface is defined as

$$F(x_s, y_s, z_s) = z_s - f_x(x_s, y_x) = 0$$

The potential flow solution will be given by $\nabla^2\phi = 0$, where $\phi = \phi_\infty + \tilde{\phi}$, $\phi_\infty = U_\infty x + W_\infty z$, and $\tilde{\phi}$ is the disturbance velocity potential ($V_\infty = 0$). The zero normal velocity boundary condition must be satisfied on the entire wing surface which, for steady flow, is given by

$$\nabla\phi\cdot\hat{n}_s = \frac{\partial f_s}{\partial x}\left\{U_\infty + \left(\frac{\partial\tilde{\phi}}{\partial x}\right)_s\right\} + \frac{\partial f_s}{\partial y}\left(\frac{\partial\tilde{\phi}}{\partial y}\right)_s - \left(\frac{\partial\tilde{\phi}}{\partial z}\right)_x - W_\infty = 0$$

where, as shown in Subsection 3.11, $\hat{n}_s = \nabla F \,|\, \nabla F\,|$. The small disturbance approximation is given by

$$\frac{|\tilde{u}|}{Q_\infty} << 1 \frac{|\tilde{v}|}{Q_\infty} << 1 \frac{|\tilde{w}|}{Q_\infty} << 1$$

and

$$\left|\frac{\partial f_s}{\partial x}\right| << 1 \left|\frac{\partial f_s}{\partial y}\right| << 1 \left|\frac{W_\infty}{Q_\infty}\right| \approx \alpha << 1 \qquad (\alpha \text{ in radians})$$

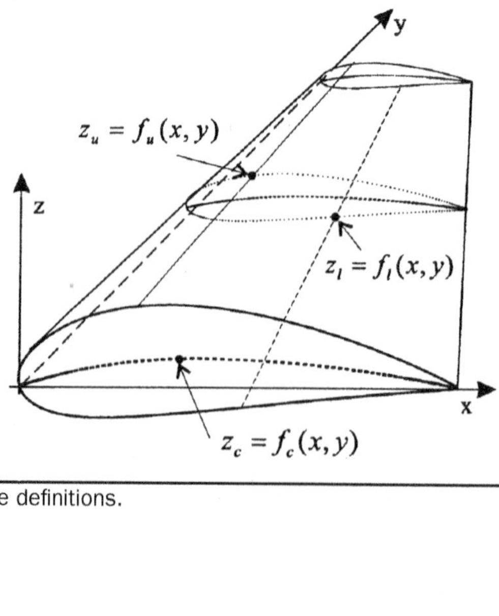

FIGURE 3.16 Airfoil surface definitions.

Using this approximation results in the following simplified form of the flow tangency condition:

$$\left(\frac{\partial \tilde{\phi}}{\partial z}\right)_s = Q_\infty \left(\frac{\partial f_s}{\partial x} - \alpha\right)$$

The left-hand side of this equation can be expanded using the Taylor series expansion. For example, application to the upper wing surface results in

$$\left(\frac{\partial \tilde{\phi}}{\partial z}\right)_u = \frac{\partial \tilde{\phi}}{\partial z}(x,y,0) + z_u \frac{\partial^2 \tilde{\phi}}{\partial z^2}(x,y,0) + O(z_u^2)$$

Linearizing the flow field means all derivatives higher than first order are ignored and the boundary condition becomes

$$\frac{\partial \tilde{\phi}}{\partial z}(x,y,0) = Q_\infty \left(\frac{\partial f_s}{\partial x} - \alpha\right)$$

Essentially, linearization has reduced the problem to one which has to find the disturbance velocity potential on the wing planform in the x–y plane rather than over the entire wing surface. As will be discussed in the following sections, many aerodynamic models utilize that fact that an airfoil or wing introduces a disturbance velocity in the z-direction on the x–y plane. This is known as downwash, which is given by

$$w(x,y,0) = \frac{\partial \tilde{\phi}}{\partial z}(x,y,0)$$

It should be noted that the small disturbance approximation is not valid near stagnation points or the leading edge.

3.13 Elementary Boundary Layer Flow

Kinetic Theory

When considering the macroscopic properties of a real fluid, we have to examine the behavior of the individual molecules that constitute the fluid. The subject which deals with the microscopic behavior of the fluid is known as kinetic theory.

The molecules which constitute any fluid are in a constant state of random motion. For a stationary fluid the molecular velocity is random in both magnitude and direction. When the fluid is moving in a particular direction the instantaneous velocity of each molecule is the vector sum of the fluid velocity and the instantaneous molecular velocity. Because the fluid velocity is superposed over the molecular velocity, it is often referred to as the ordered velocity, while the molecular velocity is known as the random velocity. This terminology also applies to other fluid properties such as its momentum and energy.

The No-Slip Condition

The surface of an airfoil is made up of molecules which leave spaces between each other of sufficient size to allow the air molecules to penetrate into them. In the 18th century Maxwell suggested that diffuse reflection (i.e., scattering in all directions) would result from the penetration of fluid molecules into the pores of the airfoil surface, where they would strike several times before escaping back into the air flow. Essentially, the air molecules are reflected from the airfoil surface irrespective of their initial direction of impact. This can cause the ordered velocity component of an individual molecule to change direction. When the average ordered velocity is taken over all the molecules, lying just above the surface, its value is found to be zero relative to the surface (i.e., after impacting the airfoil surface as many molecules drift upstream as drift downstream). This behavior is known as the noslip condition since it causes the air in direct contact with the airfoil surface to acquire the velocity of the surface.

Viscosity and Boundary Layer Formation

When the random movement of air molecules transport ordered momentum from one place to another within a moving mass of air, it is referred to as the action of viscosity. At some distance, y_e, away from the airfoil surface the air is free to move with the ordered velocity, u_e, unaware that there is a solid surface below. By considering the exchange of ordered momentum across an imaginary plane lying parallel to, and just above, the airfoil surface, it can be shown that the action of viscosity diffuses the no-slip condition out into the airflow such that the local air velocity varies from zero at the wall to u_e. This region is known as the boundary layer and thus u_e is known as the boundary layer edge velocity.

The concept of the boundary layer was first introduced by Prandtl in 1904 and is considered as the region in which the effects of viscosity are concentrated. Numerous experimental studies of boundary layer flows have revealed two distinct types of flow behavior, known as laminar and turbulent. Some of the features of these boundary layers are discussed below.

The Laminar Boundary Layer

Figure 3.17 shows that over the airfoil's leading edge the air particles move downstream in smooth and regular trajectories without appreciable mixing between different layers of air. This type of flow is known as a laminar boundary layer. The nondimensional group, which heavily influences the development of any boundary layer, is the Reynolds number, based on the surface distance from the origin of the boundary layer to the point in question. When working with boundary layer flows it is convenient to use a body-fitted (or curvilinear) coordinate system in which the x-direction is taken to represent the distance traveled along the surface and the y-direction is taken normal to the surface. Figure 3.18 illustrates the variation in local velocity within a laminar boundary layer; this variation is referred as a velocity profile.

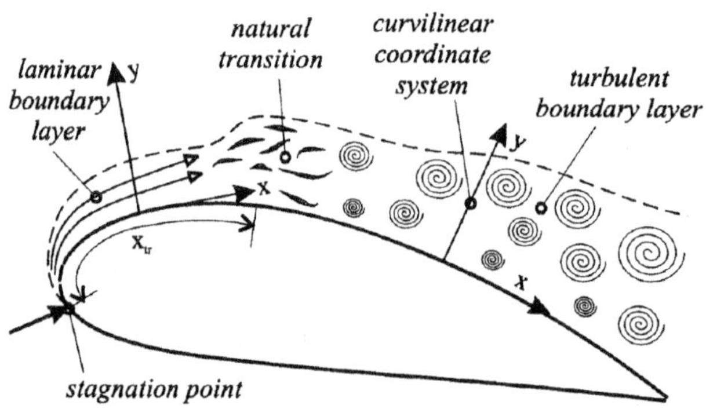

Figure 3.17 Typical boundary layer phenomena. Note that the boundary layer thickness is vastly exaggerated for illustrative reasons.

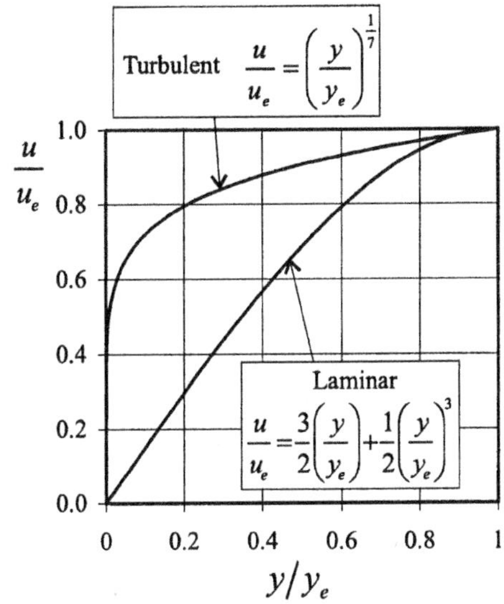

FIGURE 3.18 Velocity profiles for laminar and turbulent flow.

Skin Friction Drag

As described in Subsection 3.11, velocity gradients in a viscous fluid are always accompanied by viscous stresses. Figure 3.18 shows that there is a large velocity gradient at the airfoil surface, which induces a large viscous shear stress to act on the surface in the direction of fluid motion. When these shear stresses are integrated over the entire airfoil surface a drag force, known as the skin friction drag, is obtained.

Transition from Laminar to Turbulent Flow

The phenomena which transform the smooth laminar flow into a chaotic flow, known as a turbulent flow, are collectively known as transition. From the point of view of mathematical modeling, transition comprises of two main processes: the stability of laminar flow to small perturbations, and the amplification of these disturbances such that the transition becomes inevitable. Numerous experimental investigations of pipe flows, boundary layers, and jets have established that transition includes the following phenomena, in order of appearance:

1. Amplification of small disturbances
2. Development of isolated large-scale vortical structures
3. Formation of pockets of small-scale vortical structures known as turbulent spots
4. Growth and coalescence of turbulent spots into a fully developed turbulent flow

The exact details of transition are further complicated by factors such as free-stream turbulence, pressure gradient, wall roughness, heat transfer, and Mach number. For further information on transition see Schlichting and Truckenbrodt (1979) and White (1991).

Experimental investigations of the flow over flat plates have indicated that the maximum length of travel of a laminar boundary layer, without undergoing transition to turbulent flow, is given by

$$\mathrm{Re}_x = \frac{\rho Q_\infty x}{\mu} < 500{,}000$$

In practice, however, this value is reduced when the effects of high free-stream turbulence, surface roughness, and adverse pressure gradients are considered.

The Turbulent Boundary Layer

Figure 3.17 shows that the flow within a turbulent boundary layer consists of vortex structures known as eddies. In 1885, Reynolds suggested that the instantaneous value of any fluid property within a turbulent flow could be separated into a mean value (which stays constant over some specified time period) and a turbulence value which fluctuates about the mean value. Thus the instantaneous velocity in the x-direction is written in the form $u = U + u'$.

Figure 3.18 illustrates a typical mean velocity profile found in a turbulent boundary layer. The random eddy motion transports high-momentum air, at the outer regions of the boundary layer, toward the airfoil surface. This behavior causes the velocities close to the airfoil surface to be larger than those found in a laminar boundary layer.

The transportation of ordered momentum by eddy motion is a similar action to that of molecular viscosity, although it can be up to three orders of magnitude greater. Because of this similarity, the action of the eddies can be represented by a variable known as the eddy viscosity. Unfortunately, unlike the molecular viscosity, eddy viscosity is a variable quantity which depends on the flow field itself (plus boundary conditions) rather than a constant value fluid property. The subject which deals with relating the eddy viscosity to the mean velocity gradients (e.g., $\partial U/\partial y$) is known as turbulence modeling. For an excellent introduction to the subject Wildox (1994) should be consulted.

Figure 3.19 details the generally accepted terminology used to describe the various important regions found within a turbulent boundary layer. The inner region covers between 10% and 20% of

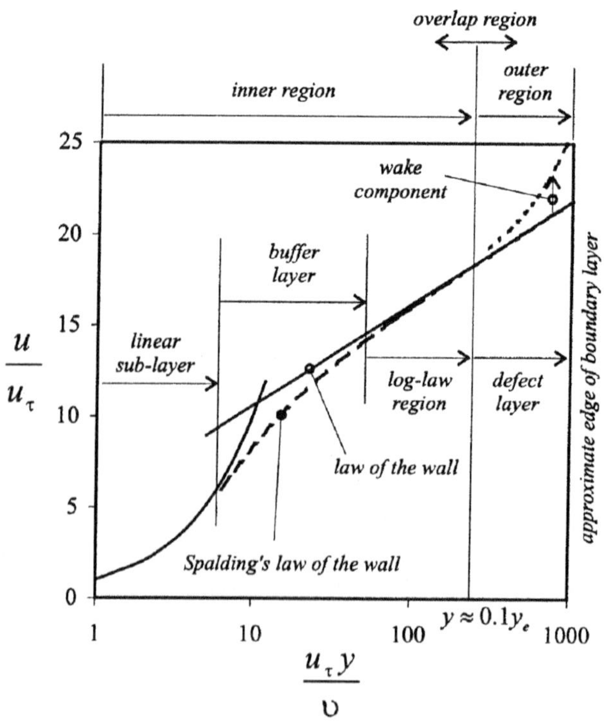

Figure 3.19 Turbulent boundary layer definitions.

the overall thickness and the total shear stress (molecular plus turbulent) is almost constant and equal to the wall value (White 1991). The inner region is further broken down into the following three sublayers:

1. The linear sublayer, where molecular viscosity dominates (the turbulent motion is restricted by the presence of the wall)
2. The buffer layer, where molecular and eddy viscosity are of similar magnitude
3. The log-law, where turbulent stresses dominate

Within the linear sublayer the viscous shear remains very close to the wall value (White 1991), which gives

$$\mu\left(\frac{dU}{dy}\right) = \tau_w \Rightarrow U = \frac{\tau_w}{\mu} y$$

In nondimensional terms this can be written as

$$\frac{U}{u_\tau} = \frac{\rho u_\tau y}{\mu} \quad \text{or} \quad u^+ = y^+ \quad \text{where } u_\tau = \sqrt{\frac{\tau_w}{\rho}}$$

where u_τ is known as the friction velocity.

As indicated in Figure 3.19, the entire velocity profile is plotted using these nondimensional groups. Moving out of the linear sublayer, Prandtl suggested the following velocity variation, known as the log-law or the law of the wall:

$$u^+ = \frac{1}{\kappa} \ln y^+ + B$$

where experimental measurements have indicated that $\kappa = 0.41$ (known as von Karman's constant) and $B = 5.0$.

Experimental data have indicated a smooth variation of u^+ with y^+ between the linear sublayer and the law of the wall, which is known as the buffer layer. Spalding's law of the wall covers both the buffer and the log-law layers and has the form

$$y^+ = u^+ + e^{-\kappa B}\left(e^{\kappa B^+} - 1 - \kappa u^+ - \frac{1}{2}(\kappa u^+)^2 - \frac{1}{6}(\kappa u^+)^3\right)$$

The outer region is often referred to as the defect layer or the law of the wake and is the region where turbulent stresses begin to decrease. The velocity variation is given by the law of the wake

$$\frac{u_e - u}{u_\tau} = -\frac{1}{\kappa} \ln\left(\frac{y}{y_e}\right) + A$$

where the value of A depends on the magnitude and direction of the pressure gradient applied to the boundary layer.

Boundary Layer Separation

Figure 3.20 illustrates the flow through a Venturi. When the flow is incompressible, the conservation of mass and energy state that the velocity must decrease and the pressure increase as the flow moves from the throat to the outlet. The increase in pressure is known as an adverse pressure gradient. The opposite case occurs between the inlet and throat and a favorable pressure gradient forms in this region. When a boundary layer is subjected to an adverse pressure gradient the slower moving fluid elements can be forced to change direction and move upstream. This is referred to as reversed

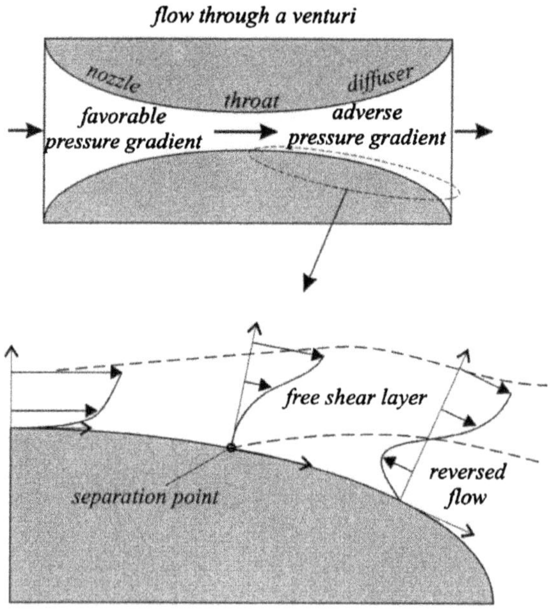

FIGURE 3.20 Boundary layer separation.

flow, and it causes the boundary layer to separate from the surface and form a free shear layer. A turbulent boundary layer is more resistant to separation than a laminar layer due to the more uniform velocity profile. Both laminar and turbulent separation have a great influence on the lift produced by an airfoil, as discussed in Subsection 3.14.

The Laminar Separation Bubble

Figure 3.21 shows a type of transition, known as free shear layer transition, which is commonly found over the leading edge of many airfoils. Under certain conditions the laminar boundary layer

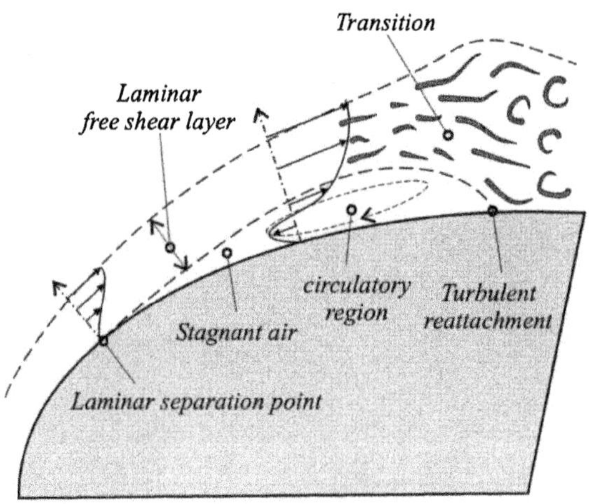

FIGURE 3.21 Flow phenomena associated with a laminar separation bubble.

will separate in the leading edge region. The resulting laminar flow free shear layer quickly undergoes transition which expands in a wedge-like shape. If the now turbulent wedge touches the airfoil surface it will reattach as a turbulent boundary layer and a laminar separation bubble will form. Ward (1963) gives an excellent review of work done on laminar separation bubbles and their effect on the stall characteristics of airfoils (see also Subsection 3.14).

The Boundary Layer Equations

The boundary layer equations are essentially the conservation equations of mass, momentum, and energy (Subsection 3.11), which have been simplified using an order of magnitude analysis. Prandtl's fundamental boundary layer assumption is that the layer is very thin in comparison to the characteristic length of the body over which it flows. Using this assumption, it can be shown that the x- and y-direction Navier–Stokes equations can be respectively reduced to the following forms:

$$\frac{\partial(\rho u u)}{\partial x}+\frac{\partial(\rho u v)}{\partial y}=-\frac{\partial p}{\partial x}+\mu\frac{\partial^2 u}{\partial y^2} \text{ and } \frac{\partial p}{\partial y}=0$$

When the flow is turbulent, steady, and incompressible the instantaneous variables (i.e, $u=U+u'$, $v=V+v'$, and $p=P+p'$) are substituted into the above equations and then time averaged. This gives the Reynolds averaged boundary layer equations in the form

$$\frac{\partial(\rho UU)}{\partial x}+\frac{\partial(\rho UV)}{\partial y}=-\frac{\partial P}{\partial x}+\mu\frac{\partial^2 U}{\partial y^2}+\frac{\partial(\tau'_{xy}/\rho)}{\partial y}$$

and

$$\frac{\partial P}{\partial y}=0$$

where $\tau'_{xy}=-\rho\overline{u'v'}$ is known as a Reynolds stress. The overbar stands for the time averaged value and is given by

$$\overline{u'v'}=\lim_{T\to\infty}\frac{1}{T}\int_0^T u'v'\,dt$$

In terms of the eddy viscosity the x-direction turbulent boundary layer equation becomes

$$\frac{\partial(\rho UU)}{\partial x}+\frac{\partial(\rho UV)}{\partial y}=-\frac{\partial P}{\partial x}+(\mu+\mu_t)\frac{\partial^2 U}{\partial y^2}$$

where

$$\tau'_{xy}=-\rho\overline{u'v'}=\mu_t\left(\frac{\partial U}{\partial y}\right)$$

The subject area of boundary layer flows has been well researched and documented. The definitive work which compiles a great deal of this work is Schlichting and Gersten (2016).

3.14 Incompressible Flow Over Airfoils

Lift Generation in Subsonic Flow

Over the years, countless experimental investigations of the subsonic flow around an airfoil have led to four fundamental observations:

1. The presence of the airfoil, within the airflow, produces significant amounts of streamline curvature as the air is forced to move around the airfoil profile.
2. In the absence of any boundary-layer separation, the upper and lower surface airstreams flow smoothly off the trailing edge—this observation holds for both cusped and finite wedge trailing-edge geometries. This observation was first made by the German mathematician Wilhelm Kutta in 1902 and used in his theoretical model of lift—it is, thus, generally referred to as the Kutta condition.
3. Figure 3.22 illustrates that, when the local static pressure is referenced to the free-stream value, regions of positive and negative gauge pressure are found to act on the airfoil surface. Lift is generated from the net resultant force exerted by this distribution of gauge pressure on the entire airfoil surface.
4. The regions of high and low gauge pressure are, respectively, accompanied by regions of negative and positive flow velocity when measured relative to the free-stream value.

Essentially the flow around an airfoil is dictated by Newton's laws of motion which are encapsulated by the Navier–Stokes equations described in Subsection 3.11. When the conservation of

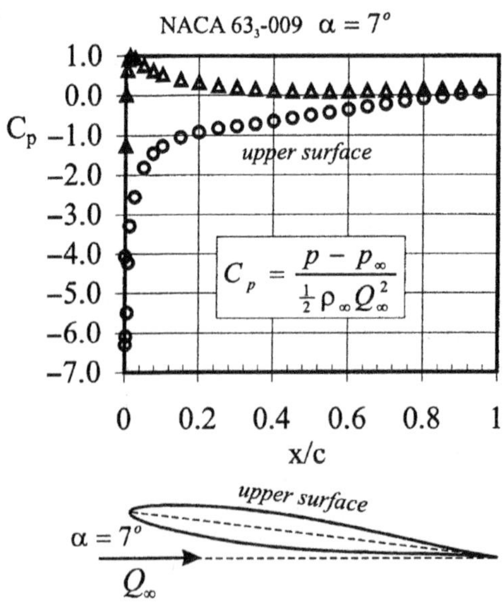

FIGURE 3.22 Typical airfoil pressure distribution (data from McCullough and Gault 1951, courtesy of NASA).

mass, momentum, and energy (if appropriate) are numerically solved for the flow around an airfoil, the resulting solution will satisfy the four fundamental experimental observations stated above. Less complex inviscid mathematical models of subsonic lift invoke the Kutta condition to introduce an amount of circulation (Subsection 3.12), which predicts lift values very close to those observed experimentally. Inviscid models have to artificially introduce a viscous boundary condition in the computation before they are capable of predicting the observed lift—the Navier–Stokes equations inherently include the effects of viscosity.

Consider the flow pictures of Figure 3.23. They show the flow generated by an airfoil at small angle of attack when the airfoil is suddenly started from rest. A similar flow is generated by an airfoil flying at a steady speed whose angle of attack is suddenly increased by a small amount. Note the appearance of a counterclockwise vortex at the sharp trailing edge that starts to separate from the trailing edge and flow downstream with the flow speed. As already discussed in Subsection 3.3, this phenomenon of vortex generation is the key to the understanding and analysis of lift generation. A similar phenomenon occurs when the angle of attack is reduced by a small amount, but the shed vortex is now rotating clockwise. With no further change in angle of attack the flow around the airfoil and the pressure distribution on the airfoil reach a steady state as soon as the shed vortex is some 20 chord lengths downstream from the trailing edge. The airfoil upper surface exhibits pressures below the free-stream pressure and the lower surface pressures above the free-stream value, as shown in Figure 3.22 for the NACA 63-009 for a positive angle of attack value of 7°. Due to this pressure difference between lower and upper surface a positive lift force is generated. It is of critical importance to note that the lift generation is caused by the vortex generation at the trailing edge. Indeed, as already noted in Subsection 3.12, airfoil lift and vortex strength are directly related by the Kutta–Joukowski lift theorem.

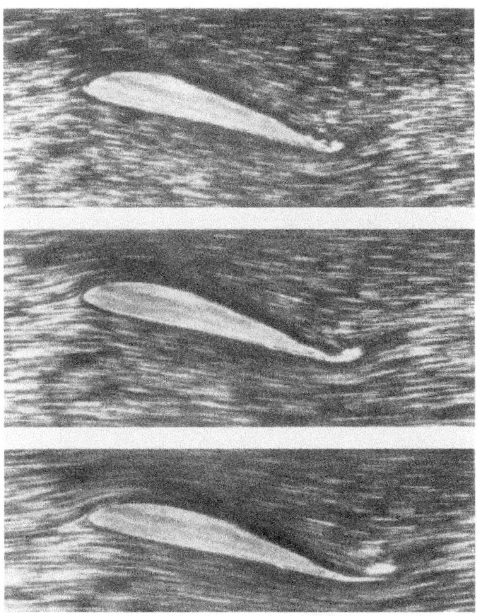

FIGURE 3.23 Trailing edge vortex generation on an airfoil starting from rest (Prandtl and Tietjens 1934).

An Overview of Mathematical Models of Lift

Nearly all mathematical models of lift use techniques which predict the velocity field around the airfoil first and then use Bernoulli's equation to obtain the pressure field. To obtain the correct velocity and pressure fields, the Kutta condition is applied in a form appropriate to the mathematical model (examples are given later). The Kutta–Joukowski lift theorem (see Subsection 3.12) is then used, in conjunction with the circulation, which accompanies the velocity field, to obtain the lift produced.

Figure 3.24 illustrates the essence of the circulation theory of lift, which can be summarized into the following steps:

1. Restrict the analysis to inviscid, incompressible, irrotational flow over an arbitrary airfoil in a free-stream flow.
2. Replace the airfoil surface with a vortex sheet of unknown variable strength. It can be argued that in a viscous flow the boundary layer (Subsection 3.13) produces the vorticity, and this can be related to the total circulation around the airfoil via Stokes theorem (Subsection 3.11).
3. Calculate the variation of vortex sheet strength subject to the following boundary condition: (i) when the vortex sheet velocity field is added to the free-stream velocity there is no normal component of velocity at every point on the airfoil surface; and (ii) that the suction and pressure surface vortex sheet strengths at the trailing edge are equal (the Kutta condition).
4. The total circulation is the net value due to the entire vortex sheet and the resulting lift per unit span is calculated using the Kuttal–Joukowski theorem.

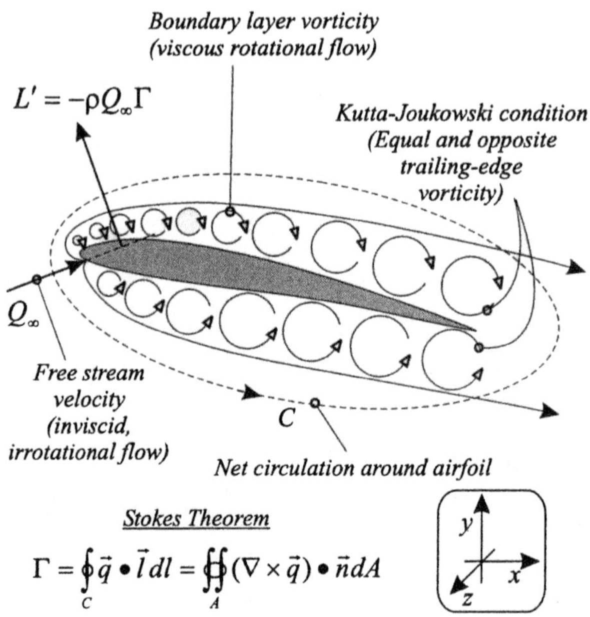

FIGURE 3.24 The circulation theory of incompressible lift. Note that the boundary layer thickness is vastly exaggerated for illustrative reasons.

Small-Disturbance Airfoil Theory

Using the principle of superposition (see Subsection 3.12), the velocity distribution around an airfoil can be decomposed into the following three separate and independent components:

1. The distribution due to the basic airfoil thickness form at zero lift
2. The distribution due to the camber line at the ideal angle of attack (explained below)
3. The distribution due to angle of attack

When the first two components are added together this is known as the basic velocity distribution. It is only dependent on the geometric properties of the airfoil. The third component is known as the additional velocity distribution and is strongly dependent on angle of attack and weakly dependent on airfoil thickness. Referring to Figure 3.25, the following points should be noted:

1. Velocity component one is denoted q_t/Q_∞ and can be found using either conformal transformation or singularity methods.
2. Velocity component two is denoted $\Delta q_c/Q_\infty$ and is obtained using thin airfoil theory. The ideal angle of attack, α_i, is defined as the angle which places a stagnation point exactly on the foremost position of the camber line. Theodorsen (1931) referred to this as "the angle of best streamlining." The lift coefficient which corresponds to the ideal angle of attack is called the design lift coefficient and is denoted $c_{l,i}$.
3. Velocity component three is denoted $\Delta q_\alpha/Q_\infty$ and is obtained using conformal transformation methods and includes the influence of airfoil thickness on the angle of attack term.

The final velocity distribution is given by

$$\frac{q}{Q_\infty} = \frac{q_t}{Q_\infty} \pm \frac{\Delta q_c}{Q_\infty} \pm \frac{\Delta q_\alpha}{Q_\infty}$$

where the plus stands for the upper surface and the negative the lower surface. Since each velocity distribution is essentially a solution to the Laplace equation, the values of $\Delta q_\alpha/q_\infty$ and $\Delta q_c/Q_\infty$ scale linearly with angle of attack and airfoil geometry respectively. For example, if the camber line ordinates are multiplied by a constant factor, the velocity distribution, ideal angle of attack and design lift coefficient all change by the same factor. Thus, Abbott and von Doenhoff (1959) tabulate the following values for NACA airfoils:

$$\frac{q_t}{Q_\infty}, \frac{\Delta q_\alpha/Q_\infty}{(c - c_{l,t})} \quad \text{and} \quad \frac{\Delta q_c/Q_\infty}{c_l}$$

As an example, consider the NACA 23012 airfoil, which has the following design parameters:

$$c_{l,i} = 0.3, \alpha_i = 1.65^\theta \quad \text{and} \quad c_{mi,/4} = -0.014$$

Using the tabulated data, given in Abbott and von Doenhoff (1959), the following velocity components were obtained at the point $x/c = 0.5$ and for a lift coefficient, c_l, equal to 0.5:

$$\frac{q_t}{Q_\infty} = 1.108, \quad \frac{\Delta q_\alpha}{Q_\infty} = 0.149(0.5 - 0.3), \quad \text{and} \quad \frac{\Delta q_c}{Q_\infty} = 0.054(0.5)$$

The quarter-chord pitching moment and the zero-lift angle are respectively given by

$$c_{m,c/4} = -0.014(c_l) \quad \text{and} \quad \alpha_{l0} = \alpha_i - \frac{57.3}{2\pi} c_{l,i}$$

Several more examples are given for other NACA airfoils in Abbott et al. (1945).

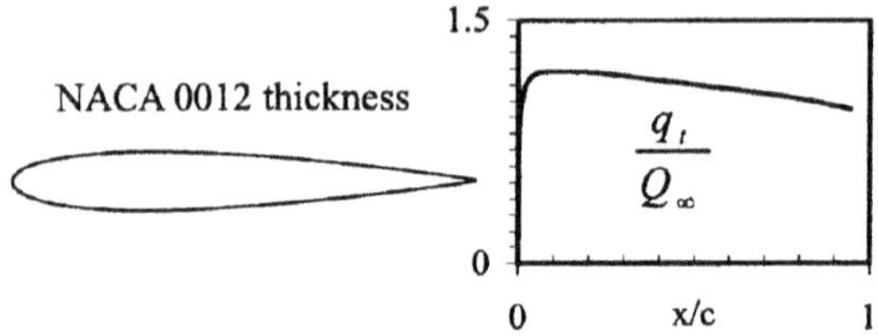

(a) Symmetrical airfoil at zero angle of attack

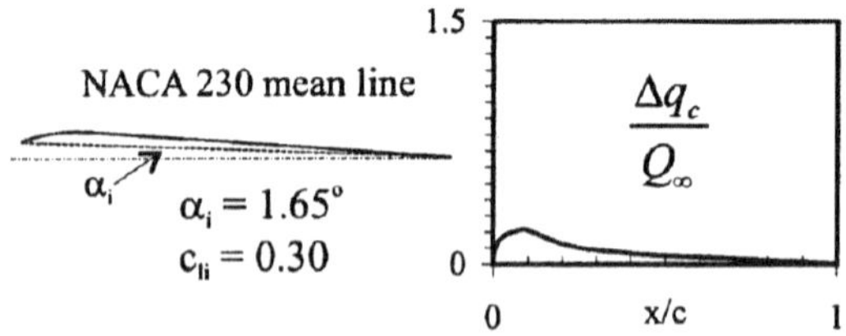

(b) Camber line at ideal angle of attack

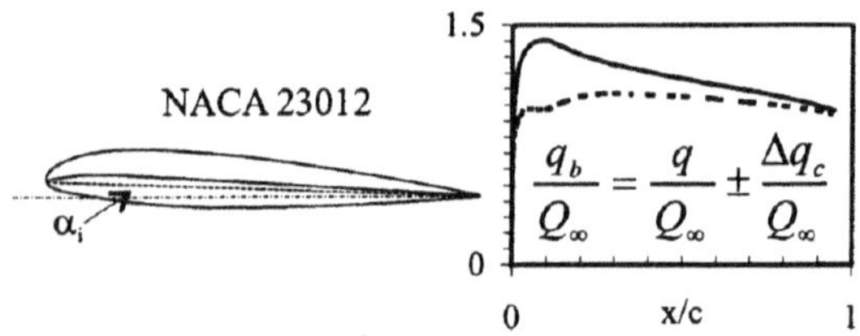

(c) The basic velocity distribution

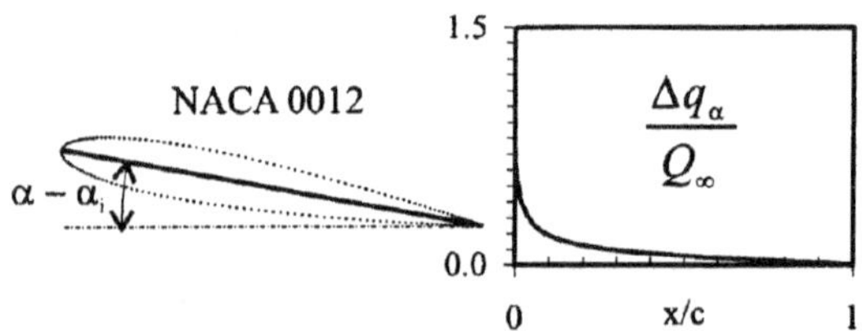

(d) The additional velocity distribution

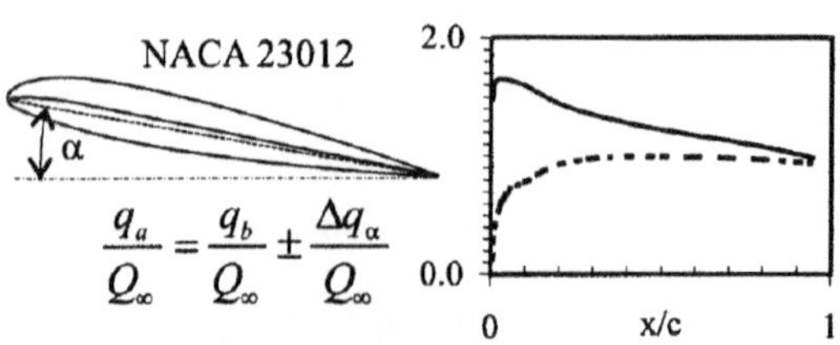

(e) The final velocity distribution

FIGURE 3.25 The velocity superposition principle used by NACA.

Conformal Transformation

(*Applications—inviscid, incompressible, irrotational flow around a family of airfoils of arbitrary thickness and camber known as Joukowski airfoils.*) Although the following analysis is applicable to a particular family of airfoils, the results obtained clearly demonstrate the following important characteristics which are common to most airfoil profiles: the influence of profile thickness on the lift curve slope is indicated along with the effect of camber on the zero-lift angle and the zero-lift pitching moment. Also, conformal transformation has played an important role in the development of the NACA family of airfoils (Theodorsen 1931).

As discussed in Subsection 3.12, the flow around a lifting cylinder can be obtained using conformal transformation techniques. The Joukowski transformation treats the lifting cylinder flow as an intermediate mapping, which is then subjected to a further transformation to obtain the flow around a cambered finite thickness airfoil. The transformation which maps a circle on the z_n-plane, of radius r_0 and center z_0, onto an airfoil on the z-plane is given by

$$z = z_n + \frac{b^2}{z_n}$$

where b is approximately equal to a quarter of the final airfoil chord and the values of z_0 and r_0 (in relation to b) control the final airfoil profile on the z-plane. Theodorsen (1931) and Theodorsen and Garrick (1932) recognized that the inverse transformation $z_n = \frac{1}{2}(z \pm \sqrt{z^2 - 4b^2})$ could be applied to an arbitrary airfoil to produce a near circle in the z_n-plane. The flow about this near circle was then related to the flow about a true circle and hence the velocity distribution around the airfoil surface was obtained.

Figure 3.26 illustrates the transformations involved in forming the lifting flow around a cambered Joukowski airfoil at an arbitrary angle of attack. The following three transformations are used:

$$w = -Q_\infty\left(z_1 + \frac{r_0^2}{z_1}\right) - \frac{i\Gamma}{2\pi}\ln\frac{z_1}{r_0}$$

$$z_n = z_1 e^{-i\alpha} + me^{i\delta}$$

$$z = z_n + \frac{b^2}{z_n}$$

A Joukowski airfoil profile, of specified thickness and camber, may be constructed using the following equations:

$$\hat{y}_{u,l} = \hat{y}_c \pm \Delta\hat{y}$$

where
$$\hat{y}_c = \hat{h}(1 - 4\hat{x}^2),$$

and
$$\Delta\hat{y} = \frac{2i}{3\sqrt{3}}(1 - 2\hat{x})\sqrt{1 - 4\hat{x}^2}$$

The maximum thickness occurs at the quarter-chord point, while the maximum camber occurs at the mid-chord position. It should be noted that $\Delta\hat{y}$ is an additional distance added to the local camber line y-coordinate in a direction perpendicular to the chord line rather than the camber line as in the generation of the NACA airfoils (see Subsection 3.9). Also, an x–y coordinate system has been used here rather than the x–z coordinate system normally used for airfoil definition and analysis, to avoid confusion with the complex number z.

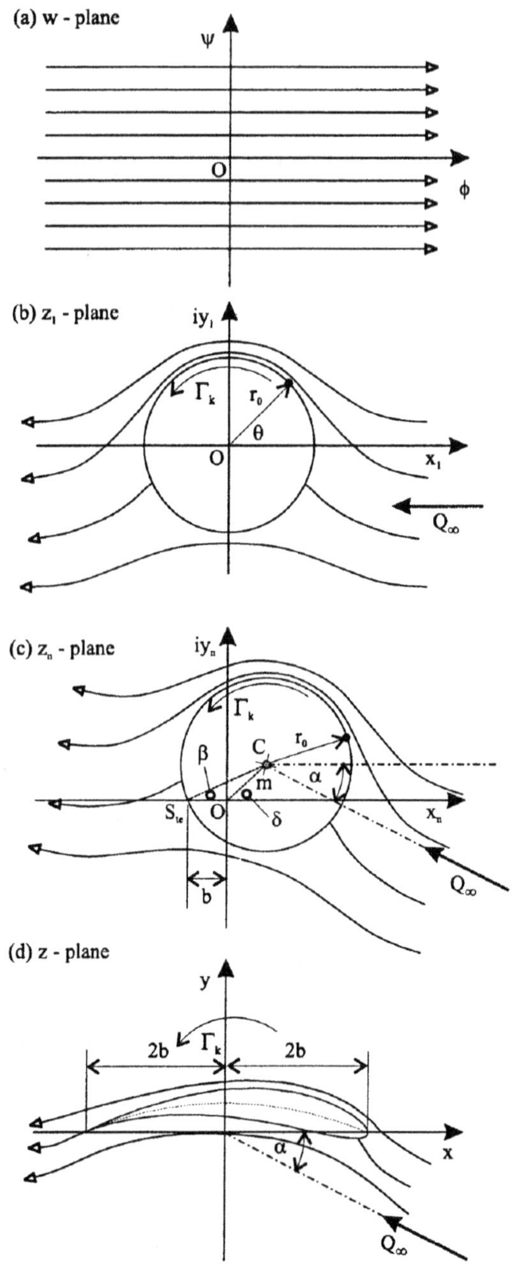

FIGURE 3.26 The Kutta–Joukowski transformation.

Referring to Figure 3.26, the complex velocity in the z-plane is given by

$$u - iv = \frac{dw}{dz} = \frac{dw}{dz_1}\frac{dz_1}{dz_n}\frac{dz_n}{dz} = \left\{-Q_\infty\left(1-\frac{r_0^2}{z_1^2}\right)-\frac{i\Gamma}{2\pi z_1}\right\}\left\{e^{-i\alpha}\right\}\left\{\left(1-\frac{b^2}{z_n^2}\right)^{-1}\right\}$$

The circulation required to place the rear stagnation point on the airfoil trailing edge (i.e., the Kutta condition) is given when $(dw/dz) = 0$ and $z_1 = r_0 e^{-i(\alpha+\beta)}$, which results in

$$\Gamma_k = 4\pi r_0 Q_\infty \sin(\alpha + 2\hat{h}) \qquad \text{(positive anticlockwise)}$$

Geometrical considerations result in

$$r_0^2 = c^2\left(\frac{1}{4}+\frac{\hat{t}}{3\sqrt{3}}\right)^2 + \hat{h}^2$$

Using the Kutta–Joukowski lift theorem, the airfoil lift coefficient is given by

$$c_l = 2\pi\left(1+\frac{4}{3\sqrt{3}}\hat{t}\right)\sin(\alpha+2\hat{h}) \approx c_{l\alpha}(\alpha-\alpha_{l0}) \tag{3.9}$$

where $c_{l\alpha} \approx 2\pi$ and $\alpha_{l0} = -2\hat{h}$.

The pitching moment around the leading edge has the form

$$c_{m,le} = -\pi(\hat{h}) - \frac{c_l}{4} = c_{m0} - \frac{c_l}{4}$$

The pitching moment around the quarter-chord position is given by

$$c_{m,c/4} = c_{m,le} + \frac{c_l}{4} = -\pi\hat{h} \tag{3.10}$$

which means that, by definition, the quarter chord locates the aerodynamic center. The variation of the center of pressure with lift coefficient can be obtained from

$$\frac{x_{cp}}{c} \approx -\frac{c_{m,le}}{c_l} = \frac{\pi\hat{h}}{c_l} + \frac{1}{4}$$

and is plotted in Figure 3.27 for a 4% cambered Joukowski airfoil.

If required, the local velocity (and hence the local pressure coefficient) at any point in the z-plane can be calculated using the complex velocity. Figure 3.28 illustrates the pressure distribution around a Joukowski airfoil at the angle of zero lift. Although this distribution produces zero life a negative pitching moment is still induced. This explains why the center of pressure, shown in Figure 3.27 tends to infinity as the lift tends to zero.

Singularity Methods (Teardrop Theory)

(*Applications—inviscid, incompressible, irrotational nonlifting flow around a symmetrical airfoil.*) Figure 3.29 shows that the effect of airfoil thickness can be modeled using a distribution of source strength along the x-axis. Applying continuity to the control volume *ABCD* and linearizing the result gives the source strength per unit length as

$$\xi(\hat{x}) = 2Q_\infty\frac{dz_u(\hat{x})}{d\hat{x}} \qquad \text{where } \hat{x} = x/c$$

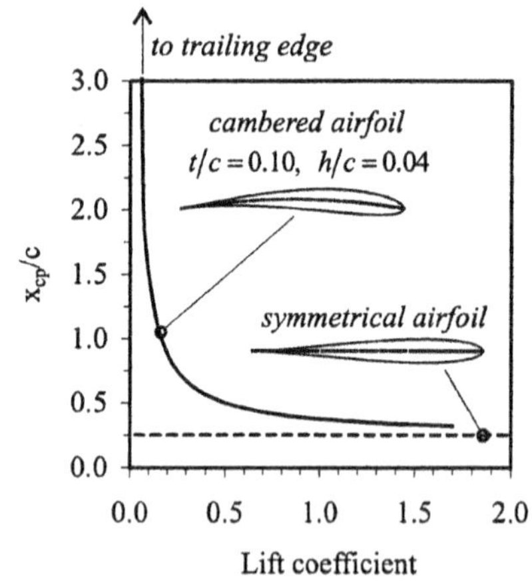

FIGURE 3.27 Variation of center of pressure with lift.

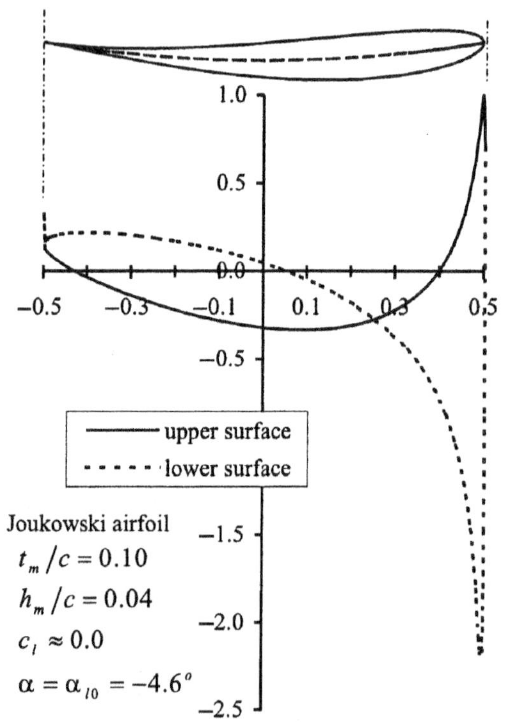

FIGURE 3.28 Zero-lift C_p distribution.

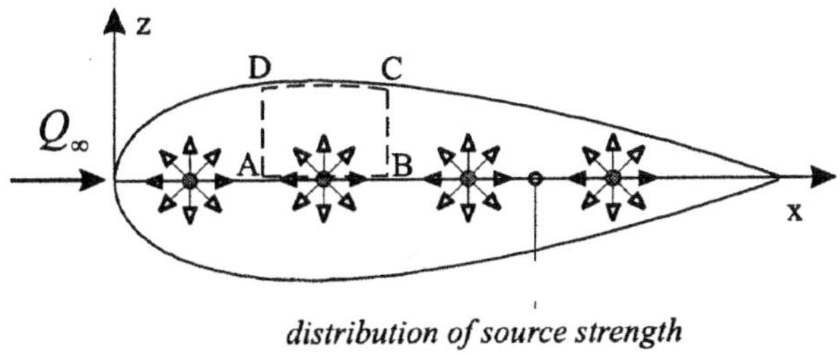

FIGURE 3.29 Thickness modeled using source distribution.

For thin airfoils, the disturbance velocity components on the airfoil surface are approximately equal to those on the x-axis, which gives

$$\tilde{u}(\hat{x}) = \frac{1}{2\pi}\int_0^1 \frac{\xi(\hat{x}_o)d\hat{x}_o}{(\hat{x}-\hat{x}_o)} \quad \text{and} \quad \tilde{w}(\hat{x}) = \pm\frac{1}{2}\xi(\hat{x})$$

The local surface velocity can now be obtained from

$$\frac{q(\hat{x})}{Q_\infty} \approx 1 + \frac{\tilde{u}(\hat{x})}{Q_\infty} = 1 + \frac{1}{\pi}\int_0^1 \frac{\xi(\hat{x}_0)d\hat{x}_0}{(\hat{x}-\hat{x}_0)}$$

This equation can be solved by first expressing the airfoil upper surface profile by a Fourier series

$$z_u(\theta) = \frac{1}{2}\sum_{n=1}^{N} b_n \sin n\theta \quad \text{where } \hat{x} = \frac{1}{2}(1+\cos\theta)$$

which results in

$$\frac{q(\hat{x})}{Q_\infty} = \frac{1}{r_i(\theta)}\left[1 + \sum_{n=1}^{N} \frac{nb_n \sin n\theta}{\sin\theta}\right]$$

where $r_i(\theta)$ is known as the Riegels factor and is given by

$$r_i(\theta) = \sqrt{1 + \frac{1}{\sin^2\theta}\left(\frac{dz_u(\theta)}{d\theta}\right)^2}$$

This procedure can be applied to any symmetric airfoil. For example, take a symmetric Joukowski airfoil whose surface contour is defined by

$$z_u(\theta) = \left(\frac{1}{2}\right)\varepsilon \sin\theta(1-\cos\theta)$$

where

$$\varepsilon = \frac{4\hat{t}}{3\sqrt{3}}$$

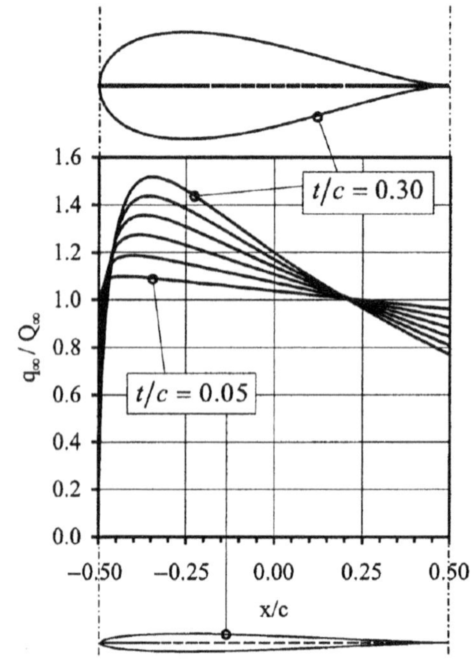

FIGURE 3.30 Effect of thickness on velocity distribution (at zero lift).

which gives the local surface velocity at any point as

$$\frac{q(\hat{x})}{Q_\infty} = \frac{1+\varepsilon(1-2\cos\theta)}{\sqrt{1+\varepsilon^2\left(\dfrac{\cos\theta-\cos 2\theta}{\sin\theta}\right)^2}}$$

Figure 3.30 displays the surface velocity distribution for various thickness ratios (in steps of 0.05). As discussed in Subsection 3.17, the maximum surface velocity is highly important when considering the behavior of an airfoil at high subsonic velocities.

Thin Airfoil Theory

(*Applications—inviscid, incompressible, irrotational flow around airfoils of thickness less than 12% chord and camber less than 2% chord.*) Figure 3.31 shows that to make the camber line of a thin airfoil a streamline of the flow, the upwash, at all points along the camber line, due to the angle of attack has to be equal to the downwash induced by the entire vortex sheet. This criterion gives the fundamental equation of thin airfoil theory:

$$\int_0^c \frac{\gamma(x)dx}{2\pi(x_0-x)} = Q_\infty\left(\alpha - \frac{dz_c(x_0)}{dx}\right) \tag{3.11}$$

To solve equation (3.11), the coordinate transformation $x=\left(\frac{1}{2}\right)c(1-\cos\theta)$ is first used to give the following form of the fundamental equation:

$$\int_0^c \frac{\gamma(\theta)\sin\theta d\theta}{2\pi(\cos\theta-\cos\theta_0)} = Q_\infty\left(\alpha - \frac{dz_c(\theta_0)}{dx}\right)$$

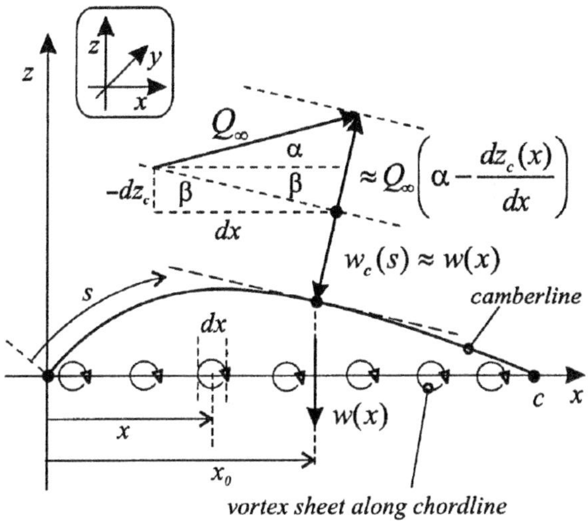

FIGURE 3.31 Thin airfoil mathematical model.

The solution to this equation can be expressed in the form

$$\gamma(\theta) = 2Q_\infty\left(A_0 \frac{1+\cos\theta}{\sin\theta}\right) + 2Q_\infty \sum_{n=1}^{\infty} A_n \sin n\theta \tag{3.12}$$

The fundamental form of this vorticity distribution can be obtained using conformal transformation methods applied to flat plate and circular arc airfoils (Houghton and Brock 1970). It should be noted that equation (3.12) satisfies the Kutta condition of smooth trailing-edge flow since the vortex sheet strength tends to zero when $\theta = 180°$. Using this vorticity distribution reduces equation (3.11) into the form

$$\frac{dz_c(\theta_0)}{dx} = (\alpha - A_0) + \sum_{n=1}^{\infty} A_n \cos n\theta_0$$

Since this is a Fourier series, the coefficients are now given by

$$A_0 = \alpha - \frac{1}{\pi}\int_0^\pi \frac{dz_c(\theta_0)}{dx} d\theta_0 = \alpha - \alpha_i$$

$$A_n = \frac{2}{\pi}\int_0^\pi \frac{dz_c(\theta_0)}{dx} \cos n\theta_0 \, d\theta_0$$

For any given airfoil camber line, the A-coefficients can be obtained using either analytical or numerical integration. The value α_i is known as the ideal angle of attack. When the free-stream velocity is set at the ideal angle of attack, $A_0 = 0$ and the infinite leading edge vorticity (and hence infinite local velocity), inherent in equation (3.12), is avoided.

The total circulation around the airfoil is given by

$$\Gamma = \int_0^c \gamma(x)\,dx = \int_0^\pi \gamma(\theta)\frac{1}{2}c\sin\theta \, d\theta = \pi c Q_\infty\left(A_0 + \frac{1}{2}A_1\right)$$

Using the Kutta–Joukowski theorem gives the lift coefficient, and the lift curve slope respectively as (α in radians):

$$c_l = 2\pi(A_0 + \frac{1}{2}A_1) = c_{l\alpha}(\alpha - \alpha_{l0}) \tag{3.13}$$

where

$$c_{l\alpha} = 2\pi \text{ and } \alpha_{l0} = -\frac{1}{\pi}\int_0^{\pi} \frac{dz_c(\theta_0)}{dx}(\cos\theta_0 - 1)d\theta_0$$

The lift coefficient corresponding to the ideal angle of attack is known as the ideal or design lift coefficient and is given by

$$c_{l,i} = 2\pi\left(\frac{1}{2}A_1\right) = 2\pi(\alpha_i - \alpha_{l0}) \tag{3.14}$$

Pitching moment characteristics can be obtained from

$$c_{m,le} = -\frac{1}{2}\pi(A_0 + A_1 - \frac{1}{2}A_2)$$

$$c_{m,c/4} = \frac{1}{4}\pi(A_2 - A_1)$$

Since A_1 and A_2 are independent of α, the quarter-chord point locates the aerodynamic center (see Subsection 3.9) for all thin airfoils. The center of pressure is given by

$$x_{cp} \approx -\frac{c_{m,le}c}{c_l} = \frac{c}{4}\left[1 + \frac{\pi}{c_l}(A_l - A_2)\right]$$

All thin symmetric airfoils thus have the center of pressure (Subsection 3.9) at the quarter chord point since $A_1 = A_2 = 0$ for this type of airfoil.

Information regarding the pressure difference between the upper and lower surface of a thin airfoil can be obtained by considering Bernoulli's equation with a velocity perturbation along an arbitrary stream line.

$$p_\infty + \frac{1}{2}\rho_\infty Q_\infty^2 = p + \frac{1}{2}\rho_\infty(Q_\infty + \tilde{u})^2 \Rightarrow p \approx p_\infty - \rho_\infty Q_\infty \tilde{u}$$

The pressure difference across the thin airfoil is thus given by

$$\Delta p(\theta) = (p_l - p_u) = \rho_\infty Q_\infty(\tilde{u}_u - \tilde{u}_l) = \rho_\infty Q_\infty \gamma(\theta)$$

Figure 3.32 illustrates the pressure difference across a symmetrical thin airfoil, which can be written in the form

$$\frac{\Delta p}{\frac{1}{2}\rho_\infty Q_\infty^2 \alpha} = 4\left(\frac{1+\cos\theta}{\sin\theta}\right) = \sqrt{\frac{1-x/c}{x/c}} \tag{3.15}$$

where α is in radians.

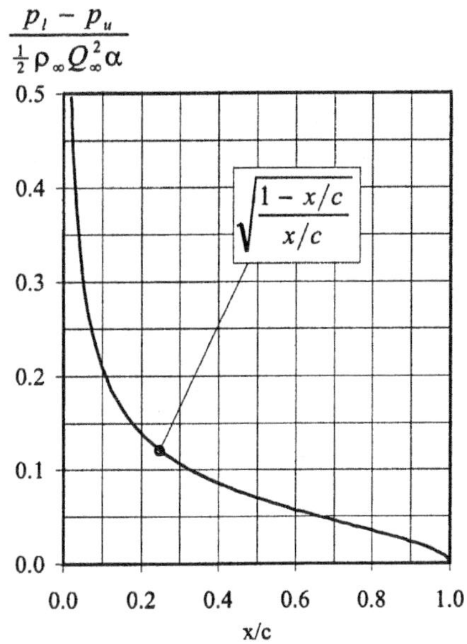

FIGURE 3.32 Pressure difference across a symmetrical airfoil.

The Lumped Vortex Model and the Rear Aerodynamic Center

Figure 3.33 shows that, from a far field point of view, the continuous vortex sheet, used in thin airfoil theory, can be replaced by a single vortex placed at the one quarter-chord position. This single bound vortex requires only one control point where the flow tangency boundary condition is satisfied. Denoting this point by kc gives

$$\frac{-\Gamma}{2\pi\left(kc - \frac{1}{4}c\right)} + Q_\infty \alpha = 0$$

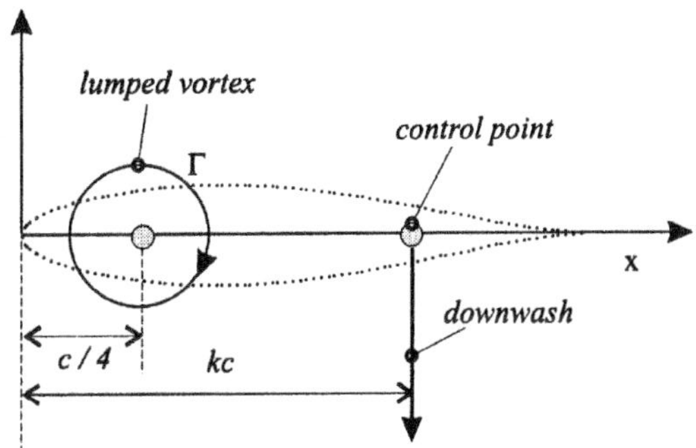

FIGURE 3.33 The lumped vortex model.

The value kc can be calculated for any cambered airfoil by using the thin airfoil result for the total circulation. For simplicity, a symmetrical airfoil is used, which gives

$$\Gamma = \pi c Q_\infty \alpha \quad \text{and} \quad \text{hence } k = \frac{3}{4}$$

Katz and Plotkin (1991) demonstrate that the lumped vortex model is an excellent method for estimating the effect on the lift due to close proximity of other airfoils (e.g., biplanes) or solid surfaces (e.g., airfoils in ground effect).

For an airfoil of zero thickness, thin airfoil theory shows that the slope of the camber line at the three quarter chord point is equal to the zero-lift angle (in radians). For example, the camber line of a circular arc airfoil can be represented by the equation $z_c = h \sin^2 \theta$, which results in the following:

$$\alpha_{l0} = -2\hat{h} = (dz_c/dx)_{3c/4}$$

When a similar calculation is carried out on the 12% thick NACA 23012, the following results are obtained:

$$\alpha_{l0} = -1.09^\circ \quad \text{and} \quad (dz_c/dx) = -1.26^\circ$$

Thus, to a first approximation, for any thin airfoil

$$\alpha_{l0} \approx \left(\frac{dz_c}{dx}\right)_{3c/4}$$

Because of the two important properties described above, the three quarter chord point is often referred to as the rear aerodynamic center. As discussed in Subsection 3.15, the rear aerodynamic center is utilized in certain finite wing models to aid in the prediction of spanwise lift distributions over swept wings.

Vortex Panel Methods

(Applications—inviscid, incompressible, irrotational flow around airfoils of arbitrary thickness and camber.) Figure 3.24 illustrates that the boundary layer can be thought of as a vorticity layer wrapped around the airfoil contour. Figure 3.34(a) shows that the airfoil surface can be replaced by a vortex sheet (see also Subsection 3.12) whose variation in strength can be related to the free-stream velocity by using the airfoil surface flow tangency boundary condition in the form

$$\vec{Q}_\infty \cdot \vec{n}_i + \frac{1}{2\pi} \oint_c \gamma_j \frac{\partial \theta_{ij}}{\partial n_i} dl_j = 0 \tag{3.16}$$

Figure 3.34(b) illustrates that this integral equation may be numerically solved by decomposing the airfoil surface into a number of vortex panels over which the vorticity is piecewise constant. This is known as a first order vortex panel method, and equation (3.16) takes the form

$$\vec{Q}_\infty \cdot \vec{n}_i + \sum_{j=1}^{N} \frac{\gamma_j}{2\pi} \int_{l_j} \frac{\partial \theta_{ij}}{\partial n_i} dl_j = 0 \tag{3.17}$$

Applying equation (3.17) to $N - 1$ panels produces $N - 1$ simultaneous equations for N unknown values of γ_j. The Nth equation is obtained by applying the Kutta–Joukowski in the form $\gamma_1 = \gamma_{N+1}$. The lift per unit span is then obtained using

$$L' = \rho_\infty Q_\infty \sum_{j=1}^{N} \gamma_j l_j$$

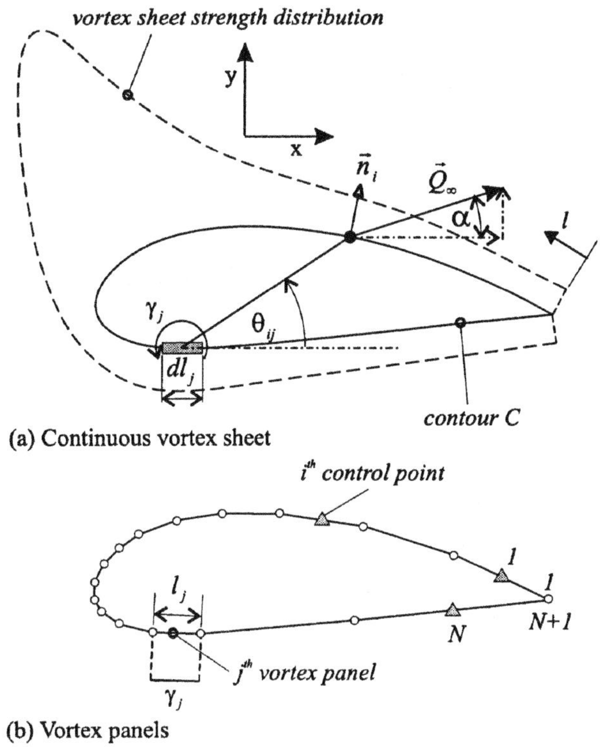

FIGURE 3.34 The vortex panel method.

For further information on vortex panel methods and other numerical models of aerodynamics Katz and Plotkin (1991) should be consulted.

Low-Speed Airfoil Stalling Characteristics

Airfoil stall is defined as the flow conditions which accompany the first peak in lift coefficient. Figure 3.35 illustrates the dominant boundary layer phenomena associated with the various types of subsonic airfoil stall. Figure 3.36 shows the variation in lift coefficient with angle of attack which accompany each stall type. Additional information regarding the various boundary layer phenomena can be found in Subsection 3.13. A short description of each stall is given below.

1. *Thin airfoil stall:* This stall is sometimes referred to as long bubble stall because it involves the formation of a laminar separation bubble which grows in length toward the trailing edge with increasing values of angle of attack. The small discontinuity in the lift curve, indicated in Figure 3.36, occurs when the bubble reaches a size large enough to initiate a considerable reduction in the leading edge suction peak (see Figure 3.37(a)). As the bubble grows in length, the lift starts to gradually decrease. Maximum lift is low relative to other stall types and occurs when the bubble reattachment point reaches the trailing edge. Any further attempts to increase the angle of attack results in bubble thickening, followed by bursting, with a rapid loss of the remaining lift.
2. *Leading edge stall:* This stall is also known as short bubble stall because it involves the formation of a laminar separation bubble which decreases in length as the lift is increased.

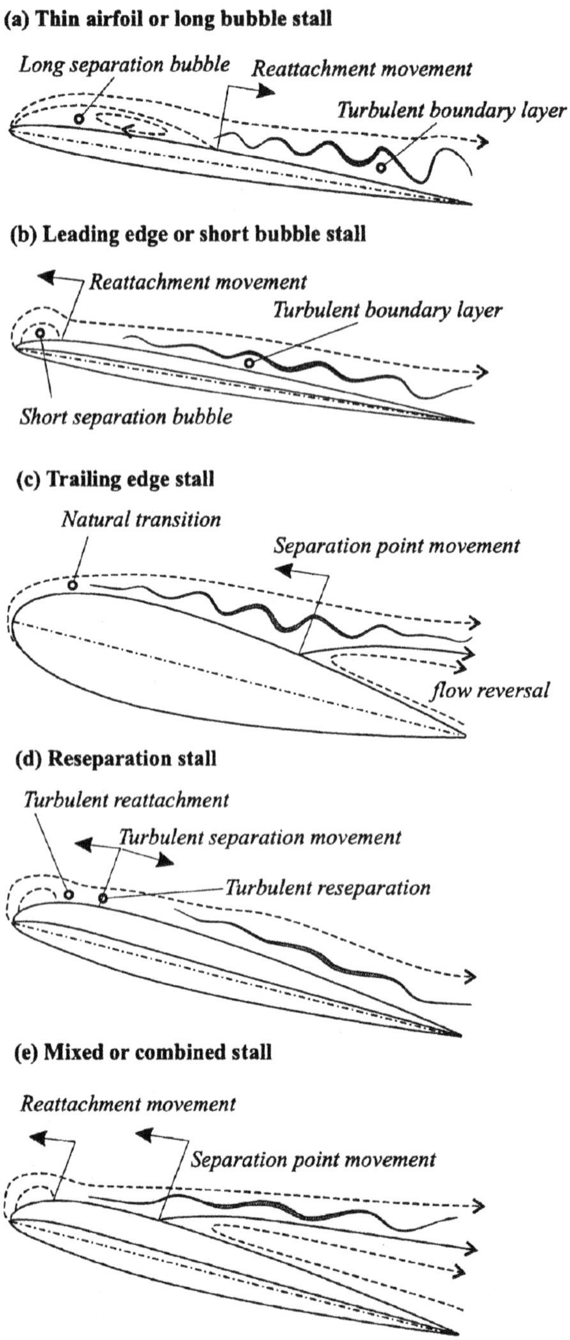

FIGURE 3.35 Types of airfoil stall.

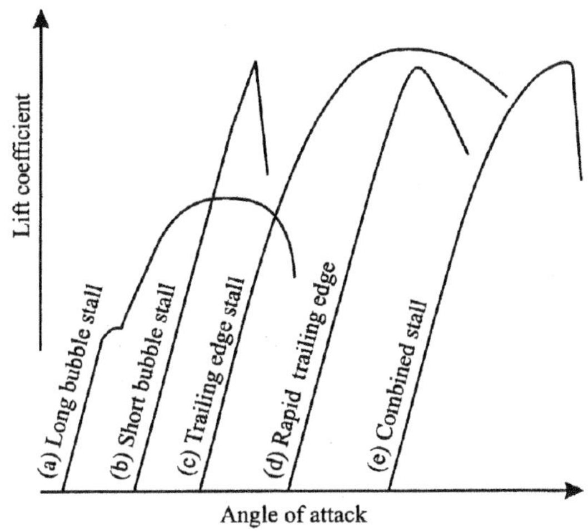

FIGURE 3.36 Types of subcritical airfoil stall.

Eventually turbulent reattachment fails to take place and the bubble bursts resulting in a catastrophic loss of lift. Since the bubble, prior to bursting, is small, it has little effect on the leading edge suction peak (see Figure 3.37(b)), and thus the maximum lift is larger than that associated with a long bubble stall.

3. *Trailing edge stall:* This stall involves turbulent boundary layer separation which starts at the trailing edge and progresses toward the leading edge as the angle of attack is increased. The rate of forward movement of the separation point is dependent on the airfoil, Reynolds number, and angle of attack. This can occasionally result in a lift curve which would normally be attributed to a leading edge stall. Figure 3.37(c) shows that trailing edge separation induces a constantpressure region to form over the rear of the airfoil, which causes a reduction in the leading edge suction peak.
4. *Reseparation stall:* This type of stall involves the sudden separation of the turbulent boundary layer just downstream of the reattachment point of a short laminar separation bubble. This behavior causes the bubble to burst. Evans and Mort (1959) have provided evidence which suggests that reseparation and failure to reattach after transition are different phenomena.
5. *Combined stall:* Sometimes referred to as a mixed stall, this type of stall can be thought of as a race between bubble bursting and trailing edge separation for the determination of maximum lift. For example, a noticeable amount of trailing edge separation may form prior to bubble bursting, resulting in a rounding of the lift curve preceding an abrupt loss of lift (Figure 3.36).

Gault (1957) studied the stall characteristics of 150 airfoils (obtained by numerous investigators) over a range of Reynolds number. He found a very useful correlation could be made between the type of stall and the upper surface ordinate at the 1.25% chord. Figure 3.38 displays this correlation, and Table 3.6 gives the relevant ordinate for most of the NACA series airfoils. The correlation is only strictly valid for airfoils with aerodynamically smooth surfaces, no high-lift, devices and tested in low-turbulence free-stream flows.

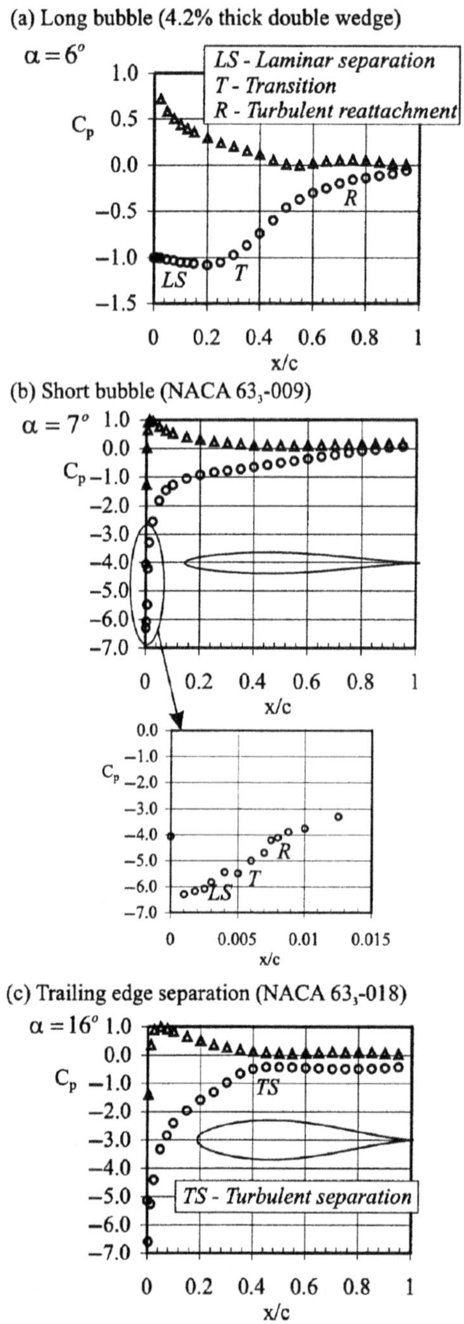

FIGURE 3.37 Typical pressure distributions associated with various boundary layer phenomena (data from McCullough and Gault 1951, courtesy of NASA).

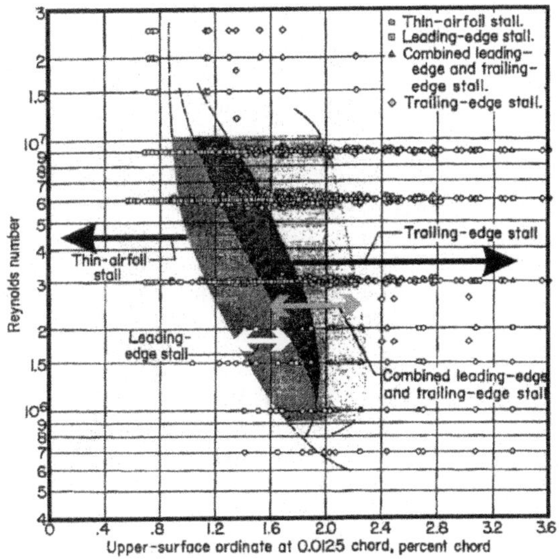

Figure 3.38 Gault's low-speed stall correlation (Gault 1957, courtesy of NASA).

NACA series	Thickness ratio (% chord)									
	0	6	8	9	10	12	15	8	21	24
00xx	0.00	0.95	1.26	1.42	1.58	1.89	2.37	2.84	3.31	3.95
24xx	0.12	1.11	1.44	1.62	1.78	2.15	2.71	3.28	3.87	4.44
44xx	0.25	1.25		1.81		2.44	3.07	3.76	4.45	5.2
230xx	0.36	1.42		2.02		2.67	3.34	4.09	4.87	5.65
16-0xx	0.00	0.65		0.97		1.29	1.61	1.94	2.26	
63-0xx	0.00	0.77		1.15	1.27	1.52	1.88	2.22	2.53	
63-2xx	0.11	0.89		1.29	1.41	1.69	2.08	2.46	2.78	
63-4xx	0.21	1.02		1.44		1.85	2.30	2.76	3.10	
64-0xx	0.00	0.75	1.01	1.13	1.25	1.49	1.84	2.18	2.52	
64-2xx	0.11	0.88	1.14	1.26	1.40	1.65	2.03	2.49	2.77	
64-4xx	0.21	1.01		1.41		1.83	2.25	2.70	3.06	
65-0xx	0.00	0.72	0.95	1.06	1.17	1.39	1.70	2.01	2.30	
65-2xx	0.11	0.84		1.19	1.30	1.54	1.88	2.21	2.54	
65-4xx	0.21	0.97		1.32	1.47	1.70	2.07	2.43	2.78	
66-0xx	0.00	0.69	0.92	1.03	1.14	1.36	1.67	1.95	2.24	
66-2xx	0.11	0.81		1.16	1.28	1.51	1.85	2.15	2.46	
66-4xx	0.21	0.93		1.30		1.67	2.03	2.36	2.68	
63A0xx	0.00	0.75	1.00		1.25	1.49	1.84			
64A0xx	0.00	0.74	0.98		1.23	1.46	1.81			
65A0xx	0.00	0.72	0.95		1.18	1.41	1.75			

Data taken from Gault 1957, courtesy of NASA.

Table 3.6 Upper Surface 1.25% Ordinate for Various NACA Series Airfoils

Types of Incompressible Flow Drag over Airfoils

When the flow is two-dimensional and incompressible there are two types of drag:

1. *Pressure drag:* This rearward facing force, sometimes referred to as form drag, arises from boundary layer thickness and separation effects which do not allow the trailing edge pressures to recover fully to those found in the leading edge region.
2. *Skin friction drag:* As described in Subsection 3.13, skin friction drag arises from viscous shear stresses which act on the airfoil surface.

A boundary layer cannot exist if the entire flow field is deemed to be inviscid, and thus both the pressure and skin friction drag terms will be zero.

3.15 Incompressible Flow Over Finite Wings

Prandtl's Lifting Line Model

(*Applications—inviscid, incompressible, irrotational, flow over unswept, tapered and twisted wings with aspect ratios greater than* 3.) From 1912 to 1918, Ludwig Prandtl and his colleagues in Germany developed an incompressible theory of finite wing aerodynamics which could be split into two parts; the study of two-dimensional flow around a wing section (an airfoil); and the modification of each span-wise airfoil flow to account for the three-dimensional flow which occurs over a finite wing. The strength of Prandtl's model lies in the fact that the airfoil characteristics can be obtained either from theory (see Subsection 3.14) or from wind tunnel testing.

Figure 3.39 shows that since the tip of a finite wing cannot sustain a differential pressure between the upper and lower surfaces, the lift, and hence the circulation, must reduce to zero. As

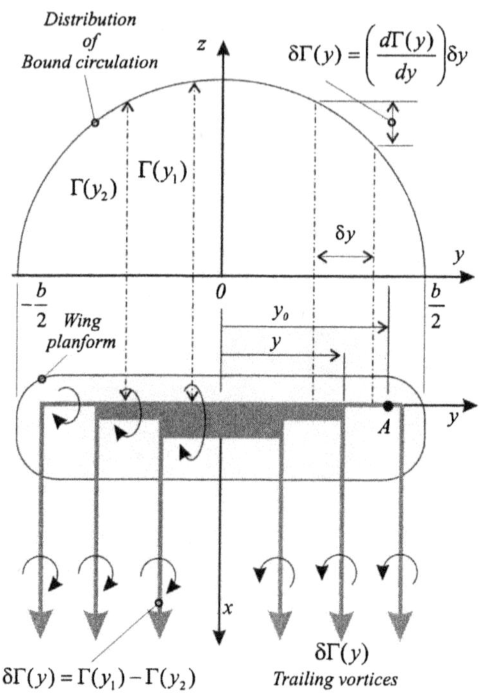

FIGURE 3.39 Prandtl's lifting line model.

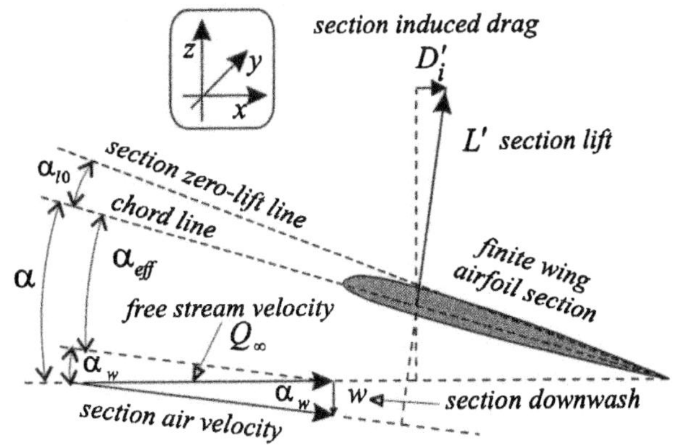

FIGURE 3.40 Effect of finite wing trailed wake.

the bound vortex reduces in strength, Helmholtz's laws (Subsection 3.12) state that the difference between the old and new circulation must be shed downstream as a trailed vortex filament. Thus, Prandtl's lifting line model replaces the wing with a bound vortex, located at the one-quarter chord position, and a wake consisting of an infinite number of trailed vortex filaments.

Using the Biot–Savart Law (see Subsection 3.12), Figure 3.39 illustrates that the downwash (velocity component in the z-direction), due to the entire trailed wake, at an arbitrary point along the bound vortex is given by

$$w(y_0) = -\frac{1}{4\pi}\int_{-b/2}^{b/2}\frac{\{d\Gamma(y)/dy\}dy}{y_0 - y} \tag{3.18}$$

Figure 3.40 shows that the downwash reduces the wing section angle of attack and cants the local lift vector rearward, which gives rise to a drag component known as the induced drag.

Using Figure 3.39 to define the relationship between the effective and geometric angles of attack and combining equation (3.18) with the Kutta–Joukowski lift theorem and the definition of the section lift coefficient results in the formulation of Prandtl's simple lifting line equation:

$$\alpha(y_0) = \frac{2\Gamma(y_0)}{Q_\infty c(y_0)c_{l\alpha}(y_0)} + \alpha_{l0}(y_0) + \frac{1}{4\pi Q_\infty}\int_{-b/2}^{b/2}\frac{(d\Gamma(y)/dy)dy}{y_0 - y} \tag{3.19}$$

To solve this equation, we first use the coordinate transformation $y = -\frac{1}{2}b\cos\theta$ along with the following general circulation distribution:

$$\Gamma(\theta) = \frac{1}{2}Q_\infty c_r c_{l\alpha,r}\sum_{n=1}^{N} A_n \sin n\theta \tag{3.20}$$

This procedure results in the following form of the Prandtl's lifting line equation:

$$\alpha(\theta) = \frac{c_r c_{l\alpha,r}}{c(\theta)c_{l\alpha}(\theta)}\sum_{n=1}^{N} A_n \sin n\theta + \alpha_{l0}(\theta) + \frac{c_r c_{l\alpha,r}}{4b}\sum_{n=1}^{N}\frac{nA_n \sin n\theta}{\sin\theta} \tag{3.21}$$

To numerically solve for the lift distribution over an arbitrary wing planform, equation (3.21) is applied at a chosen number of spanwise locations, i.e., at different values of θ. This will result in N equations in N unknown coefficients $A_1, A_2, \ldots, A_N$. All coefficients will be involved when the lift distribution is asymmetric, while only the odd numbered coefficients will be required when the distribution is symmetrical. For symmetrically loaded rectangular wings it is normally only necessary to retain only the first three or four coefficients. Once the *A*-coefficients have been obtained the following quantities can be calculated:

1. Section induced angle of attack:

$$\alpha_w(\theta) = -\frac{w(\theta)}{Q_\infty} = \frac{c_r c_{l\alpha,r}}{4b} \sum_{n=1}^{N} \frac{a A_n \sin n\theta}{\sin\theta} \tag{3.22}$$

2. Section lift coefficient:

$$c_l(\theta) = \frac{L'(\theta)}{\frac{1}{2}\rho_\infty Q_\infty^2 c(\theta)} = \frac{c_r c_{l\alpha,r}}{c(\theta)} \sum_{n=1}^{N} A_n \sin n\theta \tag{3.23}$$

3. Lift per unit span:

$$L'(\theta) = \frac{1}{2}\rho_\infty Q_\infty^2 c(\theta) c_l(\theta) = \rho_\infty Q_\infty \Gamma(\theta) \tag{3.24}$$

4. Wing lift coefficient:

$$C_L = \frac{L}{\frac{1}{2}\rho_\infty Q_\infty^2 S} = \frac{2}{Q_\infty S}\int_{-b/2}^{b/2} \Gamma(y)\,dy = \frac{c_r c_{l\alpha,r}}{4b}\frac{b^2}{S}\pi A_1 = \frac{c_r c_{l\alpha,r}}{4b} A_R \pi A_1 \tag{3.25}$$

5. Induced drag coefficient:

$$C_{D,ti} = \frac{D_i}{\frac{1}{2}\rho_\infty Q_\infty^2 S} = \frac{2}{Q_\infty S}\int_{-b/2}^{b/2} \Gamma(y)\alpha_i(y)\,dy \tag{3.26}$$

$$= \frac{C_L^2}{\pi A_R}\left(1 + \sum_{n=2}^{N} \frac{a A_n^2}{A_1^2}\right) = \frac{C_L^2}{\pi A_R}(1+\delta)$$

Figures 3.41 and 3.42 present spanwise lift distributions obtained for various tapered and twisted wings. These data were obtained by solving Prandtl's lifting line equation using the four spanwise locations, $\theta = \pi/2, \pi/8, \pi/2, 3\pi/8$, to solve for four symmetric *A*-coefficients.

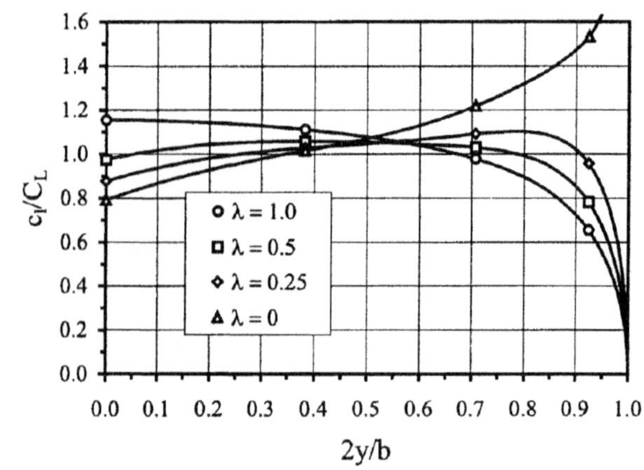

FIGURE 3.41 Effect of taper on spanwise lift distribution.

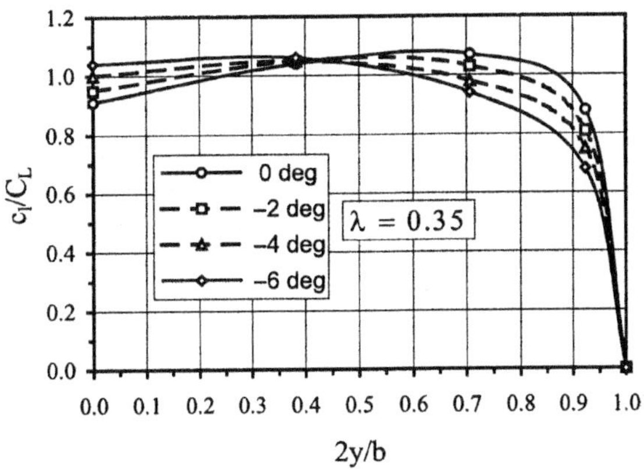

FIGURE 3.42 Effect of twist on the spanwise lift distribution of a tapered wing.

Figure 3.43 presents the variation of δ with taper ratio for untwisted wings of various aspect ratio. It is worth noting that for an untwisted wing the induced drag has a minimum value at a taper ratio of around 0.35.

The Elliptical Wing

Rather than solving for the circulation distribution associated with a particular wing planform (the direct problem), an elliptical spanwise distribution of circulation can be assumed and the wing planform and characteristics then calculated (the indirect problem). An elliptical distribution of circulation is defined by

$$\Gamma(y) = \Gamma_r \sqrt{1 - \left(\frac{2y}{b}\right)^2} \quad \text{or} \quad \Gamma(\theta) = \frac{1}{2} Q_\infty c_r c_{l\alpha,r} A_1 \sin n\theta$$

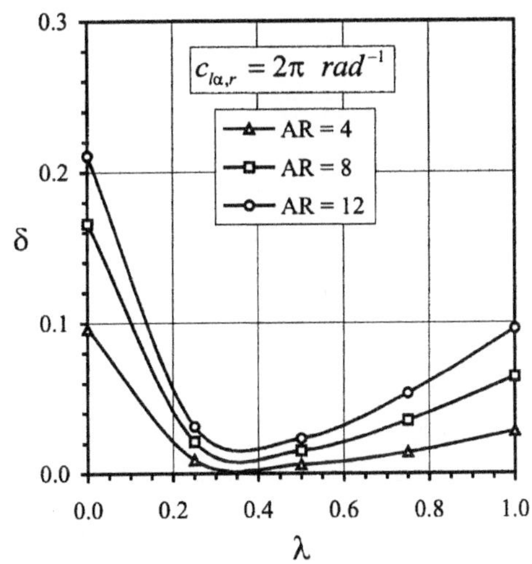

FIGURE 3.43 Variation of δ with taper ratio for an untwisted wing.

Since only one A-coefficient has been stipulated the wing must have no aerodynamic twist. Substituting the elliptical value of A_1 into equations (3.22), (3.23), and (3.25) results in a constant spanwise induced angle of attack and a constant spanwise distribution of section lift coefficient, c_l, equal to that of the wing lift coefficient, C_L. Equation (3.24) thus states that the wing must have an untwisted elliptical spanwise variation in chord given by $c(\theta) = c_r \sin\theta$. Equation (3.26) gives $\delta = 0$, and thus an untwisted elliptical wing has the lowest induced drag coefficient.

The Wing Lift Curve Slope

Consider a wing of arbitrary planform and aerodynamic twist. The wing life coefficient is given by

$$C_L = C_{l\alpha}(\alpha_r - \alpha_{L0})$$

and the root section lift coefficient is given by

$$c_{l,r} = c_{l\alpha,r}(\alpha_r - \alpha_{w,r} - \alpha_{l0,r})$$

Combining these two equations gives the following relationship:

$$C_{L\alpha} = \frac{c_{l\alpha,r}}{1 + \dfrac{c_{l\alpha,r}}{\pi A_R}(1+\tau)} \tag{3.27}$$

where

$$(1+\tau) = \frac{\pi A_R}{c_{l\alpha,r}}\left[\frac{c_{l,r}}{C_L} + \frac{c_{l\alpha,r}}{C_L}(\alpha_{w,r} + \alpha_{l0,r} - \alpha_{L0}) - 1\right]$$

For an untwisted, elliptic wing with constant airfoil section $c_{l\alpha,r} = c_{l\alpha}$ and $\tau = 0$ and equation (3.27) reduces to the form frequently used for incompressible flow.

Figure 3.44 displays the variations in τ with taper ratio for untwisted wings of various aspect ratio. As before, these data were obtained by solving Prandtl's lifting line equation using four spanwise locations.

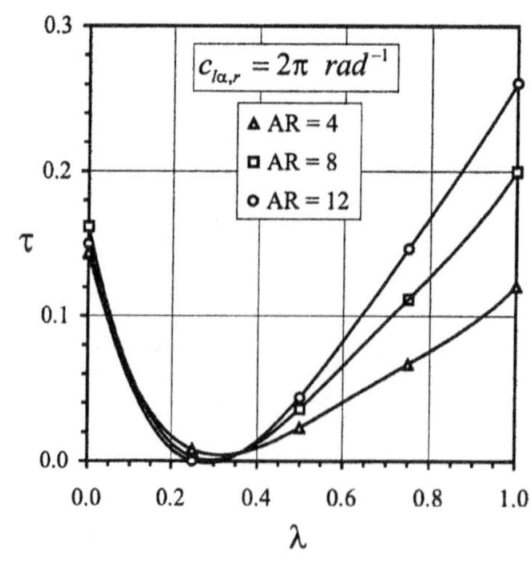

FIGURE 3.44 Variation of τ with taper ratio for an untwisted wing.

The Monoplane Wing Equation

The general circulation distribution is sometimes given as

$$\Gamma(\theta) = 2Q_\infty b \sum_{n=1}^{N} A_n \sin n\theta \tag{3.28}$$

which results in the following form of Prandtl's lifting line equation:

$$\alpha(\theta) = \frac{4b}{c(\theta)c_{l\alpha}(\theta)} \sum_{n=1}^{N} B_n \sin n\theta + \alpha_{l0}(\theta) + \sum_{n=1}^{N} \frac{nB_n \sin n\theta}{\sin\theta} \tag{3.29}$$

Equations (3.28) and (3.29) are equivalent mathematical statements of the lifting line model, as previously expressed by equations (3.20) and (3.21), where the A and B coefficients are related in the following manner:

$$B_n = \left(\frac{c_r c_{l\alpha,r}}{4b}\right) A_n = d_r A_n$$

Equation (3.29) is occasionally written in the following way, which is known as the monoplane wing equation:

$$d(\theta)\{\alpha(\theta) - \alpha_{l0}(\theta)\}\sin\theta = \sum_{n=1}^{N} B_n \sin n\theta(nd(\theta) + \sin\theta)$$

where

$$d(\theta) = \frac{c(\theta)c_{l\alpha}(\theta)}{4b}$$

Extended Lifting Line and Lifting Surface Theories

(Applications—inviscid, incompressible, irrotational flow over swept, tapered, twisted, and yawed wings.) Prandtl's lifting-line model assumes that the effects of camber and thickness are only governed by the local two-dimensional flow field around any particular wing section. This assumption is inappropriate when the wing is swept, and more sophisticated models have to be developed. As discussed in Subsection 3.18, the use of wing sweep brings many benefits during high-speed subsonic and supersonic flight, and thus its effect on low-speed flight has to investigated.

In general, the collection of swept-wing methods, known as extended lifting-line and lifting-surface models, all adopt the following solution methodology;

1. Distribute vorticity over the projection of the wing planform and the trailed wake onto the x–y plane.
2. Apply the Kutta condition in an appropriate form.
3. Use the Biot–Savart law to apply the surface flow tangency boundary condition at specified control points to produce a system of simultaneous algebraic equations for the unknown vorticity values.
4. Solve for the unknown vorticity values.

Differences in the various methods arise due to the following:

1. The manner of distributing the vorticity
2. The mathematical expression used to describe the vorticity distributions
3. The position and number of control points
4. The precise mathematical procedure used to obtain the solution

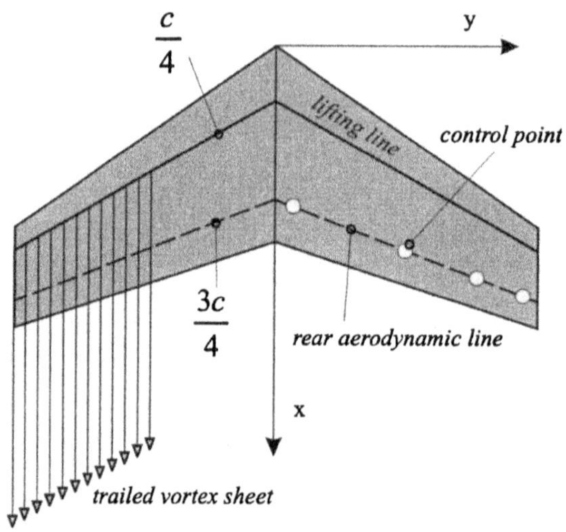

FIGURE 3.45 Weissinger's three-quarter chord method.

Thin airfoil theory (Subsection 3.14) shows that to account properly for the effects of airfoil camber, the vorticity must be distributed along the chord line. However, this theory also indicated that an approximate method could be used which calculated a value for the vorticity at the quarter chord based on the satisfaction of the surface flow tangency condition at the three-quarter chord point (the rear aerodynamic center).

As illustrated in Figure 3.45, extended lifting-line theory (also known as the three-quarter-point method) places a lifting line at the quarter-chord point, which produces a continuous sheet of trailing vortices. The surface flow tangency condition is then enforced at the rear aerodynamic center. Further details of this type of method are given by Weissinger (1947). An alternative to the extended lifting line methods is the vortex lattice model (Falkner 1943), where, as shown in Figure 3.46, the

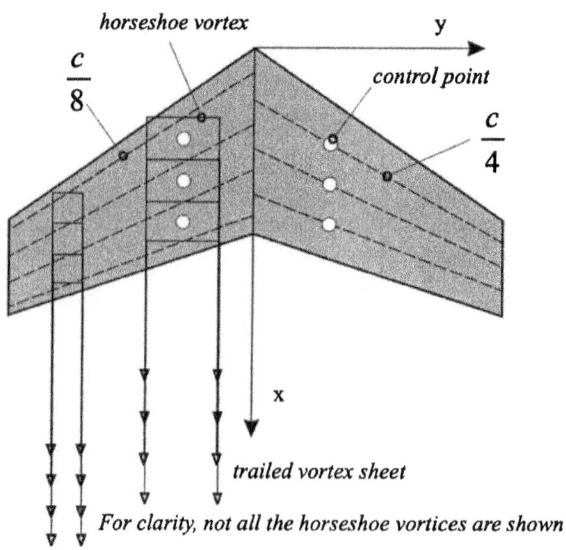

FIGURE 3.46 The vortex lattice method.

vorticity is distributed over the entire wing surface in the form of a finite number of elemental horseshoe vortices (a single vortex filament formed into a horseshoe shape). A chosen number of control points are now specified along the camber line rather than just one at the three-quarter chord point. The influence of all the elemental horseshoe vortices on each control point is found using the Biot–Savart law.

The vortex lattice method was the forerunner of the lifting surface model, which utilizes a continuous distribution of vorticity in both the chordwise and spanwise directions. Application of the Biot–Savart law yields a surface integral equation for the local downwash which has to be solved for the unknown distribution of vorticity subject to the surface flow tangency condition. For more information on vortex lattice and lifting surface methods Katz and Plotkin (2001) should be consulted.

Semiempirical Methods—Diederich's Method

(*Applications—compressible subcritical flow over swept, tapered, and twisted wings.*) Lifting-line, vortex lattice, lifting-surface, and panel methods all rely on the use of numerical algorithms to obtain the final solutions. A method is now described below which is highly amenable to spreadsheet analysis. This method is termed semiempirical because it utilizes correlations based on the results from the more sophisticated numerical models.

The following equations describe a semiempirical method developed by Diederich (1952). This method can be applied to subcritical compressible flow, which is described in more detail in Subsections 3.16, 3.17, and 3.18. It has been placed in this section for two reasons: first, because it is an excellent alternative to the numerical methods described above, and secondly, because it is used below to aid the discussion on the stalling characteristics of finite wings.

In general, the lift distribution of an arbitrary wing can be considered to be the superposition of two independent components:

1. An additional lift distribution, which depends on wing planform and angle of attack
2. A basic lift distribution, which gives zero wing lift and depends on camber and aerodynamic twist

This can be written in the basic form

$$c_l(y) = c_{la}(y) + c_{lb}(y) = \left(\frac{c_{la}(y)}{C_l}\right) C_L + c_{lb}(y)$$

$$= c_{la1}(y) C_{L.\alpha}(\alpha_L - \alpha_{L0}) + c_{lb}(y)$$

The lift distribution is often written in the following nondimensional parameters (Anderson 1936):

$$\frac{c_l(\hat{y})c(\hat{y})}{c_g} = L_a(\hat{y})C_L + L_b(\hat{y})\left(\frac{\varepsilon_{a,t}c_{l\alpha}}{E}\right)$$

where $\hat{y} = 2y/b$ and $L_a(y)$ and $L_b(y)$ are described below. The additional lift distribution is given by

$$L_a(\hat{y}) = \frac{c_l(\hat{y})c_{la}(\hat{y})}{c_g C_L} = C_1 \frac{c(\hat{y})}{c_g} + C_2 \frac{4}{\pi}\sqrt{1-\hat{y}^2} + C_3 f(\hat{y})$$

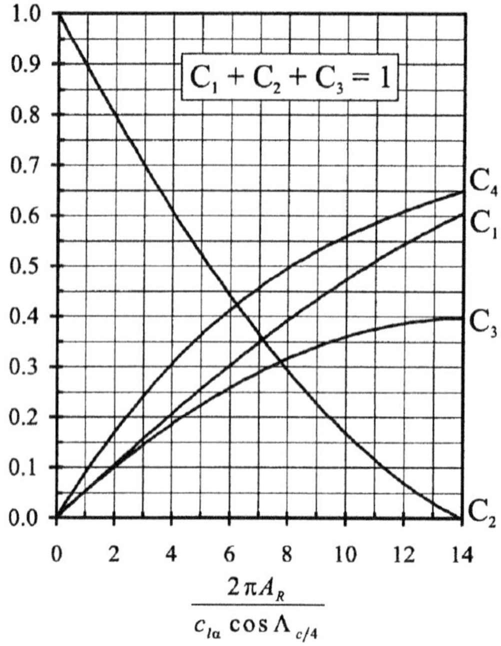

FIGURE 3.47 Factors in Diederich's method (data taken from Diederich 1952, courtesy of NASA).

The values of C_1, C_2, and C_3 depend on the aspect ratio and sweep of the wing and are given in Figure 3.47, while the function $f(\hat{y})$ is given in Figure 3.48 and depends only on the effective sweep (defined in Subsection 3.18) as

$$\cos\Lambda_i = \frac{\beta_0 \cos\Lambda}{\sqrt{1-M_\infty^2 \cos^2\Lambda}}$$

where $\beta_0 = \sqrt{1-M_\infty^2}$. The basic lift distribution is given by

$$L_b(\hat{y}) = \frac{c(\hat{y})c_{lb}(\hat{y})}{c_g}\left(\frac{E}{\varepsilon_{a,t}c_{l\alpha}}\right) = C_4 E\beta_0 \cos\Lambda_\beta \left(\frac{\varepsilon_a(\hat{y})}{\varepsilon_{a,t}} + \alpha_{01}\right)L_a(\hat{y})$$

where C_4 is given in Figure 3.46 and

$$E = 1 + \frac{2\lambda}{A_R(1+\lambda)}$$

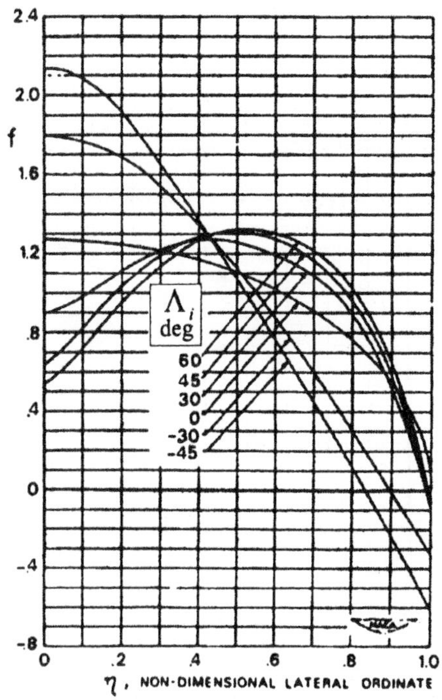

FIGURE 3.48 Function used in Diederich's method (Diederich 1952, courtesy of NASA).

which is known as Jones's edge factor (Jones 1941). The factor α_{01} is given by

$$\alpha_{01} = \frac{(\alpha_{L,0} - \alpha_{l0,r})}{\varepsilon_{at}} = -\int_0^1 \frac{\varepsilon_a(\hat{y})}{\varepsilon_{a,t}} L_a(\hat{y})\, d\hat{y}$$

and, for unswept linearly tapered wings, has the simplified form of

$$\alpha_{01} = -C_1\left(\frac{1+2\lambda}{3(1+\lambda)}\right) - (C_2 + C_3)\frac{4}{3\pi}$$

Figure 3.49 illustrates distributions of $c_{la1}(\hat{y})$ and $c_{lb}(\hat{y})$ for a linearly tapered wing with the following characteristics: $A_R = 15$, $\lambda = 0.6$, $\Lambda = 0$, $\varepsilon_{a,t} = -2.2°$, and $\beta_0 \approx 1$. The zero lift angle of the wing (see Subsection 3.10) is given by

$$\alpha_{L0} = \alpha_{l0,r} + \alpha_{01}\varepsilon_{a,t}$$

and the lift curve slope is given by

$$C_{L\alpha} = \frac{c_{l\alpha}\cos\Lambda_i}{\sqrt{\dfrac{\beta_0^2}{\cos^2\Lambda_i} + \left(\dfrac{c_{l\alpha}}{\pi A_R}\right)^2} + \dfrac{c_{l\alpha}}{\pi A_R}}$$

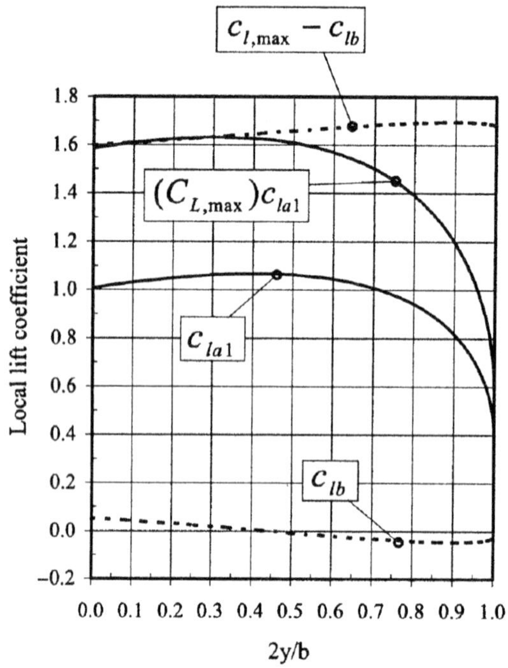

FIGURE 3.49 Spanwise lift distribution from Diederich's method.

The wing lift coefficient can now be obtained from

$$C_L = C_{L\alpha}(\alpha_r - \alpha_{l0})$$

where the wing angle of attack, α_r, is defined as the angle between the root chord and the free-stream velocity.

The wing pitching moment about the aerodynamic center of the wing (see Subsection 3.10), $C_{M,AC}$, is composed of two parts, one due to the spanwise distribution of section camber and one due to the combined effect of twist and sweep. Using the approximation that the section aerodynamic center lies at the quarter chord position, the wing pitching moment can be calculated from the following equation:

$$C_{M,AC} = C_{M,\text{camber}} + C_{M,\text{twist}}$$

where

$$C_{M,\text{camber}} = \frac{b}{S\bar{c}}\int_0^1 (c_{m,c/4})c^2 d\hat{y}$$

$$C_{M,\text{twist}} = -\frac{b^2}{2S\bar{c}}\int_0^1 (c_{lb})c\hat{y}\tan\Lambda_{c/4}\, d\hat{y}$$

The position of the wing aerodynamic center can be found using the method described in Subsection 3.10 and Figure 3.8.

The Stalling Characteristics of Finite Wings

The maximum wing lift coefficient will be reached when the local lift coefficient, at any spanwise position, reaches the maximum value appropriate to that location. When this occurs the wing is said to be stalled. Prior to the attainment of this condition an amount of trailing edge separation will have formed, on the wing upper surface, at this particular spanwise position. As the wing angle of attack is increased the trailing edge separation spreads over a region referred to as a stall cell. When considering the stability and control of an aircraft, it is important that this stall cell does not initially occur over any of the wing control surfaces, in particular the ailerons.

Figure 3.49 illustrates the spanwise variation of $(C_{l,\max} - c_{lb})$, which, for a given wing twist distribution, depends only on the airfoil sections which compose the wing. The maximum wing lift coefficient is given by the smallest value of $(C_{l,\max} - c_{lb})/c_{la1}$. The spanwise position at which this condition is satisfied identifies the location where the first stall cell will appear. The wing, used in this example, had a maximum wing lift coefficient of 1.54 and the stall cell first appeared at 26% of the span.

In general, a wing is tapered to reduce the wing root structural loads due to weight. However, as shown by Figure 3.41, taper increases the values of c_l toward the wing tips, which in turn increases the chance of flow separation over the ailerons. By introducing negative aerodynamic twist at the wing tip the tip values of c_l are reduced and the stall cell is forced inboard toward the wing root. In theory, this also makes the spanwise lift distribution more elliptical, which should reduce the induced drag. As discussed in Subsection 3.18, wing sweep is used to increase the drag divergence Mach number. For an aft-swept wing this has a similar effect on the spanwise distribution of c_l as increasing the taper. However, again this can be controlled by the use of aerodynamic twist. Chappell (1968) documents various empirical correlations which allow the stalling characteristics of various wings to be estimated.

3.16 Shock Wave Relationships

Isentropic Flow

Combining the first law of thermodynamics for a closed-system noncyclic process with the definitions of entropy, work done, and a perfect gas gives two equations for the entropy change between two end states:

$$s_2 - s_1 = c_v \ln\frac{T_2}{T_1} + R\ln\frac{v_2}{v_1} \tag{3.30}$$

and

$$s_2 - s_1 = c_v \ln\frac{T_2}{T_1} - R\ln\frac{p_2}{p_1} \tag{3.31}$$

When the flow is regarded as both reversible and adiabatic there can be no change in entropy and the flow is said to be isentropic. Setting $s_2 - s_1$ in equation (3.31) results in the isentropic flow relationship

$$\frac{p_2}{p_1} = \left(\frac{T_2}{T_1}\right)^{\gamma/\gamma-1}$$

Many compressible flow fields can be regarded as isentropic since the effects of viscous diffusion and heat transfer are often small in the free stream. The opposite is true, however, within the boundary layer, which is a strong source of entropy generation. As indicated later, this is also the case for shock waves.

Speed of Sound and Mach Number

The speed of sound through an arbitrary fluid can be calculated from $a = \sqrt{\gamma RT}$, where T is the local fluid temperature. The Mach number is defined the ratio of the local fluid velocity to the speed of sound, i.e., $M = q/a$.

The Stagnation State

The stagnation value, of a fluid property is defined as the value it would ascertain if the flow were isentropically brought to rest without any work transfer. Under these conditions, the conservation of energy reduces to

$$h + \frac{q^2}{2} = \text{constant} = h_0$$

The stagnation temperature and pressure are respectively given by

$$\frac{T_0}{T} = 1 + \frac{\gamma - 1}{2} M^2 \quad \text{and} \quad \frac{p_0}{p} = \left(\frac{T_0}{T}\right)^{\gamma/\gamma-1}$$

Combining these two equations results in a useful relationship among the stagnation pressure, the static pressure, and the Mach number given by

$$\frac{p_o}{p} = \left(1 + \frac{\gamma - 1}{2} M^2\right)^{\gamma/\gamma-1} \tag{3.32}$$

Incompressible Flow Limit

The stagnation density of a moving perfect gas is given by the isentropic relation

$$\frac{\rho_o}{\rho} = \left(1 + \frac{\gamma - 1}{2} M^2\right)^{1/(\gamma-1)}$$

For air, with $\gamma = 1.4$, the difference between the stagnation and the static densities is less than 5% for Mach numbers less than 0.32. As a result, it is generally accepted that the flow should be treated as compressible above Mach numbers of 0.3. Since the local maximum velocity over an arbitrary airfoil is approximately three times the free-stream value, the entire flow field can only be regarded as completely incompressible for free-stream Mach numbers of around 0.1. For low Mach numbers, equation (3.31) may be reduced to Bernoulli's equation using the binomial theorem, i.e.,

$$p_o = p + \frac{1}{2}\gamma M^2 p, \quad \text{which gives} \quad p_0 = p + \frac{1}{2}\rho q^2$$

Mach Wave Formation

Any object moving through a fluid will cause pressure waves to propagate through the surrounding fluid at the local speed of sound. Figure 3.50 illustrates the spread of pressure waves from a

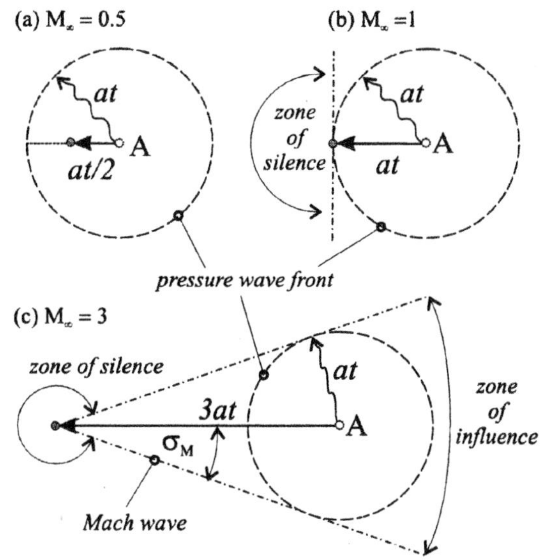

FIGURE 3.50 Mach wave formation.

small body traveling at subsonic and then supersonic speeds. For the supersonic case only the fluid that lies within the cone indicated is aware of the presence of the body. This is known as the zone of influence and the vertex angle, σ_M, is known as the Mach angle and is related to the Mach number as follows:

$$\sigma_M = \sin^{-1}(1/M_\infty)$$

Shock Wave Formation

To illustrate how a shock wave forms, consider the acceleration from rest of a piston within a long cylinder. The initial movement of the piston causes a Mach wave to travel downstream, which leaves the air behind with a slightly increased pressure and temperature. This, in turn, increases the local speed of sound so subsequent Mach waves, produced by the moving piston, travel faster downstream. A continual series of Mach waves will eventually merge to produce a strong shock wave across which the local pressure, temperature, velocity, and entropy will abruptly change. Although the flow through a shock wave is nonisentropic, the flow ahead and aft can often be considered to be isentropic.

Normal Shock Wave Relations

Applying the conservation of mass, momentum, and energy to a control volume which lies across a normal shock wave results in the following equations which are often referred to the Rankine–Hugoniot relationships.

$$M_2^2 = \frac{(\gamma-1)M_1^2+2}{2\gamma M_1^2-(\gamma-1)}$$

$$\frac{p_2}{p_1} = \frac{2\gamma M_1^2-(\gamma-1)}{(\gamma+1)}$$

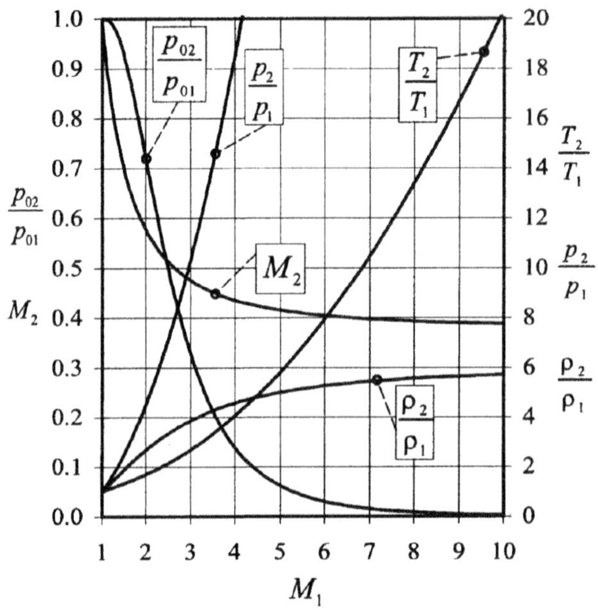

FIGURE 3.51 Normal shock wave properties ($\gamma = 1.4$).

$$\frac{p_{02}}{p_{01}} = \left[\frac{(\gamma+1)M_1^2}{2+(\gamma-1)M_1^2}\right]^{\gamma/(\gamma-1)} \left[\frac{2\gamma M_1^2-(\gamma-1)}{(\gamma+1)}\right]^{-1/(\gamma-1)}$$

$$\frac{T_2}{T_1} = \frac{[2\gamma M_1^2-(\gamma-1)][(\gamma-1)M_1^2+2]}{(\gamma+1)^2 M_1^2}$$

$$\frac{\rho_2}{\rho_1} = \frac{(\gamma+1)M_1^2}{2+(\gamma-1)M_1^2}$$

$$s_2 - s_1 = c_p \ln\left\{\frac{[2\gamma M_1^2-(\gamma-1)][(\gamma-1)M_1^2+2]}{(\gamma+1)^2 M_1^2}\right\} - R\ln\left\{\frac{2\gamma M_1^2-(\gamma-1)}{(\gamma+1)}\right\} \tag{3.33}$$

Figure 3.51 graphically illustrates the ratios of the various flow variables as a function of the upstream Mach number. A number of sets of tables and graphs are available which give similar information, e.g., Houghton and Brock 1975; Ames Research Staff 1947, 1953. Equation (3.33) indicates that the flow through a Mach wave ($M_1 = 1$) is isentropic while there is an increase in entropy through a shock wave ($M_1 > 1$).

Oblique Shock Wave Relations

Figure 3.52 defines the flow geometry associated with an oblique shock wave where σ is known as the shock wave angle and δ is referred to as the turning (or deflection) angle. Since there is only a change in the velocity normal to the oblique shock wave, we can use the normal shock wave relations defined above. This results in the following equations:

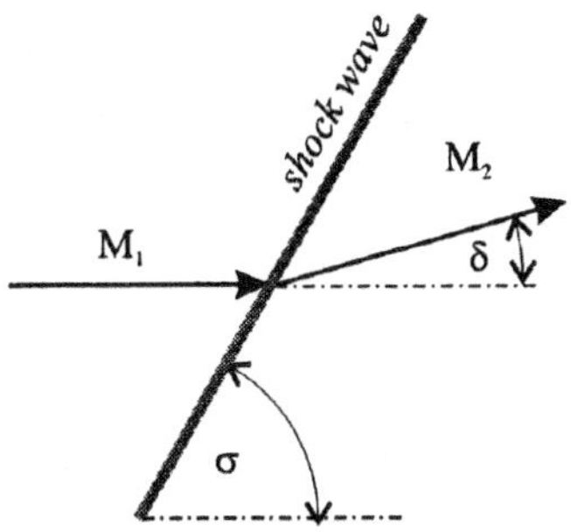

FIGURE 3.52 Oblique shock wave angles.

$$M_2^2 \sin^2(\sigma-\delta) = \frac{(\gamma-1)M_1^2 \sin^2\sigma + 2}{2\gamma M_1^2 \sin^2\phi - (\gamma-1)}$$

$$\frac{p_2}{p_1} = \frac{2\gamma M_1^2 \sin^2\sigma - (\gamma-1)}{(\gamma+1)} \tag{3.34}$$

$$\frac{T_2}{T_1} = \frac{[2\gamma M_1^2 \sin^2\sigma - (\gamma-1)][(\gamma-1)M_1^2 \sin^2\sigma + 2]}{(\gamma+1)^2 M_1^2 \sin^2\sigma}$$

$$\frac{\rho_2}{\rho_1} = \frac{(\gamma+1)M_1^2 \sin^2\sigma}{2+(\gamma-1)M_1^2 \sin^2\sigma}$$

It can be shown using the principle of increasing entropy that for an oblique shock $M_1 \sin\sigma \geq 1$. Equation (3.34) shows that the ratio of p_2/p_1 tends to unity as the shock wave angle tends to the Mach angle. Hence, the limits on the shock wave angle are $\sin^{-1}(1/M_1) \leq \sigma \leq 90°$. It should also be noted that the flow behind an oblique shock wave can be supersonic since $M_2 \sin(\sigma-\delta) \leq 1$. In order to utilize the above relationships, the relation between δ, σ, and M_1 has to known. From geometrical considerations, it can be shown that

$$\tan\delta = \frac{2(M_1^2 \sin^2\sigma - 1)/\tan\sigma}{2+M_1^2(\gamma+\cos 2\sigma)} \tag{3.35}$$

It is worth noting that the turning angle is 0 when $\sigma = 90°$ (normal shock) and also when $\sigma = \sin^{-1}(1/M_1)$ (Mach wave). A typical oblique shock wave solution procedure would be first to obtain σ for a given M_1 and δ and then to use the value of $M_1 \sin\sigma$ along with normal shock wave tables. The value of M_2 is given by the normal shock wave value divided by $\sin(\sigma-\delta)$.

Figures 3.53 and 3.54 show graphical representations of equation (3.35). For a given value of M1, equation (3.35) may be differentiated, which gives the maximum turning angle to occur when

$$\sin^2\sigma_{\max} = \frac{(\gamma+1)M_1^2 - 4}{4\gamma M_1^2} + \frac{(\gamma+1)^{1/2}}{4\gamma M_1^2}[16+8(\gamma-1)M_1^2 + (\gamma+1)M_1^4]^{1/2}$$

For a given Mach number $\sigma_{\max}$ can be calculated and thus the maximum turning angle, $\delta_{\max}$, can be obtained.

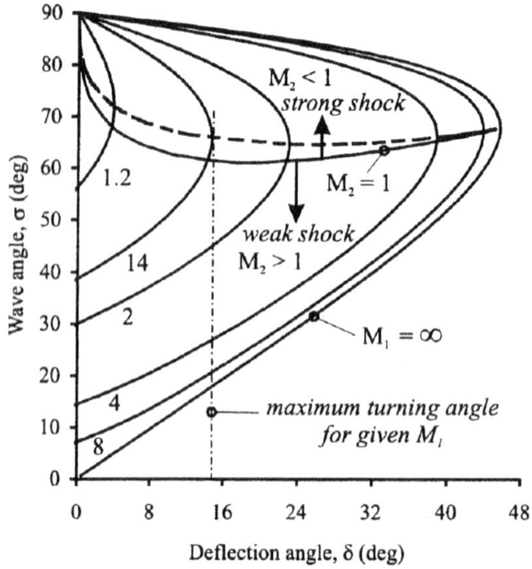

Figure 3.53 Oblique shock wave graph and definitions.

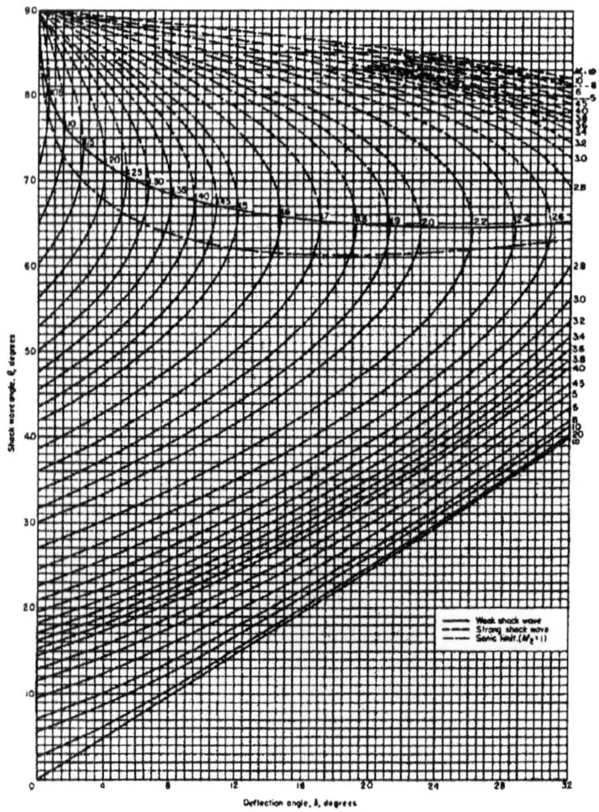

Figure 3.54 Standard oblique shock wave chart.

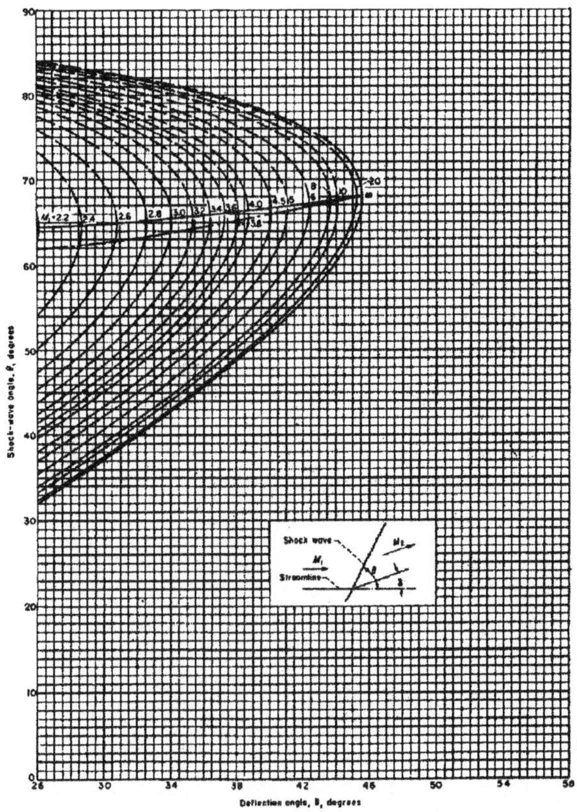

FIGURE 3.54 *(Continued)*

When a body forces the flow to deflect an amount greater than the maximum deflection angle, for the particular Mach number, a detached curved shock wave will form. For any deflection angle less than the maximum, there is a low and a high wave angle, which are respectively known as the weak and strong shock solutions. Experiments have shown that the observed shock angle nearly always corresponds to the weak solution.

Also indicated in Figure 3.53 is the locus of turning and wave angles which result in $M_2 = 1$, which are related by the following equation:

$$\sin^2\sigma_s = \frac{(\gamma+1)M_1^2 - 3 + \gamma}{4\gamma M_1^2} + \frac{(\gamma+1)^{1/2}}{4\gamma M_1^2}[9+\gamma+2(\gamma-3)M_1^2+(\gamma+1)M_1^4]^{1/2}$$

Expansion Waves (Prandtl-Meyer Flow)

Consider the flow through a Mach wave which has been produced by the flow turning through a small deflection, $d\delta$ (positive in anticlockwise direction). From geometrical considerations, and using the fact that the velocity component tangential to the wave remains constant, it can be shown that

$$\frac{dq}{q} = \frac{-d\delta}{\sqrt{M^2-1}} \tag{3.36}$$

Combining the steady flow energy equation with the speed of sound gives the following results:

$$\frac{dp}{p} = \frac{\gamma M^2 d\delta}{\sqrt{M^2 - 1}} \tag{3.37}$$

$$\frac{d\rho}{\rho} = \frac{M^2 d\delta}{\sqrt{M^2 - 1}} \tag{3.38}$$

Using these differential changes in pressure and density with equations (3.30) and (3.31) results in the flow across the Mach wave being isentropic. Thus, a small deflection produces the following changes in the flow field:

$$dM \propto -d\delta,\ dq \propto -d\delta,\ dp \propto d\delta,\ d\rho \propto d\delta, \text{ and } ds = 0$$

Figure 3.55 shows the flow around a sharp convex corner to consist of an infinite number of Mach waves, each turning the flow through a differentially small angle. For negative changes in wall angle the Mach waves diverge and the flow remains isentropic throughout the entire fan. Such flows are called either centered expansion fans or Prandtl–Meyer flows. Differentiating the definition for the Mach number gives

$$\frac{dq}{q} = \frac{dM}{M} + \frac{da}{a} = \frac{dM}{M} + \frac{1}{2}\frac{dp}{p} - \frac{1}{2}\frac{d\rho}{\rho}$$

Using equations (3.35)–(3.37), the total turning angle is given from

$$-\int_0^\delta d\delta = \int_1^M \frac{2\sqrt{M^2 - 1}}{2 + (\gamma - 1)M^2} \frac{dM}{M}$$

Carrying out the integration gives the total turning angle equal to

$$-\delta = \sqrt{\frac{\gamma + 1}{\gamma - 1}} \tan^{-1}\left\{\sqrt{\frac{\gamma + 1}{\gamma - 1}(M^2 - 1)}\right\} - \tan^{-1}\left\{\sqrt{(M^2 - 1)}\right\} = \upsilon(M) \tag{3.36}$$

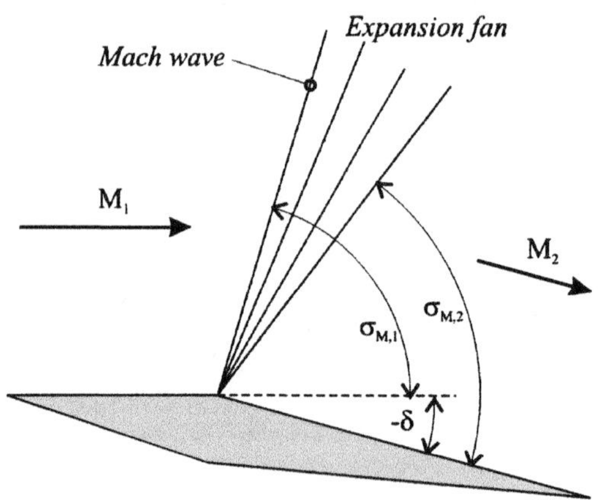

FIGURE 3.55 Prandtl–Meyer expansion fan.

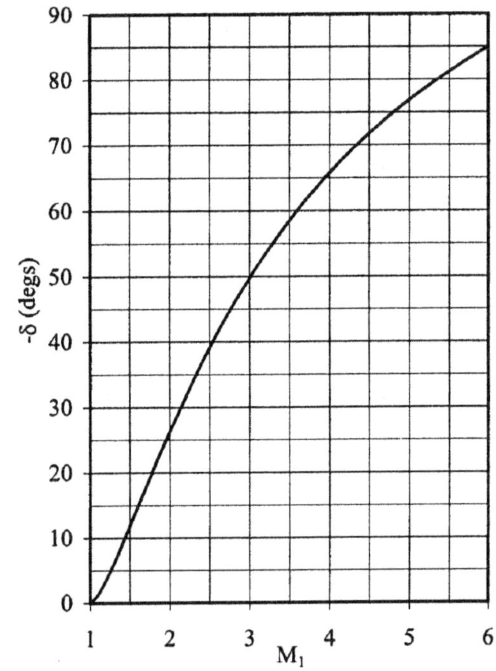

FIGURE 3.56 The Prandtl–Meyer function.

$v(M)$ is sometimes referred to as the Prandtl–Meyer function and is equal to 0 when the Mach number is unity. Knowing the upstream Mach number, M_1 and the deflection angle, δ, the downstream Mach number, M_2 can be calculated as follows:

1. From equation (3.36), Figure 3.56, or tables (Houghton and Brock 1975; Ames Research Staff 1947, 1953), obtain the value of δ_1 corresponding to M_1.
2. Calculate $\delta_2 = \delta - \delta_1$ and then find M_2 from tables, graphs, or equation.
3. Any other downstream property can be obtained from isentropic relationships or isentropic flow tables (noting that the stagnation pressure remains constant across the fan).
4. Calculate the fan boundaries using $\sigma_{M,1} = \sin^{-1}(1/M_1)$ and $\sigma_{M,2} = \sin^{-1}(1/M_2)$.

Further Compressible Flow Phenomena

There are many more important and interesting compressible flow phenomena which have not been covered here. For further information Anderson (1990) and Oosthuizen and Carscallen (1997) should be consulted.

3.17 Compressible Flow Over Airfoils

Shock-Expansion Theory of Lift

(*Applications—two-dimensional, inviscid, isentropic, irrotational, supersonic compressible flow past a thin airfoil at low angles of attack. The airfoil must be made up of straight line segments and the deflection angles must be small enough not to induce a detached shock.*) Consider the supersonic flow past a flat plate at

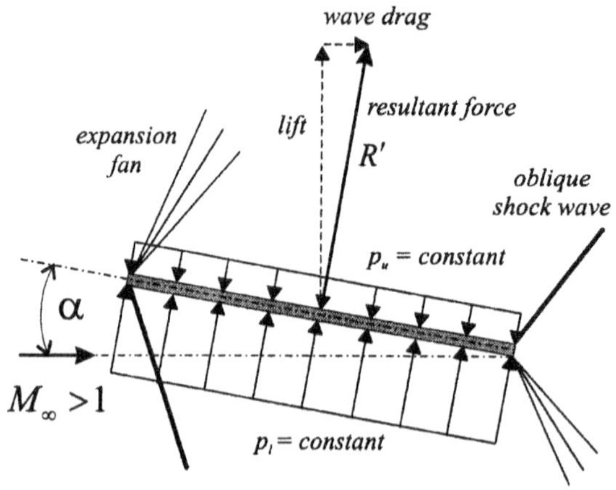

Figure 3.57 The shock-expansion theory of lift.

some angle of attack as depicted in Figure 3.57. As indicated in the figure, the upper and lower surfaces develop uniform pressure distributions which are due to the system of oblique shock waves and centered expansion fans. These pressures may be calculated using the relevant techniques described in Subsection 3.16 and thus the resultant aerodynamic force can be computed; see Anderson (2016) for numerical examples.

Linear Theory for Perturbated Compressible Flow

(Applications—two-dimensional, inviscid, isentropic, irrotational, compressible flow past a thin airfoil at low angles of attack. The free-stream flow can be subsonic or supersonic, but not transonic or hypersonic.) We define a disturbance (or perturbation) velocity potential as follows:

$$u = Q_\infty + \hat{u} \qquad v = 0 \qquad w = \tilde{w}$$

$$\phi = Q_\infty x + \tilde{\phi} \qquad u = \frac{\partial \phi}{\partial x} = Q_\infty + \frac{\partial \tilde{\phi}}{\partial x} \quad w = \frac{\partial \phi}{\partial z} = \frac{\partial \tilde{\phi}}{\partial z}$$

Starting from the continuity equation and utilizing Euler's equation (Subsection 3.11) and the isentropic speed of sound relation, we obtain the disturbance velocity potential equation:

$$\left[a^2 - \left(Q_\infty + \frac{\partial \tilde{\phi}}{\partial x}\right)^2\right]\frac{\partial^2 \tilde{\phi}}{\partial x^2} + \left[a^2 - \left(\frac{\partial \tilde{\phi}}{\partial z}\right)^2\right]\frac{\partial^2 \tilde{\phi}}{\partial z^2} - 2\left(Q_\infty + \frac{\partial \tilde{\phi}}{\partial x}\right)\left(\frac{\partial \tilde{\phi}}{\partial z}\right)\frac{\partial^2 \tilde{\phi}}{\partial x \partial z} = 0$$

Expressing the disturbance potential in terms of the disturbance velocities, substituting in the energy equation, and excluding both the transonic flow ($0.8 < M_\infty < 1.2$) and hypersonic flow ($M_\infty > 5$) regimes, we obtain:

$$(1 - M_\infty^2)\frac{\partial^2 \tilde{\phi}}{\partial x^2} + \frac{\partial^2 \tilde{\phi}}{\partial z^2} = 0 \tag{3.39}$$

This linear partial differential equation can be solved with the following boundary conditions (also see Subsection 3.12):

1. Free-stream condition: $\tilde{u} = \tilde{w} = 0$ or $\tilde{\phi} = constant$.
2. Body flow tangency condition: $\tilde{w} = \partial\tilde{\phi}/\partial z = Q_\infty(\alpha - dz_s/dx)$.

If the fluid is assumed to be a perfect gas, the linearized pressure coefficient is given by

$$C_p = \frac{p-p_\infty}{\frac{1}{2}\rho_\infty Q_\infty^2} = \frac{p-p_\infty}{\frac{1}{2}\gamma M_\infty^2 p_\infty} = -\frac{2\tilde{u}}{Q_\infty} \tag{3.40}$$

Subsonic Compressibility Correction Methods

(*Applications—two-dimensional, inviscid, isentropic, irrotational, subsonic compressible flow past a thin airfoil at low angles of attack.*) Subsonic compressibility correction methods relate the subsonic compressible flow past a particular airfoil to the incompressible flow past a second airfoil which is geometrically related to the first through an affine transformation. An affine transformation changes all the coordinates in a given direction by a uniform ratio. These methods generate expressions which are known as similarity laws. There are four methods in this class: Gothert's rule, the Prandtl–Glauert rule, Laitone's rule, and the Karman–Tsien rule. The elegance of the compressibility correction methods lies in the fact that compressible flow airfoil characteristics can be predicted by modifying the incompressible data obtained from either the methods described in Subsection 3.14 or from low-speed wind tunnel tests.

Gothert's rule essentially covers the fundamental transformation technique which is now described. Consider the following transformation:

$$\hat{x} = \zeta_x x, \quad \hat{z} = \zeta_z z, \quad \hat{\phi}(\hat{x},\hat{y}) = \zeta_\phi \tilde{\phi}(x,z)$$

Substituting these variables into equation (3.39) gives

$$\frac{\partial^2 \hat{\phi}}{\partial \hat{x}^2} + \frac{\partial^2 \hat{\phi}}{\partial \hat{z}^2} = 0$$

if we stipulate that

$$\frac{\zeta_z}{\zeta_x} = \sqrt{1-M_\infty^2} = \beta_0$$

Since $\hat{\phi}$ is a solution to the Laplace equation, it must represent the disturbance velocity in incompressible flow. The incompressible and compressible airfoil profiles are geometrically related in the following manner:

$$(dz_s/dx)_{\text{inc}} = \beta_0 (dz_s/dx), \quad \hat{t}_{\text{inc}} = \beta_0 t \quad \text{and} \quad \hat{h}_{\text{inc}} = \beta_0 \hat{h}$$

The airfoil surface flow tangency condition on the incompressible plane is given by

$$\frac{\partial \hat{\phi}}{\partial \hat{z}} = Q_\infty \left\{ \left(\frac{\zeta_\phi}{\zeta_z}\right)(\alpha)_{\text{inc}} - \left(\frac{\zeta_\phi \zeta_x}{\zeta_z^2}\right)\frac{d\hat{z}_s}{d\hat{x}} \right\}$$

which means that the flow boundary conditions are satisfied on both the compressible and incompressible planes when

$$(Q_\infty)_{\text{inc}} = Q_\infty, \quad \alpha_{\text{inc}} = \beta_0 \alpha \quad \text{and} \quad \zeta_z^2 = \zeta_x \zeta_\phi$$

The compressible pressure coefficient is given by

$$C_p = \frac{p-p_\infty}{\frac{1}{2}\rho_\infty Q_\infty^2} = -\frac{2\tilde{u}}{Q_\infty} = -\frac{2}{Q_\infty}\frac{\partial \tilde{\phi}}{\partial x} = -\frac{\zeta_x}{\zeta_\phi}\frac{2}{Q_\infty}\frac{\partial \hat{\phi}}{\partial \hat{x}} = \frac{1}{\beta_0^2}(C_p)_{\text{inc}}$$

The Prandtl–Glauert rule utilizes the fact that for affinely related airfoils in incompressible flow the local pressure coefficient at corresponding points is approximately proportional to the thickness ratio, the camber ratio, and the angle of attack. Consider two airfoils in incompressible flow with the following geometric relationship:

$$\hat{t}_{\text{inc},2} = \frac{1}{\beta_0}\hat{t}_{\text{inc},1}, \quad \hat{h}_{\text{inc},2} = \frac{1}{\beta_0}\hat{h}_{\text{inc},1}, \quad \text{and} \quad \alpha_{\text{inc},2} = \frac{1}{\beta_0}\alpha_{\text{inc},1}$$

Then it follows that

$$(C_p)_{\text{inc},2} \approx \frac{1}{\beta_0}(C_p)_{\text{inc},1}$$

This result is now combined with Gothert's rule to relate the compressible and incompressible flow fields in the following manner

$$\frac{\hat{t}}{\hat{t}_{\text{inc},2}} = \frac{\hat{h}}{\hat{h}_{\text{inc},2}} = \frac{\alpha}{\alpha_{\text{inc},2}} = 1 \quad \text{and} \quad C_p = \frac{1}{\beta_0}(C_p)_{\text{inc},2}$$

Therefore, the compressible flow pressure coefficient can be obtained from the incompressible value for the same airfoil profile as

$$C_p = \frac{(C_p)_{\text{inc}}}{\sqrt{1-M_\infty^2}} \tag{3.41}$$

where $(C_p)_{\text{inc}} = (C_p)_{\text{inc},2}$.

Since the lift and moment coefficients are obtained from the integrated pressure coefficient, we obtain

$$c_l = \frac{(c_l)_{\text{inc}}}{\sqrt{1-M_\infty^2}} \quad \text{and} \quad c_m = \frac{(c_m)_{\text{inc}}}{\sqrt{1-M_\infty^2}}$$

Laitone first hypothesized that the local Mach number (rather than the free-stream value) should be used to calculate β_0. The isentropic relations were then utilized to express the local Mach number in terms of the free-stream value to obtain the following modification to the Prandtl–Glauert rule:

$$C_p = \frac{2\beta_0(C_p)_{\text{incomp}}}{2\beta_0^2 + \frac{1}{2}\{2M_\infty^2 + (\gamma-1)M_\infty^4\}(C_p)_{\text{incomp}}} \tag{3.42}$$

The Prandtl–Glauert rule assumes that the local speed of sound does not vary from point to point around the airfoil. This approximation was taken into account by Karman and Tsien, which resulted in

$$C_p = \frac{(C_p)_{\text{incomp}}}{\beta_0 + \frac{1}{2}\{1-\beta_0\}(C_p)_{\text{incomp}}} \tag{3.43}$$

The two pressure coefficients in the Karman–Tsien rule do not strictly refer to the same airfoil profile since the transformation used distorts the geometry on the incompressible plane. However, this effect is small and the rule is commonly used for a fixed airfoil geometry.

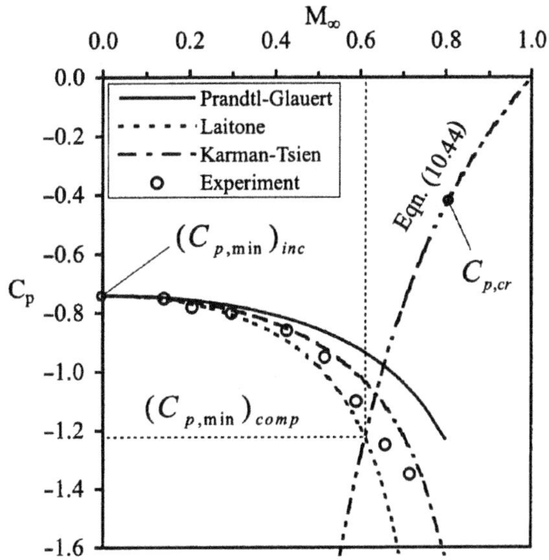

FIGURE 3.58 Compressibility correction methods (data from Stack et al. 1938, courtesy of NASA).

Figure 3.58 displays a comparison between the three compressibility correction equations and wind tunnel data. In this example, the minimum experimental pressure coefficient was used at each Mach number and Laitones criterion gives a critical Mach number of just above 0.6 for this airfoil.

Critical Mach Number

The critical Mach number is defined as the free-stream Mach number at which the local Mach number on the airfoil surface becomes unity. The local pressure coefficient which coincides with this point is given, for isentropic flow by

$$C_{p,cr} = \frac{2}{\gamma M_{cr}^2}\left[\left(\frac{1+\frac{1}{2}(\gamma-1)M_{cr}^2}{1+\frac{1}{2}(\gamma-1)}\right)^{\gamma/\gamma-1} - 1\right] \approx \frac{-2}{\gamma+1}\frac{1-M_{cr}^2}{M_{cr}^2} \tag{3.44}$$

Figure 3.58 shows that the critical Mach number may be estimated using either experimental data or one of the compressibility corrections along with equation (3.44). Laitone's correction intersected equation (3.44) at the lowest Mach number; which resulted in a critical Mach number of 0.6.

Linear Theory for Perturbated Supersonic Flow

(*Applications—two-dimensional, inviscid, isentropic, irrotational, supersonic compressible flow past a thin airfoil at low angles of attack.*) The following method was formulated by Ackeret (1925) and is often referred to as Ackeret's first order (or linear) theory. For Mach numbers between 1.2 and 5.0, equation (3.39) takes the form

$$\beta_1^2 \frac{\partial^2 \tilde{\phi}}{\partial x^2} - \frac{\partial^2 \tilde{\phi}}{\partial z^2} = 0$$

where

$$\beta_1 = \sqrt{M_\infty^2 - 1}$$

This equation has the classical wave equation form and thus has the general solution

$$\tilde{\phi} = \tilde{\phi}_1(x - \beta_1 z) + \tilde{\phi}_2(x + \beta_1 z)$$

By first considering $\tilde{\phi}_2 = 0$ and then $\tilde{\phi}_1 = 0$, it can be shown that lines of constant $\tilde{\phi}$ are in fact Mach waves respectively emanating from the upper and lower airfoil surfaces. Application of the flow tangency boundary condition to an arbitrary thin airfoil at a low angle of attack the velocity disturbances on the upper and lower surfaces are given by

$$\tilde{u} = \pm \frac{Q_\infty}{\beta_1}\left(\alpha - \frac{dz_s}{dx}\right) \quad \text{and} \quad \tilde{w} = \mp Q_\infty \left(\alpha - \frac{dz_s}{dx}\right)$$

where the upper sign applies to the upper surface. Thus, the local pressure coefficient is given from equation (3.40) as

$$C_p = -\frac{2\tilde{u}}{Q_\infty} = \mp \frac{2}{\beta_1}\left(\alpha - \frac{dz_s}{dx}\right) \tag{3.45}$$

If the upper and lower surface coordinates are written in terms of the camber line geometry and a thickness distribution (i.e., $\hat{z}_{u,l} = \hat{z}_c \pm \hat{z}_t$), the following relationships can be formulated:

$$\Delta C_p = \frac{p_l - p_u}{\frac{1}{2}\rho_\infty Q_\infty^2} = \frac{4}{\beta_1}\left(\alpha - \frac{d\hat{z}_c}{d\hat{x}}\right)$$

$$c_l = \int_0^1 \Delta C_p \, d\hat{x} = \frac{4\alpha}{\beta_1} \quad c_{l\alpha} = \frac{4}{\beta_1} \quad \alpha_{l0} = 0$$

$$c_{m,le} = -\int_0^1 \Delta c_p \hat{x} \, d\hat{x} = -\frac{2\alpha}{\beta_1} - \frac{4}{\beta_1}\int_0^1 \hat{z}_c \, d\hat{x}$$

$$c_{m,c/2} = c_{m,le} + \frac{1}{2}c_l = -\frac{4}{\beta_1}\int_0^1 \hat{z}_c \, d\hat{x}$$

$$\frac{x_{cp}}{c} \approx \frac{c_{m,le}}{c_l} = \frac{1}{2} - \frac{1}{2\alpha}\int_0^1 \hat{z}_c \, d\hat{x}$$

It is interesting to note that the following supersonic airfoil characteristics:

1. The lift coefficient is independent of airfoil shape.
2. The zero-lift angle is always zero.
3. The aerodynamic center is always located at the mid-chord position.

Also, drag is produced even though the flow is inviscid. It is known as wave drag, and the drag coefficient is given by

$$c_{d,w} = \frac{4\alpha^2}{\beta_1} + \frac{4}{\beta_1}\left[\int_0^1 \left(\frac{d\hat{z}_c}{d\hat{x}}\right)^2 d\hat{x} + \int_0^1 \left(\frac{d\hat{z}_t}{d\hat{x}}\right)^2 d\hat{x}\right] \tag{3.46}$$

$$= c_{d,i} + c_{d,t}$$

The first term of equation (3.46) is known as the wave drag due to lift and the second two terms taken together are known as the wave drag due to thickness. Equation (3.46) shows that the flat plate has the lowest wave drag and can therefore be regarded as the best supersonic airfoil profile. For a diamond profile (double-wedge) the wave drag has the form

$$c_{d,w} = \frac{4\alpha^2}{\beta_1} + \frac{4\hat{t}^2}{\beta_1}$$

Figure 3.59 illustrates a comparison between Ackeret's theory and wind tunnel data obtained for a biconvex airfoil. The difference between the prediction and the data can be addressed by using higher-order mathematical models such as that of Busemann; further information on higher-order supersonic theories can be found in Hilton (1951).

Drag Divergence Mach Number

As shown in Figure 3.58, it was relatively easy to calculate the critical Mach number for a given airfoil. Figure 3.60 illustrates the variation in drag coefficient, along with the changes in flow field around an airfoil, as the free-stream Mach number is increased past the critical value. For free-stream Mach numbers which lie between the critical and sonic values, the drag coefficient starts to increase rapidly (diverge) due to the formation of shock waves which terminate local pockets of supersonic flow on both upper and lower surfaces of the airfoil. The value of free-stream Mach number at which this phenomena occurs is known as the drag divergence Mach number. Nitzberg and Crandall (1952) observed that as M_∞ was further increased the shock waves grew in strength and eventually gave rise to boundary layer separation. This phenomenon was referred to as shock stall since the lift coefficient started to decrease rapidly at this point. As M_∞ approaches unity, the drag coefficient can easily increase by a factor of 10 or more. When the free-stream flow becomes supersonic (past the sound barrier) the drag coefficient drops due to formation of a system of shock waves which do not induce large amounts of flow separation.

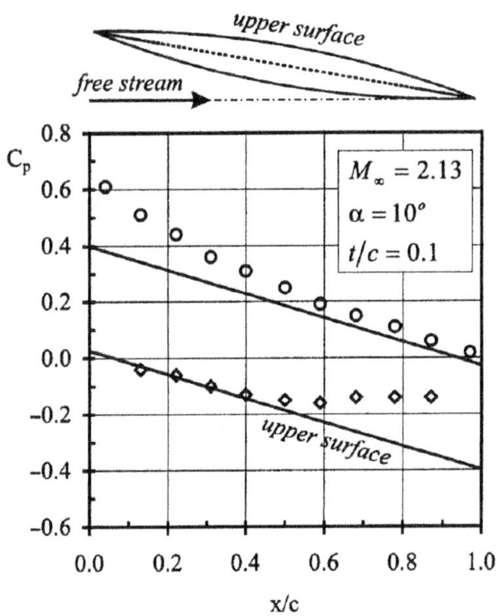

FIGURE 3.59 Ackeret's linear theory (data from Ferri 1940, courtesy of NASA).

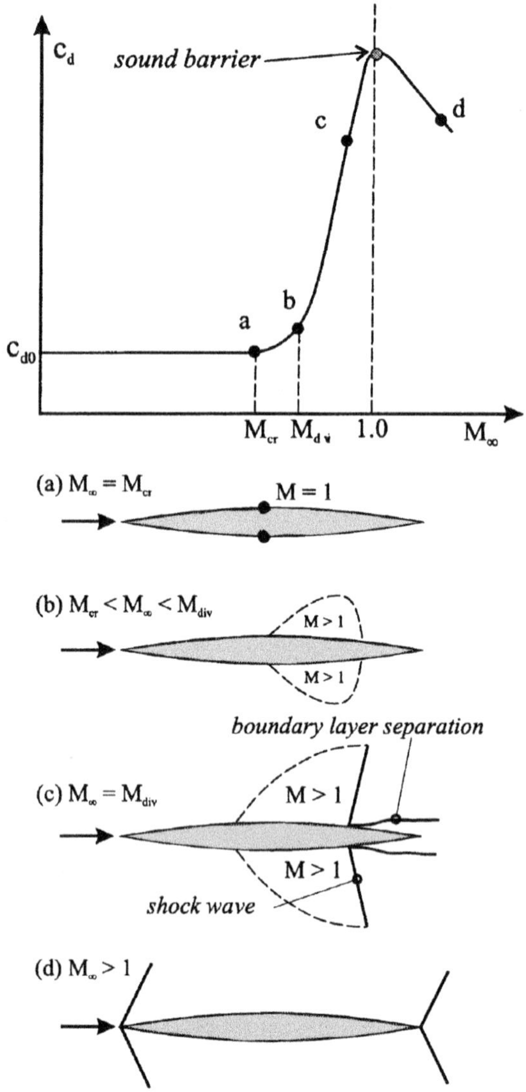

FIGURE 3.60 Drag divergence flow field changes.

The Supercritical Airfoil

Wind tunnel testing has shown that decreasing the airfoil thickness significantly increases M_{cr} (a similar effect can be achieved by using a swept wing, as discussed in Subsection 3.18). However, there are many wing design factors which prohibit the use of very thin airfoil sections. When $M_\infty < M_{cr}$, no shock waves form and the airfoil is termed subcritical. An airfoil profile which has been designed to produce a large upper surface region of low supersonic flow which can be terminated by a weak shock wave is referred to as supercritical and was originally developed by

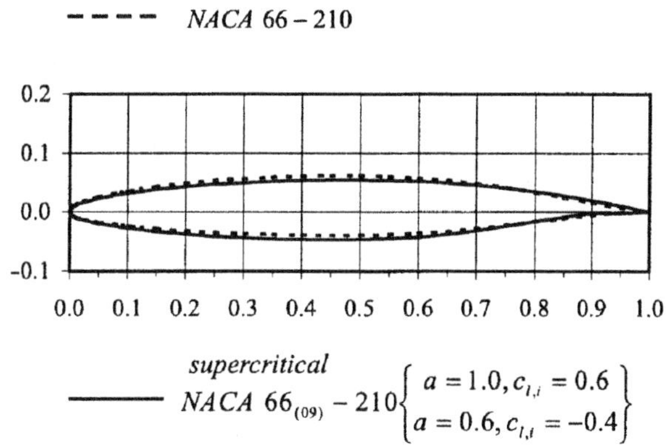

Figure 3.61 The supercritical airfoil.

Whitcomb and Clark (1965). Although initially designed to encourage laminar flow, the NACA 6 series airfoils turned out to have high critical Mach numbers. These profiles have been successfully modified to produce families of supercritical airfoils, as demonstrated by von Doenhoff et al. (1947) and Loftin and Cohen (1948). Figure 3.61 shows that a typical supercritical airfoil has a low upper surface curvature and a concave lower surface at the trailing edge. Figure 3.62 illustrates that, at low angles of attack, the lift of a supercritical airfoil is produced by two mechanisms: a low pressure supersonic region spread over a large portion of the upper surface (known as a roof-top pressure distribution), and a high pressure region over the concave lower surface (known as a rear loading).

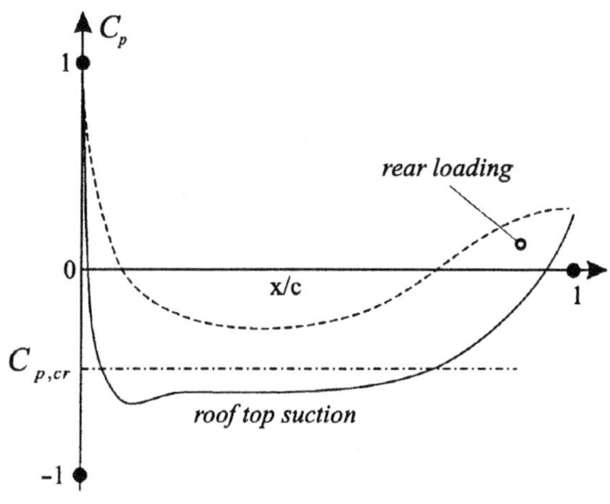

Figure 3.62 Typical supercritical pressure distribution.

3.18 Compressible Flow Over Finite Wings

Subsonic Compressibility Correction Methods

(*Applications—three-dimensional, inviscid, isentropic, irrotational, subsonic compressible flow past a finite wing of arbitrary aspect ratio, thin airfoil section and at low angles of attack.*) Gothert's rule, as described in Subsection 3.17, can be easily extended to three dimensions with the following affine transformation:

$$\hat{x} = \zeta_x x, \quad \hat{y} = \zeta_y y, \quad \hat{z} = \zeta_z z, \quad \hat{\phi}(\hat{x}, \hat{y}) = \zeta_\phi \tilde{\phi}(x, z)$$

The governing equation for the linearized disturbance velocity potential becomes

$$\frac{\partial^2 \hat{\phi}}{\partial \hat{x}^2} + \frac{\partial^2 \hat{\phi}}{\partial \hat{y}^2} + \frac{\partial^2 \hat{\phi}}{\partial \hat{z}^2} = 0$$

if we stipulate that $\zeta_y = \zeta_z$ and $\zeta_z / \zeta_x = \sqrt{1 - M_\infty^2} = \beta_0$.

Flow boundary conditions are satisfied on both the compressible and incompressible planes when

$$(Q_\infty)_{\text{inc}} = Q_\infty, \quad \alpha_{\text{inc}} = \beta_0 \alpha, \quad \text{and} \quad \zeta_z^2 = \zeta_x \zeta_\phi$$

Normally the transformation uses $\zeta_x = 1$, $\zeta_y = \zeta_z = \beta_0$, and $\zeta_\phi = \beta_0^2$. The compressible and incompressible wings are then geometrically related as follows:

$$\frac{\hat{t}_{\text{inc}}}{\hat{t}} = \frac{\hat{h}_{\text{inc}}}{\hat{h}} = \beta_0, \; (c_t / c_r)_{\text{inc}} = c_t / c_r, \; (A_R)_{\text{inc}} = \beta_0 A_R$$

$$\tan \Lambda_{\text{inc}} = \frac{1}{\beta_0} \tan \Lambda, \text{ and } (A_R)_{\text{inc}} \tan \Lambda_{\text{inc}} = A_R \tan \Lambda$$

Shapiro (1953) points out that, for finite wing flows, it is not possible to formulate an equivalent three-dimensional version of the Prandtl–Glauert rule since the local pressure coefficients for two affinely related wings are not just proportional to the airfoil thickness ratio. Thus, the wings in the incompressible and compressible flows will not be of identical geometry. The pressure, lift, and moment coefficients are related as follows:

$$\frac{C_p}{(C_p)_{\text{inc}}} = \frac{C_L}{(C_L)_{\text{inc}}} = \frac{C_m}{(C_m)_{\text{inc}}} = \frac{1}{\beta_0^2} \quad \text{and} \quad \frac{C_{L.\alpha}}{(C_{L\alpha})_{\text{inc}}} = \frac{1}{\beta_0}$$

The wing lift curve slopes are related in the following manner (Hilton 1951):

$$\frac{(C_{L\alpha})}{(C_{L\alpha})_{\text{inc}}} = \frac{\pi A_R + (C_{l\alpha})_{\text{inc}}(1+\tau)}{\pi A_R \beta_0 + (C_{l\alpha})_{\text{inc}}(1+\tau)}$$

It is worth noting that the tail surfaces of an aircraft become less effective than the main wing at high Mach numbers, due to their smaller aspect ratio, and thus may have an impact on the stability of the aircraft.

Mach Cone Flow Classification

As discussed in Subsection 3.16, a source of pressure disturbance in a supersonic flow can only influence the flow within a downstream conical volume extending from the source and whose curved surface is defined by Mach waves. In a similar fashion, the source of disturbance can only be influenced by the flow within a Mach cone facing upstream. Identifying the critical points on a wing from which Mach cones emanate is an important part of all supersonic finite wing theories. This method is used to identify whether a portion of the wing lies in either a subsonic or a supersonic flow and whether it behaves as though the flow is two dimensional or, as discussed below, conical (or cone symmetric). Figure 3.63, in conjunction with Table 3.7, documents examples of this technique.

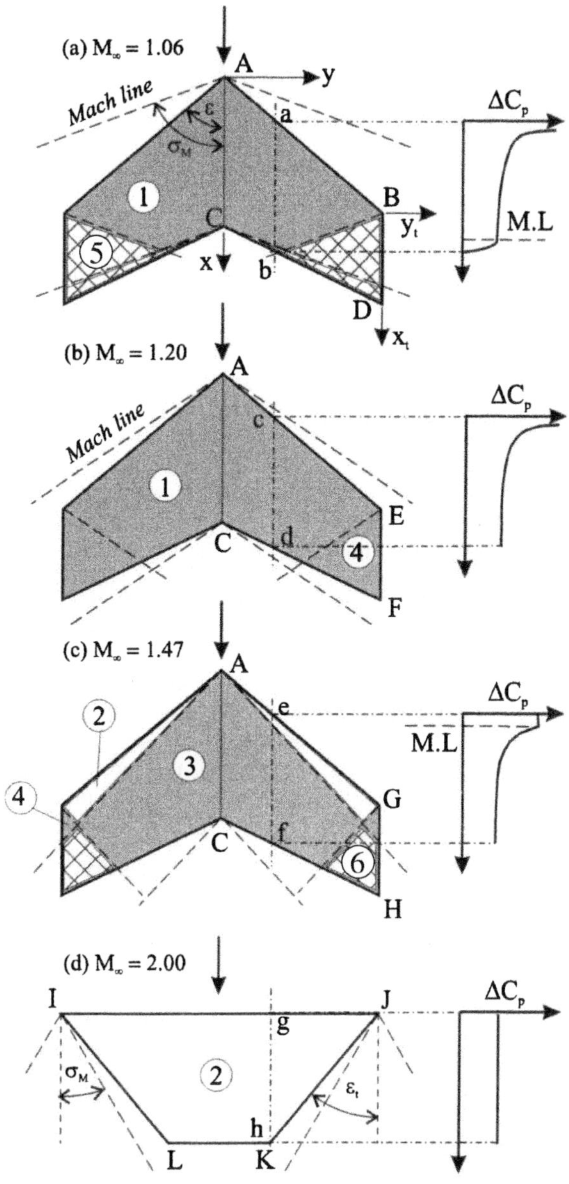

FIGURE 3.63 Examples of Mach cone zoning.

Type of flow	Zones
Subsonic conical flow	1, 3, 4
2D supersonic flow	2
Superposed flow	5, 6

TABLE 3.7 Mach Cone Zoning

Any edge of the wing perimeter can be classified as subsonic or supersonic based on whether the component of the free-stream velocity normal to the edge is respectively less than or greater than the speed of sound. Thus, subsonic edges will lie behind a Mach line, while supersonic edges will lie in front. Referring to Figure 3.63, the following edge number can be defined:

$$m = \frac{\tan \varepsilon}{\tan \sigma_M} = \beta_0 \tan \varepsilon$$

where subsonic edges have $m < 1$ and supersonic edges have $m > 1$. As illustrated in Figure 3.63, subsonic leading and trailing edges have high and zero lift loadings respectively, whilst supersonic edges have finite lift loadings.

The Method of Supersonic Singularities

(Applications—three-dimensional, inviscid, isentropic, irrotational, supersonic compressible flow past a finite wing of arbitrary planform and aerodynamic twist, thin airfoil section and set at low angles of attack.) The linearized governing equations for the perturbation velocity potential in both subsonic and supersonic flow can be solved using Green's theorem along with the concept of source, doublet, and vortex potential functions. When the flow is supersonic, the influence a singularity has on the rest of the flow field is confined to the volume within the Mach cone. Some of the solutions contained within this section were obtained using these methods (see Heaslet and Lomax 1955 for further details).

Conical (or Cone-Symmetric) Flow

(Applications—three-dimensional, inviscid, isentropic, irrotational, supersonic compressible flow past a finite wing whose planform perimeter consists of straight line segments, no aerodynamic twist, thin airfoil section and set at low angles of attack.) The conical flow field was originally conceived by Busemann (1947) and is defined as a flow in which all the flow properties are uniform along rays emanating from a single point. By using the conical flow model along with a suitable coordinate transformation, Busemann converted the linearized supersonic perturbation velocity potential equation into the Laplace equation in polar coordinates. As with the method of singularities, many of the solutions contained within this section were obtained using conical flow methods. Further details can be found in Lagerstrom (1950).

Effect of Wing Sweep

Figure 3.64 shows part of a infinite wing which has been swept back by an angle Λ. There are two types of swept-back wing: bent-back, where the wing is rotated such that the selected airfoil profile remains in a plane normal to the leading edge, and sheared-back, where the chosen profile remains parallel to the free-stream velocity. The effect of sweep relies on the fact that it is only the velocity component normal to the leading edge ($M_\infty \cos \Lambda$) which controls the pressure distribution over the wing section (Jones 1945). Thus, even though the free-stream Mach number may be equal to unity, the value of ($M_\infty \cos \Lambda$) may be small enough to avoid shock wave formation. It is

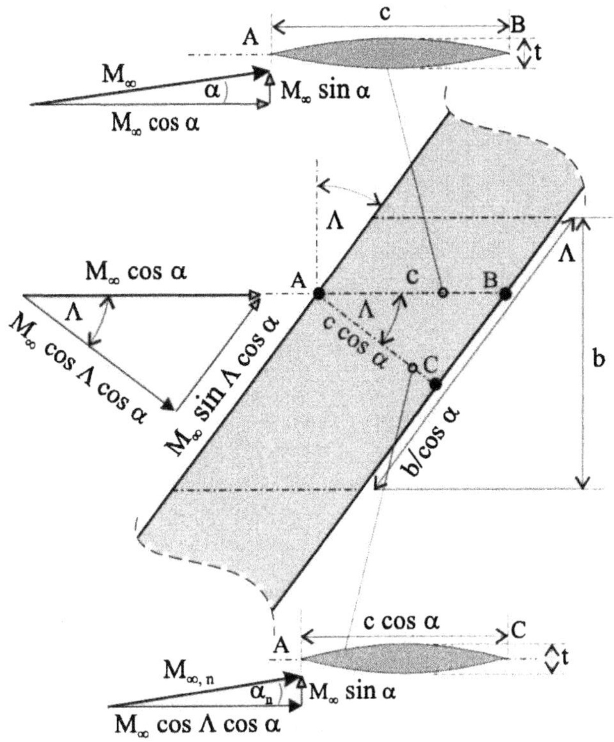

Figure 3.64 Airflow components over a swept wing.

worth noting that shear-back increases the airfoil thickness normal to the leading edge and is thus not as effective as bend-back. Referring to Figure 3.64, the following relationships are obtained for an infinite wing:

$$M_{\infty,n} = M_\infty \sqrt{1-\sin^2 \Lambda \cos^2 \alpha} \approx M_\infty \cos\alpha \tag{3.47}$$

$$\alpha_n = \arctan\left(\frac{\tan\alpha}{\cos\Lambda}\right) \approx \frac{\alpha}{\cos\Lambda} \tag{3.48}$$

$$c_l = (c_l)_n \left(\frac{M_{\infty,n}}{M_\infty}\right) \approx (c_l)_n \cos^2 \alpha \tag{3.49}$$

$$c_{l\alpha} = (c_{l\alpha})_n \cos\alpha$$

$$c_d = (c_d)_n \cos\Lambda \left(\frac{M_{\infty,n}}{M_\infty}\right)^2 + c_{d,f} \tag{3.50}$$

$$\approx (c_{d,p})_n \cos^3 \alpha + c_{d,f}$$

$$\hat{t}_n = \hat{t} / \cos\Lambda \tag{3.51}$$

When the free-stream flow is subsonic, an equation can be derived (Schlichting and Truckenbrodt 1979) which includes the effects of both compressibility and sweep on the local pressure coefficient and is given by

$$C_p = \left(\frac{\cos \Lambda}{\sqrt{1 - M_\infty^2 \cos^2 \Lambda}} \right) (C_{p,n})_{\text{inc}} \tag{3.52}$$

where $(C_{p,n})_{\text{inc}}$ is the incompressible flow pressure coefficient obtained for the airfoil profile normal to the wing leading edge at an angle of attack equal to $\alpha/\cos \Lambda_i$ and referenced to Q_∞. Gothert's rule for a swept infinite wing gives the relationship

$$\cos \Lambda_i = \frac{\beta_0 \cos \Lambda}{\sqrt{1 - M_\infty^2 \cos^2 \Lambda}}$$

Using the Prandtl–Glauert compressibility rule along with equations (3.44) and (3.52) allows the effect of sweep to be clearly illustrated in Figure 3.65.

The effect of sweep when the flow is supersonic can be illustrated by considering an infinite sheared-back wing which has a diamond airfoil profile. Combining Ackeret's theory with equations (3.47) to (3.51) results in the following relationships:

$$c_l = \frac{4\alpha \cos \Lambda}{\sqrt{M_\infty^2 \cos^2 \Lambda - 1}}$$

$$c_d = \frac{4(\alpha^2 + \hat{t}^2) \cos \Lambda}{\sqrt{M_\infty^2 \cos^2 \Lambda - 1}} + c_{d,f}$$

Using these equations, the beneficial effect sweep has on the lift-to-drag ratio is clearly shown in Figure 3.66, where $c_{d,f}$ = 0.006. The main advantage of wing sweep in supersonic flow is the reduction in wave drag due to the reduction in oblique shock wave strength.

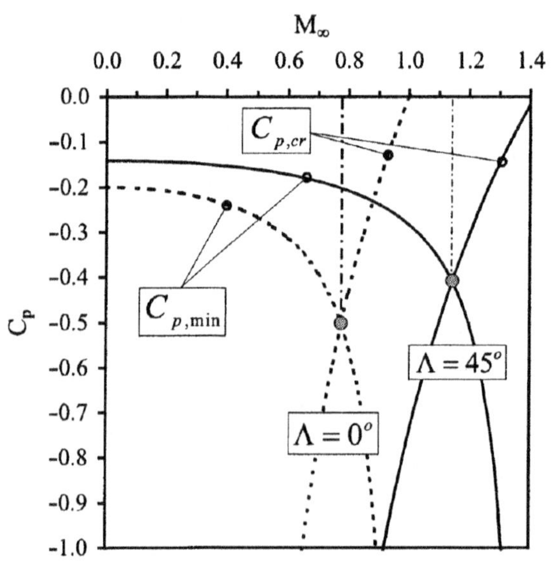

FIGURE 3.65 Swept wing critical Mach number.

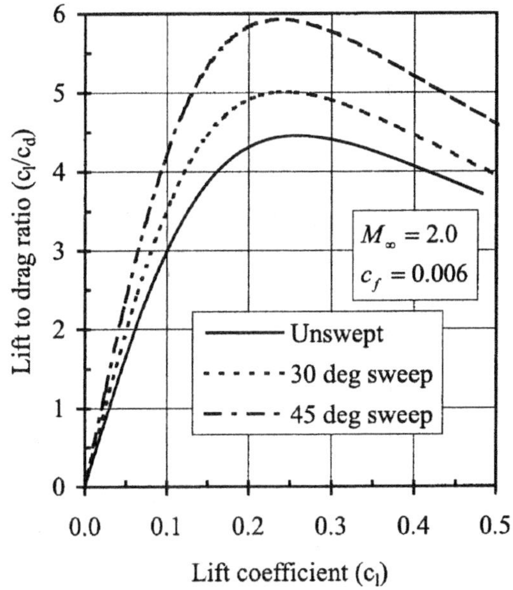

FIGURE 3.66 Effect of sweep on supersonic lift to drag ratio.

Surface Pressure Distributions—Zone 1 Subsonic Leading Edge ($m < 1$)

Figure 3.67 shows the flow over a wing with a flat plate profile and subsonic leading edges. The spanwise pressure distribution over the upper surface can be computed using conical flow theory and is given by

$$\frac{C_p}{C_{p,pl}} = \frac{m}{E(m)} \frac{1}{\sqrt{1-r_0^2}}$$

where

$$r_0 = \frac{y}{x \tan \varepsilon} = \frac{\tan \theta}{\tan \varepsilon}$$

and

$$C_{p,pl} = \mp \frac{2\alpha}{\beta_1} \quad \text{(negative denotes upper surface)}$$

$$E(m) = \int_0^{\pi/2} \sqrt{1-(1-m^2)\sin^2\theta}\, d\theta$$

$$= \frac{\pi}{2}\left\{1 - \left(\frac{1}{2}\right)^2 (1-m^2)^2 - \left(\frac{1\times 3}{2\times 4}\right)^2 (1-m^2)^4 - \cdots\right\}$$

$E(m)$ is known as an elliptic integral of the second kind. The mean value of pressure over the span is given by

$$\frac{\bar{C}_p}{C_{p,pl}} = \frac{m}{E(m)} \frac{\pi}{2}$$

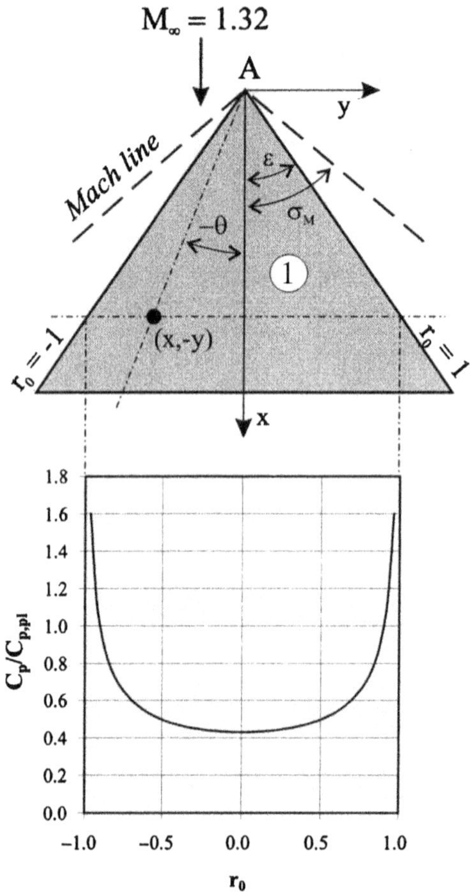

FIGURE 3.67 Subsonic leading edge spanwise distribution of pressure coefficient.

Surface Pressure Distributions—Zone 2 Supersonic Leading Edge (*m* > 1)

Figure 3.68 illustrates the flow over a supersonic edge. The pressure distribution over a swept-back inclined flat plate can be computed directly from Ackeret's theory taking into account the sweep angle, i.e.,

$$C_p = \mp \frac{2\sin\varepsilon}{\sqrt{M_\infty^2 \sin\varepsilon - 1}}\alpha$$

With reference to the flow over an unswept flat plate, the pressure distribution over a swept supersonic edge has the form

$$\frac{C_p}{C_{p,pl}} = \frac{m}{\sqrt{m^2 - 1}} \tag{3.53}$$

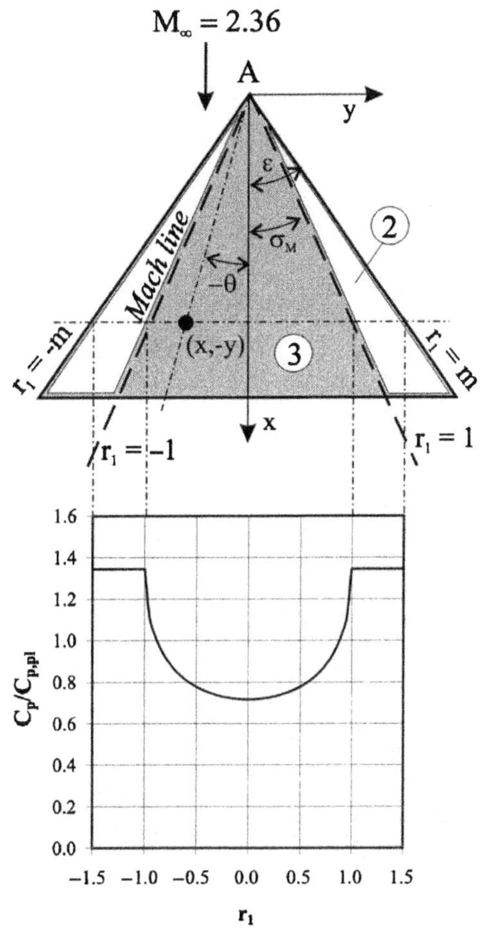

FIGURE 3.68 Supersonic leading edge spanwise distribution of pressure coefficient.

Surface Pressure Distributions—Zone 3 Supersonic Leading Edge, Region behind Mach Line ($m > 1$)

Figure 3.68 illustrates the variation in pressure coefficient after the Mach wave which is downstream of a supersonic leading edge. The variation is given by

$$\frac{C_p}{C_{p,pl}} = \frac{2}{\pi}\frac{m}{\sqrt{m^2-1}}\left(\arccos\sqrt{\frac{1-r_1^2}{m^2-r_1^2}}\right) \tag{3.54}$$

where

$$r_1 = \frac{y}{x\tan\sigma_M} = \frac{\tan\theta}{\tan\sigma_M}$$

Surface Pressure Distributions—Zone 4 Subsonic Side Edge Bounded by a Supersonic Region

A side edge is defined as an edge, on the wing perimeter, which is parallel to the free-stream flow, e.g., side *BD* in Figure 3.63. When the wing leading edge (flat plate profile) is swept, the pressure coefficient is given by

$$\frac{C_p}{C_{p,pl}} = \frac{1}{\pi}\frac{m}{\sqrt{m^2-1}}\left\{\arccos\left(1+2r_3\frac{m+1}{m-r_3}\right)\right\}$$

where

$$r_3 = \frac{y_t}{x_t \tan\sigma_M} = \frac{\tan\theta_t}{\tan\sigma_M}$$

When the side edge is part of an unswept wing, the pressure coefficient becomes

$$\frac{C_p}{C_{p,pl}} = \frac{1}{\pi}\arccos(1+2r_3)$$

Surface Pressure Distributions—Nonconical Zones

Zones which are influenced by more than one Mach cone are indicated by the crosshatched regions in Figure 3.63. These regions involve nonconical flow fields and thus are not covered by the solutions discussed above. However, solutions can be found using a superposition procedure as described by Jones and Cohen (1957).

Airloads—Delta Wing with Subsonic Leading Edge

Once the surface pressure distributions are known, the aerodynamic forces and moments can be calculated. For a delta wing with subsonic leading edges the wing lift coefficient is given by

$$C_L = \frac{m}{E(m)}\frac{2\pi}{\beta_0}\alpha$$

The drag of a wing with a subsonic leading edge is composed of a component due to the lift and the suction force at the leading edge. The drag coefficient is given by

$$C_D = \frac{C_L^2}{\pi A_R}\{2E(m)-\sqrt{1-m^2}\}$$

Airloads—Delta Wing with Supersonic Leading Edge

Interestingly, the mean spanwise pressure coefficient, as obtained from equations (3.53) and (3.54), turns out to be exactly $C_{p,pl}$, which results in both the lift and drag coefficients being equal to the two-dimensional results below:

$$C_L = c_l = \frac{4}{\beta_1}\alpha$$

$$C_D = c_d = \frac{4}{\beta_1}\alpha^2 = \frac{mC_L^2}{A_R}$$

Airloads—Rectangular and Trapezoidal Wings

Figure 3.63(d) illustrates a wing with a trapezium planform which is described as possessing raked tips. The values of lift and wave drag coefficients, expressed as a ratio of Ackeret's two-dimensional values, are given by

$$\frac{C_L}{c_l} = \frac{C_{D,w}}{c_{d,w}} = 1 - \frac{1 - r_4}{2A_R\beta_1}$$

where $r_4 = (\tan \varepsilon_t)/(\tan \sigma_M), c_l = 4\alpha/\beta$, and $c_{d,w}$ is given by equation (3.54). This formula is only applicable if the two side edge Mach cones do not overlap, i.e., when $A_R\beta_1 > 2$.

References

Abbott, I. H. and von Doenhoff, A. E. 1959. *Theory of Wing Sections,* Dover, New York.

Abbott, I. H., von Doenhoff, A. E. and Stivers, L. S. 1945. *Summary of Airfoil Data,* NACA Rpt. 824.

Ackeret, J. 1925. *Air Forces on Airfoils Moving Faster than Sound Velocity,* NACA TM-317.

Ames Research Staff. 1947. *Notes and Tables for Use in the Analysis of Supersonic Flow,* NACA TN-1428.

Ames Research Staff. 1953. *Equations, Tables and Charts for Compressible Flow,* NACA Rpt. 1135.

Anderson, J. D. 1990. *Modern Compressible Flow,* McGraw-Hill, New York.

Anderson, J. D., 1991. *Fundamentals of Aerodynamics,* McGraw-Hill, New York.

Anderson, J.D. 2016. *Fundamentals of Aerodynamics,* 6th edition, McGraw Hill, New York.

Anderson, R. F., 1936. *Determination of the Characteristics of Tapered Wings,* NACA Rpt. 572.

Busemann, A. 1947. *Infinitesimal Conical Supersonic Flow,* NACA TM-110.

Chappell, P. D., 1968. "Flow Separation and Stall Characteristics of Plane Constant-Section Wings in Subcritical Flow," *Journal of the Royal Aeronautical Society,* vol. 72.

Diederich, F. W., 1952. *A Simple Approximate Method for Calculating Spanwise Lift Distributions and Aerodynamic Influence Coefficients at Subsonic Speeds,* NACA TN-2751.

Eskinazi, S. 1967. *Vector Mechanics of Fluids and Magnetofluids,* Academic Press, New York.

Evans, W.T. and Mort, K.W. 1959. *Analysis of computed Flow Parameters for a Set of Sudden Stalls in Low-speed Two-dimensional Flow,* NACA TN D-85.

Falkner, V. M. 1943. *The Calculation of Aerodynamic Loading on Surfaces of Any Shape,* R&M-1910, British A.R.C.

Ferri, A. 1940. *Experimental Results with Airfoils Tested in the High Speed Tunnel at Guidonia,* NACA TM-946.

Gault, D. E. 1957. *A Correlation of Low-Speed, Airfoil-Section Stalling Characteristics with Reynolds Number and Airfoil Geometry,* NACA TN-3963.

Heaslet, M. A. and Lomax, H. 1955. "Supersonic and Transonic Small Perturbation Theory," in *General Theory of High Speed Aerodynamics,* ed. W. R. Sears, High Speed Aerodynamics and Jet Propulsion 6, Princeton University Press, Princeton, NJ.

Hilton, W. F. 1951. *High Speed Aerodynamics,* Longmans, London.

Houghton, E.L. and Brock, A.E. 1970. *Aerodynamics for Engineering Students,* St. Martin's Press, New York.

Houghton, E.L. and Brock, A.E. 1975. *Tables for the Compressible Flow of Dry Air,* Edward Arnold, London.

Jones, R. T. 1941. *Correction of the Lifting-Line Theory for the Effect of the Chord,* NACA TN-817.

Jones, R. T. 1945. *Wing Plan Forms for High-Speed Flight,* NACA Rpt. 863.

Jones, R. T. and Cohen, D. 1957. "Aerodynamics of Wings at High Speeds," in High Speed Aerodynamics and Jet Propulsion 7, ed. A. F. Donovan and H. R. Lawrence, *Aerodynamic Components of Aircraft at High Speeds,* Princeton University Press, Princeton, NJ.

Katz, J. and A. Plotkin. 1991. *Low-Speed Aerodynamics: From Wing Theory to Panel Methods,* McGraw-Hill, New York.

Katz, J. and Plotkin, A. 2001. *Low Speed Aerodynamics: From Wing Theory to Panel Methods,* 2nd edition, McGraw Hill, New York.

Karamcheti, K. 1980. *Principles of Ideal-Fluid Aerodynamics,* Krieger, Malabar, FL.

Lagerstrom, P. A. 1950. *Linearized Supersonic Theory of Conical Wings,* NACA TN-1685.

Loftin, L. K., Jr. and Cohen, K. S. 1948. *An Evaluation of the Characteristics of a 10 Percent Thick NACA 66-series Airfoil Section with a Special Mean Camber Line Designed to Produce a High Critical Mach Number,* NASA TN-1633.

Lugt, H. J. 1995. *Vortex Flow in Nature and Technology,* Krieger, Malabar, FL.

Nitzberg, G. E. and Crandall, S. M. 1952. *A Comparative Examination of Some Measurements of Airfoil Section Lift and Drag at Supercritical Speeds,* NACA TN-2825.

Niven, A. J., 1988. "An Experimental Investigation into the Influence of Trailing-Edge Separation on an Aerofoil's Dynamic Stall Performance" (Ph.D. Thesis, University of Glasgow).

McCullough, G. B. and Gault, D. E. 1951. *Examples of Three Representative Types of Airfoil-Section Stall at Low Speed,* NACA TN-2502.

Oosthuizen, P. H. and Carscallen, W. E. 1997. *Compressible Fluid Flow,* McGraw-Hill, New York.

Prandtl, L. and O. G. Tietjens. 1934. *Fundamentals of Hydro and Aerodynamics,* Dover, New York.

Schlichting, H. 1987. *Boundary Layer Theory.* McGraw-Hill, New York.

Schlichting, H. and Gersten, K. 2016. *Boundary Layer Theory*, 8th edition, McGraw Hill, New York

Schlichting, H. and Truckenbrodt, E. 1979. *Aerodynamics of the Airplane,* McGraw-Hill, New York.

Shapiro, A. H. 1953. *Compressible Fluid Flow,* Ronald Press, New York.

Sommerfeld, A. 1950. *Mechanics of Deformable Bodies,* Academic Press, London.

Stack, J., Lindsey, W. F., and Littell, R. E. 1938. *The Compressibility Burble and the Effect of Compressibility on Pressures and Forces Acting on an Airfoil,* NACA Rpt. 646.

Theodorsen, T. 1931. *Theory of Wing Sections of Arbitrary Shape,* NACA Rpt. 411.

Theodorsen, T. and Garrick, I. E. 1932. *General Potential Theory of Arbitrary Wing Sections,* NACA Rpt. 452.

Vallentine, H. R. 1959. *Applied Hydrodynamics*, Butterworths, London.

Von Doenhoff, A. E., Stivers, L. S., and O'Connor, J. M. 1947. *Low-Speed Tests of Five NACA 66-series Airfoils Having Mean Lines Designed to Give High Critical Mach Numbers,* NASA TN-1276.

Ward, J. W., 1963. "The Behaviour and Effects of Laminar Separation Bubbles on Aerofoils in Incompressible Flow," *Journal of the Royal Aeronautical Society,* vol. 67.

Weissinger, J. 1947. *The Lift Distribution of Swept-Back Wings,* NACA TM-1120.

Whitcomb, R. T. and Clark, L. R. 1965. *An Airfoil Shape for Efficient Flight at Supercritical Mach Numbers,* NASA TMX-1109.

White, F. 1979. *Fluid Mechanics,* McGraw-Hill, New York.

White, F. 2017. *Fluid Mechanics,* 8th edition, McGraw Hill, New York.

White, F. 1991. *Viscous Fluid Flow,* McGraw-Hill, New York.

Wilcox, D. C., 1994. *Turbulence Modeling for CFD,* DCW Industries Inc., La Cañada, CA.

PART 3

Aerodynamics of Low-Aspect-Ratio Wings and Bodies of Revolution

Max F. Platzer

For flows over bodies of revolution, low-aspect-ratio wings, or wing-body configurations with wings of low aspect ratio, methods are available which require little computational effort, yet yield good first estimates of the aerodynamic forces and moments provided the assumptions of small perturbation theory are satisfied.

3.19 Incompressible Inviscid Flow Over a Low-Aspect-Ratio Wing at Zero Angle of Attack

Consider a symmetric low-aspect-ratio wing at zero angle of attack whose thickness distribution is described by

$$z = h(x, y)$$

To be considered a low-aspect-ratio wing, $c > B$ (Figure 3.69).

The flow over this type of body can be modeled by using a distribution of sources in the wing's chord plane $z = 0$. Using the low-aspect-ratio condition, Keune and Oswatitsch (1953) showed that the perturbation velocity potential can be written as

$$\varphi(x,y,z) = U/\pi \int_{-b(x)/2}^{+b(x)/2} h'(x,\eta)\ln\sqrt{(y-\eta)^2+z^2}\,d\eta + \Phi sp$$

where

$$Q(x) = 2\int_{-b(x)/2}^{+b(x)/2} h(x,y)dy$$

$$\text{and}\quad h'(x,y) = \frac{\partial h(x,y)}{\partial x}$$

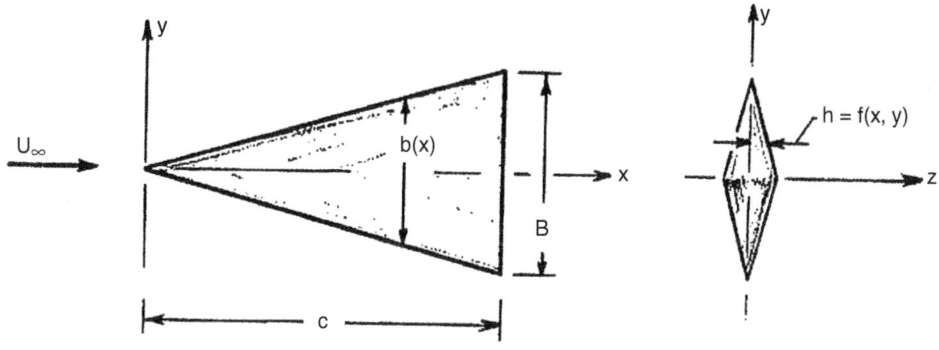

FIGURE 3.69 Low-aspect-ratio wing.

is the cross-sectional area of the wing at the streamwise station x

$$\varphi(x,y,z) = U/\pi \int_{-b(x)/2}^{+b(x)/2} h'(x,\eta) \ln \sqrt{(y-\eta)^2 + z^2}\, d\eta + \Phi sp \tag{3.55}$$

$$\Phi sp = -(U/4\pi) \int_0^c Q'(\xi)/[(x-\xi)^2 + y^2 + z^2]^{1/2} d\xi - (U/2\pi)\, Q'(x) \ln (y^2 + z^2)^{1/2}$$

The first term in the above equation has the form of a two-dimensional source. Therefore, this term represents a distribution of two-dimensional sources in the spanwise direction. These sources produce a cross flow in each y–z plane. Hence, this term is called the cross-flow potential. The other two terms give the influence of those parts of the wing (or wing-body combination or body of revolution) situated further upstream and downstream from the y–z plane being considered. Hence, these terms give the "spatial influence," which we denote as Φsp. It is important to note that the spatial influence depends only on the total cross-sectional area distribution $Q(x)$. It makes no difference how the area is distributed over any given cross section. The result depends only on how the total cross-sectional area varies in the streamwise direction.

The spatial influence can be further simplified by noting that $y^2 + z^2 = r^2$ and evaluating the spatial influence for vanishingly small r, yielding the spatial influence in the form

$$\Phi sp(x) = (U/4\pi)\, Q'(x) \ln(1/4x(c-x)) - (U/4\pi) \int_0^c (Q'(\xi) - Q'(x))/|x-\xi|\, d\xi$$

3.20 Wave Drag

A similar analysis can be performed for supersonic flow. It shows that the cross-flow potential remains unchanged and the spatial influence becomes

$$\Phi sp(x) = (U/2\pi)\left[Q'(x) \ln \beta/2x - \int_0^x [Q'(\xi) - Q'(x)]/(x-\xi)\, d\xi \right]$$

where $\beta^2 = M^2 - 1$.

While in inviscid incompressible or subsonic flow there is no finite drag (d'Alembert's paradox), the integration of the pressure distribution yields a finite drag in supersonic flow, i.e., the wave drag. It turns out that the wave drag is a function *only* of the spatial influence Φsp which is a *function only of the total cross-sectional area distribution* $Q(x)$.

3.21 Equivalence Rule or Area Rule

Keune (1952) and Keune and Oswatitsch (1953) derived the above result using the linearized potential equation for purely subsonic or supersonic flow. Oswatitsch (1952) showed that this conclusion also holds for transonic flow using the transonic small perturbation equation, see also Oswatitsch and Keune (1955).

The *equivalence rule* can be stated in the following form (Ashley and Landahl 1965):

- Far away from a general slender body the flow becomes axisymmetric and equal to the flow around the equivalent body of revolution
- Near the slender body, the flow differs from that around the equivalent body of revolution by a two-dimensional constant-density cross-flow part that makes the flow tangency condition satisfied

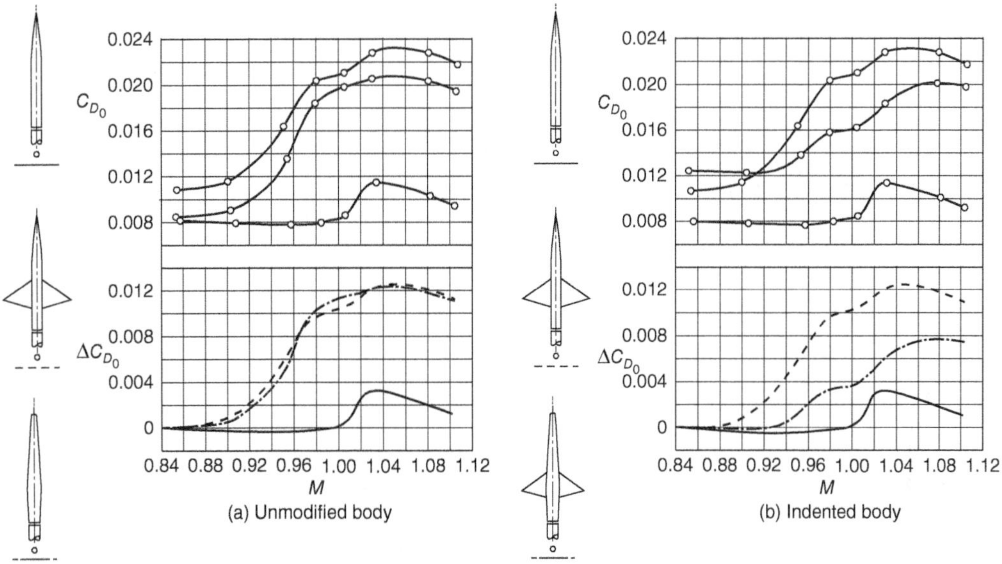

FIGURE 3.70 Whitcomb's measurements (Whitcomb 1956).

- The wave drag is equal to that of the equivalent body of revolution whenever:
    - Either the body ends with an axisymmetric portion
    - Or the body ends in a point or the body ends in a cylindrical portion parallel to the free stream

The importance of the streamwise cross-sectional area variation was first recognized by Otto Frenzl in 1943 in wind tunnel tests carried out at the German Junkers airplane company and documented in the German Patent No. 932410, dated March 21, 1944 (Meyer 2010). Richard Whitcomb independently found this rule in careful wind tunnel tests at the NACA Langley Research Center shown in Figure 3.70. It is now generally referred to as the area rule.

3.22 Bodies of Revolution at Small Angle of Attack

Consider subsonic, transonic or supersonic flow past bodies of revolution at small angle of attack. In the preceding analysis of the flow over nonlifting slender wings and bodies it was necessary to distinguish between subsonic and supersonic flow and to omit any analysis of transonic flow because of the difficulty to solve the nonlinear transonic small perturbation equation. In contrast, the case of lifting flow over slender wings and bodies can be treated with a single theory.

Consider the flow over the body of revolution shown in Figure 3.71. A coordinate system is attached to the nose of the body such that the x-coordinate coincides with the body axis, $r = R(x)$ is the body radius (which is a function of x), and R_B is the base radius.

Assume the body to be moving to the left with velocity U at angle of attack α. Consider a plane *fixed in space* and normal to the flight direction. Figure 3.72 shows the body at three different times as it passes through this plane. Observe that the cross section grows with time and moves downward.

Visualize the flow pattern which is generated by the body *in the fixed plane*. The air must move radially outward in order to accommodate the changing body cross section. Furthermore, the body

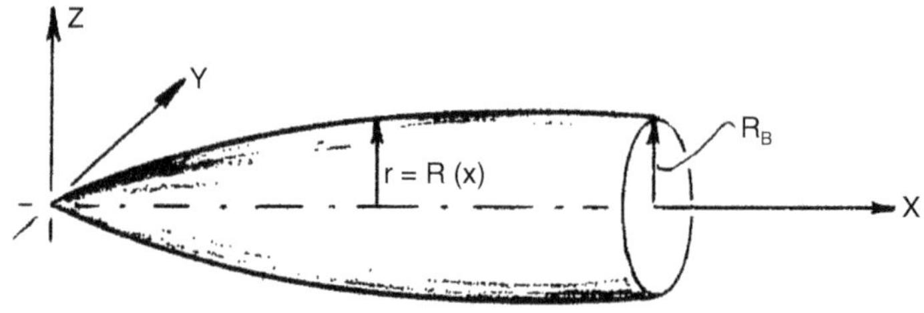

FIGURE 3.71 Body of revolution.

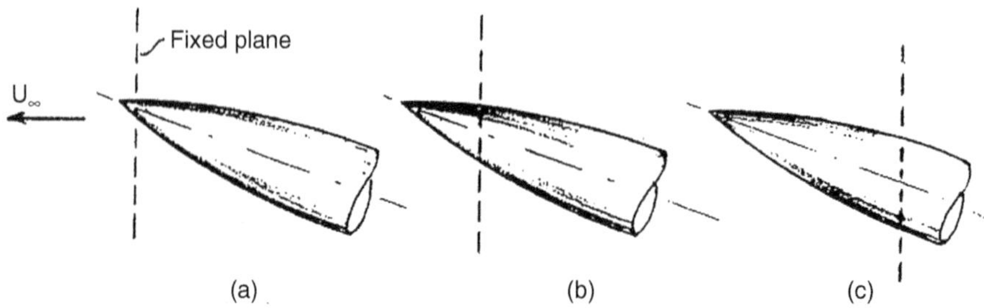

FIGURE 3.72 Body of revolution penetrating plane fixed in space.

axis is seen to move downward with the velocity w_a, (the subscript "a" denoting axis). The velocity w_a is determined by the flight speed and the angle of attack. An observer located in the fixed plane will see a flow pattern generated by a cylinder (of length dx), which is moving with the velocity w_a and whose diameter is growing or shrinking.

Since the angles of attack are assumed to be small and the bodies to be slender the velocities w_a are quite small, even though the flight speed may be supersonic. Hence, the flow problem generated in the fixed plane can be treated as an incompressible flow problem. The preceding analysis of nonlifting flow over slender bodies or wings has already shown the existence of a cross flow and a spatial influence. For the case of lifting flow the spatial influence vanishes. The reason for this simplification becomes clear if one recalls that lifting flows are modeled by distributing doublets instead of sources. Doublet solutions are obtained from source solutions by differentiation in the direction of the doublet axis (in this case the z-axis). Differentiation of equation (3.55) with respect to z leads only to a cross flow (with a doublet distribution) because the spatial influence depends only on the x-coordinate and hence vanishes.

3.23 Cross-Flow Analysis for Slender Bodies of Revolution at Small Angle of Attack

The cross flow is seen to be the same as that generated by a two-dimensional doublet with its axis pointed in the z-direction. Therefore, the required cross-flow analysis merely requires to obtain the solution for incompressible inviscid two-dimensional flow over a cylinder. This flow problem is well known, e.g., Ashley and Landahl (1965). It leads to the following result for the local lift on the body of revolution

$$L'(x) = \rho_\infty U \partial/\partial x \, [R^2(x) \, w_a(x)]\pi$$

and introducing the local air mass $m(x) = \rho_\infty \pi R^2(x)$ the local lift can be expressed as

$$L'(x) = U\partial/\partial x\,[m(x)\, w_a(x)]$$

It is apparent that the local lift is caused by the "convection effect" as the body moves with the speed U through the cross-flow plane. If the body flies at a steady angle of attack α, then w_a is $U\alpha$ for small α. However, if the body is maneuvering or deforming, then w_a is also a function of time and this local rate of change has to be taken into account by writing

$$L'(x, t) = (\partial/\partial t + U\partial/\partial x)\,[m(x)\, w_a(x, t)]$$

Hence

$$L'(x, t) = (d/dt)[m(x)\, w_a(x, t)]$$

One recognizes that the local lift is given by the total rate of change of the momentum given by the expression $m(x)\, w_a(x, t)$ and therefore reflects Newton's second law. The body imparts a change in momentum to a certain amount of air and, as a reaction, the body experiences a lift force. For a body of revolution this amount of air is exactly equal to the air it displaces. This air mass is usually denoted as "apparent" or "virtual" mass.

Integration from nose to tail yields the total lift on the body of revolution at small angle of attack

$$L = \int L'\,dx = \pi\rho_\infty U \int_{\text{nose}}^{\text{tail}} \partial/\partial x[R^2(x) w_a(x)]dx$$

Inserting

$$w_a(x) = U\alpha$$

the total lift becomes

$$L = \pi\rho_\infty U^2\alpha\,[R^2(\text{tail}) - R^2(\text{nose})]$$

If the body has a pointed nose and a flat tail of base radius R_B, the total lift is

$$L = \rho_\infty U^2 \pi\alpha R_B^2$$

If a lift coefficient is introduced using the base area as the reference area, then division by the dynamic pressure $\rho_\infty U^2/2$ and the base area πR_B^2 yields the simple formula

$$C_l = 2\alpha \tag{3.56}$$

This result, first derived by Munk (1924) in his analysis of low-speed flow over airships, leads to the following interesting observations:

- No lift will be generated if the body has a constant diameter.
- No lift will be generated if the body has a pointed nose and tail, although an unstable pitching moment will occur.

Since the development of the above slender body theory has been based on small perturbation theory, the bodies have to be "slender," meaning that a length-to-diameter ratio of 10 or more is desired. Equally importantly, all slopes have to be small and have to change gradually. For example, slender body theory would be applicable to a projectile with a parabolic-arc nose, but not to a cone-cylinder missile.

3.24 Lift on a Slender Wing

R. T. Jones (1946) recognized in 1946 that Munk's analysis also applies to high-speed flows and he extended it to the determination of the lift on a slender low-aspect-ratio wing (Figure 3.73).

The equation for the local lift is obtained as before by the equation

$$L' = U\partial/\partial x\,[m(x)\,w_a(x)]$$

For a wing flying at a small steady angle of attack

$$w_a(x) = U\alpha$$

The "virtual mass" $m(x)$ can be obtained by transforming the wing cross section (a flat plate in this case) into a circle by means of conformal mapping. One finds that the virtual mass is that contained in the circumscribed circle (whose diameter is equal to the local wing span $b(x)$). Hence,

$$m(x) = \rho_\infty \pi b^2(x)/4$$

Assuming a pointed wing, integration from tip to trailing edge then yields the total lift

$$L = \rho_\infty U^2 \pi \alpha B^2/4$$

or, using the definition of the aspect ratio $AR = B^2/S$, where B is the span at the trailing edge and S is the wing area, one obtains

$$L = \rho_\infty U^2 \pi \alpha S\, AR/4$$

Dividing this expression by the dynamic pressure and the wing area S yields the lift coefficient

$$C_l = \pi \alpha AR/2 \tag{3.57}$$

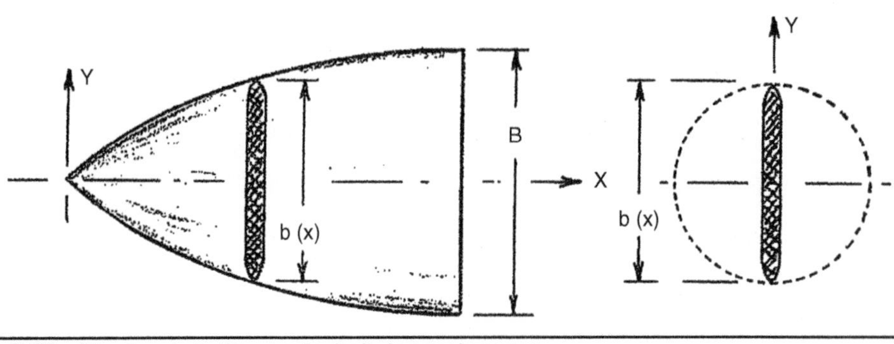

FIGURE 3.73 Slender wing with monotonically increasing span.

The comparison of this equation for lift with experimental results shows that this simple approach gives good results for aspect ratios below unity. For larger aspect ratios one must use lifting surface theory.

A similar analysis for wing-body combinations with a pointed nose, where R and B are the base radius and the wing span at the trailing edge, as previously defined, yields

$$L = \rho_\infty U^2 \pi\alpha\,[B^2/4 - R^2 + 4R^4/B^2]$$

Setting $R = 0$ reduces this equation to the previous wing-alone result. Similarly, setting $R = B/2$ gives the body-alone result.

3.25 Low-Aspect-Ratio Wing-Body Combinations at Large Angle of Attack

On low-aspect-ratio delta wings and other highly swept wings the flow starts to separate along its entire leading edge as the angle of attack is increased and two distinct vortices originate at the wing apex, as shown in Figure 3.74. These vortices generate a low-pressure region on the upper surface and therefore the delta wing experiences a lift, which is significantly greater than predicted by linear theory. However, at a certain angle of attack the vortices start to "burst," as shown in Figure 3.75. This phenomenon is usually referred to as "vortex breakdown." It is accompanied by a rapid drop in lift.

A similar flow phenomenon occurs on bodies of revolution flying at high angle of attack. However, in this case the shape and position of the flow separation line depends on the body's geometry and the specific flow conditions. The resulting flow phenomena and structures are extremely complex even though the configuration may be a simple delta wing or body of revolution. The flows become even more complex when one considers the interactions between the vortices shed from canard-wing-body combinations as is typical for missile configurations. Fortunately, modern computational aerodynamics makes it possible to analyze these highly complex flows, as outlined in the next chapter.

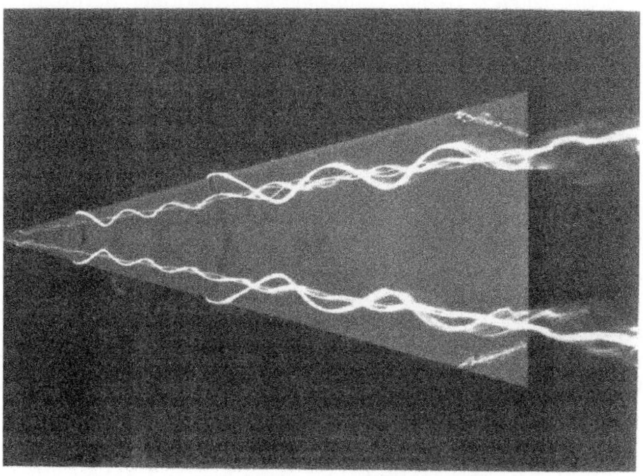

FIGURE 3.74 Flow over delta wing at large angle of attack (van Dyke 1982).

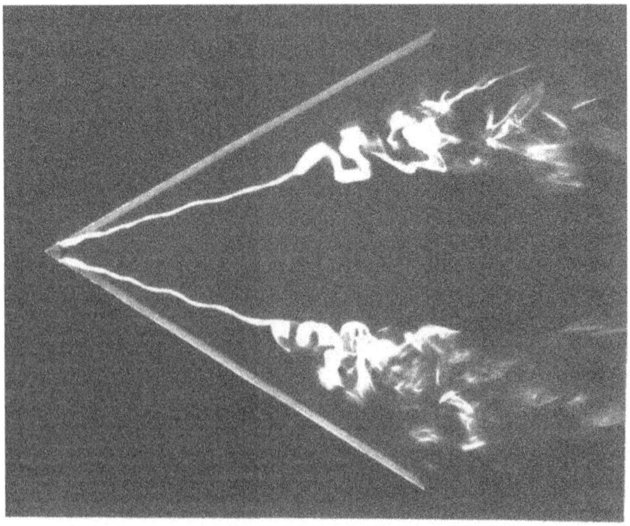

FIGURE 3.75 Vortex burst on delta wing at very large angle of attack (van Dyke 1982).

The reader is referred to Moore (2000) for a compilation of the available empirical and semi-empirical aerodynamic data. Additional comprehensive information can be found in Hemsch (1992), Mendenhall (1992), Rom (1992), Jones (1990), Cebeci (1999), and Cebeci (2005).

References

Ashley, H. and Landahl, M. 1965. *Aerodynamics of Wings and Bodies*, Addison-Wesley Publishing Company. Reading, Massachusetts.

Cebeci, T. 1999. *Modeling and Computation of Boundary-Layer Flows*, Springer. Berlin, Heidelberg, New York.

Cebeci, T. 2005. *Computational Fluid Dynamics for Engineers*, Springer. Berlin, Heidelberg, New York. Berlin, Heidelberg, New YorkHemsch, M. J. (ed.) 1992. *Tactical Missile Aerodynamics: General Topics*, vol. 141, *Progress in Astronautics and Aeronautics*, American Institute of Aeronautics and Astronautics. Washington, D.C.

Jones, R. T. 1946. *Properties of Low-Aspect Ratio Pointed Wings at Speeds below and above the Speed of Sound*, NACA Report 835.

Jones, R. T. 1990. *Wing Theory*, Princeton University Press. Princeton, N.J.

Keune, F. 1952. *Low Aspect Ratio Wings with Small Thickness at Zero Lift in Subsonic and Supersonic Flow*, KTH-AERO TN 21, Royal Institute of Technology, Stockholm, Sweden, Jun.

Keune, F. and Oswatitsch, K. 1953. *Nicht angestellte Koerper kleiner Spannweite in Unter- und Ueberschallstroemung*, Zeitschrift fuer Flugwissenschaften, vol. 1, no. 6, pp. 137–145, Nov.

Mendenhall, M. R. (ed.) 1992. *Tactical Missile Aerodynamics: Prediction Methodology*, vol. 142, Progress in Astronautics and Aeronautics.Washington, D.C.

Meyer, H. U. (ed.) 2010. *German Development of the Swept Wing 1935–1945*, American Institute of Aeronautics and Astronautics.Reston, VA.

Moore, F. G. 2000. *Approximate Methods for Weapon Aerodynamics*, Progress in Astronautics and Aeronautics, vol. 186, American Institute of Aeronautics and Astronautics. Washington, D.C.

Munk, M. M., 1924. *The Aerodynamic Forces on Airship Hulls*, NACA Report 184.

Oswatitsch, K. 1952. *The theoretical Investigations on Transonic Flow in the Aeronautics Department of the Royal Institute of Technology, Stockholm, Sweden*, Proc. 8th International Congress on Theoretical and Applied Mechanics (1952), vol. 1, pp. 261–262, Istanbul.

Oswatitsch, K. and Keune, F. 1955. *Ein Aequivalenzsatz fuer nicht angestellte Fluegel kleiner Spannweite in schallnaher Stroemung, Zeitschrift fuer Flugwissenschaften*, vol. 3, no. 2, pp. 29–46.

Rom, J. 1992. *High Angle of Attack Aerodynamics*, Springer. Berlin, Heidelberg, New York.

van Dyke, M. 1982. *An Album of Fluid Motion*, The Parabolic Press, Stanford, California.

Whitcomb, R. T. 1956. *A Study of Zero-Lift Drag Rise Characteristics of Wing-Body Combinations near the Speed of Sound*, NACA Report 1273.

PART 4

Computational Aerodynamics

John A. Ekaterinaris

Computational fluid dynamics (CFD) has been evolved over the years as a stand-alone discipline of computational sciences. The genesis of CFD can be traced back to the early 1900s. CFD started evolving more rapidly with the increase of the computational power (Figure 3.76) in the 1980s. Since then the computational power keeps increasing year after year and the numerical algorithms are improving. The introduction of parallel computing and computer clusters in the late 1990s made available more and cheaper computational capabilities not only to large research establishments but also to industries and universities. As a result, CFD became a readily available tool for design and for investigation of new concepts in fluid dynamics. There are many review articles and books for CFD. In this short review, the numerical algorithms and basic discretization approaches developed for incompressible and compressible flow of importance in low- and high-speed aerodynamics are presented. The interested reader can find more details in the literature given in the bibliography (including most of the classical texts in CFD) and references therein.

3.26 Governing Equations

The Navier–Stokes (NS) equations are macroscopic descriptions of the conservation laws for mass momentum and energy. The main objective of CFD methods for compressible flow is to use discretization approaches and techniques to obtain numerical solution of the time dependent NS equations. For compressible flow, e.g., for flows at Mach number M greater than 0.3, and in the absence of body forces the NS equations can be derived (Landau and Lifshitz 1987; Bachelor 2000; Panton 2013). For a stationary control volume Ω with respect to Eulerian reference frame with boundary $\partial\Omega$ enclosing the volume Ω, and outward normal vector $\mathbf{n}$ to $\partial\Omega$ the NS equations are

$$\frac{\partial}{\partial t}\iiint_{\Omega} \rho \, d\Omega + \oiint_{\partial\Omega} \mathbf{n}\cdot\rho\mathbf{u}\, dS = 0 \tag{3.58}$$

$$\frac{\partial}{\partial t}\iiint_{\Omega} \rho\mathbf{u} \, d\Omega + \oiint_{\partial\Omega} \mathbf{n}\cdot\rho\mathbf{u}\otimes\mathbf{u}\, dS = \oiint_{\partial\Omega} \mathbf{n}\cdot\vec{\vec{\tau}}\, dS \tag{3.59}$$

$$\frac{\partial}{\partial t}\iiint_{\Omega} \rho e \, d\Omega + \oiint_{\partial\Omega} \mathbf{n}\cdot\rho e\mathbf{u}\, dS = \oiint_{\partial\Omega} \mathbf{n}\cdot(\mathbf{u}\cdot\vec{\vec{\tau}} - \mathbf{q})\, dS \tag{3.60}$$

This form of the equations governing compressible flow is the control volume form of the governing equations and it is the basis for the finite-volume methods that will be presented later.

The differential form of the governing equations is obtained after application of the Gauss divergence theorem to replace the surface integrals in equations (3.58)–(3.60) with volume integrals. The differential form of the NS equations is the starting point for the application of finite-difference and finite element discretizations. The vector and tensor forms of the NS equations for compressible flow are

$$\frac{\partial \rho}{\partial t} + \nabla\cdot\rho\mathbf{u} = 0 \qquad \frac{\partial \rho}{\partial t} + \frac{\partial \rho u_j}{\partial x_j} = 0 \tag{3.61}$$

$$\frac{\partial \rho\mathbf{u}}{\partial t} + \nabla\cdot(\rho\mathbf{u}\otimes\mathbf{u} - \vec{\vec{\tau}}) = 0 \qquad \frac{\partial \rho u_i}{\partial t} + \frac{\partial}{\partial x_j}(\rho u_i u_j - \tau_{ij}) = 0 \tag{3.62}$$

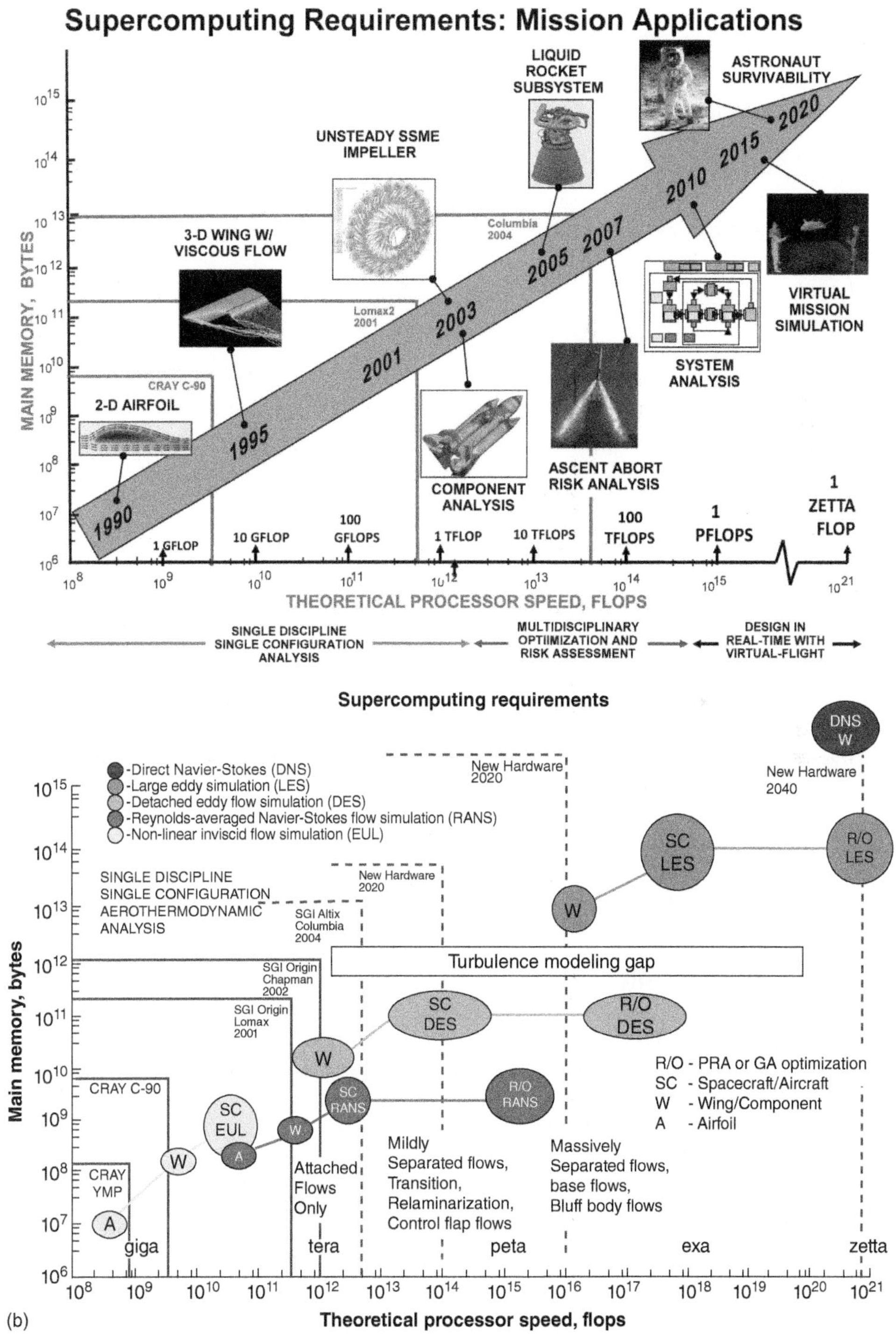

FIGURE 3.76 (*a*) Increase of computing power in the early years; (*b*) current and projected capabilities.

$$\frac{\partial \rho e}{\partial t}+\nabla\cdot(\rho e\mathbf{u}-\mathbf{u}\cdot\vec{\vec{\tau}}+\mathbf{q})=0 \qquad \frac{\partial \rho e}{\partial t}+\frac{\partial}{\partial x_j}((\rho e+p)u_j-u_i\tau_{ij}+q_j)=0 \tag{3.63}$$

where ρ is the fluid density, $\mathbf{u}=(u,v,w)$ or $u_j=(u_1,u_2,u_3)$ is velocity, $\mathbf{q}$ is heat transfer obtained by Fourier's law for heat conduction $\mathbf{q}=-\kappa\nabla T$, and $\vec{\vec{\tau}}$ is the stress tensor

$$\vec{\vec{\tau}}=-(p+\frac{2}{3}\mu\nabla\cdot\mathbf{u})I+\mu(\nabla+\nabla^T)\mathbf{u} \quad \text{or} \quad \tau_{ij}=-p+\mu S_{ij},\, S_{ij}=\frac{\partial u_i}{\partial x_j}+\frac{\partial u_j}{\partial x_i}-\frac{2}{3}\delta_{ij}\frac{\partial u_k}{\partial x_k}$$

Far from solid walls the inviscid flow approximation is valid and the inviscid flow equations, known as Euler equations, are obtained when the viscous terms in the momentum and energy equations are ignored.

The incompressible flow equations can be obtained from the compressible flow equations employing the constant density and internal energy assumptions and discarding the energy equation. The vector and tensor forms of the incompressible flow equations are

$$\nabla\cdot\mathbf{u}=0 \qquad \frac{\partial u_j}{\partial x_j}=0 \tag{3.64}$$

$$\frac{\partial \mathbf{u}}{\partial t}+\nabla\cdot(\mathbf{u}\otimes\mathbf{u})=-\frac{1}{\rho}\nabla p+\frac{\mu}{\rho}\nabla^2\mathbf{u}=-\frac{1}{\rho}\nabla p+\nu\nabla\cdot(\nabla\mathbf{u}) \qquad \frac{\partial u_i}{\partial t}+u_j\frac{\partial u_i}{\partial x_j}=-\frac{1}{\rho}\frac{\partial p}{\partial x_i}+\nu\frac{\partial^2 u_k}{\partial x_k^2} \tag{3.65}$$

The incompressible flow equations must be used to obtain numerical solutions for problems in low speed aerodynamics such as wind turbine rotor aerodynamics and flapping wings of micro air vehicles. The derivation of the Navier–Stokes equation for compressible and incompressible flow can be found in fluid mechanics textbooks (Landau and Lifshitz 1987; Bachelor 2000; Panton 2013).

Due to the nonlinearity of the governing equations the few available analytical solutions are for one- or two-dimensional simple problems of limited interest to practical applications. Therefore, the governing equations must be discretized on a numerical mesh before the application of any methods. As a result grid generation in nontrivial geometries of interest to aerodynamic applications is the first task before the discretization of governing equation with any available method.

3.27 Grid Generation

The governing equations, Euler, or NS must be discretized on a numerical mesh and a domain with finite extent. In addition, at the boundaries of the domain appropriate boundary conditions must be specified. The numerical meshes can be structured or unstructured. The Cartesian mesh (Figure 3.77, top) is the simplest type of mesh and it can be used for a limited number of simple canonical problems, such as the driven lid cavity problem. Even for this problem it is beneficial to use some type of grid stretching in the near wall region (Figure 3.77, bottom) in order to better resolve the near wall steep flow gradients. Cartesian and Cartesian-type stretched meshes can be easily constructed for simple two and three-dimensional problems of little interest to practical applications. For Cartesian meshes finite difference (FD) and finite volume (FV) discretizations are more efficient and this feature has been exploited in immersed boundary methods (Sotiropoulos and Yang 2014) that can be used to obtain numerical solutions over complex aerodynamic bodies at moderate Reynolds numbers.

The bulk of the available CFD software uses structured or block structured body-fitted (Thompson 1999) non-Cartesian meshes (Figure 3.78), referred to as generalized coordinate meshes.

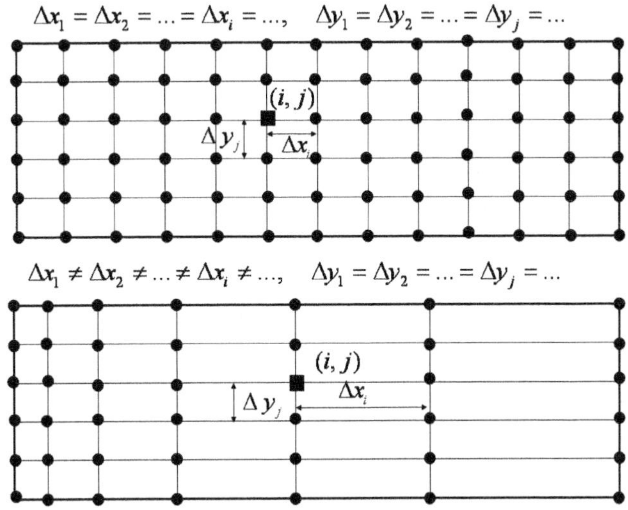

FIGURE 3.77 Cartesian mesh (top) stretched quadrilateral mesh (bottom).

For finite difference methods, the solution variables are considered on the nodes of the mesh, and in order to make possible the numerical solution of the governing equations on body-fitted meshes of the physical domain, generalized transformations of coordinates are applied and the numerical solution is performed on the transformed canonical domain (Pulliam and Steger 1980), as shown in Figure 3.79. Generation of body-fitted meshes over complex configurations, referred to as grid generation (Thompson 1999), is a nontrivial task more like an art or a science. Grid generation, which is the first step before the application of any CFD approach, is carried out with free, for example, Gmsh

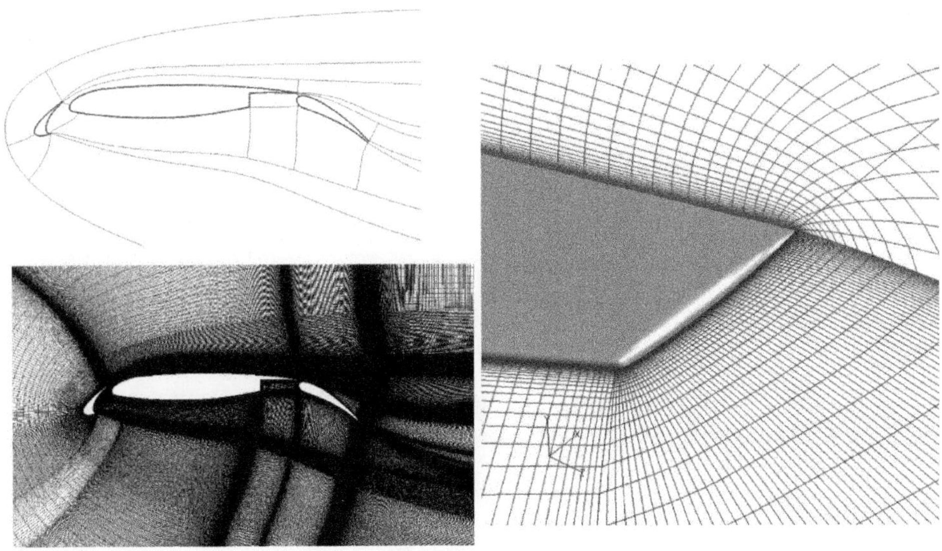

FIGURE 3.78 Block structured body-fitted mesh for NS solutions over a multi-element airfoil (left); detail of the wing tip mesh of single block structured mesh over a wing (right).

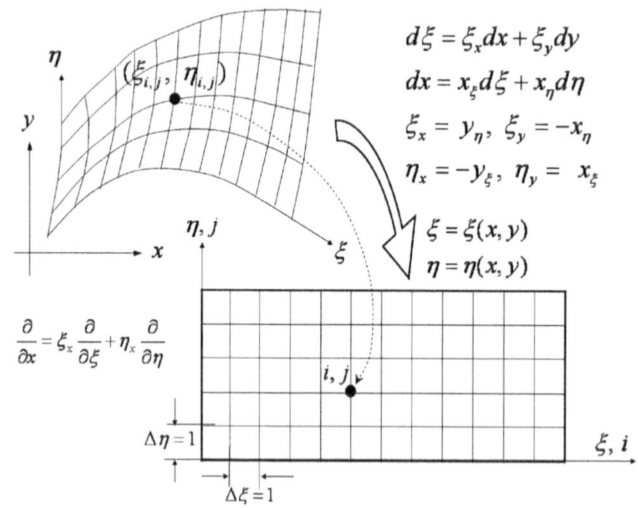

FIGURE 3.79 Generalized coordinate transformations in two dimensions. The derivatives are expressed in the canonical transformed domain using the chain rule and the numerical solution is performed in the transformed Cartesian domain.

http://gmsh.info/, or commercial software; see, for example, http://www.pointwise.com and https://www.betacae.com/.

Mesh generation over complex configurations of interest to aerodynamics (Figure 3.80) is facilitated with unstructured meshes.

Unstructured meshes offer better distribution of computational cells, and they are well suited for adaptive mesh refinement, which can be used to reduce the cost of large-scale computations without compromising numerical accuracy. For example, Figure 3.81 shows that the lambda-shock structure over the ONERA M6 wing can be better resolved with an unstructured mesh.

For structured body-fitted meshes, the computational elements are hexahedra and the solution variables are at the vertices of the hexahedra. Unstructured meshes can support different element types (tetrahedra, prisms, pyramids, and hexahedra) and they can be of mixed type, e.g., prismatic-tetrahedral. The solution variables for finite volume methods on structured or unstructured meshes can be the centers of the elements (Figure 3.82, left) or the vertices of the elements (Figure 3.82, right).

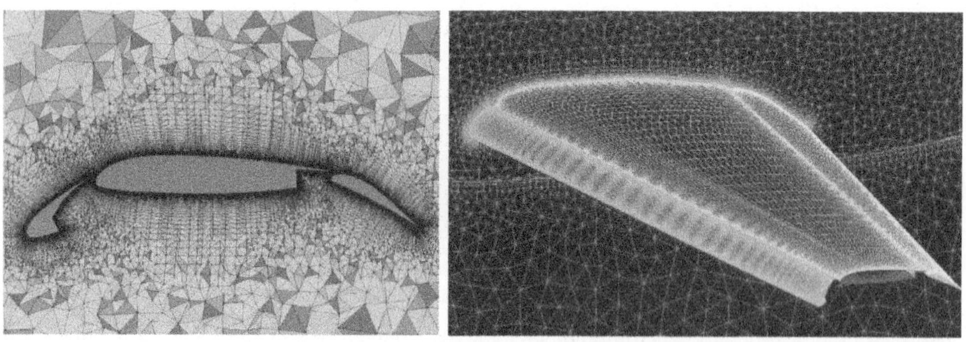

FIGURE 3.80 Unstructured mesh over a multi-element airfoil (left) and a multi element wing (right).

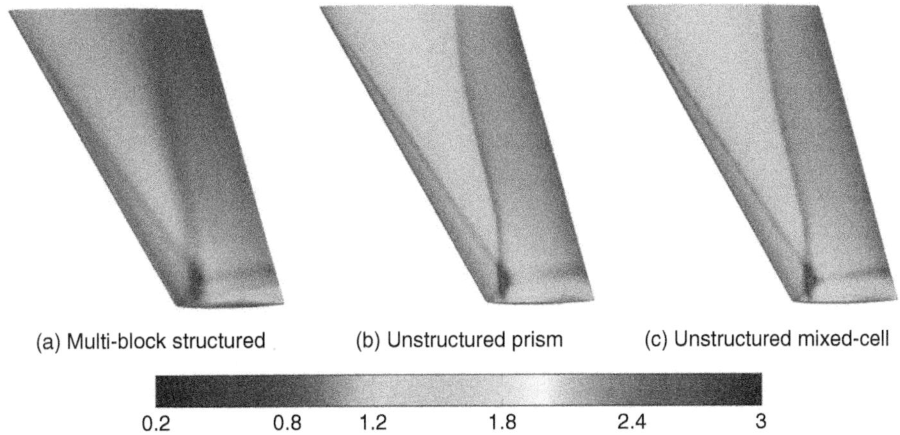

Figure 3.81 Surface pressure distributions over the ONERA M6 wing obtained with different meshes.

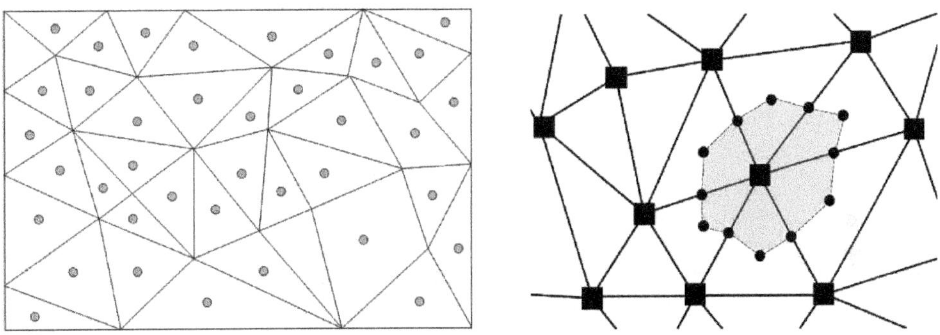

Figure 3.82 Cell-centered (left) and vertex based (right) examples of a triangular mesh.

3.28 CFD Methods for the Compressible Navier-Stokes Equations

There exists a large number of freely available and commercial CFD software for the numerical solution of the compressible NS equations. The majority of the available CFD software is based on the finite volume discretization approach. For a number of reasons the finite volume (FV) methods gained popularity over the finite difference (FD) or the finite element (FE) methods. The finite volume method is applied to the control volume form of the governing equations and it is by construction conservative both on the cell level and globally. This property is important for shock capturing and many shock capturing schemes were developed for the finite volume discretization. The FV is well suited for unstructured meshes, where it is possible to apply adaptive mesh refinement by subdividing cells in the areas of interest without excessively increasing the total number of cells, as for example in structured meshes with FD methods.

Finite Difference Methods

Finite difference methods use the differential form of the governing equations expressed on an arbitrary curvilinear coordinates space (ξ, η, ζ, τ), shown in Figure 3.79, for two dimensions. The transformed equations are not more complicated than the Cartesian form of equations (3.61)–(3.63), they retain the strong conservation law form, and in nondimensional form are written as

$$\partial_\tau \hat{\mathbf{q}} + \partial_\xi(\hat{\mathbf{E}}^{\mathrm{i}} - \hat{\mathbf{E}}^{\mathrm{v}}) + \partial_\xi(\hat{\mathbf{F}}^{\mathrm{i}} - \hat{\mathbf{F}}^{\mathrm{v}}) + \partial_\xi(\hat{\mathbf{G}}^{\mathrm{i}} - \hat{\mathbf{G}}^{\mathrm{v}}) = 0 \tag{3.66}$$

where

$$\hat{\mathbf{q}} = \frac{1}{J}\begin{bmatrix} \rho \\ \rho u \\ \rho v \\ \rho w \\ \rho e \end{bmatrix}, \quad \hat{\mathbf{E}}^{\mathrm{i}} = \frac{1}{J}\begin{bmatrix} \rho U \\ \rho u U + \xi_x p \\ \rho v U + \xi_y p \\ \rho w U + \xi_z p \\ (\rho e + p)U - \xi_t p \end{bmatrix}, \quad \hat{\mathbf{E}}^{\mathrm{v}} = \frac{1}{J}\begin{bmatrix} 0 \\ \xi_x \tau_{xx} + \xi_y \tau_{xy} + \xi_z \tau_{xz} \\ \xi_x \tau_{xx} + \xi_y \tau_{xy} + \xi_z \tau_{xz} \\ \xi_x \tau_{xx} + \xi_y \tau_{xy} + \xi_z \tau_{xz} \\ \xi_x \beta_x + \xi_y \beta_y + \xi_z \beta_z \end{bmatrix} \tag{3.67}$$

with analogous definitions for the other viscous and inviscid fluxes, see, e.g., Pulliam and Steger 1980, Plecher et al. 2013, and Lomax et al. 2011 for the definitions of other flux vectors, the metric terms ξ_x, η_x, ζ_x, etc., and the Jacobian of the transformation J.

In these equations, $U = \xi_t + \xi_x u + \xi_y v + \xi_z w, V, W$ are the contravariant components and τ_{xx}, τ_{xy}, etc., are the components of the stress tensor in Cartesian coordinates. For example, $\tau_{xy} = \tau_{yx} = \mu(u_x + v_y)$, where the Cartesian derivatives are obtained via the chain-rule relations $u_x = \xi_x u_\xi + \eta_x u_\eta + \zeta_x u_\zeta$ and for the equally spaced $\Delta\xi = \Delta\eta = \Delta\zeta = 1$ transformed domain (ξ, η, ζ) the derivatives are evaluated with simple central difference, second-order accurate formulas.

$$\left.\frac{\partial u}{\partial \xi}\right|_{i,j} = \frac{u_{i+1,j,k} - u_{i-1,j,k}}{2(\Delta\xi)} + 0(\Delta x^2)$$

The main reason that finite difference discretizations were successful in obtaining solutions for three-dimensional flow problems (Pulliam and Steger 1980) was that implicit time stepping became possible through efficient implementation in the vector machines of the 1980s of the approximate factorization algorithm of Beam and Warming (1976). The factorized Beam–Warming algorithm is

$$\begin{aligned}&\left(I + h\delta_\xi \widehat{A} - \varepsilon_{\mathrm{impl}} J^{-1}\nabla_\xi\Delta_\xi J\right)\times\left(I + h\delta_\eta \widehat{B} - \varepsilon_{\mathrm{impl}} J^{-1}\nabla_\eta\Delta_\eta J\right)\times\left(I + h\delta_\zeta \widehat{C} - \varepsilon_{\mathrm{impl}} J^{-1}\nabla_\zeta\Delta_\zeta J\right)\\ &(\hat{q}^{n+1} - \hat{q}^n) - (\Delta t)\left(\delta_\xi \hat{E}^n + \delta_\eta \hat{F}^n + \delta_\zeta \hat{G}^n\right) - \varepsilon_{\mathrm{expl}} J^{-1}\left[(\nabla_\xi\Delta_\xi)^2 + (\nabla_\eta\Delta_\eta)^2 + (\nabla_\zeta\Delta_\zeta)^2\right] J\,\hat{q}^n\end{aligned} \tag{3.68}$$

where $\widehat{A}$, $\widehat{B}$, $\widehat{C}$ are the flux Jacobian matrices resulting from time linearization of the flux vectors and they are derived in detail in Hirsch (2002) and they are given by

$$\begin{matrix}\widehat{A},\\ \widehat{B},\\ \text{or}\,\widehat{C}\end{matrix} = \begin{bmatrix} k_0 & k_1 & k_2 & k_3 & 0 \\ k_1\varphi^2 - u\theta & k_0 + \theta - k_1(\gamma-2)u & k_2 u - k_1(\gamma-1)v & k_3 u - k_1(\gamma-1)v & k_1(\gamma-1) \\ k_2\varphi^2 - v\theta & k_1 v - k_2(\gamma-1)u & k_0 + \theta - k_2(\gamma-2)v & k_3 v - k_2(\gamma-1)w & k_2(\gamma-1) \\ k_3\varphi^2 - w\theta & k_1 w - k_3(\gamma-1)u & k_2 w - k_3(\gamma-1)v & k_0 + \theta - k_3(\gamma-2)w & k_3(\gamma-1) \\ \theta[2\varphi^2 - \gamma e] & k_1[\gamma e - \varphi^2] - (\gamma-1)u\theta & k_2[\gamma e - \varphi^2] - (\gamma-1)v\theta & k_2[\gamma e - \varphi^2] - (\gamma-1)w\theta & k_0 + \gamma\theta \end{bmatrix}$$

where $\varphi^2 = 0.5(\gamma-1)(u^2 + v^2 + w^2)$, $\theta = k_1 u + k_2 v + k_3 w$ and k_0, k_1, k_2, k_3 stand for metric terms, e.g., for $\widehat{B}$, $k_0 = \eta_t$, $k_1 = \eta_x$, $k_2 = \eta_y$, $k_3 = \eta_z$. The flux Jacobian matrices $\widehat{A}$, $\widehat{B}$, $\widehat{C}$ have real eigenvalues and a complete set of eigenvectors, they can be diagonalized as $\widehat{A} = \hat{T}\,\Lambda\,\hat{T}^{-1}$, and they are used in this form for the construction of upwind schemes.

The Beam–Warming algorithm allows to overcome stability limitations imposed by explicit time marching schemes, such as explicit Runge–Kutta methods, and in the form of equation (3.68) contains second-order implicit $\varepsilon_{\text{impl}} J^{-1}\nabla_\xi \Delta_\xi J$ and fourth-order $\varepsilon_{\text{expl}} J^{-1}\left[(\nabla_\xi \Delta_\xi)^2 + (\nabla_\eta \Delta_\eta)^2 + (\nabla_\zeta \Delta_\zeta)^2\right] J\,\hat{q}^n$ explicit dissipations with constant coefficients $\varepsilon_{\text{impl}}$, $\varepsilon_{\text{expl}}$, to stabilize the calculations and avoid unphysical overshoots at steep gradients and to eliminate numerical overshoots due to the Gibbs phenomenon at solution discontinuities. The dissipation coefficients, $\varepsilon_{\text{impl}}$, $\varepsilon_{\text{expl}}$, for finite volume methods, were modified by Jameson et al. (1981) and they are spatially varying depending on the computed solution as

$$\varepsilon_{\text{impl}} = \varepsilon^{(2)}_{i+1/2,j} = \kappa^{(2)} \max(\nu_{i+1,j}, \nu_{i,j}), \quad \varepsilon_{\text{expl}} = \varepsilon^{(4)}_{i+1/2,j} = \max\left(0, (\kappa^{(4)}, \varepsilon^{(2)}_{i+1/2,j})\right)$$

$$\nu_{i,j} = \frac{|p_{i+1,j} - 2p_{i,j} - p_{i-1,j}|}{|p_{i+1,j}| + |2p_{i,j}| + |p_{i-1,j}|}, \quad \kappa^{(2)} = 1/4, \quad \kappa^{(4)} = 1/265$$

The Beam–Warming (BW) algorithm with variants, such as the diagonalized BW algorithm and the BW algorithm with Steger–Warming (1981) flux vector splitting, has been the workhorse of numerical simulations of complex configurations carried out at NASA. It has been implemented in the OVERFLOW code (Nichols and Buning 2010) https://overflow.larc.nasa.gov/ that is still in use in many fields of aerodynamics. The OVERFLOW code includes Chimera Grid Tools that facilitate meshing over very complex geometries by allowing overset grids, shown in Figure 3.83. According to the Chimera approach developed by Steger structured meshes are constructed over components of the complex geometry, such as the parts of the fuselage, the tail, and flaps. These meshes can intersect with neighboring meshes and a background Cartesian-type mesh. The computations are carried out on each individual mesh and at the intersection information is exchanged by interpolating data from the neighboring mesh and imposing it as boundary condition. Recent addition of adaptive mesh refinement capabilities (Yee 1985) in the OVERFLOW code make it more suitable for computations of flows with embedded complex flow features, such as wing tip vortices, that require high-mesh resolution.

Finite difference central second-order accurate discretizations of the conservation law form of equation (3.67) require artificial dissipation. However, a second or higher order of accuracy can be achieved by using upwind schemes (Yee 1985) and higher than second-order evaluation of the

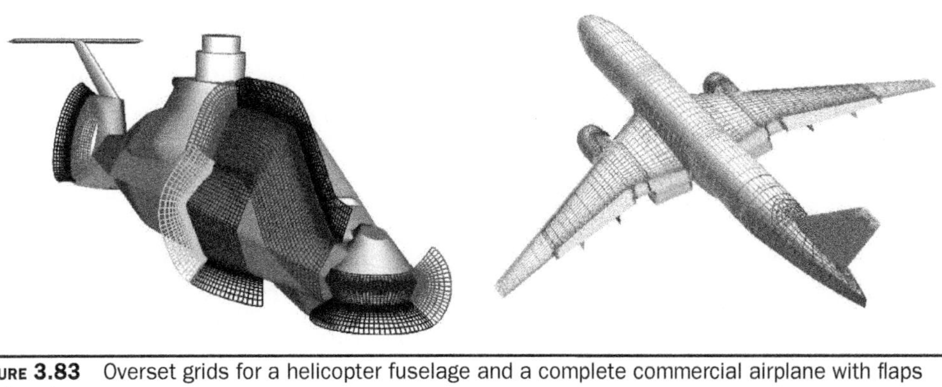

Figure 3.83 Overset grids for a helicopter fuselage and a complete commercial airplane with flaps deployed.

convective fluxes without adding artificial dissipation. The upwind schemes employed in FD are analog of the upwind flux methods developed for finite volume methods and are used in many FD codes. Another way to obtain high-order accuracy is to use higher-order accurate, explicit, finite-difference (FD) formulas, such as fourth- or sixth-order central differences that have a five- and seven-point stencil, respectively. In order to avoid the wide stencils of explicit FD formulas and obtain better resolution in wave space (less dispersion) for the same order of accuracy, compact difference schemes can be used (Visbal and Gaitonde 2002; Ekaterinaris 2005). Compact differentiation schemes evaluate the derivative globally along a line using tridiagonal matrix inversion for fourth- and sixth-order schemes. When the derivatives are evaluated with higher-order explicit or compact FD formulas it is not straightforward to construct higher than fourth order explicit dissipation. For higher-order compact schemes, the explicit dissipating is replaced by the compact (spectral-type) filters introduced by Visbal and Gaitonde (1998, 2002) and implemented in the FDL3DI Code (Gaitonde and Visbal 1998) of the Air Force Research Laboratory (AFRL). Compact filters cannot be used at discontinuities. However, the FDL3DI code has been extensively used for subsonic computations without shocks and for large eddy simulations (LES) of compressible flows.

Another approach to obtain high-order accuracy for finite difference discretizations in generalized coordinates is to use the essentially nonoscillatory (ENO) or weighted ENO (referred to as WENO) scheme. ENO and WENO schemes (Barth and Deconinck 1999; Ekaterinaris 2005) have been developed for finite volume discretizations. The WENO scheme can be used in the finite difference context for equally spaced meshes and for generalized coordinate transformations must be applied for the equally spaced transformed domain. A finite difference WENO scheme can be viewed as a central difference scheme plus a dissipative part (Ekaterinaris 2005). As a result, the dissipative part of the WENO scheme can be used to stabilize high-order central difference discretizations. A more systematic approach for stabilizing high-order (fourth-order or higher) central difference discretizations has been introduced by Yee et al. (1999) with the so-called characteristic-based filters. These characteristic-based filters have been successfully demonstrated for a wide range of problems with discontinuities and the adaptive numerical dissipation control mechanism they provide is very general and can be used in the finite volume or the finite element context.

Finite Volume Methods

The basic idea of the finite volume (FV) discretization (LeVeque 2004) applied to systems of nonlinear conservation laws $\mathbf{u}_t + \nabla \cdot \mathbf{f}(\mathbf{u}) = 0$, can be demonstrated for the one-dimension system $\mathbf{u}_t(x,t) + \mathbf{f}_x(\mathbf{u}(x,t)) = 0$. Consider the cell average

$$\overline{u}_i = \overline{\mathbf{u}}(x_i, t) = \frac{1}{\Delta x_i} \int_{x_i - 1/2}^{x_i + 1/2} \mathbf{u}(x,t)\,dx \tag{3.69}$$

over the cell i centered at x_i for $x_{i-1/2} \le x \le x_{i+1/2}$ and integrate $\mathbf{u}_t(x,t) + \mathbf{f}_x(\mathbf{u}(x,t)) = 0$ over the cell i of length Δx_i to obtain.

$$\frac{d\overline{u}_i(t)}{dt} = -\frac{1}{\Delta x_i}\left[f\left(u\left(x_{i+1/2}, t\right)\right) - f\left(u\left(x_{i-1/2}, t\right)\right)\right] \tag{3.70}$$

For FV discretizations the values $u^-_{i+1/2}$ (interior) and $u^+_{i+1/2}$ (exterior) to the cell interface are not the same and the physical flux function at the interface, $f_{i+1/2}$, is replaced with a continuous numerical flux, $\hat{f}_{i+1/2}$. The numerical flux $\hat{f}_{i+1/2} = h(u^-_{i+1/2}, u^+_{i+1/2})$ must be consistent with the physical flux. The local Lax–Friedrichs (LF) flux, $h_{LF}(a,b) = 0.5[f(a) + f(b) - \lambda(b-a)]$, where λ is the spectral radius of the flux Jacobian, $\lambda = \max_u |\partial f(\mathbf{u})/\partial \mathbf{u}|$, is the simplest flux. An extensive presentation on numerical fluxes developed over the years can be found in Toro (2009).

In order to obtain higher than first order accuracy, the values are not set to $u_{i+1/2}^{+} = \bar{u}_{i+1}$ and $u_{i+1/2}^{-} = \bar{u}_i$ but they must be reconstructed from cell averages nearby, e.g., $u_{i+1/2}^{-} = r(\bar{u}_{i-1}, \bar{u}_i, \bar{u}_{i+1})$, to obtain a higher order of accuracy. For multidimensional problems with structured meshes, polynomial-type reconstruction can be carried out on a direction per direction basis. For unstructured meshes gradient-type, k-exact, or any other form of reconstruction, such as ENO (Ekaterinaris 2005), must be performed. Then the resulting numerical fluxes $\hat{f}_{i+1/2} = h(u_{i+1/2}^{-}, u_{i+1/2}^{+})$, $\hat{f}_{i-1/2} = h(u_{i-1/2}^{-}, u_{i-1/2}^{+})$ yield the following higher-order, conservative, finite volume scheme:

$$\frac{d\bar{u}_i(t)}{dt} = -\frac{1}{\Delta x_i}\left[\hat{f}_{i+1/2} - \hat{f}_{i-1/2}\right] \tag{3.71}$$

The finite volume (FV) method for the Navier–Stokes is based on the integral form of the governing equations which in compact notation is

$$\frac{\partial}{\partial t}\iiint_{\Omega} q\, d\Omega + \iint_{\partial\Omega} \mathbf{F}(\mathbf{q})^{\text{inv}} \cdot d\mathbf{s} = \iint_{\partial\Omega} \mathbf{F}(\mathbf{q}, \nabla\mathbf{q})^{\text{vis}} \cdot d\mathbf{s} \tag{3.72}$$

FV methods can be applied with centered numerical fluxes plus artificial dissipation, as it was proposed by Jameson et al. (1981), or by using upwind flux functions. Most current implementations of FV methods are based on upwind fluxes (LeVeque 2004; LeVeque 2009), such as Roe's flux, the HLLC flux, and others (there is extensive literature on upwind fluxes; see Yee 1999, Hirsch 2002, and LeVeque 2004 and references there for more details).

After finite volume discretization, e.g., by employing a dual grid with control volumes constructed using a median-dual vertex-based scheme, the following semi-discrete form for each control volume is obtained:

$$\frac{\partial}{\partial t}\iiint_{\Omega_i} \mathbf{q}\, d\Omega + \underbrace{\sum_{j\in\aleph(i)} \left(\tilde{F}_{ij}^{inv} + \tilde{F}_{ij}^{vis}\right)\Delta S_{ij}}_{R_i} = 0$$

or

$$\frac{\partial}{\partial t}\iiint_{\Omega_i} \mathbf{q}\, d\Omega + R_i(\mathbf{q}) = 0 \tag{3.73}$$

where $\mathbf{q}$ is the vector of the state variables, $\tilde{F}_{ij}^{inv}$ and $\tilde{F}_{ij}^{vis}$ are the numerical approximations of the inviscid and viscous fluxes ΔS_{ij} is the area of the face associated with the edge ij, and $\aleph(i)$ is the set of the neighboring nodes to the node i.

The semi-discrete form of equation (3.73) can be advanced in time with explicit or implicit methods. In the absence of viscous terms, the Euler equations can be advanced in time efficiently, especially when explicit time marching is combined with multigrid acceleration. Jameson and Yoon (1987) proposed an implicit factored algorithm for time advancement of the Euler equations on structured meshes. This method is known as the LU-SGS scheme and has the form $LD^{-1}U\,\Delta\mathbf{q} = -(\Delta t)\,\hat{R}$, where $\hat{R}$ is the finite volume discretization of $\hat{R} = (\hat{E}_{i+1/2,j,k} - \hat{E}_{i-1/2,j,k}) + (\hat{F}_{i,j+1/2,k} - \hat{F}_{i,j-1/2,k}) + (\hat{G}_{i,j,k+1/2} - \hat{G}_{i,j,k-1/2})$.

$$\begin{aligned} &LD^{-1}U\,\Delta\mathbf{q} = -\Delta t\,\hat{R} \\ &L = I + \Delta t(D_{\xi}^{-}\hat{A}^{+} - \hat{A}^{-} + D_{\eta}^{-}\hat{B}^{+} - \hat{B}^{-} + D_{\zeta}^{-}\hat{C}^{+} - \hat{C}^{-}) \\ &D = I + \Delta t(\ \hat{A}^{+} - \hat{A}^{-} + \ \hat{B}^{+} - \hat{B}^{-} + \ \hat{C}^{+} - \hat{C}^{-}) \\ &U = I + \Delta t(D_{\xi}^{-}\hat{A}^{-} - \hat{A}^{+} + D_{\eta}^{-}\hat{B}^{-} - \hat{B}^{+} + D_{\zeta}^{-}\hat{C}^{-} - \hat{C}^{+}) \end{aligned} \tag{3.74}$$

where the matrices $\hat{A}^{\pm}, \hat{B}^{\pm}, \hat{C}^{\pm}$ denote the flux Jacobian matrices with positive (+) and negative (−) eigenvalues that are obtained after similarity transformations $\hat{A}^{\pm} = \hat{T}_{\xi} \Lambda^{\pm} \hat{T}_{\xi}^{-1}$ or $\hat{A}^{\pm} = [\hat{A} \pm \rho(\hat{A})I]$, where $\rho(\hat{A})$ is the spectral radius of flux Jacobian. Implementation of the LU-SGS scheme on finite volume codes combined with multigrid acceleration significantly accelerated convergence to steady state.

For fully unstructured FV discretizations time marching is obtained by advancing in time the full system as follows. The equations are linearized in time about the current state

$$\left(\frac{\Omega_i}{\Delta t_i^n} \delta_{ij} + \frac{\partial R_i(\mathbf{q}^n)}{\partial \mathbf{q}_j} \right) \Delta \mathbf{q}_j^n = -R_i(\mathbf{q}^n) \tag{3.75}$$

and use a Krylov subspace method or implement a Jacobian free Newton Krylov subspace approach, that does not require explicit evaluation and storage of $\partial R_i(\mathbf{q}^n)/\partial \mathbf{q}_j$, for solving this system.

Finite volume methods or unstructured meshes are well suited for simulation of subsonic, transonic and supersonic flows with strong shocks over complex aerodynamics configurations. In contrast to finite difference methods, it is not straightforward to design FV methods with higher than second-order accuracy. Higher-order reconstructions for unstructured meshes, such as ENO or WENO, are intensive in terms of memory requirements and computational time. However, second-order accurate in space finite volume methods with different upwind schemes and total variation diminishing (TVD) limiters, which make possible resolution of strong shocks without numerical oscillations, have been the workhorse of computational aerodynamics for the past few decades. Such methods have been implemented in many commercial codes and in research codes such as FUN3D (https://fun3d.larc.nasa.gov/) and the freely available SU2 code (Economon 2016) (http://su2.stanford.edu/).

High-Order Finite Element Methods

Finite element (FE) methods that can achieve a high order of accuracy in unstructured meshes were developed in recent years for the time-dependent *N–S* equations. FE methods for the *N–S* equations start from the differential form of the governing equations. The tensor form of the *N–S* equations (3.61)–(3.63) in short-hand notation is

$$\partial_t \mathbf{q} + \partial_j F_{ij}^{inv}(\mathbf{q}) - \partial_j F_{ij}^{vis}(\mathbf{q}, \nabla \mathbf{q}_p) = 0 \tag{3.76}$$

where $\mathbf{q} = [\rho, \rho u_i, \rho e]^T$ is the conservative variable vector, the vector valued functions $F_{ij}^{inv}(\mathbf{q})$ and $F_{ij}^{vis}(\mathbf{q}, \nabla \mathbf{q}_p)$ represent the inviscid and viscous fluxes, respectively, and the components of the tensor $\nabla \mathbf{q}_p = Q_{ij} = \partial_j(\mathbf{q}_p)_i = \partial_j(u, v, w, T)^T$ are used to form the viscous stress tensor τ_{ij}.

The starting point of FE discretization is the weak formulation of the governing equations obtained by multiplying the conservative form (3.76) with a weighting function $w_k(\mathbf{x})$ and integrating over the element Ω_m. Then after approximating the solution in each element with $\mathbf{q}^e = \sum_{i=0}^{N_p} c_i(t) b_i(\mathbf{x})$, where $b_i(\mathbf{x})$ is a basis function (a polynomial of degree *k*) and integrating by parts one obtains the following discrete form

$$\begin{aligned} \partial_t \int_{\Omega_m} w_m q_m^e \, d\Omega = & \int_{\Omega_m} \partial_j w_m F_{mj}^{inv}(\mathbf{q}^e) d\Omega - \oint_{\partial\Omega_m} w_m \widehat{F_{mj}^{inv}(\mathbf{q}^e)} n_j dS \\ & - \int_{\Omega_m} \partial_j w_m F_{mj}^{vis}(\mathbf{q}^e, \nabla \mathbf{q}_p^e) d\Omega + \oint_{\partial\Omega_m} w_m \widehat{F_{mj}^{vis}(\mathbf{q}^e, \nabla \mathbf{q}_p^e)} n_j dS \end{aligned} \tag{3.77}$$

where $n_j = \mathbf{n}$ is the outward normal vector to the faces of the element. For Galerkin methods the expansions (or bases) functions $b_i(\mathbf{x})$ and the weighting functions $w_i(\mathbf{x})$ belong to the same polynomial space. Continuous finite element discretizations correspond to a centered-type discretization of the conservation laws and require some form of dissipation to stabilize the calculations.

In the classical finite element discretizations the bases functions are polynomials or other functions that are continuous at the element interfaces (C^0 interface continuity requirement). Such a discretization, however, corresponds to a centered-like discretization and in order to prevent nonlinear instability resulting from the convective terms in regions with steep flow gradients and flow discontinuities some form of dissipation must be added. As a result, FE discretizations have global character (result into large stiffness matrix) and thus require solution of large systems of equations, and without modifications are not well suited for flows with strong shocks that are of interest for applications in aerodynamics. In order to overcome this difficulty other methods that employ local expansions for the element only, such as the spectral volume, the spectral difference, and the discontinuous Galerkin method (Ekaterinaris 2005), have been developed. The discontinuous Galerkin (DG) method (Ekaterinaris 2005; Cocburn 2000; Wang 2007) in particular, which is a mixture of the finite element and the finite volume methods, has been extensively used not only for the high-order accurate numerical solutions of high-speed flows with strong shocks but in many other fields of computational mechanics (Shu 2016).

For the DG discretization continuity restrictions at the element interfaces are relaxed and the bases functions are polynomials (often tensor product of hierarchical polynomials) defined within the element without imposing continuity restrictions at the element interfaces. Therefore, the term $\widetilde{F_{mj}^{inv}(\mathbf{q}^e)n_j}$ in equation (3.77) can be replaced with a suitable numerical flux at the element interface as in the FV methods. Furthermore, for the DG method, since the piecewise constant approximation within the element is not employed any more as in FV methods, there is no need to resort to reconstruction of the interface values to a higher order of accuracy as in equation (3.71). It has been proven that numerical solutions obtained with DG discretizations that employ expansion bases functions, which are tensor products of polynomials $P^k(x_i)$ of degree k, along each direction i, achieve a $(k+1)$ global order of accuracy.

The DG method is very well suited to adaptive refinement because nonconforming elements (elements with hanging nodes, see Figure 3.84) can be treated in a natural manner due to the local character of the expansion bases. Parallelization of DG discretizations is facilitated because each element requires information only from the faces of the neighboring elements in contrast to the large number elements (layer of several elements for WENO discretization for example) that are required for other high order methods. Superior shock-capturing capabilities have been also developed for the DG method (Figure 3.85) either with the TVB limiters (Panourgias and Ekaterinaris 2016) or with other approaches that result in discontinuity capturing within the cell (see Panourgias and Ekaterinaris (2016) and references therein). Application of adaptive mesh refinement locally (h-type) refinement and the increase of resolution for selected elements through the increase of polynomial expansion (p-type of refinement) (Wang 2011) are of particular interest to large eddy simulations of compressible flows. It appears that DG codes with hp refinement capabilities that are emerging will be used in direct simulations and adaptive LES.

Turbulence and Transition Modeling

The objective of CFD numerical solutions is to simulate accurately complex flows over full helicopter or aircraft configurations (see, e.g., Figure 3.83) for realistic Reynolds numbers in the order of 10 million. For this range of Reynolds numbers, the spatial and temporal scales of turbulence are very small and the mesh requirements are very large to perform LES or direct numerical simulation (DNS) even for the largest available supercomputers. Therefore, the CFD solutions required for design and testing of new concepts have to rely on the Reynolds averaged

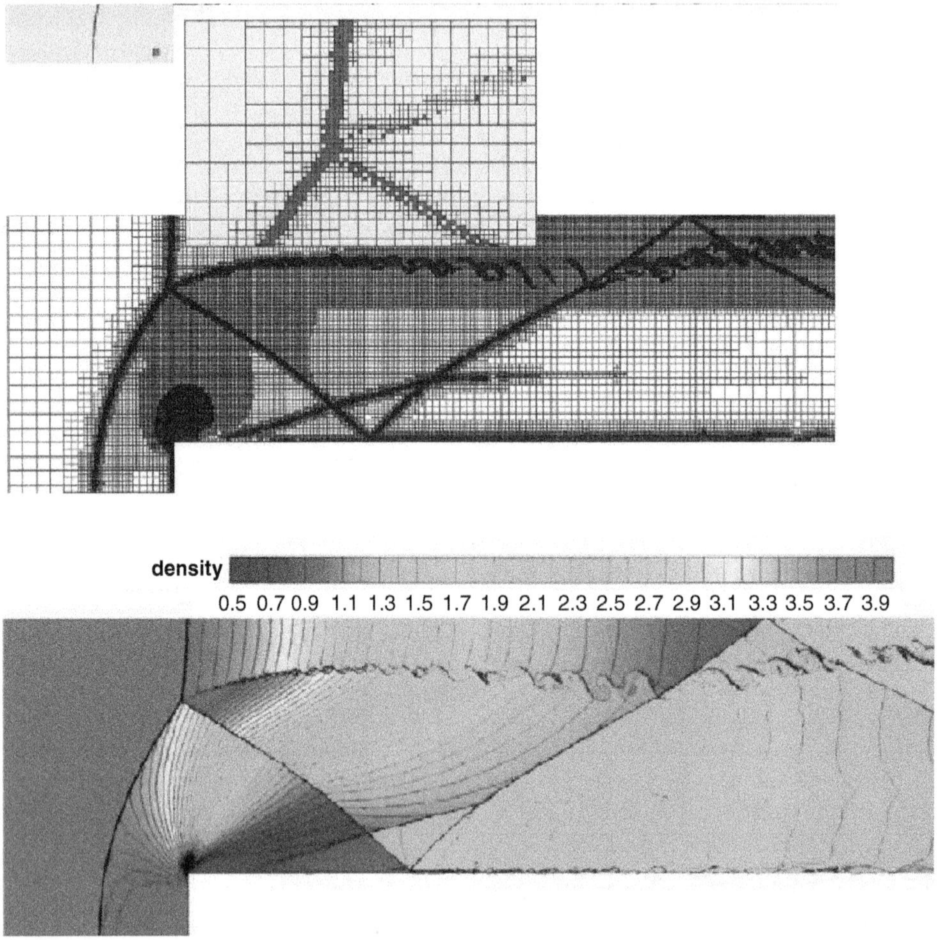

FIGURE 3.84 Adaptive mesh refinement and computed solution with nonconforming elements (from Panourgias and Ekaterinaris 2016).

Navier–Stokes (RANS) equations and use turbulence models developed over the years starting from simple algebraic models, to one- and two-equation turbulence models even to seven equations Reynolds-stress turbulence models. In the RANS equations the physical eddy viscosity, μ, which is a function of temperature for compressible flow and constant for incompressible flow, is replaced by the turbulent eddy viscosity $\mu + \mu_{tur}$. The turbulent eddy viscosity μ_{tur} is not constant even for incompressible flow but varies in space. It must be emphasized that once the RANS equations with turbulence models are used the dynamics of turbulence is lost and the effect of Reynolds stresses on the mean flow is obtained only through the turbulence model. A detailed presentation of turbulence models currently in use for CFD simulations is given by Wilcox (2006). Most of the finite-difference, finite-volume, or finite-element CFD codes use a one-equation turbulence model (Spalart and Allmaras 1992) or a two-equation turbulence model such as the $k - w$ turbulence model Menter (1994) in order to evaluate the eddy viscosity μ_{tur} that

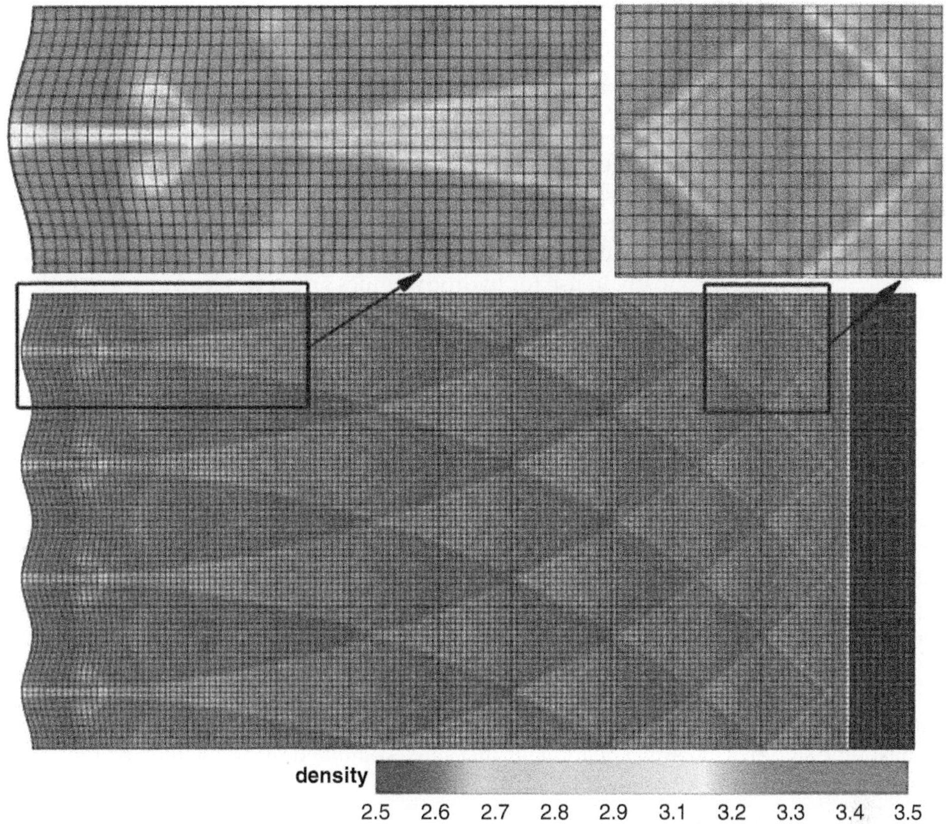

FIGURE 3.85 Normal shock reflecting from a wavy wall; sub-cell shock capturing with P5 expansions and use of nonlinear filter (from Panourgias and Ekaterinaris 2016).

must be added to the physical viscosity μ. All turbulence models in use in aerodynamics have been developed and calibrated to predict attached flows or mildly separated incompressible flows. In fact, for simulations of flows with significant compressibility effects compressibility corrections have been applied. For separated flows, free sear flows, and vortex dominated flows these models initially developed for attached wall-bounded flows had to be modified. For example, corrections for rotational flows were applied to the one equation Spalart–Almaras (1992) turbulence model in order to improve predictions of vortex dominated flows, such as flows over delta wings and helicopter rotor wakes.

It has been documented in a number of experimental investigations that many external aerodynamic flows evolve from laminar to transitional and then to fully turbulent. Due to the importance of transition the first attempts to incorporate transitional flow effects in aerodynamic calculations can be traced to boundary-layer methods (see, e.g., the XFOIL code freely available at http://web.mit.edu/drela/Public/web/xfoil/).

The effect of transition was incorporated for turbulence models (Ekaterinaris and Platzer 1998; Menter et al. 2006; Langtry and Menter 2009) and improved predictions were obtained by taking into account transitional flow effects for a number of important applications.

CFD Methods for the Incompressible *N–S* Equations

The basic discretizations techniques and turbulence models employed for compressible flows are used for incompressible flow calculations, where the density is constant and the pressure changes are not related any more to density through the equation of state. For incompressible (constant density) flows, the equation of energy is not required. However, the numerical methods must guarantee that the kinetic energy, which is obtained by multiplying the momentum equations with the flow velocity, must be conserved at the discrete level. Finite difference discretizations of the incompressible flow equations are used with immersed boundary methods (Sotiropoulos and Yang 2014). Finite volume methods (Patankar 1980; Ferziger and Peric 2002) with second-order accuracy in space are mostly in use for hydrodynamics and low-speed aerodynamics (see, e.g., the OpenFoam free access code http://www.openfoam.com/). Finite element discretizations (Hughes 2000; Deville et al. 2004; Zienkiewicz et al. 2005; Karniadakis and Sherwin 2005) have also been applied for incompressible flows. There are freely available finite element codes such as Nektar http://www.nektar.info/, an h/p high-order platform suitable for LES and DNS of incompressible flows in complex domains. Finite element discretization (Hughes 2000) is also performed in SimVascular http://simvascular.github.io/, a computational framework developed for biomedical flows.

A significant difference between the incompressible flow equations and the compressible flow equations is the elliptic-type of constraint imposed by incompressibility $\nabla \cdot \mathbf{u} = 0$ and the pressure that does not appear in the continuity equation but appears only in the momentum equations. As a result, during time advancement of the momentum equations the pressure must be adjusted accordingly so that incompressibility (divergence free condition for the velocity field) is achieved at the next time step. This is achieved either with fractional time step methods (Karniadakis and Sherwin 2005) or by solving the Poisson equation for pressure with the velocities computed from the previous, or an intermediate time step, in order to guarantee that incompressibility is enforced at the next time step. The Poisson equation for pressure is obtained by taking the divergence of the momentum equations and invoking continuity:

$$\nabla^2 p = \nabla \cdot ((\mathbf{u} \cdot \nabla)\mathbf{u})$$

The Poisson equation for pressure is linear. However, the numerical solution of the Poisson equation involves numerical solution of a large system of linear equations of the form $Ax = b$. The matrix A is large and quite sparse even when high-order space discretization is employed. However, due to the linearity of the Poisson equation multigrid methods are very effective and they are employed to accelerate convergence of the numerical solution.

The artificial compressibility approach, which introduces the pseudocompressibility parameter β to add the pressure in the continuity equation $\partial_t p + \beta^2 \nabla \cdot \mathbf{u} = 0$ has been also employed to overcome difficulties with the numerical solution of the incompressible flow equations. The addition of pressure in the continuity makes the incompressible flow equations similar in form to the compressible flow equations and the same numerical methods developed for compressible flow are used in the artificial compressibility approach.

LES and DNS of Turbulence

Large eddy simulations (LES) (Sagaut 2006; Garnier 2009) capture the large scales of turbulence and model the small scales with subgrid scale (SGS) models. This separation of scales in LES is possible because for most turbulent flows the dynamics of the smaller scales are universal, independent of the flow geometry, and can be modeled by SGS models. The separation between large (resolved) scales and small (modeled) scales in LES is not associated with statistical averaging. This scale separation is formalized on the mathematical level through the application of a frequency low-pass filter to the exact numerical solution. Application of the filter to the spatially and temporally varying velocity component $u(\mathbf{x}, t)$, for example, yields the resolved part, $\overline{u}(\mathbf{x},t)$, of this velocity component as a convolution of the velocity with a kernel G characteristic of the filter used.

$$\overline{u}(\mathbf{x},t) = G * u = \int_{-\infty}^{+\infty}\int_{-\infty}^{+\infty} \overline{u}(\tilde{\mathbf{x}},\tilde{t})G(\mathbf{x}-\tilde{\mathbf{x}},t-\tilde{t})d\tilde{t}\, d^3\tilde{\mathbf{x}} \tag{3.78}$$

In addition to the spectral filter (that often is not explicitly applied for LES obtained with second-order accurate methods), some form of filtering is inevitably performed by the numerical method itself and by the mesh used for discretization.

In the Fourier space, $(\mathbf{k},\omega)$, the spectrum $\hat{u}(\mathbf{k},\omega)$ of the velocity $u(\mathbf{x},t)$ is related to the spectrum of the filter kernel $\hat{G}(\mathbf{k},\omega)$ by

$$\hat{\overline{u}}(\mathbf{k},\omega) = \hat{u}(\mathbf{k},\omega)\hat{G}(\mathbf{k},\omega)$$

According to these definitions the unresolved part of $u(\mathbf{x}, t)$ that is denoted as $u'(\mathbf{x}, t)$ is

$$u'(\mathbf{x},t) = u(\mathbf{x},t) - \overline{u}(\mathbf{x},t) = (1-G) * u$$
$$\widehat{u'}(\mathbf{x},t) = \hat{u}(\mathbf{x},t) - \hat{\overline{u}}(\mathbf{x},t) = (1-\hat{G})\hat{u}$$

with these definitions the filter can be applied to the incompressible (Sagaut 2006) or the compressible (Garnier 2009) Navier–Stokes equations in the physical or the spectral space. After application of the filter to the incompressible flow equations obtain

$$\begin{aligned} &\frac{\partial \overline{u_i}}{\partial x_i} = 0 \\ &\frac{\partial \overline{u_i}}{\partial t} + \frac{\partial(\overline{u_i u_j})}{\partial x_j} = -\frac{\partial p}{\partial x_i} + \nu\frac{\partial \overline{u_k}}{\partial x_{kk}} \end{aligned} \tag{3.79}$$

The nonlinear term $\overline{u_i u_j}$ that appears in the filtered momentum equations is not known. In order to discretize these equations this term must be expressed as a function of $\overline{u_i}$ (resolved scales) and $u_i' = u_i - \overline{u_i}$ (unresolved or modeled scales) which are the only unknowns in the LES decomposition. There is a number of decompositions that allow to express $\overline{u_i u_j}$ in terms of known quantities. A common decomposition of $\overline{u_i u_j}$ is Leonard's double decomposition $\overline{u_i u_j} = \overline{\overline{u_i}\,\overline{u_j}} + \overline{\overline{u_i} u_j'} + \overline{\overline{u_j} u_i'} + \overline{u_i' u_j'}$. If in addition one sets $\overline{\overline{u_i}\,\overline{u_j}} = \left(\overline{\overline{u_i}\,\overline{u_j}} - \overline{u_i}\,\overline{u_j}\right) + \overline{u_i}\,\overline{u_j}$, the term $L_{ij} = \overline{\overline{u_i}\,\overline{u_j}} - \overline{u_i}\,\overline{u_j}$ representing the interactions among the large scales is called Leonard tensor, the triple Leonard decomposition (Sagaut 2006) of the filtered momentum equations is obtained

$$\frac{\partial \overline{u_i}}{\partial t} + \frac{\partial}{\partial x_j}\left(\overline{u_i}\;\overline{u_j}\right) = -\frac{\partial \overline{p}}{\partial x_i} + \nu\frac{\partial}{\partial x_j}\left(\frac{\partial \overline{u_i}}{\partial x_j} + \frac{\partial \overline{u_j}}{\partial x_i}\right) - \frac{\partial \tau_{ij}}{\partial x_j} \tag{3.80}$$

where $\tau_{ij} = \overline{u_i u_j} - \overline{u_i}\,\overline{u_j}$ is the subgrid stress tensor that needs to be modeled.

Invoking the Boussinesq hypothesis the subgrid model for the deviatoric part τ_{ij}^d can be written as

$$\tau_{ij}^d = \tau_{ij} - \frac{1}{3}\delta_{ij}\tau_{kk}, \qquad \frac{\partial \tau_{ij}^d}{\partial x_j} = \frac{\partial}{\partial x_j}\left(\nu_{sgs}\left(\frac{\partial \overline{u_i}}{\partial x_j} + \frac{\partial \overline{u_j}}{\partial x_i}\right)\right) = \frac{\partial}{\partial x_j}\left(-2\nu_{sgs}(\overline{S_{ij}} - \frac{1}{3}\delta_{ij}\overline{S_{kk}})\right)$$

A number of models have been proposed for ν_{sgs} in the literature (see Sagaut 2006; Garnier et al. 2009; and references therein). A simple model is the Sagoririnsky model that gives $\nu_{sgs}(\mathbf{x},t) = \left(C_s\overline{\Delta}\right)^2 |S(\mathbf{x},t)|$, where for local equilibrium and assuming the Kolmogorov spectrum $C_s \sim 0.18$, $\overline{\Delta} = (\Delta x \times \Delta y \times \Delta z)^{1/3}$ for a Cartesian mesh, and $|S(\mathbf{x},t)| = \left(2\overline{S}_{ij}\,\overline{S}_{ij}\right)^{1/2}$. The Sagoririnsky model has been criticized as too dissipative. Improved predictions can be obtained with the dynamic Smagorinsky model (Sagaut 2006) that computes a locally varying constant C_s invoking Germano's identity to account for the backscatter of energy from unresolved to resolved scales.

The main advantage of LES is that they capture directly the large scales of turbulence and they are therefore particularly suitable and successful for separated flow, free shear flows, such as jets, and combustion. They require, however, an almost isotropic and sufficiently fine mesh capable of capturing the resolvable by the LES scales, and as a result demand much larger computational resources than RANS. For high Reynolds number wall bounded flows the resolution requirements of LES become very large. In addition, it is well known that non-universal small scales exist in the near wall region and modeling of these subgrid scales is not straightforward. In order to overcome these difficulties and to keep the computational resources down to reasonable levels the so-called DES and hybrid RANS/LES approaches (Sagaut et al. 2013) have been proposed. In these approaches the near-wall flow region is essentially computed with a turbulence model using RANS resolution. Then the flow away from the wall is computed on a denser mesh with LES, in the hybrid RANS/LES approach, or with a modified form of the turbulence model, in the DES approach, but again on a mesh with LES-like resolution.

The required mesh density for achieving the required resolution in LES is not known a priori. On the other hand, mesh refinement in LES, which must be applied in all three directions is prohibitively expensive. As a result, LES can often be under-resolved. In addition, when explicit filtering is not applied it is not easy to reach mesh-independent solutions. In order to overcome the resolution issues related with anisotropic meshes and limitations existing in modeling subgrid scales, especially for compressible flow simulation where more terms that appear in the filtered equations, need to be modeled, the monotone implicit LES (MILES) approach (Grinstein 2011) was employed. In the MILES approach, there is no separation of scales through explicit filtering, and no attempt is made to model subgrid scales. It is assumed, however, that the numerical diffusion of the high resolution upwind scheme acts as a subgrid scale model and therefore the simulation of the Navier–Stokes equations on a fine isotropic mesh effectively reproduces the results of an LES simulation. The MILES approach has been successful for free shear flows such as jets.

Direct numerical simulations (DNS) of turbulent flows do not assume any separation of scales and do not involve any type of modeling. DNS solve numerically the Naveir–Stokes preferably with a high-order scheme (often with a spectral method) on a mesh that is capable of resolving all scales of turbulence down to the Kolmogorov scale. Clearly, DNS simulations require enormous resources even for moderate Reynolds numbers, but when they are performed correctly they have the same reliability as a high-quality experiment.

LES and DES is currently in use in many industrial applications, such simulations in aeronautics and turbomachinery, simulations of flow over cars, trucks and locomotives, simulation in urban terrains, and weather prediction to name a few. In general, LES predictions are in better agreement with measurements than RANS predictions. Instantaneous realizations of LES (see Figure 3.86) closely resemble experimental flow visualization. On the other hand, direct numerical simulations (see Figure 3.87) are still used to investigate the dynamics of turbulence and study the turbulent flow structures in order to obtain a better understanding of turbulence to improve subgrid LES models, and turbulence modes for RANS.

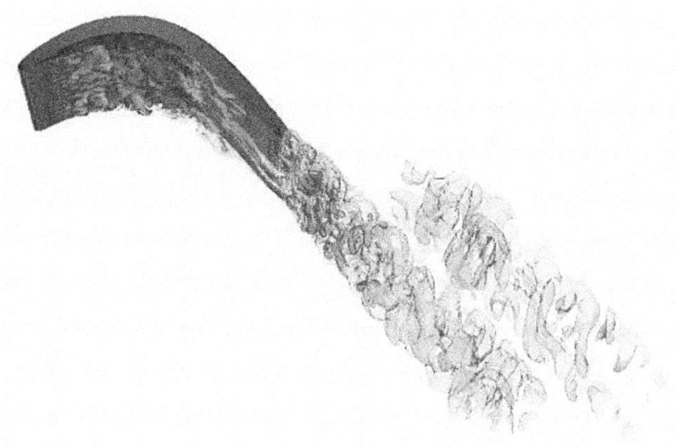

FIGURE 3.86 LES over the lower surface of a turbomachinery blade.

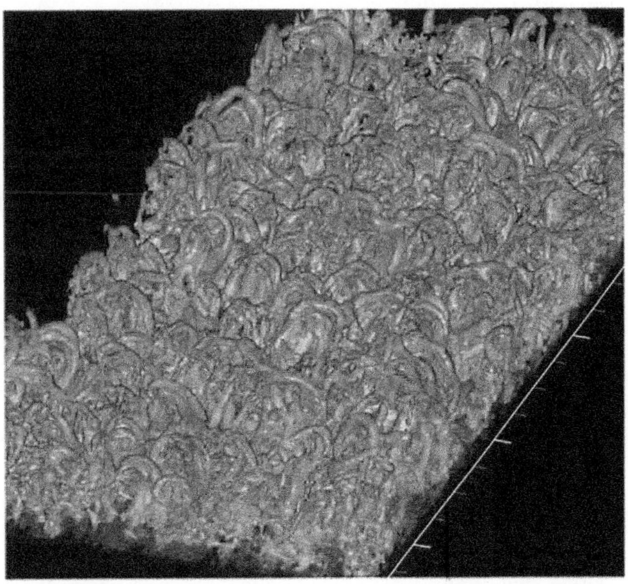

FIGURE 3.87 DNS of a turbulent boundary layer. Computed flow structure shown by isosurfaces of the second invariant of the velocity gradient tensor.

References

Bachelor, G. K. 2000. *An Introduction to Fluid Dynamics*, Cambridge University Press.

Barth, T. J. and Deconinck, H. 1999. *High-Order Methods for Computational Physics* (lecture notes in computational science and engineering), Springer. New York, N.Y.

Beam, R. and Warming, R. F. 1976. "An Implicit Finite-Difference Algorithm for Hyperbolic Systems in Conservation-Law-Form," *Journal of Computational Physics*, vol. 22, pp. 87–110.

Buning, P. G. and Pulliam, T. H. 2016. "Near-Body Grid Adaption for Overset Grids, 46th AIAA Fluid Dynamics Conference," Washington, D.C., Jun. 13–17.

Cocburn, B., Karniadakis, G. E., and Shu, S.-W. (eds) 2000. *Discontinuous Galerkin Methods: Theory, Computation and Applications* (lecture notes in computational science and engineering), Springer New York, N.Y.

Deville, M. O., Fisher, P. F., and Mund, E. H. 2004. *High-Order Methods for Incompressible Fluid Flow*, Cambridge University Press.Cambridge, U.K.

Economon, T. D., Palacios, F., Copeland, S. R., Lukaczyk, T. W., and Alonso, J. J. 2016. "SU2: An Open-Source Suite for Multiphysics Simulation and Design," *AIAA Journal*, vol. 54 (3), pp. 828–846.

Ekaterinaris, J. A. 2005. "High-Order Accurate, Low Numerical Diffusion Methods for Aerodynamics," *Progress in Aerospace Sciences*, vol. 41, pp. 192–300.

Ekaterinaris, J. A. and Platzer, M. F. 1998. "Computational Prediction of Airfoil Dynamic Stall," *Progress in Aerospace Sciences*, vol. 33 (11–12), pp. 759–846.

Ferziger, J. H. and Peric, M. 2002. *Computational Methods for Fluid Dynamics*, 3rd ed., Springer.

Gaitonde, D. and Visbal, M. 1998. "High-Order Schemes for Navier-Stokes Equations: Algorithm and Implementation into FDL3DI," Technical Report AFRL-VA-WP-TR-1998-3060, Air Force Research Laboratory, Wright-Patterson AFB, Dayton, Ohio.

Garnier, E., Adams, N., and Sagaut, P. 2009. *Large Eddy Simulation of Compressible Flows*, Springer, New York, N.Y.

Grinstein, F. F., Margolin, L. G., and Rider, W. J. 2011. *Implicit Large Eddy Simulation: Computing Turbulent Fluid Dynamics*, Cambridge University Press, Cambridge, U.K.

Hirsch, C., 2002. *Numerical Computation of Internal and External Flows*, vols. 1–2, John Wiley & Sons, New York, N.Y.

Hughes, T. J. R. 2000. *The Finite Element Method*, Dover Publications, Mineola, N.Y.

Jameson, A., Schmidt, W., and Turkel, E. 1981. "Numerical Solution of the Euler Equations by Finite Volume Methods Using Runge-Kutta Time-Stepping Schemes," AIAA Paper AIAA-1981-1259.

Jameson, A. and Yoon, S. 1987. "Lower-Upper Implicit Schemes with Multiple Grids for the Euler Equations," *AIAA Journal*, vol. 25, pp. 929–935.

Karniadakis, G. E. and Sherwin, S. J. 2005. *Spectral/hp Element Methods for CFD*, Oxford University Press, Oxford, U.K.

Landau, L. D. and Lifshitz, E. M. 1987. *Fluid Mechanics*, 2nd ed., Elsevier. New York, N.Y.

Langtry, R. B. and Menter, F. R. 2009. "Correlation-Based Transition Modeling for Unstructured Parallelized Computational Fluid Dynamics Codes," *AIAA Journal*, vol. 47 (12), pp. 2894–2906.

LeVeque, R. L. 2004. *Finite Volume Methods for Hyperbolic Problems*, Cambridge.

Lomax, H., Pulliam, T. H., and Zingg, D. W. 2011. *Fundamentals of Computational Fluid Dynamics*, Springer New York, N.Y.

Menter, F. R. 1994. "Two-Equation Eddy-Viscosity Turbulence Models for Engineering Applications," *AIAA Journal*, vol. 32 (8), pp. 1598–1605.

Menter, F. R., Langtry, R. B., and Volker, S. 2006. "Transition Modelling for General Purpose CFD Codes," *Flow Turbulence and Combustion*, vol. 77 (1), pp. 277–303.

Nichols, R. H. and Buning, P. G. 2010. *OVERFLOW User's Manual*, Version 2.2, NASA Langley Research Center, Hampton, VA, Aug.

Panourgias, K. T. and Ekaterinaris, J. A. 2016. "A Nonlinear Filter for High-Order Discontinuous Galerkin Discretizations with Discontinuity Resolution within the Cell," *J. of Computational Physics*, vol. 326, pp. 234–257.

Panourgias, K. T. and Ekaterinaris, J. A. 2016. "A Discontinuous Galerkin Approach for High-Resolution Simulations of Three Dimensional Flows," *Computer Methods in Applied Mechanics and Engineering*, vol. 299, pp. 254–282.

Panton, R. L. 2013 *Incompressible Flow*, 3rd ed., John Wiley & Sons. New York, N.Y.

Patankar, V. S. 1980. *Numerical Heat Transfer and Fluid Flow*, Taylor and Francis.

Plecher, R. H., Tannehill, J. C., and Anderson, D. A. 2013. *Computational Fluid Mechanics and Heat Transfer*, 3rd ed., CRC Press, New York, N.Y.

Pulliam, T. H. and Steger, J. L. 1980. "Implicit Finite-Difference Simulations of Three-Dimensional Compressible Flow," *AIAA Journal*, vol. 18, pp. 159–167.

Sagaut, P. 2006. *Large Eddy Simulation of Incompressible Flows*, Springer.

Sagaut, P., Deck, S., and Terracol, M. 2013. *Multiscale and Multiresolution Approaches in Turbulence—LES, DES, and Hybrid RANS/LES Methods: Applications and Guidelines*, 2nd ed., World Scientific Publishing Company. Hackensack, N.J.

Shu, C. W. 2016. "High-Order WENO and DG Methods for Time-Dependent Convection-Dominated PDEs: A Brief Survey of Several Recent Developments," *J. of Computational Physics*, vol. 316, pp. 598–613.

Sotiropoulos, F. and Yang, X. 2014. "Immersed Boundary Methods for Simulating Fluid-Structure Interaction," *Progress in Aerospace Sciences*, vol. 65, pp. 1–21.

Spalart, P. R. and Allmaras, S. R. 1992. "A One-Equation Turbulence Model for Aerodynamic Flows," AIAA Paper 92-0439, Jan.

Steger, J. L. and Warming, R. F. 1981. "Flux Vector Splitting of the Inviscid Gasdynamic Equations with Application to Finite Difference Methods," *J. of Computational Physics*, vol. 40 (2), pp. 263–293.

Thompson, J. F., Bharat, B. K., and Weatherill, N. P. 1999. *Handbook of Grid Generation*, CRC Press. New York, N.Y.

Toro, E. F. 2009. *Riemann Solvers and Numerical Methods for Fluid Dynamics*, Springer.

Visbal, M. R. and Gaitonde, D. V. 2002. "On the Use of Higher-Order Finite-Difference Schemes on Curvilinear and Deforming Meshes," *Journal of Computational Physics*, vol. 181, pp. 155–185.

Wang, Z. J. 2007. "High-Order Methods for the Euler and Navier–Stokes Equations on Unstructured Grids," *Progress in Aerospace Sciences*, 43 (1–3), pp. 1–41.

Wang, Z. J. 2011. *Adaptive High-Order Methods in Computational Fluid Dynamics*, World Scientific Publishing, Hackensack, N.J.

Wilcox, D. C. 2006. *Turbulence Modeling for CFD, DCW Industries*, 3rd ed. Lake Arrowhead, California, New York, N.Y.

Yee, H. C. 1985. "On the Implementation of a Class of Upwind Schemes for Systems of Hyperbolic Conservation Laws," NASA Technical Memorandum 86839.

Yee, H. C., and Sandham, N. D., and Djomehri, M. J. 1999. "Low-Dissipative High-Order Shock Capturing Methods Using Characteristic Based Filters," *J. Comp Physics*, vol. 150 (1), pp. 199–238.

Zienkiewicz, O. C., Taylor, R. L., and Nithirasu, P. 2005. *The Finite Element Method for Fluid Dynamics*, 6th ed., Elsevier. New York, N.Y.

PART 5

Aeronautical Measurement Techniques[1,2]

Muguru S. Chandrasekhara

3.29 General

Aerospace testing requires a facility that can simulate the various conditions of flight or flow of interest. Most often, a wind tunnel is used for this although, water tunnels also serve the purpose admirably.

A wind tunnel (or a water tunnel) is a device for moving a steady uniform stream over a model placed in its working section. To move air over a model needs more power than to move the model through air. However, such an arrangement is more convenient to make measurements.

There are two main types of wind tunnels:

1. Open Circuit Wind Tunnel
2. Closed Circuit Wind Tunnel

In open circuit wind tunnels, fresh air is drawn continuously and discharged. Because, the air simply passes once through the test section (where models are mounted) the power requirement of open circuit tunnels tends to be smaller. But, they are more susceptible to drafts, gusts, storms, and other disturbances and so the flow quality in the test section may change unpredictably.

Closed circuit tunnels continuously recirculate the air through a return circuit. The flow passes over *corner vanes* as it returns, which adds to losses. Note that losses are proportional to the square of the velocity V. To reduce local velocity, additional components and designs will be needed. The energy input to the stream causes the air temperature to rise by 5–10°C before stabilizing to a smaller rate of increase afterward. So, the flow temperature has to be monitored, particularly when temperature sensitive instrumentation like a hot wire anemometer is used. It helps to provide a vent so that the static pressure in the tunnel does not drift as the air heats up during a run. It is also useful to keep the test section pressure slightly above atmospheric pressure so that outside air does not gush into the test section and affect the flow quality.

3.30 Major Components of a Wind Tunnel

- Blower or compressor: for air source.
- Settling chamber with honey combs and screens for flow calming.
- Contraction: to accelerate the flow and to generate a uniform flow; a large contraction (inlet to exit area) ratio enables achieving a very low turbulence level at the test section.
- Test section: to place the model and conduct the studies.
- Diffuser: to recover the kinetic energy in the flow before discharging or feeding it back to the blower.

Generally, the turbulence level in good subsonic tunnels tends to be less than 0.1%. In supersonic tunnels, it will be likely larger. A necessary instrument in all wind tunnels is a means to measure the flow speed, which usually is accomplished using a pitot-static tube or a combination of a pitot total pressure tube and a wall static pressure port. In either case, appropriate calibrated read out devices or software are necessary.

3.31 High-Speed Tunnels

The power required to run a wind tunnel is proportional to the cube of the velocity. Most high-speed wind tunnels are typically run as intermittent facilities. Because of this,

- They are simpler to design and less costly to build.
- A single drive may be sufficient to run several tunnels.
- Starting the tunnel is faster and so the starting loads are less severe.

Some issues to be dealt with are as follows:

They need faster instrumentation.

1. They do not give a long run time and thus, controlling flow conditions is difficult.
2. In supersonic flows, the starting loads (during flow establishment) are significantly high and the models must be designed to withstand these loads.

There are two types of intermittent tunnels.

1. Blow-down tunnels, where high-pressure air is usually discharged to the atmosphere
2. In-draft tunnels, where the air is drawn from the atmosphere and discharged to a vacuum

Transonic flow testing requires wind tunnels that are capable of providing for wall interference effects. The extreme sensitivity of transonic flow to wave interactions makes it a difficult task. Wall adaptation is commonly used either through physical wall shape change or through bleeding and reintroducing a portion of the air suitably (usually based on potential theory).

At higher supersonic and hypersonic Mach numbers, a combination of the pressure-vacuum system is necessary. Also, at these speeds, condensation and liquefaction of air can occur and so, the designs must include provisions to prevent these through drying, preheating the air. To achieve independent variation of Mach number, Reynolds number and dynamic pressure, cryogenic fluids (such as nitrogen) are used, which permits studies of the effects of compressibility, friction, deformation, etc.

Some major, well-known national and international facilities include

- The cryogenic wind tunnel at NASA Langley, VA
- The world's largest wind tunnel—80 ft × 120 ft tunnel—at NASA Ames, Moffett Field, CA
- The AEDC Aeropropulsion Systems Test Facility (high altitude, high Mach number flight engine testing) at Tullahoma, TN
- The German-Dutch (DNW) wind tunnels, a complex of 11 tunnels in five separate locations
- European transonic wind tunnel in Cologne, Germany

3.32 Specialized Wind Tunnels

For special purpose studies such as boundary layer research including transition phenomenon, unsteady aerodynamics, etc., special wind tunnels are needed. In the former, long working sections with a flow that is extremely quiet (low turbulence intensity) is necessary. The long test section enables generating a thick boundary layer that will be easier to explore. Special provisions may be needed to manipulate the longitudinal flow pressure gradient. Unsteady flow studies require facilities that can reproduce appropriate dynamic similarity parameters such as relevant range of the degree of unsteadiness (the reduced frequency) in addition to Mach number and Reynolds number scaling. Data acquisition in unsteady flows also requires phase-locking and ensemble averaging techniques to be followed. As such, the facility must be suitably instrumented to provide the required information via encoders or other electronic timing devices.

3.33 Flow Measurement Techniques

The particular measurement technique used depends on the type and nature of data being acquired. For example, velocity or pressure at a point or across a whole plane, or other flow aspects such as density field, surface shear stress, etc. However, for most new problems, flow visualization can provide insight into the key flow aspects that may help select an appropriate measurement technique. Although flow visualization is generally perceived to be qualitative in nature, many newer methods also can provide quantitative flow information.

Flow Visualization

Most flow visualization methods provide the streamline (or streakline) pattern of the flow. Water tunnels serve as an excellent medium for an environmentally acceptable and colorful flow visualization. Higher Reynolds numbers can also be simulated in water tunnels also due to the lower kinematic viscosity of water. Commercially available food coloring diluted appropriately is introduced from dye-ports on models at selected locations. The ports are similar to static pressure ports. The resulting flow pattern is imaged under adequate lighting. The technique has been used to generate flow details in a number of flows ranging from simple streaklines to more complex flows such as over delta wings, aircraft in ground effect, turbomachinery, missile bodies, etc. Remarkable flow details can be captured, which can uncover interesting flow phenomena as seen in the image (Figure 3.88) above. Here, both near and off-surface flow past a UCAV 1303 model can be seen. It shows tip-stall occurring at a low angle of attack attributable to vortex bursting at the cranked trailing edge.

Other techniques using fluorescent dye or hydrogen-bubble can also be used, however, these require more apparatus.

Wind tunnel air flow can be visualized using tufts or smoke at low speeds for which a smoke generator is needed along with a suitably designed model and accessories. Smoke diffusion due to turbulence can be a major problem and hence, a large contraction ratio helps. Surface shear stress behavior can be studied by spraying a mixture of a tracer material such as titanium dioxide and oleic acid in what is known as oil-film method. Even color images can be obtained by using artists' pigment. Excellent flow details such as vortex imprints, lines of separation or reattachment, nodal and saddle points can picked up if care is taken to ensure proper mixture composition, usually achieved through trial and error.

Both the above tend to be largely qualitative although some quantitative details may also be obtained. Some techniques to be discussed later will discuss quantitative approaches.

Measurement of Pressure

By far the most commonly and easily measured quantity in wind tunnel studies is the wall static pressure. It is also the most useful because integration of static pressures over the model provides the force acting on it. The following details will be of use here:

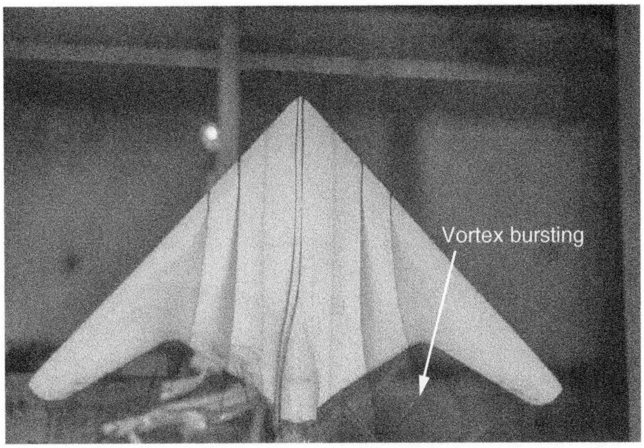

Figure 3.88 Dye-flow visualization over a UCAV 1303 wing depicting surface flow pattern and low angle of attack tip-stall (red dye on star-board wing tip) due to vortex bursting.

The **total or stagnation pressure p_o** is defined as the pressure recorded when a streamline is brought to rest isentropically. The stagnation pressure is constant in the flow in most cases, unless there are shocks present across which it drops. Stagnation pressure is simply measured by connecting a pitot-total tube to a manometer or a pressure sensor. Measuring the stream temperature can permit accounting for any variations if changes are large.

The **static pressure p** of a stream is defined as the pressure at which the local stream flows. When static pressure data is required, the model must be equipped with static pressure ports on its surface. These must be connected a sensor such as a Scanivalve or similar sensor. It is important to have a stable and reliable reference against which the static pressures are measured. Many a times, the atmospheric pressure is used, but it may drift when a storm approaches. Freestream static pressure serves as an excellent reference pressure.

The **dynamic pressure** is the difference between the two values as can be seen by applying Bernoulli equation between two points 1 and 2 in the flow:

$$\frac{p_1}{\rho g}+\frac{V_1^2}{2g}+z_1=\frac{p_2}{\rho g}+\frac{V_2^2}{2g}+z_2=p_o \tag{3.81}$$

Since in most cases, there is no height difference, the above reduces to

$$p_o=p+\frac{1}{2}\rho U_\infty^2 \tag{3.82}$$

which states that for a flow moving with U_∞ as the freestream velocity, the dynamic pressure = ½ ρU_∞^2, where ρ is the density of the fluid.

The wind tunnel speed can be measured using a commercially available pitot-static (P-S) tube, which contains both a pitot total pressure tube and static pressure tube using the above relation. When the flow is compressible, the above should be replaced with

$$\frac{p_o}{p}=\left[1+\frac{(\gamma-1)}{2}M^2\right]^{\frac{\gamma}{(\gamma-1)}} \tag{3.83}$$

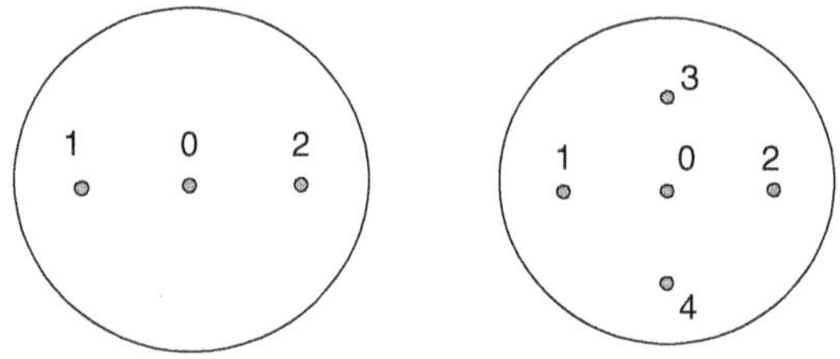

FIGURE 3.89 Multihole probe: (*a*) three-hole (yaw) probe, (*b*) five-hole (pitch and yaw) probe.

Here, care should be taken to ensure that no shocks were present. In case of transonic or supersonic flows, shocks will likely be present and hence, p_0 measured by the P-S tube may be **after** the shock and the flow will not be brought isentropically to rest. Hence, additional measurements will be needed. So, both p_0 and wall static pressure p where the static port of the P-S tube is located should be measured against a known reference (atmospheric). If the ratio of $p/p_0 > 0.528$, no shock is present and so, p_0 measured is the true total pressure. If $p/p_0 < 0.528$, then the flow has become supersonic, so a shock has occurred and thus, p_0 as measured is **after** the shock. Appropriate relations must then be used to get p_0 **before** the shock and then, the stream Mach number.

In cases when flow angularity is present (vortical flows), then multiple-hole probes will be needed. Three, five, and even more number of tubes are used. Figure 3.89 shows a schematic of such probes. Typically, these probes are about 1/8–¼ inch in diameter, depending upon the manufacturer, slight differences can be seen. In Figure 3.89(a), which is a three-hole probe, hole-0 serves to provide the stagnation pressure at zero-degree yaw, holes 1 and 2 together provide the flow yaw angle. It should be calibrated for a known range of positive and negative yaw angles before use. Figure 3.89(b) shows a five-hole probe, where in addition to the yaw measurement capability, holes 3 and 4 are added to provide the flow pitch angle as well. Once again, a thorough parametric calibration for different pitch and yaw angles should be completed before use. Use of commercially available systems and services save time and effort. These probes are satisfactory up to about 45° in pitch and yaw before larger errors begin appear. The leads from the various ports can be connected to a scanning system and the probe itself mounted on a computer controlled traverse for automated data acquisition and subsequent processing.

Measurement of unsteady surface pressures is often desired. A preferred approach is to use unsteady pressure transducers mounted on the surface where data is needed. Additional electronics including power supplies, amplifiers and filtering (signal conditioners) will be needed which adds significantly to the cost of measurements. They also have to be calibrated against known pressures in steady flows (for want of a known and calibrated unsteady pressure source). It is a challenging task in general, and special care must be taken to account for instrumentation zero drifts and temperature effects. If the pressure probes discussed earlier are used, then some means of correcting for the lag effects in the pressure leads will be needed. These are generally derived assuming one-dimensional unsteady flow analysis in the pressure tubing. Data acquisition involves time stamping of the samples at appropriate sampling frequency and subsequent ensemble averaging.

Measurement of Loads

Although loads and moments can be derived from the above discussed pressure measurements many times, it is of interest to measure them directly. Typically one can achieve this using a strain-gage-type internal balance. The balance consists of sets of electric resistance strain gages for each moment that are arranged in a Wheatstone bridge pattern. Typically, a strain gage unit is bonded to the specimen with insulating cement under no-load conditions. A load is applied to it, which produces a deformation in both the specimen and the resistance element. This deformation causes a change in resistance which can be used to measure strain. In all gages, moments are the actually measured quantities, but the forces along the three axes can be derived knowing the balance design details. Both static and dynamic loads can be measured.

The balance is normally a ½ in or up to 1 in diameter, few inches long cylindrical body and the strain gages suitably arranged on orthogonally machined flats on the body. It is housed inside the model and should be attached to it securely to ensure that all loads are directly transferred to the balance and not to any other component. A standard approach is to use it in the rear sting mount although other methods may be possible. Any change of alignment between the balance and the model gives rise to erroneous data and may also result in the loss of model. One of the methods of attachment is to use a locking taper of 1 in 12. This is to ensure that the force required to separate the model from the balance is positive, which has to be applied through the disassembly screw. The forward end of the balance is designed to house an attachment screw to assemble the model and lock the two together. Another consideration is to keep all components located inside the model where air circulation is a minimum. This will keep the temperature constant and reduce erratic changes in bridge output. All power supply and signal wires to the strain gage bridges should be well shielded from the air/water flow also.

Gage factor is a parameter that is used in the selection of the balance and to assess the sensitivity of the gages used. It is defined as the percentage change in resistance of the gage to the percentage strain. It can be shown that only the change in resistance is needed to determine the strain and hence, loads. Typically the gage factor depends upon the material and is around 2 for most wire-type strain gages. Since the system is sensitive to temperature, temperature compensated gages must be used. However, commercially available units are generally temperature compensated. Standard power supplies and signal conditioner units are used to energize the gages and process the signals prior to reading them routinely in the data acquisition (DAQ) systems.

Calibration of the gages has to be specifically and carefully performed with loading applied in the exact way they are used to identify the normal force, side force, and drag force along with the respective moments. The instrumentation is linear and, generally, the calibration can be repeatedly used.

Measuring lift force and pitching moment, side force and yawing moment is fairly straight forward. Rolling moment requires a special design inside the balance that the gage is only sensitive to this moment. Drag force is the smallest component of force and so is the hardest to measure accurately in most normal wind or water tunnel studies. Most balances that measure drag have a specially designed unit that permits it.

Measurement of Velocity and Its Fluctuations

The Hot Wire Anemometer

A hot wire anemometer is an instrument used to measure velocity and its fluctuations in fluid flow. The velocity sensing element a very fine wire, about 5 μm in diameter and about 0.5–2.0 mm long, strung between two needles known as probe supports. It is called the hot wire because it is kept hotter than the surrounding fluid, whose velocity is being measured. Along with the associated electronic hardware the instrument becomes known as the *Hot Wire Anemometer* (HWA). The type of wire and its material is generally chosen depending upon the number of velocity components measured and the fluid in which it is used from a manufacturer's list. The wire is set up as one arm of a Wheatstone bridge circuit with the bridge initially in balance. The wire is heated to a temperature above that of the flowing fluid which cools it. The resulting imbalance of the bridge circuit then becomes a measure of the fluid velocity. Feedback electronics act at a high frequency to supply the bridge with additional current to restore the temperature to the original value and hence the name constant temperature anemometer (CTA). The technique is used in this mode of operation. Thus, a hot wire does **not directly** measure the velocity. Since, the technique is based on forced convection heat transfer, it is imperative that its **thermal inertia** be small, with a small size, it is indeed possible to keep the inertia small and thus, one can use it in measurement of turbulent flows including to measure its kinetic energy spectrum and dissipation characteristics.

A hot wire sensor (in general, most sensors) should satisfy the following requirements:

It should have the smallest volume possible in order to be able to respond to all the fine scale structures in the flow. This will also ensure adequate high-frequency response. A hot wire sensor should be short to maximize spatial resolution, and aerodynamic disturbance, but long to minimize conduction losses to supports, provide a more uniform temperature distribution along its length (for proper response) and minimize support interference. The sensor diameter should be small to eliminate output noise due to separated flow around the sensor, be capable of high frequency response, and improve signal-noise ratio. A large diameter is required from strength considerations. One should ensure to reduce contamination effects from dirt and other particles in the flow. Thus, conflicting requirements are evident and a compromise is sought. Typically, 2.5–5-μm-diameter sensors, with length-to-diameter ratios $2l/d$ of 100–600 are used. When ruggedness is important, hot-film sensors are recommended, which operate identically, with a slightly reduced frequency response.

The hot wire technique is based on rules of convective heat transfer as expressed in King's law. According to it, the Nusselt number in the flow is related to the flow Reynolds number through the equation

$$Nu = A + B\,Re^{0.5} \tag{3.84}$$

Here, A and B are constants determined through calibration and the exponent n ranges between 0.45 and 0.55; with 0.5 as the value for most calibrations. Based on the curve fit to the calibration data, slight adjustments are made. Calibration of a hot wire is the process of establishing the *actual* relationship between the output voltage of a hot wire anemometer and the velocity, in a low-turbulence flow. It is performed in a clean flow such as in the potential core of a free jet or the free-stream of wind tunnel. The procedure consists of incrementally increasing the flow velocity from a small value to the maximum value and reading the output voltage of the HWA. A typical hot wire calibration is a nonlinear curve and the sensitivity decreases with increasing velocity. In view of King's law, we can write the voltage of the bridge circuit is

$$E^2 = A + BU^n \tag{3.85}$$

Here, E is the bridge output voltage as recorded and U is the calibration velocity. With $n = 0.5$ for most flows, it becomes

$$E^2 = A + B\sqrt{U} \tag{3.86}$$

When equation (3.86) is plotted, a linearized calibration results with A as the y-intercept and B as the slope of the line. To avoid any linearization issues and calibration errors, one can simply list the E and U values in a lookup table and the DAQ software can interpolate each instantaneous voltage value build a time series of velocity and compute the relevant statistical values for the velocity.

Orientation of a hot wire probe is critical for proper measurements since it relies on heat transfer from the wire to the flow, which occurs regardless of the probe orientation. Proper probe orientation can be determined using the *cosine law*, according to which the maximum heat transfer occurs when the flow is directed normally to the wire. If an infinitely long hot wire is considered, then the angle sensitivity of the hot wire is given by the cosine law. Thus, it is necessary to determine the particular orientation needed for each component of velocity. In Figure 3.90, the probe is

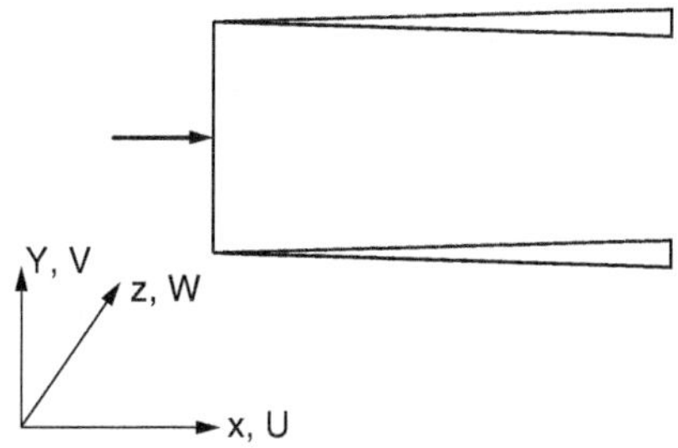

FIGURE 3.90 Orientation requirements of a hot wire.

oriented to measure the x (or U)-component of velocity, which is normal to the wire body. However, one sees that the z (or W)-component of velocity is also normal. Thus, to obtain satisfactory measurements of U-component velocity, the W-component must be either 0 or very small. The heat transfer is only minimally sensitive to the tangential V-component (see below). The following equations provide a basis here.

$$U_{\text{eff}} = U\sqrt{\cos^2\alpha + k_T{}^2\sin^2\alpha} \tag{3.87}$$

In equation (3.87)

U_{eff} is the effective velocity seen by the wire (= $U\cos\alpha$ for an infinitely long wire)

α is the flow angle to the probe axis

k_T is an empirical factor (not always a constant) for tangential velocity sensitivity

For $0 < \alpha < 60°$ for $(l/d) = 100$, $k_T = 0.2$

It tends to 0 as (l/d) approaches 300

Additional guidance can be found in the literature for when W is large. It is

$$U_{\text{eff}} = \sqrt{U^2 + k_T{}^2V^2 + k_N{}^2W^2} \tag{3.88}$$

K_N tends to be large (1.0–1.2) and hence, it is necessary to orient the probe such that W is small.

A major limitation of the hot wire is rectification of the output. The technique is *insensitive* to flow direction which is a significant limitation in separated and reversed flows and in very highly turbulent flows in which the flow direction may vary rapidly.

Despite these restrictions, the hot wire technique can successfully yield multicomponent velocity data. Two-component velocity can be measured using an "x-wire," which has a design that makes it look like the letter x. However, many designs exist and usually, one selects them from the manufacturer's catalog and then orients it appropriately based on the design and expected flow velocities and angularity. The calibration must also be accordingly carried out. Since most measure a sum or difference of two velocities, individual components of velocity must be suitably resolved. Three component measurements are also possible. The calibration and data reduction effort required grows rapidly in this case.

Before proceeding with the calibration and actual measurements, it is necessary to tune the system to the highest flow frequency based on estimated flow scales. Without this step, the measurements, especially in highly turbulent flows will be inaccurate and more importantly, the feedback circuit may become unstable and start to oscillate causing excessive current flows and damage the probes. Some commercial units may already be tuned so that this step may not be necessary.

Particle Image Velocimetry

Particle image velocimetry (PIV) is a particle tracking velocity measurement technique in which flow velocities are measured by recording multiple images of the same "tracer" particles. The particles are typically 0.2–0.22 μm (oil) fog mist and are distributed uniformly. The region is illuminated using a light source which is flashed multiple times in a certain small time window. The actual particle displacement in the images for a known time window (of a few nanoseconds to microseconds) is measured (calculated) using sophisticated correlation techniques from which the velocity is calculated simply as distance traveled in unit time. Note that this approach gives the velocity vector seen in the plane of the light sheet, which can be oriented in the traditional X, Y, or Z planes and the vector resolved to get the velocity components.

A typical PIV setup, shown in Figure 3.91, consists of powerful lasers that can be pulsed at precisely timed instants, some optics to produce a light sheet, a device for seeding the flow with particles, a high-quality digital camera to record the images and a frame grabber along with sophisticated software for evaluating and extracting the particle displacement information from these images. It is important that the plane of the light sheet is normal to the camera and lens plane. If not, the use of the Scheimpflug technique offers a way to mitigate the optical distortion and effects of nonorthogonality. Laser pulses of up to 500 mJ energy levels are used. The exposure durations are very low, order of a few nanoseconds. The time between pulses is a function of the flow speed and ranges from milliseconds to nanoseconds. The velocity values computed are generally averaged over up to 100 images.

Data processing involves *interrogating* a group of particles (say in a 32 × 32 pixel area) and through a statistical correlation of a sampled number of image fields, the particle group displacement is arrived at by cross-correlating the data suitably and thus, its velocity. It should be noted that even though a group of particles is followed during this *interrogation process*, the velocity is deemed to be determined at *a point* in the flow field. This point is the *centroid* of the interrogation window.

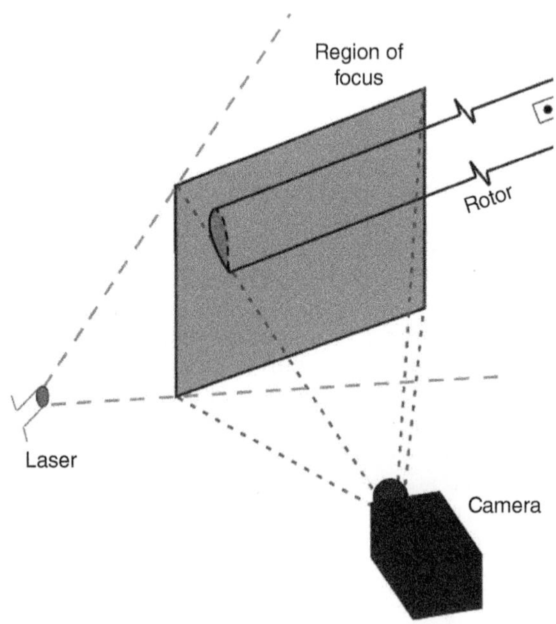

FIGURE 3.91 Schematic of the PIV system optics.

The use of correlation techniques enables determination of particle displacements regardless of whether the images are independently recorded on CCD cameras or whether they are recorded on a single photographic plate. This step is also referred to as *image correlation* in digital PIV technique. The fact that the particles move off the sampled region, disappear through three dimensional motion out of the plane of the laser sheet, a large number of particles that cause just enough scatter to appear only as noise in the background, etc., add to the measurement uncertainty.

Several uncertainty sources are inherent to the technique and should be carefully minimized. These include particle image density, velocity gradients within the interrogation window, uncertainty due to different particle size imaging, effects of using different interrogation window sizes, etc.

The technique can provide global velocity fields in short test times making it valuable for a number of aerodynamic applications.

3.34 Density-Based Optical Flow Field Measurement Methods

Optical methods refer to flow measurement techniques that use a measurable change in the flow property for flow diagnostics. In this way, they are different from other flow visualization techniques. Several optical methods have been used in aerodynamics, but here we concern ourselves with the following three most commonly used density-based techniques:

1. Schlieren technique: based on the gradient of fluid density.
2. Shadowgraphy technique: based on the second derivative of fluid density.
3. Interferometry: based on the density change itself.

There are many advantages to these methods. The most important is that no probe is introduced into the flow and hence, it is not disturbed in any way. Since a property change is used, the response is instantaneous. Hence, the flow transients can be detected accurately and thus, very detailed flow information becomes available. Also, the whole flow field is measured *at once*. This means that the experiment time can be very short. These techniques are most frequently used in intermittent, high speed tunnels, where the run time is very short. The techniques can be both qualitative and quantitative.

The principle of optical methods lies in the following:

A light beam passing through a fluid is disturbed due to the inhomogeneous distribution of refractive index in it. Note that the change in refractive index is brought about by the changing local densities in the flow. The two important aspects of this effect are

1. Light is deflected from its original path, *always toward the higher density* region of the flow.
2. The phase of the disturbed light wave is shifted with respect to that of the undisturbed wave.

The starting point for all methods is the Gladstone-Dale equation which relates the refractive index n of the fluid medium and its density ρ as,

$$\left(\frac{n-1}{\rho}\right) = C = \text{constant} \tag{3.89}$$

Here, C is known as the Gladstone–Dale constant. It depends strongly on the gas and weakly on the wavelength λ.

A schlieren system consists of a light source located at the focal point of a mirror to produce a parallel beam of light. This beam passes through the test section, experiences deflections based on the local density and emerges from the other side. It is then focused on to a knife edge (usually a

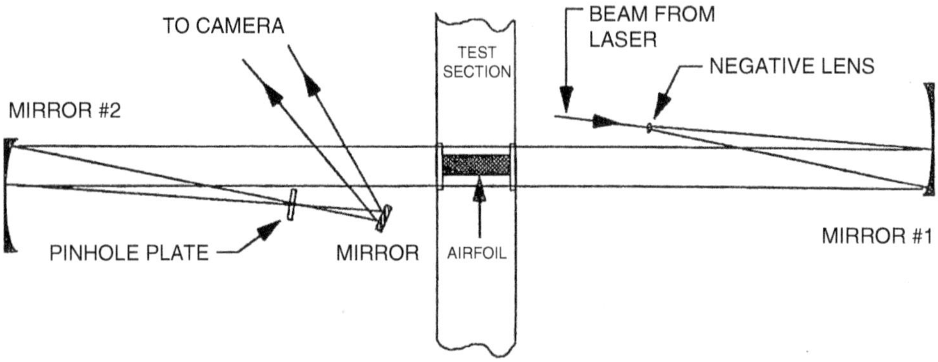

FIGURE 3.92 Optical arrangement of the schlieren system (and also of a point diffraction interferometer).

razor blade) by another mirror. This arrangement is known as the Z-type schlieren system and it introduces a slight degree of astigmatism, which is not a major issue since the technique is largely qualitative only.

In Figure 3.92, the knife edge takes the place of the photographic plate which is used for the interferometry technique. The knife edge allows the determination of the direction of the density gradient. The sensitivity of the system can be tuned by adjusting the knife edge. It can be effectively set up to accentuate the density gradient to achieve better flow visualization (to change the image contrast also). Imaging optics can then be set up to obtain a magnified image for capture by a camera. Also, a horizontal knife edge yields the vertical density gradient information and vice versa.

A shadowgraph is obtained by simply setting up a camera on the opposite side of the first mirror in Figure 3.92, which records the light beam as it exits the test section. In a shadowgraph system, the linear deflection of the disturbed light beam directly produces an image unlike its angular deflection as in the schlieren system. A major limitation of the shadowgraph is the fact that since it responds to the second gradient of density, the light beam deflection is not linear. It varies locally very strongly and through regions of different density changes. This also makes interpretation of the shadowgraph images somewhat more difficult. One should already have an idea of the flow before the images are interpreted (for example, the location of the shocks) making the use of shadowgraphy difficult over curved surfaces.

Interferometry uses the principle of interference of light waves to produce fringes that correspond to changes in flow density. Typically, there is an undisturbed light wave, known as *reference beam* that is not affected by the flow and another, the *signal beam* or *test beam* that is affected by the flow. The signal beam is deflected by the flow density changes and hence, it travels a different path length and arrives at the image plane with a different phase when compared to the reference beam. When these two wave fronts reach the image plane, they interfere with each other and the path length difference produces interference fringes that represent a definite change of density from fringe to fringe.

If $\overline{\Delta PL}$ is the path length difference between the signal and the reference beams and Δ is the phase difference between the two beams due to it, then,

$$\frac{\Delta}{2\pi} = \frac{\overline{\Delta PL}}{\lambda} \tag{3.90}$$

where λ is the wave length of the light used. If $\frac{\overline{\Delta PL}}{\lambda} = 0$ or an integer, then there is no path length difference and hence, constructive interference occurs, leading to uniform brightness. If $\frac{\overline{\Delta PL}}{\lambda}$ is a half integer (½, 3/2, 5/2, 7/2, etc.), then there is destructive interference and a uniformly dark region results. For gases, in view of the Gladstone–Dale relation, we could obtain the density by knowing the fringe number using

$$\varepsilon = \frac{\overline{\Delta PL}}{\lambda} = \frac{C}{\lambda}\int(\rho - \rho_{\text{ref}})dZ \tag{3.91}$$

where Z is along the beam path (through the test section only) and ρ_{ref} is the density of the undisturbed medium. Note that the above relationship is valid for path length differences less than the coherence length of the light source. With the use of lasers, this is of the order of 10–15 cm.

The well-known interferometer that is often well described is the Mach–Zehnder interferometer. However, its implementation is extremely complex. In the recent past, a very simple to implement technique known as point diffraction interferometry has been developed and will be described below.

The technique is based on the principle of interference between a flow aberrated (due to flow density changes) signal beam passing around a pinhole and the diffracted reference beam originating at the pinhole, Figure 3.93. Light from a point source is directed through a flow field and focused on to a pinhole in a partially transmitting plate as shown in Figure 3.92. This one beam of light essentially becomes *two* beams with the portion of the beam passing through a pre-created spot (pin-hole) as the reference beam and that passing around it as the signal (object) beam. The object wave is attenuated because of the transmittance of the pinhole (point-diffractor) plate, but it retains all the phase information introduced during its passage through the test section. If the pinhole is sufficiently small, it acts as a local point source by "cleaning up" the distortions in the beam. Thus, light passing through it is diffracted as a reference wave. At the image plane, the object and the reference wave fronts interfere producing fringes, which are contours of constant optical path length difference corresponding to definite density changes in the flow field being investigated. Quantitative flow field data can be recovered by simply counting the fringes using the equation (3.92) and tracing the fringe centerline for flow mapping:

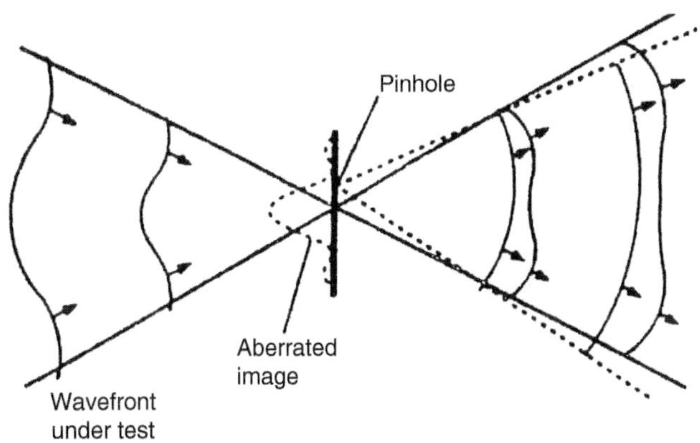

FIGURE 3.93 Principle of point diffraction interferometry.

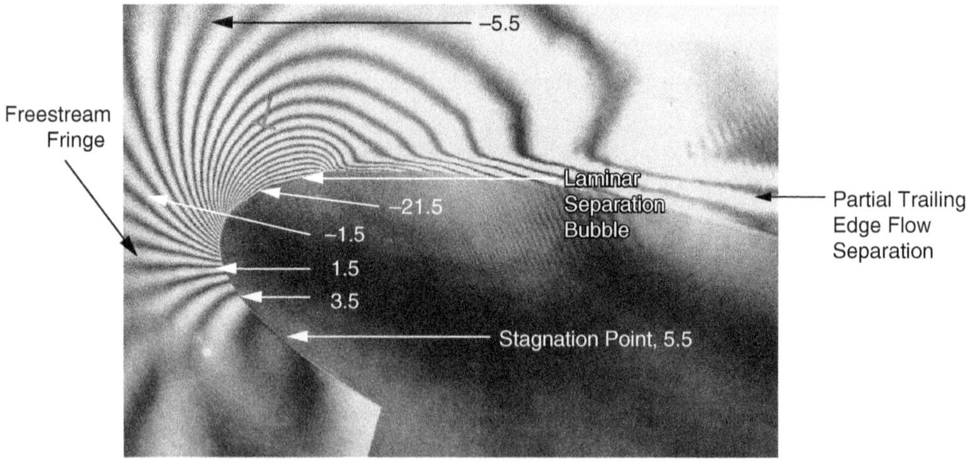

FIGURE 3.94 A point diffraction interferogram image, $M = 0.3$, $\alpha = 17°$.

A point diffraction interferogram image showing the laminar separation bubble near an airfoil leading edge is shown in Figure 3.94.

$$\rho - \rho_{\text{ref}} = \left(\frac{\lambda}{n_0 - 1}\right)\left(\frac{\varepsilon \rho_0}{L}\right) = A\varepsilon \tag{3.92a}$$

$$\frac{\rho}{\rho_0} = \frac{\rho_{\text{ref}}}{\rho_0} + \frac{A\varepsilon}{\rho_0} \tag{3.92b}$$

where A is a facility-dependent constant that involves the wavelength of the laser used also.

Since $\frac{\rho_{\text{ref}}}{\rho_0}$ is a function of the free stream Mach number only, $\frac{\rho}{\rho_0}$ can be determined by knowing the fringe number. In this method, a positive fringe number represents deceleration and vice versa.

3.35 Other Flow Field Measurement Methods

The field of aerodynamic measurements includes several other methods such as liquid crystal technique for surface shear stress measurements, Pressure sensitive paint (PSP) for surface pressure distribution measurement and so on. These are fairly complex imaging methods and their implementation is fairly involved. Hence, are not discussed.

References

Barlow, J. B., Rae, Jr., W. H., and Pope, A. 1999. *Low Speed Wind Tunnel Testing*, 3rd ed., John-Wiley & Sons, Inc. ISBN 0-471-55774-9. New York, N.Y.

Brock, N., Chandrasekhara, M. S., and Carr, L. W. 1991. "A Real Time Interferometry System for Unsteady Flow Measurements," *ICIASF'91 RECORD*, IEEE Publication 91CH3028-8, pp. 423–430.

Goldstein, R. J. 1983. *Fluid Mechanics Measurements*, Hemisphere Publishing Corp. ISBN 0-89116-244-5. New York, N.Y.

Merzkirch, W. 1987. *Flow Visualization*, 2nd ed., Academic Press, Inc. ISBN 0-12-491351-2. Cambridge, Mass.

Raffel, M., Willert, C. E., and Kompenhans, J. 1998. *Particle Image Velocimetry: A Practical Guide*, Springer. ISBN 3-540-63683-8. New York, N.Y.

Smits, A. J. and Lim, T. T. (eds) 2000. *Flow Visualization: Techniques and Examples*, Imperial College Press. ISBN 1-86094-193-1. London.

PART 6

Fast Response Pressure Probes

Anthony J. Gannon

Fast response pressure probes are useful where high fidelity data of a rapidly changing flow field is required. Examples of this type of field would be the pressure variation on a gas-turbine engine casing due the blades rotating or transient data over a body during rapid maneuvering. This high frequency data is often very useful in gaining insight into complex transient flow fields or as a validation tool for transient simulations.

Some attention will be given to the operational principles of these probes but the main focus of this discussion will be their practical use. This will include their installation, an extremely important consideration if accurate transient data is to be obtained. Depending on what data is desired, single installations or arrays of these probes can be used simultaneously. Probe limiting performance and the required data acquisition systems will be discussed. The post processing of the data beyond just the raw pressure traces can yield useful flow characteristics and some of these techniques will be mentioned.

The limiting sampling frequency is usually governed by the natural frequency of the sensor.

- Motivation; why we need these types of sensors.
- What will be covered here.
- Typical performance possible with these sensors. (Give broad stroked at this point.)
- Examples where they would be useful.

3.36 Probe Types and Ranges

Most fast response pressure probes are based on a semiconductor construction and rely on the mechanical strain of some element in response to a change in pressure. These elements are usually very small allowing for a fast response in a change to pressure. Without delving into the different manufacturer approaches, a full bridge-type arrangement usually allows for this strain to be measured.

Depending on the specific application a probe of the correct pressure and frequency response needs to be chosen. As will be described the two are somewhat interrelated and both characteristics must be considered when choosing a probe for a particular application. Probes are usually rated by their maximum differential pressure range or the range within which they can be expected to give a linear or predictable reading. There is usually a certain amount of margin above which the probe will not be damaged but will not give accurate readings. Also contained within the data sheet of the probe is the natural frequency of the sensor. Obviously if the pressure field of interest has a frequency nearing or above this natural frequency the probe sensor element may be excited and deliver meaningless results; typically the higher the pressure rating of the probe the higher its natural frequency due to the stiffer measuring elements required. This data is available on the technical specifications sheet of a probe.

So-called *microphone* probes are usually the most sensitive probes available with a sensitivity range of 100–194 dBa with a mechanical resonant frequency of 125 kHz. On the other end of the scale, high-pressure probes can measure up to maximum pressure of over 5 MPa and resonant frequency of 1,000 kHz.

Usually the core element of a probe is configured to measure differential pressures between two pressures of interest. Figure 3.95 shows a 2-mm-diameter basic barrel-type fast response probe with the sensor face and rear reference pressure tube connection shown. Also shown, is a

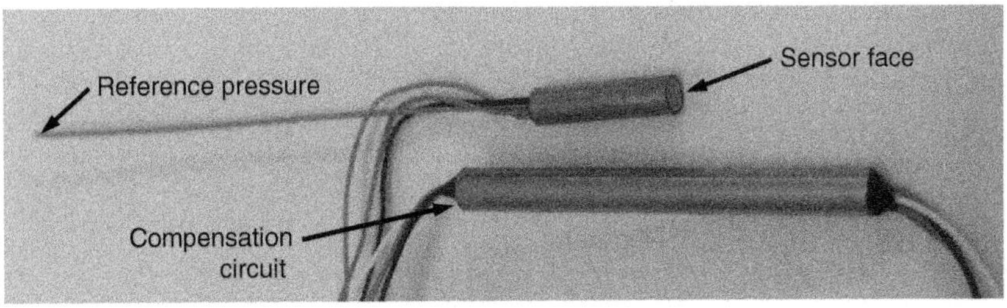

FIGURE 3.95 Basic fast response probe with compensation circuit.

compensation circuit, usually to account for temperature which simplifies the probe calibration. So-called surface mount probes are also available for situations where a low profile mount is desired. Usually probes are configured in the way shown in that the differential pressure between two levels of interest is being measured. It is up to the user to fix the pressure on one side of the probe with the other being used to measure the pressure field of interest. As mentioned most probes are temperature compensated so a calibration at one temperature will be valid over the specified temperature range of the probe. Probes that operate away from ambient temperatures are available with both high temperature (>400°C) and cryogenic types available from various vendors. They usually have additional heat shielding over the signal wires and modified temperature compensation circuits.

Fast response pressure probes tend to be small but are available in a range of sizes. The sampling areas are usually circular and range from 1.5 to 4 mm in diameter. The effect of the size choice will be discussed later but in general the smaller the probe the more fragile it is and can be more difficult to mount.

3.37 Probe Mounting

The performance of a probe is highly dependent on its mounting and proximity to the flow. Ideally the probe should be mounted flush with the surface where the pressure field is to be measured. Sometimes the probes are slightly recessed to avoid damage in harsh environments, a typical one being over the blade tips of turbomachines. This type of installation could be used in an engine to give useful data but it is also desirable to make sure the probe has a useful life. Mounting of probes in cavities causes two problems; an attenuation and lagging of the pressure signal. Both of these errors can, however, be accounted for through a calibration method outlined by Kupferschmied et al. (2000). This allows accurate amplitudes and phases to be measured while still allowing for some protection of the probe.

This attenuation of the pressure amplitudes can be useful in flow fields where a periodic pressure is being measured as it is often desired to know the mean pressure. Using a very long port high frequencies can be damped out and this mean pressure used as the reference pressure to the rear of the probe. This allows the pressure about the mean to be measured and lowers the chance of overpressuring the probe through the use of a fixed reference pressure.

Figure 3.96 shows some basic probe mounts. Example (*a*) has an externally referenced pressure while (*b*) has a reference static port close to the probe that can be used as a reference port if so desired. In the case of (*b*) a separate measurement of the mean pressure from the static port would be required if absolute pressures were desired. In this second situation, care must be taken that the static tube is long enough to act as a filter. Physically the probes are usually smooth on the outside and so some sort of bonding or clamping is required with care taken not to damage the probe during this process.

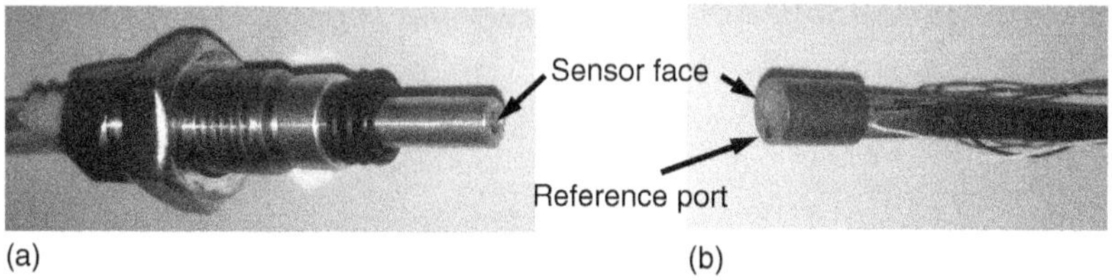

FIGURE 3.96 Mounted probes (*a*) externally referenced pressure (*b*) passively references pressure.

3.38 Measuring Considerations

Obviously, the type of flow field being measured will affect the probe choice with the pressure and temperature ranges being the most influential factors. As mentioned, the frequency of the field must also be considered to ensure that it is below the natural frequency of the probe. Often neglected is the size of the flow features that are being measured. On full-sized bodies, this is less of a consideration but often high-speed pressure data from scale models in wind-tunnels is desired.

Figure 3.97 shows an example where the size of the probe is large relative to the object being investigated (Gannon et al. 2005). The single-blade tip shown passes the pressure probe at a speed of 396.2 m/s (1,300 ft/s) with the blades passing at a frequency of 9.9 kHz, resulting in a rotating shock structure that clearly required high-speed pressure measurements to resolve it. It can be seen in the figure that the blade leading edge size is smaller than that of the probe. This results in the time averaged pressure over an area around the blade leading edge being measured as it passes the probe. If this effect is borne in mind the results are still useful. When comparing the experimental data to simulations for example it must be ensured that the values extracted from the simulation undergoes the same averaging. As mentioned in the case above, shock waves were present and the physical probe size will always have the effect of smearing this phenomenon out.

3.39 Multisensor Probes

Fast response probes that are analogous to multihole pitot-type probes are also used. Many are custom manufactured for particular applications while some commercial products are available. Once calibrated these types of probes allow the transient pressure and velocities at a particular point to be measured.

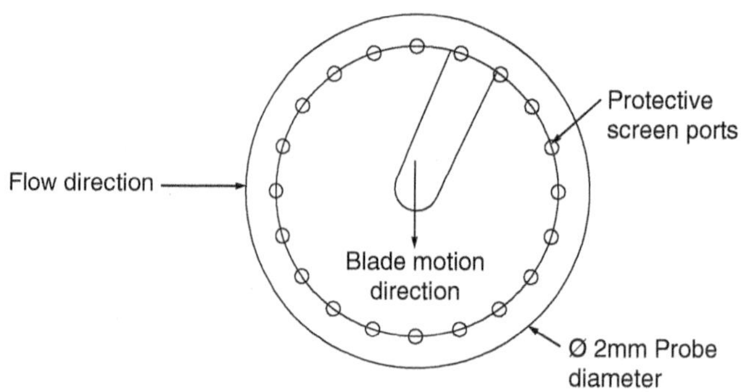

FIGURE 3.97 Probe size relative to transonic rotor leading edge.

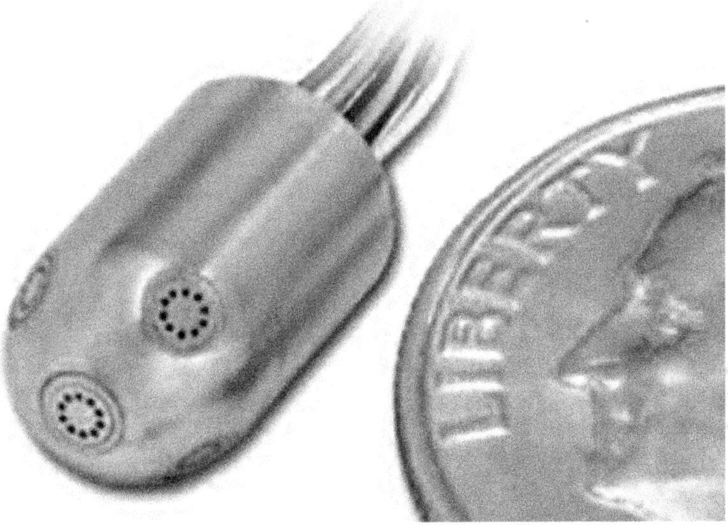

FIGURE 3.98 A Multisensor fast response probe (NASA 2013).

Kupferschmied et al. (2000) show a number of these types of probes and a technical brief sponsored by NASA (2013) shows a slightly different variation (Figure 3.98). The workings of these probes and their calibration is similar to that of traditional multihole probes, but they have the capability of taking fast measurements of the flow field. The issues mentioned previously also apply to these types of probes; the probe must be small enough to resolve the flow structure that is being measured and if the sensors are recessed the lag and attenuation must be taken into account.

3.40 Data Acquisition

Most fast response probes can be sampled using the same techniques as those for a Wheatstone bridge using a full-bridge amplifier. Most modern sampling is done using digital methods and when choosing a suitable system certain capabilities are desired. The choice and performance of a data acquisition system will greatly affect the quality and usefulness of the data captured from an array of fast response probes. As transient phenomena are usually being measured it is usually desirable to sample the probes simultaneously which leads to demanding requirements. A summary of these is as follows:

- A high resolution of at least 16 bits
- A high sampling rate per channel
- Ample data storage and bandwidth to allow for continuous sampling

Ample digital resolution is required to discern small changes in pressures and a 16-bit resolution is usually possible. A compromise between a high resolution and fast sampling rate is sometimes made as it is simpler to sample at lower bit resolutions than high ones.

A fast sampling rate per channel is required and caution should be taken when purchasing sampling systems. Many specify the total sampling rate of the device which is then distributed among the channels available. An inherent problem with this type of device is often that a single analogue-to-digital integration circuit is shared among the channels. If there is a large change in sample voltages from one channel to the next this can result in so-called *cross-talk*, where the voltage sampled from the one channel can affect the voltage sampled from the next. It is best to use devices that dedicate an analogue-to-digital integration circuit to each channel and ensure that the device explicitly states what the simultaneous data acquisition rate is. This type of system also has the advantage that the data points are all taken at exactly the same time. Sampling frequency capabilities are constantly improving with time. For example, around the year 2000, a 16-bit, 200-kHz system with 16 simultaneous channels was one of the fastest commercially available products. In 2014, a-16 bit, 1-MHz system with 64 simultaneous channels was available.

Finally, the length and frequency of data sampling will often put strains on the storage systems. The bandwidth of the system must be able to capture the data at the rate at which it is generated and then store it in some way. Solid-state systems are often faster than spinning storage media and should be considered. In the following section, it is noted that observation of the data in the frequency domain is sometimes desired. Depending on the frequencies that are found in the flow field a sampling frequency of at least double the highest frequency in the flow field to avoid aliasing of the signal is required. Long samples may also be required if low frequency signals are of interest even though the flow field may be dominated by high-frequency signals. This is often the case in turbomachinery.

3.41 Postprocessing

Once the data has been captured from probes it must usually be postprocessed in some way. The raw data are usually in the form of a digitally stored voltage, which using the calibration curve of the probes is converted to a pressure. As mentioned earlier, arrays of high-speed probes are often used as these can give a transient picture of a particular flow field, which might lead to better physical insights of the flow structure or be useful in investigating the performance of numerical simulations. Investigation within the time domain and frequency domain are both useful and an example of each is presented here. The case of flow over a transonic compressor rotor is used as an example but the techniques remain valid for other types of flow.

Time Domain Analysis

Figure 3.99 shows an array of fast response pressure probes positioned over a transonic compressor rotor. An external once-per-revolution trigger was also sampled simultaneously to allow for synchronization of the data. Knowing the physical locations of the probes and accurately recording times of the samples it is possible to create a contour plot of the pressure field projected on the case wall.

Figure 3.100 shows raw data traces from high-speed probes for varying conditions through a transonic rotor blade corresponding to the data sampled from probe number 3 in Figure 3.99.

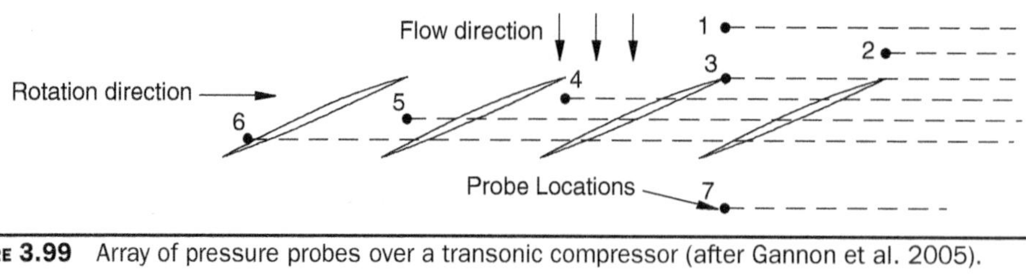

FIGURE 3.99 Array of pressure probes over a transonic compressor (after Gannon et al. 2005).

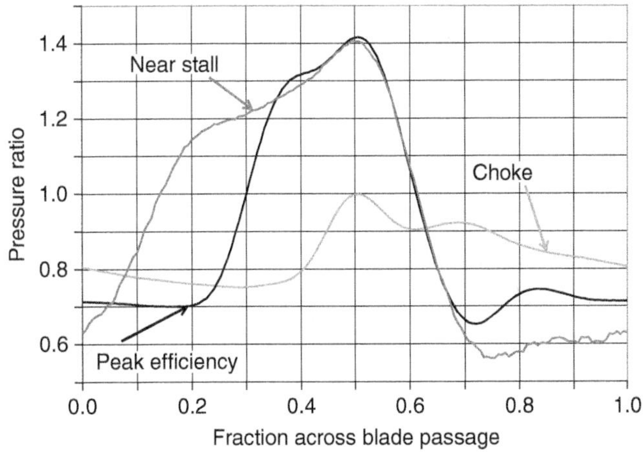

FIGURE 3.100 Time traces from various flow conditions.

Taking all the traces simultaneously it is then possible to build a contour plot of the pressure field that is projected onto the case wall as the blades pass.

Figure 3.101 shows the projected pressure field with the position of the blades visible and an oblique shock waves attached to their leading edges. This type of comparison is useful in gaining insight into flow phenomena as well as evaluating numerical models. In this case, it can be seen in the experimental data (*a*) that the shock wave extend significantly upstream. This highlighted a problem in the simulation that the inlet boundary was attenuating the shock wave, leading to changes being made in future simulations.

Frequency Domain Analysis

With certain types of flow fields such as in rotating flows associated with gas turbines or flapping propulsion it is often useful to investigate pressure data in the frequency domain. The transformation of data sampled in the time domain into the frequency domain is usually performed using a fast Fourier transformation (FFT), sometimes called discrete Fourier transforms (DFT). This is an extremely efficient method for application to discrete data sets (Cooley and Tukey 1965). Most software packages have an implementation of these transforms with the most common being that of Frigo and Johnson (2005). A free online resource is available and references within the paper.

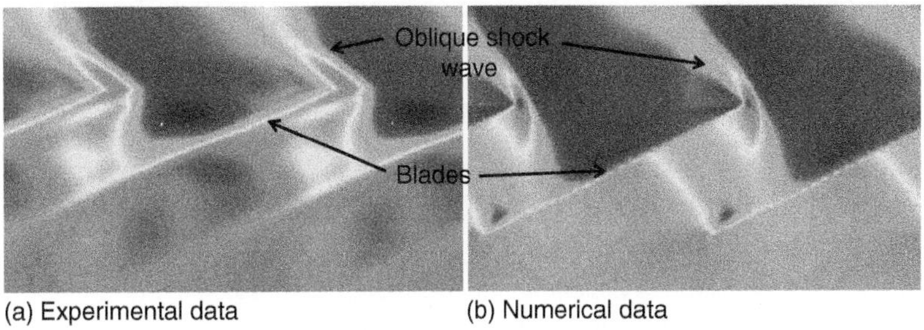

FIGURE 3.101 Pressure contours projected onto the case wall captured from high-speed probes compared to a numerical simulation (after Gannon et al. 2005).

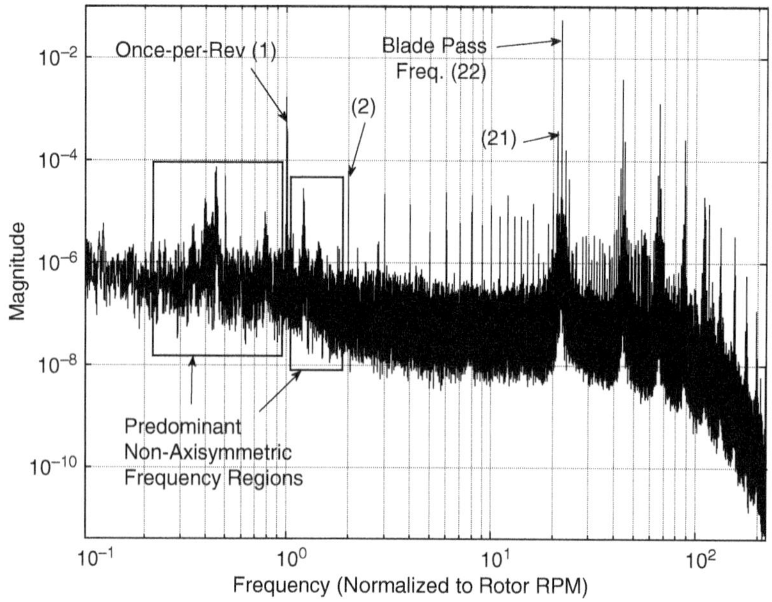

FIGURE 3.102 High-speed pressure probe data transformed into the frequency domain (Gannon et al. 2012).

Figure 3.102 shows high-speed data sampled from a fast response probe and transformed into the frequency domain using a discrete FFT. Depending on what is being analyzed, it is often useful to scale the frequency to some useful reference, in this case the rotational frequency of the rotor. The dominant frequencies are that of the blade passing and the associated harmonics, which is to be expected as this is the strongest forcing function. The rotor had 22 blades in this case so the main peak occurs at 22 due to the scaling. The next largest frequency is at the rotor speed, which is due to differences in flow between one passage and the next. This was surprising as it is generally assumed that the flow in all passages is the same during normal operation. An investigation of the data in the time domain would not yield this result as easily.

The most surprising result was the observation of frequencies below the rotor rotational speed. This indicated that there were periodic flow patterns that move from one flow passage to the next at a frequency less than the rotor rotation frequency. This clearly violates the assumption that the flow in each compressor passage is the same. This observation would have been difficult to find if the data was left in the time domain. In addition, it required a high-sampling speed to avoid aliasing of the blade passing frequencies and harmonics and a long sample time to allow the low frequencies to be observed within the frequency domain.

References

Cooley, J. W. and Tukey, J. W. 1965. "An Algorithm for the Machine Computation of the Complex Fourier Series," *Mathematics of Computation*, vol. 19, Apr., pp. 297–301.

Frigo, M. and Johnson, S. G. 2005. "The Design and Implementation of FFTW3," Proc. IEEE 93 (2), pp. 216–231.

Gannon, A. J., Hobson, G. V., and Shreeve, R. P. 2005. "Measurement of the Unsteady Case-Wall Pressures over the Rotor of a Transonic Fan and Comparison with Numerical Predictions," ISABE 2005, 17th International Symposium on Airbreathing Engines, Munich, Sep. 4–9.

Gannon, A. J., Hobson, G. V., and Davis, W. L. 2012. "Axial Transonic Rotor and Stage Behavior near the Stability Limit," *Journal of Turbomachinery*, Jan. vol. 134, p. 011009-1.

Kupferschmied, P., Köppel, P., Gizzi, W., Roduner, C., and Gyarmathy, G. 2000. "Time-Resolved Flow Measurements with Fast-Response Aerodynamic Probes in Turbomachines," *Measurement Science Technology*, vol. 11, pp. 1036–1054.

Ned, A., Kurtz, A., Shang, T., Goodman, S., and Giemette, G. 2013. "Fully Integrated, Miniature, High-Frequency Flow Probe Utilizing MEMS Leadless SOI Technology," NASA Technical Briefs, April, Document ID: 20130012640.

PART 7

Fundamentals of Aeroelasticity

Jonathan Cooper

3.42 Aeroelasticity

Aeroelasticity is the study of the interaction of aerodynamic forces on elastic bodies. It is most famously characterized by Collar's Aeroelastic Triangle (Figure 3.103), which shows the interdependence of aerodynamic, elastic, and inertial forces. For example, if an aerodynamic load is applied to, say, a wing, this will cause the wing to deflect. However, this deflection will alter the manner in which the aerodynamic forces act on the wing, and so on. Most aeroelastic effects are not desirable and in some cases can lead to structural failure. Consequently, the study of aeroelasticity is very important for the design of aerospace structures as well as other structures such as racing cars, bridges, chimneys, power cables, etc.

The various aeroelastic phenomena can be categorized as to whether they are static or dynamic, involve an attached or separated air flow, or behave in a linear or nonlinear fashion. We are particularly interested in the nature of each phenomenon and the deflections and loads that occur. For aerospace structural design it is important to determine the critical air speed at which aeroelastic phenomena occur and the behavior just before any instability occurs.

The aeroelastic equations of motion for an n-DOF system, such as an aircraft, in terms of coordinate system $\mathbf{y}$ can be written as a second-order differential equation such that

$$\mathbf{A}\ddot{\mathbf{y}} + (\rho V\mathbf{B} + \mathbf{D})\dot{\mathbf{y}} + (\rho \mathbf{V}^2\,\mathbf{C} + \mathbf{E})\mathbf{y} = \mathbf{0}$$

where $n \times n$ matrices **A, D,** and **E** are, respectively, the inertial, structural damping, and stiffness matrices as before (note the change in notation). Compared to the equations described earlier in this section, extra terms are included to represent the aerodynamic damping ($\rho V\mathbf{B}$) and aerodynamic stiffness ($\rho V^2\mathbf{C}$), which depend upon the density ρ (and hence the altitude) and airspeed V. The damping term reflects the effective change in incidence due to a vertical motion of velocity y' in an airflow of velocity V. The stiffness term reflects the change in incidence due to structural rotations. Thus, the modal parameters of an aeroelastic system vary with the flight condition. Note that the aerodynamic terms also depend upon the frequency of vibration, due to the so-called unsteady aerodynamic behavior, and these effects must be accounted for in a complete dynamic aeroelastic analysis.

Divergence

Divergence is a static aeroelastic phenomenon that results in structural failure. When an aerodynamic load or moment is applied, say to a wing, then there is a resultant deflection, and once equilibrium has been achieved the aerodynamic forces and moments are balanced by the structural restoring forces and moments. Classically, divergence occurs when the aerodynamic moment overcomes the structural restoring force and the structure fails.

For example, consider the simple rigid aerofoil in Figure 3.104 with initial angle of incidence α, eccentricity ec between the lift acting on the aerodynamic center (at the ¼ chord) and flexural axis, chord c, 2D lift curve slope a_1, unit span and torsional stiffness k_θ. The moment due to the aerodynamic lift (considered to be proportional to the dynamic pressure q) is balanced by the spring restoring moment, thus,

$$\text{Lift} \times \text{eccentricity} = qc^2 a_1(\alpha + \theta)e = k_\theta \theta \Rightarrow \theta = \frac{qc^2 a_1 e\alpha}{k_\theta - qc^2 a_1 e}$$

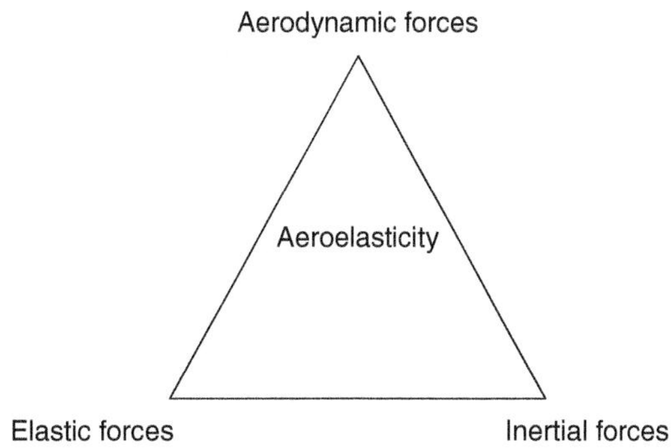

Figure 3.103 Collar's aeroelastic triangle.

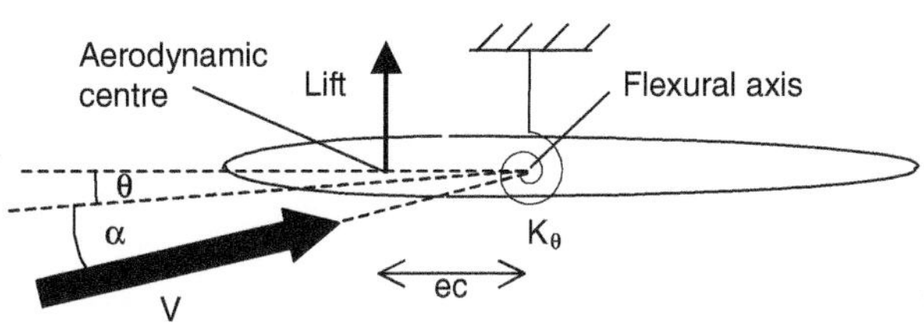

Figure 3.104 Divergence example.

As the airspeed (and hence dynamic pressure) increases, the angle of twist θ increases. At the critical divergence speed, $k_\theta = qc^2a_1e$ and structural failure occurs.

Control Effectiveness/Reversal

The flexibility of wings means that the application of the control surfaces can have a different effect, depending upon the flight speed. With zero control surface angle, the lift occurs at the aerodynamic centre at the ¼ chord, causing a nose-up pitching moment. However, if a control angle of β is applied, the resultant extra lift occurs somewhere around the 2/3 chord point, causing a nose-down moment to be applied. If the aerofoil were rigid, no torsional deflection would occur; however, in practice the aerofoil will rotate downward. This rotation reduces the lift obtained through application of the control angle. Figure 3.105 shows how the effectiveness (here defined as the ratio between the lift for flexible and rigid wings) varies with airspeed. The effectiveness drops with increasing speed until it reaches the reversal speed, when it becomes zero. At this speed, application of the control surface has no effect. Beyond this speed, the effectiveness becomes negative and application of the control will have the opposite effect to that intended. Although not disastrous, this phenomenon is undesirable, as control of the aircraft is poor around the reversal speed.

Flutter

Flutter is a violent unstable oscillation that results in structural failure. It is the most important aeroelastic phenomenon, and considerable effort is spent in the design and prototype testing stages of aircraft to ensure that it cannot occur. Flutter classically occurs when two modes (wing bending and torsion) interact at a certain flight condition and effectively extract energy from the airflow.

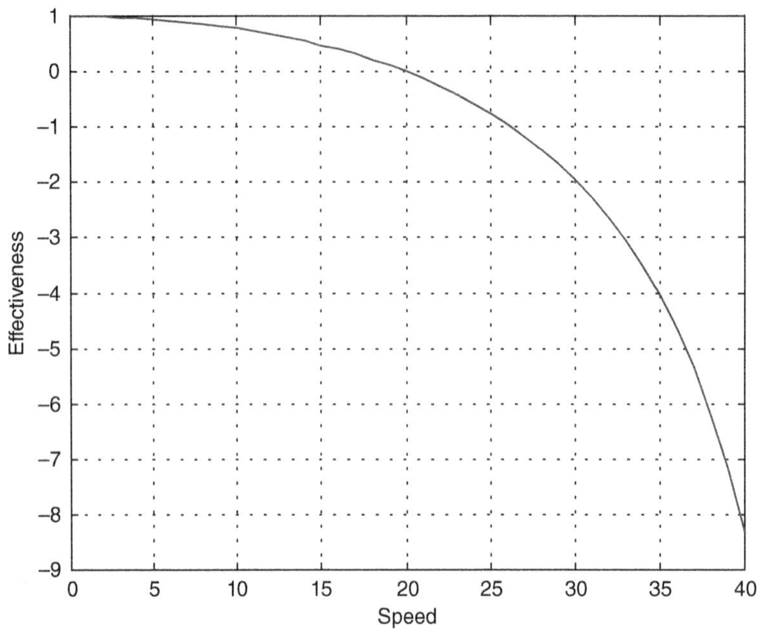

FIGURE 3.105 Typical control effectiveness versus speed.

Figure 3.106 shows how the frequency and damping ratio of a binary aeroelastic system change with speed. The frequencies move closer together but do not necessarily join together. One of the damping ratios gets very large, whereas the other eventually becomes negative. The point where one of the damping ratios becomes zero is the critical flutter speed. Beyond this speed the damping ratio becomes negative and any small disturbance will cause an unstable vibration with disastrous consequences. Although the flutter analysis of an aircraft contains many more modes, the critical flutter mechanism is nearly always binary in nature.

Nonlinear Aeroelastic Effects

If the system contains either structural, aerodynamic, or control nonlinearities, it is possible that the flutter oscillations become limited to some constant amplitude. Such an effect is known as a limit cycle oscillation (LCO), see Figure 3.107. Traditional linear analysis techniques are unable to predict LCO, which, although undesirable, is not immediately catastrophic. Other nonlinear aeroelastic effects with limited amplitude oscillations include the transonic phenomenon of control surface buzz, where a vibration is caused by the movement of a shock over the control surface. Stall flutter occurs when an aerofoil reaches an angle of attack such that the flow separates and lift is lost. This results in the angle of attack reducing and the flow reattaching and cause the lift to increase the angle of attack, and so on.

Buffet/Buffeting

A further important aeroelastic phenomenon is buffeting, where turbulent separated flows (buffet) from one part of a structure impinge on another part, causing a vibration known as buffeting. It rarely produces an instantaneous catastrophic failure, but the loads can be severe, resulting in reduced fatigue lives. This effect is currently a severe problem for twin-finned military aircraft.

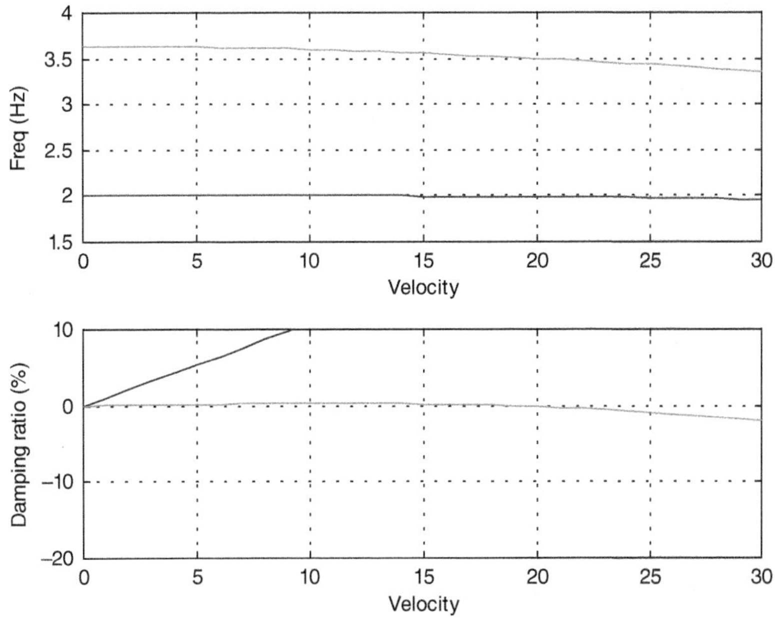

Figure 3.106 Frequency and damping trends for binary flutter system.

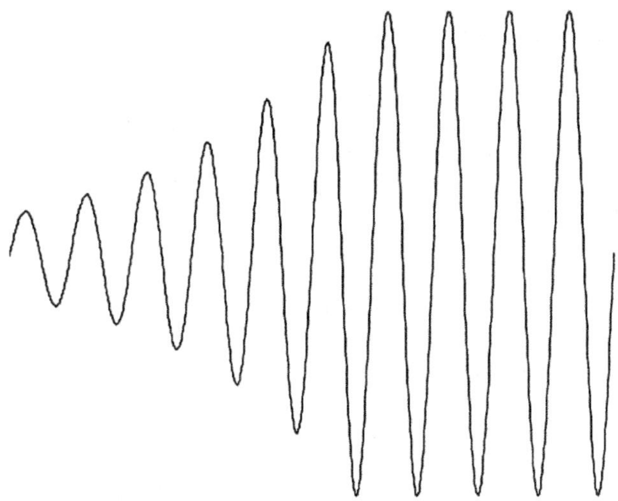

Figure 3.107 Typical limit cycle oscillation.

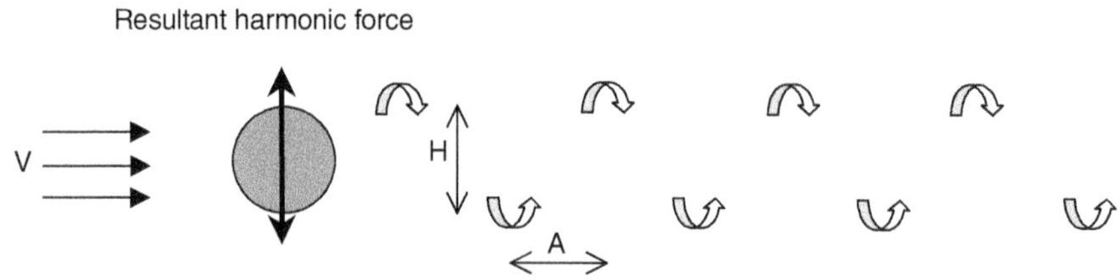

FIGURE 3.108 Vortex shedding.

Vortex Shedding

Vortex shedding is very important effect for the design of chimneys and buildings. Figure 3.108 shows how under certain Reynolds numbers the flow around a cylinder results in a Von Karman vortex street, whereby two streams of alternating vortices form downstream of the body. These vortices give rise to a sinusoidal force perpendicular to the flow. The frequency of the shedding ω is related to the air speed V by the Strouhal number

$$S = \frac{\omega D}{V}$$

where D is the diameter of the cylinder.

Should the frequency of the force correspond to one of the natural frequencies of the structure, then large deflections can result causing fatigue problems. Solutions to this include the use of helical shrouds on the upper one-third of a vertical chimney to break up the vortex formation.

Negative Damping—Galloping

Galloping is a phenomenon associated with power transmission cables, where in strong winds vertical vibrations of 10 m in a span of 150 m have been observed. A purely circular cross-section cannot gallop, but certain cross-sections are prone to this phenomenon either through the formation of ice on the cable or certain configurations of cable winding (e.g., the Severn suspension bridge). These cross-sections lead to negative aerodynamic damping, resulting in increasing oscillations.

3.43 Aircraft Airworthiness Certification

Aircraft manufacturers must demonstrate to the airworthiness authorities that all new aircraft are safe to fly. The airworthiness regulations cover the full range of operational aspects and possible types of failure (stress, fatigue, etc.). Here we shall be concerned with items relating to structural dynamics.

Once the dynamic mathematical model of the aircraft has been determined, it is possible to predict the response to the many dynamic loads that may be encountered, e.g., maneuvers, takeoff/landing, store release, etc. The aeroelastic behavior, e.g., flutter boundaries and response to gusts, may also be predicted, following the addition of an aerodynamic model. It is mandatory for certification purposes that testing must be performed in order to validate the models, and there are two major vibration tests that must be undertaken: ground vibration testing and flight flutter testing.

Ground Vibration Testing

Ground vibration testing (GVT) is performed to measure the aircraft's modal characteristics—natural frequencies, damping ratios, and mode shapes. These results are then used to validate the dynamic model (usually determined using finite elements) and can be used as the basis for adjusting (updating) the model.

The aircraft must be freely supported so that the natural frequencies of the support do not overlap with those of the aircraft. Bungees, inflatable airbags, and semideflated tires are all methods that are commonly used as supports. For very large aircraft, such as the A380, the fundamental vibration frequencies are so low that it will not be possible to support the structure at a lower frequency. It is likely, in this case, that the mathematical model will have to include the support mechanism.

The aircraft is excited using electromechanical shakers that can be controlled to give the required input of prescribed frequency and amplitude. It is usual to use at least four shakers, although for large aircraft it is more likely that up to four may be used. The response to the excitation is measured using accelerometers, with typically upward of 500–1,000 being used for the test of a large civil aircraft. The whole data-acquisition process is computer controlled, the exact procedure used depending upon which type of analysis method is to be employed.

The traditional (phase separation) approach is to excite the structure with broadband random, or stepped sine, signals and to calculate the frequency response functions (FRFs) from the measured data. The FRFs are then curve-fitted using system identification methods to estimate the natural frequencies, damping ratios, and mode shapes. This approach is likely to produce complex mode shapes which, while not being erroneous, can lead to problems when comparing with the finite-element model (invariably based upon proportional damping and given real modes). Consequently, the aerospace industry has traditionally employed the force appropriation (phase resonance) approach, whereby the structure is excited at each individual frequency and the amplitude and phase of each shaker are adjusted until only the normal mode at that frequency is excited. It is straightforward to compare the resultant normal modes with the finite-element model.

Flight Flutter Testing

Once an aerodynamic model is added to the validated structural model, it is possible to predict the frequency and damping behavior against speed or Mach number (e.g., Figure 3.106). Flight flutter testing is performed to demonstrate that the aircraft is flutter free throughout the design flight envelope. Figure 3.109 illustrates the typical flight envelope clearance procedure that is used. There are three steps to the process:

1. The aircraft is flown at some constant flight condition and excited using one of a number of approaches: aerodynamic vanes, control surfaces, explosive devices, eccentric masses, or simply atmospheric turbulence. The response to this excitation is measured using accelerometers in the same way as the GVT, except that far fewer are used.
2. The measured excitation and response data are curve-fitted using system identification methods to determine frequencies and damping ratios.
3. The decision is made to move to the next flight test point, traditionally based upon damping ratio versus speed trends, although it is also possible to predict the flutter speed using the frequency and damping ratios obtained during the tests.

This procedure is repeated until the entire envelope is cleared. Extrapolation of results is used to establish a safety margin, typically of 20%.

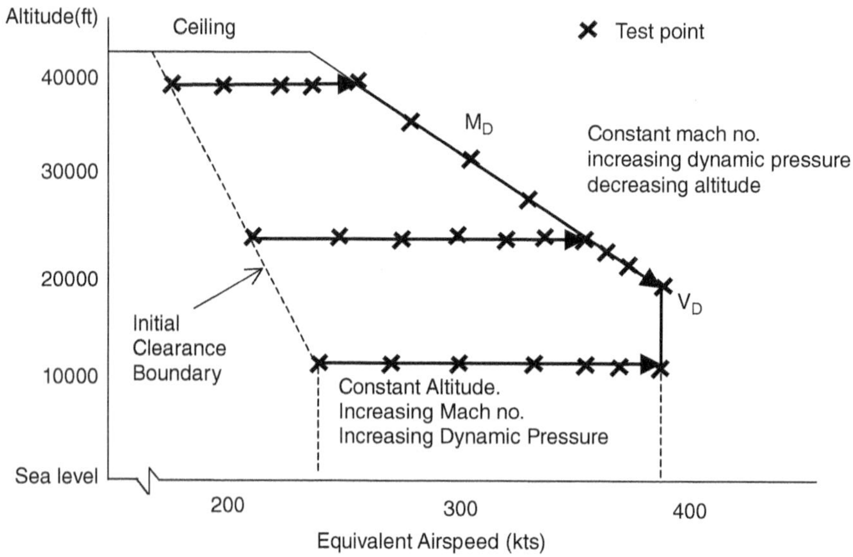

FIGURE 3.109 Typical flight envelope clearance.

3.44 Aeroelastic Design

Wing flutter usually occurs due to the interaction of two modes, one bending and the other torsion. However, this simplistic behavior becomes much more complicated with the addition of engines, stores, and control surfaces. Ideally, the designer should aim to place the inertial and flexural axes on the aerodynamic centre (¼ chord), in which case it is impossible for flutter to occur. However, whereas it is not too difficult to place the flexural axis close to the aerodynamic center, it is not as easy to do so with an inertial axis which lies aft of the aerodynamic center.

In order to keep the flutter speed as high as possible, the following general rules can be followed:

- The distance between the flexural axis and the aerodynamic center must be kept as small as possible.
- The "wind-off" natural frequencies should be kept well separated.
- An increase in the torsional stiffness will increase the flutter speed.
- Control surface flutter can be avoided by mass-balancing the control surfaces.

Should the aeroelastic prediction show that the flutter speed is less than the desired design speed, some alteration of the structure is required. Traditionally this has been achieved by adding mass to move the inertial axis forward, though this process is by no means obvious due to the interaction between all of the modes. Care must be taken to ensure that while curing a problem between two modes, a different instability is not caused instead.

Further Reading

E.H. Dowell, et al., A modern course in aeroelasticity, 5th ed., Springer New York 2015.

PART 8

Computational Aeroelasticity

Guru P. Guruswamy

The modern era of computational aeroelasticity (CA) began in the late 1970s with advancements in computational fluid dynamics (CFD) based on the transonic small perturbation theory coupled with modal structures. In the mid-1980s, development of methods using the Euler and Navier–Stokes equations-based CFD and the modal/finite element model (FEM) equations-based computational structural dynamics (CSD) began. Methods based on Lagrange's equations of motion were formulated with forcing aerodynamic data computed from CFD. Integration of solutions between the fluid and structural dynamics equations are accomplished using the time domain (coupled) and frequency domain (uncoupled) approaches. Time domain equations are solved time-accurately using mostly the linear acceleration method, also known in the literature as the Newmark time integration scheme, retaining fluids and structural solvers in separate computational domains. Several attempts have been made to solve the fluid and structural dynamics equations monolithically in a single computational domain with mixed results. Uncoupled equations are solved for stability boundaries using the artificial damping approach known as U-g method. Alternate faster methods, such as indicial and harmonic methods, though they have limitations, were also developed to generate aerodynamic data for uncoupled analysis. Later developments include the reduced order method, an alternate of the indicial approach, for the Euler equations. Efforts have also been made to include active controls and the effects of thermal loads in CA. With advancements in computational resources, particularly massively parallel computers, the fidelity of both CFD and CSD has significantly increased, and advanced FEM codes such as NASTRAN are used for CSD. In this respect, aeroelastic codes such as ATRAN3S, XTRAN3S, CAPTSD, ENSAERO, STAR, ENS3DE, HiMAP, OVERFLOW, and FUN3D were developed at NASA and DoD. These codes are applied for improved understanding of the physics associated with nonlinear flows, including moving shocks and flow separation coupled with structures for aircraft, rotorcraft, and spacecraft. They enabled successful prediction of important phenomena such as transonic-flutter-dip, vortex-induced aeroelastic oscillations; active control-surface reversals; vertical tail buffet; load coupling in the transonic regime for flexible launch vehicles; and blade-vortex interactions of rotating blades. All of these were observed in experiments but beyond the limits of linear aerodynamic methods. This chapter, mostly focused on time-accurate methods, gives a history of the developments, their current status, and the future directions of CA.

3.45 Beginning of Transonic Small Perturbation Theory

Back in the 1970s, aerospace designers got—and are still getting—a lot of mileage out of linear aerodynamics methods based aeroelastic computations by using the aerodynamic influence coefficients formulation (Appa and Somashekar 1969) that couples aerodynamics equations with finite element (FE) structural equations. However, linear aerodynamic theories have challenging limitations in modeling nonlinear flows. In order to overcome these limitations, researchers began developing, in the late 1970s, computational aeroelasticity (CA) methods based on computational fluid dynamics (CFD) and computational structural dynamics (CSD). The development work, as reported in the introduction of Guruswamy (1980), a joint NASA, Air Force, and academic effort, was led by William Ballhaus (director, NASA Ames Research Center), James Olsen (chief scientist, U.S. Air Force Flight Dynamics Laboratory), and Henry Yang (dean of engineering, Purdue University).

The breakthrough development of the unsteady, transonic small perturbation theory (TSP)–based code LTRAN2 (Ballhaus and Goorjian 1980) triggered the start of CA. Based on a domain decomposition approach (see appendix in this part), a scheme for time-accurate coupling of CFD-TSP and lumped (Ballhaus and Goorjian 1980) CSD systems for two degrees of freedom (2DOF) airfoils was then developed in the code ATRAN2S (Guruswamy and Yang 1981). Due to the complexities of CFD boundary conditions, the simpler CSD was kept as a separate computational domain. Fluid structure interaction (FSI) is simple because the coupling involves only 1D forces. It was validated with both the frequency domain approach and the kernel function aerodynamic theory (Guruswamy 1980; Guruswamy and Yang 1981). Using ATRAN2S, the transonic dip in the flutter speed of airfoils that was seen in experiments (Farmer and Hanson 1976) was successfully predicted (Yang et al. 1981). Results from aeroelasticity codes LTRAN2/ATRAN2S significantly contributed to addressing new transonic aeroelasticity phenomena reported in a classic 1980 paper by Ashley at Stanford University (Ashley 1980).

The LTRAN2 code, with the use of Jameson's rotated difference scheme (Jameson 2014) and the NASA Ames sheared grid method (Guruswamy and Goorjian 1985), led to the development of 3D codes such as XTRAN3S (Borland 1989), ATRAN3S (Guruswamy 1985), and CAPTSD (Bennett 1988), that used 3D TSP equations along with a 2D modal form of the CSD equations. FSI involved the exchange of 2D surface data such as pressures to CSD and transverse deflections to CFD, and is explained in a review paper by this author edited by Klaus-Jürgen Bathe MIT (Guruswamy 2002). TSP-based methods served as the foundation for CA, as shown in the complexity-fidelity ladder in Figure 3.110. These CA codes were capable of predicting the onset of the transonic flutter

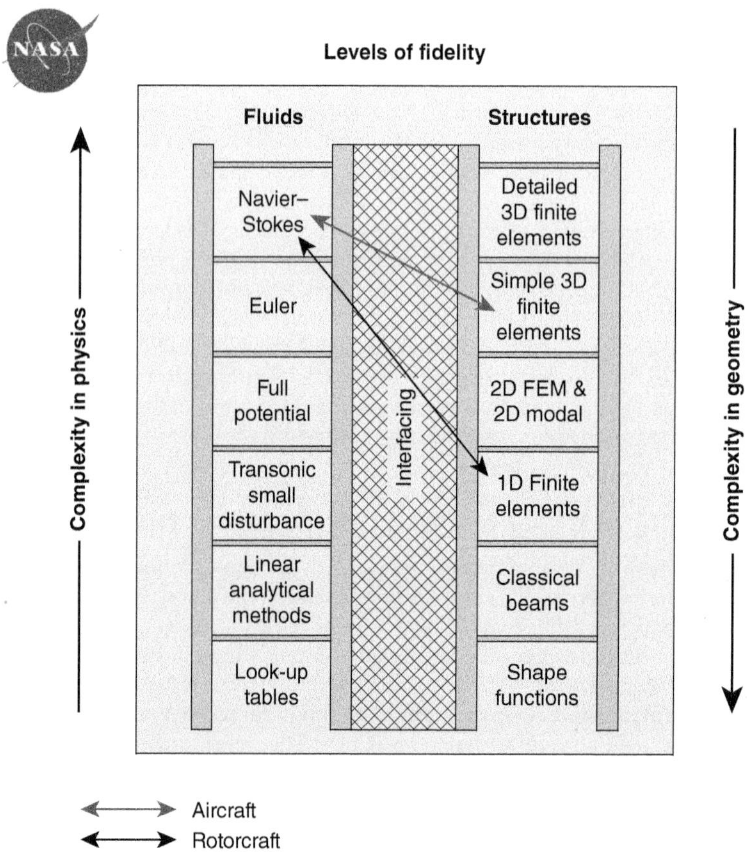

FIGURE 3.110 Fluid and structural domains in computational aeroelasticity.

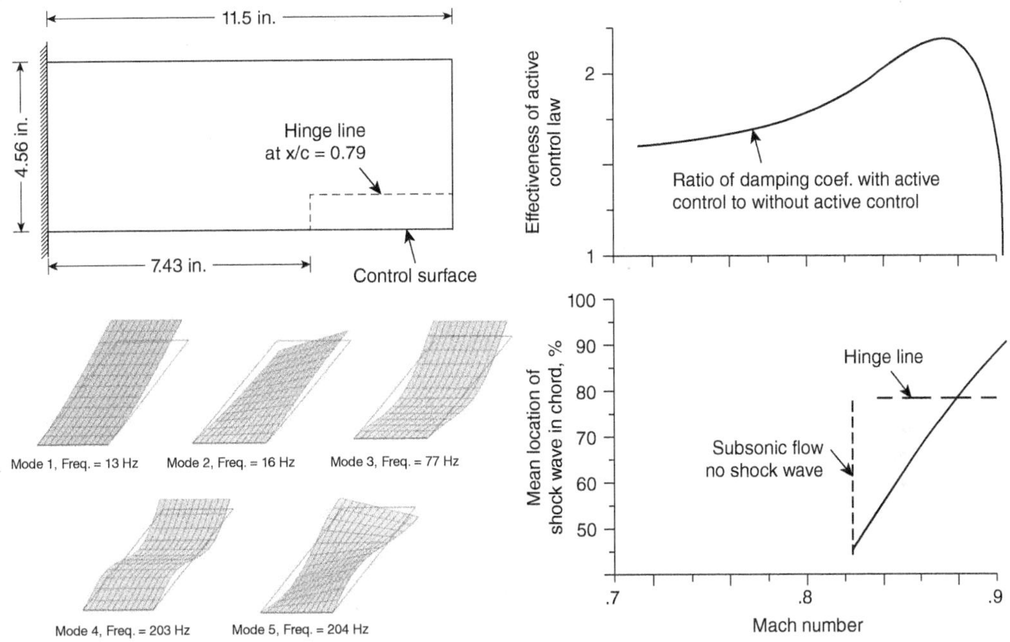

FIGURE 3.111 Effect of shock wave location on the effectiveness of active controls.

dip phenomenon (Farmer and Hanson 1976), which was beyond the limits of linear aerodynamic theories. More development efforts followed to include control surface effects for 2D airfoils, leading to a 3DOF aeroelasticity model (Yang and Chen 1982). The development and application of the ATRAN3S code for clean wings led to the discovery of new physics, such as active control law reversal in the transonic regime (Guruswamy 1989). Figure 3.111 (taken from Guruswamy (1989)) shows reversal of control surface effectiveness as the shock wave crosses the hinge line. Since the TSP approach models displacement with induced velocities without moving the grid, it served as a robust CA tool for several years.

3.46 Development of Euler and Navier-Stokes-Based Computational Aeroelasticity Tools

By the early 1980s, finite difference algorithms based on the diagonal form of the Beam-Warming central-difference scheme (Beam and Warming 1978) with algebraic turbulence models were used to solve the unsteady Euler and Reynolds–averaged Navier–Stokes (RANS) equations (Peyret and Viviand 1975) on rigid bodies. Around the mid-1980s, a first-of-its-kind code, ENSAERO (Guruswamy 1990) (cited in the classic review paper on aeroelasticity by Dugundji (2003) at MIT) was developed to time-accurately couple the 3D Euler equations with the 2D modal CSD equations. Patched-structured grids were used for the flow solver. The major challenge was to embed a moving grid for flexible configurations into an Euler/RANS solver. This author established a first-of-its-kind development for 3D CA (Guruswamy 1990) that was validated with two

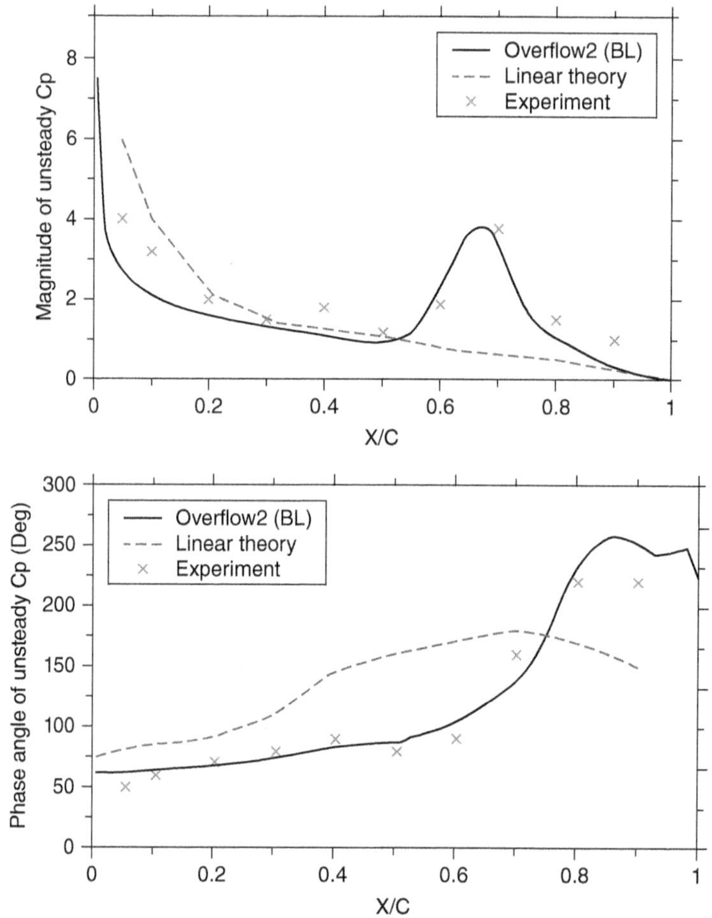

FIGURE 3.112 Jump in phase angle in transonic regime at $M_\infty = 0.90$ (NASA TND344).

well-known experiments (Lessing et al. 1960; Dogget et al. 1959), as shown in Figures 3.112 and 3.113. Figure 3.112 from Guruswamy (1990) shows the jump in phase angle as captured by computations and Figure 3.113 shows the computation of the transonic-dip in flutter speed.

The arrival of advanced mainframe computers such as the 1985 Cray-2 (Bailey 1990) expedited ENS-based CA development. Other researchers continued with 3D full-potential (Malone and Sankar 1985) and 2D Euler (Bendiksen 1987) approaches, but these were mostly of academic interest. As a follow-up to XTRAN3S, the U.S. Air Force funded the development of the ENS-based aeroelastic code ENS3DE (Schuster 1990), and a group at NASA Dryden (now Armstrong) Flight Research Center developed the STARS code (Gupta 1997). Similar codes were developed in European and Asian countries. The boundary conditions of ENS, which are more extensive than those of TSP, needed more sophisticated FSI techniques; details are reported in a review paper by this author (Guruswamy 2002). The most advanced FSI approach that conserves the work done by fluid and structural forces is reported in Guruswamy and Byun (1995).

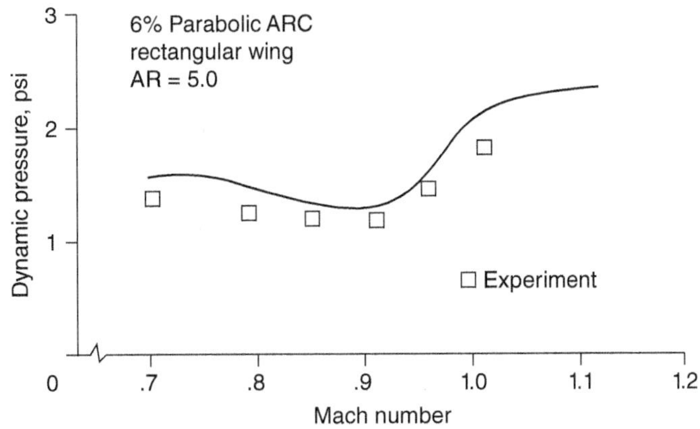

FIGURE 3.113 Transonic-dip phenomenon captured by a CA method (NASA TMX-79).

Computations using ENSAERO helped to explain the unconventional aeroelastic phenomenon of leading vortex-induced aeroelastic oscillations of a swept-wing aircraft (Dobbs and Miller 1985) caused by the coupling of lateral motion of the vortex core with bending motion (Guruswamy 1992). This phenomenon was discussed in detail by aeroelasticity legends such as Ashley (Stanford), Yoshihara (Boeing), Platzer (Naval Postgraduate School) and others noted in (Guruswamy 1992). These new findings continue to significantly impact the design of future high-speed civil transport systems.

Modeling moving and active controls in the transonic regime is a major challenge in CA. In the early 1990s, moving control surfaces were successfully modeled with sheared grids and validated for the Northrup Grumman F-5 fighter wing using ENSAERO (Obayashi and Guruswamy 1994). The code was then applied by aerospace industry to design active controls for an active-flexible-wing (Yeh 1995). Despite attempts to use large expensive and inadequately validated gap-filled grids, as discussed in Potsdam et al. (2011), the validated and numerically efficient shearing grid approach is still a useful design tool (Guruswamy 2014) today.

ENS-based codes were further improved in the mid-1990s by adding upwind solver options, along with a more efficient procedure to model static aeroelasticity (Obayashi and Guruswamy 1995). During the same timeframe, efforts were made to add a finite element option to augment the existing modal CSD option. Stress computations using a wing-box finite-element model (FEM) (MacMurdy 1994) were performed for a typical wing. In this period, CA computations were routinely used for advanced configurations such as the Rockwell X-31 aircraft, shown in Figure 3.114.

Efforts, though limited, were also made to couple the ENS equations with flight dynamics and aeroservoelasticity (Appa et al. 1996; Raveh et al. 2001).

The time step needed for FSI plays a leading role in the computer time required for CA computations. The size of the time step needed for numerical integration between CFD and CSD solutions is small since it is constrained by CFD requirements. Hence Newmark's single-step time integration (Guruswamy 1990) is adequate. However, in order to use larger time steps, attempts were made to introduce a staggered time integration scheme (Farhat et al. 2006; Silbaugh and Baeder 2008) in which CSD solutions are subiterated, using Newton's method. This procedure does not account for changes in the CFD grid during CSD (Silbaugh and Baeder 2008) subiterations and adds more bookkeeping. It also misses flow details such as jumps in phase angles near shock waves.

Figure 3.114 Surface pressures over a Rockwell X-31 aircraft at $M_{\infty} = 0.31$ and 20° angle of attack.

3.47 Computational Aeroelasticity in Rotorcraft

The development of CA for rotorcraft lagged behind that for aircraft (see Figure 3.110). Though complexities involved in rotorcraft dynamics is part of the reason for this, the main reason can be attributed to a heavy dependence on comprehensive codes (CC) that use a detailed structural model but a simple linear and/or lookup tables and/or empirical aerodynamics. The first notable rotorcraft CA effort, by Caradonna et al. in 1986 (Tung et al. 1986), involved correction of the airloads of CC by using full-potential-theory based CFD. Both CC and CFD codes are run independently without any direct coupling. This approach, known as "loose coupling" (LC), was later renamed "delta coupling" (Datta and Chopra 2004). The LC method came into use three decades back when running CFD codes was very expensive, but such a hybrid approach is still heavily used by the rotorcraft community (Datta and Chopra 2004; Jain et al. 2015) today. This is in spite of observations by independent users, such as this one, taken from Jung et al. (2012), "However, the predicted accuracy seriously relied on the selection of empirical parameters associated with the wake modeling. Once the correlation has been made with the detailed CFD/CSD approach, the tuned parameters in a CSD approach would yield reliable and efficient aeroelastic analysis solutions for a rotor in a given flight condition."

(Guruswamy 2010, 2013) showed that ENS based CA could be used for rotorcraft without depending on the non-time-accurate hybrid LC approach. This was accomplished by embedding the ENSAERO FSI solver module in the structured overset-grid based ENS code OVERFLOW2 (Guruswamy 2013). Because of the limitations of overset methods in modeling the deforming grids, only blade flexibility could be modeled. A similar effort was reported later for a proprietary configuration by the French national aerospace research center, ONERA (Sicot et al. 2014).

Figure 3.115 shows the validation of a time-accurate approach for a public domain full rotorcraft, the higher-harmonic aeroacoustic rotor (HART II) (Guruswamy 2013). It is obvious that computational results for HART II by others using an LC approach that pre-tunes (by reverse engineering) the trim angles based on thrust measured in the experiment, may show results closer to the experiment.

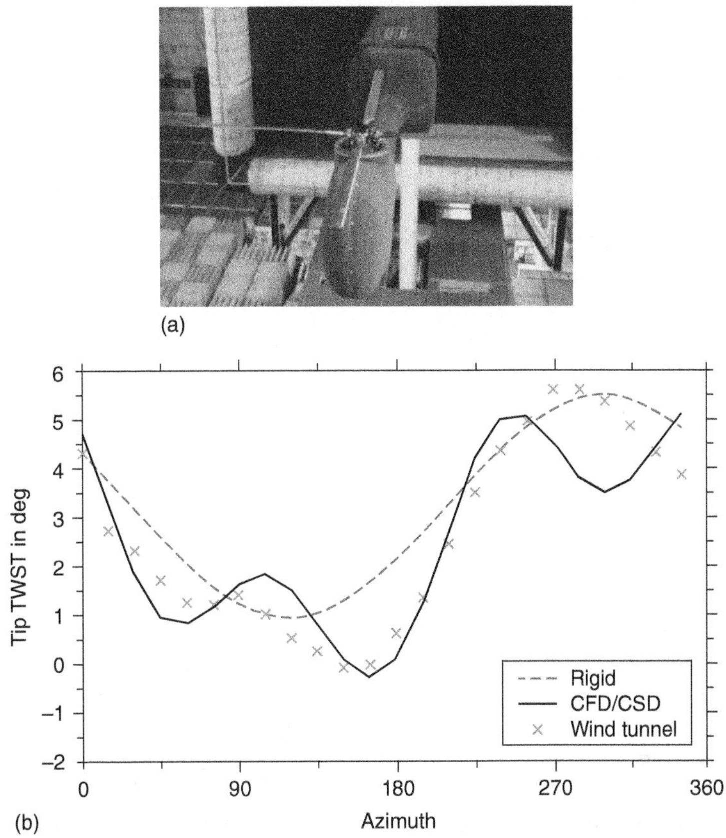

FIGURE 3.115 (*a*) Wind tunnel model HART II rotorcraft; (*b*) CFD-based time-accurate tip responses for HART II without using CC code (Guruswamy 2013).

3.48 Impact of Parallel Computers and Development of Three-Level Parallel Solvers

Computational aeroelasticity requires large computer resources in terms of both memory and CPU time due to use of the ENS and FEM equations. In the early 1990s, parallel computers started emerging, and NASA selected the ENSAERO code for further multidisciplinary development using parallel computers. Basic methods to make CFD and FEM computations parallel were initiated under NASA's former High Performance Computing and Communications (HPCC) Program (Holst 1992) (established in 1999). This resulted in the high-fidelity multidisciplinary analysis process (HiMAP), a first-of-its kind software in which CFD and CSD, including controls, run in parallel within and among themselves (Byun et al. 1999; Guruswamy 2000), as shown in Figure 3.116. HiMAP included a capability to run multiple cases in parallel. Use of higher fidelity CSD models, such as an improved wing-box finite element model (Bhardwaj et al. 1998) and NASTRAN (Eldred 1998), were implemented in HiMAP. This code was successfully

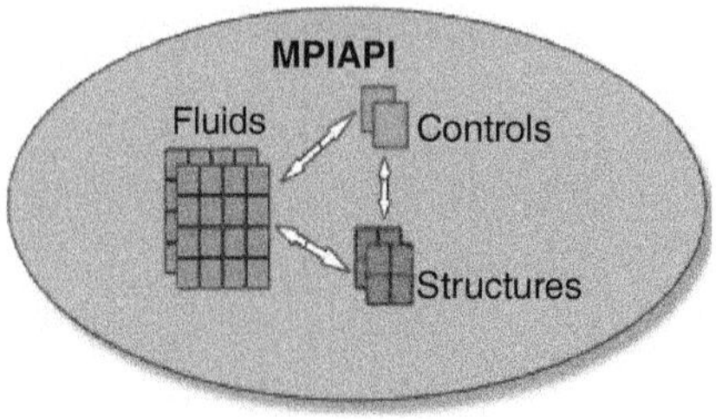

FIGURE 3.116 Multilevel parallel process in NASA's HiMAP software (Guruswamy 2000).

FIGURE 3.117 Pressure distribution of an aeroelastically deformed Lockheed L-1011 TriStar aircraft model at $M_\infty = 0.88$.

used for the analysis of both civil and military aircraft configurations, such as high-speed civil transport (HSCT) (Bhatia 2003), advanced Subsonic Transport (AST) (Goodwin 1999), F-18 vortex-induced vertical tail oscillations (Findlay 2000), and the F18-A's abrupt wing stall (Jones 2002). Using HiMAP, the first-of-its-kind aeroelastic computations on parallel computers were performed for the full Lockheed L-1011 TriStar aircraft (with 30 million flow grid points and 5 structural modes with 700 degrees of freedom) (Guruswamy 2007). Results for the L-1011 aircraft are shown in Figure 3.117. Because of weak viscous flow modeling capability and the need for an extremely small time step, efforts to use unstructured grids for the L1011 were not fully successful (Goodwin et al. 1999).

Several parallel computer-based CA codes such as those from a U.S. Department of Defense effort (Morton 2015) are becoming available elsewhere in the United States and in European and Asian countries.

With the rapid growth of parallel computers, thousands of cores are readily available at the users' disposal. However, the challenge is to use them efficiently and productively. Often, users resort to brute-force techniques—such as manually submitting multiple jobs and "babysitting" them—to exploit the use of large numbers of cores. Efficient tools such as RUNDUA, a dual-level parallel protocol that creates a single job environment for multiple jobs running on multiple cores (Guruswamy 2013) was developed in 2013 to increase user productivity. Figures 3.118(a) and (b) shows the computation of a flutter boundary during atmospheric reentry (Mach number decreasing from 5.5 to 0.5) of a hypersonic transport, using the RANS equations with 30 million grid points. The complete flutter boundary using indicial time responses at 100 flow conditions were computed in 17.5 hours of wall clock time on 4,000 cores of NASA's Pleiades supercomputer (Guruswamy 2016).

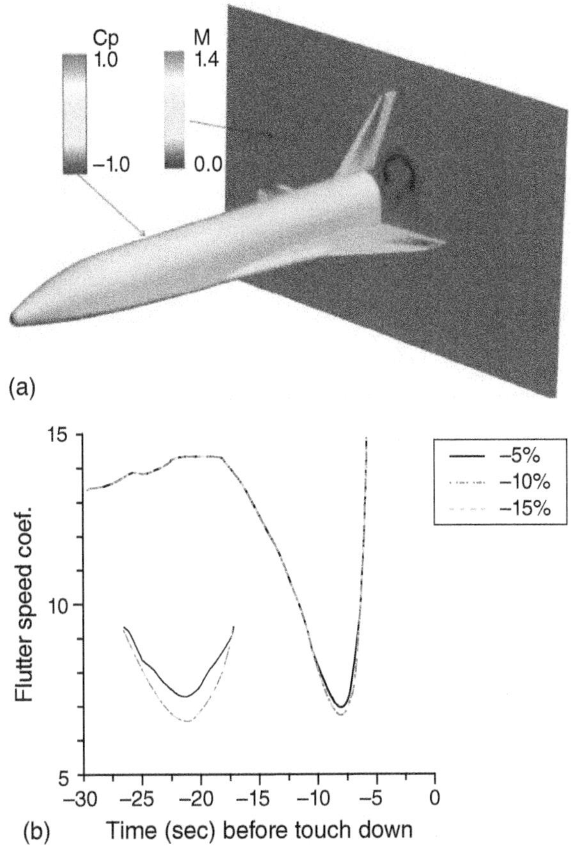

FIGURE 3.118 (*a*) Surface pressures at $M_\infty = 0.95$; (*b*) flutter boundary during atmospheric entry ($M_\infty = 5.5$ to 0.5, $\alpha = 12.0–2°$).

3.49 Conclusion

A summary of the development, validation, and applications of time-accurately coupled CFD/CSD methods to model computational aeroelasticity is presented for this handbook on aerospace engineering. Given the availability of current high-performance computer hardware, computational time is no longer an issue, and there is no need to use CFD in a hybrid mode to tune existing linear analysis codes.

Although efforts to date have involved elaborate validations with existing experiments, new experiments more suitable for validation—particularly in the rotorcraft area—are required. In the course of researching this study, it was found that use of a staggered time integration method—which adds more bookkeeping rather than improving the solutions—is not warranted. Structured patched grids are more suitable for aeroelastic analysis of complex configurations than structured overset and unstructured grids. Robust codes that combine the efficient procedure of modeling deformations using patched structured grids and multibody movements using overset structured grids are needed. Use of unstructured and Cartesian grids for CA may have to wait till they reach numerically similar efficiency as structured grids.

3.50 Appendix: Domain Decomposition Approach

Often we come across configurations with highly flexible components (wings) physically connected with highly rigid components (fuselage). Because of ill-conditioned global stiffness matrix, solving the system in a monolithic formulation is not practical. As a result, a substructure or zonal approach (Zienkiewicz 1977) is needed where each component is solved in a separate computational domain and linked at the boundaries.

A similar situation occurs when solving CFD and CSD together. As a result, a domain decomposition (DD) approach was developed (Zienkiewicz 1977) and extended to full aircraft configurations (Guruswamy 2007). The Lagrangian approach for structures, in which the mesh moves with deformations, and the Eulerian approach, where fluid moves through the mesh, are used. This particular DD method, which instantaneously couples Eulerian and Lagrangian motions, is also known in some later literature as the arbitrary Lagrangian–Eulerian (ALE) method. The main feature of the DD method is the capability to mix methods in a simulation, allowing a Lagrangian, fixed-mesh body to move through and interact with a surrounding Eulerian flow. With DD, aeroelastic phenomena can be simulated with full coupling between the fluid and the solid body. In spite of the success of the DD method, efforts to monolithically solve the CFD and CSD equations that result in very slow convergence (Felker 1993) are still being revisited (Sankaran et al. 2009).

References

Appa, K., Argyris J. H., Guruswamy G. P., and Martin, C. A. 1996. "Synergistic Aircraft Design Using CFD Air Loads," AIAA-96-4057, 6th AIAA/NASA/USAF MDO Symposium, Sep.

Appa, K. and Somashekar, B. R. 1969. "Application of Matrix Displacement Methods in the Study of Panel Flutter," *AIAA J.*, vol. 7, no. 1, Jan., pp. 50–53.

Ashley, H. 1980. "Role of Shocks in the 'Sub-Transonic' Flutter Phenomenon," *J. of Aircraft*, vol. 17, no. 3, Mar., pp. 187–197.

Bailey, R. F. 1990. "NASA's Supercomputing Experience," NASA Technical Memorandum 102890, Dec.

Ballhaus, W. F. and Goorjian, P. M. 1980. "Implicit Finite-Difference Computations of Unsteady Transonic Flows about Airfoils," *AIAA J.*, vol. 15, no. 679, Jun.

Beam, R. M. and Warming, R. F. 1978. "An Implicit Factored Scheme for the Compressible Navier–Stokes Equations," *AIAA J.*, vol. 16, no. 4, pp. 393–402.

Bendiksen, O. 1987. "Transonic Flutter Analysis Using the Euler Equations," AIAA 1987-911, 28th Structures, Structural Dynamics and Materials Conference, Apr.

Bennett, R. M., Batina, J. T., and Cunningham, H. J. 1988. "Wing Flutter Calculations with the CAP-TSD Unsteady Transonic Small Disturbance Program," NASA TM 10058.

Bhardwaj, M., Kapania, R., Reichenbach, and Guruswamy, G. P. 1998. "Computational Fluid Dynamics/Computational Structural Dynamics Interaction Methodology for Aircraft Wings," *AIAA J.*, vol. 36, no. 12, Dec., pp. 2179–2186.

Bhatia, K. 2003. "Airplane Aeroelasticity: Practice and Potential," *J. of Aircraft*, vol. 40, no. 6, Nov.–Dec.

Borland, C. J. 1989. "Additional Development of the XTRAN3S Computer Program," NASA CR-181743.

Byun, C., Farhangnia, M., and Guruswamy, G. P. 1999. "Aerodynamic Influence Coefficient Computations Using Euler/Navier-Stokes Equations on Parallel Computers," *AIAA J.*, Nov., vol. 37, no. 11, pp. 1393–1400.

Datta, A. and Chopra, I. 2004. "Validation and Understanding of UH-60A Vibratory Loads in Steady Level Flight," *J. of the American Helicopter Society*, vol. 49 (3), Jul., pp. 271–287.

Dobbs, S. K. and Miller, G. D. 1985. "Self-Induced Oscillation Wind Tunnel Test of a Variable Sweep Wing," AIAA Paper 85-0739, Apr.

Dogget, R. V., Rainey, A. G., and Morgan, H. G. 1959. "An Experimental Investigation on Transonic Flutter Characteristics," NASA TMX-79, Nov.

Dugundji, J. 2003. "Personal Perspective of Aeroelasticity during the Years 1953–1993," *J. of Aircraft*. vol. 40, no. 5, Sep.–Oct.

Eldred, L. B., Byun, C., and Guruswamy, G. P. 1998. "Integration of High Fidelity Structural Analysis into Parallel Multidisciplinary Aircraft Analysis," AIAA98-2075, Apr.

Farhat, C., van Der Zee, K., and Geuzaine, P. 2006. "Provably Second-Order Time-Accurate Loosely-Coupled Solution Algorithms for Transient Nonlinear Computational Aeroelasticity," *Comp. Methods in Applied Mech. Eng.*, vol. 195, pp. 1973–2001.

Farmer, M. G. and Hanson, P. W. 1976. "Comparison of Supercritical and Conventional Wing Flutter Characteristics," NASA TM X-72837.

Felker, F. F. 1993. "Direct Solution of Two-Dimensional Navier-Stokes Equations for Static Aeroelasticity Problems," *AIAA J.*, vol. 31, pp. 148–153.

Findlay, D. 2000. "Numerical Analysis of Aircraft High Angle of Attack Unsteady Flows," AIAA 2000-1946, Jun.

Goodwin, S. A., Weed, R. A., Sankar, L. N., and Raj, P. 1999. "Toward Cost-Effective Aeroelastic Analysis on Advanced Parallel Computing Systems," *J. of Aircraft*, vol. 36, no. 4, Jul.–Aug., pp. 710–715.

Gupta, K. K. 1997. "STARS—An Integrated General-Purpose Finite Element Structural, Aeroelastic, and Aeroservoelastic Analysis Computer Program," NASA TM-4795, May.

Guruswamy, G. P. 1980. "Aeroelastic Stability and Time Response Analysis of Conventional and Supercritical Airfoils in Transonic Flow by the Time Integration Method," Ph.D. Thesis, Purdue University, Dec.

Guruswamy, G. P. 1985. "ATRAN3S: An Unsteady Transonic Code for Clean Wings," NASA-TM 86783, Dec.

Guruswamy, G. P. 1989. "An Integrated Approach for Active Coupling of Structures and Fluids," *AIAA J.*, vol. 27, no. 6, Jun., pp. 788–793.

Guruswamy, G. P. 1990. "Unsteady Aerodynamic and Aeroelastic Calculations for Wings Using Euler Equations," *AIAA J.*, vol. 28, no. 3, Mar. 1990, pp. 461–469 (also AIAA Paper 88-2281).

Guruswamy, G. P. 1992. "Vortical Flow Computations on a Flexible Blended Wing-Body Configuration," *AIAA J.*, vol. 30, no. 10, Oct., pp. 2497–2503.

Guruswamy, G. P., 2000. "HiMAP—A Portable Super Modular Multilevel Parallel Multidisciplinary Process for Large Scale Analysis," *Advances in Engg. Software*, Elsevier, vol. 31, nos. 8–9, Aug.–Sep.

Guruswamy, G. P. 2002. "A Review of Numerical Fluids/Structures Interface Methods for Computations using High Fidelity Equations," *Computers and Fluids*, vol. 80, pp. 31–41.

Guruswamy, G. P. 2007. "Development and Applications of a Large Scale Fluids/Structures Simulation Process on Clusters," *Computers & Fluids*, vol. 36, Mar. 2007, pp. 530–539.

Guruswamy, G. P. 2010. "Computational-Fluid-Dynamics and Computational-Structural-Dynamics Based Time-Accurate Aeroelasticity of Helicopter Blades," *J. of Aircraft*, vol. 47, no. 3, May–Jun., pp. 858–863.

Guruswamy, G. P. 2013. "Dual Level Parallel Computations for Large Scale High-Fidelity Database to Design Aerospace Vehicles," NASA TM-2013-216602, Sep.

Guruswamy, G. P. 2013. "Time-Accurate Aeroelastic Computations of a Full Helicopter Model using the Navier-Stokes Equations," *International J. of Aerospace Innovations*, vol. 5, no. 3+4, Dec., pp. 73–82.

Guruswamy, G. P. 2014. "A Modular Approach to Model Oscillating Control Surface Using Navier-Stokes Equations," 5th Decennial AHS Aeromechanics's Specialists Conference, San Francisco, Jan.

Guruswamy, G. P. 2016. "Dynamic Stability Analysis of Hypersonic Transport during Reentry," Paper AIAA-2016-0280, AIAA Atmospheric Flight Mechanics Conference, San Diego, CA, Jan.

Guruswamy, G. P. and Byun, C. 1995. "Direct Coupling of the Euler Flow Equations with Plate Finite Element Structures," *AIAA J.*, vol. 33, no. 2, Feb. (also AIAA-93-3087 including shell finite-element).

Guruswamy, G. P. and Goorjian, P. M. 1985. "Efficient Algorithm for Unsteady Transonic Aerodynamics of Low-Aspect-Ratio Wings," *J. of Aircraft*, vol. 22, no. 3, Mar., pp. 193–199.

Guruswamy, G. P. and Yang, T. Y. 1981. "Aeroelastic Time Response Analysis of Thin Airfoils by Transonic Code LTRAN2," *Computers and Fluids*, vol. 9, no. 4, Dec., pp. 409–425.

Holst, T. L., Salas, M. D., and Claus, R. W. 1992. "The NASA Computational Aerosciences Program—Toward Teraflops Computing," AIAA 92-0558AIAA Aerospace Meeting and Exhibit, Jan., Reno, NV.

Jain, R., Lim, J., and Jayaraman, B. 2015. "Modular Multisolver Approach for Efficient High-Fidelity Simulation of the HART II Rotor," *J. of the American Helicopter Society*, vol. 60 (3), Jul.

Jameson, A. 2014. "Transonic Flow Calculations," Princeton University, MAE Report #1651, Mar. 22.

Jones, K., Platzer, M., Rodriguez, D., and Guruswamy, G. P. 2002. "On the Effect of Area Ruling of Transonic Abrupt Wing Stall," Symposium Transsonicum IV, Gottingen, Sep.

Jung, S. M., You, Y. H., Kim, J. W., Sa, J. H., Park, J. S., and Park, S. H. 2012. "Correlation of Aeroelastic Response and Structural Loads for a Rotor in Descent," *J. of Aircraft*, vol. 49, no. 2, Mar.-Apr., pp. 398–406.

Lessing, H. C., Troutman, J. L., and Menees, G. P. 1960. "Experimental Determination of the Pressure Distribution on a Rectangular Wing Oscillating in the First Bending Mode for Mach Numbers 0.24 to 1.30," NASA TN D-344, Dec.

MacMurdy, D., Guruswamy, G. P., and Kapania, R. 1994. "Aeroelastic Analysis of Wings Using Euler/Navier-Stokes Equations Coupled with Improved Wing-Box Finite Element Structures," AIAA 94-1587, Apr.

Malone, J. and Sankar, L. 1985. "Unsteady Full Potential Calculations for Complex Wing-Body Configurations," AIAA 1985-4062, 3rd Applied Aerodynamics Conference, Jun.

Morton, S. A. 2015. "Kestrel Current Capabilities and Future Direction for Fixed Wing Aircraft Simulations," AIAA 2015-0039, 53rd AIAA Aerospace Sciences Meeting, Jan.

Obayashi, S. and Guruswamy, G. P. 1994. "Navier-Stokes Computations for Oscillating Control Surfaces," *J. of Aircraft*, vol. 31, no. 3, May-Jun., pp. 631–636.

Obayashi, S. and Guruswamy, G. P. 1995. "Convergence Acceleration of a Navier-Stokes Solver for Efficient Static Aeroelastic Computations," *AIAA J.*, vol. 33, no. 6, Jun., pp. 1134–1141.

Peyret, R. and Viviand, H. 1975. "Computation of Viscous Compressible Flows Based on Navier–Stokes Equations," AGARD-AG-212.

Porter, L. "Physics Based Fundamental Aerodynamics," http://www.nasa.gov/about/highlights/porter_bio.html.

Potsdam, M., Fulton, M. V., and Dimanlig, A. 2011. "Multidisciplinary CFD/CSD Analysis of the SMART Active Flap Rotor," Proceedings of the 66th Annual Forum of the American Helicopter Society, Phoenix, Arizona, May.

Raveh, D., Levy, Y., and Karpel, M. 2001. "Efficient Aeroelastic Analysis Using Computational Unsteady Aerodynamics," *J. of Aircraft*, vol. 38, no. 3, May–Jun.

Sankaran, V., Sitaraman, J., Flynt, B., and Farhat, C. 2009. "Development of a Coupled and Unified Solution Method for Fluid-Structure Interactions," *Computational Fluid Dynamics*, 2008. Springer, Berlin. ISBN: 978-3-642-01272-3.

Schuster, D., Atta, E., and Vadyak, J. 1990. "Static Aeroelastic Analysis of Fighter Aircraft Using a Three-Dimensional Navier–Stokes Algorithm," *J. of Aircraft*, vol. 27, no. 9. Jan., pp. 820–825.

Sicot, F., Gomar, A., and Dufour, G. 2014. "Time-Domain Harmonic Balance Method for Turbomachinery Aeroelasticity," *AIAA J.* vol. 52, no. 1, Jan.

Silbaugh, B. and Baeder, J. 2008. "Coupled CFD/CSD Analysis of a Maneuvering Rotor Using Staggered and Time-Accurate Coupling Schemes," AHS International Specialists' Conference on Aeromechanics, San Francisco, CA, Jan.

Tung, C., Caradonna, F. X., and Johnson, W. 1986. "The Prediction of Transonic Flows on an Advancing Rotor," *J. of the American Helicopter Society*, vol. 31, no. 3, Jun., pp. 4–9.

Yang, T. Y. and Chen, C. H. 1982. "Transonic Flutter and Response Analyses of Two Three Degree of Freedom Airfoils," *J. of Aircraft*, vol. 19, Oct., pp. 875–884.

Yang, T. Y., Guruswamy, G. P., Striz, A. G., and Olsen, J. J., 1981. "Reply by Authors to H.P.Y. Hitch," *J. of Aircraft*, vol. 18, no. 2, Feb., pp. 159–160.

Yeh, D. T., "Aeroelastic Analysis of a Hinged-Flap and Control Effectiveness Using Navier-Stokes Equations," AIAA-95-2263, Jun. 1995.

Zienkiewicz, O. C. 1977. *The Finite Element Method*, McGraw-Hill Company (UK) Limited, London.

PART 9

Acoustics in Aerospace: Predictions, Measurements, and Mitigations of Aeroacoustics Noise

Nesrin Sarigul-Klijn

3.51 Introduction

Acoustics is considered as a wave motion and is different from optics in that acoustic waves require an elastic medium to travel. In the aerospace field, acoustics mostly deals with predictions of sound generation mechanisms on a vehicle system due to air flow, its effects at a receiver location, and noise mitigation methods of active or passive types.

Acoustic pressure disturbances are small amplitude changes to an ambient state. The ambient state of a fluid is characterized by its pressure, density, and velocity when no disturbances in flow are present. These ambient state variables satisfy the fluid dynamic equations. If the medium is homogeneous, then all ambient quantities become independent of position. The equations governing fluid dynamics are widely applied in the field of aerodynamics. Although sound obeys the same laws as aerodynamics, most acoustic phenomena involve only small changes in pressure values in the fluid. However, aerodynamics is not affected by these small changes. These effects are an order of magnitude smaller than the total motion of the fluid. As a result, most aerodynamic simulations do not include acoustics. In this part, we first cover the past, present, and ongoing studies involved in theoretical, computational, and experimental approaches to model noise sources and sound propagation to a receiver site. Then, we provide new techniques that combine theory, experiments, and computations in order to develop robust acoustics prediction methods for aerospace systems. Finally, examples from numerical and experimental aeroacoustics studies are presented.

3.52 Aeroacoustics Theoretical Background

Aeroacoustics involves the study of aerodynamically generated sound. When the sound has adverse effects or is at an unwanted frequency and loudness level it is referred to as "noise." Air transportation systems generated noise caused environmental concerns as early as 1960s as described by Ffowcs Williams (1963). Today, the noise generated by airborne systems is still a great concern for communities near airports, in particular during the take-off and approach-to-land phases of flight. Major noise sources are from the advancement of throttle to full power setting "propulsive" and/or the deployment of high lift devices, "nonpropulsive" types. In general, the aeroacoustics field also encompasses the study of acoustics signatures associated with complex fluid structure interactions of vehicles at varying Mach numbers, such as reusable launch vehicles during ground or air launch, and operations of unmanned aerial vehicles (UAVs) with their current rapidly increasing usage in urban environments. The aeroacoustics field involves theoretical, numerical, and experimental research to better understand the physics for relating flow field pressure fluctuations to the turbulent dynamics of flows at subsonic to hypersonic ranges of Mach numbers. For high-speed flows, formulations must include the effects of aerothermal-chemical reactions. Discoveries from aeroacoustics studies either contribute to new quieter designs (Dowling and Hynes 2006) or modifications of existing designs, i.e., "acoustics afterthoughts" of aerospace systems that meet environmental noise standards (Sarigul-Klijn et al. 2011; Sarigul-Klijn 2012).

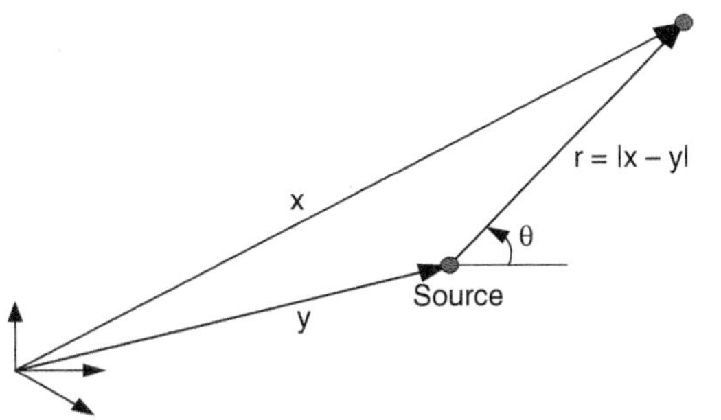

Figure 3.119 Representation of the source at near-field and the receiver at far-field in an observer-based reference system (r source receiver distance, θ directivity).

Turbulent flow generated noise, called *source*, can be in the boundary layer of the aerospace vehicle or in regions of complex geometries as a result of flow separation, or be part of the incoming flow at high speeds. The goal is to predict the effects at a receiver location and a two-step solution is usually considered in aeroacoustics simulations (Figure 3.119). The first step is the identification of the sound source via Navier–Stokes computations of the flow field to determine the time-dependent pressure near the solid boundary. The second step, the transmission from the known sound source, is treated as an independent problem. Figure 3.120 depicts this two-step process.

Governing Equations for the Near-Field of an Acoustic Source

Determination of the pressure fluctuations near the vehicle surface requires accurate knowledge of the behavior of the boundary layer. It is known that the potential theory-based boundary layer corrections are insufficient to describe the time-dependent behavior. Full Navier–Stokes computations that will allow the boundary layer to be resolved accurately must be incorporated in the solution. The fluid motion can be described by using the equations of conservation of mass, momentum, and energy. These equations are written as follows:

$$\frac{\partial}{\partial t}\begin{pmatrix} \rho \\ \rho\vec{v} \\ \rho E \end{pmatrix} + \vec{\nabla}\cdot\begin{pmatrix} \rho\vec{v} \\ \rho\vec{v}\otimes\vec{v} + p\bar{\bar{I}} - \bar{\bar{\tau}} \\ \rho\vec{v}H - \bar{\bar{\tau}}\cdot\vec{v} - \kappa\nabla T \end{pmatrix} = \begin{pmatrix} 0 \\ \rho\vec{f}_e \\ W_f + q_H \end{pmatrix} \tag{3.93}$$

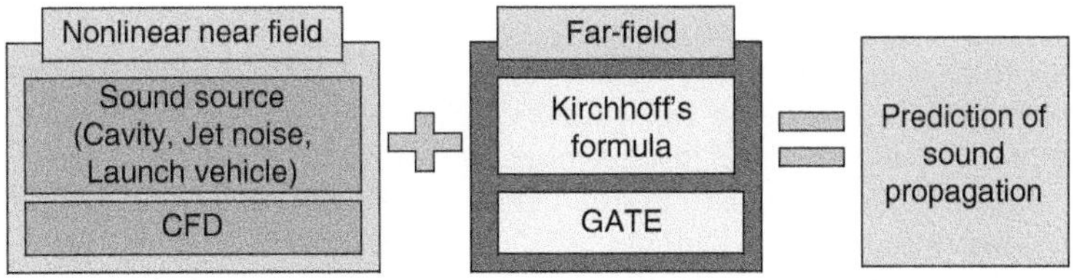

Figure 3.120 Two-step solution: near-field and far-field (Sarigul-Klijn et al. 2001; Sarigul-Klijn, 2012).

where ρ is the density, $\vec{v}$ is the velocity vector, E is the internal energy, H is the enthalpy, p is the pressure, $\bar{\bar{\tau}}$ is the stress tensor, $\bar{\bar{I}}$ is the identity matrix, κ is the thermal conductivity of the fluid, and q_H represents the contribution of heat sources. External forces represented by $\vec{f}_e$ while W_f represents the work done by external forces, $W_f = \rho \vec{f}_e \cdot \vec{v}$. The coefficients of the stress tensor, $\bar{\bar{\tau}}$, for a Newtonian fluids are given by

$$\tau_{ij} = \mu(T,p)\left[\left(\partial_i v_j + \partial_j v_i\right) - \frac{2}{3}(\vec{\nabla}\cdot\vec{v})\delta_{ij}\right] \tag{3.94}$$

where μ is the absolute viscosity of the fluid and δ_{ij} is the Dirac delta function.

If the sources are known, then we have unknown thermodynamic variables ρ, u, v, w, T, p, s, and h or e. h and e are functions of two other variables, p and T. Rewriting the energy equation,

$$\rho\frac{\partial e}{\partial t} = -p(\vec{\nabla}\cdot\ \vec{v}) + \vec{\nabla}\cdot(\kappa\nabla T) + \varepsilon_v + q_H \tag{3.95}$$

or static enthalpy (h)

$$\rho\frac{\partial h}{\partial t} = \frac{dp}{dt} + \vec{\nabla}\cdot(\kappa\nabla T) + \varepsilon_v + q_H \tag{3.96}$$

where ε_v is the dissipation term,

$$\varepsilon_v = (\bar{\bar{\tau}}\cdot\nabla)\cdot\vec{v} \tag{3.97}$$

We can then remove e and h from the energy equation by substituting one of the relationships given in equation (3.98) to equation (3.95) or (3.96):

$$de = c_v dT \qquad \text{or} \qquad dh = c_p dT \tag{3.98}$$

where c_v and c_p are the specific heats of the fluid at constant volume and pressure, respectively. Under the perfect gas assumption, we have the equation of state: $p = \rho RT$. We now have six equations and six unknowns. The enthalpy is decoupled, and can be obtained independently of the flow solution from basic thermodynamics:

$$Tds = dh - \frac{dp}{\rho}$$

Turbulence Modeling

Turbulence models were developed by writing the fundamental equations in time-averaged form in terms of their mean and the added variables representing fluctuating values. These quantities are computed to describe the turbulent energy in the domain without needing refined discretization to model the fluid motion accurately. There are two types of turbulence models: algebraic or transport equation models and one- or two-equation models.

Algebraic turbulence models assume a linear relationship between the turbulent stress and the mean stress tensor thereby creating a tight link between the mean value and the turbulence. The turbulent kinetic energy and the turbulent viscosity are the unknowns. A second equation is written for the turbulent viscosity based on dimensional analysis. Transport equation models attempt to utilize turbulence properties.

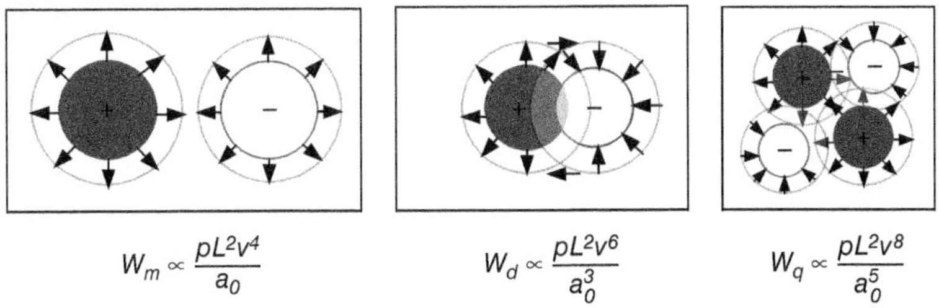

FIGURE 3.121 Idealized sources, their radiated-power (*W*) to flow-velocity (*v*) dependency: (*L*) monopole, (*C*) dipole, and (*R*) quadrupole.

One equation models include the Baldwin–Barth and Spalart–Allmaras and two equation turbulence models include the k-ε, k-ω, and q-ω models. The k-ε model is the most popular and uses the mean viscous dissipation of the turbulent kinetic energy as the transported variable.

Idealized Noise Source Modeling

The mechanisms of sound generation in fluid can be explained by three types of noise source models: monopole, dipole, and quadrupole (Figure 3.121). The radiation from a monopole source is equivalent to a pulsating sphere having both the amplitude and phase of the pressure as symmetric. A dipole consists of two equal monopole radiators 180° out of phase with a very small separation as compared to the wavelength. Quadrupole radiation is produced by Reynolds stresses in a turbulent flow with no solid obstacles. It is known to be the primary sound source in supersonic Mach number flows.

Sound Transmission to the Far-Field

There are different approaches for the prediction of the acoustic field at a receiver located in the far-field. This sub-subsection first covers commonly used approaches and then provides a sample from advanced techniques.

High-Fidelity Computational Fluid Dynamics-Based Solution

The near-field flow time-dependent pressure history around the acoustic source is generated using a high-fidelity computational fluid dynamics solution. It is possible to use computational fluid dynamics solutions everywhere in the domain only if the region of the fluid is not too large in terms of acoustic wavelengths. However, currently the treatment is done in two steps. There are numerical techniques developed which use the flow values calculated near the source as an input to a Kirchhoff integral formulation for acoustic far-field predictions. The nonlinear acoustics field is accurately modeled using computational fluid dynamics in the near-field around the source where the compressibility effects are large.

Acoustic Analogy and Surface Integral-Based Methods

The primary analytical methods used to determine the acoustic field are Lighthill's acoustic analogy and Kirchhoff's surface integral method. Lighthill's equations are complete because they are derived from the Navier–Stokes equations without approximations (Lighthill 1952; Lighthill 1954). Developed in the 1950s, the equations are still used today, including jet aeroacoustics studies. Later, Curle (1955) added solid stationary boundary effects and Ffowcs Williams and Hawkings added moving surface effects. Lighthill described the fluid medium as a stationary field of quadrupoles of finite volume. Curle's improved work showed that this field is equivalent to a volume quadrupole field and a surface dipole field, resulting in the Lighthill–Curle equation. This equation, provided below, allows the acoustic pressure at a point in space emitted by a fluid-solid interface to be evaluated in terms of the pressure fluctuations in the control volume and at the surface.

$$\rho-\rho_o=\frac{1}{4\pi c_o^2}\frac{\partial^2}{\partial x_i \partial x_j}\int V\frac{T_{ij}\left(\mathbf{y},t-\frac{r}{c_o}\right)}{r}d\mathbf{y}-\frac{1}{4\pi c_o^2}\frac{\partial}{\partial x_i}\int S\frac{P_i\left(\mathbf{y},t-\frac{r}{c_o}\right)}{r}dS(\mathbf{y}) \tag{3.99}$$

where P_i is the force per unit area along the x_i direction, and $T_{ij}=\rho v_i v_j+p_{ij}-c_o^2\rho\delta_{ij}$ with c_o as the speed of sound in the fluid at rest, p_{ij} as the compressive stress tensor and v_i, v_j are the velocity components in the i and j directions. The last term of the T_{ij} equation $c_o^2\rho\delta_{ij}$ represents the stresses in the uniform acoustic medium at rest. The second term in equation (3.99) is a dipole distribution of strength P_i. It is shown that for low-speed flows sound generated in the fluid in the absence of a boundary is negligible, the first term on the right side. Then, this leaves an expression for density in terms of the pressure at the surface. Since the intensity is given by the mean square variation in density, it can be determined using equation (3.99) and known pressure fluctuations on the surface.

Instead of solving the full nonlinear flow equations including the far-field, it is possible to extend the nonlinear near-field acoustic sources to the linear far-field sound field through the traditional Lighthill's acoustic analogy or surface integral methods, such as Kirchhoff's or Ffowcs Williams–Hawking's method. Kirchhoff's integral method applies the wave equation between a known acoustic surface S and a point in space within a uniform field with given free-stream velocity. A Kirchhoff's surface is assumed to include all the nonlinear effects and sound sources. Outside the Kirchhoff's surface the acoustic flow field is linear and is governed by the wave equation.

$$\frac{1}{c^2}\frac{\partial^2 p}{\partial t^2}-\nabla^2 p=0 \tag{3.100}$$

A surface S containing all acoustic sources and nonlinear effects is first selected. The acoustic pressure p due to the sources contained by S can be evaluated at any point in space x, using Kirchhoff's integral formula:

$$p(\mathbf{x},t)=\frac{1}{4\pi}\int_S\left[\frac{p}{r_o^2}\frac{\partial r_o}{\partial n_o}-\frac{1}{r_o}\frac{\partial p}{\partial n_o}+\frac{1}{cr_o\beta^2}\frac{\partial p}{\partial t}\left(\frac{\partial r_o}{\partial n_o}-M_\infty\frac{\partial x_o}{\partial n_o}\right)\right]_{t_r}dS \tag{3.101}$$

where n is the outward normal to S. The subscript o indicates the Prandtl–Glauert transformation of the variable, $x=x_o$, $y_o=\beta y$, $zo=\beta z$, and $\beta=\sqrt{1-M_o^2}$. This transformation corrects for the path taken by the sound in a uniform flow. For a given receiver site time t, all values are computed at the retarded time $t_r=t-t_a$, where t_a is the time for the sound to reach the receiver.

The first term in equation (3.101) represents simple geometric spreading. The second and third terms represent the waveform. The distortion of the sound field by the uniform flow is also contained in the third term.

3.53 Computational Aeroacoustics and Future Directions

Computational aeroacoustics (CAA) deals with the simulation of pressure pulsations generated by unsteady flows and their interactions with solid objects using numerical techniques that can solve acoustics problems without needing expensive full-scale experiments. It is also possible to use CAA to guide an experiment to reduce cost, especially at reduced scales. However, in spite of the continued advances in formulations, algorithms, and computer hardware, it is still prohibitively expensive and hence not yet feasible to conduct direct numerical simulations to predict the sound signature at a receiver directly. Techniques such as the Lighthill, Ffowcs Williams, and Hawkings

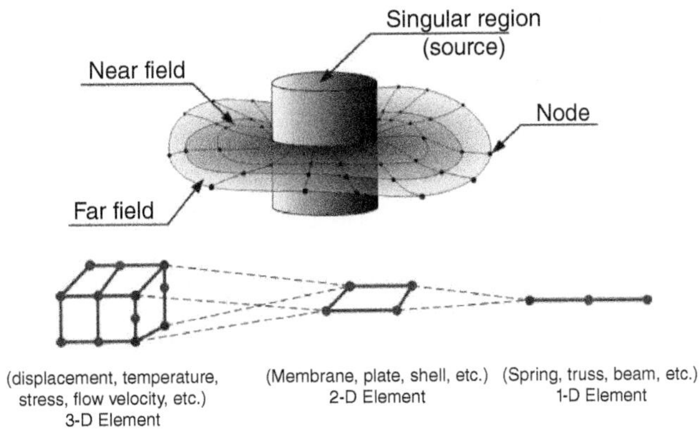

FIGURE 3.122 Unique-aeroacoustics technique involving high order transfinite elements (Sarigul-Klijn 1997).

acoustic analogy, and Kirchhoff surface integral methods which predict the far-field sound field based on near-field inputs have been a possible means of solution.

There are, however, new developments that one day may help eliminate the prohibitive nature of direct simulations, and one such technique is described in the references (Sarigul-Klijn 2012; Kuo and Sarigul-Klijn 2012). This approach incorporates the gradient adaptive transfinite elements (GATE) that allow seamless computation of the near- and far-field sound (Figure 3.122). It allows the discretization of the domain containing high-order three-dimensional finite elements to compute the flow field and sound source as well as two- and one-dimensional elements to solve the transmission problem.

At this stage of the development, this approach solves the Navier–Stokes equations in the near-field and the far-field solution is obtained from Kirchhoff integral formulation evaluated on a control surface surrounding the nonlinear near and mid fields with sources. A method is also developed in determining where to place the Kirchhoff surface to produce an accurate far-field solution (Figure 3.123). Applications using this method include cavity acoustics and launch vehicle noise predictions as summarized in the applications sub-subsection of this subsection.

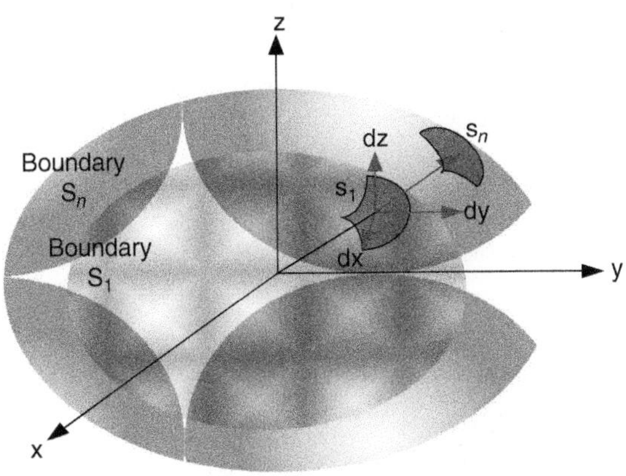

FIGURE 3.123 Placement of surface integral boundary S_1 to S_n to improve far-field solution. The energy passing through the boundary surface $S_1 \cdots, S_n$ remains constant.

3.54 Noise Measurements: Anechoic Chamber Experiments

Simultaneous computation and experimental studies are important in order to advance the field of aeroacoustics. In reference (Sarigul-Klijn et al. 2008) a scaled testing apparatus is described which allows to identify noise sources and measure acoustics signature from aircraft during approach. The measured values are compared to computational scaled simulations prior to numerical simulation of full scale systems. This apparatus serves as an anechoic wind tunnel at low Mach numbers which can also be used to investigate levels of external damage, such as bird strikes on aircraft wings using acoustic signature measurements. Positions 1, 2, and 3 show the standard microphone locations (Figure 3.124). Microphone types depend on the sound field. If it is plane, directional microphones can be used. However, if the field is diffused omnidirectional microphones are needed. Once the acoustics signature of the scaled structure with and without damage is captured and the results are validated with the results of the computational high fidelity aeroacoustics simulation the full scale simulations using the virtual-reality after simultaneous scaled testing-simulation validations (VAST) architecture, shown in Figure 3.125, can begin (Sarigul-Klijn 2008).

3.55 Applications

The studies of cavity noise, acoustic signatures from air and ground launched vehicles, high-lift device noise control via design change using deployable microdevices and environmental impact studies are important applications.

Cavity Acoustics

References Sarigul-Klijn et al. (2001) and Sarigul-Klijn (2008) provide examples of computational and experimental studies of cavity acoustics at speeds from Mach 0.26 to 0.672. The numerical simulations were conducted using the software packages OVERFLOW and CFD-FASTRAN.

Cavities with aspect ratios between 0.5 and 2.5 were considered. The dependence of the sound pressure level (SPL) on Mach number is plotted in Figure 3.126 for a cavity with an aspect ratio of 0.5. These values compare well with experimental data taken from Reference (Ahuja and Mendoza 1995), as marked on the same figure, thus demonstrating that the combined computational fluid dynamics-Kirchhoff surface integral technique yields good agreement with experimental data for complex problems.

High Lift Device Noise Reduction via Deployable Micro Devices

Simulations described in references (Sarigul-Klijn et al. 2011; Sarigul-Klijn 2012; Kuo and Sarigul-Klijn 2012) indicate that it is possible to control aircraft nonpropulsive noise during the approach-to-land phase of flight via use of deployable microdevices. The additional lift generated by these deployable microdevices is adequate to offset the lift loss from the lower setting angle of the

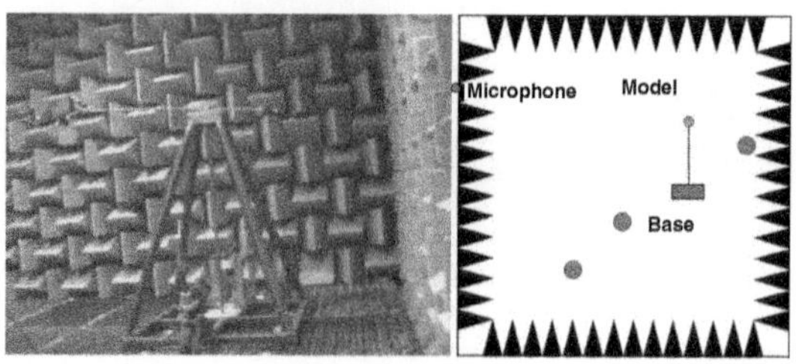

Figure 3.124 (Left) Anechoic Wind Tunnel; (right) typical microphone locations.

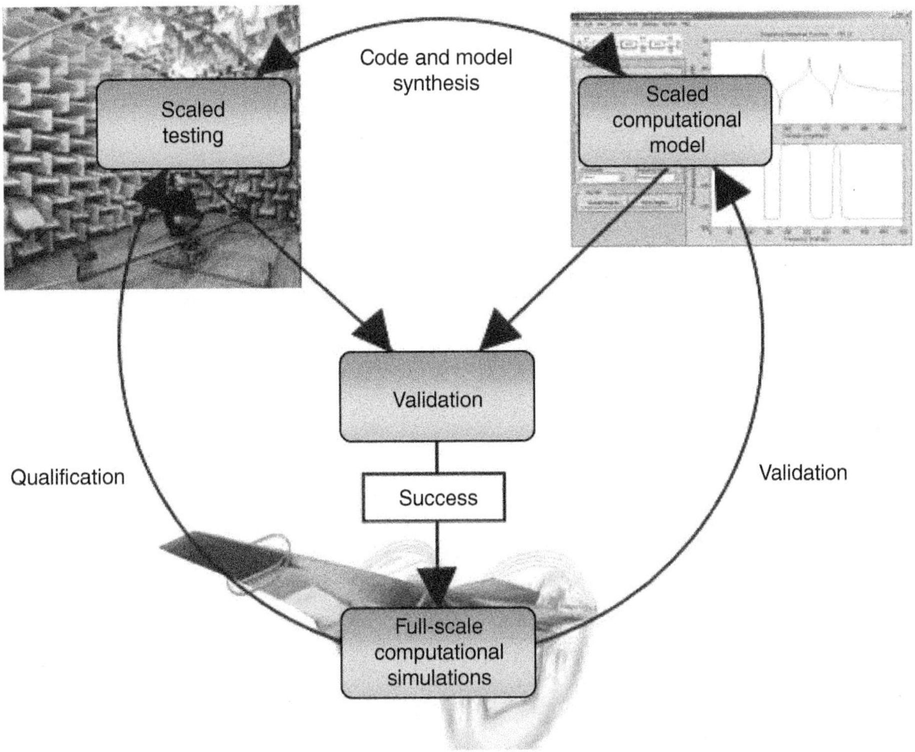

FIGURE 3.125 Acoustics based VAST architecture (Sarigul-Klijn 2008).

high-lift devices. Time-accurate Reynolds averaged Navier–Stokes simulations and the FW&H acoustic analogy were used to study the three-dimensional unsteady flow field and acoustic components around a three-element high-lift wing with and without microdevices. Deployable microdevices are designed to be attached to the pressure side of the high-lift surface near its trailing edge to help reduce the noise generated. The analysis revealed that with the deployment of the microdevice, along with reduced high-lift device setting angles, an overall airframe noise reduction of 2–5 dB is obtained over the entire frequency range, Figure 3.127.

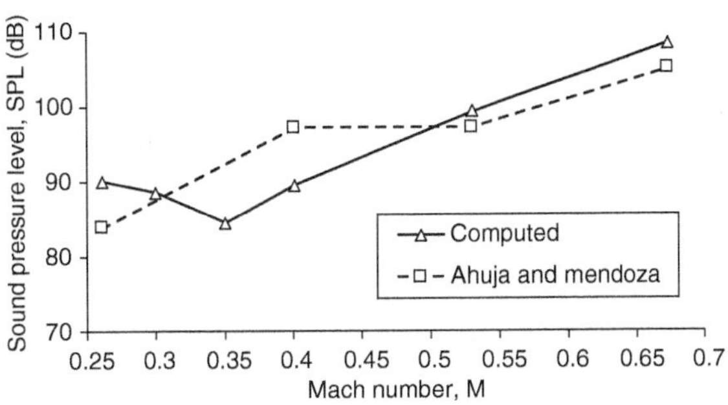

FIGURE 3.126 Sound pressure level versus Mach number for cavity aspect ratio = 0.5 aspect ratio of 0.5 driven cavity.

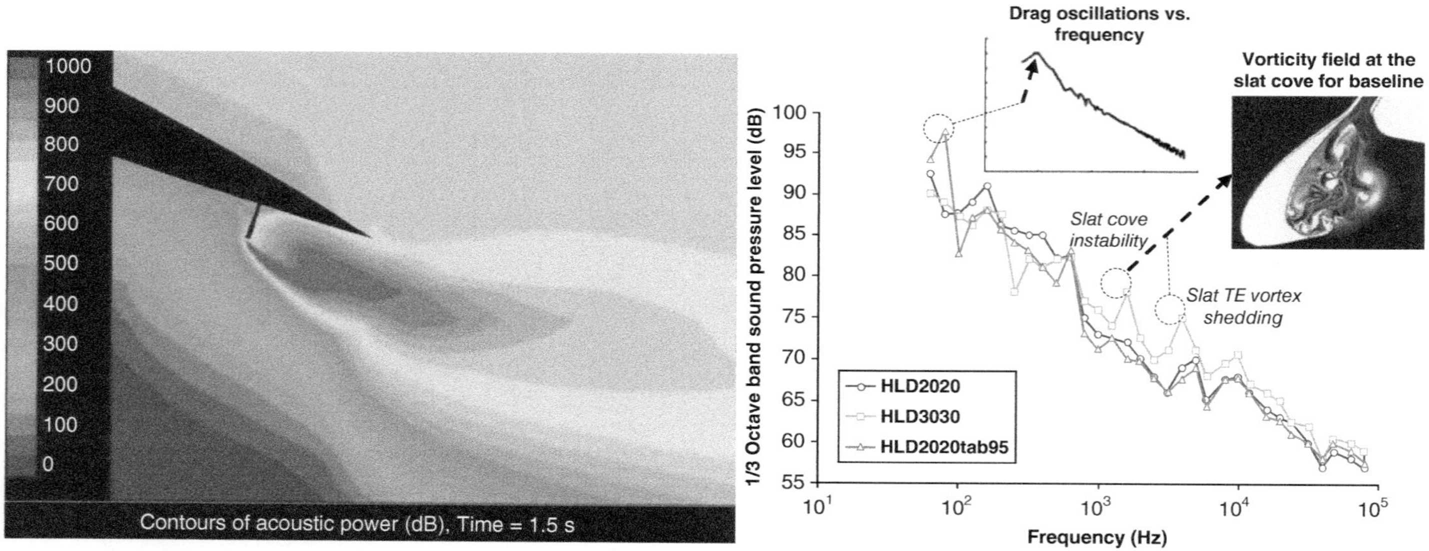

Figure 3.127 (Left) Acoustic power contour at the deployable micodevice region, (right) comparison between the baseline configuration and the reduced HLD settings from an observer at 270°.

Noise reduction in the mid-frequency range, which is the most sensitive range for human hearing, was particularly evident, thus demonstrating its potential for application on commercial airliners as well as aerial platforms for noise reduction during the approach-to-landing phase of flight.

Air and Ground Launched Vehicle Acoustics

The first paper on rocket noise characterization was published by McInerny et al. (1997). As a rocket or launch vehicle ascends, the highest sound pressure levels are experienced in the far-field. The sound generation mechanisms and propagation process is extremely complex, because measured noise data depends on a wide range of rocket types and thrusts.

Rocket exhaust plume parameters and a schematic of the rocket plume are given in Figure 3.128. The rocket plume consists of a laminar and a supersonic core. The former has a length between 16 and 20 exit nozzle diameters and the latter a length between 25 and 35 diameters. The major sound sources are the turbulence in the transition region at the edge of the supersonic core and the fully developed turbulence at the end of transition region. Typical rocket nozzle exhaust Mach numbers are between 3.0 and 3.5 and the temperature of a rocket exhaust is in the range of 1800–2100 K. Peak angles of directivity are between 50° and 60° based on ground test data and close to 70° if inferred from launch data. The angle of maximum radiation relative to the exhaust axis increases as the speed of sound increases in the flow. The directional characteristics of various types of jets and rockets are illustrated in Figure 3.129 for sound pressure levels (SPL) (NASA SP-8072 1971).

During the launch and the initial phase of ascent of a rocket or space vehicle, sound is generated by the release of high velocity engine exhaust gases, and by subsonic or supersonic vehicle movement through the atmosphere. A comparison of air versus launch noise on the ground using the previously described methods can be found in references (Sarigul-Klijn 2012; Kuo and Sarigul-Klijn 2012).

Environmental Impact and Sound Mitigation

The U.S. noise standards are defined in the Code of Federal Regulations (CFR) Noise Standards: Aircraft Type and Airworthiness Certification (14 CFR Part 36). The FAA regulates the maximum noise level that each nonmilitary aircraft can emit through requiring aircraft to meet certain noise certification standards. These standards designate changes in maximum noise level requirements. The FAA also incorporates the model of the International Civil Aviation Organization (ICAO) which sets global aircraft noise standards known as *stages*. At present, the FAA mandates that nearly all aircraft that fly within the United States comply with Stage 3 requirements.

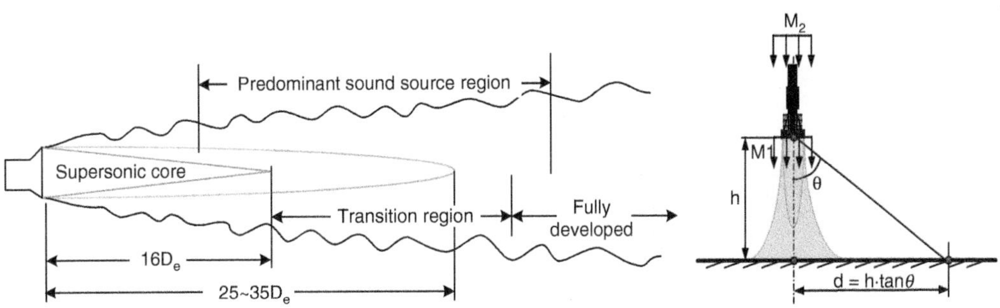

FIGURE 3.128 (Left) Undeflected rocket plume: dominant sound source regions, (right) directivity θ on vertical launch.

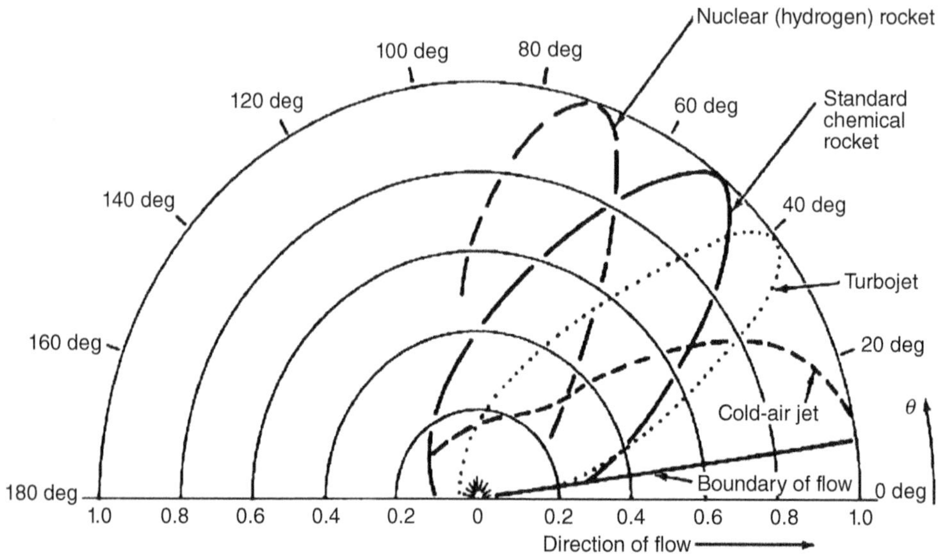

FIGURE 3.129 Far-field directivity of SPL for different types of jet flow (from NASA SP-8072 1971).

The noise generated by air transportation vehicles has been a major environmental issue, in particular for urban areas. Aircraft noise is a major complaint when dealing with noise pollution. Adverse health and other effects caused by environmental noise include high-blood pressure, stress, and insomnia. The increase in volume of air traffic means that the overall community noise has not decreased, notwithstanding the advent of more advanced noise control technologies. Therefore, the development of airport noise reduction strategies and airport noise monitoring methods become increasingly important. The effects of short-duration extreme noise such as jet noise near airports are different as compared with constant noise. In addition to the use of noise mitigation design changes or aeroacoustics afterthoughts such as chevrons and deployable microdevices (Figure 3.130),

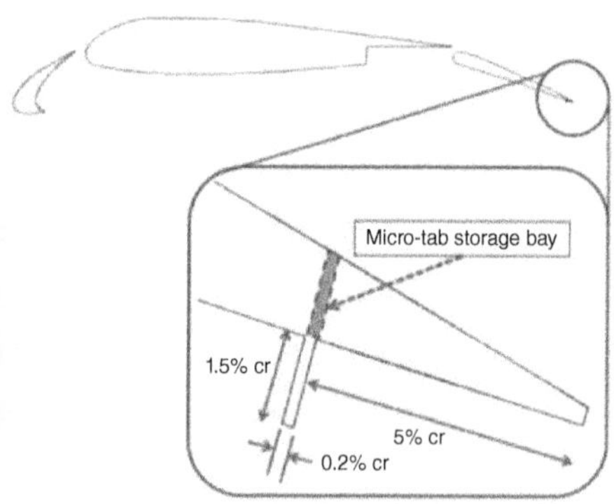

FIGURE 3.130 Sample noise mitigation systems for existing designs "aeroacoustic afterthoughts": (left) Chevrons for propulsive noise (Courtesy of NASA), and (right) deployable microdevices for high lift device noise (courtesy of TNCC).

noise emissions have prompted certain airports to establish stringent noise abatement procedures such as Bob Hope Airport in Southern California and London Heathrow Airport.

Despite the efforts of advanced schemes and measurements (Lilley 1958; Lilley 1974; Dowling and Hynes 2006; James et al. 2010; Karabasov 2010; Sarigul-Klijn et al. 2011; Sarigul-Klijn 2012; Kuo and Sarigul-Klijn 2012) aeroacoustics performance predictions of engine noise still present a challenge, especially when there are noise mitigation changes with chevrons to the design. For space launch vehicles, at the initial phases sound is generated by the release of high velocity engine exhaust gases (Sarigul-Klijn et al. 1997) and by subsonic or supersonic vehicle movement through the atmosphere. The fluctuating pressures associated with acoustic energy during launch can cause vibration of structural components and can be harmful to the environment.

Basic Terms

c = speed of sound, m/s (ft/s)

λ = wavelength, m (ft)

(c and λ values at 15°C: in air; c: 340 m/s (1225 ft/s), 0.3 m (1 ft) and in water; 1500 m/s (5,400 ft/s), 0.15 m (5 ft), f = frequency, Hz)

(The typical range of frequency of the human ear is: 20–20,000 Hz. The maximum sensitivity of the ear is around 3000 Hz or 3 kHz.)

SPL = sound pressure level (dB, reference to 20 μPa)

$$\text{SPL} = 20 \log \frac{P_{\text{rms}}}{p_{\text{ref}}} \text{DB}_{\text{ref}}\ 20\ \mu\text{Pa}$$

(The reference pressure level of 20 μPa corresponds to the onset of hearing at 1,000 Hz for an average human while the onset of pain of 140 dB SPL corresponds to pressure fluctuations of 200 Pa.)

References

Ahuja, K. K. and Mendoza, J. 1995. "Effects of Cavity Dimensions, Boundary Layer and Temperature on Cavity Noise with Emphasis on Benchmark Data to Validate Computational Aeroacoustic Codes," NASA Technical Report N95-24879.

Bridges, J. and Brown, C. A. 2004. "Parametric Testing of Chevrons on Single Flow Hot Jets," NASA/TM-2004-213107, Glenn Research Center, Cleveland, OH.

Curle, N. 1955. "The Influence of Solid Boundaries upon Aerodynamic Sound," *Proc. R. Soc. Lond., Series A*, vol. 231, pp. 505–514.

Dowling, A. P. and Hynes, T. 2006. "Towards a Silent Aircraft," *Aero J. R. Aero Soc.*, vol. 110, pp. 487–494.

Ffowcs Williams, J. E. 1963. "The Noise from Turbulence Convected at High Speed," *Phil. Trans. R. Soc. Lond. Series A*, vol. 255, pp. 469–503. doi:10.1098/rsta.1963.0010.

Goldstein, M. E. 1976. *Aeroacoustics*, McGraw-Hill, New York, 1976.

Goldstein, M. E. 2003. "A Generalized Acoustic Analogy," *J. Fluid Mech.*, vol. 488, pp. 315–333. doi:10.1017/S0022112003004890.

Howe, M. S. 2003. *Theory of Vortex Sound*, Cambridge University Press, Cambridge, UK, Cambridge Texts in Applied Mathematics.

James, M. M., et al. 2010. "Aircraft Jet Source Noise Measurements of an F-22 Using a Prototype Near-Field Acoustic Holography Measurement System," ASA, 2010.

Karabasov, S. A. 2010. "Understanding Jet Noise," *Phil. Trans. R. Soc. Series A*, vol. 368, pp. 3593–3608, 2010. doi:10.1098/rsta.2010.0086.

Kuo, B. C. and Sarigul-Klijn, N. 2012. "Conceptual Study of Micro-Tab Device in Airframe Noise Reduction: (II) 3D Computation," *Aerospace Science and Technology*, vol. 17, pp. 32–39.

Lighthill, M. J. 1952. "On Sound Generated Aerodynamically. I. General Theory," *Proc. R. Soc. Lond. Series A*, vol. 222, pp. 564–587. doi:10.1098/rspa.1952.0060.

Lighthill, M. J. 1954. "On Sound Generated Aerodynamically II. Turbulence as a Source of Sound," *Proc. R. Soc. Lond. Series A*, vol. 222, pp. 1–32.

Lilley, G. M. 1958. "On the Noise from Air Jets," *Aeronaut. Res. Council Rep. Mem.*, vol. 20, p. 376.

Lilley, G. M. 1974. "On the Noise from Jets," *Noise Mechanisms*, AGARD-CP-131, pp. 13.1–13.12.

McInerny, S. A., et al. 1997. "The Influence of Low-Frequency Instrumentation Response on Rocket Noise Metrics," *J. Acoust. Soc. Am.*, vol. 102, p. 2780.

Mohan, N. K. D., et al. 2015. "Acoustic Sources and Far-Field Noise of Chevron and Round Jets," *AIAA Journal*, vol. 53. pp. 2421–2436. ISSN 0001-1452.

Morris, P. J. 2009. "A Note on Noise Generation by Large Scale Turbulent Structures in Subsonic and Supersonic Jets," *Int. J. Aeroacoustics*, vol. 8, pp. 301–316. doi:10.1260/147547209787548921.

NASA SP-8072. 1971. Acoustic loads generated by propulsion system.

Panda, J. 2008. "The Sources of Jet-Noise: Experimental Evidence," *J. Fluid Mech.*, vol. 615, pp. 253–992. doi:10.1017/S0022112008003704.

Rossiter, J. E. 1964. "Wind-Tunnel Experiments on the Flow over Rectangular Cavities at Subsonic and Transonic Speeds," Aeronautical Research Council Reports & Memoranda No. 3438, London, U.K.

Sarigul-Klijn, N. 1997. *Gradient Adaptive Transfinite Elements (GATE) Family: Steep Gradients and Finite Element Method*, NSK, Davis. ISBN 10-9643757-1-0.

Sarigul-Klijn, N., et al. 1997. "Vehicle Aerodynamic Noise Characterization," TNCC TR 1999003.

Sarigul-Klijn, N., Dietz, D., Karnopp, D., and Dummer, D. 2001. "A Computational Aeroacoustic Method for Near and Far Field Vehicle Noise Predictions," AIAA-2001-0513.

Sarigul-Klijn, N. 2008. "'Smart' Monitoring of Structural Health in Flight-Environment," ICEE2008 conference (invited Conference key note address), Japan.

Sarigul-Klijn, N. 2012. "How to Predict Near and Far Field Acoustics: A Unique Aeroacoustics Approach with Designs to Control Noise of Launch Vehicles and Transport Aircraft (Invited Seminar) at Advanced Modeling & Simulation Seminar Series," NASA Ames RC, California, U.S.

Sarigul-Klijn, N. 2015. "Acoustics: Experiments, Theory, Computations and Noise Mitigations," DynaaTECC, TR20130803, University of California Davis.

Sarigul-Klijn, N., et al. 2008. "Scaled Acoustics Experiments and Vibration Prediction Based Structural Health Monitoring," Paper No. IMECE2008-68613, pp. 363–371. doi:10.1115/IMECE2008-68613.

Sarigul-Klijn, N., et al. 2011. "In-Flight Deployable Micro Devices in Noise Control: Design and Evaluation," Paper No. IMECE2011-62294, pp. 129–136. doi:10.1115/IMECE2011-62294.

SECTION 4

Aircraft Performance, Stability, and Control

Section Editors: Trevor M. Young, Douglas G. Thomson, and Rafał Żbikowski

PART 1

Aircraft Performance

Trevor M. Young

Notation

a	speed of sound
a	acceleration
a_c	centripetal acceleration
A_R	aspect ratio
c_t	specific fuel consumption of turbojet or turbofan airplane, defined in terms of mass flow rate
c_t'	specific fuel consumption of turbojet or turbofan airplane, defined in terms of weight flow rate
c_p	specific fuel consumption of piston or turboprop airplane, defined in terms of mass flow rate
c_p'	specific fuel consumption of piston or turboprop airplane, defined in terms of weight flow rate
C_L	lift coefficient
C_D	drag coefficient
C_{Di}	lift-dependent drag coefficient
C_{D0}	lift-independent drag coefficient (zero-lift drag coefficient)
CAS	calibrated airspeed
CS	certification specification
D	drag (force)
e	Oswald efficiency factor
EAS	equivalent airspeed
EASA	European Aviation Safety Agency
f_{acc}	acceleration factor (for climb or descent)
FAR	Federal Aviation Regulation
g	acceleration due to gravity
h	height
h_{SC}	screen height
H	geopotential height
ICAO	International Civil Aviation Organization
ISA	International Standard Atmosphere
K	lift-dependent drag factor
L	lapse rate
L	lift (force)
LRC	long-range cruise (speed)
m	mass
M	Mach number
MRC	maximum range cruise (speed)
MTOW	maximum takeoff weight
n	load factor

N rotational speed of engine
p pressure
P power
P_D drag power (power required)
P_S shaft power
P_T thrust power (power available)
q dynamic pressure
Q mass of fuel burned per unit time (fuel flow rate)
Q' weight of fuel burned per unit time (fuel flow rate)
r radius of turn
r_a specific air range (SAR)
R range
R gas constant
ROC rate of climb
ROD rate of descent
ROS rate of sink
s ground distance
S wing reference area
SAR specific air range
SFC specific fuel consumption
t time
T thrust
T temperature
TAS true airspeed
TOW takeoff weight
USC United States Customary (units)
V true airspeed (TAS)
V_e equivalent airspeed (EAS)
V_S stall speed
V_1 decision speed in takeoff
W weight (force)
x still air distance

Greek

δ relative pressure (pressure ratio)
ϕ angle of bank
γ flight path angle
γ ratio of specific heats of air
λ ground effect factor
η_p propeller efficiency
μ_B coefficient of braking friction
μ_R coefficient of rolling friction
θ relative temperature (temperature ratio)
ρ density
σ relative density (density ratio)
Ω rate of turn

4.1 Standard Atmosphere and Height Measurement

International Standard Atmosphere (ISA)

The *International Standard Atmosphere* (ISO 2533 1975) is an idealized model of the atmosphere which by international agreement is used for aircraft performance analysis and operation. The ISA describes a hypothetical vertical distribution of *temperature, pressure,* and *density,* which greatly simplifies numerical analysis. The ISA is identical to the International Civil Aviation Organization Standard Atmosphere (ICAO 1993) and the U.S. Standard Atmosphere for heights up to 32 km. For aviation purposes, the ISA may be defined in two regions:

- The *troposphere* extends from the ISA datum height (ISA sea level) to the tropopause, at a geopotential height of 11,000 m (36,089.24 ft). The temperature is assumed to be exactly 15°C at the sea-level datum and to decrease linearly with altitude at a lapse rate of 6.5°C per 1000 m (exactly). The ISA datum is not defined by a geographical height, but rather by a reference pressure of 101,325 N/m^2 (2,116.217 lb/ft^2).
- The lower *stratosphere* is the region above the tropopause to a geopotential height of 20,000 m (65,616.80 ft) in which it is assumed that the temperature is a constant –56.5°C (exactly).

It is customary to define the air temperature (T), pressure (p), and density (ρ) as ratios of the standard sea-level (ISA datum) values and to use the subscript 0 to denote the standard sea-level conditions. Standard values are given in Table 4.1. The following definitions are used:

Relative temperature (temperature ratio):

$$\theta = \frac{T}{T_0} \quad \text{(absolute temperatures)} \tag{4.1}$$

Standard values	Symbol	SI units	Equivalent
Temperature at sea level	T_0	288.15 K	518.67°R
Lapse rate in troposphere	L	6.5 K per 1,000 m	1.9812 Kper 1,000 ft 3.56616°R per 1,000 ft
Temperature of tropopause	T^*	216.65 K	389.97°R
Height of tropopause	H^*	11,000 m	36,089.24 ft
Sea-level pressure	p_0	101,325 N/m^2 1013.25 hPa	29.92126 in. Hg 2,116.217 lb/ft^2
Sea-level density	ρ_0	1.225 kg/m^3	0.002376892 slugs/ft^3
Gravitational acceleration	g_0	9.80665 m/s^2	32.17405 ft/s^2
Speed of sound at sea level	a_0	340.294 m/s	1,116.450 ft/s 661.4786 kt
Gas constant	R	287.05287 m^{2}s^{-2}K^{-1}	3,089.811 ft^{2}s^{-2}K^{-1} 1,716.562 ft^{2}s^{-2}°R^{-1}
Ratio of specific heats of air	γ	1.40	

Table 4.1 Standard Values of the ISA

Relative pressure (pressure ratio):

$$\delta = \frac{p}{p_0} \tag{4.2}$$

Relative density (density ratio)

$$\sigma = \frac{\rho}{\rho_0} \tag{4.3}$$

Temperature, density, and pressure are not independent, as is evident from the perfect gas law, which for arbitrary conditions can be written as:

$$\frac{p}{\rho} = RT \tag{4.4}$$

where R is the gas constant. It follows that the ratios σ, δ, and θ are related by the following expression:

$$\sigma = \frac{\delta}{\theta} \tag{4.5}$$

Temperature, Pressure, and Density in the Standard Atmosphere

The pressure at any altitude can be determined by the following equation, as the sea-level conditions and the temperature are known:

$$\int \frac{1}{p} dp = -\int \frac{g}{RT} dh \tag{4.6}$$

The temperature is a linear function of height in the troposphere, and it has a constant value in the stratosphere. The integral is evaluated for these two regions by setting the gravitational acceleration equal to the standard sea-level value. For the troposphere, the integration is performed from sea level to altitude H, yielding the pressure ratio. In the stratosphere the integration is performed from the tropopause to the altitude H. The density ratio is determined from the perfect gas law (equation [4.4]). The resulting equations are given in Table 4.2. Tabulated values of the ISA are given in Table 4.3 as a function of geopotential height in feet, the unit of measure used

	In the troposphere	Tropopause	In the stratosphere
Relative temperature	$\theta = 1 - \frac{L}{T_0} H$	$\theta^* = 0.751865$	$\theta = \theta^*$
Relative pressure	$\delta = \left[1 - \frac{LH}{T_0}\right]^{g_0/RL}$	$\delta^* = 0.223361$	$\frac{\delta}{\delta^*} = e^{-(g_0/RT^*)(H-H^*)}$
Relative density	$\sigma = \left[1 - \frac{LH}{T_0}\right]^{(g_0/RL)-1}$ Alternatively, if θ and δ are known, use $\sigma = \frac{\delta}{\theta}$	$\sigma^* = 0.297076$	$\frac{\sigma}{\sigma^*} = e^{(-g_0/RT^*)(H-H^*)}$ Alternatively, if θ and δ are known, use $\sigma = \frac{\delta}{\theta}$

Note: The conditions at the tropopause are denoted by superscript asterisk (*).

Table 4.2 Equations for the ISA

H, ft	*T*, K	θ	*p*, N/m^2	δ	ρ, kg/m^3	σ	a, m/s
–2 000	292.11	1.013 75	108 866	1.074 42	1.298 3	1.059 85	342.63
–1 800	291.72	1.012 38	108 092	1.066 78	1.290 8	1.053 74	342.39
–1 600	291.32	1.011 00	107 322	1.059 19	1.283 4	1.047 66	342.16
–1 400	290.92	1.009 63	106 557	1.051 64	1.276 0	1.041 61	341.93
–1 200	290.53	1.008 25	105 797	1.044 13	1.268 6	1.035 59	341.69
–1 000	290.13	1.006 88	105 041	1.036 67	1.261 2	1.029 59	341.46
–800	289.73	1.005 50	104 289	1.029 25	1.253 9	1.023 62	341.23
–600	289.34	1.004 13	103 541	1.021 87	1.246 7	1.017 68	341.00
–400	288.94	1.002 75	102 798	1.014 54	1.239 4	1.011 76	340.76
–200	288.55	1.001 38	102 059	1.007 25	1.232 2	1.005 87	340.53
0	288.15	1.000 00	101 325	1.000 00	1.225 0	1.000 00	340.29
200	287.75	0.998 62	100 595	0.992 79	1.217 8	0.994 16	340.06
400	287.36	0.997 25	99 869	0.985 63	1.210 7	0.988 35	339.83
600	286.96	0.995 87	99 147	0.978 51	1.203 6	0.982 56	339.59
800	286.57	0.994 50	98 430	0.971 43	1.196 6	0.976 80	339.36
1 000	286.17	0.993 12	97 717	0.964 39	1.189 6	0.971 06	339.12
1 200	285.77	0.991 75	97 008	0.957 39	1.182 6	0.965 35	338.89
1 400	285.38	0.990 37	96 303	0.950 43	1.175 6	0.959 67	338.65
1 600	284.98	0.989 00	95 602	0.943 52	1.168 7	0.954 01	338.42
1 800	284.58	0.987 62	94 905	0.936 64	1.161 8	0.948 38	338.18
2 000	284.19	0.986 25	94 213	0.929 81	1.154 9	0.942 77	337.95
2 200	283.79	0.984 87	93 525	0.923 02	1.148 1	0.937 19	337.71
2 400	283.40	0.983 50	92 840	0.916 26	1.141 3	0.931 64	337.47
2 600	283.00	0.982 12	92 160	0.909 55	1.134 5	0.926 10	337.24
2 800	282.60	0.980 75	91 484	0.902 88	1.127 7	0.920 60	337.00
3 000	282.21	0.979 37	90 812	0.896 24	1.121 0	0.915 12	336.77
3 200	281.81	0.978 00	90 143	0.889 65	1.114 3	0.909 66	336.53
3 400	281.41	0.976 62	89 479	0.883 09	1.107 7	0.904 23	336.29
3 600	281.02	0.975 25	88 819	0.876 58	1.101 1	0.898 82	336.06
3 800	280.62	0.973 87	88 163	0.870 10	1.094 5	0.893 44	335.82
4 000	280.23	0.972 50	87 511	0.863 66	1.087 9	0.888 09	335.58
4 200	279.83	0.971 12	86 862	0.857 26	1.081 4	0.882 75	335.34
4 400	279.43	0.969 75	86 218	0.850 90	1.074 9	0.877 45	335.11
4 600	279.04	0.968 37	85 577	0.844 58	1.068 4	0.872 16	334.87
4 800	278.64	0.967 00	84 940	0.838 29	1.062 0	0.866 90	334.63
5 000	278.24	0.965 62	84 307	0.832 05	1.055 5	0.861 67	334.39
5 200	277.85	0.964 25	83 678	0.825 84	1.049 2	0.856 46	334.16
5 400	277.45	0.962 87	83 053	0.819 67	1.042 8	0.851 27	333.92
5 600	277.06	0.961 50	82 431	0.813 53	1.036 5	0.846 11	333.68
5 800	276.66	0.960 12	81 814	0.807 44	1.030 2	0.840 97	333.44

TABLE 4.3 International Standard Atmosphere (–2,000 to 60,000 ft)

H, ft	T, K	θ	p, N/m^2	δ	ρ, kg/m^3	σ	a, m/s
6 000	276.26	0.958 75	81 200	0.801 38	1.023 9	0.835 86	333.20
6 200	275.87	0.957 37	80 589	0.795 35	1.017 7	0.830 77	332.96
6 400	275.47	0.956 00	79 983	0.789 37	1.011 5	0.825 70	332.72
6 600	275.07	0.954 62	79 380	0.783 42	1.055 3	0.820 66	332.48
6 800	274.68	0.953 25	78 781	0.777 51	0.999 16	0.815 64	332.24
7 000	274.28	0.951 87	78 185	0.771 63	0.993 04	0.810 64	332.00
7 200	273.89	0.950 50	77 594	0.765 79	0.986 95	0.805 67	331.76
7 400	273.49	0.949 12	77 005	0.759 98	0.980 89	0.800 72	331.52
7 600	273.09	0.947 75	76 421	0.754 21	0.974 85	0.795 80	331.28
7 800	272.70	0.946 37	75 840	0.748 48	0.968 85	0.790 90	331.04
8 000	272.30	0.945 00	75 262	0.742 78	0.962 87	0.786 02	330.80
8 200	271.90	0.943 62	74 689	0.737 12	0.956 92	0.781 16	330.56
8 400	271.51	0.942 25	74 118	0.731 49	0.951 00	0.776 33	330.32
8 600	271.11	0.940 87	73 551	0.725 90	0.945 11	0.771 52	330.08
8 800	270.72	0.939 49	72 988	0.720 34	0.939 24	0.766 73	329.84
9 000	270.32	0.938 12	72 428	0.714 81	0.933 41	0.761 96	329.60
9 200	269.92	0.936 74	71 872	0.709 32	0.927 60	0.757 22	329.36
9 400	269.53	0.935 37	71 319	0.703 87	0.921 82	0.752 50	329.11
9 600	269.13	0.933 99	70 770	0.698 45	0.916 06	0.747 81	328.87
9 800	268.73	0.932 62	70 224	0.693 06	0.910 34	0.743 13	328.63
10 000	268.34	0.931 24	69 682	0.687 70	0.904 64	0.738 48	328.39
10 200	267.94	0.929 87	69 143	0.682 38	0.898 97	0.733 85	328.14
10 400	267.55	0.928 49	68 607	0.677 10	0.893 32	0.729 24	327.90
10 600	267.15	0.927 12	68 074	0.671 84	0.887 70	0.724 66	327.66
10 800	266.75	0.925 74	67 545	0.666 62	0.882 11	0.720 09	327.42
11 000	266.36	0.924 37	67 020	0.661 43	0.876 55	0.715 55	327.17
11 200	265.96	0.922 99	66 497	0.656 28	0.871 01	0.711 03	326.93
11 400	265.56	0.921 62	65 978	0.651 16	0.865 51	0.706 53	326.69
11 600	265.17	0.920 24	65 463	0.646 07	0.860 02	0.702 06	326.44
11 800	264.77	0.918 87	64 950	0.641 01	0.854 57	0.697 61	326.20
12 000	264.38	0.917 49	64 441	0.635 98	0.849 14	0.693 17	325.95
12 200	263.98	0.916 12	63 935	0.630 99	0.843 73	0.688 76	325.71
12 400	263.58	0.914 74	63 432	0.626 03	0.838 36	0.684 37	325.46
12 600	263.19	0.913 37	62 932	0.621 09	0.833 01	0.680 01	325.22
12 800	262.79	0.911 99	62 436	0.616 20	0.827 68	0.675 66	324.97
13 000	262.39	0.910 62	61 943	0.611 33	0.822 38	0.671 33	324.73
13 200	262.00	0.909 24	61 453	0.606 49	0.817 11	0.667 03	324.48
13 400	261.60	0.907 87	60 966	0.601 69	0.811 87	0.662 75	324.24
13 600	261.21	0.906 49	60 482	0.596 91	0.806 64	0.658 49	323.99
13 800	260.81	0.905 12	60 001	0.592 17	0.801 45	0.654 24	323.75

TABLE 4.3 International Standard Atmosphere (–2,000 to 60,000 ft) (*Continued*)

H, ft	*T*, K	θ	p, N/m²	δ	ρ, kg/m³	σ	a, m/s
14 000	260.41	0.903 74	59 524	0.587 45	0.796 28	0.650 02	323.50
14 200	260.02	0.902 37	59 049	0.582 77	0.791 14	0.645 83	323.26
14 400	259.62	0.900 99	58 578	0.578 12	0.786 02	0.641 65	323.01
14 600	259.22	0.899 62	58 110	0.573 50	0.780 93	0.637 49	322.76
14 800	258.83	0.898 24	57 644	0.568 90	0.775 86	0.633 35	322.52
15 000	258.43	0.896 87	57 182	0.564 34	0.770 82	0.629 24	322.27
15 200	258.04	0.895 49	56 723	0.559 81	0.765 80	0.625 14	322.02
15 400	257.64	0.894 12	56 266	0.555 31	0.760 81	0.621 07	321.77
15 600	257.24	0.892 74	55 813	0.550 83	0.755 84	0.617 01	321.53
15 800	256.85	0.891 37	55 363	0.546 39	0.750 90	0.612 98	321.28
16 000	256.45	0.889 99	54 915	0.541 97	0.745 98	0.608 96	321.03
16 200	256.05	0.888 62	54 471	0.537 58	0.741 09	0.604 97	320.78
16 400	255.66	0.887 24	54 029	0.533 23	0.736 22	0.600 99	320.53
16 600	255.26	0.885 87	53 590	0.528 90	0.731 37	0.597 04	320.29
16 800	254.87	0.884 49	53 155	0.524 60	0.726 55	0.593 11	320.04
17 000	254.47	0.883 12	52 722	0.520 32	0.721 76	0.589 19	319.79
17 200	254.07	0.881 74	52 292	0.516 08	0.716 99	0.585 30	319.54
17 400	253.68	0.880 36	51 864	0.511 86	0.712 24	0.581 42	319.29
17 600	253.28	0.878 99	51 440	0.507 67	0.707 52	0.577 57	319.04
17 800	252.88	0.877 61	51 019	0.503 51	0.702 82	0.573 73	318.79
18 000	252.49	0.876 24	50 600	0.499 38	0.698 14	0.569 91	318.54
18 200	252.09	0.874 86	50 184	0.495 28	0.693 49	0.566 12	318.29
18 400	251.70	0.873 49	49 771	0.491 20	0.688 87	0.562 34	318.04
18 600	251.30	0.872 11	49 360	0.487 15	0.684 26	0.558 58	317.79
18 800	250.90	0.870 74	48 952	0.483 12	0.679 68	0.554 84	317.54
19 000	250.51	0.869 36	48 548	0.479 13	0.675 13	0.551 12	317.29
19 200	250.11	0.867 99	48 145	0.475 16	0.670 59	0.547 42	317.04
19 400	249.71	0.866 61	47 746	0.471 21	0.666 08	0.543 74	316.79
19 600	249.32	0.865 24	47 349	0.467 30	0.661 60	0.540 08	316.54
19 800	248.92	0.863 86	46 955	0.463 41	0.657 13	0.536 44	316.28
20 000	248.53	0.862 49	46 563	0.459 54	0.652 69	0.532 81	316.03
20 200	248.13	0.861 11	46 174	0.455 71	0.648 28	0.529 20	315.78
20 400	247.73	0.859 74	45 788	0.451 89	0.643 88	0.525 62	315.53
20 600	247.34	0.858 36	45 404	0.448 11	0.639 51	0.522 05	315.28
20 800	246.94	0.856 99	45 023	0.444 35	0.635 16	0.518 05	315.02
21 000	246.54	0.855 61	44 645	0.440 61	0.630 83	0.514 97	314.77
21 200	246.15	0.854 24	44 269	0.436 90	0.626 53	0.511 45	314.52
21 400	245.75	0.852 86	43 896	0.433 22	0.622 25	0.507 96	314.26
21 600	245.36	0.851 49	43 525	0.429 56	0.617 99	0.504 98	314.01
21 800	244.96	0.850 11	43 157	0.425 93	0.613 76	0.501 02	313.76

Table 4.3 International Standard Atmosphere (–2,000 to 60,000 ft) (*Continued*)

H, ft	T, K	θ	p, N/m²	δ	ρ, kg/m³	σ	a, m/s
22 000	244.56	0.848 74	42 791	0.422 32	0.609 54	0.497 58	313.50
22 200	244.17	0.847 36	42 428	0.418 73	0.605 35	0.494 16	313.25
22 400	243.77	0.845 99	42 068	0.415 18	0.601 18	0.490 76	312.99
22 600	243.37	0.844 61	41 709	0.411 64	0.597 03	0.487 37	312.74
22 800	242.98	0.843 24	41 354	0.408 13	0.592 91	0.484 00	312.48
23 000	242.58	0.841 86	41 001	0.404 64	0.588 80	0.480 65	312.23
23 200	242.19	0.840 49	40 650	0.401 18	0.584 72	0.477 32	311.97
23 400	241.79	0.839 11	40 301	0.397 74	0.580 66	0.474 01	311.72
23 600	241.39	0.837 74	39 956	0.394 33	0.576 62	0.470 71	311.46
23 800	241.00	0.836 36	39 612	0.390 94	0.572 60	0.467 43	311.21
24 000	240.60	0.834 99	39 271	0.387 57	0.568 61	0.464 17	310.95
24 200	240.20	0.833 61	38 932	0.384 23	0.564 63	0.460 92	310.70
24 400	239.81	0.832 24	38 596	0.380 91	0.560 68	0.457 70	310.44
24 600	239.41	0.830 86	38 262	0.377 62	0.556 75	0.454 49	310.18
24 800	239.02	0.829 49	37 930	0.374 34	0.552 84	0.451 29	309.93
25 000	238.62	0.828 11	37 601	0.371 09	0.548 95	0.448 12	309.67
25 200	238.22	0.826 74	37 274	0.367 86	0.545 08	0.444 96	309.41
25 400	237.83	0.825 36	36 949	0.364 66	0.541 23	0.441 82	309.15
25 600	237.43	0.823 99	36 627	0.361 48	0.537 40	0.438 69	308.90
25 800	237.04	0.822 61	36 307	0.358 32	0.533 59	0.435 59	308.64
26 000	236.64	0.821 23	35 989	0.355 18	0.529 81	0.432 50	308.38
26 200	236.24	0.819 86	35 673	0.352 07	0.526 04	0.429 42	308.12
26 400	235.85	0.818 48	35 360	0.348 97	0.522 30	0.426 37	307.86
26 600	235.45	0.817 11	35 049	0.345 90	0.518 57	0.423 33	307.61
26 800	235.05	0.815 73	34 740	0.342 85	0.514 87	0.420 30	307.35
27 000	234.66	0.814 36	34 433	0.339 83	0.511 19	0.417 30	307.09
27 200	234.26	0.812 98	34 129	0.336 82	0.507 52	0.414 30	306.83
27 400	233.87	0.811 61	33 826	0.333 84	0.503 88	0.411 33	306.57
27 600	233.47	0.810 23	33 526	0.330 88	0.500 26	0.408 37	306.31
27 800	233.07	0.808 86	33 228	0.327 94	0.496 65	0.405 43	306.05
28 000	232.68	0.807 48	32 932	0.325 02	0.493 07	0.402 51	305.79
28 200	232.28	0.806 11	32 639	0.322 12	0.489 51	0.399 60	305.53
28 400	231.88	0.804 73	32 347	0.319 24	0.485 96	0.396 70	305.27
28 600	231.49	0.803 36	32 058	0.316 38	0.482 44	0.393 83	305.01
28 800	231.09	0.801 98	31 770	0.313 55	0.478 93	0.390 97	304.75
29 000	230.70	0.800 61	31 485	0.310 73	0.475 45	0.388 12	304.48
29 200	230.30	0.799 23	31 202	0.307 94	0.471 98	0.385 29	304.22
29 400	229.90	0.797 86	30 921	0.305 16	0.465 54	0.382 48	303.96
29 600	229.51	0.796 48	30 642	0.302 41	0.465 11	0.379 68	303.70
29 800	229.11	0.795 11	30 365	0.299 67	0.461 70	0.376 90	303.44

TABLE 4.3 International Standard Atmosphere (–2,000 to 60,000 ft) (*Continued*)

H, ft	T, K	θ	p, N/m^2	δ	ρ, kg/m^3	σ	a, m/s
30 000	228.71	0.793 73	30 090	0.296 96	0.458 31	0.374 13	303.17
30 200	228.32	0.792 36	29 817	0.294 27	0.454 94	0.371 38	302.91
30 400	227.92	0.790 98	29 546	0.291 59	0.451 59	0.368 65	302.65
30 600	227.53	0.789 61	29 277	0.288 94	0.448 26	0.365 93	302.38
30 800	227.13	0.788 23	29 010	0.286 30	0.444 95	0.363 22	302.12
31 000	226.73	0.786 86	28 745	0.283 69	0.441 65	0.360 53	301.86
31 200	226.34	0.785 48	28 482	0.281 09	0.438 38	0.357 86	301.59
31 400	225.94	0.784 11	28 220	0.278 51	0.435 12	0.355 20	301.33
31 600	225.54	0.782 73	27 961	0.275 96	0.431 88	0.352 66	301.07
31 800	225.15	0.781 36	27 704	0.273 42	0.428 66	0.349 93	300.80
32 000	224.75	0.779 98	27 449	0.270 90	0.425 46	0.347 31	300.54
32 200	224.36	0.778 61	27 195	0.268 40	0.422 28	0.344 72	300.27
32 400	223.96	0.777 23	26 944	0.265 92	0.419 11	0.342 13	300.01
32 600	223.56	0.775 86	26 694	0.263 45	0.415 97	0.339 56	299.74
32 800	223.17	0.774 48	26 447	0.261 01	0.412 84	0.337 01	299.47
33 000	222.77	0.773 11	26 201	0.258 58	0.409 73	0.334 47	299.21
33 200	222.37	0.771 73	25 957	0.256 17	0.406 63	0.331 95	298.94
33 400	221.98	0.770 36	25 715	0.253 78	0.403 56	0.329 44	298.68
33 600	221.58	0.768 98	25 474	0.251 41	0.400 50	0.326 94	298.41
33 800	221.19	0.767 61	25 236	0.249 06	0.397 46	0.324 46	298.14
34 000	220.79	0.766 23	24 999	0.246 72	0.394 44	0.321 99	297.87
34 200	220.39	0.764 85	24 764	0.244 40	0.391 44	0.319 54	297.61
34 400	220.00	0.763 48	24 531	0.242 10	0.388 45	0.317 10	297.34
34 600	219.60	0.762 10	24 300	0.239 82	0.385 48	0.314 68	297.07
34 800	219.20	0.760 73	24 070	0.237 55	0.382 53	0.312 27	296.80
35 000	218.81	0.759 35	23 842	0.235 30	0.379 60	0.309 87	296.54
35 200	218.41	0.757 98	23 616	0.233 07	0.376 68	0.307 49	296.27
35 400	218.02	0.756 60	23 392	0.230 86	0.373 78	0.305 13	296.00
35 600	217.62	0.755 23	23 169	0.228 66	0.370 90	0.302 77	295.73
35 800	217.22	0.753 85	22 948	0.226 48	0.368 03	0.300 43	295.46
36 000	216.83	0.752 48	22 729	0.224 32	0.365 18	0.298 11	295.19
36 000	216.65	0.751 87	22 632	0.223 36	0.363 92	0.297 08	295.07
36 200	216.65	0.751 87	22 512	0.222 17	0.361 98	0.295 50	295.07
36 400	216.65	0.751 87	22 296	0.220 05	0.358 52	0.292 67	295.07
36 600	216.65	0.751 87	22 083	0.217 94	0.355 09	0.289 87	295.07
36 800	216.65	0.751 87	21 872	0.215 86	0.351 70	0.287 10	295.07
37 000	216.65	0.751 87	21 663	0.213 79	0.348 33	0.284 35	295.07
37 200	216.65	0.751 87	21 455	0.211 75	0.345 00	0.281 63	295.07
37 400	216.65	0.751 87	21 250	0.209 72	0.341 70	0.278 94	295.07
37 600	216.65	0.751 87	21 047	0.207 72	0.338 43	0.276 27	295.07
37 800	216.65	0.751 87	20 846	0.205 73	0.335 19	0.273 63	295.07

Table 4.3 International Standard Atmosphere (–2,000 to 60,000 ft) (*Continued*)

H, ft	T, K	θ	p, N/m²	δ	ρ, kg/m³	σ	a, m/s
38 000	216.65	0.751 87	20 646	0.203 76	0.331 98	0.271 01	295.07
38 200	216.65	0.751 87	20 449	0.201 81	0.328 81	0.268 42	295.07
38 400	216.65	0.751 87	20 253	0.199 88	0.325 66	0.265 85	295.07
38 600	216.65	0.751 87	20 059	0.197 97	0.322 55	0.263 30	295.07
38 800	216.65	0.751 87	19 867	0.196 08	0.319 46	0.260 79	295.07
39 000	216.65	0.751 87	19 677	0.194 20	0.316 41	0.258 29	295.07
39 200	216.65	0.751 87	19 489	0.192 34	0.313 38	0.255 82	295.07
39 400	216.65	0.751 87	19 303	0.190 50	0.310 38	0.253 37	295.07
39 600	216.65	0.751 87	19 118	0.188 68	0.307 41	0.250 95	295.07
39 800	216.65	0.751 87	18 935	0.186 87	0.304 47	0.248 55	295.07
40 000	216.65	0.751 87	18 754	0.185 09	0.301 56	0.246 17	295.07
40 200	216.65	0.751 87	18 574	0.183 32	0.298 67	0.243 81	295.07
40 400	216.65	0.751 87	18 397	0.181 56	0.295 82	0.241 48	295.07
40 600	216.65	0.751 87	18 221	0.179 83	0.292 99	0.239 17	295.07
40 800	216.65	0.751 87	18 046	0.178 10	0.290 18	0.236 88	295.07
41 000	216.65	0.751 87	17 874	0.176 40	0.287 41	0.234 62	295.07
41 200	216.65	0.751 87	17 703	0.174 71	0.284 66	0.232 37	295.07
41 400	216.65	0.751 87	17 533	0.173 04	0.281 93	0.230 15	295.07
41 600	216.65	0.751 87	17 366	0.171 39	0.279 24	0.227 95	295.07
41 800	216.65	0.751 87	17 200	0.169 75	0.276 57	0.225 77	295.07
42 000	216.65	0.751 87	17 035	0.168 12	0.273 92	0.223 61	295.07
42 200	216.65	0.751 87	16 872	0.166 51	0.271 30	0.221 47	295.07
42 400	216.65	0.751 87	16 711	0.164 92	0.268 70	0.219 35	295.07
42 600	216.65	0.751 87	16 551	0.163 34	0.266 13	0.217 25	295.07
42 800	216.65	0.751 87	16 392	0.161 78	0.263 59	0.215 17	295.07
43 000	216.65	0.751 87	16 236	0.160 23	0.261 07	0.213 11	295.07
43 200	216.65	1	16 080	0.158 70	0.258 57	0.211 08	295.07
43 400	216.65	0.751 87	15 927	0.157 18	0.256 09	0.209 06	295.07
43 600	216.65	0.751 87	15 774	0.155 68	0.253 64	0.207 06	295.07
43 800	216.65	0.751 87	15 623	0.154 19	0.251 22	0.205 08	295.07
44 000	216.65	0.751 87	15 474	0.152 71	0.248 81	0.203 11	295.07
44 200	216.65	0.751 87	15 326	0.151 25	0.246 43	0.201 17	295.07
44 400	216.65	0.751 87	15 179	0.149 81	0.244 08	0.199 25	295.07
44 600	216.65	0.751 87	15 034	0.148 37	0.241 74	0.197 34	295.07
44 800	216.65	0.751 87	14 890	0.146 95	0.239 43	0.195 45	295.07
45 000	216.65	0.751 87	14 748	0.145 55	0.237 14	0.193 58	295.07
45 200	216.65	0.751 87	14 607	0.144 16	0.234 87	0.191 73	295.07
45 400	216.65	0.751 87	14 467	0.142 78	0.232 62	0.189 90	295.07
45 600	216.65	0.751 87	14 328	0.141 41	0.230 40	0.188 08	295.07
45 800	216.65	0.751 87	14 191	0.140 06	0.228 19	0.186 28	295.07

Table 4.3 International Standard Atmosphere (–2,000 to 60,000 ft) (*Continued*)

H, ft	*T*, K	θ	*p*, N/m²	δ	ρ, kg/m³	σ	a, m/s
46 000	216.65	0.751 87	14 056	0.138 72	0.226 01	0.184 50	295.07
46 200	216.65	0.751 87	13 921	0.137 39	0.223 85	0.182 73	295.07
47 400	216.65	0.751 87	13 788	0.136 08	0.221 71	0.180 98	295.07
46 600	216.65	0.751 87	13 656	0.134 77	0.219 59	0.179 25	295.07
46 800	216.65	0.751 87	13 525	0.133 49	0.217 48	0.177 54	295.07
47 000	216.65	0.751 87	13 396	0.132 21	0.215 40	0.175 84	295.07
47 200	216.65	0.751 87	13 268	0.130 94	0.213 34	0.174 16	295.07
47 400	216.65	0.751 87	13 141	0.129 69	0.211 30	0.172 49	295.07
47 600	216.65	0.751 87	13 015	0.128 45	0.209 28	0.170 84	295.07
47 800	216.65	0.751 87	12 891	0.127 22	0.207 28	0.169 21	295.07
48 000	216.65	0.751 87	12 767	0.126 00	0.205 30	0.167 59	295.07
48 200	216.65	0.751 87	12 645	0.124 80	0.203 33	0.165 99	295.07
48 400	216.65	0.751 87	12 524	0.123 60	0.201 39	0.164 40	295.07
48 600	216.65	0.751 87	12 404	0.122 42	0.199 46	0.162 82	295.07
48 800	216.65	0.751 87	12 286	0.121 25	0.197 55	0.161 27	295.07
49 000	216.65	0.751 87	12 168	0.120 09	0.195 66	0.159 72	295.07
49 200	216.65	0.751 87	12 052	0.118 94	0.193 79	0.158 20	295.07
49 400	216.65	0.751 87	11 937	0.117 80	0.191 94	0.156 68	295.07
49 600	216.65	0.751 87	11 822	0.116 68	0.190 10	0.155 18	295.07
49 800	216.65	0.751 87	11 709	0.115 56	0.188 28	0.153 70	295.07
50 000	216.65	0.751 87	11 597	0.114 46	0.186 48	0.152 23	295.07
50 200	216.65	0.751 87	11 486	0.113 36	0.184 70	0.150 77	295.07
50 400	216.65	0.751 87	11 376	0.112 28	0.182 93	0.149 33	295.07
50 600	216.65	0.751 87	11 268	0.111 20	0.181 18	0.147 90	295.07
50 800	216.65	0.751 87	11 160	0.110 14	0.179 45	0.146 49	295.07
51 000	216.65	0.751 87	11 053	0.109 08	0.177 73	0.145 09	295.07
51 200	216.65	0.751 87	10 947	0.108 04	0.176 03	0.143 70	295.07
51 400	216.65	0.751 87	10 843	0.107 01	0.174 35	0.142 32	295.07
51 600	216.65	0.751 87	10 739	0.105 98	0.172 68	0.140 96	295.07
51 800	216.65	0.751 87	10 636	0.104 97	0.171 03	0.139 61	295.07
52 000	216.65	0.751 87	10 534	0.103 97	0.169 39	0.138 28	295.07
52 200	216.65	0.751 87	10 434	0.102 97	0.167 77	0.136 95	295.07
52 400	216.65	0.751 87	10 334	0.101 99	0.166 16	0.135 64	295.07
52 600	216.65	0.751 87	10 235	0.101 01	0.164 57	0.134 35	295.07
52 800	216.65	0.751 87	10 137	0.100 04	0.163 00	0.133 06	295.07
53 000	216.65	0.751 87	10 040	0.099 087	0.161 44	0.131 79	295.07

Table 4.3 International Standard Atmosphere (–2,000 to 60,000 ft) *(Continued)*

H, ft	*T*, K	θ	*p*, N/m²	δ	ρ, kg/m³	σ	a, m/s
53 200	216.65	0.751 87	9 943.9	0.098 139	0.159 90	0.130 53	295.07
53 400	216.65	0.751 87	9 848.8	0.097 200	0.158 37	0.129 28	295.07
53 600	216.65	0.751 87	9 754.6	0.096 270	0.156 85	0.128 04	295.07
53 800	216.65	0.751 87	9 661.3	0.095 349	0.155 35	0.126 82	295.07
54 000	216.65	0.751 87	9 568.8	0.094 437	0.153 86	0.125 60	295.07
54 200	216.65	0.751 87	9 477.3	0.093 534	0.152 39	0.124 40	295.07
54 400	216.65	0.751 87	9 386.6	0.092 639	0.150 93	0.123 21	295.07
54 600	216.65	0.751 87	9 296.8	0.091 753	0.149 49	0.122 03	295.07
54 800	216.65	0.751 87	9 207.9	0.090 875	0.148 06	0.120 87	295.07
55 000	216.65	0.751 87	9 119.8	0.090 005	0.146 64	0.119 71	295.07
55 200	216.65	0.751 87	9 032.6	0.089 144	0.145 24	0.118 56	295.07
55 400	216.65	0.751 87	8 946.1	0.088 292	0.143 85	0.117 43	295.07
55 600	216.65	0.751 87	8 860.6	0.087 447	0.142 48	0.116 31	295.07
55 800	216.65	0.751 87	8 775.8	0.086 610	0.141 11	0.115 19	295.07
56 000	216.65	0.751 87	8 691.8	0.085 782	0.139 76	0.114 09	295.07
56 200	216.65	0.751 87	8 608.7	0.084 961	0.138 43	0.113 00	295.07
56 400	216.65	0.751 87	8 526.3	0.084 148	0.137 10	0.111 92	295.07
56 600	216.65	0.751 87	8 444.8	0.083 343	0.135 79	0.110 85	295.07
56 800	216.65	0.751 87	8 364.0	0.082 546	0.134 49	0.109 79	295.07
57 000	216.65	0.751 87	8 284.0	0.081 756	0.133 20	0.108 74	295.07
57 200	216.65	0.751 87	8 204.7	0.080 974	0.131 93	0.107 70	295.07
57 400	216.65	0.751 87	8 126.2	0.080 200	0.130 67	0.106 67	295.07
57 600	216.65	0.751 87	8 048.5	0.079 432	0.129 42	0.105 65	295.07
57 800	216.65	0.751 87	7 971.5	0.078 672	0.128 18	0.104 64	295.07
58 000	216.65	0.751 87	7 895.2	0.077 920	0.126 95	0.103 64	295.07
58 200	216.65	0.751 87	7 819.7	0.077 174	0.125 74	0.102 64	295.07
58 400	216.65	0.751 87	7 744.9	0.076 436	0.124 54	0.101 66	295.07
58 600	216.65	0.751 87	7 670.8	0.075 705	0.123 34	0.100 69	295.07
58 800	216.65	0.751 87	7 597.4	0.074 981	0.122 16	0.099 726	295.07
59 000	216.65	0.751 87	7 524.7	0.074 263	0.121 00	0.098 772	295.07
59 200	216.65	0.751 87	7 452.7	0.073 553	0.119 84	0.097 827	295.07
59 400	216.65	0.751 87	7 381.4	0.072 849	0.118 69	0.096 891	295.07
59 600	216.65	0.751 87	7 310.8	0.072 152	0.117 56	0.095 964	295.07
59 800	216.65	0.751 87	7 240.9	0.071 462	0.116 43	0.095 046	295.07
60 000	216.65	0.751 87	7 171.6	0.070 778	0.115 32	0.094 137	295.07

Table 4.3 International Standard Atmosphere (–2,000 to 60,000 ft) (*Continued*)

internationally for aircraft operations. Note that the ISA is defined in terms of SI units and values in U.S. Customary (USC) units are obtained by conversion which can lead to small rounding error discrepancies.

Off-Standard Atmospheric Conditions

Whereas the ISA is essential for calculations of airplane performance prediction, flight tests and aircraft operations will generally not be conducted at atmospheric conditions that comply exactly with the ISA model. The evaluation of flight test data must, therefore, take into account actual flight conditions. Frequently it is required to determine the density ratio, given that the ambient temperature (T_{test}) was measured at a pressure height of H_{test}. The procedure is as follows:

1. From ISA tables, determine δ at the pressure height H_{test}.
2. Determine the relative temperature:

$$\theta = \frac{T_{test}}{T_0} \quad \text{(absolute temperatures)}$$

3. Determine σ from equation (4.5).

Height Scales

In airplane performance work it is necessary to distinguish between different height scales that are used.

Geometric Height (h)

Geometric height is the true vertical distance of a point from a datum plane, which in this case is the mean sea-level datum. It is used to define the height of buildings and terrain, for example.

Geopotential Height (H)

Geopotential height is the height in a hypothetical uniform gravitational field that would give the same potential energy as the point under consideration in the actual, variable gravitational field. Note that the integration of equation (4.6) to give the pressure and density variations with height in the ISA is performed using the standard value of g. Consequently, ISA tables are defined in terms of geopotential height. The difference between geometric and geopotential height for typical airplane cruise altitudes is small and is usually ignored (for example, at 40,000 ft the difference is less than 0.2%).

Pressure Height (h_p)

The ISA defines a unique relationship between pressure and geopotential height, as is evident from the equations in Table 4.2. The height in the standard atmosphere can thus be considered as a scale of pressure. By definition, the pressure height at a point in any atmosphere (standard or off-standard) is the height in the ISA which has the same pressure.

Flight Level (FL)

For airplane operations, altitude is specified in terms of a flight level, which is pressure height expressed in hundreds of feet. FL 350, for example, represents a pressure height of 35,000 ft. The standard sea-level setting of 1013.2 hPa (29.92 in. mercury) is used as the reference pressure.

Density Height (h_ρ)

This height scale is not as widely used as pressure height. *Density height* is the equivalent height for a given density in the ISA.

Altimeter Settings

An altimeter is essentially a pressure gauge that works on the principle of differential pressure between the inside and outside of a sealed chamber. The relationship between pressure and height is defined by equations based on those given in Table 4.2, where the constants correspond to those of the ISA. An altimeter thus provides the pilot with a reading of pressure height (i.e., altitude). For an altimeter to be a useful instrument for day-to-day operations, where varying atmospheric pressure conditions would be encountered, it must be possible to adjust the zero height reading. The pilot accomplishes this by rotating the subscale knob on the instrument, in effect changing the reference pressure used to determine the height. For standard aircraft operations, altitude is measured in feet. There are three useful settings that may be used by the pilot.

Standard Setting

By agreement, all aircraft traffic above a specified transition height use the standard sea-level setting of 1013.2 hPa (29.92 in. mercury) as the reference pressure. Because all aircraft operate by the same rules, there is no danger of collision even if the ambient conditions depart substantially from standard ISA conditions. This is the standard adopted for flight level operation.

QNH Setting

If the pilot selects a reference pressure that results in the altimeter correctly reading the elevation of the local airport (when the aircraft is on the runway), this is called a QNH setting. It is used for takeoff and climb-out to the transition altitude (at which point the pilot switches over to the standard setting), as well as for landing (where the setting is based on the destination airport). For operations below the transition height, QNH can be used throughout the flight.

QFE Setting

The pilot sets the subscale knob such that the altimeter reads zero on the ground, irrespective of the airport's actual elevation, for a QFE setting. During flight the altimeter will indicate the height above that airport.

Further Reading

Further details of the Standard Atmosphere can be found in ISO 2533 (1975), ESDU 68046 (1992), ESDU 72018 (1972), and ICAO (1993), including tables with height in meters and off-standard atmospheric properties. Lowry (1991), Swatton (2008), and Young (2017) provide more information on altimeters and the use of pressure height for flight operations.

4.2 Airspeed and Airspeed Measurement

Speed of Sound (*a*)

The speed of sound in the atmosphere is a function of the ambient conditions and is given by:

$$a = \sqrt{\gamma R T} \tag{4.7}$$

where γ is the ratio of specific heats of air, R is the gas constant, and T is the ambient temperature (absolute). This equation can also be written as a function of the temperature ratio and the standard sea-level value of the speed of sound, i.e.,

$$a = a_0 \sqrt{\theta} \tag{4.8}$$

Note that equation (4.8) is correct for any temperature, corresponding to a standard or off-standard day. In the ISA the temperature decreases linearly with altitude to the tropopause and thus the speed of sound will also decrease, although not linearly, but as a function of $\sqrt{\theta}$. In the stratosphere the speed of sound is constant.

True Airspeed (V)

The true airspeed (TAS) is the speed of the airplane relative to the surrounding air mass.

Mach Number (M)

An airplane's flight Mach number is defined as the ratio of its TAS to the speed of sound in the ambient air.

$$M = \frac{V}{a} \tag{4.9}$$

Dynamic Pressure (q)

The *dynamic pressure* is defined as:

$$q = \frac{1}{2}\rho V^2 \tag{4.10}$$

For high-speed flight Mach number and not TAS is used as the measure of speed. For this reason it is convenient to express the dynamic pressure in terms of Mach number and pressure ratio, i.e.,

$$q = \frac{1}{2}\rho_0 \delta a_0^2 M^2 \tag{4.11}$$

Airspeed Indication (Incompressible Flow)

The Bernoulli equation for incompressible flow states that the sum of the static pressure and dynamic pressure along a streamline is constant. It can be represented as:

$$p_t = p + \frac{1}{2}\rho V^2 = \text{constant} \tag{4.12}$$

where p_t is the total pressure and p is the static pressure.

A *Pitot-static* system is used for measuring airspeed. It has two pressure openings: a *total pressure* port and a *static pressure* port. A small Pitot tube is aligned approximately in the direction of the incoming air; this port senses the total pressure of the air as it momentarily comes to rest. The second port senses the static air pressure at the side of the tube, or on the side of the fuselage. A mechanical measuring device is used to measure the difference in air pressure between the two ports. This, by the application of Bernoulli's equation, is a measure of the dynamic pressure, and, provided air density is known, may be used to indicate the airspeed for incompressible flow. For airspeeds above about 200 kt, compressibility effects become important and the form of the Bernoulli expression given above in equation (4.12) is not satisfactory. The practical measurement of airspeed on actual aircraft introduces specific difficulties, which are discussed below under calibrated airspeed.

Ground Speed (V_g)

The *ground speed* is the actual speed of the airplane relative to the ground. It is given by the sum of the TAS vector and the wind velocity.

Equivalent Airspeed (V_e)

The *equivalent airspeed* (EAS) is the equivalent speed which the airplane would have at sea level (air density ρ_0) if it developed the same *dynamic pressure* as it does moving at its TAS at the altitude concerned (air density ρ). Mathematically this can written as:

$$q = \frac{1}{2}\rho_0 V_e^2 = \frac{1}{2}\rho V^2$$

Thus,

$$V_e = \sqrt{\frac{\rho}{\rho_0}}\,V = \sqrt{\sigma}\,V \tag{4.13}$$

Equivalent airspeed is a very useful parameter for engineering analysis. However, for flight operations reference is made to *calibrated airspeed.*

Calibrated Airspeed (V_C)

The *calibrated airspeed* (CAS)—very occasionally called the *rectified airspeed*—is the airspeed reading on a calibrated airspeed indicator connected to a pitot-static system that is assumed to be entirely free of error. It is common practice in aircraft operations to write KCAS for *knots calibrated airspeed.* A better representation of the operation of an airspeed indicator than that given above is provided by the following equation, which applies to *compressible* airflow:

$$\left(\frac{p_t}{p}\right)_{\text{compressible}} = \left(1 + \frac{\gamma - 1}{2}M^2\right)^{\gamma/(\gamma-1)} \tag{4.14}$$

As the pitot-static system provides a measurement of $(p_t - p)$, this term will be isolated on the left-hand side of the equation, and noting that $\gamma = 1.4$ and $M = V/a$, the equation becomes equation (4.15):

$$(p_t - p) = p\left\{\left[1 + 0.2\left(\frac{V}{a}\right)^2\right]^{3.5} - 1\right\} \tag{4.15}$$

Equation (4.15) can be manipulated into a format that can be used to determine airspeed based on acquired pressure data. This is done by selecting standard sea-level values for pressure and speed of sound on the right hand side of the equation and by defining the resulting speed as calibrated airspeed. The left-hand side of equation (4.15) is provided by the pitot-static system. From this equation it is evident that an airspeed indicator can be correctly calibrated for any Mach numbers at sea level, where CAS will always equal EAS. However due to the use of sea-level values, CAS will be greater than EAS by a small amount that increases with Mach number and altitude.

Compressibility Correction for CAS

The difference between CAS and EAS is called the *compressibility correction* and is designated as ΔV_C. By definition:

$$V_C = V_e + \Delta V_C \tag{4.16}$$

and the magnitude of ΔV_C is given by:

$$\Delta V_C = a_0\left\{\sqrt{5\{[\delta\{[1 + 0.2M^2]^{3.5} - 1\} + 1]^{1/3.5} - 1\}} - M\sqrt{\delta}\right\} \tag{4.17}$$

The compressibility correction factor is negligibly small for operations below about 10,000 ft and 200 kt CAS. A useful equation for converting CAS to Mach number, accounting for compressibility correction, is (Young, 2017):

$$M = \sqrt{5\left\{\left[\frac{1}{\delta}\left\{\left[1+0.2\left(\frac{V_c}{a_0}\right)^2\right]^{3.5}-1\right\}+1\right]^{1/3.5}-1\right\}} \tag{4.18}$$

Indicated Airspeed (V_I)

The indicated airspeed (IAS) is the reading of an actual airspeed indicator. Most frequently the speed is indicated in knots (nautical miles per hour), but units of miles per hour and km per hour are also used. During flight the IAS may differ from the CAS because of an instrument calibration error or an error arising from the inability of the Pitot-static system to measure the correct total and static pressures accurately. The error associated with an individual instrument may be corrected using charts supplied by the manufacturer; modern instruments are very accurate and the instrument error is usually negligible. Total pressure recovery and static pressure errors are very small on well-designed systems, and the instrument can be calibrated to reduce these errors. On a modern airliner the correction is taken into account by the air data computer, and for all practical purposes the IAS will then equal the CAS.

Further Reading

Consult Lan and Roskam (1981), Lowry (1999), Eshelby (2000), or Young (2017) for further details on airspeed measurement.

4.3 Drag and Drag Power (Power Required)

Drag Components

For the purposes of performance analysis the airplane's overall drag coefficient can be divided into two components, i.e.,

$$C_D = C_{D0} + C_{Di} \tag{4.19}$$

where C_{D0} is the *zero-lift* drag coefficient (i.e., the drag coefficient when C_L equals zero) and C_{Di} is the *lift-dependent* drag coefficient (often called induced drag). C_{Di} is mainly wing *trailing vortex drag,* which is proportional to C_L^2 but also includes additional lift-dependent interference drag due to the effect of the fuselage, nacelles, etc., on the wing planform, and a small contribution due to the increase of boundary layer (profile) drag with angle of attack.

Drag Polar

The characteristic C_D versus C_L relationship for an airplane is commonly referred to as a drag polar. It is convenient to represent the drag polar by a mathematical model. To an acceptable approximation, the lift-dependent drag coefficient of the whole aircraft can be taken to be proportional to C_L^2 and the drag polar written as:

$$C_D = C_{D0} + KC_L^2 \tag{4.20}$$

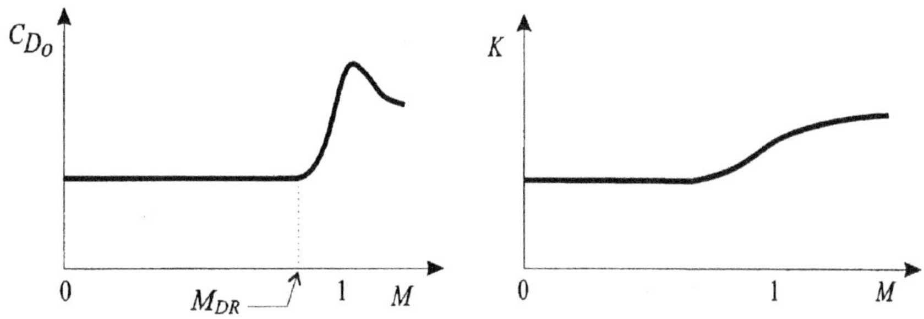

FIGURE 4.1 Influence of Mach number on drag parameters.

where K is the *lift-dependent* drag factor. For initial performance evaluation of an airplane, it is often useful to express K in terms of a span efficiency factor. The equation is then written as:

$$C_D = C_{D0} + \frac{1}{\pi A_R e} C_L^2 \tag{4.21}$$

where A_R is the wing aspect ratio and e is the Oswald factor, which is generally in the range of $0.65 \le e \le 0.90$. High aspect ratio wings are seen from this equation to result in low induced drag. The use of this idealized *parabolic* drag relationship greatly simplifies calculations and is sufficiently accurate for most performance work, providing the following factors are noted.

- *Mach number:* At a speed known as the *drag rise Mach number* (M_{DR}) the drag starts to rise rapidly due to compressibility effects (Figure 4.1). This drag increase is called *wave drag*. The drag rise Mach number thus sets an upper limit to the validity of the low-speed drag polar.
- *Aircraft configuration:* During flight the pilot may change the drag characteristics of the aircraft in several ways, such as by deploying air brakes or flaps or lowering the undercarriage. Each of these factors will change the drag polar.
- *Ground effect:* Operation of the aircraft in close proximity to the ground results in a change in the trailing vortex sheet and a reduction in the lift-dependent drag. This effect depends on the aircraft type and its proximity to the ground.

Actual Drag Polars

Drag polars derived from experimental results will usually differ slightly from the idealized parabolic form given by equation (4.20). The effect of substantial camber on a wing, for example, will result in the minimum drag ($C_{D_{\text{min}}}$) point occurring not at $C_L = 0$, as would be expected from equation (4.20), but at some small positive value of the lift coefficient ($C_{L_{\text{min drag}}}$). The drag polar in such cases can be approximated by:

$$C_D = C_{D_{\text{min}}} + K(C_L - C_{L_{\text{min drag}}})^2 \tag{4.22}$$

For most calculations, however, the difference between $C_{D_{\text{min}}}$ and C_{D_0} is small and equation (4.20) can thus be used.

High-Speed Drag Polars

For any airplane it is possible to represent its low-speed flight characteristics in terms of a single drag polar. For aircraft that operate in the transonic flight regime, it is necessary to take into account

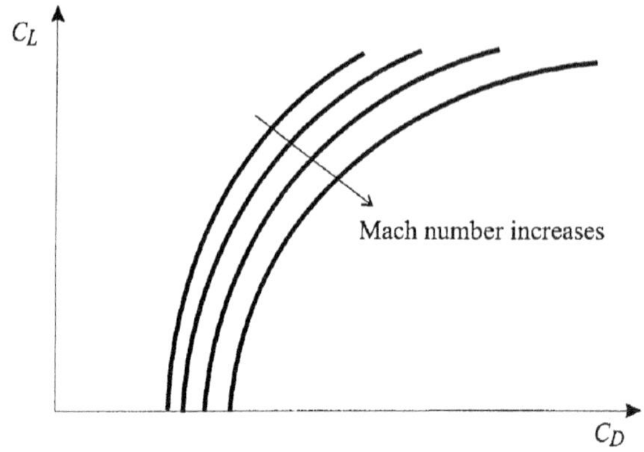

FIGURE 4.2 High-speed drag polars.

the compressibility drag rise. It is seen from Figure 4.1 that above M_{DR} there is a unique drag polar for each Mach number (Figure 4.2).

Drag versus EAS Relationship

It is convenient in many applications to write the drag in terms of V_e, based on the parabolic drag polar:

$$D = \left[\frac{C_{D_0}\rho_0 S}{2}\right]V_e^2 + \left[\frac{2KW^2}{\rho_0 S}\right]\frac{1}{V_e^2} \tag{4.23}$$

or alternatively:

$$D = A_1 V_e^2 + B_1 V_e^{-2} \tag{4.24}$$

where

$$A_1 = \frac{C_{D_0}\rho_0 S}{2} \quad \text{and} \quad B_1 = \frac{2KW^2}{\rho_0 S}$$

The substitution has introduced constants A_1 and B_1 to simplify the mathematics. It is important to note that A_1 and B_1 will change if the airplane configuration or weight changes. Equation (4.23) shows that there are two contributions to the drag of the airplane. The first, the lift-independent drag (D_0) contribution, is proportional to V_e^2, whereas the second contribution, the lift-dependent drag (D_i) term, is inversely proportional to V_e^2. At low speeds D_i is the dominant part, whereas at high speeds D_0 is dominant as shown in Figure 4.3.

Minimum Drag Condition

The speed at which the airplane's drag will be a minimum can be determined by differentiating equation (4.23) with respect to V_e and setting the resultant equal to zero.

$$V_{e_{md}} = \left[\frac{2W}{\rho_0 S}\sqrt{\frac{K}{C_{D_0}}}\right]^{1/2} \tag{4.25}$$

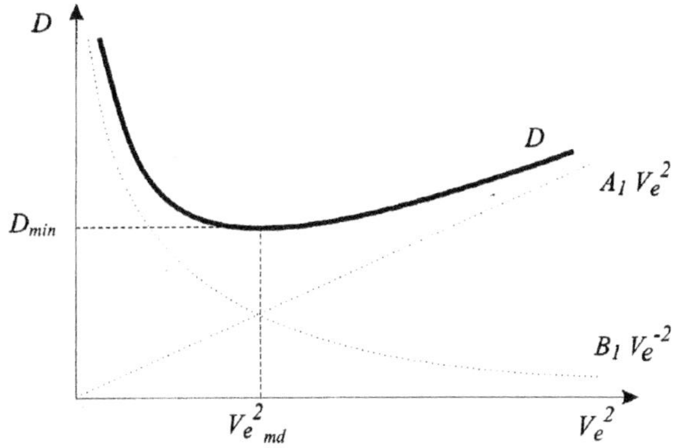

FIGURE 4.3 Drag function.

where: $A_1 = \dfrac{C_{D_0} \rho_0 S}{2}$ and $B_1 = \dfrac{2KW^2}{\rho_0 S}$

At this speed there are equal contributions to the drag from D_0 and D_i and the minimum drag is given by:

$$D_{min} = 2W\sqrt{KC_{D0}} \tag{4.26}$$

Lift-to-Drag Ratio

The lift-to-drag ratio is a measure of an aircraft's *aerodynamic efficiency*. For an aircraft in *steady level flight*, as shown in Figure 4.4, the thrust (T) will equal the drag (D) and the lift (L) will equal the weight (W). By combining these equations it is seen that the thrust required to sustain steady (i.e., unaccelerated) level flight for a given airplane weight depends on the lift-to-drag ratio, i.e.,

$$T = \frac{1}{(L/D)} W \tag{4.27}$$

An aerodynamically efficient airplane, with a high (L/D) ratio, will require a lower thrust to maintain steady level flight than a comparable airplane of the same weight.

Maximum Lift-to-Drag Ratio

Although the value of (L/D) will change during flight, each aircraft has a maximum value which it cannot exceed. The maximum lift-to-drag ratio is a figure of merit, widely used to assess

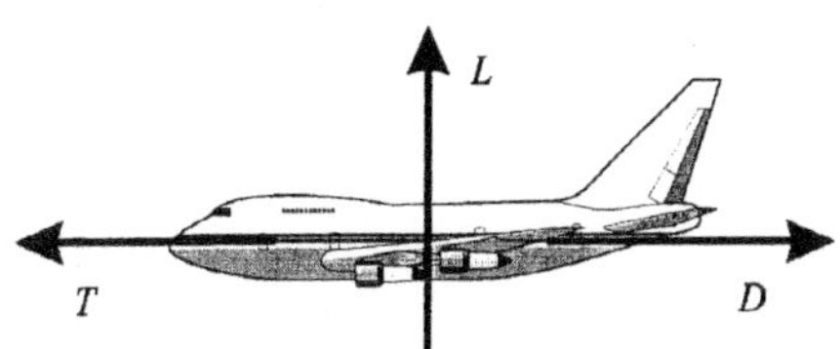

FIGURE 4.4 Steady level flight.

aerodynamic efficiency. The value of $(L/D)_{max}$ can be determined graphically by drawing a line tangent to the drag polar through the origin. Based on the parabolic drag polar, the maximum lift-to-drag ratio occurs at the conditions associated with *minimum drag*. The minimum drag lift coefficient ($C_{L_{md}}$) for a parabolic drag polar is:

$$C_{L_{md}} = \sqrt{\frac{C_{D0}}{K}} \tag{4.28}$$

and the lift-to-drag ratio corresponding to this condition is:

$$\left(\frac{L}{D}\right)_{max} = \frac{1}{2}\sqrt{\frac{1}{C_{D_0}K}} \tag{4.29}$$

Power Required and Power Available

The product of an airplane's total drag and its speed is a useful concept for performance analysis. By definition:

$$\text{Drag power} = P_D = DV \tag{4.30}$$

Because the thrust produced by the aircraft's engine(s) is required to overcome the drag, the drag power can be considered as the power required for flight. The thrust power is by definition the net propulsive force multiplied by the speed; it is thus the power available to propel the aircraft due to the engine(s). By definition:

$$\text{Thrust power} = P_T = TV \tag{4.31}$$

These two quantities, the drag power (power required) and the thrust power (power available), are equal in steady (i.e., unaccelerated) level flight, as the thrust is equal to the drag (see Figure 4.4). However, in climbing, descending, or accelerated flight this will generally not be true. An excess of *thrust power* to *drag power* will enable the aircraft to perform a sustained climb and a deficit will result in the aircraft losing height.

Minimum Power Condition

The drag power (power required) can be expressed as a function of the EAS based on the parabolic drag polar. Using equations (4.13) and (4.23), it follows that:

$$P_D = \left[\frac{C_{D0}\rho_0 S}{2\sqrt{\sigma}}\right]V_e^3 + \left[\frac{2KW^2}{\sqrt{\sigma}\rho_0 S}\right]\frac{1}{V_e} \tag{4.32}$$

This function has a minimum at a speed that is a little slower than the minimum drag speed. The flight condition is identified by the subscript *mp* for *minimum power.* At the speed for minimum power ($V_{e_{mp}}$) the drag will be slightly greater than D_{min}. Expressions for the EAS, drag, lift-to-drag ratio, and C_L corresponding to the minimum power condition, derived using the parabolic drag polar, are given in Table 4.4.

Further Reading

Consult Anderson (1999), Mair and Birdsall (1992), Filippone (2012), or Young (2017) for further details on aircraft drag. Methods for estimating drag polars for airplanes are provided by Lan and Roskam (1981), Torenbeek (1982), and Raymer (2012).

Drag:	Drag power:
$D = \left[\frac{C_{D_0}\rho_0 S}{2}\right]V_e^2 + \left[\frac{2KW^2}{\rho_0 S}\right]\frac{1}{V_e^2}$	$P_D = DV = D\frac{V_e}{\sqrt{\sigma}} = \left[\frac{C_{D0}\rho_0 S}{2\sqrt{\sigma}}\right]V_e^3 + \left[\frac{2KW^2}{\sqrt{\sigma}\rho_0 S}\right]\frac{1}{V_e}$
At the minimum drag condition:	**At minimum drag power condition:**
$C_{D_i} = C_{D_0}$	$C_{D_i} = 3C_{D_0}$
$D_{min} = 2W\sqrt{KC_{D_0}}$	$D_{mp} = \frac{4}{\sqrt{3}}W\sqrt{KC_{D_0}}$
$D_{min} = \frac{1}{(L/D)_{max}}W$	$D_{mp} = \frac{2}{\sqrt{3}}D_{min} = 1.15D_{min}$
$C_{L_{md}} = \sqrt{\frac{C_{D_0}}{K}}$	$C_{L_{mp}} = \sqrt{\frac{3C_{D_0}}{K}}$ $C_{L_{mp}} = \sqrt{3}C_{L_{md}} = 1.73C_{L_{md}}$
$V_{e_{md}} = \left[\frac{2W}{\rho_0 S}\sqrt{\frac{K}{C_{D_0}}}\right]^{1/2}$	$V_{e_{mp}} = \left[\frac{2W}{\rho_0 S}\sqrt{\frac{K}{3C_{D_0}}}\right]^{1/2}$ $V_{e_{mp}} = \frac{1}{\sqrt[4]{3}}V_{e_{md}} = 0.760\ V_{e_{md}}$
$\left(\frac{L}{D}\right)_{max} = \frac{1}{2}\sqrt{\frac{1}{KC_{D_0}}}$	$\left(\frac{L}{D}\right)_{mp} = \frac{1}{4}\sqrt{\frac{3}{KC_{D_0}}}$ $\left(\frac{L}{D}\right)_{mp} = \frac{\sqrt{3}}{2}\left(\frac{L}{D}\right)_{max} = 0.866\left(\frac{L}{D}\right)_{max}$

TABLE 4.4 Summary of Performance Parameters Based on the Parabolic Drag Polar

4.4 Engine (Powerplant) Performance

Installation Considerations

The analysis of an aircraft's performance is based on the *net installed engine thrust*. For a jet engine this takes into account the inlet pressure recovery, power and bleed air extraction, and drag contributions associated with the propulsion system. The airplane's drag is a function of the entire flowfield around the airplane, and this flowfield is influenced by the engine inlet and exhaust stream-tubes. For example, if the pilot throttles back, spillage will occur around the lip of the inlet, resulting in an increment in drag. It is conventional to regard these components of drag, whose magnitudes depend on the position of the throttle, as decrements of thrust rather than as increments of airframe drag. The accurate prediction of an aircraft's performance thus requires a consistent definition of what constitutes *propulsion system thrust* and what constitutes *propulsion system drag*. A general rule that is often used regards all fore and aft components of force that depend on throttle setting as increments or decrements of thrust.

Turbofan Thrust Variation

The following functional relationship describes the net thrust in terms of the dominant parameters:

$$\frac{T}{\delta} = f_1\left(\frac{N}{\sqrt{\theta}}, M\right) \tag{4.33}$$

where δ and θ are the ambient pressure and temperature ratios respectively and N is the rotational speed of the engine (ESDU 70020 1970). In the case of multiple-shaft engines, the speed is usually defined as the speed of the low-speed compressor. For a particular engine setting (i.e., N is constant) the thrust is a function of the atmospheric conditions and the Mach number. It is useful to consider the effect of these variables for various flight conditions. For an airliner the cruise will generally be at a constant Mach number; if the height does not change very much, then, from equation (4.33), it is seen that the thrust will be constant. In the stratosphere, where temperature is constant, the thrust for a given Mach number will decay approximately linearly with pressure as the height increases, i.e.,

$$\frac{T}{\delta} = \text{constant} \tag{4.34}$$

In the troposphere there will be a steadily decreasing thrust as the air pressure is reduced. Mair and Birdsall (1992) indicate that for turbofan engines with high bypass ratios, the thrust will decay as an approximate power law function of the air density, i.e.,

$$\frac{T}{T_{SL}} \propto \sigma^n \tag{4.35}$$

where T_{SL} is the sea-level reference thrust. The exponent $n = 0.6$ has been shown to provide a reasonable approximation to actual engine data (Mair and Birdsall 1992).

Thrust Ratings

Jet engine manufacturers use several different techniques to set (control) the thrust on their engines. One popular method uses an engine shaft speed as the index, and another, widely used, method uses an *engine pressure ratio* (EPR) as the thrust setting parameter. The EPR is essentially the ratio of the pressure in the exhaust flow to the pressure of the flow just ahead of the compressor. To avoid exceeding engine design limitations and to achieve the maximum life of the turbine, the pilot needs to adhere to specified thrust ratings during each stage of the flight (Young, 2017, provides details on turbofan engine thrust ratings).

Specific Fuel Consumption for Turbojet/Turbofan Engine

The rate at which fuel is burnt in the engine is usually expressed as a specific fuel consumption (SFC) rather than in absolute terms. The SFC of turbojet and turbofan engines is commonly known as the *thrust specific fuel consumption* (TSFC). It can be defined in two ways—a source of much confusion. SFC is either the *mass* of fuel (m_f) burned per unit time, divided by the thrust (convenient when working in SI units), or the *weight* of fuel (W_f) burned per unit time, divided by the thrust (convenient when working in USC units). The SFC is a figure of merit used to assess an engine's efficiency in converting fuel into thrust. The symbol that is widely used is c. Various subscripts/superscripts are used to distinguish between different expressions of SFC. Using the first definition:

$$\text{SFC} = c_t = \frac{-(dm_f/dt)}{T} = \frac{Q}{T} \quad \text{(convenient for SI units)} \tag{4.36}$$

where Q is the mass fuel flow. The reason for the minus sign is that the rate of change of aircraft fuel mass is negative but SFC is positive. In SI units SFC can be measured in $kgN^{-1}s^{-1}$; however, traditional engineering practice has been to quote SFC in terms of $kgN^{-1}h^{-1}$ (or $mgN^{-1}s^{-1}$). Alternatively, SFC can be expressed in terms of weight flow, i.e.,

$$\mathrm{SFC} = c_t' = \frac{-(dW_f/dt)}{T} = \frac{Q'}{T} \quad \text{(convenient for USC units)} \tag{4.37}$$

The customary units used in the industry are lb $lb^{-1}h^{-1}$.

The following functional relationship (ESDU 70020 1970) describes the fuel flow in terms of the dominant parameters:

$$\frac{Q}{\delta\sqrt{\theta}} = f_2\left(\frac{N}{\sqrt{\theta}}, M\right) \tag{4.38}$$

Combining this equation with (4.33) and (4.36) leads to the functional relationship for SFC:

$$c_t = \frac{Q}{T} = \sqrt{\theta} f_3\left(\frac{N}{\sqrt{\theta}}, M\right) \tag{4.39}$$

This equation indicates that the SFC will depend on pressure height and Mach number for a given engine speed.

SFC Models for Turbojet/Turbofan Engine

It is usually not possible to obtain expressions for the functions f_1, f_2, and f_3, and as a result simple algebraic expressions are used which allow the SFC to be represented with reasonable accuracy over a limited speed range. Several methods are described in ESDU 73019 (1982).

1. The variation of SFC over segments of the cruise is often small, and the use of a mean constant value will usually yield satisfactory results for cruise analysis.
2. A more accurate method is to assume a law of the form:

$$c_t = c_1\sqrt{\theta}M^n \tag{4.40}$$

   With suitably chosen values of the constants c_1 and n, this expression is reported to provide an accurate approximation to measured SFC figures for turbofans within the limited range of N, θ, and M values associated with subsonic cruising flight.
3. A third model that takes into account the variation of SFC with M is:

$$c_t = c_2 + c_3M \tag{4.41}$$

   This can provide a reasonable approximation to the manufacturers' data (at constant height and engine speed).

Turbojet/Turbofan Idealizations

For turbojet/turbofan engines, the following idealizations are often made to simplify the cruise analysis and enable performance problems to be solved analytically:

1. Thrust in the cruise is assumed to be independent of speed.
2. The thrust is assumed to be proportional to ambient pressure for altitude variations.
3. The SFC is assumed to be independent of speed and altitude.

The assumption that thrust is invariant of forward speed greatly simplifies analytical analyses of airplane performance. Whereas, this is a reasonable approximation for turbojet engines, it does not accurately represent the thrust characteristics of modern high bypass ratio jet engines at all altitudes.

Turboprop Engine Performance

In a turboprop engine a gas turbine drives the propeller, which generates the thrust. The product of the propeller thrust (T_p) and the forward velocity (V) is by definition the thrust power (power available). A propeller is never 100% efficient in converting the shaft power (P_S) to *thrust power* and it is thus necessary to introduce a propeller efficiency (η_p), i.e.,

$$\eta_p P_S = T_p V \tag{4.42}$$

The analysis of a turboprop engine is complicated by the fact that there is a small amount of jet thrust provided by the residual energy in the exhaust gases. The total thrust (T) must include the residual jet thrust (T_J), i.e.,

$$T = T_p + T_J = T_p\left(1 + \frac{T_J}{T_p}\right) \tag{4.43}$$

It is convenient to rate turboprop engine output in terms of an *equivalent power*, rather than in terms of thrust. The equivalent power (P_e) is equal to the shaft power plus the equivalent power contribution of the jet thrust. In other words, the equivalent power is the hypothetical power that would be needed to drive the propeller (at the same propeller efficiency) to produce the same total thrust, i.e.,

$$\eta_p P_e = TV \tag{4.44}$$

With this definition and referring to equations (4.42) and (4.43), the equivalent power can be written as:

$$P_e = P_S\left(1 + \frac{T_J}{T_p}\right) \tag{4.45}$$

Although the ratio T_J/T_p does change during flight, the impact of this variation on the ratio P_e/P_S is small and for preliminary calculations the ratio P_e/P_S can be assumed to be constant.

Power Models for Turboprop

The shaft power of a *flat rated* engine at either the climb or cruise power rating is essentially constant over a range of speeds, from sea level up the rated height, typically more than 10,000 ft. Above this height, the power output will decay with height in much the same way as described above for the turbojet/turbofan engine. The ratio of maximum shaft power to the sea-level reference power $(P_S)_{SL}$ is given by:

$$\frac{P_S}{(P_S)_{SL}} \propto \sigma^n \tag{4.46}$$

which is analogous to equation (4.35). Furthermore, the shaft power will tend to increase with Mach number because of an increase in ram pressure in the inlet. The variation of power with speed at a fixed height has been shown (Mair and Birdsall 1992) to follow an approximate power law relation:

$$\frac{P_S}{(P_S)_{SL}} = AM^n \tag{4.47}$$

where A and the exponent n (between 0 and 1) are selected to suit the engine data.

Specific Fuel Consumption for Turboprop

There are a number of ways of defining SFC for a turboprop engine. It is given the symbol c_p, and when working in SI units it is convenient to define SFC as the *mass* of fuel burned per unit time (Q) divided by the equivalent power (P_e).

$$\text{SFC} = c_p = \frac{\left(-\dfrac{dm_f}{dt}\right)}{P_e} = \frac{Q}{P_e} \quad \text{(convenient for SI units)} \tag{4.48}$$

From a theoretical perspective it is correct to base SFC on the *equivalent power.* However, engine specifications are frequently based on *shaft power,* and it is thus necessary to check this before using the data. In SI units, SFC for a turboprop engine can be expressed in terms of kg $W^{-1}s^{-1}$. However, common engineering practice is to use kg $W^{-1}h^{-1}$ or μgJ^{-1}. The alternative definition of SFC is *weight* of fuel burned per unit time divided by the equivalent power, i.e.,

$$\text{SFC} = c_p' = \frac{\left(-\dfrac{dW_f}{dt}\right)}{P_e} = \frac{Q'}{P_e} \quad \text{(convenient for USC units)} \tag{4.49}$$

Engineering practice for many years has been to express the SFC of turboprop engines in units of lb $hp^{-1}h^{-1}$. If the SFC is based on shaft power, it is usually written as lb $shp^{-1}h^{-1}$.

Turboprop Idealizations

For turboprops in the range of speeds used for subsonic cruising, shaft power varies with Mach number and altitude, and simple idealizations are not possible. The variation of SFC with height and Mach number is very small and for preliminary calculations can be ignored.

Piston Engine Performance

The power produced by a reciprocating piston engine is directly proportional to the *mass flow* of the air in the intake manifold. Two factors influence the mass flow: the *air density* and the *manifold pressure.* For normally aspirated engines (i.e., not supercharged) the intake manifold pressure will be equal to ambient pressure or a little greater than ambient at high speed due to the influence of ram air pressure in the inlet. As modern piston engine aircraft fly at low Mach numbers, the ram effect is small and can often be neglected. A piston engine suffers a considerable drop in power as it climbs. To prevent this loss of power, the intake manifold pressure can be boosted by means of a mechanical air compressor. Superchargers and turbochargers are capable of maintaining sea-level pressure in the intake manifold to heights of over 10,000 ft (i.e., the altitude above which pilots require an oxygen supply in unpressurized aircraft).

Piston Engine Thrust Power

The product of the propeller thrust and the forward velocity is equal to the thrust power (power available). The engine power (P) is related to the thrust power (P_T) by the propeller efficiency.

$$P_T = \eta_P P \tag{4.50}$$

Specific Fuel Consumption for Piston Engine

The SFC of a piston engine is defined in a similar way to that of a turboprop engine, and the same notation is used. In SI units it is convenient to define SFC as the *mass* of fuel burned per unit time (Q) divided by the engine power (P), while in USC units it is common to use the *weight* of fuel burned per unit time divided by the power. Thus,

$$\text{SFC} = c_p = \frac{\left(-\dfrac{dm_f}{dt}\right)}{P} = \frac{Q}{P} \quad \text{(convenient for SI units)} \tag{4.51}$$

or

$$\text{SFC} = c'_p = \frac{\left(-\dfrac{dW_f}{dt}\right)}{P} = \frac{Q'}{P} \quad \text{(convenient for USC units)} \tag{4.52}$$

Piston Engine Idealizations

The following idealizations are often made in order to analyze the airplane's cruise performance numerically.

1. For a given throttle setting and altitude, the power is assumed to be independent of speed (for the range of speeds normally used for cruising).
2. The shaft power is assumed to be proportional to ambient density for altitude variations. (For engines that are supercharged or turbocharged, sea-level power will be maintained to the rated height.)
3. The SFC is assumed to be independent of speed and altitude.

Further Reading

For further information on turbine engine performance consult Mair and Birdsall (1992), Mattingly et al. (2003), or Young (2017). Lan and Roskam (1981) and Lowry (1999) describe the performance of piston engines; both references provide details on propeller performance and efficiency.

4.5 Level Flight Performance

Level Flight Turbojet/Turbofan Performance

In this analysis the airplane is considered to be flying straight (i.e., not turning) and level (i.e., not climbing or descending) at constant velocity. A very useful analysis technique is to superimpose on a single graph the thrust (T) and drag (D) variations as functions of V_e. For the idealized turbojet/turbofan engine (see Subsection 4.4) operating at a set throttle position, the thrust decreases with altitude and is independent of V_e. The family of T curves is drawn as horizontal lines, with the greatest thrust at sea level and reducing with increasing height. The D curve for a given aircraft weight is independent of altitude if the function is plotted against EAS rather than TAS. A single curve will thus represent the drag at any altitude. The drag acting on the airplane can be modeled by equation (4.23), derived from the parabolic drag polar. The intersection points of the T and D curves represent a series of steady state level flight conditions (Figure 4.5). A low thrust line on the graph will have two points of intersection with the drag function, one at a speed less than

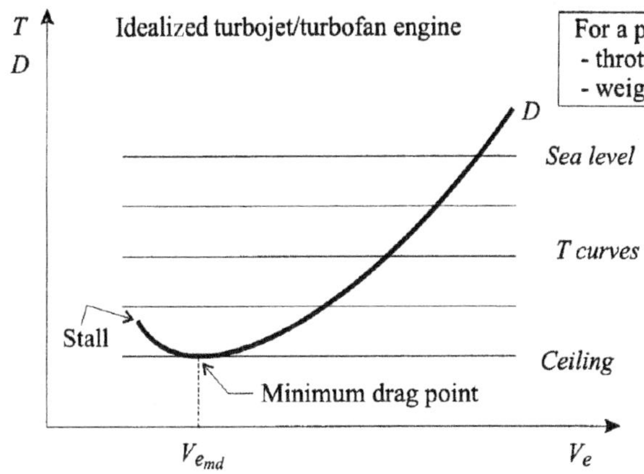

FIGURE 4.5 Thrust and drag for the idealized turbojet/turbofan engine airplane.

minimum drag speed ($V_{e_{md}}$) and one at a speed greater than $V_{e_{md}}$. The thrust functions for which this will be true correspond to flight at high altitude where the thrust has been substantially reduced, or alternatively at any altitude, but with a reduced throttle setting.

Level Flight Piston Engine Performance

For the idealized piston engine the T curves are rectangular hyperbolas, as thrust power (P_T) is constant. For a set throttle position, the thrust variation with altitude is shown superimposed on the drag versus EAS graph in Figure 4.6. The lowest thrust curve that will intersect the drag curve, giving the absolute ceiling, does so not at the $V_{e_{md}}$, as was the case for the turbojet/turbofan engine, but at the minimum power speed ($V_{e_{mp}}$). From a practical perspective, it is not convenient to represent piston engine performance in terms of thrust. Piston engines are rated in terms of power, and changes in the throttle setting will result in changes in the thrust power. The thrust can be replaced

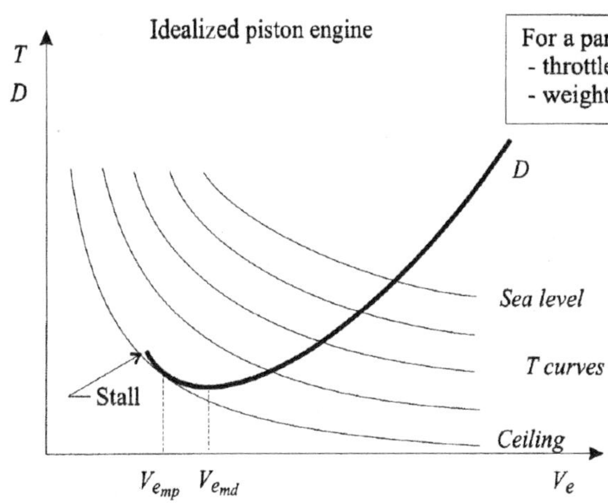

FIGURE 4.6 Thrust and drag for the idealized piston engine airplane.

by an equivalent group of terms which includes the engine power and propeller efficiency. Based on equations (4.42) and (4.13) the thrust can be expressed as:

$$T = \frac{\eta_p P\sqrt{\sigma}}{V_e} \tag{4.53}$$

Maximum Level Speed

It is apparent that the construction of T and D versus V_e graphs for the maximum throttle setting will give, at their intersection, the maximum level speed condition at any altitude. For a piston engine aircraft (flying at speeds below any compressibility effects), the *maximum level speed* corresponding to a maximum power (P_{max}) can be determined using the parabolic drag polar. As the aircraft is in steady level flight, thrust power equals drag power and hence from equation (4.32), it can be deduced that:

$$\eta_p P = \frac{C_{D_0}\rho_0 S}{2\sqrt{\sigma}} V_e^3 + \frac{2KW^2}{\rho_0 S\sqrt{\sigma}} V_e^{-1} \tag{4.54}$$

Equation (4.54) applies to any steady level flight condition. Note that the equation does not provide a closed-form solution for V_e and must be solved by iteration. For the maximum level speed, it is possible to obtain an approximate solution by noting that the second of the two terms is a function of V_e^{-1} and this term becomes very small (in comparison to the first term), as speed increases. Hence,

$$V_{e_{max}} \approx \sqrt[3]{\frac{2\eta_p P_{max}\sqrt{\sigma}}{C_{D_0}\rho_0 S}} \tag{4.55}$$

For turbojet/turbofan-powered aircraft the maximum speed is usually within the drag rise Mach regime, and the low-speed drag polar is not applicable, rendering the approach given above inaccurate due to the difficulty in accounting for the wave drag.

Speed Stability—Turbojet/Turbofan Airplane

In Figure 4.7 the thrust for a selected throttle setting of a jet airplane is superimposed on the drag relationship. If the aircraft is flying straight and level at a speed V_{e_1}, it will be operating at a

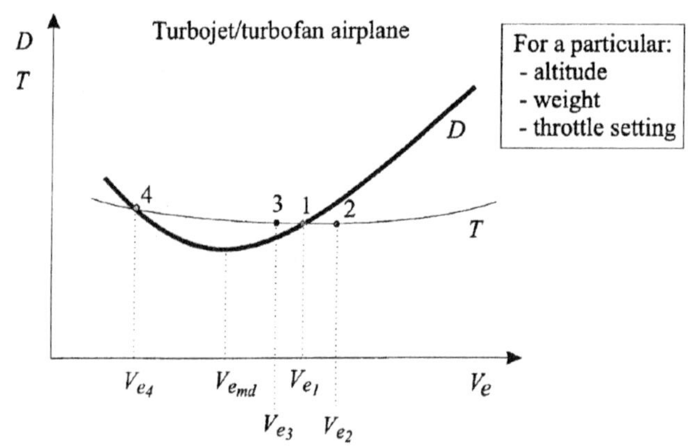

FIGURE 4.7 Speed stability for turbojet/turbofan engine airplane.

condition of *equilibrium*. If the aircraft speeds up a little (without the pilot changing the throttle position) to a speed V_{e_2}, then T will be less than D and the aircraft will tend to slow down to the original speed. By a similar argument, if the speed decreased to V_{e_3}, this would cause the aircraft to accelerate, as the drag would have decreased and T will be greater than D. At the slow speed equilibrium point V_{e_4} the situation is different. If the aircraft slows down, the drag will increase, causing the aircraft to slow down further. It can thus be concluded that a jet airplane is unstable with regard to speed changes for operation below the minimum drag speed ($V_{e_{md}}$). This region is called the back end of the drag curve. If the aircraft slows down from $V_{e_{md}}$ there will be a thrust deficit and the aircraft will start to sink, requiring a corrective action by the pilot.

Speed Stability—Piston Engine Airplane

For piston engine aircraft, a similar but not identical deduction to that presented for the jet airplane can be made. Although the form of the drag function is unchanged, the thrust curves are rectangular hyperbolas (for the idealized engine). The curve corresponding to the lowest power setting that will permit steady level flight touches the drag curve at the minimum power speed ($V_{e_{mp}}$). For speeds less than $V_{e_{mp}}$ the aircraft will be unstable with regard to speed changes. From a performance perspective, there are two important differences regarding speed instability between piston engine and jet aircraft: the first is that because $V_{e_{mp}} = 0.76V_{e_{md}}$ (based on the parabolic drag polar), speed instability affects a smaller speed range for the piston engine airplane, and the second is that because piston engines produce high thrust at low speed, the thrust deficit is comparatively small. For these reasons, pilots of piston engine light aircraft are usually not aware of the condition.

Absolute Ceiling—Turbojet/Turbofan

As the thrust produced by a jet engine decays with altitude, there will be a maximum altitude (for every thrust setting) beyond which it will not be possible for the airplane to climb. The absolute ceiling is defined as the maximum altitude at which level flight can be maintained with maximum available thrust. An aircraft can fly higher than its absolute ceiling by a zoom maneuver, in which kinetic energy is exchanged for potential energy, but these altitudes cannot be sustained. The absolute ceiling is given by the intersection of the lowest thrust curve with the lowest point on the drag curve on Figure 4.5. Steady flight is possible only at one speed; based on the parabolic drag polar and the idealized thrust relationship, this speed is $V_{e_{md}}$. Flight at the absolute ceiling is difficult to sustain and largely of theoretical interest.

Service Ceiling—Turbojet/Turbofan

A service ceiling is a practical upper operational limit, which represents the greatest altitude at which a given rate of climb (e.g., 100 ft/min or 300 ft/min) can be achieved at a particular gross weight, thrust setting, air temperature, and speed (Raymer 2012; Young 2017).

Single-Engine Inoperative Performance Requirements

In the event of an engine failure of a multiple-engine aircraft, the service ceiling will be significantly reduced. En route flight planning over mountainous terrain is based on an airplane's one-engine-inoperative performance, defined in terms of a minimum climb performance potential that is achievable with the remaining engines set at maximum continuous thrust. The so-called *net flight* path represents the airplane's gross (or actual) flight path diminished by a gradient penalty of 1.6% for four engine aircraft, 1.4% for a three engine aircraft, and 1.1% for a two engine aircraft (FAR Part 25.123). Airlines plan flights in such a way that following an engine failure at any stage during the flight, the airplane will be able to drift down along a predefined corridor with the net flight path clearing all terrain and obstacles by at least 2,000 ft.

Ceiling—Piston Engine

For the idealized piston engine aircraft the absolute ceiling is achieved by flying at $V_{e_{mp}}$. For piston engine airplanes the absolute ceiling is largely of theoretical significance because most of these aircraft are unpressurized and will usually not be flown higher than about 10,000 ft (a height at which oxygen is required by the crew and passengers).

Further Reading

For further details on level flight performance, speed stability, and aircraft ceilings, consult Mair and Birdsall (1992), Anderson (1999), Lowry (1999), Raymer (2012), Filippone (2012), or Young (2017).

4.6 Climbing and Descending Flight

Climb Speed Schedule

The initial part of a typical climb for an airliner will be at constant CAS. The implication of this is that the TAS will increase as the air density drops (Figure 4.8). The combined effect of an increase in TAS and a reduction in the speed of sound results in a rapid increase in the Mach number. To avoid a substantial increase in drag as the speed approaches the drag rise Mach number, a change to the climb schedule will usually be required at some point. The climb speed schedule may, for example, require the pilot to climb at 300 kt (CAS) until Mach 0.82 is reached and then to hold the Mach number constant for the remainder of the climb. Climbing at constant Mach number implies that there will be a slight decrease in TAS up to the tropopause, but in the stratosphere it will be at constant TAS.

Climb Analysis

Figure 4.9 shows an aircraft performing a climb in still air with a climb angle of γ. For a *steady* climb the aircraft is at a constant TAS and is in a state of equilibrium at all points along the flight path. The sum of the forces acting along the flight path in this case is zero. In flight this is seldom the situation and very often the aircraft will *accelerate* along the flight path, as is the case for a constant CAS climb. Under these flight conditions the airplane's flight path is not exactly straight but will have a

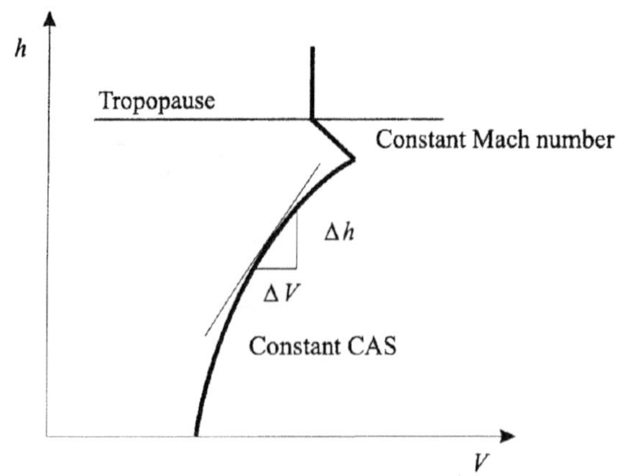

FIGURE 4.8 Typical airliner climb schedule: constant CAS, followed by constant Mach number climb.

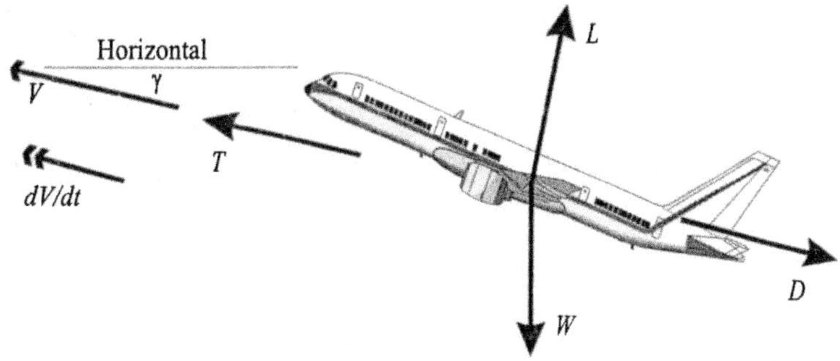

FIGURE 4.9 Climbing flight.

slight curve. The curvature, however, is very small and the centripetal acceleration is approximately zero and, for a typical climb analysis, can be ignored. For most problems involving climbing or descending flight, the angle between the thrust line and the flight path is relatively small and can be ignored.

Angle of Climb

By summing the forces acting on the aircraft in Figure 4.9, it can be shown that the climb angle (γ) relates to the thrust (T), weight (W), and drag (D) by the following relationship:

$$\sin\gamma = \frac{\frac{T}{W} - \frac{D}{W}}{\left(1 + \frac{V}{g}\frac{dV}{dh}\right)} \tag{4.56}$$

where h is the height. This equation is valid for a climb (γ positive) or descent (γ negative) and is applicable to an aircraft performing an accelerated climb or descent. For steady flight the rate of change of speed with respect to height is zero and the equation may be simplified, i.e.,

$$\sin\gamma = \frac{T}{W} - \frac{D}{L}\cos\gamma \tag{4.57}$$

The angle of climb during a typical steady-speed climb is small, permitting a small angle approximation to be used (where γ is measured in radians):

$$\gamma \approx \frac{T}{W} - \frac{D}{L} \tag{4.58}$$

Equation (4.58) provides a graphical method for the determination of the climb angle. A plot of the ratios (T/W) and (D/L) for a selected aircraft weight, superimposed on the same graph, as a function of V_e is prepared. The angle of climb (measured in radians) is given by the difference between the two curves. This is illustrated for the idealized turbojet/turbofan in Figure 4.10.

Climb Gradient

The climb gradient is given by $\tan\gamma$. The climb gradient represents the ratio of the gain in height to the horizontal distance flown and is usually expressed as a percentage.

Best Angle of Climb Speed for Turbojet/Turbofan Aircraft

The flight conditions that will give the maximum angle of climb are of interest in clearing obstacles after takeoff. For the idealized turbojet/turbofan it is evident from Figure 4.10 that the best angle of climb speed, at any altitude is $V_{e_{md}}$.

Best Angle of Climb Speed for Piston Engine Aircraft

The expression for the angle of climb is:

$$\sin\gamma = \frac{\eta_p P\sqrt{\sigma}}{WV_e} - \frac{D}{W} \tag{4.59}$$

In the case of the idealized piston engine, the thrust curves are rectangular hyperbolas and the speed that will achieve the maximum angle of climb will be between the stall speed and $V_{e_{mp}}$, but will depend on the altitude. It is possible to obtain an expression for the best angle of climb speed for a piston engine aircraft. However the expression does not have a closed-form solution and must be solved by iteration. Using the parabolic drag polar, the equation that will give the best angle of climb speed for a piston engine aircraft is:

$$C_{D_0}\rho_0 SV_e^4 + \eta_p P\sqrt{\sigma}V_e - \frac{4KW^2}{\rho_0 S} = 0 \tag{4.60}$$

The solution predicted by this equation is based on the parabolic drag polar and the assumptions of the idealized piston engine/propeller combination. At the very low speeds associated with this flight condition, the accuracy of these idealizations is reduced, resulting in a poor speed estimation. If actual data of power, drag and propeller efficiency are available, then a superior approach would be to use equation (4.59) and to solve for the optimum speed using either a numerical or graphical technique.

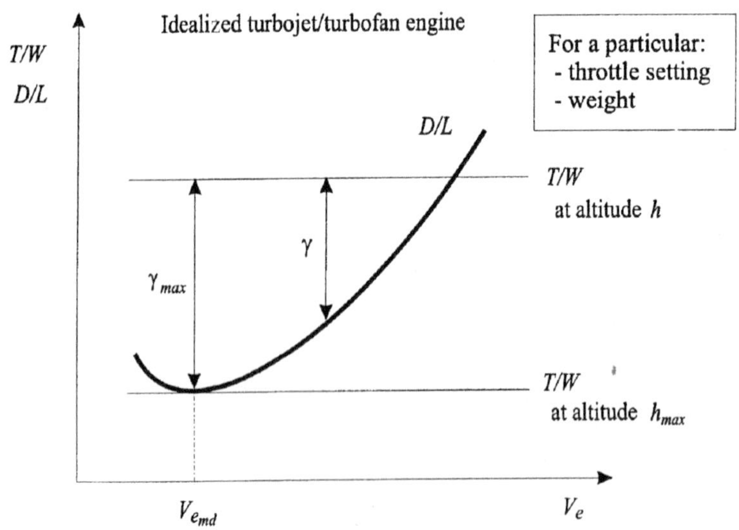

FIGURE 4.10 Angle of climb for the idealized turbojet/turbofan engine airplane.

General Equation for Rate of Climb (ROC)

The rate of climb is often written as R/C, but this notation has been the source of some confusion, particularly when written in an equation, and for this reason the abbreviation ROC will be used here. The rate of climb in the absence of updrafts is the change of height with respect to time. From equation (4.56) the ROC can be deduced:

$$\text{ROC} = \frac{dh}{dt} = V\sin\gamma = \frac{\left(\frac{T}{W} - \frac{D}{W}\right)V}{\left(1 + \frac{V}{g}\frac{dV}{dh}\right)} \tag{4.61}$$

The usual method of evaluating equations (4.56) and (4.61) is to introduce an acceleration factor, defined as:

$$f_{\text{acc}} = \frac{V}{g}\frac{dV}{dh} \tag{4.62}$$

hence,

$$\text{ROC} = \frac{\left(\frac{T}{W} - \frac{D}{W}\right)V}{\left(1 + f_{\text{acc}}\right)} \tag{4.63}$$

The significance of this factor is illustrated in Figure 4.8. The slope of the curve is (dh/dV), which changes with height. For a constant CAS climb the *acceleration factor* will increase as the altitude increases. This is evident from the fact that the TAS increases and (dV/dh), which is the reciprocal of the slope of the line, also increases for a constant CAS climb. The acceleration factor depends on the Mach number, the height, and the climb speed condition. Equations to determine f_{acc} are given in Table 4.5. For high-speed aircraft, it is preferable to express the ROC in terms of Mach number

The acceleration factor is by definition: $f_{\text{acc}} = \frac{V}{g}\frac{dV}{dh}$	
and its magnitude is given by: $f_{\text{acc}} = 0.7M^2\psi$	
Where for:	
Constant Mach no.	$\psi = -C$
Constant EAS	$\psi = 1 - C$
Constant CAS	$\psi = \frac{[1+0.2M^2]^{3.5} - 1}{0.7M^2[1+0.2M^2]^{2.5}} - C$
And where: C = 0.190263 (below tropopause) C = 0 (above tropopause)	

Source: Young (2017) provides derivations of these equations; consideration is also given to temperature corrections for off-standard atmospheres.

Table 4.5 Acceleration Factor for ISA Conditions

and divide both numerator and denominator by the pressure ratio (δ) to get the thrust in a form consistent with the usual presentation of thrust data for a jet engine, i.e.,

$$\text{ROC} = \frac{\left(\frac{T}{\delta} - \frac{D}{\delta}\right) M a_0 \sqrt{\theta}}{(1 + f_{\text{acc}}) \frac{W}{\delta}} \tag{4.64}$$

Rate of Climb (ROC) at Constant TAS

For a steady unaccelerated climb (i.e., constant TAS) f_{acc} is zero and equation (4.63) reduces to:

$$\text{ROC} = \frac{(T - D)V}{W} = \frac{P_T - P_D}{W} \tag{4.65}$$

The steady rate of climb is thus proportional to the excess of thrust power (power available) to drag power (power required).

Maximum Rate of Climb Speed for Turbojet/Turbofan

The optimum climb speed that will reduce the total trip fuel is very close to the maximum rate of climb speed. (Airliners are most efficient during cruise, so it is desirable to climb as fast as possible.) The maximum rate of climb is achieved at a speed higher than the minimum drag speed. An approximation of the best rate of climb speed for the idealized turbojet/turbofan at a particular altitude and throttle setting can be derived using the parabolic drag expression.

$$V_e = \left[\frac{W}{3C_{D_0}\rho_0 S} \left\{ \frac{T}{W} \pm \sqrt{\left(\frac{T}{W}\right)^2 + 12KC_{D_0}} \right\} \right]^{1/2} \tag{4.66}$$

Alternatively, a graphical method can be used to determine the speed for the best rate of climb. The latter method is particularly suitable for problems where the parabolic drag polar or the idealized thrust function is not valid.

Maximum Rate of Climb Speed for Piston Engine Aircraft

In the case of the idealized piston engine aircraft $(T/W)V_e$ lines are horizontal, as thrust power is independent of speed (Figure 4.11). It is seen that the maximum rate of climb for the idealized piston engine aircraft is at the minimum power speed. The comment made earlier (in the angle of climb analysis) regarding the validity of the parabolic drag polar and power idealization at low speeds is also valid here. A graphical approach to determine the best rate of climb speed is better if data are available.

Time to Climb

The time to climb from height h_1 to height h_2 is given by:

$$t = \int_{h_1}^{h_2} \frac{1}{\text{ROC}} dh \tag{4.67}$$

where the ROC is given by equation (4.63) or (4.65). The integration of equation (4.67) for a general problem is not easy because of the interdependency of the many variables. The best approach for determining the time to climb is to divide the climb into intervals, each interval corresponding to a change of height (Δh). The ROC is then determined at the start of the ith interval (ROC_i) and at end of the interval (ROC_{i+1}) based on the aircraft weight at the start of the interval (W_i). By making the

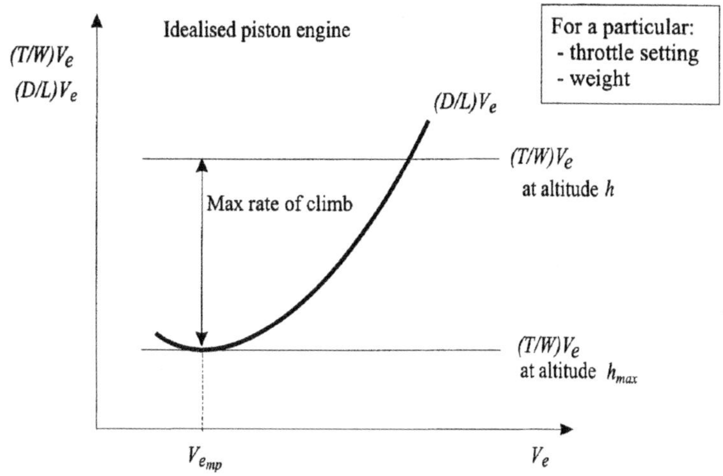

FIGURE 4.11 Rate of climb for the idealized piston engine airplane.

assumption that the ROC will change linearly across the interval, equation (4.67) may be integrated to give the increment in time (Δt_i) for the ith interval. The weight at the end of the interval (W_{i+1}) is now determined from the fuel flow and the time Δt_i. To improve the accuracy of the calculation, the ROC at the end of the interval is then recalculated using W_{i+1} and a second iteration of the increment in time is obtained. Based on a revised weight for the end of the interval, the next interval is considered.

Effect of Wind on Climb Performance

In Figure 4.12 the aircraft is shown to be flying in an air mass and the entire air mass is moving with a speed V_w. The still air angle of climb is γ. The ground speed is given by a sum of the aircraft's TAS

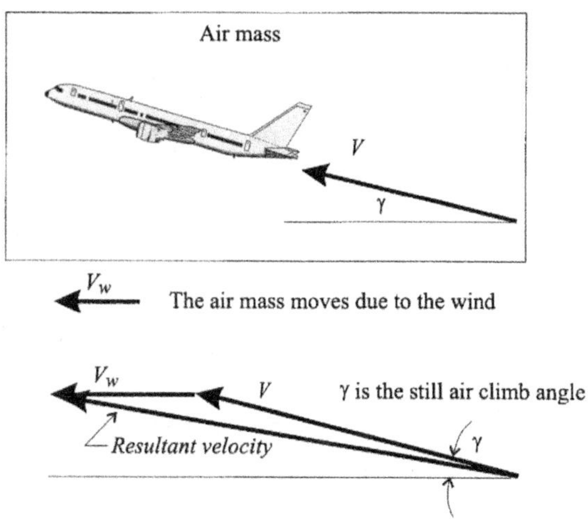

FIGURE 4.12 Effect of wind on climb performance.

vector and the wind velocity. For a tailwind the climb gradient is reduced, but the rate of climb is unchanged. Wind gradients, however, can effect an airplane's rate of climb (Young 2017).

Angle of Descent and Descent Gradient

The angle of climb and the climb gradient expressions derived above are also valid for the descent; the only difference is that the angle γ is negative. The optimum decent conditions for airline operations are those that result in the lowest trip fuel. As the descent is usually at idle thrust (lowest fuel burn), the optimum flight plan is one that results in the longest distance in descent, which implies that the descent speed should be close to the flight condition for maximum lift-to-drag ratio. For transport aircraft operations the descent may be determined by the rate of repressurization of the cabin or air traffic control requirements.

Rate of Descent (ROD)

The rate of descent, often written as R/D, may be determined directly from equations (4.63) and (4.65) derived for the rate of climb. The only difference is that the ROD is positive when (dh/dt) is negative.

Glide Angle for Unpowered Flight

Multiple engine failures are a rare occurrence. There have been a few reported cases where the engines on an airliner have failed simultaneously due to fuel starvation or the ingestion of volcanic dust. For light aircraft the incidence rate is much higher. The better an airplane can glide, the more time the pilot will have to restart the engine or find a suitable place to conduct an emergency landing. It can be shown that an aerodynamically efficient airplane with a high lift-to-drag ratio will glide very well. For a steady descent (constant TAS) with zero thrust, equation (4.58) can be rewritten as:

$$\gamma \approx -\frac{1}{L/D} \tag{4.68}$$

where γ (measured in radians) is defined positive for a climb. The smallest glide angle (which will produce the maximum range in still air) will be achieved at the flight condition of maximum lift-to-drag ratio, which implies that the aircraft should be flown at the minimum drag speed ($V_{e_{md}}$).

Sailplane Performance

For unpowered flight the term *rate of sink* (ROS) is preferable to *rate of descent*. In essence the two terms are synonymous. For steady (constant TAS flight) the ROS for unpowered flight is given by:

$$\text{ROS} = \left(\frac{D}{W}\right)V = \frac{P_D}{W} \tag{4.69}$$

In order to achieve the maximum duration in a glide from a given altitude, the aircraft must fly at a slower speed than that required for the best glide angle. From equation (4.69), it can be deduced that in still air the lowest ROS will occur at the airplane's minimum power speed ($V_{e_{mp}}$). For this class of airplane use is made of the term *glide ratio* as a figure of merit to characterize performance. The glide ratio is the horizontal distance covered divided by the loss of height, hence,

$$\text{glide ratio} = -\tan\gamma \approx \frac{L}{D} \tag{4.70}$$

This has a maximum at the same condition as the minimum glide angle. Based on the parabolic drag polar

$$\left(\frac{L}{D}\right)_{\max} = \frac{1}{2}\sqrt{\frac{\pi A_R e}{C_{D_0}}}$$

hence good gliding performance is associated with aircraft of low C_{D_0} and efficient, high aspect ratio (A_R) wings.

Further Reading

For further details on climbing and descending flight, including energy methods, consult Mair and Birdsall (1992), Anderson (1999), Eshelby (2000), or Filippone (2012). Lowry (1999) deals with light aircraft and glider performance. Young (2017) provides a detailed treatment of the performance of jet transport airplanes.

4.7 Turning Performance

Load Factor

The load factor (n) is by definition equal to the ratio of lift to weight. In straight and level flight, n equals 1; however, in a turn or a pull-up maneuver, n will be greater than 1. By definition,

$$n = \frac{L}{W} \tag{4.71}$$

During maneuvers the load factor may not exceed the structural limits of the airplane. The limiting envelope is called a *V–n* diagram and is a plot of allowable load factor versus airspeed.

Sustained Level Turn

For a sustained level turn the aircraft must maintain both speed and height in the turn. The additional lift (compared to level flight) required to provide the centripetal acceleration results in an increase in the drag. If the pilot progressively increases the angle of bank, tightening the turn, the required thrust to sustain the turn will increase. This may continue until a limiting condition is reached; the limit may be imposed by the maximum available thrust, the maximum load factor, or the maximum lift coefficient. It may not always be possible to sustain a constant speed in a level turn. If the speed is permitted to drop, the resulting instantaneous turn rate could be higher than the comparative sustained turn rate.

Angle of Bank

In a correctly banked (coordinated turn) the airplane sideslips neither inwards (overbanked) nor outwards (underbanked), so that the lift force lies in the aircraft's plane of symmetry. The component of lift acting towards the center of the turn provides the required centripetal force to accelerate an aircraft of mass m in a circular flight path of radius r, as shown in Figure 4.13. In the analysis presented below it is assumed that the thrust axis is approximately aligned with the flight direction.

Resolving the forces vertically and horizontally provides the following two equations:

$$L\cos\phi = W = mg \tag{4.72}$$

$$L\sin\phi = ma_c = m\frac{V^2}{r} = mV\Omega \tag{4.73}$$

where a_c = the centripetal acceleration
Ω = the rate of turn (units of radians per second)

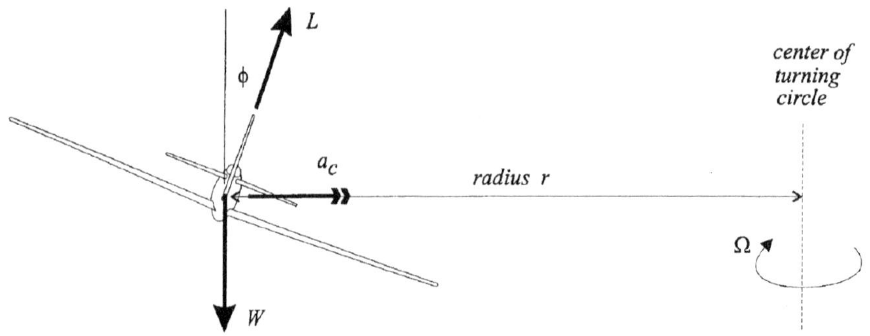

FIGURE 4.13 Turning performance.

In a turn the load factor can be obtained from equations (4.71) and (4.72).

$$n = \frac{L}{W} = \frac{1}{\cos\phi} \tag{4.74}$$

As the angle of bank is increased, it is evident that the load factor will increase. The maximum angle of bank in a sustained turn may therefore be restricted by the maximum allowable load factor.

Turn Rate in Sustained Level Turn

The turn rate is obtained from equations (4.72)–(4.74).

$$\Omega = \frac{g}{V}\sqrt{n^2 - 1} \tag{4.75}$$

For a given speed the maximum rate of turn will occur at the maximum allowable load factor. Conversely, for a given load factor, the absolute maximum rate of turn will occur at the lowest possible speed, provided that there is sufficient thrust to maintain the turn. Note that for most flights the pilot will not maneuver the airplane at turn rates anywhere near the limiting conditions.

Radius of Turn in Sustained Level Turn

The radius of turn (r) is given by:

$$r = \frac{V^2}{g\sqrt{n^2 - 1}} = \frac{V^2}{g\tan\phi} \tag{4.76}$$

The radius of turn is an important parameter for obstacle clearance after takeoff. Routine flight planning for aircraft operations out of airports with mountains in the vicinity must take into account the impact on the flight path of the loss of thrust resulting from an engine failure on multiple-engine aircraft.

Further Reading

For further details on turn performance and other maneuvers consult Mair and Birdsall (1992), Anderson (1999), Lowry (1999), Eshelby (2000), Filippone (2012), or Young (2017). Maneuver load factor design limitations are described by Torenbeek (1982) and Raymer (2012).

4.8 Stall and Spin

Stall Condition

An airplane stalls when the angle of attack exceeds the critical stall angle of attack. To enter a stall from level flight, the pilot would reduce the airspeed and, to compensate for the loss of lift, simultaneously pull the stick/yoke back, increasing the angle of attack. The stall speed is the lowest speed at which steady controllable flight can be maintained. This is preceded in most aircraft by buffeting, associated with the initial separation of airflow from the upper wing surface, striking the tailplane. A further increase in the angle of attack results in substantial separation of the flow on the upper wing surface, a reduction in lift, and an increase in drag. The aircraft loses height, and for most designs the strong nose-down pitching moment associated with the stall will rapidly reduce the angle of attack, reattaching the airflow on the wing. After initially relaxing the stick/yoke, the pilot will raise the nose and apply power to restore steady flight. It is important that the stall speeds be correctly calculated as they impact directly on the operational safety of the airplane. Apart from the obvious desire for the pilot not to inadvertently stall the aircraft without sufficient height for recovery, the importance of the stall speeds are linked to their use as reference speeds during take-off and landing.

Level Flight Stall Speed

The C_L corresponding to the stall speed (V_S) is the maximum lift coefficient ($C_{L_{max}}$), which depends on the airplane's configuration, or more precisely on the position of the high-lift devices. The purpose of leading- and trailing-edge devices, such as flaps and slats, is to increase the value of $C_{L_{max}}$ and to delay the stall. In level flight, the lift is equal to the weight and the corresponding stall speed, designated as V_{S_1}, is given by:

$$V_{S_1} = \sqrt{\frac{W}{\frac{1}{2}\rho S C_{L_{max}}}} \tag{4.77}$$

A reference stall speed V_{SR} is used as a basis (reference) to determine several operational speeds applicable to takeoff and landing. V_{SR} is a $1-g$ stall speed determined under specific conditions (FAR 25.103).

Maneuver Stall Speed

For an airplane performing a maneuver, such as a pull-up or a turn, the stall speed will be *higher* than the stall speed determined by equation (4.77). As the aircraft is not flying straight and level, the lift is equal to the weight multiplied by the load factor, hence,

$$V_S = \sqrt{\frac{nW}{\frac{1}{2}\rho S C_{L_{max}}}} = \sqrt{n}V_{S_1} \tag{4.78}$$

Factors Influencing Stall Speed

It is evident that an airplane does not have a single stalling speed, as this depends on air density, weight, aircraft configuration, and load factor. Furthermore, a change in the stall characteristics will occur when the aircraft is stalled with power on. At high angles of attack the thrust will have a significant vertical component. This will produce a small increment to the total lift, which will reduce the stall speed. For propeller-driven aircraft, with engines mounted ahead of the wings, the influence of the slipstream on the air flowing over the wings is to delay flow separation and reduce the stall speed.

Spin

Due to an asymmetry of the airflow over the wings, it is possible that one wing will stall before the other. This will result in a rolling moment because of the reduction of lift on the stalled wing, and a yawing moment because of a local increase in drag. The result is that the airplane may enter into a tight descending spiral, or spin. A pilot may deliberately put an airplane into a spin by progressively raising the nose, simultaneously reducing the engine power, and then, at the onset of the stall, deliberately yawing the aircraft using the rudder. This will result in the nose dropping abruptly, with one wing falling faster than the other and a rapid yawing of the aircraft. Depending on the airplane's aerodynamic characteristics, mass distribution, and direction of propeller motion, it may continue to spin, or it may stop yawing and just descend in a stall. If the spin becomes established, the airplane will continue to yaw at a steady rate, rapidly losing height in a motion that involves roll and sideslip. Recovery from an established spin is initiated by the pilot stopping the yawing motion by applying maximum rudder deflection in the opposite direction to the spin. The pilot then recovers from the stall condition by initially allowing the nose to drop a little, until airflow is reattached over the wings, and then pulling back on the stick/yoke, raising the nose and applying power to sustain the pullout from the dive. For some aircraft designs the relative position of the horizontal tailplane results in a wash of stalled air striking the fin in a spin, making it very difficult, and in some cases impossible, for the pilot to stop the yawing motion, particularly when the center of gravity is located in an aft position. Flight testing may require the installation of a spin parachute to assist in the recovery.

Further Reading

Consult Stinton (1996).

4.9 Range and Endurance

Fuel Consumption Definitions

The *range* relates to the *distance* that an aircraft can fly on a given quantity of fuel, whereas the *endurance* relates to the *time* that the aircraft can fly on that fuel quantity. Both parameters depend on the *rate* at which the fuel is consumed. As explained in Subsection 4.4, the fuel flow may be defined as either the *mass* of fuel consumed per unit time, which is convenient for SI units (and is given the symbol Q), or the *weight* of fuel consumed per unit time, which is convenient for USC units (given the symbol Q'). In this subsection the range and endurance equations are derived from fuel flow definitions based on the mass flow rate (see Table 4.6(a)). The alternative expressions based on the weight flow rate are summarized in Table 4.6(b). For a jet engine it is evident from equation (4.36) that the fuel flow is the product of SFC and thrust, whereas for a piston or turboprop engine, it is the product of SFC and *power*, as is evident from equations (4.48) and (4.51).

Specific Air Range (SAR)

The specific air range (r_a), also referred to as the *specific range* or the *fuel mileage,* is defined as the distance travelled per unit fuel mass consumed. Thus,

$$r_a = \left(-\frac{dx}{dm_f}\right) = \frac{\left(\frac{dx}{dt}\right)}{\left(-\frac{dm_f}{dt}\right)} = \frac{V}{Q} \tag{4.79}$$

where x = still air distance

m_f = onboard fuel mass

Q = net *mass* flow of fuel to all engines.

	Range	Endurance
Governing equation	$R = \int_{m_2}^{m_1} \frac{V}{Q} dm$	$t = \int_{m_2}^{m_1} \frac{1}{Q} dm$
Turbojet/turbofan airplane	$Q = c_t g \left(\frac{D}{L}\right) m$	
	$R = \frac{V}{c_t g}\left(\frac{L}{D}\right) \ln\left(\frac{m_1}{m_2}\right)$ for constant c_t, V, and C_L	$t = \frac{1}{c_t g}\left(\frac{L}{D}\right) \ln\left(\frac{m_1}{m_2}\right)$ for constant c_t and C_L
Piston-propeller airplane	$Q = \frac{c_p g m V}{\eta_p}\left(\frac{D}{L}\right)$	
	$R = \frac{\eta_p}{c_p g}\left(\frac{L}{D}\right) \ln\left(\frac{m_1}{m_2}\right)$ for constant η_p, c_p, and C_L	$t = \frac{\eta_p}{c_p g V}\left(\frac{L}{D}\right) \ln\left(\frac{m_1}{m_2}\right)$ for constant η_p, c_p, V and C_L

TABLE 4.6A Summary of Range and Endurance Expressions (Mass Flow Basis)

	Range	Endurance
Governing equation	$R = \int_{W_2}^{W_1} \frac{V}{Q'} dW$	$t = \int_{W_2}^{W_1} \frac{1}{Q'} dW$
Turbojet/turbofan airplane	$Q' = c_t' \left(\frac{D}{L}\right) W$	
	$R = \frac{V}{c_t'}\left(\frac{L}{D}\right) \ln\left(\frac{W_1}{W_2}\right)$ for constant c', V, and C_L	$t = \frac{1}{c_t'}\left(\frac{L}{D}\right) \ln\left(\frac{W_1}{W_2}\right)$ for constant c_t' and C_L
Piston-propeller airplane	$Q' = \frac{c_p' W V}{\eta_p}\left(\frac{D}{L}\right)$	
	$R = \frac{\eta_p}{c_p'}\left(\frac{L}{D}\right) \ln\left(\frac{W_1}{W_2}\right)$ for constant η_p c_p', and C_L	$t = \frac{\eta_p}{c_p V}\left(\frac{L}{D}\right) \ln\left(\frac{W_1}{W_2}\right)$ for constant η_p, c_p', V, and C_L

TABLE 4.6B Summary of Range and Endurance Expressions (Weight Flow Basis)

The reason for the minus sign in equation (4.79) is that the change of fuel dm_f is a negative quantity and SAR is a positive quantity.

Cruise Speeds for Jet Airplanes

The greatest possible range that an airplane may achieve (for a fixed fuel quantity) is obtained by flying at all times at the flight condition for maximum SAR. This is called the maximum range cruise (MRC) speed. The MRC speed decreases as fuel is burned (at a set altitude). In practice, airlines usually fly faster than this, sacrificing a small increase in fuel to obtain a shorter cruise time. A portion of an airline's cost is proportional to the flight time, and flying faster will reduce this. The speed that will give the lowest total trip cost for a particular set of operating costs, is called the economy (ECON) speed. This can be difficult to calculate without complete cost data, and a simpler approach is to fly at a fixed percentage faster than the MRC speed. The so-called long-range cruise (LRC) speed is typically about 2% to 4% faster than the MRC speed and has a 1% reduction in SAR, as shown in Figure 4.14.

Turbojet/Turbofan Airplane Range Equation

The change in airplane mass is equal to the change in total onboard *fuel mass*. (This is obviously true for all commercial aircraft operations, but not for military operations where weapons are released.) The still air range, R, for an airplane with initial mass m_1 and final mass m_2 is given by:

$$R = -\int_{\text{start}}^{\text{end}} r_a \, dm = -\int_{m_1}^{m_2} \frac{V}{Q} dm \tag{4.80}$$

For a jet airplane in level flight, the fuel flow may be written as:

$$Q = c_t T = c_t \left(\frac{D}{L}\right) mg \tag{4.81}$$

Hence,

$$R = -\int_{m_1}^{m_2} \frac{V}{c_t g m} \left(\frac{L}{D}\right) dm \tag{4.82}$$

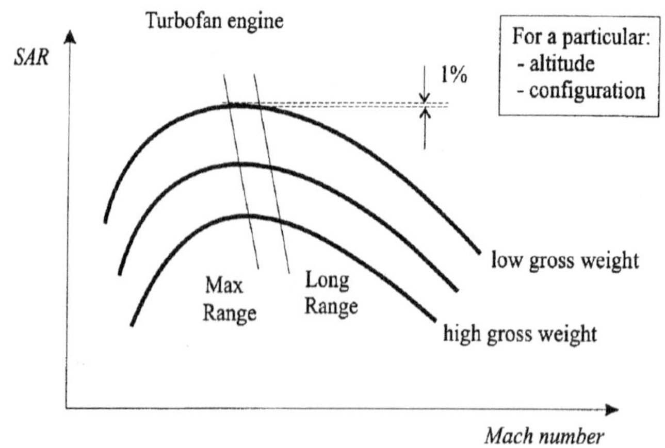

Figure 4.14 Specific air range versus Mach number.

Breguet Solution for the Turbojet/Turbofan Airplane

To evaluate equation (4.82) it is necessary to describe the variables c, V, and (L/D) during the cruise. The most widely used solution to this equation is obtained by assuming that SFC, TAS, and C_L are constant. Note that the condition of constant C_L implies that (L/D) is constant. The integration is thus straightforward.

$$R = \frac{V}{c_t g}\left(\frac{L}{D}\right)\ln\left(\frac{m_1}{m_2}\right) \tag{4.83}$$

This expression is known as the Breguet range equation (although the original equation was derived for a piston engine airplane). A number of other solutions exist to the range integral given by equation (4.82), resulting from the aircraft being flown under different constraints. Eshelby (2000) and Young (2017) present solutions for other range scenarios that assume constant C_L and altitude, and constant Mach number and altitude. The Breguet range equation is the simplest solution, and because the deviation between the results of this method and other expressions is usually not significant, it is most often used for performance estimation. For a jet airplane the Breguet range equation can be written in a slightly different way:

$$R = \frac{a_0\sqrt{\theta}}{c_t g}\left(\frac{ML}{D}\right)\ln\left(\frac{m_1}{m_2}\right) \tag{4.84}$$

It is evident from this equation that if the SFC is assumed to be constant, then the flight condition at any height that will give the greatest range for a given fuel load occurs when (ML/D) is a maximum.

Cruise-Climb

For the Breguet range equation to be valid, C_L and V must be held constant during flight. This implies that the airplane is flown in a way that ensures that the ratio W/σ remains constant. This is possible if the airplane is allowed to climb very slowly so that the relative air density decreases proportionally to the decrease in airplane gross weight. In the stratosphere, the thrust will automatically decrease as the aircraft climbs, without the throttle setting being altered. Thus, the pilot's instructions are simply to maintain a constant Mach number, allowing the aircraft to drift up as the flight progresses. This process is called a cruise-climb as altitude is not constant, and it is in fact an elegant solution for obtaining the maximum possible range. A small increase in thrust is required to maintain the climb angle, but this can be neglected for range estimation. In a cruise-climb the flight parameters for the starting condition (given the subscript 1) and the final condition (subscript 2) are related as follows:

$$\frac{W_1}{W_2} = \frac{\sigma_1}{\sigma_2} \tag{4.85}$$

Step Climb

The cruise-climb gives the greatest possible range; however, its practical use is limited by air traffic control. As a result, aircraft often fly a stepped approximation of the cruise-climb, climbing to a higher cruise altitude as fuel is burnt.

Integrated Range Method

If SAR values can be determined for the cruise, then a simple numerical integration may be performed to determine the range. The technique, usually called the integrated range method, follows directly from equation (4.80). A graph of SAR versus aircraft mass is prepared. The range is the area under the graph between the points, representing the end of the cruise (lowest weight) and the start of the cruise (highest weight).

Piston Engine Airplane Range

For a piston engine aircraft the fuel flow can be determined using equations (4.50) and (4.51) for steady level flight, i.e.,

$$Q = c_p P = \frac{c_p mgV}{\eta_p}\left(\frac{D}{L}\right) \tag{4.86}$$

The range equation (4.80) is applicable to any airplane; hence, the still air range for a piston engine airplane is given by:

$$R = -\int_{m_1}^{m_2} \frac{\eta_p}{c_p gm}\left(\frac{L}{D}\right) dm \tag{4.87}$$

Breguet Solution for Piston Engine Airplane

For an idealized piston engine the SFC and propeller efficiency are both constant and can be taken out of the integral equation (4.87). This assumption is acceptable for most applications as their variation during cruise is small. For the flight condition where C_L is constant, the integral yields:

$$R = \frac{\eta_p}{c_p g}\left(\frac{L}{D}\right)\ln\left(\frac{m_1}{m_2}\right) \tag{4.88}$$

By inspection it is seen that the maximum range will be achieved if the aircraft flies throughout the cruise at the flight condition of $(L/D)_{max}$. Equation (4.88) is the Breguet range equation for piston engine aircraft. The equation is valid for flight schedules of either constant C_L and altitude, or constant C_L and airspeed. If the altitude is constant then airspeed must be reduced to maintain a constant C_L as fuel is burned. If, on the other hand airspeed is constant, then the aircraft must fly a cruise-climb.

Payload Range Diagram

Figure 4.15 is a typical payload versus range graph for an airliner. With the *maximum* allowable payload, the amount of fuel that can be taken onboard will usually be limited not by the size of the

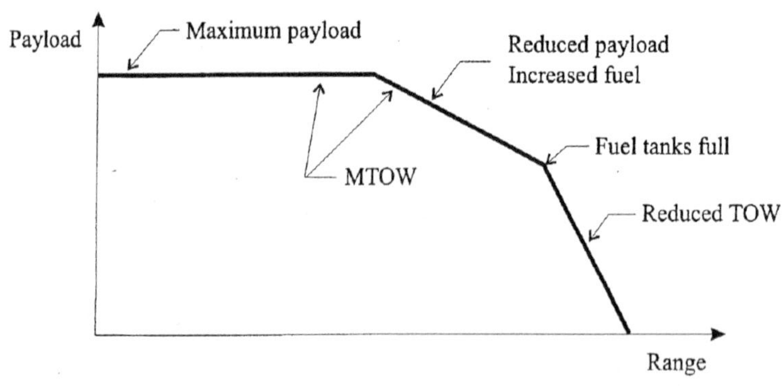

FIGURE 4.15 Typical payload-range graph for an airliner.

fuel tanks, but by the allowable takeoff weight (TOW). Under standard conditions the allowable TOW will be the maximum takeoff weight (MTOW) and the aircraft will have certain nominal range. If the nominal range is inadequate for the planned mission, then it will be necessary to reduce the payload in order to take on more fuel, but without exceeding the MTOW. Progressively longer mission lengths may be achieved by trading payload for fuel. When the point is reached when the fuel tanks are full, the only way the range can be increased is by further reducing the payload. The greatest possible range will correspond to zero payload.

Maximum Endurance for Turbojet/Turbofan Airplane

For some applications it is necessary that the airplane remain in the air for as long as possible on a given fuel load; for example, an aircraft on coastal patrol duties or an airliner holding at its destination, awaiting clearance to land. It is desirable that the airplane fly during these times at the speed for lowest fuel consumption per unit time. From equation (4.81) it is evident that if the SFC is assumed to be constant, then this occurs when (L/D) is a maximum. The airplane must therefore fly at $V_{e_{md}}$ to achieve the greatest endurance time. A plot of Q taking into account actual engine characteristics, as opposed to idealized characteristics that assume that SFC is constant, shows that the speed for minimum fuel flow is a little slower than $V_{e_{md}}$. However, this speed is in the speed instability region (see Subsection 4.5) and so for an airplane without an auto-throttle function, the speed schedule usually chosen for holding is at or very close to $V_{e_{md}}$.

Turbojet/Turbofan Airplane Endurance

The rate of change of airplane mass is equal to the fuel mass burned per unit time. Hence, if the initial mass is m_1 and final mass is m_2, the endurance time (t) is given by:

$$t = -\int_{m_1}^{m_2} \frac{1}{Q} dm = -\int_{m_1}^{m_2} \frac{1}{c_t g m}\left(\frac{L}{D}\right) dm \tag{4.89}$$

For constant c_t and C_L, the endurance time is:

$$t = \frac{1}{c_t g}\left(\frac{L}{D}\right) \ln\left(\frac{m_1}{m_2}\right) \tag{4.90}$$

This equation is valid only if the airplane is flown at a constant C_L. Thus, for flight at constant altitude, the pilot must reduce airspeed to compensate for the reduction in weight. Alternatively, if the aircraft is permitted to fly a cruise-climb, then it is possible to maintain a constant airspeed; however, this is not possible in a hold where the pilot must keep the airplane at a given altitude.

Piston Engine Airplane Endurance

It may be deduced from equation (4.86) that the lowest fuel consumption occurs when the aircraft is flown at the condition for minimum power. The endurance time for a piston engine is given by:

$$t = -\int_{m_1}^{m_2} \frac{1}{Q} dm = -\int_{m_1}^{m_2} \frac{\eta_p}{c_p g m V}\left(\frac{L}{D}\right) dm \tag{4.91}$$

With the assumptions that η_p, c_p, V, and C_L are all constant, the integral yields the following expression:

$$t = \frac{\eta_p}{c_p g V}\left(\frac{L}{D}\right) \ln\left(\frac{m_1}{m_2}\right) \tag{4.92}$$

Because the assumptions given above do not correspond to the flight condition for minimum power, the solution given by equation (4.92) does not give the greatest possible endurance time.

Further Reading

Consult Mair and Birdsall (1992), Eshelby (2000), or Young (2017) for alternative solutions to the basic range and endurance integral equations (resulting from the airplane being flown under different constraints). Anderson (1999) derives solutions to the range and endurance integrals with fuel flow based on weight rather than mass, as done here. Torenbeek (2013) and Young (2017) present details on range optimization. Lowry (1999) describes practical performance analysis of piston engine aircraft, while Smetana (2001) covers methods for assessing the en route performance of new designs.

4.10 Takeoff and Landing Performance

Takeoff

A schematic of the takeoff is shown in the Figure 4.16. The aircraft accelerates from rest to a speed that will provide sustained controllable flight, at which point the pilot will pull the stick/yoke back, causing the airplane to rotate as the tail moves downwards. A few seconds later it lifts off and climbs to clear an imaginary screen height. This height (h_{sc}) is generally 50 ft (15.2 m) for military or light aircraft and 35 ft (10.7 m) for commercial aircraft. The total takeoff distance (s) consists of a ground segment (s_g) and an air segment (s_a). The ground segment, called the *ground roll* or *ground run*, may be divided into two elements, the distance taken from rest to the point where the aircraft rotates (s_R) and the distance from the point of rotation to liftoff (s_{RL}). At speed V_R the aircraft rotates, increasing its angle of attack and lift; shortly afterwards, at a speed V_{LO}, where the aircraft has reached sufficient forward speed to generate the required lift, liftoff occurs. Methods to evaluate the takeoff distance follow.

Forces Acting on the Airplane During Takeoff

The forces acting on an aircraft during takeoff are shown in Figure 4.17. The runway has a gradient of $\tan\gamma_G$, where a positive angle of γ_G will be used to indicate an uphill takeoff. The rolling coefficient of friction is μ_R. The thrust (T) of the engine (or propeller) accelerates the aircraft. Resistance to forward motion comes from the aerodynamic drag (D), rolling friction of the tires, and, for an inclined runway, the component of weight acting parallel to the runway. The net acceleration force is given by:

$$\sum \text{forces along runway} = T - \mu_R(W\cos\gamma_G - L) - D - W\sin\gamma_G \tag{4.93}$$

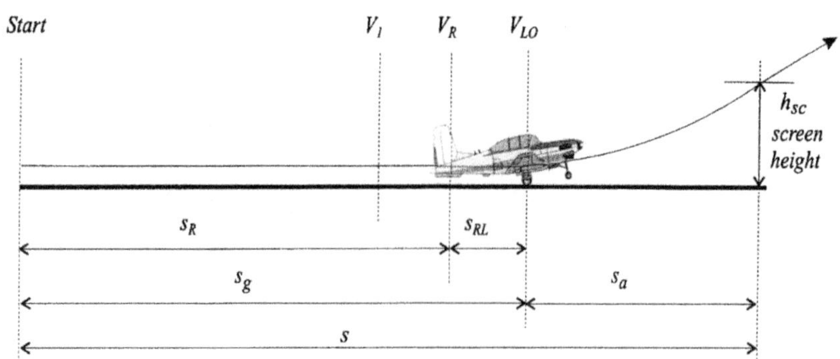

FIGURE 4.16 Takeoff profile.

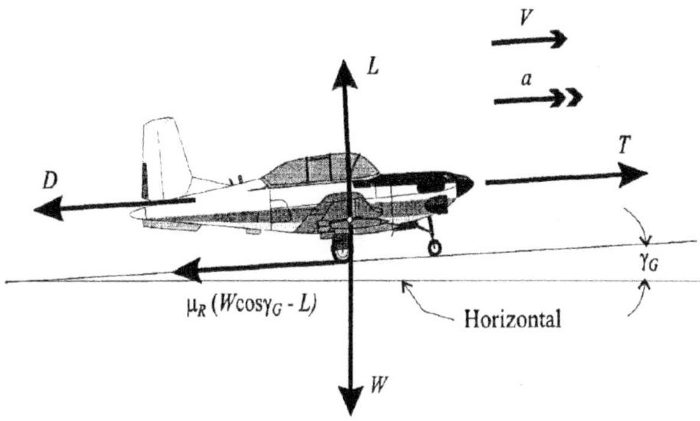

FIGURE 4.17 Forces acting on an aircraft during takeoff.

Before equation (4.93) can be used to evaluate the takeoff distance, it is necessary to describe the forces acting on the airplane during the takeoff. The weight is reduced by only a very small amount during the takeoff (due to fuel burn) and can thus be regarded as constant, but the other forces will change as the speed increases.

Lift and Drag

Because changes in the angle of attack can only result from the differential extension and contraction of the nose and main landing gear, the lift coefficient will be essentially constant up to the point of rotation. The drag coefficient will also be essentially constant. The lift and drag forces will thus vary as functions of V^2.

Ground Effect

Performance calculations in close proximity to the ground require a correction to the drag polar determined away from the ground. During the ground run the aircraft's lift-dependent drag (C_{D_i}) is reduced by a ground effect factor (λ) as a result of a reduction in the trailing vortex drag. The magnitude of λ essentially depends on the wing span and height of the wing above the ground. Torenbeek (1982) provides data to estimate the impact of this effect.

Flaps and Undercarriage

With the flaps set for takeoff and the undercarriage extended, the applicable drag polar must include terms that correct for these factors. Whereas the undercarriage would increase the value of the clean aircraft C_{D_0}, the flaps would change both C_{D_0} and C_{D_i}. Typical values may be obtained from Torenbeek (1982) or Raymer (2012).

Thrust

In general, the thrust from the engine (or propeller) depends on the atmospheric conditions and the airspeed and will vary during the takeoff run.

Rolling Friction

The value of μ_R is dependent on the tire pressure and the runway surface type and does change a little during the takeoff. However, the influence of these considerations on the ground run is very small and a mean value may be used. Because the rolling resistance on a hard dry surface is small in comparison to the other forces in equation (4.79) an approximate value of μ_R may be used without this parameter significantly affecting the calculated takeoff distance. For a hard, dry surface, the most usual value for μ_R that is used is 0.02 (ESDU 85029 1985). Other values suggested for dry concrete are 0.025 (Mair and Birdsall 1992) and 0.015 (Boeing 1989).

Analytical Evaluation of the Ground Distance (Zero Wind)

In the absence of wind, the distance to the point of rotation (s_R) is given by:

$$S_R = \int_0^{V_R} \frac{V}{a} dV \tag{4.94}$$

where s is the ground distance, V is the airspeed (equal to the ground speed in the absence of wind), and a is the acceleration. At any instant during the ground run, the acceleration may be obtained from equation (4.93) by applying Newton's second law. Because the runway gradient is always a small quantity, the approximations $\cos\gamma_G \approx 1$ and $\sin r_G \approx r_G$ (where r_G is measured in radians) can be introduced. Using the parabolic drag polar, corrected for ground effect, the acceleration can be written as:

$$a = \frac{1}{m}[T - \mu_R(W - \frac{1}{2}\rho V^2 SC_L) - \frac{1}{2}\rho V^2 S(C_{D_0} + \lambda K C_L^2) - \gamma_G W] \tag{4.95}$$

Mean Thrust

The analysis is simplified by assuming that the thrust is equal to $\overline{T}$, a mean constant value selected to give a good approximation of the takeoff distance. It has been shown (Boeing 1989) that for a jet airplane, the acceleration varies approximately linearly with V^2 from zero speed to V_R. The thrust and acceleration may thus be calculated at the speed

$$V = 1/\sqrt{2}\, V_R \approx 0.71 V_R$$

For a propeller-driven aircraft a better estimate of the ground run is obtained if the propeller thrust is calculated at a speed of $V = 0.74V_R$ (Mair and Birdsall 1992). With T taken as constant and all other variables written as functions of V^2, the integral expression (4.94) can be evaluated to give:

$$S_R = \frac{1}{2gB_2} \ln\left(\frac{A_2 + B_2 V_R^2}{A_2}\right) \tag{4.96}$$

where

$$A_2 = \frac{\overline{T}}{W} - \mu_R - \gamma_G \quad \text{and} \quad B_2 = \frac{\rho}{2(W/S)}(\mu_R C_L - C_{D_0} - K\lambda C_L^2)$$

Mean Acceleration

A popular and relatively simple method for estimating the takeoff run is based on the use of a mean acceleration ($\overline{a}$). The approach can be summarized as follows:

1. Determine the mean thrust ($\overline{T}$) at $V = 0.71V_R$ (for a jet engine) or $V = 0.74V_R$ (piston engine) from engine data.
2. Calculate $\overline{a}$ for $T = \overline{T}$ from equation (4.95) where $V = 0.71V_R$ (jet engine) or $V = 0.74V_R$ (piston engine).
3. Estimate the ground distance (s_R) from the equation for uniform acceleration, i.e.,

$$S_R = \frac{1}{2}\frac{V_R^2}{\overline{a}} \tag{4.97}$$

Effect of Wind on the Ground Distance

The component of wind acting along the runway is designated as V_w. By convention, V_w is positive for a headwind and negative for a tailwind. (Note that this is opposite to the convention usually adopted for the cruise.) At the start of the ground run, the aircraft is stationary, but the presence of the wind implies that the airspeed is equal in magnitude to V_w. In this situation equation (4.94) may be written as:

$$S_R = \int_{V_w}^{V_R} \frac{V - V_w}{a} dV \tag{4.98}$$

This expression may be evaluated as described above. The mean acceleration may be determined, as before, from equation (4.95). Based on a mean acceleration, the ground distance is:

$$S_R = \frac{1}{2\bar{a}}\left(V_R - V_w\right)^2 \tag{4.99}$$

The significance of a headwind on reducing the takeoff distance is evident from this equation.

Numerical Evaluation of the Ground Run

Equation (4.98) can be integrated numerically by dividing the takeoff run into n segments. Using the trapezoidal rule, the ground distance is given by:

$$S_R = \frac{1}{2}\sum_{i=0}^{n}\left(\frac{(V)_{i+1} - V_w}{(a)_{i+1}} + \frac{(V)_i - V_w}{(a)_i}\right)\Delta V \tag{4.100}$$

Estimation of the Rotation Distance

The rotation distance (s_{RL}) is usually small in comparison to s_R but is difficult to estimate accurately due to the changes in C_L. At the point of rotation, the pilot will pull the stick/yoke back, raising the nose and increasing the angle of attack. The time that it takes for the airplane to rotate depends on the rate that the pilot pulls the stick/yoke back and on the type of aircraft. The duration is of the order of 1 to 3 seconds; small light aircraft may rotate in 1 second or less, with large transport aircraft taking longer. An estimate of the distance s_{RL} may be obtained by multiplying the rotation time by the mean ground speed. The assumption of zero acceleration from rotation to liftoff is reasonable. In the absence of substantive data, the expression $V_{LO} = 1.2V_{S_1}$ can be used to estimate the liftoff speed.

Climb-Out to Screen Height

After the liftoff there is a transition phase in which the flight path is curved, the lift is greater than the weight, and there is a small increase in speed. After the transition the airplane will climb at an approximate constant climb angle. The point at which the screen height is reached may be either before or after the end of the curved transition. The air segment (s_a) is difficult to calculate accurately due to the variation of the governing parameters and the influence of varying pilot technique. The simplest method is to multiply an average time by the average ground speed. The time is best determined from experimental data. It would typically be between 2 and 8 seconds and is largely a function of the thrust-to-weight ratio of the airplane (Young 2017). An alternative approach to estimating s_a assumes that the flight path is a circular arc and the distance is then be calculated directly. The method is described by Mair and Birdsall (1992) and Anderson (1999).

Landing Procedure

The landing segment is shown in Figure 4.18. The aircraft descends initially along a straight glide path, which is typically at an angle of 3° to the horizontal. At the threshold (screen height), usually taken to be 50 ft (15.2 m). The pilot reduces the vertical component of the airplane's velocity by a flare.

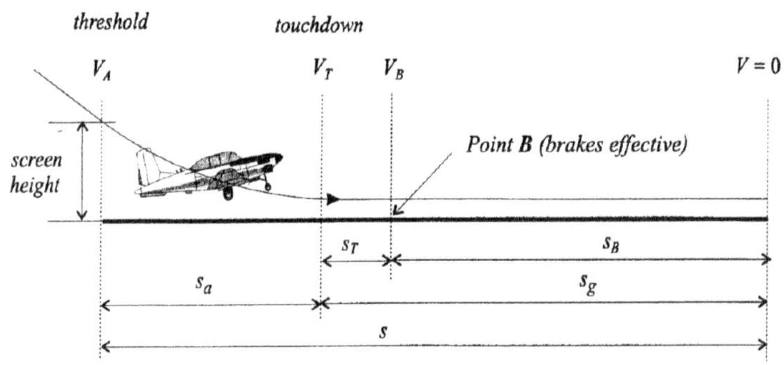

FIGURE 4.18 Landing profile.

Depending on pilot technique, there may be a short hold-off period where he or she permits the airplane to float a little, allowing the speed to reduce before it touches down. Because the landing distance depends substantially on pilot technique, analytical estimates often compare poorly with actual test data. It is common practice in theoretical analysis to assume that there is no float and that the touchdown occurs with zero vertical velocity. The speed at the touchdown (V_T) will typically be about 5% to 15% higher than the stall speed (in the landing configuration). At touchdown the nose wheel should still be well above the runway and the pilot will then allow it to descend gently onto the runway. A delay of a few seconds is typical before the pilot applies the brakes. The aircraft is brought to rest by use of the wheel brakes, sometimes assisted by lift-dumpers or spoilers and reverse thrust from the engines.

Braking Force

The analytical evaluation of the landing distance can be undertaken in an almost identical manner to that presented for the takeoff. The one significant difference is that a braking force replaces the rolling resistance. It is possible to estimate the maximum braking force, albeit with some difficulty. Under normal operations the braking force would be substantially lower than the maximum design force, which would only be required under emergency conditions. The braking force of the wheels (on a level runway) is given by:

$$\text{Braking force} = \mu_B(W - L) \tag{4.101}$$

where μ_B is the airplane braking coefficient (dimensionless). The effect of spoilers (which destroy the lift) in increasing the braking force is seen from this equation. The braking coefficient of friction is not constant and will increase as the airplane slows down. When the runway is dry, the increase is fairly small, but with a wet surface there is a large nonlinear increase. For hard, dry runways the maximum braking coefficient ($\mu_{B_{max}}$) typically increases from about 0.7 at 100 kt to about 0.8 as the speed decreases to zero, while for wet conditions $\mu_{B_{max}}$ is about 0.2 (at 100 kt) increasing to about 0.7. Maximum braking coefficient values for various runway surfaces and tyre pressures are provided by ESDU 71026 (1995). Icy slush or wet snow is a particular danger because μ_B could be reduced to less than 0.05 at the speeds associated with touchdown. The value of μ_B achieved in practice is a function of the amount of slip taking place between the tires and the runway. If the wheels are permitted to roll freely, the coefficient will equal μ_R, but as the brakes are applied the coefficient increases rapidly and then starts to reduce if the tires slip. If the brakes are manually controlled, the mean effective braking force will be about 30% to 50% of the theoretical maximum braking force (ESDU 71026 1995). Antiskid cycling by automatic braking systems protect the wheels from locking; in these cases the effective braking can be as high as 80% to 90% of the maximum value. Once a theoretical braking force has been determined, it is necessary to check if this results in the maximum permissible brake torque, or maximum brake system pressure, being exceeded. The actual braking force could thus be substantially less than that determined from a simple calculation based on a theoretical value of μ_B. Mean μ_B values of 0.3–0.5 for dry concrete or asphalt surfaces can be used for preliminary calculations (Raymer 2012; Young 2017).

Landing Distance

The total landing distance is given by:

$$s = s_a + s_T + s_B \tag{4.102}$$

An estimate of the airborne distance (s_a) can be obtained by multiplying the average time by the average ground speed. After touchdown there is a slight delay before the wheel brakes become effective, usually about 2 to 3 seconds, during which time the speed falls by a few percent. The distance s_T can be estimated from the delay time and the touchdown speed. The equations to be used for calculating the length of the ground run after the point B are essentially the same as those used for the takeoff. Equation (4.98) may be rewritten to give the braking distance, i.e.,

$$s_B = \int_{V_B}^{V_W} \frac{V - V_w}{-a}\, dV \tag{4.103}$$

where a is negative and is given by equation (4.95). The following three points are important:

1. The braking coefficient (μ_B) will replace the rolling friction coefficient (μ_R).
2. The drag coefficient will be greater than that used for the takeoff because of the greater flap angle used for landing.
3. In cases where reversed thrust is used, it is typically applied after the spoilers and brakes become effective. When thrust reversers are not available, the engines are run at idling speed and the thrust is usually small enough to be neglected.

Because the braking force cannot be represented as a function of V^2, it is often necessary to evaluate the integral by step-by-step computation. An estimate of the distance s_B may be obtained by determining the mean acceleration ($\bar{a}$) calculated for $V = 0.71V_B$. Equation (4.99), used for the takeoff analysis, may be rewritten for the landing distance:

$$s_B = \frac{-1}{2\bar{a}}\left(V_B - V_W\right)^2 \tag{4.104}$$

The determination of the required runway distance for actual aircraft operations is discussed in Subsection 4.11.

Further Reading

Further information on the takeoff and landing distance calculation is presented by Mair and Birdsall (1992), Eshelby (2000), and Filippone (2012). Lowry (1999) focuses on the performance of light aircraft and Young (2017) on jet transport airplanes. Details on the coefficient of rolling friction (μ_R) and braking coefficient (μ_B) are contained in ESDU 85029 (1985) and ESDU 71026 (1995), respectively. Stinton (1996) and Swatton (2008) describe the takeoff and landing performance from the pilot's perspective, discussing the regulatory requirements.

4.11 Airplane Operations

Regulations and Requirements

Regulations and requirements have been established to ensure that all airplanes engaged in public transport flights meet a minimum standard of safety deemed appropriate to the operation. Two complementary sets of measures contain specific details regarding the required performance of these aircraft.

The first is concerned with the *operation* of the airplane. The most important for commercial air transportation are:

- FAR 121 (Federal Aviation Regulation Part 121), *Operating requirements: Domestic, flag, and supplemental operations.*
- Commercial Air Transport Operations (OPS Part-CAT), published in European Commission regulation (EU) no. 965/2012 Annex IV.

The second set of measures pertains to the certification of new airplanes, as described in the Airworthiness Regulations/Specifications. These include:

- FAR 23 (Federal Aviation Regulation Part 23), *Airworthiness standards: Normal, utility, acrobatic, and commuter category airplanes.*
- FAR 25 (Federal Aviation Regulation Part 25), *Airworthiness standards: Transport category airplanes.*
- EASA CS-23, Certification Specifications for Normal, Utility, Aerobatic, and Commuter Category Aeroplanes.
- EASA CS-25, Certification Specifications and Acceptable Means of Compliance for Large Aeroplanes.

En Route Flight Planning—Fuel Required

The en route flight profile is divided into several parts for the purpose of flight planning, as illustrated in Figure 4.19. Specific requirements exist for the determination of the required fuel for the mission. These depend on the operator (flag or foreign), the type of airplane, the route (domestic or international), and the availability of alternate airports (if the airplane cannot land at the destination airport for any reason). For example, U.S. flag and supplemental operations on international routes where an alternate airport is specified must comply with FAR 121.645. It is stated that no person may release for flight or takeoff a turbine-engine powered airplane (not including a turboprop airplane) unless, considering wind and other weather conditions expected, unless it has enough fuel:

1. To fly to and land at the airport to which it is released; after that,
2. To fly for a period of 10% of the total time required to fly from the airport of departure to, and land at, the airport to which it was released; after that,
3. To fly to and land at the most distant alternate airport specified in the flight release; after that,
4. To fly for 30 minutes at holding speed at 1,500 ft above the alternative airport under standard temperature conditions.

The fuel required, time, and distance for each segment are determined. It is usual to calculate the trip fuel and time from brake release at the departure aerodrome to touchdown at the

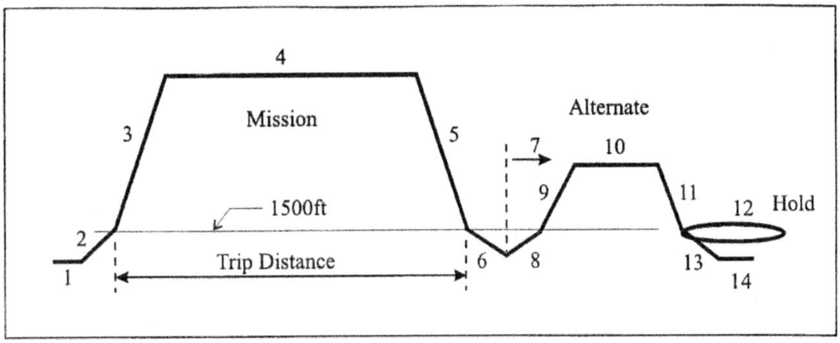

Index

1	Engine start-up, taxi	8	Climb to 1,500 ft
2	Takeoff and climb to 1,500 ft	9	Climb to alternate cruise altitude
3	Climb to cruise altitude	10	Alternate cruise
4	Cruise	11	Descent to 1,500 ft
5	Descent to 1,500 ft	12	Hold for 30 minutes
6	Approach and land	13	Approach and land
7	En route reserve	14	Taxi and shutdown

FIGURE 4.19 Typical flight profile for fuel planning.

destination aerodrome. The trip distance calculation may in some cases ignore the climb and descent below 1,500 ft (as shown in Figure 4.19). The block fuel and time includes engine start-up and taxi, and the taxi after landing.

Takeoff Reference Speeds

The takeoff is one of the critical parts of any flight. Commercial airlines are required to operate their aircraft under strict safety regulations. Definitions of the important reference speeds dealing with the takeoff operation of multiple-engine transport airplanes are given in FAR Part 25.107 and Part 25.149. The most important of these speeds are described below.

Minimum Control Speed—Ground (V_{MC_G})

This is the minimum speed on the ground at which, when the critical engine suddenly becomes inoperative and with the remaining engine(s) operating at full takeoff thrust, it is possible to recover control of the airplane with the use of primary aerodynamic controls alone (without the use of nose wheel steering) to enable the takeoff to be safely continued using normal piloting skill. The critical engine is the outboard engine that results in the most severe consequence for the takeoff.

Minimum Control Speed—Air (V_{MC})

This is the airspeed at which, when the critical engine suddenly becomes inoperative and with the remaining engine(s) operating at maximum available takeoff thrust, it is possible to recover control of the airplane using normal piloting skill and maintain straight flight with an angle of bank of not more than 5°. V_{MC} may not exceed 1.13 V_{SR}, where V_{SR} is the reference stall speed.

Takeoff Decision Speed (V_1)

This is the speed at which a multiple-engine airplane must continue the takeoff, even if one engine fails (completely). Thus, during the takeoff, up to the V_1 speed, the pilot will be able to bring the airplane safely to a stop if there is an engine failure. If there is an engine failure after V_1, the pilot shall have sufficient thrust (from the remaining engines) and sufficient remaining runway to take off safely and clear the specified screen height. The exact definition of the V_1 speed accounts for the reaction time of the pilot and is the speed of the airplane at the instant the pilot has recognized and reacted to the engine failure. The V_1 speed may not be less than V_{MC_G}.

Takeoff Rotation Speed (V_R)

This is the speed at which rotation is initiated. V_R must not be less than 1.05 times the minimum control speed (air) nor less than V_1.

Minimum Unstick Speed (V_{MU})

This is the minimum speed at which the airplane can be made to lift off the ground and continue the takeoff without displaying any hazardous characteristics. V_{MU} speeds are determined for the all-engines-operating and the one-engine-inoperative conditions.

Liftoff Speed (V_{LOF})

The liftoff speed is closely associated with the V_R speed. The all engines operating liftoff speed must not be less than 110% of V_{MU}, assuming maximum practicable rotation rate. The one engine inoperative liftoff speed must not be less than 105% of V_{MU}.

Takeoff Safety Speed (V_2)

This is a reference speed used to determine the climb performance of the airplane during the initial climb-out, with one engine inoperative. V_2 is equal to the actual speed at the 35 ft (10.7 m) height as demonstrated in flight. The airplane must be free of stall warning or other characteristics that might interfere with normal maneuvering, with asymmetric thrust, during a coordinated turn with a 30° bank angle.

Operational Field Length for Takeoff

The operational field length for a given airplane gross weight, airport elevation, ambient temperature, and wind is equal to the longest of the following three calculated distances:

1. The one-engine-inoperative takeoff distance, which is the distance required to reach a height of 35 ft for a dry runway (15 ft for a wet runway) following an engine failure, which is assumed to occur 1 second before the V_1 speed is reached.
2. The accelerate-stop (i.e., rejected takeoff) distance, which is the distance required to accelerate to V_1 and then to bring the airplane to a complete stop, using the wheel brakes only (i.e., no thrust reverse).
3. The normal all-engine takeoff distance plus a margin of 15%. This is the FAR takeoff field length for all engines operating.

Balanced Field Length

The first two distances (above) are functions of V_1. By increasing the selected V_1 speed, the calculated one-engine-inoperative takeoff distance will decrease, but the *accelerate-stop* distance will increase. There exists a unique V_1 speed, where the two distances are equal, called the balanced takeoff field length. The determination of the V_1 speed for a balanced runway is performed for a given aircraft gross weight, airport elevation, ambient temperature, and wind. There are, however, situations where the operation is planned using an unbalanced field length, i.e., the one-engine-out and accelerate-stop distances are not be equal. This may result from the use of clearways and stopways (see below).

Unbalanced Field Length—Clearways and Stopways

If the airplane is certified for an unbalanced field length takeoff and the runway has a *clearway* and/or a *stopway*, then it may be possible to increase the airplane's takeoff weight for the given runway length. A clearway is an area of prescribed width beyond the end of the runway under the control of the airport authority. Instead of reaching the height of 35 ft (10.7 m) at the end of the runway, the pilot may lift off farther down the runway and use the clearway to climb to the 35 ft (10.7 m) screen height. However, the accelerate-stop distance must still equal the available runway distance. A stopway is a hard surfaced area, aft of the runway, that may be used for braking. When a stopway is available, the additional braking distance may be taken into account to determine the accelerate-stop distance for the increased takeoff weight.

Climb-Out Gradient Requirements

The flight path after liftoff is divided into several segments (Figure 4.20). It is required that the airplane be capable of maintaining specified minimum climb gradients (as defined in FAR 25.121), with one engine inoperative during each segment. The first segment is from liftoff to the point of complete gear retraction. The airplane will have the gear extended, the flaps set for takeoff, and the engine thrust set for takeoff. The second segment starts at the point of complete retraction of the gear and ends at a height no less than 400 ft (122 m) above the runway. In this segment the airplane climbs at a speed of no less than V_2 with the gear retracted, flaps set for takeoff. The final segment extends to a height of at least 1,500 ft (457 m) or a height when the transition from takeoff to en route configuration is complete, whichever is higher. During this segment the flaps are retracted and the airplane accelerates to the en route climb speed, which is required to be 25% higher than the stall speed. The thrust will be at takeoff or maximum continuous setting. The minimum climb gradients apply after the airplane has cleaned up and accelerated along a horizontal flight path.

Obstacle Clearance Requirements

In addition to the climb gradient requirements, the airplane must be operated within the safety regulations dealing with obstacle clearance at specific airports under actual flight conditions. Buildings, trees, communication towers, etc., are all potential dangers for an aircraft that suffers an engine failure during takeoff. Furthermore, due to tailwinds, its actual flight path may be lower than that predicated by the calculated still air climb gradient. When an operator flies out of a specific airport, the net flight path must clear all obstacles within a defined flight corridor by at least 35 ft (10.7 m). The net flight path is a conservative definition that requires the actual one-engine-inoperative flight path (determined for actual headwind or tailwind conditions) to be reduced by a fixed percentage. The amount is 0.8% for two-engine aircraft, 0.9% for three-engine aircraft, and

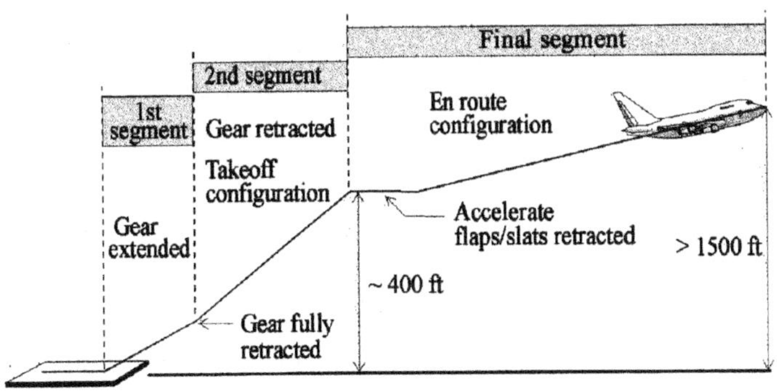

FIGURE 4.20 Climb out profile.

1.0% for four-engine aircraft (FAR 25.115). This ensures that minor errors in loading, optimistic approximations in performance predictions, or changes in wind speed or direction will not result in catastrophe when an engine failure occurs on takeoff.

Climb Requirements Following an Overshoot on Landing

One of the potentially dangerous situations that an operator has to take into account during flight planning is the possibility of an aborted landing with one engine inoperative. The airplane will approach the runway, then, after an overshoot, will need to climb, clearing all obstacles. The resulting still air climb gradient must exceed a minimum specified value for the airplane type. The minimum climb gradient is 2.1% for a two-engine airplane, 2.4% for a three-engine airplane, and 2.7% for a four-engine airplane (FAR 25.121). A second climb requirement is specified for aborted landings when all engines are operating. With the airplane in the landing configuration, all airplane types must be able to maintain a climb gradient of 3.2% (FAR 25.119).

Required Runway Length for Landing

The demonstrated landing distance is based on the airplane crossing the 50 ft (15.2 m) threshold at a speed no less than 1.23 V_{SR}. After the touchdown, the airplane will be brought to a stop by means of the wheel brakes only. This conservative assumption ensures that if thrust reversers are used, the airplane will stop within the calculated landing distance. The distance from the threshold until the airplane stops is the measured (or unfactored) landing distance. The required runway (dry) for jet airplane operations is determined by multiplying the measured landing distance by an operational reserve factor of 1.667. This factor implies that the airplane should ideally require only 60% of the runway, with the remaining 40% regarded as an *operational reserve*. For turboprop aircraft the percentage is 70%, with 30% regarded as the operational reserve. For wet runways, the landing distance is multiplied by an additional factor of 1.15. In the case of adverse runway conditions of ice or snow contamination, additional allowances are made to account for the reduced braking capability of the airplane.

Further Reading

Consult the regulations and specifications published by the FAA and EASA for complete details. Swatton (2008) and Young (2017) provide useful information on the operational implications of these regulations/specifications.

References

Anderson, J. D. 1999. *Aircraft Performance and Design,* McGraw-Hill, New York.

Boeing. 1989. *Jet Transport Performance Methods,* Boeing Commercial Airplane Company.

EASA, CS-23, Certification Specifications for Normal, Utility, Aerobatic, and Commuter Category Aeroplanes, European Aviation Safety Agency, Cologne, Germany.

EASA, CS-25, Certification Specifications and Acceptable Means of Compliance for Large Aeroplanes, European Aviation Safety Agency, Cologne, Germany.

ESDU 68046. 1992. *Atmospheric Data for Performance Calculations, amendment (d),* IHS ESDU, 133 Houndsditch, London.

ESDU 70020. 1970. *Non-dimensional Approach to Engine Thrust and Airframe Drag for the Analysis of Measured Performance Data: Aircraft with Turbo-jet and Turbo-fan Engines,* IHS ESDU, 133 Houndsditch, London.

ESDU 71026. 1995. *Frictional and Retarding Forces on Aircraft Tyres, Part II, amendment (d),* IHS ESDU, 133 Houndsditch, London.

ESDU 72018. 1972. *International Standard Atmosphere (2000 ft to 105,000 ft, data in SI units),* IHS ESDU, 133 Houndsditch, London.

ESDU 73019. 1982. *Approximate Methods for Estimating Cruise Range and Endurance: Aeroplanes with Turbo-jet and Turbo-fan engines, amendment* (c), IHS ESDU, 133 Houndsditch, London.

ESDU 85029. 1985. *Calculation of Ground Performance in Take-off and Landing,* IHS ESDU, 133 Houndsditch, London.

Eshelby, M. E. 2000. *Aircraft Performance: Theory and Practice,* Edward Arnold, London.

European Commission, Commercial Air Transport Operations (Part-CAT), Annex IV to Commission Regulation (EU) No. 965/2012, Brussels, Belgium.

FAA. FAR 23, Federal Aviation Regulation Part 23. *Airworthiness Standards: Normal, Utility, Acrobatic, and Commuter Category Airplanes.* Federal Aviation Administration, Washington, DC.

FAA. FAR 25, Federal Aviation Regulation Part 25. *Airworthiness Standards: Transport Category Airplanes,* Federal Aviation Administration, Washington, DC.

FAA. FAR 121, Federal Aviation Regulation Part 121. *Operating Requirements: Domestic, Flag, and Supplemental Operations,* Federal Aviation Administration, Washington, DC.

Filippone, A. 2012. *Advanced Aircraft Flight Performance,* Cambridge University Press, New York.

ICAO. 1993. *Manual of the ICAO Standard Atmosphere,* Doc. 7488/1, International Civil Aviation Organization (ICAO), Montreal.

ISO 2533. 1975. *Standard Atmosphere,* International Organisation for Standardization.

Lan, C.-T. E. and Roskam, J. 1981. *Airplane Aerodynamics and Performance,* Roskam Aviation and Engineering Corp.

Lowry, J. T., 1999. *Performance of Light Aircraft,* AIAA, Reston, VA.

Mair, W. A. and Birdsall, D. L. 1992. *Aircraft Performance,* Cambridge University Press, Cambridge.

Mattingly, J. D., Heiser, W. H. and Daley, D. T. 2003. *Aircraft Engine Design,* 2nd ed., AIAA, Reston, VA.

Raymer, D. 2012. *Aircraft Design: A Conceptual Approach,* 5th ed., AIAA, Reston, VA.

Smetana, F. O. 2001. *Flight Vehicle Performance and Aerodynamic Control,* AIAA, Reston, VA.

Stinton, D. 1996. *Flying Qualities and Flight Testing of the Aeroplane,* AIAA, Reston, VA.

Swatton, P. J. 2008. *Aircraft Performance Theory and Practice for Pilots,* 2nd ed., Wiley-Blackwell, Chichester, UK.

Torenbeek, E. 1982. *Synthesis of Subsonic Airplane Design,* Delft University Press, Delft.

Torenbeek, E. 2013. *Advanced Aircraft Design,* Wiley, Chichester, UK.

Young, T. M. 2017. *Performance of the Jet Transport Airplane: Analysis Methods, Flight Operations, and Regulations,* Wiley, Chichester, UK.

PART 2

Aircraft Stability and Control

Douglas G. Thomson

Notation

$\mathbf{A}$ system matrix
a_t tailplane lift curve slope, per radian
a_e elevator effectiveness, per radian
a_F lift-curve slope of fin, per radian
a_r rudder effectiveness, per radian
$\mathbf{B}$ control matrix
b wing semispan, m
$\mathbf{C}$ output matrix
$\mathbf{D}$ direct matrix
$\mathbf{E}$ gust influence matrix
C_D, C_L, C_T coefficients of drag, lift, and thrust
C_m pitching moment coefficient
$C_{D0}, C_{D\alpha}$ coefficients of empirical equation for drag coefficient
$C_{L_0}, C_{L_\alpha}, C_{L_{\delta e}}$ coefficients of empirical equation for lift coefficient
$C_{M_0}, C_{M_\alpha}, C_{M_{\delta e}}, C_{M_q}$ coefficients of empirical equation for pitching moment coefficient
$\bar{c}$ mean aerodynamic chord, m
D aircraft drag, N
g acceleration due to gravity, m/s^2
h_T offset of thrust-line from aircraft x_b body axes, m
I_{xx}, I_{yy}, I_{zz} aircraft moments of inertia about x_b, y_b, z_b body axes, kg m^2
I_{xz} product of inertia about aircraft y_b axis, kg m^2
L aircraft lift, N
L, M, N external moments, Nm
l_F distance between fin aerodynamic center and aircraft c.g., m
l_t tailplane lever arm, m
M Mach number
m aircraft mass, kg
P, Q, R angular velocities in direction of x_b, y_b, z_b body axes, rad/s
p, q, r perturbations in angular velocities in direction of x_b, y_b, z_b body axes, rad/s
p_d dynamic pressure, N/m^2
S wing area, m^2
S_f fin area, m^2
S_t tailplane area, m^2
T period of oscillation, s
T aircraft thrust, N
$\mathbf{T}$ Euler transformation matrix
t_{half} time to half amplitude, s
t_{double} time to double amplitude, s

$\mathbf{u}$	control vector
U, V, W	translational velocities in direction of x_b, y_b, z_b body axes, m/s
u, v, w	perturbations in translational velocities in direction of x_b, y_b, z_b body axes, m/s
V_f	aircraft flight velocity, m/s
V_g	gust velocity vector
V_H	tailplane volume ratio
V_v	fin volume ratio
X, Y, Z	external forces, N
$\mathbf{x}$	state vector
z_F	height of fin mean aerodynamic center above x-axis, m
x_b, y_b, z_b	body axes
x_E, y_E, z_E	earth axes
$\mathbf{y}$	output vector

Greek

α, β	incidence angles (attack and sideslip), rad
δ	control deflection, rad
ε	tailplane downwash angle, rad
Φ, Θ, Ψ	euler angles (roll, pitch, and yaw), rad
ϕ, θ, ψ	perturbations in Euler angles (roll, pitch, and yaw), rad
λ	eigenvalue
η, ζ, ξ	elevator, rudder, and aileron deflections, rad
ρ	air density, kg/m^3
σ	sidewash angle, rad

Subscripts

b	body axes set
dr	dutch roll mode
E	earth axes set
e	equilibrium condition (trim state)
e, a, r	elevator, aileron, rudder
g	gust
ph	phugoid mode
r	roll mode
s	spiral mode
sp	short-period mode

4.12 Mathematical Modeling and Simulation of Fixed-Wing Aircraft

Aircraft Nonlinear Equations of Motion

The aircraft, in its most basic form, has 6 degrees of freedom, as summarized in Table 4.7.

These consist of translational motions in the directions of the axes set fixed in the aircraft, and three rotations about these axes (Figure 4.21). The six aircraft states (or *state variables*) are (U, V, W, P, Q, R). The *body fixed frame of reference* (x_b, y_b, z_b) has its origin at the center of gravity of the aircraft, with the x_b axis pointing forwards, usually down the centerline of the fuselage, the z_b axis downwards, and the y_b axis in the starboard direction. The forces and translational velocities are

Aircraft axis	Translational			Angular			
	Motion	Velocity	Force	Motion	Displacement	Rate	Moment
x_b	Fore/aft	U	X	Roll	Φ	P	L
y_b	Sideward	V	Y	Pitch	Θ	Q	M
z_b	Heave	W	Z	Yaw	Ψ	R	N

TABLE 4.7 Slate Variables and Parameters for 6 Degree of Freedom Aircraft Model

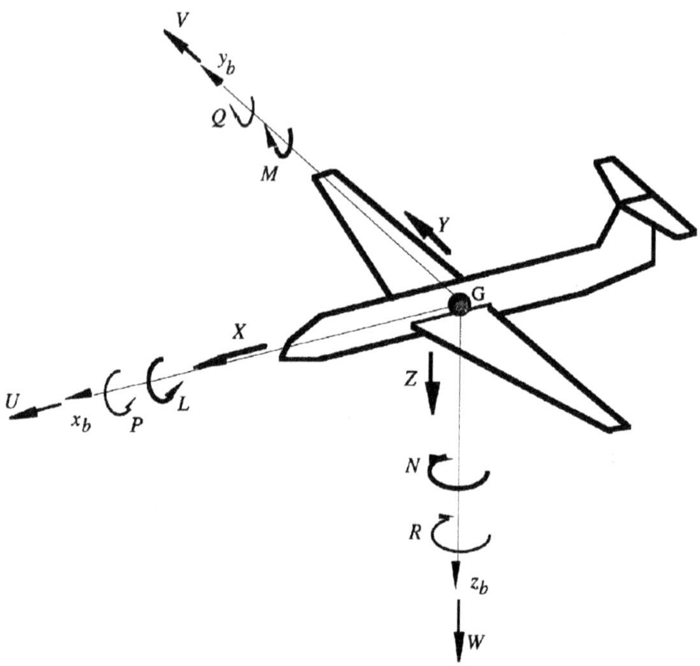

FIGURE 4.21 State variables as referred to the body axes set.

positive in these directions. The positive direction for angular quantities is determined by the right-hand rule (the right thumb is pointed in the positive direction of the axis and the direction of curl of the fingers gives the positive direction for angular quantities). Hence, a positive roll rate gives starboard wing down, a positive pitch rate gives nose up, and a positive yaw rate gives nose right.

As the aircraft is a rigid body translating and rotating in 3D space it is appropriate to apply the *Euler equations* to its motion:

$$m(\dot{U} + QW - RV) = X - mg\sin\Theta \tag{4.105}$$

$$m(\dot{V} + RU - PW) = Y + mg\cos\Theta\sin\Phi \tag{4.106}$$

$$m(\dot{W} + PV - QU) = Z + mg\cos\Theta\cos\Phi \tag{4.107}$$

$$I_{xx}\dot{P} - I_{xz}\dot{R} + QR(I_{zz} - I_{yy}) - PQI_{xz} = L \tag{4.108}$$

$$I_{yy}\dot{Q} + RP(I_{xx} - I_{zz}) + (P^2 - R^2)I_{xz} = M \tag{4.109}$$

$$I_{zz}\dot{R} - I_{xz}\dot{P} + PQ(I_{yy} - I_{xx}) - QRI_{xz} = N \tag{4.110}$$

Equations (4.105)–(4.107) are the *translational equations of motion,* derived by consideration of linear momentum. In effect they may be simply expressed as:

$$ma_x = F_x, \quad \text{where } a_x = \dot{U} + QW - RV, \quad \text{and} \quad F_x = X - mg\sin\Theta \tag{4.111}$$

$$ma_y = F_y, \quad \text{where } a_y = \dot{V} + RU - PW, \quad \text{and} \quad F_y = Y + mg\cos\Theta\sin\Phi \tag{4.112}$$

$$ma_z = F_z, \quad \text{where } a_z = \dot{W} + PV - QU, \quad \text{and} \quad F_z = Z + mg\cos\Theta\cos\Phi \tag{4.113}$$

The component accelerations (a_x, a_y, a_z) are the *absolute* (or *inertial*) accelerations of the center of gravity; recall that the terms QW, etc., occur as the frame of reference rotates as it translates. The forces F_x, F_y, and F_z are the total *external forces,* which are composed of the gravitational terms ($mg\sin\Theta$, etc.) and the aerodynamic and propulsive terms (X, Y, Z).

The rotational equations of motion are derived from the principles of angular momentum. The *external moments* (L, M, N) are due to aerodynamic and propulsive loads.

Axes Sets and the Euler Transformation

The Euler equations are referred to a frame of reference with origin located at the center of gravity of the system. As previously mentioned, this frame of reference is known as the *body fixed axes set.* This axes set moves with the aircraft and is of practical use only when referred to an *inertial frame* of reference, i.e., *the earth fixed axes set* (x_E, y_E, z_E). The origin of this axes set is nominal, but the normal convention for directions is that the x_E axis points north, the y_E axis to the east, and the z_E axis down toward the center of the earth. In practical terms the origin of this axes set is often taken as the position of the aircraft at the initiation of a simulation, with the x body and earth axes coincident. The aircraft's position and orientation in terms of the Euler or *attitude* angles (Φ, Θ, Ψ) is given relative to this axes set (Figure 4.22).

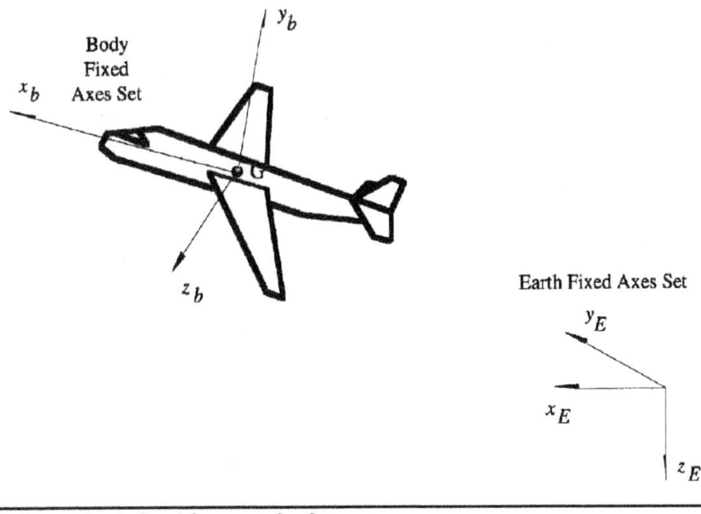

FIGURE 4.22 Earth and body fixed frames of reference.

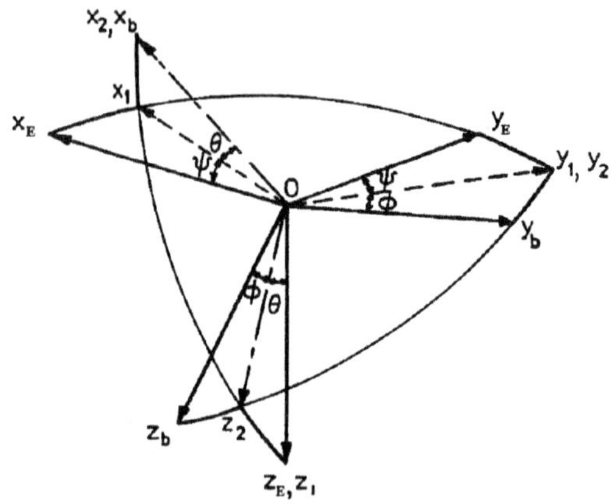

FIGURE 4.23 The Euler angle transformation.

The transformation from earth axes (O, x_E, y_E, z_E) to body axes (O, x_b, y_b, z_b) may be achieved through the action of three consecutive rotations (Figure 4.23):

1. A rotation of Ψ (the heading or azimuth angle) about Oz_E to give the intermediate frame (O, x_1, y_1, z_E)
2. A rotation of Θ (the pitch angle) about Oy_1 to give the intermediate frame (O, x_2, y_1, z_2)
3. A rotation of Φ (the roll or bank angle) about Ox_2 to give the body fixed frame (O, x_b, y_b, z_b)

Defining unit vectors in the x_b, y_b, z_b directions as (i_b, j_b, k_b) for the body axes and unit vectors in the directions x_E, y_E, z_E for the earth axes as (i_E, j_E, k_E), the transformation from the body to the earth axes frame is given by:

$$\begin{bmatrix} i_E \\ j_E \\ k_E \end{bmatrix} = \begin{bmatrix} l_1 & m_1 & n_1 \\ l_2 & m_2 & n_2 \\ l_3 & m_3 & n_3 \end{bmatrix} \begin{bmatrix} i_b \\ j_b \\ k_b \end{bmatrix} \tag{4.114}$$

where
$$\begin{aligned} l_1 &= \cos\Theta\cos\Psi \\ l_2 &= \cos\Theta\sin\Psi \\ l_3 &= -\sin\Theta \\ m_1 &= \sin\Phi\sin\Theta\cos\Psi - \cos\Phi\sin\Psi \\ m_2 &= \sin\Phi\sin\Theta\sin\Psi + \cos\Phi\cos\Psi \\ m_3 &= \sin\Phi\cos\Theta \\ n_1 &= \cos\Phi\sin\Theta\cos\Psi + \sin\Phi\sin\Psi \\ n_2 &= \cos\Phi\sin\Theta\sin\Psi - \sin\Phi\cos\Psi \\ n_3 &= \cos\Phi\cos\Theta \end{aligned}$$

or, for a general earth fixed axes vector, λ_E, for an earth to body axes transformation, we may write:

$$\lambda_E = \mathbf{T}\lambda_b \tag{4.115}$$

and to transform from body to earth axes the transpose of the matrix is used:

$$\lambda_b = \mathbf{T}^T\lambda_E \tag{4.116}$$

The matrix **T** is known as the *Euler angle transformation matrix* while its elements $l_1 \ldots n_3$ are termed the *direction cosines*.

Hence, for translational velocities where

$$\lambda_b = \begin{bmatrix} U \\ V \\ W \end{bmatrix} \quad \text{and} \quad \lambda_E = \begin{bmatrix} \dot{X}_E \\ \dot{Y}_E \\ \dot{Z}_E \end{bmatrix}$$

we would have:

$$\begin{bmatrix} \dot{X}_E \\ \dot{Y}_E \\ \dot{Z}_E \end{bmatrix} = \begin{bmatrix} l_1 & m_1 & n_1 \\ l_2 & m_2 & n_2 \\ l_3 & m_3 & n_3 \end{bmatrix} \begin{bmatrix} U \\ V \\ W \end{bmatrix} \tag{4.117}$$

and

$$\begin{bmatrix} U \\ V \\ W \end{bmatrix} = \begin{bmatrix} l_1 & l_2 & l_3 \\ m_1 & m_2 & m_3 \\ n_1 & n_2 & n_3 \end{bmatrix} \begin{bmatrix} \dot{X}_E \\ \dot{Y}_E \\ \dot{Z}_E \end{bmatrix} \tag{4.118}$$

Note that (X_E, Y_E, Z_E) is in effect the position of the aircraft relative to the earth fixed frame of reference and hence $(\dot{X}_E, \dot{Y}_E, \dot{Z}_E)$ are the component velocities in the directions of the axes. It is possible also to transform the angular velocities such that the earth fixed frame-related Euler angle rates $(\dot{\Phi}, \dot{\Theta}, \dot{\Psi})$ may be expressed in terms of their body fixed equivalents (P, Q, R):

$$P = \dot{\Phi} - \dot{\Psi}\sin\Theta \tag{4.119}$$

$$Q = \dot{\Theta}\cos\Phi + \dot{\Psi}\sin\Phi\cos\Theta \tag{4.120}$$

$$R = \dot{\Psi}\cos\Phi\cos\Theta - \dot{\Theta}\sin\Phi \tag{4.121}$$

and these expressions may be inverted to give:

$$\dot{\Phi} = P + Q\sin\Phi\tan\Theta + R\cos\Phi\tan\Theta \tag{4.122}$$

$$\dot{\Theta} = Q\cos\Phi - R\sin\Phi \tag{4.123}$$

$$\dot{\Psi} = Q\sin\Phi\sec\Theta + R\cos\Phi\sec\Theta \tag{4.124}$$

Control Variables

For a basic aircraft there are three primary flight controls: elevator, ailerons, and rudder. In mathematical terms the "δ" notation can be used, giving the control variables: δ_e, elevator; δ_r rudder; δ_a, aileron. Note that the symbols η, ζ, ξ are often adopted for elevator, rudder, and aileron.

Deflections of these three control surfaces effectively cause changes in the angular rates about the three axes of the aircraft. The mechanism and sign convention for each of them is given below.

- *Elevator* (pitch control): stick forward gives positive δ_e, elevator is depressed, thereby increasing the effective camber of the tailplane, increasing its lift, and thus producing a pitch down moment about the center of gravity (i.e., $+ve\ \delta_e$ gives $-ve\ Q$).
- *Rudder* (yaw control): Left pedal forward denotes positive δ_r, rudder is displaced to left when viewed from above (i.e., toward the port wing), the fin becomes cambered producing an increase in sideforce toward the starboard side, which produces a negative yawing moment about the center of gravity and turns the nose to the left (i.e., $+\ ve\ \delta_r$ gives $-ve\ R$).
- *Aileron* (roll control): Stick right gives positive δ_a, the port aileron is depressed increasing this wing's camber and hence lift, while the starboard aileron is raised, reducing this wing's camber and hence lift. The resulting positive rolling moment causes the aircraft to bank to the right (i.e., $+v_e\ \delta_a$ gives $+ve\ P$).

Simulation of Longitudinal Motion of a Fixed-Wing Aircraft

In this subsection a basic simulation of an aircraft is developed. The starting point is to compile a set of expressions for the external forces and moments, i.e., develop the mathematical model. For convenience the equations of motion are often split into two sets: longitudinal and lateral-directional. Here, a longitudinal simulation is presented in some detail. A full simulation is simply an extension of what is presented.

The Mathematical Model

Longitudinal motions occur in the aircraft xz-plane where the state variables are (U, W, Q) and the control variable is the elevator angle, δ_e. In effect longitudinal motions cover fore and aft motions (i.e., accelerations), climbing flight, and pitching flight. The equations used are then (4.105), (4.107), (4.109), and (4.118), with the lateral/directional variables (V, P, R, Φ, Ψ) set to zero.

$$m(\dot{U} + QW) = X - mg\sin\Theta \tag{4.125}$$

$$m(\dot{W} - QU) = Z + mg\cos\Theta \tag{4.126}$$

$$I_{yy}\dot{Q} = M \tag{4.127}$$

$$\dot{\Theta} = Q \tag{4.128}$$

In modelling the longitudinal dynamics of the aircraft it is necessary to calculate X, Z, and M as functions of the state and control variables. From Figure 4.24 we can readily see that these loads are given by:

$$X = T - D\cos\alpha + L\sin\alpha \tag{4.129}$$

$$Z = -L\cos\alpha - D\sin\alpha \tag{4.130}$$

$$M - M_A + Th_T \tag{4.131}$$

where L and D are the total aircraft lift and drag, T is the total engine thrust assumed to act a distance h_T below the x_b axis, and M_A is the aerodynamic pitching moment derived below.

The angle of attack, α, is obtained from

$$\tan\alpha = \frac{W}{U} \tag{4.132}$$

Calculation of the thrust T is dependent on the powerplant, while L, D, and M_A are obtained as follows.

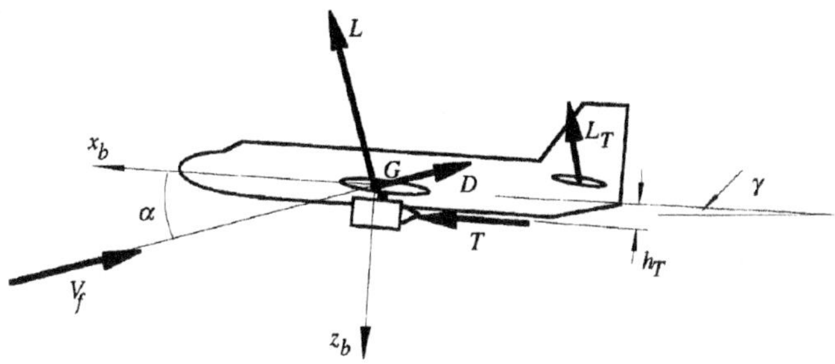

FIGURE 4.24 External forces on aircraft—longitudinal motion.

First, the forces are nondimensionalized by division by $\frac{1}{2}\rho V_f^2 S$ such that equations (4.120) and (4.121) become:

$$X = \frac{1}{2}\rho V_f^2 S(C_T - C_D \cos\alpha + C_L \sin\alpha) \tag{4.133}$$

$$Z = \frac{1}{2}\rho V_f^2 S(-C_L \cos\alpha - C_D \sin\alpha) \tag{4.134}$$

Equation (4.122) is divided by $\frac{1}{2}\rho V_f^2 S\bar{c}$ (where $\bar{c}$ is the mean aerodynamic chord) to give:

$$M = \frac{1}{2}\rho V_f^2 S\bar{c}\left(C_{M_A} + \frac{h_T}{\bar{c}}C_T\right) \tag{4.135}$$

The aerodynamic coefficients are usually obtained either from wind tunnel data or by semiempirical methods. In general terms one might write: $C_L = f(\alpha, \delta_e, M)$, where M = Mach number. Wind tunnel data may be presented in the form of a look-up table, and at any point in the simulation where α, δ_e, and M are known, C_L, C_D, and C_M are found by linear interpolation. Using semiempirical methods, typical expressions for the coefficients are:

$$C_L = C_{L_0} + C_{L_\alpha}\alpha + C_{L_{\delta e}}\delta_e \tag{4.136}$$

$$C_D = C_{D_0} + C_{D_\alpha}\alpha \tag{4.137}$$

$$C_M = C_{M_0} + C_{M_\alpha}\alpha + C_{M_{\delta e}}\delta_e + C_{M_q}\hat{q} \tag{4.138}$$

where the nondimensional pitching velocity, $\hat{q}$, is given by:

$$\hat{q} = \frac{Q\bar{c}}{V_f} \tag{4.139}$$

and

C_{D_0}, C_{L_0} = lift and drag coefficients at zero angle of attack

$C_{L_\alpha} = \frac{\partial C_L}{\partial \alpha}$ = lift-curve slope

$C_{L_{\delta e}} = \frac{\partial C_L}{\partial \delta_e}$ = change in lift coefficient due to displacement of elevator

$C_{D_\alpha} = \frac{\partial C_D}{\partial \alpha}$.

Simulation Procedure

The nonlinear differential equations of motion (4.120)–(4.123) may be solved numerically, using a Runge–Kutta scheme, for example, to give time histories of the state variables, U, W, Q, Θ in response to a deflection in the elevator angle, δ_e. The convention is to solve the equations from some initial condition representing a *trim state* of the aircraft.

Calculation of a Trim State The longitudinal trim of an aircraft is usually defined by setting the accelerations and angular velocities to zero. Equations (4.120)–(4.122) become:

$$X - mg\sin\Theta = 0 \tag{4.140}$$

$$Z + mg\cos\Theta = 0 \tag{4.141}$$

$$M = 0 \tag{4.142}$$

and hence there are three equations to satisfy for three unknowns. For a given flight velocity, V_f, altitude (hence air density, ρ) and climb angle, γ, the unknowns are the required thrust and elevator angle and the resulting fuselage pitch attitude, T, δ_e, and Θ, respectively. Hence, substituting equations (4.124)–(4.126) into (4.135)–(4.137) we have:

$$g_1(\Theta,T,\delta_e) = T - D\cos\alpha + L\sin\alpha - mg\sin\Theta = 0 \tag{4.143}$$

$$g_2(\Theta,T,\delta_e) = -L\cos\alpha - D\sin\alpha + mg\cos\Theta = 0 \tag{4.144}$$

$$g_3(\Theta,T,\delta_e) = M_A + Th_T = 0 \tag{4.145}$$

We therefore obtain a system of three nonlinear, algebraic equations g_1, g_2, and g_3, to be solved for three unknowns, T, δ_e, and Θ, which is usually solved using a Newton–Raphson (or similar) iterative scheme.

Calculation of Response to Controls Noting that $\Phi = \Psi = 0$, for longitudinal motion and rewriting the Euler transformation (4.112) accordingly to give equations (4.145) and (4.146), and recasting equations (4.120)–(4.123) provides a set of six coupled nonlinear differential equations:

$$\dot{U} = -QW + \frac{X}{m} - g\sin\Theta \tag{4.146}$$

$$\dot{W} = QU + \frac{Z}{m} + g\cos\Theta \tag{4.147}$$

$$\dot{Q} = \frac{M}{I_{yy}} \tag{4.148}$$

$$\dot{\Theta} = Q \tag{4.149}$$

$$\dot{X}_E = U\cos\Theta + W\sin\Theta \tag{4.150}$$

$$\dot{Z}_E = -U\sin\Theta + W\cos\Theta \tag{4.151}$$

which can be solved simultaneously for the six states: (U, W, Q, Θ, X_E, Z_E) in response to inputs of elevator δ_e. The elevator input feeds into the equations of motion through the lift and pitching moment (4.131) and (4.133).

4.13 Development of the Linearized Equations of Motion

Although the methods of solving the nonlinear equations computationally are well established and understood, simplified *linearized models* are far more appropriate in order to establish the stability characteristics of an aircraft.

Small-Disturbance Theory—Basic Concept

In small-disturbance theory, the aircraft's motion consists of small deviations from some reference steady flight state (a *trim state*). This assumption is valid for all of the most common flight conditions, and it is only in gross maneuvering flight (e.g., high angle of attack, high-speed maneuvering of fighter aircraft) where the linearized, small-disturbance equations are invalid and the full nonlinear equations must be applied.

Using small-disturbance theory, we assume that the instantaneous total value of each of the state and control variables is composed of two components:

$$U = U_e + u, \qquad \Theta = \Theta_e + \theta, \qquad \delta_a = \delta_{a_e} + \delta_a', \text{ etc.} \tag{4.152}$$

where the subscript e denotes the reference trim or equilibrium state of the vehicle and the lowercase denotes a perturbation from the reference state. Note that the prime notation is used for perturbations of control variables. In a similar way, it is assumed that the aerodynamic force and moments have two components: the reference value, still denoted by subscript e, and a perturbation, this time denoted by Δ:

$$X = X_e + \Delta X, \text{ etc.}$$

There are three major limitations on the use of the linearized equations of motion:

1. The linearized equations of motion are valid only for small disturbances from the reference trim state. This is a consequence of the small-disturbance assumption, and it implies that calculation of only a few seconds of disturbed (from trim) flight using the linearized equations may be valid.
2. The equations are derived for symmetrical aircraft only. This may not seem too much of a problem, but it does exclude helicopters, which are not symmetrical due to the necessity of having a tail rotor.
3. The equations are derived assuming a rigid aircraft. For small aircraft (even fighters) flying at subsonic speeds this assumption is valid because aeroelastic effects are minimal.

The Reference Trim State

It is convenient both mathematically and physically to refer a dynamic analysis of aircraft motion to a reference trim state. The following assumptions are commonly used:

1. There are no resultant accelerations on the aircraft ($\dot{U}_e = \dot{V}_e = \ldots \dot{R}_e = 0$).
2. The aircraft has no angular velocity ($P_e = Q_e = R_e = 0$).
3. The aircraft is assumed to be in wings-level ($\Phi_e = 0$), symmetric flight ($V_e = 0$).

Choice of Axes Set—Stability Axes

The most commonly used axes set for analysis are the *stability axes*. In this frame the x-axis is fixed in the aircraft in the direction of motion, i.e., the x-body axis is aligned with the relative wind, such that

$$U_e = V_f \qquad \text{and} \qquad W_e = 0$$

Choosing this axes set has the advantage that it simplifies the calculation of the external forces Figure 4.25(a) in trimmed flight to

$$X_e = T_e - D_e \tag{4.153}$$

$$Z_e = -L_e \tag{4.154}$$

Further, as $W_e = 0$, it is clear from equation (4.132) that $\alpha_e = 0$. In disturbed flight the angle between the relative wind and the x-body axis is the angle of attack, α. As the lift and drag act parallel and perpendicular to the relative wind, the external forces are obtained by resolving lift and drag through the angle α (Figure 4.25(b)):

$$X = T - D\cos\alpha + L\sin\alpha \tag{4.155}$$

$$Z = -L\cos\alpha - D\sin\alpha \tag{4.156}$$

The disadvantage of this choice is that because the aircraft will adopt a different angle of attack (and pitch attitude) for each trimmed flight speed, the x-axis will be oriented with respect to a geometrical datum at each flight speed. Consequently, the values of the moments of inertia, I_{xx}, I_{zz}, etc., will vary with reference flight speed. This is usually considered a minor effect because the angle of attack variation over the speed range may only be a few degrees.

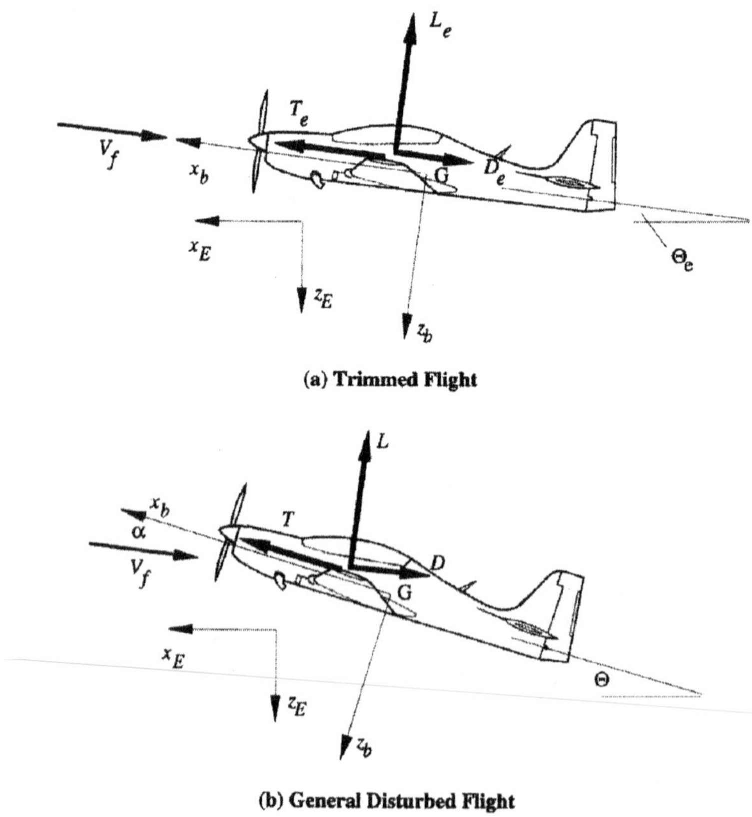

FIGURE 4.25 Definition of stability axes.

Procedure for Linearizing the Nonlinear Equations

The process of linearization can be summarized as follows:

1. Replace full nonlinear variables values by the reference (trim) plus small perturbation value (i.e., $\dot{U} = \dot{U}_e + \dot{u}$, etc.).
2. Apply appropriate trim values as listed above (i.e., $\dot{U}_e = 0, \dot{U} = \dot{u}$, etc.).
3. Make small-angle assumption for attitude perturbations (i.e., $\cos\theta \approx 1$, $\sin\theta \approx \theta$, etc.).
4. Eliminate products of perturbations (for example, it is assumed that $\dot{u} >> (qw - vr)$.
5. Eliminate the trim value of the external force or moment (for example, applying equation (4.105) at the trim state gives $0 = X_e - mg\sin\Theta_e$, etc.).

The Linearized Equations of Motion in Basic Form

Applying the procedure described above to the nonlinear equations, (4.105)–(4.110), produces the following set of linearized equations

$$m\dot{u} = \Delta X - mg\theta\cos\Theta \tag{4.157}$$

$$m\dot{v} = \Delta Y - U_e r + mg\phi\cos\Theta_e \tag{4.158}$$

$$m\dot{w} = \Delta Z + mU_e q - mg\theta\sin\Theta_e \tag{4.159}$$

$$I_{xx}\dot{p} - I_{xz}\dot{r} = \Delta L \tag{4.160}$$

$$I_{yy}\dot{q} = \Delta M \tag{4.161}$$

$$I_{zz}\dot{r} - I_{xz}\dot{p} = \Delta N \tag{4.162}$$

The expressions for the body rates in terms of the Euler angle rates, (4.119)–(4.121) are also linearized to give:

$$p = \dot{\phi} - \dot{\psi}\sin\Theta_e \tag{4.163}$$

$$q = \dot{\theta} \tag{4.164}$$

$$r = \dot{\psi}\cos\Theta_e \tag{4.165}$$

Linear Expressions for the Aerodynamic Force and Moment Perturbations

Linear expressions for the aerodynamic and propulsive force and moment perturbations are obtained by assuming that the external forces and moments are functions of the instantaneous values of the disturbance velocities, control angles and their time derivatives, i.e.,

$$X = f_1(u, \dot{u}, v, \dot{v}, w, \dot{w}, p, \dot{p}, q, \dot{q}, r, \dot{r}, \delta_e, \dot{\delta}_e, \delta_r, \dot{\delta}_r, \delta_a, \dot{\delta}_a)$$

$$\vdots$$

$$N = f_6(u, \dot{u}, v, \dot{v}, w, \dot{w}, p, \dot{p}, q, \dot{q}, r, \dot{r}, \delta_e, \dot{\delta}_e, \delta_r, \dot{\delta}_r, \delta_a, \dot{\delta}_a)$$

The method normally used to linearize the external forces and moments is to represent them by a Taylor series expansion. The Taylor series expansion for a multivariable problem can be applied to the external forces and moments to give for the X-force:

$$\Delta X = \frac{\partial X}{\partial u}u + \frac{\partial X}{\partial \dot{u}}\dot{u} + \frac{\partial X}{\partial v}v + \ldots + \frac{\partial X}{\partial \dot{\delta}_a}\dot{\delta}_a$$

where the higher-order derivatives have been neglected.

It should also be noted that the derivatives have to be evaluated at the point about which the expansion was derived—the equilibrium trim state—and hence the derivatives must be calculated using the state values from trim, denoted by the subscript e. The full set of external force and moment perturbation linearizations is:

$$\Delta X = \left(\frac{\partial X}{\partial u}\right)_e u + \left(\frac{\partial X}{\partial \dot{u}}\right)_e \dot{u} + \left(\frac{\partial X}{\partial v}\right)_e v + \ldots + \left(\frac{\partial X}{\partial \dot{\delta}_a}\right)_e \dot{\delta}_a$$

$$\vdots$$

$$\vdots$$

$$\Delta N = \left(\frac{\partial N}{\partial u}\right)_e u + \left(\frac{\partial N}{\partial \dot{u}}\right)_e \dot{u} + \left(\frac{\partial N}{\partial v}\right)_e v + \ldots + \left(\frac{\partial N}{\partial \dot{\delta}_a}\right)_e \dot{\delta}_a$$

These derivatives are known as the *stability derivatives* or the *aerodynamic derivatives*. In their full form as shown above there are six states plus three controls and their derivatives, giving 18 aerodynamic derivatives to represent each external force or moment. Clearly, if all 18 derivatives in all six equations ($6 \times 18 = 108$) were used, the equations would become large and difficult to manipulate. Fortunately, for a wide range of flight states many of the derivatives are small and may be neglected. The rationale behind neglecting certain derivatives is as follows:

1. For any condition of symmetric flight (in the xz-plane) the asymmetric forces and moments (Y, L, N) are zero. It then follows that the derivatives of the asymmetric forces and moments with respect to the symmetric variables (u, w, q, δ_e) will be zero, i.e.,
$$\frac{\partial Y}{\partial u} = \frac{\partial Y}{\partial w} = \frac{\partial Y}{\partial q} = \frac{\partial Y}{\partial \delta_e} = \cdots = \frac{\partial L}{\partial u} = \cdots \frac{\partial N}{\partial \delta_e} = 0$$
2. Similarly, the derivatives of the symmetric forces and moments (X, Z, M) with respect to the asymmetric variables ($v, p, r, \delta_r, \delta_a$) are zero, i.e.,
$$\frac{\partial X}{\partial v} = \frac{\partial X}{\partial p} = \frac{\partial X}{\partial r} = \frac{\partial X}{\partial \delta_r} = \frac{\partial X}{\partial \delta_a} = \cdots = \frac{\partial Z}{\partial v} = \cdots \frac{\partial M}{\partial \delta_a} = 0$$
3. It has also been found through experiment and experience that the derivatives with respect to acceleration are all negligible except $(\partial M)/(\partial \dot{w})$, i.e.,
$$\frac{\partial X}{\partial \dot{u}} = \frac{\partial X}{\partial \dot{v}} = \frac{\partial X}{\partial \dot{w}} = \cdots = \frac{\partial N}{\partial \dot{r}} = 0$$
4. The control rate derivatives are all negligible:
$$\frac{\partial X}{\partial \dot{\delta}_e} = \frac{\partial Y}{\partial \dot{\delta}_e} = \frac{\partial Z}{\partial \dot{\delta}_e} = \cdots = \frac{\partial N}{\partial \dot{\delta}_a} = 0$$
5. Again, through experiment and experience, the following derivatives may also be neglected
$$\frac{\partial X}{\partial q} = \frac{\partial X}{\partial \delta_e} = \frac{\partial Y}{\partial p} = \frac{\partial Y}{\partial r} = \frac{\partial Y}{\partial \delta_a} = 0$$

The Linearized Equations of Motion

The simplified expressions for perturbation aerodynamic forces and moments, $\Delta X, \ldots, \Delta N$, can be substituted into equations (4.157)–(4.162) to give:

$$m\dot{u} = \frac{\partial X}{\partial u}u + \frac{\partial X}{\partial w}w - mg\cos\Theta_e\theta \tag{4.166}$$

$$m\dot{v} = \frac{\partial Y}{\partial v}v - mU_e r + mg\cos\Theta_e\phi + \frac{\partial Y}{\partial \delta_r}\delta_r \tag{4.167}$$

$$m\dot{w} = \frac{\partial Z}{\partial u}u + \frac{\partial Z}{\partial w}w + \frac{\partial Z}{\partial q}q + mU_e q - mg\sin\Theta_e\theta + \frac{\partial Z}{\partial \delta_e}\delta_e \tag{4.168}$$

$$I_{xx}\dot{p} - I_{xz}\dot{r} = \frac{\partial L}{\partial v}v + \frac{\partial L}{\partial p}p + \frac{\partial L}{\partial r}r + \frac{\partial L}{\partial \delta_r}\delta_r + \frac{\partial L}{\partial \delta_a}\delta_a \tag{4.169}$$

$$I_{yy}\dot{q} = \frac{\partial M}{\partial u}u + \frac{\partial M}{\partial w}w + \frac{\partial M}{\partial \dot{w}}\dot{w} + \frac{\partial M}{\partial q}q + \frac{\partial M}{\partial \delta_e}\delta_e \tag{4.170}$$

$$I_{zz}\dot{r} - I_{xz}\dot{p} = \frac{\partial N}{\partial v}v + \frac{\partial N}{\partial p}p + \frac{\partial N}{\partial r}r + \frac{\partial N}{\partial \delta_r}\delta_r + \frac{\partial N}{\partial \delta_a}\delta_a \tag{4.171}$$

Linearized Equations in Compact Form

On examination, it is clear that there are two sets of equations—longitudinal equations, (4.166), (4.168), (4.170), where u, w, q are controlled by δ_e, and the lateral/directional set, (4.167), (4.169), (4.171) where v, p, r are controlled by δ_r, and δ_a. The longitudinal and lateral/directional dynamics of the aircraft may be treated separately. The above equations may be manipulated and written in a more compact form.

Including the kinematic expression (4.164), and substituting for $\dot{w}$ in (4.170) using (4.168), the longitudinal linearized equations of motion may be written in matrix form as:

$$\begin{bmatrix} \dot{u} \\ \dot{w} \\ \dot{q} \\ \dot{\theta} \end{bmatrix} = \begin{bmatrix} X_u & X_w & 0 & -g\cos\Theta_e \\ Z_u & Z_w & Z_q + U_e & -g\sin\Theta_e \\ M_u^* & M_w^* & M_q^* & M_\theta^* \\ 0 & 0 & 1 & 0 \end{bmatrix} \begin{bmatrix} u \\ w \\ q \\ \theta \end{bmatrix} + \begin{bmatrix} 0 \\ Z_{\delta_e} \\ M_{\delta_e} \\ 0 \end{bmatrix} [\delta_e] \tag{4.172}$$

where

$$X_u = \frac{1}{m}\frac{\partial X}{\partial u},\quad X_w = \frac{1}{m}\frac{\partial X}{\partial w},\quad Z_u = \frac{1}{m}\frac{\partial Z}{\partial u},\quad Z_w = \frac{1}{m}\frac{\partial Z}{\partial w},\quad Z_q = \frac{1}{m}\frac{\partial Z}{\partial q},\quad Z_{\delta_e} = \frac{1}{m}\frac{\partial Z}{\partial \delta_e},$$

$$M_u^* = M_u + M_{\dot{w}}Z_u,\quad M_w^* = M_w + M_{\dot{w}}Z_w,\quad M_q^* = M_q + M_{\dot{w}}U_e,$$

$$M_\theta^* = -M_{\dot{w}}g\sin\Theta_e,\quad M_{\delta_e}^* = M_{\delta_e} + M_{\dot{w}}Z_{\delta_e},\quad \text{and}\quad M_u = \frac{1}{I_{yy}}\frac{\partial M}{\partial u},\ \text{etc.}$$

Decoupling equations (4.169) and (4.171) and including the kinematic expressions (4.163) and (4.165) the full set of linearized lateral/directional equations may be written in matrix form as:

$$\begin{bmatrix} \dot{v} \\ \dot{p} \\ \dot{r} \\ \dot{\phi} \\ \dot{\Psi} \end{bmatrix} = \begin{bmatrix} Y_v & 0 & -U_e & g\cos\Theta_e & 0 \\ L_v^* & L_p^* & L_r^* & 0 & 0 \\ N_{sv}^* & N_p^* & N_r^* & 0 & 0 \\ 0 & 1 & \tan\Theta_e & 0 & 0 \\ 0 & 0 & \sec\Theta_e & 0 & 0 \end{bmatrix} \begin{bmatrix} v \\ p \\ r \\ \phi \\ \Psi \end{bmatrix} + \begin{bmatrix} Y_{\delta_r} & 0 \\ L_{\delta_r}^* & L_{\delta_a}^* \\ N_{\delta_r}^* & N_{\delta_a}^* \\ 0 & 0 \\ 0 & 0 \end{bmatrix} \begin{bmatrix} \delta_r \\ \delta_a \end{bmatrix} \tag{4.173}$$

where

$$Y_v = \frac{1}{m}\frac{\partial Y}{\partial v}, \qquad Y_{\delta_r} = \frac{1}{m}\frac{\partial Y}{\partial \delta_r},$$

$$L_v^* = \frac{I_{xx}I_{zz}}{I_{xx}I_{zz} - I_{xz}^2}\left(L_v + \frac{I_{xz}}{I_{xx}}N_v\right), \qquad L_p^* = \frac{I_{xx}I_{zz}}{I_{xx}I_{zz} - I_{xz}^2}\left(L_p + \frac{I_{xz}}{I_{xx}}N_p\right), \text{ etc.,}$$

$$L_v = \frac{1}{I_{xx}}\frac{\partial L}{\partial v}, \qquad L_p = \frac{1}{I_{xx}}\frac{\partial L}{\partial p}, \text{ etc.,}$$

and

$$N_v^* = \frac{I_{xx}I_{zz}}{I_{xx}I_{zz} - I_{xz}^2}\left(N_v + \frac{I_{xz}}{I_{zz}}L_v\right), \quad N_p^* = \frac{I_{xx}I_{zz}}{I_{xx}I_{zz} - I_{xz}^2}\left(N_p + \frac{I_{xz}}{I_{zz}}L_p\right), \text{ etc.,}$$

$$N_v = \frac{1}{I_{zz}}\frac{\partial N}{\partial v}, \quad N_p = \frac{1}{I_{zz}}\frac{\partial N}{\partial p}$$

The Incidence Angles Expressed in Linear Form

The nonlinear expressions for angle of attack and angle of sideslip are:

$$\tan\alpha = \frac{W}{U} \quad \text{and} \quad \sin\beta = \frac{V}{V_f}$$

By making small-angle assumptions and applying the appropriate trim information, the following linear expressions are derived:

$$\alpha = \frac{w}{U_e} \quad \text{and} \quad \beta = \frac{v}{U_e}$$

The Nondimensional Linearized Equations of Motion

The equations of motion are often used in their nondimensional form. The main advantage of this is that the derivatives (which become coefficients) of different aircraft can be directly compared. The nondimensionalizing quantities used are listed in Table 4.8. Applying the nondimensionalizing

Dimensional quantity	Divisor	Nondimensional quantity
X, Y, Z	$\frac{1}{2}\rho V^2 S$	C_x, C_y, C_z
L, M, N	$\rho V^2 S l$	C_l, C_m, C_n
U, V, W	U_e	$\hat{u}, \beta, \alpha$
P, Q, R	$\frac{1}{t^*}$	$\hat{p}, \hat{q}, \hat{r}$
$\dot{\alpha}, \dot{\beta}$	$\frac{1}{t^*}$	$D\alpha, D\beta$
$\dot{\delta}_e, \dot{\delta}_a, \dot{\delta}_r$	$\frac{1}{t^*}$	$D\delta_e, D\delta_a, D\delta_r$
m	$\rho S l$	μ
$I_{xx}, I_{yy}, I_{zz}, I_{xz}$	$\rho S l^3$	$i_{xx}, i_{yy}, i_{zz}, i_{xz}$
t	t^*	$\hat{t}$

$D = \frac{d}{d\hat{t}}$, $t^* = \frac{l}{U_e}$, and $l = \frac{\bar{c}}{2}$ (longitudinal equations), and $l = \frac{b}{2}$ (lat/dir equations).

Table 4.8 Nondimensionalizing Factors

factors given in Table 4.8 to the longitudinal set of equations (4.172) gives the following set of nondimensional equations:

$$\begin{bmatrix} 2\mu & 0 & 0 & 0 \\ 0 & 2\mu & 0 & 0 \\ 0 & -C_{m_{\dot{\alpha}}} & i_{yy} & 0 \\ 0 & 0 & 0 & 1 \end{bmatrix} \begin{bmatrix} D\hat{u} \\ D\alpha \\ D\hat{q} \\ D\theta \end{bmatrix} = \begin{bmatrix} (C_{x_u} + 2C_{L_0}\tan\theta_e) & C_{x_\alpha} & 0 & -C_{L_0} \\ (C_{z_u} - 2C_{L_0}) & C_{z_\alpha} & 2\mu & -C_{L_0}\tan\theta_e \\ C_{m_u} & C_{m_u} & C_{m_q} & 0 \\ 0 & 0 & 1 & 0 \end{bmatrix} \begin{bmatrix} \hat{u} \\ \alpha \\ \hat{q} \\ \theta \end{bmatrix} + \begin{bmatrix} 0 \\ C_{z\delta_e} \\ C_{m\delta_e} \\ 0 \end{bmatrix} [\delta_e]$$

Similarly, the lateral/directional equations in nondimensional form are:

$$\begin{bmatrix} 2\mu & 0 & 0 & 0 & 0 \\ 0 & i_{xx} & -i_{xz} & 0 & 0 \\ 0 & -i_{xz} & i_{zz} & 0 & 0 \\ 0 & 0 & 0 & 1 & 0 \\ 0 & 0 & 0 & 0 & 1 \end{bmatrix} \begin{bmatrix} D\beta \\ D\hat{p} \\ D\hat{r} \\ D\phi \\ D\Psi \end{bmatrix} = \begin{bmatrix} C_{y\beta} & 0 & -2\mu & C_{L_0} & 0 \\ C_{l\beta} & C_{lp} & C_{lr} & 0 & 0 \\ C_{n\beta} & C_{np} & C_{n_r} & 0 & 0 \\ 0 & 1 & \tan\theta_e & 0 & 0 \\ 0 & 0 & \sec\theta_e & 0 & 0 \end{bmatrix} \begin{bmatrix} \beta \\ \hat{p} \\ \hat{r} \\ \phi \\ \Psi \end{bmatrix} + \begin{bmatrix} 0 & C_{y\delta r} \\ C_{l\delta a} & C_{l\delta r} \\ C_{n\delta a} & C_{n\delta r} \\ 0 & 0 \\ 0 & 0 \end{bmatrix} \begin{bmatrix} \delta_a \\ \delta_r \end{bmatrix}$$

The nondimensional form of the derivatives are for

$$C_{x_u} = \frac{\partial C_x}{\partial \hat{u}}, \text{ etc.}$$

Note that the relationship between the compact form of the derivatives given in equations (4.172) and (4.173) do not have a direct mapping to their nondimensional counterparts.

The Equations of Motion in State Space Form

It is clear from equations (4.172) and (4.173) that the linearized equations of may be written in state space form:

$$\dot{\mathbf{x}} = \mathbf{A}\mathbf{x} + \mathbf{B}\mathbf{u} \tag{4.174}$$

where $\mathbf{x}$ = the state vector (n)
$\mathbf{u}$ = the control vector (m)
$\mathbf{A}$ = the system matrix ($n \times n$)
$\mathbf{B}$ = the control matrix ($n \times m$)

for a system with n states and m controls. For the longitudinal equations in state space form:

$$\mathbf{A} = \begin{bmatrix} X_u & X_w & 0 & -g\cos\theta_e \\ Z_u & Z_w & (Z_q + U_e) & -q\sin\theta_e \\ M_u^* & M_w^* & M_q^* & M_\theta^* \\ 0 & 0 & 1 & 0 \end{bmatrix}, \quad \mathbf{B} = \begin{bmatrix} 0 \\ Z_{\delta_e} \\ M_{\delta_e}^* \\ 0 \end{bmatrix}, \quad \mathbf{x} = \begin{bmatrix} u \\ w \\ q \\ \theta \end{bmatrix}, \quad \mathbf{u} = [\delta_e]$$

and for the lateral equations in state space form

$$\mathbf{A} = \begin{bmatrix} Y_v & 0 & -U_e & g\cos\theta_e & 0 \\ L_v^* & L_p^* & L_r^* & 0 & 0 \\ N_v^* & N_p^* & N_r^* & 0 & 0 \\ 0 & 1 & \tan\theta_e & 0 & 0 \\ 0 & 0 & \sec\theta_e & 0 & 0 \end{bmatrix}, \quad \mathbf{B} = \begin{bmatrix} Y_{\delta_r} & 0 \\ L_{\delta_r}^* & L_{\delta_a}^* \\ N_{\delta_r}^* & N_{\delta_a}^* \\ 0 & 0 \\ 0 & 0 \end{bmatrix}, \quad \mathbf{x} = \begin{bmatrix} v \\ p \\ r \\ \phi \\ \psi \end{bmatrix}, \quad \mathbf{u} = \begin{bmatrix} \delta_r \\ \delta_a \end{bmatrix}$$

The response of the aircraft in terms of variables other than the state variables can be obtained by use of an *output equation* of the form

$$\mathbf{y} = \mathbf{C}\mathbf{x} + \mathbf{D}\mathbf{u} \tag{4.175}$$

where $\mathbf{y}$ = the output vector (p)
$\mathbf{C}$ = the output matrix ($p \times n$)
$\mathbf{D}$ = the direct matrix ($p \times m$)

for an output response with p states. Further, it is possible to obtain a transfer function relating an output, y_i, from the output vector, $\mathbf{y}$, and a single control, u_j, from the control vector, $\mathbf{u}$, using the expression:

$$\frac{Y_i(s)}{U_j(s)} = \mathbf{C}_i(s\mathbf{I} - \mathbf{A})^{-1}\mathbf{B}_j + d_{ij} \tag{4.176}$$

where $\mathbf{B}_j$ is the column of the $\mathbf{B}$ matrix associated with the control u_j (the jth column of $\mathbf{B}$), $\mathbf{C}_i$ is the row of the $\mathbf{C}_i$ matrix corresponding to the output y_i (the ith row of $\mathbf{C}$), and d_{ij} is the element of $\mathbf{D}$ associated with output y_i and control u_j.

4.14 Calculation of Aerodynamic Derivatives

The linearized equations of motion given by (4.172) and (4.173) are valid for any aircraft. The main factor that determines the dynamic characteristics of a particular aircraft will be the values of its aerodynamics or stability derivatives. These derivatives can be obtained experimentally by wind tunnel testing or extracted from data recorded in flight trials. There is, of course, a need to be able to estimate the value of derivatives from basic configurational information. The following analytical expressions for derivatives are based on readily obtainable aircraft data and given in dimensional and nondimensional coefficient form.

Force Coefficients

Referring to Figure 4.25, the aircraft X and Z external forces may be expressed as:

$$X = T - D + L\alpha \qquad Z = -L + D\alpha$$

assuming that the angle of attack, α, is small. Noting that if stability axes are used the angle of attack in the trim state is zero, the trim values of the X and Z forces are:

$$X_e = -L_e \qquad Z_e = -L_e$$

and nondimensionalizing gives:

$$C_x = C_T - C_D + \alpha C_L \qquad C_Z = -C_L = -\alpha C_D$$

and

$$C_{x_e} = -C_{T_e} - C_{D_0} \qquad C_{z_e} = -C_{L_0}$$

where C_{L_0} and C_{D_0} refer to the values of these coefficients at zero angle of attack.

The Longitudinal Derivatives

1. The u-derivatives:

$$X_w = \frac{\rho S U_e}{2m} C_{x\alpha} \qquad C_{x_u} = -2(C_{D_0} + C_{L_0} \tan\Theta_e) - M_e \frac{\partial C_D}{\partial M}$$

where M_e = the Mach number in equilibrium flight.

$$Z_u = \frac{\rho S U_e}{2m}(C_{Z_u} + 2C_{Z_e}) \qquad C_{Z_u} = -M_e \frac{\partial C_L}{\partial \hat{u}}$$

$$M_u = \frac{\rho S \bar{c} U_e}{2I_{yy}} C_{m_u} \qquad C_{M_u} = M_e \frac{\partial C_M}{\partial \hat{u}} + \rho U_e^2 \frac{\partial C_M}{\partial p_d}$$

where p_d = dynamic pressure.

2. The w-derivatives:

$$X_w = \frac{\rho S U_e}{2m} C_{x\alpha} \qquad C_{x_\alpha} = C_{L_0} - C_{D_\alpha}$$

$$Z_w = \frac{\rho S U_e}{2m} C_{z\alpha} \qquad C_{z_\alpha} = -C_{L_\alpha} - C_{D_0}$$

where

$$C_{L_\alpha} = \frac{\partial C_L}{\partial \alpha} \quad \text{and} \quad C_{D_\alpha} = \frac{\partial C_D}{\partial \alpha}$$

$$M_w = \frac{\rho S U_e \bar{c}}{2I_{yy}} C_{m\alpha} \qquad C_{m_\alpha} = C_{L_\alpha}(h - h_n)$$

3. The q-derivatives—tailplane contribution only:

$$M_q = \frac{\rho S \bar{c}^2 U_e}{4I_{yy}} C_{m_q} \qquad C_{m_q} = \frac{-2V_H a_t l_t}{\bar{c}}$$

$$Z_q = \frac{\rho S \bar{c} U_e}{2m} C_{z_q} \qquad C_{z_q} = -2V_H a_t$$

where $V_H = \frac{S_t l_t}{S\bar{c}}$ = tailplane volume ratio
S_t = tailplane area
l_t = tailplane moment arm
a_t = tailplane lift curve slope = $(\partial C_{L_t})/(\partial \alpha_t)$

4. The $\dot{w}$-derivative—tailplane contribution only:

$$M_{\dot{w}} = \frac{\rho S \bar{c}^2}{4I_{yy}} C_{m\dot{\alpha}} \qquad C_{M_{\dot{\alpha}}} = -2a_t V_H \frac{l_t}{\bar{c}} \frac{\partial \varepsilon}{\partial \alpha}$$

where ε = the tailplane downwash angle.

5. The δ_e-derivatives:

$$Z_{\delta_e} = \frac{\rho S U_e^2}{2m} C_{z\delta e} \qquad C_{z\delta e} = -a_e \frac{S_t}{S}$$

$$M_{\delta_e} = \frac{\rho S \bar{c} U_e^2}{2I_{yy}} C_{m\delta e} \qquad C_{m\delta e} = -a_e V_H$$

where $a_e = (\partial C_{L_t})/(\partial \delta_e)$.

The Lateral/Directional Derivatives

1. The v-derivatives—fin contribution only:

$$Y_v = \frac{\rho S U_e}{2m} C_{y\beta} \qquad C_{y\beta} = -a_F\left(1 - \frac{\partial\sigma}{\partial\beta}\right)\frac{S_F}{S}$$

$$L_v = \frac{\rho S b U_e}{2I_{xx}} C_{l\beta} \qquad C_{l\beta} = -a_F\left(1 - \frac{\partial\sigma}{\partial\beta}\right)\frac{S_F}{S}\frac{z_F}{b}$$

$$N_v = \frac{\rho S b U_e}{2I_{zz}} C_{n\beta} \qquad C_{n\beta} = a_F V_v\left(1 - \frac{\partial\sigma}{\partial\beta}\right)$$

where a_F = lift curve slope of the fin
σ = sidewash angle
S_F = fin area
z_F = height of fin mean aerodynamic center above x-axis
V_v = fin volume ratio = $(S_F l_F)/(S_b)$

2. The p-derivatives—fin contribution only:

$$Y_p = \frac{pU_e S b}{2m} C_{y_p} \qquad C_{y_p} = -a_F\frac{S_F}{S}\left(2\frac{z_F}{b} - \frac{\partial\sigma}{\partial\hat{p}}\right)$$

$$N_p = \frac{\rho U_e S b^2}{2I_{zz}} C_{np} \qquad C_{n_p} = a_F V_v\left(2\frac{z_F}{b} - \frac{\partial\sigma}{\partial\hat{p}}\right)$$

3. The r-derivatives—fin contribution only:

$$Y_r = \frac{\rho U_e S b}{4m} C_{y_r} \qquad C_{y_r} = -a_F\frac{S_F}{S}\left(2\frac{l_F}{b} - \frac{\partial\sigma}{\partial\hat{r}}\right)$$

$$L_r = \frac{\rho U_e S b^2}{4I_{xx}} C_{l_r} \qquad C_{y_r} = a_F\frac{S_F}{S}\frac{z_F}{b}\left(2\frac{l_F}{b} - \frac{\partial\sigma}{\partial\hat{r}}\right)$$

$$N_r = \frac{\rho U_e S b^2}{4I_{zz}} C_{n_r} \qquad C_{n_r} = -a_F V_v\left(2\frac{l_F}{b} - \frac{\partial\sigma}{\partial\hat{r}}\right)$$

where l_F = distance between fin aerodynamic center and aircraft center of gravity.

4. The δ_r-derivatives:

$$Y_{\delta_r} = \frac{\rho S U_e^2}{2m} C_{y\delta_r} \qquad C_{y\delta_r} = a_r\frac{S_F}{S}$$

$$L_{\delta_r} = \frac{\rho S U_e^2 b}{2I_{xx}} C_{l\delta_r} \qquad C_{l\delta_r} = a_r\frac{S_F}{S}\frac{z_F}{b}$$

$$N_{\delta_r} = \frac{\rho S U_e^2 b}{2I_{zz}} C_{n\delta_r} \qquad C_{l\delta_r} = -a_r V_v$$

where a_r = rudder effectiveness = $(\partial C_{Y_F} / \partial \delta_r)$.

5. The δ_a-derivatives: Simple expressions for these derivatives are not available; they are normally estimated from wind tunnel tests.

4.15 Aircraft Dynamic Stability

Prediction of Stability—General Theory

The stability of any dynamic system is obtained by consideration of its free (unforced) motion. For an aircraft, unforced motion implies that there should be no control inputs (i.e., $\mathbf{u} = 0$), such that the controls remain fixed at their trim value. The state space equation (4.126) becomes:

$$\dot{\mathbf{x}} = \mathbf{A}\mathbf{x} \tag{4.177}$$

Consider the general case where the aircraft has n degrees of freedom. Equation (4.169) will have a general solution:

$$\mathbf{x} = \mathbf{x}_0 e^{\lambda t}$$

Substitution of this solution into equation (4.177) yields the familiar eigenvalue problem:

$$(\lambda \mathbf{I} - \mathbf{A})\mathbf{x}_0 = 0 \tag{4.178}$$

In this expression $(\lambda \mathbf{I} - \mathbf{A})$ is an $(n \times n)$ matrix, while $\mathbf{x}_0$ is a vector of dimension (n). Equation (4.178) has a trivial solution ($\mathbf{x}_0 = 0$), or the more useful solution that the determinant of $(\lambda \mathbf{I} - \mathbf{A})$ should be zero, giving the *characteristic equation:*

$$|\lambda \mathbf{I} - \mathbf{A}| = 0 \tag{4.179}$$

the solution of which yields n eigenvalues, λ_i, $(i = 1, n)$. If these values of λ are substituted into equation (4.178), then for the eigenvalue λ_i:

$$(\lambda_i \mathbf{I} - \mathbf{A})\mathbf{x}_0 = 0 \tag{4.180}$$

Equation (4.180) can be solved for each eigenvalue, λ_i, to give a vector of amplitudes x_0, $i = 1, n$. This is known as the eigenvector, and its value can assist in the determining which state variables are influenced by each eigenvalue.

Equation (4.179) is a polynomial function of λ which when solved gives the system eigenvalues which can have real or imaginary values upon which the stability of the system is dependent.

Imaginary Eigenvalues: $\lambda = n \pm \omega i$

The general solution for a state i takes the form:

$$x_i = x_{0_i} e^{\lambda t} = x_{0_i} e^{(n \pm \omega i)t} = x_{0_i} e^{nt} e^{\pm \omega i t} = x_{0_i} e^{nt} (\cos \omega t \mp i \sin \omega t)$$

The response is an oscillation with angular frequency ω (the imaginary part of the eigenvalue). The amplitude of the oscillation will decrease provided the real part is negative. In this case the system is said to be *dynamically stable*. Conversely, should the real part be positive, then the amplitude of the oscillations will increase and the system will be *dynamically unstable*.

Real Eigenvalues

Now the general solution for a state i takes the form:

$$x_i = x_{0_i} e^{\lambda t}$$

If λ is negative then the response will be an exponential decay and the system is then *statically stable.* If the λ is positive then the response will be an exponential growth and the system is statically unstable.

Period and Time to Half or Double Amplitude

From above it is clear that for an imaginary eigenvalue, $\lambda = n \pm \omega i$, the imaginary part, ω, determines the period of the oscillation:

$$T = \frac{2\pi}{\omega} \text{s}$$

or, more generally,

$$T = \frac{2\pi}{\text{Im}(\lambda)} \text{s}$$

The damping of the mode is usually measured by the *time to half amplitude* in the case of a convergent mode or *time to double amplitude* in the case of a divergent mode. It can be shown that:

$$t_{\text{half}} \text{ or } t_{\text{double}} = \frac{0.69}{|n|} \text{s}$$

or, more generally,

$$t_{\text{half}} \text{ or } t_{\text{double}} = \frac{0.69}{|\text{Re}(\lambda)|}$$

Note that the expressions given above are valid where λ has been calculated from the dimensional equations of motion. When the nondimensional equations are being used they must be multiplied by the factor t^* to obtain values in seconds.

Prediction of Aircraft Stability

It is possible to predict aircraft stability using the methods described at the start of this subsection. For almost all fixed-wing aircraft the longitudinal and lateral/directional modes are uncoupled and so can be treated separately. This simplifies the problem as two independent lower-order systems may be analyzed. For aircraft such as helicopters that generate nonsymmetrical loads and exhibit heavy coupling between longitudinal and lateral directional modes, much higher-order system matrices are generated.

Aircraft Longitudinal Dynamic Stability

The longitudinal stability properties of an aircraft with the controls fixed (i.e., the elevator does not move, $\delta_e = 0$) can be determined by expressing equation (4.172) in the form:

$$\begin{bmatrix} \dot{u} \\ \dot{w} \\ \dot{q} \\ \dot{\theta} \end{bmatrix} = \begin{bmatrix} X_u & X_w & 0 & -q\cos\Theta_e \\ Z_u & Z_w & Z_q + U_e & -g\sin\Theta_e \\ M_u^* & M_w^* & M_q^* & M_\theta^* \\ 0 & 0 & 1 & 0 \end{bmatrix} \begin{bmatrix} u \\ w \\ q \\ \theta \end{bmatrix}$$

This is in the same form as equation (4.177) and hence the characteristic equation can be obtained from (4.179). If it is assumed that $\Theta_e = 0$ (level trimmed flight) and that $U_e >> Z_q$, then the characteristic equation may be written in the form:

$$\lambda^4 + A\lambda^3 + B\lambda^2 + C\lambda + D = 0 \tag{4.181}$$

where $A = -(M_q^* + X_u + Z_w)$

$$B = [-M_w^* U_e + M_q^*(Z_w + X_u) + X_u Z_w - X_w Z_u]$$

$$C = [gM_u^* + U_e(M_w^* X_u - M_u^* X_w) + M_q^*(X_w Z_u - X_u Z_w)]$$

$$D = g(M_w^* Z_u - M_u^* Z_w)$$

Attempting to factorize to find λ analytically is unrealistic. In general, the longitudinal equations will give two oscillatory modes, with the characteristic equation factorizing to give:

$$(\lambda^2 + 2\zeta_{ph}\omega_{n_{ph}}\lambda + \omega_{n_{ph}}^2)(\lambda^2 + 2\zeta_{sp}\omega_{n_{sp}}\lambda + \omega_{n_{sp}}^2) = 0 \tag{4.182}$$

Most aircraft exhibit these two classical longitudinal modes: the phugoid (*ph*) and the short-period (*sp*) modes. These modes are now discussed individually.

The Phugoid Mode

The phugoid mode is characterized by lightly damped oscillations in altitude and airspeed. The period can typically be from 10 s to around 2 minutes in the case of a large airliner. Inspection of the corresponding eigenvector reveals that this mode influences the aircraft speed, u, more than its angle of attack, α. The phugoid mode is therefore a pitching oscillation at almost constant angle of attack, often lightly damped and occasionally unstable.

Although an exact analytical representation of the phugoid characteristic equation is not practical, it is possible to obtain an approximation. Assuming that the pitch acceleration ($\dot{q}$) is small, the pitching moment equation becomes one of static balance and it can be shown that the phugoid characteristic equation can be approximated by:

$$\lambda^2 + \left[-X_u - \frac{M_u^*}{M_u^*}\left(\frac{g}{U_e} - X_w\right)\right]\lambda + \frac{g}{U_e}\left(-Z_u + Z_w\frac{M_u^*}{M_w^*}\right) = 0$$

The Short-Period Mode

The short-period mode is a very fast and heavily damped oscillation in pitch. Its period can be less than a second for highly maneuverable aircraft and no more than a few seconds for larger vehicles. The variable most influenced by the short-period oscillation is angle of attack with little or no change in airspeed. This is readily confirmed by inspection of the eigenvector.

Although an exact analytical representation of the short-period characteristic equation is not practical, it is possible to obtain an approximation. Assuming that the velocity changes very little during the short-period motion, it is possible to neglect the x-equation of motion, and it can be shown that the short-period characteristic equation can be approximated by:

$$\lambda^2 - (Z_w + M_q^*)\lambda + (Z_w M_q^* - U_e M_w^*) = 0$$

Aircraft Lateral/Directional Stability

The stick-fixed stability is investigated by assuming that the aileron and rudder controls are locked ($\delta_r = \delta_a = 0$). The lateral/directional equations of motion (4.173) then become

$$\begin{bmatrix} \dot{v} \\ \dot{p} \\ \dot{r} \\ \dot{\phi} \\ \dot{\Psi} \end{bmatrix} = \begin{bmatrix} Y_u & 0 & -U_e & g\cos\Theta_e & 0 \\ L_v^* & L_p^* & L_r^* & 0 & 0 \\ N_v^* & N_p^* & N_r^* & 0 & 0 \\ 0 & 1 & \tan\Theta_e & 0 & 0 \\ 0 & 0 & \sec\Theta_e & 0 & 0 \end{bmatrix} \begin{bmatrix} v \\ p \\ r \\ \phi \\ \Psi \end{bmatrix}$$

This is in the same form as equation (4.177), and hence the characteristic equation can be obtained from (4.179). If it is assumed that $\Theta_e = 0$ (level trimmed flight), then the characteristic equation may be written in the form:

$$\lambda(\lambda^4 + A\lambda^3 + B\lambda^2 + C\lambda + D) = 0 \tag{4.183}$$

where $A = -(L_p^* + N_r^* + Y_v)$

$B = [N_v^* U_e + Y_v(L_p^* + N_r^*) + L_p^* N_r^* - L_r^* N_p^*]$

$C = [-gL_v^* + U_e(L_v^* N_p^* - L_p^* N_v^*) + Y_v(L_r^* N_p^* - L_p^* N_r^*)]$

$D = g(L_v^* N_r^* - L_r^* N_v^*)$

Attempting to factorize to find λ analytically is unrealistic. In general, the lateral/directional equations factorize to give three modes, one of which is oscillatory. The factorized characteristic equation takes the form:

$$\lambda(\lambda - \lambda_s)(\lambda - \lambda_r)(\lambda^2 + 2\zeta_{dr}\omega_{n_{dr}}\lambda + \omega_{n_{dr}}^2) = 0 \tag{4.184}$$

Most aircraft exhibit these three classical modes: the spiral mode (*s*), the roll mode (*r*), and the Dutch roll mode (*dr*). These modes are now discussed individually.

The Spiral Mode

This mode can be described as a true banked turn (i.e., without sideslip). It can be stable or unstable and is usually very slow, with time to half or double amplitude typically many seconds. When the spiral mode is stable, the turn has increasing radius, and effectively a heading change occurs. When unstable, the radius decreases and a spiral motion occurs.

Because the spiral mode is effectively a true banked turn, there is little or no sideslip. In attempting to approximate the spiral mode eigenvalue, it is therefore possible to disregard the sideforce equation of motion. Expressing the angle of bank, ϕ, in terms of the turn rate, r (for a true banked turn it can be shown that $r/\phi = g/U_e$), an approximation for the spiral mode eigenvalue is found to be:

$$\lambda_s = -\frac{\frac{g}{U_e}(L_v^* N_r^* - N_v^* L_r^*)}{(L_v^* N_p^* - N^* v L_p^*)}$$

The Roll Mode

This stable and relatively fast mode heavily influences the aircraft's roll degree of freedom. Analytically, only the eigenvector elements associated with p and ϕ are usually of significance. This mode is often modeled as a singledegree of freedom rotation about the x-axis:

$$\dot{p} = L_p^* p$$

Making the substitution $p = p_0 e^{\lambda t}$ yields the common approximation:

$$\lambda_r = L_p^*$$

The Dutch Roll Mode

This motion consists essentially of sideslip, yaw, and rolling motions in combination. As the aircraft sideslips in one direction it yaws in the other, thus maintaining an almost linear flight path. This yawing/sideslipping motion causes the aircraft to roll in the same direction as the yaw. Generally this motion is stable and relatively heavily damped. Occasionally the mode can be unstable, causing serious handling deficiencies.

By using the knowledge that sideslip and yaw mirror one another in this mode (i.e., $\Psi = -\beta$) it is possible to reduce the full lateral/directional characteristic equation to:

$$\lambda^3 - (N_r^* + L_p^*)\lambda^2 + (N_\beta^* + L_p^* N_r^* - N_r^* L_r^*)\lambda + (N_p^* L_\beta^* - L_p^* N_\beta^*) = 0$$

where $L_\beta^* = U_e L_v^*$ and $N_\beta^* = U_e N_v^*$. Note that the roll mode is a factor of this cubic.

4.16 Aircraft Response to Controls and Atmospheric Disturbances

Response Calculation—A General Approach

From the state space representation of the aircraft dynamics it is possible to derive transfer functions relating output states to controls. The Laplace transform of the state equation:

$$\dot{\mathbf{x}} = \mathbf{A}\mathbf{x} + \mathbf{B}\mathbf{u} \tag{4.185}$$

gives the general form of the transfer function:

$$\frac{X(s)}{U(s)} = (s\mathbf{I} - \mathbf{A})^{-1}\mathbf{B} \tag{4.186}$$

for a state X and control U.

Longitudinal Response to Elevator

For longitudinal response the entries into equation (4.186) are:

$$\mathbf{x} = \begin{bmatrix} u \\ w \\ q \\ \theta \end{bmatrix}, \quad \mathbf{u} = [\delta_e], \quad \mathbf{A} = \begin{bmatrix} X_u & X_w & 0 & -g\cos\Theta_e \\ Z_u & Z_w & Z_q + U_e & -g\sin\Theta_e \\ M_u^* & M_w^* & M_q^* & M_\theta^* \\ 0 & 0 & 1 & 0 \end{bmatrix}, \quad \mathbf{B} = \begin{bmatrix} 0 \\ Z_{\delta_e} \\ M_{\delta_e}^* \\ 0 \end{bmatrix}$$

which on solution will give the four transfer functions:

$$\frac{u(s)}{\delta_e(s)}, \frac{w(s)}{\delta_e(s)}, \frac{q(s)}{\delta_e(s)}, \frac{\theta(s)}{\delta_e(s)}$$

which can be used to obtain the state response to inputs of elevator. These expressions are too cumbersome to present here, but it can be appreciated by consideration of equation (4.181) that, for example, they take the form:

$$\frac{u(s)}{\delta_e(s)} = \frac{f(s)}{s^4 + As^3 + Bs^2 + Cs + D}$$

where f is a polynomial function of s (of order 4 or less) with coefficients dependent on the derivatives $X_u \ldots M^*_{\delta_e}$. The form of the response becomes apparent on consideration of equation (4.182), which allows us to deduce that:

$$\frac{u(s)}{\delta_e(s)} = \frac{f(s)}{(s^2 + 2\zeta_{ph}\omega_{n_{ph}} s + \omega^2_{n_{ph}})(s^2 + 2\zeta_{sp}\omega_{n_{sp}} s + \omega^2_{n_{sp}})}$$

It is apparent from above that the response will have two components, a short-term response related to the short-period mode and a longer-duration response related to the phugoid mode. The two characteristic motions are superimposed on one another, but because the short-period mode is very fast and heavily damped while the phugoid is of a very much longer period, the effect of the short-period motion disappears quickly, leaving only the effect of the phugoid.

Because the phugoid and short period modes are widely separated in terms of their frequencies, it is possible to decouple them to obtain approximations.

Phugoid Response—An Approximation

Extending the approximate method detailed in Subsection 4.15, it is possible to derive approximate transfer functions which define the low-frequency phugoid response:

$$\frac{u(s)}{\delta_e(s)} = \frac{\frac{g}{U_e} Z_{\delta_e} - \frac{M^*_{\delta_e}}{M^*_w}\left[S\left(X_u - \frac{g}{U_e}\right) + \frac{g}{U_e}\right]}{s^2 + \left[-X_u - \frac{M^*_u}{M^*_w}\left(\frac{g}{U_e} - X_w\right)\right]s + \frac{g}{U_e}\left(-Z_u + Z_w \frac{M^*_u}{M^*_w}\right)}$$

$$\frac{w(s)}{\delta_e(s)} = \frac{-\frac{g}{U_e}\frac{M^*_u}{M^*_w} Z_{\delta_e} - \frac{M^*_{\delta_e}}{M^*_w}\left[s^2 - sX_u - \frac{g}{U_e} Z_u\right]}{s^2 + \left[-X_u - \frac{M^*_u}{M^*_w}\left(\frac{g}{U_e} - X_w\right)\right]s + \frac{g}{U_e}\left(-Z_u + Z_w \frac{M^*_u}{M^*_w}\right)}$$

$$\frac{q(s)}{\delta_e(s)} = \frac{-\frac{Z_{\delta_e}}{U_e}\left(s - X_u + \frac{M^*_u}{M^*_w} X_w\right) - \frac{M^*_{\delta_e}}{U_e M^*_w}[s^2 - (X_u + z_w)s - X_w Z_u + X_u Z_w]}{s^2 + \left[-X_u - \frac{M^*_u}{M^*_w}\left(\frac{g}{U_e} - X_w\right)\right]s + \frac{g}{U_e}\left(-Z_u + Z_w \frac{M^*_u}{M^*_w}\right)}$$

Short-Period Response—An Approximation

Extending the approximate method detailed in Subsection 4.15, it is possible to derive approximate transfer functions which define the higher-frequency short-period response:

$$\frac{w(s)}{\delta_e(s)} = \frac{(s - M_q^*)Z_{\delta_e} + U_e M_{\delta_e}^*}{s^2 - (Z_w + M_q^*)s + (Z_w M_q^* - U_e M_w^*)}$$

and

$$\frac{q(s)}{\delta_e(s)} = \frac{(s - Z_w)M_{\delta_e}^* + M_w^* Z_{\delta_e}}{s^2 - (Z_w + M_q^*)s + (Z_w M_q^* - U_e M_w^*)}$$

Recall that the assumption is that there is no change in forward velocity during this mode ($u = 0$) and as $q = \dot{\theta}$, $q(s) = s\theta(s)$.

Lateral/Directional Response

For lateral/directional response the entries into equation (4.186) are:

$$\mathbf{x} = \begin{bmatrix} v \\ p \\ r \\ \phi \\ \Psi \end{bmatrix}, \quad \mathbf{u} = \begin{bmatrix} \delta_r \\ \delta_a \end{bmatrix}, \quad \mathbf{A} = \begin{bmatrix} Y_u & 0 & -U_e & g\cos\Theta_e & 0 \\ L_v^* & L_p^* & L_r^* & 0 & 0 \\ N_v^* & N_p^* & N_r^* & 0 & 0 \\ 0 & 1 & \tan\Theta_e & 0 & 0 \\ 0 & 0 & \sec\Theta_e & 0 & 0 \end{bmatrix}, \quad \mathbf{B} = \begin{bmatrix} Y_{\delta_r} & 0 \\ L_{\delta r}^* & L_{\delta a}^* \\ N_{\delta r}^* & N_{\delta a}^* \\ 0 & 0 \\ 0 & 0 \end{bmatrix}$$

which on solution will give 10 transfer functions:

$$\frac{v(s)}{\delta_r(s)}, \ldots, \frac{\psi(s)}{\delta_a(s)}$$

which can be used to obtain the state response to inputs of rudder and aileron. These expressions are too cumbersome to present here, but it can be appreciated by consideration of equation (4.183) that, for example, they take the form:

$$\frac{v(s)}{\delta_r(s)} = \frac{f(s)}{s(s^4 + As^3 + Bs^2 + Cs + D)}$$

where f is a polynomial function of s (of order 5 or less) with coefficients dependent on the derivatives $Y_v \ldots N_{\delta_a}^*$. The form of the response becomes apparent on consideration of equation (4.184), which allows us to deduce that:

$$\frac{u(s)}{\delta_e(s)} = \frac{f(s)}{s(s - \lambda_s)(s - \lambda_r)(s^2 + 2\zeta_{dr}\omega_{n_{dr}} s + \omega_{n_{dr}}^2) = 0}$$

Again we can see that the response of the aircraft will be made up of the component modes spiral, roll, and Dutch roll.

Roll Response to Aileron

As mentioned in Subsection 4.15, the roll motion of an aircraft can often be treated as a single-degree of freedom system. The rolling equation of motion becomes

$$\dot{p} = L_p^* p + L_{\delta_a}^* \delta_a$$

from which the roll rate can be estimated from:

$$p(t) = \frac{L_{\delta_a}^*}{L_p^*}(1 - e^{-t/\tau_r})$$

where

$$\tau_r = -\frac{1}{L_p^*} = \text{the roll time constant}$$

Dutch Roll Response to Rudder

As in Subsection 4.15, making the assumption that the aircraft center of gravity follows a straight line in a Dutch roll motion, it is possible to ignore the y-equation of motion and derive the following approximate transfer functions relating roll rate, p, and sideslip angle, β, to rudder deflection, δ_r:

$$\frac{p(s)}{\delta_r(s)} = \frac{(s^2 - N_r^* s + N_\beta^*)L_{\delta_r}^* - N_{\delta_r}^*(L_\beta^* - sL_r^*)}{s^3 - (N_r^* + L_p^*)s^2 + (N_r^* L_p^* - L_r^* N_p^* + N_\beta^*)s + (N_p^* L_\beta^* - L_p^* N_\beta^*)}$$

$$\frac{\beta(s)}{\delta_r(s)} = \frac{-N_p^* L_{\delta_r}^* - N_{\delta_r}^*(s - L_p^*)}{s^3 - (N_r^* + L_p^*)s^2 + (N_r^* L_p^* - L_r^* N_p^* + N_\beta^*)s + (N_p^* L_\beta^* - L_p^* N_\beta^*)}$$

Response of Aircraft to Atmospheric Disturbances

Consider an aircraft suddenly immersed completely in a gust of constant velocity V_g with components $-u_g$ and $-w_g$ (the negative sign simply denotes that the gust is a headwind with an upward component). (See Figure 4.26.) We can write the equations of motion in the form

$$\begin{bmatrix} \dot{u} \\ \dot{w} \\ \dot{q} \\ \dot{\theta} \end{bmatrix} = \begin{bmatrix} X_u & X_w & 0 & -g\cos\theta_e \\ Z_u & Z_w & u_e & -g\sin\theta_e \\ M_u^* & M_w^* & M_q^* & M_\theta^* \\ 0 & 0 & 1 & 0 \end{bmatrix} \begin{bmatrix} u - u_g \\ w - w_g \\ q \\ \theta \end{bmatrix} + \begin{bmatrix} 0 \\ Z_{\delta_e} \\ M_{\delta_e}^* \\ 0 \end{bmatrix} [\delta_e]$$

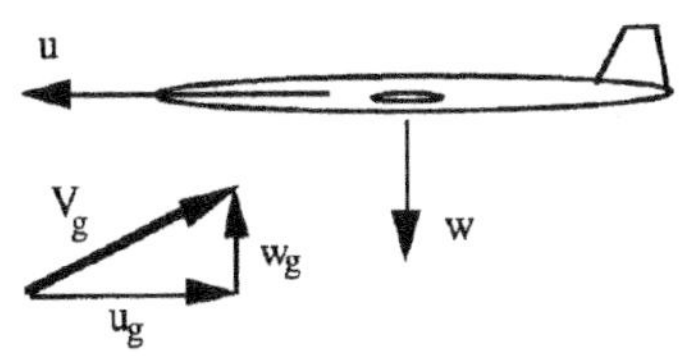

FIGURE 4.26 Aircraft acted upon by a gust.

or

$$\begin{bmatrix} \dot{u} \\ \dot{w} \\ \dot{q} \\ \dot{\theta} \end{bmatrix} = \begin{bmatrix} X_u & X_w & 0 & -g\cos\theta_e \\ Z_u & Z_w & u_e & -g\sin\theta_e \\ M_u^* & M_w^* & M_q^* & M_\theta^* \\ 0 & 0 & 1 & 0 \end{bmatrix} \begin{bmatrix} u \\ w \\ q \\ \theta \end{bmatrix} + \begin{bmatrix} 0 \\ Z_{\delta_e} \\ M_{\delta_e}^* \\ 0 \end{bmatrix} [\delta_e] + \begin{bmatrix} -X_u & -X_w \\ -Z_u & -Z_w \\ -M_u^* & -M_w^* \\ 0 & 0 \end{bmatrix} \begin{bmatrix} u_g \\ w_g \end{bmatrix}$$

which may be written as

$$\dot{\mathbf{x}} = \mathbf{A}\mathbf{x} + \mathbf{B}\mathbf{u} + \mathbf{E}V_g$$

where

$$V_g = \begin{bmatrix} u_g \\ w_g \end{bmatrix} = \text{gust vector}$$

and

$$\mathbf{E} = \begin{bmatrix} -X_u & -X_w \\ -Z_u & -Z_w \\ -M_u^* & -M_w^* \\ 0 & 0 \end{bmatrix} = \text{gust influence matrix}$$

For a sidewind of velocity $-v_g$ (i.e., from the starboard direction) we would have

$$\begin{bmatrix} \dot{v} \\ \dot{p} \\ \dot{r} \\ \dot{\phi} \\ \dot{\Psi} \end{bmatrix} = \begin{bmatrix} Y_u & 0 & -u_e & g\cos\theta_e & 0 \\ L_v^* & L_p^* & L_r^* & 0 & 0 \\ N_v^* & N_p^* & N_r^* & 0 & 0 \\ 0 & 1 & \tan\theta_e & 0 & 0 \\ 0 & 0 & \sec\theta_e & 0 & 0 \end{bmatrix} \begin{bmatrix} v \\ p \\ r \\ \phi \\ \Psi \end{bmatrix} + \begin{bmatrix} Y_{\delta_r} & 0 \\ L_{\delta r}^* & L_{\delta a}^* \\ N_{\delta r}^* & L_{\delta a}^* \\ 0 & 0 \\ 0 & 0 \end{bmatrix} \begin{bmatrix} \delta_r \\ \delta_a \end{bmatrix} + \begin{bmatrix} -Y_v \\ -L_v^* \\ -N_v^* \\ 0 \\ 0 \end{bmatrix} [v_g]$$

Note that for a large aircraft the distributed effect of the gust might impose rotations on the aircraft as well as translational velocities such that the V_g vector may include terms such as p_g and q_g.

Further Reading

Cook, M. V., *Flight Dynamics Principles,* Edward Arnold, London (1997).

Cook, M. V., *Flight Dynamics Principles: A Linear Systems Approach to Aircraft Stability and Control,* 3rd ed., Elsevier (2012).

Durham, W., *Aircraft Flight Dynamics and Control,* Wiley (2013).

Etkin, B., and Reid, L. D., *Dynamics of Flight Stability and Control,* 3rd ed., John Wiley & Sons, New York (1996).

Hancock, G. J., *An Introduction to the Flight Dynamics of Rigid Aeroplanes,* Ellis Horwood, Chichester (1995).

McCormick, B., W., *Aerodynamics, Aeronautics and Flight Mechanics,* John Wiley & Sons, New York (1979).

McLean, D. *Automatic Flight Control Systems,* Prentice Hall, London (1990).

McRuer, D., Ashkenas, I., and Graham, D. *Aircraft Dynamics and Automatic Control,* Princeton University Press, Princeton (1973).

Pallett, E. H. J., and Coyle, S. *Automatic Flight Control,* 4th ed., Blackwell Science, Oxford (1973).

Phillips, W. F., *Mechanics of Flight*, Wiley (2004).

Rolfe, K. M., and Staples, K. J. *Flight Simulation,* Cambridge University Press, Cambridge (1986).

PART 3

Computational Optimal Control

Rafał Żbikowski and Subchan Subchan

In the context of aerospace, optimal control (Bryson and Ho 1975) is the theory and practice of obtaining trajectories of vehicles by calculation of control inputs which minimize a performance criterion. Typical performance criteria are the flight time or fuel consumption and these criteria are evaluated over the whole trajectory so that optimal control problems lead to infinite-dimensional optimization problems. In engineering practice, these problems cannot be solved analytically and therefore relevant computer-based calculations must be judiciously carried out which has led to the advent of computational optimal control (Subchan and Żbikowski 2009) defined as a combination of control theory, optimization theory, numerical analysis, and software engineering applied to solve optimal control problems. Due to that combination, computational optimal control can be a demanding subject and this part aims to give a flavor of what a practitioner needs to know to become an informed user of the relevant tools.

This part is organized as follows. First, a general formulation of the optimal control problem is given in Subsection 4.17 together with characterization of the practical challenges implied by the constitutive elements of this formulation. Then, in Subsection 4.18 a brief summary of the most relevant parts of the optimal control theory are given in order to inform the presentation of the structure and capabilities of the computational optimal control tools described in the rest of the part. Subsection 4.19 is a succinct overview of the most capable numerical tools employed in the computational optimal control practice. Engineering experience of using these tools is described in Subsection 4.20 with an example of a real-world case study of trajectory shaping for a generic cruise missile.

4.17 Optimal Control Problem

Optimal control theory is a well-covered subject in books, e.g., Pontryagin et al. (1962), Athans and Falb (1966), Bryson and Ho (1975), Vinh (1981), Macki and Strauss (1982), Betts (2001), and survey papers, e.g., Pesch (1989a, 1989b), Pesch (1991), Pesch (1994), Hartl et al. (1995). Rather than treating the optimal control problem in depth, we focus here on the main elements of the problem formulation and the practical challenges implied by the presence of these elements.

The problem is to find control input as a function of time, $\boldsymbol{u} = \boldsymbol{u}(t)$, which minimizes the performance index:

$$\min_{\boldsymbol{u}\in U} J = \phi[\boldsymbol{x}(t_f), t_f] + \int_{t_0}^{t_f} \mathcal{L}[\boldsymbol{x}(t), \boldsymbol{u}(t), t]\, dt \tag{4.187}$$

with respect to the state vector functions:

$$\chi = \{\boldsymbol{x} : [0, t_f] \rightarrow \mathbb{R}^n \mid x_i, i = 1, \ldots., n, \text{ piecewise continuously differentiable}\}, \tag{4.188}$$

and the control vector functions:

$$U = \{\boldsymbol{u} : [0, t_f] \rightarrow U \subset \mathbb{R}^m \mid u_i, i = 1, \ldots, m, \text{ piecewise continuous}\} \tag{4.189}$$

subject to the following constraints:

$$\dot{x} = f(x(t), u(t)) \quad f: \mathbb{R}^{n+m} \to \mathbb{R}^n \tag{4.190}$$

$$x(0) = x_0 \in \mathbb{R}^n \quad x_0 \text{ known} \tag{4.191}$$

$$\psi(x(t_f), t_f) = \mathbf{0} \in \mathbb{R}^p \quad \psi: \mathbb{R}^n \times \mathbb{R}_+ \to \mathbb{R}^p, p \leq n, t_f, \text{ unknown} \tag{4.192}$$

$$C(x(t), u(t)) \leq \mathbf{0} \in \mathbb{R}^q \quad C: \mathbb{R}^{n+m} \to \mathbb{R}^q \tag{4.193}$$

$$S(x(t)) \leq \mathbf{0} \in \mathbb{R}^s \quad S: \mathbb{R}^n \to \mathbb{R}^s \tag{4.194}$$

Mathematically, the functions appearing in equations (4.187)–(4.194) are assumed to be sufficiently continuously differentiable with respect to their arguments. However, U allows discontinuities in controls and thus implies corners (cusps) in the states, so that x comprises piecewise smooth functions. This is a practical necessity, as many real-world applications of optimal control involve bang-bang type inputs. In other words, the optimal control solution pair $u^*(t)$ and $x^*(t)$ is not only infinite dimensional but may also be discontinuous (jumps in $u^* \in U$) or nonsmooth (cusps in $x^* \in X$).

The performance index is a scalar quality measure, e.g., fuel consumption, flight time, or other trajectory characteristics. It is worth noting that the performance index in equation (4.187) has two elements: (1) the function $\phi: \mathbb{R}^{n+1} \to \mathbb{R}$ which measures the influence of the terminal condition and (2) the integrand $\mathcal{L}: \mathbb{R}^{n+m} \to \mathbb{R}$, which measures the whole trajectory properties.

In practical aerospace problems, the presence of constraints equations (4.190)–(4.194) is essential for a meaningful problem formulation but leads to theoretical and computational challenges. The first fundamental equality constraint, equation (4.190), expresses the system's dynamics and is always present. Once the optimal control $u^* = u^*(t)$, which minimizes J is computed, the corresponding optimal state $x^* = x^*(t)$ is obtained by substituting $u^* = u^*(t)$ to equation (4.190).

Point Constraints The optimal state x^* must satisfy the boundary conditions equations (4.191) and (4.192) and these are point constraints, i.e., they act only at the extreme points of the trajectory: t_0 and t_f. By contrast, the path constraints equations (4.193) and (4.194) must be satisfied along the whole trajectory, i.e., on the time interval (t_0, t_f). One of the most remarkable features of optimal control problems is that changes in the terminal conditions equation (4.192) may have significant impact on the solution throughout the whole interval (t_0, t_f); for an example, see Figure 4.32.

Path Constraints The optimal control u^* and optimal state x^* are subject to path constraints equations (4.193) and (4.194). For mixed state-control constraints equation (4.193), either (i) $C = \mathbf{0}$ or (ii) $C < \mathbf{0}$, and the challenge is to determine the subintervals of (t_0, t_f) when the case (i) occurs. If the constraint is active at a given time t so that case (i) occurs, then equations (4.190) and (4.193) become a system of differential algebraic equations. Then the relationship between x and u is given by $C\,(x(t), u(t)) = 0$ which, in general, is an implicit equation to be solved numerically, consistently with equation (4.190). However, this is a numerically difficult problem because equation $C\,(x(t), u(t)) = 0$ then implicitly defines the state x as a function of control u, lowering the dimension of the original system of differential equations (4.190).

In the case of pure state constraints equation (4.194), either (i) $S = \mathbf{0}$, or (ii) $S < \mathbf{0}$ but the problem of active pure state constraint is significantly more challenging than for the active mixed constraint $C(x(t), u(t)) = 0$. Indeed, solving $S(x(t)) = 0$ cannot done in terms of the control $u(t)$ which makes the task of finding suitably optimal value for $u(t)$ a protracted and complex problem, see the subsubsection "Pure State Constraints." The presence of a pure state constraint equation (4.194) inevitably leads to difficult theoretical and numerical challenges and, if possible, such constraints should be avoided in the computational optimal control problem formulation.

Finally, it should be mentioned that equation (4.193) or (4.194) may be active on a subinterval of (t_0, t_f) or just at a point. In the former case, the constrained (active) subarc will be characterized by the entry time t_1 and the exit time t_2 with $t_0 \le t_1 < t_2 \le t_f$. In the latter case, the subarc collapses to a single (touch) point, $t_1 = t_2$.

Solution Strategies If problem equations (4.187)–(4.194) has a solution, it is infinite dimensional and a summary of the theory for finding such a solution is given in Subsection 4.18. In computational optimal control practice, there are two main approaches to finding the solution numerically as explained in Subsection 4.19. The indirect method approach preserves the infinite-dimensional character of the problem and reduces it to a two-point boundary value problem to be solved numerically, see the sub-subsection "Indirect Method Approach." The direct method approach approximates the continuous time interval with a set of discrete points, resulting in a large-scale finite-dimensional problem which, in principle, can be solved by the nonlinear programming methods, see the sub-subsection "Direct Method Approach."

4.18 Variational Approach to Optimal Control Problem Solution

Mathematically, the theory of optimal control (Bryson and Ho 1975) can be seen as an extension of the calculus of variations (Elsgolc 1962), where the optimal control solution $u^* = u^*(t)$ is the extremal minimizing the functional equation (4.187). Hence, the necessary conditions for the extremal are derived from the first variation of the performance index J with the constraints adjoined by the Lagrangian method. Due to the infinite-dimensional character of the problem, the Lagrange multipliers are now functions of time, $\lambda = \lambda(t)$. In the optimal control theory, these generalized Lagrange multipliers are called co-states in analogy to the system state $x = x(t)$. The necessary conditions lead to a two-point boundary value problem (TPBVP) which couples the state and co-state equations together with the initial and terminal conditions:

- State equation

$$\dot{x} = f[x(t), u(t), t] \tag{4.195}$$

- Co-state equation

$$-\dot{\lambda}^T = \frac{\partial \mathcal{L}}{\partial x} + \lambda^T \frac{\partial f}{\partial x} \tag{4.196}$$

- Stationarity condition

$$0 = \frac{\partial \mathcal{L}}{\partial u} + \lambda \frac{\partial f}{\partial u} \tag{4.197}$$

- Boundary condition

$$\left[\frac{\partial \phi}{\partial \mathbf{x}} + \mathbf{v}^T \frac{\partial \psi}{\partial \mathbf{x}} - \lambda^T\right]_{t_f} \delta x_f + \left[\frac{\partial \phi}{\partial t_f} + \mathbf{v}^T \frac{\partial \psi}{\partial t_f} + H\right] \delta t_f = 0 \tag{4.198}$$

where $H = \mathcal{L} + \lambda^T f$

If the final time and the final state are free, the boundary conditions equation (4.198) become:

$$\lambda(t_f)^T = \frac{\partial \phi}{\partial x(t_f)} + \mathbf{v}^T \frac{\partial \psi}{\partial x(t_f)}$$

$$0 = \frac{\partial \phi}{\partial t_f} + \mathbf{v}^T \frac{\partial \psi}{\partial t_f} + H(t_f)$$

In the optimal control theory, the Lagrangian approach equations (4.195)–(4.198) is conveniently recast (Pontryagin et al. 1962) in the Hamiltonian framework:

$$H(x, u, \lambda) \triangleq \mathcal{L}(x, u) + \lambda^T f(x, u) \tag{4.199}$$

$$\Phi(x, t, v) \triangleq \phi(x, u) + v^T \psi(x, u) \tag{4.200}$$

where H is the Hamiltonian, $\lambda : [0, t_f] \to \mathbb{R}^n$ are co-states, and $v \in \mathbb{R}^p$ are adjoint variables. Then, the necessary conditions (see Bryson and Ho (2000) and Pesch (1994)) become:

- differential equations of Euler-Lagrange

$$\dot{x}^T = H_\lambda = \frac{\partial H}{\partial \lambda} \tag{4.201}$$

$$\dot{\lambda}^T = -H_x = -\frac{\partial H}{\partial x} \tag{4.202}$$

- minimum principle

$$u = \arg \min_{u \in \mathcal{U}} H(x, \lambda) \tag{4.203}$$

- transversality conditions

$$\lambda^T(t_f) = \Phi_x|_t = t_f \tag{4.204}$$

$$(\Phi_t + H)|_t = t_f = 0 \tag{4.205}$$

Solving equations (4.201)–(4.205) means finding the optimal control u^* as a function of the optimal state x^* and optimal co-state λ^*:

$$u^* = u^*(x^*, \lambda^*) \tag{4.206}$$

as suggested by equation (4.203). If u appears nonlinearly in H, see equation (4.199) for the relevant definition, the solution can be computed from the equation $H_u = 0$, where H_u stands for $\partial H/\partial u$; otherwise, the general "arg min" form of equation (4.203) must be used. Finally, $H_{uu} > 0$ (positive definite Hessian) is a sufficient condition for the necessary condition $H_u = 0$ to be the optimal solution minimizing equation (4.187).

The system of equations (4.201–4.205) was obtained under the assumption that there were no path constraints, i.e., in the absence of conditions equation (4.193) or (4.194). In aerospace engineering practice, such constraints are almost always present and hence must be additionally considered. The influence of different kinds of constraints on the necessary modifications of equations (4.201)–(4.205) is summarized below in the sub-subsections "Pure Control Constraints," "Mixed State-Control Constraints," and "Pure State Constraints," arranged in the growing degree of difficulty in obtaining the optimal solution. In other words, the easiest case is the presence of pure control constraints only, see the sub-subsection "Pure Control Constraints," and the most challenging case is the presence of pure state constraints, see the sub-subsection "Pure State Constraints."

Pure Control Constraints

This case occurs when equation (4.193) simplifies to the dependence on the control $u(t)$ only:

$$C(u(t)) \leq 0 \tag{4.207}$$

As an illustrative example, consider $\boldsymbol{u}$ to be scalar ($m = 1$) and to appear linearly in H so that equation (4.199) becomes:

$$H(x, u, \lambda) = H_1(x, \lambda) + uH_2(x, \lambda) \tag{4.208}$$

It is also assumed that u obeys a simple box bounding constraint $U = [u_{\min}, u_{\max}]$ so $H_u = 0$ does not determine the optimal control solution and thus the "arg min" form of equation (4.203) must be used. If the second term $\boldsymbol{H}_2(x, \lambda)$ does not vanish identically on $[0, t_f]$, the minimum principle equation (4.203) yields:

$$u^*(t) = \begin{cases} u_{\max} & \text{when } H_2 < 0 \\ u_{\min} & \text{when } H_2 > 0 \end{cases}$$

and H_2 is the switching function, because the optimal control solution $u^* = u^*(t)$ switches between the constant values $u_{\min}$ and $u_{\max}$ according to the sign of H_2, resulting in the "bang-bang" control. In other words, for all times t when $H_2 < 0$, the optimal control solution $u^*(t)$ is held constant at $u_{\max}$, and for all times t when $H_2 > 0$, $\boldsymbol{u}^*(t)$ is held constant at $u_{\min}$.

If H_2 vanishes on a subinterval of $[0, t_f]$, the optimal control solution is a singular sub-arc (Bryson and Ho 1975, Chapter 8) computed by successive differentiation of the switching function H_2 with respect to time t until the control variable appears explicitly.

Mixed State-Control Constraints

The modification of the Hamiltonian due to the presence of mixed constraints equation (4.193) is more involved than in the pure control constraints case of the sub-subsection "Pure Control Constraints" but follows a similar approach. For illustrative purposes, it is assumed that $m = q = 1$ in equation (4.193) so that a scalar control u and scalar condition C are considered. Then the augmented Hamiltonian becomes

$$H(x, u, \lambda) = \mathcal{L}(x, u) + \lambda^T f(x, u) + \mu C(x, u) \tag{4.209}$$

The Lagrangian parameter μ is

$$\mu = \begin{cases} 0 & \text{if } C < 0 \\ \mu \geq 0 & \text{if } C = 0 \end{cases}$$

and the Euler-Lagrange equations become

$$\dot{\lambda}^T = -H_x = \begin{cases} \mathcal{L}_x - \lambda^T f_x & \text{if } C < 0 \\ \mathcal{L}_x - \lambda^T f_x - \mu C_x & \text{if } C = 0 \end{cases}$$

The control $u(t)$ along the constrained arc can be derived from the mixed constraints:

$$C(x, u) = 0 \text{ for all } t \text{ with } t_1 \leq t \leq t_2 \text{ and } t_1 < t_2 \tag{4.210}$$

and the control variable can be represented by a function

$$u = u(x) \tag{4.211}$$

if equation (4.210) can be uniquely solved for u. If $C_u \neq 0$, the multiplier μ is given by equation (4.203):

$$H_u \triangleq \mathcal{L}_u + \lambda^T f_u + \mu\, C_u \tag{4.212}$$

Pure State Constraints

The necessary modifications of equations (4.201)–(4.205) due to the presence of pure control equation (4.207) or mixed equation (4.193) constraints, summarized in the sub-subsections "Pure Control Constraints" and "Mixed State-Control Constraints," are qualitatively different from the case of pure state constraints equation (4.194) considered in this section. Both equations (4.207) and (4.193) constrained the control $\boldsymbol{u}$ directly so that the search for the thus-constrained optimal control $\boldsymbol{u}^*$ could proceed in more or less explicit manner. However, condition equation (4.194) constraints directly only the state and it is not immediately clear how the control is affected by that state constraint. Hence, the pure state constraint equation (4.194) has to be manipulated further in order to bring out explicitly its constraining influence on the control. One way of uncovering that influence is through Bryson's formulation (Bryson and Ho 1975, Section 3.11) which is illustrated for $m = s = 1$ in equation (4.194), or scalar u and S, and the constraint being active on a subinterval $[t_1, t_2]$:

$$S(x\,(t)) = 0 \quad \text{for all} \quad t \in [t_1, t_2] \subset [t_0, t_f] \tag{4.213}$$

Successive total time derivatives of equation (4.194) are taken and substituted into $\boldsymbol{f}\,(x\,(t),\, u\,(t))$, until explicit dependence on u is obtained on $[t_1, t_2]$

$$S(x) = 0,\, S^{(1)}(x) = 0, \ldots, S^{(r-1)}(x) = 0 \tag{4.214}$$

$$\text{with} \quad S^{(r)}(x, u) = 0 \tag{4.215}$$

The smallest non-negative number r such that equation (4.215) holds is called the order of the state constraint. The end-result of the successive differentiations equations (4.214)–(4.215) is that $S^{(r)}(x, u)$ becomes a mixed constraint so, by analogy with equation (4.209), the augmented Hamiltonian is

$$H(x, u, \boldsymbol{\lambda}, \mu) = \mathcal{L} + \boldsymbol{\lambda}^T \boldsymbol{f} + \mu S^{(r)} \tag{4.216}$$

and it follows that

$$\mu = \begin{cases} 0 & \text{if } S^{(r)} < 0 \\ \mu \geq 0 & \text{if } S^{(r)} \equiv 0 \end{cases}$$

The control u on the constrained arcs can be derived from equation (4.215) and μ from equation (4.203). The right-hand sides of the differential equations for the adjoint variables equation (4.202) are to be modified along $[t_1, t_2]$. However, it must be guaranteed that not only equation (4.125) but also equation (4.214) is satisfied, so the following entry conditions must be additionally fulfilled:

$$(S\,(x\,(t_1)),\, S^{(1)}\,(x\,(t_1)), \ldots, S^{(r-1)}(x\,(t_1)) = \mathbf{0} \tag{4.217}$$

As a result of the need for equation (4.217), the corresponding co-state λ generally is discontinuous at t_1 and continuous at t_2.

4.19 Numerical Solution of the Optimal Control Problem

There are two main approaches to numerical solution of the optimal control problem. The indirect method approach proceeds in two steps: (1) first an appropriate two-point boundary value problem (**TPBVP**) is formulated using the theory outlined in Subsection 4.18 and (2) then the resulting **TPBVP** is solved numerically (Ascher et al. 1995; Stoer and Bulirsch 2002). By contrast, the direct method approach sidesteps the need for derivation of the variational equations of Subsection 4.18 by directly discretizing the original formulation equations (4.187)–(4.194) and solving the resulting problem as a large-scale nonlinear programming problem (Bazaraa et al. 1993; Gill 2002). The indirect method approach is summarized in the sub-subsection "Indirect Method Approach" and the direct method approach is briefly presented in the sub-subsection "Direct Method Approach."

Indirect Method Approach

In the indirect method approach, the original problem equations (4.187)–(4.194) is not solved directly but replaced with equations (4.201)–(4.205) which constitute a two-point boundary value problem (TPBVP) to be solved numerically by multiple shooting.

A standard illustration of the TPBVP is the example of firing a shell. Given the initial barrel orientation and the initial shell speed, the corresponding trajectory can be computed by solving the corresponding initial value problem (IVP). However, if both the initial and terminal conditions are specified, then this is a TPBVP: the trajectory must be a solution of the defining differential equation, but must pass through prescribed points at both ends.

The numerical method of shooting solves the TPBVP by repeated use of well-designed (numerically convergent) IVP solvers (Stoer and Bulirsch 2002, Section 7.2). A guess of the initial point is made and the corresponding terminal point is computed and that guess is optimally modified to become the starting point for the next use of an IVP solver. This process is repeated until the shooting error is acceptably low but the numerical convergence of shooting is a challenging problem (Stoer and Bulirsch 2002, Section 7.3) despite the use of well-designed (numerically convergent) IVP solvers.

An illustration of the shooting method for a second-order equation $\ddot{x} = f(t, x, \dot{x})$ are given on Figure 4.27. The initial position $x_0 = x(t_0)$ is fixed and so is the terminal one $x_f = x(t_f)$. Thus the initial

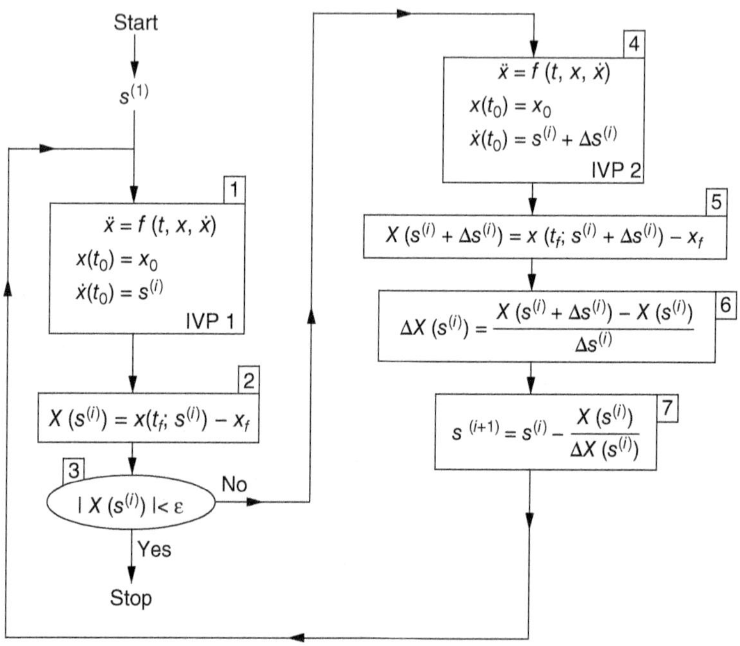

Figure 4.27 Shooting method flowchart for a second-order equation with boundary conditions $x_0 = x(t_0)$ and $x_f = x(t_f)$.

speed $\dot{x}(t_0)$ has to be iteratively modified until the end of the trajectory is within the desired accuracy ε. The first guess $s^{(1)}$ of the initial speed $\dot{x}(t_0)$ is made to start the procedure and the corresponding initial value problem (IVP 1) is solved (block 1). The error X between the obtained terminal value $x(t_f; s^{(1)})$ and the desired one $x(t_f)$ is formed (block 2) and checked against the desired accuracy ε (block 3). If the accuracy requirement is met, the desired trajectory has been found; if not, then the guess of the initial speed must be improved. The improvement is based on the idea that, ideally, the error X should be zero. In other words, we should try to find a value of the guess $s^{(i)}$ of the initial speed $\dot{x}(t_0)$ which yields $X(s^{(i)}) = x(t_f; s^{(i)}) - x_f = 0$. This is done by the well-known Newton procedure in block 7. The preceding blocks 4–6 perform the auxiliary computations: block 6 is the approximation ΔX of the derivative of X and needs the results of blocks 4 and 5. As a consequence, another IVP must be solved (block 4), so that the computation becomes more expensive.

The main drawback of the shooting method is the sensitivity to the initial guess, because of the use of Newton's iteration (block 7). To overcome this problem, the trajectory must be split up into subintervals and one must apply the same shooting method for each subinterval which results in the method of multiple shooting (Stoer and Bulirsch 2002, Section 7.3); see Figure 4.28. The multiple shooting method is the cornerstone of modern TPBVP solvers and underpins the state-of-the-art BNDSCO optimal control solver (Oberle and Grimm 1989). An additional difficulty is that in constrained optimal control problems, the jump and switching conditions on the co-state or control variables often occur so additional nodes must be inserted to improve convergence:

$$\dot{x}(t) = f_k(t, x(t)), \quad \xi_k \le t \le \xi_{k+1}, \quad 0 \le k \le s \tag{4.218a}$$

$$\boldsymbol{x}(\xi_k^+) = \boldsymbol{h}_k(\xi_k, \boldsymbol{x}(\xi_k^-)), \quad \text{for} \quad k = 1, \ldots, s \tag{4.218b}$$

$$r_i(\boldsymbol{x}(t_0), \boldsymbol{x}(t_f)) = 0, \quad \text{for} \quad 1 \le i \le n_1 \tag{4.218c}$$

$$r_i(\xi_{k_i}, \boldsymbol{x}(\xi_{k_i}^-)) = 0, \quad \text{for} \quad i = n_1 + 1, \ldots, n + s \tag{4.218d}$$

Here, ξ_k, with $k = 1, \ldots, s$, are switching points whilst $\boldsymbol{S}_j$ are the initial guesses for the $\boldsymbol{x}(t_j)$ and Ξ_j are the initial guesses for the switching points ξ_j. The approach is to compute simultaneously

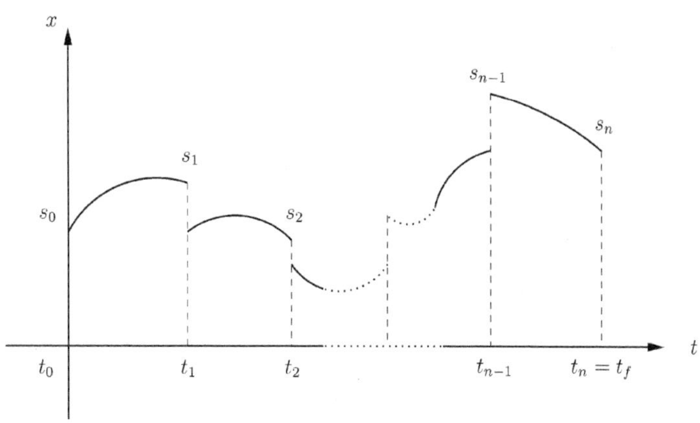

FIGURE 4.28 Multiple shooting.

the solution $x(t)$ and its switching points ξ using a modified Newton method (Deuflhard 1976). However, even the best TPBVP solver cannot overcome the fundamental problem of a narrow convergence interval inherent in TPBVP.

Direct Method Approach

The direct method approach deals with the original optimal control problem equations (4.187)–(4.194) which is discretized into a large-scale nonlinear programming (NLP) and then solved using an NLP solver (Nocedal and Wright 2006). Several discretization have been proposed, e.g., Enright and Conway (1992), Elnagr (1995), Betts (2001), Fahroo and Ross (2002). Two main approaches to the discretization of the state and/or control are briefly reviewed below.

Direct Collocation Approach

The first step in discretization (von Stryk and Bulirsch 1992) is to divide the time interval of the trajectory into subintervals by introducing nodes:

$$t_0 = t_1 < t_2 < t_3 \ldots < t_k = t_f \tag{4.219}$$

The state and control variables at each node are denoted by $x_j = x(t_j)$ and $u_j = u(t_j)$ so that the $2k$-dimensional vector Y of the NLP variables is

$$Y = [u(t_1), \ldots, u(t_k), x(t_1), \ldots, x(t_k)] \tag{4.220}$$

In order to find the values of the optimal trajectory at the time instants between the nodes equation (4.219), the controls are chosen as piecewise linear interpolating functions between $u(t_j)$ and $u(t_{j+1})$ for $t_j \le t \le t_{j+1}$ as follows:

$$u_{\text{app}}(t) = u(t_j) + \frac{t - t_j}{t_{j+1} - t_j}[u(t_{j+1}) - u(t_j)] \tag{4.221}$$

The value of the control variables in between two nodes (at the center) is given by

$$u(t_{c,j}) = \frac{u(t_j) + u(t_{j+1})}{2} \tag{4.222}$$

The piecewise linear interpolation is used to handle discontinuities in control.

The state variable $x(t)$ is approximated by a continuously differentiable and piecewise Hermite-Simpson cubic polynomial between $x(t_j)$ and $x(t_{j+1})$ on the interval $t_j \le t \le t_{j+1}$ of length q_j:

$$x_{\text{app}}(t) = \sum_{r=0}^{3} c_r^j \frac{t - t_j}{p_j} \tag{4.223}$$

$$c_0^j = x(t_j)$$

$$c_1^j = q_j f_j$$

$$c_2^j = -3x(t_j) - 2q_j f_j + 3x(t_{j+1}) - q_j f_{j+1}$$

$$c_3^j = 2x(t_j) + 2q_j f_j - 2x(t_{j+1}) + q_j f_{j+1}$$

where

$$f_j = f(x(t_j), u(t_j), t_j), \qquad q_j = t_j+1 - t_j$$
$$t_j \le t \le t_{j+1}, \qquad j = 1, \ldots, k-1$$

The value of the state variables at the center point of the cubic approximation is

$$x_{c,j} = \frac{x(t_j)+x(t_{j+1})}{2} + \frac{q(f(t_j)+f(t_{j+1}))}{8} \tag{4.224}$$

and the derivative is

$$\frac{dx_{c,j}}{dt} = -\frac{3(x(t_j)+x(t_{j+1}))}{2q} - \frac{q(f(t_j)+f(t_{j+1}))}{4} \tag{4.225}$$

In addition, the chosen interpolating polynomial for the state and control variables must satisfy the midpoint conditions for the differential equations as follows:

$$f\,[x_{\text{app}}(t_{c,j}), u_{\text{app}}(t_{c,j}), t_{c,j}] - \dot{x}_{\text{app}}(t_{c,j}) = 0 \tag{4.226}$$

Hence, equations (4.187)–(4.194) are discretized into the following NLP problem:

$$\min f(\boldsymbol{Y}) \tag{4.227}$$

subject to

$$f\,(x_{\text{app}}(t), u_{\text{app}}(t), t) - \dot{x}_{\text{app}} = 0 \tag{4.228}$$

$$x_{\text{app}}(t_1) - x_1 = 0 \tag{4.229}$$

$$\psi\,(x_{\text{app}}(t_k), (t_k) = 0 \tag{4.230}$$

$$C\,(x_{\text{app}}(t), u_{\text{app}}(t), t) \le 0 \tag{4.231}$$

$$S\,(x_{\text{app}}(t), t) \le 0 \tag{4.232}$$

where $x_{\text{app},}\, u_{\text{app}}$ are the approximation of the state and control, constituting $\boldsymbol{Y}$ in equation (4.227). Discretization equations (4.227)–(4.232) has been implemented in the DIRCOL package which uses the state-of-the-art SNOPT solver (Gill et al. 2002) as its main numerical engine.

In the NUDOCCCS package (Buskens and Maurer 2000), only the control is discretized so an NLP solver is used with respect to the discretized control only. The corresponding discretized state variables can be determined recursively using a numerical integration scheme.

One of the main advantage of DIRCOL and NUDOCCCS is that both packages calculate an approximation for the co-state variables λ. Computation of co-states is not needed in the direct method approach so this capability of DIRCOL and NUDOCCCS is an optional extra but its main advantage is that the co-state approximation can be used to improve convergence of multiple shooting solvers. It is good practice first to run DIRCOL to generate an initial approximation of u^*, x^*, and λ^*, and then use these approximation as the initial guess for BNDSCO (see the sub-subsection "Indirect Method Approach") which should then coverage well to a highly accurate solution.

Pseudospectral Method

An alternative to the direct collocation discretizations is the Legendre pseudospectral method; see Elnagar et al. (1995), Fahroo and Ross (2002), Benson (2005). This method is based on the spectral collocation in which the trajectory for the state and control variables are approximated by the Nth degree Lagrange interpolating polynomial. The value of the variables at the interpolating nodes is the unknown coefficients which in this technique are the Legendre-Gauss-Lobatto points t_i, $i = 0, \ldots, N$ distributed on the interval $\tau \in [-1, 1]$. These points can be given by $t_0 = -1$, $t_N = 1$ and for $1 \le i \le N - 1$, t_i are the zeros of $\dot{\Lambda}_N$, which is the derivative of the Legendre polynomial, Λ_N. The transformation between the LGL domain $\tau \in [-1, 1]$ and the physical domain $t \in [t_0, t_f]$ can be defined by

$$\tau = \frac{\tau_f - \tau_0}{2} t + \frac{\tau_f - \tau_0}{2} \tag{4.233}$$

The approximation for the state and control variables at the LGL points are given by the Nth degree Lagrange interpolating polynomial:

$$x(t) \approx \mathbf{X}(t) = \sum_{i=0}^{N} x(t_i) L_i(t) \tag{4.234}$$

$$u(t) \approx \mathbf{U}(t) = \sum_{i=0}^{N} u(t_i) L_i(t) \tag{4.235}$$

where $L_i(t)$ are the Lagrange interpolating polynomial of order N and is defined by

$$L_i(t) = \frac{1}{N(N+1)\Lambda_N(t_i)} \frac{(t^2 - 1)\dot{\Lambda}_N(t)}{t - t_i} = \begin{cases} 1 & \text{if } l = k \\ 0 & \text{if } l \neq k \end{cases} \tag{4.236}$$

The derivative of equation (4.234) is given by

$$\dot{x}(t_k) \approx \dot{\mathbf{X}}(t_k) = \sum_{i=0}^{N} x(t_i) \dot{L}_i(t_k) \tag{4.237}$$

where $\dot{L}_i(t_k)$ are the entries of the pseudospectral Legendre derivative matrix.

The Gauss-Lobatto quadrature rule is used to discretize the performance index equation (4.187), where w_k are the LGL weights. Finally, the boundary conditions are defined by the approximating of the state variables at $\mathbf{X}_1$ and $\mathbf{X}_N$:

$$\psi(\mathbf{X}_1, \mathbf{X}_N) = 0 \tag{4.238}$$

The pseudospectral method has been implemented in commercially available software DIDO (see Ross and Fahroo (2002)) and is (in principle) capable of producing estimates of co-states. However, the study (Benson 2005) shows that DIDO does not work for the pure state constraint case, even for the simple benchmark problem of Bryson and Ho (1975).

4.20 User Experience

The computational optimal control approaches described in the previous subsections were tested in detail by the authors on a realistic case study of trajectory shaping for a generic cruise missile (Subchan and Żbikowski 2009).

Case Study

The problem is to find the trajectory of a generic cruise missile from the assigned initial state to a final state in minimum time:

$$J = \int_{t_0}^{t_f} dt \tag{4.239}$$

The performance criterion is subject to the equations of motion:

$$\dot{\gamma} = \frac{T-D}{mV}\sin\alpha + \frac{L}{mV}\cos\alpha - \frac{g\cos\gamma}{V} \tag{4.240a}$$

$$\dot{V} = \frac{T-D}{m}\cos\alpha - \frac{L}{m}\sin\alpha - g\sin\gamma \tag{4.240b}$$

$$\dot{x} = V\cos\gamma \tag{4.240c}$$

$$\dot{h} = V\sin\gamma \tag{4.240d}$$

where t is the actual time, $t_0 \le t \le t_f$ with t_0 as the initial time and t_f as the final time. The state variables are the flight path angle γ, speed V, horizontal position x and altitude h of the missile. The thrust magnitude T and the angle of attack α are the two control variables (see Figure 4.29). The aerodynamic forces D and L are functions of the altitude h, velocity V and angle of attack α. The following relationships have been assumed (Subchan and R. Żbikowski 2009).

Axial aerodynamic force: This force is written in the form

$$D(h,V,\alpha) = \frac{1}{2}C_{dp}V^2S_{\text{ref}} \tag{4.241}$$

$$C_d = A_1\alpha^2 + A_2\alpha + A_3 \tag{4.242}$$

Note that D is not the drag force.

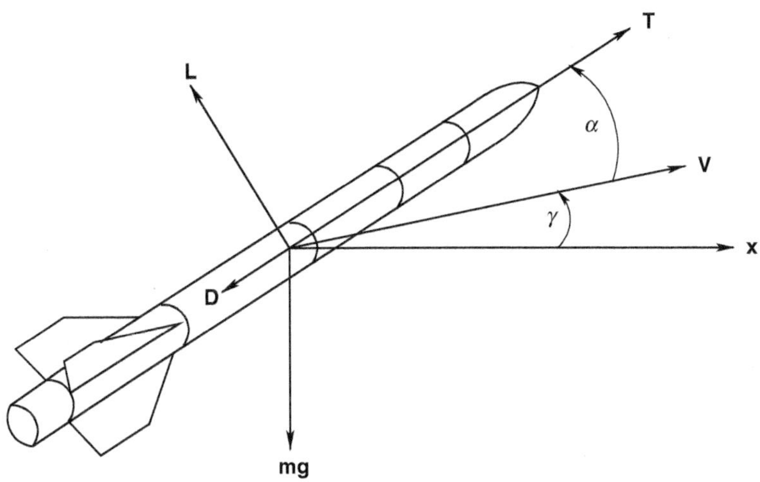

FIGURE 4.29 Definition of missile axes and angles. Note that L is the normal aerodynamic force and D is the axial aerodynamic force with respect to a body-axis frame, not lift and drag.

Normal aerodynamic force: This force is written in the form

$$L(h, V, \alpha) = \frac{1}{2} C_l \rho V^2 S_{\text{ref}} \quad (4.243)$$

$$C_l = B_1 \alpha + B_2 \quad (4.244)$$

where ρ is air density[1] given by

$$\rho = C_1 h^2 + C_2 h + C_3 \quad (4.245)$$

and S_{ref} is the reference area of the missile; m denotes the mass and g the gravitational constant; see also Table 4.9. Note that L is not the lift force.

Boundary conditions: The initial and final conditions for the four state variables are specified as follows:

$$\gamma(0) = \gamma_0, \quad \gamma(t_f) = \gamma_{tf} \quad (4.246a)$$

$$V(0) = V_0, \quad V(t_f) = V_{tf} \quad (4.246b)$$

$$x(0) = x_0, \quad x(t_f) = x_{tf} \quad (4.246c)$$

$$h(0) = h_0, \quad h(t_f) = h_{tf} \quad (4.246d)$$

In addition, constraints are defined as follows:

- State path constraints

$$V_{\min} \le V \le V_{\max} \quad (4.247)$$

$$h_{\min} \le h \quad (4.248)$$

Quantity	Value	Unit
m	1,005	kg
g	9.81	m/s^2
S_{ref}	0.3376	m^2
A_1	–1.9431	
A_2	–0.1499	
A_3	0.2359	
B_1	21.9	
B_2	0	
C_1	$3.312 \cdot 10^{-9}$	kg m^{-5}
C_2	$-1.142 \cdot 10^{-4}$	kg m^{-4}
C_3	1.224	kg m^{-3}

Table 4.9 Physical Modeling Parameters

[1]Since the cruise missile is supposed to be surface-launched only, the altitude change are expected to be small and hence a polynomial approximation of the exponential model is used.

Note that the altitude constraint equation (4.248) does not apply near the terminal condition.

- Control path constraint

$$T_{min} \le T \le T_{max} \tag{4.249}$$

- Mixed state and control constraint (see equations (4.243)–(4.245))

$$L_{min} \le \frac{L}{mg} \le L_{max} \tag{4.250}$$

where L_{min} and L_{max} are normalized; see Table 4.10

Note that equations (4.247–4.248) are pure state constraints, so the problem is expected to be challenging as explained is the sub-subsection "Pure State Constraints"; see also Maurer and Gillessen (1975), Berkmann and Pesch (1995), and Steindl and Troger (2003).

The optimal control problem considered here was to find the fastest trajectory to strike a fixed target which must be hit from above. As illustrated in Figure 4.30, the optimal trajectory has three distinct phases: level flight, climbing and diving; the level-flight phase is longest for the final speed of 250 m/s and shortest for 310 m/s. The angle of attack is one of the control variables (the other is thrust) and its optimal solution is shown in Figure 4.31. In Figures 4.30 and 4.31 the most accurate solution is the one computed with BNDSCO (see the sub-subsection "Indirect Method Approach"); in particular, the BNFSCO solution clearly captures the control variable discontinuity in Figure 4.31. The TPBVP formulations solved with BNDSCO are summarized in Figures 4.33–4.35, including the jumping and switching of the variables.

Different constraints are active during different parts of the optimal trajectories, as illustrated in Figure 4.32. During climbing, the thrust is on the maximum value for the cases of final speed 310 m/s while for the cases of final speed 250 m/s and 270 m/s during climbing the thrust switches to the minimum value. The maximum normal acceleration constraints are active only for the case 250 m/s in the middle of climbing which occurs in a few seconds. The normal acceleration and the thrust then switches to the minimum value. For the case of final speed 270 m/s the thrust switches to the minimum value at the end of climbing followed by the normal acceleration switching to the minimum value. At the start of diving, the minimum normal acceleration is active while the thrust is on the maximum value for the final speed 310 m/s case. In the middle of diving for the cases of final speed 250 m/s and 270 m/s the thrust switches back to maximum value to gain enough

Quantity	Value	Unit
$V_{min_{tf}}$	250	m/s
V_{min}	200	m/s
V_{max}	310	m/s
T_{min}	1,000	N
T_{max}	6,000	N
h_{min}	30	m
L_{min}	−4	g
L_{max}	4	g

Table 4.10 Boundary Conditions and Constraints

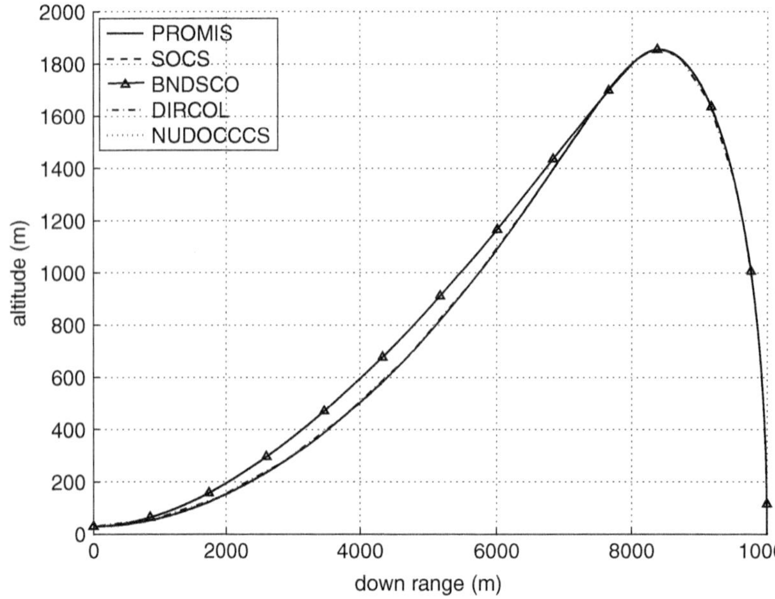

FIGURE 4.30 Comparison of PROMIS, SOCS, BNDSCO, DIRCOL, and NUDOCCCS results of the altitude versus down-range, constrained minimum time problem.

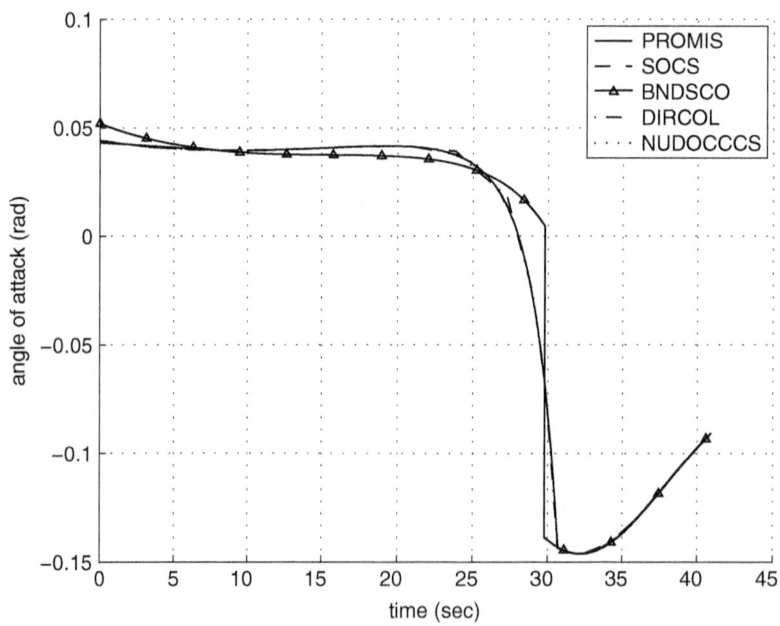

FIGURE 4.31 Comparison of PROMIS, SOCS, BNDSCO, DIRCOL, and NUDOCCCS results of the angle of attack versus time, constrained minimum time problem.

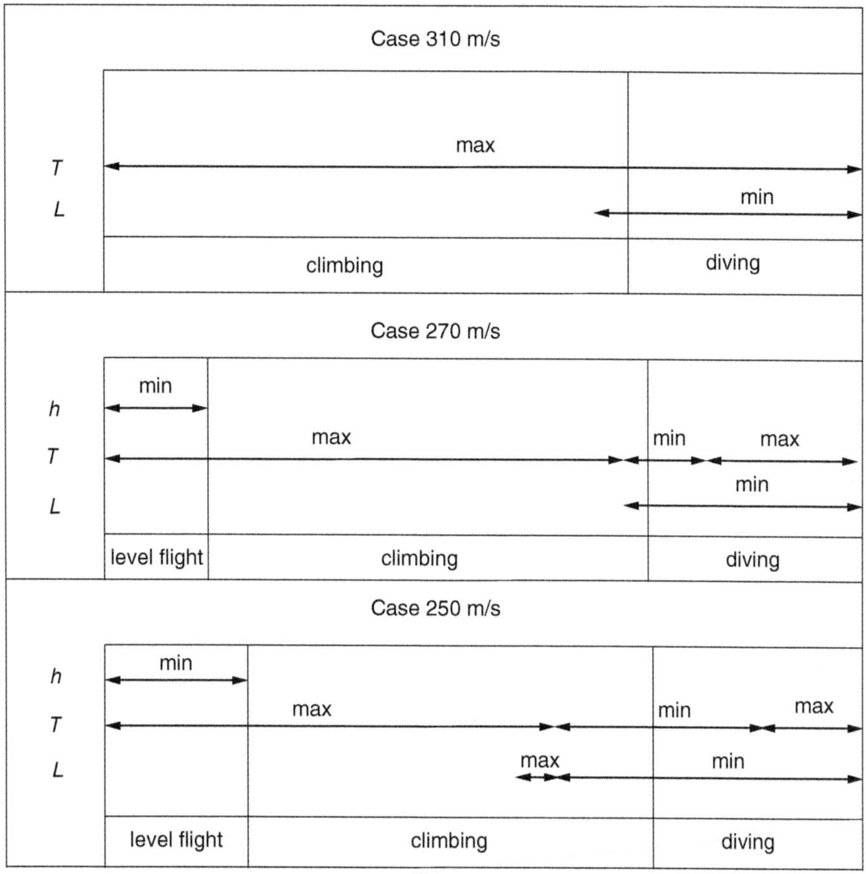

FIGURE 4.32 Switching structure of the minimum time formulation for the terminal bunt maneuver.

power to achieve the final speed while the normal acceleration is saturated on the minimum. The structure of the equations and switching time is given in Figure 4.35.

Practical Observations and Recommendations

Optimal control problems encountered in aerospace engineering practice must be solved numerically using the indirect and/or the direct method approaches. In the indirect case, the user must derive the appropriate equations for co-state variables, transversality and optimality conditions to formulate the relevant two-point boundary-value problem (TPBVP), a task requiring solid knowledge of the theory of optimal control. Numerical solution of the resulting TPBVP is the next task and requires judicious use of a well-designed solver. Using such a solver requires good understanding of the underlying numerics and a good guess of the co-state variables. In the direct case, the user does not need to perform theoretical analysis of the underlying optimal control problem and can immediately proceed to run the relevant solver. However, if the underlying problem involves several constraints which are often active, then the direct solvers are unlikely to converge. Even the convergent solutions have to be treated with caution because they produce only approximate solutions, often exhibiting artifacts due to numerical peculiarities of the

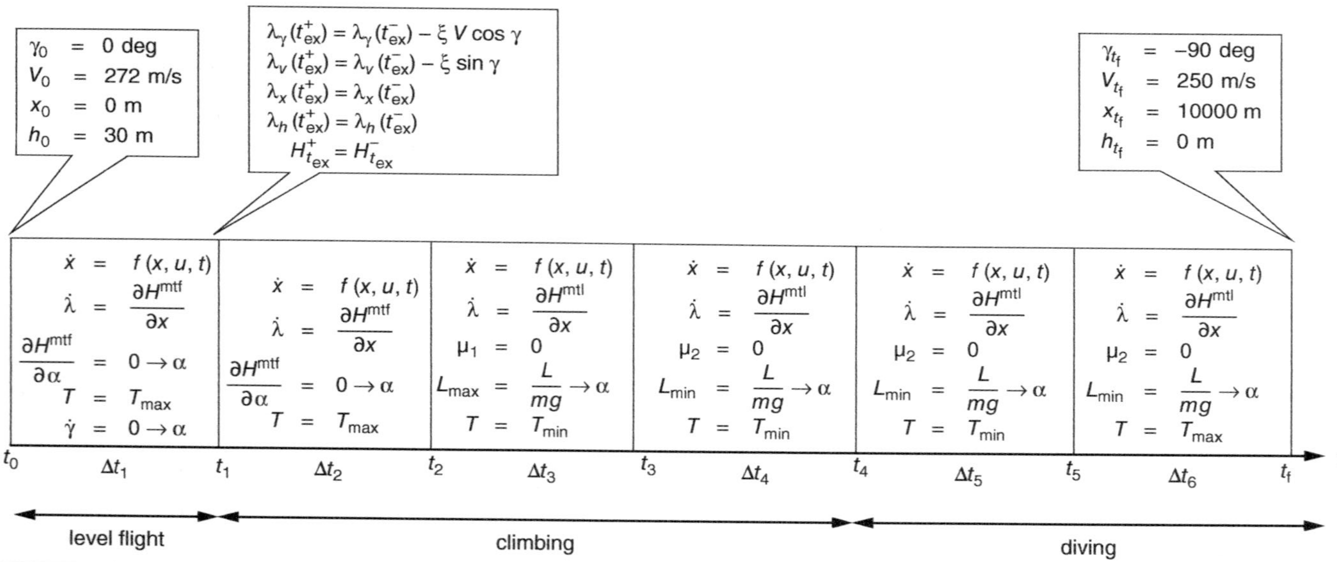

Figure 4.33 Schematic representation of the boundary value problem associated with the switching structure for the minimum time problem, case 250 m/s.

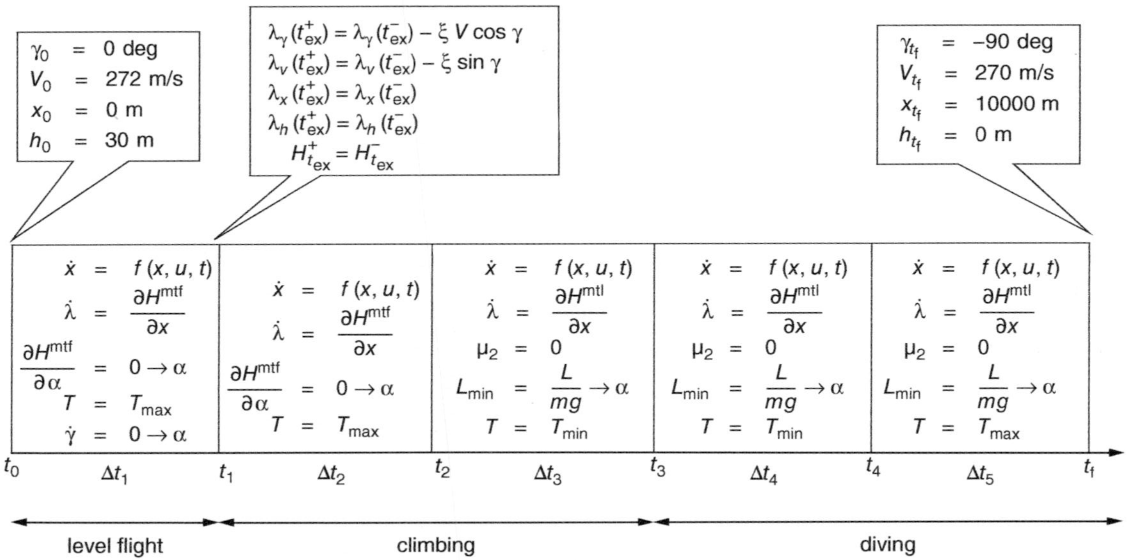

FIGURE 4.34 Schematic representation of the boundary value problem associated with the switching structure for the minimum time problem, case 270 m/s.

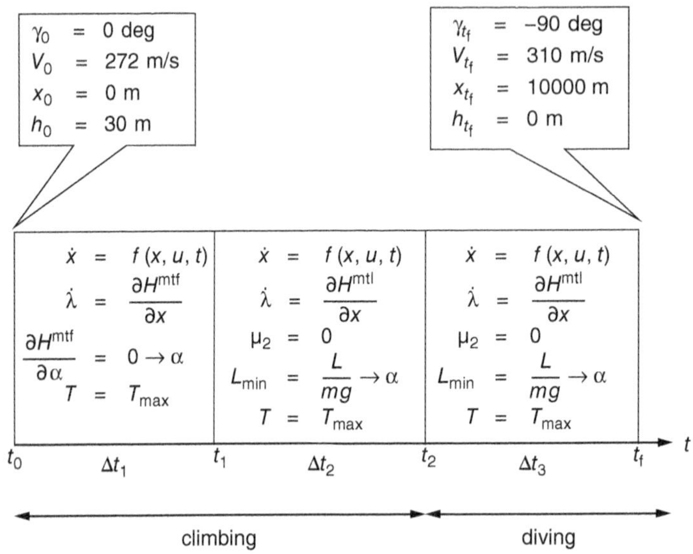

FIGURE 4.35 Schematic representation of the boundary value problem associated with the switching structure for the minimum time problem, case 310 m/s.

discretization approach adopted by the solver and the limitations of the underlying NLP numerical engine.

In practice, a combination of the direct and indirect approach (and the relevant codes) should be used (see Table 4.11), thus employing a hybrid approach. In the case study analyzed by the authors (Subchan and Żbikowski 2009). DIRCOL was used as the main direct solver and was run to discern the solution structure, including characteristic subarcs, constraints' activation and switching times. Whenever possible, DIRCOL results were compared with those of other direct solvers, NUDOCCCS, PROMIS, and SOCS. DIRCOL and NUDOCCCS codes produce initial guesses for the co-state, an essential feature to enable subsequent use of the BNDSCO code for solving the relevant two-point boundary value problem (TOBVP). The hybridization was done manually, i.e., DIRCOL, NUDOCCCS, PROMIS, and SOCS were run first, their results analysed to help formulate an appropriate TPBVP, and then the results were fed to BNDSCO as an initial guess (with co-states' guess from DIRCOL or NUDOCCCS).

Real-life optimal control problems arising in the aerospace engineering practice are too complex for automatic, deskilled solution approaches to be effective. We recommend the manual hybrid approach in which the user proceeds in three stages:

1. Direct solution (NLP via DIRCOL/NUDOCCCS/PROMIS/SOCS)
2. Analysis (optimal control theory, TPBVP formulation)
3. Indirect solution (TPBVP solution via BNDSCO)

This manual hybrid approach offers valuable insights into the problem, its solution structure, the role of constraints and boundary conditions. The insights into the influence of constraints and boundary conditions on the solution structure (e.g., the number of switching points, the number of constraints active, duration of their activation) is of significant operational and engineering value. Such insights often lead to reformulation of the original problem and suggest fruitful avenues for redesign.

Software	Documentation	Interface	Learning time	Solution quality	Running time	Ease of use	Ease of install
BNDSCO	★★	★★	★	★★★★★	★★★★★	★★	★★★★★
DIRCOL	★★★	★★	★★★	★★★	★★★★	★★★	★★★
NUDOCCCS	★★★	★★	★★★	★★★	★★★★	★★★	★★★
GESOP/PROMIS	★★★★	★★★★	★★★★	★★★	★★★	★★★★	★★★
GESOP/SOCS	★★★★	★★★★	★★★★	★★★	★★★	★★★★	★★★

★ = very bad.
★★ = bad.
★★★ = satisfactory.
★★★★ = good.
★★★★★ = very good.

Table 4.11 User Experience with Computational Optimal Control Software Packages (See also Table 4.12)

Software	Licence	Authorship/Availability	Remarks
BNDSCO	Academic	**Prof Hans Joachim Oberle** http://www.math.uni-hamburg.de	Works under both Windows and Linux
DIRCOL	Academic	**Prof Oskar von Stryk** http://www.sim.tu-darmstadt.de	1. Needs SNOPT http: //www.sbsi-sol-optimize.com 2. Works under Linux
NUDOCCCS	Academic	**Prof Christof Büskens** http://www.math.uni-bremen.de	1. Needs NAG Library www.nag.co.uk 2. Works under Linux
GESOP/PROMIS	Commercial	**Astos Solutions** http://astos.de/	Works under Windows
GESOP/SOCS	Commercial	**Astos Solutions** http://astos.de/	1. Works under Windows 2. SOCS is a Boeing FORTRAN package http://www.boeing.com/phantom/socs/

TABLE 4.12 Details of Computational Optimal Control Software Packages

Finally, it should be realized that optimal control is a challenging subject and attacking a realistic optimal control problem is not something to be undertaken lightly. To have a reasonable chance of successfully solving such a problem—usually involving nonlinear dynamics and numerous constraints—a working knowledge of the theory of optimal control is needed together with real appreciation of numerical implementation issues and software skills, preferably with FORTRAN. However, there is a large reward for studying computational optimal control because realistic (and hence practically important) problems cannot otherwise be tackled in an optimal way. Furthermore, the insights gained in deriving optimal solutions have independent value because they may lead to design improvements for the platforms considered and their mission planning. Due to high payoff in volume, or simply mission importance, aerospace application provide good justification for investing time and effort required for informed use of computational optimal control.

References

Ascher, U. M., Mattheij, R. M. M., and Russel, R. D. 1995. "Numerical Solution of Boundary Value Problem for Ordinary Differential Equations," SIAM, Philadelphia.

Athans, M. and Falb, P. L., 1966. *Optimal Control: An Introduction to Theory and Its Application*, McGraw-Hill Book Company, New York.

Bazaraa, M. S., Sherali, H. D., and Shetty, C. M. 1993. *Nonlinear Programming: Theory and Algorithms*, 2nd ed., Wiley-Interscience, New York.

Benson, D. 2005. "A Gauss Pseudospectral Transcription for Optimal Control," Ph.D. thesis, Massachusetts Institute Technology, February.

Berkmann, P. and Pesch, H. J. 1995. "Abort Landing in Windshear: Optimal Control Problem with Third-Order State Constraint and Varied Switching Structure," *Journal of Optimization Theory and Applications*, 85(1):21–57.

Betts, J. T. "Practical Methods for Optimal Control Using Nonlinear Programming," SIAM, Philadelphia, 2001.

Bryson, A. E. and Ho, Y. C. 1975. *Applied Optimal Control: Optimization, Estimation, and Control* (revised printing), Hemisphere Publishing Corporation, New York.

Büskens, C. and Maurer, H. 2000. "SQP-Methods for Solving Optimal Control Problems with Control and State Constraints: Adjoint Variables, Sensitivity Analysis and Real-Time Control," *Journal of Computational and Applied Mathematics*, 120(1–2):85–108.

Deuflhard, P., Pesch, H. J., and Rentrop, P. 1976. "A Modified Continuation Method for the Numerical Solution of Nonlinear Two-Point Boundary Value Problems by Shooting Techniques," *Numerische Mathematik*, 26:327–343.

Elnagar, G., Kazemi, M. A., and Razzaghi, M. 1995. "The Pseudospectral Legendre Method for Discretizing Optimal Control Problems," *IEEE Transactions on Automatic Control*, 40(10): 1793–1796.

Elsgolc, L. E. 1962. *Calculus of Variations*, Pergamon Press, London.

Enright, P. J. and Conway, B. A. 1992. "Discrete Approximations to Optimal Trajectories Using Direct Transcription and Nonlinear Programming," *Journal of Guidance, Control, and Dynamics*, 15(4):994–1002.

Fahroo, F. and Ross, I. M. 2002. "Direct Trajectory Optimization Pseudospectral Method," *Journal of Guidance, Control, and Dynamics*, 25(1):160–166.

Gill, P. E., Murray, W., and Wright, M. H. 2002. "SNOPT: An SQP Algorithm for Large-Scale Constrained Optimization," *SIAM Journal on Optimization*, 12(4):979–1006.

Hartl, R. F., Sethi, S. P., and Vickson, R. G. 1995. "A Survey of Maximum Principles for Optimal Control Problems with State Constraints," *SIAM Review*, 37(2):181–218.

Macki, J. and Strauss, A. 1982. *Introduction to Optimal Control Theory*, Springer-Verlag, New York/ Heidelberg/Berlin.

Maurer, H. and Gillessen, W. 1975. "Application of Multiple Shooting to the Numerical Solution of Optimal Control Problems with Bounded State Variables," *Computing*, 15:105–126.

Nocedal, J. and Wright, S. J. 2006. *Numerical Optimization*, 2nd ed., Springer, New York.

Oberle, H. J. and Grimm, W. 1989. "BNDSCO: A Program for the Numerical Solution of Optimal Control Problems," Technical Report DLR IB 515-89-22, Institute for Flight Systems Dynamics, DLR, Oberpfaffenhofen, Germany.

Pesch, H. J. 1989a. "Real-Time Computation of Feedback Controls for Constrained Optimal Control Problems, Part 1: Neighboring Extremals," *Optimal Control Applications & Methods*, 10(2):129–145.

Pesch, H. J. 1989b. "Real-Time Computation of Feedback Controls for Constrained Optimal Control Problems, Part 2: A Correction Method Based on Multiple Shooting," *Optimal Control Applications & Methods*, 10(2):147–171.

Pesch, H. J. 1991. "Offline and Online Computational of Optimal Trajectories in the Aerospace Field," In A. Miele and A. Salvetti (eds.), *Applied Mathematics in Aerospace Science and Engineering, Proceedings of a Meeting on Applied Mathematics in Aerospace Field*, Plenum Press, New York, pp. 165–220.

Pesch, H. J. 1994. "A Practical Guide to the Solution of Real-Life Optimal Control Problems," *Control and Cybernetics*, 23(1 and 2):7–60.

Pontryagin, L. S., Boltyanskii, V. G., Gamkrelidzeand, R. V., and Mishchenko, E. F. 1962. *The Mathematical Theory of Optimal Processes*, John Wiley & Sons, New York.

Ross, I. M. and Fahroo, F. 2002. "User's Manual for DIDO 2002: A MATLAB Application Package for Solving Optimal Control Problems," Technical Report AA-02-002, Department of Aerospace and Astronautics, Naval Postgraduate School, Monterey, California.

Steindl, A. and Troger, H. 2003. "Optimal Control of Deployment of a Tethered Subsatellite," *Nonlinear Dynamics*, 31(3):257–274.

Stoer, J. and Bulirsch, R. 2002. *Introduction to Numerical Analysis*, 3rd ed., Springer, New York.

Subchan, S. and Żbikowski, R. 2009. *Computational Optimal Control: Tools and Practice*, Wiley, Chichester, UK.

Vinh, N. X. 1981. *Optimal Trajectories in Atmospheric Flight*. Elsevier Scientific Publishing Company, Amsterdam.

von Stryk, O. and Bulirsch, R. 1992. "Direct and Indirect Methods for Trajectory Optimization," *Annals of Operations Research*, 37(1–4):357–373.

SECTION 5

Avionics and Air Traffic Management Systems

Section Editor: Roberto Sabatini

Acronyms

3D	Three-dimensional
4D	Four-dimensional
4DT	Four-dimensional trajectory
4-PNV	4DT planning, negotiation, and validation
AAC	Aeronautical administrative communications
ABAS	Aircraft-based augmentation system
ACARS	Aircraft communications addressing and reporting system
ACAS	Airborne collision avoidance system
ACC	Area control centers
ACT	Active control technologies/Aerodrome control towers
ADF	Automatic direction finder
ADIR	AMHS directory
ADIRS	Air data and inertial reference system
ADIRU	Air data and inertial reference unit
ADM	Aircraft dynamic model
ADS	Automatic dependent surveillance
ADS-B	Automatic dependent surveillance broadcast
ADS-R	Automatic dependent surveillance rebroadcast
AFDX	Avionics full-duplex switched ethernet
AFTN	Aeronautical fixed telecommunication network
AHRS	Attitude and heading reference system
AIAA	American Institute of Aeronautics and Astronautics
AIMS	Airplane information management system
AIS	Aeronautical information services
AIXM	Aeronautical information exchange model
AL	Alert limit
AMHS	ATS message handling system
AM(R)S	Aeronautical mobile (route) service

AMS(R)S	Aeronautical mobile-satellite (route) service
ANASTASIA	Airborne new and advanced satellite techniques and technologies in a system integrated approach
ANSP	Air navigation service provider
AOA	Angle of arrival
AOC	Airline operation center
APC	Aircraft pilot coupling
APU	Auxiliary power unit
ARINC	Aeronautical Radio Inc.
ARNS	Aeronautical radionavigation service
ARP	Aerospace recommended practice
ASAS	Airborne separation assurance systems
ASBU	ICAO's Aviation System Block Upgrades
ASDE	Airport surface detection equipment
ASEC	AMHS security
ASR	Airport surveillance radar
ASM	Airspace management
ATA	Air Transport Association
ATC	Air traffic control
ATCo	Air traffic controller
ATFM	Air traffic flow management
ATFCM	Air traffic flow and capacity management
ATK	Along track
ATM	Air traffic management
ATS	Air traffic services
ATSO	Air traffic service organizations
BC	Bus controller
BER	Bit error rate
BITE	Built-in test equipment
BM	Bus monitor
BSS	Broadcasting Satellite Service
BYDU	Backup yaw damper unit
CAA	Civil Aviation Act
CAD	Computer-aided design
CARATS	Collaborative action for renovation of air traffic systems
CAS	Calibrated air speed/Control augmentation system
CASA	Civil Aviation Safety Authority
CASR	Civil Aviation Safety Regulations
CATH	Collision avoidance threshold
CCA	Common cause analysis
CDI	Course deviation indicator
CDM	Collaborative decision making
CDMA	Code-division multiple access
CDR	Collision detection and resolution
CDTI	Cockpit display of traffic information
CFD	Computational fluid dynamics
CFR	Code of federal regulations
CG	Center of gravity
CIDIN	Common ICAO data interchange network

CIP	Current icing product
CMA	Common mode analysis
CMC	Centralized maintenance computer
CMF	Centralized maintenance function
CNDB	Customized navigation database
CNPC	Command and non-payload communications
CNS+A	Communications, navigation, surveillance, and avionics/ATM
CNSP	Communication, navigation, and surveillance providers
CONOPS	Concept of operations
COTS	Commercial-off-the-shelf
CPDLC	Controller–pilot datalink communications
CPS	Control processing station
CSMA	Carrier sense multiple access
CTA	Critical task analysis
CTAR	Critical task analysis report
CV	Collision volume
DAM	Dynamic airspace management
D-AIM	Digital Aeronautical Information Management
DCM	Direction cosine matrix
DF	Direction finder
DGNSS	Differential GNSS
DLS	Datalink service
DME	Distance-measuring equipment
DMG	Digital map generator
DR	Dead reckoning
DSS	Decision support system
DVOR	Doppler VOR
ECEF	Earth-centered, Earth-fixed
ECI	Earth-centered inertial
FDTD	Finite difference time domain
EFA	Euro fighter aircraft
EFCU	Electronic flight control unit
EFIS	Electronic flight instrument system
EHA	Electrohydrostatic actuator
EHF	Extremely high frequency
EIA	Electronics Industry Association
EICAS	Engine indicating and crew alerting system
EKF	Extended Kalman filter
ELAC	Elevator aileron computer
ELOS	Equivalent level of safety
ELT	Emergency locator transmitter
EMA	Electromechanical actuator
EMC	Electromagnetic compatibility
EMP	EM pulse
EO	Electro-optical
ESD	Electrostatic discharges
EVIGA	EKF-VIGA
FAA	Federal Aviation Administration
FAC	Flight augmentation computer

FADEC	Full authority digital engine control
FAR	Federal Aviation Regulation
FBW	Fly-by-wire
FCC	Flight control computer
FCDC	Flight control data concentrator
FCL	Flight control laws
FCS	Flight control systems
FDIR	Fault detection, isolation, and recovery
FDTD	Finite difference time domain
FEM	Finite element models
FHA	Functional hazard analysis
FIANS	Future Indian air navigation system
FIP	Forecast icing potential
FIC	Fight information center
FMEA	Failure modes and effects analysis
FMGEC	Flight management guidance envelope computer
FMS	Flight management system
FOC	Full operational capability
FOG	Fiber optic gyroscope
FRT	Fixed radius transition
FTA	Fault tree analysis
FTBP	File transfer body parts
FIXM	Flight information exchange model
FTE	Flight technical error
FWHM	Full width at half maximum
GANP	Global air navigation plan
GBAS	Ground-based augmentation system
GEO	Geostationary orbit
GFS	Global forecast system
GLONASS	Globalnaya Navigazionnaya Sputnikovaya Sistema
GML	Geography Markup Language
GMS	Ground monitoring station
GNSS	Global navigation satellite systems
GPS	Global positioning system
GRIB	General regularly distributed information in binary
GSM	Global system for mobile communications
GTG	Graphical turbulence guidance
HAL	Horizontal alert limit
HF	High frequency
HFE	Human factors engineering
HIRF	High-intensity radiated fields
HPL	Horizontal protection level
HQ	Handling qualities
HRA	Human reliability analysis
HMI2	Human-machine interfaces and interactions
HSI	Horizontal situation indicator
HTA	Hierarchical task analysis
IAP	Integrated actuator package
ICAO	International Civil Aviation Organization

IDV	Ionospheric delay in the vertical direction
IEEE	Institute of Electrical and Electronics Engineering
IF	Intermediate frequency
IFOG	Interferometric FOG
IFPCS	Integrated flight propulsion control system
IFR	Instrument flying rules
IGSO	Inclined geosynchronous orbit
IHE	IPM heading extension
ILS	Instrument landing system
IMA	Integrated modular avionics
IMU	Inertial measurement unit
INS	Inertial navigation system
IPM	Interpersonal message
IR	Infrared
IRS	Inertial reference system
ISDN	Integrated services digital network
ISL	Inter-satellite link
ISO	International Organization for Standardization
ITU	International Telecommunication Union
ITU-R	International Telecommunication Union Radiocommunication Sector
JAA	Joint Aviation Authority
JAR	Joint aviation requirements
JPRF	Jittered pulse repetition frequency
LAAS	Local area augmentation system
LAD	Local area DGNSS
LDACS	L-band digital aeronautical communication system
Loran	Long-range navigation
LF	Low frequency
LFT	Linear fractional transformation
LISN	Line impedance stabilization network
LPV	Linear parameter variant
LOP	Line of position
LOS	Line of sight
LQ	Linear quadratic
LQG	Linear quadratic Gaussian
LQR	Linear quadratic recovery
LRM	Line replaceable module
LRU	Line replaceable unit
LTR	Loop transfer recovery
MAC	Media access control
MASPS	Minimum Aviation System Performance Standards
MCDU	Multiple purpose control display unit
MEMS	Microelectromechanical systems
MET	Meteorological
METAR	Meteorological terminal air report
MF	Mid frequency
MIMO	Multi-input multi-output
MLAT	Multilateration
MLS	Microwave landing system

MS	Mode status
MSS	Mobile-satellite service
MTBF	Mean time between failure
MTI	Moving target indicator
NA	Numerical aperture
NASA	National Aeronautics and Space Administration
NASA-TLX	NASA Task Load Index
NATALIA	New Automotive Tracking Antenna for Low-cost Innovative Applications
NAVAID	Navigation aids
NAVSTAR	Navigation Signal Time and Range
NCWF	National Convective Weather Forecast
NDB	Nondirectional beacons
NED	North east down
NextGen	Next Generation Air Transportation System
NG-ADL	Next generation aeronautical datalinks
NG-FMS	Next generation flight management system
NMAC	Near mid-air collision
NOAA	National Weather Service
NSE	Navigation system error
OSD	Open space datalink
OCM	Optimal control model
OOT	Object-oriented technology
OPMET	Operational meteorological information
PAR	Precision approach radar
PBC	Performance based communications
PBN	Performance based navigation
PBO	Performance based operations
PDE	Partial differential equations
PDF	Probability distribution function
PED	Portable electric device
PF	Probability of failure
PFD	Primary flight display
PID	Proportional-integral-derivative
PIO	Pilot-induced oscillations+
PL	Protection level
PPI	Planned position indicator
PPS	Precise positioning service
PRA	Particular risk analysis
PRF	Pulse-repetition frequency
PRN	Pseudo random number
PSK	Phase shift keying
PSR	Primary surveillance radar
PSSA	Preliminary system safety assessment
PTTI	Precise time and time interval
PVA	Position, velocity, and attitude
RA	Resolution advisories
RAD	Radial (track)
RAIM	Receiver autonomous integrity augmentation
RAT	Ram air turbine

RBD	Reliability block diagram
RCS	Radar cross-section
RCP	Required communication performance
RLG	Ring laser gyroscope
RLS	Radio location service
RMS	Root mean square
RNP	Required navigation performance
RNS	Radionavigation service
RNSS	Radionavigation satellite service
RR	Reference receiver/Relative range/Radio regulations
RS	Reference station
RSS	Received signal strength
RT	Remote terminal
RTCA	Radio Technical Commission for Aeronautics
RTCM	Radio Technical Commission for Maritime Services
RVR	Runway visual range
RVSM	Reduced vertical separation minima
SA&CA	Separation assurance and collision avoidance
SAE	Society of Automotive Engineers
SAFEE	Security of Aircraft in the Future European Environment
SANDRA	Seamless aeronautical networking through integration of datalinks, radios, and antennas
SARP	Standards and recommended practices
SAS	Stability augmentation system
SBAS	Satellite based augmentation system
SCAT	Special category
SDMA	Space division multiple access
SEC	Spoiler elevator computer
SESAR	Single European Sky ATM Research
SFM	Structure from motion
SHF	Super high frequency
SIGMET	Significant meteorological information
SIL	Surveillance integrity level
SIS	Signal-in-space
SISO	Single-input single-output
SLAM	Simultaneous localization and mapping
SMGCS	Advanced surface movement guidance and control system
SMTP	Simple Mail Transfer Protocol
SNR	Signal-to-noise ratio
SoOP	Signals of opportunity
SPS	Standard positioning service
SSA	System safety assessment
SSR	Secondary surveillance radar
SST	Self-separation threshold
SSV	Self-seperation volume/surveillance state vector
STANAG	Standardization Agreement
STDMA	Self-organized TDMA
SUA	Special use airspace
SUMI	Software usability measurement inventory

SWIM	System wide information management
T&E	Test and evaluation
TACAN	Tactical air navigation
TAF	Terminal aerodrome forecast
TAS	True air speed
TBO	Time based operations/time before overhaul
TCA	Terminal control area
TCP	Trajectory change point
TCU	Terminal control units (approach)
TD	Tropospheric delay
TDOA	Time difference of arrival
TDMA	Time division multiple access
TEC	Total electron content
THS	Tail trimmable horizontal stabilizer
TID	Total ionospheric delay
TIS-B	Traffic information service—broadcast
TITAN	Thunderstorm Identification, Tracking, Analysis, and Nowcasting
TOA	Time of arrival
THS	Trimmable horizontal stabilizer
THSA	Actuator on the THS
TLS	Target level of safety
TLX	Task load index
TPSI	Time position space information
TRF	Tuned radio frequency
TSE	Total system error
TT	Transaction time
TTA	Time-to-alert
TT&C	Telemetry, tracking, and command
TTRCP	RCP transaction time
UAS	Unmanned aircraft system
UAV	Unmanned aerial vehicle
UAT	Universal access transceiver
UCS	Utilities control system
UEE	User equipment error
UERE	User equivalent range error
UHF	Ultrahigh frequency
UKF	Unscented Kalman filter
UML	Unified Markup Language
UPS	Uninterrupted power supply
UR	User receiver
URE	User range error
UT	Unscented transformation
UTC	Universal time coordinated
UTM	UAS traffic management
UV	Ultraviolet
UVIGA	UKF-VIGA
V&V	Verification and validation
VAL	Vertical alert limit
VASIS	Visual approach slope indicator system

VBA	Vibrating beam accelerometer
VBN	Vision-based navigation
VBS	Vision-based navigation sensors
VFR	Visual flying rules
VHF	Very high frequency
VHFDL	VHF datalinks
VIGA	Vision/INS/GNSS/ADM (system architecture)
VPL	Vertical protection level
VOR	VHF omnidirectional range
WAAS	Wide-area augmentation system
WAD	Wide-area DGNSS
WAIC	Wireless avionics intra-communications
WAM	Wide-area multilateration
WIDS	Weather Immediate Decision Service
WIMS	Weather information management systems
WNDS	Weather Near-Term Decision Service
WPDS	Weather Planning Decision Service
WRC	World radiocommunication conferences
WTT	Wind tunnel testing
WXXM	Weather information exchange model
XML	Extensible Markup Language
XTK	Across track
ZSA	Zonal safety analysis

PART 1

The Electromagnetic Spectrum

Florent Christophe, Subramanian Ramasamy, and Roberto Sabatini

Radio waves propagate in the vacuum or through the complex media surrounding the Earth, or other planets, and carry information from the source to the receiver. We focus in this subsection on radio waves involving man-made transmitters for avionics or astrionics applications (i.e., communications, navigation, surveillance, and radiolocation) but the case of the transmitter as a natural source may also be considered for radioastronomy or Earth observation from space.

5.1 Radio Waves in a Vacuum

Radio waves, inferred from Maxwell equations, first observed by Hertz in 1886, and applied for long range by Marconi in 1906, are a combination of an electric field E and a magnetic field H of periodic time variations, produced by electric charge displacements or electric currents. In a vacuum, far enough from those electric sources, E and H fields appear as plane waves are expressed as

$$E(r,t) = E_0 \exp(\omega t - k \cdot r) \tag{5.1}$$

$$H(r,t) = H_0 \exp(\omega t - k \cdot r) \tag{5.2}$$

where E_0 and H_0 are orthogonal vectors, their modules are linked by

$$E_0 / H_0 = 120\pi \tag{5.3}$$

The power density transported by such waves is given by

$$D = E_0^2 / 120\pi \quad \text{or} \quad D = E_0^2 / 120\pi \quad \text{or} \quad D = 120\pi H_0^2 \tag{5.4}$$

The direction of vector E_0 defines the linear polarization of the wave; ω is its frequency; and k is the wave vector indicating the direction of wave propagation, orthogonal to both E_0 and H_0. The governing relationship is given by

$$k = c / \omega \tag{5.5}$$

where c is the velocity of light in a vacuum, equal to $3 \cdot 10^8$ m/s. Vector r defines the point in space where the fields are considered. The time period and frequency can be easily derived using

$$T = 2\pi / \omega \qquad f = 1/T = \omega / 2\pi \tag{5.6}$$

as well as the spatial period or wavelength, given by

$$2\pi / k = c/f \tag{5.7}$$

Table 5.1 illustrates equation (5.7) and indicates the usual denomination of frequency bands, following a logarithmic scale. Beyond the upper extremely high frequency (EHF) limit (300 GHz), atmospheric attenuation (discussed below) makes most terrestrial applications impractical up to about 10 THz (30 μm wavelength), where the far infrared region of optics begins. The low frequencies

Frequency (MHz)	3.10^{-4}	3.10^{-3}	0.03	0.3	3	30	300	3.10^{3}	3.10^{4}
Wavelength (m)	10^{6}	10^{5}	10^{4}	10^{3}	100	10	1	0.1	0.01
Denomination	ELF	VLF	LF	MF	HF	VHF	UHF	SHF	EHF

TABLE 5.1 From Extremely Low to Extremely High Frequencies

(LF band and below) are restricted to submarine communications, where they are useful for their ability to penetrate saltwater despite their limited bandwidth and impractical wavelength-sized antennas. Most avionics or astrionics systems make use of wavelengths from a meter to a centimeter, i.e., very high frequency (VHF) to super high frequency (SHF) bands, but the selection of a frequency band for a given application is governed by bandwidth requirements and depends on antenna and wave propagation, which will be discussed now.

5.2 Antennas and Power Budget of a Radio Link

An antenna is a transducer that transforms an electric current from a transmission line into a radio wave (on transmit) or a radio wave into an electric current (on receive). This transformation should be done with maximum efficiency, but attention must be paid to the direction and polarization of the radiated wave. We will first introduce wire antennas, a classical example of which is the half-wavelength dipole. It has been shown that most of the power coming from a transmission line (like a coaxial cable often used at ultrahigh frequency (UHF) or a twin wire line for VHF and below) is radiated when it is connected to two thin collinear wires at a resonance frequency for which each wire has a length close to a quarter of the wavelength. The modulus of the electric field radiated by such an antenna is maximal in a plane orthogonal to the wires, without angular dependency in this plane, the electric field being parallel to the wires. The power efficiency of such a transducer being fair only very close to the resonance frequency, practical implementations where relative bandwidths of at least a few percent are required are based on thick dipoles.

For another classical transmission line at SHF and EHF such as the rectangular waveguide, the antenna has to adapt the guided wave progressively to a free space propagating wave (or vice versa) without major discontinuities that would make detrimental wave reflections. A natural solution is to widen the section of the waveguide progressively, making a pyramidal horn. In its aperture (the base of the pyramid) the wave is weakly coupled to the walls and thus ready to be launched to free space. Such a wave will have as its preferred beam direction the axis of the pyramid. The width of this beam in each plane (expressed in radians) may be related to the aperture of the horn antenna by approximate formulas, given by

$$\delta\phi = \lambda/a, \quad \delta\theta = \lambda/b \tag{5.8}$$

where a and b are the sides of the rectangular radiating aperture. For other aperture antennas, such as reflector antennas, for which a parabolic reflector may be illuminated by a small horn (or primary feed) located near its focus, the same formulas apply, a and b now being the dimensions of the reflector itself.

If we now consider how the radiated power P is distributed around the antenna, we may write the power density D at distance R from the antenna as if the antenna were isotropic, times a power-concentrating factor, g, i.e.,

$$D = Pg/4\pi R^2 \tag{5.9}$$

This power-concentrating factor g is known as the antenna gain[1] and is given by the approximate formula:

$$g = 4\pi S/\lambda^2 \tag{5.10}$$

where S is the surface of the aperture (i.e., $a \cdot b$), therefore:

$$g = 4\pi / \delta\phi\delta\theta \tag{5.11}$$

From the power density, the electric and magnetic fields at distance R may be derived making use of formulas (equation (5.4)) or directly from the power available at the output of a second antenna of section S' (and gain g' given by equation (5.10)) as $P' = S'D$, or

$$P' = Pgg'(\lambda/4\pi R)^2 \tag{5.12}$$

This formula governing the power transfer from one antenna to another is known as the radiocommunication equation. Compared to the equivalent thermal noise at the input of the receiver, it allows the signal-to-noise ratio, i.e., the quality of the radio link, to be inferred. Further information concerning antennas[2] may be found in Rudge et al. (1982, 1983).

5.3 Radio Wave Propagation in the Terrestrial Environment

Depending on the wavelength, many effects are likely to occur when a radio wave interacts with the Earth's surface or with the atmosphere. These effects, which are presented below, are summarized in Table 5.2. More detailed information is available in the International Telecommunication Union Radiocommunication Sector (ITU-R) Handbooks (see, e.g., ITU 1996, 1998, 2013).

Interactions with the Earth's Surface

If we first consider the ground as an homogeneous flat medium, its electrical properties (dielectric constant and conductivity) may be reduced to a refractive index ranging from 2 to 9, depending on frequency, soil composition, and moisture content. Applying Fresnel's laws of optics to an

Wavelength	>10 m	10 m-10 cm	<10 cm
Ground effects	+	+	++
Tropospheric effects	–	–	+
Ionospheric effects	++	+	–

Table 5.2 Summary of the Interactions of a Radio Wave with Earth's Environment: Reduced (–), Limited (+), or Strong (++)

[1]The notion of antenna gain, which was introduced with regard to aperture radiating antennas, also applies to wire antennas. The gain g of a wire antenna is limited to a few units (1.5 for a half-wavelength dipole), whereas it exceeds 1,000 for a $10\ \lambda \times 10\ \lambda$ aperture.

[2]The pencil beam is always associated with side lobes (either near lobes, far lobes, or even back lobes, with reference to their angular distance to the main beam), which may produce detrimental coupling effects. Parasitic side lobes might be reduced through various techniques, but at the expense of widening the main beam and reducing the gain. Those techniques, which include windowing, are similar to frequency-filtering techniques and lead to the concept of angular filtering.

electromagnetic wave incident upon the plane interface demonstrates a specularly reflected wave and a refracted wave below the surface.[3]

The reflected wave may produce an interference with the directly transmitted one, a classical situation being grazing angle at the interface where both waves with similar elevation angles cannot be separated by the antenna beam, whereas the reflection coefficient is close to −1. The resulting interference pattern exhibits at range R of a transmitter h_t above ground successive nulls at height h_n, given by

$$h_n = R\lambda(1+2n)/4h_t \tag{5.13}$$

Earth Curvature and Relief Effects

Between a transmitter and a receiver close to the Earth's surface the direct wave may be obstructed either by obstacles either coming from the local relief or by the curvature of the Earth. The limiting situation in this last case is the horizon plane tangent to the Earth's sphere, which a receiver at height h_r crosses at range R_h for a transmitter at height h_t, r_e being the Earth's radius[4]:

$$R_h = \sqrt{2r_e h_t} + \sqrt{2r_e h_r} \tag{5.14}$$

Such an obstruction does not occur at a sharp cutoff when crossing the horizon plane, due to the diffraction effect. For radio waves at wavelengths comparable to the radius of curvature of the obstacle, this diffraction may compensate most of the shadowing effect. Electromagnetic waves at MF and below are therefore suitable for propagating hundreds or thousands of kilometers near the Earth's surface.[5]

Soil Roughness Effects

Small-scale random features on the surface (either of natural origin or produced by agriculture on bare soils, but vegetation can be also accounted for, as well as wind-driven waves on the sea surface or swell) disturb the previously mentioned specular reflection, reducing the corresponding wave and creating more and more diffuse scattering as the ratio of the typical size of the irregularities to the wavelength is augmented.[6]

Interactions with the Troposphere

In this lower part of the atmosphere, less than 12 km, gaseous molecules of sufficient density, and hydrometeors (the liquid water or ice particles inside clouds, rain, snow, hail, etc.) may interact with radio waves. Molecular absorption is due to the exchange of energy between the wave and quantified levels of vibrations or rotations and appears as absorption lines. In the upper SHF and EHF bands, water vapor creates such an absorption line around 22 GHz and a rapidly increasing continuum beyond 50 GHz, whereas oxygen creates a strong absorbing continuum around 58 GHz. Those effects limit long-range surface-based applications to around 45 GHz, but radiometers take advantage of such molecular interactions for remote sensing of the atmosphere.

The major interaction with hydrometeors may be described as Rayleigh scattering from individual particles that are small with respect to the wavelength; the scattering cross-section (ratio of scattered power to incident power density) is then written for spherical water droplets of diameter d as

$$\sigma = 190d^6/\lambda^4 \tag{5.15}$$

---

[3]This refracted wave usually vanishes rapidly due to the finite conductivity of the soil.

[4]In fact, the refraction of radio waves propagating quasi-horizontally in the stratified lower troposphere bends the ray downward. In standard troposphere, this bending is usually compensated for by modifying in the horizon formula the real Earth radius of 6,400 km into a value of 8,500 km.

[5]For completeness, a creeping wave guided near the interface must also be considered in such conditions.

[6]Part of this energy is backscattered toward the transmitter, creating surface clutter for radar. See Part 4 of this section.

This relationship shows a strong increase with frequency and drop size. In a given volume of interaction where droplets of different sizes (larger than 0.1 mm in diameter, up to 7 mm for thunderstorms) are falling at different velocities, their distribution may be derived from the rain rate observed at ground level. This droplet distribution may then be used for estimating the overall scattering cross-section per volume unit, which happens to be the attenuation coefficient per length unit.

A further effect at EHF due to interaction with the lower troposphere is scintillation (rapid fluctuations of amplitude and phase) due to the transit of the wave through the time-varying heterogeneities of refractive index caused by turbulence.

Interactions with the Ionosphere

The ionosphere is mostly created, at altitudes higher than 80 km, by UV radiation and cosmic rays from the sun, ionizing the low-density air molecules. The electron density exhibits a strong dependency with solar flare activity (varying on an 11-year cycle basis), hour of the day and season, latitude (the auroral oval is the place where charged cosmic rays penetrate the high atmosphere), and altitude. The maximum electron density is observed around 350 km and varies from $N = 3{\cdot}10^{10}$ m^{-3} at night for a low-activity sun to more than $N = 10^{12}$ m^{-3} at noon for an active sun.

The propagation of a radio wave in such cold plasma, neglecting collisions, is mainly ruled by the equation giving the refractive index, given by

$$1 = \frac{1}{2}(f_p/f)^2 \tag{5.16}$$

for a frequency f which is large with respect to the plasma frequency f_p, given by (for N in m^{-3} and f_p in Hz):

$$f_p = 9\sqrt{N} \tag{5.17}$$

The maximum value for f_p is about 10 MHz. An electromagnetic wave at a frequency below or close to this value would be refracted downward before reaching the region of the maximum ionization; this effect is used for establishing over the horizon radio links in the HF band. Waves at frequencies beyond that value are able to transit through the whole ionosphere, with effects mostly due to an augmentation of the time delay. With respect to propagation in a vacuum, the path is apparently augmented by

$$\Delta R = 40.5\ \text{TEC}/f^2 \tag{5.18}$$

TEC (total electron content) is the integral of the electron density along the path. Maximum values of 10^{18} m^{-2} may be encountered for a slanting path through the whole ionosphere, giving an augmented path of 450 m at 300 MHz down to 4.5 m at 3 GHz. Such ionospheric delays cause the major errors in the budget of satellite navigation systems, most of them being corrected in dual-frequency systems. Additional effects for transionospheric systems operating at UHF and SHF comes from scintillation caused by the crossing of heterogeneities.[7]

5.4 Electromagnetic Spectrum and Its Management

To avoid interference between systems, international regulations have been agreed upon for allocating frequency bands to a single application with possible secondary applications.

[7]Anisotropy of the ionospheric plasma must also be considered due to the magnetic field of the earth, which principally affects the polarization of transionospheric electromagnetic waves.

National authorities must negotiate with their public or private users the best ways to match the limited spectral resource to an increasing demand. Among solutions for overcoming such frequency congestion bottlenecks, the design of adaptive systems more robust against interference, as well as better prediction of the effects of interference, would allow the overall number of users in the same frequency band to be increased. Also considered, despite the adverse propagation conditions in the troposphere, is the EHF band, which has large available bandwidths.

Radio frequency spectrum is a scarce natural resource with finite capacity limits and constantly increasing demands. An overview of the current aeronautical radio frequency spectrum is presented in Table 5.3. Radio Frequency Spectrum congestion imposes the need for efficient frequency spectrum management. Spectrum management is a combination of administrative and technical procedures and is necessary to ensure interference free and efficient operation of radio services (e.g., air-to-ground communications and radionavigation). The highest level of spectrum management takes place at the ITU World Radiocommunication Conferences (WRC) and are held every 4 years to discuss maintenance of the international provisions for spectrum management, which are contained in the ITU Radio Regulations (RRs). This includes maintenance of the table of frequency allocations. A consequence is that aviation frequency managers need to develop, and present a case for allocation of frequency spectrum for aviation applications.

Bands	Frequency spectrum	Aviation usages	Types of services
LF and MF	130–535 kHz	Nondirectional beacon (NDB)*	ARNS
HF	2,850–22,000 kHz	Air–ground communication (HF voice and data)†	AM(R)S
	3,023 and 5,680 kHz	Search and rescue†	AM(R)S
VHF	74.8–75.2 MHz	Marker beacon*	ARNS
	108–117.975 MHz	VOR*/ILS localizer GBAS*/VDL mode 4*	ARNS AM(R)S
	117.975–137 MHz	Air–ground and air–air communications (VHF voice and data)†	AM(R)S
	121.5, 123.1, and 243 MHz	Emergency distress frequency‡	AM(R)S
UHF	328.6–335.4 MHz	ILS glide path*	ARNS
	406–406.1 MHz	Emergency locator transmitter (ELT)‡	MSS
UHF or L	960–1,164 MHz	Distance measuring equipment (DME)*—TACAN* and future L-band digital aeronautical communication system (LDACS)	ARNS AM(R)S
	978 MHz	Universal access transceiver (UAT)†	AM(R)S
	1,020–1,040 MHz and 1,080–1,100 MHz	Secondary surveillance radar (SSR) 1090¶ extended squitter¶ automatic dependent surveillance—broadcast (ADS-B)¶ airborne collision avoidance system (ACAS)¶	ARNS
	1,164–1,215 MHz	DME/global navigation satellite system (GNSS)*	ARNS/RNSS
	1,215–1,400 MHz	Primary surveillance radar (PSR)¶	ARNS

TABLE 5.3 Overview of the Aeronautical Radio Frequency Spectrum (Adapted from IATA 2016)

Bands	Frequency spectrum	Aviation usages	Types of services
	1,525–1,559 MHz	Satellite communications (FANS)†	MSS (space-Earth)
	1,559–1,610 MHz	GNSS*	ARNS/RNSS
	1,610–1,626.5 MHz	Satellite communications (iridium)†	AMS(R)S (s-E, E-s)
	1,626.5–1,660.5 MHz	Satellite communications (FANS)†	MSS (Earth-space)
UHF or S	2,700–3,300 MHz	PSR¶ Meteorological radar	ARNS RNS/RLS
	3,400–4,200 MHz	Satellite feeder links to ATS services in Africa	
SHF or C	4,200–4,400 MHz	Radio altimeter‡ Wireless avionics intracommunications (WAIC)‡	ARNS
	5,000–5,250 MHz	Microwave landing system (MLS)* UAS CNPC†/airport surface Communication[1] (AeroMACS)	ARNS AM(R)S/ AMS(R)S
	5,350–5,470 MHz	Airborne weather radar‡	ARNS
SHF or X	8,750–8,850 MHz	Airborne Doppler radar‡	ARNS/RLS
	9,000–9,500 MHz	Precision approach radar (PAR)¶/ airborne weather radar[4]/ASDE	ARNS/RNS
SHF or Ku	13.25–13.4 GHz	Airborne Doppler radar‡	ARNS
	15.4–15.7 GHz	PAR/airborne weather radar‡/Airport Surface Detection Equipment (ASDE)¶	ARNS/RLS
SHF or K	24.25–24.65 GHz	ASDE¶	RNS
SHF or Ka	31.8–33.4 GHz	ASDE¶/airborne radar‡	RNS

AM(R)S = aeronautical mobile (route) service
AMS(R)S = aeronautical mobile-satellite (route) service
ARNS = aeronautical radionavigation service
HF = high frequency
LF = low frequency
MF = medium frequency
MSS = mobile-satellite service
RLS = radio location service
RNS = radionavigation service
RNSS = radionavigation satellite service
SHF = super high frequency
UAS CNPC = unmanned aircraft system command and non-payload communications
UHF = ultrahigh frequency
VHF = very high frequency

*ATM Infrastructure—Navigation
†ATM Infrastructure—Communication
‡Avionics and Airborne equipment
¶ATM Infrastructure—Surveillance

Table 5.3 Overview of the Aeronautical Radio Frequency Spectrum (Adapted from IATA 2016) *(Continued)*

References

International Air Transport Association (IATA). 2016. *Aviation Usages of Frequency Spectrum*, IATA, Montreal, Canada.

International Telecommunication Union (ITU). 1996. *Radiowave Propagation Information for Predictions for Earth-to-Space Path Communications*, ITU, Geneva, Switzerland.

International Telecommunication Union (ITU). 1998. *Ionosphere and its Effects on Radiowave Propagation*, ITU, Geneva, Switzerland.

International Telecommunication Union (ITU). 2013. *Radiometeorology*, ITU, Geneva, Switzerland.

Rudge, A. W., Milne, K. Olver, A. D., and Knight, P. 1982, 1983. *The Handbook of Antenna Design*, Vol. 1 (1982), Vol. 2 (1983), Peregrinus, London, U.K.

PART 2
Aircraft Environment

Marc Pélegrin

5.5 Typical Flight Profile for Commercial Airplanes

Safety

The three keywords for commercial air traffic are safety, efficiency, and environment. The local structure of the atmosphere in which the airplane flies is directly connected to safety. Accidents due to weather phenomena account for between 4% and 5% of the total number of accidents (Boeing source), and 5–7% domestic flight delays are due to meteorological causes, varying according to the season and the airport (Air France source).

For a flight of 1.5–2 hours' duration (gate to gate), accidents occur mainly during the takeoff and climb phases (more than 30%) and approach and landing phases (more than 50%). Some reasons are: at takeoff the aircraft weight could be at its maximum; *rotation*, the instant at which the pilot takes the initial climb attitude (angle of attack), corresponds to 1.3 V_{ST} (1.3 times the stall velocity); the landing phase implies a smooth junction between the airborne trajectory (altitude above terrain is related to *barometric pressure*) and the ground trajectory, which begins at the touchdown point; and atmospheric phenomena are more complex in the ground vicinity.

Until the 1950s, aircraft were considered as behaving like a rigid body; later, in flight, static deformations were included in the computation of the plane structure; the first two planes computed as deformable bodies were the B707 and the Caravelle, both introduced in 1958–1960. Nowadays, structural modes are taken into consideration at least up to the eighth first mode: two bending modes and two torsion modes of the wing, the symmetric and antisymmetric mode of the jet engine masts, first bending and first torsion mode of the fuselage. Modes of rear empennage are also considered, namely when fuel can be stored in the rear for better balance (mainly during the cruise phase), leading to fuel savings. These modes are excited by atmosphere heterogeneities or pilot actions.

Parameters Available on Board

The only measurable parameters on board linked to the atmosphere are static pressure, p_s; dynamic pressure, p_d; and total temperature, T_a. From these data true air speed (TAS) and Mach number are derived by the St. Venant or Rayleigh formula according to the Mach number. The local wind around the aircraft is derived from the true airspeed and the ground speed (if this is available on board). Precise localization systems associated with onboard inertial systems give the ground speed. The equations to be solved are listed below:

Subsonic Uncompressible Flow

Assuming that the local static temperature is available, the true airspeed is defined by

$$V_{TAS}^2 = \frac{2a^2}{\gamma-1}\left[\left(\frac{p_d-p_s}{p_s}+1\right)^{\gamma-1/\gamma}-1\right] \tag{5.19}$$

where a is the sound velocity corresponding to the local static temperature and γ is the ratio between the two heat coefficients (constant pressure, constant volume). But the static temperature is not measurable on board, and hence the value of a is not directly available.

Then a calibrated airspeed (CAS) is defined by:

$$V_{\text{CAS}}^2 = \frac{2a_0^2}{\gamma-1}\left[\left(\frac{p_d - p_s}{p_0}+1\right)^{\gamma-1/\gamma} - 1\right] \tag{5.20}$$

where a_0 and p_0 are the sound velocity and the static pressure at sea level for the standard atmosphere. The calibrated airspeed can easily be obtained on board—it is the "speed" that is shown on the panel instrument and is used by the crew to control the plane.

Compressible Flow

The Mach number is defined by

$$\frac{p_d}{p_s} = \left[\frac{(\gamma+1)^2 M^2}{4\gamma M^2 - 2(\gamma-1)}\right]^{\gamma/\gamma-1} \frac{1-\gamma+2\gamma M^2}{\gamma+1} \tag{5.21}$$

It is important to know the TAS, or at least the CAS and the Mach number, in the event of flying in a wind shear zone, in order to avoid getting out of the flight envelope of the plane. Nowadays the TAS is computed or extracted from stored tables and presented on the instrument panel on the PFD (primary flight display). The flight envelope for an A320 is represented in Figure 5.1; the envelope is graduated in V_{CAS} or M; the velocity used for piloting the aircraft is the V_{CAS}, which appears on the PFD. The V_{TAS} is computed and represented in the right corner of the PFD. In the near future, data will be automatically transmitted to ATC (air traffic control); the controllers will get ground velocities computed on board or derived from the radar tracking. In addition, atmospheric data collected and processed by planes will increase knowledge about the atmosphere.

5.6 The Atmosphere

The Standard Atmosphere

Perfect stability is assumed. The pressure is supposed to evolve according to the diagram presented in Figure 5.2. Such data are used to start the airplane computation, but the final computation should take into consideration the real atmosphere parameters, which can be extrapolated on a probability basis only. The atmosphere is divided into (Figure 5.3):

- The *troposphere*, an 8,000- to 11,000-m-high layer around the Earth, according to the latitude and the period of the year. Horizontal and vertical movements (called turbulence) occur even in clear atmosphere. Inside clouds, namely in active cumulonimbus clouds, turbulence may reach values that can compromise the safety of the flight.
- The *stratosphere*, just above the troposphere, in which air movements are mainly horizontal. Clouds may be present only in the first 5 km of thickness.
- The *tropopause*, a transition layer between the troposphere and the stratosphere. The position and thickness of this layer vary with latitude and season.

ATC works on ground velocities (not air velocities) to elaborate strategic positions of planes in a given airspace. The vertical separation is arbitrarily referenced to a barometric pressure (1,013.25 hPa): 1,000 ft in the lower space and, from 2002, in the upper space. The horizontal separations should be

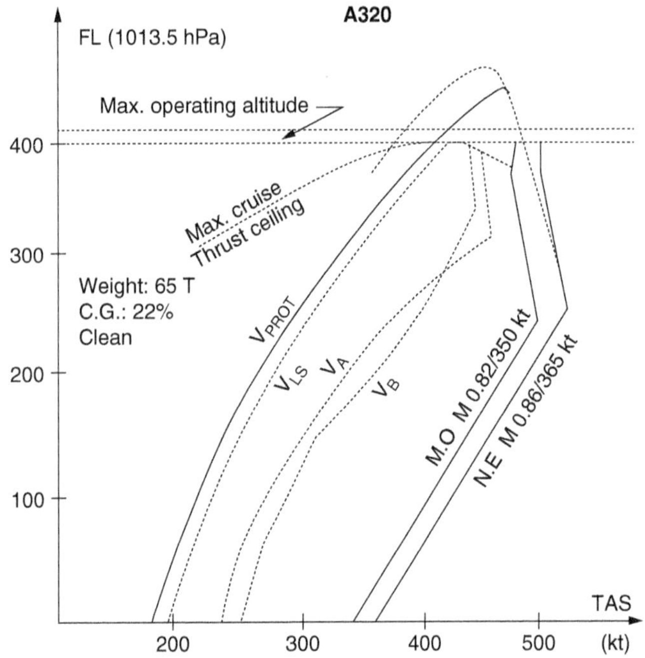

Speeds Definition

L_S: Lowest selectable speed. It corresponds to 1.13 V_S during take-off or following touch and go. It becomes 1.23 V_S as soon as any flaps/slat selection is made.

V_A: Design speed for maneuver

V_B: Design speed for maximum gust intensity and rough air speed (JAR 24.335 (d))

High Speed Protection:

MMO/VMO: M.82/350 kt
M_{MO} + 0.04/V_{MO} + 15 kt: maximum steady speed with full nose down stick.

High Angle Attack Protection:

This protection has priority over all other protections.

V_{prot}: min speed (corresponding angle of attack : alpha prot).
If alpha prot is exceeded, the angle of attack returns to and maintains alpha prot.

Figure 5.1 A320 flight envelope.

reduced in the future to cope with higher densities of planes in a given airspace. Gradients of pressure, wind, and temperature should be taken into consideration in order to guarantee a minimum separation distance in the isobaric surface on which the plane flies.

Reference values are static temperature at altitude 0 on the geoid: 1013.25 hPa, temperature 288.15 K (15°C), density 1.2922 g/dm^3. The standard atmosphere is composed of nitrogen (75.5% in mass, 78.1% in volume), oxygen (23.1%, 21%), carbon dioxide (0.053%, 0.035%), and argon

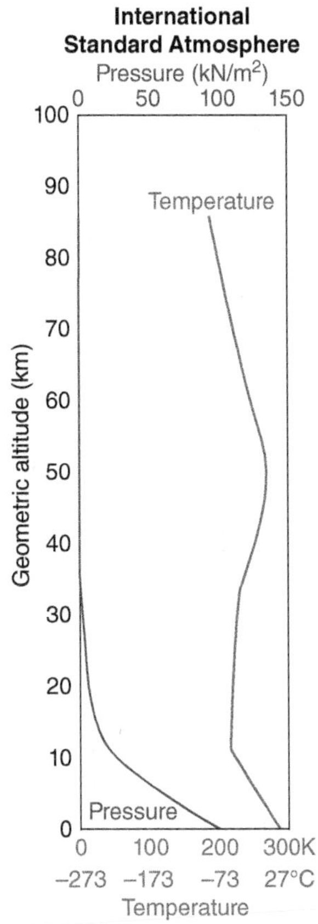

FIGURE 5.2 Standard atmosphere. (Cmglee [CC BY-SA 3.0], via Wikipedia Commons.)

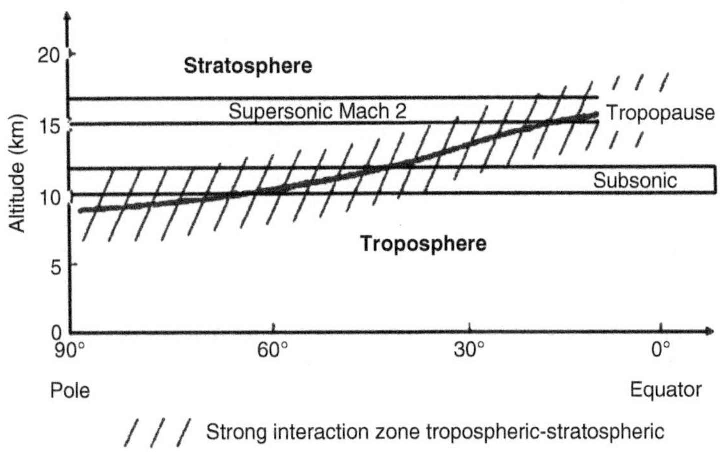

FIGURE 5.3 Troposphere, tropopause, stratosphere.

(1.28%, 0.93%). The composition is constant from 0 to 50 km in altitude. The water content can vary from 30 g/m^3 in topical zones to 1 g/m^3 in polar zones; it is the main parameter for cloud formation.

Thermal Equilibrium (Le Trent and Jancovici 2000)

The two main parameters that contribute to the Earth's temperature equilibrium and consequently to climate[8] are the energy received from the sun and the position and orientation of the Earth in its orbit.

The atmosphere interacts with both incoming energy from the sun and radiated energy from the Earth. The greenhouse effect is due to the reflection of this radiated energy; without this effect, the mean equilibrium temperature would be −18°C instead of +15°C. The energy reflected toward the Earth is due primarily to the H_2O, CO_2, CH_4, and O_3 atmosphere content.

The energy emitted by the sun also varies; the frequency of fluctuation ranges between 11 and 12 years (solar activity) and millions of years. The Little Ice Age, which occurred during the 17th and 18th centuries, was due to a deficit in solar energy. The relative stability of the climate for several hundred thousand years is only partially explained. Presently, above the atmosphere the flux received is 1,365 W/m^2, and that received on the surface of the Earth, at a global mean value and day-night periodicity, is 345 W/m^2.

As to the variation of the rotation axis of the Earth with regard to the ecliptic, the influence of planets such as Jupiter and Venus is dominant. Variation of the eccentricity of the annual cycle (main period 100,000 years), variation of the obliquity (angle between the rotation vector and the normal to the ecliptic plane; main period 40,000 years), equinoxial precession correlated with the mean distance to the sun (main period 20,000 years).

The three-atom gases H_2O, CO_2, and O_3 seem to play a dominant role in the energy balance of the planet, even though their concentration is very low. Excluding the two last centuries, the composition of the atmosphere seems to have been quite constant during the last 10,000 years: 270 ppmv (as measured from the composition of air bubbles contained into ice samples taken from Greenland and Antarctica).

Energy Flow Distribution

For the incoming solar flux:

- 30% is diffused into space (6% corresponds to an interaction between the incoming photons from the sun and the air molecules, mainly O_2; blue photons are emitted; 4% is directly reflected by the Earth surfaces, land, ocean, or ice/snow).
- 50% hits the Earth's surface and is absorbed, leading to a temperature increase; infrared radiation appears.
- 20% is directly absorbed by atmosphere; O_3 absorbs the UV radiation; the O_3 concentration is much more important above an altitude of 10–15 km, which is why the atmosphere concentration increases above the tropopause.

On the Earth's surface, the water evaporation produces a temperature drop and the condensation in the atmosphere produces a temperature rise. The main factor is the atmospheric heat. Heat transfer occurs from equatorial regions (from 30° S to 30° N) to polar regions, with a predominant transfer from tropical regions to subtropical regions (Hadley–Walker cells). The Earth's rotation (which induces the Coriolis force) makes transfers above the 30th latitudes unstable, giving rise to anticyclones and depressions with winds rotating around them.

---

[8]"Climate" comes from Greek and means "inclination" (of the rotation vector of the Earth, with regard to the ecliptic).

Oceans interact with the atmosphere with a time lag of several hundred years. In the atmosphere, air movement is fast but carries little energy. In contrast, oceans carry a high level of energy but over a long time. However, due to its pattern, the Pacific Ocean has a dominant role on a short-term basis (some years); surface currents transfer energy from equatorial zones toward polar regions in both hemispheres (El Niño); in addition, every 2–4 years, warm water is carried from west to east. For the last few years, this phenomenon has been more active, with dramatic consequences—dryness in Australia, Indonesia, and northeastern Brazil and severe rains in California, Peru, and Argentina.

The Real Atmosphere

The real atmosphere differs from the standard atmosphere because its pressure and temperature vary. Winds are caused by these differences of pressure and temperature. The main cause of instability is the daily and annual periodicities of the (apparent) sun motion; the direct consequence is the production of winds. The water content is not homogeneous even in the clear atmosphere (vapor is transparent).

Clouds

Clouds are generated by the ascending motion of moist air, the potential temperature, and the pressure inside a cloud decrease. Cloud formation depends upon the water vapor content of the atmosphere and the number and type of particles, which act as centers of condensation and possibly icing.

Clouds are either droplets of water or ice crystals; liquid droplets may exist in negative temperature (from 0° to −35°C). The basic types of clouds are (Chaon 1999):

- *Cirrus* (Ci), high-altitude (7–15 km) isolated clouds in the form of delicate filaments or white or mostly white patches or narrow bands, silken or fibrous in appearance (altocirrus: ice crystals).
- *Cumulus* (Cu), low (below 2 km) and medium (2–7 km) altitude detached clouds, generally dense; they look like a cauliflower (several kilometers in diameter) with a side wall illuminated directly by the sun or not, as the case may be; their base is relatively dark and nearly horizontal.
- *Nimbus*, mainly *nimbostratus* (Ns), low-altitude, gloomy clouds, often dark, which generate rain or snow, very often continuously.
- *Stratus* (St), generally low-altitude clouds, looking like layers or extended flat patches at low, medium, or high altitudes; if the sun is discernible no halo is produced (except at very low temperature).

From the basic clouds mentioned above, the following clouds are derived:

- *Cirrocumulus* (Cc), thin, white patch, sheet or layer of cloud without shading, composed of very small elements.
- *Cirrostratus* (Cs), transparent, whitish cloud veil of fibrous or smooth appearance, totally or partly covering the sky; these produce halo phenomena.
- *Cumulonimbus* (Cb), accumulation of big, gloomy clouds, with considerable vertical extent: such thunderclouds may generate lightnings; the upper part often spreads out in the shape of an anvil.
- *Stratocumulus* (Sc), low-level layer having a dappled or wavy structure, with dark parts.
- *Altocumulus* (Ac), white or gray patch, sheet, or layer of cloud, generally with shading, composed of laminate rounded masses, rolls.
- *Altostratus* (As), grayish or bluish cloud sheet or layer of striated, fibrous or uniform appearance, totally or partly covering the sky.

Active cumulonimbus clouds are dangerous for aircraft due to pronounced turbulence and vertical velocities of 20 m/s or more in the updraft ascending core of the air/water, severe turbulence, lightning, and ice. They can be detected by radar aboard the aircraft.

Turbulence, Wind Shear

The real atmosphere has no stability, though in many regions of the world, as in the temperate zone, the weather structure of the atmosphere evolves slowly. Weather is a consequence of air movements around the world; as for any system for which a good mathematical model exists, it should be predictable. Air movements are governed by partial differential equations and are known with reasonable certainty, but a set of accurate initial conditions (4D) is not yet available, in spite of many meteorological satellites. Meteorologists proceed by region (using ground grids with horizontal sizes varying from a few kilometers to hundreds of kilometers) and try to set coherent initial conditions for each grid. The computer then solves the equations and arrive at a correct (4/5) forecast for 48 hours. Local random motion of air within the motion of a large mass of air (which covers an area of some tens or hundreds of square kilometers) is called *turbulence* and interacts directly with the aircraft structure. The size of the turbulence ranges from several meters to several kilometers. To be certified, a plane must experience no damage (more precisely, it should stay in the elastic domain) when crossing a gust or flying in a turbulent area. Gusts are defined by specifications which vary slightly among the countries which certify planes. For example, in France, the two major conditions to be satisfied are

1. A vertical gust of "1 + cos" type (Figure 5.4), given by

$$W(t)=0 \quad \text{for} \quad t<0 \quad \text{and} \quad t>\frac{d_m}{V} \tag{5.22}$$

$$W(t)=\frac{W_n}{2}\left[1+\cos\left(\frac{2\pi Vt}{d}\right)\right] \quad \text{for} \quad 0<t<\frac{d_m}{V} \tag{5.23}$$

where W_n = gust amplitude (m/s)—according to FAR 25.341, 20 m/s for height <20,000 ft 11 m/s at 50,000 ft
d_m = gust wavelength (m) (depends on the wingspan of the aircraft)
V = aircraft velocity
t = time

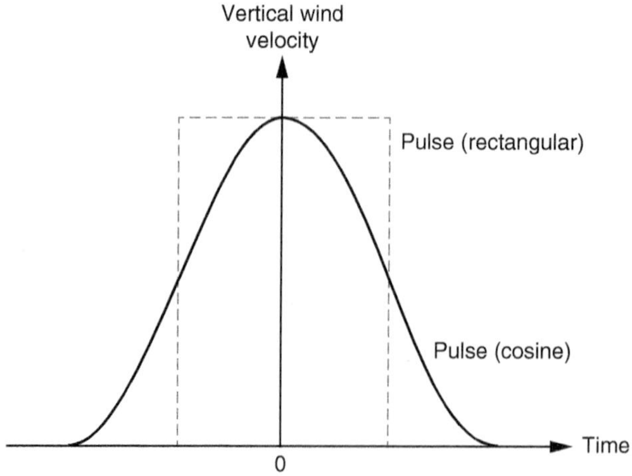

FIGURE 5.4 Vertical gust profiles.

2. A von Karman spectrum for the turbulence, given by

$$\Phi_w(\omega) = \sigma_W^2 \frac{L_W\{1+(8/3)[1.339(L_W/B)\omega]\}^2}{[1+1.339(L_W/B)\omega]^{11/6}} \tag{5.24}$$

$$\sigma_w^2 = \frac{1}{2T}\int_{-T}^{+T} \phi_w(\omega)\,d\omega \tag{5.25}$$

where L_w = turbulence length scale (m)
σ_w^2 = mean square value of velocity (m^2s^{-2})
ω = pulsation (rd/s)

Note: Numerical values must be coherent with the aircraft safety level reached at a given time. As this safety level increases, the amplitude of the gust or turbulence spectrum should be that which has a probability of occurrence of the same value as the safety level of the aircraft.

Meteorologists use four grades of turbulence, independently of the type of turbulence: *light, medium, severe,* and *extreme.*

Wind shear occurs when two layers of wind in the atmosphere have different velocities and/or directions. Due to friction between the two layers, a transition zone in between the two laminar layers is highly probable. Wind shear can exist at any altitude. Approach controllers are particularly interested in wind shear because it interacts with the safety of the landing.

Downburst

In the 1980s, a new phenomenon was identified: *downburst.* A downburst is the collapse of a mass of cold air suspended at some thousand meters of altitude by active ascending movement of air. When the collapse occurs, a downstream of saturated air may reach velocity above 40 kt (Figure 5.5; note that *downburst* is called *microburst* in the figure). As the mass of air goes down, the local temperature increases. In the upper part of the downflow there may be droplets, which may disappear by evaporation below a certain altitude (depending on the local temperature) and consequently are difficult to detect. When the stream hits the ground, a giant vortex appears. This is a very dangerous phenomenon. Predetection is difficult, its detection requires permanent real-time analysis of the structure of the local atmosphere around the airport.

Downbursts have been clearly explained by Fujita (1985): "Some aircraft accidents that occurred at low altitudes during convective activity were regarded as pilot error without blaming the weather systems as major contributing factors."

Tornadoes and Microbursts

Tornadoes consist of ascending motion of saturated warm air in a column several hundred meters in diameter. They appear on hot seas or lakes (surface temperature higher than 27°C) mainly in the afternoon. Downbursts are frequently associated with tornadoes, an additional reason to avoid tornadoes. The most spectacular phenomenon in which tornadoes and downbursts were associated occurred in July 1987 at Teton-Yellowstone (United States) and was carefully studied by Fujita (1985). The U.S. Forest Service indicated that 1 million trees were uprooted in a 61-km^2 area over a period of 26 minutes. The analysis of the orientation of the fallen trees was a powerful tool of investigation. Over an area 2.5 km wide and 39 km long, four swirl marks of spin-up vortices (tornadoes) and 72 microburst outflows were identified.

How could the tornado have maintained its fury against large frictional torque in the boundary layer over rugged terrain? Analysis of the damage caused along the trajectory of the tornado and on its sides suggests that the angular momentum of the tornado was supplied by microbursts as their outburst winds spiraled into the tornado center.

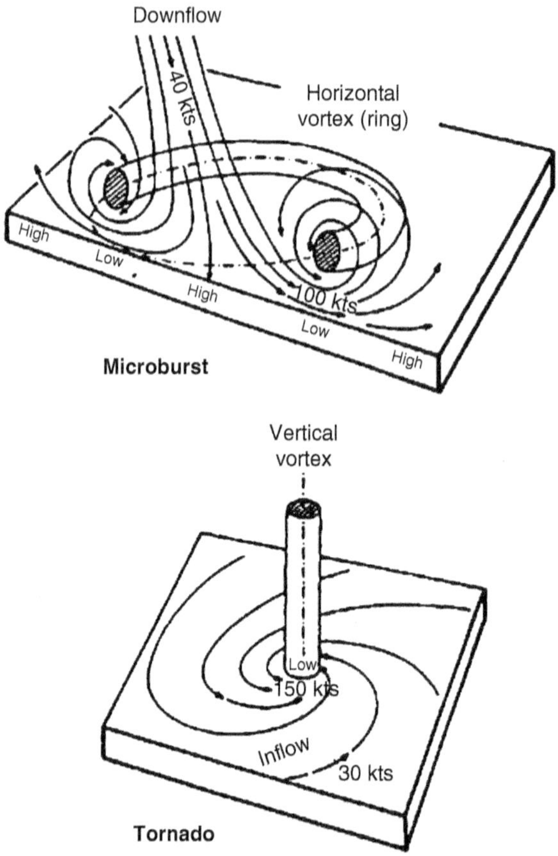

FIGURE 5.5 Downburst and tornado profiles (from Fujita 1985).

A tornado can be detected easily by radar or lidar or, most of the time, by direct observation because of the water content. However, the side microbursts which accompany the tornado are often "dry" and not directly visible; it is recommended that flying be avoided at least 2.5–5 km away from the tornado.

Jet Streams

In the lower stratosphere, jet streams are frequent. These are "tubes" of air, roughly horizontal, several hundred meters or a few kilometers in diameter. Velocities may reach 200 m/s in the center. The flow is normally clear and laminar in the core, and the transition zone is highly turbulent. The direction is usually west to east.

Lightning

Upward convective motions in the troposphere may generate concentration of electrostatic charges. In the vicinity of a cumulonimbus cloud, strong electrostatic fields (500 kV/m) may be encountered by an aircraft flying in the cloud or in its vicinity, and electromagnetic perturbations may occur. On average, long-range aircraft receive a strike once every 3,000–4,000 hours, while short-range aircraft receive a strike every 2,000–3,000 hours; damage, if any, is rarely severe. Total destruction of the plane by lightning is very rare (less than three cases during the last 50 years). However, lightning is very often accompanied by strong adverse atmospheric conditions such as severe turbulence and icing (in an accident it is very difficult to decide which phenomenon was the real cause of the accident). Lightning is a discharge between zones in which the density of electrostatic charges is high and of opposite polarity. Lightning is composed of a short-duration impulse (about 200 ps to 2 μs) with an intensity of thousands of amperes and gradients reaching 100 kA/μs, followed by another, much longer pulse (several ms) but with an intensity much lower, about 100 A. Interference with airborne radio and electronic equipment is produced by the former, and damage can be caused to an aircraft by the latter since it contains more energy. Lower-power electronic chips and the increasing use of composite materials (though with conducting material incorporated) in aircraft mean that electronic equipment will have to be studied carefully.

Optical processors and an optical data bus will replace electronic equipment in the future (around 2005–2010).

Icing

In a cumulus cloud the water content is about 2.5 g/m^3 and the diameter of droplets is between 10 μm and 40 μm. Supercooled droplets can turn into ice when they collide with an aircraft structure, forming *rime ice* if the temperature is about −30°C or, if the temperature is close to 0°C, *glaze ice*. The accretion of ice may give rise to two horns. Supercooled droplets glide on the surfaces on which they have been deposited and may turn to ice somewhere on the wing, blades, or fuselage. The laminarity, if it was present before such a cloud was entered, is destroyed, the lift coefficient is reduced, and the drag coefficient is increased.

There are two types of icing clouds. Stratus clouds extend over a large area. Their content of water is low (0.1–0.9 g/m^3), and they produce continuous icing. Cumulus clouds have a water content of about 3 g/m^3. Their size rarely exceeds a few kilometers, and they can extend from 2 to 4 km up to 12–15 km in altitude. Icing may also occur in clear air after an aircraft which has flown a long time through cool air moves into a clear, warm region. The water vapor in this warmer air condenses and freezes over the entire aircraft.

5.7 Other Atmospheric Hazards

Other hazards to aircraft due to interactions with the atmosphere are described below.

Turbulence due to the Aircraft

The main parameter to be considered is the wake vortices, which escape from the wings at their extremities. The wing-end vortex results from the difference of pressures on the suction side (above) and pressure side (below) of the wing. From the right wing, the vortex rotates in the positive direction.

Behind a plane the structure of the atmosphere is modified in such a way that another plane crossing the wake vortices or penetrating into them can be exposed to a dangerous situation. According to Thomas Heintsch of the Institute for Flight Guidance and Control, Braunschweig University, the development of the wake vortices extends up to 200–250 wingspans behind the plane, which means approximately 10 km. The shape is quite constant (intensity and extension) during the first 50 wingspans. Then the intensity decreases and the lateral extension increases. Their intensity is proportional to the mass, balance, and load factor of the plane at a given time. Separation rules were established in 1970 (ICAO). Planes are classified into three classes: heavy (H), 130 T and up (250 passengers and up); medium (M), between 130 T and 10 T (50/250 passengers); and light (L), less than 10 T (less than 50 passengers). The minimum separation distances are (planes aligned on the ILS) given by

$$H/H = 3\ \text{nm} \qquad H/M = 5\ \text{nm} \qquad H/L = 6\ \text{nm} \qquad M/L = 4\ \text{nm} \tag{5.26}$$

The separation rules are now obsolete, and real computation of the dangerous zone is possible.

The damping of the vortex is low (air viscosity); the local turbulence increases the expansion of a vortex motion. It is accepted that the decrease factor is $t^{-1/2}$ for a calm atmosphere and t^{-2} for a turbulent one.

Interference with the Ground (Puel and Saint Victor 2000)

Due to friction with the local atmosphere, the vortices go down. If a vortex encounters the ground, reflections occur and interact with the initial vortex. Let's consider a wake vortex which is descending close to the ground. Local velocities are higher than those due to the mean wind, and additional decay of the vortex appears. However, the phenomenon is more complex and the descending vortex may generate a secondary vortex (opposite in rotation). The main vortex induces a lateral flow (with regard to the axis of the vortex) which then can suck the ground boundary layer and give birth to a bulb. Due to the pressure gradient, the vorticities of the two vortices are opposite and a vortex moving toward the initial vortex may be generated (Figure 5.6). The energy stored in the second vortex comes from the first one.

The intensity of the vortex can be computed from the airplane parameters and the local characteristics of the atmosphere. A locally turbulent atmosphere damps the vortices more rapidly than a calm atmosphere. The local turbulence is slightly increased but has no structure, and the separation distance can be reduced. Nowadays the trajectories of the vortices can be estimated (position and intensity) from the aircraft parameters and wind can be measured on the airfield. For airports, equipped with lidar (or sonar), the intensity of the vortices may be roughly measured. Separation distances of planes can be computed in real time, including the lateral wind component with regard to the runway axis. The introduction of the A380 will impose real-time determination of separation between planes.

Birds

An aircraft is certified against collision with and ingestion of birds. Tests are performed on the ground: a (dead) bird of a specified mass is sent toward the cockpit windows, toward a propeller or inside a jet engine (if blades are destroyed, they should be self-contained inside the jet engine body).

The most dangerous (and frequent) case is ingestion when the plane is accelerating on the runway. If the ingestion happens before V_1 (the maximum speed after which the braking distance is higher than the length of the runway in front of the plane) and is immediately detected, the takeoff should be abandoned. In addition to the risks attached to emergency braking, the risk of a fire resulting from the ingestion is slightly higher than that due to a failure of the jet engine without any ingestion. If the ingestion appears after, or is detected after V_1, the situation is much more critical; takeoff is mandatory even if V_2 has not been reached (V_2 is the recommended speed for rotation, i.e., the speed

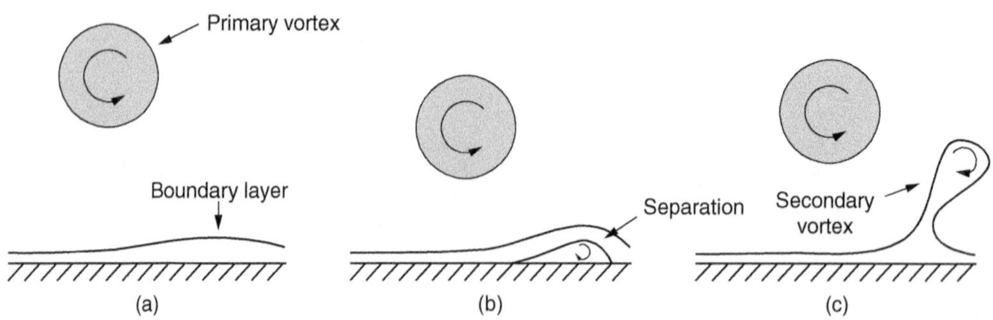

Figure 5.6 Interaction between vortices and ground (ONERA).

at which the longitudinal attitude of the plane corresponds to the initial climb). The ingestion is mainly detected by accelerometers set on the jet engine body which detect vibration of the pod.

The ingestion of birds is relatively frequent; Air France encounters some six to eight jet engine bird ingestions per year, leading to engine damages.

Meteorological Balloons

Some meteorological centers are qualified to send balloons equipped with instruments to measure atmospheric parameters across the atmosphere seven times a day (there are seven such centers in France). Balloon and nacelle are made so that collisions with aircraft are not hazardous, and the balloon is launched in accordance with present air traffic. However, some risk still exists.

Smoke from Volcanoes

There are presently 30 active volcanoes in the world, and during the last three decades about 200 volcanoes have been active. At least one fatal accident and one incident (four engines out but recovery after a while) occurred during the last 25 years due to ingestion of smoke ejected by a volcano. Even if there is no blowout, the smoke is composed of very hard dust which interacts with the engines and damages them. The largest particles of volcano smoke, most of which fall within a few days, may constitute a danger for planes flying across the volcano's plume. The consequences may be the following:

- Jet engines being turned off because particles are deposited on hot parts (600–800°C) and then form a solid state quite similar to glass
- Loss of aerodynamic data due to the pitot tube obstruction
- Erosion of front parts of the wings and opacification of glass windows
- Radio jamming due to electrical discharges encountered inside the plume
- Chemical corrosion due to acid droplets
- Fuel contamination by ash and soluble components such as Pb, Zn, and Cu

Between June 9 and 21, 1991, just after the Pinatubo eruption, nine incidents involving the replacement of 10 engines were registered.

Magnetic Storms

A magnetic storm is an ejection of charged particles coming from the sun. There is a correlation with the sun cycle (11 years). The energy involved may reach 10^{26} J within a few minutes. When the particles reach the magnetosphere, the magnetic field is modified and electromagnetic inductions appear. This is a frequent phenomenon, and the consequences are well known: temporary degradation of the position precision of satellite positioning systems may occur, or no signals may be received for many minutes or hours; satellites, mainly geostationary ones, may be partially destroyed (internal flashes, destruction of solar panels, etc.); prediction of occurrences and protection against such magnetic storms are difficult.

Traffic

The plane is not alone in the sky. The airspace is divided into six classes designated A to F. In each class a minimum of onboard equipment is mandatory. A flight can be operated under VFR (visual flying rules), where avoiding collision is the responsibility of the pilot, and IFR (instrument flying rules), where avoiding collision is the responsibility of the ground controller. In each class, IFR and VFR are possible under certain conditions, except for class A, a class in which only IFR flights are authorized. Before departure, the crew fills out a flight plan which describes the desired flight profile. According to the present traffic, the controller accepts it or modifies it. The plane is followed by ground controllers even when it flies above oceans; the crew reports the position of the plane at least every 20 minutes. Separation of planes is under the ground controller's responsibility. The situation is evolving rapidly thanks to automatic reporting systems such as ADS-B (Automatic Dependent System—Broadcast). The position of the plane, computed on board using Global Navigation Satellite Systems (GNSS) or ground systems (VOR, DME, ADF, LORAN,[9] etc.), is broadcast at a given frequency (which can be chosen from 1 second to 10 minutes). The data can be relayed by satellites and made available to ground control centers concerned with the flight.

Ingestion of Stones

During rolling on the runway or taxiways, stones can be ingested (or other objects: the Concorde accident on July 25, 2000, occurred due to ingestion of a piece of metal dropped on the runway by the plane which took off before). The structure around the pods (bottom part of the fuselage, wing, and pod itself) is sometimes modified by small ailerons in order to avoid the ingestion of stones thrown away by the wheel of the front landing gear. A small additional drag is induced.

5.8 The Ionosphere

In the atmosphere an ionized layer is situated between 60 and 1,000 km, with a maximum concentration around 400 km. Electromagnetic waves crossing this layer are more perturbed if the frequency of the crossing wave is low; HF band and below are completely reflected.

For the frequencies used in GPS (1.2 and 1.5 GHz, plus 1.1 GHz in 2006), there is a slight energy attenuation and a slight increase of the traveling time, leading to an increased distance of several meters between the satellite and the receiver. However, if two modulated frequencies are used, the perturbation may be corrected. This is why a third nonencrypted frequency will be transmitted by GPS satellites starting in 2006.

The delay Δt_1 which occurs on a frequency f_1 is related to the total electron content (TEC) encountered. If two frequencies f_1 and f_2 are used, there is a relation between the difference $\Delta t_1 - \Delta t_2$ and TEC. Then, the TEC being known, it is possible to compute the delay Δt_f which occurs on the frequency f used to determine the pseudorange (c = light velocity), given by

$$\Delta t_f = \frac{40.5}{c \cdot f^2}\,\text{TEC} \qquad (5.27)$$

Magnetic Storms

Solar activity is not constant. Solar activity disrupts the ionosphere and affects, for instance, the pseudorange in GNSS. Solar activity has a period of 11 years (high activity between 1999 and 2002). When a storm occurs, electrons and protons are ejected from the sun's surface (for large storms, up to 10^{16} g).

References

Chaon, J. P. 1999. *Cours de physique des nuages,* Meteo-France, Toulouse.

Fujita, T. 1985. *The Downburst,* University of Chicago, SMRP Research Paper 210.

Puel, F. and Saint Victor, X. de. 2000. "Interaction of Wake Vortices with the Ground," Aerospace Science and Technology, Vol. 4, issue 4, pp. 239–247.

[9]VOR = VHF omni-range, DME = distance measurement equipment, ADF = automatic direction finder, LORAN = long-range navigation.

PART 3

Electromagnetic Compatibility

J. P. Parmantier, J. P. Catani, and M. Crokaert

5.9 Introduction

Why compatibility in electromagnetics? The general answer is that electronic equipment has to operate in very different types of environments with which it has to remain compatible.

First, equipment does not have to be susceptible to the surrounding electromagnetic (EM) fields generated by the environment. The environment may be external to the entire system, as is the case with natural threats like lightning or electrostatic discharges, or human threats. Some are unintentional, but some systems, mainly military, are also concerned with intentional threats generated by EM weapons. The threat may also be internal to the system itself, generated by other pieces of equipment. Second, a piece of equipment does not have to generate EM perturbation likely to interfere with another piece of equipment. This is why compatibility with its environment is required.

In aeronautics and aerospace, electromagnetic compatibility (EMC) has been known for a long time. The reason is that all the possible EMC problems are closely connected to reliability and safety. Indeed, in the air or in space, a failure of equipment may lead to serious casualties for the system's functions and, more serious, for people. Eventually, the entire transport domain has become concerned with EMC because of the increase of electronics in all its systems. Since the advent of EM weapons in the 1970s, military systems have also been involved in EMC, with the objective of being totally hardened to EM interference. EMC has become a discipline, thoroughly accounted for by any industry, from shipbuilders to household electrical manufacturers. Now, in addition to the functional aspect, EMC is also an economic challenge. To sell their products, manufacturers have to demonstrate their compliance with EMC standards. Therefore, they have to apply protections on their equipment and optimize them in terms of price, room, and weight.

Subsection 5.10 deals with the physical process of EM coupling, which leads to the generation of EM interference. Subsection 5.11 presents the characteristics of the main EM threats and their associated standards. Subsection 5.12 introduces experimental and numerical tools commonly used in EMC design and analysis. In Subsection 5.13, engineering methods for EMC are investigated, including a discussion of the control plan, the specifications, the conception, and the installation rules. Finally, the conclusion focuses on the future of EMC.

5.10 Background of EM Coupling

Theory of EM Diffraction

The physical process that makes an interference act on a system is known as EM coupling. If an incident EM field is applied to an object, an induced current is generated on the surface of this object. This current is a potential interference for the object. Meanwhile, a scattered field is generated around the object. This field may also cause EM interference in its vicinity. On one hand, the theory of EM coupling has similarities to the theory of antennas because it is directly derived from the theory of diffraction (Stratton 1941). However, if antennas are mainly involved in far fields, EMC is mostly interested in near fields, which makes the usual approximations of antenna not applicable. On the other hand, EM coupling also has relations with circuit theory when the response of equipment connected with cable bundles is concerned or when electric protections as filters or limiters have to be considered. However, EM coupling involves additional information in terms of distributed equivalent sources induced by incident EM fields.

Considering the definition we gave of compatibility as being compatible with the ambient external EM field and as having the environment compatible with the EM-emitted field, two domains of analysis are commonly considered in EM coupling: EM susceptibility and EM emission.

EM susceptibility stands for an external stress applied onto the object under analysis. Two kinds of EM susceptibility problems are commonly distinguished: radiated problems, when an incident field is applied and conducted problems, when a current is forced on the object.

In radiated EM susceptibility, it is important to understand that the incident field E_t^{INC} is the field in the absence of the object. This condition is rigorous and comes directly from the theory of diffraction. When E_t^{INC} is applied, a current is induced on the surface in such a way that the total tangential field E_t^{TOT} on the surface verifies the following limit condition, given by

$$E_t^{\text{TOT}} = E_t^{\text{SCA}} + E_t^{\text{INC}} = Z_s \cdot J^{\text{SCA}} \tag{5.28}$$

where Z_s = surface impedance of the lossy surface
J^{SCA} = scattered current induced on the surface
E_t^{SCA} = the tangential scattered electric field

For example, if the surface is metallic, E_t^{TOT}, is zero and E_t^{SCA} verifies:

$$E_t^{\text{SCA}} = -E_t^{\text{INC}} \tag{5.29}$$

In conducted EM susceptibility, the current is directly applied on the system or forced with a generator to simulate the current induced by an incident radiated field or a perturbation generated by another part of the system. EM emission is quite similar to conducted EM susceptibility in the sense that the source is also applied on the system or comes from a piece of equipment. But, in addition, we are mainly interested in the scattered fields radiated by the surface currents.

EM Coupling Phenomenon

After the general definition of scattering on a surface, the presentation of several physical processes will help the understanding of EM coupling.

Current Redistribution on a Surface

On an external surface, the first phenomenon to consider is redistribution of currents. At very low frequency, the current scatters with respect to the different resistance paths encountered, therefore following Ohm's law. On the whole frequency range, the current follows the paths of lower impedance. Particularly when frequency increases, the impedance due to the inductance of the structure becomes more important than the resistance. Because of the inductive effect, the current lines tend to separate from each other. At very high frequency, they follow the edges of the object. In the intermediate frequency range, there is a cutoff frequency where the resistive effect balances the inductive effect. For instance, this property explains why carbon materials progressively behave like metal.

EM Penetration through a Surface

EM penetration through the surface may occur through two types of processes.

The first is EM diffusion and comes from the finite depth of the materials. It only concerns low frequencies when currents are able to penetrate into the materials. This phenomenon is also known as the skin effect: for an external excitation, the current will progressively concentrate on the external surface. The phenomenon may be understood as a generalization of the inductive effect on a finite surface applied here on the finite transverse dimension of the depth.

The second EM penetration type is due to scattering through apertures. "Apertures" is a generic name for a large variety of geometrical configurations: windows, holes, seams, junctions between panels, electromagnetic joints. The phenomenon is significantly dependent on manufacturing technology. Nevertheless, general rules may be established as a function of frequency. For instance, at low frequency, when the aperture is small compared to the wavelength, the scattered field is equivalent to the one radiated by an electric dipole and two magnetic dipoles (Degauque and Hamelin 1993; Boudenot and Laboune 1998). If the aperture is loaded with a resistive material, a cutoff frequency f_c appears on the magnetic field. Under f_c, the magnetic field penetrates, which is in agreement with a very general property in EM coupling; there is no perfect protection against the magnetic field for any kind of resistive material. Beyond f_c, the magnetic field is attenuated with a 20-dB per decade slope (see Figure 5.7). Also, the electric field is attenuated as soon as the very low frequency for almost all the resistive materials. The resistance of the junction area connecting the resistive material of the aperture to the metallic frame modifies the value of f_c (Boudenot and Labaune 1998).

Definition of EM Shields

The definition of currents induced on and through surfaces raises the important concept of EM shields. In EMC, a shield is a material that deviates the current, preventing it from running in undesired zones. Of course, the most efficient shield is metallic because all the current may be driven into it. But even if perfectly metallic, a shield is not efficient if the current is not able to flow on it. This means that the shield must be connected at both ends to ensure the derivation of the current. In addition, because currents circulate on shields, they are more likely to radiate inner scattered fields. This is why the geometry of an optimized shield defined to protect a system must tend to keep the shape of a closed enclosure (generalization of the principle of Faraday cages).

EM Coupling on Cables

Because wiring is found everywhere in electrical systems, it is the most frequent cause of EMC problems. On the one hand, coupling on cables plays a particular role because cables are receiving antenna likely to transform the incident field in generators driving interference signals propagating at the equipment input. On the other hand, this interference propagating in clean zones may radiate undesired EM fields.

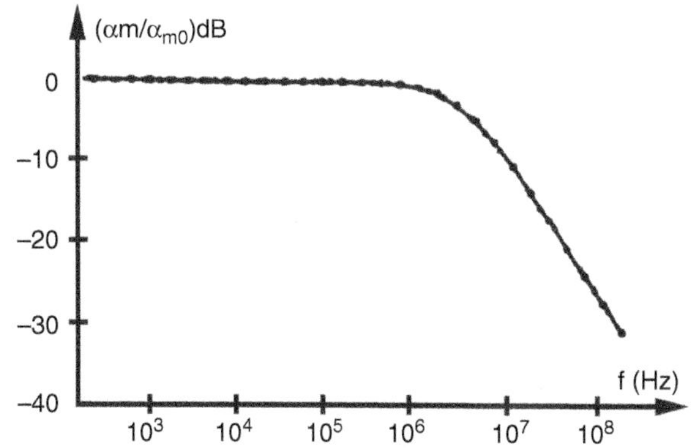

Figure 5.7 Ratio between the magnetic polarizabilities of a loaded aperture and a free aperture (from Bouderot and Labaune 1998). Magnetic/electric polarizabilities are linear coefficients defining magnetic/electric dipoles.

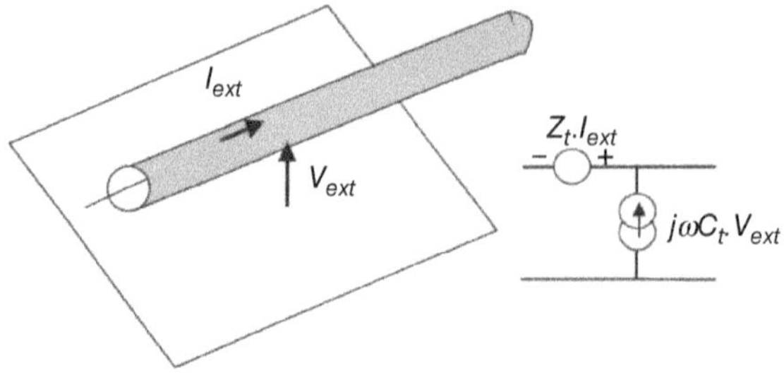

FIGURE 5.8 Equivalent model of EM coupling in a section of a shielded cable.

The other characteristic of wiring comes from its organization in bundles and branched harnesses. Therefore, there is electric and magnetic influence between wires in the same bundle. The so-called cross-coupling effect enables a perturbation on a wire to propagate from wire to wire.

To reduce radiation of cables or EM coupling of incident fields, bundles may be shielded, wrapping them in metallic screens. As seen before, because of the finite conductivity of the screen, the penetration of the magnetic field can never be totally stopped at low frequency. At higher frequency, a diffusion effect may appear in the depth of the screen. If the screen is made of a metallic coating, the scattering through small holes has also to be considered. Depending on the depth of the coating the diffusion effect may be hidden by the scattering through holes (Degauque and Hamelin 1993; Vance 1978). The penetration of the magnetic field is equivalent to applying a distributed voltage generator, linearly related to the external current I_{ext} circulating on the shield by the so-called transfer impedance Z_t. Associated with the penetration of the magnetic field is the penetration of electric field. The equivalent coupling model on the inner wires is a distributed current generator linearly related to the external common mode voltage V_{ext} developed on the shield by the so-called transfer capacitance C_t (see Figure 5.8). As seen above, the per-unit length circuit model supposes that the EM shield is correctly connected to the ground at both ends. In addition, the Z_t and C_t play a reciprocal role if one wants to determine the radiation of a shielded cable when the source of interference is on an inner wire.

5.11 EM Environment and EMC Standards

In this subsection, we consider the external environment in which the system is likely to operate and the internal environment produced by the equipment of the system itself. Hereafter, we will present the most significant threats to account for in aeronautics and aerospace.

External EM Environment

Two types of external environments may be distinguished: natural and human-made environments.

Natural EM Environment

Natural EM environments such as lightning and electrostatic discharges (ESD) are threats which man has little power to avoid. The only possible action is to control their effects.

Lightning is a serious threat capable of leading to the destruction of the system. On aircraft, it is tolerated that the system is stressed by a lightning strike. The idea of protection is to maintain the evacuation of the injected current on the outer surface of the aircraft only. For instance, on radomes, lightning protection strips are installed onto the transparent material in such a way that the current does not flow into the antenna system. In the case of space launchers, the lightning strike is not tolerated in operation to avoid accidents. In addition, the launching pads themselves are secured

Figure 5.9 Lightning protection system on Kourou's ARIANE 5 launching pad (courtesy ESA/CNES).

with protections set all around the launchers. For example, ARIANE 4 and ARIANE 5 are protected by posts that deviate the possible lightning current outside the pad (see Figure 5.9).

The tests encountered in lightning standards are formulated in time domain. Different surge waveforms have been proposed describing the different phases of the propagating current in a lightning channel (harmonization documents from the Society of Automotive Engineers, SAE-4L). The frequency spectrum of all the waveforms is lower than 50 MHz, and the maximum threat has an amplitude equal to 200 kA (waveform A).

ESD is also a serious natural EMC threat. The general process is a local increase of static potential due to charge accumulations in different locations of the system creating difference of electric potentials likely to generate sparks. On aircraft, important problems occur on cockpit canopies where charges are deposited due to triboelectricity generating sparks on the canopy and in connectors of the heating circuits. In aerospace, ESD occurring on launchers when the stages made of different materials is always a relevant problem.

In space, absolute potential voltages on satellites may generate typical current densities equal to 10 $\mu A/m^2$. Additionally, in the special case of geostationary satellites, the implantation of charges may lead to electric field surges up to 50 kV/m, with rise times lower than 10 ns. The waveform generated is variable and depends on the polarity of the discharge. Standardized tests to represent ESDs are not always available. Even if standards exist for human-origin ESD (CEI standards), there are no real identified standards for ESD on satellites or aircraft. However, on satellites the project of ISO (International Organization for Standardization) 14302 seems to be in the process of being accepted.

Human-Made Environment

The human-made EM environment is also likely to constitute a serious threat. Some threats are intentional and must be considered as generated by real weapons against which all military systems have to be protected (MIL-STD-461D, 462D, 464). Here we will mention two of the most important intentional threats: EMP and HPM. EMP stands for the *EM pulse* generated after an atmospheric or extra-atmospheric nuclear explosion (Lee 1980). The standardized threat is expressed in time domain with electric field pulses up to 100 kV/m and rise times lower than 10 ns. The frequency spectrum of this threat extends up to 100 MHz.

HPM stands for *high-power microwaves* and is associated with new types of weapons that appeared at the beginning of the 1990s. Here again, the threat is a pulse but the frequency content is larger, up to several GHz. The difference with EMP is that the generation of power requires focusing antenna and the magnitude strongly depends on the distance of the source and the capability of available technology. This is why, up to now, this threat has not been totally standardized.

Other threats are unintentional and are mainly due to high-intensity fields created by communication systems. For the civil world, such threats are standardized under the name *high-intensity radiated fields* (HIRFs). They describe the environment created by high-intensity antennas and radar as the ones likely to be encountered in the vicinity of airports or on ships. Consequently, they concern aircraft and space launchers. In common civil EMC standards, the constraint imposed on the standardized external field is as large as 200 V/m for the electric field with a frequency spectrum ranging from DC to 18 GHz (RTCA DO 160 and documents from the SAE-4R subcommittee). The demonstration of those standards is not obvious because it is quite impossible to generate a plane wave illuminating a whole object with enough level in this frequency range. Illumination of parts of the object is generally the only available solution.

Internal EM Environment

First, systems must also comply with the internal environment generated by other systems. The first elements to consider are the onboard antennas, which can generate unwanted fields in the direction of their side lobes, for example. On satellites, sensors such as altitude sensors used to work with electric fields about 10 V/m may be perturbed by fields generated by onboard antenna up to 100 V/m.

Secondly, systems that are not made to transmit fields may also be significant interference generators. This is the case with power supplies producing coherent rays on both power and ground network. We think here of uninterruptible power supplies (UPS) and regulation of chopping, widely used for the efficiency and low weight they provide. On satellites, an EMC margin is imposed on modulators and emitters to tolerate a noise of about 1 V RMS on all power inputs. In addition, to reduce common mode in the power networks, a maximum limit of the resistance of the bars is specified and is obtained by increasing the section of the bars.

Finally, the problems generated by devices external to the system itself, but likely to interact with it, must be mentioned. This is the well-known case of portable electrical devices (PEDs), for which, so far, the only solution has been to forbid their use on aircraft.

5.12 EMC Tools

Experimental Tools

Experimental tools are necessary to demonstrate the standards. For radiated susceptibility, radiated antennas are required to generate the field levels imposed. At low frequency, the size of the antenna is large but high power is made available through the use of pulsed generators. For example, in EMP the size of simulators is much greater than the size of the systems under test.

For the special case of lightning on aircraft, the technique of coaxial injection is commonly used to simulate a uniform circulation of the currents on the surface or a part of the surface. The interest of the setup is that the current injected and especially the return current can be controlled and considered as mainly symmetric. Nevertheless, it is obvious that this test modifies the circulation of the currents on the aircraft compared to a real lightning injection.

At higher frequencies, fields are generated by smaller antenna with more localized effects because the whole power cannot be applied on the whole structure. For small systems, radiated tests may be performed in anechoic chambers, but these are generally more appropriate for EM emission tests. For conducted susceptibility, common mode currents are injected into the inputs of equipment. Generally the impedance matching is provided by an LISN (line impedance stabilization network). Current injectors are transformer-like devices enabling an equivalent voltage generator to be forced in wires or bundles of wires and therefore induce a current in the equipment.

Numerical Tools

Numerical tools have become unavoidable in aeronautics and aerospace whenever EMC is concerned. Indeed, the progress made in the last 10 years by computers now makes possible the running of large computer codes requiring large memory and calculation resources. Modeling is generally used at the design phase as an efficient method for analyzing the influence of different parameters on the system's response. In that sense it also helps to optimize the tests to be performed. However, EMC computer codes are also beginning to be used for their prediction capabilities. Therefore, they may eventually replace expensive and sometimes impossible-to-achieve tests at the qualification phase. This was the case for the Airbus A340, which was too large for full standardized lightning injection tests to be carried out and for which numerical demonstrations were made.

In the following, we will present the different types of calculation techniques available in EMC.

3D EMC Computer Codes

Three-dimensional (3D) codes describe the geometry of the system and solve Maxwell's equations. Necessarily, this description is approximated with a mesh sampling the geometry and simplified because accounting for the detailed geometry is quite impossible with the capabilities of present-day computers. These computer codes are now considered fully reliable for what is called the external problem, that is, the scattering of fields and currents by the outer surface of the system. However, up to now they have not really been applicable at very high frequencies (typically frequencies larger than 1 GHz on an aircraft and satellite). Of course, it will be impossible here to investigate all the available methods, and thus only the two families commonly used are mentioned.

First are volume methods, which enable dielectric and losses in materials to be described. They require the meshing of the entire calculation volume and the simulation of limit conditions or infinite medium with absorbing conditions. These methods are frequently developed in time domain, which means they offer a wide-frequency spectrum analysis with a single pulse. The most spread-out method is the finite difference time domain (FDTD) method, valued for its robustness and simplicity of implementation (Degauque and Hamelin 1990). The problem with this method is that the mesh made of cubic cells prevents a conformal description of the surfaces. The second type of 3D methods in EMC is surface methods based on the resolution of Maxwell's equations in their integral formulation (Degauque and Hamelin 1993; Tesche et al. 1997) such as the method of moments in frequency domain.

The interest of those techniques is that the shapes of objects on which scattered currents are calculated are described precisely and only the surfaces have to be meshed. Nevertheless, the drawback of those methods is that they require large amounts of memory that limit the calculations at high frequencies.

Cable Network and Circuit Codes

The drawback of 3D codes is that they are not able to handle the complexity of EMC cable bundle problems. Since the 1980s, several computer codes based on multiconductor-transmission-line-networks techniques allow this problem to be handled (Baum et al. 1986; Parmentier and

Degauque 1996; Paul 1994). Both time domain and frequency domain techniques are available, but the latter offers the main advantages of accounting for the frequency dependence of transmission line parameters and provides models of cables independent on their length. Thanks to field-to-transmission line formalism (Lee 1980), it is possible to link those codes with 3D codes. The 3D codes calculate the distributed incident fields on the wiring path in the absence of the wiring; these wires are then used as generators for the cable network code. Compared to a 3D code, the calculation of a cable network code is fast because the network matrix is sparse. In addition, with the help of appropriate signal processing, the link between a 3D method in time domain and a cable network code in frequency domain is a very efficient technique (Parmantier and Degauque 1996).

If required, cable network codes may be linked to circuit codes to calculate the response at complex terminations. For this purpose, many SPICE-oriented computer codes, validated for a long time in other electrical domains, are available. The compaction with Thévenin equivalent (Parmantier and Degauque 1996) may be used to complete the effort to decompose the problem in subproblems as it is suggested in the theory of EM topology (Baum 1990). Therefore, the decomposition is achieved in three layers: the incident field illumination with a 3D code, the propagation on the wiring with a cable network code, and the equipment response with a circuit code.

5.13 Engineering Method

The EMC Control Plan

In industrial programs, the organization of EMC activities is under the responsibility of an architect and supervised by an EMC manager who is in charge of applying the EMC control plan. This plan covers all the activities in this domain, from the supply of equipment to the delivery of the system to the customer. The main purpose of this plan is to distribute the contract responsibilities to all the parties involved. EMC being the art of defining interfaces, it is very important to define the rules before conflicts happen. Therefore, the control plan also describes the way to manage nonconformities. Nevertheless, from the beginning of the program and the consultations for equipment purchase, it must specify the interface constraints in terms of EM emission limits, EM susceptibility, realization rules, and test methods.

EMC Specification

A General EMC Specifications document must contain all external EM constraints due to natural environment, to other systems, or to the system itself for autocompatibility.

In aeronautics, SAE-4 committees, RTCA Special Committee SC135, and EUROCAE working groups provide reference information on internal EMC on aircraft.

Up to now, there has been no real dedicated general standard for aerospace as in aeronautics. Nevertheless, in both domains EMC specifications have to be defined for each program to answer the technical clauses of the contract signed with the architect. Indeed, the architect generally has its own internal standard, coming from his experience and for which EMC specifications are mostly duplicated from one program to the other.

To maintain compatibility between systems, EMC margins are applied. By definition, this is the ratio between the susceptibility level to be demonstrated in a critical point of the system and the real perturbation level at this point. In the worst case, at least a 0-dB margin must be demonstrated, which means the perturbation level is equal to the susceptibility level. In aerospace, a 6-dB margin is commonly accepted on ground test results to account for the different conditions occurring in flight, but higher margins (20 dB) are required for sensitive equipment (pyrotechnics). The most reliable demonstration method consists of reproducing the perturbation signal in a critical point of the whole system and applying it once amplified by a factor equal to the margin required. This method has replaced old techniques consisting of the comparison of emission and susceptibility plots obtained on each piece of equipment separately. Indeed, because of the difficulty of accounting for nonlinearity and the combination of time and frequency characteristics, the incoherent summation of emission and susceptibility levels generally leads to incorrect conclusions.

EMC Test Plan

The methods for demonstrating the good behavior of the system are based on tests and analysis. They are described in the EMC test plan. Because of the market competition, the reduction of development and manufacturing costs is the constant leitmotif of the architect.

Therefore, the number of expensive tests, such as in-flight tests on aircraft, must be optimized or ground tests on ready-to-fly satellites performed only if strictly required. It is now common for no mock-up or prototype to be available to perform the design phase of a system. This is why, in aerospace, the minimum test plan is limited to the measurement of radiated fields in the launching configuration (EMC with the launcher) and the tests on electrical functions for telecommunications devices. Sometimes only a specific test for ESD susceptibility can be carried out.

Compared to aircraft, the difficulty for satellites and launchers is that it is very difficult to reproduce representative in-flight conditions. For example, the energy on a satellite is provided by solar panels. In tests, it is impossible to spread out this assembling of some 50 m^2 of photovoltaic cells and reproduce the illumination of a solar spectrum of 1,350 W/m^2. Consequently, a simpler laboratory EM source is used instead and the test is performed on a stand-alone satellite, powered by its batteries only. The test is carried out in large anechoic class 100,000 clean rooms (Figure 5.10). However, the connections between the satellite and the test bench are likely to modify the EM interface.

Additionally, for aircraft, some similar situations occur, such as for lighting. Even if, in this case, standards specify levels of current to inject in the structure, the generation of a real lightning channel connected to the structure is impossible.

Manufacturing Rules

Another document applicable to all the parties involved in a program defines the general rules for electrical manufacturing. The equipment constraints more than meet EMC needs. Some of these concern topics other than electricity, such as mechanical or thermal architecture or the integrity of onboard data transfer. Nevertheless, respect for EMC rules should theoretically avoid any redefinition. This document, written by the architect, collects all the conception and electrical manufacturing constraints. It covers very different topics, including electrical continuities, grounding of electric boxes or cable shields, installation of connectors, choice and position of cables, description of the interface circuits of power supplies, and the impedance and noise level of digital/analog data transfer devices.

On a satellite, it is frequent for the distributed power to reach 10 kW. For the near future, 20–30 kW are indicated. Primary currents overcome 200 A. The choice of the technology results from a deal between voltage losses to be maintained at the lowest level for both power saving and thermal dissipation, inductance reduction to minimize overvoltages at the switch-on and switch-off phases, and control of the magnetic moment up to a given level.

Even if aerodynamics constraints are not always relevant, all systems have to dominate their weight. Thus, whenever possible, the natural EM screen offered by the surface of the structure must be balanced with the distribution of cable shields and filters at equipment inputs.

5.14 Conclusion

In this part, we have introduced the main concepts of EMC. We have first presented the theoretical concepts required to understand the basics of EM coupling phenomenon. We have then focused on aircraft and aerospace analyzing their EMC environment and associated standards. Experimental

FIGURE 5.10 SPOT-4 satellite in an EMC test at INTESPACE (Toulouse, France, courtesy CNES/Renaut Claria).

and numerical techniques used for EMC design of systems have been mentioned. Finally, the way to account for EMC in an industrial program has been presented.

Up to now, even though EMC is identified as a requirement impossible to miss in industrial projects, accounting for it at the very beginning of the project is not so frequent. Nevertheless, industry is at a crossroads where new progress in EMC modeling is likely to reduce dramatically experimental tests and offer a wider set of configurations. Manufacturers are already working on what they call "electric mock-ups," on which EMC calculations should be systematically applied for the design of the actual system.

In this increasingly electronic world, manufacturers realize the importance of EMC for the quality of their products. In parallel, EMC should become a real electromagnetic discipline taught in engineering schools in the same way as classical disciplines such as antenna, microwaves, and electricity.

References

Baum, C. E. "The Theory of Electromagnetic Interference Control," in *Modern Radio Science 1990,* ed. J. B. Anderson, Oxford University Press, Oxford [also in *Interaction Notes,* Note 478, December 1989].

Baum, C. E., Liu, T. K., and Tesche, F. M. 1986. In *Fast Electrical and Optical Measurements,* ed. J. E. Thompson and L. H. Luessen, Nijhoff, Dordrecht, pp. 467–547.

Boudenot, J. C. and Labaune, G. 1998. *La compatabilité électromagnétique et nucléaire,* El-lipses, Paris.

Degauque, P. and Hamelin, J. 1993. *Electromagnetic Compatibility,* Oxford University Press, Oxford [French edition: *Compatabilité électromagnétique,* Dunod, Paris, 1990].

EUROCAE ED-1 D/RTCA DO 160D, *Environmental Conditions and Test Procedures for Airborne Equipment.*

Lee, K. S. H. 1980. "A Complete Concatenation of Technology for the EMP Principle: Technique and Reference Data," *Interaction Notes,* December.

MIL-STD-461D, *Requirements for the Control of Electromagnetic Interference Emissions and Susceptibility.*

MIL-STD-462D, *Measurement of Electromagnetic Interference Characteristics.* MIL-STD-464, *Electronic Environment Effects for Equipment.*

Parmantier, J. P. and Degauque, P. 1996. "Topology-Based Modeling of Very Large Systems," in *Modern Radio Science 1996,* ed. J. Hamelin, Oxford University Press, Oxford, pp. 151–177.

Paul, C. R. 1994. *Analysis of Multiconductor Transmission Lines,* John Wiley & Sons, New York.

Stratton, J. A. 1941. *Electromagnetic Theory,* McGraw-Hill, New York [French edition: *Théorie de l'électromagnétisme,* Dunod, Paris, 1961].

Tesche, F. M., Ianoz, M. V., and Karlsson, T. 1997. *EMC Analysis Methods and Computational Models,* John Wiley & Sons, New York.

Vance, E. 1978. *Coupling to Shielded Cables,* Wiley-Interscience, New York.

PART 4

Introduction to Radar

Florent Christophe

5.15 Historical Background

The capability of detecting moving objects in the environment of antennas radiating electromagnetic waves was anticipated early after the experimental evidence of such waves, thanks to Heinrich Hertz in 1886. Practical applications of radar (radio detection and ranging), i.e., the detection of vehicles and estimation of their position were demonstrated as early as the 1920s in France and Germany. The development of radar was then made possible by the availability of a microwave pulsed power oscillator, the magnetron (a derivative of which was mass produced starting in the 1970s for microwave ovens).

The importance of radar was fully recognized following its role in defending Great Britain against air attacks during World War II. Much public and industrial research funding was then invested in the United States and various European countries for solving technological issues and bringing radar to its present wide range of applications. These applications, which include geophysics, meteorology, remote sensing, air traffic control, and acquisition of military targets, come from the long-range all-weather sensing capability of radar.

5.16 Basic Principles

If the beam radiated by a directive antenna such as a parabolic dish illuminates an object, some of the incident energy is backscattered and can be detected in a receiver connected to this same antenna. The direction of the beam and the round trip time delay between the transmitted and the received waveform—usually a repetitive pulse—are used to estimate the location of the scattering object.

Radar Equation

Let P_t be the power of the transmitter, connected to a directive antenna which has a gain g_t. The power density of the field incident on a target, at range R, is given by

$$D_i = \frac{P_t g_t}{4\pi R^2} \tag{5.30}$$

The interaction with the target is described through an equivalent radar cross-section (RCS), defined as a collecting surface σ that would isotropically reradiate all the received power. This collected power is then written as:

$$P_i = \sigma D_i \tag{5.31}$$

The incident power density D_i can be used in the above equation for computing the power that has been reradiated to the receiving antenna, given by

$$P_r = P_i g_r (\lambda/4\pi R)^2 \tag{5.32}$$

where g_r is the gain of the receiving antenna. Combining equations (5.30)–(5.32) and assuming the same antenna is used for transmit and receive (i.e., $g_t = g_r = g$) result in a relationship known as the radar equation, given by:

$$P_r = \frac{P_i \lambda^2 g^2 \sigma}{(4\pi)^3 L R^4} \tag{5.33}$$

where L is a loss factor accounting for various effects such as transmitter-to-antenna or antenna-to-receiver connection losses, propagation losses, etc.

For a radar pulse of duration τ, equation (5.33) can be used to derive the received energy $P_r \cdot \tau$, and the signal-to-noise ratio after dividing this received energy by the spectral power density of the receiver noise FkT_0, where F is the noise figure with respect to the reference noise temperature T_0 (usually 300 K), k being Boltzmann's constant of value 6.02×10^{-23} J/K:

$$S/N = \frac{P_i \tau \lambda^2 g^2 \sigma}{(4\pi)^3 L R^4 F k T_0} \tag{5.34}$$

Radar Cross-Section

Electromagnetic theory allows for a better understanding of the physics over the previous definition of radar cross-section, which was simply used for computing the power flow from transmitter to target and back to the receiver.

Being a linear process, the scattering of an incident electromagnetic wave by any object of dielectric properties deviating from the surrounding medium might be characterized by a scattering matrix S connecting incident field vector E_i to scattered field vector E_s through the equation:

$$(E_s) = (S)(E_i) \tag{5.35}$$

The fields appear as vectors when projected on a polarization basis (usually vertical/horizontal, but also right circular/left circular or some other projection basis). First, diagonal terms of S correspond to copolarization scattering; second, diagonal terms to cross-polarization scattering. In the far field of the scattering object, i.e., for range:

$$R > 2D^2/\lambda \tag{5.36}$$

where D is a typical dimension of this object, E_s vanishes as $1/R$, and it can be shown that:

$$\sigma = 4\pi R^2 S_{rt}^2 \tag{5.37}$$

where S_{rt} is the term of the scattering matrix S corresponding to the transmit and receive polarizations.[10] Depending on the shape and dielectric properties of the scattering object, various methods of solving Maxwell equations are available for computing the scattering matrix in the far-field, hence the radar cross-section of targets. We will next indicate useful results for some canonical targets. For a sphere of radius r large with respect to the wavelength, a high-frequency approximation, allows the derivation of:

$$\sigma = \rho \pi r^2 \tag{5.38}$$

where ρ is the power reflection coefficient at the sphere surface: for a metal sphere, $\rho = 1$ and σ is then equal to the physical cross-section of the sphere.

---

[10] A polarimetric radar would be sensitive to more than one term of the scattering matrix (S), adding characteristic information concerning the target.

	Sphere	Circular plate	Trihedral
Peak RCS @ 1 GHz	1 m^2	140 m^2	140 m^2
peak RCS @ 10 GHz	1 m^2 Isotropic	14,000 m^2	14,000 m^2
angular behavior		HPBW 10° @ 1 GHz, 1° @ 10 GHz	Quasi-isotropic

Table 5.4 Radar Cross-Section for Three Canonical Targets of 1 m^2 Physical Cross-Section

For a metal plate of surface *S*, each dimension being large with respect to the wavelength:

$$4\pi \cdot S^2/\lambda^2 \tag{5.39}$$

when the incident wave is perpendicular to the plate (the so-called specular reflection situation) and vanishes rapidly for angles departing from perpendicular, as most of the energy is reradiated away from the incident direction. A smoother behavior, allowing for the design of passive radar calibrators, is obtained through corner reflectors. In such device, triple successive reflections combine into an outgoing ray parallel to the incident one whatever the orientation of the reflector, resulting in a quasi-isotropic behavior like that of the sphere, but at higher RCS for the same physical cross-section. Table 5.4 indicates typical values. Real targets behave quite differently from the above due to the combination of many scattering mechanisms, depending on the shape and material of the target, the wavelength, polarization, and the direction of the illuminating wave.

A simplified model for understanding complex target backscattering adds the contributions of canonical scatterers distributed across the skin of the target. The rapid change of relative phase of those scatterers with respect to frequency or angular presentation results in a noise-like appearance of respective diagrams, somehow similar to the measured ones. More accurate modeling of the backscattering mechanisms of real targets makes use of multiple interactions, creeping waves, waveguide modes for air intakes, etc.

Clutter and Other Environmental Effects

The beam radiated by a radar antenna often illuminates the natural environment, which results in backscattering energy that may compete with possible targets that the radar is searching for at the same ranges. Main sources for such so-called clutter effects are surface scattering by irregularities of the soil—either bare or vegetated—and volume scattering, mostly coming from rain cells. As opposed to coherent scattering mechanisms responsible for target backscattering, natural environment backscattering is mostly built up by superposition of a large number of individual scattering contributors with random position, i.e., making it an incoherent process in which the backscattered energy will be proportional to the surface or to the volume of the illuminated area, respectively. Since this backscattering energy can be associated with an equivalent radar cross-section through the radar equation (5.33), the ratio of such RCS to the illuminated surface or volume is defined respectively as the surface or volume reflectivity. Such reflectivity usually increases with various factors, mainly

- For ground clutter: soil roughness, vegetation density, angle of incidence, frequency
- For atmospheric clutter: rain rate, frequency

Other Environmental Aspects

The interaction of radar waves with natural media, reported as clutter when negatively affecting target detection, may also be directly used in remote sensing techniques for Earth resources or geophysical parameter characterization.

Other effects of wave propagation are to be expected on radar signals; for example, for long-range ground radar, atmospheric attenuation, tropospheric refraction, and interference of direct and ground surface-reflected rays need to be considered, or ionospheric effects for spaceborne radar observation of the Earth. Part 1 of this section gave indications and references dealing with these effects.

Principles of Clutter Rejection

In many configurations, the clutter RCS is stronger than the RCS of expected targets, resulting in the risk of blinding the radar. Improvements come from exploitation of the Doppler effect,[11] which creates a frequency shift of the backscattered signal proportional to the radial velocity of the target (i.e., the projection v_r of the velocity on line of sight) according to the formula:

$$f_d = 2v_r / \lambda \tag{5.40}$$

Since the signals backscattered by clutter are almost stationary, they can be filtered out by zero Doppler rejection, which is achieved through the moving target indicator (MTI). Target signals remain unaffected to a certain extent as far as they do not approach zero Doppler, which could be the case for targets with tangential velocity, or blind velocities resulting from undersampling of Doppler frequencies. This effect will be illustrated below.

A further opportunity for reducing clutter effects is through exploiting the polarization properties of clutter backscattering; for example, quasi-spherical rain droplets produce backscattering almost cross-polarized with respect to a circularly polarized incident wave, and therefore most radars which are likely to operate in front of strong atmospheric clutter use circular polarization.

Detection Performances

When considering the received radar signal, one of two hypotheses will apply: either it is noise only, or, the addition of noise and signal backscattered by a target.

For making a best guess, it can be shown that the optimum use of the received energy is matched filtering (i.e., processing by a filter tailored to the transmitted signal for minimal noise bandwidth) followed by thresholding. If the filter output is higher than the threshold, a target is detected. The performance of such processing is evaluated by the probability of correctly detecting a target (probability of detection) against the probability of erroneously declaring a target when in fact only noise is present (a false alarm).

The probability of a false alarm depends on the ratio between the threshold level and root mean square (RMS) noise level and the statistics of the noise, white Gaussian for thermal noise, or more complicated when clutter is added to Gaussian noise. Once the threshold is determined, the probability of detection might be computed starting from the signal-to-noise ratio and the statistics of the signal, derived from the fluctuation of RCS around its mean level.

For most situations, many successive pulses hit the targets and can contribute to improving the detection performances. The best case is that of coherent integration before thresholding, for which an improvement factor n in the signal-to-noise ratio is obtained, where n is the number of available pulses. When such coherent or Doppler processing is not feasible, noncoherent (or postprocessing) schemes are used, resulting in some degradation with respect to the optimal case.

Resolution, Accuracy, and Ambiguity of Radar Measurements

The resolution width of a radar is defined as the range of one of the parameters used for locating a target (distance, angular position, Doppler velocity when applicable) for which the output of the

[11]The Doppler effect may be understood as the phase rotation induced by enlarging or shortening the round trip from radar to target and back when the radar and/or the target is moving.

processing filter remains larger than or equal to half the peak power corresponding to the exact location of a target. With such a definition, two targets of equal RCS can be distinguished when they are separated by more than one resolution width. A resolution cell is the multidimensional patch having the resolution width along each axis. A high-resolution radar will have a resolution cell smaller than the targets (usually along the distance or Doppler velocity axis), thus allowing the analysis of target features such as its length. Information derived from such high-resolution radar is the basis for automatic target recognition.

Angular Resolution

Let $\Delta\theta$ and $\Delta\phi$ be the half-power beamwidths of the antenna in the horizontal and vertical planes. Due to the two-way effect, the half-power beamwidth of the radar—defining its angular resolution—is then $\Delta\theta/\sqrt{2}$ and $\Delta\phi/\sqrt{2}$.

Range Resolution

Along the distance axis, when dealing with a radar pulse of duration τ, advancing or delaying the processing gate by $\tau/2$ from the exact round-trip time delay of the target will result in getting half the maximum power (only half of the signal is available), and the corresponding time resolution width is therefore τ. The related range resolution width taking into account the round trip is then:

$$\Delta R = c\tau/2 \tag{5.41}$$

But we also have to consider the so-called pulse compression technique, for which the radar pulse of duration τ is modulated with a bandwidth Δf large with respect to $1/\tau$ ($k = \tau\Delta f$ is called the compression factor, being equal to 1 for nonmodulated pulse). It can be demonstrated that matched filtering results in the resolution that would give a pulse of duration τ/k.[12] A formula which remains valid for both nonmodulated and pulse compression cases is therefore:

$$\Delta R = c/2\Delta f \tag{5.42}$$

Doppler Resolution

In Fourier analysis, when duration T is available for observing signals, a frequency resolution $1/T$ is achievable. Since the Doppler frequency is related to radial velocity, the velocity resolution width can be derived as

$$\Delta v = \lambda/2T \tag{5.43}$$

where T is the duration available for coherent processing, smaller than or equal to the duration of the transmitted beam illuminating the target.

Accuracy of Radar Measurements

Some radars (e.g., tracking radars) are designed for achieving accurate location of the target; it can be shown that suitable processing, such as interpolating from overlapping resolution cells, allows for a standard deviation related to the resolution width by equation:

$$\varepsilon p = \Delta p/2\sqrt{S/N} \tag{5.44}$$

[12] This pulse compression technique allows for separating the power balance problem (a larger pulse duration gives a larger usable energy for a given available peak power from the transmitter) and the range resolution problem. Among others, classical modulations for pulse compression are swept frequency or pseudo-random biphase or polyphase modulations.

where p is one of the parameters to be measured (angle, distance, radial velocity). A signal-to-noise ratio of 13 dB, which is suitable for target detection, would then allow an improvement of a factor around 9 for estimating the parameter p when compared to the corresponding resolution width.[13]

The Ambiguities of Radar Measurements

A usual radar waveform is a pulse, the duration of which has to be short enough to overcome detrimental blinding of the receiver during transmission.[14] Targets located closer to the radar than the blind range R_b are undetectable:

$$R_b = c\tau/2 \tag{5.45}$$

But a large integration time T is also needed for Doppler separation of targets from clutter and for improved signal-to-noise. The obvious solution is to repeat the initial short pulse, with repetition period T_r, resulting in n identical pulses received during $T = nT_r$. Such repetition introduces, on one hand, the risk of erroneously referring the received signal to the last transmitted pulse when actually it comes from a previous one; the error in distance would be then a multiple of the ambiguous distance R_a, given by:

$$R_a = cT_r/2 \tag{5.46}$$

On the other hand, spectral analysis techniques conclude that sampling a signal with period T_r introduces a repetition in its spectrum of period $1/T_r$; such repetition creates an ambiguity in radial velocity (or Doppler) analysis determined by:

$$v_a = \lambda/2T_r \tag{5.47}$$

Searching for long-range high-velocity targets implies ambiguities in either distances or velocities, and postprocessing will have to overcome this situation.[15]

5.17 Trends in Radar Technology

Improved performance and reduced life-cycle costs are driving factors for injecting new technologies in radar design. Next we will discuss some of the technological issues which are specific to radar.

Transmitters

The first generation of pulsed power microwave tube, the magnetron, was able to deliver megawatts of peak power for a few microseconds, the primary energy being stored in the modulator and delivered as a high-voltage pulse. Due to its self-oscillating behavior, pulse-to-pulse coherency for Doppler processing was difficult to achieve, and next generation of tubes, such as the klystron or

[13]In tracking radars, other improvements can be used when switching from a detection to a tracking mode: larger integration time and postprocessing such as Kalman filtering. A multiport antenna with associated receivers also allows for monopulse processing, which produces the angular accuracy indicated above without negative impact of the target RCS.

[14]Some radars make use of a continuous waveform (or CW), but these need separate transmit and receive antennas with specific attention paid to their decoupling.

[15]The choice of a high pulse repetition frequency (or PRF) will make such a radar ambiguous in range, a low PRF will make it ambiguous in Doppler, whereas a medium PRF will make it ambiguous in both range and Doppler. The switching of one PRF to another within the same burst of pulses illuminating a target allows this ambiguity issue to be solved.

the traveling wave tube, are amplifiers permitting longer pulses for pulse compression together with coherent integration. Such vacuum tubes rely on electron beams interacting with microwave cavities. But they suffer from the need to handle high voltages and use heated cathodes of limited lifetime. Furthermore, the handling of high peak powers may result in the need for pressurized transmission lines to the antenna for avoiding dielectric breakdown. New-generation transmitters are therefore relying increasingly on solid state amplifiers through the association of single transistors able to deliver from a few watts to a few 100 W of peak power, depending on frequency. According to the scheme given below, specific components are to be associated with the radar transmitter for avoiding destruction of the low-noise receiver by part of the transmitted power: a circulator, which is a three-port device with a nonreciprocal ferrite core, allows transfer of more than 99% of the energy from the transmitter to the antenna, while energy coming from the antenna is transferred to the receiver. Depending on the transmitted power, further protection of the receiver might be brought by limiting diodes, possibly combined with a plasma switch. Insertion loss of those devices has to be kept as low as possible (less than 1 dB is currently achieved).

Antennas

The classical radar antenna is a parabolic dish fed by a horn at its focus, rotating around one axis for panoramic surveillance or two for target tracking. Improvements come from multiple reflectors, such as the Cassegrainian assembly for compactness (which is required for airborne radar) and sidelobe control. Close to the focal plane, multiple feeds are of interest: either for stacking beams at various elevation angles in a panoramic radar or for simultaneously receiving a single target echo in slightly separated beams for improving angular measurements in a tracking radar. Flat antennas built with slotted waveguides stacked together and suitably fed by a waveguide distribution of energy are now widely used for applications where weight or rotating inertia is constrained. In such an array antenna, each slot behaves as an individual radiating element whose contribution combined with that of the others builds a far field equivalent to what a parabolic dish of same aperture would produce.

But a breakthrough in radar antennas has been brought by electronic scanning, in which the mechanical rotation of the antenna is replaced by phase shifting of individual subsets of an array antenna. Electronic scanning might be achieved in one direction by a one-dimensional phase variation—each waveguide of the previous array antenna is fed through a phase shifter—or in two directions, which requires addressing each of the radiating elements individually. In such two-dimensional scanning arrays, the individual radiating elements can be dipoles, open waveguides, or metallic patches on a dielectric substrate. Phase shifter technology relies on either solid state diode switches or ferrite transmission lines, the insertion phase of which is modulated by an externally applied magnetic field. Insertion losses, power handling capacity, and accuracy are the main parameters for selecting these key components of an electronic scanning antenna.

An active array is an electronic scanning antenna combined with the splitting of the single high-power transmitter into many individual low-power solid state elements. These elements can be integrated into hybrid or even monolithic integrated circuits and brought together into a single transmit/receive module power amplification, phase shifting, switching, low noise amplification, and filtering for reception. Such transmit/receive modules are likely to be the core of most future radar systems as soon as low-cost production technology is available.

Receivers

The role of receivers is to transfer a very faint signal embedded into much higher parasitic signals (coming from clutter or even jammers) to a digital signal processor. Low-noise preamplification is now state of the art, with noise figures as low as 1–2 dB at frequencies ranging from 1 to 18 GHz. Frequency down-conversion can be performed with high dynamic range mixers, and bandpass filtering prepares for baseband down-conversion in both phase and quadrature for digitizing without loss of the phase. In an alternative scheme, a single channel is digitized with a residual carrier, and separation of real and imaginary parts of the complex signal is performed through numerical Hilbert filtering.

Due to the high dynamic range of signals resulting from the power −4 dependence with range—i.e., time delay referred to the transmitted pulse—compensation can be accomplished by applying an inversely varying gain prior to the final amplification or digital conversion. According to the highest clutter-to-noise or jammer-to-noise ratio which can be expected, from 1 bit to 16 bits or even more analog-to-digital converters are required; clock rates are slightly higher than twice the radar bandwidth, the exact value depending on the roll-off of the antialiasing filter. In a radar using linear shift of the carrier frequency during the pulse (also called "chirp" radar), pulse compression can be performed by a surface acoustic wave or a bulk acoustic wave device having delay-versus-frequency characteristics inverse to the transmitted pulse. In such a device operating at intermediate frequencies of 50–150 MHz and time delays up to 100 μs, the design of the piezoelectric transducers allows tailoring of the required characteristics.

Signal Processors

The digital signal processor is in charge of performing tasks as various as pulse compression if not performed in the analogic part of the receiver, digital beam forming for a phased array antenna, Doppler filtering or MTI, thresholding or plot extraction, tracking, plot-to-track association, and display management. A popular radar display is the plan polar indicator (PPI), which creates a horizontal projection of the scene surrounding the sensor, according to the azimuth rotation of the antenna and the range of the echoes detected in the beam direction.

The computing load for the most demanding of functions may range up to many hundreds of floating point operations per nanosecond (or gigaflops). The availability of general-purpose digital signal processors approaching such figures makes it less and less necessary to develop costly on-purpose processors.

5.18 Radar Applications to Aeronautics

Air Traffic Control

One of the main applications of radar to aeronautics is air traffic control (ATC). Long-range air surveillance is performed from selected national locations for controlling continental parts of major air routes. The associated radar sensors operate in the 1215–1400 MHz band, with large antennas (typical reflectors are 8 × 6 m), uniformly rotating around 6 rpm and powerful transmitters in the 100 kW range allows detection of general aviation and jetliners beyond 200 nm. Detection plots and the associated tracks built through consecutive illuminations of the target by the rotating antenna are made available to controllers by video or synthetic displays.

But beyond the cost of such a large radar (which makes it difficult to install in developing countries) there are also some major technical drawbacks:

- The accuracy of the elevation estimate at long range is much poorer than required for controlling vertical separation of routes of 2,000 or even 1,000 ft.
- The radioelectrical horizon at 200 nm is about 25,000 ft (see Part 1), which means that only jetliners close to their maximum flight level can be detected at such range.

Therefore, other sensors are required for filling those gaps, such as the so-called secondary radar. A secondary surveillance radar (SSR) can be understood as replacing the radio wave backscattering roundtrip with two single trips: on board passenger aircraft or for flying under IFR conditions, a transponder is placed which detects the incident radar signal and retransmits a similar signal. This results in a power balance much easier to achieve in each single trip, through the R^{-2} telecommunications equation (5.30) instead of the R^{-4} radar equation (5.33) for round trip. The required signal-to-noise ratio is therefore reached for lower transmit power and antenna gain, resulting in reduced cost for the radar,[16] which can be introduced in a given territory with higher density than the previously described system, which we will now call *primary radar*. Due to technological constraints, the transponder has to receive and transmit at slightly different frequencies and with additional time delay. These parameters, which need to be accounted for in the secondary radar receiver and processor, are normalized according to ICAO regulations. In addition, the retransmitted pulse can be encoded with a message indicating the flight number (A mode) and on board measured flight level (C mode) which allows accurate three-dimensional tracking from the ground.

In areas of high-density traffic and with possibly few secondary radars interrogating various transponders nearly at the same time, garbling situations may occur. Overloading either the transponders or the radar receivers with quasi-simultaneous pulses results in track losses. To overcome such situation, S mode has been introduced, in which a given transponder is activated by a selective interrogation included in the encoded received pulse. In addition, S mode operation specifies a monopulse antenna for the radar, which allows for improved azimuth accuracy.

The availability of S mode transponders on board every aircraft operating under IFR conditions, which will occur as a result of ICAO recommendations, allows the implementation of a new airborne collision avoidance system (ACAS).

Under this concept, the interrogation, reception, and processing are performed on board, thus directly providing the pilot with the necessary information concerning the surroundings for en route collision avoidance.

Other Ground-Based Radars

Approach and Ground Surveillance Radars

For accurate guidance of aircraft at landing approach and for all-weather surveillance of taxiways and runways, short-range radars are used with high enough resolution—i.e., centimeter wavelengths and beyond with medium-sized antennas—for good accuracy and detection of the various possible obstacles.

Meteorological Radar

The adverse effects of rain clutter on air target detection has been pointed out. Specific radars have been designed for meteorological purposes and are operated for detecting strong rain cells associated with thunderstorms and contributing to avoiding lightning hazards to aircraft. Of further interest for improving safety at landing is from the ability of radar to make fine Doppler analysis of the rain echoes and therefore detect specific hazards due to wind shear at final approach. For the few cases of dry wind shears, detection through the specific Doppler signature of turbulent layers at UHF is considered.

Airborne Meteorological Radar

The need for aircraft to find a safe route far away from ground radars in case of thunderstorms has led to the equipment of airborne meteorological radars protected by a radome in the nose of the aircraft. Operating at centimeter wavelengths, such radar provides the pilot with a display of heavy rain cells at ranges up to 15 nm, allowing the route to be adapted. A recently added option makes

[16] A linear antenna with azimuth-only directivity is used in secondary radar, since limited antenna gain satisfies the power balance and no direct elevation measurement is required. For increased azimuth accuracy, this linear antenna can be split into halves for monopulse processing.

this nose radar also able to detect the specific Doppler signature of wind shears at the expense of a high-quality radome for avoiding detrimental ground clutter effects induced by sidelobes resulting from radome-to-antenna interactions.

Other Airborne Radars

In preparation for landing or for terrain awareness, direct measurement of altitude above the ground is available from a radio altimeter, which is a low-power down-looking radar.

5.19 Overview of Military Requirements and Specific Developments

In addition to the need of detecting, identifying, and, if required, directing actions for shooting down targets flying beyond the altitude/velocity domain of civilian aircraft, the major specificity of air defense radars is robustness against adverse countermeasures. Therefore, ground systems have to face many challenges: modern targets of military interest may have reduced RCS (they are stealthy) and can be associated with active countermeasures (jammers) which can be carried by the target itself for self-protection, or come from standoff; passive decoys may also be encountered.

Against jammers, beam and waveform agility is required, which can be brought to the radar by the technology of active arrays. Against decoys, improved resolution may help and is also of interest against stealth targets when associated with increased radiated power.

But a drastic revision of radar design principles may be necessary for restoring detection performances in the modern countermeasures environment: low-frequency sensors (at UHF and below) and bistatic configurations (with widely separated transmitter and receiver) are among the promising techniques.

Other applications for radars on board military aircraft include

- Air-to-air detection, which is performed, for example, with a large rotating antenna protected by a lenticular radome mounted piggyback on board a large standoff platform, but is also associated with fire control in the nose radar of combat aircraft.
- Air-to-ground surveillance and reconnaissance, where synthetic aperture (which will be presented below and illustrated in Part 11 for space applications) allows for high enough resolution at long range.

Those missions where radar is the key sensor are also to be performed despite adverse countermeasures.

PART 5

Avionics Electro-Optical Sensors

Roberto Sabatini, Yixiang Lim, Alessandro Gardi, and Subramanian Ramasamy

5.20 Introduction

After a brief presentation of the physical laws necessary for the understanding of the material, this part provides a general overview of the characteristics of optoelectronic systems for aerospace applications and focuses on the main factors that affect their operational performance.

5.21 Fundamental Physical Laws

Planck's Law

The so-called "black body" is an object that absorbs fully electromagnetic energy incident, regardless of length wavelength. Planck's law on black body radiation (1900) allows one to determine the spectral composition of the radiation emitted from a black body at a given temperature (T). It can be expressed as

$$W_\lambda = \frac{2\pi hc^2}{\lambda^5} \cdot \frac{1}{e^{ch/kT\lambda} - 1} \tag{5.48}$$

where W_λ = spectral radiant emittance (W cm^{-2} μm^{-1})
λ = wavelength (μm)
h = Planck's constant (6.256×10^{-34} W s^2)
T = absolute temperature (K)
c = speed of light (2.998×10^8 m s^{-1})
k = Boltzmann's constant (1.381×10^{-23} W s K)

The terms of equation (5.48) can be conveniently rearranged to obtain the form below:

$$W_\lambda = \frac{C_1}{\lambda^5[e^{C_2/\lambda T} - 1]} \tag{5.49}$$

where C_1 is first radiation constant (3.47×10^8 W m^{-2} μm^{-1}) and C_2 is second radiation constant (1.44×10^4 μm s K).

This is illustrated in Figure 5.11, which plots the curves describing the spectral density emitted by a black body over a range of wavelengths at different temperatures.

Stefan-Boltzmann's Law

Integrating equation (5.48) with respect to λ, we obtain the expression for the radiant emittance, or the total energy emitted by a black body across all wavelengths per unit area and unit time. We thus obtain

$$W = \int_0^\infty W_\lambda d\lambda = \frac{2\pi^5 k^4}{15c^2 h^3} \cdot T = \sigma T^4 \tag{5.50}$$

where W is radiant emittance (W cm^{-2}) and σ is Stefan–Boltzmann constant (5.6697×10^{-12} W cm^{-2} K^{-4}).

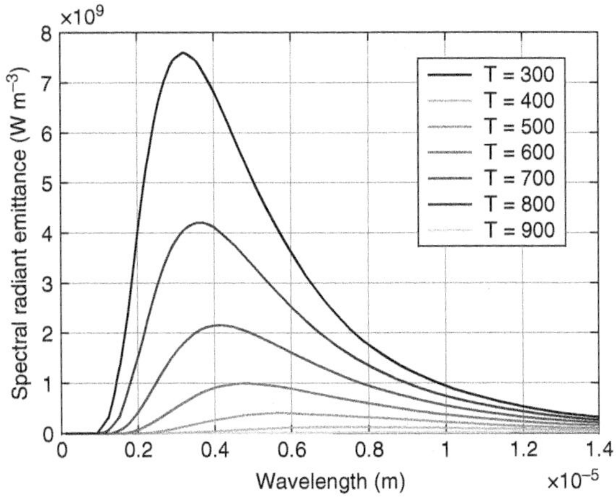

FIGURE 5.11 Black body radiation as a function of temperature and wavelength.

Wien's Displacement Law

Differentiating equation (5.48) with respect to λ and obtaining the maximum of the differential by equating it to zero, we obtain Wien's displacement law, which describes the wavelength at which maximum radiative power is obtained (i.e., the peak wavelength) at a particular temperature. The Wien's displacement law is given by

$$\lambda_p T = a \tag{5.51}$$

where λ_p is peak wavelength (µm) and a is 2897.8 µm K.

This is illustrated by the dotted line in Figure 5.12, which shows that black body radiation is shifted toward shorter wavelengths at higher temperatures. For example, at a temperature of 300 K, the maximum emission is given by

$$\lambda_p = \frac{2897.8}{300} = 9.68\ \mu\text{m} \tag{5.52}$$

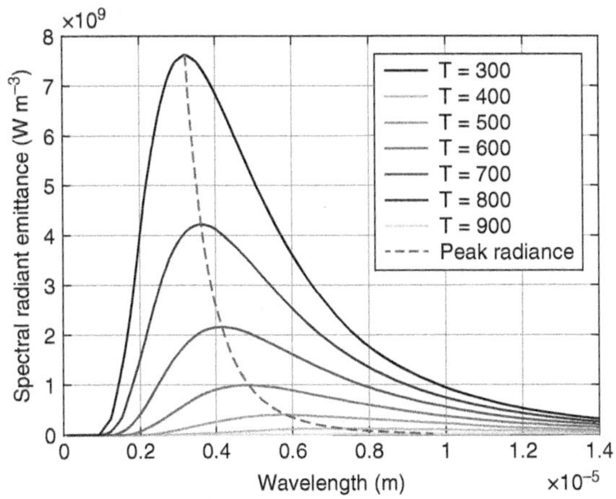

FIGURE 5.12 Emission peaks (Wien's law).

Substituting appropriate values into equation (5.51) allow the peak wavelengths emitted by the sun, plume of aircraft jets, and the human body to be obtained:

Sun: 0.5 μm

Plume: 3 μm

Human : 10 μm

5.22 IR Sensors

The sensors used in optoelectronic systems to collect infrared energy are commonly referred to as detectors. In general, it is possible to distinguish two main categories of detectors:

- Thermal detectors
- Photon detectors or quantum detectors

In thermal detectors, the incident radiation is measured by the change in the physical properties of the materials they contain. In particular, the incident radiation causes an increase in temperature of detector, which in turn determines the change of some physical parameters such as resistance or tension. The spectral response of a thermal detector is determined by the absorption characteristics of the sensor surface and, therefore, its spectral emissivity characteristics. Table 5.5 lists the most common types of thermal detectors and the physical parameters measured by them.

Unlike thermal sensors, the photon detector exploits the direct interaction between the incident photons and the electrons of the material that constitutes the detectors themselves. They are much faster and more sensitive than thermal detector. In general, photon detectors are made from semiconductor material and it is possible to group them into the following categories:

- Photoconductive detector, wherein the semiconductor substrate behaves as a variable resistor (the electrical resistance is the measured parameter). Specifically, a variation in the number of incident photons results in the variation in the number of free charges within the semiconductor, such that the conductivity (and hence the resistance) of the detector varies in proportion to the input.
- Photovoltaic detector, wherein the semiconductor assumes the characteristics of a photodiode, which produces the electronic output (current). In fact, a variation of the amount of photons incident on a *p-n* junction of the photosensitive semiconductor material (photodiode), results in voltage fluctuations at the junction and therefore the current flowing through the photodiode.
- Photoelectromagnetic detector, wherein the absorbed photons generate electrical charges, which diffuse in the semiconductor and are then separated by the application of a magnetic field. The separation of charges gives rise to a voltage, which is proportional to the incident optical signal.

Thermal detector	Measured parameter
Pyroelectric	Current
Thermopile	Tension
Bolometer	Resistance
Superconductor	Resistance

Table 5.5 The Most Common Types of Thermal Detectors

Thermal detector	Measured parameter
Photoconductive	Resistance
Photovoltaic	Current (*p-n* voltage)
Photoelectromagnetic	Tension
Photoemissive	Current (anodic)

Table 5.6 Types of Photon Detectors

- Photoemissive detector, in which the incident photons impart sufficient energy to the electrons of the photoconductive material to cause the release of the electrons themselves. These electrons are conveyed toward a collector (anode) and the associated current (proportional to the quantity and energy of the incident photons) is measured.

Table 5.6 summarizes the main characteristics of the different photon detectors. The materials most commonly used in the manufacture of photon detectors are gallium arsenide (GaAs), silicon (Si), germanium (Ge), indium antimonide (InSb) sulfide, cadmium (CdS), and mercury cadmium telluride (CdHgTe).

In order to increase the sensitivity of photon detectors, it is often necessary to provide for their cooling. In particular, this is necessary because of the relatively low energy possessed by the incident photons compared to the energy possessed by the electrons in the semiconductor material as a result of thermal agitation (thermal noise due to the Joule effect). Without going into details of physics, the photons are required to possess energy greater than a certain threshold in order to excite the electrons in the conduction band of the semiconductor so as to leave a positive hole in the valence band, given by

$$E_g = h(f - f_0) \tag{5.53}$$

where E_g = energy imparted by incident photon to electron (J)
h = Planck's constant (6.256×10^{-34} W s^2)
f = frequency of incident photon (s^{-1})
f_0 = threshold frequency of the semiconductor material (s^{-1})

5.23 Passive Optoelectronic Systems

Infrared Line Scanner

The infrared line scanner (IRLS) is a system using single or multiple imaging detectors making successive scans on rotating optics. The scans can then be processed to reconstruct the original image. Typically, the scans of an IRLS occur along the transverse axis of the aircraft on which the IRLS system is installed. The second scan axis, needed to build a two-dimensional image, is given by the movement of the aircraft along its flight path (Figure 5.13). The IRLS is one of the most effective systems for capturing images in reconnaissance missions by air. Typically, in 10 minutes of acquisition, an IRLS can cover a ground strip of about 100 NM. Scanned images can be recorded on board or transmitted in real time to a ground station connected to the aircraft via a datalink. The most important parameter in IRLS systems is the relationship between speed and altitude (V/h). It in fact determines the necessary scan-rate. Considering a detector with an instantaneous FOV (IFOV) of IFOV radians, it will follow that it observes a portion of land equal to IFOV $\cdot$ H, parallel to the direction of flight, where H is the aircraft altitude.

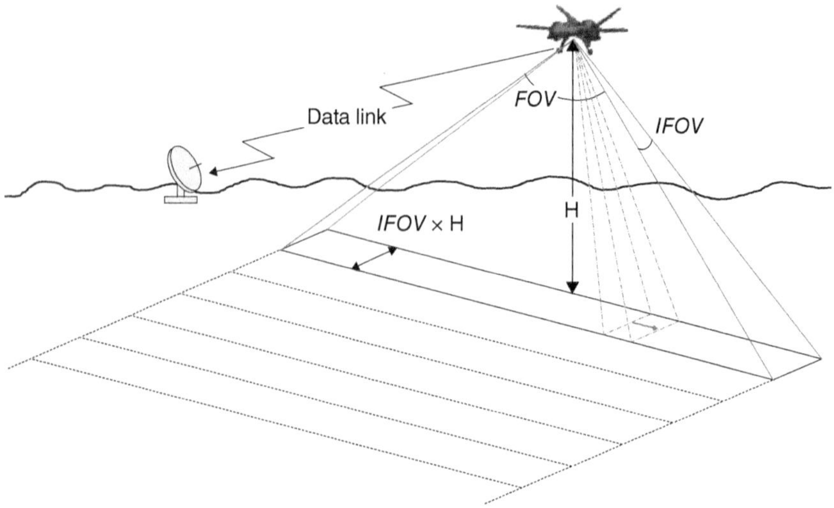

FIGURE 5.13 IRLS scan pattern and geometry.

In addition, IFOV · H is exactly the distance that the aircraft has to travel for each sweep that the IRLS makes. An aircraft traveling at a speed faster than IFOV · H creates "holes" (gapping) in the image while an aircraft traveling at a speed slower than IFOV · H creates overlaps in the images (overlapping). Therefore, the speed of the aircraft necessary to achieve a uniform scanning is given by

$$V = n\theta_{\text{IFOV}}H \tag{5.54}$$

where V = aircraft speed (m s^{-1})
n = line scan rate (s^{-1})
IFOV = instantaneous field of view (mrad)
H = aircraft altitude (m)

Therefore, the angular speed of scanning, $\dot{\text{FOV}}$, is given by

$$\dot{\text{FOV}} = n \cdot \text{FOV} = \frac{V}{H} \cdot \frac{\text{FOV}}{\text{IFOV}} \tag{5.55}$$

where $\dot{\text{FOV}}$ is angular speed of scanning (mrad s^{-1}) and FOV is scanning field of view (mrad).

A typical reconnaissance mission may provide an operating envelope of 1,000 ft/s at altitudes between 200 and 2,000 ft. This corresponds to values of V/h equal to 5 and 0.5, respectively. The need to operate at high values of V/h, especially for military systems, led to the development of complex optical scanners. An example is the split-aperture scanner, which is characterized by the use of a three- and four-sided spin mirror which is able to split the observed image. The split image is reflected to two parabolic side facets and are subsequently collected and recombined at the focal plane. This allows a reduction in the cross section of the spin mirror, such that the system can be operated at higher rotational speeds without the disadvantage of distortion from centrifugal forces. However, the design of the split-aperture sensor is somewhat complex, mainly because of the possible phenomena of narcissus strips that can occur in these systems (which occurs when the detector sees a reflected image of itself).

The dwell time (s) of the detector is the time required for it to scan a distance equal to its size in the scan direction, given by

$$t_d = \frac{\text{IFOV}}{\dot{\text{FOV}}} \tag{5.56}$$

The dwell time needed for a good reconstruction of the image is also a function of the time constant of the detector. Since the V/h ratio should vary within values defined by the operating limits, it is often necessary to adopt some compromise solution, such as the increase of the surface (or number) of the detector.

The performance of an IR imaging system is commonly characterized by its resolution. The thermal resolution of IR imaging systems can be described by the noise equivalent temperature difference (NETD), which defines the smallest temperature difference that the system is able to detect. Other performance parameters that account for not only the detector performance but also human machine interfaces include the minimum resolvable temperature difference (MRTD) and the minimum detectable temperature difference (MDTD). These describe both the thermal (sensor) resolution and spatial (display) resolution of the system and require inputs from a subjective observer in order to quantify the system performance.

The analytical derivation of the formula for the calculation of the thermal resolution of an IRLS system is rather laborious. Therefore, only the final result is presented in this brief note:

$$\text{NETD} = \frac{\pi \sqrt{A_d\, \delta f_n}\; \lambda_p\; T_b^2}{\alpha\; \beta\; A_0\; \tau_0\; D^*(\lambda_p) c_2 \int_{\lambda_1}^{\lambda_2} W_\lambda(T_b)\, d\lambda} \tag{5.57}$$

where NETD = noise equivalent temperature difference (K) (i.e., the temperature difference required for a signal-to-noise ratio of 1)

A_d = area of detector (cm^2)

δf_n = equivalent noise bandwidth (Hz)

T_b = background temperature (K)

α, β = detector angular subtenses (rad)

A_0 = effective collection area of infrared optics, including obscuration (cm^2)

$\tau_0(\lambda)$ = infrared optical transmission coefficient

$D^*(\lambda)$ = spectral detectivity (cm Hz$^{1/2}$ W^{-1})

λ_p = peak spectral response

$c_2 = 1.4388 \times 10^4$ µm K

$\int_{\lambda_1}^{\lambda_2} W_\lambda(T_b) d\lambda$ = effective spectral radiant emittance (W cm^{-2})

The equivalent noise bandwidth δf_n is given by

$$\delta f_n = k \frac{\pi (V/h)_{\max}}{2\alpha\beta} \tag{5.58}$$

where k is an equivalent filter factor.

From equations (5.55) and (5.56), we can see that NETD $\propto \sqrt{V/H}$, i.e., the dependence of NETD on the V/h factor is smaller than in other cases and as such, requires a compromise between the thermal and the angular resolution. In general, a thermal resolution of 0.5°C or better is required to produce satisfactory images.

The most important final consideration is that the performance of the IRLS is best if the system is used by a plane flying slows and low, other conditions being equal. However, it is the necessary cover that determines the flight altitude for a particular mission. In fact, a typical IRLS system with a FOV of 120° centered on the longitudinal axis of the plane has coverage of approximately 3.5 times the flight altitude. Operational considerations (rather than technical) help determine the best compromise for the groundspeed of the aircraft.

Typically IRLS systems for aeronautical use are installed inside of a pod, to be integrated on a particular aircraft as external loads. In many cases, however, there is the possibility of integrating IRLS systems directly on the aircraft.

Forward-Looking Infrared

A forward-looking infrared (FLIR) system can be likened to a television camera operating in the infrared (IR). The main function of a FLIR is to see in the dark by means of the detection and the subsequent processing of the IR electromagnetic radiation emitted from anybody. Typically, FLIR systems operate in the bands 3–5 μm (medium-wavelength infrared—MWIR) and 8–12 μm (long-wavelength infrared—LWIR). FLIR comprises of starting and scanning systems. Starting systems utilize a large number of detectors without a scanning mechanism while scanning systems use a small number of detectors that operate in a scanning pattern. Scanning systems are further decomposed into serial and parallel systems. Serial systems (Figure 5.14(*a*)) are made up of a single detector with a two-dimensional raster scanning pattern using a mirrors rotating in the vertical and horizontal planes. Parallel systems (Figure 5.14(*b*)) consist of a linear array of detectors which are scanned in a single direction, orthogonal to the arrangement of the detectors.

The operating range for the detection, recognition, and identification of imaging systems can be analyzed using the Johnson criteria, which describes the discrimination level of an imaged target based on its resolution. The resolution was measured in terms of black/white bar pairs known as cycles (Figure 5.15). An image with higher resolution has more cycles and has a higher chance of being discriminated.

The discrimination levels are divided into detection, orientation, recognition, and identification. Obviously, the number of cycles necessary for identification is higher than that of lines necessary

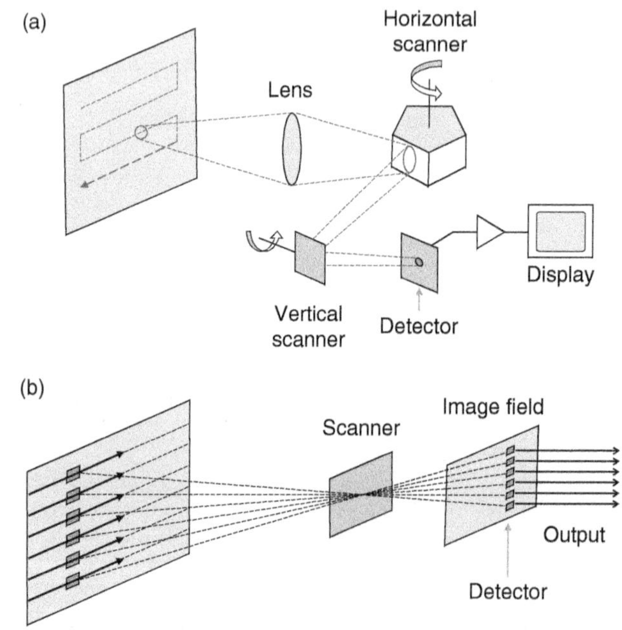

FIGURE 5.14 Scanning pattern for (*a*) serial and (*b*) parallel FLIR scanning systems.

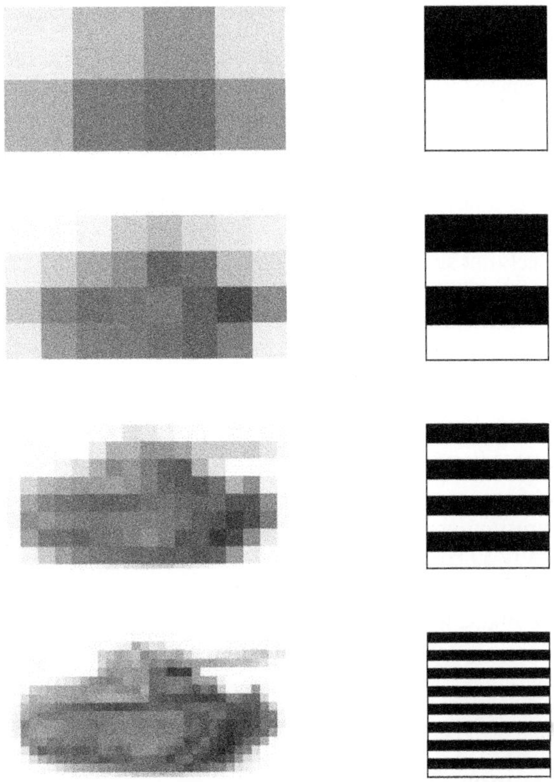

Figure 5.15 The effective resolution of an image expressed in cycles, based on the Johnson criteria.

for recognition (which in turn are higher in number than those necessary for detection). For example, detection requires 1.0 cycle whereas identification requires 6.5 cycles. The formula for calculating the operational range with a 50% probability of discriminating a target is based on the following formula, given by

$$R = \frac{h_c}{N\theta_R} \tag{5.59}$$

where R = range to target (m)
h_c = target critical dimension (m)—i.e., the smallest target dimension
N = number of black/white cycles, dependent on target type and discrimination level
θ_R = angular resolution (mrad), dependent on sensor

The angular resolution of the system is based on the number as well as the geometrical arrangement of the detectors in the array, and can vary greatly depending on the type of application.

In order to maximize the operational performance of FLIR, it is necessary to reduce the background radiation to a minimum possible level. For this purpose, the array of detectors is cooled in various ways. One of the methods most commonly used involves the use of cold finger at cryogenic cooling and special containers (Dewar) for thermal insulation from the external environment.

Infrared Search and Track

From a functional point of view, the infrared search and track (IRST) can be defined as the class of passive optoelectronic systems for military use capable of detecting, locating and tracking objects that emit in the infrared bands. IRST systems operate in spectral band of 1–15 μm, and in particular the MWIR (3–5 μm) and LWIR (5–12 μm) bands. IRST systems are a subset of IR imaging systems like FLIR. However, IRST offer a much larger field of regard and pixels per frame in IRST systems. Also, target recognition is automated in IRST systems. Because of these reasons, more complex signal processing algorithms are required to extract the information related to the target, with particular emphasis on high clutter situations. It should be noted that the IRST represents a viable alternative to many military radar systems, especially because of the "passive" nature of the discovery/tracking processes, the inherent antistealth capabilities of such systems, as well as the higher angular resolution due to the shorter wavelength used.

The main differences between the IRST systems and FLIR are as follows:

- *Field of regard (FOR):* A IRST system has a FOR greater and can therefore survey regions of the very large space (typically 360° in azimuth and 90° in elevation).
- *Presentation and frame rate:* Since an IRST sends its data to a computer, its presentation is synthetic (or semisynthetic), which allows it to have a lower frame rate (1–10 Hz) than that of a FLIR (typically 30 Hz), which is required to send image feeds in real time.
- *Number of pixels:* Typically, because of the large FOR, an IRST needs pixels/frame relationships 183 times higher compared to a FLIR and, to achieve a frame rate of 1 Hz, must be able to process a number of pixels 6 times greater than the FLIR (at a frame rate of 30 Hz) every second.

5.24 NVIS Technology Overview

The image intensifier (I^2) is the core element of NVIS systems. I^2 devices are electro-optic systems used to detect and intensify reflected energy in the visible and near infrared regions of the electromagnetic spectrum. They require some external illumination in order to operate because the image quality is a function of the reflective contrast. The performances of I^2 devices are also dependent on atmospheric and environmental conditions. Particularly, penetration through moisture can be quite effective (especially when compared to other electro-optic (EO) devices, like FLIR systems), while smoke, haze, and dust can significantly reduce I^2 performance. Signal-to-noise ratio (SNR) is the parameter commonly used to characterize I^2 systems performance.

Generation I (GEN I) NVGs were introduced into service in the mid-1960s during the Vietnam War. They used starlight scopes based on electron acceleration (i.e., no microchannel plates—MCPs). Therefore, they were characterized by high power requirements and tube gains between 40,000 and 60,000. Multiple staging, required to increase gain, often determined an increase of image distortion, and the overall systems were large/heavy (i.e., not suitable for head mount). Furthermore, GEN I systems were very susceptible to blooming and the MTBF of a typical GEN I NVG was in the order of about 10,000 hours.

Generation II (GEN II) NVGs were introduced in the late 1960s and they were small enough to be head mounted. They used electron multiplication (i.e., MCP), with increased tubes gain, reduced power requirements, and reduced size/weight. Furthermore, the new I^2 technology reduced distortion and blooming (confined to specific MCP tubules halos). Typical GEN II systems were the AN/PVS-5 ground system, and the AN/AVS-5A system modified for aircraft usage. The MTBF of typical GEN II systems was in the order of about 2,000–4,000 hours (worse than GEN I), the tube gain was approximately 10,000, and there was no inherent resolution improvement with respect to GEN I systems.

Improved photocathode performance, obtained by gallium arsenide (GaAs) components, determined a substantial improvement in spectral response with generation III (GEN III) systems. GEN III matches night sky radiation better than GEN I and GEN II systems, and can operate also in the absence of moon (starlight capability). Improved MCP performance was obtained by aluminium oxide coating, which decreases ion hits and increases MTBF (>10,000 hours). Today, GEN III systems are widely used on most ground and in aircraft applications. Figure 5.16 shows the relative responses of the GEN II/GEN III NVG systems and the human eye, together with the average night sky radiation (Johnson 1985; Ratches 1976). The improvement obtained with GEN III NVG systems is evident.

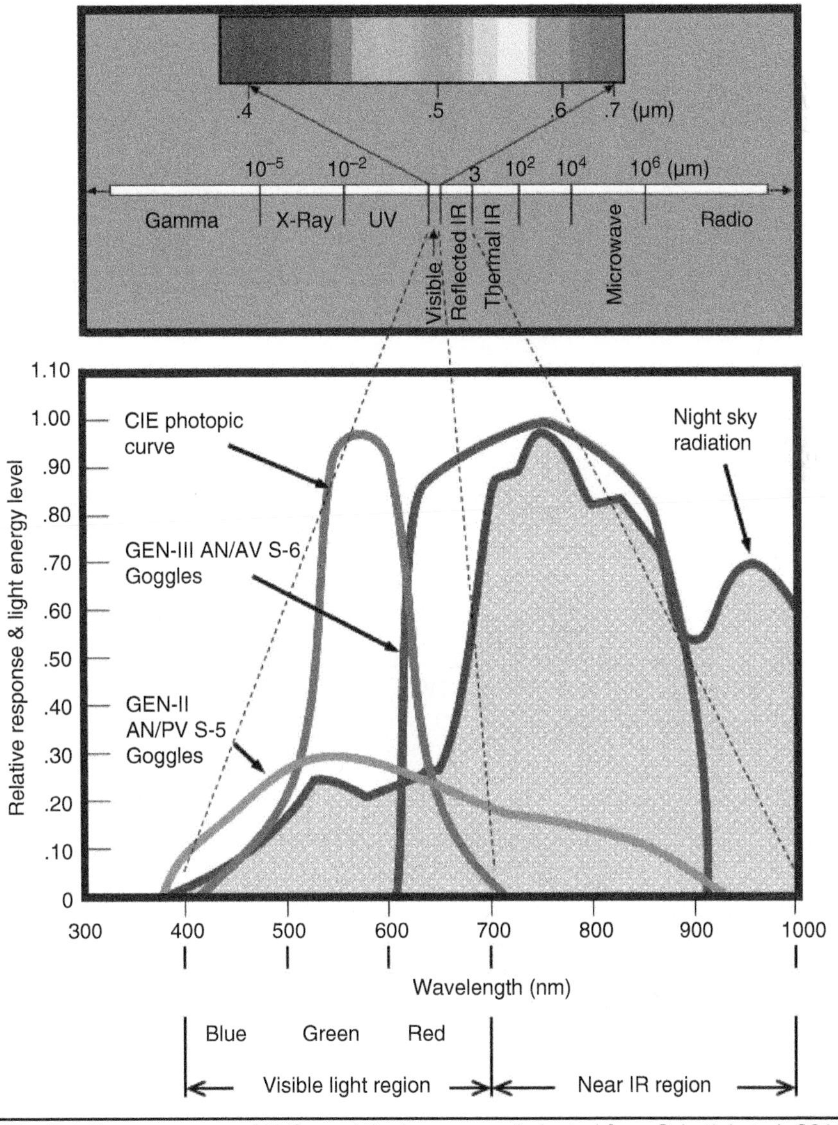

FIGURE 5.16 Relative responses of NVGs and the human eye (adapted from Sabatini et al. 2013).

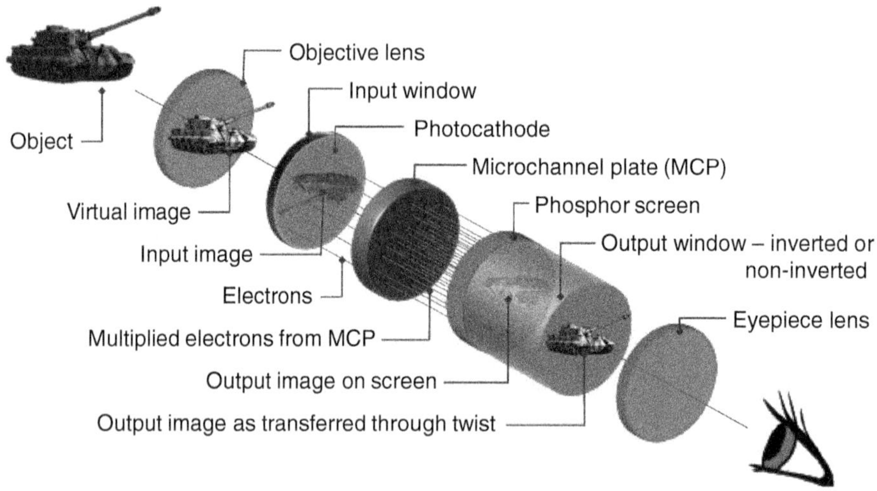

FIGURE 5.17 Architecture of an image intensifier (adapted from Sabatini et al. 2013).

As illustrated in Figure 5.17, an I^2 device is typically composed by the following elements:

- Objective lens
- "Minus blue" filter
- Photocathode
- Ion barrier film
- Microchannel plate
- Phosphor screen
- Image inverter
- Eyepiece lens

The objective lens combines the optical elements and focuses incoming photons onto the photocathode (inverted image). In most airborne NVGs, the objective lens is coated with a minus blue filter (necessary for compatible cockpit lighting). It focuses from several inches to infinity (depending on NVG). Particularly, in airborne applications, infinity focusing is used in order to obtain

- NVG external viewing
- Look under/around NVG for cockpit and instrument viewing

In airborne NVGs a minus blue filter is coated inside the objective lens. Its purpose is to reject visible light and to prevent other specific wavelengths from entering the image intensifier. Therefore, the minus blue filter allows the use of properly emitting/filtered lighting to illuminate the cockpit for viewing underneath the goggles. There are three different classes of NVG objective lens filters:

- Class A: blocks below 625 nm (blue/green)
- Class B: blocks below 665 nm (blue/green/reduced red)—allows use of color displays
- Class C (leaky green)—incorporates notch cutout to permit viewing of specific wavelength

The photocathode (PC) converts light energy (photon) to electrical energy (electrons). The PC inner surface is coated with a photosensitive material. Particularly, we list the following materials used in GEN I/II and GEN III systems:

- GEN I/II: S-20 multialkali compound, sensitive between 400 and 850 nm (peak sensitivity at 500–600 nm)
- GEN III: gallium arsenide (GaAs), sensitive from 600 to 900 nm (impact of photons cause release of electrons)

Typical PC luminous sensitivity figures are 250–550 μA/lm for GEN II systems and 1,000–1,800 μA/lm for GEN III systems. As illustrated in Figure 5.18, GEN III I^2 tubes are currently fabricated with a so-called ion barrier (IB) film. This film extends tube life (protects the PC) but reduces the system performance (i.e., degrades signal-to-noise ratio).

The MCP is a thin wafer (about 1 mm) containing various millions of glass tubes or channels (typically 4–6 million). Electrons from the PC enter the MCP tube (tube walls coated with lead compound rich in electrons) which is tilted (about 5°) to ensure the impact of the electrons with the wall. When an electron impacts the tube wall, more electrons are released resulting in a cascade process. Electrons are then accelerated toward the phosphor by an electrical potential differential (positive pole at phosphor). The ultimate output is number of electrons and their velocity. Resolution is a function of number of MCP tubes.

The phosphor screen is a thin layer of phosphor at the output of the MCP. Phosphor emits light energy when struck by electrons (electroluminescence). Light emitted by phosphor creates a visible (green) image.

The image inverter (INV) is a bundle of millions of light transmitting fibers. The bundle rotates 180° to reorient the image (fiber optic twist). It also collimates image for correct positioning at the viewer's eye. Problems in INV manufacturing and installation result in adverse image effects, such as distortion and honeycomb appearance. Some NVG designs do not incorporate a fiber optic twist for reorienting the image.

The eyepiece lens is the final optical component of the NVG. It focuses the visible image on the retina of the viewer and, generally, a limited diopter adjustment is allowed to permit some correction for individual vision variations. In general, corrective lenses must still be worn by users (the system does not correct for astigmatism). Most GEN II systems have a 15-mm eye relief and a nominal 40° FOV. GEN III systems typically have 25 mm nominal eyerelief, which also provides the 40° FOV but enhances the ability to look under/around the NVG.

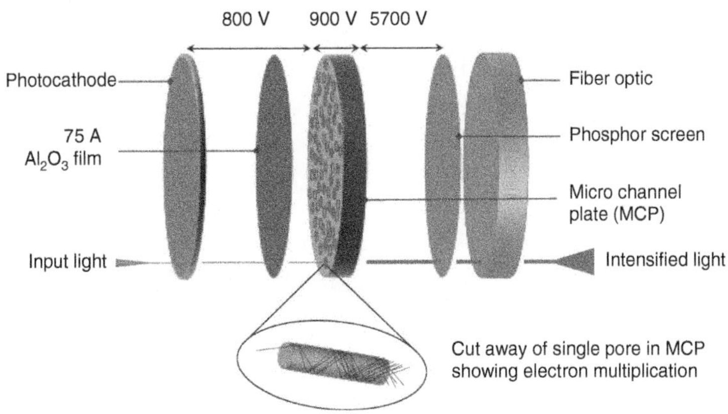

Figure 5.18 GEN III I^2 tube.

Signal-to-noise ratio (SNR) is a measure of image intensifier performance (resultant of the image intensification process). SNR for an NVG is defined as the ratio of electrons produced by ambient light (signal) to stray electrons (noise). Improved performances (larger SNRs) are produced by increasing the ambient light and/or improving the I^2 (e.g., increasing PC sensitivity and decreasing the space between the elements).

5.25 NVIS Compatibility Issues

Intensified imagery of the outside scene is of primary importance to the aircrew. Incompatible light from cockpit sources and external lights are detected by the NVG and intensified, thus reducing the NVG gain. The resulting degraded image quality may not be readily apparent to the aircrew.

NVG compatible lighting results in instruments and displays being easily read with the unaided eye at night. However, all instruments must still be readable during day. NVG compatible lighting is often invisible to the NVG, while "friendly" lighting may be visible to the goggles, but without changing the gain state of the goggle. Typically, NVG compatible instruments and displays only emit wavelengths to which the eye is most responsive (i.e., little red and no near-IR emission).

There are basically two different implementation methods which can be adopted for integrating NVG compatible lighting in the cockpit. These methods are the following:

- *Permanent lighting:* Including integral instrument/display lighting, post and bezel lighting, food lighting using existing aircraft light fixtures or LED-based light sources
- *Temporary lighting:* Including chemical light sticks and LED wiring harness

Also NVG compatible external lights can be used in order to increase mission effectiveness, increase flight safety and decrease aircraft vulnerability (IR covert mode). In this case, there are basically two different approaches possible:

- *Introducing new equipment:* Including conventional/filtered, electroluminescent and LED technologies
- *Retrofitting existing lights:* Including filtering and modifying the existing light source

Another important aspect to be considered with NVIS compatible aircraft developments is the NVG-helmet integration. Particularly, the following are the main goals to be achieved:

- Reduce the NVG-helmet moment arms.
- Reduce the weight.
- Maximize usage of the available FOV (considering eye relief, exit pupil, etc.).
- Allow use of various types of visors (including laser protection visors).

5.26 Airborne Lasers

Several search activities were performed with lasers and practical applications were investigated as early as 1950s. Since then, a number of research and development programs have concentrated on lasers, leading to a wide variety of systems, ranging from laboratory devices studying nonlinear optical emissions and propagation, to eye-safe, low size/weight and inexpensive laser-ranging binoculars. Over the last few decades, civil and military interests in airborne laser systems have been concentrated in four general areas: laser rangefinders (LRF) and target designators (LTD), laser radars (light detection and ranging, LIDAR), laser communication systems (LCS), and directed energy weapons (DEWs) (Hecht 2010; Jebur et al., 2014; Sabatini et al., 2015).

Laser Radars

LIDAR sensors can be classified into three categories, namely discrete return, full waveform, and profiling. The simplest of the three is profiling, which provides only one return. The more advanced discrete return type sensors record multiple returns while the waveform sensors record a digitized profile of the full return pulse. An improved positional accuracy is obtained from the LIDAR systems only when reliable and precise information of the aircraft location is known at both the transmission and reception times. State-of-the-art LIDAR systems are capable of capturing the reflected signal, as well as providing georeferencing of the three-dimensional coordinates of the laser returns. The basic concept of operation of a LIDAR is identical to that of conventional radar. The laser source emits a signal that is reflected by a target and then collected by the electro-optical receiver. Range to the target is determined by measuring the round-trip time of the electromagnetic impulse. Radial velocity of the target is measured by either determining the Doppler shift of the emitted wavelength or by performing multiple range measurements and calculating the rate of change in range. Similar to radar, the intensity and profile of LIDAR reflected signals vary with the beam wavelength and with the reflectance characteristics of the surface reflecting the beam. LIDARs can be categorized according to various ways. Typically, they are classified based on the type of measurement, the detection technique, the type of laser and operational wavelength, the type of interferometer employed in a coherent laser radar (where applicable), the modulation technique, the demodulation technique, the purpose, the type of data collected, or the data format. In addition, laser radars can be classed as monostatic or bistatic, depending on whether the receiver and the emitter are collocated or not.

The different types of lasers adopted for LIDAR systems and their respective carrier wavelengths summarized in Table 5.7 (Sabatini et al., 2015). The techniques employed for LIDAR are identified in Table 5.8 and its functions and measurements used are summarized in Table 5.9. The appellation is seldom sufficient to completely identify what it does and does not define the performance characteristics. The versatility of lasers is evident from their available variety. Wavelength-dependent technological

Types of lasers (typical)	Carrier wavelength
CO_2	9.2–11.2 μm
Er:YAG	2 μm
Raman shifted Nd:YAG	1.54 μm
Nd:YAG	1.06 μm
GaAlAs	0.8–0.904 μm
HeNe	0.63 μm
Frequency doubled Nd:YAG	0.53 μm

Table 5.7 LIDAR Types

Detection technique	Interferometer type	Modulation technique
Direct detection	Not applicable	Pulsed amplitude modulation (AM)
Coherent detection	Heterodyne Homodyne offset homodyne	Pulsed AM frequency modulation (FM) hybrid (AM/FM, pulse burst) none (CW)

Table 5.8 LIDAR Techniques

Functions	Measurements
Tracking	Amplitude (reflectance)
Moving target indication (MIT)	Range (time delay)
Machine vision	Velocity (Doppler shift or differential range)
Velocimetry	Angular position
Wind shear detection	Vibration spectra
Target identification	
Imaging	
Vibration sensing	

Table 5.9 LIDAR Functions and Measurements

limitations frequently prevent simple parametric extrapolation of performance from one type of system to another. These limitations make routine performance at one laser wavelength well beyond the state-of-the-art at another wavelength. Due to the physical principles, governing emission of a laser signal, tunability both in terms of wavelength and power, is very difficultly introduced. Passive optics and conventional radars—radio frequency (RF) through millimeter-wave (MMW) on the other hand—are natively capable of large tunability in their design without major changes in technology.

Receiver Detection Techniques

Direct and coherent detection types are the two types of LIDAR systems. In direct detection laser radar (Figure 5.19), the inbound radiation is focused onto a photosensitive element generating a voltage (or current) that is directly proportional to the incident energy. This process is analogous to conventional passive optical receivers.

A block diagram, of heterodyne (coherent) detection LIDAR is shown in Figure 5.20. An optical signal is generated by the laser emitter. The divergence and diameter of the laser beam are then adjusted when necessary to the rest of the system by beam-shaping optics. In a monostatic system, the transmitted laser signal enters a transmit-to-receive (T/R) switch. The T/R switch permits the LIDAR transmitter and receiver to operate through a common optical aperture. The LIDAR signal then enters the beam expander or output telescope and the scanning optics that direct the optical signal to the target. In a monostatic system, radiation reflected from the target is collected by the scanning optics and the beam expander, which now acts as an optical receiver. The T/R switch directs the received radiation to an optical mixer, where it is combined with an optical reference signal generated by the local oscillator. The combined signal is then focused onto a photosensitive

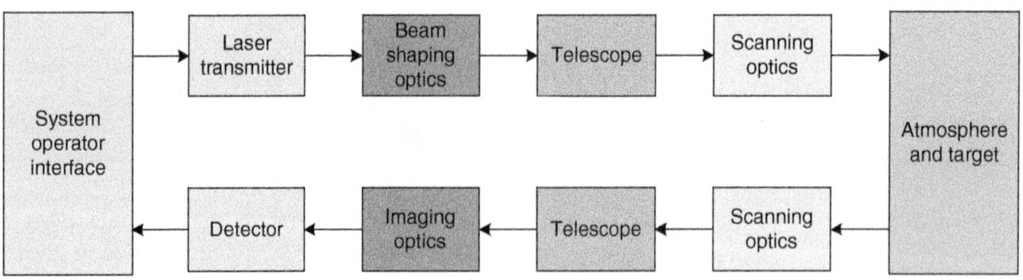

Figure 5.19 Block diagram of a direct detection LIDAR.

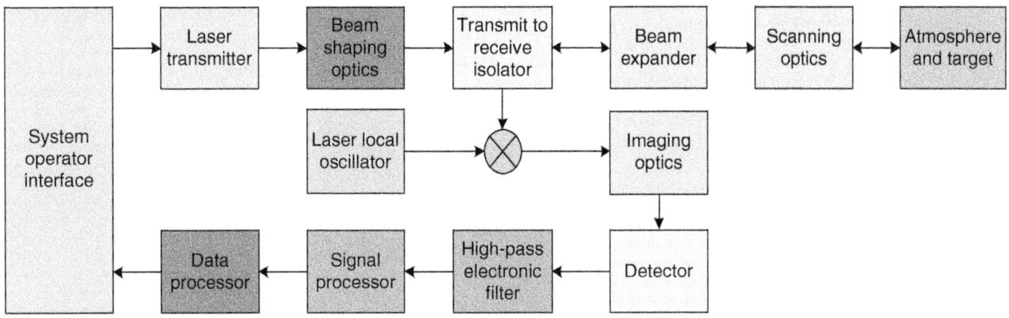

FIGURE 5.20 Block diagram of a coherent detection LIDAR.

detector by the imaging optics. The photosensitive detector generates an electrical signal in response to the received optical signal. The electrical signal is then high-pass filtered to remove any low-frequency components, such as those from background sources and from the local oscillator-induced dc signal. The high frequency components of this electrical signal contain the target information, which is then extracted from the electrical signal by signal and data processors.

In a bistatic system, the T/R switch is omitted. An additional distinction between conventional heterodyne and homodyne receivers is that while the former requires a separate laser source to serve as the local oscillator the latter uses the transmitter source for laser radiation as the local oscillator for the receiver. Offset homodyne receivers have also been developed, in which the local oscillator beam portion is frequency shifted from the transmitter beam.

Laser Range Finders

Operational range finders were introduced as early as the mid-1960s after the initial development by John D. Myers, only 5 years after Theodore Maiman presented the first working laser. Since then, a number of laser range finders (LRF) and laser target designators (LTD) have been manufactured in many countries all over the world. The high radiance and collimation of lasers makes it possible to determine distances with great accuracy. The accurate range and angle information provided by the LRF employed in modern fire control systems (FCS) is responsible for a major advance in the precision and effectiveness of weapons in battlefield conditions. A variety of laser technologies have been applied to rangefinders and Neodymium-Yttrium Aluminium Garnet (Nd:YAG) LRF, operating at a wavelength of 1,064 nm and based on the principle of pulse time-of-flight measurement, are the state of the art. LRF based on Er:fiber and Raman-shifted Nd:YAG lasers are used in cases where eye safety is fundamental. CO_2 eye-safe LRF, operating at 10.6 μm, have been developed in many configurations and they can play a significant role in conjunction with passive thermal imaging systems and other multifunctional system applications. The architecture of a typical LRF system transmitter and receiver is shown in Figures 5.21 and 5.22, respectively.

The transmitter contains an electro-optically Q-switched laser, while the radiation scattered from the target is collected by the receiver, which may be a conventional mirror or lens system. The beam divergence from the laser may be several milliradians and in order to obtain accurate target definition a simple collimating telescope has been added, which would reduce this to less than 1 mrad. The receiver may also incorporate a narrow pass-band spectral filter centered on the laser wavelength to further reduce the standing background signal which contributes to the overall system noise. The receiver electronics are illustrated in Figure 5.22 and typically include an analog section, which amplifies the return pulse whilst retaining its shape and a digital section, which performs logical timing processes and calculates the range.

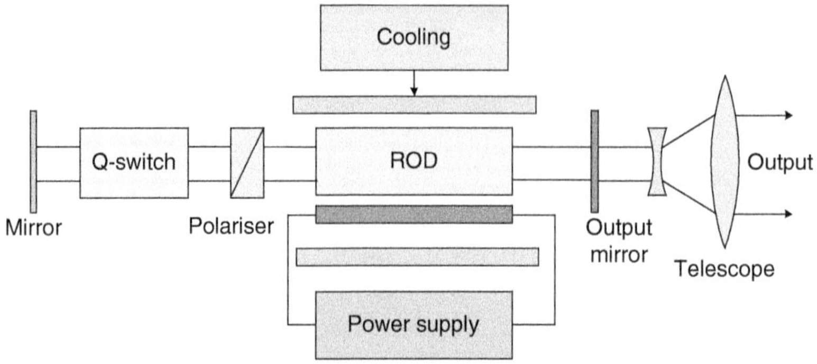

FIGURE 5.21 Typical LRF transmitter.

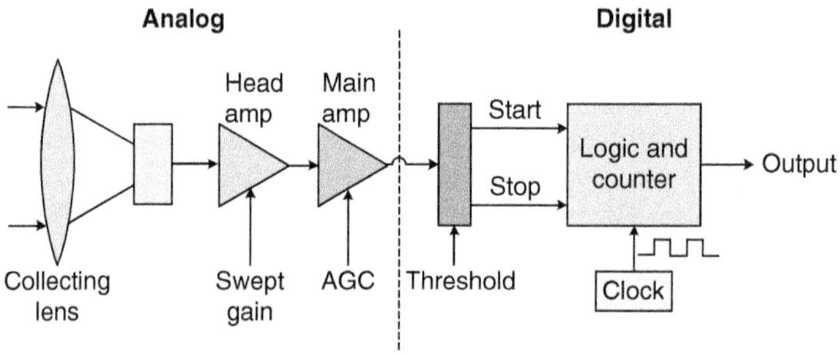

FIGURE 5.22 Typical LRF receiver.

Airborne Lasers Performance Analysis

This subsection presents the fundamental relationships to estimate the performance of airborne laser systems. These are required for design purposes as well as for experimental activities with airborne laser systems, including both developmental and operational test and evaluation in the laboratory and in flight. The generic form of the microwave radar range equation also applies to laser systems (Sabatini et al. 2015), given by

$$P_R = \frac{P_T G_T}{4\pi R^2} \cdot \frac{\sigma}{4\pi R^2} \cdot \frac{\pi D^2}{4} \cdot \tau_{\text{atm}} \tau_{\text{sys}} \tag{5.60}$$

where P_R = received signal power (W)
P_T = transmitter power (W)
G_T = transmitter antenna gain
α = transmitter beamwidth
σ = effective target cross section (m^2)
K_a = aperture illumination constant
R = system range to target (m)
λ = wavelength (m)
D = aperture diameter (m)
τ_{atm} = atmospheric transmission factor
τ_{sys} = system transmission factor

With laser systems, the transmitter antenna gain is substituted by the aperture gain, expressed by the ratio of the steradian solid angle of the transmitter beamwidth α^2 to that of the solid angle of a sphere, which is equal to the relation:

$$G_T = \frac{4\pi}{\alpha^2} \tag{5.61}$$

For laser beamwidths on the order of 1 mrad, the typical aperture gain at laser wavelengths is about 70 dB. In the far field, the transmitter beamwidth can also be expressed as

$$\alpha = K_a \frac{\lambda}{D} \tag{5.62}$$

Substituting the above expressions for transmitter aperture gain and beamwidth, equation (5.60) becomes

$$P_R = \frac{P_T \sigma D^4 \tau_{\text{atm}} \tau_{\text{sys}}}{16 R^4 \lambda^2 K_a^2} \tag{5.63}$$

Equation (5.61), obtained from the standard radar range equation, applies only in the far field of the aperture. At typical microwave bands of $\lambda = 1$ to 10^{-3} m, the far-field distances are quite short. The far-field (*Fraunhofer*) region of an aperture is typically concerned with the distance $2D^2/\lambda$ to infinity; in this vicinity, the generalized range equation applies. In certain cases, the far-field distance occurs within the feed horn assembly of a microwave antenna. As illustrated by the figure, at $\lambda = 1.064$ μm (Nd:YAG laser), a 10-cm aperture has a far-field distance of approximately 20 km. As a result, it is not unusual to operate in the near field of the optical systems; thus modifications to the range equation to account for near-field operation are required. This near-field effect modifies the beam width such that

$$\alpha = \sqrt{\left(\frac{K_a D}{R}\right)^2 + \left(\frac{K_a \lambda}{D}\right)^2} \tag{5.64}$$

Figure 5.23 shows the illustrations relative to incoherent detection and coherent detection receivers, respectively. Incoherent detection receivers at optical wavelengths are similar to video radiometers receivers (i.e., envelope detectors at microwave wavelengths). However, optical receivers have

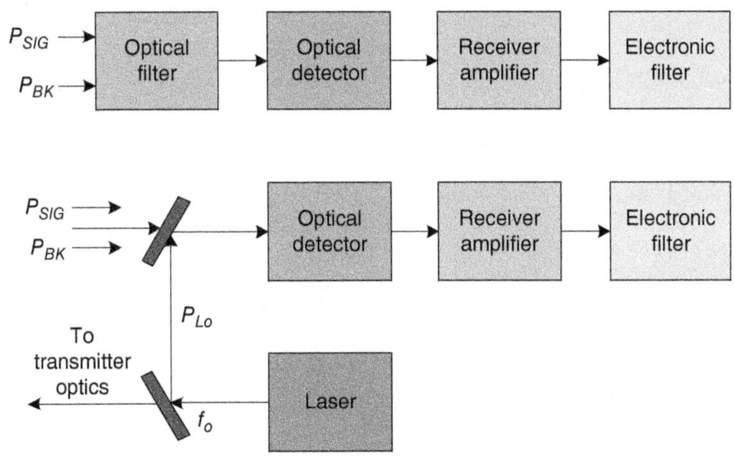

FIGURE 5.23 Laser receiver systems.

an additional term besides the signal term (P_{SIG}), the optical background power (P_{BK}) which is due to undesired signals such as sunlight, cloud reflections, flares, etc. The received optical power, after suitable filtering, is applied to the optical detector. Square law detection then occurs, producing a video bandwidth electrical signal. The coherent detection receiver is similar to the incoherent; however, a portion of the laser signal (f_o) is coupled to the optical detector via beam splitters. As a result, the optical detector has the local oscillator power (P_{LO}) in addition to the received signal power (P_{SIG}), and the competing background terms (P_{BK}).

In general, the signal-to-noise ratio (SNR) of a LIDAR system can be expressed in the form

$$\mathrm{SNR} = \frac{i_{SIG}^{\ 2}}{i_{SN}^{\ 2} + i_{TH}^{\ 2} + i_{BK}^{\ 2} + i_{DK}^{\ 2} + i_{LO}^{\ 2}} \tag{5.65}$$

where $i_{SIG}^{\ 2}$ is the mean square signal current, $i_{SN}^{\ 2}$ is the mean square shot noise current, $i_{TH}^{\ 2}$ is the mean square thermal noise current, $i_{BK}^{\ 2}$ is the mean square background noise current, $i_{DK}^{\ 2}$ is the mean square dark noise current, and $i_{LO}^{\ 2}$ is the mean square local oscillator noise current.

Laser Beam Propagation in the Atmosphere

In general, a laser beam is attenuated as it propagates through the atmosphere, mainly due to absorption and scattering phenomena. Additionally, the laser beam is often broadened, defocused, and may even be deflected from its initial propagation direction. On one hand when the output power is low, the effects are linear in behavior (absorption, scattering, and atmospheric turbulence are examples of linear effects). On the other hand, when the power is sufficiently high, new effects are observed that are characterized by nonlinear relationships (e.g., thermal blooming, kinetic cooling, bleaching, and atmospheric breakdown).

References

Hecht, J. (2010), "A Short History of Laser Development," Applied Optics, Vol. 49, Issue 25, pp. F99–F122.

Jebur, M. N., Pradhan, B., and Tehrany, M. S. (2014). "Optimization of Landslide Conditioning Factors using very High-resolution Airborne Laser Scanning (LiDAR) Data at Catchment Scale, Remote Sensing of Environment, Vol. 152, pp. 150–165.

Johnson, J. (1985). "Analysis of Imaging Forming Systems," Proceedings of the Image Intensifier Symposium (pp. 249–273). Warfare Electrical Engineering Dept.—US Army Engineering Research and Development Laboratories (Ft. Belvoir, VA, USA). Reprinted in "Selected Papers on Infrared Design," Johnson R. B., and Wolfe, W. L., (1985), SPIE Proceedings, Vol. 513, pp. 761–781.

Ratches, J. A. (1976). "Static Performance Model for Thermal Imaging Systems," *Optical Engineering*, Vol. 15, Issue 6, pp. 525–530.

Sabatini, R., Richardson, M. A., Cantiello, M., Toscano, M., and Fiorini, P., (2013). "A Novel Approach to Night Vision Imaging Systems Development, Integration, and Verification in Military Aircraft," *Aerospace Science and Technology*. DOI:10.1016/j.ast.2013.08.021

Sabatini, R., Richardson, M. A., Gardi, A., and Ramasamy, S. (2015). "Airborne Laser Sensors and Integrated Systems," *Progress in Aerospace Sciences*, Vol. 79, pp. 15–63. DOI: 10.1016/j.paerosci.2015.07.002

PART 6

Optical Fibers

Jean-Claude Mollier

5.27 Optical Fiber Theory and Applications

Basic Characteristics

Geometry (Figure 5.24)

An optical fiber is a cylindrical dielectric structure that can guide a light beam over distances ranging from tens of meters to tens of kilometers. It mainly consists of a cylindrical core (radius a, refractive index n_1) surrounded by a cladding of slightly lower refractive index n_2.

Refractive Index Profiles

The refractive index distribution in the transverse direction is given by:

$$n(r) = n_1\left[1 - 2\Delta(r/a)^g\right]^{1/2} \qquad 0 \le r \le a \tag{5.66}$$

$$n(r) = n_1\,[1 - 2\Delta]^{1/2} = n_2 \qquad r \ge a \tag{5.67}$$

where the index contrast Δ is defined as

$$\Delta = \left(n_1^2 - n_2^2\right)/2n_1^2 \tag{5.68}$$

Optical fibers are classified according to the value of the parameter g, which can be varied to get different profiles:

- Step index fiber (g ∞)
- Graded index fiber: $g = 1$ (triangular profile), $g = 2$ (parabolic profile), etc.

Spectral Loss

Depending on the application, two kinds of dielectric material are currently used to manufacture optical fibers: silica (with dopants) and plastic. The fibers with core and cladding made from silica exhibit low loss and are systematically used for long-distance applications.

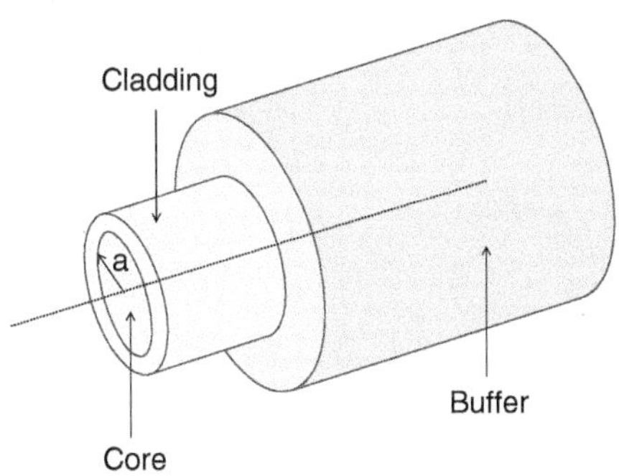

FIGURE 5.24 Optical fiber geometry.

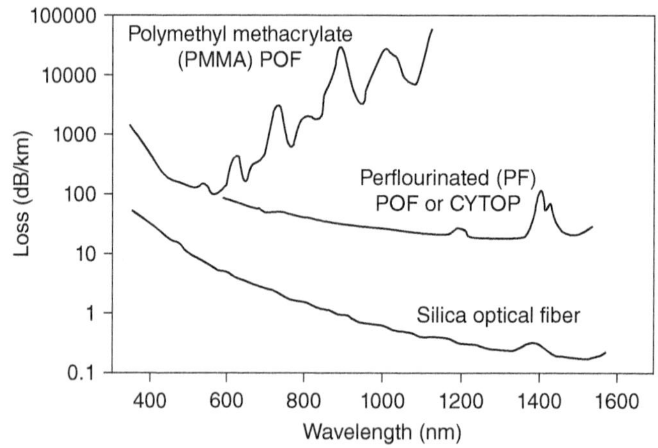

FIGURE 5.25 Spectral attenuation.

Plastic-clad fibers and all-plastic fibers are less expensive and have higher mechanical strength than silica fibers. But they are presently used for short-distance applications (less than 100 m) due to their higher losses. Typical loss spectra are represented in Figure 5.25. Table 5.10 summarizes the basic characteristics of several optical fibers.

Ray Theory of Optical Fibers

This simplified theory is valid when the core radius a is much larger than the operating wavelength λ (multimode fibers). In this geometrical optics approach, light consists of a number of rays being reflected or refracted at the interface between the core and the cladding.

Meridional and Skew Rays (Figures 5.26 and 5.27)

The rays injected into the core of the fiber are either confined to the meridian planes (meridional rays) or not confined to any plane (skew rays), depending on the injection conditions: launch angle and distance to the fiber axis.

Dielectric material	Plastic	Silica plastic	Silica Ge O_2 doped core			
Index profile	Step index	Step index	Step index		Graded index	
Core/cladding diameter (µm)	980/1000	400/480	100/140	8/125	5/125	62/125
Attenuation (dB/km)	200	10	2–5	0.2–0.5	1–3	0.9–3
Wavelength range (µm)	0.45–0.7	0.7–1	0.8–1.6	1.3–1.6	0.8–1.6	0.8–1.6
Applications	Lighting	Optical power transfer	Short-distance telecoms	Long-distance telecoms	Information networks Medium-distance telecoms	

TABLE 5.10 Optical Fiber Properties

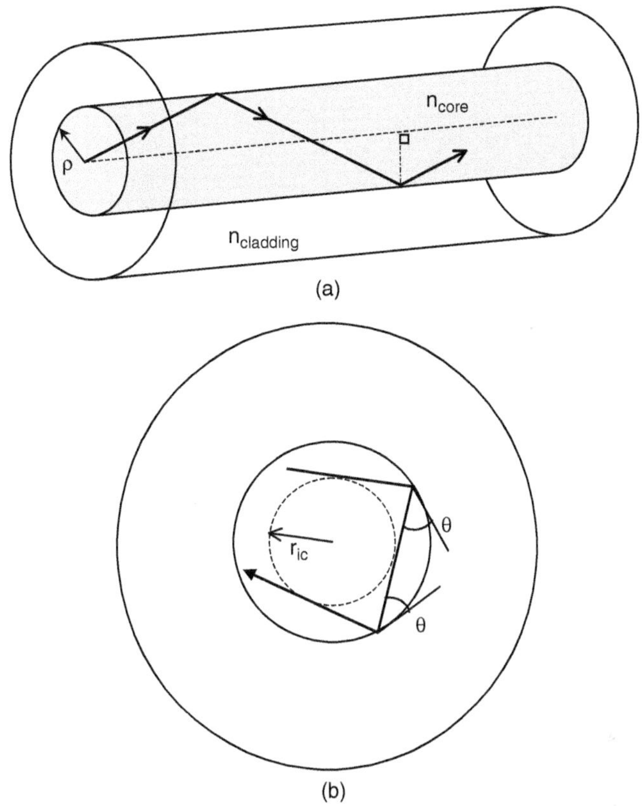

FIGURE 5.26 (*a*) Meridional ray in a step index fiber; (*b*) skew ray in a step index fiber.

Ray Paths (Figure 5.28)

The rays excited in the fiber can be classified into three different types: bound rays, refracting leaky rays, and tunneling leaky rays.

Bound rays remain guided in the core of the fiber as they undergo total reflection at the core-cladding interface.

Refracting leaky rays leak out from the fiber a very short distance.

Tunneling leaky rays leak out gradually from the core of the fiber. The attenuation distance of the corresponding optical power can vary from a few millimeters to hundreds of meters.

Numerical Aperture

Bound rays exist when the angle of incidence at the core-cladding interface is sufficiently large. It can be shown, by applying Snell's law, that the total internal reflection takes place only if the angle of incidence at the external air-core interface θ_0 is sufficiently small.

For a step index fiber, the maximum acceptance angle is given by

$$n_0 \sin\theta_0 = \sqrt{n_1^2 - n_2^2} \tag{5.69}$$

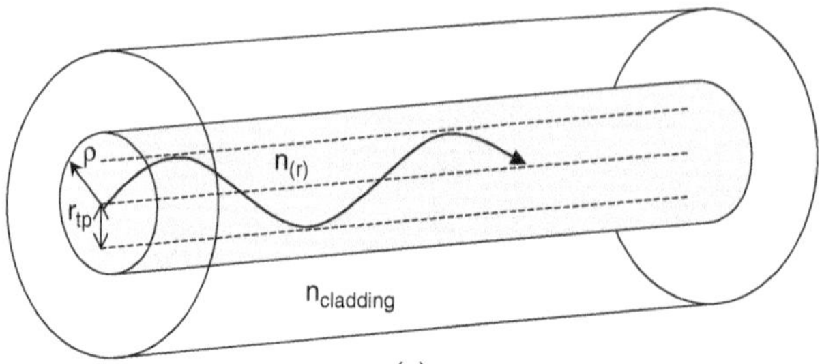

(a)

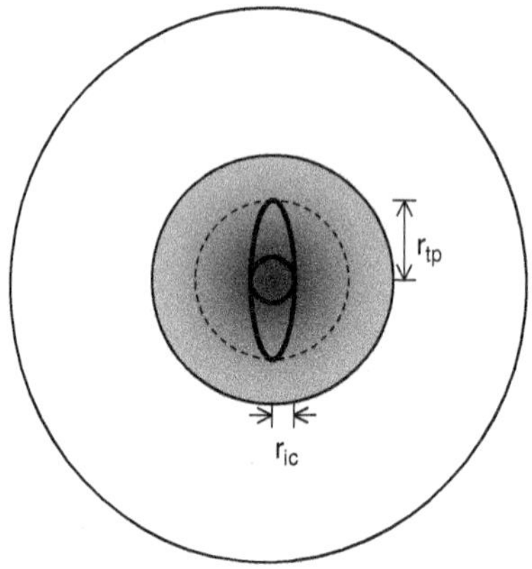

(b)

FIGURE 5.27 (*a*) Meridional ray in a graded index fiber; (*b*) skew ray in a graded index fiber.

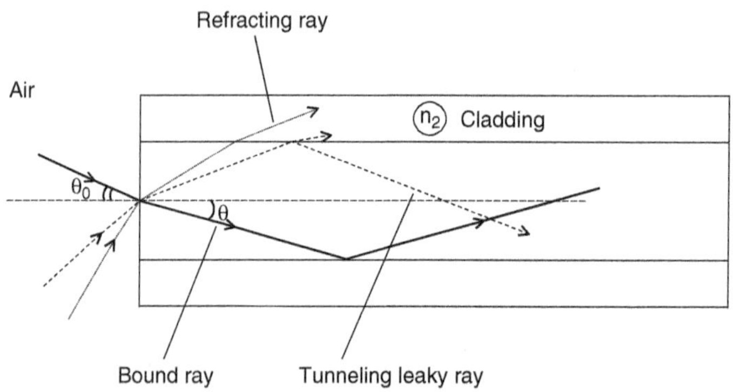

FIGURE 5.28 Various types of rays in a step index fiber.

An important parameter of the optical fiber called the numerical aperture (NA) is defined as

$$NA = \sqrt{n_1^2 - n_2^2} = n_1\sqrt{2\Delta} \tag{5.70}$$

where Δ was defined in equation (5.52).

For typical telecommunication optical fibers, NA varies from 0.1 to 0.25. For plastic fibers, NA varies from 0.4 to 0.5.

Wave Theory of Optical Fibers

To get an accurate representation for light propagation in fibers, electromagnetic analysis based on Maxwell's equations is essential. If z denotes both the fiber axis and the direction of wave propagation, the angular (Φ) and radial (r) distributions of the wave can be represented by

$$\vec{E} = \vec{E}_0(r,\Phi)\exp i(\omega t - kz) \tag{5.71}$$

$$\vec{H} = \vec{H}_0(r,\Phi)\exp i(\omega t - kz) \tag{5.72}$$

where $\vec{E}$ and $\vec{H}$ are the electric and magnetic fields, which must satisfy boundary conditions at the core-cladding interface. In addition, $\vec{E}$ and $\vec{H}$ must remain finite on the fiber axis ($r \to 0$) and decay to a negligible value outside the cladding ($r \to \infty$).

Using these boundary conditions allows the derivation of an eigenvalue equation for the propagation constant $k = \omega/v$, with v denoting the phase velocity of the wave.

Weakly Guiding Approximation

For most optical fibers, the core and cladding indices are nearly the same ($\Delta << 1$) and the eigenvalue equation can be simplified, leading to the so-called weakly guided modes: all the guided modes are linearly polarized (fixed direction of $\vec{E}$ in the fiber) and are commonly called LP_{mn} modes. The integers m and n specify the angular and radial variations of the fields (number of extrema).

In order to represent conveniently the dependence of k upon the light wavelength and the parameters of the fiber, two dimensionless parameters are introduced:

1. The normalized frequency as

$$V = ak_0\sqrt{n_1^2 - n_2^2} = ak_0 n_1 \sqrt{2\Delta} \tag{5.73}$$

where $k_0 = \omega/c = 2\pi/\lambda$ is the free space propagation constant (or wave number).

2. The normalized propagation constant b as

$$b = \frac{(k/k_0)^2 - n_2^2}{n_1^2 - n_2^2} \tag{5.74}$$

The variation of b with V forms a set of universal curves which are plotted in Figure 5.29 for a step index fiber. For a given fiber and a given operating wavelength, V is readily derived from equation (5.73). Then b is obtained from the curves which represent the guided modes LP_{mn}. Finally, k is derived from equation (5.74) for each mode of interest.

The total number of guided modes is roughly approximated by $V^2/2$ for a step index fiber and $V^2/4$ for a parabolic index profile fiber.

Cutoff Wavelengths

The condition $b = 0$, or alternatively $k = k_0 n_2$, corresponds to the cutoff of the mode. The electromagnetic wave is no longer guided in the core of the fiber and is known as an evanescent wave.

Single-Mode Fiber

As can be seen from Figure 5.30, when the condition $0 < V < 2.405$ is verified, we will have only one guided mode (LP_{01} mode). Such a fiber is referred to as a single-mode fiber and is of tremendous importance in optical communication systems.

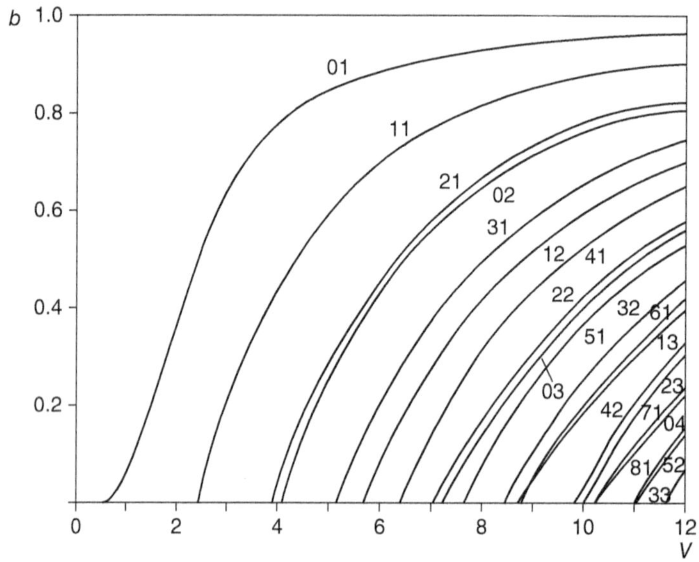

Figure 5.29 Universal curves of normalized propagation constant b for a step index fiber corresponding to guided modes LP_{mn}.

The above condition allows us to determine either the lowest operating wavelength λ_l for a given fiber or the highest value of core radius a at a fixed wavelength:

$$\lambda_l = \frac{2\pi a n_1 \sqrt{2\Delta}}{2.405} \tag{5.75}$$

$$a = \frac{2.405\ \lambda}{2\pi n_1 \sqrt{2\Delta}} \tag{5.76}$$

The core radius for single-mode silica fibers is typically in the range 3–5 μm.

Confining Factor

The confining factor $\Gamma(<1)$ is defined as the ratio of the power carried in the core to the power injected in the fiber. For a given LP mode, Γ increases markedly with V. It is essential to maintain Γ as high as possible in order to minimize the wave attenuation related to the loss mechanisms in the cladding. For a single-mode silica fiber, typical values for Γ are in the range 0.7–0.8, corresponding to $1.8 < V < 2.2$.

Dispersion in Optical Fibers

Different light components (different modes and/or different wavelengths) propagate through the fiber at different velocities. This phenomenon, called dispersion, determines the information-carrying capacity of a fiber optic communication system. When the information is coded in the form of pulses (bits), each pulse becomes larger, due to fiber dispersion, by an amount strongly depending on the refractive index profile.

Modal Dispersion

In a multimode fiber, the dominant dispersion mechanism is the intermodal dispersion: at a given operating wavelength (or given V), different modes travel with different k, i.e., different velocities (see Figure 5.29).

For a step index fiber, the full width at half maximum (FWHM) τ of the broadened pulse can be estimated as:

$$\tau \cong L \cdot n_1 \Delta / c \tag{5.77}$$

where all symbols have been defined in previous subsections.

For a parabolic graded index fiber, the FWHM τ is roughly given by:

$$\tau \cong L \cdot n_1 \Delta^2 / 8c \tag{5.78}$$

As expected, τ is proportional to the length L of the fiber. The above relations clearly show that Δ must be made as small as possible to minimize the pulse spreading.

Chromatic Dispersion

In a single-mode fiber, the dominant dispersion mechanism is the chromatic dispersion, which results from two independent phenomena:

1. *Waveguide dispersion,* determined by the difference in propagation velocity between the core and the cladding
2. *Material dispersion,* determined by the variation of propagation velocity with wavelength

In that case, the FWHM of the broadened pulse is given by

$$\tau \cong L \cdot |D_c| \cdot \Delta\lambda \tag{5.79}$$

where $\Delta\lambda$ is the spectral width of the optical signal launched into the fiber and D_c is the chromatic dispersion coefficient.

A plot showing D_c along with its two components—waveguide dispersion D_w and material dispersion D_m—is shown in Figure 5.30. Chromatic dispersion disappears at a certain wavelength. In other words, pulses can propagate without spreading and/or distortion whatever the length of the fiber is.

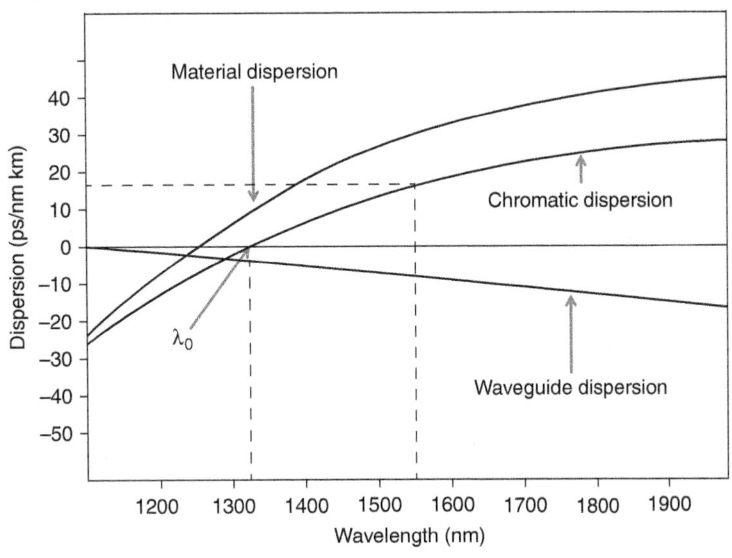

FIGURE 5.30 Chromatic dispersion, material dispersion, and waveguide dispersion for conventional single-mode fiber.

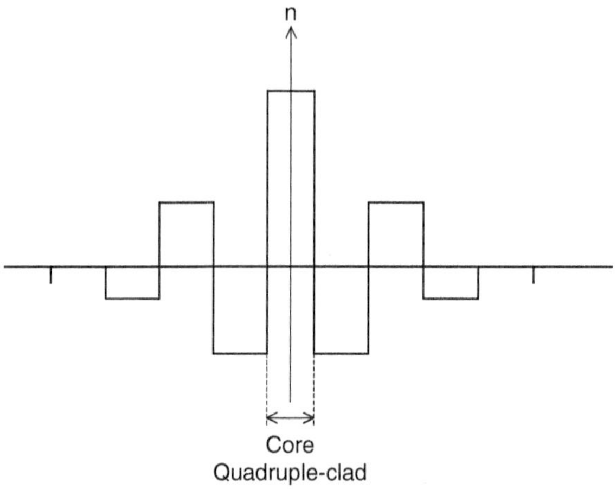

FIGURE 5.31 Index profile for dispersion-flattened fiber.

That zero-dispersion wavelength depends on fiber geometry and refractive index values in the core and the cladding. For a conventional fiber, it is approximately equal to 1.3 μm. The information transmission rate R (expressed in bits/s) is often defined as:

$$R \cong \frac{1}{2\tau} \tag{5.80}$$

and the information-carrying capacity of the fiber optic link is measured in terms of the $R \cdot L$ product, expressed in Gbps · km (symbol G means giga).

Typical values:

Single-mode silica fibers: 5–50 Gbps km; Multimode silica fibers: 0.05–0.5 Gbps km; Plastic fiber: 5–10 Mps km

Dispersion-Shifted and Dispersion-Flattened Fibers

In a silica fiber, the losses attain a minimum value of about 0.2 dB/km at 1550 nm. The shift of the zero dispersion wavelength to a region around 1,550 nm can be accomplished by slightly decreasing the core diameter and modifying the refractive index profile (triangular profile).

To achieve low dispersion in a large range of wavelengths (1.3–1.6 μm), more sophisticated index profiles are used (Figure 5.31). The resulting fiber is called dispersion-flattened fiber.

References

Buck, J. A., *Fundamentals of Optical Fibers,* John Wiley & Sons, New York, USA (1995).

Ghatak, A. and Thyagarajan, K., *Introduction to Fiber Optics,* Cambridge University Press, Cambridge, UK (1998).

Iga, K., *Fundamentals of Laser Optics,* Plenum Press, New York (1994). Lecoy, P., *Téléoms sur fibres optiques,* Hermès, Paris, France (1997).

Morthier, G. and Vankwikelberge, P. *Handbook of Distributed Feedback Laser Diodes,* Artech House, Boston, USA (1997).

Papannareddy, R., *Introduction to Lightwave Communication Systems,* Artech House, Boston, USA (1997).

PART 7

Aircraft Flight Control Systems

Dominique Brière and Sylvain Prudhomme

5.28 Foreword

Flight control is a very complex engineering challenge that encompasses many disciplines. Flight control systems (FCS) design includes understanding and modeling of the aircraft dynamics, deriving control requirements, hardware and software development and implementation, and finally in-flight validation and operation. It goes far beyond flight control laws (FCL) design, although this particular discipline is generally supported by valuable and intensive theoretical and applied research.

Historically, the technologies available for the first FCS generations and the tools used for their development were very simple. The continuous demand for enhancement in aircraft flight performance has led to progress in technology, more sophisticated FCS, and higher integration in the overall aircraft architecture. Particularly, fly-by-wire (FBW) technology revolutionized aircraft FCS in the 1980s. Several books have been published since that time that will be of interest for students or engineers interested in FCS and willing to learn more (McLean 1990; Blakelock 1991; Stevens and Lewis 1992; Pratt 2000). This part obviously cannot go into detail for all FCS design issues. It should be considered as an introduction that aims only to help the understanding of key issues. Basic information will be found on

- Flight control objectives and requirements
- Actuation principles and aircraft flight dynamics
- Control laws design process
- Technological issues for flight control implementation

When necessary, information will be particularized for civil and military FCS design. The Airbus FBW experience will be detailed and some open and challenging control problems that pertain to meeting high-level performance requirements and developing flight controls for nonconventional aircraft configurations will be presented.

5.29 Flight Control Objectives and Principles

Flight Control Objectives

In very general terms, flight control first aims to ensure safe and suitable aircraft maneuvering in the overall flight envelope. The adjectives *safe* and *suitable* have slightly different meanings for civil and military applications.

In civil aviation, *safe* must be understood in terms of airworthiness and certifiability by civil authorities (FAR 25 in the United States, JAR 25 in Europe). Aircraft manufacturers must prove the correct functioning of the FCS in all possible flight conditions, including cases of failures. Such proof can be obtained by extensive analysis and simulation and by flight testing and may be very costly. Because of these stringent safety objectives, the FCS is one of the most critical aircraft systems. In practice, *suitable* means that the FCS must provide good handling qualities (HQ), i.e., predictable dynamic response to inputs applied to the aircraft. Different requirements are set for two kinds of inputs:

- The aircraft must have a good response to commanded maneuvers (e.g., pilot demand on the stick or rudder).
- Any perturbation should be alleviated (wind shear, failure, etc.) in order to desensitize the aircraft behavior and maintain both desired trajectory and flight comfort.

The distinction between the two kinds of inputs already introduces a first consideration for the architecture of the FCL and different levels of specifications for their design. Feedback control is required to maintain stability and alleviate perturbations. Feedforward control is required to shape demands from the pilot into control signals sent to the actuators. For civil aircraft HQ, each manufacturer develops its own experience in addition to regulation rules and has specific requirements and criteria. Since the civil aircraft mission is unique, the best uniformity in dynamic behavior is generally sought over the entire flight envelope in order to provide the pilot with safe and comfortable handling. Landing and takeoff are exceptions, with specific requirements in terms of climbing, rotation, descent, approach, and flare. Higher-level requirements are set for autopiloting and flight envelope protection and monitoring.

For military aircraft, safety does not lead to the same constraints and must generally be compromised with performance and flexibility for various operational needs. However, a key safety issue is to ensure the stability of aircraft that generally have relaxed or negative natural stability margins, traits that are favorable for high-performance maneuvering. Flight control must provide the best actuation efficiency for any of the various flight configurations encountered in a mission, including weapons carriage, terrain following, ground attack, combat, and refueling. Military HQ requirements can be found in Ministry of Defence (1983) and MIL-1797-A.

Flight Control Principles

The physical principle of flight control is the generation of external forces and moments, either for maintaining aircraft equilibrium or for breaking it in order to induce dynamical motion. The only way to generate such forces is to exploit on-board available power such as thrust and hydraulic/electrical power. The latter is used for moving conventional flight control surfaces through dedicated actuators and for modifying aerodynamic forces and moments as desired on the aircraft (McCormick 1995). At high speed, aerodynamic efficiency is high and only small surface deflection is necessary to generate large forces. Low-speed flight conditions are more critical because of the need for higher aerodynamic coefficients. They generally drive the sizing of conventional moving surfaces. Some additional sophisticated active control technologies (ACT) now are included in FCS with ambitious objectives much beyond the aforementioned basic ones, mainly for military applications such as described below. Performance requirements have pushed the technology.

Development of hardware actuator technology is a specific discipline of FCS design in which tremendous progress has been made. Some advances from the Airbus experience are presented below. Other descriptions of technology can be found in Mair and Seabridge (2001). We will only retain here the following trends, which tend toward best efficiency at the lowest cost and lowest weight with the highest operational reliability:

- Transmission of power to the control surfaces by mechanical means belongs to the past.
- Hydraulic power is progressively being replaced by electrical power (all-electrical aircraft concept)
- When necessary, generation of hydraulic power is localized rather than centralized (electro-hydraulic actuator technology).

Actuators are controlled using local feedback loops and designed separately before being integrated in the aircraft as specific stability-augmented, high-bandwidth stand-alone subsystems. For this, the assumption is made that couplings with the aircraft dynamics can be neglected. Couplings may appear for specific military aircraft applications, which necessitate the integration of actuator design in overall aircraft design.

Aircraft Modeling

For sizing flight control surfaces and designing the FCL, a good understanding of the aircraft flight dynamics is necessary. As illustrated in several books dedicated to that area (Stevens and Lewis 1992; McCormick 1995; Boiffier 1998; Pratt 2000), most issues related to flight control can be studied using the general equations that govern aircraft dynamics. This subsection is not the place for going into details. We only discuss the most pertinent issues, using very few and very simple equations.

The general equations of aircraft dynamics have the following form:

$$M(\dot{X}) = F(X,U,W) \tag{5.81}$$

where X = vector of generalized aircraft internal states (from simple rigid aircraft states such as incidence, pitch angle, and rate to more sophisticated ones that describe the structural dynamics, for example)

U = vector of inputs to the actuators (thrust, moving surface deflection, etc.)

W = vector of generalized unmeasured disturbances applied to the aircraft

M = operator of generalized masses and inertia

F = operator of generalized forces and moments applied to the aircraft, some being very simple (weight is assumed concentrated at the CG), others more complex (thrust, aerodynamic forces, and moments are nonlinear functions of states X and controls U)

Actually, the construction of such equations relies on the availability of data from propulsion, structures, and aerodynamics. Permanent model updating throughout the aircraft development process positions FCS design as one of the ultimate steps, particularly concerning trimming, HQ assessment, and FCL design.

Trimming consists of finding the controls U_0 that lead to equilibrium $\dot{X} = 0$ for some flight conditions X_0 and steady disturbance W_0 by solving the following equation for U_0, given by:

$$F(X_0,U_0,W_0) = 0 \tag{5.82}$$

Trimming values X_0, U_0 are acceptable only if they lead to a stable static equilibrium, i.e., if any small instantaneous disturbance is naturally damped so that X always returns to X_0.

HQ assessment consists of studying the dynamic behavior in the neighborhood of trimmed conditions, given by

$$X = X_0 + x; \quad U = U_0 + u; \quad W = W_0 + w \tag{5.83}$$

and more precisely of analyzing the locally tangent model obtained after linearization and M inversion, given by

$$\dot{x} = A(X_0,U_0,W_0)x + B(X_0,U_0,W_0)u + E(X_0,U_0,W_0)w \tag{5.84}$$

Its eigenvalues $\lambda(A)$ are called modes and have physical meanings. Rigid modes such as short period, phugoid, and Dutch roll characterize the low-frequency aircraft dynamic behavior. They should be stable and fall within an acceptable frequency band (neither too fast nor too slow). They can be manipulated by the FCL in order to provide the pilot with suitable HQ.

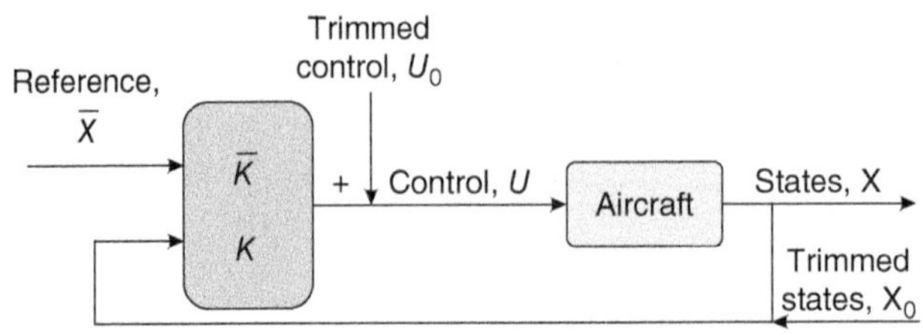

FIGURE 5.32 Generic FCL architecture.

FCL design consists of computing the gains which will be used to generate signals sent to the actuators $U = U_0 + K(X - X_0) + \bar{K}\bar{X}$ for stability augmentation and HQ enhancement via feedback gain K and for tracking pilot demand $\bar{X}$ via feedforward gain $\bar{K}$, such as illustrated in Figure 5.32.

Equations of Motion

The motion of an aircraft can be expressed in terms of a set of equations, representing the translational and angular accelerations. Applying Newton's second law to the aircraft (which is considered to be a rigid body), the following equations are obtained:

$$\vec{F} = \frac{d}{dt}\{m\,\vec{V_T}\} \tag{5.85}$$

$$\vec{M} = \frac{d}{dt}\{\vec{H}\} \tag{5.86}$$

where $\vec{F}$ = sum of all the applied forces
$\vec{M}$ = sum of applied torques
$\vec{H}$ = angular momentum
m = mass of the aircraft
$\vec{V_T}$ = total velocity vector

The sum of external forces can be subdivided into aerodynamic forces, gravitational force, and propulsive force. The forces and moments about the six degrees of freedom, are expressed as [Allerton (2009)]

$$\sum \vec{F_x} = \frac{d}{dt}\{m\vec{u}\} \tag{5.87}$$

$$\sum \vec{F_y} = \frac{d}{dt}\{m\vec{v}\} \tag{5.88}$$

$$\sum \vec{F_z} = \frac{d}{dt}\{m\vec{w}\} \tag{5.89}$$

$$\sum \vec{L} = \frac{d\vec{h_x}}{dt} \tag{5.90}$$

$$\sum \vec{M} = \frac{d\vec{h_y}}{dt} \tag{5.91}$$

$$\sum \vec{N} = \frac{d\vec{h_z}}{dt} \tag{5.92}$$

where $\vec{u}$, $\vec{v}$, $\vec{w}$ are the linear velocities along the x, y, and z axis, respectively; $\vec{F_x}$, $\vec{F_y}$, $\vec{F_z}$ are the forces acting on the aircraft along the three axes; and $\vec{h_x}$, $\vec{h_y}$, $\vec{h_z}$ are the components of angular momentum along the three axes. The effect of angular velocities on linear velocities can be easily analyzed. For example, considering the axial velocity $\dot{\vec{x}}$, the pitch rate Q increases the velocity by a factor Qz, but the yaw rate R reduces it by a factor Ry. The equations for the linear velocities are given by

$$\dot{\vec{x}} = U + Qz - Ry \tag{5.93}$$

$$\dot{\vec{y}} = V - Pz + Rx \tag{5.94}$$

$$\dot{\vec{z}} = W + Py - Qx \tag{5.95}$$

where U, V, and W are the rates in linear velocities and P is the roll rate. The above equations are differentiated to obtain:

$$\ddot{\vec{x}} = \dot{U} + \dot{Q}z + Q\dot{z} - \dot{R}y - R\dot{y} \tag{5.96}$$

$$\ddot{\vec{y}} = \dot{V} - \dot{P}z - P\dot{z} + Rx + R\dot{x} \tag{5.97}$$

$$\ddot{\vec{z}} = \dot{W} + \dot{P}y + P\dot{y} - \dot{Q}x - Q\dot{x} \tag{5.98}$$

Substituting for $\ddot{\vec{x}}$, $\ddot{\vec{y}}$, and $\ddot{\vec{z}}$ in the above equations, the below expressions are obtained

$$\ddot{\vec{x}} = \dot{U} + QW + QPy - Q^2x - RV + RPz - R^2x + \dot{Q}z - \dot{R}y \tag{5.99}$$

$$\ddot{\vec{y}} = \dot{V} - PW - P^2y - PQx + RU + RQz - R^2y - \dot{P}z + \dot{R}x \tag{5.100}$$

$$\ddot{\vec{z}} = \dot{W} + PV - P^2z + PRx - QU - Q^2z + QRy + \dot{P}y - \dot{Q}x \tag{5.101}$$

The forces acting on the element of mass are given by

$$\vec{F_x} = m\ddot{\vec{x}} \tag{5.102}$$

$$\vec{F_y} = m\ddot{\vec{y}} \tag{5.103}$$

$$\vec{F_z} = m\ddot{\vec{z}} \tag{5.104}$$

The inertia moments acting on the mass element are given by

$$M_x = \vec{F_z}y - \vec{F_y}z \tag{5.105}$$

$$M_y = \vec{F_x}z - \vec{F_z}x \tag{5.106}$$

$$M_z = \vec{F_y}x - \vec{F_x}y \tag{5.107}$$

The forces and moments acting on the aircraft can be calculated by summing the above equations over the complete aircraft mass. Assuming that the center of gravity of the aircraft coincides with the origin of the chosen axis system, the terms containing mx, my, and mz in the above equations can be ignored, since their summation over the entire aircraft yields zero. Therefore, the force equations are expressed as

$$\vec{F}_x = m\ddot{x} = m(\dot{U} + QW - RV) \tag{5.108}$$

$$\vec{F}_y = m\ddot{y} = m(\dot{V} - PW + RU) \tag{5.109}$$

$$\vec{F}_z = m\ddot{z} = m(\dot{W} + PV - QU) \tag{5.110}$$

Similarly, the moments are expressed as

$$L = I_{xx}\dot{P} + QR\left(I_{zz} - I_{yy}\right) - I_{xz}(\dot{R} + PQ) \tag{5.111}$$

$$M = I_{yy}\dot{Q} + PR(I_{xx} - I_{zz}) + I_{xz}(P^2 - R^2) \tag{5.112}$$

$$N = I_{zz}\dot{R} + PQ\left(I_{yy} - I_{xx}\right) - I_{xz}(\dot{P} - QR) \tag{5.113}$$

where I_{xx}, I_{yy}, I_{zz}, and I_{xz} are the moments of inertia and are given by

$$I_{xx} = \sum \Delta m\,(y^2 + z^2) \tag{5.114}$$

$$I_{yy} = \sum \Delta m\,(x^2 + z^2) \tag{5.115}$$

$$I_{zz} = \sum \Delta m(x^2 + y^2) \tag{5.116}$$

$$I_{xz} = \sum \Delta mxz \tag{5.117}$$

The state-of-the-art trajectory optimization algorithms employ a 3-Degrees-of-Freedom (3-DoF) or 6-Degrees-of-Freedom (6-DoF) Aircraft Dynamics Model (ADM) with variable mass. The 3-DoF Equations of Motion (EoM) describing the aircraft states and governing the translational movements along the longitudinal, lateral, and vertical axes are given by

$$\frac{d\varphi}{dt} = \frac{V\cos\gamma\sin\chi + V_{W_\varphi}}{R_M + h} \tag{5.118}$$

$$\frac{d\lambda}{dt} = \frac{V\cos\gamma\cos\chi + V_{W_\lambda}}{\cos\varphi(R_T + h)} \tag{5.119}$$

$$\frac{dh}{dt} = V\sin\gamma + V_{W_h} \tag{5.120}$$

$$\frac{dV}{dt} = \frac{\Delta f}{m} - g\sin\gamma \tag{5.121}$$

$$\frac{d\gamma}{dt} = \frac{g(n\cos\phi - \cos\gamma)}{V} \tag{5.122}$$

$$\frac{d\chi}{dt} = \left(\frac{N.g}{V}\right)\left(\frac{\sin\phi}{\cos\gamma}\right) \tag{5.123}$$

$$\frac{dm}{dt} = -\,c(P,V,h)\;T(P,V,h) \tag{5.124}$$

$$\Delta f = T(P,V,h) - D(L,V,h) \tag{5.125}$$

The above Differential Algebraic Equations (DAEs) incorporate three control variables $u = (P, n, \phi)$, where P is the engine power setting, n is the load factor, and ϕ is the bank angle. The seven state variables are $x = (m, \varphi, \lambda, h, V, \gamma, \chi)$, where m is the aircraft (variable) mass, φ is the geodetic latitude, λ is the geodetic longitude, h is the altitude, V is the true air speed, γ is the flight path angle, χ is the heading. R_M is the meridional radius of curvature, R_T is the transverse radius of curvature, W is the wind velocity, T is the thrust, D is the drag, and g is the nominal acceleration due to gravity of the Earth and c is the specific fuel consumption. This model assumes a rigid body aircraft with a vertical plane of symmetry, nil wing bending effect, a rigidly mounted aircraft engine on the vehicle body, zero thrust angle, negligible moments of inertia, varying mass only as a result of fuel consumption and uniform gravity. Wind effects are considered along the three geodetic reference axes. Earth's shape is approximated as an ellipsoid using the World Geodetic System of year 1984 (WGS 84) parameters. The lateral path is constructed in terms of segments (straight and turns) and is based on the required course change and the aircraft predicted ground speed during the turn. The specificity of the trajectory optimization and optimal control with respect to other mathematical optimization branches is the application to dynamical systems.

Control Allocation and Aircraft Trimming

Control allocation consists of finding the best correlation between available control surfaces and aircraft motions to be controlled. In the most general case, considering that all moving surfaces (and all additional control technologies) contribute to the generation of forces and moments in all directions would lead to a very complex multivariable control problem whose complexity goes beyond this Part. For conventional aircraft such as illustrated in Figure 5.33, the architecture of

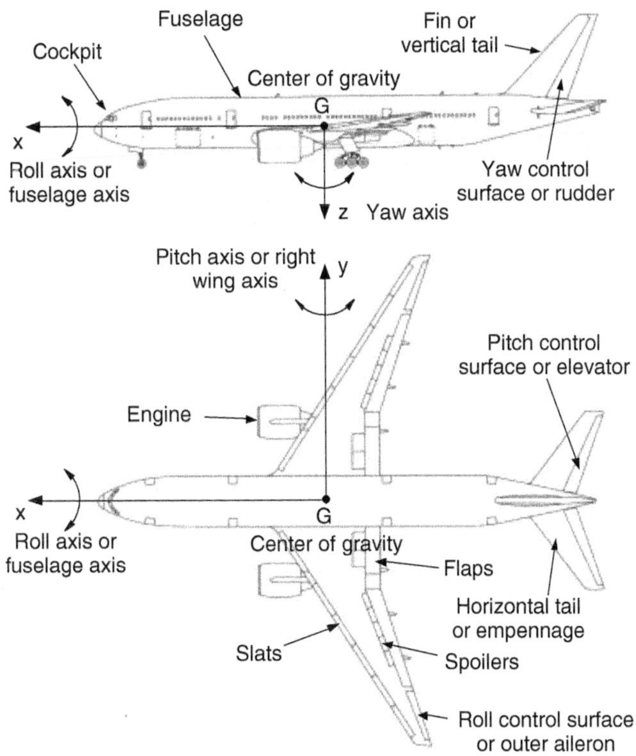

Figure 5.33 Conventional actuation architecture.

wing and tail allows natural control allocation by decoupling longitudinal and lateral actuation. For longitudinal motion, thrust generates forces against drag and allows low-frequency longitudinal axis motion control, high-lift devices (flaps and slats) generate extra lift coefficient and vertical forces which provide some low-frequency control on the vertical axis, and the horizontal tail (when trimmable) and the elevator generate moments for low and higher frequency pitch control, respectively. For lateral motion, the ailerons and rudder generate moments in roll and yaw, respectively.

For conventional aircraft, the longitudinal and lateral dynamics are decoupled because of important items:

- Mass and inertial distributions lead to the diagonal dominance of M and a natural splitting of states X into decoupled X_{lon} and X_{lat} on the left-hand side of equation (5.81).
- Conventional tail configurations allow the generation of decoupled forces and moments into F_{lon} and F_{lat} on the right-hand side.
- For most steady trimmed flight conditions in either X_{0lon} or X_{0lat}, aerodynamic coupling effects involved in cross-terms $F_{lon}(X_{lat})$ and $F_{lat}(X_{lon})$ are expressed very simply, so that equation (5.81) splits into two decoupled ones, given by

$$M_{lon}\left(\dot{X}_{lon}\right)=F_{lon}\left(X_{lon},U_{lon},W\right)+F_{lon-lat}\left(X_{0lat},W\right) \tag{5.126}$$

$$M_{lat}\left(\dot{X}_{lat}\right)=F_{lat}\left(X_{lat},U_{lat},W\right)+F_{lat-lon}\left(X_{0lon},W\right) \tag{5.127}$$

Trimming then appears as a decoupled problem, first in longitudinal for given flight conditions X_{0lon}, X_{0lat}, W_0 by solving the following equation in U_{0lon}, given by

$$F_{lon}\left(X_{0lon},U_{0lon},W_0\right)+F_{lon-lat}\left(X_{0lat},W_0\right)=0 \tag{5.128}$$

then in lateral by solving the dual equation in U_{0lat}. Another simplification is possible, but only for conventional two lifting surface configurations, which makes longitudinal trimming easier. The aerodynamic forces on the wing are not dependent on the controls on the horizontal tail. This has two major advantages:

- Equation for longitudinal trimming [see equation (5.128)] further simplifies into

$$F_{wing}\left(X_{0lon}\right)+F_{tail}\left(U_{0lon}\right)+F_{lon-lat}\left(X_{0lat}\right)=0 \tag{5.129}$$

and can be solved explicitly in U_{0lon}. Actually, the longitudinal trimming problem is itself decoupled in lift (wing), drag (thrust), and pitch (tail). This allows simple expressions of static constraints for actuator sizing, including margins to ensure static stability over a given range of mass distribution and CG location.

- Actuator sizing and wing performance are decoupled (at first order). The tail can be sized without penalizing the drag of the wing, the total drag being computed as $C_{Dwing} + C_{Dtail}$, and minimized using remaining degrees of freedom. High-speed and low-speed trimming can be solved independently without compromising performance.

Control surfaces are sized with different kinds of criteria. Among the most stringent constraints are those related to the requirement to maintain aircraft equilibrium in the presence of a maximum disturbance W_0 expected to be encountered in flight (wind, engine failure, as imposed by certification rules), mainly at landing. Some additional criteria coming from dynamical constraints (see below) may be taken into account.

The general architecture of military aircraft introduces coupling effects in both mass/inertia distribution because of compactness and force/moment generation because of actuator geometry. Control allocation is more complex than for civil aircraft. It must be considered that some control surfaces can generate forces in several directions and that in-flight dynamic allocation may be necessary either for reconfiguration in case of failure/damage or for permanent optimization of total efficiency by exploitation of redundancies.

HQ and Dynamic Stability

HQ can be assessed such as in Hodgkinson (1999) after trimming using equations. The coefficients in the time-invariant matrices A, B, E are aerodynamic derivatives $\partial F/\partial X(X_0, U_0)$ and $\partial F/\partial U(X_0, U_0)$, such as C_α, C_{mq}. For conventional aircraft configurations, the simplifications already introduced in the previous subsections apply:

- Mass and inertia distribution allow separation of x into x_{lon} and x_{lat}.
- Control allocation allows separation into longitudinal and lateral contributions.
- Terms related to aerodynamic coupling effects generally vanish when trimmed so that decoupled equations for both motions yield:

$$\dot{x}_{lon} = A_{lon}\ x_{lon} + B_{lon}\ u_{lon} + E_{lon}\ w \tag{5.130}$$

$$\dot{x}_{lat} = A_{lat}\ x_{lat} + B_{lat}\ u_{lat} + E_{lat}\ w \tag{5.131}$$

or in the Laplace domain:

$$x_{lon} = (sI - A_{lon})^{-1}(B_{lon}u_{lon}(s) + E_{lon}w(s)) \tag{5.132}$$

$$x_{lat} = (sI - A_{lat})^{-1}(B_{lat}u_{lat}(s) + E_{lat}w(s)) \tag{5.133}$$

Each equation determines specific modes, the number of which depends on the complexity of the retained modeling. In longitudinal, $\lambda(A_{lon})$ includes at least the short period and phugoid modes. In lateral, $\lambda(A_{lat})$ includes Dutch roll, spiral, and roll. Removing negligible terms leads to understandable formulas for frequencies and damping ratios, which have physical interpretation. Such analytical analysis forms the basis of HQ assessment. A good knowledge of the masses and inertias (input from structures) as well as the main aerodynamic coefficients (for example C_α, C_α for short period) is necessary. Those related to steady aerodynamics are well predicted by CFD, at least at high speed. Wind tunnel testing (WTT) or flight testing may provide updated estimates. For coefficients related to unsteady aerodynamics (C_{mq}), CFD prediction accuracy around 30% is unacceptable. Dynamic WTT can provide a better 10% accuracy.

An additional requirement is to avoid aircraft pilot coupling (APC). This phenomenon (formerly called pilot induced oscillations, PIO) is due to an implicit feedback loop introduced by the pilot. He may inject demands correlated to his feeling of the aircraft motion into the actuators, potentially destabilizing the overall closed loop if not properly phased. Such requirements are more stringent for military aircraft. Description of APC categories, criteria, and requirements can be found in Pratt (2000).

In the case where the natural behavior of the aircraft would not be good enough from a HQ point of view, different options are possible:

- Update aircraft configuration and design to make it better.
- Enhance the HQ with an automatic flight control system, such as developed in below.

5.30 Flight Control Systems Design

The FCS Design Process

The FCS design and development processes are long-term programs. There are a variety of approaches used among manufacturers based on their own experience and particular needs. Pratt (2000) and Moir and Seabridge (2001) give an overview of technical activities undertaken in the development programs and their scheduling in the standard *V*-diagram (requirement-specification-development-integration-testing process). The main concern is to meet requirements while guaranteeing certifiability at the lowest cost. This implies very structured development methodologies in order to keep traceability and control, including ground and flight testing (Bourcier 2001). The following subsections will be devoted to a more detailed presentation of the FCL design. In the first steps, models used for such design only include prediction values from structures and aerodynamics. They are refined throughout the ongoing aircraft design process. Final knowledge of the aircraft dynamics is only available after ground and flight tests, just before operation. The FCL design process must also include tools and procedures for model identification (Ljung 1987; Söderstrom and Stoica 1989) and for FCL tuning (Prudhomme and Gimonet 2001).

FCL Design Principles

The most sophisticated control design techniques may be very powerful but useless if basic aircraft stability is not understood. Before moving to modern design, this subsection gives basic physical principles of control law design (McLean 1990; Stevens and Lewis 1992). Most civil or military aircraft do not meet all the HQ requirements in open loop, generally because modes (short period or Dutch roll) are not adequately damped. Some military aircraft may even be unstable. Stability augmentation systems (SAS) provide additional artificial damping by connecting a feedback loop between measurements (generally angular rates) and the commands sent to the actuators in a single input, single output (SISO) scheme. Taking the example of longitudinal motion and assuming perfect lateral decoupling, the second order short period approximation of equations (5.130) and (5.131) gives the following open-loop transfer function between elevator deflection $u_{\text{lon}} = \delta_q$, vertical wind w, and pitch angle $x_{\text{lon}} = \theta$ in the Laplace domain:

$$\theta(s) = \frac{1}{\omega^2 + 2\zeta\omega s + s^2}\left(k\delta_q(s) + l\ w(s)\right) \tag{5.134}$$

The natural damping ratio ζ is proportional to C_{mq} and C_α and may be small or negative, i.e., unacceptable from an HQ point of view. It can be modified by a feedback loop using measurement of pitch rate $q = \dot{\theta}$ and computing:

$$\delta_q = -Kq + \bar{\delta}_q \tag{5.135}$$

This leads to the closed loop transfer function, given by

$$\theta(s) = \frac{1}{\omega^2 + (2\zeta\omega + kK)s + s^2}\left(k\bar{\delta}_q(s) + \omega(s)\right) \tag{5.136}$$

If K is correctly tuned, the closed loop damping ratio is augmented from ζ to $\zeta^x = \zeta + kK/2\omega$. It allows artificial stabilization of a naturally unstable aircraft. Similarly, using a measurement of θ enables the tuning of the disturbance rejection bandwidth by manipulation of the natural frequency ω and the θ/w transfer function. For lateral motion, the principle of SAS is similar: it consists of feeding yaw rate measurement r on the rudder actuator $u_{\text{lat}} = \delta_r$ for augmenting natural damping of the Dutch roll mode, with possible difficulties (not detailed here) due to couplings with roll and spiral modes.

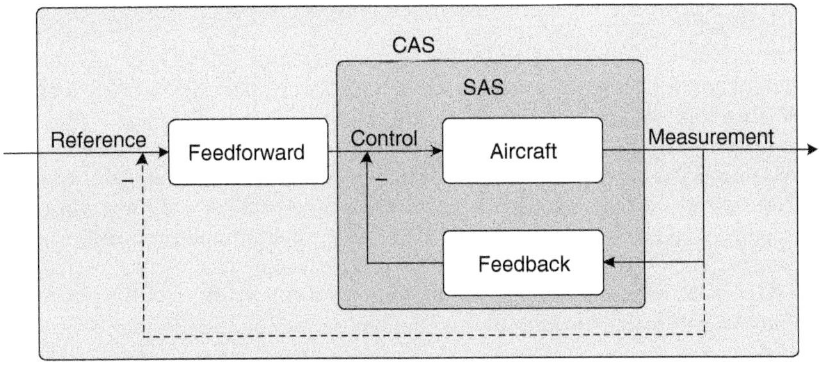

FIGURE 5.34 Illustration of SAS and CAS architecture.

In manual control, control augmentation systems (CAS) aim to provide acceptable aircraft time responses to pilot inputs. As illustrated in Figure 5.34, the laws are more sophisticated than for SAS. An additional feedforward gain K for filtering the pilot inputs are designed simultaneously with the SAS feedback. For the same example of longitudinal control, using the following control signal:

$$\delta_q = -Kq + \bar{K}(\bar{x}_{\text{lon}} - x_{\text{lon}}) \tag{5.137}$$

in equation (5.130) leads to a modified closed loop transfer function, given by

$$x_{\text{lon}}(s) = \frac{T}{\omega^2 + k\bar{K}T + (2\zeta\omega + kK)s + s^2}\left(k\bar{K}\,\bar{x}_{\text{lon}}(s) + l\,w(s)\right) \tag{5.138}$$

where $T(s)$ is the x_{lon}/θ transfer function. Both gains contribute to

- The augmentation of the closed loop damping and the manipulation of the natural frequency
- Disturbance rejection in a specified bandwidth (small x_{lon}/w for high K, $\bar{K}$)
- Additional features for the pilot, depending on the selection of reference $\bar{x}_{\text{lon}}$ (vertical acceleration N_z or pitch angle θ) and on gain structure: proportional gain ($\bar{K}$ constant) is for short-term response, proportional integral (PI) gain also offers long-term reference tracking ($x_{\text{lon}} = \bar{x}_{\text{lon}}$).

Autopiloting is the ultimate specification level. It consists of automatic tracking of target variables without any pilot input. Methods used for CAS may apply with specific selection of reference variables (altitude and speed in cruise, pitch angle or slope in climb, glide angle for landing, etc.). Gain $\bar{K}$ always include some integral terms (PI control with high gains at low frequency) for meeting the main requirement of zero steady-state error. These ultimate levels in requirements do not induce additional needs in terms of technology (actuators or measurements). Requirements generally can be met by specific FCL design. The interrelationship among flight control, guidance, and flight management is illustrated below and developed in Moir and Seabridge (2001).

For military aircraft, APC prevention consists of designing gains with values high enough for performance, but not too high, in order to avoid possible coupling between the stability-augmented aircraft and the pilot's own gain in the controlled bandwidth.

Modern FCL Design

SAS and CAS show limitations for multi-input, multi-output (MIMO) or high-order dynamic systems. Modern control design is an extension of these basic methods and opens perspectives for high-performance control of complex aircraft configurations showing high-order dynamics such as flexible aircraft or strong couplings between axes, or having redundant movable control surfaces. It provides unified tools for formulating all control objectives, including stability augmentation, tracking, disturbance rejection, etc. It has been used in the past for the integration of flight and propulsion control (FADEC) into a single MIMO integrated flight propulsion control system (IFPCS).

For MIMO control design, transfer functions are only formally used for the purpose of notation simplicity in representing generic aircraft model, given by

$$y(s)=[G_u(s)\ G_w(s)]\begin{bmatrix}u(s)\\ w(s)\end{bmatrix} \tag{5.139}$$

and two-degree-of-freedom dynamical control laws simultaneously using multidimensional measurements y and reference signals $\overline{y}$, given by

$$u(s)=\left[-K(s)\ \overline{K}(s)\right]\begin{bmatrix}y(s)\\ \overline{y}(s)\end{bmatrix} \tag{5.140}$$

Feedback loops have the same structure as in Figure 5.32, but in MIMO:

$$y(s)=(I+G_{uK})^{-1}\left[G_u(s)\ \overline{K}G_w(s)\right]\begin{bmatrix}\overline{y}(s)\\ w(s)\end{bmatrix} \tag{5.141}$$

It is clear from these equations that feedback MIMO transfer function $K(s)$ mainly deals with stability through $(I + G_{uK})$ and disturbance rejection through y/w and that feedforward gain K mainly deals with tracking through $y/\overline{y}$.

State space representation is more convenient for practical control law computation in the time domain. In its generic form for either longitudinal, lateral, or coupled motion, the aircraft model is given by

$$\dot{x}=Ax+Bu+Ew \tag{5.142}$$

$$y=Cx+Du+Jw \tag{5.143}$$

Using matrices A, B, C, D, E, J, different techniques are available for computing a generic control law under its own state space realization. For static output feedback (K constant) and assuming the aircraft is proper ($D = 0$), the obtained closed loop is given by

$$\dot{x}=(A+BKC)X+\left[B\overline{K}\ \ E+BKJ\right]\begin{bmatrix}\overline{y}\\ w\end{bmatrix} \tag{5.144}$$

$$\begin{bmatrix}y\\ u\end{bmatrix}=\begin{pmatrix}C\\ KC\end{pmatrix}X+\begin{bmatrix}0 & J\\ \overline{K} & KJ\end{bmatrix}\begin{bmatrix}\overline{y}\\ w\end{bmatrix} \tag{5.145}$$

The order of the dynamics is unchanged. Gain K directly fixes the closed loop modes and HQ. However, the number of degrees of freedom in K (number of inputs number of outputs) may be too small to meet all requirements. Dynamic feedback design offers more versatility by augmenting the closed loop dynamics. It is more challenging, however, since it requires more sophisticated tools.

For detailed information on theory and application of modern MIMO design, readers should refer to one of the many recent books. Some available techniques are based on algebraic manipulation of state space equations. One of the most powerful ones for linear systems is eigenstructure assignment (Liu and Patton 1998), which consists of assigning selected closed loop eigenvalues. It is very convenient for controlling the HQ of civil aircraft since frequencies and damping ratios are closely related to transient time responses and amplitudes of overshoots for standard step pilot demand and since it allows modal decoupling. The Airbus experience is based on this technique (Farineau 1989), as detailed below. For military applications, dynamic inversion also exploits algebraic manipulation of nonlinear control equations (Slotine and Li 1991) for constructing nonlinear control signals leading to linear closed loop behavior. Because they rely on a full description of the systems to be controlled, enhanced versions of these basic algebraic methods must be used for improving robustness. Other linear control design techniques are based on minimization of some control performance index (Maciejowski 1989), with various levels of sophistication: linear quadratic (LQ), Gaussian (LQG), loop transfer recovery (LQG/LTR), H_2 and H_∞ optimal control. Most control design problems can be formalized under the standard form of Figure 5.35. The augmented system $M(G)$ includes the aircraft model G and additional shaping functions to characterize and specify the control problem (tracking, disturbance rejection, autopiloting, etc.). Resolution consists of computing the control laws K which minimize the closed loop transfer between exogenous inputs w and controlled outputs ε in either H_2 or $H\infty$ norm. For instance, if w characterizes initial values of aircraft states, states are measured ($y = x$) and ε represents weighted states $\sqrt{Q}x$ and control effort $\sqrt{R}u$, we have a standard LQ problem. If w are stochastic noises and ε the state estimation error, we have a LQG problem. If w is a standard pilot demand and ε the distance to the outputs of a target model, we have model following control. Individual applications can be found in Magni et al. (1997). A comparison of best design practices in modern control is available in Moorhouse (2000). Modern design techniques have recently been extended to robust control by inclusion of robustness analysis using the μ theory.

The basic idea is to put aircraft model under the linear fractional transformation (LFT, also called M-Δ form) of Figure 5.36. Such modeling is a particular challenge (Ackermann 1993), but it offers capabilities for representing a continuum of models using a nominal model G_0 with an artificial feedback containing different kinds of uncertainties (parameter variations, unmodeled dynamics, etc.). Tools are available for computing robustness margins (μ-analysis), i.e., amplitude of parameter variations which lead to unstability when applied to the nominal model (Ferreres 1999) and for robust control design (μ-synthesis, linear parameter variant LPV synthesis). For nonlinear military aircraft, some nonconventional control design techniques also are available (such as QFT and neural network control) but are beyond the scope of this part.

Automatic FCL Implementation

For the sake of simplicity, for physical interpretation of models and because of tool capability, automatic FCLs generally are designed in the continuous time domain. However, onboard implementation tends to be performed on computers working in the discrete time domain. Transforming FCL from continuous to discrete is a specific automatic control problem for which solutions are given in

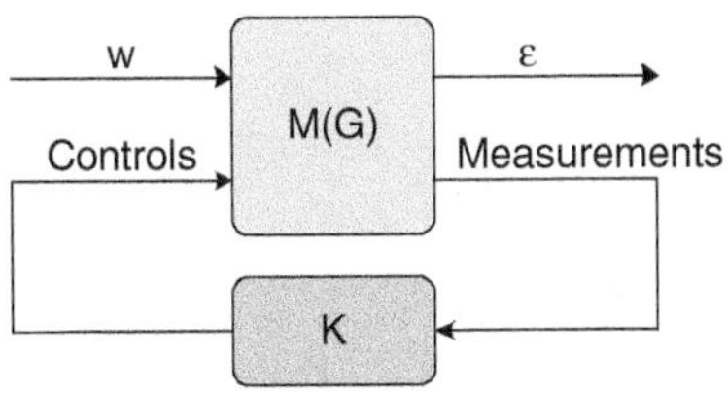

FIGURE 5.35 Standard form for control design.

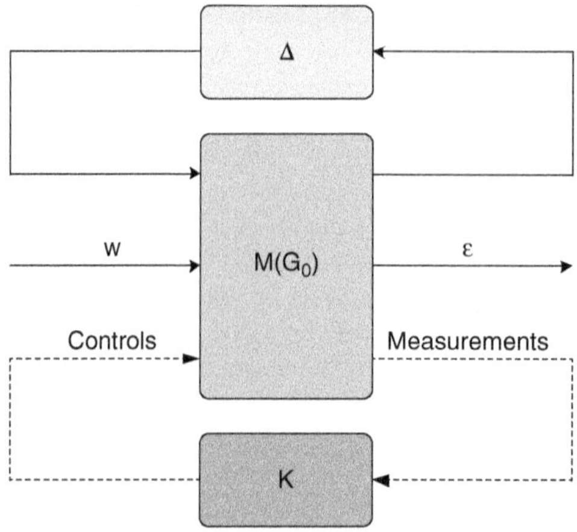

FIGURE 5.36 LFT representation for robustness analysis or robust design.

Fargeon (1989) and McLean (1990). The principle is to have sampling periods small enough so that digital control is perceived as quasicontinuous by the actuators and the aircraft. Several sampling periods may be necessary for digital control of subsystems having different bandwidths. Actuator control must be very fast (few ms sampling period) because of their high bandwidth (several hertz). In civil applications, flight control may be slower, with a piloting bandwidth of about 1 Hz and gain scheduling may be slower still for aircraft that maneuver smoothly. However, requirements are more stringent and sampling periods shorter for military aircraft maneuvering faster. FCS design teams have developed their own skills and technologies in terms of architecture, interface and redundancy. Figure 5.37 shows a typical FCS interface with other systems. Other information

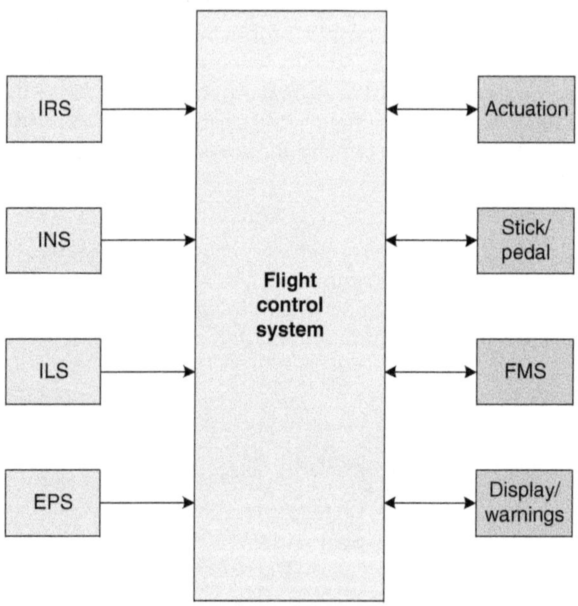

FIGURE 5.37 Typical FCS interface with other systems.

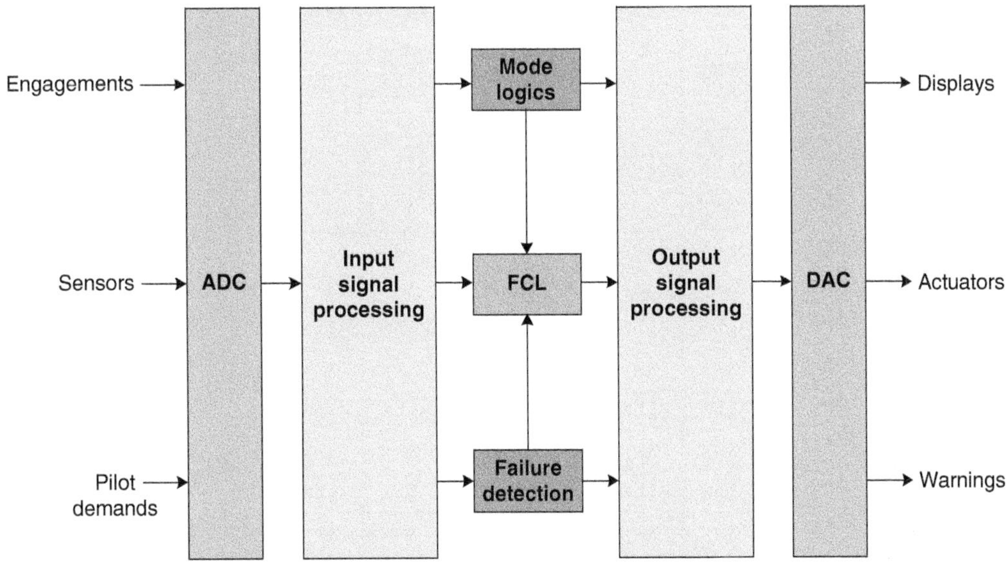

Figure 5.38 General FCC architecture.

on general system architecture is available in Pratt (2000) and Moir and Seabridge (2001). Figure 5.38 illustrates the very generic architecture of a flight control computer (FCC). The Airbus experience is presented below. Details on avionics technology can be found in Part 10 and in Spitzer (1987) and Collinson (1996).

5.31 Airbus Fly-by-Wire: An Example of Modern Flight Control

Since the A320's entry into service in 1988, the Airbus fly-by-wire (FBW) aircraft fleet has become a diverse and rapidly growing family of civil aircraft. At the end of 2001, about 2000 Airbus FBW aircraft had accumulated more than 20 million flying hours and were operated worldwide by a large variety of airlines with various cultures and backgrounds. Consequently, Airbus FBW may be considered as a legitimate reference in the domain of flight control systems. The purpose of the following subsections is to present the main flight control system features shared by this wide family, from the shorter A318 to the larger A380. The high-lift system that controls flap and slat surface extension at low speed is often designed as a secondary flight control system and is outside the scope of this presentation.

Background

Until the 1950s, flight control surfaces were mechanically linked to the cockpit controls and consequently actuated by the pilot's muscles. On such aircraft, flight control surfaces were balanced with an additional mass and horn and their rear part fitted with tabs to aerodynamically amplify or compensate pilot inputs. This technology is still successfully applied on turboprop aircraft such as the ATR 42 and ATR 72. However, aircraft size and speed could not have been increased without installing servo-controls (Figure 5.39) either to amplify the efforts generated by the mechanical linkages or to replace the linkages altogether. Direct mechanical actuation of control surfaces still exists as a complement to servo-controls or as a backup after hydraulic failure on many transport aircraft designed in the 1960s and the 1970s such as the DC9, Boeing 737, and Fokker 100. We will

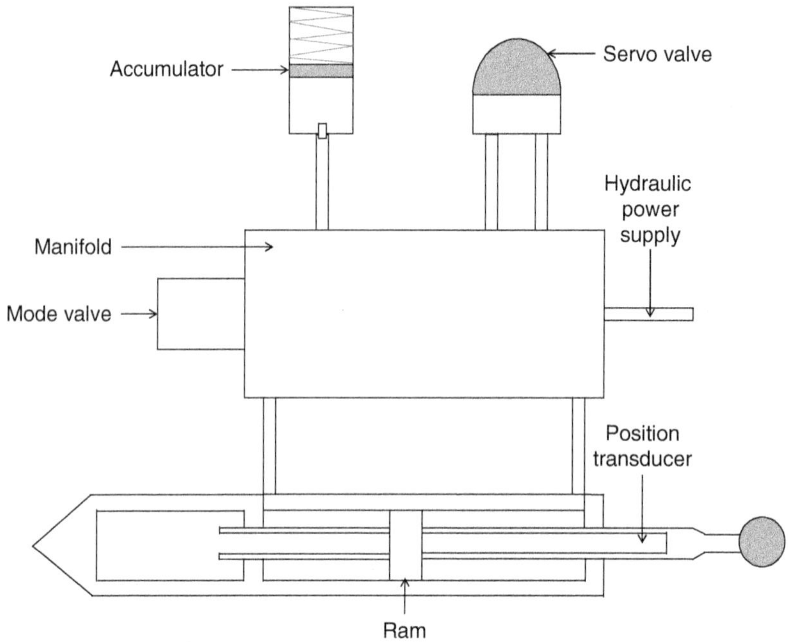

Figure 5.39 Servo control principle.

now discuss the main historical stages of the Airbus flight control systems development (first flight years are indicated in parentheses).

Caravelle (1955). This 100-seater turbofan aircraft represents the first example of a transport aircraft with permanently servo-controlled flight control surfaces. This means that on the Caravelle, mechanical linkages are no longer used to drive the surfaces directly but are instead used to transmit pilot commands to the servo-controls. The Caravelle hydraulically powered flight control system is probably the first example of a full-time critical active control system on a transport aircraft. All-weather automatic landing is another important innovation of this aircraft that opened the way for the use of electricity in critical applications. The Caravelle can be considered an early precursor of the Airbus family.

Concorde (1969). For the first time on a transport aircraft, pilot orders are transmitted to surface servo-controls by electrical wires. FBW is chosen on the Concorde to minimize the risk of HQ degradation in supersonic and transonic flight that could result from the nonlinearity (backlash, hysteresis) inherent in a mechanical transmission. A mechanical flight control system nevertheless remains installed, but only as a passive backup that will become active only after failure of the dual duplex electrical flight control systems. All the flight augmentation and flight envelope protection functions are achieved by analog computers (pitch and yaw damping, artificial feel, automatic trim, angle-of-attack protections).

A300 (1971). This aircraft launches the European Airbus story. All the flight control surfaces are permanently actuated by hydraulic servo-controls, as on the Caravelle and the Concorde, and mechanically signaled, as on the Caravelle. The flight augmentation function and envelope protection are performed by analog computers, as on the Concorde.

In the beginning of the 1980s, digital technology becomes attractive for civil aircraft application, thanks particularly to microprocessor industrialization, and quickly supersedes analog technologies. The first applications on Airbus aircraft are the digital autopilot and flight-augmentation computers; the same basic flight control architecture is maintained.

A310 (1982). FBW with digital computers appears on Airbus A310 and A300-600 with spoiler control. These upper wing surfaces provide roll control in conjunction with the ailerons when anti-symmetrically deflected on one wing and in-flight speed braking or on-ground lift dumping when symmetrically deflected on both wings. The main features of the A310 spoiler control will be applied later on the Airbus FBW family. They are summarized here:

- All the mechanical linkage and servomotors previously needed on the basic A300 spoiler control system are eliminated and all spoiler actuators electrically signaled. System weight is notably reduced and its installation simplified.
- The command/monitoring computer is the basic architecture building block. It allows detecting and neutralizing of command channel failure. Spoiler actuators are controlled by four pairs of command/monitoring units installed in two boxes.
- Each command or monitoring unit includes all the electronic components it needs to perform its function, without sharing resources with another: input and output circuits (analog, discrete, digital), processing component, electrical supply.
- All functions, including spoiler actuator servo-loops, are software based and fully under control of digital units. Actuator servo-valve command current and feedback position acquisitions are multiplexed in order to reduce drastically the electronic volume dedicated to each actuator.
- Roll control is optimized according to airspeed and flap position. Thus, spoiler contribution to roll control is significantly increased in comparison to the previous A300 mechanically signaled system. This makes it possible to eliminate the outboard aileron on A310 and A300-600 and maintain only the inboard aileron.

The EFCU also controls the automatic decoupling of the two-pitch control channels at low speed and, on pilot demand, the automatic return to zero of rudder trim. But the main step is the *A320* (1987), prepared by full-scale experimentation on Concorde (1978), then on the Airbus A300 (1983). This experimentation validated the essential evolutions further applied on the Airbus FBW families:

- Enhanced flight control laws and flight envelope protection
- Side-stick pilot controls

The A320 launched the short/medium-range Airbus single-aisle family, with the stretched *A321* (1993), then the shrunk *A319* (1995) and *A318* (2002) versions. The same flight control philosophy is maintained on the four-engine *A340* (1993) and twin-engine *A330* (1994) long-range family, their latest derivative version *A340-600* (2001) and *A340-500* (2002), the very large *A380* (target: 2004), and the military transport aircraft *A400M* (target: 2005).

Architecture Principle

In a mechanically signaled flight control system, pilot commands are sent to servo-controls by mechanical cables and rods. However, handling qualities would not be acceptable without an electrical system with computers being added to the basic mechanical system in order to perform flight-augmentation functions and flight envelope protection such as yaw damping, pilot control artificial feel, and stall warning (stick buzzer, stick pusher). The computer output is incorporated into the mechanical system by hydraulic and electrical motors. Failure of these functions is generally not hazardous as long as the basic mechanical system remains operative but handling qualities are downgraded.

The FBW term is used when the pilot (or autopilot) orders are no longer sent by mechanical cables and rods to servo-controls, but instead by computers and electrical wiring (Figure 5.40). Consequently, computers become a basic and essential part of the flight control system.

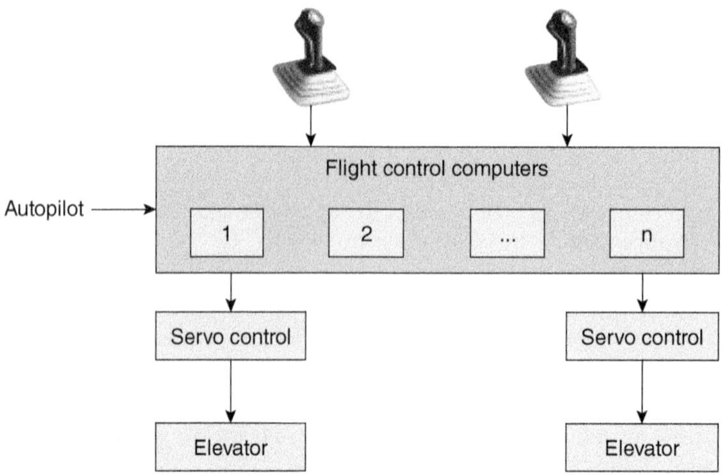

FIGURE 5.40 Airbus elevator control FBW principle.

In comparison to the previous mechanically signaled architecture, FBW enables weight savings and simplification of system installation. Evolution can be limited to this change. In addition, FBW technology offers the possibility of enhancing HQ, improving safety in emergency situations, improving cockpit ergonomy, and facilitating cross-crew qualification. This is the way selected by Airbus. All Airbus FBW aircraft share some main features:

- Side-sticks without motorization
- Enhanced flight control laws and flight envelope protection
- Architecture based on a set of highly dependable and dissimilar FBW digital computers
- Ultimate backup independent of the basic FBW system

A total flight control loss or uncontained surface runaway may lead to severe consequences. The probability of an occurrence of such failures must be extremely improbable. Some basic design principles are applied to fulfill these stringent safety objectives:

- *Redundancy:* Control channel, power supplies, surface actuators, sensors, etc., are at least duplicated so that the consequence for an aircraft of any failure or combination of failures remains acceptable according to its effect on safety. As a general rule, a single failure should have minor consequences.
- *Monitoring:* Computers must permanently monitor not only the computation they perform but also the signals they receive and the actuators they command. In addition, periodic tests may be necessary to detect hidden failures that cannot be detected by computer online monitoring.
- *Segregation:* Components, electrical routes, and hydraulic circuits must be installed to reduce the risk that a common cause, such as projections of fragments resulting from an engine burst, will lead to a total system failure.
- *Robustness:* The flight control system must be demonstrated to be immune from external aggressions such as lightning strike, electromagnetic perturbations, high and low temperature, loss of cooling air, vibrations, supply source perturbations.

- *Dissimilarity:* This additional precaution is selected by Airbus to reduce the risk that a common hardware or software defect may lead to total flight control loss (e.g., a common defect on a batch of electronic components).

Pilot Controls

Large displacement central control columns are necessary to drive mechanical systems with enough accuracy and actuators to introduce flight augmentation, flight envelope protection, and autopilot system. Hence, linked and back-driven central pilot controls are a consequence of the mechanical technology.

On a fly-by-wire aircraft, pilot control no longer drives mechanical linkages, but transducers. Large-displacement central columns are no longer suitable to sense pilot orders; lateral well-in-hand side-sticks may be used (Figure 5.41). In addition, side-sticks improve cockpit ergonomy and comfort with better visibility of instrument panels, easier accessibility, and the possibility to install a pull-out table in front of each pilot.

Controlling the aircraft with the rudder only on lateral axis and the tail trimmable horizontal stabilizer (THS) on the pitch axis was previously demonstrated with A300 flight tests. The A320 FBW takes advantage of this result and employs a mechanically signaled servo-control on the rudder and an actuator on the THS (THSA), controls activated from the rudder pedestal and pitch trim hand-wheel, respectively. This architecture allows side-stick and FBW implementation while maintaining a mechanically signaled backup.

Jamming of the cockpit controls is a rare but very severe failure case. It may be caused by any jamming of the pilot control itself or the mechanical linkage it drives, or by pilot incapacitation. If the two sticks are linked, the only way to avoid losing aircraft control if such jamming occurs is to install a decoupling device between the two sticks. A large force must be applied on the unfailed stick to declutch it from the affected one, which leads to a significant transient on the aircraft trajectory. Thus, a simple failure is capable of causing a severe situation on a mechanically signaled flight control aircraft or on a FBW aircraft with linked pilot sticks.

On Airbus FBW, side-sticks are no longer physically coupled, roll and pitch orders from each side-stick being algebraically added together by the flight control computers. Therefore, the jamming of one stick will have no effect on the other. If such a failure occurs, the opposite pilot can immediately take control of the aircraft with his unfailed stick and cancel orders of the failed stick.

FIGURE 5.41 Airbus cockpit. (*Wikimedia under Creative Commons.*)

Furthermore, he can quickly and easily neutralize orders from the failed stick by depressing a switch located on its handle.

Dual input awareness is adapted to this architecture without altering its basic advantage. Several ways are implemented to alert the crew: visual warning by flashing lights in front of each pilot (basic solution on Airbus FBW), audio warning, and now tactile warning by vibrating the handles (but not the sticks).

Autopilot orders are sent from autopilot computers to flight control computers by ARINC 429 digital buses. Servomotors are no longer useful, as the mechanical linkage has been eliminated. The active failures they were capable of introducing on flight controls are consequently eliminated. A solenoid will freeze each side-stick at neutral as soon as the autopilot mode is engaged by the crew and gives nonambiguous tactile information. It may be overridden by depressing the instinctive disconnect switch on one of the sticks or by applying a firm, but not excessive, force on each side-stick. Either of these actions will immediately disconnect the autopilot and restore manual control.

Control Laws

The flight control law objective is to achieve an accurate, stable, and safe aircraft response to pilot input at any point in the flight domain, whatever the aircraft configuration and flight phase. It is obtained on Airbus aircraft by closed loop control laws that incorporate flight envelope protection (Figure 5.42):

- Inside the normal flight domain, aircraft static stability is neutral, cross-effects between axes are automatically compensated (turn coordination, longitudinal compensation), elevators are automatically trimmed at neutral by the THSA. Aircraft pitch and roll attitudes are held as soon as the stick is at neutral and small stick corrections are sufficient to compensate for a configuration change, thrust variation, speed variation, wind gust effect, etc. The natural behavior of the lateral control law automatically compensates for a lateral dissymmetry. Hence, in case of engine flameout, the aircraft will reach a limited bank angle and sideslip while staying in the normal flight domain.

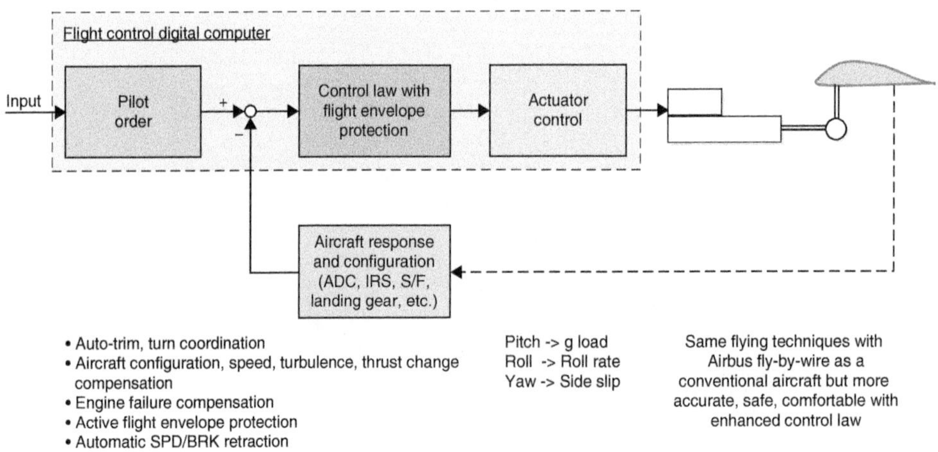

FIGURE 5.42 Airbus flight control principle.

- Outside the normal flight domain, positive stability is restored: the pilot must permanently apply a stick deflection to maintain the aircraft inside the peripheral domain. With the stick at neutral, the aircraft will come back to and stay at the limit of the normal flight domain.
- By applying full stick, the pilot will command the aircraft to fly at the limit of the peripheral domain.

The aircraft being thus protected against stalling and overstressing, the pilot can quickly and safely obtain maximum aircraft potential by a full stick deflection. Consequently, aircraft maneuverability is significantly improved.

A control law must not rely on a single resource if its failure may have severe consequences. Therefore, at least two independent sources are usually needed for each type of parameter to perform the normal control law described above (Figure 5.43). For example, three independent sources provide inertial or air data parameters on Airbus. If one fails, the normal law remains engaged. If a second one fails and is undoubtedly isolated, the control law will automatically switch to alternate mode, in which enhanced flight envelope protection is replaced by a conventional warning. If the three sources have failed, the control law will revert to simultaneously provided to the crew. Hence, the best possible handling qualities and safety level are provided, depending on the available resources.

In addition, FBW may also alleviate structural loads and consequently reduce the aircraft structural weight. This is obtained both by the basic control law and by specific load-alleviation functions:

- The basic control laws are designed with the target of guaranteeing that the pilot is able to command the maximum full-stick order without overstressing the structure. It is achieved on the pitch axis by limiting the load factor demand to 2.5 g and by avoiding exceeding the speed of the above design case (VMO/MMO protection).
- Complementary control laws may also be implemented in order to alleviate loads in specific cases. For example, on the A330/A340, the aileron and outboard spoilers are symmetrically deflected upward when the normal acceleration command exceeds 1.3 g.

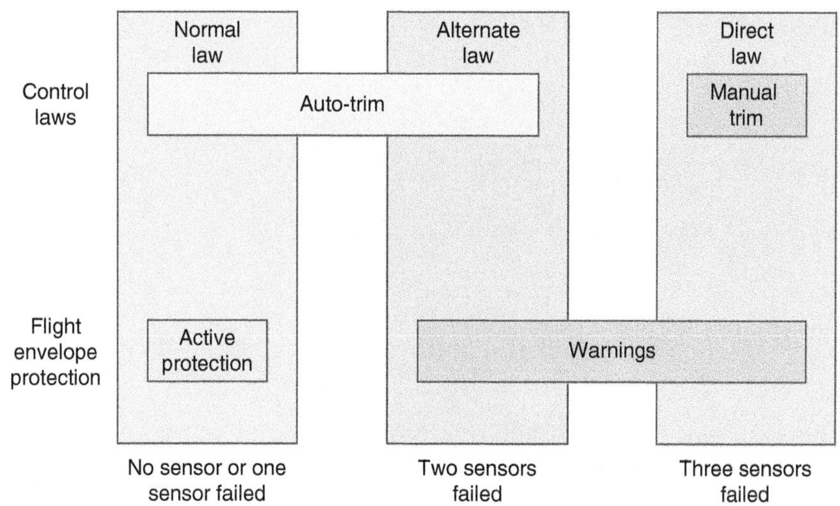

FIGURE 5.43 Control law reconfiguration (example of IRS failure).

Flight Control Computers

All the flight control functions are software based (control laws, actuator control and monitoring, failure detection, reconfiguration logic, etc.). This gives the designer greater flexibility to adapt them to aircraft characteristics and fulfill new or specific requirements.

Flight control functions are implemented in a set of redundant, dissimilar, and fail-safe digital computers. Each FCC computer is capable of detecting and isolating any failure affecting the function it performs and the actuators it controls thanks to its dual type architecture. Each computer is composed of two fully separate command and monitoring units (Figure 5.44) without hard-wired synchronization. Each one owns all the hardware resources it needs to perform all its software-based functions (analog, discrete, and digital input and output circuits, processors, electrical supplies) and to perform its self-monitoring (check-sum, watchdog, on-ground automatic self-test on electrical power-up).

The FCC automatically reconfigures the FBW system in the case of simple or multiple failures that affect themselves, the signals they receive, or the actuators they control. The reconfiguration logic is distributed between all flight control computers and follows a predefined sequence. The principle of this logic is as follows:

- In normal operation, one predefined computer is in charge of computing the control order. Digital buses to the other flight control computers dispatch this control order, and all servo-controls are slaved to the master control orders.
- If the master computer detects an internal failure, it will fall into a fail passive state by switching off all its output. Immediately, the second computer, following the predefined logic, will take over in order to compute the control order in place of the previous one.

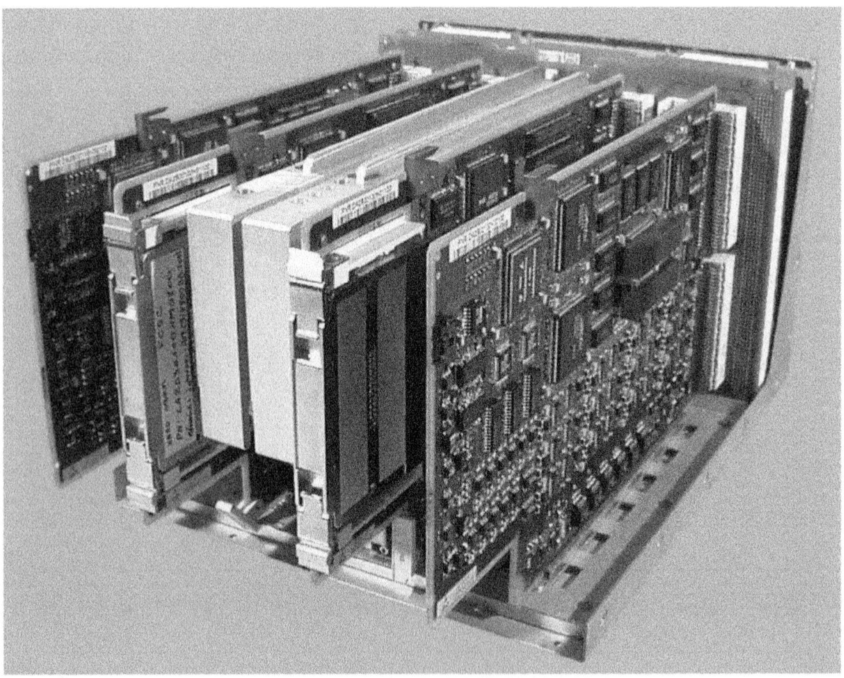

FIGURE 5.44 Airbus flight control computer (A340-600 FCC).

One computer alone is sufficient to control the aircraft, but with reduced maneuverability. For actuator control, the same principle is applied, but servo-control failure will not lead to the total loss of the computer that controls it. The actuator reconfiguration logic is also fully automatic and adapted to each type of surface. Two dissimilar types of flight control computers are used in order to prevent a single common hardware or software defect from leading to total FBW loss.

All flight control computer function, control laws, input and output monitoring, actuator control and monitoring, and reconfiguration logic are specified using a graphic language based on a well-defined library of symbols. Functional specifications are consequently accurate, complete, and unambiguous and may be automatically coded, first in order to validate them on simulation tools, then to implement them with a qualified coding tool in the flight control computers.

Flight Control Surface Actuation

On the A320 and A330/A340 family, hydraulically powered actuators control all surfaces. With the A380, a new type of electrically powered actuator will appear.

Each surface is controlled by one, two, or three actuators depending on the consequence of total surface actuation loss, taking into account that it may be combined with other failures.

Two servo-controls are installed on each aileron and elevator. In normal operation, one servo-control is active and the other one on standby. This last one will automatically become active if the first one fails. Mode change is obtained by positioning a valve controlled with a solenoid. In standby mode, both ram chambers are put into communication by the mode valve via a restrictor, and the servo-control becomes a damper. The damping coefficient is defined to avoid a detrimental surface flutter if only one servo-control remains connected to the surface in damping mode.

Each surface actuator receives its order from the flight control computer by a dedicated electrical channel. Thus, a failure of one servo-control or its control channel will not be propagated on other servo-controls of the same axis, as could be the case with mechanically signaled flight controls.

The rudder surface is powered by three synchronized servo-controls. These are mechanically signaled on the A320 and A330/A340 families and electrically signaled on A340 500/600 and later Airbus aircraft.

Each spoiler surface is controlled by a single electrically signaled servo-control. In case of control channel electrical failure, the hydraulic jack is commanded to the retracted position and the spoiler surface fully applied on the wing.

The THS is controlled by a fail-safe screw jack powered by two hydraulic servomotors. These are electrically signaled in normal operation. If all electrical channels have failed, the THSA is mechanically signaled from hand-wheels located on both sides of the center pedestal in the cockpit, one for the captain, the other for the first officer. After touchdown, the THSA is automatically positioned at a neutral position acceptable for a further takeoff. Before takeoff, the THS will be mechanically positioned at its optimum position, depending on the aircraft's center of gravity location.

System Installation and Protection

Physical segregation between the various channels of a redundant system is the basic precaution to reduce its vulnerability to a common external aggression. It is obtained on an FBW system by sending signals in separate and distant electrical routes, by installing computers in different areas in such a manner that the aircraft remains controllable in case of an external failure such as an engine burst or a local fire.

Lightning strikes and electromagnetic aggression are a concern on any aircraft. For aircraft certification, it shall be demonstrated by test that FBW may suffer the very high level required for certification with margin, not only without any damage but also without any disconnection or upset.

Maintenance Assistance

The failure isolation logic is performed by two dedicated flight control data concentrator (FCDC) units that permanently analyze all the monitoring maintenance messages that they receive from all the FCC and deduce from this analysis which line replaceable unit (LRU) has most probably failed. Maintenance assistance is provided by automatically reporting any failure to a centralized maintenance system. This system gives information on the failure environment (flight phase, time, etc.).

The FCC permanently monitors the whole FBW system, including standby functions. Consequently, the number of potential dormant failures is drastically reduced and scheduled maintenance task intervals significantly increased in comparison with mechanically signaled systems. For example, the ground spoiler function needs a weekly check on mechanically signaled A300 B2/B4 and is completely eliminated on the A320 and A330/A340. All safety checks below check A (600H on A320) have been eliminated on Airbus FBW aircraft.

In addition, all flight control equipment is specified and designed without TBO (time before overhaul).

Airbus Fly-by-Wire Advances Since the A320's Entry into Service

Airbus FBW evolves to cope with new aircraft characteristics. New and mature technologies are progressively implemented on Airbus aircraft in order to reduce weight, production cost, and maintenance cost and improve passenger comfort and safety. All this is managed with the goal of maintaining the basic Airbus principle.

A330/A340

Some significant architecture changes have been introduced.

- Yaw damper availability has been increased in order to damp Dutch roll in any situation, including ultimate mechanical control. The vertical fin is accordingly sized to reduce its area and weight and reduce aerodynamic drag. For that reason, an alternate yaw damper has been added on the rudder axis that will automatically operate in case of a total loss of the nominal yaw damping function. The backup yaw damper unit (BYDU) is electrically signaled by electronic modules fully independent from the FBW computer package, including electrical supplies.
- The two A320 flight augmentation computers have been eliminated and their flight control functions incorporated into FBW computers (control and monitoring of all yaw axis servomotors: yaw damper, rudder trim, and rudder travel limitation actuators).
- A specific control law has been added to improve comfort in turbulence by damping low-frequency vertical and lateral fuselage structural modes. Structural modes and handling modes are sufficiently distant that they can be decoupled by conventional filtering techniques.

A340-500/600

These versions are derived from the current A340, but with a longer fuselage and larger wing area in order to extend aircraft capacity and range. The first structural mode frequencies have consequently been decreased, both on longitudinal and lateral axes, and are close to HQ modes. The following significant advances of the FBW system have been developed in order to maintain good HQ and passenger comfort.

- Control laws are designed to control HQ modes and low-frequency structural mode simultaneously.
- On the rudder axis, FBW orders are sent directly to electrically signaled servo-controls. The mechanical rudder control is fully eliminated and an ultimate electrical backup is added. This backup is highly segregated from the FBW computers and aircraft electrical network.

- The inner autopilot loop is located in the FBW computers to reduce transmission delay from the AP order to the servo-control.

Perspectives for the A380

The very large A380 is presently under development. Several challenging innovations have been introduced on the A380 flight control system and presented below. They should also be applied to the A400-M transport aircraft.

Electrically Powered Actuators

Safety objectives on a modern transport aircraft require three independent power sources for primary flight control surface actuation. Up to now, hydraulics has been the only practicable solution to supply servo-controls, taking into account all the aspects: weight, performance, and cost. In other respects, the total loss of electrical sources is safety-critical on FBW aircraft obviously, but also on other transport aircraft due to the growing place of avionics. On Airbus FBW aircraft, total electrical failure is prevented by the multiplicity and diversity of supplies (engine-driven generators in normal operation, hydraulically driven generator and batteries in emergency situations) and by stringent segregation rules applied in the electrical network. Consequently, hydraulics may be at least partially eliminated, provided that electrical actuators are used in addition to hydraulic servo-controls for surface actuation. Ten years ago, a new and attractive technology of improved weight-to-power ratio, brushless electrical motors, appeared on the market. Airbus launched a trade study to evaluate their impact on FBW architecture and to select the most appropriate type of flight control actuators. Benefits of power by wire are

- Aircraft weight, reliability, production cost, and maintenance cost are reduced by eliminating the hydraulic circuit and all of its components (pumps, reservoirs, filters, pipes, selectors, etc.). This gain has been confirmed whatever the aircraft size, from a 100-seater to the largest A380.
- Aircraft safety level is improved at different levels. First, slight damage on a hydraulic circuit may lead to leakage and to its complete loss, even if hydraulic fuses are installed. With an electrical circuit, only downstream equipment is lost in the case of a circuit rupture. An electrical circuit may also be installed in an area not suitable for an hydraulic circuit and more segregated from the remaining hydraulic circuits than the hydraulic circuit it replaces. The aircraft remains controllable if a common failure mode affects all the hydraulic circuits.

New types of electrical actuators will be introduced for Airbus FBW:

- Three types of actuators have been envisaged: IAP (integrated actuator package), EMA (electromechanical actuator), EHA (electrohydrostatic actuator). For the first application, trade studies lead to selecting EHA for the following reason: EHA and conventional servo-control jacks are identical or very similar. Therefore, the experience gained with conventional servo-controls on important safety items is directly applicable to EHA: damping function integrity, lack of jamming, fatigue justification. This is not the case with EMA. Furthermore, EMA airborne experience is limited to low-speed and/or low-power actuators.
- Several EHA demonstrators have been developed by several suppliers and successfully tested in the laboratory and in flight on A320 and A340. This new technology is mature enough for new programs.

Permanent Active Stabilization

The first step has been implemented on the A330/A340 rudder control. The next will be the permanent active stabilization on the A380 elevator control. Tail horizontal stabilizer is accordingly downsized.

Full Electrical Ultimate Backup

THSA control is the last remaining mechanical control on the A340 500/600. It has been eliminated on the A380 and replaced by an electrical control. Consequently, ultimate backup becomes fully electrically signaled and will become automatically active if all flight control computers have failed, as on previous Airbus aircraft. It will be similar to the A340-500/600 rudder ultimate backup and employed on the elevator, rudder, and ailerons.

Hydraulic Pressure Increase

On all present transport aircraft, hydraulic circuits are pressurized at 3,000 psi (210 bars). This pressure will be increased to 5,000 psi (350 bars) on A380. An important weight savings will be obtained.

Autopilot Integration

The autopilot function will be integrated into the primary flight control computers. The autopilot function is being installed in 3 PRIM computers in place of two, as in the previous A320 and A330/A340 programs, and its availability will consequently be increased. In addition, the total electronics volume will be reduced, autopilot and flight control sharing the same basic resources.

AFDX Bus

The new AFDX multiplexed bus has been introduced in the A380 avionics. It will be used by FCC to communicate with other systems. Flight control critical information will continue to be transmitted with ARINC 429 networks (inertial and anemometric parameters, information exchanged between fly-by-wire computers, etc.).

Conclusion on Airbus Fly-by-Wire

Airbus has set a new standard by introducing FBW on the A320, the target being not only to reduce weight and cost, but also to improve handling qualities and safety, improve aircraft survivability in extreme situations, reduce pilot workload, and facilitate cross-crew qualification and to reduce maintenance tasks.

Nevertheless, new and mature technologies continue to be progressively implemented in new versions to improve performance and safety while minimizing costs and maintaining the basic Airbus systems philosophy.

5.32 Some Control Challenges

For illustration, we introduce three kinds of challenges in this subsection:

1. Design of FCL for aircraft having complex dynamics
2. Integration of advanced active control technologies (ACT)
3. Flight control of nonconventional configurations

FCS design engineering will also have to face many more challenges not detailed here.

Flexible Aircraft Control

Flexible aircraft control is a challenge for both military and civil aircraft. Trends toward design lighter military aircraft with new materials and larger transport aircraft lead to a more flexible structure. Difficulties in FCS design come from:

- The significant reduction of the distance between the conventional rigid aircraft flight control bandwidth and the flexible modes (sometimes overlapping)

- The high order of the total aircraft dynamics (typically 50–100 states), which now include structural modes coming from finite element models (FEM)
- The need to improve the accuracy of aerodynamics prediction
- The necessity to use sensors to extract information on the structural dynamics

Two strategies are possible for FCL design (RTO 2000). The first strategy is to try to recover (on the flexible aircraft) the rigid control performance that could be designed on the rigid aircraft by deleting the flexible contribution in the measurements used in the feedback loop. The most frequently employed technique is to introduce notch filters having zeros at flexible mode frequencies. This method, however, leads to softening of the control and degradation of the rigid performance. An alternative is to use geometrical filtering, i.e., a combination of measurements for minimizing flexible residuals in the feedback. Robustness may be poor against sensor failure, however, and it is not efficient if rigid and flexible bandwidths overlap. A more ambitious strategy is to perform simultaneous control of the rigid aircraft and of its structural dynamics. This is a fully multivariable, multiobjective control problem (Alazard et al. 1999) that can be solved after being formalized under the standard form of Figure 5.33 if models of the flexible aircraft are available with reasonable accuracy, if actuators have a high-frequency bandwidth compatible with structural modes to be controlled, and if sensors showing the contribution of the structural dynamics are available at different locations on the aircraft fuselage and wing.

Active Control Technologies

Active control technologies (ACT) generally mean nonconventional control for enhancement of military aircraft performance and maneuverability (RTO 2000). It mainly includes the popular active flow control (Gad-el-Hak 2000). However, other technologies could be considered, such as noise control, compressor instability control (Gravdahl and Egeland 1999), and combustion control, since they will all involve the same kinds of methodological issues. Most are mainly considered as isolated monodisciplinary technical challenges. However, ACT design is clearly multidisciplinary and could lead aeronautical engineering much beyond performance enhancement on given aircraft configurations. It may reopen the whole process of aircraft design. For instance, active aeroelastic wing technology opens a new design approach for aircraft wing structure (Flick et al. 1999). Integration of advanced fluidic actuation technologies may revolutionize FCS architectures (Moir and Seabridge 2001; Pratt 2001). Consequently, two main problems must be solved in terms of control design for ACT: developing control strategies for the technology itself and integrating it in aircraft FCS.

Design can be considered conventional if the two problems can be solved separately: technology is controlled with a local feedback loop and optimized separately before being integrated in the aircraft as a specific locally controlled, stability-augmented, stand-alone subsystem. Design becomes nonconventional either when the technology has a nonconventional dynamic behavior or when strong couplings appear between the dynamics involved in the technology and the aircraft's own dynamics, particularly when some attempts are undertaken to make actuators play a more ambitious role than just controlling the rigid-body flight. The two problems may need to be solved at once. Depending on complexity, it may still be possible in practice to distinguish two steps. The first step remains local control of the nonconventional technology, considering aircraft dynamics as an external input. This is a very difficult control design problem for two reasons. The first is that in most ACT, subsystems to be locally controlled have a distributed nature (vortices or shocks or separations in active flow control, wing distortion in structural control, pressure distribution in compressor instability control, etc.). Their dynamics are governed by partial differential equations (PDE), for which standard control design tools do not apply. *Local* and *distributed* refer to the aircraft scale and to the dynamics of the subsystem to be controlled, respectively. They should not be confused.

The second reason is that such local control is very demanding since it should lead to a closed loop controlled subsystem having predictable and uniform dynamic behavior for any input from aircraft dynamics. As an extension of conventional design, the nonconventional, locally controlled subsystem could then be used for aircraft flight control, with the additional difficulty of dealing with couplings in a fully multivariable control design problem.

Flight Control of Nonconventional Aircraft Configurations

Some assumptions made for modeling conventional aircraft and allowing simplified approaches for control surface sizing, HQ assessment, and FCL design are not valid for nonconventional configurations such as delta wing, tailless, three-surface, flying wing, STOL, etc., for different reasons, such as

- Couplings between the longitudinal and lateral dynamics may be significant because of different mass and inertia distributions and different aerodynamics (more compact aircraft, for instance).
- In the absence of a horizontal tail, the wing actuators have to perform adverse horizontal trimming and high-lift generation simultaneously, depending on CG location. Trimming also affects wing performance.
- Controlling lateral motion without a tail requires alternative technologies (thrust vectoring, for instance, or nonconventional actuators such as crocodile flaps).

Control surfaces sizing, HQ assessment, and FCL design for such nonconventional aircraft require the invention of new FCS design processes, including specific modeling, nonconventional requirements, and new types of compromises.

5.33 Conclusion

For the first aircraft generations, technological and methodological issues of FCS design were considered independently (McRuer and Graham 1981). Through development of modern military and civil aircraft programs, technology has been pushed by needs in performance and reliability. Airbus fly-by-wire technology is a key illustration. Sophisticated methods have been invented simultaneously, which now are part of global industrial processes. By supporting the understanding of the requirements, the specification, development, and integration of efficient and validated solutions, these tools allow the design of highly integrated FCS with high maturity, integrity, and reliability. The trend is to integrate FCS design into a global and optimized systems design process (integrated modular avionics technology, more electrical aircraft, and power optimized aircraft concepts), then to reach final integration into the overall aircraft design.

References

Ackermann, J. 1993. *Robust Control: Systems with Uncertain Physical Parameters*, Springer Verlag, Berlin.

Alazard, D., Buckarles, A., Ferreres, G. Magni, J. F., and Prudhomme, S. 1999. "Towards a Global Methodology for Flexible Aircraft Control," in *Proceedings of RTO Structural Aspects of Flexible Aircraft Control*, Ottawa, pp. 18–21.

Allerton, D. (2009), *Principles of Flight Simulation*, John Wiley and Sons Ltd, West Sussex, UK.

Babister, A. W. 1980. *Aircraft Dynamic Stability and Response*, Pergamon Press, Oxford.

Blakelock, J. H. 1991. *Automatic Control of Aircraft and Missiles*, John Wiley & Sons, New York.

Boiffier, J. L. 1998. *Dynamics of Flight: The Equations*, John Wiley & Sons, New York.

Bourcier, M. A. 2001. "Logistics Test and Evaluation in Flight Test," NATO-RTO-SCI Agardograph 10.

Collinson, R. P. G. 1996. *Introduction to Avionics*, Chapman & Hall, London.

Cook, M. V. 1997. *Flight Dynamics Principles*, Arnold, London, and John Wiley & Sons, New York.

Etkin, B. and Reid, L. D. 1996. *Dynamics of Flight*, John Wiley & Sons, New York. Fargeon, C., ed. 1989. *The Digital Control of Systems*, North Oxford Academic, Oxford.

Farineau, J. 1989. "Lateral Electric Flight Control Laws of a Civil Aircraft Based upon Eigenstructure Assignment Techniques," in *Proceedings of AIAA GNC Conference*, Boston, August.

Federal Aviation Administration (FAA). Federal Aviation Regulation 25 (FAR 25), Airworthiness Standards: Transport Category Airplanes.

Ferreres, G. 1999. *A Practical Approach to Robustness Analysis with Aeronautical Applications*, Kluwer Academic, Dordrecht.

Flick, P. M., Love, M. H., and Zink, P. S. 1999. "The Impact of Active Aerolastic Wing Technology on Conceptual Aircraft Design," in *Proceedings of RTO Structural Aspects of Flexible Aircraft Control*, Ottawa.

Gad-el-Hak, M. 2000. *Flow Control: Passive, Active and Reactive Flow Management*, Cambridge University Press, Cambridge.

Gravdahl, J. T. and Egeland, O. 1999. *Compressor Surge and Rotating Stall Modelling and Control, Advances in Industrial Control*, Springer, London.

Hodgkinson, J. 1999. *Aircraft Handling Qualities, AIAA Educational Series*, American Institute of Aeronautics and Astronautics, Reston, VA.

Joint Aviation Authorities. JAR-25, Joint Aviation Requirements for Large Aeroplanes.

Liu, G. P. and Patton, R. J. 1998. *Eigenstructure Assignment for Control System Design*, John Wiley & Sons, New York.

Ljung, L. 1987. *System Identification: Theory for the User*, Prentice Hall PTR, Upple Saddle River, NJ.

Maciejowski, J. M. 1989. *Multivariable Feedback Design, Electronic Systems Engineering Series*, Addison-Wesley, Reading, MA.

Magni, J.-F., Bennani, S., and Terlouw, J. 1997. *Robust Flight Control: A Design Challenge, Lecture Notes in Control and Information Sciences 224*, Springer, London.

McCormick, B. W. 1995. *Aerodynamics, Aeronautics, and Flight Mechanics*, John Wiley & Sons, New York.

McLean, D. 1990. *Automatic Flight Control Systems*, Prentice Hall, Englewood Cliffs, NJ.

McRuer, D. and Graham, D. 1981. "Eighty Years of Flight Control: Triumphs and Pitfalls of the System Approach," *Journal of Guidance and Control*, Vol. 4, pp. 353–362.

MIL-1797-A, Flying Qualities of Piloted Vehicles.

Minister of Defence (MoD). 1983. "Design and Airworthiness Requirements for Military Aircraft," UK MoD Definition Standards 970.

Moir, I. and Seabridge, A. 2001. *Aircraft Systems: Mechanical, Electrical, and Avionics Subsystems Integration*, American Institute of Aeronautics and Astronautics, Reston, VA.

Moorhouse, D. J., ed. 2000. "Flight Control Design: Best Practices," NATO-RTO-SCI Technical Report 29.

Nelson, R. C. 1998. *Flight Stability and Automatic Control*, McGraw-Hill, New York.

Normand-Cyrot, D., ed. 1998. Perspectives in Control, Springer, London.

Pratt, R. W., ed. 2000. *Flight Control Systems, Progress in Astronautics and Aeronautics 184*, American Institute of Aeronautics and Astronautics, Reston, Va, and Institution of Electrical Engineers, Stevenage, Herts, UK.

RTO-MP36. 2000. Proceedings of Structural Aspects of Flexible Aircraft Control.

Prudhomme, S. and Gimonet, B. 2001. "Flight Control Performance Enhancement Avoiding Aircraft Identification," paper delivered at AIAA-GNC Conference.

Slottine, J. J. E. and Li, J. W. 1991. *Applied Nonlinear Control*, Prentice Hall, Englewood Cliffs, NJ.

Söderstrom, T. and Stoica, P. 1989. *System Identification*, Prentice Hall, New York.

Spitzer, C. R. 1987. *Digital Avionics Systems: Principles and Practice*, McGraw-Hill, New York.

Stevens, B. L. and Lewis, F. L. 1992. *Aircraft Control and Simulation*, John Wiley & Sons, New York.

Tischler, M. B., ed. 1996. *Advances in Aircraft Flight Control*, Taylor & Francis, London.

PART 8

Modern Avionics Architectures

Frederic Boniol, Jack Foisseau, and Roberto Sabatini

5.34 Introduction to Avionics

In this part, computer-based technologies used for supporting more and more aircraft functions are presented. The term "avionics" (*avi*ation electro*nics*) refers to aircraft equipment serving aircraft functions such as communications, navigation, and automatic flight control. In fact between aircraft functions and embedded equipment there is an intermediate concept, called *system*, including many pieces of equipment, some of them being computers with a significant amount of software. The main top-level aircraft functions are as follows:

1. Provide and distribute communications (with air traffic management (ATM), airlines, crew, etc.).
2. Plan, generate, and control aircraft movement.
3. Provide crew, passenger, and cargo environment and services.
4. Detect and analyze aircraft conditions for flight.
5. Provide airframe movement and attachment capability.
6. Provide containment and internal support.

Each top-level function may be decomposed into subfunctions. For instance, function 2 above includes navigational guidance, aircraft performance monitoring, flight planning, thrust control, groundbreaking, aerodynamic configuration control, auto-managed and controlled flight, fuel management, etc.

Many aircraft functions performed in the past by mechanical, hydraulic, or electrical systems are now implemented as computer-based systems. A *system* is an aggregate of equipment. For instance, the flight control system of an aircraft includes equipment (like the rudder pedal, the ailerons, elevators, spoilers, hydraulic actuators, electrical and hydraulic connections, sensors, and computers inputting/outputting many signals) enabling the pilot (human or autopilot) to control the aircraft. For other systems in interface with pilot and/or crew, additional equipment types are displays, input devices, etc. Sensors and actuators are connected to computers communicating by means of data buses.

ATA Specification 100

Aircraft systems obey a standard classification for identifying aircraft equipment: ATA (Air Transport Association of America) classification 100, used by all airframers and aircraft equipment manufacturers for the preparation and distribution of technical data about systems. It divides an aircraft into major systems (Table 5.11).

ATA 100 is a numbering system providing a unique identifier for each aircraft hardware component (the first two digits identify a major aircraft system). Table 5.12 shows the subdivisions specified by ATA 100 for two major systems. With the use of digital information, a companion specification (ATA 2100) was designed to standardize the generation and distribution of ATA 100 data in electronic form.

ATA chapter	System
21	Air conditioning
22	Auto flight
23	Communications
24	Electrical power
25	Equipment/furnishings
27	Flight controls
28	Fuel
29	Hydraulic power
30	Ice and rain protection
31	Indicating/recording systems
32	Landing gear
33	Lights
34	Navigation
35	Oxygen
36	Pneumatic
38	Water/waste
42	Integrated modular avionics (IMA) and avionics data communication
49	Airborne auxiliary power
54	Nacelles/pylons
71	Power plant
76	Engine controls

Table 5.11 ATA 100 Computer-Based Systems

ATA chapter	ATA section	System/Subsystem
22		**Auto flight**
	22–10	Autopilot
	22–20	Speed attitude correction
	22–30	Auto throttle
	22–40	System monitor
24		**Electrical power**
	24–10	Generator drive
	24–20	AC generation
	24–30	DC generation
	24–40	External power
	24–50	Electrical load distribution

Table 5.12 Two Examples of ATA 100 Classification

5.35 Requirements for Avionics

In the case of avionics, three requirement types are important: real-time constraints, safety-related regulatory requirements, and economic requirements.

Real-Time Constraints

Real-time constraints are of great importance when dealing with the control loop of dynamic systems. The total time delay is important for stability and other performance aspects of closed-loop control of the flight control system, the automatic pilot system, the flight management system, etc. The time lag between data input and the corresponding output (orders to actuators, for instance) is due to (a) computing, (b) communication latency, and (c) storage time. Typically, the input is preprocessed before being used in the main computational cycle, and the output will be postprocessed after being computed. The pre- and postprocessing may involve unit conversion, reasonableness checks, comparison with inputs from other computers (voting, etc.), and integrity checks on the communication links. The time required for these activities may vary but could increase the effective lag between input and output orders above the value corresponding to the specified iteration rate. This could impact the performance of the closed-loop control system, and consequently the stability of the aircraft. Guaranteeing and verifying real-time constraints are major tasks when designing and validating avionics systems.

Regulatory Requirements

Regulatory requirements are passed to aircraft manufacturers through FARs (Federal Aviation Regulations) established by the FAA (Federal Aviation Authority). European countries have the JAA (Joint Aviation Authority), editing JARs (Joint Aviation Regulations). Compliance with these requirements must be shown before the commercial aircraft is allowed to be operated. Such requirements pertain mainly to the safety of the aircraft and its occupants. The goal of system safety is to optimize safety by identifying safety-related risks, eliminating or controlling them by design and/or procedures.

Economic Requirements

Economic requirements are often associated with direct operating costs (navigation fees, crew, maintenance, fuel and oil, etc.) and also with indirect costs driven by various factors such as use of automated system engineering tools, dispatch reliability, use of standardized computer resources, etc. Such economic requirements come from airlines and address the costs of purchasing, operating, maintaining, and servicing the aircraft.

5.36 Physical Architectures

There are three types of fundamental avionics architecture: centralized, federated, and distributed.

Centralized Architecture

A centralized architecture is characterized by signal conditioning and computations taking place in one or more computers in a line replaceable unit (LRU) located in the avionics bay with sensor and command signals transmitted over data buses. The main advantages are the computers' accessibility within the bay; a safe computer environment, simplifying qualification; and easily written and validated software. However, there are two main drawbacks: many long buses to collect and transmit data/commands, and system vulnerability to hazardous events occurring in or near the avionics bay. Centralized architectures are commonly found in rockets, missiles, and satellites.

Federated Architecture

A federated architecture is characterized by each major system sharing input and sensor data from a common set of hardware and consequently sharing their computed results over data buses. *Partitioning* (see ATA classification) is then an intrinsic feature of this architecture. One of the main advantages is that such an architecture permits independent design, configuration, and optimization of the different systems. Consequently, changes in system hardware/software are relatively easy to make. Federated architectures are commonly found in modern military aircraft (F-16, EFA, Rafale, etc.) and commercial transport aircraft (Airbus 320/330/340 Boeing B777, etc.).

Distributed Architecture

A distributed architecture is composed of multiple processors throughout the aircraft that are assigned computing and control tasks in real time. Limited signal processing may be performed in or near the sensors and the actuators. Such an architecture presents several advantages: fewer and shorter buses, faster program execution due to many computer resources, and reduced vulnerability thanks to distribution. However, such an architecture needs more processor types and then more complex software generation and validation procedures. Furthermore, some of the processors, located near sensors and actuators, may be in a more severe and less accessible environment. This could be an important drawback for maintenance costs. Distributed architecture are not yet implemented but are being examined for future avionics systems.

MIL-STD-1553B and ARINC-429 Federated Architectures

Modern military and commercial aircraft include federated avionics architectures, but based on different digital data buses.

MIL-STD-1553B-Based Architecture

This is a serial bidirectional two-wire bus based on a time-division command/response multiplex mode. Communication between different units can take place at different moments in time, and all communications between units share the same bus. Figure 5.45 shows a hypothetical architecture

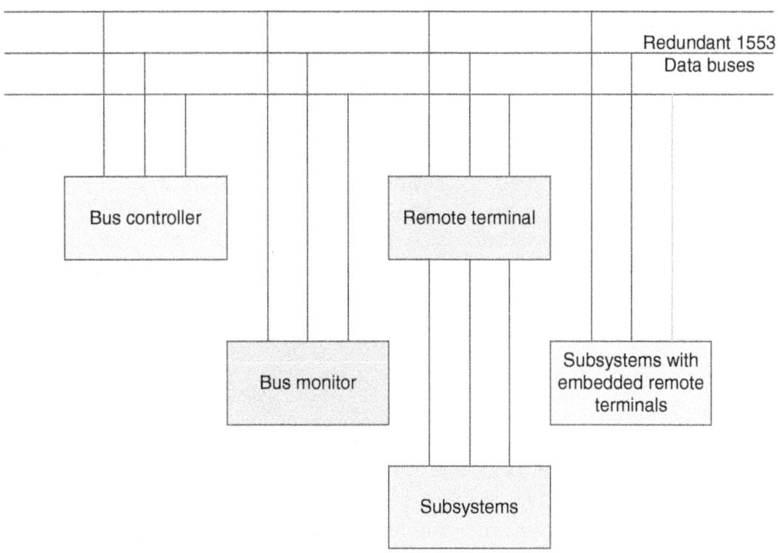

FIGURE 5.45 Typical MIL-STD-1553B-based architecture.

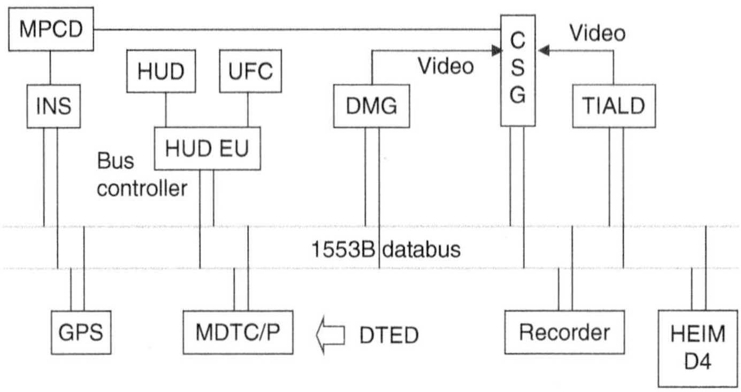

FIGURE 5.46 Example of a MIL-STD-1553B-based architecture: Jaguar GR1B avionics system.

based on three redundant 1553 data buses which operate at 1 megabit per second (Mbit/s) in a half-duplex mode using Manchester II data encoding. Hardware is the tangible portion of a 1,553 bus. The major hardware elements are the bus controller, remote terminals, and optional bus monitors. The bus controller (BC) is in charge of all data flows on the bus and initiates all information transfers. It transmits and receives data in command/response mode. This sole control of information transmission on the bus shall reside within the BC. The BC also monitors the status of the system. BCs are generally separate LRUs, but they can be part of one of the other LRUs on the bus. A bus monitor (BM) receives and stores selected bus traffic. A BM will not react to any traffic received unless the traffic is specifically addressed to it. BMs are generally used to receive and extract data for off-line purposes (flight test, maintenance, etc.). They are strictly passive and are used only as a test device. Remote terminals (RTs) are the largest fraction of units in a 1553 bus system. No RT can speak on the bus unless spoken to first by the BC and specifically commanded to transmit. RTs respond only to valid commands specifically addressed to them or to valid broadcast (all RTs simultaneously addressed) commands. Because of the address field size, there can be only up to 31 RTs on a given bus. The address of an RT is established by external wiring. This approach allows any LRU of a given type to be used in any location without internal changes to it being made. The RT can be separate from any subsystem it serves, or it can be embedded in it. The avionics architecture of the Jaguar GR1B single-seat all-weather tactical strike, ground-attack fighter aircraft, shown in Figure 5.46 is a typical 1553B-based architecture with redundant 1553 data buses.

The primary bus BC function of these three buses is provided by the fire control system. Each bus supports a top-level function: navigation and engine control, defensive avionics, etc. The EuroFighter aircraft (EFA) avionics is also based on 1553 buses. It has a flight-critical fly-by-wire control system and a utilities control system (UCS). The UCS comprises of five computers (MC 68020 microprocessors) integrated by a 1553 data bus, and it implements centralized control for the secondary electrical power, hydraulic management, fuel management, fuel gauging, environmental, etc. These systems are integrated in the same federated architecture based on the same redundant 1553 bus.

ARINC-429-Based Architecture

Similar to the MIL-STD-1553B bus, which is the basis for digital buses in modern military aircraft, ARINC-429 is the basis for digital buses in modern civil aircraft (Figure 5.47). These two standards establish vastly different approaches to data buses that reflect the differences in design philosophy and performance priorities for military and civil aircraft. Requirements for minimum weight and maximum flexibility drove 1553 to operate at 1 Mbit on a bidirectional bus, while certification requirements drove 429 to operate at either 12–14.5 or 100 kbit/s on a monodirectional bus.

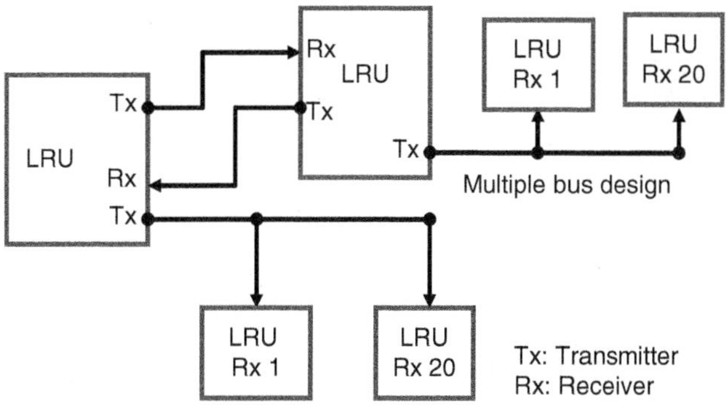

Figure 5.47 A typical ARINC 429-based architecture.

ARINC 429 is a monodirectional bus, i.e., with only one transmitter T and multiple receivers $R_0, \ldots, R_n$ ($n < 20$). If R_i needs to communicate with T, a separate bus with R_i as the source, and T as the sink is required.

This can lead to complex architecture with a great number of ARINC 429 buses. However, due to the absence of multiplex traffic, there is no need for bus controllers and remote terminals. Consequently, such architectures are more easily certified. This is a major advantage for civil aircraft.

Communications on 429 buses use 32-bit words with odd parity. ARINC specification 429 identifies two transmission rates: 100 kbit/s (high-speed bus) for transmitting large quantities of data or flight-critical information, and 12 or 14.5 kbit/s (low-speed bus) for general-purpose, low-criticality information.

ARINC 429-based architectures are commonly found in commercial aircraft designed and built in the 1980s. For instance, the avionics system of the Airbus family A320/A330/A340 is based on such digital data buses (A340 avionics architecture includes more than 300 ARINC 429 buses).

Integrated Modular Avionics

Integrated modular avionics (IMA) will make the current method of packaging line-replaceable units in electronic equipment obsolete. Functions now found in relatively large stand-alone units will be performed within a large set of hardware modules, forming a subsystem with a common chassis, common fault-tolerant processing, centralized power supplies, and flexible aircraft interfaces. From a computing point of view, IMA is basically a standard avionics architectural model containing removable components called line-replaceable modules (LRMs) communicating through multiplex digital data buses (ARINC 629, Ethernet-based, etc.). This replaces many black boxes and enables a large variety of avionics functions to share processing, memory, I/O functions, and common software such as a real-time operating system that independently processes application software while maintaining robust partitioning between software modules for the most critical flight functions. An IMA architecture example is the ARINC 629-based architecture, as in the Boeing B777. Other protocols are being studied, such as Ethernet-based deterministic protocols.

ARINC-629-Based Architecture

ARINC 629 is a multiple-transmitter digital data bus which operates at 2 Mbit/s. It uses word formats very similar to those in 1553, but without a bus controller. Each terminal has an autonomous access to the bus through a fair distributed protocol. There are two protocol schemes for ARINC 629: basic and combined. In the basic protocol, the bus operates in either the periodic or aperiodic mode. In the combined protocol, the periodic messages have priority and are guaranteed to be transmitted on a set schedule, while the aperiodic data are transmitted only when the bus is free.

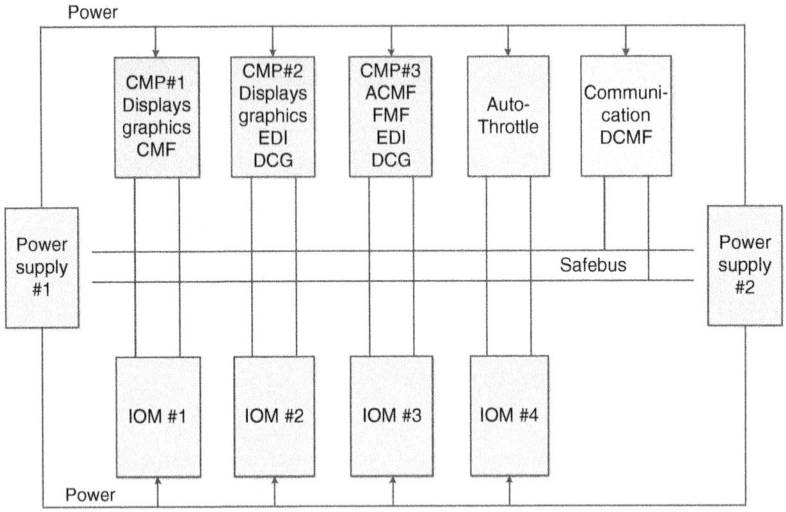

FIGURE 5.48 B777 AIMS cabinet architecture.

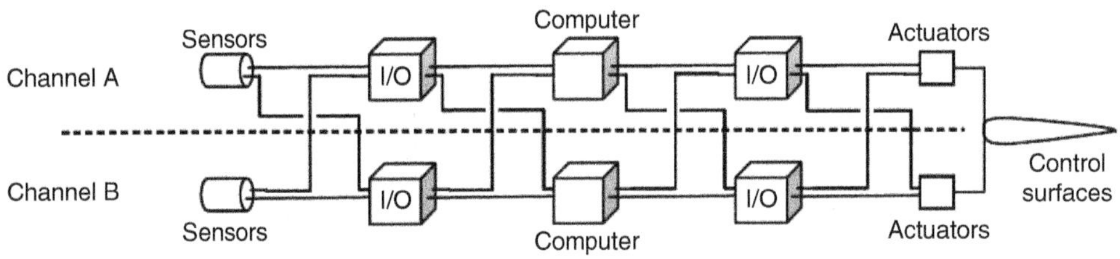

FIGURE 5.49 Typical dual logical architecture.

The Boeing B777 Airplane Information Management System (AIMS) is a typical example of ARINC 629-based IMA architecture.

The central features of AIMS, shown in Figure 5.48, are the two cabinets containing LRMs. AIMS functions performed in both cabinets include flight management and engine data interface, electronic flight instrument system (EFIS), engine indicating and crew alerting system (EICAS), central maintenance, communications management, etc.

Within each cabinet (Figure 5.49), the LRMs are interconnected by dual ARINC 659 backplane data buses. The cabinets are connected to the triplex redundant ARINC 629 system bus and to the triplex redundant fly-by-wire bus, and are also connected via ARINC 429 buses to the EFIS and EICAS displays.

5.37 Avionics Logical Architecture

Fault tolerance and maintenance are two major points of view. To meet the fault-tolerance requirements (safety and availability), avionics fault-tolerant systems generally use several types of fault tolerance, such as redundancy and dissimilarity, and have the capability for automatic, dynamic reconfiguration of the system. To meet the maintenance requirements, avionics designers have developed specific logical architecture and built-in test functions.

The Fault-Tolerance Point of View

The possible system configurations range from the simple single microcomputer with its own internal computer bus to configurations using 100 or more microcomputers and over 250 ARINC 429 data buses (e.g., the Airbus A340). Critical functions are usually implemented as fail operational (i.e., the function continues to operate in a normal manner after any single system component fails). Essential functions are usually implemented as fail passive or fail soft (i.e., a single failure will either cause loss of the function with no aircraft perturbation or cause a disturbance which is limited to a safe and acceptable level).

Avionics architectures are generally based on redundant channels configured as either independent channels (no interconnection or sharing of control signals between the parallel channels) or cross-strapped (interconnections do exist). Cross-strapping may be accomplished by either analog cross-feed or intercommunication between processors. Redundant channels can operate in either active or active-standby mode. In the active mode, all redundant channels are operating simultaneously. In the active-standby mode, some of the channels are controlling the system while the others are standing by, ready to assume control in the event one or more of the active controlling elements is declared faulty by the fault-detection logic. Some of the possible architectures are discussed in the following paragraphs.

Simplex Architecture

The simplex configuration is a single dedicated computer that may have one or more sensors/subsystems connected to input/output ports as well as actuators connected to the output ports. The principal fault-detection technique consists primarily of the data validity checks with built-in test equipment for self-monitoring.

Dual Architecture

A dual architecture comprises two identical parallel channels performing the same functions. Outputs of the two channels can be compared. Figure 5.49 shows a typical dual flight control system. In normal operation, channel A is active and channel B is in standby mode. Built-in test equipment (BITE) and software tests monitor the components of each channel. A faulty component in channel A is automatically replaced by the corresponding one in the B channel (and vice versa).

Such an architecture assumes that elementary faults can be detected by embedded functions of equipment. However, certain types of faults, e.g., latent (failure to zero), bias, slow drift, and scale factor shift, are not readily detectable. Consequently, single–dual configurations have not been widely used for critical functions.

Command/Monitor Computer

This is a particular dual configuration with cross-strapping only for output comparison. This concept has been developed and proven by Aerospatiale for Concorde and more recently for the Airbus family A320/A330/A340. Such a configuration, depicted in Figure 5.50, is composed of a pair of dissimilar channels: a command channel ensuring the function allocated to the computer (e.g., control of a control surface) and a monitoring channel ensuring that the command channel operates correctly.

Each channel includes one or more processors, their associated memories, input/output circuits, a power supply unit, and specific software (hardware and software of the two channels are dissimilar). Both channels are simultaneously active. The difference between the results computed by each channel is compared. If this difference is above an allowable threshold for a sufficiently long period of time, the channel which detects this failure cuts the links between the whole equipment (i.e., the two channels) and the outside world. System tolerances (most notably sensor inaccuracy, rigging tolerances, and computer asynchronism) must be taken into account to prevent undue failure detection, and errors which are not detectable (within the signal and timing thresholds) are assessed with respect to their handling quality and structural loads effect.

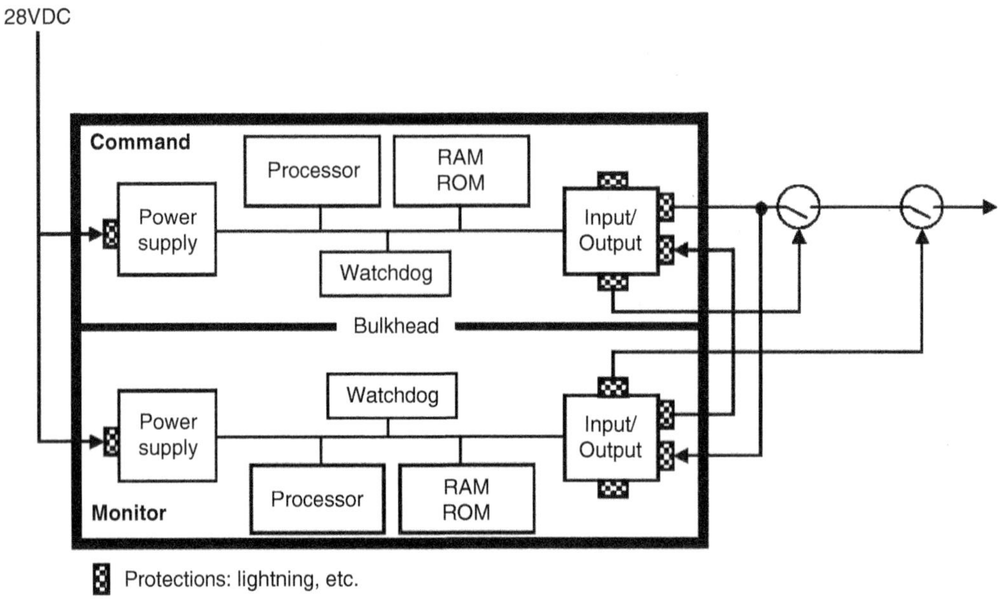

FIGURE 5.50 Command/monitor architecture.

Command/monitor computers are fail-passive (internal failures are not propagated outside the computer). This allows the design of a fault-tolerant system composed of one active equipment and N-1 standby equipment. Reconfiguration in such a system is then very simple and natural. The second equipment becomes active when the first Control/monitor digital equipment are commonly found in the essential and critical digital systems (automatic pilot, flight control system, break system, fuel management system, etc.) of Airbus built in the 1980s (see example of A320 flight control system presented below).

Dual Command/Monitor Architecture

Figure 5.51 depicts a typical dual–dual configuration.

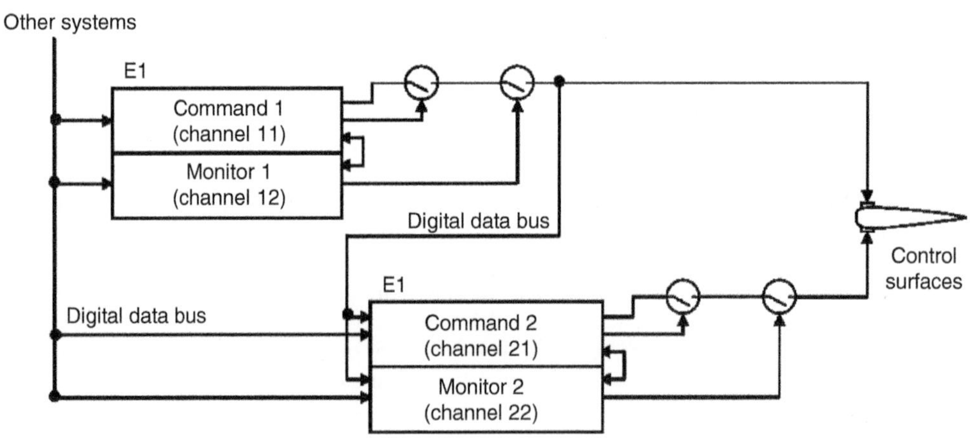

FIGURE 5.51 Dual command/monitor architecture.

While this may appear to be a four-computer system, the system is typically operated as two similar command/monitor equipment E1 and E2, each composed of two subchannels which exchange data. Data exchange also occurs between E1 and E2.

In normal operation, E1 is active and E2 is in standby mode. In case of channel 11 or channel 12 failure, the first equipment is disconnected and no data are sent to the actuators nor to the other equipment E2. After a confirmation delay, E2 then becomes active. Thus, despite one failure, the whole system remains operational. Obviously, additional equipment similar to E1 and E2 could be added to achieve any level of reliability. Such architecture can be found, for instance, in the Airbus A320 flight control system.

Nuplex Architecture

A last common redundant architecture that can be found in a great number of civil aircraft is a triplex, quadruple, and more generally Nuplex configuration. Figure 5.52 depicts the triplex configuration of the Airbus A340 Air Data and Inertial Reference System (ADIRS) computing the state data of the aircraft.

Because of its high critical level, this system is composed of three similar computers (ADIRU) simultaneously active. Consequently, the whole ADIRS system produces three altitudes, three Mach numbers, etc. These data are then compared and selected by each user system, the vote being performed in each system (FMGEC in the considered example). Two comparison monitoring systems may be used: bit-by-bit comparison and differential comparison.

One of the main differential comparison techniques is median selection requiring a minimum of three channels to tolerate one failure. Another method that is commonly used to develop the control signal in cross-strapped controllers is an averaging cross-feed. In this method, each channel accepts input from all other channels and computes their average which is then used as a signal.

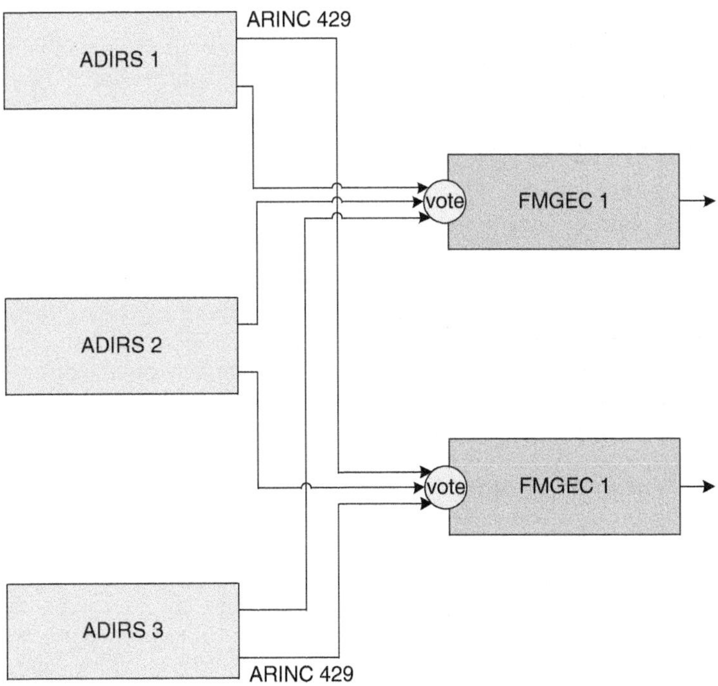

FIGURE 5.52 Typical triplex architecture: the A340 ADIRS architecture.

Only two channels are required to tolerate a single failure if autonomous false-detection techniques are used. A third channel would allow two failures.

Another way commonly found in military aircraft is bit-by-bit comparison. Contrary to differential comparisons, which do not require perfect agreement between the channels, bit-by-bit is based on the assumption that the outputs are in total agreement except in the case of a fault. In this method, the resulting signal of an Nuplex architecture is derived from a majority vote of all channel signals on a bit-by-bit basis. However, to use this method, certain criteria must be satisfied, such as time synchronism. This last constraint prevents the use of bit-by-bit comparison for most civil aircraft. Unlike military aircraft, civil avionics subsystems compute separately in an asynchronous way; parallel equipment do not share a common clock or synchronization mechanism.

Fault-Tolerant Software

Flight-critical systems require fault-tolerant software to complement the fault-tolerant hardware. Many of the concepts for fault-tolerant hardware have parallels in fault-tolerant software. The software designer must use software design concepts as adeptly as hardware concepts. Fault-tolerant software falls into three categories: multiversion programming, recovery blocks, and exception handlers (however, note that all of these techniques are subject to error if the software specification is incorrect).

Multiversion, or *N-version,* programming requires the development of two or more versions of a program that performs a specific function described in the software specification. These versions should be developed by separate software teams and may even be designed to operate on different processors. They accept a common input from an executive-level program, which in turn also compares the results of the different versions to detect faults. For instance, the two-version programming method is used for the Airbus A320 command/monitor equipment development. Channel command and channel monitor are separately developed by separate teams. *Recovery blocks* are another concept in fault-tolerant software. Acceptability checks are made on the results from a primary version of a program. If the results fail the acceptability checks, an alternative version of the program different from the primary version is invoked and the process of computation and acceptability checks is repeated. If no version produces an acceptable result, the software block is judged to have failed.

Now that a fault has occurred and been detected, it now must be recovered from. Recovery can be either backward or forward. Backward recovery is exemplified by recovery blocks, where, in the case of a fault, the executive software reinitializes the program using the same input values as used in the previous cycle and attempts to execute the program again. Forward recovery is demonstrated in N-version programming, where the outputs are compared and erroneous values generated by faulty software are ignored and only the correct value is passed to the user. Thus, even though a fault may have occurred, recovery has been made without any loss of time.

Another form of fault tolerance in software is *run-time assertions.* Watchdog timers check the time for a block of code to be executed, and if the code is not completed within the prescribed time, an error is assumed to have occurred.

An emerging concept that uses software in lieu of hardware replication to achieve fault tolerance is *analytical redundancy.* In the case of a faulty sensor, analytical redundancy combines data from the remaining functioning sensors with data from other sources in the aircraft in algorithms that compute the most probable value from the failed sensor. This computed value is then used in the same ways as a value from a functioning sensor. An equivalent concept can be applied to flight control actuators and surfaces where, if an element is lost, the remaining functioning elements can be combined in a way to offset the loss. Analytical redundancy and its companion concept for actuators are two of the cornerstones of reconfigurable flight control systems. Such analytical redundancy techniques have not yet been implemented in civil aircraft but are being examined for future military avionics systems.

Alarm and Maintenance System

Maintenance is a major factor in operating costs and therefore life-cycle costs. For this reason, built-in test equipment (BITE) and centralized maintenance function (CMF) are two main systems which impact the global architecture of the whole avionics. CMF and BITE form a distributed system composed of one or two computers CMCs (centralized maintenance computers) and a BITE function integrated in the LRUs, sensors, and actuators of the aircraft. The aim is first to recognize and correctly identify faulty elements to be changed (at least 90% of the possible faults should be recognized by the BITE components), and secondly to display and store fault data for all BITE-equipped LRUs on the aircraft. For that purpose, each BITE function integrated in each LRU tests its host LRU. In case of internal fault, or if some failure is detected on input data from other systems, the LRU's BITE component generates a message to each CMC indicating the failure. However, in order to circumvent intermittent faults, multiple successive failures over an appropriate time period are required before the BITE declares a unit failed.

In the military context, MIL-STD-2165, *Testability Program for Electronic Systems and Equipment*, requires for the BITE that 95% of all possible critical failures should be detected in less than 1 second and that 100% should be detected in less than 1 minute. Eighty-five percent of all other failures must be detected in less than 1 minute.

The CMF builds on the capabilities of the BITE to consolidate information on failures from the whole avionics system into a single database along with auxiliary information such as airspeed, altitude, time, and date when the failure occurred.

The A320 CMF, shown in Figure 5.53 is a representative example of BITE and CMF architecture in civil aircraft. It can monitor up to 70 systems, but not all of them directly, and can store the 200 most recent faults or all faults for the last 63 flight legs. The CMF also records universal time

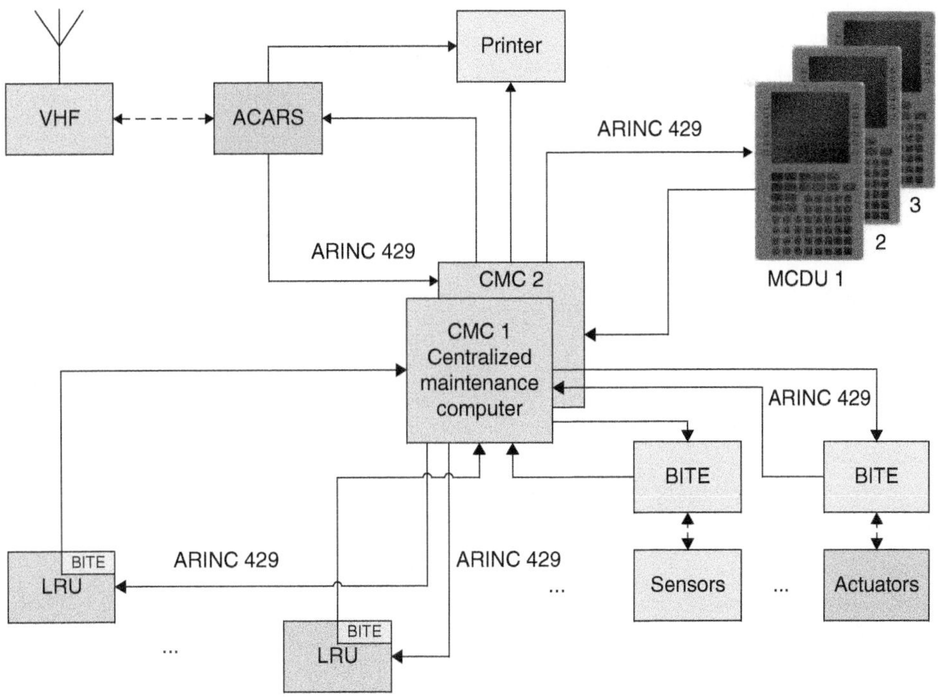

FIGURE 5.53 Airbus A320 centralized maintenance function architecture.

coordinated (UTC), flight number and phase, and any cautions and/or warnings that were in effect at the time of the fault. The interface with the CMF is through a multifunction control/display unit (MCDU) in the cockpit. All fault messages are displayed on the MCDU screen in standard formats using plain English, or can be sent to the cockpit printer and to the aircraft communications addressing and reporting system (ACARS) VHF datalink. This latter opportunity allows advance notification to maintenance personnel at the flight destination, who can begin preparation to work on the aircraft when it arrives.

5.38 Avionics Example: The Airbus A320 Flight Control System

The first electric flight control system for a civil aircraft was designed by EADS-Airbus and installed on Concorde. This is an analog, full-authority system for all control surfaces. It copies the stick commands onto the control surfaces. A mechanical backup system is provided on the three axes. All control surfaces (slats, flaps, and spoilers) are controlled electrically by high-level control laws under all circumstances.

The global architecture of the whole system is highly redundant and is based on the fail-passive command/monitoring computers.

Airbus A320 Flight Control System Architecture

Seven computers form the heart of the flight control system: two elevator aileron computers (ELACs), three spoiler elevator computers (SECS), and two flight-augmentation computers (FACs). The allocation of these computers to the flight control surfaces is shown in Figure 5.54.

The redundancy aspect is handled at the system level. The functions of the system are divided between all the computers so that each one is permanently active at least on one subassembly of its functions. For any given function, one computer is active and the others are on standby ("hot spares").

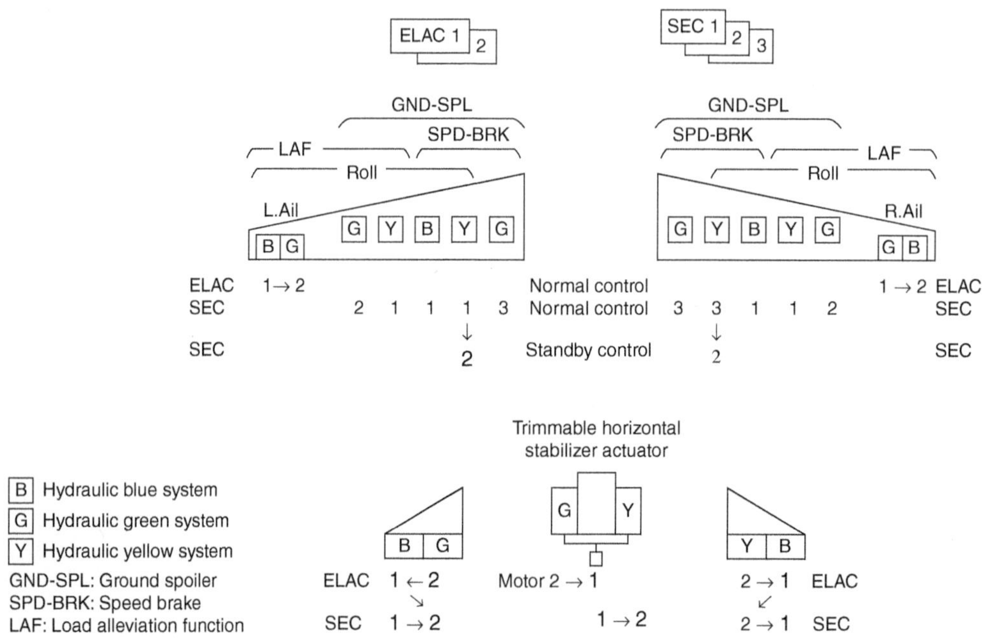

FIGURE 5.54 Airbus A320 flight control system (partial) architecture.

As soon as the active computer interrupts its operation, one of the standby computers almost instantly changes to active mode without a jerk or with a limited jerk on the control surfaces. This architecture ensures that the failure of any one ELAC, SEC, or FAC will not substantially affect the performance of the system.

The ELAC contains a pair (command and monitor part) of MC 68000 processors, each of which has approximately 40,000 words of software.

The SEC contains a pair of Intel 80186 processors, each of which has approximately 40,000 words of software. Dissimilar software is used. The functions of ELACs and SECs are identical except that ELAC control of the spoilers is via the SECs. The basic control law and all protection systems are in both ELACs and SECs. In the case of ELAC failure, the ailerons are no longer active and the aircraft is controlled by the SECs. The ELACs and the SECs are designed and manufactured by separate companies.

In addition to the ELACS and SECs of the A320, two computers are used for rudder control (FAC). However, they are not redundant with the ELACs and SECs. The FACs have the same dual structure as the ELACs and SECs. Each contains 120,000 words of software.

Redundancy Management

The Airbus A320 flight control system is highly critical and must be sufficiently dependable. This requires redundancy, dissimilarity, and reconfiguration management.

The electrical power is normally supplied by two generators, each driven by a different engine. Moreover, an auxiliary generator (APU), batteries, and a ram air turbine (RAT) are available. If the two engines shut down, the RAT is automatically extended. It then pressurizes a hydraulic system which drives a third electrical generator. The FCS computers are connected to at least two power sources.

The aircraft has three hydraulic systems [identified by G (Green), B (Blue), and Y (yellow) in Figure 5.54], one of which is sufficient to control the aircraft. Two systems are pressurized by each engine, the third being pressurized either by an electric pump or by the RAT.

The computers and actuators are also redundant. This is illustrated by the A320 pitch control (left and right elevator, plus trimmable horizontal stabilizer, THS) (Figure 5.54). Four control and monitoring computers are used, but one is sufficient to control the aircraft. In normal operation, one of the computers (ELAC2) controls the pitch, with one servo-control pressurized by the green hydraulic for the left elevator, one pressurized by the yellow hydraulic for the right elevator, and by electric motor 2 for the THS. The other computers control the other control surfaces. If ELAC2 or one of the actuators that it controls fails, ELAC1 takes over (with the servo-controls pressurized by the blue hydraulic for elevators, and with THS motor 1). Following the same failure method, ELAC1 can hand over control to SEC2. Likewise, pitch control can be passed from one SEC to the other depending on the number of control surfaces that one of these computers can handle. These priority orders are pictured by arrows in Figure 5.54. Note that three computers would be sufficient to meet the safety objectives. The additional computer is fully justified by the availability constraints: it is desirable to be able to tolerate a takeoff with one computer having failed.

Functional Reconfiguration

The high redundancy of the system permits hardware or software failures to be tolerated. However, other failures must be considered, such as air data and inertial reference unit (ADIRU) failures.

In the normal case, i.e., if the three ADIRUs are available, the pilot has full authority within a safe flight envelope provided by protections included in the control laws and by addition of protection orders to the pilot orders. Flight control is in g-load factor mode.

If only one ADIRU is available, it is partially monitored by comparison with other independent information sources (in particular, an accelerometer). The safe flight envelope is now provided by warnings, as on a conventional aircraft. Flight control is still in g-load factor mode. If all ADIRUs are lost, the flight envelope protections are also lost and the flight control law is in a degraded mode: direct mode. This law has gains which are a function of the aircraft configuration (the position of the slats and the flaps) and here again allow flight control similar to that of a conventional aircraft.

Airbus A320 Flight Control System Experience

The A320 has accrued around 3.5 million flight hours. During these revenue flights, fault-tolerance mechanisms have been activated, due, for instance, to occurrence of some hardware failures (in a consistent manner with predicted reliability), loss of both ELAC during one flight following an air conditioning failure and the subsequent abnormal temperature. In all cases, failure detection and reconfiguration were successful, including the automatic takeover by the SECs in the last case, which justifies the use of dissimilarity as a fault-tolerant mechanism.

Further Reading

Avionics Library, 1995, *Performing a Safety Certification for Avionics Components and Systems*, Avionics Communications Inc., Leesburg, VA, United States.

Avionics Library, 1995, *Principles of Avionics Data Buses*, Avionics Communications Inc., Leesburg, VA, United States.

Avionics Library, 1993, *Validating Digital Systems in Avionics and Flight Control*, Avionics Communications Inc., Leesburg, VA, United States.

Brière, D. and Traverse, P., 1993, AIRBUS A320/A330/A340 Electrical Flight Controls, a Family of Fault-Tolerant Systems, *Proceedings of Fault-Tolerant Computing Symposium*.

Electronic Industries Association, 1998, System Engineering, EIA-632.

Fishbein, S. B., 1995, Flight Management Systems—The Evolution of Avionics and Navigation Technology, Praeger, Westport, CT, United States.

Federal Aviation Administration (FAA), 2000, *System Safety Handbook*, FAA, Washington, DC, United States.

Paasch, S., 1999, Software and the Safety Assessment Process, FAA, Roston.

RTCA, DO-178B, 1992, Software Considerations in Airborne Systems and Equipment Certification, Radio Technical Commission for Aeronautics, Washington, DC, United States.

SAE, 1996, In cooperation with the FAA, Guidelines for the Certification of Highly Integrated and Complex Aircraft Systems, SAE ARP 4754, Society of Automotive Engineers (SAE), Dearborn, MI, United States.

SAE, 1996, Guidelines and Methods for Conducting the Safety Assessment Process on Civil Airborne Systems and Equipment, SAE ARP 4761, Society of Automotive Engineers (SAE), Dearborn, MI, United States.

Scott J., 1997, *Systems Engineering for Commercial Aircraft*, Ashgate, Aldershot, UIT.

Spitzer, C., 1993, *Digital Avionics Systems*, McGraw-Hill, New York, United States.

PART 9

Aeronautical Communication Systems

Roberto Sabatini, Subramanian Ramasamy, and Alessandro Gardi

5.39 Introduction

Aeronautical communication systems include air-to-ground, air-to-air, and satellite-based communications to support a number of mission-essential and safety-critical tasks, the most important of which is conflict resolution by air traffic management (ATM). This chapter outlines the essential theory and design aspects applicable to most systems.

This part specifically introduces the aeronautical communication systems and some fundamental design aspects. A general introduction and historical perspective is provided before discussing the aeronautical radio communication systems theory. Elements of communication system design include transmitter and receiver design are also provided. Various modes of aeronautical communication including VHF voice and datalink communications, military communications, ultrahigh frequency (UHF), and tactical datalinks (TDLs), as well as, long-distance mobile communications including HF and satellite links are covered. The advanced aeronautical communication systems presented in detail here include the system wide information management (SWIM) concepts.

5.40 Evolutions

The evolution of aeronautical communication systems can be segregated into legacy, current generation, and future systems.

Legacy Systems

The initial development of aeronautical communication systems can be attributed to the requirements of the First World War. The first aeronautical radio communication systems were pioneered in the 1920s. During this period, both ground-to-air and air-to-air voice communications were established. The Second World War necessitated the need for the development of new and secure radio communication systems. Tangible on-board transceivers emerged between the war years, and the main standards and practices in aeronautical VHF communications emerged in the late 1940s. These have, arguably, not significantly changed since then. This has been mainly due to very robust and proven systems (e.g., the mainstay VHF communication systems) that have served us well and is also due to the reluctance of airlines to undergo the time-consuming and cost-intensive process of retrofitting. Later, radio-controlled landing became possible as more ground-based radio navigation aids became available. The invention of the transistor paved the way for the use of modern and solid-state electronics in aeronautical communication systems. As a result of advances in electronics, a wide range of small, rugged radio and navigational equipment for aircraft were available. Furthermore, the space program began and a variety of communication satellites were also launched.

The Present—Second-Generation Systems

Today, there is a requirement to enhance the legacy of mobile communication services to provide the users with more functionality, flexibility, immunity to interference (both RF and malicious) and reliability. To an extent, this is already well underway by introducing datalink services such as aircraft communications addressing and reporting system (ACARS) and VHF datalink and aeronautical satellite services as a second-generation stop gap. The "stop gap" should be emphasized. As with many of these systems, the engineering has been "shoe-horned" into existing spectrum allocations or using proprietary technologies almost in experimental conditions. While this has bought time, the solutions are not optimized for technology, application and spectrum efficiency and are all the time aging and becoming less relevant.

The Future—Third-Generation Systems

The technology is mature for the next (third) generation of communication systems in aviation and the unit cost of this equipment is ever decreasing. The next years will see some decisive changes in aeronautical communications being driven by the availability of this technology and also by the congestion and shortfalls in the legacy systems which are becoming more exaggerated and exasperating every day. It is also clear that a rationalization of all the systems is required to simplify long-term equipage. In contrast, the terrestrial mobile communications counterparts (public mobile services) have already realized much of their third-generation systems and are already providing some fourth- and even fifth-generation system services.

Recent advances in solid-state electronics, especially the advent of micro- and nanotechnologies are supporting the introduction of lighter, smaller, cost-effective devices. The focus is shifting toward providing communication services in the communication, navigation, surveillance/air traffic management and avionics (CNS+A) context.

5.41 Aeronautical Radio Communication Types

Aeronautical communication systems provide the means to exchange data among the airborne and ground components (avionics and air traffic management), as well as, to the other external counterparts.

Any on-board communications system is essentially a transmitter and a receiver element which, in most cases, are combined into a single unit. The system typically consists of a control panel, installed in the cabin, available to the pilot or an operator (also the ground operator), one or more antennas mounted on the airframe to ensure omnidirectional coverage and the signal processing equipment. The transmitter element includes an oscillator that generates a radio frequency carrier, and a device that, through a process called modulation, places the information at base frequency (signal modulating) on the radio frequency signal. Modulation is performed to vary a characteristic of the carrier signal as a function of the modulating signal. Modulation is necessary because the basic frequencies (typically audio frequencies) do not lend themselves to be transmitted effectively in free space. The signal is then sent to the antenna, which radiates it into space. The receiver antenna captures the electromagnetic energy, providing it to the receiver which extracts the transmitted information content (Stacey 2008).

The most common forms of modulation are related to varying the amplitude, frequency, and phase of the carrier with respect to the modulating signal. After the advent of digital signals for communications, digital modulation schemes, in which a carrier signal (analog) is modulated by a discrete signal, are in practice. The digital modulation methods require the employment of digital-to-analog and analog-to-digital convertors at the transmitting and receiving ends, respectively.

Communications for aviation are predominantly provided in the high-frequency (HF) range, which covers the frequencies, 3–30 MHz, for long-distance communications, and in the very-high-frequency/ultrahigh-frequency (VHF/UHF) bands for short and medium distances. Portions of the VHF band (30–300 MHz) are for commercial purposes or allocated to particular uses. Frequencies in the range 118–154 MHz are used for communications with the air traffic authorities (VHF and amplitude modulation), while the two bands 30–88 MHz and 154–174 MHz, which use frequency modulation, are employed in land and naval applications, respectively. For military communications, air forces use the upper portion of the VHF band (225–300 MHz) and the lower portion of the UHF band (300–400 MHz), conventionally grouped under the type "UHF." Super high-frequency band (3–30 GHz) is used for applications such as radar and airborne

Type	Frequency band	Typical modulation scheme	Application
HF	3–30 MHz	Single-sideband modulation (SSB)	Communications over the radio horizon
Terrestrial band or low VHF	30–88 MHz	Frequency modulation (FM)	Communication with the ground troops
Air Force band or VHF	119–154 MHz	Amplitude modulation (AM)	Communication with the air traffic authorities
Marine band	154–174 MHz	Frequency modulation (FM)	Communications with the ships
Military UHF	225–400 MHz	Amplitude/frequency modulation (AM/FM)	Military communication
SHF	3–30 GHz	Frequency modulation (FM)	Radars, airborne ground mapping

TABLE 5.13 Aeronautical Radio Communications

ground mapping. Table 5.13 summarizes the main features of radio communications for aeronautical use.

It is clear that it is not always necessary to install all types of aircraft radio described in Table 5.13. The choice between the different possibilities is made according to specific needs (services and missions) for which it the aircraft is designed. For example, on a civil transport aircraft, a rhodium HF and VHF might be sufficient, while a military aircraft requires only the VHF and UHF. The latter also requires secure modes to support close air support and ground troops, which necessitates the requirement to install all radios except HF.

Military aircraft are generally installed with at least two communication equipment (marked with COMM1 and COMM2), each with a multiband capacity. Recently, efforts have been made in the direction of reducing the number of airborne devices and hence, single-multiband transceivers are becoming popular. To ensure maximum coverage in all flight conditions, the two communication equipment are connected to two separate antennas, one placed on the top of the fuselage and one on the bottom. Currently, aeronautical telecommunications support a variety of CNS services. Table 5.14 summarizes the currently available datalink technologies and applications for aeronautical CNS.

5.42 Aeronautical Communication System Design

Communication systems are designed to transmit information. The design of communication systems involves:

- Selection of the information–carrying waveform.
- Bandwidth and power of the waveform.
- Effect of system noise on the received information.
- Cost of the system.

Some basic definitions commonly used are:

- An Analog waveform is a function that has a continuous range of values.
- An analog communication system transfers information from an analog source to the sink.

Aeronautical Telecommunications	Communication	Air traffic control (ATC): HF and very high frequency (VHF) voice communications (radio telephone service) Communications rendered by the air traffic control institutions to the aircraft to secure the safety and mobility of the air traffic
		Aeronautical operational control (AOC): VHF voice comms, VHF open space datalink, etc. Communications by the aircraft pilot for normal navigation, etc.
		Aeronautical administrative communications (AAC): VHF open space datalink (OSD), etc. Business-use communication for airline companies, etc.
	Navigation	Radio navigation/landing: very high frequency omni directional range (VOR), distance measuring equipment (DME), tactical air navigation (TACAN), nondirectional beacon (NDB), instrument landing system (ILS), microwave landing system (MLS)
		Satellite navigation/landing system: global navigation satellite system (GNSS) and GNSS landing system (GLS)
	Surveillance	Ground Surveillance: airport surveillance radar (ASR), secondary surveillance radar (SSR), traffic collision avoidance system (TCAS), etc.
		Automatic dependent surveillance (ADS): communications to attain navigation data automatically without any human intervention, based upon the navigation data from the aircraft
		Cooperative information systems: communication to attain three dimensional air-traffic positioning data automatically without any human intervention by using multiple positioning satellites

Table 5.14 Aeronautical Datalink Technologies and Applications

- A digital waveform is defined as a function of time that can have a discrete set of amplitude values.
- A digital communication system transfers information from a digital source to the intended receiver (also called as the sink).

After the advent of digital computers and enhanced processing capabilities, digital communication systems have been predominately used in place of analog communication systems. Some key advantages of employing digital communications over their analog counterparts include cost effectiveness, security features by using data encryption algorithms, data from voice/video, and the possibility of merging data sources and transmitting over a common digital transmission system. In a digital long-distance communication system, noise dose not accumulate from repeater to repeater. Furthermore, data regeneration is possible and errors in detected data may be small, even when there is a large amount of noise on the received signal (suitable error detection and correction techniques are employed). The only disadvantages of using digital signals for communication are

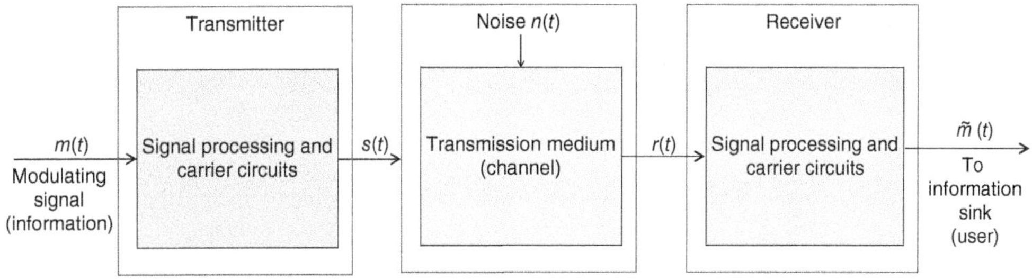

FIGURE 5.55 Components of a communication system.

more bandwidth is required than that for analog systems and synchronization must be maintained (Moir et al. 2013).

All communication systems contain three main sub systems (Figure 5.55):

- Transmitter.
- Channel.
- Receiver.

A signal processing block is used for more efficient transmission. In an analog system, the signal processor may be an analog low-pass filter to restrict the bandwidth of the information input. In a hybrid system, the signal processor may be an analog-to-digital converter (ADC) to produce digital signals that represent samples of the analog input signal. The transmitter carrier circuit converts the processed base band signal into a frequency band that is appropriate for the transmission medium of the channel.

The channel represents the path in which signals travel from transmitter to receiver. Channels may be of two types: wireless including air, space and seawater, and wire including twisted-pair telephone line, coaxial cable, waveguide, and fiber-optic cables. In general, the channel medium attenuates the signal so that the delivered information deteriorated from that of the source. The channel noise may arise from natural electrical disturbances or from artificial sources.

The receiver takes the corrupted signal at the channel output and converts it to be a base band signal that can be handled by the receiver's base band processor. The base band processor cleans up this signal and delivers an estimate of the source information to the communication system output. In digital systems, the measure of signal deterioration is usually taken to be the probability of bit error, which is also called as bit error rate (BER) of the delivered data. In analog systems, the performance measure is usually taken to be the signal-to-noise ratio (SNR) at the receiver output.

The goodness of any communication system can be measured by determining the following:

- The closeness of the estimate to the original signal dictated by quality of the transmission, SNR for analog signals, and BER for digital signals.
- The power required to transmit the signal. Lower the power, the longer is the battery life and reduced interference.
- The bandwidth required to transmit the information signal. The lower the bandwidth, the more users can share the channel. An exception to this is the spread spectrum technique, in which users use the same bandwidth.
- The quantity of information transmitted. In analog systems, information is related to the bandwidth of the signal, while in digital systems, information is expressed in bits/sec.

In analog communication systems, an optimum system might be defined as the one that achieves the largest signal-to-noise ratio at the receiver output, subject to design constraints such as channel bandwidth and transmitted power.

The dimensionality theorem for digital signaling can be defined as follows:

> Nyquist showed that if a pulse represents one bit of data, noninterfering pulses can be sent over a channel no faster than 2B pulses/sec, where B is the channel bandwidth:

$$R < B \log_2\left(1+\frac{S}{N}\right) \tag{5.146}$$

For digital communication systems, an optimum system may be defined as the system that minimizes the probability of bit error at the system output subject to constraints on the energy and channel bandwidth. According to Shannon, the probability of error would approach zero if $R < C$, where R = rate of information (bits/s)

C = channel capacity (bits/s)

B = channel bandwidth in Hz

S/N = signal-to-noise power ratio

$$C = B \log_2\left(1+\frac{S}{N}\right) \tag{5.147}$$

Capacity is the maximum amount of information that a particular channel can transmit. It is a theoretical upper limit. A number of modulation schemes can be employed for analog and digital communication system. Any type of modulated signal can be represented by:

$$v(t) = \text{Re}\left\{g(t)e^{j\omega_c t}\right\} \tag{5.148}$$

where $g(t)$ = modulated signal

f_c = carrier frequency (Hz) and $\omega_c = 2\pi f_c$

Re = real part

The formulae to express the amplitude and phase modulation schemes are given as

$$\text{Amplitude modulation: } A_c\,[1+m(t)] \tag{5.149}$$

$$\text{Phase modulation: } A_c\, e^{jm(t)} \tag{5.150}$$

where $m(t)$ is the modulating signal and A_c is the amplitude of the signal.

Two canonical forms for the generalized transmitter are the following:

> The Amplitude modulation and phase modulation (AM-PM) generation technique (Figure 5.56) involves the generation of envelope and phase functions to modulate the carrier and is given by

$$v(t) = r(t)\cos[\omega_c\, t + \theta(t)] \tag{5.151}$$

where $v(t)$ = modulated signal

$r(t)$ = resultant signal after modulation

θ = phase (rad)

The quadrature generation technique involves the use of in-phase and quadrature signals generation to modulate the carrier (Figure 5.57) and is given by

$$v(t) = x(t)\cos\omega_c\, t - y(t)\sin\omega_c\, t \tag{5.152}$$

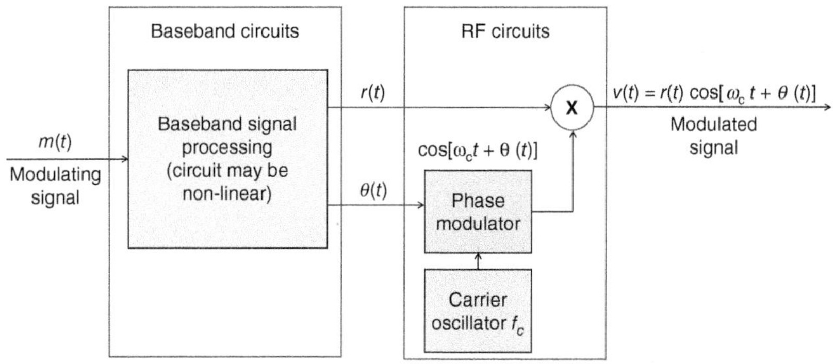

FIGURE 5.56 Generalized transmitter using the AM–PM generation technique.

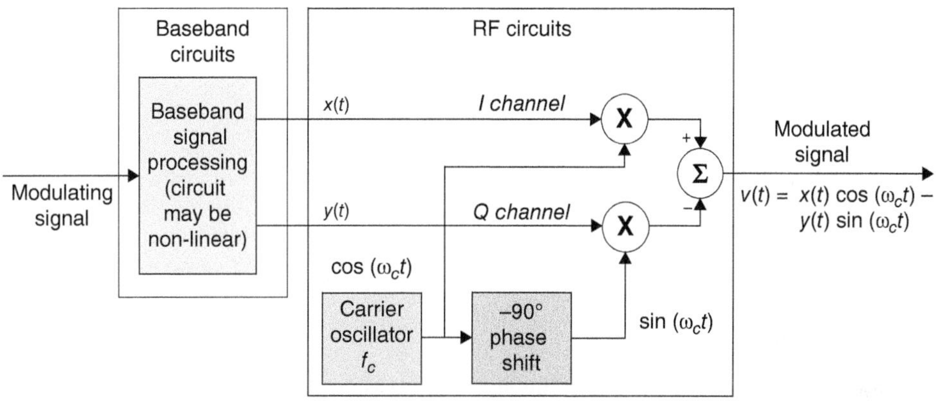

FIGURE 5.57 Quadrature generation technique.

The corresponding governing equations for the quadrature generation technique are given by

$$v(t) = x(t)\cos(\omega_c t) - y(t)\sin(\omega_c t) \tag{5.153}$$

$$g(t) = r(t)e^{j\theta(t)} = x(t)j\, y(t) \tag{5.154}$$

The two common types of receivers are the generalized transmitter using the quadrature generation technique and the superheterodyne receiver. The tuned radio frequency (TRF) receiver (Figure 5.58) is composed of RF amplifiers and detectors. There is no frequency conversion in this type of receiver and is not employed often when compared to the superheterodyne receiver. The reasons can be attributed to the difficulty in designing tunable RF stages and obtaining high gain RF amplifiers.

The superheterodyne receiver is the most popular type of radio receiver. The RF signal is downconverted to a lower intermediate frequency (IF). The main amplification takes place at the IF stage. The RF and IF frequency responses are important in providing the required reception characteristics.

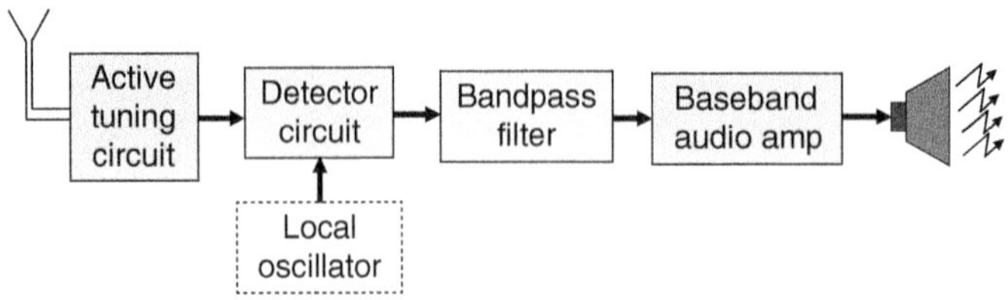

Figure 5.58 TRF receiver.

Propagation of EM Signals

The propagation characteristics of electromagnetic waves used in wireless channels are highly dependent on the frequency. Based on carrier frequency EM wave propagations can be classified as:

- Surface-wave (or ground-wave) propagation.
- Sky-wave propagation.
- Line-of-sight (LOS) propagation (direct and reflected waves).

The ground waves are the dominant mode of propagation for frequencies below 2 MHz. Diffraction of the ground causes the wave to propagate along the surface of the Earth and this propagation mode is used in AM radio broadcasting services. Diffraction of the waves in "D" layer of the atmosphere (60–90 km in altitude) supports propagation along the surface of Earth (Figure 5.59).

The sky waves are the dominant mode of propagation for EM waves in the frequency range of 2–30 MHz. A long coverage is obtained by reflection of these waves at the ionosphere and at the Earth's boundary. This mode is predominantly employed in HF band international broadcasting (shortwave radio) and aeronautical HF voice/data communications. Sky-wave propagation is caused primarily by reflection from the Flayer (90–250 miles in altitude). Waves will be bent back to Earth. Ionosphere acts as a reflector. Transmitting station will have coverage areas along the surface of Earth (Figure 5.59).

LOS is the dominant mode of propagation for EM waves above 30 MHz. This mode can be used in Satellite Communications. The disadvantage of LOS is that the signal path has to be above

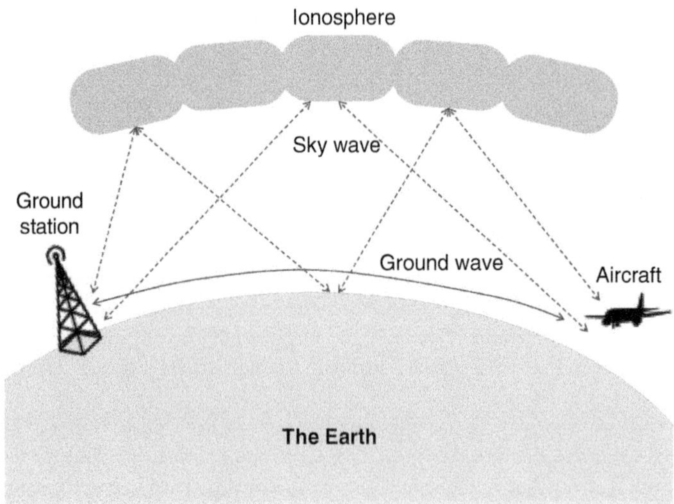

Figure 5.59 Sky and ground wave propagation.

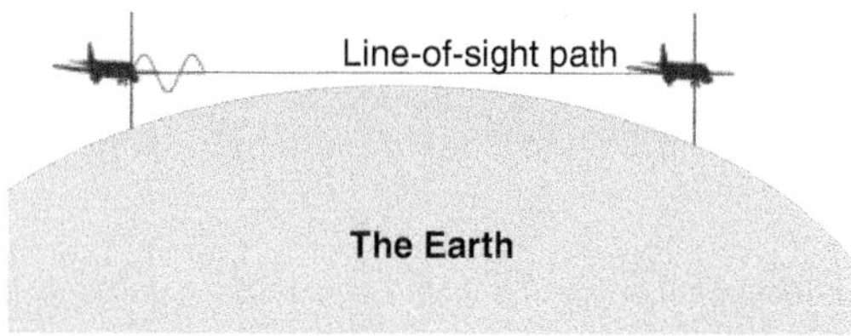

Figure 5.60 LOS propagation.

the horizon and the receiver antennas need to be placed on tall towers so that they can see each other. The LOS range for VHF transmissions and above is given by:

$$R = 1.2\sqrt{H_t} + 1.2\sqrt{H_a} \tag{5.155}$$

where R is the range (nautical miles), H_t is the height of the transmission tower (ft), and H_a is the height of the aircraft (ft). Therefore, for an aircraft flying at 35,000 ft, transmissions will generally be received by a 100 ft high radio tower if the aircraft is within a range of around 235 NM (Figure 5.60). Additionally, VHF and higher-frequency transmissions may be masked by terrain (e.g., a range of mountains).

Link Budget

Link-power budget calculations take into account all the gains and losses from the transmitter, through the medium to the receiver in a communications system. The attenuation of the transmitted signal due to propagation and the loss or gain due to the antenna is also taken into account.

The link budget parameters are:

- Transmitter power at the antenna
- Antenna gain compared to isotropic radiator
- EIRP
- Free space path loss
- System noise temperature
- Figure of merit for receiving system
- Carrier to thermal noise ratio
- Carrier to noise density ratio
- Carrier to noise ratio

Decibel Calculation

The term decibel denoted as dB is a relative unit of measurement and is used frequently in aeronautical electronic communication systems to describe power gain or loss (measure of signal change). The received power is expressed with reference to 1 W as dBW and is given by:

$$[P_R] = [EIRP] + [G_R] - [LOSSES] \tag{5.156}$$

where $[P_R]$ = received power in dBW
[EIRP] = equivalent isotropic radiated power in dBW
[GR] = receiver antenna gain in dB
[LOSSES] = total link loss in dB
dBW = 10 log10(P/(1 W)), where P is an arbitrary power in watts, is a unit for the measurement of the strength of a signal relative to 1 W.

Satellite Link

The three most important parameters of a satellite transponder are its receiver's G/T (dB/K) for the uplink, its transmitter's (EIRP) (dBW) for the downlink Eb/No(dB) and C/N(dB).

Antenna Gain

The gain of an antenna increases in power in a given direction compared to an isotropic antenna and is given by

$$G(\theta)=\frac{P(\theta)}{P_0/4\pi} \tag{5.157}$$

where $P(\theta)$ = variation of power with angle
$G(\theta)$ = gain at the direction θ
P_0 = total power transmitted

Effective Isotropically Radiated Power

An isotropic radiator is one that radiates equally in all directions. The power amplifier in the transmitter is shown as generating PT watts. A feeder connects this to the antenna, and the net power reaching the antenna will be PT minus the losses in the feeder cable (i.e., PS). The power will be further reduced by losses in the antenna such that the power radiated will be PRAD (< PT).

The received power $[P_R]$ is given by

$$[P_R] = [EIRP] + [G_R] - [LOSSES] \tag{5.158}$$

The equivalent isotropic radiated power [EIRP] is

$$[EIRP] = [P_S] + [G]\ dBW \tag{5.159}$$

where $[P_S]$ = transmit power in dBW
[G] = transmitting antenna gain in dB
$[G_R]$ = receiver antenna gain in dB

An example of antenna gain pattern (GPS directional antenna) is given by

$$G_R(dB)=7.8659\times \sin E-4.3659 \tag{5.160}$$

The corresponding antenna radiation pattern and gain is shown in Figure 5.62.

The losses are expressed as

$$[LOSSES] = [FSL] + [RFL] + [AML] + [AA] + [PL] \tag{5.161}$$

where [FSL] = free-space spreading loss in dB = PT/PR (in watts)
[RFL] = receiver feeder loss in dB

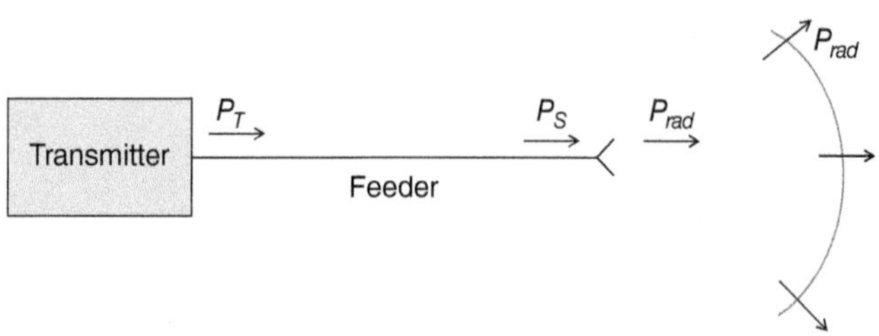

FIGURE 5.61 EIRP definition.

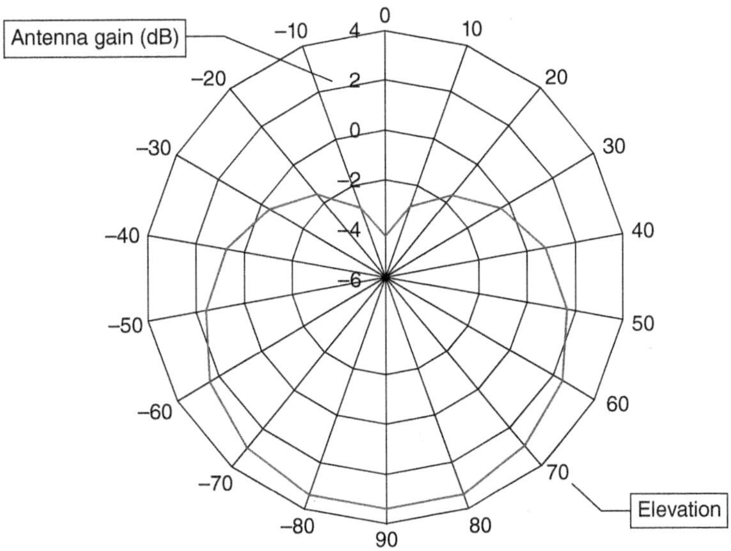

Figure 5.62 Antenna radiation pattern and gain.

[AML] = antenna misalignment loss in dB
[AA] = atmospheric absorption loss in dB
[PL] = polarization mismatch loss in dB

Demonstrated formula assumes idealized case. Free space loss only describes spherical spreading. Other effects need to be accounted for in the transmission equation:

$$P_r = \frac{P_t G_t G_r}{L_p L_a L_{ta} L_{ra} L_{\text{pol}} L_{\text{other}} L_r} \tag{5.162}$$

where L_a = losses due to attenuation in atmosphere
L_{ta} = losses associated with transmitting antenna
L_{ra} = losses associates with receiving antenna
L_{pol} = losses due to polarization mismatch
L_{other} = any other known loss (as much details as available)
L_r = additional losses at receiver (after receiving antenna)

Some intermediate variables are

$$\text{EIRP} = P_t G_t \tag{5.163}$$

where P_t is power fed into the antenna.

Hence the received power is expressed as

$$\begin{aligned} P_r &= \frac{P_t G_t G_r}{L_p L_a L_{ta} L_{ra} L_{\text{pol}} L_{\text{other}} L_r} \\ &= \frac{\text{EIRP} \times G_r}{L_p L_a L_{ta} L_{ra} L_{\text{pol}} L_{\text{other}} L_r} \\ &= \frac{P_{\text{out}} G_t G_r}{L_t L_p L_a L_{ta} L_{ra} L_{\text{pol}} L_{\text{other}} L_r} \end{aligned} \tag{5.164}$$

Therefore, there are many ways the formula could be rewritten. The user has to pick the one most suitable to each application. The transmission formula can be written in dB as

$$P_r = \text{EIRP} - L_{ta} - L_p - L_a - L_{\text{pol}} - L_{ra} - L_{\text{other}} + G_r - L_r \tag{5.165}$$

The calculation of received signal based on transmitted power and all losses and gains involved until the receiver is called "link power budget," or "link budget." The received power P_r is commonly referred to as "carrier power," C (especially in SATCOM applications).

A quick guide to perform a link budget is given below:

- First, acquire a sketch of the link path
  - Doesn't have to be artistic quality
  - Helps you find the stuff you might forget
- Next, consider the system of interest
  - Include all significant effects in the link power budget
  - Note and justify which common effects are insignificant here
- Roll-up large sections of the link power budget
  - TX power, TX ant. gain, path loss, RX ant. gain, RX losses
  - Show all components for these calculations in the detailed budget
  - Use the rolled-up results in build a link overview
- Comment on the link budget
  - Always use units on parameters (dBi, W, Hz, …)
  - Describe any unusual elements (e.g., loss caused by H_2O on radome)

5.43 VHF Voice Communications

Voice communication at very high frequency (VHF) is probably the most heavily used method of communication used by civil aircraft, although ultrahigh frequency (UHF) is generally preferred for military use. The VHF band for aeronautical applications operates in the frequency range 118.000–135.975 MHz with a channel spacing in past decades of 25 kHz (0.025 MHz). In recent years, to overcome frequency congestion, and taking advantage of digital radio technology, channel spacing has been reduced to 8.33 kHz (0.00833 MHz) which permits three times more radio channels in the available spectrum.

The VHF band also experiences limitations in the method of propagation. Except in exceptional circumstances, VHF signals will only propagate over LOS. That is, the signal will only be detected by the receiver when it has line of sight or can "see" the transmitter. VHF transmissions possess neither of the qualities of HF transmission and accordingly neither sky wave nor ground wave properties apply. This LOS property is affected by the relative heights of the radio tower and aircraft. This characteristic applies to all radio transmissions greater than 100 MHz, although the precise onset is determined by the meteorological conditions prevailing at the time.

Based on the LOS propagation, for an aircraft flying at 35,000 ft, transmissions will generally be received by a 100-ft-high radio tower if the aircraft is within a range of around 235 NM. Additionally, VHF and higher-frequency transmissions may be masked by terrain (e.g., a range of mountains). These LOS limitations also apply to equipment operating in higher-frequency bands and mean that VHF communications and other equipment operating in the VHF band or above—such as the navigation aids VOR and DME—may not be used except over large land masses, and then only when there is adequate transmitter coverage.

There are advanced techniques that may be used in sophisticated military equipment that mitigate these fundamental limitations. Such systems are said to possess an over-the-horizon (OTH) capability. Most long-range aircraft have three pieces of VHF equipment, with one usually being assigned to ARINC communications and reporting system ACARS transmissions, although not necessarily dedicated to that purpose. The requirements for certifying the function of airborne VHF equipment are given in Advisory Circular AC 20-67B, while RTCA DO-186 specifies the necessary Minimum Operational Performance Standards (MOPS).

The different VHF/UHF distress frequencies are

- 121.5 MHz and 243 MHz are the international distress frequencies (IDF) for VHF and UHF bands, respectively.
- All aircraft and air traffic control centers maintain a listening watch on the IDF 121.5 MHz.
- In addition, military controllers maintain a listening watch on 243.0 MHz in the UHF band.
- UHF receivers can also detect the harmonics of a civil VHF distress transmission and relay the appropriate details in an emergency (second harmonic of 121.5 MHz $\times$ 2 = 243 MHz).

The transmitter power level varies according to the operational region or the sector of FIR. For example, 50-W transmitters (for 200-nm range) are used on en route and oceanic regions while 20- and 10-W transmitters used in approach and TMA controls due to short range and small size of sector.

5.44 VHF Datalink Communications

The components of a VHF datalink communication system are

- Communications bearer
- Radio—VHF
- Various modulation schemes and access modes
- MODEM = ACARS/VDL terminal
- Message standard = VDL standards
- Host processing system
- Operator

A number of VHF datalinks (VHFDL) may be used. In the future, datalink rather than voice transmission will increasingly be used for air-to-ground and air-to-air communications as higher data rates may be used while at the same reducing flight crew workload. Various standards have been developed, although at present only ACARS (Aircraft Communications Addressing and Reporting System) and VDL2 (VHF Data Link 2 for Controller Pilot Data Link Communications—CPDLC) are employed operationally. ACARS is a specific variant of VHF communications operating on 131.55 MHz that utilizes a datalink rather than voice transmission. It is a digital datalink system for transmission of short, relatively simple messages between aircraft and ground stations via radio or satellite. The protocol, which was designed by ARINC to replace their VHF voice service and deployed in 1978, uses telex formats.

ACARS/VDLO/VDLA

SITA augmented their worldwide ground data network by adding radio stations to provide ACARS service. ACARS is a specific variant of VHF communications operating on 131.55 MHz that utilizes a datalink rather than voice transmission for transmission of short, relatively

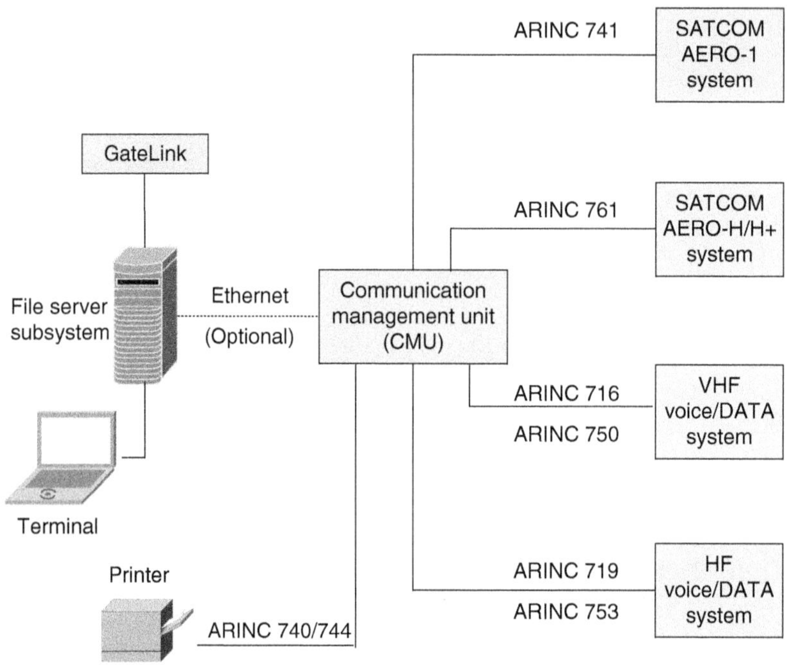

FIGURE 5.63 Typical ACARS onboard network.

simple messages between aircraft and ground stations via radio or satellite. A typical ACARS onboard network is shown in Figure 5.63. The protocol, which was designed by ARINC to replace their VHF voice service and deployed in 1978, uses telex formats. The main purpose of ACARS is to down-link operational data to the airline operational control center. The initial leg is by using VHF communications to an appropriate ground receiver, thereafter the data may be routed via land-lines or microwave links to the airline operations center. At this point it will be allowed access to the internal airline storage and management systems: operational, flight crew, maintenance, etc.

- ACARS message set includes
- Abnormal flight condition notifications
- Engine reports
- Repair and maintenance plan reports
- Manual messaging between crew and ATC (clear text format)
- Short weather reports

ICAO VDL Mode 1

The ICAO AMCP defined this Mode for validation purposes. It was the same as VDL Mode 2 except that it used the same VHF link as VHF ACARS so it could be implemented using analog radios before VHF Digital Radio implementation was completed. The ICAO AMCP completed validation of VDL Modes 1 and 2 in 1994, after which the Mode 1 was no longer needed and was deleted from the ICAO standards.

ICAO VDL Mode 2

The ICAO VDL Mode 2 is the main version of VDL. It has been implemented in a Eurocontrol Link 2000+ program and is specified as the primary link in the recently adopted EU Single Sky rule requiring all aircraft flying in Europe to be equipped with CPDLC.

In advance of CPDLC implementation, VDL Mode 2 has already been implemented in many aircraft to transport ACARS messages simplifying the addition of CPDLC. Networks of ground stations providing VDL Mode 2 service have been deployed by ARINC and SITA with varying levels of coverage. The ICAO standard for the VDL Mode 2 specifies three layers, the subnetwork, link, and physical layers. The subnetwork layer complies with the requirements of the ICAO Aeronautical Telecommunications Network (ATN) standard which specifies an end-to-end data protocol to be used over multiple air-ground and ground subnetworks including VDL. The VDL Mode 2 Link Layer is made up of two sublayers, a datalink service (DLS) and a media access control (MAC) sublayer. The datalink protocol is based on the ISO standards used for dial-up HDLC access to X.25 networks. It provides aircraft with a positive link establishment to a ground station and defines an addressing scheme for ground stations. The MAC protocol is a version of carrier-sense multiple access (CSMA). The VDL Mode 2 Physical Layer specifies the use in a 25-kHz-wide VHF channel of a digital 8-phase shift keying (8-PSK) modulation scheme, providing a data rate of 31.5 kbit/s. This is the highest data rate that can be achieved in a 25-kHz channel with a maximum range of 200 nautical miles (this required the implementation of VHF digital radios).

ICAO VDL Mode 3

The ICAO standard for VDL Mode 3 defines a protocol providing aircraft with both data and digitized voice communications (originally defined by the U.S. FAA with support from MITRE).

The VDL Mode 3 protocol is much more complex than VDL Mode 2.

The data and digitized voice packets go in time division multiple access (TDMA) slots assigned by ground stations.

The FAA implemented a prototype system around 2003 but did not manage to convince airlines to install VDL Mode 3 avionics and in 2004 abandoned its implementation.

ICAO VDL Mode 4

The ICAO standard for VDL Mode 4 specifies a protocol enabling aircraft to exchange data with ground stations and other aircraft. VDL Mode 4 uses a protocol called self-organized time division multiple access (STDMA), invented by Swedish Håkan Lans in 1988. STDMA allows VDL-4 to be "self-organizing," meaning no master ground station is required. This makes it much simpler to implement than VDL Mode 3. In November 2001, this protocol was adopted by ICAO as a global standard. Its primary function was to provide a VHF frequency physical layer for ADS-B transmissions. However, it was overtaken as the link for ADS-B by the Mode S link operating in the 1090 MHz band (selected as the primary link by the ICAO Air Navigation Conference in 2003). The VDL Mode 4 standard can also be used for air-ground exchanges. It is best used for short message transmissions between a large number of users, providing situational awareness, Digital Aeronautical Information Management (D-AIM), etc. European air traffic management modernization trials have implemented ADS-B and air-ground exchanges using VDL Mode 4 systems. However, on air transport aircraft the operational implementations of ADS-B will use the Mode S link and the implementation of CPDLC will use VDL Mode 2.

A comparison of all aeronautical datalink technologies and applications is presented in Table 5.15.

5.45 HF Communication System

High frequency (HF) covers the communications band between 3 and 30 MHz. It is a very common communications means for land, sea, and air. The utilized band is HF SSB/AM over the frequency range 2.000–29.999 MHz, using a 1-kHz (0.001-MHz) channel spacing. The primary advantage of

	ACARS (VDLO/A)	VDL1	VDL2	VDL3	VDL4
Standardization	None	None	1997	2000	2000
Operative	Since late 1970s	Obsolete	2002	Experimental since 2003	Imminent
Modulation/ air interface format/protocol	AM-ASK Unstructured	AM-ASK CSMA	8DPSK TDMA-CSMA	8DPSK TDMA	GPSK S-TDMA
Air interface speed and payload speed	2.4 kbps 300 bps	2.4 kbps 300 bps	31.5 kbps >10 kbps	31.5 kbps Up to 4 × 4.8 kbps	19.2 kbps <19.2 kbps
Prioritization	Not supported	Not supported	Not supported	Supported	Supported 4 levels
Voice services support	No	No	No	Yes	No
Service Providers	ARINC/SITA	ARINC/SITA		SITA	
Channelization	25 kHz + guard bands	25 kHz + guard bands	25 kHz + guard bands	25 kHz	25 kHz + guard bands
Deployment region	All	None	All	North America	Europe
Channel capacity		Typically 600+ per channel	Dependent on coverage volume	Theoretically up to 4,500 users	

Table 5.15 Aeronautical Datalink Technologies and Applications

HF communications is that this system offers communication beyond the line of sight. This method does, however, suffer from idiosyncrasies with regard to the means of signal propagation. With HF there are two main means of propagation, known as the sky wave and the ground wave. The sky wave method of propagation relies upon single- or multiple-path bounces between the Earth and the ionosphere until the signal reaches its intended location. The behavior of the ionosphere is itself greatly affected by radiation falling upon the Earth, notably solar radiation. Times of high sunspot activity are known adversely to affect the ability of the ionosphere as a reflector. It may also be affected by the time of day and other atmospheric conditions. The sky wave as a means of propagation may therefore be severely degraded by a variety of conditions, occasionally to the point of being unusable. The ground wave method of propagation relies upon the ability of the wave to follow the curvature of the Earth until it reaches its intended destination. As for the sky wave, the ground wave may on occasions be adversely affected by atmospheric conditions. Therefore, on occasion, HF voice communications may be corrupted and prove unreliable, although HF datalinks are more resistant to these propagation upsets. This is because HF communications are one of the main methods of communicating over long ranges between air and ground during oceanic and wilderness crossings when there is no LOS between the aircraft and ground communications stations.

For reasons of availability, most long-range civil aircraft are equipped with two HF sets with an increasing tendency also to use HF datalink if polar operations are contemplated. HF datalink (HFDL) offers an improvement over HF voice communications owing to the bit encoding inherent in a datalink message format which permits the use of error-correcting codes. Furthermore, the use of more advanced modulation and frequency management techniques allows the datalink to perform in propagation conditions, where HF voice would be unusable or incomprehensible.

An HFDL service is provided by ARINC using a number of ground stations. These ground stations provide coverage out to 2,700 NM and presently, HFDL ground stations are operating at the following locations:

- Santa Cruz, Bolivia
- Reykjavik, Iceland
- Shannon, Ireland
- Auckland, New Zealand
- Krasnoyarsk, Russia
- Johannesburg, South Africa
- Hat Yai, Thailand
- Barrow, Alaska, United States
- Molokai, Hawaii, United States
- Riverhead, New York, United States
- San Francisco, California, United States
- Al Muharraq, Bahrain
- Gran Canaria, Canary Islands
- Agana, Guam, United States
- Telde, Gran Canaria, Canary Islands

5.46 Satellite Communication System

The satellite link is probably the most basic in microwave communications since a line-of-sight path typically exists between the Earth and space. This means that an imaginary line extending between the transmitting or receiving Earth station and the satellite antenna passes only through the atmosphere and not ground obstacles. Such a link is governed primarily by free-space propagation with only limited variation due to the atmosphere. Free-space attenuation is determined by the inverse square law, which states that the power received is inversely proportional to the square of the distance. There are, however, a number of additional effects that produce a significant amount of degradation and time variation. These include rain, terrain effects such as absorption by trees and walls, and some less-obvious impairment produced by unstable conditions of the air and ionosphere.

Satellite communication is a form of radio or wireless communication and therefore must compete with other existing and potential uses of the radio spectrum. During the initial 10 years of development of these applications, there appeared to be more or less ample bandwidth, limited only by what was physically or economically justified by the rather small and low-powered satellites of the time. In later years, as satellites grew in capability, the allocation of spectrum has become a domestic and international battlefield as service providers compete among themselves, joined by their respective governments when the battle extends across borders. So, all of these factors must be considered when selecting a band for a particular application.

The general arrangement of the frequency spectrum is applied to satellite communications and other radio-communication services. Indicated are the short-hand letter designations along with an explanation of typical applications.

The most attractive portion of the radio spectrum for satellite communication lies between 1 and 30 GHz. The scale along the x-axis is logarithmic in order to show all of the satellite bands; however, observe that the bandwidth available for applications decreases in real terms the spectrum is moved toward its right (i.e., frequencies above 3 GHz).

Applications employ multiple-access systems to allow two or more Earth stations to simultaneously share the resources of the same transponder or frequency channel. These include the three familiar methods (FDMA, TDMA, and CDMA). Another multiple access system called space division multiple access (SDMA) has been suggested in the past. In practice, SDMA is not really a multiple access method but rather a technique to reuse frequency spectrum through multiple spot beams on the satellite. Because every satellite provides some form of frequency reuse (cross-polarization being included), SDMA is an inherent feature in all applications.

The required factors for an effective satellite communication system design include the performance of the satellite itself, the configuration and performance of the uplink and downlink, and the impact of the propagation medium in the frequency band of interest. Also important is the efficient transfer of user information across the relevant interfaces at the Earth users/station locations, involving such issues as the precise nature of this information, data protocol, timing, and the telecommunications interface standards that apply to the service. A proper engineering methodology guarantees that the application will go into operation as planned, meeting its objectives for quality and reliability. The RF signal in any communications link begins at the transmitting electronics and propagates from the transmitting antenna through the medium (free space and absorptive atmosphere) to the receiving antenna, where it is recovered by the receiving electronics. The carrier is modulated by a baseband signal that transfers information for the particular application. The first step in designing the communications link is to identify the overall requirements and the critical components that determine performance.

The following services are offered:

- Fixed satellite service (FSS) between Earth stations at given positions, when one or more satellites are used; the given position may be a specified fixed point or any fixed point within specified areas; in some cases this service includes satellite-to-satellite links, which may also be operated in the inter-satellite service; the FSS may also include feeder links for other services.
- Mobile satellite service (MSS) between mobile Earth stations and one or more space stations (including multiple satellites using inter-satellite links). This service may also include feeder links necessary for its operation.
- Broadcasting satellite service (BSS) in which signals transmitted or retransmitted by space stations are intended for direct reception by the general public. In the BSS, the term "direct reception" shall encompass both individual reception and community reception.
- Inter-satellite link (ISL) a service providing links between satellites.

Progress in satellite communications for aeronautical safety is realized through the revision of Standards and Recommended Practices (SARPs) and guidance material by ICAO for the aeronautical mobile satellite (route) service (AMS(R)S). Additionally, ICAO interacts with other international bodies to ensure that resources are coordinated and available. Acceptance of the applicability of datalinks to support air traffic services (ATS) as largely replacing voice communications requires assurance that all relevant elements of datalink network(s) and subnetworks (such as a satellite subnetwork) are properly coordinated and interoperable. AMS(R)S provides a global satellite subnetwork of the aeronautical telecommunications network (ATN) that provides end-to-end voice and data connectivity among end users, such as air traffic controllers, pilots, aircraft operators. Interoperability with the ATN is ensured by means of a standardized architecture for all elements of the ATN, based on ICAO SARPs and guidance material.

In general, satellite communication systems currently used include

- INMRSAT (GEO)
- MTSAT (GEO)
- IRIDIUM (LEO)

5.47 Military Aeronautical Communications

Military communications are predominantly used for communications among the different crew members. The various types of military communications in use are

- Intercom/ground crew communications
- Voice radio communications
    - HF (3–30 MHz)
    - LOW-VHF/SINCGARS (30–88 MHz)
    - VHF (AM 119–154 MHz and FM 154–174)
    - UHF (225–400 MHz) clear
    - UHF secure (Have Quick)
- Data communications using datalinks (e.g., Link 16, VMF, etc.)
- Satellite communications (e.g., SKYNET, SICRAL, MUOS, etc.)

Various types of tactical datalinks are employed and are summarized in Table 5.16. The JTIDS (joint tactical information system) or the MIDS (multifunctional information distribution system) are examples of systems that allow the exchange of tactical information relating to the mission among the military aircraft that cooperate within a mission under NATO command and control. Other communication systems use satellites (mainly in orbit geostationary) with the possibility of both regional and global coverage (using constellations of at least 3 or 4 satellites). These systems can be used effectively for both voice communications and for transmitting/receiving data (Sabatini et al. 2009).

5.48 Future Trends

Future aeronautical datalinks are intended to be used for a variety of aviation applications including: air traffic management (ATM), navigation, surveillance, and aeronautical/flight information management. A critical evaluation of datalink usage is required especially in high throughput scenarios such as in terminal maneuvering areas (TMA). As the route-based airspace design has predominantly been phased out, replaced by spacing and sequencing applications, current research is focusing on future ATM concepts for autonomous manned and unmanned aircraft operations where the next generation datalinks will become dominant players.

System Wide Information Management

A number of programs are currently addressing the challenges associated with providing a reliable, secure, integrated, and interoperable communication network. The expected result is to provide a performance-based solution, supporting time-based/performance-based operations by setting the performance requirements (e.g., availability, latency, message integrity), and enhancing the data communications bandwidth as well as the available electromagnetic spectrum.

	LINK-1	LINK-4	LINK-11	LINK-14	JTIDS/ IJMS	MIDS/LINK-16/ TADIL-J
Bearer	Point-to-point land line	Radio UHF	Radio HF (radio UHF)	HF ground/ sky wave (U/SHF)	Radio UHF	Radio UHF
Speed (bit/s)	1.20	3.80	1.80	75	28.80	57.60 or greater dependent on message packing structure and Net Design
ECM resistance	Not applicable	No	None	None	Yes	Yes
Encrypted	No	No	Yes	Yes	Yes	Yes
Nodeless	Not applicable	No	No	No	Yes	Yes
Extended LOS	Not applicable	No	Yes (HF)	Yes (HF or Satcom)	Yes (Relay)	Yes (Relay)
Secure voice	No	No	No	No	Yes	Yes
Flexible net	Not applicable	No	No	No	No	Yes
Navigation	Not applicable	No	No	No	No	Yes
Functional Area	Surveillance	Air control (LINK-4A) Fighter to Fighter (LINK-4C)	Surv. command Exchange Control and Status	Surv. eill.	Surv. eill.	Surveillance Air Control Fighter to Fighter Others as required

Table 5.16 Tactical Datalinks

The programs that are currently providing solutions for the evolution of air-ground, ground-ground and air-air communications and supporting the implementation of the System wide information management (SWIM) include

- Airborne new and advanced satellite techniques and technologies in a system integrated approach (ANASTASIA)
- Australian TAKE OFF
- Broadband Global Aeronautical Network (BGAN)
- E-enabled cabin (E-CAB)
- MOWGLY
- Net information integration services for security systems (NI2S3)
- NEtWorking the SKY (Newsky)
- New automotive tracking antenna for low-cost innovative applications (NATALIA)
- Seamless aeronautical networking through integration of datalinks, radios, and antennas (SANDRA)

- Security of aircraft in the future European environment (SAFEE)
- SWIM-Suit

High-integrity, high-throughput, secure datalink, and ground-based network infrastructure for civil/military dual usage and SWIM are required to allow a greater sharing of air traffic management information, such as weather, airport operational status, flight data, airspace status, and restrictions.

References

Moir, I., Seabridge, A., and Jukes, M., 2013, "Communications and Navigation Aids," Chapter 9, *Civil Avionics Systems*, 2nd ed., John Wiley & Sons, pp. 329–364.

Sabatini, R., Aulanier, L., Foreman, L., Martinez, M., Pour, B., Rutz, H., Snow S., et al., 2009, "Present and Future of the Multifunctional Information Distribution System (MIDS): Platform Integration, Terminal Modifications and Future Enhancements," *Proceedings of the International Data Link Society (IDLS) Annual Symposium IDLS 2009*, Vienna, Austria. More information available from https://www.eurocontrol.int/swim and http://www.faa.gov/about/office_org/headquarters_offices/ato/service_units/techops/atc_comms_services/swim.

Stacey, D., 2008, *Aeronautical Radio Communication Systems and Networks*, John Wiley & Sons.

PART 10

Ground Radio Navigation Aids

Rohan Kapoor, Alessandro Gardi, Suraj Bijjahalli, Subramanian Ramasamy, and Roberto Sabatini

5.49 Introduction

This part presents the fundamental principles of ground radio navigation systems and their application to the estimation of aircraft position and bearing. The use of ground-based navigation aids (NAVAID) to acquire a "fix" on aircraft position is gradually giving way to satellite-based navigation, i.e., the global navigation satellite system (GNSS). However, several traditional radio navigation aids remain in operation and form an important part of the communication, navigation and surveillance/air traffic management, and avionics (CNS+A) paradigm. The fundamentals of navigation using line-of-sight (LOS) positioning systems are presented first, and later subsections present specific details of radio navigation aids used in aviation.

5.50 Line-of-Sight Positioning

Terrestrial line-of-sight (LOS) systems are limited to operate within the distance to the horizon. The maximum range R (in nautical miles) at which a ground radio-based position fix can be performed is given by

$$R = 1.2\sqrt{h_T} + 1.2\sqrt{h_R} \tag{5.166}$$

where h_T and h_R are the heights of the transmitting and receiving antennas in feet. The range is constrained by the transmitter power, transmitter and receiver antenna gains, system bandwidth and noise as accounted for in the link budget of the given system. The reader is directed to Part 9 for more details on radio link budgets. Fundamentally, positioning in radio-navigation systems is achieved through determining the range and/or bearing to a ground-based station. Lines of position (LOP) can be plotted, given a range (rho) and/or bearing (theta) measurement with respect to a radio station. The intersection of two LOPs constitutes an ambiguous or unambiguous position fix. Depending on the number and combination of measurements, position fixing schemes can be rho-theta (ρ-θ), rho-rho (ρ-ρ), rho-rho-rho (ρ-ρ-ρ), or theta-theta (θ-θ), as shown in Figure 5.64. In rho-theta navigation, the bearing and range measurements are obtained with respect to one landmark with the LOPs intersecting at right angles, leading to an unambiguous position determination. Rho-rho systems measure the distances from two radio stations. Measurement of distances from three radio stations provides an unambiguous solution, a

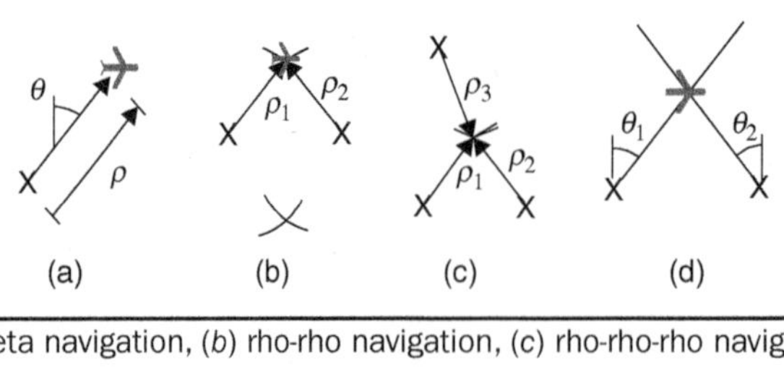

FIGURE 5.64 (*a*) Rho-theta navigation, (*b*) rho-rho navigation, (*c*) rho-rho-rho navigation, (*d*) theta-theta navigation.

concept utilized in rho-rho-rho systems. Theta-theta systems provide an unambiguous position fix by measuring bearing from two landmarks. Uncertainty in bearing is a function of the angle between the landmarks, where optimal angular separation between landmarks is 90°.

5.51 Calculation of Aircraft Position

A range measurement ρ to a single radio station can be expressed as:

$$\rho = |R - A| + \epsilon \tag{5.167}$$

where R and A are the position vectors of the radio station and the aircraft respectively; ϵ is the error in the measurement which is unique to the type of deployed system. For a rho-theta system, the 2D aircraft position is determined relative to the radio station, expressed by the vector X_{Rel} and is given by:

$$X_{\text{Rel}} = \rho \begin{bmatrix} \cos\theta \\ \sin\theta \end{bmatrix} \tag{5.168}$$

where ρ is the range measurement, and θ is the relative bearing measurement. Given the fact that the radio station location is precisely surveyed a priori, the relative aircraft position can then be computed in the desired reference frame. For a rho-rho system, equation (5.167) is expanded and applied to two range measurements. For a ground-based system and an airborne user, a single range measurement, such as from a distance measuring equipment (DME), constrains the aircraft position on a hemispherical surface. The availability of two range measurements constrains the position solution to lie on the intersection of two hemispherical surfaces (Figure 5.65).

A range measurement to a third radio station (rho-rho-rho) will provide a third equation to unambiguously solve for x_A, y_A, z_A. Given three or more range measurements, a system of equations can be formulated:

$$\rho_1 = \sqrt{(x_{R1} - x_A)^2 + (y_{R1} - y_A)^2 + (z_{R1} - z_A)^2} \tag{5.169}$$

$$\rho_2 = \sqrt{(x_{R2} - x_A)^2 + (y_{R2} - y_A)^2 + (z_{R2} - z_A)^2} \tag{5.170}$$

$$\rho_n = \sqrt{(x_{Rn} - x_A)^2 + (y_{Rn} - y_A)^2 + (z_{Rn} - z_A)^2} \tag{5.171}$$

which constitutes a set of nonlinear equations that is typically solved in practice by a least-squares approach or the implementation of a Kalman filter.

Theta measurements can be either based on a ground station providing angular information, such as in the case of aeronautical VHF omnidirectional radio range (VOR), or exploiting

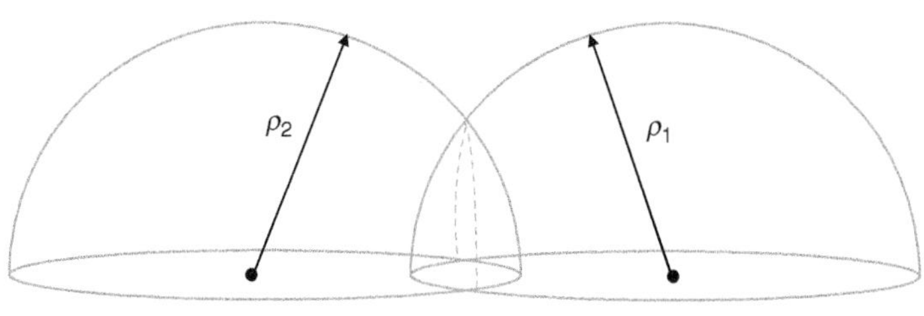

FIGURE 5.65 Rho-rho navigation system.

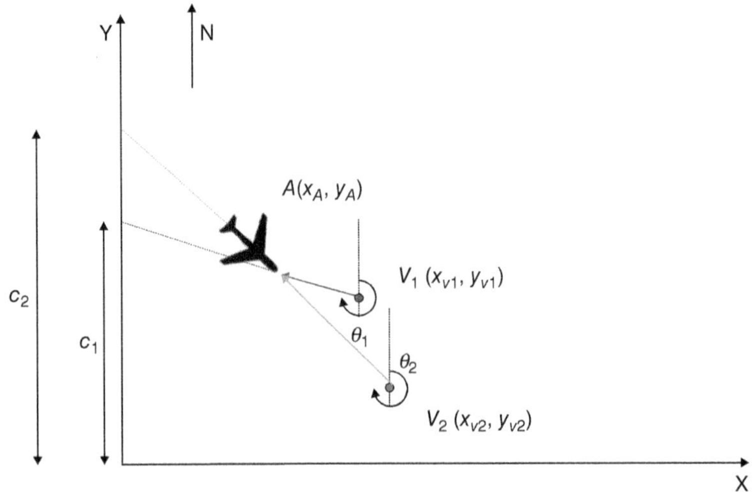

FIGURE 5.66 Computation of position using two theta-theta NAVAIDs.

angle-of-arrival (AOA) measurements. The LOPs for this type of ground-based system are straight lines radiating from the ground station, much like the spokes of a wheel. The point of intersection of the LOPs of two theta NAVAIDs is the aircraft position. A simplified scenario is illustrated in Figure 5.66, where a Cartesian coordinate reference frame describes the 2D position of two VOR stations V_1 and V_2, and the unknown coordinates of the aircraft A.

The aircraft position coordinates satisfy the equations describing the LOPs from the ground stations:

$$y_A = m_1 x_A + c_1 \tag{5.172}$$

$$y_A = m_2 x_A + c_2 \tag{5.173}$$

where m_1 and m_2 are the slopes of the two LOPs, computed from the measured bearings by

$$m_1 = \tan(270 + \theta_1) \tag{5.174}$$

$$m_2 = \tan(270 + \theta_2) \tag{5.175}$$

c_1 and c_2 are the y-intercepts of the two LOPs that can be computed using the bearing measurements and the known coordinates of the two ground stations:

$$c_1 = y_{v1} - m_1 x_{v1} \tag{5.176}$$

$$c_2 = y_{v2} - m_2 x_{v2} \tag{5.177}$$

Time of Arrival

Time of arrival (TOA) relies on range (rho) measurements. In 2D, TOA involves determining the position of the receiver based on the intersection of three or more circles whose radii being the distance of the respective transmitter from the receiver (Doukhnitch and Salamah 2008;

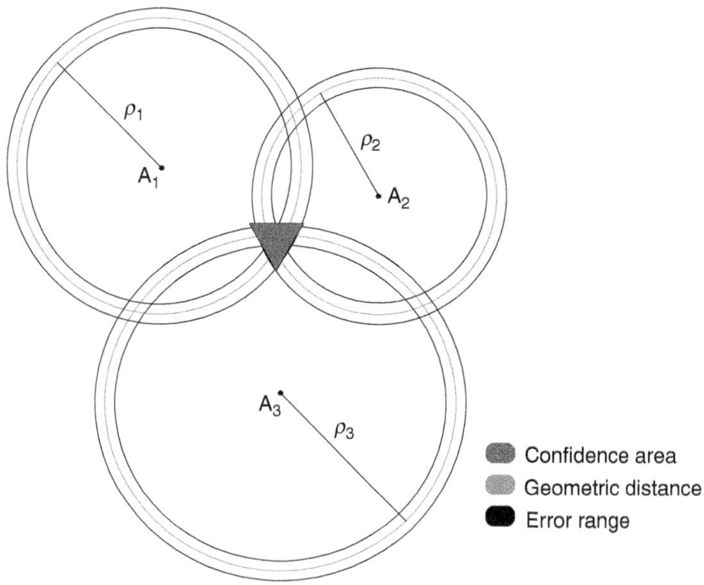

Figure 5.67 Time of arrival.

Dobbins 2011). Imperfect readings lead to a region of uncertainty in receiver location, as illustrated in Figure 5.67. Transmit time and transmitter position and movement (e.g., ephemeris) data is required to resolve the position of the receiver (Fisher 2005). Clock synchronization of transmitters and receiver is most often a challenge but can be typically resolved by adding an additional transmitter. The coordinates of the receiver can be calculated using trilateration, by solving the system of nonlinear equations by least squares algorithms for an overdetermined system. In three dimensional cases, the receiver coordinates are calculated from intersection of spheres, with the transmitters located at their respective centers.

Time Difference of Arrival

A different working principle is applied to determine position in time difference of arrival (TDOA) measurements, which is based on rho-rho-rho or multilateration. Two opposite configurations of TDOA are possible: the first involves a receiver on the platform and multiple transmitters at known fixed/moving locations. The second involves a transmitter located on the platform and multiple receivers at known fixed or moving locations. Typical examples of the first configuration include the long-range navigation (LORAN) and omega systems, whereas a typical example of the second configuration is the aeronautical multilateration (MLAT) system, used both to assist aircraft in distress and to manage traffic on the airport surface. In the first TDOA configuration, hereby called hyperbolic navigation, the receiver measures the time difference between when a signal transmitted from a primary reference transmitter (designated a "master" station) and another signal transmitted from a secondary station arrives at the receiver. The locus of points having the same value of time difference between the master transmitter M and secondary transmitters S1 and S2 forms a hyperbolic LOP as shown in Figure 5.68, and this explains the name given to this technique.

Hyperbolic positioning is used in E911 and was used in a number of ground-based long-range navigation systems till they were discontinued (Niculescu and Nath 2003). The TDOA is calculated, which, assuming no propagation delays, is a hyperbola, as illustrated in Figure 5.69. Assuming

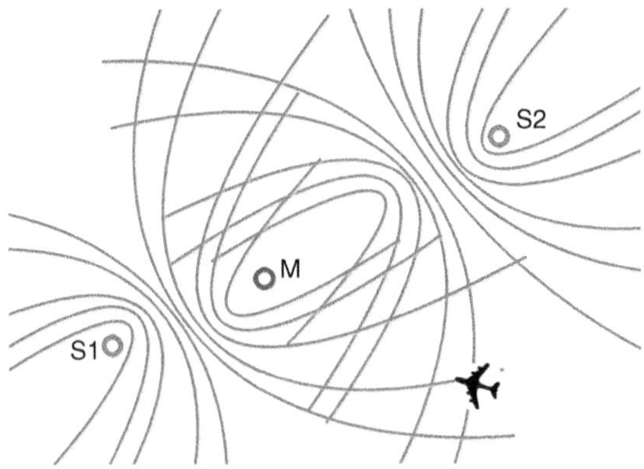

Figure 5.68 Hyperbolic LOP of M-S1 and M-S2 station pairs.

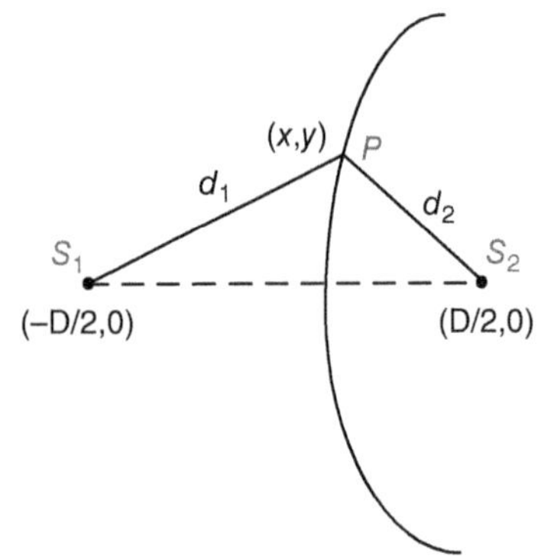

Figure 5.69 TDOA calculation geometry.

stations S1 and S2 are both located on the x-axis, the hyperbolic function can be expressed as (Niculescu and Nath 2003):

$$d_1 = \sqrt{y^2 + \left(x + \frac{D}{2}\right)^2} \tag{5.178}$$

$$d_2 = \sqrt{y^2 + \left(x - \frac{D}{2}\right)^2} \tag{5.179}$$

$$\Delta d = d_2 - d_1 = \sqrt{y^2 + \left(x - \frac{D}{2}\right)^2} - \sqrt{y^2 + \left(x - \frac{D}{2}\right)^2} \tag{5.180}$$

Equation (5.180) can be rewritten as

$$\frac{x^2}{\Delta d^2/4} - \frac{y^2}{D^2/4 - \Delta d^2/4} = 1 \tag{5.181}$$

In the second configuration (single transmitter and multiple receivers), one of the receivers is designated as a reference, which pairs with other receivers in calculation of transmitter location. So, for n available receivers, $n - 1$ pairs can be formed and emitter position is calculated from $n - 1$ hyperbolae. Usually, it requires $n + 1$ receivers to calculate transmitter location, n being the dimension of the coordinates. So, for a 2D case, three receivers are needed to calculate the location of the transmitter, provided the three receivers are not collinear (Fletcher et al. 2007). TDOA is an absolute navigation technique and requires lesser payload, does not require synchronized transmitter and receiver clocks, knowledge of transmit time or location of transmitter. However, TDOA computations require sharing of data between receivers. Research is ongoing on TDOA using signals of opportunity for various navigation applications (Fisher 2005; Yan and Fan 2008; Martin et al. 2009; Layh et al. 2014).

Angle of Arrival

Angle of arrival (AOA) is defined as the angle between the propagation direction of an incident wave and a reference direction, called orientation. Orientation is a fixed direction represented in degrees in a clockwise direction from the North. AOA is absolute when the orientation is 0° or pointing to the north, otherwise is relative. AOA measurements can be made by using an antenna array on each receiver node (Peng and Sichitiu 2007). AOA measurements can also be made using an antenna triad. Comparison of the received signal strength (RSS) to the three antennas is used to compute AOA (Fisher 2005). Localization in AOA can be solved by triangulation. Triangulation is similar to trilateration, except the receiver determines the angles of the received signals from transmitters, instead of distance measurements of transmitters from receivers in trilateration (Niculescu and Nath 2003). The triangulation problem in AOA can be reduced to trilateration by some transformations. In a 2D case, if a receiver R measures the angle to a pair of transmitters T1 and T2, it can be inferred that the position of R is somewhere on the circle determined by the measured angle and the position of the two transmitters, as illustrated in Figure 5.70.

Hence, the transformed triangulation problem involves x, y, x_0, y_0, and radius of the circle as the distance. Thus, a triangulation problem of size n is reduced to a trilateration problem of size $\binom{n}{2}$. Given (x_i, y_i), the coordinates of the ith transmitter and d_i, the distance of the ith transmitter to the receiver, the equation of the nonlinear system is

$$(x - x_i)^2 + (y - y_i)^2 = d_i^2, i = 1, \ldots, n \tag{5.182}$$

Linearizing the equation by subtracting one equation from the rest

$$2x(x_i - x_1) + 2y(y_i - y_1) = d_1^2 - d_i^2 + x_i^2 - x_1^2 + y_i^2 - y_1^2, i = 2, \ldots, n \tag{5.183}$$

The equations can be solved by least squares approximation (LSA) for an overdetermined system. Instead of using three antennas, three or more transmitters, with prior knowledge of transmitter

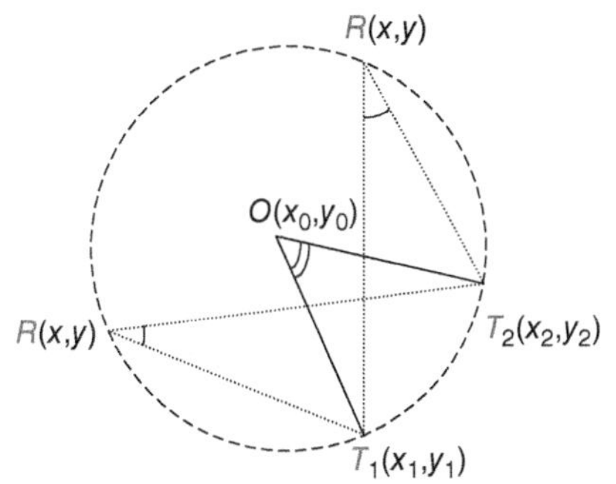

FIGURE 5.70 Angle of arrival.

power and path loss model, can be used to compute the distance of receiver from the transmitter based on RSS (Nasipuri and Li 2002). However, angular measurements are more prone to errors than TDOA measurements (Gustafsson and Gunnarsson 2003) and the accuracy of final position estimate reduces rapidly as the receivers move away from the source (Martin et al. 2009).

5.52 Air Navigation and Landing Aids

Various radio navigation techniques have been developed over the years, specifically catering to aviation requirements. Air navigation services are required for pilots to accurately determine their position with reference to ground references or geodetic data. Landing aids provide precise approach guidance to the active runway and they are categorized as nonprecision navigational aids (NDB, DME, and VOR) and precision navigational aids (ILS, MLS, and GBAS). The earliest form of electronic navigation for aviation purposes was terrestrial, or Earth-based, which relied on ground stations. Some of the ground-based radio navigation aids for aircraft are discussed below.

Direction Finder

Direction finder (DF) represents the earliest use of radio signals for navigational purposes. Direction finder can be classified as either ground-based or airborne. A loop aerial is rotated to find minimum signal strength. To eliminate ambiguity, combined signals from the loop and a whip antenna, which forms a cardioid (Figure 5.71), is used. Two orthogonally mounted loops may be used to eliminate moving parts, which, when used in conjunction with a rotating coil, forms a goniometer. Both ground waves and line-of-sight waves can be used in direction finder. The range of direction finder can vary from 200 NM at 200 kHz to 500 NM at 1.6 MHz, with accuracies of ±5°. The frequency range of direction finder for civilian use is between 118 and 156 MHz whereas, in military use, it varies from 225 to 400 MHz.

VHF Omnidirectional Range

The VHF omnidirectional range (VOR) is a terrestrial radio navigation system operating in the 108- to 118-MHz band, which is below the frequency band for VHF communications. Since its inception, the number of available channels has increased due to a reduction in channel spacing from 100 to 50 kHz. By 1946, VOR had become the U.S. standard and was later adopted by the International

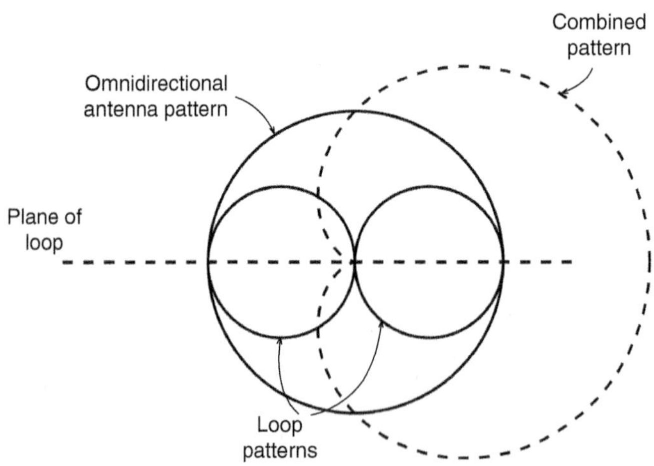

FIGURE 5.71 Loop and sense antenna patterns in direction finder.

Civil Aviation Organization (ICAO) as an international standard. The operational frequency enjoys high level of immunity from propagation and atmospheric effects. The transmitted power from ground stations is 200 W and the stations with the same frequency are placed at least 150 nm apart. Each VOR station transmits:

1. A cardioid pattern that rotates at 30 rps, resulting in a 30-Hz amplitude modulated sine wave at the aircraft receiver.
2. An 9,960-Hz omnidirectional signal that is frequency-modulated with a fixed 30-Hz tone, which serves as a reference.
3. A morse-code modulated 1,020-Hz identity tone.
4. An optional voice modulation.

The fundamental working principle is fairly simple: the relative phase between the amplitude-modulated signal and the frequency-modulated reference signal varies as a function of azimuth. The phase difference is used by airborne VOR receivers to compute heading with respect to magnetic north.

The equation for the entire VOR signal is:

$$E(t,\theta,R)=\left[\frac{E_o\cos(2\pi f_c t)}{\mathrm{R}}\right]$$
$$\{1+0.3\sin(60\pi t+\theta)+0.3\cos(19{,}920\pi t-16\cos(60\pi t))+0.1\cos(2040\pi t)\} \tag{5.184}$$

where E is the electric field intensity and f_c is the RF carrier frequency.

A schematic of the airborne VOR receiver is shown in Figure 5.72. The 30-Hz waveform is extracted and fed to the phase comparator. The 9,960-Hz data is filtered and limited to remove the

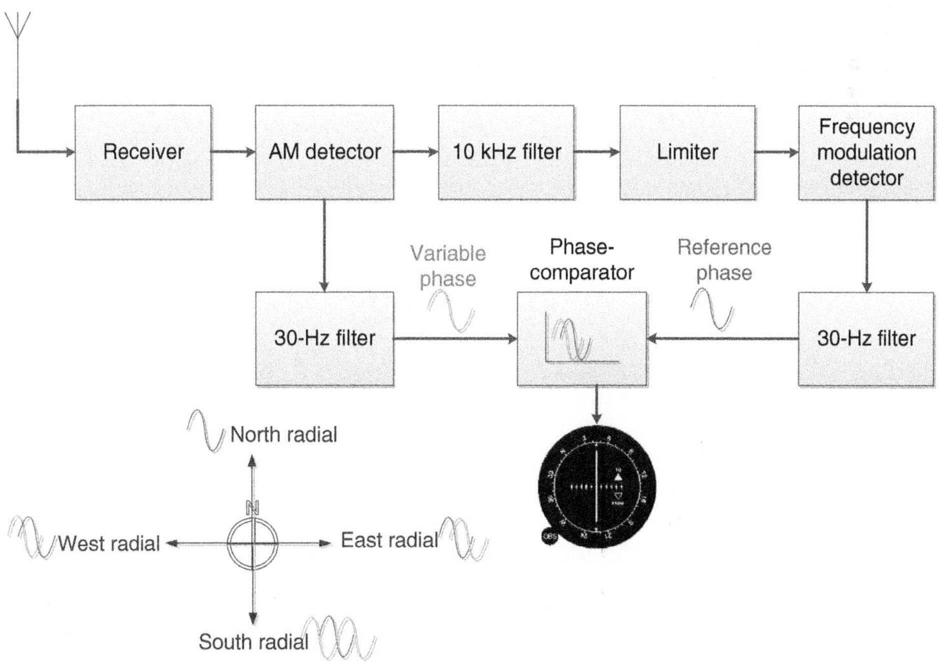

FIGURE 5.72 VOR receiver schematic.

30-Hz amplitude modulation. Feeding this signal to the frequency discriminator results in the 30-Hz reference frequency, which is then compared with the variable phase after filtering.

The phase difference is used to compute and display aircraft bearing on the instrument panel, specifically, to the course deviation indicator (CDI), or a horizontal situation indicator (HSI). The LOP of a VOR system is a radial from the ground station. The VOR station produces a total of 360 radials with a spacing of 1°. It is practically free from static interference and is unaffected by sky-waves, rendering it usable during the day and at night. The accuracy of the bearing computed from VOR is limited by errors arising from the airborne or ground-based equipment, polarization errors due to signal reflection from large obstacles, which can change the polarization of the signal from horizontal to vertical. Signal reflection and errors due to propagation are more prominent in first-generation VOR, prompting the development of Doppler-VOR (DVOR), which comprises a wide-base antenna array. The variable-phase signal in DVOR is frequency-modulated as opposed to amplitude-modulated, rendering it more robust to interference. The variable-phase signal is electronically rotated at 30 rps, inducing a Doppler shift that is measured to compute the relative bearing. Prior to the advent of GNSS, routes were defined as segments between VOR stations, or between waypoints defined at the intersection of LOPs of two VOR stations.

Distance-Measuring Equipment

Distance-measuring equipment (DME) is an ultra-high frequency (UHF) radio navigation system that measures the slant range (rho) (Figure 5.73) between the aircraft and a ground station transponder. The computation of the range is performed by the measurement of the transit time of a pulse from the aircraft to the ground station, and the subsequent transmission of an identical signal from the ground station back to the airborne receiver.

DME is essentially a form of secondary surveillance radar operating in the 962–1,213 MHz range, wherein the ground return is triggered by the aircraft. For most aerial navigation applications, the difference between the slant range and the horizontal range (plan range) is negligible. The largest error in the measurement occurs when the aircraft passes directly overhead, at which point the DME-computed range will be the aircraft altitude.

Distance is displayed digitally from 0.0 to 99.9 NM. A DME channel consists of two-carrier frequencies that are 63 MHz apart. The system operates in two modes: search mode and tracking mode. When in search mode, the airborne unit transmits a pulse pairs whose repetition frequency is random or *jittered* (jittered pulse repetition frequency—JPRF), thereby differentiating its pulse-train from those of other aircraft. The pulse rate during the search is at 150 pulse pairs per second. The ground transponder transmits an identical pulse train (with an identical pulse-repetition

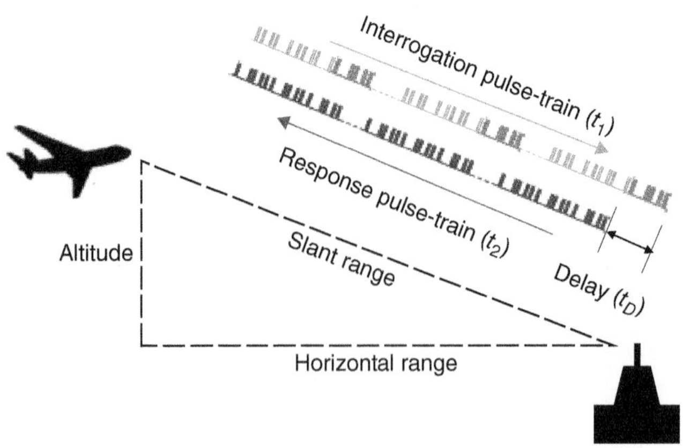

FIGURE 5.73 DME operating principle.

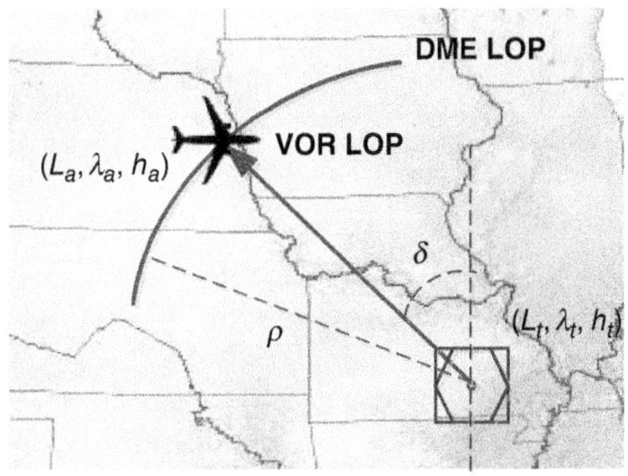

FIGURE 5.74 Collocated VOR/DME ground station.

frequency) on a carrier frequency which is either 63 MHz below or above the frequency of the wave transmitted from the aircraft. This response signal is intentionally delayed by an interval $t_D = 50$ μs from the time the interrogation pulse arrives at the ground transponder. When the matching pulses from the ground transponder is received by the airborne unit, lock-on is achieved, the pulse rate is reduced to 25 pulse pairs per second and the DME enters tracking mode. As the aircraft moves and the range to the ground station varies, its receiver-gates shift to accommodate the corresponding increase or decrease in time between transmission and arrival of the response from the ground station, and remain centered on the delay of the pulse pair.

The slant range S is easily calculated by (referring to Figure 5.74):

$$S = \frac{(t_1 + t_2 - t_D)c}{2} \tag{5.185}$$

where t_1 is the propagation time of the interrogation pulse from the airborne unit to the ground station; t_D is the delay before the ground station transmits the response pulse which is equal to 50 μs; t_2 is the propagation time of the response pulse from the ground station to the airborne transponder; c is the speed of the carrier wave $(c = 3 \times 10^8 \text{ m/s})$.

A DME station can service up to 100 aircraft at a given time, at a maximum range of 150 NM, which drops to 75 NM below 20,000 ft. Under certain circumstances, like aircraft maneuvers, change of frequency (another DME is selected) or loss of line of sight, tracking can be lost, which leads to receiver continuing to search in "nearby windows" for up to 10 seconds before reverting to search mode.

VOR and DME stations are typically collocated to form the standard ICAO-approved rho-theta NAVAID providing LOS range and bearing measurements (Figure 5.66) for an unambiguous computation of position.

Given the aircraft altitude h_a (based on the altimeter reading), the aircraft latitude and longitude (L_a, λ_a) can be approximated by:

$$L_a = L_t + \frac{\sqrt{\rho^2 - (h_a - h_t)^2}\cos(\delta + \alpha)}{R_N + h_t} \tag{5.186}$$

$$\lambda_a = \lambda_t + \frac{\sqrt{\rho^2 - (h_a - h_t)^2}\sin(\delta + \alpha)}{(R_E + h_t)\cos L_t} \tag{5.187}$$

where (L_t, λ_t, h_t) are the known latitude, longitude, and altitude of the VOR/DME station; ρ is the DME range measurement; δ is the VOR bearing; α is the magnetic declination. R_N and R_E are the meridian radius of curvature and the transverse radius of curvature, respectively, which are both functions of the latitude.

Power (W)	25	50	50–1,000	2,000+
Range (NM)	15	25	50	75–200

Table 5.17 Variation of Nondirectional Beacons Range with Power

Nondirectional Beacons

The use of low- and medium-frequency airborne direction finders has led to omnidirectional ground transmitters being designed specifically to use with them. Nondirectional beacons (NDBs) are used mainly for location of instrument landing system approach path and in nonprecision approach (Table 5.17). It has a single vertical antenna operating in 200- to 1,600-kHz range, with output power ranging from 20 W up to several kilowatts. NDB antenna has a "null cone" vertically above it that allows aircraft flying over the antenna to get a "fix," with uncertainty being a function of height. Transmitter power determines the range of beacons. Although NDBs are low cost, they suffer from reflections in mountainous areas.

Tactical Air Navigation

Tactical air navigation (TACAN) is a military rho-theta distance measurement system that uses the same pulses and frequencies as the standard DME system. In addition to a constant-duty-cycle DME beacon and antenna, a TACAN beacon comprises of additional parasitic elements rotating around the antenna for improving bearing accuracy. All DME/TACAN beacons and interrogators are interoperable among each other. Both distance and bearing in TACAN are obtained over the same radio frequency. There are nine parasitic elements mounted 40° apart, which superimpose a 135-Hz amplitude modulation on the transmitted signal. The bearing reference pulses are added to the constant-duty cycle DME beacon in the transponder. At any instant of time, TACAN system can measure distance of 110 aircraft, while allowing for unlimited azimuth measurement. A TACAN system block diagram is shown in Figure 5.75.

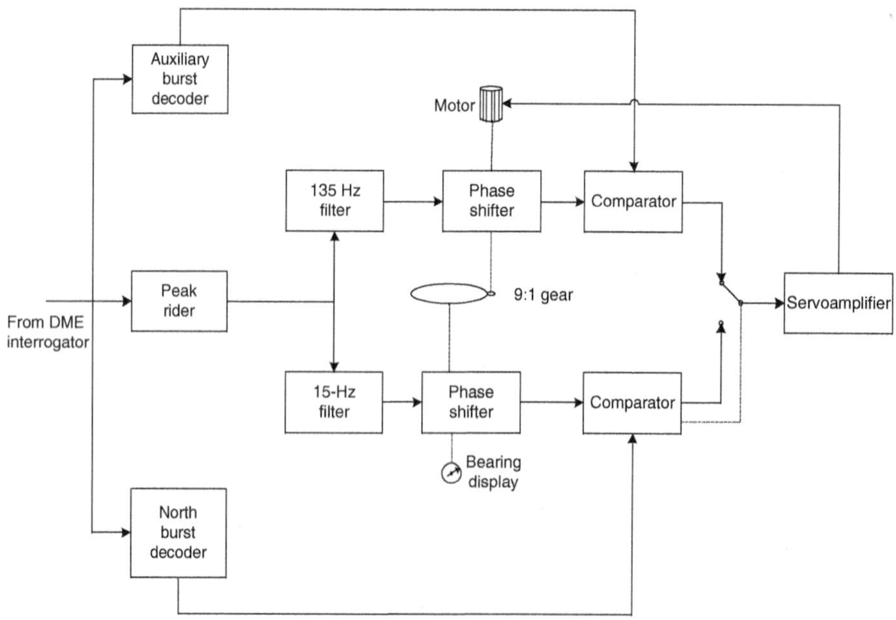

Figure 5.75 TACAN system block diagram.

Long-Range Navigation (Loran)

Loran is a hyperbolic radio-navigation system that was developed just before the outbreak of World War II in Europe. As Loran systems evolved, Loran-C was developed for navigation users and precise time and time interval (PTTI) users. Loran-C is a low-frequency radio-navigation aid with an operating frequency of 90–110 kHz. The pulse transmitted by the Loran-C stations is a 100-kHz carrier, which forms an envelope of the signal, increasing and decaying amplitude at a specified rate. The low-frequency signal propagation is affected by properties of the Earth's surface as well as the ionosphere.

Instrument Landing System

The instrumental landing system (ILS) was developed to help the pilot during landing in conditions of reduced visibility. The components of the ILS are shown in Figure 5.76. These include the following:

- Localizer provides guidance in the horizontal plane (track) along the extended centerline of the approach path to the runway. It comprises a transmitting antenna array, which is typically positioned at the far end of the runway and transmits two highly directional RF beams on the VHF band between 108.10 and 111.95 MHz, which are, respectively, modulated at 90 Hz and 150 Hz. The beams overlap such that the received signal is equally modulated along the centerline of the runway. There is a difference in depth of modulation that is proportional to the deviation from the runway centerline that drives the guidance. The total beamwidth is approximately 5° wide.
- Glide slope provides vertical guidance (flight path angle) to the runway touchdown point after the localizer signal has been acquired and the aircraft is on the approach to the runway. The radiant system is located on the side of the runway, at a distance of 125–170 m off its axis, and consists of dipole systems. The control unit is installed in the control tower. The parameters of the signals radiated are constantly tested by a triple control system that

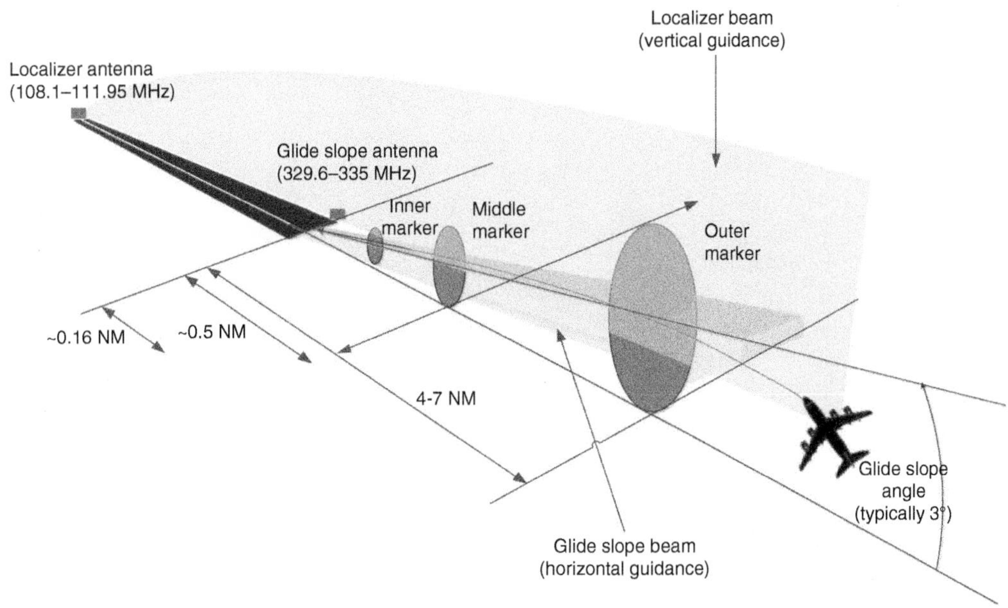

Figure 5.76 Instrument landing system.

receives the signals captured by dipoles arranged in an appropriate position. Glide slope units operate on frequencies in the band 328.6–335.4 MHz combined with the localizer frequencies. The glide path generally has an inclination of 3°, but this can be adjusted between 2° and 4° to suit the local needs of the airport/runway.

- The marker beacons provide an indication of the distance from the runway threshold. Each station includes two transmitters, a control device and a power supply grouped into one unit. The frequency of operation is 75 MHz.
- Approach lights.

Microwave Landing System

The microwave landing system (MLS) is a system of radio aid for the approach and landing of aircrafts created to replace the ILS. In fact, despite numerous upgrades that yielded a high level of affordability, ILS maintains the basic restrictions that do not they manage to meet new operational requirements that have arisen over the years. To this end and driven by voluntary flexible systems to be developed and equipped with higher performance, from 1975 the Federal Aviation Administration (FAA) americana and subsequently took into consideration the possibility of ICAO, use of systems based on the use of microwaves. Among the many identified systems was selected the MLS who possessed the following characteristics:

- low sensibility to the morphological characteristics of the site
- Low sensibility to the reflections of the surrounding environment
- 200 channels available
- High flexibility for adaptation of future needs

The MLS has much broader coverage than the ILS (Figure 5.77). In addition, the characteristics of the MLS system can pinpoint the most ideal paths down to the track (not necessarily straight), allowing the execution of both curved and segmented approaches.

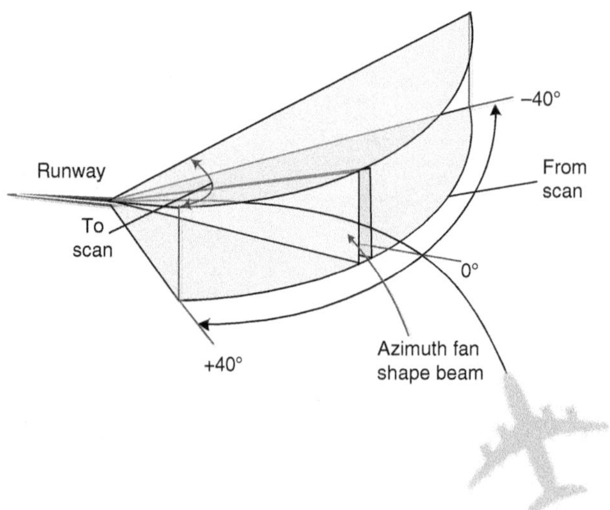

Figure 5.77 Microwave landing system.

The components of an airport MLS are

Azimuth (AZ) subsystem: The subsystem AZ is functionally similar to a localizer ILS but has a more extensive sectoral coverage for proportional guidance from a minimum of ±10° up to ±60°. For this reason, in some cases, it is possible to provide approach guidance for a single curved approach (for individual tracks or heliports).

Elevation (EL) subsystem: The function of an EL subsystem is similar to the ILS glide slope unit but unlike this, the unit allows a wide selection of downhill corners EL up to about 20°. EL signal coverage extends over the whole field azimuthal.

Precision distance measuring equipment (P/DME) subsystem: It provides continuous information similar to the discrete indications provided by marker beacons in the ILS system. The P/DME has an accuracy of ±30 m, an order of magnitude better than DME.

Back-azimuth (BK-AZ) subsystem: It is identical to the azimuth subsystem and has functions similar to that of the ILS localizer back-course. Since it is used for missed approaches and for take-off, this subsystem supports large aircraft converging on the runway in a sequence.

References

Dobbins, R., 2011, "Software Defined Radio Localization Using 802.11-Style Communications," Naval Research Laboratory.

Doukhnitch, E. and Salamah, M., 2008, "General Approach to Simple Algorithms for 2D Positioning Techniques in Cellular Networks," *Computer Communications*, Vol. 31, no. 10, pp. 2185–2194.

Fisher, K. A., 2005, "The Navigation Potential of Signals of Opportunity-Based Time Difference of Arrival Measurements," AFIT/DS/ENG/05-02, DTIC document.

Fletcher, F., Ristic, B., and Mušicki, D., 2007, "Recursive Estimation of Emitter Location Using TDOA Measurements from Two UAVs, FUSION 2007, 10th International Conference on Information Fusion.

Gustafsson, F. and Gunnarsson, F., 2003, "Positioning Using Time-Difference of Arrival Measurements," Acoustics, Speech, and Signal Processing, 2003. Proceedings (ICASSP'03), 2003 IEEE International Conference on Acoustics, Speech and Signal Processing.

Layh, T., Larson, J., Gebre-Egziabher, D., et al., 2014, "GPS-Denied Navigator for Small UAVs."

Martin, R. K., Velotta, J. S., and Raquet, J. F., 2009, "Bandwidth Efficient Cooperative TDOA Computation for Multicarrier Signals of Opportunity," *IEEE Transactions on Signal Processing*, Vol. 57, no. 6, pp. 2311–2322.

Nasipuri, A. and Li, K., 2002, "A Directionality Based Location Discovery Scheme for Wireless Sensor Networks," *Proceedings of the ACM International Workshop on Wireless Sensor Networks and Applications*.

Niculescu, D. and Nath, B., 2003, "Ad Hoc Positioning System (APS) Using AOA," INFOCOM 2003. *22nd Annual Joint Conference of the IEEE Computer and Communications*. IEEE Societies.

Peng, R. and Sichitiu, M. L., 2007, "Angle of Arrival Localization for Wireless Sensor Networks," *3rd Annual IEEE Communications Society on Sensor and Adhoc Communications and Networks*, Secon 2006.

Yan, C. and Fan, H. H., 2008, "Asynchronous Differential TDOA for Non-GPS Navigation Using Signals of Opportunity," *IEEE International Conference on Acoustics, Speech and Signal Processings*, ICASSP.

PART 11

Inertial Navigation Systems

Suraj Bijjahalli, Roberto Sabatini, and Subramanian Ramasamy

5.53 Introduction

Accurate knowledge of the aircraft state vector in terms of its position, velocity, and attitude with respect to the navigation reference frame (bank angle, pitch, and track) is crucial for aircraft safety in all weather conditions. Onboard avionic systems utilize a combination of dear reckoning (DR) and position fixing for accurate and reliable navigation. This part introduces the fundamentals of inertial measurement units (IMUs), the inertial navigation system (INS), vision-based systems, and integration architectures. Fundamental calculations of navigation are briefly described prior to explaining individual navigation sensors. A simple flat Earth representation of the range and bearing between two points is shown in Figure 5.78 (Allerton 2009).

Steering from position A to position B requires description of the distance D between the two points described here in Cartesian coordinates:

$$D = \sqrt{(x_B - x_A)^2 - (y_B - y_A)^2} \tag{5.188}$$

In practice, however, positions of the aircraft and of ground-based navigation aids and of aerodromes are specified using spherical geometry, i.e., by latitude (ϕ) and longitude (λ). The distance between the two points expressed in radians is given by

$$\delta = \sqrt{(\phi_B - \phi_A)^2 \cos\phi_A \cos\phi_B (\lambda_B - \lambda_A)^2} \tag{5.189}$$

The distance between A and B on the surface of the Earth is then

$$D = R\delta \tag{5.190}$$

where R is the radius of Earth. The relative bearing from point A to point B is given by

$$\chi = \tan^{-1}\left(\left|\frac{y}{x}\right|\right) = \tan^{-1}\left(\left|\frac{\sin\phi_B - \sin\phi_A \cos\delta}{\cos\phi_B \sin(\lambda_B - \lambda_A)\cos\phi_A}\right|\right) \tag{5.191}$$

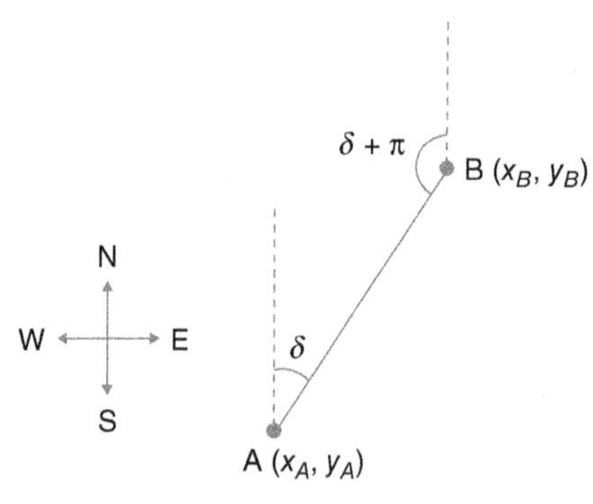

FIGURE 5.78 Range and relative bearing between two points.

5.54 Inertial Sensors

Gyroscopes

Gyroscopes exploit the property of inertia, or the resistance to a change in momentum to sense angular motion. Along with accelerometers, gyroscopes are an essential part of the attitude and heading reference system (AHRS) and inertial navigation system (INS). Spinning rotor gyroscopes have gradually been phased out owing to their mechanical complexity, poor reliability, high power consumption, and long run-up time. At present, microelectromechanical systems (MEMS) gyroscopes are widely used in industry owing to their high reliability and low cost of ownership. MEMS gyroscopes exploit Coriolis forces that are induced when a vibrating mass is subject to an angular rotation about an axis in the plane of vibration. The induced Coriolis force produces a vibration which is measured by suitable sensors to produce an output signal proportional to the applied angular rate. A schematic of a tuning fork configuration of a MEMS gyroscope is shown in Figure 5.79.

The tines of the tuning fork are subject to a driven vibration with a precise amplitude along the Y-axis by means of an oscillator. The effect of an angular rate input $\dot{\theta}$ about the X-axis induces a Coriolis response along the Z-axis, which is manifested as a torsional oscillation of the stem of the fork, wherein the amplitude of the oscillation is directly proportional to $\dot{\theta}$. The typically employed configuration is to use one tuning fork for sensing the angular rate input, and a second tuning fork, which is coupled to the first as a pick-up to produce the oscillatory output signal. The vibrating forks, their support flexures and frames are typically fabricated from wafers of single crystal piezoelectric quartz and micromachined through photolithographic processes.

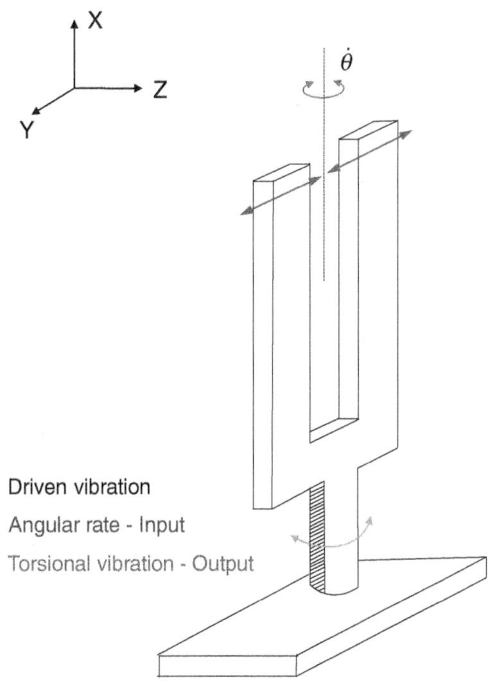

FIGURE 5.79 Schematic of a tuning-fork MEMS gyroscope.

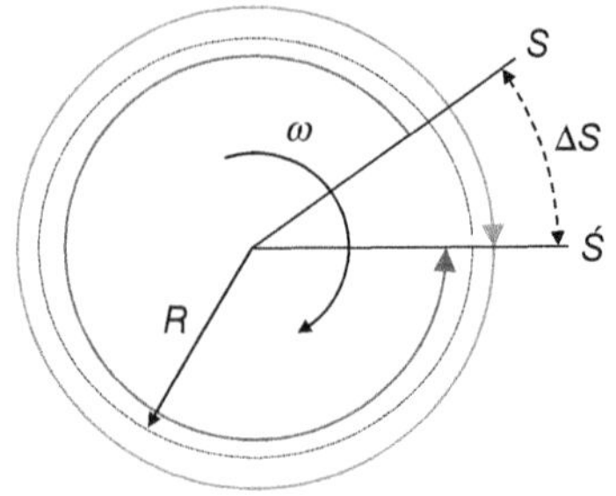

FIGURE 5.80 The Sagnac effect.

The current state of the art is represented by optical gyroscopes such as the ring laser gyroscope (RLG) and the fiber optic gyroscope (FOG), which operate on the phenomenon of the Sagnac effect, as illustrated in Figure 5.80.

Referring to Figure 5.81 a photon of light starting from point S and propagating in a clockwise direction around the circuit, and a second photon also starting from point S, but propagating in a counterclockwise direction will have the same transit time $T = 2\pi R/c$, in the absence of an input rotation rate, where R is the radius of the circular path. If an input rotation rate $\omega = \dot{\theta}$ is applied to the closed optical path, the photons will have differing transit times around the path. The Sagnac time difference ΔT between the clockwise and counterclockwise photons is presented here as

$$\Delta T = \frac{4A}{c^2}\dot{\theta} \tag{5.192}$$

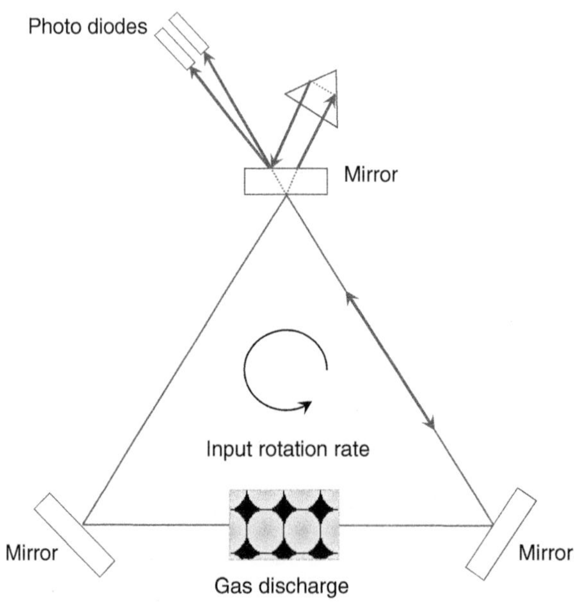

FIGURE 5.81 Ring laser gyroscope schematic.

The difference in optical path length ΔL is given by

$$\Delta L = c\Delta T = \frac{4A}{c}\dot{\theta} \tag{5.193}$$

where A is the area enclosed by the closed circuit, c is the velocity of light, and $\dot{\theta}$ is the input angular rotation rate. For a circular path, $A = \pi R^2$. The sensitivity of the sensor is, therefore, dependent on the area of the enclosed path.

The invention of the laser and the development of single mode optical fibers enabled the development of highly coherent light sources and optical paths with a large effective area, respectively. These innovations resulted in the development of the ring laser gyroscope and the fiber optic gyroscope. In an RLG, the input rotation rate is sensed by the difference in resonant frequencies of the clockwise and counterclockwise beams propagating in the laser cavity. Among the different types of FOG, the interferometric fiber optic gyroscope (IFOG) is more commonly applied in industry and operates as a Sagnac interferometer that senses the input rotation rate through measurement of the phase shift in the fringe pattern produced by interfering light beams.

A schematic of a RLG is shown in Figure 5.73. The laser cavity is formed by locating three reflecting mirrors in a triangular pattern. A helium-neon gas discharge, and a voltage source of the order of 2,000 V dc to ionize the He-Ne mixture is used to generate two laser beams propagating in opposite directions in a closed triangular path. The longitudinal mode frequency of the resonant cavity is a function of the optical path length L:

$$f = nc/L \tag{5.194}$$

where n is an integer and c is the speed of light. When an input rotation rate is applied to the laser cavity, the consequent difference in path lengths between the clockwise and counterclockwise beams causes a difference Δf in their frequencies. A fringe pattern is generated by reflecting one of the beams so that it can be combined with the other. The input rotation causes the fringe pattern to shift relative to the photodiodes at a rate and in a direction which is proportional to the difference in frequency between the beams. The outputs of the photodiodes are sinusoidal signals, which are converted to positive or negative pulses (depending on the direction of rotation) by pulse triggering circuits.

The difference in frequencies owing to a change ΔL in optical path length is

$$\Delta f = \frac{f\Delta L}{L} \tag{5.195}$$

Substituting for ΔL from equation (5.193):

$$\Delta f = \frac{4A}{\lambda L}\dot{\theta} \tag{5.196}$$

where $\lambda = c/f$ is the wavelength of the laser transition. Equation (5.196) can be rewritten as

$$\Delta f = K\dot{\theta} \tag{5.197}$$

The rotation angle θ about the gyroscope input axis over a time period T is equal to the number of pulses (positive or negative) counted in that period:

$$\int_0^T \Delta f dt = K \int_0^{\theta} d\theta \tag{5.198}$$

The laser cavity is typically machined from "CERVIT," a vitroceramic known to possess a low coefficient of thermal expansion. The mirrors are designed to have reflectivity exceeding 99.9% through a multi-dielectric coating process.

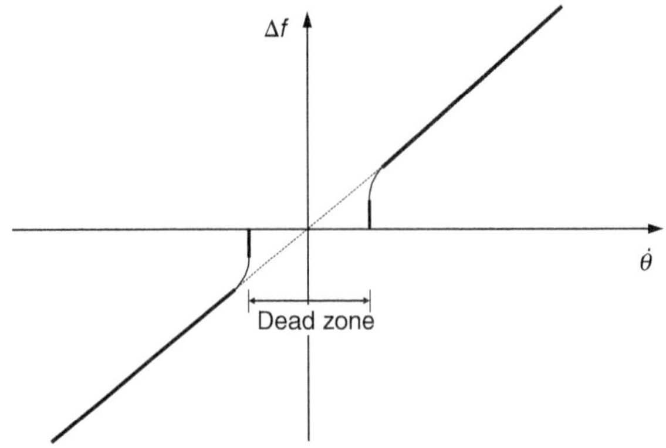

FIGURE 5.82 Lock-in zone below threshold rotation rate.

A problem to be overcome in RLGs is a phenomenon known as "lock-in." Back-scattering of a laser beam into the other due to flaws and imperfections in the mirrors can occur, which causes the frequencies of both beams to couple with one another. A null frequency difference would fail to produce a measurement of rotation rate. This problem is apparent at low input rotation rates, as shown in Figure 5.82, where rotation rates below a "lock-in rate" threshold result in a zero output.

Lock-in is commonly overcome in practice by mechanically dithering the laser cavity assembly about the input axis at a frequency of about 100 Hz with a peak velocity of about 100° per second. The amplitude of the dithering rate is chosen so as to minimize the dwell time in the lock-in zone. The dithering signal can be removed from the output through digital filtering techniques at the expense of gyroscope bandwidth. RLGs are accurate over a large dynamic range from 0.01° per hour to 400° per second, a range of approximately 10^8:1. The optical principles governing the RLG render it insensitive to acceleration. Scale factor accuracy is very high with errors occurring in the 5–10 ppm bracket. Warm-up times are negligible on start-up, and mean time between failures (MTBF) of over 60,000 hours have been demonstrated in practice.

The initial push to develop interferometric fiber optic gyroscopes (IFOG) in the 1970s was driven by the potential to deliver a cost-effective alternative to the ring laser technique. Angular motion in IFOGs is sensed by detecting the phase difference between two light beams propagating around an optical path in opposite directions. This passive interferometric technique eliminates the need for finely machined and polished reflective surfaces as in the case of RLGs. A schematic of an IFOG is shown in Figure 5.83. Light from a broad band source is passed through a beam splitter. A single optical mode from this splitter is passed through a second beam splitter and the resultant beams are propagated in opposite directions around the coil. In the absence of an input rotation

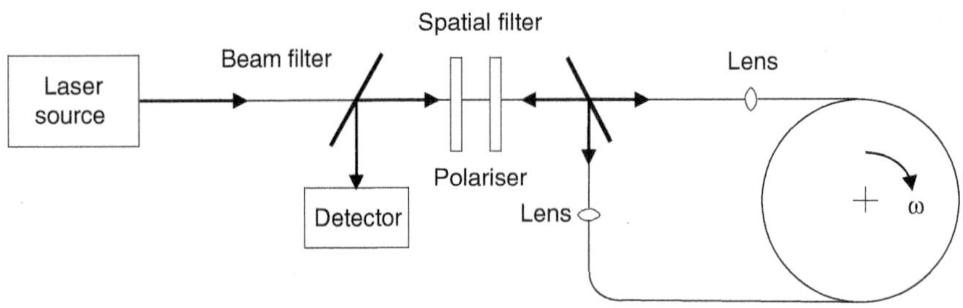

FIGURE 5.83 Schematic of an interferometric gyroscope.

rate applied to the fiber coil, the beam transit times are equal, and the fringe pattern when the light arrives back at the second beam splitter is formed by constructive interference. The intensity of the fringe pattern is detected by means of a photodiode. An input rotation rate applied to the coil causes the transit times to differ according to equation (5.193). By writing the enclosed coil area A in terms of the length L and diameter D of the coil, equation (5.193) is rewritten as:

$$\Delta T = \frac{LD}{c^2}\dot{\theta} \tag{5.199}$$

The resulting Sagnac phase shift is given by

$$\varphi = \omega\Delta T = \frac{2\pi LD}{\lambda c}\dot{\theta} \tag{5.200}$$

where $\omega = \frac{2\pi c}{\lambda}$ is the angular frequency of the light source. Referring to equation (5.200), it can be seen that the sensitivity of the IFOG is hampered owing to the occurrence of the velocity of light c in the denominator of the expression. The sensitivity is boosted by increasing the length L of the coil by increasing the number of turns. An open-loop configuration of the IFOG is simple in construction, but is particularly sensitive to any nonreciprocal effects, consequently lowering the sensitivity of the interferometer. The response of the detector in an open-loop configuration is shown in Figure 5.84.

Referring to Figure 5.84, it is readily apparent that the sensitivity of the device is low at small values of input rotation rates, owing to the "one plus cosine" curve of the response. This problem is typically addressed by incorporating a phase modulator at one end of the fiber coil to give a phase dither of 90°. Consequently, the sensitivity of the detector would then be maximum at and around small rotation rates. The response, however, would still be sinusoidal, resulting in nonlinearity for large input rotation rates. Closed-loop techniques can address the scale factor stability problems of open-loop techniques. A commonly employed method is phase-nulling, wherein a second phase-modulator operated at twice the frequency of the dither modulator is used to "back-off" the effects of the Sagnac phase-shifts caused by the rotation of the coil. The error signal in the feedback loop is the open-loop signal, which generates an additional phase-shift $\Delta\phi$, which is equal and opposite to that generated by the input rate. Thus, the device is always operated at null, which is the region of maximum sensitivity. Writing phase-shift in terms of a corresponding frequency difference, the response of the closed-loop IFOG with a single mode optical fiber with refractive index n is then given by

$$\Delta f = \frac{D}{n\lambda}\dot{\theta} \tag{5.201}$$

The velocity of light is no longer present in the response, and the output characteristic is the same as the RLG.

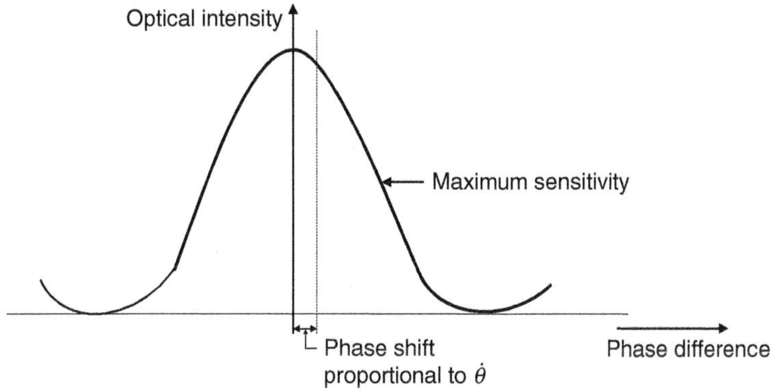

FIGURE 5.84 Open-loop detector response.

Accelerometers

Figure 5.85 is a schematic illustrating the working principle of an accelerometer. The sensor comprises a proof-mass restricted to move along the accelerometers sensitive axis by a set of springs. Deviation of the proof mass from the equilibrium position is measured by a pick-off that converts the motion into a voltage response. The assembly is enclosed by a case which is mounted on the aircraft body. The application of an external force on the case causes it to move with respect to the proof mass, thereby compressing one spring and extending the other. The case moves relative to the mass until the inertial force experienced by the mass equals the asymmetric forces applied by the springs. At this point, the acceleration of the case due to the external force matches the acceleration of the proof mass, which is proportional to the deviation from the equilibrium position.

The accelerometer measurement can be derived from the application of Newton's second law to the forces acting on the aircraft and on the proof mass. The forces acting on an aircraft of mass M comprise the following vectors: weight, propulsive thrust T, lift L, and drag D. The summation of these vectors produces a resultant acceleration a and is expressed as

$$T + L + D + Mg = Ma \tag{5.202}$$

where g is the acceleration due to gravity. Equation (5.25) can be rewritten as

$$a = \frac{T + L + D}{M} + g \tag{5.203}$$

Equation (5.203) is derived by applying a similar summation of vectors to the proof mass m of the accelerometer:

$$F_c + mg = ma \tag{5.204}$$

where F_c is the force applied by the springs to constrain the proof mass. Substituting for the acceleration from equation (5.26) and rewriting:

$$F_c = m\,\frac{(T + L + D)}{M} \tag{5.205}$$

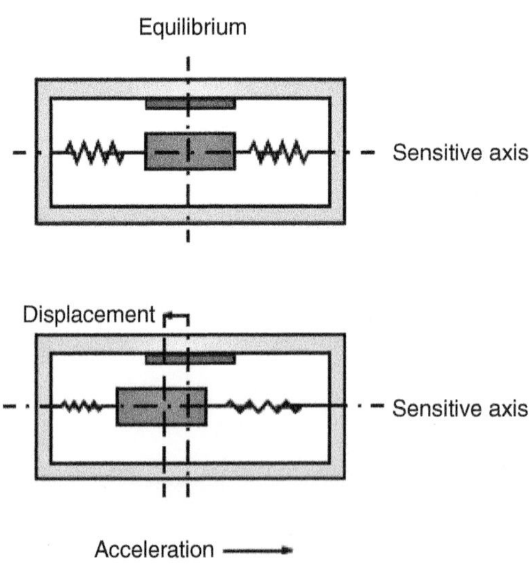

FIGURE 5.85 Schematic of an accelerometer.

The external forces per unit mass of the aircraft with the exception of gravitational force is termed the specific force, i.e.,

$$F_s = \frac{(T+L+D)}{M} \tag{5.206}$$

Thus, from equation (5.206), accelerometers can only sense the component of the specific force along its sensitive axis and not the vehicle acceleration component directly. Therefore, the magnitude and orientation of the gravitation vector relative to the accelerometer axis must be known to determine the vehicle acceleration. An exception to this rule is when the sensitive axis is exactly horizontal (i.e., there is no component of gravitational force acting along the sensitive axis), wherein the vehicles acceleration component can be directly measured. A pendulous accelerometer is shown in Figure 5.86 wherein a pendulous arm supports the proof mass, and a hinge provides damping to limit oscillation. In particular, a deflection from the null position due to an input acceleration is shown.

The spring hinge exerts a restoring torque on the pendulum which is directly proportional to the angular deflection θ. The output signal of the accelerometer is provided by the electrical pick-off measuring the deflection from the null position. The pendulosity p of the accelerometer is a parameter defined as the product of the proof mass with the distance between the hinge and the center of gravity (CG):

$$p = ml \tag{5.207}$$

The torque acting along the sensitive axis on the application of an input acceleration a is:

$$T = mal = ap \tag{5.208}$$

The restoring torque $\dot{T}$ exerted by the spring hinge is proportional to the spring stiffness and the angular deflection:

$$\dot{T} = k\theta \tag{5.209}$$

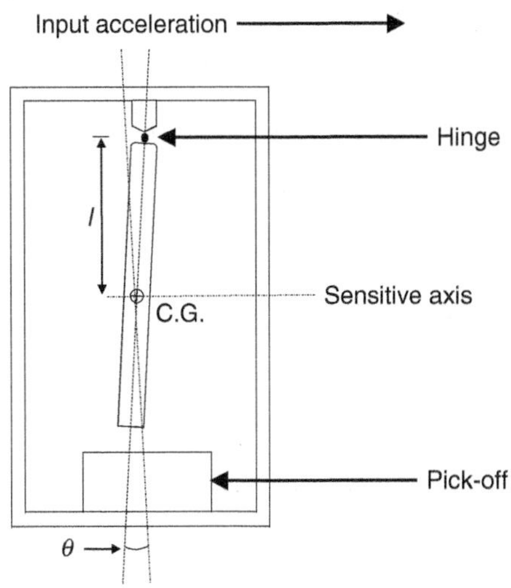

FIGURE 5.86 Pendulous accelerometer.

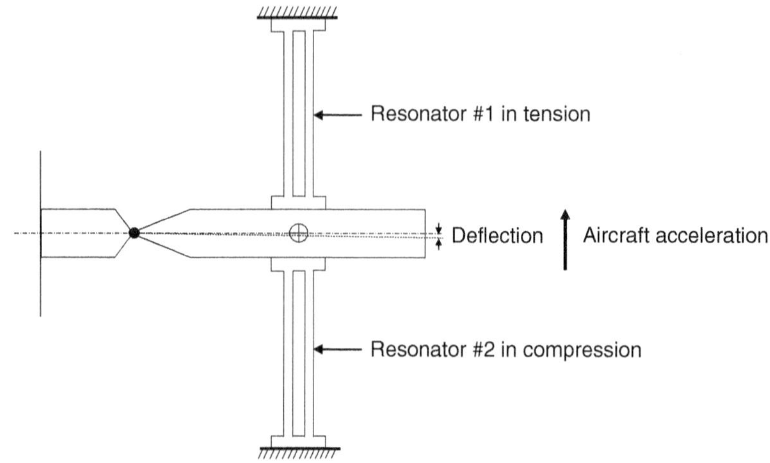

FIGURE 5.87 Schematic of a vibrating beam accelerometer.

Therefore, the input acceleration can be determined as

$$a = \frac{k}{p}\theta \tag{5.210}$$

The transfer function of the accelerometer is described by a quadratic lag filter and is given by

$$\frac{\text{Output}}{\text{Input}} = \frac{K_0}{s^2 + 2\xi\omega s + \omega^2} \tag{5.211}$$

where K_0 is the sensitivity or scale-factor of the accelerometer. ω is the undamped natural frequency of the hinge spring-proof mass system, and ξ is the damping coefficient.

Vibrating beam accelerometers (VBAs) measure acceleration based on the principle that the frequency of a vibrating beam can be altered by subjecting it to an external load that induces tensile or compressive stresses. The schematic of a VBA is shown in Figure 5.87.

A VBA typically consists of a pendulous proof mass supported by two vibrating beams that restrain the motion of the mass along the sensitive axis. An input acceleration along the sensitive axis causes a deflection of the proof mass, which in turn, loads the vibrating beams, thereby modulating the frequencies of their vibrations. A tensile stress induced in a beam increases the frequency of its vibration, whereas a compressive stress reduces it. The acceleration is determined by differencing the two frequencies. The vibrating beams are predominantly fabricated from quartz owing to its low cost and stability over time and over a range of temperature.

The accelerometers described so far are open-loop accelerometers wherein the displacement of the proof mass due to an input acceleration is measured. Inherent limitations of these designs include the high nonlinearity in the relationship between acceleration and the induced displacement (for large displacements). Additionally, the susceptibility to cross-coupling errors makes it necessary to limit deflection to about $\pm 2°$. Owing to these factors, open-loop accelerometers are suited for applications where accuracy requirements are in the order of 1–2%. These limitations are typically overcome by using closed-loop accelerometers such as the force-feedback accelerometer shown in Figure 5.88.

Contrary to open-loop accelerometers, the proof mass is maintained in a fixed position and the force necessary to maintain that position is measured. Deviation from the null position is therefore, minimal, making the sensor less dependent on pick-off linearity than open-loop accelerometers.

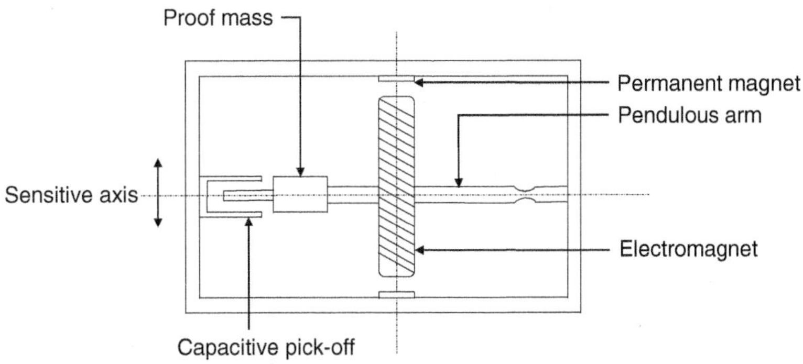

Figure 5.88 Force-feedback pendulous accelerometer.

The proof mass in the design (see Figure 5.80) is a beam fabricated from fused quartz supported by a low stiffness flexural hinge. A capacitive pick-off measures the deviation from the null position. The restoring torque required to maintain the beam in the null position is provided by a moving coil/permanent magnet torquer through feedback of the displacement measured by the pick-off. The pick-off is usually capacitive, although other sensing mechanisms include piezoresistive, electromagnetic, piezoelectric, optical, and ferroelectric.

Inertial Measurement Units and Inertial Navigation Systems

The system architecture of an IMU is shown in Figure 5.89. A single unit comprises three accelerometers and three gyroscopes to output specific force along three axes and attitude rate measurements about three axes, respectively. Unit conversion in the IMU processor is from units of potential difference or current into units of specific force and attitude rates. Parameters characterizing IMU error, including the accelerometer bias, gyroscope bias, scale factor error, and cross-coupling errors are measured during calibration across a range of temperatures and stored for compensation during operation.

All accelerometers and gyroscopes are subject to errors in specific force and attitude rate measurements: bias, scale factor error, cross-coupling error, and random noise. Laboratory calibration

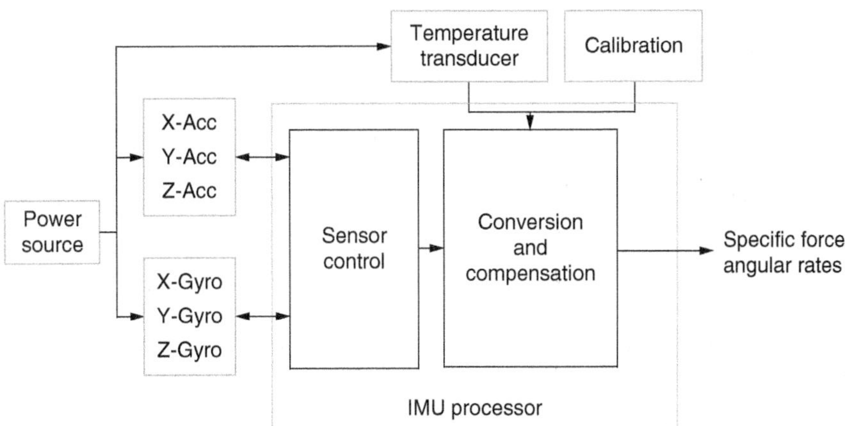

Figure 5.89 IMU architecture.

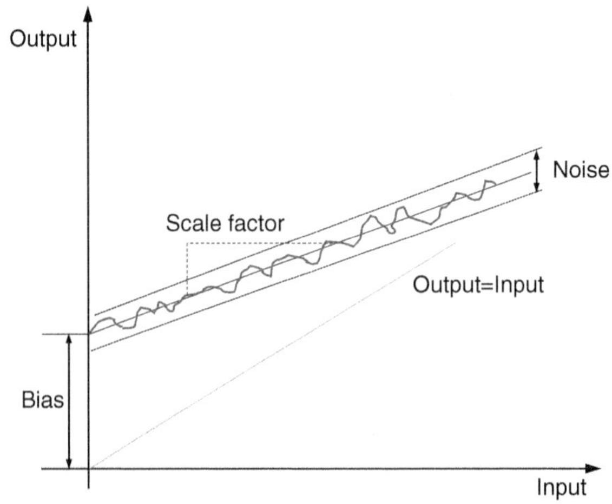

FIGURE 5.90 IMU errors.

prior to operation eliminates the fixed and temperature-dependent components of these error sources. However in-run variation of errors slowly changes during operation and cannot be corrected by calibration, but can be minimized through integration with other navigation sensors. Postcalibration errors characterize the performance of IMUs and will be briefly discussed here for each type of error. The types of error are illustrated in Figure 5.90.

The bias error is a constant error independent of the specific force and attitude rate, and typically dominates the overall error in the output. Biases can be split into a static component, which comprises the run-to-run variation, and the dynamic component, which varies during the run.

The scale factor error is the deviation of the input-output gradient from a unit slope (ideal). Cross-coupling errors arise due to a misalignment of the sensitive axes of the accelerometers and gyroscopes with respect to the body-frame axes. Consequently, sensitive axes of the sensors become coupled with one another. Electrical noise, mechanical instability and vibrations contribute to the overall random noise of the sensor output. Scale-factor errors and cross-coupling errors for an accelerometer or gyroscope may be expressed as a matrix:

$$M = \begin{bmatrix} s_x & m_{xy} & m_{xz} \\ m_{yx} & s_y & m_{yz} \\ m_{zx} & m_{zy} & s_z \end{bmatrix} \tag{5.212}$$

where the diagonal elements are the components of the scale factor error, while the off-diagonal elements are the cross-coupling coefficients. In summary, in-run IMU errors accumulate over multiple time-steps, causing the navigation solution to drift unless corrected.

Navigation Coordinate Frames

It is frequently necessary to transform vectors between coordinate frames in order to compute the navigation solution and present it in a user-friendly format. Four reference frames are briefly described herein: the Earth-centered inertial (ECI) frame, the Earth-centered, Earth-fixed (ECEF) frame, the north-east-down (NED) frame, and the body frame, followed by a description of attitude representation and coordinate frame transformation.

Earth-Centered Inertial Frame

The ECI frame is centered at the Earth's center of mass and is oriented with respect to the Earth's spin-axis and the stars. The frame does not rotate with the Earth. Although not a true inertial frame owing to the acceleration of the Earth during its revolution, the ECI is a sufficiently accurate frame for navigation. The z-axis z^i is always oriented along the Earth's spin-axis, i.e., from the center to the true North Pole. The x^i and y^i axes lie in the equatorial plane and complete the right-hand coordinate system.

Earth-Centered Earth-Fixed Frame

Similar to the ECI frame, the ECEF frame is centered at the Earth's center of mass. However, the axes x^e, y^e, and z^e remain fixed with respect to the Earth and rotate as the Earth spins. The z^e axis is oriented along the Earth's spin-axis. The x^e points from the center to the intersection of the equator with the prime meridian. The ye axis completes the right-hand coordinate system.

North-East-Down Frame

The NED frame is also a right-hand coordinate system typically centered at the aircrafts center of gravity. In scenarios where a navigation solution is sought in a small area, position can be expressed in a local NED frame, where the origin is defined at a user-specified location.

The z^n axis always points downward, i.e., it is always normal to the surface of the reference ellipsoid. The x^e axis always points in the north direction and is located in the plane tangential to the reference ellipsoid. The ye axis completes the orthogonal triad and always points in the east direction, and lies in the plane tangential to the reference ellipsoid. Figure 5.91 illustrates the ECEF and NED frames.

Body Frame

The body frame is centered at the aircrafts center of gravity and its axes are fixed with respect to the aircraft. x^b is oriented along the aircrafts longitudinal axis (points forward), y^b is orthogonal to x^b and points to the right side of the aircraft. z^b completes the right hand coordinate system and points downward.

Euler Angles and Coordinate Transformations

The attitude of the aircraft with respect to a fixed inertial reference frame is defined by three rotations applied in a particular sequence, or *Euler angles*:

1. A rotation about the z-axis of the body frame or yaw ψ
2. A rotation about the y-axis or pitch θ
3. A rotation about the x-axis or roll Φ

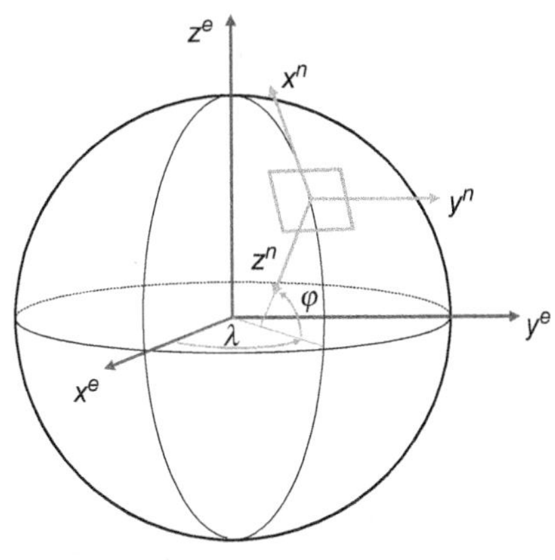

FIGURE 5.91 ECEF and NED coordinate frames.

A yaw rotation about the z-axis can be expressed as a 3×3 matrix and is given by

$$R_z = \begin{bmatrix} \cos\psi & \sin\psi & 0 \\ -\sin\psi & \cos\psi & 0 \\ 0 & 0 & 1 \end{bmatrix} \tag{5.213}$$

The pitch rotation about the y-axis is given by

$$R_y = \begin{bmatrix} \cos\theta & 0 & -\sin\theta \\ 0 & 1 & 0 \\ \sin\theta & 0 & \cos\theta \end{bmatrix} \tag{5.214}$$

The roll rotation about the x-axis is given by

$$R_x = \begin{bmatrix} 1 & 0 & 0 \\ 0 & \cos\Phi & \sin\Phi \\ 0 & -\sin\Phi & \cos\Phi \end{bmatrix} \tag{5.215}$$

The coordinate transformation matrix C_b^a, also termed the direction cosine matrix (DCM), which transforms a vector from frame b to frame a is formed by multiplying the three rotation matrices:

$$C_b^a = R_x R_y R_z \tag{5.216}$$

Coordinate transformation matrices are used in aircraft navigation to transform vectors between the ECI, ECEF, NED, and body coordinate frames. The coordinate transformation matrix can also be used to denote attitude of the aircraft with respect to a reference frame.

The gyroscopes in the strap-down IMU measure the rotation of the aircraft body referenced to an inertial frame and resolved in the body frame axes. These measurements are also termed the body rates. The Euler angle rates can be derived given from the measured body rates by

$$\begin{bmatrix} \dot{\Phi} \\ \dot{\theta} \\ \dot{\psi} \end{bmatrix} = \begin{bmatrix} 1 & \sin\Phi\tan\theta & \cos\Phi\tan\theta \\ 0 & \cos\Phi & -\sin\Phi \\ 0 & \sin\Phi\sec\theta & \cos\Phi\sec\theta \end{bmatrix} \begin{bmatrix} p \\ q \\ r \end{bmatrix} \tag{5.217}$$

where p = body rate about the roll axis or the roll rate
q = body rate about the pitch axis or the pitch rate
r = body rate about the roll axis or the roll rate

Given an initial attitude expressed in terms of Euler angles and the measured body rates, the aircraft attitude can be tracked by computing the Euler angle rates, followed by integration over the time-step interval.

Quaternions

A limitation of the use of Euler angles to express attitude lies in the fact that $\tan\theta$ and $\sec\theta$ become infinite at $\theta = 90°$ (equation (5.54)). This condition is the mathematical equivalent of "gimbal lock." Flight profiles that avoid 90° pitch angles are therefore, required to avoid mathematical singularities.

The limitation of Euler angles is addressed through the use of quaternions. The concept of quaternions is as follows:

Given a coordinate frame O with axes OA, OB, and OC, it is required to describe its attitude with respect to an inertial reference frame $\acute{O}$ with axes OA_0, OB_0, and OC_0. The coordinate frame can be moved to any required attitude by a single rotation about a suitably located axis. Let this axis make angles $\cos^{-1}\alpha$, $\cos^{-1}\beta$, and $\cos^{-1}\gamma$ with the inertial axes OA_0, OB_0, and OC_0, respectively. A single rotation μ about the axis will rotate a coordinate frame from $\acute{O}$ to O. The components of the quaternion are then expressed as

$$\begin{aligned} e_0 &= \cos\mu/2 \\ e_1 &= \alpha\sin\mu/2 \\ e_2 &= \beta\sin\mu/2 \\ e_3 &= \gamma\sin\mu/2 \end{aligned} \tag{5.218}$$

These four parameters are used to represent aircraft attitude with respect to frame $\acute{O}$. The differentials of these parameters are related to the aircraft body rates by

$$\begin{bmatrix} \dot{e}_0 \\ \dot{e}_1 \\ \dot{e}_2 \\ \dot{e}_3 \end{bmatrix} = \frac{1}{2}\begin{bmatrix} -e_1 & -e_2 & -e_3 \\ e_0 & -e_3 & e_2 \\ e_3 & e_0 & -e_1 \\ -e_2 & e_1 & e_0 \end{bmatrix}\begin{bmatrix} p \\ q \\ r \end{bmatrix} \tag{5.219}$$

The following constraint exists when using quaternions:

$$e_0^2 + e_1^2 + e_2^2 + e_3^2 = 1 \tag{5.220}$$

Equation (5.220) can be algebraically manipulated into the compact form:

$$\dot{S} = K\,S \tag{5.221}$$

where $S = \begin{bmatrix} e_0 \\ e_1 \\ e_2 \\ e_3 \end{bmatrix}$ and $K = \frac{1}{2}\begin{bmatrix} 0 & -p & -q & -r \\ p & 0 & r & -q \\ q & -r & 0 & p \\ r & q & -p & 0 \end{bmatrix}$.

By assuming a small integration time step, Δt, the body rates can be assumed to be constant over the time-step. Equation (5.58) can then be modified to relate parameter values at the end of the integration time-step with the values at the start of the time-step:

$$\begin{bmatrix} \dot{e}_0 \\ \dot{e}_1 \\ \dot{e}_2 \\ \dot{e}_3 \end{bmatrix}_{t_{n+1}} = \begin{bmatrix} 1 & -\Delta P/2 & -\Delta Q/2 & -\Delta R/2 \\ \Delta P/2 & 1 & \Delta R/2 & -\Delta Q/2 \\ \Delta Q/2 & -\Delta R/2 & 1 & \Delta P/2 \\ \Delta R/2 & \Delta Q/2 & -\Delta P/2 & 1 \end{bmatrix}\begin{bmatrix} e_0 \\ e_1 \\ e_2 \\ e_3 \end{bmatrix}_{t_n} \tag{5.222}$$

where ΔP, ΔQ, and ΔR are incremental angular rotations about the roll, pitch, and yaw axes, respectively, given by

$$\begin{aligned} \Delta P &= \int_{t_n}^{t_{n+1}} p\,dt = p\Delta t \\ \Delta Q &= \int_{t_n}^{t_{n+1}} q\,dt = q\Delta t \\ \Delta R &= \int_{t_n}^{t_{n+1}} r\,dt = r\Delta t \end{aligned} \tag{5.223}$$

The Euler angles can be computed from the quaternion parameters through:

$$\begin{aligned} \sin\theta &= 2(e_0 e_2 - e_3 e_1) \\ \tan\psi &= \frac{2(e_0 e_3 - e_1 e_2)}{e_0^2 + e_1^2 - e_2^2 - e_3^2} \\ \tan\Phi &= \frac{2(e_0 e_1 - e_1 e_3)}{e_0^2 - e_1^2 - e_2^2 + e_3^2} \end{aligned} \tag{5.224}$$

The initial values of the quaternion parameters given the initial values of Euler angles at the start of the iteration process are

$$\begin{aligned} e_0 &= \cos\frac{\psi}{2}\cos\frac{\theta}{2}\cos\frac{\Phi}{2} + \sin\frac{\psi}{2}\sin\frac{\theta}{2}\sin\frac{\Phi}{2} \\ e_1 &= \cos\frac{\psi}{2}\cos\frac{\theta}{2}\sin\frac{\Phi}{2} - \sin\frac{\psi}{2}\sin\frac{\theta}{2}\cos\frac{\Phi}{2} \\ e_2 &= \cos\frac{\psi}{2}\sin\frac{\theta}{2}\cos\frac{\Phi}{2} + \sin\frac{\psi}{2}\cos\frac{\theta}{2}\sin\frac{\Phi}{2} \\ e_0 &= \sin\frac{\psi}{2}\cos\frac{\theta}{2}\cos\frac{\Phi}{2} + \cos\frac{\psi}{2}\sin\frac{\theta}{2}\sin\frac{\Phi}{2} \end{aligned} \tag{5.225}$$

The accuracy of the numerical integration can be improved by using higher-order Runge–Kutta algorithms, and by decreasing the integration time-step size at the expense of greater computational cost. Quaternions are computationally efficient when compared to Euler angle computation, and are applicable for all attitudes (including 90° pitch angles). Quaternions can be easily converted to Euler angle form or can be directly used to form the DCM in an INS iteration. The process followed in an INS in updating aircraft PVA given initial conditions is described in the following subsection.

INS are the primary DR-based source of attitude, velocity, and position for aircraft. The navigation processor in the INS utilizes the specific force and attitude rate measurements of the IMU and integrates them over a time interval to output the Position, Velocity, and Attitude (PVA). The schematic of an INS is shown in Figure 5.92.

Computation of a navigation solution for inertial sensors is an iterative process, wherein the solution is derived by making use of the solution computed in the previous time-step. The process

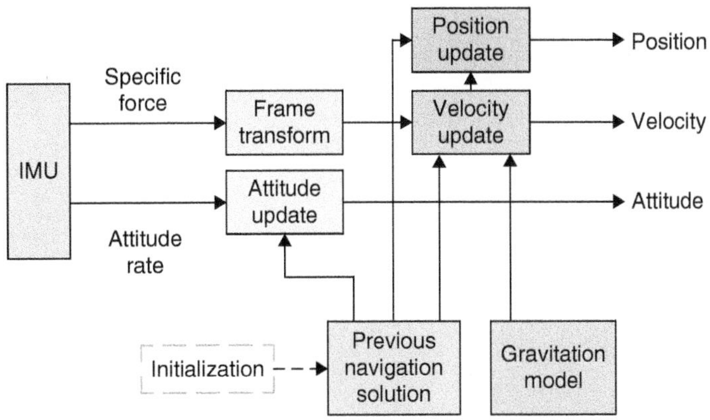

Figure 5.92 Schematic of an INS.

must be initialized using known values of PVA, or through values provided from another navigation system such as the Global Navigation Satellite System (GNSS), or terrestrial radio for the initialization of position and velocity. External sensors such as multi-antenna GNSS can also provide an initialization of the attitude. Alternatively, self-alignment comprising leveling to initialize pitch and roll, and gyrocompassing to initialize yaw (heading) is more commonly used when aviation-grade gyroscopes are available. The form or mechanization of the inertial navigation equations and their complexity depends on the frame the solution is expressed in.

However, regardless of the chosen reference frame, once initialized, the navigation computation consists of four steps:

1. *Attitude update:* The attitude rate from the IMU is used to update the attitude, expressed as the coordinate transformation from the body-fixed frame to the chosen reference frame.
2. *Frame-transformation:* The specific force measurements of the IMU must be resolved in the chosen reference frame prior to the velocity update. This step is accomplished by applying a coordinate transformation matrix that transforms the specific force measurement vector.
3. *Velocity update:* The acceleration is obtained from the specific force measurement by applying a gravitation model. The velocity is then obtained by integrating the acceleration over a discrete time-step and adding it to the velocity at the start of the time-step.
4. *Position update:* The position is obtained by integrating the velocity over the time-step and adding it to the position at the start of the time-step.

Navigation on the surface of the Earth is typically expressed in a local north-east-down coordinate reference frame, wherein the coordinates of the origin are user-defined, the horizontal plane is defined by axes pointing to the north and east, and the vertical axis points downward toward the center of the Earth.

An iteration of INS computations where the raw accelerometer measurements (specific forces) and raw gyro measurements (aircraft body rates) are used to obtain aircraft position, expressed in terms of latitude, longitude, and altitude, and aircraft attitude, expressed in terms of Euler angles, is presented.

In the following explanation, the notation ω_{ab}^{c} denotes the rotation vector from coordinate frame a to coordinate frame b, defined in frame c. The notation v_{ab}^{c} denotes velocity of frame b with reference to frame a and expressed in frame c. The notation C_a^b is a coordinate transformation matrix, alternately termed a direction cosine matrix (DCM) that transforms a vector from frame a to frame b. The rate of rotation of the Earth with reference to the inertial frame is:

$$\omega_{ie}^{e} = \begin{bmatrix} 0 \\ 0 \\ \Omega \end{bmatrix} \tag{5.226}$$

where $\Omega = 7.299{,}211{,}5 \times 10^{-5}$ rad/s is the spin rate of the Earth. When this vector is transformed to the navigation frame, it is expressed as

$$\omega_{ie}^{n} = \begin{bmatrix} \Omega \cos \varphi \\ 0 \\ -\Omega \sin \varphi \end{bmatrix} \tag{5.227}$$

where φ is the aircraft latitude. The rate of rotation of the NED frame with respect to the Earth frame, also termed the transport rate, is given by

$$\omega_{en}^{n} = \begin{bmatrix} \dfrac{v_{eb,E}^{n}}{R+h} \\ \dfrac{-v_{eb,N}^{n}}{R+h} \\ -\dfrac{v_{eb,E}^{n} \tan \varphi}{R+h} \end{bmatrix} \tag{5.228}$$

The subscripts E and N denote the east and north components of the aircraft velocity. $R = 6{,}378{,}137$ m is the nominal Earth radius.

The Earth spin rate and the transport rate are combined, and transformed to the body frame:

$$\omega_{in}^{b} = C_{n}^{b}(\omega_{ie}^{n} + \omega_{en}^{n}) \tag{5.229}$$

This quantity is subtracted from the aircraft body rates ω_{ib}^{b} sensed by the gyros:

$$\omega_{nb}^{b} = \omega_{ib}^{b} - \omega_{in}^{b} \tag{5.230}$$

This term represents the aircraft rotation rate with respect to the NED frame, which has been corrected for the effects of rotation of the Earth and the transport rate. These rotation rates are used to compute the Euler angles by computing system quaternions as described further on. The accelerometers measure specific force with respect to an inertial frame expressed in the body frame. On transforming them to the navigation frame and applying a gravitation model:

$$\begin{bmatrix} a_N \\ a_E \\ a_D \end{bmatrix} = C_{b}^{n} \begin{bmatrix} a_x \\ a_y \\ a_z \end{bmatrix} + \begin{bmatrix} 0 \\ 0 \\ g \end{bmatrix} \tag{5.231}$$

The gravity model is a function of geocentric latitude and altitude, and is approximated as:

$$g = g_e(1 + 0.005{,}288{,}4 \sin^2\varphi - 0.000{,}005{,}9 \sin^2 2\varphi)\frac{R^2}{(R+h)^2} \tag{5.232}$$

The acceleration expressed in the NED frame must be corrected to compensate for the accelerations from the rotation of the Earth and the transport rate.

Therefore, the corrected accelerations in the NED frame are given by

$$\dot{v}_N = a_N - 2v_E\Omega\sin\varphi + \frac{v_N v_D - v_E^2\tan\varphi}{R+h} \tag{5.233}$$

$$\dot{v}_E = a_E - 2v_N\Omega\sin\varphi + 2v_D\Omega\cos\lambda + \frac{v_N v_E\tan\lambda + v_E v_D}{R+h} \tag{5.234}$$

$$\dot{v}_D = a_D - 2v_E\Omega\cos\lambda - \frac{v_E^2 + v_N^2}{R+h} \tag{5.235}$$

The NED velocities are obtained by integrating the accelerations and are given by

$$\begin{aligned} v_N &= \int \dot{v}_N dt \\ v_E &= \int \dot{v}_E dt \\ v_D &= \int \dot{v}_D dt \end{aligned} \tag{5.236}$$

The velocities in curvilinear coordinates are expressed as

$$\dot{\varphi} = \frac{v_N}{R+h} \tag{5.237}$$

$$\dot{\lambda} = \frac{v_E}{\cos\lambda(R+h)} \tag{5.238}$$

The position in curvilinear coordinates is then obtained as

$$\begin{aligned} \varphi &= \int \dot{\varphi} dt \\ \lambda &= \int \dot{\lambda} dt \\ h &= \int v_D dt \end{aligned} \tag{5.239}$$

Any error in determining the tilt, or the orientation of the accelerometer sensitive axes with respect to the local vertical axis induces an error in the accelerometer measurements, which consequently results in unacceptably large horizontal velocity and position errors. It is therefore, essential to track the local vertical to minimize system tilt errors. This is accomplished through Schuler tuning, described here using the concept of the Schuler pendulum: the bob of a pendulum having a length equal to the radius of the Earth (Figure 5.93) would always remain at the Earth's center, thereby defining the local vertical regardless of the acceleration of the point of suspension of the pendulum.

The period of oscillation of such a pendulum is given by

$$T = 2\pi\sqrt{\frac{R}{g}} \tag{5.240}$$

where R is the radius of the Earth and g is the acceleration due to gravity. The nominal value of the period is computed using equation (5.240) to be approximately 84.4 minutes.

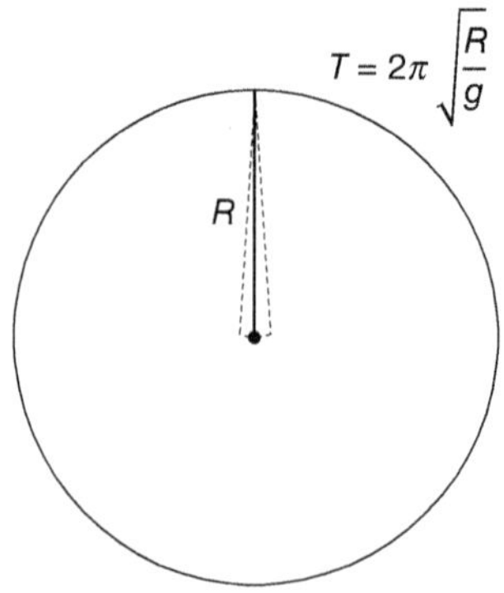

FIGURE 5.93 Schuler pendulum.

References

Allerton, D., 2009, *Principles of Flight Simulation*: John Wiley & Sons.

Cappello, F., Ramasamy, S., and Sabatini, R., 2016, A Low-Cost and High Performance Navigation System for Small RPAS Applications, *Aerospace Science and Technology*, Vol. 58, pp. 529–545. DOI: 10.1016/j.ast.2016.09.0

Collinson, R. P., 2013, *Introduction to Avionics Systems,* Springer Science & Business Media.

Groves, P. D., 2013, *Principles of GNSS, Inertial, and Multisensor Integrated Navigation Systems,* Artech house.

Sabatini, R., Cappello, F., Ramasamy, S., Gardi, A., and Clothier, R., 2015, An Innovative Navigation and Guidance System for Small Unmanned Aircraft Using Low-Cost Sensors, *Aircraft Engineering and Aerospace Technology: An International Journal,* Vol. 87, pp. 540–545.

Titterton, D. and Weston, J. L., 2004, *Strapdown Inertial Navigation Technology* Vol. 17: IET.

PART 12

Alternative Sensors and Multisensor Navigation Systems

Suraj Bijjahalli, Roberto Sabatini, and Subramanian Ramasamy

5.55 Introduction

A number of aircraft states must be computed and provided over the course of a flight to support safe and efficient navigation. The performance of a navigation system is typically assessed in terms of measurable parameters, namely accuracy, continuity, integrity, and availability. Performance requirements specified by aviation authorities as thresholds on these metrics vary depending on the phase of flight, with tighter thresholds and stringent requirements being imposed on the navigation system for the approach and landing phase, and more relaxed thresholds mandated for the cruise phase. It is common practice to employ a number of complementary sensors in an integrated navigation system to increase observability and to overcome individual sensor limitations. This part describes a few sensors that have emerged more recently in integrated navigation systems, such as vision-based sensors. Additionally, the fundamentals of sensor integration schemes are presented, along with the basics of data-fusion algorithms such as the extended Kalman filter (EKF) and the unscented Kalman filter (UKF).

5.56 Vision-Based Navigation

Vision-based navigation techniques use optical sensors (visual or infrared cameras) to extract visual features from images which are then used for localization in the surrounding environment. Cameras have evolved as attractive sensors as they help design economically viable systems with simpler sensor limitations. Vision-based systems were initially used for surveillance purposes and then integrated with the navigation and control systems. They have been implemented for various applications ranging from autonomous landing to obstacle avoidance. Another application is the augmentation of GNSS and/or INS measurements with data from the vision system using Kalman filters so as to get better estimates of the states of the vehicle.

Vision-based schemes include simultaneous localization and mapping (SLAM), the structure from motion (SFM) problem, terrain-based navigation, and navigation based on feature recognition. Vision-based navigation techniques can be classified into two categories:

1. Model-based approach.
2. Appearance-based approach.

Model-Based Approach

Model-based approaches use feature tracking in images and create a 3D model of the workspace in which robots or UAV operates. The 3D maps are created in an offline process using a priori information of the environment. Localization is carried out using feature matching between the current view of the camera and the stored 3D model. The orientation of the robot is found from 3D–2D correspondence.

The model-based approach is the most common technique implemented for vision-based navigation. However, the accuracy of this method is dependent on the features used for tracking, robustness of the feature descriptors, and the algorithms used for matching and reconstruction.

The reconstruction in turn relies on proper camera calibration and sensor noise. Knowledge of surroundings so as to develop the 3D models is also required prior to implementation which may not be case in most situations.

Appearance-Based Approach

Appearance-based (or view-based) algorithms eliminate the need for a metric model as they work directly in the sensor space. This approach utilizes the appearance of the whole scene in the image, contrary to model-based approach which uses distinct objects such as landmarks or edges. The environment is represented in the form of key images taken at various locations using the visual sensors. This continuous set of images describes the path to be followed by the platform. The images are captured while manually guiding the platform through the workspace. In this approach, localization is carried out by finding the key image with the most similarity to the current view. The platform is steered by either coding the action required to move from one key image to another or by a more robust approach using visual servoing. The modeling of the surrounding using a set of key images is more straightforward to implement compared to 3D modeling. A major drawback of this method is its limited applicability. For instance, the key images need to be recaptured each time the platform operates in a new workspace. Appearance-based approach requires a large amount of memory to store the images and is computationally costly. The following text briefly describes an appearance-based algorithm for autonomous aircraft control during the landing phase and image-processing for aircraft state estimation. The appearance-based visual navigation systems consist of three main stages:

- Learning stage: A series of consecutive frames are captured during a flight connecting the initial and target location, forming the reference visual route. The key frames are sampled and the selected images are stored in the memory to be used for guidance during autonomous navigation.
- Localization is carried out at the start of the autonomous run and consists in identifying the key image most similar to the current view. A certain number of images, preferably the ones at the start of the visual route, are compared with the current view and the key image with the least matching error is considered to be the initial position for the autonomous run.
- Autonomous run consists of the aircraft following the visual route starting from the image identified in localization. The key images of the visual route can be considered as waypoint for the aircraft to reach in sensor space. The current view is compared to the key images so as to perform visual servoing.

Figure 5.94 summarizes the vision-based navigation system using a representation of the visual route and how it is used during the autonomous run. Key image 2 has been identified as the current key image during the localization stage and is used for visual servoing. The onboard computer tracks the matching error between key images 2, 3, and its current view until the error between 2 and the current view exceeds that between 3 and the current view. At this stage, key image 3 is used to control the aircraft and the matching errors between 3 and 4 and the current view are monitored. This process is repeated until the aircraft reaches its destination. The values for the attitude rates, the difference in attitudes and deviation are then supplied to the controller.

The functional architecture of the image-processing module is shown in Figure 5.95. The image processing module uses techniques such as edge detection and optical flow to derive the state vector of the aircraft. The values for the attitude rates, the difference in attitudes and deviation are then supplied to the controller. Horizon-based attitude estimation can be used to estimate the pitch and roll of the aircraft using grayscale images provided by the onboard camera. In this approach, the horizon is assumed to be a straight line appearing as an edge in the image. A Canny edge detector can be used to identify the horizon and its pixel coordinates are used to calculate the roll and pitch and are given by

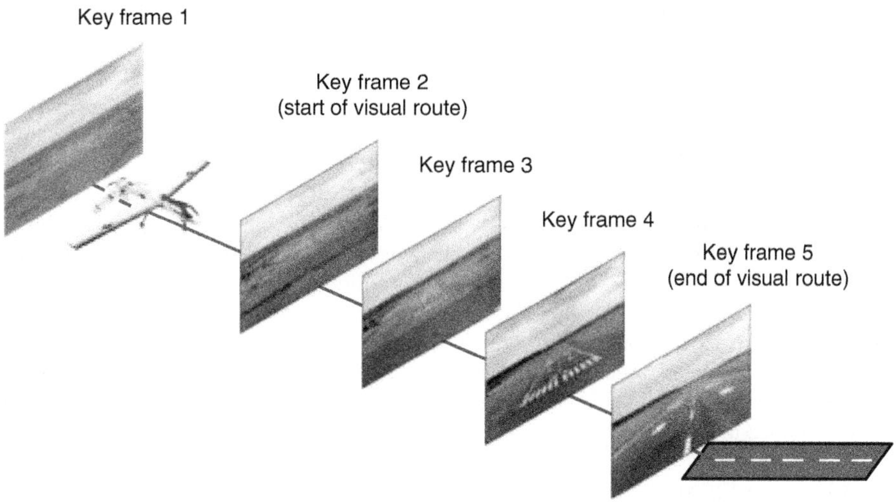

FIGURE 5.94 Appearance-based navigation (from Cappello et. al (2016)).

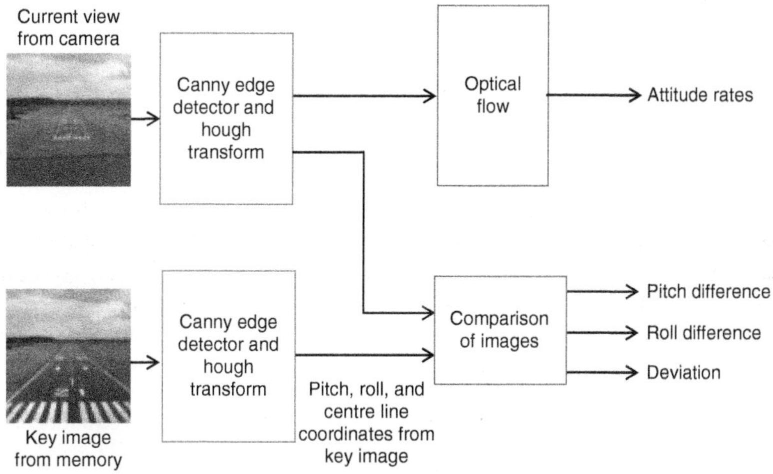

FIGURE 5.95 Functional architecture of the VBN image processing module.

$$\phi = \tan^{-1}\left(\frac{-m_y}{m_x}\right) \tag{5.241}$$

$$\theta = \tan^{-1}\left(\frac{\pm u \sin\phi + v\cos\phi}{f}\right) \tag{5.242}$$

where (m_y/m_x) gives the gradient of the horizon line. u and v are the coordinates of the horizon on the image plane. f is the focal length of the camera. Hough transform was used to detect the centerline of the runway from the image. Optical flow is typically used to calculate the body rates of the aircraft.

5.57 Integrated Navigation Systems

GNSS-based positioning solutions deliver accuracy over relatively large time windows with errors typically in the range of a few meters when using a standard, standalone configuration. Standalone GNSS equipment is however, limited in that it cannot provide attitude estimation (without resorting to a dual antenna configuration). Furthermore solution output rates are typically low, in the order of 10 Hz. Availability and continuity of the solution can also be intermittent, depending on the receiver local environment. Standalone INS solutions provide high frequency solution outputs (at least 50 Hz) of attitude, position, and velocity. However, INS errors accumulate over time, resulting in lowered accuracy and the requirement for online correction.

The complementary nature of various navigation systems can be leveraged through data-fusion schemes to provide a continuous navigation solution with short- and long-term accuracy. This subsection briefly describes commonly employed data-fusion algorithms such as the EKF, the UKF, and the particle filter. The subsection also describes integration architectures focusing on an integration of GNSS and INS for the sake of simplicity.

Extended Kalman Filter

The Kalman filter (KF) is an algorithm that outputs an optimal estimate of a system using a series of measurements from multiple sensors corrupted by noise, essentially achieving accuracy greater than would be possible using a standalone sensor. The method utilizes Bayesian inference in a recursive manner to maximize the joint probability of the states and measurements. Since the inception of the method in 1960, it has been widely applied to navigation of aerial platforms, typically implemented in a discrete time form. The algorithm runs in two steps, a prediction step that predicts the navigation state vector and its associated uncertainty by propagation through a system dynamic model, and a measurement step, where data from the sensors is incorporated to correct the prediction and output an optimal estimate. The optimality is achieved by weighting the measurements and prediction according to the associated uncertainties. Despite its obvious benefits, the original KF is subject to two assumptions that limits its practical usage. The first is that the states, process noise and measurement noise must be Gaussian. The second is that the system model and the sensor measurement models must be linear, rendering the basic algorithm inapplicable to most navigation problems. The extended Kalman filter (EKF) addresses this limitation by linearizing the models describing the system dynamics and sensor measurements. This procedure necessitates the computation of a matrix of partial derivatives (the Jacobian) for the state transition and measurement models. The EKF equations are now described beginning with the characterization of the system dynamics. The system is described in state space notation by a set of first-order nonlinear differential equations:

$$\dot{x} = f(x, w) \tag{5.243}$$

where $x \in \mathbb{R}^n$ denotes the system state vector, $f(x)$ is the nonlinear function of the states and $w \in \mathbb{R}^n$ represents a zero mean random process. The matrix of the process noise $Q \in \mathbb{R}^{n \times n}$ is given by

$$Q = E\left(ww^T\right) \tag{5.244}$$

The measurement equation is considered to be a nonlinear function of the measured states and is given by

$$z = h(x, v) \tag{5.245}$$

where $v \in \mathbb{R}^m$ is a zero mean random process described by the matrix of measurement noise $R \in \mathbb{R}^{m \times m}$:

$$R = E\left(vv^T\right) \tag{5.246}$$

It is possible to rewrite the nonlinear equation of measurements for systems with discrete-time measurements as:

$$z_k = h(x_k, v_k) \tag{5.247}$$

As the measurement equations are non-linear, it is necessary to linearize them by first-order approach to obtain the dynamic matrix of the system, F, and the measurement matrix, H. These matrices are related to the nonlinear equations and are expressed as:

$$F = \left.\frac{\partial f(x)}{\partial x}\right|_{x=\hat{x}} \tag{5.248}$$

$$H = \left.\frac{\partial h(x)}{\partial x}\right|_{x=\hat{x}} \tag{5.249}$$

where $\hat{x}$ represents the mean value. The fundamental matrix is approximated by the Taylor series expansion as follows:

$$\Phi_k = I + FT_s + \frac{F^2T_s^2}{2!} + \frac{F^3T_s^3}{3!} + \cdots \tag{5.250}$$

where T_s is the sampling time and I is the identity matrix. The Taylor series is often truncated to first-order:

$$\Phi_k \approx I + FT_s \tag{5.251}$$

The overview of the steps in an iteration of the EKF is illustrated in Figure 5.96.

The prediction of the navigation state x and its covariance P at time-step k from the previous time-step through the state-transition matrix Φ is given by:

$$\hat{x}_k^- = \Phi_k \hat{x}_{k-1}^+ \tag{5.252}$$

$$P_k^- = \Phi_k P_{k-1}^+ \Phi_k^T + Q \tag{5.253}$$

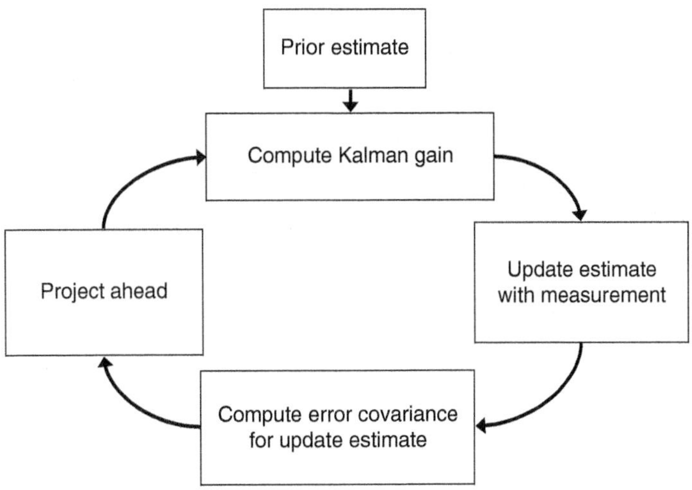

FIGURE 5.96 EKF algorithm overview.

where the (−) and (+) superscripts denote the process before and after incorporating the measurement. The ($\hat{\ }$) symbol denotes an estimate of the state. Q is the covariance of the process noise. The computation of the Kalman gain is given by:

$$K_k = P_k^- H^T \left(HP_k^- H^T + R\right)^{-1} \tag{5.254}$$

which is a function of the state covariance, the measurement noise covariance R and the measurement matrix H. The measurement is incorporated in the correction step to obtain the optimal estimate of state $\hat{x}_k^+$ and its covariance P_k^+ and they are given by

$$\hat{x}_k^+ = \hat{x}_k^- + K_k \left(z_k - h_k\left(\hat{x}_k^-\right)\right) \tag{5.255}$$

$$P_k^+ = (I - K_k H) P_k^- \tag{5.256}$$

$\hat{x}_k^+$ and P_k^+ are then used recursively in the next iteration of the filter. Despite its widespread use in aircraft navigation to integrate multiple sensors to boost system performance, the EKF is subject to limitations, prompting the development of alternative implementations of the KF. One of these methods is the unscented Kalman filter (UKF).

Unscented Kalman Filter

Although the EKF is one of the most widely used filtering methods for non-linear system estimation, this filter has two key drawbacks that make it challenging for a practical implementation (Julier and Uhlmann 1997, 2004; Lozano et al. 2008; Van Der Merwe 2004):

- Highly nonlinear state transition models causes poor performance and instability.
- The accuracy of the nonlinearly propagated mean and the covariance is limited to the first order since the filter uses a first-order truncated Taylor series linearization scheme.

The UKF was developed in order to overcome the limitations of the EKF due to first-order linearization of the nonlinear systems and provides derivative-free higher-order approximations with Gaussian distributions rather than arbitrary nonlinear functions. In comparison to the UKF, the EKF is difficult to implement and tune, and is mostly suboptimal for nonlinear systems. The unscented transform (UT) is central to the UKF, which evaluates the statistics of a nonlinear transformed random variable. A number of "Sigma Points" are generated to characterize the state vector and its uncertainty, and are propagated via the state transition model. The sigma points essentially are used to approximate the moments of the distributions representing the states. The UT process is a mathematical function that is used to estimate a given nonlinear transformation using probability distribution that is characterized only in terms of a finite set of statistics. The most common use of the UT is in the nonlinear projection of mean and covariance estimates in the context of nonlinear extensions of the KF. The UT, especially as part of the UKF, has largely replaced the EKF in many nonlinear filtering and control applications, including underwater (Challa 2011), ground and air navigation (Oh 2007), and spacecraft (Van Der Merwe 2004) applications. Generally, the UKF performs to the same order of computational expense as the EKF; however, it generally performs much better. This can be attributed to increased accuracy of the moment approximations in the UKF attributed to the UT process. In fact, the UKF achieves the same level of accuracy as the second-order EKF, which requires the derivation of Jacobian and Hessian matrices. A top-level representation of the UKF algorithm is shown in Figure 5.97.

1. Determine sigma points $\chi_{k-1}^1, \ldots, \chi_{k-1}^s$ and weights $w^1, \ldots, w^s$ to match a mean $\hat{x}_{k-1|k-1}$ and covariances matrix $P_{k-1|k-1}$.
2. Compute the transformed sigma points $\chi_k^i = f\left(\chi_{k-1}^i\right), i = 1, \ldots, s.$

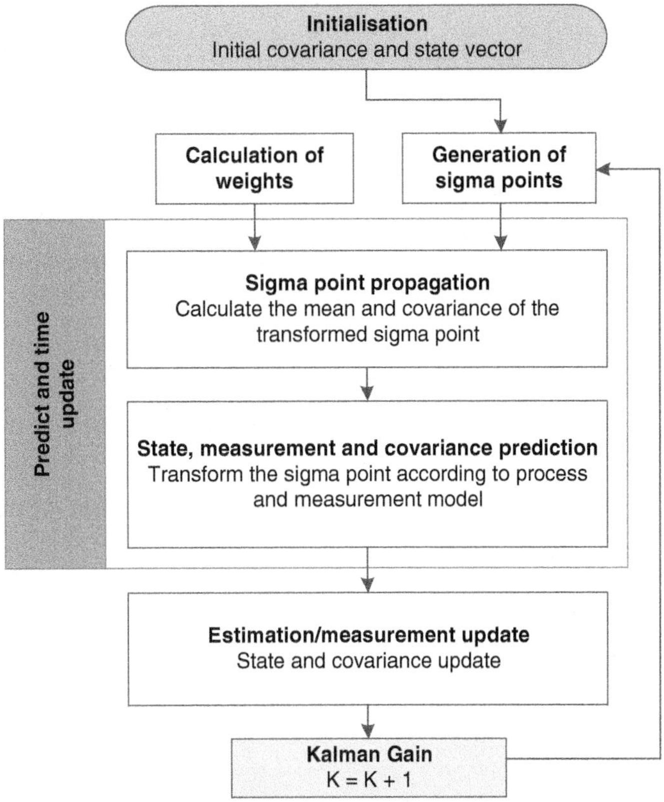

FIGURE 5.97 Top-level illustration of the UKF.

3. Compute the predicted state statistics:

$$\hat{x}_{k|k-1} = \sum_{i=1}^{s} w^i \chi_k^i \tag{5.257}$$

$$P_{k|k-1} = Q_k + \sum_{i=1}^{s} w^i \left(\chi_k^i - \hat{x}_{k|k-1}\right)\left(\chi_k^i - \hat{x}_{k|k-1}\right)^T \tag{5.258}$$

4. Determine the sigma points $\chi_k^1, \ldots, \chi_k^s$ and weights $w^1, \ldots, w^s$ to match mean $\hat{x}_{k|k-1}$ and covariance matrix $P_{k|k-1}$.
5. Compute the transformed sigma points $y_k^i = h\left(\chi_k^i\right)$, $i = 1, \ldots, s$.
6. Compute the predicted measurement statistics:

$$\hat{y}_{k|k-1} = \sum_{i=1}^{s} w^i y_k^i \tag{5.259}$$

$$S_k = R_k + \sum_{i=1}^{s} w^i \left(y_k^i - \hat{y}_{k|k-1}\right)\left(y_k^i - \hat{y}_{k|k-1}\right)^T \tag{5.260}$$

$$\Psi_k = \sum_{i=1}^{s} w^i \left(\chi_k^i - \hat{x}_{k|k-1}\right)\left(y_k^i - \hat{y}_{k|k-1}\right)^T \tag{5.261}$$

7. Compute the posterior mean and covariance matrix

$$\hat{x}^{+}_{k/k} = \hat{x}^{+}_{k/k-1} + \Psi_k S_k^{-1}\left(y_k - \hat{y}_{k|k-1}\right) \tag{5.262}$$

$$P_{k|k} = P_{k|k-1} - \Psi_k S_k^{-1} \Psi_k^T \tag{5.263}$$

Particle Filter

The particle filter overcomes the linearity and Gaussian model constraints in the KF through a representation of the posterior state probability wherein the system dynamics and sensor measurement models can be highly nonlinear, and the states and the noise terms can be non-Gaussian. The navigation state is represented by x_t, where t is time. z_t is the set of all observations. The posterior density function $p(x_t \mid z_t)$ is represented by a set of weighted samples or particles. The main idea of the PF is the estimation of the probability distribution function (PDF) of the aerial platform by a set of weighted samples $S = \{(s^{(n)}, w^{(n)}) \mid n = 1, \ldots, N\}$ such that $\sum_{n=1}^{N} \omega^{(n)} = 1$. Based on the observations, likelihood of each particle is applied in weight assignment by the equation below in each step:

$$\omega^{(n)} = p\left(z_t \mid x_t = s_t^{(n)}\right) \tag{5.264}$$

The primary steps in the algorithm are summarized below.

The PF consists of two stages, prediction and update. In the prediction state, the particles are propagated through the state transition equations. In the update stage, the weights are assigned to new particles by the following equation:

$$\omega_t^i \propto \omega_{t=1}^i \frac{p\left(z_t x_t^i\right) p\left(x_t^i \mid x_{t=1}^i\right)}{q\left(x_t^i \mid x_{t=1}^i, z_t\right)} \tag{5.265}$$

where q is the importance density function which is used in the sampling function. Posteriori density function is estimated by using the following equation:

$$p(x_t \mid z_t) = \sum_{i=1}^{N} \omega_t^i \delta(x_t - x_{t-1}) \tag{5.266}$$

The next step state vector is estimated by using the weighted average on current step particles with the following equation:

$$\hat{x}_t = \sum_{i=1}^{N} w_t^i x_t^i \tag{5.267}$$

Sequential Importance Resampling Particle Filter

One of the basic challenges in implementation of PF is appropriate choice of proposal density function. One of the common suboptimal choices of proposal density function is transitional prior distribution.

$$q\left(x_t \mid x_{t=1}^i, z_t\right) = p\left(x_t \mid x_{t=1}^i\right) = N_{x_t}\left(x_t, \Sigma_t\right) \tag{5.268}$$

By substituting $q(.)$ into each particle's weight, a new weighting can be calculated based on

$$\omega_t^i \propto \omega_{t-1}^i p(z_t \mid x_t^i) \tag{5.269}$$

In each step of the SIR algorithm, re-sampling is done after state estimation. After re-sampling, weights of all particles are equal to $\frac{1}{N}$, where N is the number of particles. Hence, the above equation is modified as follows:

$$\omega_t^i = p(z_t \mid x_t^i) \tag{5.270}$$

Initialize: the particles using prior distribution of target.
FOR i = 1:N

a. Prediction: Generate particles using dynamic model

$$\hat{x}_{t+1}^i \sim p(x_{t+1} \mid x_t^i) \tag{5.271}$$

b. Weighting: Calculate weight for particle

$$\omega_{t+1}^i \sim p(z_{t+1} \mid x_t^i) \tag{5.272}$$

End FOR
Normalize the particle weights:

$$\hat{\omega}_{t+1}^i = \frac{\omega_{t+1}^i}{\sum_{i=1}^{N} \omega_{t+1}^i} \tag{5.273}$$

Resampling: Resample the particles and obtain the new part with same weight and repeat from stage 2.

$$\left\{\left(x_{t+1}^i, \frac{1}{N}\right)\right\}_{i=1}^{N} = \text{Re-sampling}\left\{\left(x_{t+1}^i, w_{t+1}^i\right)\right\}_{i=1}^{N} \tag{5.274}$$

Integration Architectures

Integration architectures differ in the type of measurements extracted from the constituent sensors, and the method by which corrections are formed and applied to the solution. Although there is no standardized categorization of integration architectures, distinctions can be made between architectures defined as loosely coupled, and tightly coupled. A closed-loop, loosely coupled GNSS/INS integration architecture is shown in Figure 5.98.

Pseudorange measurements from the GNSS user-equipment are used to obtain a standalone GNSS solution for position and velocity. The INS equations compute a solution from IMU measurements. Both these standalone solutions are provided as inputs to the INS/GNSS Data-fusion algorithm (e.g., extended Kalman filter, unscented Kalman filter, etc.) in order to estimate INS errors, which are then fed back to correct the Inertial Navigation solution at each integration filter iteration. The corrected solution from the INS is the integrated solution. Loosely coupled architectures are relatively simple to implement, and owing to the availability of standalone GNSS and INS solutions, can support integrity monitoring through parallel-solutions. However, the overall system is still subject to the availability limitation of GNSS, i.e., a minimum of four satellites must be tracked in

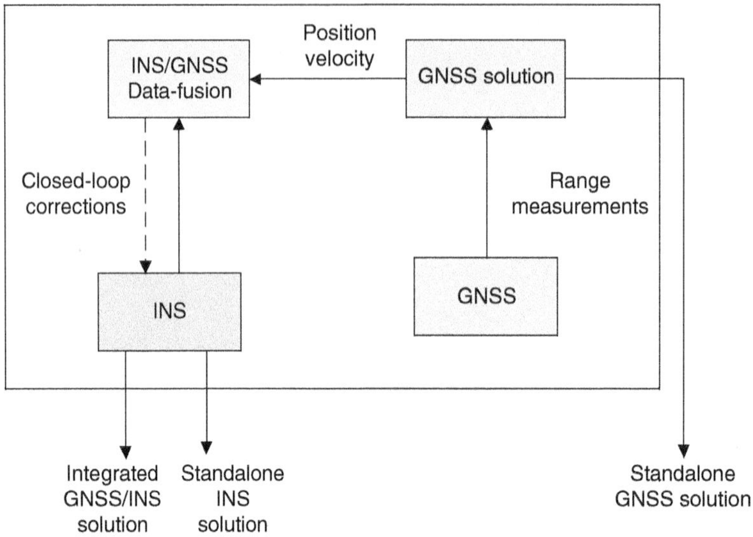

FIGURE 5.98 Loosely coupled GNSS/INS integration architecture.

order for a GNSS solution to be computed, and subsequently, for the closed-loop corrections to be formed. In general, greater accuracy and solution availability can be obtained through a tightly coupled architecture as shown in Figure 5.99. In this architecture, rather than input a standalone GNSS solution to the data-fusion algorithm, the raw range and range-rate measurements are directly fused with the INS solution to estimate errors in the INS and GNSS subsystems, which are then fed back for correction. The corrected INS solution forms the integrated solution.

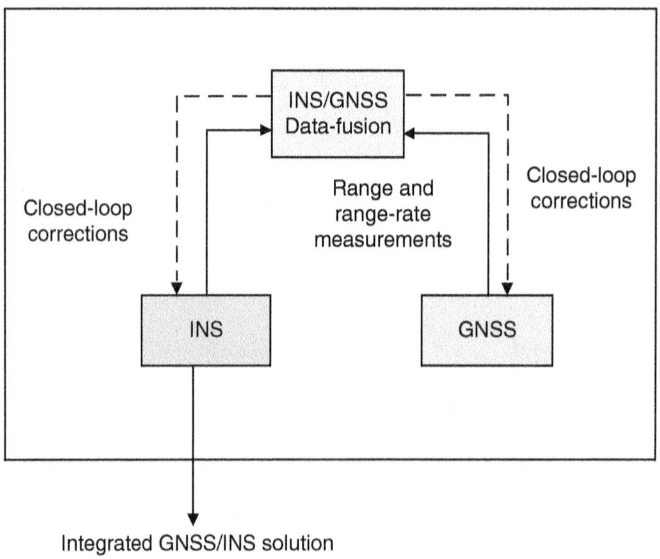

FIGURE 5.99 Tightly-coupled GNSS/INS integration architecture.

References

Challa, S., 2011, Fundamentals of object tracking, *Cambridge University Press*, Cambridge, United Kingdom.

F. Cappello, S. Ramasamy and R. Sabatini, "A Low-Cost and High Performance Navigation System for Small RPAS Applications," *Aerospace Science and Technology*, Vol. 58, pp. 529–545, 2016. DOI: 10.1016/j.ast.2016.09.002.

Julier, S. J. and Uhlmann, J. K., 1997, New extension of the Kalman filter to nonlinear systems, AeroSense'97, Proceedings Volume 3068, Signal Processing, Sensor Fusion, and Target Recognition VI; Orlando, Florida, USA.

Julier, S. J. and Uhlmann, J. K., 2004, Unscented filtering and nonlinear estimation, *Proceedings of the IEEE*, Vol. 92, no. 3, pp. 401–22.

Lozano, J. G. C., Carrillo, L. R. G., Dzul, A., et al., 2008, Spherical simplex sigma-point Kalman filters: a comparison in the inertial navigation of a terrestrial vehicle, *Proceedings of the American Control Conference*, 2008.

Oh, S. M., 2007, Nonlinear estimation for vision-based air-to-air tracking, School of Aerospace Engineering PhD Dissertation, Georgia Institute of Technology, Atlanta, Georgia, USA.

Van Der Merwe, R., 2004, Sigma-point Kalman filters for probabilistic inference in dynamic state-space models, Department of Electrical and Computer Engineering PhD Dissertation, Oregon Health & Science University, Portland, Oregon, USA.

PART 13

Global Navigation Satellite Systems

Roberto Sabatini

Various Global Navigation Satellite Systems (GNSS) are currently in service or under development. The U.S. Navigation Signal Time and Range (NAVSTAR), Global Positioning System (GPS), and the Russian GLONASS (Globalnaya Navigazionnaya Sputnikovaya Sistema) have achieved their full operational capability (FOC) back in the 1990s (Daly 1995). Other GNSS that are currently at the advanced development or deployment stages include the European GALILEO, the People's Republic of China BEIDOU, and the Indian Navic. Although these systems operate at different frequencies, employ various signal processing techniques and adopt different multiple access schemes, most of them employ or are planned to transition to code-division multiple access (CDMA) for operations at the L-band frequency of 1575.42 MHz (L1). This signal, originally introduced by GPS for Standard Positioning Service (SPS), will guarantee interoperability between the various GNSS constellations with substantial benefits to the existing vast community of GPS users, including aviation. Signal-in-space (SIS) interoperability is also being actively pursued on other frequencies, with a focus on those that are currently allocated to safety-of-life (SOL) services. Even when interoperability at SIS level cannot be guaranteed, the introduction of multi-constellation receivers can bring significant improvements in GNSS performance.

5.58 GNSS Segments

All GNSS are composed of a space segment, a control segment, and a user segment (Figure 5.100). The space segment includes the satellites required for global coverage. These satellites are predominantly in intermediate circular orbit (ICO), at nominal altitudes of 19,100 km (GLONASS), 20,184 km (GPS), and 23,222 km (GALILEO) km from the Earth's surface. BEIDOU also employs satellites in geostationary orbit (GEO) and inclined geosynchronous orbit (IGSO). The user segment includes the large variety of GNSS receivers developed for air, ground, and marine navigation and positioning applications. The control segment includes one or more control and processing stations (CPSs) connected to a number of ground monitoring stations (GMSs) and antennae located around the globe for telemetry, tracking, and command (TT&C) signals down/uplink and navigation/integrity signals uplink to the satellites. The GMS antennae passively track all GNSS satellites in view collecting ranging signals from each satellite. This information is passed on to the CPS where the satellite ephemeris and clock parameters are estimated and predicted. Additionally, satellite integrity data are analyzed and appropriate integrity flags are generated for faulty/unreliable satellites. The ephemeris/clock and integrity data are then up-linked to the satellite for retransmission in the navigation message.

The satellite clock drift is corrected so that all transmitted data are synchronized with GNSS time. The fundamental equation is the following:

$$t_{\mathrm{GNSS}} = t_s - \Delta t_s \tag{5.275}$$

where t_{GNSS} is the GNSS time, t_s is the satellite time, and Δt_s is the difference between satellite and GNSS time. The corrections are applied typically using polynomial coefficients and a relativistic correction term. The correction equation can be written as

$$\Delta t_s = a_0 + a_1 (t_{\mathrm{GNSS}} - t_{0c}) + a_2 (t_{\mathrm{GNSS}} - t_{0c})^2 + \Delta t_r \tag{5.276}$$

where a_0, a_1, and a_2 are the polynomial coefficients for phase, frequency, and age offset; Δt_r is the relativistic correction term; and t_{0c} is the time of transmission of the corrections. Typically, the

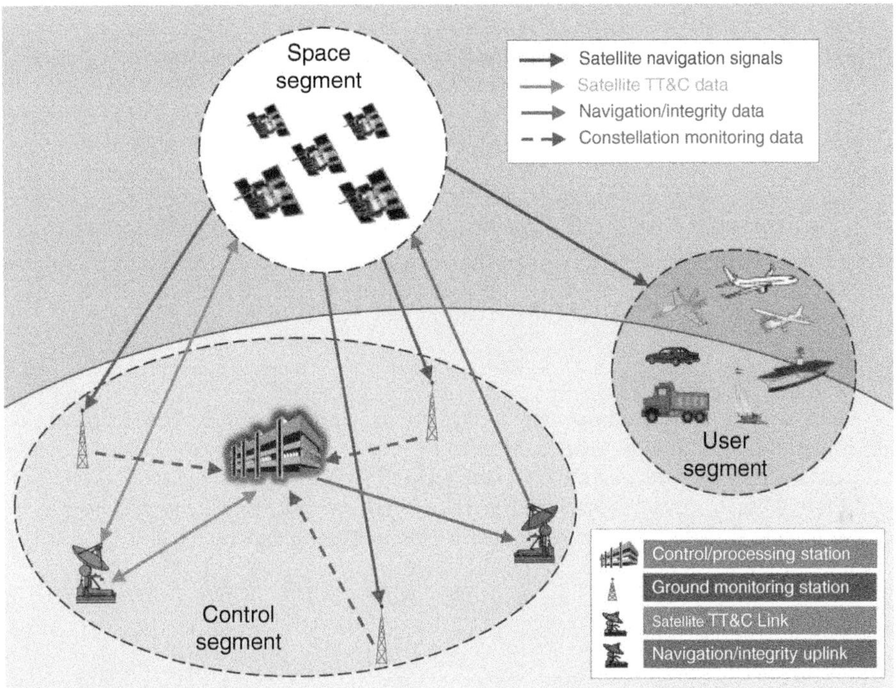

FIGURE 5.100 GNSS segments.

ephemeris corrections are obtained through the estimation of the Cartesian coordinates of the satellites along the orbits by integrating there equations of motion. For integrity purposes, suitable fault detection, isolation, and recovery (FDIR) techniques are employed. In particular, the health status of the satellite subsystems is continuously monitored (satellite payload, bus, solar arrays, battery power, and the level of propellant used for maneuvers) and any anomaly must be promptly detected and resolved. When needed, spare satellites can be activated. The signal-in-space (SIS) is also constantly monitored to guarantee the required performance standards. Despite the significant technological enhancements recently introduced in control segment integrity features, current GNSS have limited FDIR capability. In most cases, several minutes or even hours are required to provide the required integrity information (i.e., use/do not use signals) to GNSS users. Obviously, this is not acceptable for mission-essential and safety-critical aviation applications.

5.59 GNSS Observables

There are basically three types of GNSS observables that can be used in aviation GNSS receivers: pseudorange, carrier phase, and Doppler observable. Pseudoranges are commonly used in real-time aircraft navigation and can provide an accuracy ranging from about 20 m (stand-alone L1 code positioning) to about 2 m in differential GNSS (DGNSS) positioning. The carrier phases are traditionally used in high precision trajectory determination applications (e.g., TSPI systems for flight test and flight inspection) and can achieve subcentimeter accuracy. However, it is also quite common to use combinations of pseudoranges and carrier phase measurements both in real-time and post-mission applications. Moreover, it has become common practice to take advantage of various combinations of the original phase observation, such as double differences and triple differences.

Pseudorange Observable

The concept of pseudoranging is based on measuring difference between the time of transmission of the code from the satellite and the epoch of reception of the same signal at the receiver antenna. This is achieved by correlating identical pseudorandom noise (PRN) codes generated by the satellite's clock, with those generated internally by the receiver's own clock. If this time difference is multiplied by the speed of propagation of the radio wave, a range value is obtained which is the distance between the satellite and the receiver's antenna referring to the epoch of observation. Both receiver and satellite clock errors affect the pseudoranges. Therefore, they differ from the actual geometric distance corresponding to the epochs of emission and reception. The general pseudorange equation is

$$P_k^p(t_k) = \left(t_A - t^p\right) \times c \tag{5.277}$$

where P_k^p represents the actual measurement, t_k denotes the nominal time of the receiver clock k at reception, t^p denotes the nominal time of the satellite clock p at emission, and c denotes the speed of light. Equation (5.277) would correspond to the actual distance between the satellite and receiver's antenna, if there were no clock biases, the signal travelled through vacuum and there was no multipath effect. The clock drifts can be represented by the following expressions:

$$t_{r,k} = t_k + dt_k \tag{5.278}$$

$$t_k^p = t^p + dt^p \tag{5.279}$$

where the symbol r denotes the true time and the terms dt_k and dt^p are the receiver and satellite clock errors, respectively. Taking these errors and biases into account, the complete expression for the pseudorange becomes

$$P_k^p(t_k) = \left(t_{r,k} - t_r^p\right)c - \left(dt_k - dt^p\right)c + I_{k,p}{}^p(t_k) + T_k^p(t_k) + d_{k,p}(t_k) + d_{k,p}{}^p(t_k) + d_p{}^p(t_k) + \varepsilon_p \tag{5.280}$$

Therefore,

$$P_k^p(t_k) = \rho_k{}^p(t_{r,k}) - \left(dt_k - dt^p\right)c + I_{k,p}{}^p(t_k) + T_k^p(t_k) + d_{k,p}(t_k) + d_{k,p}{}^p(t_k) + d_p{}^p(t_k) + \varepsilon_p \tag{5.281}$$

where $I_{k,p}{}^p(t_k)$ and $T_k^p(t_k)$ are the ionospheric and tropospheric delays, depending on varying conditions along the path of the signal. The symbols $d_{k,p}(t_k)$ and $d_p{}^p(t_k)$ denote the receiver and satellite hardware code delays, respectively. The symbol $d_{k,p}{}^p(t_k)$ denotes the multipath of the codes, which depends on the geometry of the antenna and satellite with respect to surrounding reflective surfaces. The term ε_p denotes the random measurement noise. The term $\rho_k{}^p(t_{r,k})$ is the actual geometric distance between the receiver's antenna and the satellite at a specific epoch and therefore

$$\rho_k{}^p(t_r) = \sqrt{\left(u^p - u_k\right)^2 + \left(v^p - v_k\right)^2 + \left(w^p - w_k\right)^2} \tag{5.282}$$

The terms (u_k, v_k, w_k) are the approximate Cartesian co-ordinates of the receiver and (u^p, v^p, w^p) denote the position of the satellite at the epoch of transmission. With reference to Figure 5.101, assuming a constant receiver clock error dt_k for measurements to any satellite and omitting all other error terms in equation (5.282), the following system of equations is formed:

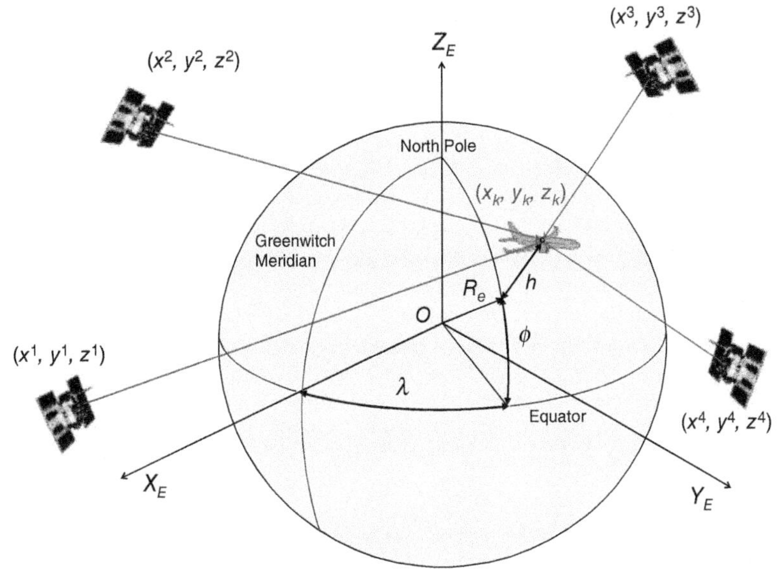

FIGURE 5.101 Navigation solution in the ECEF coordinate system (at time zero, the X_E axis passes through the North Pole, and the Y_E axis completes the right-handed orthogonal system).

$$\begin{cases} P_k^1(t) = \sqrt{(u^1 - u_k)^2 + (v^1 - v_k)^2 + (w^1 - w_k)^2} - cdt_k \\ P_k^2(t) = \sqrt{(u^2 - u_k)^2 + (v^2 - v_k)^2 + (w^2 - w_k)^2} - cdt_k \\ P_k^3(t) = \sqrt{(u^3 - u_k)^2 + (v^3 - v_k)^2 + (w^3 - w_k)^2} - cdt_k \\ P_k^4(t) = \sqrt{(u^4 - u_k)^2 + (v^4 - v_k)^2 + (w^4 - w_k)^2} - cdt_k \end{cases} \tag{5.283}$$

Therefore, the co-ordinates of the user receiver (and GNSS time) can be derived from the simultaneous observation of four (or more) satellites. If more than four satellites are visible, a least-square solution can be determined. The GNSS receiver calculates its position in an Earth-centered Earth fixed (ECEF) Cartesian coordinate system (typically using WGS84). These coordinates may be expressed to some other system such as latitude, longitude, and altitude if desired. Solution of the system requires the measurement of the pseudoranges to four different satellites. The GNSS receiver's computer may be programmed to solve directly the navigation equations in the form given above. However, the computation time required in order to solve them may be too long for many applications.

As an alternate way, these equations may be approximated by a set of four linear equations that the GPS receiver can solve using a much faster and simpler algorithm. The system of equations (5.120) can be rewritten in the form:

$$P_k^i(t) = \sqrt{(u^i - u_k)^2 + (v^i - v_k)^2 + (w^i - w_k)^2} + T, \quad i = 1,2,3,4 \tag{5.284}$$

where u^i, v^i, and w^i represent the coordinates of the ith satellite, $T = dt_k$ and the units have been chosen so that the speed of light is unity. Linearization of equation (5.284) can be performed as described in Thomin (1984) and Siouris (1993). The resulting set of linearized equations relates the pseudorange measurements to the desired user navigation information as well as the user's clock bias:

$$\left(\frac{u_n - u^i}{p_{ni} - T_n}\right)\Delta u_k + \left(\frac{v_n - v^i}{p_{ni} - T_n}\right)\Delta v_k + \left(\frac{w_n - w^i}{p_{ni} - T_n}\right)\Delta w_k + \Delta T = \Delta P_i\,, \qquad i = 1, 2, 3, 4 \tag{5.285}$$

where u_n, v_n, w_n, T_n are the nominal (a priori best-estimate) values of u_k, v_k, w_k, and T; Δu_k, Δv_k, Δw_k, ΔT are the corrections to the nominal values; p_{ni} is the nominal pseudorange measurement to the ith satellite; ΔP_i is the difference between actual and nominal range measurements. The quantities on the right-hand side are simply the differences between the actual measured pseudoranges and the predicted measurements, which are supplied by the user's computer, based on knowledge of the satellite position and current estimate of the user's position and clock bias. Therefore, the quantities to be computed $(\Delta u_k, \Delta v_k, \Delta w_k, \Delta T)$ are the corrections that the user will make to the current estimate of position and clock time bias. The coefficients of these quantities on the left-hand side represent the direction cosines of the LOS vector from the user to the satellite as projected along the Cartesian co-ordinate system.

Carrier Phase Observable

The carrier phase observable is the difference between the received satellite carrier phase and the phase of the carrier generated by the receiver oscillator. The same error sources that affect pseudoranges are responsible for the errors which determine the positional accuracy achieved with carrier phases (Leick et al. 2015). Clearly, the mathematical formulation of any specific error component is different from the pseudorange case (i.e., phase measurement errors instead of range measurement errors). Since the antenna cannot sense the number of whole carrier waves between the satellite and the receiver (called the integer ambiguity), an extra parameter is inserted in the carrier phase equation:

$$\Phi_k^p = \Phi_k(t) - \Phi^p(t) + N_k^p + I_{k,\Phi}^p(t) - \frac{f}{c}T_k^p(t) + d_{k,\Phi}(t) + d_{k,\Phi}^p(t) + d_\Phi^p(t) + \varepsilon_\Phi \tag{5.286}$$

The symbols $\Phi_k(t)$ and $\Phi^p(t)$ denote the phase of the receiver generated signal and the phase of the satellite signal respectively, at the epoch t of satellite signal reception. The symbol N_k^p is the integer ambiguity. The terms $I_{k,\Phi}^p(t)$ and $T_k^p(t)$ are the ionospheric and tropospheric delays. The ionospheric delay factor has a negative value because the carrier phase progresses when traveling through the ionosphere. Furthermore, the tropospheric factor is converted in cycles multiplying by f/c, where f is the nominal frequency and c is the speed of light in vacuum. The symbols $d_{k,\Phi}(t)$ and $d_k^p(t)$ refer to the receiver and satellite hardware delays, respectively. The symbol $d_{k,\Phi}^p(t)$ denotes the multipath effect and ε_Φ denotes the carrier phase measurement noise. Assuming synchronization of the satellite and receiver clocks, omitting other error sources (receiver phase tracking circuits, local oscillator, multipath, and measurement noise), and taking into account both the time of transmission and reception of the signal, the equation for the phase observation between a satellite i and a receiver A, can be written as (Ashkenazi 1997):

$$\Phi_A^i(\tau) = \Phi^i(t) - \Phi_A(\tau) \tag{5.287}$$

where $\Phi_A^i(\tau)$ is the phase reading (phase at receiver A of the signal from satellite i at time τ); $\Phi^i(t)$ is the received signal (phase of the signal as it left the satellite at time t); and $\Phi_A(\tau)$ is the generated signal phase (phase of the receiver's signal at time τ). If $\rho_A^i(t)$ is the range between receiver and satellite, we have

$$t = \tau - \frac{\rho_A^i(t)}{c} \tag{5.288}$$

Therefore,

$$\begin{aligned}\Phi^i(t) &= \Phi^i\left(\tau - \frac{\rho_A^i(t)}{c}\right)\\ &= \Phi^i(\tau) - \underbrace{\frac{\partial \Phi^i(t)}{dt}}_{f} \times \frac{\rho_A^i(t)}{c} + \cdots \\ &= \Phi^i(\tau) - f \times \frac{\rho_A^i(t)}{c} + \cdots\end{aligned} \tag{5.289}$$

and finally

$$\Phi_A^i(\tau) = \Phi^i(\tau) - \frac{f}{c}\rho_A^i(t) - \Phi_A(\tau) + N_A^i \tag{5.290}$$

where $\Phi_A^i(\tau)$ is the phase reading (degree or cycles); $\Phi^i(\tau)$ is the emitted signal; $\frac{f}{c}\rho_A^i(t)$ is the total number of wavelengths; $\Phi_A(\tau)$ is the generated signal; and N_A^i is the integer ambiguity. The carrier phase measurement technique typically uses the difference between the carrier phases measured at a reference receiver and a user receiver. This is, therefore, an inherently differential technique.

Doppler Observable

The equation that associates the transmitted frequency from the satellite with the received frequency is

$$f_k = \frac{f^p}{1 + \frac{r'}{c}} \tag{5.291}$$

where f_k is the received frequency, f^p denotes the emitted frequency from the satellite, r' denotes the radial velocity in the satellite-receiver direction, and c denotes the speed of light in vacuum. The Doppler frequency shift is given by the difference $f^p - f_k$. The radial velocity r' is the actual rate of change in the satellite-receiver distance and it is given by

$$r' = -\frac{f_k - f^p}{f_k} \cdot c \tag{5.292}$$

The integrated Doppler count between two epochs t_1 and t_2 is given by

$$N_{(t_1, t_2)} = \int_{t_1}^{t_2} \left(f^p - f_k\right) dt \tag{5.293}$$

More information about the Doppler observable and on some of its practical uses can be found in the literature (Gleason and Gebre-Egziabher 2009; Seeber 2003).

5.60 GPS Error Sources

The main errors affecting pseudorange and carrier phase measurements can be classified into the five broad categories listed below:

- Receiver-dependent errors: clock error; noise and resolution.
- Ephemeris prediction errors.

- Satellite-dependent errors: clock offset; group delays.
- Propagation errors: ionospheric delay; tropospheric delay; multipath.
- User dynamics errors.

Receiver-Dependent Errors

Most receivers have quartz clocks to measure the GNSS time, which are not as accurate as the atomic clocks of the satellites. Therefore, there is an offset between the receiver and satellite clocks called *receiver clock error*. This error affects both the measurement of the signal flight time and the calculation of the satellite's position at time of transmission. Measurement noise is a random error, which depends entirely on the electronic components of the receiver. Receivers for very precise measurements are designed to minimize this error component. Both noise and resolution errors can be reduced by using appropriate filtering techniques. Theoretically, receiver noise can be removed by averaging the measurements, but only over fairly long periods of observation time.

Ephemeris Prediction Errors

The ephemeris data are required for both pseudorange and phase computations. These errors are due to incorrect estimation of the satellites ephemeris at the CPS. A model for evaluating these errors (Figure 5.3) is obtained by considering the three components of the vector representing the difference between estimated and true distance: along track (ATK), across track (XTK), and radial (RAD).

The maximum error is experienced when the satellite has an elevation of 0° on the receiver horizon and the line-of-sight (LOS) user-satellite lies on the geometric plane containing ATK. In general, the error can be expressed as a function of the three components, in the form:

$$\mathrm{ERR} = \mathrm{RAD}\cos\alpha + \mathrm{ATK}\sin\alpha\,\cos\beta + \mathrm{XTK}\sin\alpha\,\sin\beta \tag{5.294}$$

where α is the angle between the LOS user-satellite and the satellite vertical and β is the angle between the ATK direction and the plane containing the LOS and the satellite vertical. Typical values for the error components are in the order of about 1-m along RAD, 7-m along ATK, 3-m along XTK. The US DoD precise ephemeris is calculated from actual observation to the satellites from the monitor stations. It is produced several days after the observation period and is available only to authorized users. Other non-DoD organizations produce precise ephemeris, both globally and locally, by suitable modeling of all forces and moments acting on the satellites. Orbit relaxation techniques can be developed within GNSS software. These techniques solve for small orbital errors in the broadcast ephemeris and produce improved relative position. Differential techniques can correct the ephemeris prediction errors.

Satellite-Dependent Errors

As discussed above, corrections to the drift of the satellite atomic clocks are computed by the CPS and then broadcasted to the users in the navigation message. The effect of satellite clock offset is negligible in most positioning applications (using the polynomial coefficients corrections computed at the CPS it is possible to reduce this error down to 1 part per 1012). The residual error is due to the fact that corrections from the CPS are periodic and not continuous. Group delays are the delays typical of the satellite electronic circuits. They are estimated on the ground before the satellites are launched and corrections are included in the navigation message.

Propagation Errors

As discussed, propagation errors include both ionospheric and tropospheric delays. As the satellite signal passes through the ionosphere, it is delayed for two reasons (Dodson 1997). First, because it travels through a nonvacuum material (propagation delay); thus the pseudorandom noise (PRN) codes are delayed, while the carrier phase is advanced when passing through the ionosphere layers. Second, because it bends due to refraction; for many applications, the error caused by the bending effect can be considered negligible if the signals are transmitted by satellites with an elevation of 15° or more.

The ionospheric delay is dependent primarily on the number of electrons that the signal encounters along its propagation path. It is therefore dependent both on the ionosphere characteristics (variable during the day and with seasons), and the path angle (elevation angle of the satellite). It is possible to approximately evaluate the ionospheric delay using the following equation (Dodson 1997):

$$\Delta\tau = 40.31\frac{\mathrm{TEC}}{cf^2} \tag{5.295}$$

where TEC is the total electron content integrated along the LOS to the satellite in units of electrons per square meter, f is the frequency in hertz, and c is the speed of light. TEC varies with time and depends on the location of the ionosphere *pierced* by the LOS to the satellite. At the L1 frequency (1,575.42 MHz), equation (5.295) can be written for a satellite that is not at the zenith (Giffard 1999):

$$\Delta\tau = 0.162\frac{F \cdot \mathrm{TEC}_v}{c} \tag{5.296}$$

where TEC_v is the TEC value for a vertical column located at the pierce point, in units of 1,016 electrons per square meter, and F is the obliquity factor. Assuming that the active region of the ionosphere can be represented by a thin shell at an elevation of 350 km, the obliquity can be approximately expressed as a simple function of the elevation angle in degrees (E) of the satellite at the receiver's antenna (Giffard 1999):

$$F = 1 + 2.74 \cdot 10^{-6}(96 - E)^3 \tag{5.297}$$

Depending on the receiver design, different models can be adopted to calculate the correction terms to be applied to the pseudorange before solving the navigation equations. Particularly, L1 code-range receivers use a sinusoidal model of the ionosphere (called Klobuchar model), which take into account the variations of the ionospheric layers (low over night, rapidly getting higher after dawn, getting slightly higher during the afternoon and rapidly getting lower after sunset). The sinusoidal parameters (amplitude and period) are transmitted in the navigation message. The relevant equations are the following (Klobuchar 1986):

$$\mathrm{IDV} = \mathrm{DC} + A_{\cos}\left[\frac{2\pi(t - \Phi)}{P}\right] \text{(day)} \tag{5.298}$$

$$\mathrm{IDV} = \mathrm{DC} \text{ (night)} \tag{5.299}$$

where IDV (ionospheric vertical delay) is expressed in nanoseconds, DC is the constant night-day offset (5 ns), A is the amplitude (whose value is between 10 and 100 ns), Φ is the constant phase offset (14.00 hours), t is the local time, and P is the period. The two factors A and P are transmitted as coefficients of a cubic equation representing a model of the ionosphere with varying latitude. As the delay also depends on obliquity of the path, elevation is included as an additional factor in the equation:

$$\mathrm{TID} = \left[1 + 16(0.53 - E_h)^3\right]\mathrm{IDV} \tag{5.300}$$

where TID is the total ionospheric delay (nanoseconds) and E_h is the elevation angle of the satellites over the horizon. As the ionosphere is dispersive at radio frequencies, two signals at frequencies the

L1 and L2 signals will be delayed by different amounts. Therefore, double frequency GNSS receivers (e.g., GPS P-code receivers) can measure the difference (Δt) between the time of reception of L1 and L2, and evaluate the delay associated with both of them. For L1 we have:

$$\Delta\tau_{L1} = \Delta T\left[\left(\frac{f_{L1}}{f_{L2}}\right)^2 - 1\right]^{-1} \tag{5.301}$$

where

$$\Delta T = \frac{40.31 \cdot \mathrm{TEC}}{cf^2}\left(\frac{1}{f_{L2}^2} - \frac{1}{f_{L1}^2}\right) \tag{5. 302}$$

The troposphere is the lower part of the atmosphere (up to about 50 km) and its characteristics depend on local humidity, temperature, and altitude. There are various expressions used to model the tropospheric delay (TD), all including the elevation angle (E). One of these equations is the following (Dodson 1997):

$$\mathrm{TD} = \mathrm{cosec}[E(1.4588 + 0.0029611N_s)] \tag{5.303}$$

where TD is the tropospheric delay and N_s is the superficial refraction index, dependent on season and latitude. In general, the tropospheric refraction can be calculated as follows:

$$N = 77.6\frac{P}{T} + 3.73\times10^5\frac{w}{T^2} \tag{5.304}$$

where P is the total pressure, T is the absolute temperature, and w is the partial water vapor pressure. The refraction of the troposphere can be separated into two components, the dry component (i.e., $77.6\,P/T$) and the wet component (i.e., $3.73\times10^5\,w/T^2$). The dry component is responsible for 90% of the total tropospheric delay and can be easily modeled. The wet component varies according to the water vapor pressure and, therefore, it is difficult to model. Table 5.18 shows typical values of the tropospheric delay for various elevation angles (Dodson 1997).

Multipath Errors

GNSS signals may arrive at the receiver antenna via different paths, due to reflections by objects along the path. Such effect is known as multipath. The reflected signal will have a different path length compared to the direct signal; therefore, it will give a biased distance measurement (Figure 5.102). Multipath depends on the environment surrounding the receiver and on the satellite geometry. Generally, multipath will be greater for low elevation satellites and code multipath is much greater than carrier phase multipath.

Multipath can be accurately modeled and removed only at static points, by taking observations at the same points and at the same hour on consecutive days. This, however, is not practical or even

Elevation angle, °	Dry component, m	Wet component, m
90	2.3	0.2
30	4.6	0.4
10	13.0	1.2
5	26.0	2.3

Table 5.18 Tropospheric Delay for Varying Elevation Angle

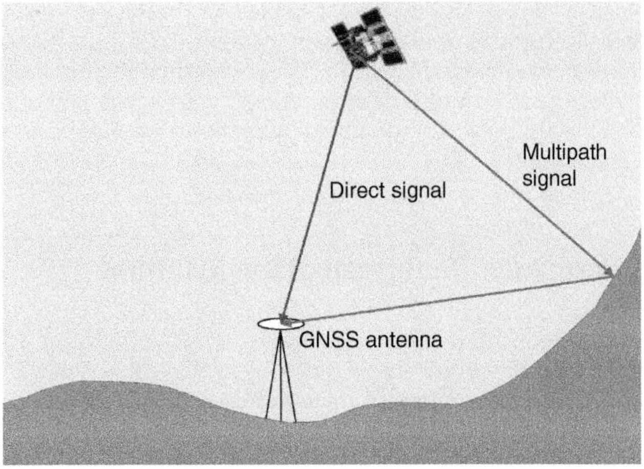

Figure 5.102 Multipath error.

impossible in a dynamic environment. Other techniques use signal-to-noise ratio (SNR) and signal phase/frequency information to detect and quantify multipath.

Despite several research efforts addressing multipath detection and mitigation techniques, this is still a major error source in GNSS applications. Techniques such as the narrow correlator (Dierendonck et al. 1992), the double delta/strobe correlator (McGraw and Braasch 1999), or the vision correlator by Fenton and Jones (2005), are not capable to eliminate multipath-related errors completely.

User Dynamics Error

There are various errors associated with a dynamic GNSS platform. These errors range from the physical masking of the GNSS antenna to the accuracy degradation caused by sudden acceleration of the antenna. If carrier phase is used, the resulting effect of "cycle slips" is the need for re-initialization (i.e., re-determination of the integer ambiguities). In general, a distinction is made between medium-low and high dynamic platforms.

5.61 UERE Vector and DOP Factors

The user equivalent range error (UERE) is an error vector along the line-of-sight user-satellite given by the projection of all system errors. The value of this vector can be reduced only by careful design of the receiver, since there is nothing the users of nonaugmented GNSS receivers can do to reduce other error sources. The UERE is generally measured in meters and is given as either a 1-sigma error (often denoted σ_r) or a 2-sigma error (95%). The portion of the UERE allocated to the space and control segments is called the *user range error* (URE) and is defined at the phase center of the satellite antenna. The portion of the UERE allocated to the user equipment is called the UE error (UEE). Specifically, the UERE is the root-sum-square of the URE and UEE. When SA was on, typical values of the UERE vector for GPS were below 10 m (95%) for P(Y) code receivers and about 33 m (95%) for C/A code receivers. With SA turned off, the UERE of C/A code receivers is typically less than 20 m, with the actual value dominated by ionospheric and multipath effects. Dual-frequency receivers (with the capability of removing almost entirely the ionospheric errors), typically experience smaller UEREs. Users of the new modernized GPS civilian signals, as well as future users of GALILEO and BEIDOU, will be able to use multiple frequencies to compensate for the ionospheric errors and thereby to achieve lower UEREs. In GNSS, the UERE values are strongly dependent on the time elapsed since the last upload from the control segment. In GPS normal operations, the time since last upload is limited to no more than 1 day. The smallest UERE and best SIS accuracy will generally occur immediately after an upload of fresh NAV message data to a satellite, while the largest UERE and worst SIS accuracy will usually be with the stalest NAV message data just prior to the next upload to that satellite.

5.62 GNSS Performance Requirements in Aviation

The aviation community has made great efforts to rationalization and standardization of the navigation performance parameters and requirements, thus specifying the so-called required navigation performance (RNP) that an airborne navigation system must achieve (ICAO 1995, 2010). Accuracy, integrity, availability, and continuity are used to describe the RNP for operations in various flight phases and within specified classes of airspace. The four parameters used to characterize the navigation systems performance are summarized (DoT/DoD 2008):

Accuracy: The accuracy of an estimated or measured position of a craft (vehicle, aircraft, or vessel) at a given time is the degree of conformance of that position with the true position, velocity and/or time of the craft. Since accuracy is a statistical measure of performance, a statement of navigation system accuracy is meaningless unless it includes a statement of the uncertainty (e.g., confidence level) in position that applies.

Integrity: Integrity is the measure of the trust that can be placed in the correctness of the information supplied by a navigation system. Integrity includes the ability of the system to provide timely warnings to users when the system should not be used for navigation.

Continuity: The continuity of a system is the ability of the total system (comprising all elements necessary to maintain craft position within the defined area) to perform its function without interruption during the intended operation. More specifically, continuity is the probability that the specified system performance will be maintained for the duration of a phase of operation, presuming that the system was available at the beginning of that phase of operation.

Availability: The availability of a navigation system is the percentage of time that the services of the system are usable by the navigator. Availability is an indication of the ability of the system to provide usable service within the specified coverage area. Signal availability is the percentage of time that navigation signals transmitted from external sources are available for use. It is a function of both the physical characteristics of the environment and the technical capabilities of the transmitter facilities.

In aviation applications, accuracy is a measure of the difference between the estimated and the true or desired aircraft position under nominal fault-free conditions. It is normally expressed as 95% bounds on horizontal and vertical position errors (ICAO 2008). In the avionics context, there are two distinguished types of accuracy that must be considered: the accuracy relative to the navigation system alone and the accuracy achieved by the combination of navigation and flight control systems. This second type of accuracy is called total system error (TSE) and is measured as deviation from the required flight trajectory. In large commercial aircraft and several military aircraft, the required flight trajectory and associated guidance information is computed by a flight management system (FMS) and constantly updated based on ATM and weather information. The navigation system estimates the aircraft state vector (i.e., position, velocity, attitude, and associated rates), determines the deviation from the required flight trajectory and sends these information either to the cockpit displays or to an automatic flight control system. The error in the estimation of the

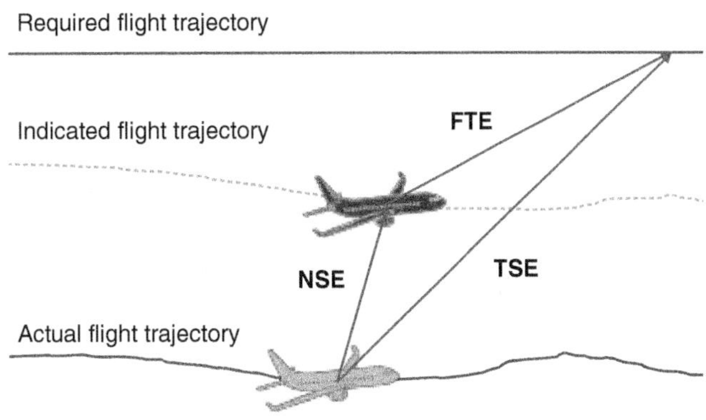

Figure 5.103 Navigation accuracy in aviation applications.

aircraft's position is referred to as navigation system error (NSE), which is the difference between the aircraft's true position and its displayed position. The difference between the desired flight path and the displayed position of the aircraft is called flight technical error (FTE) and accounts for aircraft dynamics, turbulence effects, human-machine interface, etc. As illustrated in Figure 5.103, the TSE is obtained as the vector sum of the NSE and the FTE.

The integrity risk can be conveniently defined as the probability of providing a signal that is out of tolerance without warning the user in a given period of time. It is typically derived from the high-level target level of safety (TLS) which is an index, generally determined through historic accident data, against which the calculated risk can be compared to judge whether the operation of the system is safe or not. The navigation integrity requirements in various flight phases/operational tasks have been specified by ICAO in terms of three key performance indicators: the probability of failure over time (or per single approach), the time-to-alert (TTA), and the horizontal/vertical alert limits. The TTA is the maximum allowable time elapsed from the onset of the system being out of tolerance until the equipment enunciates the alert. The alert limits represent the largest horizontal and vertical position errors allowable for safe operations. They are defined as follows (RTCA 2008):

Horizontal alert limit (HAL): The HAL is the radius of a circle in the horizontal plane (the local plane tangent to the WGS-84 ellipsoid), with its center being at the true position, which describes the region that is required to contain the indicated horizontal position with the required probability for a particular navigation mode.

Vertical alert limit (VAL): The VAL is half the length of a segment on the vertical axis (perpendicular to the horizontal plane of the WGS-84 ellipsoid), with its center being at the true position, that describes the region that is required to contain the indicated vertical position with the required probability for a particular navigation mode.

According to these definitions, in order to assess the navigation system integrity, the NSE must be determined first and checked against the alert limits (ALs) applicable to the current flight operational phase. However, since the NSE is not observable by the pilot or by the avionics on board the aircraft, another approach to evaluate the system integrity has to be adopted. The standard approach consists in estimating the worst-case NSE and confronting this value with the corresponding AL. These limits for NSE are known as protection levels (PLs) and represent high-confidence bounds for the NSE. Similarly to the positioning errors, the PLs are stated in terms of horizontal and vertical components (HPL and VPL).

5.63 GNSS Augmentation Strategies in Aviation

In aviation applications, GNSS alone does not guarantee the level of performance required in several flight phases and operational tasks. In particular, GNSS fails to deliver sufficient accuracy and integrity levels for mission-critical and safety-critical operations such as precision approach/auto-landing, avionics flight test and flight inspection. Therefore, appropriate augmentation strategies must be implemented to accomplish these challenging tasks using GNSS as the primary source of navigation data.

Accuracy Augmentation

Differential GNSS (DGNSS) techniques can be used to address accuracy requirements. Specifically, in the case of GPS, differential techniques have been developed which can provide accuracies comparable with current landing systems. A variety of local area DGNSS (LAD) and wide area DGNSS (WAD) networks have been proposed over the past 20 years and some LAD/WAD systems have achieved the levels of maturity required to enter the market. DGNSS was developed to meet the needs of positioning and distance-measuring applications that required higher accuracies than stand-alone GNSS could deliver. DGNSS involves the use of a control or reference receiver at a known location to measure the systematic GNSS errors; and, by taking advantage of the spatial correlation of the errors, the errors can then be removed from the measurement taken by moving or remote receivers located in the same general vicinity. There have been a wide variety of implementations described for affecting such a DGNSS. Our intent is to characterize various DGNSS and to compare their strengths and weaknesses. Two general categories of DGNSS can be identified: those that rely primarily upon the code measurements and those that rely primarily upon the carrier phase measurements. Using carrier phase, high accuracy can be obtained (centimeter level), but the solution suffers from integer ambiguity and cycle slips. Whenever a cycle slip occurs, it must be corrected for, and the integer ambiguity must be recalculated. The pseudorange solution is more robust but less accurate (2–5 m). As it is not affected by cycle slips, there is no need for re-initialization.

A typical DGNSS architecture is shown in Figure 5.104. The system consists of a reference receiver (RR) located at a known location that has been previously surveyed, and one or more DGNSS user receivers (URs). The RR antenna, differential correction processing system, and datalink equipment (if used) are collectively called the reference station (RS). Both the UR and the RR data can be collected and stored for later processing, or sent to the desired location in real time via the datalink. DGNSS is based on the principle that receivers in the same vicinity will simultaneously experience common errors on a particular satellite ranging signal. In general, the UR (mobile receiver) use measurements from the RR to remove the common errors. In order to accomplish this, the UR must simultaneously use a subset or the same set of satellites as the reference station. The DGNSS positioning equations are formulated so that the common errors cancel. The common errors include signal path delays through the atmosphere, and satellite clock and ephemeris errors. For PPS users, the common satellite errors are residual system errors that are normally present in the PVT solution. For SPS users, the common satellite errors also include the intentionally added errors from SA. Errors that are unique to each receiver, such as receiver measurement noise and multipath, cannot be removed without additional recursive processing (by the reference receiver, user receiver, or both) to provide an averaged, smoothed, or filtered solution.

Various DGNSS techniques are employed depending on the accuracy desired, where the data processing is to be performed, and whether real-time results are required. If real-time results are required, then a datalink is also required. For applications without a real-time requirement, the data can be collected and processed later. The accuracy requirements usually dictate which measurements are used and what algorithms are employed. In the case of differential GPS (DGPS), accuracy is independent of whether SPS or PPS is being used, although real-time PPS DGPS can have a lower

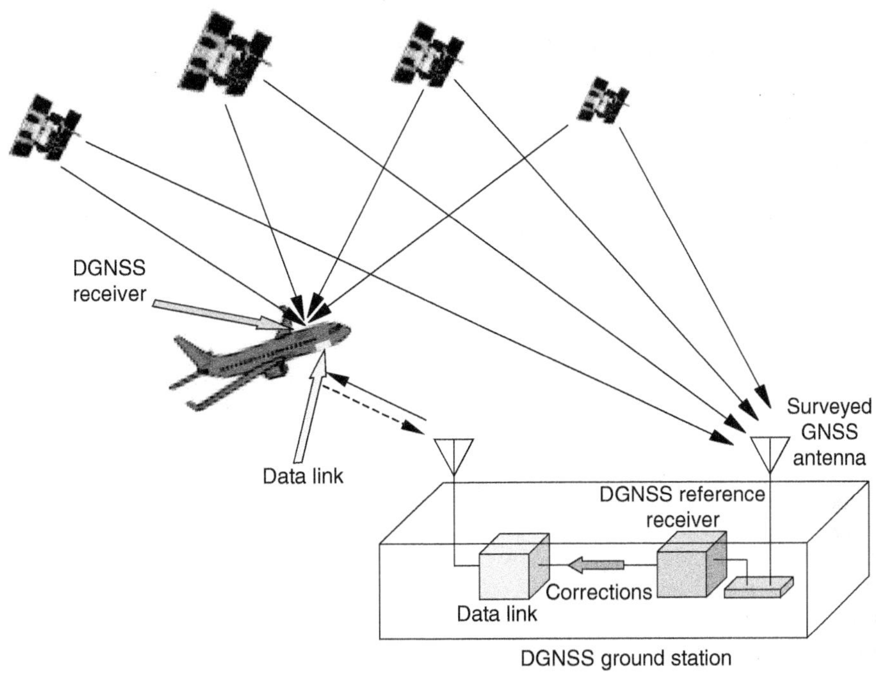

FIGURE 5.104 Typical DGNSS architecture (Sabatini and Palmerini 2008).

data rate than SPS DGPS because the rate of change of the nominal system errors is slower than the rate of change of SA. However, the user and the reference station must be using the same service (either PPS or SPS). The clock and frequency biases for a particular satellite will appear the same to all users since these parameters are unaffected by signal propagation or distance from the satellite. The pseudorange and delta-range (Doppler) measurements will be different for different users because they will be at different locations and have different relative velocities with respect to the satellite, but the satellite clock and frequency bias will be common error components of those measurements. The signal propagation delay is truly a common error for receivers in the same location, but as the distance between receivers increases, this error gradually de-correlates and becomes independent. The satellite ephemeris has errors in all three dimensions. Therefore, part of the error will appear as a common range error and part will remain a residual ephemeris error. The residual portion is normally small and its impact remains small for similar observation angles to the satellite. The first accepted standard for SPS DGPS was developed by the Radio Technical Commission for Maritime Services (RTCM) Special Committee-104 (RCTM 2011). The data interchange format for NATO users of DGPS is documented in STANAG 4392 (NATO 1998, p. 68). The SPS reversionary mode specified in STANAG 4392 is compatible with the RCTM SC-104 standards. The standards are primarily intended for real-time operational use and cover a wide range of DGPS measurement types. Most SPS DGPS receivers are compatible with the RCTM SC-104 differential message formats. DGPS standards for aeronautical use were originally developed by the Radio Technical Commission for Aeronautics (RTCA) to allow Special Category-I (SCAT-I) precision approach using range-code differential. These standards were contained in RCTA document DO-217 and intended only for limited use until an international standard could be developed for precision approach (RTCA 1993). More recently, the two separate standards were developed by RTCA for GPS wide area and local area augmentation systems (WAAS/LAAS). These standards define the minimum operational performance requirements for different WAAS/LAAS implementation types allowing WAAS operations down to CAT-I (RTCA 2006) and LAAS operations down to CAT-IIIb (Rife et al. 2004; RTCA 2004). There are two primary variations of the differential measurements and equations. One is based on ranging-code measurements and the other is based on carrier-phase measurements. There are also several ways to implement the datalink function. DGNSS can be designed to serve a limited area from a single reference station, or can use a network of reference stations and special algorithms to extend the validity of the DGNSS technique over a wide area. The result is that there is a large variety of possible DGNSS implementations using combinations of these design features.

Ranging-Code DGNSS

The ranging-code differential technique uses the pseudorange measurements of the RS to calculate pseudorange or position corrections for the UR. The RS calculates pseudorange corrections for each visible satellite by subtracting the "true" range determined by the surveyed position and the known orbit parameters from the measured pseudorange. The UR receiver then selects the appropriate correction for each satellite that it is tracking, and subtracts the correction from the pseudorange that it has measured. The mobile receiver must only use those satellites for which corrections have been received. If the RS provides position corrections rather than pseudorange corrections, the corrections are simply determined by subtracting the measured position from the surveyed position. The advantage of using position corrections is obviously the simplicity of the calculations. The disadvantage is that the reference receiver and the user receiver must use the exact same set of satellites. This can be accomplished by coordinating the choice of satellite between the RR and the UR, or by having the RS compute a position correction for each possible combination of satellites. For these reasons, it is usually more flexible and efficient to provide pseudorange corrections rather than position corrections. RTCA, NATO STANAG, and RCTM standards are all based on pseudorange rather than position corrections. The pseudorange or position corrections are time tagged with the time that the measurements were taken. In real-time systems, the rate of change of the corrections is also calculated. This allows the user to propagate the corrections to the time that they are actually applied to the user position solution. This reduces the impact of data latency on the accuracy of the system but does not eliminate it entirely. SPS corrections become fully uncorrelated with the user measurements after about 2 minutes. Corrections used after 2 minutes may produce solutions, which are less accurate than stand-alone SPS GPS. PPS corrections can remain correlated with the user measurements for 10 minutes or more under benign (slowly changing) ionospheric conditions. There are two ways of pseudorange data processing: post-mission and real-time processing. The advantage of the post-mission solution over the real-time one is that it is more accurate, because the user can easily detect blunders and analyze the residuals of the solution. On the other hand, the main disadvantage of the post-mission solution is that the results are not available immediately for navigation. The typical algorithm of the ranging-code DGNSS postprocessed solution used in flight test and inspection tasks is the double difference pseudorange.

Figure 5.105 shows the possible pseudorange measurements between two receivers (k. l) and two satellites (p, q). If pseudoranges 1 and 2 from Figure 5.97 are differenced, then the satellite clock error and satellite orbit errors will be removed. If SA is active (not the present case), it will be removed completely only if the signals utilized in each receiver are transmitted exactly at the same time. The residual error from SA is not a problem for postprocessed positioning, where it is easy to ensure that the differencing is done between pseudoranges observed at the same time. Any atmospheric errors will also be reduced significantly with single differencing.

The basic mathematical model for single difference pseudorange observation is the following:

$$P_k^p - P_l^p = \rho_k^p - \rho_l^p - (dt_k - dt_l)c + d_{k,p} - d_{l,p} + d_{k,p}^p - d_{l,p}^p + \Delta\varepsilon_p \tag{5.305}$$

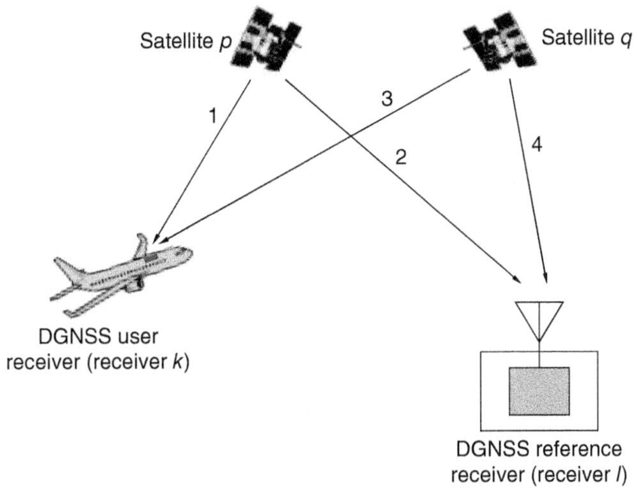

FIGURE 5.105 Pseudorange differencing.

where P_i^p is the pseudorange measurement, ρ_l^p denotes the geometric distance between the stations and satellite, dt_i denotes the receiver's clock offsets, $d_{i,p}$ denotes the receiver's hardware code delays, $d_{l,p}^p$ denotes the multipath of the codes, $\Delta\varepsilon_p$ denotes the measurement noise, and c is the velocity of light.

Equation (5.142) represents the single difference pseudorange observable between receivers. Another type of single difference apart from (5.120) is known as between-satellite single difference. There are four unknowns in equation (5.120) assuming that the coordinates of station k are known and that the difference in clock drifts is one unknown. Hence, four satellites are required to provide four single difference equations in order to solve for the unknowns. Single differences with code observations are frequently used in relative (differential) navigation. Using all pseudoranges shown in Figure 5.105, differences are formed between receivers and satellites. Double differences are constructed by taking two between-receiver single differences and differencing these between two satellites. This procedure removes all satellite dependent, receiver dependent, and most of the atmospheric errors (if the distance between the two receivers is not too large). The derived equation is

$$P_k^p - P_k^q - P_l^p - P_l^q = \rho_k^p - \rho_k^q - \rho_l^p - \rho_l^q + d_{i,p}^j \tag{5.306}$$

where $d_{i,p}^j$ denotes the total effect of multipath. There are three unknowns in equation (5.143). A minimum of four satellites is required to form a minimum of three double difference equations in order to solve for the unknowns. Using the propagation of errors law, it is shown that the double difference observables are twice as noisy as the pure pseudoranges:

$$\sigma_{DD} = \sqrt{\sigma_P^2 + \sigma_P^2 + \sigma_P^2 + \sigma_P^2} = 2\sigma_P \tag{5.307}$$

but they are more accurate, because most of the errors are removed. Note that multipath remains, because it cannot be modeled and it is independent for each receiver.

Carrier-Phase DGNSS

The carrier-phase measurement technique uses the difference between the carrier phases measured at the RR and UR. A double-differencing technique is used to remove the satellite and receiver clock errors. The first difference is the difference between the phase measurement at the UR and the RR for a single satellite. This eliminates the satellite clock error which is common to both measurements. This process is then repeated for a second satellite. A second difference is then formed by subtracting the first difference for the first satellite from the first difference for the second satellite. This eliminates both receiver clock errors which are common to the first difference equations. This process is repeated for two pairs of satellites resulting in three double-differenced measurements that can be solved for the difference between the reference station and user receiver locations. This is inherently a relative positioning technique; therefore, the user receiver must know the reference station location to determine its absolute position.

The single difference observable is the instantaneous phase difference between two receivers and one satellite. It is also possible to define single differences between two satellites and one receiver.

GNSS Integrity Augmentation

At the moment, satellite navigation systems are not certified to be used as sole means of aircraft navigation and landing. Contrary to inertial navigation systems (INS) and to the various radio navigation aids currently in use, GNSS is only certifiable for aircraft navigation if it is integrated or augmented by systems. The reason is not the accuracy but integrity. As discussed above, integrity is the ability of a system to provide timely warnings to users when the system should not be used for navigation (DoD/FAA 2008). In general, it is essential to guarantee that, with a specified probability P, either the horizontal/vertical radial position error does not exceed the prespecified thresholds (HAL/VAL), or an alarm is raised within a specified TTA interval when the horizontal/vertical radial position errors exceed the thresholds. To detect that the error is exceeding a threshold, a monitor function has to be installed within the navigation system. This is also the case within GNSS in the form of the ground segment. However, for GNSS, TTA is typically in the order of several hours, that is even too long for the cruise where a TTA of 60 seconds is required (an autoland system for zero meter vertical visibility must not exceed a TTA of 2 seconds). Various methods have been proposed and practically implemented for stand-alone GNSS integrity monitoring. A growing family of such implementations, already very popular in aviation applications, includes the so called receiver autonomous integrity monitoring (RAIM) techniques. Details about RAIM can be found in the literature (Parkinson and Spilker Jr. 1996; Brown 1992; Brown and Chin 1998; JPO 2007; Kalafus 1987; Kelly 1996; Parkinson and Axelrad 1988; Walter and Enge 1995). Regarding DGNSS, it should be underlined that it does more than increasing GNSS positioning accuracy; it also contributes to enhancing GNSS integrity by compensating for anomalies in the satellite ranging signals and navigation data message. The range and range rate corrections provided in the ranging-code DGNSS correction message can compensate for ramp and step type anomalies in the individual satellite signals, until the corrections exceed the maximum values or rates allowed in the correction format. If these limits are exceeded, the user can be warned not to use a particular satellite by placing "do-not-use" bit patterns in the corrections for that satellite (as defined in STANAG 4392 or RTCA/RTCM message formats) or by omitting the corrections for that satellite.

Step anomalies will normally cause carrier-phase DGNSS receivers to lose lock on the carrier phase, causing the reference and user receivers to reinitialize. User receiver noise, processing anomalies, and multipath at the user GPS antenna cannot be corrected by a DGNSS. These errors are included in the overall DGNSS error budget. Errors in determining or transmitting the satellite corrections may be passed on to the differential user if integrity checks are not provided within the reference station. These errors can include inaccuracies in the RS antenna location that bias the corrections, systematic multipath due to poor antenna sighting (usually in low-elevation angle satellites), algorithmic errors, receiver inter-channel bias errors, receiver clock errors, and communication errors. For these reasons, typical WAD and LAD RS designs also include integrity checking provisions to guarantee the validity of the corrections before and after broadcast. In the aviation context, various strategies have been developed for increasing the levels of integrity of DGNSS based navigation/landing systems. These include both satellite-based augmentation systems (SBAS) and ground-based augmentation systems (GBAS) to assist aircraft operations in all flight phases. These systems are effectively WAD and LAD systems that employ, respectively, DGNSS vector correction (SBAS) and scalar (pseudorange) correction techniques (GBAS). Various methods have been proposed and practically implemented for stand-alone GNSS integrity monitoring. These approaches are already very popular in aviation applications, including the so called receiver autonomous integrity monitoring (RAIM) technique. In addition to the existing SBAS, GBAS, and RAIM techniques, GNSS augmentation may take the form of additional information being provided by other on-board avionics systems. As these systems normally operate via separate principles than the GNSS, they are not subject to the same sources of error or interference. A system such as this is referred to as an aircraft based augmentation system or ABAS (Sabatini et al. 2013a and 2013b). ABAS is different from RAIM, in which the aircraft characteristics (flight dynamics, body shape, antenna location, electromagnetic compatibility, etc.) are typically not considered and various kinds of consistency check are accomplished using the GNSS signals to detect and exclude faulty/unreliable satellites.

References

Ashkenazi, V., 1997, "Principles of GPS and Observables," *Lecture Notes, Institute of Engineering Surveying and Space Geodesy (IESSG)–University of Nottingham.*

Parkinson, B. W. and Spilker Jr., J. J., 1996, "Global Positioning System: Theory and Applications—Volume I," *Progress in Astronautics and Aeronautics*, Vol. 163.

Brown, R. G., 1992, "A Baseline GPS RAIM Scheme and a Note on the Equivalence of Three RAIM Methods," *Navigation*, Vol. 39, no. 3, pp. 301–316.

Brown, R. G. and Chin, G. Y., 1998, "GPS RAIM: Calculation of Thresholds and Protection Radius Using Chi-Square Methods: A Geometric Approach," *Journal of the Institute of Navigation*, Vol. 5.

Daly, P., 1995, "GLONASS Approaches Full Operational Capability (FOC)," ION GPS-95.

Dierendonck, V. A., Fenton, P., and Ford, T., 1992, "Theory and Performance of Narrow Correlator Spacing in a GPS Receiver," *Navigation*, Vol. 39, no. 3, pp. 265–283.

DoD, 2008, "Global Positioning System Standard Positioning Service Performance Standard," *Assistant Secretary of Defense for Command, Control, Communications, and Intelligence.*

DoD/FAA, 2008, "Global Positioning System Wide Area Augmentation System (WAAS) Performance Standard," FAA, Washington, DC, United States.

Dodson, A. H., 1997, "Propagation Effects on GPS Measurements," IESSG, University of Nottingham.

DoT/DoD, 2008, *US Federal Radio/Navigation Plan*, report n., Department of Defence, Department of Transport, United States.

Fenton, P. C. and Jones, J., 2005, "The Theory and Performance of NovAtel Inc.'s Vision Correlator," *Proceedings of the ION GNSS.*

Giffard, R., 1999, "Estimation of GPS Ionospheric Delay Using L1 Code and Carrier Phase Observables," DTIC document.

Gleason, S. and Gebre-Egziabher, D., 2009, *GNSS Applications and Methods*, Artech House.

ICAO, 1995, "Report of the Special Communications/Operations Divisional Meeting," International Civil Aviation Organization (ICAO).

ICAO, 2008, *Doc 9613: Performance-Based Navigation (PBN) Manual*, ICAO, 3rd ed., International Civil Aviation Organization, Montreal, Canada.

ICAO, 2010, *Performance Based Operations Manual*, International Civil Aviation Organization, Montreal, Canada.

JPO G, 2007, *GPS Precision Positioning Service (PPS) Performance Standard*, NAVSTAR Global Positioning System Joint Program Office.

Kalafus, R., 1987, "Receiver Autonomous Integrity Monitoring of GPS," Project Memorandum DOTTSC-FAA-FA-736-1, US DOT Transportation System Center, Cambridge, MA.

Kelly, R., 1996, "Derivation of the RAIM Algorithm from First Principles with Performance Comparisons between Published Algorithms," Proceedings of the 1996 National Technical Meeting of The Institute of Navigation.

Klobuchar, J. A., 1986, "Design and Characteristics of the GPS Ionospheric Time Delay Algorithm for Single Frequency Users," PLANS'86-Position Location and Navigation Symposium.

Leick, A., Rapoport, L., and Tatarnikov, D., 2015, *GPS Satellite Surveying*, John Wiley & Sons.

McGraw, G. A. and Braasch, M. S., 1999, "GNSS Multipath Mitigation Using Gated and High Resolution Correlator Concepts," Institute of Navigation, National Technical Meeting "Vision 2010: Present and Future," San Diego, CA.

Parkinson, B. W. and Axelrad, P., 1988, "Autonomous GPS Integrity Monitoring Using the Pseudorange Residual," *Navigation*, Vol. 35, no. 2, pp. 255–274.

RCTM, 2011, "RTCM Recommended Standards for Differential GNSS Service," report n., Radio Technical Committee for Maritime Services, Washington, DC, United States.

Rife, J., Pullen, S., Pervan, B., et al., 2004, "Core Overbounding and Its Implications for LAAS Integrity," Proceedings of ION GNSS.

RTCA, 1993, "Minimum Aviation System Performance Standards DGNSS Instrument Approach System: Special Category 1 (scat-1)," Radio Technical Commission for Aeronautics (RTCA) Inc., Washington, DC, United States, RTCA DO-217.

RTCA, 2004, "RTCA DO-245A: Minimum Aviation System Performance Standards for the Local Area Augmentation System (LAAS)," SC-159, 9/12/2004, RTCA DO-245.

RTCA, 2006, *Minimum Operational Performance Standards for Global Positioning System/Wide Area Augmentation Systems Airborne Equipment*, Radio Technical Commission for Aeronautics (RTCA) Inc., Washington, DC, United States, RTCA DO-229D.

RTCA, 2008, *Minimum Operational Performance Standards for GPS/GBAS Airborne Equipment*, Radio Technical Commission for Aeronautics (RTCA) Inc., Washington, DC, United States, RTCA DO-253C.

Sabatini, R., Moore, T., and Hill, C., 2013a, A New Avionics Based GNSS Integrity Augmentation System: Part 2—Integrity Flags, *Journal of Navigation*, Vol. 66, No. 4, pp. 511–522. DOI: 10.1017/S0373463313000143

Sabatini, R., Moore, T., and Hill, C., 2013b, A New Avionics Based GNSS Integrity Augmentation System: Part 1—Fundamentals, *Journal of Navigation*, Vol. 66, No. 3, pp. 363–383. DOI: 10.1017/S0373463313000027

Sabatini, R. and Palmerini, G. B., 2008, "Differential Global Positioning System (DGPS) for Flight Testing," Flight Test Instrumentation Series, Systems Concepts and Integration Panel (SCI-135), NATO Science and Technology Organization.

Seeber, G., 2003, *Satellite Geodesy: Foundations, Methods, and Applications*, Walter de Gruyter.

Siouris, G., 1993, *Aerospace Avionics Systems*, Chapter 5, Academic Press, Inc.

Thomin, D. W., 1984, "Global Positioning System-A Modification to the Baseline Satellite Constellation for Improved Geometric Performance," report n., DTIC document.

Walter, T. and Enge, P., 1995, "Weighted RAIM for Precision Approach," Proceedings of ion GPS.

PART 14

Airborne Separation Assurance and Collision Avoidance

Subramanian Ramasamy, Alessandro Gardi, and Roberto Sabatini

5.64 Introduction

In this subsection, a short summary of the top-level components of airborne and ground surveillance systems are defined and a brief description of the lower-level services, functions, equipment, roles, and procedures under each component are presented. These top-level components as given by ICAO are conflict detection and resolution (CD&R), traffic synchronization, demand and capacity balancing, aerodrome/airspace operations, airspace and service delivery management.

5.65 Rules of AIR

CD&R algorithms take into account the same collision avoidance requirements of Rules of Air for manned aviation, delineated in the ICAO Annex 2 (FAA 2016), in order to guarantee the equivalent level of safety (ELOS). Aviation regulations, collectively known as Federal Aviation Regulation (FAR), are codified in the Code of Federal Regulations (CFR), Title 14, Chapter I (Airservices 2016a). The CFR along with supplementary material like handbooks, orders, advisory circulars, and technical standards orders issued by FAA, define appropriate standards, procedures, and practices to ensure that manufactures and operators are able to establish a minimum level of safety and reliability required for civil operations (Eurocontrol 2013). Code of Federal Aviation, Title 14, part 91 has established right-of-way rules, aircraft speed, minimum safe altitudes, equipment, instrument, and certificate requirements, maintenance, preventive maintenance, special flight operations, i.e., aerobatic flight, flight test areas, etc. Federal regulations do not permit pilots to fly below 10,000 ft or in proximity of airspace class B, C, and D at speed exceeding 250 knots. It is not allowed for aircraft to operate close to another aircraft in a way that creates a collision hazard. The pilot as well must contribute to collision avoidance by being alert to "see and avoid" other aircraft [CFR §91.111]. The CFR Title 14, Section 91 defines two types of flight rules, namely visual flight rules and instrument flight rules, which apply to all aircraft except moored balloons, kites, unmanned rockets, and unmanned free balloons, which are governed by part 101 of the same chapter, and ultralight vehicles operated in accordance with part 103 of the same chapter.

5.66 Airspace Categories and Classes

The airspace is defined as the portion of the atmosphere controlled by a country above its territory, including its territorial waters or, more generally, any specific 3D area of the atmosphere (Durand et al. 2016). According to CFR, Title 14, part 71–73, and the Aeronautical Information Manual, the national airspace is divided into two categories and four types based on the complexity or density of aircraft movements, proximity to airports, altitude, the nature of the operations conducted within the airspace, the level of safety required, and the national and public interest:

- Regulations that include airspace class A, B, C, D, E, restricted and prohibited areas
- Nonregulatory framework that includes airspace class G, military operations area (MOA), warning areas, alert areas, and controlled firing areas.

While classes are mutually exclusive, categories can be defined within any class as needed. Categories typically used to provide additional information to pilots about hazards activities common to the airspace. There are four types within these two categories:

- Controlled airspace
- Uncontrolled airspace
- Special use airspace
- Other airspace

Airspace classes A through E, ordered from more restrictive to less restrictive, corresponds to controlled airspace; however, when overlapping airspace designations apply to the same airspace, the operating rules associated with the more restrictive airspace designation apply [CFR §71.9].

5.67 Separation Standards

Separation standards refer to the minimum distance that aircraft operating in controlled airspace and at airports with an operational control tower must keep in between them (Airservices 2016b). Different separation standards apply to aircraft operating under IFR and VFR. Aircraft flying under IFR in controlled airspace up to 29,000 ft are typically separated by 1,000 ft vertically unless they are separated horizontally. When above 29,000 ft, the vertical separation increases to 2,000 ft, except in airspace where reduced vertical separation minima (RVSM) is applied. It is generally guaranteed that when aircraft are separated vertically, horizontal separation can be reduced without compromising safety. In controlled en route airspace, the horizontal separation standard between aircraft flying at the same altitude is 5 NM while in terminal area airspace, the minimum separation is 3 NM. For general aviation aircraft (noncommercial aircraft) outside of controlled airspace, separation can be as close as 500 ft vertically and 500 ft horizontally. ICAO rules divide aircraft into three main categories including light, medium, and heavy in order to establish separation between aircraft based on wake vortices. The classification is based on the maximum take-off weight of the aircraft and the separation distances are based on experience (flight tests). The ICAO classification is provided in Table 5.19 adapted from (Campos and Marques 2016). The only exception is for Airbus A380 aircraft, which has a different set of separation distance standards.

An analytical model has been proposed for calculating the safe separation distance between aircraft and is especially valuable in non-cooperative sensor–based aircraft separation. The safe separation distance is given by

$$s = \frac{A/B}{f\eta} \tag{5.308}$$

Designation	Maximum take-off weight	Aircraft example	Separation distances for a pair of aircraft
Heavy	$W > 136t$	Boeing B 747-400 (W = 396,893 kg)	Heavy-heavy 4 NM Heavy-medium 5 NM Heavy-light 6 NM
Medium	$7t < W < 136t$	Boeing B 737-300 (W = 70,080 kg)	Medium-heavy 3 NM Medium-medium 3 NM Medium-light 5 NM
Light	$W < 7t$	Cessna citation 500 (W = 5,375 kg)	Light-heavy 3 NM Light-medium 3 NM Light-light 3 NM

TABLE 5.19 ICAO Separation Standards

where s is the safe separation distance, f is the fraction of the available control moment that is used, η is the turbulent kinematic shear viscosity, and

$$A = \frac{I_f\, r^2}{2} \tag{5.309}$$

where I_f is the incident flow velocity and r is the vortex core radius. B is given by

$$B = \frac{12}{h}\,\frac{v_2}{v_1}\,\frac{S_a}{S}\,\frac{b_a}{b_2}\,\frac{a_1}{b_1}\,\frac{a_1}{c_1} \tag{5.310}$$

where h is a dimensionless factor specified by the wing planform, v_2 and v_1 are the volume loading (wing loading divided by span) of preceding and following aircraft, S_a is the aileron area, S is the wing area, b_a is the moment arm of ailerons, b is the wing span, a is the core radius of wing-tip vortices, and c is the wing chord.

5.68 Collision Detection and Avoidance

A key objective of the ATM modernization concept is the delegation of the task of separation assurance from air traffic controller to the flight crew supported by higher levels of automation. In order to be able to satisfy this goal, aircraft are required to implement automated means for airborne conflict management, which is categorized as

- Conflict detection
- Conflict prevention
- Conflict resolution

Onboard systems will need to be able to identify violation of separation minima in strategic (long term), tactical, and emergency timeframes. Resolutions for these conflicts need to be computed in order to offer the flight deck crew an alternative routing. In case coordination with other traffic participants is required, those updated trajectories need to be communicated, shared, and validated. This has to be achieved independent of ground infrastructure.

One of the key goals is to describe the system design aspects that will allow airborne separation assurance and conflict management in line with the ATM modernization as well as UAS traffic management (UTM) concepts of operations. Over the years, many efforts have been made in order to develop separation assurance and collision avoidance (SA&CA) requirements and, in 2004, the Radio Technical Commission for Aeronautics (RTCA) Special Committee 203 (SC-203) was formed (Oztekin and Wever 2015). SC-203 was handed over the task of defining Minimum Aviation System Performance Standards (MASPS) for SA&CA.

The FAA defined sense-and-avoid as "the capability of [an unmanned aircraft] to remain well clear from and avoid collisions with other airborne traffic" (FAA 2009) and identified that it consisted of two elements (Consiglio et al. 2015):

- A self-separation component that ensures a safe separation based on a variable time-based threshold. In this way, the aircraft remain "well-clear" of each other.
- A collision avoidance component that operates when the safe separation is lost and an extreme maneuver is needed to prevent a collision that is penetrating the collision volume. In fact, for the collision avoidance maneuver, a distance-based threshold is considered.

The SA&CA core and crosscutting capabilities are also identified and are here summarized in Tables 5.20 and 5.21 (Davis and Cook (2015)).

Core capability	Description
Target detection	Ability to determine presence and location of any airborne object that presents a collision threat to the unmanned aircraft covered by this SA&CA system
Target tracking	Ability to refine a detected airborne object's position (2D or 3D) and velocity based on one or more target detection(s) from a single sensor
Multisensor data fusion	Ability to refine a detected airborne object's position (2D or 3D) and velocity based on one or more target detection(s) or track(s) from multiple sensors
Object identification	Ability to discriminate type and characteristics of a detected airborne object
Threat assessment	Ability to determine risk of collision based on relative position (2D or 3D) and velocity of the detected airborne object and the unmanned aircraft covered by the SA&CA system
Alerting	Ability to communicate that a risk threshold has been met
Maneuver selection	Ability to determine the appropriate action needed to resolve identified collision risk(s) to the appropriate separation threshold
Maneuver notification	Ability to communicate the selected maneuver in order to execute within the timing window
Maneuver execution	Ability to execute the selected maneuver within the timing window
Return-to-home/ mission	Ability of UAS to return to intended or amended course after an avoidance or separation maneuver is complete

Table 5.20 Core SA&CA Capabilities

Crosscutting capability	Description
Pilot/operator presentation	Ability to provide the pilot/operator with the required information required to perform/monitor the core SA&CA capabilities
System integration	Ability to integrate the core capabilities into existing and planned manned and unmanned aircraft
Certification/qualification	Ability to show through process and evidence that the SA&CA system meets an acceptable safety level
Concept of operations (CONOPS)/ procedure validation	Ability to conform to approved CONOPS and operational procedures
Data standardization/ characterization	Ability to enable collaboration and meaningful comparisons through databases, data standards, and data integrity
Airspace/ATM/UTM	Ability to assess the impact on the airspace UAS system performance and the performance of supporting systems, such as SA&CA, may have on ATM/UTM
Traffic separation minima	Ability to determine the separation standards and the environmental or operational conditions which inform SA&CA design

Table 5.21 Crosscutting SA&CA Capabilities

In order to define a set of SA&CA standards, for the certification and operational approval of unmanned aircraft, the FAA workshop identified five subfunctions that a SA&CA system shall perform including:

- Detection of conflicting traffic
- Determination of the right of way
- Analysis of the flight paths
- Maneuvering
- Communication

According to the performance of the on board SA&CA technologies, it is possible to obtain different encounter sensing dimension. This design factor is defined as the extension of space surveyed for the conflict risks monitoring. The trajectory information of other traffic is determined after performing multisensor data fusion of information obtained from the SA&CA technologies (Figure 5.106).

Criticality analysis is carried out to prioritize (i.e., to determine if a collision risk threshold is exceeded for all tracked intruders) and to determine the action commands. If an evading action is required, the SA&CA system generates and optimizes an avoidance trajectory according to a

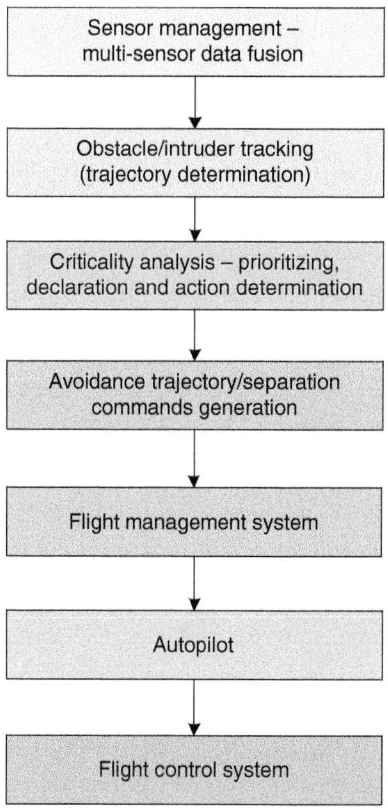

FIGURE 5.106 SA&CA system processes.

priority-based cost function based on minimum distance, fuel, time, and closure rate criteria with the aid of differential geometry algorithms to generate a smooth trajectory.

Detection of conflict risks requires the identification of potential threats in the airspace proximity close to the own ship and continuous data update and collection recording. Therefore, it is necessary to provide the system with the tools to enable the detection and identification of specific type of threats. These tools can be simple databases, e.g., for detecting ground threats, sensors set for the detection of moving threats, etc.

Considering a classical two aircraft conflict, where an aircraft on the lower segment flies at v_1 m/s, and another aircraft flies on the upper segment flies at v_2 m/s. The angle of incidence is α and D is the separation standard.

Let

$$r = \frac{v_2}{v_1} \tag{5.311}$$

If two aircraft are in conflict, the following inequality should be satisfied to avoid any conflicts.

$$(x_1 - x_2)^2 + (y_1 - y_2)^2 \le D^2 \tag{5.312}$$

The above second degree inequality is only satisfied if the below equation holds true:

$$(x_1 - x_2)^2 + (y_1 - y_2)^2 - D^2 = 0 \tag{5.313}$$

Equation (3.222) has at least one solution. Hence the discriminant is given by:

$$\Delta = D^2 \left(v_1^2 - 2v_1v_2 \cos\alpha + v_2^2\right) - \sin^2\alpha \, (l_2v_1 - l_1v_2)^2 \tag{5.314}$$

where the aircraft on the lower segment is at a distance l_1 from the intersection point and the other aircraft is at a distance l_2. Δ must be positive and the extremal points r_1 and r_2 are given by

$$r_1 = l_1 \frac{v_2}{v_1} + D \frac{\sqrt{1 + \left(\frac{v_2}{v_1}\right)^2 - 2\left(\frac{v_2}{v_1}\right)\cos\alpha}}{\sin\alpha} \tag{5.315}$$

$$r_2 = l_1 \frac{v_2}{v_1} - D \frac{\sqrt{1 + \left(\frac{v_2}{v_1}\right)^2 - 2\left(\frac{v_2}{v_1}\right)\cos\alpha}}{\sin\alpha} \tag{5.316}$$

If the length of the segment is given by $r_1 - r_2$, the function on r and α does not depend on l_1 (Granger 2002). It implies that the obtained information is sufficient and there is no uncertainty of any kind. Conflicts could be exactly predicted as soon as the position of the aircraft is known, even if they are very far away from the intersection point. A number of studies have shown that only one conflict out of three to five detected and monitored would really result in separation violation. For example, on simulated traffic on a busy day of 1999 in the French airspace above FL 320, when the uncertainties on trajectory predictions increase, the number of necessary maneuvers also increases dramatically (Granger 2002). As an example, in an analysis performed, without uncertainty, 972 conflicts occur and 1041 maneuvers are required to solve these uncertainties. When a 2% uncertainty is considered on ground speed and 5% uncertainty on climbing or descending rate, 2,461 maneuvers are necessary to solve the detected conflicts. The number of maneuvers increases to 3,881 maneuvers for a 5% uncertainty on ground speed and 15% uncertainty on vertical speed and it eventually reaches 6,819

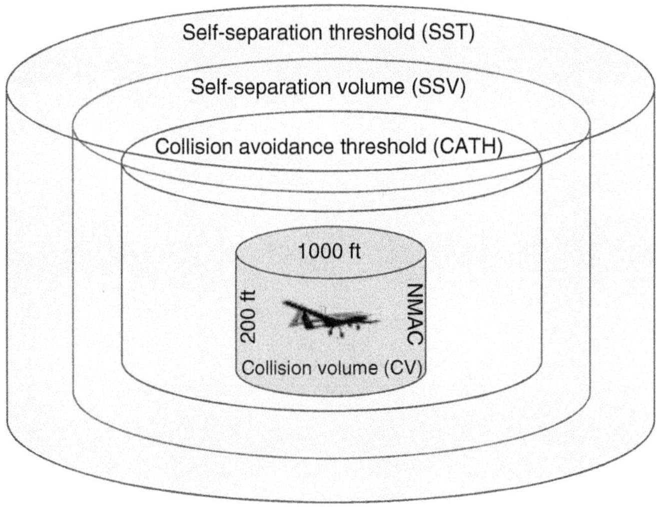

Figure 5.107 CATH, SSV, and SST boundaries.

maneuvers with 10% ground speed and 30% vertical speed uncertainties (Granger 2002). Figure 5.107 shows the different SA&CA zones and the condition of CA. As can be seen, the self-separation threshold (SST) and self-separation volume (SSV) are boundaries where there is no threat, but an aircraft detected far away can be turned into an obstacle. In controlled airspace, ATC ensures a safe distance is maintained from other aircraft, terrain, obstacles, and certain airspace not designated for routine air travel. The ATC is in charge of giving the order to reconfigure the path and avoid crossing trajectories. The second volume is called the self-separation threshold. In this situation, the ATC may have failed, or another unresolved alarm may have occurred, and the aircraft detected turns out to be a threat. The SSV could be the only function provided given that the safety analysis demonstrates that the target level of safety can be met with SSV alone. However, when all forms of SA have failed, we are now at the third volume and the CA (provided by the collision avoidance threshold (CATH) boundary) takes appropriate action to prevent a threat aircraft from penetrating the collision volume (CV). On rare occasions the UAS self-separation functions may conflict with ATC separation services, and as a general rule the UAS follows the ATC separation services.

5.69 Conflict Detection and Resolution Approaches

Conflict detection and resolution (CD&R) algorithms are implemented to monitor the airspace for both static and dynamic obstacles using onboard sensors and using information from the ground traffic management system. Cooperative systems and non-cooperative sensors are employed for object detection in the surrounding airspace. To detect and resolve a conflict, it is important to assess and compare the trajectories of the host aircraft and the detected object (Kopřiva et al. 2012). Onboard the aircraft, the trajectory prediction and computation module processes the raw sensor information and provides the trajectories of the host aircraft and the tracked objects. When multiple sensors/systems are employed for tracking, then multisensor data fusion techniques are adopted. There are three basic models of trajectory prediction methods (Figure 5.108). These can be divided based on flight profiles, state information extraction and propagation of states in the aircraft trajectory. The propagation-based approaches are commonly used. The first method is called a nominal method, in which the trajectory is predicted directly from the sensor data without considering the possibility of any uncertainty or change. The output of the nominal trajectory

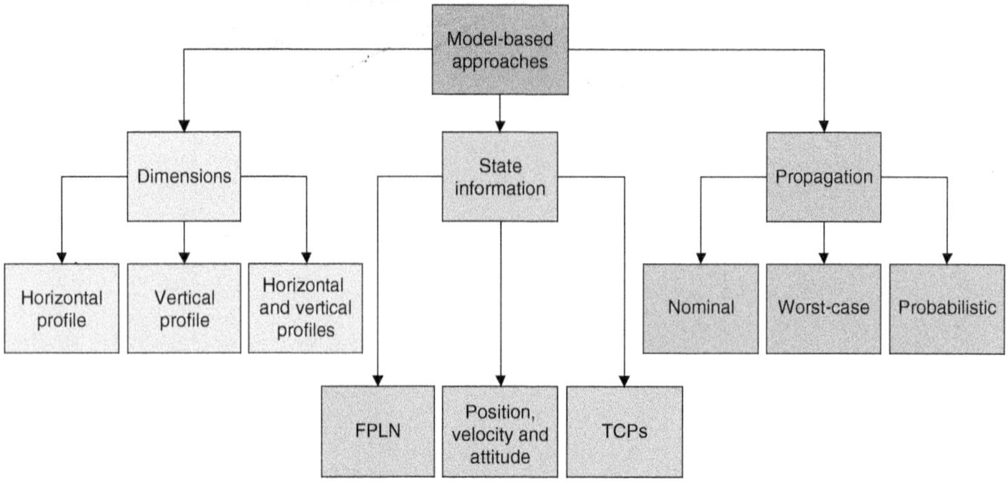

FIGURE 5.108 Trajectory prediction/estimation approaches.

predictor is a single trajectory computed from information obtained after a few time epochs. Different techniques including optimal state estimation, Taylor series based prediction, etc. can be employed for predicting the trajectory of the tracked objects. The nominal method prediction is suitable for short-term predictions, where the probability of change is relatively low. The second method is a worst case based approach and it covers the whole range of maneuvers the aircraft may perform and using a look-ahead time parameter computes the area where the aircraft may occur. This area is then considered as the predicted trajectory. In the probabilistic prediction approach, the uncertainties are used to model the potential variations in the trajectory. To construct this model, all possible trajectories are generated (like for the worst-case prediction) and each trajectory is evaluated by the probability function. The third method is a probability-based approach and is the most widely used technique, in which decisions are made based on the likelihood of the conflict. State information-based approaches employ flight plan (FPLN), position, velocity and attitude (PVA) as well as Trajectory Change Points (TCP)-based assessment techniques. Some of the key methods are rule-based, game theory, field-based, geometric, numerical optimization, combined and multiagent methods (Kopřiva et al. 2012; Melega 2014). In general, the existing conflict resolution solutions (Figure 5.109) can be divided into two parts. The first involves simple collision avoidance solutions, which are based on avoiding collisions by steering the vehicle into opposite direction using different techniques. The biggest drawback of such solutions is the intervention into the steering which may not be desirable for the mission of the UAV. These solutions do not allow a UAV to control its distance from an obstacle which may be necessary in a couple of situations (Bouabdallah et al. 2007; Roberts et al. 2007; Scherer et al. 2008; Pablo 2012). The rule-based methods use a set of prescribed rules to avoid conflict. The sets of rules are fixed during the system design phase and are inspired by the VFR.

The second part consists of localization and mapping-based solutions. Those solutions avoid collisions by mapping, positioning, and navigation within the map while the positioning and mapping is based on a complex localization algorithm. Compared to the first division, these solutions do not limit the mission, but the CA algorithm requires considerable memory and computational power compared to the already mentioned simple solutions, because of its complexity (Achtelik et al. 2008; Shen et al. 2009; Chowdhary et al. 2013).

The rule-based methods use a set of prescribed rules to avoid conflict. The sets of rules are fixed during the system design phase and are derived from the visual flight rules (VFR). In the

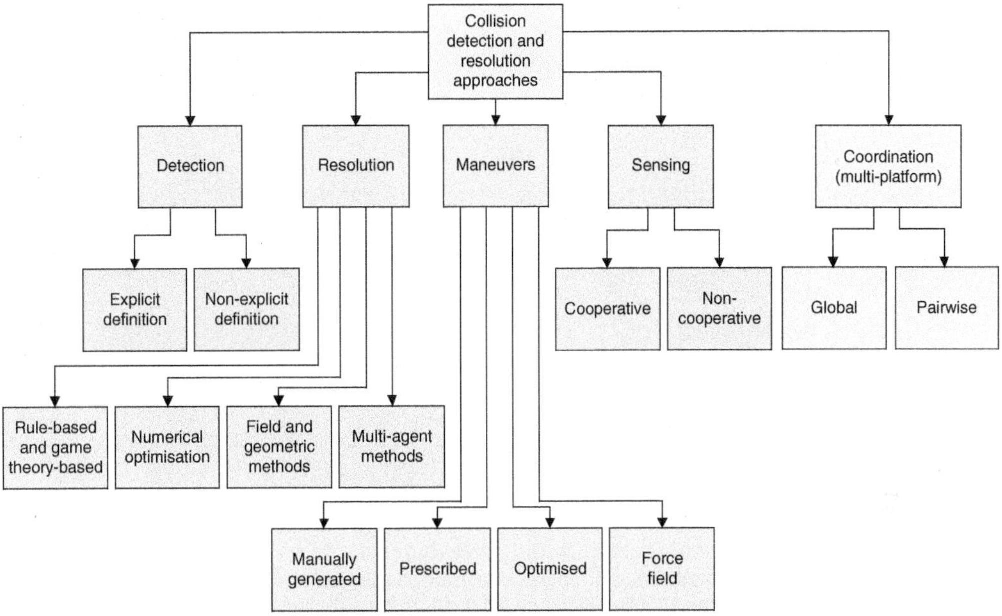

FIGURE 5.109 Collision detection and resolution taxonomy.

game theory methods, the conflict is typically modelled as a two-player differential game. The field-based methods treat each aircraft as a charged particle and are very close to a reactive control mechanism. Based on the current configuration (e.g., position of other platforms, weather condition, considered uncertainty), the field is computed. Geometrical approaches consider in many cases the whole trajectory of the airplane while it avoids observed solution. Various approaches optimize given objective functions while they are searching for the proper evasion maneuver. In many cases, these methods are considering only collision of two airplanes and for multi-collision situations they come with suboptimal approaches or sequential application of collision of two airplanes where safety is not guaranteed. Geometrical optimization is a very complex problem, especially if all available actions including change of state information is considered. Numerical optimization approaches employ a kinematic model of the aircraft and include a set of constraints and cost indexes for generating the optimised trajectories, to be flown by the aircraft. An optimal avoidance trajectory is computed, which is then flown by the aircraft to avoid any predicted conflict(s). The key benefit of employing this technique is the rigor of the mathematical formulation. Multiagent approaches employ an agent-based framework for providing conflict resolutions. In this approach, each aircraft is considered as an agent. All the aircraft in a multiagent network communicate and negotiate the conflict resolution solution with all other agents present in the network.

5.70 SA&CA Technologies

To operate in non-segregated airspace, manned and unmanned aircraft are expected to maintain safe separation from other aircraft. New self-separation and collision avoidance systems (CAS) designed for unmanned aircraft are under research and development, sponsored by many programs, both regionally and globally. To maintain airspace safety, these new systems must interoperate safely with CAS onboard manned aircraft. Therefore, the interactions of manned aircraft

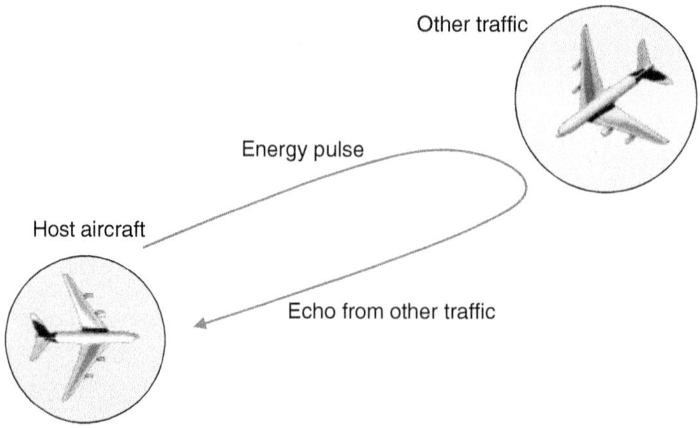

Figure 5.110 PSR surveillance principle.

collision avoidance and unmanned aircraft collision avoidance and self-separation must be understood. Many of the SA&CA systems under development for unmanned aircraft provide horizontal guidance, alerting the unmanned aircraft to turn left or right to avoid intruders. In contrast, both traffic alert and collision avoidance system (TCAS), the internationally mandated CAS required onboard all large transport aircraft, and its planned successor ACAS Xa1 (Airborne Collision Avoidance System X) issue only vertical resolution advisories such as climb or descend.

Radar surveillance can be achieved through primary surveillance radar (PSR) as well as secondary surveillance radar (SSR). PSR detects aircraft thanks to radiofrequency signals reflected by other aircraft (Figure 5.110). Therefore, PSR is described as independent, self-contained system. SSR is based on interrogation and reply by all aircraft within range, exchanging identification as well as altitude and other information (Figure 5.111). SSR is, therefore, considered to be a dependent, cooperative surveillance system. Although SSR provides more detailed information than PSR, it does not provide information on unequipped aircraft or when the SSR equipment has failed.

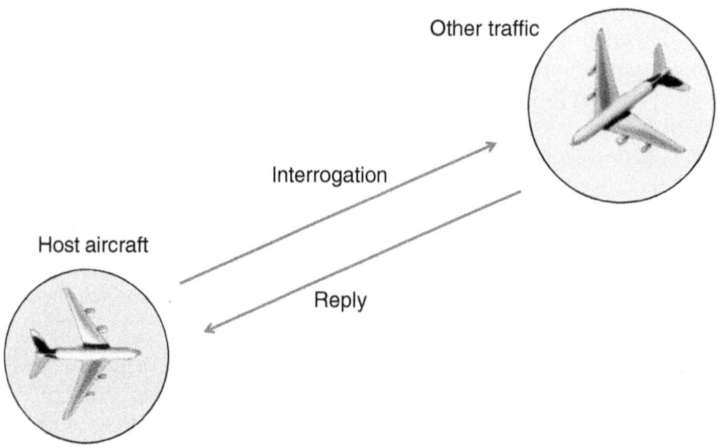

Figure 5.111 SSR surveillance principle.

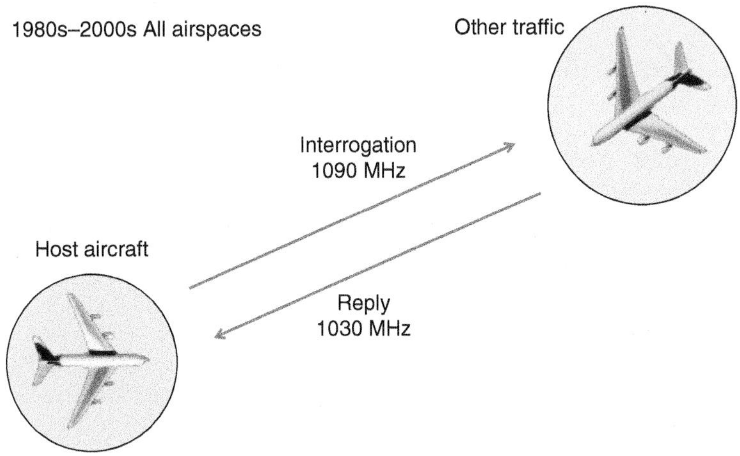

Figure 5.112 Onboard collision avoidance system—TCAS.

TCAS was developed in response to a series of mid-air collisions between civil aircraft and mandated internationally for large transport aircraft, reducing the risk of airborne collisions significantly since its introduction (Figure 5.112) (Harman 1989).

TCAS alerting logic is based on linear extrapolation of intruder trajectories and a large set of heuristic rules (Lenz and Gottstein 2005; Kuchar and Drumm 2007). There are three members of the TCAS family. TCAS I is a pilot warning indicator which displays proximate traffic and alerts the flight to other aircraft which may become potential near mid-air collision threats.

TCAS provides an effective collision avoidance function in traffic densities as high as 0.3 aircraft per square nautical mile (24 transponder-equipped aircraft within 5 NM of the TCAS-equipped aircraft). When TCAS operates in traffic densities of 0.3 transponder-equipped aircraft per square nautical mile or higher, it provides protection against collisions with other aircraft that are closing at speeds of less than 500 kt. TCAS is also capable of providing protection against aircraft closing at relative speeds of up to 1,200 kt in airspace characterized by densities of less than 0.06 transponder-equipped aircraft per square nautical mile, which is equivalent to 18 transponder-equipped aircraft within 10 MN of the TCAS aircraft.

A TCAS I-equipped aircraft can receive limited alert and advisory information about aircraft in its vicinity. TCAS I also participates in cross-link communication with TCAS II equipped aircraft. TCAS I operates by listening for aircraft transponder transmissions (replies). The replies detected may have been unsolicited transmissions from mode S transponders (squitters); replies elicited by ground station or TCAS II interrogations (passive TCAS I); or replies resulting from low-power interrogations from TCAS I (active TCAS I). Additionally, TCAS I receives traffic advisory information from TCAS II aircraft. If an intruding aircraft is equipped with TCAS II, and if that TCAS II determines that a collision hazard exists, it will transmit advisory information to TCAS I, which includes bearing (if available), range, and relative altitude of the TCAS II aircraft with respect to the TCAS I aircraft. As an option, TCAS I can be designed to provide an indication of the transmitting transponder's bearing.

In addition to the minimum standards contained in this document, several options are described herein which will allow TCAS to request and/or receive information from other mode S-equipped aircraft. These optional features provide for the exchange of information via the proposed mode S extended squitter or an air-to-air datalink. The purpose for including these options in this standard is to permit a uniform minimum capability to upgrade or enhance TCAS in the future.

The specific applications for these optional features and the applications corresponding minimum operational performance characteristics are not part of these standards.

TCAS II is referred to as the Airborne Collision Avoidance System II (ACAS II) in ICAO terminology and provides traffic advisories and if warranted, resolution advisories (RA) in the vertical plane. Nontransponding aircraft are not detected. RAs are indications given to the flight crew recommending maneuvers intended to provide separation from all threats, or restrictions to maneuvers to maintain existing separation. TCAS II operating procedures give RAs priority over ATC clearances and instructions, i.e., flight crews are instructed to follow an RA even when it conflicts with ATC guidance. Controllers have no knowledge of a TCAS RA unless notified by the flight crew (resolution advisory information is provided by TCAS II to mode S SSRs, but it is not normally presented to controllers). TCAS can all the while track up to 30 airplane in an ostensible scope of 14 NM for mode A or C targets and 30 NM for mode S targets. This data is passed along to the collision avoidance logic in the PC to decide the prerequisite for TAs or RAs. The term TCAS IV is reserved for future TCAS equipment that has additional capabilities beyond those of TCAS II equipment to generate resolution advisories in the horizontal plane (e.g., "TURN RIGHT," "TURN LEFT").

5.71 Conflict Resolution Heuristics

The following paragraphs describes how to compute and update the CRS input variables Coarse_Altitude, Fine_Altitude, Own_Corrected_Alt, Fine_Altitude_Rate, Radio_Altitude, and ZRINT. These calculations use the parameters Init_Alt, Init_Alt_Rate, Alt_Accel, Ground_Level, and Quant specified in the encounter input file of the TCAS.

At $T = 1$ for own aircraft and at the time Encounter_Start is specified for other traffic, the altitude (Alt), and altitude rate (Alt_Rate) for each aircraft are set to Init_Alt and Init_Alt_Rate, respectively. At the end of each cycle ($T = i$) Alt and Alt_Rate are updated for each aircraft for the next cycle ($T = i + 1$) as follows:

$$\text{Alt}_{i+1} = \text{Alt}_i + \text{Alt\_Rate}_i + 0.5 \times \text{Alt\_Accel}_i \tag{5.317}$$

$$\text{Alt\_Rate}_{i+1} = \text{Alt\_Rate}_i + \text{Alt\_Accel}_i \tag{5.318}$$

Calculations for Alt and Alt_Rate always use the exact value of Alt, rather than rounding to the nearest 100 ft. The CRS input variables Coarse_Altitude, Fine_Altitude, Own_Corrected_Alt, Fine_Altitude_Rate, Radio_Altitude, and ZRINT are calculated at the beginning of each cycle from these values as follows (RTCA, 2008):

$$\text{Coarse\_Altitude}_i = \text{Alt\_Own}_i \tag{5.319}$$

$$\text{Fine\_Altitude}_i = \text{Alt\_Own}_i \tag{5.320}$$

$$\text{Own\_Corrected\_Alt}_i = \text{Alt\_Own}_i \tag{5.321}$$

$$\text{Fine\_Altitude\_Rate}_i = \text{Alt\_Rate\_Own}_i \tag{5.322}$$

$$\text{Radio\_Altitude}_i = \text{Alt\_Own}_i - \text{Ground\_Level}_i \tag{5.323}$$

$$\text{ZRINT}_i = \text{Round }(\text{Alt}_{\text{Other}i}, 100) \text{ if QUANT} = 100 \tag{5.324}$$

$$\text{ZRINT}_i = \text{Round }(\text{Alt}_{\text{Other}i}, 25) \text{ if QUANT} = 25 \tag{5.325}$$

With respect to the encounter geometry, and compute and update the CRS input variables RR (relative range), BEAR (bearing), and Bear_Meas (raw bearing). These calculations use the parameters Air_Speed, Init_Range, Range_Space, and Angle specified in the Encounter Input File.

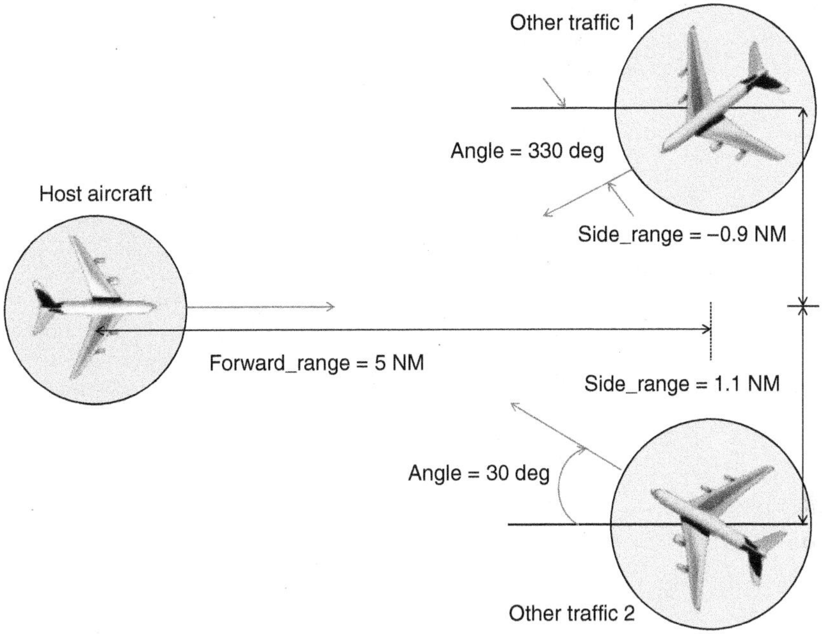

FIGURE 5.113 Example of an encounter geometry.

Figure 5.113 illustrates an example of the encounter geometry. For all encounters, own aircraft is assumed to be traveling at a constant speed in the positive x-direction. Forward_Range is the x-component of the distance of the other aircraft from own aircraft. Side_Range is the y-component of the distance of the other aircraft from own aircraft, with positive distance being measured to the right of own aircraft. Angle is the heading angle of the other aircraft measured clockwise from the x-axis. Initially, at the time Encounter_Start is specified, Forward_Range is set to Init_Range and Side_Range is set to Range_Space.

At the end of each cycle Forward_Range and Side_Range are updated for the next cycle as follows:

$$\text{Forward\_Range}_i = \text{Forward}_{\text{Range}i} - \text{Air\_Speed\_Own} - \text{Air\_Speed\_Other}_i \times \cos(\text{Angle}_{i+1}) \qquad (5.326)$$

$$\text{Side\_Range}_i = \text{Side\_Range}_i - \text{Air\_Speed\_Own} - \text{Air\_Speed\_Other}_i \times \sin(\text{Angle}_{i+1}) \qquad (5.327)$$

Air_SpeedOWN and Air_SpeedOTHER are constant throughout each test and, if the value of Angle changes, it is updated before the new Forward_Range and Side_Range are computed. The CRS input variable RR is computed at the beginning of each cycle from these values as follows:

$$\text{RR}_i = \sqrt{\text{Forward\_Range}_i^2 + \text{Side\_Range}_i^2 + \text{Relative\_Alt}_i^2} \qquad (5.328)$$

where Relative_Alt is the real altitude difference between own and other aircraft. Relative_Alt is computed as follows using the altitudes computed in the previous paragraph:

$$\text{Relative\_Alt}_i = \text{Alt\_Own}_i - \text{Alt\_Other}_i \qquad (5.329)$$

Finally, the CRS input variables BEAR and BEAR_MEAS are calculated at the beginning of each cycle as follows:

$$\text{Bear}_i = \arctan\left(\frac{\text{Side\_Range}_i}{\text{Forward\_Range}_i}\right) \tag{5.330}$$

$$\text{Bear\_Meas}_i = \arctan\left(\frac{\text{Side\_Range}_i}{\text{Forward\_Range}_i}\right) \tag{5.331}$$

and then adjusted for the quadrant in which the other aircraft is located.

TCAS has been very successful in preventing mid-air collisions over the years, but the way in which the logic was designed limits its robustness (RTCA 2008; FAA 2011). To overcome the limitations of TCAS the FAA program office began formal research on a next generation airborne collision avoidance system: ACAS X. Whereas TCAS logic is based on linear extrapolation and heuristic rules, ACAS X logic is based on a dynamic model of aircraft movement and a computer optimized lookup table of collision avoidance actions. Traffic alerts are issued to advise pilots that another aircraft is a potential threat and to prepare for a resolution advisory if necessary. A resolution advisory commands specific vertical-only maneuvers that will satisfy safety goals with minimal maneuvering. Resolution advisories include the following:

- Climb or descend
- Level off
- Maintain climb or descend
- Do not or limit climb or descent rate

Like TCAS, ACAS X avoids collisions by giving vertical guidance to an aircraft's pilot. The collision avoidance community defines a near mid-air collision (NMAC) when two aircrafts collide within 100 ft vertically and 500 ft horizontally (Kuchar and Drumm 2007; Kochenderfer et al. 2012). ACAS X is designed to avoid such NMACs. ACAS X involves both discrete advisories to the pilot and continuous dynamics of aircraft. From a table of advisories and their modeling variables, every advisory, except the clear of conflict (COC), has a vertical rate range of the form $(-\infty, \dot{h}_f]$ or $(\dot{h}_f, +\infty]$ for some vertical rate $\dot{h}_f$, which we call the target vertical velocity. We therefore model any advisory by its corresponding target vertical velocity $\dot{h}_f$, and a binary variable w, whose value is -1 if the vertical rate range of the advisory is $(-\infty, \dot{h}_f]$ and $+1$ if it is $(\dot{h}_f, +\infty]$. Note that this symbolic encoding makes it possible to represent much more advisories and is, therefore, robust to changes in the set of advisories that ACAS X allows. It is assumed that the own aircraft pilot complies with each advisory within d_p seconds and that they accelerate with at least acceleration a_r to bring the relative vertical velocity in compliance with the advisory.

This design makes ACAS X substantially easier to adapt to specific aircraft, airspace procedures, and surveillance technologies. ACAS X is a family of adaptations, two of which are more predominantly employed: ACAS Xa and ACAS Xu. Both TCAS and ACAS Xa issue resolution advisories based in large part on the projected vertical trajectories of intruder aircraft. ACAS Xa is being designed as a direct replacement for TCAS, with intent to provide improvements in safety and operational suitability. ACAS Xu is being designed for UAS, and as such it is optimized for the aerodynamic performance and surveillance systems characteristic of those platforms. ACAS Xu, unlike TCAS and ACAS Xa, is able to provide horizontal resolution advisory guidance. In this study of the interactions of vertical collision avoidance and a system providing UAS with horizontal guidance, ACAS Xu's horizontal logic fills the role of the system providing horizontal guidance. Note that ACAS Xu is also able to provide vertical RA guidance, but this analysis is focused solely on its horizontal logic. Even though ACAS Xu is a collision avoidance system, the results of this

ACAS X type	Description	Surveillance technology	Advisories
ACAS Xa (active) 	It is the general purpose ACAS X that will make active interrogations to establish the range of intruders. ACAS Xa is a successor to TCAS II system.	Active RADAR supplemented with passive	Same as current TCAS
ACAS Xo (operation) 	A mode of operation of ACAS X designed for particular operations for which ACAS Xa is unsuitable. It facilitates procedure-optimized alerting against a user-selected aircraft while providing global Xa protection for all other traffic.	Active RADAR supplemented with passive	Vertical and horizontal advisories
ACAS Xp (passive) 	It is specifically intended for unmanned aircraft. It accepts a variety of surveillance inputs and uses decision-logics that are optimized for a wide variety of performance capabilities.	Passive only	Reduced advisory set
ACAS Xu (unmanned) 	A version of ACAS X that relies solely on passive ADS-B inputs to track other traffic and does not make any active interrogation. It is intended for general aviation aircraft, which are not currently required to be equipped with TCAS II system.	Potentially radar, EO/IR, etc.	Vertical and horizontal advisories (for self-separation)

Table 5.22 ACAS X Variants

study are applicable to any CAS or self-separation system providing guidance that may cause vertical maneuvers within the collision avoidance timeframe.

Various variants of ACAS X can be employed including (Kochenderfer et al. 2012) as summarized in Table 5.22.

ACAS is different from a conventional CA system, in the sense that it does not include any autonomous resolution execution function. ACAS has characteristics and performances different from those required by a SA&CA system for UAS. One of the main reasons is that the different climb performance of various aircraft platforms (manned and unmanned) is not compatible with current TCAS requirements. More precisely, the limitations on the climb performance make the unmanned aircraft remote pilot or the avionics systems difficult to execute the suggested maneuver by TCAS advisories. Considering general aviation aircraft and aerostatic balloons, they are not usually fitted with a mode S transponder and hence in those cases, non-cooperative surveillance is required. Figure 5.114 shows the top-level architecture of ACAS. The system receives sensor measurements every second. On the basis of these sensor measurements, the system infers the distribution over the aircraft's current status. This status, or state estimation, takes into account the probabilistic dynamic model and the probabilistic sensor model. This state distribution determines where to look in the numeric logic table to determine the best action to perform, which is to issue an advisory and if so, the vertical rate to adopt. This processing chain is repeated once per second with every new sensor measurement. Critical to understanding the logic optimization process are two important concepts. The first is a Markov decision process, which is essentially the probabilistic dynamic model combined with the utility model. The second is dynamic programming, which is the iterative computational process used to optimize the logic.

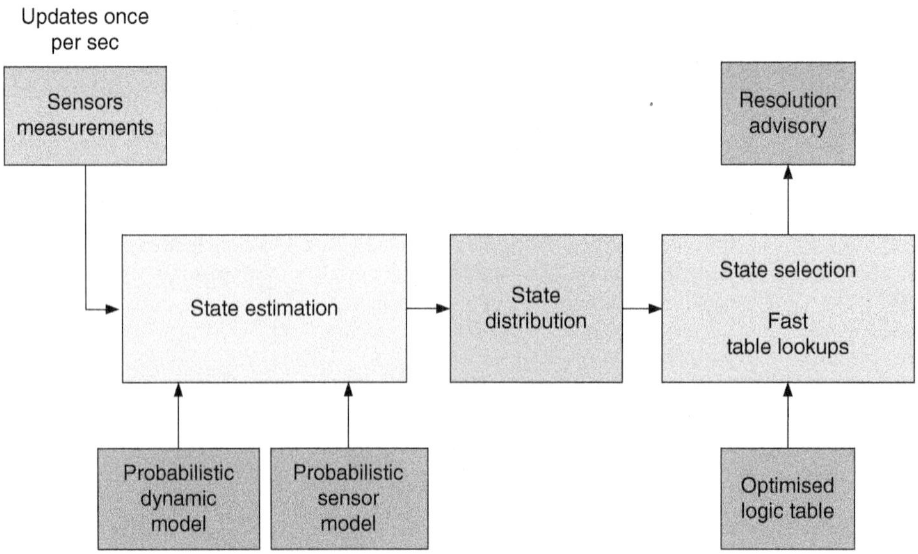

FIGURE 5.114 ACAS top-level architecture.

5.72 Automatic Dependent Surveillance

Automatic dependent surveillance—broadcast (ADS-B) is operational in most countries. In Australia, ADS-B has almost complete coverage above FL250 and reasonable coverage down to 10,000 ft. ADS-B messages are transmitted via the extended squitter mode-S transponder datalink operating at 1,090 MHz. Mode S ES capable transponders and ADS-B OUT equipage mandates are detailed in (Valovage 2007; CASA 2016). Automatic dependent surveillance—contract A (ADS-C) separation of instrument flight rules (IFR) dependent aircraft outside of radar coverage are managed procedurally, resulting in large separation standards. ADS-C relies on the automatic transmission of reports from an appropriately equipped aircraft. The content and timing of reports is described as a "contract." Digital datalink communications via satellite or VHF can be made from appropriately equipped aircraft (FANS 1/A) to enable reduced separation standards. The digital datalink enables controller pilot datalink communications (CPDLC) and automatic dependent surveillance—contract (ADS-C).

Automatic dependent surveillance—broadcast (ADS-B) has been viewed as a low-cost alternative and complement to the current primary surveillance radar. According to the definition provided by International Air Transport Association (IATA), ADS-B refers to the function on a surface vehicle or airplane which broadcasts its state vector together with other relevant information periodically. ADS-B is developed to improve situation awareness, conflict detection, conflict avoidance, runway incursion avoidance as well as surveillance in remote areas without radar covering. ADS-B is identified by RTCA as one of the key enabling technologies that make free flight possible. In addition, ADS-B integrated with other capability is expected to enable airspace with self-separation similar to visual operation today in Single European Sky ATM Research (SESAR) and NextGen programs.

All the involved vehicles, including aircraft, ground vehicles, and other objects periodically transmit their identification, positional information as well as intent information using a broadcast datalink. ADS-B is made up of the following components: a transmitting subsystem which includes a message generation as well as transmission function at the source aircraft, the propagation medium, and a receiving subsystem which is composed of message reception and report assembly functions at the receiving aircraft. Some ground-based users may have the capability to receive but not transmit. The ADS-B system consists of three functional components: receiver aircraft, transmitter aircraft, as well as ground station as illustrated in Figure 5.115. The characteristics and requirements of ADS-B make it a capable sensor for CA on small UAS in the National Airspace System. One of the key aspects of ADS-B that makes it feasible for use on small UAS is the availability of ADS-B receivers that meet the SWaP constraints of a small UAS. Another key advantage of ADS-B is the long range at which information is available. While there is a significant amount of variation in the range of ADS-B signals, the shortest expected range is 10 NM. Additionally, the long range of ADS-B is advantageous in that the quality of information transmitted over ADS-B does not degrade with range. Thus the accuracy of ADS-B is not dependent on the size, power, or range of the transmitter and receiver units. This is a significant advantage over radar and optical sensors, and makes path planning for conflict detection and separation assurance at long ranges.

However, the limitations of ADS-B system are

- The availability of GNSS information and line-of-sight ADS-B transmissions determine the performance of the ADS-B system. Typically, a multiple access scheme is specified for the ADS-B transmissions. This scheme is different for th 978 and 1,090 MHz frequency of operation. Furthermore, only a finite number of transmitters can be supported. This limitation is even more important for designing these systems for small UAS (Sahawneh, 2016).
- The cost of procuring ADS-B system.

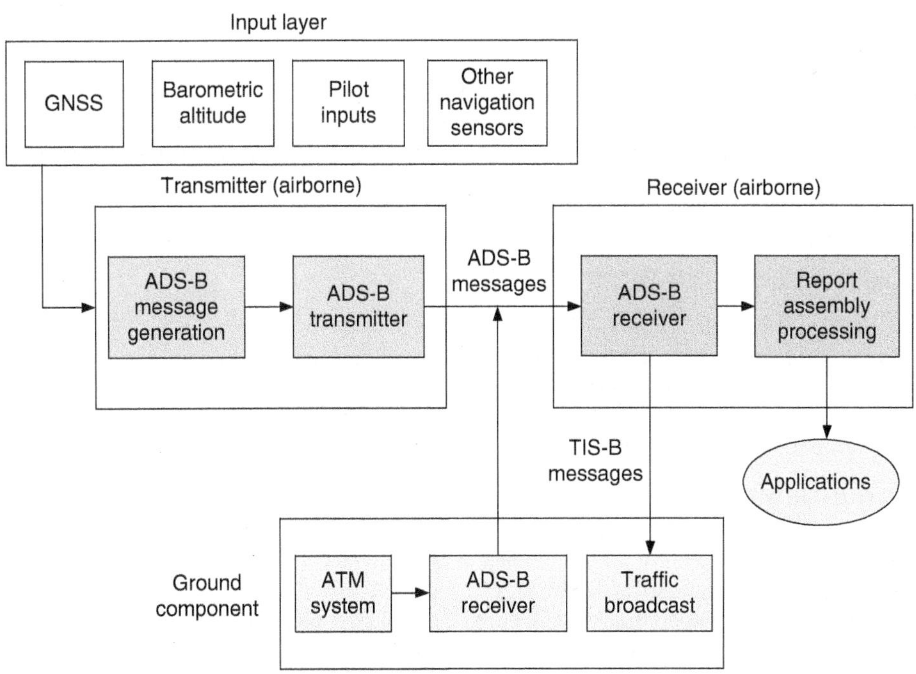

Figure 5.115 Functional component of ADS-B.

ADS-B uses three types of reports, namely, surveillance state vector (SSV) report, mode-status (MS) report, and on-condition reports. The SV report contains information regarding current kinematic states of the aircraft, which includes geometric position, horizontal and vertical velocity, and heading. In addition, the quality of SSV element is contained in the MS report. The MS message includes current operational information of the transmitting aircraft such as capability class codes, operational mode codes, and SSV quality. CC code identifies the capability of a transmitting ADS-B participant including TCAS/AS installation, cockpit display of traffic information (CDTI) capability, air referenced velocity report capability, and trajectory change report capability. OM codes indicate the current operational mode of a transmitting participant. The SSV quality codes contain the navigation accuracy category for position (NACp), navigation accuracy category for velocity (NACv), and surveillance integrity level (SIL). All ADS-B participants must provide the SSV and MS report. Upon reception of a state vector report, the system searches for a track of the same type (ADS-B, ADS-R, TIS-B) and with the same 24-bit address as the report. As an example, if no match is found, the system generates a new track containing the following parameters (RTCA, 2008):

- System time (track time of applicability)
- Position time of applicability
- Velocity time of applicability
- Track type (ADS-B, ADS-R, TIS-B)
- 24-bit address
- Position (latitude 2, longitude, altitude)
- Velocity (west/east velocity, north/south velocity, altitude rate)
- NIC
- NACP = 5
- NACV = 1
- SIL = 0
- GVA = 1 (Version 2)
- SDA = 1 (Version 2)
- State covariance matrices
- ID change flag = 0
- Index of correlated TIS-B track = null
- Time of correlation with TIS-B track = null
- Previous state (position and track time of applicability) matrix initialized as a row vector
- Estimated turn rate = null

5.73 Multilateration Systems

Multilateration systems use a network of ground-based transceivers to localize aircraft based on the difference in time of arrival of aircraft transponder transmissions. Multilateration systems are currently deployed at a number of class C CTR (for terminal area airspace surveillance and separations services within the terminal maneuvering area) and for en route and terminal area surveillance. Advanced surface movement guidance and control system (A-SMGCS) provides routing, guidance, and surveillance for the control of the surface movement of aircraft and vehicles. The system is based on a number of existing surveillance technologies to provide ATCo a situational awareness

Airspace	Conventional systems	State-of-the-art systems
Oceanic continental en route airspace with low/high-density traffic	Primary radar/secondary surveillance radar (SSR) VHF voice position reports OMEGA/ LORAN customized navigation database (CNDB)	Automatic dependent surveillance (ADS)
Continental airspace with high-density traffic	Primary radar SSR mode A SSR mode C	SSR mode A SSR mode C SSR mode S ADS
Terminal areas with high-density	Primary radar SSR mode A/C	SSR [mode A/C/S] ADS

Table 5.23 Airborne Surveillance Systems

and alerting. It is a key technology for the prevention of runway and taxiway incursions and ground collisions. Precision runway monitor is a surveillance system operational to support reduced staggered separation for aircraft on the existing parallel instrument landing system (ILS) approaches. It uses a primary radar sensor to provide accurate and high update rate surveillance of the approach and landing areas. A summary of airborne surveillance systems currently available is provided in Table 5.23.

References

Achtelik, M., Bachrach, A., He, R., Prentice S., and Roy, N., 2008, Autonomous Navigation and Exploration of a Quadrotor Helicopter in GPS Denied Indoor Environments, *Proceedings of the 1st Symposium on Indoor Flight*. Available online at: http://robot-chopper.googlecode.com/svnhistory/r479/trunk/References/2009MIT.pdf.

Airservices, 2016a, Separation Standards, Airservices Australia, Australian Capital Territory, Australia.

Airservices Australia, 2016b, Separation Standards, Australia's Air Navigation Service Provider, Canberra, Australia.

Bouabdallah, S., Becker, M., de Perrot, V., and Siegwart, R., 2007, Toward Obstacle Avoidance on Quadrotors. *Proceedings of the 12th International Symposium on Dynamic Problems of Mechanics*, Ilhabela, Brazil, pp. 1–10.

Campos, L. M. B. C. and Marques, J. M. G., 2016, On an Analytical Model of Wake Vortex Separation of Aircraft, *The Aeronautical Journal*, Vol. 120, No. 1232, pp. 1534–1565.

CASA, 2016, ADS-B mandates 2014-2017, Civil Aviation Safety Authority, Woden, Australian Capital Territory, Australia.

Chowdhary, G., Johnson, E. N., Magree, D., Wu A., and Shein, A., "GPS Denied Indoor and Outdoor Monocular Vision Aided Navigation and Control of Unmanned Aircraft," *Journal of Field Robotics*, Vol. 30, Issue 3, pp. 415–438.

Consiglio, M. C., Comstock J. R., Jr,, Ghatas, R. W., and Vincent, M. J., 2015, Unmanned Aircraft Systems (UAS) Controller Acceptability Study 3 (CAS3): Collision Avoidance, Self Separation, and Alerting Times (CASSAT): Combined PER/FER, UAS in the NAS Project Briefing to FAA and Industry, NASA Langley Research Center, Hampton, VA, United States.

Davis, K. D. and Cook, S. P., "Achieving Sense and Avoid for Unmanned Aircraft Systems: Assessing the Gaps for Science and Research," *Handbook of Unmanned Aerial Vehicles*, pp. 1841–1855, Springer, Netherlands, 2015.

Durand, N., Gianazza, D., Gotteland, J. B., and Alliot, J. M., 2016, *Airspace Management, Metaheuristics for Air Traffic Management*, John Wiley and Sons, pp. 69–83.

Eurocontrol, 2013, *European Airspace Concept Handbook for PBN Implementation*, 3rd ed., Brussels, Belgium.

FAA, 2009, Sense and avoid (SAA) for unmanned aircraft systems (UAS), Sense and Avoid Workshop, Federal Aviation Administration.

FAA, 2011, Introduction to TCAS II, Version 7.1, Federal Aviation Administration, Washington, DC, United States.

FAA, 2016, Code of Federal Regulations, CFR, Federal Aviation Administration, Washington, DC, United States.

Granger, G., 2002, Détection et Résolution de Conflits Aériens: Modélisations et Analyse, Doctoral Thesis, Informatique de l'Ecole Polytechnique.

Harman, W., 1989, TCAS-A System for Preventing Midair Collisions, *The Lincoln Laboratory Journal*, Vol. 2, Issue 3, pp. 437–457.

Kochenderfer, M. J., Holland, J. E., and Chryssanthacopoulos, J. P., 2012, Next-Generation Airborne Collision Avoidance System, *Lincoln Laboratory Journal*, Vol. 19, Issue 1, pp. 17–33.

Kopřiva, S., Šišlák D., and Pěchouček, M., Sense and Avoid Concepts: Vehicle-Based SAA Systems (Vehicle-to-Vehicle), *Sense and Avoid in UAS: Research and Applications*, ed, P. Angelov, John Wiley & Sons, Ltd, Chichester, UK, 2012. DOI: 10.1002/9781119964049.ch6

Kuchar, J. E. and Drumm, A. C., 2007, The Traffic Alert and Collision Avoidance System, *Lincoln Laboratory Journal*, Vol. 16, Issue 2.

Lenz, H. and Gottstein, J., 2015, TCAS Compatibility of Advanced Airborne Separation Assurance System Operations, *Proceedings of the AIAA Infotech at Aerospace Conference*, ISSN: 0731-5090.

Melega, M., 2014, Autonomous Collision Avoidance for Unmanned Aerial Systems, Doctoral Dissertation, Cranfield University, United Kingdom.

Oztekin A. and Wever, R., 2015, Development of a Regulatory Safety Baseline for UAS Sense and Avoid, *Handbook of Unmanned Aerial Vehicles*, Springer, Netherlands, pp. 1817–1839.

Pablo, R., Santamaria, E., Lema, J. M., Pastor E., and Barrado, C., 2012, Integration of SAA Capabilities into a UAS Distributed Architecture for Civil Applications, Sense and Avoid in UAS: Research and Applications, p. 55.

Roberts, J. F., Stirling, T. S., Zufferey, J. C., and Floreano, D., 2007, Quadrotor Using Minimal Sensing for Autonomous Indoor Flight, *Proceedings of MAV Conference (EMAV 2007)*, Toulouse, France, pp. 1–8.

RTCA, 2008, Minimum Operational Performance Standards for Traffic Alert and Collision Avoidance Systems II (TCAS II), DO-185B, Radio Technical Commission for Aeronautics, Washington, DC, United States.

Sahawneh, L. R., 2016, *Airborne Collision Detection and Avoidance for Small UAS Sense and Avoid Systems*, Doctoral Dissertation, Brigham Young University, Provo, Utah, USA.

Scherer, S., Singh, S., Chamberlain L., and Elgersma, M., 2008, Flying Fast and Low among Obstacles: Methodology and Experiments, *International Journal of Robotics Research*, Vol. 27, Issue 5, pp. 549–574.

Shen, S., Michael N., and Kumar, V., 2011, Autonomous Multi-Floor Indoor Navigation with a Computationally Constrained MAV, *Proceedings of the IEEE International Conference on Robotics and Automation*, Shanghai, China, pp. 20–25.

Valovage, E., 2007, Enhanced ADS-B Research, *IEEE Aerospace and Electronics Systems Magazine*, Vol. 22, Issue 5, pp. 35–38.

PART 15

Air Traffic Management Systems

Alessandro Gardi, Trevor Kistan, Yixiang Lim, Roberto Sabatini, and Subramanian Ramasamy

5.74 General Layout of ATM Systems

Air traffic management (ATM) is a ground-based service whose overarching mission is to prevent aircraft collisions and promote an orderly and expeditious flow of air traffic (ICAO 2001b, 2007b). This is ultimately achieved through various operational measures, typically based on amending the lateral, vertical, and longitudinal navigation of aircraft as necessary, and can be performed at various timeframes. As depicted in Figure 5.116, the provision of these measures requires suitable means to identify the conflicts, negotiate a resolution and follow the amended flight trajectory. For instrument flights in controlled airspace, these essential functions are provided by communication, navigation, surveillance, ATM (CNS/ATM), and avionics (CNS+A) technologies, but various ATM service provisions are in place for non-instrument flights and for operations in noncontrolled airspace as well. Human operators are still responsible for the majority of duties involved in the current ATM paradigm, which is largely procedural and hence more tactical than strategic in nature (Visser 1994). The ATM services are geographically distributed and are provided worldwide by a large number air navigation service providers (ANSPs) and air traffic service organizations (ATSO), to which specific portions of airspace are allocated under international agreements. To accomplish its overarching mission, the ATM organizations encompass all services to assist aircraft in safely

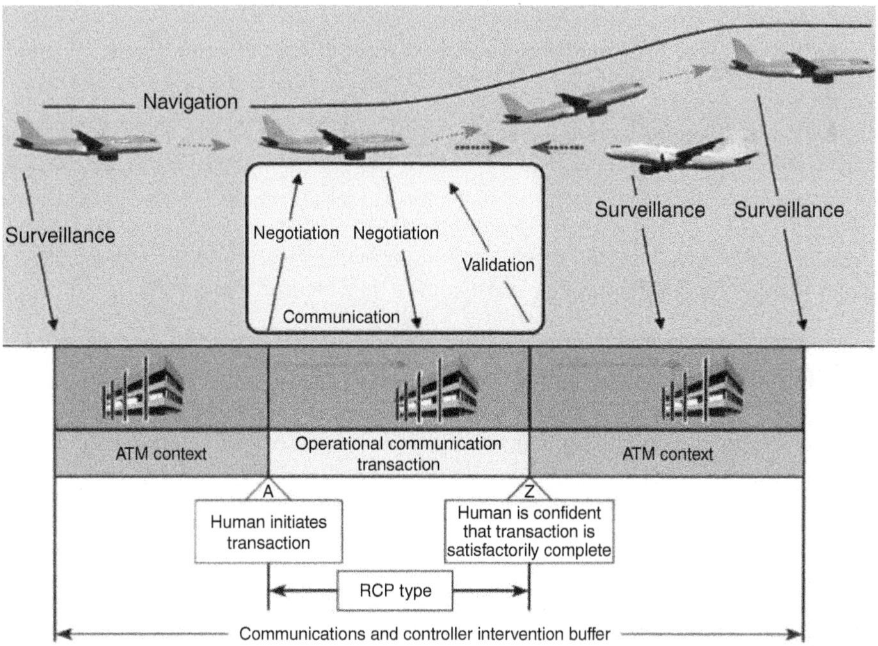

FIGURE 5.116 Overview of a deconfliction process for instrument flight in controlled airspace (Gardi et al. 2016).

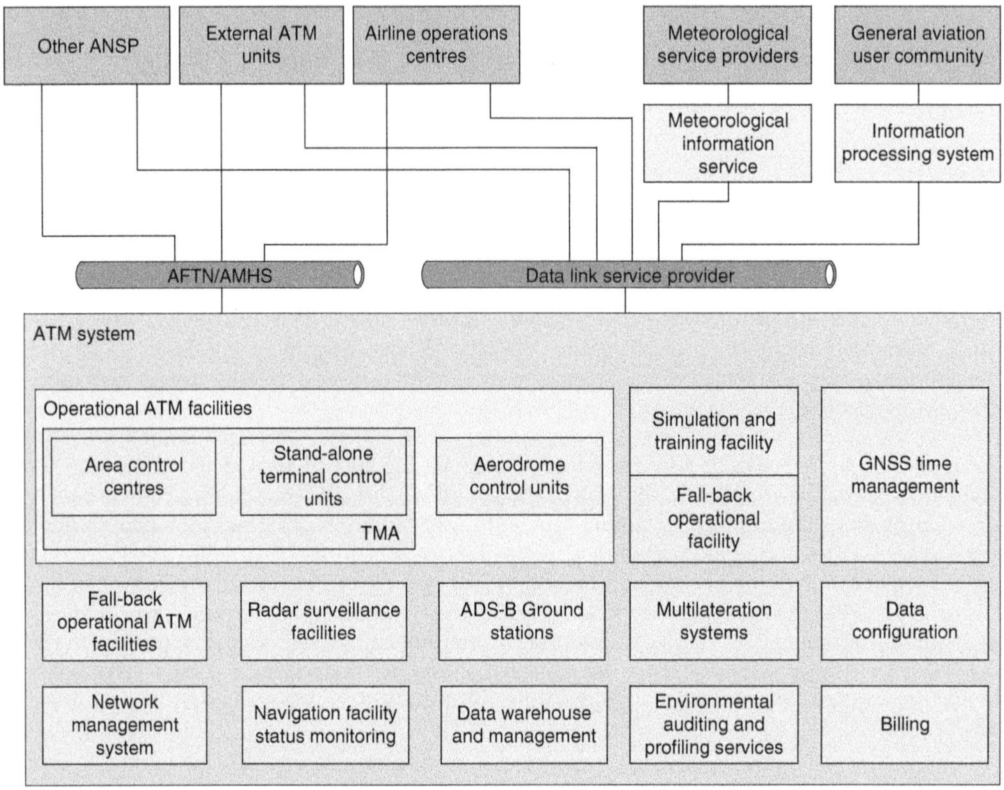

FIGURE 5.117 General layout of ATM systems.

accomplishing all their flight phases, including aeronautical meteorology, airspace management (ASM), air traffic services (ATS), and air traffic flow management (ATFM) or air traffic flow and capacity management (ATFCM). Various cyber-physical systems are involved in the provision of these services, including aeronautical communication, navigation, and surveillance infrastructures, decision support systems for planning, online management and emergency. Figure 5.117 provides an outlook of ATM systems as implemented by a medium-large ANSP.

5.75 Fundamental ATM System Design Drivers

ICAO document 9982 "Manual on Air Traffic Management (ATM) System Requirements" {ICAO, 2008 #2637} takes the move from ICAO documents 9854 "Global ATM Operational Concept" (OCD) (ICAO 2005) and 9883 "Manual on Global Performance of the Air Navigation System" (ICAO) to derive the top-level requirements to be fulfilled by the ATM service, i.e., the ATM system-of-systems, as a whole. While this document does not specifically target ATM decision support system (DSS) design, most of the fundamental requirements outlined in the document have direct implications on the DSS design. This document is, therefore, an important reference for the design, development, and deployment of emerging ATM system concepts in a global perspective. The underlying principle is to promote an integrated, harmonized and globally interoperable ATM service with a time horizon of 2025 and beyond, when a mature performance-based service is expected to be achieved. It is essential that this transition is evolutionary in nature, with backward compatibility and involving continuous and incremental performance enhancements. In this perspective, the document complements the OCD and specifies certain aspects most relevant to ATM services in greater practical detail. The document recognizes that the performance criteria that shall drive ATM service evolutions will depend to a certain extent on the local specificities, capturing the demands of differing operating environments particularly at state and region levels.

5.76 Airspace Structure

Three regulatory aspects define the airspace configuration (ICAO 2001b; Sabatini et al. 2015): ATS classification, air traffic control (ATC) sectorization, and special use airspace (SUA). The ATS classification specifies the level of service, separation provisions as well as minimum equipment and visual meteorological conditions that apply in a certain portion of the airspace. In particular, ATS classes range from alpha (A) to golf (G) in terms of decreasing levels-of-service and equipment requirements. ANSP assign the ATS classification to airspace regions based on the overall traffic volumes, the percentage of instrument flights, available CNS infrastructure and ATC resources (ICAO). According to ICAO recommendations, class "A" airspace is only accessible by instrument flight rules (IFR) traffic and require secondary surveillance and constant radio contact with ATC operators. The responsibility for deconfliction is on the ATM side. In class "G," all traffic is responsible of self-deconfliction, radio communication with ATC operators is not mandatory and no particular restrictions are introduced in terms of communication and surveillance equipment. Airspace restrictions and SUA are portions of the airspace characterized by notable regulatory limitations or by a natural or man-made hazard to aircraft operations (Sabatini et al. 2015). SUA are frequently associated with segregated operations conducted by military, government, industrial or scientific research organizations, and may be established to protect major cities or strategic and sensitive facilities from criminal threat. SUA can either be defined offline as well as during operations by means of apposite notice to airmen (NOTAM) and may have specific activation/deactivation time periods (Sabatini et al. 2015). ATC sectors reflect the organizational needs of the ANSP to provide services to large traffic volumes over a vast geographic area. Airspace is partitioned into smaller sections (ATC sectors) that are under the jurisdiction of individual air traffic controllers (ATCs). Further definitions were introduced by ICAO to guarantee global harmonization and provides a simple and efficient structure to international airspace. By definition, *control areas* are regions of airspace delineated so as to encompass sufficient airspace to contain the flight paths of IFR flights or portions thereof for which it is desired to provide the applicable parts of the ATS (ICAO 2001b). Control areas also include ATS routes (airways) and terminal control areas (TCAa), also known as terminal maneuvering areas (TMAs). Control zones complement control areas in their purpose and are specifically conceived to contain arrival and departure IFR flight paths in close proximity to the ground. For minor and isolated airports with limited IFR traffic, an appropriate combination of control zones and ATS routes is usually sufficient to contain IFR traffic safely (Gardi et al. 2016a, 2016b). This is due to the fact that tactical deconfliction by ATM operators is almost never required and occasional peaks of traffic can be successfully managed by exploiting multiple approach and departure procedures and en route holdings. Larger airports or combinations of multiple minor airports in close proximity to each other require more frequent tactical ATM interventions in terms of path stretching and separation measures. Whenever additional allowance for these tactical measures is necessary, TCAs/TMAs are established. In order to maximize safety of IFR traffic TCAs/TMAs are most commonly granted the highest levels of ATS, corresponding to airspace classes "A," "B," or less frequently "C." Due to their distinctive nature, TCAs/TMAs have become the paradigm of dense air traffic scenarios, where arriving, departing, and overflying aircraft intersect leading to frequent conflict situations. The highest levels of ATS assigned to TCA/TMA implicates that most of the responsibilities for conflict detection, separation maintenance, sequencing, and spacing lies on the ATM side. This is effectively the opposite of recreational airspace assigned classes "G" and "F" and has important consequences for UAS developers.

5.77 ATM Telecommunications Infrastructure

In terms of ATM telecommunication infrastructure, the picture is heterogeneous as various networking systems and standards coexist and are employed for different components. In general, although a number of ATM system functionalities are progressively migrated to modern IP-based connectivity on digital data networks, some interactions are still taking place via plain-text or voice messages through public switched telephone networks (PSTNs). This is especially the case for international links to neighboring ANSP in various regions of the world. Message handling systems (MHS) based on the X.400 ITU-T recommendation were selected as an inexpensive and nondisruptive evolution and are being implemented in a growing number of countries. More in detail, according to ICAO, the aeronautical fixed service (AFS) includes the following networking systems (ICAO 2013): direct ATS communications, aeronautical fixed telecommunication network (AFTN) also including the previously separated meteorological operational circuits, networks, and broadcast systems, the common ICAO data interchange network (CIDIN), the ATS message handling system (AMHS), and the inter-center communication (ICC) system. Future evolutions being developed will cater for the aeronautical/flight/weather information exchange models (AIXM/FIXM/WXXM) to fully realize the system-wide information management (SWIM).

Direct ATS Communications

Direct ATS communications including the so-called direct speech circuits and networks include both direct speech alone, direct written ("printed") communications as well as speech and datalink combined. Recommendations relating to direct communications are included in Chapter 6 of ICAO Annex 11 (ICAO 2013). Three types of direct ATS communications are outlined based on the time delay: instantaneous, 15 seconds (allowing for switchboard), and 5 minutes (record-and-retransmission) (ICAO 2001b). All telephone and voice intercom services are categorized as direct speech communications and, although not explicitly mentioned by ICAO, direct speech communications include both Voice-over-IP (VoIP) and store-and-forward voice messages. Where established by state regulators, required communication performances (RCPs) are to be fulfilled by all direct ATS communication services. Direct ATS communication facilities shall at least be established between a flight information center (FIC) and its related area control centers (ACC), approach terminal control units (TCUs) and aerodrome control towers (ACTs) as well as with all neighboring FIC. ACC shall have direct ATS communications at least with its related FIC, TCU, ACT, and neighboring ACC. Beyond ATS, FIC, and ACC shall have direct communications with international NOTAM office, relevant military units, meteorological offices, aeronautical telecommunication centers, operators offices emergency, and rescue services. Direct communications can either be established as stand alone or in combination with datalink, especially in the case of transfer of control. In this case, only instantaneous type of direct ATS communications is to be implemented. Conversely, written communications only need to meet the 5-minutes time to delivery (ICAO 2001b). Suitable facilities for automatic recording shall be provided.

Aeronautical Fixed Telecommunication Network

The aeronautical fixed telecommunication network (AFTN), which now also include the previously separated meteorological operational circuits, networks, and broadcast systems, is a character-oriented (i.e., telegraph/teletypewriter) worldwide message handling system introduced in ICAO Annex 10, as early as the 1950s (ICAO 2007a). It supports the exchange of text messages and data between all participating fixed aeronautical stations around the world, including ANSP, aviation service providers, airport authorities, and government agencies. Originally, AFTN was conceived as a teletypewriter circuit and, therefore, did not strictly fit in the OSI model. More recent implementations of AFTN can be ascribed to layers 3 and 4 of the OSI model (starting from CIDIN below), exploiting various physical and datalink layer implementations and supporting a variety of ATM applications. A considerable portion of core ATM message traffic around the world is exchanged via AFTN. The following categories of messages are exchanged through AFTN in decreasing order of priority (ICAO 2001a): distress, urgency, flight safety, meteorological, flight regularity, information, administrative, and service messages.

Common ICAO Data Interchange Network

Common ICAO data interchange network (CIDIN) was conceived in the 1980s mainly as a technological upgrade for the legacy AFTN to cope up with the increasing demands in terms of speed, capacity, and versatility. As defined by the ICAO, the CIDIN is an AFS based on X.25 recommendation by ITU-T that uses bit-oriented procedures, store-and-forward and packet switching techniques to deliver AFTN and other messages such as operational meteorological information (OPMET) (ICAO 2007a). Consequently, while the overall structure and addressing specifications remained mostly unchanged from the Annex 10-defined AFTN system, the CIDIN standard introduced a connectionless system ascribed to layers 3 and 4 of the OSI model only and that relied on X.25 recommendation by ITU-T for lower layers (ICAO 2011). This evolution introduced some flexibility in terms of physical (layer 1) and datalink (layer 2) networking infrastructures, provided they adhered to X.25 standard, which also encompasses digital networks including integrated services digital network (ISDN) to be exploited, as well as a wider range of ATM applications. CIDIN was extensively adopted in Europe and neighboring areas but had limited diffusion in other regions.

ATS Message Handling System

As the CIDIN-grade AFTN system also neared saturation in CIDIN-equipped countries and major advances were achieved in global internetworking technology, in the early 2000s the aeronautical community elected to introduce the ATS message handling system (AMHS), also called ATSMHS, as a novel stage in AFS based on proven MHS/X.400 technology over the Internet. This technology is largely similar to the common email system but lacks the capability to share binary data or graphics (ICAO 2008). Consequently, some countries elected to deviate from the X.400-based AMHS and implement the Simple Mail Transfer Protocol (SMTP) of the email system but ensuring backward compatibility with X.400. In terms of the AMHS specifications, the ICAO Doc 9705 was replaced and extended by Doc 9880–Manual on detailed technical specifications for the aeronautical telecommunication network using ISO/OSI standards and protocols (ICAO 2016), whose part IIB is entirely dedicated to AMHS. In particular, the AMHS includes the ATS message service and the ATN pass-through service (ICAO 1999). The ATN pass-through service consists in the carriage of AFTN messages over the ATN internet (i.e., exploiting X.400 or SMTP). In particular, as the purpose of legacy AFTN, CIDIN, and AMHS overlap—in the sense that each evolution expanded the scope of the former—suitable gateway systems allow the seamless retransmission of messages over PSTN, X.25, TCP/IP, SMTP, and even webmail based on the recipient and to meet local infrastructure restrictions and regulatory requirements (ICAO 2008). These gateways are outlined in ICAO (2001a) and further specified in ICAO (2016). On the other hand, the ATS message services supported by AMHS includes both basic and extended ATS message services (ICAO 1999). The extended ATS message service was introduced to cater for the following services: file transfer body parts (FTBP), IPM heading extension (IHE), AMHS security (ASEC), AMHS directory (ADIR) (EUROCONTROL 2009).

Inter-Center Communication System

The inter-center communication (ICC) system supports ATS interfacility data communication (AIDC) and shall be used to exchange the following messages between ATS centers/units over the ATN (ICAO 2001a): flight notifications, coordination, planning, transfer of control and communications, tactical flight-related data, airspace, and air traffic flow management messages. These also include notifications for flights approaching the jurisdiction boundary and coordination of boundary conditions in terms of transfer of control and of communication. AIDC shall only be used to convey tactical ATC control information and shall not be used to convey AFTN or information messages. The most up-to-date specification of AIDC is given in ICAO Doc. 9880 Part IIA (ICAO 2016). AIDC flight notification messages are dispatched by a controlling ATS unit to one or multiple downstream ATS units to disseminate information regarding changes in a particular flight's profile. AIDC flight coordination messages involve amendments in the conditions of transfer of a particular flight. AIDC transfer of control messages are used for formal control handover interactions concerning a particular flight. Similarly, AIDC transfer of communications are used to negotiate the communication handover of a particular flight. AIDC transfer of surveillance data convey surveillance information between two adjacent ATS units. Finally, AIDC general information interchange messages involve free-text/unstructured information regarding a particular flight to be transmitted (ICAO 2016).

AFS Evolutions: AIXM, FIXM, WXXM, and SWIM

The adoption of Extensible Markup Language (XML) for formatting AFS messages was eventually considered (ICAO 2008). The main advantages of XML-based exchange models are platform/technology independent, harmonized, and interoperable models that not only involve a very simple, easily understandable and better error-proof syntax, but also combine the possibility of conveying binary data with the advantage of human-readability of the raw messages. XML-based AFS protocols are now being realized with the development of the aeronautical/flight/weather information exchange models (AIXM/FIXM/WXXM). These exchange models include their associated conceptual models and XML schema. The conceptual models describe the features and properties (attributes/associations) within the exchange model domain and are specified in terms of Unified Markup Language (UML). The XML schema specifies the formal declaration of each element in the XML document and therefore can be seen as a template/prototype implementation of the conceptual model. The geography markup language (GML) standard is adopted to describe geographic entities.

5.78 ATM Surveillance Infrastructure

The ground-based surveillance infrastructure consists of a combination of primary and secondary surveillance radars (PSR/SSR), as well as automatic dependent surveillance broadcast (ADS-B) receivers and wide-area multilateration (WAM) equipment. Various portions of the airspace, especially at low altitude, are outside the coverage of most ground-based surveillance equipment and are therefore targeted by satellite-based surveillance systems gradually appearing. In terms of radar-based surveillance, wide area coverage is achieved with adequate accuracy exploiting high-power equipment with low refresh rate and limited low-altitude coverage. As a consequence, some approach-grade PSR/SSR with higher refresh rates and better low-altitude coverage capabilities are installed at major commercial airports to cater for terminal control area (TCA) needs. Due to the sensitivity of the data collected, the ground-based surveillance infrastructure is very frequently managed by designated military authorities or in conjunction/coordination with them, and this affects also the design of the related ground telecommunication infrastructure. The military authorities are also frequently involved as recipient of most/all data collected by ANSP.

5.79 Meteorological Services

RTCA DO-340 introduced an advanced concept of use for meteorological (MET) datalink services (RTCA 2012). These are specified in terms of service category, method of delivery and of the weather information involved. The MET services are classified into two categories. Category 1 services are the primary means of delivering MET information and may be exclusively relied on to support decisions without questioning their validity, while Category 2 services are useful for making noncritical decisions but should not be relied upon as the sole source of information. Category 1 services comprise both MET and Aeronautical Information Service (AIS) datalinks. MET datalinks are used onboard aircraft to provide weather information for supporting flight crew decisions. There are three types of pilot decision support services, which reflect the different planning and execution needs of the flight crew: a Weather Planning Decision Service (WPDS), a Weather Near-Term Decision Service (WNDS), and a Weather Immediate Decision Service (WIDS). WPDS provides weather information for strategic planning such as in the case of changes in routing or cruise altitude due to ATFM initiatives or destination airport closures. In such scenarios, it is assumed that the flight crew has an advance time of 20 minutes or more to comprehensively evaluate the situation and plan/validate diversions or route amendments. WNDS provides weather information for tactical decision making such as avoiding hazardous weather cells (including cumulonimbus, icing, turbulence, etc.) especially in terminal arrival/departure operations. In these cases, the flight crew has limited time for replanning and coordination with ATM operators, typically between 3 and 20 minutes. WIDS is conceived to provide weather information for freshly detected or quickly evolving weather hazards in order to allow the flight crew to initiate an emergency avoidance or abort take-off/landing. These decisions are assumed to require immediate action from within a few seconds to 3 minutes.

All the services are supported by three delivery modes: broadcast, demand, and contract. Broadcast datalink delivery service issues continuous regular transmissions of MET information to all aircraft within range. Demand and contract datalinks delivery services require two-way communication between the aircraft and ground station, where the flight crew initiates a request for specific MET information which the ground station then responds to. For the demand service, the ground station only needs to respond to an initiated request, while for the contract service, the information request is usually pre-coordinated and this requires the ground station to monitor the aircraft and provide the MET information at predefined time intervals or position.

RTCA DO-308 specifies four different categories for MET data formats (RTCA 2007). These are point data, area data, vector graphics, and gridded data. Point and area data are given in alphanumeric strings and, as their name suggest, provide weather information on either a single geospatially located point or over an area delimited by a polygonal line. Examples of point data include the conventional meteorological terminal air report (METAR) and terminal aerodrome forecast (TAF), whereas examples of area data include significant meteorological information (SIGMET). Vector graphic data are images, represented by vectors, points, lines, or other geometric entities and can be used to mark out volumes of interest. Gridded data, typically in the form of General Regularly Distributed Information in Binary (GRIB) format, consist in a four-dimensional structured grid (latitude, longitude, altitude, time) of weather data with a forecast time dimension. The amount of MET information exchanged is also dependent on the delivery method. Broadcast services always transmit the full set of weather information, while demand and contract will usually transmit a subset of information based on the initiated request and thus have shorter transmission times. The nature of the decision service also affects the type of information provided. For example, Table 5.24 shows the information provided by the three pilot decision support services.

MET service	Weather Planning Decision Service (WPDS)	Weather Near-Term Decision Service (WNDS)	Weather Immediate Decision Service (WIDS)
Time horizon	Greater than 20 min	3–20 min	Less than 3 min
Profile grid-point data	Offline and strategic online operations	Tactical online operations	Emergency operations
Airport Equivalent	METAR (incl. RVR), TAF		RVR, gusts, and wind shear

Table 5.24 MET Information Classified According to Decision Service (RTCA 2010)

As WPDS is used for route planning and optimization, the information transmitted has a significantly longer time horizon than WNDS and WIDS. The geographic extent of WPDS MET data is also significantly larger as WPDS data is conceived for strategic large-scale replanning and diversions, whereas WNDS information is used for local tactical rerouting or altitude changes and might involve higher fidelity and resolution, while these tasks are not supported by WIDS. The profile grid-point data is mainly used for arrival and departure planning and involves MET information along the planned flight profile only. The information provided in WIDS is used for more tactical operations such as airborne separation assurance systems (ASAS). Hazardous weather information is similar across all the three services, while for airport weather, WPDS and WNDS provide more comprehensive information in the form of METAR, TAF, and runway visual range (RVR), while WIDS provides only visibility and wind shear weather data. RTCA DO-324 specifies the required communication performance (RCP) for the service delivery (RTCA 2010). These are determined in terms of transaction time (TT) and can be either the RCP transaction time (TTRCP) or the normal transaction time (TT95), both expressed in seconds. A transaction is defined as the basic unit of an interaction between peer parties, which includes one or more operational messages that are transmitted from one party to the other. TTRCP is the maximum time for completion of a transaction, while TT95 is the time before which 95% of all transactions should be completed. Different TT requirements are defined for airport, terminal and en route domains. These are tabulated in Tables D7 to D15 of RTCA DO-324 (RTCA 2010).

The recommended quality of MET information is defined in ICAO Annex 3 (ICAO 2010), which lists the desirable accuracies of measurements forecasted data for a range of weather data. A partial list is provided in Table 5.25, which shows the desired quality of measured data and also the TAF data.

The rest of the document also provides recommendations regarding forecast data for general trends, take-off, and en route cases. However, these are not operational requirements but rather desirable accuracy figures to fulfil typical operational needs. RTCA DO-308 also identifies a list of candidate MET products (RTCA 2007). In particular, the global forecast system (GFS) used for

Weather data	Accuracy of measurement	Accuracy of forecast	
Wind direction	±10°	±20°	Minimum percentage within range
Mean surface wind	±1 knot up to 10 knots ±10% above 10 knots	±10 5 knots	80% of cases
Visibility	±50 m up to 600 m ±10% between 600 and 1,500 m ±20% above 1,500 m	±200 m up to 800 m ± 30% between 800 and 10 km	80% of cases
Cloud amount	±1 okta	One okta below 1500 ft Occurrence of BKN or OVC between 1,500 and 10,000 ft	80% of cases
Cloud height	±33 ft up to 330 ft ±10% above 330 ft	± 100 ft up to 1,000 ft ± 30% between 1,000 and 10,000 ft	70% of cases
Air temperature	±1°C	±1°C	70% of cases

TABLE 5.25 Recommended Quality of MET Information (ICAO 2010)

World area forecast center information	Data format	Refresh rate (hours)	Validity (hours)
Wind (latitudinal and longitudinal)	Gridded	6	36
Temperature	Gridded	6	36
Humidity	Gridded	6	36
Tropopause—height, temperature, direction	Gridded	6	36
Maximum wind—speed, direction, height	Gridded	6	36
Significant weather charts	Vector	6	13

Table 5.26 Reference Global Forecast MET Data for WPDS (RTCA 2007)

planning and offline/strategic online optimization purposes falls under the World Area Forecast Center group shown in Table 5.26.

The WPDS data described above have been sufficient for flight planning and offline/strategic online trajectory optimization applications within airborne avionics and ground-based ATM DSS, as the main factors driving the lateral/vertical planning are the 4D wind and temperature fields. Nevertheless, a significant gap is represented by the lack of information regarding weather cells and other adverse phenomena. Consequently, in terms of future CNS+A evolutions, ICAO's Aviation System Block Upgrades (ASBU) roadmap acknowledges that further improved meteorological services are required to implement advanced functionalities such as the ones involved in 4D-TBO. Long-range weather forecasts (such as GRIB and METAR) are already being supplemented by nowcasting techniques for the provision of ATM and ATFM services, particularly in the terminal area. Some of these advanced weather services are already acknowledged by RTCA in DO-308, as listed in Table 5.27. In particular, the U.S. National Center for Atmospheric Research's Thunderstorm Identification, Tracking, Analysis, and Nowcasting (TITAN) system is an example of a nowcasting system belonging to the NCWF class and used by ANSPs in the United States, Australia, and South Africa.

Nowcasting provides detailed current weather information extrapolated up to 6 hours into the future, typically with sub-kilometer spatial resolution and a temporal resolution in the order of minutes. Nowcasting systems use sophisticated algorithms to track and extrapolate individual

National Weather Service (NOAA)	Data format	Refresh rate	Validity (hours)
National Convective Weather Forecast (NCWF)	Gridded/Vector	5 min	1
Graphical turbulence guidance (GTG)	Gridded	15 min	0.25
Current Icing Product (CIP)	Gridded	1 hour	N/A
Forecast Icing Potential (FIP)	Gridded	1 hour	3
Weather Information Management Systems (WIMS) (FLYSAFE)			
WIMS thunderstorm	Gridded/vector	5 min to 6 hours	0.2–1
WIMS turbulence	Gridded/vector	6 hours	36
WIMS icing	Gridded/vector	15 min to 12 hours	0.24–24
WIMS wake vortex	Gridded/vector	1–6 hours	2–12

Table 5.27 Advanced METLINK Products for Flight Planning in the United States and Europe (RTCA 2007)

storm cells from weather radar information. Aviation weather service providers are increasingly supplementing this information with satellite imagery and ground-based sensors such as lightning detectors. The ability to downlink weather data turns datalink-equipped aircraft into mobile weather sensors able to report on the weather in their vicinity. Current-generation ATM systems are capable of receiving this data and exporting it to aviation weather service providers. Such data can contribute to a "4D weather cube," supplementing weather data along major air routes and allowing those sections of the cube that are of the most interest to aviation users to be updated more frequently and more accurately. However, the availability of a near real-time, high-resolution full 4D Weather Cube is not far off. In 2015, SESAR successfully demonstrated a web-services-enabled SWIM implementation of the FAA NEXTGEN 4D Weather Cube concept in their "Optimizing trajectories over the 4DWeatherCube" SWIM Masterclass. Further research and development is currently being carried out as part of SESAR's TOPMET project.

5.80 Trajectory Design

Specific conditions are introduced on the trajectories generated and displayed by ATM systems to ensure global interoperability. In particular, the specifications set for required navigation performance (RNP) 4D trajectory descriptors complying with RTCA DO-236C and DO-229D (RTCA 2013a, 2013b) are outlined here:

1. The fly-by transition is defined by the following equations (RTCA 2013a):

$$R = \frac{GS^2}{g \tan \mu} \tag{5.332}$$

$$Y = R \tan(0.5\ \delta\chi) \tag{5.333}$$

$$GS = \left|\vec{v}_{TAS} + \vec{v}_{wind}\right| = \begin{cases} 500\ \text{kt (low altitude)} \\ 750\ \text{kt (high altitude)} \end{cases} \tag{5.334}$$

$$\mu = \begin{cases} \min[0.5\ \delta\chi;\ 23.0°] & \text{(low altitude)} \\ 5.0° & \text{(high altitude)} \end{cases} \tag{5.335}$$

$$\text{if } Y > 20.0\ \text{nmi} \Rightarrow \begin{cases} Y = 20.0\ \text{nmi} \\ R = \dfrac{20.0}{\tan(0.5\ \delta\chi)} \end{cases} \tag{5.336}$$

2. The geometry of the fixed radius transition (FRT) is defined by the track change $\delta\chi$ and the radius R. The lead distance Y from the turn initiation to the waypoint and the abeam distance X between the waypoint and the point of the circular arc abeam the waypoint are defined based on the radius and the track change as per the following equations:

$$X = R \cdot \left(\frac{1}{\cos \delta\chi/2} - 1\right) \tag{5.337}$$

$$Y = R \cdot \tan(\delta\chi/2) \tag{5.338}$$

Context	Track change ($\delta\chi$)	Maximum radius (R)	Maximum turn initiation distance (Y)
High altitude ($h \geq 19{,}500$ ft)	$< 24.1°$	93.7 nmi	$R \cdot \tan\frac{\delta\chi}{2}$
	$\geq 24.1°$	$\frac{Y}{\tan\frac{\delta\chi}{2}}$	20.0 nmi
Low altitude ($h \geq 19{,}500$ ft)	$\leq 46.0°$	$\frac{Y}{\tan\frac{\delta\chi}{2}}$	3.65 nmi
	$> 46.0°$	8.59 nmi	$R \cdot \tan\frac{\delta\chi}{2}$
Feeder, departure, and missed approach	$\leq 46.0°$	$\frac{Y}{\tan\frac{\delta\chi}{2}}$	1.78 nmi
	$> 46.0°$	4.18 nmi	$R \cdot \tan\frac{\delta\chi}{2}$
Initial approach	$\leq 46.0°$	$\frac{Y}{\tan\frac{\delta\chi}{2}}$	1.45 nmi
	$> 46.0°$	3.41 nmi	$R \cdot \tan\frac{\delta\chi}{2}$
Intermediate approach	$\leq 46.0°$	$\frac{Y}{\tan\frac{\delta\chi}{2}}$	1.00 nmi
	$> 46.0°$	2.38 nmi	$R \cdot \tan\frac{\delta\chi}{2}$

TABLE 5.28 Geometric Characteristics of the Fly-by Theoretical Transition Area (RTCA 2013b)

3. When transitioning from one airway to another, if both require an FRT at the common waypoint, the smaller of the two radii applicable shall be selected. In the case one of the two airways does not involve an FRT, the FRT of the other shall be implemented.

The criteria for implementing the definitions above are given in Table 5.28, adapted from RTCA DO-229D (RTCA 2013b).

5.81 CNS+A Evolutions

Innovative ATM technologies are being developed by the major research initiatives in this domain including the Single European Sky ATM Research (SESAR) in Europe, the Next Generation Air Transportation System (NextGen) in the United States, OneSky in Australia, and other programs worldwide including the collaborative action for renovation of air transport systems (CARATS) in Japan, SIRIUS in Brazil, Future Indian Air Navigation (FIANS) in India, and others. Although the various countries involved are addressing national priorities and specificities in different ways (with a consequent need to address harmonization and interoperability requirements both at a regional and global level) all programs converge in promoting the evolution of ATM into a highly automated, integrated, and collaborative system, allowing a more flexible and efficient use of the airspace resources through higher levels of automation and information sharing, in line with the evolutions originally envisioned by the Future Air Navigation Systems (FANS) special committee of the International Civil Aviation Organization (ICAO) in the 1980s (Bradbury 1991). The major CNS+A innovations include

- Four-dimensional (4D) trajectory-based operations (TBO):
- Performance-based communication, navigation, and surveillance (PBC/PBN/PBS), enabling CNS performance-based operations (PBO)
- System-wide information management (SWIM)
- Enhanced line-of-sight (LOS) and beyond LoS (BLOS) aeronautical communications, involving a substantial exploitation of datalink technology
- Ground, avionics, and satellite-based augmentation systems (GBAS/ABAS/SBAS) enabling global navigation satellite systems (GNSS) as primary means of navigation in all flight phases, including precision approach and auto-landing
- Enhanced ground-based and satellite-based (LOS/BLOS) surveillance, including Automatic dependent surveillance broadcast (ADS-B) and self-separation assurance
- Collaborative decision making (CDM)
- Dynamic airspace management (DAM)
- Advances in human machine interface and interaction (HMI^2), including cognitive forms of HMI^2 adaptation

In order to deploy these enhanced capabilities, advances in integrated CNS+A systems are essential. Therefore, new ground-based and airborne CNS+A systems are being developed to improve the operational efficiency and environmental sustainability of aviation, while at the same time ensuring higher levels of safety and interoperability. These novel automated CNS+A systems allow better equipped aircraft to fly user-preferred optimal flight paths, limiting the intervention of human operators to high-level and emergency decisions. Increased airspace flexibility enables more efficient and greener flight profiles. The CNS+A systems will provide better and precise airborne automated navigation services aided by optimal aircraft separation assistance to fully exploit the available airspace resources and ground facilities. The key enabling CNS+A systems include

- Network-centric avionics and ground-based ATM systems for planning and real-time execution of 4DT, incorporating multiobjective 4DT optimization (MOTO-4D), negotiation, and validation functionalities.
- High-integrity, high-throughput, and secure LOS/BLOS next generation air-to-ground datalink (NG-ADL) system networking all the CDM stakeholders and SWIM.
- Integrated health management systems for both airborne and ground-based systems.
- CNS+A technologies for RPAS.

In order to fulfil the SESAR/Clean Sky and NextGen objectives, the online air traffic flow is optimized by the novel automated systems considering high-density airspace, rerouting, and rescheduling in real time (Sabatini et al. 2013). The overall 4DT decision making is enhanced by exploiting automated negotiation and validation schemes. This involves the exchange of constraints and operational parameters between a next generation flight management system (NG-FMS) and a 4DT planning, negotiation, and validation (4-PNV) system via the NG-ADL (Gardi et al. 2014, 2015;

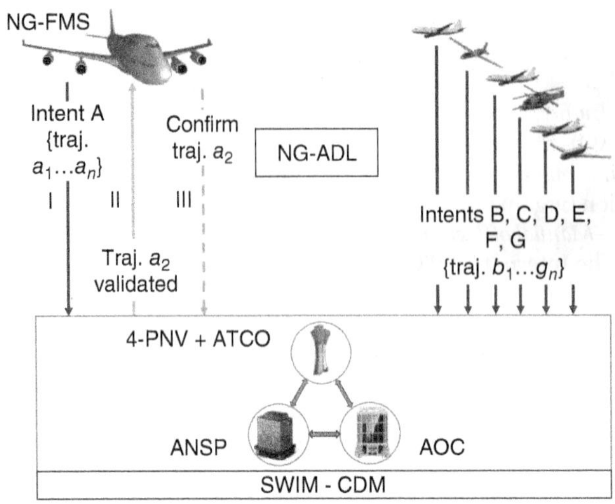

FIGURE 5.118 CNS+A enabled intent negotiation and validation scenario.

Ramasamy et al. 2014, 2015). A CNS+A enabled 4DT intent negotiation and validation context is illustrated in Figure 5.118. The innovative 4-PNV system receives 4DT intents from each manned and unmanned aircraft equipped with NG-FMS (Flow of information I in Figure 5.118). If an optimal conflict free trajectory is identified, the 4-PNV system instructs each aircraft to fly the validated trajectory (flow of information II in Figure 5.118) and the aircraft sends a confirmation to the ground (flow of information III in Figure 5.118).

When feasible trajectories cannot be identified from the NG-FMS intents, the 4-PNV system, calculates a new set of optimal trajectories based on performance weightings agreed between airline operating centers (AOC) and air navigation service providers (ANSP) and uplinks them to the respective aircraft. The onboard NG-FMS then identifies the preferred trajectory and sends a confirmation to the 4-PNV system. A subcase of this is when a new negotiation loop is initiated by the 4-PNV system (e.g., due to conflicting air traffic and/or direct inputs from the ANSP/AOC).

References

Bradbury, J. N., 1991, "ICAO and Future Air Navigation Systems," in *Automation and Systems Issues in Air Traffic Control*, Springer, pp. 79–99.

EUROCONTROL, 2009, *EUROCONTROL Specification on the Air Traffic Services Handling System (AMHS)*, report n. SPEC-136, The European Organization for the Safety of Air Navigation (EUROCONTROL), Brussels, Belgium.

Gardi, A., Ramasamy, S., Kistan, T., et al., 2016a, "UAS in the Terminal Area: Challenges and Opportunities," in R. Blockley and W. Shyy (eds), *Encyclopedia of Aerospace Engineering*, John Wiley & Sons, Chirchester, UK, Vol. UAS.

Gardi, A., Ramasamy, S., Sabatini, R., et al., 2016b, "CNS+A Capabilities for the Integration of Unmanned Aircraft in Controlled Airspace," 2016 International Conference on Unmanned Aircraft Systems (ICUAS '16), Arlington (VA), United States.

Gardi, A., Sabatini, R., Kistan, T., et al., 2015, "4-Dimensional Trajectory Functionalities for Air Traffic Management Systems," Integrated Communication, Navigation and Surveillance Conference (ICNS 2015), Herndon, VA, United States.

Gardi, A., Sabatini, R., Ramasamy, S., et al., 2014, "Real-Time Trajectory Optimisation Models for Next Generation Air Traffic Management Systems," *Applied Mechanics and Materials*, Vol. 629, pp. 327–332.

ICAO *Doc. 9426—Air Traffic Services Planning Manual*, report n., The International Civil Aviation Organization (ICAO), Montreal, Canada.

ICAO *Doc. 9883—Manual on Global Performance of the Air Navigation System*, report n., The International Civil Aviation Organization (ICAO), Montreal, Canada.

ICAO, 1999, *Doc. 9705—Manual of Technical Provisions for the Aeronautical Telecommunication Network (ATN)*, report n., The International Civil Aviation Organization (ICAO), Montreal, Canada.

ICAO, 2001a, *Annex 10 to the Convention on International Civil Aviation—Aeronautical Telecommunications, Vol. 2: Communications Procedures Including Those with PANS Status*, The International Civil Aviation Organization (ICAO), Montreal, Canada.

ICAO, 2001b, *Annex 11 to the Convention on International Civil Aviation—Air Traffic Services*, International Civil Aviation Organization (ICAO), Montreal, QC, Canada.

ICAO, 2005, *Doc. 9854—Global Air Traffic Management Operational Concept*, The International Civil Aviation Organization (ICAO), Montreal, Canada.

ICAO, 2007a, *Annex 10 to the Convention on International Civil Aviation—Aeronautical Telecommunications, Vol. 3: Communications Systems*, report n., The International Civil Aviation Organization (ICAO).

ICAO, 2007b, *Doc. 4444—Procedures for Air Navigation Services—Air Traffic Management*, report n., The International Civil Aviation Organization (ICAO), Montreal, Canada.

ICAO, 2008, *Aeronautical Communications Panel (ACP) Working Paper: Evolution of the Aeronautical Fixed Telecommunication Network (AFTN)—Progress of Work within WG-I (IPS), Including Future Work Program on IPS Based Air-Ground and Ground-Ground Applications*, report n. ACP-WGW02/WP-7, The International Civil Aviation Organization (ICAO), Montreal, Canada.

ICAO, 2010, *Annex 3 to the Convention on International Civil Aviation—Meteorological Service for International Air Navigation*, The International Civil Aviation Organization (ICAO), Montreal, Canada.

ICAO, 2011, *EUR CIDIN Manual—Sixth Edition*, report n., The International Civil Aviation Organization (ICAO)—European and North Atlantic Office (EUR), Montreal, Canada.

ICAO, 2013, *Working Document for the Aviation System Block Upgrades—The Framework for Global Harmonization*, The International Civil Aviation Organization (ICAO).

ICAO, 2016, *Doc. 9880—Manual on Detailed Technical Specifications for the Aeronautical Telecommunication Network using ISO/OSI Standards and Protocols*, The International Civil Aviation Organization (ICAO), Montreal, Canada.

Ramasamy, S., Sabatini, R., and Gardi, A., 2015, "Novel Flight Management Systems for Improved Safety and Sustainability in the CNS+A Context," Integrated Communication, Navigation and Surveillance Conference (ICNS 2015), Herndon, VA, United States.

Ramasamy, S., Sabatini, R., Gardi, A., et al. 2014, "Next Generation Flight Management System for Real-Time Trajectory Based Operations," *Applied Mechanics and Materials*, Vol. 629, pp. 344–349.

RTCA, 2007, *RTCA DO-308: Operational Services and Environment Definition (OSED) for Aeronautical Information Services (AIS) and Meteorological (MET) Data Link Services*, SC-206, 6/12/2016, RTCA DO-308.

RTCA, 2010, *RTCA DO-324: Safety and Performance Requirements (SPR) for Aeronautical Information Services (AIS) and Meteorological (MET) Data Link Services*, SC-206, 8/12/2010, RTCA DO-324.

RTCA, 2012, *RTCA DO-340: Concept of Use (CONUSE) for Aeronautical Information Services (AIS) and Meteorological (MET) Data Link Services*, SC-206, 26/9/2016, RTCA DO-340.

RTCA, 2013a, RTCA DO-236, *Minimum Aviation System Performance Standards: Required Navigation Performance for Area Navigation*, Radio Technical Committee for Aeronautics Inc. (RTCA), Washington, DC, United States.

RTCA, 2013b, *Minimum Operational Performance Standards for Global Positioning System/Satellite-Based Augmentation System Airborne Equipment*, report n. RTCA DO-229D, Radio Technical Commission for Aeronautics Inc. (RTCA), Washington, DC, United States.

Sabatini, R., Gardi, A., Ramasamy, S., et al. 2015, "Modern Avionics and ATM Systems for Green Operations," in R. Blockley and W. Shyy (eds), *Encyclopedia of Aerospace Engineering*, John Wiley & Sons.

Visser, H. G., 1994, *A 4D Trajectory Optimization and Guidance Technique for Terminal Area Traffic Management*, Delft University of Technology.

PART 16

Aerospace Systems and Software Engineering

Yixiang Lim, Eranga Batuwangala, Roberto Sabatini, Alessandro Gardi, Subramanian Ramasamy, and Trevor Kistan

5.82 Introduction

This part provides an overview of the processes involved in the aerospace systems engineering lifecycle. An aerospace system may include software, mechanical, electrical, and electronic hardware. Within the system, the various interactions between its elements interact to produce the desired behavior to satisfy predefined operational, certification, or user requirements. Figure 5.119 provides a layout of the systems development process. The development starts with a top-level requirements capture, followed by the development of systems-level requirements. These requirements are then used to define lower-level hardware (HW) and software (SW) requirements, as well as in the development of verification and validation procedures. The low level requirements are used for developing the system's software/hardware components, which are verified and validated against the higher-level requirements.

5.83 Software Life-Cycle Process

The software life cycle comprises five primary processes that determine the roles of the various stakeholders within the organization (Table 5.29):

In particular, the software development process involves the design and integration of the software into a system. This typically includes the following eight phases: requirement, design,

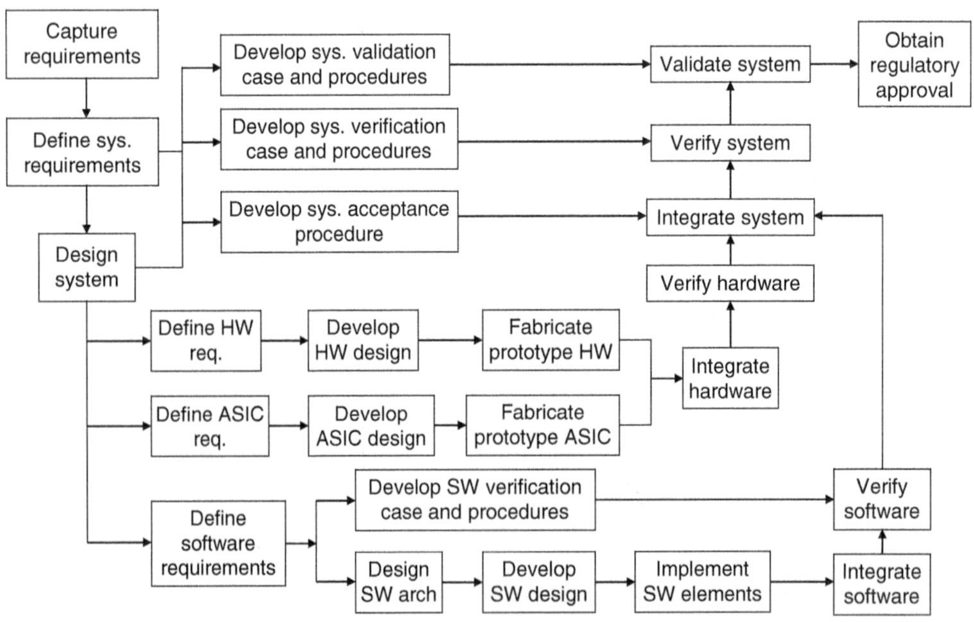

FIGURE 5.119 General layout of the system development process (ASIC—application-specific integrated circuit).

Process	Role	Stakeholder
Acquisition	Defines the activities of the acquirer—the organization that acquires a system, software product, or software service	Acquirer
Supply	Defines the activities of the supplier—the organization that provides the system, software product, or software service to the acquirer under terms of the contract	Supplier
Development	Defines the activities of the developer—the organization that defines and develops the software product during the life-cycle process	Developer
Operation	Defines the activities of operator—the organization that operates the system in its live environment for its users	Operator/user
Maintenance	Defines the activities of maintainer—the organization that provides the service of maintaining the software product	Support agency

TABLE 5.29 Software Life-Cycle Process

implementation, test, installation, operation, maintenance, and retirement, which are organized within a formal framework to provide a mechanism for communicating the software development process between the various stakeholders. These stakeholders can include the procuring agencies and contractors, as well as the system and software engineers. The common frameworks—waterfall, v-model, prototyping, and spiral models—are described below.

Waterfall

The waterfall model consists of five to seven phases of development (including the systems and software requirements, analysis, design, implantation, test, maintenance). The phases are carried out sequentially, beginning at the systems level and flowing down to the software level. The model is linear and systematic, requiring the customer's review and approval of existing requirements, design, and test documents in the current phase before proceeding to subsequent phases. As such, it is well-documented and straightforward to apply but has problems adapting if changes are introduced to design or requirements at a later stage. As such, if the customer does not have all the requirements available at the beginning of the project, the waterfall model might not be suitable.

V-Model

The V-model is an extension of the waterfall model, with a structure that associates each development activity with a test or validation at the same level of abstraction. The top-level system requirements are associated with the system integration test and evaluation (T&E) activities; subsystem (hardware and software) requirements are associated with acceptance testing; preliminary design is associated with subsystem integration; detailed design is associated with component testing; and source code is associated with unit testing. The V-model has similar characteristics as the waterfall model, with greater emphasis on test and evaluation.

Prototyping

The prototyping model allows for the quick design and development of prototype systems, and is usually useful in projects where the customer does not have a clear set of requirements at the beginning. The iterative cycle of design, prototyping, and evaluation allows early changes to be made according to customer feedback, resulting in a quicker and less costly development cycle, improved system usability and flexible adaptation to changing requirements. However, such a model also requires greater commitment from the customer in the systems test and evaluation.

Spiral

The spiral model combines the top-down, structural features of the waterfall model with the bottom-up iterative features of the prototyping model while introducing a new element of risk assessment. Each cycle of the spiral comprises four phases: determining and reviewing project objectives, alternatives, and constraints; evaluating alternatives as well as identifying and resolving risks, development, and verification of next-level product; as well as reviewing the current phase and planning the next phase. A cyclic approach allows the incremental growth of the system's degree of definition and implementation while reducing the degree of risk in the developmental cycle (e.g., the first cycle develops the concept of operations, the second develops subsystem-level requirements, the third cycle designs and verifies the subsystem components, the fourth cycle integrates and verifies the system level components, etc.). The spiral model is appropriate for large projects and provides a mechanism for supporting projects with many requirements.

Agile

The agile model is similar to the prototyping model in that systems are incrementally developed in short iterations called sprints. At the end of each sprint, a working product is demonstrated to customers, allowing continuous feedback on the product, progressive refinement and minimizing development risk. Agile models emphasize collaboration and communication with the customer, flexibility and adaptability to change, and evolving requirements. Popular methodologies used in the agile model include scrum, kanban, and lean.

5.84 Software Requirements

Requirements are defined at the start of the software development life cycle and are refined, modified, and used throughout the rest of the development process. As such, the process for managing, documenting, and maintaining requirements is an important component of the development life cycle. The various types of requirements described in Table 5.30 can be described by their purpose in the development cycle. For example, customer requirements can be used to define the operational context and constraints, such as the deployment platform, the mission scenario, or the performance parameters required to accomplish the mission.

When developing requirements, considerations must be given to the following attributes; a good requirement should satisfy the following attributes:

Specific: The requirement must address only one aspect of the system design or performance, and must be expressed in terms of the need (what is needed, why it is needed, how well it needs to be done) and not the solution (how it should be done).

Measurable: The requirement must have a measurable performance that is expressed in a manner that allows verification in an objective and preferably quantitative manner (e.g., an exact pointing requirement can be tested and thus verified prior to deployment).

Achievable: The requirement must reflect a need or objective for which a solution is technically achievable at costs considered affordable.

Relevant: The requirement must have the appropriate detail for the level of system hierarchy that it is specified for. It should not be too detailed that it constrains solutions for the current level of design (e.g., detailed requirements relating to components would not normally be in a system-level specification).

Traceable: The lower level requirements must clearly flow from and support higher level requirements. Requirements without a parent are referred to as orphans, and need to be assessed for necessity of inclusion.

Types of requirements	Description of requirements
Customer requirements	Statements of fact and assumptions that define the expectations of the system in terms of mission objectives, environment, constraints, and measures of effectiveness and suitability (measures of effectiveness/suitability).
Functional requirements	The necessary task, action, or activity that must be accomplished. Functional (what has to be done) requirements identified in requirements analysis will be used as the top-level functions for functional analysis.
Performance requirements	The extent to which a mission or function must be executed; generally measured in terms of quantity, quality, coverage, timeliness, or readiness.
Design requirements	The "build to," "code to," and "buy to" requirements for products and "how to execute" requirements for processes expressed in technical data packages and technical manuals.
Derived requirements	Requirements that are implied or transformed from higher-level requirements.
Allocated requirements	A requirement that is established by dividing or otherwise allocating a high-level requirement into multiple lower-level requirements.

Table 5.30 Different Types of Requirements

Complete: The requirement must contain all information necessary to understand the customer's needs (e.g., all mission profiles, operational and maintenance concepts, utilization environments, and constraints).

Consistent: The requirement must not conflict with other requirements. Any conflicts must be resolved up front.

The requirements engineering process is illustrated in Figure 5.120 and comprises an iterative cycle of requirements elicitation, analysis, specification, validation, and documentation, as well as modeling (which is part of the design process). Requirements are captured during the elicitation

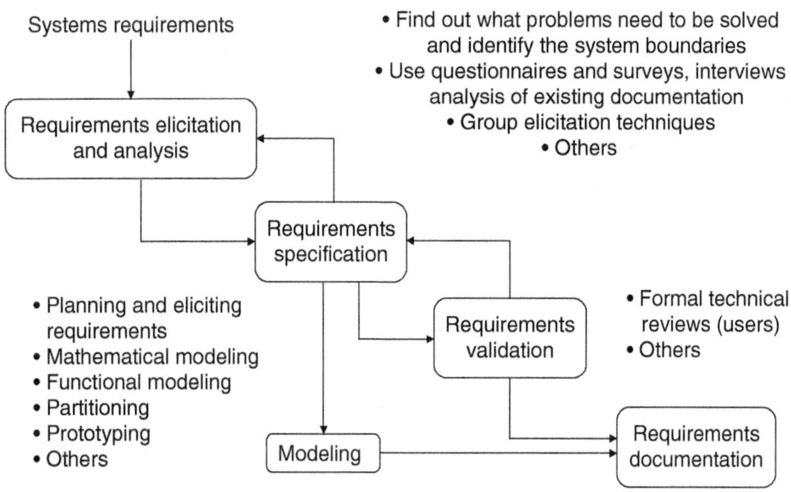

Figure 5.120 Requirements engineering process.

phase, which are captured through a number of methods such as interviews, focus groups, or team meetings. The requirements are analyzed to identify the domain constraints, as well as consistency, gaps, and ambiguity in the requirements. They can then be sorted, prioritized, and formally expressed in the specification phase (e.g., as a graphical or mathematical model). The validation phase reviews the specifications to ensure that they are consistent, correct, understandable, realistic, testable, and traceable. When these are satisfactory, the requirements and models are appropriately documented.

5.85 Software Design

Software design transforms requirements into four classes of design details or activities. The product is a software design specification describing the data, architecture, interfaces, and components.

Data design is the transformation of the information model into structures for storing data. Data structures allow the program to retrieve and store data efficiently, allow organization of the program into modules, and reduces the overall complexity of the software. The types of data structures, the links between data structures and the integrity rules of the data are specified.

Architectural design involves identifying the components and modules within the software and defining the relationships that exist between them. The three levels of detail in architecture design are *conceptual architecture*, which identifies the different software components and allocates responsibilities to each; *logical architecture*, which focuses on the design of component interfaces, interactions, connection mechanism and protocols; and *execution architecture*, which focuses on the communication, coordination, and resource allocation of software components during runtime.

Interface design is the design of the interfaces between the different components of the software/hardware, as well as between the program and the end user. It describes how the software communicates with itself, the systems that interoperate with it, and with the human user through data and control flow elements.

Component level design transforms the structural elements of the software architecture into a procedural description of the software components.

Typical languages used in the development of avionics and ATM systems include C, C++, and Ada. While C is a procedural language, C++ also supports an object-oriented programming paradigm. Ada also supports object-oriented programming and is adopted in a variety of safety critical applications. Ada possesses a number of characteristics that allow the development of reliable programs and also enhance code safety and maintainability. In recent years, Java and Python have also been adopted for development of embedded avionics and ATM systems.

Enterprise architectures describe the structure and operation of an organization and allow it to function in a cohesive and strategic manner. Architecture frameworks provide a standard basis for creating, using, and analyzing architectures. The Open Group Architecture Framework (TOGAF) is a widely used framework for designing, planning, implementing, and maintaining enterprise information technology (IT) architectures. TOGAF supports the design of enterprise architectures in four interrelated domains. These domains comprise the business architecture, which addresses the needs of the users, planners, and business management; application architecture, which addresses the needs of system and software engineers; data architecture, which address the needs of database designers, administrators, and system engineers; and technical architectures, which address the needs of hardware/software acquirers, operators, and administrators. Other frameworks include the British Ministry of Defence Architecture Framework (MODAF), Department of Defence Architecture Framework (DODAF), and NATO Architecture Framework (NAF). These architecture frameworks are used in the defence context to develop architectures for managing complex system-of-systems, ensuring interoperability within the network infrastructure, and to enable effective decision making between diverse stakeholder groups.

Model-based systems/software engineering is the application of modeling throughout different phases of the software lifecycle. Model-based systems/software engineering makes use of industry standard modeling languages such the Unified Modeling Language (UML), Systems Modeling Language (SysML), DODAF, or MODAF to support the specification, analysis, design, verification and validation of software and systems. Integrated development environments (IDEs) such as IBM's Rational Rhapsody provides a modeling environment for teams to work collaboratively on the development of complex and embedded applications.

5.86 Aerospace Software Verification and Validation

Software verification and validation (V&V) comprises two components: first, verification of the product to ensure that it performs its intended function in the absence of errors or unintended functionality, and is sufficiently robust to respond correctly to abnormal inputs and conditions; second, validation of the product to ensure that it satisfies the customer's needs as specified in the high-level requirements, that these high-level requirements have been correctly translated into lower-level requirements, and that the developed software architecture and source code meets all high and low level requirements.

Guidelines for conducting V&V can be found in the documents listed in Table 5.31—RTCA DO-178 has traditionally been the point-of-reference, and as of the writing of this subsection, DO-178B has been expanded to DO-178C (avionics-based software) and DO-278A (ground based and air traffic management software). Supplementary material, including DO-248C, DO-330, DO-331, DO-332, and DO-333, has been introduced to provide extra guidance to DO-178C and DO-278A. Further discussion of the RTCA documents can be found in (Jacklin 2012). Additionally, FAA provides a rich resource for more detailed description of software considerations—the most recent publications in this area include FAA/TC-16/4 and FAA/TC-15/57, and are also listed in Table 5.31.

The main objectives of V&V are to assess whether the system is usable in its intended operational environment, and to discover any possible defects in the system. Both static and dynamic methods are used in V&V. Static verification is conducted through software inspection to analyze the static system, and is supplemented by tool-based document and code analysis to discover problems with the software. Inspections are used to check conformance with a given specification but cannot check non-functional characteristics such as performance and usability. Tools, such as static analyzers, can be used to supplement the inspection, and work by parsing the program text to try to discover potentially erroneous conditions. The stages of static analysis are:

Control flow analysis checks for loops with multiple exit or entry points, finds unreachable code, etc.

Data use analysis detects uninitialized variables, variables written twice without an intervening assignment, variables which are declared but never used, etc.

Interface analysis checks the consistency of routine and procedure declarations.

Information flow analysis identifies the dependencies of output variables. Does not detect anomalies itself but highlights information for code inspection or review.

Path analysis identifies paths through the program and sets out the statements executed in that path.

Dynamic V&V is conducted through software testing, where the system (or a prototype of the system) is executed with test data and its operational behavior observed. Different types of tests are conducted such as defect testing, to discover system defects; as well as statistical testing, to reflect the frequency of user inputs and to estimate program reliability—the tests can reveal the presence of errors but not their absence and a successful test is one which discovers one or more errors. If defects or errors are discovered, the program undergoes debugging to locate and repair the errors.

Documents	Description
RTCA DO-178C	Provides guidelines for managing the software life-cycle processes, which include the planning, development and verification, quality assurance and certification processes for airborne software. System-level, software-level, and source code-level considerations are covered.
RTCA DO-278A	Provides guidelines which are similar to those of DO-178C, but covering ground-based CNS/ATM software: the software criticality levels in DO-178C are replaced with assurance levels in DO-278A; there is an additional section on commercial-off-the-shelf (COTS) software, and an emphasis on a system-of-systems considerations (including aspects of software communication, security, adaptability and cutover).
RTCA DO-248C	Contains background and supplementary information for further clarification of the guidance material presented in DO-178C and DO-278A. Comprises three main sections: answers to FAQs posed by industry to certification authorities; Discussion Papers generated over the course of the development of DO-178C/DO-278A—these provide greater detail in clarification; as well as rationale for the development of specific sections within the two documents.
RTCA DO-254	The hardware counterpart to DO-178, this document provides guidelines for managing the hardware development life-cycle processes.
RTCA DO-330	A standalone document for software tool developers. Provides guidelines for managing the software tool life-cycle processes. The processes are similar to those mentioned in DO-178C but have different requirements as the software covered in both documents are designed for different purposes.
RTCA DO-331	Supplements DO-178C and DO-278A when model-based development is included in the software life cycle. Model-based techniques are described, and guidance material is provided for specifying model requirements, conducting a model coverage analysis, and conducting certifiable model simulations, and model standards.
RTCA DO-332	Supplements DO-178C and DO-278A in providing assurance for the use of object-oriented technology (OOT) in the development of airborne and ground-based software. Specific OOT-based techniques are described, and guidance material is provided for the software development and verification processes, as well as for conducting vulnerability analysis and ensuring traceability.
RTCA DO-333	Supplements DO-178C and DO-278A in providing guidance for the use of formal methods (which are mathematically-based techniques) as part of the software life cycle. Formal methods are described, and guidance material is provided for the use of formal methods in the software development process, the use of formal methods for reviewing and analyzing software output, as well as using formal methods for coverage analysis and to verify source code or object code.
FAA AR-06/35, AR-06/36	A handbook (06/35) and report (06/36) to provide guidance on evaluating software development tools. The handbook provides an overview on DO-178B, the types of tools available, the criteria for qualifying these tools, and the outstanding issues which need to be addressed. The report goes into greater detail, providing perspectives on research questions posed in the problem statement. These include: the industry view on the use of software development tools, as well as the qualification, quality assessment and evaluation of these tools.

Table 5.31 List of Relevant Documents for Providing Guidelines on V&V

Documents	Description
FAA AR-06/54	A report evaluating the acceptability and suitability of structural coverage analysis tools with the objectives specified in DO-178B. The document makes references to both DO-178B and DO-248B. The different types of structural coverage are described and metrics are given for evaluating them. The requirements for a test suite are specified along with a prototype example.
FAA AR-07/19, AR-07/52, AR-07/17, AR-07/20	Handbooks (07/17 and 07/19) and reports (07/20 and 07/52) on a 3-phase research project investigating the use of Object Oriented Technology in Aviation as specified in DO-178B. Phase 1 was an industry survey and is not included in the references. Phase 2 addresses techniques to verify structural coverage of inter-component dependencies to satisfy data coupling and control coupling requirements; while phase 3 addresses structural coverage at the source and object code level to ensure the absence of unintended function. The handbooks provide a summary of the work presented in the reports.
FAA TC-16/4	A report addressing the verification and safety assurance issues for the use of adaptive systems in the Next Generation Air Transportation System. The report reviews and identifies state-of-the-art adaptive systems, the safety issues concerning the use of such systems, current certification standards, and proposes certification recommendations and requirements for verification, as well as future research needs.
FAA TC-15/57	A 3-phase report detailing the research on state-of-the-art software development and verification methods with respect to DO-178 and DO-278 (DO-178C and DO-278C were released in the middle of the project and were considered during phase 3). Phase 1 provides a review of software standards and software assurance methods used by industry and regulatory bodies; phase 2 discusses analysis techniques for providing assurance early in the design and streamlining the development process; phase 3 provides case studies of these techniques and reviews the model-based and formal methods described in DO-331 and DO-333.

Table 5.31 List of Relevant Documents for Providing Guidelines on V&V (*Continued*)

The level of detail and rigor required from the V&V is dependent on the assurance level of the software, as specified in Table 5.32—software and systems which are more safety-critical require more thorough and extensive V&V. The assurance level is determined from an initial functional hazard assessment (FHA) which identifies the criticality of the software to the functional safety of the aircraft. Based on the software assurance level, appropriate safety assessments and requirements validations are performed to determine and validate the low-level requirements at the architecture, system, source code and executable object code levels. The final implementation validation then provides confirmation that the implemented product satisfies all its intended functions and high-level requirements.

Table 5.33 provides an outline of the main components of a system safety assessment, which provides a systematic method of defining, evaluating and providing assurance for the safety requirements necessary in the development cycle of aircraft software and systems. These main components are the functional hazard assessment (FHA), preliminary system safety assessment (PSSA), system safety assessment (SSA), and common cause analysis (CCA).

Description	Failure condition	Failure probability (ARP 4761)
	Software safety level A	
Software whose anomalous behavior, as shown by the system safety assessment process, would cause or contribute to a failure of system function resulting in a **catastrophic** failure condition for the aircraft.	**Catastrophic** failure conditions would result in multiple fatalities, usually with the loss of the aircraft.	Less than $1e^{-9}$ per flight hour
	Software safety level B	
Software whose anomalous behavior, as shown by the system safety assessment process, would cause or contribute to a failure of system function resulting in a **hazardous** failure condition for the aircraft.	**Hazardous** failure conditions would result in a reduction in aircraft capability or in the flight crew's ability to cope with adverse operating conditions. These might result in: • A *large* reduction in safety margins or functional capabilities; • Physical *distress* or *excessive* workload such that the flight crew cannot be relied upon to perform their tasks accurately or completely; or • *Serious or fatal* injury to a relatively small number of the occupants other than the flight crew.	$1e^{-7}$ to $1e^{-9}$ per flight hour
	Software safety level C	
Software whose anomalous behavior, as shown by the system safety assessment process, would cause or contribute to a failure of system function resulting in a **major** failure condition for the aircraft.	**Major** failure conditions would result in a reduction in aircraft capability or the flight crew's ability to cope with adverse operating conditions. These might result in: • A *significant* reduction in safety margins or functional capabilities; • A *significant* increase in: crew workload, conditions impairing crew efficiency or discomfort to the flight crew; or • A *significant* increase in physical distress to passengers or cabin crew, possibly including injuries.	$1e^{-5}$ to $1e^{-7}$ per flight hour
	Software safety level D	
Software whose anomalous behavior, as shown by the system safety assessment process, would cause or contribute to a failure of system function resulting in a **minor** failure condition for the aircraft.	**Minor** failure conditions would not significantly reduce aircraft safety, and involve crew actions that are well within their capabilities. These might result in: • A *slight* reduction in safety margins or functional capabilities; • A *slight* increase in crew workload, such as routine flight plan changes; or • *Some* physical discomfort to passengers or flight crew.	$1e^{-3}$ to $1e^{-5}$ per flight hour

Table 5.32 Software Assurance Level Definitions and Failure Conditions, as per RTCA (2011)

Description	Failure condition	Failure probability (ARP 4761)
	Software safety level E	
Software whose anomalous behavior, as shown by the system safety assessment process, would cause or contribute to a failure of system function with **no effect** on aircraft operational capability or pilot workload.	Failure conditions that have **no effect** on safety. This means that the failure would have: • *No* effect on the operational capability of the aircraft • *No* increase in crew workload	1 to $1e^{-3}$ per flight hour

Table 5.32 Software Assurance Level Definitions and Failure Conditions, as per RTCA (2011) (*Continued*)

SSA components	Description
Functional hazard assessment (FHA)	FHA identifies the failure condition classification and the development assurance level. These allow the safety objectives to be assigned for that particular condition and subsequently, a quantitative probability requirement to be assigned.
Preliminary SSA (PSSA)	The PSSA examines the failure conditions established by the FHA on a system-by-system basis and aids in formulating the safety requirements for a given system and ensure that the system architecture incorporates those requirements. PSSA involves performing a Fault Tree Analysis (FTA) on the functions derived from the system functional architecture and calculating the probabilities of failure.
SSA	The SSA is a systematic and comprehensive evaluation of the system design using similar techniques to those employed during the PSSA activities. However, whereas the PSSA identifies the requirements, the SSA is intended to verify that the proposed design does in fact meet the specified requirements as identified during the FHA and PSSA analyses conducted previously. SSA involves calculation of reliability figures and determination of the level of redundancy needed to attain a specified safety target.
Common cause analysis (CCA)	The CCA assesses the failure modes associated with integrated systems or functions to ensure that the failure modes of the individual systems are either independent of each other, or if some dependence exists, the risks associated with that dependence are acceptable. The CCA is subdivided into three types of analysis: Zonal Safety Analysis (ZSA), Particular Risk Analysis (PRA), and Common Mode Analysis (CMA).

Table 5.33 Components of the System Safety Assessment, as Defined in SAE (1996)

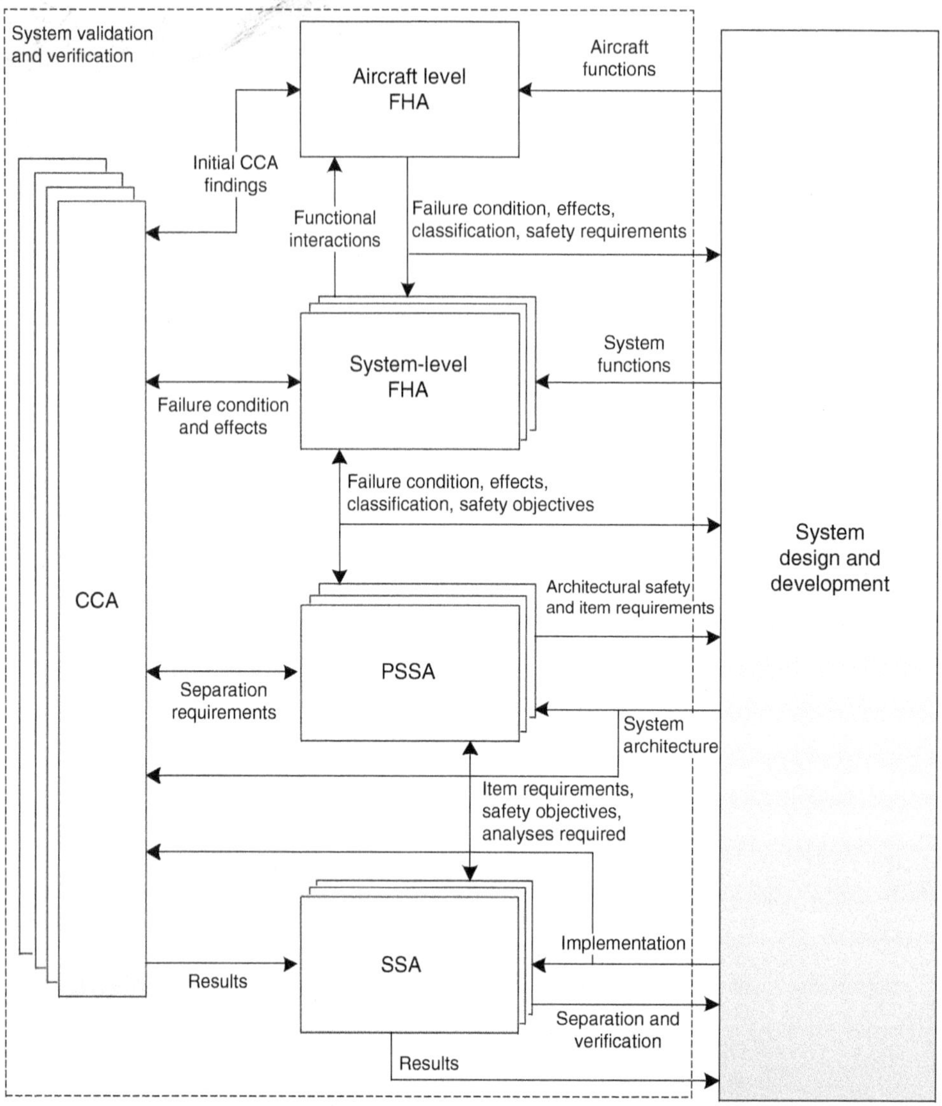

FIGURE 5.121 Safety assessment process (SAE 2010).

The relationships between the different components of the safety assessment process are illustrated in Figure 5.121, along with the interactions between system V&V, and design and development. Based on system and aircraft functions, the FHA develops the high-level requirements, which are passed to the PSSA. The PSSA develops low-level requirements and ensures that the software architecture satisfies the given high-level requirements, and the low-level requirements are passed to the SSA. The SSA is then performed to validate that the implementation of the source code satisfies the low level requirements. Throughout the process, the CCA is performed with increasing level of detail to ensure that the safety of the system-level integration is not compromised.

The review and analysis of the software development cycle covers a number of different aspects, as described in Table 5.34 and illustrated in Figure 5.122. The aspects of accuracy/consistency,

	High-level requirements	Low-level requirements	Software architecture	Source code
Compliance	The high-level requirements properly satisfy the system level (functional, performance and safety-related) requirements.	The low-level requirements satisfy the high-level requirements.	The software architecture does not conflict with the high-level requirements, especially functions that ensure system integrity such as partitioning.	The source code satisfies the low-level requirements and do not implement any undocumented functions. The source code matches the data flow and control flow as defined in the software architecture.
Accuracy/ consistency	Each high-level requirement is accurate, unambiguous, sufficient detailed, and do not conflict with the other requirements.	Each low-level requirement is accurate, unambiguous, sufficient detailed, and do not conflict with the other requirements.	A correct relationship exists between the components of the software architecture via data flow and control flow.	The source code is accurate and consistent. Includes an assessment of the: stack usage; floating-point arithmetic; resource contention and limitations; worst-case execution timing; exception handling; unused variables; and data corruption due to task or interrupt conflicts.
Compatibility	No conflicts in the areas of system response times, input/ output hardware between the high-level requirements, and the hardware/ software features of the target computer.	No conflicts in the areas of resource use, bus loading, system response times, input/ output hardware between the low-level requirements, and the hardware/ software features of the target computer.	No conflicts in the areas of initialization, asynchronous operation, synchronization and interrupts between the software architecture, and the hardware/ software features of the target computer.	

Table 5.34 Review and Analysis Aspects for Software V&V (RTCA 2011)

	High-level requirements	Low-level requirements	Software architecture	Source code
Verifiability	Each high-level requirement can be verified.	Each low-level requirement can be verified.	The software architecture can be verified (e.g., no unbounded recursive algorithms).	The source code does not contain statements and structures that cannot be verified. The code does not have to be altered to be tested.
Conformance	Software requirements standards were followed; any deviations were justified.	Software design standards were followed; any deviations were justified.	Software design standards were followed; any deviations were justified.	Software code standards were followed; any deviations were justified.
Traceability	The system level requirements were developed into the high-level requirements.	The high-level requirements were developed into the low-level requirements.		The low-level requirements were developed into source code.
Algorithm	The proposed algorithms have sufficient accuracy and behave as expected.	The proposed algorithms have sufficient accuracy and behave as expected.		

Table 5.34 Review and Analysis Aspects for Software V&V (RTCA 2011) (*Continued*)

verifiability and algorithm accuracy describe the individual characteristics of the V&V component. Compliance and traceability describe the relationships between low and high level components; conformance describes the adherence of the V&V components to design standards; and compatibility determines the hardware integration of the V&V components onto the target computer.

5.87 Tools for Safety and Reliability Assessment

This subsection describes some of the techniques used for the system safety assessment (SSA) and preliminary SSA (PSSA). The objectives of the SSA are to provide sufficient assurance for the integrity and quality of the system. It does this by showing that the failure probability of the system is within suitable thresholds, which are consistent with the specified risk categories; by providing suitable levels of redundancy; by eliminating common cause failures; by providing suitable failure warnings to operating personnel; by providing information on suitable operational and maintenance procedures; and by making allowances for changes to environmental conditions or tolerances, as well as the effects of any external or internal damage to the system.

The **parts count reliability prediction** is a widely used method in the aerospace domain for predicting the reliability of complex equipment because of its simplicity. It is primarily used for

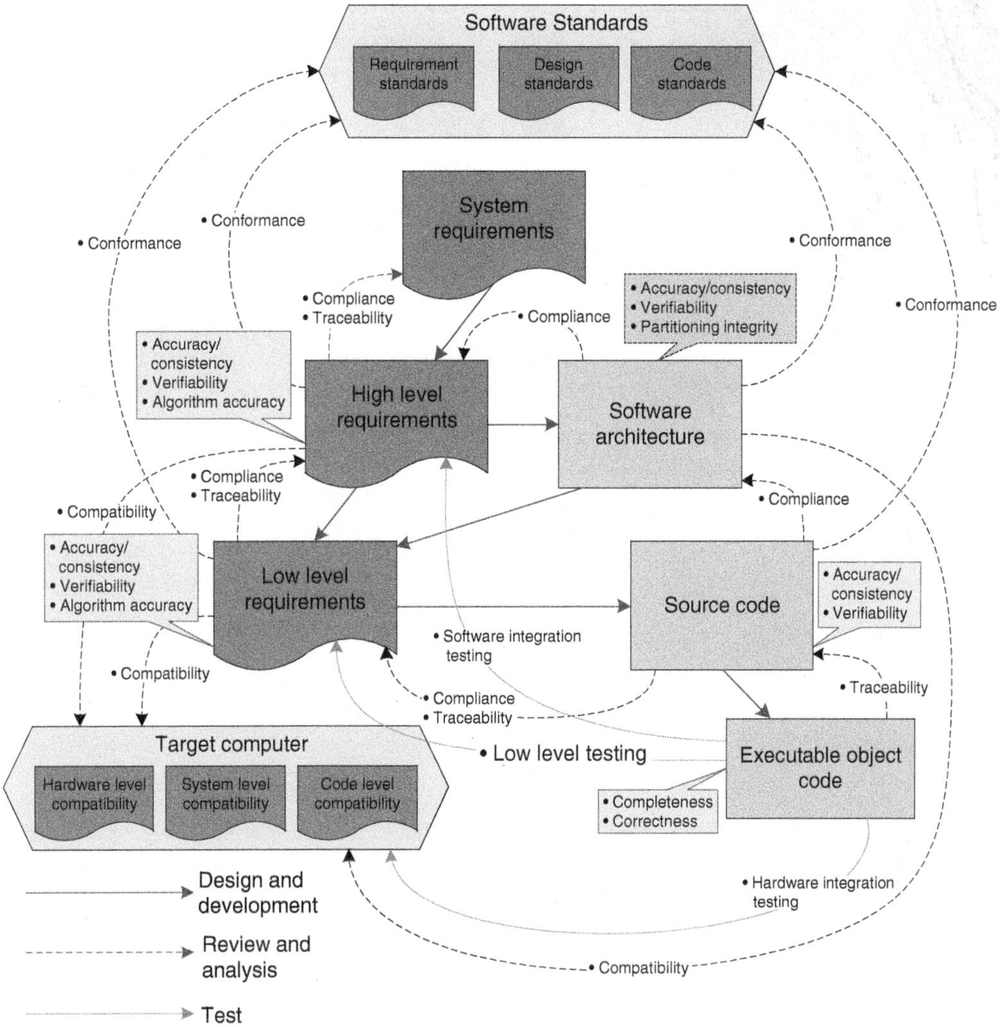

FIGURE 5.122 Illustration of the validation and verification activities as specified in (RTCA 2011).

predicting the reliability of electronic equipment—the assumption of constant failure rate might not be applicable to mechanical equipment. The overall reliability of the equipment is obtained by analyzing the reliability of its component parts. The reliability of the equipment is given by the equation

$$R_{\text{EQUIP}} = e^{-\lambda_{\text{EQUIP}} t} \tag{5.339}$$

where t is the time interval over which the reliability of the equipment applies, and λ_{EQUIP} is the failure rate of the equipment, given by

$$\lambda_{\text{EQUIP}} = \sum_{i=1}^{n} N_i (\lambda_{Bi}) \tag{5.340}$$

where n is the number of different component types, while N_i and λ_{Bi} are, respectively, the quantity and base failure rate of the ith component type used in the equipment. The base failure rates λ_{Bi} can

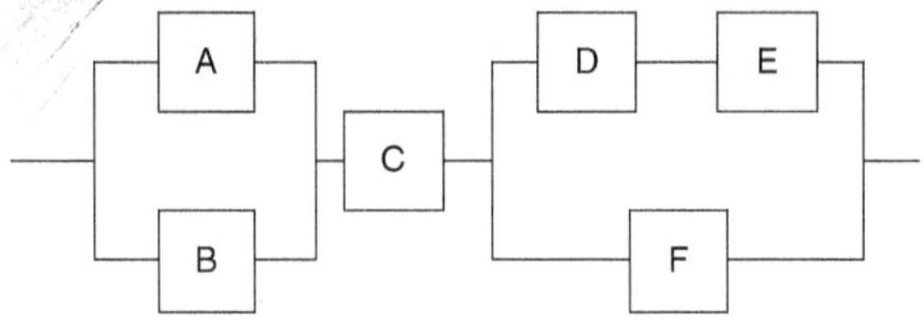

FIGURE 5.123 RBD example.

be derived from the tables in Appendix A of MIL-HDBK-217 (DOD 1991) in different operating environments: ground benign, fixed, and mobile (G_B, G_F, G_M); naval sheltered and unsheltered (N_S, N_U); airborne inhabited cargo, inhabited fighter, uninhabited cargo, uninhabited fighter and rotary wing (A_{IC}, A_{IF}, A_{UC}, A_{UF}, A_{RW}); space flight (S_F), missile flight and launch (M_F, M_L); and cannon launch (C_L).

Reliability block diagrams (RBD), also known as dependence diagrams, are also widely used in aerospace for reliability prediction, especially during the early stages of design (e.g., when conducting PSSA). The purpose of the RBD is to show by concise visual shorthand the various series-parallel block combinations that result in system success. A series combination denotes an *or*-relationship between two components (if **a** *or* **b** fails, the system fails) while a parallel combination denotes a *and*-relationship (if **a** *and* **b** fails, the system fails). For example, in Figure 5.123, the system fails if

- A and B fails, or
- C fails, or
- D and E or
- F fails

Table 5.35 provides the equations for calculating the reliability of the system, based on the component-level failure rates. If the failure rates are assumed to be constant in time, and if the components have the same failure rates, special equations can be used to further simplify the calculation.

Block layout	Reliability equation $R(t)$	Constant failure rate ($\lambda(t) = \lambda$)	
		Different failure rate for each block ($\lambda_1 \neq \lambda_2$ ... etc.)	Same failure rate for all blocks ($\lambda_1 = \lambda_2$ ... etc.)
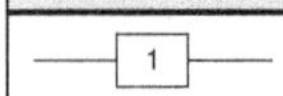	R_1	—	$e^{-\lambda t}$
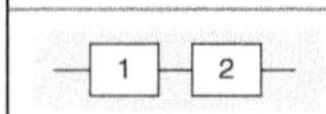	$R_1 \times R_2$	$e^{-(\lambda_1+\lambda_2)t}$	$e^{-2\lambda t}$
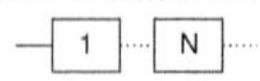	$R_1 \times R_2 \ldots R_N$	$e^{-(\lambda_1+\lambda_2 \ldots \lambda_N)t}$	$e^{-N\lambda t}$
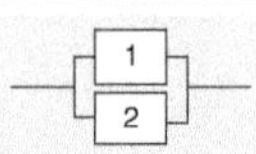	$1-[(1-R_1)(1-R_2)]$	$1-[(1-e^{-\lambda_1 t})(1-e^{-\lambda_2 t})]$	$1-(1-e^{-\lambda t})^2$
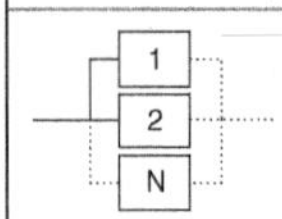	$1-[(1-R_1)(1-R_2)\ldots(1-R_N)]$	$1-[(1-e^{-\lambda_1 t})(1-e^{-\lambda_2 t})\ldots]$	$1-(1-e^{-\lambda t})^N$

TABLE 5.35 Reliability Prediction Equations for RBD Analysis

The **fault tree analysis** (FTA) provides a more formal and rigorous technique for analysis of a system's safety and reliability as compared to the RBD and parts count methodologies. The FTA is conducted to exhaustively identify the causes of an undesired (usually catastrophic) event. The FTA generates a diagram of the system and its components known as the fault tree—the main system at the top of the tree branches into lower level subsystems and components, with the relationships between different levels represented with logic gates. Table 5.36 describes some of the common logic gates used in the FTA.

Gate symbol	Name of gate	Description of gate
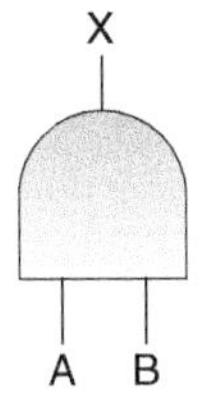	AND gate	The output fault occurs if all of the input faults occur.
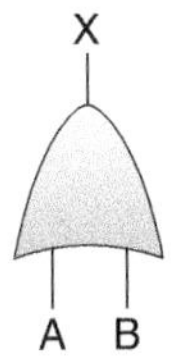	OR gate	The output fault occurs if at least one of the input faults occurs.
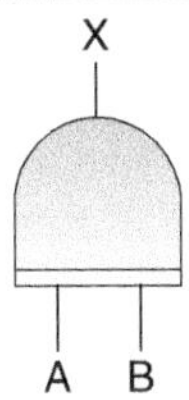	Priority AND gate	The output fault occurs if all of the input faults occur in a specific sequence.
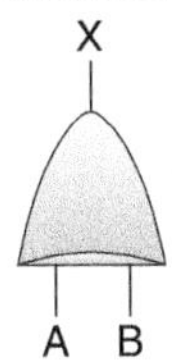	Exclusive OR gate	The output fault occurs if exactly one of the input faults occurs.
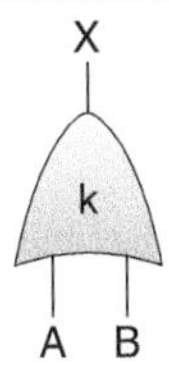	Voting OR gate	The output fault occurs if *k* or more of the input faults occur.
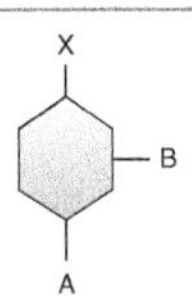	Inhibit gate	The output fault occurs if the (single) input fault occurs in the presence of an enabling condition.

Table 5.36 Description of Commonly Used Logic Gates in the FTA

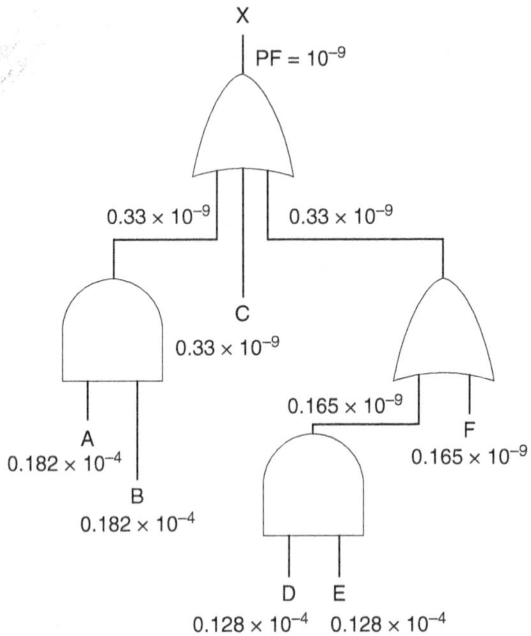

FIGURE 5.124 Equivalent FTA diagram of the example in Figure 5.115, with the PF of the main system X being propagated downward via a top down approach.

The fault tree contains the probability of failure (PF) of the various components, subsystems, and systems. The FTA utilizes both a top-down and bottom-up approach—the former approach propagates the PF of the main system (usually predefined by performance requirements) downward into subsystems and individual components (Figure 5.124) while the latter approach propagates the PF of the individual components (usually obtained through hardware/software specifications) upward through the tree to obtain the PF of the main system.

Failure modes and effects analysis (FMEA) is another formal safety assessment technique and is used to identify potential failure modes and their effects on the overall system. The analysis begins by considering the possible failure modes of each component within the system and the effects of these failures on the next higher level of the assembly. The technique proceeds upward through the assembly until the entire system is covered. A worksheet (Table 5.37) is used to document the results of the FMEA and includes information on the component name and functions; failure modes; operating/environmental conditions at time of failure; effects of failure (on the component as well as on higher-level assemblies); failure detection method (and means/ease of detection), compensating provisions; as well as severity of the failure.

Criticality analysis is often conducted alongside the FMEA to rank the potential failure modes which have been identified. The criticality accounts for the failure mode's severity and its probability of occurrence. A criticality number C_m is assigned to the mode according to the equation:

$$C_m = \beta \alpha \lambda_p t \tag{5.341}$$

where $0 \leq \beta \leq 1$ is the conditional probability of mission loss, $0 \leq \alpha \leq 1$ the failure mode ratio (which is the probability that the component will fail with the given mode; if all the failure modes are listed, the sum of all the α-values will equal one), λ_p the part failure rate and t the duration of the

Component identification	Component function	Failure modes and causes	Mission phase/ operational mode	Failure effects (local/higher/ end effects)	Means of detection	Severity class
Component 1	Function 1.1	Mode 1.1.1	Phase A	Effects 1.1.1a, 1.1.1b, 1.1.1c	Detection by X1	C
		Mode 1.1.2	Phase B	Effects 1.1.2a, 1.1.2b	Detection by X2	D
	Function 1.2	Mode 1.2.1	Phase B	Effects 1.2.1a	Detection by Y1	D
		Mode 1.2.2	Phase C	Effects 1.2.2a, 1.2.2b	Detection by Z1	C
	Function 1.3	Mode 1.3.1	Phase A	Effects 1.3.1a, 1.3.1b, 1.3.1c	Detection by Y2	B

TABLE 5.37 FMEA Worksheet

applicable mission phase, usually expressed in hours or number of operating cycles. The C_m for the component can then be found by summing up the C_m for all the failure modes, as given by the equation:

$$C_{m,\text{total}} = \sum_{i=1}^{N} C_{m,i} \tag{5.342}$$

where N is the total number of failure modes and $C_{m,i}$ is the criticality number for the ith failure mode. An example of the criticality analysis is provided in Table 5.38.

5.88 Certification Considerations for Aerospace Systems

Aviation must modernize to face sustainability challenges in an era of steady air traffic growth, with air traffic doubling its size every 15 years. Due to the inability of traditional air traffic control (ATC) systems to cope with this growth, ATC has evolved into air traffic management (ATM), with novel systems being designed and developed for communication, navigation, surveillance (CNS)

Component identification	Component function	Failure modes and causes	β (prob mission loss)	α (failure mode ratio)	$\lambda_p t$ (failures per mission)	C_m (criticality number)
Component 1	Function 1.1	Mode 1.1.1	0.01	0.18	1.2	$2.16E^{-3}$
		Mode 1.1.2	0.25	0.13	0.005	$1.625E^{-4}$
	Function 1.2	Mode 1.2.1	0.02	0.36	0.01	$7.2E^{-4}$
		Mode 1.2.2	0.10	0.09	1.25	$1.125E^{-3}$
	Function 1.3	Mode 1.3.1	0.65	0.24	0.60	$9.36E^{-2}$
					C_m total: $9.78\ E^{-2}$	

TABLE 5.38 Example of the Results from a Criticality Analysis

and avionics/ATM (CNS+A). The ever-increasing demand for capacity, efficiency, safety, and the necessity to minimize adverse effects on the environment due to fuel burn and gaseous emissions has stimulated many research efforts in the domains of avionics and ATM.

The Single European Sky ATM Research (SESAR) is a collaborative effort toward ATM modernization in the European airspace, while the Next Generation Air Transportation System (NextGen) led by the Federal Aviation Administration (FAA) is the ATM development program in the United States. SESAR and NextGen are the leaders in ATM research and development. Other ATM modernization programs include collaborative action for renovation of air traffic systems (CARATS) in Japan, SIRIUS in Brazil, OneSky in Australia and similar programs in Canada, China, India, and Russia (ICAO 2013). The International Civil Aviation Organization (ICAO) has authorized a globally coordinated plan published as the Global Air Navigation Plan (GANP) under ICAO Doc 9750 (ICAO 2013), to guide the harmonized implementation of CNS+A enhancements across regions and states.

In the CNS+A context, aircraft safety is a shared responsibility between airborne and ground-based resources. Hence this is a safety challenge requiring changes to the current regulatory framework to properly capture the nature of this shared responsibility and the concept of "integrated" CNS+A systems.

Certification of aircraft and ground equipment (hardware and software) together with organizational approvals are essential elements to ensure continued and enhanced safety. Certification also facilitates harmonization and interoperability of CNS+A systems across regions, subregions, and states. Furthermore, while new ATM and avionics technology brings with it an increased level of automation, certification would be the instrument to ensure the correct application of technology to achieve its full potential benefits. The aviation regulatory framework enforces and drives the certification process, while industry standards provide a vital link by offering methods of compliance for certification.

The current certification framework for CNS+A is evolving, and is required to keep pace with the global modernization efforts to ensure safety and sustainability of future aviation.

References

DOD, 1991, *MIL-HDBK-217F: Military Handbook—Reliability Prediction of Electronic Equipment*, report n. MIL-HDBK-217F, Department of Defense, Virginia, USA.

ICAO, 2013, 2013-2028 Global Air Navigation Plan, Doc 9750, International Civil Aviation Organization, Montreal, Canada.

Jacklin, S. A., 2012, "Certification of Safety-Critical Software under DO-178C and DO-278A," AIAA Infotech@ Aerospace Conference, Garden Grove, CA, USA.

RTCA, 2011, *RTCA DO-178C: Software Considerations in Airborne Systems and Equipment Certification*, SC-205, 13/12/2011, Radio Technical Commission for Aeronautics.

SAE, 1996, *ARP4761: Guidelines and Methods for Conducting the Safety Assessment Process on Civil Airborne Systems and Equipment*, S-18, Society of Automotive Engineers, Pennsylvania, U.S.

SAE, 2010, *ARP4754A: Guidelines for Development of Civil Aircraft and Systems*, S-18, WG-63, Society of Automotive Engineers, Pennsylvania, U.S.

PART 17

Aviation Human Factors Engineering

Yixiang Lim, Subramanian Ramasamy, Alessandro Gardi, and Roberto Sabatini

5.89 Human Performance Modeling

Early efforts sought to model the pilot as a direct controller of the aircraft using *control theory* in order to improve flight handling qualities and understand the human factor in pilot-induced oscillations (PIO)-related dynamics. Besides control theory, *human reliability analysis* (HRA) is another major domain where statistical, probabilistic, and analytical methods are used to analyze human reliability. Databases of task-related performance statistics (such as completion time and probability of success for basic operator actions) have been developed and integrated into context-specific operational frameworks to model the performance of individual operators or groups of teams. Cognitive models gained relevance as automation became increasingly more integral to both avionics and air traffic management systems. *Cognitive architectures* are used in the aviation domain to provide modeling and simulation capabilities, primarily to aid in design or for accident investigation. *Subjective ratings* have been used to support user test and evaluation through obtaining feedback from the operator on his/her self-assessed performance (NASA-TLX (Hart and Staveland 1988)) or with regards to flight handling qualities (Cooper-Harper scale (1969)) or interface usability (software usability measurement inventory—SUMI (Kirakowski and Corbett 1993)). Additionally, *physiological and psychological measures* have also been used to empirically determine operator performance (Kramer 1991; NATO 2004), and include measures of heart activity, brain evoked potential, and gaze patterns.

To evaluate the appropriateness of the assessment technique to the evaluated task or design, a list of key criteria, based on AIAA's *Guide to Human Performance Measurements* (AIAA 1993), is provided in Table 5.39.

Assessment criteria	Description
Level of detail	The metric reflects the performance of interest with sufficient detail for a meaningful analysis
Reliability	The metric is repeatable. Ideally, the same distribution of values should be obtained if the same task is repeated under identical circumstances.
Validity	The metric is appropriate for use in its intended purpose—it correctly reflects the aspect of human performance evaluated and is relevant to the task being evaluated.
Sensitivity	The metric is able to detect changes in the independent variable, and are sampled at sufficient resolution to capture the behavior of interest.
Transferability	The metric is applicable for different contexts and can be used to evaluate human performance in different settings.
Diagnosticity	The metric provides information on human performance to account for the reason for the changes in the given performance.
Nonintrusiveness	The measurement technique does not adversely affect the subject's performance when it is collecting information.
Implementation requirements	The measurement technique allows information to be easily collected, uses equipment that is easily obtainable and robust, and provides data with good quality.

Table 5.39 List of Criteria for Assessing Different Performance Measurement Techniques

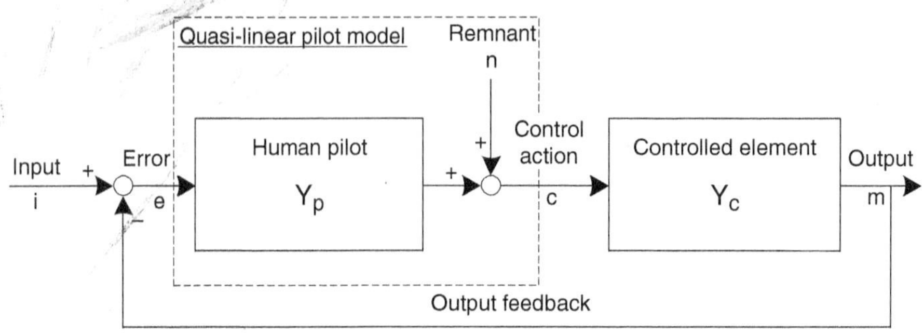

FIGURE 5.125 Single loop system for a compensatory tracking task.

Control Theory-Based Pilot Models

Control theory-based models support a deeper understanding of the effect of handling qualities on pilot performance, providing a design driver through the analysis, modeling and simulation of PIO-related pilot-aircraft interactions. The landmark controller model was the *Crossover Model,* developed and validated in *"Mathematical Models of Human Pilot Behaviour"* report (McRuer and Krendel 1974), describing a quasilinear model of a pilot-aircraft system comprising a linear component and a nonlinear remnant. The model describes a tracking task in which a human pilot has to minimize the error $e(t)$ between the system forcing function $i(t)$ and the system output $m(t)$, via a control action $c(t)$, as illustrated in by the single loop system in Figure 5.125.

The objective of the pilot-vehicle system is to bring the output $m(t)$ as close to the input $i(t)$ as possible in a compensatory tracking task. The dynamics of the human pilot is modeled by the quasilinear function Y_p, while the controlled element is represented by the transfer function Y_c. Essentially, the human operator varies his/her dynamics (Y_p) according to the platform dynamics (Y_c) such that, for most tracking tasks, the open-loop transfer function $G(j\omega)$ of the combined system near the crossover frequency ω_c is described as:

$$G(j\omega) = Y_p(j\omega)Y_c(j\omega) = \frac{\omega_c e^{-\tau_e j\omega}}{j\omega}, \quad \text{near } \omega_c \tag{5.343}$$

Here, the crossover frequency ω_c represents the open-loop gain and should be greater than the input frequency ω_i for effective tracking; and the effective delay τ_e represents the time lag due to pilot delay as well as the pilot's neuromuscular dynamics. The pilot adopts appropriate lead or lag equalization to achieve stable control such that the pilot dynamics are given as:

$$Y_p(j\omega) = K_p\left(\frac{T_L j\omega + 1}{T_l j\omega + 1}\right)e^{-s\tau_e} \tag{5.344}$$

The equalization term $\left(\frac{T_L s + 1}{T_l s + 1}\right)$ contains the lead T_L and lag T_l terms which are adjusted by the pilot to obtain a slope of −20 dB per decade for $|G(s)|$ at frequencies near ω_c. This allows the crossover model as presented to be satisfied for different types of control elements. Additionally, the gain K_p and effective time delay τ_e are adjusted by the pilot to achieve the necessary stability margins. The equalization selection and adjustment for a range of controlled elements are provided in Table 5.40.

Controlled element transfer function $Y_c(s)$	Pilot equalizer form	Pilot transfer function $Y_p(j\omega)$	Location of equalization break frequency
K_c	Lag lead	$\frac{K_p e^{-j\omega\tau_e}}{(T_I j\omega+1)}$	$T_I \gg \frac{1}{\omega_c}$
$\frac{K_c}{s}$	High-frequency lead	$K_p e^{-j\omega\tau_e}$	—
$\frac{K_c}{s^2}$	Low-frequency lead	$K_p(T_L j\omega+1)e^{-j\omega\tau_e}$	$T_L \gg \omega_c$
$\frac{K_c}{s(sT+1)}$	If $T > \tau_e$, use mid-frequency lead	$K_p(T_L j\omega+1)e^{-j\omega\tau_e}$	$T_L \doteq T$
	If $T < \tau_e$, use high-frequency lead	$K_p e^{-j\omega\tau_e}$	—
$\frac{K_c\omega_n^2}{s^2+2\zeta\omega_n s+\omega_n^2}$	If $\omega_n \ll \frac{1}{\tau_e}$, use low-frequency lead	$K_p(T_L j\omega+1)e^{-j\omega\tau_e}$	$T_L \gg \omega_c$
	If $\omega_n > \frac{1}{\tau_e}$, use lag lead	$\frac{K_p e^{-j\omega\tau_e}}{(T_I j\omega+1)}$	$T_I \gg \frac{1}{\omega_c}$

TABLE 5.40 Equalization Adjustment for Different Controlled Elements

Optimal Control-Based Pilot Models

An optimal control model (OCM) was developed as an alternative to the multiloop control models and in particular as an extension to the existing single-input/single-output systems (Stapleford et al. 1967). The system dynamics are modeled by the linearized equations of motion, given by (Kleinman et al. 1970):

$$\dot{x}(t) = Ax(t) + bu(t) + w(t) \tag{5.345}$$

where x represents the vector of vehicle states, u the vector of control inputs, and w the vector of external disturbances. The system outputs y are linear combinations of the system states, given by:

$$y(t) = Cx(t) \tag{5.346}$$

And the optimal pilot gains are obtained by minimizing the quadratic cost function, given by:

$$J(u) = E\left\{\lim_{\tau\to\infty}\frac{1}{\tau}\int_0^{\tau}\left(y^T Q y + u^T R u + \dot{u}^T G \dot{u}\right)dt\right\} \tag{5.347}$$

where Q, R, and G are the weightings on the outputs of the internal model, pilot's control action and pilot's control rate, respectively. Figure 5.126 illustrates the structure of the OCM. The human operator model is divided into two components: the filter and estimator, which provides an estimate of the delayed state, and the predictor and controller, which generates the best estimate of the current state and sends the appropriate control to the vehicle.

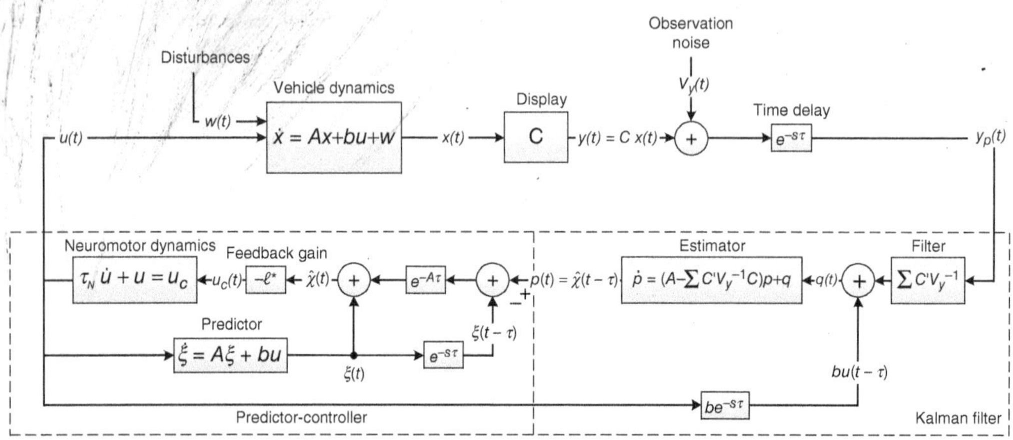

FIGURE 5.126 OCM layout.

5.90 Human Factors Engineering Program

Human factors engineering assures that the system, equipment, facility design, required human tasks and work environment are compatible with the sensory, perceptual, mental, and physical attributes of the personnel who will operate, maintain, control, and support it. The purpose for establishing a human factors engineering program is to support the design, test and integration of a particular hardware, software, system, or work environment to meet specific human-machine performance requirements, and can be accomplished by

1. First defining the required human-system performance requirements in order to accomplish specified operational or mission objectives.
2. Establishing a set of workflow procedures, which are based on an analysis of the required task division and functional allocation.
3. Designing appropriate interfaces, which enable the human operator to execute his tasks successfully. These interfaces include the hardware, software, and system components as well as the work environment.
4. Identifying critical tasks and design elements, which can potentially compromise the safety, reliability, efficiency, or effectiveness of the total human-machine performance.
5. Assessing the performance of the operator through test and evaluation, as well as obtaining feedback on the usability of the designed product. The assessment is used in subsequent redesign of the system interfaces or work procedures.
6. Evaluating the system life-cycle cost, which includes the operational, maintenance, support, and training costs.

Further details on the requirements and relevant procedures in conducting a human factors program are described in FAA/HF-STD-004 (FAA 2009b).

The human factors program comprises three main stages, namely the requirements definition, design, and evaluation stages. These three stages are typically iterated over multiple cycles, with progressive iterations providing more detail and refinement to the requirements, procedures, and system design. The key to proper management of the program is to plan the definition, design, and evaluation activities such that they can be performed concurrently, while also allowing flexibility for successive evaluations to influence the future stages. In greater detail, a human factors program comprises the following activities:

- **Requirements definition:** The requirements drive the design and evaluation process by providing a framework for identifying specific design elements, as well as for assessing the effectiveness of these elements against specific objectives. Initial requirements can be later defined in greater detail based on the assessment of previous design stages.
  - **Development of top-level requirements:** The top-level requirements are derived from the customer and specify the purpose of the product (the hardware, software, or system being designed). The designer will ensure that sufficient detail is captured in the top-level requirements, as these will form the initial technical baseline for the design and development phases.
  - **Definition of mission and operational requirements:** These requirements are used to define the mission environment, objectives, operational constraints, as well as performance and manpower requirements. An operating scenario would be developed as a representative case for evaluating the system performance and success—the scenario can be successively modified and refined based on design and evaluation needs.
  - **Capture of functional and task requirements:** The functions and tasks as well as respective subfunctions and subtasks required to accomplish the proposed mission are defined. These requirements provide details on what task an operator is required to perform as well as when the execution is required. Similar to the mission and operational requirements, these are refined over the design process.
- **Design:** Product design encompasses the design and development of procedures, software, systems, and equipment to meet the specifications established in the previous stage. Based on the top-level requirements, an initial conceptual design is developed to provide the preliminary information-flow and task sequencing processes, the basic system/subsystem layout, as well as the description/allocation of tasks and functions between the human and system components. Subsequently, the design is evaluated based on more specific requirements and design objectives and more detailed refinements are made. To ensure an efficient development cycle, human factors principles should be applied throughout the entire design process, and just not at the final stages of detailed design.
  - **Procedural development:** The procedural development includes the defining the procedures and guidelines for staffing, operations, maintenance, and training as well as the development of any associated documentation.
  - **System design:** System design includes defining the architecture as well as the information-flow and functional allocation between the system/subsystem's various modules and components. While the various subsystems can be design and tested individually, these would have to be integrated into the main system at some point in the design process.
  - **Interface design:** The human-machine interfaces and interactions (HMI^2) are developed for test and evaluation by the end user. Similar to the system design, the individual software and hardware components can be developed and evaluated independently of each other, and integrated at a later stage in the design process.
- **Evaluation:** The evaluation process assesses the effectiveness of a design in accomplishing the mission objectives against specific performance requirements. Critical areas are identified where the total performance and safety of the operator/system might be compromised, such as the identification of specific tasks which might result in operator overload or underload. Based on the findings of the evaluation process, recommendations

are made to improve specific design elements, which are reflected in the requirements capture and design stage of future iterations.

- **Analytical evaluation:** The analytical evaluation of a design employs heuristic methods to determine if the design meets performance requirements. This can involve the use of block diagrams, checklists and matrices.
- **Mock-up studies:** Product mock-ups and prototypes are used early in the evaluation process as a relatively quick method of assessing design concepts. Environment mock-ups can be built from cardboard or foam, and interface displays can be sketched on paper. These support heuristic evaluation by providing users with a feel of the final product during usability testing.
- **Computer-aided evaluation:** Computer-aided design (CAD) tools are employed to evaluate the environmental human factors by simulating virtual operators in a virtual work environment. At more mature evaluation phases, simulators are used to evaluate the performance of a real operator in a virtual work environment. Other forms of computer-aided evaluation include software validation and verification and system-level testing.
- **Human test and evaluation:** Human-in-the-loop evaluations are conducted to evaluate the performance of the system in a representative mission environment. The evaluation is conducted in stages and typically ranges incrementally from the testing of individual components/interfaces, to an integrated system, to a full field evaluation. Subjective user (workload/usability) ratings are employed to assess human performance aspects such as workload, situational awareness, and usability.
- **Trade studies:** Trade studies and risk/cost/benefit analyses are used to assess the value of deciding between particular design features and usually evaluated against metrics such as implementation and lifecycle cost, effectiveness meeting mission goals, usability, as well as operator safety and risk.

5.91 Techniques for Task Analysis

The human factors principles for the design and development of systems and equipment are specified in *FAA HF-STD-001: Human Factors Design Standard*, and require design elements to be user-centered, i.e., taking into consideration the interactions of the user with the system based on the his/her needs and requirements and incorporating these considerations into the design and development of suitable interfaces and interactions. The design of such interfaces and interactions follow from a prior assessment of the operational requirements, which comprise specific tasks or performance requirements. In general, pilot tasks can be grouped under the "Aviate, Navigate, Communicate, and Manage" categories.

Procedure Design

Procedure design aims at implementing the required safety in all operational and maintenance tasks. At the initial stage, this involves decomposing a task into its basic functions and describing the sequence of steps required to accomplish the safe completion of the given task. Tools for procedures design and development may consist of checklists, information flowcharts, decision trees, hierarchal task structures, operational sequence diagrams, or functional block diagrams. Good procedure design allows potential risks and hazards to be identified and minimized at an early stage; these can then be translated into workable systems, interfaces, and interactions.

A key component of procedural design is the task and functional analyses, which translate top-level requirements into working procedures and specific design criteria. Functions are discrete processes necessary to achieve specific objectives, and tasks are the sequential steps required to achieve a specific function; these are performed through use of facilities, equipment, software, personnel, or a combination. Kirwan and Ainsworth's *A Guide to Task Analysis* (Kirwan and Ainsworth 1992) provides a good reference on performing a task analysis.

Hierarchal Task Analysis

Hierarchal task analysis (HTA) organizes the tasks and functions in a hierarchal manner. Block diagrams are commonly used to represent the top-down relationships between objectives and tasks. Top-level objectives are decomposed into lower-level subobjectives, which can themselves be decomposed into lower-level tasks and subtasks. A complex task can thus be represented as a hierarchy of actions, which when undertaken in a sequential manner, allow the parent task to be completed. The granularity of detail (or the number of sublevels) in each task depends on a number of factors, such as the complexity or criticality of the task, the risk of human error, or the impact of a particular design on the human-machine interaction. Using the task hierarchy, a preliminary workload analysis can be conducted for the human operator and functional allocation can be performed for the system.

Cognitive Task Analysis

For more tasks with more complex human-machine interactions, it is important to understand the underlying mental processes of the human operator in order to identify and diagnose particular errors and failure modes. In these scenarios, a cognitive task analysis should be performed. This analysis can start with a list of activities, such as one obtained from a gross task analysis (Miller 1953), which describes the top-level tasks, relevant sub-tasks and user control actions. The list is then expanded and updated to model the operator's mental processes by analyzing the cognitive cues, information flow requirements and decision-making strategies in particular scenarios. As described in Salmon et al. (2010) and McIlroy and Stanton (2011), a five-phase analysis process is used to analyze the work domain (*"Where is the task performed?"*), control task (*"What are the tasks performed?"*), strategies (*"How are the tasks performed?"*), social organization and cooperation (*"Who will be performing the tasks? How can they cooperate between each other?"*), as well as worker competencies (*"What are the skills required?"*). Similar to hierarchal task analysis, various methodologies are employed in each phase for information acquisition and representation. With sufficient detail, the cognitive work analysis can eventually be used as a tool to produce formal procedural guides and requirements specifications.

One of the outputs of the cognitive task analysis is the *decision ladder* (Rasmussen 1976), describing the information processing activities involved at each stage of the decision process. The operator might skip certain steps, based on either heuristics, predefined operational procedures, or due to time constraints. These are represented as shortcuts through the ladder, illustrated by the dotted lines in Figure 5.127. A decision making strategy is represented by the set of paths through the decision ladder, which are available to the operator and can involve varying levels of situation analysis/diagnosis and action planning. The cognitive task analysis provides information on the nature of the task to be performed by the operator, identifies available cognitive strategies, and requirements, and is ultimately used to drive the systems-level definition of the related human factors requirements. These requirements include information and interface requirements, performance requirements as well as training prerequisites.

Critical Task Analysis

A critical task analysis is specified by the Federal Aviation Authority (FAA) as an approach for evaluating the human factors involved in safety- and mission-critical tasks when designing/implementing a system. The process of critical task analysis is similar to cognitive task analysis, consisting of steps to acquire data (identifying and understanding) with respect to the critical tasks, representing these tasks using tables and/or diagrams, as well as conducting an analysis to identify failure modes and

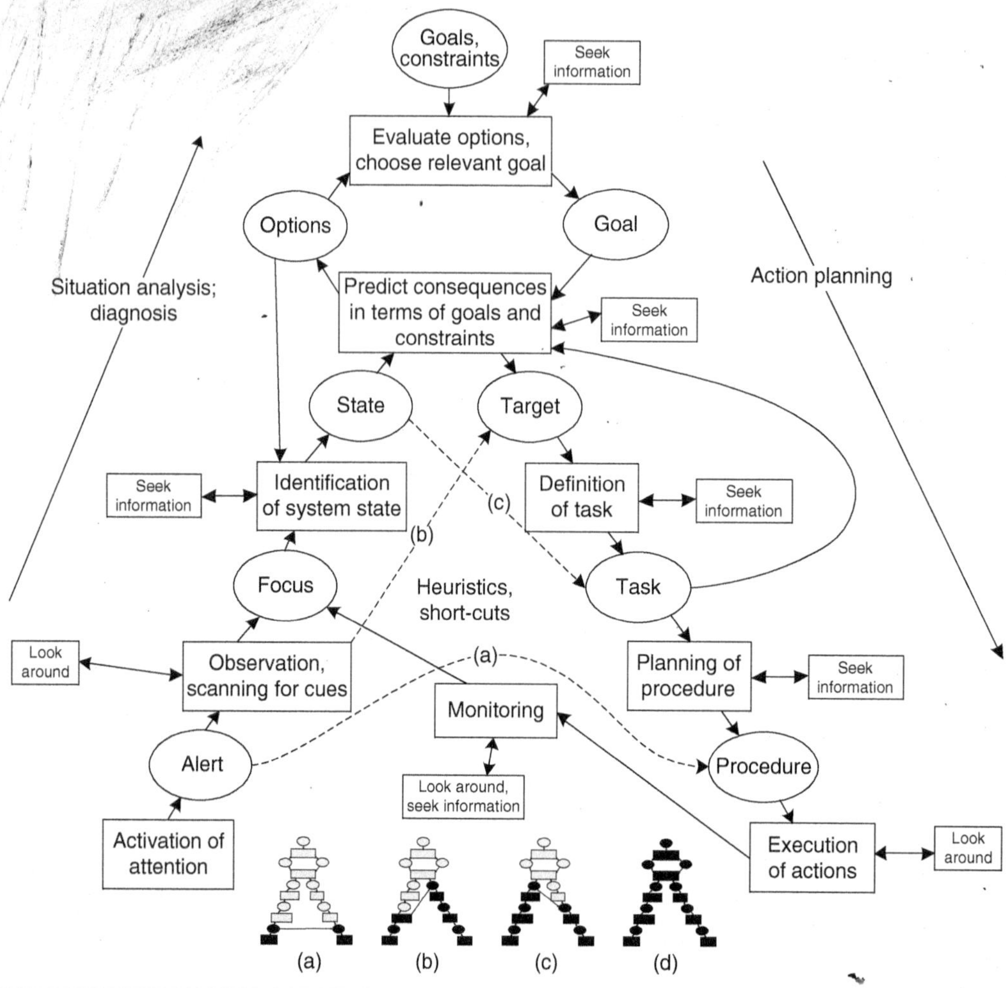

FIGURE 5.127 The decision ladder representing the operator information processing and control actions. Different decision strategies are represented at the bottom of the figure with increasing depth of information processing by the operator.

performance-influencing factors at all operational modes (including degraded and emergency modes). The results of the analysis can be used for improving system design and provide additional measures to minimize risk and improve performance. The format and content requirements for a critical task analysis report (CTAR) are specified in FAA-HF-004A (FAA 2009a). The report requires a detailed description of each critical task, containing the items described in Table 5.41.

Operational Sequence Diagram

An *operational sequence diagram* is a process flowchart describing the workflow of a particular activity. The tasks and functions, arranged sequentially in the order with which they are carried out, are divided between multiple operators and automated functions into different channels in the diagram. The flow of information between the operators and the system is indicated by unique symbols

Information category	Information description
Task-hierarchical relationships	Actions required for accomplishing the task. Probability and severity of human error. Potential for error recovery.
Information flow	Information required and available to the operator/maintainer. Feedback informing operator/maintainer of the adequacy of action(s) taken. Communications required, including type of communication. Operator/maintainer interaction.
Cognitive processes	Decision evaluation process. Decision reached after evaluation. Number of personnel required; their specialties/experience. Performance limits of personnel.
Physical processes	Actions taken for accomplishing the task. Time available for completing the task. Body movement required by action taken. Frequency and tolerance of action. Number of personnel required; their specialties/experience. Performance limits of personnel.
Workspace environment	Workspace envelope required/available to take action. Location and condition of work environment. Tools and equipment required. Job aids, training or references required. Hazards involved. Operational limits of hard/software.

TABLE 5.41 List of Content Requirements for a Critical Task Analysis Report

representing different functions or actions. Decision action, communication links, inter-relationships, task frequency, task time, and workload can be obtained from an analysis of the operational sequence diagram, and it can be used as a common reference by the human factors engineer, systems designer, and software engineer of the tasks and functions required for a particular activity. Based on the identified functions and information flows in the diagram, the respective subsystems, interfaces, and interactions can then be designed and developed.

5.92 Design Considerations

In interface design, both the content of information presented to the operator as well as the format which it is presented in shall be specified. The design of interactions involves defining the behavior of the system in response to the human operator in both nominal and off-nominal conditions.

Design of Human-Machine Interfaces

Interface design is the design of devices, applications, displays, or controls to facilitate the interactions between the human operator and the system, and requires an in-depth understanding of

Design parameters	Key considerations
Physical characteristics	• The following parameters should be adjustable to allow display readability in all operational conditions: resolution, luminance, dimming, contrast, chromaticity, grayscale. • The display should be sufficiently large to accommodate all information in a visible format. • The refresh rate should be sufficient to prevent any flickering or jitter effects. • The layout of the displays should account for the flight crew's visibility and any possible obstructions. • Displays should be grouped in a logical manner. • Visual/aural alerts should be sufficiently visible/audible to the operator.
Information characteristics	• Information should be placed in a consistent location under normal operating conditions. • The primary field of view should contain critical or frequently used displays, while the secondary field of view (requiring the pilot to turn his head) should contain ancillary and advisory information. • Pop-up text or windows should not interfere with the operator's use of the display. However, time critical warnings should appear in the operator's primary field of view. • Information display philosophy (color coding, symbology, formats, placement, etc.) should be consistent across the flight deck to minimize operator workload and error. • Information elements should be distinct and permit the operator to recognize the source of the information elements when there are multiple sources providing the same information. • Similar information elements must be grouped and arranged in a logical manner that allows the operator to identify the required information. • Alerts should be prioritized according to urgency, and given in a timely manner to reduce operator surprise. Ramping the onset and offset of warnings is recommended to avoid startling the flight crew. • Frequent false alarms or missed detections must be avoided as these increase the flight crew's workload and reduce their trust in the alerting system.
Information content	• Information specification (from the requirements capture) should be used as a guideline when designing the interface. This can include integrity, precision, and refresh rate requirements. • The precision, accuracy, level of detail and latency of displayed/annunciated information should be sufficient to allow the operator to accomplish his/her task at hand. • Failure flags should be presented in the location of the information they reference or replace. • If symbols are used, their meaning should be sufficiently clear to avoid operator misinterpretation. • If graphical elements are used, they should be presented with at a sufficient clutter level to perform the task at hand, without causing any confusion to the operator. • A consistent design philosophy should be used for alerting conditions, presentation, urgency, and prioritization.

Table 5.42 Key Design Factors and Considerations for Display and Alert (Yeh et al. 2013)

the end user's requirements, product purpose, system behavior, and operational context. Good interface design allows the operator to accomplish a given task effectively, efficiently, and consistently, requires minimal training for use, and is easy to maintain. Interface design can refer to design of the workstation environment, or alternatively the design of specific displays, alerts, or control devices.

Workstation: The user's operating environment. The workstation design considers the layout and placement of consoles and work surfaces such that the operator can perform his/her task effectively and comfortably. Design elements include consideration of user anthropometry (space clearances, physical reach, and line of sight); environmental stressors (ambient temperature, humidity, lighting, air quality, noise, vibration, and motion); as well as safety (risk of accident and threat to operator).

Displays/alerts: The set of visual, audial, or haptic devices that present information to the operator in the form of readable messages, graphs, charts, indicators, and symbols, or voice/sound/motion alerts. The design elements in the display/alert interface can be broadly classified into three design parameters: physical characteristics (the hardware-related parameters such as display size, resolution, placement, luminance, sound loudness/intensity); information characteristics (the software-related parameters such as display formats, content arrangement, alerting conditions, prioritization); and information content (accuracy, precision, resolution, color, clutter). The design considerations for these elements are listed in Table 5.42.

Controls: The set of inputs, devices, and surfaces that provide the user control of the system; these might include buttons, knobs, levers, and switches, as well as keyboards, mouse, joystick, touchpad, or voice activated systems. The key considerations in the design of a control interface are presented in Table 5.43 and include the design of control's physical characteristics (shape, dimensions, surface texture, range of motion, color); placement (location, orientation, arrangement, accessibility, visibility); function (purpose, mode of actuation, effect on the rest of the system); and user interaction (feedback, control options, method of operation).

Design of Human-Machine Interactions

Interactions design is primarily concerned with the design of task and functional allocation between the human operator(s) and the automated system(s), as well as the system's behavior and its impact on the user's cognitive processes. These cognitive processes include the user's workload, situational awareness, and vigilance. Interactions design is especially important in functions where control and decision authority is shared between both the human operator and automated system, and must ensure that the task can be completed efficiently and safely at the specified levels of automation, which can be driven either by a degradation in system performance (leading to a decrease in the level of automation), or by an increase in workload or task complexity (leading to an increase in the level of automation). The 10 levels of automation presented in Table 5.44 are based on Endsley and Kaber's taxonomy (1999) and comprise four functions: monitoring (which involves visual, audible scanning, and other means); generating options (which includes the formulation of strategies for resolving issues and achieving goals); selecting one of the options, and implementing it.

5.93 Design Evaluation

The evaluation phase assesses the designed outputs against design requirements within operational constraints to determine if the product will function according to user expectations, as well as to identify any potential hazards, failure modes, or unintended interactions as a result of

Design parameters	Key considerations
Physical characteristics	• Must be designed to prevent jamming, chafing, any form of interference or significant wear and tear. • Should be made distinguishable or predictable by differentiating its appearance through form, color, location, and labeling (e.g., push buttons should be differentiated from rotational knobs).
Placement	• Controls must be grouped and arranged in a logical and consistent manner. • The control's accessibility should be consistent with its frequency of usage and the criticality of its function. • Any pop-up text or windows must not result in unacceptable distractions, interference, or clutter.
Function	• For frequently used controls, the designer should aim to minimize the number of actions required to actuate the control. • For controls that serve multiple functions, the current function should be indicated to, or easily queried by the operator. The multifunction control must not result in unacceptable levels of workload, error rates, speed, and accuracy. Hidden functions or excessively modes complex should be avoided. • For functions that can be actuated by multiple controls, the primary and secondary control devices should be designated and clearly labeled. The use of either control should not result in confusion or any inadvertent operation.
User interaction	• Must provide to the operator any necessary information at a resolution, latency, and priority appropriate to the task at hand. • Must allow operator to provide input at an appropriate speed, precision, and accuracy based on the task at hand. • Data entry should allow the operator to easily recover from input errors. • The operator must be able to override the control authority of automated functions. • Must provide appropriate feedback to notify the operator of any changes in the control mode (e.g., in noisy conditions, aural notifications should be accompanied by other forms of feedback). • In multicrew situations, feedback should allow all crew members to be aware of any control changes. • Feedback should not result in false alarms or missed detections. • Should not result in operator confusion due to any possible inconsistencies within the interface or with other interfaces.

Table 5.43 Key Design Factors and Considerations for Control Interfaces (Yeh et al. 2013)

operator use. Methods for evaluating design outputs are summarized in Table 5.45. The product's feasibility, usability, impact on operator performance and its level of integrity/reliability is evaluated at varying levels of realism, either on the isolated component or as part of a larger integrated system. The findings of the evaluation process are then used to refine the requirements to address any potential risks and problems in the design, as well as to propose changes to areas of the design where such risks and problems have been identified.

Level of automation	Roles			
	Monitoring	Generating	Selecting	Implementing
1. Manual control	Human	Human	Human	Human
2. Action support	Human/computer	Human	Human	Human/computer
3. Batch processing	Human/computer	Human	Human	Computer
4. Shared control	Human/computer	Human/computer	Human	Human/computer
5. Decision support	Human/computer	Human/computer	Human	Computer
6. Blended decision making	Human/computer	Human/computer	Human/computer	Computer
7. Rigid system	Human/computer	Computer	Human	Computer
8. Automated decision making	Human/computer	Human/computer	Computer	Computer
9. Supervisory control	Human/computer	Computer	Computer	Computer
10. Full automation	Computer	Computer	Computer	Computer

Table 5.44 Levels of Automation as Defined by Endsley and Kaber (1999)

Assessment method	Description of method	Advantages	Limitations
Engineering analysis and evaluation • Functionality and identification testing • Critical task analysis (CTA) • Human error, reliability analysis, and risk assessment (HRA) • Failure modes and effects analysis (FMEA)	A preliminary assessment of the design output; the task workflow, functionality, as well as physical and cognitive ergonomics aspects of the design are assessed and evaluated against design requirements and/or a human factors checklist. The feasibility of the design is assessed and appropriate changes are recommended. Critical tasks and functions are identified and detailed requirements are given for further design. Key evaluation parameters are identified for measurement against the performance requirements in subsequent evaluations.	• Flexible, can be adapted to varying levels of detail. • Can be performed early in the design process to ensure good practices are followed throughout. • Can be used as an initial planning activity for more detailed assessment methods. • Relatively low requirements on cost and time.	• As the evaluation is carried out on a conceptual level, many assumptions will have to be made about the product, operating environment, and user. • Might require specialist knowledge or expertise.

Table 5.45 Methods for Evaluating a Design Output

Assessment method	Description of method	Advantages	Limitations
Part-task evaluation • Interviews/ walkthrough analysis • Usability testing	A limited evaluation of the design output, in representative conditions. A combination of interfaces, hardware and simulated systems are used to evaluate the performance of the operator when executing specific sets of tasks. Some of the human factors assessed include the visibility of the interfaces, interactions with different automation modes, availability/ correctness of presented information and operator workload, as well as task performance metrics (including error rate, accuracy, and execution time).	• Builds upon prior task analyses, such that a task does not have to be designed from scratch. • Isolated components or tasks can be assessed, allowing parallel evaluation or fault/error isolation. • The level of detail in the task can be refined as the product matures, allowing for successive iterations in the design and evaluation process.	• In complex systems, the evaluation of an isolated task might not be representative of the behavior of the larger system. Unanticipated vulnerabilities might arise from the interactions of different tasks and systems. • If a partially working system is required for assessment, the assessment might not be feasible at an early stage of the design process.
Simulation and mock-ups	Assessment of the design output in representative conditions within the confines of a laboratory environment. The realism in the assessment is varied with different levels of product integration into the host platform, the different level of detail in scenario design, the complexity in the simulation and mock-up, as well as the involvement of the human operator.	• The ability to vary the level realism allows much flexibility, from quick prototypes to detailed models and systems, and can be utilized at different stages in the design process. • Complex dependencies and interactions that are typically not captured in isolated-task evaluations might surface in a realistic simulation. • The assessment is carried out within a controlled environment (as opposed to a field test), allowing for more comprehensive data collection as well as repeatable testing. • Risk is minimized when compared to field tests.	• Complex mock-ups and simulations can be costly and time consuming to develop. • High costs associated with maintaining a simulator/laboratory. • Data collected in a simulated/mock-up environment might not be representative of actual operating conditions.

Table 5.45 Methods for Evaluating a Design Output (*Continued*)

Assessment method	Description of method	Advantages	Limitations
Field test	An assessment of the product in a highly realistic operational scenario, in an external (field) setting. The assessment is typically conducted at a mature stage of the development process (where the operational effectiveness/suitability of a finalized design is evaluated), or in the operational phase (where the performance of the product is continually monitored over its life cycle).	• The robustness/ suitability of the product is evaluated in realistic operational conditions (including users, environmental, dynamics, and organizational factors).	• High costs and time consuming as compared to lab evaluations. • Low experimental control in the external environment. • Data collection is limited by the platform, and can be affected by the external factors. • Safety and risk are more critical considerations in the field test.

Table 5.45 Methods for Evaluating a Design Output (*Continued*)

References

AIAA 1993, Guide to Human Performance Measurements (AIAA G-035A-2000), American Institute of Aeronautics and Astronautics (AIAA), USA.

Cooper, G. E. and Harper Jr, R. P. 1969, The Use of Pilot Rating in the Evaluation of Aircraft Handling Qualities, AGARD Report 567, DTIC Document.

Endsley, M. R. and Kaber, D., 1999, "Level of Automation Effects on Performance, Situation Awareness and Workload in a Dynamic Control Task," *Ergonomics*, Vol. 42, no. 3, pp. 462–492.

FAA 2009a, Critical Task Analysis Report, FAA-HF-004A, Federal Aviation Administration.

FAA 2009b, Requirements for a Human Factors Program, HF-STD-004, Federal Aviation Administration.

Hart, S. G. and Staveland, L. E., 1988, "Development of NASA-TLX (Task Load Index): Results of Empirical and Theoretical Research," *Advances in Psychology*, Vol. 52, pp. 139–183.

Kirakowski, J. and Corbett, M., 1993, "SUMI: The Software Usability Measurement Inventory," *British Journal of Educational Technology*, Vol. 24, no. 3, pp. 210–212.

Kirwan, B. and Ainsworth, L. K., 1992, *A Guide to Task Analysis: The Task Analysis Working Group*, CRC press.

Kleinman, D., Baron, S., and Levison, W., 1970, "An Optimal Control Model of Human Response Part I: Theory and Validation," *Automatica*, Vol. 6, no. 3, pp. 357–369.

Kramer, A. F., 1991, "Physiological Metrics of Mental Workload: A Review of Recent Progress," *Multiple-Task Performance*, pp. 279–328.

McIlroy, R. C. and Stanton, N. A., 2011, "Getting Past First Base: Going All the Way with Cognitive Work Analysis," *Applied Ergonomics*, Vol. 42, no. 2, pp. 358–370.

McRuer, D. T. and Krendel, E. S. 1974, Mathematical models of human pilot behavior, AD-775 905, Advisory Group for Aerospace Research and Development (AGARD).

Miller, R. B. 1953, A method for man-machine task analysis, WADC TR 53-137, DTIC Document.

NATO 2004, Operator Functional State Assessment, TR-HFM-104, North Atlantic Treaty Organization (NATO).

Rasmussen, J., 1976, "Outlines of a Hybrid Model of the Process Plant Operator," in *Monitoring Behavior and Supervisory Control*, Springer, pp. 371–383.

Salmon, P., Jenkins, D., Stanton, N., et al. 2010, "Hierarchical Task Analysis vs. Cognitive Work Analysis: Comparison of Theory, Methodology and Contribution to System Design," *Theoretical Issues in Ergonomics Science*, Vol. 11, no. 6, pp. 504–531.

Stapleford, R. L., McRuer, D. T., and Magdaleno, R. E., 1967, "Pilot Describing Function Measurements in a Multiloop Task," *IEEE Transactions on Human Factors in Electronics*, no. 2, pp. 113–125.

Yeh, M., Young, J. J., Donovan, C., et al. 2013, *Human Factors Considerations in the Design and Evaluation of Flight Deck Displays and Controls*, TC-13/44, Federal Aviation Administration, Washington DC, USA.

SECTION 6

Aeronautical Design

Michael W. Jenkins

6.1 Definitions

Aeronautics is a science centered on flight in the atmosphere. It involves the design, construction, testing, and operation of lighter-than-air and heavier-than-air aircraft. These aircraft can be either mechanically driven (jets or propellers), rocket propelled, or not driven at all (gliders or balloons). All rely upon lift for successful operation. Fixed-wing aircraft lift is generated by the dynamic action of the air on the fixed wing. Rotary-wing aircraft (helicopters and autogyros) obtain lift from rotors placed horizontally above the fuselages. A helicopter differs from an autogyro in that the rotors are turned by motor power and there is usually no auxiliary propeller for forward motion. Balloons generate lift through density differences between the balloon gas (or hot air) and the local atmosphere. Flight regimes for fixed-wing aircraft include VTOL (vertical takeoff and landing), STOL (short takeoff and landing), V/STOL (vertical and/or short takeoff and landing), CTOL (conventional takeoff and landing, and STOVL (pronounced "stoval") (short takeoff and vertical landing).

6.2 Introduction

There are three phases in an aircraft design process: conceptual design, preliminary design, and final or detail design. These phases will overlap considerably, depending on the design organization. The final design and life cycle costs are almost entirely fixed by the conceptual design results, and therefore the conceptual design process is what will be discussed in this section. In the conceptual design phase the aircraft takes shape and the influences on the design configuration of the mission and program requirements can be most readily observed. This phase traditionally relies on the use of already proven technology in the form of databases from previously successful aircraft concepts. Thus, during the conceptual studies a continual validation process is in place that lends credibility to the emerging configuration. If a radical configuration is to be developed, it is usual to precede the conceptual design with a technology base development, either through extensive wind tunnel testing or through the use of experimental aircraft specifically designed for the purpose.

6.3 Overall Approach

There are many ways to develop aircraft conceptual designs. Figure 6.1 shows one such approach. This approach relies upon the use of historical databases.

The mission requirements usually identify the aircraft type (fighter, bomber, transport), which then leads to the selection of an appropriate aspect ratio and a corresponding maximum

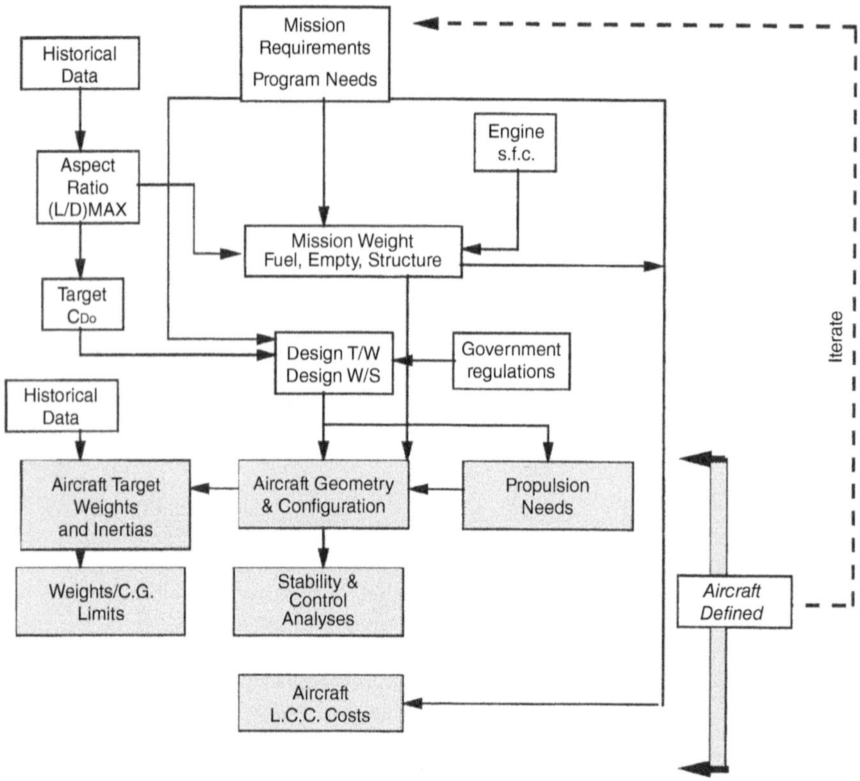

FIGURE 6.1 A conceptual design approach.

lift-to-drag ratio, L/D, from a suitable database. By assuming a parabolic drag polar (a good assumption for conceptual design), the magnitude of allowable C_{D_0} can be estimated. Application of these data to the mission requirements will, with a suitable set of engine data and through the use of standard geometric relationships, permit the configuration to emerge. Life cycle costs can be developed and the resulting aircraft can be compared with the mission and program requirements in an iterative fashion to refine a final conceptual aircraft.

This approach requires a significant amount of historical data, collected in a format for easy and quick use. Many excellent books exist on aircraft design (see References and Further Reading), containing a vast array of very useful design data. No attempt will be made in this section to duplicate these. It is understood that many other data sources exist, both private and company confidential, that can supplement or replace those identified here. Table 6.1 summarizes some characteristics of a range of fixed-wing aircraft.

Design Aims

A successful aircraft design is the result of a compromise among sometimes conflicting requirements. The external shape is usually influenced by the particular philosophy of the design organization, resulting in different configurations emerging from different design organizations designing to the same requirements. The basic design principles, however, remain fairly consistent. These are listed in Figure 6.2 in terms of desirable minima and maxima. These are postulated to promote

Manufacturer	Boeing	Dassault	Lockheed	Lockheed	Sukhoi
Name	F/A-18E/F	Rafale C	F-117A	F-22	Su 37
Gross Weight (lbs)	66,000	54,000	52,500	—	56,590
Empty Weight (lbs)	30,500	22,400	29.500	—	—
Span (ft)	44.9	33.4	43.3	44.5	48.2
Length (ft)	60.3	50	65.9	62.1	72.8
Wing Area (sq-ft)	500	490	913	740	667
Aspect Ratio	4.03	2.28	2.05	2.68	3.48
Power Plant	(2) GE F414 tf	(2) Snecma M88-2 tj	(2) GE F404 tf	(2) PW F119-PW-100 tf	(2) AL-37FU
Maximum Speed (Mach or MPH)	—	—	—	M 2.5	—
Cruise Speed (Mach or MPH)	—	M 1.8	—	—	—
Range (nm)	—	—	—	—	1,782
No. of Passengers	—	—	—	—	—
Payload (lbs)	—	—	—	—	3,090
Manufacturer	**Northrop Grumman**	**Eurofighter**	**Boeing**	**Saab**	**Mapo**
Name	F-14D	Typhoon	TAV-8B	JAS-39	MIG-31
Gross Weight (lbs)	74,349	50,264	31,000	27,557	90,000
Empty Weight (lbs)	41,780	24,239	14,223	12,900	47,000
Span (ft)	64.2/38.2	35.9	30.4	27.5	44.1
Length (ft)	61.9	52.4	50.3	46.3	74.5
Wing Area (sq-ft)	565	538	230	—	663
Aspect Ratio	7.29/2.58	2.40	4.02	—	2.93
Power Plant	(2) GE F110-GE-400 tf	(2) Eurojet 200 tf	(1) RR F402 or RR 406A tf	(1) Volvo Aero/GE RM12	(2) Soloviev D-30G
Maximum Speed (Mach or MPH)	M 2+	M 2 Class	M 0.9	M 2	M 2.8
Cruise Speed (Mach or MPH)	—	—	—	—	—
Range (nm)	—	—	—	—	1,630
No. of Passengers	—	—	—	—	—
Payload (lbs)	—	—	—	—	—

Table 6.1 Some Typical Aircraft Data

Manufacturer	Lockheed	Lockheed	Tupelov	Boeing	Antonov
Name	C-130J	C-5B	Tu 160	C-17A	An225
Gross Weight (lbs)	155,000	837,000	606,260	580,000	1,320,000
Empty Weight (lbs)	73,007	374,000	279,980	277,000	—
Span (ft)	132.6	222.8	182.6	169.8	290
Length (ft)	97.8	247.8	177.3	174	275.6
Wing Area (sq-ft)	1,745	6,200	—	3,800	9,743.70
Aspect Ratio	10.08	8.01	—	7.59	.63
Power Plant	(4) All. AE2100D3 tp	(4) GE TF39-IC tf	(4) NK-32 tf	(4) PW F117-PW-100	(6) Lotarev D-18T
Maximum Speed (Mach or MPH)	—	—	—	—	—
Cruise Speed (Mach or MPH)	340	563	M 2.3	M 0.77	430 kt
Range (nm)	—	3,000	7,640	3,000	3,800
No. of Passengers	—	270	—	3	6
Payload (lbs)	41,800	261,000	88,200	172,200	550,000
Manufacturer	**Boeing**	**Illyushin**	**Boeing**	**Sukhoi**	**Northrop**
Name	KC 135R	IL-78M	B-1B	SU-34	B-2A
Gross Weight (lbs)	322,500	462,966	477,000	99,400	336,500
Empty Weight (lbs)	119,212	204,088	186,234	—	—
Span (ft)	130.8	165.6	137	46.6	172
Length (ft)	128.8	152.8	145.8	76	69
Wing Area (sq-ft)	2,433	3,229.20	1,946	667.3	5,140
Aspect Ratio	7.03	8.49	9.64	3.25	5.76
Power Plant	(4) F108-CF-100	(4) D-30KP	(4) GE F101-GE-102	(2) AL-31F	(4) GE F118-GE-100
Maximum Speed (Mach or MPH)	—	—	—	—	—
Cruise Speed (Mach or MPH)	530 mph	M 0.80	—	M 1.8	—
Range (nm)	4,000	1,620	—	—	—
No. of Passengers	4	6	4	2	2
Payload (lbs)	—	105,820	—	17,640	40,000

Manufacturer	**Gulfstream**	**Airbus**	**Boeing Co.**	**Bombardier**	**Cessna**
Name	Gulfstream V	Airbus Corporate Jet	Boeing Business Jet	Global Express	Citation Excel
Gross Weight (lbs)	90,900	166,500	171,000	93,750	19,200
Empty Weight (lbs)	48,000	75,800	94,000	41,000	11,310
Span (ft)	93.5	111.8	112.6	94	55.7
Length (ft)	96.4	111	110.3	99.4	51.8
Wing Area (sq-ft)	1,136.5	1,318	1,342	1,022	370
Aspect Ratio	7.69	9.48	9.45	8.65	8.39
Power Plant	(2) BMW RR BR710-48	(2) CFM56-5B or IAE V2500	(2) CFM56-7B	(2) BMW RR BR700-710A2-20	(2) PWC PW545A
Maximum Speed (Mach or MPH)	M 0.89	M 0.82	M 0.82	M 0.89	M 0.77
Cruise Speed (Mach or MPH)	M 0.80	M 0.80	M 0.80	M 0.85	490 mph
Range (nm)	7,820	—	7,135	7,811	2,331
No. of Passengers	3–19	2–124	2–100	3/4–19	8–11
Payload (lbs)	—	—	—	—	—
Manufacturer	**Dassalt**	**Raytheon**	**Raytheon**	**Cessna**	**Piper**
Name	Falcon 2000	Hawker 800 XP	Beech 1900D Air-liner	Skyhawk/172R	PA46-350P Malibu Mirage
Gross Weight (lbs)	36,000	28,120	16,950	2,457	4,300
Empty Weight (lbs)	20,735	16,270	10,615	1,640	3,080
Span (ft)	63.4	51.4	57.8	36.1	43
Length (ft)	66.3	51.1	57.7	27.2	28.9
Wing Area (sq-ft)	527.6	374	310	174	175
Aspect Ratio	7.62	7.06	10.78	7.49	10.57
Power Plant	(2) CFE 738-1-18	(2) ASE TFE731-5BR	(2) PWC PT6A-67D	(1) Lyc. 10-360-L2A	1 Lyc. TI0540-AE2A
Maximum Speed (Mach or MPH)	M 0.87	M 0.80	331 mph	142	267
Cruise Speed (Mach or MPH)	M 0.80	M 0.75	331 mph	141	246
Range (nm)	3,975	2,906	1,827	668	1,547
No. of Passengers	8–19	8–15	14–21	4	6
Payload (lbs)	—	—	—	—	—

TABLE 6.1 Some Typical Aircraft Data (*Continued*)

Manufacturer	Airbus	Boeing	Boeing	Boeing	Airbus
Name	A340-600	B767-400ER	B777-200ER	MD-11	AIRBUS-A380-800
Gross weight (lb)	804,700	450,000	656,000	630,500	1,272,000
Empty weight (lb)	560,000	227,300	314,000	295,600	610,200
Span (ft)	208.7	170.3	199.9	170.6	261.7
Length (ft)	245.9	201.3	209.1	202.2	238.6
Wing area (ft^2)	4,704	2,837	4,605	2,688	9,100
Aspect ratio	9.26	10.22	8.68	7.89	7.5
Power plant	(4) RR Trent 500	(2) GE CF6-80C2B or PW4062	(2) PW4000s, GE90's or RR Trents	(3) PW4460, PW4462, or GE CF6-80C2-DFI	(4) RR Trent 970/B
Maximum speed (Mach or Mph)	M 0.86	M 0.86	M 0.87	596	M 0.89
Cruise speed (Mach or mph)	M 0.83	M 0.80	M 0.84	M 0.83	M 0.85
Range (nm)	8,600	6,469	8,723	7,660	8,200
Number of passengers	380	261–304	301–440	298–410	525 to 853
Payload (lb)	—	53,700	125,000	98,500	196,700
Manufacturer	**EMBRAER**	**Fairchild**	**British Aerospace**	**British Aerospace**	**British Aerospace**
Name	ERJ-145 LR	728 Jet	BAe-146–300	AVRO RJ 100	Concorde
Gross weight (lb)	48,501	76,850	97,500	97,500	408,000
Empty weight (lb)	26,124	44,175	54,700	56,850	173,500
Span (ft)	65.8	87.4	86.4	86.4	83.8
Length (ft)	98	84.6	101.8	101.7	203.8
Wing area (ft^2)	551	807	832	832	3,640
Aspect ratio	7.86	9.46	8.97	8.97	1.93
Power plant	(2) All. AE3007-A1	(2) GE CF34-8D	(4) ASE ALF502R-5	(4) ASE LF507	(4) RR Olympus 593 Mk. 610
Maximum speed (Mach or Mph)	M 0.78	M 0.82	M 0.70	500 mph	M 2.0
Cruise speed (Mach or mph)	575	600	466	474	1,354
Range (nm)	1,550	1,781	1,355	1,150	4,090
Number of passengers	50	80	112	128	128

Source: Aviation Week Source Book 1999.

Table 6.1 Some Typical Aircraft Data (*Continued*)

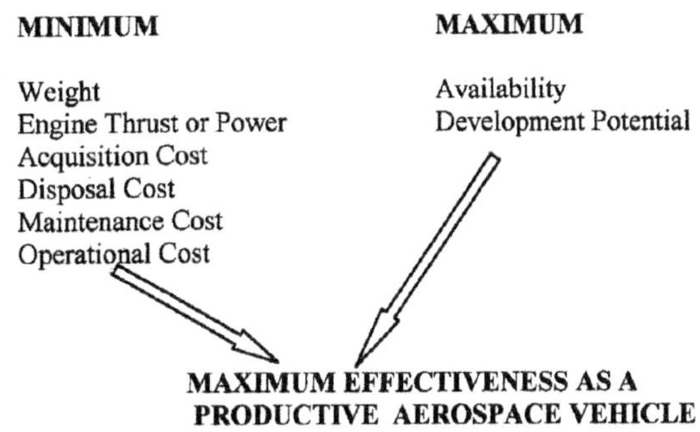

FIGURE 6.2 Generic overall design aims.

minimum acquisition cost, minimum operational cost, and maximum effectiveness as a functioning aircraft performing its mission.

Other factors specifically influence military aircraft configurations, such the desire for low infrared, low radar cross-section, and low noise signatures and cargo loading/unloading needs. Further special operational requirements also influence military aircraft configurations, see Figure 6.3, such as the need to possess VTOL, V/STOL, STOVL, STOL, aircraft carrier compatibility, in-flight refueling, and supercruise capability. For civil or commercial aircraft the corresponding factors include airport gate compatibility (influencing wing span), passenger emergency evacuation requirements, the normal loading/unloading of passengers, and low operating costs. For very large military and civil/commercial aircraft the landing gear design and layout is a function of the thickness of the paved runway surface to be used as well as airframe stowage considerations.

The need for a balanced approach to the design function is best summarized, humorously, in Figure 6.4, where bias toward one technology or another has influenced the final configuration adversely. A good design avoids such bias.

Mission Establishment

The starting point for aircraft design studies is usually a new mission requirement or the need for the application of a new technology. Mission requirements usually stem from mission area analyses; new technologies can be in the form of a new engine, new materials, or the need to explore the influence of stealth for example. A request for proposal (RFP) or similar document issued by the customer contains specifics defining the needs for the new concept.

Mission Requirements

The specifics deduced from the RFP permit design-to criteria to be established in the form of such considerations as maximum speeds, rate-of-climb, range, endurance, payload, takeoff and landing distance requirements, and costs. Also identified are the various government specifications to which the proposed concept must be designed. These contain such considerations as the allowable symmetric load factor, the gust environment through which the aircraft must fly safely, and the definition of takeoff and landing distances. Government specifications are necessary to ensure the safe design and operation of aircraft in national and international airspace.

Company Design Approach

A mission profile (Figure 6.5) is developed to reflect the major design needs. Initial sizing studies are performed, usually following the outline of Figure 6.1. The overall gross weight of the

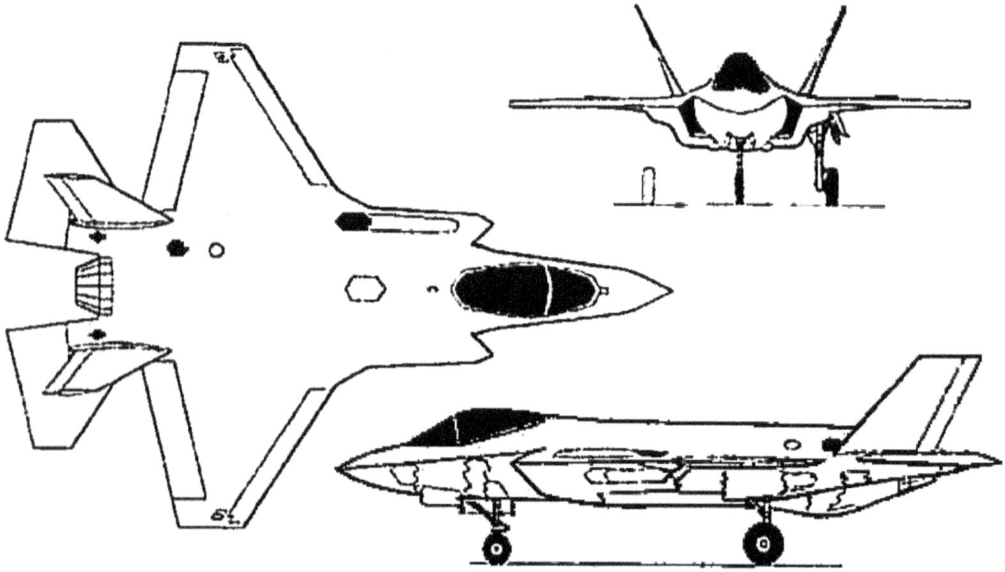

Variants	F-35A CTOL	F-35B STOVL	F-35C CATOBAR
Length ft	51.4	51.2	51.4
Wingspan ft	35.0	35.0	43.0
Wing Area ft^2	460	460	668
Empty weight lbs.	29,300	32,300	34,800
Internal fuel lbs.	18,250	13,500	19,750
Max.TO. wt. lbs.	70,000	60,000	70,000
Combat radius nml	590	450	600

Figure 6.3 Lockheed-Martin F-35A joint strike fighter variants. (*Source*: Jane's All the Worlds Aircraft, 2014–2015 edition) (F-35A three view shown) (Propulsion primarily F135–PW series engines.)

aircraft is established primarily through the use of the Breguet range and endurance equations as applied to each segment of the mission profile. Experience has shown that the combination of the minimum value of T/W and maximum value of W/S to meet all the mission requirements usually results in the minimum weight and minimum cost for the concept. This is the design aim for modern aircraft.

Customer Needs

Sometimes a design configuration is significantly influenced by special customer needs. Military cargo transports sometimes require drive-on and drive-off capabilities (Figure 6.6), resulting in fuselage floors close to the ground and wings on top of the fuselage; increasing weight of larger transports dictate unusual landing gear configurations (Figure 6.7); and VTOL needs will influence engine (Figure 6.8) or wing design (Figure 6.9).

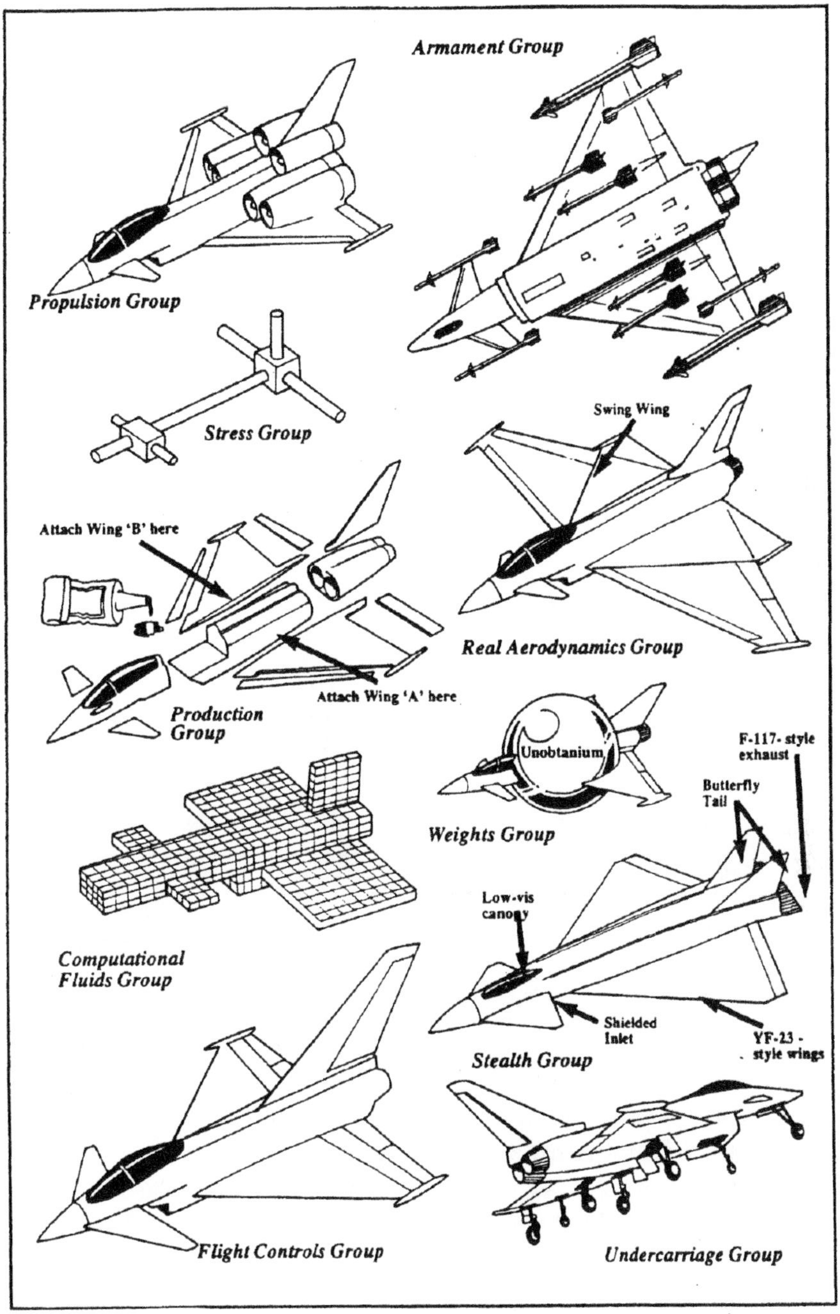

FIGURE 6.4 Light-hearted look at how specialists would have designed Eurofighter EF 2000 if given a free hand (Pete West/*Air International*).

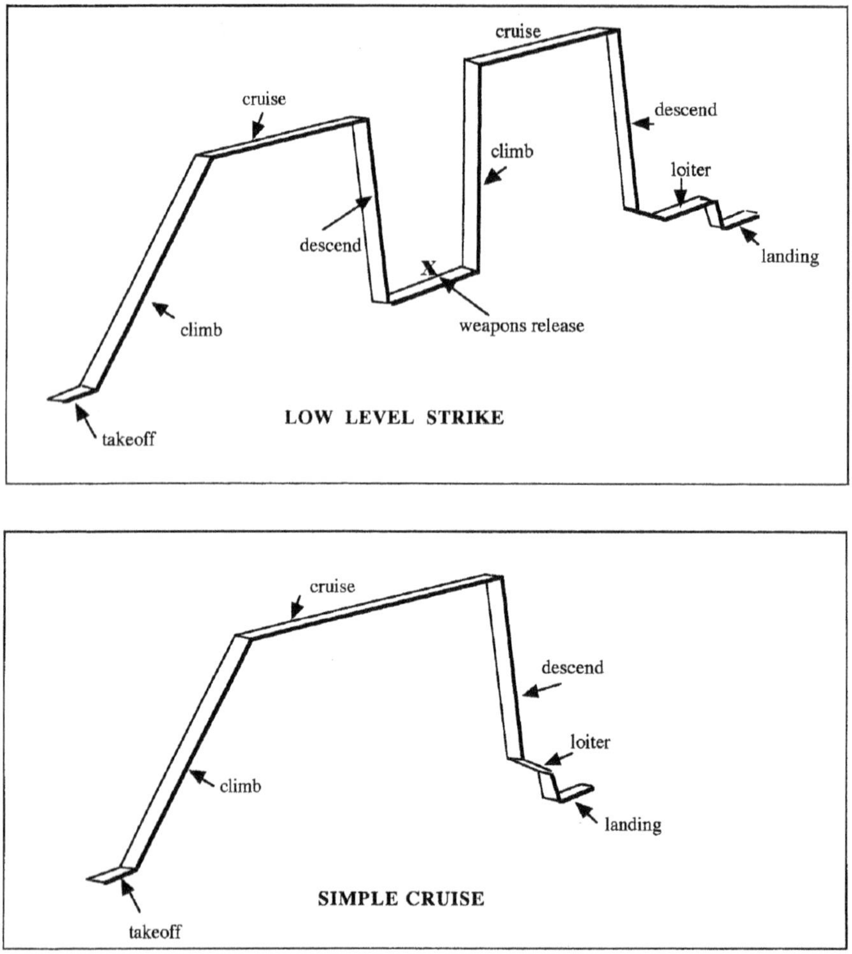

FIGURE 6.5 Some typical mission profiles.

6.4 Government Regulations

To be considered in the design process, these regulations are primarily related to the safety of the aircraft operation and of the aircraft structure. Aeronautical research and operational experience accumulated over many years have resulted in the generation of a set of requirements to which modern aircraft are now designed. The appropriate regulations define, for example, airplane maneuver loads, and stability and control requirements, which are a function of the intended use of the aircraft, and also define the gust environment through which all aircraft may eventually fly, irrespective of their intended use. Not only must the aircraft structure withstand the specified loads, but all internal equipment must be strong enough and mounted such as to withstand the resulting angular and linear accelerations. Specifically, engine gyroscopic induced loads must also be accounted for.

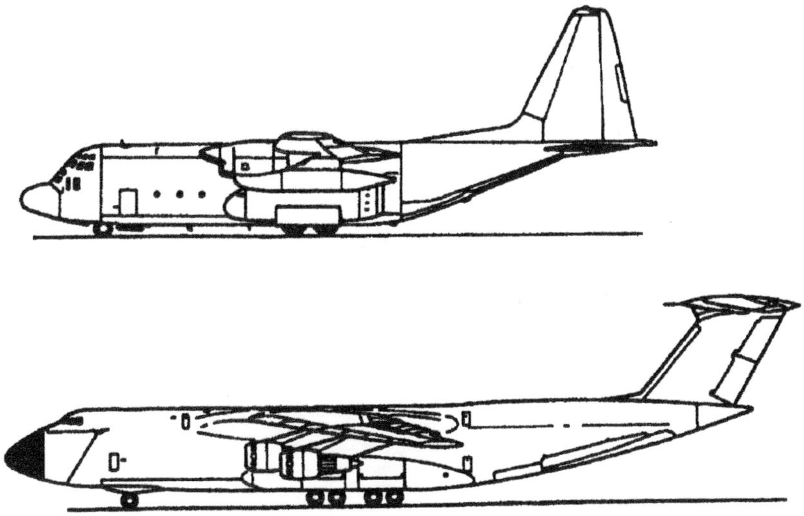

FIGURE 6.6 Military design requirements influence wing and fuselage location (Lockheed-Martin).

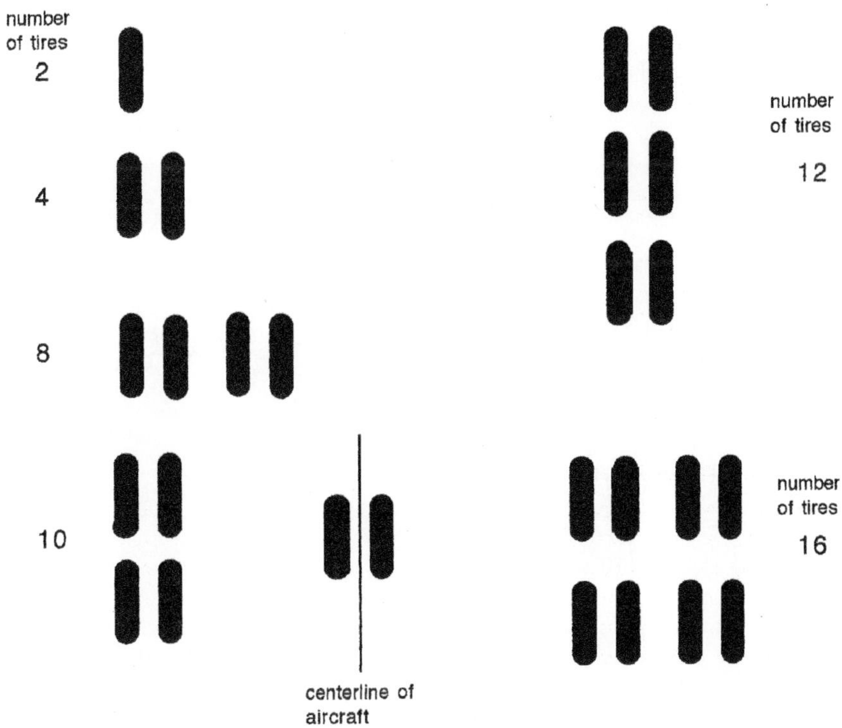

FIGURE 6.7 Some typical main gear wheel layouts per aircraft side.

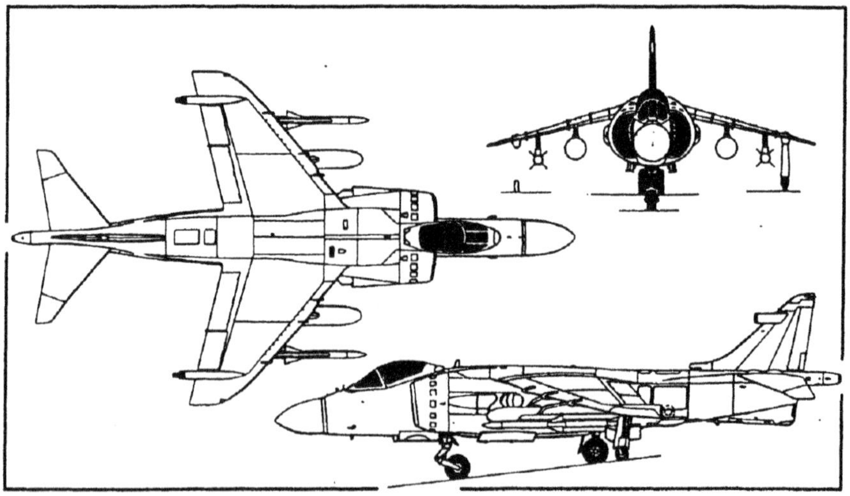

FIGURE 6.8 British Aerospace Sea Harrier (*Air International*, December 1999).

Commercial and civil aircraft are usually covered by FAR/JAR or similar regulations, and military aircraft by military specific regulations (MIL-SPECS). Figure 6.10 lists some of these. It can generally be assumed that the international civil maneuver regulations are equivalent for FAR/JAR and that the gust environment, in terms of vertical gust magnitudes, is the same.

Commercial (FAR 25) and civil (FAR 23) aircraft are separated primarily by gross weight and propulsion type (Figure 6.11).

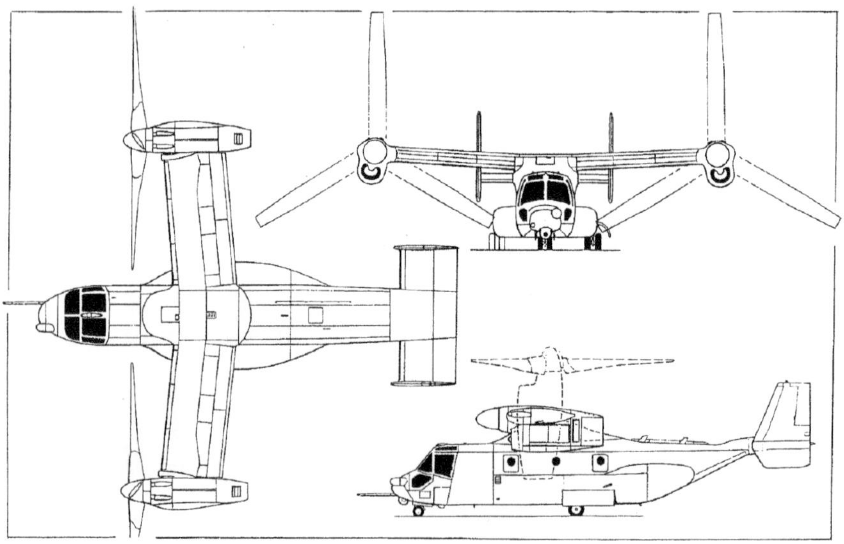

FIGURE 6.9 V-22 tilt rotor military transport (*Air International*, May 1989).

CHAPTER 1 – **FEDERAL AVIATION ADMINISTRATION**, DEPARTMENT OF TRANSPORTATION.– **FAR** – (Extract).

(FAR – Federal Aviation Regulations – U.S.)

Subchapter C – Aircraft.

Part	
21	Certification procedures for products and parts.
23	Airworthiness standards: normal, utility, acrobatic, and commuter category airplanes.
25	Airworthiness standards: traansport category airplanes.
27	Airworthiness standards : normal category rotorcraft.
29	Airworthiness standards: transport category rotorcraft.
31	Airworthiness standards: manned free balloons.
33	Airworthiness standards: aircraft engines.
35	Airworthiness standards: propellers.
36	Noise standards : aircraft type and airworthiness certification.
39	Airworthiness directives.
43	Maintenance, preventive maintenance, rebuilding, and alteration.

TYPICAL MAKE-UP OF **JOINT AIRWORTHINESS REQUIREMENTS.**

(JAR – EUROPEAN).

JAR – 25 , Large Aeroplanes.

Section 1 – Requirements.

Subpart A – General

Subpart B – Flight

Subpart C – Structure

Subpart D – Design and Construction

Subpart E – Powerplant

Subpart F – Equipment

Subpart G – Operating limitations and information

Subpart J – Gas turbine auxiliary power unit installation.

Figure 6.10 Some FAA, JAR, and U.S. Military specifications (after Nicolai 1984).

Maneuver loads and gust-induced loads are applied separately to the aircraft structure during the analysis stage of design. Upon initiation of production, at least one sample of each aircraft type is subjected to structural testing, using design loads, to validate the safety of the aircraft structure. This extensive testing is a demonstration of the structural safety of the aircraft and, because of the number of test cases involved this testing, can take up to a year to complete.

TYPICAL **MIL.SPEC. STANDARDS.** – U.S.

MIL-STD-1797A	Flying qualities of Piloted Aircraft
MIL-F-83300	Flying qualities of Piloted V/STOL Aircraft
MIL-F-9490	Flight Control Systems-Design, Installation and Test of Piloted Aircraft
MIL-S-8369	Stall/Post-Stall/Spin Flight Test Demonstration Requirements for Airplanes
MIL-C-18244	Control and Stabilization Systems: Automatic, Piloted Aircraft.
MIL-D-8708	Demonstration Requirements for Airplanes
MIL-A-8860 through 8864 and 8870.	Airplane Strength and Rigidity
MIL-P-26366	Propellers, Type Test of
MIL-I-8700	Installation and Test of Electronics Equipment in Aircraft
MIL-S-18471	Seat System, Ejectable, Aircraft
MIL-W-25140	Weight and Balance Control Data
MIL-STD-850	Aircrew Station Vision Req. for Military Aircraft
MIL-STD-757	Reliability Evaluation from Demonstration Data
MIL-C-5011	Charts; Standard Aircraft Characteristics and Performance
MIL-STD-881	Work Breakdown Structure (WBS)

FIGURE 6.10 Some FAA, JAR, and U.S. Military specifications (after Nicolai 1984). (*Continued*)

FAR 23	**FAR 25**
Propeller driven – generally	Jet or propeller driven
Less than 19 passengers	No minimum passenger limit
$W < 12{,}000$ lb generally	$W > 12{,}000$ lb
Categories	**Categories**
Normal Utility Acrobatic	Normal Cargo

FIGURE 6.11 Commercial and civil FAR divisions.

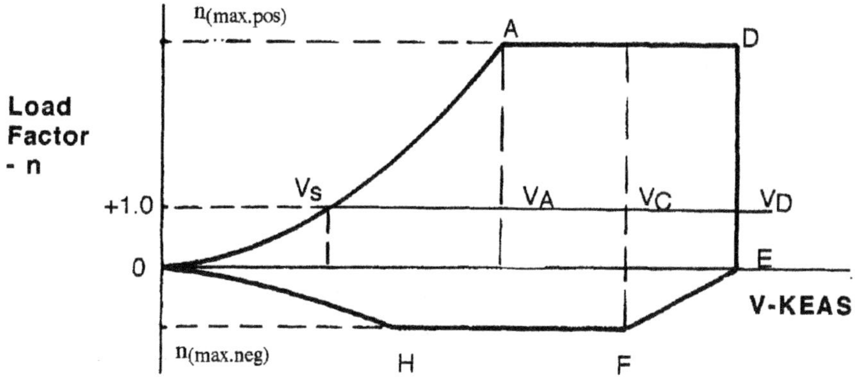

V_S = Level flight stall speed

V_A = Corner speed, or maneuver speed for military

V_C = Cruise speed,or max.level flight speed for military

V_D = Limit speed, or max. dive speed for military and civil

KEAS = Equivalent airspeed (knots)

FIGURE 6.12 Typical maneuver diagram.

V-n Diagram—Maneuver Envelope

(Figure 6.12). This defines the symmetric flight operational boundaries of an aircraft. The airplane must be structurally safe to fly on or within any point on this envelope which defines the structural, aerodynamic, and propulsion limits of the concept. The maximum value of n, for military aircraft depends upon the function or mission of the aircraft for military operations (Table 6.2). For FAR 23 aircraft n_{max} is a function of aircraft type (Figure 6.13).

For FAR 25 commercial aircraft n_{max} is also a function of maximum takeoff gross weight (MTOGW). For aircraft up to 4100 lbs, $n_{max} = 3.8$ and at higher gross weights n_{max} varies as

$$n_{max} = 2.1 + \left[\frac{24,000}{\text{MTOGW} + 10,000}\right]$$

up to a gross weight of 50,000 lb, where n_{max} becomes equal to 2.5 and thereafter remains constant for higher gross weights.

V-n Diagram—Gust Envelope

The envelope in Figure 6.14 describes the load factors imposed on a trimmed aircraft when encountering a vertical up-gust or down-gust in horizontal constant-speed flight.

A sharp-edged vertical gust is assumed to be present, and a gust alleviation factor to account for gust penetration and gust response is assumed to exist. The resultant total load factor induced as given by MIL-SPEC MIL-A-8861(AS) is

$$n = 1 \pm \frac{K_g U_{ge} V_E a}{498(W/S)}$$

USAF airplane type	USN airplane type	Positive	Negative
Fighter		8.67	–3.00
Attack	Fighter, attacker, trainer Observation	7.33 6.00	–3.00 –3.00
Trainer		5.67	–2.33
Utility	Utility	4.00	–2.00
Small bomber		3.67	–1.67
Medium bomber, assault transport	Patrol, weather, Antisubmarine, reconnaissance	3.00	–1.00
Medium transport		2.50	–1.00
Heavy bomber, heavy transport		2.00	–1.00

Source: MIL-SPEC Mil-F-8860.

TABLE 6.2 Typical Maximum Maneuver Load Factors

Normal Category

$$Positive + n_{max} = 2.1 + \left[\frac{24{,}000}{MTOGW + 10{,}000}\right]$$

$$Negative - n_{max\,neg} = 0.4(+n_{max})$$

Exceptions:–

	$+n_{max}$	$n_{max.neg}$
Utility	4.4	1.76
Acrobatic	6.0	3.0

FIGURE 6.13 Design maximum load factor for FAR 23 type aircraft (FAR 23).

where $K_g = \dfrac{0.88\mu_g}{5.3 + \mu_g}$ = gust alleviation factor

$\mu_g = \dfrac{2(W/S)}{g\,\rho a \bar{c}}$ = mass parameter

U_g = gust velocity (fps) (e.a.s.)
ρ = air density (slugs/ft^3)
(W/S) = wing loading (psf)
$\bar{c}$ = wing mean geometric chord (ft)
g = acceleration due to gravity (f/s^2)
V_E = aircraft equivalent airspeed (knots) (f/s^2)
V_E = aircraft equivalent airspeed (knots) (e.a.s.)
a = slope of airplane normal force curve (per radian)

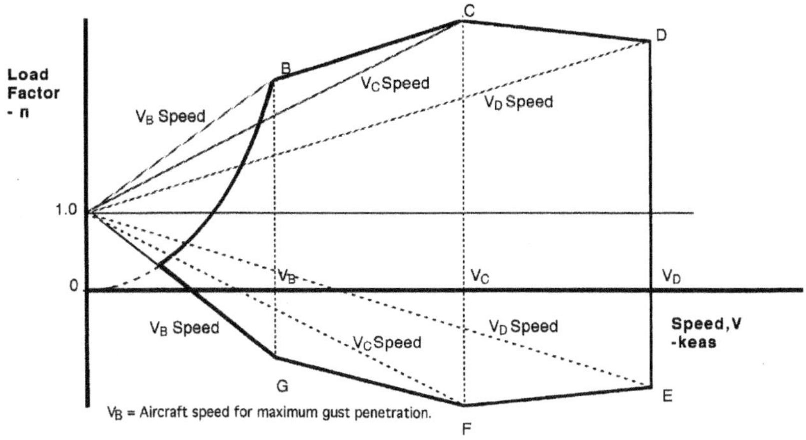

FIGURE 6.14 Typical gust design envelope.

Gust intensities (fps e.a.s.)		Maximum aircraft velocity (knots e.a.s.)
S.L. to 20,000 ft	**20,000 ft to 50,000 ft**	
±66	linear reduction to 38	V_B
±50	linear reduction to 25	V_C, V_H
±25	linear reduction to 12.5	V_D, V_L

V_B = Speed for maximum gust intensity
V_c, (V_H) = Design cruise speed (max. level flight speed at max. continuous power for military aircraft)
V_D = Design dive speed (1.25 V_C or 1.25 V_L)

Source: MIL-SPEC Mil-E-8861.

TABLE 6.3 Standard Gust Velocities

Gust velocity magnitudes and their variation with altitude are given in Table 6.3. When the symmetric maneuver and symmetric gust envelopes are superimposed, one of these will establish a maximum value of symmetric load factor. For highly maneuvering aircraft a symmetric *maneuver* load factor will be a maximum, which is the case for fighter aircraft. Most commercial aircraft will exhibit a maximum symmetric load factor as generated by the *gust* envelope. The former is called a *maneuver critical* aircraft, the latter *gust critical.*

6.5 Conceptual Design

Weight Estimation

The maximum design weight of an aircraft, W_0, can be written as follows (see Raymer 1989; Roskam 1980):

$$W_o = W_e + W_f + W_c + W_{pl} \text{ (lbs)}$$

where W_e = empty weight
W_f = fuel weight
W_c = crew weight
W_{pl} = payload weight

This can be written as

$$W_0 = \left(\frac{W_c + W_{pl}}{1 - \left(\frac{W_e}{W_0}\right) - \left(\frac{W_f}{W_0}\right)} \right) \text{(lb)}$$

where (W_e/W_0) is the empty weight fraction and (W_f/W_0) is the fuel fraction.

The empty weight fraction in conventional design is a targeted allowance for those items listed in Figure 6.15.

The likely value for this fraction depends on the function of the aircraft (see Figure 6.16). In conceptual design, empty weight and component weights are often based on aircraft with missions similar to your conceptual design (Table 6.4). This approach assumes that the structural efficiency and resulting weight of the systems for the design being developed are at least as good as those of existing aircraft.

The fuel fraction depends entirely on the mission, the anticipated L/D ratio of the aircraft, and the specific fuel consumption of the engine(s) to be used. The use of the Breguet (or similar) range and endurance equations will permit fuel fractions to be determined. For example, using the mission profile of Figure 6.17, we can postulate that the overall mission weight fraction W_{end}/W_0 is given by

$$\left(\frac{W_7}{W_0}\right) = \left(\frac{W_1}{W_0}\right)\left(\frac{W_2}{W_1}\right)\left(\frac{W_3}{W_2}\right)\left(\frac{W_4}{W_3}\right)\left(\frac{W_5}{W_4}\right)\left(\frac{W_6}{W_5}\right)\left(\frac{W_7}{W_6}\right) = \left(\frac{W_{end}}{W_{start}}\right)$$

for a seven-element mission profile, where (W_3/W_2) represents an application of Breguet's range and (W_5/W_4) the endurance equation. All other fractions represent what can be termed standard weight fractions for design purposes. These are a function of aircraft type, i.e., fighter or commercial transport.

Structure weight
- Wing weight
- Empennage weight
- Fuselage weight (including furnishings)

Propulsion weight

Systems weight
- Controls
- Environmental
- Hydraulic
- Pneumatic
- Fuel
- Electric
- Avionics
- Weapons

Landing gear weight

FIGURE 6.15 Components included in empty weight.

$$\frac{W_e}{W_0} = C \times (W_0)^D$$

W_e = empty weight (lb) W_0 = gross weight (lb)

Aircraft type	C	D
Fighters	0.6954	−0.0275
Business jets	1.0595	−0.0598
Wide-body jets	0.8342	−0.0375

FIGURE 6.16 Typical empty weight fractions (*Aerospace Source Book* 1999).

Group weight (lb)	F/A-18A	Gulfstream II	Airbus A300
Wing group	3,798	6,372	44,131
Empennage group	945	1,965	5,941
Fuselage group	4,685	5,944	35,820
Engine section	143	1,239	7,039
Landing gear group	1,992	2,011	13,611
Structure total	**11,563**	**17,531**	**106,542**
Engines & systems	6,277	6,886	22,897
Fixed equipment	5,135	11,203	35,053
Total empty weight	**22,974**	**35,620**	**164,492**
Flight design gross weight	**32,357**	**64,800**	**302,000**

Source: Roskam 1980.

TABLE 6.4 Some Typical Group Weight Values

An excellent summary (Roskam 1980) lists various values for various concepts, and it can be seen that for conceptual design purposes each of these can be assumed to be about 0.99.

$$1-(W_7/W_0) = \text{Fuel fraction} = (W_f/W_0)$$

and if a reserve percentage of fuel is required, e.g., 6% fuel reserve, then the total mission fuel fraction is given by 1.06 (W_f/W_0), which is called the total fuel fraction and represents the fuel required for the mission.

With knowledge of the crew weight and payload weight, with the total fuel fraction (W_f/W_0) and with (W_e/W_0) magnitudes, it is now possible to solve iteratively for W_0. This represents the design value for the takeoff weight, which now permits the estimation of total fuel weight and target empty weight through

$$\text{Total fuel weight} = (W_f/W_0)\cdot W_0 \text{ (lb)}$$

$$\text{Target empty weight} = (W_e/W_0)\cdot W_0 \text{ (lb)}$$

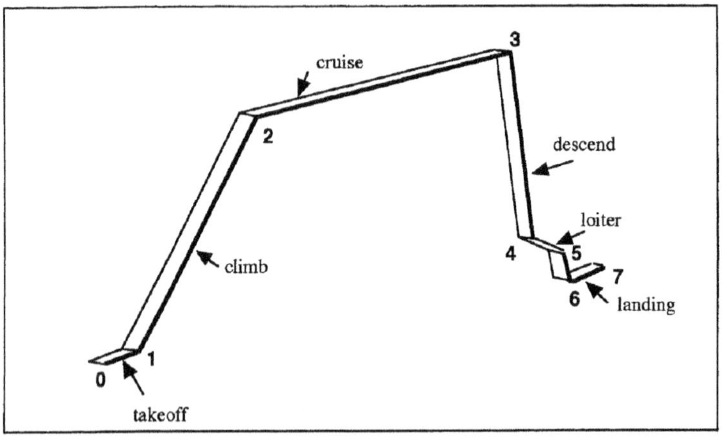

FIGURE 6.17 Typical mission profile.

Aircraft type	Wing aspect ratio	$(L/D)_{max}$	Wing taper ratio
Fighters • Subsonic	4.00–6.00	11.0	0.28–0.57
• Supersonic	2.00–3.00	9.00	0.07–0.25
Jet transports (military and civil)	7.00–9.00	17.00	0.28–0.42

For subsonic flight, $e = 0.80$ to 0.85 (clean wing).

TABLE 6.5 Likely Values of Design Parameters

Likely values of certain design parameters necessary for use in the Breguet range and endurance equations are given in Table 6.5.

Estimation of Thrust to Weight (T/W), Wing Loading (W/S), and Definition of the High Lift System Requirements

These design parameters are established by evaluations of the performance and maneuver requirements of the mission. Whereas the T/W and W/S values are generally established by in-flight performance requirements, the high-lift systems are usually dictated by airfield performance demands in terms of field length or takeoff and landing distances.

The relationship between T/W and W/S is given as follows:

$$\frac{T}{W} = \left(\frac{C_{D_0} q}{W/S}\right) + \left(\frac{n^2 (W/S)}{q.\pi \cdot A \cdot e}\right)$$

where C_{D_0} = profile drag coefficient,
q = dynamic pressure (psf)
W/S = wing loading (psf)
n = maneuver load factor
T/W = thrust to weight

all at the flight conditions being evaluated.

Maximum speed

Cruise speed

Maneuver requirements

Climb requirements – Single engine

Climb requirements – Multiengine

– All engines operating

– One engine failure

– One engine go-around

Supercruise speed

Absolute ceiling – Rate of climb = 0

Service ceiling – Rate of climb = 100 feet per minute (Mil.@ max. power)

= 500 feet per minute (Commercial jet)

Figure 6.18 Mission segments defining T/W and W/S.

By choosing appropriate values of aspect ratio (A), and induced drag factor, e, the above equation can be evaluated to show the relationship between T/W and W/S for the particular mission segment. Mission segments can be combinations of those shown in Figure 6.18. Likely values of A and e are shown in Table 6.5.

It is not unusual to assume at the conceptual design phase that the drag polar for the aircraft is of parabolic form, resulting in the generation of algebraic relationships for many design-related parameters (Table 6.6). These clean aircraft relationships can be modified when external stores or other drag-producing devices are used during parts of the mission.

The maximum value of L/D is given by

$$\frac{L}{D} = 0.5\sqrt{\frac{\pi \cdot A \cdot e}{C_{D_0}}}$$

Choosing the design value for maximum value of L/D, A and e will determine a target value of C_{D_0}.

Presenting the resulting T/W and W/S relationships in graphical format (Figure 6.19) will identify the T/W and W/S combination that will satisfy all of the mission requirements.

Figure 6.19 shows that a single combination of T/W and W/S, called the *design point*, will satisfy or exceed all mission requirements. It identifies which mission requirements define the configuration and which do not. The design point shown here is the lowest thrust-to-weight ratio and the highest wing loading necessary to meet all the requirements. The $(T/W)_{\text{min}}$ and $(W/S)_{\text{max}}$ chosen lead to the lowest installed thrust and the lightest wing weight. Any point above the two defining mission requirement lines will also satisfy the mission requirements, but probably not with the lowest cost. The area above these lines is sometimes called the design space.

The resulting airplane geometric parameters and engine requirements can then be established as follows:

$$S = (W)/(W/S)_{\text{max}} \text{ ft}^2 = \text{wing area}$$

and

$$T = (T/W)_{\text{min}} (W) \text{ lb} = \text{total performance thrust}$$

Maximum range	Maximum endurance
$C_D = 2C_{D_0}$	$C_D = 4C_{D_0}$
$C_L = \sqrt{C_{D_0}\pi Ae}$	$C_L = \sqrt{3C_{D_0}\pi Ae}$
$V = \sqrt{\frac{W}{S}\frac{2}{\rho}\frac{1}{\sqrt{C_{D_0}\pi Ae}}}$ (fps)	$V = 0.76\sqrt{\frac{W}{S}\frac{2}{\rho}\frac{1}{\sqrt{C_{D_0}\pi Ae}}}$ (fps)
$\left(\frac{L}{D}\right)_{max} = 0.500\sqrt{\frac{\pi Ae}{C_{D_0}}}$	$\left(\frac{L}{D}\right) = 0.433\sqrt{\frac{\pi Ae}{C_{D_0}}}$
For parabolic drag polar: $C_D = C_{D_0} + \frac{C_L^2}{\pi Ae}$ (Propeller-powered aircraft)	
$C_D = \frac{4}{3}C_{D_0}$	$C_D = 2C_{D_0}$
$C_L = \sqrt{\frac{C_{D_0}\pi Ae}{3}}$	$C_L = \sqrt{C_{D_0}\pi Ae}$
$V = 1.316\sqrt{\frac{W}{S}\frac{2}{\rho}\frac{1}{\sqrt{C_{D_0}\pi Ae}}}$ (fps)	$V = \sqrt{\frac{W}{S}\frac{2}{\rho}\frac{1}{\sqrt{C_{D_0}\pi Ae}}}$ ~ F.P.S.
$\left(\frac{L}{D}\right) = 0.433\sqrt{\frac{\pi Ae}{C_{D_0}}}$	$\left(\frac{L}{D}\right)_{max} = 0.500\sqrt{\frac{\pi Ae}{C_{D_0}}}$
For parabolic drag polar: $C_D = C_{D_0} + \frac{C_L^2}{\pi Ae}$ ~ Jet powered aircraft.	

Table 6.6 Some Useful Design Relationships

Use of the geometric and aerodynamic relationships of Figures 6.20–6.22 permits the aircraft configuration to develop once a wing taper ratio has been chosen.

Trade studies are performed to assess the influences on the design (primarily on W_0) of variations in certain mission parameters and in certain assumed design parameters. Included in these are range, payload, aspect ratio, specific fuel consumption, and C_{D0}, in which the effects of systematic individual variations are assessed.

Influence of Takeoff Distance Requirements on Aircraft Design

Traditionally, takeoff and landing distances are defined as being composed of a ground distance and an air distance. In each instance the air distance is measured relative to an imaginary vertical obstacle situated at the end of the runway (Figure 6.23). For civil aircraft this obstacle is 50 ft for FAR 23 category aircraft on takeoff and 50 ft for landing. For FAR 25 category aircraft this obstacle is 35 ft on takeoff and 50 ft for landing. For landing the approach speed, *VA*, is specified as $1.3(V_{SL})$, which becomes $1.2(V_{SL})$ for military land based aircraft and $1.15(V_{SL})$ for carrier-based aircraft. The design factors that are influenced are wing loading, thrust or power loading, aerodynamic drag, and lift for an RFP-dictated takeoff distance and runway surface type. The design process is

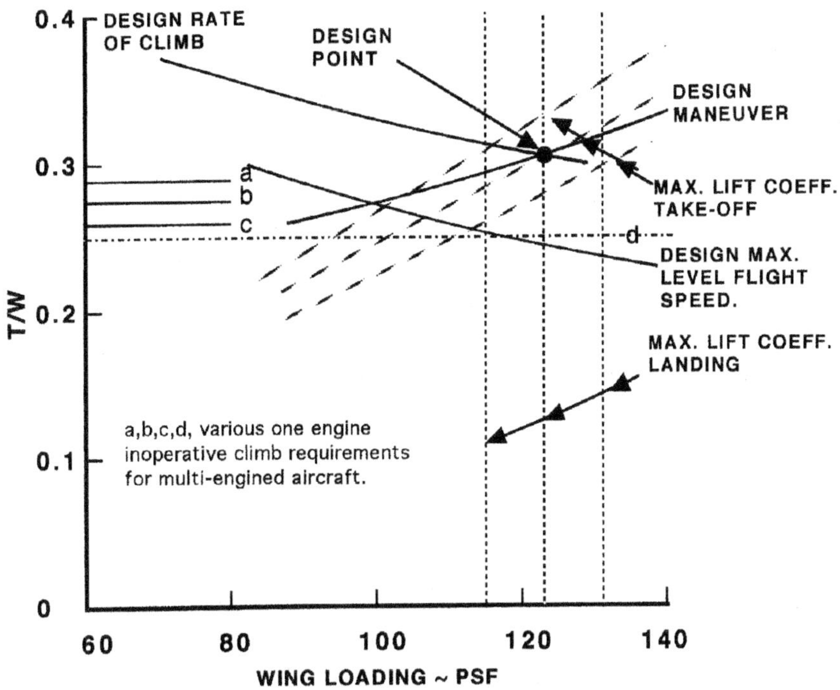

FIGURE 6.19 Typical design chart showing various mission requirements.

centered on establishing the appropriate combinations of W/S, T/W, or W/P and C_L necessary to meet the design takeoff distance. These relationships have been extensively explored (Loftin 1980, as shown in Raymer 1989).

For FAR 23 rules, the following can be used for takeoff distance estimation (Loftin 1980, as shown in Raymer 1989).

$$S_{\mathrm{TOG}} = 4.9(\mathrm{TOP}_{23}) + 0.009(\mathrm{TOP})_{23})^2 \text{ (ft)}$$

where $$(\mathrm{TOP}_{23}) = \frac{(W/S)_{TO}(W/P)_{TO}}{\sigma \cdot C_{L_{\mathrm{MAX}\,TO}}} (\mathrm{lb}^2/\mathrm{ft}^2 \cdot \mathrm{hp})$$

and $$S_{TO} = 1.66(S_{\mathrm{TOG}})$$

For FAR 25 takeoff estimates the following is applicable:

$$S_{TOFL} = \frac{37.5(W/S)_{TO}}{\{\sigma \cdot C_{L_{\max TO}}(T/W)_{TO}\}} \text{(ft)}$$

In each instance solution of the above equations will give a combination of W/S, T/W (or W/P), and $C_{L_{\max}}$ that will generate the necessary takeoff distance required. (Here P = power, σ = relative density, and subscript TO refers to takeoff conditions.)

Wing / Horizontal Tail Geometries

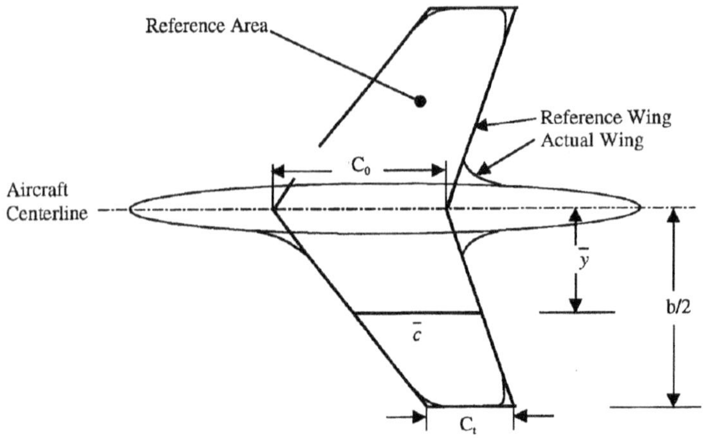

$$\bar{c} = \left(\frac{2}{3}\right)c_0\frac{(1+\lambda+\lambda^2)}{(1+\lambda)} \qquad \bar{y} = \left(\frac{b}{6}\right)\frac{(1+2\lambda)}{(1+\lambda)}$$

Typically, Wing/Horizontal Tail Aerodynamic Center $\begin{cases} = 0.25\bar{c} & (\textit{Subsonic}) \\ = 0.40\bar{c} & (\textit{Supersonic}) \end{cases}$

S = Reference Wing Area = $\left\{\frac{(c_0 + c_t)}{2}\right\} * b$

AR = Aspect Ratio = $\left\{\frac{b^2}{S}\right\}$

λ = Taper Ratio = $[c_t / c_0]$

b = Span

Given W/S, AR, and λ we can find the following:

$$b = \sqrt{AR * S} \qquad c_0 = \frac{2 * S}{b * (1+\lambda)} \qquad c_t = \lambda * c_0$$

FIGURE 6.20 Some geometric relationships for the wing and empennage.

In military aircraft design takeoff evaluations, it is usual to consider only the takeoff ground run, S_{TOG}, and (from Roskam 1980) an effective equation for this is

$$S_{TOG} = \frac{k_1 \cdot (W/S)_{TO}}{\rho \cdot \{C_{L\max TO} \cdot ((T/W) \cdot k_2 - \mu_G)_{TO} - 0.72 \cdot C_{D_0}} \text{(ft)}$$

where $k_1 = 0.0447$ for jet aircraft,

$= 0.0376$ for propeller aircraft

k_2 = a factor to allow for the reduction thrust with forward velocity on the takeoff ground run

ρ = air density, slugs/ft^3

C_{D_0} = profile drag coefficient including the effects of landing gear and flaps

Wing Sweep

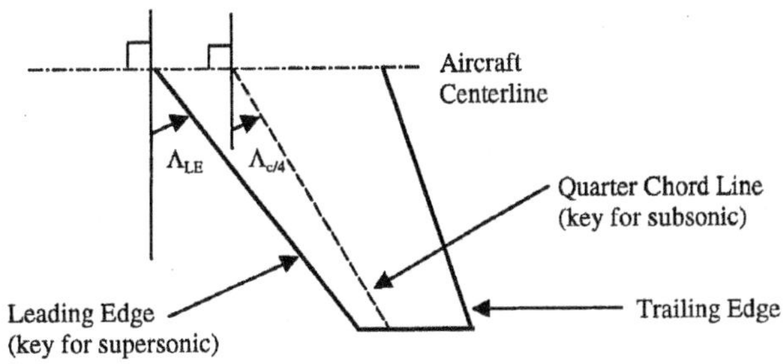

Wing and Horizontal Tail

$$TAN(\Lambda_{LE}) = TAN(\Lambda_{C/4}) + \left(\frac{[1-\lambda]}{AR[1+\lambda]}\right)$$

Vertical Tail

$$TAN(\Lambda_{LE}) = TAN(\Lambda_{C/4}) + \frac{(1+\lambda)}{2AR(1+\lambda)}$$

Vertical Tail Geometric Properties

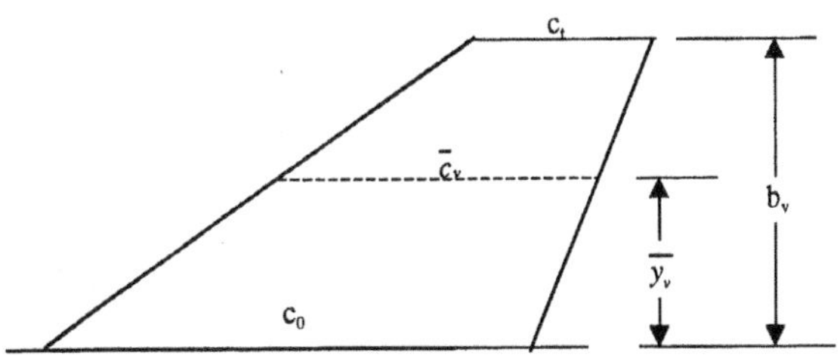

$$\lambda = c_t / c_0 \qquad S_v = [(c_t + c_0)/2] * b_v$$

$$\bar{c}_v = \frac{2}{3}\left\{\frac{c_0(1+\lambda+\lambda^2)}{(1+\lambda)}\right\} \qquad \bar{y}_v = \frac{b_v}{3}\left\{\frac{(1+2\lambda)}{(1+\lambda)}\right\}$$

Geometric Mean Chord, $\bar{c}_{geom} = \frac{c_0}{2}(1+\lambda)$ Aspect Ratio $= \frac{b_v^2}{S_v}$

FIGURE 6.21 Some geometric relationships for the wing and empennage.

μ_G = ground rolling coefficient of friction

$C_{L_{max}}TO$ = lift coefficient at the takeoff configuration

This equation will also permit the determination of combinations of W/S, T/W, and $C_{L_{max}}$ that will generate the design value of takeoff ground run. Typical results for takeoff and landing are shown in Figure 6.19.

Typical ground rolling coefficients of friction, μ_G, are given in Table 6.7.

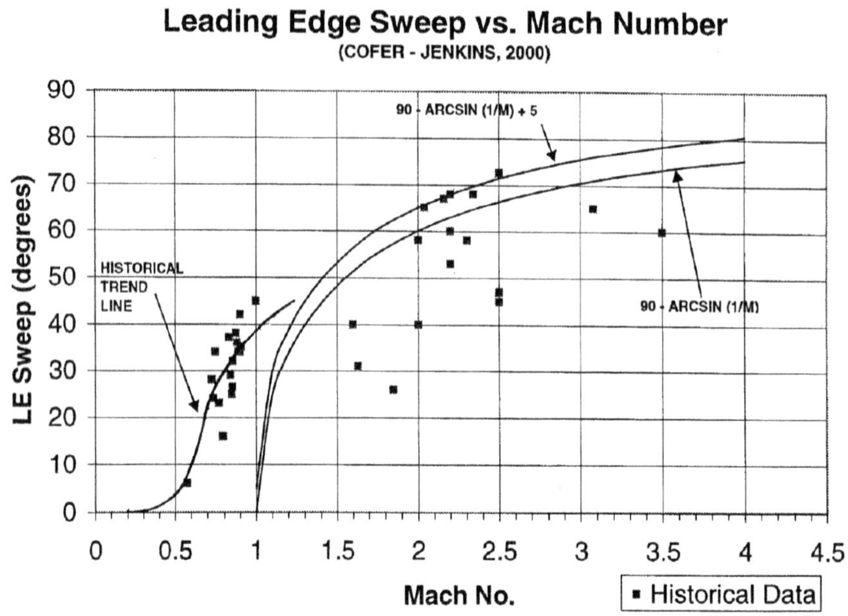

Figure 6.22 Historical leading edge sweep versus Mach number.

Influence of Landing Distance Requirements on Aircraft Design

Landing wing loading, approach speed, deceleration techniques, and pilot techniques all influence the resulting landing distance. There is usually a design landing weight specified for the type of aircraft (Table 6.8), as well as an obstacle height and a set of government requirements. Typically, for CTOL aircraft the total landing distance consists of an air distance and a ground distance. The design aim is to establish the appropriate combination of wing loading and $C_{L_{\max L}}$ necessary to meet the stated landing distance required. It can be shown that for civil/commercial design the landing distance is a function of the square of the approach speed, V_A^2, where $V_A = 1.3\ V_s$; for land-based military aircraft $V_A = 1.2\ V_s$; and for carrier-based aircraft $V_A = 1.15\ V_S$, where V_s = the appropriate stall speed in the landing configuration.

For landing design evaluations the following equations permit the generation of a range of equivalent takeoff wing loadings necessary to meet a given landing distance requirement. For FAR 23 rules, for landing,

$$(W/S)_{TO} = \frac{S_L \cdot \rho \cdot C_{L_{\max}}}{0.3606(W_L/W_0)} \text{ (psf), where } V_A = 1.3\ V_{SL}$$

For FAR 25 rules, for landing,

$$(W/S)_{TO} = \frac{S_L \cdot \rho \cdot C_{L_{\max}}}{0.3559(W_L/W_0)} \text{ (psf), where } V_A = 1.3\ V_{SL}$$

and for military aircraft design, for landing,

$$(W/S)_{TO} = \frac{S_L \cdot \rho \cdot C_{L_{\max}}}{0.3033(W_L/W_0)} \text{ (psf), where } V_A = 1.2\ V_{SL}$$

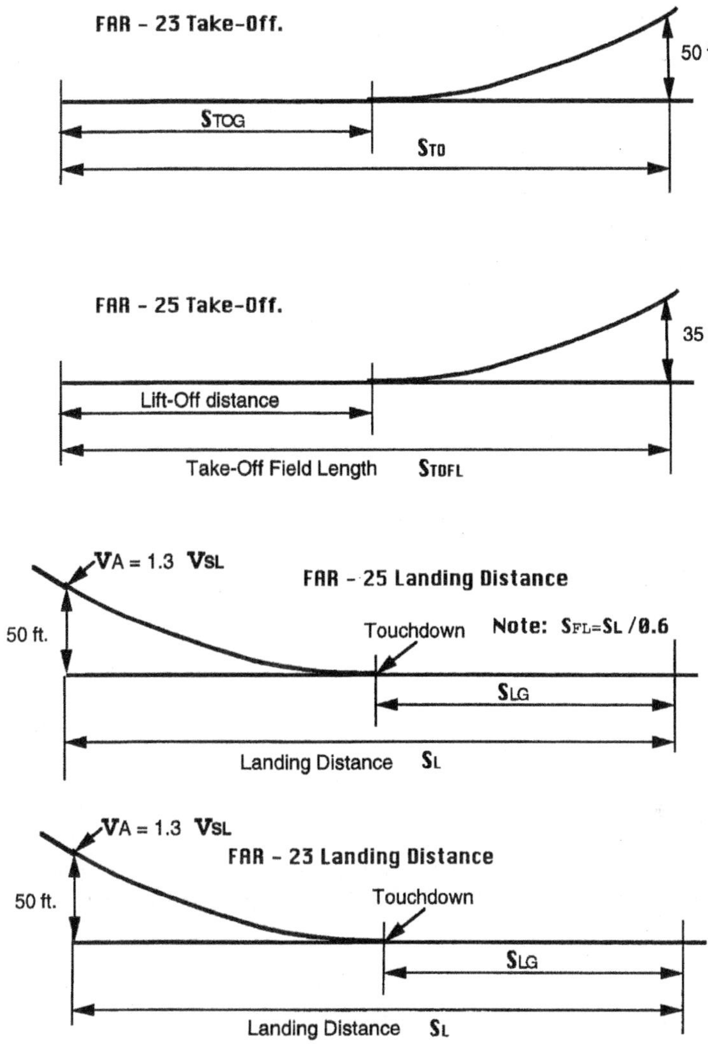

FIGURE 6.23 Takeoff and landing profiles (from Roskam 1980).

but using the FAR 25 ground rules as a good approximation (S = distance, ρ = air density). In all landing design evaluations, V_{SL} is the stall speed with the aircraft in the appropriate configuration such as landing gear down, flaps down, and the engine at the appropriate setting. S_L is defined as the desired landing distance in feet and (W_L/W_0) is the design value of landing weight to gross weight ratio (Table 6.8). No allowance is made in these distance estimates for the reduction in landing ground run due to the effects of reverse thrust, drag parachute, or spoiler application.

Geometric Layout of the Wing

Experience has shown that the magnitudes of the taper ratio (λ), aspect ratio (A), and thickness-to-chord ratio (t/c) are driven by the operational function of the aircraft. The taper ratio chosen is a compromise between structural efficiency (low values) and wing stall (high values). For example,

Surface type	μ_G
Concrete	0.02–0.03
Asphalt	0.02–0.03
Hard turf	0.05
Short grass	0.05
Long grass	0.10
Soft ground	0.10–0.30

Source: Roskam 1980.

Table 6.7 Typical Ground Rolling Coefficients of Friction

Airplane type		Average (W_L/W_0)
Propeller	• Single engine	1.00
	• Twin engine	0.99
Business Jets		0.88
Transport jets		0.84
Military trainers		0.99
Military patrol, bombers, and transports	• Jets	0.76
	• Turboprop	0.84

Source: Roskam 1980.

Table 6.8 Maximum Landing Weight Design Fractions

it can be shown that the spanwise stall point on the semispan of the wing from the centerline of the aircraft is approximately equal to the following:

$$\text{Stall point} = (1-\lambda)(b/2)\text{, where } (b/2) \text{ is the wing semispan}$$

Wing planform layout and fuselage shape are determined primarily by the performance demands of the concept, in terms of Mach number, maximum dynamic pressure, and function or purpose of the concept.

Examples of these planforms are shown in Figures 6.24–6.27.

The wing and sometimes the fuselage will be required to carry fuel. An approximate estimate of the fuel volume available in the wing is given by

$$\text{Fuel volume} = 1.07\,(\Delta c)\left(\frac{S^2}{b}\right)\left(\frac{t}{c}\right)_o \left\{\frac{1+\lambda_w(\tau_w)^{0.5}+(\lambda_w)^{2.0}\tau_w}{(1+\lambda_w)^{2.0}}\right\}\ (\text{ft}^3)$$

where $(t/c)_o$ = thickness to chord ratio of the wing at the wing root

S = wing area (ft²)

b = wing span (ft)

λ = wing taper ratio

$$\tau_w = \left[\frac{(t/c)_{\text{tip}}}{(t/c)_o}\right]\text{, for the wing}$$

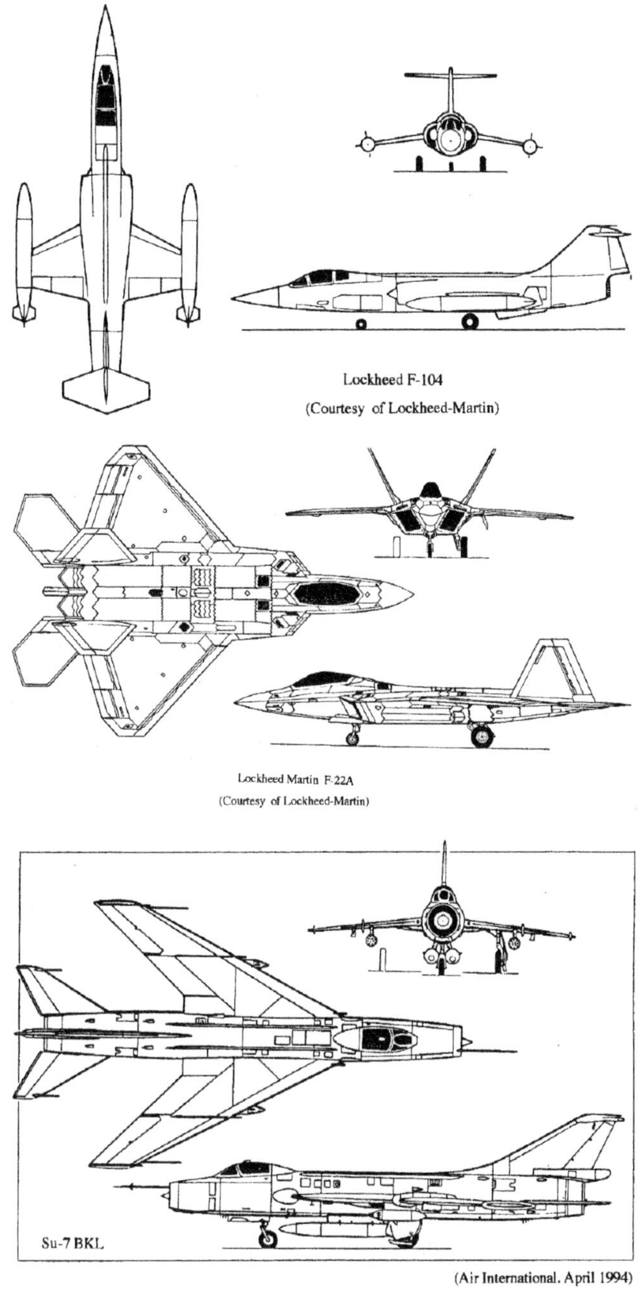

Figure 6.24 Typical fighter aircraft configurations.

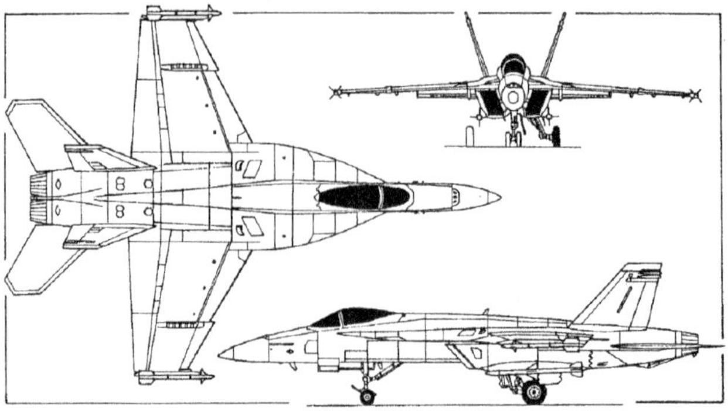

F/A-18E Super Hornet. (Mike Badrocke-Air International)

Figure 6.24 Typical fighter aircraft configurations. (*Continued*)

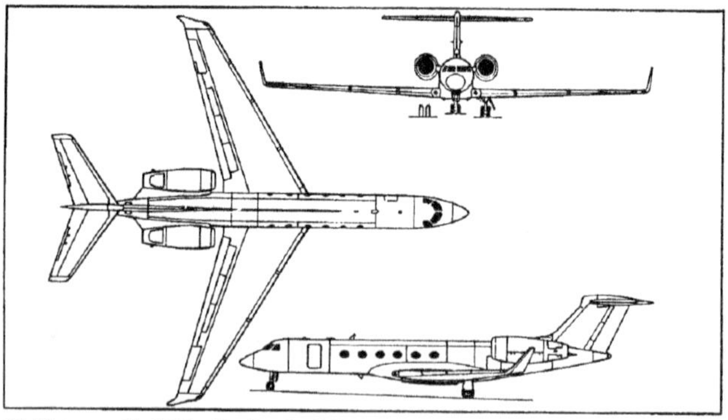

Gulfstream V. (Mike Badrocke/AI)

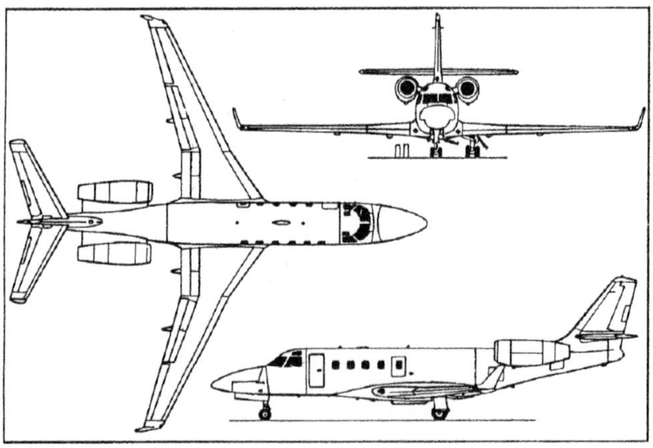

Galaxy Aerospace 1125A Astra SPX. (Mike Badrocke/AI)

Figure 6.25 Typical business jet configurations (Mike Badrocke, *Air International*, April 1994).

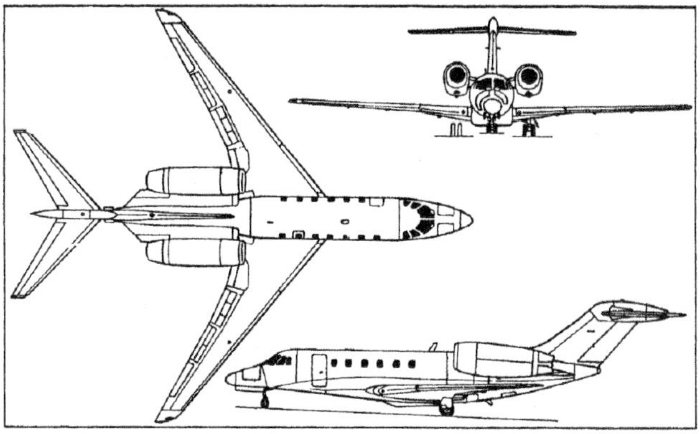

Cessna 750 Citation X. (Mike Badrocke/AI)

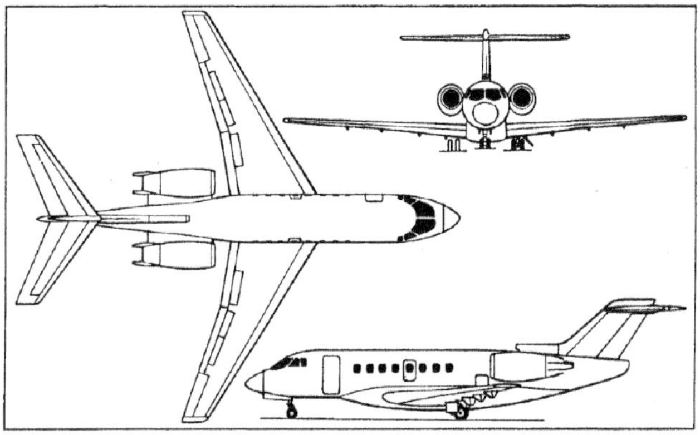

Raytheon Hawker 4000 Horizon. (Mike Badrocke/AI)

Figure 6.25 Typical business jet configurations (Mike Badrocke, *Air International*, April 1994). *(Continued)*

The magnitudes of S and b are those values covering the extent of the wing assumed to carry the fuel. Thus subscript o defines the inboard extent of the wing fuel tank, and Δ_C is the distance between the front spar and the rear spar of the wing which encloses the wing fuel. For example, $\Delta c = (0.75 - 0.25) = 0.50$ for a wing with the front spar at 25% chord and the rear spar at 75% chord. As can be seen, this volume is a function of the placement of the spars. Even if multiple spars are used, this equation is still valid as a first pass estimate.

Fuselage

Designed to house the crew, fuel, and cargo, the fuselage design must also make allowance for loading and unloading of cargo, and for single-engine aircraft the fuselage usually also houses the engine. Drag produced by the fuselage is almost always exclusively parasitic and so must be minimized. The advantage of flying wings is evident in this respect.

Historically the length and diameter of the fuselage have been shown to be a function of the mission of the aircraft, which is usually tied to the cargo or payload required. For civil and commercial aircraft the dimensions of the fuselage are closely tied to the proposed seating layout in terms of numbers of seat rows, called seat pitch, and the number of seats abreast (Figure 6.28).

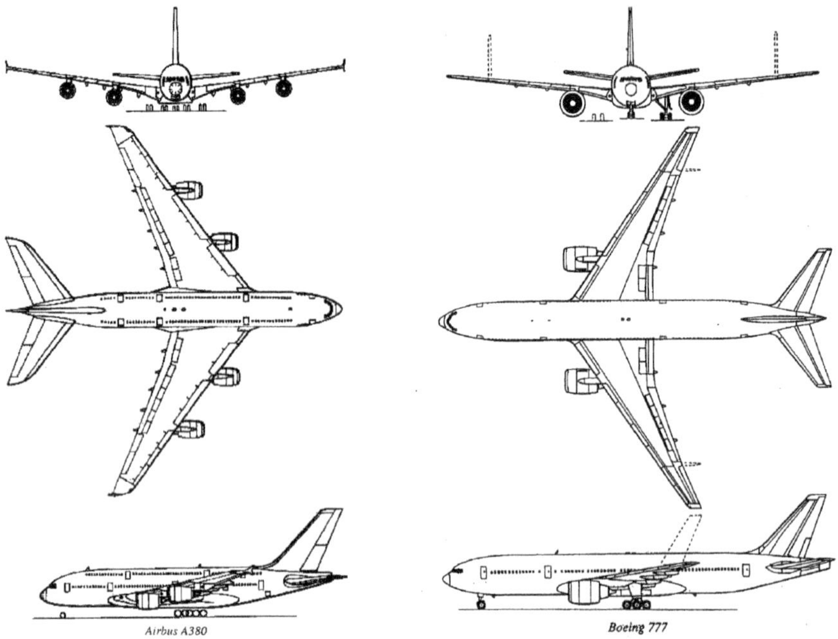

FIGURE 6.26 Typical large commercial transports (Mike Badrocke, *Air International*, April 1994).

These requirements reflect market forecasts and passenger acceptance and vary with the class of accommodations. FAA/JAR regulations determine the number of aisles required in the cabin. The maximum number of seats on either side of an aisle must not exceed three. For military cargo aircraft, loading and unloading of cargo is a paramount design parameter and usually dictates the aircraft configurational layout (Figure 6.6). For fighter/bomber type aircraft with weapons bays the fuselage requires in-flight activated fuselage doors. For low-observable aircraft these doors must open and close with great rapidity.

Lacking explicit length needs, an approximation can be obtained for a range of concepts from Table 6.9.

Empennage

Described as the combined vertical and horizontal tail, the empennage function is to provide trim, maneuver, and stability for a range of aircraft centers of gravity. The two parameters which encompass these functions are the horizontal tail volume ($\overline{V}_H$) and the vertical tail volume ($\overline{V}_V$). These tail volumes are a function of the aircraft type and aircraft engine layout (Table 6.10) and depend on the required center of gravity range to be used on the aircraft (Table 6.11).

$$\overline{V}_H = \left[\frac{S_T l_T}{S_w \overline{c}}\right]$$

where S_T = horizontal tail area

l_T = distance from aircraft center of gravity to aerodynamic center of the horizontal tail

S_W = wing reference area

$\overline{c}$ = mean aerodynamic chord of the wing

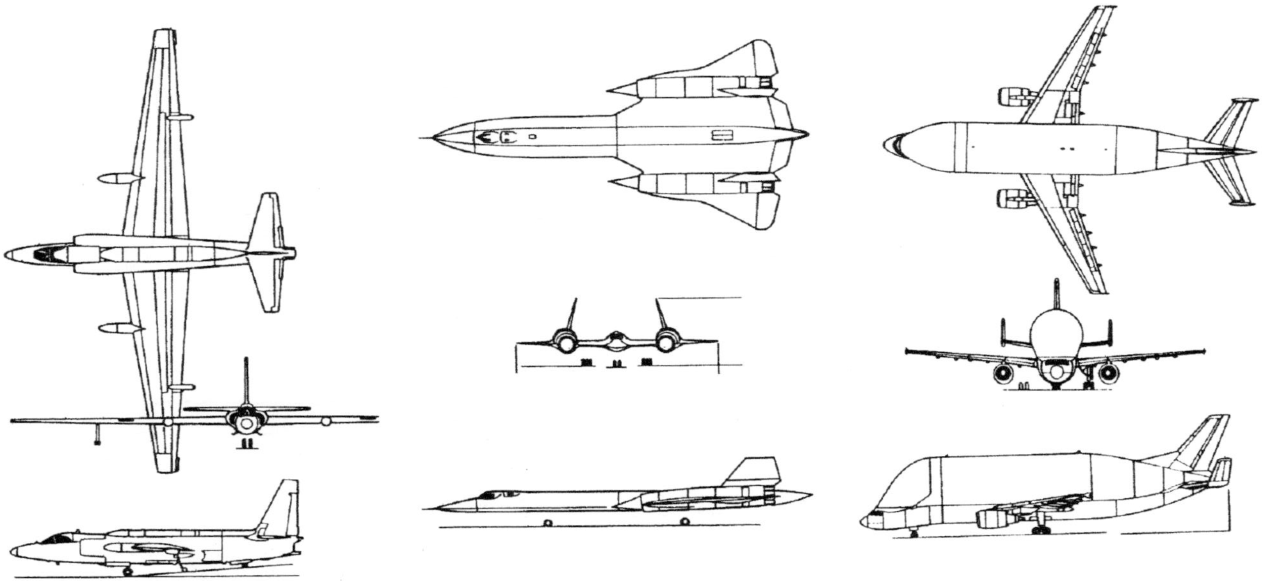

Lockheed TR-1
(Courtesy of Lockheed-Martin)

Lockheed SR-71
(Courtesy of Lockheed-Martin)

SATIC A300-608ST "Beluga" (Mike Badrocke)
(Air International, November 1997)

FIGURE 6.27 Some special-purpose configurations (Lockheed-Martin and *Air International*, November 1997).

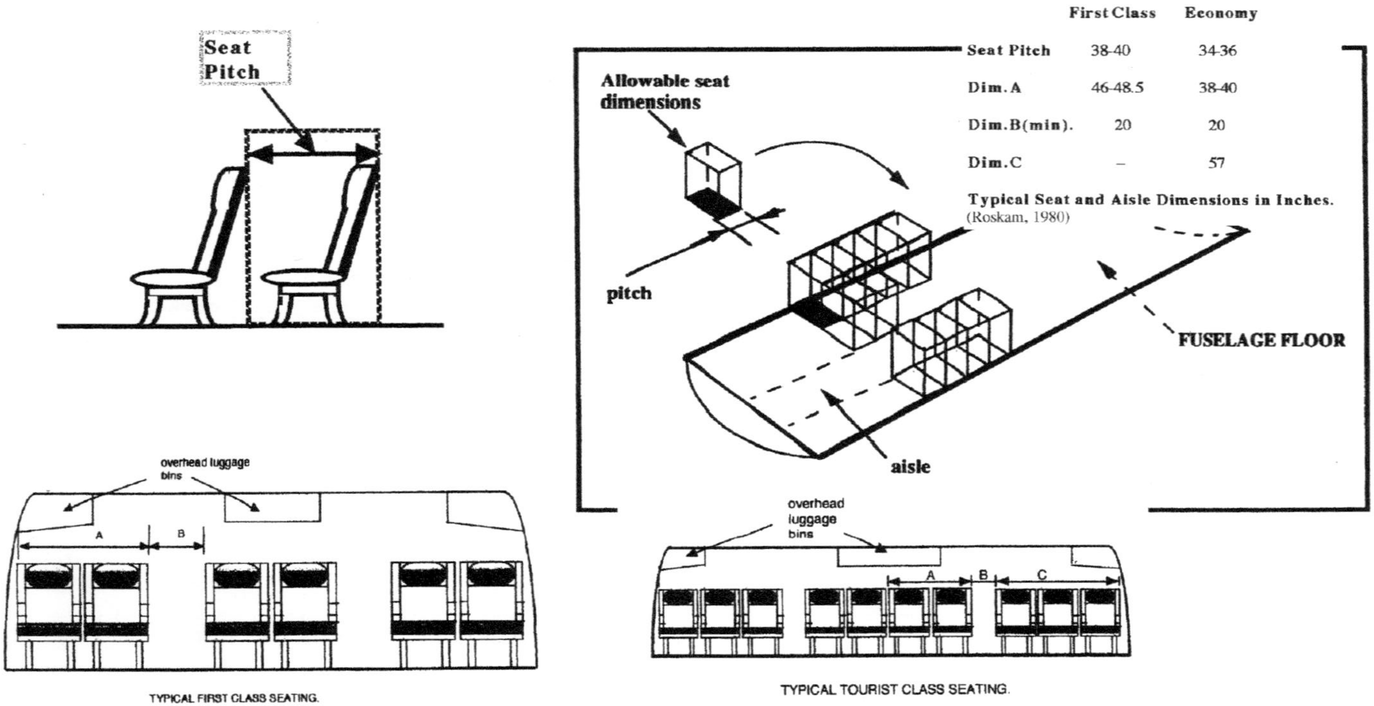

FIGURE 6.28 Passenger seat arrangement.

Fuselage length = $A \cdot W_0^B$ (ft)		
Aircraft type	**A**	**B**
Fighter	1.1106	0.3700
Business jets	1.3286	0.3738
Wide-body jets	1.9311	0.3522

Source: K. Wagoboff.

Table 6.9 Fuselage Length Relationships

Aircraft type	Typical tail volume coefficients	
	Vertical	**Horizontal**
General Aviation		
Single engine	0.04	0.70
Twin engine	0.07	0.80
Jet fighters	0.07	0.40
Military cargo/bomber	0.08	1.00
Jet transports	0.09	1.00

Source: Raymer 1989.

Table 6.10 Examples of Tail Volume Coefficients

Aircraft type	Center of gravity range (fr $\bar{c}$)	Aircraft type	Center of gravity range (fr $\bar{c}$)
Single-engine Prop.	0.06–0.27	Military trainers	0.10
Twin-engine Prop.	0.12–0.22	Fighters	0.20
Business jets	0.10–0.21	Military patrol, bombers transports	0.30
Jet transports	0.12–0.32	Flying boats, amphibious float	0.25

Source: Roskam 1980.

Table 6.11 Some Typical Center of Gravity Ranges as Fractions of Wing Mean Aerodynamic Chord, $\bar{c}$

and

$$\bar{V}_V = \left[\frac{S_V l_V}{S_w b}\right]$$

where S_V = vertical tail area
l_V = distance from aircraft center of gravity to aerodynamic center of the vertical tail
b = wingspan

The geometric relationships of Figures 6.20 and 6.21 are used to generate the remaining empennage parameters.

Landing Gear

Certain aerospace companies now specialize in the design and manufacture of landing gear units. In modern aircraft the total weight of the gear is approximately 3% to 7% of the gross weight of the aircraft, or as high as 38,000 lb (Table 6.12). Correct interface with the airframe (and the gear design itself) is necessary for a successful aircraft design. Interface entails the appropriate ground clearance in the pitch and roll axes and the necessity to avoid lateral tip-over (Figures 6.29–6.31), as well as the ability to provide effective stowage within the airframe if necessary.

A general ground rule is that if the cruise speed of the aircraft is greater than 150 knots, then retraction is recommended to reduce cruise drag.

In the gear design itself the following factors of the gear strut and tire combination need to be addressed: gear vertical, longitudinal, and lateral loads; and gear dynamics, including vertical damping and spring constants and directional stability (sometimes called shimmy). These dynamic

Aircraft type	Total landing gear weight (lb) (Typical)	% (Gear wt/W_0) (Typical)
Propeller • Single engine	191	5.6
• Twin engine	263	5.5
Business jets	659	2.8
	2,011	3.1
Transport jets	25,761	4.6
Military trainers	457	3.9
Military fighters	1,992	6.5 (U.S. Navy)
	1,848	5.9 (USAF)
Military transports • Jets	38,353	5.0
• Turboprops	5,309	3.4

Source: Roskam (1980).

Table 6.12 Total Landing Gear Weight

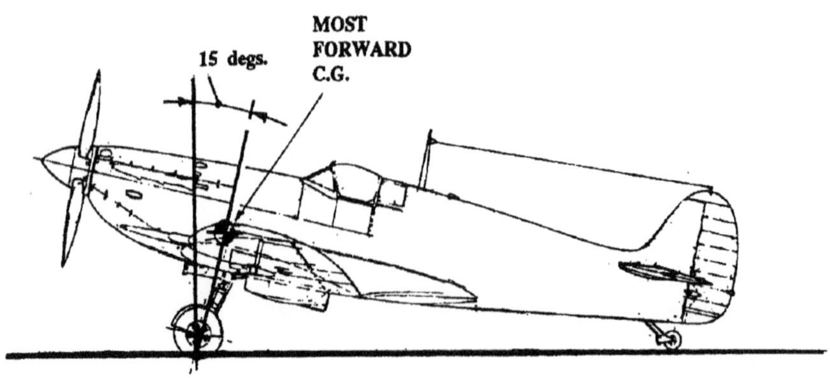

Figure 6.29 Longitudinal tip-over guideline—tail dragger.

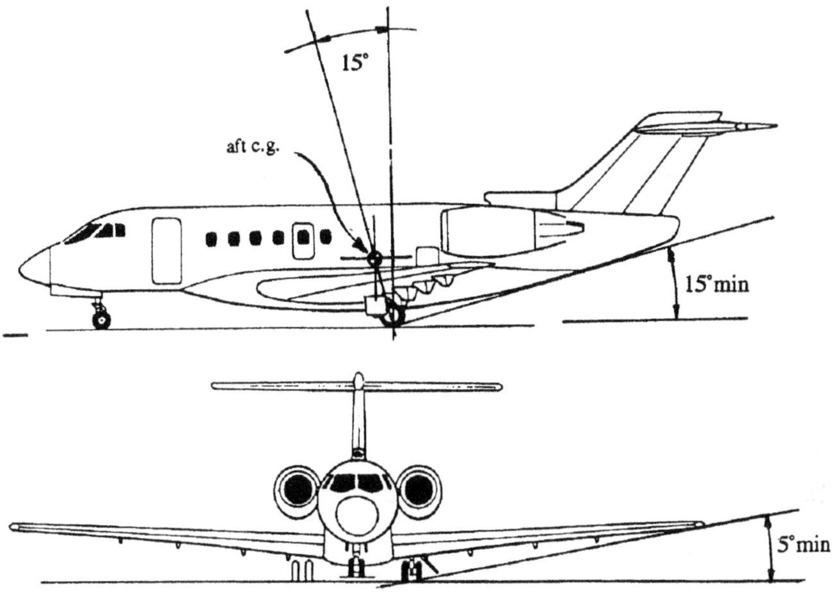

FIGURE 6.30 Lateral clearance and tip-back criteria (aircraft by Mike Badrocke, *Air International*).

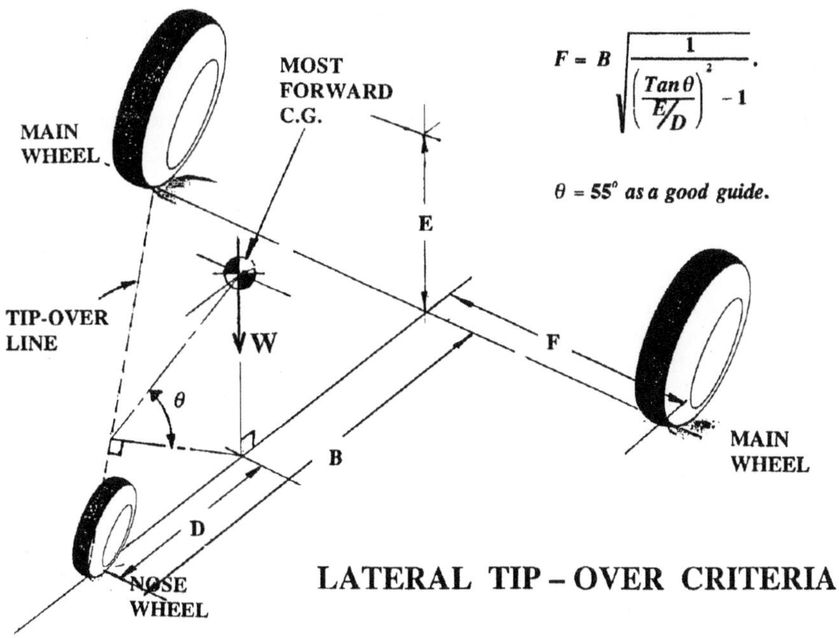

FIGURE 6.31 Lateral-directional tip-over criteria (from Roskan 1980).

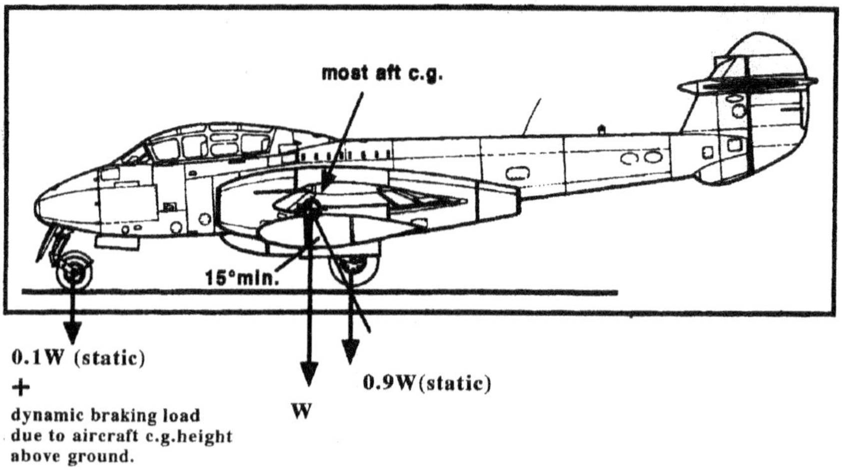

Figure 6.32 Longitudinal tip-back guidelines and wheel load guideline (after Raymer 1989).

requirements are a strong function of the design-to aircraft rate of sink. Longitudinal placement of the gear for a nose-wheel aircraft is a function of the actual most aft center of gravity of the aircraft as well as by the loads to be carried by the nose gear to ensure effective nose-wheel steering (Figure 6.32). Lateral placement of the main gear is determined primarily by the most forward center of gravity of the aircraft and by the desire to avoid lateral-longitudinal tip-over (Figure 6.31).

Sizing of the tires is a function of the tire loads and the tire operating speeds. The number and configuration of the tires on the strut has become a function of the gross weight of the aircraft, runway surface pressure (Figure 6.33), as well as the stowage location in the airframe. Gears are primarily retracted into the fuselage and/or the wings, but occasionally retraction is made into specialized fuselage pods to accommodate special requirements, usually military, and into wing pods for other needs (Figures 6.34 and 6.35).

Weight/Center of Gravity Influence on Design Layout

Every element comprising an air vehicle contributes to the total weight and resulting center of gravity (see Figure 6.36). For all air vehicles fuel burn-off will influence weight and the center of gravity; for military aircraft, weapons delivery will also influence these items. Figure 6.37 illustrates a method for estimating the longitudinal location of the center of gravity, and the following equation locates this center of gravity on the mean aerodynamic chord of the wing from the leading edge.

$$\frac{\text{c.g. location}}{\bar{c}} = \left(\frac{1}{\bar{c}}\right)\frac{\sum_{n=1}^{n=x} l_n w_n}{\sum_{n=1}^{n=x} w_n} - \left(\frac{l_A}{\bar{c}}\right) - \left(\frac{b}{4c_0}\right)\left(\frac{1+2\lambda}{1+\lambda+\lambda^2}\right)\text{Tan}\,\Lambda_{L.E.}$$

where c_o = centerline chord of the wing
λ = wing taper ratio

Vertical c.g. location can be determined by similar methods by rotating the datum line to a horizontal location beneath the aircraft.

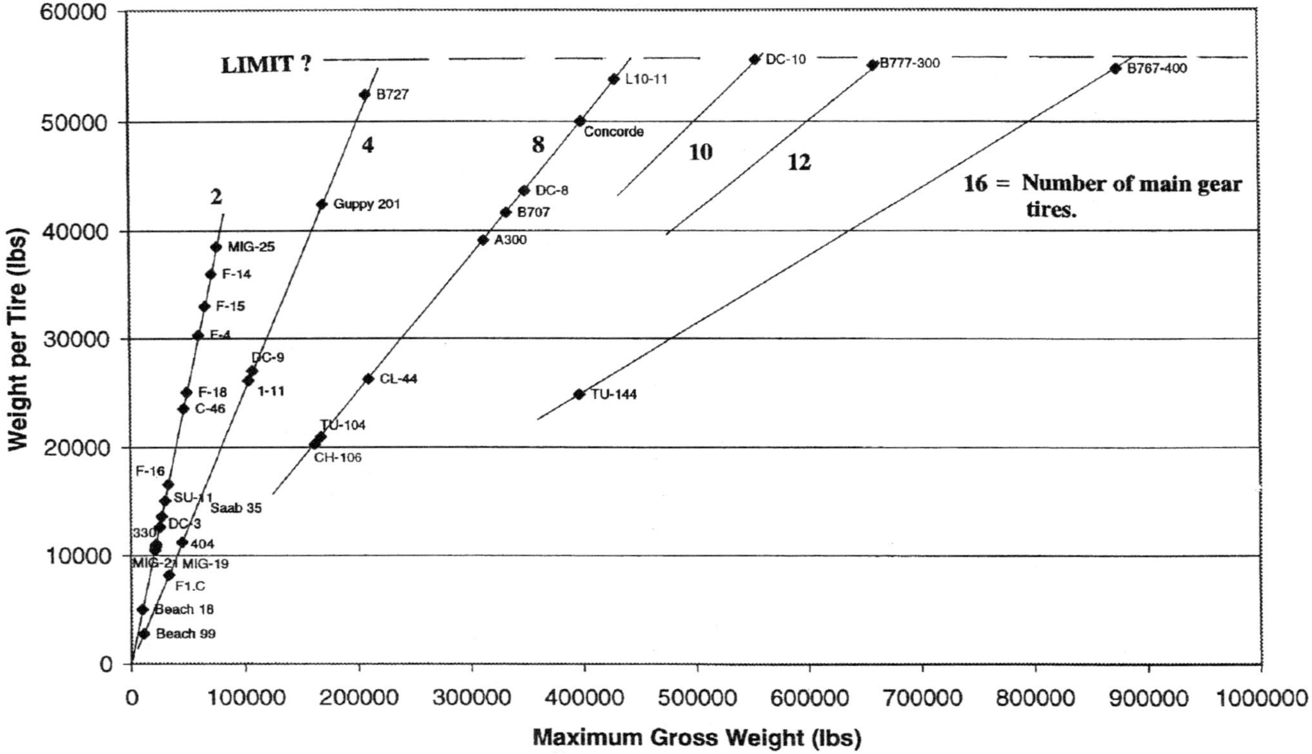

Figure 6.33 Weight per tire versus gross weight (T. Cofer).

Figure 6.34 Special landing gear pods—C 130 (Lockheed-Martin).

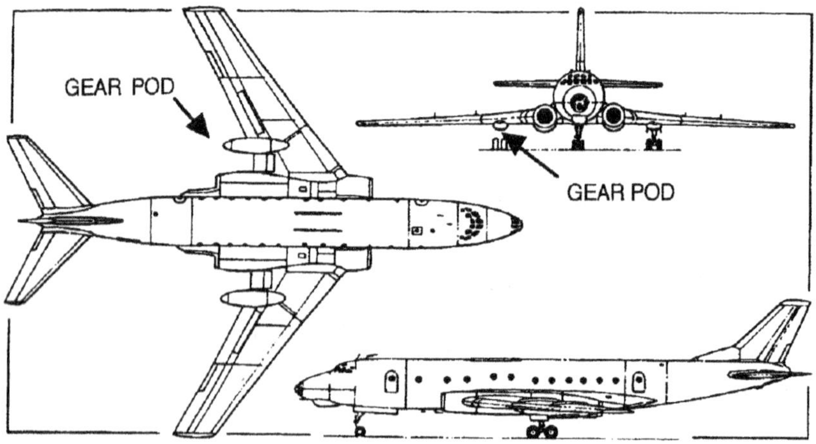

Figure 6.35 Special landing gear pods—Tu 154 (Mike Badrocke, *Air International*).

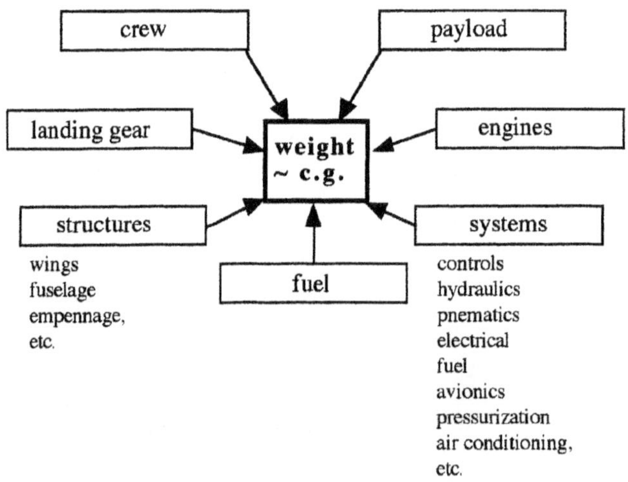

Figure 6.36 Contributors to center of gravity estimation.

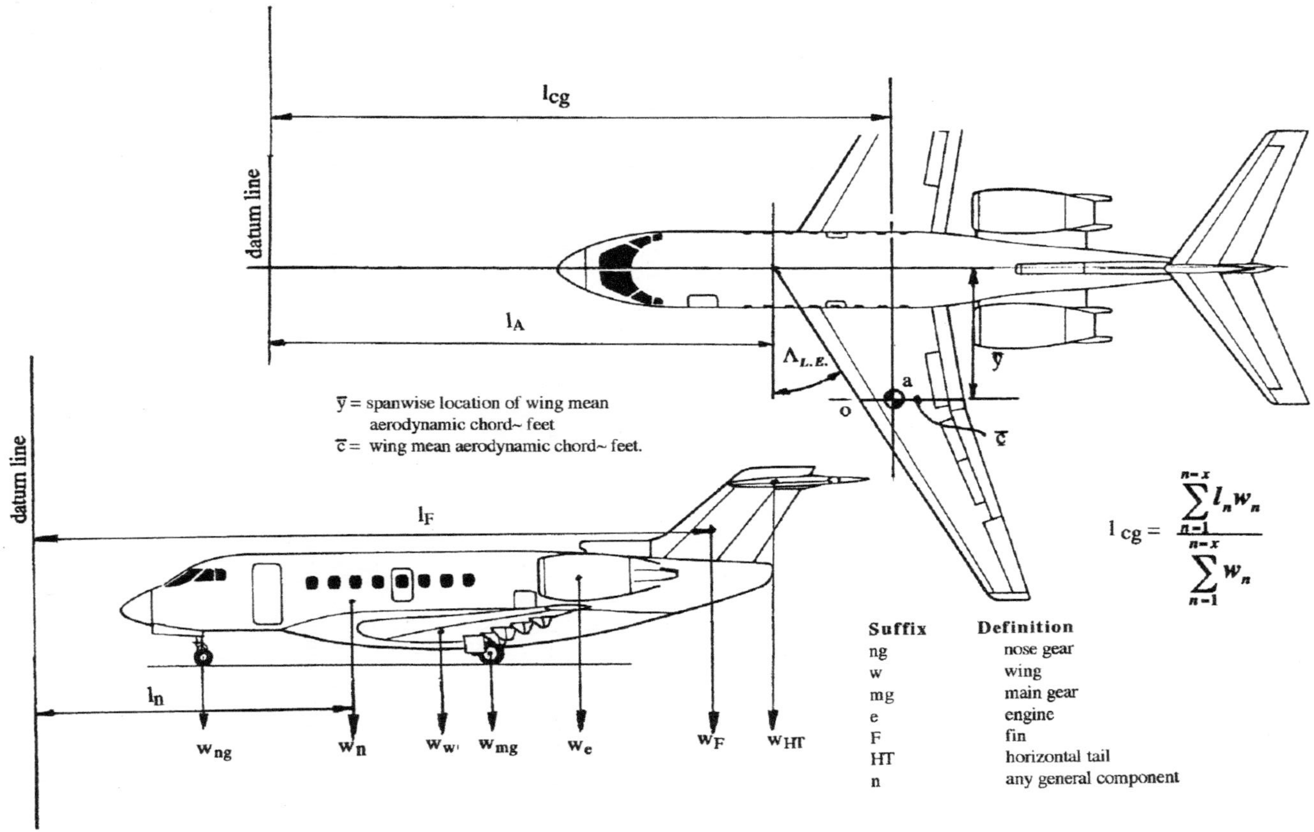

Figure 6.37 Center of gravity estimation (aircraft drawing from Mike Badrocke, *Air International*).

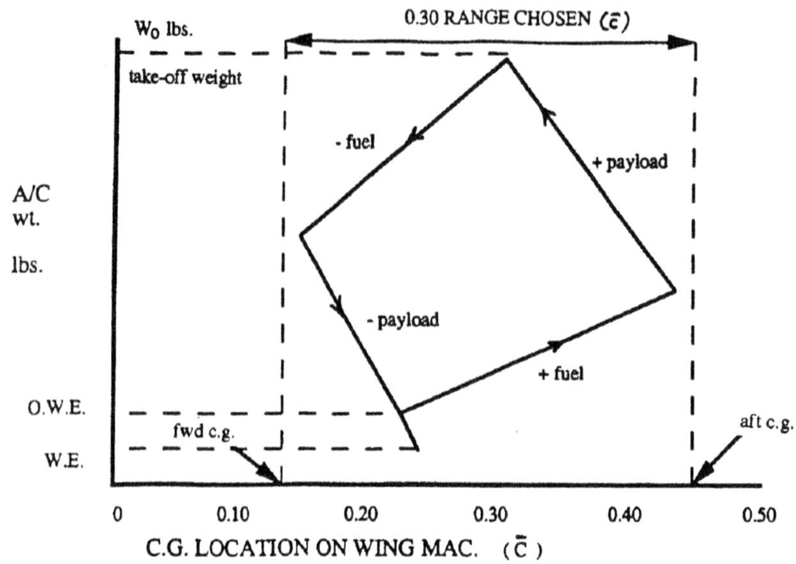

FIGURE 6.38 Typical center of gravity variation during mission.

Figure 6.38 shows a typical variation of center of gravity throughout a simple mission. The required c.g. range is shown to be $0.30\bar{c}$ as an example, and the forward (fwd) and aft c.g. limits must be sustainable by the aerodynamics generated by the horizontal tail or its equivalent.

Ground Clearance

Landing gear design also must ensure that lateral and longitudinal (Figure 6.30) ground clearances of 5° minimum and 15° minimum are observed. The latter promotes safe aircraft rotation on takeoff as well as structural clearance on landing. By observing a 15° aft c.g./main gear line relationship, inadvertent tail sinking can be avoided on rotation during takeoff.

Propulsion

The propulsive method can be propellers, with reciprocating or gas turbine power sources, or fans with gas turbines or just pure jets. The physical characteristics of typical modern gas turbines are given in Figure 6.39 for maximum thrust values up to 110,000 lb.

Typical cruise values of specific fuel consumption, sfc, in pounds of fuel per pound of thrust per hour are 0.35 to 0.38 for thrusts up to about 70,000 lb, and approximately 0.55 to 0.58 for engines with maximum thrusts from 70,000 lb to 110,000 lb. Figure 6.39 also shows the corresponding physical properties for shaft turbines up to 6,000 shaft horse power (shp). For these engines cruise sfc is between 0.45 and 0.55 lb/shp/hour. Typical fan and shaft gas turbine engines are shown in Figure 6.40, which also illustrates the techniques used to extract power to drive the various aircraft systems. Engines are usually mounted to the aircraft structure with a fixed forward mount and a flexible aft mount (Figure 6.41).

In addition to providing power (thrust) for performance needs, the power plant must also generate sufficient power to drive all the other aircraft systems necessary for the concept to perform the design mission. Some typical systems are listed in Table 6.13. Table 6.14 lists the resulting typical system power requirements. Typically, JP-4 fuel weighs 6.5 lb/gallon with a density 49 lb/cubic foot.

	GAS TURBINE	SHAFT TURBINE
overall length (in.)	$2.2941(T)^{0.3849}$	$35.459e^{0.0001.(SHP)}$
maximum diameter (in.)	$0.5649(T)^{0.4775}$	$5.0679(SHP)^{0.2310}$
weight (lb)	$0.7191(T)^{0.8642}$	$1.0287(SHP)^{0.8044}$
	Up to 110,000 lb thrust	Up to 6,000 shp

FIGURE 6.39 Typical physical characteristics of propulsion types (*Aerospace Source Book* 1999).

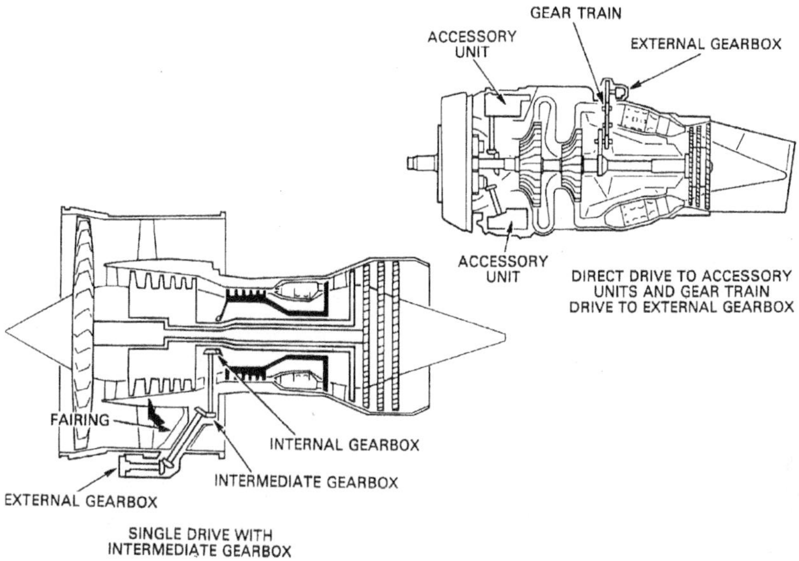

FIGURE 6.40 Typical power plants (reproduced with kind permission of Rolls-Royce plc 1986).

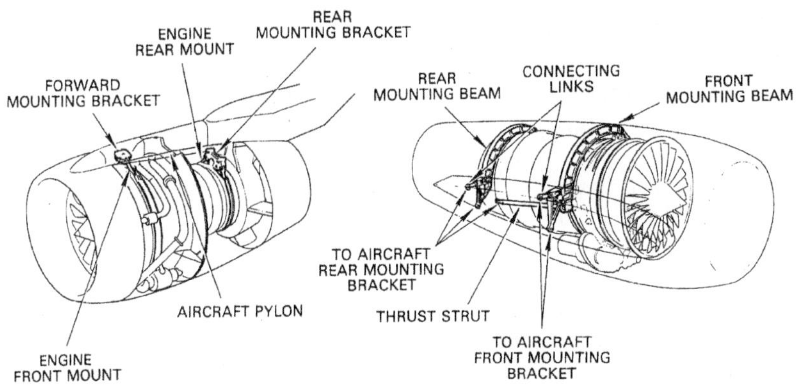

FIGURE 6.41 Engine mounting techniques (reproduced with kind permission of Rolls-Royce plc 1986).

Fuel pumps	Deicing
Hydraulic pumps	Antiicing
Cooling fans	Engine starting system
Heating system	Pressurization of fuel tanks
Cooling system	Pressurization system
radar	Electronic warfare systems

TABLE 6.13 Typical Onboard Systems

Aircraft type	Electrical, shp	Mechanical, shp	Pneumatic bleed slugs/s
Single engine, light a/c	2–4	3–5	0.01 $m_{\dot{a}}$
Twin-engine turbojets/fans	8–10	9–11	0.025 $m_{\dot{a}}$
Jet fighters—attack	100–200	100–200	0.04 $m_{\dot{a}}$
Jet transports—civil	0.0007 W_0	0.0006 W_0	0.03 $m_{\dot{a}}$
W_0 = maximum design weight (lb) $m_{\dot{a}}$ = Engine air mass flow rate (slugs/s)			

Source: Roskam 1980.

TABLE 6.14 Onboard System Typical Power Requirements

6.6 Military Aircraft Design

The basic design principles are the same as given in Subsection 6.1, with the additional need to account for the effects of specific military requirements. In general, these requirements influence the design by changes in the following:

- Engine thrust/installation/shielding
- Fuselage volume/dimensions
- External shaping of aircraft (radar reflectivity)
- External additions to the aircraft:
  - Arresting hook
  - Launch attachments (carrier)
  - Probe for in-flight refueling
- Underwing/fuselage pylons for external stores
  - Internal or external weapons
  - Folding wings
  - External surface treatment such as radar-absorbing material or finishes

All of these induce weight changes, cost increases, and performance changes, and the resulting designs are usually accomplished with the aim of minimizing these effects.

Radar cross-section values are a function of measurement azimuth and shape of the aircraft with many contributors (Figure 6.42). The rapid development of aircraft external shape/radar

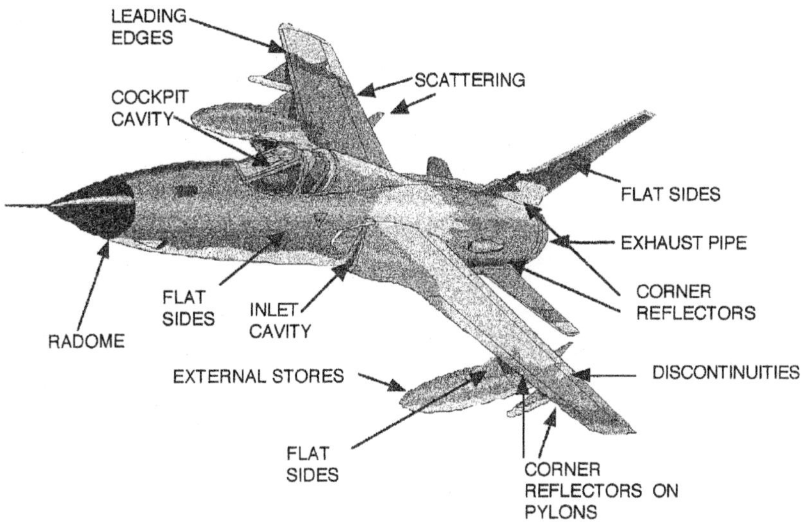

FIGURE 6.42 Major RCS contributors.

reflectivity technology is shown by comparing the early faceted approach to design (Figure 6.43) and the more modern approach (the F-22 in Figure 6.24), each aircraft design coming from the same design organization.

An indication of radar reflectivity, in terms of meters squared, and the published approximate values for a range of aircraft are shown in Table 6.15. These values are meant to be relative only and will vary with azimuth and radar frequency (*Air International*).

Minimizing radar reflectivity can be achieved by a combination of (a) absorbing incoming radar energy and (b) redirecting the reflected wave away from receiving sensors. Design centers on the development and use of radar-absorbing materials and coatings and also the appropriate external shaping and surface integrity of the aircraft. Surface integrity effects primarily involve the avoidance of long surface gaps such as weapons doors and control surface shroud lines because these reflect surface currents unidirectionally, causing reflection spikes. Aircraft external shaping is aimed at minimizing reflection spikes or on concentrating a spike in a given direction.

6.7 Commercial and Civil Aircraft Design

Aircraft that are nonmilitary and that carry passengers and or freight make up this classification. The basic design principles outlined earlier in this section apply, starting from a commercial or civil mission requirement. Since their inception, the number of passengers carried by typical commercial airliners has increased significantly (Figure 6.44). Aircraft cruise speed and cruise altitude have also increased (Figure 6.45).

The difference between military and commercial/civil aircraft empty weight is primarily due to the necessary comfort and conveniences required for the passengers in commercial/civil aircraft. These include galleys, seats, multiple toilets, overhead luggage racks, baggage storage, in-flight entertainment, and cabin pressurization. Figure 6.46 shows a typical cabin versus outside pressure comparison for a typical airliner.

Passenger seating space allowances are a function of the class of travel chosen (Figure 6.28), and are a significant factor in the passenger acceptability of a particular design. Seating space is characterized by seat pitch and seat width.

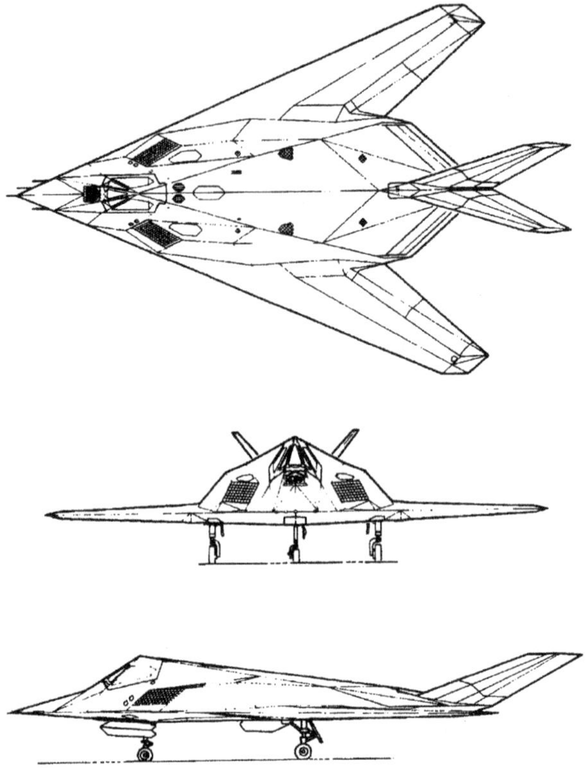

FIGURE 6.43 Lockheed-Martin F-117A (Lockheed-Martin).

Aircraft	Radar cross-section (m^2)
McDonnell Douglas F-15	40.5
Boeing B-52	99.5
Rockwell B-1A	10
Rockwell B1-B	1.02
Lockheed SR-71	0.014
Lockheed F-22A	0.0065
Lockheed F-117A	0.003
Northrop B-2	0.0014

Source: Air International Magazine 1995.

TABLE 6.15 Comparison of Typical Radar Cross-Sections

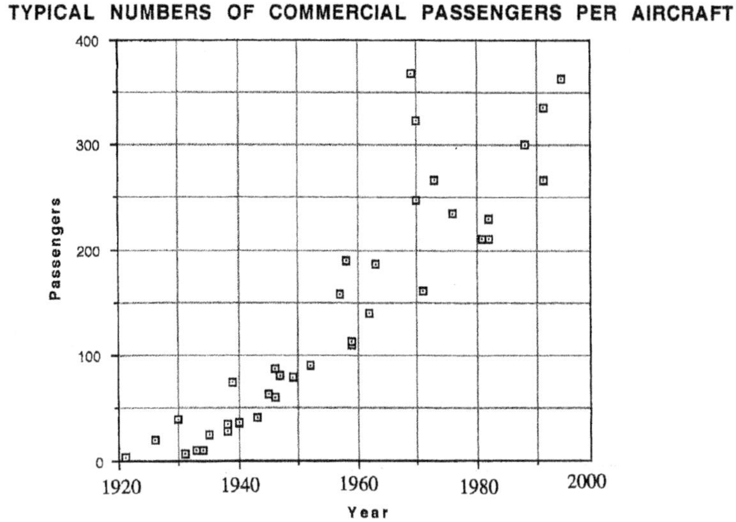

Figure 6.44 Passengers per aircraft trends.

6.8 Life Cycle Cost (LCC)

Life cycle cost is considered to be the total cost of a concept from cradle to grave (Figure 6.47). It is known that 95% of LCC is usually locked in or committed at the end of the preliminary design phase. Extensive methodology exists to estimate LCC for aircraft (see Roskam 1980).

Four elements are considered part of LCC: costs for RDT&E, acquisition and manufacturing, operations, and disposal.

In general, RDT&E costs and acquisition and manufacturing costs are composed of many contributions, each possessing the following format:

$$\Delta \text{ Costs} = f(V)^a \cdot (W^b \text{ampr}) \cdot (\bar{n})^c \cdot (\$ / \text{hr})$$

where V = maximum airspeed (knots e.a.s.)
W ampr = ampere weight (lb)
$\bar{n}$ = number of aircraft built
\$/hr = specific labor rates
exponents a, b, c, represent experience based numbers
f = numerical values based upon allowances for such things as the use of CADAM/CATIA, new materials, special observable requirements, and program security needs

The total RDT&E, manufacturing, and acquisition costs are then

$$\Sigma_1^n \Delta \text{ Costs}$$

where n = number of cost elements involved. Figure 6.48 summarizes some of these cost element sources and shows one derivation of airplane estimated price.

Other factors, such as prototype and production aircraft manufacturing rates, avionics cost, inflation, profit, cost of money, and number of aircraft produced, all influence the individual aircraft price. Figure 6.49 shows a typical variation of aircraft price with number of aircraft manufactured.

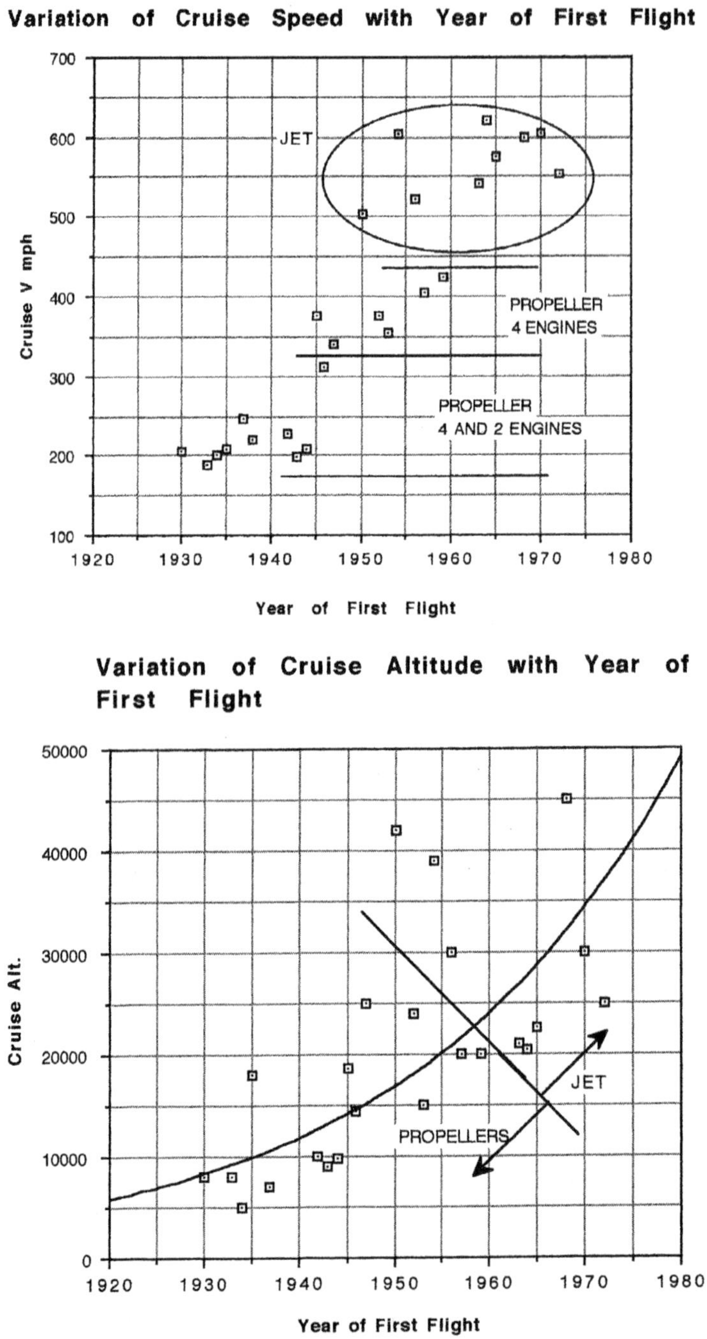

FIGURE 6.45 Cruise speed and cruise altitude trends.

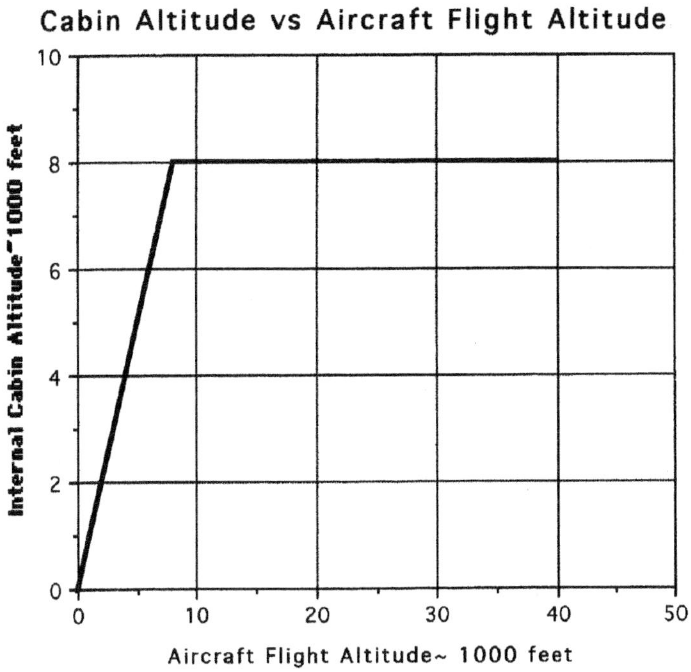

FIGURE 6.46 Typical cabin altitude variation.

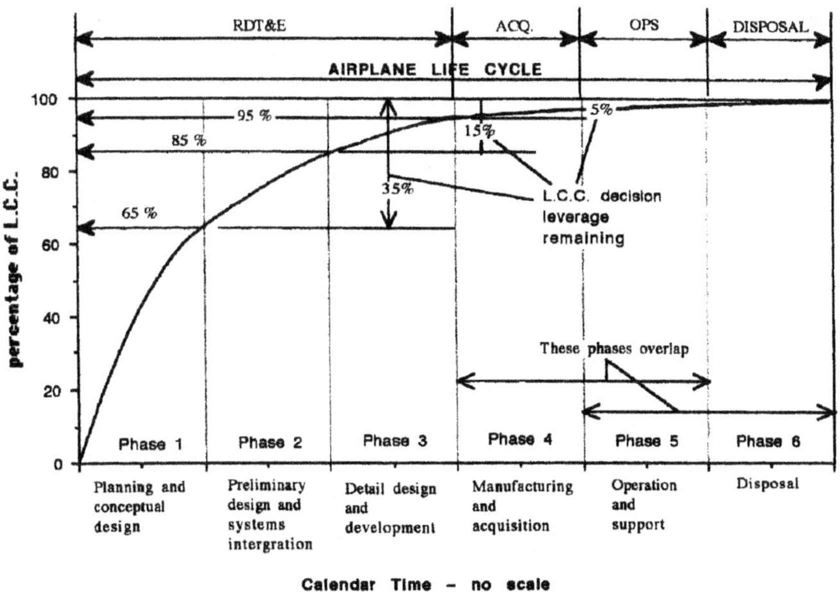

FIGURE 6.47 Overview of life cycle cost for aircraft (Roskam 1980).

$$C_{RDTE} = C_{aedr} + C_{dstr} + C_{ftar} + C_{ftor} + C_{tsfr} + C_{pror} + C_{finr}$$

= Research, development, test and evaluation cost

where C_{aedr} = Airframe engineering and design cost

C_{dstr} = Development support and testing cost

C_{ftar} = Flight test airplanes cost

C_{ftor} = Flight test operations cost

C_{tsfr} = Test and simulation facilities cost

C_{pror} = RDTE profit

C_{finr} = Cost to finance the RDTE phases

$$C_{MAN} = C_{aedm} + C_{apcm} + C_{ftom} + C_{finm}$$

= Manufacturing and acquisition cost

where C_{aedm} = Airframe engineering and design cost

C_{apcm} = Airplane program production cost

C_{ftom} = Production flight test operations cost

C_{finm} = Cost to finance the manufacturing phase

$$\textbf{Estimated airplane price} = (C_{RDTE} + C_{MAN} + C_{MANP})/(N_M)$$

where C_{MANP} = Manufacturing profit

N_M = Number of aircraft made to the production standard

each C_{XXXX} above = $\Delta \text{cost}_{XXXX}$ = a cost element

and $\sum \Delta\ \text{cost}_{XXXX}$ = Total cost

Each appropriate C_{XXXX} will also establish appropriate man hour requirements necessary for the completion of that particular task. A summation of these man hour requirements will establish the foundation for the development of program planning, manpower distribution, and headcount needs for the program.

FIGURE 6.48 Typical Cost Elements (Roskam 1980).

6.9 Commercial Aircraft Operating Costs

Total operating costs (TOC) are composed of the sum of the indirect operating costs (IOC) and the direct operating costs (DOC). These costs can be expressed as either \$ per nautical mile or \$ per block time. (Sometimes ¢ or £ are substituted for \$.) Block time is defined as flight time plus 25 minutes for long haul or flight time plus 15 minutes for short haul (Jenkinson et al. 1999).

The magnitude of DOC is influenced in the design phase of an aircraft, whereas in general the magnitude of IOC is not.

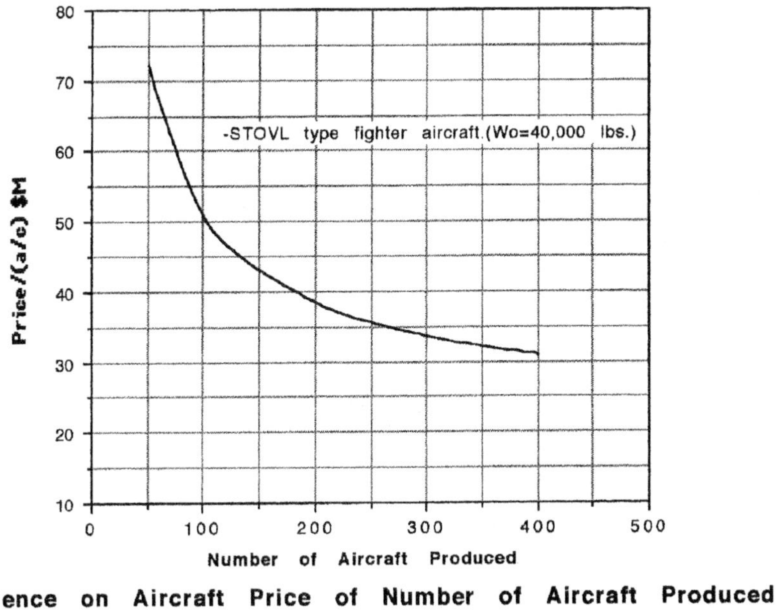

Figure 6.49 Typical influence of production run on aircraft cost.

Elements of each cost component are summarized below (see Figure 6.50).

- Direct operating costs:
- Flight costs:
  - Crew costs (not including flight attendants)
  - Petroleum, oil, and lubricants
  - Insurance costs:
    - Airframe ground and flight risk
    - Passenger liability
    - Third-party liability
    - Cargo damage
- Maintenance costs:
  - Airframe:
    - Labor costs
    - Material costs
  - Engine:
    - Labor costs
    - Material costs
  - Maintenance burden
- Depreciation costs:
  - Airframe depreciation costs, less engine, avionics, and spares
  - Engine depreciation costs

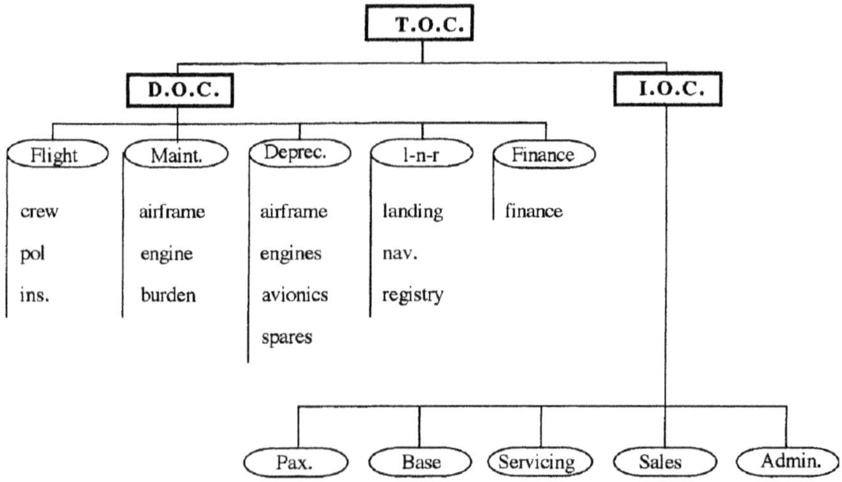

FIGURE 6.50 Elements of total operating costs (TOC) for commercial aircraft.

- Avionics depreciation costs
- Depreciation of spare parts:
  - Airframe
  - Engines
- Landing fees, navigation and registry taxes:
  - Landing fees
  - Navigation fee per aircraft
  - Registry taxes: function of aircraft size and particular state or country
- Finance DOC costs:
  - Function of operators' philosophy in financing fleet
- Indirect operating costs:
  - Passenger services (PAX): meals, cabin attendants, passenger insurance, sales and reservations, baggage handling, security
  - Maintenance (and depreciation) of ground equipment and base facilities
  - Airplane freight and traffic servicing
  - Sales: travel agent commissions; publicity; entertainment
  - Administrative: administrative and accounting and requirements for corporate staffers and their facilities

Equations for the estimation of the contributions of the above components to the DOC and IOC are extensively covered in Roskam (1980). The calculations rely upon historical data for the many classes of aircraft available in the commercial field and are shown to be fairly accurate in comparison with published TOCs.

All estimations of TOC other than those performed by the actual aircraft designers and operators can only be approximate and comparative between concepts. This is because actual inputs and actual mathematical models are by necessity commercially most sensitive and not available in the public domain.

6.10 Unmanned Air Vehicles

Generically, UAVs are flying vehicles that rely upon wing, rotor, or gas bag methods to generate lift and that do not carry a human. Historically, UAVs have acquired changing acronyms starting as remotely piloted vehicles (RPVs) through UAVs to the currently popular uninhabited aerial vehicles. All have a variety of control or navigation techniques such as direct control from the ground, relying upon line-of-sight control, through preprogrammed systems to semiautonomous and autonomous navigation systems. Launch and recovery techniques are ground based or air based, with the former using runways similar to manned aircraft or zero-zero assisted launch and recovery. Catapults, launch rails, and rocket assists are methods used for assisted takeoff, and nets or arresting wires for landing. In-flight parachute recovery and capture by specially modified manned aircraft have also been used. Propellers and jets form standard propulsion methods, with solar cells being constantly evaluated.

Air vehicle design of RPVs follows the principles outlined for manned aircraft, with appropriate (empty weight to gross weight) ratios, with suitable *CDo*'s, and with correct specific fuel consumption appropriate to the type of propulsion used. Table 6.16 shows characteristics of a range of typical RPVs.

UAV missions are many and varied and are constantly under development (see Table 6.17).

The takeoff gross weight of RPVs can range from 90 lb up to 26,000 lb. For weights up to 1,000 lbs, a good design value of $(W_e/W_0) = 0.4095\ (W_0)^{0.0648}$, and for weights greater than 1,000 lb lightweight aircraft design guidelines can be used initially. Propulsion types vary according to mission. For propeller-driven UAVs, the required shaft horsepower can be correlated with W_0, and this shows shp = $0.4109\ (W_0)^{0.7513}$, and this relationship can be used for conceptual design for W_0 up to 2000 lb.

For jet-driven vehicles propulsion can be sized following the manned aircraft design approach. The variety of potential applications for civil and military UAVs results in the choice of electrical or I.C.E. (Internal Combustion Engines) as a major propulsion system. Analysis of production UAVs shows the following maximum mission parameters achieved for each class of propulsion used.

Mission Parameters:

ENDURANCE (hours): Electric 9.0, I.C.E. 45.0

RANGE (km): Electric 200, I.C.E. 2,272

PAYLOAD (kg): Electric 37, I.C.E. 2,948

CEILING (feet): Electric 18,000, I.C.E. 60,000

(These mission parameters are not achieved by any one UAV, but rather are individual maxima achieved in the general class of UAVs shown.)

Airframe Configurations

A range of airframe configurations exist in the current production UAVs. These include fixed wing, blended wing-body, tail sitter, multi-rotor from one rotor through quadcopter to octocopter. These multi-copter concepts are primarily electric motor UAVs.

Unmanned Aerial Vehicles and Drones							
Manufacturer	**Aerovironment Inc.**	**BAI Aerosystems Inc.**	**BTA Automatic Piloting International**	**Bombardier Inc.**	**CAC Systems**	**Finmeccanica**	**Flight Refueling Ltd.**
Designation	HILINE	Exodrone	Sheddon Mk3	CL 327 Guardian	Fox TX	Mirach 20	Falconet
Country	USA	USA	Israel	Canada	France	Italy	England
Mission	High alt Long-end recon.	Multipurpose	Recon/surv/ target acq.	Recon/surv/ target acq.	EW	Surv. Target, acq.	Target
Overall Length (ft)	19	6.1	8.6	6	9.1	11.8	12
Maximum Overall Span (ft)	50	8.1	13.5	13.3	12	13.6	10
Launch Weight (Less Booster) (lbs)	770	9.1	90	770	250	374	450
Power Plant	(1) Akkeman eng. recip.	(1) 8 hp 2-stroke	(1) 100 cc, 10 hp 2-stroke	(1) Williams WTS 117-5	(1) Limbach 22 hp. recip.	(1) 26 hp.	(1) Microturbo TJA24
Payload type	Autop. data link, nav. comp.	TV, IR, EW	TV	Various	Autop., nav., GPS	TV or FLIR/ ELINT	—
Speed (MPH or Mach)	120	100	45–80	98	60–120	138	M 0.72
Endurance	1–2 days	2.5 hr.	6 hr	6.5 hr.	5 hr.	4 hr. +	1 hr.
Maximum Altitude	40,000	—	15,000	18,000	10,000	11,500	—

Manufacturer	General Atomics	Israel Aircraft Industries	Meggitt Defence Systems Ltd.	Oerlikon Contraves AG	Lockheed-Martin Skunk Works	A.S. Yakoviev Design Bureau	Teledyne Ryan Aeronautical
Designation	I-GNAT	Searcher	Spectre	Ranger	DeskStar	Yak-061 (Shimel)	Global Hawk
Country	USA	Israel	England	Switzerland	USA	Russia	USA
Mission	Recon./surv.	Recon./surv/target acq.	Surveillance, EW	Recon./surv./target acq.	Acq./recon/surv.	Recon.	Recon.
Overall Length (ft)	20.8	16.7	8.9	15.1	15	9.1	44.4
Maximum Overall Span (ft)	42.2	23.7	10.7	18.7	65	10.6	116.2
Launch Weight (Less Booster) (lbs)	1,400	818	—	605	8,600	285	25,600
Power Plant	(1) Rotax 914 gasoline	(1) Limbach 550 47 hp.	(1) TTL WAE 342	(1) Goebler-Hirt 42 hp. 2-stroke	(1) Williams Rolls EJ44 tf	(1) Kuznetsov P-032 re	(1) All. AE3007H tf
Payload	EO, IR, SAR	TV & FLIR	CCD camera, FLIR, EW	EO/IO sensor	SAR	TV or IR	SAR/MTI, EO, IR
Speed (MPH or Mach)	161	121	77–150	60–137	288+	87	454
Endurance	40 hr. +	12 hr.	3–6 hr.	5 hr.	8 hr. +	2 hr.	42 hr.
Maximum Altitude	23,000	15,000	23,000	14,765	45,000 +	9,840	65,000

Table 6.16 Typical Unmanned Aerial Vehicles, Drones, and Rotary Wing Aircraft

Rotary-Wing Aircraft							
Manufacturer	**Eurocopter**	**Kamov**	**Bell Helicopter Textron**	**The Boeing Co.**	**Finmeccanica**	**MIL**	**Bell Boeing**
Name	AS365 N4	Ka-29	OH-58D	AH-64D	A-129	MI-28N	MV – 22B
Gross Weight (lbs)	10,580	24,230	5,500	17,650	11,243	25,300	57,000
Empty Weight (lbs)	5,564	—	3,300	11,150	6,989	—	33,140
Rotor Diameter (ft)	41.3	52.1	35	48	39	56.4	38
Max. Length Unfolded Blades (ft)	41.7	52.1	40.7	58.1	47.7	67.3	57.3
Plower Plant	(2) Tur. Amiel 2C ts	(2) TV3-117VK	(1) All. 250-C30R	(2) GE T700-GE-701	(2) LHTEC T800 ts	(2) TV3-117VMA	(2) RR-AE 1107C
Maximum Speed (Mach or MPH)	190.00	160	150	226	185	200	296
Hover Ceiling	12,480	14,100	10,000	14,300	13,600	4,593	5,400
Still-Air Range (nm)	—	255	256	290	305	270	2,416
No. of Passengers	13	16	—	—	—	—	24
Manufacturer	**The Boeing Co.**	**Eurocopter**	**MIL**	**EH Industries Ltd.**	**Hindustan Aeronautics Ltd.**	**Mitsubishi Heavy Industries**	**United Technologies Corp.**
Name	CH-47D	AS332L 1	MI-26TC	EH101	ALH	MH-2000	UH-60L
Gross Weight (lbs)	50,000	18,960	123,000	32,188	9,900	9,920	16,994
Empty Weight (lbs)	23,455	9,920	62,040	19,861	5,500	5,510	11,284
Rotor Diameter (ft)	60	51.2	105	61	43.3	40	53.7
Max. Length Unfolder Blades (ft)	99	61.4	110.7	74.8	52.1	45.9	64.8
Plower Plant	(2) ASE T55-L-712 ts	(2) Tur. Makila 1A1	(2) Progress D-136	(3) GE CT7-6A	(2) Tur TM 333 of LHTEC T-800s	(2) MG5-100 ts	(2) GE T700-Ge-700/701C ts
Maximum Speed (Mach or MPH)	187	173	155	173	180	170	184
Hover Ceiling	10,100	10,663	4,922	12,500	9,850	7,700	9,500
Still-Air Range (nm)	1,450	516	416	864	460	435	373
No. of Passengers	33	24	82	30	12	8	11

Source: Aerospace Source Book 1999.

TABLE 6.16 Typical Unmanned Aerial Vehicles. Drones, and Rotary Wing Aircraft (*Continued*)

Small-area surveillance	Reconnaissance
Large-area surveillance	Command and control
Strike	Deception
Aerial spraying	Electronic intelligence
Communications relay	Jamming
Atmospheric sampling	Air-to-air combat
Monitoring ground sensors	Electronic relay
Aircraft research	Systems research

Table 6.17 Some Typical Potential Missions for UAVs.

6.11 Lighter-Than-Air Vehicles (LTA)

Generically called Aerostats, there are three classes of LTA: nonrigid, semirigid, and rigid airships. All rely on buoyancy for lift, and modern airships use inert helium as the gas. (Early airships used inflammable hydrogen as gas: $\rho = 0.00532$ lb/ft^3 for hydrogen and $\rho = 0.01056$ lb/ft^3 for helium.) Some nonrigid aerostats use hot air for buoyancy; these are called hot air balloons. All airships use ballast weight and gas release for vertical ascent and descent (Figure 6.51). Horizontal thrust from jet engines or propellers controls the forward motion of the airship, and tailfins provide stability and airship angular control. More modern airship designs use vectored thrust from swiveling engines to assist in angular orientation and vertical ascent and descent. All modern airships enclose the gas in sealed gas bags with a controllable vent valve. Ascent speed, operating altitude, and payload weight depend on the percentage of available gas volume in the gas bag to the volume available.

$$\text{Available lift } = F \cdot V(\rho_a - \rho_g) - W_e - W_f - W_c \text{ (lb)}$$

where F = fraction of gas used to volume available
V = volume available in gas bag (ft^3)
ρ_a = density of air at local altitude (lb/ft^3)
ρ_g = density of gas at same condition as ρ_a (lb/ft^3)
W = weight (lb)
$_e$ = empty weight (lb)
$_f$ = fuel weight (lb)
$_c$ = crew weight (lb)

Empty weight is composed of weight for the necessary structure, operating systems, ballast, and propulsion systems. The gas and air in the buoyancy system behave according to Boyle's law and Charles' law. For airships with partially inflated gas bags, ascent causes gas expansion at constant lift until the gas totally fills the available volume. That altitude is called pressure altitude, and further ascent will require venting of the excess gas volume caused by the increased altitude (see Figure 6.52). This venting will reduce the available lift. Operating altitude is reached when the available lift is equal to the total weight of the airship.

Nonrigid Airship

A nonrigid airship is primarily a balloon whose shape is maintained by internal gas pressure. Elongated shapes are used to minimize air resistance in forward flight, and fins, if used, provide directional and longitudinal stability and control. Ventral gondolas are supported by a system of gas bag internal wires which distribute the gondola weight along the dorsal part of the gas bag.

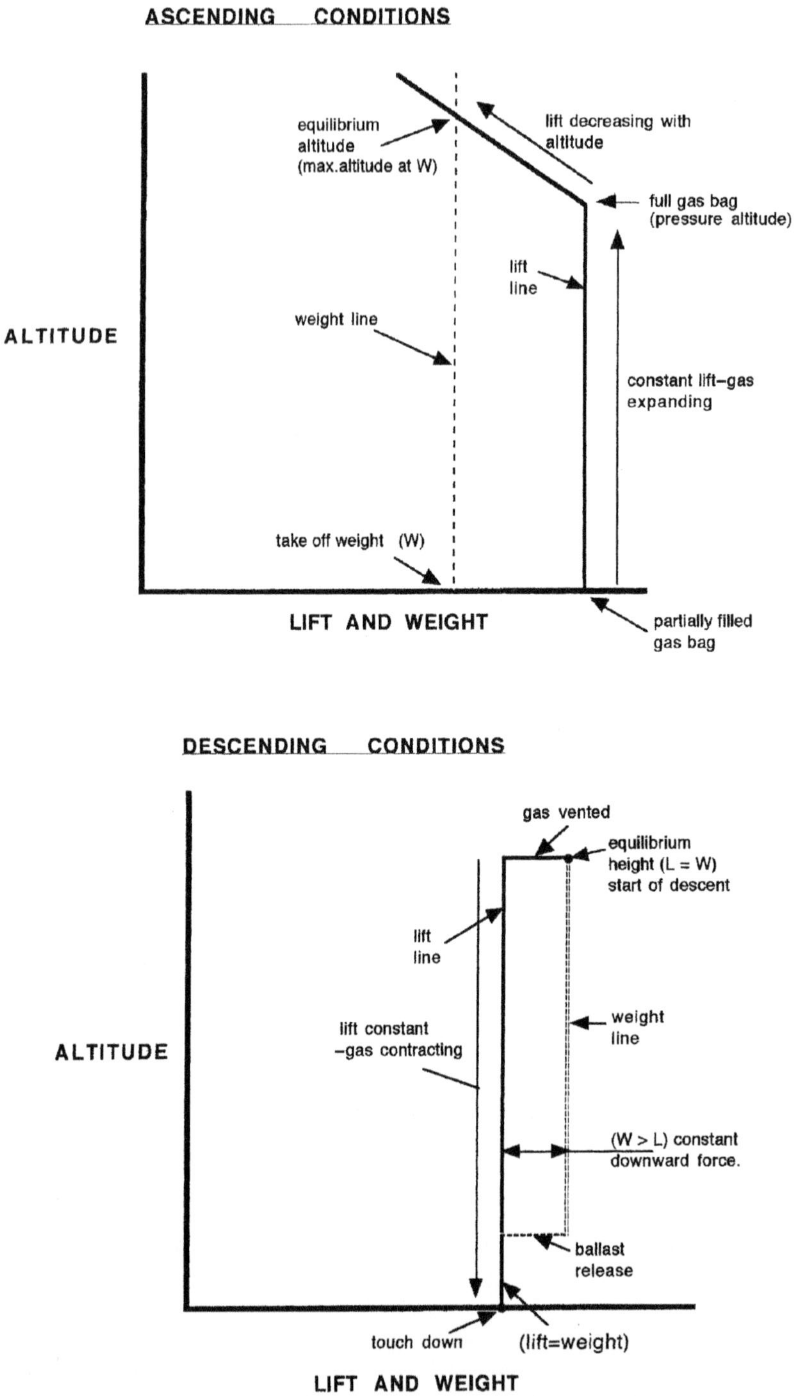

FIGURE 6.51 Ascending and descending conditions for LTA vehicles.

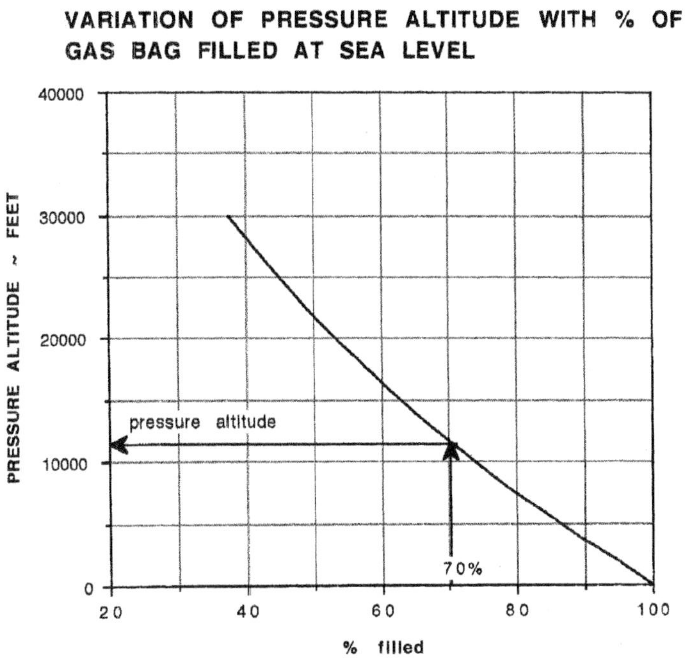

FIGURE 6.52 Pressure altitudes for LTA vehicles.

Fore and aft air-filled ballonets on the lower side of the hull are pressurized to retain external shape at sea level. With increasing altitude, gas expansion causes the air-filled ballonets to be compressed and to vent, and pressure altitude is reached when all ballonets are empty of air. Longitudinal stiffeners are placed on the nose of nonrigid airships to retain nose shape at forward velocities and provide some rigidity for mooring.

Semirigid Airship

A girder-like ventral keel forms a payload and gas bag mounting structure on semirigid airships. The shape of the gas bag is maintained by its pressure. Aft-mounted fins provide directional and longitudinal stability and angular control.

Rigid Airship

The external shape consists of a series of circular frames connected by longitudinal structural members extending from nose to tail of the hull (Figure 6.53). The hull is covered by a layer of fabric material. This shape, designed to minimize drag, is filled with individual (up to 15) gas bags. The structure supports payload compartments, propulsive system, control systems, and crew areas. Internal to the structure are walkways, passages, and ladders permitting access to all critical flight components and to mooring areas.

6.12 V/STOL Air Vehicles

This covers a vast array of configurations, with the principle discriminators being shown in Figure 6.54. Power assist is defined as the integration of thermodynamic units with the lift system. These units can be either the main forward propulsive engines or lift-dedicated engines.

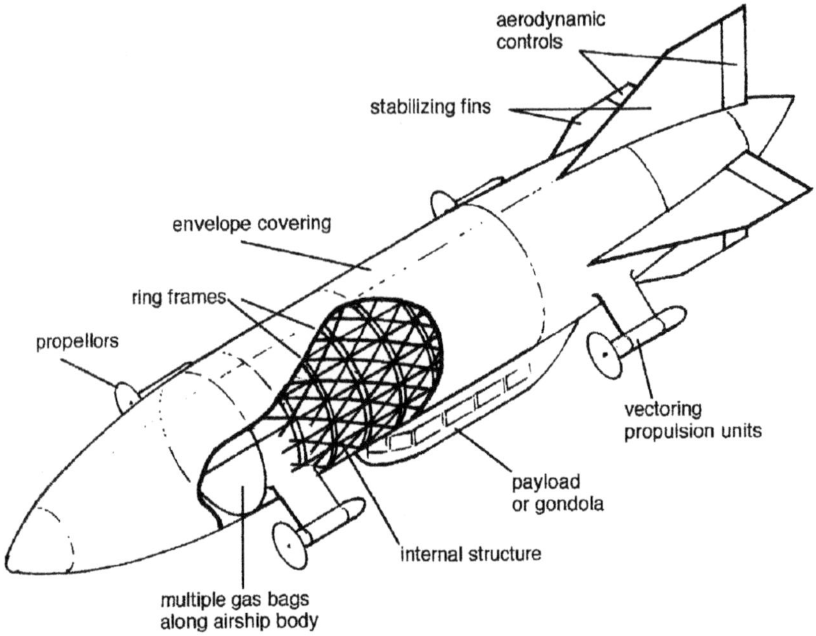

Figure 6.53 Cutaway of typical rigid airship.

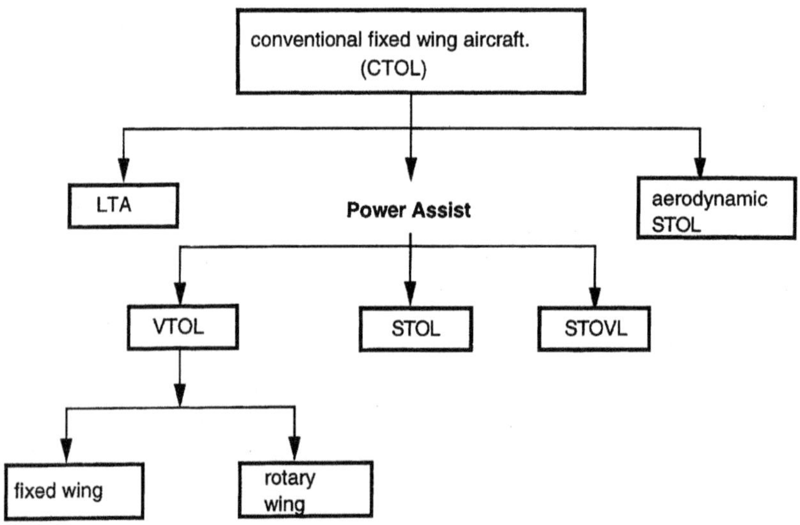

Figure 6.54 CTOL–STOL–VTOL relationships.

Aerodynamic STOL entails the application of leading-edge and trailing-edge devices and the general use of low wing loading. Power-assist STOL and STOVL concepts require the discrete airframe impingement of or vectoring of the main propulsive jet efflux (vectored thrust). Fixed-wing VTOL concepts use either vectored thrust or a combination of vectored thrust and direct-lift thrust generators. The latter can be either direct-lift fans or dedicated direct-lift engines. Rotary-wing air vehicles are designated as either helicopters or autogyros. The former use engines to drive the main lift rotor and tail rotor (Figure 6.55). The latter use a forward-thrusting engine to establish a forward velocity and generate lift by autorotation of the main lift rotor.

Too many combinations of configuration variables exist in the design of V/STOL air vehicles to cover here. The general guidelines for fixed-wing aircraft, however, still apply for their design. These are primarily minimum weight, maximum availability, and minimum acquisition and operating cost. In wing-supported conventional flight of V/STOL concepts, fixed-wing performance and design guidelines apply. In hover and in transition from hover to wing-supported flight, other performance and design requirements need to be met for V/STOL air vehicles.

To provide for vertical linear excursions in hover, it is necessary that the vertical component of thrust, T, be greater than the air vehicle operating weight, W, with T/W on the order of 1.1 as a typical value. Linear and angular control about all three axes is required in hover, with angular control requirements being quoted in terms of angular accelerations at time t = zero. These control requirements are mission dependent and are usually specified by the procuring agency.

Transition is the stage of flight during which the primary lift is gradually transferred from the hover type to the wing-supported type. (Control methods must also be similarly transitioned.) Transition can occur in a climbing, descending, or level flight trajectory. In a level flight transition a primary design goal is to ensure that lift equals weight throughout. This can be achieved by controlled or programmed variations in angle of attack and in pitch angle throughout the transition. Such variations, coupled with corresponding propulsive lift changes, result in what is known as a

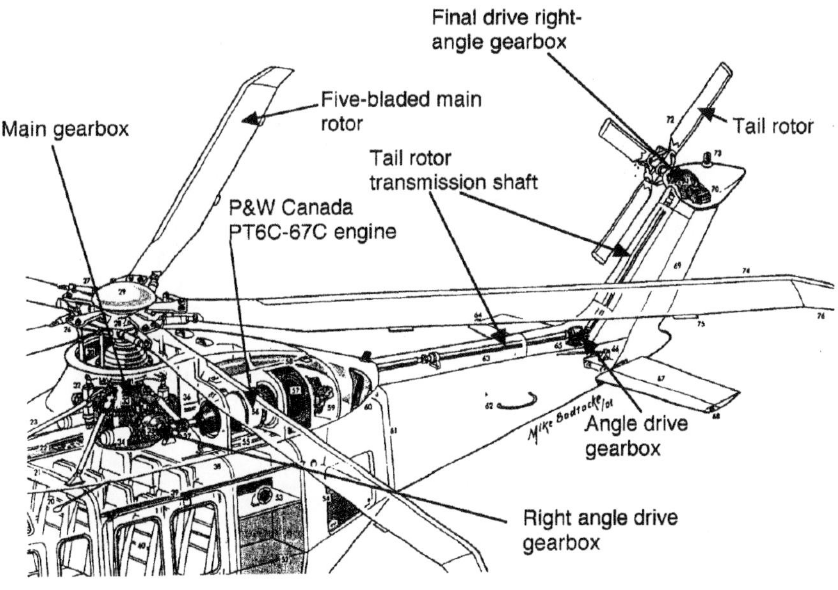

FIGURE 6.55 Typical helicopter power train (after Mike Badrocke, *Air International*).

transition corridor. There are many likely transition corridors for each air vehicle. It is usual to minimize transition time.

Helicopters use the main rotor to generate lift, forward velocity, pitch, lateral, and roll control. These quantities are varied through the use of cyclic pitch control of the rotor and the use of flapping hinges and lag mechanisms on the rotor blades. The tail rotor generates directional or yaw control (Figure 6.56). Multiple rotors are possible, and multiple blades are possible on each rotor (Figure 6.57), with blades in excess of four per rotor being generally reserved for heavy-lift vehicles. Performance capability depends upon the relative magnitudes of drag and power (or thrust) available from the main rotor (Figure 6.58).

The total power required is the sum of the required induced, profile, and parasite powers. Induced power is referred to as the power required to produce lift; profile power is that required to drag the lift-producing rotor blades through the air; parasite power is that necessary to drag the fuselage through the air in forward flight. In hover, at zero forward velocity, induced and profile power requirements predominate, and, as shown in Figure 6.58, in forward flight parasite power increases. The maximum horizontal speed is defined where the power available equals the total power required.

At lower speeds the excess power can be used to establish horizontal acceleration, establish a rate of climb, or set up a combination of both. Figure 6.59 shows a typical rate-of-climb variation with speed for a helicopter. Speeds for maximum endurance and maximum range can be established through the identification of minimum power-required and the tangential point on the power required curve, respectively, as also shown in Figure 6.59.

A criterion for establishing rotor hovering efficiency can be established through the use of the rotor figure of merit, M. Dimensionally,

$$M = \left[\frac{1}{\sqrt{2}}\right]\cdot\left\{\frac{T}{P}\right\}\cdot\sqrt{\frac{T}{\rho\,\pi\,R^2}}$$

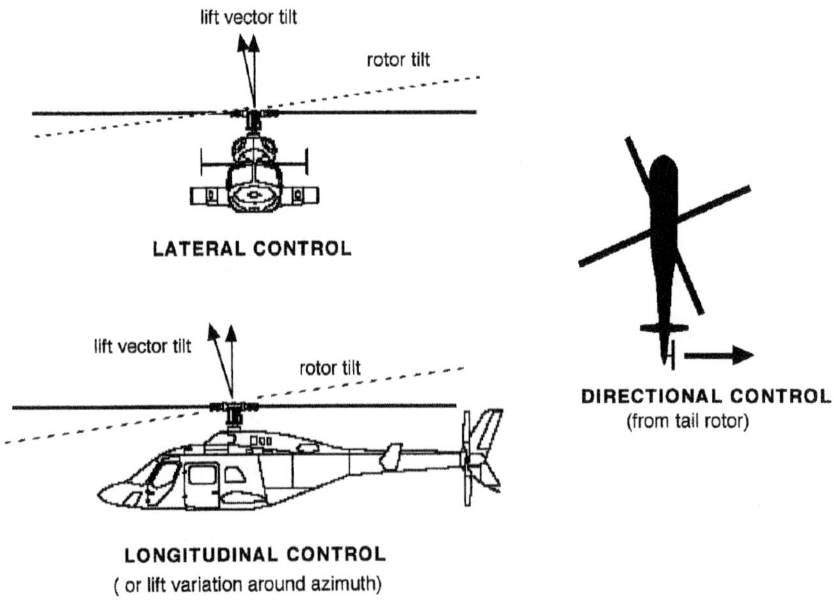

FIGURE 6.56 Typical helicopter control methods.

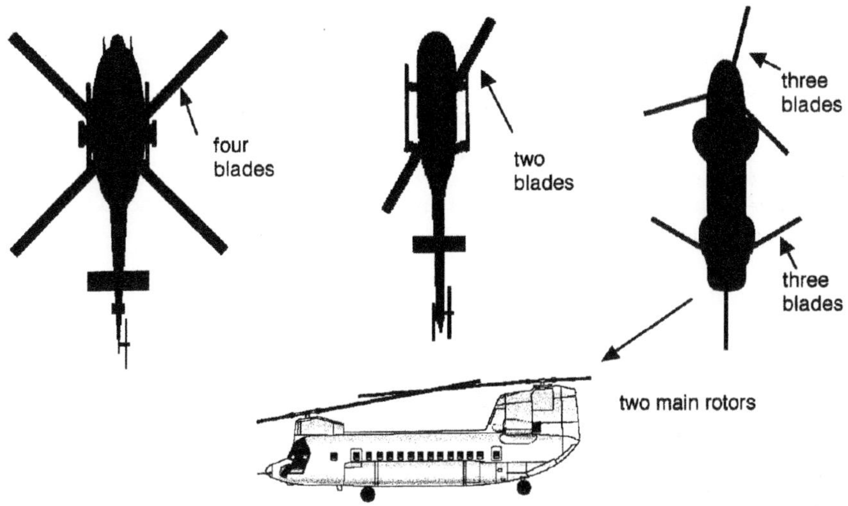

FIGURE 6.57 Typical rotor blade configurations.

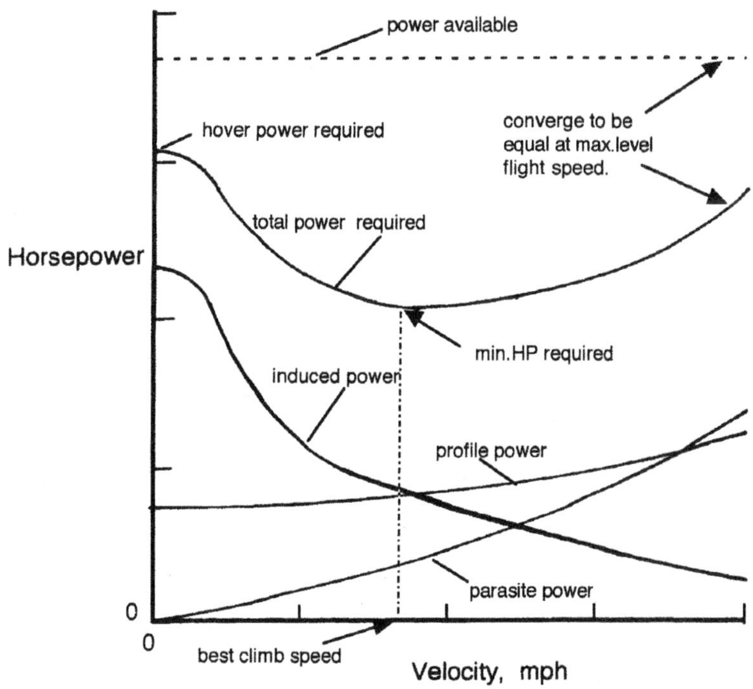

FIGURE 6.58 Helicopter power requirements.

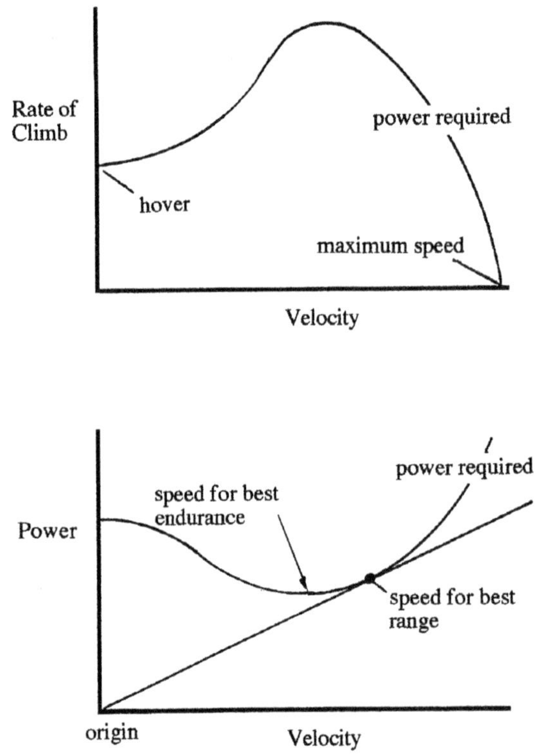

FIGURE 6.59 Some performance aspects of helicopters.

where T = hover thrust (lb)
P = horsepower
R = rotor diameter (ft)
ρ = air density (slugs/ft^3)

For sea-level conditions, this reduces to

$$P.L. = 37.93\left[\frac{M}{\sqrt{D.L.}}\right]$$

where $P.L.$ = power loading (lb/horsepower)
$D.L.$ = disk loading (psf)

An ideal rotor will have $M = 1.00$. This is a rotor with zero profile drag power, minimum induced power (constant inflow), and zero rotational and tip losses. Such a rotor represents an upper limit to the hovering performance of any helicopter. Figure 6.60. shows the variation of $D.L.$ and $P.L.$ for a range of M.

The gross takeoff weight, W_0, of helicopters can be defined as follows:

$$W_0 = W_c + W_{pl} + W_F + W_S$$

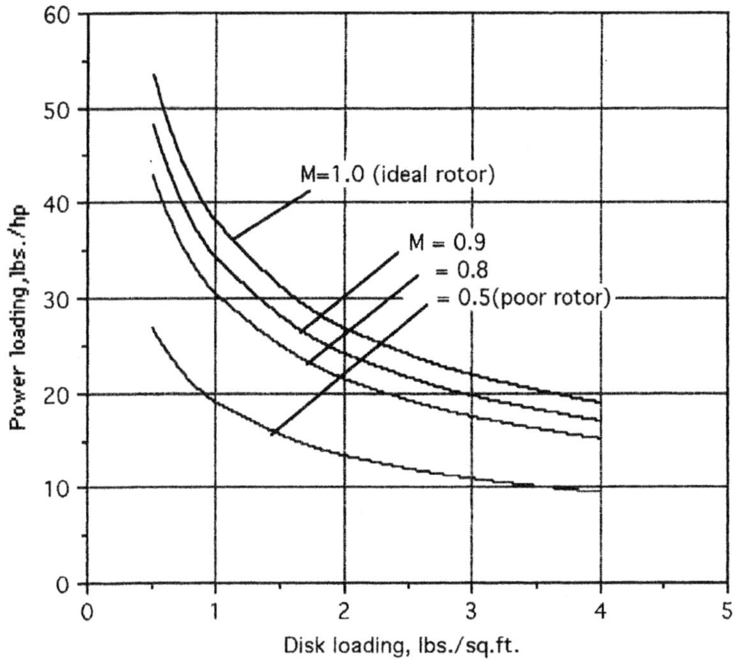

FIGURE 6.60 Power loading and disk loading relationships.

where subscript c = crew weight
subscript $p.l.$ = payload weight
subscript F = fuel weight
subscript S = structure weight

$$\text{Power required, } HP_r = HP_o + HP_i + HP_p + HP_{acc} + HP_{tr}$$

where subscript o = profile power
subscript i = induced power
subscript p = parasite power
subscript acc = accessory power
subscript tr = tail rotor power

If special on-board mission oriented payloads are required, then an additional power requirement can be added.

Helicopter design then centers on the establishment of suitable inputs to these two equations to reflect current technology. The main driver in the design of the helicopter is the rotor aerodynamics, structure, mechanics, and hub. Typical helicopter design data are shown in Table 6.16.

For pure vectored thrust V/STOL concepts, without exhaust impingement on the airframe, it is possible to shorten the takeoff ground roll considerably by distributing the total thrust between a lift component and a forward air vehicle accelerating component. Figure 6.61 shows the assumed system of forces acting on an aircraft with thrust deflected through an angle measured from the horizontal. If ground rolling friction and aerodynamic drag are ignored, a good approximation can be obtained of the effectiveness of vectored thrust in reducing ground roll in a rolling takeoff. The equations of Figure 6.61, developed from the force diagram, are graphed in Figure 6.62 which

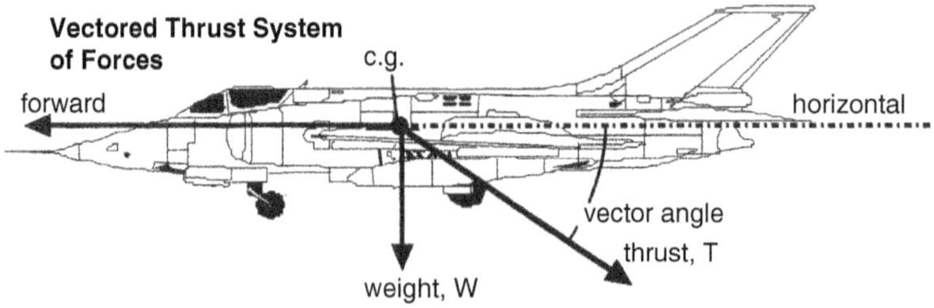

$$\left(\frac{S_{g\theta}}{S_g}\right) = \left(\frac{1}{\cos\theta_T}\right) - \left(\frac{T}{W}\right)\tan\theta_T$$

$$\left(\frac{S_{g\theta}}{S_g}\right)_{opt.} = \sqrt{1-\left(T/W\right)^2} = \cos\theta_T$$

Where θ_T = thrust deflection angle measured from the horizontal,

$\left(\frac{T}{W}\right)$ = thrust to weight ratio,

S = ground roll dist., subscript, θ implies dist. with thrust deflected through angle θ_τ

S_g = dist. with thrust undeflected

FIGURE 6.61 Vectored thrust system of forces.

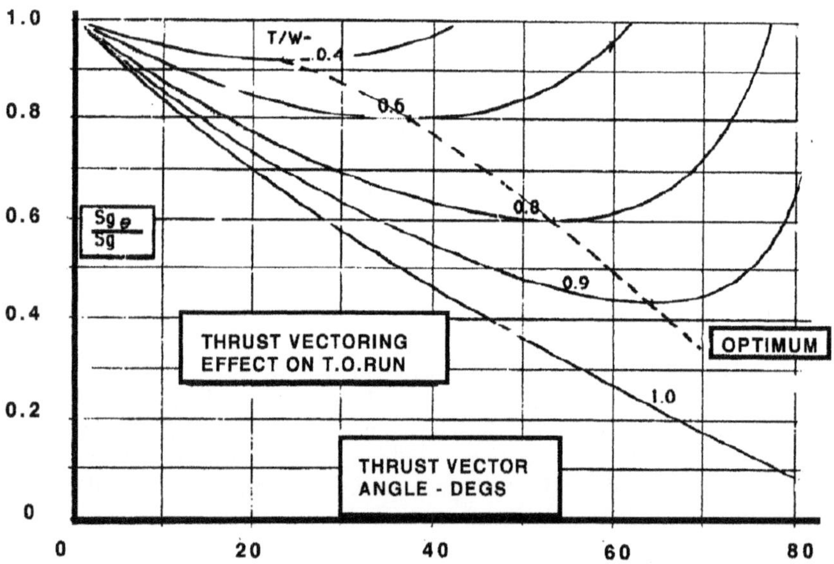

FIGURE 6.62 Takeoff ground run with vectored thrust.

illustrates the effectiveness of vectoring with a constant vector angle. The optimum relationship between T/W and thrust vector angle is also shown.

6.13 Performance

Axes System

For the purposes of analysis and design, a system of orthogonal axes is assumed to exist in the airframe (Figure 6.63). Aircraft rotational motions are then measured relative to these axes and are termed *longitudinal,* about the y-axis; *directional,* about the z-axis; and *rolling,* about the x-axis. Performance measures are usually established in the longitudinal plane, which involves motion fore and aft, along the x-axis, and in the pitching plane, that is, nose-up and nose-down rotation about the y-axis. Under these rules it is usual to establish a force diagram (Figure 6.64) from which a series of equations can be set up to establish performance capability.

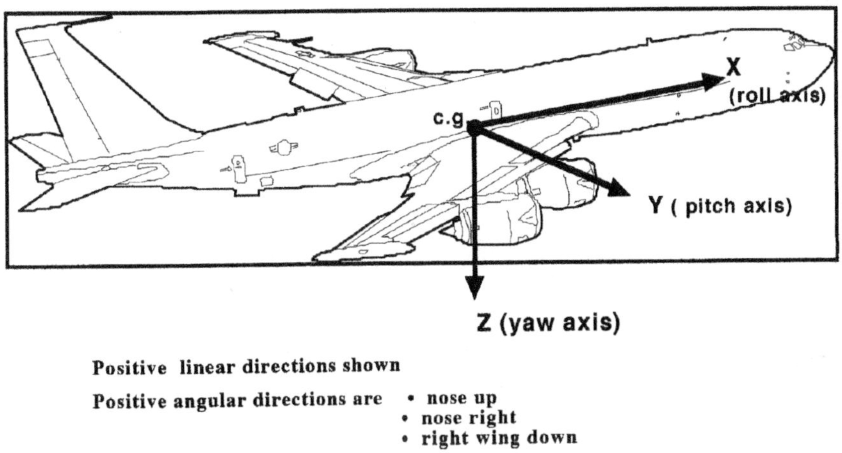

FIGURE 6.63 Axes system.

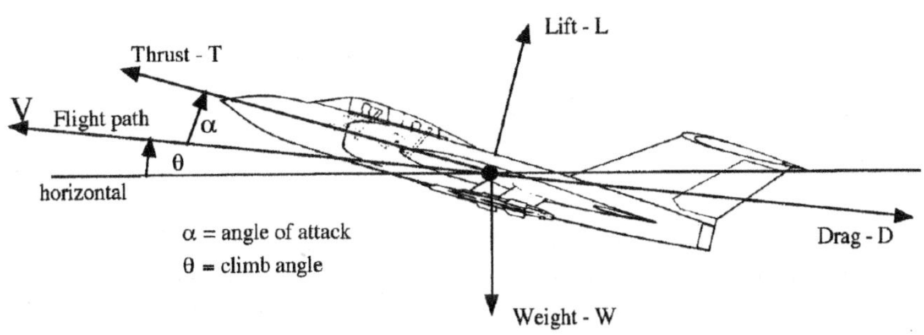

FIGURE 6.64 Force diagram.

For equilibrium in straight flight the following equations can be established from Figure 6.64:

$$F_x = T\cos\alpha - D - W\sin\theta, \text{ force parallel to the flight path}$$

$$F_z = W\cos\theta - L - T\sin\alpha, \text{ force perpendicular to the flight path}$$

which for steady, level, unaccelerated flight, when a small angle of attack is also assumed, becomes

$$T = D \text{ and } L = W$$

where n = load factor
θ = climb angle
α = angle of attack

Drag

For the complete aircraft, drag is composed of three major components: profile, induced, and wave drag. Profile is made up of skin friction drag and form or pressure drag. Induced drag is generated by the backward inclination of the lifting surface vector, caused by tip vortices. Lifting surface definition includes wing and empennage surfaces; the latter drag is sometimes referred to as trim drag. Wave drag is caused by the formation of shock waves and is sometimes referred to as compressibility drag.

For subsonic aircraft, the total drag coefficient, *CD*, is given by

$$C_D = \frac{\text{Profile drag}}{q \cdot s} + \frac{\text{Induced drag}}{q \cdot s}$$

or

$$C_D = C_{Do} + \frac{C_L^2}{\pi A \cdot e}$$

where C_{Do} is the profile drag coefficient and C_{Di} is the induced drag coefficient. Figure 6.65 shows a typical variation of total drag with airspeed. Minimizing profile drag is a key factor in establishing maximum performance. Figure 6.66 shows typical values of C_{Do} for a range of aircraft, and these remain essentially invariant with Mach number until the formation of shock waves (near mach number = 1.0), at which point C_{Do} begins to increase. Delaying this increase to the highest possible Mach number is a good design aim. The onset of this increase in drag is a function of the cross-sectional area distribution of the complete aircraft. The addition of external stores to military aircraft will increase C_{Do}.

Figure 6.67 shows how the total drag will vary with altitude, and, as can be seen, minimum drag magnitude does not vary with altitude, but the speed for minimum drag increases with increasing altitude.

Thrust

Thrust can come from a jet engine or from a propeller. In each case thrust will vary with altitude and airspeed. For jet engines the aircraft basic performance capabilities can be assessed by assuming thrust is invariant with velocity and varies only with altitude. (Graphical methods can be used to evaluate performance for engines with significant thrust–speed dependency.) Figure 6.68 shows a typical variation of thrust with altitude for a high bypass ratio engine.

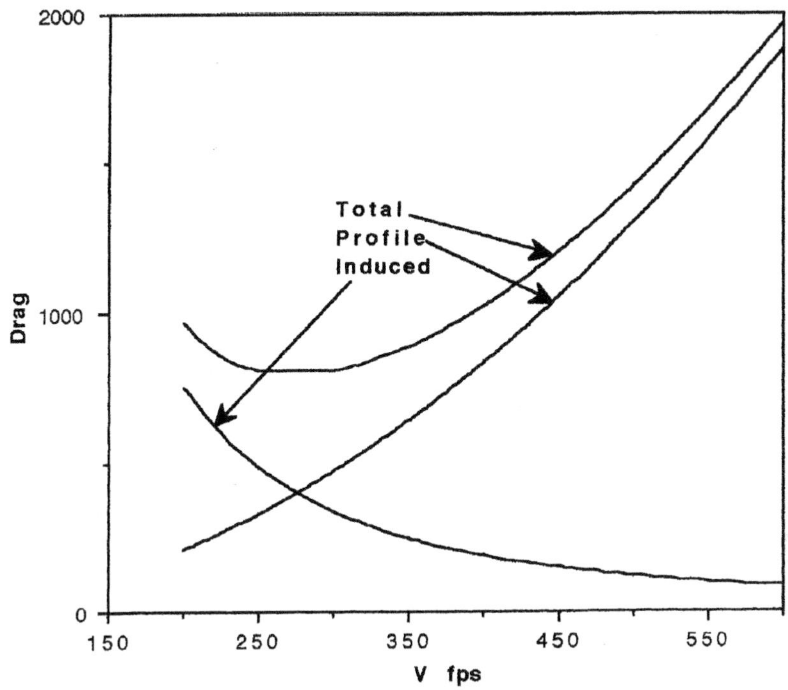

Figure 6.65 Aircraft drag elements.

For propeller-driven aircraft, power can also be assumed to be independent of forward velocity with the same caveat for power–speed dependency as for jet engines.

Airplane Performance

The relative interplay between thrust available and drag constitutes the available performance of an aircraft. Figure 6.69 illustrates a performance estimation sequence showing the two major elements, air performance and field performance. It is usual to estimate the performance of aircraft using a set of standard atmosphere tables to describe the environmental conditions at the appropriate mission segment. This permits performance comparisons between concepts to be made effectively.

Steady flight or unaccelerated level flight conditions are usually evaluated, and under these conditions the performance is said to be static performance. For some high-performance aircraft, such as modern-day fighters, accelerated flight conditions are taken into account in estimating the maneuverability capability of the concept.

Both the thrust available and drag vary with altitude and Mach number. Therefore, the performance capability of a concept becomes a function of speed and altitude. The thrust variation is a function of the engine type and depends on engine bypass ratio, whereas the drag variation is a function of the aircraft shape.

An overview of performance capability can be obtained from a graphical representation of thrust and drag at a given altitude (Figure 6.70). Here maximum and minimum speeds occur at $(T - D) = 0$, minimum drag speed is identified, and at speeds where $(T - D) > 0$, excess power is available for use in climb or in level flight acceleration. As altitude increases overall, $T - D$ over the speed range decreases (Figure 6.71), eventually reaching zero or less than zero for all speeds.

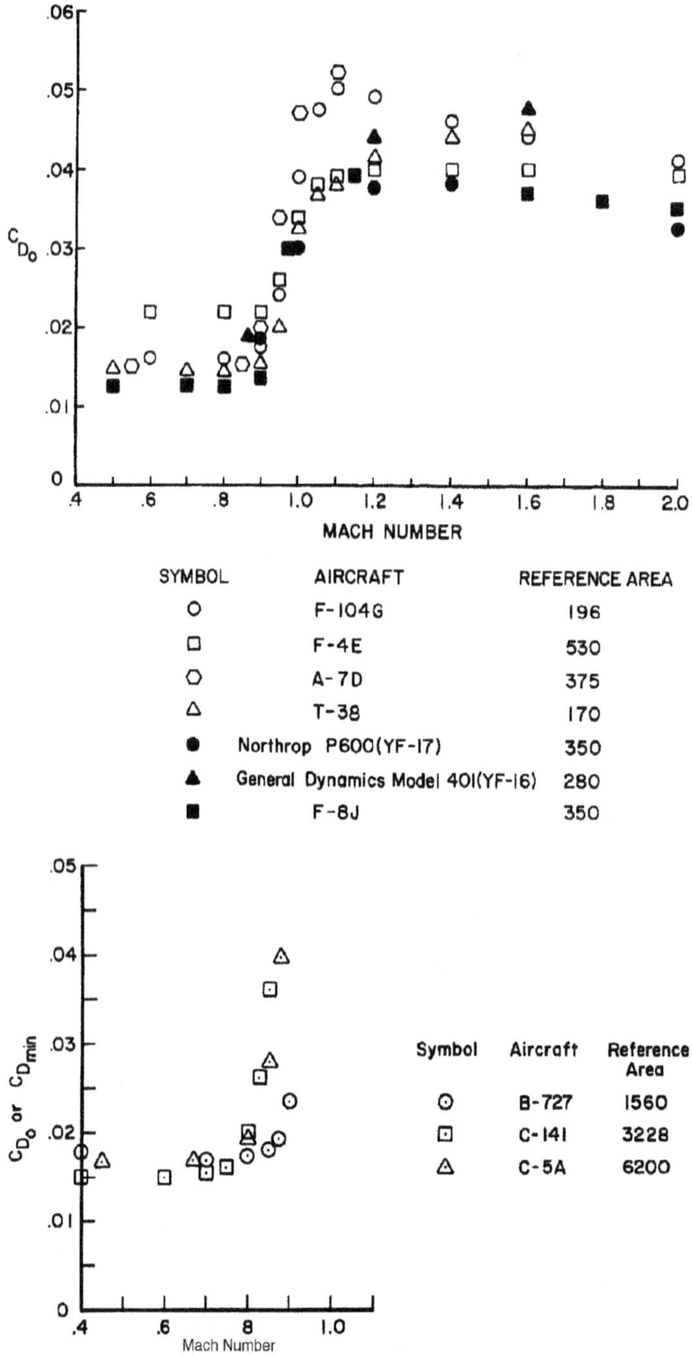

FIGURE 6.66 Measured profile drag coefficients (Nicolai 1984).

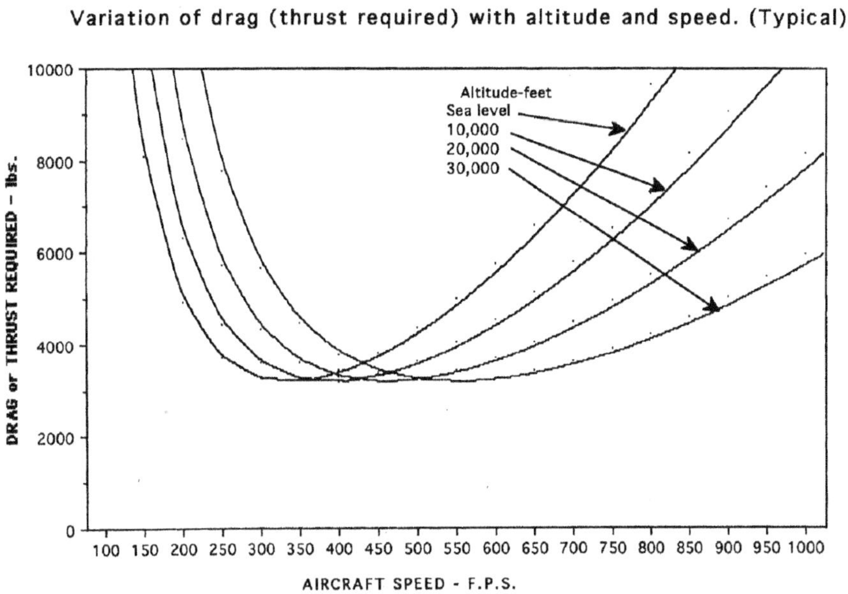

FIGURE 6.67 Typical drag variation with altitude.

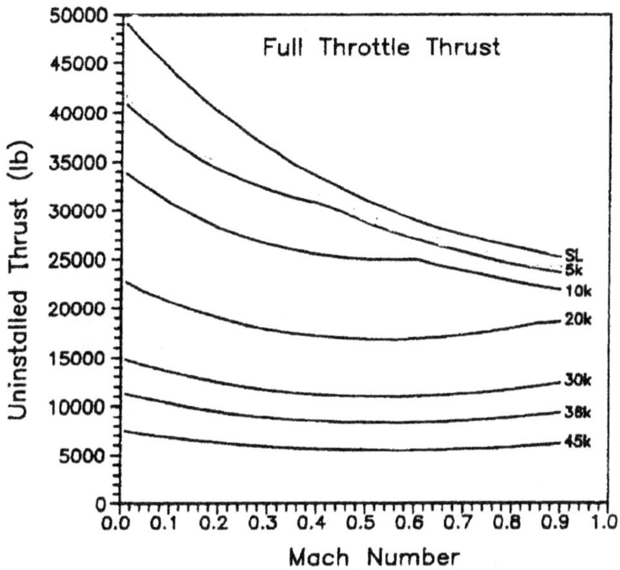

FIGURE 6.68 Typical thrust variation with altitude (Raymer 1989).

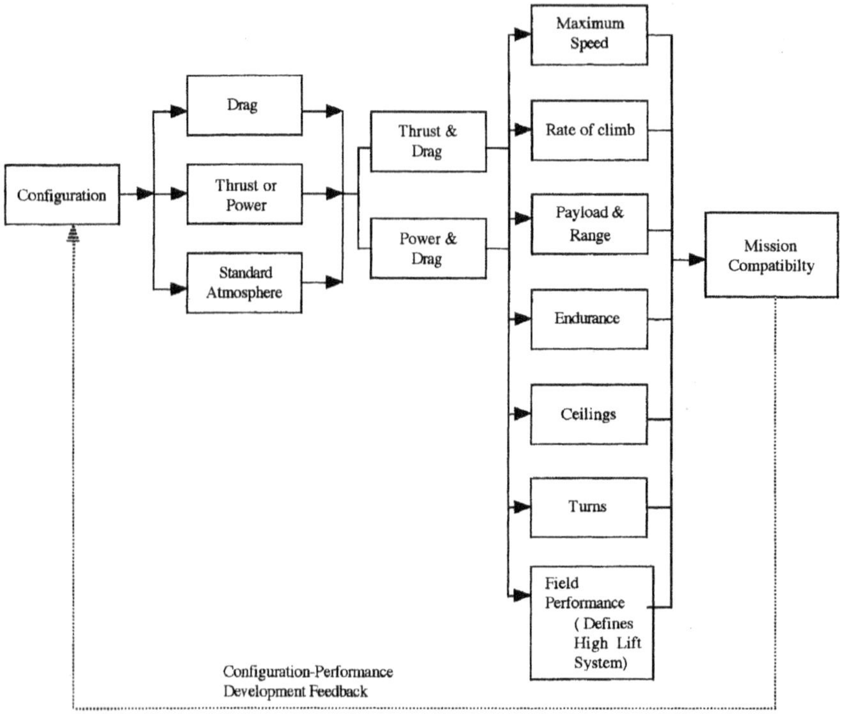

Figure 6.69 Overview of performance estimation sequence.

Absolute ceiling is reached when $(T - D) = 0$ and theoretically the aircraft can fly at one speed only, which is that speed where the thrust available curve is tangential to the drag curve. Service ceilings are a more practical measure of operating altitudes, and the rate-of-climb conditions for these are tabulated in this design section for commercial and military aircraft.

Much simplification in performance estimations can be achieved by assuming that the drag polar is parabolic. This results in the development of many performance equations which eloquently display the influence of many design parameters. The parabolic assumption is made for the performance methods presented herein.

Rate of Climb (R/C)

Defined as $V(T - D)/W$, the rate of climb will vary with altitude (see Figure 6.72), with a maximum value occurring for each altitude at a speed given by

$$V = \left\{ \frac{(T/W)\left(\frac{W}{S}\right) \pm \sqrt{\left(\frac{T}{W}\right)^2 \left(\frac{W}{S}\right)^2 + 12\left(\frac{W}{S}\right)^2 \left(\frac{C_{Do}}{\pi Ae}\right)}}{3C_{Do}\rho} \right\}^{0.5} \quad \text{(fps)}$$

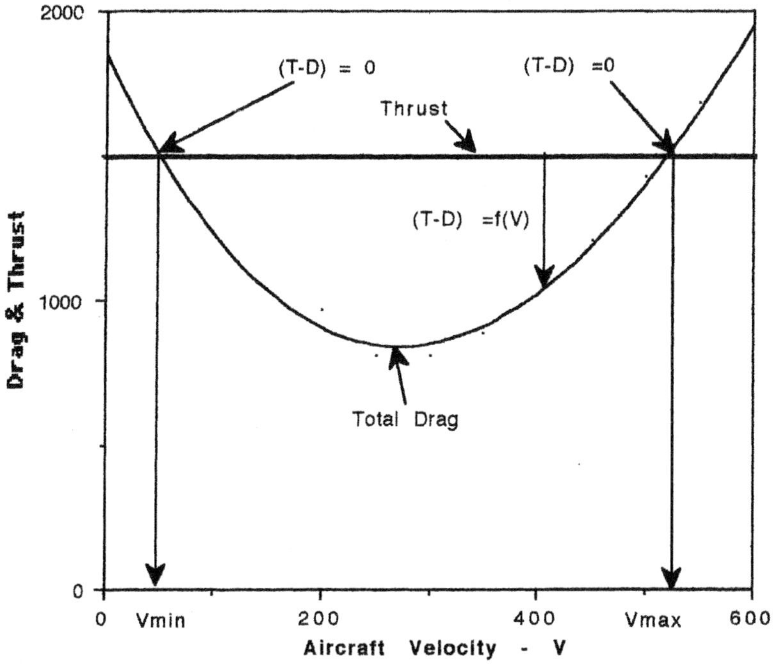

FIGURE 6.70 Typical variation of thrust and drag with velocity.

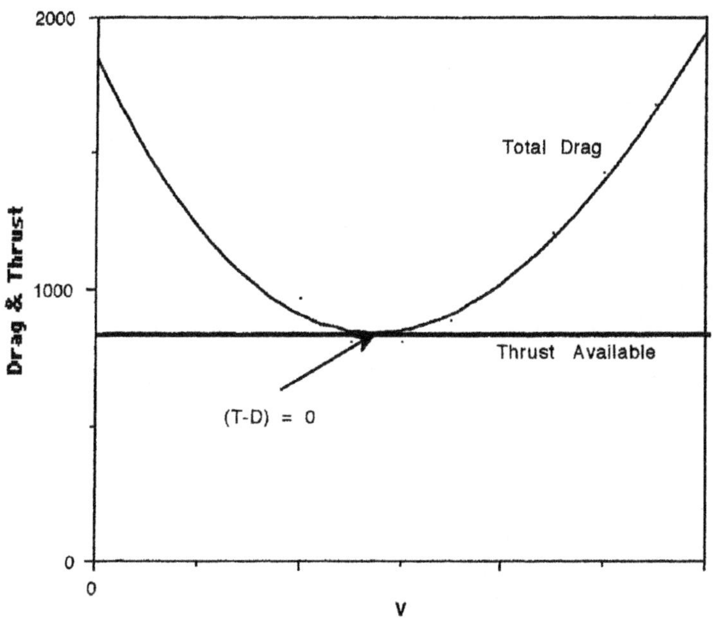

FIGURE 6.71 Typical variation of thrust and drag relationships at ceiling.

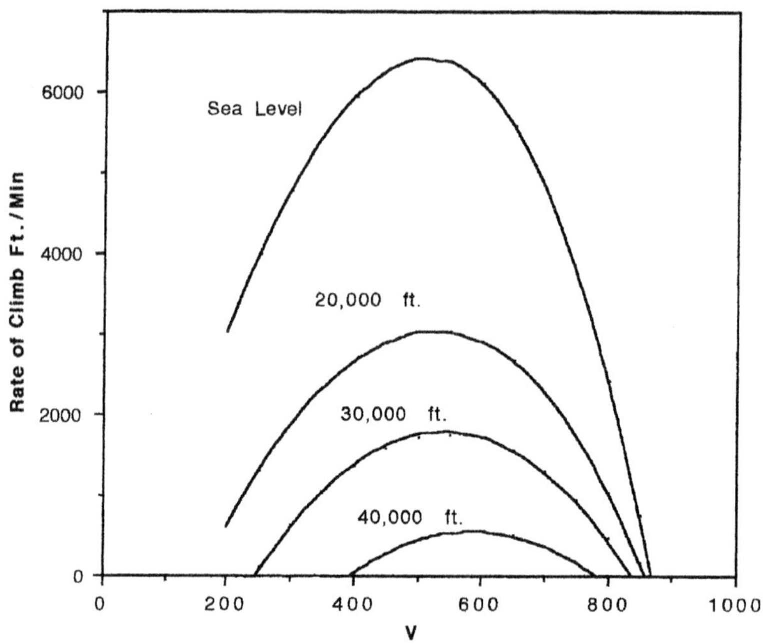

FIGURE 6.72 Typical rate of climb versus altitude.

The climb angle corresponding to maximum rate of climb, $(R/C)_{max}$, is given by

$$\Theta = \sin^{-1}\left(\frac{(R/C)_{max}}{V}\right) \text{(degrees)}$$

The maximum climb angle is given by

$$\Theta_{max} = \sin^{-1}\left(\frac{T}{W} - \frac{1}{L/D}\right) \text{(degrees)}$$

at a speed of

$$V = \sqrt{\frac{2}{\rho}\left(\frac{1}{C_{D_0}\pi Ae}\right)^{0.5} \frac{W}{S} \cos\Theta_{max}} \text{ (fps)}$$

All of the above are summarized in the climb hodograph of Figure 6.73.

Gliding Flight

In power-off flight the aircraft will glide and descend (Figure 6.74). The maximum straight-line ground range in a glide from height h is given by

$$\text{Max. horizontal distance, } l, = h\left(\frac{L}{D}\right)_{max}$$

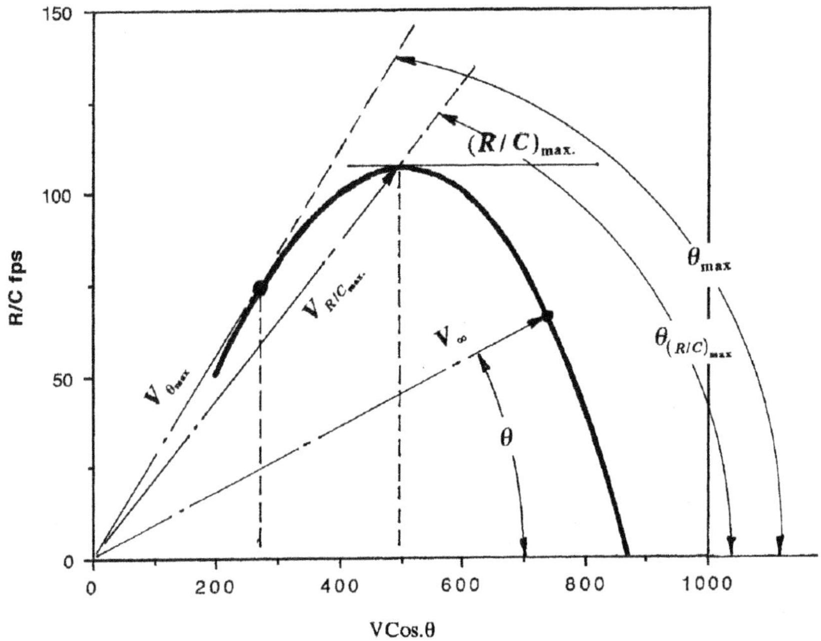

FIGURE 6.73 Climb hodograph.

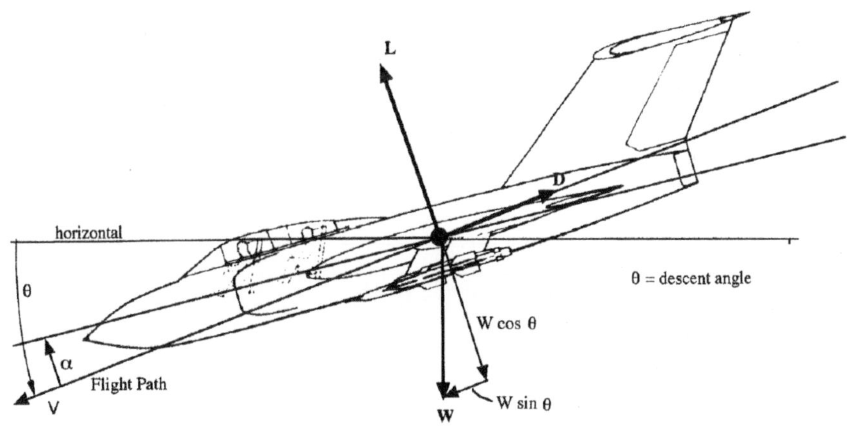

FIGURE 6.74 Forces in power-off flight.

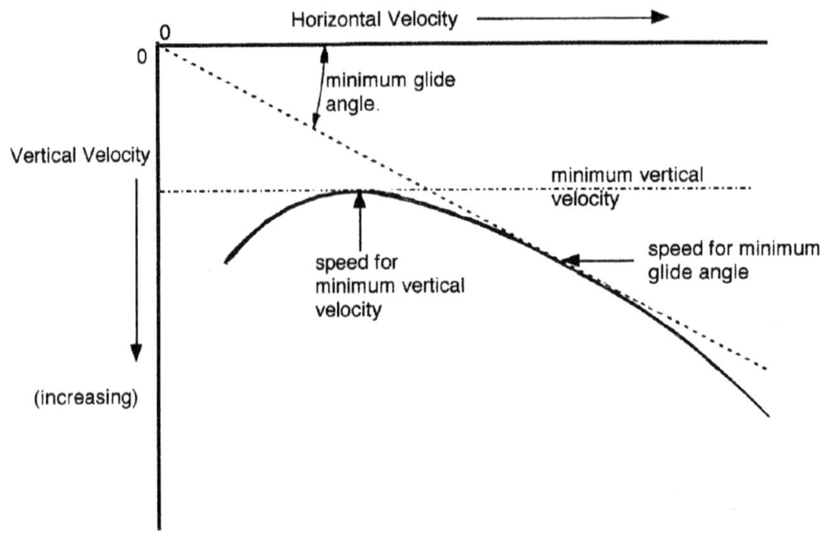

FIGURE 6.75 Glide hodograph.

and it will occur at an aircraft speed given by

$$V = \sqrt{\frac{W}{S}\left(\frac{2}{\rho}\right)\frac{1}{\sqrt{C_{Do}\pi Ae}}}\ \text{(fps)}$$

and at a descent angle given by

$$\Theta_{\min} = \tan^{-1}\left(\frac{1}{(L/D)_{\max}}\right)\text{(degrees)}$$

Minimum sink speed (vertical velocity) in a glide is given by

$$V_v = \sqrt{\frac{2}{\rho\left((C_L)^3/(C_D)^2\right)}\left(\frac{W}{S}\right)}\ \text{(fps)}$$

Figure 6.75 summarizes these glide equations on a glide hodograph.

Maximum Range and Maximum Endurance

The flight conditions for these performance parameters are specified by aircraft speeds; that is, a speed for maximum endurance and a speed for maximum range. Maximum endurance occurs at a flight condition appropriate to the longest period of *time* spent in the air for a given quantity of fuel, and maximum range corresponds to the maximum *distance* traveled on a given quantity of fuel. Table 6.18 summarizes the flight conditions for propeller and jet aircraft, and Figure 6.76 compares the maximum range and endurance speeds for two identical aircraft, one powered by a jet and the other by a reciprocating engine propeller combination. The maxima on each curve of the various aerodynamic parameters indicate "best" or maximum conditions. Table 6.19 summarizes the Breguet range and endurance formulae for propeller and jet aircraft.

Propulsion type	Maximum range	Maximum endurance
Propeller	V such that C_L/C_D is a maximum	V such that $C_L^{3/2}/C_D$ is a maximum
Jet	V such that $C_L^{1/2}/C_D$ is a maximum	V such that C_L/C_D is a maximum

Table 6.18 Speed, Maximum Range, Maximum Endurance Relationships

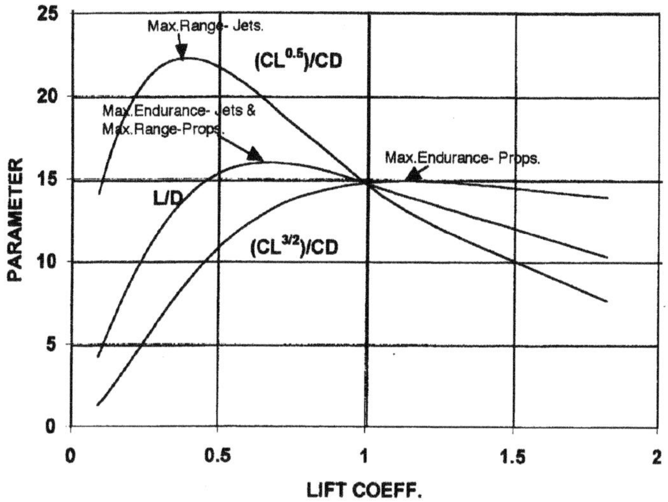

Figure 6.76 Performance comparisons—jet and propeller.

Propulsion type	Maximum range (ft)	Maximum endurance (h)
Propeller	$R = \frac{n}{c}\frac{C_L}{C_D}\ln\frac{W_0}{W_1}$	$E = \frac{n}{c}\frac{C_L^{3/2}}{C_D}(2\rho S)^{1/2}\{W_1^{1/2} - W_0^{-1/2}\}\left(\frac{1}{3600}\right)$
Jet	$R = 2\sqrt{\frac{2}{\rho S}}\cdot\left(\frac{1}{c}\right)\cdot\left(\frac{C_L^{1/2}}{C_D}\right)\cdot\{W_0^{1/2} - W_1^{1/2}\}$ at constant altitude	$E = \left(\frac{1}{3600\ c}\right)\frac{C_L}{C_D}\ln\frac{W_0}{W_1}$
where c = specific fuel consumption is lb/lb thrust/s for jets, lb/ft lb/s/s for reciprocating engines ρ = air density (slugs/ft^3) S = wing area (ft^2) W_0 = weight at start of range (endurance) segment (lb) W_1 = weight at end of range (endurance) segment (lb)		

Table 6.19 Maximum Range, Maximum Endurance Relationships

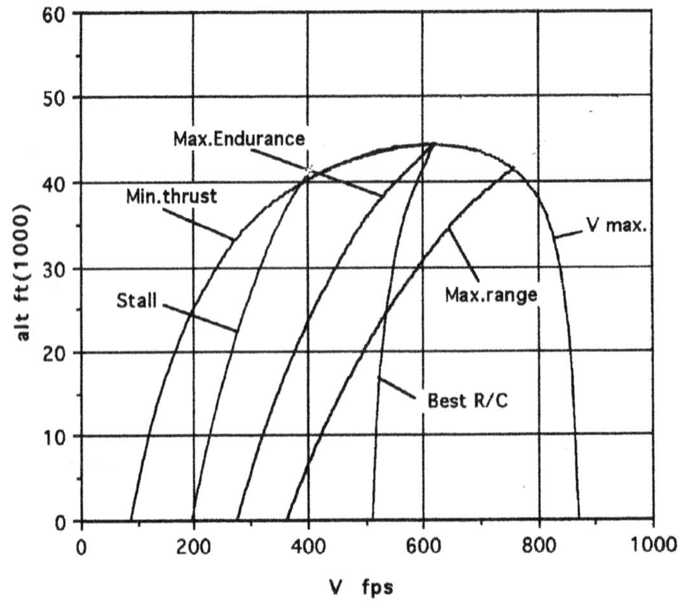

FIGURE 6.77 Typical speed-altitude limits.

Table 6.6 summarizes the analytical expressions for the maximum range and maximum endurance speeds for both jet-propelled and propeller-driven aircraft based on an assumption that the drag polar can be assumed parabolic in nature. Also shown are the corresponding C_L, C_D, and L/D values.

Figure 6.77 summarizes a typical variation of the above speeds with altitude for a jet-propelled aircraft and includes the stall speed based upon the following equation:

$$V_{\text{stall}} = \sqrt{\left(\frac{W}{S}\right)\left(\frac{2}{\rho}\right)\left(\frac{1}{C_{L\max}}\right)} \text{ (fps)}$$

where ρ = air density (slugs/ft^3)
$C_{L\max}$ = maximum lift coefficient
W/S = wing loading (psf)

Pull-up, Push-over, and Horizontal Turns

The magnitude of $T - D$ can also be used for maneuver in the form of horizontal and vertical flight path changes. In the horizontal level flight turn the airspeed and altitude are assumed to remain constant. The corresponding load factor and bank angle are defined as

$$n = \left(\frac{L}{W}\right)$$

$$\text{Bank angle, } \phi = \sec^{-1}\left(\frac{1}{n}\right)$$

All of these maneuvers are initiated from trimmed, straight, and level flight. In a pull-up the forces acting on the aircraft are shown in Figure 6.78 and the pull-up is evaluated at the instant of

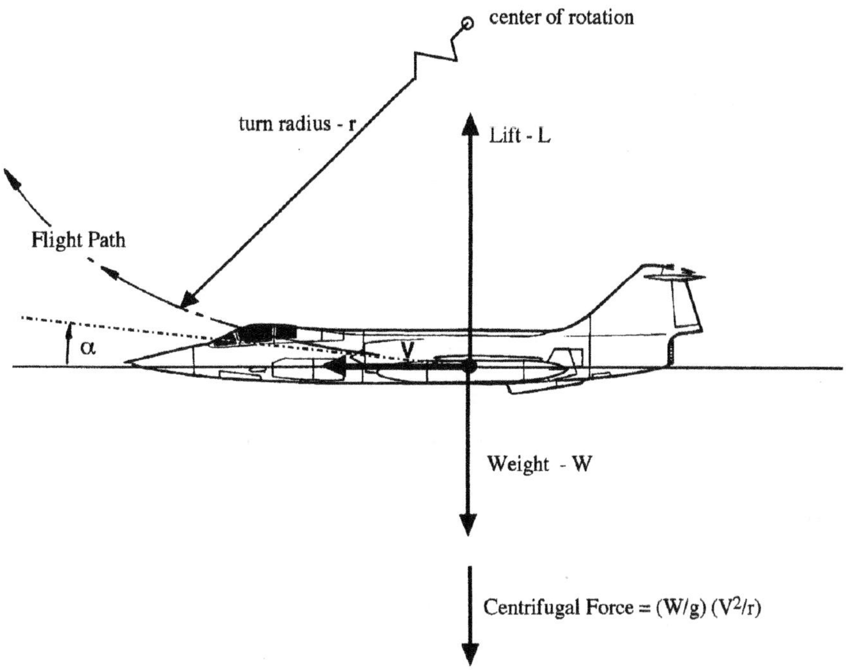

FIGURE 6.78 Forces in a pull-up.

maneuver initiation, with the resulting performance being strictly applicable to that instant. In a pull-down the aircraft is rolled inverted and a pull-down maneuver is initiated, with the resulting performance measures being applicable to the instant that pull-down is initiated. In this case a similar system of forces acts on the aircraft with, in this instance, lift and weight acting together in a downward direction to oppose centrifugal force, which is acting upwards.

The system of forces acting in a horizontal turn is shown in Figure 6.79, and in this case these forces can be assumed to be valid throughout the turn provided sufficient additional thrust is added. In this maneuver the bank angle, φ, and V are assumed constant throughout the turn.

All of these maneuvers depend on the application of normal load factor, n, primarily through an increase in angle of attack. The resulting turn performance measures are the available radius of turn, R ft, and resulting rate of turn, w, radians per second. The equations to evaluate pull-up, pull-down, and horizontal turns are given in Table 6.20, where the applicable load factor, n, and corresponding speed, V, are derived from the design maneuver flight envelope.

Takeoff and Landing

This aspect of performance is concerned with estimating the distance required to clear a given obstacle height (Figures 6.80 and 6.81). This distance consists of ground, transition, and air distance elements for takeoff, and an air distance and ground roll for landing (Anderson 1999).

In general, the ground roll distance, s_g, for takeoff can be shown to be approximated by the following equation, which serves to illustrate the influence of certain aircraft parameters:

$$S_g = \frac{1.21(W/S)}{g \cdot \rho \cdot C_{L\max}\left(\frac{T}{W} - \frac{D}{W}\right)} \text{(ft)}$$

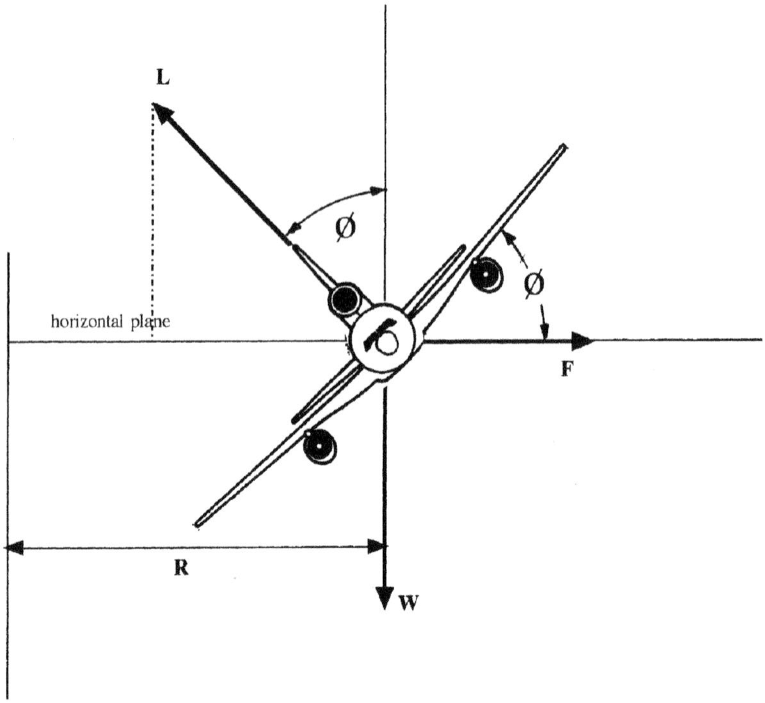

Figure 6.79 Forces in a horizontal turn.

where T/W = aircraft thrust-to-weight ratio
D = aircraft total drag

and where 1.21 corresponds to the assumption that liftoff speed, V_{LO}, occurs at $(1.1\ V_s)^2$. To achieve short takeoff distances, it is necessary that high values of $C_{L\ max}$ and $T - D$ be present, coupled with low values of W/S.

	Pull down	Pull up	Horizontal turn
R (ft)	$\frac{V^2}{g(n+1)}$	$\frac{V^2}{g(n-1)}$	$\frac{V^2}{g\sqrt{n^2-1}}$
ω (rads/s)	$\frac{g(n+1)}{V}$	$\frac{g(n-1)}{V}$	$\frac{g\sqrt{n^2-1}}{V}$
t (s for 180° turn)	—	—	$\frac{\pi}{\omega}$
$\sec \varphi = \left(\frac{L}{W}\right) = n$ (load factor) φ = bank angle (deg s)			

Table 6.20 Horizontal Turns, Pull-up, and Pull-down Relationships

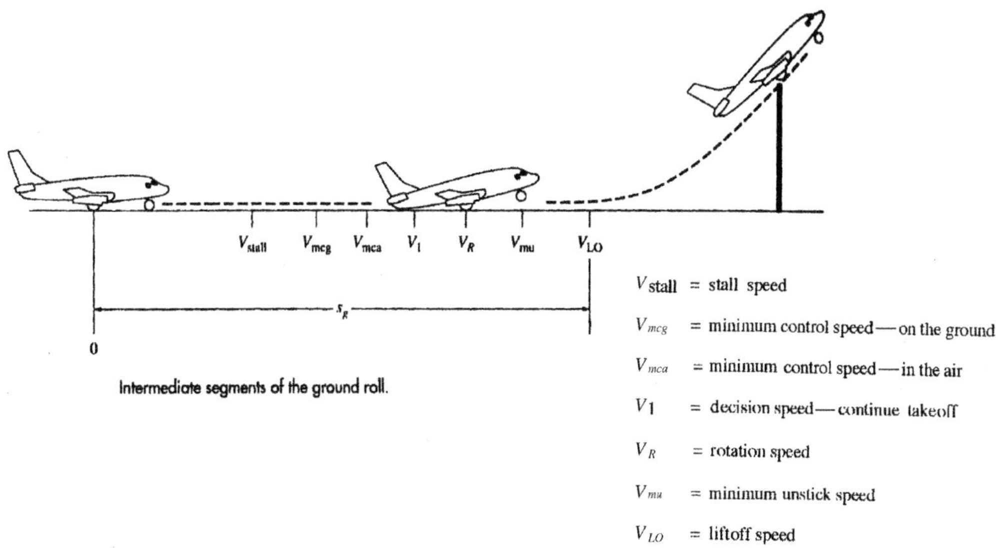

FIGURE 6.80 Interim segments of ground roll—takeoff (Anderson 1999).

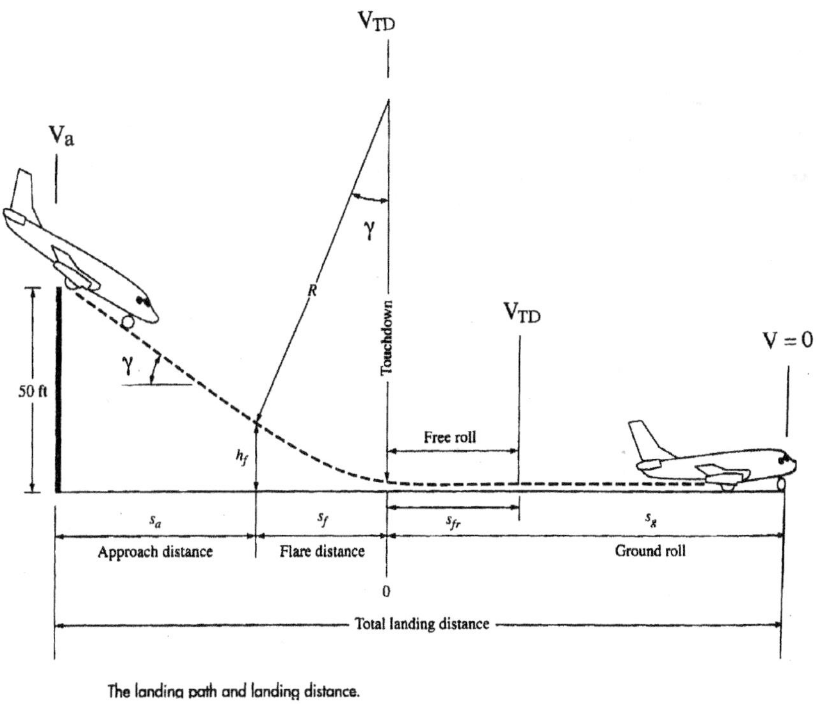

FIGURE 6.81 Interim segments of ground roll—Landing (Anderson 1999).

The following equation can be used to estimate the landing ground roll distance, s_L, for modern jet transports (Anderson 1999):

$$S_L = \frac{1.69\,W^2}{g \cdot \rho \cdot S \cdot C_{L\max}[T_R + (D + \mu_G(W - L))]} \text{ (ft)}$$

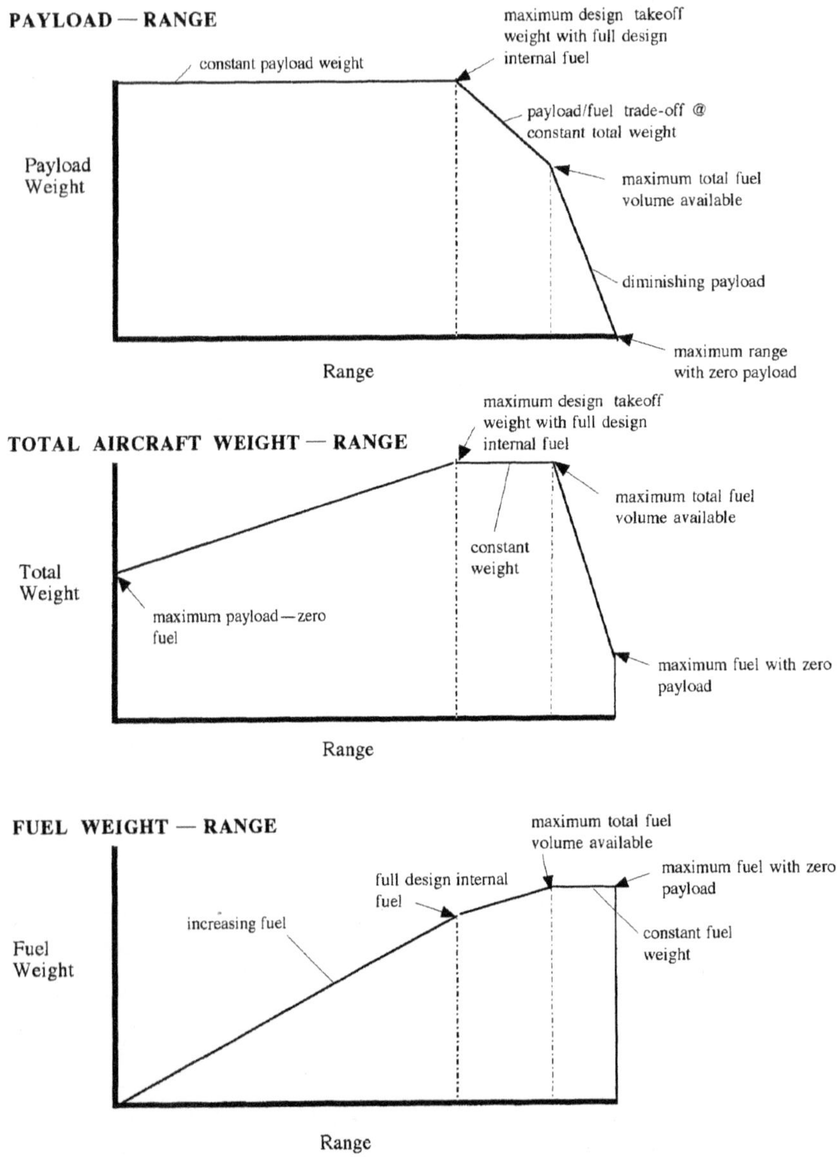

Figure 6.82 Definition of payload range and corresponding fuel and total weight.

where VT is the touchdown velocity = $1.3(V_{stall})$ (fps)

$g = 32.2$ ft/s^2

ρ = air density (slugs/ft^3)

S = wing reference area (ft^2)

$C_{L\,max}$ = maximum lift coefficient on landing

T_R = reverse thrust (lb)

D = total drag (lb)

μ_B = braking coefficient of friction (usually = 0.40)

W = aircraft landing weight (lb)

L = residual lift on the wing (lb) with a proviso that all the terms in the [ ] brackets are evaluated at a speed equal to 0.70 (V_T)

To minimize s_L, it is necessary to increase the design terms in the denominator and reduce the landing weight. The residual lift on the wing can be minimized or made equal to zero by deploying wing-mounted upper surface spoilers at the start of the ground run on landing. This in turn reduces the magnitude of s_L.

The remaining elements of the takeoff and landing distance estimates can be established through methods developed in Anderson (1999).

Payload Range

A measure of the mission effectiveness of a cargo or passenger carrying aircraft is displayed by the payload range diagram, which shows what payload weight can be carried for what range. Figure 6.82 shows a typical payload range diagram and how an increase in range, beyond that for maximum payload, is possible by trading additional fuel for payload. The corresponding total mission weight variation with range is also shown, and this assumes that only enough fuel is loaded to achieve the chosen range. The fuel weight required to establish the corresponding range concludes the figure. In all instances additional fuel above the design full internal fuel can be accommodated by using temporary fuselage fuel cells which can be located inside the cargo space in place of the normal cargo.

References

Aerospace Source Book. 1999. *Aviation Week & Space Technology*, McGraw-Hill, New York.

Air International. 1989–2001, Vol. 36–60, Key Publishing LTD, Stamford, Lincolnshire, UK.

Anderson, J. D., Jr. 1999. *Aircraft Performance and Design*, WCB/McGraw-Hill, Boston.

Jenkinson, D., Simpson, P., and Rhodes, D. *Civil Jet Aircraft Design*, AIAA Education Series, Washington, DC.

Loftin, L. K., Jr. 1980. *Subsonic Aircraft: Evolution and the Matching of Size to Performance*, NASA Reference Publication 1060.

Nicolai, L. M. 2010. *Fundamentals of Aircraft Design*, AIAA Education Series, Washington, DC.

Raymer, D. P. 2013. *Aircraft Design—A Conceptual Approach, 5th ed.*, AIAA Education Series, Washington, DC.

Rolls-Royce plc. 1986. *The Jet Engine*, Rolls-Royce plc, Derby.

Roskam, J. 1980. *Aircraft Design*, Parts I through VIII, DAR Corporation, Lawrence, KS.

Further Reading

Avallone, E. A. and Baumeister, T., III, eds., *Marks' Standard Handbook for Mechanical Engineers*, 10th ed., McGraw-Hill, New York (1996).

Gessow, A. and Myers, G. C., Jr., *Aerodynamics of the Helicopter*, Macmillan, New York (1952).

Huenecke, K., *Modern Combat Aircraft Design*, Naval Institute Press, Annapolis, MD (1987).

Khoury, G. A. and Gillett, J. D., *Airship Technology,* Cambridge Aerospace Series 10, Cambridge University Press, Cambridge (1999).

Mair, W. A. and Birdsall, D. L., *Aircraft Performance,* Cambridge Aerospace Series 5, Cambridge University Press, Cambridge (1992).

McCormick, B. W., Jr., *Aerodynamics of V/STOL Flight,* Academic Press, New York (1967).

Schaufele, R. D., *The Elements of Aircraft Preliminary Design,* Aires, Santa Ana, CA (2000).

Shapiro, J., *Principles of Helicopter Engineering,* McGraw-Hill, New York (1955).

Smetana, F. O., *Flight Vehicle Performance and Aerodynamic Control,* AIAA Education Series, Washington, DC (2001).

Stinton, D., *The Anatomy of the Airplane-Second Edition,* Blackwell Science, Oxford, and AIAA, Washington, DC (1998).

Stinton, D., *The Design of the Airplane,* Van Nostrand Reinhold, New York (1983).

Stinton, D., *Flying Qualities and Flight Testing of the Airplane,* AIAA Education Series, Washington, DC (1996).

SECTION 7

Spacecraft Systems

Section Editor: Brij N. Agrawal

PART 1

Space Missions

Brij N. Agrawal

7.1 Introduction

The Soviet Sputnik satellite was the first to orbit earth, launched on October 4, 1957 (Figure 7.1). It was a 58.4-cm (23-in) and 83.4-kg (184-lb) metal ball. It carried a thermometer, a battery, and a radio transmitter, which changed the tone of its beeps to match the temperature changes. The interior was pressurized with nitrogen gas. On the outside, four-whip antennas transmitted on shortwave frequencies around 27 MHz. After 92 days, drag took over and it burned in earth's atmosphere. A space race between Soviet and the United States ensued.

The first U.S. earth satellite, named Explorer I, was developed by JPL and launched on February 1, 1958 (Figure 7.2). Its mass was 14 kg; orbit: $354 \times 2{,}515$km at 33.2° inclination. During an experiment directed by Dr. James A. Van Allen, Explorer I discovered a radiation belt around the earth, known as the Van Allen belts.

Currently, satellites are an integral part of civilian and military life. Satellites are used for communications, navigation, imaging, reconnaissance, weather prediction, remote sensing, and astronomy. Initially, space missions were funded by governments. Today, commercial companies are often major contributors for funding and use of satellites.

Earth satellites offer significant capability improvement compared to terrestrial techniques in many fields. The fundamental advantage of a satellite is its ability to obtain a global look at a large portion of the earth's surface. A system of three satellites in geosynchronous orbit can almost cover earth's entire surface. For astronomy, satellites provide unique advantages in comparison to

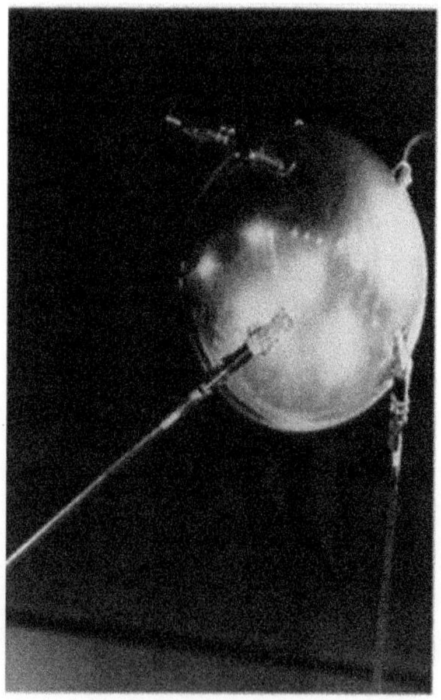

FIGURE 7.1 Soviet Sputnik satellite.

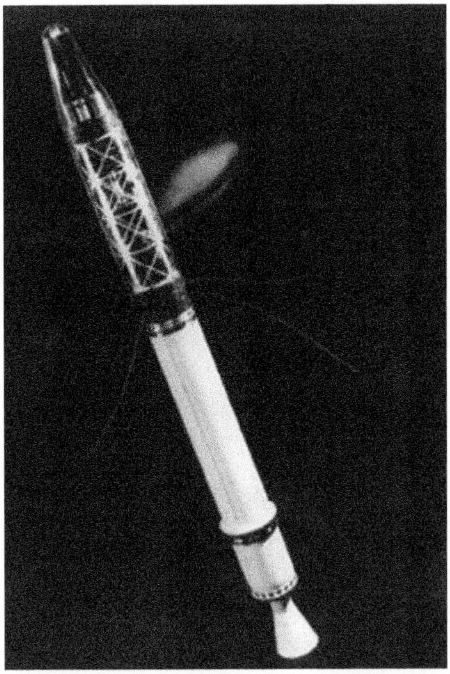

FIGURE 7.2 Explorer I.

ground telescopes. Interplanetary satellites can view planets with close proximity. It has led to the application of satellites in several areas, as follows:

- Communications
- Navigation
- Meteorological
- Earth resource
- Astronomy
- Commercial imaging
- Science
- Human space flight
- Military missions

Commercial communications satellites are the major portion of satellites both in numbers and revenues. By the end of 2013, nearly 1,200 satellites were operational: 40% are commercial communications satellites, 13% are government communications satellites, 13% are remote sensing, 12% are R&D, 8% are navigation, 7% are military surveillance, 5% are scientific, and 3% are meteorological satellites. The growth is significant with the introduction of small satellites.

The satellite industry has had significant revenues. In 2015, total revenues were $208 billion, consisting of $128 billion for satellite services, $60 billion for ground equipment manufacturing, $17 billion for satellite manufacturing, and $5.5 billion for launch services. There is great interest in using satellites for the Internet. In this chapter, we introduce some of the missions.

A space system consists of three parts: space segment, launch segment, and ground segment. Space segment represents in-orbit satellites. Launch segment puts the satellites into orbit. Ground segment is used for telemetry, command and ranging during launch, health monitoring of satellites, and sending/receiving data from the satellites for the missions.

As shown in Figure 7.3, a spacecraft consists of two basic parts: the mission payload and the spacecraft bus. The payload performs the missions of the spacecraft, such as communications by radio links, earth images for weather forecasting, and high-resolution photography for assessment of visible earth resources. The spacecraft bus supports the payload by providing the required orbit and attitude control, electric power, thermal control, mechanical support, and a two-way command and datalink to the ground.

A spacecraft bus can be divided into six subsystems as shown in Figure 7.3. A brief description of these subsystems is provided here.

Attitude Control

The attitude control subsystem maintains the spacecraft's attitude, or orientation in space, within the limits allowed. It consists of sensors for attitude determination and actuators, such as thrusters and/or angular momentum storage devices, for providing corrective torques. The attitude disturbance torques come from many sources, such as solar pressure, the gravity gradient, and any misalignment of the thrusters. Performance requirements of the attitude control subsystem depend significantly on the mission. For example, imaging satellites require challenging jitter and pointing performance. In contrast, the requirements for communications satellites are less challenging, as radio frequency has much higher wavelengths than visible for imaging satellites.

Propulsion

The propulsion subsystem injects the spacecraft into the desired orbit and keeps the orbital parameters within the limits allowed. The disturbance forces that will alter the spacecraft's orbit arise from the gravitational forces of the sun and moon and the earth's ellipticity.

Electrical Power

The electrical power subsystem provides electric power to the spacecraft during all mission phases. The primary power is provided through the conversion of light energy from the sun into electrical energy by the use of solar cells. During the eclipse period, when there is no sunlight, electrical power is provided by rechargeable batteries. The batteries are charged by the solar array during the sunlit period. Power control electronics control the bus voltage.

Thermal Control

The thermal control subsystem keeps the temperatures of spacecraft equipment within specified ranges by using the correct blend of surface coatings, insulation, and active thermal control devices. In a communications satellite, the major tasks for the thermal control subsystem are to maintain the temperatures of the batteries within narrow limits, to prevent the hydrazine fuel from freezing, and to radiate into space the large amount of heat generated by the traveling-wave tubes (TWTs). For an imaging satellite, dimensional stability of the telescope is most challenging.

Structure

The structural subsystem provides the mechanical interface with the launch vehicle, gives mechanical support to all spacecraft subsystems, sustains launch loads, and provides precise alignment where needed, such as for antennas and thruster jets.

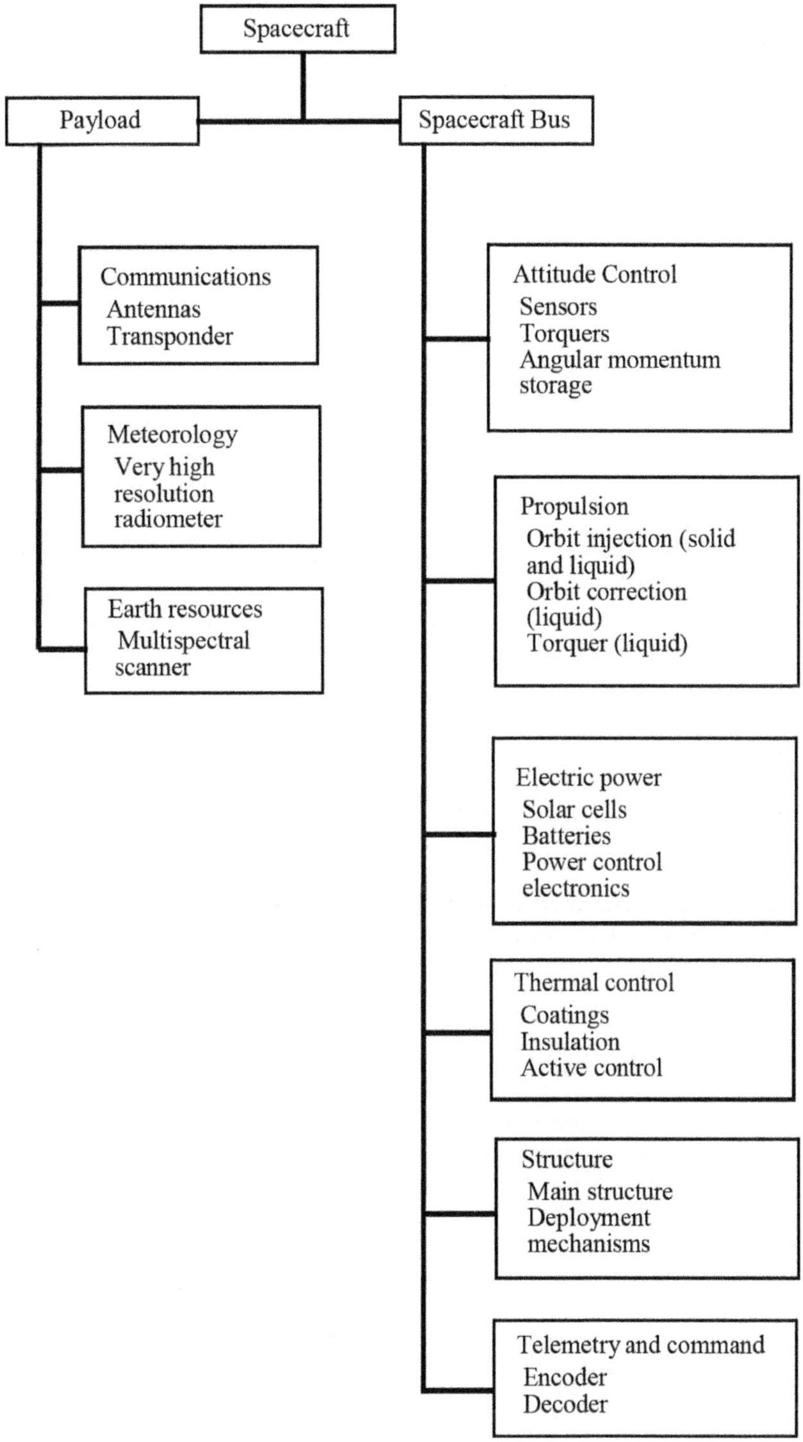

FIGURE 7.3 Spacecraft block diagram.

Telemetry and Command

The telemetry and command subsystem maintains the two-way command and datalink to the ground. The command portion receives the command data, decodes them, and carries out the stored command sequences. The data-handling portion accepts data from the subsystem sensors and, after encoding them, telemeters them to the ground.

7.2 Orbits

Before we discuss satellite missions, it is desirable to review different orbits.

Geosynchronous Orbit

These orbits are circular orbits with an altitude of 35,786 km. Its angular orbit equals earth's rotation rate. Orbits with zero inclination are called geostationary orbits. For this orbit, satellite appears stationary over a point on the earth. Therefore, the ground stations can be nontracking, fixed antennae. Three, equally spaced satellites will provide full earth coverage, except near poles.

Low Earth Orbit

Altitude for these orbits range from 100 to 2,500 km. This altitude keeps the satellites out of the Van Allen belts. Orbit period is approximately 90 minutes. This orbit has been used for weather, imaging, and communications satellites. The International Space Station is in this orbit. Normally, this orbit will require a large number of satellites for full earth coverage.

Medium Earth Orbit

Altitude for these orbits range from 800 to 20,000 km. In this orbit, satellites are exposed to large amounts of radiation from the Van Allen belts. This orbit requires fewer satellites than low earth orbit (LEO) for full earth coverage. Global Positioning Satellite constellation uses this orbit.

High Elliptical Orbit

High elliptical orbit (HEO) is also known as Molniya. It was developed by Russians for communications with the northern regions, as an alternate to GEO. Apogee is placed over area of interest. Three satellites provide full coverage of northern latitudes.

Sun-Synchronous Orbit

For sun-synchronous orbit, parameters are chosen such that the orbital plane processes with nearly the same period as planet Earth's solar orbit period. In this orbit, spacecraft keeps the same orientation with respect to the sun. Spacecraft crosses perigee at the same time every orbit. It is widely used for imaging satellites as they rely on certain lighting condition. This orbit can also simplify spacecraft thermal design.

Polar Orbit

Polar orbits are 90° inclination orbits, useful for spacecraft that carry out mapping or surveillance operations. Planet rotates below a polar orbit, allowing the spacecraft access to virtually every point on the surface.

Escape Orbit

Escape orbit is when a trajectory takes the spacecraft away from earth on a path to another body, generally into hyperbolic orbits.

7.3 Satellite Missions

Communications

The microwave frequency range (1–30 GHz) is best suited for carrying the large volume of communications traffic existing today. The signals of this frequency band are not appreciably deflected by the earth's ionosphere, limiting the line of sight to about 100 km. In addition, these signals will not travel through earth. Over land and over short water links, microwave systems may be constructed using repeater stations approximately every 50 km with small dish-type antennas that receive the beamed signal and using a second antenna to retransmit it toward the next station in the chain. The placing of such a chain across one of the world's major oceans, such as the Atlantic Ocean, presents insurmountable engineering difficulties. Another approach to transmitting a microwave signal across the Atlantic Ocean would be by erecting a repeater tower nearly 800 km high in the center of the ocean. This project would pose even greater engineering difficulties. As these microwave frequency become congested due to heavy use, addition spectrum above 30 GHz in the Q/V bands and even W bands are now being explored for use in new satellite systems.

In the space age, it is possible to have a repeater in the sky without building the tower. A satellite launched into a circular orbit of an altitude 35,786 km above the equator, called the geosynchronous orbit, would appear to be stationary from a point on the earth. A satellite in this position can view about one-third of the earth's surface. Any earth station in that coverage area directed toward that satellite can potentially communicate with any other earth station. The first well-known article on communications satellites was written in 1945, by Arthur C. Clark. He showed that three geosynchronous satellites, powered by solar energy converted into electricity using silicon cells, could provide worldwide communications. It took major advances in rocket technology for launching satellites and communications equipment to translate Clark's concept into reality.

The first human-made active communications satellite was SCORE, Signal Communications by Orbiting Receiving Equipment, and was launched in December 1958. It was designed to operate for 21 days of orbital life. The operating life was, however, limited to 12 days due to the failure of the batteries. During the late 1950s and early 1960s, the relative merits of passive and active communications satellites were often discussed. Passive satellites merely reflect incident signal, whereas active satellites have equipment that receives, processes, and retransmits incident signal. Project ECHO used two large spherical passive satellites, one 30.48 m (100 ft) in diameter and the second 41.5 m (135 ft) in diameter. They were launched in 1960 and 1964, respectively. The initial orbits were $1{,}520 \times 1{,}688$ km with 48.6° inclination and $1{,}032 \times 1{,}316$ km with 85.5° inclination, respectively. The satellite that paved the way for commercial communications satellites was Telstar. Telstar was a low-earth orbit active repeater satellite. It was soon followed by a geostationary active repeater satellite, Syncom. The Syncom satellites were spin stabilized. Syncom 1 was launched in February 1963. The intended orbit was at the geosynchronous altitude with a 32° inclination. The satellite, however, failed during apogee motor fire. Syncom 2 was successfully launched in July 1963 by a Delta launch vehicle.

Communications satellites can be divided into three categories on the basis of their applications: fixed service, broadcast, and mobile. In the fixed satellite service (FSS), the earth station locations are fixed geographically and provided a fixed-path, two-way communications link (receive/transmit) to the satellite. These satellites are further divided into international, regional, and domestic service satellites, depending on the area of use.

A traditional international communication space link is shown in Figure 7.4. Individual telephone calls are processed at local telephone exchanges and relayed to the satellite earth station via a microwave link. At the earth station, these signals are stacked in frequency to form a composite baseband signal that frequency modulates the radio carrier wave sent up to the spacecraft. This is called frequency-division multiplexing. The spacecraft amplifies the received signal, translates its frequency, and sends it back to the receiving earth station, where it is passed into the local

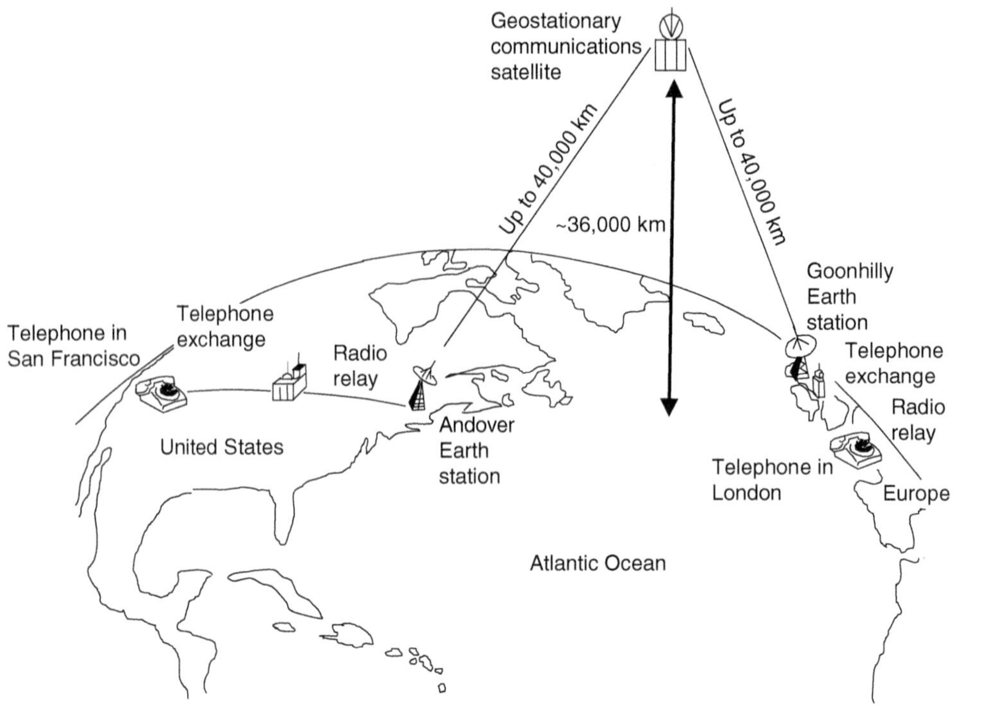

FIGURE 7.4 Network for international communications.

terrestrial telephone system for transmission to users after demultiplexing. The current trend in space communications is to eliminate a major portion of the terrestrial system in the link. This has been made possible by the increase in the power of satellite signals, to the level where smaller antennas on customer premises can receive and transmit signals to the satellite.

In broadcast satellites, the satellites receives the TV/radio signal from a central station and transmits it back, after very high signal power amplification, to small, receive-only antennas, usually of less than 1 m diameter, on the roofs of private houses. In mobile satellites, the mobile terminals on the sea, air, and ground, such as ships, airplanes, and automobile/trucks, communicates with a fixed or mobile terminal through the mobile satellite, as shown in Figure 7.5.

The Intelsat System

The creation of Intelsat was preceded by the formation of Comsat with the U.S. Communications Satellite Act of 1962, enacted in August 1962 during President Kennedy's term, which authorized the formation of the Communications Satellite Corporation (Comsat) to establish international commercial communications satellite system. Comsat was incorporated on February 1, 1963. Intelsat (International Telecommunications Satellite Organization) was founded in July 1964 and subsequently grew to over 100 member countries. Comsat represented the interests of the United States in Intelsat. Now Intelsat has become a commercial company and Comsat was bought by Lockheed Martin.

The major decision to be made in 1963 was whether the first Intelsat satellite system should be based on satellites in a medium-altitude orbit of 10,000–13,000 km above the earth or in a geosynchronous orbit at 35,786 km (22,236 mi). The available experience at that time favored the medium-altitude system since Telstar had been successful and Syncom 1 had failed during launch in

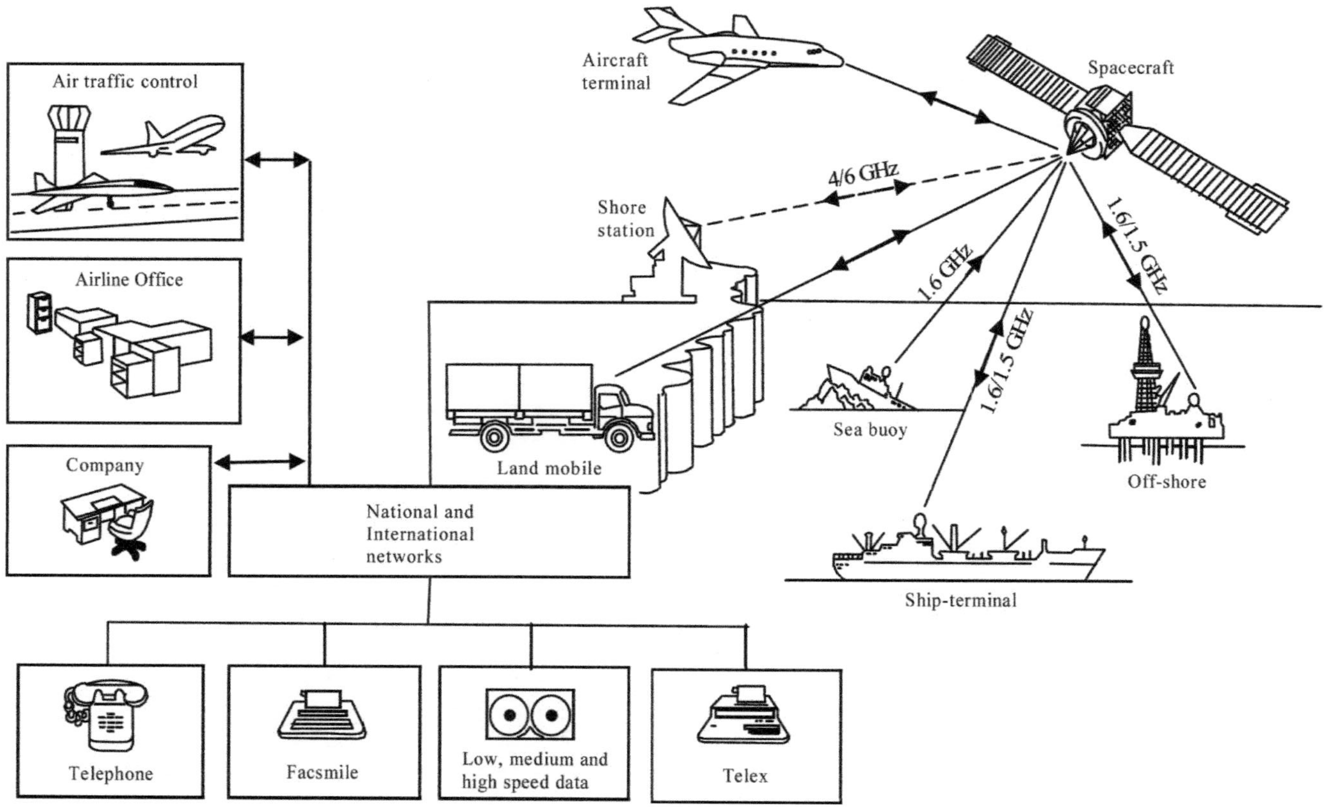

FIGURE 7.5 Mobile network system.

February 1963. However, it was decided that whereas a successful launch of one or two medium-altitude satellites would, at best, allow only experimental use of the system (since 18–24 of such satellites would be needed for a fully operable commercial system), the successful launch of a single geosynchronous satellite would permit immediate operational use on one ocean region. Accordingly, a contract was awarded in early 1964 to the Hughes Aircraft Co. for the construction of a geosynchronous communication satellite (Early Bird). When Early Bird successfully attained its geosynchronous orbit, it was renamed Intelsat I. This was a very important decision and the majority of the communications satellites operate in geosynchronous orbit. Globalstar and Iridium were developed for low orbit with many satellites, but were not commercial competitive. Recently, there is interest in low orbit for Internet services.

The first commercial satellite, Intelsat I, built by Hughes (now Boeing), initially known as the Early Bird, was launched in April 1965. It weighed only 38 kg in orbit and was single spin stabilized. It had a fixed, omnidirectional antenna which transmitted in a full 360° arc. It carried 240 two-way telephone circuits or one transatlantic TV channel. It did not, however, allow multiple access; only one pair of earth stations could operate with the satellite at any one time. It provided services across the North Atlantic, between Europe and North America.

The Intelsat II series of satellites, also built by Hughes, were single spin stabilized and weighed 86 kg in orbit. They carried 240 two-way telephone circuits or one TV channel. An important innovation in the Intelsat II satellite series was the multiple-access capability, enabling many pairs of earth stations to be connected simultaneously by the same satellite. The first successful Intelsat II launch took place in October 1966 and with these satellites, Intelsat coverage was eventually provided over both the Atlantic and Pacific Ocean areas.

The Intelsat III series satellites, built by TRW (now Northrop Grumman), provided a dramatic increase in capacity, from 240 circuits for Intelsat I and II to 1,500 circuits plus a TV channel. They incorporated a mechanically despun antenna, which always pointed toward the earth, so the transmission power could be focused in the direction of the earth instead of radiating some of the power into space. They weighed 152 kg in orbit. Intelsat III satellites were placed over all three major ocean areas: the Atlantic, the Pacific, and the Indian. So in 1969, The Intelsat system achieved full global coverage for the first time; it was a translation of Clark's concept into reality.

The Intelsat IV series satellites were built by Hughes and launched by the Atlas-Centaur vehicle. They weighed 832 kg in orbit and were dual-spin-stabilized spacecraft. Intelsat IV provided 3,750 two-way telephone circuits and two TV channels. The major advanced of Intelsat IV over its predecessors was the use of additional "zone-beam" antennas, which covered selected small portions of the visible earth. The resulting concentration of radiated power in these zone beams contributed to the increased capacity. Intelsat IV satellites also had four global coverage horns. The first Intelsat IV was launched in January 1971. The Intelsat IV satellite was bandwidth limited rather than power limited, the first time that had ever happened to a communications satellite.

The Intelsat V satellites series, built by Ford Aerospace (now Space Systems Loral) were the first three-axis-stabilized commercial satellites. Currently, three-axis-stabilized is the preferred design for communications satellite because of high power requirements. The design of Intelsat V is discussed in more details in the following section.

Intelsat V: Three-Axis Stabilization

Intelsat V, as shown in Figure 7.6, was a three-axis-stabilized spacecraft. It weighs approximately 1,900 kg at launch and was designed to be compatible for launch by Atlas-Centaur, STS-PAM-A, or Ariane. Depending on the operational configuration, each satellite can carry up to 12,000 two-way telephone circuits, and two color-TV transmissions. The communications subsystem provides an RF bandwidth capability of 2,137 MHz. It is accomplished by extensive frequency reuse of 4 and 6 GHz by both spatial and polarization isolation and by introducing the 14/11-GHz bands to international traffic. Polarization isolation is provided between the hemi and zone beams of the

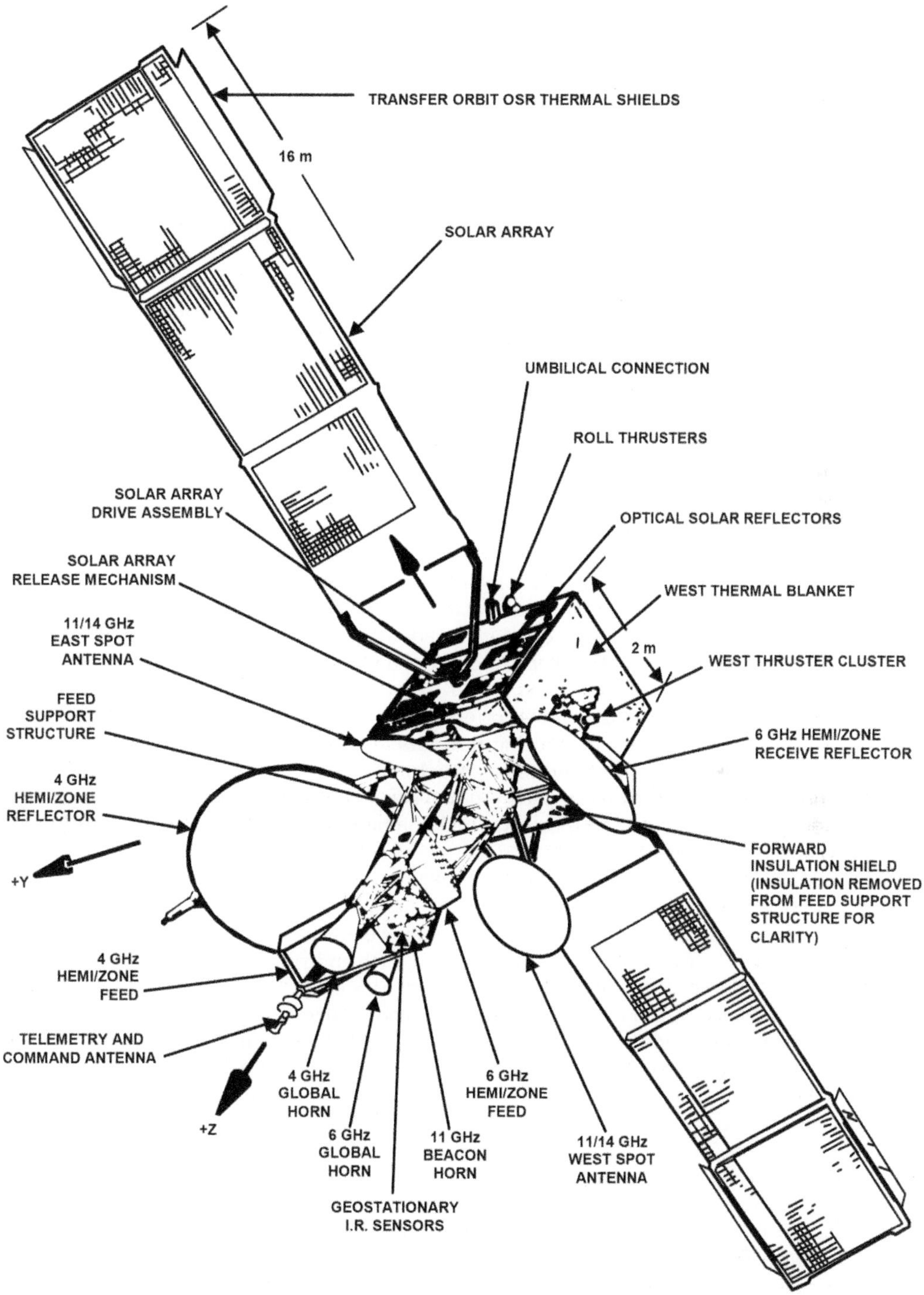

FIGURE 7.6 Intelsat V spacecraft configuration. (Courtesy of Intelsat and FACC.)

6/4-GHz coverages by using right-circular polarizations for the hemi beams and left-circular polarization for the zone beams.

The spacecraft antennas consist of communications and telemetry/command antennas. The communications antennas consist of 4-GHz transmit and 6-GHz receive hemi/zone antennas, 14/11-GHz east and west spot antennas, the 4-GHz transmit, and the 6-GHz receive global antennas and 11-GHz beacon antennas. The hemi/zone transmit and receive antennas consist of single offset-fed parabolic reflectors of diameters 2.244 m and 1.54 m, respectively, illuminated by clusters of square feed horns. Each 14/11-GHz spot beam antenna consists of a nominal 1-m diameter single offset-fed reflector illuminated by a conical corrugated feed horn. The two (4-GHz transmit and 6-GHz receive) global antennas are circularly polarized conical horns. A switching network provides considerable interconnectivity between the various coverage areas as well as channel allocation between hemi, zone, and global coverages. The communications subsystems uses graphite/epoxy material extensively for antenna feeds, waveguides, contiguous output multiplexers, and input channel filters.

The communications subsystem consists of 27 independent transponder channels, of which 24 are at least 72 MHz wide. Three types of traveling-wave tube amplifiers (TWTAs) are used. At 4-GHz, 4.5- and 8.5-W output TWTAs are used for zone and hemi/global use, respectively. At 11 GHz, 10-W output power TWTAs are used for spot beam channels.

Intelsat V is spin stabilized during transfer orbit. The stabilization is provided by active nutation control using thrusters. In geosynchronous orbit, the spacecraft is three-axis-stabilized. Pitch attitude is controlled by a fixed-momentum-wheel reaction torque. The roll attitude is controlled by thrusters. The yaw attitude is passively maintained because of roll-yaw coupling due to the angular momentum. The pitch unbalanced solar pressure disturbance torque, a magnetic roll torque compensation scheme using a dipole aligned with the spacecraft yaw axis is included. This reduces the number of daily thruster firings around the roll axis from 200 to about 17 during the solstices.

The station keeping mode occurs during corrections for north-south or east-west station keeping. These corrections are implemented by firing thrusters in pairs. This also results in disturbance torques due to thrust misalignment and imbalance. In this mode, the momentum wheel is either operational or commanded to a preset speed. An active yaw control is also provided by using finite digital sun sensors and yaw thrusters. The expected pointing accuracy (3σ) is $\pm 0.12°$ in the roll and pitch axes and $\pm 0.33°$ in the yaw axis.

The Intelsat V propulsion system consists of a solid-propellant apogee motor and a combined catalytic/electrothermal monopropellant hydrazine subsystem. Two 22.2-N (5-lb_f) thrusters are used during geosynchronous transfer orbit for orientation correction and active nutation damping. Spacecraft spin/despin, east-west station keeping and pitch and yaw are performed by the 2.67 N (0.6 lb_f) thrusters. These thrusters also serve as a backup to the 0.3-N (0.07-lb_f) electrothermal thrusters normally used for north-south station keeping. Such thrusters can deliver an average specific impulse of about 304 seconds by heating hydrazine propellant to about 2,200°C prior to ejection. The mass saving of hydrazine over 7 years is on the border of 20 kg compared to the use of catalytic hydrazine thrusters, which deliver a specific impulse of about 230 seconds. The propellant is kept in two screen-type surface-tension propellant/pressurant tanks which are fed to two redundant sets of thrusters.

The electrical power system for Intelsat V is designed to accommodate a continuous spacecraft primary load of approximately 1 kW for a 7-year orbit lifetime. Primary power is provided by two sun-oriented solar array wings. During any period of insufficient solar array power, such as during an eclipse, power is supplied by two nickel-cadmium batteries. Each solar array wing consists of a deployment mechanism, three rigid panels, and a solar array drive mechanism which provides a rotation of 1 revolution per day to keep the array pointed at the sun. The total panel area is 18.12 m^2 and is covered with 17,568 solar cells. During transfer orbit, the array is stowed in such a way that one outer panel per wing provides load support and battery charging.

Each battery consists of 28 series-connected 34-Ah nickel-cadmium cells. The allowable depth of discharge is less than 55%. The power control electronics consist of a power control unit (PCU) and shunt dissipators. The output of each solar array wing is independently regulated to 42 ± 0.5 V dc by use of a sequential linear partial shunt regulator.

The structural configuration consists of three main elements: antenna module, communications module, and spacecraft bus. The spacecraft bus structure and communications module structure are U-shaped and when joined together, form a rigid box of dimensions 1.65 × 2.01 × 1.77 m. The box is additionally stiffened by a central tube which supports the apogee motor. Aluminum is the main material of these structures, in contrast to the antenna module, which uses graphite/epoxy extensively in order to limit thermal distortion.

Thermal control of the Intelsat V spacecraft is accomplished by using conventional passive techniques. The passive thermal design is augmented with heater elements for components having relatively narrow allowable temperature ranges. High thermal dissipators, such as the TWTAs, are located on the north and south panels of the main body so that they may efficiently radiate their energy to space via heat sinks and optical solar reflector radiators. The east and west panels, antenna deck, and aft surfaces are covered with multilayer insulation to minimize the effect of solar heating on equipment temperature during a diurnal cycle. Thermal control for the antenna module tower is achieved by the use of a three-layer thermal shield. Thermal control of the antenna reflectors, positioners, feeds, and horns is obtained by using thermal coatings and insulation. Heater elements are used on the propellant tanks, lines, valves, apogee motors, and batteries to maintain temperature above the minimum allowable levels.

Most of the satellites built in the 1970s, such as the Intelsat V, are three-axis-stabilized in geosynchronous orbit with similar configuration. However, recent satellites are much heavier; launch mass up to 6,000 kg, electric power up to 16 kW, and communications capacity several orders of magnitude higher, in the range 140 Gbit/s. This is possible with the use of advanced technology in the majority of the subsystems. Communication payloads have used deployable antennas up to 16 m in diameter, high level of frequency reuse, higher frequencies, and complex feed network. Attitude control systems provide order of magnitude higher pointing accuracy by using fine attitude sensors, such as star tracker and fiber gyros and modern control techniques. Propulsion systems use bipropellant and electric propulsion to increase specific impulse, resulting in saving of propellant mass. Structures are made of carbon fiber epoxy resulting in lower mass, high stiffness, and dimensional stability. Current satellites use high power and large number of TWT amplifiers resulting in higher power dissipation without significant increase in radiator size. In thermal control, heat pipes and deployable radiators are used. There is dramatic increase in electric power requirements. Current satellites use high efficiency triple junction solar cells and nickel hydrogen and lithium ion batteries with higher power mass density. Communications satellites are used for several different applications as discussed in the following section.

Fixed Satellite Services

Fixed satellite services (FSS) are offered essentially in C band (6/4 GHz), Ku band (14/12 GHz) and, now, Ka band (30/20 GHz), while defense fixed services are primarily in UHF band [320 MHz/240 MHz, X band (8/7 GHz), and Ka band 30/20 GHz]. This service is between fixed antennas with a clear line of sight to the satellite and 3–6 dB of link margin. Largely GEO but now several new systems in MEO and LEO are planned.

Mobile Satellite Services

Mobile satellite services (MSS) are offered in the lower VHF, UHF bands (137 MHz, 400/432 MHz, 1.6/1.5 GHz, 2.0 GHz, 2.5/2.6 GHz bands) because mobile terminals may be shadowed or partially blocked. They need to have frequencies that are more tolerant of not always having a clear line of sight to the satellite. They also need higher link margins (10–22 dB) because of blockage by foliage, power line poles, buildings, hillsides, etc. They also need more power (power flux density) to be received especially by small portable or mobile user terminals (i.e., transceivers).

Broadcast Satellite Services

A direct broadcast satellite (DBS) receives TV and radio programs from the central earth stations and transmits them back, after very high signal power amplification, directly to homes for reception on parabolic antennas of less than 1 m diameter. The feasibility of direct broadcast services was demonstrated by ATS-6 in 1974 through experiments in India for educational purposes.

Broadcast satellite services (BSS) are offered essentially 18/12 GHz bands allocated by the ITU. DirecTV and Dish in the United States, BskyB and Eutelsat Hotbird in Europe, NHK's Broadcasting Satellites in Japan, Nimiq in Canada are some BBS Services around the world—All GEO. The size of user terminals for BBS today is in the 0.8–0.35 m range.

The DBS satellites differ from conventional satellites, mainly in terms of transmitted power. For conventional satellites, TWTAs typically operate at 10 W. However, on DBS satellites, TWTAs operate nominally at 200 W. This exceptional high power level influences spacecraft design, specifically temperature control.

Internet Satellite Services

Recently, there is great interest in providing Internet services by satellites. Several high-throughput satellites in GEO and several satellites in LEO are planned for Internet. Emphasis on the areas not covered currently by terrestrial services.

Some of the challenging recent communications satellites are discussed in the following section.

EchoStar T1 In 2009, Space System Loral (SSL) launched EchoStar T1 (formerly TerreStar-1), at that time the world's largest commercial satellite (Figure 7.7). Built for TerreStar Networks, EchoStar T1 has an 18-m antenna reflector that enables voice, data, and video communications to be transmitted to mobile devices the size of a typical smartphone using 2 GHz spectrum. The spacecraft uses a two-way ground-based beam forming (GBBF) system, making it capable of generating hundreds of spot beams covering the continental United States, Canada, Alaska, Hawaii, Puerto Rico, and the U.S. Virgin Islands. EchoStar T1 is located at a geostationary orbital slot at 111.0° West longitude (Griffin and French 1991). It had a launch mass of 6,910 kg (15,230 lb) launched by Ariane.

Figure 7.7 EchoStar T1.

High-Throughput Satellite High-throughput satellite (HTS) is a classification for communications satellites that provide at least twice, though usually by a factor of 20 or more, the total throughput of a classic FSS satellite for the same amount of allocated orbital spectrum, thus, significantly reducing cost-per-bit. ViaSat-1 does provide more than 100 Gbit/s of capacity, which is more than 100 times the capacity offered by a conventional FSS satellite. When it was launched in October 2011, ViaSat-1 had more capacity (140 Gbit/s) than all other commercial communications satellites over North America combined at that time.

The significant increase in capacity is achieved by a high-level frequency reuse and spot beam technology which enables frequency reuse across multiple narrowly focused spot beams (usually in the order of hundreds of kilometers), as in cellular networks, which both are defining technical features of high-throughput satellites. By contrast, traditional satellite technology utilizes a broad single beam (usually in the order of thousands of kilometers) to cover wide regions or even entire continents. In addition to a large amount of bandwidth capacity, HTS are defined by the fact that they often, but not solely, target the consumer market. In the last 10 years, the majority of high-throughput satellites operated in the K_a band, however, this is not a defining criterion. At the beginning of 2017, there were at least 10 K_u band HTS satellites projects, of which three were already launched and seven were in construction.

HTS are primarily deployed to provide broadband Internet access service (point-to-point) to regions unserved or underserved by terrestrial technologies. Here, they can deliver services comparable to terrestrial services in terms of pricing and bandwidth. While many current HTS platforms were designed to serve the consumer broadband market, some are also offering services to government and enterprise markets, as well as to terrestrial cellular network operators who face growing demand for broadband backhaul to rural cell sites. For cellular backhaul, the reduced cost per bit of many HTS platforms creates a significantly more favorable economic model for wireless operators to use satellite for cellular voice and data backhaul.

ViaSat-1 ViaSat-1, shown in Figure 7.8, is a high throughput communications satellite owned by ViaSat Inc. and Telesat Canada. Launched on October 19, 2011, aboard a Proton rocket, it holds the Guinness record for the world's highest capacity communications satellite with a total

FIGURE 7.8 ViaSat-1.

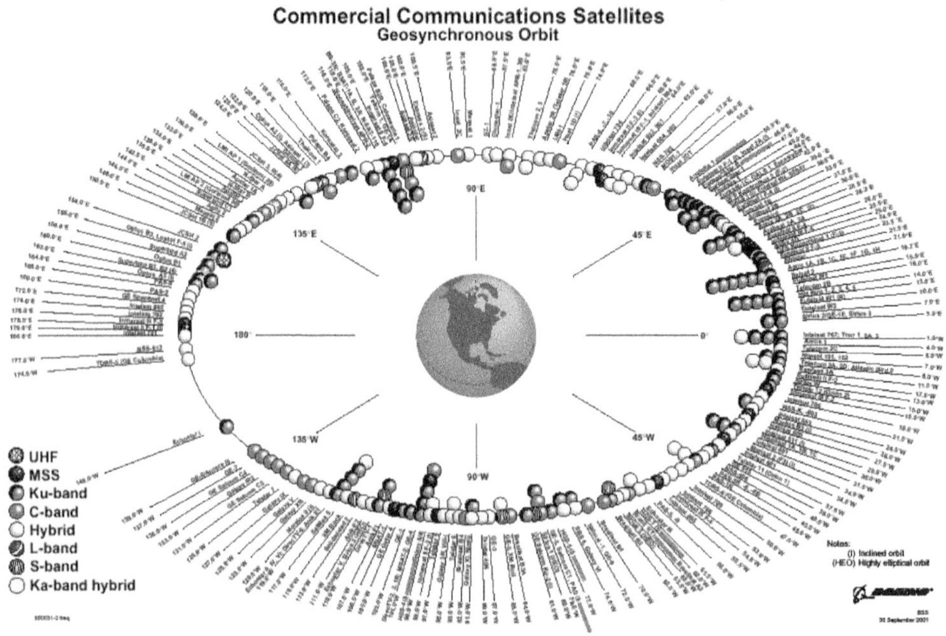

FIGURE 7.9 Commercial communications satellites in geosynchronous orbit.

capacity in excess of 140 Gbit/s, more than all the satellites covering North America combined, at the time of its launch. ViaSat-1 is capable of two-way communications with small dish antennas at higher speeds and a lower cost-per-bit than any satellite before. The satellite was positioned at the Isle of Man registered 115.1° west longitude geostationary orbit point, with 72 Ka-band spot beams; 63 over the United States (eastern and western states, Alaska and Hawaii), and nine over Canada.

The Canadian beams are owned by satellite operator Telesat and will be used for the Xplornet broadband service to consumers in rural Canada. The U.S. beams will provide fast Internet access called Exede, ViaSat's satellite Internet service. Its launch mass is 6,740 kg, launched by Proton and built by Space System Loral (SSL).

Geosynchronous orbit is very crowded with commercial communications satellites as shown in Figure 7.9.

Navigation Satellite

Navigational satellites are used to determine position of a user in three dimensions (e.g., latitude, longitude, altitude) by measuring the instantaneous range from three spacecraft. Velocity in three dimensions is determined by examining changes in position over time. Satellite ranging and triangulation provide an effective way to create a position "fix."

The first satellite navigation system, Transit, used by the U.S. Navy, was tested in 1960. It used a constellation of five satellites and could provide a navigational fix approximately once per hour. During the Cold War arms race, the nuclear threat was used to justify the cost of providing a more capable system. These developments led eventually to the deployment of the Global Positioning System (GPS). The U.S. Navy required precise navigation to enable submarines to get an accurate fix of their positions before they launched their SLBMs. The USAF had requirements for a more accurate and reliable navigation system, as did the U.S. Army for geodetic surveying for which

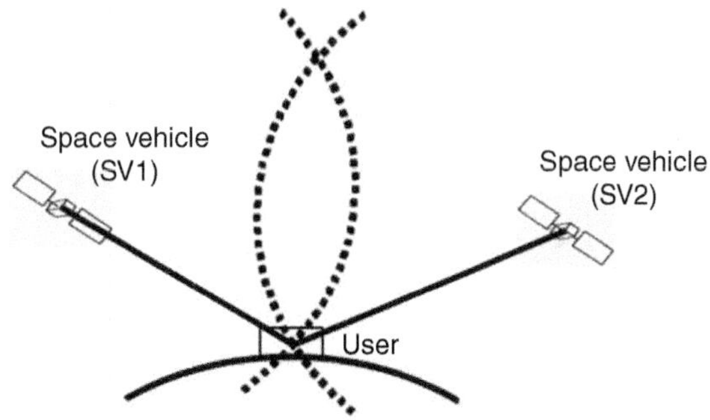

Figure 7.10 Two-dimensional example of determination of user's position.

purpose they had developed the SECOR system. SECOR used ground-based transmitters from known locations that sent signals to satellite transponder in orbit. A fourth ground-based station, at an undetermined position, could then use those signals to fix its location precisely. The last SECOR satellite was launched in 1969.

In 1978, the first experimental Block-I GPS satellite was launched and by December 1993, GPS achieved initial operational capability (IOC), indicating a full constellation (24 satellites) was available and providing the Standard Positioning Service (SPS). Full operational capability (FOC) was declared by Air Force Space Command (AFSPC). Theory behind GPS is as follows:

Two-Dimensional Example

Range from a single satellite puts the user on a circle. Range from a second satellite positions the user at one of two locations identified by the intersection of two circles, as shown in Figure 7.10. How do we tell which location is correct? One of the two locations is unreasonable and we ignore it.

To produce these results, the user must be able to answer these key questions: (1) Where are the satellites located (to define the center of each sphere)? and (2) How far away is each satellite (to define the radius of each sphere)?

Position of the spacecraft in three dimensions is given by SV_x, SV_y, and SV_z and position of the user in three dimensions is given by U_x, U_y, U_z.

$$\text{Range} = \sqrt{(SV_x - U_x)^2 + (SV_y - U_y)^2 + (SV_z - U_z)^2}$$

Since communication signals travel at the speed of light ($c = 3 \times 10^8$ m/s),

$$\text{Range} = c(t_{\text{received}} - t_{\text{sent}})$$

The equation becomes

$$\sqrt{(SV_x - U_x)^2 + (SV_y - U_y)^2 + (SV_z - U_z)^2} = c(t_{\text{received}} - t_{\text{sent}})$$

What is known is satellite position (SV_x, SV_y, SV_z) and time that the communication signal is sent by the satellite, t_{sent}. What is not known is user position, U_x, U_y, and U_z. One strategy that can be used to solve the problem is to obtain communication signals from three different spacecraft. Rely on the user's clock to identify the time that each communication signal is received so that three range measurements can be created. Three measurements create a unique solution for the three unknown components of user position. However, the problem is that the user's clock must match the clocks used by the satellites. A timing error of 20 ns produces a range error of 6 m.

Only atomic clocks can provide this accuracy (10–20 ns of UTC). To get desired accuracy, the user must have an atomic clock. Atomic clocks are expensive, so providing each user with an atomic clock is not feasible. What can the user do instead? Create an additional range measurement from a fourth spacecraft to estimate errors in the user's clock. Accuracy of the system is now limited by random errors present in the system. Some of these errors can be reduced by using special techniques.

GPS was designed for and is funded and controlled by the U.S. Department of Defense, but its use has been extended to commercial users. GPS permits land, sea, and airborne users to instantaneously determine their three-dimensional position, velocity, and time 24 hours a day, under all weather conditions, anywhere in the world. The system consists of orbiting satellites, a ground control system, and the individual receivers of the users. The system is used by aircraft, ships, spacecraft, land-vehicles, and handheld users. GPS is now used worldwide for aid in auto drive. Figure 7.11 shows GPS nominal configuration.

Global Navigation Satellite System Global navigation satellite system (GLONASS) is a Russian system that is similar to U.S.-based GPS. The system consists of 24 satellites in three 19,100 km circular orbit planes. It is designed to have five satellites in view of a user at any time.

Galileo Galileo is a proposed European navigation satellite constellation, expected to be fully operational in 2019. It will be interoperable with GPS and GLONASS. It will consist of 30 satellites at 24,000 km in circular orbits inclined at 55–60°. It will have 27 operational and 3 active spares.

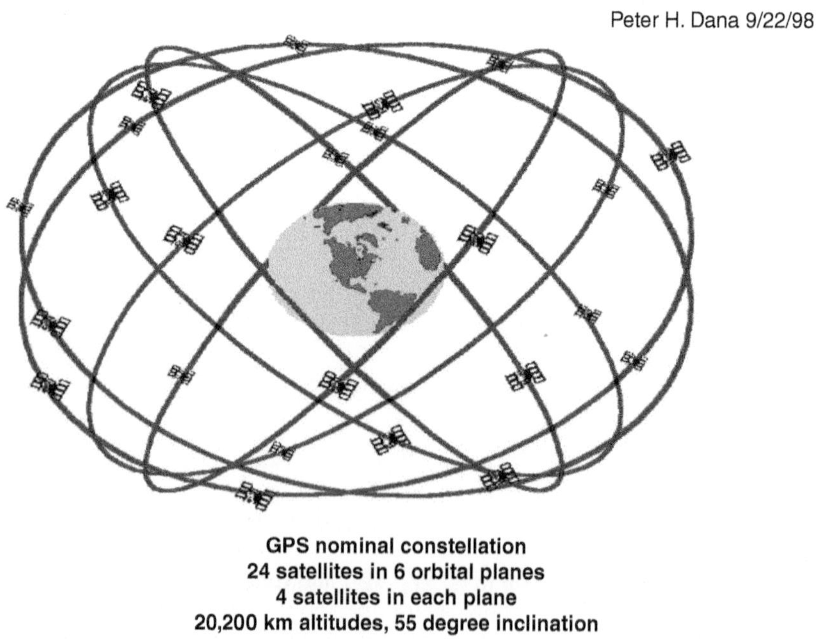

FIGURE 7.11 GPS nominal configuration with 12 hours orbit.

Meteorological Satellites

Conventionally, used ground-based or airborne weather observation systems include radiosonde stations, meteorological radars, weather ships, and so on. However, these do not provide global coverage and suffer from serious limitations in making certain types of observations. In contrast, satellite observation of meteorological phenomena and parameters can provide nearly continuous global coverage and it overcomes the problems of the installation and maintenance of, and the transmission of data from, meteorological sensing instruments in remote land areas, oceans, or high altitudes in the atmosphere.

Satellites obtain a full map of the coverage region using a radiation wavelength in either the visible or infrared range. Radiation in the visible range is absorbed or reflected differently by atmospheric gases, masses of water vapor, cloud, snow, land, or water. The visible images, however, can be obtained only when the underlying region of the earth is illuminated by the sun. This limitation is surmounted by using infrared (IR) images. By taking visual and IR images of all of earth, the motion of a cloud formation over a large portion of the globe can be measured.

There are two types of meteorological satellites: low-altitude sun-synchronous and high-altitude geosynchronous orbit satellites. A typical low-altitude sun-synchronous satellite uses a polar orbit with a 1.92-hour period and cycles back to cover the same earth surface every 12 hours. On the other hand, the geosynchronous orbit satellite has a continuous view of one area of the earth's surface. The optical and spacecraft attitude control requirements for viewing the thin (19 km) tropospheric/stratospheric shell of the earth are quite formidable for the geosynchronous-altitude spacecraft, but are within the present state of the art. For example, to achieve a ground resolution of 1 km in the visible light range for an imaging system corresponds to a field of view of 0.2 mrad and a maximum allowable jitter of the optical control and, thus, an attitude control of no more than 2.5 μrad during the picture-taking sequence. Silicone photodiodes (solid state) are good for the visible light range and require no cooling. For infrared detectors to work efficiently, cryogenic temperatures are required. Hence, infrared detectors require special coolers.

The first meteorological satellite was Tiros 1, launched on April 1, 1960. Tiros, and the later Nimbus and ESSA satellites, are all low-altitude meteorological satellites. ATS I, a spin-stabilized satellite, launched on December 7, 1966, was geosynchronous and provided for the first time a full-disk photograph of the earth and its cloud cover.

The sixth World Meteorological Congress 1971 urged the member countries of the World Meteorological Organization (WMO) to conduct comprehensive atmospheric and meteorological observations in their geographical regions. The primary function of WMO is to plan, produce, and distribute worldwide weather data. Five satellites form the geosynchronous satellite portion of this network: West Europe's Meteosat is positioned at 0°, the Soviet Union's GOMS at 70 to 80°E, Japan's GMS at 140°E, and the U.S. GOES at 75° and 135°W longitudes.

GOES-16, as shown in Figure 7.12, is an American geosynchronous weather satellite, which is part of the Geostationary Operational Environmental Satellite (GOES) system operated by the U.S. National Oceanic and Atmospheric Administration. It is the first of the next generation of geosynchronous environmental satellite. It provides atmospheric and surface measurements of the earth's Western Hemisphere for weather forecasting, severe storm tracking, space weather monitoring, and meteorological research.

The GOES-16 spacecraft uses A2100 bus, is three-axis-stabilized, and designed for 10 years of on-orbit operation preceded by up to 5 years of on-orbit storage. It is built by Lockheed Martin and launch mass is 5,192 kg and dry mass 2,857 kg. Its dimension is 6.1 × 5.6 × 3.9 m and 4 kW power. It provides near-continuous observations as well as vibration isolation for the earth-pointed optical bench and high-speed spacecraft-to-instrument interfaces designed to maximize data collection. The cumulative time that GOES-16 science data collection (including imaging) is interrupted by momentum management, station-keeping and yaw flip maneuvers will be under 120 min/year. This represents a nearly two order of magnitude improvement compared to older GOES satellites.

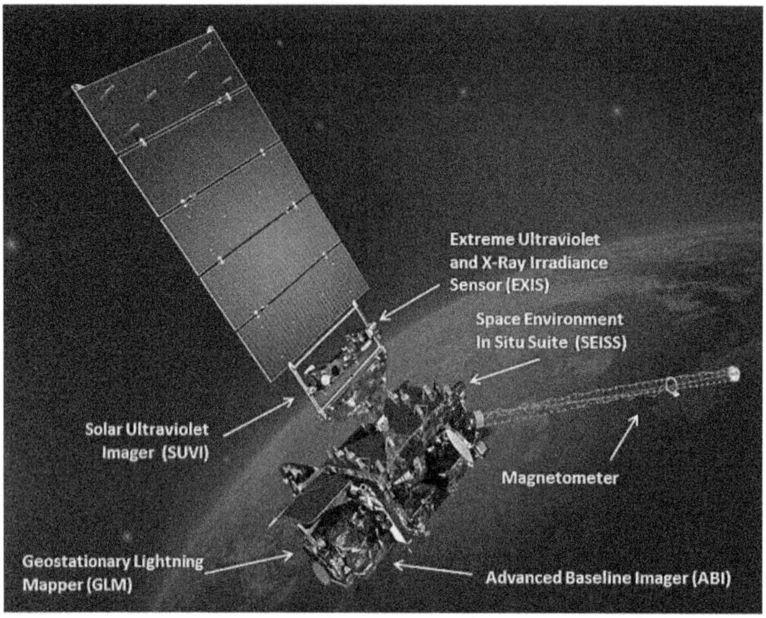

FIGURE 7.12 GOES-16 satellite.

The GOES-16 instrument suite includes the following three types of instruments: Advanced Baseline Imager (ABI), Geostationary Lightning Mapper (GLM), and sensing, solar imaging, Solar Ultraviolet Imager (SUI). The ABI is the primary instrument on GOES-16 for imaging earth's weather, climate, and environment. ABI is able to view the earth across 16 spectral bands, including two visible channels, four near-infrared channels, and 10 infrared channels. It will provide three times more spectral information, up to four times the spatial resolution (depending on the band), and more than five times faster coverage than earlier satellites. Forecasters will be able to use the higher resolution images to track the development of storms in their early stages. GLM takes continuous day and night measurements of the frequent intracloud lightning that accompanies many severe storms, and does so even when the high-level cirrus clouds atop mature thunderstorms may obscure the underlying convection from the imager. GLM's potential for improvement is in tornado warning lead time and false alarm rate reduction. Solar Ultraviolet Imager (SUVI) is a telescope that observes the sun in the extreme ultraviolet (EUV) wavelength range. SUVI will observe and characterize complex, active regions of the sun, solar flares, and the eruptions of solar filaments that may give rise to coronal mass ejections. Depending on the size and the trajectory of solar eruptions, the possible effects to the earth's environment, referred to as space weather, include the disruption of power utilities, communication and navigation systems, and possible damage to orbiting satellites and the International Space Station. SUVI observations of flares and solar eruptions will provide an early warning of possible impacts to the earth environment and enable better forecasting of potentially disruptive events.

Earth Resource Satellites

The work on the study of the earth's resources by satellites began in 1960 with the initiation of the Earth Resources Technology Satellite (ERTS) Program, later called Landsat. The first satellite under this program, Landsat 1, was launched in July 1972. It was equipped with a multispectral scanner (MSS), which transmitted data on water, soil, vegetation, and minerals to earth stations for interpretation and analysis by scientists in the United States and 30 other countries. Landsat 2 and 3, launched in 1975 and 1978, respectively, have sent data to more than 100 nations participating in the program. The new satellite, Landsat-D, provides further improvements in the scanning capability with its thematic mapper (TM) sensor. The data collected by Landsat are used in a wide range of disciplines, including agricultural crop classifications, forest inventories, geological structure identifications, water resource determinations, oceanographic resource locations, and environmental impact studies.

The Landsat 1, 2, and 3 satellites orbit in the sun-synchronous orbit, which has an 81° inclination to the equator and at an altitude of 919 km. The orbit period is 103 minutes. The near-polar orbit is sun synchronous, always crossing the equator from north to south at 9:30 A.M. suntime. Each successive orbit is shifted westward by 2,875 km at the equator. The 14 orbits each day are shifted 159 km (at the equator) westward relative to the 14 orbits on the preceding day. After 18, the orbital cycles repeat. Thus, it is possible to map the earth completely in 18 days and to generate fairly high resolution images of 185×185 km areas in a variety of spectral bands. Landsat-D operates at an altitude of 705 km.

The multispectral scanner consists of a vibrating mirror, a telescope, and a bank of 24 detectors to give four colored images in four different spectral bands: green, red, and two infrared bands. A single image is approximately 185×185 km^2, and made up of 7.5 million picture elements, each being 80 m square. Each picture element is quantized into one of 64 intensity levels and the intensity value is transmitted to the ground. One full frame is sent every second.

The first three satellites, Landsat 1, 2, and 3, used a Nimbus-type bus, developed originally for meteorological satellites. Landsat-D uses the Multi-Mission Spacecraft bus. It has two sensors, and MSS and a TM. The TM is a new sensor that provides images of 30-m resolution.

The TM is essentially bigger and better than MSS. It images in seven spectral bands versus four for the scanner. The mapper has 100 imaging detectors compared with 24 for the multispectral scanner. Landsat-D has the capacity to generate 800 frames a day (550 MSS and 250 TM), compared to 190 MSS frames per day for Landsat 1, 2, and 3.

Landsat 8

Landsat 8, as shown in Figure 7.13, is an American earth observation satellite launched on February 11, 2013. It is the eighth satellite in the Landsat program. Originally called the Landsat Data Continuity Mission (LDCM), it is collaboration between NASA and the United States Geological Survey (USGS).

Figure 7.13 Landsat 8.

The Landsat 8 spacecraft was built by Orbital Sciences Corporation, under contract to NASA, and uses Orbital's standard LEOStar-3 satellite bus. The spacecraft launch mass is 2,623 kg and dry mass 1,512 kg. It is in sun synchronous orbit with an altitude of 702 km and inclination of 98.22° with orbit period of 98.8 minutes. All components, except for the propulsion module, are mounted on the exterior of the primary structure. A single deployable solar array generates power for the spacecraft components and charges the spacecraft's 125 amp-hour nickel-hydrogen (Ni-H_2) battery. A 3.14-terabit solid state data recorder provides data storage aboard the spacecraft and an X-band antenna transmits OLI and TIRS data either in real time or played back from the data recorder. It has two imagers, Operational Land Imager (OLI), built by Ball Aerospace and Thermal Infrared Sensor (TIRS), built by Goddard Space Flight Center. The OLI and TIRS are mounted on an optical bench at the forward end of the spacecraft. Resolution is 15/30/100 m for panchromatic/multispectral/thermal images.

Landsat 8's Operational Land Imager (OLI) improves on past Landsat sensors. The OLI instrument uses a push broom sensor instead of whiskbroom sensors that were utilized on earlier Landsat satellites. The push broom sensor aligns the imaging detector arrays along Landsat 8's focal plane allowing it to view across the entire swath, 115 mi (185 km) cross-track field of view, as opposed to sweeping across the field of view. With over 7,000 detectors per spectral band, the push broom design results in increased sensitivity, fewer moving parts, and improved land surface information. OLI collects data from nine spectral bands. Seven of the nine bands are consistent with the TM and Enhanced Thematic Mapper Plus (ETM+) sensors found on earlier Landsat satellites, providing for compatibility with the historical Landsat data, while also improving measurement capabilities. Two new spectral bands, a deep blue coastal/aerosol band and a shortwave-infrared cirrus band, will be collected, allowing scientists to measure water quality and improve detection of high, thin clouds.

The Thermal Infrared Sensor (TIRS) conducts thermal imaging and supports emerging applications such as evapotranspiration rate measurements for water management. The TIRS focal plane uses GaAs Quantum Well Infrared Photodetector arrays (known as QWIPs) for detecting the infrared radiation—a first for the Landsat program. The TIRS data will be registered to OLI data to create radiometrically, geometrically, and terrain-corrected 12-bit Landsat 8 data products. Like OLI, TIRS employs a push broom sensor design with a 185-km swath width. Data for two long wavelength infrared bands will be collected with TIRS. This provides data continuity with Landsat 7's single thermal IR band and adds a second.

RADARSAT-1

RADARSAT-1 is Canada's first commercial earth observation satellite (Figure 7.14). It utilized synthetic aperture radar (SAR) to obtain images of the earth's surface to manage natural resources and monitor global climate change. RADARSAT-1 was launched on November 4, 1995, from Vandenberg

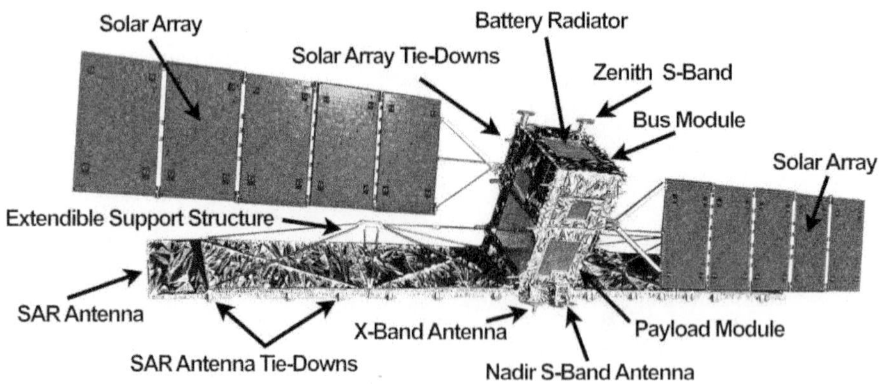

Figure 7.14 RADARSAT-1.

AFB in California, into a sun-synchronous orbit above the earth with an altitude of 798 km (496 mi) and inclination of 98.6°. Orbit period is 100.7 minutes and descending node 6.00 hours and ascending node 18.00 hours. Developed under the management of the Canadian Space Agency (CSA) RADARSAT-1's images are useful in many fields, including agriculture, cartography, hydrology, forestry, oceanography, geology, ice and ocean monitoring, arctic surveillance, and detecting ocean oil slicks.

RADARSAT-1 used a synthetic aperture radar (SAR) sensor to image the earth at a single microwave frequency of 5.3 GHz, in the C band (wavelength of 5.6 cm). The SAR support structure was designed and manufactured by Northrop Grumman Astro Aerospace and deployed to 15 m (49 ft) in length on orbit. Unlike optical satellites that sense reflected sunlight, SAR systems transmitted microwave energy toward the surface and recorded the reflections. Thus, RADARSAT-1 imaged the earth, day or night, in any atmospheric condition, such as cloud cover, rain, snow, dust or haze.

RADARSAT-1 was manufactured by MDA formerly SPAR and Ball Aerospace. Its launch mass was 2,750 kg and power 2,100 W. Its planned mission duration was 5 years, but lasted more than 17 years. Transmitter peak power is 5 kW and average power 300 W.

Each of RADARSAT-1's seven beam modes offered a different image resolution. The modes included Fine, which covers an area of 50 × 50 km (31 × 31 mi) [2,500 km^2 (970 mi^2)] with a resolution of 10 m (33 ft); Standard, which covered an area of 100 × 100 km (62 × 62 mi) (10,000 km^2 [3,900 mi^2)] and had a resolution of 30 m (98 ft); and ScanSAR wide, which covered a 500 × 500-km (310 × 310-mi) [250,000 km^2 (97,000 mi^2)] area with a resolution of 100 m (330 ft). RADARSAT-1 also had the unique ability to direct its beam at different angles.

Astronomy

Following are the space programs related to astronomy.

- Hubble Space Telescope
- Chandra X-ray Observatory
- Advanced X-ray Astrophysics Facility (AXAF)
- Extreme Ultraviolet Explorer (EUVE)
- Far Ultraviolet Spectroscopic Explorer (FUSE)
- Compton Gamma Ray Observatory (CGRO)
- Rossi X-ray Timing Explorer (XTE)
- Solar and Heliospheric Observatory (SOHO)

Hubble Space Telescope

NASA's Hubble Space Telescope was the first astronomical observatory to be placed into orbit around earth with the ability to record images in wavelengths of light spanning from ultraviolet to near-infrared. Launched on April 24, 1990, aboard the Space Shuttle Discovery, Hubble is currently located about 340 mi above earth's surface where it completes 15 orbits per day—approximately one every 95 minutes. The satellite moves at the speed of about 8 km/s (5 mi/s), fast enough to travel across the United States in about 10 minutes.

Hubble Space Telescope, as shown in Figure 7.15, is 13.2 m long and 4.26 m wide at the back, where the scientific instruments are housed. Weighing nearly 1,134 kg, the telescope is approximately the same size and weight as a school bus. The observatory is powered by two solar arrays that convert sunlight into electrical energy that is stored in six large batteries. The batteries allow the observatory to operate during the shadowed portions of Hubble's orbit when earth blocks the satellite's view of the sun. In the middle of the spacecraft, near its center of gravity, are four 45-kg reaction wheels used to reorient the observatory. The observatory uses high-precision gyroscopes (gyros) to detect its rate and direction of motion. In addition to gyros, Hubble has

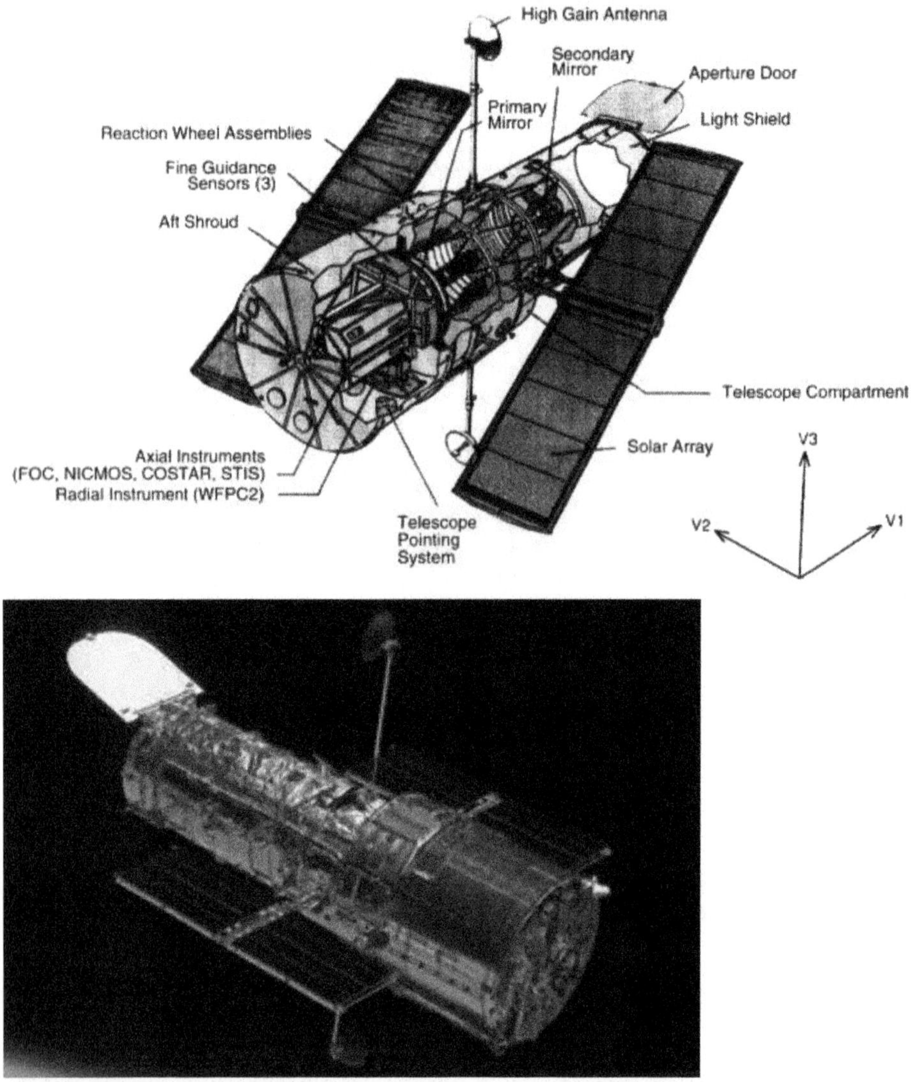

FIGURE 7.15 Hubble Space Telescope.

three Fine Guidance Sensors (FGS) that act within the spacecraft's overall pointing and control system to keep the telescope virtually motionless while observing. Hubble jitters less than 7 milliarcseconds (mas) in a 24-hour period when locked on its target. This is equivalent to shining a laser on a dime 200 mi away for this period. Commands and data are transmitted between the spacecraft and the control center through two high-gain antennas that communicate through NASA's Tracking Data and Relay Satellite System that are in geosynchronous orbit. The science data is then forwarded from the control center to the Space Telescope Science Institute via a wide-area network for processing, dissemination, and archiving.

Hubble telescope is classified as a Cassegrain, named after a 15th century French cleric who was among the first to suggest this basic optical design. Light hitting the telescope's main or primary mirror is reflected to a smaller, secondary mirror suspended above the primary. The secondary, in turn, reflects the light back through a hole in the primary where it enters Hubble's instruments (cameras and spectrographs) for final focus before it hits their detectors. Hubble's primary mirror is not only exquisitely polished, but at 2.4 m in diameter, collects an immense amount of light. Hubble can detect objects that are 10 billion times fainter than the unaided eye can see. High above the blurring effects of earth's atmosphere, Hubble also gets a much clearer view of the cosmos than do telescopes located on the ground. The space telescope can distinguish astronomical objects with an angular diameter of a mere 0.05 arcsec—the equivalent to discerning the width of a dime from a distance of 86 mi. This resolution is about 10 times better than the best typically attained by even larger, ground based telescopes. High resolution enables Hubble to locate such objects as dust discs around stars or the glowing nuclei of extremely distant galaxies.

Also because it circles above the atmosphere, Hubble can view astronomical objects across a wider range of the electromagnetic spectrum than ground-based telescopes which are limited by atmospheric absorption at various wavelengths. This gives astronomers using Hubble a fuller view into the energetic processes that create the radiation seen and measured. Finally, Hubble's observations are predictably consistent. The telescope's seeing conditions do not change from day to day or even orbit to orbit. Astronomers can revisit targets with the expectation that they will be imaged at the same high quality each time. This optical stability is critical for detecting tiny motions or other small variations in celestial objects. Such is not the case for ground-based observatories, where observing conditions vary with weather and directly affect the quality of the images acquired.

Commercial Imaging Systems

Several imaging satellites have been developed in the last few decades. Table 7.1 gives performance of some of these satellites.

	French satellites			US satellites				
	SPOT 1-4	SPOT 5	Pleiades	Ikonos II	QuickBird 2	WorldView 1	GeoEye 1	WorldView 2
Mean Wavelength, λ_{mean}(μm)	0.615	0.615	0.655	0.675	0.675	0.675	0.675	0.675
Aperture diameter, D(m)	0.31	0.31	0.65	0.70	0.6	0.6	1.1	1.1
λ_{mean}/D, (μr)	1.98	1.98	1.01	0.96	1.13	1.13	0.61	0.61
Pixel pitch, p(μm)	13	4.6	9.12	12	12	8	8	8
Focal length, f(m)	1.08	1.08	12.905	10	8.84	8.00	13.3	13.3
IFOV, (μr)	12.0	4.3	0.7	1.2	1.4	1.0	0.6	0.6
F#	3.48	3.48	19.85	14.29	14.73	13.33	12.09	12.09
Optical Q	0.165	0.466	1.426	0.804	0.829	1.13	1.02	1.02
Eff. angular res., α_{eff} (mr)	12.8	5.3	1.4	1.8	2.1	1.8	1.0	1.0
Altitude, h (km)	828	828	693	682	451	497	685	773
Resolution, $\Delta = r \cdot \text{IFOV}$ (m)	10.0	3.5	0.49	0.82	0.61	0.50	0.41	0.46
Eff. ground res., Δ_T (m)	10.6	4.4	1.00	1.24	0.94	0.88	0.70	0.78

Table 7.1 Commercial Imaging Satellites

Figure 7.16 WorldView-4 imaging satellite.

The following section provides details of WorldView-4, highest resolution commercial imaging satellite at the time of its launch in 2016.

WorldView-4 WorldView-4, as shown in Figure 7.16, previously known as GeoEye-2, is a third generation commercial earth observation satellite launched on November 11, 2016. The spacecraft is operated by DigitalGlobe. With a maximum resolution of 31 cm (12 in), WorldView-4 provides similar imagery as WorldView-3, the highest resolution commercially available at the time of its launch. The spacecraft's telescope is called the GeoEye Imaging System-2 also known as SpaceView 110, which was designed and built by ITT Corporation (later ITT Exelis and Harris). The telescope mirror is 1.1 m (3.6 ft) in diameter. It provides panchromatic images at a highest resolution of 31 cm/pixel (12 in/pixel) between 450 and 800 nm, and multispectral images at 124 cm/pixel (49 in/pixel) in red, green, blue, and near-infrared channels (655–690, 510–580, 450–510, and 780–920 nm, respectively). It is sun synchronous orbit with altitude 610 km, inclination 98°, and orbit period 97 minutes. It was built by Lockheed Martin.

Scientific Satellites

There have been several scientific satellites for interplanetary missions. Some of the satellites are discussed in the following section.

Mars Global Surveyor

The Mars Global Surveyor, as shown in Figure 7.17, arrived on Mars on September 11, 1997, with a mission to map the Martian surface. Spacecraft employed aerobraking to enter orbit. Improperly deployed solar array caused change in aerobraking schedule and steerable high gain antenna stuck

Figure 7.17 Mars Global Surveyor.

in fixed position. JPL is in charge of the mission and Lockheed Martin Astronautics built the spacecraft. Primary mission ended January 31, 2001.

Cassini

Cassini, as shown in Figure 7.18, was launched in October 1997. Huygens probe will investigate moon Titan. It is a cooperative effort between many agencies: NASA, ESA, ASI, and various individual institutions. JPL built the spacecraft bus and ESA built the probe.

Human Space Flight

Soviet Mir Space Station

The Soviet Mir Space Station was complex and made up of six different modules; the first was the Mir module (Figure 7.19). There were six docking stations on the module: five on the transfer compartment and one at the opposite end for resupply. In addition, a separate Docking Module was attached to facilitate and provide proper clearances for the docking of the US Space Shuttle between 1995 and 1998. 31 spacecraft and 64 supply vessels docked. Soyuz-TM and Space Shuttles docked for personnel changes; Progress-M spacecraft dock for resupply.

International Space Station

International Space Station (ISS) (1998–present) (Figure 7.20). Zarya was the first component launched on November 20, 1998; there are currently six partners: United States, Russia, Canada,

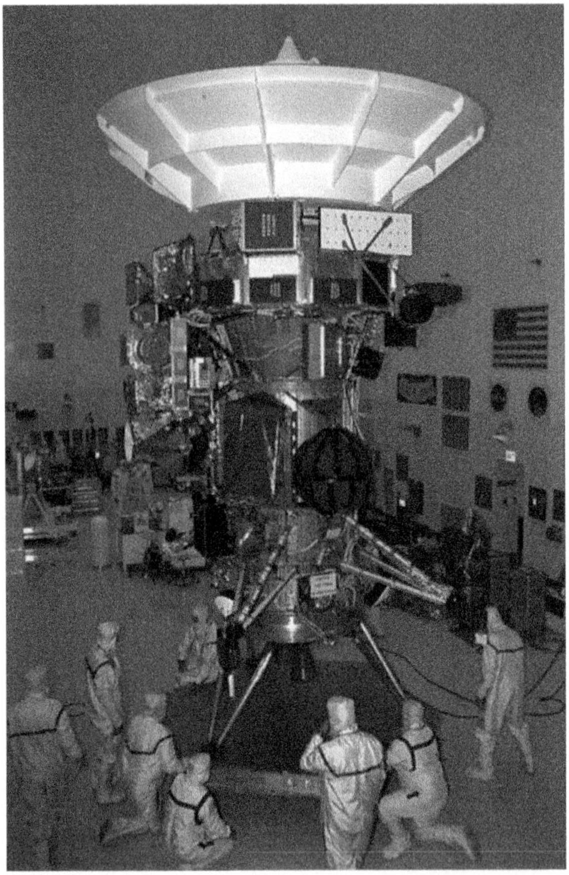

FIGURE 7.18 Cassini.

Japan, ESA, and Brazil. Crews as large as seven can be supported with the addition of the habitation module for 90-day stays. Mass of the station is 187,016 kg; orbit: 360 km, inclination 51.63°, period 92 min; living volume: 425 m^3.

Military Satellites

The first military use of satellites was for reconnaissance. In the United States the first formal military satellite program, Weapon System 117L, was developed in the mid-1950s. Within this program, a number of subprograms were developed, including Corona. Satellites within the Corona program carried different code names. The first launches were code named Discoverer. This mission was a series of reconnaissance satellites, designed to enter orbit, take high-resolution photographs and then return the payload to earth via parachute. The Corona program continued until May 25, 1972. Corona was followed by other programs. They remain classified.

The Soviet Union began the *Almaz* (Russian: Алмаз) program in the early 1960s. This program involved placing space stations in earth orbit as an alternative to satellites. Three stations were launched between 1973 and 1976: Salyut 2, Salyut 3 and Salyut 5. Following Salyut 5, the Soviet Ministry of Defense judged in 1978 that the time consumed by station maintenance outweighed the benefits relative to automatic reconnaissance satellites.

Figure 7.19 Soviet Mir Space Station.

Figure 7.20 International Space Station.

Communications

Communications satellites have been found to be of great military use in the following applications: long-distance communications to isolated areas; high-data-rate transmission for intelligence applications; rapid extension of circuits into new areas; and mobile communications to moving platforms, such as aircrafts, ships, and submarines. The design of a military satellite is driven by survivability against physical attack, electronic countermeasures, security for sending secret messages, and flexibility for mobile system and remote-area access. These design considerations result in an increase in complexity and cost. Communications satellites are used for military communications applications. Typically military satellites operate in the UHF, SHF (also known as X-band) or EHF (also known as K_a band) frequency bands.

The U.S. Armed Forces maintains international networks of satellites with ground stations located in various continents. Signal latency is a major concern in satellite communications, so geographic and meteorological factors play an important role in choosing teleports. Some of the major military activities of the U.S. Army are in foreign territories; the U.S. government needs to subcontract satellite services to foreign carriers headquartered in areas with favorable climate.

Military Strategic and Tactical Relay, or Milstar, is a constellation of military satellites managed by the United States Air Force. Six spacecraft were launched between 1994 and 2003, of which five are operational, with the sixth lost in a launch failure. They are deployed in geostationary orbit and provide wideband, narrowband, and protected military communication systems. Wideband systems support high-bandwidth transfers. Protected systems offer more sophisticated security protection like antijam features and nuclear survivability, while narrowband systems are intended for basic communications services that do not require high bandwidth.

The United Kingdom also operates military communication satellites through its Skynet system. This is currently operated with support from Astrium Services and provides near worldwide coverage with both X band and ultrahigh frequency (UHF) services. Skynet 5 is the United Kingdom's most recent military communications satellite system. There are four Skynet satellites in orbit, with the latest launch completed in December 2012. The system is provided by a private contractor, Astrium, with the U.K. government paying service charges based on bandwidth consumption. Following subsections discuss U.S. Navy's Mobile User Objective System (MUOS).

Mobile User Objective System

The Mobile User Objective System (MUOS), as shown in Figure 7.21, is a UHF (300 MHz–3 GHz frequency range) SATCOM system, primarily serving the United States Department of Defense (DoD). International allies use is under consideration. The MUOS replaces the legacy UHF follow-on (UFO) system before that system reaches its end of life to provide users with new capabilities and enhanced mobility, access, capacity, and quality of service. Intended primarily for mobile users (e.g., aerial and maritime platforms, ground vehicles, and dismounted soldiers), MUOS will extend users' voice, data, and video communications beyond their lines-of-sight.

MUOS is an array of geosynchronous satellites that will provide global satellite communications (SATCOM) narrowband connectivity for communications use by the United States at data rates up to 384 kbit/s. The program consists of five satellites, four ground stations, and a terrestrial transport network. Lockheed Martin is the prime system contractor and satellite designer for MUOS. First MUOS satellite, MUOS-1, was launched into space successfully on February 24, 2012, carried by an Atlas V rocket.

WCDMA System

The MUOS operates as a global cellular service provider to support the war fighter with modern cell phone-like capabilities, such as multimedia. It converts a commercial third generation (3G) wideband code division multiple access (WCDMA) cellular phone system to a military UHF SATCOM radio system using geosynchronous satellites in place of cell towers. By operating in the UHF frequency band, a lower frequency band than that used by conventional terrestrial cellular networks,

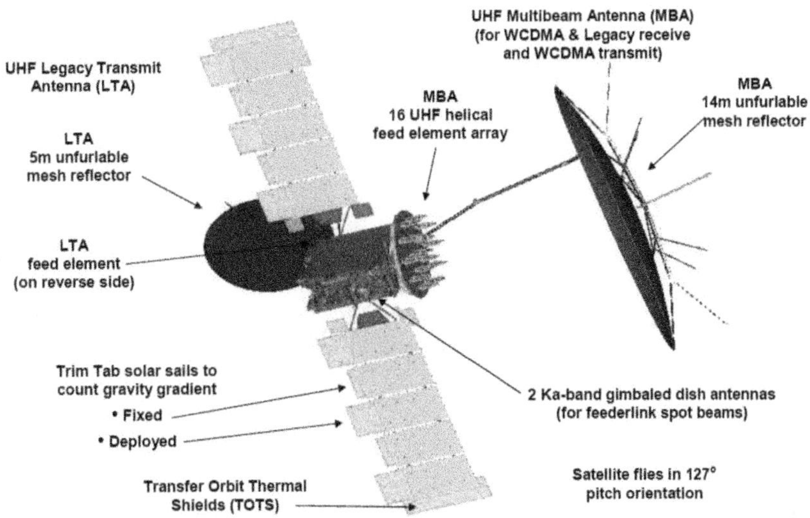

FIGURE 7.21 Mobile user objective system (MUOS).

the MUOS provides warfighters with the tactical ability to communicate in "disadvantaged" environments, such as heavily forested regions where higher frequency signals would be unacceptably attenuated by the forest canopy. The MUOS constellation consists of four operational satellites and one on-orbit spare. MUOS provides military point-to-point and netted communication users with precedence-based and preemptive access to voice, data, video, or a mixture of voice and data services that span the globe. Connections may be set up on demand by users in the field, within seconds, and then released just as easily, freeing resources for other users. In alignment with more traditional military communications methods, preplanned networks can also be established either permanently or per specific schedule using the MUOS' ground-based Network Management Center.

Legacy Payload

In addition to the cellular MUOS WCDMA payload, a fully capable and separate UFO legacy payload is incorporated into each satellite. The "legacy" payload extends the useful life of legacy UHF SATCOM terminals and enables a smoother transition to MUOS.

7.4 Launch Vehicles

A typical flight profile used to obtain geosynchronous orbit is shown in Figure 7.22. The satellite is first injected into a low-altitude circular parking orbit, the low altitude being about 200 km. The satellite, with its upper stages still attached, coasts in the parking orbit until it is above the equator, at which point the first of the upper stages, also called the perigee kick motor (PKM), injects the satellite into a geosynchronous transfer orbit (GTO). The transfer orbit is a highly elliptical orbit with maximum altitude equal to that of the geosynchronous orbit, 35,786 km. The inclination angle of the orbit depends on the latitude of the launch site and launch azimuth. A due-east launch, with azimuth equal to 90°, results in a transfer orbit inclination equal to the latitude of the launch site; any other azimuth angle of the launch will increase the inclination. Since a zero inclination angle is required for fully geostationary satellites, it is advantageous to have the launch site as close to the equator as possible. The satellite is injected into the geosynchronous orbit by the second upper stage, the apogee kick motor (AKM), which not only increases the transfer orbit apogee velocity to match the geosynchronous orbit velocity, but reduces the orbit inclination to zero, as well.

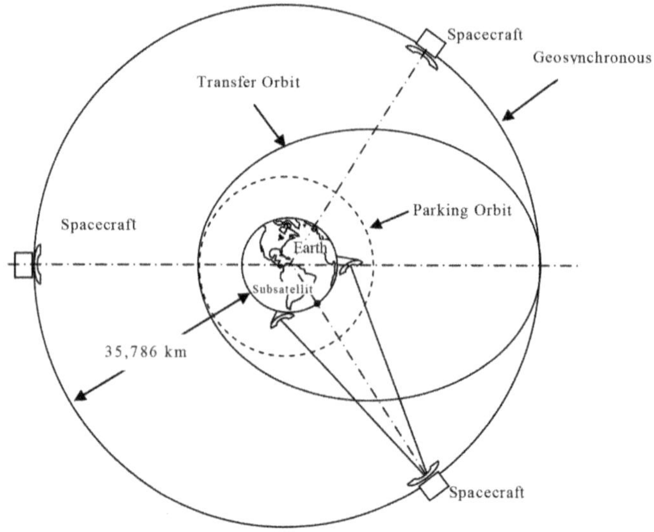

FIGURE 7.22 Flight profile for geosynchronous orbit.

For a geosynchronous orbit satellite, the launch vehicle selected will determine whether the PKM and AKM are part of the spacecraft or of the launch vehicle. Launch vehicles impose several constraints on spacecraft designs. The most obvious constraints are the mass and volume available to the spacecraft. Other launch vehicle characteristics that influence spacecraft design include mechanical and electrical interfaces, radio link performance through the fairing (required for certain spacecraft ground tests while they are enclosed in the fairing), launch vibration environment, nutation angle at separation from spin-stabilized stages, and thermal environment. Therefore, spacecraft designers need to know much more about the launch vehicle than the available mass and volume. This additional information is usually available in the user's manual of the launch vehicle. As an example for launch vehicle, details of Ariane V launch vehicle is given in the following section.

Ariane V

Ariane V, as shown in Figure 7.23, is a European heavy lift launch vehicle that is part of the Ariane rocket family, an expendable launch system used to deliver payloads into geostationary transfer orbit (GTO) or LEO. Two satellites can be mounted using a SYLDA carrier. Up to eight secondary payloads, usually small experiment packages or mini satellites, can be carried with an ASAP. Its height is 46–52 m, diameter 5.4 m, and mass 777,000 kg. Launch site is Guiana Space Center. Payload launch capability is 16,000 kg in LEO and 6,950 kg in GTO. It consists of two stages; details of stages are provided as follows. Ariane V has several versions.

Main Stage

Ariane V cryogenic H173 main stage is called the EPC. It consists of a large tank 30.5 m high with two compartments, one for liquid oxygen and one for liquid hydrogen, and a Vulcain 2 engine at the base with a vacuum thrust of 1,390 kilonewtons (kN) [310,000 pounds-force (lbf)]. The H173 EPC weighs about 189 tonnes, including 175 tonnes of propellant. After the main cryogenic stage runs out of fuel, it can reenter the atmosphere for an ocean splashdown. Attached to the sides are two P241 solid rocket boosters, each weighing about 277 tonnes full and delivering a thrust of about 7,080 kN

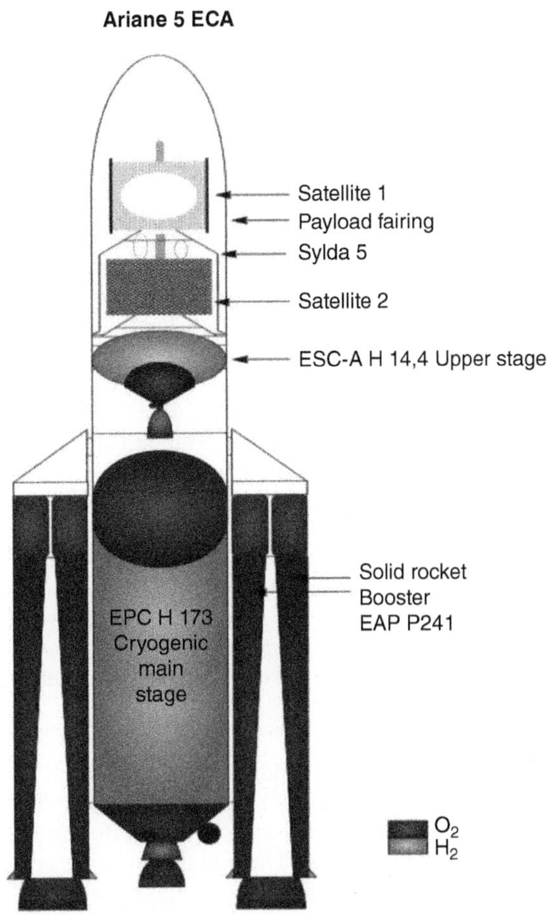

FIGURE 7.23 Ariane 5 ECA launch vehicle.

(1,590,000 lbf). They are fueled by a mix of ammonium perchlorate (68%) and aluminum fuel (18%) and HTPB (14%). They each burn for 130 seconds before being dropped into the ocean.

Second Stage

The second stage is on top of the main stage and below the payload. The Ariane V G used the EPS, which is fueled by monomethylhydrazine (MMH) and nitrogen tetroxide. It also has 10 tonnes of storable propellants. Ariane V ECA uses ESC, which is cryogenic upper stage that is fueled by liquid hydrogen and liquid oxygen.

7.5 Ground Segment

The ground segment consists of satellite operation center (SOC) and several receiving and transmitting earth stations located around the world. SOC is used during satellite launch and monitoring the health of satellites through telemetry, tracking, and also sending command through the earth stations. User of satellite services will have their own earth stations for getting data from satellite payload. In this section, we will discuss two large ground segment systems, Air Force Satellite Control Network (AFSCN) and NASA Deep Space Network (DSN).

Air Force Satellite Control Network

The Air Force Satellite Control Network (AFSCN) provides support for the operation, control, and maintenance of a variety of U.S. Department of Defense and some non-DoD satellites. This involves continual execution of Telemetry, Tracking, and Commanding (TT&C) operations. In addition, the AFSCN provides prelaunch checkout and simulation, launch support, and early orbit support while satellites are in initial or transfer orbits and require maneuvering to their final orbit. The AFSCN provides tracking data to help maintain the catalog of space objects and distributes various data such as satellite ephemeris, almanacs, and other information.

The AFSCN consists of satellite control centers, tracking stations, and test facilities located from around the world. SOCs are located at Schriever Air Force Base near Colorado Springs, Colorado, and various other locations throughout the continental United States. These SOCs are manned around the clock and are responsible for the command and control of their assigned satellite systems. The SOCs are linked to remote tracking stations (RTSs) around the world. Space vehicle checkout facilities are used to test launch vehicles and satellite platforms to ensure that the onboard systems operate within specifications. The RTSs provide the link between the satellites and the SOCs as some satellites, especially in geostationary orbit, never come within view of their control center. Each antenna at an RTS is referred to as a "side." Previously, side A typically included an 18-m diameter (60-ft) dish antenna. Side B typically included a 14-m (46-ft) antenna. At some sites, the B side included a 10-m-diameter (33 ft.) antenna. Over time, however, as the network upgraded and/or replaced the antennas, the old conventions no longer apply.

NASA Deep Space Network

The NASA Deep Space Network (DSN) is a worldwide network of U.S. spacecraft communication facilities, located in the United States (California), Spain (Madrid), and Australia (Canberra), that supports NASA's interplanetary spacecraft missions. It also performs radio and radar astronomy observations for the exploration of the solar system and the universe, and supports selected earth-orbiting missions. DSN is part of the NASA Jet Propulsion Laboratory (JPL). Similar networks are run by Europe, Russia, China, India, and Japan.

DSN currently consists of three deep-space communications facilities placed approximately 120° apart around the earth. They are the Goldstone Deep Space Communications Complex outside Barstow, California; the Madrid Deep Space Communication Complex Madrid, Spain; and the Canberra Deep Space Communication Complex (CDSCC) in Australia. Each facility is situated in semi-mountainous, bowl-shaped terrain to help shield against radio frequency interference. The strategic placement with nearly 120° separation permits constant observation of spacecraft as the earth rotates, and helps to make the DSN the largest and most sensitive scientific telecommunications system in the world. The antennas at all three DSN complexes communicate directly with the Deep Space Operations Center (also known as Deep Space Network operations control center) located at the JPL facilities in Pasadena, California.

The DSN supports NASA's contribution to the scientific investigation of the Solar System: It provides a two-way communications link that guides and controls various NASA unmanned interplanetary space probes, and brings back the images and new scientific information these probes collect. All DSN antennas are steerable, high-gain, parabolic reflector antennas. The antennas and data delivery systems make it possible to acquire telemetry data from spacecraft, transmit commands to spacecraft, upload software modifications to spacecraft, track spacecraft position and velocity, perform very long baseline interferometry observations, measure variations in radio waves for radio science experiments, gather science data, and monitor and control the performance of the network.

FIGURE 7.24 A 70-m antenna at Goldstone.

Each complex consists of at least four deep space terminals equipped with ultra-sensitive receiving systems and large parabolic-dish antennas. There are one 34-m (112-ft) diameter high efficiency antenna (HEF), two or more 34-m (112-ft) beam waveguide antennas (BWG) (three operational at the Goldstone Complex, two at the Robledo de Chavela Complex near Madrid, and two at the Canberra Complex), one 26-m (85-ft) antenna, and one 70-m (230-ft) antenna (Figure 7.24).

References

Agrawal, B. N. 1986. *Design of Geosynchronous Spacecraft*, Prentice-Hall, New Jersey.

Ariane User's Manual, *European Space Agency*. www.arianespace.com/wp-content/uploads/.../Ariane5_Users-Manual_October2016.pdf

Brown, C. D. 2002. *Elements of Spacecraft Design*, AIAA, Washington, DC.

Griffin, M. D. and French, J. R. 1991. *Space Vehicle Design*, AIAA, Washington, DC.

Wertz, J. R., Everett, D. F., and Puschell, J. J. 2011. *Space Mission Engineering: The New SMAD*, Microcosm Press, Hawthorne, CA.

Wertz, J. R. and Larson, W. J. 2010. *Space Mission Analysis and Design*, Springer, NY.

https://www.sslmda.com/html/satexp/terrestar1.html

https://en.wikipedia.org/wiki/High-throughput_satellite

https://en.wikipedia.org/wiki/ViaSat-1

https://en.wikipedia.org/wiki/GOES-16

https://landsat.gsfc.nasa.gov/landsat-8/landsat-8-overview/

https://en.wikipedia.org/wiki/Radarsat-1

https://en.wikipedia.org/wiki/Hubble_Space_Telescope

https://en.wikipedia.org/wiki/WorldView-4

https://en.wikipedia.org/wiki/Mir

https://en.wikipedia.org/wiki/Mobile_User_Objective_System

https://en.wikipedia.org/wiki/Ariane_5

https://en.wikipedia.org/wiki/Air_Force_Satellite_Control_Network

https://deepspace.jpl.nasa.gov/

PART 2

Test and Product Certification of Space Vehicles

Louis L. Maack

The focus of this subsection is on the test program role in the product certification of space vehicles. The test program anchors with concrete evidence the validity of the analytical predictions, which support the space vehicle development, qualification, and flight acceptance process. This section addresses the test basics: why we test, what we test, when we test, how we test, where we test.

Space vehicles can be most simply described as strong, lightweight, stiff, thermoelastically stable containers, with deployable appendages, housing the electronic and mechanical components that satisfy mission requirements in unforgiving environments. They are unique in that they are very expensive, low-volume procurements, requiring a significant amount of hand touch labor during the hardware integration and extensive test program performed to verify and validate by test that the space vehicle can fulfill mission requirements.

There is a significant difference between mission "stated" requirements and "real" requirements. Stated requirements are those provided by the customer at the beginning of a system procurement effort either in a request for information or proposal statement of work (SOW). Real requirements are those that reflect the verified needs of users for a particular system or capability that is agreed on during the system requirements definitization negotiation. Identifying the real requirements and the methodology for verification and validation requires an interactive and iterative requirements development activity, supported by effective practices, processes, mechanisms, methods, techniques, and tools as illustrated in Figure 7.25.

Product certification is both validation and verification of these documented requirements. The NASA Systems Engineering Handbook states that verification consists of proving that a system complies with its requirements, whereas validation consists of proving that the total system accomplishes its purpose.

1. Validate the product can accomplish all mission objectives: did we build the right thing? To validate requirements is to prove that it is possible to satisfy them, to build the right system, ensure that the system does what it is supposed to do and guarantee correctness of an end product, compliance of the system with the customer's needs, and completeness of the system.
2. Verify the product can satisfy all specification requirements: did we build the thing right? System verification, on the other hand, is a process of proving that a system meets its requirements, build the system right, ensure that the system complies with its design requirements, guarantee the consistency of the product at the end of each phase, with itself and with the previous prototypes and enable the smooth transition from model, to prototype, to preproduction unit and ultimately to the production unit.

7.6 Validation Basics

Validation uses the tool set consisting of analysis, test, demonstration, and inspection individually and in combination to ensure that we are building the right system in compliance with the customer's needs, and completeness of the system. Validation ensures that the system does what it is supposed to do, throughout all phases of the mission life, including off-nominal conditions. As we will show later, "*test like you fly*" process is a powerful tool for system validation.

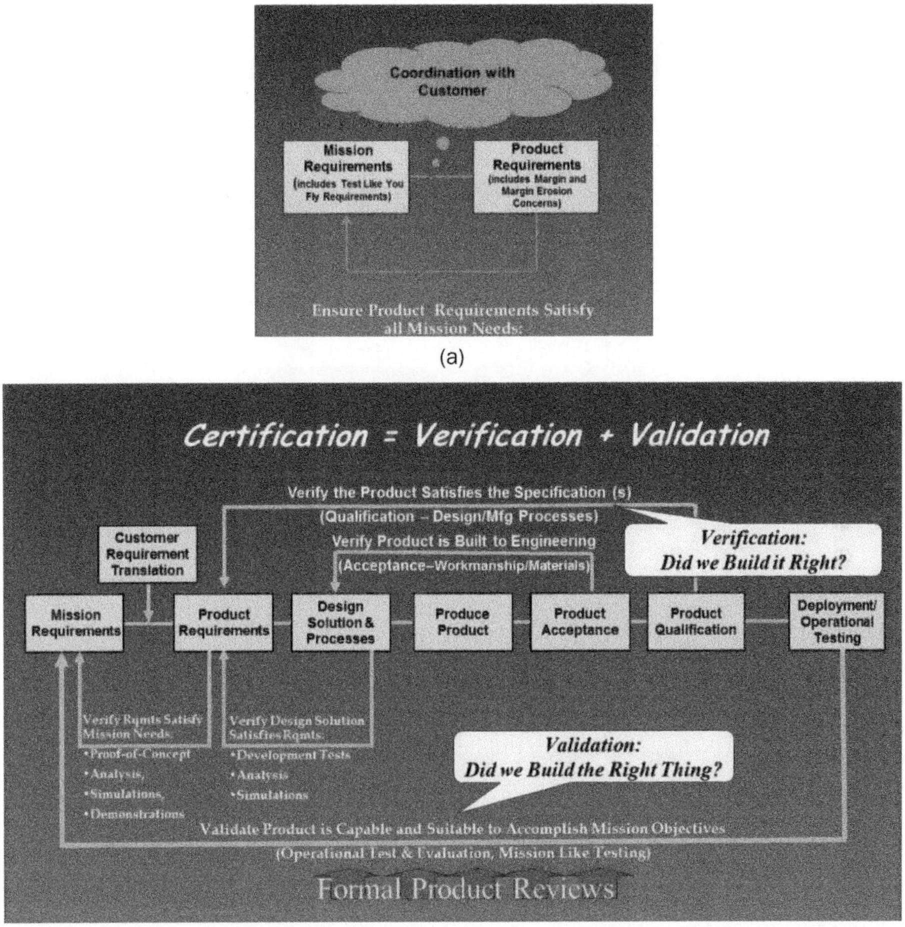

(a)

(b)

FIGURE 7.25 Space vehicle certification process.

7.7 Verification Basics

Verification, similarly, applies the same tool set individually and in combination to verify that the system performance, for the duration of the mission, meets all specified requirements.

- Analysis—test—inspection—analysis
- Analysis—demonstration—analysis
- Analysis—inspection

This includes verification of positive margin with respect to design and maximum operating environments and parameters.

- *Ultimate limit:* Laws of physics, thermodynamics, etc. (e.g., pressurize a propellant tank till it bursts)

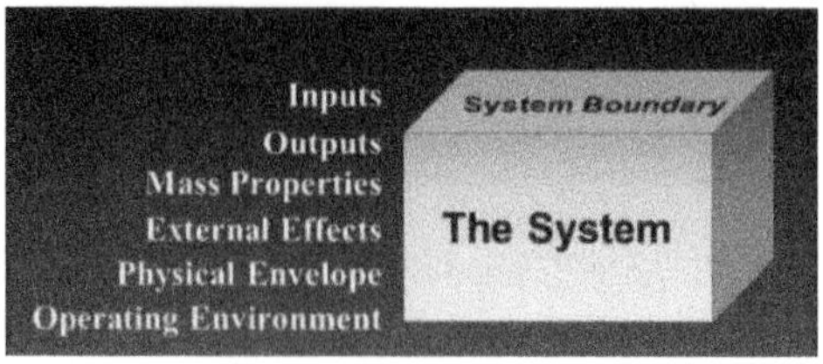

FIGURE 7.26 Requirements addressed at system boundaries.

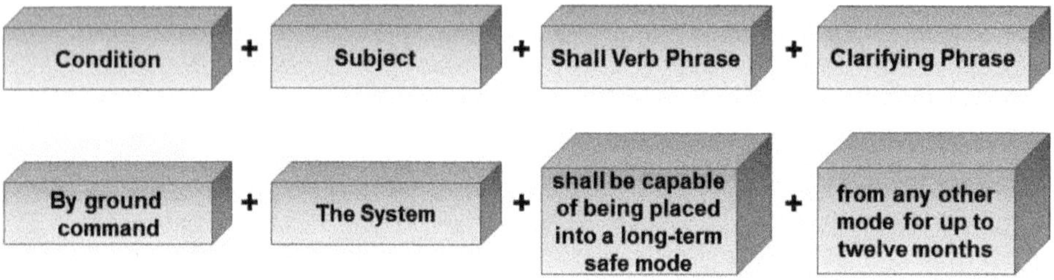

FIGURE 7.27 Requirements writing basics.

- Design limit, test limit, maximum predicted operation, normal operational condition
- *Trend for margin erosion:* When flight article is used to verify margin, minimize how much useful life of the product is consumed by test

7.8 Requirements Development Basics

It is critical to ensure requirements are addressed at the system boundary and that they are externally observable and verifiable as pictured in Figure 7.26.

Basic requirements must be verifiable and precisely stated. Requirements must be concise, clear, unambiguous, and complete. It is important to ensure that there are no missing requirements and to address margin including margin erosion criteria.

A simple rule for formulating each requirement involves four parts as shown in Figure 7.27: a condition statement, a subject, a "shall" verb phrase and a clarifying statement.

7.9 Certification Requirements and Test Plan Development

Requirements development is enabled by adopting "systems—thinking—competencies" as presented by R. Valerdi and W. B. Rouse, "When Systems Thinking Is Not a Natural Act," 5th IEEE Systems Conference, San Diego, CA, April, 2010. The seven system competencies:

- Ability to define the "universe" appropriately—the system operates in this universe
- Ability to define the overall system appropriately—defining the right boundaries
- Ability to see relationships—within the system and between the system and universe
- Ability to see things holistically—within and across relationships

- Ability to understand complexity—how relationships yield uncertain, dynamic, nonlinear states
- Ability to communicate across disciplines—to bring multiple perspectives to bear
- Ability to take advantage of a broad range of concepts, principles, models, methods and tools—because any one view is inevitably wrong

The requirements plan defines how the hardware and software requirements evolve and how to address validation and verification activities in the system life cycle as illustrated in Figure 7.27.

1. Identify the stakeholders and gain an understanding of customers' and users' needs for the planned system and their expectations of it.
2. Identify and clarify requirements by stating them in simple sentences and providing them as a set.
3. Analyze the requirements and define them to ensure that they are well understood and mean the same thing to all stakeholders.
4. Specify the requirements to detail in a specification document.
5. Prioritize the requirements to address the highest priority first and possibly release a version of a product later that addresses lower priorities.
6. Derive, partition, allocate, tracking, and manage requirements.
7. Test and verify requirements to ensure designs, code, test plans, and system products comply with the system requirements.
8. Validate requirements to confirm that the real requirements are implemented.

7.10 Verification Methods

1. Verification by analysis—is a statistical or quantitative analysis or computer simulation that confirms the product design satisfies the specified functional, performance, interface, or design requirements.
2. Verification by test—is a quantitative measurement of performance parameters or functions through the application of functional, electrical, or mechanical stimuli that may include exposure to natural or simulated environments to verify compliance with specified requirements.
3. Verification by demonstration—is the exercise of the system to ensure that specified functions can be performed.
4. Verification by inspection—is the examination of a product's physical characteristics to ensure compliance with workmanship, quality, physical condition, and dimensional tolerance requirements stated in a product specification or design documentation.

Why Do We Emphasize Verification by Test?

In Test We Trust—testing offers the most credible evidence of validation, of the four methods discussed. Test, most realistically, places the design solution relative to its use and involves a very controlled stimulus–response exercise that exposes detailed operation characteristics for clear evaluation.

Test identifies flaws in the overall life cycle validation and verification process including: requirements, analysis, design, parts, and production. In addition, test identifies process variability, the source of defects.

- It is the primary method of test like you fly.
- It is the ultimate proof.
- It is key to mission success.
- All customers require it.

7.11 Test Basics

Testing involves an organized process of stimulating a system (or representation of the system) and measuring responses using special test equipment. The test plan commonly includes a sequence of events, each with a predefined setup and an anticipated response. Test events are structured to prove that a particular design concept offers a viable solution to the item requirements to conform to a documented test process as shown in Figure 7.28.

Upon completion of the test program, we can conclude with confidence that we either can or cannot satisfy the requirements with current design concepts and technology. There are cases where testing is challenging, impossible, or so expensive as to be prohibitive. For example, it is very difficult to test large items in a zero- or microgravity condition. There are drop towers that can sustain the zero-gravity condition for seconds and aircraft that can fly an arc and sustain this condition for several minutes. But these environments cannot handle very large items nor for very long time intervals. The zero-G performance of a whole space station cannot be tested on earth because of its size and must rely on analysis as the principle method of verification.

Testing Categories

The following are the category of tests that can be applied at the unit/component, subsystem, and/or system level of assembly.

1. *Development testing:* Informal tests conducted on representative articles to characterize engineering parameters, gather data, and validate the design approach for the system to perform the mission.

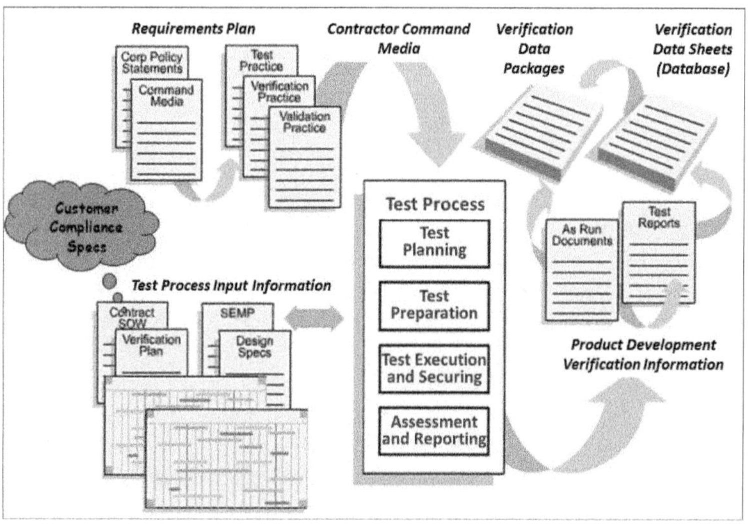

FIGURE 7.28 Typical test process.

2. *Qualification testing:* Formal tests conducted on sacrificial flight units to demonstrate proof of design, verify manufacturing process, and maximum tested margin (these units do not fly unless refurbished and recertified).
3. *Protoqualification/protoflight testing:* Formal tests conducted on flight units to demonstrate limited proof of design tests, verify manufacturing process and workmanship and the maximum tested margin that minimizes consumption of useful life.
4. *Acceptance testing:* Formal workmanship conducted on flight units to verify with margin to maximum and normal operations.

7.12 Compliance Documents

Compliance documents are the road map. They continue to evolve based on lessons painfully learned. Examples of compliance documents are

- Mil-Std (Military Standards) especially Mil-Std-1540 the original space vehicle test compliance document
- NASA test requirements specifications
  - GSFC-STD (NASA Goddard Space Flight Center developed)
  - CEQATR
  - Jet propulsion laboratory design principles
- Prospective customer compliance documents
- Contractor compliance documents and command media
- Standards from the European Space Agency (ESA): European Cooperation on Space Standardization (ECSS) for general industrial standards and European Space Components Cooperation (ESCC) for electrical, electronic and electromechanical parts, including a European Preferred Parts List for space missions.
- Standards for the Japanese Space Exploration Agency (JAXA) and for numerous countries active in space vehicle missions.

Mil-Std-1540 Evolution

This subsection provides a historical context of the "gold standard of testing" Mil-Std-1540 authored by the Aerospace Corporation, which became the primary U.S. government testing standard for space hardware:

April '74—Mil-Std-1540 "Proof Copy," titled: "Test Requirements for Space Vehicles"

May '74—Mil-Std-1540A released

Oct '82—Mil-Std-1540B released with Notices 1 (Jul '89), 2 (Feb '91) and 3 (Feb '91) also released

Sep '94—Mil-Std-1540C released, retitled: "Test Requirements for Launch, Upper-Stage and Space Vehicles" introduced during the acquisition reform/single process initiative (SPI) era.

Jan '99—Mil-Std-1540D released, retitled: "Product Verification Requirements for Launch, Upper Stage and Space vehicles." It relied heavily on Mil-Hdbk-340. It was prepared during acquisition reform. Process oriented, as part of SMC's current Systems Engineering Revitalization program it was found to be inadequate for the current plan of reestablishing key specifications and standards as part of SMC's technical baseline.

Dec '02—Mil-Std-1540E TOR 2003(8583)-1 released. It combined the best of versions B&C, with limited distribution to industry for review and comment.

Jan '04—Mil-Std-1540E TR-2004(8583)-1 released. It was published to support near-term government acquisitions with the intent of using it as a compliance document and to provide the basis for the updated Military Standard 1540 (revision E) with unlimited distribution. It was considered an "interim mil-std" and was also released as SMC-TR-04-17.

September '06—Mil-Std-1540E TR-2004(8583)-1 Rev: A released with minor revisions and increased references to aerospace TORs (e.g., software TOR); it was also released as SMC-TR-06-11.

13 June '08—also released as SMC-S-016 as an SMC Standard.

NASA Test Specifications

GSFC-STD-7000—General Environmental Verification Specification (GEVS) for GSFC Flight Programs and Projects:

- NASA Goddard Space Flight Center (GSFC) developed.
- Provides requirements and guidelines for environmental verification programs for GSFC payloads, subsystems, and components and describes methods for implementing those requirements. It contains a baseline for demonstrating by test or analysis the satisfactory performance of hardware in the expected mission environments, and that minimum workmanship standards have been met. It elaborates on those requirements, gives guideline test levels, provides guidance in the choice of test options, and describes acceptable test and analytical methods for implementing the requirements.

CEQATR—Constellation Program Environmental Qualification and Acceptance Testing Requirements

- NASA Johnson Space Center (JSC) developed. This document defines the baseline environmental test requirements to be applied for the qualification and acceptance of Constellation (Cx) flight hardware, systems, and major assemblies. It also presents alternative strategies and associated test requirements when appropriate approval to deviate from the baseline qualification/acceptance program has been granted. Qualification testing is a risk mitigation strategy for potential design deficiencies. The testing stresses the hardware and systems beyond the design operating and nonoperating conditions to ensure positive margins exist for design requirements, material, and process variability. Acceptance testing is a risk mitigation strategy for verifying that the manufacturing and assembly process has been accomplished in an acceptable manner and the product performs within specified parameters.

Jet Propulsion Lab (JPL)—Design Principles

Other Approaches/Terminologies

Some other terms/approaches typical for a production test program that you may encounter

- HALT—highly accelerated stress testing
- HASS—highly accelerated stress screening

- HAST—highly accelerated stress testing
- HASA—highly accelerated stress audit
- ESS—environmental stress screening
- ESSEH—ESS of electronic hardware
- TAF—test analyze and fix

HALT and HASS testing is a method which is used for improving the reliability of hardware. It was originally developed for commercial industries where the actual "real-world" environments are unknown, and proved to be an effective technique for reducing field failures and warranty claims. The philosophy of HALT is to apply severe environments to a prototype of the hardware in order to rapidly precipitate potential failures, and then correct them before production commences. The philosophy of HASS is to apply environments to production hardware in order to precipitate manufacturing defects.

Validation and Verification Methodology

In physics, an operator is a function that acts on the space of physical states. As a result of its application on a physical state, another physical state is obtained, very often along with some extra relevant information.

Leveraging "systems thinking," we view the test program as a planned sequence of operators acting singly or in combination on the spacecraft and its subsystems, leading to flight worthiness certification. The sequenced aggregate of all these operators acting on the space vehicle constitutes the Formal Test Program shown in Figure 7.29.

- T: Power-on-time
- Mk: Mission environments (all expected operating environments addressed M1: launch acoustic, M2 launch vibration, M3 shock, M4 thermal, …, etc.)
- A: Ambient environment

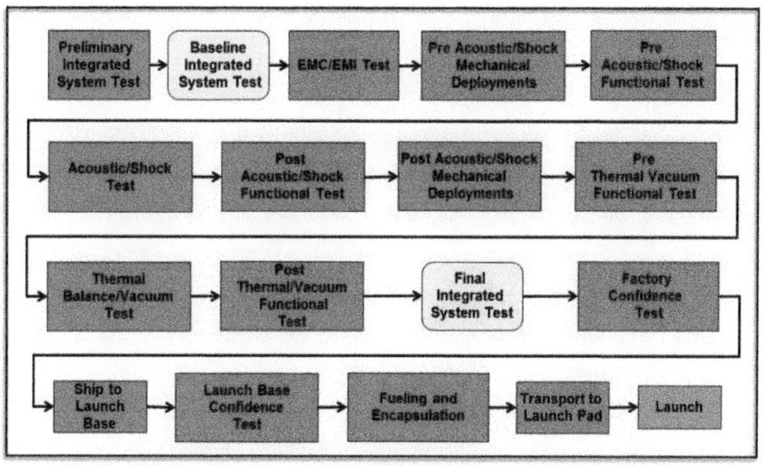

Figure 7.29 Typical test program.

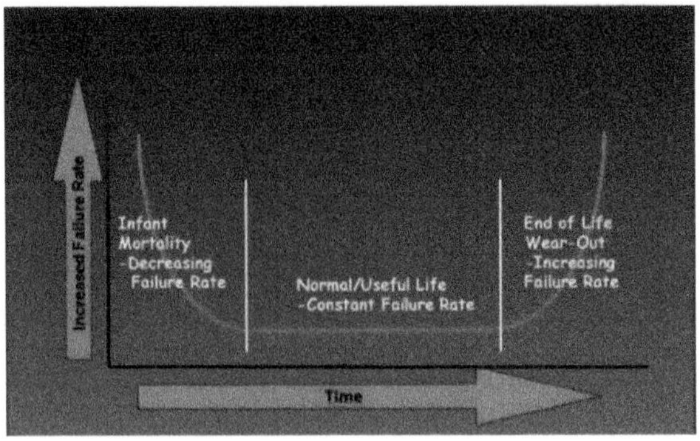

FIGURE 7.30 Bath tub curve.

- Pm: Performance tests (P1: Power tests, P2: Propulsion tests, …, etc.)
- Dn: Demonstrations (D1: TLYF days/weeks-in-Life demo, …, etc.)
- Ip: Inspections (I1: Structure and fastener inspection, I2: Faraday cage inspection, …, etc.)

Operator Examples

Power-on-time operator T, with its "bath tub" curve, acts on spacecraft electronic components to screen out "infant mortality" and ensure reduced failure rate in-orbit. T is the optimal power on or "burn in" time to eliminate early failures without adversely using up the design life of the component as shown in Figure 7.30.

Mission thermal environment operator M_k subjects the space vehicle components individually and the integrated space vehicle to mission thermal or thermal vacuum cycles over temperature extremes with margin as shown in Figure 7.31.

Launch environment operator M_l subjects the space vehicle components individually and the integrated space vehicle to launch environment levels with margin as shown in Figure 7.32.

Detailed structural testing of space vehicles is provided in supporting Subsection 7.29, Test Verification, Qualification and Flight Acceptance.

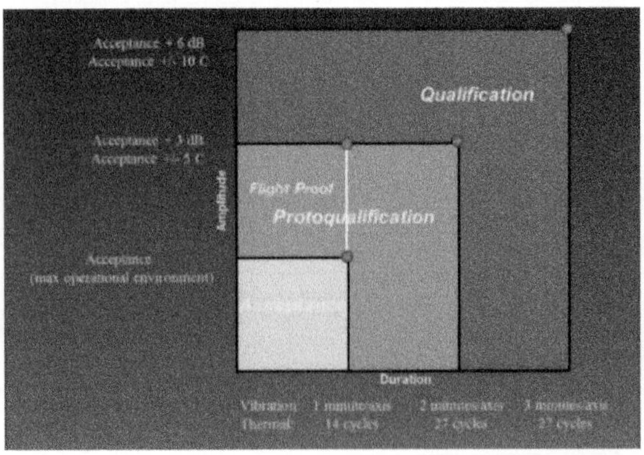

FIGURE 7.31 Thermal/thermal vacuum cycling.

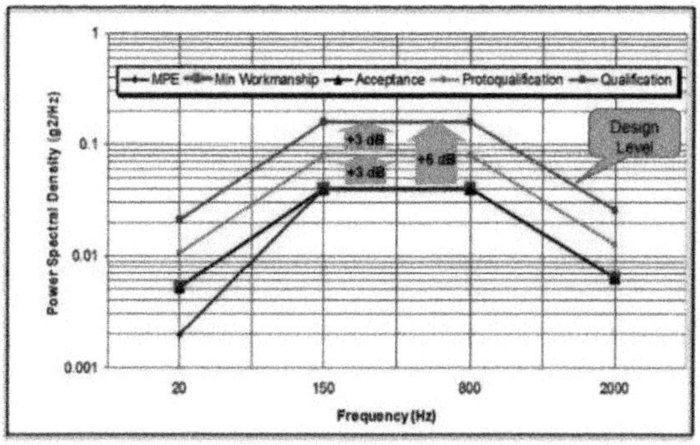

FIGURE 7.32 Notional launch environment test levels.

Why Test Like You Fly?

As smart as we all are, especially if we include the best in the aerospace industry ... even if we use all of the technology, tools and lifetimes of lessons learned expertise ... a complex system of systems will always find a way to outsmart all of us ... that is why "test as you fly (TLYF)" is a both a journey and a work in progress. TLYF is a key to both mission success and space vehicle product certification.

The TLYF discussion that follows has been excerpted, from AEROSPACE REPORT NO. TOR-2010(8591)-6 (Test Like You Fly: Assessment and Implementation Process) with the permission and encouragement of the author, Julie White of the Aerospace Corporation. This Aerospace report, with cover page shown in Figure 7.33, is approved for public release with unlimited distribution.

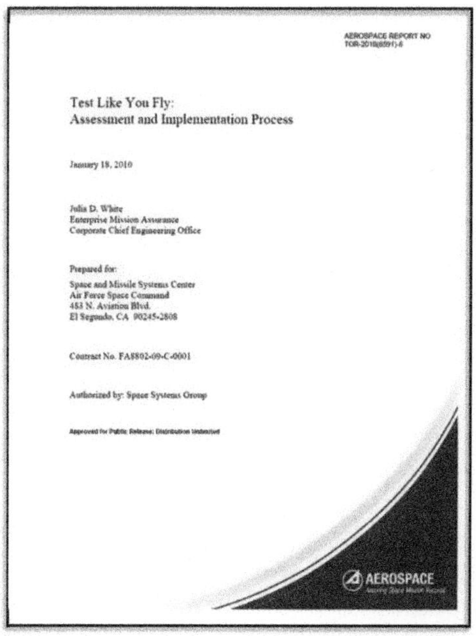

AEROSPACE REPORT NO
TOR-2010(8591)-6

Test Like You Fly:
Assessment and Implementation Process

January 18, 2010

Julia D. White
Enterprise Mission Assurance
Corporate Chief Engineering Office

Prepared for:
Space and Missile Systems Center
Air Force Space Command
483 N. Aviation Blvd.
El Segundo, CA 90245-2808

Contract No. FA8802-09-C-0001

Authorized by: Space Systems Group

Approved for Public Release: Distribution Unlimited

AEROSPACE

FIGURE 7.33 Cover of TLYF TOR.

7.13 TLYF Overview

TLYF is an approach that provides a unique assessment process that focuses on determining the "mission-related" or "like you fly" risks associated with potential flaws in our space systems. It encompasses much more than "test."

TLYF can be any of the following:

- A systems engineering methodology that focuses on more than verifying requirements, but focuses on the validation of a system's ability to perform its mission.
- A process to assess the mission concepts for testability and to assess the risk for those concepts that are not readily testable.
- A mission assurance mission validation tool to ensure that the acquired systems can accomplish the intended mission.
- A test technique for mission operability at all levels of assembly. This has an "end-to-end" aspect, meaning that it crosses interface boundaries. Ends are truly the ultimate ends during the total operations chain test (TOCT), but are "brought in" for early validation of segments and lower levels of assembly.
- A specific readiness test: total operations chain (space + ground) days-in-the-life (DITL)/ weeks-in-the-life (WITL) operability test.

Principles and Tenets of TLYF

First: The system should never experience expected operations, environments, stresses, or their combinations for the first time in flight.

Second: Do only smart things with the space system.

Third: TLYF is a complement to other forms of performance and functional testing, NOT a replacement for other perceptive testing (e.g., vibration testing with electronics powered and active).

Fourth: When you cannot test like you fly—worry (or do risk management).

TLYF Drivers

Before anyone can test like you fly, it is necessary to know how the mission will be flown. The process to assess the mission concepts for testability flows from that knowledge.

- What is feasible and practical to test?
- What needs to be available (documentation, hardware, software, procedures, trained personnel) to conduct feasible, practical tests in a flight-like way?

Like you fly testing is driven by mission operations concepts, flight constraints, flight conditions, and mission considerations. TLYF has an "operability" aspect and an "end-to-end" aspect, even if the ends are not very far apart. The prime "like you fly" characteristics are those in the time domain: continuous clock, timing, duration, order/sequence of events, including an appropriate set of initial conditions, a set of time-ordered events that include transactions and interactions among the elements between and at the ends, and any and all mission characteristics are applied (where possible).

TLYF Working Definition

TLYF is a prelaunch systems engineering approach that examines all applicable mission characteristics and determines the fullest practical extent to which those characteristics can be applied in testing.

Reducing Risk in Key Mission Areas

The TLYF approach focuses on reducing risk in key mission areas. Lessons from a number of catastrophic failures have taught us that it is necessary to include TLYF tests that are "mission operations centric," where the focus is to demonstrate pre-launch the capability of an integrated items or system to perform the mission. Demonstrating that hardware survives an environment, although a necessary prerequisite, is not the same thing as showing that the integrated hardware, software, processes, and procedures work.

First: Programmatic and physical limitations and constraints need to be considered upfront.

Second: Flight HW is not exposed to known damaging test configurations or environments. However, if the execution of a TLYF test reveals a flaw that damages the HW unknowingly, from a TLYF perspective, this is viewed as a successful test, to preclude that same flaw on orbit with detrimental effects.

Third: Recognize the value of other tests and the importance of TLYF as a tool in ensuring mission success.

Fourth: Account for the risks derived from not conducting LYF tests.

TLYF Implementation

The implementation of test you fly involves an assessment process shown in Figure 7.34. It starts with an analysis of the entire mission and the development of "like you fly" tests.

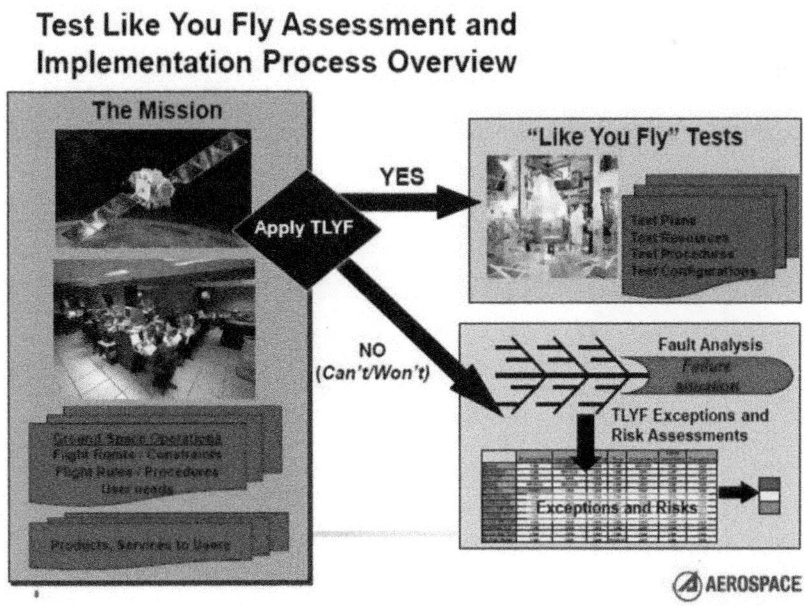

FIGURE 7.34 Test like you fly implementation process.

Time Component and TLYF

Much of our established testing is based on test specifications (e.g., MIL-STD-1540) and ignores the time component, which is a vital factor in executing a mission.

TLYF tests show the ability to properly transition from one activity to the next, to sustain duty-cycle driven activities, to demonstrate timing interactions between asynchronous activities, and allow for error growth and discovery in software execution.

At all levels of assembly, use a realistic mission timeline to include all first-time, mission-critical, and mission-objective events, and sustained activities appropriate for the level of assembly and in context of that assembly's contribution to the mission.

A propulsion system, for example, can only be tested in a very limited way once integrated into a space vehicle, due to facility and contamination issues. Mission conditions and characteristics need to be applied at the lower levels of assembly to be able to expose flaws that are only able to be seen under those conditions.

TLYF is ultimately a specific mission readiness test of the total operations chain, end-to-end (space + ground) days-in-the-life/weeks-in-the-life mission operability test as shown in Figure 7.35. This critical TLYF is called the "total operations chain test" (TOCT). Experience from integration and test organizations is that this test finds defects that cannot be detected in any previous testing, including full electrical, functional, and performance testing. The TOCT demonstrates a specific mission duration, whether it be a day, days, or weeks in the life of the system.

Failing to Test Like You Fly

Postmortem failure analysis of failed missions shows consistent violations of the test like you fly approach to be significant contributors of loss of mission as listed in Figure 7.36.

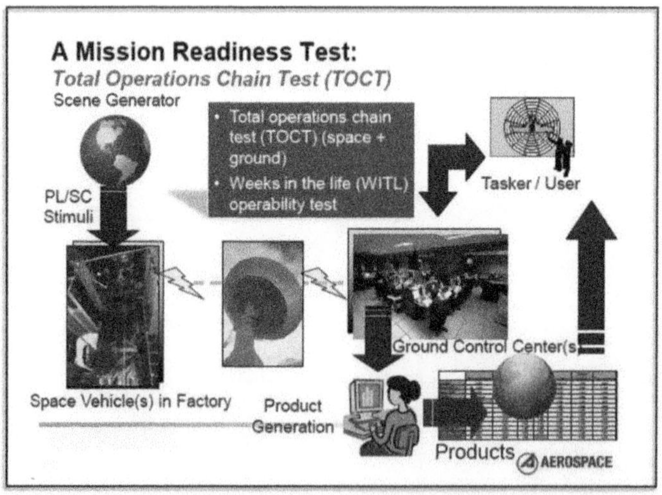

FIGURE 7.35 TLYF mission readiness total operations chain test.

Vehicle	Year	Mission Critical Anomaly Root Cause	TLYF Violation	Result
Titan CT-2 launch	1990	Miswiring prevented satellite separation: Ground test using non-flight software did not identify the problem.	Test what you fly	Loss of mission
ESEX payload	1999	Exploded in space after leaking battery electrolyte caused short circuit; Bettery qualified in non-flight condition.	Test like you fly	Loss of mission
Mars polar lander	1999	Faulty touchdown sensor logic caused vehicle to crash; Test not rerun with hardware and software after modification.	Test like you fly	Loss of mission
Mars climate observer	1999	English-Metric units error crept into modified software; Being deemed non-critical, was never tested.	Test like you fly	Loss of mission
Wire	1999	Start-up transient in pyrocontroller caused premature telescope cover deployment allowing coolant to escape; GSE power supply hid the problem.	Test what you fly	Loss of mission
Terriers	1999	Torquer coil installed upside down; Hardware and software never tested together.	Test like you fly	Loss of mission
Milstar 2-F1 launch	1999	Improper filter coefficient loaded into flight software; No validation of filter constants actually flown.	Test what you fly	Loss of mission
Genesis return capsule	2004	Four deceleration switches installed backwards causing parachute failure; Never tested in flight configuration.	Test like you fly	Significant loss of science product

FIGURE 7.36 Test like you fly approach violations that impacted missions.

PART 3

Space Safety Engineering and Design

Joseph N. Pelton

This chapter seeks to address the topic of space safety. It first seeks to define what is meant by the term "space safety" and explain that this is actually a quite broad term. The definition used here includes the safety design and operation of automated spacecraft as well as spacecraft and launcher systems with crew aboard. It also includes the safety of people and the Earth environment against dangers that might ensue from various space-related programs and operations. This includes but is not limited to the safe return of objects from outer space, the safe launch of objects into outer space, sub-orbital flights and operations in the so-called "Proto-zone"—or the area above commercial air space and below outer space. It certainly includes issues such as coping with the problem of orbital space debris, cosmic hazards, and even the area now known as the sustainability of space.

The detailed discussion that then follows seeks to address these various areas of space safety. This discussion highlights some of the most important issues, problems and possible solutions to the safety, reliance or resilience concerns noted in this broad definition. It notes how the concept of space safety that began with a narrow interest on designing safe spacecraft and crewed vehicles has expanded over time to include considerations such as orbital debris, safe re-entry of space objects, cosmic hazards and more.

Finally there is some discussion of how space systems analysis techniques can be applied to these expanded areas of space safety. The key is to find ways to "optimize" reliability and safety when considering a wide and expanding number of space safety issues and problems. These issues range from designing and implementing new systems in space to such concepts as using space systems in areas such as climate change, cosmic hazards, or reduction of orbital space debris. This chapter concludes with a discussion of how current and future trends that include new commercial space activities, new activities in "near space" that involve spaceplanes and hypersonic transport, as well as concerns over space debris, have dramatically expanded the scope of space safety. In the process, the field of "space safety" has increasingly become relevant and important to topics that involve much more than astronaut safety and now embrace topics key to the future sustainability of space and all human life on the Planet.

7.14 Introduction

At first blush it might seem that space safety engineering is a somewhat narrow and well-focused field, but this is not really the case. This topic really covers not only the safety of space mission equipment in all aspects of its design and operation. It, of course, particularly includes the safety of the crew if it is a manned mission. Then there is the safety of the launch site and the surrounding area. Space safety even covers the earth, air traffic safety, and space environment. Finally, there is the issue of environmental pollution that particularly involves the stratosphere and the emissions from launchers, spaceplanes, and even hypersonic transportation that in current design concepts are more like rocket launchers than airplanes. Further there are space safety concerns due to the hazards that space missions can create in earth orbit, such as the mounting problem of orbital space debris that could ultimately deny future access to space.

In short, space safety involves all aspects of mission and related flight safety for individual rocket launches, but it also covers the long-term sustainability of space for which the U.N. Committee on the Peaceful Uses of Outer Space has created a special working group on this topic. In this sense, space safety concerns orbital space debris and planetary defense against comets, asteroids, solar radiation flares, coronal mass ejections, and even changes to the earth's magnetic field (i.e., shifting of the earth's magnetic poles) that affects the natural protective shield of the Van Allen belt that saves us from various types of solar storms and cosmic radiation.

One attempt to define space safety in this broad sense is provided below:

> "Space safety refers to space mission hazards and relevant risk avoidance and mitigation measures. The space mission hazards include threats to human life, loss of space systems, and pollution of the Earth environment. Space safety, in a wider sense, encompasses the safeguard of critical and/or high-value space systems and infrastructures, as well as the protection of orbital and planetary environments. Space safety is a necessity for the sustainable development of space activities, whether targeted to minimize hazards for human spaceflight or when directed toward protecting vital space assets so vital to human commerce and safety here on Earth. Space safety is devoted to protecting space infrastructure from growing human-made orbital debris and even the risks posed to Earth population from re-entering objects." (Pelton et al., "Space Safety," 2005)

When viewed from this perspective the field of "space safety" thus becomes quite broad and can entail many different areas of expertise such as materials research, combustion, and propulsion engineering, flight safety and air and space traffic management and control, systems engineering, biological sciences and life support systems for space systems, radiation standards, orbital mechanics, electrical-magnetic engineering, chemical engineering, mechanical engineering, computer sciences and programming, etc.

The International Association for the Advancement of Space Safety has attempted to define what is meant by "space safety" and what activities are included in this field. The list of these activities, as provided on the IAASS webpage, is as follows:

1. Ensure that citizens of all nations are equally protected from the risks posed by overflying space systems and objects during launch and reentry/return operations.
2. Ensure that space systems are developed, built, and operated according to common minimum ground and flight safety rules, which reflect the status of knowledge and the accumulated experience of all space-faring nations.
3. Seek to prevent collisions or interference with other aerospace systems during launch, on-orbit operation, and reentry.
4. Ensure the protection of the ground, air and on-orbit environments from chemical, radioactive, and debris contamination related to space operations.
5. Ensure that mutual aid provisions for space mission safety emergencies are progressively agreed, developed, and made accessible without restriction anywhere on the earth and in outer space. (IAASS, 2016)

To make this chapter as useful as possible and to assist readers with a particular interest in one aspect of space safety or another, the text will be divided into several sections as follows:

- Unmanned systems design and engineering
- Crewed systems design and engineering
- Combustion and materials engineering and safety
- Suborbital flight systems, spaceplanes, and hypersonic transport
- Launch site design and safety standards

- Licensing and safety controls for various types of launcher systems
- Air and space traffic control and management
- Atmospheric and environmental pollution
- Orbital mechanics and orbital debris concerns
- Cosmic hazards and planetary defense and safety
- Systems engineering and safety
- Future trends in space safety engineering and design

These sections will note what the safety and reliability concerns are, plus the general approach taken and where appropriate some examples given. The brief chapter cannot provide in-depth details but relevant webmail references will be provided when relevant and appropriate.

The remaining part of the chapter will be devoted to the last two sections on systems engineering and safety and future trends in space safety engineering and design where the greatest emphasis will be given.

For the reasons already given, the study and development of the "field of space safety" presents a number of problems. First of all there is the daunting "scope" issue. Space safety really touches on every aspect of space from astrophysics to orbital mechanics, from electrical and magnetic engineering to life sciences to traffic control systems, from reliability testing to probability analysis to combustion research. No one can truly be an "expert" in safety engineering because it requires a synoptic overview of the entire field of space. This means that most systems design engineering involving satellites, launchers, space vehicles, or any other aspect of space starts out with a system being envisioned and designed in a preliminary manner and then "safety experts" or reliability analysts asked to sign off on the design. Safety engineers are often seen as the people to say "no" to a system or component design because it is not reliable enough or has not been tested for sufficient duration, or it has not met its specifications.

In most space agencies "safety" comes down to reliability testing, "fixes" of system design to meet a specification or getting a "variance" approved by a high-level management official to let the problem with meeting a specification be over ridden. The Columbia Shuttle and the Challenger Shuttle both had thousands of such "variances" approved by NASA Management. Some of these were incredibly small and trivial and others such as the "foam insulation" and "O-ring" problems were the source of the Shuttle failures. The real key issue to understand is that safety engineering—to be truly successful—must start with the design engineers and their conception of a space vehicle, system, or component. The key is to start with a design that maximizes reliability and consistent performance over a long term.

One of the unique aspects of space system designer Burt Rutan, who designed SpaceShipOne and literally thousands of space systems, is that he has that unique ability to think of designs that are as efficient, reliable, simple, and ultimately safe as possible. English philosopher Thomas Ockham is famous for his "Ockham's Razor" theory. This rule of logic goes: "If there are two solutions to a problem choose the simplest." It could be argued that he hit on the principle that is at the core of effective design of systems that perform the best and turn out to be the most reliable. The ceramic tile design for the thermal protection system for the Space Shuttle program should have been discarded from the outset as too complicated and thus too unsafe to provide totally reliable performance over the longer term. There are almost countless examples in the case of space systems design that if a "safety engineer" had been a part of the design team from the start, plans would have been scrapped as too complicated and thus likely to have been too unreliable.

7.15 Unmanned Space Systems Design and Engineering

Most rocket launchers, most satellites, and most things that go into space are unmanned and thus have no crew aboard. Telecommunications and broadcasting satellites, remote sensing satellites, surveillance satellites, meteorological satellites, space navigation satellites, defense and military satellites, experimental satellites to collect data and measure phenomena in space, and most rocket launchers have no crew aboard. This is significant in that rigor of the design specifications can be relaxed, perhaps by several orders of magnitude. Subsystem and system testing can be relaxed. A very good design can allow for testing of the first few components in a manufacturing process, but relaxing the degree of testing once consistent test results are obtained. Unmanned satellites and spacecraft can be much smaller and simpler if there is no need to create and operate a very, very reliable life support system and also if the satellite does not have to return safely to earth.

Further, new technology is rapidly changing the manufacturing process and the swift development of new tools such as 3D printing that can allow the rapid and quality production of components and subcomponents for spacecraft. These developments are changing the nature of satellite design and production. Ironically quality, reliability, and safety of spacecraft can be increased while costs are decreased due to automation and use of 3D printing of components. This has given rise to new systems architecture for satellite communications systems, remote sensing, and other space applications. Instead of increasing performance through always designing larger and more sophisticated spacecraft with larger aperture multibeam antennas and higher performance power systems, some space networks have opted to create networks with larger constellations that employ smaller and less complex spacecraft, typically deployed in lower orbits.

There are some ironic consequences in that some of these architectural network design changes substitute one type of risk for another. The substitution of a low earth orbit constellation for telecommunications that deploys a large number of smaller spacecraft with multiple spares involves different considerations as to safety and reliability. The large number of satellites in the constellation provides service robustness and presumably increased reliability of service. The large size of the constellation, however, increases the risk of orbital collisions—not so much with other satellites in the constellation but debris or small satellites deployed in similar orbits.

Table 7.2 below indicates some of the more clear-cut differences in the reliability and "safety" concerns with respect to applications satellites deployed in GEO orbit, versus those deployed in MEO or LEO orbits. For a more detailed discussion of low earth orbits (LEOs), medium earth orbits (MEOs) and geosynchronous earth orbits (GEOs) see other relevant books (e.g., Pelton, *Basics of Satellite*, 2006).

7.16 Crewed Space Systems Design and Engineering

Astronaut safety is what most people think about when they hear the term "space safety." The enormous publicity associated with the Apollo I fire and the Challenger and Columbia Space Shuttle disasters has reached billions of people via radio, television, the print media and the Internet. The design approach that was used in the earliest astronaut and cosmonaut flights was to develop very high reliability standards for each and every subsystem within the launcher and spacecraft of the overall design and then do very extensive testing. This testing involved lifetime durability, shaker table testing, and thermal and vacuum testing. If the performance or safety standard used in the design of rocket launchers or spacecraft in the past had been perhaps "x" or "y," then the performance or reliability standard was now set 10 times higher. The difficulty of meeting those standards often led to the granting of variances or official "waivers" being granted by the Office of Space Safety and Mission Assurance, or in some cases by high-level management when there were questions as to whether a "waiver" can be safely granted. In the cases of both the Challenger and

Differences in reliability and safety concerns between GEO application satellites and LEO constellations		
Safety issue or process	**GEO networks**	**LEO or MEO constellation**
Lifetime	GEO satellites designed for 15–18 year lifetimes	LEO satellites designed for 7–10 year lifetime. They are harder to maintain in orbit due to gravitational pull and atmospheric drag during SolarMax.
Sparing	Very few spares are needed for global system. One or two spares can back up global network.	Large number of spares are deployed and can be used to restore service.
End of Life disposal	Satellites can be lifted to parking orbit, where they can stay for millions of years.	Controlled deorbit necessary under 25 year rule and risks of safe deorbit rise as more and more satellites deployed in LEO. Especially if constellations of thousands of satellites are launched.
Concerns about orbital debris	The controls for GEO have assisted in eliminating new debris from upper stage launchers. Biggest problem is possible satellite conjunctions or "zombie" satellites.	40% of orbital debris is now in low earth orbit and in a relatively small band in terms of total volume. Voluntary guidelines adopted by UN COPUOS are not adequate to new plans for so-called "mega-LEO" systems for communications.
Cross-links	These are typically not used for GEO satellites because of huge distances between regions in GEO.	Cross links can be used in LEO constellations to work around failed satellites.

Source: Chart provided by the author with all rights reserved.

Table 7.2 Comparing the Safety/Reliability Aspects of GEO versus LEO Satellites

Columbia missions, thousands of waivers were granted. And the failures could be traced back to waivers that were officially granted. In the case of the Challenger launch the weather conditions were too cold and this led to the shrinkage of the famous "O" ring that let fuel leak through to the hot rocket exhaust. In the case of Columbia, the problem of the foam insulation hitting the thermal insulation tiles was detected in earlier launches and the problem was allowed to go unfixed (NASA Technical Standard 2012).

Demanding reliability standards and avoiding the granting of "waivers" to meeting these standards are clearly key steps to be taken to create reliable launchers and spacecraft. But one can also build safety into the design in the first place. In the report that was done to review Shuttle safety at George Washington University after the Columbia accident, the list of recommendations that were assembled included

1. ***Design of a launcher systems that is much simpler in component parts:*** The Shuttle with some 30,000 plus parts was much too complex to achieve truly high levels of safe operation over sustained periods of time and also could never be highly cost efficient in addition to safety concerns (Joseph N. Pelton et al., *Vulnerabilities*).

2. *Design of a system that is more resilient:* The ceramic tiles in the thermal protection system were very fragile and could never have met the requirement for rapid turn-around for relaunch as promised to the U.S. Congress.
3. *Design of a system that is capable of update and could be easily refurbished:* The nearly 5 km of wiring in the Shuttle, that were largely hidden and not available for rewiring or maintenance, were yet another key vulnerability for a space vehicle that was conceived as having many trips to orbit over time.
4. *Design of a system that is not the result of political intervention:* There was significant intervention in the design of the Shuttle to make it less costly from the White House and by industrial interests to include solid rocket boosters in the design. This redesign of the Shuttle to add the solid boosters due to political intervention increased the risk factors, reduced astronaut escape options, and actually led to the Columbia failure. After nearly 3 years of down time for reworking the STS design to improve the thermal foam insulation between the Shuttle and the booster rocket and after nearly $2 billion of expenses the problem was never truly solved.
5. *Design of crew transport system to allow for astronaut escape and shut down of the liquid fueled rocket system:* The problem with solid fuel rocket systems are that they cannot be shut down after ignition. Hybrid systems such as those used in the Dreamchaser or Virgin Galactic are "throttleable," but these systems are also quite "dirty" in terms of particulate emissions.
6. *Systems that are for cargo transport only should be automated and separated from astronaut launchers that should have higher level of safety and escape capacity:* Cargo launchers that do not have to include life support systems and meet safety and reliability requirements associated with astronaut launch can be designed, manufactured and operated at lower cost.

The full report provides many more examples of the ways in which safety design could have created a space transportation system that had a higher degree of safety and reliability for sustained use in space launch missions (Pelton et al, *Vulnerabilities* 2005).

The space transportation system was able to provide low earth access for nearly 30 years and support the construction of the International Space Station as well as the deployment and servicing of the Hubble Space Telescope, but this design, from the perspective of safety design, clearly had flaws (Figure 7.37) (NASA Space Shuttle 2016).

And the safety concerns with the design of future crewed space missions by NASA in the form of the Space Launch System (SLS) and the Orion capsule remain today. The latest safety concerns were reported in a recent article in *Aerospace America* that reports on an outside review of NASA current crew launcher systems by the Aerospace Safety Advisory Panel. The first issue was the fact that NASA accepted in 2010 the safety standard of 1 loss of crew in 270 missions for low earth orbit missions. More recently when the concept of using the SLS and Orion for a mission to Mars was assessed by NASA the results were surprisingly low. "NASA determined the loss-of-crew rate on that type of mission would be 1 in 75." This is a number lower than assigned to the Shuttle at the end of its mission life (Werner and Zak 2015).

At this point the Aeronautical Space Advisory Panel has indicated their disappointment with the level of safety for such an important and high-profile mission. Their report on the subject in 2014 expressed concern in the following manner: "It was the ASAP's hope that the inherently safer architecture of the SLS and Orion, as compared to the Space Shuttle, including full abort capability separation of the energetics from the crew module, and parachute reentry instead of aerodynamics, would greatly improve inherent safety." (Werner and Zak 2015).

It was the hope that the lessons from the faults in the safety design of the Space Shuttle would have been learned and applied in the design of the next generation of launch systems, but this has

FIGURE 7.37 The Space Shuttle design failed key safety design criteria (graphic courtesy of NASA).

not been the case. The next generation of launch systems has continued to combine both liquid and solid fueled launchers, and the objective of safety performance has been set too low according to the advisory safety panel (SLS Architecture 2016) (Figure 7.38).

7.17 Combustion and Materials Engineering and Safety

The issue of materials and combustibles that might be safely or at least prudently used in spacecraft design is a difficult issue. New materials and combustibles as well as electronic ion propulsion systems are constantly under development. Despite this changing array of construction materials and fuels that might be used to develop launchers, spacecraft, or particular payload instrumentation, knowing what has been employed in the past and safety standards that might have been

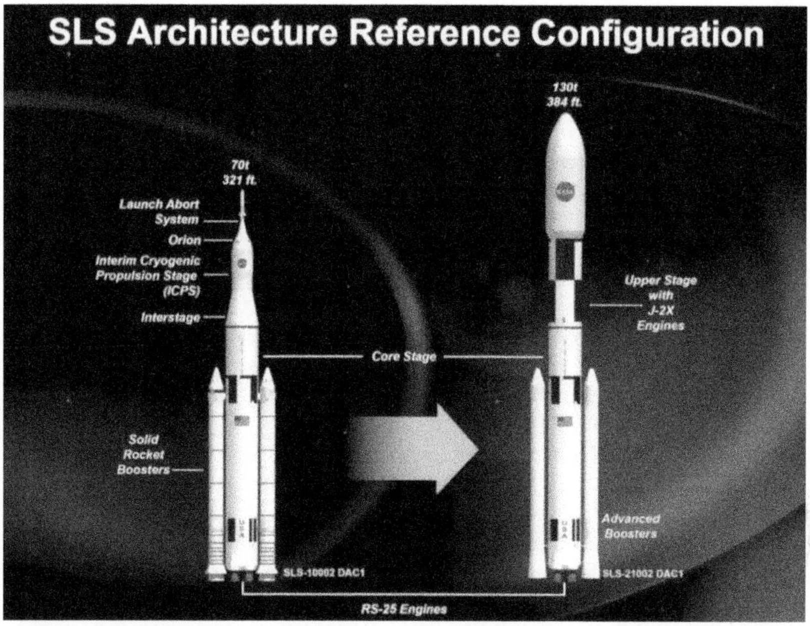

FIGURE 7.38 The conceptual design for the space launcher system that retains solid boosters (graphic courtesy of NASA).

already effectively tested and used is a prudent place to start. Thus, design engineers and safety specialists should always start by considering the status of tests and safety standards that have evolved over a number of decades. There are many published standards of safety related to materials, rocket and station-keeping fuels. In some cases one must look at several dimensions of safety such as those relating to hypergolic fuels where one must take safety precautions with regard to handling of fuel when it is loaded, when fuel tanks are designed, when the hydrazine jets are fired and when exhausting this fuel at end of life prior to deorbit.

In the new era of large-scale manufacturing of satellites and components using 3D printing, the choice of materials may not only involve strength and low-mass density, but also suitability for use in new manufacturing systems. In the area of launching system fuels, there may be new questions not only about propulsive performance and "throttleability" but also the degree of atmospheric pollution to the stratosphere that comes from solid fuel particulates. It is easier and less costly to check on the performance and reliability testing that has been carried out by space agencies or aerospace companies than to set up one's own testing program.

7.18 Suborbital Flight Systems, Spaceplanes, Hypersonic Transport, and New Uses of the "Protozone" or "Near Space"

For many years there was a clear divide between those involved in outer space related activities and those who were involved in aviation. In recent years, there have been a growing number of activities that involve the area above commercial air space (i.e., 21 km and above) and the area wherein satellites or spacecraft are able to maintain earth orbit (i.e., 160 km and above). This area is sometimes called "subspace," "near space," "protospace," or the "protozone." Only recently have organizations such as the International Telecommunication Union (ITU) and the International Civil Aviation Organization (ICAO) started to consider the issues that increasing use of this region might entail. This is a region where spaceplanes, hypersonic transport, high altitude platform systems, dark sky stations, stratospheric balloons and dirigibles, plus military surveillance craft might pass and at greatly varying speeds that might range from essentially stationary to Mach 6 or 7. Currently there are no air traffic management or control systems governing the use of this area. This is a very significant safety issue that will be discussed further in the section on space traffic control.

In addition there are the safety issues involved with the design and operation of spaceplanes and hypersonic transport. These suborbital craft might fly at velocities in the range of Mach 2 to Mach 6 and as such represent much less difficult technological demands than orbital launchers that involve much greater velocities and thermal environments.

At one time, the U.S. and European approaches to regulating the safety of these craft were moving in much different directions. The European Aviation Safety Agency (EASA) was proceeding to create safety certification standards for all "winged" craft including spaceplanes. In contrast, the U.S. Federal Aviation Administration Office of Commercial Space Transportation (FAA-AST) was proceeding to grant experimental licenses. Under the new U.S. Commercial Space Act of 2015, this process is now to continue until 2025. EASA has now shifted its regulatory approach to be more of a case-by-case authorization and deferred to the various European space agencies. Ultimately if point-to-point hypersonic transport is to offer transoceanic services and there are regular spaceplane flights, some form of certification of "safety to fly" standards will need to be established. One suggestion is that safety and performance standards first be established for subsystems. This might be for such things as avionic systems, life support systems, thermal protection, etc.

The difficulty with safety certification for spaceplanes is the short duration of the burn of rocket engines. In the case of jet aircraft engines these motors operate over long periods of time and thus can be certified by long lifetimes. In contrast, rocket engines are typically short-burst chemical explosive events. This dilemma of how to create safety certification standards for rocket engines is one of the major safety challenges that exist today. Presumably this issue will be addressed and resolved within the next decade as the current case-by-case licensing domain under the aegis of the FAA-AST runs out. If the U.S. and European regulatory agencies can reach a common agreement on this approach, then others that are working in this area, such as Japan, China, India, and Russia, will likely also agree to a safety oversight and certification standards process that is common on a global scale.

It also seems likely that the ITU will provide international coordination with regard to the radio frequency allocations process in this area. Likewise it seems that the International Civil Aviation Organization will play a key role with regard to coordination of safety certification or regulatory oversight as to safety standards that might be adopted. There are other matters related to radiation health standards that might involve the World Health Organization. Further there are concerns about atmospheric pollution from stratospheric spaceplane and hypersonic transportation systems that may also include efforts at global regulation that might involve the World Meteorological Organization (WMO) and the United Nations Environment Programme (UNEP).

Comparing airplanes, jets, suborbital space planes and rockets-to-orbit				
Comparative factor	**Airplane**	**Jet**	**Spaceplane**	**Rocket to low earth orbit**
Velocity (m/s)	250	500	1,600	7,800
Height (km)	Up to 10	Up to 20	Up to 120	200+
Specific energy (J/kg)	0.13	0.7	14.5	324

Source: Table by permission of author.

TABLE 7.3 Difference in Performance Factors for Different Types of Craft

7.19 Launch Site Design and Safety Standards

Currently the only specific regulatory role that the U.S. FAA-AST has in the oversight of the new spaceplane industry is to protect the safety of the general public against accidents. This means that it inspects and certifies the safety of spaceports to make sure that they are as safe as possible against accidents that might endanger those living or working in proximity to the launch facilities. Currently those that are signing up for spaceplane flights as "space tourists" are required to sign agreements acknowledging that they are taking high risk experimental flights. Thus, they must agree not to hold the government responsible for any accidents with one written agreement. They are also required to sign another agreement to hold the spaceplane operator harmless as well. This type of "experimentally licensed" flight would appear to be something that can be used for a time for "space tourism" flights but would not be adequate for regulatory point-to-point flights for hypersonic transport.

Some of the spaceplane companies such as Virgin Galactic, Swiss Space Systems (S3), etc. are moving to develop business plans to create a business based partially on the launch of small satellites using their craft to assist with such operations. The safety regulations associated with such ventures where people are not involved can thus be less stringent.

7.20 Licensing and Safety Controls and Management for Various Types of Launcher Systems

The world of space used to be much simpler. There was a time when it was essentially for missions by governmental space agencies or a few commercial space applications companies for telecommunications or remote sensing who were under the strict control of governments. Today in the world of "new space" there is an ever-growing diversification of space industries with many more activities now planned. The world of "new space" envisions spaceplane rides, hypersonic transport, high-altitude carrier planes for launch of satellites, a variety of small, medium, and larger launchers for various types of payloads, large-scale commercial satellite constellations for communications, IT services, remote sensing, high-altitude platforms for various applications, stratospheric dirigible excursions, and even space mining. The safety oversight and certification of all of these activities will become increasingly difficult and will have many dimensions. There are issues of space traffic control, due diligence against the creation of new orbital debris, the safe deorbit of satellites and the safe operation of private launches, especially when they may involve reusable vehicles that are intended to come back and land at launch sites.

In the past launch sites had a range control officer that was responsible for the safety of people at the launch site and nearby populations and had control of a flight termination system so that an errant launch vehicle could be blown up before a flight from Kennedy Space Center, for instance, could crash into Orlando or Miami, Florida. Today, with a growing number of private space ventures and the operation of spaceplanes, hypersonic transport, high-altitude platform systems, the safety and control of all of these activities has become much more difficult and complex. Oversight of environmental impact statements with regard to every launch, ability to terminate a dangerous flight, or control of satellite constellations to avoid in-orbit collisions and major space debris are increasingly real space safety concerns.

7.21 Air and Space Traffic Control and Management

The increasing focus of space safety concerns is how to implement in a practical, comprehensive and cost-effective way what is called *space traffic control*. This is in many ways a complicated issue. For starters, the area of concern actually involves the areas above commercial air space that is currently up to 20 km (12.5 mi). Thus, there is first a matter of definition of what is meant by space traffic control and who is in charge. Many countries claim sovereignty and control of space over their country and territorial waters way up into outer space but do not have the ability to enforce it. In the case of territorial waters, there are various zones of control, including an "economic zone" that extends to 100 km. No such agreements about zones of control or guaranteed free passage currently apply beyond the regulation of air space up to 20 km. In the United States, Europe, and a number of other countries, there are requirements for environmental impact statements associated with rocket launches but this is not a universal requirement around the world.

The existing radar systems for air traffic control and Global Navigation Satellite Systems (GNSSs) and related software are not optimized to cope with safety and precise resolution above today's regulated airspace. This is to say there is not only a need to establish new regulatory agreements, probably by extending the authority provided by the Chicago Convention of 1945 under which ICAO operates, but there are likely new technical abilities needing to be established as well. The new S band radar "Space Fence" that is being established by the United States in the Micronesia region might provide additional infrastructure needed in this regard, but this facility was established to monitor space objects and not monitor operations in the protozone.

Currently, the way rocker launches are handled is that air traffic is diverted away from the launch area. This process can affect the diverting of hundreds or even thousands of aircraft flights. As the volume of flights by spaceplanes and hypersonic transport increases, this total segregation of air traffic and space traffic will become more and more inefficient. Thus, there will be pressure to create a system whereby all types of craft flying in air space, the protozone, or even outer space might be subject to a uniform control process. Conceptually, this may sound quite simple, but in practice there are many complicated technical, legal and regulatory issues to be sorted out at the national and international level. These are sufficiently complicated that they may take decades to be resolved. And based on precedent, it may take a major accident to create sufficient pressure to force resolution of these issues and to create sufficient impetus for the needed investment in radar hardware and to develop the software required to provide the necessary accuracy in GNSS systems for air and space traffic control that is needed by the middle of the 21st century.

7.22 Atmospheric and Environmental Pollution

Perhaps the most overlooked and most important aspect of space safety relates to the issue of atmospheric and environmental pollution. The few launches by chemical rockets and by solid fuel missile systems were sufficiently small that no one focused on these launches as an air pollution issue to worry about. The "new space" revolution that is leading us toward a new environment whereby there may be a very high volume of spaceplane and hypersonic transport vehicles operating in the stratosphere should be registered as a danger signal as to the environmental effects of this new space transport revolution.

The first thing to understand is that the stratosphere is a much different environment than air at sea level. The density of atmosphere can be 100 times less dense at the higher levels of the stratosphere. Further not all "rocket fuels" are the same. A solid fuel rocket can spew particulates that are 100 times more polluting than a liquid fueled rocket using liquid oxygen and liquid hydrogen as a fuel. Even with these systems water vapor is a pollutant to the stratosphere.

The biggest issue is thus the climate-changing potential of the emissions from rocket launchers. But there are other environmental concerns with a significant safety implication. One of these is the dangerous and poisonous fuels such as hydrazine and other hypergolic fuels. These are dangerous to load into satellites and should be vented at the end of life to avoid creation of ruptured tanks or the return of this dangerous gas back to earth. Even greater safety importance is that of nuclear fuel and how it is handled to be as safe as possible. International guidelines have been adopted that provides safety restriction and processes with regard to radioisotopes and nuclear power sources (international guidelines for nuclear reactors in space).

There are scores of these types of power systems that have been launched into space and some have even crashed back into earth creating radioactive pollution. In January 1978, a Russian satellite, namely Kosmos 954, crashed into Canada and in the process scattered radioactive debris over a large area, estimated to be equivalent to the size of Austria. Under the international liability agreement for space activities, Russia was unquestionably at fault and immediately negotiated a settlement with the Canadian government. This is perhaps the most well-publicized instance of a nuclear activity involving radioactive power generators in spacecraft, but there have been other incidents as well (Galloway 1979).

7.23 Orbital Debris Concerns and Tracking and Sensor Systems

Back in the 1970s and 1980s, a NASA scientist named Dr. Donald Kessler suggested that if care was not taken with regard to defunct satellites, upper stage rockets and other debris elements, this could lead to a future problem with regard to orbital space debris. At the time there was a much greater problem with natural space debris in the form of micrometeorites, meteorites, and bolides and the warning was not taken seriously. Yet over time, more and more satellites were launched the amount of space debris accumulated in earth orbit. Figure 7.39 shows the ever-increasing accumulation of orbital space debris that has been accumulating in low earth orbit (LEO) and highlights the increase in "trackable" debris that occurred with the shooting down of the Fengyun derelict meteorological satellite in 2007 and the collision between the Russian Kosmos satellite and the Iridium satellite that followed (Orbital Debris 2008).

The build-up of this debris is now beginning to be considered an increasingly severe problem. There are now some 6 metric tons of debris in earth orbit and over 40%, or some 2.7 metric tons, is

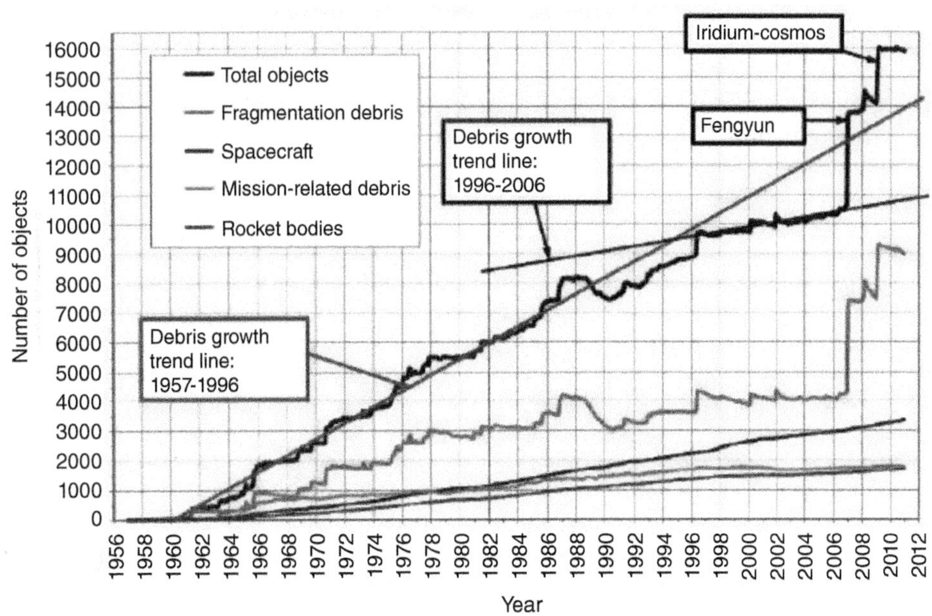

FIGURE 7.39 The mounting orbital debris—particularly in LEO—since the 1960s (chart courtesy of NASA).

in the LEO orbits—particularly in the very congested polar or near-polar sun-synchronous regions. The collision between the Iridium satellite and the Russian derelict Kosmos weather satellite produced well over 2000 new trackable, space debris elements and the Chinese missile strike against an out-of-commission Chinese meteorological satellite created over 2,000 new "trackable" items. These two "accidents" represent a sizeable portion of the 22,000 debris elements being actively monitored at this time.

Increasing study is being devoted to the possibility of active debris removal because of the potential of further debris bringing what is now known as the "Kessler syndrome." This is a condition where there is sufficient debris elements that an on-going avalanche of debris elements creates a truly hostile space environment in which it is not possible to launch new satellites and existing satellites would be increasingly put at risk.

In addition there are now plans for the future launch of so-called "Mega-LEO" constellations with hundreds if not thousands of satellites deployed to support Internet optimized low-latency services. At this time, OneWeb (with nearly 1,000 operational satellites plus spares) and LeoSat (with several hundred satellites plus spares) are now specifically declared and are either in manufacture or final design formulation. Further there have been indications from SpaceX founder Elon Musk that his company may seek to design, build and deploy a constellation with up to 4,000 small satellites in a giant LEO constellation as well. These initiatives raise several questions as to whether there is market demand for such Internet optimized systems, and equity issues as to whether these quickly deployed networks might preclude others from making similar use of outer space, etc.

From the simple perspective of space safety, there are several questions. These include whether such a large constellation can be safely managed to avoid spacecraft collisions? Can such large Mega-LEO systems be successfully designed to avoid interference with GEO satellite networks that have guaranteed high priority of service over non-GEO networks? And does the deployment of such space networks raise significantly the possible collision with an uncontrolled piece of space debris in the already most crowded area of space in proximity to earth? Should the deployment of such Mega-LEO constellations trigger a deadly cascade, then this could endanger hundreds if not thousands of satellites and make it virtually unsafe to launch new satellites safety for some time to come.

7.24 Cosmic Hazards and Planetary Defense and Safety

There is unfortunately even greater danger to satellite networks that may be in the process of evolving due to changes related to the earth's magnetic poles. Research satellites such as the Swarm Project by the European Space Agency (ESA) have confirmed that the magnetic poles of the planet are migrating. The North magnetic pole is shifting southward and the South magnetic pole is shifting northward. The magnetic poles are key to the formation of the Van Allen belts. There have been models developed at the Neil Bohr Institute that suggest that the protective shielding of earth by the Van Allen belts could be reduced to 15% of that constituted by these belts when fully formed when the poles are furthest apart.

These now changing magnetic field conditions, which have not changed for tens of thousands of years, are now changing fairly rapidly. Under these new conditions our electric power systems, our satellites, our computer networks, and supervisory control and data acquisition (SCADA) networks that control much of our urban infrastructure are seemingly going to be at much greater risk. The bottom line is that the sun gives off radiation that is sometimes emitted as powerful flares. The sun also produces explosions of ion particles that are known as coronal mass ejections (CMEs) that often occur together with these flares. These CMEs can create natural electromagnetic pulses (EMPs) that are the most dangerous events to ground and space infrastructure. The Carrington Event of 1859 set telegraph offices on fire and brought the "Northern Lights" down to Cuba and Hawaii. An event like this today, with the Van Allen belts protective shielding greatly diminished, could endanger our power grids, or SCADA networks, our telecommunications and IT networks and most certainly our vital satellites for communications, remote sensing, weather satellites, and timing and navigation ("A Super Solar Flare" 2016).

And in more recent times, there was the Montreal event of 1989 that resulted in the loss of the electrical power systems from Chicago to Montreal. In 2003, there was the "Halloween Event" that resulted in power losses in Scandinavia. In space, there are warnings that can be sounded because the flares travel at the speed of light, but the ions travel at only a few million kilometers per hour. Thus, application satellites can be powered down to save them from the worst of these solar storms (Halloween Storm 2016).

In terms of space infrastructure safety and protection, there are serious concerns about what these natural EMPs can cause in coming years. Since the magnetic shift of the poles take a long time—perhaps two centuries or more—the challenge to our vital space infrastructure will be posed for a very long time.

The strategies for protecting against violent CME events are limited. One can design spacecraft to be quickly powered down on very short notice. Further heavy duty circuit breakers can be designed into satellites to protect the on-board electronics. There are studies that are underway to see if there could be a way to create an electromagnetic shielding system at the Lagrangian point 1 that might be created to screen earth and orbiting satellites from the most violent solar storms.

7.25 Systems Engineering and Space Safety

The above discussion serves, if nothing else, to emphasize the great breadth of scientific and engineering information and knowledge that is needed to cope with the range of space safety and reliability issues that are now of interest and concern. It turns out that there are range of skill sets and scientific and engineering capabilities that involve the safety design and reliability of space systems for unmanned and manned missions. There are an even greater amount of technical abilities and engineering skills associated not with the safe design of satellite systems, but rather safety issues that arise from space related activities. These include such areas as pollution from rocket launches, the impact of solar storms on satellites and human infrastructure, the problems associated with orbital debris and meteorites and what might be done to address these problems, and so on. The range of skills sets is very broad indeed.

Systems engineering involves a structured way at looking at a problem and defining its various dimensions and finding a way to optimize a design or safety concept to meet specifically defined goals. This analytic process can apply to many types of space safety problems and issues. In the design of spacecraft, systems engineer analysis can assess structural integrity, propulsion efficiency, aerodynamic shape effectiveness, but it can also be applied to safety design optimization. These analytic concepts involve looking at whether the design is (1) as simple and fault free as possible; (2) provides the least possible single points of failure; (3) provides for ease of update, service, and retrofit for improvements; (4) allows for lifetime and resilience testing prior to launch into space; and (5) provides for redundancy in key areas of vulnerability, especially in the case of space systems with crew. (Thus in human crewed systems there would typically be a provision for emergency escape systems for astronauts. However, before this provision is included, there would be efforts to design in the highest level of safety that would be possible with a well-crafted design without driving costs to unreasonable levels.)

Systems design is a demanding process of trade-offs and optimization. It can be applied to safety optimization as well. In this case, the prime objective or design challenge becomes to create a design that substitutes safety and reliability rather than performance and cost-efficiency as the prime objective.

When the issue of space safety specifically translates to developing fuels that are less polluting, or safer ways to deorbit satellites, or the creation of systems to protect critical ground infrastructure or satellites from cosmic hazards that threaten, the same basic concepts of systems design can likewise be applied. The system design engineering approach is often thought of as optimizing performance and cost-efficiency, but it can certainly be adapted to include reliability, safety, or even beauty and aesthetics.

In addition to applying systems engineering and optimization techniques to safety design and operations, there is another important aspect to consider. This is to apply the very large data base of knowledge that has been acquired over the more than half a century of human space activity. A good deal of information has actually been translated into safety standards. In the case of materials, pyrotechnics, life support systems, thermal controls and engineering, power systems and much more the acquired information has been assembled and translated into space safety and resilience standards. These standards are assembled and available in many books and websites (Sgobba and Pelton 2010). There are many other tools and structured approaches to space safety engineering and design. These include data bases where engineers, scientists, and technicians can enter information and experiences with regard to testing or recommendations with regard to safety improvements. There are number of safety training courses offered by space agencies, universities, and organizations such as the IAASS as well as technical consultants that specialize in space safety.

In the "tug of war" that sometimes emerges between those seeking to pursue safety, resilience, and reliability versus those that are pursuing "mission assurance" and completion of a program, there is a mechanism that is often a very problematic aspect of any failure analysis. This mechanism is the so-called granting of a "Waiver" when a programmatic goal and a safety inspection or safety-related issue comes into conflict. It seems significant that in a NASA listing of definitions of key terms that the exposition on "waivers" appears to be the longest and most complicated in that it seems to have several different aspects and grounds for differing interpretation as follows:

NASA official waiver definition: [1] A written authorization to depart from a specific directive requirement. [2] A documented authorization releasing a program or project from meeting a requirement after the requirement is put under configuration control at the level the requirement will be implemented. [3] A written authorization granting relief from an applicable requirement and documenting the acceptance of any associated risk. For NASA ELV payload projects, waivers typically are approved for a single mission and have a specific duration. However, a waiver identified early in the design or specification/requirement review(s) may apply throughout the project or to multiple missions that use a common upper stage and/or a common spacecraft bus. [4] A variance that authorizes departure from a specific safety requirement where an increase in risk, due to the fact that the requirement is not satisfied, has been documented and accepted by the appropriate authority. [5] A variance that authorizes departure from a specific safety requirement (NASA Technical Standard, 2012).

The facts show that all three of the NASA fatal accidents can be traced back to granted waivers. In the case of Apollo I, this was allowing paper to be in a pure oxygen environment. In the case of Challenger this was an O-ring thermal condition exception. In the case of Columbia, this was allowing foam insulation pieces hitting the ceramic tiles at high velocity and not proceeding with a safety review investigation. This history and the record of thousands of waivers being granted to maintain program schedules seems to suggest that this is a fundamental safety issue. It certainly leads credence to the criticism that because NASA, ESA, JAXA, and other space agencies are exclusively responsible for their safety oversight, this represents a programmatic weakness. It also seems to suggest that the granting of waivers appears to be a key and valid point that deserves to be reexamined and seriously addressed. Some suggest that the fault could in part lie with creating too demanding safety or resilience standards in the first instance and without sufficient focus, rather than setting critical safety standards in a more targeted manner with the view to solving deficiency to key standards problems rather than overcoming them with waivers. This is not merely a problem of safety standard administration and organization, but of safety culture.

7.26 Future Trends in Space Safety Engineering, Design, and Study

The initial concepts as to the meaning of space safety began with ways to build reliable spacecraft and rockets, including those that were "man rated" to support the launch of astronauts into space. These concepts generally focused on reasonable standards of safety and reliability that can be verified by life or stress testing and other forms of qualification and quality assessment. These concepts were largely developed by the various civilian and defense-related space agencies around the world and the aerospace companies that supported these efforts. Over time, however, the field of "space safety" has grown in scope and complexity. There were added areas of expertise, when exploratory missions were sent to other celestial bodies and there were concerns about biological contaminants to earth. Then there were concerns about nuclear power supplies and large space objects returning to earth and doing damage. There were concerns about the safety of launch sites and spaceports and injury to people that operated these sites or lived nearby. More recently there have been new concerns about orbital space debris, cosmic hazards, changes to the protective shielding of earth by the Van Allen belts, and even the implications of the emergence of the so-called "new space" industries. These new space entities are involved in developing suborbital space tourism flights and low-cost launchers (i.e., Virgin Galactic, Blue Origin, Sierra Nevada, etc.). Others are engaged in developing new commercial launch systems to outer space (Space X, Sierra Nevada, Bristol Aerospace, Reaction Engines Ltd., etc.). Yet others are developing new ideas such as hypersonic transcontinental transport, high-altitude platform systems (HAPS), stratospheric remotely or robotically piloted freighter transport, or dark sky stations. All of these new ventures are not only posing new safety issues in terms of space and protospace traffic control and management, but also developing new and better ways to access space more cost effectively, more reliably, and with greater levels of safety. Perhaps the most profound change of all is the emerging recognition that "space safety" is not just about building safe space systems, but increasingly it involves the sustainability of earth and space and protective humanity in such terms as planetary defense against cosmic hazards such as asteroids, comets, solar storms, orbital debris, and even at the level of assisting with systems to cope with climate change. This is a profound shift in terms of who is being protected from what threat, but also a grand challenge in terms of the scope of technologies, disciplines, and expertise that are needed to cope with these entirely new and growing number of "space safety" concerns.

7.27 Conclusions

Within NASA, there is an Office of Safety and Mission Assurance. If one examines the processes that this office uses and the training courses that are offered through this unit of NASA, it is clear that its mission is that of the traditional qualification and quality office that does independent verification and validation of space safety standards and copes with "waivers" when those standards are not or cannot be met. It is clear that much the same approach to "space safety" is taken by most of the other space agencies around the world, whether it is the European Space Agency (ESA), the Japanese Space Agency (JAXA and ISAS), the Russian Space Agency (Roscosmos), the Indian Space Research Organization (ISRO), the Chinese National Space Agency (CNSA), etc. The emergency of "new space" industrial activities, the growing threat of orbital space debris, the increasing plans for industrial use of protospace, the new recognition of various cosmic threats that includes asteroids, comets, solar storms, changes to the earth magnetic fields, and climate change, however, are all serving at least one new purpose. That purpose is to redefine the very meaning, scope, and importance of what is now encompassed by "space safety."

In today's new world of space safety, everybody from astrophysicists to orbital mechanics engineers, from astrobiologists to robotics engineers, are involved in this ever-expanding effort. One of the key next steps forward will be to determine which international agency (such as perhaps ICAO?) will take on the topic of space and protospace traffic management and especially the control of space planes and hypersonic transport. Another key question is whether legislative bodies around the world, such as space agencies or some other scientific or regulatory agencies, will take on such issues as planetary defense, protective systems against natural or man-made electromagnetic pulses (EMPs), and even space-based systems to cope with solar storms, earth's changing magnetic field systems and climate change. At one time, space agencies might have assumed such responsibilities, but today it may be the time for other national or international entities to take on these tasks with a clearly defined mandate and appropriate financial and technical resources adequate to these daunting tasks that could involve the survival of billions of people, if not the entire planet.

References

"A Super Solar Flare." 2016. http://science.nasa.gov/science-news/science-at-nasa/2008/06may_carringtonflare/ (last accessed April 20, 2016).

Galloway, Eileen. Winter 1979. "Nuclear Powered Satellites: The U.S.S.R. Cosmos 954 and the Canadian Claim," *Akron Law Review*, https://www.uakron.edu/dotAsset/4acf6020-02f6-4cd8-aa19-416a0303ad10.pdf (last accessed on April 21, 2016).

"Halloween Storm of 2003 Still the Scariest." 2016. http://www.nasa.gov/topics/solarsystem/features/halloween_storms.html

International Association for the Advancement of Space Safety (IAASS) Space Safety Objectives. 2016. www.iaass.space-safety.org (last accessed April, 2016).

NASA Space Shuttle Lot. 2016. https://www.google.co.za/search?q=space+shuttle+pictures+nasa&biw=1024&bih=475&tbm=isch&imgil=0MhLvfG7fAwXwM%253A%253BRpXzPuSkzjKePM%253Bhttp%25253A%25252F%25252Fwww.fanpop.com%25252Fclubs%25252Fnasa%25252Fimages%25252F27326777%25252Ftitle%25252Fnasa-space-shuttle-lot-photo&source=iu&pf=m&fir=0MhLvfG7fAwXwM%253A%252CRpXzPuSkzjKePM%252C_&usg=__DG489IPlFIPIuPufAW35EdM7atY%3D&ved=0ahUKEwje-tqlsprMAhUHWhoKHb5GD8gQyjcIJw&ei=9_oVV56jLYe0ab6NvcAM#imgrc=0MhLvfG7fAwXwM%3A (last accessed, April 20, 2016).

NASA Technical Standard, NS870922, Rev. 2012. http://www.hq.nasa.gov/office/codeq/doctree/NS870922.pdf2

Orbital Debris History of Orbital Fragmentation, 18th Edition, NASA. 2008. https://www.google.co.za/?gfe_rd=cr&ei=0qUYV4OTHOuo8went4XQDQ&gws_rd=ssl#q=history+of+orbital+fragmentations+18th+edition

Pelton, J. 2006. *Basics of Satellite Communications*, 2nd ed., International Engineering Consortium. Chicago, Illinois.

Pelton, J., Sgobba, T., and Trujillo, M. 2015. "Space Safety," Chap. 11, in Kai-UweSchrogl, Peter L. Hays, Jana Robinson, Denis Moura, and Christina Giannopapa (eds.), *Handbook of Space Security*, Springer Press, New York.

Pelton, J., Smith, D., Helm, N., MacDoran, P., Caughran, P., and Logsdon, J. 2005. *Vulnerabilities and Risk Reduction in U.S. Human Space Flight Programs*, Space and Advanced Communications Research Institute (SACRI) George Washington University Research Report.

Sgobba, T. and Pelton, J. 2010. "Space Safety Standards," in J. Pelton and R. Jakhu (eds.), *Space Safety Standards and Regulation*, Elsevier Publishing, New York.

SLS Architecture Reference Configuration. 2016. https://www.google.co.za/search?q=picture+of+NASA+Space+launch+system&biw=1024&bih=475&tbm=isch&imgil=fhBFWJCZp0kNmM%253A%253BhGaHHAoPV9VfoM%253Bhttp%25253A%25252F%25252Fwww.spaceflightinsider.com%25252Fmissions%25252Fhuman-spaceflight%25252Fnasas-sls-program-manager-talks-block-1b-and-beyond-part-three%25252F&source=iu&pf=m&fir=fhBFWJCZp0kNmM%253A%252ChGaHHAoPV9VfoM%252C_&usg=__VDXIssJ25dl50sobMlTnRVPomAk%3D&ved=0ahUKEwjD75bzuJrMAhXEnRoKHY8-CmYQyjcIMw&ei=5AEWV8PtMcS7ao_9qLAG#imgrc=fhBFWJCZp0kNmM%3A (last accessed on April 20, 2016).

Werner, D. and Zak, A. 2015. "Maximizing Safety," *Aerospace America*, October, pp. 18–25.

PART 4

Spacecraft for Human Operation and Habitation

Gary H. Kitmacher

7.28 Introduction

Several new spacecraft designed for human habitation and operations are being developed in the early 21st century as spaceflight enjoys a resurgence. The new vehicles may be used for Earth orbital servicing of the International Space Station (ISS), which is in orbit today, or for future lunar and planetary missions. Spacecraft have been carrying humans for more than 50 years at the time of this writing. They have been developed by more than a dozen nations to several different standards.

These spacecraft carry their human crew on designated missions. The design of the spacecraft is influenced by the specific environment through which it will fly and the functions the crew will be expected to perform. This includes its starting place on the earth, its traverse through the earth's atmosphere, the regions of space through which it will fly, and the attributes of any world to which it may approach or land on, including the earth.

7.29 Premium Placed on Mass and Volume

A spacecraft usually coasts in a nearly frictionless environment and to change velocity, whether direction or speed, it must expend mass or energy. The quantity of mass or energy, which has to be expended, is directly related to the mass of the spacecraft. The mass is dependent on the spacecraft's volume. Because spacecraft all involve moving mass at high velocity, following specific trajectories, and maintaining and changing orientation, of necessity spacecraft are as low mass as is practical to construct them.

A primary constraint on the mass and design of any spacecraft is the capacity of its launch vehicle. Because of constraints on mass and volume, many of the systems are designed to be as efficient as possible, often using miniaturized components.

Spacecraft travel faster than any other vehicles that have ever carried humans. While some are designed to fly at high velocity through the atmosphere and are therefore designed to be aerodynamic, because some spacecraft travel only in a vacuum, they may not be streamlined at all and do not look particularly fast. Figure 7.40 illustrates the configuration and respective sizes of all manned spacecraft flown between 1961 and 2016. Table 7.4 provides pertinent parameters for each.

7.30 Common Attributes of Manned Spacecraft

A spacecraft that is designed for humans must be designed:

- To protect the inhabitants from the natural space environment
- To protect the inhabitants from induced environments resulting from the spacecraft's operation (Figure 7.41)
- To fulfill the missions for which the spacecraft is intended

It must provide protection for its crew against the natural atmospheric pressure, temperature, radiation, and micrometeorites or other particulate debris in the environment in which it operates. The spacecraft must carry an artificial environment that is conducive to life. Humans and systems

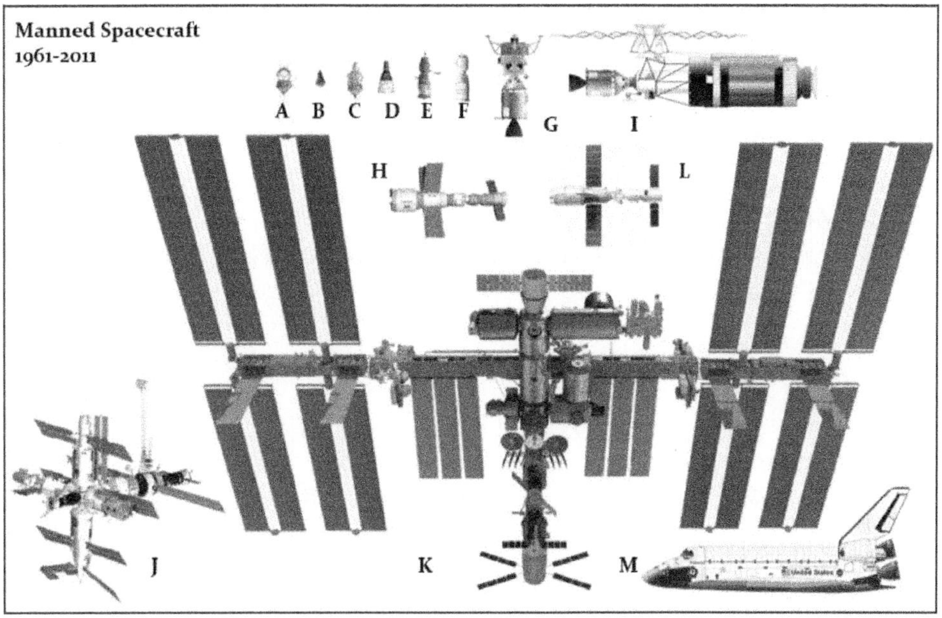

FIGURE 7.40 Relative sizes of all manned spacecraft flown between 1961 and 2016. A, Vostok; B, Mercury; C, Voskhod; D, Gemini; E, Soyuz; F, Shen Zhou; G, Apollo; H, Salyut; I, Skylab; J, Mir; K, ISS; L, Tian Gong; M, Space Shuttle. (*G. Kitmacher.*)

operating in the environment of the spacecraft will require electrical energy. They will create waste thermal energy which must be dissipated. Humans will require water and food for consumption and they will create effluents and waste which must be disposed of or utilized. The spacecraft will require sensors to enable its orientation and location in space, and mechanical or propulsive systems to reorient and relocate itself.

Space-rated human systems tend to be expensive because human life must still be protected even when systems fail. Therefore, systems must be redundant for life-critical support functions. All critical systems essential for crew safety must be designed to be two-fault tolerant. When this is not practical, systems have to be designed so that no single failure can cause the loss of the crew. In-flight maintenance can be considered as a third leg of redundancy as long as mission operations, appropriate tools, spares, and logistics support it (Figures 7.42 through 7.45).

Hardware and software reliability is verified by integrated system-level testing prior to flight. Verification based on in-flight system monitoring continues throughout the flight phase and prior to reflight. A successful, comprehensive test program validates analytical models, verifies the environmental envelope in which the vehicle must perform, and provides a performance database prior to flight. Vehicle reliability is verified by testing at the integrated system level prior to flight. It is reverified by monitoring during flight.

The on-board flight crew should have the capability to control their vehicle and its systems, including overriding any automated controls. The crew must be provided with insight that is adequate to enable operation of the vehicle systems. Autonomous operation without reliance on ground support should also be supported.

Potential risks to the crew, the vehicle, the payload, and the mission are identified throughout the design process. Candidate mission architectures, operations, elements, and equipment should be optimized together with crew capabilities to maximize system performance and minimize risk.

Spacecraft	Dimensions	Pressurized volume	Mass	Crew size	Years operating	Cabin at inosphere	Longest* flight	Recovery	Max G
Earth Orbital, Earth launch and landing									
A. Vostok	5 × 2.5 m	3 cu m	4700 kg	1	1961–1963	14·7 psi air	5 days	land, pilot ejection	9 G
B. Mercury	3 × 2 m	1.6cu m	1400 kg	1	1959–1963	5 psi O_2	34 hrs	ocean, chute	12 G
C. Voskhod	5 × 2.5 m	4·5 cu m	5700 kg	2, 3	1964–1965	14·7 psi air	2 days	land, chute, rocket	9 G
D. Gemini	6 × 3 m	2.3 cu m	3800 kg	2	1964–1966	5 psi O_2	14 days	ocean, chute	8 G
E. Soyuz	8 × 2.7 m	6 cu m	7200 kg	1, 2, 3	1966–	14·7 psi air	18 days	land, chute, rocket	8 G
F. ShenZhou	9·3 × 2.8 m	14 cu m	7800 kg	1, 3	1999–	14·7 psi air	15 days	land, chute, rocket	8 G
Lunar Trajectory, Moon Landing and Launch									
G. Apollo									
CSM	11 × 4 m	6 cu m	28800 kg	1, 3	1966–1975	5 psi O_2	12.5 days	ocean, chute	4 G
LM	7 × 9.4 m	4·5 cu m	16400 kg	2	1969–1972	5 psi O_2	3 days	none	1 G
Space Stations							Days Occupied	No. of visiting manned flights	Longest Duration
H. Salyut 1	20 × 4 m	99 cu m	18500 kg	3	1971	14·7 psi air	24	2	24
Salyut 3	14·5 × 4 m	90 cu m	18500 kg	3	1974–1975	14·7 psi air	15	1	16
Salyut 4	15.8 × 4 m	90 cu m	18500 kg	2	1974–1977	14·7 psi air	92	2	63
Salyut 5	14·5 × 4 m	100 cu m	19000 kg	2	1976–1977	14·7 psi air	67	2	49
Salyut 6	15.8 × 4 m	90 cu m	19800 kg	2	1977–1982	14·7 psi air	683	16	184
Salyut 7	16.4 × 4 m	90 cu m	18900 kg	2, 3	1982–1991	14·7 psi air	816	11	236
I. Skylab	25 × 17 m	361 cu m	77,111 kg	3	1973–1974	5 psi O_2	168	3	84
J. Mir	28 × 31 m	350 cu m	129700 kg	2.3	1986–2001	14·7 psi air	4592	39	437
K. ISS	100 × 93 m	915 cu m	430000 kg	3, 6	1998–current	14·7 psi air	341 days	ongoing	340
L. Tiangong 1	10.4 × 3.3 m	15 cu m	8506 kg	3	2012–2013	14·7 psi air	25 days	2	15
Space Plane									
M. Space Shuttle	24 × 37 m	132 cu m	110000 kg	2 to 8	1981–2011	14·7 psi air	17 days	gliding flight	3·5 G

*longest independent flight; Soyuz and Apollo spacecraft dock in a passive mode to space stations for several months.

TABLE 7.4 Manned Spacecraft Flown between 1961 and 2011

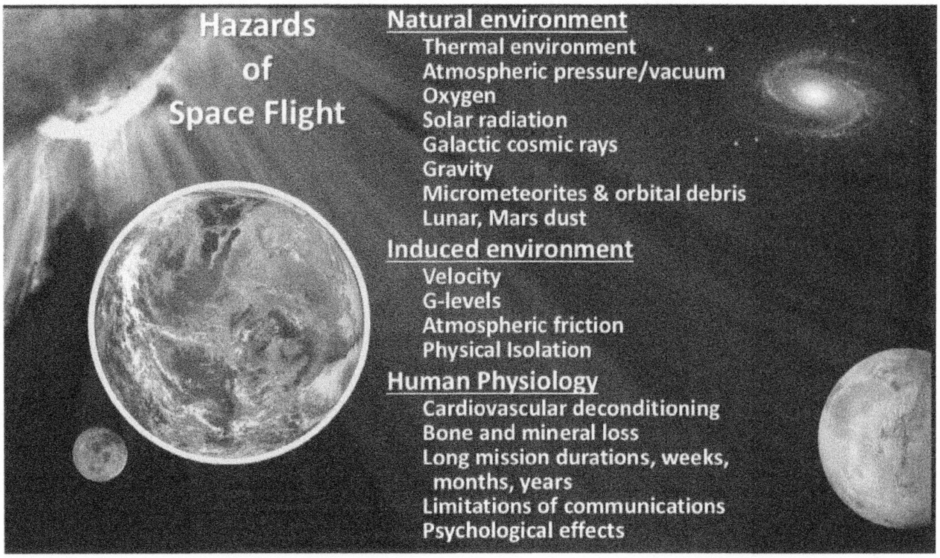

FIGURE 7.41 Hazards of human space flight stem from both natural and induced environments. (*G. Kitmacher.*)

FIGURE 7.42 The crew compartment of the U.S. Space Shuttle was a separate insulated and isolated, pressurized structure within the fuselage. (*NASA.*)

FIGURE 7.43 Pressure vessels of the ISS modules are aluminum shells with exterior layers of insulation and micrometeorite protection. (*NASA.*)

The crew should always be aware of risks imposed by the systems, vehicle operations, and environments throughout all mission phases.

Any spacecraft, with or without crew, which operates in proximity or while docking or berthing with a crewed spacecraft must be designed and operated to eliminate hazards to both vehicles. Provisions should enable abort, breakout, and separation by either vehicle without violating the design and operational requirements of either vehicle. Crews on-board the manned vehicle must have the capability to command safety critical functions on the unmanned vehicle.

FIGURE 7.44 The continuously closed environment of a spacecraft like the ISS can lead to contaminants. Here astronauts wear portable breathing masks to protect themselves from ammonia used in the cooling system. (*NASA.*)

FIGURE 7.45 The ISS cupola provides 3D viewing and serves as the flight deck, but windows must be protected against extreme temperatures, direct sunlight and overheating and orbital debris with special insulated shutters. (*NASA*.)

If one space vehicle interacts with another, there are interrelationships which must be sufficiently understood. The interface between elements must be specified so that performance is controlled in a complementary fashion and so overall performance is optimized.

For earth-to-orbit (ETO) vehicles, a crew escape system should be provided for safe crew recovery from in-flight failures across the flight envelope from prelaunch to landing.

For beyond-earth-orbit (BEO) missions, spacecraft and propulsion systems should have sufficient energy to fly abort trajectories and provide power and critical consumables to enable crew survival (Figure 7.46).

(a) (b) (c)

FIGURE 7.46 (*a*) A crew trains to land the Apollo Lunar Module on the moon. The two astronauts stood, strapped and velcroed to the floor, because seats weighed too much and had been removed in the final design. Even with main engine firing, G-level never got beyond one earth gravity. One astronaut focused on the lunar surface, looking for landmarks and manually flying the spacecraft, while the other astronaut kept his eyes inside the cockpit, and focused on instrumentation and computer readings. (*NASA*.) (*b*) Inside the ISS cupola. The cupola provides viewing of an entire hemisphere surrounding the ISS. (*NASA*.) (*c*) Inside the ISS cupola, using laptop computers, hand controllers, along with direct viewing out the windows, two crewmembers maneuver the Japanese logistics vehicle, HTV, to an ISS berthing port. (*NASA*.)

7.31 Optimization of Humans with Machines

The spacecraft is a complex machine. But the most complex, highly developed systems aboard the spacecraft are the astronauts. The spacecraft provides specialized capabilities in power, propulsion, orientation, and guidance. Computers can provide computational and control capabilities. The spacecraft computers can manage equipment and systems devoting untiring attention to monotonous tasks. Human intelligence, however, can interpret a wide variety of unrelated information to deal with unforeseen situations.

Spaceflight is challenging for the human organism. The space environment is foreign and threatening. Life relies on proper functioning of spacecraft and human systems. Physical and environmental conditions affect crew performance and safety.

The spacecraft design and facilities should address the full range of human needs, beyond just the need for human survival. It is important for the spacecraft to facilitate human habitation functions. It is important that the spacecraft be designed to facilitate information communication to the crew. It is important that the system of humans + machine be integrated and optimized. Technology for remote or autonomous operation, and controls and displays for virtual and telepresence capabilities can have a significant impact on the ability of the crew to rapidly acquire, assess, and use information related to all operations. Implications of advanced human/machine systems operations are seen in

- Task allocation between crew and machine
- Intelligent system design
- Automation capabilities
- Simulation requirements
- Crew experience, training, and skill
- System safety

In addition to monitoring and controlling the status of various systems, the crew of a spacecraft should be able to communicate with mission control on earth with regard to mission and system status. But, the on-board crews' ability to communicate with earth can vary depending on the mission, mission phase, and operation. For missions in low earth orbit (LEO), communication between earth and spacecraft may be continuous and near instantaneous. For earth-to-orbit and return vehicles, there may be periods of black-out or periods during which control functions must be made more quickly than humans can react. For some beyond-earth-missions, communications between earth and on-board crew may be delayed to the point that the on-board crew must operate independently with messages sent and received only after a delay of hours (Figures 7.47 through 7.49).

7.32 Human Spacecraft Configuration

Spacecraft come in a variety of forms, tailored for a given payload and mission. Configuration is influenced by the environments in which the spacecraft operate.

The earliest human carrying spacecraft were often referred to as "capsules" because they were small and self-contained, often designed around their one-person human crew. The early spacecraft were as compact as their designers could make them, including minimal volume for the astronaut. They were designed exclusively for operation and constrained by their launch vehicles. The early spacecraft were launched either in place of the nose cone of a launch vehicle or within the aerodynamic fairing that formed the nose cone. The use of a compact, relatively simple, symmetric configuration-reduces aerodynamic loads and minimizes flight control complexity. The compact vehicle's reduced mass, minimizes fuel requirements.

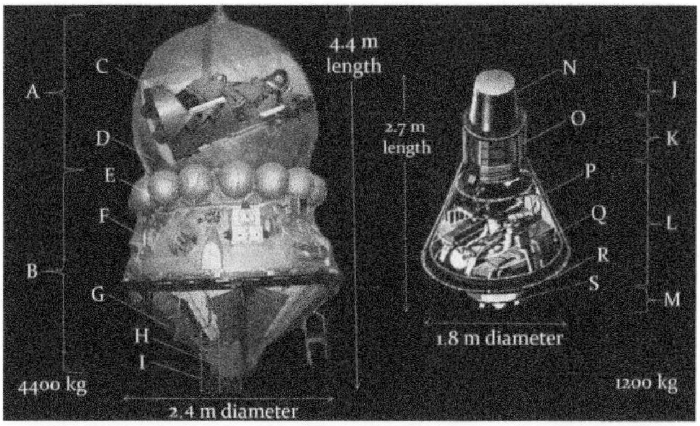

FIGURE 7.47 On the left, a cutaway of the Russian Vostok spacecraft, the first spacecraft to carry a human. The cosmonaut reclines in (A) the ball-shaped capsule, seated in (C) an ejection seat which served for emergency escape and for routine landings; (D) the capsule was protected by a resin heat shield during return from orbit. (B) A service module housed most systems and consumable supplies: (E) oxygen and nitrogen tanks, (F) orientation rockets, (G) thermal radiators, (H) retrorocket, (I) telemetry antenna. On the right is the U.S. Mercury spacecraft. Its major sections: (J) instrument compartment, (K) parachute compartment, (L) crew and systems compartment, (M) retrorocket module. Features: (N) horizon scanner and drogue parachute, (O) main and backup parachutes, (P) astronaut and controls, (Q) environmental, communications, and power systems, (R) resin heat shield, (S) retro- and posigrade rockets. (*NASA.*)

A shape that provides maximum volume with minimum surface area reduces structural stress, atmospheric leakage, and protects better against the potential hazards of radiation and orbital debris or micrometeorites. Generally, the simpler the shape, such as spherical, cylindrical, or conical reduces the amount of required structure and reduces stresses caused by internal pressurization while maximizing volume. A spherical ball is optimal (Figures 7.47 and 7.48).

FIGURE 7.48 The Voskhod was based on Vostok. Three cosmonauts rode in very tight quarters without spacesuits. An additional module was added on top of the capsule containing a backup retrorocket. The Voskhod is in preparation for launch, being encapsulated in a nose cone shroud which will be jettisoned before reaching orbit. (*RSC Energia.*)

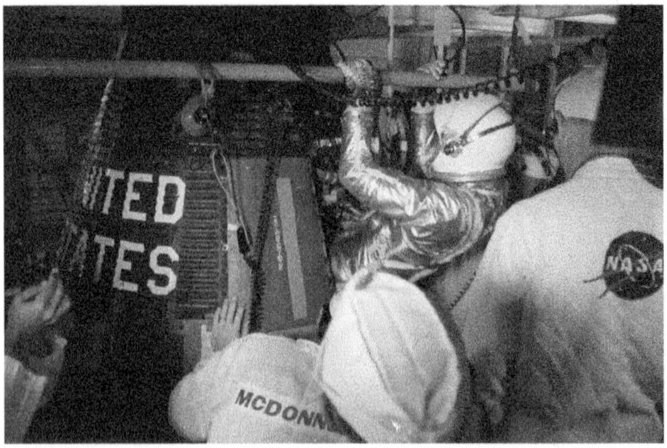

Figure 7.49 The Mercury spacecraft was a U.S. contemporary of the Vostok. The tiny capsule, 2 m diameter and 3 m long, carried one astronaut. Size and height restrictions meant all Mercury astronauts were of small stature. (*NASA*.)

Construction processes for manned spacecraft have been borrowed from the aircraft industry, in large measure because many of the same companies producing spacecraft were previously producing aircraft. In many ways, requirements for spacecraft and aircraft were similar; although there was not a knowledge base for designing spacecraft, both air and space vehicles had to be reliable and operate in demanding and hostile environments (Figures 7.49 through 7.51). Early spacecraft were designed with aluminum semi-monocoque structures. In addition to providing overall rigidity, a primary mechanical function of the spacecraft structure was to support subsystem components, particularly during the launch, where stresses tend to be highest; and to allow the craft to be handled on the ground during assembly and transportation. More recent manned spacecraft designs use the same semi-monocoque philosophy but with CAD-CAM and 3D printing using waffle pattern or isogrid, can now be used to manufacture a unitized spacecraft structure that doubles as the pressure vessel (Figure 7.52).

Figure 7.50 Engineers work on the structure of an early prototype Mercury capsule in 1959. (*NASA*.)

Figure 7.51 Work on a production Mercury spacecraft in a clean room at the McDonnell Aircraft Corporation in 1962. (*NASA*.)

(a) (b) (c)

Figure 7.52 Stages in the assembly of the Orion spacecraft. (*a*) Image shows the isogrid pressure vessel for the spacecraft prior to the integration of systems hardware. (*NASA*.) (*b*) Image shows systems hardware installed. (*NASA*.) (*c*) Image shows the completed command module heat shield that covers the systems installations. (*NASA*.)

Because of mass constraints and the need to operate beyond the earth's atmosphere, new technologies and new lightweight materials have been used to provide protection against the space environment. Typical surface materials used are aluminum alloys, honeycomb panels, resins, carbon fiber and lightweight metallized polyesters, plastics, polyimides, polycarbonates and synthetic fibers for thermal insulation, and protection against meteorites and orbital debris.

The two foremost designers of manned spacecraft, Russia and the United States, have somewhat different design philosophies (Figure 7.53). Russian designs have evolved more slowly with newer systems based heavily on earlier systems. The early differences in design were due in part to launch capacity. The United States possessed a smaller nuclear weapon and therefore the nuclear weapon delivery system, the Atlas intercontinental ballistic missile (ICBM) had a limited payload capacity. The Russians had a larger nuclear bomb and their ICBM, the R-7 had to have a larger carrying capacity. The Russians could therefore afford a larger irregularly shaped spacecraft which launched inside of an aerodynamic shroud. Their spacecraft was large enough to carry an earth-like,

FIGURE 7.53 Russian and U.S. launch vehicles used for manned spacecraft. From left to right: 1961 Vostok/R-7, 1964 Voskhod/R-7, 1967 Soyuz/R-7, 1961 Mercury/Redstone, 1962 Mercury/Atlas D, 1965 Gemini/Titan II, 1968 Apollo/Saturn 1b, 1968 Apollo/Saturn V, 1981 Space Shuttle, proposed Orion/Space Launch System. (*G. Kitmacher.*)

higher pressure atmosphere of nitrogen and oxygen, and many of their systems were air-cooled inside the pressurized compartment. For spacewalks, the Russians added a separate airlock compartment. The entire pressurized compartment was protected during reentry in the earth's atmosphere by a larger and heavier spherical heat shield that encompassed the spacecraft. By comparison, the first three U.S. spacecraft designs, Mercury, Gemini, and Apollo, were conical in shape, designed to serve as the nose cones of the Redstone, Atlas, Titan, and Saturn launch vehicles, which boosted them into space (Figure 7.54). The conical shape was a tight fit for the astronauts who would sit at the broad end; it permitted the placement of a relatively small circular heat shield, which would protect the spacecraft from heating while providing an aerodynamic face during their reentry into the earth's atmosphere. The reduced launch vehicle capacity meant the United States used a lower pressure pure oxygen atmosphere. The lower pressure permitted a lighter structure. There was no room for an airlock, so for spacewalks the entire spacecraft would be depressurized.

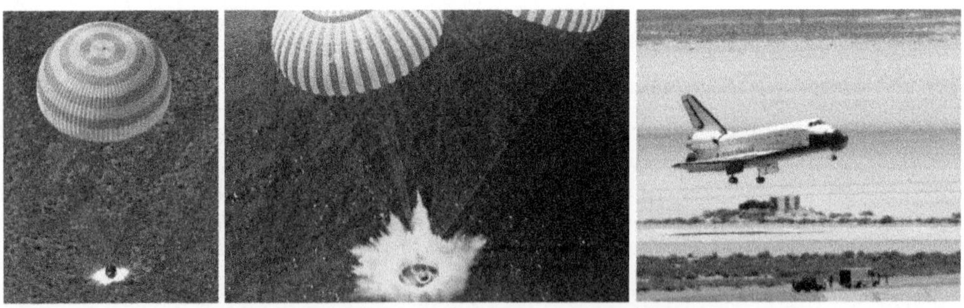

(a) (b) (c)

FIGURE 7.54 Methods of landing. (*a*) Soyuz descends under single parachute and uses a solid fuel retrorocket that fires at 1 m above the ground. (*NASA.*) (*b*) Apollo used three parachutes and landed in the ocean. (*NASA.*) (*c*) Space Shuttles glided to a runway landing. (*NASA.*)

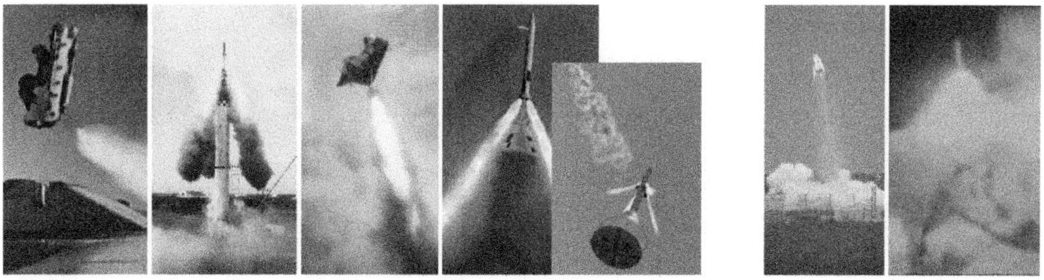

(a) (b) (c) (d) (e) (f) (g)

FIGURE 7.55 Launch escape. (*a*) Vostok ejection seat. All six Vostok cosmonauts ejected as a normal part of their mission. (*RSC Energia*.) (*b*) Mercury escape rocket. (*NASA*.) (*c*) Space Shuttle ejection seat. (*NASA*.) (*d*) Apollo escape rocket. (*NASA*.) (*e*) Orion escape rocket. (*NASA*.) (*f*) Dragon escape rocket. (*NASA*.) (*g*) Only actual use of an escape rocket during a crew emergency occurred in 1983, when Soyuz T-10 caught fire on the launch pad during a night launch. Automated systems failed due to cable burn through. The escape rocket fired after 20 seconds in response to ground personnel manual commands. The two-man crew survived without injury while the launch vehicle exploded beneath them. (*RSC Energia*.)

Equipment and electronics could not be air-cooled but instead relied upon active thermal control requiring hardware to be mounted on cold plates with coolant pumped through to externally mounted radiators (Figures 7.55 and 7.56).

Because of Soviet concerns about domestic security at the height of the cold war, an outgrowth of the Soviet experience during World War II, the Soviets placed their primary launch site in a land-locked area on the Kazakh steppe. This necessitated designing the Soviet and Russian spacecraft primarily for land landing and for water landings as a contingency. The United States, by comparison had the world's largest naval force and placed its launch test centers along the coast. This necessitated designing the early manned spacecraft primarily for water landings.

The first manned spacecraft, the Soviet (Russian) Vostok was spherical in shape. The spherical return capsule had an off-center center-of-gravity, which passively maintained the spacecraft's

(a) (b) (c) (d)

FIGURE 7.56 Heat protection is required for vehicles returning to earth from an orbital velocity of 28,000 km/h or a lunar return velocity of 40,000 km/h due to heating from atmospheric friction. (*a*) Vostok's ball had a resin heat shield which burned and melted away during atmospheric entry. The resin was covered with hexagons of metallic foil to reflect sunlight in space. The base which entered the atmosphere first is to the right; the metallic hexagons are burned away. Hexagons remain on the opposite, cool side of the spacecraft. (*RSC Energia*.) (*b*) The Apollo Command Module had a heat shield similar in design to Vostok, with a resin heat shield. The areas exposed to sunlight during the mission had an aluminized mylar covering for solar reflection, visible in the second image. (*NASA*.) (*c*) The honeycomb pattern of the Apollo heat shield is visible. Each honeycomb cell was filled with resin. (*NASA*.) (*d*) The nose of a Space Shuttle. The nose cone to the left is a high-temperature reinforced carbon-carbon ceramic. Much of the surface of the Shuttle was covered with lightweight ceramic tiles that were glued to the underlying aluminum airframe and surface skin. (*NASA*.)

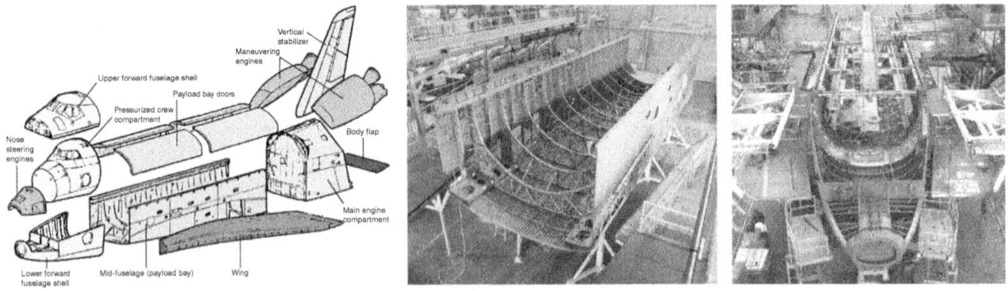

FIGURE 7.57 Structural components of the Space Shuttle during assembly. Much of the airframe was aluminum. (*NASA*.)

orientation while returning through the atmosphere without using active orientation which would have necessitated using control thrusters and active automated or crew piloting. The biconic shape of the U.S. capsules required more sophisticated orientation and control systems, requiring active orientation, control, and piloting (Figure 7.57).

Design philosophies were also different because of the political and social structure of the two countries and these effects on their respective manufacturing industries during most of the 20th century. Continuity of personnel under the Soviet government, who had jobs assured for their lifetimes, led to long tenures in working positions. Under the Soviets, manufacturers were state-owned and a relatively small number of specialized manufacturing companies led Russian designs to maintain design continuity and hardware commonality from one spacecraft to the next, and to evolve slowly as technologies improved. The United States proceeded with new starts and changes of direction every time a major new contract was issued and as new personnel and new manufacturers proposed new designs, processes, and materials. As fall-out from these different philosophies, Russian spacecraft are integrated at the module level; U.S. spacecraft hardware is usually designed at the major component and systems level, often with different subsystems designed and built by separate subcontractor companies and subsequently integrated into the spacecraft.

These differences had an effect on the definition of requirements, interfaces, and documentation. The Soviets often internalized requirements and developed minimal documentation because their hardware was designed and developed internally with closely established supervision from concept through flight. The United States, depending on a large number of external hardware suppliers and with large workforce distributed at far-flung geographic locations, required more extensive specification of materials and processes, documentation, control and verification of hardware during the manufacture, integration, and test and flight process.

The U.S. Space Shuttle was a winged, aerodynamic hypersonic flying vehicle. It was designed and built like an airplane with active aerodynamic control systems: wings, vertical stabilizer, elevons, rudder, speed brakes, as well as with rocket thrusters and systems that would function in the weightlessness of orbital flight. Major components were built by aerospace manufacturers around the country. Its aluminum airframe, not any different from the typical large jet aircraft, was covered with ceramic thermal protection tiles to protect against the several thousand degree temperatures of reentry. The Soviet Buran shuttle, following the U.S. shuttle by about a decade, was similar in design, using similar technologies and similar materials (Figures 7.42 and 7.57).

The Lunar Module (LM) was the spacecraft designed to land astronauts on the moon during the Apollo missions. The LM was designed to be enclosed inside of an aerodynamic fairing during launch. In many respects it was the world's first true spacecraft in that it could only operate in the vacuum environment of space. It had two stages. The lower and more massive descent stage had spindly legs, oxygen tanks and batteries to support the astronaut on the moon's surface, and large spherical fuel tanks and a large rocket engine for the descent to the moon, all packaged in a boxy octagon.

The upper stage had a variety of appendages, protruding antennae and docking targets, bulbous multifaceted compartments containing the fuel tanks used to launch from the moon, and a cylindrical crew compartment with large triangular windows.

Weight was particularly critical in landing on the moon. Every ounce translated to fewer seconds the astronauts would have to select and maneuver to a landing location. In order to reduce and eliminate every spare ounce from the LM, the manufacturer, Grumman aircraft, offered cash prizes for every idea that eliminated unneeded weight. The LM had to operate in the environment of the lunar surface, where temperatures went abruptly from as high as 120°C (250°F) in sunlight to –157°C (–250°F) in shade. The design strategy pioneered during Apollo was to wrap the LM in insulation so that the functional systems would be thermally isolated. Solid exterior panels were eliminated in favor of multilayer metallic mylar foils and polyimide films that were taped in place. For the pressure shell of the cabin, aluminum ingots were chemically milled to be so thin that a misplaced kick by an astronaut's boot might easily puncture the shell and depressurize the cabin. The hatch through which the astronauts would emerge to walk on the moon was so thin that it noticeably bulged when the cabin was pressurized. Seats were eliminated and instead the astronauts stood erect, shoulder-to-shoulder, velcroed, and tethered in place. In the 1G experienced while the main engine was firing, or the 1/6G environment of the moon, it was not an inconvenience. It was described as feeling "like standing in an elevator" on earth.

The LM was structurally and aerodynamically incapable of operation in flight in the earth's atmosphere and was never intended to return to earth. Every LM flown on ten Apollo missions was a new vehicle, never previously flown or tested. Although the LM could not be tested in the actual operating environment before its mission, every system was tested on the ground at the component, assembly, subsystem, and the fully integrated levels prior to flight (Figure 7.58).

Space stations, like the International Space Station (ISS) that is in orbit today, are intended for operation only in space. ISS modules were intended by their designers at the outset to be used as a basic element for future manned space activities. Modules first used for the ISS were intended to eventually provide permanent habitats in lunar orbit, on the moon's surface, and for vehicles traveling to Mars (Figure 7.59).

From the outset, the U.S. Space Shuttle was intended to be the launch vehicle and assembly platform for the ISS. So, the Shuttle's design heavily influenced the design of the ISS modules. The U.S. modules had to be carried in the Shuttle's payload bay. This permitted a maximum diameter of 460 cm (15 ft) and a length of 1,830 cm (60 ft). Designers wanted the ISS modules to be as volumetrically large as possible, but the modules also were constrained by the capacity of

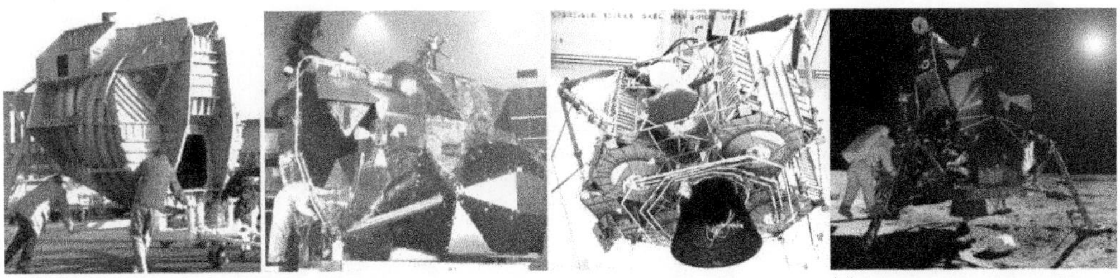

(a) (b) (c) (d)

Figure 7.58 The Apollo Lunar Module (LM) during fabrication and on the moon. (*a*) The forward ascent stage was the main crew cabin; it was a cylindrical structure. Its forward face accommodated a hatch and two large triangular windows. The materials were as thin as could be made while still protecting the crew, and in places a misplaced astronaut boot could have easily ruptured the pressure vessel. (*Grumman.*) (*b*) A completed ascent stage prior to flight. (*Grumman.*) (*c*) The LM descent stage cruciform structure which held the large engine for descending to the moon and tanks for fuel, oxygen, and water. (*Grumman.*) (*d*) The Apollo 11 LM-5, nicknamed Eagle, on the moon. (*NASA.*)

(a) (b)

Figure 7.59 (*a*) ISS module proposed for use for the First Lunar Outpost. (*NASA.*) (*b*) ISS modules proposed for use for the Earth-to-Mars vehicle. (*Boeing.*)

their launch vehicles. The modules had to be sized not only for volumetric constraints but for Shuttle's lift capacity to the station orbit. So the modules had to be cut back to no more than 1,130 cm (37 ft) long because of mass.

7.33 Space Vehicle Architecture

The ISS provides a habitable environment for its crew. The ISS had to be designed to sustain the habitat environment for decades. The environmental control system (ECS) and supporting systems had to be accessible to the on-board crew to maintain and repair. Modularity permits system components to be replaced; components are launched to the station, operate for a period, and later return for refurbishment (Figure 7.60).

Each U.S. module of the ISS was launched by the Space Shuttle. Mainly the Shuttle did the assembly, attaching each module to the ISS already assembled in orbit. Each additional module provided new components of subsystems. Systems integrated across modules support the overall vehicle. The modules of the ISS were intended from the outset as habitats, and the diameter of the modules and the 460 cm (15 ft) wide opening of the Shuttle's payload bay were the critical

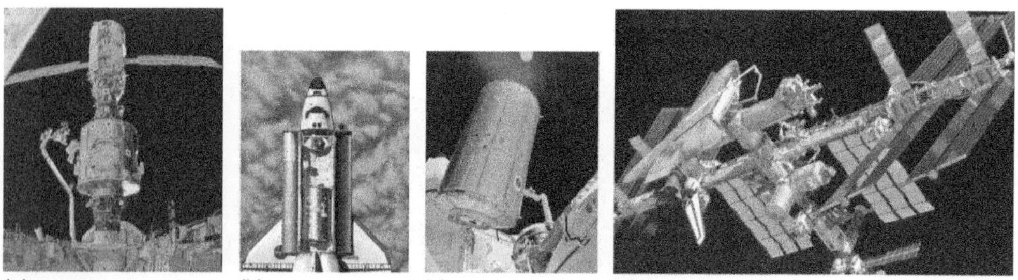

(a) (b) (c) (d)

Figure 7.60 ISS assembly. (*a*) The first two modules of the ISS are connected by the Shuttle. The module at the top, the Russian built FGB, could function independently and autonomously and provided electrical power, propulsion, maneuvering capability, and several of the systems required for the vehicle to survive in orbit without the Shuttle. (*NASA.*) (*b*) A Shuttle approaches ISS carrying one of the long ISS laboratories in its payload bay. (*NASA.*) (c) Robotic arms on the Shuttle and ISS help to maneuver the U.S. Lab module into place. (*NASA.*) (*d*) One of the last Shuttle flights is shown docked to the modules of the completed ISS. The original FGB module, in orbit for 14 years at the time of this image, is to the lower right. (*NASA.*)

(a) (b) (c) (d) (e)

Figure 7.61 Space stations prior to the ISS. Figures (*a–d*) show the progression in design of the Russian Long Duration Orbital Stations (DOS). (*a*) Salyut 1 was used by a single crew but the crew suffocated when their Soyuz spacecraft departed after a month long mission. (*b*) Salyut 4, occupied by two crews, one staying for 2 months. Salyuts 3 and 5 were "Almaz" stations, designed for military purposes. (*c*) Salyut 6 and 7 with a TKS space ferry. These stations were serviced by Progress and TKS ferry vehicles and had docking ports on both ends. Salyut 6 operated for 5 years with 16 visiting manned spacecraft. The longest stayed more than 6 months. Salyut 7 was used for about 5 years by 10 crews. Salyut 7 was the first modular space station using TKS modules. The TKS was originally designed to carry cosmonauts to military space stations but was never used in that capacity. It became the basis for modules like the FGB. (*d*) Mir Orbital Station was composed of a Salyut-style core module, 4 TKS-type modules and two unique modules. Mir remained in service for 15 years and hosted 39 visiting crews. Crews stayed as long as 1 year and one cosmonaut stayed for 437 days. Ten Space Shuttles flew to Mir and six U.S. astronauts lived and worked on board. Two of the FGB modules were outfitted by NASA as part of the first phase of the ISS program. (*e*) U.S. Skylab Orbital Workshop, hosted three crews for stays of 1, 2, and 3 months. (*G. Kitmacher.*)

dimensions, specifically sized to provide interior ISS compartments that were large enough to be outfitted for habitation.

The ISS was designed, from the outset, for in-space assembly of separate modules and elements, most of which could not function independently. The overall size, shape, and arrangement of the ISS was modularized so that changes to the overall station design could and were made many times during the multidecadal program. Some modules were eliminated and some were added. As international agreements established new percentages of "ownership," bartered in trade for Shuttle launches and additional crew members, equipment from one country or another was moved into new locations in the different country's modules (Figures 7.61 and 7.62). Individual modules were optimized for their functions. The United States, ESA, and Japanese laboratories

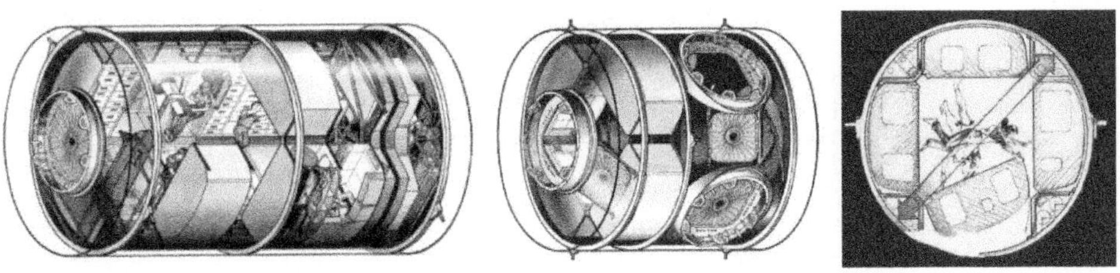

(a) (b) (c)

Figure 7.62 (*a*) Cutaway view of the ISS U.S. Lab Module, Destiny, shows one berthing port and hatch on one endcone. The opposite endcone has the same. A total of six racks can be located on each wall, floor and ceiling for a total of 24 racks. (*NASA.*) (*b*) Cutaway view of the ISS Node 1, showing berthing port and hatch on each endcone and four additional berthing ports/hatches around the circumference. Six modules can be attached at the berthing ports. Four racks fit at one end of Node 1. Nodes 2 and 3 are longer than Node 1 and each can fit 8 racks. (*NASA.*) (*c*) Cross section of a U.S. segment ISS module showing the four racks on ceiling, walls, and floor. Each is attached to a corner stand-off and can be pivoted away from the pressure vessel wall for access. (*Boeing.*)

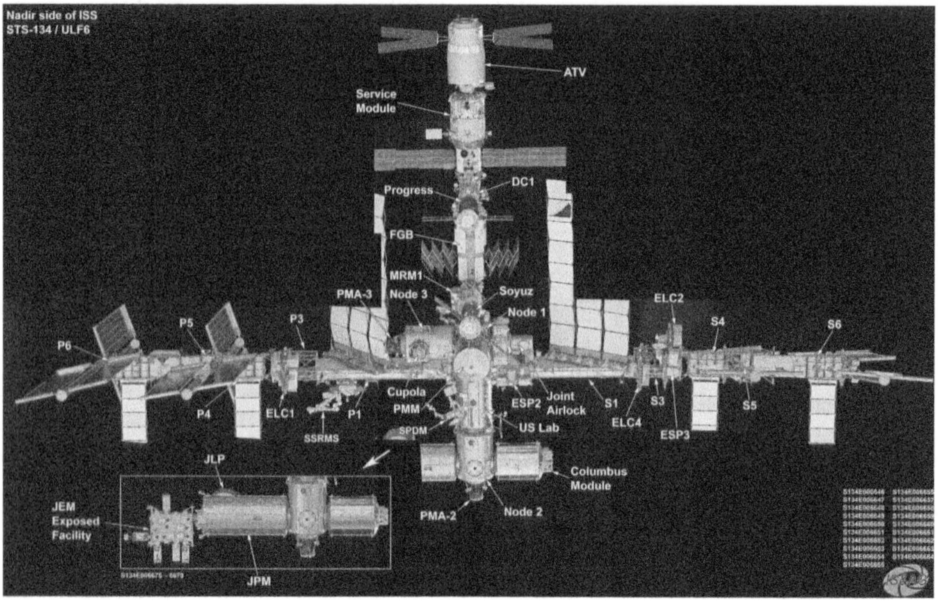

	length	width	mass*	pres vol	features
NASA/RSA FGB Zarya	13 m	4 m	19300 kg	72 m^3	3 docking ports, 2 solar arrays 3kW, propulsion
RSA Zvezda Service Mod.	13 m	4 m	20300 kg	85 m^3	4 docking ports, 2 solar arrays 3kW, propulsion 14 windows
US Node 1 Unity	6 m	4 m	11900 kg	25 m^3	4 racks, 6 berthing ports
US Lab Destiny	9 m	4 m	24000 kg	42 m^3	24 racks, 2 berthing ports
US Node 2 Harmony	7 m	4 m	14790 kg	31 m^3	8 racks, 6 berthing ports
US Node 3 Tranquility	7 m	4 m	14790 kg	31 m^3	8 racks, 6 berthing ports
Joint Airlock Quest	5 m	4 m	6100 kg	34 m^3	
NASA/ESA Cupola	2 m	3 m	1900 kg	5 m^3	
US Press Mating Adapter 1,2,3	2 m	2 m	1200-1600 kg	6 m^3 (total)	
ESA Lab Columbus	7 m	4 m	19300 kg	32 m^3	10 racks
JAXA Lab Kibo	11 m	4 m	15900 kg	52 m^3	23 racks, 2 berthing ports, equipment airlock
JAXA Logistics Mod	4 m	4 m	4200 kg	18 m^3	12 racks
NASA Leonardo Multipurpose	7 m	5 m	4428 kg	31 m^3	16 racks
RSA Piers Docking Mod.	5 m	3 m	3800 kg	13 m^3	
RSA Poisk Docking Mod.	5 m	3 m	3800 kg	15 m^3	
RSA Rassvet Research Mod.	6 m	2 m	5100 kg	17 m^3	
CSA ISS Remote Manipulator	18 m	.4 m	1497 kg		
CSA Dextre Manipulator	4 m	.9 m	1662 kg		
JAXA Manipulator	10 m	.4 m			
JAXA Fine Arm	2 m				
JAXA Exposed Facility	6 m	5 m	4000 kg		
Z1 Truss	5 m	4 m	8760 kg		Control moment gyros, 2 berthing mechanisms
S0 Truss	13 m	5 m	13970 kg		
S1 Truss	14 m	5 m	14120 kg		Thermal Control Radiators
S3/4 Truss	73 m	11 m	15820 kg		Solar arrays, a rotary joint
S6 Truss	73 m	11 m	15820 kg		Solar arrays
P1 Truss	14 m	5 m	14000 kg		Thermal Control Radiators
P3/4 Truss	14 m	5 m	15820 kg		Solar arrays, a rotary joint
P6 Truss	73 m	11 m	15820 kg		Solar arrays

FIGURE 7.63 ISS principal elements. (*NASA.*)

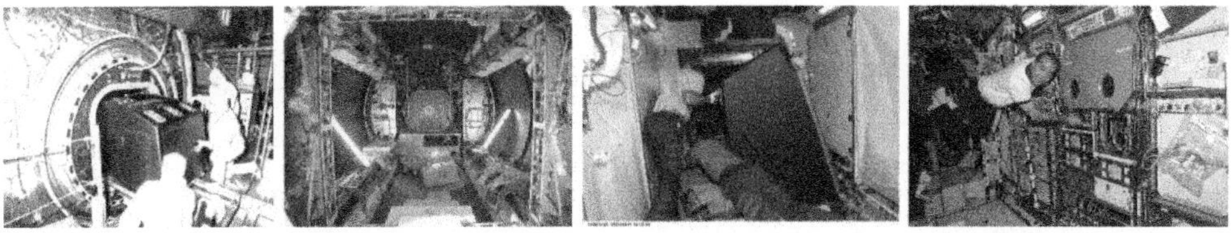

(a) (b) (c) (d)

Figure 7.64 (*a*) Before flight, a rack is taken through the hatch for installation in the U.S. Lab. (*NASA.*) (*b*) Interior of the U.S. Lab prior to the installation of racks, shows the four corner stand-offs that serve as utility runs. (*NASA.*) (*c*) Crew in orbit prepare to install a rack. (*NASA.*) (*d*) Several racks installed along one wall of the U.S. Lab in orbit. (*NASA.*)

were long cylinders with hatches and berthing ports on only one or two ends. The United States and European-built nodes had six hatches and berthing ports and were intended to link as many as six modules together. Other modules were stripped down for carrying logistics to and from the station; another was designed for use as an airlock (Figures 7.63 and 7.64). Although the partially assembled station was always a functional spacecraft, after 13 years, 37 individual Space Shuttle launches, and four Russian rocket launches, ISS was deemed to have reached "assembly complete." However, 5 years later additional modules were still being added for technology demonstrations and to replace earlier failing elements.

Earlier space stations, like the U.S. Skylab or Soviet Almaz and Salyut, were monolithic. They were orbited on the top of a rocket as a single integrated fully functional system ready to support their astronaut crews. Once in orbit, appendages and modules would spring into place and lock into position. Later stations, like the Russian Mir Orbital Station and the ISS required in-space assembly because of their intended size and enhanced capabilities.

7.34 ISS Crew Compartment Design

Throughout the early station design, an Architectural Control Document defined environmental and habitability requirements. The NASA Standard 3000, Man-Systems Integration Standards established design considerations and trades required to support the morale, comfort, and health of crews in space. "Man-systems," today more commonly referred to as "human-systems," was recognized as one of nine primary systems of the space station, along with electrical power, data management, thermal control, communications, guidance, extravehicular activity, and environmental control. Man-systems defined architectural parameters, including orientation, personnel movement, viewing, work location design, nomenclature and markings, displays and controls, and facility management. Man-systems human factors engineers designed 15 subsystems: crew quarters, restraints and mobility aids, operational and personal equipment, emergency provisions, workstations, galley/food, personal hygiene, illumination, wardroom, stowage, trash/waste management, internal close outs, on-orbit maintenance, inventory management, and crew health care. Crew health care in turn was comprised of health maintenance, exercise, and environmental health (Figure 7.65).

Racks

For the module interiors, trade studies looked at a building block approach and considered the most efficient block size. At one end of the spectrum was the potential to change-out entire modules; at the opposite, small modular elements, the size of Space Shuttle lockers 0.06 m^3 (2 ft^3) were considered.

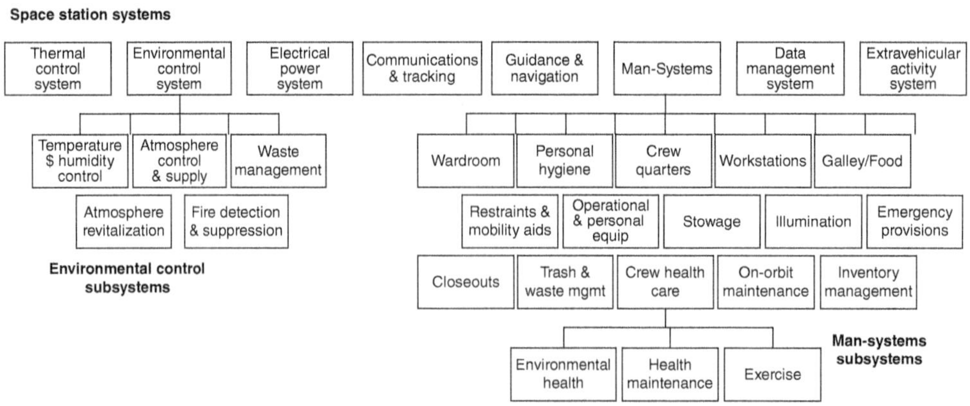

FIGURE 7.65 Work breakdown structure space station systems definition. (*G. Kitmacher.*)

Refrigerator-sized equipment racks were deemed to be the most efficient solution for housing most major systems hardware. Their standard width was based on earth-based electronic equipment racks.

These *"standard racks"* became the design unit into which most systems hardware was built. The racks could be carried to and from orbit in logistics modules. Hatches in the U.S. segment of the station were sized to permit the racks to be transferred throughout the station. There are four major categories of racks. Avionics racks house computers, fans, power converters, air conditioners, and other equipment that keep the vehicle functioning and crew alive. Payload racks house the various science facilities and experiments. Stowage racks house food, clothing, maintenance hardware, experiment stowage and other loose items. Crew support racks or *functional units* contain the galley, and commode/waste management compartment. Some vacant rack volumes are outfitted into individual crew sleep compartments.

Design concepts looked at the lay-out of systems in the module interiors. Astronauts who had flown on earlier space stations asked that a relatively constant ceiling and floor arrangement be maintained, and they recommended that the floor be oriented toward earth. Though systems hardware would be designed into the avionics racks, utility runs had to accommodate plumbing for fluids, airflow, electrical and data lines throughout the interior volume of the station. These were placed into "stand-offs." The utility run stand-offs run longitudinally along the module cylinder walls spaced evenly at the "four corners," and standard racks between the stand-offs form a floor, ceiling, port, and starboard. Down the center of each module is a generally open volume through which the crew can move.

Lights and air vents are installed in the stand-offs on either side of the ceiling racks. Air returns are located in the stand-off on either side of the floor racks. The standoffs are also where the racks attach to the module. At the base of each rack, pivot pins provide a hinge on which each rack rotates away from the cylinder of the pressure vessel. This gives the crew easy access behind the racks and to the pressure shell. Although there is no up or down in zero-G, the constant orientation of the lights and vents in the upper stand-offs and air return in the bottom establish an artificial up and down in the modules (Figures 7.65 and 7.66).

Far more stowed gear and consumables, which include food and clothing, must be stowed than can fit into stowage racks. Most is stowed in white collapsible fabric rectangular bags called cargo transfer bags (CTB). Stowage space is at a premium on ISS, and the CTBs are stowed in vacant areas throughout the modules.

While the Russian segment has a similar interior arrangement of ceiling, floor and walls with the floor toward earth, Russian systems are mounted directly to the module walls or structure. Smaller hatches in the Russian segment preclude the transfer of standard racks from the United States into the Russian segment.

Key to the Near-Earth Space Environment

Studies of the space environment in the vicinity of earth up to 20,000 nmi altitude show that the following geophysical effects are of prime importance in the design of spacecraft: gravitation, atmosphere, radiation, and micrometeoroids. Some variations of density, temperature, and gravitational effects as a function of altitude are shown in the accompanying Figure 7.66.

Gravitational effects for operations near the earth or other celestial bodies are normally represented by the point source gravitational parameter μ from which the velocity required for circular and escape orbits can be computed as $(\mu/r)^{1/2}$ and $(2\mu/r)^{1/2}$, respectively (r = distance from the gravitational source). For near-earth orbits of long duration, and for precise space trajectory calculations,

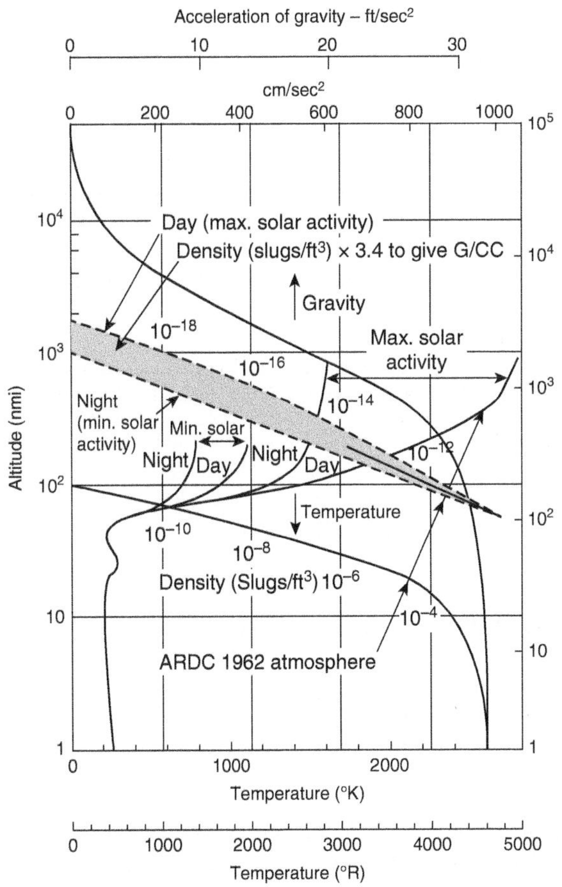

FIGURE 7.66 Characteristics of the near-earth orbital space environment. (*TRW Space Data*.)

it is also necessary to account for the nonspherical shape of the earth, by use of a potential function such as the following:

$$P(r,\theta) = \frac{\mu}{r}\left[\begin{array}{c} 1 + J\left(\frac{R}{r}\right)^2 (1 - 3\cos^2\theta) + \frac{K}{2}\left(\frac{R}{r}\right)^3 (5\cos\theta^3 - 3\cos\theta) \\ + \frac{D}{35}\left(\frac{R}{r}\right)^4 (35\cos^4\theta - 30\cos\theta^2 + 3) \end{array}\right]$$

where μ = GM the gravitational parameter

r = geocentric radius

Θ = latitude angle measured from North Pole

R = equatorial radius of earth, and J, K, D are constants characteristic of the earth's mass distribution; i.e., $J = 1624 \times 10^{-6}$, $K = 2.3 \times 10^{-6}$, $D = 9.27 \times 10^{-6}$

Atmospheric characteristics of the earth and other celestial bodies include the variation of density and pressure with altitude due to gravitational effects; the temperature resulting from solar radiation and atmospheric convection; and the chemical constituents of the atmosphere. Provisions necessary to overcome these atmospheric effects are pressurized cabins with adequate oxygen supply, temperature control, selected bearing materials and lubricants, aerodynamic drag and lift devices for reentry, heat shielding, dielectric protection for electronic apparatus, etc.

Radiation effects, such as degradation of electronics and reduction of solar array power output, result from the natural electron, and proton environments. These effects require the provision of adequate shielding to protect against injury or damage to personnel or apparatus from bombardment of radiation particles, and attitude controls suitable for overcoming the torque due to solar pressure. Galactic cosmic rays can cause single event upsets (SEU) in spacecraft electronics.

Micrometeoroids/debris are prevalent in sufficient quantities that their size, momentum, frequency, and penetration effects need to be considered. Provisions for protection against damage due to normal or abnormal micrometeoroid density include reserve wall thickness and puncture-sealing provisions.

7.35 Systems

A manned spacecraft must be designed to provide a habitable living and working environment. Because of the harsh natural environment's extremes of lighting, temperatures, external debris and the lack of atmospheric pressure, the spacecraft must be built to withstand and protect both the astronauts and the supporting systems. This chapter focuses on the systems unique for human habitation and environmental control.

Environmental Control System

A habitable spacecraft is a self-contained, isolated chamber, and must provide all of the necessities to support and protect life. Human interfaces are critical in an environment designed for habitation and do not necessarily have the same interface requirements as in earth's gravity. Gases, fluids, particulates, and temperature flow do not behave in a weightless environment like they do in earth's gravity. Human physiology sometimes reacts differently in the weightlessness of spaceflight than in 1-G gravity. Because the spacecraft environment may be closed, the potential exists for contaminants in the air or water not anticipated from ground based operation. The cabin of a spacecraft must maintain an atmosphere that is healthy and comfortable.

Because the operation of the environmental control system is considered life critical, every failure must be considered possible and the system has to be designed to accommodate multiple simultaneous failures in order to ensure crew survivability.

Spacecraft environmental control encompasses four specific functions. A fifth function is shared with the thermal control system (TCS).

- The system maintains the atmosphere aboard a spacecraft.
- The system accommodates and maintains the water supply, which includes water collection, storage, purification, and distribution.
- The system provides for human waste collection, storage, processing, and disposal.
- The system provides for fire detection and suppression.

In conjunction with the TCS, the environmental control system regulates temperatures inside the spacecraft.

Other functions that must be supported which relate to environmental control include the storage and provisioning of food, crew exercise, radiation shielding, and hygiene and the maintenance of sanitary conditions, and the disposal of garbage and waste.

Atmospheric Pressurization and Composition

On ISS the spacecraft's cabin atmosphere is continually cleaned and reconstituted to ensure the right mixture of different constituent gases is maintained. Oxygen is replenished. Carbon dioxide and water vapor are removed.

Earth's atmosphere is a mixture of 78% nitrogen, 21% oxygen, 0.5% water vapor, and small amounts of carbon dioxide and several other trace gases. At sea level, atmospheric pressure is 14.7 lb/in^2, or 760 mm of Hg.

We depend on the correct mixture of gases and the pressure of our atmosphere to be able to breathe. Because in the weightlessness of spaceflight there is no convection, and because individual gases can originate in particular locations, it is important that the air circulation be maintained at all times in order to prevent a build-up of hazardous gases in any one place.

The Near-Earth Space Environment

On earth, atmospheric pressure depends upon altitude. As determined through tests since the late 1700s, a minimum atmospheric pressure of at least 0.9 psi is required to prevent the vaporization of body fluids. The human body is specifically dependent on oxygen in the atmosphere. Without an adequate supply of oxygen, we would lose consciousness and die within minutes. The human body consumes oxygen and therefore the environmental control system must replenish oxygen as it is consumed. Nitrogen is not critical to human functioning; however, the body has evolved to use the atmosphere in which oxygen is diluted. The partial pressure of oxygen in the atmosphere at sea level is approximately 3.5 lb/in^2. If the partial pressure of oxygen falls below 2.5 psi, then there is serious danger of slipping into unconsciousness (Figures 7.66 and 7.67).

Undiluted oxygen at full sea level pressure can be toxic after several hours. However, toxic effects are reduced if the pressure of oxygen is reduced. There are no toxic effects if one breathes pure oxygen at a pressure similar to the partial pressure of oxygen at sea level, 3.5 psi, for multiple weeks. There is evidence that breathing pure oxygen even at a low pressure may be toxic after several weeks. Therefore, a physiologically inert gas is added to the atmosphere for mission durations of 2 weeks or longer to prevent oxygen toxicity.

Pure oxygen readily oxidizes with other materials, causing a flammability hazard. Therefore, dilution of oxygen with an inert gas like nitrogen is important for reducing flammability risk.

Through the body's metabolism and respiration from the lungs, carbon dioxide is exhaled into the atmosphere. A small amount of carbon dioxide in the atmosphere is harmless. However, when it is concentrated, carbon dioxide is toxic. High carbon dioxide levels increase heart rate and respiration rate and cause problems with the acid-base balance in the body. High CO_2 levels will cause

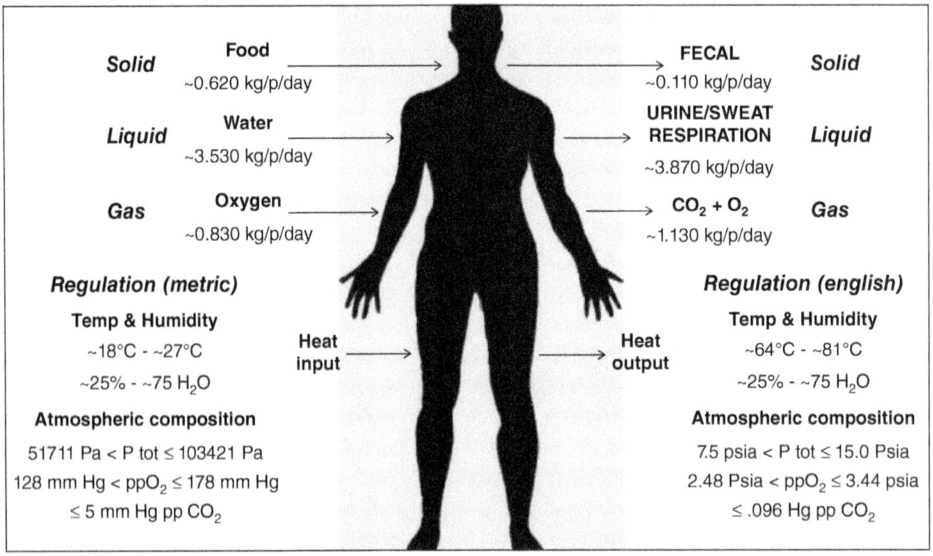

FIGURE 7.67 Mass and energy balance for a human being. (*Johnson Space Center Crew and Thermal Systems Division.*)

drowsiness, headache, dizziness, visual and hearing dysfunction, eventually unconsciousness, and ultimately death. Human breathing continuously produces carbon dioxide. It is important that air circulation moves CO_2 away from the crewmember's head, and that filtration removes carbon dioxide from the atmosphere of a manned spacecraft and keep its concentration below 0.3% or 3 mm Hg.

The amount of water vapor in the atmosphere, measured by humidity, is highly variable. Life is highly tolerant of wide extremes in humidity in the earth's atmosphere. Humidity is important for crew comfort and can also have deleterious effects on the spacecraft, contributing to the degradation of materials and increased microbial growth. Water vapor is continuously expelled by the crew through respiration and perspiration. Therefore, a manned spacecraft must have the equipment to condense and collect water vapor from the cabin atmosphere.

A habitable spacecraft must have a system to monitor and maintain a healthy atmosphere in the cabin. This spacecraft environmental control system (ECS) has to replenish the oxygen that the crew consumes, remove water vapor and carbon dioxide, identify and remove other toxic materials, maintain the overall cabin pressure, and maintain the temperature in the cabin at a comfortable level. If the crew is expected to live in the spacecraft for periods exceeding a few weeks, then the cabin atmosphere should contain a diluent gas like nitrogen. The types and concentrations of materials, particularly oxygen, are critical to human life and yet the crew cannot readily sense the concentration of oxygen or many other materials in the air they breathe. Therefore, the crew must rely upon instruments to verify that the materials in the atmosphere are being maintained in a safe balance.

Open-Cycle versus Closed-Cycle ECSs

It is always desirable to provide an interior habitat atmosphere similar in pressure and composition to the atmosphere on earth. On earth, atmospheric pressure is about 14.7 psi; the composition contains a significant proportion of about 78% nitrogen and 21% oxygen. The large proportion of nitrogen reduces the likelihood of oxygen poisoning which can occur if the crew breathes oxygen at too high a partial pressure. The dilution with nitrogen reduces flammability hazards.

The type of ECS used in a spacecraft varies dependent primarily on mission duration. An open-cycle system usually makes little or no effort to recycle environmental consumables. Open-cycle systems are usually simple and reliable. System equipment weighs less but resources required for life support are linearly dependent on flight time. Open-cycle systems are most commonly used in short-duration vehicles such as those used to transport crews between earth and space.

A closed-cycle ECS recycles life support resources. Waste products are recovered and processed into consumables suitable for reuse. A closed-cycle system is used on long-duration missions in order to reduce resupply requirements. Closed-cycle systems rely on sophisticated machinery for resource processing. The equipment weighs more initially and requires additional power and cooling than an open-cycle system. The increased complexity of the equipment lessens reliability and increases requirements for maintenance. Closed-cycle ECS systems are most commonly used in long-duration spacecraft such as a space station.

We will examine two different types of environmental control systems for comparison.

Open-Cycle ECS, Apollo Lunar Module Example

The Apollo Lunar Module (LM) landed two men on the moon. The LM was designed for a 2 day, two person lifespan using pure oxygen at a cabin pressurization of 0.38 atm (5.5 psi). Its ECS was designed to be simple, lightweight, and reliable. An open cycle, single gas, reduced pressure, pure oxygen atmosphere was used.

The ECS system pressurized the crew cabin, provided breathable oxygen for the astronauts, cooled electrical and electronic equipment, provided oxygen and water for the astronauts' spacesuits, and provided water for drinking, food preparation, and fire extinguishment. Most of the system was located inside the crew cabin. Oxygen and water supply tanks were outside the crew compartment. The system was fully disposable, meant to be discarded when the two day mission of the LM was over, and was not designed for in-flight maintenance (Figures 7.68 and 7.69).

Oxygen The amount of oxygen required was determined by adding crew metabolic consumption, cabin leakage, and repressurization requirements. Metabolic consumption varies but was estimated at 2 lb/crewman/day. Each time astronauts went outside to walk on the moon, the cabin air was dumped overboard. Oxygen was required to repressurize the cabin after each moon walk. Enough O_2 was provided for four repressurizations. Leakage from the cabin was about 0.02 kg (0.05 lb) per hour, or less than 1.4 kg (3 lb) for the mission duration. Oxygen was also diverted from the LM to the astronauts' spacesuits.

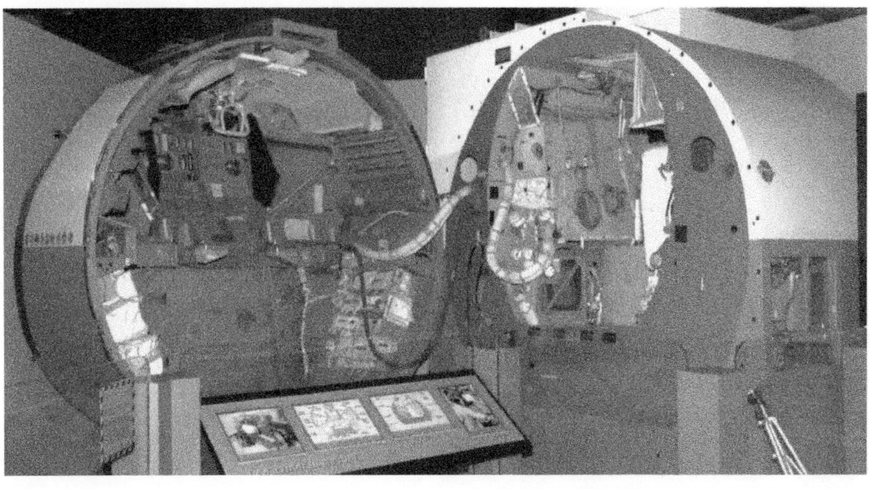

FIGURE 7.68 Lunar Module interior. This is the LM mission simulator used for astronaut training. To the left is the forward cockpit from which the commander and pilot flew the vehicle. To the right is the aft cockpit in which most of the ECS controls and components are located.

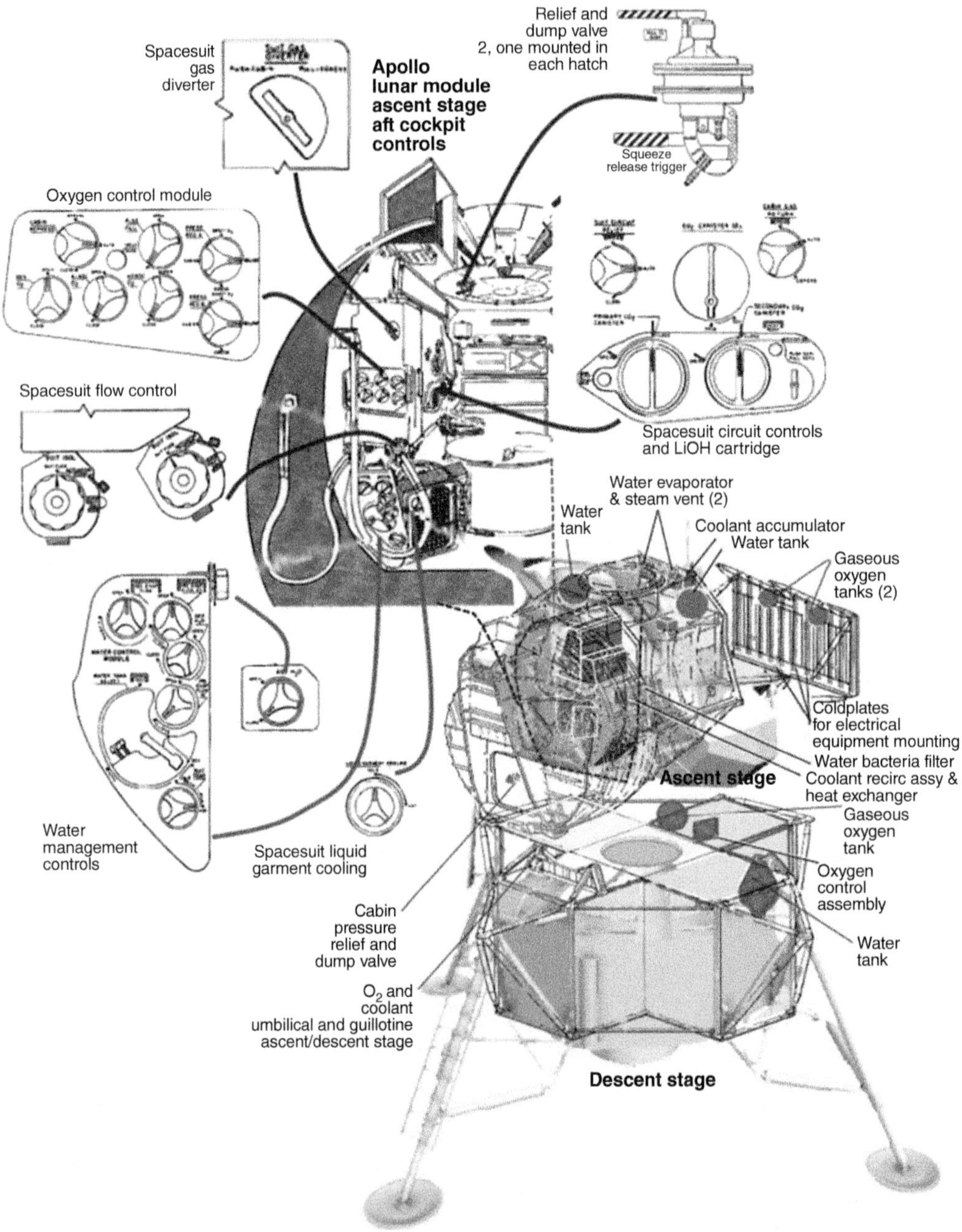

FIGURE 7.69 Apollo Lunar Module environment control system.

Oxygen was stored as a high-pressure gas in three tanks at 177 atm (2,800 psi). One tank held 22 kg (48 lb) of O_2 and two tanks, each held 1 kg (2.4 lb). Storage of O_2 as a high-pressure gas is the simplest, though heaviest, method of oxygen storage.

Astronaut Metabolism Astronauts consumed oxygen and produced carbon dioxide and water. Each astronaut produced about 0.9 kg (2 lb) of carbon dioxide and 0.23 kg (0.5 lb) of respired water per day. This amount of water is very small compared the total amount of water the body gains and loses each day. Water is lost through respiration, perspiration, urination, and in fecal waste. Typical water use is about 2.3 kg (5 lb) per person per day and about 1.4 kg (3 lb) is released into the atmosphere. The amount of water used can vary widely depending on crew activity, temperature, and humidity in the air.

The carbon dioxide and water that enters the air from the crew was removed by the ECS.

Carbon Dioxide Removal In the LM, cabin air was routed through a canister. In the canister, the air passed through a debris trap, then a lithium hydroxide (LiOH) bed, and then an activated charcoal filter. The debris trap caught particulate contaminants. The lithium hydroxide removed the carbon dioxide (CO_2) by combining chemically and producing water:

$$CO + 2LiOH \rightarrow Li_2CO_3 + H_2O$$

About 2.5 lb of lithium hydroxide are required per person per day. For short missions, lithium hydroxide is the lightest method of CO_2 removal. Then, the air passed through an activated charcoal filter that absorbed odors.

Redundant fans ensured that airflow circulation remained adequate throughout the cabin, around the astronauts, through the astronauts' spacesuits, and into the LiOH canister.

From the canister, the air passed through a sublimator to a heat exchanger. Air temperature was reduced and water vapor was condensed to liquid when heat was absorbed by a coolant. Then the air and water flowed into centrifugal water separators which spun to remove the water from the air. The water was diverted into water tanks. The dried oxygen went back to the cabin. The cabin pressurization system added O_2 to the cabin air as needed to maintain the cabin pressure.

Closed-Cycle ECS, ISS Example

The International Space Station (ISS) has been in orbit since 1998. Assembly in orbit was completed in 2011 and ISS is expected to stay in operation through the mid-2020s. Crew size is typically 6. Each astronaut usually stays on board for about 6 months and as three of the crew return they are replaced by three new crewmembers.

Inside the ISS, the ECS maintains the atmosphere and water supply. On long missions, like the ISS or on future flights to the planets, considerable weight can be saved if consumables such as oxygen and water are regenerated by the ECS. This requires a sophisticated closed-cycle ECS. The ECS hardware is much larger, weighs more, uses more electrical power and requires ongoing maintenance, all in contrast to the simple open-cycle ECS of the Apollo LM. But when the ISS ECS works, the use of consumables, like air, oxygen, nitrogen, or water, is minimized and a small supply can last a long time. One primary purpose of the ISS is to develop and prove the technology for the closed-cycle ECS.

The ISS is divided into a Russian segment and a U.S. segment. The U.S. and Russian segments each have separate environmental control systems that operate independently of one another. However, normally hatches are open between the segments and air moves between the segments. The air, water vapor, and contaminants are all shared. Fresh supplies of atmospheric gases and water are sometimes shared by manual transfer of gas or water containers.

In the early phases of the ISS program, U.S. and Russian ECS systems depended more on one another for redundancy. As the program has progressed, additional systems and improved systems have been used to improve the environmental recycling capabilities.

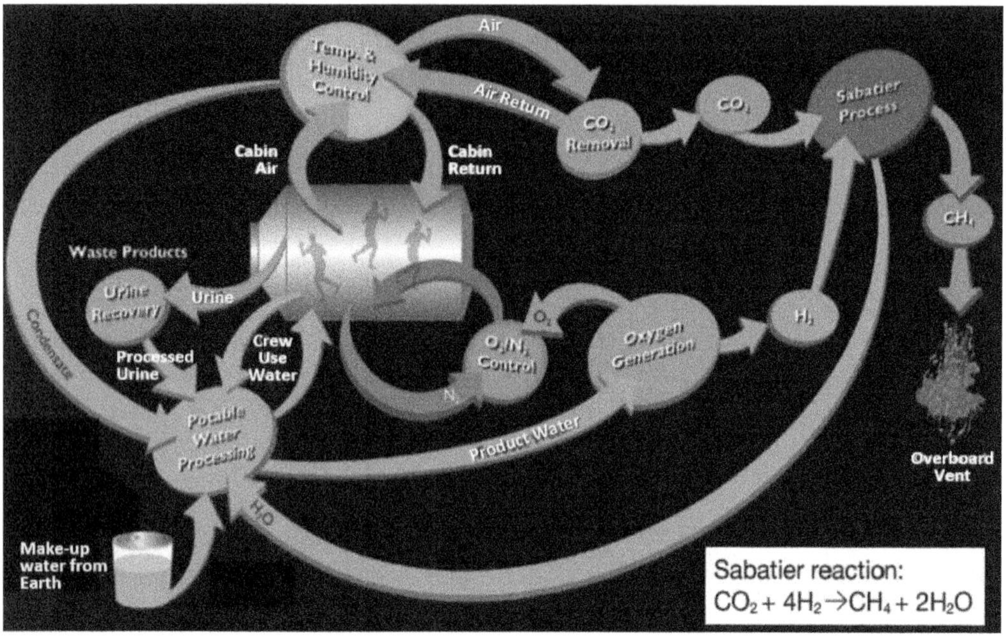

FIGURE 7.70 Closed-cycle ISS environmental control functions.

The ISS ECS was designed for minimum risk of failure at the system level. At the individual module level the failure tolerance for many of the ECS functions is zero; a function is lost if a component fails. However, across the entire ISS habitable compartment, there is redundancy for all critical functions (Figure 7.70).

The philosophy used for designing the U.S. ECS was to minimize the use of consumable supplies by using methods which would regenerate consumables where feasible. Waste products which are recycled include water produced and used by the crew, moisture from the air, urine, and CO_2. These products can be captured and processed to provide potable water and oxygen. As much mass is recycled as possible.

U.S. and Russian ECS Design Philosophies The U.S. and Russian ECS design philosophies are different. This is because the U.S. and Russian systems were designed for two different programs. The Russian system has been developed over the course of several earlier space stations from Salyut 1 through Mir and the ECS is very similar to the system used on the Mir Orbital Station. The Russian and U.S. space station programs were merged late in the U.S. ECS design process.

The Russian approach to ensuring that a capability is maintained, even in the case of malfunctions, tends toward having backup capabilities provided by a different type of system. For example, to maintain the supply of breathing oxygen, if the Elektron electrolytic O_2 generator fails, the backup is either oxygen stored in pressurized tanks or chemical candles which are burned to create oxygen. In the case of malfunctions, the primary system is restored to functionality by the system's replacement on a regular resupply mission. There are some cases of redundant components; for instance there are two air circulation fans in each Russian module for redundancy (Figure 7.71).

The U.S. approach to ensuring that a capability is maintained even in the case of malfunctions tends toward using redundant equipment; i.e., having two identical units with one reserved for emergency use, or alternatively, operating multiple units at less than their full capacity. This leads

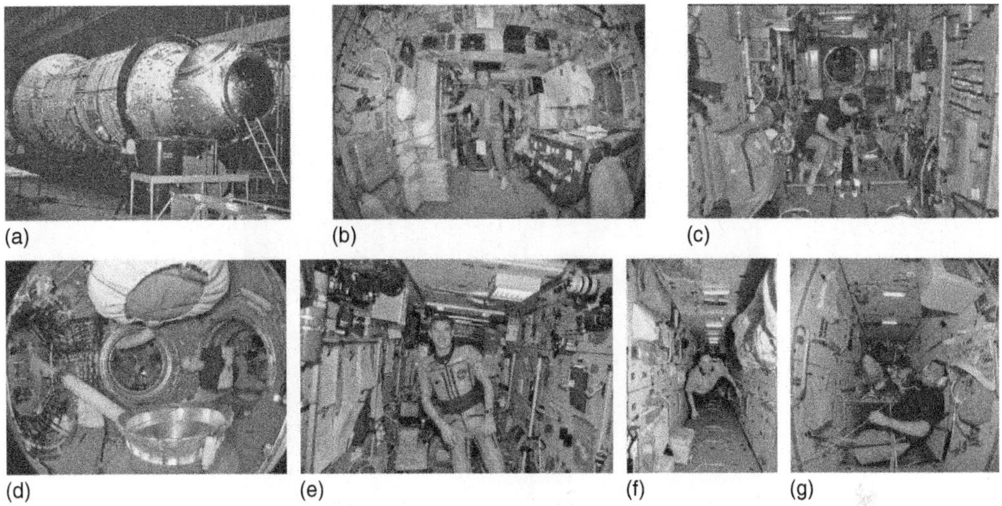

Figure 7.71 Russian ISS modules. (*a*) Service Module (SM, Zvezda) structure prior to completion. (*b*) Interior of SM looking forward. (*c*) SM interior aft, showing several windows in the floor. (*d*) Interior of SM docking node. (*e*) FGB, Zarya interior. (*f*) FGB is narrow. (*g*) Wall panels are hinged for access to mounted system equipment and stowed equipment behind.

to having two CO_2 removal units in two separate modules. There are exceptions to this approach. For instance, there is only a single U.S. water processor and one U.S. commode.

In addition to oxygen (O_2), another by-product of electrolysis is hydrogen (H_2). H_2 can be a safety concern because of flammability. The H_2 is dealt with in different ways by the U.S. and Russian segments. The U.S. system is designed to ensure that the quantity of combustible gas is kept to a minimal, negligible amount. The Russians used a different approach. The electrolyzer operates inside a pressurized N_2 jacket. The N_2 flushes O_2 and H_2 from the lines. The design precludes the leakage of hazardous gas to the atmosphere. But this design precludes individual components from being accessible for replacement (Figure 7.72).

Another example of the effect of different philosophies is the design of the oxygen generation systems (OGS). The U.S. and Russian systems both use electrolysis of water for generating oxygen, but there are significant design differences. The U.S. approach was to design hardware to be serviceable in orbit. Components of the U.S. OGS are designed as orbital replaceable units (ORU) and are readily accessible for replacement. The Russian approach does not require that components be individually replaceable. However the entire unit can be removed and replaced.

Another difference between the U.S. and Russian systems relates to analysis versus operation of ground hardware in order to predict problems in space. The Russians maintain identical systems on the ground concurrently with the flight unit. One ground unit is operated in the same manner as the flight unit, though slightly in advance of the flight unit. If problems occur with the ground unit, this allows problems in orbit to be anticipated and corrective actions to be taken. A second unit is maintained on the ground to serve as the replacement for the on-orbit unit in case of irreparable problems in orbit. Although the U.S. maintains a test unit on the ground, operation of U.S. systems is planned through analysis rather than by maintaining or operating an exact duplicate on the ground.

Atmosphere At present, the ISS consists of 15 pressurized modules. The U.S. segment is composed of seven U.S. modules, two Japanese modules, and one European module. The Russian segment is composed of five Russian modules. Nonpermanent modules are added at times for

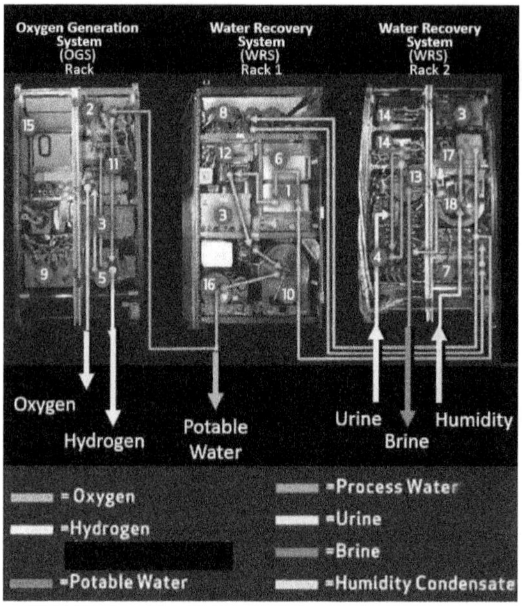

1. Catalytic reactor
2. Deionizer beds
3. Digital controller
4. Distillation assembly
5. Electrolysis cell stack
6. Gas separator
7. Multifiltration beds
8. Particulate filter
9. Power supply
10. Product water tank
11. Pumps and valves
12. Reactor health sensor
13. Storage tanks
14. Urine processor pumps
15. CO_2 reduction system
16. Water processor delivery pump
17. Water processor pump and separator
18. Wastewater tank

Figure 7.72 ISS closed-cycle ECS racks. (*NASA.*)

logistics, technology demonstration, or science purposes. As many as four sortie vehicles have visited the ISS at a time. The U.S. system is capable of supporting about seven people. The Russian system is sized for three people. The systems are partially redundant so that if a portion of one system is not functioning the other system can temporarily takes its place. The total pressurized volume is about 900 m^3 (32,900 ft^3).

Atmospheric pressure, temperature, and other environmental attributes are maintained within set limits on-board the ISS. The ISS crew compartment is maintained at a pressure between 0.95 atm (14.0 psi) and 1.01 atm (14.9 psi). Pressure between adjacent isolated modules can be equalized within a few minutes. In addition to crew metabolic requirements, some systems on-board the ISS require air cooling and require a minimal level of air pressure or need to be turned off. The air on ISS is composed of nitrogen, oxygen, small amounts of CO_2 and other trace gases; relative amounts of each gas are monitored and controlled.

Temperature within the cabin is monitored continuously. Typically the temperature set-point is decided by the crew commander within a narrow range and the temperature is set, monitored and maintained by mission control in the U.S. and Russia. The crew has the capability to override the temperature setting using the on-board portable computer system. In the U.S. Lab, the temperature is typically set at 22°C (72°F). Temperature is often quite variable from one module to another and may necessitate a change of clothing in order for the crewmember to remain comfortable. In addition to crew comfort, temperature levels are maintained within a narrow range because of the potential for moisture condensation and resulting mold formation.

Humidity is monitored and water is removed from the cabin air as humidity condensate at the approximate rate of 1.5 kg/person/day (3.3 lb/person/day).

Air is circulated throughout the interior pressurized volume. Within the U.S. segment modules, outlet vents are mounted along both sides of the ceiling stand-offs between the ceiling mounted standard racks. Returns are mounted in the floor stand-offs; air flows from ceiling to floor.

Fire The likelihood of a fire is reduced by careful design, selection and location of potential ignition sources and strict control of flammable materials. Fire risk locations are identified in advance. In the case of a fire being detected, an alarm sounds, and power and airflow circulation is removed in the local area and between adjacent connecting modules. A goal is to identify fires early and to fight fires in a localized area. Fires are suppressed by the crew using portable fire extinguishers.

Atmospheric and Fluid Contaminants Several kinds of unanticipated contaminants are possible in the closed environment of a spacecraft. Contaminants can include undesirable solids, liquids, or gases. These can originate in materials used during manufacture or off-gassing after manufacturing or from decomposition from aging. Contaminants can result from maintenance, leaked propellants, leaked cooling fluids, by-products of experiments or payloads, or from over-heated motors or overheating of nonmetallics. Contaminants can result from crew effluents or detritus, or from hygiene consumables, food, or other unanticipated sources. The spacecraft maximum allowable concentrations (SMACs) identify toxic effects and allowable concentrations of specific foreseen substances.

Space station environments are closed and sealed for years. Gases and fluids can be reused over and over again. Contaminants can remain in the cabin atmosphere and become concentrated over time. On Earth, contaminants are often ignored because of ventilation; at worst they usually only cause temporary discomfort. In a long duration spacecraft, as contaminants become concentrated, they can create unanticipated difficulty with the functioning of environmental control hardware. As the contaminant concentrations increase, changes to the system design may be required to preclude danger to the crew.

During long-duration operations on the Mir Orbital Station and the ISS, several contaminating materials and compounds were unanticipated. It was long known that crewmember's bones demineralized and weakened in zero gravity. The minerals exited the crewmember as elevated calcium levels in the urine. What was unanticipated by environmental control system designers was that the calcium precipitated as calcium sulfate. The calcium sulfate interfered with proper operation of the water distillation and purification system. Other unforeseen contaminants, the origins of some which are still uncertain, have also required environmental control system changes.

U.S. spacesuits were originally designed for use on short-duration Space Shuttle missions and would be returned to the manufacturer after every flight for cleaning and refurbishment. On the ISS, the same space suits remain in orbit for years. Facilities are not available on the ISS for complete cleaning or refurbishment and over time, contaminants can increase. In one instance, increased particulates slowed the circulation of environmentally controlled fluids. Fluids backed up in the suit and became a breathing hazard for the astronaut during a spacewalk.

Years of testing, including tests in similar operational environments, including zero gravity and multiple-year durations using identical system and personnel consumables, may be required in order to fully understand fluid chemistry effects and contaminants, some of which may be present in minimal and almost untraceable quantities but which can become concentrated and a significant concern and design issue during long-duration operations in space.

Contaminant Identification and Control Some fluid contaminants may require chemical treatment. Aerosol contaminants are most easily removed by filtration. Many gases are easily removed through absorption by activated charcoal. Activated charcoal cannot remove methane, hydrogen, and carbon monoxide. Minor concentrations of some of these gases are toxic and explosive; their removal can be accomplished by use of a catalytic burner which will oxidize them into water and carbon dioxide, which in turn can be removed by the other equipment.

On the ISS, the major constituent analyzer (MCA) is a mass spectrometer that analyzes N_2, O_2, CO_2, H_2, and H_2O levels in the air. In the U.S. Lab, the volatile organic analyzer (VOA) uses a gas chromatograph/ion mobility spectrometer (GC/IMS) to identify violations of spacecraft maximum allowable concentration (SMAC) limits. In the case that limits are exceeded, the crew is alerted and all ventilation is commanded to off or standby.

The trace contaminant control system (TCCS) removes and disposes of gaseous contaminants. It consists of an activated charcoal adsorption bed and thermal catalytic oxidizer with postsorbent bed. The activated carbon is impregnated with phosphoric acid for ammonia removal. Ammonia is a concern because of its toxicity and its use in the ISS thermal control system. The high-temperature catalytic oxidizer converts carbon monoxide, methane, hydrogen, and other low molecular weight compounds to carbon dioxide, water or other compounds which can be more easily utilized. A sorbent lithium hydroxide (LiOH) filter removes acidic products of catalytic oxidation.

The immediate response in the case of atmospheric contamination that is harmful to the crew is the donning of breathing masks. Masks with a 15-minute oxygen supply are provided in multiple locations in each lab, node, and airlock. Additional breathing masks with 1-hour oxygen supplies are provided through fixed ports in each module.

In the case of a major contamination event, toxic contaminants are removed most expeditiously by venting of the cabin atmosphere from a module into space. It takes approximately 24 hours to completely vent the atmosphere from a module. It would take approximately 75 hours to repressurize the U.S. Lab from gaseous oxygen and nitrogen sources.

Rapid Decompression Unanticipated air leaks are not normal but have occurred. Separate alarms identify rapid pressure change as well as low pressure. Instrumentation will detect off-nominal pressure before the crew is able to. When a rapid decompression is detected, external vents are closed air revitalization systems are switched off; if pressure goes lower than a specified level, systems, and hardware that rely on air cooling will be commanded off. If there is a leak in the hull, in the U.S. segment most of the cabin pressure wall is accessible by moving standard racks and other equipment away from a leak location. Repair kits are available for the manual sealing of leaks from the interior.

Air Revitalization

Air Use Oxygen is used metabolically by the ISS crew during respiration; the rate of oxygen use is driven by the ISS crew metabolic demand, which correlates to crew member activity level and size.

Air Loss Air is normally lost from the ISS in several ways. Some air is lost when visiting vehicles undock from ISS. The air is lost as a result of depressurization of the vestibule (the volume between the ISS and the visiting vehicle hatches).

Air is lost when U.S. or Russian airlocks are depressurized for extravehicular activity (EVA) or spacewalks. The Quest U.S. segment airlock loses less atmospheric gas overboard than the Russian airlocks because in the Quest most air is pumped from the airlock back into the ISS cabin; however, the pump can only be operated to 0.14 atm (2 psi), so the remaining air is vented overboard. All of the air from the Russian airlocks is vented overboard.

Some systems on-board the ISS vent air overboard. The carbon dioxide (CO_2) removal assembly (CDRA) in the U.S. segment of ISS and Vozdukh from the Russian side of ISS, both remove CO_2 from the cabin and vent it overboard. The Russian micropurification unit (БМП) removes atmospheric contaminants. All of these lose some air overboard during their normal operation.

Some payloads or experiments are pressurized with cabin air and then vent the air overboard through the experiment vacuum exhaust system (VES).

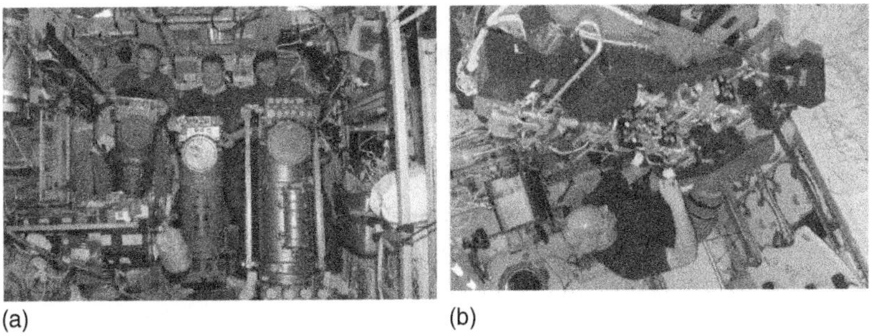

Figure 7.73 (*a*) Used, new electron units in the service module. (*NASA.*) (*b*) U.S. CO_2 CDRA system undergoing on-orbit maintenance. (*NASA.*)

Air leaks from the ISS modules at a rate of about 0.45 kg/day (1 lbm/day). The amount of oxygen lost to leakage is small in comparison with that used by the crew (Figures 7.73 and 7.74).

Maintenance of Air Pressure and Composition To maintain the cabin total pressure and the correct relative proportions of each gas, oxygen and nitrogen have to be added back to the ISS cabin. Oxygen is introduced to the cabin atmosphere at a nominal rate of 0.84 kg/person/day (1.84 lb/person/day). Oxygen partial pressure is maintained at 0.2 atm (2.83 psi). Several methods are used to maintain pressure and composition. (Figure 7.73)

The primary method of oxygen replenishment in the Russian segment of ISS uses the Elektron system, and in the U.S. segment, the oxygen generation assembly (OGA). Both generate oxygen through the electrolysis of water. The water molecules are split into oxygen and hydrogen. The oxygen is produced at ambient pressure and vented into the cabin. Similar technology is used on naval submarines.

Most vehicles that visit the ISS, carrying logistics, supplies or crew, carry air in their cabins. Most also carry nitrogen and oxygen stored in tanks. Russian progress resupply vehicles carry 170 kg (375 lb) each of oxygen and nitrogen in 206 atm (3,027 psi) tanks. Both cabin air and stored gases are used to replenish the ISS atmosphere.

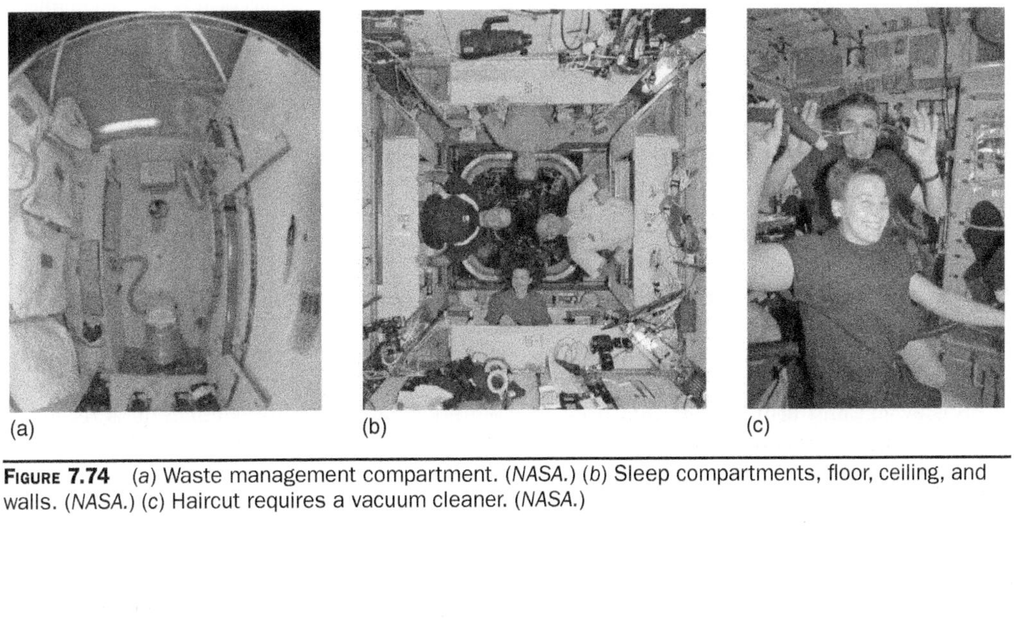

Figure 7.74 (*a*) Waste management compartment. (*NASA.*) (*b*) Sleep compartments, floor, ceiling, and walls. (*NASA.*) (c) Haircut requires a vacuum cleaner. (*NASA.*)

On the U.S. segment, the ISS has high-pressure oxygen and nitrogen tanks mounted to the exterior of the Quest airlock. The tanks provide a reserve of stored gaseous nitrogen and oxygen on the station that is used after other resources have been expended. The 91 kg (200 lb) of either oxygen or nitrogen is stored at 3,000 psi. The Quest tanks can be refilled from reserves brought up by visiting vehicles.

On the Russian segment of the ISS, a backup option for oxygen replenishment uses solid fuel oxygen generation (SFOG) canisters. These are solid fuel lithium or sodium perchlorate candles, which are burned to create gaseous oxygen. Each candle can supply enough oxygen for one crewmember for one day.

Temperature and Humidity Control The temperature and humidity control (THC) subsystem ensures that temperature and humidity levels are maintained within specification limits.

The metabolic activity, respiration, and perspiration from the crew generates heat and humidity. Most electrical equipment, including motors, lights, and electronics also generate heat. The heat from most equipment is dissipated using coolant flowing through cold-plates to exterior radiators. Heat in the air is dissipated through airflow and ventilation.

Within avionics racks, avionics air assemblies (AAA) blow warm air into module ducting. Within most modules, cabin air assemblies (CAAs) blow warm air into a heat exchanger, where the air is cooled and water is condensed back to a liquid. A centrifugal liquid separator removes condensed liquid water from the air stream.

Cooled air is ventilated through the two ceiling stand-offs into the module cabin. High-efficiency particulate air (HEPA) filters in the return air ducts in the floor stand-offs capture particulates and microorganisms. Crew members manually vacuum the filters periodically in order to remove particulates.

Carbon Dioxide Removal When ISS was first placed in orbit, CDRAs were relied upon to remove metabolically produced carbon dioxide from the air. The CDRAs use a zeolite and silica gel molecular sieve to desiccate the incoming air and adsorb the CO_2. Air and water are returned to the cabin. Two zeolite beds are used in alternating sequence. As CO_2 from the first zeolite bed is desorbed and vented overboard into space, a second zeolite bed is used to absorb CO_2. The venting of the CO_2 to space results in a net loss of O_2.

Later a Sabatier CO_2 methanation reactor was developed and placed on-board. With the Sabatier reactor, the CO_2, produced by the astronauts' metabolic processes, reacts with H_2 from the oxygen generation assembly in a Sabatier reactor to produce methane and water according to the chemical equation:

$$CO_2 + 4H_2 \rightarrow CH_4 + 2H_2O$$

Further closure and conservation of consumables may be realized at some point in the future if waste methane (CH_4) can be separated through a Bosch reactor, which would use a chemical reaction of carbon dioxide and hydrogen to produce carbon and water according to the reaction equation:

$$CO_2 + 2\,H_2 \rightarrow C + 2H_2O$$

Water Recovery and Processing A greater weight of water is needed than any other consumable. Water is needed for crew consumption for drinking, to reconstitute food, for personal hygiene and for equipment cooling. Therefore, an adequate supply of water must be a primary consideration in planning a spacecraft.

Wastewater and urine is collected from the commode and humidity condensate is received from the temperature and humidity control subsystem. Wastewater is collected from all of the U.S. segment modules and sent via wastewater plumbing to a condensate tank, where it is temporarily stored. The condensate tank can store up to 68 kg (150 lb) of wastewater.

In the Service Module (*Zvezda*) in the Russian segment, the water recovery system processes wastewater and water vapor from the atmosphere. The processed water is primarily used in the Elektron system to produce oxygen but can be used for crew consumption if needed.

In Node 3 (*Tranquility*) in the U.S. segment, the water recovery system (WRS) can process water vapor collected from the atmosphere, wastewater and urine into water that is intended for drinking.

The WRS is housed in ECS racks. Three ECS racks are used for processing and reclaiming water and air. The WRS consists of a urine processor and a water processor. The urine processor uses a low-pressure vacuum distillation process that uses a centrifuge to separate liquids and gasses. It is designed to handle a load of 9 kg (20 lb) per day, corresponding to a 6-person crew. About 70% of the water can be recovered from the urine. A significant and unanticipated finding from the ISS WRS operation is that considerably more calcium sulfate must be processed out of the urine in orbit than the amount in urine collected on earth. The difference corresponds to the loss of the mineral from the astronauts' bones during long-duration zero-G spaceflight.

Water from the urine processor and from other wastewater sources feed the water processor. Filter beds and a high-temperature catalytic reactor filter out gasses and solids in the water processor. Tests verify that the water is suitable for human consumption before the water reenters the stored supply. If water is found to be unacceptable, it is cycled back through the water processor assembly (Figure 7.72).

Waste Collection and Management

There are two waste and hygiene compartments (WHCs) on the ISS. One is located in the Russian segment in close proximity to the electron water electrolyzer and the second in the U.S. Tranquility Node in close proximity to the water recovery system. The WHC is an example of a rack designed for crew entry (Figure 7.74).

The waste management system (WMS) is designed to collect and isolate liquid and solid metabolic waste from the crew. Because there is no gravitational force to assist in moving waste away from the human and through the system, a suction system driven by a fan drives airflow and carries away noxious or contaminated air, urine, and feces to their respective collection containers.

For urination, the crew person uses a custom sized funnel at the end of a vacuum hose. The urine flows through the hose to a holding and stabilization chamber.

For solid waste, the crewmember uses a commode with a form fitting seat at chair height. Positioning of the astronaut's posterior to the seat is critical in order to form a good seal. A proper seal aids both in solid waste transport and in reducing the likelihood of leaks or contaminants.

The commode seat normally has a lid in place when not in use. For use the lid is lifted, and a blower which creates a vacuum starts. The suctioned cabin air passes through a carbon filter and returns to the cabin.

A replaceable bag is mounted to the commode seat prior to use. The bag is perforated to permit airflow while in use. After use the entire bag is detached and sealed and suction pulls the bag into a solid waste container below the seat. Periodically, as they are filled, the filled solid waste containers are placed into a logistics vehicle for storage and disposal.

Personal Hygiene

Astronauts on the ISS have available toothbrushes, toothpaste, razors, brushes and combs, a variety of soaps, cleaners and sanitizers, and a variety of wet and dry wipes. Most of these are designed for use without water. Towels and a vacuum cleaner are used to collect waste products and to dry the crew member. There is no shower on the ISS at present as it was deemed unneeded after the experience of using showers on the earlier Skylab and Mir space stations. The showers on those spacecraft were intended for temporary set-up and this required careful and extensive cleaning and drying between uses.

Clothing

The total amount of clothing carried per astronaut is minimized to reduce logistics and stowage requirements. At the present time clothes are not washed on the ISS.

Clothing that is worn directly on the skin is usually changed every few days. Outerwear clothing is usually worn for multiple weeks and may be used throughout a 6-month-long stay on the ISS. Many astronauts will recycle their clothing, using clothes initially for daily wear, and subsequently for exercise.

Soiled clothing is stored and disposed of in logistics vehicles.

Food

Food on the ISS comes in several forms including rehydratable, thermostabilized, natural form, and irradiated. Astronauts are free to select from a wide variety of menu items and will typically stay on a 1-week to 10-day cycle after which menu items are repeated. Nutritionists ensure that the menus meet nutritional requirements. Much of the food is available commercially, "off-the-shelf" with little modification except for packaging. Most food is provided by the United States and Russia, both of which maintain laboratories for the development, testing and packaging of foods. Several other space agencies also develop and provide specialty foods for their astronauts. Most food is individually packaged for single servings. Experimental units have been used for growing vegetables and these have been consumed during the course of some space missions (Figure 7.75).

Two galleys, one in the U.S. segment and the other in the Russian segment, provide tables for food and utensil restraint. There is no oven or stove for cooking; however, food warmers are used for heating food and drinks. Although refrigerators and freezers are not provided specifically for food, units intended for supporting experiments have often been used for cooling foods and drinks.

Exercise

Because of the physiological effects of zero gravity, crewmembers are required to exercise for about 2.5 hours per day. Several exercise devices are provided, including treadmills, bicycle ergometers, and a resistive exercise unit which is used like a weight lifting machine on earth but which uses

FIGURE 7.75 Crewmembers gather in Node 1 Galley. (*NASA.*)

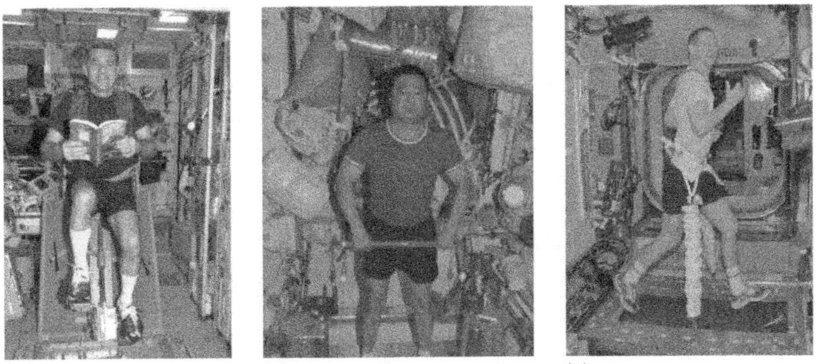

(a) (b) (c)

Figure 7.76 Exercise includes (*a*) bicycle ergometer (*NASA*); (*b*) weight lifting (using springs) (*NASA*); and (c) treadmill (*NASA*).

springs to provide resistance in place of gravity. Most exercise equipment is isolated from the spacecraft structure in order to minimize vibration and acceleration forces.

Exercise is considered a critical life support function on the ISS. If adequate exercise cannot be provided for all crewmembers in case of equipment breakdown, then within a specified time period consideration is given to returning crewmembers to earth (Figure 7.76).

Habitability

Spaceflight is a challenging activity for the human organism. The human is isolated. The space environment is foreign and threatening. Life relies on proper functioning of spacecraft and systems at all times. Physical and environmental conditions can induce physiological, biomedical, and psychosocial stressors on the astronaut that can affect crew performance and safety.

The spacecraft design and facilities must address the full range of human needs, beyond just the need for human survival. Systems and utilities should eliminate sensory issues associated with lighting, noise, vibration, smell, and air circulation. Accommodations should be provided that are similar to analogous earth-bound facilities for personal and group functions, fully accommodating privacy, personal space, hygiene, sleep, housekeeping, storage, eating, and meeting.

In the case of the ISS, in the early years of the station's design, there was considerable emphasis on design for habitability which considered psychological and sociocultural factors. The U.S. Habitation Module was to have included private sleep quarters for each astronaut, a fully enclosed shower and hygiene compartment, and a complete galley with refrigerators, freezers, and systems for cooking and eating, entertainment, and meeting facilities. This module and its facilities were to have provided separation from the environmental control system hardware with its attendant noise, vibration, and potential for noxious odors and fluid leaks.

Later in the ISS program, costs grew, and in response major elements and systems were eliminated or indefinitely deferred. The Habitation Module and all of its facilities were eliminated. A permanent and dedicated storage module was eliminated. The Cupola control and observation module was eliminated. Much of the hardware for the Habitation Module, Storage Module, and Cupola had previously been designed and manufactured but they were cancelled in order to reduce integration, outfitting, and launch costs. Eventually the Storage and Cupola elements were added back into the program. The Habitation Module was never reintroduced. After years of operation, out of necessity, some habitability features, such as private sleep compartments and a second waste management compartment were reintroduced. These habitation accommodations were added in modules not originally intended for routine or permanent habitation.

For long-duration habitation, consideration must be given to the full range of human activities and functions. All functions must be accommodated; all should be designed into the system from the outset. For early, short-duration spacecraft, the emphasis was on engineering, mission operability, mass constraints, and cost mitigation. Early spacecraft were designed to be operated, not lived in. For long-duration space habitats, human-centered habitability functions must be prioritized and cannot be ignored.

Micrometeorites and Orbital Debris

The ISS flies at an altitude between 320 km (200 mi) and 390 km (240 mi). Since the beginning of the space age, low earth orbit (LEO) has become littered with the remnants of satellites and rocket stages. The debris comes in all sizes from microscopic to entire multi-ton derelict satellites and rocket stages. The ISS flies at an orbital velocity of about 8 km/s (5 mi/s) and any object in earth orbit flies at a similar velocity, depending on altitude, but often in very different trajectories. At such high velocities, almost any impact would have the force of a small explosion. The impact of a 10-cm (4-in) object on the surface of the ISS would have an explosive force equivalent to 7 kg (15 lb) of TNT. The impact of larger debris could potentially destroy the ISS. Exterior surfaces of the ISS show the results of such impacts as craters, dents, and holes. Debris in the ISS orbit is one of the greatest risks to the vehicle and crew.

The U.S. Department of Defense (DoD) Space Surveillance Network tracks orbital debris as small as 5 cm (2 in). In order to mitigate the risk of a collision between debris this size or larger and the ISS, NASA, and the DoD work together to identify space junk which can pose a threat to the ISS and periodically the ISS is maneuvered out of the way of potential collisions.

The habitable pressurized ISS modules, including the temporary sortie or logistics vehicles, are all designed to protect against the impact of small 1 cm (0.4 in) diameter or smaller debris. The shielding consists of a metal outer aluminum bumper mounted at a distance from the inner pressure shell, and with several intervening layers of multilayer insulation and blankets of Nextel and Kevlar. The multiple layers help to insulate the pressure shell from the extremes of heat and cold on the day and night sides of earth, and also serve as the debris shield. When a piece of debris strikes the outer bumper debris panel, it vaporizes and dissipates kinetic energy (Figure 7.77).

Some debris with higher energy could penetrate the outer bumper. As a piece of debris travels through each layer, it will lose energy and fracture into smaller pieces. The fragments will spread out from the point of penetration and will either not make it to the primary pressure shell, or will strike the pressure shell but not penetrate.

Debris that is more massive or more energetic than can be stopped by the shielding may cause a penetration of the ISS pressure hull. This would result in an overboard air leak and depressurization. If the hole can be found before the internal ISS air pressure drops too low, the crew can

Figure 7.77 MLI blankets which include aluminized Mylar, Nextel, and Kevlar layers insulate the pressure shell and serve as a debris barrier for micrometeorites and orbital debris energy dissipation. (*NASA.*)

use a kit to patch an accessible leak. If the air pressure drops too rapidly, the crew may have to close hatches and isolate other parts of the ISS from a leaking module. It is critical in such a case that all of the crew stay on the side of the isolated module that permits access to their earth-return vehicles.

The standard racks enhance the ability to deal with orbital debris or micrometeorites. They provide additional hardware layers on the interior that protect the crew and ease access to the vehicle pressure shell to enable astronauts to access and repair leaks in cases where the shell is breached.

Repair kits allow the crew to repair holes up to approximately 1.25 cm (0.5 in) in diameter if the hole can be accessed. The crew would work in a module until the cabin pressure got below 0.6 atm (9.5 psi). Below that pressure the crew would either isolate the depressurizing module or take refuge in their return vehicle to prepare to depart the ISS.

Windows

The ISS has a large number of windows. Because of mass and structural considerations, windows are primarily placed according to experiment and operations requirements. The ISS was designed in a period before digital imagery had become commonplace and at that time it was felt that windows were required for the crew to control nearby vehicles, robotics, and to observe and assist during spacewalks (EVA).

There are 13 windows in the Russian Service Module, most of which are earth facing. Each U.S. segment hatch has a window in it. Most windows in closed hatches are covered with thermal blankets or internal stowage. The U.S. Lab has a single, large-diameter window used primarily for earth observation experiments. The Japanese laboratory has two large-diameter windows, overlooking the Japanese robotic arm. The Cupola is an observation module with seven large windows. It is essentially the flight deck of the ISS and is used by the crew for earth viewing and to monitor the arrival and departure of U.S. cargo vehicles, robotics, and EVAs.

Windows for science experiments are specially ground, polished, and coated to specification to ensure specified optical quality. The optical quality windows are protected with shutters to ensure protection against external contaminants, micrometeorites or debris, and excessive heat and sunlight.

All of the windows consist of at least two structural panes of glass that provide redundant pressure barriers. Most windows also usually have thinner interior and exterior protective panes that are designed for ease of removal and replacement in the case of scratches or contamination (Figure 7.78).

Space Robotic Systems

Five robotic arms are used on the exterior of the ISS. The first robotic arm system on ISS was mounted on the station the first year that astronauts took up residence and was used throughout the ISS

(a) (b)

Figure 7.78 (*a*) Best view is from the ISS Cupola. (*NASA*.) (*b*) Optical-quality windows are found in several modules. (*NASA*.)

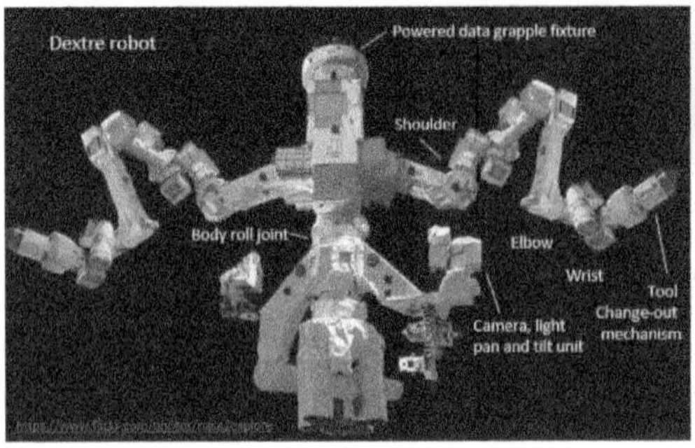

(a)

(b)

FIGURE 7.79 (*a*) Dextre Canadian robot is used outside ISS (*NASA*) and (*b*) Robonaut 2 performs functions inside the ISS (*NASA*).

assembly process. The Canadian Space Station Remote Manipulator System (SSRMS), named Canadarm II, is a second-generation system based on the smaller and less sophisticated robot arm carried on the U.S. Space Shuttle. Canadarm II operates in conjunction with a mobile base and mobile transporter system (MTS). The MTS is a small railroad car, which travels along rails mounted on the football field length truss. The robot arm can stay mounted on the MTS or can "walk" end-over-end from fixtures at several locations on the exterior of station modules (Figure 7.79).

Astronauts on the ISS operate the Canadarm II and Mobile Transporter from robotics workstations inside the ISS. The workstations provide the data processing, video equipment, and hand controllers used by the astronaut to control the system. Some operations are performed from the Cupola, where the crew can watch the arm while looking out the windows. Cameras help to increase the crew's situational awareness.

The special purpose dexterous manipulator (SPDM), called Dextre, looks very much like a robot. It provides a smaller set of arms that have a resemblance to human shoulders, arms, elbows, and wrists with a large set of tools can be operated independently or mounted on Canadarm II to conduct fine or delicate operations.

The Japanese laboratory module, Kibo (Hope) has a large experiment airlock for deploying payloads outside the ISS. The Japanese Experiment Module Remote Manipulator System Main Arm (JEM RMS MA) is mounted outside the airlock to assist in those deployments. A smaller arm designed for delicate interface work is called the JEM Small Fine Arm (JEM SFA).

The European Robotic Arm (ERA) is the last robotic arm to arrive on the ISS, awaiting delivery on one of the last Russian modules, the multipurpose laboratory. ERA will work on the exterior of the Russian segment, moving end-over-end on prefixed base points. The arm will be used for external payload deployments, and support of astronauts during spacewalks. Astronauts will be able to operate the arm either from inside or outside the station.

Thermal Control

A spacecraft's TCS must provide a comfortable cabin environment by establishing thermal equilibrium in the cabin at the proper temperature. Heat that is added to the cabin interior must be balanced by rejecting heat through the TCS. Equipment throughout the interior and exterior of the vehicle must be kept within a specified temperature range.

For most spacecraft the primary role of the TCS is to reflect, collect, and reject heat. However, there are usually localized areas of the spacecraft in cold locations that require heaters to maintain temperature. The ISS has more than 300 heaters. The heaters are inside of modules, and prevent condensation from forming, and outside the vehicle on fluid lines and hardware installations and keep hardware operating within temperature specifications.

Temperatures outside of the ISS range from −184° to 149°C (±300°F). The thermal balance of the ISS structures, external equipment, and modules are maintained within an allowable temperature range through both passive and active systems.

Passive thermal control measures include vehicle exterior color, surface texture, and the use of multilayer insulation (MLI). These can limit environmental heat from solar radiation from transferring into the vehicle. MLI can serve multiple purposes. MLI provides a barrier between the heat on the vehicle's exterior and the inside of the pressure vessel and helps prevent heat from entering or escaping. It controls heat transfer rates and minimizes temperature gradients. An aluminized cloth layer outside some components provides protection from the erosional effects of atomic oxygen and can provide micrometeoroid orbital debris (MMOD) protection. An inner layer of MLI, immediately adjacent to the module pressure hull, can provide protection during handling and installation of the intermediate and outer MLI layers. MLI is sometimes used as a protective measure for astronaut safety to prevent crew contact with extreme temperatures. The outer surface coatings and paints must be compatible with atomic oxygen and radiation found in the station orbit. MLI is used outside the module pressure shells, on truss segments, and on orbital replacement units (ORUs).

Active TCS systems circulate coolant fluids inside and outside the vehicle. The coolant fluids carry heat to external radiators. External radiators on the Russian segment of ISS are mounted on the bodies of the Russians modules. Integrating radiators as part of the vehicle structure can offer weight savings and assist in the protection against MMOD. On the U.S. segment of ISS, radiators are separate structures on rotating mounts on the truss and exposes both sides of the radiator to space and can be oriented for optimal rejection of heat energy. Between the U.S. and Russian segments of the ISS there are no operational interfaces with the sole exception of airflow.

Within the modules of the ISS, heat load originates with crew metabolic heat. The amount of heat depends on the number of crew and level of activity of the crewmembers. All powered equipment generates heat. Forced air ventilation provided by the ECS helps the flow of heat; however, the ability to control temperature adequately requires the active TCS.

Powered equipment is located primarily in the standard racks. The equipment is attached to cold plates. Heat from the equipment is conducted into the cold plate. Cooling water from the active internal TCS flows through the cold plate and carries the heat away. There are more than 100 cold plates within the U.S. Lab module.

The coolant water from the internal TCS flows into heat exchangers mounted on the forward end-cone of the U.S. Lab. The heat exchangers are shared between the ISS internal and external TCS fluid loops. The external TCS uses ammonia to carry the heat. Ammonia was chosen because of its high thermal capacity and wide range of operating temperatures. Ammonia was not used inside the habitable modules because of its toxicity to humans.

Ten heat exchangers, located in the U.S. Lab, Node 2 and Node 3, Columbus, and Kibo have active, standalone internal TCS systems. The end-cone-mounted heat exchangers interface with the external ECS. Other modules like the Cupola, Quest airlock, and Node 1 must be connected to the other modules' internal TCS.

From the module heat exchangers, the ammonia flows through cold plates cooling avionics on the ISS truss and then to six 23 m (75 ft) long pointable radiators. The heat is radiated into space. Total heat rejection capability is 70 kW.

The Russian segment has a similar system although the working fluids are different. Inside the crew compartment, the coolant is triol that is similar to ethylene glycol. The external TCS uses polymethyl siloxane. The radiators for the Russian segment are mounted on the exterior of the Zvezda and Zarya modules.

Spacesuits

One of the most iconic images of human space flight is that of the space suited astronaut in orbit floating separately from the spacecraft. Early spacesuits were worn as emergency devices intended to protect the astronaut in case of an unexpected loss of cabin pressure. Apollo astronauts relied on their spacesuits to allow them to walk on the moon's surface. On the ISS, the spacesuit became a necessity for high-altitude construction. One hundred and sixty-four spacewalks and more than a thousand hours of extravehicular activity (EVA) by astronauts from the Space Shuttle and the ISS were required to assemble all of the components.

Three different types of spacesuits are worn for ISS missions. The Russian Sokol suit is worn by astronauts launching to and returning from the ISS in the Soyuz spacecraft. It is a protective measure against inadvertent loss of cabin pressure. The Russian Orlan suit and U.S. Extravehicular Mobility Unit (EMU) are worn when astronauts exit the ISS pressurized modules to conduct EVA or spacewalks.

A spacesuit like the Sokol is used only inside the spacecraft. It is different from the EVA suit in several respects. When inflated, the protective device need only provide sufficient mobility for the astronaut to maintain position in the spacecraft and operate the controls. The suit is tailored and fitted to the shape of the astronaut in the recumbent Soyuz-seated position. With the suit pressurized, the astronaut can scan the display panels, reach the necessary controls, and communicate with other astronauts and mission control. The Sokol suit remains connected to the spacecraft's environmental control system (ECS). Either breathing air from the spacecraft cabin, or pure oxygen from pressurized bottles provides breathing air into the suit. The suit operates at 0.38 atm (5.8 psi) above the ambient cabin atmospheric pressure. The suit is intended to be worn in a pressurized cabin for 30 hours or in a vacuum for 2 hours.

Russian and US EVA suits are used outside the pressurized cabin of the spacecraft. The suits provide their own portable ECS mounted to the astronaut's back. The suit ECS provides oxygen for breathing, carbon dioxide removal, and water circulating through the underwear for cooling. When outside, the astronaut remains attached to the spacecraft by a simple cable. Outside of the spacecraft the astronaut can maneuver large elements of the ISS in the weightlessness environment. Astronauts must be able to use their hands and fingers to use tools; dexterity must be enhanced and for this reason MLI insulation is reduced. In some cases electric heaters are used in the glove's fingers. The gloves are specially designed to maximize dexterity while protecting the crewmember from the space environment.

The EVA suits are designed to be worn in the extreme hot and cold of space. Multilayer insulation (MLI) in the suit lining minimizes the amount of heat entering the suit from space. Nevertheless, heat from external sources together with the metabolic heat produced internally by the astronaut, must be removed from the suit by the portable life support system (PLSS). For this reason a liquid cooling garment (LCG) is worn as an undergarment and has flexible tubes that circulate cooling water. The cooling water conducts heat away from the skin by direct contact. The circulating water is cooled in a heat exchanger in the backpack-mounted portable life support system and then reused.

Early spacesuits attempted to use circulating breathing oxygen for cooling and this proved unsuccessful. Joints in the pressurized suit resist bending and motion. The restriction of movement caused by the inflated suit requires that the astronaut expend much greater energy than if they were only wearing ordinary clothing. Because oxygen is used at a low pressure in the suit, its effectiveness in cooling is reduced and is inadequate for carrying away perspiration and heat. The inadequacy of convection cooling using low-density oxygen means that the astronaut tends to perspire more while wearing the suit. This can result in dehydration. For this reason a small drink container is mounted within the suit for use by the astronaut during the spacewalk (Figure 7.80).

Remote Operation

The data management system of the ISS makes many of the systems on ISS operable remotely from the mission control centers located in several places around the world. This permits the crew to focus on the on-board mission and science activities. It increases dependence on the on-board computer system. Such an approach can create issues if, for instance, a computer is plugged into a

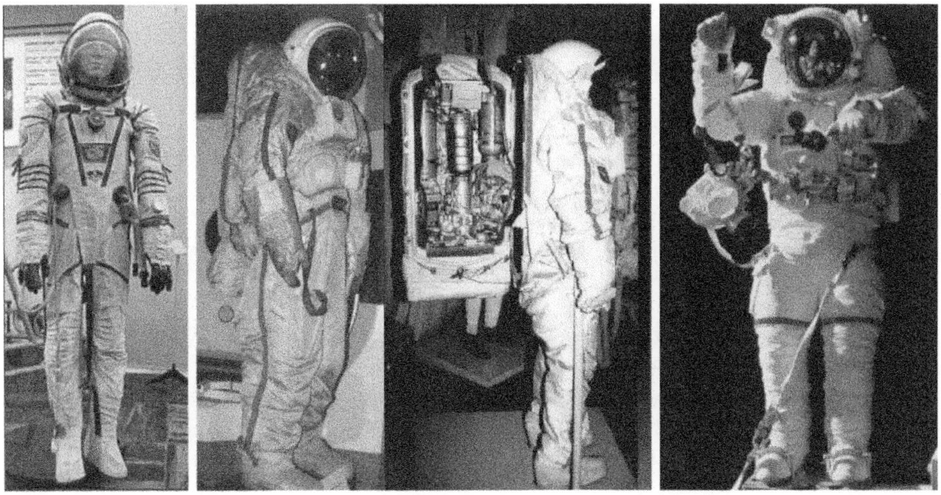

FIGURE 7.80 (*a*) Russian Sokol suit. (*NASA.*) (*b*) Russian Orlan EVA suit and open PLSS. (*NASA.*) (*c*) U.S. Extravehicular Mobility Unity (EMU) suit. (*NASA.*)

power circuit that the same computer controls. The ISS was also designed to rely upon systems management by ground control. This places a reliance on telemetry provided by the spacecraft to ground control and on commands that are provided by ground control to the spacecraft. In many cases, the astronaut is only relied upon to control ISS system operations infrequently. For future missions in which communications delays or communications is lost periodically, more reliance may need to be placed on on-board computer control, robotics or the human crew.

7.36 Summary

This chapter focused on topics of specific interest for the design of space vehicles in which astronauts live and operate. While today's International Space Station is an example of current space facility design and technology, human space flight vehicles have been evolving since the beginning of the space program more than 50 years ago and several examples of spacecraft types and configurations during that period were described.

The capabilities, configurations, and systems of manned spacecraft have varied widely depending on their missions, the era in which they were developed, the countries that were responsible for their development, and the goals of their respective programs. Spacecraft have come in a variety of forms, tailored for a given payload and mission. Configuration was influenced by the environments in which the spacecraft operated. However, there are many common features of manned spacecraft. Ultimately, the manned spacecraft must protect its human crew from the environment in which it operates. It has to provide a breathable atmosphere and all of the consumables required for the human to not only survive but to operate in an effective manner for the length of the mission.

Design requirements should be levied on the spacecraft that delineate human rating requirements, crew health requirements, habitation requirements, and systems standards. Specific design requirements are imposed for specific types of missions or operations.

Spacecraft in which humans live and operate must be designed to protect human life throughout all mission phases and even when major systems fail. All systems critical for crew safety must be robust and redundancy must be provided for all life-critical functions.

References

Chapanis, Alphonse. 1996. *Human Factors in Systems Engineering*. Wiley. New York.

Connors, Mary M., Harrison, Albert A., and Akins, Faren R. 1985. *Living Aloft: Human Requirements for Extended Spaceflight*. NASA.

Faget, Max. 1965. *Manned Space Flight*. Holt, Rinehart and Winston, New York.

Grumman, *Lunar Module Documentation*.

https://www.hq.nasa.gov/alsj/alsj-LMdocs.html.

Kitmacher, Gary H. 2002. *Design of the Space Station Habitable Modules*. 53rd International Astronautical Congress.

http://www.spacearchitect.org/pubs/IAC-02-IAA.8.2.04.pdf.

Kitmacher, Gary H. 2007, 2010. *Reference Guide to the International Space Station*. NASA. Washington, DC.

https://www.nasa.gov/pdf/508318main_ISS_ref_guide_nov2010.pdf.

Larson, Wiley and Pranke, Linda. 2007. *Human Spaceflight*. McGraw Hill. New York.

Purser, Paul, Faget, Max, and Smith, Norman. 1964. *Manned Spacecraft, Engineering Design and Optimization*. Fairchild. New York.

SECTION 8

Astrodynamics

Roy Y. Myose and Atri Dutta

Notation

a	semimajor axis length (defines the orbit size)
E	eccentric anomaly angle (used for time of flight in elliptic orbit)
F	hyperbolic eccentric anomaly angle (used for time of flight in hyperbolic orbit)
H	angular momentum per unit mass
L	Lagrange point (for restricted three-body problem, with subscript) or latitude (for field of view)
m	mass
p	semilatus rectum (defines the orbit size)
r	radial distance (measured from center of gravitational body)
T	orbital period (for circular and elliptic orbits)
t	time since periapsis passage
TOF	time of flight
U	mechanical energy per unit mass (sum of kinetic and potential energies per unit mass)
v	velocity
γ	flight path angle (direction of velocity vector measured above horizon line)
Γ	rocket thrust vector
ε	eccentricity (defines the orbit shape)
θ	true anomaly angle (measured in direction of travel from periapsis)
μ	gravitational parameter
Ω	angular velocity (for restricted three-body problem, in rad/s units) or right ascension of ascending node (for orbital elements, in rad units)
$\dot{\Omega}$	regression of nodes (for orbital precession, in deg/s units)

Subscripts

- a apoapsis (location in elliptic orbit farthest from gravitational body)
- c condition corresponding to circular orbit
- p periapsis (location in orbit closest to gravitational body)
- ∞ condition, where $r = \infty$ (for hyperbolic and parabolic orbits)

8.1 Orbital Mechanics

Introduction

Over the past 50 years, there have been significant achievements in the field of astronautics—astronauts have visited the moon, cosmonauts have stayed up in space for more than a year, and robotic probes have visited most of the planets in the solar system. Closer to home, there are hundreds of communication satellites circling the globe today. These satellites provide communication links between distance countries and help to support the Internet-driven global economy of the 21st century. The ability to accomplish even the simplest of these space missions depends on a fundamental understanding of orbital mechanics. The foundation for the modern-day study of orbital mechanics was laid by Johannes Kepler and Isaac Newton some four centuries ago. Many of the terms used in orbital mechanics, however, originate from ancient Greek studies of celestial mechanics and analytical geometry. A list of nomenclature is given above as a convenient reference to aid readers who are new to the study of orbital mechanics. Although there are many standard textbooks covering the topic of orbital mechanics, some of the notation used differs from reference to reference. Thus, the nomenclature list also helps to clarify the notation used in this section of the handbook.

Terminology and Orbit Types

The orbital mechanics part of the space mission begins with the delivery of a spacecraft into orbit around a gravitational body such as the earth. Depending on the amount of energy (and angular momentum) provided by the launch vehicle, the spacecraft will follow a specific flight path. The shape of this flight path can be circular, elliptic, parabolic, or hyperbolic. Closed orbits around the earth follow circular or elliptic flight paths, while parabolic or hyperbolic flight paths are followed by spacecraft leaving the earth's gravitational field, as shown in Figure 8.1. With the exception of a circular orbit, there is one location in the orbit which is closest to the gravitational body. For the generic gravitational body, this location is called periapsis and has radius r_p, as shown in Figure 8.2. For specific gravitational bodies, the suffix is changed; the point of closest approach to the earth is called *perigee*, while the closest approach to the sun is called *perihelion*. The periapsis location is used as the reference starting point for measuring the true anomaly angle θ so that $\theta = 0$ at the periapsis. Another important location in the orbit is the semilatus rectum p, which is defined as the location where $\theta = 90°$, as shown in Figure 8.2. For elliptic orbits, there is one location in the orbit which is farthest from the gravitational body. This location, called the *apoapsis*, corresponds to $\theta = 180°$ and has radius r_a. The periapsis and apoapsis are used as two points, which define the so-called line of apsides. In addition, the periapsis and apoapsis are the end points of the major axis of the ellipse. Thus, for an elliptic orbit the semimajor axis length a is given by

$$a = \frac{r_p + r_a}{2} \tag{8.1}$$

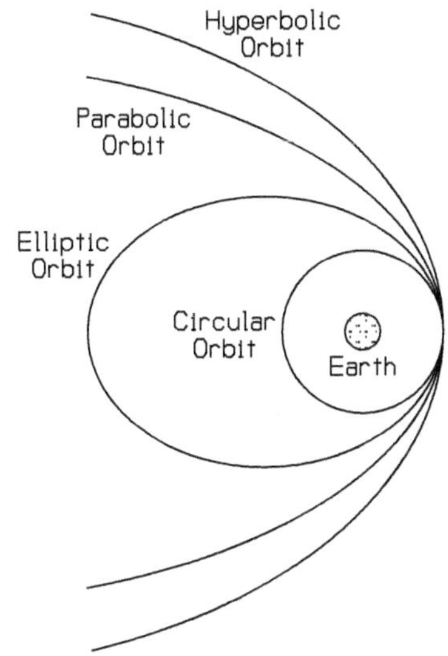

FIGURE 8.1 Different orbit types (adapted from Myose 2001).

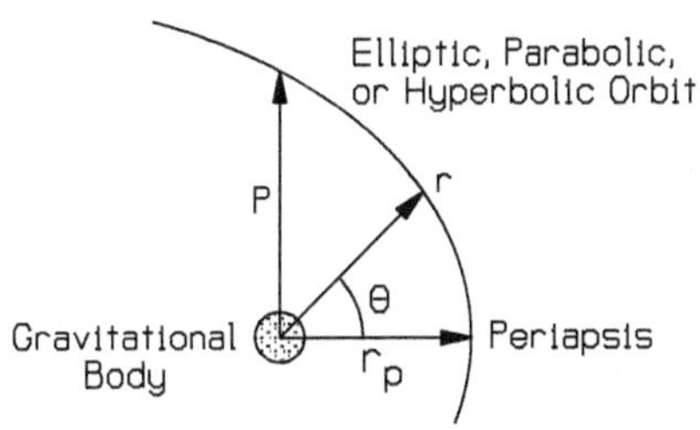

FIGURE 8.2 Definition of periapsis, semilatus rectum, and true anomaly angle (adapted from Myose 2001).

Scalar Relationships for Position, Energy, and Velocity

For circular orbits, the radius r_c is constant everywhere which means that for this special case:

$$r_c = r_p = p = r_a = a \tag{8.2}$$

The four different orbit shapes (circular, elliptic, parabolic, and hyperbolic) are all geometric shapes which can be formed by sectioning off a cone, as shown in Figure 8.3. Consequently, the motion of a spacecraft is governed by the well-known equation for a conic section, which is given by:

$$r = \frac{p}{1 + \varepsilon \cos \theta} \tag{8.3}$$

where r is the radial distance measured from the center of the gravitational body, p is the semilatus rectum, ε is the eccentricity, which is a measure of the orbit's shape, and θ is the true anomaly angle.

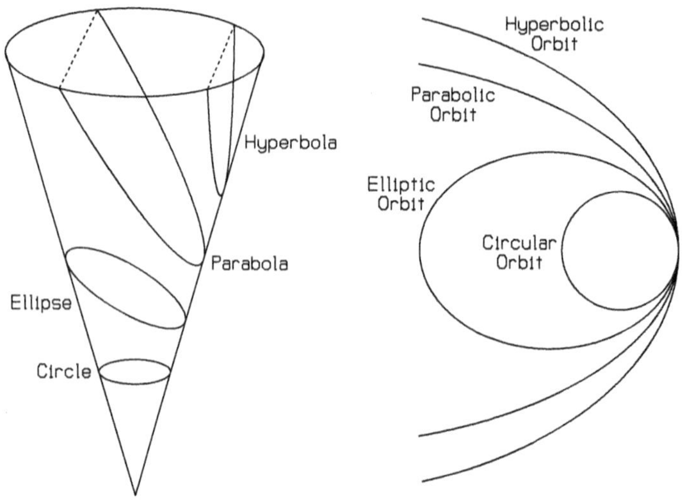

FIGURE 8.3 Conic section orbital shapes (adapted from Myose 2001).

Equation (8.3), called the orbit equation, describes the functional relationship between r, the radial distance from the gravitational body, and θ, the true anomaly angle. Since cosine is an even function, equation (8.3) shows that the radius at $+\theta$ is the same as the radius at $-\theta$. The value of the eccentricity is $\varepsilon = 0$ for a circular orbit, $0 < \varepsilon < 1$ for an elliptic orbit, $\varepsilon = 1$ for a parabolic orbit, and $\varepsilon > 1$ for a hyperbolic orbit.

One unique parameter which is often of interest in hyperbolic orbits is the asymptote angle θ_∞, which is shown in Figure 8.4. This angle is measured between the asymptote line and the line of apsides. The asymptote line is simply the extension of the spacecraft's direction of

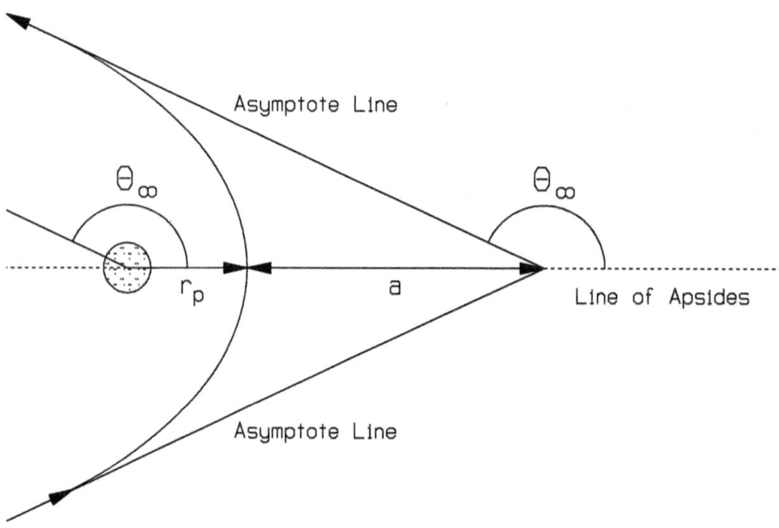

FIGURE 8.4 Hyperbolic orbit (adapted from Myose 2001).

travel when it escapes the planet's gravity well. A second asymptote line can be drawn from a mirror image along the line of apsides. This second asymptote line is an extension of the direction of travel for a spacecraft which starts outside the planet's gravity well and is captured by the planet's gravity well. Since the spacecraft must start off or end up outside the planet's gravity well, the radial distance is ∞ in this case. Substituting $r = \infty$ into the orbit equation results in an asymptote angle of

$$\theta_\infty = \cos^{-1}\left(\frac{-1}{\varepsilon}\right) \tag{8.4}$$

The orbit equation given in equation (8.3) only describes the flight path taken by the spacecraft and not the energy or angular momentum required to place the spacecraft into such an orbit. For the nonmaneuvering case associated with the basic orbital mechanics problem, the mechanical energy and the angular momentum of the spacecraft must remain constant. The conservation laws for energy and angular momentum are derived from Newton's laws. If **r** represents the vector measured from the gravitational body and pointed toward the spacecraft, then application of Newton's second law together with Newton's law of gravitation results in (see Bate et al. 1971; Hale 1994; Wiesel 1997; Myose 2001):

$$\frac{d^2\mathbf{r}}{dt^2} = -\left(\frac{G(m_1 + m_2)}{r^3}\right)\mathbf{r} \tag{8.5}$$

where G is the universal gravitational constant. For typical astrodynamics problems, the mass m_2 of the spacecraft is significantly less than the mass m_1 of the gravitational body. In this case, $G(m_1 + m_2) \approx Gm_1 = \mu$, where μ is called the gravitation parameter. Typical values for the gravitational parameter as well as other data (based on Beatty and Chaikin 1990; Brown 1992; Hale 1994) are given in Tables 8.1 and 8.2. Using the gravitational parameter μ, the modified form of Newton's law given in equation (8.5) can be written as

$$\frac{d^2\mathbf{r}}{dt^2} = -\left(\frac{\mu}{r^3}\right)\mathbf{r} \tag{8.6}$$

It should be noted that the spacecraft mass is not a part of equation (8.6). Thus, the conservation laws are related to the energy per unit mass and the angular momentum per unit mass.

Celestial body	μ (km³/s²)	Equatorial radius of surface (km)
Sun	1.3271×10^{11}	6.960×10^5
Mercury	2.2032×10^4	2.439×10^3
Venus	3.2486×10^5	6.051×10^3
Earth	3.9860×10^5	6.378×10^3
Moon	4.9028×10^3	1.783×10^3
Mars	4.2828×10^4	3.393×10^3
Jupiter	1.2671×10^8	7.1492×10^4
Saturn	3.7940×10^7	6.0268×10^4
Uranus	5.7816×10^6	2.5559×10^4
Neptune	6.8713×10^6	2.4764×10^4
Pluto	1.0209×10^3	1.500×10^3

TABLE 8.1 Physical Data

Celestial body	Semimajor axis length of orbit (km)	Eccentricity	Inclination (deg.)
Mercury	5.7910×10^7	0.2056	7.004
Venus	1.0820×10^8	0.0068	3.394
Earth	1.4960×10^8	0.0167	0.000
Moon	3.8440×10^5	0.05	18.3–28.6
Mars	2.2794×10^8	0.0934	1.850
Jupiter	7.7833×10^8	0.0483	1.308
Saturn	1.4270×10^9	0.0560	2.488
Uranus	2.8710×10^9	0.0461	0.774
Neptune	4.4971×10^9	0.0097	1.774
Pluto	5.9135×10^9	0.2482	17.148

TABLE 8.2 Orbital Data

The conservation of energy can be derived by taking a dot product between the radius vector **r** and the modified form of Newton's law given by equation (8.6) (see Bate et al. 1971; Hale 1994; Wiesel 1997; Myose 2001). The result is a scalar equation involving the spacecraft's velocity v and is given by

$$U = \left(\frac{1}{2}\right)v^2 - \frac{\mu}{r} = \text{constant} = -\frac{\mu}{2a} \tag{8.7}$$

where U is the total mechanical energy per unit mass, $\frac{1}{2}v^2$ is the kinetic energy per unit mass, and $-(\mu/r)$ is the potential energy per unit mass. It should be noted that $r = \infty$ is the reference location for zero potential energy in astrodynamics rather than the surface of the earth, which is commonly used in physics textbooks. Thus, a spacecraft has negative potential energy inside the earth's gravitational well. It requires a significant amount of kinetic energy to climb out of the gravity well and leave the earth's gravitational field. The total energy for circular and elliptic orbits is negative because the semimajor axis lengths are finite positive values. Thus, spacecraft in circular and elliptic orbits remain trapped inside the gravity well. A spacecraft in parabolic orbit has just enough energy to climb out of the gravity well with zero residual velocity at $r = \infty$. Thus, the total energy for a parabolic orbit is zero and the semimajor axis length is ∞. Finally, a spacecraft in hyperbolic orbit has positive residual velocity at $r = \infty$. This means that the total energy for a hyperbolic orbit is greater than zero and the semimajor axis length is negative. The orbital characteristics for the different orbit types are summarized in Table 8.3.

Orbit type	Total energy	Semimajor axis length	Eccentricity
Circular	$U < 0$	$a > 0$	$\varepsilon = 0$
Elliptic	$U < 0$	$a > 0$	$0 < \varepsilon < 1$
Parabolic	$U = 0$	$a = \infty$	$\varepsilon = 1$
Hyperbolic	$U > 0$	$a < 0$	$\varepsilon > 1$

TABLE 8.3 Orbital Characteristics

The conservation of energy given by equation (8.7) can be rearranged to obtain a general equation for velocity as follows:

$$v = \sqrt{\frac{2\mu}{r} - \frac{\mu}{a}} \tag{8.8}$$

In the special case of a circular orbit, which has a constant radius, the semimajor axis length is given by $a = r_C$. Substituting this result into equation (8.8) results in a circular orbit velocity of

$$v_C = \sqrt{\frac{\mu}{r_C}} \tag{8.9}$$

Another velocity of interest is the escape velocity, which is the special case where a spacecraft has just enough energy to escape the gravity well. In this case, the spacecraft has zero excess energy when it reaches $r = \infty$. Since energy must be conserved, this means that the total energy of this orbit is zero, which corresponds to the case of a parabolic orbit. Substituting $a = \infty$ into equation (8.8) results in an escape velocity of

$$v_{\text{escape}} = \sqrt{\frac{2\mu}{\mathbf{r}}} \tag{8.10}$$

When equation (8.10) is applied to determine the escape velocity from the surface of the earth using μ and r from Table 8.1, the well-established result of $v_{\text{escape,earth}} = 11.2$ km/s is obtained.

Vector Relationships for Angular Momentum and Orientation of the Orbit

The conservation of angular momentum can be derived by taking a cross-product between the radius vector **r** and the modified form of Newton's law given by equation (8.6) (see Bate et al. 1971; Hale 1994; Wiesel 1997; Myose 2001). The result is a vector equation given by

$$\mathbf{H} = \mathbf{r} \times \mathbf{v} = \text{constant} \tag{8.11}$$

where **H** is the angular momentum per unit mass. The radius and velocity vectors are the two lines which define the orbital plane while the angular momentum is perpendicular to the orbital plane as shown in Figure 8.5. If no maneuvers are performed, the spacecraft remains in orbit within this

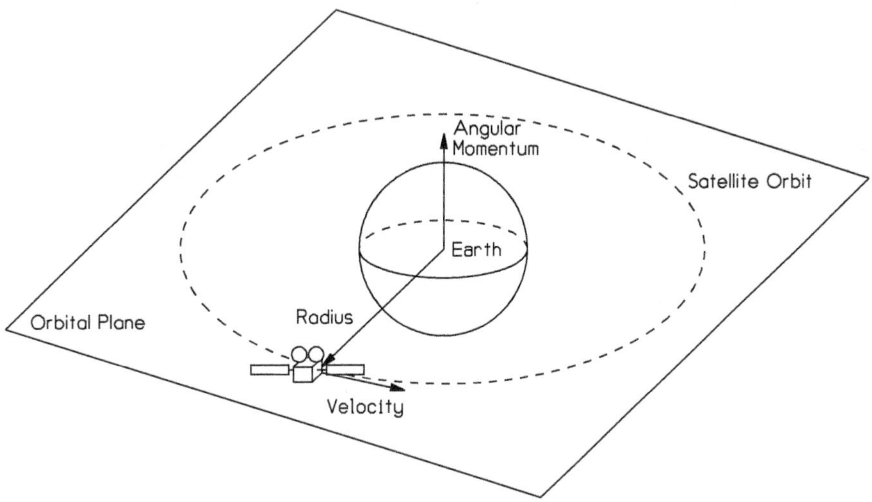

FIGURE 8.5 Orbital plane and angular momentum vector (from Myose 2001).

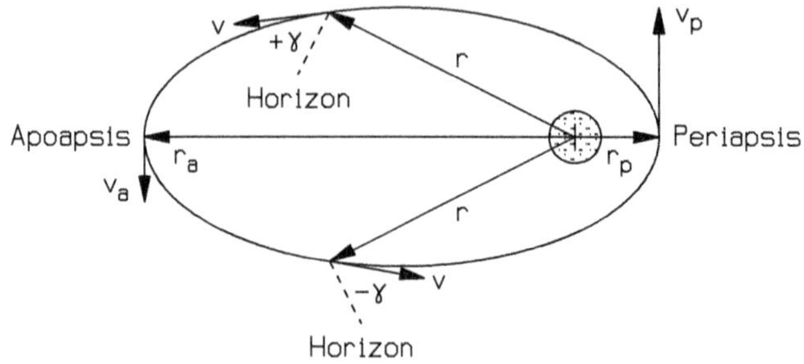

FIGURE 8.6 Flight path angle (from Myose 2001).

plane and the magnitude as well as the direction of the angular momentum vector remains constant according to equation (8.11).

For the circular orbit shown in Figure 8.5, the velocity vector is always perpendicular to the radius vector. In general, the velocity does not have to be perpendicular to the radius. Instead, the velocity must always be tangent to the flight path. For noncircular orbits, this means that the velocity vector is typically oriented above or below the horizon line as shown in Figure 8.6. The angle between the velocity vector and the horizon line is called the flight path angle γ. When the spacecraft is moving away from the periapsis, $\gamma>0$, while $\gamma<0$ when the spacecraft is moving toward the periapsis. At the periapsis and apoapsis, the velocity and radius are perpendicular to each other which mean that the flight path angle is equal to zero. When the idea of flight path angle is applied to equation (8.11), a scalar form of the conservation of angular momentum results:

$$H = rv\cos\gamma \tag{8.12}$$

For the special case when the spacecraft is located at the periapsis and apoapsis, equation (8.12) becomes

$$H = r_p v_p = r_a v_a \tag{8.13}$$

Since the angular momentum vector is perpendicular to the orbital plane, the vector **H** can be used as part of a three-axis reference system which defines the orientation of an orbit in three dimensions. Within the orbital plane, a two-axis system can be constructed using the periapsis and semilatus rectum directions. Determining the periapsis direction requires a quantity called the eccentricity vector ε which is pointed in the periapsis direction and has a magnitude equal to the eccentricity. Starting with the cross-product between the modified form of Newton's law given by equation (8.6) and the angular momentum, it can be shown that (see Bate et al. 1971; Hale 1994; Wiesel 1997; Myose 2001):

$$\mathbf{v}\times\frac{\mathbf{H}}{\mu} - \mathbf{r}/r = \varepsilon \tag{8.14}$$

If a dot product is taken between **r** and the eccentricity vector equation given by equation (8.14), a slightly different form of the orbit equation is obtained (see Bate et al. 1971; Hale 1994; Wiesel 1994; Myose 2001):

$$r = \frac{H^2/\mu}{1 + \varepsilon\cos\theta} \tag{8.15}$$

One consequence of this result is that the orbits of all celestial objects must be conic sections in shape. This was deduced by Johannes Kepler in the 16th century, who stated in his first law that all planetary orbits are elliptic orbits with the sun at one focus.

Conic Section Relationship and Orbital Elements

A one-to-one relationship between total energy and semimajor axis length was given in equation (8.7). A similar one-to-one relationship between angular momentum (magnitude) and semilatus rectum is obtained by comparing the two different forms of the orbit equation given by equations (8.3) and (8.15). If the orbit equation is evaluated at the periapsis, then a relationship between semilatus rectum and periapsis radius is also obtained. Furthermore, the definition of semimajor axis length given by equation (8.1) can be used to relate the semilatus rectum to the semimajor axis length. The result is an equation relating the angular momentum to the various conic section parameters (see Bate et al. 1971; Hale 1994; Wiesel 1997; Myose 2001):

$$\frac{H^2}{\mu} = p = r_p(1+\varepsilon) = a\,(1-\varepsilon^2) \tag{8.16}$$

Equation (8.16) shows that any pair of (independent) conic section parameters is sufficient to define a spacecraft's angular momentum and total energy. For example, the semimajor axis length and eccentricity can uniquely determine a spacecraft's angular momentum and total energy except in the case of a parabolic orbit.

The angular momentum and total energy are two dynamical quantities, which can be used to define a spacecraft's orbital characteristics, also known as orbital elements. Since these two dynamical quantities are directly related to conic section parameters, two geometric quantities can also be used to define the characteristics of an orbit. Oftentimes the semimajor axis length and eccentricity are chosen as the two orbital elements which define the in-plane orbital characteristics. The current position of the spacecraft can be defined by the true anomaly angle (or time since periapsis passage). Three other angles are used to define the orientation of the orbit in three dimensions. The complete set of orbital elements can then be formed from these six different parameters.

Time of Flight Relationships

One consequence of the conservation laws (for energy and for angular momentum) is that a spacecraft moves quickly at periapsis and slowly at apoapsis. This means that the circumferential distance traveled by the spacecraft in a given amount of time is large at periapsis and small at apoapsis. Suppose the area swept out by the radius vector in a given amount of time is considered. Due to the conservation laws, the area swept out near the periapsis would be comparable to the area swept out near the apoapsis as shown in Figure 8.7. In fact, it can be shown from first principles that the area A swept out by the radius vector in a given amount of time is related to the angular momentum according to (see Bate et al. 1971; Hale 1994; Wiesel 1997; Myose 2001):

$$\frac{dA}{dt} = \frac{H}{2} \tag{8.17}$$

Since the angular momentum is constant, equation (8.17) proves Kepler's second law, which states that the radius vector connecting the spacecraft and the gravitational body sweeps out equal areas in equal time.

The orbital period T, which is the time it takes to complete one full orbit, can be determined if the area for the closed orbit is known. For elliptic orbits, the area is given by πab and the semiminor

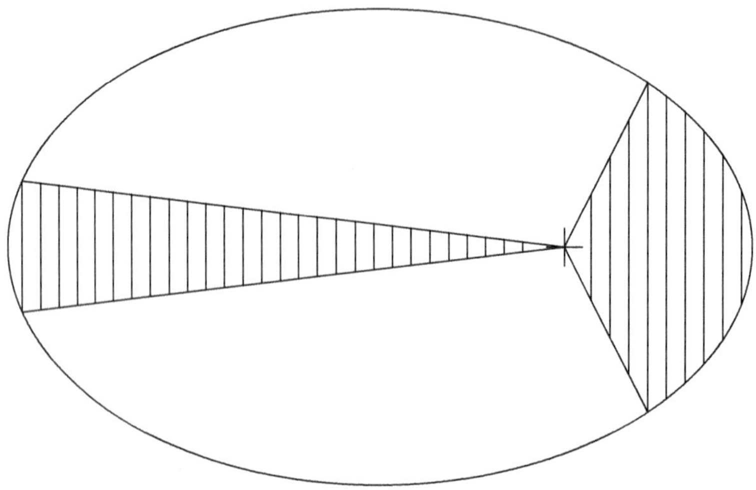

FIGURE 8.7 Area swept out by radius vector near apoapsis (left) and periapsis (right).

axis length b follows the relationship $b^2 = a^2(1 - \varepsilon^2)$. This means that the orbital period is given by (see Bate et al. 1971; Hale 1994; Wiesel 1997; Myose 2001):

$$T = 2\pi\sqrt{\frac{a^3}{\mu}} \tag{8.18}$$

Equation (8.18) is Kepler's third law, which states that the orbital period squared is proportional to the semimajor axis length cubed.

Many problems of interest require determining the time of flight (TOF) between the periapsis and a location on the orbit, where the true anomaly angle θ is known. In order to determine this time since periapsis passage in an elliptic orbit, a concept called the auxiliary circle must be used. The auxiliary circle is a circle with a radius equal to the semimajor axis length of the ellipse which just touches the periapsis and apoapsis, as shown in Figure 8.8. If a line perpendicular to the line of apsides is extended from the location on the orbit to the auxiliary circle, the resulting angle on the auxiliary circle is given by the eccentric anomaly angle E, as shown in Figure 8.8. Using relationships from analytical geometry, it can be shown that the true anomaly and eccentric anomaly angles are related as follows (see Bate et al. 1971; Hale 1994; Wiesel 1997; Myose 2001):

$$\cos E = \frac{\varepsilon + \cos\theta}{1 + \varepsilon\cos\theta} \text{ and } \cos\theta = \frac{\cos E - \varepsilon}{1 - \varepsilon\cos E} \tag{8.19}$$

The time since periapsis passage t is given by (see Bate et al. 1971; Hale 1994; Wiesel 1997; Myose 2001):

$$t = \sqrt{\frac{a^3}{\mu}}(E - \varepsilon\sin E) \tag{8.20}$$

It should be noted that the eccentric anomaly angle E used in equation (8.20) must be in units of radians. One interesting property of all orbits is that the orbital behavior is mirror imaged along the

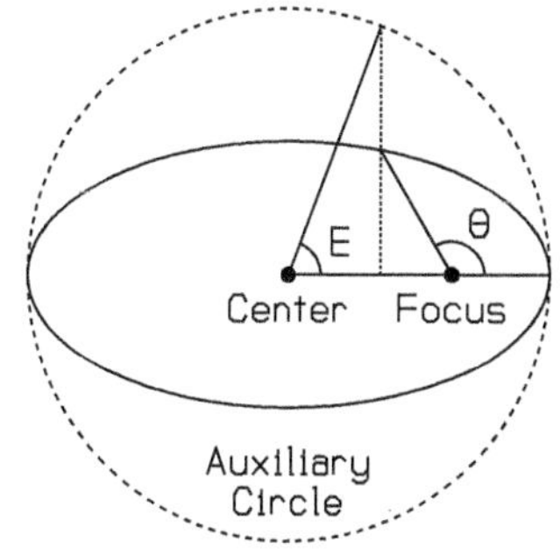

FIGURE 8.8 Auxiliary circle (adapted from Myose 2001).

line of apsides. This means that the time since periapsis passage for a location corresponding to a true anomaly angle of θ is exactly the same as the time it takes to travel from a location corresponding to a true anomaly angle of $-\theta$ to the periapsis.

Although the time since periapsis passage given by equation (8.20) is valid only for an elliptic orbit, a similar equation is available for a hyperbolic orbit. In this case, the hyperbolic eccentric anomaly angle must be defined. The hyperbolic eccentric anomaly angle F is related to the true anomaly angle as follows (see Bate et al. 1971; Hale 1994; Wiesel 1997; Myose 2001):

$$\cosh F = \frac{\varepsilon + \cos\theta}{1+\varepsilon\cos\theta} \quad \text{and} \quad \cos\theta = \frac{\cosh F - \varepsilon}{1-\varepsilon\cosh F} \tag{8.21}$$

The time since periapsis passage for a hyperbolic orbit is given by (see Bate et al. 1971; Hale 1994; Wiesel 1997; Myose 2001):

$$t = \sqrt{\frac{-a^3}{\mu}}(\varepsilon \sinh F - F) \tag{8.22}$$

Example of Geotransfer Orbit

A very useful and often crowded orbital location is the geosynchronous orbit. Many geosynchronous orbit-bound satellites are first brought to a low earth parking orbit by a rocket launch vehicle before final delivery to geosynchronous orbit. The transfer orbit between the low earth orbit (LEO) and the final geosynchronous earth orbit (GEO) is an elliptic orbit whose perigee is equal to the LEO radius and whose apogee is equal to the GEO radius. As an example, we will determine the characteristics of this transfer ellipse if LEO corresponds to 100 nautical mile altitude (i.e., 185 km above the earth's surface) and GEO corresponds to a radius of 42,164 km.

Problem statement: given $h_p = 185$ km and $r_a = 42{,}164$ km, determine the orbital characteristics of this transfer ellipse and the TOF from perigee to apogee.

Problem solution: the radius of the earth's surface, from Table 8.1, is $r_{\text{earth}} = 6{,}378$ km, which means that $r_p = r_{\text{earth}} + h_p = 6{,}563$ km. From equation (8.1), the semimajor axis length of the transfer ellipse would be $a = (r_p + r_a)/2 = 24{,}363.5$ km. Based on the conservation of energy, the total energy would be $U = -\mu/(2a) = -8.18\ \text{km}^2/\text{s}^2$. Using the conic section relationship, equation (8.16), the eccentricity of this transfer ellipse would be given by $\varepsilon = 1 - (r_p/a) = 0.7306$. Also from the conic section relationship, the semilatus rectum and angular momentum would be $p = r_p(1 + \varepsilon) = 11{,}358$ km and $H=\sqrt{\mu p}$ = 67,285 km²/s. The orbital period is given by equation (8.18), $T=2\pi\sqrt{a^3/\mu}$ = 37,846 s. Since the orbit is a mirror image along the line of apsides, the TOF from perigee to apogee is the same as the TOF from apogee to perigee. This means that the TOF from perigee to apogee is half the orbital period, or $T/2$ = 18,923 s. An alternative approach would be to determine the eccentric anomaly angle at θ = 180° (the answer is $E = \pi$) and then substitute it into equation (8.20).

Circular Restricted Three-Body Problem

The circular restricted three-body problem investigates the motion of a spacecraft in the gravitational field of two point masses m_1 (larger of the two) and m_2 (smaller of the two) referred to as primary masses. It is assumed that the primary masses move in circular orbits about their mutual center of mass. Note that the center of mass of the two primary bodies would be unaccelerated, i.e., it is either at rest or moves with a uniform speed in a straight line. The mass m of the spacecraft is considered to be small compared to the primary masses and, therefore, does not affect the motion of the primary masses. These assumptions allow the motion of the primary masses to be determined by solving the two-body problem. We are interested in describing the motion of the mass m. Note that, unlike the two-body problem, the circular restricted three-body problem does not have a closed-form analytical solution.

In order to analyze the motion of the mass m, we consider a noninertial reference frame (see top inset of Figure 8.9) at the center of mass C of the primary masses. The x-axis of the reference frame is along the $m_1 - m_2$ line pointed away from m_1, the z-axis is normal to the plane of motion of m_1 and m_2. The y-axis is in-plane and follows the right-hand rule. Consider that the unit vectors along the three axes of the reference frame are given by $\boldsymbol{i}$ along the x-axis, $\boldsymbol{j}$ along the y-axis, $\boldsymbol{k}$ along the z-axis. The angular velocity (vector) $\boldsymbol{\Omega}$ of the noninertial reference frame is given by

$$\Omega = \sqrt{\frac{G(m_1+m_2)}{r_{12}^3}}k \tag{8.23}$$

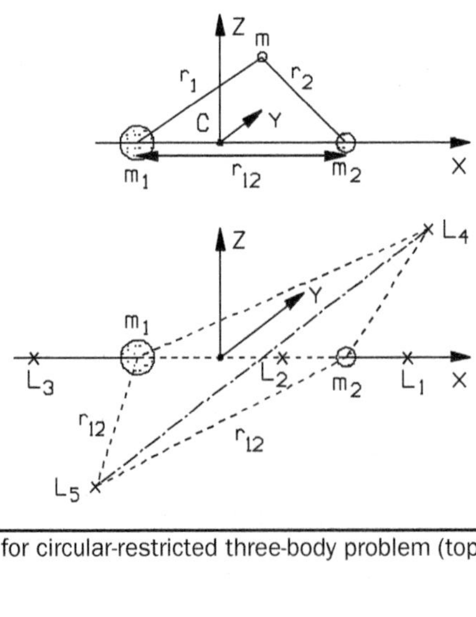

FIGURE 8.9 Reference frame for circular-restricted three-body problem (top) and Lagrange points (bottom).

We consider that the position of the spacecraft is given by coordinates x, y, and z, and then the equations of motion in the $\boldsymbol{i}, \boldsymbol{j}, \boldsymbol{k}$ directions, respectively, are given by (see Prussing and Conway 2013):

$$\ddot{x} = 2\Omega\dot{y} - \Omega^2 x = -\frac{\mu_1}{r_1^3}\left(x + \frac{m_2}{m_1+m_2}r_{12}\right) - \frac{\mu_2}{r_2^3}\left(x - \frac{m_1}{m_1+m_2}r_{12}\right) \tag{8.24a}$$

$$\ddot{y} + 2\Omega\dot{x} - \Omega^2 y = -\frac{\mu_1}{r_1^3}y - \frac{\mu_2}{r_2^3}y \tag{8.24b}$$

$$\ddot{z} = -\frac{\mu_1}{r_1^3}z - \frac{\mu_2}{r_2^3}z \tag{8.24c}$$

These equations cannot be solved analytically, i.e., it permits only numerical solution. However, we can obtain key insight into the dynamics of the mass m without solving the equations. For instance, for the equilibrium points of this dynamic system, all time derivatives of the positional coordinates should go to zero and the differential equations reduce to a set of nonlinear algebraic equations, as given below:

$$-\Omega^2 x = -\frac{\mu_1}{r_1^3}\left(x + \frac{m_2}{m_1+m_2}r_{12}\right) - \frac{\mu_2}{r_2^3}\left(x - \frac{m_1}{m_1+m_2}r_{12}\right) \tag{8.25a}$$

$$-\Omega^2 y = -\frac{\mu_1}{r_1^3}y - \frac{\mu_2}{r_2^3}y \tag{8.25b}$$

$$0 = -\frac{\mu_1}{r_1^3}z - \frac{\mu_2}{r_2^3}z \tag{8.25c}$$

One of the immediate conclusions from the third equation is that the Z-coordinate has to be zero. This means that the equilibrium points for the system must lie in the X–Y plane containing the two primary masses. The first two sets of equations result in five equilibrium points, referred to as Lagrangian points. As evident from the second equation, some of these equilibrium points will correspond to $Y = 0$. This means that these Lagrange points lie on the line collinear with the primary masses. There are three such points and they are referred to as L_1, L_2, and L_3 Lagrange points (see bottom inset of Figure 8.9). The remaining two Lagrange points occupy the vertex of possible equilateral triangles that can be formed, with the length of the sides of the equilateral triangle being the distance r_{12} from the two primary masses. These are referred to as the L_4 and L_5 Lagrange points. These points also turn out to be stable equilibrium points, and examples of stable orbits can be found in nature. For instance, the Trojan asteroids are roughly at 60° to both Sun and Jupiter, suggesting that they are near the L_4 and L_5 Lagrange points of the Sun-Jupiter system.

8.2 Orbital Maneuvers

Introduction

An important orbit mentioned in the example of the previous subsection is the geosynchronous earth orbit (or GEO). A large number of communication satellites as well as some of the earth observation weather satellites are located in GEO. The idea of using GEO for communication relays was first proposed in 1945 by Arthur C. Clarke, who was trained as an engineer before becoming a science fiction writer. The Early Bird satellite made by Hughes (which is now a part of the Boeing Company) was the first commercial communication satellite to be put into GEO. Twenty years later, in 1984, the number of commercial communication satellites in GEO had grown to 75. In 2001, that number increased threefold—to 223 commercial communication satellites in GEO. To put this into perspective, a large communication satellite relays as many as 100,000 simultaneous telephone or Internet connections. Thus, there has been a tremendous increase in global communication traffic in 35 years.

Three conditions must be met in order for a satellite to be located in GEO. First, the satellite must have an orbital period equal to the rotation rate of the earth. Contrary to popular belief, the rotation rate of the earth is not one revolution in 24 hours. A mean solar day, which corresponds to 24 hours, is the time it takes for the sun to be located at the same position when viewed by a ground observer. During this 24-hour period, the earth not only rotates about its own axis, but it also moves in its orbit around the sun. This motion around the sun contributes a 360° rotation in 365¼ days, i.e., an additional degree-per-day rotation. Thus, when this additional contribution is taken into account, the rotation rate of the earth is determined to be one revolution every 23 hours 56 minutes and 4 seconds, which is called a sidereal day. A satellite with an orbital period equal to a sidereal day would have a semimajor axis length of 42,164 km based on equation (8.18). The second requirement for a GEO satellite is that it must be in a circular orbit, i.e., an eccentricity of $\varepsilon = 0$. Since the satellite travels at a constant rate in a circular orbit, there is no eastward or westward shift of the satellite from the perspective of a ground observer. The third requirement for a GEO satellite is that the orbit must lie in the equatorial plane. This prevents any north-south drift of the satellite during its orbit. Since the satellite motion is synchronized with the rotation of the earth, a GEO satellite will appear to be located at a fixed location above the equator for a ground observer. One consequent advantage of GEO is that satellite tracking is not required, which makes it easy to use fixed ground-based antennas. Another advantage for GEO satellites is continuous observation of a fixed area of the earth, which is a desirable quality for an earth observation weather satellite.

Regardless of whether the spacecraft is intended to be a GEO communication satellite or a scientific probe to another planet, a typical spacecraft is delivered to a low earth parking orbit by a rocket launch vehicle. This subsection focuses on the issue of how to maneuver a spacecraft from one orbit to another, in particular the velocity changes required to accomplish the orbit change. This may entail not only an increase (or decrease) in the spacecraft's velocity within the plane of the orbit, but also a change in the spacecraft's orbital plane.

Hohmann Transfer

The most basic problem in orbital maneuvers is a transfer from one circular orbit to another circular orbit, e.g., from LEO to GEO. For this situation, the most fuel-efficient transfer orbit is half of an ellipse, which is tangent to the two circular orbits—in other words, an ellipse whose perigee corresponds to the low circular orbit radius and apogee corresponding to the high circular orbit radius as shown in the left-hand illustration of Figure 8.10. Such an elliptic transfer orbit is called the

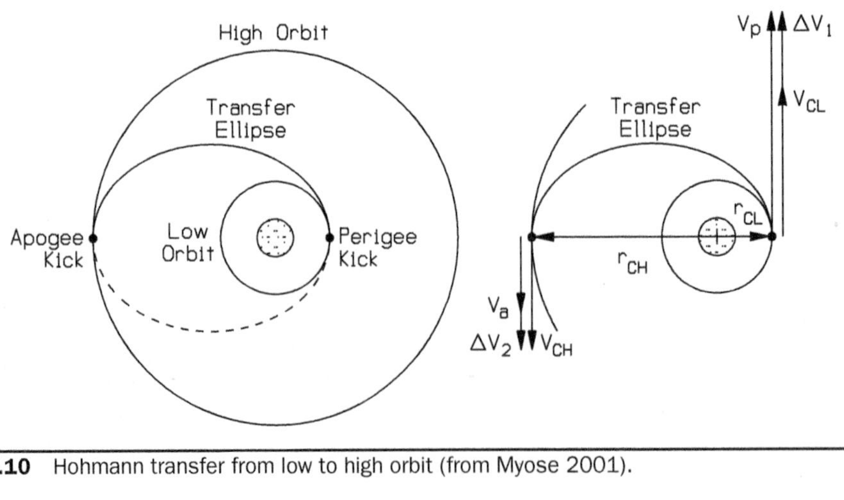

FIGURE 8.10 Hohmann transfer from low to high orbit (from Myose 2001).

Hohmann transfer, which was named after Walter Hohmann, who discovered this maneuver in 1925. In order to go from low to high orbit, the velocity must be increased at perigee and at apogee. These two engine burns are often called the perigee and apogee kicks. The velocity increment associated with each of these engine burns is simply the difference in the final and initial velocities. The initial and final velocities are the circular orbit velocity given by equation (8.9) and the perigee or apogee velocity on the transfer ellipse, which is given by equation (8.8). At perigee, the initial velocity is the low circular orbit velocity (v_{CL}) while the final velocity is the perigee velocity (v_p) for the Hohmann elliptic transfer orbit. Thus, the first velocity increment (Δv_1) is given by

$$\Delta v_1 = v_p - v_{CL} \tag{8.26a}$$

$$v_{CL} = \sqrt{\frac{\mu}{r_{CL}}} \tag{8.26b}$$

$$v_p = \sqrt{\frac{2\mu}{r_{CL}} - \frac{\mu}{a}} \tag{8.26c}$$

$$a = \frac{r_{CL} + r_{CH}}{2} \tag{8.27}$$

At apogee, the final velocity is the high circular orbit velocity (v_{CH}) while the initial velocity is the apogee velocity (v_a) for the Hohmann elliptic transfer orbit. Thus, the second velocity increment (Δv_2) is given by

$$\Delta v_2 = v_{CH} - v_a \tag{8.28a}$$

$$v_{CH} = \sqrt{\frac{\mu}{r_{CH}}} \tag{8.28b}$$

$$v_a = \sqrt{\frac{2\mu}{r_{CH}} - \frac{\mu}{a}} \tag{8.28c}$$

where equation (8.27) gives the semimajor axis length a for the Hohmann transfer orbit.

As an example, consider a Hohmann transfer from 6,563 km radius circular LEO to GEO which corresponds to a circular orbit radius of 42,164 km. Based on equation (8.27), the semimajor axis length of the Hohmann transfer is $a = 24{,}363.5$ km. An earth orbit means that the gravitational parameter is $\mu_{earth} = 3.986 \times 10^5$ km^3/s^2 based on Table 8.1. The circular orbit, perigee, and apogee velocities determined using equations (8.26b), (8.26c), (8.28b), and (8.28c) are $v_p = 10.252$ km/s, $v_{CL} = 7.793$ km/s, $v_{CH} = 3.075$ km/s, and $v_a = 1.596$ km/s. This means that the required velocity increments, based on equations (8.26a) and (8.28a), are $\Delta v_1 = 2.459$ km/s and $\Delta v_2 = 1.479$ km/s. The total velocity increment (Δv_{total}), which relates to the fuel expenditure associated with the perigee and apogee kicks, is given by the sum of the two velocity increment magnitudes, i.e., $\Delta v_{total} = 3.938$ km/s. The TOF was determined in the example of the previous subsection to be TOF = $T/2 = 18{,}923$ s, where the orbital period T was given by equation (8.18).

Fast Transfers

The TOF in a Hohmann transfer is equal to half the period of an elliptic orbit because a Hohmann transfer is formed from half of an ellipse. If a reduction in the TOF is desired, then the transfer orbit must be changed from an ellipse, which is tangent to the low and high orbits, to another conic section with a larger eccentricity. Such a transfer orbit is called a fast transfer. The advantage of a fast transfer is its reduced TOF, which comes at the cost of significantly increased fuel expenditure.

The conservation laws discussed in the previous subsection showed that a spacecraft moves quickly at perigee and slowly at apogee. This means that truncating the ellipse to exclude the apogee area can decrease the TOF. The type I fast transfer shown in Figure 8.11 is such a transfer orbit. A type I fast transfer must have its perigee at the low orbit radius, while the high orbit is intercepted at a true anomaly angle of $\theta_2 < 180°$. A type II fast transfer, shown in Figure 8.12, must start from the low orbit radius with a true anomaly angle past perigee ($\theta_1 > 0$), while its apogee is located at the high orbit radius. A type III fast transfer, shown in Figure 8.13, is the general case with $\theta_1 > 0$ and $\theta_2 < 180°$.

The relevant velocity vectors are aligned with each other if the transfer orbit intercepts the circular orbit at either the perigee or apogee. For the type I fast transfer, the first velocity increment will then be given by equation (8.26) as indicated by Figure 8.11. For the type II fast transfer, the second velocity increment is given by equation (8.28) as indicated by Figure 8.12. At locations other than perigee and apogee, the transfer orbit's velocity is oriented at a nonzero flight path angle. This means that the velocity increment is not aligned with the other velocities in the case of Δv_2 for type I

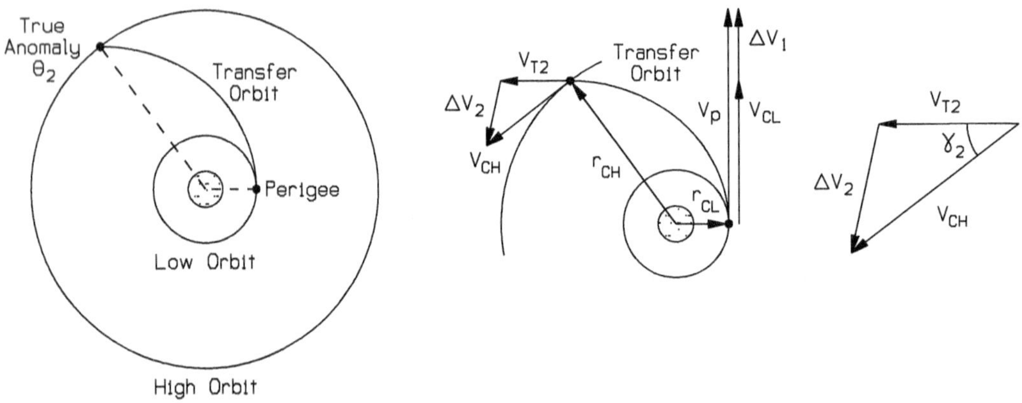

FIGURE 8.11 Fast transfer type I: starting from perigee (adapted from Myose 2001).

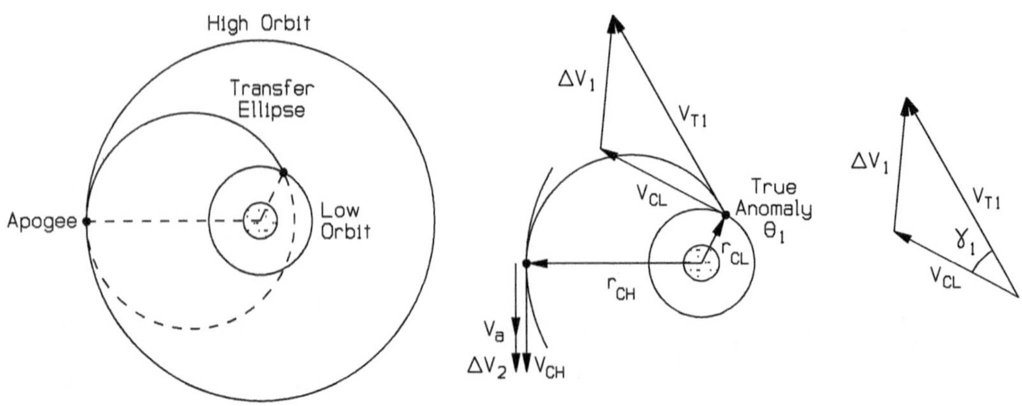

FIGURE 8.12 Fast transfer type II: starting from apogee (adapted from Myose 2001).

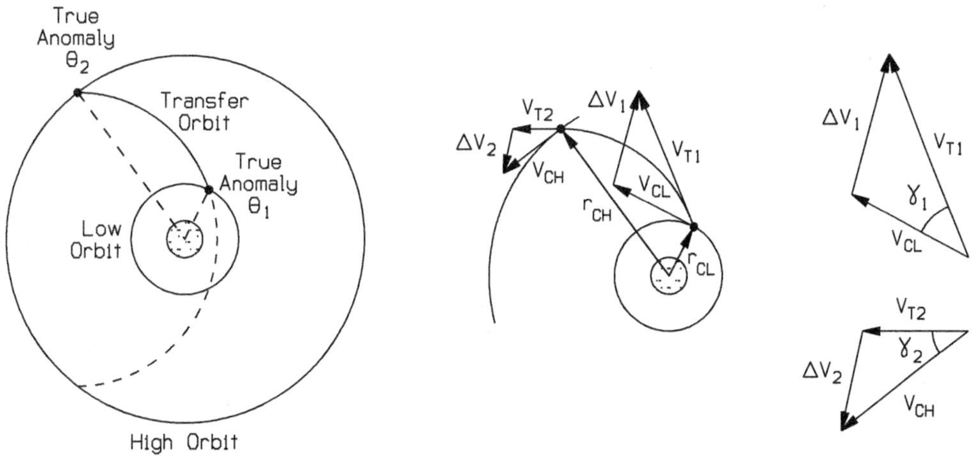

FIGURE 8.13 Fast transfer type III: general case (adapted from Myose 2001).

(Figure 8.11), Δv_1 for type II (Figure 8.12), and both Δv_1 and Δv_2 for type III (Figure 8.13). In such a situation, the cosine law must be used to determine the velocity increment as follows:

$$\Delta v_1 = \sqrt{v_{T1}{}^2 + v_{CL}{}^2 - 2 v_{T1} v_{CL} \cos \gamma_1} \tag{8.29}$$

$$\Delta v_2 = \sqrt{v_{T2}{}^2 + v_{CH}{}^2 - 2 v_{T2} v_{CH} \cos \gamma_2} \tag{8.30}$$

As an example, consider transfer from 6,563 km radius circular LEO to GEO, which corresponds to a circular orbit radius of 42,164 km. Suppose a type I fast transfer with a high orbit intercept at a true anomaly angle of $\theta_2 = 120°$ is used. The characteristics of this type I fast transfer can be determined from the orbit equation, which is given by either equation (8.3) or (8.15). Since the semilatus rectum p and angular momentum H are fixed for the fast transfer, the orbit equation can be rearranged to determine the eccentricity as follows:

$$r = \frac{H^2/\mu}{1+\varepsilon\cos\theta} \Rightarrow r_{CL}(1+\varepsilon\cos\theta_1) = \frac{H^2}{\mu} = r_{CH}(1+\varepsilon\cos\theta_2) \Rightarrow$$

$$\varepsilon = \frac{r_{CH} - r_{CL}}{r_{CL}\cos\theta_1 - r_{CH}\cos\theta_2} \tag{8.31}$$

Substituting the relevant values result in an eccentricity of $\varepsilon = 1.2878$ for this fast transfer, which means that this transfer orbit is hyperbolic. Since $r_{CL} = r_p$ for a type I fast transfer, the semimajor axis length of $a = -22{,}805$ km can be determined from the conic section relationship given by equation (8.16). An earth orbit means that the gravitational parameter is $\mu_{earth} = 3.986 \times 10^5$ km^3/s^2 based on Table 8.1. Substituting these values into equation (8.26a, b, c) results in $v_p = 11.788$ km/s, $v_{CL} = 7.793$ km/s, and $\Delta v_1 = 3.995$ km/s. This first velocity increment is 1.6 times the first velocity increment for the Hohmann transfer found in the previous example.

The transfer orbit's velocity at high orbit, v_{T2}, can be found from the general velocity equation given by equation (8.8). Substituting the relevant values of $\mu_{earth} = 3.986 \times 10^5$ km^3/s^2, $r_{CH} = 42{,}164$ km, and $a = -22{,}805$ km results in $v_{T2} = 6.032$ km/s. Since the perigee conditions are known, the angular momentum can be found using equation (8.13) to be $H = r_p v_p = 77{,}362\ \text{km}^2/\text{s}$. The scalar form of the conservation of angular momentum given by equation (8.12) can then be used to determine the flight path angle at the time of high orbit intercept. That is, $\gamma_2 = \cos^{-1}(H/r_{CH} v_{T2}) = 72.29°$. Based on equation (8.28b), the required final velocity in GEO is $v_{CH} = 3.075$ km/s. Using the cosine law relationship given in equation (8.30) results in a second velocity increment of $\Delta v_2 = 5.878$ km/s. This second velocity increment is four times the second velocity increment for the Hohmann transfer found in the previous example. Finally, the total velocity increment for this fast transfer is $\Delta v_{total} = 9.873$ km/s, which is 2½ times the total velocity increment for the Hohmann transfer.

Substituting $\theta_2 = 120°$ and $\varepsilon = 1.2878$ into equation (8.21) results in a hyperbolic eccentric anomaly angle of $F = 1.4317$ rad. This value along with $\mu_{earth} = 3.986 \times 10^5\ \text{km}^3/\text{s}^2$ and $a = -22{,}805$ km gives a TOF of 6,053 s based on equation (8.22). In comparison, the Hohmann transfer TOF found in the previous example was about three times longer at 18,923 s.

Orbital Inclination and Launch Site Latitude

The geosynchronous orbit is one important orbit mentioned earlier in the introduction. One requirement for GEO is that the orbit must lie in the equatorial plane. The initial low earth parking orbit, on the other hand, oftentimes does not lie in the equatorial plane. Figure 8.14 shows such a case where the inclination angle i is defined to be the angle between the angular momentum vector and the earth's North Pole direction. Orbits with an inclination angle of $i = 0$ are called equatorial orbits. Orbits with an inclination angle in the range $0 < i < 90°$ are called prograde orbits because its direction of motion has an eastward component which is the rotation direction of the earth. Orbits with an inclination angle of $i = 90°$ are called polar orbits because they fly over both poles. Orbits with an inclination angle in the range $90° < i < 180°$ are called retrograde orbits because their direction of motion is toward the west, which is counter to the rotation direction of the earth.

The smallest inclination angle for the low earth parking orbit is obtained when the rocket is launched toward the east. In this case, the resulting inclination angle is exactly equal to the latitude angle of the launch site because the cross-product between the launch site's radius vector and the due east velocity vector results in an inclined angular momentum vector, as shown in Figure 8.14.

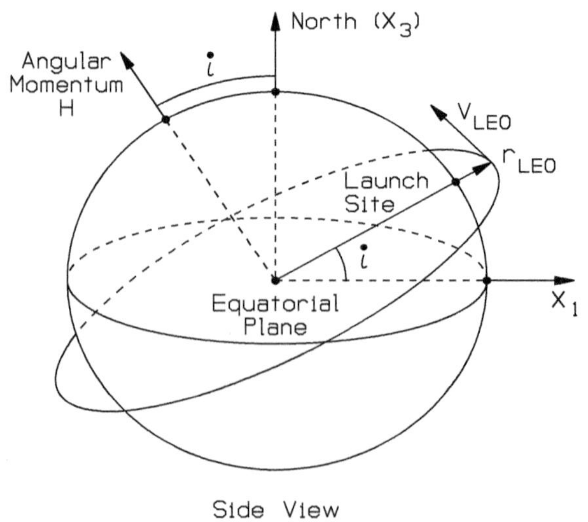

FIGURE 8.14 Launch latitude and orbital inclination (from Myose 2001).

Country	Launch site	Location	Latitude
China	Xichang	China	28.2° N
Europe	Kourou	French Guiana	5.2° N
India	Sriharikota	India	13.9° N
Japan	Tanegashima	Japan	30.2° N
Russia	Baikonur	Kazakhstan	45.6° N
Russia	Plesetsk	Russia	62.8° N
United States	Cape Canaveral (KSC)	Florida, United States	28.5° N
United States	Vandenberg AFB	California, United States	34.7° N

TABLE 8.4 Launch Site Latitudes

If the rocket is launched in any other direction, then the resulting inclination angle of the low earth parking orbit is greater than the latitude angle of the launch site. This is because northeastwardly launches would mean that the rocket would continue to travel farther north while southeastwardly launches would indicate that the rocket originates from a higher more northerly latitude.

A launch site located on or near the equator has two advantages. First, the inclination angle of the low earth parking orbit would be relatively small if the rocket was launched toward the east. This means that very little inclination change is required to transfer from the parking orbit to GEO. Second, a launch site located on the equator would obtain a free boost of 0.464 km/s due to the rotation of the earth. However, this boost in velocity diminishes with increasing latitude. Table 8.4 lists the latitudes of some of the major launch sites (from Brown 1992; Isakowitz 1995). Most launch sites are located within the territorial limits of the national entity launching the rocket. Consequently, there are no ground-based rocket launch facilities located on the equator. One notable exception (although no longer in operation) was the multinational Sea Launch rocket developed by the Boeing Company, which used a floating base located at the equator in the Pacific Ocean to launch the Russian Zenit rocket launch vehicle.

Pure Inclination Change

Table 8.4 shows that most launch sites are not located on the equator. This means that the low earth parking orbit is typically inclined. If the desired final destination is located at a different inclination angle, then an inclination change maneuver must be performed. Figure 8.15 depicts such a situation with an inclined initial orbit and an equatorial orbit such as GEO for the desired final orbit. The inclination can only be changed at the intersection points of the two orbits. There are two possible intersection points called nodes—the ascending node, where the spacecraft is headed northward, and the descending node, where the spacecraft is headed southward. A line connecting the two nodes is called the line of nodes.

If only the inclination of the orbit is to be changed without affecting any other orbital characteristic, then the total energy as well as the magnitude of the angular momentum must remain the same. Such a maneuver, shown in Figure 8.15, is called a pure inclination change. In this case, the radius and velocity magnitudes are the same before and after the inclination change. Based on the bisection of a right triangle, the velocity increment associated with a pure inclination change is given by

$$\Delta v_{\text{pure inclination}} = 2v \sin(\Delta i/2) \tag{8.32}$$

Pure inclination changes are extremely fuel-expensive maneuvers to perform. As an example, consider a spacecraft which is initially in a 28.5° inclination 6,563 km radius circular low earth

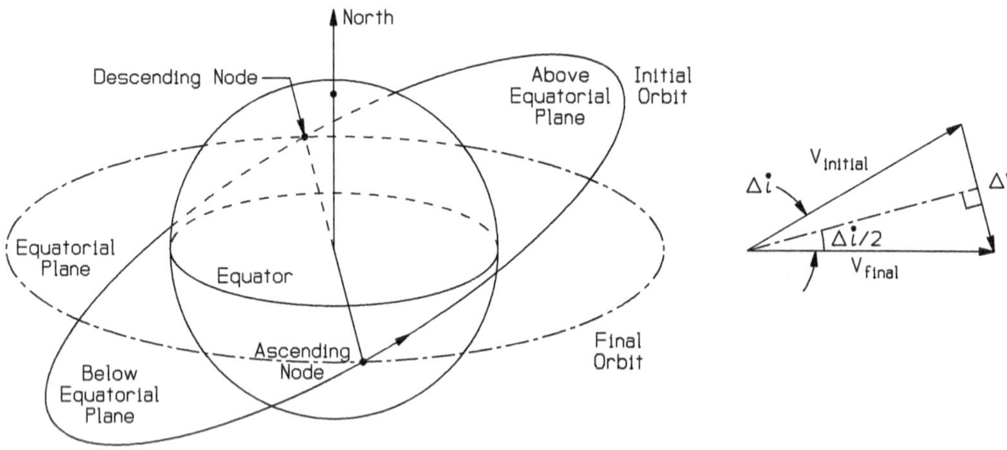

FIGURE 8.15 Intersecting nodes and pure inclination change (from Myose 2001).

parking orbit. If an equatorial orbit with the same circular orbit radius is desired, then $v = v_{\mathrm{CL}} = 7.793$ km/s and $\Delta i = 28.5°$ would give a pure inclination change velocity increment of $\Delta v_{\text{pure inclination}} = 3.837$ km/s. In comparison, a Hohmann transfer, which delivers a spacecraft from 6,563 km radius to GEO at 42,164 km, requires a velocity increment of $\Delta v_{\text{Hohmann}} = 3.938$ km/s. If, on the other hand, the pure inclination change is performed at the high orbit, a much smaller fuel expenditure is involved. In this case, $v = v_{\mathrm{CH}} = 3.075$ km/s, so that the pure inclination change velocity increment is $\Delta v_{\text{pure inclination}} = 1.514$ km/s.

Combined Hohmann Transfer and Inclination Change Maneuver

The example cited above involved a three-engine burn maneuver. Perigee and apogee kicks for the Hohmann transfer within the plane of the transfer orbit and a separate engine burn to change the inclination of the orbit. In an actual LEO to GEO transfer, the out-of-plane maneuver is combined with the two in-plane maneuvers. Figure 8.16 depicts such a combined Hohmann transfer and inclination change maneuver. Suppose the change in inclination angle during the first engine burn is Δi while the total change in inclination is i. Based on the cosine law, the velocity increments at perigee and apogee are given respectively by

$$\Delta v_1 = \sqrt{v_p^2 + v_{\mathrm{CL}}^2 - 2v_p v_{\mathrm{CL}} \cos(\Delta i)} \tag{8.33}$$

$$\Delta v_2 = \sqrt{v_a^2 + v_{\mathrm{CH}}^2 - 2v_a v_{\mathrm{CH}} \cos(i - \Delta i)} \tag{8.34}$$

The total velocity increment is the sum of these two velocity increments. That is,

$$\Delta v_{\text{total}} = \sqrt{v_p^2 + v_{\mathrm{CL}}^2 - 2v_p v_{\mathrm{CL}} \cos(\Delta i)} + \sqrt{v_a^2 + v_{\mathrm{CH}}^2 - 2v_a v_{\mathrm{CH}} \cos(i - \Delta i)} \tag{8.35}$$

Based on the results of the previous example, one might conclude that all of the inclination change should be performed at the high orbit because the velocity involved is smallest at that location. However, this is not the case. The optimum solution is the minimization of the total velocity increment, which is given by equation (8.35). Since v_p, v_{CL}, v_a, v_{CH}, and i are fixed constants, the only

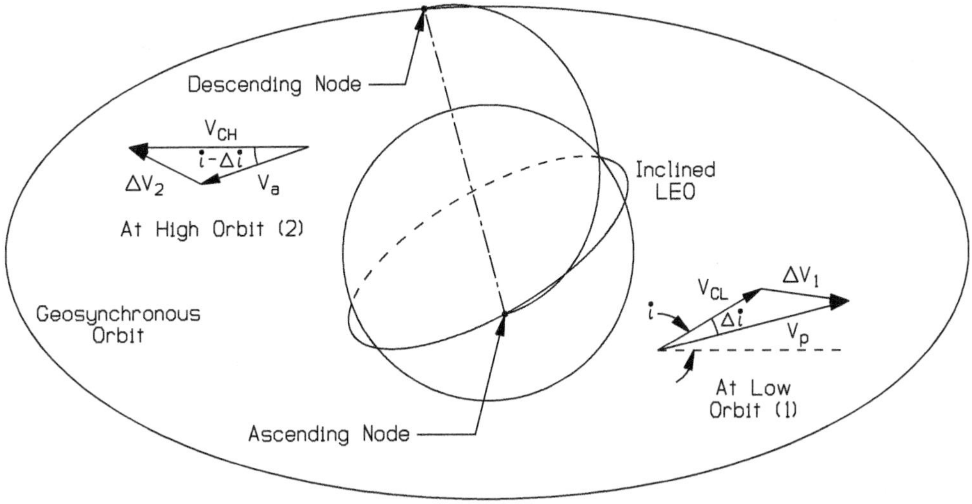

FIGURE 8.16 Combined Hohmann transfer and inclination change maneuver (from Myose 2001).

variable in equation (8.35) is the amount of inclination change Δi during the first engine burn. This means that the minimum total velocity increment is obtained if $\partial(\Delta v_{\text{total}})/\partial(\Delta i) = 0$. Taking a partial derivative of equation (8.35) results in

$$\frac{v_p v_{\text{CL}} \sin(\Delta i)}{\sqrt{v_p^2 + v_{\text{CL}}^2 - 2v_p v_{\text{CL}} \cos(\Delta i)}} - \frac{v_a v_{\text{CH}} \sin(i - \Delta i)}{\sqrt{v_a^2 + v_{\text{CH}}^2 - 2v_a v_{\text{CH}} \cos(i - \Delta i)}} = 0 \tag{8.36}$$

The optimum solution is dependent on the specific values for v_p, v_{CL}, v_a, v_{CH}, and i. This means that equation (8.36) must be solved numerically to determine the optimum inclination change and total velocity increment. The optimum solution is obtained at small nonzero values of Δi because the first and second terms involve the sine (and cosine) of two different angle values.

For the 28.5° inclination 6,563 km radius circular low earth parking orbit to GEO example cited earlier, the optimum inclination change angle turns out to be $\Delta i = 2.2°$. This results in a total velocity increment of $\Delta v_{\text{combined}} = 4.273$ km/s. This compares to an in-plane Hohmann followed by a pure inclination change at high orbit requiring a total velocity increment of $\Delta v_{\text{Hohmann}} + \Delta v_{\text{pure inclination}} =$ 3.938 km/s + 1.514 km/s = 5.452 km/s.

Continuous Thrust Transfer

Chemical propulsion onboard a spacecraft results in a large amount of thrust so that the thrusters need to be operated for a short time in order to provide a certain velocity change. The implicit assumption with high-thrust transfers is that a velocity change occurs instantaneously, allowing the transfer to be modeled as a set of impulses. The Δv for an impulse can then simply be computed by taking a difference of the two velocities that the spacecraft has on the two orbits, the one before the impulse and the one after. Modern solar-electric propulsion systems that have been used in deep space propulsion and recently on all-electric satellites generate low levels of thrust, several orders of magnitude smaller than their chemical counterparts. If the spacecraft employs a low-thrust propulsion system, the thrusters have to operate for a significantly long time in order to make appreciable changes to the orbit. Hence, these transfers can no longer be modeled as impulsive, and the spacecraft dynamics under the action of the low thrust need to be considered in order to analyze the low-thrust spacecraft maneuver. The equation of motion of the spacecraft, modified from equation (8.6), is given by

$$\frac{d^2\mathbf{r}}{dt^2} = -\left(\frac{\mu}{r^3}\right)\mathbf{r} + \frac{\mathbf{\Gamma}}{m} \tag{8.37}$$

where $\mathbf{r}$ is the radius vector of the spacecraft, μ is the gravitational parameter, $\mathbf{\Gamma}$ is the thrust vector, and m is the mass of the spacecraft. If the thrust is zero, then the equation is that of the two-body problem given in equation (8.6), whose solution is a conic section (orbit). In the presence of a nonzero thrust, the additional acceleration term would be small compared to the local gravitational perturbation, because of the low thrust generated by the spacecraft thrusters. One way of modeling the low-thrust transfer is to consider that the low thrust only makes infinitesimal changes to the orbit instantaneously, and the cumulative action over a long time results in a considerable change in the orbit.

The analysis of low-thrust transfers is challenging and usually do not result in analytical relationships like in two-impulse transfers, except in the case of simpler transfers. For instance, in the case of a circle-to-circle transfer of a spacecraft whose orbital elements (referred to as osculating elements) vary slowly enough to permit the consideration of the osculating orbit remaining almost circular, simple relationships for the transfer time t_f, and the velocity change Δv can be derived:

$$t_f = \frac{m_0}{\Gamma}\left(\sqrt{\frac{\mu}{r_0}} - \sqrt{\frac{\mu}{r_f}}\right) \tag{8.38}$$

$$\Delta v = \sqrt{\frac{\mu}{r_0}} - \sqrt{\frac{\mu}{r_f}} \tag{8.39}$$

where r_0 and r_f are the initial and final orbit radius, m_0 is the initial mass of the spacecraft, and Γ is the magnitude of the engine thrust. These relationships are derived by considering the variation of the energy of the spacecraft during the transfer and equating it to the rate of work done by the onboard thruster (note that the energy otherwise stays constant in the absence of any thrusting). Other key considerations in the derivation of the above simplified relationships are that the thrust is tangential and the mass is assumed to be constant during the transfer; in the real case, the spacecraft's mass will decrease over time due to depletion of propellant. A detailed analysis of the quasicircular continuous thrust transfer can be found in Wiesel (2010) and in Prussing and Conway (2013). The velocity change for a continuous thrust transfer is greater than the corresponding Hohmann transfer because part of the thrust is used to overcome the gravitational force, thereby resulting in gravity losses.

Example of Electric Orbit-Raising

Let us consider the continuous thrust transfer of a 1,500-kg spacecraft from a 6,600-km radius circular earth orbit to the geosynchronous orbit of 42,164 km radius. We assume that the thrust provided by the electric propulsion system onboard the spacecraft is 0.3 N and the specific impulse is 1,800 s. Then, the velocity change required for the low-thrust transfer can be computed as $\Delta v = 7.7714 - 3.0747 = 4.6967$ km/s. The time required for the transfer is therefore given by $t_f = (1{,}500/0.3) \times 4.6967 \times 1{,}000 = 23{,}483{,}500$ s, which is a little more than 9 months. This is many times more than the transfer time of 18,945 s, which corresponds to a few hours that would result from a Hohmann transfer between the two circular orbits. Yet, an electric transfer does have an advantage over a traditional chemical transfer. To see this advantage, let us compute the final mass of the spacecraft at the geosynchronous orbit using the rocket equation. Using $\Delta v = 4.6967$ km/s, the spacecraft final mass deployed will be 1,150 kg if it performs the electric transfer, while its mass will be around 402 kg if it performs a Hohmann transfer (assuming a specific impulse of 300 s); the reason for the huge savings in the fuel expenditure is the superior propellant management of the electric propulsion system, as reflected by its specific impulse being several times higher than chemical rockets. Note that the consideration of constant mass actually results in a higher estimate of the transfer time; as the spacecraft becomes lighter by spending fuel, the spacecraft acceleration increases over time as the engine thrust remains constant. However, the transfer time is still in the order of months, and is one of the major challenges of its acceptability throughout the satellite industry.

8.3 Earth Orbiting Satellites

Introduction

The previous subsection covered the basic aspects of orbital maneuvers, e.g., transferring a satellite from an inclined LEO to GEO. This subsection focuses on additional issues which are relevant to earth orbiting satellites, such as the field of view, ground tracks, orbital rendezvous, orbit determination, special types of earth orbits, the effect of earth oblateness, and orbital perturbations.

Field of View

One critical piece of information with regard to mission specification is the issue of the satellite's field of view (FOV). The FOV determines the amount of area on the earth's surface, which is visible from the satellite. This information is critical for determining the number of satellites required to provide continuous global coverage in the case of a satellite constellation such as the NAVSTAR Global Positioning System located in medium earth orbit (MEO). From the perspective of a ground observer, the FOV determines the north and south latitude limits from which the observer can sight a satellite such as the INTELSAT communication satellite located in GEO.

The upper and lower limits of a satellite's FOV are defined by lines which are tangent to the earth's surface. If a satellite is located infinitely far away, then the lines are parallel and the FOV is 50% of the earth's surface area as shown in the left-hand illustration of Figure 8.17. If a satellite is a finite distance away, then the FOV is less than 50% of the earth's surface area as shown to the right of Figure 8.17. The northern and southern latitude limits of the FOV (i.e., the angle L in Figure 8.17) are given by the relationship (see Brown 1992; Hale 1994; Myose 2001):

$$\cos L = \frac{r_{\text{earth}}}{r} \tag{8.40}$$

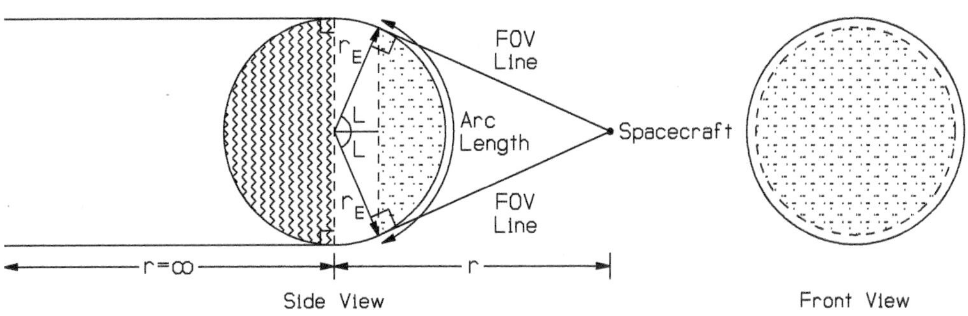

FIGURE 8.17 Field of view (adapted from Myose 2001).

where r is the orbital radius of the satellite. The FOV fraction, which is the fraction of the earth's surface area visible by the satellite, is given by (see Brown 1992; Hale 1994; Myose 2001):

$$\text{FOV fraction} = \frac{1}{2}\left(1 - \frac{r_{\text{earth}}}{r}\right) \tag{8.41}$$

The circumferential arc length visible by the satellite is given by (see Brown 1992; Myose 2001):

$$\text{FOV arc length} = 2Lr_{\text{earth}} \tag{8.42}$$

The FOV for a GEO satellite can easily be determined from equations (8.40) to (8.42). Since $r_{\text{earth}} = 6{,}378$ km and $r = r_{\text{GEO}} = 42{,}164$ km, the FOV latitude limits are $L = \pm 81.3°$ (i.e., 81.3° north to 81.3° south latitude), FOV fraction = 0.424 or 42.4%, and FOV arc length is 18,100 km. It should be noted that a minimum of three GEO satellites is required to obtain global coverage because the FOV fraction for each GEO satellite is 42.4%.

A theoretical latitude limit of 81.3° was obtained for the FOV of GEO satellites in the example above. In practice, however, the latitude limit of the FOV is less than this theoretical value because even a small vertical obstruction would get in the way of the line of sight for a ground observer located at 81.3° north or south latitude. This further reduction in the FOV is not a significant problem for most communication satellite applications because the major population centers are not located in the far north or south. For countries such as Russia, however, this is a problem because a large part of their land mass is located in the far north, especially the undeveloped areas, which have poor communication lines. Russian engineers devised an ingenious solution called the Molniya orbit, which involves a large eccentricity ($\varepsilon = 0.69$) and a high inclination angle ($i = 63.435°$). The Molniya orbit, shown in Figure 8.18, results in a very large hang time near its apogee.

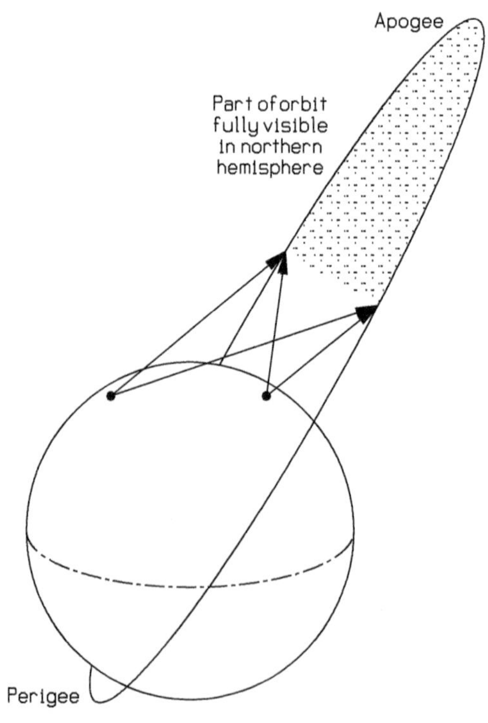

Figure 8.18 Molniya orbit (adapted from Myose 2001).

This allows the Molniya satellite to be visible in the northern hemisphere during 8 hours of its 12-hour orbital period. A constellation of three Molniya satellites therefore provides continuous coverage. A major drawback of the Molniya system, however, is that satellite tracking is required.

Ground Tracks

One distinct character of a GEO satellite is that it appears to be stationary to a ground observer, hovering over a fixed point above the equator. The main reason is because the orbital period is the same as the earth's rotation rate, and consequently the satellite moves in sync with the earth. If the satellite is located in the equatorial plane but at a radius which is less than (or greater than) 42,164 km, then it will no longer move in sync with the earth. In this case, the satellite would appear to fly over the equator. The flight path taken by the satellite traced onto the earth's surface is called the ground track.

When a satellite is located in an inclined orbit, it follows a flight path, as shown in Figure 8.19. The satellite starts at the ascending node labeled A1 for the first orbital pass, reaches its northernmost location, labeled C, crosses the equator again, but this time at the descending node, and so on. The resulting ground track for a circular LEO is a sinusoidal path as shown in Figure 8.20. Three

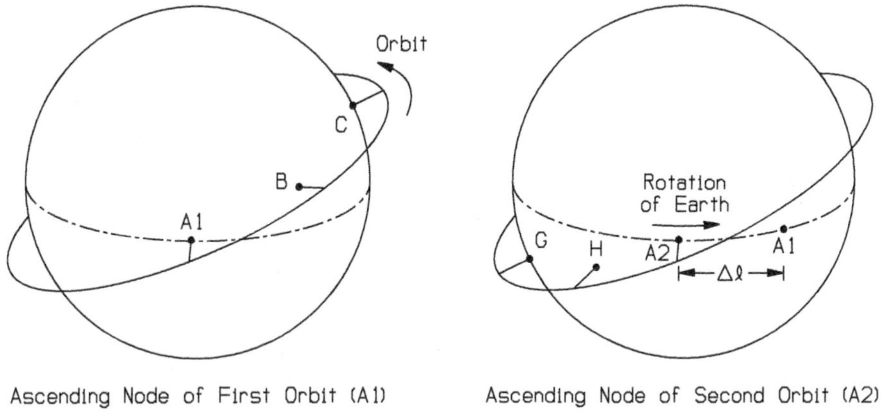

FIGURE 8.19 Ground track of a satellite in a prograde orbit (adapted from Myose 2001).

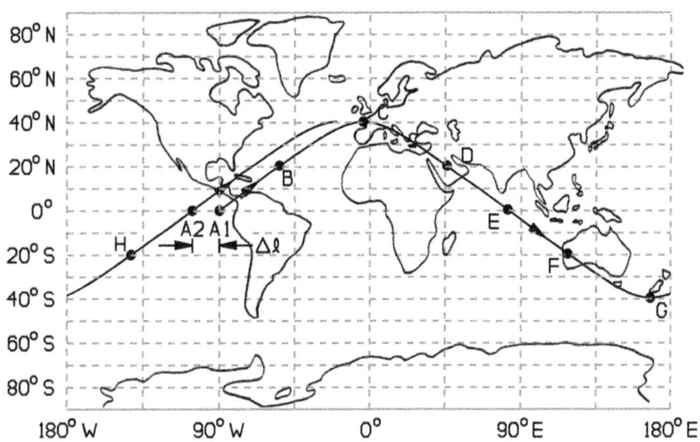

FIGURE 8.20 Ground track of a satellite in 40° inclination circular orbit (adapted from Myose 2001).

interesting pieces of information can be obtained by studying the satellite's ground track. First, the northernmost and southernmost latitudes correspond to the inclination of the orbit. Thus, the inclination of the orbit for the ground track shown in Figure 8.20 would be 40°. Second, the symmetry of the sinusoidal ground track pattern indicates that the orbit type is circular. If the orbit is noncircular, the resulting ground track lacks some form of symmetry in the sinusoidal pattern due to the difference in speed at perigee and apogee. See Myose (2001) and Hale (1994) for additional details on this issue. The third interesting piece of information which can be obtained from the ground tracks is the orbital period. Suppose the satellite's ground track is traced out starting from the beginning of the first orbital pass, which is labeled A1 in Figures 8.19 and 8.20. During the time it takes for the satellite to complete one orbit, the earth does not stand still but moves to the east. Since the satellite is left behind by the earth, the ground track appears to shift west, from the perspective of a ground observer, to the location labeled A2 in Figures 8.19 and 8.20. Subsequent orbits result in further westward shifts in the ground track. Since the earth's rotation rate is 360° in one sidereal day (which is 86,164 s), each longitudinal shift $\Delta\ell$ in the ground tracks is given by

$$\Delta\ell = t\left(\frac{360^\circ}{86{,}164}\right) \quad \text{or} \quad \Delta\ell = \frac{t}{239.34} \tag{8.43}$$

where the time t which is relevant to the present situation is the orbital period T. As an example, the longitudinal shift in the ground track for 6,563 km radius circular LEO is 22.11° (to the west) because its orbital period is 5,291.3 s based on equation (8.18).

Orbital Rendezvous

Satellites in geosynchronous orbits are assigned to specific orbital slots corresponding to particular longitude locations on the equator. Longitudes are defined toward the east or west starting from 0° longitude, corresponding to the meridian location of Greenwich, England. Longitude values range from 0° to 180° in both the eastward and westward direction, as shown in Figure 8.20. In order to distinguish east longitudes from west longitudes, we will define east longitudes as positive angle values and west longitudes as negative angle values. For example, Wichita State University in Kansas is located at 97.3° west longitude, so its longitude will correspond to −97.3°.

The geosynchronous satellite placement problem is illustrated in Figure 8.21. The goal is to place a satellite at a particular orbital slot; for example, a final destination corresponding to a longitude of −97.3°. Suppose the transfer orbit involves a true anomaly angle difference of $\theta_2 - \theta_1$ between high orbit intercept (at θ_2) and transfer initiation (at θ_1). For a Hohmann transfer, which involves a perigee to apogee half ellipse, the true anomaly angle difference is $\theta_2 - \theta_1 = 180°$, whereas it would be less than 180° for a fast transfer. An initial glance at the problem might suggest that the Hohmann transfer should be initiated (i.e., its perigee should be located) 180° west of (i.e., clockwise from) the final destination longitude. However, this ignores the rotation of the earth during the time it takes to complete the Hohmann transfer. Based on equation (8.43), the rotation of the earth $\Delta\ell$ during the TOF is given by

$$\Delta\ell = \frac{\text{TOF}}{239.34} \tag{8.44}$$

This means that the final destination longitude is situated differently depending on the phase of the transfer, i.e., the start or end of the transfer orbit, as shown in the right-hand illustration in Figure 8.21. The lead angle κ_l is the angle difference between the transfer initiation longitude and the final destination longitude at the time the Hohmann transfer is initiated at perigee. Based on Figure 8.21, the lead angle is given by

$$\kappa_l = (\theta_2 - \theta_1) - \Delta\ell \tag{8.45}$$

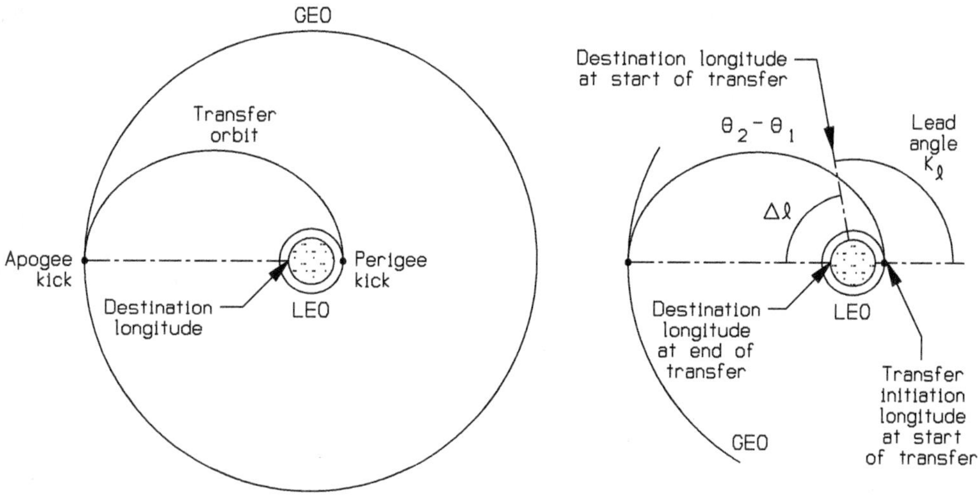

FIGURE 8.21 Transfer initiation for a geosynchronous satellite (adapted from Myose 2001).

where $(\theta_2 - \theta_1) = 180°$ for a Hohmann transfer. Since the required transfer initiation longitude is west of the desired final destination longitude by an angle κ_l, the lead angle value must be subtracted from the final destination longitude:

$$l_{\text{transfer initiation}} = l_{\text{final destination}} - \kappa_l \tag{8.46}$$

As an example, consider placing a satellite in GEO (with 42,164 km radius) at an orbital slot of 97.3° west longitude starting from a 6,563-km radius 28.5° inclination circular LEO. The semimajor axis length of the Hohmann transfer is $a = 24{,}363.5$ km based on equation (8.1) or (8.27). Using the orbital period relationship given by equation (8.18), $\text{TOF} = T/2 = 18{,}923$ s. Based on equation (8.44), the earth rotates $\Delta\ell = 79.06°$ during the Hohmann transfer. Since the true anomaly angle difference is 180° for a Hohmann transfer, the lead angle is given by equation (8.45) to be $\kappa_l = 180° - 79.06° = 100.94°$. Based on equation (8.46), the required transfer initiation longitude is $l_{\text{transfer initiation}} = -97.3° - 100.94° = -198.24°$. Since longitudes are only defined between ±180°, the required transfer initiation longitude is $l_{\text{transfer initiation}} = -198.24° + 360° = +161.76°$ longitude (where + means east).

The discussion in the orbital maneuvers subsection showed that the combined Hohmann transfer and inclination change maneuver must be performed at the node locations. The nodes for a low earth parking orbit depend on the orbital radius, and information about their locations is usually given by the organization providing the launch service. However, a rough idea about the process involved in determining the node locations can be illustrated by a simple example. Suppose a due east launch is performed from Cape Canaveral, which is located at 80.5° west longitude. The resulting orbit will have a 28.5° inclination, and the first crossing of the equator, which is the descending node, occurs one quarter of an orbit later, or 90° east of the launch site. However, there is a 22.11° westward shift per orbit due to earth rotation for a 6,563-km radius LEO. This means that there is a westward shift of about 22.11°/4 = 5.53° during this quarter orbit. Thus, the first descending node occurs at roughly −80.5° + 90° − 5.53° = +3.97° (east) longitude. The first ascending node will occur 180° to the east, less the shift due to earth rotation, which is 22.11°/2 = 11.06° for this half of the orbit. This means that the first ascending node occurs at 3.97° + 180° − 11.06° = +172.91° (east) longitude. Neither the descending nor the ascending node for this first orbit corresponds to the required launch initiation longitude. The ascending nodes for subsequent orbits are shifted 22.11° to the west. Thus, the ascending nodes for the second and subsequent orbits are 150.8°, 128.69°,

106.58°, 84.47°, 62.36°, 40.25°, 18.14°, −3.97°, and so on. A further refinement of this process requires knowledge of the actual rocket launch trajectory to determine the initial node location and the inclusion of earth oblateness effect, which is discussed in the later part of this subsection.

None of the orbits in the example above have ascending nodes corresponding to the required transfer initiation longitude. The closest longitude is the ascending node for the first orbit, i.e., 172.91°. This is 11.15° ahead (i.e., farther east) than the required transfer initiation longitude of 161.76°. If the transfer is initiated at the ascending node of the first orbit, the satellite will arrive 11.15° ahead of the desired final longitude. Thus, some type of longitude correction must be performed once the satellite arrives at the high orbit radius. This type of problem, shown in the left-hand illustration in Figure 8.22, is called the secondary intercept problem. Since the satellite is currently ahead of the desired longitude, an initial thought would be to slow down and allow the earth to catch up. However, this braking maneuver would cause the satellite to be placed into a secondary intercept ellipse whose orbital period is less than the GEO period of 86,164 s, which is synchronized with the earth's rotation. This will cause the satellite to increase its longitudinal error angle after one full orbit is completed, as shown in the middle illustration of Figure 8.22. Therefore, the correct secondary intercept maneuver is initially to increase the speed, which would place the satellite into an ellipse whose orbital period is greater than the GEO period. The required secondary intercept period and semimajor axis length are given by the following relationship:

$$2\pi\sqrt{\frac{a^3}{\mu_{\text{earth}}}} = T = 86{,}164\left(1+\frac{\Delta\ell_{\text{error}}}{360^\circ}\right) \tag{8.47}$$

where $\Delta\ell_{\text{error}}$ is positive if the satellite is ahead (i.e., east) of the desired final longitude and negative if it is behind (i.e., west) of the desired final longitude. The velocity increment required to inject the satellite into this secondary intercept ellipse is given by

$$\Delta v_1 = v_{\text{secondary intercept}} - v_{\text{CH}} \tag{8.48a}$$

$$v_{\text{secondary intercept}} = \sqrt{\frac{2\mu_{\text{earth}}}{r_{\text{CH}}} - \frac{\mu_{\text{earth}}}{a}} \tag{8.48b}$$

$$v_{\text{CH}} = \sqrt{\frac{\mu_{\text{earth}}}{r_{\text{CH}}}} \tag{8.48c}$$

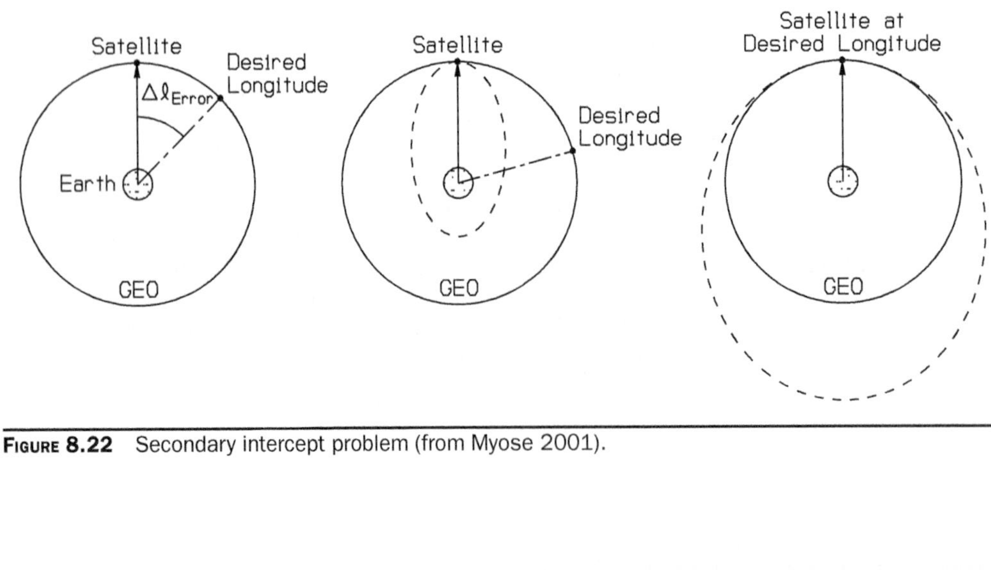

FIGURE 8.22 Secondary intercept problem (from Myose 2001).

where r_{CH}, the high circular orbit radius, is the 42,164 km for GEO. After one full orbit on the secondary intercept ellipse is completed, a second velocity increment which slows the speed down and places the satellite back onto GEO is required. This second maneuver is exactly the opposite of the first secondary intercept maneuver, so that

$$\Delta v_2 = -\Delta v_1 \tag{8.49}$$

Fuel expenditure is associated with any type of engine burn, whether it increases the velocity or decreases it. Thus, the total velocity increment associated with the initiation and completion of the secondary intercept is given by the sum of the absolute values of these velocity increments. That is,

$$\Delta v_{total} = |\ \Delta v_1\ | + |\ \Delta v_2\ | \tag{8.50}$$

As an example, consider the situation described above where the satellite arrives at the GEO radius 11.15° ahead of the desired longitude location. In this case, $\Delta\ell_{error} = +11.15°$ so the required secondary intercept orbital period and semimajor axis length based on equation (8.47) are $T = 88{,}833$ s and $a = 43{,}030.3$ km, respectively. Using equations (8.48)–(8.50), the required velocities are found to be $v_{SI} = 3.106$ km/s, $v_{CH} = 3.075$ km/s, $\Delta v_1 = 0.031$ km/s, $\Delta v_2 = -0.031$ km/s, and $\Delta v_{total} = 0.062$ km/s.

In practice, the longitudinal error between the actual and required transfer initiation longitude is dealt with in a slightly different manner. A GEO-bound satellite is placed in a Hohmann transfer orbit with an apogee slightly larger (or smaller) than the GEO radius (Blevis and Stoolman 1981). The subsequent apogee kick places the satellite into a high orbit with an orbital period, which is slightly different than the GEO period. The satellite is then allowed to drift until the final destination longitude is reached, at which point small velocity adjustments are made.

Orbit Determination and Orbital Elements

Delivering a GEO satellite to the assigned orbital slot requires a matchup between the required transfer initiation longitude and the actual node location for the low earth parking orbit. Otherwise, some sort of secondary intercept maneuver is required to correct the longitude error. Either method requires a precise determination of the satellite's orbital characteristics in the low earth parking orbit and in the high orbit before any type of maneuver is performed. A total of six parameters, called orbital elements, is required to define the characteristics of an orbit in three dimensions. The reason is because the governing equations of motion, the modified form of Newton's law given in vector form by equation (8.6), are three separate second-order differential equations. Since two constants of integration are required for each second-order differential equation, a total of six constants of integration or orbital elements are involved in three-dimensional orbits.

In the basic orbital mechanics discussion, two parameters were mentioned as the required elements for determining the characteristics of an orbit within the plane of the orbit. They can come in different forms (dynamic or geometric), but the most commonly used parameters are the semimajor axis length a and the eccentricity ε. One additional in-plane parameter determines the current position of the satellite. Typically, the time since perigee passage is used. However, the true anomaly angle θ will be used for the purposes of the discussion which follows. There is indeed a one-to-one correspondence between the true anomaly angle and the time since perigee passage, as can be seen from equation pairs (8.19) and (8.20) for elliptic orbits and (8.21) and (8.22) for hyperbolic orbits. However, equations (8.20) and (8.22) are transcendental, and no analytical solutions are available to determine the eccentric anomaly angle. Determining the anomaly angles from the time information therefore requires an iterative numerical technique.

Three other orbital elements besides the semimajor axis length, eccentricity, and true anomaly angle are required to define the characteristics of an orbit. The remaining three orbital elements

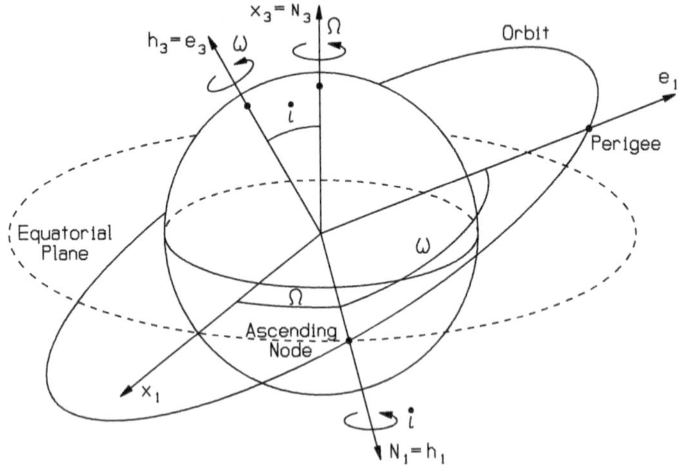

FIGURE 8.23 Orbital elements (from Myose 2001).

define the orientation of the orbit in three dimensions measured with respect to a three-axis geocentric (or earth-based) coordinate system. This earth-based coordinate system **X** illustrated in Figure 8.23 has its origin at the center of the earth. The X_3 axis is pointed in the North Pole direction and the X_1 axis is pointed toward the vernal equinox direction. The vernal equinox is the location where the sun crosses the equator on the first day of spring. Although not shown, the X_2 axis lying in the equatorial plane completes the geocentric coordinate system, and it is simply orthogonal to the other two axes according to the usual right-hand rule involving cross-products, where $X_2 = X_3 \times X_1$. The orbit-based coordinate system with its origin at the center of the earth has already been introduced in the orbital mechanics subsection. The e_1 axis corresponds to the perigee direction, the e_2 axis to the semilatus rectum direction, and the e_3 axis to the angular momentum direction.

A set of three angles called Euler rotation angles separates the geocentric and orbit-based axes. Euler rotation angles are involved whenever a coordinate system undergoes a series of rotations to obtain another coordinate system. Starting from the geocentric coordinate system **X**, the three rotations are (1) the right ascension of ascending node **Ω**, (2) the inclination angle i, and (3) the argument of perigee ω. The first rotation is made about the north pole direction X_3, toward the east by an angle **Ω**, until the one axis is pointed in the ascending node direction. This results in the formation of a new set of axes called the node crossing coordinate system **N** with N_1 pointed toward the ascending node, N_3 in the north pole direction, which makes it coincident with X_3, and N_2 completing the three-axis orthogonal coordinate system. The second rotation is made about the ascending node direction N_1, by the inclination angle i, until the three axis is pointed in the orbit's angular momentum direction. This results in the formation of another set of axes **h**, where h_1 points toward the ascending node and is therefore coincident with N_1, h_3 is pointed in the angular momentum direction, and h_2 completes the three-axis orthogonal coordinate system. The third and final rotation is made about the angular momentum direction h_3, in the direction of travel in the orbit by an angle ω, until the one axis is pointed in the perigee direction. The result of this final rotation is the orbit-based coordinate system **e**.

During a typical satellite mission, multiple sightings of the satellite are obtained through ground-based visual observations or radar measurements. These geocentric measurements must be converted into orbit-based parameters such as a, ε, θ, Ω, i, and ω. That is, a set of dynamical information such as the radius and velocity vectors at one point on the orbit must be converted into the geometric orbital elements. Given **r** and **v**, the orbital elements can be determined from the relationships which follow (see Bate et al. 1971; Hale 1994; Myose 2001). The semimajor axis length a can be determined from equation (8.7), the conservation of energy, which after some rearrangement is given by

$$a = \frac{1}{\left(\frac{2}{r} - \frac{v^2}{\mu}\right)} \tag{8.51}$$

The eccentricity can be determined from the magnitude of the eccentricity vector $\boldsymbol{\varepsilon}$. Substituting equation (8.11) into equation (8.14) and then applying the vector triple product rule, the eccentricity vector is given by

$$\underline{\boldsymbol{\varepsilon}} = \frac{1}{\mu}\left[\left(v^2 - \frac{\mu}{r}\right)\underline{\mathbf{r}} - (\underline{\mathbf{r}} \cdot \underline{\mathbf{v}})\underline{\mathbf{v}}\right] \tag{8.52}$$

The true anomaly angle is the angle between the perigee and the radius vector. Since the eccentricity vector $\boldsymbol{\varepsilon}$ is pointed in the perigee direction $\mathbf{e}_1$, the true anomaly angle can be found from:

$$\theta = \cos^{-1}\left(\frac{\underline{\boldsymbol{\varepsilon}} \cdot \underline{\mathbf{r}}}{\varepsilon\ \mathrm{r}}\right) \tag{8.53}$$

Since cosine is an even function, two solutions are possible. One possible technique for determining the correct true anomaly angle is to observe the sign of the dot product $\mathbf{r} \cdot \mathbf{v}$, which is the second parenthetical term in equation (8.52). If $\mathbf{r} \cdot \mathbf{v} > 0$, then the true anomaly angle is in the range $0 < \theta < 180°$. If $\mathbf{r} \cdot \mathbf{v} < 0$, then the true anomaly angle is in the range $180° < \theta < 360°$. The inclination angle lies between the angular momentum vector and the North Pole direction. Thus,

$$i = \cos^{-1}\left(\frac{\underline{\mathbf{H}} \cdot \underline{\mathbf{X}}_3}{H}\right) \quad \text{where } \underline{\mathbf{H}} = \underline{\mathbf{r}} \times \underline{\mathbf{v}} \tag{8.54}$$

Quadrant correction is not required in the case of the inclination angle, since it is only defined between 0° and 180°. Before the right ascension of ascending node and argument of perigee angles can be determined, the node crossing direction N_1 must first be determined. The node crossing direction must be perpendicular to both the angular momentum direction h_3 and the North Pole direction X_3 as indicated in Figure 8.23. Thus,

$$\underline{\mathbf{N}}_1 = \left(\frac{\underline{\mathbf{X}}_3 \times \underline{\mathbf{H}}}{H}\right) \Bigg/ \left\|\frac{\underline{\mathbf{X}}_3 \times \underline{\mathbf{H}}}{H}\right\| \tag{8.55}$$

It should be noted that $(\mathbf{X}_3 \times \mathbf{H})/H$ is not a unit vector, so that this resultant vector must be normalized by its magnitude in order to make the $\mathbf{N}_1$ axis direction given in equation (8.55) a unit vector. Based on the definition for the right ascension of ascending node and the argument of perigee,

$$\Omega = \cos^{-1}(\underline{\mathbf{X}}_1 \cdot \underline{\mathbf{N}}_1) \tag{8.56}$$

$$\omega = \cos^{-1}\left(\frac{\underline{\mathbf{N}}_1 \cdot \underline{\boldsymbol{\varepsilon}}}{\varepsilon}\right) \tag{8.57}$$

Quadrant correction may be required since cosine is an even function, the details of which are given in Myose (2001) and Bate et al. (1971).

Since large numerical values are involved in astrodynamics problems, measurement data used to determine the orbital elements are oftentimes normalized by units of measure called canonical units. For earth orbiting satellites, the normalization for distance is made with respect to the radius of the earth's surface, where 1 distance unit (DU) is 6,378 km. Also, the gravitational parameter for earth is used as a reference, i.e., 1 distance unit cubed divided by a time unit squared (DU^3/TU^2)

is 3.986×10^5 km^3/s^2, which means that a time unit (TU) is equivalent to 806.786 s. For interplanetary missions a different set of references is used to define the canonical units. In this case, the earth's orbital radius around the sun (1 DU = 1.496×10^8 km) and the gravitational parameter for the sun (1 DU^3/TU^2 = 1.327×10^{11} km^3/s^2) are used as the reference normalization values. Canonical units are used simply as a matter of tradition and convenience, just as the problems discussed thus far used kilometers rather than the SI standard of meters.

To illustrate the use of the relationships discussed above, consider the following example from Myose (2001). Problem statement: determine the six orbital elements given that the radius and velocity vectors written in terms of geocentric coordinates are

$$\mathbf{r} = (-0.73993\mathbf{X}_1 + 0.73993\mathbf{X}_2 + 0.60415\mathbf{X}_3)\ \text{DU}$$

$$\mathbf{v} = (-0.77264\mathbf{X}_1 - 0.64637\mathbf{X}_2 + 0.05155\mathbf{X}_3)\ \text{DU/TU}$$

Solution: the radius and velocity magnitudes are $r = 1.2083$ DU (i.e., $r = 1.2083 \times 6378 = 7706.5$ km) and $v = 1.0087$ DU/TU (i.e., $v = 1.0087 \times 6378/806.786 = 7.974$ km/s). Based on equation (8.51), the semimajor axis length is $a = 1.5679$ DU. Keeping in mind that $\mu = 1$ in canonical units, the first term in equation (8.52) is $(v^2 - \mu/r)\mathbf{r} = -0.14044\mathbf{X}_1 + 0.14045\mathbf{X}_2 + 0.11467\mathbf{X}_3$. The parenthetical part of the second term is $\mathbf{r} \cdot \mathbf{v} = +0.12458$, so that $(\mathbf{r} \cdot \mathbf{v})\,\mathbf{v} = -0.09625\mathbf{X}_1 - 0.08052\mathbf{X}_2 + 0.00642\mathbf{X}_3$. Finally, the eccentricity vector is $\boldsymbol{\varepsilon} = -0.04419\mathbf{X}_1 + 0.22097\mathbf{X}_2 + 0.10825\mathbf{X}_3$, which means that its magnitude is $\varepsilon = 0.25$. Since $\mathbf{r} \cdot \mathbf{v} > 0$, the true anomaly angle based on equation (8.53) is $\theta = 30°$ rather than 330°. From equation (8.11), $\mathbf{H} = \mathbf{r} \times \mathbf{v} = (0.42865\mathbf{X}_1 - 0.42865\mathbf{X}_2 + 1.04996\mathbf{X}_3)$ DU^2/TU, which means that its magnitude is $H = 1.21239$ DU^2/TU. The inclination angle $i = 30°$ based on equation (8.54). The node crossing direction, which is a unit vector, is given by $\mathbf{N}_1 = 0.70711\mathbf{X}_1 + 0.70711\mathbf{X}_2 + 0.0\mathbf{X}_3$ based on equation (8.55). Finally, the right ascension of ascending node and the argument of perigee from equations (8.56) and (8.57) are $\Omega = 45°$ and $\omega = 60°$, respectively.

In the discussion above, the radial position and velocity vectors in terms of geocentric coordinates were used as the starting basis for determining the orbital elements. Determining the orbital elements based on more rudimentary measurements such as position and time is beyond the scope of this discussion. In addition, there are reference coordinate systems besides the geocentric and orbit-based systems described above. The interested reader is referred to Bate et al. (1971), Chobotov (1991), Escobal (1976), and Vallado (2001).

Orbital Precession due to Earth Oblateness

The westward shift of the ground track discussed above is only an apparent shift in the ground track viewed from the perspective of the ground observer. On the other hand, the oblateness of the earth results in an actual precession or wobble in the orbit. The reason for this effect is illustrated in Figure 8.24 where the Earth bulges out at the equator by about 20 km, radially. Because of this bulging effect, the satellite experiences a nonradial gravitational force at the maximum and minimum latitude locations in its orbit. This nonradial force causes the angular momentum vector of the orbit to precess in a manner similar to the wobbling of an earth-based gyroscope. This precession motion causes the ascending node to shift physically westward in prograde orbits and eastward in retrograde orbits.

As discussed earlier, the right ascension of ascending node Ω is the angle measured from a reference axis on the equator (called the vernal equinox) to the ascending node. The rate $\dot{\Omega}$ at which the node regresses westward is given by the relationship (see Prussing and Conway 1993; Griffin and French 1991; Kaplan 1976):

$$\dot{\Omega} = \left[\frac{3}{2} J_2 \sqrt{\frac{\mu}{a^3}} \left(\frac{r_{\text{equator}}}{p}\right)^2 \cos i\right]\left(\frac{180}{\pi}\right) \tag{8.58}$$

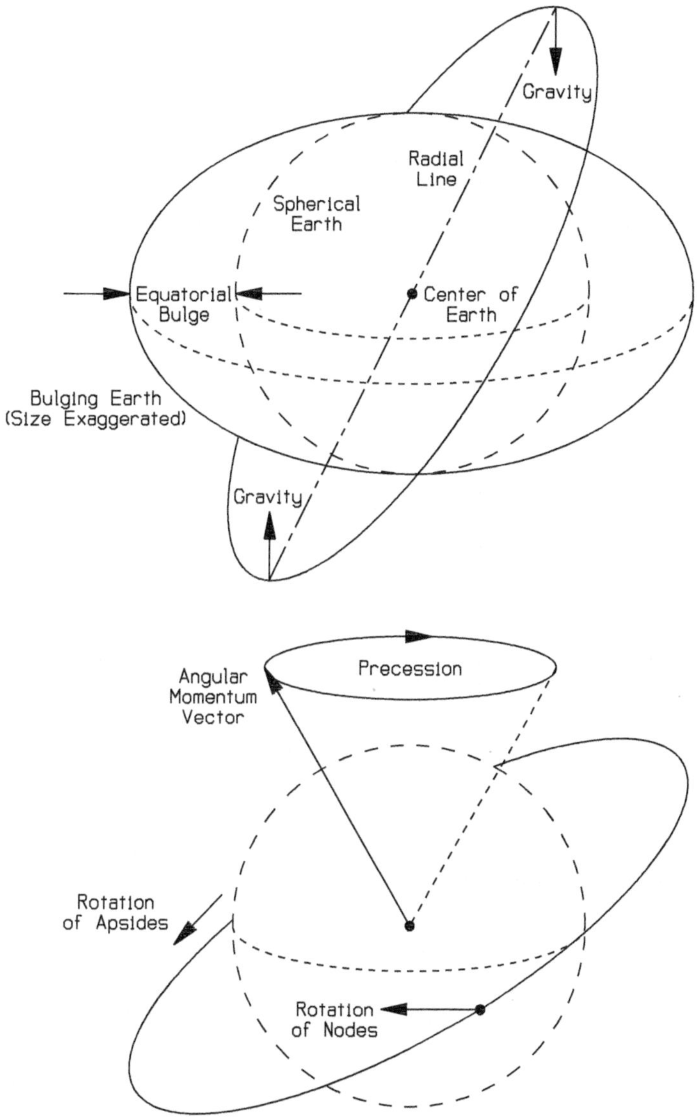

FIGURE 8.24 Earth oblateness effect (adapted from Myose 2001).

where $\dot{\Omega}$ is given in units of degrees per second after the $180/\pi$ conversion factor, $r_{equator}$ is the equatorial radius of the planet given in Table 8.1, and J_2, the zonal harmonic, for the earth is 1.0826×10^{-3}. Brown (1992) provides the zonal harmonic coefficients for many of the other planets in the solar system. Contrary to some references, a sign convention of positive for westward shift and negative for eastward shift is used in equation (8.58) for the sake of consistency with the earlier discussion on longitudinal shifts in the ground track.

As an example, consider a 6,563 km radius 28.5° inclination circular LEO. Since this is a circular earth orbit, the semimajor axis length and semilatus rectum are given by $a = p = 6{,}563$ km and the gravitational parameter from Table 8.1 is $\mu_{earth} = 3.986 \times 10^5$ km^3/s^2. Based on equation (8.58), the regression of nodes would then be $\dot{\Omega}$ = 9.170 × 10^{-5} deg/s. Since the orbital period is 5,291 s based on equation (8.18), the westward shift in the ascending node after one orbit would be $\Delta\Omega = \dot{\Omega}T = 0.49°$. This compares to the 22.11° westward shift in the ground tracks due to earth rotation, discussed earlier.

The effect of precession due to earth oblateness can be used as an advantage in a certain type of orbital application called sun synchronous orbits. Earth resource observation satellites such as the American Landsat or French Spot require comparable lighting conditions from one day to the next, and this requirement can be met using a sun synchronous orbit. The idea is to match the precessional rotation of the orbit with the earth's orbital motion around the sun. This means that the precession rate must be 360° in 365¼ days, which corresponds to the earth completing one orbit around the sun in 1 year. One important issue, however, is that this precession must be oriented toward the east, i.e., a negative regression of nodes. Thus, $\dot{\Omega} = -360°/365\frac{1}{4}$ days $= -0.9856$ deg/day or a regression of nodes of about 1° per day toward the east. This value for the regression of nodes can be obtained from a multitude of orbital radius and inclination angle pairs. The Landsat 4, for example, is in a 7,083 km radius 98.2° inclination circular LEO (Damon 1995).

For noncircular orbits, there is an additional earth oblateness effect called the rotation of apsides, where the perigee is shifted due to a physical rotation of the orbit within the orbital plane. As discussed earlier, the argument of perigee ω is the angle measured from the ascending node to the perigee. The rate $\dot{\omega}$ which the argument of perigee is rotated is given by the relationship (see Kaplan 1976; Griffin and French 1991; Prussing and Conway 1993):

$$\dot{\omega} = \left[\frac{3}{2} J_2 \sqrt{\frac{\mu}{a^3}} \left(\frac{r_{equator}}{p}\right)^2 (2 - 2.5\sin^2 i)\right]\left(\frac{180}{\pi}\right) \tag{8.59}$$

Here, positive values indicate a rotation of apsides in the direction of travel (i.e., same rotation direction as the orbit's angular momentum vector) while negative values indicate a rotation of apsides opposite to the direction of travel. The highly eccentric Molniya orbit described earlier is located at a unique inclination angle (i.e., i = 63.435°), where the parenthetical term (2 – 2.5 $\sin^2 i$) = 0. This means that the Molniya orbit does not entail any rotation of apsides. This ensures that the apogee is always located in the northern hemisphere, where the largest hang time (or coverage) is required.

Orbital Perturbations due to Other Effects

The orbit equation representing the dynamics of a spacecraft assumed a point mass for the central body, and ignored any nongravitational force and the gravitational forces due to other bodies. In reality, other bodies affect the motion of the spacecraft, and aerodynamic drag force may be encountered if the spacecraft travels within the atmosphere of the central body. For a spacecraft orbiting around the earth, the effect of the sun and the moon, nongravitational forces such as atmospheric drag (provided the perigee of the orbit is sufficiently close to the surface of the earth), and the (nonspherical) oblateness of the earth must be considered in order to truly describe the dynamics of the spacecraft. Figure 8.25 provides a schematic description of these effects, which are only secondary compared to the gravitational force due to the central body, and is therefore modeled as perturbations that cause the spacecraft to deviate from its orbit as predicted by the orbit equation. In the immediately preceding portion of this subsection, the effect of the earth's oblateness was already considered. We now consider some of the other important perturbations for a spacecraft in an LEO.

- *Atmospheric drag:* Owing to the presence of a dissipative force, the spacecraft loses energy; the loss of energy means a decrease in the semimajor axis length of the orbit. If the spacecraft is in a circular orbit, then small changes in the orbital radius (i.e., semimajor axis length) is given by (Prussing and Conway 2013):

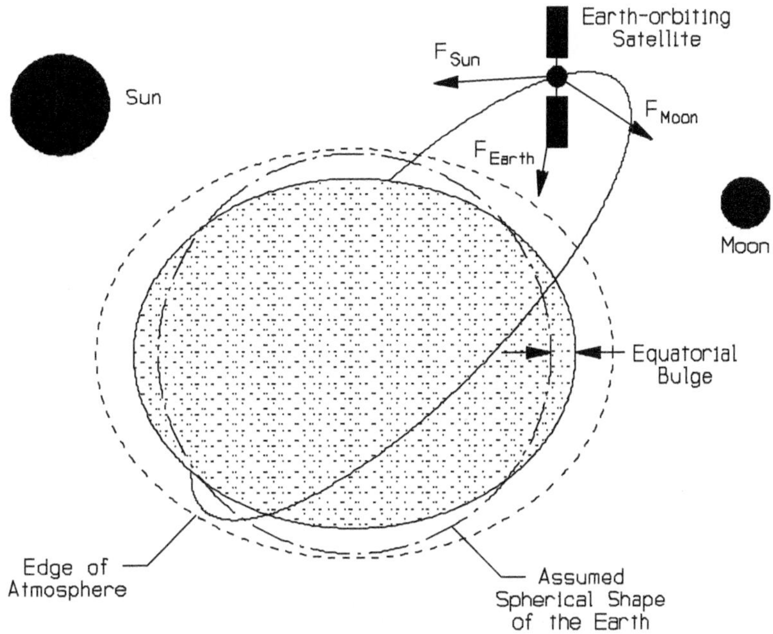

FIGURE 8.25 Perturbing forces due to third and fourth bodies on a spacecraft in an earth orbit, with the shaded shape demonstrating the oblateness of the earth (objects not to scale).

$$\sqrt{a_f} - \sqrt{a_i} = -\frac{1}{2}\sqrt{\mu}\rho(BC)\Delta t \tag{8.60}$$

where a_i is the initial semimajor axis length, a_f is the semimajor axis length after time Δt, μ is the gravitational parameter, ρ is the atmospheric density, and BC is the ballistic coefficient of the spacecraft which is given by $C_D A/m$ with C_D being the drag coefficient, A being the (frontal) surface area, and m being the spacecraft mass. If the spacecraft is in an elliptic orbit, the atmospheric drag near the perigee is more pronounced than at the apogee owing to the closer distance to the surface of the earth. Hence, the spacecraft is slowed down more at the perigee resulting in a lowering of the apogee of the spacecraft. Hence, the eccentricity of the orbit is reduced, meaning that the atmospheric drag results in circularizing the orbit of the spacecraft. With sufficient long exposure to the atmospheric drag, the spacecraft will deorbit resulting in an eventual reentry.

- *Perturbation due to the sun and the moon:* The perturbations caused by the lunar and solar forces of attraction capture the gravitational effects that are neglected in the consideration of only two bodies to determine the orbit equation. The attraction of a third body results in no secular change in the semimajor axis length in a geocentric circular orbit; however, the remaining orbital elements undergo secular variations. Details of these variations can be found in the work of Cook (1962).
- *Solar radiation pressure:* Electromagnetic radiation possesses momentum, and the reflection of the radiation off the spacecraft results in a change in momentum. This change in momentum generates a small perturbing force on the spacecraft directly proportional to the area-to-mass ratio of the spacecraft. In particular, the effect of solar radiation pressure is prominent for geosynchronous orbiting spacecraft having large solar arrays.

8.4 Interplanetary Missions

Introduction

The basic interplanetary mission is a truly complex problem involving multiple gravitational bodies. There is the earth where the spacecraft originates from, the sun during transit across the solar system, and the destination planet. The simple two-body problem consisting of a spacecraft and a gravitational body such as the earth is easy to solve, and its governing equations were discussed in Subsection 8.1. However, there is no solution available for the general *N*-body problem, which is associated with interplanetary missions. Instead of an exact solution, an approximate technique called the method of patched conics will be used. In this technique, the interplanetary mission is broken up into phases according to the primary gravitational body that influences the spacecraft. The first phase (I) is the hyperbolic earth departure, the second phase (II) is the heliocentric (i.e., sun-centered) transit across the solar system, and the third phase (III) is the hyperbolic arrival at the destination planet. These orbits or conic sections around different primary gravitational bodies are then patched together to complete the full interplanetary mission. Another complex problem associated with interplanetary missions is the issue of proper launch timing in order to intercept the destination planet, which is a moving target. This issue will be addressed in the final part of this subsection. Keeping these complexities in mind, it is indeed an accomplishment to visit other planets that are trillions of kilometers away.

For the purposes of this discussion, a number of simplifying assumptions will be made. First, a Hohmann transfer will be used for the heliocentric transit from the earth to the destination planet. Interested readers should see Myose (2001) and Hale (1994), which discuss the issue of fast heliocentric transfers. The second and third assumptions are as follows: planetary orbits around the sun are circular and planetary orbits are in the same orbital plane as the earth. With the exception of Mercury and Pluto, these are reasonable assumptions to make according to Table 8.2. Brown (1992) discusses the issue of non-coplanar interplanetary transfers as well as planetary ephemeris, which specify the planetary positions within their elliptic orbits around the sun.

Sphere of Influence and Method of Patched Conics

As discussed earlier, the interplanetary mission must be broken up into separate phases according to the primary gravitational bodies that influence the spacecraft. This means that the boundary, or, in the case of three dimensions, the sphere of influence for each gravitational body, must be determined. Since gravitational force is inversely proportional to the distance squared, spacecraft located far away from a planet would primarily be under the influence of the sun. On the other hand, spacecraft located close to a planet are influenced primarily by the planet. The sphere of influence (SOI) radius which defines this demarcation was determined by Laplace to be

$$r_{\text{SOI}} = R_{\text{planet}} \left(\frac{\mu_{\text{planet}}}{\mu_{\text{sun}}} \right)^{0.4} \tag{8.61}$$

where R_{planet} is the planet's orbital radius around the sun, which should not be confused with r_{planet}, the equatorial radius for the surface of the planet. For planets in circular orbit around the sun, as was assumed in this subsection, R_{planet} would correspond to the orbit's semimajor axis length. Table 8.5 lists the SOI radius for the various planets based upon this assumption. It should be noted that the SOI radii for Mercury and Pluto deviate from the average value much more than the other planets since their orbits are relatively eccentric.

A spacecraft located inside the SOI boundary, i.e., $r < r_{\text{SOI}}$, is viewed to be in orbit around the planet, and all values such as the gravitational parameter, radius, and velocity must be referenced with respect to the planet. On the other hand, a spacecraft located outside the SOI is viewed to be

Planet	Semi-major axis length (a) for heliocentric orbit (km)	μ (km^3/s^2)	r_{SOI} (km)	r_{planet}, equatorial radius of surface (km)	r_{SOI}/r_{planet}
Mercury	5.7910×10^7	2.2032×104	1.124×10^5	2.439×10^3	46.1
Venus	1.0820×10^8	3.2486×105	6.162×10^5	6.051×10^3	101.8
Earth	1.4960×10^8	3.9860×105	9.247×10^5	6.378×10^3	145.0
Mars	2.2794×10^8	4.2828×104	5.772×10^5	3.393×10^3	170.1
Jupiter	7.7833×10^8	1.2671×108	4.821×10^7	7.1492×10^4	674.3
Saturn	1.4270×10^9	3.7940×107	5.456×10^7	6.0268×10^4	905.3
Uranus	2.8710×10^9	5.7816×106	5.172×10^7	2.5559×10^4	2023.7
Neptune	4.4971×10^9	6.8713×106	8.681×10^7	2.4764×10^4	3505.6
Pluto	5.9135×10^9	1.0209×103	3.359×10^6	1.500×10^3	2239.7

TABLE 8.5 Sphere of Influence Radius for Planets in the Solar System

in orbit around the sun. In this case, the gravitational parameter, radius, and velocity must be referenced with respect to the sun. In order to distinguish between the different phases of an interplanetary mission, upper- and lowercase lettering will be used for radius and velocity. During phase II, radius and velocity values are measured with respect to the sun, and they will be denoted with uppercase lettering. For phases I and III, lowercase lettering will be used for radius and velocity measured with respect to the earth or the destination planet.

Whenever the SOI boundary is crossed, the reference system must be changed from the earth to the sun or from the sun to the destination planet. In each case, the gravitational parameter corresponding to the new gravitational body must be used. Since the SOI radius is several orders of magnitude less than the heliocentric orbital radius, as shown in Table 8.5; it is reasonable to assume that the radius at the start and end of phase II are given by R_{earth} and R_{planet}, respectively. Finally, the velocity after crossing over the SOI boundary can be determined from the following relationships:

$$\mathbf{V}_{T1} = \mathbf{v}_{\infty,earth} + \mathbf{V}_{earth} \tag{8.62}$$

$$\mathbf{v}_{\infty,in} = \mathbf{V}_{T2} - \mathbf{V}_{planet} \tag{8.63}$$

In equations (8.62) and (8.63), the subscripts $T1$ and $T2$ refer to the start and end of the phase II heliocentric transfer orbit, respectively. Subscripts "∞,earth" and "∞,in" refer to the hyperbolic excess velocity measured with respect to the earth and the destination planet in phases I and III, respectively. This is the spacecraft's velocity when it is located infinitely far away from the earth or the planet. In this subsection, planetary orbits around the sun were assumed to be circular. Thus, V_{earth} and V_{planet} can be determined from equation (8.9), which in the present situation can be written as

$$V_{earth} = \sqrt{\mu_{sun}/R_{earth}} \tag{8.64a}$$

$$V_{planet} = \sqrt{\mu_{sun}/R_{planet}} \tag{8.64b}$$

The phase II heliocentric transfer orbit velocities can be determined from the conservation of energy, equation (8.7) or (8.8), to be

$$V_{T1} = \sqrt{\frac{2\mu_{\text{sun}}}{R_{\text{earth}}} - \frac{\mu_{\text{sun}}}{a}} \tag{8.65a}$$

$$V_{T2} = \sqrt{\frac{2\mu_{\text{sun}}}{R_{\text{planet}}} - \frac{\mu_{\text{sun}}}{a}} \tag{8.65b}$$

Since the phase II heliocentric transfer orbit is a Hohmann transfer, the semimajor axis length is given by equation (8.1) or (8.27), which in the present case can be written as

$$a = \frac{R_{\text{earth}} + R_{\text{planet}}}{2} \tag{8.66}$$

Since R_{earth} corresponds to the perihelion in the present Hohmann transfer case, the eccentricity for the phase II heliocentric transfer orbit can be determined from the conic section relationship, equation (8.16), to be

$$R_{\text{earth}}(1+\varepsilon) = a(1-\varepsilon^2) \text{ or } R_{\text{earth}} = a(1-\varepsilon) \text{ or } \varepsilon = 1 - \frac{R_{\text{earth}}}{a} \tag{8.67}$$

Figure 8.26 graphically illustrates the change in reference frame for the case of a Hohmann transfer to an outer planet. In equations (8.62) and (8.63), all the velocities ($\mathbf{V}_{T1}$, $\mathbf{v}_{\infty,\text{earth}}$, $\mathbf{V}_{\text{earth}}$, $\mathbf{V}_{T2}$, $\mathbf{v}_{\infty,\text{in}}$, and $\mathbf{V}_{\text{planet}}$) are given in vector form. For the present case of a Hohmann transfer, however, the equations can be written in scalar form since the velocities are all co-linear and aligned with each other, as illustrated in Figure 8.26. Although all three phases are shown in the figure, the focus is the phase II

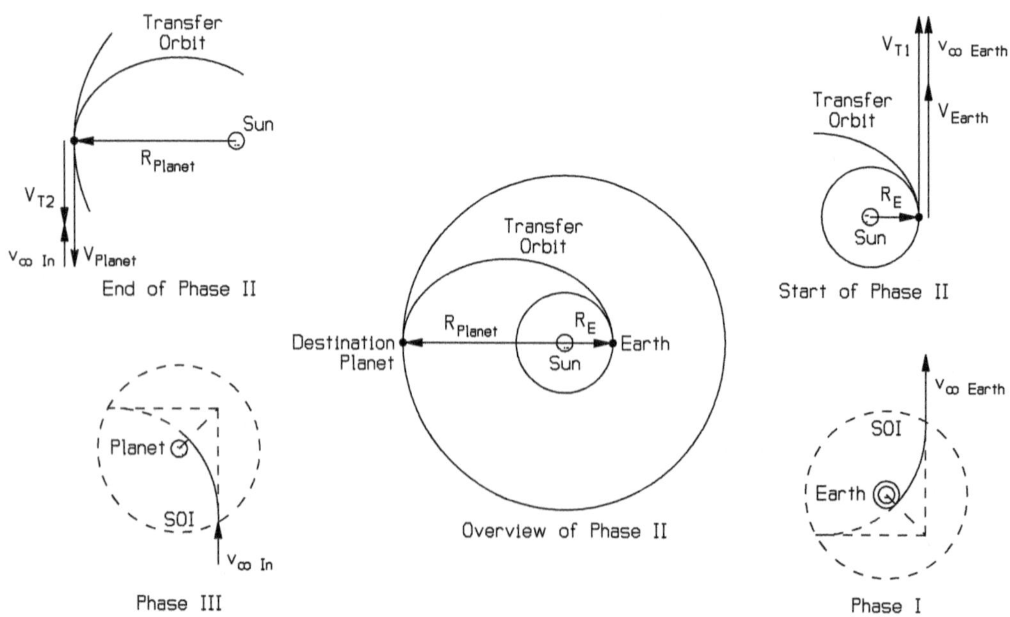

FIGURE 8.26 Interplanetary mission using the method of patched conics for a Hohmann transfer to an outer planet (from Myose 2001).

heliocentric portion. Thus, the phase II perihelion is located to the right of the origin (i.e., the sun) in the middle illustration just as the +X axis would normally be oriented toward the right. Consequently, the perigee and periapsis for phases I and III are skewed at an angle in the lower right and lower left SOI illustrations. Although the hyperbolic orbits within the SOI of the earth and the destination planet are shown in the figure, they will be discussed later in more detail. The focus of the following example, just as in the figure, will be on the phase II heliocentric transfer orbit, which is the first design step associated with an interplanetary mission.

As an example, consider sending a spacecraft from earth to Jupiter using a Hohmann transfer. In this case, the perihelion would correspond to $R_{\text{earth}} = 1.496 \times 10^8$ km and the aphelion would correspond to $R_{\text{Jupiter}} = 7.7833 \times 10^8$ km, according to Table 8.5. This means that the semimajor axis length for the phase II heliocentric transfer orbit is $a = 4.6397 \times 10^8$ km based on equation (8.66). Using equation (8.67), the eccentricity of the phase II heliocentric transfer orbit is determined to be $\varepsilon = 0.6776$, which is an ellipse as expected for a Hohmann transfer. The gravitational parameter for the sun is $\mu_{\text{sun}} = 1.3271 \times 10^{11}$ km^3/s^2, according to Table 8.1. The planetary velocities, based on equation (8.64), are $V_{\text{earth}} = 29.784$ km/s and $V_{\text{Jupiter}} = 13.058$ km/s. Based on equation (8.65), the heliocentric velocities at the start and end of phase II are $V_{T1} = 38.577$ km/s and $V_{T2} = 7.415$ km/s, respectively. The hyperbolic excess velocity required at the end of phase I, based on equation (8.62), is $v_{\infty,\text{earth}} = 8.793$ km/s. Finally, the hyperbolic excess velocity of the spacecraft upon arrival at Jupiter's SOI is $v_{\infty,\text{in}} = -5.643$ km/s based on equation (8.63). Note that the minus sign simply indicates that the incoming velocity in phase III (i.e., $v_{\infty,\text{in}}$) is oriented in a direction opposite to V_{Jupiter}, as shown in Figure 8.26. For a Hohmann transfer to an outer planet such as Jupiter, the hyperbolic excess velocity $v_{\infty,\text{earth}}$ at earth departure is oriented in the same direction as V_{earth}, while the hyperbolic excess velocity $v_{\infty,\text{in}}$ upon arrival at the destination planet is oriented opposite to V_{planet}. The reverse is true for a Hohmann transfer to an inner planet such as Mercury or Venus. That is, $v_{\infty,\text{earth}}$ is opposite to V_{earth} while $v_{\infty,\text{in}}$ is in the same direction as V_{planet} upon arrival at an inner planet.

Earth Departure

When an interplanetary mission is planned, the phase II part of the mission is analyzed first in order to determine the required phase I hyperbolic excess velocity. Once this is found, the earth departure condition can be determined from basic orbital mechanics relationships. In a typical mission, the spacecraft is initially placed in a circular low earth parking orbit by a rocket launch vehicle. Then, the spacecraft is injected into the earth departure hyperbolic orbit at the perigee of the hyperbola as illustrated in Figure 8.27. The orientation chosen for the figure is such that the perigee of the hyperbola is toward the right of the origin, which is the center of the earth. The radius and velocity at this injection point will be denoted with a subscript i rather than with a subscript p in order to avoid confusion with the perihelion of the phase II transfer orbit or with the planet's heliocentric orbital radius and velocity. An assumption will be made that the spacecraft's radial distance at the SOI boundary is large enough so that its contribution to the potential energy is negligible, i.e., $\mu_{\text{earth}}/r_{\text{SOI}} << v_\infty^2/2$ contributes only a small error, on order of about 1%. The required injection velocity v_i at the perigee of the earth departure hyperbola is then a much simpler relationship that can be determined from the conservation of energy, equation (8.7), to be

$$\frac{1}{2}v_i^2 - \frac{\mu_{\text{earth}}}{r_i} = U = \frac{1}{2}v_\infty^2 - \frac{\mu_{\text{earth}}}{r_{\text{SOI}}} \approx \frac{1}{2}v_\infty^2 \tag{8.68a}$$

$$v_i = \sqrt{\frac{2\mu_{\text{earth}}}{r_i} + v_\infty^2} \tag{8.68b}$$

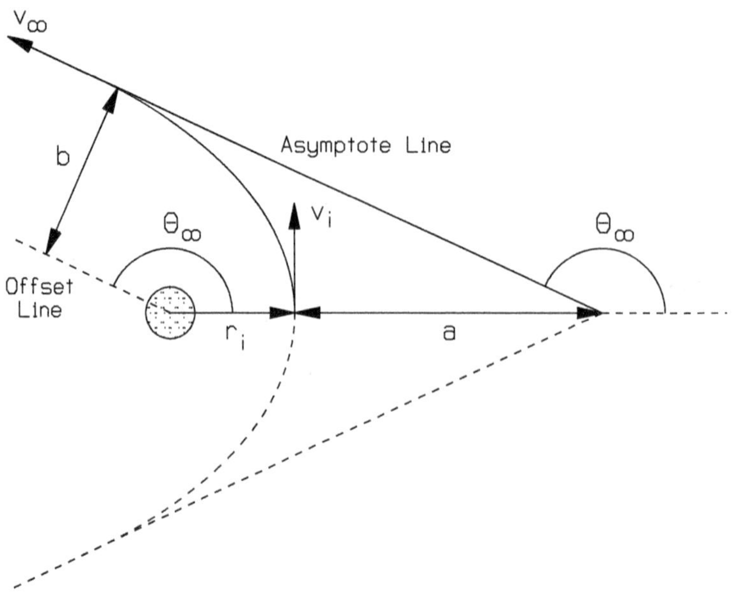

FIGURE 8.27 Phase I hyperbolic earth departure condition (adapted from Myose 2001).

For a circular orbit, the velocity vector is perpendicular to the radius vector. Since the spacecraft is injected into the earth departure hyperbola starting from a circular parking orbit, the injection velocity v_i is in the same direction as the initial circular parking orbit velocity v_C. This means that the velocity increment Δv would simply be the difference between these two velocities. That is,

$$\Delta v = v_i - v_C \tag{8.69a}$$

$$v_C = \sqrt{\frac{\mu_{\text{earth}}}{r_i}} \tag{8.69b}$$

Since the total energy is directly related to the semimajor axis length, the earth departure hyperbola's semimajor axis length is found to be

$$\frac{1}{2}v_\infty^2 = U = -\frac{\mu_{\text{earth}}}{2a} \quad \text{or} \quad a = -\frac{\mu_{\text{earth}}}{v_\infty^2} \tag{8.70}$$

The remaining characteristic of the earth departure hyperbola, the eccentricity ε, can be found from the conic section relationship, equation (8.16), to be

$$r_i(1+\varepsilon) = a(1-\varepsilon^2) \quad \text{or} \quad r_i = a(1-\varepsilon) \quad \text{or} \quad \varepsilon = 1 - \frac{r_i}{a} \tag{8.71}$$

The equation for the asymptote angle, equation (8.4), was given earlier in the basic orbital mechanics subsection, and is presented here once again:

$$\theta_\infty = \cos^{-1}(-1/\varepsilon) \tag{8.72}$$

The offset distance b, which is also the semiminor axis length, is given by $b^2 = a^2(1 - \varepsilon^2)$. Substituting equations (8.71) and (8.72) into this semiminor axis length relationship results in three additional relationships for the offset distance (see Brown 1992; Myose 2001):

$$b = \sqrt{a^2(\varepsilon^2 - 1)} = r_i\sqrt{\frac{\varepsilon+1}{\varepsilon-1}} = r_i\sqrt{\frac{-2a}{r_i}+1} = r_i\sqrt{\frac{2\mu_{earth}}{r_i v_\infty^{\;2}}+1} \tag{8.73}$$

For a Hohmann transfer, the hyperbolic excess velocity $v_{\infty,earth}$ must be oriented in the same direction as V_{earth} in the case of transfer to an outer planet and opposite to V_{earth} in the case of transfer to an inner planet. Thus, V_{earth} must be oriented along the offset line that is shown as a dashed line in Figure 8.27 in the case of a Hohmann transfer to an outer planet.

As an example of the earth departure condition, consider the Earth-to-Jupiter Hohmann transfer discussed earlier. The required hyperbolic excess velocity at the end of phase I was found to be $v_{\infty,earth} = 8.793$ km/s. Suppose a spacecraft is injected into this departure hyperbola starting from a circular low earth parking orbit of radius $r_i = 6{,}563$ km. Based on equation (8.68), the required injection velocity is $v_i = 14.099$ km/s. Since the circular parking orbit velocity is $v_C = 7.793$ km/s from equation (8.69b), the velocity increment required to inject the spacecraft into the earth departure hyperbola, based on equation (8.69a), is $\Delta v = 6.306$ km/s. This is a substantial Δv requirement in light of the fact that the original rocket booster raised the velocity from zero on the ground to a value of $v_C = 7.793$ km/s for the parking orbit. Based on equations (8.70) and (8.71), the characteristics of the earth departure hyperbola are semimajor axis length of $a = -5155.4$ km and eccentricity of $\varepsilon = 2.273$, i.e., $a < 0$ and $\varepsilon > 1$, as expected for a hyperbolic orbit. It should be noted that the semimajor axis length and eccentricity for phase I are quite different from those for phase II that were found in the previous example. Care should therefore be taken to avoid confusing the two sets of values in light of the fact that the same notation (a and ε) is used. The asymptote angle of $\theta_\infty = 116.1°$, found using equation (8.72), specifies the injection location measured relative to the earth's velocity vector V_{earth}. Based on equation (8.73), the offset distance is $b = 10{,}523$ km. This value is several orders of magnitude smaller than the earth's orbital radius around the sun (i.e., $R_{earth} = 1.496 \times 10^8$ km). Since $b << R_{earth}$, the heliocentric radius at the start of phase II being equal to R_{earth} is a reasonable assumption to make for the method of patched conics.

Planetary Arrival

When the spacecraft reaches the SOI boundary of the destination planet, the spacecraft can possibly arrive at four different quadrants. For the Hohmann transfer case to an outer planet, this is illustrated in Figure 8.28. The orientation chosen for the figure is consistent with Figure 8.26, where the phase II perihelion is located to the right of the origin, which is the sun. If the phase II aphelion radius is less than R_{planet}, then the spacecraft arrives at the sunny side of the planet's SOI as shown in the two lower middle illustrations. If the phase II aphelion radius is larger than R_{planet}, then the spacecraft arrives at the dark side of the planet's SOI (i.e., the side farther away from the sun) as shown in the lower left and lower right illustrations. Furthermore, the spacecraft's arrival at the planet's SOI can be timed so that it is ahead of the planet, as shown in the two lower left illustrations, or behind the planet, as shown in the two lower right illustrations. It is evident from the two lower-right illustrations that it is not possible for the spacecraft to penetrate the planet's SOI from behind, because the spacecraft moves away from the planet. Therefore, arrival in front of the planet is required for SOI penetration in the case of a Hohmann transfer to an outer planet. The situation is reversed in the case of a Hohmann transfer to an inner planet whereby arrival behind the planet is required for SOI penetration. The interested reader should see Myose (2001) for the penetration and nonpenetration quadrants of a fast transfer, which are more complicated.

Another issue that must be considered is whether the spacecraft impacts the planet or successfully enters an orbit around the planet. If the spacecraft's phase II aphelion radius is exactly equal

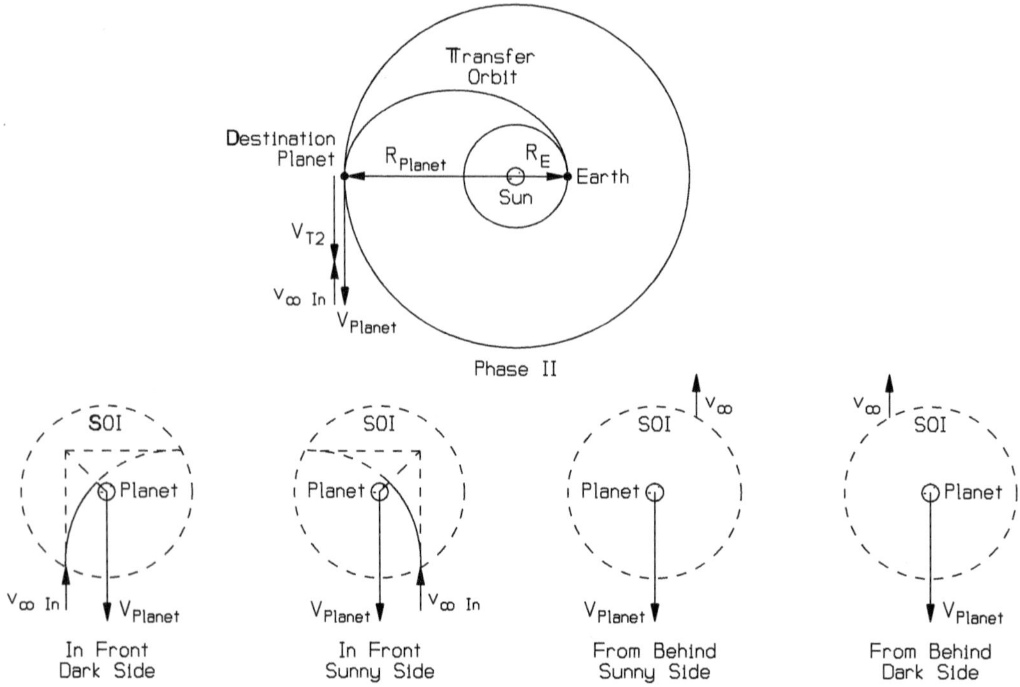

FIGURE 8.28 Encounter with an outer planet after a Hohmann transfer (from Myose 2001).

to R_{planet}, then the spacecraft will be headed directly into the planet and it will obviously impact the planet. Due to the geometry associated with hyperbolic orbits, offset distances much larger than the planet's surface radius will still result in planetary impact. Figure 8.29 illustrates the situation where the spacecraft's periapsis just touches the planet's surface. The orientation chosen for the figure is such that the periapsis of the arrival hyperbola is toward the right of the origin, which is the center of the destination planet. Using the appropriate gravitation parameter and substituting the planet's surface radius r_{planet} for the periapsis radius r_i into equation (8.73) results in an impact distance d of

$$d = r_{planet}\sqrt{\frac{2\mu_{planet}}{r_{planet}v_{\infty}^{2}} + 1} \tag{8.74}$$

If the offset distance b for the arrival hyperbola is less than or equal to the impact distance d (i.e., $b \leq d$), then the spacecraft will impact the planet.

Finally, if a closed orbit such as a circular parking orbit around the planet is to be established, then a braking maneuver must be performed at the periapsis of the arrival hyperbola. The process is essentially the reverse of the earth departure procedure outlined in equations (8.68)–(8.73), with the exception that the appropriate gravitational parameter for the planet must be substituted for μ_{earth}. It should be noted, however, that the arrival hyperbola's characteristics, such as the periapsis radius r_i and the offset distance b, are the desired target values. If the actual offset distance b is different from the desired value, then a midcourse correction must be performed (see Brown 1992).

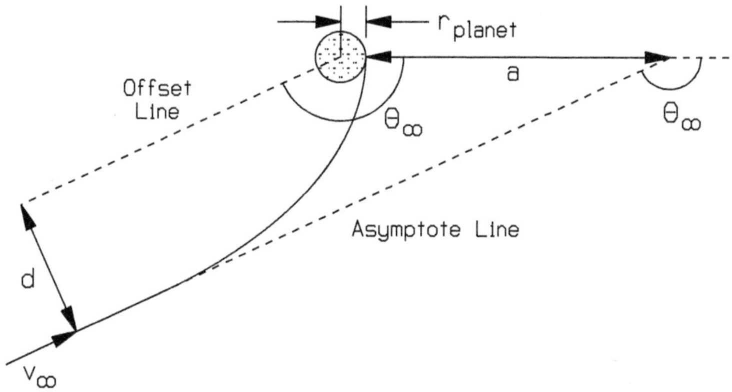

FIGURE 8.29 Impact distance (adapted from Myose 2001).

Gravity Assist Maneuver

As discussed earlier, it is possible for a spacecraft to penetrate the SOI after a Hohmann transfer to an outer planet by arriving in front of the planet on either the sunny side or the dark side. The resulting flyby of the planet is illustrated in Figure 8.30 for the sunny-side case and Figure 8.31 for the dark-side case. The orientation chosen for Figure 8.30 is such that the periapsis of the arrival hyperbola is toward the right of the origin, which is the center of the destination planet. On the other hand, the orientation chosen for Figure 8.31 is such that V_{planet} is oriented in the same direction as V_{planet} in Figure 8.30. The orientation chosen for both of these figures is different from the orientation of the same situations shown in the two lower-left illustrations of Figure 8.28. Notice that the hyperbolic orbit resulting from a sunny-side penetration shown in Figure 8.30 results in a counterclockwise flyby. On the other hand, the hyperbolic orbit resulting from a dark-side penetration shown in Figure 8.31 results in a clockwise flyby. Both of these orbits, however, are trailing-side

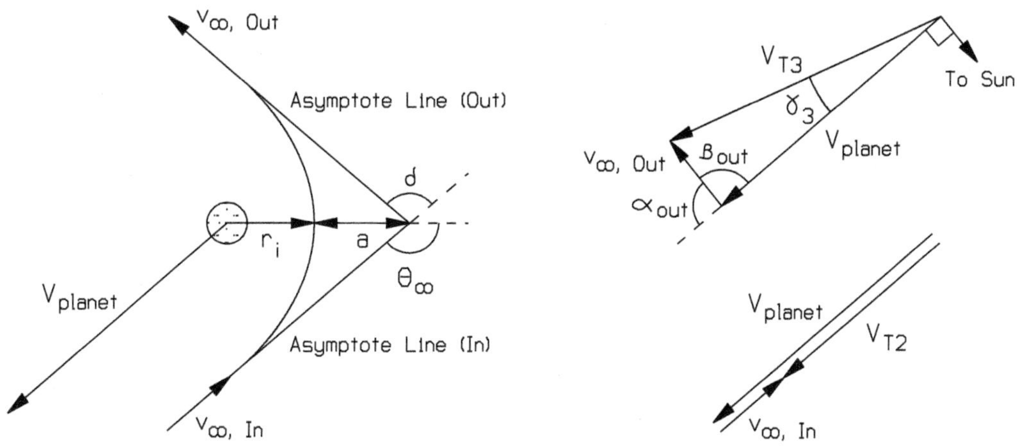

FIGURE 8.30 Counterclockwise planetary flyby after a Hohmann transfer to an outer planet (adapted from Myose 2001).

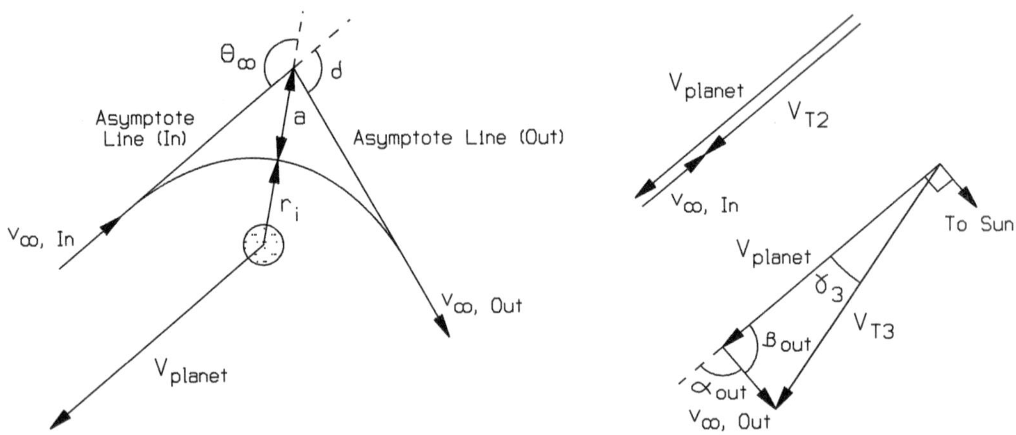

FIGURE 8.31 Clockwise planetary flyby after a Hohmann transfer to an outer planet (adapted from Myose 2001).

flybys because the orbital path never cuts across V_{planet}. In general, trailing-side flybys result in an increase in the spacecraft's heliocentric velocity, i.e., its velocity after exiting the planet's SOI increases. This behavior of velocity increase can be explained from two different physical perspectives. One perspective is the gravity assist view, where gravity acts to "pull" the spacecraft as it flybys the rear of the planet. This "pulling" motion is in the same direction as the spacecraft's heliocentric travel direction, and therefore increases the spacecraft's heliocentric velocity. Another perspective is the "slingshot" view, where the hyperbolic excess velocity's direction is changed. In this view, $v_{\infty,in}$ opposes the heliocentric travel direction (i.e., opposite to V_{planet}) when it penetrates the SOI, but it is turned by the time it departs so that $v_{\infty,out}$ is less opposed. In comparison to trailing-edge flybys, described above, leading-side flybys cut across V_{planet}, resulting in a reduction in the spacecraft's heliocentric velocity. Such a situation can occur from a planetary flyby after a Hohmann transfer to an inner planet or after a fast transfer. The interested reader should see Myose (2001) and Hale (1994), which address the Hohmann transfer to an inner planet case and the fast transfer case.

A number of different angles associated with the hyperbolic flyby and the heliocentric velocities need to be defined in order to explain the change in velocity resulting from a planetary flyby. From a "slingshot" perspective, the important parameter that must be determined is the amount of turning accomplished by the hyperbolic flyby. As can be seen from Figures 8.30 and 8.31, this turning angle δ is related to the asymptote angle according to

$$2\theta_{\infty} = \delta + 180° \tag{8.75}$$

Substituting equation (8.72) and then using a trigonometric identity results in

$$\delta = 2\sin^{-1}\left(\frac{1}{\varepsilon}\right) \tag{8.76}$$

The flight path angle γ is defined to be the angle between the velocity of interest and the horizon line, where the horizon line is defined to be perpendicular to the radius vector. Normally, the radius vector is drawn from the center of the gravitational body of interest. In the case of the heliocentric velocity V_T, it is evident that flight path angle γ should be measured with respect to the sun. In the case of the hyperbolic excess velocity v_∞, the appropriate gravitational body is not so clear.

The normal perspective would be that v_∞ is the hyperbolic excess velocity measured with respect to the destination planet (or the earth when appropriate). Another perspective, however, is that v_∞ is the difference between the heliocentric velocity V_T and the planet's orbital velocity around the sun V_{planet}. In this case, the appropriate gravitational body would be the sun, and the horizon line should be drawn with respect to the sun. In order to avoid confusion, α will be used to denote the heliocentric flight path angle for the hyperbolic excess velocity as shown in Figures 8.30 and 8.31. Finally, β will be defined as the supplementary angle to α, so that

$$\alpha + \beta = 180° \tag{8.77}$$

In the case of a Hohmann transfer to an outer planet, $v_{\infty,\text{in}}$ opposes V_{earth} so that $\alpha_{\text{in}} = 180°$, which means that $\beta_{\text{in}} = 0°$. The heliocentric flight path angle α_{out} associated with the outgoing hyperbolic excess velocity $v_{\infty,\text{out}}$ will depend upon the type of flyby. For a counterclockwise flyby,

$$\alpha_{\text{out}} = \alpha_{\text{in}} - \delta \tag{8.78}$$

For a clockwise flyby,

$$\alpha_{\text{out}} = \alpha_{\text{in}} + \delta \tag{8.79}$$

where flight path angles greater than 180° correspond to the situation where the velocity lies below the horizon line. Thus, the range of values for α (as well as β) should be limited to ±180°. Once the outgoing heliocentric flight path angle α_{out} is determined within this range, the supplement angle β_{out} can be found using equation (8.77). Since no engine burns are performed during the flyby, the outgoing hyperbolic excess velocity $v_{\infty,\text{out}}$ is exactly the same magnitude as the incoming hyperbolic excess velocity $v_{\infty,\text{in}}$. Only the direction has been changed. Once the spacecraft completes the flyby and exits the planet's SOI, the spacecraft is in a new conic section patch, i.e., phase IV, which is a heliocentric orbit. The new heliocentric velocity V_{T3} and its flight path angle γ_{T3} corresponding to the start of phase IV can be determined from the cosine and sine laws as follows:

$$V_{T3} = \sqrt{v_\infty^{\ 2} + V_{\text{planet}}^{\ 2} - 2v_\infty V_{\text{planet}} \cos\beta_{\text{out}}} \tag{8.80}$$

$$\gamma_{T3} = \sin^{-1}\left(\frac{v_\infty}{V_{T3}} \sin\beta_{\text{out}}\right) \tag{8.81}$$

As an example, consider a Jupiter flyby where the periapsis of the hyperbolic orbit is r_i = 72,000 km. Note that this periapsis radius is greater than Jupiter's surface radius of 71,492 km listed in Table 8.5. In an earlier example, the hyperbolic excess velocity at Jupiter arrival and Jupiter's orbital velocity were found to be $v_{\infty,\text{in}} = 5.643$ km/s and $V_{\text{Jupiter}} = 13.058$ km/s, respectively. Although the equations outlined in the earth departure condition will be used, the appropriate gravitational parameter, from Table 8.5, is $\mu_{\text{Jupiter}} = 1.2671 \times 10^8$ km³/s² in this case rather than μ_{earth}, which is written in the equations. Substituting μ_{Jupiter}, $v_{\infty,\text{in}}$, and r_i into equations (8.70)–(8.73) results in $a = -3.979 \times 10^6$ km, $\varepsilon = 1.018$, $\theta_\infty = 169.2°$, and $b = 7.6 \times 10^5$ km. These values are the characteristics required for the arrival hyperbola in order to obtain a periapsis of 72,000 km. It should be noted that the offset distance of $b = 7.6 \times 10^5$ km is well within Jupiter's SOI radius of 4.821×10^7 km and is substantially smaller than Jupiter's orbital radius around the sun of $R_{\text{Jupiter}} = 7.7833 \times 10^8$ km given in Table 8.5. Substituting $\varepsilon = 1.018$ into equation (8.76) or $\theta_\infty = 169.2°$ into equation (8.75) results in a turning angle of δ = 158.4°. This is a very large amount of turning, which leads to a substantial increase in the heliocentric velocity, as will be shown below. As discussed earlier, α_{in} = 180° and $\beta_{\text{in}} = 0$ in the case of a Hohmann transfer to an outer planet. If the SOI penetration occurs in front on the sunny side (i.e., $R_{T2} = R_{\text{Jupiter}} - b$) as shown in the lower middle-left illustration of Figure 8.28, then a counterclockwise flyby will result, and $\alpha_{out} = 21.6°$ according to equation (8.78). In this case, β_{out} would then be 158.4° based on equation (8.77). If the SOI penetration occurs in front on the dark side (i.e., $R_{T2} = R_{Jupiter} + b$) as shown in the lower-far left illustration in Figure 8.28, then a clockwise flyby will result, and $\alpha_{out} = 338.4° = -21.6°$, according to equation (8.78). In this case, the outgoing supplement angle would be $\beta_{out} = 201.6° = -158.4°$ based on equation (8.77). Since cosine is an even function, the heliocentric velocity V_{T3} at the start of phase IV does not depend upon the sign of the outgoing supplement angle β_{out}. This means that the heliocentric velocity V_{T3} = 18.422 km/s for both counterclockwise and clockwise flybys based on equation (8.80). This compares to a heliocentric velocity of V_{T2} = 7.415 km/s that the spacecraft had before entering Jupiter's SOI. Based on equation (8.81), the heliocentric flight path angle at the start of phase IV is $\gamma_{T3} = 6.475°$ for counterclockwise flyby and $\gamma_{T3} = -6.475°$ for clockwise flyby. A positive flight path angle indicates that the velocity vector is oriented above the horizon line and the spacecraft therefore moves farther away from the sun. On the other hand, a negative flight path angle indicates that the velocity vector is oriented below the horizon line and the spacecraft moves closer to the sun.

Once the heliocentric velocity and flight path angle at the start of phase IV are found, the characteristics of this phase IV heliocentric orbit can be determined. The appropriate gravitational parameter is $\mu_{sun} = 1.3271 \times 10^{11}$ km^3/s^2, and the heliocentric radius at the start of phase IV, i.e., upon exiting Jupiter's SOI, is $R_{Jupiter} = 7.7833 \times 10^8$ km. The angular momentum and total energy based on equations (8.12) and (8.7) are

$$H = R_{Jupiter} V_{T3} \cos \gamma_{T3} = 1.425 \times 10^{10} \text{ km}^2/\text{s}$$

$$U = \frac{1}{2} V_{T3}^{\ 2} - (\mu_{Sun}/R_{Jupiter}) = -0.8210 \text{ km}^2/\text{s}^2 = -\mu_{Sun}/(2a) \text{ or } a = 8.082 \times 10^{10} \text{ km}$$

Using the conic section relationship given by equation (8.16), the semilatus rectum, eccentricity, and perihelion for the phase IV heliocentric orbit are found to be $p = 1.529 \times 10^9$ km, $\varepsilon = 0.9905$, and $R_{perihelion} = 7.68 \times 10^8$ km. It should be noted that the Jupiter flyby has increased the heliocentric eccentricity from a moderately eccentric one at $\varepsilon = 0.6776$ for phase II to a highly eccentric one with $\varepsilon = 0.9905$ for phase IV. Based on equation (8.1), the aphelion for the phase IV heliocentric orbit is $R_{aphelion} = 1.609 \times 10^{11}$ km, which is well beyond the average radial position of Pluto since its semi-major axis length is 5.9135×10^9 km according to Table 8.5. Both the counterclockwise flyby and the clockwise flyby result in the same overall orbital characteristics for phase IV. The only difference between the two is that they are located at different points on the orbit. Using the orbit equation, equation (8.3) or (8.15), the true anomaly angle upon exiting Jupiter's SOI is found to be $\theta_{T3} = 13°$ for the counterclockwise flyby, whereas $\theta_{T3} = -13° = 347°$ for the clockwise flyby. That is, the spacecraft is headed toward the aphelion after a counterclockwise flyby, whereas it would be headed toward the perihelion after a clockwise flyby.

Intercept Problem

In the previous example, the heliocentric orbit was substantially energized by a Jupiter flyby to the point that the phase IV heliocentric orbit is nearly parabolic with ε = 0.9905 and its aphelion is beyond the outer edges of the solar system. It would not be possible to obtain a significantly more energetic phase IV heliocentric orbit from a single Jupiter flyby alone, since the hyperbolic flyby's periapsis is already close to Jupiter's surface. If multiple flybys are used, however, it is possible to increase the orbital energy and then completely escape the solar system. This was accomplished by the Voyager spacecraft through a series of gravity assists from Jupiter, Saturn, Uranus, and Neptune. The planetary alignment required to obtain such a sequence of planetary flybys is very rare, and the technical issue of timing the spacecraft launch is extremely complex. Although only the most basic case will be considered, this timing issue is the subject of the final part of this subsection.

The key issue associated with the intercept problem is the difference in orbital period between the spacecraft and the planets. Kepler's third law, given by equation (8.18), shows that the orbital period is proportional to the semimajor axis length (raised to the 3/2 power). This means that the smaller the orbital radius, the faster an object travels in its orbit. If the spacecraft is delivered to an outer planet using a Hohmann transfer, the earth is located in the lowest orbit since $R_{earth} < a_{Hohmann\ transfer} < R_{planet}$. This means that it takes less time for the earth to complete one lap in its orbit, i.e., traveling 360° of true anomaly compared to the spacecraft in the Hohmann transfer or the destination planet. To view this from a different perspective, suppose the TOF is fixed to the time it takes for the spacecraft to travel from the earth to the outer planet. During this time, the spacecraft travels 180° of true anomaly in the Hohmann transfer. In comparison, the destination planet travels less than 180°, say $\theta_{planet} = 180° - \kappa_1$, because it moves at the slowest rate while the earth travels more than 180°, say $\theta_{earth} = 180° + \kappa_2$, because it moves at the fastest rate. The lead angle κ_1 is the angle difference between the earth and the destination planet at the time the spacecraft is launched from the earth. The arrival difference angle κ_2 is the angle difference between the earth and the destination planet at the time the spacecraft arrives at the destination planet. These angle differences are shown graphically in Figure 8.32.

The equations associated with the intercept problem are simply an extension of the TOF relationships discussed in Subsection 8.1. If the semimajor axis length of the Hohmann transfer orbit is known, then its TOF can be determined from equation (8.18) because the TOF for a Hohmann transfer is one-half its orbital period. For the earth and the destination planet, which are in circular orbits, the angular velocity is a constant and is given by the relationship:

$$n_{earth} = \sqrt{\frac{\mu_{sun}}{(R_{earth})^3}} \tag{8.82a}$$

$$n_{planet} = \sqrt{\frac{\mu_{sun}}{(R_{planet})^3}} \tag{8.82b}$$

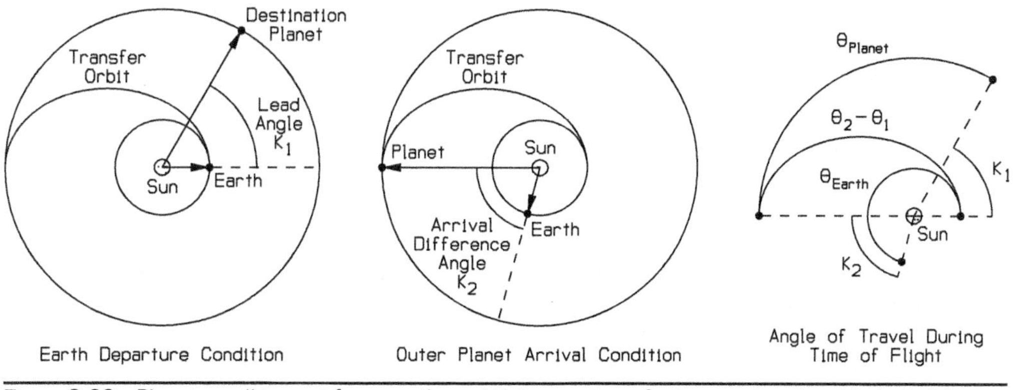

FIGURE 8.32 Planetary alignment for an outbound Hohmann transfer to an outer planet (adapted from Myose 2001).

The angle traveled by the earth and the destination planet, given in units of radians, would then be the product of the angular velocity n and the TOF. Consequently, the lead angle κ_1 and the arrival difference angle κ_2 would be given by

$$\kappa_1 = (\theta_2 - \theta_1) - \theta_{\text{planet}} = (\theta_2 - \theta_1) - n_{\text{planet}} \cdot \text{TOF} \tag{8.83a}$$

$$\kappa_2 = \theta_{\text{earth}} - (\theta_2 - \theta_1) = n_{\text{earth}} \cdot \text{TOF} - (\theta_2 - \theta_1) \tag{8.83b}$$

where $\theta_2 - \theta_1 = 180° = \pi$ in the case of a Hohmann transfer orbit.

The fact that the earth has traveled ahead of the destination planet poses a problem for a return-type mission. If a Hohmann transfer is to be used in the return mission, then the earth must be located behind the outer planet. Thus, proper timing associated with a return mission is a more complicated problem involving some waiting time. The interested reader should see Myose (2001), Hale (1994), Brown (1992), and Wiesel (1997) for further details.

References

Bate, R. R., Mueller, D. D., and White, J. E. 1971. *Fundamentals of Astrodynamics*, Dover, New York.

Beatty, J. K. and Chaikin, A. 1990. *The New Solar System*, Cambridge University Press, Cambridge.

Blevis, Z. O. and Stoolman, L., eds. 1981. *Hughes Aircraft Company Geosynchronous Spacecraft Case Histories*, Hughes Aircraft Company, El Segundo, CA.

Brown, C. D. 1992. *Spacecraft Mission Design*, American Institute of Aeronautics and Astronautics, Washington, DC.

Chobotov, V. A., ed. 1991. *Orbital Mechanics*, American Institute of Aeronautics and Astronautics, Washington, DC.

Cook, G. E., 1962. "Luni-Solar Perturbations of the Orbit of an Earth Satellite." *The Geophysical Journal of the Astronomical Society*. vol. 6, no. 3, pp. 271-291.

Damon, T. D. 1995. *Introduction to Space*, Krieger Publishing, Malabar, FL.

Escobal, P. R. 1976. *Methods of Orbit Determination*, Krieger Publishing, Malabar, FL.

Griffin, M. D. and French, J. R. 1991. *Space Vehicle Design*, American Institute of Aeronautics and Astronautics, Washington, DC.

Hale, F. J. 1994. *Introduction to Space Flight*, Prentice-Hall, Englewood Cliffs, NJ.

Isakowitz, S. J. 1995. *International Reference Guide to Space Launch Systems*, American Institute of Aeronautics and Astronautics, Washington, DC.

Kaplan, M. H. 1976. *Modern Spacecraft Dynamics and Control*, John Wiley & Sons, New York.

Myose, R. Y. 2001. *Fundamentals of Astronautical Engineering*, Aerospace Engineering 415 Class Notes Version 0.6, Copyright 2001, Department of Aerospace Engineering, Wichita State University, Wichita, KS.

Prussing, J. E. and Conway, B. A. 1993, 2013. *Orbital Mechanics*, Oxford University Press, Oxford.

Vallado, D. A. 2001. *Fundamentals of Astrodynamics and Applications*, Microcosm Press, Malabar, FL.

Wiesel, W. E. 1997. *Spaceflight Dynamics*, McGraw-Hill, New York.

Wiesel, W. E. 2010. *Spaceflight Dynamics*, McGraw-Hill, New York.

SECTION 9

Rockets and Launch Vehicles[1]

Jerry Jon Sellers and Michael J. Sekerak

Rockets take us where we need to go in space. They form the core of the propulsion systems found on everything from fireworks to Space Shuttles to the Starship *Enterprise*. Propulsion systems:

- Get us into space
- Move us around once we get there
- Change our attitude (the direction we're pointing)

Once a spacecraft is delivered to space by a launch vehicle, its propulsion subsystem provides the ΔV needed to take it to the final mission orbit and then make orbital corrections and maneuvers throughout the mission lifetime.

Propulsion is also essential for attitude control. These thrusters are either the sole method of attitude control (like the Shuttle's reaction-control system) or they complement the primary system for large slewing maneuvers or to provide a means for momentum dumping. In this chapter, we will peel back the mysteries of rocket science to see how rockets work and how propulsion systems are put together for both spacecraft and launch vehicles.

9.1 Rocket Science

A rocket is simply a system that takes mass plus energy and converts them into a force to move a vehicle. The input mass for a rocket is generally called *propellant*. The force produced by a rocket is called *thrust*.

Our examination of rocket systems begins by looking at the output—thrust. This requires us to dust off Newton's laws to see how high-speed exhaust going in one direction pushes a vehicle in another. Next, we will see how this thrust, over time, produces a velocity change for the vehicle, and—most important for mission planning—how to calculate this effect and ensure we have enough propellant to get our vehicle where we want it to go. We will then turn our attention to the process at the heart of a rocket: how it converts mass plus energy into the high-speed exhaust. We will tie all these concepts together by looking at the simplest example of a rocket—cold-gas thrusters—to see how varying some of the inputs and design variables changes their thrust and overall system efficiency.

[1]Adapted from *Understanding Space: An Introduction to Astronautics* by Dr. Jerry Jon Sellers (Teaching Science and Technology Inc.) with contributions by Dr. Michael J. Sekerak (NASA).

Thrust

A rocket ejects mass at high speed in one direction so a vehicle can go in the other. The simplest example of this is a balloon. All of us have blown up a toy balloon and let go of the stem to watch it fly wildly around the room. What makes the balloon go? Recall Newton's third law: *For every action there is an equal but opposite reaction.*

When you blow into a balloon, you force air into it, making the rubber skin tighten, increasing the internal air pressure, and storing mechanical energy like a spring. When you let go of the stem, the air pressure has an escape route, so the skin releases, forcing the air out under pressure. Following Newton's third law, as the air, which has mass, is forced out in one direction (the action), an equal force pushes the balloon in the opposite direction (the reaction).

Let's look at this action/reaction situation in a bit more detail to see where the force comes from. Consider an astronaut perched in a wagon armed with a load of rocks. If he is initially at rest and begins to throw the rocks in one direction, because of Newton's third law, an equal but opposite force will move him (and the wagon load of rocks) in the opposite direction.

To throw the rocks, the astronaut has to apply a force to them. This force is identical in magnitude, but opposite in direction, to the force applied to the astronaut and thus, the wagon. However, remember the concept of conservation of linear momentum. It tells us the change in speed of the rock (because it has less mass) will be greater than the change in speed of the wagon.

We will say the rocks leave the wagon at a rate called the *mass flow rate*, $\Delta m/\Delta t = \dot{m}$ (kg/s). Recall that linear momentum is mass times velocity. If the exit speed of the rocks is V_{exit}, the ejected mass has a momentum of $\dot{m}V_{exit}$. But remember, momentum is always conserved! So as the momentum of the rocks goes in one direction, the momentum of the wagon goes in the other direction. This is the basic principle that produces rocket thrust. A rocket expends energy to eject mass out one end at high velocity, pushing it (and the attached vehicle) in the opposite direction.

Dropping the vector notation to look at magnitudes only, we can say

$$\dot{P}_{rocket} = \dot{P}_{exhaust} \tag{9.1}$$

$$\dot{P}_{rocket} = \dot{m}V_{exit}$$

where $\dot{P}_{rocket}$ = time rate of change of the rocket's momentum (N)
$\dot{P}_{exhaust}$ = time rate of change of the exhausted mass momentum (N)
$\dot{m}$ = mass flow rate of the exhaust products (kg/s)
V_{exit} = exit velocity of the exhaust (m/s)

Notice this momentum change has the same units as force. This is the force on the rocket we defined to be the thrust. As we will see later, depending on the type of rocket used, the effective thrust delivered may be slightly different than $\dot{m}V_{exit}$. For this reason, we define a more comprehensive term called *effective exhaust velocity*, C, so we can express the rocket thrust as:

$$F_{thrust} = \dot{m}C \tag{9.2a}$$

where F_{thrust} = rocket's total thrust (N)
C = effective exhaust velocity (m/s)
$\dot{m}$ = mass flow rate (kg/s)

This relationship should make sense from our astronaut example. He can increase the thrust on the wagon either by increasing the rate at which he throws the rocks (higher $\dot{m}$) or by throwing the rocks faster (higher C). Or he can do both. For example, if he were throwing bowling balls he could achieve high $\dot{m}$ but with lower velocity than if he were throwing small pebbles.

Of course, exhaust velocities for typical rockets are much, much higher than anyone can achieve by throwing rocks. For typical chemical rockets, like the ones used by the Space Shuttle, the exhaust velocity can be as high as 3 km/s. Because these high velocities are hard to visualize, it is useful to think about the raw power involved in a rocket engine. Recall that kinetic energy is:

$$KE = 1/2\, mV^2$$

We define *power* as energy expended per unit time. Thus, the power in the jet exhaust of a rocket is

$$P_J = 1/2\, \dot{m}C^2 \tag{9.2b}$$

where P_J = Jet power in a rocket ($J/s = W$).

At lift-off, the Space Shuttle's three main engines plus its solid rocket boosters produce 26.6 billion watts of power. That is equivalent to over 13 Hoover Dams! We will see the effect of all that power next.

The Rocket Equation

To understand better how we use the thrust produced by rockets to get a vehicle where we want it to go, we must first introduce a new concept—impulse. Impulse will help us understand the total velocity change rockets deliver.

Impulse

So a rocket produces thrust that pushes on a vehicle. Then what happens? If you push on a door, it opens. If you hit a ball with a bat, it flies out to left field. Returning to our astronaut in the wagon, realize that to give the rocks their velocity, he has to apply a force to them over some length of time. Force applied to an object over time produces an *impulse*. Dropping vector notation, we express this law as

$$F = \frac{\Delta p}{\Delta t}$$

If we multiply both sides by Δt, we get

$$F\Delta t = \Delta p$$

The left side of this equation represents a force, F, such as the force we would generate if we hit a baseball with a bat, over some time, Δt. When our bat hits that fastball, it seems like the impact is instantaneous, but the bat actually stays in contact with the ball for a fraction of a second, applying its force to the ball during that time. The result of that force acting over time is seen on the right side of the equation, where Δp represents the resulting change in momentum.

From this relationship, realize that to change momentum you can either apply a large force acting over a short time (like a bat hitting a ball) or a smaller force acting over a longer time (like an ant slowly moving a bread crumb). We define total impulse, I, to be force times time, or change in momentum

$$I \equiv F\Delta t = \Delta p \tag{9.3}$$

where I = total impulse (N/s)
F = force (N)
Δt = time (s)
Δp = momentum change (N/s)

Impulse works the same way for rockets as it does for baseballs. We want to change the velocity and hence the momentum of our rocket, so we must apply some impulse. This impulse comes from the rocket thrust acting over some time interval. But as we saw, we can produce the same impulse from a rocket by applying a small thrust over a long time or a large thrust over a short time.

Although total impulse is useful for telling us the total effect of rocket thrust, it does not give us much insight into the rocket's efficiency. To compare the performance of different types of rockets, we need a new parameter we call *specific impulse*. Specific impulse, I_{sp}, is the ratio of the total impulse to the propellant weight required to produce that impulse (how much "bang" you get for your buck).

$$I_{sp} \equiv \frac{1}{\Delta W} \tag{9.4}$$

where I_{sp} = specific impulse (s)
I = total impulse (N/s)
ΔW = change in propellant weight (N) (using weight rather than mass to calculate I_{sp} is simply a convention established by the founders of rocket science and, as a result, I_{sp} has the unusual units of seconds). Substituting for total impulse

$$I_{sp} = \frac{F_{thrust}\Delta t}{\Delta W}$$

$$I_{sp} = \frac{F_{thrust}}{\dot{W}} \tag{9.5}$$

where I_{sp} = specific impulse (s)
F_{thrust} = force of thrust (N)
$\dot{W}$ = propellant weight flow rate (N/s)

I_{sp} represents the ratio of what we receive (momentum change) to what we spend (propellant). So the higher the I_{sp}, the more efficient the rocket. To get a more useful expression for I_{sp}, we replace the weight flow rate with the mass flow rate, $\dot{m}$, times the gravitational acceleration at sea level (a constant).

$$I_{sp} = \frac{F_{thrust}}{\dot{m}g_o} \tag{9.6}$$

where $\dot{m}$ = mass flow rate
g_o = gravitational acceleration constant = 9.81 m/s^2 (sea level)

Earlier, we found the force of thrust in terms of the mass flow rate and the effective exhaust velocity. By substituting equation (9.2) into equation (9.6), we get another useful expression for I_{sp}.

$$I_{sp} = \frac{C}{g_o} \tag{9.7}$$

where C = effective exhaust velocity (m/s).

Notice g_o is a constant value representing the acceleration due to gravity at sea level. This means no matter where we go in the universe, we humans will use the same value of g_o to measure engine performance.

As a measure of engine performance, I_{sp} is the same as the miles per gallon (mpg) rating given for cars. The higher the I_{sp} is for a rocket, the more ΔV it will deliver for a given mass of propellant. Another way to think about I_{sp} is that the faster a propulsion system can expel propellant, the more efficient it is.

Realize that we express I_{sp} in terms of propellant mass. For some space missions, especially those involving small satellites, conserving volume can be just as important, or even more important, than conserving mass. Therefore, comparing only the I_{sp} of two systems may not tell us the whole story. For this reason, we define another useful term called *density-specific impulse*, I_{dsp}, which we find by multiplying the rocket's I_{sp} times the average density of the propellants, δ_{av}.

$$I_{dsp} = \delta_{av} I_{sp} \tag{9.8}$$

where δ_{av} = average density of propellants (kg/m^3).

By comparing the mass and volume between different system options, mission planners can do more realistic trade-offs.

Velocity Change

When you take a long trip in your car, you have to make sure you will have enough gas in the tank to get there. This concern is equally important for a trip into space. But how do we determine how much propellant, we need for a given mission?

Naturally, some rockets are more efficient than others. For example, one rocket may need 100 kg of propellant to change velocity by 100 m/s while another may need only 50 kg. To figure out how much propellant we need for a given trip, we need a relationship between the velocity change and the amount of propellant used. We find this relationship by setting thrust equal to momentum change.

$$F_{thrust} = \dot{m}c = \frac{\Delta p_{rocket}}{\Delta t}$$

where
F_{thrust} = effective thrust from the rocket (N)
$\dot{m}$ = propellant mass flow rate (kg/s)
c = effective exhaust velocity (m/s)
$\frac{\Delta p_{rocket}}{\Delta t}$ = the rocket's rate of change of momentum (N)

From this relationship we can derive the *ideal rocket equation*. It tells us how much propellant we need for a given ΔV.

$$\Delta V = C \ln\left(\frac{m_{initial}}{m_{final}}\right) \tag{9.9}$$

where
ΔV = velocity change (m/s)
C = effective exhaust velocity (m/s)
ln = natural logarithm of the quantity in the parenthesis
$m_{initial}$ = vehicle's initial mass, before firing the engine (kg)
m_{final} = vehicle's final mass, after firing the engine (kg)

Equation (9.9) is one of the most useful relationships of rocket propulsion. Armed with this equation, we can determine how much propellant we need to do anything, from stopping the spin of a spacecraft in orbit, to launching a probe to another solar system. Notice that we are taking the natural logarithm of the ratio of initial to final mass. *The difference between initial and final mass represents the amount of propellant used.* ΔV is also a function of the effective exhaust velocity. This relationship should make sense because, as the propellant leaves the rocket faster, momentum changes more, and the rocket goes faster.

We can substitute the definition of I_{sp} into the rocket equation (9.9) to compute the ΔV for a rocket if we know the I_{sp}, as well as the initial and final rocket mass.

$$\Delta V = I_{sp} g_o \ln\left(\frac{m_{\text{initial}}}{m_{\text{final}}}\right) \tag{9.10}$$

where I_{sp} = propellant's specific impulse (s)
g_o = gravitational acceleration at sea level (9.81 m/s^2)

Rockets

Now that we've seen what rockets do—expel high-speed exhaust in one direction so a space vehicle can go in the other—let's take a closer look at how they do it. For purposes of discussion, we can break this process into two steps. First, energy must be *transferred* to the propellant in some form. Second, the energized propellant must be *converted* into high-speed exhaust. There are only two basic types of rockets currently in use. Their classification depends on the form of energy that is transferred to the propellant and converted to high-speed exhaust. These are:

- Thermodynamic rockets—rely on thermodynamic energy (heat and pressure)
- Electrodynamic rockets—rely on electrodynamic energy (charge, electric and magnetic fields)

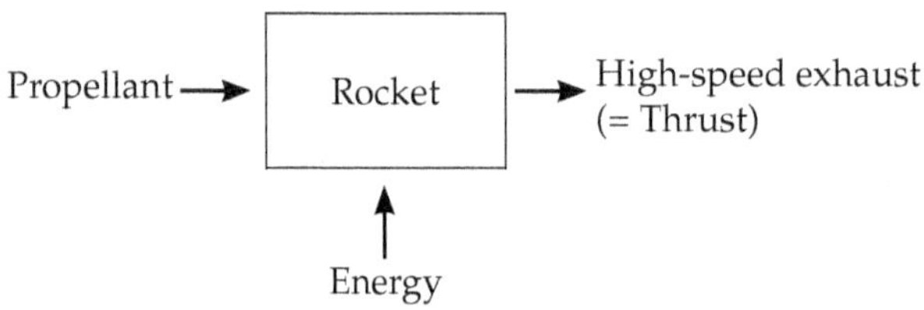

Thermodynamic energy is in the form of heat and pressure. We are all familiar with this. A covered pot of water on the stove reaches high temperature and produces high-pressure steam. We have all seen how the thermodynamic energy in steam can be used to drive trains or in power plants to produce electricity. In a *thermodynamic rocket*, thermodynamic energy is transferred to the propellant in the form of heat and pressure. Heat is either liberated from the propellant by a chemical reaction or supplied directly from electrical, solar, or nuclear sources. Gaseous or liquid propellants are delivered to the rocket under pressure, supplying additional thermodynamic energy. However, for now the result is the most important thing. Once energy is transferred to the propellant, we end up with a high-temperature, high-pressure gas, in other words, a gas with lots of thermodynamic energy. Air in a toy balloon or the result of burning liquid hydrogen and liquid oxygen inside the Shuttle main engines are two extreme examples.

Later in this section we will look at the simplest type of rocket, a cold gas thruster, that relies on gas under pressure alone as its source of thermodynamic energy. In Subsection 9.2 we will look at other, more complex and efficient types of thermodynamic rockets.

Electrodynamic rockets rely on electrodynamic energy. *Electrodynamic energy* relates to the energy available from charged particles moving in electric and magnetic fields. To understand this, we need to use the concept of *charge*.

Charge is a fundamental property of matter, like mass, and can be either positive or negative. Like charges repel each other and opposite charges attract. Typically, a molecule of propellant has the same number of protons and electrons, making it electrically neutral. However, if one or more electrons can be "stripped off," the resulting molecule will be left with a net positive charge,

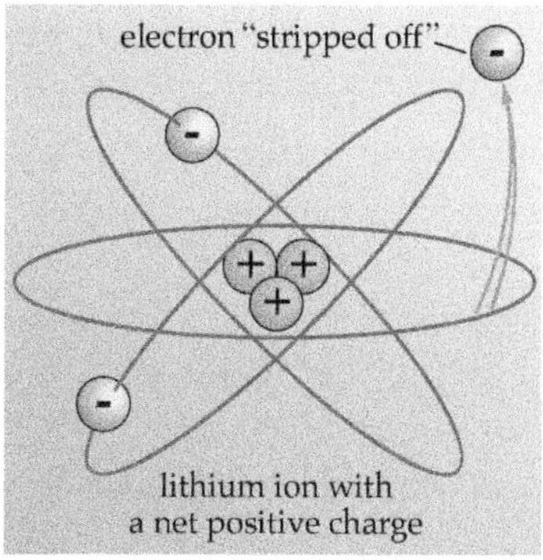

FIGURE 9.1 The ion.

making it an *ion*, as shown in Figure 9.1. To create the ion, some electrodynamic energy must be supplied by the electrical power subsystem (EPS). In a thermodynamic rocket the inherent energy of the energized propellant is quite high; however, the inherent energy of an ion is relatively low. However, once a particle is charged, it can be easily accelerated to very high velocities using additional electrodynamic energy. Later in this section we will look at how this is done inside an electrodynamic rocket.

The form of energy transferred to the propellant determines how it can be converted to high-speed exhaust. In the rest of this section we will look at the two ways of doing this:

- Thermodynamic expansion—using nozzles
- Electrodynamic acceleration—using electric and magnetic fields

Thermodynamic Expansion—Nozzles

By far the most commonly used types of rockets rely on nozzles. *Nozzles* convert the thermal energy produced by chemical, nuclear, or electrical sources into kinetic energy through thermodynamic expansion. To understand how nozzles do this, we must first understand fluid mechanics. Once we understand how fluids behave, we can look at nozzles to see how they convert low-speed, high-temperature gases into high-velocity exhaust. We will then look at how we can predict and measure the performance of thermodynamic rockets and look at a simple example of a cold-gas rocket to see how all these principles fit together.

Fluid Mechanics Let's start by looking at one of the simplest examples, the air in a balloon. Assuming the air in the balloon behaves as a perfect gas, we can relate the pressure, density, and temperature of the gas using the perfect-gas law:

$$P = \rho RT \tag{9.11}$$

where P = pressure (N/m^2)
ρ = density (kg/m^3)
T = temperature (K)
R = specific gas constant (J/kgK)
$= Ru/M$
Ru = universal gas constant (= 8314.41 J/kmol K)
M = molecular weight of the gas (kg/kmole)

It is important to review some basic assumptions that will make our discussion valid. A perfect gas flow has the following properties:

- No heat transfer into/out of the fluid—this is known as an *adiabatic flow.*
- The flow is reversible, meaning total energy is conserved. Together, these two assumptions are called *isentropic flow.*
- Flow in one dimension—two and three dimensions get more complicated.
- "Frozen flow"—meaning all chemical reactions are completed inside the combustion chamber.
- Steady flow—meaning mass flow rate, energy, and momentum are constant.

Earlier we discussed mechanisms for transferring energy to the propellant. In most cases the propellant arrives in the combustion chamber under pressure, so it already has some mechanical energy. As we said earlier, the most common means of adding energy is through heat, such as that produced in chemical reactions (combustion). We describe the total energy in fluid systems in terms of their *specific enthalpy, h.*

$$h = u + Pv \tag{9.12}$$

where h = specific enthalpy (J/kg)
u = internal energy (J/kg)
P = pressure (N/m^2)
v = specific volume (m^3/kg)

Computing enthalpy allows us to separate the energy in the exhaust due to the internal energy (heat) from the mechanical energy (pressure). To see the usefulness of this concept, we need to put the fluid in motion. Let's connect our balloon to a pipe and watch the gas flow as shown in Figure 9.2.

Assuming mass flow rate is constant, we can compute that rate by multiplying the density of the exhaust by the flow velocity and the cross-sectional area of the pipe:

$$\dot{m} = \rho V A \tag{9.13}$$

where $\dot{m}$ = fluid's mass flow rate (kg/s)
ρ = fluid's density (kg/m^3)
V = velocity of flow (m/s)
A = pipe's cross-sectional area (m^2)

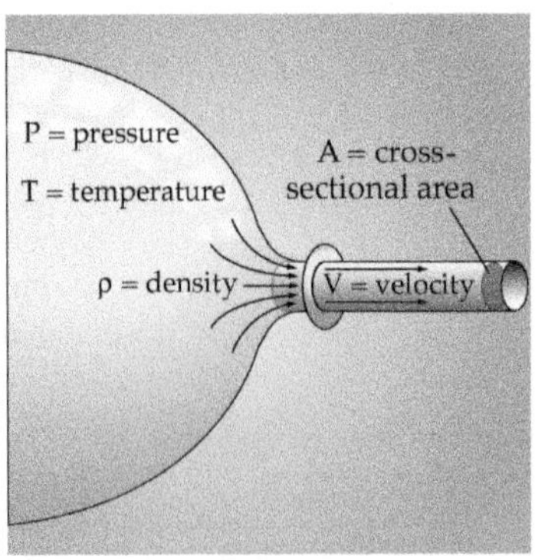

Figure 9.2 Fluid in motion.

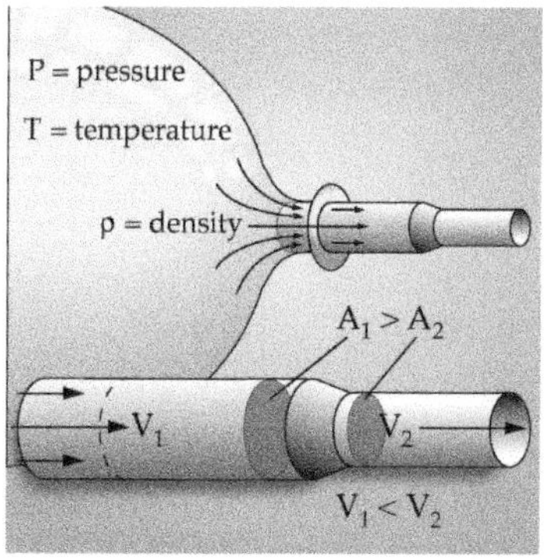

FIGURE 9.3 The Venturi effect.

Now what happens when we vary the pipe's cross-sectional area? If we reduce the pipe's area, the flow velocity *increases* to maintain a constant mass flow rate. This increase in flow velocity to maintain constant mass flow rate due to a constriction in area is called the *Venturi effect*. Looking at it the other way, if we *increase* the pipe's cross-sectional area, the flow velocity decreases to maintain the same mass flow rate. This is illustrated in Figure 9.3.

These effects are common ones, so they are fairly intuitive. However, it only works for low speed flows. For very high-speed flows, the opposite effect takes place. As the area increases, the flow speeds up. For the case of steady flow, with no heat transfer to or from the system, the sum of the specific enthalpy, h, and one-half the square of the flow velocity, $\frac{1}{2}V^2$, is constant.

$$h + \frac{1}{2}V^2 = \text{constant} \tag{9.14}$$

where

$$h = u + Pv$$

Note that the second term is the same as the specific kinetic energy (independent of the mass) of the flow. Thus, the internal energy and gas pressure can be traded for kinetic energy. In other words, the *velocity can increase at the expense of enthalpy and vice versa*. This relationship is the *Bernoulli principle*, named after its discoverer, Swiss mathematician Daniel Bernoulli (1700–1782). This is one of the most important concepts in science. It helps us explain the dynamics of weather and how birds and planes fly.

But this still does not immediately explain how the flow speed increases when the cross-sectional area expands for very high-speed flow. To understand this, we need to delve a little deeper into the behavior of high-speed gases by looking at the speed of sound.

In 1947, Chuck Yeager, piloting the Bell X-1 rocket plane, became the first person to break the sound barrier by traveling faster than the speed of sound. The speed of sound represents the velocity at which pressure disturbances move through a medium. In other words, if this book falls on the floor, it will create a pressure disturbance (sound) that will travel out from the source at a specific speed that depends on characteristics of the air in the room. You will hear the sound a fraction of a second before someone on the other side of the room hears it. For short distances like these, it seems almost instantaneous. But if you have ever witnessed a thunderstorm, you have seen flashes of lightning in the distance and heard the thunder a few seconds later. By counting the seconds between seeing the lightning and hearing the thunder, and by knowing the speed of sound, you can get a good estimate of the distance to the storm. We can find the speed of sound, a_o, from:

$$a_o = \sqrt{\gamma RT} \tag{9.15}$$

where a_o = speed of sound (m/s)
γ = ratio of specific heats (dimensionless)
R = specific gas constant (J/kgK)
T = temperature (K)

Notice in equation (9.15) we introduce the ratio of specific heats, γ. This parameter is constant for a particular gas (or gas mixture) at a given temperature and pressure. We can compute it using various gas-modeling techniques that are beyond the scope of our discussion here, or we can measure it experimentally. As we will see later, γ is an extremely useful parameter for calculating rocket efficiency and other characteristics. Applying equation (9.15) to air at 20°C, we get a speed of sound of 346 m/s.

The ratio of the velocity of the flow (or the velocity of the Bell X-1) to the speed of sound is called the Mach number, M_a.

$$M_a = V/a_o \tag{9.16}$$

From equations (9.14) and (9.16), we can derive a relationship between Mach number, M_a, the change in area, dA, and change in flow velocity, dV.

$$dA/A = \left(M_a^2 - 1\right)dV/V \tag{9.17}$$

where M_a = Mach number
dA = infinitesimal area change (m²)
A = pipe's cross-sectional area (m²)
dV = infinitesimal velocity change (m/s)
V = velocity of the flow (m/s)

Looking closely at this relationship, we can see that for subsonic flow ($M_a = 1$) a positive change in area (expansion) leads to a negative change in velocity (decrease). However, for supersonic flow ($M_a > 1$) the opposite must occur, a positive area change (expansion) leads to a positive velocity change (increase). Figure 9.4 illustrates these relationships. This flip-flop in relationships between area and velocity change above and below the speed of sound is due to conservation of energy (the trade-off between enthalpy and velocity) and conservation of mass (constant flow rate).

Without this effect, rockets as we know them would not be possible. We take advantage of this principle to convert the enthalpy of gases, in the form of heat and pressure in the combustion chamber, into kinetic energy using nozzles. Just as with our balloon example, the process begins in the combustion chamber where high-pressure (and usually high-temperature) exhaust products are created with little or no velocity ($M_a << 1$). From the combustion chamber, the products flow first into a converging section of the nozzle where we know the velocity of the flow increases. The narrowest portion of the nozzle is called the throat. In the throat, the flow velocity reaches the speed of sound ($M_a = 1$) and we say the flow is "choked." As the nozzle expands beyond the throat the velocity again increases to supersonic before being exhausted at the nozzle exit. Figure 9.5 shows a simplified combustion chamber and nozzle diagram.

Nozzles The more we expand the exhaust, the higher the velocity. But there are practical limits. Equation (9.17) showed us that the increase in velocity comes at the expense of the enthalpy of the gases. Recall, enthalpy is a measure of the gases' internal energy (heat) plus mechanical

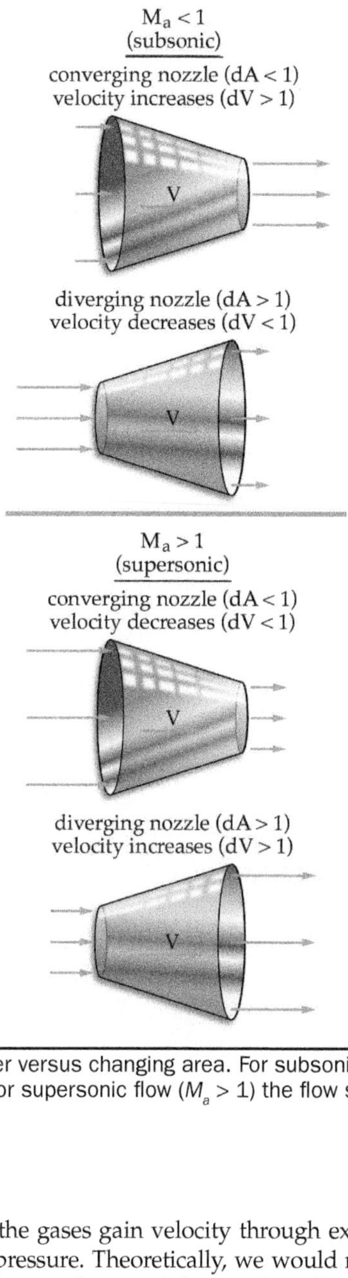

FIGURE 9.4 Changing Mach number versus changing area. For subsonic flow ($M_a < 1$) the flow velocity increases when area decreases. For supersonic flow ($M_a > 1$) the flow speed increases when area increases.

energy (pressure). Therefore, as the gases gain velocity through expansion in the nozzle, they lose enthalpy, both temperature and pressure. Theoretically, we would need an infinitely long nozzle to expand the exhaust to zero exit pressure (vacuum). In practice, of course, this is not possible. Instead, rocket scientists design nozzles that are long enough for the conditions in which they operate. As we will see next, the most important condition to consider is the outside air pressure.

Earlier, we used our astronaut in the wagon example to show that rocket thrust equals the mass flow rate, $\dot{m}$, times the effective exhaust velocity, C.

$$F = \dot{m}C \tag{9.18}$$

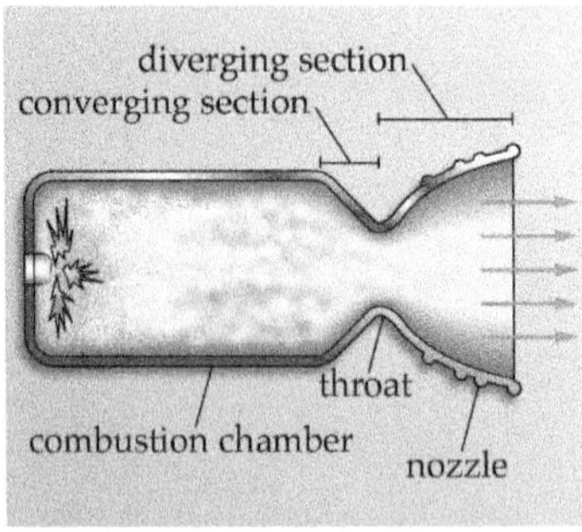

Figure 9.5 Standard combustion chamber and nozzle configuration.

However, this represents only the thrust produced from momentum change, called the *momentum thrust*. But for rockets using nozzles to convert thermal energy into kinetic energy, the thrust due to the momentum of the exhaust is only part of the story.

As we learned, unless a nozzle is infinitely long, the exhaust will have some exit pressure, P_{exit}. This pressure also contributes to the rocket's thrust. To see this, consider an imaginary "control volume" drawn around a rocket as shown in Figure 9.6. Acting on the boundaries of this volume, we've drawn the atmospheric pressure $P_{atmosphere}$ acting on all sides except at the nozzle exit. At the nozzle exit, the pressure is P_{exit}, drawn inward for consistency. Notice that due to symmetry, $P_{atmosphere}$ cancels everywhere except in the direction parallel to momentum thrust over an area equal to the nozzle exit area, A_{exit}. The net force exerted on the rocket from this pressure differential is called the *pressure thrust*. It equals the difference between exit pressure, P_{exit}, and atmospheric pressure, $P_{atmosphere}$, times the exit area, A_{exit}. Its magnitude is

$$F_{pressure\ thrust} = A_{exit}(P_{exit} - P_{atmosphere}) \tag{9.19}$$

where $F_{pressure\ thrust}$ = pressure thrust (N)
A_{exit} = nozzle's exit area (m^2)
P_{exit} = exit pressure (N/m^2)
$P_{atmosphere}$ = atmospheric pressure (N/m^2)

By adding together momentum thrust and pressure thrust, we express the magnitude of the total thrust on the rocket, using the total rocket thrust equation

$$F_{thrust} = \dot{m}V_{exit} + A_{exit}(P_{exit} - P_{atmosphere}) \tag{9.20}$$

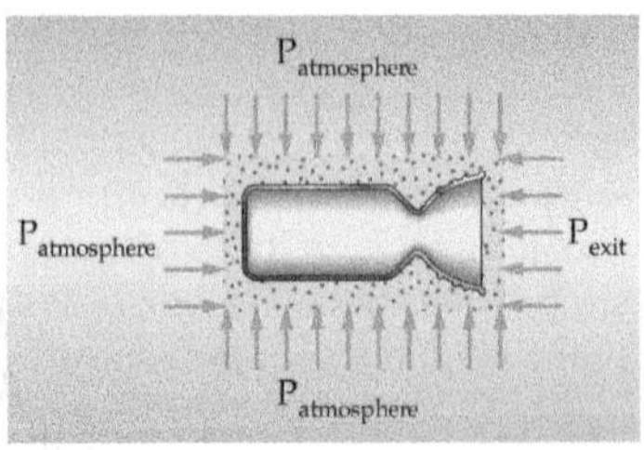

Figure 9.6 Pressure thrust.

where F_{thrust} = rocket's total thrust (N)
V_{exit} = exit velocity of exhaust products (m/s)

To further simplify this equation, we can now fully define the *effective exhaust velocity, C,* to be

$$C = V_{exit} + \frac{A_{exit}}{\dot{m}}(P_{exit} - P_{atmosphere}) \tag{9.21}$$

So we are back to where we started with

$$F = \dot{m}C$$

Only now we know a lot more about where C comes from and how nozzle expansion effects it.

A casual glance at the total rocket thrust in equation (9.2) may lead us to conclude we'd want to make $P_{exit} >> P_{atmosphere}$ to maximize total thrust. Although this would appear to increase the amount of thrust generated, a big loss in overall efficiency would actually reduce the effective thrust. Recall that for supersonic flow, as the gases expand they *increase* in velocity while, due to the Bernoulli principle, they decrease in pressure. Thus, the higher the V_{exit}, the lower the P_{exit}. For the ideal case, the pressure thrust should be zero ($P_{exit} - P_{atmosphere} = 0$), which means the exit pressure exactly equals the atmospheric pressure ($P_{exit} = P_{atmosphere}$). In this case the exit velocity and thus the momentum thrust is maximized.

But what happens when $P_{exit} \neq P_{atmosphere}$? When this happens, we have a rocket that is not as efficient as it could be. We can consider two possible situations:

- *Over expansion*: $P_{exit} < P_{atmosphere}$. This is often the case for a rocket at lift-off. Because most launch pads are near sea level, the atmospheric pressure is at a maximum. This atmospheric pressure can cause shock waves to form just inside the nozzle. These shock waves represent areas where kinetic energy turns back into enthalpy (heat and pressure). In other words, they rob kinetic energy from the flow, lowering the exhaust velocity and thus decreasing the overall thrust.
- *Under expansion*: $P_{exit} > P_{atmosphere}$. In this case, the exhaust gases have not expanded as much as they could have within the nozzle and thus there is a loss in the sense that we have not converted all the enthalpy we could have into velocity. This is the normal case for a rocket operating in a vacuum, because P_{exit} is always higher than $P_{atmosphere}$ ($P_{atmosphere} = 0$ in vacuum). Unfortunately, you would need an infinitely long nozzle to expand the flow to zero pressure, so in practice we must accept some loss in efficiency.

In Subsection 9.3 we will see how we deal with this problem for launch-vehicle rocket engines. Figure 9.7 illustrates these cases of expansion.

The total expansion in the nozzle depends, of course, on its design. We define the nozzle *expansion ratio, ε,* to be the ratio between the nozzle exit area, A_e, and the throat area, A_t:

$$\varepsilon = A_e/A_t \tag{9.22}$$

Later in this section we will see how varying expansion ratio can affect engine performance.

Characteristic Exhaust Velocity We know rocket thrust depends on effective exhaust velocity, C, the rocket's output. But how do you measure C? Unfortunately, you can't just stick a velocity transducer into superheated rocket exhaust. Therefore, when doing rocket experiments, we need to have some other, more measurable parameters available. We know we can vary rocket performance (thrust, I_{sp}, etc.) by changing the inputs: $\dot{m}$, of the propellant going into the combustion chamber, and the resulting pressure in the combustion chamber, P_c. In addition to these dynamic parameters, there are also important physical dimensions of the rocket that can be varied, such as the area of the

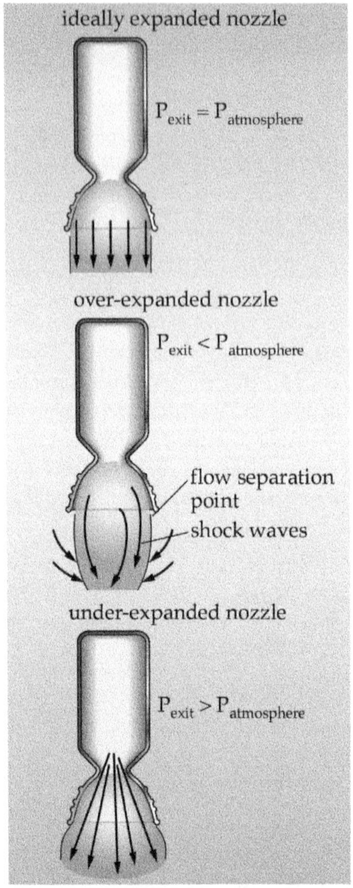

FIGURE 9.7 Nozzle expansion.

nozzle throat, A_t, and the expansion ratio, ε. To understand the relationship between these design variables, we define another important rocket performance parameter called *characteristic exhaust velocity*, C^*, in terms of chamber pressure, P_c, mass flow rate, $\dot{m}$, and throat area, A_t.

$$C^* = P_c A_t / \dot{m} \tag{9.23}$$

The nice thing about C^* is that not only can we easily measure it experimentally, we can also compute it by modeling combustion and flow characteristics. Rocket scientists use a variety of computer codes to predict characteristic exhaust velocity. These techniques are beyond the scope of the discussion here, but for a complete description see Humble et al. (1995). One of the most important parameters to find using these modeling techniques is the ratio of specific heats, γ, for the products in the combustion chamber. Knowing this value, we can compute C^* using:

$$C^* = a_o / \Gamma \tag{9.24}$$

where a_o = speed of sound in the fluid (m/s)
$\Gamma = \gamma(2/(\gamma+1))^{[(\gamma+1)/(2\gamma-2)]}$ (unitless)

By comparing C^* from experiments with C^* from our predictions, we can determine how efficiently the rocket transfers the available energy to the propellants relative to an ideal value. This is especially useful as it allows us to measure the performance of the combustion chamber (energy transfer) independent from the nozzle (energy conversion). No rocket is perfect. If the measured C^* is over 90% of predicted ideal C^*, it is considered good performance. As we learned earlier, absolute performance is measured by I_{sp} and I_{dsp}, both of which are a function of effective exhaust velocity, C. While it is fairly straight forward to determine I_{sp} experimentally by measuring delivered thrust and the total mass of propellant used, it is also important to be able to predict it for a given type of rocket. Fortunately, we can compute it directly from C^* and the ratio of specific heats for a given reaction.

As one would expect, one of the primary goals of rocket design is to maximize performance. We express rocket performance most often in terms of mass efficiency (I_{sp}). How do we maximize I_{sp}? We have just seen that effective exhaust velocity, and hence specific and density-specific impulse are all functions of C^*. From equation (9.24), we know C^* depends on the speed of sound in the combustion chamber, a_o, which depends on the gas constant, R, and the combustion temperature, T. Looking at the equation, we can see that the higher the temperature, the higher the speed of sound and thus the higher the I_{sp}. What about R? Remember, R is *inversely* proportional to the propellant's molecular weight. *Molecular weight* is a measure of the weight per molecule of propellant. Thus, to improve I_{sp} for thermodynamic rockets, we try to maximize the combustion temperature while minimizing the molecular weight of propellant. We can express this relationship more compactly as:

$$I_{sp} \propto \sqrt{\frac{T_{combustion}}{M}} \tag{9.25}$$

where I_{sp} = specific impulse (s)
$T_{combustion}$ = combustion temperature (K)
M = molecular weight (kg/mole)
[*Note:* the symbol "∝" means proportional to]

As a result, the most efficient thermodynamic systems operate at the highest temperature with the propellants having the lowest molecular weight. For this reason, hydrogen is often the fuel of choice because it has the lowest possible molecular weight and achieves high temperatures during combustion. Unfortunately, the low molecular weight also means low density. Thus, while hydrogen systems achieve high I_{sp}, they often do it at the expense of I_{dsp}. Designers must trade-off mass versus volume efficiency depending on the mission requirements.

Finally, we need to know what total thrust our rocket produces. We can relate characteristic exhaust velocity, C^*, to effective exhaust velocity, C, through yet another parameter called the *thrust coefficient*, C_F.

$$C_F = C/C^* \tag{9.26}$$

We can also relate the thrust coefficient to the thrust, F, chamber pressure, P_c, and throat area, A_t, through

$$C_F = F/(P_c A_t) \tag{9.27}$$

In this way, we can compare the measured rocket thrust to the ideal thrust from theoretical modeling to determine how well the nozzle converts enthalpy into kinetic energy. Like characteristic exhaust velocity, thrust coefficient gives us a way to determine the performance of the nozzle, independent from the combustion chamber. Again, no nozzle is perfect. However, a well-designed nozzle, for the correct expansion conditions, should achieve 95% or better.

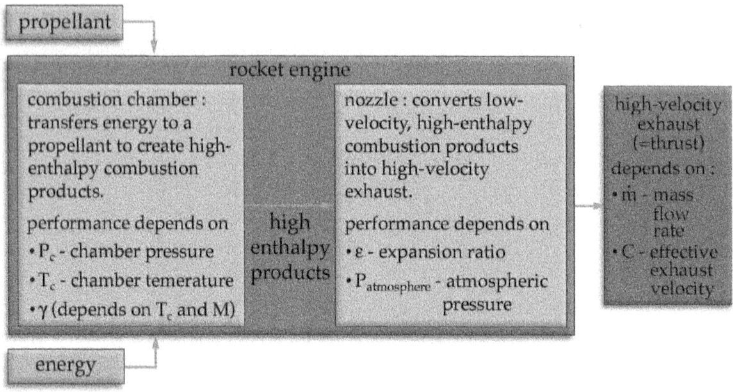

Figure 9.8 Expanded systems view of a thermodynamic rocket.

Summary Let's review what we have discussed about thermodynamic rockets. Figure 9.8 further expands our systems view of a thermodynamic rocket and summarizes important performance parameters. Recall, there are two important steps to the rocket propulsion process—energy transfer and mass acceleration. These two steps take place in the combustion chamber and nozzle, respectively. The most important output is the thrust that moves the vehicle from point *A* to point *B*.

Now let's put all these principles together by looking at one specific example—the simplest type of rocket in use, a cold-gas thruster.

Cold-Gas Rockets A *cold-gas rocket* uses only mechanical energy in the form of pressurized propellant as its energy source, similar to the toy balloon example we talked about at the beginning of the chapter. While spacecraft designers do not send balloons into orbit, the basic principles of cold-gas rockets are not that different. A coiled spring stores mechanical energy that can be converted to work, such as running an old-fashioned wind-up watch. Similarly, any fluid under pressure has stored mechanical energy that can be used to do work. Any rocket system containing fluids under pressure (and virtually all do) uses this mechanical energy in some way. As we will see, usually this energy is a minor contribution to the overall energy of the propellant. However, for cold-gas rockets this is the only energy the propellant has.

Table 9.1 summarizes basic principles and propellants used by cold-gas rockets. There is a small volume upstream of the nozzle where the gas collects prior to expulsion. This is similar to the combustion chamber needed for chemical rockets.

Cold-gas rockets are very reliable and can be turned on and off repeatedly, producing very small, finely controlled thrust pulses (also called impulse bits)—a desirable characteristic for attitude control. A good example of them is on the manned maneuvering unit (MMU) used by Shuttle astronauts. The MMU uses compressed nitrogen and numerous small thrusters to give astronauts complete freedom to maneuver.

Unfortunately, due to their relatively low thrust and I_{sp}, we typically use cold-gas systems only for attitude control or limited orbital maneuvering on small spacecraft. Even so, they can serve as a good example of trading off some of the basic rocket parameters we have talked about in this section.

Figure 9.9 presents the results of analyzing five variations of the same basic cold-gas rocket. From the baseline design using nitrogen as the propellant at room temperature (298 K) and 5 bar (72.5 psi) chamber pressure, the figure shows the effects of increasing P_c or T_c and reducing ε or

Operating principle	Uses the mechanical energy contained in a compressed gas and thermodynamically expands the gas through a nozzle producing high-velocity exhaust
Propellants	Helium (He), nitrogen (N_2), carbon dioxide (CO_2), or virtually any compressed gas
Advantages	• Extremely simple • Reliable • Safe, low-temperature operation • Short impulse bit (thrust pulses)
Disadvantages	Low I_{sp}, I_{dsp} compared to chemical rockets
Example	UoSAT-12 cold-gas thrusters: Propellant = N_2, Pc = 4 bar, thrust = 0.1 N

TABLE 9.1 Summary of Cold-Gas Rockets

lowering the molecular weight (and γ) of the propellant by switching to helium. From this analysis, we can draw some general conclusions about basic trade-offs in rocket design:

- Increasing chamber pressure increases thrust (with little or no effect on specific impulse or density-specific impulse).
- Increasing chamber temperature increases specific impulse and density-specific impulse with a slight decrease in thrust.
- Decreasing expansion ratio (underexpanded condition) decreases specific impulse, density-specific impulse, and thrust.
- Decreasing propellant molecular weight increases specific impulse at the expense of decreasing density-specific impulse and thrust.

Electrodynamic Acceleration

We have spent some time discussing thermodynamic expansion and acceleration of exhaust using nozzles to convert propellant with thermodynamic energy into high-speed flow. But there is a second method for propellant acceleration currently gaining wider use in spacecraft—electrodynamic acceleration. To take advantage of this method, we must start with a charged propellant. Recall, the force of attraction (or repulsion) between charges depends on the strength of the charges involved and the distance between them. We expressed this as Coulomb's law:

$$F = KQq/R^2\mathbf{R} \tag{9.28}$$

where F = electrostatic force on charge 1 (N)
K = constant (9×10^9)
Q = value of charge 1 (coulombs, C)
q = value of charge 2 (coulombs, C)
R = distance between charges (m)
$\mathbf{R}$ = unit vector in direction of force

An *electric field* exists when there is a difference in charge between two points, i.e., there is an imbalance between positive and negative charges in a confined region. We call the energy an

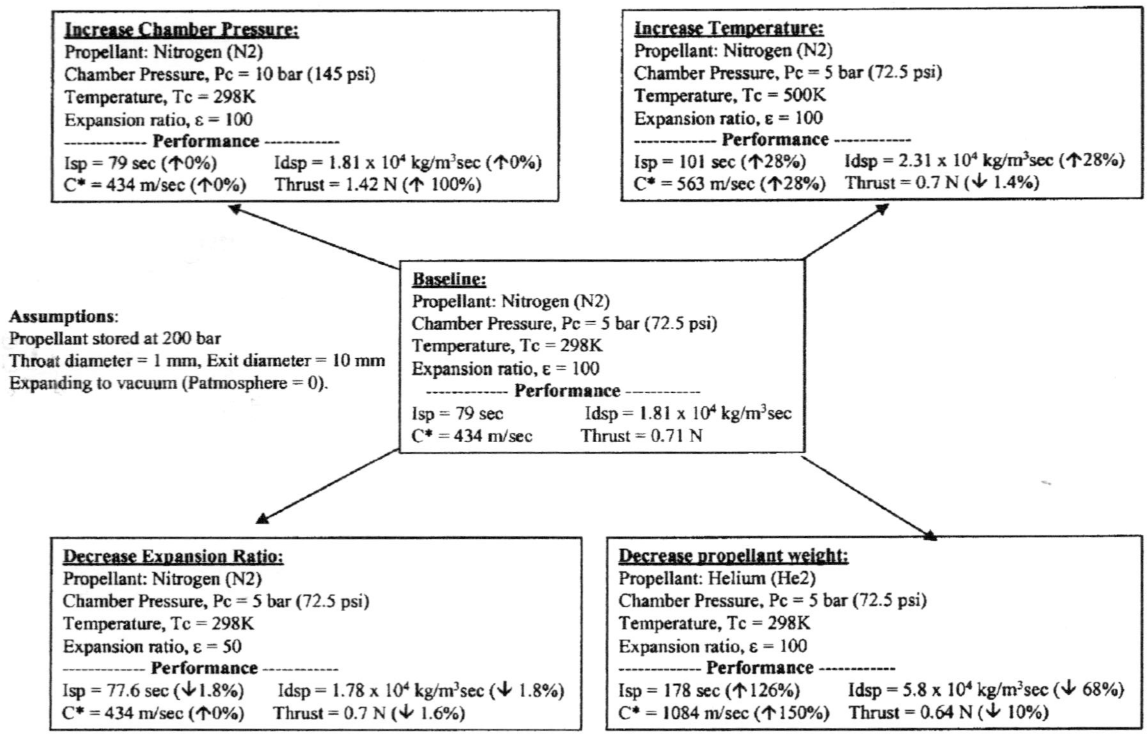

FIGURE 9.9 Cold-gas rocket trade-offs. This figure illustrates various trade-offs in rocket design by looking at the simplest type of rocket, a cold-gas thruster. From the baseline case using nitrogen as the propellant, the specific impulse, density-specific impulse, characteristic exhaust velocity, C^*, and thrust are affected by changing chamber pressure, chamber temperature, expansion ratio, and the molecular weight of propellant. (Analysis courtesy of Johnson Rocket Company, Inc.)

electric field can transmit to a unit charge the *electrical potential*, described in terms of volts/m. The resulting force on a unit charge is called *electrostatic force*.

A simple example of electrostatic force in action can be seen if you have ever rubbed a balloon through your hair and stuck it to a wall. As you rub the balloon it picks up a net positive charge. When it is placed against the wall, initially neutral, the positive charges on the surface are pushed away leaving a net negative charge. The opposite charges attract each other creating a force strong enough to keep the balloon in place. The strength of the electrostatic force is a function of the charge and the electric field.

$$\mathbf{F}_i = m_i \mathbf{a}_i = q_i \mathbf{E} \tag{9.29}$$

where $\mathbf{F}_i$ = electrostatic force vector (N)
m_i = mass of charged particle (kg)
$\mathbf{a}_i$ = acceleration of charged particle (m/s^2)
q_i = charge on particle (coulombs, C)
$\mathbf{E}$ = electric field vector (electric potential) (V/m)

This is illustrated in Figure 9.10. Notice that the direction of the force is parallel to the electric field.

Electrodynamic rockets take advantage of this principle to create thrust. In the simplest application, they only need some charged propellant and an electric field. As with any rocket, the two things we are most interested in are thrust, F, and specific impulse, I_{sp}. From equation (9.2) we know thrust depends on the mass flow rate, $\dot{m}$, and effective exhaust velocity, C:

$$F = \dot{m}C \tag{9.30}$$

and from equation (9.7) we know specific impulse, I_{sp}, is directly related to C by

$$C = I_{sp} g_o$$

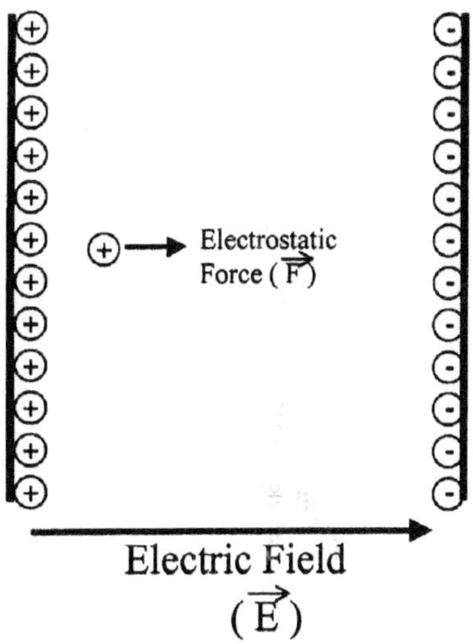

FIGURE 9.10 Electrostatic force. An electric field exists when there is an imbalance between positive and negative charges in a confined region. This field will impart an electrostatic force on a charged particle within the field, making it accelerate.

In an electrodynamic rocket, high $\dot{m}$ is achieved by having a high density of charged propellant. High exhaust velocity comes from having a strong electric field and/or applying the electrostatic force for a longer time. We can summarize these effects on performance as follows:

- Higher charge density → higher $\dot{m}$ → higher thrust
- Stronger electric field → stronger electrostatic force on propellant → higher acceleration → higher exhaust velocity → higher I_{sp}

Thus, by varying the charge density and the applied field, a wide range of thruster designs can be created. Naturally, there are practical design issues that limit how high you can increase each of these parameters. Let's start with charge density.

Charge density is limited by the nature of the propellant and how it is charged. Earlier, we defined an ion to be a positively charged propellant molecule that has had one or more electrons stripped off. Ions are handy in that they are simple to accelerate in an electric field. Unfortunately, when you try to pack lots of positive ions into a small confirmed space, they tend to repel each other. This creates a practical limit to the charge density you can achieve with ions.

One way around this is to create a *plasma* with the propellant instead. A plasma is an electrically neutral mixture of ions and free electrons. A plasma can be seen inside a common florescent lamp or neon light. When a gas, such as neon, is placed in a strong electric field, the electrons become only weakly bound to the molecules creating a "soup" of ions and free electrons. The glow that is given off is the result of electrons jumping back and forth between energy states within the molecule. Because it is electrically neutral, a plasma can contain a much higher charge density than simply a collection of ions alone.

So far we have only considered the acceleration effect from an applied electric field. However, whenever an electric field is applied to a plasma, a magnetic field is also created (or said to be *induced*). Charged particles are also accelerated by magnetic fields but at right angles to the field, instead of parallel to it. To determine the combined effect of electric and magnetic fields on a charged particle we must look at the cross-product of their interaction.

$$F_{em} = m_i a_{em} = q_i(\mathbf{E} + V_i \times \mathbf{B}) \tag{9.31}$$

where F_{em} = electromagnetic force on a particle (N)
a_{em} = electromagnetic acceleration (m/s²)
$\mathbf{E}$ = electric field vector (V/m)
V_i = velocity of particle (m/s)
$\mathbf{B}$ = Magnetic field vector (tesla)

Some types of electrodynamic rockets rely on this combined effect to produce thrust. However, for most cases the electrostatic force is dominant and we can ignore the effect of the magnetic field for simple analysis of performance. When we consider the second parameter that limits thruster performance, the strength of the electric field, we can focus mainly on the practical limits of applied power. From equation (9.2b) the jet power of a rocket is found from:

$$P_J = \frac{1}{2}\dot{m}C^2$$

Thus, for a given charge density, the exhaust velocity increases with the square root of the power. As you would expect, there are practical limits to the amount of power available in any spacecraft, limiting the ultimate performance of electrodynamic rockets.

While exhaust velocity (and I_{sp}) go up with power, there is a trade-off between thrust and exhaust velocity, as illustrated by the following relationship:

$$F = 2P/C \tag{9.32}$$

where F = thrust (N)
P = power (W)
C = effective exhaust velocity (m/s)

Therefore, when designing an electric thruster, you can have high exhaust velocity or high thrust, but not both at the same time. In Subsection 9.2 we will look at some specific examples of electric thrusters and compare their performance.

9.2 Propulsion Systems

Subsection 9.1 gave us a look at rockets as a system. But rockets, as important as they are, comprise only one part of an entire propulsion system. In this section, we will concentrate less on rocket theory and more on propulsion system technology to learn what essential components we need, and how they're put together.

Figure 9.11 shows a block diagram for an entire propulsion system. To design a specific system, we start with the desired thrust, usually at some very specific time. The propulsion system controller manages these inputs and formulates commands to send to the propellant management actuators to turn the flow of propellant on or off. For some systems, the controller also manages the energy input to the rocket. For example, in an electrodynamic rocket or a thermoelectric rocket, the system has to interface with the spacecraft's electrical power subsystem (EPS) to ensure it provides the correct power level. The controller uses sensors extensively to monitor the temperature and pressure of the propellant throughout.

Regardless of whether you are using a thermodynamic or an electrodynamic rocket, you need to store and handle some propellant. In this section, we will start by looking at propellant management, how to store liquid or gaseous propellants and supply them to the rocket as needed. We will then review in detail most of the various thermodynamic and electrodynamic rocket technologies currently in use or on the drawing boards. Following this discussion, we will look briefly at important factors for selecting and testing propulsion systems. Finally, since rocket scientists are always striving to improve propulsion system performance, we will look at what exotic concepts may one day take spacecraft to the stars.

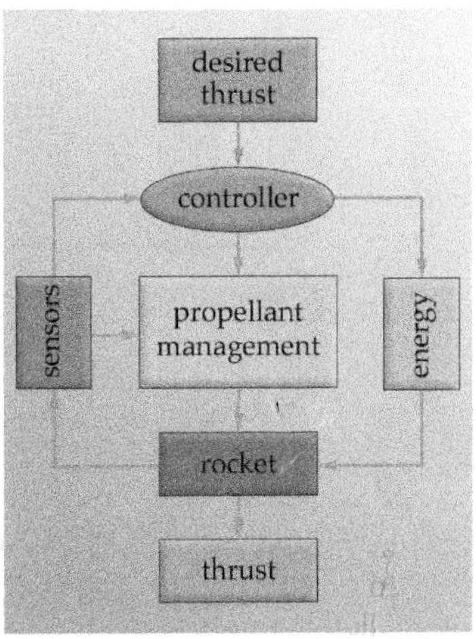

FIGURE 9.11 Block diagram of a complete propulsion system.

Propellant Management

All rockets need propellant. The job of storing propellant and getting it where it needs to go at the right time is called *propellant management*. The propellant management portion of a propulsion system has four main tasks:

1. Propellant storage
2. Pressure control
3. Temperature control
4. Flow control

We will look briefly at the requirements and hardware for each of these tasks.

Gaseous propellants, such as nitrogen for cold-gas rockets, are normally stored in tanks under high pressure to minimize their volume. Typical gas storage pressures are 200 bar (3000 psi) or more. Unfortunately, we cannot make a liquid propellant denser by storing it under pressure. However, depending on how we pressurize the liquid propellant for delivery to the combustion chamber, the storage tanks may have to be designed to take high pressure as well. In any case, propellant tanks are typically made from aluminum, steel, or titanium and designed to withstand whatever pressure the delivery system requires.

As we learned in Subsection 9.1, combustion chamber pressure is an important factor in determining rocket thrust. This pressure depends on the delivery pressure of the propellants. Pressurizing the flow correctly is another function of propellant management. There are two approaches to achieving high pressure flow: pressurefed systems and pump-fed systems.

Figure 9.12 shows how a *pressure-fed propellant system* relies on either a gaseous propellant stored under pressure, or a separate tank attached to the main tank, filled with an inert, pressurized gas such as nitrogen or helium to pressurize and expel a liquid propellant. The high-pressure gas squeezes the liquid propellant out of the storage tank at the same pressure as the gas.

To minimize volume, the storage pressure of the gas is typically much higher than the pressure needed in the combustion chamber. To reduce or regulate the high pressure in the gas storage tank

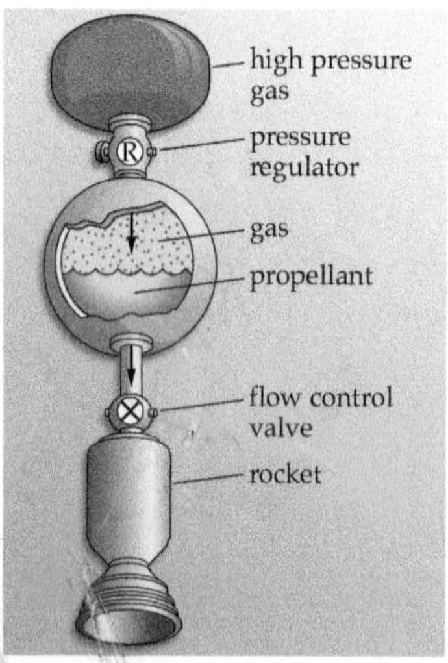

FIGURE 9.12 Pressure-fed propellant system.

to the lower pressure for propellant delivery, mechanical regulators are typically used. As high-pressure gas flows into a *regulator*, the gas pushes against a carefully designed diaphragm. The resulting balance of forces maintains a constant flow rate but at a greatly reduced output pressure. For example, a gas stored at 200 bar may pass through a regulator that reduces it to 20 bar before it goes into a liquid propellant tank. Pressure regulators are common devices found in most rocket plumbing systems. SCUBA tanks use regulators to reduce high-pressure air stored in the tank to a safe, lower-pressure for breathing.

The main drawback of pressure-fed systems is that the more liquid propellant in the tank, the more pressurizing gas is needed. For very large propulsion systems, such as on the Space Shuttle, enormous quantities of high-pressure propellant must be delivered to the combustion chamber each second. To do this using a pressure-fed system would require additional large, high-pressure gas tanks, making the entire launch vehicle larger and heavier. Instead, most launch vehicles use pump-fed delivery systems.

Pump-fed delivery systems rely on pumps to take low-pressure liquid and move it toward the combustion chamber at high pressure, as shown in Figure 9.13. Pumps impart kinetic energy to the propellant flow, increasing its pressure. On the Space Shuttle, massive turbo-pumps burn a small amount of H_2 and O_2 to produce mechanical energy. This energy is used to take the liquid propellants normally stored at a few bar and boost the feed pressure to over 480 bar (7,000 psi) at a flow rate of 2.45×10^5 L/s (6.5×10^4 gal/s). Spinning at over 30,000 rpm, the Shuttle propellant pumps could empty an average-size swimming pool in only 1.5 s!

Regardless of the propellant-delivery system used, the pressure of propellants and pressurizing gases must be constantly monitored. *Pressure transducers* are small electromechanical devices used to measure the pressure at various points throughout the system. This information is fed back to the automatic propellant controller and sent to ground controllers via telemetry.

Temperature control for propellant and pressurant gases is another important propellant-management function. The ideal gas law tells us that a higher gas temperature causes a higher pressure and vice versa. The propellant-management subsystem must work with the spacecraft environmental control and life support subsystem (ECLSS) to maintain gases at the right temperature and prevent liquid propellants from freezing or boiling. In the deep cold of outer space, there is a danger of propellants freezing. For instance, hydrazine, a common spacecraft propellant, freezes at 0°C. Usually the spacecraft ECLSS maintains the spacecraft well above

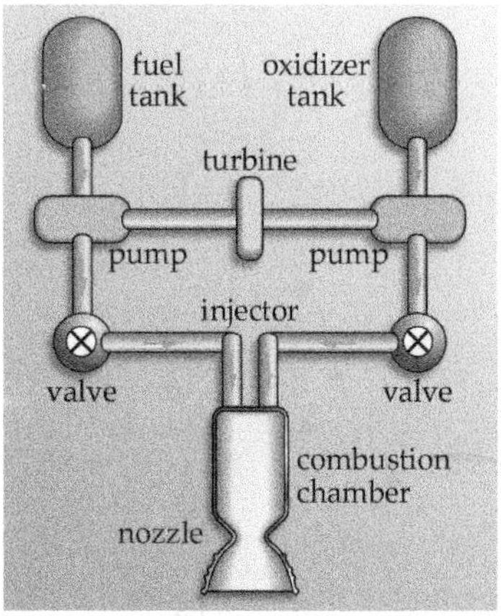

FIGURE 9.13 Pump-fed propellant management.

this temperature, but in some cases exposed propellant lines and tanks may need heaters to keep them warm.

On launch vehicles, propellant thermal management often has the opposite problem. It must maintain liquid oxygen (LOX) and liquid hydrogen at temperatures hundreds of degrees below zero Centigrade. Using insulation extensively helps control the temperature. However, some boil-off of propellants prior to launch is inevitable and must be planned for.

Finally, the propellant management system must control the flow of gases and liquids. It does this using valves. Valves come in all shapes and sizes to handle different propellants, pressures, and flow rates. Fill and drain valves are needed to fill the tanks prior to launch (and drain them in emergencies). Tiny, electrically controlled low-pressure valves pulse cold-gas thrusters on and off to deliver precise micro-amounts of thrust. Large pyrotechnic valves mounted below liquid-propellant tanks keep them sealed until ignition. When the command is sent, a pyrotechnic charge fires, literally blowing the valve open, allowing the propellant to flow. Of course, these valves are good for only one use. To protect against overpressure anywhere in the system, *pressure relief valves* automatically release gas if the pressure rises above a preset value. Check-valves are designed to allow liquid to flow in only one direction, preventing backflow in the wrong direction. Other valves throughout the system ensure propellant flows where it needs to when the system controller sends the command. Some of these other valves lead to redundant lines that ensure the propellant flows even when a main valve malfunctions.

Let's briefly review the components needed for propellant management. Propellants and pressurant gas are stored in tanks. Below the tanks, valves control the flow throughout the system and regulators reduce the pressure where needed. Pressure and temperature are measured at various points in the system using transducers and other sensors. Figure 9.14 gives an example schematic of a simple cold-gas rocket system showing the necessary components for propellant management.

Thermodynamic Rockets

As we learned in Subsection 9.1, thermodynamic rockets convert thermodynamic energy (heat and pressure) transferred to a propellant into high-speed exhaust using nozzles. There are a wide

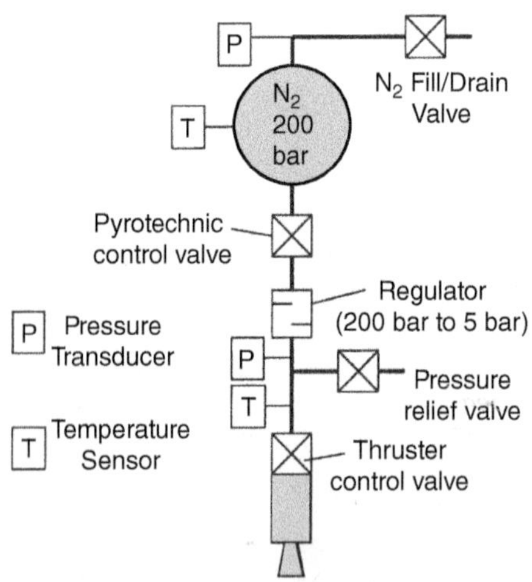

FIGURE 9.14 Example schematic for a cold-gas rocket system.

variety of other thermodynamic rockets currently available or being considered. We can classify these based on their source of energy:

- Cold-gas—use mechanical energy of a gas stored under pressure
- Chemical—rely on chemical energy (from catalytic decomposition or combustion) to produce heat
- Solar thermal—use concentrated solar energy to produce heat
- Thermoelectric—use the heat produced from electrical resistance
- Nuclear thermal—use the heat from a nuclear reaction

Because we examined simple cold-gas rockets in detail in the last section, here we will review the other four types and compare their relative performances.

Chemical Rockets

The vast majority of rockets in use today rely on chemical energy. When you strike a match, the flame represents a combustion process taking place. The fuel—the wood in the match—is chemically combining with the oxygen in the air to form various chemical by-products (CO, CO_2, water, etc.) and, most importantly, heat. In *chemical rockets*, the energy is stored in the chemical bonds of the propellants. In the Shuttle main engines, liquid hydrogen (H_2) and liquid oxygen (O_2) are combined in the most basic of all chemical reactions:

$$2H_2 + O_2 \rightarrow 2H_2O + \text{Heat}$$

All combustion reactions must have a *fuel* (such as hydrogen) plus an *oxidizer* (such as oxygen). These two combine, liberating a vast amount of heat and creating by-products that form the exhaust. The heat is transferred to the combustion products, raising their temperature. The place where this chemical reaction and energy transfer takes place is called the *combustion chamber*. Although the propellants arrive in the combustion chamber under pressure, delivered by the propellant management system, this mechanical energy is small compared to the thermal energy released by the chemical reaction.

Chemical rockets generally fall into one of three categories:

- Liquid
- Solid
- Hybrid

Let's briefly review the operating principles and performance parameters of each.

Liquid Chemical Rockets Liquid chemical rockets are usually one of two types, bipropellant or monopropellant. As the name implies, *bipropellant* rockets use two liquid propellants. One is a fuel, such as liquid hydrogen (LH_2), and the other is an oxidizer, such as liquid oxygen (LOX). Brought together under pressure in the combustion chamber by the propellant management system, the two compounds chemically react (combust) releasing vast quantities of heat and producing combustion products (these vary depending on the propellants). To ensure complete, efficient combustion, the oxidizer and fuel must mix in the correct proportions. The *oxidizer/fuel ratio* (*O/F*) is the proportion, by mass, of oxidizer to fuel.

Some propellant combinations, such as hydrogen and oxygen, will not spontaneously combust on contact. They need an igniter, just as your car needs a spark plug, to get started. This need, of course, increases the complexity of the system somewhat. Propellant chemists thus strive to find combinations that react on contact. We call these propellants *hypergolic* because they do not need a separate means of ignition. Hydrazine (N_2H_4) plus nitrogen tetroxide (N_2O_4) is an example of a hypergolic fuel and oxidizer combination.

Another important feature in selecting a propellant is storability. Although the liquid hydrogen and liquid oxygen combination used in the Space Shuttle main engines offers high performance (specific impulse around 455 s), it must be super-cooled to hundreds of degrees below zero Centigrade. Because of their low storage temperature, we call these propellants *cryogenic*. Unfortunately, it is difficult to maintain these extremely low temperatures for long periods (days or months). When the mission concept calls for long-term storage, designers turn to *storable propellants* such as hydrazine and nitrogen tetroxide that remain stable at room temperature for a very long time (months or even years).

The Titan, an early intercontinental ballistic missile (ICBM), used hypergolic, storable propellants because the missiles stayed deep in underground silos for many years. These propellants are also used in the Shuttle's orbital-maneuvering engines and reaction-control thrusters as well in virtually all spacecraft using liquid chemical rockets. The penalty paid for the extra convenience of spontaneous combustion and long-term storage is a much lower performance than the cryogenic option (specific impulse around 300 s). In addition, current hypergolic combinations are extremely toxic and require special handling procedures to prevent propellant release. Table 9.2 summarizes key points about bipropellant rockets.

As the name implies, monopropellant chemical rockets use only a single propellant. These propellants are relatively unstable and easily decompose through contact with the correct catalyst.

Hydrogen peroxide (H_2O_2) is one example of a monopropellant. You may have used a low-concentration (3%), drug store variety of this compound to disinfect a bad scrape or to bleach your hair. Rocket-grade hydrogen peroxide, also called high-test peroxide (HTP), has a concentration of 85% or more. It is relatively safe to handle at room temperatures, but when passed through an appropriate catalyst (such as silver) it readily decomposes into steam (H_2O) and oxygen, releasing significant heat in the process. Typical HTP reactions exceed 630°C. This relatively high temperature, combined with the molecular weight of the reaction products, gives HTP monopropellant rockets an I_{sp} of about 140 s. These types of thrusters were successful used on the X-15 rocket plane and the Scout launch vehicle.

By far the most widely used monopropellant today is hydrazine (N_2H_4). It readily decomposes when exposed to a suitable catalyst, such as iridium, producing an I_{sp} of about 180 s. The main disadvantage of hydrazine is that it is higly toxic. This means specialized handling procedures and equipment are needed during all testing and launch operations.

The biggest advantage of monopropellant over bipropellant systems is simplicity. The propellant-management system needs to maintain only one set of tanks, lines, and valves. Unfortunately, there is a significant penalty in performance for this added simplicity (two-thirds the I_{sp} of a bipropellant system or less). However, for certain mission applications, especially station keeping and attitude control on large communication satellites, this trade-off is worth it. The benefit grows when we use

Operating principle: A liquid oxidizer and a liquid fuel react together in a combustion process liberating heat and creating exhaust products that are thermodynamically expanded through a nozzle.	
Advantages: • High I_{sp} • Can be throttled • Can be re-started	**Disadvantages:** • Must manage two propellants • Intense heat of combustion creates thermal control problems for chamber and nozzle

Table 9.2 Bipropellant Rockets

<table>
<tr><td colspan="2">Operating principle:
A single propellant is decomposed using a catalyst releasing heat and creating by-products that are thermodynamically expanded through a nozzle.</td></tr>
<tr><td>Advantages:
• Simple, reliable
• Single propellant to manage</td><td>Disadvantages:
• Lower I_{sp} than bipropellant</td></tr>
</table>

TABLE 9.3 Monopropellant Rockets

hydrazine as the fuel with nitrogen tetroxide in a large bipropellant rocket for initial orbital insertion and then, by itself in a smaller, monopropellant rocket for station keeping. Such "dual-mode" systems take advantage of the flexibility offered by hydrazine to maximize overall system performance and simplicity. Table 9.3 summarizes key points about monopropellant rockets.

Solid Chemical Rockets The fireworks we watch on the Fourth of July are a good example of solid rockets at work. Solid rockets date back thousands of years to the Chinese, who used them to confuse and frighten their enemies on the battlefield. In modern times these rockets create thrust for intercontinental ballistic missiles as well as space launch vehicles.

Just as liquid bipropellant rockets combine fuel and oxidizer to create combustion, solid rockets contain a mixture of fuel, oxidizer, and a binder, blended in the correct proportion and solidified into a single package called a *motor*. A typical composite solid rocket fuel is powdered aluminum. The most common oxidizer is ammonium perchlorate (AP). Together, the fuel and oxidizer comprise about 85–90% of the rocket motor mass, with an oxidizer/fuel ratio of about 8:1. The remaining mass of the motor consists of the binder, which holds the other ingredients together and provides overall structural integrity. Binders are usually hard, rubberlike compounds such as hydroxyl terminated polybutadiene (HTPB). During combustion, the binder also acts as additional fuel.

As we learned in Subsection 9.1, rocket thrust depends on mass flow rate. In a solid rocket motor, this rate depends on the propellant burn rate (kg/s) and the burning surface area (m^2). The faster the propellant burns and the greater the burning surface area, the higher the mass flow rate and the higher the resulting thrust. Propellant burning rate depend on the type of fuel and oxidizer, their mixture ratio, and the binder material. The total burning area depends primarily on the inside shape of the solid propellant. During casting, designers can shape the hollow inner core of the solid propellant to adjust the surface area available for burning so they can control burning rate and thrust. The Space Shuttle's solid rocket motors, for example, have a star-shaped core, specifically tailored so the thrust decreases 55 s into the flight to reduce acceleration and the effects of aerodynamic forces.

Because solid motor combustion depends on exposed propellant surface area, manufacturers must carefully mold the propellant mixture to prevent cracks. Burning occurs on any exposed surface, even along undetected cracks in the propellant grain. Investigators linked the Space Shuttle *Challenger* accident to an improperly sealed joint between solid motor segments. This open seal exposed the motor case to hot gases, burning it through and causing the accident.

The *Challenger* disaster highlighted another drawback of solid motors—once they start, they are very difficult to stop. With a liquid rocket, we simply turn off the flow of propellant to shut off the engine. Solid motors burn until all the propellant is gone. To stop one prior to that would require us to blow off the top or split it open along its side, releasing internal pressure and thus, inhibiting combustion. That is not a very practical solution on the way to orbit!

Despite their drawbacks, solid motors are used on a variety of missions because they offer good, cost-effective performance in a simple, self-contained package that does not require a

Operating principle: An oxidizer and fuel are blended together in a single, solid grain. Combustion takes place along any exposed surface, producing heat and byproducts that are thermodynamically expanded through a nozzle.	
Advantages: • Simple, reliable • No propellant management needed • High I_{dsp} compared to bipropellant	**Disadvantages:** • Susceptible to cracks in grain • Cannot be restarted • Difficult to stop • Modest I_{sp}

Table 9.4 Solid Rockets

separate propellant management system. One important use of solid motors is to augment liquid engines on launch vehicles. Without the solid rocket boosters, the Space Shuttle could not get off the ground. Several expendable launch vehicles use various combinations of strap-on solid motors to give users a choice in payload-lifting capacity, without the need to redesign the entire vehicle. For example, three, six, or nine solid motors can be added to the Delta II launch vehicle, depending on the payload mass. Solid motors also provide thrust for strapon upper-stages for spacecraft needing a well-defined ΔV to go from a parking orbit into a transfer orbit.

A solid rocket motor's I_{sp} performance depends on the fuel and oxidizer used. After the chemicals are mixed and the motor is cast, however, I_{sp} and thrust profile are fixed. I_{sp} for typical solid motors currently in use ranges from 200 to 300 s, somewhat more than for a liquid, monopropellant rocket but slightly less than for a typical liquid bipropellant engine. Their big performance advantage is in terms of I_{dsp}. For example, the Shuttle's solid rocket boosters (SRBs) have an I_{dsp} 6% less than the I_{dsp} of the liquid main engines (SSMEs), even though the I_{sp} for the SSMEs is almost 70% higher. This makes solid motors ideal for volume-constrained missions needing a single, large ΔV. Table 9.4 summarizes key points about solid rocket motors.

Hybrid Chemical Rockets *Hybrid propulsion systems* combine aspects of both liquid and solid systems. A typical hybrid rocket uses a liquid oxidizer and a solid fuel. The molded fuel grain forms the combustion chamber and the oxidizer is injected into it as illustrated in Figure 9.15.

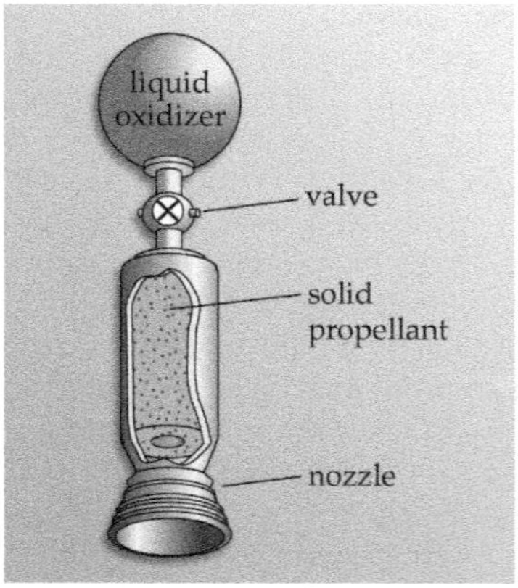

Figure 9.15 Hybrid rocket motor.

Operating principle: Hybrid rockets typically use a liquid oxidizer with a solid fuel. The oxidizer is injected into a hollow port (or ports) within the fuel grain where combustion takes place along the boundary with the surface.	
Advantages:	**Disadvantages:**
• Only one liquid propellant to manage • Safe, cannot create an explosive combination of fuel/oxidizer • Environmentally friendly products	• Limited heritage • Modest I_{sp}

Table 9.5 Hybrid Rockets

A separate sparking system or a superheated oxidizer initiates combustion. The hybrid combustion process is similar to burning a log in the fireplace. Oxygen from the air combines with the log (fuel) in a fast oxidation process and burns. If we take away the air (throttle the oxidizer), the fire goes down. If we use a bellows or blow on the fire, we increase the flow of air and the fire grows.

A properly designed hybrid rocket can offer the flexibility of a liquid system with the simplicity and density of a solid motor. Hybrids are safe to handle and store, similar to a solid, but can be throttled and restarted, similar to a liquid engine. Their efficiencies and thrust levels are comparable to solids. For example, one interesting hybrid configuration uses high-test peroxide (HTP) oxidizer with polyethylene (plastic) fuel. At an O/F of 8:1 this system offers an I_{sp} of around 290 s and I_{dsp} of around 3.8×10^5 kg/m^{3}s. It has the added advantage that the HTP can be used alone as a monopropellant making it a dual-mode system. Unfortunately, at this time hybrid rocket research and applications lags far behind research on liquid and solid systems. Table 9.5 summarizes key points about hybrid rockets.

Chemical Rocket Summary Table 9.6 compares the I_{sp} and I_{dsp} of the thermodynamic rockets we've discussed in this section and compares their performance and key features.

Solar Thermal Rockets

In chemical rockets, the heat is a by-product of a chemical reaction. But heat can also be produced in other ways, then transferred directly to the propellant using conduction and/or convection. One convenient source of heat is the sun. By concentrating solar energy using mirrors or lenses, we can create extremely high temperatures (2,000 K or more) on a focused point. By passing a propellant, such as hydrogen, through this point, it can directly absorb the heat, reaching very high temperatures before being expanded through a nozzle to achieve high velocity exhaust. In this way, *solar thermal rockets* use the limitless power of the sun to produce relatively high thrust with high I_{sp}.

The natural advantage of a solar thermal rocket is the abundant source of solar energy, saving the need to produce the energy on the spot or carry it along as chemical energy. It can use virtually any propellant. The best I_{sp}, of course, comes from using hydrogen. Theoretical and experimental results indicate a liquid-hydrogen, solar-thermal rocket could achieve a specific impulse of about 700 s. Thrust levels are limited by the basic engineering problems of efficiently transferring heat between the thermal mass that absorbs solar energy and the propellant. However, thrusts in the several-Newton range should be achievable. Another important operational challenge for solar thermal rockets is deploying and steering the large mirrors needed to collect and focus the solar energy. Several concepts for solar thermal rockets have been proposed, such as the Solar Orbital Transfer Vehicle, SOTV. However, up to now, none have been tested in orbit. Table 9.7 summarizes key features of solar thermal rockets.

Thermoelectric Rockets Of course, the solar energy is only available when the sun is shining. A spacecraft in eclipse, or far from the sun, needs another heat source. If you hold your hand up next

Type	Propellant combinations (O/F) [specific gravity]	I_{sp} (s)	I_{dsp} (s kg/m³)	Advantages	Disadvantages
Liquid	—	—	—	• Can be re-started • Can be throttled	• Propellant management needed • Intense heat of combustion creates thermal control problems for chamber and nozzle
Bipropellant	LO_2/LH_2 (5:1) [1.15:0.07] LO_2/Kerosene(RP-1) (2.25:1) [1.15:0.8] N_2O_4/N_2H_4 (1.9:1) [1.43:1.0]	477 370 334	462 385 429	• High I_{sp} • "Green" propellants • Storable fuel • Good I_{dsp} • Storable propellants • Good I_{sp}	• Cryogenic fuel/oxidizer difficult to store • Cryogenic oxidizer • Toxic propellants
Monopropellant	N_2H_4 (hydrazine) [1.00] H_2O_2 (hydrogen peroxide) [1.37]	245 181	246 247	• Simple • One set of propellant management • Moderate I_{sp} • Dual-mode capable • Large flight heritage • "Green" propellant	• Low I_{sp} • Toxic • Little flight heritage
Solid	NH_4ClO_4(AP)/Al (includes a binder, e.g., HTPB) [1.95:1.26]	300	539	• Simple • High I_{dsp} • No propellant management needed	• Sensitive to cracks in propellant grain • Difficult to stop • Cannot restart
Hybrid	H_2O_2(85%)/PE(8:1) [1.37:0.9]	333	437	• One set of propellant management • Safe operation (cannot explode) • "Green" propellants	• Limited heritage • Modest I_{sp}

LOX = Liquid oxygen; LH_2 = liquid hydrogen; RP-1 = rocket propellant-1, Kerosene; N_2O_4 = nitrogen tetroxide; N_2H_4 = hydrazine; HTP = high-test hydrogen peroxide (85% H_2O_2, 15% H_2O); HTPB = hydroxyl terminated polybutadiene (rubber); PE = polyethylene (plastic).

TABLE 9.6 Thermodynamic Rockets

Operating principle:	
Solar energy is concentrated using lenses or mirrors onto a heat transfer chamber. A propellant, such as liquid hydrogen, flows through the chamber, healing up, and is then expended through a nozzle.	
Advantages:	**Disadvantages:**
• Limitless energy supply can be refueled and reused	• Less effective at the outer planets.
• Potentially very high I_{sp} (~700 s with H_2)	• Must carefully point a large mirror or lens
• Environmentally friendly products	• No flight heritage

Table 9.7 Solar Thermal Rockets

to a conventional light bulb, you will feel the heat given off by the resistance of the filament in the bulb. For space applications, the energy source is the electrical energy provided by the spacecraft's electrical power subsystem (EPS). By running electricity through a simple resistor, or by creating an arc discharge as you would with a spark plug, we can create heat. *Thermoelectric rockets* transfer this heat to the propellant by conduction and convection.

One of the simplest examples of a thermoelectric rocket is a *resisto-jet*. As we see in Figure 9.16, electrical current flows through a metal heating element inside a combustion chamber. The resistance (or electrical friction) in the metal causes it to heat up. As propellant flows around the heating element, heat is transferred to it via convection, increasing its temperature before it expands through a nozzle.

This simple principle can be applied to virtually any propellant (NASA even investigated using urine on the Space Station as a propellant). The resisto-jet concept can significantly increase the specific impulse of a conventional cold-gas rocket, making it, in effect, a "hot-gas" rocket with twice the I_{sp} (from 60 s to more than 120 s for nitrogen). Resisto-jets improve the performance of a conventional hydrazine mono-propellant rockets by heating the exhaust products, thus boosting their I_{sp} by about 50% (from 200 s to over 300 s). The direct benefit of a resisto-jet rocket comes from adding heat to the propellant, so, the hotter it gets, the higher its I_{sp} and I_{dsp}. Hydrazine resisto-jets are gaining wide use as mission designers become increasingly able to trade extra electrical power for a savings in propellant mass. The Iridium spacecraft, for example, rely on hydrazine resisto-jets to take them from their initial parking orbit to their final mission orbit.

Another method for converting electrical energy into thermal energy is by using a spark or electric arc. To form an arc, we create a gap in an electrical circuit and charge it with a large amount of electricity. When the electrical potential between two points gets high enough, an arc forms (during a thunderstorm we see this as dazzling displays of lighting). An *arc-jet rocket* passes propellant through a sustained arc, increasing its temperature. Arc-jet systems can achieve relatively high I_{sp} (up to 1000 s) with small but significant thrust levels (up to 1 N). Like resisto-jets, arc-jet rockets can

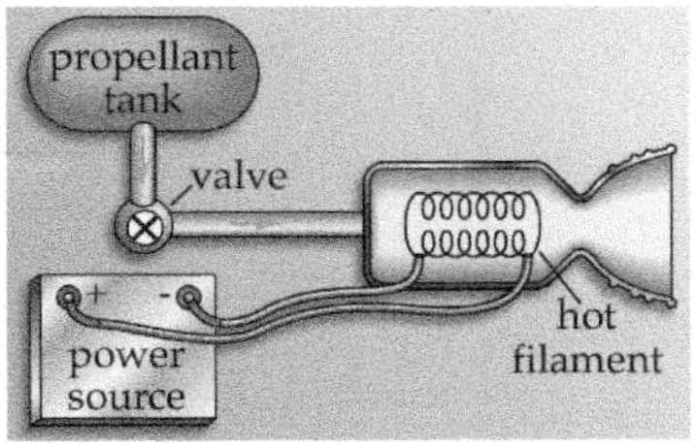

Figure 9.16 Resisto-jet.

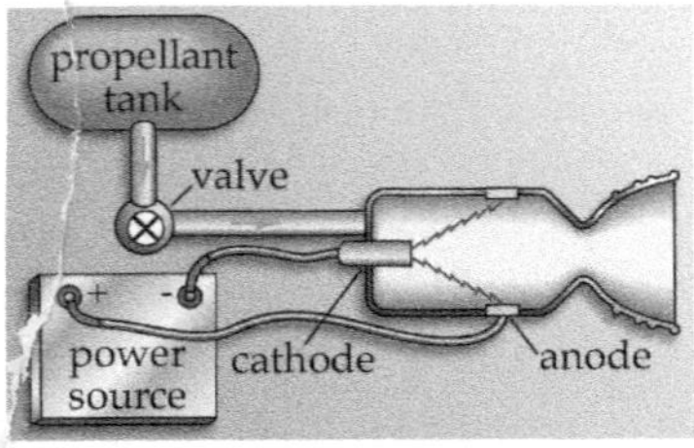

FIGURE 9.17 Arc-jet thruster.

use almost any propellant. Current versions use hydrazine, liquid hydrogen, or ammonia. A schematic for an arc-jet system is shown in Figure 9.17. The ARGOS spacecraft, for example, was launched in 1999 to test a 25-kW ammonia arc-jet, producing a thrust of around 0.8 N with a specific impulse over 800 s.

As you would expect, the primary limitation on thermoelectric rocket thrust and efficiency is the amount of power available. In Subsection 9.1 we saw a simple relationship among input power, thrust, and specific impulse.

$$F = 2P / I_{sp} g_o$$

where F = thrust (N)
P = power (W)
I_{sp} = specific impulse (s)
g_o = gravitational acceleration = 9.81 m/s^2

Using this simple equation, we can fine-tune the design of a thermoelectric thruster, trading off thrust versus power versus I_{sp}. For example, if we double the power input we can increase thrust by a factor of 4 for the same I_{sp}. Table 9.8 summarizes key features of thermoelectric rockets.

Nuclear Thermal Rockets

Another potentially useful heat source in space is nuclear energy. On Earth, nuclear reactors harness the heat released by the fission of uranium to produce electricity. In much the same way, a *nuclear thermal rocket* uses its propellant, such as liquid hydrogen, to flow around the nuclear core, absorbing the thermal energy while cooling the reactor. As you can see in Figure 9.18, propellant enters the reaction chamber, where it absorbs the intense heat from the nuclear reaction.

Operating principle: Heat is produced through electric resistance or spark discharge inside a heat transfer chamber. A propellant, such as water or ammonia (NH_3), flows through the chamber, heating up, and is then expended through a nozzle.	
Advantages: • Simple, reliable • Can be used as an add-on to conventional monopropellant rocket to boost I_{sp} ~50% • High power arc-jet offers very high I_{sp} (>800 s with NH_3)	**Disadvantages:** • Requires large amounts of onboard electrical power • Relatively low thrust (<1 N)

TABLE 9.8 Thermoelectric Rockets

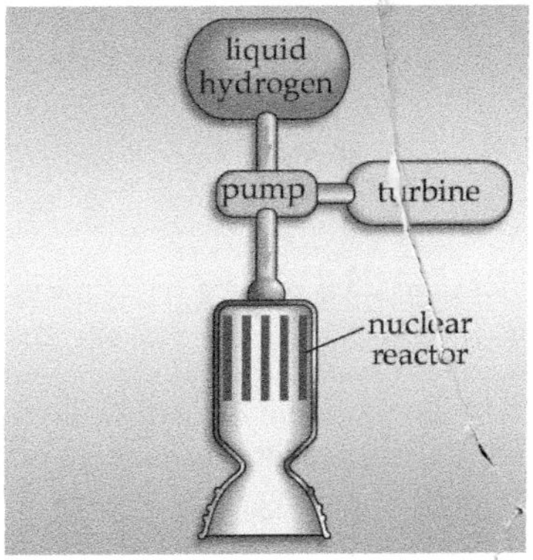

FIGURE 9.18 Nuclear thermal rocket.

From there, thermodynamic expansion through a nozzle produces high thrust (up to 106 N) and high I_{sp} (up to 1000 s using hydrogen).

Because of their relatively high thrust and better efficiencies, nuclear thermal rockets offer a distinct advantage over chemical systems, especially for manned planetary missions. These missions must minimize transit time to decrease the detrimental effects of free fall as well as exposure to solar and cosmic radiation on the human body. Ironically, future astronauts may escape the danger of space radiation by using the energy from a nuclear reactor to propel them to their destination faster. Extensive research into nuclear thermal rockets was done in the United States. in the 1960s as part of the NERVA program. Additional work was done in the 1980s when great theoretical advances occurred in heat transfer. Unfortunately, environmental and political concerns about safe ground testing of nuclear-thermal rockets (let alone the potential political problems of trying to launch a fully fueled nuclear reactor) have severely reduced research into this promising technology. Table 9.9 summarizes key features of nuclear thermal rockets.

Electrodynamic Rockets

While thermodynamic rockets offer relatively high thrust over a very wide range—10^{-1} to 10^6 N—basic problems in heat transfer pose practical limits on the specific impulse that they can ultimately reach (up to 1,000 s or so for nuclear rockets). To achieve the higher efficiencies demanded by future, more challenging interplanetary and commercial missions, we need to take a different approach—electrodynamic rockets.

Operating principle: Heat is produced through nuclear fission inside a reactor. A propellant, such as liquid hydrogen, flows through the reactor, heating up, and is then expended through a nozzle.	
Advantages: • Long-term energy supply, can be refueled and reused • Potentially very high I_{sp} (~1,000 s with H_2) • High thrust (~10^6N)	**Disadvantages:** • Environmental and political problems with testing and launching nuclear reactors • No flight heritage

TABLE 9.9 Nuclear Thermal Rockets

As discussed in Subsection 9.1, electrodynamic rockets rely on electric and/or magnetic fields to accelerate a charged propellant to very high velocities (more than 10 times the exhaust velocity and I_{sp} of the Shuttle main engines). However, this high I_{sp} comes with a price tag—high power and low thrust. Recall from equation (9.2) the relationship among power, exhaust velocity, and thrust.

$$F = 2P/C$$

Power, of course, is always a limited commodity on a spacecraft—especially if you are trying to use an electrodynamic thruster. Given a finite amount of power, equation (9.2) tells us we can have high exhaust velocity only at the expense of thrust. As a result, practical limits on power availability make electrodynamic thrusters unsuitable for launch vehicles or when a spacecraft needs a quick, large impulse, such as when it brakes to enter a capture orbit. Even so, because of their very high I_{sp}, mission planners are increasingly willing to sacrifice power and thrust (and the extra time it will take to get where you need to go) in order to save large amounts of propellant mass.

As we indicated in Subsection 9.1, there are many ways to use electric and/or magnetic fields to accelerate a charged propellant. Here we will focus on the two primary types of electrodynamic rockets currently in use operationally:

- Electrostatic rockets—use electric fields to accelerate ions
- Electromagnetic rockets—use electric and magnetic fields to accelerate a plasma

Electrostatic Rockets

Ion Thrusters

An *ion thruster* (also called an *electrostatic thruster*) uses an applied electric field to accelerate an ionized propellant. Figure 9.19 illustrates the basic operating principles of an ion thruster. First, the thruster ionizes propellant by stripping off the outer shell of electrons, making positive ions. It then accelerates these ions by applying a strong electric field. If the engine ejected the positive ions without neutralizing them, the spacecraft would eventually accumulate a negative charge over time due to the leftover electrons. To prevent this, as Figure 9.19 illustrates, it uses a neutralizer source at the exit plane to eject electrons into the exhaust, making it charge-neutral.

The ideal ion thruster propellant is easy to ionize, store, and handle. Early ion thruster research used mercury and cesium, both metals that are easy to ionize. Unfortunately, they are also toxic, making them difficult to store and handle. Currently, the most popular propellant for ion thrusters is xenon. Xenon is a safe, inert gas that stores as a dense liquid (4.56 times the density of water)

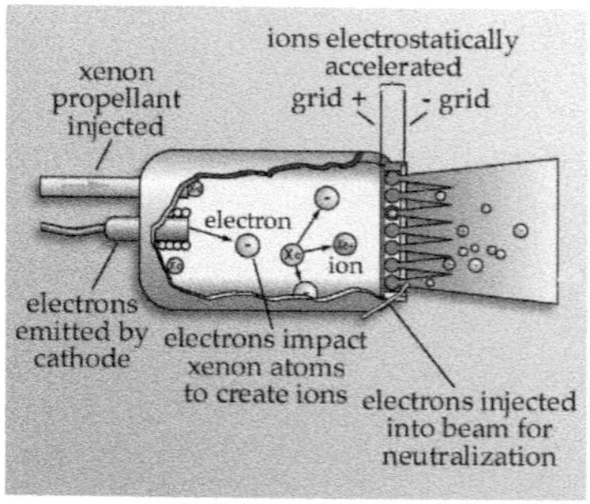

FIGURE 9.19 Simple ion thruster.

under a moderate pressure of 58.4 bar at room temperature. This high-density propellant also gives ion thrusters excellent density-specific impulse.

Ion thrusters offer an electrically efficient (~90%, meaning 90% of the power goes to accelerating propellant) propulsion with very high specific impulse (as high as 10,000 s). As we discussed earlier, thrust and I_{sp} in electrodynamic rockets are limited only by available power. Ion thrusters have been used on a variety of space missions. Perhaps their most exciting application is on interplanetary missions. Dawn mission used an ion rocket to explore asteroids.

Hall-Effect Thrusters

A Hall-effect thruster (HET) is another type of electrostatic thruster. HETs take advantage of a unique effect called a "Hall current" that occurs when we apply a radial magnetic field to a conducting plasma. The interaction of the applied magnetic field and resulting electric field causes the electrons to circle the discharge channel in a continual Hall current.

The electrons in the Hall current collide with the incoming neutral gas to ionize the atoms and create the plasma. The much more massive ions do not feel the magnetic field as much, and are accelerated by the electric field as illustrated in Figure 9.20. Note that the ions that produce thrust are accelerated electrostatically similar to ion thrusters. The magnetic field only confines the electrons to ionize the neutral gas to create the plasma. The electric field is not created by charge separation, but by the resistance of the plasma across the strong magnetic field. Russian scientists

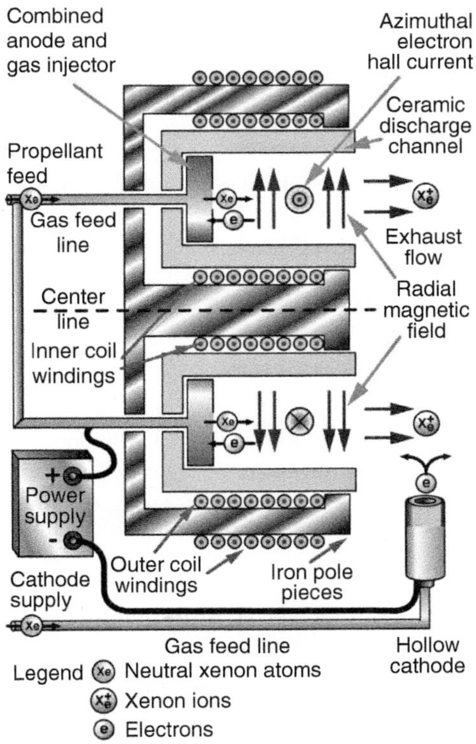

FIGURE 9.20 Hall-effect thruster (HET).

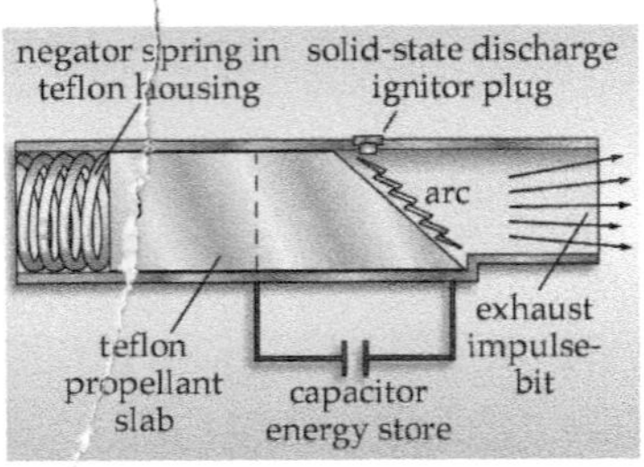

FIGURE 9.21 Pulsed-plasma thruster (PPT).

pioneered many of the modern advances in HETs with their stationary plasma thrusters (SPTs) series of HETs being the most widely used. The first Russian HETs were launched in the early 1970s for station-keeping applications and they are very common on GEO communication satellites. The first U.S.-designed HET was launched in 2006 and HETs are now used for orbit raising as well as station keeping. As with ion thrusters, xenon is also the most widely used propellant. Recent advancements in Hall thruster design include nested channel thrusters where two or more Hall thrusters are concentric in one thruster body. This allows for significantly higher ratios of thrust to propulsion system mass because they share some elements of the magnetic circuit.

Electromagnetic Rockets

Unlike all other types of rockets, which operate continuously, pulsed-plasma thrusters (PPTs) operate in a noncontinuous, pulsed mode. And in contrast to ion and plasma thrusters, PPTs use a solid propellant, usually Teflon (PTFE). A high-voltage arc pulses over the exposed surface of the propellant. This intense electric field vaporizes it, creating a plasma. The resulting induced magnetic field accelerates the plasma. Figure 9.21 shows a schematic for a simple PPT. A number of missions have used PPTs for spacecraft station keeping. Their advantage is their precisely controlled, low thrust levels. Because they operate in a pulsed mode, they don't need continuous high power. Instead, they can gradually store electrical energy in a capacitor for release in high power bursts (the same technique used in a camera flash). This low-power, pulsed operating mode makes them suitable for many small satellite applications.

Compared to electrostatic thrusters, PPTs are relatively low in energy-conversion efficiency (20%). However, they provide respectable I_{sp} (700–1500 s) but with low thrust (10^{-3} to 10^{-5} N). Their biggest potential advantage is in ease of integration. Because they don't require any additional propellant management, they can be built as simple, self-contained units that, in principle, we can easily bolt onto a spacecraft. Table 9.10 summarizes key information about the electrodynamic rockets we've discussed.

Due to their low thrust levels, EP systems must operate continuously far longer than traditional chemical systems. For example, the 30-cm ion engine on Deep Space 1 operated for 16,265 h during the mission and ion engines have been tested on the ground for over 40,000 h of continuous operation. Due to these long operation times, erosion of components eventually leads to failure. The erosion is caused by ions unintentionally striking the thruster, causing material to sputter away. In ion engines, erosion of the grids is the most common failure mechanism. In Hall thrusters, the discharge channels are typically made of ceramic such as boron nitride and the most common failure mechanism is erosion of the channels from ion bombardment. For both systems, the cathode that supplies electrons can also erode until failure.

Type	Propellant	Operating principle	Electrical efficiency	Thrust (N)	I_{sp} (s)
Ion (or electrostatic) thruster	xenon	Applied electric field is used to accelerate ionized propellant	90%	0.1–1.0	2000–10,000
Stationary plasma thruster (SPT)	xenon	Combined electric and magnetic fields produce a "Hall effect" that ionizes the propellant, which is then accelerated by the electric field	60%	0.1–1.0	~2000
Pulsed plasma thruster (PPT)	Teflon (PTFE)	An electric arc is discharged in a pulsed mode over a solid propellant, vaporizing it and creating a plasma. Interaction between the applied electric field and resulting magnetic field accelerates the plasma.	20%	10^{-5}–10^{-3}	~1500

Adapted from SPAD.

TABLE 9.10 Electrodynamic Rockets

A recent development for Hall thrusters is magnetic shielding, which offers the promise of almost unlimited operation time. In this design the magnetic field lines have been carefully designed to minimize plasma contact with the discharge channel walls, thus limiting erosion.

System Selection and Testing

So far, we looked at all the pieces that make up propulsion systems and many of the various rocket technologies options available. But many questions about propulsion system applications remain.

- How do mission planners go about selecting the best technology from this large menu? How do researchers decide which is the best technology to pursue for future applications?
- How are new or improved systems tested and declared fit for flight?

As with most technology decisions, there is rarely one best answer for any given application. Sometimes, as in the case of our *FireSat* example, the severe constraints on volume, power and mass, coupled with the modest ΔV requirements leave only a few realistic options—cold-gas thrusters, or possibly a monopropellant system. Even when we narrow the field, the choice of the right propulsion system for a given mission depends on a number of factors that we must weigh together.

One way to trade-off various rocket options is to select one with the lowest total cost. But here, cost represents much more than simply the engine's price tag. The *total cost* of a propulsion system includes at least eight other factors, in addition to the bottom-line price tag, that we must consider before making a final selection:

1. Mass performance—measured by I_{sp}
2. Volume performance—measured by I_{dsp}
3. Time performance—how fast it completes the needed ΔV, measured by total thrust
4. Power requirements—how much total power the EPS must deliver
5. Safety—how safe the system (including its propellant) is and how difficult it is to protect engineers working with the system
6. Logistics requirements—how difficult it will be to transport the system and propellant to the launch site and service it for flight
7. Integration cost—how difficult the system is to integrate and operate with other spacecraft subsystems and the mission operations concept
8. Technical risk—new, untried systems carry more risk to the mission than proven technology

Different missions (and mission planners) naturally place a higher value on some of these factors than others. Other missions, such as a complex commercial mission, may place a high priority on reducing technical risk. For them, a new type of plasma rocket engine, even if it offers lower mass cost, may be too risky when they consider all other factors. When asking what the best option is for a given mission, "it depends" is usually the best answer!

Once a system is selected, engineers must conduct a rigorous testing and qualification process before it can be declared safe for use. New rocket development usually progresses from relatively crude engineering model testing under atmospheric conditions to more elaborate testing of flight models under high-altitude or vacuum conditions. Of course, for specialized systems such as electromagnetic thrusters (e.g., ion thrusters or SPTs), testing can only be done under vacuum conditions using highly accurate thrust stands to measure micronewtons (10^{-3} N) of thrust. During experimental testing, rocket scientist carefully measure mass flow rates, chamber pressures, temperatures, and other parameters and compare them to predicted values based on thermochemical and other models.

Because rockets typically involve high pressures, high temperatures, high voltages, and hazardous chemicals, safety issues are a primary concern. These concerns carry through from initial development of new rockets to servicing of proven systems in preparation for flight. In the case of launch vehicle propulsion, human lives may depend on safe, reliable operation. As discussed earlier, special loading procedures and equipment are used to ensure safe handling of hazardous propellants.

Ensuring system reliability involves a complex series of ground tests measuring performance over a wide range of conditions. These can range from relatively simple tests to ensure the system does not leak at flight pressure, to complicated tests that require widely varying O/F ratios and expansion conditions. In addition to performance, all the typical space environment testing done for other subsystems, such as thermal and vacuum testing, must also be accomplished for the propulsion subsystem.

Exotic Propulsion Methods

Chemical rockets have given us access to space and taken spacecraft beyond the solar system. Electrostatic rockets offer a vast increase in mass efficiency making exciting, new missions possible. However, to really open space to colonization and allow humans to challenge the stars, we need new ideas. Exotic propulsion systems are those far out ideas still on the drawing boards. While there are many exotic variations to the rockets we have already discussed (such as using a high-energy density or meta-stable chemicals, nuclear fusion, or antimatter to create superheated products), here we will focus on even more unconventional types of propulsion—ones that produce thrust *without* ejecting mass:

- Solar sails
- Tethers

We will see how these far out concepts can be used to give us even greater access to the solar system. Then we will go beyond that to look at one of the unique challenges of interstellar flight.

Solar Sails

Just as a sail is used to harness the force of the wind to move a ship, a *very* large *solar sail* can be used to harness the force of solar pressure to prepell a spaceship without ejecting mass. Of course, the farther it goes from the sun, the less solar pressure it can collect, so a solar sail would work best inside Mars's orbit.

How large would a sail need to be? This force can be determined from

$$F = F_s/cA_s(1+r)\cos I \tag{9.33}$$

where F_s = solar constant = 1358 W/m^2 at Earth's orbit around the sun
c = speed of light = 3×10^8 m/s
A_s = surface area (m^2)
r = surface reflectance (where $r = 1$ for a perfect reflector)
I = incidence angle to the sun (deg)

To produce just 5 N of thrust near Earth (about 1 lb), a 1-km^2 sail would be needed. To achieve escape velocity from a low-Earth orbit (assuming a total spacecraft mass of only 10 kg), this force would have to be applied for more than 17 years. Of course, a solar sail uses no propellant, so the thrust is basically "free." As long as no one is in a hurry, a solar sail offers a cheap way to get around. Solar sails have been proposed to maneuver mineral-rich asteroids closer to Earth to allow for orbital mining operations.

Tethers

Another imaginative means of propulsion that does not need propellant uses very long cables called *tethers*. Typically, these booms are only a few meters long. Using a small mass at the end of a very long tether, tens or even hundreds of kilometers long, produces the same stabilizing effect. But even more interesting effects become possible as well.

Picture a large spacecraft such as the Shuttle in a circular orbit. Now imagine a small payload deployed upward (away from Earth), from the Shuttle at the end of a very long tether as shown in Figure 9.22.

Typically, when we compute orbital velocities we assume we were dealing with point masses effected only by gravity. From an orbital mechanics standpoint, this point mass assumption is valid only at the center of mass of the Shuttle/payload system. If the payload mass is small with respect to Shuttle's mass, the system's center of mass will not move significantly when it deploys. Thus, the orbital velocity of the system will stay about the same. What does this mean for the payload? Secured by the tether, it is being pulled along in orbit at the Shuttle's orbital velocity. But the payload is above the Shuttle. Orbital velocity depends on distance from the center of the Earth. Therefore, because the payload is higher than the Shuttle, its proper circular, orbital velocity would be somewhat *lower* than the velocity it maintains due to the tether. Or, said another way, the tether forces it to travel *faster* than orbital mechanics would dictate for its altitude.

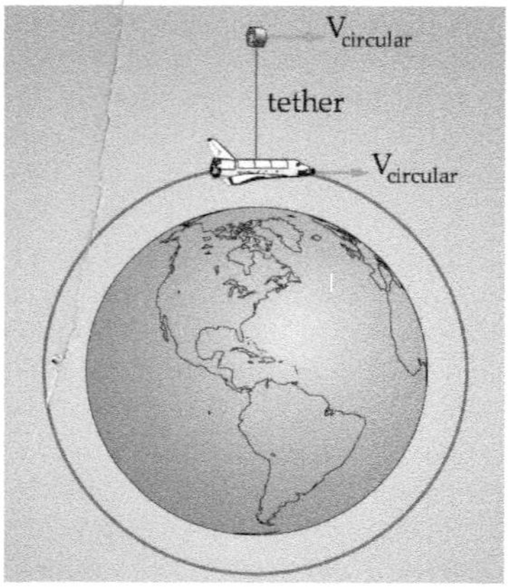

FIGURE 9.22 Space tether deployment.

Now, what happens if we suddenly cut the tether? Orbital mechanics would take over and the payload would suddenly find itself at a velocity too fast for a circular orbit at that altitude. The situation would be as if its velocity were suddenly increased by firing a rocket. It would enter an elliptical orbit with a higher apogee one-half orbit later. Analysis indicates this new apogee altitude would be seven times the length of the tether higher than the original circular orbit. In other words, if the original altitude of the payload were 300 km and the tether length were 10 km, the new elliptical payload orbit would be 310 × 370 km, as illustrated in Figure 9.23.

If the payload were deployed downward instead of upward, the opposite would happen. Its orbit would shrink, so that half an orbit after the tether released, the payload would reach perigee. This technique was used by the Small Expendable-tether Deployment System (SEDS) mission in 1993 to successfully deorbit a small payload (Humble et al. 1995).

Of course, tether propulsion is not completely "free." We still need to add the mass of the tether and its deployment motors and gears. And we need extra electrical power to operate the

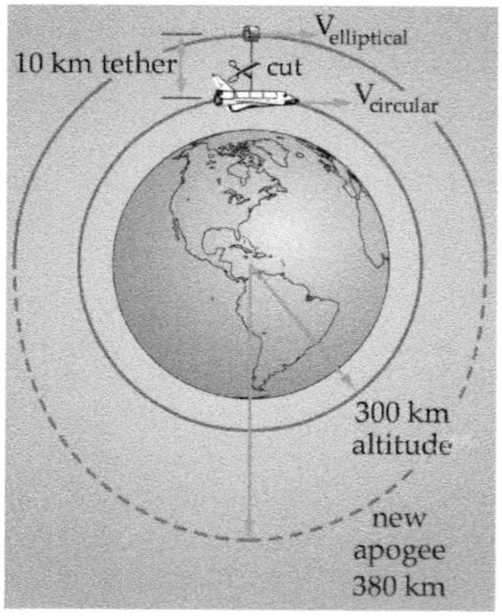

FIGURE 9.23 Tether orbit boost.

tether-deployment mechanisms. However, once we put these systems in place, we could conceivably use the tether system over and over again to boost or deorbit payloads.

Space Shuttle astronauts have performed a number of experiments to investigate the exciting possibilities of tethers. So far, these experiments have focused on the practical problems of deploying, controlling, and reeling in a small payload at the end of a long tether. Future applications for tethers are truly unlimited. A series of rotating tether stations could be used to sling-shot payloads, passing them from one to the other all the way from low-Earth orbit out to the Moon. Another exiting use of tethers is for power generation. A conducting tether passing through the Earth's magnetic field each orbit can generate large amounts of electrical power (Forward and Hoyt, 1999).

Interstellar Travel

The ultimate dream of space exploration is someday to travel to other star systems. Actually, the first human-built starships are already on their way out of the solar system. Launched in 1972 and 1973, NASA's Pioneer 10 and Pioneer 11 probes became the first spacecraft to leave our local planetary neighborhood and begin their long journey to the stars. Unfortunately, at their present velocities, they are not expected to pass near another stellar body in over 2 million and 4 million years, respectively!

Obviously, these travel times are far too long to be useful for scientists who want to be around to review the results from the mission. Hollywood's version of rocket science can take advantage of hyperspace and warp drive to allow roundtrip times to nearby stars in the space of a single episode. Unfortunately, real-world rocket science is far from these amazing means of propulsion.

Assuming we could develop efficient on-board energy sources, such as fusion or antimatter, and rely on ion or other extremely efficient types of rockets to achieve very high specific impulse, there is still the limit imposed by the speed of light. If a rocket could thrust continuously for several years, even at a very low thrust level, it would eventually reach very high velocity.

One aspect of Albert Einstein's theory of relativity says that as an object's velocity approaches light speed, its perception of time begins to change relative to a fixed observer. This time adjustment leads to the so-called twin paradox. To visualize this concept, imagine a set of twins, a sister and brother. If the sister leaves her brother and sets off on a space mission that travels near the speed of light, when she returns, she'll find her brother much older than she is! In other words, while the mission seemed to last only a few years for her, tens or even hundreds of years would have passed for her brother.

We express this *time dilation* effect, sometimes called a tau (τ) factor, using the Lorentz transformation:

$$\tau = \frac{t_{\text{starship}}}{t_{\text{Earth}}} = \sqrt{1 - \frac{V^2}{c^2}} \tag{9.34}$$

where t_{starship} = time measured on a starship

t_{Earth} = time measured on Earth

V = starship velocity

c = speed of light = 300,000 km/s

The tau factor, τ, tells us the ratio of time aboard a speeding starship compared to Earth time. As the spacecraft's velocity approaches light speed, τ gets very small, meaning that time on the ship passes much more slowly than it does on Earth. While this may seem convenient for readers thinking about a weekend journey to the star Alpha Centauri (4.3 light years away), Einstein's theory also places a severe speed limit on would-be space travelers. As a spacecraft's velocity increases, its effective mass also increases. Thus, as the ship's velocity approaches light speed, it needs more thrust than it did at lower speeds to get the same velocity change. To attain light speed, it would need an infinite amount of thrust to accelerate it's infinite mass. For this reason alone, travel at or near the speed of light is well beyond current technology.

For years, scientists and engineers said travel beyond the speed of sound, the so-called sound barrier, was impossible. But in October 1947, Chuck Yeager proved them all wrong while piloting the Bell X-1. Today, jet planes routinely travel at speeds two and three times the speed of sound. Perhaps by the 23rd century some future Chuck Yeager will break another speed barrier and take a spacecraft beyond the speed of light.

9.3 Launch Vehicles

Now that we have seen the types of rockets available, let's see how they are used to solve perhaps the most important problem of astronautics—getting into space. Launch vehicles come in many different shapes and sizes, from the mighty Space Shuttle to the tiny Pegasus. In this section, we will start by examining the common elements of modern launch vehicles. Looking at launch vehicles as systems, we will review the various subsystems that work together to deliver a payload into orbit and focus on the unique requirements for the massive propulsion systems needed to do the job. Finally, we will look at staging to see why a launch vehicles are broken into pieces that are used and discarded on the way to orbit.

Launch Vehicle Systems

A launch vehicle needs most of the same subsystems to deliver a payload (the spacecraft) from the ground into orbit. The two biggest differences between a launch vehicle and a spacecraft are the total operation time (about 10 min versus 10^+ years) and the total ΔV needed (>10 km/s versus 0–1 km/s). Let us start by looking at the challenges of launch vehicle propulsion to see how we must adapt the technologies discussed earlier in this chapter to the challenging launch environment. Then we will briefly review the other subsystems needed to support these large rockets and safely deliver spacecraft (and people) into space.

Propulsion Subsystem

The launch vehicle propulsion subsystem presents several unique challenges that sets it apart from the same subsystem on a spacecraft. These include:

- Thrust-to-weight ratio—must be greater than 1 to get off the ground
- Throttling and thrust-vector control—may need to vary the amount and direction of thrust to decrease launch loads and allow for steering
- Nozzle design—nozzles face varying expansion conditions from the ground to space

Let us go through each of these challenges in more detail.

Thrust-to-Weight Ratio To get a rocket off the ground, the total thrust produced must be greater than the weight of the vehicle. We refer to the ratio of the thrust produced to the vehicle's weight as the *thrust-to-weight ratio*. Thus, a launch vehicle's propulsion system must produce a thrust-to-weight ratio greater than 1. For example, the thrust-to-weight ratio for the Atlas launch vehicle is about 1.2 and the Space Shuttle's about 1.6.

Even though chemical rockets are not as efficient as some options discussed in the last section, they offer very high thrust and, more importantly, very high thrust-to-weight ratios. For this important reason, only chemical rockets are currently used for launch vehicle applications.

Throttling and Thrust Vector Control For virtually all spacecraft applications, rocket engines are either on or off. There is rarely a need to vary their thrust by *throttling* the engines. However, for launch-vehicle applications, throttling is often needed, greatly adding to the complexity (and cost!) of launch vehicle propulsion systems.

One reason for throttling has to do with the high launch forces that are generated on the vehicle as it flies through the atmosphere. Within the first minute or so of launch, the vehicle's velocity increases rapidly while it is still relatively low in altitude where the atmosphere is still fairly dense. The effect of passing through this dense atmosphere at high velocity produces *dynamic pressure* on a vehicle. Without careful attention to design and analysis, these launch loads could literally rip the vehicle apart. During design, some maximum value is assumed, based on extensive analysis of expected launch conditions, that cannot be exceeded without risking structural failure. Prior to each launch, engineers carefully measure and analyze the winds and other atmospheric conditions over the launch site to ensure the vehicle won't exceed it's design tolerances. In many cases, this is based on an assumed thrust profile for the vehicle that decreases, or throttles down, during peak dynamic pressure. The Space Shuttle, for example, reduces the main engines' thrust from 104% to 65% and the solid rocket boosters are likewise tailored to reduce their thrust during this phase of flight to keep dynamic pressure below a predetermined, safe level.

Another reason for throttling is to keep total acceleration below a certain level. An astronaut strapped to the top of a launch vehicle feels the thrust of lift-off as an acceleration or *g-load* that pushes him or her back into the seat. From Newton's laws, we know the total acceleration depends on the force (thrust) and the total mass of the vehicle. If the engine thrust is constant, the acceleration will gradually increase as the vehicle gets lighter due to expended propellant. This means the acceleration tends to increase over time. To keep the overall g-load on the Space Shuttle under 3 g's, the main engines throttle back about six minutes into the launch to decrease thrust to match the propellant that is used.

Some vehicles also need throttling for landing. The decent stage engine in the Lunar Excursion Module (LEM) used during the Apollo missions allowed an astronaut to throttle the engine over a range of 10–100% so they could make a soft touch down on the lunar surface.

Finally, launch vehicle rockets often have the unique requirement to be able to vary their thrust direction to allow for steering. This *thrust vector control* (*TVC*) is usually managed by gimbaling the entire engine to point the thrust in the desired direction. The Space Shuttle, for example, can vary the thrust direction for each main engine by ±10°. Of course, the mechanical gears and hydraulic actuators needed to move massively thrusting rocket engines can be quite complicated. Earlier rockets used more simple methods of thrust vector control. The V-2 rocket, for example, used large, movable ablative vanes stuck into the exhaust to change direction. Other launch vehicles use separate steering rockets or direct injection of gases into the exhaust flow to change thrust direction.

Nozzle Design In Subsection 9.1, we discussed the importance of external pressure and nozzle expansion ratio to overall engine performance. We prefer not to have a rocket nozzle either overexpanded or underexpanded, but instead designed for ideal expansion. In comparison, spacecraft rocket engines always work within a vacuum and designers simply use the greatest expansion ratio possible for the best performance. For launch vehicle rocket engines, the choice of expansion ratio is not so simple.

During launch, the external pressure on the first stage engines goes from sea level (1 bar or 14.7 psi) to near zero bars (vacuum) in just a few minutes. Ideally, we would like the nozzle to increase its expansion ratio throughout the trajectory to change the exit pressure as atmospheric pressure decreases. Unfortunately, with current technology, the weight of the hardware to do this is far too great. Instead, we design the nozzle to achieve ideal expansion at some design altitude about two-thirds of the way from the altitude of engine ignition to the altitude of engine cutoff.

For example, if we design a rocket to go from sea level to 60,000 meters, a reasonable choice for the desired exit pressure would be the atmospheric pressure at about 40,000 meters altitude. As a result, our rocket would (by design) be over-expanded below 40,000 meters and under-expanded above 40,000 meters. As we see in Figure 9.24, a nozzle designed in this way offers better overall performance than one designed to be ideally expanded only at sea level.

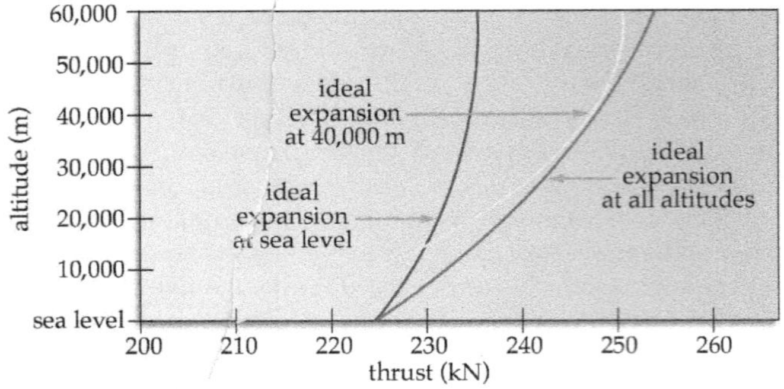

FIGURE 9.24 Thrust versus altitude for different nozzle designs.

Guidance, Navigation, Control Subsystem

A launch vehicle must deal with the issue of attitude determination and control in a much more dynamic environment than that faced by a spacecraft. The guidance, navigation, and control (GNC) subsystem keeps the launch vehicle aligned along the thrust vector to prevent dangerous side loads, keeps the thrust vector pointed according to the flight profile, and ensures the vehicle reaches the correct position and velocity for the desired orbit.

As with all control systems, the GNC subsystem has actuators and sensors. The primary launch vehicle actuators are the main engines, which make use of TVC and throttling to get the rocket where it needs to go. GNC sensors are typically include accelerometers and gyroscopes to measure acceleration and attitude changes. Even though the accuracy of these sensors drifts over time, they are usually sufficiently accurate for the few minutes needed to reach orbit. Future launch vehicles will use the Global Positioning System (GPS) for position, velocity, and attitude information.

Communications and Data Handling

Throughout launch, the vehicle must stay in contact with the Launch Control Center. There, flight controllers constantly monitor telemetry from the launch-vehicle subsystems to ensure they are functioning nominally. To do this, the vehicle needs a communication and data handling subsystem to process on-board data and deliver telemetry to the ground. Launch-vehicle data handling is very similar to a spacecraft's. Computers process sensor information and compute commands for actuators, as well as monitor other on-board processes. On expendable vehicles, these subsystems can be relatively simple because they need to work for only a few minutes during launch and will not be exposed to long periods of space radiation.

Communication equipment is also very similar in concept to that found on spacecraft. However, for safety reasons, an independent means of tracking the launch vehicle's location on the way to orbit is needed. In the Launch Control Center, Range Safety Officers monitor a launch vehicle's trajectory using separate tracking radar, ready to send a self-destruct command if it strays much beyond the planned flight path to endanger people or property.

Electrical Power

Launch vehicle electrical power requirements are typically quite modest compared to a spacecraft's. Launch vehicles need only enough power to run the communication and data handling subsystems, as well as sensors and actuators. Again, because of their limited lifetimes, expendable launch vehicles typically rely on relatively simple batteries for primary power during launch. The Space Shuttle uses fuel cells powered by hydrogen and oxygen.

Structure and Mechanisms

Finally, we must design the launch vehicle's structures and mechanisms to withstand severe loads and perform the numerous mechanical actuations and separations that must happen with split-second timing. A typical launch vehicle can have tens or even hundreds of thousands of individual nuts, bolts, panels, and load-bearing structures that hold the subsystems in place and take the loads and vibrations imposed by the engines' thrust and the atmosphere's dynamic pressure. Since the majority of a launch vehicle's volume is propellant tanks, these tend to dominate the overall structural design. Often the tanks become part of the primary load bearing structure. For the Atlas launch vehicle, used during the Mercury program, the thin-shelled tank literally inflated with a small positive pressure to create the necessary structural rigidity.

In addition to the problem of launch loads and vibrations, hundreds of individual mechanisms must separate stages and support other dynamic actions throughout the flight. These mechanisms tend to be larger than similar mechanisms on spacecraft. During staging, large sections of the vehicle's structure must literally break apart, usually by explosive bolts. Gimbaling the massive engines to change their thrust direction requires large hinges, hydraulic arms, and supporting structure.

Launch vehicle designers have the challenge of carefully integrating all of these structures and mechanisms with the engines, tanks, and other subsystems to create a compact, streamlined vehicle. Sadly, for expendable vehicles, all the painstaking design and expensive construction and testing to build a reliable launch vehicle burns up or drops in the ocean within 10 minutes after launch.

Staging

Getting a payload into orbit is not easy. As we learned in Subsection 9.2, the state of the art in chemical rockets (the only type currently available with a lift-to-weight ratio >1) can only deliver a maximum I_{sp} of about 450 s. Given the ΔV needed to get into orbit, and the hard realities of the rocket equation, that means most of a launch vehicle will be taken up by propellant. In fact, over 80% of a launch vehicle's lift-off mass is propellant. All of this propellant must be contained in large propellant tanks, that also add mass. Of course, the larger the mass of propellant, tanks, and other subsystems, the less mass is available for payload. One way of reducing the vehicle's mass is to somehow get rid of mass that is no longer needed. After all, why carry all that extra tank mass along when the rocket engine empties the tank steadily during launch anyway? Instead, why not split the propellant into smaller tanks and then drop them as they empty? Fighter planes flying long distances use this idea in the form of drop tanks. These tanks provide extra fuel for long flights and are dropped as soon as they are empty, thus lightening and streamlining the plane. This is the basic concept of staging.

Stages consist of propellant tanks, rocket engines, and other supporting subsystems that are discarded to lighten the launch vehicle on the way to orbit. As each stage burns out, it is dropped off and the engines of the next stage ignite (hopefully) to continue the flight into space. As each stage drops off, the vehicle's mass decreases, meaning a smaller engine can keep the vehicle on track into orbit.

Table 9.11 gives an example of how staging can increase the amount of payload that can be delivered to orbit. For this simple example, notice the two-stage vehicle can deliver more than twice the payload to orbit as a similar-sized, single-staged vehicle with the same total propellant mass—even after adding 10% to the structure's overall mass to account for the extra engines and plumbing needed for staging. This added payload-to-orbit capability is why all launch vehicles currently rely on staging.

In Table 9.11, for both cases, the size of the payload delivered to orbit compared to the weight of the entire launch vehicle is pretty small—5% or less. About 80% of a typical vehicle is propellant. The other 15% or so is made up of structure, tanks, plumbing, and other subsystems. Obviously, we could get more payload into space if only our engines were more efficient. However, with engines

	Launch vehicle	Parameters	Payload to orbit
Single Stage 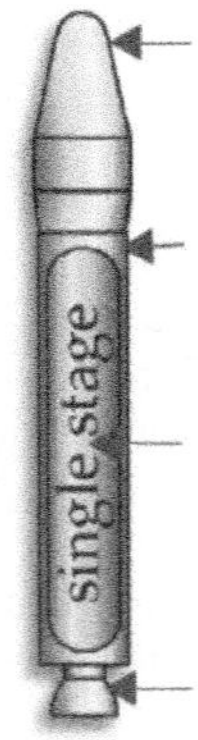 	$m_{payload}$ = 25 kg m_{inert} = 1,051 kg $m_{propellant}$ = 8,899 kg engine I_{sp} = 430 s	ΔV_{design} = 9,310 m/s Inert Mass Fraction = 0.1056,685 **Totals** m_{inert} = 1,051 kg $m_{propellant}$ = 8,899 kg $m_{vehicle}$ = 10,000 kg	$m_{payload}$ = 25 kg
Two Stage 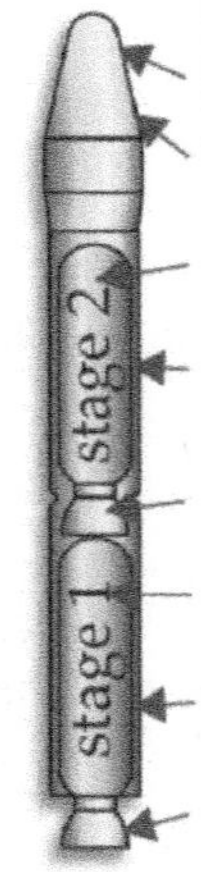 	$m_{payload}$ = 600 kg m_{shroud} = 25 kg $m_{propellant}$ = 1,729 kg m_{inert} = 204 kg engine I_{sp} = 430 s $m_{propellant}$ = 6,655 kg m_{inert} = 786 kg engine I_{sp} = 430 s	ΔV_{design} = 9,310 m/s Inert Mass Fraction = 0.1056,685 **Totals** m_{inert} = 990 kg $m_{propellant}$ = 8,384 kg $m_{vehicle}$ = 10,000 kg	$m_{payload}$ = 600 kg

Comparing Single-stage and Two-stage Launch Vehicles. These examples try to make an apples-to-apples comparison between a single stageto-orbit vehicle and a two-stage vehicle. For this example, identical total Delta-V_{design}, rocket engine I_{sp}, shroud mass, and inert mass fractions were assumed. Inert mass fraction is defined to be the ratio of stage inert mass (dry mass minus payload) divided by total stage mass. Both examples assume an inert mass fraction of 0.1056,685. Payload mass on small vehicles such as these are EXTREMELY sensitive to inert mass fraction, especially for the single stage option. This low inert mass fraction is beyond the state of practice for LOX/LH_2 launch vehicles currently in operation but within reach of current technology. However, the key result from this example is the clear advantage of the use of staging over single-stage-to-orbit.

Table 9.11 Comparing a Single-Stage and Two-Stage Launch Vehicle

operating at or near the state of the art, the only other option, as the examples show, is to shed empty stages on the way into orbit.

Now, let's see how we use the rocket equation to analyze the total ΔV we get from a staged vehicle. We start with

$$\Delta V = I_{sp} g_o \ln\left(\frac{m_{\text{initial}}}{m_{\text{final}}}\right)$$

Recognize that for a staged vehicle, each stage has an initial and a final mass. Also, the I_{sp} may be different for the engine(s) in different stages. To get the total ΔV of the staged vehicle, we must add the ΔV for each stage. This gives us the following relationship for the ΔV of a staged vehicle with n stages.

$$\Delta V_{\text{total}} = \Delta V_{\text{stage 1}} + \Delta V_{\text{stage 2}} + \ldots + \Delta V_{\text{stage } n} \tag{9.35}$$

$$\Delta V_{\text{total}} = I_{\text{sp stage 1}}\, g_o \ln\left(\frac{m_{\text{initial stage 1}}}{m_{\text{final stage 1}}}\right) + I_{\text{sp stage 2}}\, g_o \ln\left(\frac{m_{\text{initial stage 2}}}{m_{\text{final stage 2}}}\right) + \cdots$$
$$+ I_{\text{sp stage } n}\, g_o \ln\left(\frac{m_{\text{initial stage } n}}{m_{\text{final stage } n}}\right) \tag{9.36}$$

where ΔV_{total} = total ΔV from all stages (m/s)
$I_{\text{sp stage } n}$ = specific impulse of stage n (s)
g_o = gravitational acceleration at sea level (9.81 m/s²)
$m_{\text{initial stage } n}$ = initial mass of stage n (kg)
$m_{\text{final stage } n}$ = final mass of stage n (kg)

What is the initial and final mass of stage 1? The initial mass is easy; it is just the mass of the entire vehicle at lift-off. But what about the final mass of stage 1? Here we have to go to our definition of final mass when we developed the rocket equation. Final mass of any stage is the initial mass of that stage (including the mass of subsequent stages) less the propellant mass burned in that stage. So for stage 1,

$$m_{\text{final stage 1}} = m_{\text{initial vehicle}} - m_{\text{propellant stage 1}}$$

Similarly, we can develop a relationship for the initial and final mass of stage 2, stage 3, and so on.

$$m_{\text{final stage 2}} = m_{\text{final stage 1}} - m_{\text{structure stage 1}}$$

$$m_{\text{final stage 2}} = m_{\text{initial stage 2}} - m_{\text{propellant stage 2}}$$

Overall, staging has several unique advantages over a single-stage vehicle. It

- Reduces the vehicle's total weight for a given payload and ΔV requirement
- Increases the total payload mass delivered to space for the same-sized vehicle
- Increases the total velocity achieved for the same-sized vehicle
- Decreases the engine efficiency (I_{sp}) required to deliver a same-sized payload to orbit

All of these staging advantages come with some drawbacks. These include

- Increased complexity because of the extra sets of engines and their plumbing
- Decreased reliability because we add extra sets of engines and the plumbing for the upper stages
- Increased total cost because more complex vehicles cost more to build and launch

Another interesting limitation of staging has to do with the law of diminishing returns. So far, you may be ready to conclude that if two stages are good, four stages must be twice as good. But this is not necessarily the case. Although a second stage significantly improves performance, each additional stage enhances it less. By the time we add a fourth or fifth stage, the increased complexity and reduced reliability offset the small performance gain. That is why most launch vehicles have only three or four stages.

Engineers working on single-stage to orbit (SSTO) concepts must overcome many technical challenges in propulsion and materials to make this idea feasible. However, the great promise of a completely reusable launch vehicle that can operate like an airplane offers the potential for operations far more cost effective than current expendable vehicles.

References

Forward, R. L. and Hoyt, R. P. 1999. "Space Tethers," *Scientific American*, February, pp. 66–67.

Humble, R. J., Gary, H. N., and Larson, W. J. 1995. *Space Propulsion Analysis and Design*, McGraw-Hill, New York.

SECTION 10

Earth's Environment and Space

Section Editor: Michael J. Rycroft

PART 1

The Earth and Its Atmosphere

Michael J. Rycroft

10.1 The Earth in Space

The Universe began with the Big Bang, about 14 billion years ago. About 4.6 billion years ago, the Sun formed during the gravitational collapse of an enormous cloud of gas and dust. Rotating around the Sun was a disk of material out of which the planets grew. These planets now orbit the Sun in the ecliptic plane. Our Earth is the third planet away from the Sun.

The different materials within the Earth gradually separated, with the densest (iron) forming an electrically conductive core. Dynamo action in the core generates the geomagnetic field. Above is the mantle, a region which convects very slowly. The Earth has a relatively thin crust, a few tens of kilometers thick, compared with the Earth's equatorial radius of 6,378 km. The crust and uppermost mantle are made up of 12 tectonic plates, hundreds of kilometers or so thick, which move relative to each other at up to 1 cm/year.

The Earth's atmosphere has evolved through outgassing from the interior via volcanoes. Oxygen was generated by dissociation of water vapor; the hydrogen that was also formed gravitationally escaped into space. Oxygen is also generated via photosynthesis by plants.

The Earth's hydrosphere (oceans), atmosphere, biosphere, and solid surface interact with each other in many complex ways. This area of science is now termed Earth Systems Science (Ernst 2000).

10.2 Properties of the Earth's Atmosphere

This subsection gives an overview of the physical properties of the atmosphere—its temperature, pressure, density, and composition—and how parameters expressing these vary with height (termed profiles). It also shows average wind distributions through the atmosphere.

The variation of temperature with height divides the atmosphere into different regions, as shown in Figure 10.1. The troposphere, the region of the Earth's weather, reaches up to about 14 km altitude in the middle latitudes, up to 18 km in the tropics, and only 9 km in the polar regions. The temperature decreases with increasing height; the lapse rate is typically 6.5 K/km. At the top of the troposphere, the tropopause, the temperature begins to rise with increasing height. This region is called the stratosphere. It is a stably stratified layer, with considerably less rapid vertical mixing than in the troposphere. Above the stratopause is the mesosphere, about which little is known. The mesopause is the coldest region of the atmosphere, typically at a temperature below 180 K. It is coldest in the local summer, paradoxically. Above lies the thermosphere.

Figure 10.1 shows, using a logarithmic scale, the density and also approximate pressure as a function of altitude. The pressure at 100 km altitude is about one millionth of its value at the surface (close to 1,010 hPa, on average). Some 90% of the Earth's atmosphere resides in the troposphere, and 99.5% in the troposphere and stratosphere. The upper atmosphere is tenuous indeed.

Conventional meteorological measurements of temperature over the Earth's land and sea surface are complemented by thermistor measurements aboard hydrogen or helium-filled balloons, plus values derived from satellite instruments using infrared radiation to determine the temperature field in three dimensions. An average over longitude, the zonal mean temperature, in kelvins is shown in Figure 10.2 (Andrews 2000). The cold summer mesopause is evident. Noctilucent clouds of tiny ice crystals often form here then.

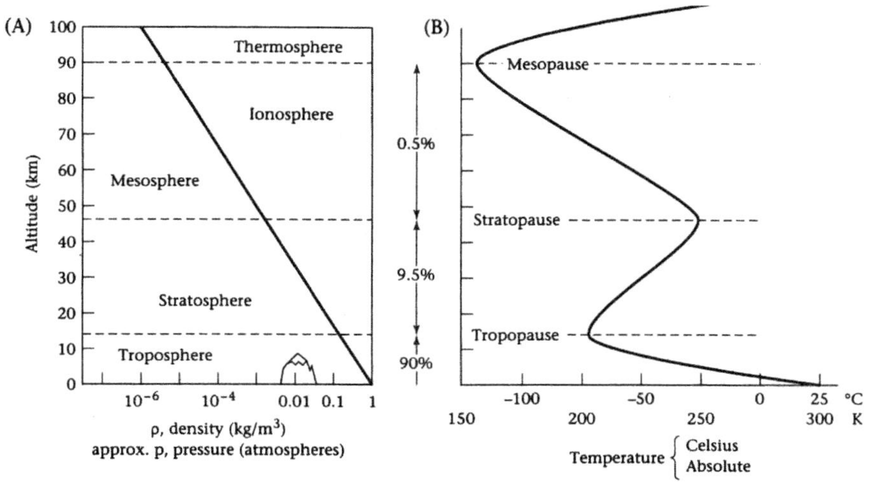

FIGURE 10.1 Variation of the Earth's atmospheric density and approximate pressure as a function of altitude in km. The density is 1.17 kg/m³ at the surface, where the pressure is 1 atm or 1,010 hPa (or mb) (equivalent to a weight of 10.1 kg/m²). The temperature of the atmosphere (shown in degrees Celsius (centigrade) and also in degrees absolute, or Kelvin) varies considerably; it is higher where the Sun's energy is absorbed. At 0°C the absolute temperature is 273.15 K (from Ernst 2000).

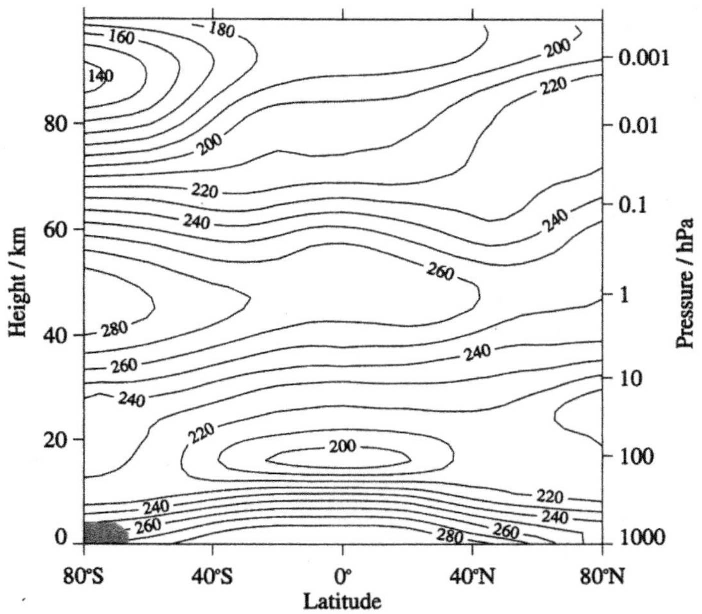

FIGURE 10.2 Zonal mean temperature (K) for January (from Fleming et al. 1990 and Andrews 2000).

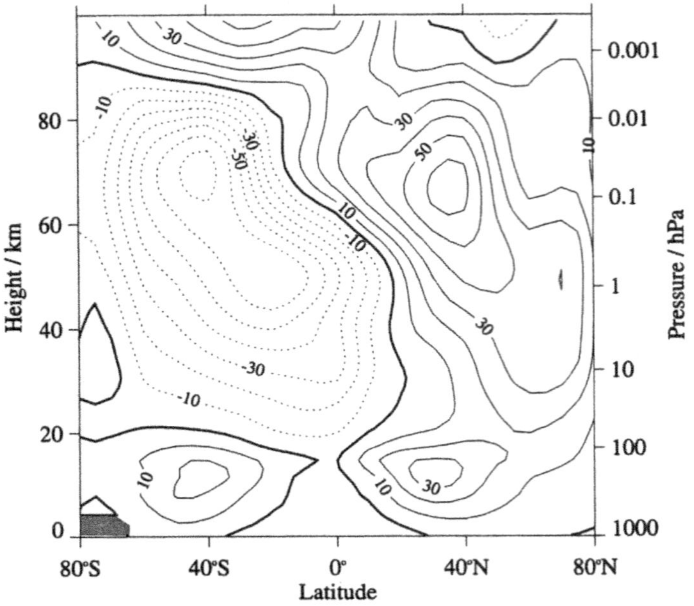

FIGURE 10.3 Zonal mean wind (ms^{-1}) for January (from Fleming et al. 1990 and Andrews 2000). Thin solid lines: eastward winds (westerlies); thick solid lines: zero winds; dashed lines: westward winds (easterlies).

The equator-to-pole temperature gradient drives the atmospheric heat engine, the weather machine. Winds blow over the Earth, attempting to reduce the temperature differences that exist between different places. Figure 10.3 shows average zonal winds for the month of January (Andrews 2000). At altitudes below 100 km the atmospheric composition is essentially the same everywhere. The predominant gas is nitrogen (N_2), 78% by volume. Next comes oxygen (O_2), 21% by volume. Argon, an inert (or noble) gas (A), is present at 0.93% by volume. Carbon dioxide (CO_2) now constitutes 0.036%, or 360 parts per million by volume (ppmv). In 1800, it was at 280 ppmv, and in 1900, 295 ppmv, the increase being due to the burning of fossil fuels. Water vapor (H_2O) is present in amounts ranging from 0 up to 0.04% (Wallace and Hobbs 1977). Methane (CH_4), other inert gases, hydrogen (H_2), and ozone (O_3) are all at the 10^{-4}% level. Trace gases, at the 10^{-6}% level, include the oxides of nitrogen, carbon monoxide, sulphur dioxide (from volcanoes), and human-made chlorofluorocarbons (CFCs), which are also termed freons.

To put matters into perspective, the total mass of the Earth is 6×10^{24} kg, the mass of the oceans is 1.3×10^{21} kg, and the mass of the atmosphere is 5×10^{18} kg (Wallace and Hobbs 1977).

10.3 How the Earth's Atmosphere Works

This subsection discusses the underlying physical mechanisms, namely the absorption of solar radiation and gravity, which determine the profile of important atmospheric parameters.

The atmosphere becomes hotter where radiation from the Sun is absorbed. Visible light is absorbed at the Earth's surface, and the tropical regions become especially hot. Solar ultraviolet radiation is absorbed by molecular oxygen (O_2) and by ozone (O_3) in the stratosphere. X-rays from the Sun's

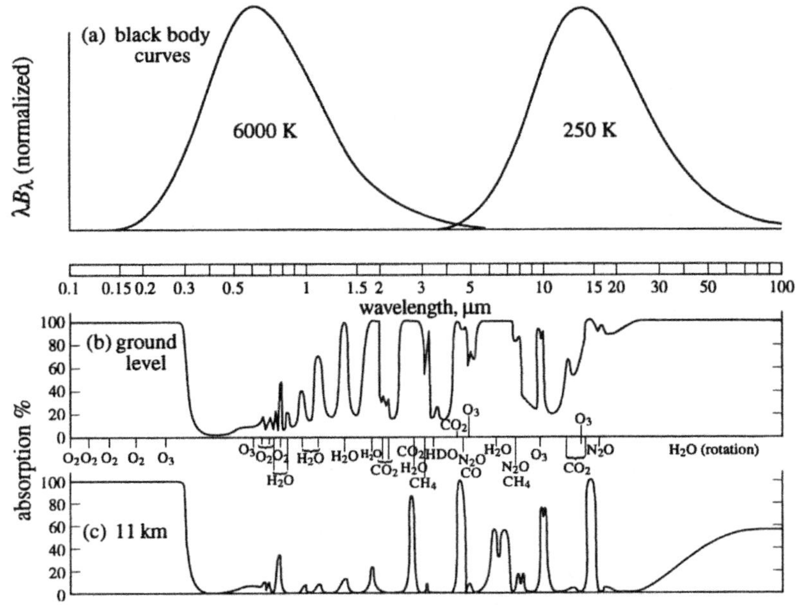

FIGURE 10.4 (a) Blackbody curves for 6000 K and 250 K, the Sun and the Earth, respectively; (b) atmospheric absorption spectrum for solar radiation reaching the ground; (c) the same for radiation reaching the temperate tropopause. The areas beneath the curves in (a), proportional to the energy fluxes, are the same over the globe for a year (from Goody 1995).

corona (its outer atmosphere) are absorbed in the thermosphere. At 400 km altitude the temperature rises to ~800 K during solar minimum conditions and to ~2,000 K at solar maximum conditions. The interval between one solar maximum and the next (occurring in 1957, 1968, 1979, 1990, and 2001) is about 11 years. This is termed the solar cycle.

The temperature of the Sun's surface, the photosphere, is close to 6,000 K. It radiates a blackbody spectrum (shown in Figure 10.4(a)), which peaks in the visible part of the spectrum at a wavelength of 0.6 μm (600 nm), in the yellow. Figure 10.4(b, c) (Goody 1995) shows that most of this radiation reaches ground level, with shorter wavelength radiation (wavelengths below 0.3 μm, the ultraviolet) being absorbed by O_2 and O_3 in the stratosphere.

Molecular gases such as water vapor, carbon dioxide, methane, and ozone absorb infrared radiation with wavelengths greater than 0.7 μm. In the far infrared, peaking at a wavelength between 10 and 20 μm, is radiation emitted into space by the Earth-atmosphere system at an average temperature near 250 K. Most of this terrestrial radiation is absorbed by molecular trace gases in the atmosphere. Reradiated both down and up, this mechanism accounts for the greenhouse effect. It maintains the Earth–atmosphere system at a temperature of 288 K, 33°C warmer than it would otherwise be. Thus, there is no doubt that the greenhouse effect is a good thing for humans living on the Earth. What could be a bad thing, at least for many, is an increasing greenhouse effect. Termed *global warming*, this is due to increasing amounts of gases such as CO_2, which are infrared active gases, produced by the burning of fossil fuels (Houghton 1997; Houghton et al. 2001). Other mechanisms also contribute to global warming.

Thermodynamics plays an important role in our understanding of atmospheric phenomena. The ideal gas law or equation of state relates the pressure p, density ρ, and temperature T, as

$$p = \rho RT$$

where R is the gas constant for 1 kg of gas.

For a mass of gas equal to its molecular weight in kg (28 for N_2, 32 for O_2), which contains 6×10^{26} molecules (N_A, Avogadro's number), the universal gas constant is $R^* = 8300$ J K^{-1} kilomole^{-1}. With the volume of this amount of gas being V, the ideal gas law becomes

$$pV = R^*T$$

For one molecule of gas, the universal gas constant is Boltzmann's constant, $k = 1.38 \times 10^{-23}$J K^{-1}. Thus $R^* = N_A \times k$. Hence,

$$p = \frac{N_A}{V} kT = NkT$$

where N is the number density of molecules, each of mass m.

At sea level, $N = 2.7 \times 10^{25}$ m^{-3}. The reader can check that the atmospheric pressure at sea level, with $T = 288$ K, is $p_0 = 10^3$ hPa.

Going up into the atmosphere, the atmospheric pressure decreases. The atmospheric pressure is due to the mass of air above unit area (1 m^2). For the equilibrium of a slab of air of thickness dz,

$$\frac{dp}{p} = -g\frac{dz}{RT}$$

Integrating, for an isothermal atmosphere at constant temperature T, from p_0 up to the height z, where the pressure is p,

$$p = p_0 \exp\left(-\frac{gz}{RT}\right) = p_0 \exp\left(-\frac{z}{H}\right)$$

The pressure decreases exponentially with increasing height z above the Earth's surface. In a distance

$$H = \frac{RT}{g} = \frac{kT}{mg}$$

p becomes p_0/e. The pressure falls to $p_0/2.72$ over a height equal to the scale height. Inserting T = 288 K for near the Earth's surface, the scale height is 7 km.

Over a height range of 2.3-scale height, about 15 km, the pressure decreases by a factor of 10. Over twice that distance the pressure decreases a hundredfold. This theory thus explains the variation of pressure with height shown in Figure 10.1(a).

The mass of atmosphere per unit area (1 m^2) of the Earth's surface is in fact equal to the mass of atmosphere of uniform density at sea level N over a scale height H. This is about 10^4 kg, namely 10 tonnes, and explains the sea-level atmospheric pressure of 10^5 Pa (10^3 hPa).

The theory given here can be extended, with greater algebraic complexity, to consider a nonisothermal, and hence more realistic, atmosphere.

At heights above 110 km, termed the turbopause, the gas composition starts to vary with altitude. This is because lighter gases, e.g., atomic oxygen (O) and hydrogen (H_2), float above the heavier gases such as molecular oxygen (O_2). This process is called diffusive, or gravitational, separation.

Thus, at a height of 300 km the neutral gas scale height is about 70 km, an order of magnitude greater than at sea level because the absolute temperature is five times greater and the molecular weight is halved.

At around this height the exosphere begins. This is where the atmosphere is so thin that the frequency of collisions between atoms/molecules is so small that the mean free path between collisions becomes comparable with the scale height. It is likely that an upward-moving hydrogen atom (H), produced from water vapor or methane in the stratosphere, will not collide with another particle. If its temperature is above 5,200 K, the thermal velocity of a hydrogen atom exceeds the velocity required to escape altogether from the Earth's gravitational field. This is a mechanism by which the Earth loses mass.

The Earth gains mass, ~10^5 kg per day, by the influx of meteoroids to the top of the atmosphere. A meteor glows for a few seconds in the upper atmosphere (~100 km altitude) as it burns up. More massive meteoroids can reach ground level; these are called meteorites.

10.4 Atmospheric Dynamics and Atmospheric Models

The atmosphere is warmed where solar radiation is absorbed. The atmosphere moves—winds below—to try to remove these temperature (or pressure) differences. Because the Earth is rotating (at an angular velocity Ω), a moving parcel of air is subjected to the Coriolis force. In the northern hemisphere, this makes the air parcel veer (to the right) of the direct line from high pressure to low pressure. This leads, for the northern hemisphere, to an anticlockwise circulation of air around a low-pressure system, storm, or cyclone (Andrews 2000; James 1994; Wallace and Hobbs 1977).

Applying Newton's second law of motion, the acceleration of a parcel of air of unit mass is equal to

$$F\,\underline{v} \wedge \underline{k} - \frac{\Delta p}{p} + \underline{F}_{fr}$$

The first term is the Coriolis force, the second the pressure gradient force, and the third the frictional force at the surface. Here $F = 2\Omega \sin \phi$, where ϕ is the geographic latitude and k is a unit vector in the vertical direction ($+z$).

For what is termed geostrophic flow, with steady state conditions and

$$\underline{F}_{fr} = 0, \qquad \frac{\Delta p}{p} = \underline{v} \wedge \underline{k}$$

Thus, the wind blows at right angles to the pressure gradient.

This is the explanation of Buys-Ballot's law: in the northern hemisphere, the low-pressure region is on the left if one's back is to the wind. The winds are stronger when the isobars (pressure contours on a weather map) are closer together.

The circulation of the atmosphere is shown in Figure 10.5 (Ernst 2000). It can be understood in terms of the theoretical notions presented. In the stratosphere/mesosphere, there is one convective cell (like a Hadley cell) from summer high latitudes to winter mid-latitudes.

In Figure 10.5, three large convecting cells of air (shown in cross-section on the left-hand side of the globe) define the circulation of the lower atmosphere in each hemisphere. The surface components of each atmospheric cell form the zonal wind belts that drive the surface circulation of the ocean. The limbs of the atmospheric cells include (1) zones of rising moist air, low pressure (L), and

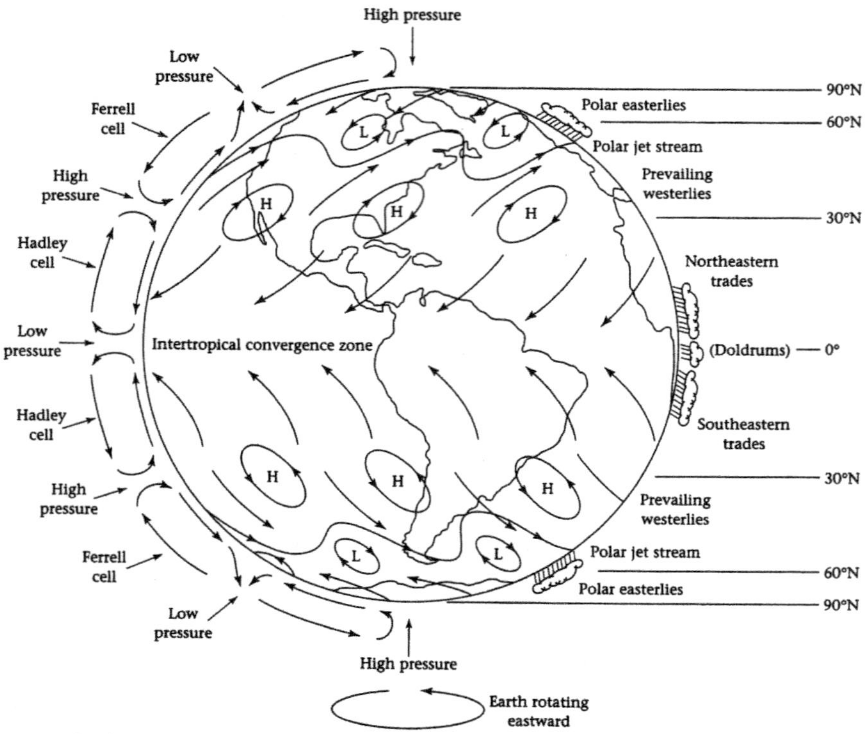

FIGURE 10.5 Schematic illustration of global atmospheric circulation and surface wind patterns (arrows on the Earth's surface) (from Ernst 2000).

high rainfall in the equatorial zone and along the polar fronts (at 50° to 60°N and S), and (2) zones of descending dry air, high pressure (H), and low rainfall over the polar regions and in the mid-latitudes (at approximately 30°N and S). Eastward rotation of the Earth and the Coriolis effect cause surface winds to veer to the right of their motion in the northern hemisphere and to the left of their motion in the southern hemisphere. The polar jet streams are not surface winds but rather flow eastward at high tropospheric altitude along the polar fronts in wave-like patterns. They influence the positions of the individual high- and low-pressure systems north and south of these fronts.

Models of atmospheric motion are needed to produce weather forecasts. The laws of conservation of mass, momentum, and energy are applied to describe and predict (using observed initial conditions) the dynamics of the atmosphere. Solar energy is absorbed at the Earth's surface and through the atmosphere.

The exchange of energy and water vapor between the surface and the atmosphere must be described (Houghton 1997). Water vapor is important because of its latent heat, which it gives out when it condenses, resulting in cloud formation. Latent heat has to be supplied to evaporate water from the oceans.

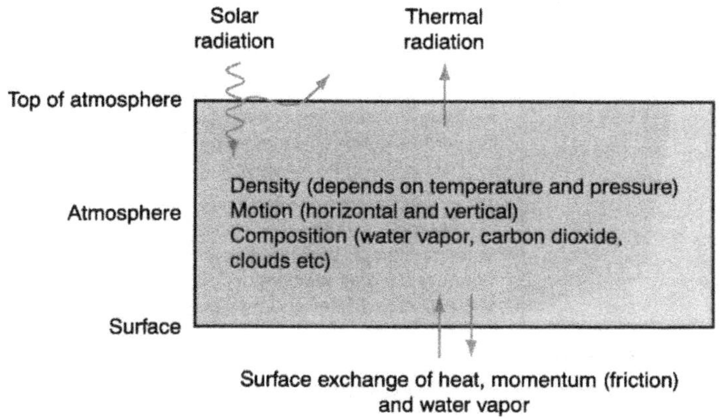

FIGURE 10.6 Schematic diagram illustrating the parameters and physical processes involved in atmospheric models (from Houghton 1997).

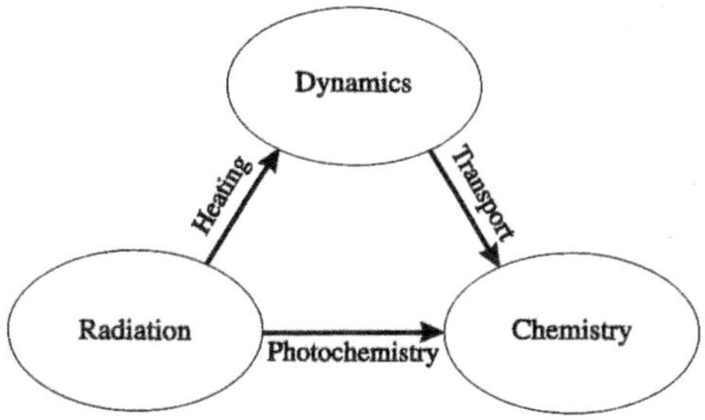

FIGURE 10.7 Diagram showing some of the interactions among dynamics, radiation, and chemistry in the atmosphere (from Andrews 2000).

The atmospheric processes involved in a model of atmospheric behavior are illustrated in diagrammatic form in Figure 10.6, from Houghton (1997). The interactions among atmospheric dynamics, radiation, and chemistry (which is especially important in the stratosphere), dealing with ozone (O_3), are illustrated in Figure 10.7, from Andrews (2000).

Ranging from short- and long-wave radiation to volcanoes, ice sheets, and both gas exchange and heat transfer with the ocean, Houghton (1997) summarizes the physical processes involved in the Earth's weather and the climate system. These processes can be well modeled using powerful digital computers today. The anthropogenic impact, i.e., the effects of human activities on the climate (the long-term average of weather), is a great cause of concern today (Houghton et al. 2001). The changes are both natural and human-caused, which are of especial importance. Figure 10.8, from Houghton et al. (2001), highlights these concerns diagrammatically.

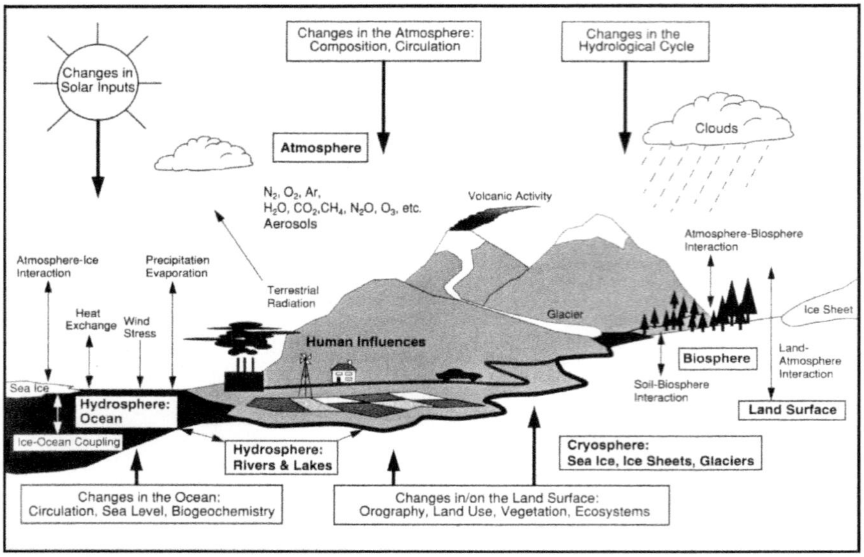

FIGURE 10.8 Schematic view of the components of the Earth's climate system (bold), their processes and interactions (thin arrows), and some aspects that may change (bold arrows) (from Houghton et al. 2001).

10.5 Electrical Phenomena in the Atmosphere

Thunderstorms are at the high-energy end of the spectrum of atmospheric phenomena. Lightning discharges are dramatic events indeed, which can cause significant damage.

Within a thundercloud there are strong upward convective motions, with the lower part of the cloud containing water droplets, and the upper part ice crystals/particles. Heavy ice particles fall as hail, and falling particles collide with other droplets, causing electric charges to be produced.

An upward electric current in a thundercloud carries ~100°C of positive charge to the top, leaving −100°C near the bottom (MacGorman and Rust 1988). The potential difference between the top and bottom of the cloud may reach 100 MV or more. Electrical breakdown of the atmosphere can occur, leading to a lightning discharge to Earth, a cloud-to-ground discharge carrying either negative or positive charge to ground. Alternatively, a discharge may occur within a cloud, an intracloud discharge, or from one cloud to the next, an intercloud discharge.

The electric dipole moment within the cloud or that formed by its electrical image below the ground is destroyed in a short time (~μs). This acts as a strong impulsive source of radio signals at frequencies up to many MHz. These signals propagate in the Earth–ionosphere waveguide.

Some electric current continues up from the top of the thundercloud, charging the ionosphere to a potential of ~ +250 kV with respect to the Earth. The good-conducting ionosphere is almost an equipotential surface. Far away from thunderstorm regions, a fair weather current ~2 $\mu A\ m^{-2}$ flows downward. The global atmospheric electric circuit is completed by currents flowing through the land and sea, and finally by point discharge currents below the thundercloud.

Changes to the electrical conductivity of the atmosphere associated with changes in the flux of cosmic rays or energetic charged particles from the Sun or the magnetosphere may modify the properties of the atmospheric electric circuit (Rycroft et al. 2000).

Sprites are upward electrical discharges from the top of a particularly energetic thundercloud to the ionosphere; they are especially likely to occur after a positive cloud-to-ground strike. They glow strongly, at altitudes between about 70 and 90 km, for only a few milliseconds.

The audio-frequency components of an atmospheric discharge (sferic, for short) can propagate into the ionosphere and be guided by ducts of enhanced ionization along geomagnetic field lines to the opposite hemisphere. Traveling through a dispersive plasma (an electrically charged gas), they emerge as descending frequency tones and are termed *whistlers*.

The lowest radio frequency component of a lightning discharge at ~8 Hz excites the fundamental resonance of the dielectric shell of atmosphere between the good-conducting Earth and ionosphere. These Schumann resonances occur when the wavelength is comparable with the Earth's circumference.

References

Andrews, D. G. 2000. *An Introduction to Atmospheric Physics.* Cambridge University Press, Cambridge.

Ernst, W. G., ed. 2000. *Earth Systems: Processes and Issues,* Cambridge University Press, Cambridge.

Fleming, E. L., Chandra, S., Barnett, J. J., and Corney, M. 1990. "Zonal Mean Temperature, Pressure, Zonal Wind and Geopotential Height as Functions of Latitude," *Advances in Space Research,* vol. 10, Chap. 12, pp. 11–59.

Goody, R. 1995. *Principles of Atmospheric Physics and Chemistry,* Oxford University Press, New York.

Houghton, J. T. 1997. *Global Warming: The Complete Briefing,* 2nd ed., Cambridge University Press, New York.

Houghton, J. T., Ding, Y., Griggs, D. J., Noguer, M., van der Linden, P. J., Dai, X., Maskell, K., and Johnson, C. A., eds. 2001. *Climate Change 2001: The Scientific Basis,* Cambridge University Press, Cambridge.

James, I. N. 1994. *Introduction to Circulating Atmospheres,* Cambridge University Press, Cambridge.

MacGorman, D. R. and Rust, W. D. 1998. *The Electrical Nature of Storms,* Oxford University Press, New York.

Rycroft, M. J., Israelsson, S., and Price, C. 2000. "The Global Atmospheric Electrical Circuit, Solar Activity and Climate Changes," *Journal of Atmospheric and Solar-Terrestrial Physics,* vol. 62, pp. 1563–1576.

Wallace, J. M. and Hobbs, P. V. 1977. *Atmospheric Science: An Introductory Survey,* Academic Press, New York.

PART 2

The Near-Earth Space Environment

Michael J. Rycroft

10.6 Background

The material presented in a section on the near-Earth space environment can be organized in several ways. First, descriptions of the several different aspects of the space environment can be given; this approach is followed here and by Mitchell (1994), Skrivanek (1994), and Tascione (1994). Detailed works on particular regions, such as the ionosphere (Kelley 1989; Rishbeth and Garriott 1969; Schunk and Nagy 2000) or the plasmasphere (Lemaire and Gringauz 1998), or on charged particles trapped by the geomagnetic field (Walt 1994), have been published. Alternatively, the Sun–Earth connection, or solar–terrestrial physics, now popularly termed space weather, has been considered by Freeman (2001), Hargreaves (1992), Kivelson and Russell (1995), Gombosi (1998), Suess and Tsurutani (1998), and Song et al. (2001). This important subject is also discussed in Part 6.

Another approach, which might appeal to readers of this handbook, is to consider the impacts that the space environment has on satellites in orbit around the Earth or on spacecraft. This approach is adopted in DeWitt et al. (1993), Dyer et al. (2000), Fortescue and Stark (1991, Chap. 2), Hastings and Garrett (1996), Holmes-Siedle and Adams (1994), Wertz and Larson (1999, Chap. 8), and Tribble (1995). Further detailed information is available in all these references.

To understand the nature of the near-Earth space environment, some basic knowledge of plasma is required. Plasma is the fourth state of matter. The first state is a solid (e.g., ice); when it is heated, a liquid is formed (water). When a liquid is heated (i.e., given extra energy), a gas, or vapor (steam), is formed. When a gas is heated, an electron is detached from a significant fraction of the molecules to create an overall electrically neutral gas, a mixture of positively charged ions and negatively charged electrons. This partially ionized gas is termed a plasma if three conditions are met (see below).

The three conditions for plasma behavior are:

1. The number of electrons N_D in a sphere of radius equal to λ_D must be very much greater than 1.
2. The typical dimension of the problem of interest must be much greater than λ_D.
3. The electron plasma frequency must be greater than the electron-neutral collision frequency, so that plasma waves will not be damped out.

Under the action of an applied electric field, the electrons move with respect to the much more massive ions. The electrons oscillate at the electron plasma frequency, whose value is f_{pe} (in Hz) = $9\sqrt{N_e(\text{in m}^{-3})}$, where N_e is the electron density.

An electric charge moving at velocity $\underline{v}$ perpendicular to a magnetic field $\underline{B}$ experiences a Lorentz force perpendicular to both $\underline{v}$ and $\underline{B}$; $\underline{F} = q\,\underline{v} \wedge \underline{B}$. Thus, an electron whose negative charge has magnitude e gyrates in a circle of radius (the Larmor radius) equal to mv/eB about the magnetic field direction. The gyro-frequency, or cyclotron frequency, is $f_{Be} = eB/2\pi m$, where m is the electron mass; f_{Be} (in Hz) = 28 B (in nT). The electron gyroradius (in km) is about $100\sqrt{W(\text{energy, in keV})}/B$ (in nT).

The Debye length is the distance in a plasma over which the electric field due to one particular positive charge is appreciable. It is given by λ_D (in m) = $\sqrt{T(\text{in K})/N_e(\text{in m}^{-3})}$. At distances greater than λ_D, electrons shield the remainder of the plasma from the effect of this particular ion.

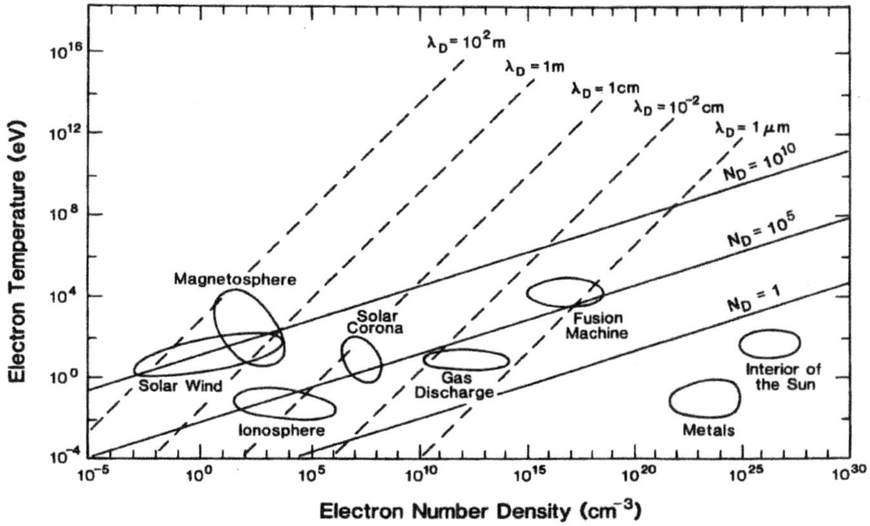

FIGURE 10.9 Values of the Debye length λ_D and the number of particles in the Debye sphere N_D for various plasmas having different electron temperatures and electron densities (from Kivelson and Russell 1995).

Figure 10.9 shows typical electron density and temperature values for different types of plasmas. Satellite-borne instruments can make satisfactory measurements in the magnetosphere and solar wind without affecting the plasma system being investigated, i.e., these three conditions are properly met.

10.7 The Plasma Environment

Gas in the thermosphere is ionized by ultraviolet and X-radiation from the Sun. The partially ionized gas, the plasma, so formed is termed the ionosphere. Representative profiles of the electron density as a function of height, by day and night, are shown in Figure 10.10. The specific radiations causing different ionospheric layers to be formed—the C, D, E, F1, and F2 layers—are also shown.

At 100 km altitude during the day, N_e is ~10^{11} m^{-3} and the number density of neutrals is ~10^{19} m^{-3}. Only 1 particle in 10^8 is charged, so the ionosphere is a very weakly ionized plasma there. At 300 km, about 1 particle in 10^3 is ionized, still a weakly ionized plasma.

At the top of the ionosphere, the strength of the ionizing radiation is great but there are only a few particles to ionize. Conversely, no ionizing radiation reaches down to the mesophere, but the gas concentration is large there. Thus, a layer of ionization, with a characteristic profile termed a Chapman layer, is produced between these two heights.

Figure 10.11 (Gombosi 1998) shows theoretical Chapman functions for an over-head Sun (S_0, for solar zenith angle $\chi = 0°$) and for greater solar zenith angles.

$$S = S_0 \exp\left[1 - \frac{z}{H} - \sec\chi \exp\left(-\frac{z}{H}\right)\right]$$

where the altitude z is given in units of scale height of the neutral species being ionized, normalized to the altitude of greatest rate of production of ionization for an overhead Sun.

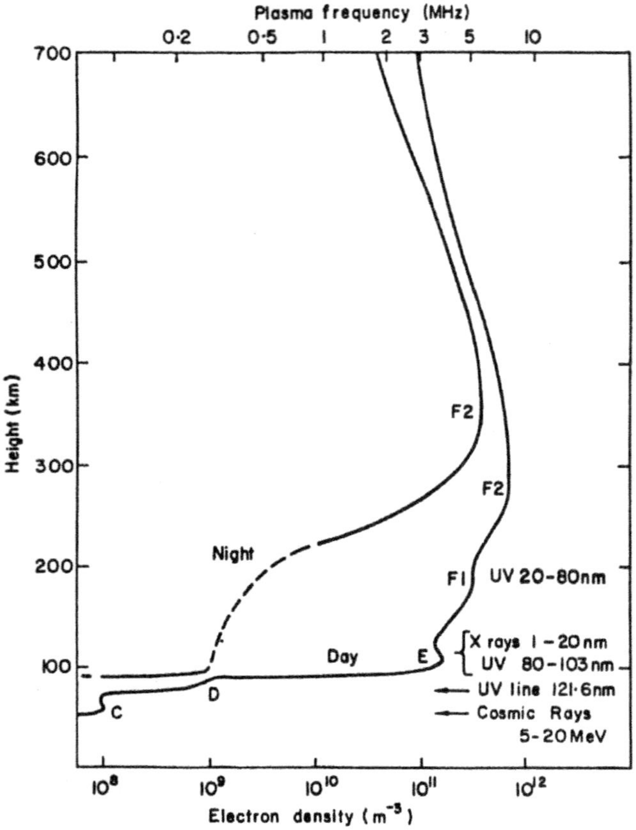

FIGURE 10.10 Typical ionospheric electron density particles, by day and night, and the different radiations responsible for the different layers.

Of the various ionospheric layers, the E layer obeys the Chapman formalism most closely. Transport effects—motions due to neutral winds and/or to electric fields—are strong in the F layer. During magnetic storms, the neutral gas composition at each particular height varies from its usual composition, and so the ionization is also affected in a complicated way. Operating near the equator, a fountain effect produces the so-called Appleton anomaly of larger electron densities at ~20° latitude than at the equator. In the equatorial and auroral regions, plasma irregularities with scale widths less than 1 km but elongated up to ~100 km along the geomagnetic field are created by plasma instabilities.

The ionosphere and its irregularities affect the propagation of radio waves used for space communications and/or navigation to a certain extent. In order to propagate through the ionosphere from the ground to a satellite or vice versa, the radio frequency used must exceed the plasma frequency of the F_2 layer (~10 MHz).

High-frequency (HF, 3–30 MHz) radio waves are refracted (bent) by the ionosphere as they propagate through it. Actually two modes of propagation exist in a plasma, termed ordinary (O) and extraordinary (X) modes. At HF, the O and X modes travel at slightly different velocities, both of which differ a little from the velocity of light in free space (c, which is nearly 3×10^5 km s^{-1}).

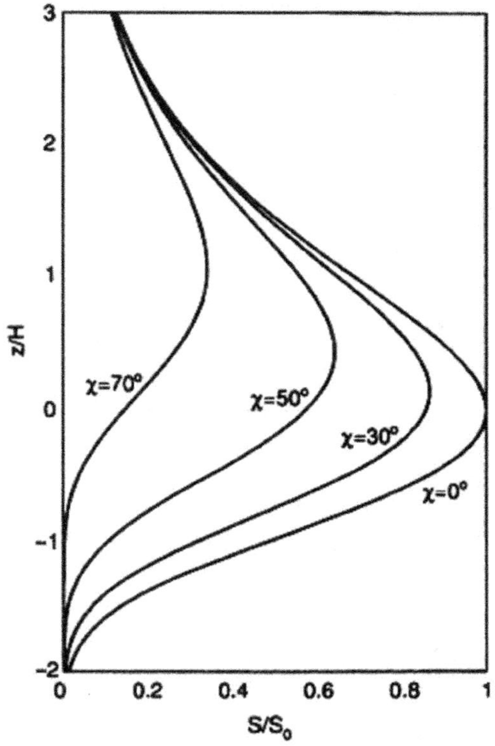

FIGURE 10.11 The normalized Chapman ionization function (from Gombosi 1998).

Further, HF waves are scattered by ionospheric irregularities, giving rise to scintillations. These are rapid (less than 1 s) variations of signal amplitude (fading) and/or phase.

At higher frequencies (VHF, 30–300 MHz, and UHF, 300 MHz–3 GHz), both the refraction and scattering effects are less than at HF. This explains why radio signals in the UHF band, or even higher frequencies, are ideal for communication with satellites or for communications systems via satellites. Account has to be taken of both refraction and scattering effects for navigation methods using global positioning by satellites (GPS).

These effects also have to be accounted for when interpreting satellite altimetry results obtained using the radar principle. The necessary correction depends upon the electron density integrated along the ray path, which is termed the total electron content (TEC). Synthetic aperture radar (SAR) signals may also be affected by the ionosphere through which they have traveled.

Because the ionosphere is a birefringent medium, i.e., the O and X modes have slightly different refractive indices linked to their different velocities, the plane of polarization of a radio signal received from a satellite rotates. Due to the Faraday effect, such observations can be interpreted in terms of spatial variations of the ionospheric electron density. The principles of tomography are now being applied to study latitudinal variations of the ionosphere by observing radio signals from satellites in polar orbit.

Due to the Doppler effect, the frequency of the radio signal observed on the ground differs slightly from that transmitted from the orbiting satellite. Using two frequencies, the differential Doppler effect can be observed and used to derive some properties of the ionosphere.

The interaction between the ionospheric (or interplanetary) plasma and a satellite (or spacecraft) can lead to charging of the satellite/spacecraft. High-voltage systems aboard the satellite/spacecraft may discharge—or arc—to the surrounding plasma. The satellite/spacecraft is likely to sustain permanent damage in that event.

Such charging effects are much more prevalent at geostationary orbit (GEO, at a geocentric distance of 6.6 Earth radii) than in low Earth orbit (LEO, at heights typically between 300 and 1000 km). This is because at GEO the electron density is so low, usually between 10^6 and 10^7 m^{-3} (Lemaire and Gringauz 1998).

At altitudes above a few hundred kilometers, the ionospheric plasma is constrained to move along geomagnetic field lines from one hemisphere to the other. On field lines out to a geocentric distance of four Earth radii, the plasma density is relatively high, greater than 10^9 m^{-3}. Then the plasma density decreases dramatically, by a factor of 10 to 100, at the plasmapause. This field-aligned surface is usually on a field line crossing the equatorial plane at a geocentric distance between four and five Earth radii, termed $L = 4–5$.

10.8 The Neutral Gas Environment

Even though the atmosphere is very thin in LEO, it still exerts a drag force on a satellite and on other particles in orbit. The density of the Earth's atmosphere above 110 or 120 km altitude changes as the atmosphere expands in response to extra ultraviolet and X-radiation from the Sun with increasing solar activity.

Above this altitude, the atmosphere expands as it becomes hotter at times near solar maximum. This is evident from Figure 10.12. At 500 km altitude, the temperature is nearly 2000 K at solar maximum, but only ~800 K at solar minimum; the neutral gas density is almost a hundred times larger at solar maximum than at solar minimum.

The drag force on a satellite

$$F_D = \frac{1}{2}\rho A C_D v^2$$

depends crucially on the neutral gas density ρ, the cross-sectional area of the object A presented as it moves at velocity v; C_D is the aerodynamic drag coefficient. For objects in LEO, v is almost 8 km s^{-1}.

As satellites or other objects in LEO reenter the Earth's atmosphere, their total energy decreases, their kinetic energy increases, and their orbital period τ decreases as time t progresses. Application of the law of the conservation of energy to satellites in a circular orbit (Hargreaves 1992) shows that

$$\frac{d\tau}{dt} = \frac{3\pi_\rho A C_D (z + R_E)}{M_s}$$

where z is the altitude, R_E is the radius of the Earth, and M_s is the mass of the satellite.

This equation was used in the early days of space research, the 1960s, to obtain the first information on $\rho(z)$, the neutral gas density as a function of height z, under different solar activity conditions.

At heights between 200 and 600 km the main atmospheric constituent is atomic oxygen (O). This is much more chemically reactive than molecular oxygen (O_2). It oxidizes the surface materials of satellites. For example, a front-silvered mirror becomes gray and its optical, thermal, and mechanical properties are degraded (Tribble 1995).

In the ram direction the surface of a satellite glows. This is because the kinetic energy of the satellite is sufficient to excite the atmospheric atoms or molecules. These then radiate when they fall back to their ground state.

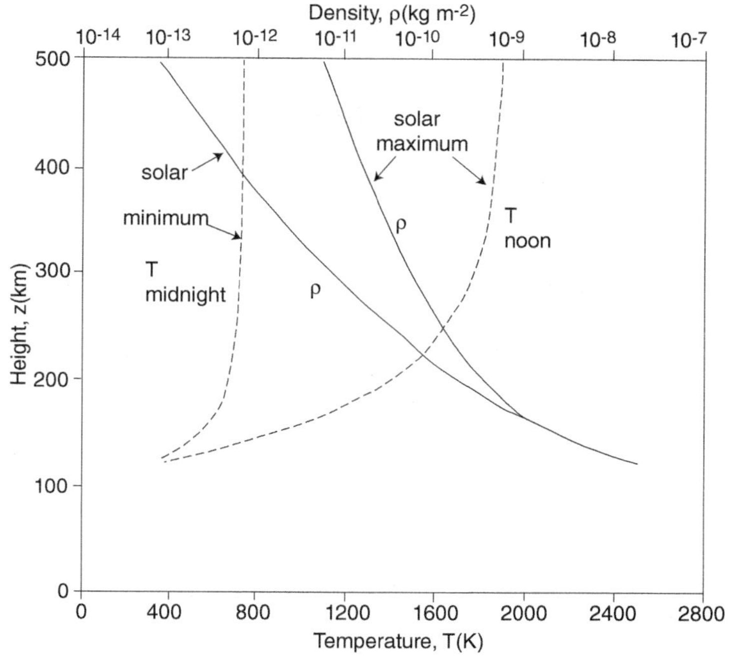

FIGURE 10.12 Variation with height (z) above 120 km of the atmospheric density (ρ), solid lines, upper scale, and the atmospheric temperature (T), dashed lines, lower scale, for both solar maximum and minimum conditions (based on CIRA 1965).

10.9 The Vacuum Environment

At ~400 km altitude, the pressure is ~10^{-6} Pa, 11 orders of magnitude less than at the Earth's surface. It is hard to make such a good vacuum on the Earth's surface, and so space is a good vacuum laboratory. The Space Shuttle wake shield facility utilizes this situation.

Under such conditions, lubricants between moving metal surfaces may not work well. Gas particles stuck on satellite surfaces outgas and may then stick on sensitive surfaces such as the lens of a telescope or thermal control surfaces, thereby reducing their performance.

The Sun's radiation is bright in space—there is no absorption of its ultraviolet radiation by ozone. This degrades the mechanical or thermal properties of some surface materials and may affect the performance of an optical instrument. Integrated across the spectrum, the solar constant of 1.368 kW m^{-2} and the terrestrial infrared radiation, heat a satellite. In the absence of gas to carry away heat by convection, excess heat can only be conducted through the satellite to a radiator that radiates energy away into the blackness of space (actually at 2.7 K, due to the cosmic microwave background radiation arising from the Big Bang).

10.10 The Radiation Environment

Charged particles with an energy much greater than the thermal energy of the plasma (less than 1 eV, equivalent to 11,600 K) constitute the radiation environment of space. Ions with an energy ~0.1 keV impinging upon the surface of a satellite eject atoms from the surface. Called sputtering, this process, over time, can remove a thin coating applied to part of a satellite's surface. Much higher energy charged particles can penetrate solar cells and microelectronic chips inside a

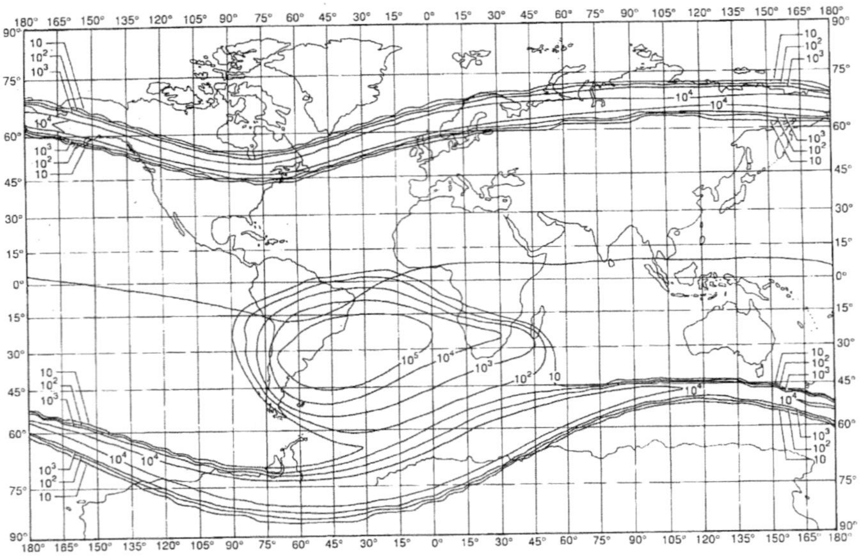

Figure 10.13 Contours of electron fluxes at >1 MeV, in units of cm^{-2} s^{-1}, at 500 km altitude, showing effects due to the outer Van Allen radiation belt and the South Atlantic geomagnetic anomaly (from Holmes-Siedle and Adams 1994).

satellite, causing them to change their state or even destroying them completely. Holmes-Siedle and Adams (1994) discuss many details. The charged-particle environment of near-Earth space, especially strong in the vicinity of the South Atlantic geomagnetic anomaly (see Figure 10.13), is hazardous to astronauts and cosmonauts carrying out extravehicular activities (EVAs).

Here some facts about the radiation environment are given. Deeper explanations are given in Part 6. The most energetic charged particles (electrons of greater than 1 GeV), galactic cosmic rays, come from outside the solar system. When the Sun is especially active, it can emit bursts of charged particles, protons, helium ions, and electrons (of greater than 1 MeV). These enter the magnetosphere at high latitudes. The usual interaction between the solar wind and the magnetosphere causes it to be populated by energetic ions and electrons (of greater than 1 keV); charged particles trapped on geomagnetic field lines are termed the Van Allen radiation belts.

An electron with an energy of some tens of keV, typical of the Van Allen belt population, undergoes three types of motion. These are indicated in Figure 10.14. The most rapid component of the electron's motion is gyration about the geomagnetic field line, on a time scale of less than 1 ms. The electron bounces from one hemisphere to the other on a time scale of seconds. The third component of motion, with a time scale of hours, is a longitudinal drift around the Earth, with electrons drifting eastward and protons westward. This constitutes a westward-directed ring current which reduces the geomagnetic field at the Earth's surface on the equator (nominally 31,000 nT) by some tens or hundreds of nT.

In meridional cross-section, contours of the fluxes of electrons (above 0.5 MeV), in cm^{-2} s^{-1}, are illustrated in the lower part of Figure 10.15. The inner and outer belts, separated by a slot region, are clearly seen. The contours come closest to the Earth's surface near 60° geomagnetic latitude, to account for the high flux values at all longitudes evident in Figure 10.13.

The upper part of Figure 10.15 shows the proton fluxes, in units of cm^{-2} s^{-1}. While it is not apparent from these figures, the Van Allen belts are electrically neutral overall. The mechanism believed to be responsible for the proton belt is the decay of neutrons produced when cosmic rays hit the top of the atmosphere. This is termed the cosmic ray albedo neutron decay (CRAND) mechanism.

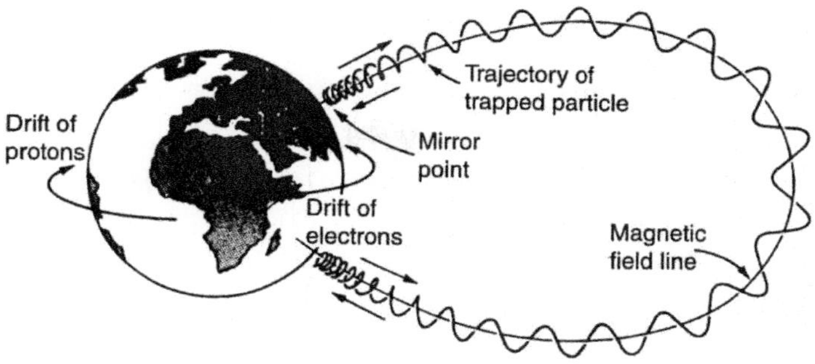

FIGURE 10.14 The motion of a charged particle trapped in the Earth's magnetic field (from Gombosi 1998).

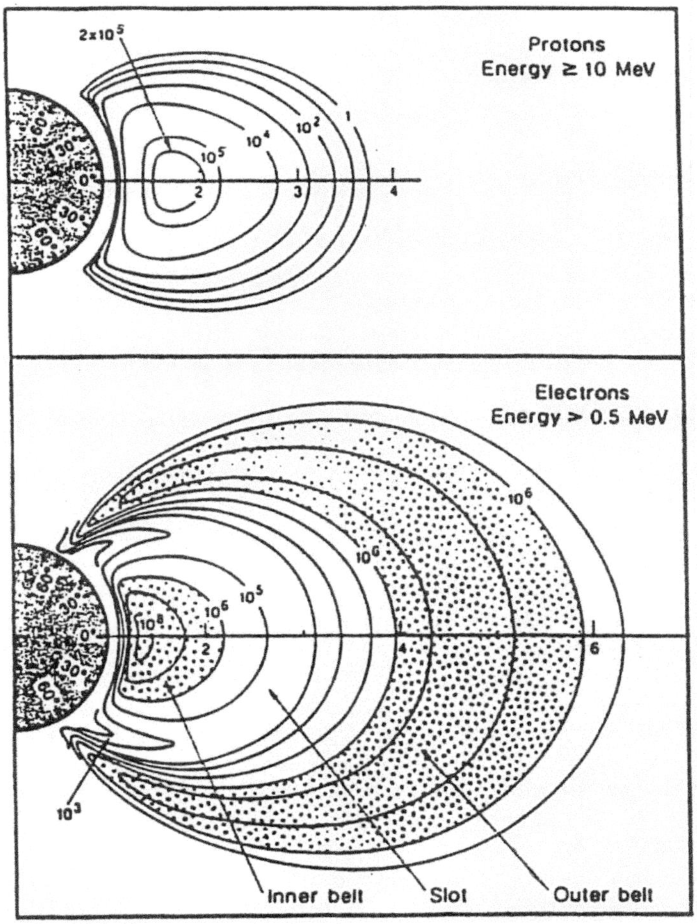

FIGURE 10.15 Fluxes of protons and electrons in the Van Allen radiation belts, in meridional cross-section.

At any particular time, the charged particle fluxes may vary from these typical values by up to two orders of magnitude. In other words, the radiation belts, especially the outer one, are very dynamic in their response, on a time scale of minutes, to solar wind changes.

10.11 The Micrometeoroid and Space Debris Environment

The relative velocities, and hence the kinetic energy involved, in a collision between an orbiting satellite and a piece of debris are so large that a small particle (more than a millimeter across) can seriously damage the structure of a satellite. Particles with sizes larger than ~0.1 m can be observed from the ground, using either radars or optical telescopes.

The subject is of such importance that it warrants extensive discussion in Part 7.

References

COSPAR International Reference Atmosphere (CIRA). 1965. North-Holland, Amsterdam.

DeWitt, R. N., Duston, D., and Hyder, A. K., eds. 1993. *The Behavior of Systems in the Space Environment*, Kluwer Academic Publishers, Dordrecht.

Dyer, C. S., Truscott, P. R., Sanderson, C., Watson, C., Peerless, C. L., Knight, P., and Mugford, R. 2000. "Radiation Environment Measurements from CREAM & CREDO during the Approach to Solar Maximum," *IEEE Transactions on Nuclear Science*, vol. 47, no. 1.

Fortescue, P. W. and Stark, J. P. W. eds. 1991. *Spacecraft Systems Engineering*, John Wiley & Sons, Chichester.

Freeman, J. W. 2001. *Storms in Space*, Cambridge University Press, Cambridge.

Gombosi, T. J. 1998. *Physics of the Space Environment*, Cambridge University Press, Cambridge.

Hargreaves, J. K. 1992. *The Solar-Terrestrial Environment*, Cambridge University Press, Cambridge.

Hastings, D. and Garrett, H. 1996. *Spacecraft-Environment Interactions*, Cambridge University Press, Cambridge.

Holmes-Siedle, A. and Adams, L. 1994. *Handbook of Radiation Effects*, Oxford University Press, Oxford.

Kelley, M. C. 1989. *The Earth's Ionosphere: Plasma Physics and Electrodynamics*, Academic Press, San Diego.

Kivelson, M. G. and Russell, C. T. 1995. *An Introduction to Space Physics*, Cambridge University Press, Cambridge.

Lemaire, J. F. and Gringauz, K. I. 1998. *The Earth's Plasmasphere*, Cambridge University Press, Cambridge.

Mitchell, D. G. 1994. "Space Systems Engineering," in *Fundamentals of Space Systems*, ed. V. L. Pisacane and R. C. Moore, Oxford University Press, Oxford.

Rishbeth, H. and Garriott, O. K. 1969. *Introduction to Ionospheric Physics*, Academic Press, New York.

Schunk, R. W. and Nagy, A. F. 2000. *Ionospheres: Physics, Plasma Physics, and Chemistry*, Cambridge University Press, Cambridge.

Skrivanek, R. A., ed. 1994. *Contemporary Models of the Earth's Environment*, AIAA, Washington, DC.

Song, P., Singer, H. J., and Siscoe, G. L., eds. 2001. *Space Weather*, AGU, Washington, DC.

Suess, S. T. and Tsurutani, B. T., eds. 1998. *From the Sun: Auroras, Magnetic Storms, Solar Flares, Cosmic Rays*, AGU, Washington, DC.

Tascione, T. F. 1994. *Introduction to the Space Environment*, Krieger, Malabar, FL.

Tribble, A. C. 1995. *The Space Environment: Implications for Spacecraft Design*, Princeton University Press, Princeton, NJ.

Walt, M. 1994. *Introduction to Geomagnetically Trapped Radiation*, Cambridge University Press, Cambridge.

Wertz, J. R. and Larson, W. J. 1999. *Space Mission Analysis and Design*, Kluwer Academic Publishers, Dordrecht.

PART 3

The Solar System

Michael J. Rycroft

10.12 Physical Properties of the Planets

In this subsection the physical characteristics and properties of the bodies in the solar system are reviewed. Moving in almost circular orbits about the Sun, under the action of its gravitational force, are the planets and asteroids. Moving in elliptical orbits about the Sun are Pluto, comets from the Kuiper belt, and more distant comets from the Oort cloud. Asteroids and comets are remnants of the early solar system (Beatty et al. 1999; Lewis 1997; Lodders and Fegley 1998; Mendell 1999).

All the planets except Pluto orbit the Sun in the same plane, the ecliptic plane. Starting from closest to the Sun, the planets are Mercury, Venus, Earth, and Mars (rocky planets, termed the terrestrial planets), the asteroids (rocky material which did not form a planet) then Jupiter, Saturn, Uranus, and Neptune (large gaseous planets). The different planetary orbits are shown, together with their symbols, in Figure 10.16 (from Lewis 1997).

Table 10.1 summarizes, generally to two significant figures, numerical information about the Sun, the planets, and their motions. Information on one planetary satellite, the one orbiting the Earth—namely the Moon—is included. The distance at which a planet orbits the Sun is given in astronomical units (1 AU, the Sun-Earth distance, is almost 1.5×10^{11} m). The mass of the body is given in kg, its radius in km, and its density in kg m^{-3}. It is evident that the density of the inner (terrestrial planets) is several times greater than that of the outer (giant gaseous planets). The eccentricity of the orbit shows by how much the orbit departs from a circular orbit; the eccentricity is zero for an exactly circular orbit. The period for one orbit around the Sun is given in years. The period for one revolution of the planet around its axis of rotation is shown in days. It is remarkable that the larger outer planets rotate much more rapidly than the smaller inner (terrestrial) planets. Finally, the acceleration due to gravity on the surface of the planet is given in g; for the Earth, $1\,g = 9.8$ ms^{-2}.

All these values have to be known to great accuracy in order to navigate a spacecraft to rendezvous with a planet or asteroid.

Table 10.2 highlights the major satellites of the planets in the solar system, and ring systems. For each stated satellite its radius is given in km.

New planetary satellites and ring systems discovered by instruments aboard spacecraft are considered in the next subsection.

10.13 Space Age Discoveries

One of the brightest planets in the sky, Mercury, was imaged directly by Mariner 10 in 1974 and 1975. Half its surface has been photographed. Like the Moon, it is heavily cratered. Because its spin axis is very close to perpendicular to its orbital plane, there are no seasons on Mercury. Its surface temperature is low (~100 K) at night, but high (~700 K) during the day (Lodders and Fegley 1998). The atmosphere of Mercury is very thin, the surface pressure being less than 10^{-9} hPa. The most abundant species is believed to be argon, with a number density less than 3×10^{13} m^{-3} (Beatty et al. 1999). There is a weak magnetic field near its surface, whose properties are tabulated in Table 10.3. As discussed in Part 6, the magnetic dipole moment is given as a multiple of the Earth's magnetic moment, the surface field at the magnetic equator is given in nanoteslas (nT), the angle between the rotational and magnetic axes is shown in degrees (°), and a typical distance from the center of the planet to its magnetopause is given in units of planetary radius. The magnetopause is the boundary between the

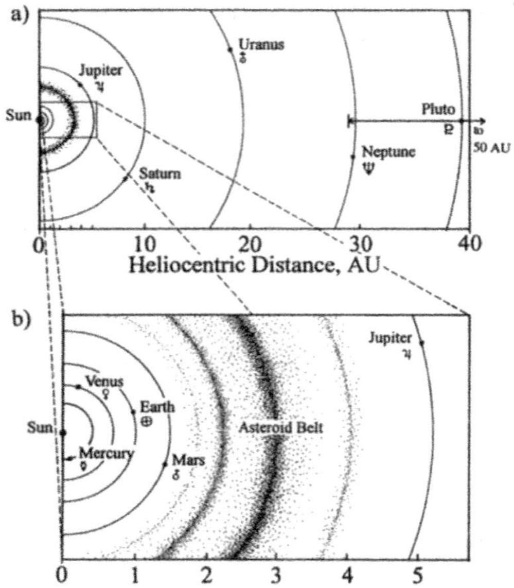

FIGURE 10.16 Diagram, to scale, of (a) the orbits of the outer planets and (b) the inner planets and asteroids (from Lewis 1997).

planet's magnetic field and the interplanetary magnetic field carried away from the Sun by the solar wind plasma.

The main difference between Venus and Earth is the dryness of Venus. Its atmosphere of 96.5% carbon dioxide (CO_2) has a surface pressure some 90 times greater than Earth's. Since CO_2 is a very effective greenhouse gas, the surface temperature is about 750 K (Beatty et al. 1999; Marov and Grinspoon 1998). The thick clouds of aqueous sulfuric acid droplets at heights between 45 and 70 km prevent the Venusian surface from being seen in the visible part of the spectrum. However, the atmosphere is transparent to microwaves, and its surface has been impressively mapped by the radar instrument of the Magellan mission in the early 1990s. Both impact craters and volcanoes are very evident. Optical images of flat basaltic rocks on the surface of Venus were obtained by the Venera 14 lander in 1982. Venus has a significant ionosphere. No planetary magnetic field has been detected by magnetometers aboard spacecraft in the vicinity of Venus.

The origin of the Earth and the Moon is discussed by Canup and Righter (2000). The rather well-known properties of the Earth and the Moon are summarized, in the context of solar system studies, by Beatty et al. (1999) and Lodders and Fegley (1998). Detailed information on the Moon, as an object worthy of detailed study by both robotic and crewed space missions, is given in Part 4.

Mars, the red planet, may have harbored forms of life in past eons and may still be an environment where life exists. Because of the great interest in the search for life on Mars using space missions, the planet warrants a separate discussion in Part 5. Excellent images of Mars have been taken by the Hubble space telescope in orbit around the Earth. In the winter and springtime on Mars, carbon dioxide frost and a (water) ice cap are seen (Fischer and Duerbeck 1998). The atmospheric pressure on the Martian surface is only 6 hPa. The magnetic field of Mars is very small (see Table 10.3).

Images of some asteroids have been taken by the Hubble space telescope (Fischer and Duerbeck 1998) and by the Galileo spacecraft. The Near Earth Asteroid Rendezvous (NEAR Shoemaker)

Object	Distance from Sun (AU)	Mass (kg)	Radius (km)	Density (10^3 kgm^{-3})	Orbital eccentricity	Orbital period (year)	Rotation period (day)	Surface gravity (g)
Sun	—	2.0×10^{30}	7.0×10^5	1.4	—	—	25–32	28
Mercury	0.39	3.3×10^{23}	2.4×10^3	5.4	0.21	0.24	59	0.39
Venus	0.72	4.9×10^{24}	6.0×10^3	5.2	0.007	0.62	243	0.90
Earth	1.00	6.0×10^{24}	6.4×10^3	5.5	0.017	1.00	1.00	1.00
Moon	1.00	7.4×10^{22}	1.7×10^3	3.3	0.055	1.00	28	0.17
Mars	1.52	6.4×10^{23}	3.4×10^3	4.0	0.093	1.88	1.03	0.38
Asteroids	2.2–3.3	Total ~10^{22}	>1	Few	~0.15	3.3–6	Fraction	<0.001
Jupiter	5.2	1.9×10^{27}	7.1×10^4	1.3	0.048	11.9	0.41	2.6
Saturn	9.5	5.7×10^{26}	6.0×10^4	0.69	0.056	29	0.43	1.1
Uranus	19	8.7×10^{25}	2.6×10^4	1.3	0.047	84	0.70	0.9
Neptune	30	1.0×10^{26}	2.5×10^4	1.6	0.009	165	0.74	1.1
Pluto	39	10^{22}	10^3	2	0.25	248	6.4	0.07
Comets	$1–5 \times 10^4$	Total ~10^{27}	<10	<1	0.5–1.0	3–10		Negligible

Sources: Lewis 1997; Lodders and Fegley 1998; Mendell 1999.

Table 10.1 Numerical Information on Bodies in the Solar System

Planet	Satellite	Radius (km)
Mercury		
Venus		
Earth	Moon (see Part 4)	1.7×10^3
Mars	Phobos, Deimos	~10, ~6
Jupiter	16, including Io, Europa and Ganymede	1.8×10^3, 1.6×10^3, 2.6×10^3
Saturn	18, largest Titan; rings	2.6×10^3
Uranus	5, largest Titania; rings	790
Neptune	2, larger Triton	1.4×10^3

TABLE 10.2 Planetary Satellites Known before Direct Space Observation

Planet	Magnetic dipole moment	Equatorial surface field (nT)	Angle between axes (°)	Magnetopause distance (Planetary radii)
Mercury	7×10^{-4}	300	14	1.5
Venus	$<10^{-4}$	—	—	—
Earth	1.00	31,000	11	10
Mars	$<2 \times 10^{-4}$	< 30	—	—
Jupiter	2×10^4	430,000	10	80
Saturn	600	22,000	< 1	20
Uranus	50	23,000	60	20
Neptune	25	14,000	47	25

Source: Based on Beatty et al. 1999.

TABLE 10.3 Properties of Planetary Magnetic Fields

mission has obtained images of a few asteroids. At the end of its mission in February 2001 it landed on the asteroid Eros. The properties of asteroids are summarized by Kowal (1996).

The giant planets are gaseous, mainly hydrogen and helium, but with methane and ammonia at the 0.1% level by volume. If the mass of Jupiter were 13 times its actual mass, it would be classified as a brown dwarf (Beatty et al., 1999). It is therefore more of a "failed star" than a planet. Jupiter radiates 1.7 times the amount of solar radiation that it receives (Lodders and Fegley 1998).

Voyager 1 and 2 images of Jupiter's surface exhibit zonal bands of clouds, signifying strong jet streams. Weather systems are also clearly evident, the largest of which is the very long-lived Great Red Spot. Bursts of lightning, as well as whistlers, have been observed on Jupiter. Rings of light around the two magnetic poles of Jupiter, signifying its auroral zones, have been seen with the Hubble space telescope (Fischer and Duerbeck 1998) and by ground-based infrared telescopes. Jupiter's magnetosphere is enormous in scale.

In July 1994, pieces of the comet Shoemaker-Levy 9 hit Jupiter, causing material from below the clouds to rise up (Spencer and Mitton 1995). Comets are believed to be "dirty snowballs," carbon-rich ices left over from the early days of the solar system.

Figure 10.17 Voyager 1 image of Saturn taken in November 1980 of Saturn, its ring system, and its moons Tethys (above) and Dione (below) (NASA image). The shadows cast by the ring system and by Tethys onto the tops of the clouds account for the dark features on the planet's surface in this image.

When the Voyager 1 spacecraft reached Jupiter in 1979, Jupiter's ring system was discovered in the vicinity of the four innermost satellites (Lodders and Fegley 1998). Of the next four satellites (the Galilean satellites), Io is known to be the most volcanically active body in the solar system. Both the Voyager and Galileo spacecraft have observed sulfur emitted by volcanoes. Europa and Ganymede have water ice on their surface and oxygen and ozone trapped in the ice. Spacecraft magnetometer observations near Europa indicate the presence of conducting oceans below the ice. Such an environment may be conducive to life (Beatty et al. 1999).

Saturn is a somewhat smaller version of Jupiter; its surface exhibits a similar pattern of zonal winds and clouds. Ammonia and phosphine appear as absorption features in Saturn's infrared spectrum. Saturn's rings are the planet's most distinctive and dramatic feature (Figure 10.17). They are thought to be formed of ice and rocks, plus some carbonaceous material (Lodders and Fegley 1998). The particles forming the rings range in size from submicron ($<10^{-6}$ m) to some meters.

Both Voyager spacecraft visited Jupiter and Saturn, as is evident from their trajectories shown in Figure 10.18 (Rothery 1992). The Cassini mission was launched in 1997 to arrive at Saturn in 2004. It carries the Huyghens probe, which will parachute down to the surface of Titan (which is a little larger than the planet Mercury). The Sun's light reflected by Titan shows absorption features in the red, which are ascribed to methane present at the level of up to 10%. The dominant gases in Titan's atmosphere are inferred to be nitrogen (N_2) and argon (A).

The Voyager 2 spacecraft journeyed on to Uranus, reaching it in January 1986 (Miner 1998). Models of the interior that are consistent with Voyager 2 findings have been developed.

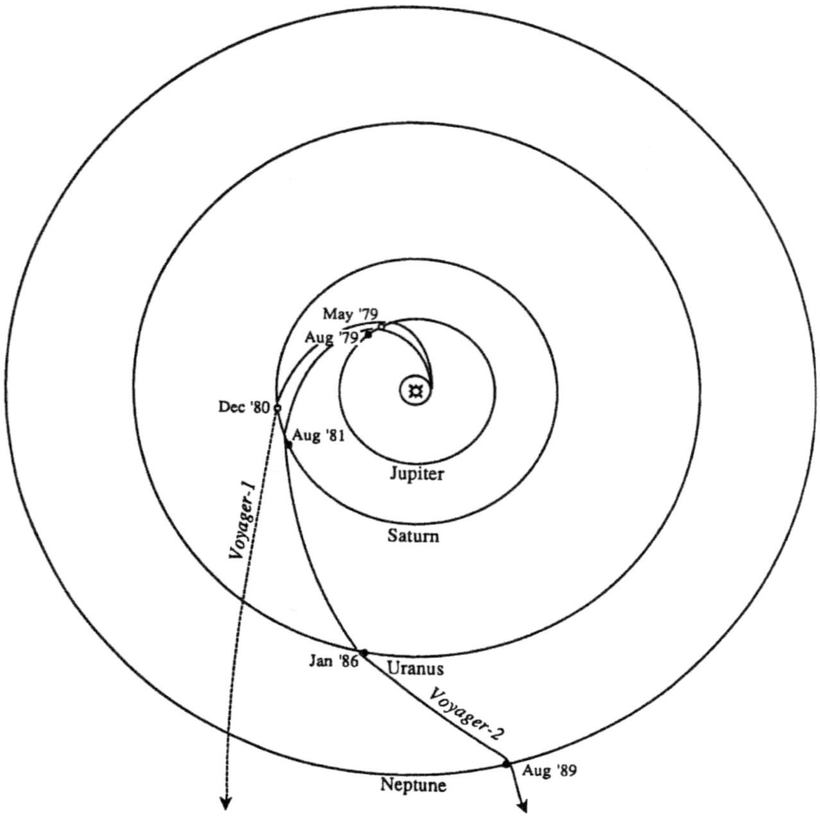

FIGURE 10.18 Trajectories of the two Voyager spacecraft. Voyager 1's encounter with Saturn took it out of the ecliptic plane. After Neptune, Voyager 2 continued on a course heading below the plane of the solar system (from Rothery 1992).

Its characteristic blue-green color is due to methane (CH_4) in the cloudy atmosphere. The effective temperature of Uranus is 59 K.

As indicated in Figure 10.19 (from Miner 1998), the unique feature of the planet Uranus is that its axis of rotation is almost in the ecliptic plane. Uranus rotates in a retrograde (east to west) direction. The best-fit magnetic dipole axis makes a large angle (60°) with the rotational axis. The magnetic field is probably generated by dynamo activity in a partially fluid core extending out to 0.7 radius of Uranus. The rings of Uranus were observed in detail for the first time in January 1986. Ten satellites of Uranus were also discovered in Voyager 2 images (Miner 1998); their radii vary from 10 to 80 km.

Observed by Voyager 2 and the Hubble space telescope, Neptune is not very dissimilar to Uranus. It has a cloudy atmosphere, but at the cloud tops water is not present because it is too cold (Lodders and Fegley 1998). Dynamo action in a conducting interior produces a magnetic field, which is offset from Neptune's center by 0.55 radius of Neptune and tilted at 47° to the rotational axis. Six small satellites were discovered by Voyager 2, which imaged its rings in August 1989 (Lodders and Fegley 1998).

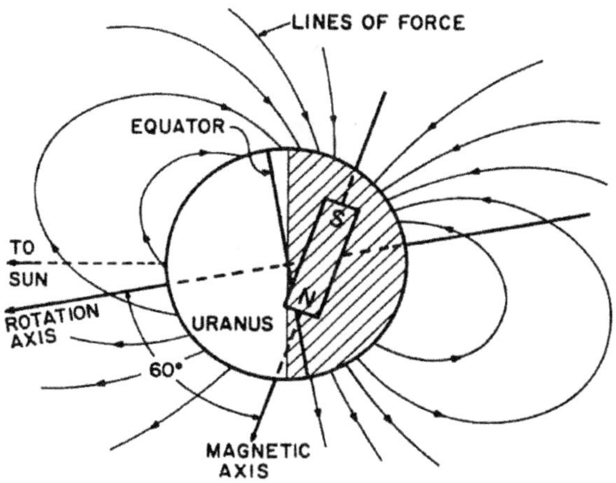

FIGURE 10.19 The simplest representation of the magnetic field of Uranus is a magnetic dipole tilted at 60° to the rotation axis and offset by 0.3 Uranian radii toward the dark north polar region (from Miner 1998).

Neptune's largest satellite, Triton, was also observed then. Its average surface temperature is only 38 K. Its nitrogen-rich atmosphere has a surface pressure of 1.6 Pa.

Pluto and its satellite Charon (whose radius is 0.52 radius of Pluto) are icy worlds. Pluto has predominantly nitrogen ice with a surface temperature of 40 K, whereas Charon's surface is of water ice, also at a temperature of 40 K (Lodders and Fegley 1998; Stern and Mitton 1998).

Comets have a nucleus, coma, dust tail, and plasma tail. The Giotto probe visited Halley's comet in 1986, taking the first close-up image of a comet's nucleus subliming. The nucleus is a 10-km-sized conglomerate of ice and rocky dust particles (Lewis 1997; Lodders and Fegley 1998). Several other instrumented spacecraft intercepted the orbit of Halley's comet in 1986, and other comets have been investigated directly by other space missions.

References

Beatty, J. K., Collins, C., and Chaikin, A., eds. 1999. *The New Solar System,* Sky, Cambridge, MA.

Canup, R. M. and Righter, K., eds. 2000. *Origin of the Earth and Moon,* University of Arizona Press, Tucson, AZ.

Fischer, D. and Duerbeck, H. 1998. *Hubble Revisited,* Copernicus, New York.

Kowal, C. T. 1996. *Asteroids: Their Nature and Utilization,* Wiley-Praxis, Chichester.

Lewis, J. S. 1997. *Physics and Chemistry of the Solar System,* Academic Press, San Diego.

Lodders, K. and Fegley, B., Jr. 1998. *The Planetary Scientist's Companion,* Oxford University Press, New York.

Marov, M. Y. and Grinspoon, D. H. 1998. *The Planet Venus,* Yale University Press, New Haven, CT.

Mendell, W. 1999. In *Keys to Space: An Interdisciplinary Approach to Space Studies,* ed. A. Houston and M. Rycroft, McGraw-Hill, New York.

Miner, E. D. 1998. *Uranus: the Planet, Rings and Satellites,* Wiley-Praxis, Chichester.

Rothery, D. A. 1992. *Satellites of the Outer Planets: Worlds in their Own Right,* Clarendon Press, Oxford.

Spencer, J. R. and Mitton, J. 1995. *The Great Comet Crash: The Collision of Comet Shoemaker-Levy 9 and Jupiter,* Cambridge University Press, Cambridge.

Stern, A. and Mitton, J. 1998. *Pluto and Charon: Ice Worlds on the Ragged Edge of the Solar System,* John Wiley & Sons, Chichester.

PART 4

The Moon

Peter Eckart

10.14 Origin of the Moon

The origin of the Moon has been pondered, modeled, studied, and endlessly debated, but still the mystery has not been ultimately solved. There are three classical theories: the coaccretion model, the fission model, and the intact capture model.

The coaccretion model assumes that Moon and Earth developed close to each other as separate bodies. This model is based on the dominant model for the origin of the solar system, suggesting that it condensed out of a cloud of hot gas, called the solar nebula. In such a model the planets are small pieces of the condensed material left over after star formation. Thus, our Moon and the moons of the other planets would be similar to the planets that form around a star, a moon being the left-over bits of a planet formation. One of the major problems of this model is the fact that the densities of Moon and Earth are so different. This appears unlikely if they were formed at the same time, in the same region, and from the same original matter. Also, there is a problem in accounting for the angular momentum of the Earth and Moon, explaining why material around Earth stayed in orbit rather than falling on Earth.

The fission model assumes that the Moon was initially a part of Earth. This theory is based upon the assumption that initially the original planet had a high rotational rate, and led to the suggestion that the Moon was flung off Earth very early in Earth's history. According to calculations, this offspring should have a mass of about 1–2% of the original body's mass, and the ratio of the diameters of the resulting spherical bodies should be about 1:4—these are quite exactly the ratios of Earth and Moon. However, there are problems explaining how the Earth was spinning fast enough to fling off a blob of material from which the Moon formed. Because no explanation for this aspect could be found, the fission model had to be rejected.

The intact capture model assumes that the Moon's and Earth's flight paths crossed early in the history of the solar system, resulting in the Moon being captured by Earth. But, according to this model, the minimum distance of Earth and Moon would have to be less than 60,000 km. Also, the fact that Moon's orbit around Earth is almost circular and the symmetry of the Earth–Moon system make this theory appear very unlikely. Moreover, the probability of capturing a fully formed Moon is considered extremely tiny.

Thus, none of the three classical models of lunar origin are particularly successful at explaining all the properties of the Moon and all the orbital characteristics of the Earth–Moon system. Therefore, the giant impact (or Big Whack) model was created, which combines essential aspects of the three classical models. The idea that the Moon formed as a consequence of a giant impact of a planetoid on Earth, originally proposed in 1976, is currently accepted wisdom in planetary science. It assumes that during the last phase of Earth's accretion, after its core had formed and while it was still molten, an object the size of Mars smashed into this proto-Earth at an oblique angle (capture model). The resulting monumental explosion deposited large quantities of material in Earth orbit, and the primitive Moon formed from this material (fission model). Although the giant impact model can explain most of the current phenomena of the Earth–Moon system, including the geochemical properties of the Moon and the orbital dynamics of the Earth–Moon system (coaccretion model), it is far from certain that the problem of lunar origin has been solved, and more data will be required either to prove or to reject it (Spudis 1996).

10.15 Orbital Parameters

The Earth–Moon-System

Viewed from above the North Pole of the Earth, the Moon travels counterclockwise in a slightly elliptical path around the Earth. The eccentricity of this orbit oscillates between 0.044 and 0.067 (mean value 0.0549). Measured from center to center, the Earth–Moon distance ranges from 356,410 km at the Moon's closest approach (perigee) to a maximum value of 406,697 km at furthest separation (apogee). The mean value of the semimajor axis is 384,400 km. As the Moon orbits the Earth, it shows different phases, or lighting conditions of its surface, at various times of the month. The Moon's orbit around the Earth is synchronous, i.e., its rotational period is the same as its period of revolution around the Earth. Therefore, it always keeps the same hemisphere, the near side, pointing toward the Earth. Conversely, the other hemisphere, the far side, is always turned away from the Earth (Heiken et al. 1991).

When the Earth–Moon system is orbiting the Sun and the Moon is orbiting the Earth, the time interval between two successive new moon phases is 29.53059 days; this is a synodic month or a lunar day. For one complete orbit around the Earth, the Moon needs 27.32166 days, a sidereal month. During these 27.3 days, the center of gravity of the Earth–Moon system advances 27° on its way around the Sun. To be in alignment again with the Earth and the Sun, the Moon has to move another 27° on its orbit around Earth. A summary of lunar characteristics is given in Table 10.4 (Spudis 1996).

Lunar Librations

The plane of the lunar orbit is inclined to the plane of the ecliptic by 5°9′, and the Moon's polar axis is inclined 1°32′ to the plane of the ecliptic. This results in an inclination of 6°41′ of the Moon's rotation axis relative to a line perpendicular to the lunar orbital plane (see Figure 10.20). This geometric

Property	Moon	Earth	Moon-to-Earth ratio
Mass	7.353×10^{22} kg	5.976×10^{24} kg	1/81 (0.0123)
Mean radius	1.738 km	6.371 km	1/3.7 (0.2727)
Surface area	3.80×10^{7} km²	5.10×10^{8} km²	1/13.5 (0.07429)
Flattening[a]	0.0005	0.0034	1/7.5 (0.147)
Mean density	3.34 g/cm³	5.517 g/cm³	1/1.65 (0.605)
Gravity at equator	1.62 m/s²	9.81 m/s²	1/6 (0.165)
Escape velocity at equator	2.38 km/s	11.2 km/s	1/4.7 (0.2125)

Parameter	Value
Mean distance from Earth	384.400 km
Mean eccentricity	0.0549
Mean inclination of the ecliptic	5°09′
Mean inclination to the lunar equator	6°41′
Length of sideric month	27.3217 days
Length of synodic month	29.5306 days

[a]Flattening = (equatorial diameter – polar diameter)/equatorial diameter.
Source: Heiken et al. 1991.

TABLE 10.4 Lunar Characteristics

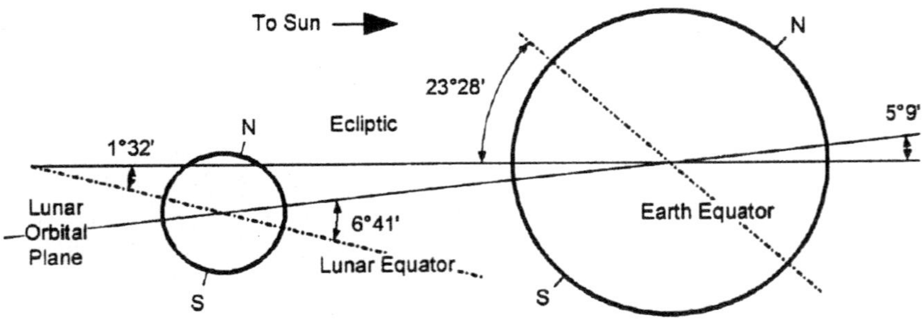

Figure 10.20 Schematic of lunar orbital characteristics (angular relationships have been exaggerated) (Heiken et al. 1991).

relationship causes what is called the latitudinal libration. Due to this libration it is possible to see an additional 6.7° of the lunar surface at the North Pole and South Pole, over a two-week interval. The Moon's longitudinal libration is caused by the slight noncircularity of its orbit around the Earth. The Moon revolves at a constant rate, so that after one quarter of the rotational period it has rotated through 90°. At the same time, however, it has moved through 97° of its orbit, so that seen from Earth the Moon's central meridian would appear to be displaced by about 7°. Because of this libration, it is possible to see an additional 7° along both eastern and western limbs of the lunar disc. Finally, there is a daily or diurnal libration, a parallax effect, due to the axial rotation of the Earth itself. The position of an observer on the Earth's equator will move laterally about 13,000 km during a 12-hour period. This gives a libration of 1° at the western edge of the Moon upon rising and 1° at the eastern edge when it sets. Because of these combined librations, a total of about 60% of the lunar surface can be seen from Earth. As a result of the lunar librations, all near side lunar sites can expect continuous communication with Earth; a site located near the terminator will see the Earth set below the lunar horizon, thus allowing direct communication only part of the time. Sites on the far side of the Moon cannot communicate with Earth without the aid of relay satellites or relay antennas on the lunar surface (Heiken et al. 1991).

Lunar Diurnal Cycle

A lunar day lasts about 29.5 days (see above). Therefore, at the equator and at all latitudes, except very close to the poles, a 14-day-night-14-day-light cycle results. Due to the Moon's orbital parameters (see above), the Sun's elevation at the poles is only ±1°32′, hence a ½-year-light-½-year-night-cycle results. As a result, some craters in the South polar region are permanently shaded (estimated 2% of the lunar surface) (Heiken et al. 1991).

10.16 Lunar Geography

The main features of the lunar surface are the highlands, maria, and craters. The light-colored highlands, or terrae, are the original primordial crust of the Moon. They have been continuously penetrated by meteoroids which created their very rough landscape. The dark maria or lowlands, which were formed by immense lava flows, are the most obvious and smoothest of all lunar features (see Figures 10.21 and 10.22). They are easily visible with the naked eye. Most lunar craters have a circular outer shape, sunken floors, a central peak, and rims rising to modest heights above the outer surface. It is generally accepted that meteorite impact is the predominant crater-forming mechanism. Craters appear all over the Moon, but there are fewer in number the younger they are (Heiken et al. 1991; Moore 1981).

Figure 10.21 The maria of the lunar near side (Heiken et al. 1991).

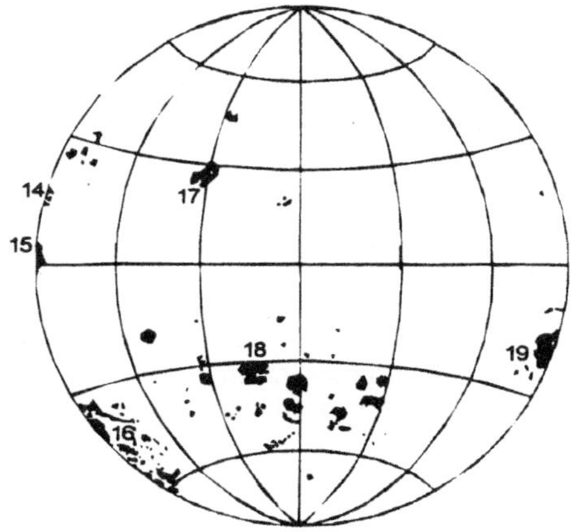

Figure 10.22 The maria of the lunar far side (Heiken et al. 1991).

In order to be able to locate particular points on the lunar surface, lunar maps follow Earth convention, with North at the top and the prime meridian referenced to crater Mösting A on the Moon's near side. Before 1974, lunar longitude was expressed in terms of 0°–180° East (+) and 0°–180° West (–) of the prime meridian. This convention is also used here because it is used in most literature. After 1974, a lunar longitude system of 360°, proceeding counterclockwise (increasing to the East) from the prime meridian, was adopted by international agreement for future cartography (Heiken et al. 1991).

10.17 Lunar Geology

Like the Earth, the Moon is not a uniform, homogeneous planet. It consists of different rocks, formed in different ways at different times. To learn about the Moon and its history, one has to learn something about lunar rocks—what they are made of and how they are classified. Some of the

Moon's rocks are familiar. For example, the basalt lava flows that cover the dark maria formed in the same way as many of the lava flows on Earth—by melting deep in the planet, followed by ascent of molten rock to the surface and eruption through fractures and vents. Within any group of rocks, e.g., basaltic lavas, the rocks are not identical. There are differences in mineral composition, in mineral shapes and sizes, and in each rock's chemical composition. The rocks of the Moon fall into four distinct groups: basaltic volcanic rocks, pristine rocks, breccias and impact melts, and lunar soil.

The basaltic volcanic rocks of the maria were produced inside the Moon by partial melting of the lunar mantle, followed by the buoyant rise of molten rock to the lunar surface and its eruption. Two types of volcanic rocks have been erupted to the lunar surface—lava flows and pyroclastic (volcanic ash) rocks. The volcanic plains that form the maria are rich in iron (Fe), and many are rich in titanium (Ti), most of which is in the mineral ilmenite ($FeTiO_3$). The common way of distinguishing mare basalts is to classify them according to their TiO_2 content into high-Ti basalts (>9 wt-% TiO_2), low-Ti basalts (1.5–9 wt-% TiO_2), and very low-Ti basalts (<1.5% TiO_2). Pristine rocks from the highlands are rocks that have original lunar compositions, uncontaminated by impact mixing. Pristine highland rocks are divided into three major groups: ferroan anorthosites (low-Na plagioclase, low-Mg pyroxene), Mg-rich rocks (plagioclase, high-Mg pyroxene), and KREEP rocks (K (potassium), rare Earth elements, phosphorus). However, most lunar rocks have been shattered, crushed, and recompacted by meteoroid impacts (breccias) and/or formed by melting of older lunar rocks during micrometeoroid impacts (impact melts). Breccias containing a single rock type are called monomict breccias, while multiple rock types are called polymict breccias. Polymict breccias are the most common of lunar breccias. There are seven major groups: fragmental breccias, glassy melt breccias, crystalline melt breccias, clast-poor impact melt rocks, granulitic breccias, dimict breccias, and regolith breccias. Finally, there is the lunar soil, which is composed of fragmental (<1 cm), unconsolidated debris within the lunar regolith that covers the lunar surface.

Lunar Regolith

The lunar regolith is the source of virtually all our information about the Moon. *Regolith* is a general term for the layer or mantle of fragmental and unconsolidated rock material, whether residual or transported and of highly varied character, that nearly everywhere forms the surface of the land and overlies or covers bedrock. It includes rock debris of all kinds, including volcanic ash. On the airless and lifeless Moon, the lunar regolith basically results from two uniquely different processes—the continuous impact of large and small meteoroids on the lunar surface and the steady bombardment of the lunar surface by charged atomic particles from the stars and the Sun. All of the lunar landings and all of the photographic investigations show that the entire lunar surface consists of a regolith layer that completely covers the underlying bedrock, except perhaps on some very steep-sided crater walls and lava channels, where there may be exposed bedrock. Studies of returned samples have shown that the bulk of the lunar regolith consists of particles less than 1 cm in size, although larger cobbles and boulders, some as much as several meters across, are commonly found at the surface. Although lunar soil is lexicographically synonymous with lunar regolith, the term *lunar soil* usually refers to the sub-cm fraction of the lunar regolith. Lunar soil is a somewhat cohesive, dark gray to light gray, very fine-grained, loose, clastic material derived primarily from the mechanical disintegration of basaltic and anorthositic rocks. The mean grain size of analyzed soils ranges from about 40 mm to about 800 mm, averaging 60–80 mm. Individual lunar soil particles are mostly agglutinates, as well as various rock and mineral fragments. Agglutinates are individual particles that are aggregates of smaller lunar soil particles (mineral grains, glasses, and even older agglutinates) bonded together by vesicular, flow-banded glass. Agglutinate particles are small (usually less than 1 mm) and contain minute droplets of Fe (much of which is very fine-grained single-domain FeO), and troilite (FeS). Agglutinates make up a high proportion of many lunar soils, about 25–30% on average, although their abundance may range from a rare 5% to about 65%.

Overall, lunar soils range in composition from basaltic to anorthositic, and they include a small (<2%), meteoritic component. Although the chemical compositions of lunar soils show considerable variation, physical properties such as grain size, density, packing, and compressibility are rather uniform. One manifestation of such uniform physical properties is that seismic velocities in regolith materials at both mare and highland sites are quite similar, varying between 92 m/s and 114 m/s. The current consensus is that the regolith is generally about 4–5 m thick in the mare areas but may average about 10–15 m in older highland regions, based on orbital radar data and modeling. Beneath this true regolith is a complex zone that probably consists of large-scale ejecta and impact-fractured, brecciated bedrock. Because of its surficial, unconsolidated, and fine-grained nature, it is likely that the regolith will be an important raw material to be used for lunar base construction and shielding, mining, road building, and resource extraction.

Soil Composition

It may appear incongruous for a planet that has no atmosphere, but oxygen is overwhelmingly the most abundant element in lunar surface materials. Table 10.5 presents the elements in their order of abundance, both in atom-% and weight-%. Thanks to the implantation of ions from the solar wind into the grains of lunar soil, even elements such as hydrogen (H), carbon (C), nitrogen (N), oxygen (O), phosphorus (P), and sulfur (S), as well as some noble gases, are available on the surface of the Moon. The solar wind is a plasma of chemical elements, expelled as ionized atoms from the atmosphere of the Sun. The principal element of the solar wind is H, the second most abundant element is helium (He), followed by C, N, and O. The concentration of H in typical lunar soil is about 50 g/kg. This corresponds to about 1.5 million liters of water per km^2 to a depth of 2 m, or 100 g/m^3 of H, assuming a soil density of 1.75 g/cm^3. So the problem is not that H and the other elements are scarce on the Moon, but whether they are economically accessible (Heiken et al. 1991).

Water Ice in the Lunar Polar Areas?

The very small inclination of the Moon's equator to the ecliptic led to the most interesting speculation that there may be water ice preserved in the lunar polar areas, for example in permanently

Element	Average abundance (Atom-%)	Average abundance (Mass-%)	Sites
O	60%	45%	Maria/highlands
Si	10.5%	21%	Maria/highlands
Al	16%	13%	Highlands
	5%	5%	Maria
Ca	5%	10%	Highlands
	4.5%	8%	Maria
Mg	5%	5.5%	Maria/highlands
Fe	2.5%	6%	Highlands
	6%	15%	Maria
Ti	<1%	<1%	Maria/highlands
Na	<1%	<1%	Maria/highlands

Source: Heiken et al. 1991.

TABLE 10.5 Average Abundance of Elements in Lunar Soil

shadowed craters of the South Pole. This water ice may originate from comets that struck the Moon long ago. Debate over the presence or absence has developed over several years, but it still remains unclear whether or not the ice exists. In 1998, the Lunar Prospector mission provided further evidence for the existence of water ice, but still no ultimate proof (Heiken et al. 1991).

10.18 Physical Surface Properties

The physical properties of the lunar surface discussed in this subsection cover mainly the geotechnical properties, i.e., the surface properties needed to evaluate engineering problems, in particular the mechanical properties of lunar regolith. The experience of the first human being to set foot on the Moon, Neil Armstrong, gives an initial impression of these properties:

I am at the foot of the ladder. The LM [Lunar Module] footpads are only depressed in the surface about 1 or 2 inches, although the surface appears to be very, very fine-grained, as you get close to it, it's almost like a powder; down there, it's very fine . . . I'm going to step off the LM now. That's one small step for [a] man. One giant leap for mankind. As the—The surface is fine and powdery. I can—I can pick it up loosely with my toe. It does adhere in fine layers like powdered charcoal to the sole and sides of my boots. I only go in a small fraction of an inch. Maybe an eighth of an inch, but I can see the footprints of my boots and the treads in the sandy particles.

The ranges of geotechnical properties of lunar materials are less than those that occur in surficial materials on Earth. This is because of the following factors:

- The familiar terrestrial geologic processes of chemical weathering, running water, wind, and glaciation are absent on the Moon. These processes tend to produce well-sorted sediments with uniform grain sizes.
- The three main constituents most likely to produce unusual or problem soils on Earth are absent on the Moon. There is no water, and therefore no clay minerals or organic materials.
- The variety of minerals in lunar soil is much less than that found on Earth. Many soil particles are simply fragments of rocks, minerals, and glass stuck together with glass (agglutinates).

The most significant variable of all geotechnical properties is the relative density. To a certain extent, it also controls the other physical properties discussed in this subsection (Heiken et al. 1991).

Particle Size Distribution

The particle size distribution in an unconsolidated material, such as lunar soil, is a variable that controls to various degrees the strength and compressibility of the material, as well as its optical, thermal, and seismic properties. The majority of lunar soil samples fall in a fairly narrow range of particle size distributions. In general, the soil is a well-graded (or poorly sorted) silty sand to sandy silt. The median particle size is 40–130 mm, with an average of 70 mm, i.e., approximately half of the soil, by weight, is finer that the human eye can resolve. Roughly 10–20% of the soil is finer than 20 mm, and thus a thin layer of dust adheres electrostatically to everything that comes in contact with the soil—space suits, tools, equipment, and lenses. Housekeeping is thus a major challenge for operations on the lunar surface (Heiken et al. 1991).

Specific Gravity

The specific gravity, G, of a soil particle is defined as the ratio of its mass to the mass of an equal volume of water at 4°C. Many terrestrial soils have a specific gravity of 2.7, i.e., the density of the individual particles is 2.7 g/cm^3, or 2.7 times that of water (1 g/cm^3). Values for lunar soils mostly range from 2.3 to > 3.2 (agglutinate/glass particles: 1.0 to >3.32; basalt particles: >3.32; breccia particles: 2.9–3.1). A value of 3.1 is recommended for general scientific and engineering analyses of lunar soils (Heiken et al. 1991).

Bulk Density and Porosity

The bulk density, ρ, of soil is defined as the mass of the material contained within a given volume, usually expressed in g/cm^3. The porosity, n, is defined as the volume of void space between the particles divided by the total volume. Bulk density, porosity, and specific gravity, G, are interrelated as:

$$\rho = G\rho_w(1-n)$$

with the density of water, $\rho_w = 1\,\text{g/cm}^3$. The *in situ* bulk density of lunar soil is a fundamental property. It influences bearing capacity, slope stability, seismic velocity, thermal conductivity, electrical resistivity, and the depth of penetration of ionizing radiation. Consequently, considerable effort has been expended over the years in obtaining estimates of this important parameter. Taking into account all of the measurements, approximations, and analyses of the returned Apollo core samples and other measurements, the best estimate for the average bulk density of the upper 60 cm of lunar soil is 1.66 g/cm^3. For regions below 3 m no direct tactical data about the density of the lunar regolith exist. But it is known that the density approaches a maximum value at about 50 cm depth and increases very slowly beyond that (Heiken et al. 1991).

Shear Strength

The shear strength of a granular soil is typically defined in terms of the classic Mohr–Coulomb equation:

$$\tau = c + \sigma \tan \Phi$$

where τ = shear strength (kPa)
c = cohesion (kPa)
σ = normal stress (kPa)
Φ = friction angle

The shear strength therefore consists of two components: a cohesive component that is independent of applied stress, and a frictional component that is directly proportional to the normal stress. The shear strength governs such important engineering properties as ultimate bearing capacity, slope stability, and trafficability. As a result, estimates of lunar soil cohesion and friction angle have been the object of intensive research. Based on a variety of data sources, including the Apollo missions, best-estimate values for cohesion and friction angle have been developed as indicated in Table 10.6 (Heiken et al. 1991).

Bearing Capacity

The bearing capacity describes the ability of a soil to support an applied load, such as an astronaut, a vehicle, or a structure. Usually the topic of bearing capacity is divided into two categories: ultimate bearing capacity and allowable bearing capacity. Each of these is then subdivided further into static and dynamic quantities. The ultimate bearing capacity defines the maximum possible load that can be applied without causing gross failure, such as the overturning of a structure. The allowable bearing capacity defines a lesser load that can be applied without exceeding a given amount of settlement. A settlement limit is usually imposed either for structural or operational requirements.

Location	Cohesion c (kPa)	Friction angle ϕ (°)
In general	0.1–1	30–50
Crater wall (inner)	0.17–1.0	45–25
Crater slope (outer)	0.52–2.7	45–25
Horizontal ground	0.34–1.8	45–25
Intercrater areas		41–43
0–15 cm	0.44–0.62	44–47
0–30 cm	0.74–1.1	
30–60 cm	2.4–3.8	52–55
0–60 cm	1.3–1.9	48–51

Source: Heiken et al. 1991.

TABLE 10.6 Best-Estimate Values for Cohesion and Friction Angle

The static ultimate bearing capacity, q_{ult}, can be estimated on the basis of plasticity theory. It is therefore controlled by the soil density, its shear strength, and the size of the footing. Using the *in situ* bulk density estimates and the *in situ* shear strength estimates given earlier in this subsection, the static ultimate bearing capacity versus footing width can be calculated (see Figure 10.23). This means that the ultimate load (stress × area) for a circular or square footing is proportional to the cube of its width. Consequently, the ultimate bearing capacity of the lunar surface is more than sufficient to support virtually any conceivable structure.

The dynamic ultimate bearing capacity defines the maximum resistance to impact loading. This dynamic capacity is always greater than the static capacity because of the inertial resistance of the soil. Even after the experience from several manned lunar landings, many scientists and engineers

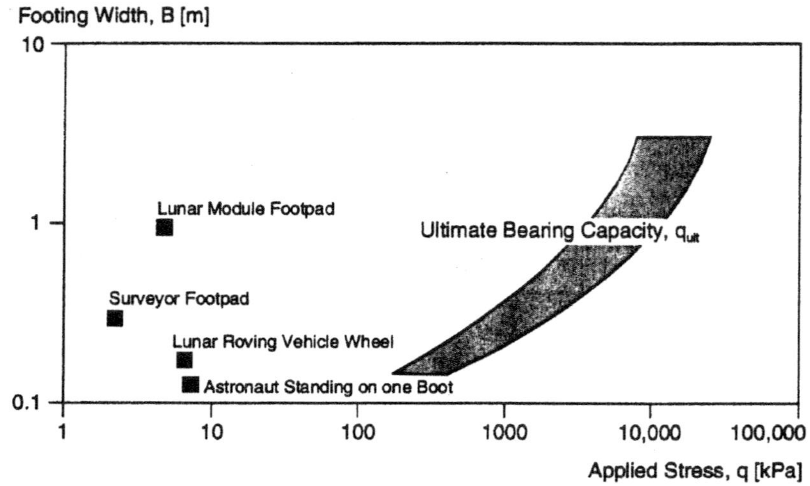

FIGURE 10.23 Static ultimate bearing capacity data for lunar surface material (Heiken 1991).

still believed that an astronaut could hammer a rod or core tube into the lunar surface material to almost any depth. In fact, the practical limit of the Apollo core tubes was only about 70 cm, and it typically required about 50 hammer blows to reach this depth. An analysis showed that, if energy losses were neglected, the number of hammer blows required to reach a given depth would increase with the square of the depth. If energy losses were included, then there would be a depth beyond which no amount of hammering would drive a rod or core tube farther in (Heiken et al. 1991).

Slope Stability

Many numerical methods have been developed to evaluate the stability of a soil slope, i.e., its ability to stand without support. On the Moon, the absence of water greatly simplifies the analysis of slope stability. The factor of safety, *F.S.*, against slope failure can be reduced to the expression:

$$F.S. = N\frac{\rho g_m h}{c}$$

where ρ = density of soil
c = cohesion of soil
g_m = acceleration of gravity on the Moon = 1.624 ms^{-2}
h = height of slope
N = stability number, which is a function of friction angle, Φ, and slope angle, θ

A constructed slope could be either an excavation, a compacted embankment, or a dumped pile (see Figure 10.24). The safe depth, i.e., the depth up to which no slope failure will occur, of an excavation in an intercrater area can be calculated by combining values for the *in situ* density with the *in situ* shear strength. Using a factor of safety of 1.5, which is more than adequate for design

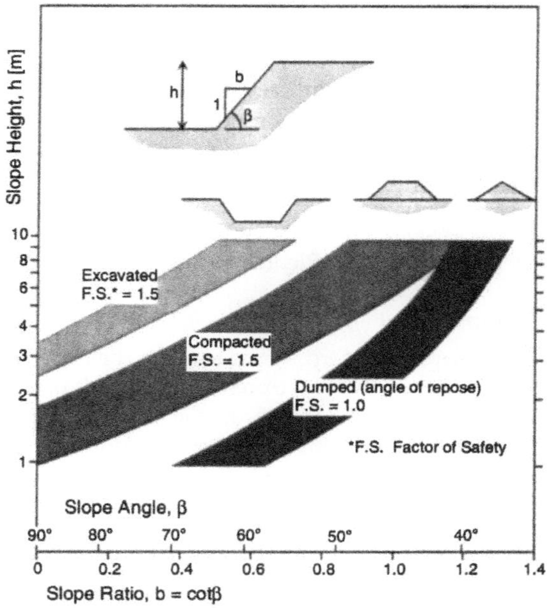

FIGURE 10.24 Calculated stability of artificial slopes constructed from lunar surface material (Heiken 1991).

purposes, calculations show that a vertical cut could be made in lunar soil to a depth of about 3 m and a slope of 60° could be maintained to a depth of about 10 m. In order to construct an embankment, the soil must first be excavated, then transported, placed, spread, and compacted. As discussed above, *in situ* lunar soil is very dense, with a greater density than could be produced with mechanical compaction equipment. The processes of handling and manipulating the lunar soil would loosen it considerably, and it would not then be possible to compact the soil back to its original, undisturbed density. As a result, the density of the soil in a compacted embankment would be less than its original density, and the maximum possible slope angle of the embankment would be less than that of an excavation in undisturbed soil. Assuming a compacted relative density of 65–75%, a 10-m-high slope could be constructed at an angle of about 45°. If lunar soil were simply dumped in a pile, it would attain a relative density of about 30–40% and the factor of safety would be 1.0. The pile could be raised to a height of 10 m at an angle of nearly 40°. Not much is known about the stability of natural lunar slopes. The very limited cone penetrometer data obtained by both human and robotic missions have established that the soil on the slopes is actually somewhat weaker than the soil in the flatter intercrater areas, at least to a depth of 70 cm (Heiken et al. 1991).

Seismic Activities

The Apollo passive seismometers monitored the Moon's seismic activity for almost eight years. It was found that the release of seismic energy from the Moon is about seven orders of magnitude lower than that of Earth. The sources of seismicity on the Moon are:

- Monthly deep-focus moonquakes caused by Earth–Moon tidal stresses.
- Shallow moonquakes that are fewer but stronger and may be due to tectonic processes. They account for most of the seismic energy released in the Moon.
- Thermal moonquakes that may be due to thermal degradation of young lunar surface features. They last for about an hour.
- Moonquakes caused by meteoroid impacts, which vary widely in energy. Meteoroid impacts of all energies tend to be most common when meteoroid showers peak, particularly among the largest meteoroid impacts that tend to occur in the months of April to July. The largest recorded impacts, in July 1972 and May 1975, represented meteoroids of about 5 tonnes. Overall, seven meteoroid impacts of 1 tonne or more were observed within 5 years during lunar seismic monitoring.

Overall, the Moon can be considered seismically and tidally stable, with a magnitude of the largest events recorded being below 5 on the Richter scale (Heiken et al. 1991; Moore 1981).

Thermal Properties of Lunar Soil

Since the Moon is a small planetary body, there is good reason to believe that it has cooled considerably during its 4.6-billion-year history. Most of the present heat flux is probably generated by radioisotopes, mainly ^{40}K, ^{232}Th, ^{235}U, and ^{238}U, present in the interior to a depth of about 300 km. Experiments installed on the Moon during the Apollo missions provided extensive information on the temperature and thermal properties of the lunar surface layer to a depth of 3 m, including surface temperature variations, near-surface thermal properties, subsurface temperature variations, and thermal conductivity. At the Apollo sites, mean temperatures 35 cm below the surface are 40 to 45 K above those at the surface. This is primarily due to the fact that the upper 1 to 2 cm of the lunar surface have an extremely low thermal conductivity, with this conductivity being temperature-dependent. At a depth of about 2 cm, the conductivity increases greatly to values five to seven times greater than the surface value. This increase of conductivity appears to be mainly due to a large increase in the soil compaction and grain boundary contacts with depth. In lower layers of the

Surface material	Parameter γ $\left[\frac{cm\sqrt{s}K}{J}\right]$	Density ρ $\left[\frac{kg}{m^3}\right]$	Specific heat c $\left[\frac{J}{kgK}\right]$	Conductivity k $\left[\frac{W}{mK}\right]$
Total range	5.97–334	500–3000	755–1007	2.14×10^{-3}–1.13
Range for particulate material heavily mixed with blocks	57.2–119	1200–2000	837	7.12×10^{-3}–1.8×10^{-2}
Blocks (rocks)	7.2	2500	837	9.22×10^{-1}
Range, excluding blocks	95.5–238	500–1100	837	4.18×10^{-3}–1.17×10^{-2}
Average maria	95.5–191	800–1500	837	4.18×10^{-3}–8.8×10^{-2}

Source: Heiken et al. 1991.

TABLE 10.7 Summary of Lunar Surface Thermal Property Values

lunar surface, the thermal conductivity of the lunar soil is on the order of 1.4–3.0 10^{-4} W/cm K. This is approximately a factor of 10 higher than the conductivity at the surface. Lunar surface thermal property values are summarized in Table 10.7 (Heiken et al. 1991).

Electrical and Electromagnetic Properties of Lunar Soil

The electrical properties of the lunar surface materials are those of silicates characterized by extremely low loss and low electrical conductivity. In the total absence of water, the DC electrical conductivity ranges from 10^{-14} mho/m for lunar soil to 10^{-9} mho/m for lunar rocks at 300 K in darkness. Upon irradiation with sunlight, there is a more than 10^6-fold increase in conductivity of both lunar soils and rocks. The relative dielectric permittivity k' for lunar materials as a function of bulk density, ρ, is approximately:

$$k' = 1.9\,\rho$$

The relative dielectric permittivity is controlled by bulk density and is independent of chemical or mineralogical composition, variations of frequency above 1 MHz, and temperature variations within the range of lunar surface temperatures. The loss tangent for high-frequency electromagnetic loss in lunar soil is given by:

$$\tan\theta = 10^{[0.038(\%\,TiO_2+\%\,FeO)+0.312\rho-3.260]}$$

In both equations, ρ is the bulk density in g/cm³. The extremely low electrical conductivities and low loss tangents indicate that lunar materials are very transparent to electromagnetic waves. For example, radio transmissions should readily penetrate through the lunar soils to a depth of about 10 m. As a result, radio communications on the lunar surface need not necessarily be by direct line of sight, but may penetrate low hills. The low conductivity and low loss are also responsible for the fact that lunar materials are readily chargeable and will remain electrically charged for long periods of time. The large photoelectric change in electrical conductivity at lunar sunrise and sunset can charge surface soil particles to the point that they will levitate and move. Such charged soils and mobile particles could readily coat surfaces and be hazardous to visibility and equipment operations during the lunar night.

10.19 Lunar Surface Environment

Lunar Surface Temperature

The temperature of the lunar surface is determined by the amount of radiation absorbed by the Sun, as well as the heat load from the interior. Assuming constant thermal properties for the surface material, the surface temperature curve depends only on a single parameter, the thermal inertia, defined as:

$$\gamma = \frac{1}{\sqrt{k\rho c}}$$

where k = thermal conductivity
ρ = bulk density
c = specific heat

The thermal inertia of the lunar surface layer does vary with temperature and depth, but including these complications does not significantly increase the fidelity of the temperature model for engineering purposes. The lunar thermal inertia is so low that the daytime surface temperature is essentially in thermal equilibrium with absorbed incident solar radiation. At the equator this translates to a subsolar point temperature of about 390 K. The predawn temperature falls to approximately 110 K at the equator and to lower values at higher latitudes (see Table 10.8).

In general, the lunar surface temperature increases by about 280 K from just before lunar dawn to lunar noon. The temperature at lunar noon varies throughout the year because of varying distance from the Sun. The noon temperature increases about 6 K from aphelion (greatest distance from the Sun) to perihelion (smallest distance to the Sun). There is a large difference in mean temperature, i.e., the temperature averaged over a complete day-night cycle, just below the lunar surface. Estimated average surface temperatures and temperature extremes for different areas of the Moon are presented in Table 10.8. The temperature at the poles is basically unknown, but it might be as low as 40 K in some permanently shaded areas, i.e., inside craters. This might be the case for about 2% of the lunar surface (Heiken et al. 1991).

Lunar Albedo

The albedo is defined as the fraction of light or electromagnetic radiation reflected by a body or particular surface. The average albedo of the Moon is very low (0.09), i.e., only 9% of the light received by the Moon is reflected back to space or to objects on the surface. For lunar surface materials, the brightness of any area, observed from the Earth at full Moon, yields a value that is

	Shadowed polar craters	Other polar areas	Front equatorial	Back equatorial	Limb equatorial	Typical midlatitudes
Average temperature	40 K	220 K	254 K	256 K	255 K	220 K < T < 255 K
Monthly range	none	10 K	140 K	140 K	140 K	110 K

Source: Heiken et al. 1991.

TABLE 10.8 Estimated Average Surface Temperatures and Temperature Extremes for Different Areas of the Moon

Regions	Minimum	Normal albedo maximum	Average
Front side			
Mare	0.06	0.09	0.07
Highland	0.11	0.20	0.15
Entire face	0.07	0.10	0.09
Far side	—	—	0.22

Source: Smith 1971.

TABLE 10.9 Normal Albedo Values of the Lunar Near and Far Side

virtually that of the normal (perpendicular) albedo for that area. The normal albedo is defined as the normal reflectance of a surface element which is illuminated normally. The value of the normal albedo for an area of lunar surface will depend on the local chemical and mineralogical composition, particle size, and packing density. In general, crater ray systems are the brightest features on the Moon, and highland areas are brighter than maria. Table 10.9 presents normal albedo values of the lunar near and far side.

Lighting Environment of the Lunar Surface

Natural lighting on the lunar surface is very dependent on location. Any spot on the lunar surface goes through the same light cycle during every lunar day. The incident light angle depends on the latitude of the lunar surface position. In addition to sunlight, some natural light is available in the form of earthshine, i.e., sunlight reflected from the Earth. Since the Earth is bigger than the Moon, and the Earth's global average albedo (0.39) is greater than the Moon's (0.09), the brightness of full earthshine as seen from the Moon is 58 times greater than the brightness of full Moon as seen from the Earth. Brightness is a function of albedo times radius squared. The earthshine light is site-dependent because the same side of the Moon always faces the Earth. Sites on the far side of the Moon never see any earthshine, while sites on the near site see the Earth go through all phases, but with varying efficiencies. Earthshine overlapped by sunshine has no benefit. Sites near the Earth terminator receive only minimal earthshine when not in sunlight. Overall, natural lighting will not be available for entire lunar cycles at most sites.

With no atmosphere to scatter light, the ambient light level of the Moon drops to only what is reflected by objects and the ground. There are three situations where the direction of the available light causes visibility problems that are attributed to the lack of ambient light:

1. Shadowing—low light angles (below 30° elevation) cause extremely long shadows that can hide obstacles and craters.
2. Washout—high light angles (within 30° of the normal) do not allow reflected light to reach the eye. An object simply does not appear until one is right next to it. There is also a lack of shadows on most of the lunar landscape, which causes objects to be less defined.
3. Back Lighting—in case of moving down-light, i.e., when the Sun is right behind the observer, an effect similar to washout results, but only in one direction. The surface appears featureless when looking about 10° to either side of the down-Sun or down-light direction.

During the Apollo missions astronauts found that their space suits could be used to reflect light into the shadows when a little extra light was needed.

The Lunar Atmosphere

The lunar atmosphere is very tenuous and not very well characterized. The undisturbed gas concentration is only about 2×10^5 molecules/cm^3 during the lunar night, falling to perhaps 10^4 molecules/cm^3 during the lunar day. This is about 14 orders of magnitude less than for the Earth's atmosphere, a difference so extreme that the Moon is often said to have no atmosphere at all. The major constituents of the ambient lunar atmosphere are neon, hydrogen, helium, and argon. Neon and hydrogen are derived from the solar wind. Helium is mostly derived from the solar wind, but about 10% may be radiogenic and lunar in origin. Argon is mostly ^{40}Ar that is derived from the radioactive decay of lunar ^{40}K. Table 10.10 lists the most probable abundances of these and other species in the undisturbed lunar atmosphere (Heiken et al. 1991).

Since the lunar atmosphere can be considered as a planetary exosphere, all neutral atoms move along ballistic or escaping trajectories in the gravitational field with no collisions between them. This implies that any localized injection of new particles from a landing spacecraft or from a lunar base will quickly spread all around the Moon, in a time on the order of the ballistic free flight time of the molecules, i.e., in less than three hours for oxygen atoms which have a mean thermal speed of 0.6 km/s. Light gases will all escape from the lunar gravitational field. For example, atomic H disappears in less than 120 minutes from the sunlit lunar hemisphere where the temperature is 400 K and the thermal speed of H is 2.5 km/s, while the escape velocity of the Moon is only 2.38 km/s (ESA 1992).

The Lunar Magnetic Field

The magnetic field of the Moon is essentially negligible. During the Apollo missions, an extremely weak magnetic field of 3–300 nT (nanotesla) was measured. Additionally, an external field due to the solar wind of 5–10 nT is present when the Moon crosses the Earth's magnetic tail during four days per orbit. Although the Moon has no global magnetic field, its surface is dotted with small zones where the crust is strongly magnetized. These magnetic zones may be caused by unusual impact conditions or subsurface geology (Spudis 1996).

Radiation Environment of the Lunar Surface

On the lunar surface, there are two kinds of incoming radiation—electromagnetic radiation and ionizing radiation. Basically all of the electromagnetic radiation in the solar system is emitted by

	Abundances [Molecules/cm^3]		Scale heights	
Species	Daytime	Nighttime	Daytime	Nighttime
^{20}Ne	4×10^3 to 10^4	10^5	100 km	25 km
He	8×10^2–4.7×10^3	4–7×10^4	511 km	128 km
H_2	2.5–9.9×10^3	10^4–1.5×10^5	1,022 km	256 km
Ar	2×10^3	$<10^2$	55 km	—
CH_4	1.2×10^3	—	—	—
CO_2	10^3	—	—	—
NH_3	4×10^2	—	—	—
OH + H_2O	0.5	—	—	—

Source: Heiken et al. 1991.

TABLE 10.10 Most Probable Abundances of Lunar Atmosphere Elements

the Sun. The solar electromagnetic radiation at a distance of 1 AU from the Sun has an average energy density of about 1,368 W/m^2. The most important aspect of electromagnetic radiation on the lunar surface is its potential use for solar power production.

Ionizing radiation consists mainly of protons, electrons, and some heavier nuclei. These particles interact with the Moon in different ways, depending on their energy and composition, resulting in penetration depths that vary from μm (micrometers) to m. Any kind of lunar surface habitat will have to be protected from the three different kinds of ionizing radiation in space:

1. Solar wind
2. Solar cosmic rays (SCRs)
3. Galactic cosmic radiation (GCR)

Meteoroid Environment

The term *meteoroid* is used for a naturally occurring solid body, traveling through space, that is too small to be called an asteroid or a comet. Meteoroids with diameters less than about 1 mm are commonly classified as micrometeoroids. The term *meteorite* is used for meteoroids that have fallen upon a planet and have been recovered. Almost all lunar rock surfaces that were exposed to space contain numerous microcraters. Studies of lunar rocks have revealed the average meteroid flux during the past several hundred million years. The average annual cumulative meteoroid model estimates for the lunar surface are:

For $10^{-6} < m < 10^{6}$:

$$\log Nt = -14.597 - 1.2131 \log m$$

For $10^{-12} < m < 10^{-6}$:

$$\log Nt = -14.566 - 1.5841 \log m - 0.063 (\log m)^2$$

where Nt = number of particles/m^2/s of mass m or greater
m = mass (in g)

The velocities of meteoroids that hit the Moon can be calculated to range from 13 to 18 km/s. The meteoroid flux at the lunar surface shows a significant enhancement from small meteoroids (<1 mm), traveling toward the Moon from the Sun, and a somewhat lesser enhancement of large particles (>1 mm), arriving from the direction in which the Earth is traveling. Whichever side of the Moon is facing into the direction of the Earth's motion in its orbit around the Sun will be more exposed to the larger and more hazardous meteoroids (Heiken et al. 1991).

Environmental Impact of Lunar Surface Activities

The Moon, in its pristine condition, serves as an important, well-preserved fossil of the solar system. Therefore, any lunar surface activity, in particular the installation of a lunar base and mining of lunar resources, will cause environmental concern. Most lunar scientific activities require that the unique lunar environment be preserved. Lunar base operations and mining might affect this environment in adverse ways. Specific potential environmental impacts include increased atmospheric pressure, which would compromise astronomical observations, and increased radio frequency background through lunar communication relay satellites, which could affect the use of the far side of the Moon for radio telescopes. Extensive mining efforts could scar the lunar surface irreversibly, which could destroy some of the potential for geological discoveries.

References

Eckart, P. 1996. *Spaceflight Life Support and Biospherics,* Kluwer Academic Publishers, Dordrecht; Microcosm, Torrance, CA.

Eckart, P. 1999. *The Lunar Base Handbook,* McGraw-Hill, New York.

ESA. 1992. ESA Report No. SP-1150, ESA, Noordwijk.

Heiken, G., Vaniman, D., and French, B. 1991. *Lunar Sourcebook: A User's Guide to the Moon,* Cambridge University Press, Cambridge.

Landis, G. 1990. "Degradation of the Lunar Vacuum by a Moon Base," *Acta Astronautica,* vol. 21, no. 3, pp. 183–187.

Mendell, W., ed. 1984. *Lunar Bases and Space Activities of the 21st Century,* Lunar and Planetary Institute, Houston.

Moore, P. 1981. *The Moon,* Mitchell Beazley, London.

Spudis, P. D. 1996. *The Once and Future Moon,* Smithsonian Institution Press, Washington, DC.

PART 5

Mars

G. Komatsu

10.20 Orbital Characteristics

Mars is the fourth planet from the Sun. It has a metallic core and a rocky mantle and crust. Its orbital characteristics and geometrical properties are summarized in Table 10.11. The main difference between the orbits of Earth and Mars is the eccentricity of Mars's orbit, 0.0934, giving the planet a larger range of solar illumination than Earth experiences.

A particularly interesting orbital property is obliquity. Because of the lack of large satellites stabilizing Mars, the obliquity is in chaotic motion and it can change up to 60°; some orbits can change dramatically in less than 45 Myr (Lasker and Robutel 1993). The large range of obliquity could have a profound impact on the geomorphic processes on the Martian surface.

10.21 Solid Geophysical Properties and Interiors

The geophysical parameters of solid Mars are summarized in Table 10.12. Mars is a terrestrial-type planet smaller than Earth and Venus but larger than the Moon and Mercury. The surface area of Mars is approximately equivalent to Earth's land area. Like other planets in the Solar System, Mars was probably born from the gas surrounding the young Sun. Like the other planets formed near the Sun, Mars did not collect enough gas, and so it is rocky. The absence of extensive surface chemical measurement and seismological data limits detailed discussion of the bulk composition and interior structure of Mars. However, based on the mean density and the moment of inertia, it is estimated that Mars should have a core.

Mars Global Surveyor observations found that the magnetic field of Mars does not encompass the entire planet, implying that the Martian dynamo driven in the metallic core is essentially extinct. Instead, magnetic field sources of multiple scales, strength, and geometry were discovered in the ancient southern highland crust (Acuña et al. 1999). They are observed as East–West bands of alternating polarity, the longest extending for over 2,000 km. These are perhaps remnant magnetization formed during early Martian history when the dynamo was active.

Because Mars has no sea level, the level where the average atmospheric pressure is about 6.1 mbar (hPa), the triple point of water, has been used as a datum. The MOLA (Mars Orbiter Laser Altimeter) on board the Mars Global Surveyor acquired topographic information about Mars with high accuracy (Smith et al. 1999b). Dominant features of Mars include southern highlands, the low northern plains, and the Tharsis province (Figure 10.25). The northern depression has been proposed to have been formed by an internal mechanism such as convective mantle overturn and crustal thinning which resulted from a thermal event, or by a giant impact.

The gravity field of Mars has also been measured by the Mars Global Surveyor (Smith et al. 1999a). The rough, elevated southern hemisphere appears to be in a state of near-isostatic compensation, whereas the smooth, low northern plains display a wider range of gravitational anomalies that indicates a thinner, but stronger, surface layer than in the South.

Semimajor axis	1.52366 AU
Eccentricity	0.0934
Inclination	1°.8504
Longitude of ascending node	49°.59
Longitude of perihelion	335°.94
Mean orbital period	686.98 Earth days, 669.60 Mars solar days
Obliquity	25°.19
Length of day	88642.663 ± 0.002 s (sidereal)
	99775.2 s (solar)

Source: Based on Kieffer et al. 1992a.

Table 10.11 Orbital and Rotational Parameters of Mars

Parameter	Value
Mass	6.4185×10^{23} kg
Mean radius	3.389.508 m*
Volume	1.6318×10^{22} m^3
Mean density	3.9335 ± 0.0004 g cm^{-3}
Surface gravity	3.758 ms^{-2} (at pole), 3.711 ms^{-2} (at equator)
Total surface area	1.4441×10^{14} m^2

Source: Based on Kieffer et al. 1992a. excerpt * from Smith et al. 1999b.

Table 10.12 Geophysical Parameters of Solid Mars

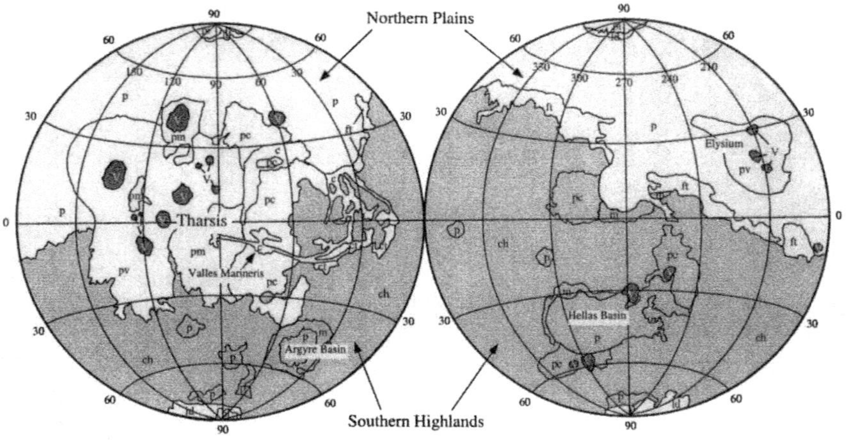

Figure 10.25 Simplified physiographic map of Mars, showing ancient (ch: cratered highlands, m: mountainous terrain), modified (p: plains, c: channel and canyon floors, ct: chaotic terrain, ft: fretted terrain), volcanic (V: volcanic constructs, pv: volcanic plains, pm: moderately cratered plains, pc: cratered plains), and polar (pi: permanent ice, ld: layered deposit), areas (modified from Mutch and Head 1975).

10.22 Surface and Subsurface

Surface Materials

The global surface composition of Mars has been studied by Earth-based telescopes and spacecraft instruments. The early telescopic observations revealed that the albedo of Mars is about 0.15 and that the Martian surface has a red color and is divided into lighter and darker regions. This difference is explained by the degree of oxidation of basaltic rocks. The primary reason for the red color of the Martian surface has been generally attributed to the presence of iron oxide minerals, with the amount of the iron oxides varying depending on the minerals. In addition, clays have been detected in the bright regions and dust. It is also possible that materials in the bright regions are largely amorphous or poorly crystalline, and palagonite may be the main composition because of excellent spectral matching. Palogonite can be formed when volcanic eruptions occur in water-rich environments or during subsequent alteration of basaltic ash. The TES (Thermal Emission Spectrometer) on board the Mars Global Surveyor identified a few regions of high crystalline hematite concentration (Christiensen et al. 2000). Based on the detection of absorption features near 1 μm and 2 μm, the presence of pyroxene has been inferred in the dark regions of Mars, consistent with mafic composition of the surface estimated from the geomorphology of volcanoes. TES examined the dark regions and concluded that the highlands are basaltic whereas the northern plains appear to be more of andesitic composition (Bandfield et al. 2000). Carbonate deposits have been searched for extensively since they are closely linked with the presence of water on Earth. However, neither ground-based telescopic observation nor TES has identified an appreciable amount of carbonate on the Martian surface, within their detection limits.

Based on direct *in situ* analyses by Viking lander XRFS (X-ray fluorescence spectrometer) and supplementary information from the Martian meteorite Shergotty, a representative chemical composition of Mars soil has been derived (Banin et al. 1992; Table 10.13). Soil samples measured at the two Viking sites are remarkably similar to each other in terms of chemical composition despite the great distance between them. This may indicate that a thin semihomogeneous blanket soil material covers much of the Martian surface. There is no known terrestrial soil similar to the Martian soil, which is characterized by a very high concentration of sulfur.

The Mars Pathfinder APXS (alpha proton X-ray spectrometer) measurements of surface soils and rocks added another piece of information about the composition of the Martian surface. The soils at the Pathfinder landing site are similar to those at the Viking sites (Rieder et al. 1997). The measurement of rocks at the Pathfinder landing site yielded compositions close to andesite, a type of lava common on continental margins on Earth. However, it is not known whether the measured samples are igneous, sedimentary, or metamorphic rocks.

Eolian Processes

On present-day Mars, eolian (wind) processes are the most active geological force in shaping landforms. But there are many unanswered questions about these processes. Winds are capable of eroding and transporting a large quantity of materials over a geologically short time scale. Despite the tenuous atmosphere on Mars compared with that on Earth, there exist landforms such as yardangs and dunes, representative features formed by wind erosion and transport, respectively. Dunes are globally distributed, as revealed in the new Mars Global Surveyor images, but the main dune fields are found at high latitudes in both hemispheres and in craters of the southern highlands. These dunes are probably made of sand-sized material, but the source of the sand remains controversial (Greeley et al. 1992). How active are the eolian landforms on Mars? For example, observations made on a dune field over a 20-Earth-year period indicate that the upper limit of the dunes' motion is 200 times less than on Earth. Even if they are still active, grain movement is slow.

Constituent	Selected average concentration (%)	Source
SiO_2	43.0	Direct soil analysis by Viking XRFS
Al_2O_3	7.2	Direct soil analysis by Viking XRFS
Fe_2O_3	18.0	Direct soil analysis by Viking XRFS
MgO	6.0	Direct soil analysis by Viking XRFS
CaO	5.8	Direct soil analysis by Viking XRFS
TiO_2	0.6	Direct soil analysis by Viking XRFS
K_2O	0.2	Analyses of Martian meteorite
P_2O_5	0.8	Analyses of Martian meteorite
MnO	0.5	Analyses of Martian meteorite
Na_2O	1.3	Analyses of Martian meteorite
Cr_2O_3	0.2	Analyses of Martian meteorite
SO_3	7.2	Direct soil analysis by Viking XRFS
Cl	0.6	Direct soil analysis by Viking XRFS
CO_3	< 2	Estimated from Label Release simulation
H_2O	0.1	Variable: based on direct soil analysis on Viking GC-MS

Source: From Benin et al. 1992.

TABLE 10.13 Representative Chemical Composition of Martian Soil

If this is the case, it is possible that many of the dunes are relict features formed during periods when the climate was somewhat different—for example, with a denser atmosphere. So-called wind streaks are bright, or dark, coneshaped surface features emanating from obstacles such as impact craters. These features are considered to have formed through the interaction of wind, sediments, and obstacles. Therefore, the directions of the wind streaks probably imply the direction of prevailing wind in the area. In fact, an excellent correlation has been found between the orientations of some wind streaks and wind directions derived from a GCM (Global Circulation Model).

Dust on Mars is globally distributed; it has been moved all over the planet by dust storms. There is a question of how the tenous Martian atmosphere raises dust particles. According to theoretical and experimental studies, the optimum size for a particle to be moved by wind on Mars is about 100 μm (Greeley et al. 1992). The minimum wind velocities required to move dust-sized particles (<20 μm) are much larger than the wind velocities observed on Mars. Alternative dust-raising mechanisms include dust devils, the hitting of dust particles by saltating sands, and the suspension of dust aggregates and later pulverization into fine grains. In fact, dust devils have been observed on Mars, but it is not clear if they raise enough dust to initiate dust storms.

Impact Cratering

Impact cratering is a major geological process affecting the surface of Mars. The impact craters on Mars exhibit a very wide range of morphology due to the diversity of target materials, the presence of an atmosphere, and postimpact erosional processes. For example, many Martian impact craters

	Crater density	Absolute age
Amazonian-Hesperian boundary	400 > 2 km crater/10^6 km^2	1.8–3.55 Gyr
	67 > 5 km crater/10^6 km^2	
Hesparian-Noachian boundary	200 > 5 km crater/10^6 km^2	3.5–3.8 Gyr
	25 > 16 km crater/10^6 km^2	

Source: Based on Tanaka et al. 1992.

TABLE 10.14 Mars Time-Stratigraphic System

are characterized by a lobate ejecta blanket, which is sometimes a overlapping set of lobes. This morphology has been attributed by many researchers to the presence of a volatile-rich layer in the target, but other researchers have suggested that ejecta interaction with the atmosphere is responsible for formation of the lobes.

The surface chronology of Mars has been defined mostly by the relative age based on the impact crater population density. For example, the southern highlands are the oldest, dating back to the end of the heavy asteroid/comet bombardment that formed Mars. The northern plains are much younger than the southern highlands; topographically they are generally flat. Image observations suggest that their surfaces are covered by young lava flows and sediments. The Martian history is divided into three periods—Noachian, Hesperian, and Amazonian, from old to young (Table 10.14). The absolute ages of the surface types are estimated by applying an inferred impact flux at the Mars orbit, the likely value ranging from six times the lunar rate to the same as the lunar rate. However, this approach is not yet established well enough for us to have great confidence in the ages derived. The big problem is that we do not know the variation of the influx of projectiles, such as asteroids and comets, through time. The resolution of this problem must await the results of the sample return mission, which should provide the absolute age of calibration sites. Impact craters on Mars exhibit a wide range of degradational morphology, which implies that geological processes have been actively operating on the planet. However, how active these processes have been, and when, are still a matter of debate.

Impact cratering may have played an important role in the atmosphere of early Mars. It has been suggested that very large basin-forming objects could have eroded the atmosphere to its present-day tenuous state (Melosh and Vickery 1989).

Vulcanology

Mars is a world of giant volcanoes. The largest is Olympus Mons, which is over 600 km in diameter and reaches more than 25 km above the surrounding plains. These giant volcanoes are called shield volcanoes because they have very small height-to-diameter ratios; this implies that the volcanoes were made primarily by eruptions of low-viscosity lava flows such as basalt. Due to the lack of Earth-type plate tectonics, the locations of the Martian shield volcanoes with respect to the magma chambers may have been fixed, and this may explain the extraordinary sizes of the volcanoes. Large shield volcanoes are concentrated in three regions. The largest one is Tharsis that is centered on the equator and 105°W. Tharsis is a 6,000 km wide volcano-tectonic bulge overriding both the southern highlands and northern plains. The second is Elysium, which is centered on 25°N, 210°W. There are also eroded volcanic edifices distributed around the Hellas Basin in the southern hemisphere. Other major volcanic landforms include volcanic plains, very flat volcanic edifices called paterae, and tholi, which may be made of lava and ash layers. Recent Mars Global Surveyor observations of the Martian surface have revealed extensive coverage of lava flows in some parts of the northern plains. Their ages, based on crater counting, are on the order of 10 million years (Hartmann and Berman 2000), extremely young by Martian standards. This implies that Mars is practically still alive volcanically.

Tectonics

Global-scale tectonics on a planet are an indication of how the planet loses its heat generated internally. Earth loses heat through active plate tectonics and Venus seems to do it through plume activities. Smaller planetary bodies, such as the Moon and Mercury, are considered to have lost their heat primarily by conduction; they are practically dead internally. Mars does not have characteristic topographic features that indicate Earth-type plate tectonics, such as rifts, subduction zones, and collisional mountain chains. Because Mars is smaller than Earth, it is possible that Mars has already lost the majority of its internal heat.

For regional tectonics, the gigantic Tharsis bulge shows a wide range of structural features. In planetary geology, tectonic structures are used to infer the stress field acting in the region. For example, the Tharsis province is characterized by a wide range of structures such as grabens and wrinkle ridges, indicating tensional and compressional stress fields, respectively. The origin of Valles Marineris, the 400 km long canyon system, is often explained as being where the crust is spread by tensional stress. Many compressional landforms are observed associated with Olympus Mons, implying that the volcano is pressing down the crust because of its loading.

Polar Caps and Ground Ice

The polar regions of Mars are characterized by complex geology of ice caps, layered deposits, and peripheral deposits including dune fields. Polar ice is classified into two components, summer residual ice and CO_2 frost. The northern residual cap is made of H_2O, whereas the southern residual appears to be CO_2, but it is not known if the southern residual is underlain by H_2O ice. The layered deposits consist of pairs of light and dark bands, but the composition of the bands is poorly constrained. They were probably formed as the result of cyclic climate change causing the deposition of dust and ice. The dunes around the northern polar cap seem to have some sources within the layered deposits, implying a genetic link between them.

According to theoretical models, very thick layers of ground ice can exist in the subsurface of Mars, at the depths of 1–3 km near the equator and 3–8 km in the polar regions (see, e.g., Fanale 1976). This type of modeling attempts to determine the upper boundary of ground ice layers, based on a diffusion process, by changing crustal temperature, which is related to surface temperature. If crustal temperatures are above the frost point of atmospheric water vapor, it is possible that ice may be lost by diffusion over time. Assuming that there was no replenishment of water since the early history of Mars, the upper several hundred meters of Martian ground ice at low latitudes should have been desiccated by now, unless the ice was protected from diffusion by some mechanism. In reality, we observe a number of landforms on the Martian surface at low latitudes, which can be attributed to processes due to shallow ground ice. These landforms include lobate debris aprons and thermokarst. This implies that there may have been hydrologically active events to replenish ice later in Mars history.

Hydrology

Water on Mars is one of the most exciting topics about the red planet. The issue can be addressed from two aspects—present conditions and past conditions. Present-day Mars is extremely dry; the atmosphere has 0.03% water content. The majority of the surface water inventory is concentrated in the polar caps, and a large quantity of water may lie underneath the surface as ground ice. The modern surface condition of Mars does not allow liquid water to flow freely, but new data from the MGS suggest that liquid water may have come out onto the surface very recently in geological terms to produce small-scale landforms such as gullies (Malin and Edgett 2000). The Mars explorations conducted mainly by the Mariner 9 and Viking orbiters have identified rich evidence of surface water activities and ground ice melting in the planet's past.

The discovery of numerous valleys on the Martian surface is the strongest evidence that liquid water played an important role in shaping the landscapes on the planet. Compared with their terrestrial counterparts, these valleys are less densely populated with water and have smaller drainage in general. Many of the valleys have short, stubby tributaries and a theater-headed terminus, characteristics similar to terrestrial sapping valleys. Sapping valleys are formed primarily by groundwater processes that undermine the rocks and cause the collapse of roof rocks. Although surface runoff cannot be excluded, sapping processes appear to be the dominant mechanism for the formation of valley networks on Mars. In any case, replenishing of the water to the source areas would be required, and hydrogenated circulation and precipitation have been proposed. For the latter, the past climate history of Mars would be different.

Channels on a grand scale exist on the surface of Mars. Called outflow channels, they were formed by ancient catastrophic floods, perhaps of water, with discharge rates a few orders of magnitude greater than that of the Amazon River. The majority of the outflow channels originate from the collapsed ground called chaotic terrain. It is considered that the ground ice layers became unstable due to magmatic heating, and this caused a catastrophic release of water to the surface. There are also pieces of evidence indicating active ice-related processes that modified the outflow channels. There are many surface features indicative of past ponding of water, including shorelines such as wave-cut terraces and deltas, and sediments. The biggest may have been the northern plains ocean (see, e.g., Parker et al. 1989), but ponding perhaps also occurred in smaller topographic basins, inside crater rims, and in parts of the Valles Marineris. How often water filled the topographic lows is not well understood. A diverse group of landforms distributed widely in mid to high latitudes has been interpreted to be glacial (Kargel and Strom 1992) and periglacial. Some of these interpretations have been opposed, but nonetheless ice seems to have played a major role in the geomorphology of Mars. The implications of ice processes on Mars are significant. In particular, glaciation requires a continuous supply of water to the accumulation zone in order to maintain the ice movement. And this means a hydrological cycle.

On Earth, hydrological elements such as oceans, lakes, rivers, and surface and subsurface ices are closely linked to each other. Water transport through the atmosphere is also very important. Whether this is also true for Mars is not known, and this point needs further investigation. Baker et al. (1991) proposed a hypothesis that connects many pieces of observational evidence of water-related landforms and envisions episodic global hydrological cycles relatively late in Mars history. According to the hypothesis, the episodic hydrological cycles were short-lived and therefore did not produce the range of erosional landforms seen on Earth. The hypothesis suggests that rapid formation of northern plain oceans by outbursts of water from below ground created the gigantic outflow channels. Perhaps the atmosphere was temporarily warmed due to greenhouse gases released at the same time as the water outflows, and precipitation may have occurred to supply water to valley networks and glaciers in the southern highlands.

10.23 Atmosphere

Mars's tenous present-day atmosphere is mostly made of CO_2 and is characterized by its low water content (Table 10.15). The global annual average atmospheric pressure at the surface is 5.6 mbar (hPa), but this value varies seasonally and spatially. The variation of atmospheric pressure is controlled by the sublimination and condensation of CO_2 on the polar caps. The ground temperatures at the equator range mostly from about 160 K to 180 K at night, to a peak temperature of about 260–280 K during the day (Carr 1996). The atmospheric dynamics are strongly affected by conditions unique to Mars, including the behavior of CO_2 described above, the radiative effect of dust loading, and variations of

Gas	Abundance
CO_2	95.32%
N_2	2.7
^{40}Ar	1.6
O_2	0.13
CO	0.07
H_2O	0.03
$^{36+38}Ar$	5.3 ppm
Ne	2.5 ppm
Kr	0.3 ppm
Xe	0.08 ppm
O_3	0.04–0.2 ppm

Source: Based on Owen 1992.

TABLE 10.15 Composition of the Martian Lower Atmosphere

insulation due to the orbital eccentricity and the phasing of perihelion (Zurek et al. 1992). The average sunlight Mars receives is less than half the Earth receives. But the sunlight falling on the Martian surface varies greatly due to the planet's large orbital eccentricity and dust in the atmosphere.

The thermal structure of the Martian atmosphere is very different from the Earth's in some aspects. The major differences are caused by the lack of sufficient ozone and the variable dust loading in the Martian atmosphere (Zurek 1992). The ozone abundance in the Martian atmosphere is less than one three-hundredth of that in the terrestrial atmosphere, allowing strong UV light from the Sun to reach the surface. Because of the low ozone abundance, the Martian atmosphere lacks the stable, temperature-increasing layer equivalent to the stratosphere on Earth. Airborne dust in the Martian atmosphere plays the role of ozone by absorbing the Sun's radiation; hence the dusty Martian lower atmosphere somewhat resembles the terrestrial stratosphere. Obviously the dust loading in the Martian atmosphere is a strong function of season and region. Thus, the thermal structure of the Martian atmosphere can vary tremendously.

The most predominant large-scale atmospheric phenomenon is the dust storm. The largest of the observed regional dust storms and all of the planet-encircling dust storms have occurred almost exclusively in the spring and summer of the southern hemisphere (Kahn et al. 1992).

A number of pieces of observational evidence indicate that the past climate on Mars was not always the same as today's. For example, landforms such as valleys require the action of liquid water on or near the surface, implying that the atmosphere was once much thicker and warmer than now. Determining exactly when and how the climate was different is not an easy task without any of the proxy data that are still available in the case of Earth but not for Mars. There are two main lines of thought concerning past climatic variations. One is that the landforms indicative of liquid water date back from the early Mars, immediately after the formation of the planet. Therefore, the early history of Mars had a thick, warm, and wet atmosphere, enough to allow liquid water (Pollack et al. 1987). The other asserts that Mars has experienced episodic climatic changes even in its late history; the hydrological processes occurring during such episodes produced the range of water-related landforms observed today (Baker et al. 1991).

Although the Martian magnetic field is not dipolar, but is crust remnant magnetization, sources with an equivalent magnetic moment as large as 13×10^{17} amp-m^2 exist (Acuña et al. 1999).

	Phobos	Deimos
Radii of triaxial ellipsoid:	a: 13.3 ± 0.3 km	7.6 ± 0.5 km
	b: 11.1 ± 0.3 km	6.2 ± 0.5 km
	c: 9.3 ± 0.3 km	5.4 ± 0.5 km
Volume	5,680 ± 250 km^3	1,052 ± 250 km^3
Mass	1.08 × 10^{16} kg	1.8 × 10^{15} kg
Mean density	1.905 ± 0.053 g cm^{-3}	1.7 ± 0.5 g cm^{-3}
Semimajor axis	9,378.5 km	23,458.8 km
Eccentricity	0.0152	0.0002
Inclination	1°.03	I°.83
Orbital period	0.31891023 day	1.2624407 day
Length of day	Synchronous	Synchronous

Source: Based on Kieffer et al. 1992a.

TABLE 10.16 Parameters of the Martian Satellites

The ionized atmosphere, the ionosphere, of Mars extends to high altitudes over the strong crustal magnetic fields. The crustal magnetic field is sufficiently strong to produce an asymmetric, time-varying obstacle to the flow of the solar wind around Mars.

10.24 Satellites

The two moons of Mars have general characteristics similar to those of other small bodies of the Solar System (Table 10.16), particularly the asteroids. Spectral studies indicate that Phobos and Deimos have a surface composition resembling that of carbonaceous asteroids, but with little bound or absorbed water. Phobos and Deimos have very different surface geologies. Phobos is heavily cratered and characterized by linear grooves that are related to the impact crater Stickney, whereas Deimos has a much smoother surface with subdued craters.

10.25 Search for Life on Mars

Mars has been the planet of choice when it comes to the image of extraterrestrial life in the minds of many people. Mars has very extreme environmental conditions in terms of its suitability for the survival of life, but many of these conditions are still the most similar in the Solar System to the Earth's. Studies of terrestrial life have increased our knowledge about the range of conditions in which life can survive (Rothschild and Mancinelli 2001). Life on Earth flourishes in hot springs, on sea floors, deep in subsurface rocks, and even in hypersaline lakes. This greatly increases the survivability of present or ancient life, if it ever existed, on Mars. The report of evidence of ancient life from a Martian meteroite ALH84001 (McKay et al. 1996), although not fully supported by the entire scientific community, stirred strong interest among scientists and the general public about life on Mars. However, the surface condition of Mars is extremely lethal to terrestrial-type organisms due to strong UV light from the Sun and the oxidizing conditions. Therefore, future investigations will focus on deep subsurface and hydrothermal systems as good candidates in the search for life on Mars. The earliest life on Earth is commonly considered to be hyperthermophile (Nisbet and Sleep 2001).

Whether life originated on Mars is another issue. For life, early Mars could have had a relatively safe environment, with oceans and a warm, wet atmosphere. However, the lack of knowledge on how life begins makes it difficult to assess Mars as a birthplace of life.

An interesting possibility exists for the transportation of life between planetary bodies. There is a group of meteorites called SNC meteorites that are considered to have come from Mars. These meteorites were ejected from Mars at a time in its history when giant impacts were common and so the inner planets were constantly exchanging their surface materials. If microorganisms were shielded in the debris, it is possible that they survived the space trip between the planets. Therefore, life may have come from Mars to the Earth, or indeed vice versa.

10.26 Exploration

Before the Space Age, the main means of observing Mars was to use ground-based telescopes. The history of Mars robotic exploration began with Mars 1 (Table 10.17). Important early results were obtained from missions such as Mariner 9 and the Viking orbiters and landers. Recently, Mars Pathfinder and Mars Global Surveyor have started a series of new explorations that is supposed to pave the way to eventual human flight to the red planet. The Mars Pathfinder lander was equipped with a panoramic camera, a meteorological device, and an alpha particle/proton/X-ray spectrometer for chemical analysis. The Mars Global Surveyor orbiter was equipped with a camera with high spatial resolution, 1.5 m. Other major instruments included a laser altimeter for a topographic survey and a thermal emission imaging spectrometer for mapping surface minerals. Future missions will focus more on detailed analyses of various environmental and geological properties and exobiological investigations of the planet. For example, the European Mars Express mission will conduct a survey of the upper several kilometers of the Martian crust using a ground-penetrating radar called MARSIS (Mars Advanced Radar for Subsurface and Ionospheric Sounding). Future rovers will have a long-range capability, up to some tens of kilometers, to enable flexible investigations of the geology in the landing areas. Sample return missions are also under consideration.

One of the major objectives in the upcoming Mars exploration program is the search for present life or evidence of ancient life on Mars. Many future missions will examine areas of recent water seepage (Malin and Edgett 2000), hydrothermal vents around volcanoes, and sedimentary basins and related water ponding.

Humans visiting Mars has been a dream for many people for generations. It has been envisioned in various science fiction books and movies and discussed extensively among space

Name	Country	Launch date	Accomplishments
Mars 1	U.S.S.R.	Nov. 1, 1962	First flyby in 1963
Mariner 4	U.S.A.	Nov. 28, 1994	First photographs
Mars 3	U.S.S.R.	May 28, 1971	First landing
Mariner 9	U.S.A.	May 30, 1971	Photographed extensively
Viking 1	U.S.A.	Aug. 20, 1975	Successful orbiter and lander mission
Viking 2	U.S.A.	Sep. 5, 1975	Successful orbiter and lander mission
Phobos 2	U.S.S.R.	July 12, 1988	Orbital investigation of surface
Mars Pathfinder	U.S.A.	Dec. 4, 1996	Lander and rover
Mars Global Surveyor	U.S.A.	Nov. 7, 1996	Extensive mapping of surface

Table 10.17 List of Major Spacecraft Missions to Mars

scientists and engineers. Without a political will such as that which drove the Apollo project during the Space Race, the possibility of human missions to Mars in the near future is slim. But other factors, such as the future discovery of life on Mars or major technological advancements, may change this situation.

The main technological obstacles to human flights to Mars are the long duration, the harsh space and surface environments, and fuel. But all of these are soluble, considering the ingenuity of the human race. The real question is when and why we will undertake the venture. The Apollo project was a huge success, and it created a view that the Earth is an oasis floating in the barren Universe. What kind of vision will human flights to Mars create? We now live in a more globally connected world than in 1969, the year of the Moon Landing. The journey to Mars may thus be conducted with the joint will of people worldwide rather than under conditions of competition between two nations. If this is so, the journey of humans to Mars will signify the arrival of a new way of human thinking.

References

Acuña, M. H., Connerney, J. E. P., Ness, N. F., Lin, R. P., Mitchell, D., Carlson, C. W., McFadden, J., Anderson, K. A., Réme, H., Mazelle, C., Vignes, D., Wasilewski, P., and Cloutier, P. 1999. "Global Distribution of Crustal Magnetism Discovered by the Mars Global Surveyor MAG/ER Experiment." *Science,* vol. 284, pp. 790–793.

Baker, V. R., Strom, R. G., Gulick, V. C., Kargel, J. S., Komatsu, G., and Kale, V. S. 1991. "Ancient Oceans and Ice Sheets and the Hydrological Cycle on Mars," *Nature,* vol. 352, pp. 589–594.

Bandfield, J. L., Hamilton, V. E., and Christensen, P. R. 2000. "A Global View of Martian Surface Compositions from MGS-TES," *Science,* vol. 287, pp. 1626–1630.

Banin, A., Clark, B. C., and Wänke, H. 1992. "Surface Chemistry and Mineralogy," in *Mars,* ed. Kieffer et al., University of Arizona Press, Tucson, pp. 594–625.

Carr, M. H. 1996. *Water on Mars,* Oxford University Press, New York.

Christensen, P. R., Bandfield, J. L., Clark, R. N., Edgett, K. S. Hamilton, V. E., Hoefen, T., Kieffer, H. H., Kuzmin, R. O., Lane, M. D., Malin, M. C., Morris, R. V., Pearl, J. C., Pearson, R., Roush, T. L., Ruff, S. W., and Smith, M. D. 2000. "Detection of Crystalline Hematite Mineralization on Mars by the Thermal Emission Spectrometer: Evidence of Near-Surface Water," *Journal of Geophysical Research,* vol. 105, pp. 9623–9642.

Fanale, F. P. 1976. Martian Volatiles: Their Degassing History and Geochemical Fate," *Icarus,* vol. 28, pp. 179–202.

Greeley, R., Lancaster, N., Lee, S., and Thomas, P. 1992. "Martian Eolian Processes, Sediments and Features," in *Mars,* ed. Kieffer et al., University of Arizona Press, Tucson, pp. 730–766.

Hartmann, W. K. and Berman, D. C. 2000. "Elysium Planitia Lava Flows: Crater Count Chronology and Geological Implications," *Journal of Geophysical Research,* vol. 105, pp. 15,011–15,026.

Kahn, R. A., Martin, T. Z., Zurek, R. W., and Lee, S. W. 1992. "The Martian Dust Cycle," in *Mars,* ed. Kieffer et al., University of Arizona Press, Tucson, pp. 1017–1053.

Kargel, J. S. and Strom, R. G. 1992. "Ancient Glaciation on Mars," *Geology,* vol. 20, pp. 3–7 (1992).

Kieffer, H. H., Jakosky, B. M., and Snyder, C. W. 1992a. "The Planet Mars: From Antiquity to the Present," in *Mars,* ed. Kieffer et al., University of Arizona Press, Tucson, pp. 1–33.

Kieffer, H. H., Jakosky, B. M., Snyder, C. W., and Matthews, M. S. 1992b. *Mars,* University of Arizona Press, Tucson.

Lasker, J. and Robutel, P. 1993. "The Chaotic Obliquity of the Planets," *Nature,* vol, 361, pp. 608–612.

Malin, M. C. and Edgett, K. S. 2000. "Evidence for Recent Groundwater Seepage and Surface Runoff on Mars," *Science,* vol. 288, pp. 2330–2335.

McKay, D. S., Gibson, E. K., Thomas-Keprta, K. L., Vali, H., Romanek, C. S., Clemett, S. J., Chillier, X. D. D., Maechling, C. R., and Zare, R. N. 1996. "Search for Life on Mars: Possible Relic Biogenic Activity in Martian Meteorite ALH84001," *Science,* vol. 273, pp. 924–930.

Melosh, H. J. 1993. "Blasting Rocks off Planets," *Nature,* vol. 363, pp. 498–499.

Melosh, H. J. and Vickery, A. M. 1989. "Impact Erosion of Primordial Atmosphere of Mars," *Nature,* vol. 338, pp. 487–489.

Mutch, T. A. and Head, J. W. 1975. "The Geology of Mars: A Brief Review of Some Recent Results," *Reviews of Geophysics and Space Physics,* vol. 13, pp. 411–416.

Nisbet, E. G. and Sleep, N. H. 2001. "The Habitat and Nature of Early Life," *Nature,* vol. 409, pp. 1083–1091.

Owen, T. 1992. "The Composition and Early History of the Atmosphere of Mars," in *Mars.* ed. Kieffer et al., University of Arizona Press, Tucson, pp. 818–834.

Parker, T. J., Saunders, R. S., and Schneeberger, D. M. 1989. "Transitional Morphology in West Deuteronilus Mensae, Mars: Implications for Modifications of the Lowland/Upland Boundary," *Icarus,* vol. 82, pp. 111–145.

Pollack, J. B., Kasting, J. F., Richardson, S. M., and Poliakoff, K. 1987. "The Case for a Wet, Warm Climate on Early Mars," *Icarus,* vol. 71, pp. 203–224.

Rieder, R., Economou, T., Wänke, H., Turkevich, A., Crisp, J., Brückner, J., Dreibus, G., and McSween, H. Y., Jr. 1997. "The Chemical Composition of Martian Soil and Rocks Returned by the Mobile Alpha Proton X-ray Mode," *Science,* vol. 278, pp. 1771–1774.

Rothschild, L. J. and Mancinelli, R. L. 2001. "Life in Extreme Environments," *Nature,* vol. 409, pp. 1092–1101.

Smith, D. E., Sjogren, W. L., Tyler, G. L., Balmino, G., Lemoine, F. G., and Konopliv, A. S. 1999. "The Gravity Field of Mars from Mars Global Surveyor," *Science,* vol. 286, pp. 94–97.

Smith, D. E., Zuber, M. T., Solomon, S. C., Phillips, R. J., Head, J. W., Garvin, J. B., Banerdt, W. B., Muhleman, D. O., Pettengill, G. H., Neumann, G. A., Lemoine, F. G., Abshire, J. B., Aharonson, O., Brown, C. D., Hauck, S. A., Ivanov, A. B., McGovern, P. J., Zwally, H. J., and Duxbury, T. C. 1999b. "The Global Topography of Mars and Implications for Surface Evolution," *Science,* vol. 284 pp. 1495–1503.

Tanaka, K. L., Scott, D. H., and Greeley, R. 1992. "Global Stratigraphy," in *Mars,* ed. Kieffer et al., University of Arizona Press, Tucson, pp. 345–382.

Zimbelman, J. R. 2000. "Non-active Dunes in the Acheron Fossae Region of Mars between the Viking and Mars Global Surveyor Areas," *Geophysical Research Letters,* vol. 27, pp. 1069–1072.

Zurek, R. W. 1992. Comparative Aspects of the Climate of Mars: An Introduction to the Current Atmosphere," in *Mars,* ed. Kieffer et al., University of Arizona Press, Tucson, pp. 799–817.

Zurek, R. W., Barnes, J. R., Haberle, R. M., Pollack, J. B., Tillman, J. E., and Leovy, C. B. 1992. "Dynamics of the Atmosphere of Mars," in *Mars,* ed. Kieffer et al., pp. 835–933.

PART 6
The Sun–Earth Connection

D. N. Baker

10.27 Introduction

Humans have been fascinated since time immemorial by the Sun and its relationship to the Earth. Every civilization has speculated about the place of Earth in the cosmos. Solar and lunar eclipses have instilled a sense of awe from ancient times, and structures such as Stonehenge have shown the compelling inspiration that results as people try to understand Sun–Earth connections. Certainly in the past few hundred years, since Copernicus and Galileo, we have come to new scientific understandings of the Sun and its broader place in the realm of the stars. We have also come to a clearer picture of the Sun as a driver of magnetic disturbances at Earth. Especially in the past few decades, scientists have arrived at a much clearer picture of solar activity and its effects on Earth and on human technological systems.

The Sun reaches an activity peak every 11 years or so, during which it exhibits a wide variety of powerful disturbances. The shock waves and clouds of hot, ionized gas (plasma) emitted by the Sun can be extremely energetic and can accelerate interplanetary charged particles to millions or even billions of electron volts. If such disturbances happen to strike the Earth and its magnetosphere, the result can be violent geomagnetic storms that can have profound effects on the near-Earth environment that we call *geospace*.

As understanding of the global Sun–Earth system has evolved, human technological systems have also grown more sophisticated. A complex web of communication links connects virtually all points on the Earth's surface, power grids criss-cross the continents, and spacecraft pass through much of the volume in the near-Earth space above us. The "cyberelectric cocoon" that has been spun around the Earth has greatly enhanced global communications, navigational accuracy, national security, and weather prediction. However, reliance on these systems has also put human society into danger. Every technological system that utilizes radio waves or conducting wires or computer processors can be adversely affected by powerful disturbances in the solar–terrestrial system. Given the growing human dependence on advanced technologies, it is likely that the Sun–Earth connection will be increasingly important in the coming decades.

This portion of this section discusses the connected Sun–Earth system. The Sun is examined as the basic driver of the system, and the interplanetary medium is briefly discussed. The near-Earth space environment is described with an eye toward understanding the ultimate effects of solar variations on the atmosphere. Finally, near-Earth technologies are considered in the light of new geospace understandings.

10.28 The Sun and the Heliosphere

As illustrated in Figure 10.26, the Sun is a complex and highly structured object. At the center is the hydrogen-burning core. The nuclear fusion proceeding within this core gives rise to temperatures in excess of 10 million degrees Kelvin (K). Outside the core is the radiation zone, where energy is transported rapidly toward the outer layers of the Sun. In the convection zone, as shown in Figure 10.26, heated solar plasma percolates toward the surface and cools somewhat, thereby setting up large-scale circulatory eddies. Such convective motions produce powerful currents and related magnetic fields.

The visible surface of the Sun, called the photosphere, has a temperature of about 6,000 K. There are, however, regions in the photosphere of somewhat cooler plasma which appear slightly darker.

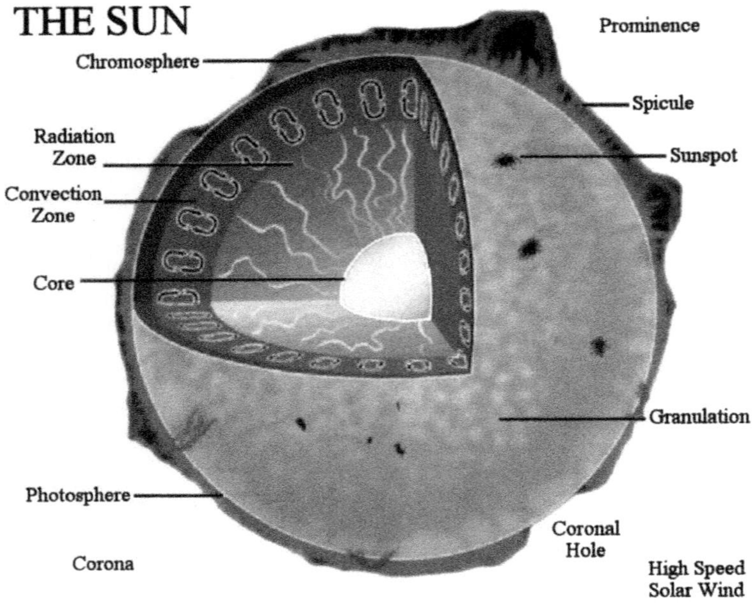

FIGURE 10.26 A schematic diagram showing the principal features of the Sun, including its interior and its outer atmosphere (adaptation of figure courtesy of P. Dusenbery, Space Science Institute).

Called sunspots, these are broad regions of strong, emergent magnetic field. The number of sunspots varies quite substantially over the course of a period of 10–12 years (average value 11 years) that is known as the solar cycle. At the time of sunspot minimum, the photosphere appears rather homogeneous and uniform (aside from small-scale granulations) and solar activity is very low. At the time of solar maximum, however, the number of sunspots is high, solar activity is very prevalent, with many bright, intense regions of X-ray and ultraviolet emissions, and there are frequent eruptive disturbances with powerful large-scale prominences.

The solar layer above the photosphere is called the chromosphere. It is normally visible from Earth only during solar eclipses. The hot, outermost layer of the Sun (directly above the chromosphere) is called the solar corona. There is a rapid temperature transition in the Sun's atmosphere, from 6,000 K at the surface to over 1 million K in the corona. The hot gases of the corona escape outward toward the cold void of interplanetary space and in so doing form the solar wind. The solar wind is a highly ionized gas plasma formed mostly of protons and electrons with an important minor proportion of helium and other trace constituent elements up to iron. Quite importantly, the solar wind has embedded within it the interplanetary magnetic field, which is "frozen" into the hot plasma flow.

From the base of the corona to a few solar radii heliocentric distance, the solar wind is accelerated to supersonic speeds. How this occurs, in detail, is still an essentially unsolved mystery. Near the poles of the Sun, the solar wind emerges from regions called coronal holes. In these regions of weak X-ray emissions, the magnetic flux tubes extend far out into interplanetary space and hence are called "open" field lines. The solar wind from coronal holes is observed to have high flow speeds (600–800 km/s) and is rather cool. At lower solar latitudes, the magnetic flux tubes are generally "closed" in that they emerge from the solar surface and then return in looping arcs back to the surface. The solar wind plasma from the lower latitude portion of the Sun has lower

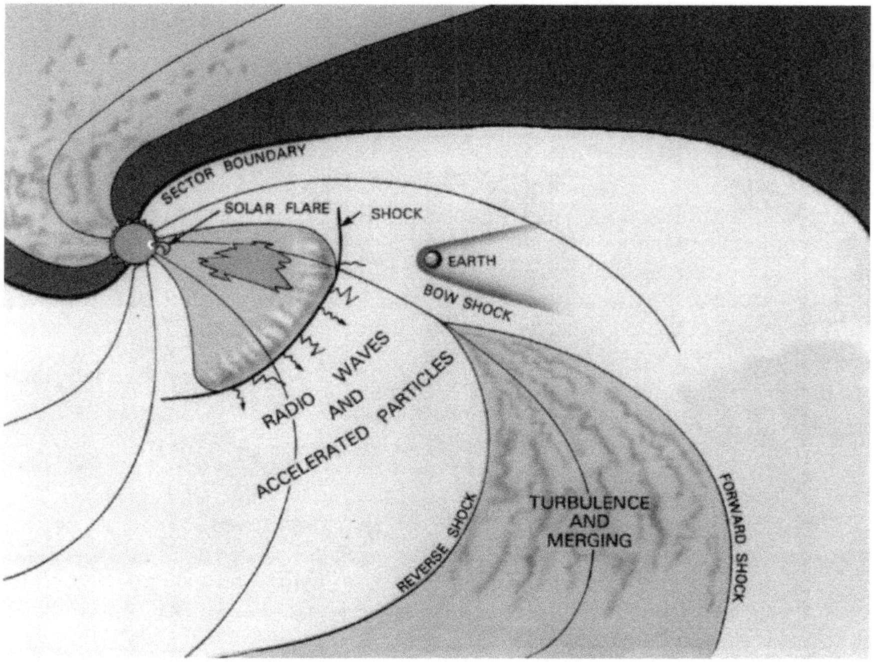

FIGURE 10.27 A schematic diagram of the heliosphere and some of its structural elements near the ecliptic plane (adaptation of diagram from NASA/GSFC).

characteristic speeds (300–500 km/s) and normally is hotter and denser than the plasma flowing out of coronal holes.

During certain portions of the solar activity cycle, especially in the approach to sunspot minimum conditions, coronal holes can extend to low solar latitudes in limited longitudinal sectors of the Sun. The high-speed flows from these regions can give rise to solar wind streams near the ecliptic plane. Such streams can have a significant effect upon the Earth as the streams propagate outward from the Sun (shown by the dark-shaded regions in Figure 10.27). On the other hand, at solar maximum conditions, solar flares and large expulsions of coronal material (coronal mass ejections, CMEs) can occur on the Sun, as also illustrated schematically in Figure 10.27. The flares and CMEs tend to have a major influence throughout broad regions of the heliosphere, the huge volume of the Sun's extended plasma influence.

Figure 10.27 is an ecliptic-plane cut of the three-dimensional heliosphere, which has been mapped by such space missions as Ulysses (which has flown over the poles of the Sun at ~1 AU heliocentric distance) and by the Pioneer and Voyager missions, which are presently many tens of AU away from the Sun. Observations and modeling suggest that the heliosphere extends outward to perhaps 80–100 AU, where the solar wind flow ends in a termination shock. Beyond this is the boundary of the Sun's effective plasma influence, called the heliopause. Depending upon the speed of the Sun's motion through the local interstellar medium (LISM), i.e., whether it is supersonic or subsonic, there may also be a standing bow shock wave in front of the heliosphere.

As suggested by Figure 10.27, the magnetic and plasma structure of the heliosphere plays an important role in driving terrestrial magnetospheric processes. The time variations associated with solar flares, CMEs, and interplanetary shock waves mingle with the more stable features such as high-speed streams and sector boundaries. This entire pattern of structures and transient features tends to rotate with the 27-day period of the Sun. The Earth is bathed in many of these solar and heliospheric features in an ever-changing pattern of plasma variations and energy fluctuations.

10.29 Structure and Dynamics of the Magnetospheric System

Each of the planets in our solar system with an intrinsic magnetic field (Mercury, Earth, Jupiter, Saturn, Uranus, and Neptune) has a region around it called the planet's magnetosphere. (Even the unmagnetized planets Venus and Mars have magnetosphere-like regions around them.) Figure 10.28 is a schematic diagram of the Earth's magnetosphere: this serves as a comparative basis for all the known planetary systems. On the sunward side—extending outward about 10 Earth radii (1 R_E = 6,372 km, mean)—the Earth's magnetic field stands off the solar wind flow. The boundary between the solar plasma and the terrestrial influence is called the magnetopause. The solar wind flow distorts the Earth's field on the antisolar side into an elongated magnetotail. Inside the magnetotail, near the central plane, is a region of hot, dense plasma called the plasma sheet. At the center of the plasma sheet is a region of magnetic field reversal called the neutral sheet. On either side of the plasma sheet are huge cylindrical regions of strong magnetic fields and very low plasma densities called the tail lobes. The tail lobe magnetic fields are produced by solenoidal currents of large scale that flow across the width of the plasma sheet (i.e., the cross-tail currents) and then close over the top and bottom surfaces of the magnetotail to form the nightside magnetopause currents.

The magnetic flux tubes of the outer magnetosphere all tend to funnel back toward the Earth and connect to the uppermost part of the atmosphere (see Figure 10.28). The ionized upper layers of the atmosphere are called the ionosphere. There are powerful electrical currents flowing into, out of, and within the ionosphere. These field-aligned currents and electrojet currents provide a major avenue of energy coupling between the magnetosphere and the ionosphere and lead to strong Joule heating and energy dissipation.

The inner part of the Earth's magnetosphere has a very strong, essentially dipolar magnetic field configuration. The strongly confined plasmas in this region drift under the magnetic field

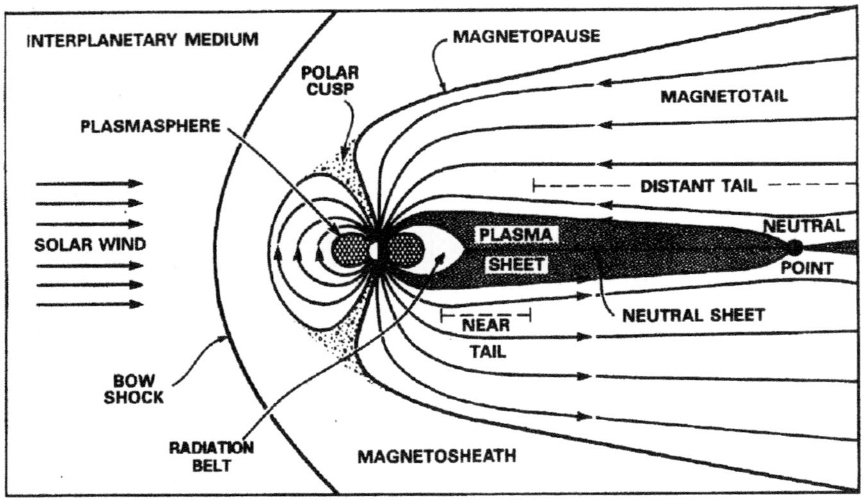

FIGURE 10.28 A noon–midnight cross-sectional view of the Earth's magnetosphere.

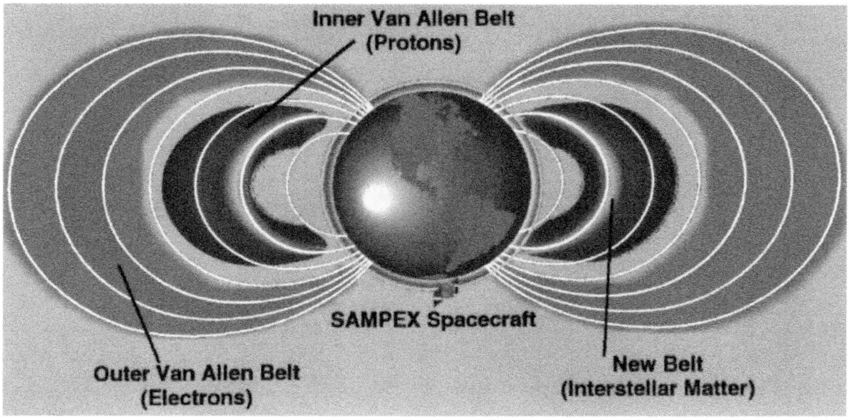

Figure 10.29 A cross-sectional diagram of the Earth's radiation belts (adaptation of figure courtesy of R. Mewaldt).

gradients and curvature forces to form the extraterrestrial ring current. This is also a region of high, steady plasma density, called the plasmasphere. As shown more explicitly in Figure 10.29, the part of the magnetosphere closest to the Earth also is the region of very energetic charged particle confinement, called the Van Allen radiation belts. The Van Allen belts contain highly time variable fluxes of relativistic electrons and high-energy protons. The inner belt (Figure 10.29) is mainly composed of very energetic protons, while the outer belt is composed principally of relativistic electrons.

10.30 The Solar-Terrestrial Energy Chain

The energy that drives the geospace environment has its origins at the Sun (as illustrated schematically in Figure 10.26). Nuclear fusion operating within the core subsequently drives convection in the outer layers of the Sun, and the associated dynamo action produces intense magnetic fields in the Sun's outer atmosphere. Eventually, emerging magnetic flux and hot, high-pressure gas are carried outward in the Sun's corona to form the expanding solar wind flow. This magnetized solar wind (after a transit time of several days) interacts with the outer parts of Earth's magnetic field and imparts—in a highly variable way—the energy to drive all forms of magnetospheric, auroral, and ionospheric disturbances (see Figure 10.30). These disturbances can be manifested by changes in auroral luminosity and by the range of phenomena covered by the general term *geomagnetic activity*.

As Figure 10.30 makes obvious, the ultimate deposition of energy from the solar wind occurs in the atmosphere. It is clear that solar–terrestrial coupling extends at least into the lower thermosphere and the mesosphere. However, mounting evidence suggests that solar–terrestrial coupling extends deeply into the stratosphere as well. Thus, in many ways the energy flowing from the Sun plays a key role in many aspects of life on (and near) the Earth's surface.

As noted above in the discussion of Figure 10.27, there can be another chain of connection between the Sun and the Earth, apart from the usual solar wind link: this alternate route is the one associated with transient events on the Sun. Large solar flares and CMEs can produce copious quantities of energetic charged particles, X-rays, gamma rays, and ultraviolet (UV) light bursts, as well as fast solar wind flows that can have the most extreme effects on the terrestrial environment.

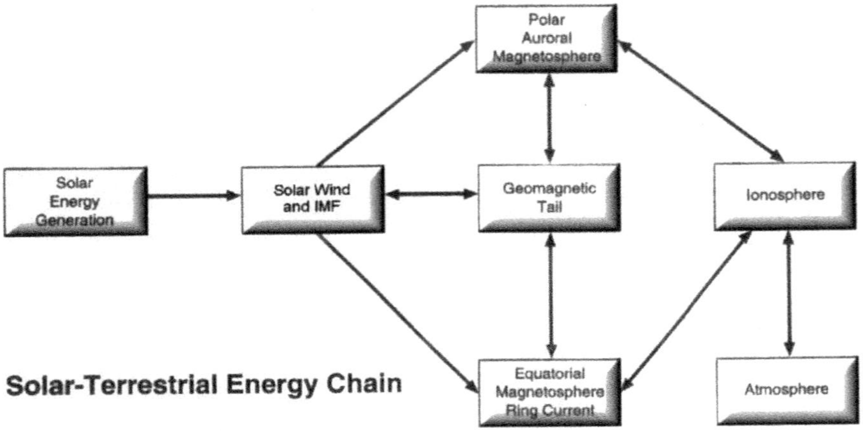

FIGURE 10.30 Schematic diagram showing the flow of energy from the Sun into the Earth's ionosphere–atmosphere system.

It is often these extreme solar events that are of the greatest concern in terms of severe consequences to human life and human technological systems.

10.31 Dynamics of the Magnetosphere-Ionosphere-Atmosphere System

The elementary episode of energy coupling from the solar wind to the magnetosphere and the subsequent dissipation of this energy is called a *magnetospheric substorm.* Initial phenomenological descriptions of substorms have been followed in subsequent decades by many debates and controversies about the nature and causes of these events. Figure 10.31 shows the flow of energy into and through the magnetosphere-ionosphere system during substorms. The energy to drive substorms comes from solar wind input, which is increased when the interplanetary magnetic field (IMF) turns southward. The left side of Figure 10.31 shows the substorm features related to the magnetotail. There is rather broad agreement that the magnetosphere exhibits a "growth phase" and thus stores energy in the magnetotail prior to substorm onset. The right side of Figure 10.31 shows the substorm processes occurring in the ionosphere. The directly driven aspects of substorms are manifested in the ionosphere by dissipation associated with the global convection process. As this convective dissipation proceeds, the magnetotail is concurrently loaded with energy. This makes the storage process (the substorm growth phase) an integral part of the driven portion of substorms.

An essential aspect of substorms shown in Figure 10.31 is the explosive dissipation that constitutes the unloading process. This occurs in the substorm expansion phase and is localized initially on the nightside of the Earth. There are many specific dissipation processes that can be identified during the typical energy unloading event, including plasma sheet heating, ring current injection, energetic charged particle bursts, and plasmoid formation and release. In addition to plasmoids that carry magnetotail energy back downtail to the solar wind, some portion of the stored energy may also be so far tailward that it is not geoeffective. Explosive dissipation is strikingly obvious in the ionosphere, as well as in the magnetotail, during substorm expansion phases. There is substantial energy dissipation in a substorm for a typical period of 1–2 hours. The ionospheric energy dissipation is predominantly in the form of Joule heating and charged particle precipitation. However, there is also energy dissipation through auroral luminosity and kilometric radiation.

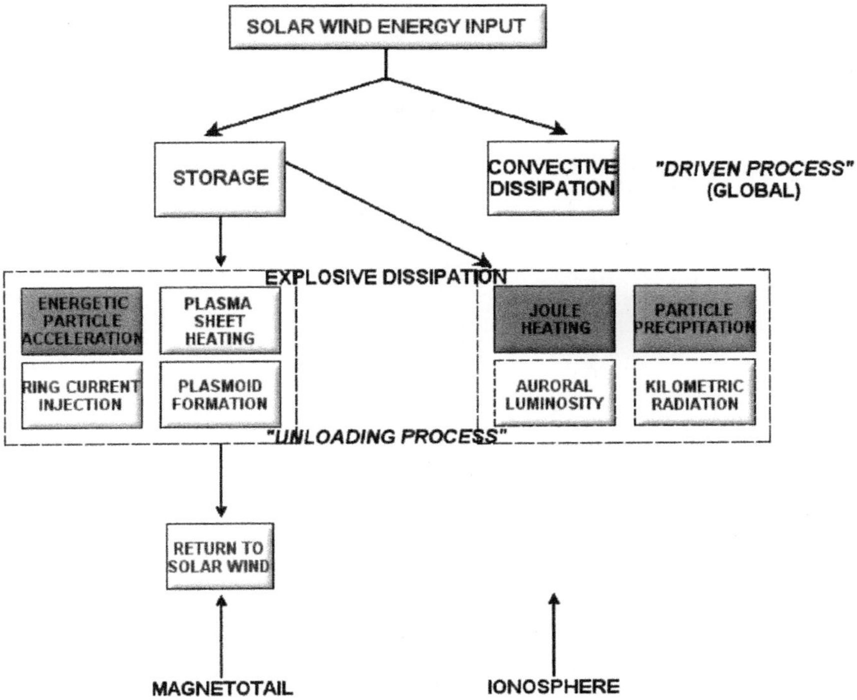

Figure 10.31 A detailed flow diagram of energy in the magnetosphere–ionosphere system during magnetospheric substorms.

Figure 10.32 shows the sequence of magnetotail states that occur during the course of a magnetospheric substorm. The various panels illustrate the noon–midnight meridian plane. Magnetic field lines are sketched, and the gray-shaded region in each panel represents the plasma sheet. The diagram shows how during a substorm the magnetotail: (a) occupies an initial or ground state; (b) undergoes a stress-developing growth phase; (c) exhibits an explosive onset of tail magnetic reconnection; (d) diverts cross-tail current during the expansion phase and pinches off much of the plasma sheet to form a plasmoid; and (e) eventually relaxes and recovers to something approaching the initial state. The overall timescale for the sequence depicted in Figure 10.32 to occur is typically 2–3 hours.

As described above, the passage of solar wind containing southward interplanetary magnetic field imparts energy into the near-Earth space environment and produces dynamical responses of the magnetosphere, ionosphere, and atmosphere. A great deal is known about the responses of geospace to these inputs as a result of past satellite and ground-based observing programs. Episodes of southward IMF drive strongly enhanced convection of plasmas and magnetic fields within the magnetosphere; a strong extraterrestrial ring current is formed as plasma sheet particles drift deep into the inner magnetosphere on open drift paths. This strong ring current is a defining feature of geomagnetic storms. The drift inward takes some 3–4 hours. During this time the solar wind drivers have usually subsided, the convection has consequently decreased, and the charged particles have begun to move along closed drift paths. The symmetric ring current decays slowly through collisions with the hydrogen geocorona. Ring current ions also experience wave-particle

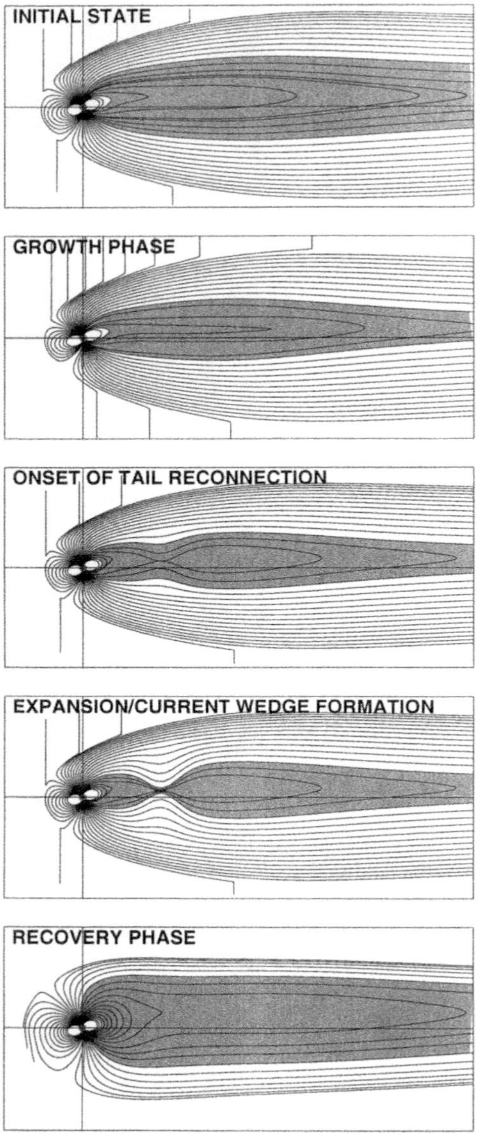

Figure 10.32 Diagram illustrating the sequence of steps occurring in the Earth's magnetotail during substorms.

interactions, which cause the recovery to quiet conditions. Figure 10.33 shows a rather typical geomagnetic storm interval in which the Dst index (a commonly used index of ring current strength and geomagnetic storm development) is plotted along with concurrent solar wind and IMF data. The September 1998 storm portrayed developed in a few hours and decayed over a several-day interval.

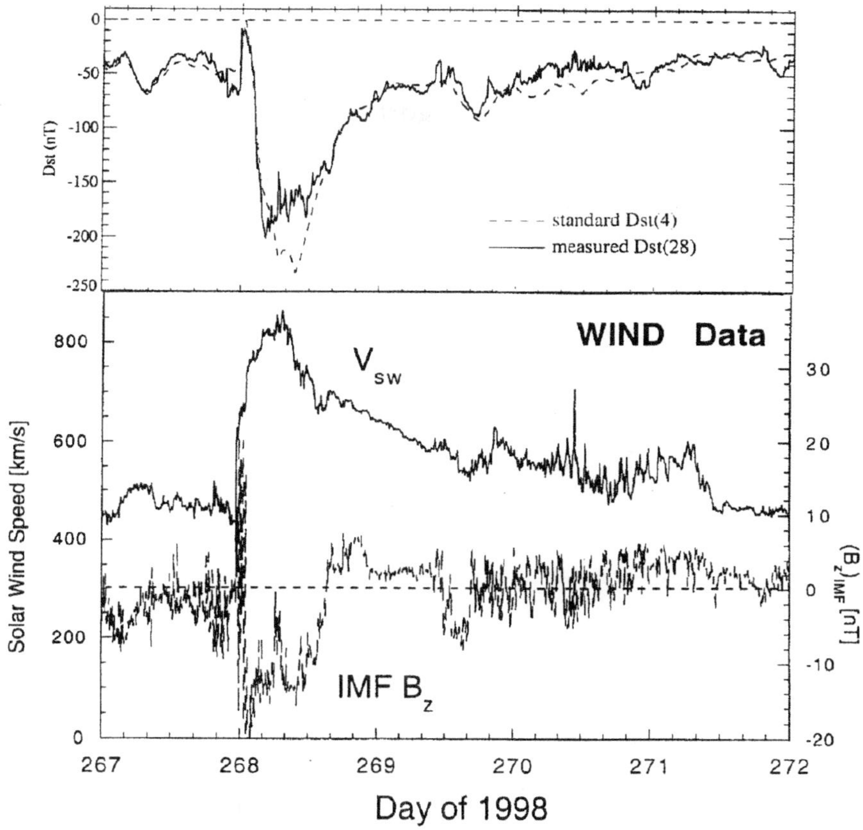

FIGURE 10.33 A typical geomagnetic storm development as seen in the Dst index (during late September 1998). Solar wind speed (V_{SW}) and southward component of the IMF data are also shown.

Enhanced convection stresses the magnetosphere to the point that instabilities are common both in the near-tail and in the mid-tail regions, producing dramatic reconfigurations of the tail's basic structure (as shown in Figure 10.32). In the neartail, strong currents are diverted through the high-latitude ionosphere and produce bright auroral displays. In the mid-tail, field lines merge, accelerating plasma earthward at high velocity and ejecting plasmoids downtail into the solar wind. The earthward flows from the mid-tail are thought to trigger the current disruption in the near-tail. Dipolarization of the magnetic field during substorms generates intense induction electric fields, which accelerate charged particles to high energies in the near-Earth magnetotail. Plasma sheet material moves into, around, and through the inner magnetosphere, modulating the intensity and composition of the ring current. Plasma composition measurements inside 30 R_E indicate that the plasma sheet contains substantial amounts of both solar wind and ionospheric plasma.

Radiation belt particles respond adiabatically to the magnetic field perturbations associated with the building stormtime ring current. If an enhanced solar wind dynamic pressure moves the dayside magnetopause to small radial distances as the storm develops, the radiation belt particles drifting rapidly on closed trajectories may encounter the magnetopause and be lost. The radiation belts rebuild to much higher levels in association with high-speed streams (if they occur) following the storm maximum. The exact mechanism or collection of mechanisms by which magnetospheric charged particles are accelerated to relativistic energies are unknown at present, although several promising candidates exist. It appears that ULF waves in the magnetosphere which are driven by the high-speed solar wind streams may play a key role in the acceleration of substorm particle populations to relativistic energies. If a ring current develops while the radiation belts are enhanced, the associated magnetic field distortion scatters the radiation belt particles to small pitch angles where they encounter the high densities at the foot of the magnetic field lines and are lost. Dramatic losses from the inner belts may also result from the interaction of whistlers (generated by lightning) with the relativistic electrons there; this mechanism introduces an interesting direct coupling with weather systems in the lower atmosphere.

10.32 Importance of Atmospheric Coupling

The neutral atmosphere is not a passive element in the coupled near-Earth system. It is heated at high latitudes by both auroral currents and charged particle precipitation associated with the substorm current systems, thereby expanding to higher altitudes and modifying its composition. The temperature and composition changes move in waves from the conjugate auroral regions and meet at the equator. The changing neutral atmosphere modifies the ionospheric plasma. These variations alter the conductivity of the ionospheric plasma and thus actively modulate the current flow between the ionosphere and magnetosphere. The heating of the high-latitude upper atmosphere in response to auroral inputs during active times can exceed solar inputs. The low-altitude ionosphere is enhanced in the auroral region through impacts of precipitating magnetotail particles on the neutral atmosphere. Large currents (>10^6A) flow along the auroral oval, subauroral electric fields are generated, and for short intervals these can extend all the way down to the equator. Thus, magnetosphere–atmosphere coupling is an important last leg of the Sun–Earth connection issue.

During the last three decades, important efforts have been underway to develop an understanding of variations in middle-atmospheric ozone (O_3) due to natural processes and to the effects of humankind. It is now reasonably well understood that the global balance of O_3 is governed in part by the balance between the production of O_3 and its destruction by reactions within the NO_y, Cl_y, HO_y, and O_y chemical families. One of the most important catalytic cycles in this global balance is that due to the following reactions:

$$\begin{array}{c} NO + O_3 \rightarrow NO_2 + O_2 \\ NO_2 + O \rightarrow NO + O_2 \\ \hline O_3 + O \rightarrow 2O_2 \end{array}$$

The bulk of the odd nitrogen (i.e., NO_y, oxides of nitrogen) in the stratosphere is formed from the oxidation of N_2O by O (^1D), forming NO. It is known, however, that middle-atmospheric NO_y is formed by ion chemistry initiated by the precipitation of energetic charged particles into the upper atmosphere (see Figure 10.34).

The ion and neutral odd-nitrogen chemistry initiated by energetic charged particles produces secondary electrons, e*, which in turn ionize and dissociate the major atmospheric species. This ionization is followed by a series of recombination reactions involving nitrogen and its ions, which produce additional atomic nitrogen. The resulting atomic nitrogen may be in either the ground level or an excited level. Nitric oxide is formed by the reaction of atomic nitrogen with O_2. Reactions involving excited nitrogen are faster than those with the ground-state N atom. The destruction of odd nitrogen proceeds through the reaction of N atoms with NO to produce molecular nitrogen and atomic oxygen (Figure 10.34). In the sunlit atmosphere the photodissociation of NO is important; however, in the fall, winter, and spring in the polar region, the photolytic reaction is negligible

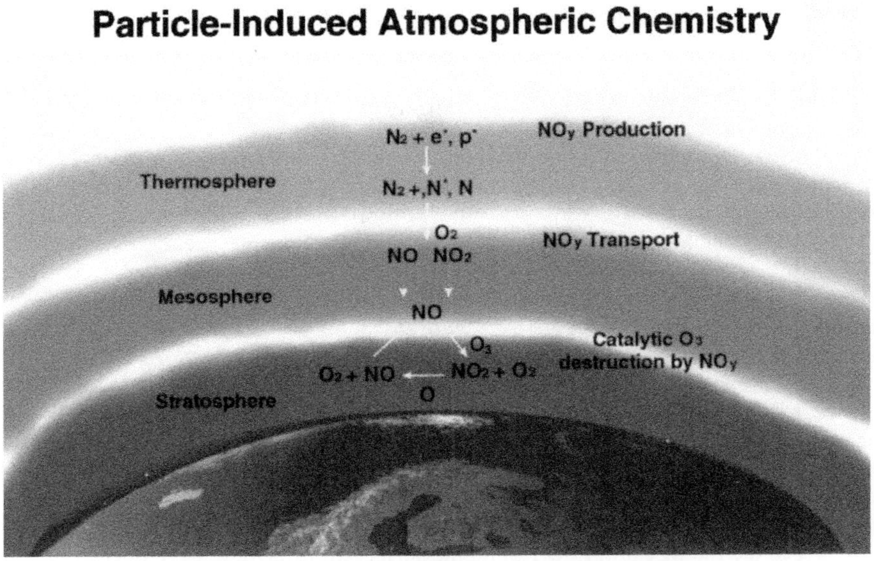

FIGURE 10.34 Charged particle-induced chemical changes in the various layers of the Earth's atmosphere and the catalytic destruction of ozone (O_3).

and the resulting NO lifetime is sufficient for downward transport to bring the NO into the mesosphere and stratosphere.

Global atmospheric models driven by the measured flux of precipitating energetic electrons indicate that column densities of NO_y can increase by 20–40% near 25 km due to the electron inputs. This in turn can lead to O_3 reductions at those altitudes by up to ~20% compared to ozone levels expected without energetic electron inputs. These results imply that solar-wind-magnetosphere-atmosphere coupling is a key determinant of the overall chemistry of the middle atmosphere. This charged particle linkage therefore represents a solar-terrestrial coupling mechanism which, for the last solar cycle, was as important to stratospheric O_3 as were solar UV flux variations. Thus, such recent work shows that the solar-terrestrial energy coupling chain extends down into the stratosphere, just above the troposphere, that part of the Earth's atmosphere which determines weather and climate.

10.33 Sun–Earth Connections and Human Technology

The effects of the Sun on the near-Earth space environment often are considered most significant in terms of their influences on human technology and humans in space. The National Space Weather Program strategic plan of the United States speaks about the wide-ranging impacts of space weather on satellites, power grids, communications systems, and many other technologies. Certainly, astronauts in space and airline passengers and crew on high-flying transpolar routes can be adversely affected by both solar disturbances and geospace responses to these.

Solar flares and solar energetic charged particles can damage solar cells and thereby greatly shorten satellite lifetimes. The heating of the upper atmosphere that occurs due to solar flares substantially increases atmospheric drag on low-altitude satellites. Ionospheric disturbances cause radio signal scintillations and "blackouts" in certain radio frequency ranges. Some of the largest solar-induced magnetic storms disrupt electrical power grids (as occurred in the Canadian province of Quebec in March 1989). All of these can have paralyzing large-scale consequences for human activities.

As another example, the Galaxy 4 spacecraft was a heavily used communications satellite at geostationary orbit; its sudden failure in May 1998 caused the loss of pager service to some 45 million customers as well as numerous other communications outages. Analysis by operators and builders continues as to the exact cause of the Galaxy 4 failure. Using a wide array of space data sets, researchers analyzed the magnetospheric and solar wind conditions during the April–May 1998 period. There was large solar and magnetic activity in early May. Strong evidence was found that the fluxes of highly relativistic electrons were substantially elevated above average conditions for a period of about 2 weeks prior to the May 19 failure of Galaxy 4. Thus, evidence was presented that the internal (deep-dielectric) charging mechanism acting on electronic chips may have played an important role in the Galaxy 4 failure.

There were several other failures and spacecraft anomalies in early to mid-May 1998. Whether the Galaxy spacecraft failure incident was or was not due to a space weather effect, it clearly showed the vulnerability of modern society to individual spacecraft failures. The large number of users affected by the loss of just the one Galaxy spacecraft shows how dependent society has become on space technology and how fragile modern communication systems can be. The Galaxy failure had a large impact because the spacecraft was optimally located over the central United States and could best handle digital pager signals. Therefore, 80% of all pager traffic was directed through it. Increasingly, phones, TV, radio, bank transactions, newspapers, credit card systems, etc., all depend upon satellites for some part of their communications links rather than being all ground-based. It seems unwise to have complex, societally significant systems susceptible to single-spacecraft failures.

10.34 Summary

Study of the near-Earth space environment and its relationship to the Sun has come a tremendous distance in recent decades. Today, owing to a wealth of new observational tools and modeling methods, there is a much more profound understanding of the Sun, of major solar disturbances, and of how these disturbances affect the Earth. Over the course of the next solar cycle (see Figure 10.35), we can expect even further improvements in our understanding of Sun–Earth connections.

Study of the Sun–Earth system has improved because of the number and quality of observing platforms that are available compared to earlier times. The ability to observe the Sun nearly continuously, and especially to observe the most active regions on the Sun, is important for space weather predictions. When a major solar disturbance has occurred, it is also important to track the subsequent interplanetary motion of solar outbursts and to model the motion of solar charged particles and fields. Finally, as the solar outputs reach the vicinity of Earth, there is a need to use the large array of observation platforms available and the global models of the magnetosphere-ionosphere-atmosphere system to predict and to assess their geo-space consequences. All this is developing now and should, in the next solar cycle, improve quite dramatically (Figure 10.35). We can envision a program of Sun–Earth research with a thorough underpinning of basic physical understanding. Through a combination of data assimilation, numerical modeling and better definition of user needs, solar-terrestrial research may soon be at a stage of specification and forecasting of the near-Earth space environment that rivals modern tropospheric weather predictions.

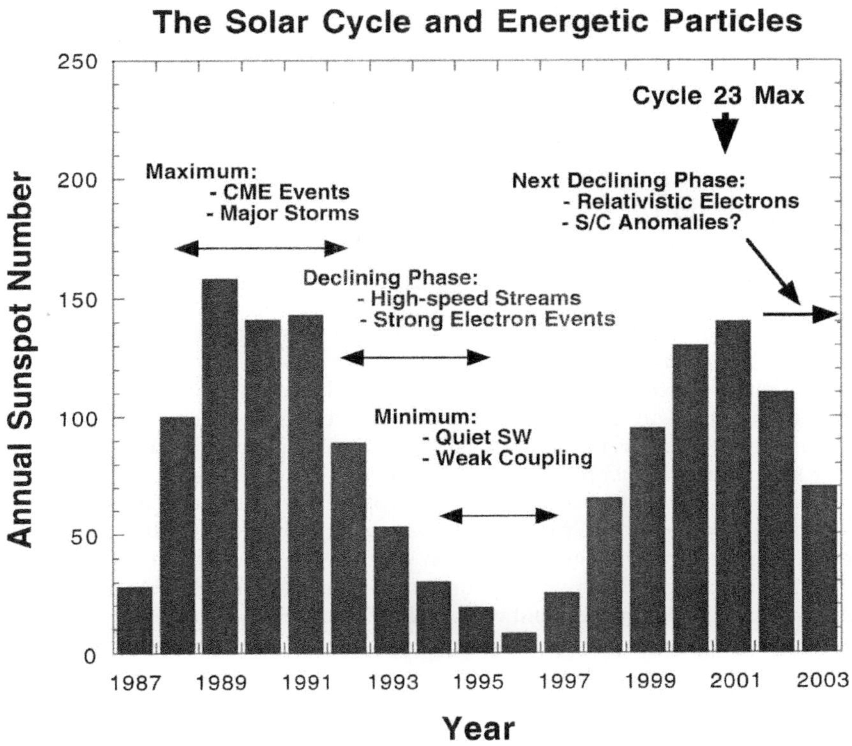

FIGURE 10.35 Portions of two solar activity cycles (1987–2003) as seen in sunspot number and showing their near-Earth consequences; after the maximum of the current solar cycle in 2001, will there be many examples of anomalous—dangerous—events to spacecraft?

Further Reading

Baker, D. N., "Critical Issues in Space Plasma Physics," *Physics of Plasmas,* vol. 6, pp. 1700–1708 (1999).

Baker, D. N. "Effects of the Sun on the Earth's Environment," *Journal of Atmospheric and Solar-Terrestrial Physics,* vol. 62, pp. 1669–1681 (2000).

Barth, C. A., "Nitric Oxide in the Lower Thermosphere," *Planetary and Space Science,* vol. 40, pp. 315–336 (1992).

Burlaga, L. F., Interplanetary Magnetohydrodynamics, Oxford University Press, New York (1995).

National Space Weather Program Strategic Plan, Office of the Federal Coordinator for Meteorological Services, NOAA, Silver Spring, MD (1995).

Parker, E. N., "Dynamics of the Interplanetary Gas and Magnetic Fields." *Astrophysical Journal,* vol. 128, pp. 664–676 (1958).

PART 7
Space Debris

Rüdiger Jehn

10.35 Introduction

Space debris can be defined as any man-made Earth-orbiting or reentering object which is nonfunctional with no reasonable expectation of assuming or resuming its intended function, including fragments and parts thereof.

Since the launch of Sputnik on October 4, 1957, with more than 4,000 launches about 5,000 satellites have been placed in orbit. In total about 27,000 larger objects—satellites, rocket upper stages, mission-related objects like telescope covers or bolts, and fragments from in-orbit explosions have been observed by ground-based radar and telescopes. About 18,500 of these have burnt up in the atmosphere, leaving about 8,500 larger objects currently in Earth orbit. Figure 10.36 illustrates the steady increase of objects in Earth orbit.

Radar, optical, and infrared telescopes of the U.S. Space Surveillance Network track more than 1,000 objects every day, update their orbital elements, and maintain the orbital data in a publicly available catalogue. Of these catalogued objects only 600–700 are operational satellites (7%). Nearly half of the catalogued objects are fragments from explosions in space (43%). As of today, more than 160 explosions in space are known. Intentional explosions have not been performed for more than 10 years, but fragmentations of rocket upper stages or propulsion units are still taking place four to five times a year. The reason is often residual propellant, which can ignite even after many years in space. There are about 1,350 discarded upper stages from launch vehicles (16% of the catalogued objects), 1,900 defunct satellites (22%), and more than 1,000 mission related objects like ejected covers or screwdrivers dropped by an astronaut (12%). Their number is steadily increasing, despite a reduced number of rocket launches per year (see Figure 10.37).

In low Earth orbit, objects down to a size of about 10 cm are maintained in the catalogue. In geostationary orbit, objects have to have a size of about 1 m to enter into the catalogue. The boundaries are not precisely determined. Whether an object can be detected depends on its material properties as well as on the observing geometry (elevation over the horizon, Sun illumination, number of observations, etc.). The sensitivity of radar and optical telescopes is shown in Figure 10.38.

As can be seen in Figure 10.36, the number of catalogued objects increases by about 210 per year on average. Temporary declines are correlated with solar activity. At high solar activity the high atmosphere is heated, increasing the local air density and thus increasing the air drag of satellites. This, in turn, causes the satellites to lose altitude more rapidly and finally to burn up in the atmosphere.

This is the only natural mechanism which cleans space. However, it is only effective in low Earth orbit. At altitudes above 1,000 km, the annual decrease in orbital altitude is very small. For a typical satellite in a 1,000 km circular orbit, it will take about 2,000 years until air drag pulls it down into the dense atmosphere, where it will finally burn up. This means that space activities at these altitudes will inevitably lead to a continuous increase in man-made objects.

Operational satellites can be damaged or even destroyed by a collision with space debris. Since the relative velocities are very high—9–13 km/s on average in low Earth orbit—centimeter-sized particles can produce considerable damage. In addition, new debris is generated during such collisions, in turn increasing the collision probability.

The biggest danger is due to untrackable objects. The number of these is much higher than the number of trackable, or catalogued, objects. The orbits are widely unknown and the spatial distribution can only be described statistically. The number of objects larger than 1 cm is estimated to be about 350,000.

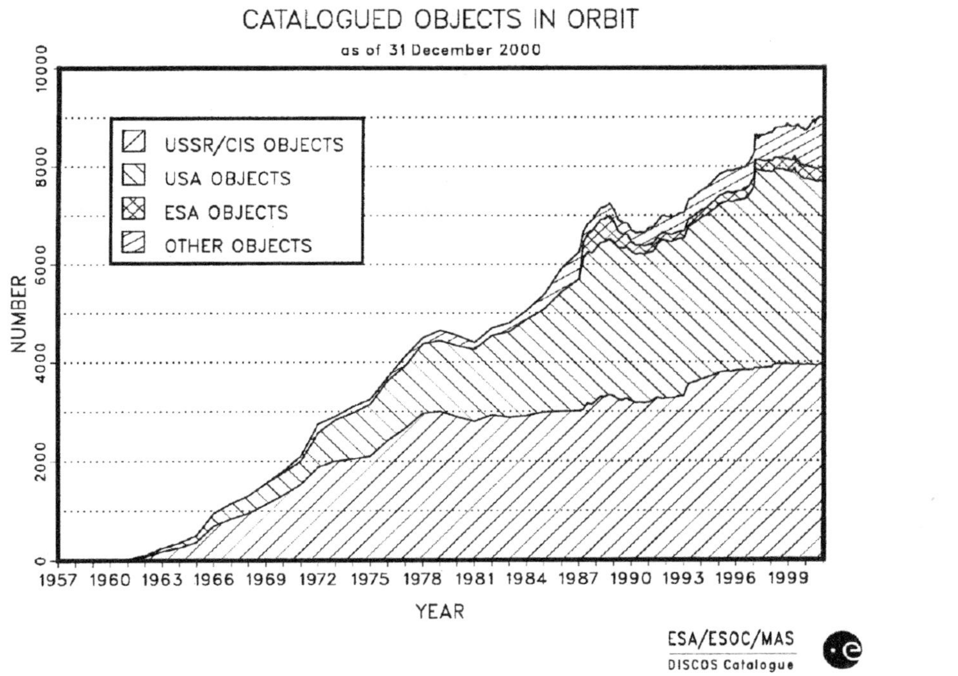

FIGURE 10.36 Catalogued objects in orbit. These are objects of at least 10 cm size in low Earth orbit and at least 1 m size in geostationary orbit.

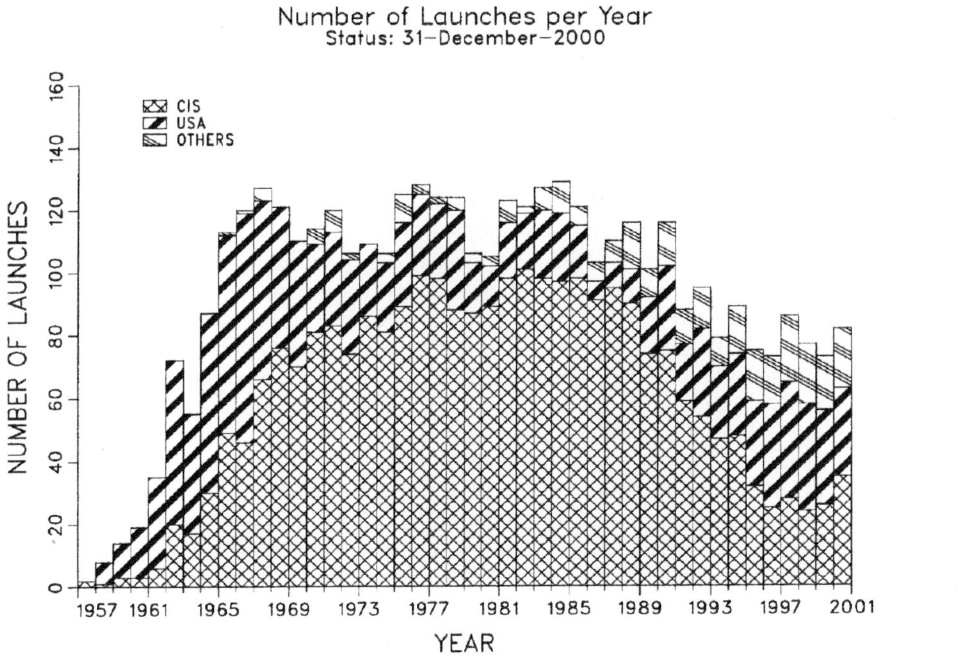

FIGURE 10.37 Launch record. The number of successful launches per year is listed (launch rate).

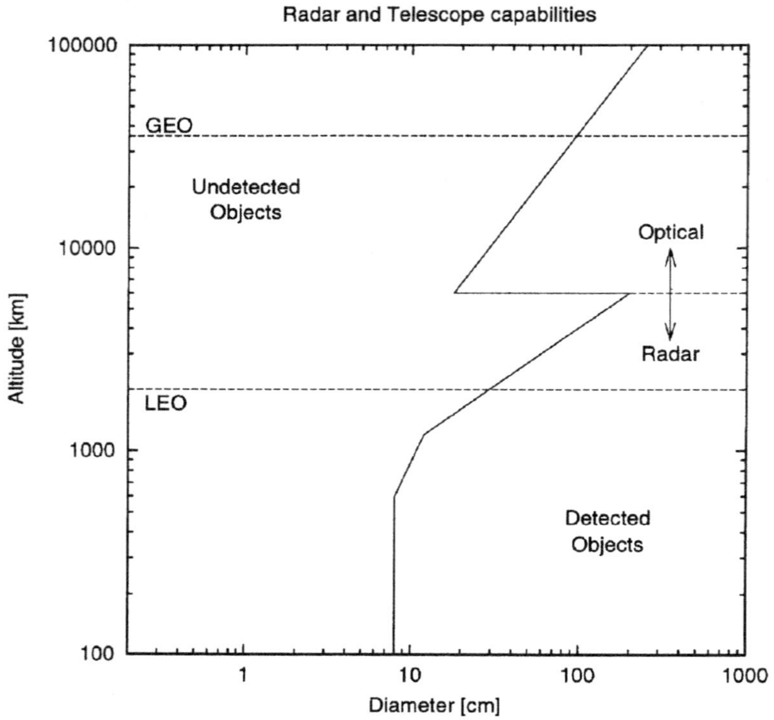

FIGURE 10.38 Sensitivity of the U.S. Space Surveillance Network. Objects to the left of the solid line are too small to be routinely tracked. Radar detect objects of about 10 cm in LEO, and telescopes are used to track objects of about 1 m in the geostationary orbit.

10.36 Spatial Distribution of Space Debris

The operational orbit of a satellite is chosen with consideration of various criteria such as mission objective, orbital altitude, orbital perturbations, requirements on orbit control, launch costs, radiation dose, etc. Some orbits are of much higher interest than other orbits. Today, most spacecraft reside in low-eccentricity orbits below 1,500 km altitude. Near-polar and Sun-synchronous orbits are preferred for Earth observations. Eccentric 12-hour orbits with a critical inclination of 64.3° have been used for many years. The geostationary orbit is also becoming populated with more and more satellites.

The spatial distribution of the trackable space debris is by no means homogeneous. It is closely linked to the orbits of operational spacecraft. For instance, in cases of highly eccentric orbits, the apogees are nearly all located in the northern hemisphere. Figure 10.39 shows the spatial distribution of the trackable space debris population as a function of altitude.

The spatial density has maxima between 900 and 1,200 km altitude and around 1,500 km altitude. Higher up the density continuously decreases, with intermediate peaks at 20,000 km altitude (12-hour orbits, e.g., GPS and GLONASS) and at 35,800 km (geostationary satellites).

The approximate measured particle flux of space debris in low Earth orbit is shown in Figure 10.40 as a function of particle diameter. The dashed curve shows the meteoroid flux, which is of the same order of magnitude as the debris flux for particle sizes between 10 μm and 1 mm. At smaller sizes and also at larger sizes, the space debris flux predominates.

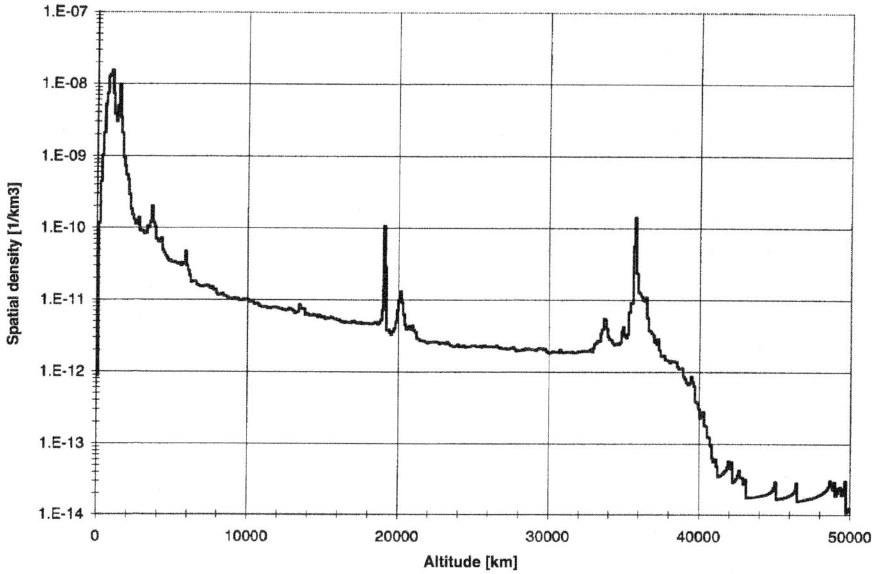

Figure 10.39 Spatial density of catalogued objects (as of January 2000; source: N. Johnson).

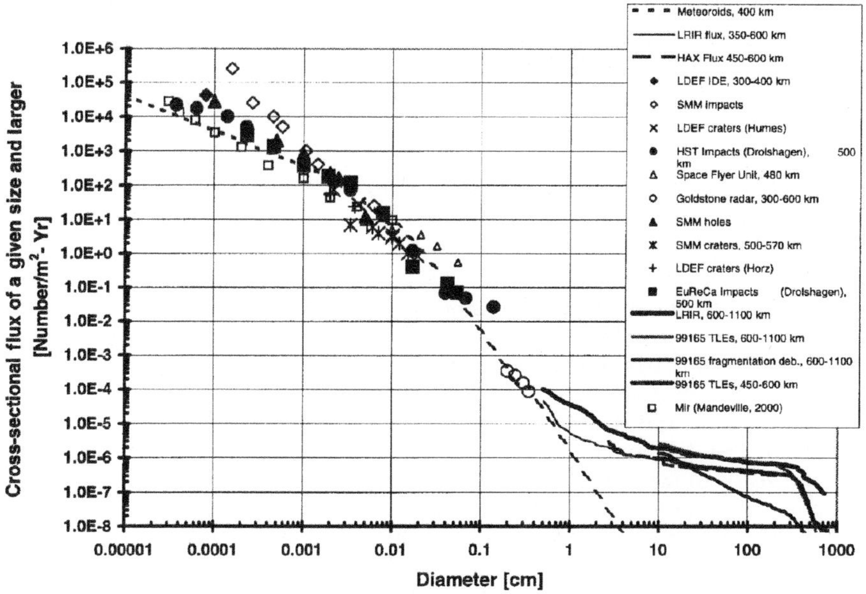

Figure 10.40 Approximate measured debris flux in low Earth orbit (source: N. Johnson).

One source of 1-10-μm-sized space debris is aluminium oxide, which is used as catalyst in solid rocket motors (SRM). During each of the more than 1,000 SRM burns which have been made up to today, trillions of such Al_2O_3 objects are released in space, and current models predict their number in orbit to be on the order of 10^{17}.

Space debris resides mainly in the neighborhood of orbits which are highly frequented by operational satellites. Therefore, these orbits are exposed to the highest collision risk. To estimate the collision risk of individual orbits the MASTER model was developed at ESA (Klinkrad et al. 2000). It describes the spatial distribution of space debris and meteoroids from low Earth orbit to the geostationary altitude for objects larger than 1 μm. The model population is based on:

- The catalogued population
- The simulation of more than 150 explosions in space
- The simulation of more than 1,000 SRM firings
- The release of 150 kg of sodium-potassium droplets
- The release of surface material (e.g., paint flakes)

In total, 350,000 objects larger than 1 cm are estimated to be in orbit, with 110,000 on average being below 2,000 km altitude.

10.37 The Collision Risk

A simple method to estimate the collision probability is presented here. Assuming that a debris model provides the particle flux F and the satellite has a cross-sectional area of A, then the average number of impacts on the surface A during time T is:

$$N = \int_0^T F A\, dt$$

If we further assume that an impact is a rare event and follows a Poisson distribution, then the probability of at least one impact during time T is:

$$P = 1 - e^{-N} \approx N$$

For a constant flux F and cross-section A, the collision probability during time T is:

$$P = N = F\ A\ T$$

The collision risk in low Earth orbit is studied below for two examples: the European remote sensing satellite ERS and the International Space Station.

ERS is in a Sun-synchronous orbit at an altitude of 780 km and an inclination of 98.5°. Spacecraft in such an orbit face the highest collision risk. The reasons are twofold: the spatial density of debris reaches a maximum at this altitude, and most of the possible collision partners will impact at ±20° around the flight direction of ERS. The reason for this special collision geometry is the high inclination, which leads to many near-collisions over the poles with debris in the same orbit, but with ascending nodes about 180° different from the node of ERS. The probability of colliding with an object of 1 cm is $5 \times 10^{-5}\ m^{-2}\ y^{-1}$ and the most probable collision velocity is 15 km/s. Figure 10.41 illustrates the distribution of collision velocities for three different orbital inclinations. The higher the inclination, the higher the average collision velocity.

The collision geometry as well as the total collision probability in the ISS orbit are quite different. Due to the ISS-inclination of 51°, most of the impacts occur at an azimuth between +30° and +60° and −30° and −60° (see Figure 10.42). The azimuth is measured in the local horizontal

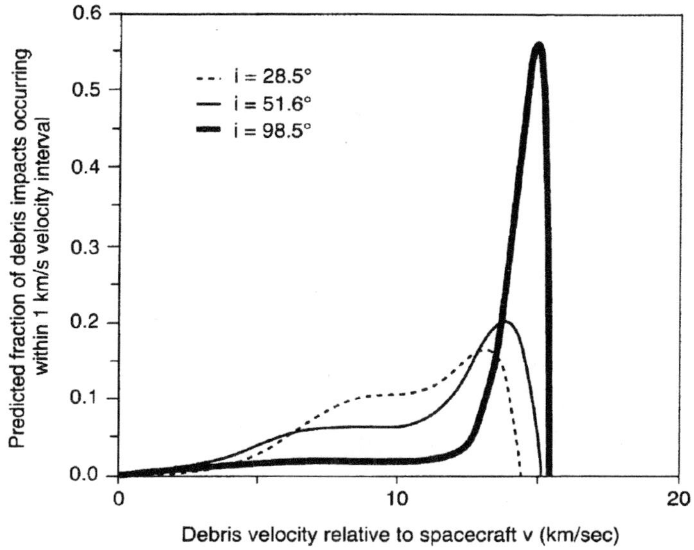

FIGURE 10.41 Calculated collision velocity distribution.

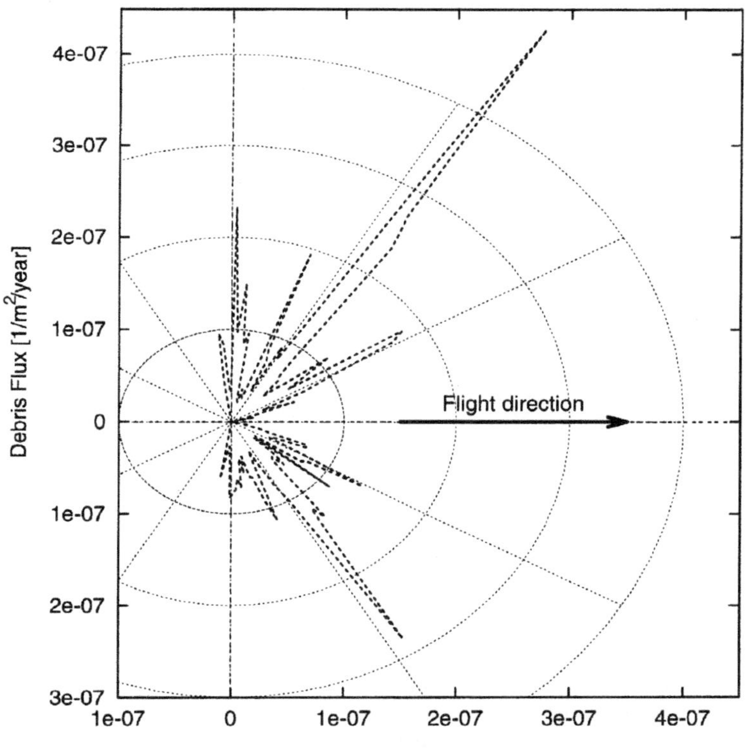

FIGURE 10.42 Space debris flux on the International Space Station.

plane with 0° indicating the flight direction. This means the ISS needs the most effective debris shields around 45° on its "left" and "right" sides (left and right in the sense of a passenger in an aircraft). Little debris is coming from "above" and nearly no debris is coming from "below." The total 1 cm debris flux is $0.7 \times 10^{-5}\ m^{-2}\ y^{-1}$, which is seven times less than the flux in the ERS orbit. Assuming a surface of 100 m^2 for an ISS module and a projected lifetime of 20 years, the probability that the module will be hit by 1 cm debris is 1.4%. The average impact velocity is 10 km/s (see Figure 10.41).

10.38 The Geostationary Orbit

Since 1963, about 650 satellites have been launched into geostationary orbit. The majority of them were boosted from GTO into a nearly synchronous orbit by an apogee kick motor. For satellites equipped with solid rocket motors, the motors are usually ejected at the end. A considerable number of satellites are directly inserted into the geostationary orbit by the launch vehicle, which itself remains in the vicinity of the geostationary orbit. In total about 900 objects are catalogued near the geostationary orbit, 300 of them operational and under orbit control (Hernández and Jehn 2000).

The Sun's, Moon's, and Earth's oblateness lead to long-period perturbations of the inclination of geostationary satellites. The period is 53 years and the maximum inclination is 15°. Thus, uncontrolled satellites build up an inclination of 15° after 27 years. They keep crossing the geostationary orbit twice a day with a velocity of 800 m/s with respect to the controlled geostationary satellites.

In order to eliminate this collision risk, GEO satellites are moved out of the geostationary orbit at the end of their missions. Three burns are recommended to raise the orbit by about 300 km, which is considered a safe distance to avoid future interference with active GEO spacecraft. The change in velocity that is required to raise the semimajor axis by 300 km is 11 m/s, and the propellant requirements correspond to propellant for three months of station-keeping. This means spacecraft operators have to stop operations three months before the spacecraft runs out of fuel and give up considerable revenue in order to reorbit their spacecraft. This is currently the only possibility to preserve the unique resource of the geostationary orbit.

However, only one-third of the aging satellites are properly reorbited into the recommended graveyard orbits 300 km above GEO. In recent years one-third were abandoned without any end-of-life measures, adding further debris to the already 125 objects librating from East to West and back with periods of up to 10 years. About one-third are moved out of GEO just high enough to free their orbital slots to make them available to new satellites.

There is a further collision risk when a couple of satellites occupy the same orbital slot (colocation). The orbital slots typically have extensions of 0.2° (140 km) in longitude and latitude. Since the GEO resource is limited, satellites have to share the same longitude slot (e.g., seven Astra satellites at 19.2° east). Special station-keeping strategies are necessary to eliminate the collision risk among the controlled satellites. The disadvantage is a slightly increased propellant consumption.

10.39 Long-Term Evolution of the Space Debris Environment and Mitigation Measures

All major space-faring nations have developed models to simulate the long-term evolution of the space debris environment. Currently, the steady increase in the number of objects in space is mainly due to satellite launches and explosions in space (five per year on average). There is an average increase of 200 catalogued objects per year. The solid line in Figure 10.43 shows the predicted number of 10 cm objects over the next 200 years if space activities continue as today ("business-as-usual"). The population will double about every 75 years.

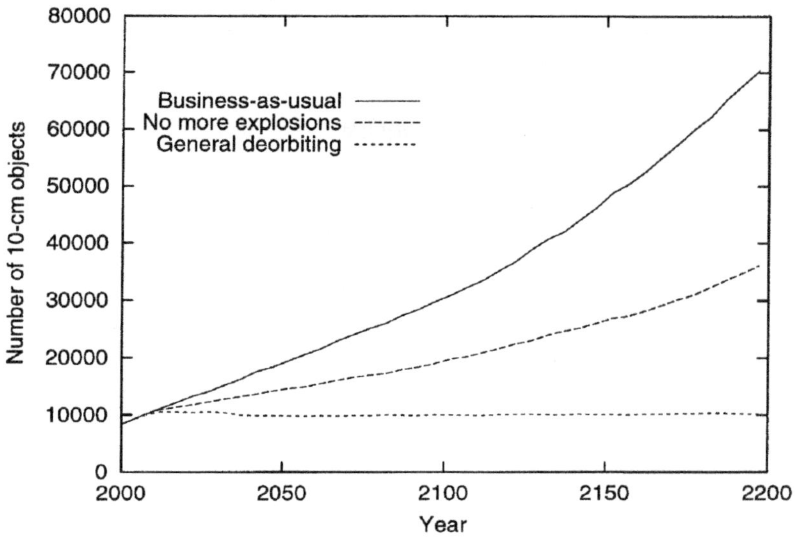

FIGURE 10.43 Simulated number of 10 cm objects in low Earth orbit depending on future space debris mitigation measures. Only if all explosions are prevented and if the deorbiting of old satellites and rocket bodies is introduced can the continuous increase of small objects in space be stopped (source: A. Rossi).

At some point in time the increase will no longer be dominated by launches and explosions. If a certain number of objects is reached (which means that a certain collision probability is also reached), the increase will be dominated by collisions. Simulations by Anselmo et al. (1999) predict that in the business-as-usual case the rate of catastrophic collisions will increase from 0.1/year in the year 2000 to about 5/year within the next 200 years. Kessler (1991) introduced the term *critical density*. The spatial density at a given altitude is called critical if the number of fragments created by collisions exceeds the number of objects removed from this altitude by air drag. If the critical density is reached, the number of debris particles will increase even if all satellite launches are suspended. Today the critical density is reached at altitudes of 800–1,000 km and at 1,500 km. Space debris mitigation measures are especially necessary at these altitudes.

As immediate steps, two measures must be taken: (1) prevention of in-orbit explosions, and (2) avoidance of mission-related objects. The first measure has the biggest impact on the future environment since about half of the catalogued objects in space are fragments from explosions. To prevent the breakup of rocket upper stages, they have to be made harmless after separation of the payload(s). The U.S. Delta rocket upper stage now performs a burn to depletion, and the Ariane upper stage and the Japanese H-1 second stage release their residual fuel to avoid a later explosion. Also, spacecraft batteries and critical items with a potential to explode have to be made safe. The second measure requires minimizing the number of objects released during spacecraft operations. Typical mission-related objects are adapters between two satellites of a dual launch, telescope covers, bolts, yo-yos, etc. Spacecraft designers and operators must design the mission such that objects stay attached to the spacecraft and do not form additional space debris. If no objects are released after the year 2005 and if no more explosions happen after 2010, then the number of 10 cm objects will increase moderately and, within 100 years, reach only half of the value of the business-as-usual scenario.

However, to stop the ever-increasing amount of debris and stay below the critical space debris density, more ambitious mitigation measures need to be taken. In the long run, spacecraft and rocket stages have to be returned to Earth after completion of their mission. Currently, there are international efforts to come to a world-wide agreement to limit the orbital lifetime of retired spacecraft in low Earth orbit to a few decades (25 years is the limit in the NASA and European space debris mitigation standards). When such an agreement is applicable, all low Earth orbit spacecraft will be required to lower their orbits such that atmospheric drag will cause their decay within a specified time. And when solar electric propulsion has replaced chemical propulsion as the principal means of orbital transfer, the deorbiting of spacecraft in high Earth orbits up to the geostationary orbit will become possible without excessive propellant penalties. The dotted line in Figure 10.43 shows the number of 10 cm objects if all spacecraft and rocket bodies are deorbited after end-of-mission (which last 10 years in the simulation) starting in the year 2010.

References

Anselmo, L., Rossi, A., and Pardini, C. 1999. "Updated Results on the Long-Term Evolution of the Space Debris Environment," *Advances in Space Research,* vol. 23, no. 1, pp. 201–211.

Hernández, C. and Jehn, R. 2000. *Classification of Geostationary Objects—Issue 3,* ESA/ESOC, Darmstadt, Germany.

Kessler, D. J. 1991. "Collisional Cascading: The Limits of Population Growth in Low Earth Orbit," *Advances in Space Ressearch,* vol. 11, no. 12, pp. 63–66.

Klinkrad, H., Bendisch, J., Bunte, K. D., Krag, H., Sdunnus, H., and Wegener, P. 2000. "The MASTER-99 Space Debris and Meteoroid Environment Model," 33rd COSPAR Scientific Assembly, Warsaw, Poland, July.

Further Reading

Flury, W., and Klinkrad, H., eds., *Space Debris,* 32nd COSPAR Scientific Assembly, Nagoya, Japan, *Advances in Space Research,* vol. 23, no. 1 (1999).

International Academy of Astronautics (IAA), *Position Paper on Orbital Debris,* IAA, Paris (1995).

Johnson, N. "The World State of Orbital Debris Measurements and Modelling," in *Space Safety and Rescue 1997,* ed. G. W. Heath, AAS Science and Technology Series 96, Univelt, San Diego, pp. 121–128 (1999).

National Research Council, *Orbital Debris: A Technical Assessment,* National Academy Press, Washington, DC (1995).

Office of Science and Technology Policy, *Interagency Report on Orbital Debris,* Washington, DC (1995).

Proceedings of the Third European Conference on Space Debris, ESA SP-473, Darmstadt, Germany (2001).

Technical Report on Space Debris, UN Doc. A/AC.105/720 (1999).

SECTION 11

Spacecraft Subsystems

Section Editor: Brij N. Agrawal

PART 1

Attitude Dynamics and Control

Sachin Agrawal

11.1 Introduction

An attitude control subsystem determines and controls the orientation, or the attitude, of a spacecraft about its center of mass. It should be noted that orbit dynamics and control deals with the motion of the center of mass of a spacecraft. On the other hand, attitude dynamics and control is concerned with the rotational motion of the spacecraft body about its center of mass. Depending on the mission requirements of the spacecraft, there are several attitude control approaches that have been developed. Generally, imaging satellites have higher attitude control requirements. Some of the control approaches are passive, some active, and some a combination of both. This chapter provides a basic understanding of these attitude control approaches.

As shown in Figure 11.1, the basic elements of an attitude control system are spacecraft attitude dynamics, attitude sensors, control laws, actuators, and disturbance torques. Attitude dynamics can be based on rigid or flexible bodies, depending on the spacecraft. The sensors provide information on the attitude of the spacecraft. Sensors could be earth sensors, sun sensors, star trackers, rate gyros, and magnetometers. The sensor outputs are used by control laws to determine the required control torques. Control torques are provided by the actuators, which can be reaction wheels, momentum wheels, control moment gyros, thrusters, and magnetic torquers. Disturbance torque tends to degrade spacecraft performance and can be from earth's gravitational field, aerodynamic drag, solar radiation pressure, earth's magnetic field, and internally generated torque by the spacecraft.

11.2 Rigid-Body Dynamics

The attitude dynamics of a spacecraft is based primarily on rigid-body dynamical equations. These equations, however, are modified for flexibility in the case of flexible spacecraft, requiring high-attitude accuracies. The attitude motions are normally represented in the spacecraft body reference frame, which is rotating and accelerating.

Hence, the subject of relative motion and transformation between reference frames plays an important role in attitude dynamics. This section deals with the general motion of a particle in a rotating reference frame and derives the equations of motion of a rigid body.

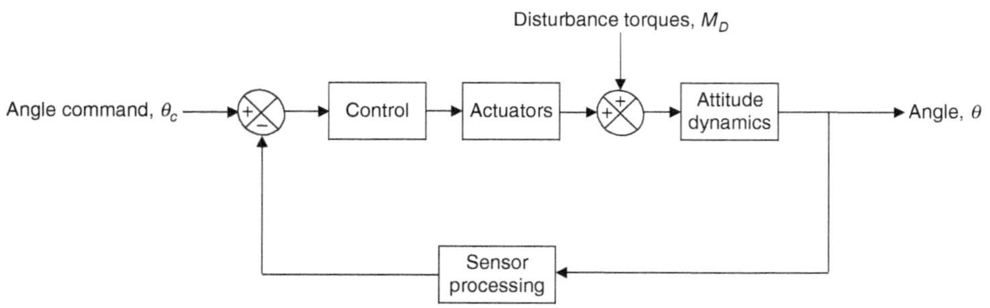

FIGURE 11.1 Block diagram of a attitude control system.

Derivative of a Vector in a Rotating Reference Frame

In many situations, it is convenient to refer to a vector with respect to a moving reference frame. In such cases, the time derivative of the vector consists of two parts: one due to the rate of change of the vector relative to the translating reference frame and the other due to the rotation of the frame. Let $\hat{\mathbf{n}}_x, \hat{\mathbf{n}}_y, \hat{\mathbf{n}}_z$ define a set of orthonormal basis vectors fixed in an inertial frame, N. Also, $\hat{\mathbf{b}}_x, \hat{\mathbf{b}}_y, \hat{\mathbf{b}}_z$ define a set of orthonormal basis vectors in reference frame B, rotating relative to frame N with angular velocity vector ${}^N\vec{\omega}^B$. We can define ${}^N\vec{\omega}^B$ as

$$ {}^N\vec{\omega}^B = \omega_x \hat{\mathbf{b}}_x + \omega_y \hat{\mathbf{b}}_y + \omega_z \hat{\mathbf{b}}_z \tag{11.1} $$

The origin point O of both reference frames are coincident as shown in Figure 11.2. The position vector from the point O to the point P can be written in the form

$$ \vec{\mathbf{r}}^{P/O} = r_x \hat{\mathbf{b}}_x + r_y \hat{\mathbf{b}}_y + r_z \hat{\mathbf{b}}_z \tag{11.2} $$

The time derivative of $\vec{\mathbf{r}}^{P/O}$ or point P's velocity in frame N is simply

$$ {}^N\vec{\mathbf{v}}^{P/O} = \frac{{}^N d}{dt}(\vec{\mathbf{r}}^{P/O}) = \dot{r}_x \hat{\mathbf{b}}_x + r_x \dot{\hat{\mathbf{b}}}_x + \dot{r}_y \hat{\mathbf{b}}_y + r_y \dot{\hat{\mathbf{b}}}_y + \dot{r}_z \hat{\mathbf{b}}_z + r_z \dot{\hat{\mathbf{b}}}_z \tag{11.3} $$

It can be shown (Thomson 1961) that the time derivatives of the unit vectors are

$$ \frac{{}^N d}{dt}(\hat{\mathbf{b}}_x) = {}^N\vec{\omega}^B \times \hat{\mathbf{b}}_x, \quad \frac{{}^N d}{dt}(\hat{\mathbf{b}}_y) = {}^N\vec{\omega}^B \times \hat{\mathbf{b}}_y, \quad \frac{{}^N d}{dt}(\hat{\mathbf{b}}_z) = {}^N\vec{\omega}^B \times \hat{\mathbf{b}}_z \tag{11.4} $$

Moreover, the relative velocity or time derivative of the position vector $\vec{\mathbf{r}}^{P/O}$ with respect to frame B is

$$ {}^B\vec{\mathbf{v}}^{P/O} = \frac{{}^B d}{dt}(\vec{\mathbf{r}}^{P/O}) = \dot{r}_x \hat{\mathbf{b}}_x + \dot{r}_y \hat{\mathbf{b}}_y + \dot{r}_z \hat{\mathbf{b}}_z \tag{11.5} $$

where ${}^B\vec{\mathbf{v}}^{P/O}$ is recognized as the time rate of change of $\vec{\mathbf{r}}^{P/O}$ relative to the reference frame B. Substituting equations (11.4) and (11.5) into equation (11.3), we get

$$ {}^N\vec{\mathbf{v}}^{P/O} = {}^B\vec{\mathbf{v}}^{P/O} + {}^N\vec{\omega}^B \times \vec{\mathbf{r}}^{P/O} \tag{11.6} $$

where ${}^N\vec{\omega}^B \times \vec{\mathbf{r}}^{P/O}$ can be thought of as the velocity seen due to the frame B's rotation with respect to frame N.

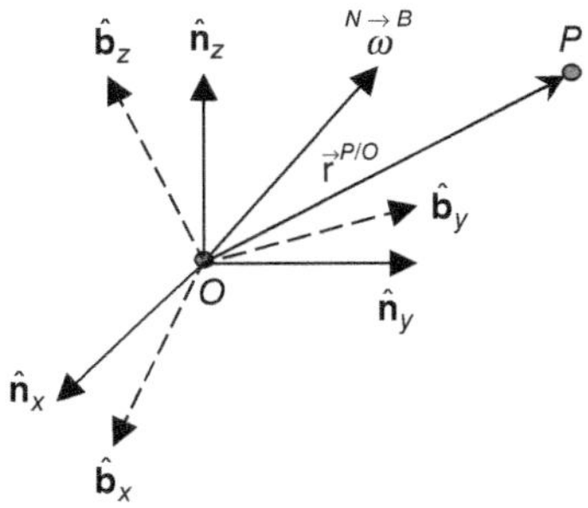

FIGURE 11.2 Derivative of vector in a rotating frame.

General Motion of a Particle

This approach can be used to develop expressions for the general motion of a particle. Generally, frame B is capable not only of rotation relative to N, but also of translation, as shown in Figure 11.3.

Denoting the origin of reference frame B by point C, the position vector of a particle P relative to frame N can be written in the form

$$\vec{\mathbf{r}}^{P/O} = \vec{\mathbf{r}}^{C/O} + \vec{\mathbf{r}}^{P/C} \tag{11.7}$$

where $\vec{\mathbf{r}}^{C/O}$ is the position vector from O to C and $\vec{\mathbf{r}}^{P/C}$ is the position vector of P relative to the frame B. Differentiating equation (11.7) with respect to time, we obtain the absolute velocity of P in the form

$${}^{N}\vec{\mathbf{v}}^{P/O} = \frac{{}^{N}d}{dt}(\vec{\mathbf{r}}^{P/O}) = \frac{{}^{N}d}{dt}(\vec{\mathbf{r}}^{C/O}) + \frac{{}^{N}d}{dt}(\vec{\mathbf{r}}^{P/C}) \tag{11.8}$$

Noting that ${}^{N}\vec{\mathbf{v}}^{C/O} = \frac{{}^{N}d}{dt}(\vec{\mathbf{r}}^{C/O})$ is the velocity of the point C with respect to frame N and recalling equation (11.6), we obtain the absolute velocity of P as follows:

$${}^{N}\vec{\mathbf{v}}^{P/O} = {}^{N}\vec{\mathbf{v}}^{C/O} + {}^{B}\vec{\mathbf{v}}^{P/C} + {}^{N}\vec{\omega}^{B} \times \vec{\mathbf{r}}^{P/C} \tag{11.9}$$

where ${}^{B}\vec{\mathbf{v}}^{P/C} = \frac{{}^{B}d}{dt}(\vec{\mathbf{r}}^{P/C})$ is the velocity of P with respect to frame B. The absolute acceleration of P can be obtained by differentiating equation (11.9) so that

$${}^{N}\vec{\mathbf{a}}^{P/O} = \frac{{}^{N}d}{dt}({}^{N}\vec{\mathbf{v}}^{P/O}) = \frac{{}^{N}d}{dt}({}^{N}\vec{\mathbf{v}}^{C/O} + {}^{B}\vec{\mathbf{v}}^{P/C} + {}^{N}\vec{\omega}^{B} \times \vec{\mathbf{r}}^{P/C}) \tag{11.10}$$

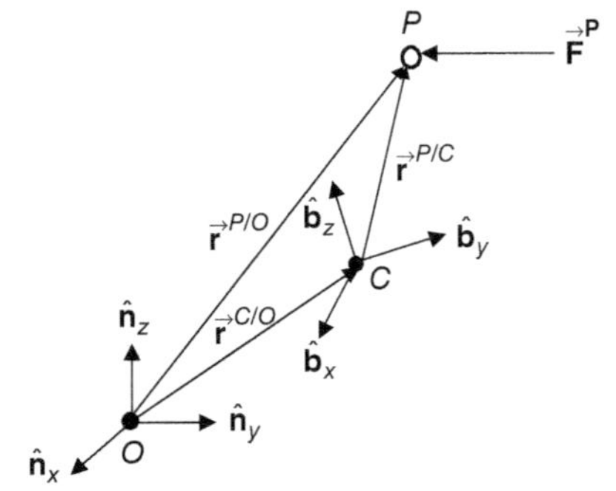

FIGURE 11.3 General motion of a particle.

But ${}^{N}\vec{\mathbf{a}}^{C/O} = \frac{{}^{N}d}{dt}({}^{N}\vec{\mathbf{v}}^{C/O})$ is the acceleration of C. Also, ${}^{N}\vec{\alpha}^{B} = \frac{{}^{N}d}{dt}({}^{N}\vec{\omega}^{B})$ is the angular acceleration of reference frame B relative to frame N. Moreover, recalling equation (11.6), we can show

$$\frac{{}^{N}d}{dt}({}^{B}\vec{\mathbf{v}}^{P/C}) = \frac{{}^{B}d}{dt}({}^{B}\vec{\mathbf{v}}^{P/C}) + {}^{N}\vec{\omega}^{B} \times {}^{B}\vec{\mathbf{v}}^{P/C}$$
$$= {}^{B}\vec{\mathbf{a}}^{P/C} + {}^{N}\vec{\omega}^{B} \times {}^{B}\vec{\mathbf{v}}^{P/C} \tag{11.11}$$

where ${}^{B}\vec{\mathbf{a}}^{P/C} = \frac{{}^{B}d^2}{dt^2}(\vec{\mathbf{r}}^{P/C})$ is the acceleration of particle P, relative to frame B. Substituting equation (11.11) into equation (11.10) and using equation (11.6), we obtain the absolute acceleration of P:

$${}^{N}\vec{\mathbf{a}}^{P/O} = {}^{N}\vec{\mathbf{a}}^{C/O} + {}^{B}\vec{\mathbf{a}}^{P/C} + {}^{N}\vec{\alpha}^{B} \times \vec{\mathbf{r}}^{P/C} + {}^{N}\vec{\omega}^{B} \times ({}^{N}\vec{\omega}^{B} \times \vec{\mathbf{r}}^{P/C}) + 2({}^{N}\vec{\omega}^{B} \times {}^{B}\vec{\mathbf{v}}^{P/C}) \tag{11.12}$$

where $2({}^{N}\vec{\omega}^{B} \times {}^{B}\vec{\mathbf{v}}^{P/C})$ is known as the Coriolis acceleration.

Momentum of a Rigid Body

As shown in Figure 11.4, for a rigid body B we can define B_p to be an arbitrary point fixed on B and B_{cm} to be the center of mass. Let dm be a differential element of mass at B_p.

The linear momentum of a rigid body B can be written in the form

$${}^{N}\vec{\mathbf{P}}^{B} = \rho \int_{B} {}^{N}\vec{\mathbf{v}}^{B_{cm}/O}\, dV = \int_{B} {}^{N}\vec{\mathbf{v}}^{B_{cm}/O}\, dm \tag{11.13}$$

where $dm = \rho dV$ and $\int_B = \iiint_B$ is shorthand notation to express the volume integral over body B. Using equation (11.9) with axes $\hat{\mathbf{b}}_x, \hat{\mathbf{b}}_y, \hat{\mathbf{b}}_z$ fixed in the rigid body such that ${}^{N}\vec{\mathbf{v}}^{B_p/B_{cm}} = 0$, we obtain

$${}^{N}\vec{\mathbf{P}}^{B} = \int_{B} {}^{N}\vec{\mathbf{v}}^{B_p/O}\, dm$$
$$= m\,{}^{N}\vec{\mathbf{v}}^{B_{cm}/O} + {}^{N}\vec{\omega}^{B} \times \int_{B} \vec{\mathbf{r}}^{B_p/B_{cm}}\, dm \tag{11.14}$$

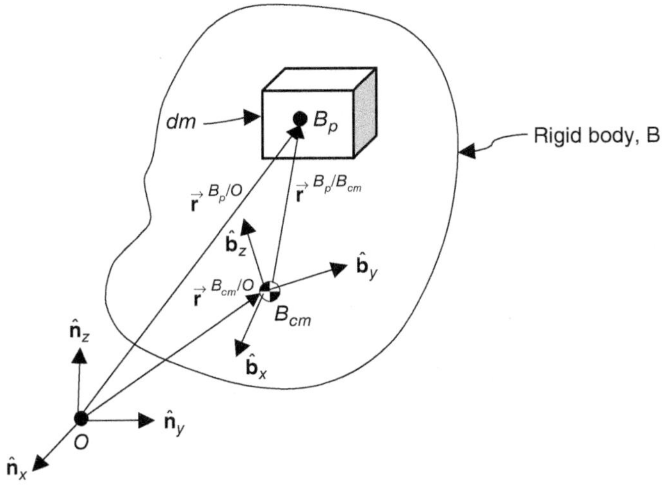

FIGURE 11.4 System with rigid body B.

Equation (11.14) can be simplified because the origin of the rotating reference frame B coincides with the center of mass of the rigid body such that $\int_B \vec{\mathbf{r}}^{B_p/B_{cm}}\,dm = 0$ (Thomson 1961). It follows that

$$ {}^N\vec{\mathbf{P}}^B = m\,{}^N\vec{\mathbf{v}}^{B_{cm}/O} \tag{11.15}$$

or the linear momentum of rigid body B is equal to the product of its mass and the velocity of its center of mass.

Using the same approach as we did for linear momentum, the angular momentum of the rigid body B, in reference frame N about B_{cm}, can be written in the form

$$\begin{aligned} {}^N\vec{\mathbf{H}}^{B/B_{cm}} &= \int_B \vec{\mathbf{r}}^{B_p/B_{cm}} \times {}^N\vec{\mathbf{v}}^{B_p/O}\,dm = \int_B \vec{\mathbf{r}}^{B_p/B_{cm}} \times ({}^N\vec{\mathbf{v}}^{B_{cm}/O} + {}^N\vec{\omega}^B \times \vec{\mathbf{r}}^{B_p/B_{cm}})\,dm \\ &= \left(\int_B \vec{\mathbf{r}}^{B_p/B_{cm}}\,dm\right) \times {}^N\vec{\mathbf{v}}^{B_{cm}/O} + \int_B \vec{\mathbf{r}}^{B_p/B_{cm}} \times ({}^N\vec{\omega}^B \times \vec{\mathbf{r}}^{B_p/B_{cm}})\,dm \end{aligned} \tag{11.16}$$

Once again, since the origin of the frame B coincides with the center of mass of the rigid body (the mass center is fixed), equation (11.16) reduces to

$$ {}^N\vec{\mathbf{H}}^{B/B_{cm}} = \int_B \vec{\mathbf{r}}^{B_p/B_{cm}} \times ({}^N\vec{\omega}^B \times \vec{\mathbf{r}}^{B_p/B_{cm}})\,dm \tag{11.17}$$

Writing the angular velocity vector ${}^N\vec{\omega}^B$ in terms of its components

$$ {}^N\vec{\omega}^B = \omega_x\hat{\mathbf{b}}_x + \omega_y\hat{\mathbf{b}}_y + \omega_z\hat{\mathbf{b}}_z \tag{11.18}$$

and recalling equations (11.18) and (11.17) yields

$$ {}^N\vec{\mathbf{H}}^{B/B_{cm}} = [I_{xx}\omega_x - I_{xy}\omega_y - I_{xz}\omega_y]\,\hat{\mathbf{b}}_x + [I_{yy}\omega_y - I_{xy}\omega_x - I_{yz}\omega_z]\,\hat{\mathbf{b}}_y + [I_{zz}\omega_z - I_{xz}\omega_x - I_{yz}\omega_y]\,\hat{\mathbf{b}}_z \tag{11.19}$$

In equation (11.19), I_{xx}, I_{yy}, I_{zz} can be computed with

$$ I_{xx} = \int_B (r_y^2 + r_z^2)\,dm, \quad I_{yy} = \int_B (r_x^2 + r_z^2)\,dm, \quad I_{zz} = \int_B (r_x^2 + r_y^2)\,dm \tag{11.20}$$

where I_{xx}, I_{yy}, I_{zz} are the moments of inertia of the body about $\hat{\mathbf{b}}_x, \hat{\mathbf{b}}_y, \hat{\mathbf{b}}_z$, respectively. Also, $I_{xy} = \int_B r_x r_y\,dm$, $I_{xz} = \int_B r_x r_z\,dm$, $I_{yz} = \int_B r_y r_z\,dm$ are the products of inertia. The axes about which products of inertia are zero are called the principal axes of the moment of inertia.

Equation (11.19) can be written in the matrix form

$$ {}^N\vec{\mathbf{H}}^{B/B_{cm}} = \begin{bmatrix} H_x \\ H_y \\ H_z \end{bmatrix} = \begin{bmatrix} I_{xx} & -I_{xy} & -I_{xz} \\ -I_{xy} & I_{yy} & -I_{yz} \\ -I_{xz} & -I_{yz} & I_{zz} \end{bmatrix} \begin{bmatrix} \omega_x \\ \omega_y \\ \omega_z \end{bmatrix} \tag{11.21}$$

Kinetic Energy of a Rigid Body

The kinetic energy of a rigid body B in N is defined as

$$ {}^N T^B = \frac{1}{2}\int_B {}^N\vec{\mathbf{v}}^{B_p/O} \bullet {}^N\vec{\mathbf{v}}^{B_p/O}\,dm \tag{11.22}$$

Since, we are considering the body to be rigid then ${}^{B}\vec{\mathbf{v}}^{B_p/B_{cm}} = 0$. Using equation (11.9) with equation (11.22), we have

$$ {}^{N}T^{B} = \frac{1}{2} m \, {}^{N}\vec{\mathbf{v}}^{B_{cm}/O} \cdot {}^{N}\vec{\mathbf{v}}^{B_{cm}/O} + {}^{N}\vec{\mathbf{v}}^{B_{cm}/O} \cdot ({}^{N}\vec{\omega}^{B} \times \int_{B} \vec{\mathbf{r}}^{B_p/B_{cm}} dm) + \frac{1}{2} \int_{B} ({}^{N}\vec{\omega}^{B} \times \vec{\mathbf{r}}^{B_p/B_{cm}}) \cdot ({}^{N}\vec{\omega}^{B} \times \vec{\mathbf{r}}^{B_p/B_{cm}}) dm \tag{11.23} $$

The first integral in equation (11.23) is zero because frame B coincides with the center of mass such that

$$ {}^{N}T^{B} = \frac{1}{2} m({}^{N}v^{B_{cm}/O})^2 + \frac{1}{2}\left[I_{xx}\omega_x^2 + I_{yy}\omega_y^2 + I_{zz}\omega_z^2 - 2\,\omega_x\omega_z I_{xz} - 2\,\omega_y\omega_z I_{yz} - 2\,\omega_x\omega_y I_{xy}\right] \tag{11.24} $$

where ${}^{N}v^{B_{cm}/O}$ is the magnitude of ${}^{N}\vec{\mathbf{v}}^{B_{cm}/O}$. The first term in equation (11.24) represents the kinetic energy of the body as if it were in pure translation and the remaining terms constitute the kinetic energy of the body as if it were in pure rotation about B_{cm}. The kinetic energy of rotation, ${}^{N}T^{B}_{\text{rot}}$, can be expressed in matrix form as

$$ {}^{N}T^{B}_{rot} = \frac{1}{2}[\omega_x \omega_y \omega_z] \begin{bmatrix} I_{xx} & -I_{xy} & -I_{xz} \\ -I_{xy} & I_{yy} & -I_{yz} \\ -I_{xz} & -I_{yz} & I_{zz} \end{bmatrix} \begin{bmatrix} \omega_x \\ \omega_y \\ \omega_z \end{bmatrix} \tag{11.25} $$

Equations of Motion

The equations of motion are based on Newton's second law, which states that the rate of change of the linear momentum of a particle is equal to the force acting on the particle. It should be noted that in applying Newton's second law one must measure the motion relative to an inertially fixed frame.

Let us consider a particle P that has mass m and is moving under the action of a force ${}^{N}\vec{\mathbf{F}}^{P}$ as shown in Figure 11.3. According to Newton's second law:

$$ {}^{N}\vec{\mathbf{F}}^{P} = m \frac{{}^{N}d}{dt}({}^{N}\vec{\mathbf{v}}^{P/O}) = m \, {}^{N}\vec{\mathbf{a}}^{P/O} \tag{11.26} $$

The angular momentum of the particle P relative to an arbitrary point C is

$$ {}^{N}\vec{\mathbf{H}}^{P/C} = \vec{\mathbf{r}}^{P/C} \times m \, {}^{N}\vec{\mathbf{v}}^{P/O} \tag{11.27} $$

The time derivative of the angular momentum is

$$ \frac{{}^{N}d}{dt}({}^{N}\vec{\mathbf{H}}^{P/C}) = \frac{{}^{N}d}{dt}(\vec{\mathbf{r}}^{P/C}) \times m \, {}^{N}\vec{\mathbf{v}}^{P/O} + \vec{\mathbf{r}}^{P/C} \times m \frac{{}^{N}d}{dt}({}^{N}\vec{\mathbf{v}}^{P/O}) \tag{11.28} $$

Using equations (11.9) and (11.26), we obtain

$$ \begin{aligned} \frac{{}^{N}d}{dt}({}^{N}\vec{\mathbf{H}}^{P/C}) &= \frac{{}^{N}d}{dt}(\vec{\mathbf{r}}^{P/C}) \times m \, {}^{N}\vec{\mathbf{v}}^{P/O} + \vec{\mathbf{r}}^{P/C} \times {}^{N}\vec{\mathbf{F}}^{P} \\ &= {}^{N}\vec{\mathbf{v}}^{P/C} \times m({}^{N}\vec{\mathbf{v}}^{C/O} + {}^{N}\vec{\mathbf{v}}^{P/C}) + \vec{\mathbf{r}}^{P/C} \times {}^{N}\vec{\mathbf{F}}^{P} \\ &= {}^{N}\vec{\mathbf{v}}^{P/C} \times m \, {}^{N}\vec{\mathbf{v}}^{C/O} + \vec{\mathbf{r}}^{P/C} \times {}^{N}\vec{\mathbf{F}}^{P} \end{aligned} \tag{11.29} $$

Using the cross-product property $\vec{\mathbf{a}} \times \vec{\mathbf{b}} = -(\vec{\mathbf{b}} \times \vec{\mathbf{a}})$, we have

$$\frac{{}^N d}{dt}({}^N\vec{\mathbf{H}}^{P/C}) = -({}^N\vec{\mathbf{v}}^{C/O} \times m\ {}^N\vec{\mathbf{v}}^{P/C}) + \vec{\mathbf{r}}^{P/C} \times {}^N\vec{\mathbf{F}}^{P} \tag{11.30}$$

By definition (Agrawal 1986), however, the moment of the force ${}^N\vec{\mathbf{F}}^P$ about point C is

$${}^N\vec{\mathbf{M}}^{P/C} = \vec{\mathbf{r}}^{P/C} \times {}^N\vec{\mathbf{F}}^{P} \tag{11.31}$$

Substituting equation (11.31) into equation (11.30), we obtain

$${}^N\vec{\mathbf{M}}^{P/C} = \frac{{}^N d}{dt}({}^N\vec{\mathrm{H}}^{P/C}) + {}^N\vec{\mathrm{v}}^{C/O} \times m\,{}^N\vec{\mathrm{v}}^{P/C} \tag{11.32}$$

The force and moment equations were derived for a single particle but they can be extended to the motion of rigid bodies. As shown in Figure 11.4, for rigid bodies we can define B_p to be an arbitrary point fixed on body B and B_{cm} to be body B's center of mass. Recognizing there is no motion relative to the frame B and using equation (11.12), we can write the force equation of the motion as

$${}^N\vec{\mathbf{F}}^{\mathrm{B}} = \int_{\mathrm{B}} {}^N\vec{\mathbf{a}}^{B_p/O}\, dm = \int_{\mathrm{B}} \left[{}^N\vec{\mathbf{a}}^{B_{cm}/O} + {}^N\dot{\vec{\omega}}^{B} \times \vec{\mathbf{r}}^{B_p/B_{cm}} + {}^N\vec{\omega}^{B} \times ({}^N\vec{\omega}^{B} \times \vec{\mathbf{r}}^{B_p/B_{cm}}) \right] dm \tag{11.33}$$

where ${}^N\vec{\mathbf{F}}^{\mathrm{B}}$ is the resultant of the external forces acting on the rigid body B and dm is an infinitesimal mass element. Note that the mutual attraction forces between any two particles that are internal to the rigid body cancel out in pairs (Schaub and Junkins 2003). If B_{cm} coincides with the center of the mass of the rigid body, then by definition $\int_{\mathrm{B}} \vec{\mathbf{r}}^{B_p/B_{cm}}\, dm = 0$ and equation (11.33) reduces to

$${}^N\vec{\mathbf{F}}^{\mathrm{B}} = m\,{}^N\vec{\mathbf{a}}^{B_{cm}/O} \tag{11.34}$$

The moment equations of motion can be obtained in an analogous manner. Using equation (11.6), we can write the moment equation about B_{cm} as

$$\begin{aligned} {}^N\vec{\mathbf{M}}^{\mathrm{B}/B_{cm}} &= \frac{{}^N d}{dt}({}^N\vec{\mathbf{H}}^{\mathrm{B}/B_{cm}}) \\ &= \frac{{}^B d}{dt}(\vec{\mathbf{H}}^{\mathrm{B}/B_{cm}}) + {}^N\vec{\omega}^{B} \times \vec{\mathbf{H}}^{\mathrm{B}/B_{cm}} \end{aligned} \tag{11.35}$$

where ${}^N\vec{\mathbf{M}}^{\mathrm{B}/B_{cm}}$ is the resultant of the external moments about B_{cm} and ${}^N\vec{\mathbf{H}}^{\mathrm{B}/B_{cm}}$ is the angular momentum of the rigid body about B_{cm}. Therefore, the moment of the external forces about the center of the mass of a rigid body is equal to the time rate of change of the angular momentum of the body about the center of the mass. We can conveniently express the angular momentum in terms of components along $\hat{\mathbf{b}}_x, \hat{\mathbf{b}}_y, \hat{\mathbf{b}}_z$ in the form

$${}^B\vec{\mathbf{H}}^{\mathrm{B}/B_{cm}} = H_x \hat{\mathbf{b}}_x + H_y \hat{\mathbf{b}}_y + H_z \hat{\mathbf{b}}_z \tag{11.36}$$

Equations (11.34) and (11.39) can be written in each component as follows:

$$F_x = m\,a_x \tag{11.37a}$$

$$F_y = m\,a_y \tag{11.37b}$$

$$F_z = m\,a_z \tag{11.37c}$$

$$M_x = \dot{H}_x + \omega_y H_z - \omega_z H_y \tag{11.37d}$$

$$M_y = \dot{H}_y + \omega_z H_x - \omega_x H_z \tag{11.37e}$$

$$M_z = \dot{H}_z + \omega_x H_y - \omega_y H_x \tag{11.37f}$$

where H_x, H_y, and H_z are given by equation (11.21). In the special case where the rotating frame B coincides with the principal axes of moment of inertia then the components of the angular momentum are reduced to

$$H_x = I_{xx}\,\omega_x \tag{11.38a}$$

$$H_y = I_{yy}\,\omega_y \tag{11.38b}$$

$$H_z = I_{zz}\,\omega_z \tag{11.38c}$$

If we substitute equation (11.38a–c) into equation (11.37d–f), we get

$$M_x = I_{xx}\,\dot{\omega}_x + \omega_y\,\omega_z\,(I_{zz} - I_{yy}) \tag{11.39a}$$

$$M_y = I_{yy}\,\dot{\omega}_y + \omega_x\,\omega_z\,(I_{xx} - I_{zz}) \tag{11.39b}$$

$$M_z = I_{zz}\,\dot{\omega}_z + \omega_x\,\omega_y(I_{yy} - I_{xx}) \tag{11.39c}$$

which are known as Euler's moment equations.

11.3 Orientation Kinematics

One important aspect of attitude dynamics and control is the relative orientation between reference frames. In many applications the orientation is usually defined between a moving reference frame or "body frame" (as shown by reference frame B, previously) and a fixed reference frame or "inertial frame" (as shown by reference frame N, previously). There are three common representations of the orientation between reference frames. They are (1) direction cosines, (2) Euler angles, and (3) quaternions.

Direction Cosines

As shown in Figure 11.5, we can define a reference frame A by a set of right-handed, unitary orthogonal bases vectors $\hat{\mathbf{a}}_1, \hat{\mathbf{a}}_2, \hat{\mathbf{a}}_3$ and another arbitrary reference frame B by a set of right-handed, unitary orthogonal bases vectors $\hat{\mathbf{b}}_1, \hat{\mathbf{b}}_2, \hat{\mathbf{b}}_3$.

The direction cosines are defined as the cosine of the angle between the basis vectors

$$^{A}R^{B}_{ij} = \hat{\mathbf{a}}_i \bullet \hat{\mathbf{b}}_j = \left\|\hat{\mathbf{a}}_i\right\| \left\|\hat{\mathbf{b}}_j\right\| \cos\theta_{ij} \qquad \text{for } (i, j = 1, 2, 3) \tag{11.40}$$

where θ_{ij} is the angle between $\hat{\mathbf{a}}_i$ and $\hat{\mathbf{b}}_j$, which can be found using

$$\theta_{ij} = \text{acos}\left(\frac{\hat{\mathbf{a}}_i \bullet \hat{\mathbf{b}}_j}{\left\|\hat{\mathbf{a}}_i\right\| \bullet \left\|\hat{\mathbf{b}}_j\right\|}\right) \tag{11.41}$$

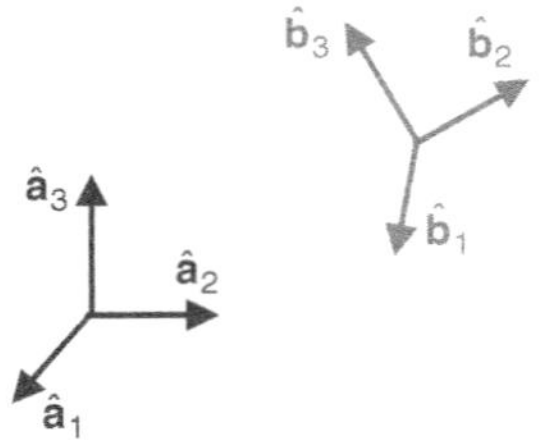

FIGURE 11.5 Right-handed, orthonormal bases $\hat{\mathbf{a}}_1, \hat{\mathbf{a}}_2, \hat{\mathbf{a}}_3$ and $\hat{\mathbf{b}}_1, \hat{\mathbf{b}}_2, \hat{\mathbf{b}}_3$.

The relative orientation from frame A to frame B can be described by a 3×3 rotation matrix with nine direction cosines called the direction cosine matrix (DCM):

$$ {}^{A}R^{B} = \begin{bmatrix} {}^{A}R^{B}_{11} & {}^{A}R^{B}_{12} & {}^{A}R^{B}_{13} \\ {}^{A}R^{B}_{21} & {}^{A}R^{B}_{22} & {}^{A}R^{B}_{23} \\ {}^{A}R^{B}_{31} & {}^{A}R^{B}_{32} & {}^{A}R^{B}_{33} \end{bmatrix} \tag{11.42} $$

Since the unit basis vectors $\hat{\mathbf{b}}_1, \hat{\mathbf{b}}_2, \hat{\mathbf{b}}_3$ are orthogonal, we can express $\hat{\mathbf{a}}_1, \hat{\mathbf{a}}_2, \hat{\mathbf{a}}_3$ in terms of them using the DCM in equation (11.42):

$$ \hat{\mathbf{a}}_1 = {}^{A}R^{B}_{11}\hat{\mathbf{b}}_1 + {}^{A}R^{B}_{12}\hat{\mathbf{b}}_2 + {}^{A}R^{B}_{13}\hat{\mathbf{b}}_3 \tag{11.43a} $$

$$ \hat{\mathbf{a}}_2 = {}^{A}R^{B}_{21}\hat{\mathbf{b}}_1 + {}^{A}R^{B}_{22}\hat{\mathbf{b}}_2 + {}^{A}R^{B}_{23}\hat{\mathbf{b}}_3 \tag{11.43b} $$

$$ \hat{\mathbf{a}}_3 = {}^{A}R^{B}_{31}\hat{\mathbf{b}}_1 + {}^{A}R^{B}_{32}\hat{\mathbf{b}}_2 + {}^{A}R^{B}_{33}\hat{\mathbf{b}}_3 \tag{11.43c} $$

which in matrix form is

$$ \begin{bmatrix} \hat{\mathbf{a}}_1 \\ \hat{\mathbf{a}}_2 \\ \hat{\mathbf{a}}_3 \end{bmatrix} = {}^{A}R^{B} \begin{bmatrix} \hat{\mathbf{b}}_1 \\ \hat{\mathbf{b}}_2 \\ \hat{\mathbf{b}}_3 \end{bmatrix} \tag{11.44} $$

An important property of direction cosine matrices is ${}^{b}R^{a} = ({}^{a}R^{b})^{-1} = ({}^{a}R^{b})^{T}$. Three elementary rotations about the first, second, and third axes of the reference frame A can be described by the following rotation matrices:

$$ R_1(\theta_1) = \begin{bmatrix} 1 & 0 & 0 \\ 0 & \cos(\theta_1) & \sin(\theta_1) \\ 0 & -\sin(\theta_1) & \cos(\theta_1) \end{bmatrix} \tag{11.45a} $$

$$R_2(\theta_2) = \begin{bmatrix} \cos(\theta_2) & 0 & -\sin(\theta_2) \\ 0 & 1 & 0 \\ \sin(\theta_2) & 0 & \cos(\theta_2) \end{bmatrix} \tag{11.45b}$$

$$R_3(\theta_3) = \begin{bmatrix} \cos(\theta_3) & \sin(\theta_3) & 0 \\ -\sin(\theta_3) & \cos(\theta_3) & 0 \\ 0 & 0 & 1 \end{bmatrix} \tag{11.45c}$$

where $R_i(\theta_i)$ denotes the direction cosine matrix R of a rotation about the ith axis of A with an angle θ_i.

The angular velocity ${}^A\vec{\omega}^B = \omega_1\hat{\mathbf{b}}_1 + \omega_2\hat{\mathbf{b}}_2 + \omega_3\hat{\mathbf{b}}_3$ can be expressed in terms of the derivatives of ${}^AR^B_{ij}$:

$$\omega_1 = {}^AR^B_{13}\,{}^A\dot{R}^B_{12} + {}^AR^B_{23}\,{}^A\dot{R}^B_{22} + {}^AR^B_{33}\,{}^A\dot{R}^B_{31} \tag{11.46a}$$

$$\omega_2 = {}^AR^B_{11}\,{}^A\dot{R}^B_{13} + {}^AR^B_{21}\,{}^A\dot{R}^B_{23} + {}^AR^B_{31}\,{}^A\dot{R}^B_{33} \tag{11.46b}$$

$$\omega_3 = {}^AR^B_{12}\,{}^A\dot{R}^B_{11} + {}^AR^B_{22}\,{}^A\dot{R}^B_{21} + {}^AR^B_{32}\,{}^A\dot{R}^B_{31} \tag{11.46c}$$

Euler Angles

Euler angles are the most common set of attitude parameters. We can describe the orientation of frame B with respect to frame A through three successive rotation angles $(\theta_1, \theta_2, \theta_3)$ called Euler angles. The order of the three successive rotations is important. We can define a sequence of rotations as the following:

$$\begin{bmatrix} \hat{\mathbf{a}}'_1 \\ \hat{\mathbf{a}}'_2 \\ \hat{\mathbf{a}}'_3 \end{bmatrix} = \begin{bmatrix} \cos(\theta_3) & \sin(\theta_3) & 0 \\ -\sin(\theta_3) & \cos(\theta_3) & 0 \\ 0 & 0 & 1 \end{bmatrix} \begin{bmatrix} \hat{\mathbf{a}}_1 \\ \hat{\mathbf{a}}_2 \\ \hat{\mathbf{a}}_3 \end{bmatrix} = R_3(\theta_3) \begin{bmatrix} \hat{\mathbf{a}}_1 \\ \hat{\mathbf{a}}_2 \\ \hat{\mathbf{a}}_3 \end{bmatrix} \tag{11.47a}$$

$$\begin{bmatrix} \hat{\mathbf{a}}''_1 \\ \hat{\mathbf{a}}''_2 \\ \hat{\mathbf{a}}''_3 \end{bmatrix} = \begin{bmatrix} \cos(\theta_2) & 0 & -\sin(\theta_2) \\ 0 & 1 & 0 \\ \sin(\theta_2) & 0 & \cos(\theta_2) \end{bmatrix} \begin{bmatrix} \hat{\mathbf{a}}'_1 \\ \hat{\mathbf{a}}'_2 \\ \hat{\mathbf{a}}'_3 \end{bmatrix} = R_2(\theta_2) \begin{bmatrix} \hat{\mathbf{a}}'_1 \\ \hat{\mathbf{a}}'_2 \\ \hat{\mathbf{a}}'_3 \end{bmatrix} \tag{11.47b}$$

$$\begin{bmatrix} \hat{\mathbf{b}}_1 \\ \hat{\mathbf{b}}_2 \\ \hat{\mathbf{b}}_3 \end{bmatrix} = \begin{bmatrix} 1 & 0 & 0 \\ 0 & \cos(\theta_1) & \sin(\theta_1) \\ 0 & -\sin(\theta_1) & \cos(\theta_1) \end{bmatrix} \begin{bmatrix} \hat{\mathbf{a}}''_1 \\ \hat{\mathbf{a}}''_2 \\ \hat{\mathbf{a}}''_3 \end{bmatrix} = R_1(\theta_1) \begin{bmatrix} \hat{\mathbf{a}}''_1 \\ \hat{\mathbf{a}}''_2 \\ \hat{\mathbf{a}}''_3 \end{bmatrix} \tag{11.47c}$$

where A' and A'' are two intermediate reference frames with the basis vectors $\hat{\mathbf{a}}'_1, \hat{\mathbf{a}}'_2, \hat{\mathbf{a}}'_3$ and $\hat{\mathbf{a}}''_1, \hat{\mathbf{a}}''_2, \hat{\mathbf{a}}''_3$, respectively.

We can combine this sequence of rotations to get the rotation matrix or direction cosine matrix in terms of the 3-2-1 Euler angles such that

$$\begin{bmatrix} \hat{\mathbf{b}}_1 \\ \hat{\mathbf{b}}_2 \\ \hat{\mathbf{b}}_3 \end{bmatrix} = {}^B R^A \begin{bmatrix} \hat{\mathbf{a}}_1 \\ \hat{\mathbf{a}}_2 \\ \hat{\mathbf{a}}_3 \end{bmatrix} \tag{11.48}$$

where

$$\begin{aligned} {}^B R^A &= R_1(\theta_1) R_2(\theta_2) R_3(\theta_3) \\ &= \begin{bmatrix} {}^B R^A_{11} & {}^B R^A_{12} & {}^B R^A_{13} \\ {}^B R^A_{21} & {}^B R^A_{22} & {}^B R^A_{23} \\ {}^B R^A_{31} & {}^B R^A_{32} & {}^B R^A_{33} \end{bmatrix} \\ &= \begin{bmatrix} c\theta_3 c\theta_2 & c\theta_2 s\theta_3 & -s\theta_2 \\ s\theta_1 s\theta_2 c\theta_3 - c\theta_1 s\theta_3 & s\theta_3 s\theta_2 s\theta_1 + c\theta_3 c\theta_1 & c\theta_2 s\theta_1 \\ c\theta_1 s\theta_2 c\theta_3 + s\theta_1 s\theta_3 & c\theta_1 s\theta_2 s\theta_3 - s\theta_1 c\theta_3 & c\theta_1 c\theta_2 \end{bmatrix} \end{aligned} \tag{11.49}$$

and $c\theta_i \triangleq \cos(\theta_i)$ and $s\theta_i \triangleq \sin(\theta_i)$.

The inverse transformation from the rotation matrix to the 3-2-1 Euler angles is

$$\theta_1 = \mathrm{a}\tan 2\left({}^B R^A_{23}, {}^B R^A_{33}\right) \tag{11.50a}$$

$$\theta_2 = -\mathrm{a}\sin\left({}^B R^A_{13}\right) \tag{11.50b}$$

$$\theta_3 = \mathrm{a}\tan 2\left({}^B R^A_{12}, {}^B R^A_{11}\right) \tag{11.50c}$$

In terms of 3-1-3 Euler angles, the rotation matrix from A to B is

$$\begin{aligned} {}^B R^A &= R_3(\theta_3) R_1(\theta_2) R_3(\theta_1) \\ &= \begin{bmatrix} c\theta_3 c\theta_1 - s\theta_3 c\theta_2 s\theta_1 & c\theta_3 s\theta_1 + s\theta_3 c\theta_2 c\theta_1 & s\theta_3 s\theta_2 \\ -s\theta_3 c\theta_1 - c\theta_3 c\theta_2 s\theta_1 & -s\theta_3 s\theta_1 + c\theta_3 c\theta_2 c\theta_1 & c\theta_3 s\theta_2 \\ s\theta_2 s\theta_1 & -s\theta_2 c\theta_1 & c\theta_2 \end{bmatrix} \end{aligned} \tag{11.51}$$

and the inverse transformation from the rotation matrix to the Euler angles are

$$\theta_1 = \mathrm{a}\tan 2\left({}^B R^A_{31}, -{}^B R^A_{32}\right) \tag{11.52a}$$

$$\theta_2 = \cos^{-1}\left({}^B R^A_{33}\right) \tag{11.52b}$$

$$\theta_3 = \mathrm{a}\tan 2\left({}^B R^A_{13}, {}^B R^A_{23}\right) \tag{11.52c}$$

For 3-2-1 Euler angles the singularity occurs for $\theta_2 = \pm 90°$ and for 3-1-3 Euler angles the singularity occurs for $\theta_2 = 0°$ or $\theta_2 = 180°$. The complete set of 12 transformations between the various Euler angle sets and direction cosine matrices can be derived. Note that each of the 12 possible sets of Euler angles has a geometric singularity.

The angular velocity can be written in terms of frame B as

$$^{A}\vec{\omega}^{B} = \omega_1\hat{\mathbf{b}}_1 + \omega_2\hat{\mathbf{b}}_2 + \omega_3\hat{\mathbf{b}}_3$$

and we can express the B frame rotation in terms of Euler angles (ϕ, θ, ψ) and Euler angle rates ($\dot{\phi}$, $\dot{\theta}$, $\dot{\psi}$) using the basis vectors for intermediate reference frames A' and A''

$$\begin{aligned} ^{A}\vec{\omega}^{B} &= \dot{\psi}\hat{\mathbf{a}} + \dot{\theta}\hat{\mathbf{a}}' + \dot{\phi}\hat{\mathbf{a}}'' \\ &= R_1(\phi)R_2(\theta)R_3(\psi)\begin{bmatrix} 0 \\ 0 \\ \dot{\psi} \end{bmatrix} + R_1(\phi)R_2(\theta)\begin{bmatrix} 0 \\ \dot{\theta} \\ 0 \end{bmatrix} + R_1(\phi)\begin{bmatrix} \dot{\phi} \\ 0 \\ 0 \end{bmatrix} \end{aligned} \tag{11.53}$$

and the kinematic differential equation for the 3-2-1 Euler angles is

$$\begin{bmatrix} \omega_1 \\ \omega_2 \\ \omega_3 \end{bmatrix} = \begin{bmatrix} 1 & 0 & -s\theta \\ 0 & c\phi & s\phi c\theta \\ 0 & -s\phi & c\phi c\theta \end{bmatrix}\begin{bmatrix} \dot{\phi} \\ \dot{\theta} \\ \dot{\psi} \end{bmatrix} \tag{11.54}$$

the inverse of equation (11.54) is

$$\begin{bmatrix} \dot{\phi} \\ \dot{\theta} \\ \dot{\psi} \end{bmatrix} = \frac{1}{c\theta}\begin{bmatrix} c\theta & s\phi s\theta & c\phi s\theta \\ 0 & c\phi c\theta & -s\phi c\theta \\ 0 & s\phi & c\phi \end{bmatrix}\begin{bmatrix} \omega_1 \\ \omega_2 \\ \omega_3 \end{bmatrix} \tag{11.55}$$

Similarly, the kinematic differential equation for the 3-1-3 Euler angle is

$$\begin{bmatrix} \omega_1 \\ \omega_2 \\ \omega_3 \end{bmatrix} = \begin{bmatrix} s\psi s\theta & c\psi & 0 \\ c\psi s\theta & -s\psi & 0 \\ c\theta & 0 & 1 \end{bmatrix}\begin{bmatrix} \dot{\phi} \\ \dot{\theta} \\ \dot{\psi} \end{bmatrix} \tag{11.56}$$

and the inverse relationship is

$$\begin{bmatrix} \dot{\phi} \\ \dot{\theta} \\ \dot{\psi} \end{bmatrix} = \frac{1}{s\theta}\begin{bmatrix} s\psi & c\psi & 0 \\ c\psi s\theta & -s\psi s\theta & 0 \\ -s\psi c\theta & -c\psi c\theta & s\theta \end{bmatrix}\begin{bmatrix} \omega_1 \\ \omega_2 \\ \omega_3 \end{bmatrix} \tag{11.57}$$

Again for 3-2-1 Euler angles the kinematic differential equation has a singularity for $\theta_2 = \pm 90°$ and for 3-1-3 Euler angles the singularity occurs for $\theta_2 = 0°$ or $\theta_2 = 180°$.

Quaternions

The orientation of frame B in A can be described by aligning $\hat{\mathbf{b}}_i$ with $\hat{\mathbf{a}}_i$ (for i = 1, 2, 3) and then subjecting B to a right-handed rotation in A of amount θ about a skew unit vector $\hat{\lambda}$ whose direction in both A and B remains constant throughout the orientation. The orientation of B in A can be defined in terms of this skew-vector rotation, where $\lambda_1, \lambda_2, \lambda_3$ are defined as

$$\lambda_1 = \hat{\lambda} \bullet \hat{\mathbf{a}}_1 = \hat{\lambda} \bullet \hat{\mathbf{b}}_1 \tag{11.58a}$$

$$\lambda_2 = \hat{\lambda} \bullet \hat{\mathbf{a}}_2 = \hat{\lambda} \bullet \hat{\mathbf{b}}_2 \tag{11.58b}$$

$$\lambda_3 = \hat{\lambda} \bullet \hat{\mathbf{a}}_3 = \hat{\lambda} \bullet \hat{\mathbf{b}}_3 \tag{11.58c}$$

where the unit vector $\hat{\lambda}$ is also known as the Euler axis or the Eigenaxis. So far we have seen that most methods for describing the orientation of two frames possess mathematical singularities. We can avoid this using a set of four scalars q_1, q_2, q_3, q_4 called the quaternions (or Euler parameters). Using the skew-vector rotation scalars, we can now define the quaternions as

$$q_1 = \lambda_1 \sin\left(\frac{\theta}{2}\right) \tag{11.59a}$$

$$q_2 = \lambda_2 \sin\left(\frac{\theta}{2}\right) \tag{11.59b}$$

$$q_3 = \lambda_3 \sin\left(\frac{\theta}{2}\right) \tag{11.59c}$$

$$q_4 = \cos\left(\frac{\theta}{2}\right) \tag{11.59d}$$

where q_1, q_2, q_3, q_4 must satisfy the constraint:

$$q_1^2 + q_2^2 + q_3^2 + q_4^2 = 1 \tag{11.60}$$

The orientation of B in A by quaternions q_1, q_2, q_3, q_4 results in the following direction cosine matrix:

$$\begin{bmatrix} \hat{\mathbf{a}}_1 \\ \hat{\mathbf{a}}_2 \\ \hat{\mathbf{a}}_3 \end{bmatrix} = \begin{bmatrix} 1-2q_2^2-2q_3^2 & 2(q_1q_2-q_3q_4) & 2(q_1q_3+q_2q_4) \\ 2(q_1q_2+q_3q_4) & 1-2q_1^2-2q_3^2 & 2(q_2q_3-q_1q_4) \\ 2(q_1q_3-q_2q_4) & 2(q_1q_4+q_2q_3) & 1-2q_1^2-2q_2^2 \end{bmatrix} \begin{bmatrix} \hat{\mathbf{b}}_1 \\ \hat{\mathbf{b}}_2 \\ \hat{\mathbf{b}}_3 \end{bmatrix} \tag{11.61}$$

If the orientation of B in A is described by the quaternions ${}^Aq_1^B, {}^Aq_2^B, {}^Aq_3^B, {}^Aq_4^B$ then

$${}^Aq_1^B = -{}^Bq_1^A \tag{11.62a}$$

$${}^Aq_2^B = -{}^Bq_2^A \tag{1162b}$$

$${}^Aq_3^B = -{}^Bq_3^A \tag{11.62c}$$

$${}^Aq_4^B = {}^Bq_4^A \tag{11.62d}$$

When the orientation of B in A is described by the quaternions ${}^Aq_1^B, {}^Aq_2^B, {}^Aq_3^B, {}^Aq_4^B$ and the orientation of C in B is described by the quaternions ${}^Bq_1^C, {}^Bq_2^C, {}^Bq_3^C, {}^Bq_4^C$ then the quaternions describing the orientation of C in A are

$$\begin{bmatrix} {}^Aq_1^C \\ {}^Aq_2^C \\ {}^Aq_3^C \\ {}^Aq_4^C \end{bmatrix} = \begin{bmatrix} {}^Aq_4^B & -{}^Aq_3^B & {}^Aq_2^B & {}^Aq_1^B \\ {}^Aq_3^B & {}^Aq_4^B & -{}^Aq_1^B & {}^Aq_2^B \\ -{}^Aq_2^B & {}^Aq_1^B & {}^Aq_4^B & {}^Aq_3^B \\ -{}^Aq_1^B & -{}^Aq_2^B & -{}^Aq_3^B & {}^Aq_4^B \end{bmatrix} \begin{bmatrix} {}^Bq_1^C \\ {}^Bq_2^C \\ {}^Bq_3^C \\ {}^Bq_4^C \end{bmatrix} \tag{11.63}$$

and

$$\begin{bmatrix} {}^Aq_1^C \\ {}^Aq_2^C \\ {}^Aq_3^C \\ {}^Aq_4^C \end{bmatrix} = \begin{bmatrix} {}^Bq_4^C & {}^Bq_3^C & -{}^Bq_2^C & {}^Bq_1^C \\ -{}^Bq_3^C & {}^Bq_4^C & {}^Bq_1^C & {}^Bq_2^C \\ {}^Bq_2^C & -{}^Bq_1^C & {}^Bq_4^C & {}^Bq_3^C \\ -{}^Bq_1^C & -{}^Bq_2^C & -{}^Bq_3^C & {}^Bq_4^C \end{bmatrix} \begin{bmatrix} {}^Aq_1^B \\ {}^Aq_2^B \\ {}^Aq_3^B \\ {}^Aq_4^B \end{bmatrix} \tag{11.64}$$

We can find the orientation of C in B or the difference between quaternions ${}^Aq_1^B, {}^Aq_2^B, {}^Aq_3^B, {}^Aq_4^B$ and ${}^Aq_1^C, {}^Aq_2^C, {}^Aq_3^C, {}^Aq_4^C$ using

$$\begin{bmatrix} {}^Bq_1^C \\ {}^Bq_2^C \\ {}^Bq_3^C \\ {}^Bq_4^C \end{bmatrix} = \begin{bmatrix} {}^Aq_4^B & {}^Aq_3^B & -{}^Aq_2^B & -{}^Aq_1^B \\ -{}^Aq_3^B & {}^Aq_4^B & {}^Aq_1^B & -{}^Aq_2^B \\ {}^Aq_2^B & -{}^Aq_1^B & {}^Aq_4^B & -{}^Aq_3^B \\ {}^Aq_1^B & {}^Aq_2^B & {}^Aq_3^B & {}^Aq_4^B \end{bmatrix} \begin{bmatrix} {}^Aq_1^C \\ {}^Aq_2^C \\ {}^Aq_3^C \\ {}^Aq_4^C \end{bmatrix} \tag{11.65}$$

The angular velocity of B in A can be expressed in terms of the quaternion derivatives $\dot{q}_1, \dot{q}_2, \dot{q}_3, \dot{q}_4$

$$\begin{bmatrix} \omega_1 \\ \omega_2 \\ \omega_3 \\ 0 \end{bmatrix} = \begin{bmatrix} q_4 & q_3 & q_2 & -q_1 \\ -q_3 & q_4 & q_1 & -q_2 \\ q_2 & -q_1 & q_4 & -q_3 \\ q_1 & q_2 & q_3 & q_4 \end{bmatrix} \begin{bmatrix} \dot{q}_1 \\ \dot{q}_2 \\ \dot{q}_3 \\ \dot{q}_4 \end{bmatrix} \tag{11.66}$$

For small, slow rotations the angular velocity components can be approximated as $\omega_1 \cong 2\dot{q}_1$, $\omega_2 \cong 2\dot{q}_2, \omega_3 \cong 2\dot{q}_3$.

The kinematical equations are

$$\begin{bmatrix} \dot{q}_1 \\ \dot{q}_2 \\ \dot{q}_3 \\ \dot{q}_4 \end{bmatrix} = \frac{1}{2} \begin{bmatrix} q_4 & -q_3 & q_2 & -q_1 \\ q_3 & q_4 & -q_1 & q_2 \\ -q_2 & q_1 & q_4 & q_3 \\ -q_1 & -q_2 & -q_3 & q_4 \end{bmatrix} \begin{bmatrix} \omega_1 \\ \omega_2 \\ \omega_3 \\ 0 \end{bmatrix} \tag{11.67}$$

which can be written as

$$\begin{bmatrix} \dot{q}_1 \\ \dot{q}_2 \\ \dot{q}_3 \\ \dot{q}_4 \end{bmatrix} = \frac{1}{2} \begin{bmatrix} 0 & \omega_3 & -\omega_2 & \omega_1 \\ -\omega_3 & 0 & \omega_1 & \omega_2 \\ \omega_2 & -\omega_1 & 0 & \omega_3 \\ -\omega_1 & -\omega_2 & -\omega_3 & 0 \end{bmatrix} \begin{bmatrix} q_1 \\ q_2 \\ q_3 \\ q_4 \end{bmatrix} \tag{11.68}$$

This is often written more compactly as

$$\dot{\vec{\mathbf{q}}} = \frac{1}{2}(q_4 \, {}^N\vec{\omega}^B - {}^N\vec{\omega}^B \times \vec{\mathbf{q}}) \tag{11.69a}$$

$$\dot{q}_4 = -\frac{1}{2}({}^N\vec{\omega}^B)^T \vec{\mathbf{q}} \tag{11.69b}$$

where

$$\vec{\mathbf{q}} = \begin{bmatrix} q_1 \\ q_2 \\ q_3 \end{bmatrix}$$

For small, slow rotations the kinematical equations can be approximated as $\dot{q}_1 \cong \frac{1}{2}\omega_1$, $\dot{q}_2 \cong \frac{1}{2}\omega_2, \dot{q}_3 \cong \frac{1}{2}\omega_3$. The values $\dot{q}_1, \dot{q}_2, \dot{q}_3, \dot{q}_4$ must satisfy the following motion constraint:

$$q_1\dot{q}_1 + q_2\dot{q}_2 + q_3\dot{q}_3 + q_4\dot{q}_4 = 0 \tag{11.70}$$

Converting between Orientation Methods

It is often convenient to convert from one orientation method to another. The quaternions associated with a set of 3-2-1 Euler angles (successive body-fixed rotations in the order of $\psi\hat{\mathbf{b}}_3, \theta\hat{\mathbf{b}}_2, \phi\hat{\mathbf{b}}_1$ can be calculated by the following equations:

$$q_1 = \sin\frac{\phi}{2}\cos\frac{\theta}{2}\cos\frac{\psi}{2} - \cos\frac{\phi}{2}\sin\frac{\theta}{2}\cos\frac{\psi}{2} \tag{11.71a}$$

$$q_2 = \cos\frac{\phi}{2}\sin\frac{\theta}{2}\cos\frac{\psi}{2} + \sin\frac{\phi}{2}\cos\frac{\theta}{2}\sin\frac{\psi}{2} \tag{11.71b}$$

$$q_3 = \cos\frac{\phi}{2}\cos\frac{\theta}{2}\sin\frac{\psi}{2} - \sin\frac{\phi}{2}\sin\frac{\theta}{2}\cos\frac{\psi}{2} \tag{11.71c}$$

$$q_4 = \cos\frac{\phi}{2}\cos\frac{\theta}{2}\cos\frac{\psi}{2} + \sin\frac{\phi}{2}\sin\frac{\theta}{2}\sin\frac{\psi}{2} \tag{11.71d}$$

Using equation (11.60), we can calculate q_1, q_2, q_3, q_4 from the DCM:

$$q_1 = (R_{23} - R_{32})/(4q_4) \tag{11.72a}$$

$$q_2 = (R_{31} - R_{13})/(4q_4) \tag{11.72b}$$

$$q_3 = (R_{12} - R_{21})/(4q_4) \tag{11.72c}$$

$$q_4 = \pm\frac{1}{2}\sqrt{1 + R_{11} + R_{22} + R_{33}} \tag{11.72d}$$

provided that $|q_4| = \frac{1}{2}\sqrt{1 + R_{11} + R_{22} + R_{33}}$ is not small.

Quaternions are an advantageous method for representing orientation between two reference frames because computations are faster and there are no singularities.

Spacecraft Orientation

The reference frames typically used in spacecraft attitude control are shown in Figure 11.6. An inertial reference frame N defined by orthogonal unit basis vectors $\hat{\mathbf{n}}_x, \hat{\mathbf{n}}_y, \hat{\mathbf{n}}_z$ is used to determine the orbital position of the spacecraft. This is sometimes also known as the earth centered inertial frame. The attitude motion of a spacecraft is most commonly described in terms of ϕ, θ, and ψ or roll, pitch, and yaw, respectively.

The nominal roll axis $\hat{\mathbf{o}}_x$ is along the orbit velocity vector; the nominal yaw axis $\hat{\mathbf{o}}_z$ is along the vector from the center of mass of the spacecraft to the center of mass of the earth; and the nominal pitch axis $\hat{\mathbf{o}}_y$ is normal to the orbit plane in such a way that the $\hat{\mathbf{o}}_x, \hat{\mathbf{o}}_y, \hat{\mathbf{o}}_z$ reference frame is a

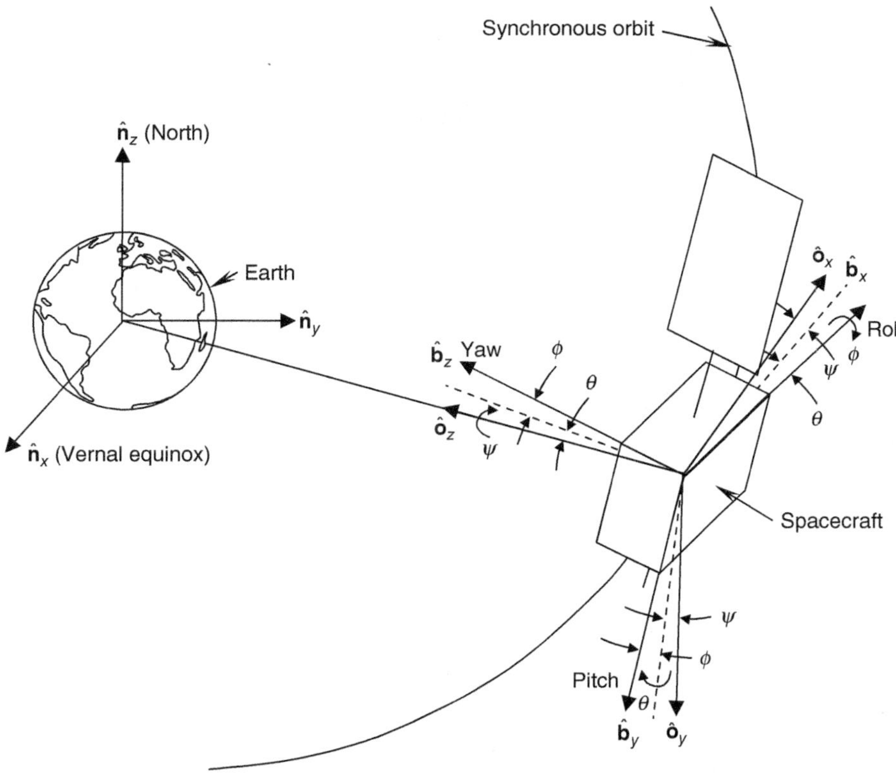

FIGURE 11.6 Reference frames in attitude control.

right-handed mutually orthonormal frame. The reference frame O defined by orthogonal unit basis vectors $\hat{\mathbf{o}}_x, \hat{\mathbf{o}}_y, \hat{\mathbf{o}}_z$ is also called the orbit reference frame. The origin of the orbit reference frame is at the center of mass of the spacecraft and is rotating with respect to the inertial reference frame $\hat{\mathbf{n}}_x, \hat{\mathbf{n}}_y, \hat{\mathbf{n}}_z$ at the orbit angular rate ω_o, which is one revolution per day for a geosynchronous spacecraft.

The perturbed attitude of the spacecraft body reference frame $\hat{\mathbf{b}}_x, \hat{\mathbf{b}}_y, \hat{\mathbf{b}}_z$ is obtained from the nominal attitude by the following rotations: ψ about $\hat{\mathbf{o}}_z$, θ about the once-displaced $\hat{\mathbf{o}}_y$, and ϕ about the twice-displaced $\hat{\mathbf{o}}_x$ axis. This orientation of the frame O to frame B is the 3-2-1 Euler rotation sequence given by

$$
{}^{B}R^{O} = \begin{bmatrix} c\theta c\phi & c\theta s\phi & -s\theta \\ s\phi s\theta c\psi - c\phi s\psi & c\psi c\phi + s\psi s\theta s\phi & c\theta s\phi \\ c\phi s\theta c\psi + s\phi s\psi & s\psi s\theta c\phi - c\psi s\phi & c\theta c\phi \end{bmatrix} \tag{11.73}
$$

11.4 Attitude Stabilization

Spacecraft attitude control systems can be classified into two broad categories: spin stabilization and three-axis stabilization. Spin stabilization is based on gyroscopic stiffness, which is the ability of the spinning spacecraft to resist internal or external disturbance torques that affect pointing.

There are two subsets of this class of spacecraft: single-spin stabilization and dual-spin stabilization. In single-spin stabilization, the whole body rotates about the axis of maximum or minimum principal moment of inertia. However, as will be shown, for spacecraft with energy dissipation elements, only the maximum moment of inertia is the stable spin axis. Early communications satellites, such as Syncom and Intelsat I, were single-spin stabilized. The primary limitations of these satellites were that they could not use earth-oriented antennas, thus resulting in the requirement for an omnidirectional antenna with a very low antenna gain toward the earth. This limitation is overcome with a dual-spin stabilized spacecraft, which is divided into two parts: the platform and the bus, each part rotating at different rates. The platform, consisting of antennas and communications equipment, orients toward the earth by rotating at one revolution per day for a geosynchronous orbit. The bus rotates at a higher spin rate, nominally at 60 rpm, to provide gyroscopic stiffness.

Spin Stabilization of a Rigid Spacecraft

In the absence of external moments, Euler's equations for a rigid body are

$$I_{xx}\,\dot{\omega}_x = (I_{yy} - I_{zz})\,\omega_y\,\omega_z \tag{11.74a}$$

$$I_{yy}\,\dot{\omega}_y = (I_{zz} - I_{xx})\,\omega_x\,\omega_z \tag{11.74b}$$

$$I_{zz}\,\dot{\omega}_z = (I_{xx} - I_{yy})\,\omega_x\,\omega_y \tag{11.74c}$$

where the body frame B is aligned along the axes of principal body axes of the spacecraft. It can be seen from equation (11.38a) through equation (11.38c) that pure rotation about any of the principal axes of inertia is possible. The question is which of the principal axes are the axes of stable spin. Let us assume that the spacecraft spins uniformly about the $\hat{\mathbf{b}}_y$ axis, the pitch axis, and allow a small perturbation from that motion to occur. The pitch axis is the preferred spin axis by spacecraft designers because it can be fixed inertially. The initial motion can be described by $\omega_y = \omega_S, \omega_x = \omega_z = 0$. The perturbed motion can be described by $\omega_y = \omega_S + \varepsilon, \omega_x \neq 0, \omega_z \neq 0$, where ω_x and ω_z are small. The resulting linearized equations are

$$I_{xx}\,\dot{\omega}_x = (I_{yy} - I_{zz})\,\omega_S\omega_z \tag{11.75a}$$

$$I_{yy}\dot{\varepsilon} = 0 \tag{11.75b}$$

$$I_{zz}\,\dot{\omega}_z = (I_{xx} - I_{yy})\,\omega_S\omega_x \tag{11.75c}$$

From equation (11.74b), we concluded that ε remains constant. Differentiating equations (11.74a) and (11.74c), and substituting $\dot{\omega}_z$ and $\dot{\omega}_x$ from equations (11.74c) and (11.74a), respectively, we obtain

$$\ddot{\omega}_x + \frac{(I_{yy} - I_{zz})(I_{yy} - I_{xx})}{I_{xx}I_{zz}}\,\omega_S^2\,\omega_x = 0 \tag{11.76a}$$

$$\ddot{\omega}_z + \frac{(I_{yy} - I_{xx})(I_{yy} - I_{zz})}{I_{xx}I_{zz}}\,\omega_S^2\omega_z = 0 \tag{11.76b}$$

For stability, the coefficient of ω_x and ω_z in equation (11.75a, b) must be positive. Hence, the stability conditions are (1) $I_{yy} > I_{xx}, I_{zz}$, in which case the pitch axis is the axis of the maximum moment of inertia, or (2) $I_{yy} < I_{zz}, I_{xx}$, in which case the pitch axis is the minimum moment of inertia. Hence, for a perfectly rigid spacecraft, stable spin can take place about the axis of the maximum moment of inertia or about the axis of the minimum moment of inertia. The axis of the intermediate moment of inertia is unstable.

Let's consider an axisymmetric spacecraft such that $I_{yy} = I_S$ and $I_{xx} = I_{zz} = I_T$. Then equation (11.76a, b) can be reduced to

$$\ddot{\omega}_x + \left(\left(\frac{I_S}{I_T} - 1\right)\omega_S\right)^2 \omega_x = 0 \tag{11.77a}$$

$$\ddot{\omega}_z + \left(\left(\frac{I_S}{I_T} - 1\right)\omega_S\right)^2 \omega_z = 0 \tag{11.77b}$$

From equation (11.77a, b), the solution to the second-order differential equations is of the form of undamped oscillators such that

$$\omega_x = A_1 \cos(\lambda t) + B_1 \sin(\lambda t) \tag{11.78a}$$

$$\omega_z = A_2 \cos(\lambda t) + B_2 \sin(\lambda t) \tag{11.78b}$$

where

$$\lambda = \left(\frac{I_S}{I_T} - 1\right)\omega_S = (\sigma - 1)\omega_S \tag{11.79}$$

and A_i and B_i $(i = 1, 2)$ are the amplitudes, σ is the inertia ratio I_s/I_T. Let $\omega_{x,0}$ and $\omega_{z,0}$ be the initial body angular velocity, then the constants A_1 and A_2 must be

$$A_1 = \omega_{x,0} \qquad A_2 = \omega_{z,0} \tag{11.80}$$

Differentiating equation (11.78a, b) and substituting into equation (11.76a, b), we find

$$B_1 = A_2 \qquad B_2 = -A_1 \tag{11.81}$$

We can find a closed-form solution to the body angular velocity for the axially symmetric, torque free case:

$$\omega_x(t) = \omega_{x,0}\cos(\lambda t) + \omega_{z,0}\sin(\lambda t) \tag{11.82a}$$

$$\omega_y(t) = \omega_{y.0} \tag{11.82b}$$

$$\omega_z(t) = \omega_{z,0}\cos(\lambda t) - \omega_{x,0}\sin(\lambda t) \tag{11.82c}$$

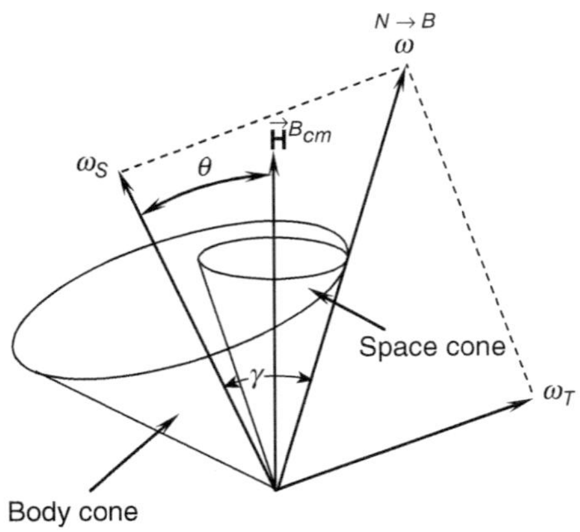

FIGURE 11.7 Nutational motion for a disk-shaped body, $I_S/I_T > 1$.

Nutational motion can be represented in terms of the body cone and the space cone, as shown in Figures 11.7 and 11.8. The body cone is fixed in the body and its axis coincides with the spin axis. The space cone is fixed in space and its axis is along the angular momentum vector.

For a disk-shaped body, $I_S/I_T > 1$, the inside surface of the body cone rolls on the outside surface of the space cone, as shown in Figure 11.7. For a rod-shaped body, $I_S/I_T < 1$, the outside surface of the body cone rolls on the outside surface of the space cone as shown in Figure 11.8. The nutation angle, θ, and the angle, γ, are given by

$$\theta = \text{a}\sin(I_T\omega_T, H) \quad \text{and} \quad \gamma = \text{a}\tan 2(\omega_T, \omega_S) \tag{11.83}$$

The nutational frequency is

$$\omega_{NUT} = \frac{I_S}{I_T}\omega_S \tag{11.84}$$

In the inertial reference frame, the transverse component of the angular velocity, ω_T, is rotating with the angular velocity, ω_{NUT}, because the $\hat{\mathbf{b}}_x$ and $\hat{\mathbf{b}}_z$ axes are rotating with angular velocity ω_S.

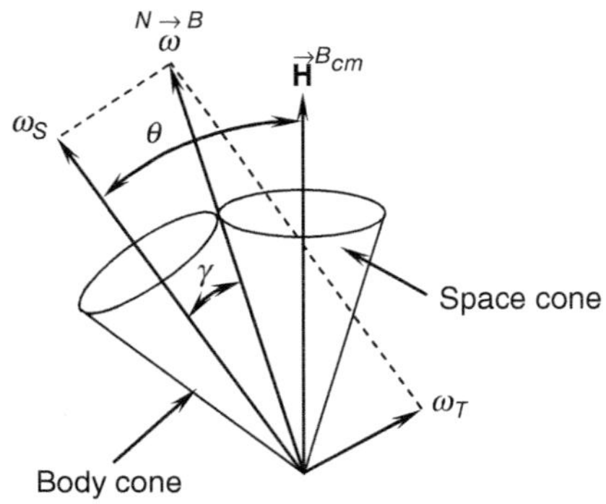

FIGURE 11.8 Nutational motion for a rod-shaped body, $I_S/I_T < 1$.

Because the angular momentum vector is fixed in space, the angular momentum component $I_S\,\omega_S$ is also rotating with the angular velocity ω_{NUT}. The spacecraft motion consists of the spacecraft rotation about its spin axis and the spin axis rotating about the angular momentum vector with angular velocity ω_{NUT}. This latter motion is called "nutational motion."

11.5 Spin Stabilization of an Energy-Dissipating Spacecraft

Single-Spin Stabilization

In single-spin stabilization, the entire spacecraft is spinning about a principal moment of inertia axis. For a rigid spacecraft, stable spin can take place about the axis of either the maximum or the minimum moment of inertia. However, a spacecraft is likely to have several components experiencing relative motion, so the satellite is not rigid. In the absence of external torques, the angular momentum of the spacecraft is constant, but the internal energy dissipation will cause the spacecraft to move toward a minimum-energy state. Assuming that the potential energy is constant, the spacecraft will tend to reach a minimum kinetic-energy state. The kinetic energy of spacecraft T is related to the magnitude of angular momentum H and to the moment of inertia I_S about the spin axis by

$$T = \frac{H^2}{2\,I_S} \tag{11.85}$$

For a constant angular momentum H, the kinetic energy is minimum when the moment of inertia I_S is maximum. Hence, for a nonrigid spacecraft there is only one axis of stable spin, namely, the axis of the maximum moment of inertia.

In the case of energy dissipation, the nutation angle θ is no longer constant. The rate of change of the angle θ can be related to energy dissipation by assuming that the moments of inertia do not vary significantly as a result of the relative motion within the spacecraft and that the angular momentum of the relative motion is negligible compared to the rigid angular momentum of the motion. The kinetic energy T and the angular momentum magnitude H are given by

$$T = \frac{1}{2}\,[I_T(\omega_x^2 + \omega_z^2) + I_S\omega_S^2] \tag{11.86}$$

$$H^2 = I_T^2(\omega_x^2 + \omega_z^2) + I_S^2\omega_S^2 \tag{11.87}$$

The nutation angle θ is given by equation (11.83) and is no longer constant. Inserting equations (11.86) and (11.87) into equation (11.83), we obtain

$$\sin^2\theta = \frac{I_T}{(I_S - I_T)\,H^2}\,(2\,I_S\,T - H^2) \tag{11.88}$$

Differentiating equation (11.88) with respect to time, we have

$$\dot{\theta} = \frac{1}{\sin 2\theta}\,\frac{2\,I_T\,I_S}{(I_S - I_T)}\,\frac{\dot{T}}{H^2} = \frac{2\,I_T\,\omega_{NUT}\,\dot{T}}{\sin 2\theta\,\lambda\,H^2} \tag{11.89}$$

Equation (11.89) gives the rate of change of the nutation angle in terms of the energy dissipation. Because $\dot{T}$ is negative, the nutation angle decreases only if $I_S > I_T$. This confirms that for a nonrigid spacecraft, stable spin takes place only about the axis of maximum moment of inertia. The analysis is sometimes known as the energy-sink approach.

Dual-Spin Stabilization

The limitation of single-spin stabilization is overcome in a dual-spin-stabilized spacecraft, which consists of a rotor providing gyroscopic stabilization and a platform pointing toward the earth. Iorillo (1965) presented a stability criterion for dual-spin-stabilized spacecraft in 1965. In a dual-spin-stabilized spacecraft, the stable spin axis can be the axis of minimum moment of inertia if the rate of energy dissipation in the platform is higher than that in the rotor by a certain factor. The spin axis for the dual-spin stabilized spacecraft, Intelsat IV, is the axis of minimum moment of inertia. Likins (1967) and other investigators have proved the stability criterion of Iorillo in a rigorous manner.

11.6 Three-Axis Stabilization

The control torques for a three-axis stabilized system are provided by a combination of momentum wheels, reactions wheels, control moment gyros, magnetic torquers, and thrusters. Broadly, however, there are two types of three-axis stabilized systems: a momentum biased system and a zero momentum system.

In a momentum biased system, the angular momentum along the pitch axis provides gyroscopic stiffness. In these systems, the pitch and roll axes are controlled directly and the yaw axis is controlled indirectly due to gyroscopic coupling of yaw and roll errors, thus eliminating the need for a yaw sensor. In zero-momentum systems, all three axes are controlled independently, thus requiring a yaw sensor also. For earth-pointing geosynchronous spacecraft, the momentum biased system has been commonly used. A linearized analysis of the three-axis stabilization system is given in this section. The linearized equations of motion of a three-axis stabilized spacecraft are derived first. The angular velocities of the spacecraft can be expressed in terms of the orbital rate ω_0 and the Euler angles ψ, θ, and ϕ, which are now known as yaw, pitch, and roll *errors*, respectively. The angular rates of the three-axis stabilized system are

$$^{N}\vec{\omega}^{B} = {}^{N}\vec{\omega}^{O} + {}^{O}\vec{\omega}^{B} \tag{11.90}$$

Using the direction cosine matrices we derived in the subsection "Direction Cosines," we have

$$\begin{bmatrix}\omega_x\\ \omega_y\\ \omega_z\end{bmatrix} = R_1(\phi)\begin{bmatrix}\dot{\phi}\\ 0\\ 0\end{bmatrix} + R_1(\phi)R_1(\theta)\begin{bmatrix}0\\ \dot{\theta}\\ 0\end{bmatrix} + R_1(\phi)R_2(\theta)R_3(\psi)\begin{bmatrix}0\\ 0\\ \dot{\psi}\end{bmatrix} + R_1(\phi)R_2(\theta)R_3(\psi)\begin{bmatrix}0\\ \omega_0\\ 0\end{bmatrix}$$

$$= \begin{bmatrix}1\\ 0\\ 0\end{bmatrix}\dot{\phi} + \begin{bmatrix}0\\ \cos\phi\\ -\sin\phi\end{bmatrix}\dot{\theta} + \begin{bmatrix}-\sin\theta\\ \sin\phi\cos\theta\\ \cos\phi\cos\theta\end{bmatrix}\dot{\psi} - \begin{bmatrix}\cos\theta\sin\psi\\ \cos\phi\cos\psi + \sin\phi\sin\theta\sin\psi\\ -\sin\phi\cos\psi + \cos\phi\sin\theta\sin\psi\end{bmatrix}\omega_0$$

Using the small-angle approximation such that $\sin\alpha = \alpha$, $\cos\alpha = 1$, and ignoring the coupling terms we have

$$\begin{bmatrix}\omega_x\\ \omega_y\\ \omega_z\end{bmatrix} \cong \begin{bmatrix}\dot{\phi} - \omega_0\,\psi\\ \dot{\theta} - \omega_0\\ \dot{\psi} + \omega_0\,\phi\end{bmatrix} \tag{11.91}$$

Let's consider a system S that consists of the spacecraft rigid body and a actuator body. The angular momentum of the system S can be written in the form

$$^{B}\vec{\mathbf{H}}^{S/S_{cm}} = H_i\hat{\mathbf{b}}_i + \sum_{k=1}^{n}\vec{\mathbf{h}}_k \bullet \hat{\mathbf{b}}_i \quad (\text{for } i = x, y, z) \tag{11.92}$$

where the angular momentum of the spacecraft body is

$$H_i \hat{\mathbf{b}}_i = I^N \vec{\omega}^B \bullet \hat{\mathbf{b}}_i = [\, I_{xx}\, \omega_x \quad I_{yy}\, \omega_y \quad I_{zz}\, \omega_z \,]^T \tag{11.93}$$

The angular momentum of the momentum wheels is

$$\sum_{k=1}^{n} \vec{\mathbf{h}}_k \bullet \hat{\mathbf{b}}_i = [\, h_x \quad h_y \quad h_z \,]^T \tag{11.94}$$

where n is number of momentum wheels, $h_i = J\omega_i$ (for $i = x, y$, and z) is the angular momentum of momentum exchange device in the B frame, and J is the spin axis inertia of the momentum wheel.

Using (11.35) the equation of motion for the system, S, with momentum wheels is

$$\vec{\mathbf{M}}_C \bullet \hat{\mathbf{b}}_i = (I^N \vec{\alpha}^B + {}^N\vec{\omega}^B \times I^N \vec{\omega}^B - \vec{\tau}) \bullet \hat{\mathbf{b}}_i \;\text{ for } (i = x, y, z) \tag{11.95}$$

where $\vec{\mathbf{M}}_C$ is the external control torque from thrusters or torquers on system, S, and $\vec{\tau}$ is the torque produced by the momentum wheels onto system S. Using equation (11.95) in conjunction with equations (11.91) and (11.92) while ignoring nonlinear terms, we obtain the equations of motion:

$$\begin{bmatrix} M_{C_x} \\ M_{C_y} \\ M_{C_z} \end{bmatrix} \cong \begin{bmatrix} \dot{h}_x + I_{xx}\,(\ddot{\phi} - \omega_0\, \dot{\psi}) + h_z\,(\dot{\theta} - \omega_0) - h_y\,(\dot{\psi} + \omega_0\, \phi) \; - (I_{zz} - I_{yy})\, \omega_0\,(\dot{\psi} + \omega_0\, \phi) \\ \dot{h}_y + I_{yy}\, \ddot{\theta} + h_x\,(\dot{\psi} + \omega_0\, \phi) - h_z\,(\dot{\phi} - \omega_0\, \psi) \\ \dot{h}_z + I_{zz}\,(\ddot{\psi} - \omega_0\, \dot{\phi}) + h_y\,(\dot{\phi} - \omega_0\, \psi) - h_x\,(\dot{\theta} - \omega_0) - (I_{yy} - I_{xx})\, \omega_0\,(\dot{\phi} - \omega_0\, \psi) \end{bmatrix} \tag{11.96}$$

11.7 Disturbance Torques

Disturbance torques or external torques must be considered when sizing and designing the attitude control system of a spacecraft. The main disturbance torques on a spacecraft are due to the environment effects from the gravity gradient, solar radiation pressure, magnetic field, aerodynamic drag which we can express as $\vec{\mathbf{M}}_{GG}$, $\vec{\mathbf{M}}_{SRP}$, $\vec{\mathbf{M}}_{MAG}$, and $\vec{\mathbf{M}}_{DRAG}$, respectively. The total disturbance torque on the spacecraft body B taken with respect to its center of mass B_{cm} is

$$\vec{\mathbf{M}}_D = \vec{\mathbf{M}}_{GG} + \vec{\mathbf{M}}_{SRP} + \vec{\mathbf{M}}_{MAG} + \vec{\mathbf{M}}_{DRAG} \tag{11.97}$$

The equations for the gravity-gradient, solar radiation pressure, and magnetic field torques will be derived in the following subsections.

Gravity-Gradient Disturbance Torque

A spacecraft body B experiences a gravity-gradient torque due to the variation of distances between the spacecraft center of mass and the center of mass of the earth. The gravity gradient has been used on early-low-orbit satellites to maintain the earth pointing of antennas or other instruments.

The net force on body B, due to the gravitational potential of a primary body E, is given by (Schaub and Jenkins 2003)

$$\vec{\mathbf{F}}_{GG}^{B} = -\mu_E \int_B \frac{\vec{\mathbf{p}}^{B_{dm}/E_{cm}}}{\left|\vec{\mathbf{p}}^{B_{dm}/E_{cm}}\right|^3} dm \tag{11.98}$$

where $\vec{\mathbf{p}}^{B_{dm}/E_{cm}} = \vec{\mathbf{r}}^{B_{cm}/E_{cm}} + \vec{\mathbf{r}}^{B_{dm}/B_{cm}}$.

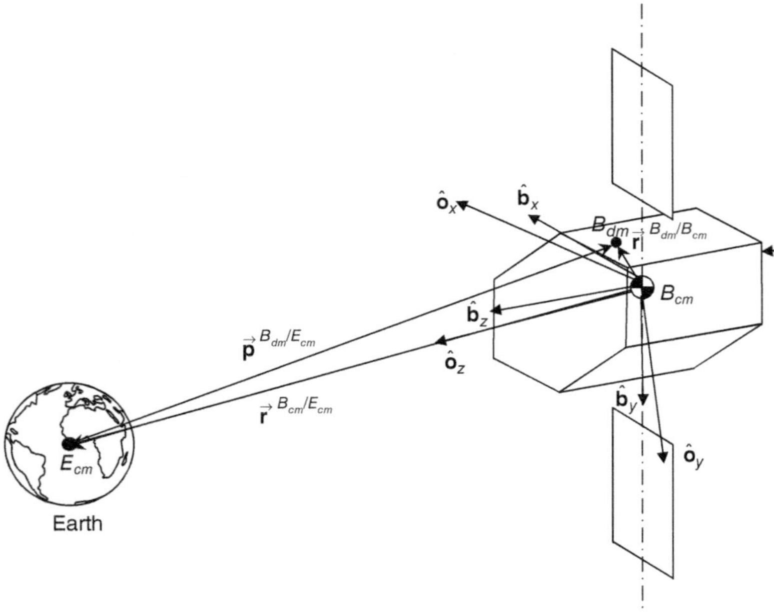

FIGURE 11.9 Gravity force on infinitesimal mass element.

As shown in Figure 11.9, $\vec{\mathbf{p}}^{B_{dm}/E_{cm}}$ is the position vector from the primary body E's center of mass to the infinitesimal mass element dm, $\vec{\mathbf{r}}^{B_{cm}/E_{cm}}$ is the position vector from the center of mass of E to the center of mass of B, and $\vec{\mathbf{r}}^{B_{dm}/B_{cm}}$ is the position vector from the center of mass of B to the infinitesimal mass element dm on the spacecraft body. If the body E is the earth then, $\mu_E = GM_E = 3.986\times 10^{14}\, m^3 s^{-2}$, where G is the gravitational constant and M_E is the mass of the earth.
The gravity gradient torque is

$$\vec{\mathbf{M}}_{GG} = -\mu_E \int_B \vec{\mathbf{r}}^{B_{dm}/B_{cm}} \times \frac{(\vec{\mathbf{r}}^{B_{dm}/B_{cm}} + \vec{\mathbf{r}}^{B_{cm}/E_{cm}})}{\left|\vec{\mathbf{p}}^{B_{dm}/E_{cm}}\right|^3} dm \tag{11.99}$$

where we can use the binomial expansion up to first-order terms on the denominator of equation (11.99) to obtain

$$\vec{\mathbf{M}}_{GG} = -\mu_E \int_B \left[\vec{\mathbf{r}}^{B_{dm}/E_{cm}} \times \left(\frac{\vec{\mathbf{r}}^{B_{dm}/B_{cm}} + \vec{\mathbf{r}}^{B_{cm}/E_{cm}}}{r_E^{\,3}} \left(1 - 3\frac{\vec{\mathbf{r}}^{B_{dm}/B_{cm}} \bullet \vec{\mathbf{r}}^{B_{cm}/E_{cm}}}{r_E^{\,2}} \right) \right) dm \right] \tag{11.100}$$

where $r_E = \left\| \vec{\mathbf{r}}^{B_{cm}/E_{cm}} \right\|$. Carrying out the integral in equation (11.100), we obtain the gravity-gradient torque:

$$\vec{\mathbf{M}}_{GG} = \frac{3\mu_E}{r_E^{\,5}} \vec{\mathbf{r}}^{B_{cm}/E_{cm}} \times \left(\begin{bmatrix} I_{xx} & I_{xy} & I_{xz} \\ I_{yx} & I_{yy} & I_{yz} \\ I_{zx} & I_{zy} & I_{zz} \end{bmatrix} \vec{\mathbf{r}}^{B_{cm}/E_{cm}} \right) \tag{11.101}$$

If we align the orbit frame O as done previously in the subsection "Direction Cosines" then the position vector $\vec{\mathbf{r}}^{B_{cm}/E_{cm}}$ is

$$\vec{\mathbf{r}}^{B_{cm}/E_{cm}} = r_E \hat{\mathbf{o}}_z \tag{11.102}$$

We can use ${}^{B}R^{O}$ defined in equation (11.73) to get $\vec{\mathbf{r}}^{B_{cm}/E_{cm}}$ in terms of the spacecraft body frame:

$$\vec{\mathbf{r}}^{B_{cm}/E_{cm}} = -r_E \sin\theta \hat{\mathbf{b}}_x + r_E \sin\phi \cos\theta \hat{\mathbf{b}}_y + r_E \cos\phi \cos\theta \hat{\mathbf{b}}_z \tag{11.103}$$

Substituting equation (11.103) into equation (11.101), the gravity gradient torque in terms of 3-2-1 Euler angles:

$$\begin{bmatrix} M_{GG_x} \\ M_{GG_y} \\ M_{GG_z} \end{bmatrix} = \frac{3\mu_E}{r_E^3} \begin{bmatrix} (I_{zz} - I_{yy}) \sin\phi \cos\phi \cos^2\theta \\ (I_{zz} - I_{xx}) \sin\theta \cos\theta \cos\phi \\ (I_{xx} - I_{yy}) \sin\theta \cos\theta \sin\phi \end{bmatrix} \tag{11.104}$$

where r_E is the orbit radius and the orbital angular rate is given by

$$\omega_o = \sqrt{\frac{\mu_E}{r_E^3}} \tag{11.105}$$

Inserting equation (11.105) into equation (11.104) and assuming that the angles θ and ϕ are small, the gravity-gradient torque is reduced to

$$\begin{bmatrix} M_{GG_x} \\ M_{GG_y} \\ M_{GG_z} \end{bmatrix} \cong 3\,\omega_0^2 \begin{bmatrix} \phi\,(I_{zz} - I_{yy}) \\ \theta\,(I_{zz} - I_{xx}) \\ 0 \end{bmatrix} \tag{11.106}$$

It should be noted that if the moment of inertia about the yaw axis I_{zz} is the minimum inertia axis, then the gravity gradient will provide a stabilizing torque along the roll and pitch axes. However, there is no gravity gradient torque along yaw axis. Therefore, for a spacecraft that is passively stabilized by gravity gradient torques, we need to actively control along the yaw axis.

Solar Radiation Pressure Disturbance Torque

The impingement of solar radiation on an opaque surface generates pressure, which can exert a force on the spacecraft body. If the flux created by impinging photons on the spacecraft surface is arrested by the material then a fraction ρ_s of the impinging photons will be specularly reflected. Another fraction ρ_d will be diffusely reflected, and another fraction ρ_a will be absorbed by the surface. For an opaque surface:

$$\rho_s + \rho_d + \rho_a = 1 \tag{11.107}$$

Figure 11.10 shows the solar radiation force due to the absorbed photons, which can be computed as (Agrawal 1986)

$$\begin{aligned} \vec{\mathbf{F}}^{B}_{SRP_a} &= \int_A \rho_a P(\hat{\mathbf{s}} \cdot \hat{\mathbf{n}})\hat{\mathbf{s}}\, dA \\ &= \int_A \rho_a P \cos(\theta_i)\hat{\mathbf{s}}\, dA \end{aligned} \tag{11.108}$$

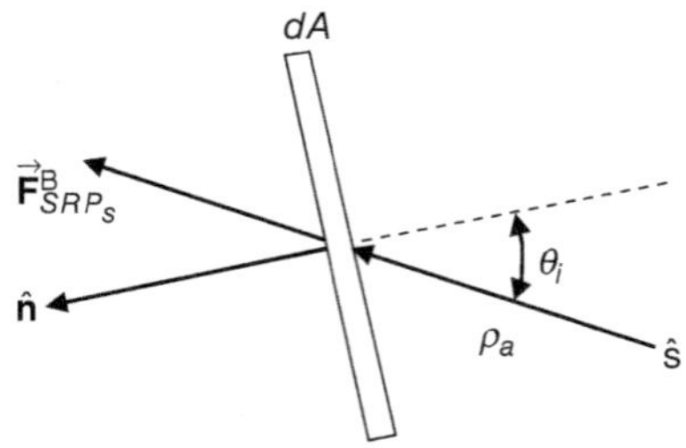

FIGURE 11.10 Solar radiation pressure due to absorbed photons.

where $P = W/c = 4.644 \times 10^{-6} Nm^2$ is the pressure that acts on the surface from the sun, W is the flux density for the sun, and c is the speed of light

dA = the infinitesimal surface area on the spacecraft

$\hat{\mathbf{s}}$ = the unit vector from the spacecraft to the sun

$\hat{\mathbf{n}}$ = the unit vector normal to the infinitesimal surface area

$\hat{\mathbf{s}} \bullet \hat{\mathbf{n}} = \|\hat{\mathbf{s}}\|\|\hat{\mathbf{n}}\| \cos\angle(\hat{\mathbf{s}} \bullet \hat{\mathbf{n}}) = \cos(\theta_i)$, where θ_i is the incident angle from the photons

The radiation force due to specularly reflected photons as shown in Figure 11.11 can be calculated with

$$\vec{\mathbf{F}}^{B}_{SRP_s} = \int_A 2\rho_s P \cos^2(\theta_i)\hat{\mathbf{n}}\, dA \tag{11.109}$$

As shown in Figure 11.12, a fraction ρ_d of the incoming photons is assumed to be diffusely reflected. The incoming photon's momentum may be considered as stopped at the surface and sequentially radiated uniformly into the hemisphere.

Integrating the momentum corresponding to the hemisphere, it can be seen that the tangential component of outgoing momentum is canceled due to symmetry, leaving the component normal to the surface. The total force caused by diffusely reflected photons is

$$\vec{\mathbf{F}}^{B}_{SRP_d} = \int_A \rho_d P \cos(\theta_i)\left(\hat{\mathbf{s}} + \frac{2}{3}\hat{\mathbf{n}}\right) dA \tag{11.110}$$

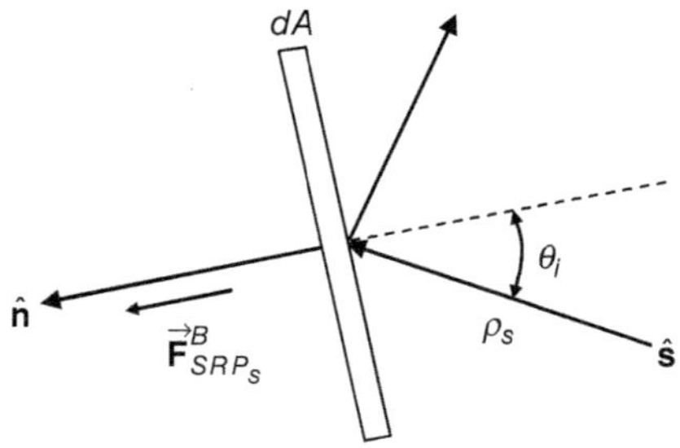

FIGURE 11.11 Solar radiation pressure due to specular photons.

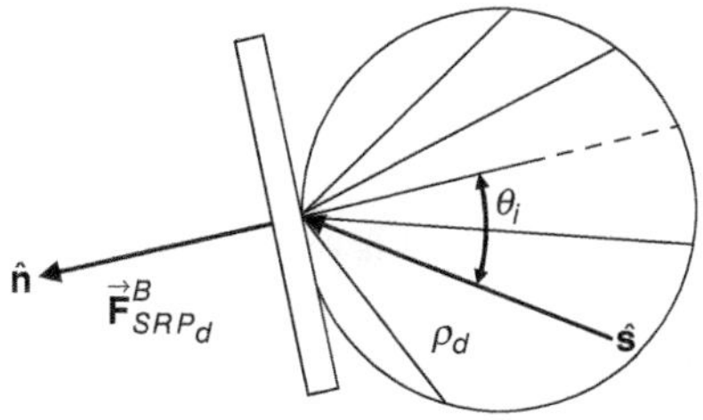

FIGURE 11.12 Solar radiation pressure due to diffusely reflected photons.

Combining equations (11.108), (11.109), and (11.110), the total force due to solar radiation pressure is

$$\vec{\mathbf{F}}^{B}_{SRP} = \int_A P\cos(\theta_i)\left[(\rho_d+\rho_a)\hat{\mathbf{s}}+\left(2\rho_s\cos(\theta_i)+\frac{2}{3}\rho_d\right)\hat{\mathbf{n}}\right]dA \tag{11.111}$$

Noting that the surface area A is flat, the solar radiation force is

$$\vec{\mathbf{F}}^{B}_{SRP} = P\cos(\theta_i)\left[(1-\rho_s)\hat{\mathbf{s}}+\left(2\rho_s\cos(\theta_i)+\frac{2}{3}\rho_d\right)\hat{\mathbf{n}}\right]A \tag{11.112}$$

If we assume that the spacecraft is lumped into a single surface A, whose orientation $\hat{\mathbf{n}}$ is fixed in the inertial frame, then the moment due to the solar radiation force is

$$\begin{aligned}\vec{\mathbf{M}}_{SRP} &= \vec{\mathbf{r}}^{B_{cp}/B_{cm}} \times \vec{\mathbf{F}}^{B}_{SRP}\\ &= \vec{\mathbf{r}}^{B_{cp}/B_{cm}} \times \left\{P\cos(\theta_i)\left[(1-\rho_s)\hat{\mathbf{s}}+\left(2\rho_s\cos(\theta_i)+\frac{2}{3}\rho_d\right)\hat{\mathbf{n}}\right]A\right\}\end{aligned} \tag{11.113}$$

where $\vec{\mathbf{r}}^{B_{cp}/E_{cm}}$ is the position vector from the center of mass of the spacecraft to the center of pressure of area A.

Magnetic Disturbance Torques

The magnetic disturbance torque on a spacecraft is influenced by the orbital altitude, the magnetic dipole moment, and earth's magnetic field. Assuming the magnetic field of the earth is due to a magnetic dipole vector $\vec{\mathbf{m}}$ it can be calculated as (Wertz 1978)

$$\vec{\mathbf{B}}_o = \frac{\mu_D}{r_C^3}\left(3(\hat{\mathbf{m}}\cdot\hat{\mathbf{r}})\hat{\mathbf{r}}\cdot\hat{\mathbf{m}}\right) \tag{11.114}$$

where $\mu_D = 7.943\times 10^{15}\,\mathrm{Nm^2A^{-1}}$ is the magnetic dipole strength of the earth
r_C = the radius of the circular orbit measured from the center of earth
$\hat{\mathbf{m}}$ = the unit vector in the direction of the magnetic dipole
$\hat{\mathbf{r}}$ = the unit vector at which point the field is desired

The earth's magnetic field components with respect to a circular orbit reference frame are given by

$$\begin{bmatrix} B_{o_x}\\ B_{o_y}\\ B_{o_z}\end{bmatrix} = \frac{\mu_D}{r_C^3}\begin{bmatrix}\cos(\upsilon)[\cos(\varepsilon)\sin(i)-\sin(\varepsilon)\cos(i)\cos(u)]-\sin(\upsilon)\sin(\varepsilon)\sin(u)\\ -\cos(\varepsilon)\cos(i)-\sin(\varepsilon)\sin(i)\cos(u)\\ 2\cos(\upsilon)[\cos(\varepsilon)\sin(i)-\sin(\varepsilon)\cos(i)\cos(u)]+2\cos(\upsilon)\sin(\varepsilon)\sin(u)\end{bmatrix} \tag{11.115}$$

where υ = true anomaly of the spacecraft
ε = magnetic dipole tilt from the North Pole
i = orbit inclination
u = right ascension angle of the magnetic dipole with respect to right ascension of the orbit normal

The earth's magnetic field vector in body frame B is given by

$$\begin{bmatrix} B_x \\ B_y \\ B_z \end{bmatrix} = {}^{B}R^{O} \begin{bmatrix} B_{o_x} \\ B_{o_y} \\ B_{o_z} \end{bmatrix} \tag{11.116}$$

The interaction between the magnetic moment created by the spacecraft and earth's magnetic field will create a magnetic disturbance torque given by

$$\vec{\mathbf{M}}_{MAG} = \vec{\mathbf{M}}_{SCMAG} \times \vec{\mathbf{B}} \tag{11.117}$$

where $\vec{\mathbf{M}}_{SCMAG}$ is the spacecraft magnetic moment that can be controlled by magnetorquers and $\vec{\mathbf{B}}$ is the earth's magnetic field in the body frame.

Aerodynamic Drag Disturbance Torque

For low-earth orbit spacecraft, one of the main disturbance torques is due to the aerodynamic drag force, which tends to decrease the orbit altitude. When applied to the spacecraft body in a circular orbit, the drag force is (Vallado 2001)

$$\vec{\mathbf{F}}^{B}_{DRAG} = -\frac{1}{2}\frac{C_D A}{m}\rho\|\vec{\mathbf{v}}_{rel}\|\vec{\mathbf{v}}_{rel} \tag{11.118}$$

where $\vec{\mathbf{v}}_{rel}$ = the relative velocity of the spacecraft with respect to the rotating atmosphere
ρ = local atmospheric density model
A = is the spacecraft cross-sectional area
C_D = coefficient of drag on the spacecraft
m = coefficient of drag on the spacecraft

If the spacecraft is in a circular orbit and the atmosphere has a mean motion due to earth's rotation then $\vec{\mathbf{v}}_{rel}$ is

$$\begin{aligned} \vec{\mathbf{v}}_{rel} &= \vec{\mathbf{v}}^{B_{cm}} - \vec{\mathbf{v}}^{D} \\ &= \vec{\mathbf{v}}^{B_{cm}} - \omega_E \hat{\mathbf{d}}_y \times \vec{\mathbf{r}}^{B_{cm}/E_{cm}} \end{aligned} \tag{11.119}$$

where $\vec{\mathbf{v}}^{B_{cm}}$ = velocity of the spacecraft body
$\vec{\mathbf{v}}^{D}$ = velocity of earth's atmosphere in the atmosphere frame D, which is rotating at the rotational rate of the earth
$\omega_E = 7.2921150 \times 10^{-5}$ rad/s, which is earth's rotational rate
$\vec{\mathbf{r}}^{B_{cm}/E_{cm}}$ = position of the spacecraft from the center of the earth to the spacecraft's center of mass taking into account it's local orbital position.

The disturbance torque from the aerodynamic forces is

$$\vec{\mathbf{M}}_{DRAG} = \vec{\mathbf{r}}^{B_{cp}/B_{cm}} \times \vec{\mathbf{F}}^{B}_{DRAG} \tag{11.120}$$

where $\vec{\mathbf{r}}^{B_{cp}/B_{cm}}$ is the position vector from the system's center of mass to the center of pressure.

11.8 Spacecraft with a Fixed Momentum Wheel and Thrusters

Many three-axis stabilized communications satellites have used an attitude control system consisting of a momentum wheel and thrusters (or magnetic torquers). The momentum wheel provides gyroscopic stiffness for the spacecraft. The attitude dynamics of the fixed momentum wheel system is similar to that of dual-spin stabilized spacecraft. The major difference lies in the fact that the angular momentum of a dual-spin stabilized spacecraft is significantly higher than that of a three-axis stabilized spacecraft with a fixed momentum wheel. Hence, the rate of attitude error buildup due to disturbing moments is higher for a fixed-momentum-wheel system. For a dual-spin stabilized spacecraft, the attitude errors are corrected periodically by ground commands or autonomously. In a fixed momentum wheel system, however, the thrusters are fired by the onboard software to correct the attitude errors automatically. The system is shown in Figure 11.13. The pitch and roll errors are detected by earth sensors.

Geosynchronous Satellite with a Momentum Wheel

For a geosynchronous satellite with a momentum wheel that is nominally aligned with the pitch axis, the components of the angular momentum of the wheel in the spacecraft body frame B are

$$\begin{bmatrix} h_x & h_y & h_z \end{bmatrix}^T = \begin{bmatrix} 0 & -h & 0 \end{bmatrix}^T \tag{11.121}$$

where $h = J\,\Omega$ is the magnitude of the angular momentum of the wheel, J is the moment of inertia of the momentum wheel about the spin axis, and Ω is magnitude of the angular velocity of the momentum wheel about the spin axis. Inserting equations (11.106) and (11.121) into equation (11.96), we obtain

$$\begin{bmatrix} M_{C_x} \\ M_{C_y} \\ M_{C_z} \end{bmatrix} = \begin{bmatrix} I_{xx}\,\ddot{\phi} + [h - \omega_0\,(I_{xx} - I_{yy} + I_{zz})]\,\dot{\psi} + [4\,\omega_0^2\,(I_{yy} - I_{zz}) + \omega_0\,h]\,\phi \\ I_{yy}\,\ddot{\theta} + 3\,\omega_0^2\,\theta\,(I_{xx} - I_{zz}) - \dot{h} \\ I_{zz}\,\ddot{\psi} - [h - \omega_0\,(I_{xx} + I_{zz} - I_{yy})]\,\dot{\phi} + [\omega_0^2\,(I_{yy} - I_{xx}) + \omega_0\,h]\,\psi \end{bmatrix} - \begin{bmatrix} M_{D_x} \\ M_{D_y} \\ M_{D_z} \end{bmatrix} \tag{11.122}$$

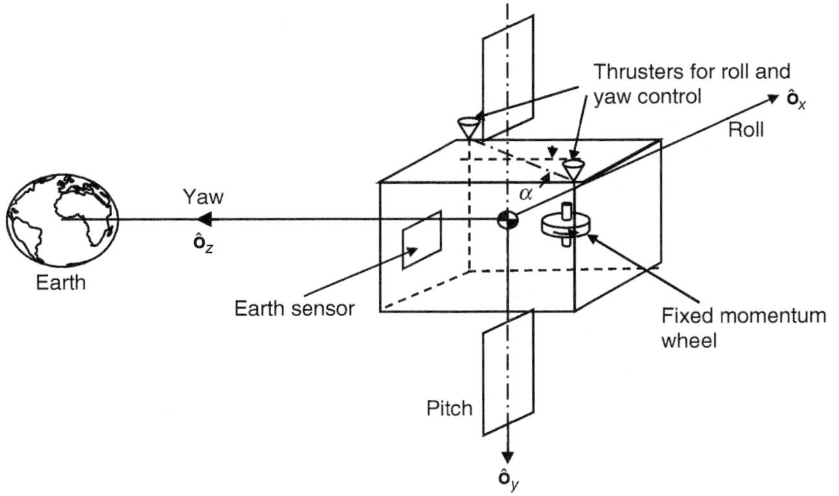

FIGURE 11.13 Fixed momentum wheel system with thrusters.

where $\vec{\mathbf{M}}_C$ is the control torque and $\vec{\mathbf{M}}_D$ is the external disturbance torque on the spacecraft body. Assuming that

$$h \gg \max\left[4\omega_o^2(I_{yy}-I_{zz}),\ \omega_0(I_{xx}-I_{yy}+I_{zz}),\ \omega_o^2(I_{yy}-I_{xx})\right] \tag{11.123}$$

the equations of motion become

$$\begin{bmatrix} M_{C_x} \\ M_{C_y} \\ M_{C_z} \end{bmatrix} = \begin{bmatrix} I_{xx}\,\ddot{\phi}+\omega_0\,h\,\phi+h\,\dot{\psi} \\ I_{yy}\,\ddot{\theta}+3\,\omega_0^2\,\theta\,(I_{xx}-I_{zz})-\dot{h} \\ I_{zz}\,\ddot{\psi}+\omega_0\,h\,\psi-h\,\dot{\phi} \end{bmatrix} - \begin{bmatrix} M_{D_x} \\ M_{D_y} \\ M_{D_z} \end{bmatrix} \tag{11.124}$$

From equation (11.124), it is clear that for small attitude errors, the equation about the pitch axis is uncoupled from the equations about the roll and yaw axes. The equations about the roll and yaw axes are coupled by the angular momentum of the wheel, h. Because of this, we can discuss the pitch axis and roll-yaw axes controls separately.

Pitch Control of a Geosynchronous Satellite with a Momentum Wheel

From equation (11.124), the linearized pitch axis equation is

$$M_{C_y}(t)+M_{D_y}(t)=I_{yy}\,\ddot{\theta}(t)+3\,\omega_0^2\,(I_{xx}-I_{zz})\,\theta(t)-\dot{h}(t) \tag{11.125}$$

If $I_{xx}=I_{zz}$, then equation (11.125) reduces to

$$M_{C_y}(t)+M_{D_y}(t)=I_{yy}\,\ddot{\theta}(t)-\dot{h}(t) \tag{11.126}$$

The pitch error can be controlled by applying a torque proportional to the pitch position error and pitch rate error. In this case, this control is provided by changing the angular momentum of the wheel according to

$$\dot{h}(t)=-K_\theta(\tau_\theta\,\dot{\theta}(t)+\theta(t)) \tag{11.127}$$

where K_θ is the gain and τ_θ is the lead time constant. Substituting equation (11.127) into equation (11.126) and assuming that the external control torque $M_{C_y}(t)=0$, the pitch equation becomes

$$M_{D_y}(t)=I_{yy}\,\ddot{\theta}(t)+K_\theta\,\tau_\theta\,\dot{\theta}(t)+K_\theta\,\theta(t) \tag{11.128}$$

Taking the Laplace transform of equation (11.128) with zero initial conditions and considering the homogenous part of the equation, we get

$$s^2+\frac{K_\theta\,\tau_\theta}{I_{yy}}s+\frac{K_\theta}{I_{yy}}=0 \tag{11.129}$$

For K_θ, $\tau_\theta > 0$, (11.129) is the equation of the motion of a damped second-order system with the characteristic equation:

$$s^2+2\xi_\theta\omega_\theta s+\omega_\theta{}^2=0 \tag{11.130}$$

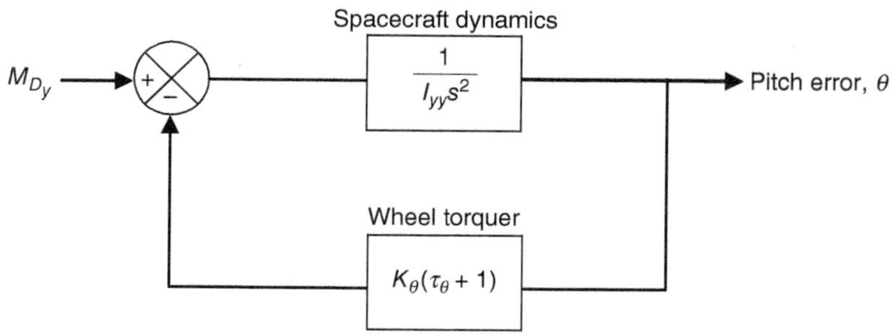

FIGURE 11.14 Block diagram for pitch axis control.

and comparing equations (11.129) and (11.130), the natural frequency ω_θ and damping ratio ζ_θ are given by

$$\omega_\theta = \sqrt{\frac{K_\theta}{I_{yy}}} \qquad \zeta_\theta = \frac{\tau_\theta}{2}\sqrt{\frac{K_\theta}{I_{yy}}} \tag{11.131}$$

The control block diagram is shown in Figure 11.14. The momentum wheel is intended to apply a moment to the spacecraft to null the disturbance torque.

The transfer functions for pitch control are

$$G(s) = \frac{1}{I_{yy}s^2} \tag{11.132}$$

and

$$H(s) = K_\theta(\tau_\theta\, s + 1) \tag{11.133}$$

yielding

$$G(s)\,H(s) = \frac{K_\theta\,(\tau_\theta s + 1)}{I_{yy}\,s^2} \tag{11.134}$$

Hence, the open-loop transfer function $G(s)H(s)$ has two poles at $s = 0$ and one zero at $s = -1/\tau_\theta$. The root locus for the pitch control is shown in Figure 11.15.

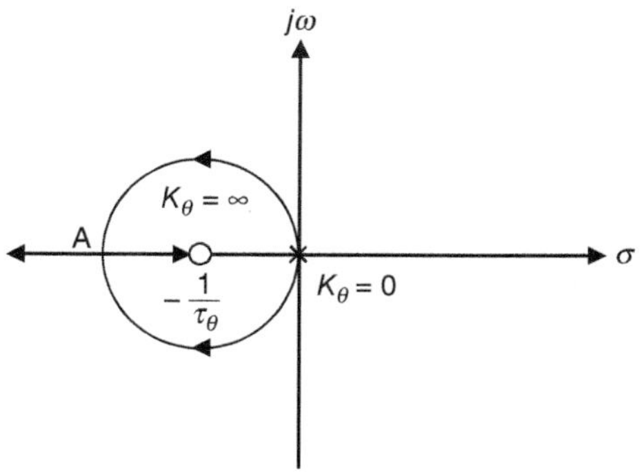

FIGURE 11.15 Root-locus plot for pitch axis control.

Point A, which is generally the design point, corresponds to a critically damped system $\zeta_\theta = 1$. Using equation (11.131), we conclude that for a critically damped system:

$$\tau_\theta = 2\sqrt{\frac{I_{yy}}{K_\theta}} \tag{11.135}$$

and the closed-loop transfer function is

$$\frac{\theta(s)}{M_{D_y}(s)} = \frac{1}{I_{yy}\,(s+\sqrt{K_\theta/I_{yy}})^2} \tag{11.136}$$

The pitch error is introduced from several sources, such as initial pitch error, impulse moments from a thruster during desaturation period (to keep the wheel speed within normal allowable range), during station keeping because of misalignment of the thrusters, and a cyclic moment due to the solar radiation pressure disturbance. The disturbance torque during station keeping is generally large enough so that control by the wheels is abandoned in favor of thruster control. In this case, the wheel speed is kept constant while the thrusters are used to provide attitude control torque. The parameters of the control system, such as K_θ and τ_θ, are selected such that the attitude errors due to theses disturbances are within allowable limits.

If M_{D_y} is an impulse moment such that $M_{D_y}(t) = M_0\,\delta(t)$ and using the inverse Laplace transform on equation (11.136) yields

$$\theta(t) = \frac{M_0}{I_{yy}}\,t\,e^{-t/\tau} \tag{11.137}$$

where $\tau = \sqrt{\frac{I_{yy}}{K_\theta}}$. The maximum pitch error occurs at $t = \tau$ and has the value

$$\theta_{\max} = \frac{M_0\,\tau}{I_{yy}}\frac{1}{e} \tag{11.138}$$

For a cyclic disturbing moment of the type

$$M_{D_y}(t) = M_0 \cos\omega_0 t \tag{11.139}$$

such as in the case of solar radiation pressure moment, the steady-state response is

$$\theta(t) = \theta_{\max}\cos(\omega_0\,t - \varphi) \tag{11.140}$$

where the amplitude is given by

$$\theta_{\max} = \frac{M_0\tau^2}{I_{yy}}\frac{1}{\left[(1-\tau^2\,\omega_0^2)^2 + (2\tau\omega_0)^2\right]^{1/2}} \tag{11.141}$$

and the phase angle by

$$\varphi = \tan^{-1}\frac{2\tau\omega_0}{1-\tau^2\omega_0^2} \tag{11.142}$$

For $\tau\omega_0 \ll 1$, equation (11.140) can be reduced to

$$\theta(t) = \frac{M_0\,\tau^2}{I_{yy}} \cos\omega_0\, t \tag{11.143}$$

The gain K_θ is selected on the basis of the steady-state error and the time constant of the system.

Noting that the control torque on the spacecraft is provided by the momentum wheel, we can use equation (11.127) and obtain

$$J\,\dot{\Omega}(t) = -K_\theta(\tau_\theta\,\dot{\theta}(t) + \theta(t)) \tag{11.144}$$

The Laplace transformation of equation (11.144) for an initial angle $\theta(0) = 0$ is

$$\Omega(s) = -\frac{K_\theta}{J\,s}(\tau_\theta s + 1)\,\theta(s) + \frac{\Omega(0)}{s} \tag{11.145}$$

where $\Omega(0)$ denotes the initial angular velocity of the wheel. Considering a critically damped system, $\zeta_\theta = 1$ and $\tau_\theta = 2\,\tau$, we can use equations (11.144) and (11.145) and write

$$\Omega(s) = -\frac{2\tau s + 1}{J\,s\,(\tau s + 1)^2} M_{D_y}(s) + \frac{\Omega(0)}{s} \tag{11.146}$$

For an impulsive disturbance of magnitude $M_{D_y}(t) = M_0\delta(t)$, the Laplace transform is $M_{D_y}(s) = M_0$, and the momentum wheel speed response is

$$\Omega(t) = -\frac{M_0}{J}\left[1 + \left(\frac{t}{\tau} - 1\right)e^{-t/\tau}\right] + \Omega(0) \tag{11.147}$$

The steady-state response, obtained by taking $\lim_{t\to\infty}\Omega(t)$, is simply

$$\Omega_{ss} = -\frac{M_0}{J} + \Omega(0) \tag{11.148}$$

It is clear from equations (11.147) and (11.148) that the long-term effects of an impulsive disturbance moment are that the spacecraft attitude does not change but the speed of the momentum wheel changes as if the disturbance moment acted on the wheel directly. The negative sign in equation (11.148) is due to the fact that the angular momentum is along the negative pitch axis.

For a cyclic disturbing torque, such as that given by equation (11.139), the Laplace transform of the wheel speed response is

$$\Omega(s) = -\frac{(2\tau s + 1)\,s\,M_0}{J\,s\,(\tau s + 1)^2\,(s^2 + \omega_0^2)} + \frac{\Omega(0)}{s} \tag{11.149}$$

Noting that for $\tau\,\omega_0 \ll 1$, the steady-state response can be reduced to

$$\Omega_{ss} = -\frac{M_0}{J\,\omega_0} \sin\omega_0\, t + \Omega(0) \tag{11.150}$$

A secular (unidirectional nonzero) disturbing torque will lead to an infinite wheel speed unless an external moment along the pitch axis is applied, such as by thrusters or magnetorquers, to reduce the change in wheel speed. Thus, the secular torques will require periodic thruster firings to keep the wheel speed within allowable limits. This procedure is called the momentum wheel desaturation or momentum unloading. Cyclic disturbance moments result in a cyclic wheel speed with an average value of $\Omega(0)$. The momentum wheel is generally designed in such a way that the wheel speed variation due to cyclic disturbance moments is within allowable limits.

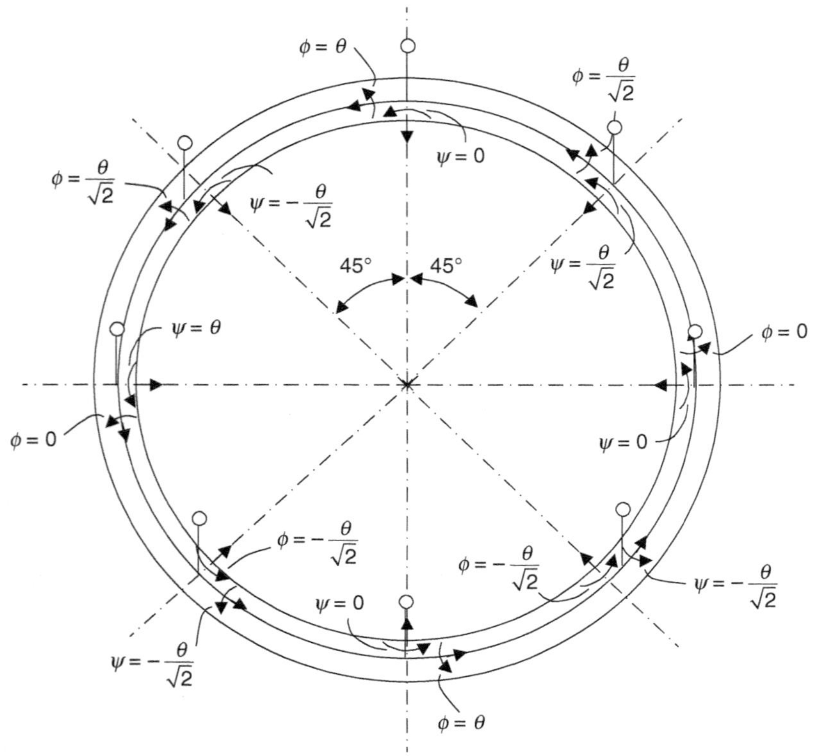

FIGURE 11.16 Yaw and roll coupling.

As shown in Figure 11.16, the control torque is proportional to the pitch error $\theta(t)$ and the pitch error rate $\dot{\theta}(t)$. The pitch error can be determined by attitude sensors, such as earth sensors. However, to determine the error rate, rate gyros may be required.

Roll-Yaw Control of Geosynchronous Satellite with a Momentum Wheel

Using equation (11.124), the Laplace transform of the equation of motion is the roll-yaw axes

$$\begin{bmatrix} M_{C_x} \\ M_{C_z} \end{bmatrix} = \begin{bmatrix} I_{xx}\, s^2 + \omega_0\, h & h\, s \\ -h\, s & I_{zz}\, s^2 + \omega_0\, h \end{bmatrix} \begin{bmatrix} \phi(s) \\ \psi(s) \end{bmatrix} - \begin{bmatrix} M_{D_x} \\ M_{D_z} \end{bmatrix} \tag{11.151}$$

For an uncontrolled spacecraft $M_{C_x} = M_{C_z} = 0$, the motion is the same as that of a dual-spin stabilized spacecraft. In the symmetric case $I_{xx} = I_{zz}$, the transfer function between the yaw angle and the yaw disturbance torque is

$$\frac{\psi(s)}{M_{D_z}(s)} = \frac{I_{xx}\, s^2 + \omega_0\, h}{(I_{xx}\, s^2 + \omega_0\, h)(I_{xx}\, s^2 + \omega_0\, h) + h^2\, s^2} \tag{11.152}$$

so that the characteristic equation can be written in the form

$$s^4 + (2\,\omega_0\,\omega_n + \omega_n^2)\, s^2 + \omega_0^2\,\omega_n^2 = 0 \tag{11.153}$$

where $\omega_n = h/I_{xx}$ is the nutation frequency. The roots of the characteristic equation are

$$s^2 = -\frac{\omega_n^2}{2}\left[\left(1+2\frac{\omega_0}{\omega_n}\right) \pm \sqrt{1+4\frac{\omega_0}{\omega_n}}\right]$$

$$= -\frac{\omega_n^2}{2}\left[\left(1+2\frac{\omega_0}{\omega_n}\right) \pm \left(1+2\frac{\omega_0}{\omega_n} - 2\left(\frac{\omega_0}{\omega_n}\right)^2 + \cdots\right)\right] \tag{11.154}$$

Note that when $\omega_0/\omega_n \ll 1$, the poles are $s = \pm j\,\omega_n$ and $s = \pm j\,\omega_0$. Hence the force-free motion consists of two periodic motions. The first motion is the nutational motion with frequency equal to ω_n and the second motion consists of coupled yaw and roll motions at the orbital rate ω_0. Figure 11.17 shows the yaw-roll coupling with zero nutation.

The angle θ between the orbit normal and the angular momentum vector appears as yaw error ψ at point A. After a quarter period and at point B, the angle θ appears as roll error alone. Hence yaw and roll error interchange every quarter of the orbit period. It is necessary only to sense and control one error directly, either yaw or roll, as the other error is controlled indirectly due to this coupling effect. Generally, for communications satellites the roll error has a greater impact on the antenna pointing accuracy and is easier to sense than the yaw error. Therefore, for a fixed momentum wheel system, the roll error is sensed and controlled directly and the yaw error is controlled indirectly. The following methods can be used to control roll-yaw errors.

The WHECON Control System (Agrawal 1986)

In this method, the control torque is applied along the roll axis thrusters and it is proportional to the roll error ϕ and the roll error rate $\dot{\phi}$. Thrusters historically have been commanded on or off. So, a modulator is necessary to transform the control signal to the on-off signal for thrusters. The common practice is to use a pseudorate modulator, as shown in Figure 11.17. A Schmitt trigger with hysteresis and real pole feedback produces a train of pulses whose average is proportional to the error and the error rate. This process is sometimes called a "derived-rate increment system" because it offers a method of synthesizing angular rate when direct measurement is not possible. The thrusters are offset so as to provide control torque about the yaw axis as well. The external control torques along the roll and yaw axes are

$$M_{C_x} = -K\cos\alpha(\tau\,s+1)\,\phi(s) \tag{11.155}$$

$$M_{C_z} = K\sin\alpha(\tau\,s+1)\,\phi(s) \tag{11.156}$$

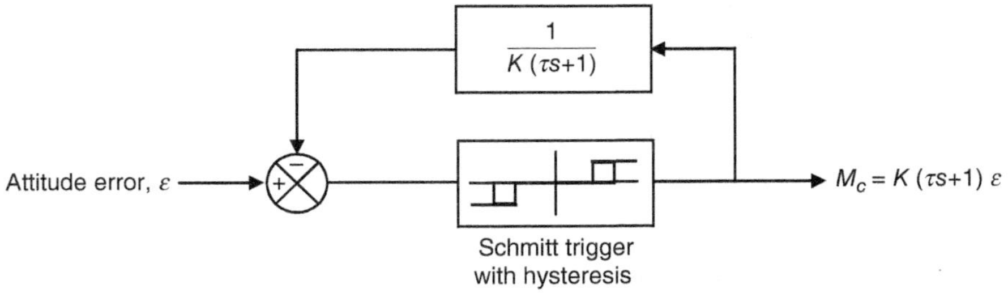

Figure 11.17 Block diagram of pseudorate modulation.

where K is the gain, determined by the thrust force and moment arm, α is the offset angle, and τ is the lead time constant of the pseudorate circuit. Substituting equations (11.155) and (11.156) into equation (11.151), we obtain

$$\begin{bmatrix} M_{D_x} \\ M_{D_z} \end{bmatrix} = \begin{bmatrix} I_{xx}\, s^2 + \omega_0\, h + K\cos\alpha(\tau\, s+1) & h\, s \\ -K\sin\alpha(\tau\, s+1) - h\, s & I_{zz}\, s^2\, \omega_0\, h \end{bmatrix} \begin{bmatrix} \phi(s) \\ \psi(s) \end{bmatrix} \tag{11.157}$$

The characteristic equation corresponding to equation (11.157) is

$$\begin{aligned} & s^4(I_{xx}\, I_{zz}) + s^3(K\,\tau\, I_{zz}\cos\alpha) + \ldots \\ & s^2(h\,(\omega_0\, I_{xx} + \omega_o\, I_{zz} + h + K\,\tau\,\sin\alpha) + K\,\cos\alpha\, I_{zz}) + \ldots \\ & s(h(K\,\tau\,\omega_0\,\cos\alpha + K\,\sin\alpha)) + 1(\omega_0^2\, h^2 + \omega_0\, h\, K\,\cos\alpha) = 0 \end{aligned} \tag{11.158}$$

According to Routh's criterion, a necessary condition for stability is that all coefficients of the characteristic equation be positive. The coefficients are stable for positive values of K, τ, α, and h. Therefore, in a fixed-momentum-wheel system, the angular momentum of the wheel is kept along the negative pitch axis.

Magnetic Control and Desaturation

Magnetic torque-rods generate a magnetic dipole moment whose interaction with the earth's magnetic field produces torques that are used to remove excess momentum. For satellites in low earth orbit, torque rods have been used as a means of magnetic control. Momentum unloading logic responds to accumulated momentum due to torques that create angular momentum components that are secular, while ignoring any periodic components since that are nominally stored by the reaction wheels. This momentum offset can computed as

$$\vec{\mathbf{H}}_{UNL} = (\vec{\mathbf{H}} - \vec{\mathbf{H}}_N) \bullet \hat{\mathbf{b}}_i \tag{11.159}$$

where $\vec{\mathbf{H}}_{UNL}$ = excess momentum from the wheels to be unloaded
$\vec{\mathbf{H}}$ = actual momentum wheel momentum
$\vec{\mathbf{H}}_N$ = nominal or periodic momentum stored on the reaction wheels

Using equation (11.159), we can derive the basic control law using the magnetic torquers as (Sidi 1997)

$$\vec{\mathbf{M}}_C = k\vec{\mathbf{H}}_{UNL} = \vec{\mathbf{M}}_{SCMAG} \times \vec{\mathbf{B}} \tag{11.160}$$

where k is the unload control gain. After taking the cross product of both sides of equation (11.160) with $\vec{\mathbf{B}}$ and using the vector triple product expansion to obtain

$$\vec{\mathbf{B}} \times k\vec{\mathbf{H}}_{UNL} = \vec{\mathbf{B}} \times (\vec{\mathbf{M}}_{SCMAG} \times \vec{\mathbf{B}}) = B^2\vec{\mathbf{M}}_{SCMAG} - \vec{\mathbf{B}} \bullet (\vec{\mathbf{M}}_{SCMAG} \bullet \vec{\mathbf{B}}) \tag{11.161}$$

Assuming that $\vec{\mathbf{M}}_{SCMAG}$ is perpendicular to $\vec{\mathbf{B}}$ and solving for $\vec{\mathbf{M}}_{SCMAG}$ in equation (11.161) we get

$$\vec{\mathbf{M}} = -\frac{k}{B^2}(\vec{\mathbf{B}} \times \vec{\mathbf{H}}_U) \tag{11.162}$$

such that

$$\begin{bmatrix} M_{SCMAGx} \\ M_{SCMAGy} \\ M_{SCMAGz} \end{bmatrix} = -\frac{k}{B^2}\begin{bmatrix} B_y H_{UNL_z} - B_z H_{UNL_y} \\ B_z H_{UNL_x} - B_x H_{UNL_z} \\ B_x H_{UNL_y} - B_y H_{UNL_x} \end{bmatrix} \tag{11.163}$$

where B_x, B_y, B_z are measured as earth magnetic field strength components in the body frame by a magnetometer. When the magnetic moment produces a magnetic torque not precisely proportional to the momentum to be unloaded then the control torque is

$$\vec{\mathbf{M}}_C = -\frac{k}{B^2}\left[B^2\vec{\mathbf{H}}_{UNL} - \vec{\mathbf{B}}(\vec{\mathbf{B}}\cdot\vec{\mathbf{H}}_{UNL})\right] \tag{11.164}$$

Physically, if the excess momentum to be unloaded is parallel to the magnetic field $\vec{\mathbf{B}}$ then the momentum cannot be unloaded.

Using classical control techniques as shown previously, the attitude control law can be

$$\begin{bmatrix} M_{C_x} \\ M_{C_y} \\ M_{C_z} \end{bmatrix} = \begin{bmatrix} -k_\phi(\phi + \tau_\phi\,\dot{\phi}) \\ -k_\theta(\theta + \tau_\theta\,\dot{\theta}) \\ -k_\psi(\psi + \tau_\psi\dot{\psi}) \end{bmatrix} \tag{11.165}$$

where k_ϕ, k_θ, k_ψ are the control gains for roll, pitch, and yaw τ_ϕ, τ_θ, τ_ψ are the lead time constants for roll, pitch, and yaw.

11.9 Three-Axis Reaction Wheel System

A three-axis reaction wheel system can be considered as a combination of three independent pitch, roll, and yaw control systems as shown in Figure 11.18. Reaction wheels differ from momentum wheels in that the spin rate is increased and decreased while the momentum wheels usually have

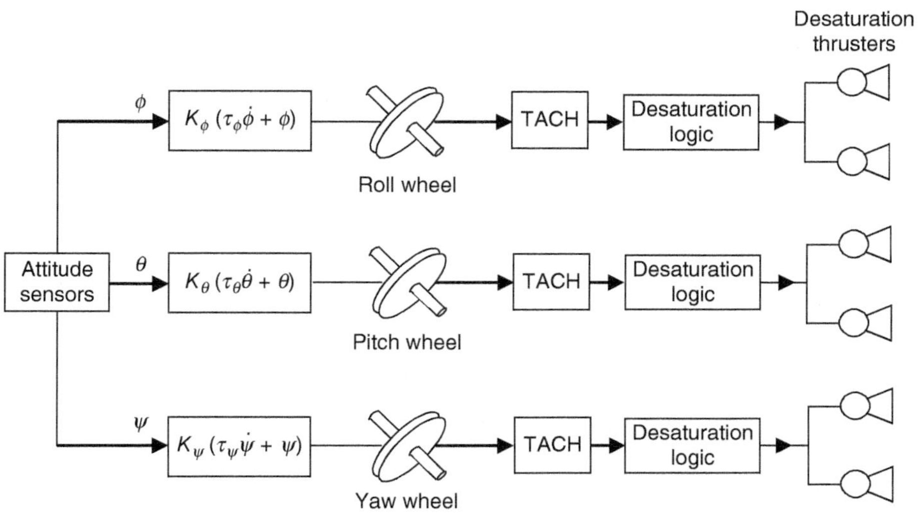

FIGURE 11.18 Block diagram of the three-axis reaction control system.

constant spin rate. Each axis is controlled by varying the speed of the reaction wheel in response to the attitude error. The system requires a reaction wheel with zero nominal angular momentum and an attitude sensor for each axis (i.e., roll, pitch, and yaw).

Reaction Wheel Configuration

Suppose n axially symmetric reaction wheels, RW_1, RW_2, …, RW_n are connected to spacecraft body B such that the wheel axes are fixed in B and are parallel to unit vectors $\hat{\mathbf{w}}_1, \hat{\mathbf{w}}_2, \ldots, \hat{\mathbf{w}}_n$. The total angular momentum of the wheels can be expressed as

$$\sum_{k=1}^{n} h_k \hat{\mathbf{w}}_k = h_1 \hat{\mathbf{w}}_1 + h_2 \hat{\mathbf{w}}_2 + \ldots + h_n \hat{\mathbf{w}}_n \tag{11.166}$$

The reaction wheel assembly can be configured such that it consists of four wheels RW_1, RW_2, RW_3, RW_4 whose spin axes are along unit vectors $\hat{\mathbf{w}}_1$, $\hat{\mathbf{w}}_2$, $\hat{\mathbf{w}}_3$, and $\hat{\mathbf{w}}_4$. The wheels are offset such that the angle β between $\hat{\mathbf{b}}_x$ and $\hat{\mathbf{b}}_z$ as shown in Figure 11.19.

We can express unit vectors $\hat{\mathbf{w}}_1$, $\hat{\mathbf{w}}_2$, $\hat{\mathbf{w}}_3$, and $\hat{\mathbf{w}}_4$ in terms of $\hat{\mathbf{b}}_x$, $\hat{\mathbf{b}}_y$, and $\hat{\mathbf{b}}_z$

$$\hat{\mathbf{w}}_1 = -\sin\beta \hat{\mathbf{b}}_y - \cos\beta \hat{\mathbf{b}}_z \tag{11.167a}$$

$$\hat{\mathbf{w}}_2 = -\sin\beta \hat{\mathbf{b}}_y - \cos\beta \hat{\mathbf{b}}_x \tag{11.167b}$$

$$\hat{\mathbf{w}}_3 = -\sin\beta \hat{\mathbf{b}}_y + \cos\beta \hat{\mathbf{b}}_z \tag{11.167c}$$

$$\hat{\mathbf{w}}_4 = -\sin\beta \hat{\mathbf{b}}_y + \cos\beta \hat{\mathbf{b}}_x \tag{11.167d}$$

Using equations (11.167a–d), we can develop equations relating h_1, h_2, h_3, h_4 to h_x, h_y, h_z:

$$\begin{bmatrix} h_x \\ h_y \\ h_z \end{bmatrix} = \begin{bmatrix} 0 & -\cos\beta & 0 & \cos\beta \\ -\sin\beta & -\sin\beta & -\sin\beta & -\sin\beta \\ -\cos\beta & 0 & \cos\beta & 0 \end{bmatrix} \begin{bmatrix} h_1 \\ h_2 \\ h_3 \\ h_4 \end{bmatrix} \tag{11.168}$$

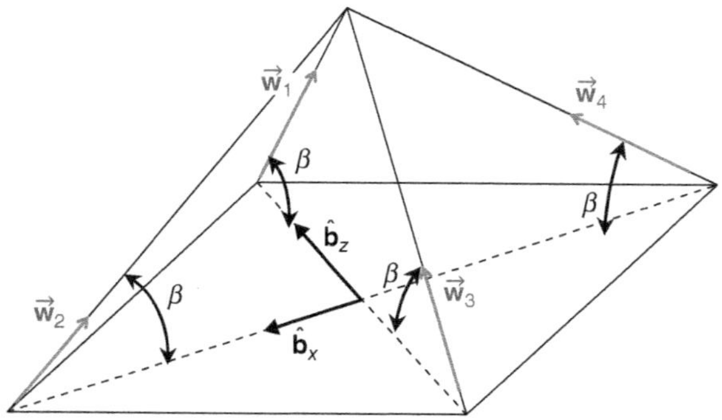

FIGURE 11.19 Tetrahedral reaction wheel configuration.

The resulting equation of motion in terms of the Euler body angles for a three-axis reaction wheel system we have

$$M_{C_x} + M_{D_x} = I_{xx}\,\ddot{\phi} + [4\,\omega_0^2\,(I_{yy} - I_{zz}) - \omega_0\,h_y] - [h_y + \omega_0\,(I_{xx} - I_{yy} + I_{zz})]\,\dot{\psi} + \dot{\theta}\,h_z - \omega_0\,h_z + \dot{h}_x \quad (11.169a)$$

$$M_{C_y} + M_{D_y} = I_{yy}\,\ddot{\theta} + 3\,\omega_0^2\,\theta\,(I_{xx} - I_{zz}) + \omega_o\,h_x\,\phi - h_z\,\dot{\phi} + \omega_0\,h_z\,\psi + h_x\,\dot{\psi}_x + \dot{h}_y \quad (11.169b)$$

$$M_{C_z} + M_{D_z} = I_{zz}\,\ddot{\psi} + [\omega_0^2\,(I_{yy} - I_{xx}) - \omega_0\,h_y]\,\psi + [h_y + \omega_0\,(I_{xx} + I_{zz} - I_{yy})]\,\dot{\phi} - h_x\,\dot{\theta} + \omega_0\,h_x + \dot{h}_z \quad (11.169c)$$

Equations (11.169a–c) are coupled but since h_x, h_y, h_z, and ω_0 are small then the coupling terms are small. If the coupling terms are neglected, the equations of motion about the roll, pitch, and yaw axes become independent, and hence they can be controlled independently as shown in Figure 11.18.

The control torques are applied by letting the time derivative of the angular momentum of the reaction wheels have the form of a classical lead compensator:

$$\dot{h}_x(t) = K_\phi(\tau_\phi\,\dot{\phi} + \phi) \quad (11.170a)$$

$$\dot{h}_y(t) = K_\theta(\tau_\theta\,\dot{\theta} + \theta) \quad (11.170b)$$

$$\dot{h}_z(t) = K_\psi(\tau_\psi\,\dot{\psi} + \psi) \quad (11.170c)$$

The control of each axis is very similar to the pitch axis control dynamics of the fixed momentum-wheel system. The secular or inertial disturbing moments will lead to unacceptably high wheel speeds unless the thrusters or other devices are used to apply an external moment along the wheel axis to reduce the wheel speed. Therefore, the secular torques will require desaturation of the reaction wheels. Periodic disturbing torques result in periodic wheel speeds and do not require desaturation if the variation in the speed is within allowable wheel speed limits.

Quaternion Control

Quaternions are useful for feedback control on large-angle orientation spacecraft maneuvers and especially amenable to real-time computation. A quaternion feedback control can be of the form

$$\vec{\mathbf{u}} = -\mathbf{K}\vec{\mathbf{q}}_e - \mathbf{K}_d\,{}^N\vec{\omega}^B \quad (11.171)$$

where $\mathbf{K}$ and $\mathbf{K}_d$ are the position control gains and rate control gains, respectively.

In equation (11.171), $\vec{\mathbf{q}}_e$ is the error quaternion or difference between $q_{1d}, q_{2d}, q_{3d}, q_{4d}$, which is the desired quaternion, and q_1, q_2, q_3, q_4, the quaternion in the spacecraft body frame. Using equation (11.65), the error quaternion is

$$\vec{\mathbf{q}}_e = \begin{bmatrix} q_{1e} \\ q_{2e} \\ q_{3e} \\ q_{4e} \end{bmatrix} = \begin{bmatrix} q_{4d} & q_{3d} & -q_{2d} & -q_{1d} \\ -q_{3d} & q_{4d} & q_{1d} & -q_{2d} \\ q_{2d} & -q_{1d} & q_{4d} & -q_{3d} \\ q_{1d} & q_{2d} & q_{3d} & q_{4d} \end{bmatrix} \begin{bmatrix} q_1 \\ q_2 \\ q_3 \\ q_4 \end{bmatrix} \quad (11.172)$$

In order to accurately propagate the quaternions q_1, q_2, q_3, q_4, one needs to numerically integrate the first-order kinematical equation given by equation (11.67). There are a variety of methods including approximate Taylor Series expansion, Runge-Kutta, and rotation vector algorithms. Each method has a tradeoff in terms of algorithm complexity and accuracy. The interested reader can refer to Wie and Barba (1989) for more information.

Trapezoidal Slew Maneuvers

Let us consider a rigid spacecraft that is required to maneuver about an inertially fixed axis as fast as possible but within the saturation limits of the reaction wheels and the rate gyros. Assume a trapezoidal maneuver profile: constant acceleration, α, for t_{accel} seconds up to the maximum velocity, ω_{max}, a coast phase t_{coast} at maximum angular rate ω_{max}, and a deceleration interval t_{decel} back to zero velocity. This is depicted in Figure 11.20.

Using Figure 11.20, we can calculate the total slew angle as

$$\theta_{slew} = \tfrac{1}{2}\alpha t_{accel}^{2} + \omega_{max} t_{coast} + \tfrac{1}{2}\alpha t_{decel}^{2} \tag{11.173}$$

and the corresponding time to perform the slew maneuver is

$$t_{slew} = t_{accel} + t_{coast} + t_{decel} \tag{11.174}$$

where $t_{accel} = t_{decel}$ and $\omega_{max} = \alpha t_{accel}$. Depending on the parameters that are available and the constraints of the hardware we can create the following cases to perform a trapezoidal slew maneuver.

Case 1: Given ω_{max}, θ_{slew}, and t_{slew} then solve for α. If $\frac{1}{2}\omega_{max} t_{slew} > \theta_{slew}$ then the maneuver is triangular (no coast phase required) such that

$$t_{accel} = \frac{t_{slew}}{2};\ t_{decel} = t_{accel};\ t_{coast} = 0;\ \alpha = \frac{4\theta_{slew}}{t_{slew}^{2}} \tag{11.175}$$

If $\omega_{max} t_{slew} > \theta_{slew}$ then the maneuver is trapezoidal (coast phase at max rate required) such that

$$t_{accel} = t_{slew} - \frac{\theta_{slew}}{\omega_{max}};\ t_{decel} = t_{accel};\ t_{coast} = t_{slew} - 2t_{accel};\ \alpha = \frac{\omega_{max}}{t_{accel}} \tag{11.176}$$

otherwise the maneuver cannot be achieved within given constraints.

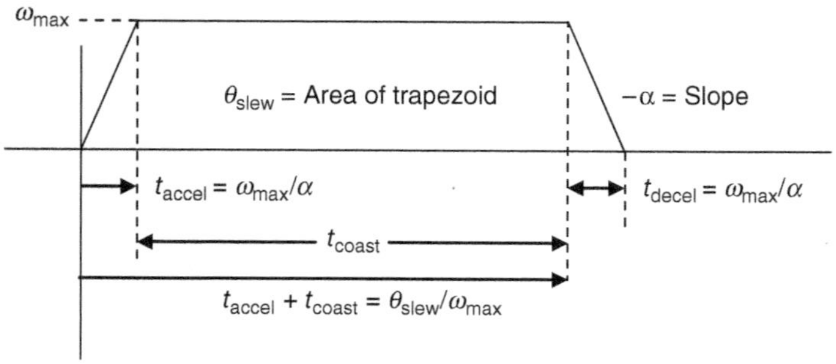

FIGURE 11.20 Trapezoidal slew profile.

Case 2: Given ω_{max}, θ_{slew}, and α then solve for t_{slew}. If $\alpha\sqrt{\frac{\theta_{slew}}{\alpha}} < \omega_{max}$ then the maneuver is triangular (no coast phase required) such that

$$t_{accel} = \sqrt{\frac{\theta_{slew}}{\alpha}}; \quad t_{decel} = t_{accel}, t_{coast} = 0; t_{slew} = 2t_{accel} \tag{11.177}$$

otherwise the maneuver is trapezoidal (coast phase at max rate is required) such that

$$t_{accel} = \frac{\omega_{max}}{\alpha}, t_{decel} = t_{accel}, t_{coast} = \frac{\theta_{slew}}{\omega_{max}} - t_{accel}, t_{slew} = 2t_{accel} + t_{coast} = \frac{\theta_{slew}}{\omega_{max}} + \frac{\omega_{max}}{\alpha} \tag{11.178}$$

Case 3: Given t_{slew}, θ_{slew}, and α then solve for ω_{max}. If $\alpha\left(\frac{t_{slew}}{2}\right)^2 > \theta_{slew}$ then the maneuver is trapezoidal (triangular in limiting case) such that

$$\omega_{max} = \left(\frac{\alpha t_{slew}}{2}\right) - \sqrt{\left(\frac{\alpha t_{slew}}{2}\right)^2 - \alpha\theta_{slew}}; \; t_{accel} = \frac{\omega_{max}}{\alpha}; \quad t_{decel} = t_{accel}; \; t_{coast} = t_{slew} - 2t_{accel} \tag{11.179}$$

otherwise the maneuver cannot be achieved within given constraints.

11.10 Control Moment Gyroscope

A control moment gyroscope contains a rotor that has a constant spin rate. The rotor is mounted on a gimbal (or a set of gimbals). Torqueing the gimbal motor results in a reaction torque on the spacecraft body that is orthogonal to both the rotor spin axis and gimbal axis. The CMG can be considered a torque amplification device because a small gimbal torque input can produce a large control output on the spacecraft. Because the CMGs are capable of generating large torque, they are often used for large spacecraft with any or all of the following requirements: fast slewing, large control torque, and precise pointing.

There are two basic types of control moment gyros: (1) single gimbal control moment gyros (SGCMG) and (2) double gimbal control moment gyros (DGCMG). The SGCMGs are considerably simpler than DGCMGs from hardware point of view, resulting in significant cost, power, and weight advantages. However, the gimbal control laws are much simpler for DGCMGs because of extra degree of freedom per device.

Single Gimbal Control Moment Gyro Steering Law

Let us consider a system S that consists of the spacecraft rigid body B, and the CMG bodies $G_1, \ldots, G_k$. The angular momentum of the system S about it's center of mass can be written in the form

$$^{B}\vec{\mathbf{H}}^{S/S_{cm}} = \left(I\,^{N}\vec{\omega}^{B} + \sum_{k=1}^{n}\vec{\mathbf{H}}^{G/G^{k}_{cm}}\right) \cdot \hat{\mathbf{b}}_i \;\; (\text{for } i = x, y, z) \tag{11.180}$$

where $\sum_{k=1}^{n}\vec{\mathbf{H}}^{G/G^{k}_{cm}}$ is the angular momentum of the CMG, $G_1, \ldots, G_k$, with respect to its center of mass.

The torque of the CMGs can be found using

$$\vec{\tau}_G = \frac{^{B}d}{dt}\left(\sum_{k=1}^{n}\vec{\mathbf{H}}^{G/G^{k}_{cm}}\right) + {}^{N}\vec{\omega}^{B} \times \sum_{k=1}^{n}\vec{\mathbf{H}}^{G/G^{k}_{cm}} \tag{11.181}$$

The equations of motion for the system S with CMGs and gravity-gradient disturbing torque is

$$\vec{\mathbf{M}}_C \cdot \hat{\mathbf{b}}_i = (I\,{}^N\vec{\alpha}^B + {}^N\vec{\omega}^B \times I\,{}^N\vec{\omega}^B - \vec{\tau}_G - \vec{\mathbf{M}}_{GG}) \cdot \hat{\mathbf{b}}_i \quad \text{for } (i = x, y, z) \tag{11.182}$$

where $\vec{\mathbf{M}}_C$ = external control torque from thrusters or torque rods onto system S
$\vec{\mathbf{M}}_{GG}$ = disturbance torque due to the gravity gradient onto system S
$\vec{\tau}_G$ = torque produced by the CMGs onto system S

We can find the scalars for the CMG torques in terms of the body components

$$\vec{\tau}_G \cdot \hat{\mathbf{b}}_x = \dot{h}_x + \omega_y h_z - \omega_z h_y \tag{11.183a}$$

$$\vec{\tau}_G \cdot \hat{\mathbf{b}}_y = \dot{h}_y + \omega_z h_x - \omega_x h_z \tag{11.183b}$$

$$\vec{\tau}_G \cdot \hat{\mathbf{b}}_z = \dot{h}_z + \omega_x h_y - \omega_y h_x \tag{11.183c}$$

Rearranging equations (11.183a–c), the desired CMG momentum rate is given as

$$\dot{h}_x = \tau_{G_x} - \omega_y h_z + \omega_z h_y \tag{11.184a}$$

$$\dot{h}_y = \tau_{G_y} - \omega_z h_x + \omega_x h_z \tag{11.184b}$$

$$\dot{h}_z = \tau_{G_z} - \omega_x h_y + \omega_y h_x \tag{11.184c}$$

The CMG angular momentum is a function of the CMG gimbal angles $\delta_1, \ldots, \delta_n$. Taking the partial differential with respect to gimbal angles, we get

$$\begin{bmatrix} \dot{h}_x \\ \dot{h}_y \\ \dot{h}_z \end{bmatrix} = \mathbf{A} \begin{bmatrix} \dot{\delta}_1 \\ \vdots \\ \dot{\delta}_n \end{bmatrix} \tag{11.185}$$

where $\mathbf{A}$ is the Jacobian matrix defined as

$$\mathbf{A} = \begin{bmatrix} \dfrac{\partial h_x}{\partial \delta_1} & \cdots & \dfrac{\partial h_x}{\partial \delta_n} \\ \vdots & \ddots & \vdots \\ \dfrac{\partial h_z}{\partial \delta_1} & \cdots & \dfrac{\partial h_z}{\partial \delta_n} \end{bmatrix} \tag{11.186}$$

The CMG steering control law is used to find gimbal rates $\dot{\delta}_1, \ldots, \dot{\delta}_n$ that can generate the commanded $\dot{h}_x$, $\dot{h}_y$, and $\dot{h}_z$ while avoiding geometric singularities and meeting constraints such as rate limits and hardware stops. Computing the gimbal rates requires the inverse of $\mathbf{A}$. Since $\mathbf{A}$ is a $m \times n$ matrix, where $n > m$, we can find the gimbal rates using a pseudoinverse $\mathbf{A}^{\dagger}$:

$$\begin{bmatrix} \dot{\delta}_1 \\ \vdots \\ \dot{\delta}_n \end{bmatrix} = \mathbf{A}^{\dagger} \begin{bmatrix} \dot{h}_x \\ \dot{h}_y \\ \dot{h}_z \end{bmatrix} = \mathbf{A}^T (\mathbf{A}\mathbf{A}^T)^{-1} \begin{bmatrix} \dot{h}_x \\ \dot{h}_y \\ \dot{h}_z \end{bmatrix} \tag{11.187}$$

In order to avoid singularities, **A** must be full rank or $rank(\mathbf{A}) = m$. For certain sets of gimbal angles the pseudoinverse does not exist and the pseudoinverse steering law encounters singular states. The singularity condition can also be stated in terms of matrix determinant as

$$\det(\mathbf{A}\mathbf{A}^T) = 0 \tag{11.188}$$

One can think of singular situation occurring in practice when all the CMG torque vectors are perpendicular to the commanded torque vector. This situation is also known as "gimbal lock."

CMG Configuration

In order to get the optimal redundancy, several CMG configurations have been developed. The configuration of the assembly should be made such that the total angular momentum of the CMG is zero. Figure 11.21 shows four SGCMGs in a pyramid configuration used to provide torques about all three axes and also provide redundancy.

The gimbal axes are orthogonal to the pyramid faces causing the rotor angular momentum vector to rotate about each gimbal axis. Each face is included with a skew angle of β from horizontal plane or platform. Each CMG, generally, has the same angular momentum about its spin rotor axis. The total angular momentum for the CMGs shown in Figure 11.21 is

$$\begin{bmatrix} h_1 \\ h_2 \\ h_3 \end{bmatrix} = \mathrm{H}_1 \begin{bmatrix} -\cos\beta\sin\delta_1 \\ \cos\delta_1 \\ \sin\beta\sin\delta_1 \end{bmatrix} + \mathrm{H}_2 \begin{bmatrix} -\cos\delta_2 \\ -\cos\beta\sin\delta_2 \\ -\sin\beta\sin\delta_2 \end{bmatrix} + \mathrm{H}_3 \begin{bmatrix} \cos\beta\sin\delta_3 \\ -\cos\delta_3 \\ \sin\beta\sin\delta_3 \end{bmatrix} + \mathrm{H}_4 \begin{bmatrix} \cos\delta_4 \\ \cos\beta\sin\delta_4 \\ \sin\beta\sin\delta_4 \end{bmatrix} \tag{11.189}$$

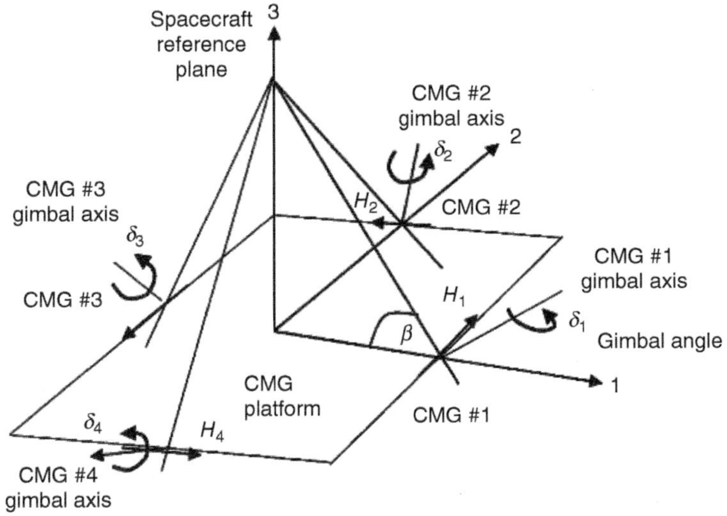

FIGURE 11.21 Pyramid CMG configuration (Wie 1988).

Assuming that all the four CMGs have same spin angular momentum such that $H = \mathbf{H}_1 = \mathbf{H}_2 = \mathbf{H}_3 = \mathbf{H}_4$, the time derivative of the CMG angular momentum is

$$\begin{bmatrix} \dot{h}_1 \\ \dot{h}_2 \\ \dot{h}_3 \end{bmatrix} = \mathrm{H}\mathbf{A} \begin{bmatrix} \dot{\delta}_1 \\ \dot{\delta}_2 \\ \dot{\delta}_3 \\ \dot{\delta}_4 \end{bmatrix} \tag{11.190}$$

where

$$\mathbf{A} = \begin{bmatrix} -\cos\beta\cos\delta_1 & \sin\delta_2 & \cos\beta\cos\delta_3 & -\sin\delta_4 \\ -\sin\delta_1 & -\cos\beta\cos\delta_2 & \sin\delta_3 & \cos\beta\cos\delta_4 \\ \sin\beta\cos\delta_1 & \sin\beta\cos\delta_2 & \sin\beta\cos\delta_3 & \sin\beta\cos\delta_4 \end{bmatrix} \tag{11.191}$$

This configuration with a $\beta = 54.73°$ has been extensively studied in the literature and the momentum envelope becomes nearly spherical. Other mounting arrangements have been used for SGCMGs. As an example, the six parallel mounted SGCMGs have been successfully used to control the MIR Space Station.

11.11 Effects of Structural Flexibility

In the previous subsections, we assumed the system to contain only rigid bodies; however, most spacecraft have some structural flexibility that needs to be considered when designing control laws. Structural flexibility is introduced from many sources, such as the solar arrays, antennas, and antenna support structure. In the future, spacecraft are expected to become much more mechanically flexible, due to large deployable antennas and solar arrays. These spacecraft will require advanced attitude control concepts. It is beyond the scope of this part to provide a detailed discussion of these concepts. However, a simple example will be discussed to introduce the fundamentals of structural flexibility and attitude control interactions.

Solar Array Flexibility along the Pitch Axis

Let us consider that the torsional flexibility of a solar array in the pitch axis is represented by a torsional spring and a disk, as shown below in Figure 11.22.

The equations of motion for the system are

$$I_b\,\ddot{\theta}_b(t) + K\,(\theta_b(t) - \theta_s(t)) = M(t) \tag{11.192}$$

$$I_s\,\ddot{\theta}_s(t) + K\,(\theta_s(t) - \theta_b(t)) = 0 \tag{11.193}$$

where I_b and I_s are the pitch axis inertias of the central body and the solar array, θ_b and θ_s are the pitch angles of the body and the solar array, K is the torsional stiffness of the solar array, and $M(t)$ is the external torque along the pitch axis. Let

$$\theta_r(t) = \theta_s(t) - \theta_b(t) \tag{11.194}$$

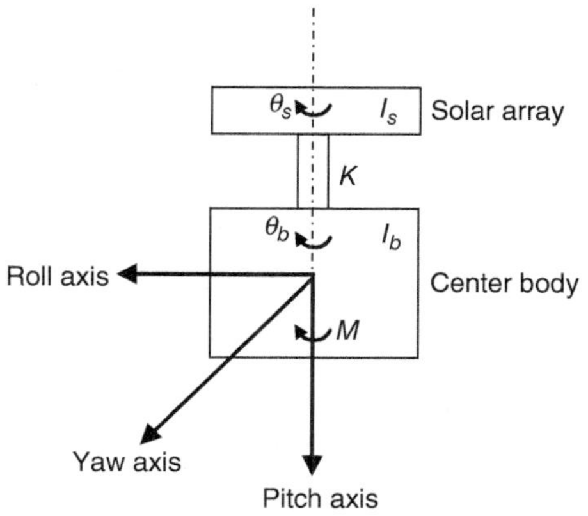

FIGURE 11.22 Model of solar array flexibility.

Adding equations (11.192) and (11.193) and using equation (11.194), we get

$$I_T\,\ddot{\theta}_b(t) = M(t) - I_s\,\ddot{\theta}_r(t) \tag{11.195}$$

$$\ddot{\theta}_r(t) + \omega_n^2\,\theta_r(t) = -\,\ddot{\theta}_b(t) \tag{11.196}$$

where $I_T = I_b + I_s$ is the total pitch inertia and

$\omega_n = \sqrt{K/I_s}$ is the fixed base natural frequency of the solar array.

Taking the Laplace transform of equations (11.195) and (11.196) with initial conditions equal to zero, θ_b can be expressed as

$$I_T\,s^2\,\theta_b(s) = M(s) + \frac{I_s\,s^4\,\theta_b(s)}{s^2 + \omega_n^2} \tag{11.197}$$

The transfer function in this form is shown in Figure 11.23. The flexibility effects are introduced as a feedback torque.

By rearranging the terms in equation (11.197), we get

$$\frac{\theta_b(s)}{M(s)} = \frac{s^2 + \omega_n^2}{s^2\,I_b\,(s^2 + \Omega_n^2)} \tag{11.198}$$

where Ω_n is the free-free natural frequency of the spacecraft and is given by

$$\Omega_n = \omega_n\sqrt{1 + \frac{I_s}{I_b}} \tag{11.199}$$

By factoring the denominator of equation (11.198), the transfer function is

$$\frac{\theta_b(s)}{M(s)} = \frac{1}{I_T\,s^2} + \frac{I_s}{I_b\,I_T(s^2 + \Omega_n^2)} \tag{11.200}$$

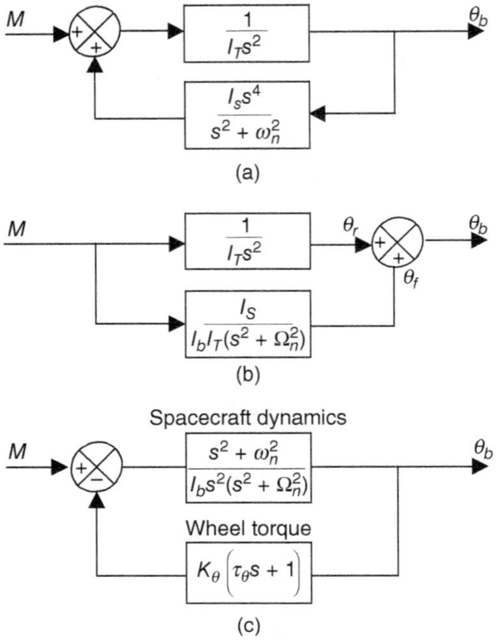

FIGURE 11.23 Block diagrams for pitch axis with solar array flexibility.

The transfer function in the form of equation (11.200) is represented in Figure 11.23(b). Here the pitch angle of the body consists of two components. The first component θ_r is obtained by assuming the entire spacecraft to be a rigid body and the second component θ_f is due to the flexibility of the solar array. By using the transfer function from equation (11.200), the pitch axis control is modified as shown in Figure 11.23(c). Here, the base fixed natural frequency of the solar array ω_n is the zero of the open-loop transfer function and the pole is the free-free natural frequency of the spacecraft Ω_n.

The root locus of the pitch control is shown below in Figure 11.24. The system is stable, although the flexible modes are lightly damped. To avoid interaction of the flexibility with the attitude control, it is desirable to have wide separation between the control frequencies and structural frequencies.

Example Pitch Control Determine the parameters for the pitch attitude control system consisting of a fixed momentum wheel with thrusters for a spacecraft whose mass properties are: $I_{xx} = 2700\ \text{kgm}^2$, $I_{yy} = 1360\ \text{kgm}^2$, $I_{zz} = 2200\ \text{kgm}^2$. Wheel desaturation thruster torque is 0.422 Nm with a pulse of 0.2 seconds. The permissible attitude control errors are 0.05° in pitch. Find the thruster and control time constant for the pitch and roll-yaw control.

Solution The pitch control design can be based on the maximum allowable pitch error during wheel desaturation. The desaturation torque impulse M_{D_y} is

$$M_{D_y} = 0.422 \times 0.2 = 0.0844\ \text{Nms}$$

The pitch error due to desaturation impulse is given by equation (11.137). From equation (11.138), the time constant of the system is given by

$$\tau = \frac{\theta_{\max} I_{yy} e}{M_y}$$

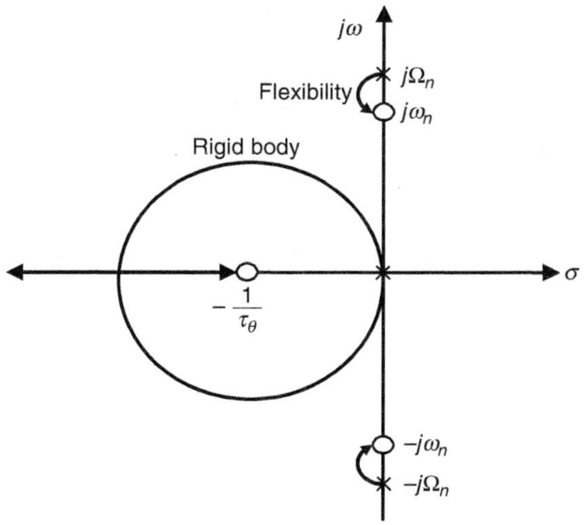

FIGURE 11.24 Root locus for pitch axis control with flexibility.

Assuming that $\theta_{max} = 0.04°$ and a 0.01° error from other sources then

$$\tau = \frac{0.04 \times \pi \times 1360 \times 2.718}{180 \times 0.0844} = 31 \text{ s}$$

The time constant in equation (11.136) is defined as

$$\tau = \sqrt{\frac{I_{yy}}{K}} \quad \text{or} \quad K = \frac{I_{yy}}{\tau^2}$$

Substituting the numerical values in the equation above yields

$$K = \frac{1360}{(31)^2} = 1.41 \text{ Nm(rad)}^{-1}$$

The lead-time constant τ_θ from equation (11.135) is given by

$$\tau_\theta = 2\sqrt{\frac{I_{yy}}{K_\theta}} = 2\,\tau = 2 \times 31 = 62 \text{ s}$$

Mode Stabilization Using Classical Control Techniques

Using classical control techniques one can begin control design by mode stabilizing the rigid-body mode and subsequently gain/phase stabilizing the unstable flexible modes. We can analyze the contribution of the rigid body and flexible modes separately for a spacecraft where the single-axis plant model is

$$G(s) = \frac{\theta(s)}{M(s)} = \underbrace{\frac{1}{Is^2}}_{\text{rigid body mode}} + \underbrace{\frac{K}{I\left(s^2 + 2\xi_f\omega_f + \omega_f^2\right)}}_{\text{flexible mode}} \tag{11.201}$$

where K = flexible modal gain
ω_f = flexible mode frequency
ξ_f = passive damping ratio

The proportional integral derivative (PID) controller is

$$u(s) = k_p + k_d s + \frac{k_i}{s} = k_\theta\left(1 + \tau_d s + \frac{\tau_i}{s}\right) \tag{11.202}$$

Initially, we consider only the rigid body mode in equation (11.201) and let $M(s) = -u(s)$, then the characteristic equation of a continuous-time PID in closed loop is

$$s^3 + \frac{1}{I}(k_d s^2 + k_p s + k_i) = 0 \tag{11.203}$$

Factoring this as a dominant second-order character with an integrator root yields (Wie 1988)

$$(s^2 + 2\zeta_r\omega_r s + \omega_r^2)\left(s + \frac{1}{\tau_i}\right) = 0 \tag{11.204}$$

One can select gain values so that the dominant second-order poles of this system would match that of a second-order system. Acceptable performance requirements are (1) 90-second settling time within 2% of final steady state response (i.e., 4 time constants), (2) a damping ratio of $\zeta_n = 0.707$, and (3) integrator time constant of $\tau_i = 10\tau$ can be used. The initial parameter selection to meet performance requirements can be summarized below:

$$T_{settling} = 4\tau = \frac{4}{\zeta_r\omega_r}, \qquad \tau_i = 10\tau = \frac{10}{\zeta_r\omega_r} \tag{11.205a}$$

$$k_p = I\left(\omega_r^2 + \frac{2\zeta_r\omega_r}{\tau_i}\right),\ k_i = I\left(\frac{\omega_r^{\ 2}}{\tau_i}\right),\ k_d = I\left(2\zeta_r\omega_r + \frac{1}{\tau_i}\right) \tag{11.205b}$$

If the sample frequency is much faster than the bandwidth of the control loop, then the values calculated for the continuous design can be used directly as starting values in the discrete implementation (Wie 1988).

After the unstably interacting flexible structural modes have been identified, the initial filter coefficients can be selected. In their continuous-time forms, the filters chosen have the transfer functions:

$$F_1(s) = \frac{\omega_1^2}{s^2 + 2\zeta_1\omega_1 + \omega_1^2},\ F_2(s) = \frac{s^2 + 2\zeta_{2n}\omega_2 s + \omega_2^2}{s^2 + 2\zeta_{2d}\omega_2 s + \omega_2^2} \tag{11.206}$$

The roll-off filter $F_1(s)$ was designed to have damping ζ_1 and a break frequency at ω_1, which is about halfway between the crossover frequency and the worst-case solar array mode. High-frequency structural modes are often gain-stabilized by a steep roll-off filter at a frequency above the control bandwidth and well below the first structural mode frequency. This filter most often will not attenuate the structural mode(s) and so an additional filter $F_2(s)$ can be designed to give a notch at ω_2, which is the frequency at which the structural mode in question will still rise above the 0 dB line. The magnitude or "depth" of a notch filter is given by

$$K_{\text{depth}} = 20\log_{10}\left(\frac{\zeta_{2n}}{\zeta_{2d}}\right) \tag{11.207}$$

where the values $\zeta_{2n} < \zeta_{2d}$ are selected to give adequate depth. The filter coefficients must be mapped to a discrete implementation. The initial parameters should be properly simulated to determine that robustness margins and performance requirements are satisfied.

11.12 Attitude Determination

For a three-axis stabilized spacecraft, whose main body is not spinning, pitch and roll attitudes are easily determined by earth sensors. The yaw attitude, however, is difficult to determine and is usually estimated.

Earth Sensors

There are several types of earth sensors, such as scanning, balanced radiation, and edge tracking types. We will focus on the scanning type. The scanning sensor scans the earth's 14 to 16 μm infrared radiance profile with a resonant torsion-bar suspended mirror. The scanning movement is provided by a closed-loop feedback system. Since the spacecraft is not rotating, scanning of the sensor becomes necessary. The sensor provides two scanning beams separated by an angle which is less than that subtended by the earth. The sensor operation is shown in the following Figure 11.25.

Scanning is performed in the east to west direction. For a nominal spacecraft orientation, the sensor is centered about the earth's center with one beam scanning across the northern half of the earth disk and the other across the southern half. Attitude is determined by comparing the angle between the first horizon crossing and a center reference point to the angle from the reference to the second horizon crossing. Let H_1 be the angle of scan at the entrance to the radiance image, H_2 be the angle of scan at the departure to the radiance image, and H_0 is the reference angular position, normally at mirror center position. One way to determine these angles is to have an optical encoder mounted with a mirror so as to generate a pulse for each incremental angle of the scan motion and to generate a center reference pulse at the mirror center position. For pitch sensing, a count-up of the encoded pulses is started at H_1 until the center reference pulse is reached and then a count-down is started until H_2 is reached. The output count is proportional to the pitch error and will be zero for zero pitch error. For roll attitude

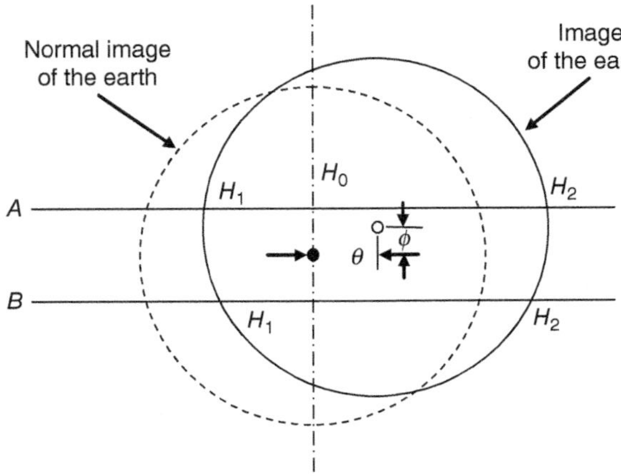

FIGURE 11.25 Earth sensor operation.

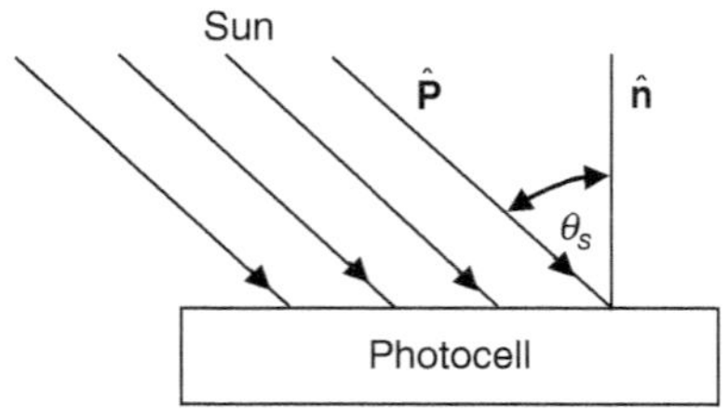

FIGURE 11.26 Sun sensor operation.

sensing, a count-up is started at H_1 and terminated at H_2. The difference in the count for the north and south scan is proportional to the roll error. The pitch attitude error θ is given by

$$\theta \cong \frac{(H_1 + H_2)}{2} - H_0 \text{ at A or B} \tag{11.208}$$

The roll attitude error ϕ is given by

$$\phi \cong (H_1 - H_2)_A - (H_1 - H_2)_B \tag{11.209}$$

Sun Sensors

Sun sensors are frequently called cosine sun sensors because a common type is based on the sinusoidal variation of the output current of a silicon cell with sun angle. This concept is shown by the following relationship and accompanying Figure 11.26

$$\vec{\mathrm{I}} = \mathrm{I}\hat{\mathbf{P}} \bullet \hat{\mathbf{n}} = \mathrm{I} \bullet \cos\theta_s \tag{11.210}$$

Digital Sun Sensors

A digital sensor determines the angular deviation of the sun line from the sensor optical axis. Figure 11.27 shows the measurement component of the sensor.

The sun image is reflected by a material with an index of refraction n, which may be unity. The sun image illuminates a pattern of slits and the slits are divided into a series of rows with a photocell beneath each row. Four classes of rows are illustrated: (1) an automatic threshold adjust (ATA), (2) a sign bit, (3) encoded bits, and (4) fine bits.

Because the photocell voltage is proportional to $\cos\theta_s$ (θ_s = sun angle), a fixed threshold is inadequate for determining the voltage at which a bit is turned on. This is compensated for by using the ATA slit, which is half the width of the other slits. A bit is turned "on" if its photocell voltage is greater than the ATA photocell voltage. The sign bit, or most significant bit, determines which side of the sensor the sun is on. The encoded bits provide a discrete measure of the linear displacement of the sun image relative to the sensor center line or null. A Gray code, named after the inventor, is most widely used. The fine bits can be used to provide increased resolution.

Rate Gyroscopes

Rate gyros are important devices that measure the angular motion of the spacecraft including the angular rates and attitude. One disadvantage of rate gyros is the sensor noise characteristic of these devices. It is beyond the scope of this part to provide rate gyro noise models as shown in Lam et al. (2003). Instead, we will describe the different types of rate gyroscopes that have been used for

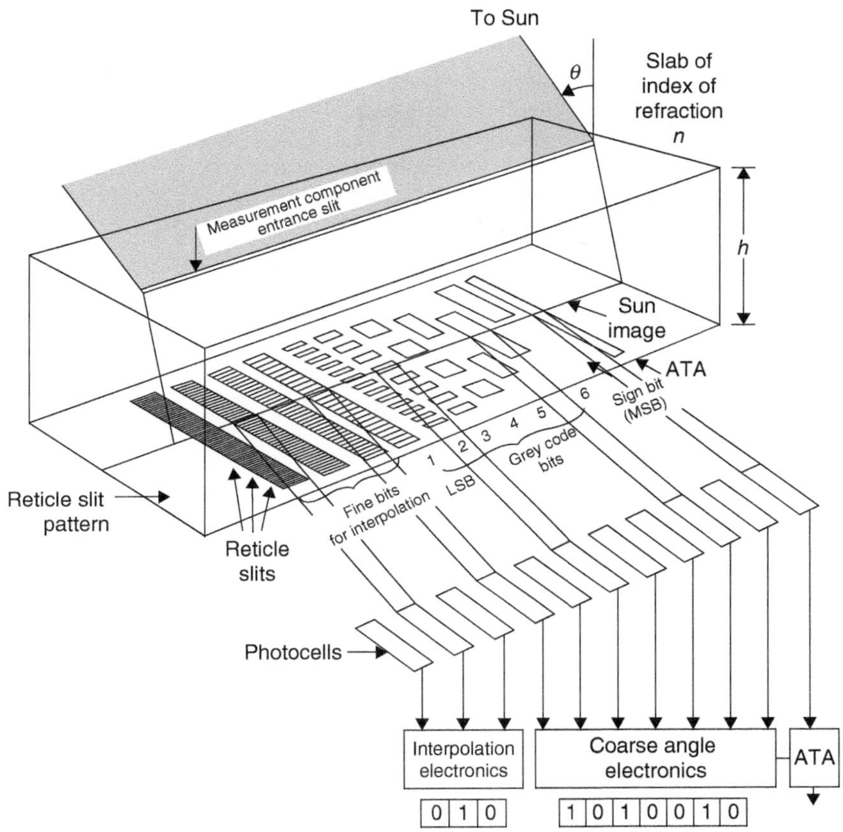

FIGURE 11.27 Digital sun sensor operation.

attitude determination: (1) spinning mass, (2) laser, (3) ring laser, (4) fiber optic, and (5) hemispherical resonator:

1. Spinning mass gyroscope has a mass spinning steadily with a free movable axis (gimbal). When the axis is titled the gyroscopic effect causes precession (motion orthogonal to the tilt sense) on the rotating mass axis, therefore letting you know that the angle moved. Because of mechanical constraints, the idea came up for fixing the axis with springs where the spring tension is proportional to the precession speed. By integrating the spring tension one can get angle.
2. Laser gyroscopes splits the beam of laser light into two beams in opposite directions through narrow tunnels in a closed optical circular path around the perimeter of a temperature stable triangular glass block with reflecting mirrors placed in each corner. When the gyro is rotating at some angular rate, the distance traveled by each beam becomes different—the shorter path being opposite to the rotation. The phase-shift between the two beams is measured by an interferometer and is proportional to the rate of rotation (Sagnac effect).

3. Ring laser gyroscopes uses interference of laser light within a bulk optic ring to detect changes in orientation and spin. It is an application of the Sagnac effect.
4. Fiber optic gyroscope is a device that uses the interference of light to detect mechanical rotation. The sensor is a coil with as much as 5 km of optical fiber. Two light beams travel along the fiber in opposite directions. Due to the Sagnac effect, the beam travels against the rotation experiences, a slightly shorter path than the other beam. The resulting phase shift affects how the beams interfere with each other when they are combined. The intensity of the combined beam then depends on the rotation rate of the device.
5. Hemispherical resonator gyroscopes use a standing wave induced in a brandy snifter, then the snifter is tilted, and the waves continue in the same plane of movement and do not fully tilt with the snifter. This trick is used to measure angles. Instead of brandy snifters the system uses hollow globes made of piezoelectric material. The electrodes used to induce and sense the standing waves are evaporated directly onto the material. This system has almost no moving parts and is very accurate.

Star Trackers

A star tracker is a device that measures the direction of a star in the spacecraft body frame. By comparing these measurements with known star directions stored in a star catalog, the orientation of the spacecraft is determined. Star trackers are the most accurate means of attitude determination, with accuracies down to arc seconds. The computer hardware/software to process the star tracker data is more complex and extensive than any other attitude sensor. One drawback is that star trackers are subject to occultations and interference from the sun, earth, and other bright light sources. But in spite of these drawbacks, their accuracy and versatility has resulted in their use in a variety of spacecraft. A star tracker commonly has the following basic components: (1) sun shade, (2) optical system, (3) image definition device, (4) detector, and (5) electronic assembly. Sun shades are designed to improve sensor performance by protecting the optical system from sunlight and scattered light reflected by dust particles, thruster exhaust particles, and other parts of the spacecraft. In spite of protection provided by the sun shade, star sensors are usually unusable within 30–60° from the sun. The optical system consists of mainly of a lens. This lens projects the star image onto a focal plane. The image deflection device selects a small portion of the sensor's FOV (called instantaneous field of view, IFOV), which contains the star image. This can either be a reticle on an opaque plate or an image dissector tube. In an image dissector tube, the IFOV electronically scans the FOV. The detector (such as photomultiplier) transforms the optical signal (light) to an electronic signal. The electronics assembly receives the signal from the detector and processes it before sending it to the attitude determination software. More information on star trackers can be found in Sidi (1997).

References

Acord, J. D. and Nicklas, J. C. 1964. "Theoretical and Practical Aspects of Solar Pressure Attitude Control for Interplanetary Spacecraft," in *Progress in Astronautics and Aeronautics*, vol. 13; *Guidance and Control II*, R. C. Langford and C. J. Mundo, eds., Academic Press, New York.

Agrawal, B. N. 1986. *Design of Geosynchronous Spacecraft*, Prentice-Hall, New Jersey.

Canon, R. H., Jr. 1962. "Some Basic Response Relations for Reaction—Wheel Attitude Control," *ARS Journal*, Jan.

D'Azzo, J. J. and Houpis, C. 1966. *Feedback Control System Analysis and Synthesis*, McGraw-Hill, New York.

Dougherty, H. J., Scott E. D., and Rodden, J. J. 1968. "Analysis and Design of WHECON – An Attitude Control Concept," Paper no. 68-461, AIAA 2nd Communications Satellite Systems Conference (8-10 April, San Francisco).

Dougherty, H. J., Lebsock, K. L., and Rodden, J. J. 1971. "Attitude Stabilization of Synchronous Communications Satellites Employing Narrow-Beam Antennas," *Journal of Spacecraft and Rockets*, vol. 8, no. 8, pp. 834–841, Aug.

Franklin, G. F., Powell, J. D., and Enaemi-Naeini, A. 1994. *Feedback Control of Dynamics Systems*, Addison Wesley, Massachusetts.

Iorillo, A. J. 1965. "Nutation Damping Dynamics of Axisymmetric Rotor Stabilized Satellites," ASME Winter Meeting, Chicago, Nov.

Iwens, P. P., Fleming, A. W., and Spector, V. A. 1974. "Precision Attitude Control with a Single Body-Fixed Momentum Wheel," AIAA Mechanics and Control of Flight Conference, Paper No. 79-894.

Kaplan, M. H. 1976. *Modern Spacecraft Dynamics and Control*, Wiley, New York.

Lam, Q. M., Stamatakos, M., Woodruff, C., and Ashton, S. 2003. "Gyro Modeling and Estimation of Its Random Noise Sources," AIAA Guidance, Navigation, and Control Conference, Paper No. 2003-5562.

Likins, P. W. 1967. "Attitude Stability Criteria for Dual-Spin Spacecraft," *Journal of Spacecraft and Rockets*, vol. 4, no. 12, pp. 1638–1643, Dec.

Meirovitch, L. 1970. *Methods of Analytical Dynamics*, McGraw-Hill, New York.

Mitiguy, P. 2013. *Advanced Dynamics and Motion Simulation*, Prodigy Press, California.

Sabroff, A. E. 1968. "Advanced Spacecraft Stabilization and Control Techniques," *Journal of Spacecraft and Rockets*, vol. 5, no. 12, pp. 1377–1393, Dec.

Schaub, H. and Junkins, J. L. 2003. *Analytical Mechanics of Space Systems*, AIAA, Virginia.

Sidi, M. J. 1997. *Space Dynamics and Control: A Practical Engineering Approach*, Cambridge University Press, New York.

Thomson, W. T. 1961. *Introduction to Space Dynamics*, Wiley, New York.

Vallado, D. A. 2001. *Fundamentals of Astrodynamics and Applications*, Microcosm, Inc., California.

Wertz, J. R. 1978. *Spacecraft Attitude Determination and Control*, Microcosm, Inc., California.

Wie, B. 1988. *Space Vehicle Dynamics and Control*, AIAA, Virginia.

Wie, B. and Barba, P. M. 1989. "Quaternion Feedback for Spacecraft Large Angle Maneuvers," *Journal of Guidance, Control, and Dynamics*, vol. 12, no. 3, pp. 375–380.

PART 2

Observation Payloads

Jeffery J. Puschell

11.13 Overview

Space-based observation payloads collect data (e.g., imagery, spectral radiance, distance) on remote objects such as the surface of the earth, relatively nearby objects such as another satellite or distant astrophysical objects that can be processed into information (e.g., temperature, reflectance, chemical composition, topography) to provide insight into physical characteristics of these remote objects for scientific, weather forecasting, military, and policy-making purposes—just to name a few. Observation payloads operate across the entire electromagnetic spectrum from gamma rays and X-rays to radio frequencies (RFs). In addition, observation payloads include cosmic ray and other directional particle sensors used to probe the sun, supernovae, active galactic nuclei, and other unknown astrophysical objects.

Observation payload missions have dramatically changed the way we see our world by providing vitally important imagery and remote sensing data on life-threatening hurricanes and other severe weather, surveillance and reconnaissance data of critical importance to military planners and national decision makers and stunning imagery and measurements of distant astrophysical objects at wavelengths not accessible from the earth's surface. Future space-based observation missions are expected to provide continuous video surveillance of the earth, improve weather forecasting by constraining forecast models with more accurate and detailed measurements, contribute to more complete understanding of climate change, and help resolve some of the deepest cosmological mysteries of our time related to dark matter and dark energy.

Observation payloads can be divided into two basic types: so-called passive payloads such as visible-infrared and microwave imagers that observe intrinsic emission from the scene, whether it be reflected sunlight, thermal emission or gamma rays and active payloads such as lidars and radars that supply their own source of light either from the same platform as the receiver or another cooperative illumination source to enhance signal-to-noise ratio and spatial resolution and enable specific types of measurement such as 3D imaging or trace chemical detection. Passive sensor examples include visible-infrared imagers like the Visible Infrared Imaging Radiometer Suite (VIIRS) for the NOAA/NASA Joint Polar Satellite System (JPSS) that observes reflected sunlight and thermal emission from the scene and passive microwave radiometers like the NASA/JAXA Global Precipitation Measurement (GPM) Microwave Imager (GMI) that detect thermal emission from the earth's surface and atmosphere. Radars and laser-based sensors like DLR's TerraSAR-X and NASA's ICESAT are examples of active systems.

Earth observations occur in three primary spectral regions known as solar reflectance, thermal infrared, and microwave or radio frequency. Space astronomy payloads like NASA's Chandrasekhar X-Ray Observatory and Compton Gamma Ray Observatory have observed objects at X-ray and gamma ray wavelengths, in addition to these four primary spectral regions. Schowengerdt (2009) is an excellent source for detailed discussion of solar reflectance and thermal infrared systems. Sharkov (2003) presents a comprehensive treatment of passive microwave observation systems. Pillai et al. (2008) provide extended discussions of active space-based RF systems.

Solar reflectance systems operate from the near ultraviolet (~0.35 μm) to the shortwave infrared (SWIR) (~2.5 μm) by observing reflected sunlight. These systems offer the potential for higher spatial resolution than other spectral regions because of their relatively short wavelengths, but typically operate only in daylight, unless a laser or other source of active scene illumination is used (cf. Nischan et al. 2003) or the system has special design features to detect very low levels of light from reflected moonlight, artificial lighting, or airglow (cf. Elvidge et al. 2013).

Thermal infrared or emissive systems operate in the spectral region dominated by thermal emission from commonly observed objects on earth (~3–300 μm). Objects of interest include the surface of the earth (~300 K), the upper atmosphere (~220 K), biomass fires (~600–1,200 K), and missile launches (~300–1,600 K). The thermal infrared region is affected by absorption of light by molecules in the atmosphere. Observations of the earth's surface are possible only within certain wavelength ranges known as atmospheric windows, which occur in between strong molecular absorption bands. These strong absorption bands provide a means for probing temperature structure of the atmosphere along with the 3D distribution of trace gases such as water vapor and ozone. Thermal infrared sensors can operate both day and night because the detected signal is from thermal emission by the scene itself and does not rely on reflected sunlight.

Microwave/RF systems operate at millimeter and centimeter wavelengths. Passive microwave radiometers operate chiefly at millimeter wavelengths (20–200 GHz). Spatial resolution of these systems is three to five orders of magnitude worse than the visible wavelength sensors with the same aperture size, but they are capable of collecting unique information over large areas. Active RF systems also known as radars provide their own illumination of the scene in the centimeter to millimeter bands. The reflected signals can be processed to identify physical features in the scene. Radar systems can be designed to penetrate most clouds because the water droplets and ice particles inside clouds are too small to scatter longer radar wavelengths effectively.

11.14 Observational Payload Types

Space-based observation payloads can be categorized into the following major system types: passive solar reflectance, emissive or thermal infrared, passive microwave, X-ray, gamma ray, active electro-optical (EO), and active RF, including real aperture and synthetic aperture systems. Table 11.1 summarizes advantages and disadvantages of each type along with missions generally addressed by each kind of system. The discussion below describes these observational payload types in detail.

Passive solar reflectance systems observe that part of the electromagnetic spectrum dominated by reflected sunlight—the ultraviolet (~0.3 μm) through the SWIR at ~2.5 μm. These systems offer the potential for relatively high spatial resolution from space because their shorter operating wavelengths diffract less than longer wavelengths. However, because they rely on reflected sunlight for earth observations, most of these systems operate effectively only during daytime. Observation payloads in this category include imagers like the DigitalGlobe commercial imagers that specialize in collecting high spatial resolution imagery in relatively few spectral bands, multispectral (~20 spectral bands with $\lambda/\delta\lambda$ ~10 typically) synoptic imagers like the geosynchronous Advanced Baseline Imager (ABI) onboard GOES-16 and polar orbiting environmental imagers like VIIRS onboard the Suomi National Polar-Orbiting Partnership (NPP) satellite that provide routine global-scale observations (cf. Figure 11.28 and Table 11.2), hyperspectral (~100 spectral band with $\lambda/\delta\lambda$ ~100 across a broad contiguous spectral region) imagers such as ARTEMIS and Hyperion that provide more detailed spectral measurements over limited spatial regions, low-light-level imagers like the VIIRS Day-Night Band that provide panchromatic imagery of moonlit, artificially, or airglow-lit scenes on earth at night (Figure 11.29) and polarimeters such as POLDER that provide measurements of linear polarization of spectral radiance in different spectral bands across the entire solar reflectance spectral region.

Observation payload type	Passive solar reflectance	Passive emissive	Passive microwave	Passive X-ray and gamma ray	Active electro-optical	Active RF–real aperture radar	Active RF–synthetic aperture radar
Features/ Benefits	Mature, efficient technology, high spatial resolution even with relatively small apertures, rich phenomenology across many missions	Mature, efficient technology, rich phenomenology across many missions, effective day or night	Mature technology, rich phenomenology, effective day or night and through virtually all weather conditions	Enables detection of high energy events on Earth and throughout the Universe	Day/night operation, capability to map surfaces, clouds, winds, trace gas concentrations in 3-D	Day/night, all weather capability for detecting precipitation, mapping clouds, characterizing large scale features and altimetry	Day/night, all weather performance with spatial resolutions comparable to much shorter wavelength systems because of the large synthetic aperture
Concerns	Less effective at night for Earth observations except with special low light level technology, does not penetrate dense clouds in planetary atmospheres or dust in nebulae and elsewhere in galaxies	High sensitivity systems require cryogenic cooling, achievable spatial resolution is not as good as solar reflectance systems for same system aperture	Relatively coarse spatial resolution even for large apertures	Limited angular resolving power and aperture efficiency, technology less commonly available than for other passive systems	Long lived reliable lasers for operation in space remain a challenge, area coverage rate limited by laser pulse repetition frequency and system sensitivity, more complex than passive EO systems and require more power	Coarse spatial resolution, because of practical limits on aperture collector size and relatively long wavelength, long lived reliable radar transmitters remain a challenge, more complex than passive RF systems and require more power	High spatial resolution available with limited area coverage, significant processing required for creating synthetic aperture, long lived reliable radar transmitters remain a challenge, more complex than passive RF systems and require more power

Observation Payload Missions							
Operational Environmental Remote Sensing/Earth Science	Land, ocean color, coastal water, clouds, aerosols imaging for weather forecasting, climate change understanding and forecasting, air pollution monitoring	Land surface temperature, sea surface temperature, cloud top heights, cloud properties, temperature and water vapor sounding for weather forecasting, climate change understanding and forecasting, air pollution monitoring, biomass fire detection and monitoring	Temperature and water vapor soundings, quantitative precipitation measurements, sea ice characterization, soil moisture		Aerosol, cloud and trace gas characterization, surface topography, flood plain mapping, sea ice characterization, vegetation canopy characterization, altimetry, wind velocity, underwater remote sensing	Cloud and precipitation mapping, low horizontal resolution altimetry	Detailed mapping of topography, vegetation, deformation in Earth surface, subsurface structures, ocean waves, ocean wind, sea ice, soil moisture,
Military Reconnaisance/ Surveillance	Damage assessment, mission planning, intelligence, camouflage counter-measures	Damage assessment, mission planning, intelligence, camouflage countermeasures, fire detection and monitoring	Soil moisture measurement, all weather, day/ night	Detection of nuclear device detonation	3-D imaging, altimetry, camouflage mitigation		All weather day/ night high resolution surveillance and reconnaisance, foliage and ground penetrating radar, detection and tracking of moving vehicles, detection and tracking of ocean vessels, soil moisture maps
Missile Warning/ Defense	Detection of missile launches and tracking of missiles in flight	Detection of missile launches and tracking of missiles in flight along with physical characterization of plume and reentry vehicles			Very accurate measurements of velocity and location of missiles, plumes and reentry vehicles		All weather, day/ night strategic reconnaisance and surveillance of known and possible missile launch sites

Table 11.1 Observational Payload Types: Features/Benefits, Concerns, and Sample Missions

Observation payload type	Passive solar reflectance	Passive emissive	Passive microwave	Passive X-ray and gamma ray	Active electro-optical	Active RF–real aperture radar	Active RF–synthetic aperture radar
Planetary Science	Imaging of planets, moons and planetary systems to determine physical characteristics and composition	Imaging of planets, moons and planetary systems to characterize atmosphere and surface temperature	Imaging of planets, moons and planetary systems to characterize atmosphere and surface temperature		Characterization of topography and atmosphere of other planets, moons and planetary systems	Low resolution altimetry, atmospheric mapping	Detailed mapping of topography, deformation in planetary surfaces, subsurface structures and moisture through thick clouds and without sunlight
Observational Astrophysics	Imaging of distant astrophysical objects in our Galaxy and beyond	Imaging of distant astrophysical objects in our Galaxy and beyond	Detection of molecules in distant astrophysical objects	Characterization of high energy events in the Sun, supernova remnants, center of our Galaxy, active galactic nuclei and unknown objects			
Commercial Remote Sensing	High spatial resolution imaging of urban, agricultural and forest areas	High spatial resolution imaging of urban areas to monitor energy usage in residential and industrial areas	Commercial weather forecasting, continuous tracking of ships at sea		Altimetry, vegetation characterization	Cloud and precipitation mapping, low horizontal resolution altimetry	All weather day/night high resolution monitoring of urban areas including highways, airports and harbors, foliage and ground penetrating radar, detection and tracking of moving vehicles, detection and tracking of ocean vessels, soil moisture maps, detailed topography mapping

Table 11.1 Observational Payload Types: Features/Benefits, Concerns, and Sample Missions (*Continued*)

FIGURE 11.28 Spectacular solar reflectance natural color image of Hurricane Darby and Tropical Storm Emily collected on July 17, 2016, by the VIIRS on the Suomi NPP satellite. (*NASA*.)

Emissive or thermal infrared systems observe that part of the electromagnetic spectrum dominated by thermal emission from the scene itself in atmospheric transmission windows from approximately 3 to 300 μm in wavelength. In some cases, these systems operate outside atmospheric transmission windows to enable less cluttered observations high above the surface of the earth. These systems operate effectively day or night because they do not rely on sunlight to create an observable signal. Like solar reflectance systems, emissive systems include multispectral imagers like VIIRS and ABI (Table 11.3) that was preceded by the nearly identical Advanced Himawari Imager for Japan on Himawari-8 and Himawari-9 (Figure 11.30), hyperspectral systems like the Thermal Emission Spectrometer that flew on Mars Global Surveyor to map mineralogy on Mars and search for evidence of water (Figure 11.31) and ultraspectral (~1,000 spectral band with $\lambda/\delta\lambda$ ~1,000 across a broad contiguous spectral region) systems like the Atmospheric Infrared Sounder onboard the NASA Aqua satellite, the Cross-track Infrared Sounder for JPSS and the Infrared Atmospheric Sounding Interferometer onboard the EUMETSAT MetOp satellites that provide detailed spectra of the earth's atmosphere and surface to enable 3D instantaneous maps of temperature distribution on the surface and in the atmosphere along with maps of water vapor and other trace gases.

Passive microwave systems observe thermal emission at much longer wavelengths than emissive infrared systems. The microwave part of the electromagnetic spectrum extends from approximately 1 mm to 1 m in wavelength. Normally, microwave radiation is referenced to frequency rather than wavelength. In frequency, the microwave spectral region ranges from approximately 0.3 to 300 GHz. Like emissive infrared systems, passive microwave systems operate effectively day or night. At frequencies less than 10 GHz, microwave radiation is relatively unaffected by clouds, so that these longer wavelength systems can operate under virtually all weather conditions. For higher frequency microwave systems, clouds and fog along

	Band No.	Wavelength (μm)	Horiz sample interval (km downtrack × crosstrack)		Driving EDRs	Radiance range	Ltyp or Ttyp
			Nadir	End of scan			
VIS/NIR FPA Silicon PIN Dioders	M1	0.412	0.742 × 0.259	1.60 × 1.58	Ocean Color	Low	44.9
					Aerosols	High	155
	M2	0.445	0.742 × 0.259	1.60 × 1.58	Ocean Color	Low	40
					Aerosols	High	146
	M3	0.488	0.742 × 0.259	1.60 × 1.58	Ocean Color	Low	32
					Aerosols	High	123
	M4	0.555	0.742 × 0.259	1.60 × 1.58	Ocean Color	Low	21
					Aerosols	High	90
	I1	0.640	0.371 × 0.387	0.80 × 0.789	Imagery	Single	22
	M5	0.672	0.742 × 0.259	1.60 × 1.58	Ocean Color	Low	10
					Aerosols	High	68
	M6	0.746	0.742 × 0.776	1.60 × 1.58	Atmospheric Corr'n	Single	9.6
	I2	0.865	0.371 × 0.387	0.80 × 0.789	NDVI	Single	25
	M7	0.865	0.742 × 0.259	1.60 × 1.58	Ocean Color	Low	6.4
					Aerosols	High	33.4
CCD	DNB	0.7	0.742 × 0.742	0.742 × 0.742	Imagery	Var.	6.70E-05
S/MWIR PV HgCdTe (HCT)	M8	1.24	0.742 × 0.776	1.60 × 1.58	Cloud Particle Size	Single	5.4
	M9	1.378	0.742 × 0.776	1.60 × 1.58	Cirrus/Cloud Cover	Single	6
	I3	1.61	0.371 × 0.387	0.80 × 0.789	Binary Snow Map	Single	7.3
	M10	1.61	0.742 × 0.776	1.60 × 1.58	Snow Fraction	Single	7.3
	M11	2.25	0.742 × 0.776	1.60 × 1.58	Clouds	Single	0.12
	I4	3.74	0.371 × 0.387	0.80 × 0.789	Imagery Clouds	Single	270 K
	M12	3.70	0.742 × 0.776	1.60 × 1.58	SST	Single	270 K
	M13	4.05	0.742 × 0.259	1.60 × 1.58	SST	Low	300 K
					Fires	High	380 K
LWIR PV HCT	M14	8.55	0.742 × 0.776	1.60 × 1.58	Cloud Top Properties	Single	270 K
	M15	10.763	0.742 × 0.776	1.60 × 1.58	SST	Single	300 K
	I5	11.450	0.371 × 0.387	0.80 × 0.789	Cloud Imagery	Single	210 K
	M16	12.013	0.742 × 0.776	1.60 × 1.58	SST	Single	300 K

Table 11.2 VIIRS Spectral Bands with Horizontal Sampling Interval and Typical Scene Spectral Radiance in Units of W m^{-2} μm^{-1} sr^{-1} or Typical Scene Brightness Temperature in K for Each Radiance Range (Puschell et al)

Figure 11.29 Composite image of the continental United States at night assembled from data acquired by the "Day-Night Band" of the VIIRS onboard the Suomi NPP satellite in April and October 2012. (*NASA Earth Observatory/NOAA NGDC.*)

with absorption by water and oxygen molecules in the atmosphere can influence the signal from the surface. Well-known passive microwave systems include GMI, AMSU, SSM/I, and SSMI/S (Figure 11.32). GMI is a multichannel, conical-scanning, microwave radiometer to measure precipitation from space. GMI has 13 microwave channels ranging in frequency from 10 to 183 GHz. In addition to carrying channels similar to those on the Tropical Rainfall Measuring Mission (TRMM) Microwave Imager (TMI), the GMI carries four high frequency, millimeter-wave, channels at about 166 and 183 GHz. AMSU is used primarily for mapping temperature and water vapor structure in the atmosphere. The system flying today is the most recent version of systems that have operated onboard the TIROS/POES satellites since 1978. SSM/I is used primarily for near-surface wind speed, total column water vapor, total column cloud liquid water, and precipitation. It has been onboard DMSP satellites since 1987, but has heritage back to the Nimbus 7 and Seasat missions launched in 1978. SSM/I operates in seven channels ranging from 19 to 89 GHz in frequency. AMSU is divided into two parts called A (used mostly for temperature sounding) and B (used mostly for water vapor sounding), which together operate in 20 spectral channels ranging from 23.8 to 183.3 GHz in frequency and with different spectral resolutions for some frequencies. AMSU-A has a spatial sample size at nadir of about 45 km. The higher frequency AMSU-B has a spatial sample size of about 15 km at nadir. SSM/I operates with spatial resolutions ranging from 13 km to 69 km, depending on frequency. The SSMI/S instrument is the follow-on instrument to SSM/I and measures microwave energy at 24 discrete frequencies from 19 to 183 GHz with a swath width of 1,700 km at 12.5 to 75 km spatial resolution at nadir. The synoptic solar reflectance and emissive infrared systems used in conjunction with these passive microwave systems operate with approximately 1 km spatial

GOES imager (ABI) band	Wavelength range (μm)	Central wavelength (μm)	Nominal subsatellite IGFOV (km)	Sample use
1	0.45–0.49	0.47	1	Daytime aerosol over land, coastal water mapping
2	0.59–0.69	0.64	0.5	Daytime clouds fog, insolation, winds
3	0.846–0.885	0.865	1	Daytime vegetation/burn scar and aerosol over water, winds
4	1.371–1.386	1.378	2	Daytime cirrus cloud
5	1.58–1.64	1.61	1	Daytime cloud-top phase and particle size, snow
6	2.225–2.275	2.25	2	Daytime land/cloud properties, particle size, vegetation, snow
7	3.80–4.00	3.90	2	Surface and cloud, fog at night, fire, winds
8	5.77–6.6	6.19	2	High-level atmospheric water vapor, winds, rainfall
9	6.75–7.15	6.95	2	Midlevel atmospheric water vapor, winds, rainfall
10	7.24–7.44	7.34	2	Lower-level water vapor, winds, and SO_2
11	8.3–8.7	8.5	2	Total water for stability, cloud phase, dust, SO_2 rainfall
12	9.42–9.8	9.61	2	Total ozone, turbulence, and winds
13	10.1–10.6	10.35	2	Surface and cloud
14	10.8–11.6	11.2	2	Imagery, SST, clouds, rainfall
15	11.8–12.8	12.3	2	Total water, ash, and SST
16	13.0–13.6	13.3	2	Air temperature, cloud heights, and amounts

TABLE 11.3 Summary of the Wavelengths, Resolution, and Sample use of the ABI Bands (The minimum and maximum wavelength range represent the full width at half maximum (FWHM or 50%) points) [The Instantaneous Geometric Field of View (IGFOV)] (Schmit et al)

resolution using much smaller aperture sizes, because of the diffraction performance advantage of shorter wavelengths.

X-ray imagers observe thermal and nonthermal emission from astrophysical objects including the sun, neutron stars, supernovae, the center of our Galaxy, and distant unknown objects at wavelengths ranging from less than 3 pm to 10 nm, corresponding to frequencies ranging from above 10 EHz (1 EHz is 10^{18} Hz) down to 30 PHz (1 PHz is 10^{15} Hz). X-rays are often referenced in units of electron volts (eV) rather than wavelength or frequency. In eV, the X-ray spectral region extends from about 0.1 to 400 keV. The first X-ray observations from space were made with Geiger counters that flew onboard converted V-2 rockets starting in 1949. UHURU also known as SAS-1 launched in 1970 was the first satellite with an X-ray observation payload. The Chandra X-ray Observatory, launched in 1999, observes the 0.1 to 10 keV spectral region, with 2.4 μrad (0.5 arcsec) angular resolution using a 1.2-m diameter X-ray telescope. The satellite is in a highly elliptical

FIGURE 11.30 First full disk emissive infrared image in the 13.3-μm band taken by the imager onboard Japan's Himawari-8 satellite on December 18, 2014. (*Japan Meteorological Agency.*)

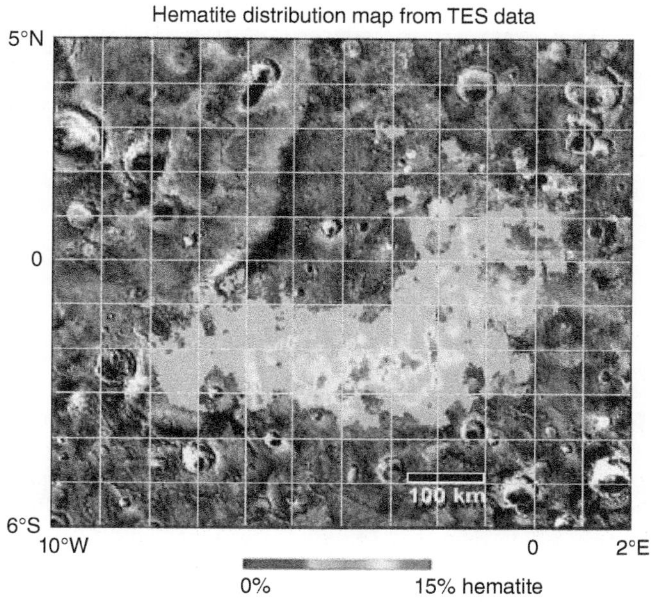

FIGURE 11.31 Analysis of hyperspectral data from the infrared spectrometer TES onboard Mars Global Surveyor found evidence for past surface water on Mars by mapping the spectral signature of hematite, an indicator of dried-up lakes. (*Christensen et al. 2001.*)

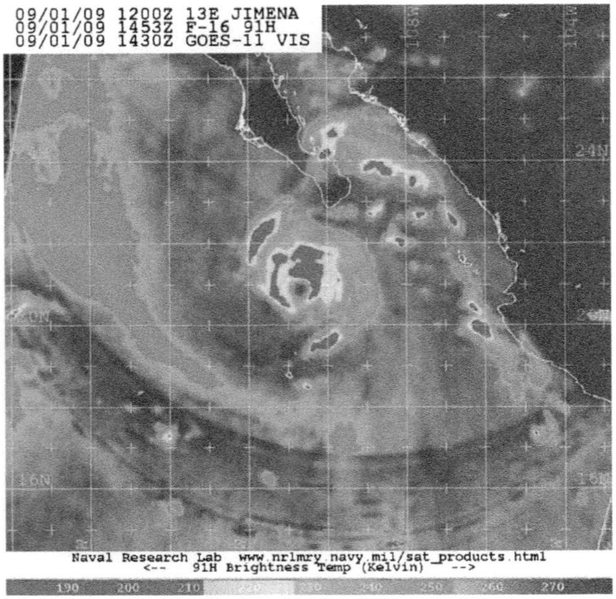

FIGURE 11.32 91-GHz brightness temperatures (in K) measured by the SSMI/S passive microwave sensor on DMSP F-16, on September 1, 2009. The spiraling patterns near the center indicate thunderstorms associated with Hurricane Jimena. Courtesy of the U.S. Air Force and the Naval Research Laboratory. (*U.S. Air Force and the Naval Research Laboratory.*)

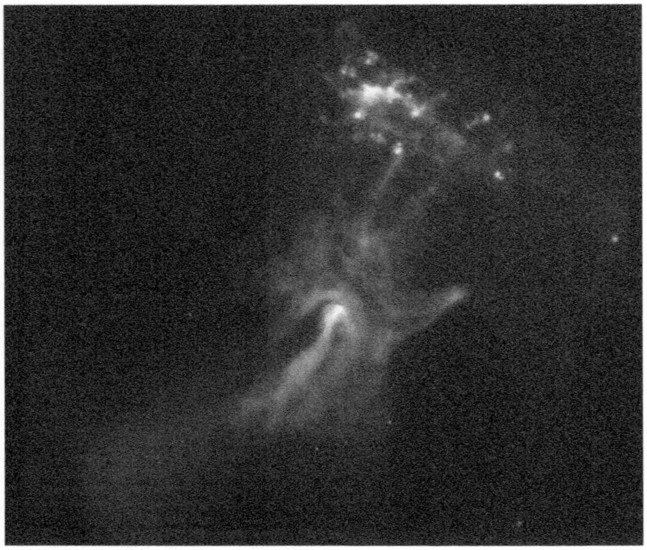

FIGURE 11.33 Chandra X-ray observatory image of the region near the pulsar PSR B1509-58. This rapidly spinning neutron star, which was discovered by the Einstein X-ray observatory in 1982, is approximately 17,000 light years from earth. The nebula surrounding the pulsar is about 150 light year in extent. (*NASA.*)

orbit that enables the observatory to make X-ray measurements (Figure 11.33) from well out the earth's Van Allen belts. The effective collection area for the Chandra Observatory is only 0.04 m^2 at 1 keV, despite the much larger telescope diameter. Conventional optical approaches are not effective at X-ray wavelengths, because no suitable refractive material is available to build X-ray lenses and standard reflector telescopes do not work because X-rays are either absorbed or transmitted at near normal incidence to reflecting mirrors. Therefore, X-ray instruments use either a coded aperture method or a grazing incidence telescope also known as Wolter telescope to collect and focus X-rays. In Wolter telescopes, the angle of reflection from the mirror is very low—typically 10 arc minutes to 2 deg. Chandra uses a Wolter telescope that consists of nested cylindrical paraboloid and hyperboloid surfaces coated with iridium or gold. Chandra uses four pairs of nested mirrors. The relatively thick substrate (2 cm) and very careful polishing allowed a very precise optical surface, which is largely responsible for Chandra's unprecedented and unmatched angular resolution. However, thickness of the substrates limits the fill factor of the aperture, leading to the low-effective collection area.

Gamma ray payloads observe emission of the highest energy photons from astrophysical objects including the sun, supernovae, the center of our Galaxy, and distant unknown objects. Gamma rays do not penetrate earth's atmosphere very well so that direct astronomical observations of gamma rays must occur from high in the atmosphere or from space. Wavelengths for gamma rays are shorter than 3 pm, corresponding to frequencies greater than 10 EHz (1 EHz is 10^{18} Hz). The first gamma ray observations of the sun and distant galactic and extragalactic objects were made by OSO-3 launched in 1967. Shortly thereafter, mysterious gamma ray bursts from very distant, but still unidentified objects were discovered by the Vela satellites designed to detect gamma rays from nuclear explosions on earth. Recent gamma ray observation payloads include the instruments onboard the Fermi Gamma Ray Telescope, a successor to the Compton Gamma Ray Observatory. These instruments detect gamma rays in two broad spectral ranges using scintillation detectors that detect small flashes of light created by gamma rays passing through a crystal and pair production techniques at the highest energies that create electron-positron pairs created by incoming gamma rays.

Active EO systems like lidars, ladars, and altimeters use active illumination of the scene by a laser to create a measurable signal that can be read out rapidly over time to remotely sense backscattered light from water droplets, aerosols and trace gases in the atmosphere as a function of distance from the system and for range measurements and altimetry, when looking at the earth, moon, and other space objects. The first lidar in space was the Lidar In-space Technology Experiment (LITE) carried onboard the Space Shuttle in 1994. LITE measured clouds with a backscatter lidar based on a Nd:YAG laser operating in three different wavelengths: 1,064, 532 (second harmonic), and 355 nm (third harmonic). Figure 11.34 illustrates the mission concept for LITE. CALIPSO, launched as part of the A-train of Earth Science satellites in 2006, is also a backscatter lidar based on a Nd:YAG laser. CALIPSO transmits both the fundamental (1,064 nm) and second harmonic (532 nm) in a nadir beam. The vertical resolution is 30 m and horizontal resolution 333 km. ICESAT launched in 2003 carried a Nd:YAG-based laser altimeter used primarily for measuring thickness of Arctic ice sheets, but also for providing global topography and vegetation data. The main challenge for building a laser-based observational space payload is producing a reliable laser for extended duration use in space—previous instruments like ICESAT have experienced failure of some laser modules after only a few months of operation. Lessons learned from ICESAT benefit future missions such as the upcoming AEOLUS mission with the ALADIN lidar, planned for the Atmospheric Dynamics mission. ALADIN is a wind-measuring lidar, based on the third harmonic of a Nd:YAG laser. ALADIN will measure wind speed along the lidar line of sight by measuring the Doppler shift of the backscattered light using two methods: coherent (heterodyne) detection by mixing the return signal with a local oscillator and a direct direction method using a narrowband etalon.

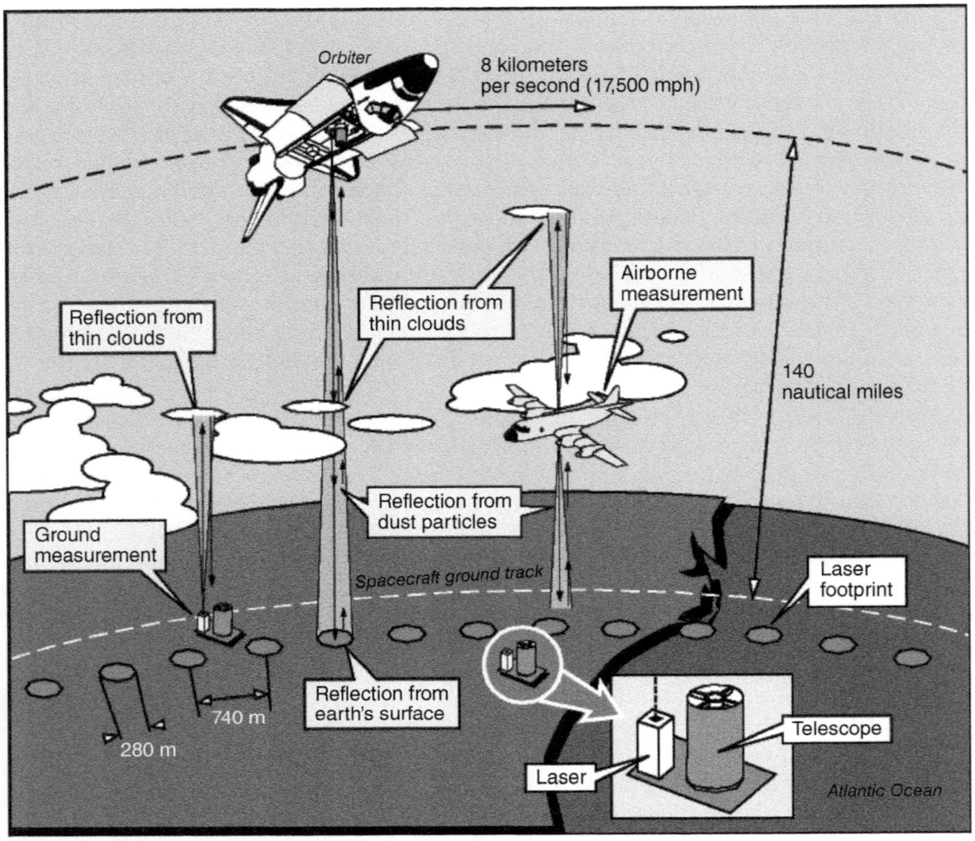

FIGURE 11.34 Mission concept for the shuttle-based LITE lidar experiment. Laser light from the shuttle illuminates thin clouds, dust particles, and the earth's surface. Light reflected back to LITE's telescope is measured over with high temporal resolution to measure high and extent of clouds and particles above the earth. Measurements from the ground and from aircraft provided ground truth information. (*NASA*.)

These techniques will be applied to measuring Doppler shifts of light returned to receiver from Mie (aerosol) and Rayleigh (molecular) scattering, respectively. While useful as a pathfinder for future winds lidar missions for providing information not available today to numerical weather forecast models, AEOLUS will only measure a single wind component.

Active RF systems like real aperture radars and synthetic aperture radars (SARs) use active illumination of the scene by a coherent RF transmitter to create a signal that can be detected and read out rapidly in time to produce detailed range resolved RF imagery of precipitation in clouds, surface wakes on the ocean, and detailed topography of earth and other planetary solid surfaces. Like passive microwave systems, active RF systems operate at much longer wavelengths than visible and infrared systems that are routinely used for imagery—typically at centimeter wavelengths. Consequently, for practical payload sizes, active RF systems tend to operate at much coarser spatial resolution than visible and infrared systems—but they can operate through almost any possible weather condition. An example is the precipitation radar (PR) onboard the TRMM satellite. The TRMM PR launched in 1997 was the first spaceborne radar designed to measure the vertical structure of rainfall precipitation in the lower atmosphere. It operated at 13.8 GHz (2 cm in wavelength) using a 128-element electronically scanned array to allow fast and complex cross-track scanning over a swath width of 215 km with a cross-range spatial resolution of about 4.3 km and a range resolution of 250 m. As shown in Figure 11.35, the TRMM PR operated in conjunction with other observational payload systems—a passive microwave imager and a visible-infrared imager. The TRMM PR was replaced by a similar PR in the Core Observatory of the GPM mission launched in 2014. TRMM-like systems are very useful, but they cannot provide spatial resolution approaching visible-infrared systems with feasible collector sizes. A different type of radar takes advantage of the relatively relaxed tolerances at these long RF wavelengths for coherent phase and Doppler shift measurements of a radar return signal. SAR processes coherent measurements of phase and Doppler shift across the wide beam width of a RF system to synthesize a virtual aperture with a size equivalent to the distance the actual antenna moves while the scene location remains illuminated in the beam. The corresponding spatial resolution in the along-spacecraft track direction approaches resolutions achievable with visible and infrared systems. SEASAT launched in 1978 was the first known SAR in space. It operated in L-band at a frequency of 1.275 GHz or 23.5 cm in wavelength. SEASAT operated for only about 105 days, failing when a short circuit in the satellite's electrical system ended the mission. Nevertheless, the mission was very successful and laid the processing foundation for following systems. Since SEASAT, SARs have

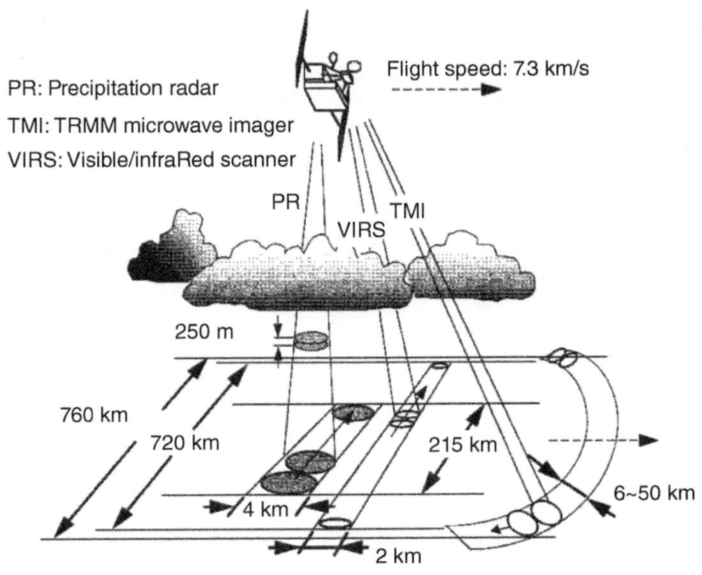

FIGURE 11.35 The TRMM PR observed rainfall in conjunction with a passive microwave imager, the TMI, and a visible infrared imager called VIRS. While the PR and VIRS made cross-track scans, the TMI made a conical scan, similar to the SSM/I and SSMI/S. (*Kozu et al. 2001.*)

Figure 11.36 TerraSAR-X multitemporal high-resolution spotlight image of Sydney, Australia Harbor. (*Weringhaus and Buckreuss 2010.*)

been developed and flown by Canada, Japan, Europe, and Russia. TerraSAR-X, an X-band (9.65 GHz) SAR developed by the German Aerospace Center (DLR) and launched in 2007 has produced remarkable imagery that is especially useful for change detection in urban landscapes (cf. Figure 11.36). Along with its partner satellite, TanDEM-X, launched in 2010, high-resolution digital elevation maps are being produced for flood plain maps and traffic monitoring among many other applications (cf. Figure 11.37).

FIGURE 11.37 TanDEM-X image showing Salar de Uyuni, the largest salt flats in the world, located next to a volcanic region in the Atacama Desert. Darkest areas show lowest lying parts of the salt flats.

11.15 Observational Payload Performance Figures of Merit

Regardless of spectral region or mode of operation, performance of an observational payload is described by figures of merit that depend on the observation geometry, effective sensor aperture size, integration time, detector efficiency and noise characteristics, spectral bandwidth, transmission through the atmosphere, spatial sample size, and spectral radiance from the scene itself.

Signal-to-noise ratio (SNR) is the most widely used observational payload performance figure of merit. For randomly varying signal, SNR is defined as the mean signal divided by standard deviation of the signal statistical distribution. SNR value depends on where signal is measured in the system and how it is detected. For example, SNR measured at the detector front surface is different than SNR measured at the detector electronics output, because the detector itself and associated electronics have noise and possibly gain factors that affect SNR value. Similarly, an observational payload system that detects signal by directly converting photons to photocarriers such as electron-hole pairs will have a different SNR than one that uses heterodyning techniques to mix a signal with a local coherent optical source prior to photodetection. Typically, SNR is calculated at the detector electronics output, but is referenced to the front end of the detector so that the signal and all noise sources are expressed as equivalent amplified photocarriers.

Direct detection photocarrier-based observational payloads are most commonly used in EO space systems. SNR for such a payload is given by

$$\mathrm{SNR} = n_s < G > / (\Sigma \sigma_i^2)^{1/2}$$

where n_s, the mean number of signal photocarriers created in the detector over an integration time t_{int} is given by

$$n_s = \eta R_s A \Omega \tau \Delta\lambda t_{int}$$

and η is detector quantum efficiency in converting photons to carriers, R_s is input signal photon spectral radiance defined as spectral radiance defined by band averaged photon energy, A is payload collector area, Ω is instantaneous solid angle field of view for a single detector, τ is optical transmittance of the payload, $\Delta\lambda$ is the measurement spectral bandwidth, and $<G>$ is the average detector gain. The sum of the individual variances σ_i^2 in the denominator is given by

$$\Sigma\sigma_i^2 = <G^2> (n_s + n_b + n_d) + \sigma_q^2$$

where $<G^2>$ is the average of the square of the gain, n_b is the number of photocarriers created by background light inside the payload during t_{int}, n_d is the number of equivalent photocarriers resulting from detector dark noise, and σ_q^2 is the variance due to all other noise sources in units of equivalent amplified detector photocarriers.

Other well-known figures of merit include

- Noise equivalent spectral radiance (NESR or NEDN) or minimum detectable signal: mean input signal that produces SNR = 1 given by

$$\text{NESR} = (<G^2> (n_s + n_b + n_d) + \sigma_q^2)/(<G> \eta A \Omega \tau \Delta\lambda t_{int})$$

- Sensitivity: signal corresponding to the minimum SNR required for acceptable measurements for a specific mission.
- Noise equivalent difference in temperature (NEDT): temperature uncertainty corresponding to system noise for a specified blackbody scene temperature and optical frequency or wavelength given by

$$\text{NEDT} = \text{NESR}/(\partial B(\lambda, T_{scene})/\partial T)$$

  for Planck function $B(\lambda, T_{scene})$ and scene temperature, T_{scene}

- Modulation transfer function: image quality performance defined by ability of observational payload to transfer object contrast to the observed image
- Uncertainties in mission specific data products such as geophysical parameters (e.g., sea surface temperature, ozone column density) that are affected by algorithm noise such as errors in atmospheric characteristics or effect of approximations used in data analysis in addition to observational payload measurement sensitivity
- Calibration: stability, repeatability, and absolute accuracy of observational payload measurements

The following example illustrates how some of these figures of merit are used in observational payload design for a specific mission.

Example: The mission is to measure earth surface temperature globally twice a day using a single satellite with an observational payload that provides 1 km spatial resolution or better using emissive infrared spectral bands at 10.8 and 12.0 μm. Spectral bandwidth is defined by $\Delta\lambda/\lambda = 0.1$ in each band. Required sensitivity in each band corresponds to NEDT of 0.1 K or better for a 300 K scene. Before designing the payload, we need to determine some basic mission parameters, starting with the orbit. Polar sun-synchronous orbits (PSSOs) that pass over each earth location at nearly the same time of day appear to be ideal for this mission because they can provide twice daily global coverage using a wide swath that extends almost from horizon to horizon. Observing this wide swath requires either an extremely wide field of view instrument with a large number of detectors for each spectral band to cover the wide swath like a pushbroom or an instrument with a gimbaled scan mirror or telescope that moves the line of sight of a narrow field of view across the surface with a single detector or possibly a few detectors per band. At emissive infrared wavelengths, a pushbroom imager is much larger than an actively scanned instrument, because the wide field of view requires very large optical elements compared with the size of the scanner. Therefore, a gimbaled telescope approach is selected for this mission to minimize instrument size and to minimize pixel distortion as a function of scan angle versus other scanning approaches. The selected spectral bands are in the 8 to 13 μm window of atmospheric transmission to the surface near the maximum wavelength for room temperature emission, which makes them a good choice for measuring earth surface temperature. These spectral bands should prove useful for other missions, such as measuring cloud top temperature, detecting fires and possibly measuring fire temperature. Selected spatial resolution and NEDT requirements are similar to current systems such as VIIRS, MODIS, and AVHRR. PSSOs with inclinations within 15° of the pole are possible for altitudes up to about 2,000 km. However, above about 1,000 km in altitude, the orbit is well within the inner Van Allen radiation belt, which exposes the payload to life limiting electron and energetic proton radiation. At altitudes below about 750 km, adjacent swaths may not overlap orbit-to-orbit depending on the minimum acceptable satellite elevation angle as seen from earth surface, resulting in coverage gaps that do not meet the global coverage mission requirement. An orbital altitude of 800 km is selected for this mission, because it is high enough to provide global coverage every 12 hours with no gaps and low enough to avoid serious radiation exposure to the payload. For this orbit, the swath width is ±56.6° or about 2,980 km on the surface for a minimum satellite elevation angle of 20°. Swath to swath overlap is greater than about 1.5° in scan angle at all latitudes. As the payload line of sight is scanned out to its maximum scan angle, the projected detector instantaneous field of view (IFOV) onto the surface grows according to the measurement geometry. Projected IFOV on earth grows differently for along-satellite track and cross-track directions, with the cross-track projection at end of scan, producing a much larger ground sample size because the earth surface curves away in the scan direction. At the end of scan, the cross-track IFOV needs to be 194 μrad to provide a 1-km pixel, based on simple geometric considerations. Similarly, the along-track IFOV needs to be about 565 μrad to provide a 1-km pixel at end of scan. Minimum aperture size for this payload is determined by diffraction at the maximum wavelength of 12.0 μm for the smallest IFOV, which is the 194 μrad cross-track IFOV. For a circular aperture, the angular full width of the central diffraction peak is given by $2.44 \times \lambda_{max}/(\text{aperture diameter})$, which corresponds to the smallest IFOV. Solving for aperture diameter results in a minimum aperture diameter of about 0.150 m. Is this aperture diameter large enough to collect enough light during the integration time to meet the NEDT requirement of 0.1 K in both spectral bands at 300 K? Before we can answer that question, another critical measurement parameter constrained by mission measurement geometry and orbital characteristics needs to be determined: integration time. Integration time is the dwell time of the cross-track IFOV over a spot on the earth. This dwell time is determined by the scan rate, which is required to cover the entire swath without gaps as the satellite orbits the earth. Simple orbital dynamics calculations show that the effective ground speed of the satellite for this orbit is 6.62 km s^{-1}. Assuming the scanning telescope needs to rotate through 360° or 2π rad in scan angle during the orbital dwell time of the projected along-track IFOV at nadir, the scan rate is given by

$$2\pi \text{ rad} \times 6.62 \text{ km s}^{-1}/(800 \text{ km} \times 565\ \mu\text{rad}) = 92.0 \text{ rad s}^{-1} = 14.6 \text{ revolutions s}^{-1}$$

which results in an integration time of 194 μrad/92.0 rad s^{-1} or 2.11 μs. Assuming plausible payload performance values for this integration time ($n_b = 200{,}000$, $n_d = 7{,}500$, $\sigma_q^2 = 20{,}000$, $\langle G \rangle = 1$, $\langle G^2 \rangle = 1$, $\eta = 0.7$, $\tau = 0.7$), the number of signal photocarriers at 12.0 μm collected during an integration time is

$$n_s = 0.7(5.41 \times 10^{20} \text{ phot s}^{-1} \text{ m}^{-2} \text{ sr}^{-1} \text{ μm}^{-1})\ (\pi/4\ (0.150 \text{ m})^2)\ (194 \text{ μrad} \times 565 \text{ μrad})\ 0.7 \times 1.2 \text{ μm} \times 2.11 \text{ μs}$$
$$= 1.30 \times 10^6 \text{ photocarriers}$$

with R_s calculated as a photon spectral radiance using the Planck blackbody function for 12.0 μm in wavelength and 300 K scene temperature. SNR is then

$$\text{SNR} = 1.0 \times 1.30 \times 10^6/((1.0\ (1.30 \times 10^6 + 200{,}000 + 7{,}500) + 20{,}000))^{1/2} = 1050$$

which leads to NEDT = 0.071 K at 300 K. Similarly, NEDT for the 10.8 μm band can be shown to be 0.069 K for this set of assumptions. These NEDT values meet mission requirements, but with relatively little margin in performance against these requirements. In general, it is not considered good design practice to proceed at a very early stage of design without substantial margin, because delivered hardware performance may not meet expectations for a variety of reasons. A prudent margin factor in early design is 1.4× against the most challenging requirement, which is NEDT at 12.0 μm. Margin could be built into this system by increasing the aperture size by about 1.4×, but that would result in a much larger instrument that would likely be much more expensive to build, host on a spacecraft and launch. Other design options to increase margin that are expected to be less expensive to implement than increasing instrument size include adding a detector to each spectral band in the along-track direction to make it possible to slow down the scan rate by 2× and increase the integration time by 2× and adding a detector to each spectral band in the cross-track direction to provide two independent detector samples of each scene element and enable using a technique known as sample averaging or software time delay and integration to improve SNR and NEDT by the square root of the total number of detectors or 1.4×. Yet another option is to proceed at risk for NEDT at end of scan, while noting that multiple detector samples can be combined at smaller scan angles to improve NEDT by a factor that would vary as a function of scan angle. For example, at nadir, SNR at 12.0 μm for a 1-km pixel would be about 3.5× better than the value of 1050 for a single detector sample corresponding to a 155 × 452-m ground sample, resulting in 5× margin with respect to the mission requirement. All of these options and others would need to be explored by the design team with the mission customer to determine the best course of action for this specific mission.

References

Elvidge C. D., Baugh K., Zhizhin M., and Hsu F. C. 2013. "Why VIIRS data are superior to DMSP for mapping nighttime lights," *Proceedings of the Asia-Pacific Advanced Network*, v. 35, pp. 62–69. http://dx.doi.org/10.7125/APAN.35.7.

Kozu, T. et al. 2001. "Development of Precipitation Radar Onboard the TRMM Satellite," IEEE Transactions on Geoscience and Remote Sensing, vol. 39, pp. 102–116.

Nischan M. L., Joseph R. M., Libby J. C., and Kerekes J. P. 2003. "Active Spectral Imaging," *Lincoln Laboratory Journal*, vol. 14, pp. 131–144.

Pillai, S. U., Li, K. Y., and Himed, B. 2008. *Space-Based Radar*, McGraw-Hill Education.

Puschell, J. J., Ardanuy, P. E., and Schueler, C. F. 2016. "System Engineering of the Visible Infrared Imaging Radiometer Suite (VIIRS): Improvements in Imaging Radiometry Enabled by Innovation Driven by Requirements," *SPIE Proceedings*. doi:10.1117/12.2242302.

Schmit, T. J., Gunshor, M. M., Menzel, W. P., Gurka, J. J., Li, J., and Bachmeier, A. S. 2005. "Introducing the Next-Generation Advanced Baseline Imager on GOES-R," *Bulletin of the American Meteorological Society*, doi:10.1175/BAMS-86-8-1079.

Schowengerdt, R. A. 2009. *Remote Sensing: Models and Methods for Imaging Processing*, Elsevier Science & Technology.

Sharkov, E. A. 2003. *Passive Microwave Remote Sensing of the Earth: Physical Foundations*, Springer Praxis Books.

Weringhaus, R. and Buckreuss, S. 2010. "The TerraSAR-X Mission and System Design," IEEE Transactions on Geoscience and Remote Sensing 48: 606–614.

PART 3

Spacecraft Structures

Kenneth R. Hamm, Jr., and Constantinos Stavrinidis

The objectives of this part are to provide an overview of processes and techniques involved in the design, manufacturing, and verification of spacecraft structures and to show the interaction of spacecraft structures with other spacecraft parts or subsystems and equipment.

11.16 Role of Spacecraft Structures and Various Interfaces

Structures form the skeleton of all spacecraft and support key spacecraft components in desirable locations where various constraints need to be adequately considered, for example, thermal control, fields of view for antenna and sensors, and lengths and weights of cables. The stowed configuration must fit within the launch vehicle's (LV's) payload envelope, meet the weight and center of gravity requirements, and provide access for installing and maintaining components.

Structures need adequately to protect spacecraft components from dynamic environments during ground operations, launch deployment, and mission operations. They need to successfully resist the deployment of antennas and sensors and provide adequate structural stability during their operation. Structures must of course be sufficiently light to allow useful payload mass.

The structural stiffness needs to cover adequately LV frequency constraints in order not to interfere with the dynamics of the LV during flight. Similarly, the structural frequency characteristics of the spacecraft in its deployed configuration must not interfere with the spacecraft's own control system.

The materials used must survive ground, launch, and on-orbit environments, e.g., time-varying applied forces, pressure, humidity, radiation, contamination, thermal cycling, and atomic particles, without rupturing, collapsing, excessively distorting, or contaminating critical components. The structure materials need to support thermal control, and in some cases electrical conductivity.

The manufacturing, handling, and storage processes must be selected/conceived so as not to introduce life-limiting factors.

Structure categories (Figure 11.38) are as follows:

- Primary structure is the backbone, or major load path, between the components of the spacecraft and the LV. It may also provide the static reference plane for sensitive optical measurements and may require tight tolerances on position to control alignment of the instruments during flight operations and measurement.
- Secondary structures might include support beams, booms, trusses, and solar panels. These usually attach directly to the primary structure in some fashion either by bolting, bonding, or some combination.
- Tertiary structures refer to the smallest structures such as enclosure boxes that house electronics and brackets that support electrical cables.

Secondary and tertiary structures are sometimes combined under the secondary category, such as in many of the NASA Standards, where specific requirements for loads/load factors are spelled out for primary and secondary structure.

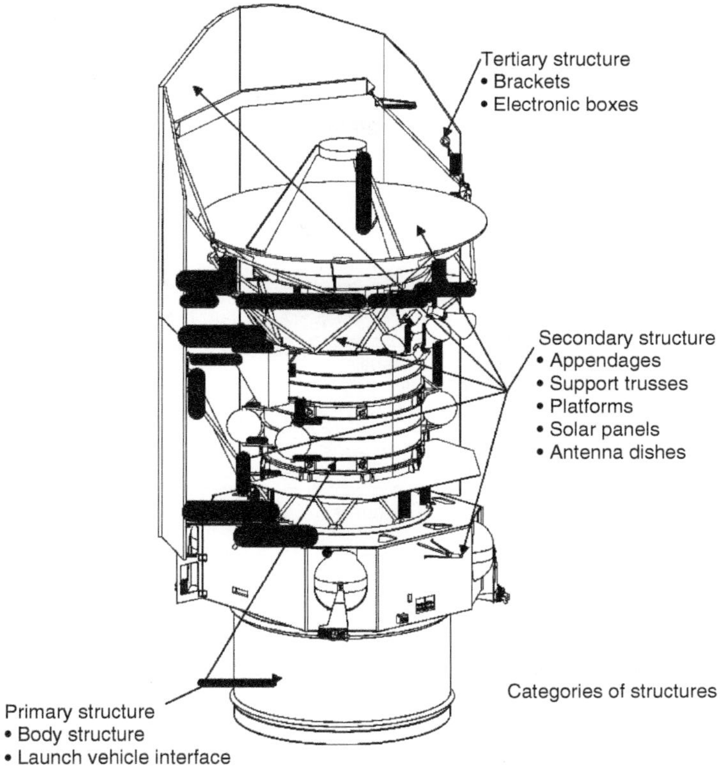

Figure 11.38 Categories of structures. HERSCHEL infrared telescope. (*European Space Agency.*)

Typical characteristics of these categories are

- The primary structure is usually designed for stiffness, or natural frequency, though certain local areas may need to provide adequate strength, especially at interface and/or attachment locations. The primary structure must also be designed to survive steady-state accelerations and transient loading during launch. Another major characteristic of the primary structure is to interface adequately with the launcher.
- The design of secondary structures is influenced significantly by steady-state accelerations and transient loading during launch and is therefore highly affected by the mass and inertia of the components they support. However, loads generated by mission operations, on-orbit thermal cycling, and acoustic pressure during launch are often a more severe environment for lightweight and flexible deployable appendages.
- Stiffness and structural stability over a lifetime including thermal cycling are other driving requirements for primary, secondary, and tertiary structures.

Many times metallic structure can be used. Analysis and overall testing can then be used to verify/validate the design. Composite design processes in many cases require the development and test of many smaller subcomponents in what is referred to as the "Building Block" approach.

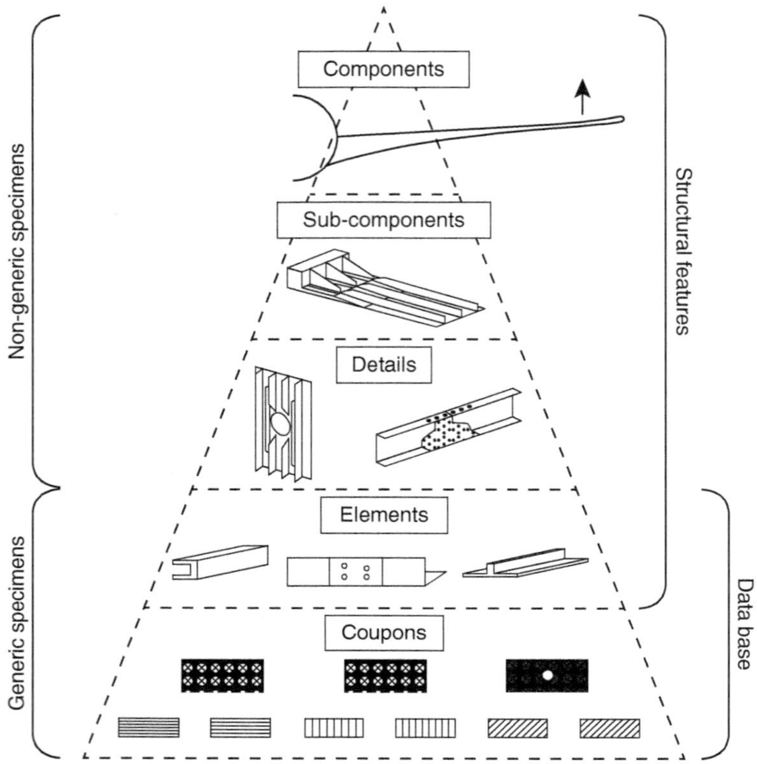

FIGURE 11.39 Generic building block approach for aircraft/spacecraft. (*FAA.*)

Starting with material coupons, the design works up through detailed pieces, sub components, components, and finally to the full scale article, all of which are tested to evaluate behavior and strength, as shown in Figure 11.39.

Interfaces

The structural subsystem interfaces on the one hand with the launcher (launcher/payload adapter) and on the other hand with the payload and the other spacecraft (S/C) subsystems.

Being the physical support medium for all hardware, the structural subsystem is usually designed before all other S/C elements, and therefore interdependence of the structure and other S/C elements necessitates close interaction between structures engineers and other engineers of the spacecraft team. The structural layout and design many times needs to be finalized at the preliminary design timeframe so as to allow time for detailed design of other spacecraft components and instruments that interface to the primary structure.

11.17 Mechanical Requirements

Mechanical requirements are defined, and typical requirement sources are now identified, for spacecraft structures.

Strength is the amount of load applied to a structure that must not cause rupture, collapse, or deformation significant enough to jeopardize the mission. This basic structural requirement for all space programs applies to all life-cycle events. Strength inadequacy can lead to spectacular failures.

Structural life is the number of loading cycles a structure can withstand before its materials fatigue (rupture from cyclic loading), or the duration of sustained loads a structure can withstand before its materials creep (deform excessively from sustained loading) or crack from stress corrosion. Again, this applies to all life-cycle events.

Structural response attenuation is the magnitude of vibration in response to external loads which must be contained so as to avoid damaging critical components or the LV. This is controlled by the inherent damping in a structure which is described below.

Natural frequency is the frequency at which a structure will vibrate, when excited by a transient load or cyclic load input, and then left undisturbed. It depends on mass properties and stiffness. Each structure has an infinite number of natural frequencies, corresponding to different mode shapes of vibration. Natural frequencies must usually be above a given value or outside a certain range. The lowest fundamental frequency of a structure is of particular concern and corresponds to its first mode of operation. In the stowed configuration the natural frequencies of the combined spacecraft/LV system must not interfere with the control system of the LV or result in excessive loads. Likewise, for on-orbit configuration, vibrations must not affect the control system of the spacecraft.

Stiffness is often prescribed for substructures to achieve the required natural frequency for a larger assembly, or alternatively, to provide the necessary structural stability for a sensor or antenna.

Damping is the dissipation of energy during vibration; it is a structural characteristic that limits the magnitude and duration of response to input forces. It is designed to control loads and ensure that any vibration decays before influencing the control system of the spacecraft.

Mass properties include mass, center of mass, and mass moments and products of inertia. These are imposed by the LV and allocated to all substructures to achieve the required natural frequencies of a larger assembly.

Dynamic envelope is the physical space the spacecraft or substructure must stay within while deflecting under loads to avoid contact between the spacecraft and the payload fairing of the LV or between parts of the spacecraft.

Structural stability is the ability to maintain location or orientation within a certain range. Typical concerns are thermoelastic distortions, material yielding, and shifting of mechanical joints. The intention is to ensure that critical instruments, such as antennas and sensors, will find their targets, since inadequacy in this area may result in performance degradation.

Mechanical interfaces include surface and geometrical position features, such as flatness or bolt hole locations. It may also include stiffness requirements or thermal stability requirements. It is derived from designs of mating structures to ensure proper fit and alignment and avoid excessive deformations and loads.

11.18 Space Mission Environment and Mechanical Loads

Environments on earth, during launch, and in space are the design drivers for spacecraft structures. When applicable, on-orbit performance of spacecraft structures such as pointing accuracy and structural stability are further important design drivers. Activities related to space mission environments and mechanical loads are identified:

- Structures must not only survive the environments to which the spacecraft is subjected but also protect the spacecraft instruments and control system components and allow them to function properly during flight.
- The selected materials, nonstructural as well as structural, must not degrade before and during the mission.
- Ground testing needs to envelop mission environments with margin. As a result, test environments need to be defined at an early stage and structures need to be designed adequately to cover them.

- Mechanical loads can be static or dynamic. Mechanical loads can be external (e.g., engine thrust, sound pressure, and gusts of wind during launch) or self-contained (e.g., mass loading of a vibrating satellite during environmental testing or in space after the force that caused the excitation is removed).

The design of a spacecraft and its subsystems must cater to all the loads which will be experienced. Typical loading events are

- Testing (tailored to cover adequately flight, in-orbit, and ground transportation environment)
- Ground handling and transportation
- Liftoff
- Stage separations
- Stage ignition
- Stage or main engine cut-off
- Maximum aerodynamic pressure
- Random vibrations
- Acoustic (noise) loading
- Spin-up and deployments
- Attitude control system firings
- Reentry
- Emergency landing (for shuttle space transportation system)

The above events induce accelerations, shock, and vibration to the structure. Careful attention is needed to cover these effects adequately. In general, the limiting factors in the structural design are set by the dynamic effects rather than by steady-state accelerations. Some examples are

- Solid rocket engines have combustion chambers which run the length of the fuel column, and they burn continuously and cannot be throttled. The thrust output changes over time and the subsequent loss of mass results in a dynamic loading of the structure.
- Liquid fuel rockets have a combustion chamber fed by separate fuel tanks; they can be (but rarely are) throttled to reduce thrust loads at critical aerodynamic phases of the flight and to give a more controlled trajectory. The tanks of liquid fuel act to some extent as damper and attenuate vibrations from the engines. However, the liquid nature of the fuel can sometimes become excited by external flight forces and induce large inertial loads in the vertical direction of flight (POGO) or lateral to flight (Slosh). These must also be accounted for and designed against.

The launcher agency will issue standard guidelines for the design and qualification of spacecraft. These apply to the mounting interfaces at its base and are concerned with both steady and dynamic forces. They will be stated as flight limit loads, i.e., levels which one would not expect to be exceeded in 99% of launches. The launcher agency will, however, require the spacecraft designer to demonstrate (by a combination of test and analysis) that the design can withstand these levels with significant margins, differing from those of aircraft systems, and which typically are

- Flight limit loads: 1.0
- Flight acceptance: 1.1
- Design qualification: 1.25–1.4

For composite structures, the factors can be somewhat higher

- Flight acceptance: 1.2
- Design qualification: 1.5–2.0

Every designer should consult the applicable standards and or requirements document relating to the mission profile which is being flown and determine which types of load factors or safety factors may be required based on the design concept for that particular flight vehicle or payload.

Therefore, test levels for flight acceptance are set for structures and equipment which have previously (in prototype form) passed tests at the design qualification level. The implication here is that if a one-model program is followed (protoflight approach, i.e., the prototype is actually flown), then the model must clear the higher qualification test. Separate factors for the materials may also be specified by the launch agency. Typically these are 1.1 at yield and 1.25 at ultimate stress for metals, and up to 2.0 for composites. Therefore, a structure clearing acceptance test levels plus material safety factors will have a healthy margin over actual flight loads.

- Margin of safety: terms used in strength analysis are identified below.
- Load factor, n: a dimensionless multiple of g's (gravitational acceleration) that represents inertia force. It is often used to define limit load.
- Limit load (or design limit load): the maximum acceleration, force, or moment expected during the mission or for a given event, at a statistical probability defined by the selected design criteria (usually between 99% and 99.87%). Typically 2σ or 3σ statistical events are used to envelope maximum load limits.
- Limit stress: the predicted stress level corresponding to limit load.

The allowables and factors of safety are somewhat different from those discussed for aircraft.

- Yield failure: permanent deformation. In the case of composites there is usually little yielding of the material prior to ultimate failure, therefore margins for composite are usually written against an ultimate criteria.
- Ultimate failure: rupture or collapse. This is the point at which structural failure of a component is expected or predicted to occur.
- Allowable load (or stress): the minimum strength (load or stress) of a material or a structure at a statistical probability defined by the selected design criteria (usually 99% probability at 95% confidence level).
- Allowable yield load (or stress): the highest load (or stress) that, based on statistical probability, will not cause yield failure.
- Allowable ultimate load (or stress): the highest load (or stress) that, based on statistical probability, will not cause ultimate failure. For common aerospace metals, these values are readily available. However, with the large number of variables associated with composites, significant test programs are required to develop A-basis and B-basis values for strength.
- Yield factor of safety, FS_y: a factor applied to the limit load or stress for the purpose of decreasing the chance of detrimental deformation; usually between 1.0 and 2.0 for flight structures, depending on the option, whether personnel safety is at risk, and how sensitive the mission is to small deformations.
- Ultimate factor of safety, FS_u: a factor applied to the limit load or stress to decrease the chance of ultimate failure; usually between 1.25 and 3.0 for flight structures, depending on the test option and whether people are at risk.

- Design yield load: limit load multiplied by the yield factor of safety; this value must be no greater than the allowable ultimate load.
- Design ultimate load: limit load multiplied by the ultimate factor of safety; this value must be no greater than the allowable ultimate load.
- Design yield stress: predicted stress caused by design yield load; this value must not exceed the allowable yield stress.
- Design ultimate stress: predicted stress caused by the design ultimate load; this value must not exceed the allowable ultimate stress.
- Yield margin of safety, MS_y:

$$MS_y = \frac{\text{Allowable yield load (as stress)}}{\text{Design yield load (or stress)}} - 1$$

- Ultimate margin of safety, MS_u:

$$MS_u = \frac{\text{Allowable ultimate load (as stress)}}{\text{Design ultimate load (or stress)}} - 1$$

11.19 Project Overview: Successive Designs and Iterative Verification of Structural Requirements

The assessment of load distribution within a spacecraft structure is largely an iterative process starting with generalized launcher predictions. These are used as a basis for the design of the initial spacecraft concept and therefore provide subsystem target specifications according to their location.

The first step in structural design is to convert the mission requirements into a spacecraft concept and specify the underlying parameters with as much detail as can be expected at the concept stage. This may be quite mission specific, for example, contrasting the thermal dissipation of a communication satellite with the precision requirements of large antennas and telescopes.

In general, the following requirements need to be covered from an early stage:

- Overall configuration meeting mission objectives
- Accommodation for the payload and spacecraft systems
- Ability to withstand launch loads

Selection of material system

- Stiffness
- Provision of environmental protection
- Alignment
- Thermal and electrical paths
- Accessibility

At this stage the configuration effort will dominate as the distribution of masses becomes established. The concept of a load path from the interface with the LV (from which all accelerations are imparted) through the spacecraft structure to the mounting points of individual systems or units is followed. At each mounting point individual sets of interface requirements (alignment, thermal, field of view, screening, connections, and accessibility) will be generated which therefore set some of the constraints to be applied to the design of the structure. Subsequently, the problem of providing a structure to meet the specification might be well outlined. However, the need for high mass efficiency and reliability presents challenges. The selection of materials is often dominated by stiffness and structural dynamics rather than the stress levels. Testing as a means of validation is an inherent part of the design, by full test or by limited testing supported with analysis and modeling. The final design stage requires a coupled loads analysis which combines the characteristics of the full assembly of the spacecraft with those of the LV.

In the case of large structures, such as a space station, reconfigurations well beyond the initial concept will occur on orbit during its 25-year lifetime. Addition of modules and long beams will change the very low-frequency dynamic characteristics during attitude maneuvers and orbit boosting, presenting an additional design criterion. The design methodology is summarized in Figure 11.40.

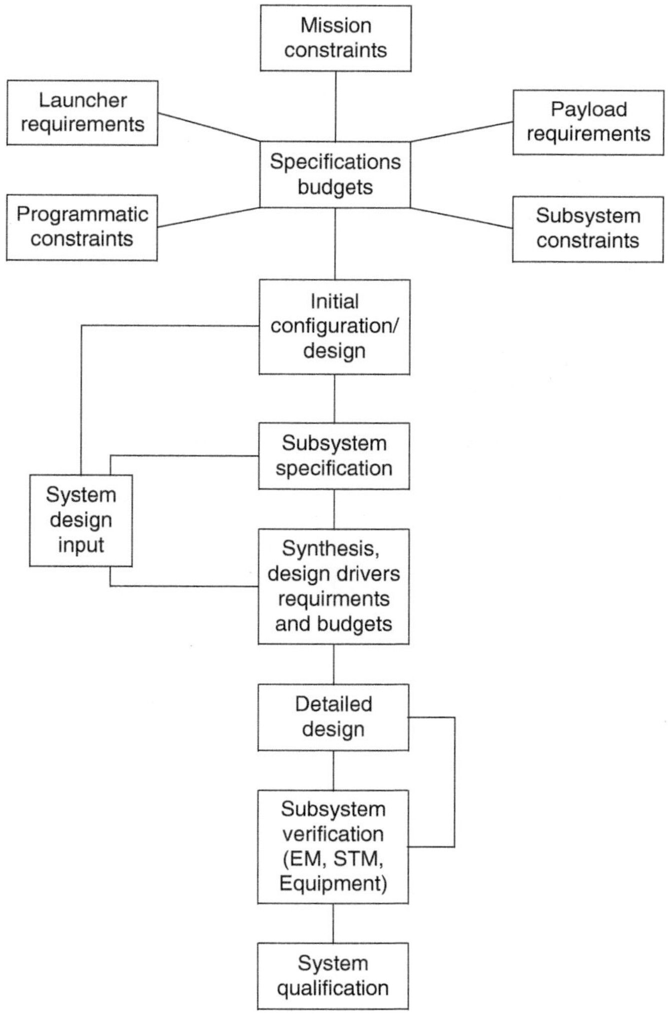

FIGURE 11.40 Design methodology.

11.20 Analytical Evaluations

With the development of modern fast-access large-memory computers it is now possible, with a readily available commercial finite-element code, to model in detail satellite mechanical systems and to examine their behavior under various static or time-dependent load conditions. The finite-element method was outlined in Part 3. Commercial packages are available (NASTRAN, ABAQUS, ANSYS, and other specialty software packages) which enable a spacecraft structure to be evaluated under static, dynamic, and thermal loads. The natural frequencies and modes of vibration are determined as a first step in dynamic evaluations to assess the characteristics of the structure. These are employed to evaluate the compatibility of the satellite with launcher requirements and for modal response analysis of the structure excited by external loads to determine the satellite response at different locations. The finite-element model of the satellite system is employed in coupled loads dynamic analysis (CLDA) with the launcher to examine the launch assembly configuration and determine representative test excitation (tailoring of test profile and level). Figure 11.41 shows a highly three-dimensional satellite and its finite element model.

In such analytical evaluations, it is particularly important to assess qualitatively and quantitatively the analytical predictions. A number of qualitative checks are presented here which assist practicing engineers to identify mathematical errors.

Many times the subsystem synthesis and detailed design/analysis will be repeated over in a series of "design/analysis cycles" in which various trades and or design refinements can be investigated. This leads to the final design cycle, where the final spacecraft or payload item is analyzed or assessed in full and safety margins written against all components.

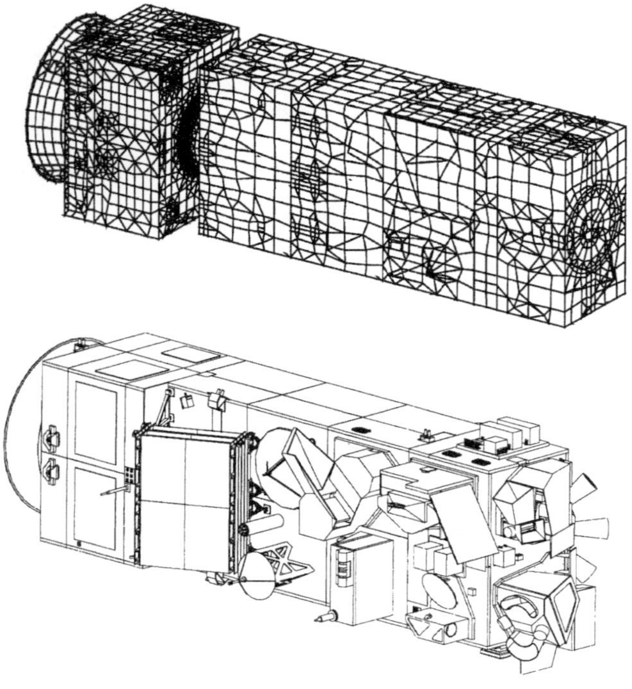

Figure 11.41 The ENVISAT satellite and FE model. (*European Space Agency.*)

The underlying physics of the problem should be respected in analytical modeling, and these assist in qualitative checks of the finite-element mathematical model. For example:

- No stress/loads should be generated due to rigid body translations and rotations of a structure.
- The mass matrix of the structure (often simplified for dynamic analysis) should represent the total mass and inertia properties of the structure.
- No stress/loads should be generated in a homogeneous structure free to expand when subject to a uniform temperature field.

11.21 Test Verification, Qualification, and Flight Acceptance

The objective of a structural test is to engender confidence in the analytical predictions which support satellite development and ultimately to support the qualification and flight acceptance of the satellite system. The types and purpose of the different tests are now presented.

The *static test* is achieved by subjecting the structure to discrete loads or by centrifuge test to simulate inertia loading. The static loads can be imparted with a known weight distribution (sandbags), discrete application points (wiffle-tree with actuators), a centrifuge as stated before, or with vibration (sine burst testing). Sine burst deserves some explanation and cautions. This approach uses a vibration table to impart dynamic loads into a structure at frequencies much lower than the natural frequency of the structure undergoing test. In this case, the structure should behave in a quasistatic nature, responding based on the mass/inertia distribution of the stack dictates. However, this test is highly dependent on the table input and this input cannot be determined a priori. So one must gradually creep up on the estimated load value and make sure the structure is behaving linearly until the final g loading is reached. If there are any nonlinearities (table stiction, malfunction, improper settings for input loads) significant damage can be done to the test article, and if the protoflight approach is being used, damage to the actual flight vehicle. Therefore, great care should be used when attempting to test via sine burst for static qualification loading of flight articles.

In the end, the static test provides insight on the validity of the stiffness matrix but is used mainly to qualify the strength adequacy of the primary structure and critical structural interfaces, e.g., satellite/launcher and satellite/payload interfaces.

The *modal survey test* is achieved by exciting the structure with small exciters to determine the natural frequencies and modes of vibration of the satellite. Since the applied excitation forces at resonance are compensated only by damping, high responses can be achieved with small excitation forces. Modal survey tests identify the satellite natural frequencies, modes of vibration, and damping to determine the dynamic compatibility of the satellite with the launcher and support the verification of the finite element model which is used in LV/spacecraft (S/C) coupled loads dynamic analysis (CLDA) in the loads cycle assessment and to tailor the vibration test level.

The *shaker vibration (sine) test* supports the verification of the mathematical model used in forced frequency response predictions and is particularly useful to determine the amplification of the excitation input from the L/V–S/C interface to various elements of the satellite (amplification factor Q output/input). The excitation input may be notched (filtered) to avoid exceeding limit loads. The main purpose of the shaker vibration sine test is to qualify the adequacy of the secondary structures when subjected to dynamic environment, and also to verify the adequacy of the satellite system by performing functional tests after satellite system qualification and flight acceptance shaker tests.

The *shaker vibration random test* supports the verification of satellite units subjected to random dynamic environment which might be experienced during flight. The latter usually results from acoustic excitation of structural interfaces. The random vibration loads are usually applied in three orthogonal axes.

The *shock test* supports the verification and qualification of the satellite structure and instruments subjected to shock environment due to pyro and latching loads, e.g., release of the L/V–S/C interface clampband, and release of booms, solar panels, antennas, etc.

11.22 Satellite Qualification and Flight Acceptance

Classical Approach

In the classical approach, structural, in some cases engineering, and flight models are tested prior to satellite flight. The purpose of these models and the tests which are performed are summarized below.

Prototype: A separate test article (not to be flown) is subjected to the loading required to qualify the structure. The actual flight article then undergoes acceptance level testing.

Model	Tests
Structural model (representative of satellite flight structure, units might be mechanically representative or with electronics represented with mass simulators)	• Static/centrifuge test at qualification load level to check and qualify the structural strength in particular primary structure and critical interfaces. • Modal survey test identifies vibration, natural frequencies, natural modes, and damping to support verification of mathematical model which is used in loads cycle and coupled loads dynamic analysis of the satellite with the launcher. • Shaker sine vibration test at qualification test input level to qualify satellite secondary amplification from S/C–L/V interface to various parts of the spacecraft. • Random and acoustic loads at qualification levels and duration.
Flight model	• Modal survey test is in general not performed with the flight model unless significant design changes have been introduced or there are built standard differences with the structural model. • Shaker sine vibration test at flight level. • Acoustic test at flight level. • Shock test to determine compatibility of the satellite system with shock environment in particular payload units. • Random and acoustic loads at acceptance levels.
Engineering model	• The engineering model is usually not structurally representative and is employed for electrical purposes. In a number of cases the structural model may be employed as engineering model after the mechanical qualification tests.

Protoflight Approach

The major difference between the protoflight approach and the classical approach is that the flight hardware is subjected to qualification loads. (There is no prototype structure.) The employment of the hardware model and the test which are performed is summarized below.

Structural model (this is refurbished after qualification test as the flight model—units might be mechanically representative or with electronics represented with mass simulators)	• Static/centrifuge test at qualification load level • Modal survey test • Shaker sine vibration test at qualification test input level • Acoustic and/or random test at qualification level • Shock test
Flight model (structural model refurbished to flight model standard after qualification test)	• Shaker sine vibration test at flight level • Acoustic and/or random test at flight level (qualification level acceptance duration) • Shock test

11.23 Materials and Processes

The selection of appropriate materials for spacecraft structures applications requires a knowledge of the way each material property can best be used and where various limitations must be adequately recognized. Selection criteria include but are not limited to the specific strength, specific stiffness, stress corrosion resistance, thermal parameters, sublimation and erosion, ease of manufacture and modification, material cleanliness, and compatibility.

Ground and launch environments have to be considered when selecting materials for handling, moisture absorption/desorption, shocks and quasistatic and dynamic loads. The space environment in particular influences the selection of materials. Among the most important factors are

- Atomic oxygen affects mostly polymeric materials (important exception: Ag).
- Charged particles require conductive materials to prevent local electrical charging (ITO conductive coating).
- Thermal radiation and uneven temperatures lead to thermal distortions (particularly multilayer composite materials).
- Vacuum facilitates outgassing (important selection criterion).
- Micrometeoroids and debris.

Analysis of the possible impact of an environment aspect and verification of suitability/acceptability of selected materials is performed using a structured and documented approach. In addition, costs, availability of data, material availability, previous experience, and ability to be easily repaired can affect the choice of materials.

Two classes of materials are suitable for spacecraft structures: metal alloys and composites. Metals are homogeneous and isotropic. Composites are inhomogeneous and orthotropic. The basic properties are summarized in Table 11.4. The most expensive metallic alloy is that of beryllium, which is justified for satellite structures in view of its outstanding specific stiffness.

Typical applications of metals and composites:

- Aluminum: skins, shells, truss members, face sheets, and core of sandwich structures
- Titanium: attachments fittings, fasteners, pressure vessels

	E(GPa)	(UTS) (MPa)	e_u (%)	p (Mg/m³)	E/p	UT S/p
Aluminum	71	482	3.0	2.80	25	172
Titanium	110	1,000	2.5	4.43	25	226
Steel	207	1,723	7	7.78	26	221
E glass	70	2,200	~4.5	2.54	27.6	866
Kevlar (aramid)	130	2,900		1.45	90	2,000
Carbon	380	2,700	~1	1.86	204	1,450
E glass*	40	840		1.80	22.2	467
Kevlar*	82	1,500		1.39	59	1,080
Carbon*	200	1,500		1.55	129	970

*Denotes fiber volume fraction of 60%.

TABLE 11.4 Mechanical Properties of Metallic and Polymer Composites

- Beryllium: stiffness-critical appendages, support structures for optical equipment
- Steel: fastening hardware, ball bearings

Composites are used in stiffness-driven applications, such as for reflectors or, in the case of Kevlar, where transparency to electromagnetic waves is required (Figure 11.42).

Jointing is important and relies basically on three approaches: mechanically fastened joints, bonded joints, and welding. In general, mechanical joints lead to stress concentrations, which require special attention for load introduction, in particular for composite materials which are susceptible to such concentrations. Bonding is the most elegant approach as it guarantees a better continuity of the load transfer. However, it requires proper shaping of the parts to be assembled to have a progressive load transfer so as to minimize through-thickness stresses. In composites attention shall be given to avoid or limit the peeling forces.

The tolerance between parts and cleanliness of the surfaces is also very important. Welding is reserved to metals and provides excellent tightness. Fusion associated with the welding process introduces weak zones in the joint, particularly when the heat-affected zone is not small, as in the tungsten inert gas process. High-performance welded joints are required for pressure vessels and can be obtained by electron beam welding or laser welding.

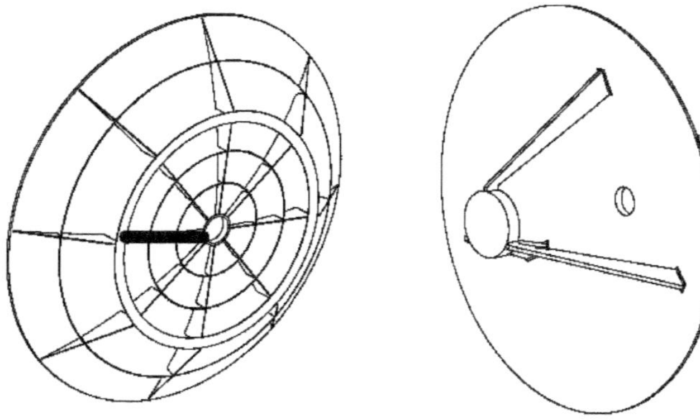

FIGURE 11.42 Composites reflector.

Two applications are leading the search for improved materials: high stability of scientific instruments of antennas and high-temperature materials for reentry vehicles. For high stability, new materials like carbon fiber-reinforced cyanate ester resins are considered due to their low moisture absorption/desorption. For high temperatures, carbon-carbon and carbon-silicon-carbide offer good mechanical strength under high thermal fluxes leading to temperatures well above 1,000°C. However, oxidation of these materials under severe heat fluxes requires the development of adequate coatings.

11.24 Manufacturing of Spacecraft Structures

Manufacturing of metallic or composites parts is fundamentally different. Because the processing of composites starts from raw materials and not semifinished products as for metals, it is possible to produce fairly complex parts as a single piece (unitization), saving on manufacturing and assembly costs (see Figure 11.43).

Metals

Machining is the process of removing material with cutting or grinding tools. Many machining operations are automated to some extent; although for low-volume production as typical of spacecraft manufacturing, manual processes are often more cost effective.

Forming is one of the most economical methods of fabrication. The most limiting aspect of designing formed parts is the bend radius, which must be quite large to limit the amount of plastic strain in the material. Super-plastic forming and diffusion bonding, at temperatures up to 1,000°C, can produce complex components but only for titanium alloys since others are prone to surface oxidization, which inhibits the diffusion bonding. Spin forming has been successfully applied for manufacturing of aluminium pressure vessels.

Forging, in which structural shapes are produced by pressure, is well suited for massive parts such as load introduction elements. Where large sheets have to be manufactured (as for the skin of main thrust cylinders), chemical milling is frequently used to reduce the thickness of the sheet where possible. This process is more reliable than machining when processing very thin elements, but adequate tolerances have to account for various inaccuracies as thickness variations of the original piece of material are achieved by masking.

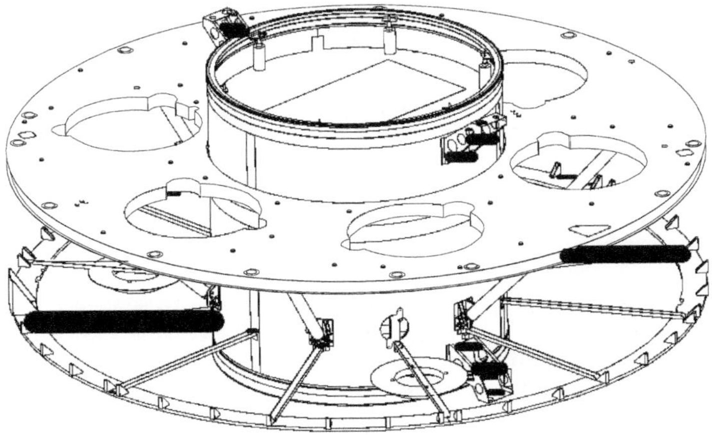

FIGURE 11.43 CLUSTER spacecraft structural layout. (*European Space Agency.*)

Casting can be used to produce parts of complex shapes. However, the quality of a casting is difficult to control because as the material solidifies, gas bubbles can form, resulting in porosity. Material strength and ductility are not as high as with most other processes.

One of the newest methods is actual three-dimensional (3D) printing of parts, either from metal particles or plastic. New printing techniques can take complex digital models of a part and produce a final component in a relatively short time frame. Material properties of these 3D printed part are usually lower (strength/stiffness) as a result of the printing process so this must be taken into account during the design/analysis phase if 3D parts are planned for flight application.

11.25 Composites

Composite aerospace structures have been around for more than 30 years. We shall mention only the most common type, where very fine fibers are embedded in a resin, forming unidirectional laminae (pre-preg). The lamina is then laid up in a chosen stacking sequence to suit the designer, forming a laminate. The most popular combination has been carbon fibers in an epoxy resin which is cured under elevated temperature and pressure (thermosets) and undergoes a physical and non-reversable chemical change within the resin. Less popular, but gaining more traction are thermoplastic resins which have the characteristic of being solid at a low temperature but more liquid/viscous at higher temperatures. Composites made from thermoplastic can be reformed over again to change shape and or configuration.

Carbon is a brittle material which does not suffer dislocations but is sensitive to the most minor scratch or flaw. The fibers are only microns in diameter, and it has been found that a typical unflawed length is on the order of 15 to 30 diameters. Flaws in adjacent fibers will never coincide, so the resin is able to provide a load path bypassing the fiber fracture. This is the basis for laminated composite structures.

Table 11.4 shows the values for strength and stiffness for pure fibers with no flaws and for unidirectional composites having fiber volume fractions of 60%. Average values for typical aerospace metallics are included in the table for comparison. Strengths for composite materials are higher than for those of metals, while the stiffnesses are comparable. Where composites excel, of course, is in their low density. With the exception of glass composites, specific stiffness and strength figures are much higher than those found in metals. Of the three composites, the highest value of specific stiffness is for carbon composite which is better by a factor of 4 than that for any metal listed. It is stiffness which controls the buckling performance of structures, and more than 70% of the primary aircraft structure is governed by buckling. Modern high-performance military aircraft need stiffness for their aeroelastic margins, and most could not have been built without using carbon composites. However, even more important than high specific stiffness is the ability to tailor properties as required. Other fiber systems have unique properties which may be of use in certain applications. For example, glass fibers have larger strain to failure values which may be necessary in applications where larger displacements are needed without causing rupture of the fiber.

Composites have poor transverse strength perpendicular to the fibers, and for most structures where the stress field is bidirectional, the designers will choose to assemble each ply into a laminate with a stacking sequence carefully chosen for the local stress field. Lay-ups vary from quasi-isotropic $(+45, -45, 0, 90°)_{ns}$ to cross-ply $(0, 90)_{ns}$, and angle ply $(+45, -45)_{ns}$. The subscript n denotes the number of sublaminates, and the subscript s denotes a symmetric sequence of plies. Symmetrical lay-ups are most often utilized since they do not distort when curing. However, clever nonsymmetrical combinations have been used for aeroelastic tailoring, where wing flexure can be accompanied by beneficial twisting, which would not happen in homogenous metals.

Weak transverse fiber strength is equally important in the through-thickness direction of the laminate when the individual laminae can easily delaminate by failure of the resin matrix. This is the Achilles' heel of laminated composite structures, and currently several solutions to this problem are being evaluated. Textile technology such as weaving or stitching is possible for the dry fibers, which are then impregnated by infusion of the resin. While effective, this process results in a reduction of strength in the in-plane directions. Recent advances in 3D weaving technology have enabled more development of single piece complex geometry with relatively good strengths in all three directions. Z-pinning is an option for laminated composites already cured. Here small pins of carbon, aramid, or metal are driven dynamically into the laminate from the outer surface. Some in-plane strength is sacrificed by the process, but the through-thickness strength and fracture toughness increase dramatically. The other alternative is to design a structure which has no through-thickness stresses, but this is virtually impossible near any 3D features such as joints and panel stiffeners. Also, low-velocity impact damage is a threat which can cause delamination and some 70% loss in the residual compressive strength.

Thermoplastic resins are an alternative which does not require a chemical change during manufacture, just application of heat to a flat laminated sheet. At the moment these appear to be too expensive, but they are much quicker to mass produce since a thermoset composite spends 1½–2 hours in an autoclave. Their properties are better, particularly the through-thickness fracture toughness, where the thermoplastic matrix exhibits modest ductility. The Airbus 380, for example, uses thermoplastic in the leading edge, and these must withstand bird-strike without excessive losses in stiffness or strength.

Composite structures do have several other virtues. Their fatigue life is exceptional. Corrosion is not much of a problem, although moisture absorption in the matrix is of concern. It is also easier to design a stealthy aircraft with a small radar cross-section using composites than it is in metallic structures. The coefficient of expansion for carbon is virtually zero. The high-temperature performance (strength and creep) for thermoset materials is better than for aluminum alloys.

The structural properties of stiffness and strength of laminated composite structures are not formed until the manufacturing process is complete. They can be theoretically deduced from the unidirectional lamina properties, knowing the stacking sequence, but this is beyond the scope of this book. Several texts on composite structures are available (Hull 1981; Agarwal and Broutman 1990; Niu 1992).

11.26 Composite Structures

Laminated composite structures do seem to have everything going for them when we judge the specific strength and stiffness properties summarized in Table 11.4. As mentioned, for military aircraft, where performance has been the main driver, particularly aeroelastic efficiency, there has been little choice other than to use carbon/epoxy composites. As affordability becomes a more significant driver for military as well as civil aircraft, we might expect composites to become less attractive. However, composite structures have made great strides in reducing cost and part count. The unitization concept, which involves reducing dramatically the separate number of parts, is now being exploited for composite structures. The degree of automation in manufacturing has advanced to the use of automated tow or tape-laying machines whose head can move with five degrees of freedom so that double curvature and any choice of stacking sequence and orientation is now possible. It also looks as if techniques borrowed from textile technology will drastically reduce manufacturing costs with a small loss of performance. These techniques use dry or tacky fiber preform, often a woven fabric, and then infuse the resin under pressure and heat. Resin transfer molding (RTM) introduces the resin at selected points, and resin infusion methods deploy film or tiles which can be sandwiched between layers of the preform and then infused as the temperature increases. One of the reasons for high costs is the necessary use of an autoclave for thermosetting resins. We expect to see increasing use of room temperature cure, probably using vacuum-assisted RTM. So if we look at the use of carbon fiber composites in civil aircraft (where the acquisition cost certainly has to be justified), they are used, for example, in the Airbus A340-600 (long-range version) for the fin, the fuel-carrying tail-plane, rudder and elevator, ailerons, wing-fuselage fairings, floors, engine cowlings, nose and main undercarriage doors, and so on. The Boeing 787 has advanced the state of the art by including composites in major fuselage, wing components, and other primary structure. And recent designs in spacecraft bus architecture have included one-piece "monocoque" modules made from composite laminate or honeycomb construction that can be stacked together to produce various size/dimension spacecraft from simple submodules.

The fatigue performance of carbon fiber composites is exceptional. All this is not excessive conservatism. In fact, the wings of the single-aisle long-range Boeing and Airbus aircraft will almost certainly be composites, and cured in one piece to avoid heavy bolted joints. Herein lies the weakness of laminated composites. They are vulnerable to stresses transverse to the fiber direction where failure values in tension or shear may be only 4% of the in-plane strength. As stress concentrations give rise to local stress gradients, and if there is a gradient in stress along the direction of the fibers, the equations of equilibrium show that there will be an interlaminar shear, and if these also have a gradient there will be a through-thickness compression or tension (peeling) stress. Industry is well aware of this and has useful design rules to alleviate the problem. For example, when laminates are tapered a minimum taper ratio of 1:20 is aimed for. However, this does not solve the problem of complex parts such as any joint, whether bonded or bolted. It has to be admitted that highly 3D bolted joints are designed very conservatively and are probably too heavy. However, the ability of structural designers to recognize, analyze, and design for these stress concentrations will undoubtedly improve.

Finally, a novel hybrid composite material called Glare (GLAssfibre REinforced) is being introduced for newer aircraft like the Airbus 380. This material is a sandwich of unidirectional glass fiber and very thin sheets of aluminum alloy, produced in any stacking sequence that the customer needs. The main reason for this hybrid was to exploit the fiber bridging capability of high-strain glass fiber if the aluminum alloy developed a crack. This it does admirably, although the basic cost of Glare is high. However, when it is manufactured then all the very thin aluminum sheets can be laid down in a staggered fashion so that their edges are not on top of each other. This means that very much larger sheets can be made than that of the basic aluminium alloy. The very large fuselage of the Airbus 380 can then be constructed using only one or two sheets of Glare fore and aft of the wing support frames. This saves more manufacturing costs than the basic increase in the cost of the material. It should also be welcomed by airline operators since its fatigue performance is without equal in homogenous metallic structures (Figures 11.44 and 11.45).

Manual lay-up is a costly technique which is well adapted to production of single parts or very small quantity of parts. When the number of identical parts to be produced grows, filament winding, RTM, and braiding are cost-effective alternatives.

Filament winding consists of wrapping bands of continuous fiber or strands or rovings over a mandrel in a single machine-controlled operation. A number of layers of the same or different patterns are placed on the mandrel. The fibers may be impregnated with the resin before winding (wet winding), preimpregnated (dry winding), or post-impregnated. The first two winding sequences are analogous to wet or dry lay-up in the reinforced plastic fabrication methods. The process is completed by curing the resin binder and removing the mandrel. Curing is normally conducted at elevated temperature without pressure. Finishing operations such as machining or grinding are usually not necessary.

RTM is closed-mold low pressure process. The fiber reinforcement is placed into a tool cavity, which is then closed. The dry reinforcement and the resin are combined within the mold to form the composite part. This process allows the fabrication of composites ranging in complexity from simple, low-performance small parts to complex elements of large size.

In braiding operation, a mandrel is fed through the center of a braiding machine at a uniform rate and the fibers or yarns from the carriers are braided around the mandrel at a controlled angle. The machine operates like a maypole, the carriers working in pairs to accomplish the

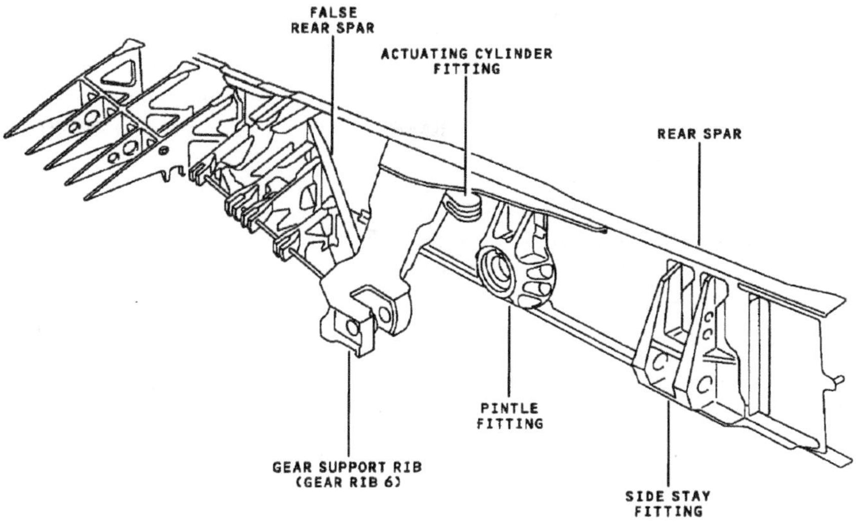

FIGURE 11.44 Attachment points of landing gear to rear spar on Airbus 340. (*Airbus UK Ltd.*)

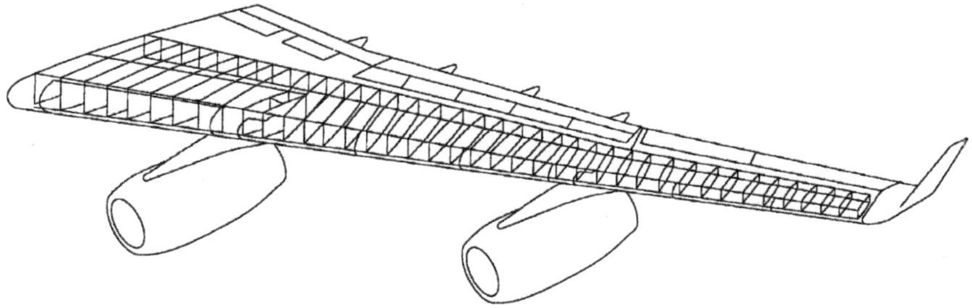

FIGURE 11.45 Layout of ribs in Airbus 340. (*Airbus UK Ltd.*)

over-and-under braiding sequence. Parameters in the braiding operation include strand tension, mandrel feed rate, braider rotational speed, number of strands, width, and the perimeter being braided. Interlaced fibers result in stronger joints. Applications include lightweight ducts for aerospace applications.

References

Agarwal, B. D. and Broutman, L. J. 1990. *Analysis and Performance of Fiber Composites*, 2d ed., John Wiley, New York.

Bruhn, E. F. 1965. *Analysis and Design of Flight Vehicle Structures*, Tri-State Offset, Cincinnati.

Curtis, H. D. 1997. *Fundamentals of Aircraft Structural Analysis*, Times-Mirror HE Group, Los Angeles.

Dowell, E. H. 1995. *A Modern Course in Aeroelasticity*, Kluwer Academic, Dordrecht.

Foreman, R. G., Kearney, V. E., and Engle, R. M. 1967. "Numerical Analysis of Crack Propagation in Cyclically Loaded Structures," *ASME Transactions, Journal of Basic Engineering*, vol. 89(D), pp. 459–464.

Hull, D. 1981. *An Introduction to Composite Materials*, Cambridge University Press, Cambridge.
Kuhn, P. 1956. *Stresses in Aircraft Shell Structures*, McGraw-Hill, New York.
Megson, T. H. 1997. *Aircraft Structures for Engineering Students*, 3d ed. Edward Arnold, London.
Miner, M. A. 1954. "Cumulative Damage in Fatigue," *ASE Transactions, Journal of Applied Mechanics*, vol. 67, p. A159.
Niu, M. C. Y. 1988. *Airframe Structural Design*, CONMILIT, Hong Kong.
Niu, M. C. Y. 1992. *Composite Airframe Structures*, CONMILIT, Hong Kong.
Paris, P. C., Bucci, R. J., Wessel, E. T., Clark, W. G., and Mager, T. R. 1972. "Extensive Study of Low Fatigue Crack Growth Rates in A533 and A508 Steels," *ASTM STP* 513, pp. 141–176.
Rivello, R. M. 1969. *Theory and Analysis of Aircraft Structures*, McGraw-Hill, New York.
Rooke, D. P. and Cartwright, D. J. 1976. *Compendium of Stress Intensity Factors*, HMSO, London.
Timoshenko, S. and Gere, J. M. 1961. *Theory of Elastic Stability*, McGraw-Hill, New York.

PART 4

Satellite Electrical Power Subsystem

Abbas A. Salim

11.27 Introduction

The electrical power subsystem (EPS) provides power to the satellite during all mission phases, beginning with the switchover to internal batteries during lift-off on through the on-orbit phase. It generates, stores, conditions, and distributes electrical power as required by the various satellite loads. The selection of a suitable EPS architecture/topology and the selection and sizing of its components starts with a set of performance requirements derived from the satellite mission requirements. The satellite performance requirements are generally specified in the satellite system-level performance specification. The EPS performance specification, which is often derived from the satellite system-level performance specification, includes the EPS design, performance, interface, and test requirements. A requirements compliance matrix is generally a part of the same specification. The selection of the topology, bus voltage, power management and distribution (PMAD) technique, and component technologies are heavily dependent on the nature of the mission and duration, orbit, and the payload. The payload could be a set of instruments deployed in the low earth orbit (LEO) to monitor the earth's environment or imaging the parts of the earth. It could be a global positioning system (GPS) in the medium earth orbit (MEO) or it could be a set of transponders in a geosynchronous earth orbit (GEO) for telecommunication, such as television broadcast. The spacecraft/satellites are generally classified by the nature of the payload it carries such as earth observation system (EOS), navigation system, meteorological or weather monitoring system or telecommunication system. Additionally, there are planetary and interplanetary probes carrying a large set of instruments to explore the various planets in the solar system such as Mars, Jupiter, etc., and exoplanet detection. The focus of this part is communication satellites located in GEO and are utilized for telecommunication. The EPS design principles and component technologies discussed here are applicable to all other types of satellites in general.

EPS Mass and Cost

As shown in Figures 11.46 and 11.47, the EPS mass and cost are the major satellite bus mass and cost drivers. Batteries and solar arrays are the heaviest components within EPS. However, with the introduction of the lithium-ion batteries, the mass of the batteries is significantly lower than nickel-hydrogen batteries. On the cost side, the solar arrays are the major satellite bus cost driver.

EPS Components

Communication satellites are generally designed for a maximum of 15-year mission life. However, many of them exceed the design mission life depending upon the amount of fuel available for station-keeping and de-orbiting. A typical EPS functional block diagram is shown in Figure 11.48. The main power source is solar arrays. Nickel hydrogen (NiH_2) and lithium-ion (Li-ion) batteries are most common for energy storage. However, the Li-ion battery technology has taken over NiH_2 batteries in most satellite applications due to their superior performance and lower mass and cost. Power regulation and control electronics may include a series, shunt, or a peak-power tracking solar array regulator, battery charge/discharge regulators/electronics, and command, control, and telemetry electronics. A combination of sequential switching shunt regulators (S3Rs) and battery charge/discharge (buck/boost) regulators is most common in the communication satellite EPS.

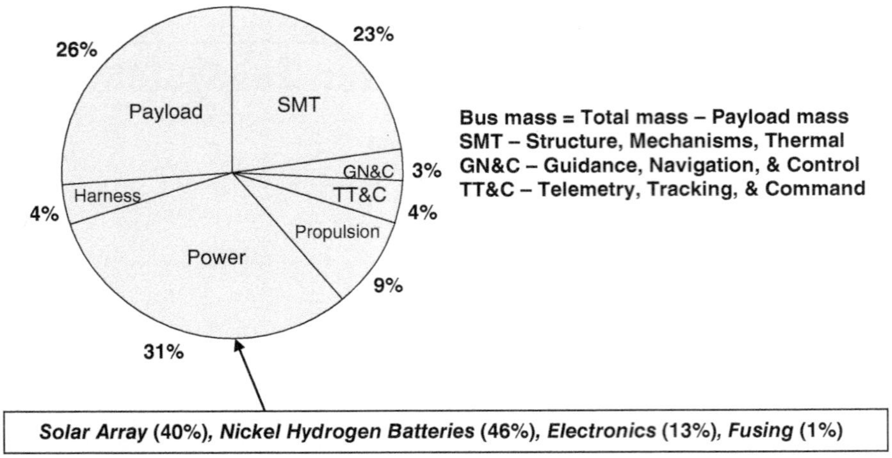

FIGURE 11.46 A typical communication satellite mass breakdown.

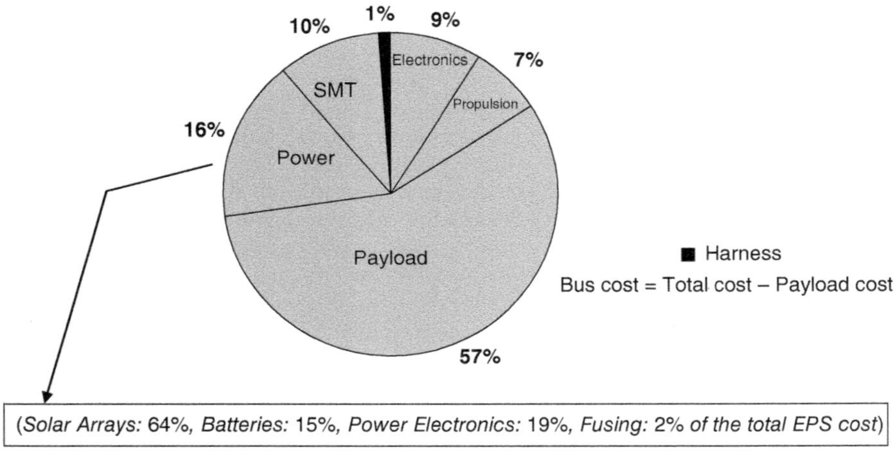

FIGURE 11.47 A typical communication satellite cost breakdown.

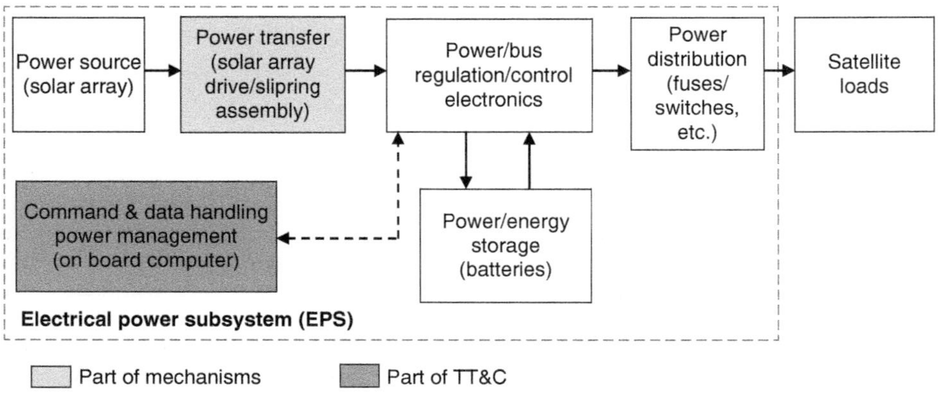

FIGURE 11.48 Functional block diagram—communication satellite EPS.

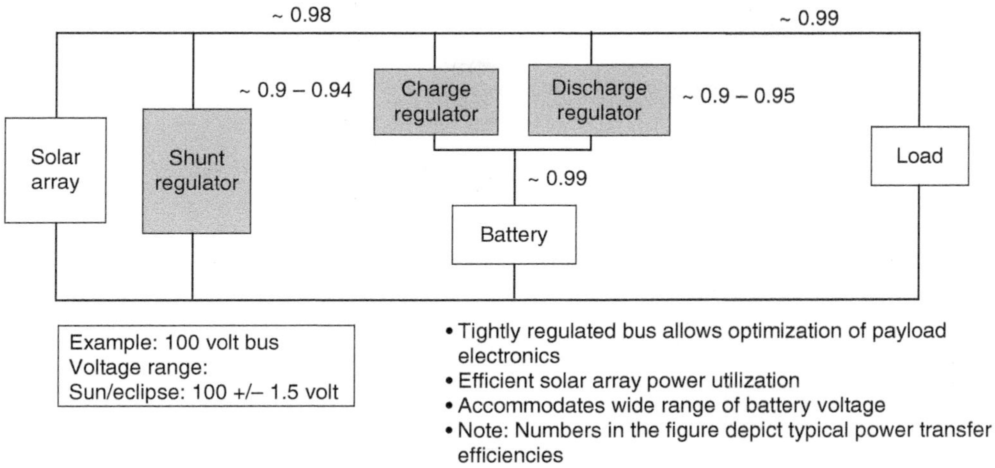

FIGURE 11.49 EPS topology—fully regulated bus.

For satellite with multi-kilowatt payloads, a fully regulated bus as shown in Figure 11.49 is the most efficient solution. The payload power requirement for most communication satellites is in the range of 5 to 20 kW, which is dependent on the type, strength, and number of transponder channels. Nearly 80% of the power is consumed by the payload and around 20% by the various subsystem equipment.

EPS Design

The EPS design starts with a set of pre-defined requirements flowed down from the satellite-level performance specification as listed in Table 11.5.

The higher-level design requirements result in a set of derived requirements that govern the design of the satellite power subsystem. Typical derived requirements are listed in Table 11.6.

EPS Sizing

Typical satellite loads requiring power for a GEO satellite are shown in Figure 11.50. The power subsystem component sizing and technology selection starts with the development of a power budget that needs to take into consideration the key drivers as listed in Table 11.7. A typical power budget format is shown in Table 11.8.

Selection of Satellite Bus Voltage

The selection of the bus voltage is dependent upon several factors including total satellite load power and operating voltage of available payload and bus equipment. Higher bus voltage results in efficient transfer of power from source to loads and considerable power harness mass savings. It is advisable to design high voltage busses for large multikilowatt satellites. Common industry standard bus voltages include 35, 50, 70, and 100 V (Ref. 1). Satellite busses with greater than 100 V have also been designed and flown. With the advent of solar electric propulsion for orbit raising and station-keeping as opposed to chemical propulsion, the demand for solar power continues to grow. Therefore, the processing and conditioning of power at higher bus voltages result in significant mass savings and lower heat dissipation.

Satellite System-Level Requirements
Mission imposed • Duration – Low earth orbit (LEO): 1–5 years typical, 10 years extended – High earth orbit (HEO): 7–15 years (medium and geostationary) • Orbit – Type: circular or elliptical – Location: Sun, Earth, other – Orbital parameters • Period time • Sun angle range • Fractional sun • Maximum eclipse duration
Environment Imposed • Radiation – Near-earth – Geostationary – Van Allen belt (for MEO and extended transfer orbit) • Thermal – Near-earth – Geostationary – Interplanetary • Micrometeoroids • Mechanical – Vibration – Shock
Satellite and Customer Imposed • Attitude/orientation – Payload – Communications • Mass • Volume/form factor • Power profile • Voltage regulation/distribution • Launch • Integration and test • Cost

TABLE 11.5 Satellite System-Level Requirements Affecting Power Subsystem Design

Subsystems Interfaces

The EPS interfaces with practically all subsystems on the satellite. It provides power to the various subsystem components, monitors the load current, and protects the satellite power bus from any load faults. Figure 11.51 provides the details of the subsystem interfaces.

• Generate, store, condition, and distribute electrical power as required by the various satellite loads throughout the mission life in required modes of operation in the given orbit or orbit location.
• Perform these functions over the entire orbital design life, in the presence of all specified design environments.
• Designed, manufactured, and tested to meet worst-case power requirements for all modes of operation.
• No single point failure in the power system shall keep the subsystem from satisfying the satellite load requirements.
• Support satellite transfer orbit demands and station-keeping requirements, if any.
• Provide sufficient telemetry for on-orbit performance monitoring.
• Sufficient end of life load margin.
• The design of the satellite shall allow removal and replacement of batteries without affecting the acceptance status of the rest of the satellite.
• Separate wiring circuits to power redundant satellite loads.
• Protection to prohibit permanent degradation of the power subsystem in the event of a load failure.
• Automatic load shedding to prevent battery over-discharge.
• Power subsystem compatibility with all satellite modes of operation.

TABLE 11.6 Power Subsystem Design Requirements—Derived

There are a number of interfaces between the EPS components and the satellite that require careful attention for the survival and satisfactory performance of the EPS components as illustrated in Figure 11.52.

The following sections provide the details of the three key components of the power subsystem, namely, energy source (solar arrays), energy storage (batteries), and PMAD electronics.

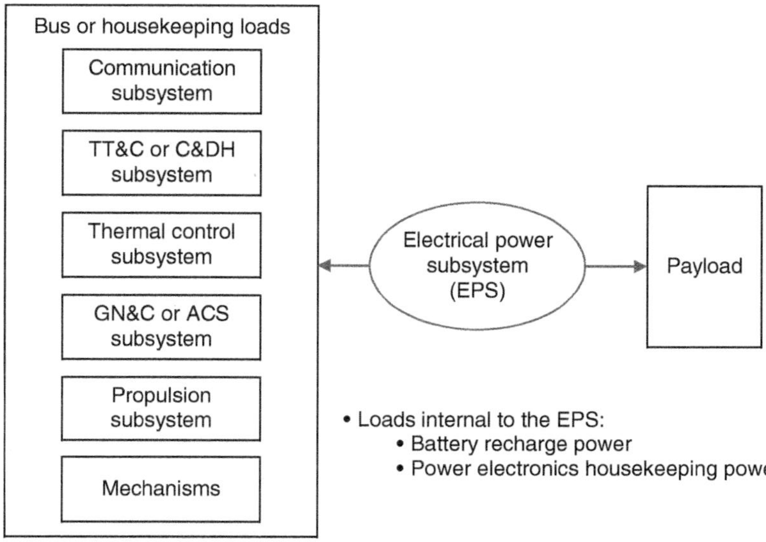

FIGURE 11.50 Typical satellite power subsystem loads.

Satellite loads • Beginning of life (BOL) and end of life (EOL) • Transfer orbit • On orbit (solstice and equinox/sunlight and eclipse) • Orbital average—for solar array and battery sizing • Peak—for fuse and harness sizing • Load margin considerations (1–20% based on design maturity)
Satellite orbital location
Satellite mission life

TABLE 11.7 Satellite Power Budget Considerations

Satellite Loads	Transfer Orbit	Equinox Eclipse	Autumnal Equinox Sunlight	Winter Solstice Sunlight	Vernal Equinox Sunlight	Summer Solstice Sunlight	Station Keeping: Winter Solstice Sunlight	Station Keeping: Summer Solstice Sunlight
Payload Payload margin Payload harness loss								
Payload Total								
Bus TT&C baseband & RF GN&C EPS Propulsion Thermal Bus margin Bus harness								
Bus Total								
Battery recharge power								
Total System Load								
Solar array EOL capability Available load margin Required load margin Battery depth of discharge (DOD)—nominal Battery DOD (with failed cells per battery)								

TABLE 11.8 Typical GEO Satellite Power Budget Summary

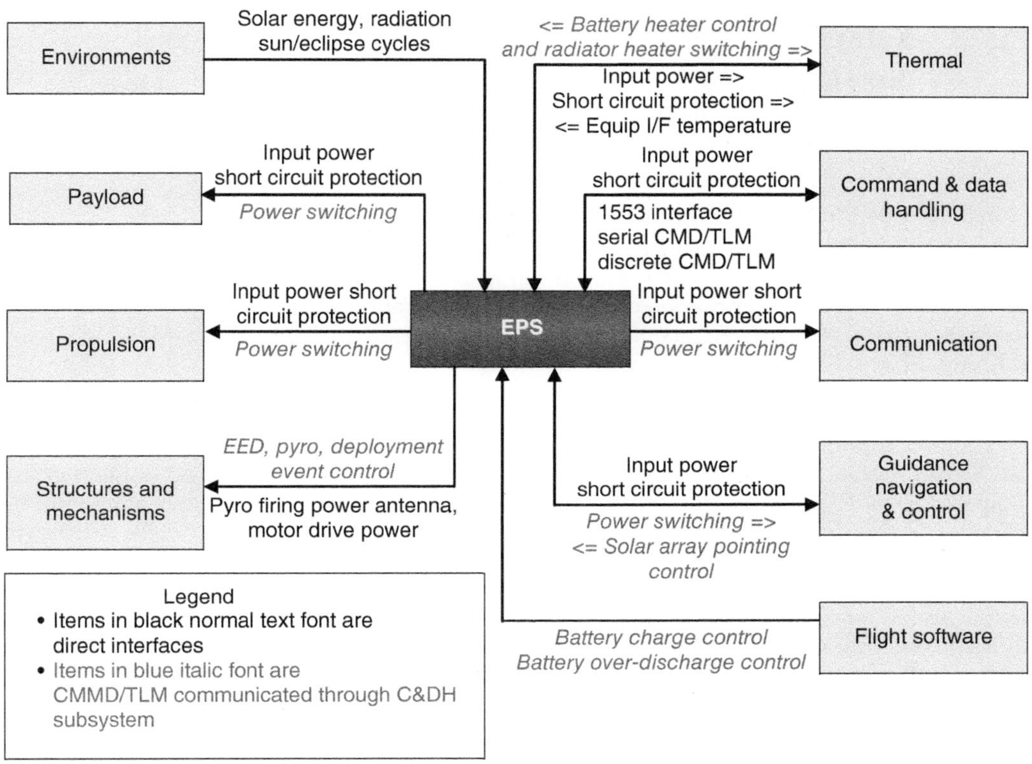

FIGURE 11.51 Typical interfaces—power subsystem and other satellite subsystems.

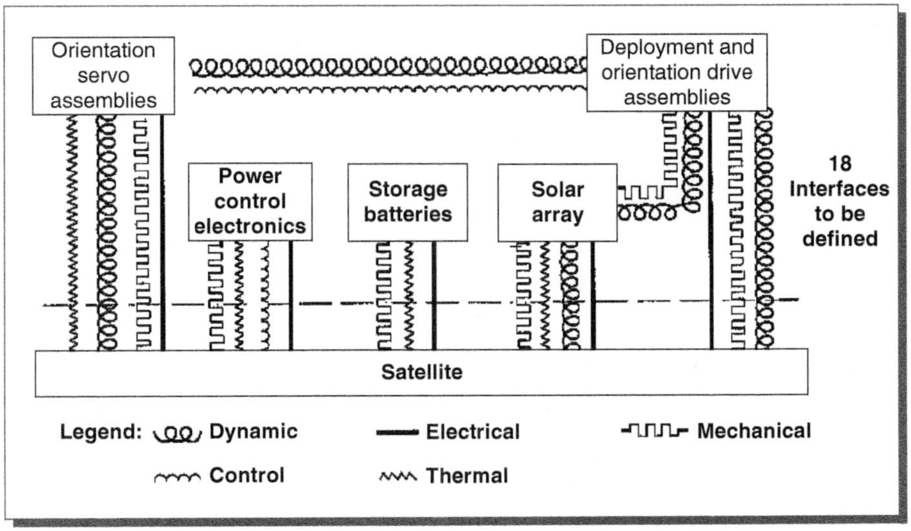

FIGURE 11.52 Satellite and EPS component interfaces.

11.28 Solar Arrays

Giang Lam

Discussion in this section is focused on photovoltaics for the space environment in earth orbit only (Refs. 2 to 7).

Conversion of Photons into Electrical Energy

Earlier satellite solar arrays, during the 1980s and into the early 90s, made use of lower efficiency single-junction solar cells that utilize silicon (Si) as the base photovoltaics material. The single-junction cells used a smaller portion of the light spectrum, approximately 300 to 1200 nm, to convert into electrical power. The single-junction solar cell efficiency peaked at around 14 to 16% (High Efficiency Si from Sharp Corp.) with a voltage output in the range of 1.0 to 1.2 V per cell. The photovoltaics technology used on the International Space Station (ISS) is based on the Si cells developed during the 1980s.

Gallium arsenide (GaAs) solar cells promised higher efficiency and greater radiation tolerance that helped minimize efficiency losses and retain more power at a mission's end of life (EOL) as compared to Si. Higher EOL power translated into a smaller solar array at the beginning of life (BOL). Typical early single-junction GaAs cells had BOL efficiency in the range of 17% to 19% with a voltage output in the range of 2.0 V and higher.

Because of improved manufacturing techniques developed in the semiconductor industry using vapor deposition for depositing microlayers of circuitry materials, more junctions can be added to the solar cell to improve its conversion efficiency by utilizing more of the solar spectrum. Multi-junction solar cells increase their efficiencies by turning more of the different wavelengths into electrical energy. Figure 11.53 illustrates the different portions of the light spectrum that are converted by the different layers of the triple-junction solar cell. The higher wavelength infrared region, 900 to 1700 nm, is absorbed by the germanium (Ge) layer, the lower wavelength ultraviolet, 300 to 700 nm, by the gallium-indium-phosphate (GaInP) layer, and the in-between visible wavelengths by the gallium-arsenide (GaAs) layer. As more efficient multi-junction solar cells are developed, wider ranges of the light spectrum are utilized to increase their efficiencies. (Note: The area under the curve in Figure 11.53, represents energy captured from the sunlight.)

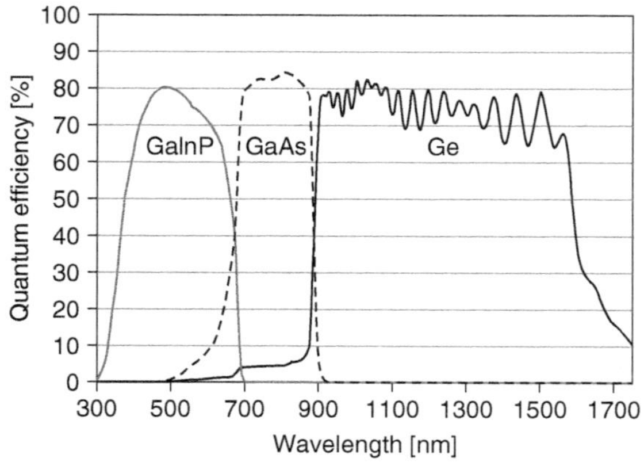

FIGURE 11.53 Triple junction solar cell quantum efficiency versus wavelength.

Solar Cell Size

Solar cell production starts with a 4-in diameter semiconductor wafer and the solar cell's final size is cut from the single wafer into smaller cells per design need. Smaller individual cell size (for example, 2 cm × 4 cm cell) can be packaged more efficiently onto a solar panel, but the cumulative cell cost is higher because of the larger quantity of cells required and the manufacturing cost at the next assembly due to higher part count. Packaging factor describes the percentage of active cell area to available substrate area on a solar array panel. A packaging factor of 90% or higher is a very efficient package due to less wasted space and weight. However, the trade-off is in overall cost. Higher part counts in assembling the solar panel results in higher labor cost associated with cell welding, bonding, wiring, testing, etc. A large-size cell is less efficient in packing factor (empty unusable spaces on the panel) but results in lower assembly cost. The current largest cell size, qualified to operate in the space environment, cut from a 4-in wafer, is a single 60 cm^2 square cell with an I_{mp} of approximately 1 A.

Present Solar Cell Efficiency

Present solar cell efficiencies utilized for today's mission are advanced triple junction (sometimes referred to as MJ for multijunction) GaAs solar cells with 27 to 29% BOL efficiency in 1 air mass zero (AM0). These MJ solar cells are commonly used on the majority of today's commercial and military satellites, interplanetary probes, and classified missions. They are currently an establish technology to manufacture with manageable production yield and the risk of insertion into solar array for mission design is low. Also, available as an emerging developing technology are the four-junction inverted metamorphic multi-junction (IMM) solar cells being qualified for space with efficiency in the range of 30 to 33% BOL. Cost of top-of-the-line IMM cells is a driver for mission planning because of the limited production yields. The nonrecurring cost of qualifying IMM cells for use in a particular mission's environment would also have to be considered as a major factor in mission planning. However, the trade-off is that the IMM cell can meet a mission's power requirement with a smaller size and lighter weight solar array whereas the MJ cell may not.

Future Efficiency Projection

Future research showing potential of greater than 40% with multiple junctions IMM solar cells, greater than four junctions, where more of the light wavelengths are utilized. Solar cells with recorded efficiencies greater than 40% are available in the form of development multijunction cells and solar concentrators, however, in limited production quantities because of production yield issues.

Solar Array Design Principles

Key solar array design principles are listed in Table 11.9.

Electrical Characteristics

A solar cell *I–V* (current–voltage) curve, as shown in Figure 11.54, has the following three significant points: I_{sc}, short-circuit current; P_{mp}, maximum power output point; and V_{oc}, open-circuit voltage. Corresponding to P_{mp}, there is a maximum power current, I_{mp}, and a maximum power voltage, V_{mp}. The output voltage of the cell decreases as the current increases from no load. The rate of change is indicative of a low source impedance (<1.0 Ω). As illustrated by Figure 11.55, the solar cell acts as a transistor where current flow is from negative to positive with solar input. When the maximum current capability of the solar cell is approached, the solar cell then acts as a current-limited source with a high source impedance (>1000 Ω).

Basic solar cell characteristics:
• Each cell has certain voltage (function of material, no. of junction) – Independent of cell size – Each cell has a given current (directly proportional to area of cell) – Cell is specified at a given temperature and illumination level • Normally 28 C and 135.3 mW/sq. cm • Solar cells/array almost never used under those conditions • Specific cell choices are up to EPS engineer and solar array designer • Most cell power is available at Maximum Power Point—MPP (less everywhere else)
Solar cell "real" application conditions
• Almost always hotter (voltage—negative temperature coefficient, current—positive temperature coefficient) • Sun angle may not be normal—illumination proportional to COS (angle) • Operating voltage may not be at V_{mpp} • Coverglass coatings, thickness affect cell performance • The above conditions affect cell voltage, current, power, even at BOL
End of life (EOL) operation further degrades cells
• Radiation degrades cell voltage, current, power
Highly dependent upon orbit and orbital location
Solar cell coverglass thickness and panel shielding affect radiation performance
Degradation of thermal surfaces may alter array temperatures
Other issues
• Solar array is almost always the costliest element of EPS – Solar cells account for around 75% of array cost – Higher efficiency cells often cost more per cell, but need fewer cells – Fewer cells may mean lower overall cell cost, and reduced solar array assembly cost (fewer cells to lay down) • Solar array area and mass proportional to number of cells and amount of power required • Solar array inertia is proportional to array size and affects GN&C performance • Deployed arrays usually need balanced arrays on both sides of vehicle
Solar wind effects
Solar arrays must be stowed for launch
• Usually need some area visible to sun in stowed configuration • Have to figure out how to do deployment – More complex = more expensive and riskier
Flexible arrays useful for very high-power applications
• Much more complex, expensive (lots of overhead for small arrays) – Flexible blankets are thin. Provide much less radiation shielding – May or may not be lighter than rigid panels – Flatness problems may cause illumination angle problems – Only certain type of cells can be used with flexible arrays – Cannot supply power during chemically propulsive transfer orbit if needed
Spinning/Rotating satellites with body mounted cells or curved panels cause illumination problems
• 1/Pi effective area—power only when array pointed at sun
Array has an "area fill factor" (active cell area/total array area)
• Smaller for bigger, nonrectangular cells (but array may be cheaper)

TABLE 11.9 Solar Array Design Principles

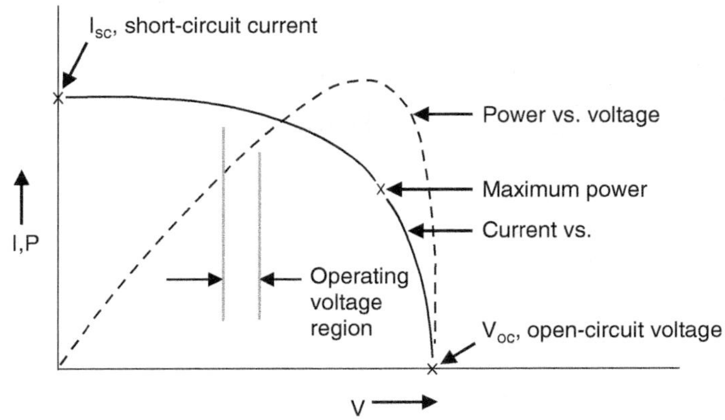

FIGURE 11.54 Typical solar cell *I–V* (current–voltage) curve.

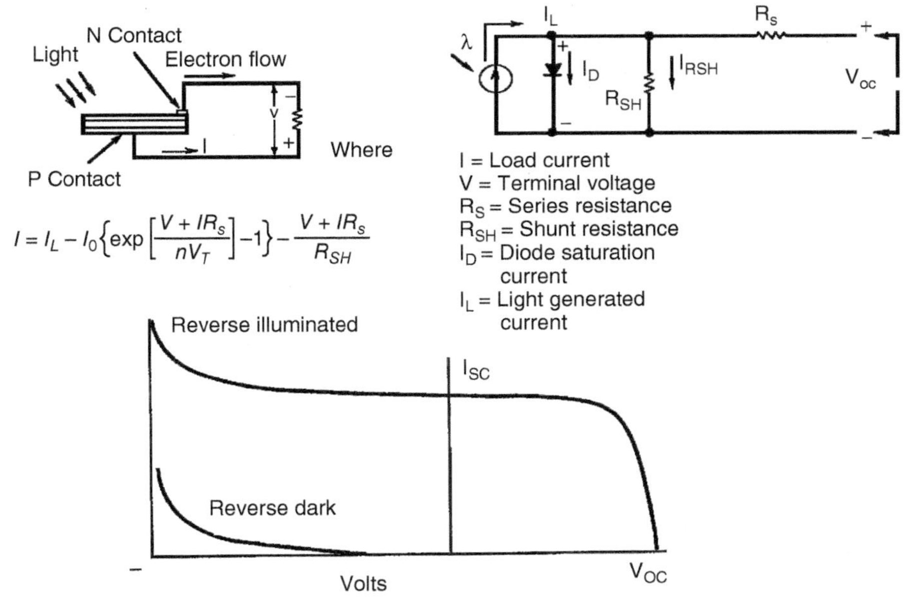

FIGURE 11.55 Schematic of the solar cell.

Operating Voltage

The V_{oc} of a cell is driven by the solar cell material with typical voltage in the 2.0-V range for today's multijunction GaAs cells. The solar cells are connected in series into electrical strings to increase the voltage to meet the minimum satellite bus voltage needs. High-power communication satellites in geosynchronous orbit (GEO) operate in the 70 V to 100 V to minimize harness losses and circuit complexity. With the higher voltage multijunction cells, 40 to 60 cells in series are required for a 70- to 100-V bus. The electrical string, sized for the minimum bus voltage at operating temperature, wiring losses and blocking diode forward voltage drop, should have its voltage operating point to the left of the P_{mp} to maintain current margin. The slope of the available current drops off quickly when the electrical string approaches open circuit voltage (to the right of the P_{mp}). The solar cell string voltage is per the following equation where V_{cell} is the independent variable (at temperature), I is the dependent variable and can be modeled based on the modified Brown model (Ref. 2), $R_{harness}$ = the array overall wiring losses, and V_{diode} = the string blocking diode (for reverse current protection) forward voltage drop.

$$V_{string} = \text{No. of cell in series} \times V_{cell} - I^{*}R_{harness} - V_{diode}$$

Temperature

Both current and voltage respond to temperature variation as illustrated in Figure 11.56. As the temperature increases, charge carrier energy increases and results in a longer minority carrier lifetime. Current increases as a result. Conversely, as the solar cell temperature increases the solar cell voltage decreases. The voltage decreases significantly more than the current increases with temperature. The magnitude of the change is affected by radiation exposure.

The normalized temperature performance is computed by

$$\text{Normalized current or voltage} = 1 + \beta(T_{cell} - 28°\text{C})$$

where β = voltage or current temperature coefficient for a particular cell type (radiation dependent) in %/°C and T_{cell} is the cell temperature.

Circuit Current

The I_{mp} of a cell is proportional to the individual cell's area. For example, a 2 cm × 4 cm size solar cell produces half as much current as a 4 cm × 4 cm size, and so on. Electrical strings are combined in parallel into circuits per Figure 11.57 to increase the operating current. The maximum current is limited by the satellite's power system de-rated current capability and is usually sized at BOL before on-orbit environment degradation.

Solar Array Sizing

Key steps necessary for the design of a mass and cost-efficient solar array are listed in Table 11.10.

Mission Power

The power generating requirements of a solar array-powered mission are

1. The satellite's load power requirement including contingency margin, redundancies, losses
2. Battery recharge power

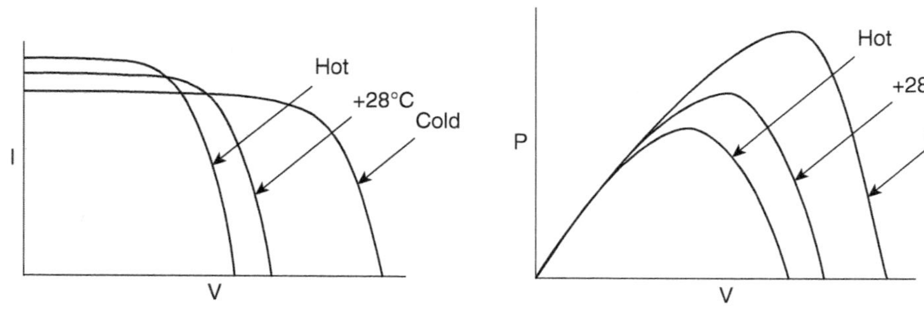

Figure 11.56 Operating temperature impact on solar cell I–V characteristic.

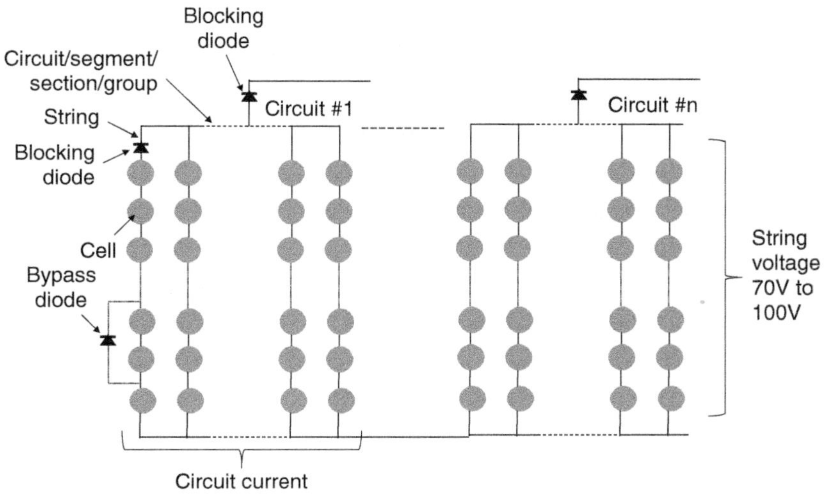

FIGURE 11.57 Typical solar array circuit.

Step 1	Find orbit, orbit location, mission life, EOL margin, etc. (Source: RFP/Proposal, S/C Perf. Spec., etc.)
Step 2	Find solar array launch, transfer orbit, in-orbit configurations. Any Deployment, area, mass, cost, and shadowing constraints (Source: SMT Eng., Systems Eng., PMO, etc.)
Step 3	Determine S/C load power requirements including design contingency, margin, losses and battery recharge power as well as operating voltage requirements (Source: S/C Power Budget & Mass Properties Documents, Systems Eng., PHSL, etc.)
Step 4	Must know approximate operating temperature, illumination, total radiation dose, etc. (Source: SMT Eng., Product Center, analysis, etc.)
Step 5	Determine candidate cell technology(s), coverglass thickness, etc., that are available and will work (Source: Product Center, Vendors, etc.)
Step 6	Perform sizing analysis (Source: Product Center, analysis tools, etc.) • # series cells = Operating voltage, Vmpp of cells at temp/radiation • # parallel cell strings = Current required, current per cell at temp/radiation/illumination, shunt current May need a lot more cells if spinner, or array doesn't point normal to sun, etc.
Step 7	Iterate mass, cost, performance for the best overall design (EPS Eng., Product Center)
	• Remember: Array design is affected by selected EPS architecture. Want to operate as near MPP as possible at EOL (extract as much power as possible from array) • Remember: Design must be amenable to easily accommodate changing requirements, mass/cost/launch constraints, etc.

TABLE 11.10 Key Sizing Steps

3. Bus operating voltage
4. Maximum circuit current
5. Dedicated electric propulsion power
6. EOL power (typical 15 years for a GEO commercial communication mission)

BOL power is verified by test (BOL power), and EOL power is verified by analysis (EOL power) based on a variety of factors described in more details in the following sections.

The factors that are involved in the solar array power sizing analysis include the operating environment such as temperature, radiation degradation over mission life, micrometeoroid damage, contamination, and solar intensity. Other factors also to be considered are development risks of inserting new technology solar cells, the mechanical deployment structure, and the overall cost. In today's aerospace environment and budget realities, cost (solar cell costs, development/qualification costs, production costs, etc.) is usually the first or second driver for a solar array design.

Operating Environments

The operating environment of a satellite's mission provides the requirements for the overall solar array sizing design. A major factor for the overall array size is the radiation degradation of the solar cell's semiconductor material over the lifetime in earth's orbit. The percentage of degradation is based on the total exposed radiation dosage (expressed as damage equivalent 1 MeV fluence), a function of the orbit altitude and time of exposure and time in orbit of the mission. Reference 3 describes in detail the shielding analysis and power degradation analysis standard used industry wide. The remaining factors that influence the overall solar array size, and ultimately cost, are listed and explained below.

1. Operating temperature: The solar cell operating temperature, as a function of its power conversion efficiency, thermal optical property, solar intensity as per time of the year, thermal conductivity of the supporting mechanical substrate, and orientation of the satellite versus time.
2. Solar intensity: The solar intensity in earth's orbit varies with the season (Figure 11.58) due to variation in earth-sun distance throughout the year and directly impact the solar cell's output current (Figure 11.59). One sun at AM0 is equivalent to 1353 W/m^2. See Ref. 3 for relative intensity changes throughout the year.
3. Current mismatch: Variation in individual solar cell's output current within the same string can cause power losses as much as 2%. As a good practice, cells grouped within an electrical string are matched to each other to within 5 mA.

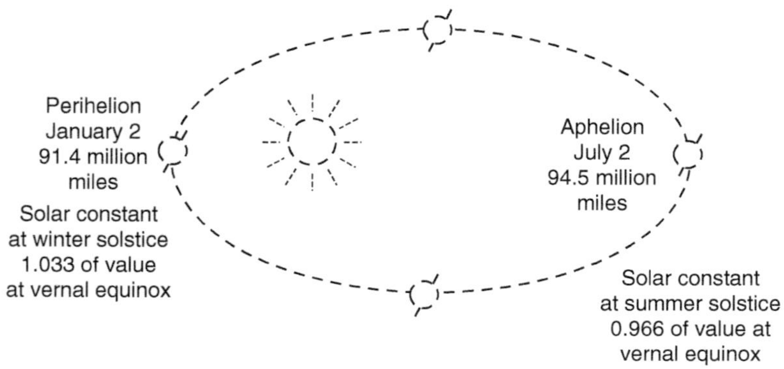

FIGURE 11.58 Solar intensity variation within year.

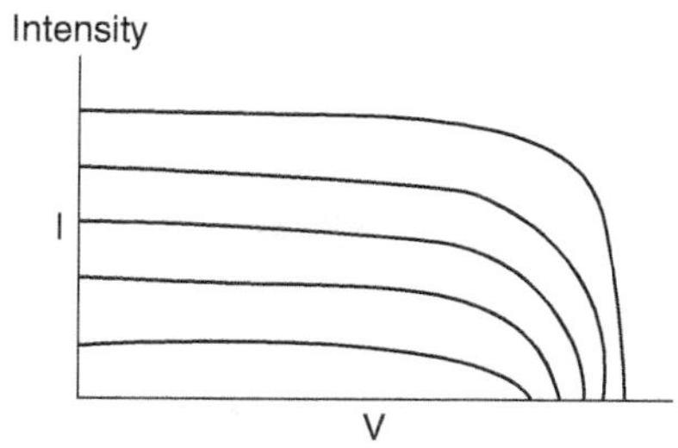

FIGURE 11.59 Current output versus solar intensity.

4. Shadowing: Sweeping shadow across the solar array surface throughout orbit, due to large deployed reflectors or the satellite body, can cause temporary partial loss of power and should be considered during system design.
5. Contamination: UV darkening of the coverglass bonding adhesive usually occur immediately during the first year in orbit and the current loss can be up to 2%.
6. Micrometeoroids: Potential particle damage, expressed statistically, damaging solar cells, and power wires that can lose power from at least an electrical string or circuit or combination of both. Power margin of one extra circuit or more are usually factored in the array size.
7. Electrostatic discharge mitigations in charge particles environment (trapped electrons and protons in earth's orbit): Surface charge buildup on the solar array can lead to electrical discharge to the nearest available metal surface when the surface potential builds up to greater than or equal to 500 V. The discharges by themselves are high current but are temporary and self-extinguish; however, in the vicinity of power generating solar cells, the cells can power and sustain the discharges as secondary discharges that can pyrolize the dielectric layer and short-circuit the solar array circuits to ground which reduces the available power to the satellite. See Ref. 5 for more detailed information on electrostatic discharge.
8. Attitude control's thruster plume: Plumes from the satellite's orbit attitude control thrusters can degrade the solar array power performance by ablating the antireflective coating from the coverglass of the solar cells. In certain conditions, the thruster plumes can initiate ESD events and cause short-circuits and loss of power from the solar cells in direct exposure to the thruster plumes.

Mission Life (Number of Thermal Cycles, Metallurgical Joints Fatigue, Reliability)

The mission life and orbit attitude affects the reliability of the solar array due to number of eclipse cycles (number of hot/cold thermal cycles), the temperature extremes reached and the different materials' coefficient of thermal expansion (CTE) that fatigues the various metal-to-metal joints that connect electrical strings and circuits together. The typical metallic joints associated with solar arrays are

1. Solar cell string joints, at the n and p junctions to interconnects
2. String blocking diode joints
3. Solar cell string termination to round wire joints
4. Round wire to connector contact joints

Qualification by test of the solar array design, using Ref. 7 as a guideline, with sufficient sample size of the above-mentioned joints in the operating life thermal cycling environment is required. One or more sub-size qualification solar array coupons can be used to show compliance to the mission life requirement. Metallic joints by solder is usually the least reliable method because of the brittleness of the joint, low melting point, low fatigue strength, and concern with dendrites associated with the tin and lead components of solder. Brazing and welding are the preferred metallic joining methods for its high reliability operating in the extremes of the space environment.

Mechanical Properties

Solar cells require packaging on a mechanical support structure to stow against the satellite to survive the launch acoustic and dynamics environment and deploy upon command once in orbit, to generate power and sun-tracking as per Figure 11.60. The deployed solar arrays are usually the largest appendage on the satellite and their center of mass are far enough from the satellite that maintaining attitude control is not trivial. The reliability of the deployment mechanisms to deploy the solar arrays to power the satellite is also critical to the satellite mission.

The deployed solar array wings are usually oriented normal to the orbit plane (±Y axis) in order to maximize solar exposure at any point in orbit. Each solar array wing tracks the sun via a 360° solar array drive mechanism, which continuously orients the panel surface normal to the sun at any position in orbit, and in most cases, open loop tracking during eclipse.

The followings provide a brief description of the various mechanical factors associated with the solar arrays:

1. Minimum stowed and deployed frequencies: The steadiness required of the satellite to perform its mission determines the stiffness of the deployed solar arrays. The stiffness of the deployment hinges is the "softest" component in the solar array that drives the first mode natural frequency of the array.
2. Deployment temperatures (hinges, dampers, harnesses): The temperature of the deployment mechanisms determines the available torque margin to drive out the solar array. The electrical power harness produces resistive torque that the hinges have to overcome, and the resistive torque is directly proportional to temperature. The dampers that limit the deployment rate and end-of-travel lockup loads on the hinges are also temperature dependent. The damping material is usually sealed silicone fluid, which can leak at hot temperature and contaminate the surface of the adjacent solar cells and impact electrical power.
3. Deployment method: Reliability of the deployment mechanisms determines the probability of a successful deployment critical to the mission. As discussed in the following sections, rigid solar arrays on honeycomb structure have the simplest and most reliable deployment method using spring-driven hinges and dampers. Flexible solar arrays have more complex mechanisms required to successfully deploy a solar array.
4. Launch loads and acoustic environment: The launch vibration environment produces the dynamics load that can amplify the stresses that the solar array structure experiences during launch while stowed inside the launch vehicle fairings.

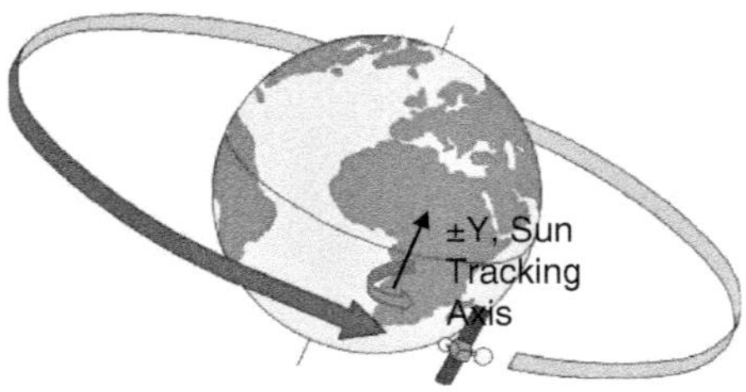

FIGURE 11.60 Solar array orientation.

Deployable Solar Array Technologies

Different concepts of deploying solar arrays, supporting photovoltaics, in space include rigid honeycomb structure, flexible film carrier, and solar concentrator, where the solar intensity is magnified onto smaller number of solar cells to produce relatively same amount of power. Another developing technology is the combined use of solar array for the satellite loads, regardless of type, and electric propulsion, where a dedicated portion of the solar array directly powers the plasma electric propulsion that lowers the overall launch mass.

Rigid Honeycomb Substrate The most common type of deployable arrays in use on the majority of satellite is the deployable rigid honeycomb substrates, which employ stored-energy mechanisms to deploy on-orbit upon command. The solar cell circuits are bonded onto large area honeycomb panels, graphite facesheets, and aluminum honeycomb core, which have relatively high structural strength and rigidity in addition to their light mass. An electrical insulation layer, usually Kapton™, isolates the solar cell circuits from the conductive graphite facesheet. Figure 11.61 shows the stack up of the different construction material layers.

The panels are folded up and stowed against the satellite structure (Figures 11.62 and 11.63), which limits the solar panel size to be generally no larger than the satellite panel that they are stowed

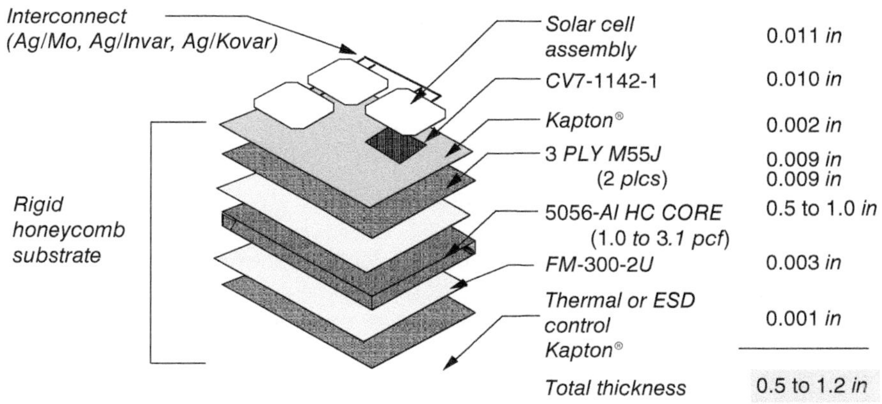

FIGURE 11.61 Typical rigid honeycomb substrate array design.

FIGURE 11.62 Satellite with stowed rigid honeycomb panels, single and multi-panel solar array wings shown.

Figure 11.63 Deployed rigid honeycomb solar array wing.

against. The maximum number of panels that can be deployed per solar array wing is limited by its deployed length and the satellite's attitude control capability due to the solar array's center of mass being further away from the satellite's center of mass. Launch vehicle fairing space is another limiting factor. The stiffness of the deployment mechanisms after deployment lock-up is usually the main driver that determines a solar array wing's minimum deployed natural frequency.

Flexible Substrate Solar Arrays The advantage of flexible solar arrays, where solar cell circuits are packaged on a thin film flexible substrate material like Kapton or similar material, where it can be stowed for launch into a small envelope or volume and deployed into a large area generating from 10 to 25 kW of electrical power per single solar array wing, using today's available photovoltaic technologies. The trade-off is a higher starting launch mass as compared to rigid arrays, and higher recurring cost associated with the deployment structures required of flexible solar array in order to attain the high generating power. The cross-over point where the advantages of flexible solar array starts to gain over rigid honeycomb solar arrays is approximately 12 to 16 kW per solar array wing (Figure 11.64). Beyond this power level, additional solar array power can be added to a flexible solar array with minimal addition in structural mass and cost by simply adding more flexible solar panels into a single wing. Adding more rigid panels carry more mass and cost penalty in order to gain power because of its inherent large size and mass.

The two flexible solar array types that are being pursued by different organizations include rectangular fold-up blankets shown in Figure 11.65 that extend accordion style, and circular fold-up blankets shown in Figure 11.66 that extend into a circular fan style.

Flexible Fold-Up Blankets Unlike large area rigid honeycomb substrates, solar cell strings and circuits are bonded onto flexible thin film substrate like Kapton that can fold like paper for more compact stowage. An extendible structure then stretches out upon command to extend the flexible panels into a large surface area solar array to collect more solar energy. Figure 11.65 shows an example of two flexible fold up blankets being deployed by a single extendible mast in the middle on the ISS.

Circular Fold-Up Blankets Figure 11.66 shows two circular Ultraflex™ solar arrays, from Orbital-ATK, that have flown on the Cygnus commercial resupply mission to the ISS. Each solar array stows segments of the circle comprising of solar cell strings and circuit into a flat sandwich package

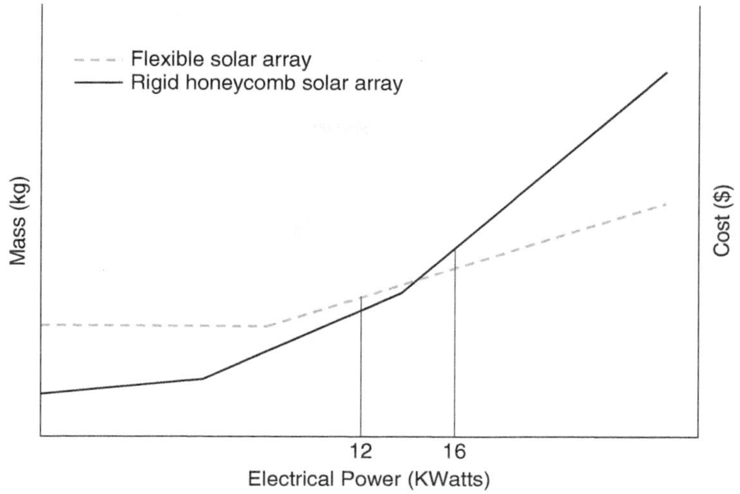

Figure 11.64 Flexible versus rigid solar array trade.

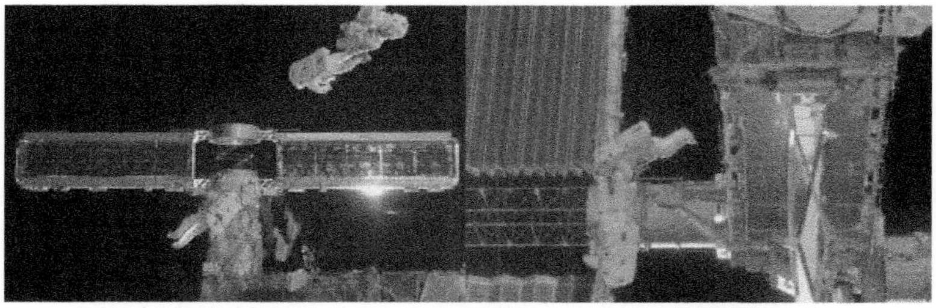

Figure 11.65 Flexible solar array wing on the International Space Station (ISS).

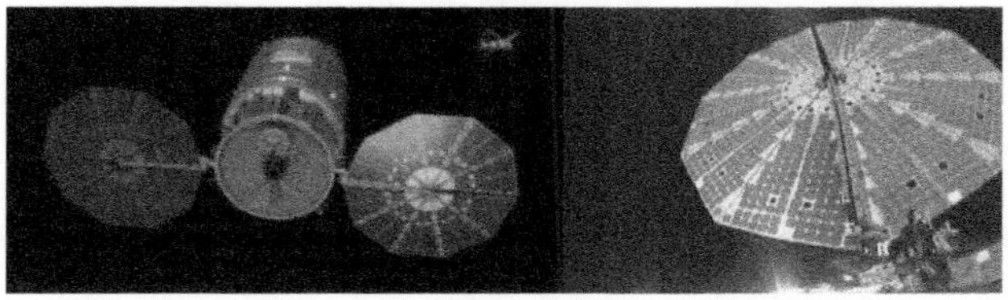

Figure 11.66 Circular fold-up blankets, Ultraflex, by Orbital-ATK.

for launch. Upon command, the sandwich releases from the vehicle and a motor-driven system pulls the segments out into a circular fan.

Solar Electric Propulsion

With solar electric propulsion (SEP), large high-power solar cell arrays convert collected sunlight energy to electrical power to directly power, instead of through the EPS regulated power, more fuel-efficient electric thrusters that provide gentle but nonstop thrust, as opposed to impulses of conventional chemical propulsion, throughout the mission. Several electric thrusters can be combined to increase the power of an SEP satellite. Such a system can potentially provide enough force over a period of time to move cargo and perform orbital transfers (Ref. 5). The advantage of electric propulsion thrusters is its low propellant mass required for a mission, which translates into lower overall satellite mass even after accounting for a larger solar array to power the thrusters. However, the available thruster power to accelerate the satellite is smaller than heritage chemical propulsion power so more time is needed for orbital transfer.

The electric propulsion thruster operates at high voltages, from 100 V to greater than 1000 V, so the design of the solar array and the EPS with the higher operating voltage becomes more complex. Some GEO commercial communication satellites currently in-orbit utilize SEP for both GEO transfer orbit and on-orbit station keeping.

Solar Concentrator

Solar concentrator arrays work by focusing the solar intensity onto a smaller solar cell area to increase the power output through the use of concentrating lens. The benefits of solar concentrators are fewer solar cell quantity required for relatively the same amount of power. However, the required development effort to successfully deploy a concentrator in space from a stowed configuration is still a high technology risk. More development efforts are being expended to reevaluate solar concentrator for deep space missions where the solar intensity is exponentially lower than in earth orbit.

11.29 Batteries

Joseph Troutman

Overview

For a satellite, rechargeable batteries are used during the periods when the electric power from the solar array is not available (that is in eclipse) or is insufficient (that is during stationkeeping or to support the peak power demand). Additionally, the batteries will have to meet the following operational conditions. During ground power switch-off and launch, the housekeeping functions of the satellite are maintained from the batteries until the fairing is jettisoned and the solar array panels are exposed to the sunlight. Even then, in many satellite missions the solar array (if not deployed) in transfer orbit will deliver power only intermittently, requiring a battery to supplement the power. Another vital function of the battery prior to final acquisition on station is to deliver the electric power for various pyrotechnic actuators, such as for the ignition of the apogee boost motor or the deployment of the solar array.

The type and design of the battery is dependent upon the operational mission parameters and the specific orbit the satellite will reside in. Battery systems for low earth orbit (LEO) operating satellites typically have a maximum life period of 5 to 10 years due to the number of charge/discharge cycles it will experience. A typical LEO battery system will undergo 16 charge/discharge cycles per day (24-hour period) due to the number of eclipse periods it will experience. The duration of those eclipse periods may vary depending upon the orbit inclination and the earth orbit seasonal effects. A battery system designed for a geosynchronous orbit (GEO) will experience much fewer eclipse periods than a LEO orbit due to its distance from the earth. A LEO satellite is much closer to the earth (typically 400 to 750 km above the earth surface) and will remain in the shadow of the earth through each orbit, whereas a GEO battery system (35,786 km above the earth surface) will experience much fewer eclipse periods as the satellite does not remain in the shadow of the earth during much of the year. In a GEO orbit, the battery system will only experience 90 eclipse cycles per year (two periods of 45-day eclipse periods centered around the vernal and autumnal equinoxes) with the eclipse time ranging from 0 to 72 minutes maximum). This is a less stressful operation condition on the battery system and therefore the battery system can perform for a much longer operational period than an LEO orbit battery system. Typical GEO battery systems will operate for a time period of 10 to 15 years with more and more satellites reaching for the 20-year operational scenario.

Chemistry

The development of lithium ion rechargeable battery cells in the late 1970s set the stage for various lithium ion technologies used today for space applications. The initial chemistry used, lithium cobalt oxide (LCO) is still used today in many applications. Lithium ion is the most common battery cell chemistry used today and advancement in the lithium technology is continuing. Lithium ion chemistry is a leap in improvements over nickel hydrogen technology with its high-energy density, high discharge rate characteristics, higher average cell operating voltage (3.7 V against 1.25 V for nickel-hydrogen), and compact packaging features. There are several lithium ion chemistries used today but two chemistries are primarily used for space batteries. They are the LCO and lithium nickel cobalt aluminum (NCA) oxide chemistries. An LCO technology has been the baseline chemistry for most of the space flight batteries in orbit today. The newly developed NCA chemistry provides an improvement over the LCO chemistry by providing high discharge power capability and even longer life, and low-capacity fade capabilities. Other chemistries used today include lithium nickel manganese cobalt oxide, lithium sulfur, lithium iron phosphate, and lithium titanate. Each chemistry has its benefits and is used in various applications but today's space batteries are relying primarily on LCO and NCA technologies. Both, LCO and NCA have performed well in ground life test demonstrations which take on the order of 5 to 7 years to accomplish. These demonstrations are an important part of reducing mission risk prior to implementing a new battery chemistry on a major satellite program.

Variations in cell chemistry result in varied cell energy densities. Energy densities for lithium iron phosphate chemistry cells are in the range of 90 to 100 Wh/kg, whereas the energy density for an NCA chemistry cell can be in the 265 Wh/kg range. Energy density is a key factor when selecting a cell for a GEO spacecraft as the energy capability and mass of the cell are critical requirements for each mission. However, high energy density cells do not come without their potential issues. Typically, the cell vendor will thin down the cell walls to minimize the mass of the cell, which may introduce possible side wall ruptures in the event of a thermal runaway situation. Although energy density is extremely important when selecting the proper cell for a GEO mission, all performance characteristics of the cells must be taken into account, which would include charge/discharge rate capability, capacity fade over mission life, mass, volume, and internal resistance.

Battery Depth-of-Discharge

Another factor affecting the operational life of a lithium-ion battery system is the operational depth-of-discharge (DOD) or energy removed from the battery system during the discharge period of the orbit. Lithium ion battery chemistry (like other battery chemistries) will have a reduced operational life when experiencing high DOD levels. This is due to several factors but mainly due to degradation of the electrodes (aging resulting in capacity fade) and increased internal resistance over life. Knowing this the battery designer must select the proper battery size (capacity) to ensure the DOD level achieved is such that the capacity fade over the mission life and increased internal resistance of the cells still allows for full mission power delivery from the battery system. For LEO orbit battery systems to be able to perform for a period of 5 to 7 years a low DOD level is required. A typical LEO DOD may be in the range of 5% to 20%, whereas a high DOD for a LEO orbit would be in the 30% to 40% range. For GEO battery systems, the eclipse period varies over a 45-day period twice a year with the eclipse duration beginning at zero minutes and reaching a maximum of 72 minutes at day 22 then decreasing down to zero minutes at day 45. This cycle occurs twice a year during spring and fall equinox seasons. There are no eclipse periods and hence no battery cycling during summer and winter solstice seasons of the year when the satellite is never in the shadow of the earth. However, the batteries might experience a few charge/discharge cycles during station keeping in solstice seasons. Figure 11.67 presents the effect of DOD level on Li-ion

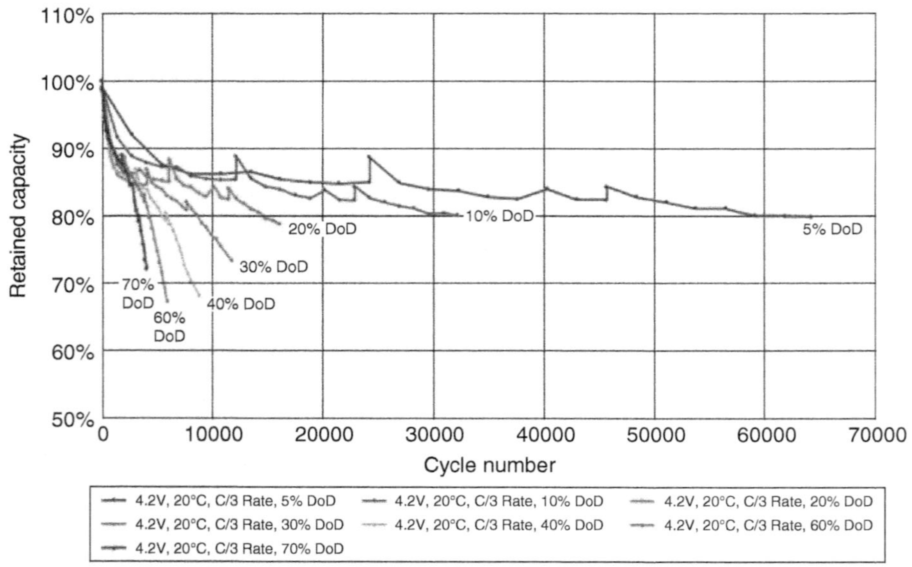

FIGURE 11.67 Effects of DOD on Li-ion battery cycle life (based on Sony Hard Carbon 18650 cell test data).

retained capacity over cycle life with a fixed C/3 charge/discharge rate and a constant operating temperature of 20°C. From the figure, it can clearly be seen that operating at a higher DOD level reduces the number of operating cycles over life.

Operating Temperature

The operating temperature is another driving factor for the expected electrical performance of the lithium-ion battery system. The operating temperature of the battery system has a similar effect on the battery performance as does the DOD levels. Typically, with lithium ion chemistries, higher operating temperatures result in shorter cycle life due to decomposition of electrolyte in the battery cells. It should be noted, however, that operating at too low of a temperature will result in minimal available capacity from the battery due to increased internal resistance, and operating beyond the cell suppliers recommended temperatures (both high and low) could result in an operational safety issue. The safety issue may include potential cell thermal runaway (increasing rise in cell temperature) which may result in venting of the battery cells when the battery system is operating in an extreme high temperature condition (>80°C). The safety issue on the extreme low-end operating temperature is potential battery cell internal shorting resulting from dendrite growth (generation of whisker which can create a current path between the electrodes). Extreme low operating temperature also results in high internal resistance that causes heating up of the cell resulting in a potential battery cell venting condition. Typical thermal operating conditions for lithium ion chemistries are 0 to 40°C. The optimum thermal operating condition within that specified range is dependent upon the lithium ion chemistry selected. Figure 11.68 presents cycling data demonstrating the effects of operating temperature on cycle life of lithium ion battery cells.

End-of-Charge Voltage Conditions

The end-of-charge voltage (EoCV) can also have an impact on the cycle life of a battery system. Battery systems are designed and configured to meet the mission requirements at the end of

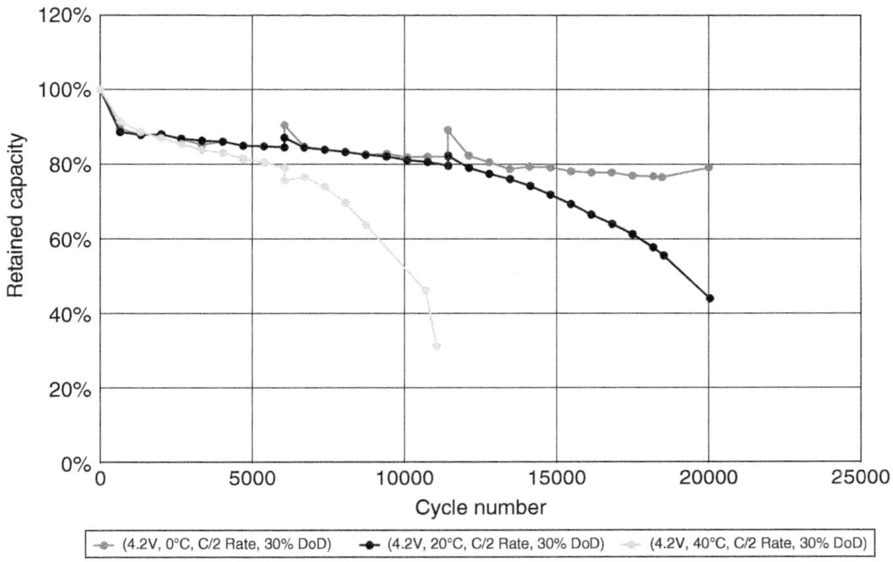

Figure 11.68 Operating temperature effects on cycle life (based on Sony Hard Carbon 18650 cell test data).

designed mission life (EOL) of a satellite. As such the battery system will have excess energy capability at the BOL. To take advantage of this excess energy capability and to minimize capacity fade due to charging the battery to 100% state of charge (SOC) at BOL, the satellite operators typically charge the battery to a lower EoCV value during the earlier mission years. With the excess energy (capacity) at BOL and using the mission designed DOD level, the end of discharge voltage (EoDV) is much higher at BOL as there is no degradation of capacity or increase in internal resistance of the battery cells due to minimal cycling. Reducing the EoCV and not charging the battery system to its 100% SOC voltage level results in a lower capacity fade or degradation during the early years of the satellite mission. As the number of cycle increases over the mission years and the EoDV level begins to approach the minimum EoDV value, the EoCV is increased to keep the bus voltage level (minimum battery voltage level) above the minimum EoDV. Figure 11.69 provides test data demonstrating the cycle life capability of a battery system at various EoCV values.

Battery Cell Selection

Selecting the proper (optimum) battery cell for a satellite application is dependent upon many factors. Besides, operational DOD level and operating temperature, there are other mission-specific requirements, which may also drive the selection of the battery cell chemistry/cell vendor. Mission-specific requirements may include pulse power delivery capability, high overall energy density (Wh/kg) to minimize mass and volume, mission life and associated required cycle life, and cost. Table 11.11 provides a capabilities overview of a few available lithium ion cells and their electrical/mechanical benefits, which need to be considered when evaluating battery cell selection. Table 11.12 provides an overview of some commercial off-the-shelf (COTS) and custom-made cells and chemistries. The selection of the cell packaging format and cell chemistry is directly related to satisfying the mission-specific requirements of a program. Many of these cell technologies are currently in use in unmanned and manned space programs.

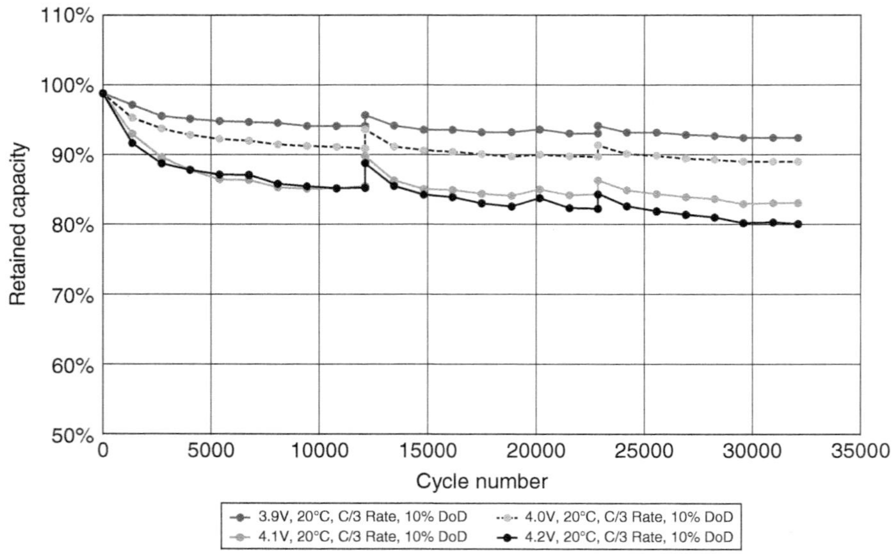

FIGURE 11.69 Effects of EoCV on cycle life of a battery (based on Sony Hard Carbon 18650 cell test data).

Battery Management System

All satellite missions using rechargeable lithium-ion batteries require some sort of battery management system (BMS). The BMS has several functions, which are described below:

1. Cell voltage monitoring: Individual or virtual cell (two or more cells connected in parallel) voltage monitoring is performed to assess the health of the battery cells, detect overvoltage and undervoltage conditions, and to provide guidance to the electrical power system on the state of the battery system with respect to energy within the cells.
2. Total battery voltage monitoring: The monitoring of the total battery voltage is used at the satellite system level to determine the voltage state of the spacecraft/battery bus. The voltage of the battery is used to determine situations where a safe mode condition may need to be implemented if the voltage is at a very low state. Additionally, a high battery voltage may provide information on the battery state of charge or an over charge condition.
3. Cell charge voltage balancing: Some battery cells may have different internal resistance values, varied self-discharge rates, or may be at a different operating temperature than other cells within a battery pack. These factors may cause the voltage of an individual cell to be different than the other cells causing a voltage divergence (imbalance) between cells. This could result in a cell operating in a more strenuous condition resulting in a higher capacity or voltage fade over the cycle life. Cell voltage divergence may also lead to one cell reaching the maximum charge voltage prior to the remaining cells, resulting in a battery that would not get fully recharged as the charge current would need to be terminated due to one cell reaching its maximum charge voltage limit. To prevent this situation an individual/virtual cell voltage balancing methodology is used to ensure all cells remain within a specific voltage tolerance during the recharge period.

There are three basic cell voltage balancing methods used today. They are a simple resistive bypass circuitry placed around each individual/virtual cell, an active DC/DC converter connected

	Panasonic NCR-18650A	LG Chem INR18650MJ1	Sony US – 18650HCM	E-Moli One IHR18650C
Nominal Capacity - Typical (Ah)	3.1	3.5	1.5	2
Nominal Voltage (V)	3.6	3.6	3.6	3.6
Maximum Voltage (V)	4.2	4.2	4.2	4.2
Minimum Voltage (V)	2.5	2.5	2.5	2
Maximum Discharge Current	2C	2.9C	1.3C	10C
			High Rate	
Energy Density (Wh/Kg)	245.27	257.14	130.75	160.00
Diameter - Maximum (mm)	18.6	18	18	18.4
Height - Maximum (mm)	65.2	65	65	65.2
	High Energy			
Weight (g)	45.5	49	41.3	45
Discharge Operating Temperature (°C)	–10 to +60	–20 to +60	–20 to +60	–20 to +60
Charge Operating Temperature (°C)	0 to 60	0 to 45	0 to 40	0 to 45
Cycle Life	80% of minimum capacity after 300 cycles of 0.3C charge/1C discharge	High Cycles	80% of minimum capacity after 2000 cycles of 0.66C charge/0.47C discharge	80% of minimum capacity after 350 cycles of C charge/10C discharge
Anode	Hard Carbon	Hard Carbon	Hard Carbon	Hard Carbon
Cathode	Lithium Cobalt Oxide	Lithium Cobalt Oxide	Lithium Cobalt Oxide	Lithium Cobalt Oxide

Table 11.11 COTS Battery Cell Characteristics for Cell Selection

around each individual/virtual cell, or a "common" bus technique where the cells are switched on and off on a common bus line to allow higher voltage cells to share current to lower voltage cells charging them up to the voltage level of the higher voltage cells. All balancing techniques have their advantages and disadvantages, which include high costs, complexity to the battery system high thermal dissipation, higher amount of energy needed to power the BMS, extra mass, and volume. Figure 11.70 shows the general configuration of the cell voltage balancing electronics and an example of the simple resistive shunt balancing method.

Cell Packaging Form Factor Selection

There are several mechanical packaging formats available for lithium ion cells in the market today. One very popular standard is the cylindrical 18650 cell format where the diameter of the cell is 18 mm and the height of the cell is 65 mm. 26650 is also a popular cylindrical design format where the cell diameter is 26 mm and the height is 65 mm tall. These cells are typically COTS cells, mass

Chemistry and cell identification		Cell format	Capacity	Maximum discharge current	Energy density
Lithium Cobalt Oxide					
	Panasonic 18650A	18650	3.1 Ah	6A	245 Wh/kg
	LG Chem ICR18650B2	18650	2.6 Ah	3.9A	202 Wh/kg
	Sanyo UR18500F	18650	1.7 Ah	3.4A	160 Wh/kg
	Samsung ICR18650-26F	18650	2.6 Ah	5.2A	204 Wh/kg
	Sony US18650HCM	18650	1.5 Ah	1.95A	130 Wh/kg
	E-Moli One IHR18650C	18650	2.0 Ah	20A	160 Wh/kg
	E-Moli One ICR18650J	18650	2.4 Ah	4.8A	192 Wh/kg
	E-Moli One ICR18650M	18650	2.8 Ah	5A	217 Wh/kg
Lithium Manganese and Nickel-Manganese-Cobalt Oxide					
	IHR18650C	18650	2.0 Ah	20A	160 Wh/kg
Lithium Nickel Cobalt Aluminum Oxide					
	QL18650	18650	1.5 Ah	1.5A	118 Wh/kg
	LG Chem INR18650MJ1	18650	3.5 Ah	10A	257 Wh/kg
	QL075KC	Large Prismatic	72 Ah	75A	142 Wh/Kg
Lithium/Florinated carbon					
	Panasonic BR-AAK	AA	2.5 Ah	500 mA	465 Wh/kg
	Panasonic BR-C	C	5 Ah	1 A	355 Wh/kg
	Quallion D Cell	D	15 Ah	1.5 A	600 Wh/kg
	Quallion QC0025B	Small cylindrical	25 mAh	100 uA (continuous) 10 mA (pulse)	165 Wh/kg

TABLE 11.12 Cell Packaging Format and Chemistry Examples with Electrical Characteristics

produced by cell vendors such as Sony Corporation, E-One Moli Energy, Panasonic, etc., and are used in commercial terrestrial products but can also be tested and screened for use in launch vehicle and satellite applications. In general, COTS cells of the same lot from an automated assembly line are expected to pose fewer cell-to-cell variations when obtained from a reputable manufacturer with a high production rate. Qualifying a COTS cell for space applications requires extensive testing in the following area; environmental (thermal, vibration, shock, radiation), safety (overcharge, over-discharge, shorting), chemical (multicell testing for ensuring the same chemical and mechanical materials are used for uniformity across all cells produced), and life characteristics (cycle life at various DOD levels, temperatures, and charge/discharge rates). This testing can become quite extensive and costly but has shown to be a very viable process for using COTS cells for long life satellite projects. COTS 18650 cells range in available capacity values of 1.5 to 3.5 Ah and offer both high energy and high power (high current) performances.

The other mechanical form factors used today are the large cylindrical and prismatic or pseudo prismatic cells. These large format cells are available in a much higher capacity than an 18650 or

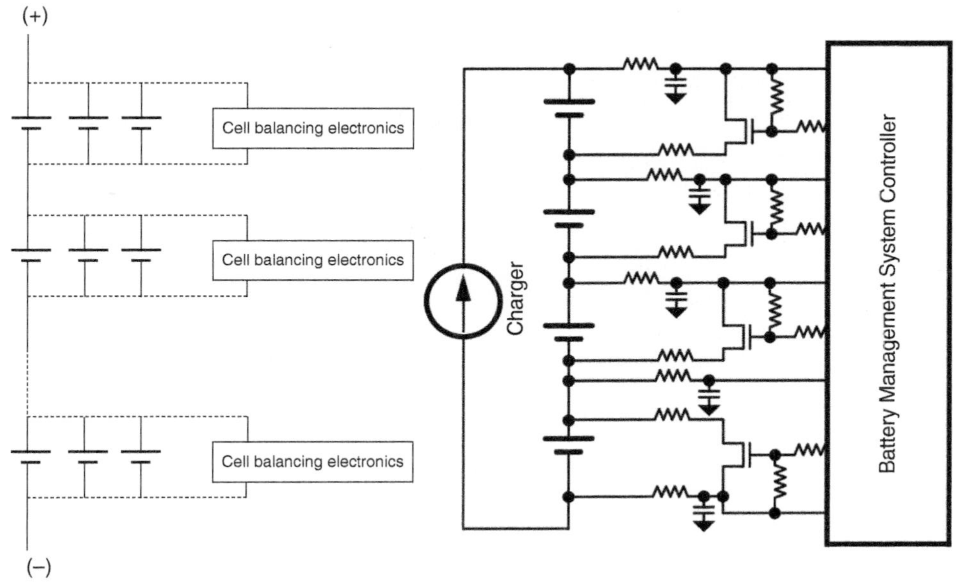

FIGURE 11.70 Cell voltage balancing diagrams.

26650 small cylindrical cell. Large format cell capacity typically ranges from 20 to 200 Ah. The selection of a large mechanical format cell versus the small mechanical format cell is mission specific and must be weighed against all the program requirements (mass, volume, safety, cost, schedule, electrical performance, and heritage). The packaging (mechanical) form factor of the cell does not have any effect on the battery cell's electrical performance assuming it is operated within the recommended thermal operating temperature. Figure 11.71 shows the performance curve for a large prismatic (rectangular) space cell and a space qualified 18650 format cylindrical cell. The electrical performance for the two mechanical formats is identical when using the identical cell chemistry and acceptable cell operating temperature environment as shown in Figure 11.71.

The selection of a small cylindrical cell or a large prismatic format cell (Figure 11.72) for a specified mission is very much mission dependent as both technologies would be able to meet the overall mission requirements. Considerations when deciding to select a small cylindrical cell or large prismatic cell are volume of an integrated battery design (one format may have an improved packing factor resulting in a smaller battery), mass (one format may require additional mechanical features to constrain the cells resulting in added mass), complexity of the thermal design needed to reject heat and maintain temperature uniformity, cost (one format may require more touch labor to assemble than the other resulting in additional costs and one format may result in an overall increase in cell costs for the battery design).

Battery Design Configurations

Space batteries can be electrically configured in one of two ways. There is the series before parallel electrical connections and the parallel before series electrical connections. The selection of one of these configurations is dependent upon the mission-specific requirements. The series before parallel configuration is an industry standard. Extensive cell space qualification process and subsequent individual cell screening and matching process ensures that each cell selected for the space battery will have identical electrical performance characteristics. Furthermore, individual cell screening

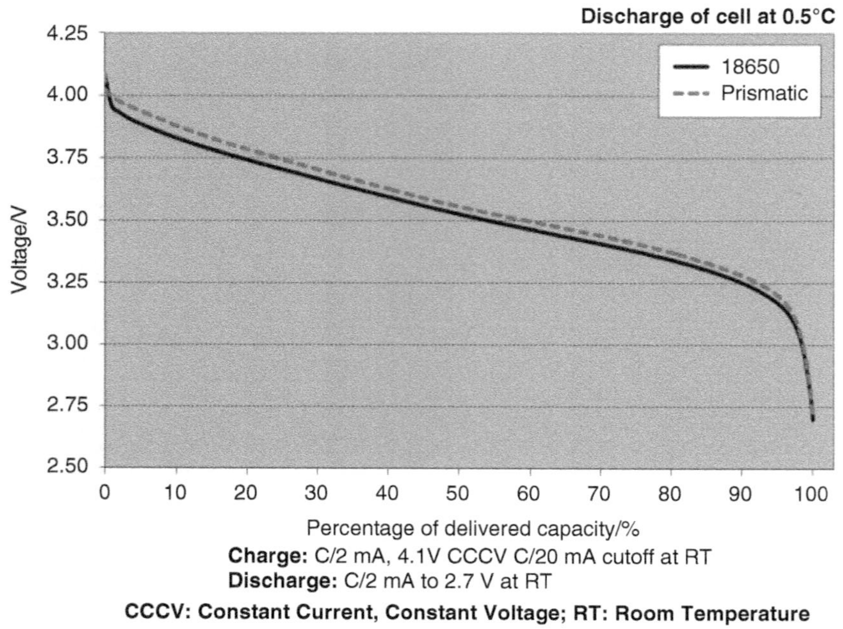

FIGURE 11.71 Comparison of small and large mechanical format cells.

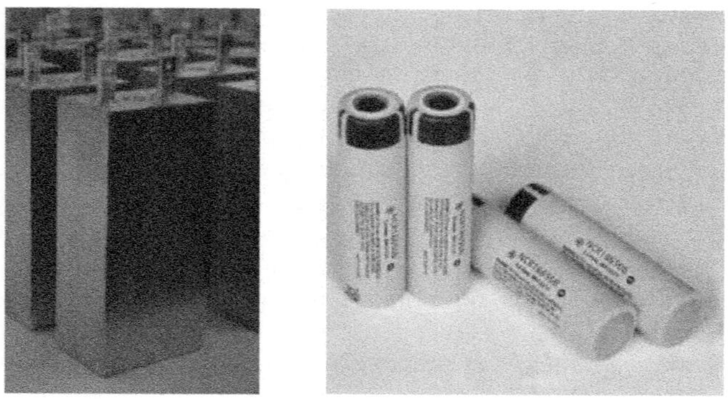

FIGURE 11.72 Large prismatic cells and 18650 cell formats.

and matching minimizes the cell voltage divergence over life. The parallel before series configuration is selected when there is a mission requirement for individual cell voltage monitoring or virtual cell voltage monitoring. A virtual cell is multiple cells placed in parallel to create a higher capacity value but remaining within the voltage limits of a single battery cell. Individual cell voltage monitoring is always required on all space manned missions and any high-value military or commercial satellite program. In this case, the series before parallel configuration would not be a viable solution as each individual battery cell would need to be monitored and would require a multitude of telemetry wires resulting in added cost and complexity to the battery system, especially in a high-power payload satellite where thousands of cells may be needed. The diagram (Figure 11.73) shows the comparison between the two configurations.

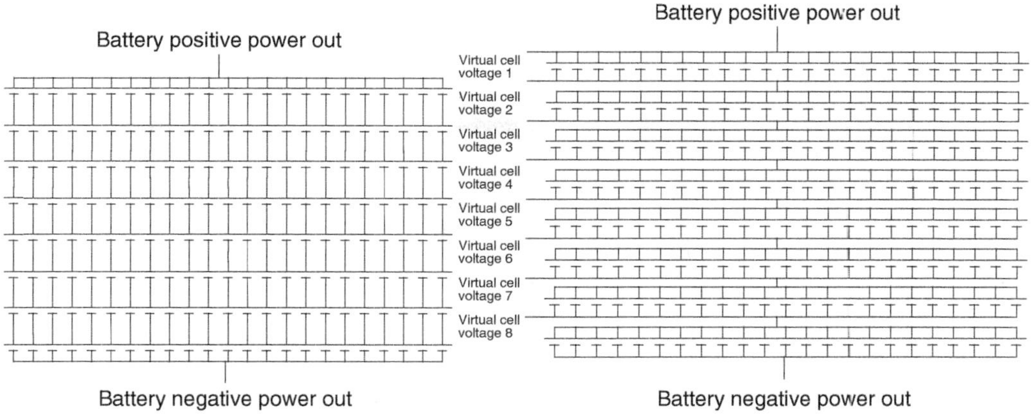

FIGURE 11.73 Series then parallel and parallel then series electrical connection configurations.

The number of cells in series in a configuration determines the operating battery voltage and the number of strings in parallel determines the energy or total capacity of the battery.

Geosynchronous Orbit Battery Operational Performance

A battery system in a geosynchronous orbit provides the satellite power during the eclipse portions of the satellite orbit and also the peak load demand, which is above the power generation capability of the solar array system. The variable eclipse period occurs only 90 days per calendar year in a GEO orbit and only then is the battery typically required to provide the power to the satellite bus. During the two solstice seasons throughout the year the battery is held at a specific SOC in the event it is needed for any pulse load capability, station keeping or in the event the satellite loses sun lock. Assuming the satellite maintains sun lock (solar arrays always pointed at the sun) the spacecraft can operate off the power delivered from the solar arrays alone and will not need any power from the battery system. During these nonrequired operational battery periods, the battery is stored at a reduced SOC value to minimize capacity fade. Storing a battery system at full or 100% SOC will increase the capacity fade of the battery system. During the extended solstice periods, the battery system may be charged to an SOC value of 50% to minimize the capacity fade. Figure 11.74 shows the effect of storage SOC on battery capacity demonstrating the lower storage SOC results in a lower capacity fade through life. The same protocols should be used prelaunch: satellite batteries should be stored at about 50% SOC in a cool environment prior to launch to safeguard against impedance growth and irreversible performance losses.

Zero-Volt Technology

The current lithium-ion cell technology can suffer permanent performance degradation if the cell is discharged below approximately 2.5 V. This makes the recovery of the satellite from a dead bus situation impossible. Furthermore, U.S. government satellite power systems are often required to provide some type of contingency device that both protects the lithium battery from a deep discharge while insuring the power system is revived in the event that power is later regained from the solar arrays. These devices add significant complexity to the power system design.

As a result, a new electrochemistry for lithium ion cells has been developed that enables the battery systems to discharge to a zero-volt level without damage or degradation to the overall battery performance for the life of the satellite system. Satellites have a control system which constantly points the satellite solar arrays toward the sun to provide the necessary power to the satellite

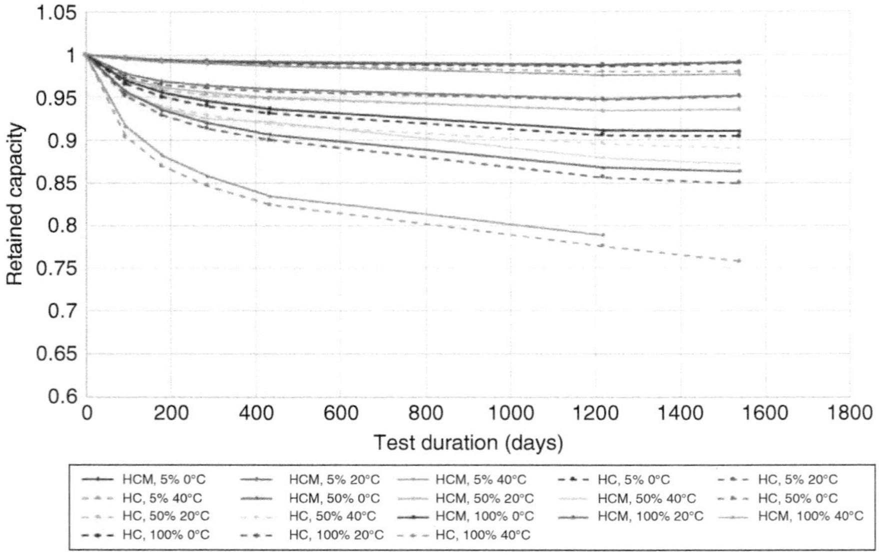

FIGURE 11.74 Storage or non-operational SOC effects on battery capacity.

during sunlight periods within the orbit. At times, there may be an anomaly within the spacecraft control system, which results in the off-pointing of the solar array from the sun. This situation will result in the battery system taking control of the spacecraft bus and providing the necessary power to keep the satellite operational. However, it may take an extended period for the satellite to regain a sun lock condition and the battery may get to a state where it could be completely discharged resulting in a "dead" bus scenario. With the zero-volt technology, the battery would be able to be discharged to a zero-volt or 100% energy discharge state and then recover its full performance following a regain of sun lock by the satellite. When a regain of the sun lock situation occurs, the satellite would automatically begin to power the critical loads on the satellite and begin to recharge the battery system. Recharging of the battery system will be executed through a low-level charge of C/200 until the battery system voltages reaches a value of 2.5 V per cell at which time a typical C/2 charge rate could be applied to bring the battery up to full SOC without any degradation in continued battery performance throughout the satellite mission life. Figure 11.75 is a summary of test data for recently produced cells undergoing a zero-volt storage period followed by charge/discharge cycling. The data in Figure 11.75 demonstrate that zero-volt technology is able to store the battery system at a zero-volt state for 40 months followed by recharge to its typical operational state and electrical cycling without any degradation in the battery cycle life performance.

11.30 Power Control Electronics

Power conditioning is an essential function of the EPS design for the various reasons as illustrated in Figure 11.76.

As briefly discussed in Subsection 11.27, power conditioning electronics generally consists of a voltage regulator to manage the power from the source (typically a solar array) and the combination of a battery charge and discharge regulators. The topologies of these regulators vary widely and are selected based on the nature of the payload and its voltage and power requirements.

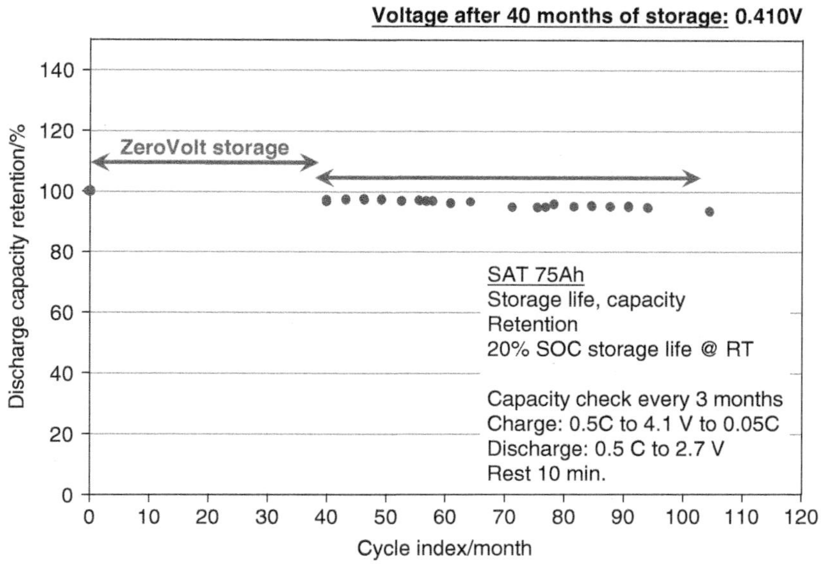

FIGURE 11.75 Zero-volt cell technology performance capability.

The selection is further influenced by the variation in the available solar array power and battery charge/discharge requirements.

The simplest configuration for an electric power subsystem would be to connect the battery, the solar array, and the load, all in parallel. The battery will discharge into the load during eclipse, and during the sunlight period the array will supply power to the load and to the battery for recharging. However, based on the solar array and the battery electrical characteristics, the need for power conditioning electronics between the source and the load becomes unavoidable.

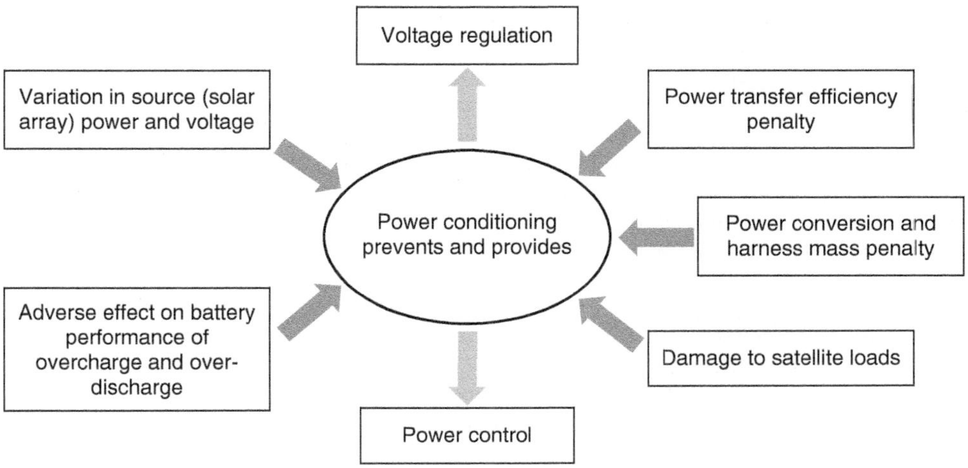

FIGURE 11.76 Need for power conditioning.

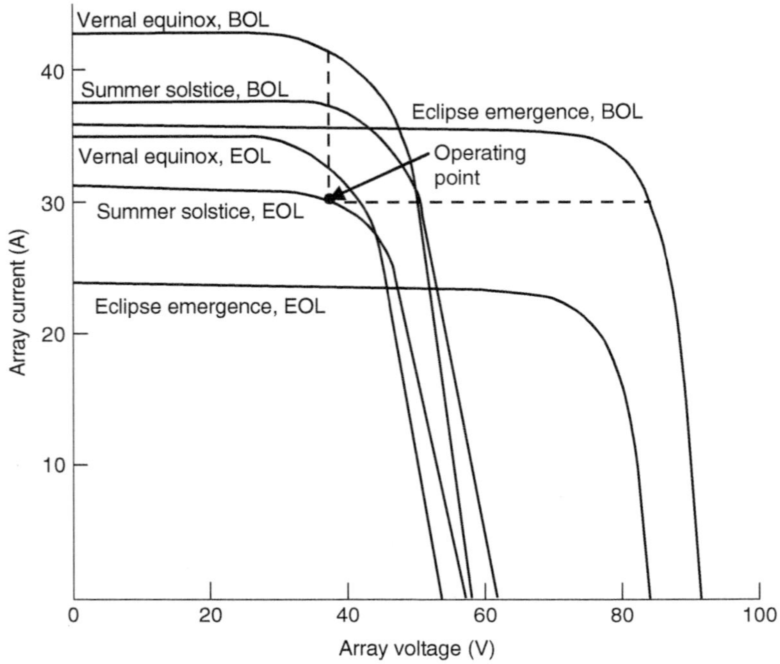

FIGURE 11.77 Solar array electrical characteristics.

The solar array output is described by a plot of current versus voltage (*I–V* curve) in Figure 11.77. The *I–V* curve changes both due to seasonal variation in the array temperature and solar intensity, and due to radiation degradation of the solar cells, as shown in Figure 11.77. The solar array voltage is maximum when the satellite comes out of eclipse due to the very low temperature of the solar cells. The solar arrays are sized to meet the load requirements at the end of life at summer solstice with the operating point slightly left of the maximum power point on the *I–V* curve. As a result, the solar array provides excess power that needs to be managed in order to avoid an overvoltage condition on the bus. The simplest solution to regulate/manage the excess solar array power would be a variable resistance (linear shunt) in parallel with the real loads. If the load decreased, the resistance of the shunt would decrease to draw more current. The total demand on the solar array would be the same and the voltage would be steady.

Because of the solar array size, a simple linear shunt would be enormous. It would have to have the capability to dissipate the output of the entire array. This would also generate a lot of waste heat. The problem of managing the enormous amount of heat is alleviated by employing a scheme of switched shunts as depicted in Figure 11.78.

A switched shunt is basically a short circuit across the array output. The switch is generally a power hexfet. For the tight regulation of the bus voltage, multiple solar cell strings are divided in smaller circuits or segments and a combination of multiple switched shunts and a small linear shunt is employed. The switched shunts are turned on and off sequentially and hence this scheme is generally known as sequential switching shunt regulator (S3R). When a switch is closed, it draws the array segment's short circuit current. Isolation diodes prevent the switches from pulling down the entire bus. The voltage drop in the switch is quite low and hence the dissipation. The *I–V* curves with a number of shunts constantly ON are too far apart to tightly regulate the bus (e.g., within 1.0 V) on any one segment. To maintain the bus voltage for a given current demand, the

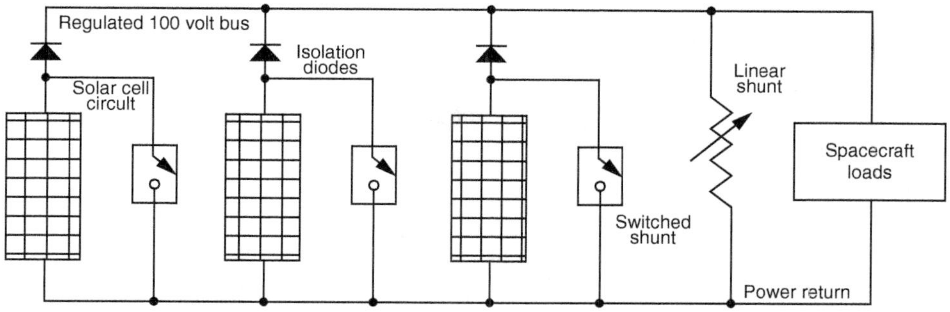

FIGURE 11.78 Solar array switch shunts.

shunt control electronics operates one shunt on and off with a defined duty cycle. When the control electronics releases a shunt, the bus voltage starts to rise. The voltage ramps up until it reaches the upper limit of an error detector. The control electronics then turns the switch back on and the bus voltage ramps down. The duty cycle depends on the amount of current drawn by the loads. If the current is half way between two *I–V* curves, the duty cycle will be 50%. The peak to peak variation in the bus voltage is less than ±0.1 V. The frequency is about 1 kHz in this example. As the array output decreases (such as during eclipse), the last active shunt will spend more and more time off until it turns off altogether and the next one in the sequence starts to cycle. The shunt power electronics utilizes the various techniques for switching shunt, such as limit cycle, pulse width modulation (PWM), etc.

Other solar array access power control techniques include array segment bus-connect/disconnect, series regulators, linear sequential full and partial shunts (Ref. 8), peak power trackers, array off-point-control, etc., depending on the mission, array size, space environment, and other system-level requirements.

Another problem is in the charging of the battery. The battery connected in parallel across the array will act as a shunt element and will force the system to operate at its charge voltage, accepting the difference between the load current and the available array current as charge current. Therefore, at the beginning of life, very high charge currents will go into the battery. These high charge currents will result in battery overheating and a reduction in the battery cycle life. Hence, a regulator to control the charging current is required. The battery discharge voltage changes as a function of the discharge rate, temperature, and depth of discharge, as discussed in the preceding section. To keep the discharge voltage constant for a regulated bus, some type of battery discharge regulator will be required.

For the battery charge control a switching regulator normally referred to as a buck converter is very common. The battery voltage in the most common EPS architectures is generally lower than the bus voltage and hence, the need for a buck converter. The buck converters are generally designed to accept a wide-ranging battery input voltage and are modularized to accommodate different satellite load power requirements and to achieve scalability. They are also designed to accept a range of battery charge rate commands and EOC voltage levels initiated by the satellite on board computer (OBC) in response to the battery state-of-charge, temperature, and voltage. The power management software in the OBC includes the algorithms for the battery charge control, including for lithium-ion batteries, charge current tapering at the pre-selected EOC voltage level.

For the battery discharge voltage regulation, a boost regulator or a converter is required. Like the buck converter, a boost converter is also modularized to accommodate the varying satellite load requirement and to achieve scalability. Both, buck and boost converters are generally greater than 95% efficient at or near full load. Sometimes both functions are combined into a common converter, generally known as a bidirectional power converter (BPC).

11.31 Subsystem Design

Overview

This subsection provides an overview of the power subsystem design process. The EPS is designed to meet satellite requirements that are stated in terms of load profile, orbit, mission life, and system configuration. Initially, different design concepts are analyzed to achieve an optimum design of EPS for given system requirements. The trade-offs are performed among mass, cost, and reliability. There are several issues, such as bus voltage, single versus dual bus, unregulated versus regulated bus, solar cells and solar panel design, and fault protection, which are normally decided on the basis of past experiences of the manufacturer and the preference of the customer. The design process for a typical EPS is outlined in Figure 11.79. The design tradeoffs are further detailed in Table 11.13.

The power requirements for each load element of the satellite system are evaluated for each major phase of the mission, beginning with switchover from ground supply to internal battery power during lift-off, through the on-orbit operation phases. The electric power system for a communications satellite is designed primarily for synchronous orbit load requirements. As shown in Table 11.8, typical breakdown of power requirements for a three-axis-stabilized communications satellite are given for three critical times during the year: autumnal equinox, summer solstice, and eclipse. In this case, the communications subsystem is fully operational during eclipse. The batteries that provide power during eclipses constitute approximately 40% of the electrical power subsystem mass. To reduce the EPS mass, some satellites are designed to have only a fraction of the communications subsystem on during eclipse. The heater power requirements for the satellite generally vary with the season in order to maintain the equipment within their desired operating temperature ranges. The battery charging load also varies because of the requirements of a higher

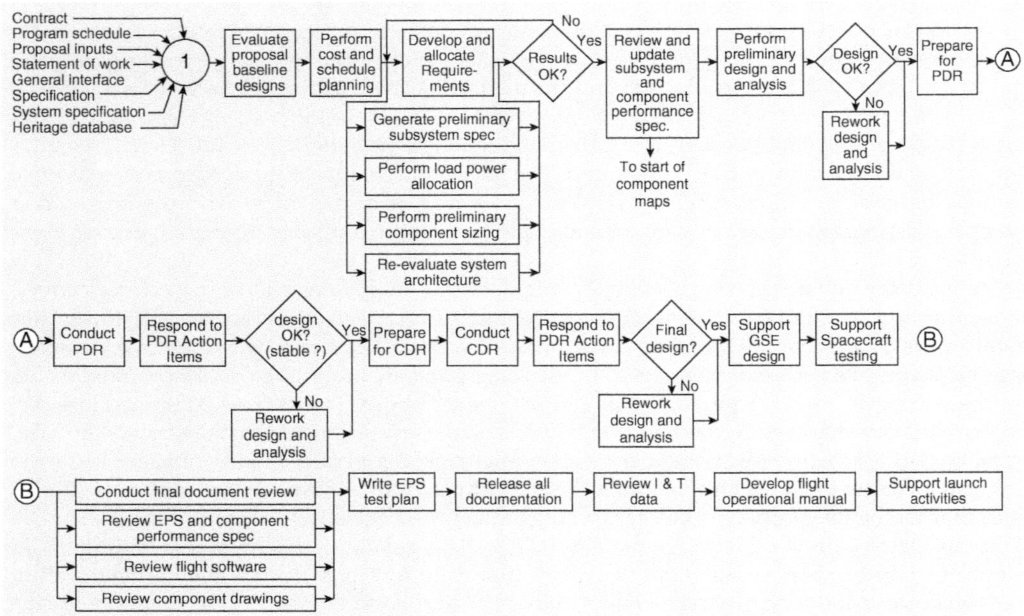

FIGURE 11.79 The EPS design process.

Power subsystem design tradeoffs
Power system • Configuration • Component selection
Battery • Number of series cells • Number of parallel batteries • Number of packs/battery • Depth of discharge capability • Thermal control • Charge control • Capacity, weight, volume, cycle life capability
Solar array • Series-parallel arrangement • Materials • Coverglass • Cell size • Shunt tap location • Dipole reduction • Thermal • Orientation • Weight, area, operating temperature, mission life
Power electronics • Power handling capability • Weight, volume

TABLE 11.13 Power Subsystem Design Tradeoffs Comparisons and Selection

charge rate during equinox to fully charge the battery following an eclipse discharge, and a lower charge rate during solstice seasons for trickle charging. During the design phase, it is a common practice to use a 5% load contingency to take into account the uncertainty in the calculation of equipment load. For solar arrays, an extra 10% design margin is used for uncertainty in the radiation degradation and other power prediction factors. During autumnal equinox, the load requirements are higher than that in the summer solstice. The extra power requirements during equinox usually do not pose a problem because the available solar array power during equinox is higher than that at solstice.

The selection of bus voltage is frequently based on the desire to use existing equipment, which has been proven on another satellite. The main power-switching semiconductors are limited to approximately 100 V by present technology, except for a few which operate up to 400 V. They normally set the upper limit for the bus voltage. The minimum voltage is determined by the distribution losses that can be tolerated. Low voltages, requiring high currents, will result in high losses. In the 1960s, bus voltages of 20 to 30 V were common. In the 1970s and early 1980s, bus voltages were of the order of 40 to 50 V. Bus voltages of 70 and 100 to 120 V are now common for a majority of satellites.

The choice of the bus voltage level and range has a major effect on the power control electronics. In a regulated bus configuration, with separate solar array and battery regulators, the array and battery voltages may be chosen independently of each other. To simplify the battery charge and discharge regulators, the voltages should permit the use of either buck or boost regulators, rather than buck-boost types. For an unregulated bus, the designs of load regulators will be simplified if the voltage range of the battery discharge lies within that of the solar array. In general, load power conditioning will be eased if the solar array EOL voltage is made equal to the battery EOL discharge voltage.

For a lightweight solar array on a three-axis-stabilized satellite, the array voltage at eclipse exit may be more than double the normal array voltage and it decays in about 5 to 20 minutes as the array warms up. If a high level (e.g., 50 V) is chosen for the nominal array voltage, the eclipse exit voltage spike of 130 V can require careful attention to the load regulator design.

In a single-bus system all the loads are connected to a single bus. In a dual-bus system, the steady-state loads are divided between two independent satellite power buses to maintain equal battery depth of discharge and to provide additional reliability, and protection against single point faults. In general, two complete sets of housekeeping equipment are installed with redundant units of the same types connected to opposite main buses. The failure of one unit will require use of the redundant unit, which is provided from the other power bus. In a dual-bus system, no single fault in the power subsystem or in the loads can affect more than half of the satellite system. A slight disadvantage of the dual-bus system is the constraint on the use of smaller redundant units to maintain power balance.

The current trend for higher-power communications satellite has resulted in greater emphasis on the optimization of electric power subsystem mass. An electric power subsystem consists of three main components: solar array, battery, and power control electronics. The solar arrays have been made lighter by using higher efficiency and thinner solar cells and lightweight deployable solar panels. The mass of the batteries is reduced by using new battery technology, namely, lithium-ion, which offers higher energy density, improved specific weight, etc., as highlighted in Subsection 11.29. The power control electronics mass is reduced by using higher switching frequencies and advanced power control concepts and devices.

Fault Protection

The power subsystems for large GEO satellites with significant business plans are designed to be single point failure free. There are three types of failures: (1) failures of power subsystem components, (2) failures in load components, and (3) harness faults.

Failures in Power Subsystem Components

The solar cells in a solar array can fail open, short, or short to ground. The design practice is to diode couple sections consisting of one or more solar cells strings into the main bus. Usually there are a large number of these sections. A single failure will be limited to the loss of one of these sections in the worst case.

The batteries have more complex failure modes. Because of the mass of batteries, cell-rather than battery-level redundancy must be used. A single cell in a battery can fail open, short, or short to ground. The case can rupture due to the evolvement of hydrogen gas. This failure mode can occur when the cell is overcharged and its voltage becomes too high or a cell's voltage is actually reversed. To protect against these failures, cells require individual monitoring. One control approach that has been used is the individual cell bypass circuit. The bypass circuit will bypass the cell if a cell develops an open circuit condition. Battery over discharge is usually managed by shedding the loads that is triggered by the load shed macros built into the power management flight software.

Failures in Load Components

The failure in load components can be short or open. If the load component has an open-circuit failure mode there is no danger to the power subsystem. However, a short-circuit failure could endanger the entire power subsystem. One of the simplest approaches is to equip each load with parallel redundant fuses. Alternatively, an undervoltage detector for each load equipment will preclude any danger to the equipment and is also used to sequentially unload the main bus in case of power deficit.

Harness Failures

Short-circuit failures in a wiring harness can be corrected by one of the following: dual or multiple bus, diode isolation of sources, and double insulation system.

Analysis

Detailed subsystem-level analyses are necessary to ensure full compliance with the EPS design/ performance requirements. These are listed in Table 11.14.

Telemetry

For the power subsystem in-orbit performance evaluation and for fault diagnosis, it is important to continuously monitor certain critical performance parameters. These parameters typically include

- Bus voltage
- Battery voltage (cell and battery level)
- Load currents
- Solar array input currents
- Solar array return currents

Types of analyses
• Launch and ascent • Energy balance – Transfer orbit – Operational orbit (normal and Station keeping) • Lockup analysis • Component dissipation • Stability and transient response
Typical conditions • Beginning of life and end of life – Minimum load – Maximum sun – Nominal and failure • Failures
• Analysis tools • Typical EPS analysis programs • Multiple battery simulation • Solar array synthesis • Energy balance • Solar array shadowing analysis • Bus dynamic performance analysis

Table 11.14 Typical Power Subsystem Performance Analyses

- Solar array temperatures
- Solar cell open circuit voltage
- Solar cell short circuit current
- Battery charge/discharge currents
- Battery temperature
- Others as needed

This part of the handbook is a top-level summary of steps necessary to design a reliable and compliant power subsystem for a given satellite application. This part concentrates on the GEO satellite power subsystem design. However, the steps described in this part are equally applicable to the design of the satellite power subsystems launched in other earth orbits and for the exploration of other planets.

Acknowledgments

The author greatly appreciates the support of the following reviewers and contributors:

- James Haines, Retired, Electrical Power System Section Head, European Space Agency, Noordwijk, The Netherlands
- Dr. Mukund Patel, Professor, U.S. Merchant Marine Academy, Kings Point, New York
- Giang Lam, Engineer Sr. Staff, Lockheed Martin Space Systems, Sunnyvale, California
- Joseph Troutman, Senior Manager, EnerSys Advanced Systems, Longmont, Colorado
- Robert Danielak, Senior Manager, Lockheed Martin Space Systems, Denver, Colorado
- Dr. Judith Jeevarajan, Research Director, Underwriters Laboratories Inc., Houston, Texas
- Dr. Margot Wasz, Senior Scientist, Aerospace Corporation, El Segundo, California
- Dr. Thomas Barrera, President, LIB-X Consulting, Long Beach, California

References

1. Salim, A. A. 2000. "In-Orbit Performance of Lockheed Martin's Electrical Power Subsystem for A2100 Communication Satellites," AIAA-2000-2809.
2. www.pveducation.org/pvcdrom/
3. Rauschenbach, H. S. 1976. *Solar Cell Array Design Handbook*, JPL SP 43-38 vol. I, Jet Propulsion Laboratory, California Institute of Technology, Pasadena, CA Oct.
4. http://www.pvmeasurements.com/Quantum-Efficiency-Measurements/qex10-quantum-efficiency-measurement-system.html
5. Hoeber, C. F. et al. 1998. "Solar Array Augmented Electrostatic Discharge in GEO," AIAA 98-1401.
6. https://www.nasa.gov/mission_pages/tdm/sep/index.html
7. 2013. "Standard: Qualification and Quality Requirements for Electrical Components on Space Solar Panels," AIAA-S-112, revision A.
8. Abbas. A. Salim, 1975. "A Simplified Minimum Power Dissipation Approach to Regulate the Solar Array Power in a Satellite Power Subsystem," 11th IECEC.

PART 5

Systems Engineering, Requirements, Independent Verification and Validation, and Software Safety for Aerospace Systems

Thomas M. Hancock III

There are two truths in aerospace: (1) systems will fail and (2) people will die. It is our job to mitigate the first and eliminate the second.

History has shown that avionics and software are the No. 1 cause of spacecraft failure and the No. 3 cause of launch vehicle failure. As an example, on December 11, 1998, the Mars Climate Orbiter was lost when it entered the atmosphere of Mars and impacted the surface because of a unit of measurement error between the ground and spacecraft teams. On January 3, 1999, the Mars Polar lander was lost when it also impacted the surface of Mars from an altitude of 20 ft due to a faulty software switch that activated when the landing legs deployed, shutting down its terminal descent engines.

Analysis of both missions showed a common cause of failure, a lack of systems engineering and no application of independent verification and validation (IV&V).

In addition, during development of the International Space Station a decision was made to reuse software from the U.S. Laboratory module for U.S. Node 2. During testing, the IV&V activity discovered a significant software error. If IV&V had not been applied, U.S. Node 2 would have been lost once it was removed from the Space Shuttle cargo bay and docked to U.S. Node 1 (see Figure 11.80). Basically, this software error would have prevented the internal thermal control system from being activated within the maximum 2-hour limit before the module would freeze up. (Note: Once power to U.S. Node 1 is disconnected from the Space Shuttle as it is removed from the payload bay, the unpowered node has less than 2 hours to power up and activate internal thermal control before it is lost.)

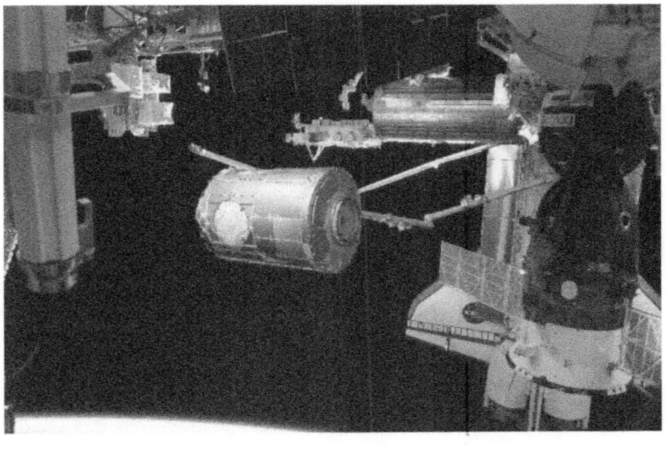

Figure 11.80 U.S. Node 2 suspended between the Space Shuttle and ISS. Mating to the ISS must take place within 2 hours before it suffers irreversible damage.

11.32 Developing Software for Aerospace Systems

First, the mission concept and high-level requirements are established by the customer. From the high-level requirements and mission concept, the systems engineering team begins to decompose high-level requirements into functional areas. At this time, the IV&V team is introduced into the process. The IV&V team will do an assessment of the software development plan and processes, the high-level requirements, the functional decomposition of the requirements and the system architecture. In addition, the IV&V team will (in addition to other things) conduct an organizational assessment of the software development team and its past experience in the software language, development process, tools, and systems. These conclusions are documented in a start-up assessment report that will discuss the project in detail, the IV&V teams' understanding of it and justification for the level of technical rigor to be applied to the program.

Key factors in this report include

Intensity: Intensity is based on project complexity, or project risks associated with meeting a specific capability.

Rigor: There are cases where a standard set of IV&V work instructions may not apply to different development methods.

Project risks: Project risks can affect both intensity and rigor.

In all cases, the level of technical rigor can be tailored to meet the unique nature of a project. The type of IV&V to be applied to a project is dependent on the software criticality of the program and the level of risk identified during the start-up assessment process. The final product at this phase is the IV&V project plan. The project plan is basically an agreement between the program, the IV&V organization, and the funding body on what IV&V activity will be conducted, how it will be done and with whom and at what level they will interface with the program.

Taking place at the same time, the software safety team is beginning its assessment of the program. Analysis of requirements and identification of hazards and risks is under way. Inputs from the safety team to the baselined requirements (recommending safety-related requirements and additional requirements as driven by the human rating standard) are discussed. Definitions and an agreed to understanding on the meaning of requirements are documented. Controls for each hazard are developed. In some instances the controls will require additional or redefined requirements.

Having these processes take place early will limit the impact of errors, misinterpretation of requirements, and cost to the program. It is also necessary to make sure that the difference between reliability and safety is fully communicated to the software development team, systems engineers, and program managers. Basically, think of reliability as that 1 in 10,000 chances something could go wrong and safety the ability to do something about it when it does.

An example on how systems engineering is necessary to guide the software development considers that during the Ares I launch vehicle development program, the coders and related management decided the instrumentation program and command list (IPCL) was not flight critical and should be controlled as class C software (the IPCL is the definition and bit assignment values of all valid commands and status for the flight software). An active discussion by systems engineers demonstrated the need to make and control the IPCL as class A software product. Because you can have the best written software modules in the world but if the IPCL is wrong, then what's the point.

As part of a continues process, system engineers, working with the IV&V and safety teams, are modifying requirements and design documents to introduce recommended changes. While neither IV&V nor safety teams can force requirements to change, they can refuse to provide flight certification, leaving program management in the difficult position of trying to justify why safety or IV&V recommendations were ignored before launch or first flight.

When correctly implemented, the IV&V and safety teams become imbedded elements of the development organizations and are included at each milestone review and test activity. Early, active and open sharing of information will lead to a safer, more maintainable, and cost-effective software program.

Systems Engineering

Systems engineering is the creation of an idea expressed by a model known as the requirements that specifically defines what the code is to do. It all starts with an operational concept. The operational concept outlines the mission and high-level functions of a system: the "what and why." The "what and why" is decomposed into other requirements that further define the system into the "how and when." The "how and when" become detailed requirements that are interpreted into code.

Software Requirements

Software requirements represent the bases for an aerospace systems software design and architecture (see Table 11.15). They also serve as the contractual obligation of the development organization to the customer. Requirements specify the performance, function and other needs of an aerospace vehicle systems and subsystems. Poorly written requirements are a leading cause of code errors, development schedule slips, critical in flight software problems, testing difficulties, and sustainability issues. In some instances developers merge design and requirements documents. In others significant requirement detail is left to design documents where not one shall is found. By better defining, breaking down and controlling requirements, a significant reduction in risk, along with cost and schedule saving can be achieved for NASA, DoD, and commercial missions.

Requirements type	Definition
High-Level Requirements	Software requirements developed from analysis of system requirements, safety requirements, and system architecture.
Derived Requirements	Additional requirements resulting from the software development processes, which may not be directly traceable to higher level requirements. These must be tagged, and verified. If a derived requirement is from a high-level requirement, then low-level requirement(s) must be traced to satisfy it.
Low-Level Requirements	Software requirements are decomposed from high-level requirements, derived requirements, and design constraints from which source code is directly implemented without further information. These requirements are developed during the software design phase and capture the algorithm and other requirements generated as part of the design process. These must be tagged, traced, and verified.
Safety Requirement	A requirement at any system indenture level that is decomposed or derived from higher functional and physical hazards and their mitigation. Any design feature, probability, level of rigor assignment, check/ inspection, test, assumption, etc., which is used to support Space flight certification criteria is a potential safety requirement.

Table 11.15 Requirements Types and Definitions

Analysis was conducted by reviewing the established requirements for different software elements of crewed spacecraft (Environmental Control Life Support Subsystem, Thermal, Command and Data Handling, Guidance Navigation and Control, Communications), and robotic missions. Part of this analysis identified the number of complex requirements, conducted analysis of those requirements and the problems identified by different development and review organizations during each phase (design, code, and test) and during space flight operations.

Requirements Analysis

Requirements analysis is an evaluation of the complexity of a given requirement. Requirements complexity can be best defined as

- Number of shall statements in a given requirement
- Number of parts, steps, or elements in a given requirement
- Number of states a given requirement can impact
- Number of conditions set in the requirement
- How requirements are assigned in complex algorithms (a logical brake down of a complex algorithm)
- How a requirement interacts with other requirements
- Evaluating the ease of coding a given requirement
- Assessing the testability of a complex requirement

From this, it can be stated that requirements with multiple shall statements have not been sufficiently decomposed and will result in highly complex code and significant test and interface issues. Those requirements with a significant number of steps or parts can result in developers inventing code to support steps and interfaces not called for in the requirements or interface control documents. In addition, testers may not be able to accurately test the code according to the requirements due to an inability to stimulate a sufficient number of steps in the code with simulation tools. This leads to an understanding that requirements that manage multiple states, conditions, or share operations between multiple states, risk not being able to correctly manage the states or conditions.

11.33 Impact of Poorly Written Requirements

Analysis has shown that poorly written requirements can have a significant impact on code and test activities and can result in

- A loss of requirements detail
- A loss of requirements definition
- The software design and requirements documents not reflecting the actual flight software
- Overly complex and untestable code
- Code spanning multiple modules (unclean code)
- "Invention" by coders (coding actions not documented in the requirements and coding implied functions)

- Guessing by testers (testing issues have resulted from test team members testing according to the code and not the requirements or even design)
- Ground commanding and operation error that can result in a loss of science data, instruments, or missions.

11.34 Benefit of Requirements Analysis

By conducting analysis during the requirements phase you are able to anticipate potential problems during code and test phase (by creating a list of requirements to watch). From this list you will be better able to monitor the software development effort and reduce risk and cost to a mission. By compiling a list of requirements to watch (preloss analysis, preloss analysis is developing a list of requirements and other key elements of the overall development effort), you are giving developers and testers a heads up of potential issues to look for. In addition, you can help the project to accurately document the requirements, code, and test products.

Requirement Analysis Questions

When conducting requirements analysis ask the following questions as a guide in your efforts.

- Are the requirements clean?
- Can the requirement be contained in a single module of code?
- Do the requirements represent a logical decomposition?
- Evaluate the number of states the requirement can impact.
- Does the requirement call for unique actions for more than one state?
- Are states interdependent?
- Do conditions that impact states cover more than one code module?
- If not interdependent, can they be logically broken out?
- Is the requirement complete?
- Is everything (all actions) defined in the requirement? (Complete requirements should be reviewed to see if they can be further decomposed.)
- Can code be easily and cleanly developed from the requirements?
- When thinking about how the requirement would be implemented do holes in the requirements appear?
- Can test cases be developed from the requirements?
- Can a logical test case (created from the requirement) be developed that will test each condition or element of the complex requirement?
- Are any elements of the requirement not testable?
- Does the requirement span more than one capability?

  Example: Fault protection algorithms strung between multiple capabilities of a subsystem.

11.35 Application of Independent Verification and Validation

After the first X-43A hypersonic research, aircraft was destroyed (the X-43A failed because the vehicle control system design was deficient for the trajectory flown due to inaccurate analytical models which overestimated the system margins), major independent reviews were conducted to determine how and why the mission was lost and what could be done to mitigate the reason for the failures.

This loss mitigation data was then applied to future missions. Again, two major threads emerged: little or no systems engineering was applied and independent verification and validation of flight and ground software, requirements, design, and interfaces was not done. By not applying IV&V, an opportunity to identify and resolve the primary reason for each failure was missed. In addition, system testing was cut back or skipped in place of analysis.

Development organizations that apply IV&V techniques can expect better software and system performance, higher confidence in software reliability, and a better chance for meeting program acceptance criteria. From a management perspective you can expect better visibility into development, better decision criteria, reduced maintenance cost, and reduced frequency of operational changes to the code.

Simply put, the application of IV&V techniques revolve around two key goals:

1. Making sure the right system is built
2. Making sure the right system is built the right way

Application of IV&V Techniques

IV&V is a systems engineering process employing rigorous methodologies for evaluating the correctness and quality of the software product throughout the software life cycle. IV&V techniques are adapted to the unique characteristics of a program. When applying IV&V techniques, the development organization uses a systems' perspective, reviewing the needs, systems use, interfaces and limitations of the software and development team. When preparing to use IV&V techniques the development organization should start by addressing the following self-assessment criteria:

- Determine the consequences of failure.
- Determine the likelihood of failure.

Typically, IV&V is applied to a system at a set level of rigor and are defined by four categories:

1. Complete—independent testing, requirements analysis, code and test analysis
2. Full—requirements analysis, code and test analysis
3. Limited—requirements analysis
4. None

The level of rigor can be tailored to the specific needs of a program, but in general the activity defined for each category is conducted.

11.36 Consequences of Failure

Consequence of failure is an evaluation of what will happen if a subsystem develops a fault (recoverable or unrecoverable). When evaluating the consequences of failure, concentrate on creating a watch list of requirements and code for the most critical functions.

Example criteria for determining the consequence of failure include

- Potential for loss of life
- Potential for serious injury
- Potential for catastrophic mission failure

- Potential for partial mission failure
- Potential for loss of equipment or subsystems
- Potential for waste of resources
- Potential for adverse visibility
- Potential effect on operations

11.37 Likelihood of Failure

This information is then evaluated against the likelihood of failure. Likelihood of failure is an estimate for the potential of the type of consequence identified above to occur. This assessment represents a real-world view of the state of the development organization.

Likelihood or failure can include factors such as

- Level of integration
- Requirements maturity
- Number of software lines of code
- Schedule pressure
- Software team experience
- Maturity and utilization of software processes
- Management team experience

While evaluating, these factors rely on people who are independent of the parent development team (from a different project or section of the organization).

11.38 General IV&V Techniques

IV&V techniques are built around the software life cycle and divided into four areas:

1. Requirements analysis
2. Design analysis
3. Code analysis
4. Test analysis

IV&V Requirements Analysis

Requirements analysis is involved in verifying system and software requirements are correct, complete, traceable, and testable. Typically, this will include the analysis of requirements specifications and interface control documents. Development teams can conduct independent reviews (audits) of the documents in question and provide that information to project management for disposition.

Design Analysis

Design analysis is intended to verify the design that satisfies the levied requirements and does not contain any characteristics that will cause failures under operational scenarios/conditions and to validate the architecture and correctness of algorithms. Development teams can utilize systems engineers to repeatedly conduct system-level analysis throughout the life of the project, making sure the original (also new and changed) requirements are being implemented by the design. This type of IV&V analysis was used in identifying the problems that lead to the loss of the Mars Polar Lander and the first X-43A. Each problem had a high likelihood of being identified prior to flight if IV&V had been utilized.

Code Analysis

Code analysis involves verifying the code is free of implementation errors, reflects the design, and is correct. Development organizations can benefit by establishing small teams to conduct informal and formal reviews of code and documentation outside of the standard review process. Even if this is only done for select subsystems and code modules the opportunity for finding code errors (some possibly significant) is enhanced. Review teams should include someone from every organization developing code (if ground software is developed by a different organization than the spacecraft, include people from both teams). This type of combined team review could have identified the software problems that resulted in the loss of the Mars Climate Orbiter.

Test Analysis

Test analysis is designed to ensure all requirements are tested and every critical function is fully exercised. Development organizations that conduct an independent assessment of the test program should ensure no holes are left in different levels of testing and that regression testing is standard, repeatable, and satisfactorily conducted.

Other IV&V Techniques

There are other IV&V techniques that can be applied by development organizations. These include

- Preloss analysis
- Requirements complexity analysis.

Preloss Analysis

Preloss analysis covers identifying high-risk elements and functions, the application of new technology and reuse from past mission or related subsystems. As an example, preloss analysis on the International Space Station was conducted on requirements and code reused from the U.S. Laboratory Module for the U.S. Node 2 Module. By creating a list of existing requirements and code errors identified in U.S. Laboratory flight software, they were able to keep a watch on similar requirements and code (based on reuse) in U.S. Node 2. This resulted in the early detection of significant errors, prior to test and reduced overall risk, and cost to the program.

Requirements Complexity Analysis

Requirements complexity analysis is an evaluation of the complexity of a given software requirement. Requirements complexity is a leading cause of serious software errors. These problems escalate as the percentage of complex requirements increases for a mission.

Requirements complexity (overly complex requirements) can be best defined as

- Number of shall statements in a given requirement
- Number of parts, steps, or elements in a given requirement programmer of states a given requirement can impact.
- Number of conditions set in the requirement
- How requirements are assigned in complex algorithms (a logical brake down of a complex algorithm)
- How this complex requirement interacts with other requirements

- Evaluating the ease of coding a given complex requirement
- Assessing the testability of a complex requirement

Analysis has shown a saving can be achieved by correctly implementing complex requirements and accounting for complexity as a risk issues. This type of analysis can represent a real savings and reduction of risk for NASA, DoD, or commercial space missions.

11.39 Software Safety

Safety results from the safety assessment process and includes functional, integrity, and reliability requirements as well as design considerations. The system safety assessment process determines and categorizes the failure conditions of the system. Within the safety assessment process, safety-related requirements are defined to ensure the integrity of the system by specifying the desired immunity from and system responds to these failures conditions. The requirements are identified for hardware and software to preclude or limit the effects of faults and may provide fault detection, fault tolerance, fault removal and avoidance.

The software safety processes are responsible for the refinement and allocation of system requirements to software as determined by the system architecture. Safety-related requirements are developed and refined into system requirements that are verified by the software verification process activities. These requirements and the associated verification establish what the software performs and its intended functions under any foreseeable operating conditions. Basically, until proven otherwise, all software within a safety critical system shall be assumed to be safety critical.

NASA has 10 rules for the development of safety critical software:

1. Restrict all code to very simple control flow constructs. Do not use GOTO statements, setjmp or longjmp constructs, or direct or indirect recursion.
2. All loops must have a fixed upper bound. It must be trivially possible for a checking tool to statically prove that a preset upper bound on the number of iterations of a loop cannot be exceeded. If the loop-bound cannot be proven statically, the rule is considered violated.
3. Do not use dynamic memory allocation after initialization.
4. No function should be longer than what can be printed on a single sheet of paper (in a standard reference format with one line per statement and one line per declaration). Typically, this means no more than about 60 lines of code per function.
5. The assertion density of the code should average a minimum of two assertions per function. Assertions must always be side effect-free and should be defined as Boolean tests.
6. Data objects must be declared at the smallest possible level of scope.
7. Each calling function must check nonvoid function return values, and the validity of parameters must be checked inside each function.
8. Preprocessor use must be limited to the inclusion of header files and simple macro definitions. Token pasting, variable argument lists (ellipses), and recursive macro calls are not allowed.
9. The use of pointers should be restricted. Specifically, no more than one level of dereferencing is allowed. Pointer dereference operations may not be hidden in macro definitions or inside typedef declarations. Function pointers are not permitted.

10. All code must be compiled, from the first day of development, with all compiler warnings enabled at the compiler's most pedantic setting. All code must compile with these setting without any warnings. All code must be checked daily with at least one—but preferably more than one—state-of-the-art static source code analyzer, and should pass the analyses with zero warnings.

Failure Conditions

Failure conditions describe system functions in terms of failure states of the function or as "a condition having an effect on the aerospace system and/or its occupants, either direct or consequential, which is cause or contributed to by one or more failures or errors, considering flight phase and relevant adverse operational or environmental conditions or external events." Failure conditions are related to states of systems as opposed to any specific equipment failure modes.

Hazards

Hazards are a real or potential condition that can cause injury, illness, or death to personnel; damage to or loss of a system, equipment or property; or damage to the environment. All "failure conditions" are considered hazards.

Safety Analysis

Safety analysis is a general term used to describe an evaluation of a system. An analysis can be made up of various descriptive or qualitative discussions or specific analyses. The term is used in safety process as a systematic, comprehensive evaluation of the implemented system to show that the relevant safety requirements are met.

Safety Assessment

Safety assessment is a compilation, to show that system failure conditions were identified and shown to comply with the regulations (quantitative and/or qualitative probability). It includes all safety analyses accomplished, as well as, supporting development testing and analysis results that support the safety analyses or safety requirements. The safety assessment report adds the element of hazard control, risk assessment, and acceptability of residual risks.

Software Hazard Analysis

Software hazard analysis is a system-level analysis of the software requirements, a first step in the process to assess risk and is performed by looking at the information flow from the overall systems perspective. Software hazard analysis is a black box process and does not consider the actual implementation of the software. Hazard analysis should also be performed whenever changes are made to the software.

Software Hazard Criticality Index

A number of factors influence the level of rigor specified for software hazards. Level of rigor is based upon the criticality of the hazards [also referred to as the software hazard criticality index (SHCI)]. As hazard severity and level of software autonomy increases, the corresponding level of rigor for verification also increases. The level of rigor will be specified in the software safety management plan, and be reflected in requirements verification plans.

An SHCI assists the program in allocating resources for the software safety verification effort. The SHCI is established using the hazard categories for the columns and the software control categories for the rows. The SHCI is completed by assigning values to each element just as hazard risk index (HRI) numbers are assigned in the hardware hazard risk assessment matrix. Unlike the hardware-related HRI, a low index number does not mean that a design is unacceptable. Rather, it indicates that greater resources (higher level of rigor) are needed for the analysis and testing of the software to satisfy safety verification requirements.

Software Hazard Controls

Software hazard controls are used to mitigate or eliminate the effect of any identified hazard (see Table 11.16).

Software Safety Criticality Classifications

NASA classifies software criticality into seven categories (classes A–G), with class A software being safety critical (class A software requires the highest level of rigor), and class B being mission critical. The remaining classes of software cover scenarios that are unlikely to have acute catastrophic consequences resulting from software malfunction (loss of crew or mission). The distinction between the two is that class A software systems are required to ensure human safety and mission viability, whereas class B is responsible for mission viability, or equivalently, no loss of mission.

NASA Software Classifications

Class A—human rated space systems

Class B—mission critical non-human-rated space systems or large scale aeronautical vehicles

Class C—mission support software or aeronautic vehicles or major engineering or research facilities

Class D—basic science/engineering design and research technology software systems

Category	Category description
AT (I) Autonomous–Time Critical	SW that exercises control over potentially hazardous subsystems or components without the possibility of real time human intervention to preclude occurrence of the hazard. SW failure directly contributes to the potential occurrence of the hazard.
AN (IIa) Autonomous–Not Time Critical	SW that exercises control over potentially hazardous subsystems or components, but that allows sufficient time for human or Independent safety Intervention to prevent hazard occurrence. The human or Independent safety Intervention is not sufficient for safety. Corrective action(s) may be required.
IT (IIb) Information–Time Critical	SW that provides Information display requiring an Operator(s) to take an Immediate action to prevent hazard occurrence. IT SW failure may contribute to, or fail to prevent, a hazard.
OC (IIIa) Operator Controlled	SW that requires human action to complete a function that commands potentially hazardous subsystems or components.
SR (IIIb) Information–Safety Critical	SW generates information of a safety critical nature used to make safety critical decisions. There are several, redundant, independent safety measures for each hazardous event.
NSC (IV) Not Safety Critical	No safety function is controlled by the SW. Used to bound and provide rationale for safety analyses.

Table 11.16 Software Hazard Definitions

Class E—design concepts and research technology software
Class F—general purpose computing, business IT software (multicenter/multiproject)
Class G—general purpose computing, business IT software (single center)

These definitions have been generally adopted by industry.

Safety Critical Software

Key actions and requirements that must be implemented for safety critical software include

- Software is always initialized to a known safe state.
- Software only transitions between predefined known states.
- Termination activity of critical functions is performed to a known safe state.
- Operator overrides require at least two independent actions.
- Software rejects commands received out of sequence.
- Software detects inadvertent memory modification and recovers to a known safe state.
- Software performs integrity checks on inputs and outputs to/from the software system.
- Software performs prerequisite checks prior to the execution of safety-critical software commands.
- No single event or action is allowed to initiate an identified hazard.
- Software responds to an off nominal condition within the time needed to prevent a hazardous event.
- Software provides error handling of critical functions.
- Software has the capability to place the system into a safe state.
- Safety-critical elements (requirements, design elements, code components, and interfaces) are uniquely identified as safety-critical.
- Requirements are incorporated in the coding methods, standards, and/or criteria to clearly identify safety-critical code and data within source code comments.

Safety Approval

Safety approval allows use of an element in its proposed launch or reentry application after examination of the safety element's performance characteristics. The performance characteristics of the safety element were already provided as part of the safety approval. However, the developer needs to show evidence that the safety element is suitable for the particular launch or reentry application being proposed.

The use must be consistent with launch or reentry safety and fall within the operating limits of its approval. The NASA or FAA will evaluate whether its use does not exceed the limits of the safety approval. In addition, a safety approval does not relieve the developer from demonstrating the safety of any portion of the applicant's launch or reentry not already covered by the safety approval.

In order to receive a safety approval, the developer must verify to the NASA or FAA's satisfaction that acceptable performance criteria have been met.

A developer may be required to:

- Address potential hazards and risks to public safety posed by use of the approved safety element.
- Provide engineering and safety analyses, system tests, quality assurance procedures, manufacturing processes, and test plans and results.

- Validate the adequacy and reliability of the various analyses and procedures used in the safety element's demonstration.
- Submit test results that show a measure of proficiency and experience for personnel involved in training.

The NASA or FAA will verify and validate performance to acceptable criteria before issuing a safety approval. As part of the verification process the developer may be required to:

- Develop a plan that identifies the methods of verification, which include demonstration analysis, inspection, and testing.
- Develop procedures or reports documenting verification methods and results.
- Conduct verification.
- Submit verification reports and results.

The developer may need to identify additional criteria applicable to a particular software element.

Safety Approval Usage

The use must be consistent with launch or reentry safety criteria and fall within the operating limits of its approval. The FAA or NASA will evaluate whether its use does not exceed the limits of the safety approval. In addition, a safety approval does not relieve the developer from demonstrating the safety of any portion of the applicant's launch or reentry capability not already covered by the safety approval.

Safety Approval Reuse

The developer may apply to reuse an existing safety approval by submitting to the NASA or FAA:

- The developer must describe any proposed changes in the approved systems or services and provide any additional information necessary to support the fitness of the proposed changes to meet appropriate standards.
- The NASA or FAA will conduct reviews required for a safety approval to determine whether the safety approval may be renewed or reused for an additional term. The NASA or FAA may request a description of how the element has been used, including its success or failure rates.
- The NASA or FAA may amend the expiration date of an existing safety approval or issue a new safety approval after conducting the required review. Additional or revised terms and conditions necessary to protect public health and safety or safety of property may be imposed.

If the NASA or FAA denies reuse, the developer may correct any deficiencies identified and resubmit for reconsideration.

Safety Analysis

Safety Analysis is a general term used to describe an evaluation of a system. An analysis can be made up of various descriptive or qualitative discussions or specific analyses. The term is used in safety process as a systematic, comprehensive evaluation of the implemented system to show that the relevant safety requirements are met.

Safety Assessment

Safety Assessment is a compilation, to show that system failure conditions were identified and comply with the regulations (qualitative and/or qualitative probability and level of rigor). It includes all

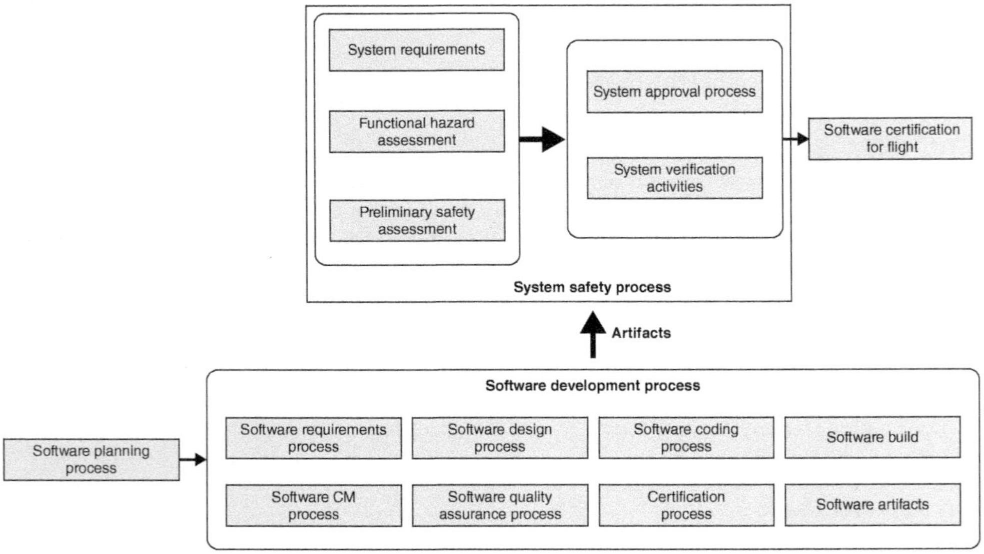

FIGURE 11.81 Software certification process.

safety analyses accomplished, as well as, supporting development testing and analysis results that support the safety analyses or safety requirements (see Figure 11.81). The Safety Assessment Report (SAR) adds the element of hazard control, risk assessment, and acceptability of residual risks.

Inclusion of COTS

The inclusion of COTS into a space system can offer the opportunity to leverage existing software certification. However, COTS are not a guarantee of cost savings. The overall complexity of software in space systems is increasing because of the requirements for autonomy in long term missions. This trend will accelerate because more and more mission software requirements are being assigned to the spacecraft. COTS software obsolescence in mission critical systems is caused by the length of time it takes to develop software technology for a project. Once software is developed for a flight critical system it is difficult and even impractical to upgrade to the current software technology base. The biggest barriers to COTS software acquisition is the lack of verification artifacts. If a vendor's supports the availability of verification artifacts and its product has been used in a similar system, then the barriers are low. If not, the cost may equal to new development efforts. COTS products that lack artifacts must be fully tested to the same level of rigor as new development. Overall software affordability is a big driver of program cost because of its role in the implementation of mission and system requirements.

Legacy Systems and COTS Space Flight Certification

It can be difficult to meet all of the lifecycle process for systems that have already been developed for COTS systems.

There are some strategies to deal with these types of systems:

1. Architect the system so the previously developed components or COTS has minimal safety effect.

2. Perform reverse engineering of the previously developed components or COTS. This approach may not be practical for SHCI level A, B, or C, or for complex systems where few of the lifecycle artifacts exist.
3. NASA or FAA may prioritize the lifecycle activities on a risk assessment for COTS and legacy systems.
4. Lack of data for the system (i.e., missing requirements documentation, missing structural coverage test cases, etc.) may necessitate the development of new artifacts that meet the required Process Through partitioning, it may be shown that previously developed COTS components require a less rigorous SHCI level than other software in the system; however, sufficient evidence (data, analyses, testing, etc.) must be provided to show the SHCI level is suitable for the previously developed or COTS item.

Waivers

Waivers are a method for requirements, capabilities, functions, and related software to be knowingly excluded form a software build or release. Waivers are used to identify contractually obligated requirements of the system to be delayed or excluded by agreement of the end user, the NASA or FAA. Waivers require an identified work around, steps or methods to assure the waiver does not create an unacceptable risk to the operation of the software. Specifically, the waiver needs to identify the requirements that are excluded. The exact shortfall in system capability and defined actions and how to work around the waiver. Each software thread given a waiver needs to be fully documented and all interdependences identified. Interfaces and related data definitions are examined to assure any risk is understood and avoided. All software waivers should be considered safety critical.

11.40 Certification

Flight certification and qualification can only be achieved by up-front planning. Defined and agreed to software life cycle process objectives are applied to each system undergoing development or modification. However, the implementation of all life cycle process objectives may not be required. To determine what software life cycle process objectives are to be applied, a "level of rigor" is determined by systems engineers, IV&V and safety team based on requirements developed from the system architecture and the system safety processes.

The life cycle process that must be implemented for development and qualification of each is determined by its role in the system architecture, as well as, its contribution to any associated hazards.

If flight certification and safety are not part of the initial software development process, it may not be feasible to perform the sufficient software design analysis, generate the data, and perform the flight certification and qualification analysis required to obtain flight approval. This will significantly increase both program costs and safety risks.

PART 6

Thermal Control

Brij N. Agrawal

11.41 Introduction

A spacecraft contains many components which will function properly only if they are maintained within specified temperature ranges. The temperature limits for typical equipment in a spacecraft are given in Table 11.17. The temperatures of these components are influenced by the net thermal energy exchange with the spacecraft thermal environment. The spacecraft thermal environment is determined by the magnitude and distribution of radiation input from the sun and the earth. Component temperatures are established by the heat radiated from external surfaces to the space sink, and internal equipment heat dissipation, together with the characteristics of the conduction and radiation heat-transfer paths between these "sources and sinks." The objective of

	Thermal design temperature Limits (°C), min/max	
Subsystems/equipment	**Nonoperating/turn-on**	**Operating**
Communications		
Receiver	–30/+55	–10/+45
Input multiplex	–30/+55	–10/+30
Output multiplex	–30/+55	–10/+40
TWTA	–30/+55	–10/+55
Antenna	–170/+90	–170/+90
Electric power		
Solar array wing	–160/+80	–160/+80
Battery	–10/+25	–0/+25
Shunt assembly	–45/+65	–45/+65
Attitude control		
Earth/sun, sensor	–30/+55	–30/+50
Angular rate assembly	–30/+55	–1/+55
Momentum wheel	–15/+55	–1/+45
Propulsion		
Solid apogee, motor	+5/+35	–
Propellant tank	+10/+50	+10/+50
Thruster catalyst bed	+10/+120	+10/+120
Structure		
Pyrotechnic mechanism	–170/+55	–115/+55
Separation clamp	–40/+40	–15/+40

Table 11.17 Typical Temperature Limits

thermal control design is to provide the proper heat transfer between all spacecraft elements so that the temperature-sensitive components will remain within their specified temperature limits during all mission environmental conditions, including prelaunch, launch, transfer orbit, and synchronous orbit phases.

A typical spacecraft thermal control design process is shown in Figure 11.82. Initially, system trade-offs are performed to determine the spacecraft configuration. The objective is to meet the requirements of all subsystems within the launch vehicle constraints. The equipment locations are based on a trade-off between structural and thermal requirements. Preliminary analyses are

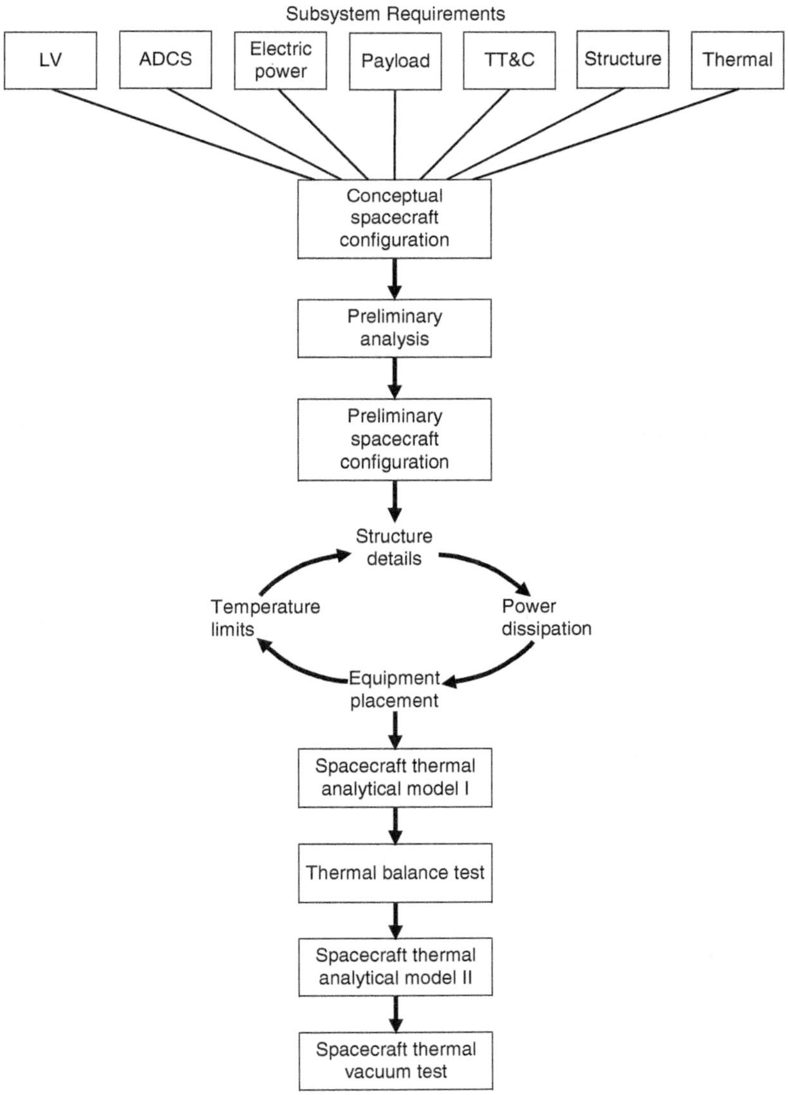

FIGURE 11.82 Thermal control design process.

performed to determine the adequacy of the thermal design concept. Next, detailed subsystem designs and trade-offs are performed. In thermal design, the interaction between structural design, equipment power dissipation, equipment location, and the desired component temperature limits are important considerations. A trade-off is performed in terms of thermal control parameters such as thermal coatings, insulation, metal thickness, heat pipes, and louvers, to establish the best thermal design for a spacecraft. After the trade-off studies and preliminary design are completed, an analytical thermal model is developed to predict temperatures. If the equipment temperatures are not within the allowable temperature limits, the thermal design is modified and the process is repeated.

The accuracy of the analytical thermal model can be improved and verified by performing thermal balance tests on a suitable thermal test model. Equipment heat dissipation is simulated by resistive heaters. Absorbed external flux may be simulated by a variety of means, such as lamps which approach the sun in spectral and intensity properties, warm-plate infrared sources, or embedded film-resistive heaters. This test serves as the verification of the spacecraft thermal control subsystem. The flight spacecraft itself is later subjected to thermal vacuum qualification and acceptance tests where test conditions and equipment temperature (expected in orbit) are imposed on the spacecraft. Typically, design margins of 10°C and 5°C beyond those predicted for orbit are used for qualification and acceptance testing, respectively. This part provides the basics of spacecraft thermal control.

11.42 Heat Transfer

There are three modes of heat transfer: conduction, convection, and radiation. Conduction in a solid is the transfer of heat from one part to another under the influence of a temperature gradient without appreciable displacement of particles. Conduction involves the transfer of kinetic energy from one molecule to an adjacent molecule. The same process takes place within liquids and gases. Convection involves the transfer of heat by mixing one part of a fluid with another. The fluid motion may be entirely the result of the difference of density resulting from temperature differences (i.e., natural convection), or the motion may be produced by means of mechanical fluid movers (i.e., forced convection). Solid bodies, as well as liquids and gases, are capable of radiating and absorbing thermal energy in the form of electromagnetic waves. Both conduction and convection rely on a heat-transfer medium, whereas radiation can occur in a vacuum. In a spacecraft, heat is mostly transferred by conduction throughout the solid parts of the spacecraft and radiated across interior volumes and into space from external surfaces.

Conduction

Heat transfer by steady unidirectional conduction is given by the following relation of Fourier proposed in 1822:

$$Q = -KA\frac{dT}{dx} \tag{11.211}$$

where Q = rate of heat conduction along the x axis, W

K = thermal conductivity, a physical property of the body, $W/m{\cdot}K$

A = cross-sectional area of the path normal to x axis, m^2

$\frac{dT}{dx}$ = temperature gradient along the path, K/m

The thermal conductivities of typical materials are given in Table 11.18.

Materials	Thermal conductivity, *K*, at 25°C *(W/m·K)*
Aluminum	210
Aluminum alloys	117–175
Magnesium	157
Magnesium alloys	52–111
Titanium	21
Stainless steel	16.2
Teflon	0.25

TABLE 11.18 Thermal Conductivity of Spacecraft Materials

Let us consider a plane wall with thickness X and cross-sectional area A, whose two surfaces are kept in steady state at different but constant temperatures T_1 and T_2. Using equation (11.211), the heat-transfer rate from T_1 to T_2, Q, is given by

$$Q = \frac{K}{X} A (T_1 - T_2) \tag{11.212}$$

The amount of heat flow through a unit area of the surface per unit time, the specific heat flux, is given by the equation

$$q = \frac{Q}{A} = \frac{K}{X}(T_1 - T_2) \tag{11.213}$$

Equation (11.213) can be written as

$$Q = \frac{T_1 - T_2}{R_c} \tag{11.214}$$

where

$$R_c = \text{thermal resistivity, k/w}$$

$$= \frac{X}{KA} \tag{11.215}$$

Fourier's laws for the heat conduction process are similar to Ohm's law for electric current. The heat flow Q corresponds to current; the driving force for heat flow; temperature difference $(T_1 - T_2)$ corresponds to voltage E; and thus the thermal resistivity R_c is directly analogous to electrical resistivity.

Radiation

Radiation is the transmission of energy by electromagnetic waves. The electromagnetic spectrum and the names given to radiation transmitted in various ranges of wavelengths are shown in Figure 11.83. The frequency of radiation depends on the nature of its source. For example, a metal bombarded by high-energy electrons emits X-rays, high-frequency electric currents generate radio waves, and a body emits thermal radiation by virtue of its temperature. Radiation in the

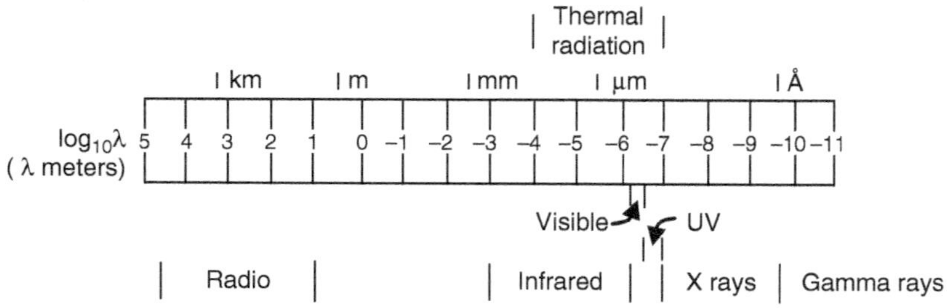

FIGURE 11.83 Spectrum of electromagnetic waves. (*Gray and Muller 1972.*)

wavelength range 0.1–100 μm is called thermal radiation. The radiation within the wavelength band of 0.38–0.76 μm is visible.

When radiation falls on a body, a fraction (α) of it is absorbed, a fraction (ρ) is reflected, and the remainder (τ) is transmitted through the body. These fractions are related by

$$\alpha + \rho + \tau = 1 \tag{11.216}$$

where α = absorptivity
ρ = reflectivity
τ = transmissivity

Most solid materials absorb practically all radiation within a very thin surface layer, usually less than 1 mm thick. For these opaque materials, $\tau = 0$ and $\alpha + \rho = 1$.

In the analysis of radiative heat transfer, it is useful to introduce the concept of a blackbody. A blackbody can be defined as one that (1) absorbs all radiation incident upon it (and reflects and transmits none) and (2) emits at any particular temperature the maximum possible amount of thermal radiation.

Stefan-Boltzmann Law

The rate at which energy is radiated from a blackbody is proportional to the fourth power of its absolute temperature. This law was discovered experimentally by Stefan in 1879 and was deduced theoretically 5 years later by Boltzmann using classical thermodynamics. The law can be stated as

$$E_b = \sigma T^4 \tag{11.217}$$

where E_b = rate at which energy is radiated from a unit area of the surface of a blackbody to the hemisphere of the space above it, W/m^2
σ = Stefan-Boltzmann constant = $5.67 \times 10^{-8}\ W/m^2\ K^4$
T = absolute temperature, K

Figure 11.84 shows the radiative flux as a function of wavelength for a blackbody at a number of temperatures. The analytical expression, derived by Planck from quantum theory, is

$$E_{b\lambda} = \frac{C_1 \lambda^{-5}}{e^{(C_2/\lambda T)} - 1} \tag{11.218}$$

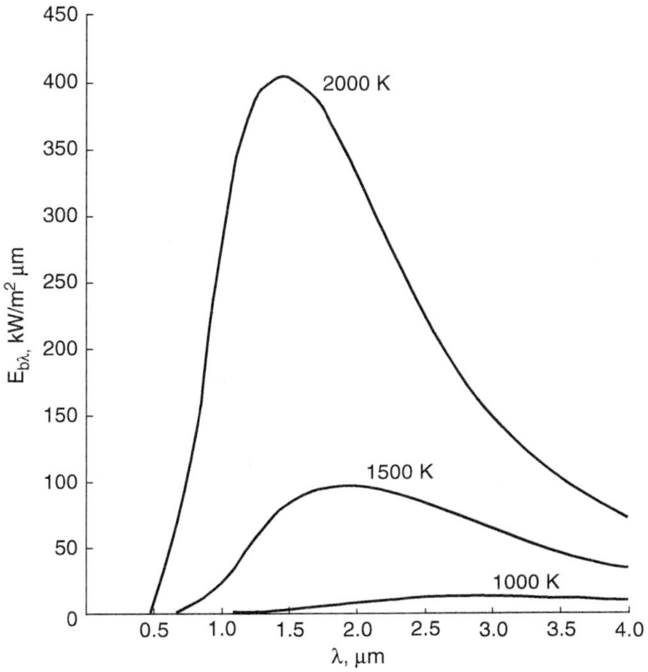

FIGURE 11.84 Spectral distribution of radiation emitted by a blackbody. (*Gray and Muller 1972.*)

where $E_{b\lambda}$ = monochromatic emissive power for a blackbody
T = absolute temperature, K
$C_1 = 3.74 \times 10^{-19}$ kW · m²
$C_2 = 1.439 \times 10^{-2}$ m · K
λ = wavelength, m

From Figure 11.84 it can be seen that in addition to the amplitude increase with temperature, the maximum moves toward the shorter wavelengths. If the temperature of a hot body is less than 500°C (773 K), virtually none of the radiation will fall within the band of the wavelength corresponding to visible light. Even at 2,500°C, the temperature of an incandescent lamp, only about 10% of the energy is emitted in the visible range. The wavelength at which maximum flux is emitted (λ_{max}) is inversely proportional to the absolute temperature (T), a relationship known as Wein's displacement law:

$$\lambda_{max} T = 2.9 \times 10^{-3}\,\text{m} \cdot \text{K} \tag{11.219}$$

Intensity of Radiation

The intensity of radiation in the direction normal to the emitting surface is related to the blackbody emissive power by

$$I_n = \frac{E_b}{\pi} \tag{11.220}$$

When the direction of emission is at an angle ϕ to the normal of the surface, the projected area of emission is $\cos\phi$ and not 1. In this case, the intensity of emission is

$$I_\phi = \frac{E_b \cos\phi}{\pi} \tag{11.221}$$

Equation (11.221) is a statement to the Lambert cosine law.

Kirchhoff's Law

Kirchhoff's law states that for a given wavelength λ, the absorptivity and emissivity are equal:

$$\alpha_\lambda = \varepsilon_\lambda \tag{11.222}$$

For a given range of the spectrum limited by the wavelengths λ_1 and λ_2:

$$\alpha(\lambda_1, \lambda_2) = \varepsilon(\lambda_1, \lambda_2) = \frac{\int_{\lambda_1}^{\lambda_2} \alpha_\lambda I_\lambda \, d\lambda}{\int_{\lambda_1}^{\lambda_2} I_\lambda \, d\lambda} \tag{11.223}$$

where I_λ is the monochromatic intensity of the radiation at the wavelength.

For a blackbody, $\alpha_\lambda = \varepsilon_\lambda = 1$ and, therefore, $\alpha = \varepsilon = 1$. A gray body is defined as one whose emissivity is constant and does not vary with wavelength, $\alpha = \varepsilon =$ constant over the entire spectrum range.

For realistic coatings, however, α_λ (or ε_λ) is a function of λ. The absorptivity α_λ or ε_λ of white paint, black paint, and second surface mirror surfaces are shown in Figure 11.85 as a function of wavelength. In practice, the absorptivity α and emissivity ε are different in most cases, as they

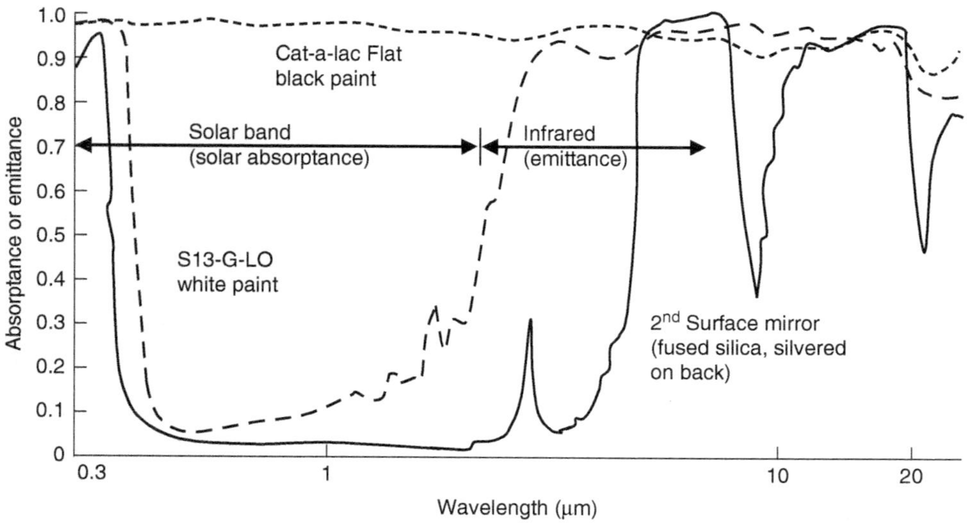

FIGURE 11.85 Actual representation of three material properties.

depend on the range of the spectrum corresponding to the absorbed and emitted radiation. In spacecraft thermal design, normally α refers to solar absorption α_s and ε to infrared emittance, as discussed below.

Solar Absorptivity

The solar spectral irradiance is given in Figure 11.86. Since only 2% of the total irradiance corresponds to the range below 0.22 μm wavelength and 3% above 2.7 μm, a good approximation of α_s is given by

$$\alpha_s = \frac{\int_{0.2}^{2.8} \alpha_\lambda I_\lambda \, d\lambda}{\int_{0.2}^{2.8} I_\lambda \, d\lambda} \tag{11.224}$$

Infrared Emissivity

The relation between radiation intensity, temperature, and wavelength indicates that emissivity is a function of temperature of the radiating surface. For the range near room temperature (20°C), the effective spectrum may be defined as between 5 and 50 μm. Thus, the infrared emissivity ε is very closely approximated by

$$\varepsilon = \frac{\int_{5}^{50} \alpha_\lambda I_\lambda \, d\lambda}{\int_{5}^{50} I_\lambda \, d\lambda} \tag{11.225}$$

Solar absorptivity α_s and infrared emissivity ε for some coatings currently used for spacecraft thermal control are given in Table 11.19. All thermal coatings degrade to a certain extent in the space environment due to an exposure to ultraviolet radiation. Table 11.19 also gives expected degraded values after 7 years in space.

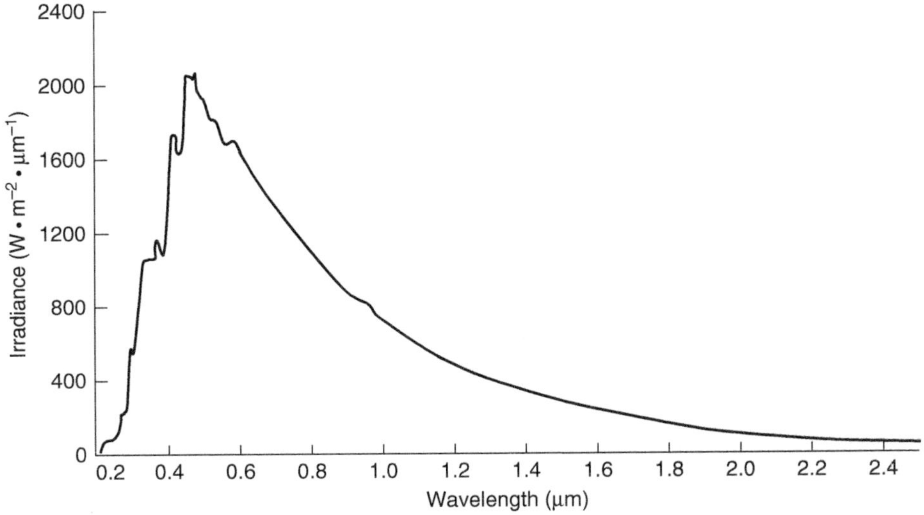

FIGURE 11.86 Solar spectral irradiance curve.

Surface	Typical application	Solar absorptance, α,		Emittance, ε	
		BOL[a]	EOL[b]	BOL[a]	EOL[b]
Black paint	Interior structure	0.9	0.9	0.9	0.9
White paint	Antenna reflector	0.2	0.6	0.9	0.9
Optical solar reflector	North and south panel radiators	0.08	0.21	0.8	0.8
Graphite/epoxy	Solar panel and antenna structure	0.84	0.84	0.85	0.85
Aluminized kapton	Thermal insulation	0.35	0.50	0.6	0.6
Tiodized titanium	Apogee motor thermal shield	0.6	0.6	0.6	0.6
Aluminum, aluminum tape, deposited aluminum	Propellant insulation	0.12	0.18	0.06	0.06
Anodized aluminum	Interior structure	0.2	0.6	0.8	0.8
Solar cells	Solar panels	0.65–0.75	0.65–0.75	0.82	0.82
Gold		0.2–0.3		0.03–0.06	

[a]BOL, Beginning of life; [b]EOL, end of life, 7 years.

Table 11.19 Thermal Properties of Surfaces

View Factors

The view factor from one surface to another is defined as the fraction of the total radiation emitted by one surface which is directly incident on the other. If a closed system is composed of n surfaces, the following equation for view factors for the ith surface applies to all of the n surfaces:

$$\sum_{j=1}^{n} F_{ij} = 1 \tag{11.226}$$

Reflections

The surfaces of some solids which are highly polished and smooth behave like mirrors to thermal radiation (i.e., the angle of reflection equals the angle of incidence). Such reflection is called specular. Most commercial materials have rough surfaces; that is, their surface irregularities are large compared to the wavelength of radiation. Reflection of thermal radiation from this kind of surface occurs indiscriminately in all directions. Such reflection is called diffuse.

Radiative Coupling

The net heat exchange between two black surfaces i and j is

$$\begin{aligned} Q_{ij} &= A_i F_{ij} E_{bi} - A_j F_{ji} E_{bj} \\ &= A_i F_{ij}(E_{bi} - E_{bj}) \\ &= A_i F_{ij} \sigma (T_i^4 - T_j^4) \end{aligned} \tag{11.227}$$

From equation (11.227) it can be seen that an analogy may be drawn between radiative heat transfer and the flow of electric current through a resistor. The voltage is analogous to blackbody emissive power E_b, the current to net heat exchange Q, and the electrical resistance to the reciprocal exchange area $1/A_iF_{ij}$. The radiative coupling between two black bodies i and j is given by

$$R_{ij} = A_i F_{ij} \tag{11.228}$$

In real situations, we are usually concerned with the transfer among more than two nonblack surfaces, which involves reflection.

External Heat Flux

The major heat flux on an earth orbiting spacecraft consists of solar flux, albedo flux, and the thermal radiation of the earth. Other types of incident flux that have an influence on the spacecraft thermal behavior only during short periods are the infrared flux coming from the internal wall of the fairing before the fairing is jettisoned, the aerodynamic heating after the fairing is jettisoned, which will depend on velocity and altitude, and plume heating coming from the third-stage motor, apogee motor, and so on.

Solar Flux

The yearly average solar flux outside the earth's atmosphere is approximately 1,353 W/m^2. The solar constant is defined as the flux existing at a distance of one astronomical unit from the sun and is closely approximated by this previous quantity. Solar flux varies as the inverse square of the distance from the sun, which causes a variation of 3% during the year due to the eccentricity of the earth's orbit. The maximum (1,399 W/m^2) occurs at the perihelion of the earth's orbit around January 3, and minimum (1,309 W/m^2) occurs at aphelion around July 4. Solar flux is assumed to be independent of orbit altitude, from near-earth to geosynchronous. Due to the great distance between the sun and the spacecraft, it is assumed that the solar flux impinges on the spacecraft with parallel rays. Hence the solar intensity (in units of power) on a surface not shadowed by another body is obtained simply as the product of the solar flux times the projected area of the surface on the plane perpendicular to the sun vector. Hence, the solar flux incident on a surface is given by

$$Q = S\mu_i A \tag{11.229}$$

where Q = incident solar intensity, W
S = solar flux, W/m^2
A = total area of the surface, m^2
μ_i = ratio of projected (effective) area of the surface/total area
= $\cos\theta$ for a plane surface where θ is
= $\frac{1}{4}$ for a sphere
= solar aspect coefficient

Albedo Flux

The albedo flux is the fraction of the total incident solar radiation on the earth which is reflected into space as a result of scattering in the atmosphere and reflection from clouds and the earth surfaces. The albedo flux constant ϕ_a is given by

$$\phi_a = Sa \tag{11.230}$$

where S is the solar flux constant and a is albedo coefficient.

The recommended annual mean value of a is 0.30 ± 0.02. The value of a varies from 0.1 to 0.8. The variation is related to the increase in the average cloud cover with distance from the equator and the high reflectance of snow- and ice-covered surfaces in high-latitude regions. Low values of mean albedo ($a < 0.3$) exist in the latitude range from 30°N to 30°S.

In the calculation of incident albedo flux on a spacecraft, it is commonly assumed that the albedo coefficient is constant over the earth's surface and that the earth's surface reflects diffusely and obeys Lambert's law. However, the calculation is still complex because the albedo flux depends on the position of the spacecraft, orientation of the sun, and the spacecraft altitude. For a simple case of a spherical spacecraft and averaging over the spacecraft surface's and sun's orientation, the albedo flux is

$$\phi_A = \frac{Sa}{8}\left(1-\sqrt{1-\frac{R_e^2}{\delta^2}}\right) \tag{11.231}$$

where R_e is the radius of the earth and δ is the distance of the spacecraft from the center of mass of the earth.

Thermal Radiation of the Earth

A portion of the incident solar radiation is absorbed as heat by the earth and its atmosphere and is reemitted as thermal radiation according to the Stefan-Boltzmann law. The mean annual value of thermal radiation near the earth's surface is

$$\phi_r = 237 \pm 7 \text{ W/m}^2 \tag{11.232}$$

It is usually assumed that earth-emitted thermal radiation is constant over the earth's surface and that the earth's surface emits diffusely and obeys Lambert's law. Flux incident (ϕ_T) on a spacecraft surface is a function of altitude and for a spherical spacecraft is given by

$$\begin{aligned}\phi_T &= \frac{1}{2}\sigma T_e^4\left(1-\sqrt{1-\frac{R_e^2}{\delta^2}}\right)\\ &= \frac{1}{2}\phi_r\left(1-\sqrt{1-\frac{R_e^2}{\delta^2}}\right)\end{aligned} \tag{11.233}$$

where T_e = mean earth black-body temperature, K
ϕ_r = earth's thermal radiation, W/m^2
R_e = radius of the earth, km
δ = distance of the spacecraft from the earth center of mass

11.43 Thermal Analysis

Spacecraft temperatures are computed from solutions of simple heat balance equations of the form

$$\text{heat stored} = \text{heat in} - \text{heat out} + \text{heat dissipated} \tag{11.234}$$

For a spacecraft in orbit above a planetary atmosphere, the heat that is absorbed by the spacecraft includes absorbed sunlight, reflected sunlight (albedo), and planet-emitted radiation. Heat is produced within the spacecraft by power dissipated primarily by electrical and electronic components. Heat is rejected from the spacecraft by infrared radiation from external surfaces. Heat is also exchanged among spacecraft component parts by radiation and conduction.

Let us consider an infinitely conductive, therefore isothermal, spherically shaped spacecraft in geosynchronous orbit where the albedo flux and the earth's radiation are negligible. The spacecraft is assumed to have no dissipative equipment. The temperature of the spacecraft is a direct function of its surface thermo-optical properties: the absorptivity α_S and emissivity ε.

The heat-balance equation is

$$\alpha_s S\pi R^2 = \varepsilon\sigma T^4 4\pi R^2 \tag{11.235}$$

where R is the radius of the sphere; S is the solar flux intensity (say, 1,353 W/m^2); α_S and ε are the absorptivity and emissivity of the surface, respectively; σ is the Stefan-Boltzmann constant = 5.67×10^{-8} W/m^2·K^4; and T is the equilibrium temperature.

The equilibrium temperature T is therefore

$$T = \left(\frac{\alpha_S S}{4\varepsilon\sigma}\right)^{1/4} \tag{11.236}$$

If the spacecraft surface is painted white with $\alpha_S = 0.2$ and $\varepsilon = 0.9$ (Table 11.19), the equilibrium temperature is equal to –83°C. White paint is sometimes termed a "cold" coating because it absorbs very little solar flux (just like white summer clothing) but has a high emissivity. For a spacecraft surface painted black, $\alpha_S = \varepsilon = 0.9$, the equilibrium temperature is 5°C. Black paint is called a "mean" coating. For a gold surface $\alpha_S = 0.25$ and $\varepsilon = 0.045$, the equilibrium temperature is 154°C. Gold is called a "warm" coating.

Isothermal Radiator

During spacecraft configuration study, preliminary thermal analyses are performed to estimate the average temperature of the spacecraft and the required radiator size. For such analyses the spacecraft is assumed to be isothermal. The calculation of the average spacecraft temperature is a valuable tool in the configuration study, even though it will be supplemented later by a more detailed calculation of the temperatures at many points in the spacecraft.

The heat-balance equation for an isothermal spacecraft is

$$mc_p\frac{dT}{dt} = \text{external heat flux} + \text{equipment heat dissipation} - \text{heat radiated in space} \tag{11.237}$$

where m = total spacecraft mass, kg
C_p = specific heat, W·s/K·kg
T = absolute temperature, K

Neglecting the albedo flux and the thermal radiation of the earth for a geosynchronous orbit satellite, the external heat flux from the sun is given by

$$\text{external heat flux} = \alpha_S S\mu_i a \tag{11.238}$$

where α_S = solar absorptance of the spacecraft surface
S = solar flux intensity, W/m^2
a = surface area, m^2
μ_i = solar aspect coefficient
= projected (effective) area of the surface/a

$$\text{heat radiated in space} = \varepsilon A \sigma T^4 \tag{11.239}$$

where ε = emissivity

σ = Stefan-Boltzmann constant = 5.67 × 10^{-8} W/m²·K⁴

A = radiator area, m²

T = absolute temperature, K

Hence the thermal balance equation becomes

$$mc_p \frac{dT}{dt} = \alpha_S S \;{}_i a + P - \varepsilon A \sigma T^4 \tag{11.240}$$

where P is the equipment heat dissipation. For a steady-state condition, $dT/dt = 0$. The steady-state temperature is given by

$$T_E = \frac{P + \alpha_S S \;{}_i a}{\varepsilon A \sigma}^{1/4} \tag{11.241}$$

In a geosynchronous orbit, a spacecraft is subject to two eclipse seasons with a maximum eclipse period of 72 minutes per day. During these eclipse periods, $S = 0$ and temperatures drop. To ensure that the equipment is able to withstand eclipse periods, it is necessary to calculate the minimum temperature to be experienced during eclipses. It is also important to know the temperature-time profile during the return to the steady-state conditions.

Equation (11.240) can be rewritten in terms of steady-state temperature T_E (from equation (11.241)) as

$$\frac{dT}{dt} = \frac{T_E^4 - T^4}{4\tau T_E^3} \tag{11.242}$$

where the time constant

$$\tau \equiv \frac{mc_p}{4\varepsilon \sigma A T_E^3} \tag{11.243}$$

The solution of equation (11.243) depends on whether the temperature is decreasing, such as during an eclipse period, or the temperature is increasing, such as immediately after eclipse. The steady state temperatures for these two periods T_E will be different.

For the case of radiative cooling, no solar influx, and $T > T_E$, the solution can be written as

$$\frac{t}{\tau} + C = 2 \coth^{-1}\frac{T}{T_E} - \cot^{-1}\frac{T}{T_E} \tag{11.244a}$$

For the case of radiative heating, with solar flux and $T < T_E$, the solution is given by

$$\frac{t}{\tau} + C = 2 \tanh^{-1}\frac{T}{T_E} + \tan^{-1}\frac{T}{T_E} \tag{11.244b}$$

where C is the constant of integration and is determined from initial conditions, temperature at $t = 0$, which is known. It should be noted that the temperature cannot be written explicitly as a function of time.

Example 11.1 In a three-axis stabilized geosynchronous spacecraft, the communications equipment is mounted on the north- and south-facing panels. The total thermal dissipation on each panel is 300 W. For the equipment to operate properly, the allowable temperature range for the radiator on the panel exterior is 5°/+37°C. The mass of the radiator including the mounted equipment is 85 kg and its specific heat c_p is 900 W·s/kg·K.

Determine

(a) Radiator size for communications equipment.

(b) Equinox/eclipse temperature for the cases: (1) batteries provide full power to communications equipment during eclipse, resulting in no change in thermal dissipation; (2) batteries provide power for only a part of the communications equipment, such that the dissipation is halved. For the second case, determine the radiator temperature at the end of the eclipse.

(c) Determine the time required for the radiator to reach the minimum operating temperature of 5°C after the eclipse for the second case. The radiator heat dissipation during this period can be assumed to be 375 W by turning on the heaters instead of TWTs.

Solution

(a) The (south-facing) radiator is sized for the hottest condition, winter solstice at EOL. The optical solar reflector (OSR) radiator is assumed to be isothermal and to take into account the solar reflection and IR radiation from the solar array and the antennas, the efficiency of the radiator is assumed to be 90% during noneclipse period. The heat-balance equation is

$$\varepsilon\sigma T^4\eta A = \alpha_S AS\sin\theta + P$$

In this problem, ε = emittance of the radiator = 0.8 from Table 11.19, $\sigma = 5.67\times10^{-8}$ W/m²·K⁴; η = efficiency = 0.9; A = area of the radiator to be determined; T = radiator temperature in K = 310 K, maximum allowable temperature; α_S = solar absorptance at EOL = 0.21 from Table 11.19; S = solar intensity at winter solstice = 1,397 W/m²; θ = solar aspect angle = 23.5° at winter solstice; and P = 300 W. Substituting these values in the heat-balance equation, the area of the radiator is given by

$$A = \frac{300}{0.8\times5.67\times10^{-8}\times(310)^4\times0.9 - 0.21\times1397\sin 23.5^\circ}$$

$$= \frac{300}{377-117.14} \simeq 1.16\text{ m}^2$$

(b) At equinox during the non-eclipse period, $\theta = 0$ and the heat-balance equation is

$$\varepsilon\sigma T^4\eta A = P$$

or

$$T = \left(\frac{P}{\varepsilon\sigma\eta A}\right)^{1/4}$$

$$= \left(\frac{300}{0.8\times5.67\times10^{-8}\times0.9\times1.16}\right)^{0.25}$$

$$= 282\text{ K} = 9^\circ\text{C}$$

So the temperature during the equinox non-eclipse period is within the prescribed limits. If the temperature was too low, auxiliary heaters would have been necessary. This concept is termed as an augmented passive or semi-active thermal design.

In the first case where the batteries provide full power to the communications equipment during eclipse, the radiator temperature will remain approximately the same during eclipse as during the noneclipse period at equinox.

In the second case, with batteries providing partial power, heat dissipation on 150 W, the equilibrium temperature T_E during eclipse is given by

$$T_E = \left(\frac{P}{\varepsilon\sigma A}\right)^{1/4}$$
$$= \left(\frac{150}{0.8 \times 5.67 \times 10^{-8} \times 1.16}\right)^{1/4} = 231\ \text{K}$$

The time constant τ from equation (11.243) is given by

$$\tau = \frac{mc_p}{4\varepsilon\sigma A T_E^3}$$

Substituting the values of the parameters in the above equation, the time constant τ is given by

$$\tau = \frac{85 \times 900}{4 \times 0.8 \times 1.16 \times 5.67 \times (231)^3 \times 10^{-8}}$$
$$= 2.949 \times 10^4\ \text{s}$$

For the case of radiative cooling, the temperature from equation (11.244a) is given by

$$\frac{t}{\tau} + C = 2\left(\coth^{-1}\frac{T}{T_E} - \cot^{-1}\frac{T}{T_E}\right)$$

at $t = 0$, $T = 282$ K, and $T_E = 231$ K. Substituting these values in the equation, the integration constant C is equal to 0.936. The equation above is solved by assuming temperature T and determining time t corresponding to the temperature.

At $T = 273\ \text{K} = 0°\text{C}$, the time $t = 71$ minutes.

At $T = 272\ \text{K} = -1°\text{C}$, the time $t = 80$ minutes.

The maximum eclipse period is 72 minutes. Using linear interpolation, the temperature at 72 minutes is 272.9 K.

(c) We want to determine the time required to reach 278 K after eclipse. The radiator heat dissipation is 375 W.
The equilibrium temperature T_E is given by

$$T_E = \left(\frac{375 \times 10^8}{0.8 \times 5.67 \times 0.9 \times 1.16}\right)^{1/4}$$
$$= 298.3\ \text{K}$$

The time constant τ is given by

$$\tau = \frac{85 \times 900}{4 \times 0.8 \times 0.9 \times 1.16 \times 5.67 \times (298.3)^3 \times 10^{-8}}$$
$$= 1.52 \times 10^4\ \text{s}$$

For this case of radiative heating, the temperature is given from equation (11.244b) as

$$\frac{t}{\tau} + C = 2\left(\tanh^{-1}\frac{T}{T_E} + \tan^{-1}\frac{T}{T_E}\right)$$

at $t = 0$, $T = 272.9$ K, and $T_E = 298.3$ K. Substituting these parameters in the equation, the integration constant C is determined to be 4.595. Next, we would like to determine the time for the radiator temperature to reach 5°C or 278 K. Substituting the values of τ, C, T, and T_E in the equation, t is 64 minutes, the "warm-up period," after maximum period of eclipse:

11.44 Thermal Control Techniques

The typical evolution of a spacecraft thermal design can be considered in three stages: the conceptual design, the preliminary design, and the detailed design. The main task of the thermal designer during the conceptual design stage is to influence the spacecraft configuration in such a way as to achieve effective thermal control. Feasibility studies, mission planning, and gross design trade-offs also occur at this stage.

After the finalization of the conceptual design, the preliminary design begins. System-level trade-offs are performed on the basis of spacecraft or mission requirements. A more detailed configuring of the spacecraft, such as the overall electronics packaging layout, begins. Preliminary definitions of subsystem thermal requirements and characteristics are obtained and power and mass budgets established. Analyses are performed to verify the adequacy of the conceptual thermal design. If warranted, modifications to the thermal design are made.

The foregoing process is continued in greater depth in the detailed design stage. To assist a detailed thermal analysis of spacecraft design, extensive use is made of digital computers.

The thermal control techniques can be divided into two classes: passive thermal control and active thermal control.

Passive Thermal Control

A passive thermal control system maintains the component temperature within the desired temperature range by control of the conductive and the radiative heat paths through the selection of the geometrical configurations and thermo-optical properties of the surfaces. Such a system does not have moving parts, moving fluid, or electric power input other than the power dissipation of spacecraft functional equipment. Passive thermal control techniques include thermal coatings, thermal insulations, heat sinks, and phase-change materials (PCMs).

Thermal Coating

External surfaces of a spacecraft radiatively couple the spacecraft to space, the only heat sink available. Because these surfaces are also exposed to external sources of energy, their radiative properties must be selected to achieve a balance at the desired temperature between internally dissipated and external sources of power and the heat rejected into space. The two properties of primary importance are the emittance of the surface ε and the solar absorptance α_s. Table 11.19 gives the properties of different thermal coatings. Two or more coatings can be combined in an appropriate pattern to obtain a desired average value of α_s and ε (e.g., a checkerboard pattern of white paint and polished metal).

For a radiator, low α_s and high ε ($\alpha_s/\varepsilon \ll 1$) are desirable to minimize solar input and maximize heat rejection to space. For a radiator coating, both the initial values of α_s and ε, and any changes in these values during a mission lifetime, are important. For long-life missions, 7–10 years, degradation can be large for all white paints. For this reason, the use of a second surface mirror coating system is preferred. Such a coating is vapor-deposited silver on typically 0.2-mm-thick fused silica, called an OSR.

The values of the radiative properties are subject to uncertainties arising from four sources: (1) property measurement errors; (2) manufacturing reproducibility; (3) contamination before, during, and after launch; and (4) space environment degradation.

Degradation of the thermal coating in the space environment results from the combined effects of high vacuum, charged particles, and ultraviolet radiation from the sun. The last two factors vary with the mission trajectory, and orbit degradation data are obtained continually from flight tests and laboratory measurements.

Thermal Insulation

Thermal insulation is designed to reduce the rate of heat flow per unit area between two boundary surfaces at specified temperatures. Insulation may be single homogeneous material, such as low-thermal-conductivity foam, or an evacuated multilayer insulation (MLI) in which each layer acts as low-emittance radiation shield and is separated by low-conductance spacers. MLIs are widely used in the thermal control of spacecraft and components to (1) minimize heat flow to or from the component, (2) reduce the amplitude of temperature fluctuations in components due to time-varying external radiative heat flux, and (3) minimize the temperature gradients in components caused by varying directions of incoming external radiative heat.

MLI consists of several layers of closely spaced radiation-reflecting shields which are placed perpendicular to the heat-flow direction. The aim of radiation shields is to reflect a large percentage of the radiation the layer receives from warmer surfaces. To avoid direct contact between shields, low-conductivity spacers are commonly used. Sometimes embossing or crinkling of the shields produces small contact areas whose thermal joint conductance is low. The space between shields is evacuated to decrease gas conduction. For space applications, proper venting of the interior insulation is provided to avoid undue pressure loads on the shields during ascent and to pass outgassing products from the insulation material for long-term on-orbit missions. An evacuated MLI provides, for a given mass, insulation which is orders of magnitude greater than that furnished by more conventional materials such as foams and fiberglass batting.

For typical applications, the MLIs are shown in Figure 11.87. For normal temperatures, the outer skin is 25-μm aluminized Kapton and the remaining layers are made of aluminized Mylar separated by Dacron mesh to provide low thermal conductivity. The outer aluminized Kapton layer provided an outer covering for handling a moderate α_s/ε ratio and blanket protection from

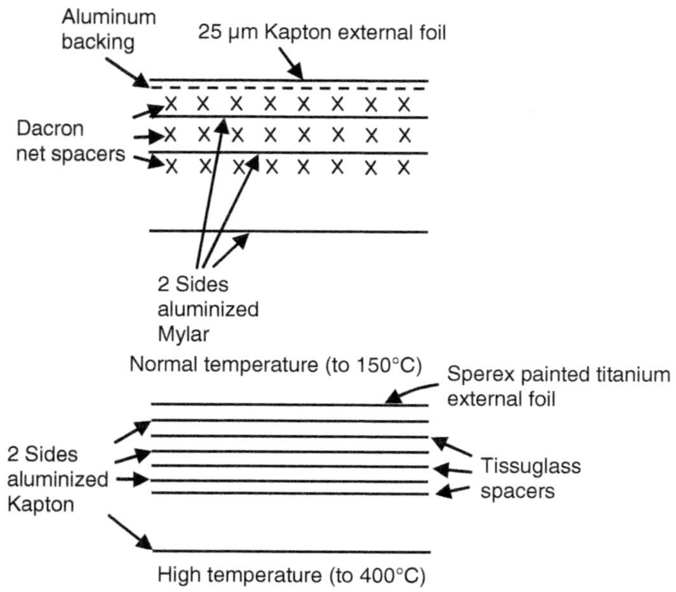

FIGURE 11.87 Typical multilayer blanket composition.

the space environment for a 7- to 10-year mission. The Kapton provides high-temperature protection since its useful maximum temperature is 343°C compared to 121°C for Mylar. The insulation blanket that is subject to apogee motor plume heating has an outer layer of titanium or stainless material. These materials can withstand the high temperature (1,400°C) of the engine exhaust gases while protecting the remaining blanket assembly. The outer layer for such an insulation blanket is painted black to provide a high emittance in order to radiate the absorbed apogee motor heating.

Heat Sinks/Thermal Doublers

Heat sinks are materials of large thermal capacity that are placed in thermal contact with the component whose temperature is to be controlled. When heat is generated by the component, the temperature rise is restricted because the heat is conducted into the sink. The sink will then dispose of this heat to adjacent locations through conduction or radiation. The heat sink can serve the same function in reverse; that is, heat sinks can prevent severe cooling during the periods of low heat absorption or generation. Heat sinks are commonly used to control the temperature of those items of electronic equipment which have high dissipation or a cyclical variation in power dissipation. The equipment and structure of the spacecraft itself usually provides a heat sink.

PCMs

Solid-liquid PCMs present an attractive approach to spacecraft passive thermal control when the incident orbital heat fluxes or on-board equipment dissipation changes widely for short periods.

The PCM thermal control system consists primarily of a container which is filled with a material capable of undergoing a chemical phase change. When the temperature of the spacecraft surface increases, the PCM will absorb the excess heat when it melts. When the temperature decreases, the PCM solidifies. PCMs used for temperature control are those whose melting point is close to the desired temperature of the equipment. Then the latent heat associated with the phase change provides a large thermal inertia when the temperature of the equipment is passing through the melting point. However, the PCM cannot prevent a further temperature rise when all the material is melted.

Active Thermal Control

Passive thermal control may be adequate and efficient in terms of spacecraft added mass for the applications where the equipment has a narrow specified temperature range and there is a great variation in equipment power dissipation, surface thermo-optical properties due to space degradation, and solar flux during the mission. In such cases, temperature sensors may be placed at the critical equipment location. When critical temperatures are reached, mechanical devices are actuated to modify the thermo-optical properties of surfaces or electrical power heaters come on or off to compensate for the variation in the equipment power dissipation. For spacecraft with high-power-dissipation equipment, such as high-power TWTA, it may be more efficient in terms of added mass to use heat pipes to increase thermal conductivity in place of heat sinks.

This section provides a brief review of active control elements, such as heat pipes, louvers, and electric heaters.

Heat Pipes (Chi 1976; Dunn and Reay 1976)

A heat pipe is a thermal device that provides efficient transfer of a large amount of thermal energy between two terminals with a small temperature difference. Heat pipes provide an almost isothermal condition and can be considered an extra high-thermal-conductivity device.

A heat pipe, as shown in Figure 11.88, consists of a closed tube whose inner surfaces are lined with capillary wick. Heat at the evaporation portion of the heat pipe vaporizes the working fluid. The resulting difference in pressure drives vapor from the evaporator to the condenser, where it condenses and release the latent heat of vaporization. The loss of liquid by

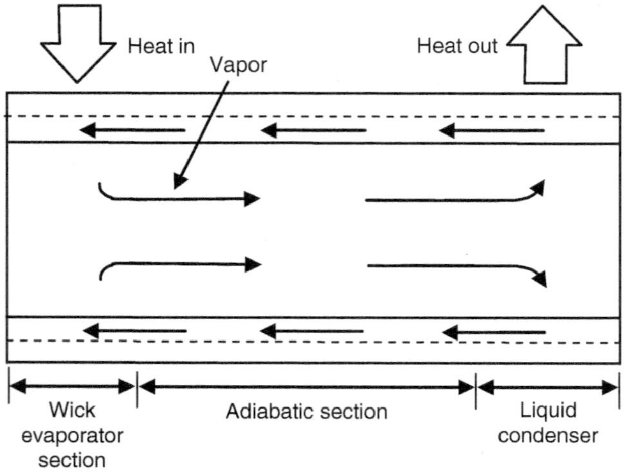

FIGURE 11.88 Schematic diagram of heat pipe.

evaporation results in the liquid-vapor interface in the evaporator entering the wick surface, and a capillary pressure is developed there. This capillary pressure pumps the condensed liquid back to the evaporator for reevaporation. So the heat pipe transports the latent heat of vaporization continuously from the evaporator section to the condenser without drying out the wick, provided that the flow of working fluid is not blocked and a sufficient capillary pressure is maintained. The amount of heat transported as latent heat of vaporization is usually several orders of magnitude higher than that which can be transmitted as sensible heat in a conventional convective system.

For the heat pipe to work, the maximum capillary pressure must be greater than the total pressure drop in the pipe. Neglecting the pressure drops in the evaporator and the condenser, the condition can be stated as

$$(P_c)_{\max} \geq \Delta P_l + \Delta P_v + \Delta P_g \tag{11.245}$$

where $(P_c)_{\max}$ = maximum capillary pressure
ΔP_l = pressure drop required for the liquid to flow from the condenser to the evaporator
ΔP_v = pressure drop required for the vapor to flow from the evaporator to the condenser
ΔP_g = pressure head due to gravity which is near zero in space

If equation (11.245) is not satisfied, the wick will dry out in the evaporator region and the pipe will not operate.

Variable-Conductance Heat Pipes (VCHP)

The purpose of a VCHP is to control the operating temperature of given equipment against variations in the heat dissipations of the equipment and/or in the thermal environment.

The change in the conductance of the heat pipe may be exerted by controlling the mass flow using either of the following techniques:

1. Interruption of the fluid flow, either the liquid flow in the wick or the vapor flow in the core, by means of thermostatically controlled valves
2. Reduction of the condensation rate

The second technique is most widely used. It is normally achieved by the introduction of a noncondensible gas in the condenser, partially displacing the condensible gas.

Louvers (Kelly et al. 1976)

For a spacecraft in which the changes in internal power dissipation or external heat fluxes are severe, it is not possible to maintain the spacecraft equipment temperatures within the allowable design temperature limits unless the α_s/ε ratio can be varied. A very popular and reliable method which effectively gives a variable α_s/ε ratio is through the use of thermal louvers. When the louver blades are open, the effective α_s/ε ratio is low (low α_s, high ε) and when the blades are closed, the effective α_s/ε ratio is high (low α_s, low ε). The louvers also reduce the dependence of spacecraft temperatures on the variation of the thermo-optical properties of the radiators.

Electrical Heaters

Electrical heaters (resistance elements) are used to maintain the temperature above minimum allowable levels. The heater is typically part of a closed-loop system that includes a temperature-sensing element and an electronic temperature controller. In some applications, bimetallic thermostats are used. Electrical heaters are used in an on-off control mode, a ground controllable mode, a proportional control mode, or simply in a continuous-on mode.

11.45 Spacecraft Thermal Design

A spacecraft thermal design is highly dependent on the mission and the type of attitude stabilization system and orbit. In a dual-spin-stabilized spacecraft, the spinning solar array drum equalizes the solar flux and provides a comfortable temperature to the internal equipment. For geosynchronous spacecraft, the north face can radiate freely from the spacecraft except for the obstruction caused by antennas. The south face normally contains the apogee motor nozzle and is usually covered by the shield or the MLI blanket.

In geosynchronous three-axis-stabilized spacecraft, the main body is basically nonspinning. It rotates about N/S axis at one revolution per day in order that the antennas continually orient toward the desired area on the earth surface. This results in a diurnal variation of solar flux on the earth, anti-earth, west, and earth surfaces of the spacecraft. These surfaces are normally covered with MLI to avoid the extreme daily temperature variation. The north and south surfaces are affected only by seasonal variation of solar incidence angle of $\pm 23.5°$ and shadowing by spacecraft appendages. The thermal control design of a three-axis-stabilized spacecraft is normally more difficult than that of dual-spin-stabilized spacecraft.

As examples, this section provides a review of the thermal designs of a dual-spin stabilized spacecraft, INTELSAT IV and three-axis-stabilized spacecraft, Intelsat V. Both are geosynchronous spacecraft.

Dual-Spin-Stabilized Spacecraft

Figure 11.89 shows the thermal control elements of Intelsat IV, which is dual-spin-stabilized spacecraft. The despun section consists of an antenna/communications subsystem. The remaining subsystems, propulsion, attitude control, electric power, and so on, are in the spun section.

Since the antenna/mast assembly is despun, the sun makes one complete revolution relative to it in a 24-hour period. Although the antenna assembly is not generally sensitive to temperature extremes, the temperature gradients cause structural distortion, which results in antenna pointing error. To reduce temperature gradients, MLI blankets are used extensively.

The TWTs and their power supplies provide the majority of the internal power dissipation of the spacecraft. Hence it is necessary to control the local temperature and the bulk temperature in the despun compartment. The bulk temperature control requires rejecting the power into space,

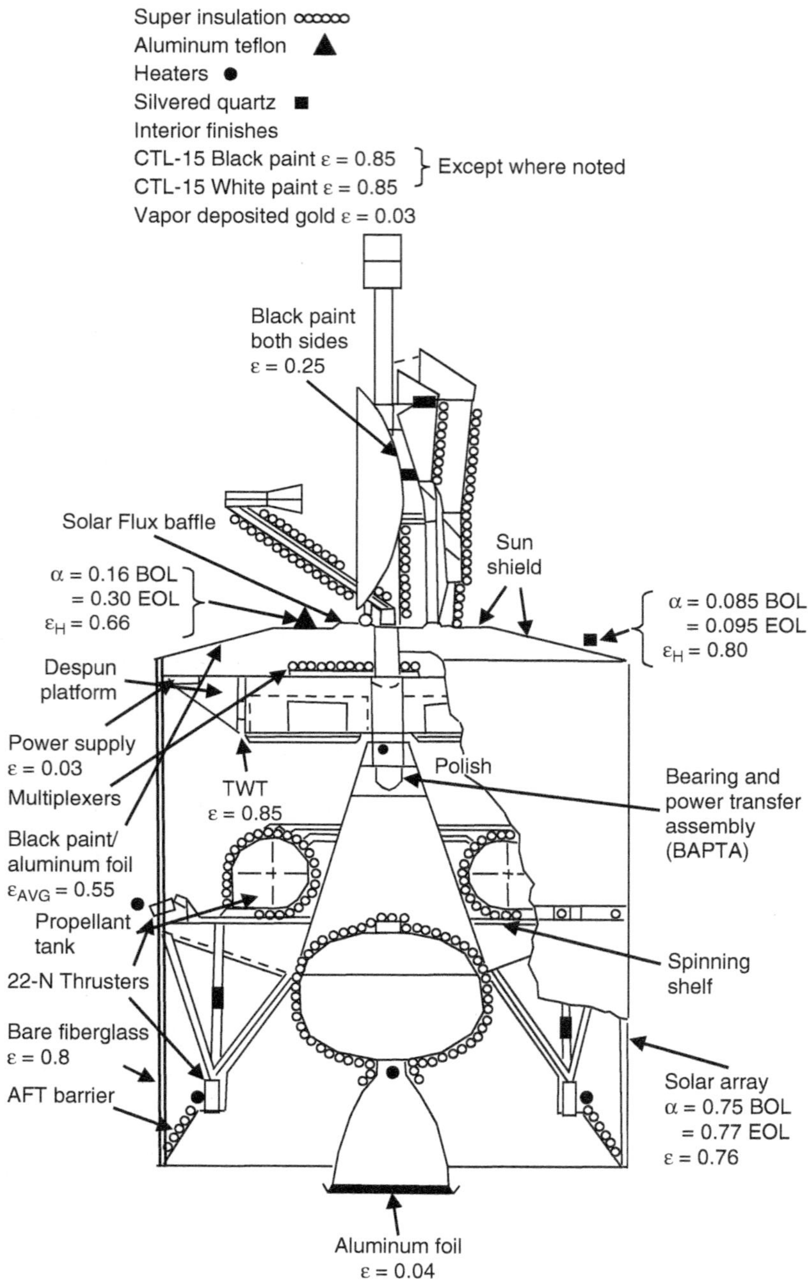

Figure 11.89 Intelsat IV thermal control. (Robinson 1972.)

necessitating radiators with low α_s/ε ratio. The forward-end sunshield and spinning solar array perform this function. The forward sunshield has two sections: (1) the in-board section is covered on the space side with aluminized Teflon ($\alpha_s = 0.16$, $\varepsilon = 0.66$) and (2) the outboard section (spinning) is covered with silver quartz ($\alpha_s = 0.085$, $\varepsilon = 0.80$). The internal thermal coupling of the sunshield with the despun compartment is provided by black paint ($\varepsilon = 0.85$) on the compartment shelf and a combination of black paint and aluminum-foil strip ($\varepsilon = 0.55$) on the inside of the sunshield.

The local temperature requires the simultaneous sinking of the dissipated power of the equipment to the mounting surface and protection from eclipse cooling. To improve the contact surface condition, TWTs and all electronic boxes are mounted with RTV filler. The communication receiver and certain other electronic boxes are gold plated ($\varepsilon = 0.03$) for required radiation isolation during eclipse. Output multiplexers are covered with MLI to decouple them from sunshield cooling during eclipses. The remaining boxes are painted with high-emittance white paint ($\varepsilon = 0.85$).

Positioning and Orientation Subsystem

The temperature control in the positioning and orientation (P&O) subsystem is designed to prevent hydrazine from freezing in orbit. The propellant tanks are covered with MLI. The lines and valves are conductively decoupled from the spacecraft structure with low-conductivity spacer materials and wrapped with low-emittance aluminum-foil tape ($\varepsilon = 0.85$) to reduce radiation cooling during eclipse. Heaters are used around the lines and valves. The thruster chamber is enclosed in a canister consisting of two concentric shells of stainless steel separated by refrasil batting to shield the spacecraft from very high thruster firing temperatures (800°C).

The temperature of the apogee motor, propellant, and nozzle throat area is kept in the temperature range 4–32°C throughout the transfer orbit in order that the apogee motor may ignite properly. The apogee motor case is covered with MLI and the heaters heat the nozzle throat on command. Postapogee motor firing heat soakback to the spacecraft is minimized by MLI and the conduction isolation at the mounting ring.

The solar panel provides a very suitable environment for the spun section except during eclipse. The aft thermal barrier is used to minimize heat loss out of the aft end and shields the spacecraft from apogee motor exhaust plume heating and contamination. The thermal barrier is composed of a stainless sheet with black paint on the outside and gold coating on the inside, and nickel-foil layers. The low-power-dissipating equipment, the spacecraft housekeeping, is mounted on the spacecraft structure in the spun section.

Three-Axis-Stabilized Spacecraft

As discussed earlier, in a three-axis-stabilized spacecraft, the main body is nonspinning. This results in the sun making a complete revolution with respect to the east/west surfaces in a 24-hour period and the north/south surfaces see the variation of solar incidence angle of only ± 23.5° in a 1-year period. Therefore, the solar flux variation is significantly smaller on the north/south surfaces in comparison to the east/west surfaces. Considering the solar flux variation on the spacecraft surfaces, the thermal control design for a typical three-axis-stabilized spacecraft consists of locating high-power-dissipation equipment, such as TWTs, on the north/south surfaces where the heat can be radiated directly to space. The east/west surfaces are covered with MLI to minimize the effect of solar flux variation on the equipment.

Intelsat V

The thermal configuration for Intelsat V, a three-axis-stabilized spacecraft, is shown in Figures 11.90 and 11.91 for a transfer orbit and a synchronous orbit, respectively. The thermal control of Intelsat V spacecraft is accomplished using passive techniques, including a selective location of the power-dissipating equipment, a selective use of surface finishes, and the regulation of thermal paths. The passive design is augmented with heater elements for the batteries,

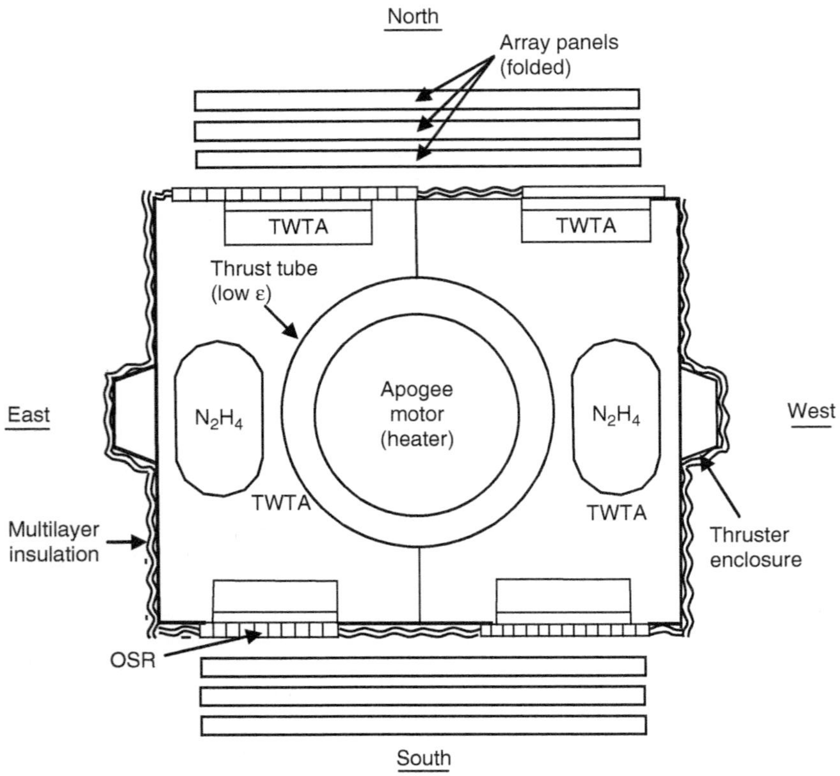

FIGURE 11.90 Preoperational-phase thermal configuration. (*Courtesy of INTELSAT and FACC.*)

hydrazine-propellant lines and tanks, and the apogee motor to keep the temperatures above minimum allowable levels. Heater elements are employed in the catalytic thrust chambers and electrothermal thrusters to optimize performance and useful lifetime. The thermal control for three major parts of the spacecraft—main body, antenna module, and solar arrays—is achieved, essentially, individually with heat transfer between these parts minimized.

Main Body Thermal Control

The overall thermal control of the main body is achieved by (1) the heat dissipation of components in the communications and support subsystem modules, (2) absorption of solar energy by the OSR radiator on the north/south panels, and (3) the reemission to space of infrared energy by the OSR radiator. High thermal dissipators, such as TWTAs, are located on the north and south panels so that they may efficiently radiate their energy to space via the heat sinks and the OSR radiators. The east and west panels, antenna deck, and anti-earth surfaces are covered with MLI to minimize the effect of solar incidence on equipment temperature control during a diurnal cycle. Most of the equipment is directly attached to the interior surface of the north and south radiator panels, thereby maximizing the thermal efficiency of transporting heat from the source to the radiator. The requirements for some equipment are best satisfied, however, by utilizing other locations. As an example, receivers are mounted on the earthward panel to minimize waveguide lengths and to achieve a more stable temperature level than would occur in locations near to TWTAs.

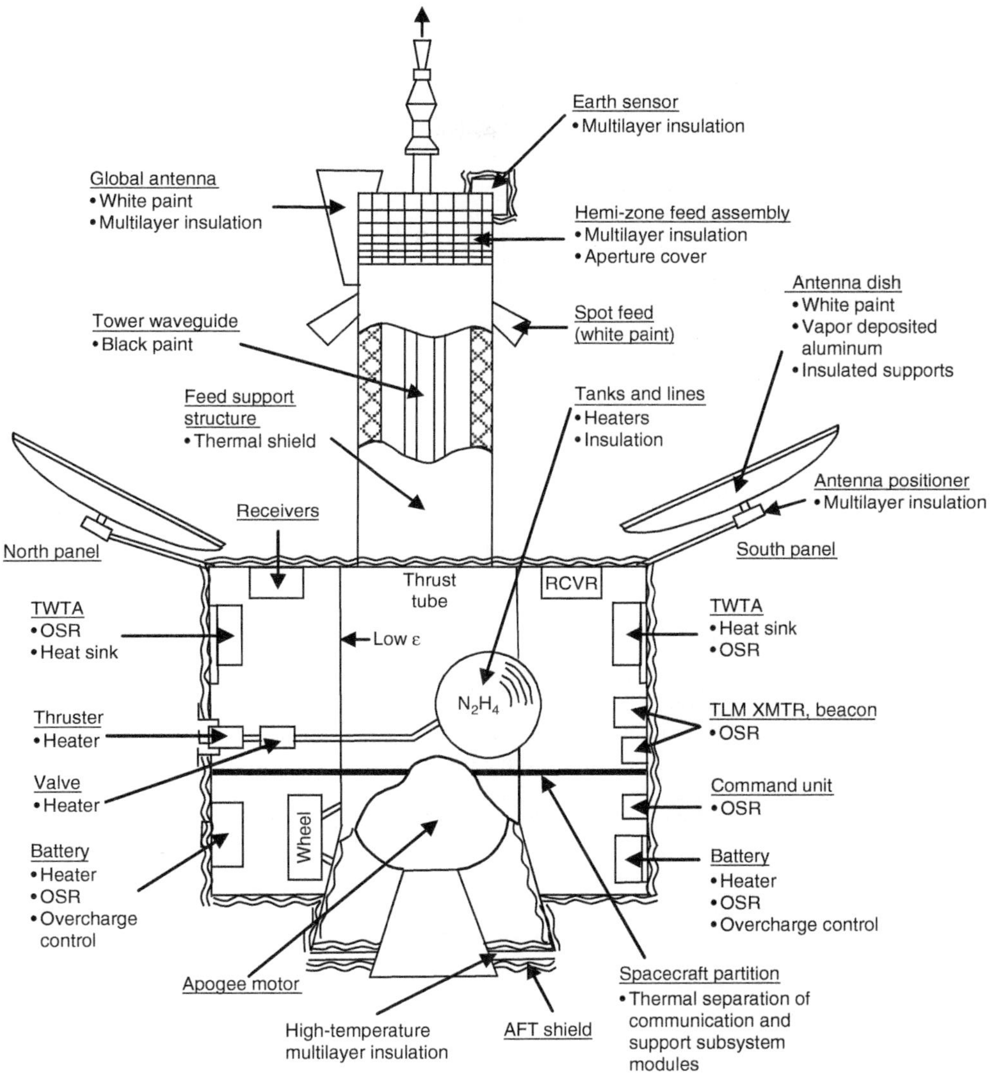

Figure 11.91 Thermal subsystem operational configuration features (solar array and shunt limiters not shown). (*Courtesy of INTELSAT and FACC.*)

The propulsion subsystem components are mounted on or near to the east and west panels to accommodate thruster location requirements and to minimize propellant line lengths. The concentrated heat load of the TWTA collectors is distributed by means of high-conductivity ASTM 1050 tempered aluminum heat sinks between the TWTA units and their mounting panels.

The conductive heat transfer with the antenna module has been minimized primarily to limit the large diurnal temperature effect of the antenna module on the main body. This is achieved by using MLI between the antenna module and the main body, graphite/epoxy for the tower legs, and low-thermal-conductance spacers.

The propellant tanks, lines, and valves are covered with insulation, and heaters are provided to keep the temperature above the freezing temperature of hydrazine. To protect against plume heating during the apogee motor firing, high-temperature insulation blankets, which consist of a titanium outer layer to permit a temperature of up to 500°C, are used on the aft side of the spacecraft. The heat-soak-back effects of the apogee motor are minimized by using a low-emittance surface inside the thrust tube and on the motor casing ($\varepsilon = 0.25$) and by conductively decoupling the apogee motor. The high-temperature insulation blankets minimize the plume heating effects.

Antenna Module

For spacecraft in a synchronous orbit, the sun appears to travel completely around the antenna module once each day. The various parts of the module are subject to full sun and complete shadowing; the shadowing is caused by the main body, the module elements, and by the earth during equinox. Therefore, the antenna module is subject to a rather severe environment, considering that minimal heat is dissipated within the module itself.

The thermal control of the tower structure and tower waveguide is achieved mainly by the use of a three-layer thermal shield around the tower structure. The shield is painted black on the outside to prevent solar reflections from concentrating on antenna reflectors. The interior surface of the shield is bare Kapton, enhancing radiative heat transfer within the tower. The shield serves to minimize the diurnal temperature variation within the tower. The thermal control of reflectors is normally achieved by the use of white paint on the concave surfaces of the reflectors and a MLI blanket on the convex surface.

Solar Array Thermal Control

The primary elements of a solar array are the solar array panels, the shunts, and the yoke. Thermal control of the solar array panels in geosynchronous orbit is achieved by the absorption of solar fluxes on the solar cells on the front surfaces of the panels and reemission of infrared energy from the front and graphite/epoxy back surface of the panels.

Shunts with beryllium heat sinks are located on the anti-sun surfaces of each of the two inboard panels. The front side of the solar panel opposite the shunt is painted white. The shunts and beryllium heat sinks are covered with aluminum tape on the anti-sun surface. The white paint reduces absorbed solar energy. The aluminum tape lowers the shunt's radiative coupling to space, raising the minimum shunt temperature experienced at the end of the equinox eclipse. Thermal control of the yoke is achieved by painting the yoke white.

Further Reading

Agrawal, B. N. 1986. *Design of Geosynchronous Spacecraft*, Prentice-Hall, New Jersey.

Brown, C. D. 2002. *Elements of Spacecraft Design*, AIAA, Washington, DC.

Chi, S. W. 1976. *Heat Pipe Theory and Practice*, McGraw-Hill, New York.

Comsat Technical Review, vol. 2, no. 2, 1972.

Dunn, P. D. and Reay, D. A. 1976. *Heat Pipes*, Pergamon Press, Elmsford, New York.

Gilmore, D. 2002. *Spacecraft Thermal Control Handbook Volume I: Fundamental Technologies*, Aerospace Press, El Segundo.

Gray, W. A. and Muller, R. 1972. *Engineering Calculations in Radiative Heat Transfer*, Pergamon Press, Elmsford, New York.

Griffin, M. D. and French, J. R. 1991. *Space Vehicle Design*, AIAA, Washington, DC.

Groll, M. 1978. "Heat Pipe Technology for Spacecraft Thermal Control," *Spacecraft Thermal and Environmental Control Systems*, ESA SP-139, Nov.

Howell, J. R. and Siegal, R. 1969. *Thermal Radiation Heat Transfer*, NASA SP-164.

Hughes Aircraft Company Geosynchronous Spacecraft Case Histories, Jan. 1981.

Hwangbo, H., Hunter, J. H., and Kelly, W. H. "Analytical Modelling of Spacecraft with Active Thermal Control System," AIAA 8th Thermophysics Conference, Palm Springs, Calif., Jul. 16–18, 1973, AIAA Paper No. 73-773.

Kelly, W. H. and Reisenweber, J. H., Jr. 1981. "Optimization of a Heat Pipe Radiator for Spacecraft High-Power TWTAs," IV International Heat Pipe Conference, Sep. 7–10, The Royal Aeronautical Society, London.

Kelly, W. H. Reisenweber, J. H., Jr., and Flieger, H. W. 1976. "High Performance Thermal Louver Development," AIAA 11th Thermophysics Conference, San Diego, Calif., Jul. 14–16, AIAA Paper No. 76-460.

McAdams, W. H. 1954. *Heat Transmission*, McGraw-Hill, New York.

Robinson, J. A. Fall 1972. "The Intelsat IV Spacecraft—Thermal Control," *COMSAT Technical Review*, vol. 2, no. 2.

Spacecraft Thermal Control, NASA SP-8105, May 1973.

Van Villet, R. M. 1965. *Passive Temperature Control in the Space Environment*, Macmillan, New York.

Wertz, J. R. and Larson, W. J. 2010. *Space Mission Analysis and Design*, Springer, New York.

Wertz, J. R. Everett, D. F., and Puschell, J. J. 2011. *Space Mission Engineering: The New SMAD*, Microcosm Press, Hawthorne, CA.

Wiebelt, J. A. 1966. *Engineering Radiation Heat Transfer*, Holt, Rinehart and Winston, New York.

PART 7

Communications

Gerald Lo and Brij N. Agrawal

11.46 Introduction

This part presents an overview of the communications subsystem, commonly referred to as the payload for communications satellites. To understand the communications subsystem of a spacecraft, one begins with a communication link. A communication link starts with the source signal. The source signal could be a human voice that has been converted into an electrical analog waveform by the microphone, or it could be the signal generated by a color TV camera or it could be a series of binary digits (ones and zeros) generated by teletype machines or data terminals. For transmission over long distances, it is inconvenient and expensive to keep the source signal in its original form. Usually, this electrical source signal, or as it is commonly called by communications engineers, the "baseband signal," travels only a short distance down the cable before it reaches a center where the signal is processed for further transmission. (The term "baseband" is also used by communications engineers in a wider sense than that described here.) Generally, it refers to signals of relatively low frequencies. As an example, a voice signal leaving the telephone handset reaches the local switching office as its first destination. If it is a local call, the signal is switched to the receiving party's telephone and there will be no further signal processing. However, if it is a long-distance call, the signal has to be processed and amplified along the way using the most economical means of transmission. If the signal is carried over the air via microwave or satellite transmission, it is more convenient to use radio frequencies for transmission. Radio frequencies are at much higher frequencies than the source signal. For example, a voice signal spreads from about 300 Hz to about 4,000 Hz, and a microwave signal can be operating in the 4×10^9 Hz or 4 GHz band. Thus a means of carrying the source signal in the microwave frequency band is necessary. This process of converting the source signal into a signal suitable for transmission over air in a microwave frequency band is in fact modulation and upconversion. This type of signal processing is common to both terrestrial microwave transmission and satellite microwave transmission.

A simplified block diagram of an active satellite repeater is shown in Figure 11.92. Frequency translation from the uplink to the downlink and signal amplification is shown in this figure.

11.47 Basic Units and Definitions in Communications Engineering

One of the most basic quantities in communications engineering is signal power. This is usually expressed in the well-known units of watts or milliwatts. Since the communications engineer has to multiply and divide quite often, it is more convenient to express quantities in a logarithmic form or in the form of the logarithmic of a ratio. Thus, if the signal power S is expressed in the logarithmic form with respect to 1 W, then

$$S(\text{dBW}) = 10 \log_{10} \frac{\text{signal power}}{1 \text{ watt}} \tag{11.246}$$

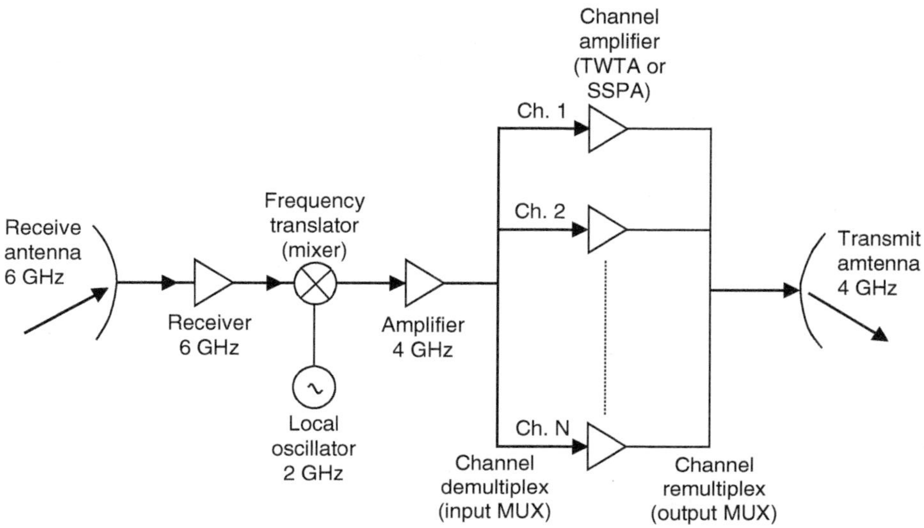

FIGURE 11.92 Simplified block diagram of a satellite communication subsystem (payload).

Similarly, if the signal power S is expressed in the logarithmic form with respect to 1 mW, then

$$S(\text{dBm}) = 10 \log_{10} \frac{\text{signal power}}{1 \text{ mW}} \tag{11.247}$$

Thus one can convert dBW to dBm by adding 30 dB to the figure, or conversely, one can convert dBm to dBW by subtracting 30 dB from the figure.

Another useful quantity that communications engineers use is power flux density (PFD). For a propagating plane radio wave, this is a measure of the amount of electromagnetic energy passing through a given aperture or "window." Usually, this is expressed in watts per square meter or more conveniently in dBW/m^2.

11.48 Frequency Allocations and Some Aspects of the Radio Regulations

Since all countries on earth share the same radio frequency spectrum, the International Telecommunications Union (ITU) Radio Regulations provide the requisite coordination and regulations to ensure that everyone's need is met on an equitable basis. Thus, the radio frequency spectrum is heavily segmented so that each segment or frequency band of the spectrum is allocated by mutual consent for a specific purpose on a coordinated basis. Most of the frequency bands are allocated for more than one type of service and precautions in coordinations are therefore necessary to avoid mutual interference.

Since the satellite transmitters may share the same frequency bands with other terrestrial services, a maximum PFD limit at the earth's surface (e.g., –148 dBW/m^2/4 kHz) is usually imposed on the satellite transmitters so that the satellites do not interfere with existing terrestrial microwave links. Thus, satellite designers often do not have complete freedom to increase the satellite capacity by increasing the power transmitted by the satellite. However, there are certain bands reserved primarily for broadcasting satellites and for these bands the allowable PFDs are usually somewhat higher than those of the fixed-satellite services. These bands are usually reserved for direct-to-home satellite broadcasting services. For further details on this subject, the reader is referred to the ITU Radio Regulations (1982).

Communications engineers have used acronyms to designate various frequency bands. Following is a list of these.

Low frequency (LF)	30–300 kHz
Medium frequency (MF)	300 kHz–3 MHz
High frequency (HF)	3–30 MHz
Very high frequency (VHF)	30–300 MHz
Ultrahigh frequency (UHF)	300 MHz–3 GHz
Super high frequency (SHF)	3–30 GHz
Extremely high frequency (EHF)	30–300 GHz

1 Hz = 1 Hz = one cycle per second

1 kHz = 10^3 Hz (k = kilo)

1 MHz = 10^6 Hz (M = Mega)

1 GHz = 10^9 Hz (G = Giga)

In addition to these band designations, communications engineers commonly use letters to designate bands in the microwave spectrum. These designations have no official international standing, and various engineers have used limits for the bands than those listed below.

Band destination	Frequency band (approx.)
L Band	400 MHz–1.5 GHz
S Band	1.5–3.9 GHz
C Band	3.9–6.2 GHz
X Band	6.2–10.9 GHz
K Band	10.9–36.0 GHz
Q Band	36.0–46.0 GHz
V Band	46.0–56.0 GHz

Frequently, satellite engineers call the 4/6-GHz band for fixed satellite services the C band, the 11/14-GHz band the Ku band, and the 20/30-GHz band the Ka band.

The following are some popular bands allocated for satellites communications, extracted and simplified from ITU Radio Regulations (1982).

Service	Band	Usage
Mobile satellite	1.535–1.559 GHz	Downlink
	1.6265–1.6605 GHz	Uplink
Broadcast satellite	2.5–2.69 GHz	Downlink
Fixed satellite	3.4–4.2 GHz	Downlink
	4.5–4.8 GHz	Downlink
	5.85–7.075 GHz	Uplink
	7.25–7.75 GHz	Downlink
	7.9–8.025 GHz	Uplink
	8.025–8.4 GHz	Uplink + downlink
	10.7–11.7 GHz	Downlink
Fixed + broadcast satellite	11.7–12.75 GHz	Downlink
Fixed satellite	12.75–13.25 GHz	Uplink
	14.0–14.5 GHz	Uplink
Broadcast satellite	14.5–14.8 GHz	Uplink
Fixed satellite	17.3–17.7 GHz	Uplink
Fixed satellite	17.7–18.1 GHz	Uplink + downlink
	18.1–21.2 GHz	Downlink
Inter satellite	22.55–23.55 GHz	Intersatellite
Fixed satellite	27.5–31.0 GHz	Uplink
Inter satellite	32.0–33.0 GHz	Intersatellite
Fixed satellite	37.5–40.5 GHz	Downlink

To conclude this subsection on frequency bands and frequency allocations, it is necessary to point out that there is a free-space wavelength associated with every radio frequency. It is sometimes more convenient to refer to the wavelength rather than the frequency of the radio waves. In free space (or air), the relationship is as follows:

$$\lambda = \frac{c}{f} \tag{11.248}$$

where λ is the wavelength, f is the frequency in hertz, and c is the velocity of light and is approximately equal to 3×10^8 m/s.

11.49 Electromagnetic Waves, Frequency, and Polarization Selection for Satellite Communications

Basic Concepts of Electromagnetic Waves

Electromagnetic waves are traveling waves of the electric and magnetic fields that vary sinusoidally with time and with distance in the direction of propagation. They are analogous to the acoustic traveling waves in air. This acoustic analogy does not extend to the vectorial (or directional) nature of the electric and magnetic fields; that is, the electric or magnetic fields have orientations in space due to the "lines of force" nature of these fields. These lines of force are quite evident when static electric or magnetic fields are examined. Thus, in general, electric or magnetic fields can be resolved into three orthogonal components in space, for example, in Cartesian coordinates, E_x, E_y, and E_z and H_x, H_y, and H_z, where E and H represent the amplitude of the electric and magnetic fields, respectively. In vectorial form:

$$\mathbf{E} = E_x\mathbf{i} + E_y\mathbf{j} + E_z\mathbf{k} \tag{11.249}$$

$$\mathbf{H} = H_x\mathbf{i} + H_y\mathbf{j} + H_z\mathbf{k} \tag{11.250}$$

Fundamental electromagnetic theory tells us that if the propagation of the plane wave is in the z direction, $E_z = H_z = 0$ (i.e., the fields are transverse to the direction of propagation). In free space, E and H are related by a simply constant:

$$E = Z_0 H$$

where $Z_0 = \sqrt{\mu/\varepsilon}$ is the characteristic impedance of free space and it is 377 Ω, and μ and ε are, respectively, the permeability and permittivity of the propagation medium. Thus, most of the time, communications engineers will only have to work with the E or the H field in free space. Now choosing to express the electromagnetic waves in the E fields only, for a plane wave propagating in the z direction:

$$\mathbf{E} = E_x\mathbf{i} + E_y\mathbf{j} \tag{11.251}$$

and both E_x and E_y have sinusoidal variations with time. If $E_y = 0$, the wave is said to be horizontally polarized, and if $E_x = 0$, the wave is said to be vertically polarized. The communications engineer has the choice of sending the signal via horizontally polarized or vertically polarized electromagnetic waves. E_x and E_y components are orthogonal to each other and they are mutually independent of each other. Figure 11.93 illustrates the transmission of the horizontally and vertically polarized waves. Both the vertically and the horizontally polarized waves are known as linearly polarized (LP) waves. This is only one of the many possible orthogonal sets of polarized electromagnetic waves. Another very popular orthogonal set consists of circularly polarized waves (Figure 11.94). If

$$\mathbf{E} = E_0 \sin\left(\omega t + \frac{\pi}{2}\right)\mathbf{i} + E_0 \sin(\omega t)\mathbf{j} \tag{11.252}$$

where t is time and $\omega = 2\pi f$, f being the frequency, and if one looks into the direction of propagation (z direction), the resultant E field (or vector) has a constant magnitude and it rotates clockwise as it progresses down the z direction. This is known as a right-hand circularly polarized wave (RHCP). Equation (11.252) shows that an RHCP wave can be generated if both E_x and E_y are present and equal in magnitude but with E_x leading E_y by 90° in time phase. The left-hand circularly polarized wave (LHCP) is similarly generated by having E_y leading E_x by 90° in time phase, that is

$$\mathbf{E} = E_0 \sin(\omega t)\mathbf{i} + E_0\left(\omega t + \frac{\pi}{2}\right)\mathbf{j} \tag{11.253}$$

Thus, LHCP is "orthogonal" to RHCP. In the most general situation, the E_x component is not equal to the E_y component, and these two components may have a difference in time phase other than 90°. In that case, the wave will be elliptically polarized. The orientation of the major axis of the polarization ellipse (as traced out by the tip of the rotating E vector) as well as the ellipticity of the ellipse is dependent on the relative time phase and the relative magnitude of the E_x and E_y components. The axial ratio of an elliptically polarized wave is defined by the power ratio between

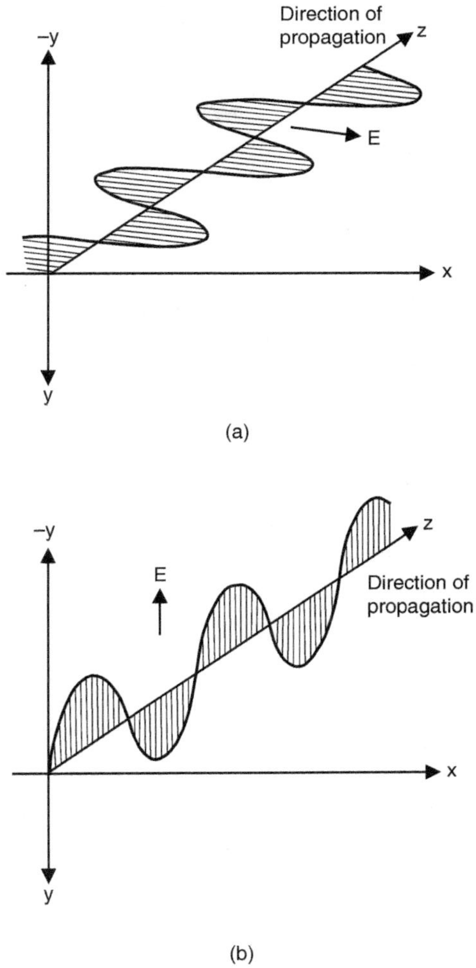

FIGURE 11.93 (a) Horizontally polarized wave; (b) vertically polarized wave.

the major and minor axes of the polarization ellipse, usually expressed in decibels. It can now be seen that linear polarization is a special case of elliptical polarization where the ellipse becomes infinitely long, and circular polarization is another special case where the ellipse becomes perfectly round.

Propagation Effects

Since the satellite orbit is above the earth's atmosphere, line-of-sight satellite communication paths must traverse the entire atmosphere at various incident angles (or elevation angles for the earth stations). At low frequencies, the ionosphere literally prevents radio waves from penetrating it. At high frequencies, the atmosphere gives rise to loss mechanisms that result in propagation fade. This is especially true in the presence of precipitation (e.g., rain) in the communications path. Figure 11.95

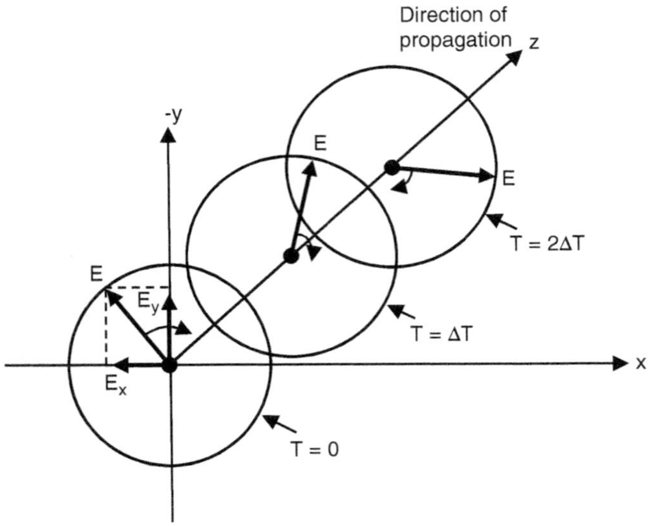

FIGURE 11.94 Right–hand circular polarization (RHCP).

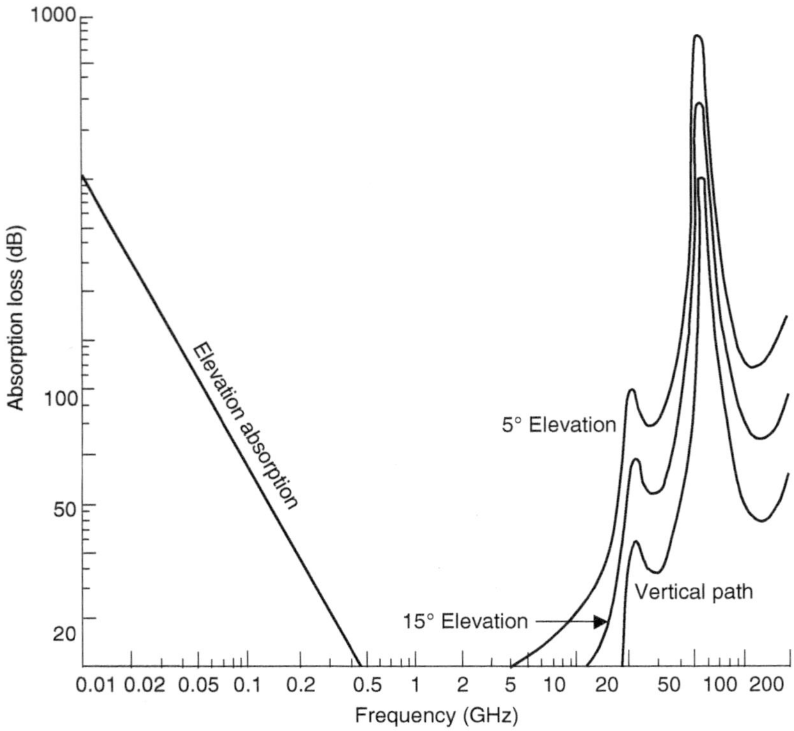

FIGURE 11.95 Absorption in the atmosphere caused by electrons, molecular oxygen, and uncondensed water vapor (Ref. 11).

shows the absorption loss of the clear sky atmosphere for radio waves. It can be seen that the most useful window of the spectrum for satellite communication is between 100 MHz and 30 GHz.

In addition to signal fades, precipitation in the atmosphere causes depolarization of an otherwise perfectly polarized electromagnetic wave. For example, a perfect LP wave may become elliptically polarized after traversing a rainstorm. Or a perfect RHCP wave may contain an LHCP component in addition to loss in signal strength after propagating through the atmosphere containing precipitation. In less likely circumstances, a sandstorm can also give rise to fade and depolarization. The ratio of the wanted polarization component to the unwanted orthogonal component is called cross-polar discrimination ratio (XPD) and is usually expressed in decibels. It will become clear later why depolarization is important when spectrum reuse via orthogonal polarization is discussed.

At frequencies below the X band, Faraday rotation, caused by the ionosphere, is another propagation phenomenon of considerable importance. While the ionosphere does not appreciably attenuate the electromagnetic wave at frequencies above 100 MHz, the *E* field of an LP wave is twisted or rotated as it passes through the ionosphere without being depolarized to any large extent.

11.50 Link Consideration

Concepts of Antenna Radiation Pattern and Antenna Gain

A microwave radiator (or antenna) does not necessarily radiate energy with equal intensity in all directions, but if it does, it is called an isotropic radiator. The gain of an isotropic radiator is defined to be unity (i.e., 0 dB). If, however, an antenna concentrates its radiated energy in a particular direction within a small solid angle, the antenna will have gain over the isotropic antenna in that direction. Figure 11.96 illustrates the polar radiation diagram of an isotropic antenna and also that of an antenna with high gain in a certain direction. The peak gain of the antenna is defined as the ratio of maximum radiated power intensity of that antenna (in a certain direction)

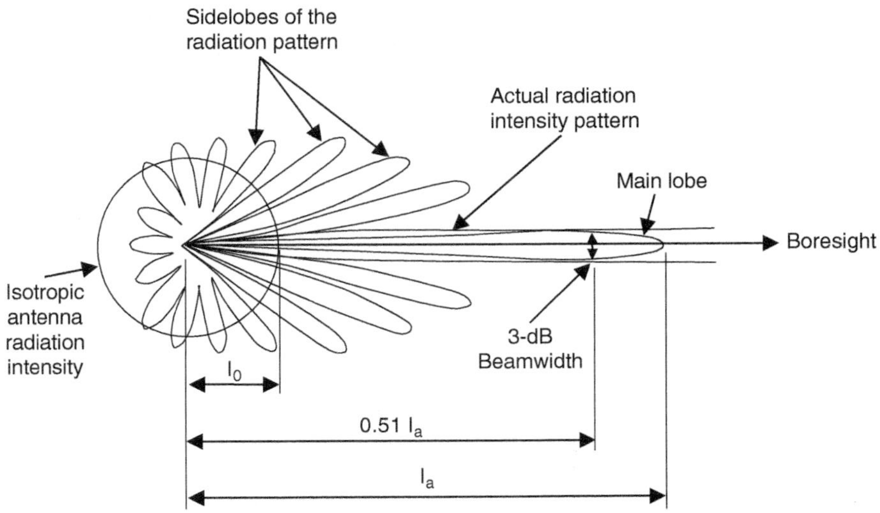

FIGURE 11.96 Polar radiation pattern of an antenna.

to the radiated power intensity of a hypothetical isotropic antenna fed with the same transmitting power, that is,

$$\text{antenna peak gain}\,|_{\text{dB}} = 10\log_{10}\frac{I_a}{I_0} \tag{11.254}$$

where I_a is the maximum radiated power intensity and I_0 is the power intensity of an isotropic antenna.

This definition of the gain of an antenna can be generalized to include the gain $G(\theta, \phi)$ of the antenna in directions other than that of maximum radiation, that is,

$$G(\theta,\phi)\,|_{\text{dB}} = 10\log_{10}\frac{I(\theta,\phi)}{I_0} \tag{11.255}$$

where θ and ϕ define the arbitrary direction in a polar (r, θ, ϕ) coordinate system and $I(\theta, \phi)$ is the radiated power intensity in the direction of (θ, ϕ). The peak gain, sometimes called the boresight gain of an antenna, is a measure of the antenna's ability to confine as much of the radiated power within as small as a solid angle as possible (i.e., the higher the gain, the smaller is the solid angle). A good way to quantify this solid angle is the 3-dB beamwidth of the antenna. The 3-dB beamwidth of the antenna is defined by the two points on either side of the boresight in the polar radiation pattern where the gains are 3 dB (i.e., half power) below the maximum boresight gain of the antenna. This is illustrated in Figure 11.96. Note that 3 dB below the maximum is equivalent to a radiation intensity where the E field is 0.707 times the maximum E field at boresight. The E field is expressed in volts per meters rather than in dBW/m^2, which is the unit used to described radiated power intensity. Also, the 3-dB beamwidth (θ_1) in the θ direction may not necessarily be the same as the 3-dB beamwidth (ϕ_1) in the ϕ direction. A useful empirical formula that relates antenna gain to the two bandwidths is

$$\text{boresight gain} = \frac{30{,}000}{\theta_1 \times \phi_1} \tag{11.256}$$

where gain here is expressed in real numbers (not in decibels) and θ_1 and ϕ_1 in degrees. Equation (11.256) represents a fairly efficient antenna (i.e., $\eta \geq 70\%$; see the definition of η in the following section).

Figure 11.96 also illustrates the concepts of the "main beam" and "sidelobes" of an antenna. The main beam of an antenna is the main lobe of the radiation pattern centered around the boresight of the antenna. For example, the solid angle bounded by the 3 dB points can be considered the main beam of the antenna. Fundamental antenna theory shows that the peak gain (or boresight gain) of a microwave antenna can be predicted from the aperture area A of the antenna. Here A is the projected area of the antenna reflector in a plane perpendicular to the direction of maximum radiation (i.e., the direction of the boresight).

An ideal antenna would have the following relationships:

$$G_{\max} = \frac{4\pi A}{\lambda^2} \tag{11.257a}$$

where λ is the wavelength and $G_{\max}$ is the maximum boresight gain, and for a circular aperture:

$$G_{\max} = \left(\frac{\pi D}{\lambda}\right)^2 \tag{11.257b}$$

where D is the diameter of the antenna. G_{max} can, of course, be expressed in decibels above isotropic [if ten times the log to the base 10 of the right-hand side of equation (11.257a) or (11.257b) is taken]. For a practical antenna, the boresight gain or the maximum gain will be less than that shown in equation (11.257b). The boresight gain G_0 can be expressed as

$$G_0 = \frac{4\pi A}{\lambda^2}\eta = \frac{4\pi A_{eff}}{\lambda^2} \tag{11.258}$$

where η is the efficiency factor and A_{eff} is the effective area of the antenna. Usually, η is somewhere between 40% and 85% for a typical well-designed microwave antenna.

Power Transfer from a Transmitting Antenna to a Receiving Antenna

To assess the quality of the signal, the communications engineer must calculate the signal power that is being transferred from a transmitting antenna to a receiving antenna. Let G_T and G_R be the boresight gains of the transmitting and receiving antennas, respectively. Assume that the transmitting and receiving antennas are properly aligned such that their boresights are collinear (i.e., they are looking at each other on-axis), as shown in Figure 11.97.

Let P_T be the total power radiated by the transmit antenna. If the transmit antenna were isotropic, the PFD ϕ at a distance r from the transmit antenna would be

$$\phi = \frac{P_T}{4\pi r^2} \tag{11.259}$$

Since the transmit antenna has a gain of G_T over the isotropic antenna, the PFD at the receiving antenna is actually

$$\phi = \frac{P_T G_T}{4\pi r^2}$$

From equation (11.258), the effective area A_R of the receiving antenna with an antenna gain of G_R is

$$A_R = \frac{\lambda^2 G_R}{4\pi}$$

Since A_R is the capture area of the receive antenna, the power P_R received by the receiving antenna is ϕA_R, that is,

$$P_R = \frac{P_T G_T G_R \lambda^2}{(4\pi r)^2} \tag{11.260}$$

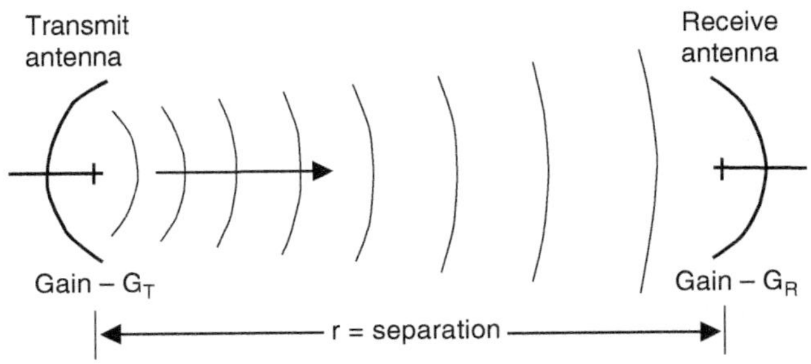

FIGURE 11.97 Power transfer from transmit antenna to receive antenna.

Equation (11.260) is the fundamental equation governing power transfer between two antennas. The factor $(4\pi r/\lambda)^2$ is commonly called "path loss," P_L, by systems engineers. Although the well-known inverse-square law is frequency independent, this definition of path loss is frequency dependent, that is,

$$P_L = \left(\frac{4\pi r}{\lambda}\right)^2 \tag{11.261}$$

Thus, if equation (11.260) is written in the logarithmic form in decibels, then

$$\frac{P_R}{P_T}\bigg|_{\text{dB}} = G_T\big|_{\text{dB}} + G_R\big|_{\text{dB}} - P_L\big|_{\text{dB}} \tag{11.262}$$

Example 11.1 The earth station antenna has an on-axis gain of 42.5 dB and the satellite antenna has a gain of 30 dB. The separation between the geosynchronous satellite and the earth station is 39,500 km. If the satellite delivers 10 W of signal power at 4 GHz, what is the received signal level at the earth station when atmospheric losses are neglected?

Solution At 4 GHz, the wavelength:

$$\lambda = \frac{3\times 10^8}{4\times 10^9}\,\text{m} = 0.075\,\text{m}$$

Path loss P_L is

$$P_L\big|_{dB} = 20\log_{10}\frac{4\pi\times 3.95\times 10^7}{0.075} = 196.4\,\text{dB}$$

Transmitted power is 10 W or 10 dBW. Then according to equation (11.262), the received power P_R is

$$\begin{aligned} P_R &= 10 + 42.5 + 30 - 196.4\ \text{dBW} \\ &= -113.9\ \text{dBW or} - 83.9\ \text{dBm} \end{aligned}$$

The received signal power is approximately 4 pW. This example is fairly typical of a domestic satellite downlink for TV-receive-only earth stations using a 4.5-m receiving antenna.

Equivalent Isotropically Radiated Power

Systems engineers find it convenient to express the product of the transmit antenna gain and the total radiated power and call it equivalent isotropically radiated power (EIRP), that is, power that appears to have radiated from an isotropic antenna. This is a very common term in satellite communications and is given by

$$\text{EIRP}\big|_{\text{dBW}} = \text{antenn again}\big|_{\text{dB}} + P_T\big|_{\text{dBW}} \tag{11.263}$$

Thus, in the numerical example above, the satellite EIRP is 10 + 30 = 40 dBW. Once the EIRP of a satellite transmitter is known, it is a simple matter to calculate the PFD at the surface of the earth. For geosynchronous satellites, the PFD in dBW/m² at the earth's surface can be written as

$$\text{PFD} = \text{EIRP of satellite}\big|_{\text{dBW}} - 162\ \text{dB} \tag{11.264}$$

at the subsatellite point, if atmospheric losses are neglected.

Concepts of Thermal Noise, G/T, and C/N

Noise could potentially corrupt the transmitted signal in a communication link. It must be emphasized at the outset that thermal noise is only one of the chief contributors to satellite link degradation. There are other contributors, such as distortion in the transmission equipment, nonlinearities in active devices, and interference from other transmissions. However, thermal noise has been shown to be one of the major contributors to link degradation in satellites communications. The name "thermal noise" suggests that the noise is of thermal origin. Indeed, the origin of thermal noise stems from random motions of electrons within the devices in the transmission path. A quantum mechanical model leads to the following equation for the thermal noise power density P_N in watts per hertz bandwidth (Figure 11.98):

$$P_N(f) = \frac{hf/2}{e^{hf/(kT)} - 1} \text{ W/Hz} \tag{11.265}$$

where f is the frequency in hertz, h is Planck's constant $= 6.62 \times 10^{-34}$ J·s, k is the Boltzmann constant $= 1.38 \times 10^{-23}$ J/K, and T is the absolute temperature of the device or the noise source in kelvins. Equation (11.265) gives the available noise power and this power is the actual noise contribution if the source and load are conjugate matched (i.e., maximum transfer of power from source to load). Also, at microwave frequencies and at moderate temperatures, $hf \ll kT$. Equation (11.265) reduces to

$$P_N(f) = \frac{kT}{2} \text{ W/Hz} \tag{11.266}$$

and if the communication system does not distinguish between negative and positive frequencies (as is the case):

$$P_N(f) = kT \text{ W/Hz}$$

For a finite filter bandwidth or receiver bandwidth of B hertz, the total noise power P_N received is

$$P_N = kTB \tag{11.267}$$

Equation (11.267) is the fundamental approximate noise equation used by communications engineers to calculate the noise power in a certain communication channel. It is emphasized here that equation (11.267) cannot be used to calculate noise power in optical links. Thermal noise is of secondary importance in optical devices, where quantum noise and shot noise dominate.

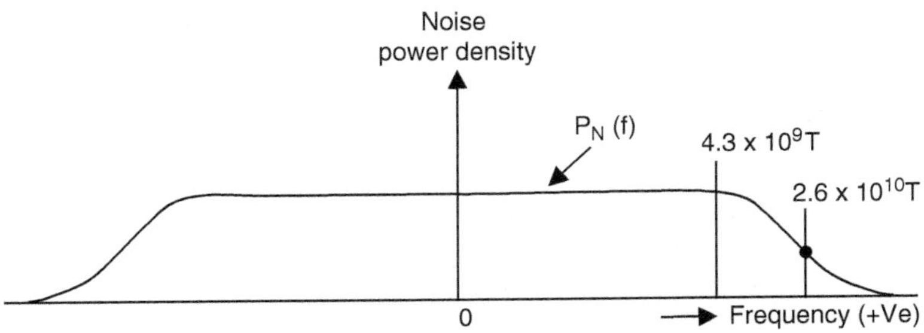

FIGURE 11.98 Thermal noise power density versus frequency.

If thermal noise produced by different devices is uncorrelated, they can be summed on the basis of a simple power addition. For example, in a receiving system, the effective antenna thermal noise can be added directly to the effective receiver amplifier thermal noise. It is therefore convenient to develop the concept of effective thermal noise temperature T_e of microwave devices.

Let us assume the antenna effective noise temperature is 60 K and that T_s of the receiver is 100 K, the total noise power P_N in a bandwidth of 30 MHz is

$$P_N = 1.38 \times 10^{-23} \times (60 + 100) \times 30 \times 10^6 \text{ W}$$
$$\text{i.e. } P_N = -228.6 + 22 + 74.8 \text{ dBW}$$
$$\text{or } P_N = -131.8 \text{ dBW where } P_N = N$$

The carrier level C was determined in the example to be −113.9 dBW; then the carrier-to-noise ratio C/N is

$$\frac{C}{N} = -113.9 - (-131.8) = 17.9 \text{ dB}$$

This is a realistic C/N ratio for a TV signal delivered by a typical C-band satellite and received on the 4.5-m antenna and amplified by a good solid-state microwave receiver.

Finally, readers should observe that all the discussions above on link considerations apply equally well for any pair of transmitting and receiving microwave stations, satellite-earth stations or terrestrial microwave repeater stations.

11.51 Communications Subsystem of a Communications Satellite

Having discussed the fundamental concepts and relationships between the various quantities used to characterize a microwave link, this subsection is devoted to the design and design concepts of the communications system of a typical communication satellite.

Spacecraft Antennas and the Reuse of the Allocated Spectrum (Bandwidth)

There are certain significant differences between the design of a satellite microwave antenna and the design of an earth station microwave antenna. The satellite antenna must be light and survive in the space environment, where the temperature extremes are usually encountered; whereas the weight of an earth station antenna is not a significant factor and the earth station survives in a completely different environment. The satellite antenna is usually designed to provide communications "coverage" over an area or a certain landmass, but the earth station antennas in most instances need to "cover" only a single point (i.e., the satellite in the sky). The satellite antenna designer is very often constrained by fairing envelopes of the launch vehicles and sometimes may have to resort to stowed and deployable designs to meet the various launch vehicle constraints. However, the earth station antenna designer is very much concerned with the sidelobe levels (see Figure 11.96) because of the fear of adjacent satellite interference. Until recently, satellite antenna designers have concentrated only on the antenna (gain) performance within the required main beam coverage area. Now, spectrum reuse techniques via spatial beam isolation have forced satellite antenna designers to come up with low-sidelobe designs as well. The technique of spectrum reuse via spatial beam isolation is illustrated in Figure 11.99.

Figure 11.99(a) shows a satellite antenna designed to cover the entire visible surface of the earth all at once. This allows the allocated spectrum (by ITU) to be used only twice (i.e., via two orthogonal polarizations). If the required coverage can be split into subareas, a multibeam antenna can be

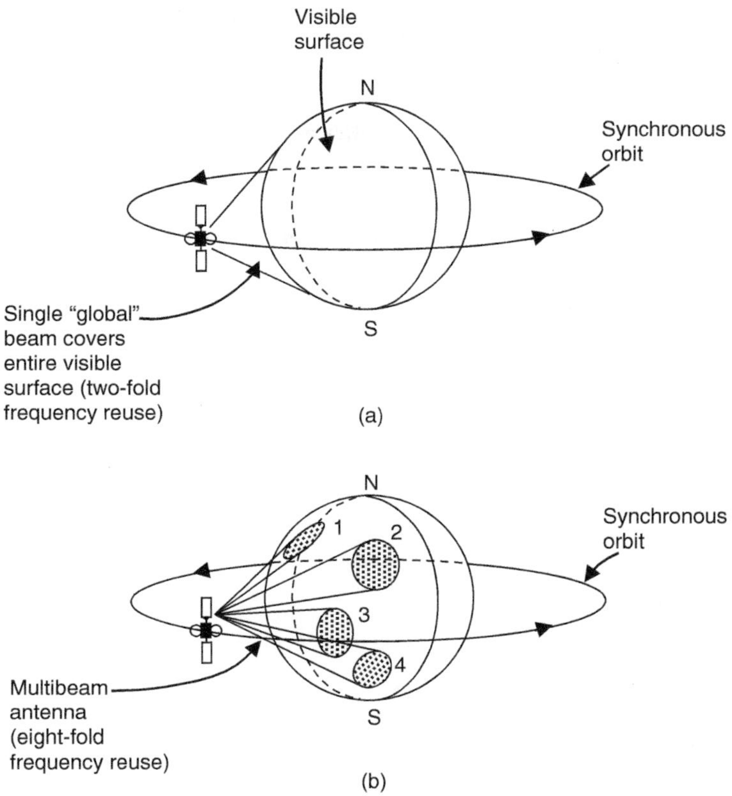

FIGURE 11.99 (a) Global beam coverage; (b) spot beam coverages.

designed so that the allocated spectrum is reused a number of times. In Figure 11.99(b), the spectrum is reused eight times (four beams each with two orthogonal polarizations), assuming, of course, that sufficient mutual isolations between the four antenna beams are kept low enough such that effectively each of the main beams of the antenna covers each of the service areas and practically nowhere else.

Basic Types of Microwave Antennas for Satellites

As indicated earlier, the electromagnetic spectrum available for satellite communication stretches from about 100 MHz to about 30 GHz. Such a wide range of frequencies certainly offers the possibility of many different design approaches for the antenna designers.

Below 1 GHz, the antennas could be of dish (reflector) type or it could be the wire type most often seen on rooftops of a lot of houses. Because Faraday rotation of the LP E field is significant below 4 GHz, communications engineers normally use circularly polarized waves rather than LP waves for these frequencies. A very popular antenna that generates circularly polarized waves is the helical antenna shown in Figure 11.100.

Above 1 GHz, the most popular antenna designs are the parabolic-dish-type reflector antennas. These designs offer efficient and mechanically simple solutions to the antenna problem. Figure 11.101 shows that these designs are derived from the well-known designs of optical

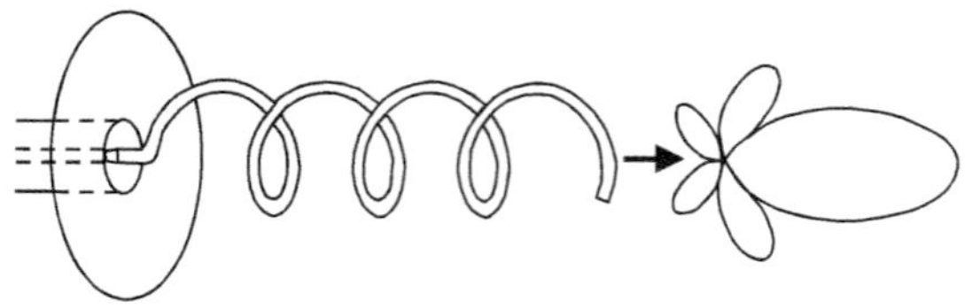

FIGURE 11.100 Helical antenna and its radiation pattern.

telescopes. For the purpose of eliminating blockage by the antenna feed itself, offset-fed designs have become the mainstay of satellite antennas. In particular, the simple offset-fed paraboloid antenna has become the most popular design for microwave satellite antennas. This simple design avoids feed blockage and does not require the additional subreflector that is present in some other designs. In contrast, earth station antennas up until now have mostly adopted the symmetrical Cassegrain design or the symmetrical Gregorian design. This divergence in design approaches for satellite and earth station antennas is largely due to (1) the large difference in the antenna sizes (i.e., earth station antennas are typically many times bigger than the satellite antenna), and (2) the simplicity that is required in the satellite antenna in order to survive the launch, the deployment in orbit, and the space environment during the lifetime of the spacecraft.

It can be observed that in all the designs shown in Figure 11.101, the reflector (or dish) is actually fed by a primary source antenna. The primary source antenna is in fact another antenna in itself. This leads to the discussion of some simple and popular horn antenna designs used for feeding dishes or sometimes used directly as the radiating device. Figure 11.102 shows four common types of horn antennas.

While the reflector antennas offer relatively high antenna gains efficiently, the horn antenna is a simple way of achieving low-gain radiation. Typically, horn antenna 3-dB beamwidths vary from 10° to about 120°, whereas reflector antennas are seldom used if the 3-dB beamwidth is greater than about 5–10°. Reflector antennas are used mostly when the 3-dB beamwidths are in the order of 5° or less. Microwave energy above 1 GHz is very efficiently "piped" from one point to another via waveguides or hollow metallic tubes of either rectangular or circular cross section. These waveguides are the high-frequency equivalents of a twisted pair of wires or coaxial cables commonly used for lower frequencies. A flared section of the simple rectangular or circular waveguide forms a horn radiator. A horn antenna can be visualized as a transition from the guided mode of propagation of the electromagnetic energy within the "pipe" to the free-space unimpeded mode of propagation of the electromagnetic energy.

Satellite Transponder Design

Having discussed the basic design concepts for communication satellite antennas, it is now appropriate to turn to the microwave repeater or transponder that processes the signals received by the receive antenna and transmit the processed signal via the transmit antenna. Figure 11.93 gave a simple diagram of the communication payload. The transponder portion of that block diagram is now expanded in Figure 11.103 to give more details. One of the main functions of the satellite transponders is to transpose the uplink frequency downlink frequency. This function is accomplished in the frequency translator or mixer, item 3 in Figure 11.103. A conventional C-band communications satellite would have the uplink in the band 5.925–6.425 GHz and the downlink in the band 3.7–4.2 GHz. Thus, a stable on-board local oscillator must provide the difference frequency of 2.225 GHz to downconvert the uplink frequencies to downlink frequencies. The example in Figure 11.103 shows a "single"-conversion system; that is, the conversion from uplink frequencies to downlink frequencies is achieved in one step.

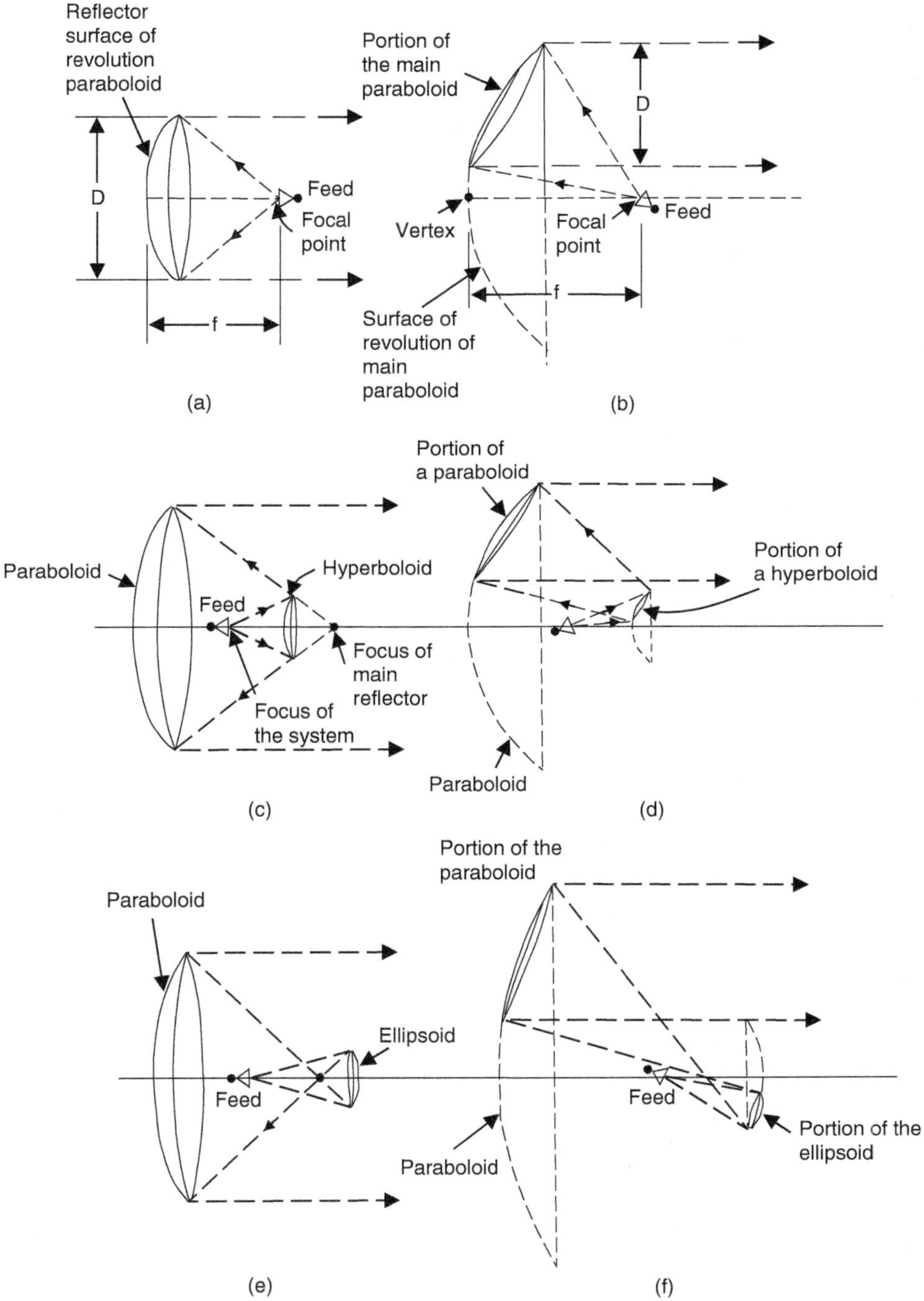

FIGURE 11.101 (a) Focal–fed symmetrical parabolic antenna; (b) offset–fed paraboloid (no feed blockage); (c) center–fed symmetrical Cassegrain antenna; (d) offset–fed Cassegrain antenna; (e) center–fed symmetrical Gregorian antenna; (f) offset–fed Gregorian antenna.

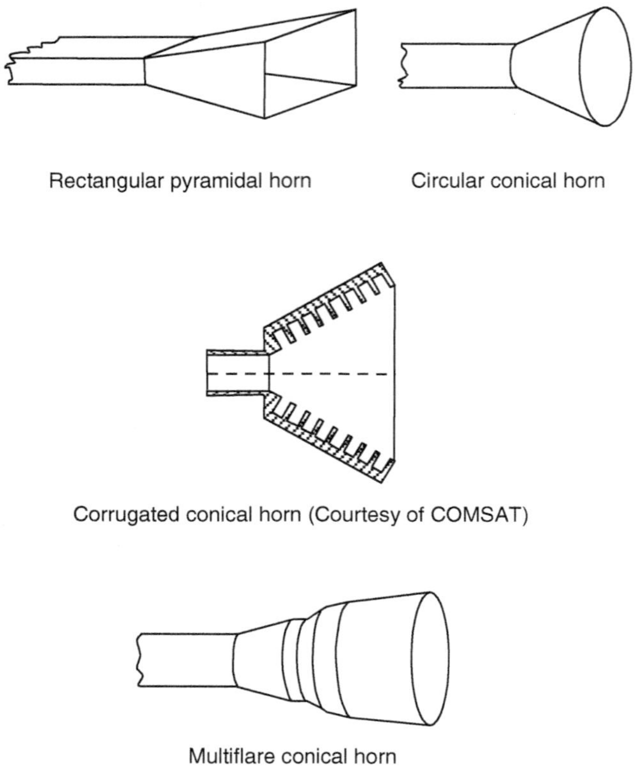

FIGURE 11.102 Feed Horn Antennas.

Following the flow of the microwave signal through the communications subsystem, the bandpass filter (item 1) preceding the receiver (item 2) passes only the uplink signals received by the antenna and rejects out-of-band signals that may cause degradation in the receiver. The receiver amplifies the very weak signals to a certain convenient level so that downconversion can be affected without much degradation to the signal. Next, another amplifier (item 4) boosts the signal to a higher level before the various RF channels are demultiplexed into individual channels in the input demultiplexer, often referred to as the "input mux." More will be said later concerning the reasons behind this arrangement of having individual RF channels. Each individual channel is now amplified by a channel driver (item 6) and a high-powered transmitter (item 7). Sometimes an automatic level control circuit (ALC) is incorporated into the channel driver so that a constant signal level is presented to the input of the high-powered transmitter.

Two types of high-powered transmitters are commonly used, the traveling-wave-tube amplifier (TWTA) and the solid state power amplifier (SSPA). The use of SSPAs in space is a recent development. Most C-band SSPAs are of the type that uses gallium arsenide field-effect transistors. SSPAs promise good linearity, better reliability, and lower mass. Although the situation may change, TWTAs are still the most common high-powered transmitters used in present-day communications satellites. This is due mainly to the high RF power-to-dc power efficiency (≈40%) offered by the TWTAs and also due to the accumulated knowledge of the behavior of space TWTAs.

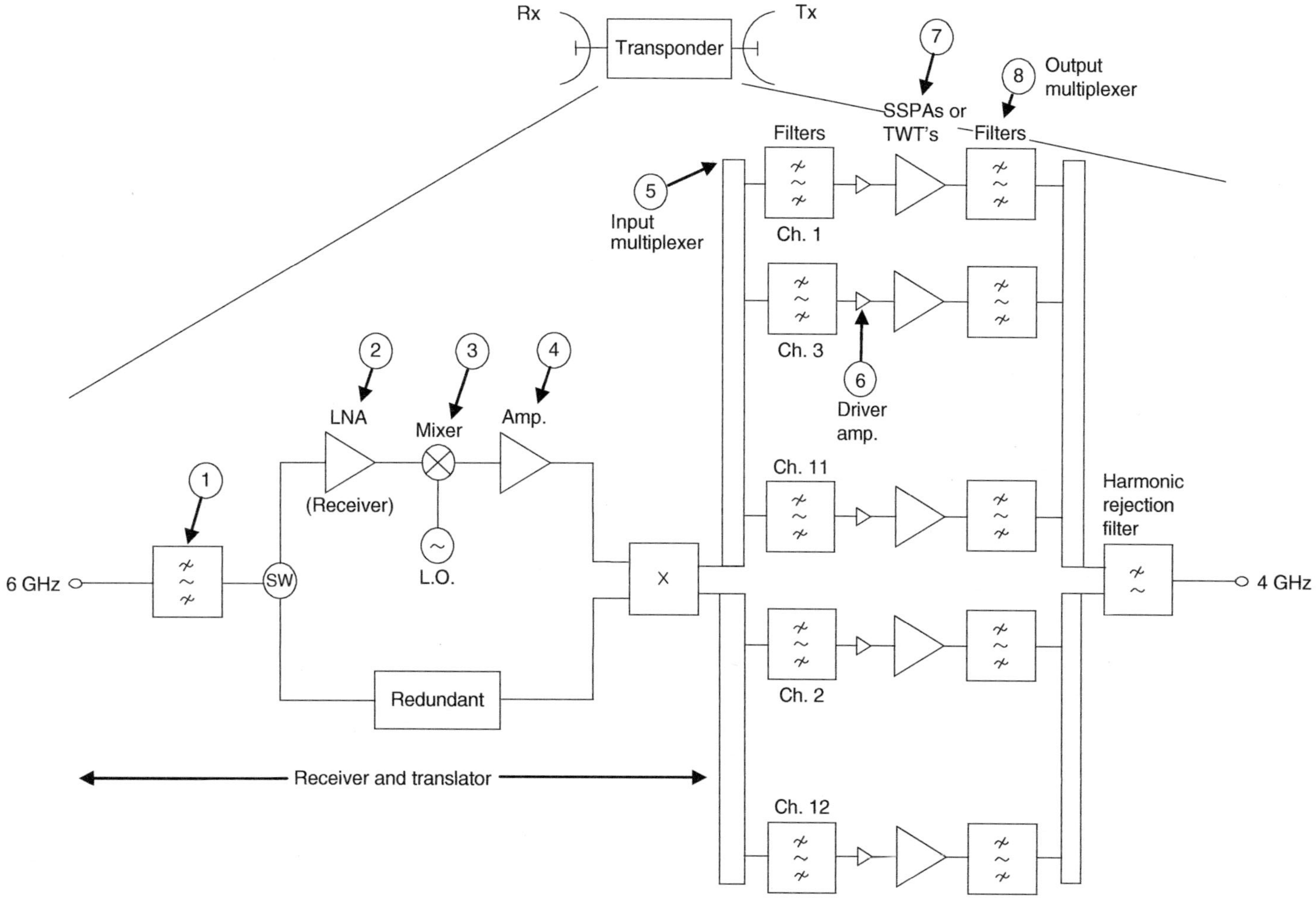

Figure 11.103 Typical satellite transponder.

The output signals at the outputs of all the high-powered transmitters are combined together in the output multiplexer (item 8, output mux) so that a single transmit antenna now transmits all the signals back to the earth.

11.52 Some Common Modulation and Access Techniques for Satellite Communications

As shown in the following equation, a sinusoidal microwave signal $\upsilon(t)$ cannot transfer information unless some of its parameters are varied. A sinusoidal signal can be expressed as

$$\upsilon(t) = A \sin(\omega t + \phi) \tag{11.268}$$

where A is the amplitude of the signal, $\omega = 2\pi f$ is the angular frequency, f is the frequency in hertz, t is time, and ϕ is the time phase of the sinusoid with respect to some reference. There are three quantities associated with the sinusoidal wave: the amplitude A, the frequency f (or ω), and the phase ϕ. Any of the three quantities can be varied or modulated by the source information so that the microwave sinusoidal "carrier" signal can now convey information.

If A is modulated by the source information, the modulation process is called amplitude modulation (AM). If the frequency f (or ω) is modulated, the modulation process is called frequency modulation (FM), and if the phase ϕ is modulated it is called phase modulation. AM is not commonly used for satellite communications. However, FM has been extensively used on early satellites. So we will discuss FM in more details.

In FM technique, it was pointed out that the instantaneous frequency ω of an RF carrier is the derivative of the phase θ of the RF carrier, that is,

$$\omega = \frac{d\theta}{dt} \tag{11.269}$$

If the modulating signal or source signal consists also of a sinusoidal waveform $\sin \omega_m t$ and if k is the peak angular frequency deviation in the FM process, then

$$\frac{d\theta}{dt} = \omega_c - k \sin \omega_m t \tag{11.270}$$

where ω_c is the steady unmodulated carrier frequency of the RF. Integrating both sides with respect to t gives

$$\theta = \omega_c t + \frac{k}{\omega_m} \cos \omega_m t \tag{11.271}$$

The use of $-k \sin \omega_m t$ rather than $k \cos \omega_m t$ to represent FM is a matter of convenience. Thus the resultant RF waveform $\upsilon(t)$ is

$$\begin{aligned}\upsilon(t) &= A \cos\theta = A \cos\left(\omega_c t + \frac{k}{\omega_m} \cos \omega_m t\right) \\ &= A \cos(\omega_c t + X \cos \omega_m t)\end{aligned} \tag{11.272}$$

where $X = k/\omega_m$ is modulation index.

Carson's rule for bandwidth for FM signal is

$$B = 2(f_k + f_m) \tag{11.273}$$

where $k = 2\pi f_k$ and $2\pi f_m = \omega_m$, by definition. Equation (11.273) is the well-known Carson's rule bandwidth for FM signals, where f_k is the peak frequency deviation in hertz and f_m is the highest modulating source frequency, also in hertz. Since k (or f_k) is a free parameter, link performance can usually be improved by increasing k (or f_k) at the expense of using up bandwidth. Thus a trade exists between bandwidth and available RF power. This can be seen in Figure 11.104, which shows the relationship between predetection signal C/N and postdetection basebound signal S/N for a typical FM modulator.

It will be noticed that the trade-off between bandwidth and available RF carrier power exists only as long as the carrier to noise ratio (C/N) at the input to the FM modulator is above a certain threshold, as shown in Figure 11.104. Once the input C/N drops below the threshold, the output signal to noise ratio S/N drops very rapidly. Most communication links attempt to maintain the input C/N several decibels above the threshold to ensure high-quality transmission.

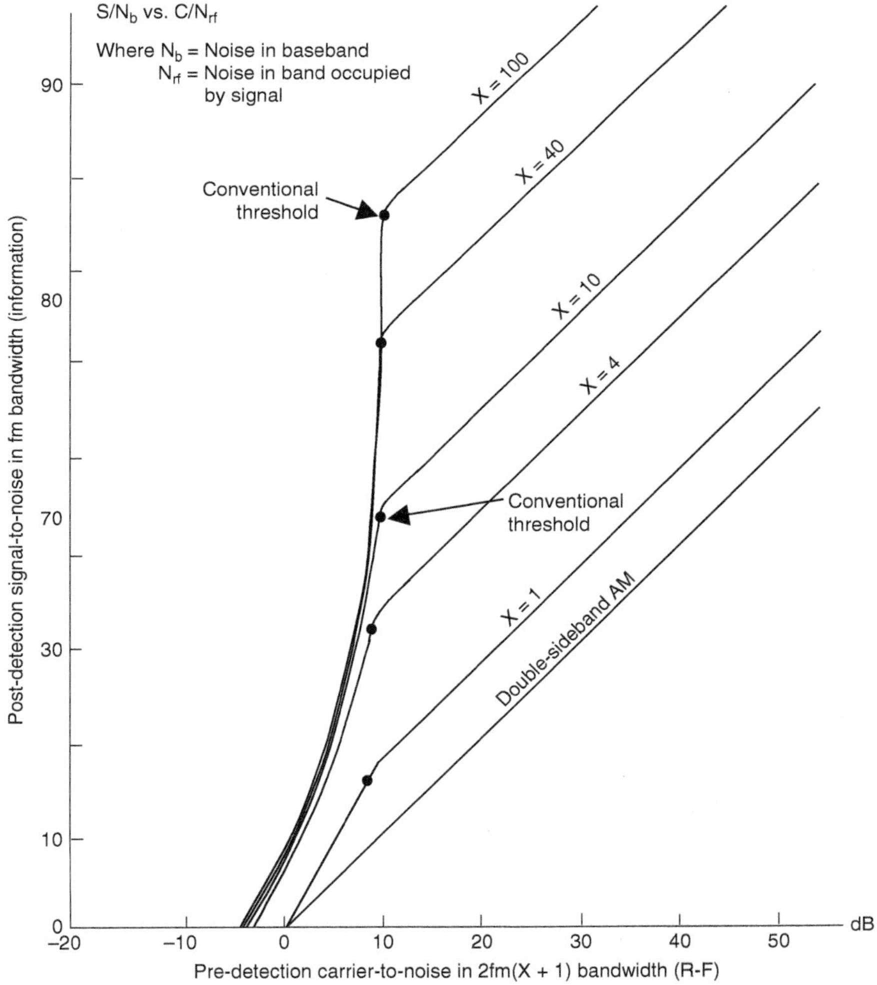

FIGURE 11.104 Signal-to-noise ratio in $2f_m(X + 1)$ bandwidth.

To conclude the discussion on AM and FM techniques, it can be said that AM technique is usually unsuitable for satellite transmission. However, Carson bandwidth limited FM technique will result in an almost constant signal level and, thus, amplitude linearity is not critical. Furthermore, FM offers the advantage of a trade between bandwidth and carrier power. Until a few years ago, FM had been the principal modulation technique used in satellite communications.

Digital Modulation

Due to the widespread introduction of digital systems in terrestrial networks, digital modulation techniques have become rather common for satellite communications. The rest of this subsection deals with one of the most common digital modulation techniques, the quaternary-phase phase-shift-keyed system (QPSK), also known as 4ϕ-PSK. Before QPSK modulation can be discussed, it is necessary to introduce the binary digital waveform, called the nonreturn to zero (NRZ) waveform, and discuss how analog signals can be converted to this NRZ format.

Digital Encoding of Analog Signals

Pulse Code Modulation (PCM)

This is a relatively well-established and common technique developed primarily for telephony. The idea is rather simple. If an analog voltage waveform generated by the microphone in the telephone handset could be sampled periodically, the voltages of these samples could be encoded in a digital binary word of a certain length. Figure 11.105 illustrates this technique.

In the particular example shown in Figure 11.105, each voltage sample is encoded by a 4-bit binary word of ones and zeros. The 4-bit word would allow $2^4-1 = 15$ possible discrete voltage levels in the encoding process. The encoder must choose between one word and another if the sampled voltage is not exactly the same as one of the selected voltage levels. All voltages in between two fixed levels will be encoded by the same digital word. Thus, there will be quantization error in this process of encoding. At the receiving end, the electronic circuitry generates the voltage levels according to the information in the 4-bit words, and with the help of a bandpass filter, the original analog waveform is more-or-less reproduced to within the quantization errors introduced. In addition to the number of levels in the encoding process, the frequency of sampling plays an important role. Nyquist (1924) showed that for good reproduction of a waveform that contains frequencies up to f_{max}, the sampling rate must at least twice f_{max}. For example, a telephone voice circuit contains

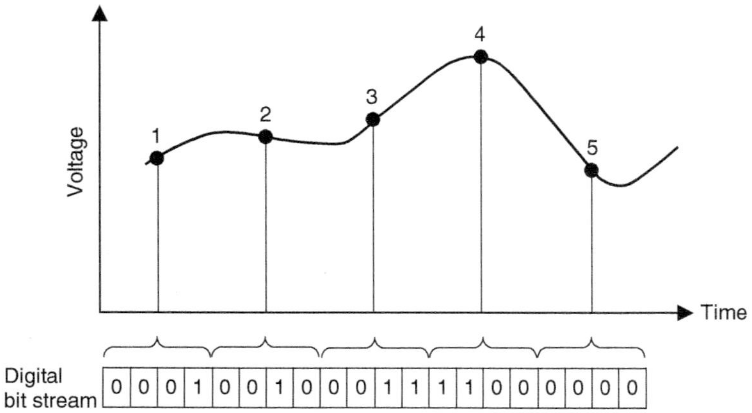

FIGURE 11.105 PCM encoding with a 4-bit word for each sample.

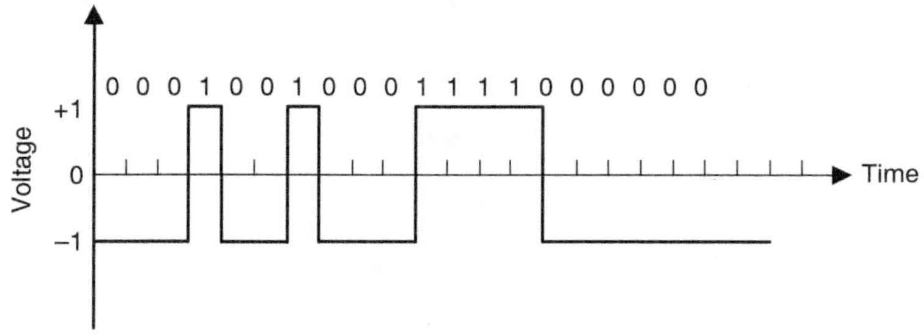

FIGURE 11.106 NRZ waveform for the digital bit stream.

frequencies up to about 3,400 Hz; thus, 8 kHz is a good sampling frequency according to Nyquist. Studies have also shown that an 8-bit word for encoding each voltage sample ensures good reproduction for the waveform (i.e., quite acceptable quality to the human ear). A conventional PCM encoding system would, therefore, require $8 \times 10^3 \times 8 = 64$ kbits/s for each voice circuit. This is indeed the bit rate adopted by most countries as the standard bit rate for one PCM voice circuit. Thus the PCM technique converts the analog waveform into a series of ones and zeros. If one is represented by +1 V and zero is represented by −1 V, the resulting digital waveform is called a NRZ waveform (Figure 11.106).

Delta Modulation (Δ-mod)

This is another encoding technique whereby an analog waveform is turned into a digital bit stream of ones and zeros. Instead of encoding the magnitude of each sampled voltage, sampling is done much more frequently and the encoder stores the previous sample and compares it with the current sample. The encoder sends out a one if the current sample exceeds or equals the previous sample and sends out a zero if the current sample is less than the previous sample.

Differential PCM (DPCM)

To combine the virtues of the techniques of PCM and Δ-mod, there exists yet another encoding technique, called DPCM. In this case, the current sample is compared with the previous sample and only the difference between the two samples is encoded.

Modulation

Having discussed digital encoding, now we discuss on how we use digital modulation.

A Fourier analysis of the time-domain NRZ waveform (Figure 11.106) would yield a $(\sin x/x)^2$ spectrum in the frequency domain as shown in Figure 11.107. The spectrum is continuous and without discrete components if the digital sequence is truly random (i.e., there are no correlated bits in the NRZ bit sequence). The NRZ spectrum is also infinite, although the sidebands would become insignificantly small as the frequency becomes infinite.

One technique is amplitude shift keying. In this technique, as shown in Figure 11.108, amplitude shifts from bit 1 to 0. Second technique is called frequency shift keying. In this technique, as shown in Figure 11.109, frequency shifts from bit 1 to 0. Third technique is called phase shift keying. In this technique, as shown in Figure 11.110, phase shifts from bit 1 to 0. Phase shift keying is quite popular.

QPSK Modulation

In this technique, four possible phases of the RF carrier are formed, resulting in transmission of two bits/symbol.

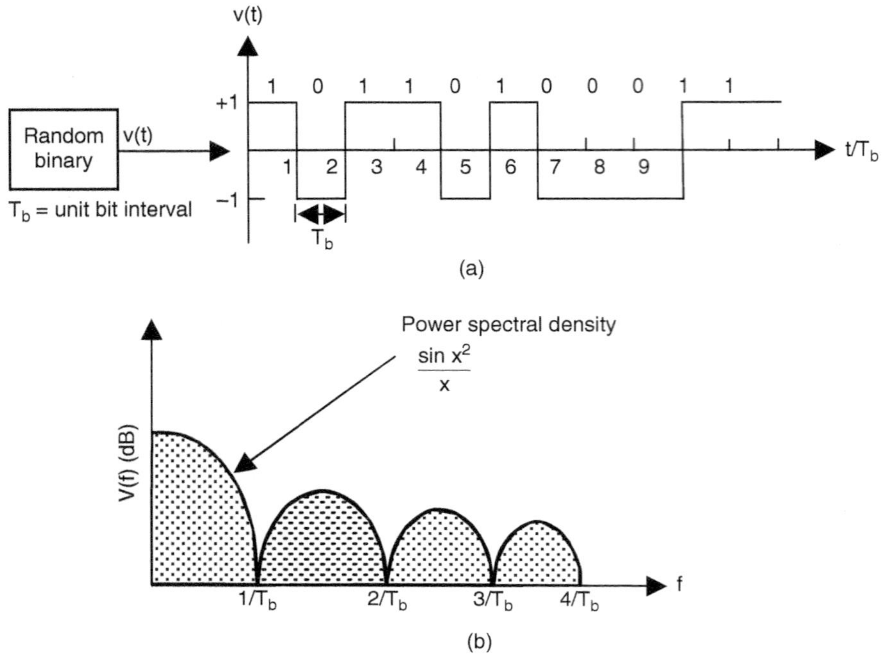

FIGURE 11.107 Power spectrum of an NEZ waveform: (a) time-domain representation; (b) frequency-domain representation.

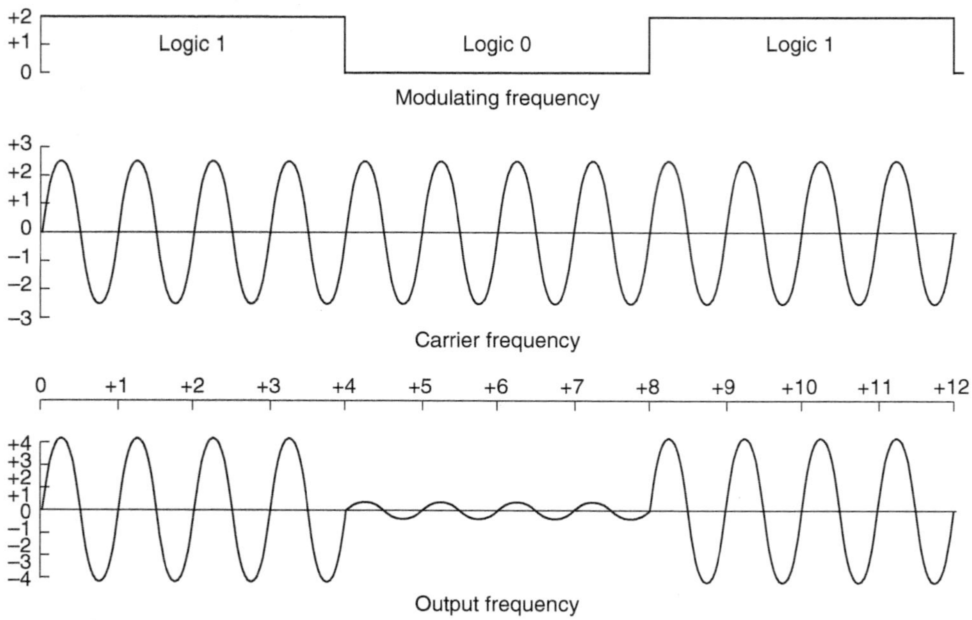

FIGURE 11.108 Amplitude Shift Modulation.

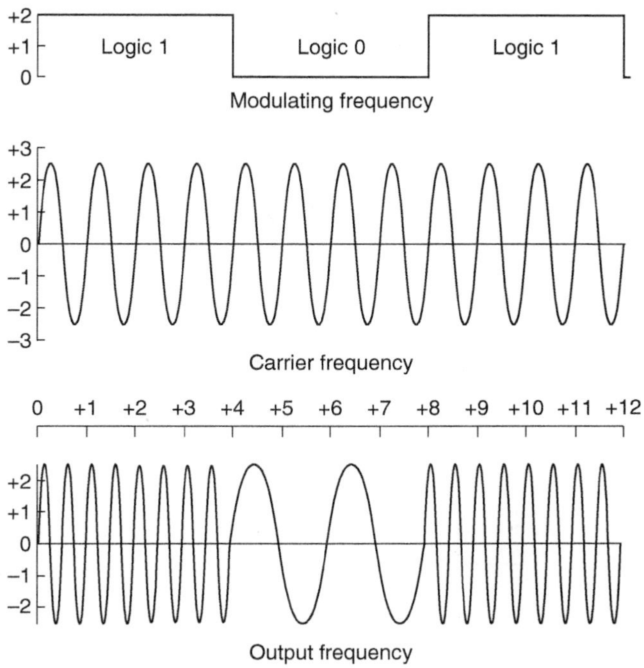

FIGURE 11.109 Frequency Shift Keying.

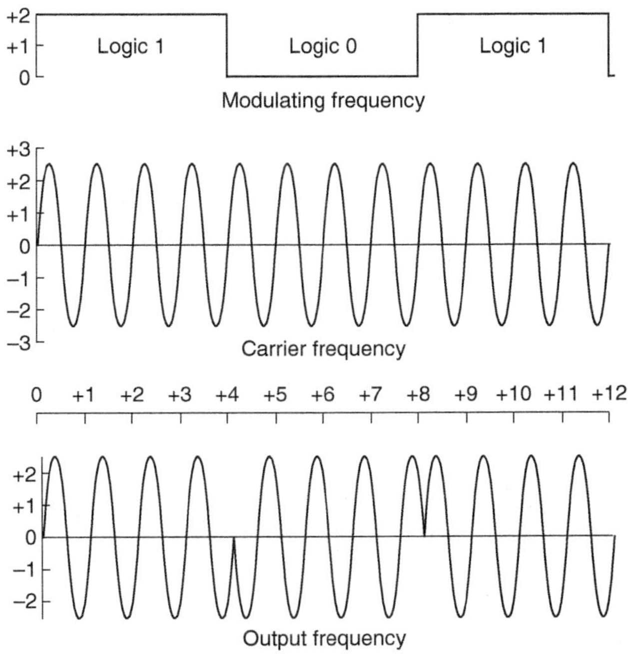

FIGURE 11.110 Phase Shift Keying.

As the name implies, QPSK is in reality the superposition of two biphase PSK signals in time quadrature. Figure 11.111 shows the vector diagram of an RF carrier that is four-phase-PSK-modulated by two independent uncorrelated but synchronous NRZ inputs, commonly called the I and the Q channels (in-phase and quadrature channels). The I-channel NRZ biphase modulates the in-phase component of the RF carrier and the Q-channel NRZ independently biphase modulates the quadrature component of the RF carriers. These two signals are superimposed on each other to form the four-phase PSK-modulated RF carrier shown in Figure 11.111(c). The phase of the RF carrier changes at $T, 2T, 3T, \ldots$, and can assume only one of the four phases shown in Figure 11.111(c).

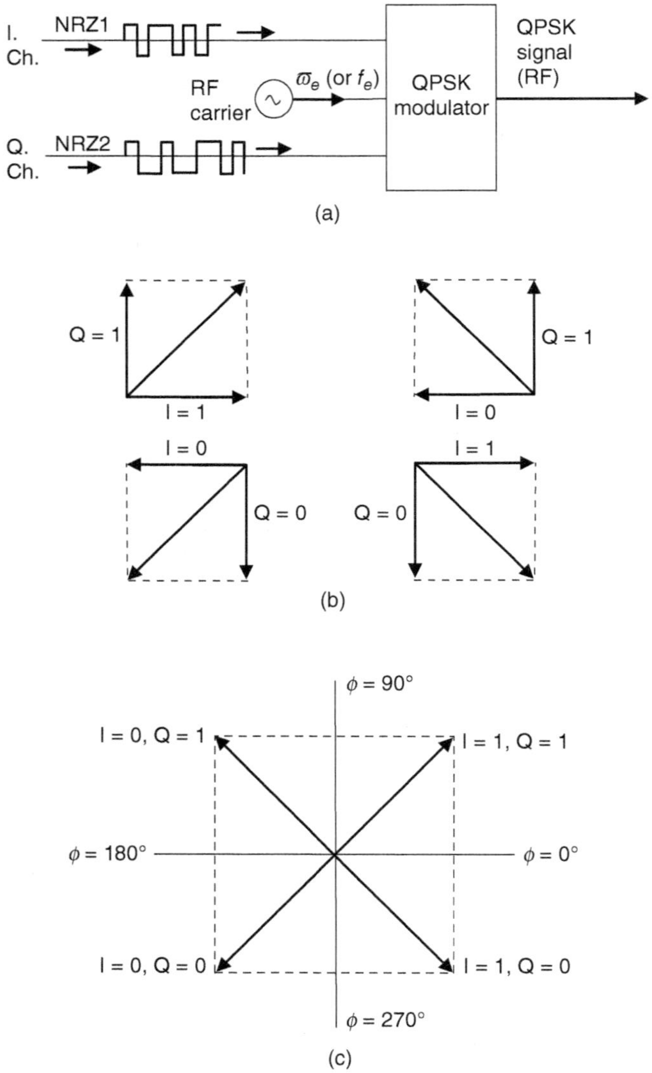

FIGURE 11.111 (a) Simplified QPSK modulator; (b) formation of the four possible phases of the RF carrier; (c) four phases of the QPSK modulated RF carrier, I & Q channels superimposed.

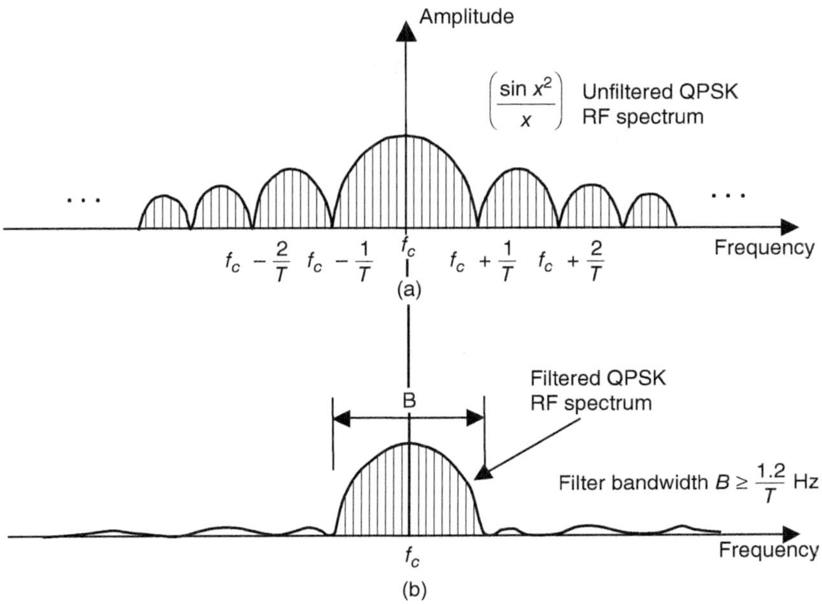

Figure 11.112 Unfiltered and filtered spectrum of a QPSK signal.

It can be shown that the RF spectrum of a QPSK modulated signal retains the shape around the carrier frequency but with no carrier component. This is shown in Figure 11.112(a). As in the case of FM modulation, sideband truncation is needed in practice to limit the bandwidth of the signal to a finite value. This is shown in Figure 11.112(b). Not too much degradation is incurred if the bandwidth B of the filter is greater than or equal to $1/T \times 1.2$, and contains most of the main lobe of the spectrum that lies between $f_c + 1/T$ and $f_c - 1/T$.

Thermal noise and other interference in the link will cause the carrier phase to jitter, and eventually at some instance there will be enough phase shift ($\geq \pi/2$) to cause an error. Theoretical bit error rate (BER) versus link C/N performance is shown in Figure 11.113. It can be seen that at BER $\leq 10^{-4}$, the BER improves at the approximate rate of one decade per decibel in C/N improvement. A BER worse than 10^{-4} is considered an outage in most links. This is similar to the FM threshold in an FM link. It should be noted the higher order modulation sends more bits/symbol and therefore uses less spectrum at a given data rate. BSPK uses 1 bit per symbol. QPSK uses 2 bits per symbol. 8PSK uses 3 bits per symbol. Higher order has higher data rate/spectral efficiency, but requires higher E_b/N_0 bit energy noise ratio for the same.

With the conclusion of this discussion of QPSK modulation, the two most common modulation techniques in satellite communications, FM and QPSK, have been described.

Ratio of Bit Energy to Power Noise Density $E_b/P_N(f)$

Digital systems need a further conversion to find the ratio of energy per bit, E_b, to noise density $P_N(f)$. The carrier power C is equal to the energy per bit multiplied by the bit rate R (bits per second). So $C = E_bR$. Therefore, the decibel equation for the ratio of bit energy to power noise density is

$$E_b/P_N(f) = C/P_N(f) - 10\log_{10} R \quad \text{(dB)} \tag{11.274}$$

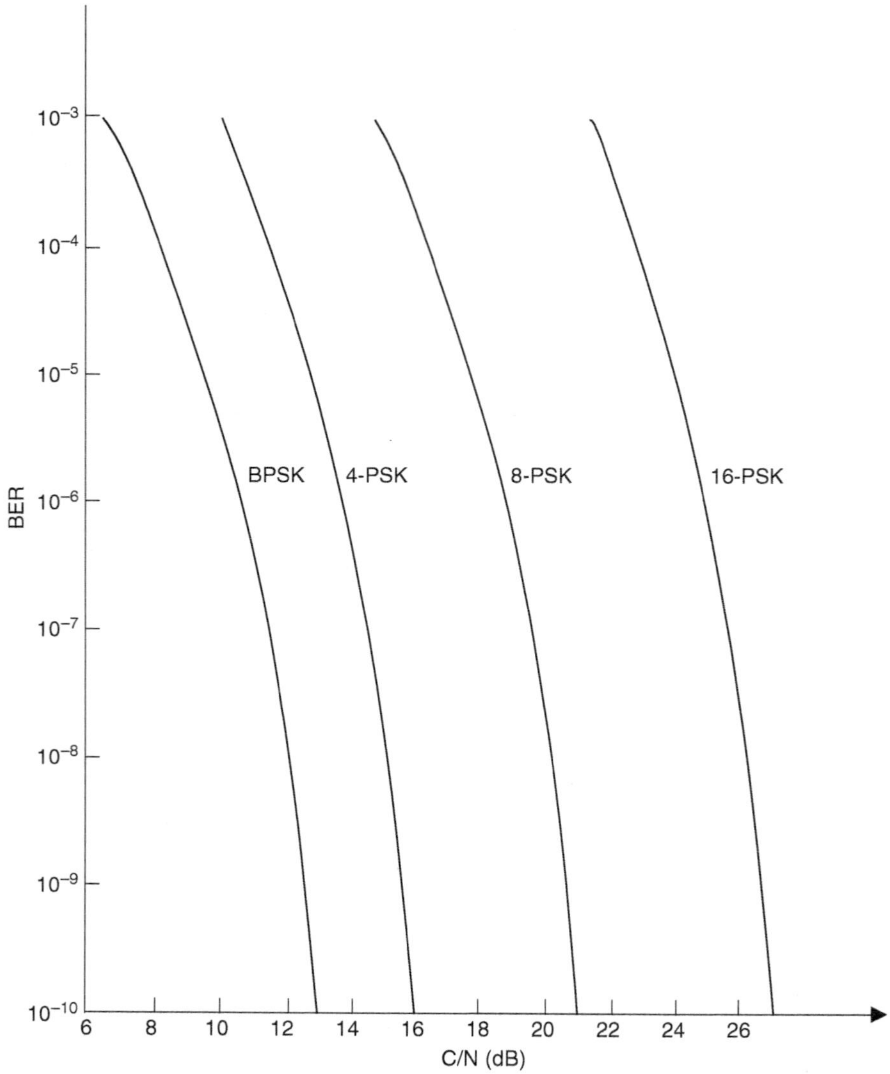

FIGURE 11.113 Theoretical BER performance of ideal M-ary PSK coherent systems. The rms C/N is specified in the double-sided minimum (Nyquist) bandwidth.

Example 11.2 The TV source signal from the camera has a highest component frequency of 4.5 MHz. If FM peak deviation of 4 MHz is used, what is the required RF bandwidth?

Solution Using equation (11.273), the Carson bandwidth will be the required bandwidth and it is

$$B = 2(4.0 + 4.5) = 17 \text{ MHz}$$

Example 11.3 The symbol rate of a digital source signal is 60 Msymbols/s. If two such digital sources are to be combined to form a QPSK modulated RF carrier, what is the RF bandwidth required, and what is the required minimum C/N for a BER of 10^{-6}?

Solution Figure 11.107(a) shows that the main lobe of the $(\sin x/x)^2$ spectrum resulting from such a modulation technique is $2/T$ Hz, where T is the symbol rate. Here, $T = (60\times10^6)^{-1}$ s. Therefore, $2/T$ Hz $= 60\times10^6\times2 =$ 120 MHz. However, Figure 11.112(b) shows that after appropriate filtering, the minimum required bandwidth in practice should be $1/T\times1.2 = 72$ MHz. Thus, a bit rate of $2\times60 = 120$ Mbits/s can be transmitted in an RF bandwidth of 72 MHz when QSPK is used as a modulation technique. This is indeed the case for Intelsat's time-division multiple access (TDMA) system.

Figure 11.113 shows that a BER of 10^{-6} would require a minimum overall C/N of 13.5 dB, where there is no further distortion introduced in the satellite. In practice, with a typical satellite transponder and a typical pair of modems (modulator and demodulator), the C/N required is approximately 16.0 dB, that is, an extra 2.5 dB is required to account for the nonlinearities and distortions in the transmission path.

Access Techniques

This section deals with the problem of accessing the satellite from the ground and the problem of sharing the satellite's available capacity with other users of the same satellite. It was pointed out earlier that (ITU) allocated bandwidths for satellite communications were already predetermined. Very few individual users can, in fact, use up the entire 500 MHz of bandwidth on a full-time basis. Some sharing schemes are therefore necessary to allow access of the satellite by more than one user.

The block diagram in Figure 11.104 already suggested one form of sharing arrangement (i.e., the allocated 500 MHz bandwidth is channelized to suit the user's needs). A common channelization scheme for C-band downlink is shown in Figure 11.114. With a channel bandwidth of 36 MHz, depending on the modulation index chosen, either one or two TV signals can be transmitted in one channel or a 60-Mbit/s data stream can be transmitted in the channel, assuming, of course, that sufficient satellite power is used to overcome the thermal noise. Even 36 MHz may prove to be much more than required for small users. For example, 24 voice channels may require a bandwidth of less than 1.5 MHz. Thus, any of the channels can be further subdivided on a frequency basis to allow small users to access the satellite directly. Figure 11.115 shows the details of how some of the RF channels might be used. This subdivision of the channel is done by the earth stations using small carriers; it is not done in the satellite. This process of sharing a satellite channel by subdividing the channel on the frequency basis is called frequency-division multiple access (FDMA).

Another way of sharing the channel is on the basis of time; each user uses the entire channel, but for only a fraction of the time. To allow time sharing to happen, time is divided into consecutive frames and each user occupies the channel for only a fraction of the time in each of the frames. Figure 11.116 shows this arrangement for four users, A, B, C, and D, sharing channel 11 in Figure 11.115. This process of sharing by multiple users on a time basis is called TDMA.

Another way is using code division multiple access (CDMA). In CDMA, all users share the same time and frequency band of the satellite repeater channel. Signals are separated by using

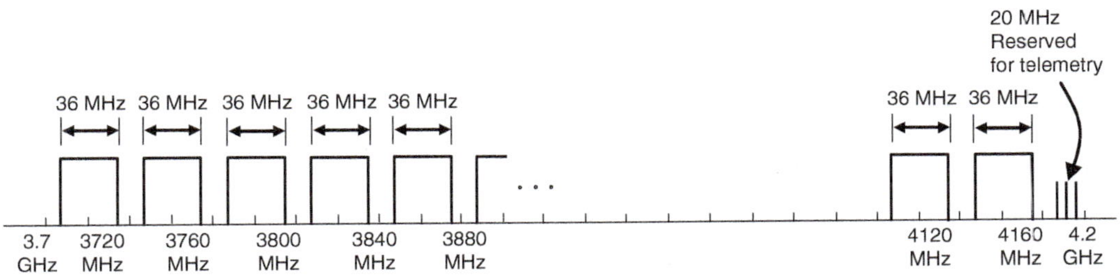

FIGURE 11.114 Typical C-band domestic satellite channelization (for the downlink).

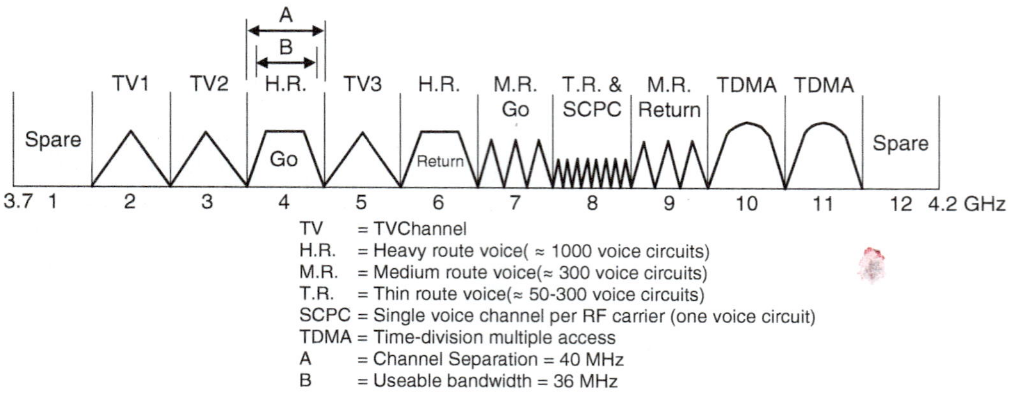

FIGURE 11.115 Example of channel utilization (domestic satellites).

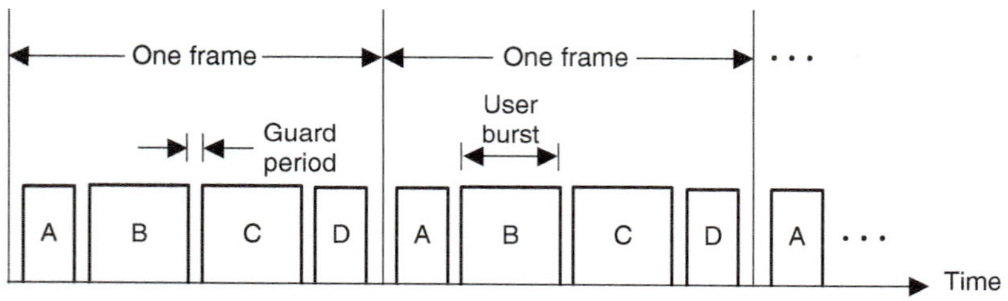

FIGURE 11.116 TDMA burst-time plan for four users, A, B, C, D.

unique codes. Codes must be orthogonal so that user A does not respond to a code intended for user B. Code is combined with the useful information at each transmitter. The signal can only be demodulated by a receiver that has all the codes and location of the code. Transmission of useful information in CDMA will require greater radio frequency bandwidth than to transmit information alone from other techniques like FDMA and TDMA. For this reason, it is also called spread spectrum transmission. Since the bandwidth is so much greater than is necessary to handle the data, a jammer can come with any point within the spread spectrum signal and the data is not affected. Another technique in spread spectrum is frequency hopping. Single frequency operation on a narrowband channel is easy to jam. Frequency hopping allows you to hop off the jammed frequency to others. The more you hop, the more frequency band is used up.

11.53 Satellite Capacity and the Sizing of Satellites

Capacity Consideration

In a thermal-noise-limited environment where the link degradation is due only to thermal noise that enters into the communication link, Shannon (1948) showed the following:

$$H = B \log_2 \left(1 + \frac{C}{N}\right) \tag{11.275}$$

where H = maximum information capacity possible in the link
B = available bandwidth
C = signal power
$N = P_N$ = thermal noise power

It can be seen that to achieve a certain communication capacity H, bandwidth B can be traded off with signal power C if thermal noise N is to remain constant for that particular link. As discussed earlier, AM techniques do not allow such a trade-off, and also, sufficient linear power must be available to establish a communication link. On the other hand, FM techniques with the modulation index as a free parameter do allow such a trade between H and C and furthermore, nonlinear power (i.e., amplitude nonlinearities) can be used if only one carrier is transmitted by the transmitter.

For digital communications, Shannon-Hartley equation (11.275) can be written as follows:

$$\frac{H}{B} = \log_2\left(1 + \frac{E_b}{N_o}\frac{H}{B}\right) \tag{11.276}$$

The signal power C can be written as energy per bit E_b times the bit rate per second that is channel capacity. So C is equal to E_bH. N_o is noise density. So N is equal N_oB. Substituting these values in equation (11.275), we get equation (11.276). As discussed earlier, by increasing power, we can increase E_b/N_o and we can use the higher order phase shift keying method, so increasing data rates, bits per second. Therefore, for digital link there is also a trade-off between signal power and channel capacity. The ratio H/B is spectral efficiency of the communications link, the ratio of bit rate to the bandwidth of the channel. For $H > B$ is called bandwidth limited. For $H < B$ is called power limited.

Example 11.4 A domestic geosynchronous satellite system is required for TV program delivery. The C band has been chosen, and at least 24 TV channels must be available at an EIRP of 36 dBW at EOL. The size of the projected landmass is approximately 12(2 × 6) square degrees. If 100% eclipse protection is required, what is the approximate size of the satellite?

Solution The first step is to evaluate the total dc power requirement and the size of the main antenna of the satellite. Since 500 MHz is available in the C band (3.7–4.2 GHz), it is convenient to use orthogonal polarization to double the 500 MHz to a 1-GHz bandwidth. A 12-RF-channel-per-500-MHz channelization scheme (see Figure 11.114) would be suitable. This will allow a total of 24 RF channels, each having a 36-MHz bandwidth. Equation (11.256) is for an elliptical beam $\theta_1^\circ \times \theta_2^\circ$. We have a rectangular area of size 6° × 2°. In addition, if a pointing error of ± 0.1° is assumed, the peak gain of a shaped beam antenna is approximately

$$\begin{aligned} G_{\max} &= \frac{30{,}000}{6.2\times 2.2}\left(\frac{\pi\times 3.1\times 1.1}{6.2\times 2.2}\right) \\ &= 1727 \approx 32.4\ \text{dBi} \end{aligned}$$

correcting for the difference in the coverages between an elliptical beam and a rectangular beam, and allowing for pointing errors by enlarging the beam dimensions by 0.2°. Note that the area of an ellipse is πAB, where A and B are, respectively, the semi-major and semi-minor axes.

The edge gain would be 32.4 − 3 = 29.4 dB. Assume that the loss between the antenna input and the transmitter output is 1.5 dB and allow 1 dB for the aging of the transmitter. The BOL RF power per channel is

$$P_1 = 36 + 1.0 + 1.5 - 29.4 = 9.1\ \text{dBW} = 8.1\ \text{W}$$

Assume a dc-to-RF efficiency of 35% for the transmitter. The total dc power required for all the transmitters (24 channels) would be = $24\times 8.1\times 1/0.35 = 555$ W. Allow 150 W for housekeeping and the other electronic and microwave active devices. The total dc power requirement is then approximately 700 W.

The eclipse can last 72 minutes, so the batteries must be sized to support 700 W through at least 72 minutes with a depth of discharge compatible with the reliability requirements. Similarly, the solar array should be sized to support the total power requirements plus the power necessary for battery charging during eclipse seasons at the end-of-life.

The antenna size can be estimated as follows. The $6° \times 2°$ rectangular coverage is best approximated by three contiguous 2° component beams. The size of the reflector necessary for the component beam can be estimated from equations (11.256) and (11.258):

$$\eta\left(\frac{\pi D}{\lambda}\right)^2 = \frac{30{,}000}{2\times 2} = 7500$$

Solving for D and assuming that $\eta = 0.7$ and $\lambda = 7.5$ cm, we have

$$D = \sqrt{\frac{7500}{0.7}} \times \frac{7.5}{\pi} \approx 250 \text{ cm} \text{ or } 2.5 \text{ m}$$

The exact size of the reflector in an optimum configuration needs sophisticated computer simulation before it can be determined. Therefore, determination of the optimum reflector focal length and diameter is beyond the scope of this subsection.

The example above illustrates the approximate size of a typical domestic satellite that is capable of delivering either 24 simultaneous TV programs or about 20,000 simultaneous telephone one-way circuits (10,000 duplex circuits).

Further Reading

Agrawal, B. N. 1986. *Design of Geosynchronous Spacecraft*, Prentice-Hall, Inc., Englewood Cliffs, NJ.

CCIR Recommendations 465-1 and *CCIR Report 391-4*, vol. IV, part 1: Fixed Satellite Service, Geneva, 1982.

Hannan, P. W. 1961. "Microwave Antennas Derived from Cassegrain Telescope," *IRE Transactions on Antennas and Propagation*, pp. 140–153, Mar.

ITU Radio Regulations, ITU, Geneva, 1982.

Jayant, N. S. 1976. *Waveform Quantization and Coding*, IEEE Press, IEEE, New York.

Lane, S. O., Caulfield, M. F., and Taormina, F. A. 1984. "INTELSAT VI Antenna System Overview," AIAA Conference on Communications Satellites, Orlando, FL, Mar.

Martin, J. 1978. *Communications Satellite Systems*, Prentice-Hall, Inc., Englewood Cliffs, NJ.

Nyquist, H. 1924. "Certain Factors Affecting Telegraph Speed," *Bell System Technical Journal*, vol. 3, pp. 324–326, Apr.

Potter, P. D. 1963. "A New Horn Antenna with Suppressed Sidelobes and Equal Beamwidths," *Microwave Journal*, pp. 71–76, Jun.

Ruze, J. 1966. "Antenna Tolerance Theory," *Proc. IEE*, vol. 54, pp. 633–640.

Shannon, C. E. 1948. "A Mathematical Theory of Communications," *Bell System Technical Journal*, vol. 27, pp. 379–423 and pp. 623–651.

Wozencraft, J. M. and Jacobs, I. M. 1965. *Principles of Communication Engineering*, Wiley, New York.

SECTION 12

Spacecraft Design

Section Editor: Brij N. Agrawal

PART 1

Design Process and Design Example

Brij N. Agrawal

12.1 Spacecraft Design Process

Spacecraft design is performed in many stages. It is a multidisciplinary process, requiring experts in many fields working together, as discussed in Part 2. Four organizations play important roles in the process. First is the user group that will use the products of the satellites, such as communications, images, weather prediction, and remote sensing. Second is the spacecraft operators group, such as commercial, government, or military who, based on input from the user groups for current and future requirements, come up with the mission requirements, acquire the satellites, and operate and deliver the product to the users. Third group is spacecraft manufacturers who design, build, test, and get the satellite ready for launch. Fourth group is launch vehicle providers who launch the satellites in the desired orbits.

Spacecraft design process starts when the spacecraft operator decides need for new spacecraft. It could be the need to replace an old spacecraft, additional needs from the user group, or development of new technologies that will greatly improve performance of the spacecraft. Spacecraft operator will perform the following steps in the process of acquisition of a new spacecraft:

1. Develop mission requirements for the new spacecraft. For communication satellites, the requirements will be coverage area, data rates, and EIRP. For imaging satellites, it will be desired resolution, spectral bands, revisit time, and geopositioning accuracy.
2. Based on the mission requirements, perform conceptual design and trade-off, as discussed in Part 2. It uses Excel-based simple empirical formulas based on previous spacecraft. This will provide spacecraft conceptual design without detailed analyses.
3. Evaluate the mission requirements based on the results from the conceptual design study. Modify the mission requirements and repeat conceptual design study, if necessary. It should be noted that this is the early stage of spacecraft design and cost of the effort is low, but cost reduction opportunities are high.
4. Perform a spacecraft feasibility study that will include simple analysis, selection of components, etc. The spacecraft design example given in Subsection 12.2 can be considered a spacecraft feasibility study. This is the time to get feedback from the spacecraft manufacturers. Spacecraft manufacturers generally develop standard spacecraft buses. Therefore, a new spacecraft may need minor modifications to the spacecraft bus, but main effort remains in the payload design.
5. Finalize mission requirements and acquisition process and issue request for proposal (RFP) if mission requirements are met by spacecraft manufacturers.

Next step in spacecraft design process is for the spacecraft manufacturers to perform the preliminary spacecraft design for the proposal. It will involve experienced subsystem and system engineers. The proposal spacecraft design should prove to the spacecraft operator in sufficient details and analyses that the spacecraft contractor will be able to meet the mission requirements within the proposed schedule and cost.

Spacecraft operator will then evaluate the proposals based on technical performance, past experience, and cost. Once the spacecraft manufacturer is selected, spacecraft design process moves to spacecraft contractor. The contract requires several design reviews to evaluate the progress in spacecraft design and development. In general, design reviews facilitate communication between the design team, management, and the customer. Some of the main design reviews are discussed as follows.

Preliminary Design Review

The preliminary design review (PDR) demonstrates that the preliminary design meets all system requirements with acceptable risk and within the cost and schedule constraints and establishes the basis for proceeding with detailed design. It will show that the correct design options have been selected, interfaces have been identified, and verification methods have been described.

Critical Design Review

The critical design review (CDR) is an intermediate design review that occurs after the detail design is complete, prior to the fabrication of prototypes or preproduction models. This review is conducted to evaluate the design against the detailed requirements. It has many of the components as in PDR, including the provision of assumptions and calculations used in the design, project progress, and risk management. A production assessment is often included.

Final Design Review

The final design review (FDR) is conducted after prototypes or preproduction units have been through verification testing. Problems encountered during this testing and the respective solutions are examined. Any necessary changes to the product with respect to performance, cost, reliability, and manufacturing issues are agreed upon prior to the initiation of full-scale production.

12.2 Spacecraft Design Example

This subsection provides an example of electro-optic infrared imager spacecraft design. This work was performed by students of the Naval Postgraduate School (NPS) under capstone course AE 4871, Spacecraft Design and Integration, in 2008. Professor Brij Agrawal was instructor with CAPT Daniel Bursch and CAPT Al Scott as other faculty advisors. Dr. Jason Wilkenfield, Operational Responsive Space (ORS), was sponsor of the project. The students were CDR Michael Tsutagawa, LCDR Nathan Walker, LCDR J. Alan Blocker; LT Thomas Childers, LCDR Joseph Cascio, LT Mark Gerald, LCDR Troy Hicks, LCDR Chance Litton, LCDR Tanya Lehmann, Capt. Margaret Sullivan, and LT Scott Williams.

Mission Requirements

The objective of the project was to design an electro-optic infrared (EOIR) imager with a short development timeline and a 1–2 year mission life. Furthermore, the design of the spacecraft bus must be within ORS design philosophies of smaller, cheaper, and faster using modular, multimission buses with standardized interfaces, and consistent with integrated system engineering team (ISET) recommendations. ORS office also released broad agency announcements for similar mission requirements. The design study by NPS students became a government design study.

The mission requirements (MR) for the spacecraft were given as follows in Table 12.1.

The iSat design team approached the design of the satellite in two phases. The payload and orbital parameters were developed to meet the Mission Requirements as outlined by the ORS office. The spacecraft bus, however, was intended to be a multimission ORS bus based on the ISET standards. This design falls within the class A ISET mission dataset for a LEO precision pointer with some deviation due to some of the unique aspects of a high-resolution mid-wave IR imager. The overall design philosophy of the bus was to incorporate modularity, simplicity, and ease of integration while meeting as many of the mission and ISET General Bus Standard (GBS) requirements as possible.

Requirement	Threshold	Objective
Mission		
Revisit approx. hours	24	12
Images approx. per day	30	60
Task-to-collect approx. hours	12	4
Task-to-product approx. hours	13	4.25
Operational availability	Night	Night/day
Data storage orbits	2	
ADCS		
Geo-loc accuracy, meters	20	10
Imager		
Footprint approx., km	25	50
Max slew angle approx., degrees	30	45
Spacial resolution approx., meters	1.5	1
Spectral range approx., μm	3.0–5.0	Vis/MWIR
Thermal control	Cryo-cooled	Cryo-cooled
Absolute temp. approx., °F	15	10
Relative temp. approx., °F	3	2
Mass/Size		
Mass approx., kg	485	450
Bus diameter approx., meters	1.8	1.3
Payload diameter, meters	1.8	1.3
Pri Mirror Diameter, meters	1	0.75
Overall Length, meters	5	3
Launch Vehicle	Min IV	Min 1
Orbit		
Sun-synchronous Time	10:00 PM	Twilight
Altitude-km	300	800
Multi-Mission		
MissionType	IR imager	Vis-IR imager
Modularity		
ISET	Bus	Payload-Bus
Plug-n-Play (PnP)	Bus	Payload-Bus
TT&C		
Operations	MMSOC, AFSCN	MMSOC, AFSCN
Imagery downlink	CDL, DCGS	CDL, DCGS

TABLE 12.1 Mission Performance Requirements

Based on these requirements, each subsystem engineer developed a list of derived design requirements to accomplish both MR and ISET GBS requirements. A preliminary mass budget by subsystem engineers was developed and commercially available components were evaluated based on the mission requirements, mass constraints, and flight heritage. Each subsystem engineer then engaged in trade analysis based on their derived system requirements and commercially available components. The iSat team then used The Aerospace Corporation's Concept Design Center database and program to refine the initial mass, power, and thermal requirements for the spacecraft bus. A baseline equipment list was established and iSat engineers began a more detailed analysis of the performance of each subsystem. Trade studies within each subsystem were made and system level decisions were made.

Table 12.2 provides comparison of ISET standard requirements and iSat mission requirements. The driving requirement for the bus was to remain under 250 kg.

System Summary

Orbit

Orbit altitude selection for iSat was primarily a function of image resolution with consideration given to orbital drag and thermal stability. Because the primary mirror was constrained in size and mass, to achieve 1 meter resolution required the satellite to fly at an altitude of approximately 300 km. Orbital drag at that altitude requires a large propulsion system to overcome atmospheric drag.

Mission orbit	Units	iSet Leo	iSat req.
Altitude	km	350–705	300–600
Inclination	°	0–98	97.2
Mass (maximum)			
Bus wet mass	kg	250	250
Payload	kg	175	250
Comm. rates and data storage			
Telemetry (downlink)/command (uplink)	kpbs	2,000/2	100
Theatre downlink/uplink	Mpbs	250/0.2	80
Data storage	GB	2	2.25
Power			
Bus orbital average	W	220	220
Payload orbital average	W	400	400
Peak power	W	700	1,690
Voltage	V	28 ± 6	28
Pointing			
Pointing accuracy	°	0.05	0.01
Slew rate	°/s	2	2
Absolute position knowledge	m	20	8
ΔV (minimum)	m/s	175	175

Table 12.2 Example ISET Standard Requirements and iSat Mission Requirements

The iSat team proposed a compromised elliptical orbit with perigee at 300 km and apogee at 600 km, which reduces the propellant requirements. While resolution will suffer at apogee, there will be increased area coverage.

A sun-synchronous orbit was chosen for iSat based on MRs, but it is also advantageous for this particular mission. A sun-synchronous, 0600/1800 orbit (or terminator orbit) was chosen primarily for MWIR image quality due to thermal crossover; the thermal cycling for the spacecraft is also significantly reduced. With a nadir pointing space vehicle, one face of the satellite will always face the sun and one will always face deep space. While there will be a significant amount of heat that must be distributed, the result is a fairly stable thermal balance problem. For the mid-wave IR imager, the focal plane array must be cryo-cooled to approximately 100 K and must be tightly controlled. The stability offered by the terminator orbit will assist in the stability. An additional benefit for the terminator orbit is that a fixed solar array can be pointed at the sun almost continuously.

Launch Vehicle Selection

The Falcon 1, Falcon 1e, Minotaur I, and Minotaur IV were the primary launch vehicles considered. Of these four, only the Minotaur series has successfully placed a payload into orbit at the time of the design study. The basic bus was designed to meet the Falcon 1 envelope for possible future ORS missions; however, iSat was notionally designed for a September 2009 launch which required a Minotaur I or IV. Initial estimates of payload mass (250–350 kg) and size (4 m long) precluded a Minotaur I launch. Early in the program design the Minotaur IV was chosen as the launch vehicle for iSat.

Configuration Trade Studies

The ideas of modularity and simplicity in integration and testing were a key concern in the configuration of the space vehicle. The system took inspiration from a number of sources, including TACSATs 2 and 4 and a number of commercial imagery satellites as well as the ISET requirements. Looking to orbiting imaging satellites, Worldview 1 demonstrated a bus design where most of the satellite components were mounted externally with a bus structure of four load bearing posts. The externally mounted components, while not necessarily modular, seemed to ease the integration of the satellite and, assuming standard interfaces and plug and play technology is employed, could allow future updates to the iSat bus. This seemed an attractive feature and influenced the overall configuration of the bus. The only systems that were designed with modularity at the forefront were the attitude control subsystem and the propulsion system. The propulsion tanks would be mounted internally with the thrusters mounted near the payload interface close to the center of mass of the integrated space vehicle. The attitude control system was deemed to be a significant source of vibration and was mounted on the antinadir end of the bus. One of the advantages of the terminator orbit was that it allowed fixed solar arrays to sun soak with little reliance on the attitude control system. With the understanding that another orbit would require a different solar array design, the iSat team chose to design the solar arrays for this particular mission. This is an example of a design decision that was made based on the mission vice the ISET standards. There was concern that the equipment mounted externally would be subject to unacceptable acoustic loads during launch. The design of the solar arrays allowed them to essentially wrap around the bus in the stowed configuration which will provide some protection to the bus during launch.

Final Spacecraft Configuration

In its stowed configuration, see Figure 12.1, iSat fits inside the dynamic envelope of a Minotaur IV with a maximum diameter of 1.95 m and height of 3.0 m from the launch vehicle adapter. The deployed configuration is shown in Figure 12.2.

Spacecraft Mass Budgets

Minimizing the mass of the bus and not sacrificing required performance to meet the mission was a challenging aspect of this project and drove the design of the propulsion, attitude control,

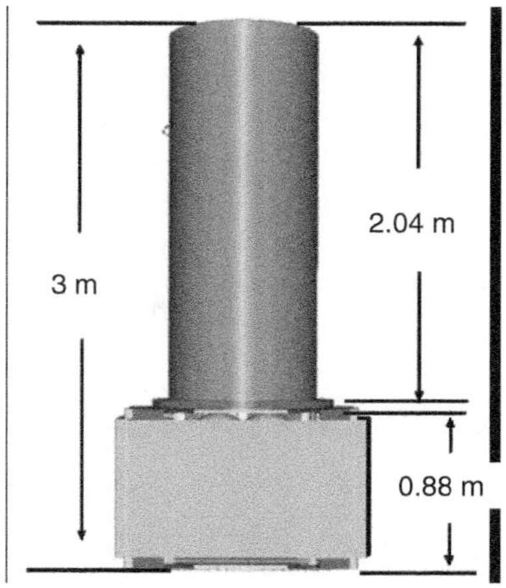

Figure 12.1 iSat in stowed configuration.

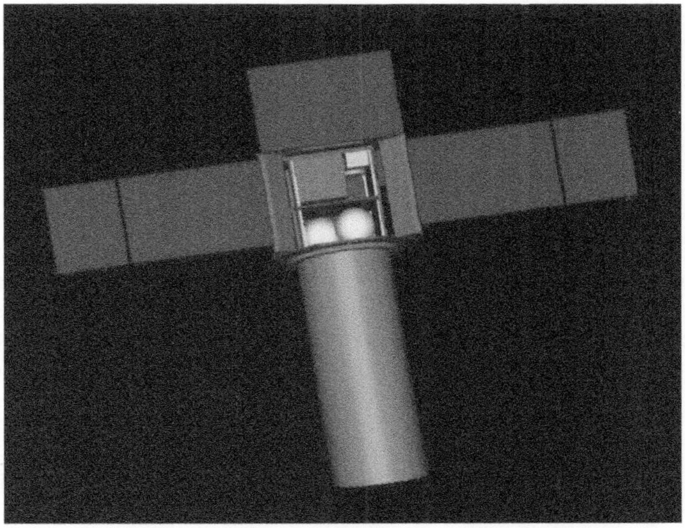

Figure 12.2 iSat in deployed configuration.

electrical power systems, and orbit choice as discussed above. Initial mass estimates by subsystem were made based on *Space Mission Analysis and Design* (SMAD) by Larson and Wertz for small satellites. The estimates were altered slightly because the bus was essentially being designed separately from the payload and the mass distributions from the SMAD are based on the entire spacecraft. Once the initial mass budget was promulgated and initial component selection completed, the concept design center software provided by the Aerospace Corporation was used to refine the design and estimate power requirements. The final estimated mass and the original mass budget is

	Original budget (kg)	Final budget (kg)
Structure	46.4	61.9
Power	70.7	41.3
ADCS	50.8	54.1
Prop (dry)	8.8	7.7
Thermal	6.6	6.1
TT&C	5.0	7.5
C&DH	10.5	10.0
Dry bus	**198.8**	**188.6**
Propellant	29.1	28.1
Margin	22.1	33.3
Bus total	**250.0**	**250.0**
Payload	235.0	184.0
Margin	0.0	27.6
TOTAL	**485.0**	**461.6**

TABLE 12.3 iSat Mass by Subsystem

shown in Table 12.3. The initial budget included a 10% margin, but was considered too low for the maturity of this design. The final estimate includes a 15% margin for the bus (assuming a 250-kg wet bus) and a 15% margin for the payload.

Payload

The mission of this payload was to provide NIIRS 5, infrared imagery in the 3–5-μm wavelength range. In other words, the goal was to provide photographic quality images built on photons emitted from average temperature targets on the ground. This mission results in several challenging demands that were balanced against each other including ground resolution, primary mirror diameter and mass, noise equivalent delta temperature (NEDT), and mission altitude and profile.

Trade Studies

The selection of a primary wavelength for the sensor within the 3–5 μm band was driven by three priorities. First, to receive information from the ground, the atmosphere must be largely transparent to the particular wavelength chosen. There are two windows of interest, from between 3.5 and 4 μm and between 4.5 and 4.9 μm. Initially, the lower wavelength window seemed more appealing because it reduced the aperture size needed to obtain resolution, as described in the next section. However, higher wavelengths enable a better NEDT. NEDT is essentially the temperature resolution of the image based on a signal-to-noise ratio of 1. The higher band allows for a more precise NEDT and produces higher quality thermal images.

Ground Resolution and Aperture Sizing

The first consideration of image resolution was the physical ground sample distance (GSD) framed by each of the focal plane array's (FPA) pixels. GSD is a function of pixel pitch, altitude, and the effective focal length of the optical system. Figure 12.3 illustrates the equation, where o is the GSD, d is altitude, f is focal length, and i is pixel pitch.

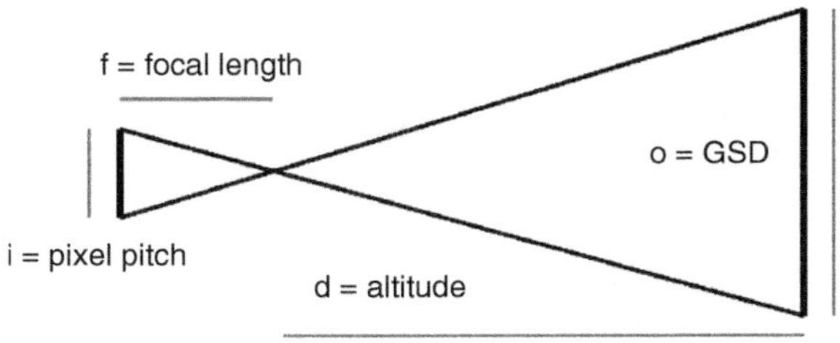

Figure 12.3 Ground sample distance relationship.

The equation for the relations is

$$\frac{0}{d} = \frac{i}{f} \tag{12.1}$$

As a GSD of 1 m is desired at an orbit altitude of 300 km and the best pitch available in our wavelength band is 18 μm, the necessary focal length of our system can be determined as follows:

$$f = \frac{i.d}{o} = \frac{18\ \mu\text{m} \cdot 300\ \text{km}}{1\ \text{m}} = 540\ \text{cm} \tag{12.2}$$

Therefore, a 5.4-m effective focal length is required with an 18-μm pitch FPA in order to achieve a 1-m GSD. In order to find resolution, we have to consider diffraction limits.

The Rayleigh criterion is an analytical method of determining the approximate minimum angular distance between two point sources must be from each other at the target in order to be distinguishable from one another at the image. For a circular aperture, the Rayleigh criterion is as follows:

$$\theta = 1.22\frac{\lambda}{D} \tag{12.3}$$

where θ is the angular resolution, λ is the wavelength of the incoming light, and D is the diameter of the aperture. Recognizing that most of the valuable information lies within the full width half max (FWHM) portion of the central lobe rather than to the first absolute minimum, allows the aperture to be further reduced. However, the 1.22, which is derived from the spherical Bessel functions, can be dropped to estimate the FHWM distance, vice the distance between first-order minima. Assume Q = 1.8, number of pixels lies within FHWM, the diameter of aperture is determined from the following equation,

$$D = \frac{\lambda f}{Qd} \tag{12.4}$$

Substituting the values in the right-hand side of equation (12.4), telescope aperture diameter is 80 cm.

Design Description

The optical payload consists of an 80-cm Cassegrain telescope supported by a quartz rod cluster within a 2-m barrel and an electronics sensing, processing, and cooling suite located behind the primary mirror. This entire structure is attached to the bus via vibration damping struts and electrical signal and power cabling to facilitate the modularity between bus and payload desired for the mission architecture. The last 20 cm of the outer end of the barrel is cut at a 17° angle to shield the interior of the barrel from incident sunlight while imaging. The secondary mirror is situated at the base of this cut; therefore, the distance between the two major optical surfaces is 1.6 m, also allowing for 20 cm behind the primary mirror for the mirror's body and the payload's electronic suite.

Focal Plane Array The primary decision factors in selecting the specific FPA for the instrument sensor were the ability to detect infrared light in the 4.8 μm wavelength, a small pitch to enable a more precise GSD and a large imaging axis, and a large array to achieve a larger imaging area, and, therefore, a larger swath. Most commercially available MWIR FPAs have 24 and 32 μm pitch pixels and are typically not larger than 640 pixels square. One particular FPA currently available in 2008 that exceeds these typical values is the Hawaii-2RG (H2RG) built by Rockwell Scientific. It is CMOS sensor with 18 μm pitch and is 2,048 by 2,048 pixels. This FPA was selected.

Primary and Secondary Mirrors The primary considerations in mirror selection were mass, structural strength, and thermal stability. The limiting factor is thermal stability and the glass ceramic Zerodur, manufactured by Schott Optiks, provides the best thermal stability of space tested materials. It has a temperature coefficient of expansion of less than 0.2×10^{-6}/K. At a density of 2.53 g/cm^3 and the size of the mirror, a solid Zerodur mirror would weigh in the neighborhood of 120 kg. However, Zerodur also has the necessary strength to allow significant honeycombing in order to light weight the mirror by almost 50%.

Electronics The processing structure is provided by a card box produced in-house by SAIC. The key cards in this system are the sensor board, housing, and reading the FPA, the analog control electronics (ACE) board, which submits read orders to the sensor board, and the Frame Summer and Frame Buffer boards, which control the summing process, allowing enough information to be produced with short frame speeds.

Thermal Control

It is necessary to maintain the temperature of the optical train at or below 250 K and of the FPA at 110 K to minimize thermal photon noise on board the spacecraft and to minimize thermal transients to reduce optical distortion. This is accomplished by a number of material and operational methods. First, while the sun-synchronous terminator orbit selection results in the highest temperatures due to the constant exposure to sunlight, it also provides a constant face toward deep space and largely eliminates thermal transients due to the orbital period. Second, the telescope barrel is coated with a highly reflective white paint to minimize solar absorption. Third, a combination of mirror coatings and bandpass filters reduces the incident energy from imaging operations onto the FPA by up to 99%. Fourth, a static radiator is affixed to the deep space side of the payload to reject much of the heat produced by the electronics suite behind the primary mirror. Finally, a small cryo-cooler is used to supercool the FPA.

Mass and Power Budgets

The mass and power budgets of the payload are shown in Tables 12.4 and 12.5.

Structure Subsystem

Overview/Requirements

The structural subsystem was designed to survive the payload launch environments of the Minotaur I and IV, while meeting four main requirements: (1) provide sufficient support for all spacecraft subsystem components; (2) ensure the dimensions of spacecraft are within the payload

Parts & Masses			
Optics Assembly	Primary Mirror	Zerodur (LW)	74 kg
	Secondary Mirror	Zerodur (LW)	2 kg
	Relay Optics	Zerodur	3 kg
	Barrel	Carbon Fiber	30 kg
	Mirror Supports	Quartz	18 kg
	Door	Carbon Fiber	5 kg
Electronics Deck	Electronics Boxes	Carbon Fiber/Alum	9 kg
	Focal Plane Array and		1 kg
	structure		6 kg
	Cryocooler & Support		16 kg
	Deck Plate		6 kg
	Connectors		6 kg
	Cabling		
Total Mass			**176 kg**

TABLE 12.4 Payload Mass Budget

Power requirements		
Imaging	FPA and Sensor Board	1 W
	Analog Control Electronics	19 W
	Frame Summer	20 W
	Frame Buffer	15 W
	Camera Interface Board	8 W
	Cryocooler	75 W
	Spacecraft I/O Board	6 W
	Power Supply Board	38 W
	BackPlane	0 W
		182 W
Standby	FPA and Sensor Board	0 W
	Analog Control Electronics	0 W
	Frame Summer	0 W
	Frame Buffer	0 W
	Camera Interface Board	0 W
	Spacecraft I/O Board	0 W
	Cryocooler	31.5 W
	Power Supply Board	0 W
	BackPlane	0 W
		31.5 W

TABLE 12.5 Payload Power Budget

fairing size of the launch vehicle, sized from the Orbital Sciences Minotaur I to the Minotaur IV; (3) provide adequate support during launch and orbit insertion against both axial and lateral accelerations; and (4) minimize subsystem mass leading to an overall spacecraft mass of less than 250 kg.

The iSat structure was designed to survive the sum of steady and dynamic accelerations for the worst case events, stage 3 ignition (axial) and stage 2 ignition (lateral) for the Minotaur IV, with a designed safety factor of 1.3.

Design Consideration

The initial design concept was for iSat to be of a modular design type. To this end, the main structure was chosen to be rectangular instead of cylindrical, which allowed for the increased feasibility of a modular design and the availability of additional mounting surfaces. The upper and lower structure shelf caps were chosen to be octagonal in order to maximize the structure area inside the payload fairing and increase the usable surface in the event that more space was needed for mounting additional components. In order to meet thermal requirements, the primary heat-producing components, such as the battery box and the integrated avionics units (IAUs), were place on the +Y panel, the permanent deep-space facing panel, to take advantage of the thermal dissipation benefits presented. To leverage the modular design, most individual subsystem components are placed within proximity to one another. This expedites spacecraft manufacturing and testing of various subsystems. Unfortunately, the TT&C and C&DH subsystems, due to the number of antennae required and the ADCS subsystem, due to the requirement to have the star tracker as close to the primary mirror as possible, are unable to conform to the close component proximity desired.

Material Selection

The structural subsystem design consists of a single rectangular support system with top and bottom octagonal shelf caps. Figure 12.4 illustrates the structural design.

Initially, solid panels made from AL-7075 (aluminum alloy 7075) were used for all of the panels and shelves. The calculations showed that the mass for the four vertical panels alone would be 58 kg too high for structural mass budget. To provide the requisite strength but a lighter mass, aluminum honeycomb was used instead.

Mass Budget

The structure subsystem has an overall mass of 55.8 kg. The four vertical panels have a total mass of 7 kg; with the three horizontal shelves, a total of 35.4 kg. The remaining mass was for the four vertical columns (2.8 kg), eight horizontal columns (2.24 kg), and miscellaneous supports (8.4 kg).

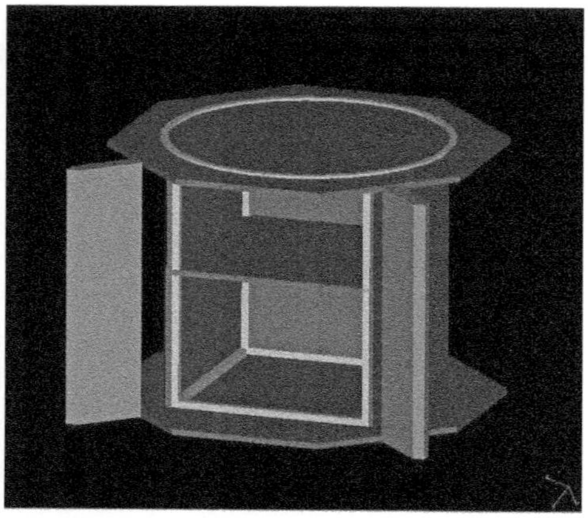

FIGURE 12.4 iSat structural design.

Performance

Analysis of the structural subsystem was done by using finite element method in IDEAS software. An acceleration loading analysis was conducted followed by modal analysis. The acceleration loading analysis estimated the performance of the stowed spacecraft under maximum loading conditions. Maximum acceleration loading condition of 8.82 g axial and 5.27 g lateral are used. These loading conditions are taken from the launch vehicle Minotaur IV load factors and safety factor of 1.3 and represent the worst case for each vehicle. In reality, this situation could not occur due to the fact that the maximum axial and lateral values are derived from different stages of the Minotaur IV, but are designed for in order to err on the side of safety. Initial estimate for the spacecraft response was a maximum displacement of 1.1 mm and a maximum stress of 13.4×10^6 Pa, both occurring in the octagonal shelves. This gives a margin of safety for allowable stress of 28×10^6 Pa.

Modal analysis was performed to demonstrate if the modal response of the spacecraft is within the specifications of the launch vehicles. The components were again modeled as lumped masses with even distribution over the surface(s) on which they were mounted. The first lateral mode occurred at 58 Hz, and the first axial mode occurred at 64 Hz. Both frequencies are higher than the first fundamental frequency requirement of both launch vehicles and ISET, therefore, are within parameters.

Telemetry, Tracking, and Control Subsystem

Overview and Requirements

The basic requirements for the communications subsystem was to securely communicate with the Air Force Satellite Control Network (AFSCN) for telemetry and commanding and have a method for downlinking data into the distributed common ground station (DCGS) network without affecting payload operations. The satellite must have two possible paths for all the information to travel and meet the data rates of at least 250 Mbps data downlink and 200 kbps command uplink for in-theater, and 2 Mbps data downlink and 2 kbps command uplink with the command and control network.

Trade Studies

Several factors were considered in the trade space for the system to be used on iSat. Besides the aforementioned requirements, the entire system must be a small volume, low mass, and power efficient to be commensurate with the small satellite model and provide the other systems increased margins. It must have a flexible configuration capable of easily integrating our data pipes and very reliable in order to justify integrating single components into the design. For these reasons, the components selected were the AeroAstro S-Band transponder as well as their combination S/X-Band transponder, as one unit. They are flight proven, Space Ground Link System (SGLS) transponders with interfaces to an MCU-110 crypto unit. These transponders meet all the required factors and consist of three separate units: a receiver, an interface, and a transmitter, which provide additional flexibility in the placement within the satellite as well as expandability and interchangeability options, depending on the ultimate communications requirements. They also provide the greatest flexibility in transmitter power by allowing the RF output to be set as low as 1 W with software control versus the 3–5 W minimum RF output from the comparable L-3COM transponders, which are not modular and are larger than the AeroAstro units. Additionally, while the use of the common data link (CDL) protocol is the expected standard protocol for connectivity from the satellite to the ground, the limited equipment selection and high costs were enough to deter its selection and find a different frequency to pass the data to the DCGS using X band.

With the rigid requirements placed on the ADCS system and payload capabilities, the primary requirement for the TT&C subsystem for iSat was for constant connectivity for command and telemetry and the rapid download of data as to not interfere with the mission requirements.

To meet these requirements, small, lightweight antennas with beam widths of 90° were selected and fixed to the solar array panels for system simplicity and mass considerations. With the selected Antenna Development Corporation micropatch and microstrip antennas, the system is reliable, lightweight, and robust without impacting or restricting the other subsystems.

The link budget was calculated using the most conservative data. The smallest dish size in the AFSCN is 10 m and 1 m was used for the in-theater command node. A satellite antenna efficiency of only 15% was estimated. Conservative numbers in every part of the link budget yielded healthy margins for all downlinks above the required 3 dB. These margins indicate robust communication capabilities for the stated requirements.

Command and Data Handling Subsystem

Overview/Requirements

The Command and Data Handling (C&DH) subsystem is an integrated C&DH system. These are the primary functions for subsystem spacecraft command processing, data acquisition assistance, and satellite system management: command processing, data acquisition assistance, Tracking, Telemetry, and Communications (TT&C), attitude determination and control system (ADCS) processing, interface with all ADCS components, electrical power system management, battery charging control, power distribution, and command and data storage.

Since the payload will handle all image processing functions, the C&DH subsystem is only required to store that data and assist in acquisition and transmission when necessary.

The general required amount of storage for a satellite of this size compared to the storage needed to meet the mission requirements were just about equal, 2 GB versus approximately 2.25 GB. Each image was estimated to require the same storage as an 8 bit per pixel RGB image, which equates to 12.5 MB per image for a 2 × 2-km IR image. The system required enough storage for 3 days of images with 60 images taken per day.

Trade Studies

Only two companies were found to carry products that were close to satisfying all the necessary requirements of the subsystem. Therefore, only the modular units produced by SEAKR Engineering and Broad Reach Engineering were compared for this mission. SEAKR Engineering offers a modular solution with a compact PCI (cPCI) or VME backplane in 3U or 6U form factors. There are numerous central processing units (CPUs) available as well as ADCS and minimal power supply cards. The modules are space qualified and can hold from 8 to 12 cards depending on the size selected. Broad Reach offers their integrated avionic unit (IAU), a PCI 3U form factor. They also provide a number of CPU options, as well as ADCS and EPS capabilities. Their unit can hold eight cards that can be filled with any of their available cards. The IAU weighs less than 5 kg and requires approximately 40 W of power. While both products can meet all the requirements, the Broad Reach IAU can be acquired with no changes to its design or configuration, while the SEAKR units would require at least minimal modifications to their COTS design. Redundancy and overall reliability were the primary considerations in the design of the C&DH architecture. These are primarily for fault tolerance, but also provide for operation in a limited manner if there is a single component failure. Having two separate processors in a star configuration with all other components allows them to individually process data and then compare the independent calculations, thereby preventing any catastrophic events. This configuration was selected.

Propulsion Subsystem

The major function of propulsion system is to compensate for atmospheric drag. The selected orbit is elliptical, between 300 and 600 km altitude. To estimate the required ΔV for this orbit, a simple spreadsheet was constructed. At each altitude given in the SMAD, an estimated percentage of an

Alt.	% of orbit	Solar min.		Solar max.	
300	0.1397188	154	21.5167	747	104.3699
350	0.1407649	43.7	6.151427	311	43.77789
400	0.1418110	13.6	1.92863	140	19.85354
450	0.1428571	4.55	0.65	66.6	9.514286
500	0.1439033	1.64	0.236001	32.9	4.734417
550	0.1449494	0.659	0.095522	16.8	2.435149
600	0.1459955	0.303	0.044237	8.9	1.29936
Total ΔV (m/s) req'd/year:		**30.62251**		**185.9846**	

TABLE 12.6 Orbit Drag Calculation (Solar Min. and Max.)

orbit was calculated to determine approximately how much time the satellite would spend at or near that altitude. These percentages were then multiplied by the values from the SMAD for annual atmospheric drag compensation required at each altitude, and the sums of those products were then calculated to determine the total annual ΔV requirement for both solar cycle minimum and maximum. Those totals are also shown in Table 12.6.

Because solar maximum only occurs every eleven years, it was decided that the propulsion system would be designed for only 1 year's operation, vice two. There is some risk in this decision, but it was deemed acceptable in return for the mass and cost savings. Following the ORS guidance, a value of 80% of solar max., or 155 m/s, was determined to be the final value to design to; however, because the ISET standards for the bus design mandate a minimum of 175 m/s ΔV from the propulsion system, that's the value to which the system was eventually sized.

Trade Studies

The first trade study conducted considered electric versus liquid chemical propulsion. Electric propulsion was initially considered attractive due mainly to its propellant mass savings. Taking the L-3 Communications 8-cm xenon ion propulsion system (XIPS) as a representative example, an Isp of 2,800 seconds was used to calculate the propellant requirement. This engine could provide the required drag compensation for only 3 kg of propellant per year, but it carries two major penalties. First, it would require very long thrusting times. This is acceptable in long-life missions to the asteroid belt and outer planets; not so in a short-lived mission centered on frequent imaging. Second, the power requirements for the engine were prohibitively large. When combined with the long thrusting time, the power generation system was deemed incapable of supporting this engine choice. When comparing liquid chemical propulsion systems, there are two main types to consider: bipropellant and monopropellant. Initially the bipropellant option was preferred due to its low system mass. With an Isp of 290 seconds, a hydrazine bipropellant system (N_2H_4 fuel, N_2O_4 oxidizer) would be capable of supplying the required ΔV for iSat with 29 kg of propellant. Compared to the cold gas thruster, with an Isp of only 220 seconds and a propellant requirement of 38 kg, the bipropellant system is clearly the better choice. However, the resistojet, which combines an electric heater to superheat the propellant gives an Isp of 300 s, and is therefore able to provide the same ΔV with only 28 kg of propellant. The downside to the resistojet is a mass penalty in the thrusters themselves. Since bipropellants are hypergolic, they require no external heat input to ignite; simply bringing the fuel and oxidizer together in the combustion chamber is sufficient to light the system. Monopropellant requires external energy to achieve an Isp of 300 seconds, and therefore, the thrusters are heavier by nearly 100%. Taken as a whole, the resistojet system is slightly more massive, but the return for that mass is a simpler, more reliable propulsion system. For these reasons, the sponsor chose to pursue the resistojet system. The final trade study was very brief, and concerned diaphragm tanks versus surface-tension tanks. Since surface-tension tanks provide no slosh protection, they were deemed inappropriate for this design. When deciding between polymer or metal diaphragm tanks, the metal diaphragm options, while attractive due to their near-zero slosh, were deemed too heavy and expensive.

Design Description

The thruster chosen for this system is the Aerojet MR-502A Improved Electrothermal Hydrazine Thruster (IMEHT). It has a nominal thrust of 0.8 N and a nominal Isp of 299 seconds. For redundancy, a second thruster will be mounted with its nozzle as close as possible to that of the primary thruster. The tank design process initially yielded a single spherical vessel of 43 cm diameter; however, this was too large to fit into the allotted height of the tank shelf on the bus. A spherical/cylindrical and spherical/elliptical design was also explored, but it was desirable to design a system with redundancy. For this reason, two smaller spherical tanks were chosen for reasons of packaging and reliability. Titanium was determined to be the best compromise between weight and cost, therefore, a tank similar to the ATK-PSI 80290 would need to be constructed. The requisite tanks would be constructed of 6Al-4V titanium, and each would be 34.2 cm in diameter. Positive fuel expulsion is accomplished via a polymer diaphragm retained (welded in) at the sphere mid-plane. An overview of the system's basic components and design is shown in Figure 12.5.

The total system mass of 35.9 kg is allocated as follows:

- Propellant/pressurant: 28.2 kg
- Tanks: 5.0 kg
- Lines/valves/fittings: 0.9 kg
- Thrusters (2): 1.8 kg

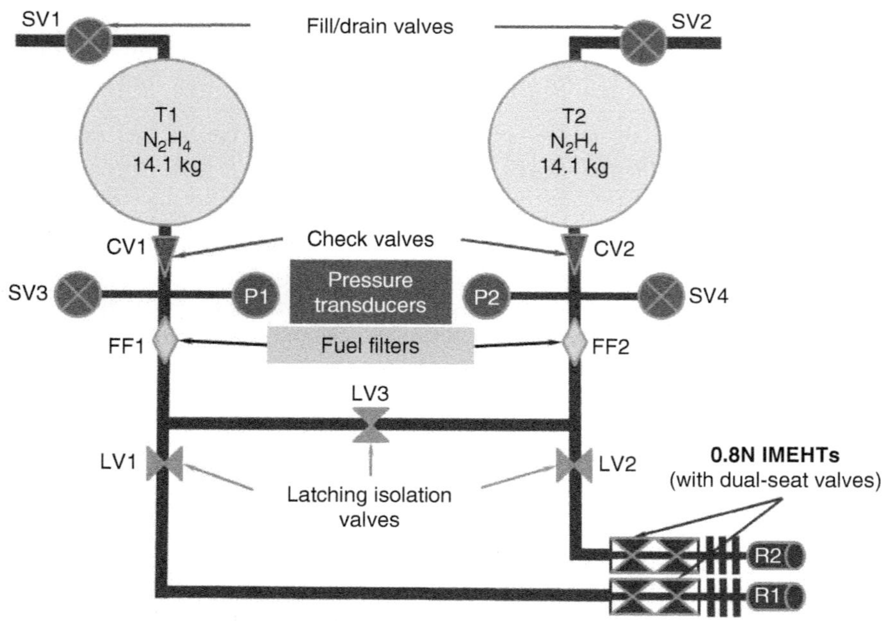

Figure 12.5 Propulsion system schematic.

The continuous maintenance power level is approximately 6 W and the power required during thrusting is 900 W for the augmentation heater and valve actuations.

Attitude Determination and Control System

Overview/Requirements

The attitude determination and control subsystem (ADCS) acts to stabilize the satellite and control it through all of its mission operations and day-to-day housekeeping maneuvers. The design consists of a zero momentum, three axis-stabilized systems driven by payload imaging requirements. Those requirements are

- Reference maneuver of three 60° slews allowing for four images in 452 seconds
- Geolocation accuracy of ±10 m (objective), ±20 m (threshold)

The reference maneuver was developed from the requirement for 60 images per day, and by ensuring four pictures per pass we obtain the desired number of images. The remaining requirement is a characteristic of similar imaging satellites. Before components could be selected for the subsystem design, calculations of the environmental disturbance torques and spacecraft moment of inertia (MOI) were determined. With these calculations, the derived requirements were found:

- Reference maneuver torque of 0.1 N·m
- Reference maneuver momentum storage of 17 N·m·s
- Pointing accuracy of 0.0025°
- Torque dipole of 95 A·m^2

The torque and momentum calculations were done for the reference maneuver of three 60° slews in 452 seconds, including time for settling and imaging. The necessary pointing accuracy was determined from the required geolocation accuracy. The torque dipole was calculated from the worst case environmental disturbance torque of 3.81E-03 N·m.

Trade Studies

The key trade study revolved around whether or not to use a reaction wheel system (RWS) or a control moment gyro (CMG) system as the primary actuators. CMGs deliver much more performance than reaction wheels, but at increased cost and complexity. The initial proposed payload demanded a high slew rate from the ADCS system to accomplish a nodding maneuver during imaging. This would allow the payload to point at a target for a short period of time while frame summing as part of the image processing procedure. The slew rate of 1.6°/s at 300 km and short time to build up that slew rate exceeded the capabilities of all reaction wheels investigated that were small enough to be installed in the spacecraft's ADCS compartment. Later in the design process, this payload requirement was lifted; ultimately, reaction wheels were chosen over CMGs primarily due to the nonavailability of a suitable CMG system.

Design Description

The selection of components shown in Figure 12.6 was driven by accuracy and agility requirements levied to the subsystem by the payload. The availability issues for CMGs led to an easy decision to choose reaction wheels as our primary actuators. Due to the high pointing accuracy requirements of the imaging system, a star tracker was the only choice for the primary attitude sensor. The rest of the component selection decisions were made in accordance with our particular orbit and mission.

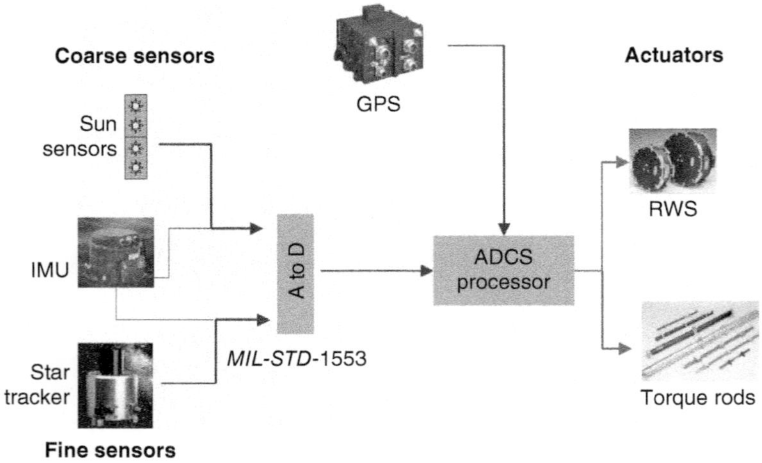

Figure 12.6 ADCS system diagram.

Reaction Wheels Two different size wheels were chosen for the satellite. Since the roll axis would be tasked most heavily, a TW35E1000 1 N·m Goodrich wheel was employed. For the remaining two axes, the smaller TW16B200, 0.2 N·m wheel was chosen. The three wheels were placed in an orthogonal configuration, maximizing the application of torque at the expense of giving up redundancy.

Magnetic Torque rods were selected for reaction wheel desaturation and to act as a backup means of attitude control. Sizing of the torque rods began with calculating each of the worst case disturbance torques and finding an overall worst case torque of 3.81E-03 N·m. From this value a required torque dipole of 95.31 A·m^2 was calculated. Three Goodrich TR100CAR torque rods were chosen with a linear moment of 110 A·m^2. Each rod has a mass of 3.5 kg and requires 2 W.

Star Tracker The Ball Aerospace CT-602 star tracker was chosen to provide the high pointing accuracy required by the imaging spacecraft. This star tracker has an accuracy of 3 arcsec and can be improved to 1 arcsec by employing a Kalman filter and postprocessing. The mass of the star tracker is 5.4 kg and requires 9 W of power.

Inertial Measurement Unit A Honeywell Miniature Inertial Measurement Unit (MIMU) working in conjunction with the CT-602 star tracker should provide the spacecraft with constantly updated high-accuracy attitude information. The MIMU is a Ring Laser Gyro (RLG) assembly with an extremely low bias of 0.005°/hour. System mass is estimated at 4.7 kg. The peak power for the unit is 32 W, but typically draws 22 W.

Global Positioning System Receiver The Monarch Global Position System (GPS) receiver manufactured by General Dynamics was chosen to provide high accuracy position and timing information for the satellite. This signal is designed to be accurate within 20 ns and provides position accuracy within 8.5 m. The typical position accuracy was listed as 3 m. The unit mass is 3.7 kg and requires a peak power of 25 W and steady-state power of 19 W.

Sun Sensors Five AeroAstro MSS-01 Medium Sun Sensors provide coarse attitude knowledge. The sensors are located on five of the six external bus panels. The accuracy of each sensor is dependent on the FOV, providing 1° accuracy at an 18° half-angle FOV and 4° at a 30° half-angle FOV.

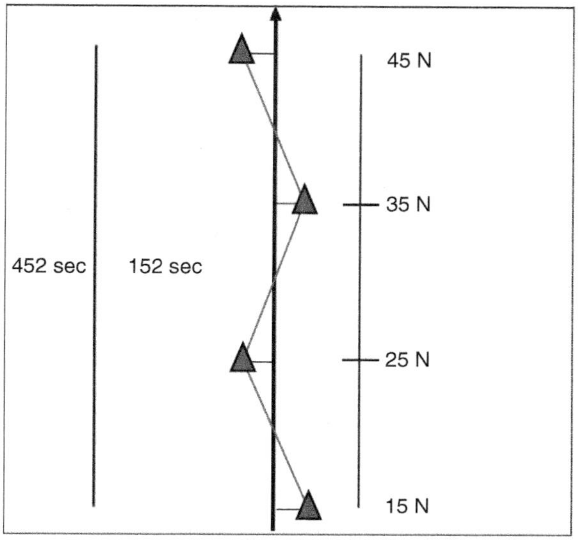

FIGURE 12.7 Reference maneuver.

Matlab/Simulink Analysis

A Simulink model analysis was completed to make sure that the chosen ADCS configuration would feasibly meet the mission requirements and not exceed the specifications of the individual components. The simulated environmental torques in the model were aerodynamic, solar, and gravity gradient. Also, note that a rigid body assumption is made for the dynamics modeling. The controller used in the simulation was a gain-scheduled, proportional integral derivative (PID). The first maneuver considered was the reference maneuver stated in the mission requirements (Figure 12.7). In order to achieve 60 images per day, this maneuver was created to ensure that between the latitudes of 15° N and 45° N, we would be able to capture four images. The primary AORs of interest are situated within these lines of latitude. With 16 revolutions per day at our chosen orbit, we will be able to get the required number of images per day.

It is shown that the reference maneuver called for four images to be taken in 452 seconds could be taken in less than 300 seconds. The required torque remains below the 1 N·m maximum value provided by the TW 35E1000 wheel positioned on the roll axis. It should also be noted that this maneuver utilizes only the roll axis motion. Two additional maneuvers were evaluated in relation to actual targets. The simulation results for both showed that, like in the reference maneuver, iSat is capable of capturing the required images.

Mass and Power Budgets

Tables 12.7 and 12.8 display the mass and power budgets for the ADCS subsystem, respectively.

Thermal Control Subsystem

Introduction

Thermal control subsystem is designed to maintain all spacecraft components within their normal operating temperatures under all design circumstances encountered during the spacecraft's design life, while also adhering to the requirements outline in the ORS ISET standards. iSat's design orbit

Component	Weight each (kg)	Weight total (kg)
Reaction wheels	13.6/8	29.6
Torque rods	3.5	10.5
Star tracker	5.4	5.4
IMU	4.7	4.7
GPS	3.7	3.7
Sun sensors	0.036	0.18
Total	—	54.1

Table 12.7 ADCS Mass Budget

Component	Power SS (W)	Power peak (W)
Reaction wheels	68	925
Torque rods	5.1	8.1
Star tracker	9	9
IMU	22	32
GPS	19	25
Sun sensors	—	—
Total	123.1	999.1

Table 12.8 ADCS Power Budget

is a sun-synchronous 6:00 a.m.–6:00 p.m., 300 × 600-km altitude orbit inclined at approximately 98°. The orbit period is 93.5 minutes and depending on the time of year, little to no eclipse. The LEO orbit required analysis for heating not only from the sun, but also albedo and earth infrared heating effects. Due to iSat's orientation, the radiator is always looking at deep space, while the rest of the spacecraft exposed to the sun is covered with multilayer insulation (MLI) to minimize heating and cooling of the spacecraft during daylight and eclipse. During normal simulated operations, iSat produced approximately 255 W of heat energy. During safe mode, this dropped to 166 W.

Requirements

The top level requirements for this design were governed by the following ISET standards:

- At the payload/bus interface, the bus side must remain between −30°C and +55°C.
- The payload side of the interface must remain between −30°C and +55°C to ensure 10 W conductive heat transfer.
- The spacecraft bus surfaces shall not radiate more than 10 W to the payload and the payload shall not radiate more than 60 W to the spacecraft bus.

The derived requirements were based upon the type of equipment that was chosen by each subsystem. The battery and propellant are the most temperature limiting components on iSat. The lithium-ion battery's allowable temperature range is 15–35°C while the propellant must be maintained above 2°C. The rest of the electronic components have a wider temperature range for operation: −20 to +50°C.

Trade Studies

The following trades were considered in the initial design of the iSat thermal control subsystem: heat pipes, thermal doublers, or radiator mounted components; louvered, flat plate or honeycomb type radiator; paint coatings or multilayer insulation; thermostat and heaters or heat switches; and variable emissivity coatings versus paint coatings or other material with high emissivity and low absorptivity on the radiators. Due to the relatively low-heat loads, and one surface always looking into deep space, thermal control became simple. Heat pipes, louvers, and variable emissivity material were not needed. Radiator with silver Teflon surface finish, heaters, MLI, and paints were the main thermal control elements.

Thermal Design

Preliminary thermal analysis indicated approximately 0.70 m^2 of radiating area is required to dissipate the worst case heat load and approximately 170 W of heater power would be required to keep critical components of the spacecraft and fuel warm in cold case. The radiator is divided into three sections, each radiating heat from different equipment. The transponders are mounted to the section of the radiator opposite the payload and center section of the radiator dissipates the heat generated by the reaction wheels and other internal components. The batteries and the avionics boxes are mounted to the section of the radiator that is located near the payload. The radiator has 10 mils silvered Teflon surface finish with 966 acrylic adhesives attaching it. This particular finish provides the high emissivity and low solar absorptivity necessary for the radiator to work properly. The remaining areas of the spacecraft as well as the payload/bus interface are covered in MLI with allowances made for protruding instrumentation. The MLI consists of 25 layers of 5-mil aluminized Kapton, coated with silicon oxide with approximately 8 mils polyester Dacron netting in between each layer. The layers are sandwiched between two Kevlar/polyamide trilaminate sheets. The aluminized layers as well as the Dacron netting are provided by Sheldahl Corporation. The interior of the spacecraft bus is painted black with Chemglaze Z306 to facilitate radiant heat transfer among the internal components.

The heaters in the spacecraft are lightweight heater strips manufactured by Tayco Engineering. The heater strips provide 1 W/cm^2 and are lightweight at 1.5 g/cm^2. For temperature control, the 271 series thermal switches manufactured by Honeywell with a mass of only 8 g each and rated for 100,000 cycles were used. Because of tight temperature control, batteries were thermally insulated from the rest of the spacecraft with separate radiator and heater control. The payload was modeled primarily to analyze the interface between itself and the spacecraft bus, as per the ISET requirements. The exterior of the payload is painted white with Chemglaze A276. This analysis indicated that in conjunction with the cryogenic cooler, two radiators approximately 0.232 m of radiating area is required to dissipate the worst case heat load and maintain the focal plane at −163°C (110 K). These radiators will consist of a stiffened flat aluminum plate coated in silvered Teflon. For thermal control, the calculated masses are 0.367 kg for MLI, 5.62 kg for radiators, 0.070 kg for patch heaters and thermostat, and 0.5 kg for coatings and conductive fillers. Total mass is 6.557 kg. Power for heaters is 174 W.

Modeling with I-DEAS TMG

iSat was modeled using I-DEAS TMG thermal analysis software. The thermal analysis proceeded in a measured manner, increasing in complexity and fidelity until a satisfactory model was achieved with credible results. Next, the satellite bus was modeled with the solar arrays and all of the satellite's subsystem equipment attached using the same model used in the structural analysis. The payload model built by the payload engineer was utilized in the I-DEAS thermal model. Since the primary concern for thermal control of the payload was the temperature of the focal plane and the size of the radiators, it was determined that the electronics suite and the cryogenic cooler would be modeled as nongeometric units. For simplicity, it was assumed that both the electronics suite and cryogenic cooler would be attached to the circular mesh in the center of the electronics deck.

Each material in the spacecraft was defined by its thermal and optical characteristics. For simplicity only, aluminum, honeycomb aluminum, MLI, titanium (propellant tanks), carbon fiber (payload), and solar cell material were modeled. Then the entire structure was meshed component-by-component, face-by-face until all parts of the satellite had been meshed. Thermal couplings between components were also performed. This was necessary to ensure that conductive heat transfer was simulated between connected components. Finally, the orbit was defined in I-DEAS TMG. A sun-synchronous orbit from 300 to 600 km was created. Thermal heat loads were calculated 24 times per orbit.

Results for 6 A.M.–6 P.M. Orbit

The results of the hot case simulation are shown in Figure 12.8. The batteries operated near the upper end of their temperature range and were the most challenging component to keep within its limits. The heater requirement was 150 W for the batteries, and appeared to be continuously required. The hydrazine tanks maintained a temperature of approximately 5°C with a combination of 10 W from the heaters and 2.5 cm of double aluminized Mylar wrapped around each tank. The focal plane temperature reached 109 K during orbital maintenance, but it was not a concern since iSat would not be imaging at that time and the focal plane temperature decreases back down to 110 K during imaging time intervals. All other components were able to be maintained within their required temperature bands. Although the temperatures at the payload interface are much lower than that required by ISET, the difference between the bus side temperature and the payload side temperature is less than $\Delta T = 85°C$, therefore, there is still less than 10 W conductive heat transfer.

Similar simulations were made for cold case and the temperatures for the equipment were found to be within allowable limits.

Electric Power Subsystem

Overview/Requirements

The function of electric power subsystem is to provide power up to EOL power for all operating modes of satellite. The electrical power subsystem (EPS) is designed to produce an average beginning of life (BOL) power of 773 W and an average end of life (EOL) power of 735 W for a 2-year design mission. The solar array area required is 3.015 m^2, using 1,116 solar cells. Lithium ion batteries provide a 150 Ah or 540 Wh energy capacities that will be used for peak power loading conditions. The power management and control will be performed by the IAU, and will monitor

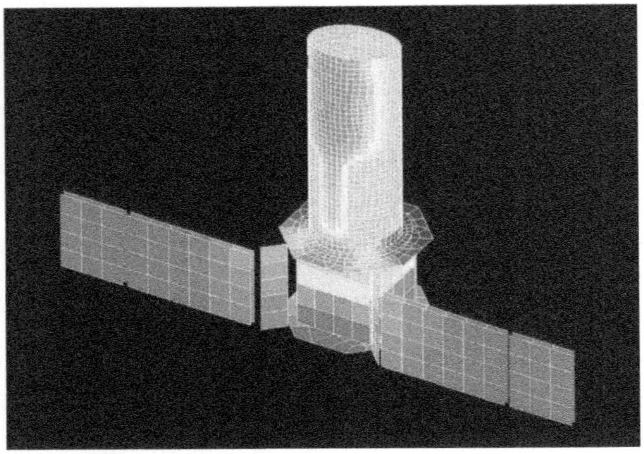

FIGURE 12.8 Hot case exterior temperature profile.

battery status and distribute power to subsystems up to 23 outputs. A separate battery charger was incorporated in the design to provide individual cell charging for safety reasons. This system was designed to handle seven modes of operations (modes). The orbit maintenance mode was found to consume the greatest power of all these missions. Recharge time of a 50% DOD battery during the worst case solar season (summer solstice) will take 8.2 orbits.

Engineering Design

The engineering design for the EPS was broken into the categories of power generation, storage, regulation and control, and distribution.

Power Generation—Solar Arrays

Selected Solar Cells The drivers for choosing the solar cells were primarily based on high solar cell efficiency, space-flight qualified heritage, and the manufacturer's reputation. The solar cell chosen for the design was Emcore's 28.5% BTJ Triple-Junction solar cell, primarily due to its highest fully space-qualified efficiency. Solar cell efficiency decreases with increased temperature. Since the efficiency for the solar cells are given with a 28°C operating temperature, design calculations were made to predict the actual temperature that the solar cells in orbit would experience and account for the reduced efficiency.

Selected Array Deployment Component The Starsys AH-9060 powered hinge system was chosen for its light weight of 0.29 kg each and actuator extension performance of 100 seconds/90° after a 90-second warm-up period.

Solar Array Design for iSat The solar arrays were designed to an area of 3.015 m^2 with 4.03 m^2 of available surface area. The remaining area would house shunts for the purpose of power conditioning from power coming directly from the arrays to the subsystems. The solar arrays also incorporated patch antennas for TT&C. Figure 12.9 depicts the solar array design.

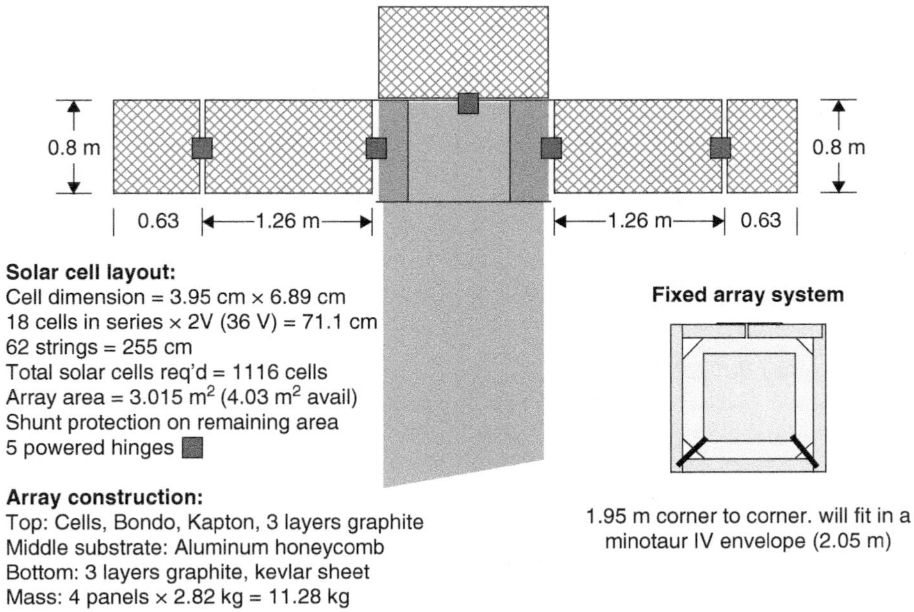

FIGURE 12.9 iSat solar array design.

The EPS was designed with a bus voltage of 28 V dc. To meet this requirement and account for losses over the life of the satellite, the solar arrays were designed to generate 36 V dc. The solar flux at winter solstice was used to size the solar cell strings with each cell estimated to produce 2.05 V. Each string was designed with 18 cells in series to produce 36 V dc per string. The individual cell height of 3.95 cm stacked 18 times produced the 71.1 cm array height requirement. The individual cell length of 6.89 cm for 62 strings produced the 255 cm length requirement. A total of 1,116 cells are required to fill the 3.015 m^2. The arrays were designed with an aluminum honeycomb substrate and three sheets of graphite on both faces. The front side has a Kapton layer for insulating purposes with the solar cells bonded to the Kapton. The back side has a Kevlar sheet on top of the graphite. A cover glass is typically on top of the cells for radiation fluency purposes, but is not included due to the low proton fluency at 300 km LEO orbit and the short 2-year mission.

Power Storage—Batteries

Battery Selection The battery selection drivers were long life cycles, high energy capacity, mass, size, DOD less than 60% (ISET requirement), space-flight qualified heritage, and the manufacturer's reputation. The battery chosen is the Quallion QL015KA lithium ion battery due to its high energy capacity of 15 Ah and low mass of 0.38 kg each. Though the manufacturer's specification sheet advertises greater than 100,000 cycles, the actual tested number of cycles to date is currently 18,000. The Quallion QL075 KA battery was also considered, but was not chosen due to its mass of 1.82 kg each, which is almost 4.8 times more mass than the QL015KA.

In order to provide a battery bus voltage of 28 V, 10 of the 3.6 V batteries would be connected in series to provide 36 V dc. Again, due to losses and degradation over time, it is better to overdesign the bus voltage and have the power regulator manage the step down to 28 V. One concern of using lithium ion batteries is the fire potential due to overcharging. Individual battery charging will be required to prevent these situations as will be discussed in the next section.

Battery Charger Selection The selected battery charger design component is the Mitsubishi 360 W/16 output battery charger. The unit provides individual direct cell charging in the range of 3.0–4.5 constant voltage. Though its mass of 7.9 kg is a trade-off, the battery charger is required for safety purposes to prevent battery fires.

Power Control and Regulation—Computer

The power control unit selection drivers consist of the ability to regulate battery charging from power produced by the solar arrays, the ability to control power distribution to all spacecraft subsystems, reasonable mass/size, and space-flight qualified heritage. The IAU from Broadreach Engineering was chosen for its multisystem avionics package at only 5 kg per box. The IAU-boasted EPS, ADCS, and TT&C management with ability to have 2 GB of storage in its CASI board for its 400+ MIPS processor board. The power distribution solution was provided by the LASI boards for the Broadreach IAU box shown.

EPS Estimated Mass Budget

This design met and exceeded the initial mass budget of 70 kg, as seen in Table 12.9, due to the improved technology with solar cell efficiency and lithium ion batteries.

Design Simulation and Analysis

The EPS power profiles were modeled in Simulink for seven modes. These modes are

1. Mode 1—Transit, nominal power state with no imaging, transmitting, and orbit maintenance
2. Mode 2—Transit (receive and transmit for 9 minutes)
3. Mode 3—Image (slew 5 minutes and no transmit)

Solar arrays	11.3 kg
Deploy mechanism (5 × 0.29 kg)	1.45 kg
Batteries (10 × 0.38 kg + case)	4.0 kg
Battery charger	7.9 kg
IAU (2 × 5.0 kg)	10.0 kg
Wiring (4% of 422 kg)	16.9 kg
Total Mass	**41.3 kg**

Table 12.9 EPS Mass Budget

4. Mode 4—Image (slew 5 minutes and transmit 9 minutes)
5. Mode 5a—Orbit maintenance (burn every day)
   - Mode 5b—Orbit maintenance (burn every 7 days)
   - Mode 5c—Orbit maintenance (burn every 14 days)
   - Mode 5d—Orbit maintenance (burn every 21 days)
6. Mode 6—Safe mode (receive only)
7. Mode 7—Deployment mode (battery power only)

iSat was designed for constant sun conditions of the 6 a.m.–6 p.m. Terminator orbit and the battery were primarily required for peak power loading conditions during the various modes. It was found that orbit maintenance required the greatest power consumption. For this reason, four different orbit maintenance power profiles were created to provide the operator the boundaries to work within for orbit maintenance.

Cost Analysis

The cost goal by the sponsor, operationally responsive space office, for a production unit in a multiple buy is $40 million for each space vehicle.

Estimation Method

iSat Bus A parametric cost model was the primary cost estimation method employed for the iSat bus. Parametric cost models are based on equations, or cost estimating relationships (CERs), relating cost as a function of one or more spacecraft parameters such as mass or power. The CERs that comprise a parametric cost model are developed using cost data from previous satellite programs. The accuracy of a parametric cost estimate depends on the similarity of the system being assessed to the systems that were used to develop the cost model. Parametric cost estimates are often employed in the early stages of spacecraft design when detailed designs and development schedules are unavailable. Since vendor cost data was not available for every component and detailed knowledge of manufacturing and assembly processes are beyond the scope of this course, a parametric cost estimate was chosen for estimating the cost of the iSat bus. The most applicable cost model available was the Small Satellite Cost Model 2007 (SSCM07). The SSCM07 is based on small satellites from 10 to 1,000 kg that rely on existing technology with reduced redundancy and complexity, simplified systems engineering, and shorter lifetimes. The iSat bus fits within these criteria. The SSCM07 can be used to estimate the cost of developing and producing one spacecraft, not including concept development, operations, launch vehicles, upper stages, and associated ground equipment. Contractor fees and incentives, government costs, and concept development costs are

Item	Cost FY09 ($ million)
iSat bus initial acquisition cost	52.9
Average cost of first 5 buses	30.1
Average cost of first 10 buses	26.4
Total cost of 5 bus buy	150.6
Total cost of 10 bus buy	264.2

Table 12.10 iSat Bus Cost Estimate

not included in the estimate. Furthermore, the SSCM07 does not estimate payload, or payload software costs, except for the cost of payload-to-bus integration.

IR Payload A hybrid engineering build-up/analogy cost estimate was the primary estimate used for the 80-cm IR payload. An engineering build-up estimate, often called a bottoms-up estimate, is obtained by accumulating costs from the lowest level of detail within the work breakdown structure (WBS). SAIC provided a bottoms-up/analogy estimate for a 75-cm IR payload that SAIC was designing. The cost difference between an 80-cm payload and a 75-cm payload was deemed negligible.

Projected Cost

iSat Bus Cost Estimate Since the iSat bus is compatible with any payload that meets ISET standards, the bus may be purchased without the 80-cm IR payload. For this reason, the bus cost estimates are shown separately in Table 12.10.

The $52.9 million initial acquisition cost in Table 12.10 is the 80th percentile estimate obtained from the SSCM07 risk analysis with an added 10% contractor fee. The low and high percentages in the SSCM Cost Risk table were set to zero, except for the ADCS and IA&T estimates. Pointing control for the iSat bus is 0.001°, which is an order of magnitude more accurate than all of the spacecraft in the SSCM07 database. Vendor cost data was collected on iSat's ADCS components and it was determined that material alone would most likely cost around $4 million, which was equal to the SSCM07 estimate for ADCS recurring costs, including labor. For that reason, the ADCS most likely estimate was increased by 5%, the low estimate was set to 5% below the most likely estimate, and the high estimate was set to 25% above the most likely estimate to create a triangular probability distribution. The IA&T cost element was also a concern because the ORS concept relies on increased acceptable risk and decreased testing. As a result, the low estimate for IA&T was set to 30% below the most likely estimate to create a triangular probability distribution. A 95% learning curve was applied to the 80th percentile bus recurring cost, resulting in a decrease in the cost of each successive iSat bus.

80-cm IR Payload Cost Estimate The IR payload initial acquisition cost, excluding software development, is estimated at $21.05 million. This estimate is a result of a Monte Carlo simulation performed on the payload elements of the SAIC cost estimate using Crystal Ball with 100,000 trials (Figure 12.10). The 80th percentile estimate was chosen and a 10% contractor fee was added.

A 95% learning curve was applied to the 80th percentile payload recurring cost, resulting in a decrease in the cost of each successive IR payload. The recurring costs of the payload were estimated at 57.8% of the total cost provided by SAIC.

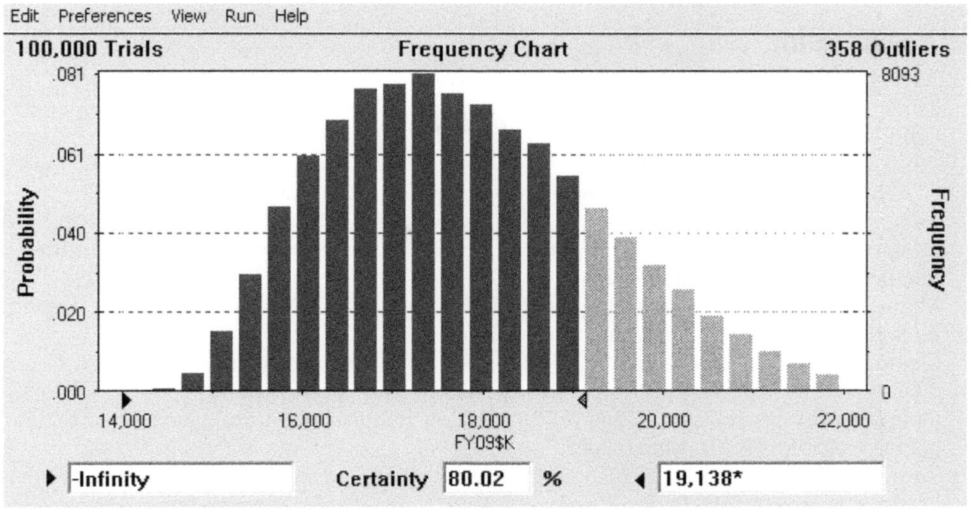

Figure 12.10 Result of Crystal Ball Monte Carlo Simulation on IR payload estimate.

Item	Cost (FY09, $ million)
BAA requirement	40.0
iSat initial acquisition cost	71.9
Average cost of first 5 satellites	42.3
Average cost of first 10 satellites	37.4
Total cost of 5 iSat buy	211.5
Total cost of 10 iSat buy	373.6

Table 12.11 iSat Cost Estimate Summary

iSat Cost Estimate Table 12.11 displays a summary of the iSat cost estimate. The initial acquisition cost includes both recurring and nonrecurring costs, including development costs for one iSat with an 80-cm IR payload. The $71.9 million initial acquisition cost is the 80th percentile cost resulting from a 100,000 trial Monte Carlo simulation performed using Crystal Ball. The inputs to the simulation included each element from the SSCM07 iSat bus cost estimate and each element from the SAIC payload estimate. The elements were correlated using the SSCM correlation coefficients of the SSCM07 manual, and a correlation coefficient of 0.2 for each payload element to the other payload and bus element.

A 95% learning curve was applied to the 80th percentile recurring cost, resulting in a decrease in the cost of each successive satellite. If development and other nonrecurring costs are included, the average cost per satellite when 5 satellites are purchased is about $42.3 million. Due to the learning curve, when 10 satellites are purchased, the average cost per satellite decreases to about $37.4 million, meeting the $40 million requirement.

Further Reading

1. Agrawal, B. N. 1986. *Design of Geosynchronous Spacecraft*, Prentice-Hall, New Jersey.
2. Brown, C. D. 2002. *Elements of Spacecraft Design*, AIAA, Washington, DC.
3. Gilmore, D. 2002. *Spacecraft Thermal Control Handbook*, vol. I: *Fundamental Technologies*, Aerospace Press, El Segundo.
4. Griffin, M. D. and French, J. R. 1991. *Space Vehicle Design*, AIAA, Washington, DC.
5. Kellams, G. and Leonard, C. "Small Satellite Producibility, Affordability Approaches and TacSat 4 Design for Manufacturing and Assembly Results," 22nd Annual AIAA/USU Conference on Small Satellites, 2008 (SSC08-I-6).
6. Mahr, E. M. 2007. Small Satellite Cost Model 2007 User's Manual, Aerospace Report No. ATR-2007(8617)-5, The Aerospace Corporation.
7. Operationally Responsive Space (ORS) General Bus Standard (GBS) ORSBS-002, December 2007.
8. Operationally Responsive Space (ORS) Mission Requirements and Concept of Operations (CONOPS) ORSBS-001, March 2007.
9. Wertz, J. R. and Larson, W. J. 2010. *Space Mission Analysis and Design*, Springer, New York.
10. Wertz, J. R., Everett, D. F., and Puschell, J. J. 2011. *Space Mission Engineering*: *The New SMAD*, Microcosm Press, Hawthorne, CA.

PART 2

Concurrent Engineering

Robert Stevens

12.3 Introduction

Aerospace engineers are responsible for designing, integrating, and testing complex systems with many subsystems and interfaces. Ultimately the goal is to create a system comprised of individual elements that meets mission objectives within cost and schedule constraints. Because subsystem behaviors are highly coupled with each other, design decisions made for one subsystem will impact the behavior of all the others thus affecting total system performance. When conceiving of a system, how do aerospace engineers comprehend all the design options, make decisions, and develop a system holistically that achieves acceptable performance characteristics? Just like writers will outline their main points before writing an essay and artists will rough in shapes on canvas before fleshing out a detailed painting, systems engineers will model whole system performance before diving into the detailed design. One approach is to start with a baseline configuration guess given the mission objectives, then refine subsystem concept designs sequentially. When one subsystem design is estimated, the engineer will pass key design parameters to the next subsystem engineer to estimate their subsystem characteristics, and so forth. Continuing to iterate sequentially will eventually converge on a design solution (hopefully!). The problems with this sequential approach are as follows:

- It can take a long time to converge to a design solution.
- Since changes in one part of the system affect the performance of all the others, the process can get stuck in a repeating loop without converging.
- Requirements may evolve throughout the process, so the engineers developing the various system elements may be using different sets of requirements.
- The focus is on individual elements and not the whole system.

An alternative approach is to develop the whole system design simultaneously. Instead of developing subsystems independently and throwing key design parameters over the fence to the next subsystem engineer, all the subsystems may be developed in parallel with a common set of requirements and a shared system model. Some advantages to this approach are as follows:

- Fast convergence to feasible design.
- Focus is on whole system, not individual piece parts.
- Captures design "ripple effects" throughout the whole system in near real time.
- All subsystem engineers and models share the same common centralized model of the whole system so all are working to the same requirements, constraints, and assumptions.

What Is Concurrent Engineering?

Concurrent engineering (CE) is a method by which multidisciplinary teams of experts work together simultaneously to rapidly create and evaluate total system concepts that meet stakeholder needs. Teams are often comprised of engineers, scientists, payload specialists, customers, cost analysts, and other stakeholders. They often use a connecting tool that links together each of the customized subsystem tools to enable near real-time system model updates as the concept design is matured. Using this parallel engineering approach enables rapid and comprehensive concept design.

Why Is Concurrent Engineering Valuable?

Using a concurrent engineering method helps engineers choose the right system design solution to meet mission goals, and forces key conversations with stakeholders early in the life cycle of a project to avoid problems later. Ultimately, spending a little time up front in developing concepts using CE will save cost and/or enhance performance in the long run. When defining the mission goals and the initial concept of an aerospace system, engineers are often tempted to breeze through analyzing alternatives and jump right into a design solution. In practice, however, it is necessary for us to step back and assess our options because most of the total life cycle costs depend on getting it "right" at the start. Once a program has started marching down a selected path, opportunities to make significant cost-saving or performance-enhancing decisions dwindle away. In fact, it has been said that of all decisions affecting the life cycle costs of complex mechanical systems, over 70% are made during the concept definition phase (Ullman 1997; A3 Study 1983). Using CE early to define a system helps to rapidly explore trade options, select the right one, and provide analysis to support decisions makers (Figure 12.11). Holistic system engineering is woven into the very fabric of the concurrent engineering method thereby reducing the risk of having an assemblage of well-designed individual subsystems that will not work together. Although the concurrent engineering process may be applied in later phases as well, it is most often used in this early phase.

Common Uses

Blank Sheet Design

When considering a new spacecraft concept, it is common for engineers to quickly estimate total system size, mass, power, and cost based on parametric or analogous relationships obtained from historical satellite data. Sometimes, however, it is necessary to start a new concept. Estimates based on parametric or legacy spacecraft depend heavily on the relevance to the current mission and the accuracy of the historical data. These methods fall short when the goal is to generate a concept

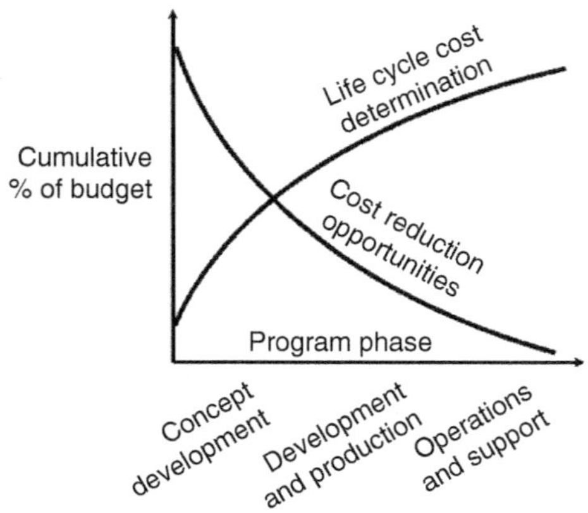

FIGURE 12.11 Cost curve across system life cycle.

design for a system where historical data is insufficient or irrelevant, mission requirements are not well defined, or the intent is to break from tradition and develop something new. Furthermore, these approaches do not produce supporting comprehensive design analysis like a custom "made to order" system model would. In this case, the CE team will develop a system design from the bottom up, often to support one or more of the following activities:

- Determining concept feasibility
- Generating, refining, and validating key mission requirements
- Writing proposals or requests for proposals
- Identifying key design and cost drivers
- Identifying technologies that need to be matured to support new missions

Note that "blank sheet" design is somewhat of a misnomer. It is perfectly fine and even encouraged to use a similar class spacecraft design as an initial guess to jump start the iterative design process.

Trade Study

Considering the astronomical costs associated with the design, development, production, operation, and disposal of many satellite systems, it is no wonder that systems engineers and program managers have a deep interest in exploring many alternative designs. By "white boarding" the concepts and alternatives, decision makers and engineers are able to mature the system's concept of operations (ConOps) and requirements. Establishing clear and comprehensive mission requirements is critical to avoid leaping too far into the design process, just to have to redesign the system later to accommodate new or rewritten requirements that had been overlooked or poorly considered. These brainstorming activities early in the life of a satellite system require methods and tools capable of quickly comparing candidate solutions and providing analysis to back them up. Because the nature of the CE process is to develop quick concepts, it is possible to evaluate many options in a short amount of time. To achieve this response time, most models are simple and produce adequate fidelity results nearly instantaneously rather than high fidelity solutions that can take hours or even days to achieve. The level of modeling fidelity should only be detailed enough so that trade options may be compared against each other with a speed that keeps pace with concepts as they are conceived. This helps avoid "paralysis by analysis" that can often bog down a project.

Design Analysis

The design solutions resulting from the CE process need to be feasible and defendable, therefore it is imperative that this methodology yield supporting analyses. Fortunately, by nature of the bottom up design approach, many of the mathematical models used to generate key parameters shared by the subsystem tools may also be used to analyze key system performance metrics. Analyses come in the form of communications link budgets, payload performance charts, day in the life power system performance, etc.

Engineering Changes

In addition to exploring new concepts, CE also provides a method to rapidly assess the impact of various design excursions on future series systems. Legacy systems like the GPS series may require upgrades, have requirements changes, and contain components that are obsolete. The CE process enables engineers to run "what if" scenarios to provide program managers with the information necessary to make decisions about future development of the series. What if new technologies offer the opportunity to reduce size and power of a satellite? What if shrinking the size of a single satellite presents opportunities to share a ride to orbit saving launch costs? These are the types of questions that CE can help answer.

12.4 Concurrent Engineering Methodology

The interconnected nature of various system elements and the impact they have on system performance requires that tools also be interconnected so that we may observe the ripple effects on the whole system resulting from changes to a single element, requirement, or function. Since tools do not operate by themselves, a CE process is also needed to capture the impacts that our design decisions have on the whole integrated system. This section addresses the tool and the process that comprise the concurrent engineering methodology.

The Tool

To perform a holistic integrated system design using a CE method, a tool is needed to link together each element's individual model, and enable an exchange of information between them through a central parameter repository. This tool may be as simple as a single spreadsheet, in which case a single experienced engineer could conceivably perform a design, or it may consist of an integrated data exchange architecture that stores shared model data in a database accessible by all the members of a CE team. Either way, each subsystem model receives input parameters and requirements from a common parameter exchange which the designer uses to develop their subsystem, and then sends their model outputs to the parameter exchange for use by all the other subsystem tools. Fortunately, a common set of subsystems exists for most spacecraft that can serve as a checklist to ensure all the spacecraft functions are addressed and that the tool is adequately prepared for a mission study. The following are some common subsystems:

- Structure and mechanisms
- Payload and payload communication subsystem
- Attitude determination and control subsystem (ADACS)
- Propulsion
- Electrical power subsystem (EPS)
- Thermal control subsystem (TCS)
- Command and data handling (C&DH)
- Telemetry, tracking, and command or control (TT&C)
- Guidance, navigation, and control (GNC)

In addition to the spacecraft subsystem models, the CE tool may include models from other disciplines, such as

- Astrodynamics
- Ground and ground software
- Flight software (FSW)
- Launch vehicle
- Cost
- Risk and reliability
- Environmental control and life support (ECLS) for human spaceflight

Note that although this list is representative of space systems, with minor changes it could also serve for aeronautical systems. In order to perform a system design integrating all these elements, tool developers need to create the models necessary to produce exchange parameters that will be shared among them. These models often need to estimate size, mass, and power (SMaP) and performance of a subsystem. Estimation may be accomplished using one or more of the following methods: bottom up, parametric, or analytical.

Bottom Up

Just as we decomposed the whole system into subsystems for our blank sheet design, we can further decompose a subsystem into its constituent parts. Estimates for component SMaP and performance are gleaned from catalogs, databases and vendor specification sheets. Although this straightforward method requires more inputs and can be more time consuming than other methods, it yields a custom subsystem solution, often with supporting analysis. This is the best choice when you want to explore new design options that deviate from the norm or that are not well represented by existing systems. Subsystem models commonly represented using this estimation method include payload, EPS, ADACS, TT&C, C&DH, propulsion, and ground. The tool has the task of summing component mass and load power values from the bottom up to estimate the subsystem SMaP which feeds into the total system SMaP. For the most often used subsystem components, the tool can include table lookup values making it easier to observe trades between component selections in real time during a CE team study.

Analytical

When available, analytical equations are useful when relating various design parameters. For instance, the rocket equation relates maneuvering capability in terms of ΔV with propellant mass and rocket performance parameters like the specific impulse. Model elements that use this estimation method include propulsion, EPS, and astrodynamics. Notice that some of the subsystems on this list also use other estimation methods. Methods are not mutually exclusive. For instance the propulsion system model uses the rocket equation to determine the propellant mass, but it might estimate the propellant tank and plumbing mass using another method. The down side to analytical relations is that they are not always available or are too cumbersome for concept design. Modeling some of the more integrated subsystems can involve unwieldy (or nonexistent) simulations requiring a detailed model of the complete system, which we do not have in the concept development phase. The thermal control subsystem often falls in this category, so detailed simulations will have to wait. Not to worry though, there is still one other estimation method that is useful for cases like this.

Parametric

The parametric method estimates subsystem characteristics using empirical formulas derived from historical satellite data. For example, many large satellite systems have a structure dry mass fraction of about 20–30%. It provides quick estimation and can be accurate when the data set is relevant to the system being developed and under similar assumptions. As a word of caution, one must be sure to understand the data that is behind the parametric relationships used for estimation. For example, if designing a microsatellite, you would want to avoid using a parametric estimation method for a subsystem that is based on historical data of large satellites. If you seed your parametric estimation tools with large satellite data only, you will likely end up with a large satellite. Subsystems and elements that cannot be sized rapidly or accurately using another method are scaled empirically from historical trends. These may include structure (mass fraction), TCS (scaled from power dissipation), and software (scaled from complexity). Although not listed above, cost may be estimated using the same three methods.

One useful diagram that helps us understand the parameters shared between subsystem models is the System Model N^2 Diagram shown in Figure 12.12. The diagram lists the major model elements (N of them) along the diagonal of the N-by-N matrix. To observe model exchange parameters of a specific model element, first find its cell along the diagonal. The input parameters to this element are in the same vertical column. The element's outputs share the same horizontal row.

Power			Dissipated power, temp range				Available power	Mass
Load	**C&DH**	Data rate	Heat load					Mass
Load	Ground station access	**Comm**	RF power out	Beam width				Mass, Ant FOV, EMI
Load			**Thermal**				Payload shield, radiator	Mass, radiator req
Load, pointing strategy		Pointing accuracy	Mag torque rod/coil heat	**ADACS**	Momentum dump req		Accuracy and slew agility	Mass, thruster locations
Load			Propellant temp range	Thrust point req	**Prop**		Thruster loc (plume)	Mass, thruster locations
Radiation, Orbit, eclipse	Period, radiation environ	Max slant range, orbit	Solar beta angle, eclipses	Atm density, orbit	Delta V and maneuver schedule	**Astro**	Mission orbit parameters	Radiation environ-ment
Load, Op Modes	Data rate, Op mode	Data rate req	Temp range req	Agility and accuracy		Sensor req	**Payload**	Mass, FOV, baffles, etc.
Load, dry mass, array dimensions	Radiation shielding	Available antenna place area	Avail surface area (radiators)	MOI, drag area, CM, CP	S/C dry mass	Drag area, mission mass		**Structure**

FIGURE 12.12 System model N^2 diagram.

Each off-diagonal cell of the matrix contains the parameters that are exchanged between two element models that are along the diagonal. The purpose of the diagram is not to illustrate physical or data interfaces, but rather to show the key requirements and design parameters that are created and shared by the various element models. This helps the tool developers create the models necessary to transform the given inputs into outputs required by other modelers. For example, the "Astro" model will output a *DV* budget based on the launch and mission ConOps, which the propulsion subsystem model will use as input to the rocket equation to determine the required propellant mass, which will alter the mass properties, which may impact decisions for the attitude control subsystem, and so on. High level mission requirements like payload performance and mission lifetime serve as input to the whole process. If we desire to model cost, ground stations, and other external interfaces, we could add those too; however, the focus of this discussion will remain on the space vehicle design.

The Process

The CE process for concept development of an aerospace system follows a systems engineering approach that is tailored to be collaborative, comprehensive, and quick. This process involves some or all of the following activities:

- Establish mission goals and requirements
- Identify system functions

- Brainstorm configuration options
- Prepare the tools
- Evaluate options and select concept(s)
- Derive subsystem requirements
- Develop selected concept(s)

Establish Mission Goals and Requirements

As with the systems engineering process, the CE process starts with defining requirements, therefore engineers and other stakeholders must work together to determine mission objectives. The initial mission goals do not need be overly defined or comprehensive. We are mainly interested in high level requirements that significantly drive the system size, mass, power, cost, and performance. We can wait on detailed design until after the winning concepts are selected. At the concept design phase, typical mission requirements will address performance in terms of how well, how often, how much, and for how long. Requirements are "soft," i.e. tradeable, during concept development. It is far better to evaluate the cost and performance impacts of requirements decisions up front, before charging down a design path with "hard" requirements and getting locked into an untenable design demanding herculean correction efforts during later development phases like integration and test (I&T). During this fluid stage, computer tools may not even be necessary. An open discussion with the all the experts in the room may suffice. One of the first tasks is to establish the system boundaries. For a space system, are we just focusing on a single satellite, or does our complete system include ground stations, multiple vehicles in a constellation, multiple mission types (like communications relays, imagers, etc.), launch vehicles, and end users? The latter would constitute a space architecture system. For architectures, we would need to consider launch vehicles, constellation performance, and ground station network availability. Although the CE process is applicable to any system domain definition, for the discussion that follows we will focus on a single satellite vehicle system.

Example Suppose we have an objective to develop a satellite mission to estimate growth and deforestation rates of selected forests in the midlatitudes. We might write some initial mission requirements and constraints that look like this:

- Estimate tree density for selected forests between 60°S and 60°N latitude and monitor changes (within ±50 trees per acre per week) for 15 years
- Monitor up to four forests at a time per week
- Use a rideshare to get to orbit instead of a dedicated launch
- Use single ground station for command and control (C2) and data downlink at U.S. Naval Academy, Annapolis, MD

Notice the first two requirements answer key questions that will drive design decisions without providing the solution. The CE team must resist the urge to charge straight into design solutions by first fully comprehending the problem. We have addressed the basic questions of how well the system must perform, how much, how often, and for how long. Based on this minimal set of requirements, we can already recognize the direct implications for the mission orbit, communications ConOps, data storage, structural configuration (for rideshare constraint), payload sensors, and mission life. A requirement not explicitly stated here, but implied nonetheless, is to keep costs low ("rideshare" is probably a cost-driven constraint). Remember that in the early concept development phase, requirements are not written in stone. They are negotiable and therefore we are free to change them as we perform the concept design to save cost or enhance performance. This aspect is unique to the early phases of system development since requirements must solidify as the system design matures and opportunities to make changes diminish. Requirements that are changed in later phases can often lead to "requirements creep," which impact project cost and schedule.

Identify System Functions

Once the initial set of mission requirements have been established, we can identify the functions that the space system must perform to achieve them. In the context of a systems engineering process, this is called functional analysis (NASA 2007). Identifying these functions will help us to do three tasks; identify configuration options, derive specific requirements needed for subsystem design, and prepare the tools with adequate models that relate our design choices with system performance. The list below shows a common set of functions that many spacecraft and launch systems must perform, and trade options to consider for each. Notice that all the functions are expressed as action verbs.

- *Stow:* Solar arrays, antennas, and other external and deployable appendages need to be stowed against the sides of the bus, so the dimensions must be consistent. The stowed vehicle needs to fit inside the payload fairing of the launch vehicle. Other stowage related options include multi-manifesting versus single vehicle per launch which will have implications on spacecraft size, launch vehicle selection and total launch costs.
- *Launch:* The launch vehicle must be capable of transporting the satellite mass to the mission orbit, parking orbit, or transfer orbit. The spacecraft may have to maneuver itself into position if the launch vehicle that it rides in does not carry it all the way to its destination orbit. Ride sharing can offer a less expensive ride to orbit than a dedicated launch since the cost is shared with others; however, this may mean the launch vehicle deposits the spacecraft in a non-optimal orbit for your mission.
- *Maneuver to mission orbit:* As previously mentioned, the launch vehicle may not drop the satellite off in its final orbit. If this is the case, then the satellite will need to execute a series of orbital thrust maneuvers such as
  - Altitude and/or plane changes.
  - Orbit phasing maneuvers, useful for populating an orbit plane. In this case, one can trade shorter time to station for additional propellant.

  Other trades include using conventional versus alternative propulsion, like electrical propulsion (EP) or a hybrid. The decision is between time to orbit versus propellant mass and, in the case of EP, required power. When maneuvering to GEO using EP propulsion, consider the time spent in the Van Allen radiation belts and the resulting degradation to sensitive electronics and solar cells. In LEO, it is possible to take advantage of natural orbit dynamic perturbations (precession) to achieve desired parameters if the additional time to reach mission orbit is acceptable.
- *Deploy:* Dimensions of deployment mechanisms and structures for arrays, antennas, and booms need to be considered in both the stowed state as well as the deployed state to account for shadowing on arrays and field of view for antennas.
- *Command and control the spacecraft:* Commanding the spacecraft involves receiving the command and control (C2) signals from the ground, then disseminating and executing the commands on board the spacecraft using its TT&C and C&DH subsystems. Operators on the ground send low data rate C2 signals to the satellite telling it what to do. Trades include
  - *Continuous positive control versus non-continuous.* If continuous, then non-GEO satellites will require crosslinks to relay signals, otherwise they will only receive direct uplinks when over command ground stations.

- *RF band.* These C2 signals usually contain little actual data, thus low data rate. We may use a UHF or S-band, so for a bottom up estimate, hardware should be chosen accordingly.
- *Robustness.* More robust systems may include multiple receivers, processors, and antennas for redundancy to reduce risk of losing one system and subsequently losing control of the satellite with a single point failure (redundancy versus single string). An antenna option may be to use omnidirectional or large beam width antennas for better signal reception coverage for various spacecraft orientations which comes at the expense of reduced gain. Thus if the spacecraft loses attitude control, operators can still communicate with it.

- *Downlink health/status:* The spacecraft will send information about itself to the ground operators who monitor its health and status (e.g., voltages and temperatures). The TT&C subsystem is responsible for performing this function. Like the C2 uplink, the data rate requirements for this downlink are usually low and the same trade options apply.
- *Manage spacecraft power:* There are several methods of powering the spacecraft subsystems. All involve generating, storing, and distributing electrical energy, which are actually three subfunctions of the function to "manage spacecraft power." Some of the trade options are listed below.
  - *Solar power versus primary battery.* Very short life missions may use a primary non-rechargeable battery as a power source; otherwise solar power is usually the primary source unless the spacecraft flies beyond the reach of useable sunlight.
  - *Alternative power sources.* For missions where solar power is not a good option, interplanetary missions for instance, alternative sources like radioisotope thermoelectric generator (RTG) should be considered.
  - *Solar collection control strategy.* The solar power collection strategy depends largely on the vehicle steering ConOps. If the payload does not need to be pointed in a particular direction all the time, then the solar arrays may be placed in a "sun bathing" mode to recharge the batteries at a maximum rate during the sunlit portion of the orbit. On the other hand, if the payload needs to face a particular direction most of the time, such as toward earth nadir, then designers need to decide if they can rotate the spacecraft about the pointing axis or gimbal the arrays to achieve better solar power collection (such as yaw tracking). If not, then the fixed arrays will need to be sized to accommodate non-optimal array orientations. This would be the case if a satellite had to maintain a fixed local vertical, local horizontal (LVLH) orientation.
- *Control angular momentum:* This function should be disaggregated into several subfunctions since they vary significantly between missions; therefore, our model needs to account for them. The spacecraft must estimate its orientation, stabilize its attitude, maintain attitude, point, and unload momentum accumulated in momentum exchange devices due to perturbation torques. Sensors such as sun sensors, star trackers and earth sensors are used to estimate orientation. Component selection usually depends on pointing knowledge requirements. There are several methods of establishing and maintaining attitude stability, including momentum bias control, 3-axis control, gravity gradient, and spinning the vehicle. The choice depends on pointing accuracy, stability, and slewing requirements. Mechanisms that may be used to accomplish these functions include momentum exchange devices (e.g., reaction wheels and control moment gyros) and/or inertial controllers (thrusters and magnetic torque rods). If momentum exchange devices are used for attitude control, then an inertial controller must be used to unload the momentum buildup using externally generated torque.

- *Maintain station:* Throughout its lifetime on station, a satellite will drift due to orbit perturbations resulting from atmospheric drag in LEO or multiple gravitating body dynamics in GEO. In LEO, drag can cause the satellite to lose altitude over time depending on the size and mass of the vehicle and the atmospheric density. Station keeping may be accomplished using conventional thrusters (at the expense of propellant), electric propulsion (at the expense of power and time), or other nontraditional means (like electrodynamic tethers in LEO).
- *Perform mission:* This is the sole purpose of launching and operating the spacecraft in the first place. Missions vary widely, but typically fall into one or more of the following categories; science, reconnaissance, weather, navigation, or communications. In general, the functions will include sensing, storing, relaying, transmitting, and processing. It would be difficult to list here all the possible functions and options for the mission types. The mission payload will perform functions that impose requirements on the rest of the system. For example a payload function may be to cool a component to cryo temperatures, requiring a cryo-cooler which may impact radiator design, vibration isolation mounts, and so on. The tool that links together all the subsystems should have the capability to characterize payload performance and relate it to the overall system design. Another function of payloads that collect data is to transmit the data to the ground station. Some of the trades include
    - *Data downlink latency.* Does the data collected by the payload need to be relayed to the ground nearly real time, or may the data be stored and relayed at a later time? This is a big decision for the designer since it impacts the data storage capacity of the C&DH subsystem, and determines if cross links are required.
    - *On board versus off board processing.* Mission planners need to determine if the mission data users will perform processing on raw data downlinked from the satellite, or if the satellite will perform its own processing then downlink the processed data. This decision impacts the payload processor selection and the payload communications capabilities. On board processing (OBP) will require that the spacecraft have more processing horsepower, but it may also reduce the volume of data that needs to be downlinked, thus reducing the demands on the payload communications subsystem.
    - *Separate payload communications versus combined TT&C and payload communications.* In some instances the same communications subsystem can perform the functions of both the TT&C and payload data downlink. If it is desirable to maintain separation of function with the bus, then payloads will contain their own dedicated processing and communication subsystems to downlink their data.
- *Maintain bus and payload temperature within allowable limits:* Several components may have relatively small operating temperature ranges such as batteries and propellant tanks and plumbing. The TCS can maintain this range through active or passive means. Active methods include heaters, cryo-coolers, and louvers. Passive means include MLI blankets, surface finishes, and material selection. A radiator surface is also required to radiate absorbed and generated heat into space. For concept design purposes, a complete bottom up TCS estimate is not possible; however, rough estimates based on similar systems or mass and power fractions are useful.
- *Dispose of the system (deorbit or maneuver to disposal orbit):* Current space policy dictates that a spacecraft shall either deorbit or maneuver to a disposal orbit within 25 years to reduce orbital debris. Satellites in LEO that are subject to atmospheric drag will most likely burn up upon reentering the atmosphere. If the time to reentry is expected to be greater than 25 years, then the satellite will need to actively accelerate the reentry at the EOL.

For satellites above LEO, the best option may be to maneuver the spacecraft into a disposal (or junkyard) orbit where the chance of crashing into an active satellite is remote. Trade options include

- *Conventional thruster versus alternative.* Systems using conventional thrusters for deorbiting require a reserve amount of propellant at EOL for this final maneuver. Electric propulsion systems use less propellant and are lighter, but they require more power and for a much longer time. This maneuver would be at the EOL though when other power loads are light, presumably with the payload turned off. Alternatives that can save propellant mass for satellites that are subject to drag include drag enhancing devices like deployable tethers, drogues and inflatables.
- *Controlled versus uncontrolled reentry (LEO).* If the satellite reentry at EOL needs to be controlled then the final deorbit maneuver will require an impulsive ΔV, probably delivered by a conventional thruster. This maneuver should plunge the orbit perigee much deeper into the upper atmosphere to an altitude of about 50 km to better control the atmospheric reentry point. If controlling the reentry is not critical, then a less demanding perigee lowering maneuver is acceptable to save EOL propellant mass.

Each function on our list may be decomposed into subfunctions, but for purposes of developing a CE model for concept design, we only need to decompose these functions to achieve the desired level of modeling fidelity. For example, "Maneuver to mission orbit" is a function performed by our propulsion subsystem. The propulsion subsystem model uses the rocket equation to convert the total maneuver DV into a propellant mass output, which was our chief purpose for this parameter. No further function decomposition is necessary since it has served its purpose. If estimating the necessary characteristics of a subsystem would be significantly improved by fragmenting it further into its subfunctions, then the tool should be modified to address these details.

ForestSat Example Listed below are some options that we will consider for our ForestSat mission. The configuration options are shown in the trade tree in Figure 12.13.

- *Payload type:* Two methods of forest monitoring to be considered are visible and hyperspectral imaging. The concept would be to image desired forests over multiple passes. Using a visible imager with enough resolution to distinguish large groupings of trees, we could employ algorithms to estimate population in selected areas. To distinguish types of trees and other vegetation we could use a hyperspectral imager, an improvement which would come at increased cost and complexity.

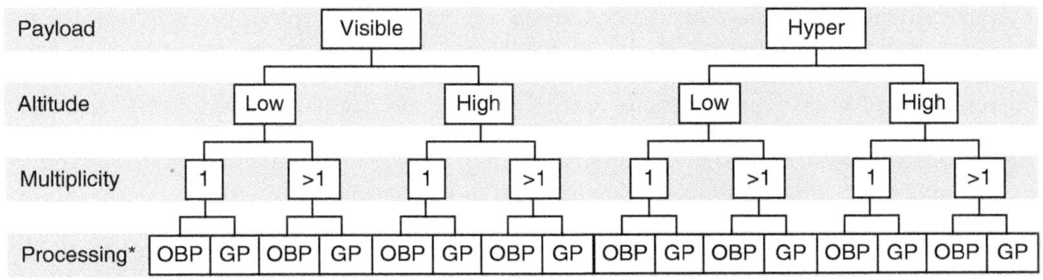

FIGURE 12.13 Trade tree for ForestSat.
*OBP = on board processing; GP = ground processing.

- *Low versus high mission altitude:* High altitudes are characterized by long-dwell times over any given region, but the ride to orbit might be expensive and the imager will be big. At low altitudes the satellite would collect higher resolution images; however, it would pass over target areas and ground stations relatively quickly leaving shorter time to collect and downlink data possibly causing a communications bottleneck if there is lots of data to downlink.
- *Single 15 year vehicle vs. multiple shorter life satellites:* Long-life vehicles will require more design redundancy and will need to factor in greater component degradation (like solar cells), but we would only need to build one satellite that would launch once. Shorter life vehicles launched in intervals to replace older ones may be smaller and cheaper per unit, but they require multiple launches.
- *On board processing versus ground processing:* Processing the forest data on board using image processing may significantly reduce the data that must be downlinked to the ground stations. The cost though is in the flight software development, testing, and to a lesser extent the processing hardware. Storing the raw forest data and then downlinking while overhead a ground station can save on flight software costs, but there will be a greater data volume in the downlinks.

Brainstorm Configuration Options

When exploring the trade space, engineers will find that system performance is at odds with the cost. At this point, cost may be understood in terms of SMaP. Tradeable mission design parameters include mission altitude, downlink ConOps, and satellite class. The CE team may choose to develop a trade tree as shown in Figure 12.13, where the various options are shown for each key design parameter.

Prepare the Tools

We need to verify that we are modeling all the physical subsystems and components that will perform our list of functions, and that our models will relate design decisions to performance and cost. This translation of the functional architecture into a physical architecture is called design synthesis. The common physical architecture elements (subsystems) were already discussed in the subsection, The Tool. Most of the common spacecraft functions on our list are accomplished by at least one of the subsystems in the spacecraft physical architecture. For those functions that are not represented by hardware, we will need to identify some. Again, we only need to synthesize to the desired level of system concept design fidelity. We do not need to create a physical architecture decomposed down to every circuit board.

Evaluate Options and Select Concept(s)

It would be cumbersome or impossible to explore every possible combination of options; therefore, it is often instructive to choose a small set of option combinations to explore, preferably representing different regions of the trade space for completeness (i.e., "pruning" the trade tree). The team can then evaluate this smaller, more manageable set of configuration options. This down selection process may involve developing various concept designs and comparing, or it may simply be a discussion among the CE team members. Although it is wise to avoid moving too quickly into a design solution possibly missing better alternatives, it is also essential to avoid spending too much time exploring trade space and succumbing to paralysis by analysis.

ForestSat Example

- *Payload type:* There is clearly a tug-of-war between payload cost and performance. A high-resolution hyperspectral system would certainly meet requirements, but it may bump the concept up to a larger vehicle size with more complex software. We will instead select a lower resolution visible imager to keep launch and development costs down. The performance risk incurred by making this decision will be mitigated thorough analysis showing that it is "good enough." If it is determined that requirements will not be met then we will revisit the requirements.

- *Low versus high altitude:* We will select a LEO orbit for this mission. There are no driving persistence requirements that would necessitate long dwell times over a region. The ConOps will be to schedule regional mappings up to four regions per pass. The lower revisit rate over each region is not a major concern given that the satellite has weeks to capture regional images.
- *Single 15 year vehicle versus multiple shorter life satellites:* Since sharing a ride as a secondary launch payload to orbit is a requirement, it would be wise to consider standardized rideshare-friendly satellites, like CubeSats. Because of their standard interfaces with host launch vehicles and deployer devices, there are ample ride sharing opportunities, which open the door to the "multiple launch" option. Choosing this path would cause us to forgo some design redundancy (thus lifetime) and performance. It would, however, keep unit costs low and spread out risk over multiple vehicles and launches since all the eggs would not be in one basket. For ForestSat, we will plan on multiple CubeSats launched every 2–4 years to achieve the mission lifetime goal while spreading launch and deployment risks out over time. Budget permitting, this would also allow for incremental design improvements with each subsequent build. If through the CE process we discover that the mission cannot be achieved using a CubeSat, then we will either change our original requirements to accommodate this option (we are allowed to do this!), or redesign our satellite to achieve the requirements as written.
- *On board processing versus ground processing:* To keep flight software and testing costs low, we will put the data processing burden on the ground. We will mitigate the data throughput problem using on board data compression and include plenty of data storage space. On board communications hardware will need to be sized to relay data at rates consistent with access times with the U.S. Naval Academy ground station.

Derive Subsystem Requirements

As we did with the mission requirements, we allocate requirements specific to each of the functions (function allocation) by establishing how well they must be performed, and how long, how often, how fast, etc. When subsystem-specific performance ranges are established, then CE team members will be prepared to select their components, design their subsystems, and perform initial analysis. The following example illustrates some of the derived requirements that will be used to develop ForestSat.

ForestSat Example The requirements in the following table were derived from the mission goals and requirements, and they provide more details that the engineers will need to design the system (Table 12.12).

Develop Selected Concept(s)

After the CE team has down selected and generated requirements, the team can perform more detailed design and analysis on the selected system concept(s). Starting with the requirements set, the team iterates through a procedure of receiving input parameters, updating, and generating output for use by the other subsystems. Design changes to one subsystem, even seemingly innocuous ones, often cause ripple effects throughout the whole design. Iteration is complete when the system design converges to a feasible solution. Depending on the complexity of the payload, the maturity of the requirements, the number of configurations, and the desired level of confidence in the design, and the experience of the team, this iterative phase can take a few hours or a few days.

Element	Functions	Concept design requirements
Astro	Launch, maneuver to and maintain station	Launch to orbit with inclination ≥60° to cover 60°N to 60°S latitude range, and altitude range between 500 km (orbit decay limitation) and 1,000 km (deorbit capability limited <25 years)
Structures	Stow, deploy	Conform to CubeSat specifications
Payload	Collect, store and downlink data	Image with GSD ≤30 m, compress data 2×, store data for up to 8 orbit revs (max time between ground station passes), downlink stored data in ≤5 minutes (average overhead pass at 600 km)
ADACS	Stabilize, point and unload momentum	Maintain three-axis stability and pointing with ≤1° accuracy, slew to up to four target areas in ≤5 minutes, unload momentum using magnetic torque rods
Propulsion	Maneuver to and maintain station, then deorbit at EOL	Perform reentry into earth's atmosphere at EOL before 25 year limit (uncontrolled)
EPS	Manage satellite power	Generate and store energy during worst case power conditions (max. eclipse = 35 minutes), provide power for critical functions in degraded tumbling state (multiple cells for robustness)
TCS	Maintain temp limits	Use passive control methods except possibly a battery heater, place radiator surface on anti-sun side to remove waste heat
C&DH	Command system activities	Provide processor to command bus and payload activities along with memory to store health and status data for up to 8 revs
TT&C	Uplink C2, downlink health	Provide redundant C2 comm systems and antennas that provide connectivity with the ground in any orientation (for robustness)
Ground	Receive and process data	Constraints: Ground station EIRP = 24 dBW, Gr/T = 0 dBi/K Assure ground station access (Annapolis, MD latitude 39°N)

Table 12.12 ForestSat Requirements

Models

The CE process provides the framework with which a team may rapidly and concurrently develop concept designs, and the tool provides the mechanism by which the various elements of the aerospace system are linked. The various elements of the interconnected tool, such as the subsystems and other disciplines, require mathematical models to relate with the whole system. A model is only a representation of reality and therefore can never be 100% accurate. As British statistician George Box famously put it, "All models are wrong, but some are useful." Although models can often be created to represent reality very accurately, for concept design we only need the model detailed enough to meet the concept study objectives (e.g., determine feasibility and identify design drivers). There are many good references containing models useful for subsystem size, mass and power (SMaP) estimation (Wertz et al. 2011; Pisacane and Moore 1994). The purpose of the describing models in this section is not to restate those existing models, but rather to show how they connect together to support the holistic CE process. This list of models is not an exhaustive one addressing all possible spacecraft types and configurations, but it does support bottom-up concept design of a great number of earth orbiting satellites to demonstrate the application of a concurrent engineering methodology.

Payload

A vehicle concept design starts with identifying the payloads, and estimating their size, mass, and power requirements. The spacecraft bus will be designed to support the payloads since this is its sole function. If a payload equipment list is available, then a bottom up system build would provide an accurate estimate for payload size and mass. Mission ConOps would drive payload duty cycle and power loads that must be accommodated by the EPS. If a similar payload exists with known characteristics, then this analogous system would serve well as a starting point. If instead there are no relevant analogous systems, then a parametric estimation approach may serve us well. It would be impossible to list all payload types and parametric relationships between mass, power, performance, and cost for each. As an example however, we will examine ForestSat with one such relationship.

ForestSat Example ForestSat requires an optical system with a GSD of 30 m. The focus of this CE example will remain on the bus design, so we will not discuss the design of the payload, however it is worth pointing out that GSD is proportional to $\frac{\text{range}}{\text{aperture diam}}$, so at higher altitudes a camera would need a larger aperture diameter to maintain the same GSD performance as a camera at a lower altitude. Larger aperture diameter translates to more mass, more power, and more cost. For this reason, we will assume the satellite will operate at the lower end of our feasible altitude range. We will select a COTS payload camera that meets our requirement at 600 km. The camera spec sheet indicates that it includes its own recording and processing as well as 256 GB of memory to store the data. The payload electronics compresses the data per our requirements. The mass and orbit average load power are 0.5 kg and 1.25 W, respectively.

Electrical Power Subsystem

For our concept design, we will use solar cells to provide the primary source of power, and rechargeable batteries as the secondary power source to support spacecraft electronics while the vehicle transits through the earth's eclipse. The electrical power subsystem (EPS) model should evaluate the electrical energy balance between loads and sources to ensure that a concept is feasible, i.e., has power margin. It also needs to size the battery with enough capacity to provide energy during eclipse. Inputs to the model will be the independent electric loads supplied by the other subsystem models. We start by calculating the energy generated during sunlight. The instantaneous power generated by an array during the sunlit portion of the orbit is given by $P_{\text{inst}} = P_{\text{peak}} \cos\alpha$, where α is the sun incidence angle relative to the solar array normal (maximum sunlight when $\alpha = 0$). The peak output power at EOL, P_{peak}, collected by an array pointing directly at the sun with an effective area A_e (including cell packing factor), is given by $P_{\text{peak}} = SA_e\eta$, where S is the solar flux and η is the array efficiency at EOL. For quick estimation, it is useful to compute generated power on an orbit average basis, where the orbit average power (OAP) generated is given as

$$OAP = P_{\text{peak}} c_{\text{avg}} \frac{t_s}{\mathcal{P}} = P_{\text{peak}} \hat{P}$$

where c_{avg} is the average $\cos\alpha$ during the sunlit portion of the orbit, $\mathcal{P}$ is the orbital period, and t_s is the time in the sun. It is assumed that P_{peak} does not itself change quickly on the time scale of an orbit. In general, however, $\cos\alpha$ does change during a single revolution and depends on how the array is controlled. An array may be fixed on a satellite that rotates relative to the sun, or it may rotate about a gimbal axis relative to a spacecraft that is fixed in a particular reference frame. For an array that points directly at the sun for two-thirds of its orbital period around the earth, then $\hat{P} = 2/3$, where $\hat{P}$ is the normalized OAP. Some systems may be allowed to point the arrays to the sun 100% of the time while in sunlight, either by steering the spacecraft or controlling array gimbals, but this may be undesirable because of limitations on spacecraft control or gimbal complexity.

To ensure that we have a positive orbit average power margin we build a system such that the OAP generated is greater than the orbit average load in sunlight and eclipse, P_{Ls} and P_{Le} respectively, such that

$$\text{OAP} > \frac{1}{\mathcal{P}}\left(\frac{P_{Ls}t_s}{\eta_s} + \frac{P_{Le}t_e}{\eta_e}\right)$$

where t_e is the time in eclipse, and η_e and η_e are the efficiencies of paths from arrays to loads and arrays to batteries to loads, respectively. Detailed loads are summed after our first design iteration, but an initial condition is required to start us off. If we have a rough idea of our payload load power, then we can guess a system power based on power fractions derived from satellite data (more to come on estimated mass and power fractions based on average spacecraft data in the section, System).

The satellite EPS will need to generate enough OAP to meet the load with some margin during worst case conditions (see System subsection for discussion on margins). After array properties are defined and we know P_{peak}, we then determine $\hat{P}$ to calculate the orbit average power as OAP $= P_{\text{peak}}\hat{P}$. The following discussion describes models for $\hat{P}$ representing various array orientations and controls.

Arrays That Are Fixed in the LVLH Frame Instead of a "sunbathing" array as previously described, suppose that a satellite must remain in a fixed local vertical, local horizontal (LVLH) orientation with fixed arrays, perhaps body mounted. In this case, the ram and nadir faces must remain fixed in this rotating frame and the sun incidence angle on fixed arrays in these directions will vary as shown in Figure 12.14. The zenith facing array for instance will collect zero power over the terminators when $\nu = -\pi/2$ and $\nu = \pi/2$ and will collect maximum power at the local noon position when $\nu = 0$. The instantaneous cosine factor for the fixed zenith facing array is given as $c_z = \cos B \cos\nu$, where B is the solar beta angle defined in Figure 12.15 and ranges from 0 to $\pi/2$. The normalized

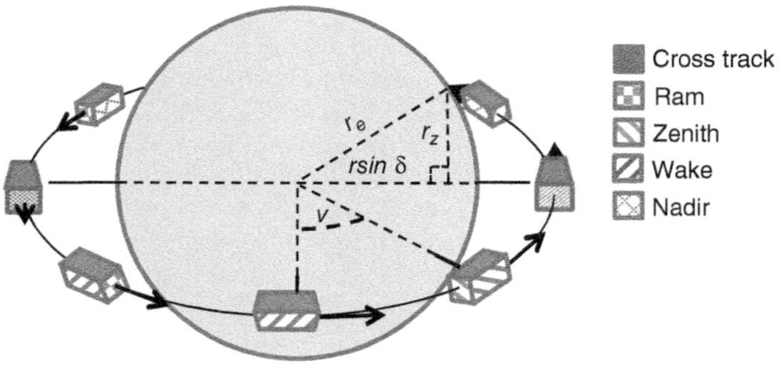

FIGURE 12.14 Sun exposure on a vehicle flying in a fixed LVLH orientation. View is shown from the perspective of the sun.

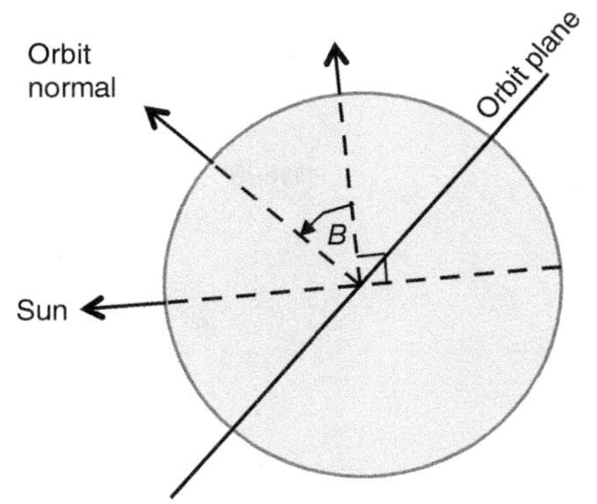

FIGURE 12.15 Solar beta angle B.

OAP is calculated by integrating the cosine factor during sunlight between the terminators. For circular orbits, this may be written as

$$\hat{P}_z = \frac{1}{2\pi}\int_{-\pi/2}^{\pi/2} \cos B \cos v \, dv = \frac{\cos B}{\pi}$$

It is assumed that the solar beta angle does not change much during a single orbit and may therefore be removed from the integrand. Although B will actually vary throughout the lifetime in general due to sun-earth geometry and orbit precession, it only does so on a much longer time scale than a single revolution, so this is a reasonable simplifying approximation. When the orbit perpendicular is directed toward the sun, i.e., $B = \pi/2$, then no sunlight hits the zenith array, i.e., $\hat{P} = 0$. Similar analysis for a fixed array facing the ram direction may be performed by integrating between satellite sunrise and local noon ($-\pi + \delta \leq v \leq 0$), where δ is the angle that spans local midnight to the terminator (Figure 12.16). The eclipse half angle δ is derived by recognizing the right triangle formed by the three sides shown in Figure 12.14, where $r_z = r \sin B \cos \delta$, r is the orbit radius, and r_e is the radius of the earth. With some manipulation, the relationship between the orbit radius, eclipse half angle, and solar beta angle is given by

$$\cos^2\delta \cos^2 B = 1 - R^2$$

whenever $\sin B < R < 1$ (the eclipse season condition) and where $R = r_e / r$. When the orbit is inclined enough relative to the sun such that $R \leq \sin B$, then there are no eclipses, the satellite is always in the sun and $\delta = 0$. The instantaneous cosine factor for the fixed ram face is $c_r = \cos B \sin v$. Carrying out the integration yields the normalized OAP value for the ram face

$$\hat{P}_r = \frac{1}{2\pi}\int_{-\pi+\delta}^{0} -\cos B \sin v \, dv = \begin{cases} \dfrac{\cos B + \sqrt{1-R^2}}{2\pi} & R > \sin B \text{ (eclipse)} \\ \cos B/\pi & R \leq \sin B \text{ (no eclipse)} \end{cases}$$

Full derivations for various arrays facing all the other cardinal directions in the LVLH frame are provided in Stevens (2015). These models are useful for relating spacecraft pointing requirements

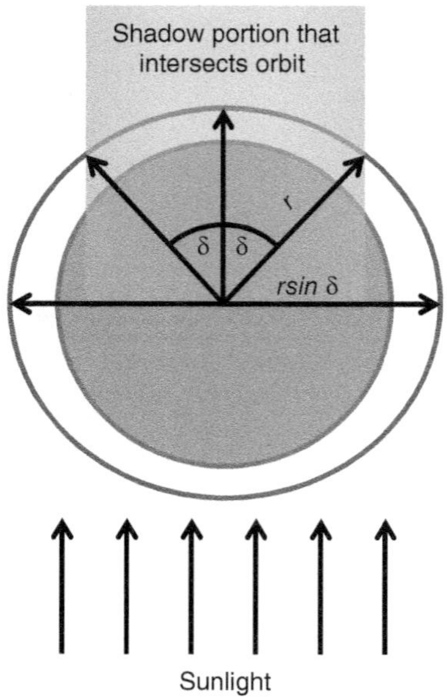

FIGURE 12.16 Eclipse geometry.

(like nadir pointing satellites), solar beta angle conditions, array placement, and array control. Here are a couple considerations.

- *Array placement, orientation, and control:* Although batteries must be sized to accommodate the maximum expected eclipse duration, the $B = 0$ max eclipse case does not always represent the worst case from a solar power collection viewpoint. Depending on array placement sometimes even low B conditions offer better power collection than higher B conditions. Consider a zenith array in Figure 12.14 which favors times when solar B is low. Array placement and control may be arranged to generate a more consistent power output in various beta angle conditions. Care must be taken though to avoid placing arrays where they may be shadowed by deployables, including other arrays.
- *Array performance for orbits with limited beta angle ranges:* Some mission orbits have limited solar beta angle ranges such as sun synchronous or low inclination orbits, therefore the array performance ranges will be constrained as well. For non-sun synchronous orbits, the maximum beta angle is driven by the orbit inclination i and the tilt of the earth with respect to the ecliptic plane (23.5°). The maximum beta angle is

$$B_{max} = \begin{cases} i + 23.5^\circ & 0 < i \le 66.5^\circ \\ 90^\circ & 66.5^\circ < i \le 156.5^\circ \end{cases}$$

  In reality this maximum may drift a little over time, but for estimation purposes, this will suffice.

It is worth reviewing the mission requirements to see if it is possible to yaw, roll, or pitch the array allowing it to seek the sun while still maintaining a specific spacecraft pointing orientation (nadir, ram, wake, or cross track). If this is permissible, then the spacecraft may use attitude control or implement an array gimbal to optimize the array's orientation with respect to the sun.

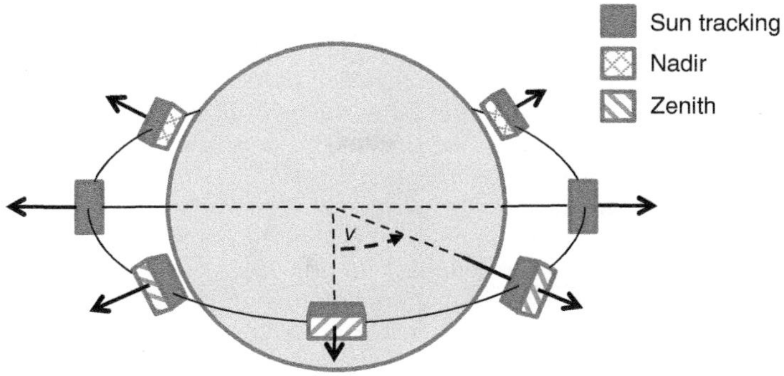

Figure 12.17 Sun exposure on a nadir pointing satellite with yaw steering. View is shown from the perspective of the sun.

Yaw Steered Arrays Suppose that the satellite is not constrained to be fixed in the LVLH frame as previously described, but rather only one side must remain fixed facing toward earth (e.g., nadir as shown in Figure 12.17). Alternatively suppose that the spacecraft must be fixed in the LVLH frame, but is able to control an array allowing it to rotate to an optimal sun pointing orientation. Both these situations permit a sun seeking array a degree of control freedom about the local yaw axis (zenith-nadir axis) to achieve the maximum solar power given the nadir pointing constraint. In this yaw-controlled sun-seeking array case, the instantaneous cosine factor while in sunlight c_{yaw} is given as $c_{\text{yaw}} = \sqrt{1-\cos^2 B \cos^2 v}$, therefore for circular orbits the normalized OAP, $\hat{P}$, is (Stevens 2015)

$$\hat{P}_{\text{yaw}} = \frac{1}{2\pi}\int_{-\pi+\delta}^{\pi-\delta} \sqrt{1-\cos^2 B \cos^2 v}\; dv, \qquad R > \sin B \qquad \text{(during eclipse season)}$$

The $\hat{P}$ equation may be solved relatively easily using numerical integration. During seasons when there are no eclipses ($R \le \sin B$), the eclipse half angle $\delta = 0$. Notice that when the orbit is tilted 90° with respect to the sun, i.e., $B = \pi/2$, then $c_{\text{yaw}} = 1$, meaning that the array can point directly at the sun all the time with no eclipse. Also, when the satellite is over a terminator ($v = -\pi/2$ or $\pi/2$) then the cosine factor at that instant is one regardless of beta angle, meaning that the array can point directly at the sun for that instant.

Roll Steered Arrays There may be other cases where the satellite must point one face toward the ram or wake direction, but there is no restriction on rolling the spacecraft about the in-track vector. In this case, the sun tracking array would achieve an instantaneous cosine factor $c_{\text{roll}} = \sqrt{1-\cos^2 B \sin^2 v}$, and normalized OAP of

$$\hat{P}_{\text{roll}} = \frac{1}{2\pi}\int_{-\pi+\delta}^{\pi-\delta} \sqrt{1-\cos^2 B \sin^2 v}\, dv, \qquad R > \sin B \qquad \text{(during eclipse season)}$$

This form is very similar to the yaw sun tracking case and also applies to spacecraft fixed in the LVLH frame that have a gimbaled sun tracking array free to rotate along the in-track axis. Notice that when the satellite is over the local noon position (i.e., when $v = 0$) that the cosine factor at that instant is unity indicating that the array is able to perfectly point at the sun, regardless of solar beta angle. Incidentally the integral for $\hat{P}_{\text{roll}}$ has a closed form solution since it may be put into the form of an elliptic integral, but numerical integration is perhaps preferable if your tables for incomplete elliptic integrals of the second kind are not handy.

Pitch Steered Arrays If the satellite operates in situations where the beta angle range is low (i.e., orbit plane is nearly edge-on viewed from the sun), then it could be advantageous to have a pitch-controlled array such that the control axis is in the orbit perpendicular direction. In this case, the only loss would be the cosine loss due to the tilt of the orbit plane with respect to the sun. This appears in cases of low inclination orbits like geosynchronous orbits or certain sun synchronous orbits. The instantaneous cosine factor is simply $c_{\text{pitch}} = \cos B$.

The equations for cosine factors and normalized OAP for various array control situations are summarized in Table 12.13, and their solutions are plotted in the nomographs shown in Figures 12.18 and 12.19 for various altitudes, solar beta angles, and pointing situations. During EPS design, the nomographs are useful for simultaneously relating the mission pointing ConOps with array performance. For instance, body mounted arrays facing the zenith direction on a spacecraft that must remain fixed in the LVLH frame will deliver the most power during times of low solar beta angles, but are weak at higher solar beta angles. One design option to consider would be to add a fixed arrays facing the cross-track directions, which will boost power generation during times of low beta angle. Gimballing a deployable array about an axis may also be a design solution at the expense of added complexity. Not all solar array configuration and control cases are covered here, however these estimates serve well in many situations. Also, although they were derived assuming circular orbits, they are also good approximations for orbits with small eccentricities.

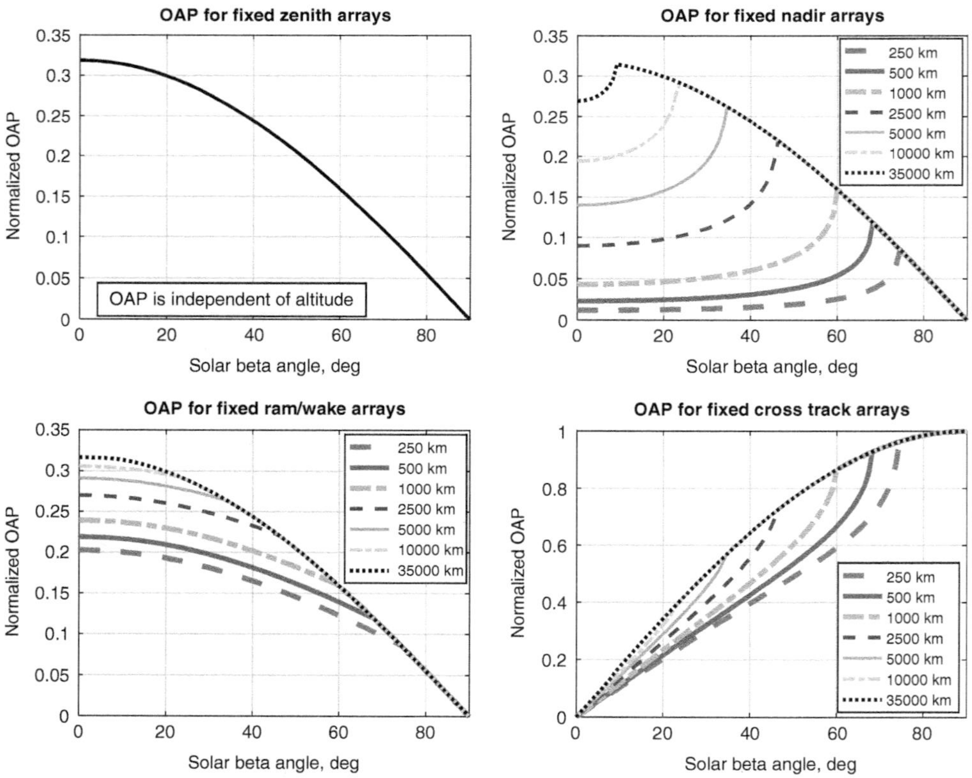

FIGURE 12.18 Solar array performance for arrays fixed in LVLH orientations. The nomographs relate orbit average power (OAP) performance with array location on the satellite and solar beta angle conditions.

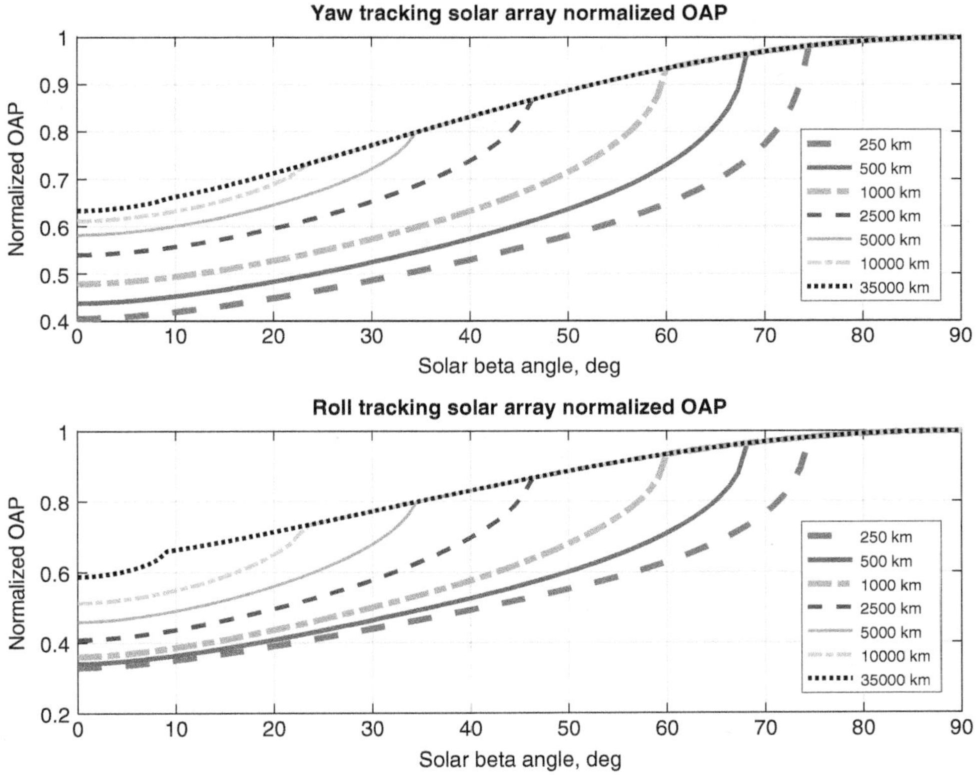

FIGURE 12.19 Solar array performance for steered arrays. The nomographs relate orbit average power (OAP) performance with array control (yaw or roll steering) and solar beta angle conditions.

To estimate the mass of the arrays, multiply the computed array area by the area density that includes cells, blankets and harnessing. Some area density estimates obtained from Reddy (2003) are listed here.

- Rigid planar array: 3.2 kg/m^2
- Flexible planar array: 2.3 kg/m^2
- Flexible thin film array: 0.7 kg/m^2

It might be tempting to use the lightest array option; however, slew requirements must be considered. Rigid planar arrays should be used on highly agile systems performing rapid slews and/or systems with tight pointing requirements, where wing vibration must be minimized. If using deployable arrays, yokes and gimbals must be accounted for as well if applicable.

Battery Design Other functions of the EPS include storing energy during sunlight and dispensing electric energy to the subsystems during eclipse, assuming a primary rechargeable battery. The bottom up approach to battery design may be achieved by selecting individual battery cells and stringing them together in series to compute total voltage ($V_1 + V_2 + \cdots + V_{N\text{cells}} = V_{\text{total}}$). Then connect these strings of cells, each with capacity C in Amp-hours, in parallel to compute the total battery capacity ($C_1 + C_2 + \cdots + C_{N\text{strings}} = C_{\text{total}}$). The cell specification sheets will indicate the values for their cells (for

Array control	Instantaneous cosine factor in sunlight	Limits of sunlight exposure on arrays	$\hat{P}$ for orbits with eclipses $R > \sin B$	$\hat{P}$ (no eclipses) $R \le \sin B$
Fixed zenith	$\cos B \cos v$	$-\pi/2 \le v \le \pi/2$	$\frac{\cos B}{\pi}$	$\frac{\cos B}{\pi}$
Fixed nadir	$-\cos B \cos v$ $\cos B \cos v$	$-\pi + \delta \le v \le -\pi/2$ $\pi/2 \le v \le \pi - \delta$	$\frac{\cos B - \sqrt{R^2 - \sin^2 B}}{\pi}$	$\frac{\cos B}{\pi}$
Fixed wake	$\cos B \sin v$	$0 \le v \le \pi - \delta$	$\frac{\cos B + \sqrt{1 - R^2}}{2\pi}$	$\frac{\cos B}{\pi}$
Fixed cross track	$\sin B$	$-\pi + \delta \le v \le \pi - \delta$	$\sin B\left[1 - \frac{1}{\pi}\cos^{-1}\left(\frac{\sqrt{1-R^2}}{\cos B}\right)\right]$	$\sin B$
Fixed ram	$-\cos B \sin v$	$-\pi + \delta \le v \le 0$	$\frac{\cos B + \sqrt{1 - R^2}}{2\pi}$	$\frac{\cos B}{\pi}$
Yaw sun tracking	$c_{yaw} = \sqrt{1 - \cos^2 B \cos^2 v}$	$-\pi + \delta \le v \le \pi - \delta$	$\frac{1}{2\pi}\int_{-\pi+\delta}^{\pi-\delta} c_{yaw}\, dv$	$\frac{1}{2\pi}\int_{-\pi}^{\pi} c_{yaw}\, dv$
Roll sun tracking	$c_{roll} = \sqrt{1 - \cos^2 B \sin^2 v}$	$-\pi + \delta \le v \le \pi - \delta$	$\frac{1}{2\pi}\int_{-\pi+\delta}^{\pi-\delta} c_{roll}\, dv$	$\frac{1}{2\pi}\int_{-\pi}^{\pi} c_{roll}\, dv$
Pitch sun tracking	$\cos B$	$-\pi + \delta \le v \le \pi - \delta$	$\cos B\left[1 - \frac{1}{\pi}\cos^{-1}\left(\frac{\sqrt{1-R^2}}{\cos B}\right)\right]$	$\cos B$

Table 12.13 Mathematical Models for Performance of Solar Arrays in Various Fixed Orientations and Control Modes (given the solar beta angle B such that $0 \le B \le \pi/2$, eclipse half angle δ such that $0 \le \delta \le \pi/2$, and altitude ratio R)

instance, many lithium ion cells are ~3.7 V). The battery voltage will determine the bus voltage, so components will need to be compatible with this voltage. If step up or down voltage electronics are required to make voltages compatible, the mass will be accounted for in the power conditioning and distribution electronics (stay tuned). Capacity is generally proportional to the size and mass of the cells, so it is important to derive a good requirement for the total battery capacity. This required battery capacity (in terms of energy in Watt-hours for instance) is determined as follows:

$$C_{\mathrm{req}} = \frac{P_{Le} t_e}{\mathrm{DoD}\,\eta_{\mathrm{batt}}}$$

where DoD is the battery depth of discharge, and η_{batt} represents the battery and discharge regulator efficiencies. The total number of parallel strings of cells will need to yield a capacity that meets this requirement with some margin. The sum of all the individual cell masses, along with any spares, will yield the total battery mass.

An alternative to the bottom up battery estimation method is a parametric method using a cell's energy density in Watt-hours/kg. Computing the required energy capacity in Amp-hours is accomplished by dividing C_{req} by the average battery voltage during discharge (the bus voltage would be a good start). To approximate the battery weight, simply divide the required energy capacity by the energy density. Battery types and DoD values may be obtained from the subsection, Power Storage—Batteries.

Power Conditioning and Distribution (PCAD) and Harnesses Although a bottom-up estimation for the PCAD is possible, for concept design it is often simpler to use a parametric relationship. We can estimate the PCAD mass as roughly 15% of the total EPS mass. The data used to create this relationship comes from 27 satellites that have a total EPS mass between 10 and 200 kg. The estimate varies in accuracy with a coefficient of determination $R^2 = 0.76$. Once again, exercise caution when estimating systems outside the range of this data set.

ForestSat Example We will estimate the solar arrays, battery, and supporting electronics using a combination of analytical and parametric estimation methods.

Solar arrays: We first need to estimate the orbit average load. Without having the detailed roll up of all the subsystem loads, we will take an initial guess based on the payload orbit average load of 1.63 W (including a generous 30% margin). To obtain a total system orbit average load we recognize that on average, for small satellites, the payload power load represents about a third of the total spacecraft load (see the System subsection for description of average mass and power fractions). Our initial orbit average load estimate is therefore roughly 5 W. As we iterate through the design, we will obtain better and better subsystem load estimates, but this is good enough to get us started. If the vehicle were to continuously operate in a sun bathing mode, then body mounted solar cells alone would be enough to produce the requisite power. Given that the satellite will point the camera toward nadir while the attitude controller yaws about the nadir-zenith axis, our arrays will only collect sunlight at an oblique angle in general. To find out how much the OAP degrades due to cosine losses, refer to Figure 12.19 for the worst case conditions (in this case it is $B = 0$), and observe that the normalized orbit average power is $\hat{P} = 0.45$. The required array peak power at EOL is therefore $P_{peakReq} = 5/\hat{P} \approx 11$ W. This is too much power for body mounted cells alone to produce, so we will need to include a deployable 3U solar panel (30 cm × 10 cm) that can fold along the side of the CubeSat bus wall when stowed. Using COTS cells, we can populate one 3U body face and one deployable panel with 12 × 1 W cells. The deployable 30 × 10 cm panel with an area density corresponding to rigid arrays will weigh just under 100 g.

Battery: To provide enough energy to support a 5-W load during the maximum 36 minute eclipse, the battery will be comprised of Li-ion cells with a cumulative capacity of

$$C_{req} = \frac{P_{Le} t_e}{\text{DoD}\,\eta_{batt}} = \frac{5 \cdot 36}{60 \cdot 0.2 \cdot 0.9}\ 17\ \text{W} \cdot \text{h}$$

where we assumed a DoD of 20% and battery discharge efficiency of 90%. Many CubeSat electronics operate using 5–7 V, so we will assume a 5-V electric bus, therefore $C_{req} = 17\ \text{W} \cdot \text{h}/5\ \text{V} = 3.4\ \text{A} \cdot \text{h}$. We find that a particular COTS Li-ion cell weighing 50 g provides 3.7 V and 2.2 A·h capacity. Two cells in a series string will provide enough voltage for our 5 V electric bus, and two strings in parallel will provide more than enough capacity. We estimate the total mass of the four cell battery as 200 g.

PCAD. Using the rule of thumb that PCAD is about 15% of the total EPS mass (including the PCAD itself), we obtain a PCAD mass of $(200 + 100)(0.15)/0.85 \approx 50$ g. We need to be careful using this estimate since the data used to create this rule of thumb were obtained from larger satellites. This value will serve as our initial guess, but as we refine the design through successive iterations we may want to build up a PCAD using actual hardware mass values. In this case, the estimate so far seems to fit a typical 40–50 g printed circuit board (PCB) used in many CubeSats, so we are probably not too far off. Note that larger systems would need to include a harness mass as well [Wertz et al. (2011) contains estimation methods for harnesses], however for CubeSats the harnesses consist of small jumper wires and connectors that are included in the PCAD estimate.

Attitude Determination and Control Subsystem

The attitude determination and control subsystem (ADACS) functions we identified in the CE process are accomplished using components that fall into two categories: sensors and actuators. Attitude sensors are almost always less massive and less power hungry than the actuators; however, it is important to consider the placement on the vehicle, field of view, and measurement accuracy. Sensors must be chosen to meet pointing knowledge, accuracy, and stability requirements. For systems requiring high accuracy and stable pointing, such as for laser communications and some imaging missions, high precision star trackers may be required along with a high precision inertial measuring unit (IMU) to smooth out attitude estimation between samples. A description of the various sensor options, their uses and performance characteristics are listed in SMAD (Wertz et al. 2011). The second component category consists of actuators. These include reaction wheels, control moment gyros (CMGs), thrusters, and magnetic torquers. We will specifically focus on reaction wheels due to their ubiquitous use; however, some of the models described here also apply to the other actuators.

Reaction Wheel Performance and Mass Estimation The reaction wheel assembly (RWA) will provide control authority needed for spacecraft fine pointing. We will use a combination of an analytical method and a bottom up method to estimate the mass and power of the RWA. There are two key reaction wheel performance characteristics to consider in the concept design: the torque capability and angular momentum capacity. The torque and momentum requirements are in turn driven by the slewing requirements and the perturbation torques that must be overcome. Slewing requirements may be driven by several required functions such as tracking a surface target, perhaps a communication site on the surface of the earth as the satellite passes overhead, or pointing the satellite at multiple separated targets in a short window of time, or performing a yaw flip maneuver for thermal or power management. Once these required functions have been identified, it is necessary to compute the required wheel torque and momentum capabilities. To accomplish this task using simple mathematical models, we make the following assumptions:

- Wheel sizing is based on a single wheel with a spin axis that is parallel to the desired spacecraft rotation axis. This assumption reduces the Euler moment vector equations to much simpler scalar equations, yet still captures the limiting cases.
- All slew maneuvers are considered "rest-to-rest," so that the vehicle starts in one orientation with zero angular velocity and finishes in another orientation after a slew maneuver with zero angular velocity.
- All maneuvers are either "bang-bang" (torque limited wheel) or "bang-coast-bang" (momentum limited wheel).
- Rigid body motion. Flex modes and jitter are not explicitly considered in the model.
- The whole spacecraft body moves to orient the payload when slewing. The payload is not gimbaled on a stabilized bus platform.

If slew rate and acceleration requirements are not specified, then slew performance may be characterized as achieving a specific slew angle θ within a specified time t. A common estimation technique is to assume a reaction wheel will perform "bang-bang" control to achieve the slew maneuver, by producing maximum torque for half of the required maneuver time, then switching over to produce negative torque for the remaining half maneuver time to generate angular deceleration. This technique, however, assumes that the wheels have sufficient momentum capacity to achieve the maneuver. If not, then a coast period is added to the middle of the slew maneuver, during which the wheel operates at constant speed and produces no torque. Therefore when selecting wheels, it is important to consider both the torque capability T, and momentum capability H. These requirements may be derived using

$$T = \frac{4I\theta}{t^2 - t_c^2}, \quad 0 \le t_c < t \tag{12.5}$$

$$H = \frac{2I\theta}{t + t_c} \tag{12.6}$$

where I is the moment of inertia (MOI) about the desired rotation axis (usually chosen to be the maximum inertia), and t_c is the "coast" time when no torque is applied and the spacecraft rotates at a constant rate. These equations allow for "bang-coast-bang" maneuvers, allowing for smaller wheels that may have high torque capability, but low momentum capacity. A plot showing the locus of minimum torque and momentum requirements to achieve a particular desired maneuver in the range $0 \le t_c < t$ appears in Figure 12.20. For instance, a particular maneuver may require the spacecraft to slew to point its camera, settle and dwell on the target region, then slew to a new region within the limited time that it is passing overhead. Having analytically derived minimum specifications for wheel selection, we now seek actual wheels that can do the job, and then use a bottom up approach to estimate mass and power. Wheels with torque and momentum characteristics obtained from their specification sheets are plotted as markers on the Torque-Momentum plot

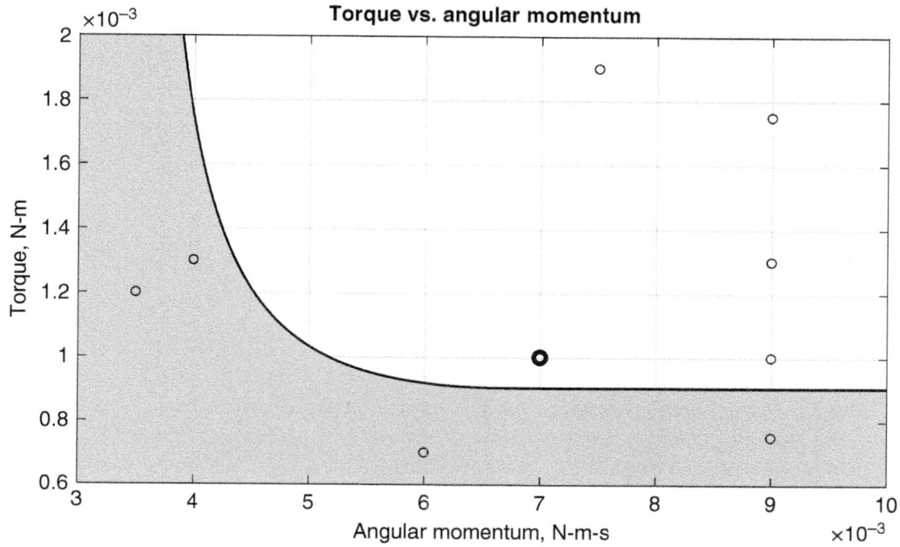

FIGURE 12.20 Slew requirements and reaction wheel selection. The curve represents the minimum torque-momentum requirements derived from the ForestSat slew requirement. Markers represent different reaction wheels with their associated torque and momentum capacity values.

in Figure 12.20. Wheels represented by points above and to the right of the requirement curve signify a feasible design solution. Those below and left of the curve would not meet requirements. Wheels that have lower momentum and/or torque capacity are usually less massive and less power hungry (all else equal), so feasible points closer to the "knee" in the curve are often desirable. Using the CE process, the MOI input comes straight from the Structures model (refer to the system model N^2 diagram in Figure 12.12).

For redundancy and enhanced performance, a fourth wheel is often added to the RWA. Arranged so that the four wheel rotation axes are in a pyramid formation, more torque can be generated along an arbitrary axis than with a single wheel, and momentum may be shared among the wheels to increase total momentum storage capability. Additionally, should one wheel fail, the remaining three could still provide the capability for three-axis attitude control even though the spin axes are not along the body x, y, and z directions. With four wheels, the torque and momentum improvement along one of these cardinal directions is improved by a factor that depends on the pyramid angle. Assuming the four identical reaction wheels have axes that form a symmetric right pyramid, the pyramid angle α is measured from the base to a spin axis. Two opposite wheels, each with a maximum torque of T_{max}, could operate in unison to achieve a torque along their shared base line direction with a magnitude up to $2T_{max}\cos\alpha$, an improvement factor of $2\cos\alpha$ over a similar three-wheel RWA. This same improvement factor applies to momentum capacity.

When estimating the load power for the wheels, it is important to consider what operations they will be doing and for how long. While slewing, a wheel (or wheels) will be operating closer to its peak torque capacity, thus drawing a peak load power. On the other hand, wheels that simply maintain a fixed orientation, nadir-pointing for instance, a rule of thumb is to use 1/3 of the max speed power load. The load power values for various modes of operation may be obtained from the selected wheel specifications sheet.

Momentum Unloading While maintaining stable spacecraft attitude, the wheels in the RWA are slowly but constantly spinning up to null small, but persistent, external torques. The momentum buildup in the wheels eventually saturates their momentum capacity and they need to be unloaded using external torque provided by thrusters or magnetic torque rods (using the earth's magnetic field). Although these disturbance torques are extremely small, they can have a significant impact on the attitude control over time. Disturbance torques are highly dependent on the satellite's orbit regime. Also, these torques are not always constant, but can be periodic, e.g., torqueing one way for half of an orbit revolution, then changing direction and torqueing the other way for the other half. The following list describes common causes and behaviors of perturbation torques, and how we address them in our CE model.

- *Gravity gradient:* This torque results from the fact that the vehicle center of mass and the center of gravity are not collocated. This torque is modeled as

$$T_g = \frac{3\mu}{2r^3}\left|I_{min} - I_a\right|\sin 2\theta$$

  where μ is the earth's gravitational constant, r is the orbit radius, I_{min} is the minimum inertia axis aligned with the nadir-zenith direction, I_a is the axisymmetric axis ($I_{min} \ll I_a$), and θ is the angle between the I_{min} axis and the local vertical. This model uses MOI estimates derived from the Structures model and the orbit radius from Astro. The maximum required off-nadir pointing angle would also contribute to this torque. For satellites not intending to use gravity gradient stabilization, this can be a secular torque if the vehicle points off nadir for an extended period of time. For torque estimation, the angle θ may be chosen as the edge of the field of regard of the earth facing sensor or signal line of sight as applicable.
- *Magnetic dipole interaction with the earth's magnetic field:* The satellite will have residual magnetic dipoles due to magnetized components and current loops in circuits. This torque

is modeled as the vector equation $\boldsymbol{T}_m = \boldsymbol{M} \times \boldsymbol{B}$, where $\boldsymbol{M}$ is the internal magnetic dipole vector and $\boldsymbol{B}$ is the local magnetic field vector of the earth with maximum magnitude of $B_{max} \approx \gamma/r^3$ near the equatorial plane, and $B_{max} \approx 2\gamma/r^3$ near the poles, where $\gamma \cong 8 \times 10^{15}$ tesla-m^3 and 1 tesla $= 1N/(A \cdot m)$. Maximum magnetic torque occurs when $T_m = MB$, so this will serve as a worst magnetic torque disturbance estimate. If using magnetic torque rods for momentum desaturation, then they should each have a dipole greater than the expected magnitude of the satellite's intrinsic dipole $\boldsymbol{M}$ with some margin. Astro provides the orbit radius input for this model.

- *Atmospheric drag:* A constant drag torque caused by an offset between the drag center of pressure and the center of mass r_a can cause momentum accumulation in the RWA. For example, a LEO satellite that is nadir pointing traveling with velocity V with a large deployable array on one side facing into the "wind" will likely be subject to a constant drag torque

$$T_d = \frac{1}{2} C_D A_D \rho V^2 r_a$$

where A_D is the average drag area provided by the Structures model, ρ is the atmospheric density provided by the Astro model, and C_D is the coefficient of drag which may be estimated as 2.0 for a cube (Reynerson) since fluid dynamics are not applicable in the thin atmosphere. A decent estimate for r_a in some cases is to assume that the center of pressure of the structure is roughly the center of the cross-sectional area of the whole structure as viewed by the velocity direction (a box plus an array panel, payload, or antenna on one side perhaps). The distance from this point to the computed vehicle center of mass from the Structures model projected in the same plane will yield the offset estimate. If the vehicle is inertially pointed or sun pointed, then there may not be any secular buildup of momentum due to drag since the torque magnitude would cycle with the orbital period. In this case the maximum drag torques occur every quarter revolution before reversing direction, so each reaction wheel would need to absorb a quarter rev's worth of momentum accumulation to avoid constantly having to dump momentum.

- *Solar radiation pressure:* Unlike the other disturbances, torque due to solar radiation pressure is independent of the orbit radius and therefore is the dominant perturbation torque when a satellite with large reflective surfaces is far from earth, e.g., geosynchronous orbit. Similar to the atmospheric drag torque, the maximum torque due to solar radiation pressure is determined as the product of the solar radiation pressure force and the center of solar radiation pressure offset from the center of mass r_s, i.e.

$$T_s = \frac{SA_s}{c}(1 + \rho_s) r_s$$

where S is the solar flux, A_s is the effective area exposed to the sun, c is the speed of light (3×10^8 m/s), and ρ_s is the solar reflectance, where $0 \le \rho_s \le 1$. Estimates for A_s and r_s are input from the Structures model. Vehicles with constant orientation with respect to the sun or that are inertially fixed will be subject to secular growth of momentum in the wheels. Ram or nadir pointing vehicles will most likely only be subject to cyclic torques. Total disturbance torque may be assumed to be the dominant torque alone. A more conservative estimate would be to take the root sum square of all the contributing torques.

Magnetic Torque Rod Estimate For vehicles low enough to use earth's magnetic field to generate magnetic control torque, the most common unloading actuators are torque rods or torque coils. Magnetic torquers are simple, require no propellant and they may be used to continuously unload wheel momentum. Using a bottom up method, the designer can select three rods (x, y, and z), each with a dipole value that is sufficient to null the secular torques and exceed the spacecraft's intrinsic magnetic dipole (perhaps with 10× safety factor). The vendor spec sheets will provide the dipole performance range and the mass. The generated magnetic dipole moment is proportional to the control current (and thus power) which is useful for our power load estimate. If the maximum disturbance torque is due to drag T_d, then the torquers must generate a magnetic dipole $M > T_d r^3/\gamma$. Unfortunately the earth's magnetic field will not always be aligned optimally to generate torque about the desired axis to compensate for drag. Consider equatorial orbits where a drag torque about the pitch axis cannot be nulled using torque rods since the earth's magnetic dipole vector is closely aligned with the pitch axis, thus no control torque can be generated along the pitch axis. It is better to include a penalty factor $f < 1$ for the nonoptimal alignment so that $M > T_d r^3/(\gamma f)$. For inclinations of 30° and greater, $f \approx 0.7$ is a safe estimate.

ACS Thruster and Propellant Estimate Thrusters may also be used to dump momentum and they are not beholden to the earth's magnetic field, so they would work in any orbit. The downside, however, is that they require propellant and a more complex design with nozzles, plumbing, and tankage. The thrusters must compensate for a lifetime of momentum accumulation due to external disturbance torques by dumping momentum each time wheels approach saturation. To estimate propellant mass, choose a desired time interval between unloading maneuvers t based on the ConOps. Compute required thruster torque as $T_A = T_D \Delta t/t_{\text{on}}$, where t_{on} is the thruster "on" time. The applied ACS torque T_A nulls the accumulated disturbance torque T_D using a pair of thrusters on opposite sides of the spacecraft. For a given moment arm distance from the vehicle center of mass to an attitude control thruster r_t, each thruster in a pair must achieve a thrust of $F_A = T_A/(2r_t)$ to keep the wheels from saturating. Over the course of the satellite lifetime, this unloading maneuver will need to be accomplished $N = t_{\text{life}}/\Delta t$ times. From the thrust equation, we compute the propellant burn rate as $\dot{m} = T_A/(I_{sp} g_o r_t)$ and the lifetime ACS propellant mass as $m_p = \dot{m} N t_{\text{on}}$. Thruster performance properties like the specific impulse and thrust are inputs from the Propulsion model.

ForestSat Example We will use a bottom up estimation for the ADACS.
Sensors: The pointing requirements do not require the accuracy of a star tracker. We will use five coarse sun sensors and an earth nadir sensor for attitude control and sun tracking. A three-axis magnetometer will be included to provide magnetic field estimates used to control the torque rods. Additionally, a 1-W GPS receiver with a patch antenna on the zenith face will be included for navigation.

Actuators: Our driving torque and momentum requirements are derived from our need to slew and image four different target regions during a 5 minute overhead pass. Assume a 1-minute settle and dwell on each of the four targets, which leaves 1 minute total for four slew maneuvers, or 15 seconds for each maneuver. The most stressing case would be if all the targets were on opposite horizons relative to the ground track when the satellite would have to slew back and forth. We derive the maximum slew angle as $\theta = 2\text{asin}(r_e/r)$, which is twice the earth's angular radius (maximum satellite nadir angle η_s when $\varepsilon = 0°$ in Figure 12.21). For our 600 km mission altitude, $\theta = 132°$. We do not have an accurate mass properties value at this point, but we will start with a 3-kg 3U CubeSat box with a solar array wing and calculate a maximum MOI of $I \approx 0.02\ \text{kg}\cdot\text{m}^2$. Assuming a rest to rest maneuver time of t = 15 seconds, and using Eqs. (12.5) and (12.6), we derive the slew requirement curve shown in Figure 12.20. We find a commercial off the shelf (COTS) 90 g reaction wheel with torque and momentum capacity of 0.01 N·m and 0.007 N·m·s, respectively (indicated as the bold marker in Figure 12.20). This wheel meets our slew requirements since it is in the feasible region. We will use four reaction wheels in a pyramid to reduce risk of losing three-axis attitude control if a wheel fails during the 5 year lifetime. Using a 35° pyramid angle, we also achieve a 60% improvement over the maximum torque and momentum limits of our selected reaction wheels since $2\cos(35°) \approx 1.6$.

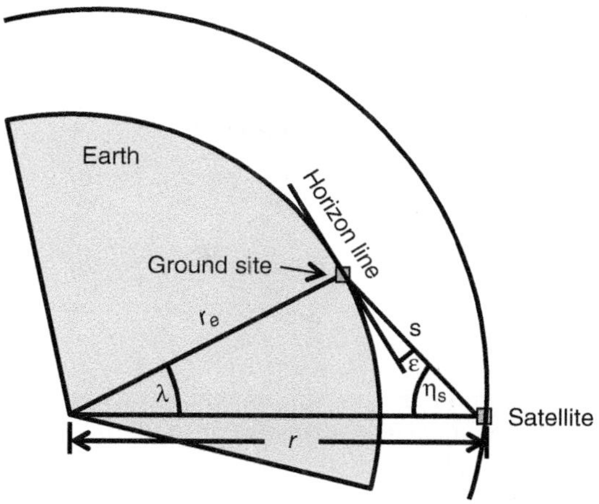

FIGURE 12.21 Satellite, earth, and ground site geometry.

We will include three magnetic torque rods to dump the momentum accumulated in the reaction wheels. After computing the disturbance torques, the dominant one is determined to be due to magnetic torque assuming a spacecraft dipole of 0.01 A·m^2. Selecting three COTS torque rods, each with a dipole 10× this value would be more than adequate. Besides the sensors and actuators, we will also include a 40-g ADACS processor.

Astrodynamics Model

Referring to the N^2 diagram in Figure 12.12, it is clear that the Astrodynamics model produces many outputs that are used as requirements and constraints for all the subsystems. This model does not itself represent a subsystem; however, it does provide some key outputs such as altitude, inclination, required ΔV, and other orbit geometry parameters. Some of the functions that the Astro model supports are "maneuver to mission orbit," "maintain station," and "dispose of system at EOL."

Maneuver to Mission Orbit Establishing the initial mission orbit can be costly from a propulsion system perspective if the launch vehicle does not directly insert the satellite into the desired orbit. The CE team will need to decide what combination of launch vehicle and satellite propulsion capabilities best suits the mission. Supposing that the launch vehicle will deposit the satellite in a parking or transfer orbit, we will a model a combined Hohmann transfer and inclination change maneuver to compute the ΔV requirements. The details of a Hohmann transfer may be found in just about any book on spacecraft dynamics and control.

Orbit Phasing Maneuver If the spacecraft must maneuver to a different location within its circular orbit plane (same altitude, different phase) then a phasing maneuver will be required. This is useful if populating a plane of co-altitude satellites in a constellation and you need the satellites to be spaced apart on the ring. The maneuver is executed in two parts. The first burn slightly changes the orbit semimajor axis which makes the new transfer period slightly shorter (if orbit was lowered) than the desired orbit. This is executed just like the first burn of a Hohmann transfer to get into a phasing transfer orbit to "rendezvous" with the co-altitude target position that is an angular distance of ϕ from the current position. The target orbit mean motion is $n = \sqrt{\mu/a^3}$ while the "interceptor" will be in a smaller, faster elliptical transfer orbit with mean motion $n_T = \sqrt{\mu/a_T{}^3}$. The target traverses an arc distance of $2\pi N_{rev} - \phi$ in the same amount of time that the interceptor traverses $2\pi N_{rev}$ in its transfer orbit, where N_{rev} is the number of revolutions.

$$\frac{2\pi N_{rev} - \phi}{n} = \frac{2\pi N_{rev}}{n_T}$$

The transfer orbit's semimajor axis is therefore

$$a_T = a\left(1 - \frac{}{2\pi N_{rev}}\right)^{2/3}$$

Using a_T, the first impulsive burn of a Hohmann transfer may be calculated to determine the required ΔV and propellant mass for the phasing maneuver. The second burn will be the same magnitude as the first, but in the opposite direction to recircularize the orbit once the satellite is at the desired station. When planning the ConOps, the designer must trade between maneuver time and propellant mass. The shorter the time allowed for the maneuver (i.e., less revs), the more propellant is required. From this equation we can derive a few rules of thumb. Adjusting to a GEO orbit slot requires ~ $6\,\frac{\text{mps}}{\text{deg/day}}$. For example, to shift at a rate of 1° per day in GEO, you would need 6 m/s of ΔV. A drift rate of 2° per day requires 12 mps, and so on. For LEO orbits (~300–2,000 km) the value is closer to $1\,\frac{\text{mps}}{\text{deg/day}}$.

Maintain Station The station keeping models depend on the mission orbit regime, LEO, MEO, or GEO.

LEO station keeping: At low altitudes where perturbations due to drag dominate, a maneuver is accomplished at regular intervals to maintain a station within a maximum in-track deviation distance D with a reference position in the center (Chao 2005).

$$D = \left(-\frac{da}{dt}\right)\frac{3}{4}nt_m^2$$

where a is the orbit semimajor axis, n is the orbit mean motion, and t_m is the time between maneuvers. The orbit decay rate $-da/dt$ depends on the satellite's ballistic coefficient $B^* = C_D A_D/m$, the atmospheric density ρ_o, the rotation rate of earth ω_e, the orbit inclination i, and is approximated as

$$\frac{da}{dt} \approx -B^* \rho_o n a^2 \left(1 - \frac{\omega_e}{n}\cos i\right)^2 \tag{12.7}$$

where A_D and C_D are the drag area and coefficient of drag (~2.0), respectively. The estimated time between maneuvers and the ΔV required per maneuver are given, respectively, as

$$t_m = 4\sqrt{\frac{D}{-3n\,da/dt}}$$

$$\Delta V = -\frac{1}{2}\frac{da}{dt}nt_m$$

Total lifetime ΔV requirements for LEO station keeping is computed as $\Delta V_{life} = \Delta V t_{life}/t_m$.

MEO station keeping: This orbit regime is relevant to GPS satellites and some others that are beyond the reach of the atmosphere. Orbits in the MEO regime are affected by different perturbations than those in LEO. Satellites in MEO, like GPS, have a period of roughly 12 hours and due to resonance effects of the earth gravity harmonics tend to cause drift in the longitude of the nodes. The drift motion of the longitude of nodes tends be pendular about stable longitudes near 27.8° and 27.8° + 180°. Orbits that do not have a node line at these points will require station keeping maneuvers if it is required to maintain station. The reader is referred to Chao (2005) for the derivation of these equations.

GEO station keeping: Geostationary orbits (zero inclination, zero eccentricity) require longitude (east-west) station keeping as well as inclination (north-south) station keeping. Minimum east-west station keeping ΔV requirements are given as (Agrawal 1986)

$$\Delta V_{E-W} \approx 1.74 \sin 2\Delta\lambda \quad \text{mps/year}$$

where Δ is the difference between the actual and desired mean longitude. If it is desired to control the inclination, then north-south station keeping requires approximately

$$\Delta V_{N-S} = 47 \text{ mps/year}$$

It is advisable to include approximately 20% margin on these values to account for misaligned thrusters, attitude error, etc. These ΔV requirements (BOL stationing, station keeping during mission, and EOL disposal) are fed into the propulsion model which uses the rocket equation to compute the propellant mass.

Electric Propulsion Electric propulsion is a fuel efficient way to continuously keep station, at the expense of a large load power. The ΔV requirement to change the inclination comes from the Edelbaum equation (Ruggiero et al. 2011)

$$\Delta V = V\sqrt{2 - 2\cos\frac{\Delta i\pi}{2}}$$

where Δi is the inclination tolerance.

Dispose of System at EOL Satellites have a 25-year life limit to meet space debris mitigation requirements, so unless a satellite is in a LEO orbit that will deorbit naturally within this time, they will likely need to deploy a drag inducing device or carry some EOL ΔV to for a deorbit maneuver. A final ΔV maneuver would lower perigee into the atmosphere. In higher altitudes, a Hohmann transfer is used to move to a junkyard orbit.

A summary of some rules of thumb for calculating station keeping ΔV is shown in Table 12.14. Following the CE process, all the ΔV values derived in this Astro model are output to the Propulsion model for propulsion system sizing.

Orbit	LEO	MEO	GEO
Altitude, km	300–2,000	26,560	35,786
$\Delta V_{N\text{-}S}$, mps/year	0	0	40–50
$\Delta V_{E\text{-}W}$, mps/year	2*	0.2	2.5
$\Delta V_{phasing}$, mps/deg/day	0.9–1.1	3.6	6

*Depends on altitude, ballistic coefficient, and predicted atmospheric density.

TABLE 12.14 Rules of Thumb for Calculating Station Keeping ΔV in Different Orbit Regimes

ForestSat Example Once ForestSat is on station, there are no explicit requirements to maintain a particular altitude or position in the orbit. Let us evaluate the ΔV requirements to remain on orbit by compensating for drag and for a final deorbit maneuver. First, determine the ballistic coefficient B^*. At this point we do not know the spacecraft mass, but we can still estimate B^*. We will assume a shape of a box with two solar array "wings" that are attached along two edges of the box that are the same dimension as the sides of the box upon which they are stowed during ascent. For our small satellite we will assume a mass density of 1 kg/U (original CubeSat standard). For the average drag area A_D we will estimate the average between the maximum and minimum cross sectional areas facing the "wind" as 0.02 m^2/U. The ballistic coefficient is therefore estimated as $B^* = 2(0.02)/1 = 0.04\ \text{m}^2/\text{kg}$. To determine the deorbit lifetime of the satellite, we can integrate Eq. (12.7) or use software that performs orbit decay analysis. The atmospheric density varies exponentially with altitude. Using a NRLMSISE-00 atmospheric model, our satellite in a 600-km circular orbit can expect to deorbit naturally in about 14–17 years (depends on launch date). This is roughly the highest altitude that a typical CubeSat can achieve without having to carry a deorbit propulsion system or a deployable deorbit device to meet the 25 year disposal requirement at EOL. In the spirit of simplicity, ForestSat will forgo a propulsion system since we can meet our other mission requirements without one. We will add a compact deorbiting tether package (100 g) on one side of the spacecraft that will be deployed at EOL to ensure ForestSat can deorbit within the 25 year limit, even at higher altitudes.

Structures and Thermal Control

Performing a bottom up estimate for the spacecraft structures and thermal control subsystems is difficult to do in the dynamic CE environment since this process is fluid by nature and there is not much design detail at this point. A parametric estimation approach is frequently used to determine the SMaP for these subsystems. Typical structure mass fractions can range from 9% of the system dry mass for some CubeSats to 25% for some large satellites. When sizing the bus, the internal dry mass density provides an indicator of the feasibility of your bus size relative to the mass. For reference consider that the average large satellite (1,000$^+$ kg) has an internal dry mass density of about 200–250 kg/m^3. On the other end of the size spectrum, a CubeSat by definition has a mass density of 1,000–1,300 kg/m^3 and many everyday consumer electronics like smart phones can have a density of about 1,600 kg/m^3. The internal dry mass density only includes the mass of bus subsystems and payload that are inside the confines of the body and does not include external appendages, solar arrays, antennas, etc. The volume used in calculating this density is internal volume less the propellant tank volume. The shape and size of the bus body will subsequently affect the thermal radiator performance as well as the dimensions of solar arrays that fold against the body walls when stowed. Other key parameters that the Structures model needs to estimate are the center of mass (CM), moment of inertia (MOI) and for some LEO satellites, drag area and the center of pressure (CP). These values contribute to computing disturbance torques in the ADACS model.

Like the Structures model, the thermal control subsystem is often estimated using a parametric approach rather than a bottom up approach. The method expressed here describes how relevant historical data may be used to generate parametric equations. The key word is "relevant" and the reader is cautioned to use data relevant to their systems of interest. The example values used here are representative of a particular sample set and included for instructional purposes.

TCS Mass Estimation The mass fraction approach provides a rough estimate for TCS mass and power based on the payload power, payload mass, and overall spacecraft dry mass. The TCS mass m_{TCS} may be estimated using a mass fraction relating to the uncontingencied (i.e., no margin) spacecraft dry mass m_{sc} as follows:

$$m_{\text{TCS}} = aP_{PL} + bm_{sc}^2 \qquad (12.8)$$

where P_{PL} is the uncontingencied power of the payload, $a = 0.013$ kg / W and $b = 1 \times 10^{-4}$ kg^{-1}. This parametric equation is based on a sample of twelve LEO microsatellites less than 200 kg and has a correlation $R^2 = 0.46$. Because the spacecraft mass in this formula includes the TCS mass (the value we are trying to compute), this equation in its present form can lead to circular reference errors in the CE tool. To avoid this we sum all N_{sub} subsystem masses *except* for the TCS mass such that the spacecraft dry mass may be expressed as

$$m_{sc} = \sum_{i=1}^{N_{\text{sub}}-1} m_i + m_{\text{TCS}}$$

Substituting into the original equation and solving the resulting quadratic equation yields an equation for the TCS mass that is completely independent the total system dry mass, thus avoiding circular reference errors.

$$m_{\text{TCS}} = \frac{1}{2b}\left[1 - 2b\tilde{m} \pm \sqrt{(2b\tilde{m}-1)^2 - 4b(b\tilde{m}^2 + aP_{PL})}\right] \tag{12.9}$$

where $\tilde{m} = \sum_{i=1}^{N_{\text{sub}}-1} m_i$. Choosing the solution to Eq. (12.9) such that $m_{\text{TCS}} \ll m_{sc}$ will yield the desired TCS mass estimate.

TCS Power Estimation The power estimate may also be achieved using a parametric relationship. Using a sample set of seven microsats less than 200 kg, the following formula will estimate the TCS load power, P_{TCS}, with a correlation $R^2 = 0.95$

$$P_{\text{TCS}} = cP_{PL}^2 m_{PL}$$

where m_{PL} is the uncontingencied mass of the payload and $= 1.55 \times 10^{-4}$ W^{-1}kg^{-1}.

ForestSat Example We will use a 10% dry mass fraction to estimate our 3U CubeSat structure mass, so that equates to 300 g of our estimated 3 kg system. ForestSat will use passive thermal control with the exception of a small 10 g heater for the Li-ion battery to keep it warm during cold conditions. For illustrative purposes, had we used the parametric relationship shown in Eq. (12.8), we would have gotten

$$m_{\text{TCS}} = \left[(0.013)(1) + \left(1 \times 10^{-4}\right)3^2\right] \times 1{,}000 \approx 14 \text{ g}$$

The power for this heater is almost negligible as it is used only as a precaution in cold conditions. It is not used in a normal mode, so we will not bother with the heater power in our ForestSat normal mode power budget.

Propulsion

The orbit maneuvering requirements directly drive the propulsion system design with the ΔV inputs being supplied by the Astro model. Additionally, the Astro model will supply the launch and stationing ConOps that include the various phases of initial stationing maneuvers (i.e., plane changes, altitude changes, etc.) and station keeping so that the interim satellite mass may be determined at each successive phase. The rocket equation is used to determine the propellant mass m_p at each phase based its interim post burn "final" mass m_f and is given as

$$m_p = m_f\left(e^{\Delta V / I_{sp} g_0} - 1\right)$$

Off the shelf thrusters are selected to achieve the desired thrust and specific impulse I_{sp} performance. Electric propulsion may be used for either maneuvering, if long thrusting times are not an issue,

Propulsion system	Propellant density, kg/m^3
Mono N_2H_4	1,010
H_2O_2	1,450
Biprop MMH N_2O_4	878
Biprop N_2H_4 N_2O_4	1,010
Cold gas	1.2
Xenon HET	1,858
Xenon resistojet	1,858
Xenon ion engine	1,858

Table 12.15 Density for Selected Rocket Propellants

or for constant station keeping. Although extremely efficient, these systems are power hungry, so it is critical to feed back the load power to the EPS model to properly capture the impact on the solar array and battery sizing.

For conventional propulsion systems, we need estimate the tanks to store the propellant and, in the case of regulated pressure systems, a pressurant tank is also required. Given the total propellant mass requirement, a propellant type may be selected. The required propellant volume is simply computed as $V_p = m_p/\rho_p$, where ρ_p represents the propellant density. Density values for some common propellants are shown in Table 12.15. An appropriately sized tank may be selected from a list of commercially available units to determine the mass, however during the design iteration process the total satellite mass, and thus propellant mass, can sometimes change significantly. This method can be burdensome in the dynamic CE environment. An alternative method is to estimate tank size parametrically by using empirical tank data to form estimating equations. The following is the general form of a linear equation that may be used to estimate tank mass (units in parentheses).

$$m_{\text{tank}}(\text{kg}) = a \times [V_T(m^3) \times P_{\text{burst}}\ (Pa)] + b \tag{12.10}$$

Values for coefficients a and b are shown in Table 12.16 for various tank types. The burst pressure P_{burst} may be calculated as the BOL operating pressure of the propellant multiplied by a safety factor.

Propellant tank type	a	b
Elastomeric diaphragm	1.44e-5	2.68
Metal diaphragm	2.63e-5	1.71
Prop management device (PMD)	1.38e-5	3.06
No PMD metal	1.21e-5	0.90
No PMD COPV	1.18e-5	3.60
Metal pressurant tanks	7.61e-6	1.65
COPV pressurant tanks	5.34e-6	1.27

Table 12.16 Coefficients Used in Eq. (12.10) to Estimate Different Propellant and Pressurant Tank Types

Propulsion system type	Blowdown BOL, kPa	Blowdown EOL, kPa	Regulated kPa
Monopropellant	2,240	690	1,550
Bipropellant	2,760	970	1,790
Cold gas	20,680	690	17,240
Electric propulsion	20,680	690	20,680

BOL = beginning of life; EOL = end of life.

Table 12.17 Tank Pressures for Various Propulsion Systems at BOL and EOL

Typical BOL pressures are listed in Table 12.17 for several propellant categories. The tank required tank volume V_T in Eq. (12.10) depends on the type of pressure feed system, regulated or blowdown. A regulated system uses a pressurized inert gas stored in a separate pressurant tank that provides constant regulated pressure to feed propellant from the propellant tank to the thrusters via valves. Since roughly 5% of the propellant in the tank is unused (ullage), the total propellant tank volume for a regulated feed system is $V_T \approx V_p/0.95$ (regulated). In a blowdown system however, the pressurant gas and propellant share the volume of a single tank, thus at BOL

$$V_T = V_p + V_{g\text{BOL}} \tag{12.11}$$

where the p and g subscripts indicate propellant and pressurant gas, respectively. Blowdown propellant feed systems are more volume efficient than regulated systems, but as propellant is expended throughout the lifetime, pressure decreases and so does thruster performance. Because the pressurant gas stays on board (hopefully no leaks), its mass at BOL is equivalent to its mass at EOL, thus for constant temperature and assuming an ideal gas

$$P_{\text{BOL}} V_{g\text{BOL}} = P_{\text{EOL}} (V_T - V_{\text{residual}}) \tag{12.12}$$

where the residual gas volume is $V_{\text{residual}} \approx 0.02V_p$. Combining equations (12.11) and (12.12) and solving for tank volume yields

$$V_T = \frac{V_p(P_{\text{BOL}} - 0.02P_{\text{EOL}})}{P_{\text{BOL}} - P_{\text{EOL}}} \quad \text{(blowdown)} \tag{12.13}$$

ForestSat does not need a propulsion system, but the following example illustrates how to use these equations for tank sizing. To estimate the mass of a tank with a blowdown feed system storing 10.1 kg of hydrazine monopropellant, N_2H_4, we first derive the propellant volume as 0.01 m^3 given that the density is 1010 kg/m^3 (see Table 12.15). From Table 12.17 we discover that the BOL and EOL pressure is respectively 2240 kPa and 690 kPa, thus using Eq. (12.13) we obtain a tank volume V_T of 0.014 m^3. We will use a safety factor of 1.5 to determine the burst pressure $P_{\text{burst}} = 1.5(2,240) = 3,360$ kPa. Using Eq. (12.10) and the appropriate coefficients from Table 12.16, we estimate the tank mass as $m_{\text{tank}} = 1.44 \times 10^{-5}\ [0.014 \times 3.36 \times 10^6] + 2.68 = 3.36$ kg.

The following are some tank considerations for different satellite and propellant types.

- Satellites that are three-axis stabilized require tanks with a Propellant Management Device (PMD) to ensure propellant is properly fed to the thrusters. Two types of PMD tanks may be used. Diaphragm PMD tanks are the less expensive option, however surface tension PMD tanks are typically less massive.

- Metal surface tension tanks are typically used to store liquid propellants. However, if propellant sloshing must be kept to a minimum (for tight CG or MOI control for instance), then diaphragm tanks might be a better choice at the expense of higher dry tank mass. There are metal and elastomeric diaphragms. Metal will be heavier, all else equal, but elastomeric diaphragms have compatibility issues with oxidizers, therefore they are used for monoprop systems.
- A composite overwrapped pressure vessel (COPV) tank is a light weight alternative to a similar metal tank for storing liquid propellants or pressurants, but they are usually more expensive.
- Tankage for biprop systems must accommodate fuel and oxidizer.

The tanks must fit within or on the spacecraft which itself fits in the launch vehicle's payload fairing. It is useful to assign a member of the team to track the dimensions of the tanks and other large components to ensure they will fit (a CAD tool is often helpful). Sometimes multiple tanks or tanks of different shapes (cylindrical for instance) must be used to meet physical space constraints.

Telemetry, Tracking, and Command Subsystem

All satellites have a telemetry, tracking and command (TT&C) subsystem to communicate with the ground so there are many analogous systems with which we could start our design, but for illustrative purposes we will focus on models supporting a bottom up approach. We will need to estimate the size, mass, power, and performance of the transmitter, receiver, amplifier, encryption devices, antennas, and other communications hardware that are capable of transmitting signals to and receiving signals from the intended link participants and/or relays. To ensure a communication subsystem design meets data rate and other system performance requirements, we perform a link budget analysis for both the uplink and downlink given the ground station locations, capabilities, and link geometry.

Communication connectivity with the various nodes is determined by evaluating the signal strength to noise ratio for a given set of hardware at the maximum expected separation distance between the satellite and the ground station (or other satellite). This is achieved by summing all the gains and losses in the communications link. A transmitter's effective isotropic radiated power (EIRP) is a measure of its signal output power, including amplification, P_t coupled with the gain of its antenna G_t and given as (Gordon and Morgan 1993):

$$\text{EIRP} = 10 \log_{10} P_t - L_t + G_t \text{ (dBW)} \tag{12.14}$$

where L_t represents line and pointing losses. For the uplink link budget, the ground station EIRP would need to be known (provided by the ground site) or derived from its antenna characteristics. The C2 uplink signal is subject to a loss due to path length between transmitter and receiver, atmospheric absorption or scattering, and rain among other factors. The space loss is given as

$$L_s = 20 \log_{10} S + 20 \log_{10} f + 32.44 \text{ (dB)} \tag{12.15}$$

where S is the path length between transmitter and receiver in km and f is the frequency in MHz. The measure of a receiver's capability to receive a signal is characterized by its G_r/T ("gee over tee," or figure of merit), which is a function of its receiver gain G_r, and the thermal noise temperature that it is exposed to, T in Kelvin, and is given as

$$\frac{G_r}{t} = G_r - 10 \log_{10} T \text{ (dBi/K)}$$

The noise associated with this thermal noise temperature interferes with the signal entering the receiver. Noise temperature ranges from 30 K for receivers staring into space, perhaps while crosslinking with other satellites, to 290 K when they are aimed toward earth receive the uplink. Typical $G_r T$ values for spacecraft systems range from –20 to 10 dBi/K (see Gordon, G., Morgan 1993). The measure of link performance is captured in the carrier power to noise density ratio C/N_o, and written as

$$\frac{C}{N_o} = \text{EIRP} - L_s + \frac{G_r}{T} + 228.6 \quad (\text{dBHz}) \tag{12.16}$$

where N_o is the noise power in a bandwidth of 1 Hz. For digital communications a different, but related, performance metric is used; the energy per bit to noise density ratio E_b/n_o. Because the carrier signal power C is equal to the bit rate R times the energy per bit E_b, i.e., $C = E_b R$, then the decibel form of the energy per bit to noise density ratio equation may be written as

$$\frac{E_b}{N_o} = \frac{C}{N_o} - 10\log_{10} R \tag{12.17}$$

where the second term on the right side is the rate loss due to bit rate R in bits per second. A positive 3 dB margin is a typical value used to ensure link closure. If using forward error correction (FEC) then some additional coding gain may be included.

For the downlink budget, the satellite transmitter EIRP would be computed in the same manner as described for the ground station. If selecting a transmitter-antenna package, then the EIRP may be provided in the vendor provided spec sheet. Otherwise, the system must be broken down into its components and "built up." The ground station receiver G_r/T should be used to complete the link budget computation.

Typical data rates for the TT&C subsystem are small, measured in single digit kbps. If in addition you are designing a payload communications system, the data rates will probably be much larger on the downlink, perhaps measured in Mbps, but the same process may be applied. Once components are selected that meet the link performance requirements, the ConOps are considered to determine the load power to output to the EPS tool. For instance, if the downlink transmitter only needs to be "on" for a maximum time overhead a ground station (provided by the Astro tool), say 10 minutes in a 100-minute period, then the orbit average load duty cycle should reflect this.

ForestSat Example ForestSat requires full connectivity in any orientation, so we will try to close the link budget using a pair of ¼ wave dipole omnidirectional antennas. The cost of using these is that there will be no antenna gain. There is also no need for continuous contact with the ground, so a crosslink relay is not required. For communication redundancy, we will select a pair of small S-band transceivers to communicate with the ground. One will operate using a receiver duty cycled 1/8th of the time and will transmit when commanded overhead the ground station. The fraction of the circular orbit that the satellite is in contact with the ground station, assuming it passes straight overhead, is equivalent to double the earth central angle λ, where $\lambda = \pi/2 - \eta_s - \varepsilon$, and η_s is the nadir angle shown in Figure 12.21. The satellite nadir angle is computed as $\eta_s = a\sin(r_e \cos\varepsilon/r)$, where the elevation angle ε is the angle from the ground station horizon line to the satellite, r_e is the earth's radius (6,378 km), and r is the orbit radius (6,978 km). We will use $\varepsilon = 15°$ to ensure that the line of sight between satellite and ground station clears obstacles like buildings and trees to compute $2\lambda \approx 26°$. The maximum fraction of the circular orbit that the transmitter must be on is therefore $26/360 \approx 1/14$, which we will use to compute the transmitter orbit average load (corresponds to about 7 minutes at this altitude).

To find a small radio that suits our needs, we will perform a downlink budget to estimate the required RF power out requirement based on our data rate. Suppose we determine that downlinking 60 MB of forest data during an average overhead time of about 5 minutes would meet our mission requirements, then we would need to downlink data at an average rate of approximately 1.6 Mbps (recall there are 8 bits in a byte). We will combine telemetry and payload data on the same downlink where either radio would be capable of each function, but payload data throughput will dominate so we will use its rate for the link budget. To achieve this data rate using QPSK with a bit error rate (BER) less than 10^{-6}, we estimate that we would need an E_b/N_o greater than or equal to 10.8 dB (without code gain). See Gordon, G., Morgan 1993 for relating BER with E_b/N_o. We will implement some FEC to achieve a code gain of 6.1 dB (Viterbi soft decision), bringing the required E_b/N_o to 4.7 dB. This computation for the *required* E_b/N_o is only a function of the modulation type, bit rate, and FEC code used. Now we take the physical properties of our system into account to perform the link budget using Eqs. (12.14) through (12.17) to determine the EIRP required from the satellite's TT&C subsystem as follows:

1. Determine the space loss using Eq. (12.15). We compute the slant range S from Figure 12.21 as

$$S = r_e[(r^2/r_e^2 - \cos^2\varepsilon)^{1/2} - \sin\varepsilon] = 1{,}626 \text{ km}$$

   We will assume an S-band frequency of 2,100 MHz, so the space loss in decibels is $L_s = 163$ dB. Other losses such as rain and atmospheric polarization losses also contribute to space loss, but are small by comparison so we will ignore these.

2. Determine C/N_o using Eq. (12.17) provided the desired data rate $R = 1.6$ Mbps and required $E_b/N_o = 4.7$ dB.

$$C/N_o = 4.7 + 10\log_{10}(1.6\times10^6) = 66.7 \text{ dB}$$

3. Determine the satellite's required EIRP provided the ground station's receiver G_r/T (0 dBi) using Eq. (12.16). The satellite's EIRP is computed as

$$\text{EIRP}_{\text{sat}} = \frac{C}{N_o} + L_s - \frac{G_r}{T} - 228.6 = 1.1 \text{ dB}$$

4. Find the required satellite transmitter power required to accomplish this EIRP using an antenna gain for an omnidirectional whip antenna ($G_t = 0$ dB) and Eq. (12.14). Assume line and pointing losses are $L_t = 3$ dB. This would require a transmitter RF power output of

$$P_t = 10^{(\text{EIRP}+L_t-G_t)} = 2.57 \text{ W}$$

   We find a COTS transceiver that fits well with CubeSat dimensions with modest load power requirements with an RF output power of 2 W. Not quite enough to close the link using an omnidirectional antenna. We have the option of reconsidering our original requirement of 60 MB/pass average downlink rate, or we can switch to a patch antenna with a 5 dB gain to close the link. This comes at the cost of losing total omnidirectional coverage. Given that we still have a second omnidirectional antenna that meets our "contact with ground in any orientation" requirement, we will change the downlink ConOps to point the patch antenna toward the ground station during downlink passes. This COTS radio has a mass of 90 g, a transmitter peak load power of 10 W and a receiver load of 1/3 of a watt. Given the orbit duty cycles previously calculated (1/10th and 1/8th for the transmitter and receiver, respectively), the orbit average loads are 1 W and 42 mW, respectively.

C&DH

Estimating the C&DH subsystem using parametric relationships is challenging because the processing requirements vary widely for different missions. If there is a spacecraft with similar complexity and data handling requirements with known C&DH characteristics, then they would provide a good starting estimate for our concept. If this is not available, then we may use a parametric or a bottom up approach. One consideration is to decide if the C&DH subsystem performs payload processing and handling functions, or if it is only responsible for bus functions. If the C&DH subsystem records and processes payload data, then a rough parametric estimate for C&DH mass and power may be expressed as

$$m_{\text{CDH}} = 8.94 + 0.078 m_{PL} \quad (R^2 = 0.59)$$

$$P_{PL} = 12.9 + 8.57\sqrt{m_{\text{CDH}}} \quad (R^2 = 0.58)$$

where m_{PL} and P_{PL} represent the payload mass in kilograms and load power in watts, respectively. A word of caution: The empirical data used to generate these relationships were from satellites over 50 kg and may not be applicable to smaller spacecraft. Performing a bottom up estimate will be more accurate and the following components should be considered.

- Flight processors
- Data recorders
- I/O boards
- Power management boards
- Chassis
- Remote interfaces

There are also integrated systems that perform all the functions of these components. Integrated boards are especially useful for micro and picosats (like our ForestSat). When choosing a recorder, consider its rate and capacity. This may be a function of the payload's data collection rate and volume if the payload does not perform these functions on its own. Radiation hardening is important to reduce the risk of single event upsets (SEUs). For longer life missions, consider a redundant C&DH system to reduce risk of single string failure.

ForestSat Example The imaging payload includes its own data recording and processing capability, therefore the C&DH subsystem only needs to perform bus functions. We will use a COTS integrated processor with a mass of 70 g and peak power of 5 W that will operate with a duty cycle of approximately 10% (constant average load of 500 mW).

System

At the system level, we sum the subsystem masses and orbit average loads to arrive at the total system current best estimate (CBE) values. We must recognize, however, that at this early phase of development we are unable to account for every bracket and wire. These items will certainly go along for the ride to space so we need to account for their mass and load, but we do not need to (nor could we) itemize down to this level. We account for this mass and load with a system margin that will be added to the CBEs. The margin amount depends on how well we know the subsystems. Well defined systems will not need as much margin as those that are less mature. Typical system margins for both mass and power at this early phase of development range from 20% to 30% of the respective CBE values. For ForestSat, we will use 20% system margin for both mass and power since we used actual space-rated COTS hardware for many of our subsystem estimates and have reasonable confidence in them. Incidentally, we could have assigned individual margins for each subsystem based on their maturity in terms of technology readiness level (TRL) and placed a system margin on top of those. For simplicity we did not do this, but further guidelines that address when and how to apply these margins are detailed in GSFC-STD (2009).

The examples thus far have focused on the initial guess for our ForestSat concept design. The next step in the process is to sum the estimated subsystem masses and load power values in a bottom up fashion to achieve a more refined total. These new mass and power estimates will require members of the team to change their subsystem design or ConOps. The team iterates the design in this fashion until they converge on a final feasible concept design. In other words, the design closes. The "concurrent" part of concurrent engineering is due to the fact that this iterative process of updating model inputs and producing new outputs occurs simultaneously in a collaborative design environment.

ForestSat Example After summing the loads, it is determined that the CBE average load is not 5 W, but actually closer to 6 W. Although our single deployable 6-W solar wing along with the 6-W body-mounted array would produce a total peak power of 12 W when directly pointing at the sun, in actual operation they will only produce a fraction of this amount on average since we have required one face to remain nadir pointing. As described in the EPS section, the arrays are fixed (i.e., no control gimbals) so the spacecraft rotates about the yaw axis to minimize solar cosine losses. Recalling our normalized OAP for yaw tracking solar arrays in worst case conditions, $\hat{P} = 0.45$ (solar $B = 0$), we determine the orbit average power to be OAP $= 12 \times 0.45 = 5.4$ WOAP without losses, so not enough to support our load. We can either redefine our ConOps requirements to allow a sun bathing mode during low solar beta conditions (thus sacrificing some imaging opportunities), or add another solar array to meet power requirements without interrupting the imaging mission. In this case, we will add another deployable array bringing the total orbit average power to 8.1 WOAP before losses. Even with 90% array transmission losses, the new array configuration meets the load requirement. Our four cell battery pack still meets these updated power requirements; however, the PCAD mass will be a little higher. To mitigate the risk of losing power if the spacecraft loses attitude control, we will include a single cell on other faces as well to enable operation in a degraded mode.

These modifications to the EPS will impact the rest of the system design and performance. When we add a new wing and update the mass properties, other subsystems are affected. The ADACS must slew a spacecraft with more inertia. The EOL deorbit performance may change due to fact that the ballistic coefficient has changed. Structure mass will be impacted. Using the CE process, we can observe the impact of this change as it ripples through the rest of the system design. After a few design iterations, we achieve the final ForestSat concept design as shown in Figure 12.22 with a summary of mass and load power values shown in Table 12.18.

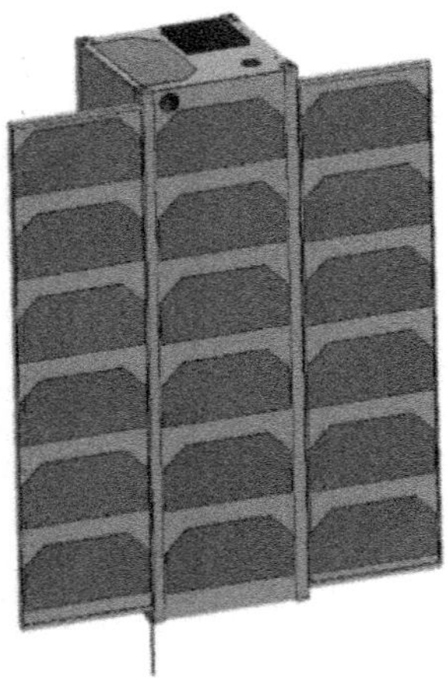

Figure 12.22 ForestSat concept design. (*Courtesy of University of Southern California's Aero Design Team.*)

Subsystem	Mass, kg	% Mass	Power, W	% Power
Payload*	0.70	22.1	1.25	20.2
ADACS	0.66	20.8	1.58	25.5
C&DH	0.08	2.4	0.50	8.0
EPS	0.61	19.2	0.70	11.3
Structure	0.32	10.0	0.00	0.0
TT&C	0.21	6.7	1.04	16.8
TCS	0.07	2.1	0.10	1.6
System margin	0.53	16.7	1.03	16.6
Total	3.18	100.0	6.20	100.0

*Includes mass of deorbit device.

Table 12.18 ForestSat Mass and Orbit Average Load Summary

It is sometimes helpful to compare the design results against the average mass and power fractions for similar spacecraft that have flown in the past. Average mass and power fractions for large and small satellites are shown in Tables 12.19 and 12.20. The results of a CE concept design do not need match, or even be close to, these average values. No two satellites are exactly alike. For large deviations, however, there should be some explanation. Common exceptions include unusually large payloads, agile spacecraft with aggressive slewing requirements, or missions with payloads that require cryo-cooling. ForestSat, for instance, has no propulsion system so this obviously is a key difference. CubeSats do not need the same proportional amount of structure as larger satellites. Also, ForestSat's pointing requirements were fairly aggressive, so we would expect the ADACs mass and power to be higher than average. Performing an alternative configuration using only

Subsystem	Average mass fraction, %	STDEV, %	Average power fraction, %	STDEV, %
Payload	24.8	13	33.6	23
ADACS	7.9	5	15.2	11
C&DH	4.7	2	14.7	12
EPS	19.8	7	11.2	15
Propulsion	11.1	9	5.4	7
Structure	24.4	10	0.0	0
TT&C	4.3	3	11.5	13
TCS	3.0	2	8.5	9

*Total dry mass 200–500 kg.
Mass fractions are based on spacecraft dry mass. Power fractions are based on total OAP (bus + payload). Data was compiled from 48 missions.

Table 12.19 Average Mass and Power Fractions for Small Satellites*

Subsystem	Average mass fraction, %	STDEV, %	Average power fraction, %	STDEV, %
Payload	36.4	16	50.5	11
ADACS	5.0	2	10.0	3
C&DH	3.4	2	8.9	10
EPS	19.8	7	5.9	5
Propulsion	4.7	4	0.4	1
Structure	24.6	9	0.4	1
TT&C	2.6	3	8.9	8
TCS	3.5	2	15.0	8

*Total dry mass >500 kg.

Mass fractions are based on spacecraft dry mass. Power fractions are based on total OAP (bus + payload). Data was compiled from 25 missions, although not all subsystem data was available for all missions.

Table 12.20 Average Mass and Power Fractions for Large Satellites*

three reaction wheels instead of four reveals that ForestSat's total mass would drop approximately 4% and total power load would drop approximately 1%. This estimate includes all the other subsystem adjustments necessary to accommodate the change. Slew performance would still meet requirements, albeit with slimmer margin and there would be no backup in case a reaction wheel should fail during the 5 year mission life. The systems engineer and other stakeholders can use this information to determine if the increased performance and lifetime risk is worth the cost savings, resulting from choosing a simpler system.

12.5 Summary

A concurrent engineering methodology is useful for performing rapid holistic system concept design. Using models that estimate size, mass, power, and performance, a concurrent engineering team can collectively and iteratively converge on feasible design solutions. This rapid process is useful for performing trade studies, blank sheet designs, engineering changes on existing systems, and design analysis. The models presented here and in the references may be implemented in tools that capture the integrated nature of complex systems and show the effect that specific design choices have on the whole system.

References

Agrawal, B. 1986. *Design of Geosynchronous Spacecraft*, Prentice-Hall Inc., Englewood Cliffs, NJ.

Chao, C. 2005. *Applied Orbit Perturbation and Maintenance*, The Aerospace Press, El Segundo, CA.

Gordon, G. and Morgan, W. 1993. *Principles of Communication Satellites*, John Wiley and Sons, Inc., New York.

GSFC-STD-1000 Rev E, Rules for the Design, Development, and Operation of Flight Systems, Greenbelt, MD, 2009.

NASA/SP-20007-6105, NASA Systems Engineering Handbook, Washington DC, 2007.

Pisacane, V. and Moore, R. 1994. *Fundamentals of Space Systems*, Oxford University Press, New York.

Reddy, M. 2003. "Space Solar Cells—Tradeoff Analysis," *Solar Energy Materials and Solar Cells*, vol. 77, no. 2, pp. 175–208, May 15.

Reynerson, C. 2011. "Aerodynamic Disturbance Force and Torque Estimation for Spacecraft and Simple Shapes Using Finite Plate Elements, Part 1," *The Phoenix Index, Inc.*, U.S.

Ruggiero, A., Pergola, P., Marcuccio, S., and Andrenucci, M. 2011. "Low-Thrust Maneuvers for the Efficient Correction of Orbital Events," 32nd International Electric Propulsion Conference, Wiesbaden, Germany, Sep. 11–15.

Stevens, R. 2015. "Concurrent Engineering Methods and Models for Satellite Concept Design," IEEE Aerospace Conference, Big Sky Montana.

"The Affordable Acquisition Approach Study (A3 Study), Part KKI, Final Briefing," Headquarters Air Force Systems Command, Andrews AFB, MD, 1983.

Ullman, D. G. 1997. *The Mechanical Design Process*, McGraw-Hill, New York.

Wertz, J., Everett, D., and Puschell, J. 2011. *Space Mission Engineering: The New SMAD*, Microcosm Press, Hawthorne, CA.

PART 3

Small Spacecraft Overview

Stevan M. Spremo, Alan R. Crocker, and Tina L. Panontin

12.6 Introduction

Small spacecraft have been an integral part of space study, exploration, and commercialization since humankind's first steps into low earth orbit with Sputnik-1 and Explorer-1. As defined in Table 12.21, "small spacecraft" are considered to be those with wet masses below 500 kg. Although the 500 to 1,000 kg mass range is defined technically as a "medium" spacecraft class, it is common to omit reference to medium classifications and refer to a spacecraft as either "small" or "large."

Development of small satellite platform standards have created new opportunities in the nanosatellite market. The standard "one unit" (1U) cubesat spacecraft, illustrated in Figure 12.23, is a cube 10 cm on each side with a mass of approximately 1 kg. [1]. Cubesats can be composed of a single cube (a "1U" cubesat) or several cubes combined forming, for instance, 3U, 6U, or even 27U integrated units. Representative configurations and sizes are compared in Figure 12.24.

Most cubesat masses qualify them as nanosatellites; however, cubesats less than 1 kg may be considered pico spacecraft. Practical femtosats have not yet entered the marketplace, but attempts to develop and launch such "ChipSats" systems are underway (Jones 2016).

Although their mass and volume constraints can significantly limit power, propulsion, and communication subsystem sizes, small spacecraft do provide certain advantages. With generally lower complexity and fewer payloads, small satellites can be produced at an increased cadence than larger ones. Rapid development provides organizations with more agility in meeting requirements and adapting to new technologies. For instance, the Department of Defense Operationally Responsive Space (ORS) program pursues small satellites to meet its primary goal to *rapidly* assemble, test, and launch satellites in support of warfighters. Rapid development can also drive costs down such that use of small satellites may allow more frequent research missions and technology demonstrations. Low-cost developments can also provide opportunities for engineering and project management education and capability maturation. Cubesats, in particular, are often used as educational and technology demonstration platforms.

<table>
<tr><th colspan="2">Spacecraft class definition</th><th>Wet mass, kg</th><th></th></tr>
<tr><td colspan="2">Large spacecraft</td><td>≥1,000</td><td></td></tr>
<tr><td rowspan="7">Small spacecraft</td><td>Medium spacecraft</td><td>500–1,000</td><td></td></tr>
<tr><td>Mini spacecraft</td><td>100–500*</td><td></td></tr>
<tr><td>Micro spacecraft</td><td>10–100</td><td rowspan="2">} Cubesats</td></tr>
<tr><td>Nano spacecraft</td><td>1–10</td></tr>
<tr><td>Pico spacecraft</td><td>0.01–1</td><td></td></tr>
<tr><td>Femto spacecraft</td><td>0.01–0.01**</td><td></td></tr>
</table>

*Organization-dependent upper thresholds vary from 180 to 500 kg.
**Organization-dependent upper thresholds vary from 0.01 to 0.1 kg.

Table 12.21 Spacecraft Classification Based on Wet Mass

FIGURE 12.23 Typical 1U cubesat.

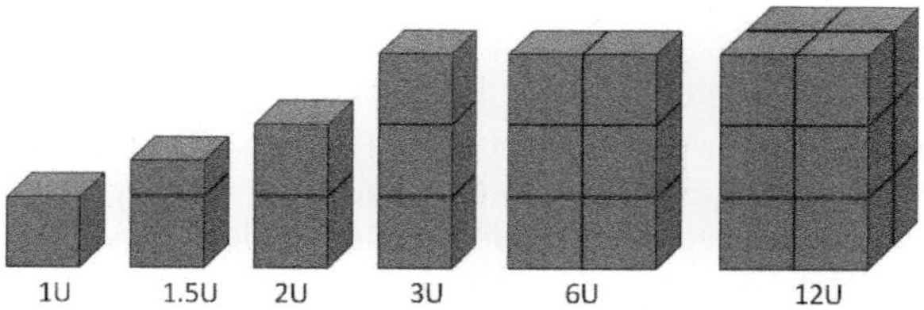

FIGURE 12.24 Comparative size of 1U, 3U, and 6U cubesats.

12.7 History and Evolution of Small Spacecraft

Small spacecraft are not a new phenomenon but rather are the original class of space vehicle and a cornerstone of the United States space program. The first artificial satellites—the Soviet Union's Sputnik 1 and The United States' Explorer 1—were small satellites. More than 50 small spacecraft followed as part of NASA's Explorer Program—missions investigating earth science, astronomy and heliophysics (NASA 2017). Though small in size, complexity, and cost, NASA's Explorer spacecraft were well-engineered and highly reliable. Many continued to operate for 5 or more years; the longest-lived, IMP-8, operated for 33 years (NASA Space Science Data Coordinated Archive 2017).

Small spacecraft missions were not limited to just the Explorer Program. Many of the early interplanetary Mariner and Pioneer spacecraft may be categorized as small spacecraft. NASA's Apollo Program deployed two small lunar satellites as part of the Apollo 15 and 16 missions of the early 1970s. Organizations outside of NASA, including the Department of Defense, academia, and foreign governments, successfully developed and operated small spacecraft as well.

NASA began the Small Explorer Program (SMEX), in 1989 as a follow-on to the Explorer Program. Cost capped at $120 million (FY17), with a primary goal of utilizing small spacecraft to conduct mission in astrophysics, space physics, and upper atmospheric science, the SMEX program was also to usher in a new generation of engineers through apprenticeship with experienced spacecraft developers. A decade later, NASA initiated the University-class explorer ("UNEX")

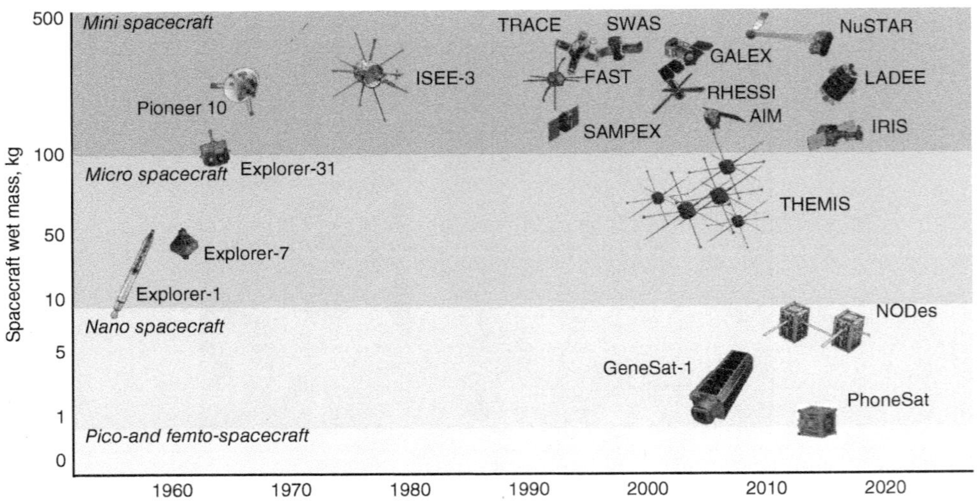

FIGURE 12.25 Example small spacecraft and mass trends over the past six decades.

program to conduct missions at a per mission cost of less than $15 million (FY17) that provides even more "hands-on" training ground for future spacecraft developers. Figure 12.25 shows the growing variety of NASA's small spacecraft over many decades.

There are many examples of smallsats that provide cost-effective platforms for astrophysics, heliophysics, and earth science research. One is the 200 kg Interface Region Imaging Spectrograph (IRIS) spacecraft that provides important insights into our sun. Another is the 385 kg Stardust mission, launched in 1999, that collected comet dust samples and returned those samples to earth. The Cyclone Global Navigation Satellite System (CYGNSS), launched in 2016, placed eight microsatellites in low earth orbit to study the formation and intensity of tropical cyclones and hurricanes (NASA 2016). The Time History of Events and Macroscale Interactions during Substorms (THEMIS) mission launched a constellation of five smallsats to study earth's magnetosphere. Military and commercial interest in smallsats has grown as well. In 2005, the 100 kg U.S. Air Force XSS-11 spacecraft demonstrated the ability to rendezvous with and repair another satellite (David 2005).

In 1999, the cubesat standard revolutionized the secondary payload industry, replacing nonfunctional "ballast mass" with small, low-cost spacecraft and creating a new, diverse community of spacecraft developers. Cubesats provide low-cost access to space, enabling research and educational projects that might not otherwise be possible using "traditional" methods. Their low cost make cubesat educational programs accessible from the elementary school to university levels. Cubesat design, launch, and operation programs are a growing part of STEM curricula in K-12 schools, community colleges, and universities.

A cubesat dispenser design, the Poly-Picosat On-Orbit Deployer (P-POD) shown in Figure 12.26, emerged in parallel with the cubesat standard to enable easy integration with existing launch vehicles. The mechanically robust dispenser accommodated a pack of three 1U spacecraft until commanded to eject the cubesats with a spring-actuated foot. Today, multiple cubesat dispenser options are available as open design standards and proprietary commercial solutions.

FIGURE 12.26 P-POD cubesat launcher.

Beyond educational access to space, cubesats provide a low-cost, yet effective, means to demonstrate new technologies. For example, in 2012, NASA's PhoneSat spacecraft demonstrated the application of COTS technologies for affordable spacecraft. NASA has also utilized cubesats for space biology experimentation since 2006. The U.S. Army's Operational Nanosatellite Effect (ONE) program began investigating the use of cubesat platforms in 2011 (Brinton 2011). Cubesats may also begin to play support roles in larger missions, for example, NASA's InSight mission to Mars will use two Mars Cube One (MarCO) spacecraft to provide communications relay functions during InSight's entry, descent and landing (Matousek 2014).

Low-cost earth imaging is an emerging commercial market that benefits from small satellite capabilities. Several new companies plan to leverage the simplicity, modularity and affordability of cubesats to create satellite constellations (Bandyopadhyay et al. 2015). One such constellation already includes over 100 operating 3U cubesats. From low earth orbit, these satellites may provide earth imaging at the maximum resolution allowed under U.S. export law.

Government cubesat mission costs vary from $3 to $30 million and are typically determined by the scope of scientific capability, complexity, propulsion, control system type, and radiation tolerance requirements. A summary of early NASA cubesats based on a standard "*.Sat" cubesat bus is provided in Table 12.22 (Lee et al. 2015).

Figure 12.27 illustrates the quickly growing number and diversity of missions utilizing nano- and microsatellites. Earth observation and science areas have grown significantly since 2013. Overall, the number of cubesat launches have seen large increases, as shown in Figure 12.28. The rate has remained high since 2013—a record 103 small satellites were launched simultaneously on an Indian Polar Satellite Launch Vehicle in Feb. 2017. The increased cubesat launch rate reflects the expanding viability of these platforms for research and commercial activities, aided in part by the availability of commercially produced standard small spacecraft buses. Projections suggest a total launch demand of about 2,400 satellites weighing 1–50 kg over the next 5 years (Doncaster et al. 2017). Recent reductions in this 5 year projection may be the result of prolonged launch delays and a growing backlog of small satellite missions yet to launch.

Mission	Facts	Launch	Launch vehicle	Status
GeneBox	• First nanosat to integrate a 1U bus to a 2U payload • First nanosat to integrate onto the first privately launched prototype space station (Genesis 1) • First nanosat fluidic/optical contained payload system	July 12, 2016	Dnepr	Successful
GeneSat 1	• First nanosat to be integrated onto a Minotaur 1 • First biological payload in a 3U nanosat configuration • First to fly a Microhard radio • First university-run mission operations for a NASA nanosat	December 16, 2006	Minotaur 1	Successful
PreSat	• First nanosat with a biological payload to be integrated onto a Falcon 1 • First NASA Ames mission integrating 2 nanosat spacecraft at once	June 1, 2008	Falcon 1	Launch vehicle did not reach orbit
NanoSail-D 1	• First nanosat solar sail to integrate onto a Falcon 1 • First nanosat joint effort by NASA Ames Research Center and NASA Marshall Space Flight Center • First nanosat to be viewed from the ground	June 1, 2008	Falcon 1	Launch vehicle did not reach orbit
PharmaSat 1	• First principal investigator–led science mission in a 3U nanosat configuration • First peer-reviewed publication from a nanosat mission • First nanosat pharmaceutical experiment • First NASA Ames full science mission in a 3U nanosat	May 19, 2009	Minotaur 1	Successful
O/OREOS	• First nanosat to integrate onto a Minotaur 4 • First 3U nanosat to have 2 payloads running 3 experiments over a 6-month period • First university-run mission operations to track 2 nanosats simultaneously	November 19, 2010	Minotaur 4	Successful
NanoSail-D 2	• First 3U nanosat to be deployed from a primary satellite • First solar sail to be deployed and used as a deorbit mechanism • First nanosat to be optically tracked from the ground	November 19, 2010	Minotaur 4	Re-entered Fall 2011

Table 12.22 Early NASA Cubesat Missions Based on the *.Sat Cubesat Bus

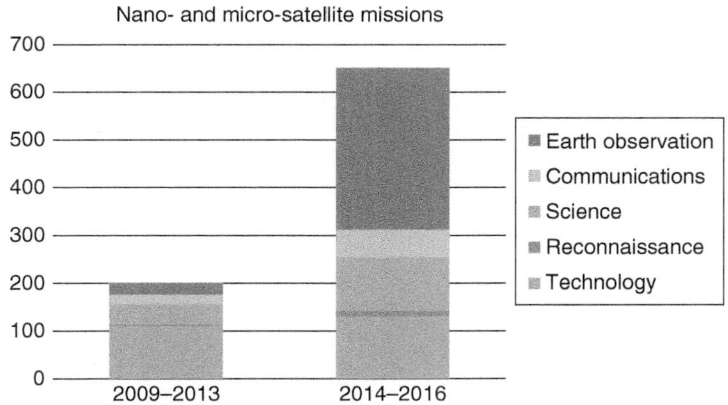

FIGURE 12.27 Increasing volume of nano- and microsatellite missions. (Doncaster et al. 2017.)

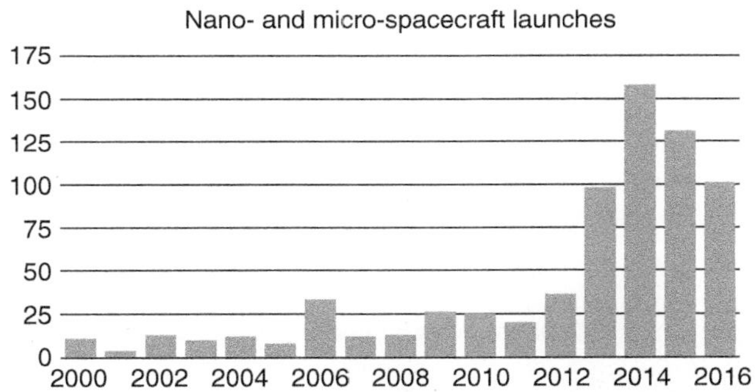

FIGURE 12.28 Annual nano- and microsatellite launch rates. (Doncaster et al. 2017.)

12.8 Programmatic Considerations

Early satellites may have been small due to the limited capabilities of early launch vehicles, and spacecraft masses generally increased to keep pace with the increasing payload mass capabilities of new launch vehicles. More options are now available to the spacecraft designer, and the choice to implement a project as a small spacecraft will consider factors such as mission objectives, available budgets, launch mass allocation, propellant/delta v mass fraction requirements, technology readiness, risk tolerance, and long-term programmatic organizational goals.

The key feature of small spacecraft—flight system affordability—can have fundamental impacts on several programmatic concerns. Cost estimation can be challenging, especially as new, standardized platforms change the scale and complexity of these satellites. The growing smallsat market has increased competition over scarce launch opportunities, causing prolonged launch delays. Finally, risk management approaches for smallsat missions may differ from those of much larger missions.

Cost

The introduction of standardized designs and a diverse set of COTS small spacecraft buses have lowered spacecraft development costs to a level accessible to most. Entry-level spacecraft development kits are priced below U.S. $10,000, and corresponding ground communications systems may be assembled from similarly priced components. Cost models, such as the Aerospace Corporation Small Spacecraft Cost Model (SSCM) provide cost estimation capabilities for mini and medium spacecraft projects but do not generally support cubesat cost estimation (Mahr 2016).

The per-kilogram payload launch cost is often the driving factor in overall mission cost. For many years, this measure held steady at \$100,000/kg to low earth orbit (LEO) and at least \$225,000 per kilogram to lunar orbit. Ridesharing allows cubesat project to achieve launch costs of \$30,000 kg to \$50,000/kg. As commercial access to space continues to grow, these costs are expected to decrease.

Launch Opportunities

Launch frequency can vary with the success of a launch vehicle product line. Typically, a government or commercial entity may seek a launch service provider that can launch every 3 months. High reliability launch vehicles—those that launch on schedule and successfully deliver their payloads—provide the best opportunity for all spaceflight projects, but typically at higher cost. Small spacecraft projects that seek lower-cost launch services may do so at increased risk of launch delays or launch failure, and ridesharing leaves secondary small satellite projects prone to delay when primary spacecraft issues force launch slips. A recent history of launch failures across many separate commercial launch service providers has reduced the overall smallsat launch rate.

The NASA CubeSat Launch Initiative (CSLI) was formed in 2008 to match cubesat projects with ride-share opportunities across the launch industry. CSLI subsidizes launches for non-profit, educational, and government organizations that could not otherwise acquire launch services. As illustrated in Figure 12.29, organizations across the U.S. have benefited from these CSLI-provided launch opportunities. Universities, high schools, and even an elementary school have flown

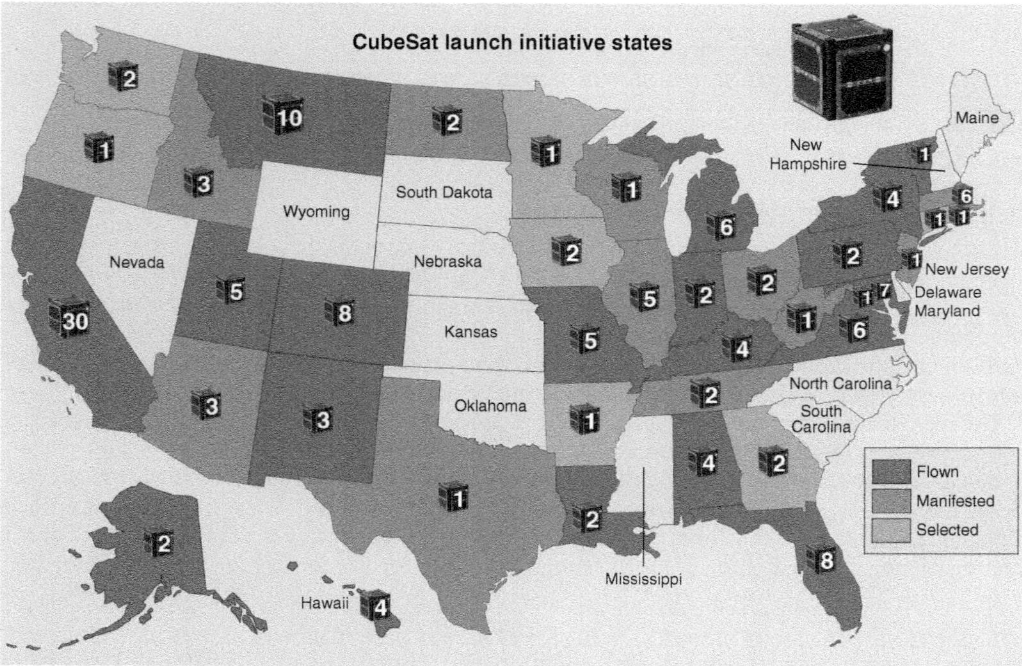

Figure 12.29 CubeSat Launch Initiative Recipients by State. (NASA CubeSat Launch Initiative Website 2017.)

science, technology, engineering and mathematics (STEM) cubesats that promote the education of the next generation of U.S. aerospace workers.

Risk Management

The shorter schedules and lower development costs associated with small satellites can raise concerns of an increased mission failure likelihood, but small spacecraft projects can leverage cost effective risk mitigation strategies. Decreased complexity and use of heritage systems can mitigate the risks associated with the typically reduced testing and technical oversight investments in small spacecraft projects. At the same time, reliance on COTS parts, particularly in higher radiation environments, is a recognized risk area.

Even when higher failure likelihood remains, the overall risk (likelihood × consequence) may remain much smaller than the risks associated with larger spacecraft projects. The risk posture for a small spacecraft development should be determined during the proposal and/or formulation phase of the mission. Stakeholders must all be in agreement in order to maintain cost, schedule, and risk discipline.

NASA's methods for characterizing and managing risk are documented in NASA Procedural Requirements (NPR) 8705.4, *NASA Payload Risk Management*, in which all NASA spacecraft and instruments are considered payloads of launch systems or other carrier vehicles. Table 12.23

Characterization	Class A	Class B	Class C	Class D
Priority (criticality to agency strategic plan)	High priority	High priority	Medium priority	Low priority
National significance	Very high	High	Medium	Low to medium
Complexity	Very high to high	High to medium	Medium to low	Medium to low
Mission lifetime (primary baseline mission)	Long, >5 years	Medium, 2–5 years	Short, <2 years	Short, <2 years
Cost	High	High to medium	Medium to low	Low
Launch constraints	Critical	Medium	Few	Few to none
In-flight maintenance	N/A	Not feasible or difficult	May be feasible	May be feasible and planned
Alternative research opportunities or reflight opportunities	No alternative or reflight opportunities	Few or no alternative or reflight opportunities	Some or few alternative or reflight opportunities	Significant alternative or reflight opportunities
Examples	HST, Cassini, JIMO, JWST	MER, MRO, Discovery payloads, ISS Facility Class Payloads, Attached ISS payloads	ESSP, Explorer payloads, MIDEX, ISS complex subrack payloads	SPARTAN, GAS Can, technology demonstrators, simple ISS, express middeck and subrack payloads, SMEX

Table 12.23 NASA Payload Risk Classification

summarizes the document's risk classifications. For each class, NASA defines design, test, and mission assurance philosophies and provides further guidance in of areas that directly affect cost, schedule, and risk, such as single point failures [redundancy, electrical, electronic, and electromechanical (EEE) parts, test levels, etc.]. Note that safety requirements do not vary between mission classes; these requirements cannot be waived or amended if it is proven that valid safety risk exists.

Class A missions are the most costly given the requirements levied on the project for mission success. As the least costly, and presumably lower priority, missions, class D missions are assigned less stringent requirements. Small spacecraft are typically defined as class C or class D missions. It is typically assumed that class D missions are "higher risk" as compared to class A missions, However, there is a lower inherent consequence of class D mission failures, given their low priority, significance, and investment level (e.g., development cost). Thus, the total risk (consequence × likelihood) of mission failure may be no higher than that of an average spacecraft.

Small satellite development strategies that can achieve reasonable costs while supporting mission success include

- Capability driven (design-to-cost) approaches rather than the standard requirements driven approach
- Use of heritage hardware (build-to-print) and high technology readiness (TRL) solutions
- Applying large margins in the initial design phase to provide flexibility and parallel development (e.g., mass constraints can drive the design requirements of a spacecraft to a considerable extent—the need for optimizing a design will increase the cost of a development).
- Verification by test rather than reliance on increasing analysis sophistication
- Choosing the level of integration for verification based on function criticality
- Use of modular, layered architectures
- Standardized interface, components, and software for multiuse satellite bus developments
- Reduction of radiation-induced parts failure likelihood through limiting mission lifetime or the added shielding
- Use of redundant spacecraft—such as cubesat swarms—rather than through redundant subsystems to achieve reliability goals

Additional discussion of similar low-cost development strategies may be found in the RAND Corporation's 1998 report "The Cosmos on a Shoestring."

Because it is one of the most important drivers of both risk and cost, complexity is an important variable to attend to in development of low-cost, small spacecraft. Methods to reduce complexity include

- Careful reuse of software
- Minimized use of pyrotechnics and complex mechanical systems
- Separation of critical functions to prevent cascade failures
- Streamlined organizational interfaces and communication
- Use of a small, focused core team throughout the project (e.g., avoid fractional FTE)
- Allowance for significant systems engineering resources

Systems engineering and mission assurance efforts are an important and cost-effective part of small spacecraft project risk management. Examination of typical costs distribution to subsystems and processes in a spacecraft development in Figure 12.30 shows that systems engineering (~4% of total) and mission assurance (~2%) costs are not large portions of the overall cost. However, they

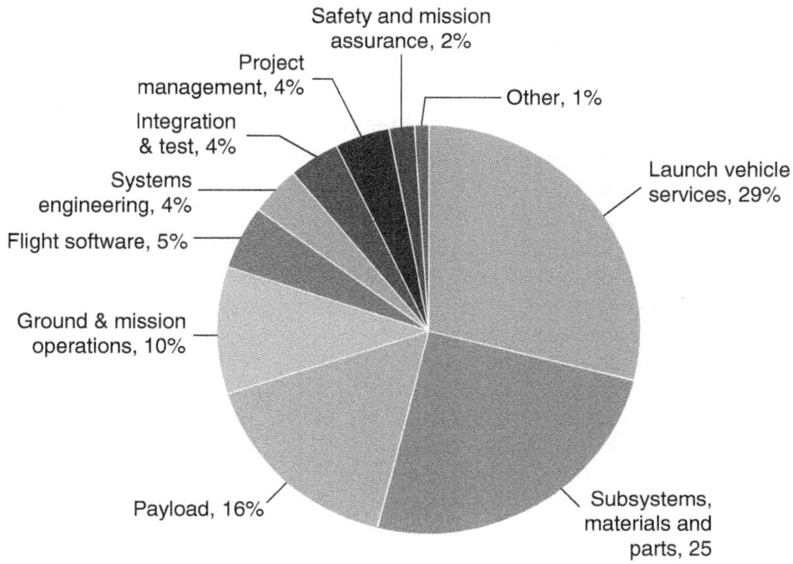

FIGURE 12.30 Example small spacecraft costs by category.

are critically important to ensuring success of a mission. Cost-cutting in these areas saves little and can significantly increase the probability of failure.

12.9 Life Cycle Considerations

Small satellite development and operation involve the same activities and phases that are associated with larger spacecraft.

Integration and Test

Integration and test of spacecraft is an essential process for small spacecraft builders seeking to reduce the risk of failure in during launch and operation. An understanding of how to tailor testing approaches for each application is key to keeping testing costs to a minimum. A logical testing flow, including adequate configuration management process of documents and hardware, is also necessary to make sure test costs are not unnecessarily increased. It is important not to under-test or over-test a spacecraft. Guiding test documents already adopted in the industry such as Goddard Environmental and Verification Specification (GEVS) have been written with past lessons learned in mind. Shock, vibration, acoustic, electromagnetic interference (EMI), electromagnetic compatibility (EMC), thermal vacuum, and other general environmental stress screening tests are common for small spacecraft.

A qualification and acceptance program is considerably more costly but allows for design modifications if a design flaw is found in the qualification model. Qualification systems are tested with higher loads to ensure a flight system should predictably pass an acceptance test once fully integrated based up the qualification unit. This is due to the higher margin factors placed on qualification unit versus acceptance units. Design flaws are usually witnessed in the higher stressed qualification program eliminating significant risk that the flight unit will experience failure in acceptance level stressing conditions. Proto-flight approaches condense the schedule taking more risk that only one article with be tested at higher loads but assumes more risk for failure on the flight article.

Test programs can add unnecessary risk if the system is over tested. It is not uncommon for a team to feel obligated to retest an entire system if a piece of the configuration is changed. Sending an entire spacecraft back for acceptance vibe retest may be a flawed approached if it originally passed its primary test. A calculated risk must be taken and weighed since retest may be more risky to the hardware than no test. Elimination of a simple subsystem vibe acceptance test in this case may spare potential harm to the overall spacecraft system.

Launch and Deployment

Historically, government-subsidized launches have been a main pathway for most of these secondary spacecraft. Recently, there has been a shift in LEO based missions with new commercial based launch opportunities. Four general categories of launch option exist for small satellites—dedicated launch, rideshare, hosting, and deferred deployment.

Mini- and medium-spacecraft may use dedicated launch vehicles. The growing market for small spacecraft has spurred the development of a class of smaller launch vehicles that support small spacecraft needs. For example, IRIS and NuSTAR were both carried to orbit by Orbital ATK Pegasus XL launch vehicles that air-launch from carrier aircraft.

For smaller cubesats, launch and deployment opportunities typically leverage mass and volume margin from a larger mission's launch vehicle. Today, three launch options are available to cubesat customers.

Ridesharing arrangements provide an affordable meant for cubesats to "ride along" on and deploy directly from a launch vehicle. Often, cubesat deployers are mounted to the Evolved Expendable Launch Vehicle Secondary Payload Adapter (ESPA) ring that interfaces the launch vehicle to the primary payload. A typical configuration for such secondary payload mounting is shown in Figure 12.31. The Indian Space Research Organization Polar Satellite Launch Vehicle (PSLV) deployed a record 104 satellites—the 680-kg, medium-class CartoSat-2 mapping satellite and 103 nanosatellites. However, such secondary payload launch opportunities have been limited and inconsistent over the last several years. One hundred and fifty-eight smallsats (1–50 kg) were launched in 2014, 131 in 2015, and 101 in 2016 (Foust 2017, launch woes diminish demand for small satellites). This decline was due primarily to delays in launches caused by launch vehicle failures and other setbacks. Some of this pressure on ride shares may eventually be alleviated by the

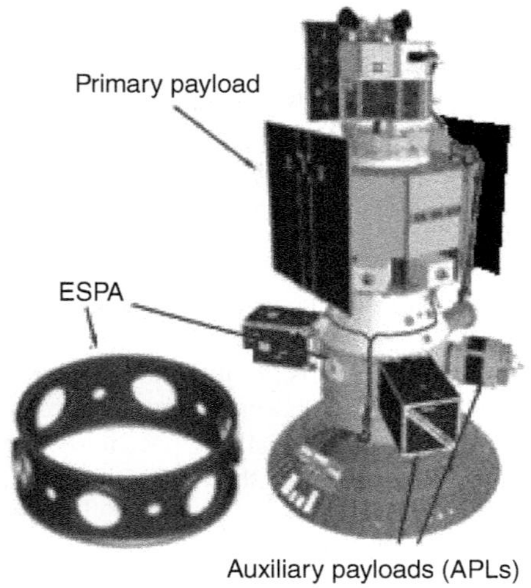

Figure 12.31 Cubesat secondary payloads installed on an ESPA ring.

emergence of a new generation of dedicated small launch vehicles. Until then, the full advantage of fast development times for small satellites will be difficult to realize due to the challenge in securing timely, cost-effective launches.

Cubesats may also be integrated into large satellites for deployment after the host spacecraft arrives on orbit. The 150-kg Fast, Affordable Science and Technology Satellite (FASTSAT), itself a mini-spacecraft launched from an ESPA ring, deployed the NanoSail-D cubesat after it separated from its launch vehicle (Boudreaux 2013).

Finally, cubesats may be ferried to an orbital platform (typically the International Space Station) for deployment at a later date. Commercial providers now offer services to transport cubesats to the ISS and deploy them. Figure 12.32 illustrates a typical cubesat deployment from the ISS.

By whatever means, launch generally represents the most severe vibrational and shock environment for the spacecraft and the greatest opportunity for spacecraft mechanical failure. The duration and levels will vary, but NASA's General Environmental Verification Specification (GEVS) provides a comprehensive approach for testing to the anticipated environments for most available launch vehicles. Launch phase durations can vary in the number of minutes but in general, most LEO launches are around approximately 10 minutes in duration. Differences in thermal coefficient of expansion (TCE) between dissimilar materials can also lead to failure after repeated thermal cycles or even after just one major change in thermal conditions (rate dependent in some cases).

Small spacecraft may be launched powered on or powered off. There are advantages to both approaches, but launch costs are typically lower for a spacecraft that is launched powered off. Launching powered on brings additional safety considerations—additional, redundant inhibits are needed to prevent premature initiation of propulsion systems, for instance.

Small spacecraft deployment may be accomplished through several means. Most common for Cubesats is with a dispenser in which a door is opened and a spring pusher foot ejects that spacecraft. For non-cubesat small spacecraft, ESPA rings and payload adapter systems are used with spring load separation systems. After deployment a spacecraft will immediately begin a safe mode routine after powering up or if already powered, the spacecraft will typically begin a detumbling routine and start working through a sequence of GN&C events with a ground-based mission operations team.

FIGURE 12.32 Cubesat deployment from ISS.

Ground Data Systems and Operations

Ground data systems should be designed/built in parallel with the spacecraft. Requirements for supporting ground data systems can vary based on spacecraft complexity, mission complexity, and destination. Complex spacecraft and missions may demand near-continuous communication with ground support functions. This generally necessitates the use of multiple ground-based communications terminals provided by communications networks such as the Near Earth Network (NEN). Simpler missions that only require brief daily or less frequent communications sessions may use single ground stations.

Small spacecraft with deep space destinations (lunar orbit or beyond) may require use of Deep Space Network communications assets, while earth orbiting small spacecraft may use a range of communications options from the NEN to single independent ground stations. Multiple universities have developed and fielded such ground stations, and turnkey COTS ground station systems are marketed as solutions for cubesat developers.

Mission operations can take many forms, as dictated by the same needs that define ground data system requirements. Complex spacecraft and missions may require larger operations teams and continuous uplink of command sequences, while simple and/or highly automated spacecraft may require little more that periodic data transfer and simple monitoring. In general, mission operations teams are sized and trained to support at least once-a-day communications sessions.

Decommissioning

Another area of concern for small satellites, particularly cubesats and their constellations, is their potential contribution to orbital debris. Depending on which orbit they travel to by way of rideshare, cubesats could be in orbit from months to up to 25 years. If spacecraft are in orbits that will exceed 25 years, then deorbit systems must be used. There have been advancements in drag-based systems that deploy and increase drag to meet these requirements. Other methods require propulsion systems to lower the orbit—these are often mass, volume, and cost prohibitive. Without specific requirements and corresponding solutions for deorbiting, very small spacecraft may contribute to debris hazard for years and the problem could exponentially increase as space-based commercial industry grows.

12.10 Small Spacecraft Technologies

In general, small spacecraft must provide the same, albeit appropriately proportioned, functional capabilities as any spacecraft. Smallsat lifetime and level of subsystem redundancy, however, are typically more limited than those of larger spacecraft. Mass and volume constraints significantly affect tend to limit power, propulsion, and communication subsystem performance. These constraints, combined with typically low project budgets, encourage innovative solutions. The use of commercial off the shelf (COTS) technologies and components from other industries in small satellites has become increasingly common as a means to simplify system development efforts and achieve the low cost goals typical of such projects.

While early smallsat development involved the creation of new and unique spacecraft bus designs, more recent smallsat projects have successfully leveraged a growing number of off-the-shelf bus designs. The increasing quantity and variety of small spacecraft leverages a growing availability of commercial and open standards. Commercial smallsat buses, such as Orbital Sciences' LEOSat-2, Surrey Satellite Technology Ltd.'s SSTL-70, Millennium Space Systems' Altair, and Sierra Nevada Corporation's SN-100, allow development of mini satellites. Commercially produced kits simplify the development and integration for academic spacecraft projects. Use of other COTS technologies and components can enable faster, lower cost development of novel cubesat architectures. For example, NASA's PhoneSat leveraged a commercial smartphone as the spacecraft command data handler.

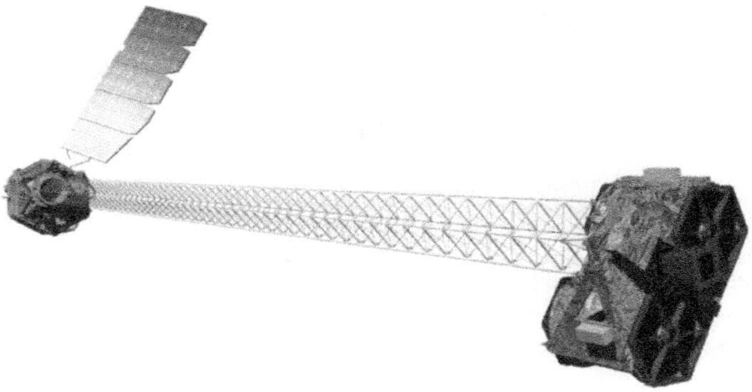

Figure 12.33 NuSTAR in deployed configuration.

Deployable structures, although common in all spacecraft classes, can be particularly important in small spacecraft. Such structures allow smallsats to achieve much larger dimensions to support power generation, communications, and instrument needs. The NuSTAR spacecraft, shown in Figure 12.33, deployed 10 m mast structure to accomplish on-orbit assembly of an X-ray telescope. The Dellingr spacecraft will similarly deploy magnetometers on the folded boom structure shown in Figure 12.34. These techniques allow small launch packages to become much larger scientific instruments. Other novel mechanical designs hold promise to provide larger support structures for solar arrays, communications antennae, and other key components for future small spacecraft.

Small spacecraft electrical power systems typically rely on photovoltaic solar cells and lithium batteries to generate and store energy. While low-cost single junction photovoltaic cells may be used for very low-power spacecraft, most small spacecraft designs leverage space-rated, high-efficiency triple junction photovoltaic cells. Early cubesats integrated flush-mounted solar arrays into the spacecraft sides, but increased power needs have inspired novel designs for larger deployable solar arrays. While traditional, single-axis hinge mechanisms allow the deployment of cubesat solar array "wings," newer mechanisms employ multi-hinged umbrella-style solar arrays.

Traditionally, small spacecraft have employed chemical mono- or bipropellant propulsion systems for recurring tasks such as primary trajectory maneuvers, trajectory correction maneuvers, attitude maneuvers, and momentum management (such as "momentum dumping" used to manage reaction wheels). Such systems, however, are typically complex and demand significant fractions of overall spacecraft mass and volume. Traditional chemical propulsion systems also introduce safety concerns. Cold gas propulsion systems offer simpler, and generally safer, design alternatives. The low mass of small spacecraft can make electric propulsion techniques, including Hall effect thrusters, plasma thrusters, and ion propulsion systems, reasonable alternatives. NASA's iSat mission will demonstrate practical use of an Iodine Hall thruster to maneuver a cubesat in low earth orbit (Dankanich 2014). Solar sail propulsion technologies may prove to be viable for cubesats as well. The 6U Lunar Flashlight and Near Earth Asteroid Scout (NEA-Scout)

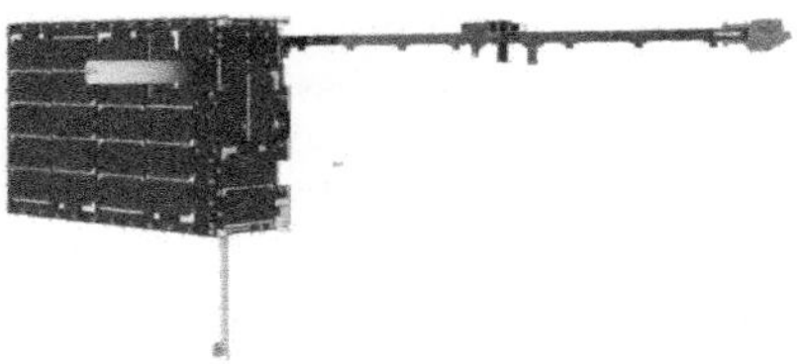

Figure 12.34 Dellingr spacecraft with magnetometer boom and antennae extended.

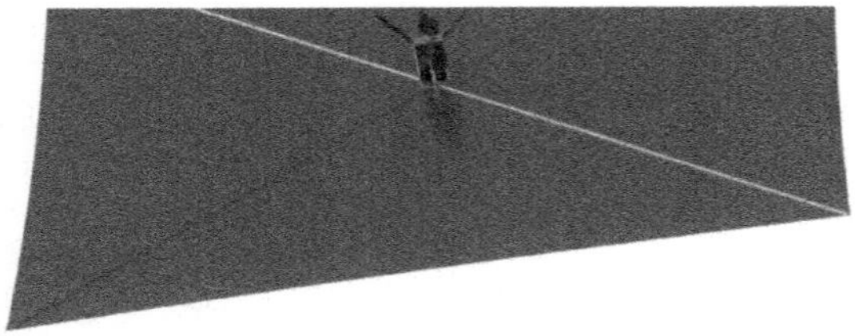

Figure 12.35 Solar sail configuration for Lunar Flashlight and NEA Scout spacecraft.

spacecraft, shown in Figure 12.35, will each deploy 9 × 9-m solar sails to provide propulsion (Castillo-Rogez 2014).

Like larger spacecraft, smallsat guidance, navigation, and control (GNC) subsystems may employ star trackers, reaction wheels, sun sensors, earth limb sensors, inertial measurement units, torque coils, torque rods, Global Positioning System (GPS) receivers, and gyroscopes as GNC sensors and effectors driven by software- and/or hardware-driven control systems. Key drivers in GNC performance are antenna pointing accuracy requirements, payload pointing, and attitude restrictions defined by thermal environment constraints. Many LEO cubesats require only general pointing, on the order of ±1°, to adequately point solar arrays toward the sun and antennae toward communications terminals. For these systems, hysteresis rods and permanent magnets that interact with the earth's magnetic field may be adequate. Cubesats with such passive attitude control are ideal for achieving pristine microgravity environments in LEO for genetic based experiment. Beyond LEO, however, active attitude control is necessary due to the absence of a magnetic field.

Thermal control system implementation options for small spacecraft are limited due to mass and power constraints. Passive thermal control methods such as multilayered insulation (MLI), surface coating (coatings, paints, and tape), heat pipes, and conductive thermal pathways/straps, provide low-cost and low-mass options that are particularly attractive in pico- and nanosat applications. Active cooling mechanisms, such as electrical patch heaters, thermoelectric coolers (TECs), and pump-driven liquid thermal conditioning loops, provide greater control over the spacecraft's internal thermal environment at the cost of additional complexity, higher spacecraft mass, and increased electrical power requirements. Recent demonstrations indicate that use of a pressurized bus container can bound temperature extremes experienced by spacecraft avionics, thereby reducing the necessary thermal system design effort and reducing development cost.

Small spacecraft communication systems have employed at least 25 radio frequency (RF) bands including VHF, UHF, L-band, S-band, X-band, and Ka bands (NASA perspective on Cubesats and Highlighted Activities 2015). The recent development of software-defined radios, in which a radio can be tuned to specific frequency throughout a dynamic range using software commanding, has introduced new options for smallsat communications. Smallsats may use patch, whip, evolvable (computer-generated using genetic algorithms), parabolic reflector, or deployable antennas. Laser-based optical communications systems with very high data transfer rates will likely play an important role in future missions, but cubesat mass, power, and volume constraints will likely limit performance on these platforms.

Small spacecraft command and data handling (CDH) designs typically trade high processor performance for low mass, low power, and robustness. Additionally, the need to control costs tends to drive small spacecraft projects to use COTS components and interface standards. For example, COTS processors (such as the BAE systems' RAD-750 radiation hardened single board processor) and standard interfaces such as the serial bus, universal serial bus (USB), and I^2C protocols have been used in small spacecraft. Satellites flown within the Van Allen belts (commonly low earth orbit missions) are subject to a comparatively benign radiation environment. Satellites flown through or beyond the 60,000 km edge of the Van Allen belts, however, experience much higher radiation doses and require design features—either radiation hardening or radiation shielding—to handle this exposure.

Radiation hardening entails designing an electronic chip with radiation dosing as a driving requirement. The primary objective of radiation hardening is to ensure the performance of device operates as intended and extends the useful life of the device while experiencing radiation and post radiation dosing. In space a number of radiation pathways may be incident upon an electronic device while in, passing through, and travel beyond earth's radiation belts/magnetosphere. Material selection and physical design of gate geometry of solid state electronics can be a key factor. In some cases, larger electronic gate sizes are preferred in radiation environments as they provide greater reliability over newer, smaller gate designs. Electronics with slow performance and high reliability often are chosen over processing speed. For missions requiring radiation hardening, processing speed may be traded for overall functional reliability. When the cost of radiation hardened parts is too high, some missions turn to alternate systems architectures utilizing items such as "watch dog" timer functions that recover from processor latch-ups before such events become total permanent failures.

Radiation shielding may be used to protect devices not inherently radiation hardened. A design can mitigate the amount of radiation that reaches electronic circuitry by way of layering radiation materials that have an effective linear energy transfer stopping characteristic. Heavy materials such as lead are effective in stopping radiation transmission but are often too heavy for use in small spacecraft. Instead, a clever layering of lighter materials, each with good linear energy transfer stopping characteristics, can provide effective shielding.

12.11 Case Studies

Several recent NASA projects demonstrate emerging trends in smallsat development. The GeneSat-1 mission represents NASA first use of a cubesat for astrobiology research. The COTSAT spacecraft illustrates a novel design approach to simplify spacecraft design, and the Lunar Atmosphere Dust Environment Explorer (LADEE) mission illustrates the use of a mini-satellite bus to execute planetary exploration.

GeneSat-1

GeneSat-1, NASA's first cubesat satellite, was a joint effort involving NASA Ames Research Center, Santa Clara University, California Polytechnic University at San Luis Obispo, and Stanford University. This spacecraft launched successfully in December 2006 on a Minontaur-1 launch vehicle as a rideshare with the TacSat-2 mission.

NASA leveraged high reliability, radiation tolerant, low-power electronics, coupled with the recent introduction of high-efficiency triple-junction solar cell technology, to advance cubesat technology readiness levels (TRLs). The 3U spacecraft bus, illustrated in Figure 12.36, consumed

FIGURE 12.36 GeneSat-1.

an average of just 3–5 W and a peak of 5 W—a feat enabled by both clever system engineering, recent improvements in microprocessor technologies, and the identification of low-cost COTS communication technologies that were compatible with the space environment. Despite its low-power consumption, GeneSat-1 operated its biological experiments within a benign payload temperature of 27 ± 0.5°C.

Given the minimal technology investment associated with cubesat development, COTS parts are used extensively and the predicted probability of flawless on-orbit operation was considered low. To ensure that GeneSat-1 would not negatively impact the primary mission, the required mechanical integrity and reliability of the cubesat system and its dispenser were considered to be very high. This approach is now known at the secondary payload "do no harm campaign approach." Many of the early cubesats such as GeneSat-1 had very large structural margins added to ensure the mechanical integrity of the system would never be a concern. This system was intentionally overdesigned for mechanical strength to alleviate concerns of structural failure damaging a neighboring environment on the launch system. The team tested the system to strict NASA mechanical environmental testing requirements and proved the system could not negatively impact TacSat-2's success.

Cost Optimized Test of Spacecraft Avionics and Technologies

NASA Ames Research Center developed the 250-kg Cost Optimized Test of Spacecraft Avionics and Technologies (COTSAT-1) as a rapid prototype, low-cost spacecraft for science experiments and technology demonstration. COTSAT-1 demonstrated significant spacecraft design cost reduction through methods and technologies that maximized reuse of previously developed spacecraft hardware, software, and related technology on future missions. The spacecraft platform, shown in Figure 12.37, was designed to accommodate low-cost access to space for various remote-sensing payloads while allowing future expansion for potential biological payloads. Though much larger in mass and volume than a cubesat, COTSAT used similar philosophies in COTS parts selection and low cost targets. The system's size enabled it to carry, for a similar cost, a range of payload sizes beyond that possible using the Cubesats standard.

COTSAT-1 leveraged the use of the one atmosphere pressurized structure to house spacecraft components, a design feature similar to that previously used in Soviet and Russian spacecraft. This artificial environment makes it feasible to incorporate a wide array of prebuilt hardware, including

Figure 12.37 COTSAT.

commercial off-the-shelf (COTS), modified off-the-shelf (MOTS), and government off-the-shelf (GOTS) hardware. Hybridizing a one-atmosphere pressure vessel with current COTS technologies provides a significant subsystem cost reduction, in most cases, by orders of magnitude by elimination of many of the space hardening steps that would be needed for hardware directly exposed to space vacuum. By using COTS hardware, the spacecraft program can utilize technology investments already made by commercial vendors (Swank 2009).

COTSAT-1 also incorporated industry data interface standards such as USB 2.0 and Ethernet to reduce development and integration time. Where inexpensive COTS hardware solutions are not readily available, subsystems are designed and developed in-house. This includes reaction wheels, star tracker, and electrical power systems. The one atmosphere avionics bus structure encases a PC-104 CDH and a custom electrical power system. The COTSAT-1 project used open source software, including GNU/Linux and existing software libraries and device drivers to reduce software development and integration time.

Beyond LEO, radiation exposure still remains a challenge for this platform although programs in low-cost shielding are underway and NASA has a solution to this issue in development. This means a system of this low cost could also be a candidate for a new class of spacecraft that is low cost for travel beyond the Van Allen belts. Resolution of radiation limitations in commercial electronics would pave the way for future low-cost exploration platforms beyond low earth orbit.

Commercial organizations have begun licensing COTSAT patents, bringing the derived technologies to the marketplace and into service in low earth orbit.

Lunar Atmosphere Dust Environment Explorer

Another example of low-cost approach to building deep space vehicle was demonstrated successfully on the Lunar Atmosphere Dust Environment Explorer (LADEE) Mission. The LADEE spacecraft, shown in Figure 12.38, was designed, developed and integrated at NASA Ames research Center for a cost of $282 million (including the launch vehicle). The LADEE system was 383 kg

Figure 12.38 LADEE.

system when fully fueled. The mission included three science instruments [Neutral Mass Spectrometer (NMS), Ultraviolet Spectrometer (UVS), and Lunar Dust Experiment (LDEX)], to characterize the exosphere of the moon while in a low-altitude retrograde elliptical orbit. The spacecraft also carried the Lunar Laser Communications Demonstration (LLCD) payload, the first demonstration of earth-to-space laser communication.

Many of the systems used on LADEE were COTS or modified COTS. LADEE launched on a Minotaur V rocket on September 6, 2013 and successfully performed its mission requirements in addition to a technology breakthrough with the successful demonstration of a Laser Communications system achieving a 622 Mbps data transmission rate from lunar orbit to an Earth base ground receiving station.

12.12 Conclusion

Small spacecraft support a broad spectrum of needs in space exploration and commercialization. From their humble beginnings of early spaceflight, the opportunities and available options for the application of small satellite technologies makes spaceflight more accessible to government, commercial, and academic institutions. Today, the key challenge in expanding the small spacecraft market is the provision of sufficient affordable launch opportunities.

As humankind's brief space history has shown, small spacecraft platforms will continue to serve an important role in space exploration. They likely will be the genesis and proving ground for future space technology advancements. They also will continue to serve as the training ground for future space innovators and as a stepping stone leading to large spacecraft flagship exploration missions. One can only wonder what the seed of small spacecraft technology development will yield in the future.

Summary

Small spacecraft play a major role in earth, lunar, planetary, stellar, and interstellar discoveries. As technologies improve, instruments scale down in size, and their advantages in reduced cost and development time continue to attract investment, small satellites will play an even more important role. Today, the growth rate of small spacecraft utilization is limited by the availability of affordable launch opportunities.

References

Bandyopadhyay, S., Subramanian, G. P., Foust, R., Morgan, D., and Chu, S. J. 2015. A Review of Impending Small Satellite Formation Flying Missions. 53rd AIAA Aerospace Sciences Meeting. Kissimmee, FL: American Institute of Aeronautics and Astronautics.

Boudreaux, M. 2013. The Fast, Affordable, Science and Technology Satellite (FASTSAT) Mission. 27th Annual AIAA/USU Conference on Small Satellites. Logan, UT: American Institute of Aeronautics and Astronautics.

Brinton, T. 2011, January 17. Retrieved from SpaceNews: spacenews.com/us-army-cubesat-completes-monthlong-demonstration-2-more-may-launch-year/

Castillo-Rogez, L. J. et al. 2014. *Near Earth Asteroid Scout Mission*. Retrieved from: www.lpi.usra.edu/sbag/meetings/jul2014/presentations/0930_Thu_Castillo_NEAScout.pdf

CubeSat Program. 2014. CubeSat Design Specification. California Polytechnic University.

David, L. 2005, July 22. Military Micro-Sat Explores Space Inspection, Servicing Technologies. Retrieved from Space.com: www.space.com/1336-military-micro-sat-explores-space-inspection-servicing-technologies.html

Doncaster, B., Shulman, J., and Williams, C. *2017 Nano/Microsatellite Market Forecast*. Spaceworks Enterprises, Inc. 2017. Atlanta, GA.

Foust, J. 2010. Emerging Opportunities for Low-Cost Small Satellites in Civil and Commercial Space. Proceedings, 24th Annual AIAA/USU Conference on Small Satellites. Logan, UT: American Institute of Aeronautics and Astronautics.

Foust, J. 2017, February 2. "Launch woes diminish demand for small satellites". SpaceNews. Retrieved from http://spacenews.com/launch-woes-diminish-demand-for-small-satellites/

Hille, K. B. 2011. *Growing Up at Goddard: Shuttle Small Payloads Launched Careers of Many.* Greenbelt, MD. Retrieved from https://www.nasa.gov/centers/goddard/news/features/2011/small-payloads.html

John, W. and Dankanich, K. A. 2014. The iodine Satellite (iSAT) Hall Thruster Demonstration Mission Concept and Development. *50th AIAA/ASME/SAE/ASEE Joint Propulsion Conference.* Cleveland, OH: American Institute of Aeronautics and Astronautics.

Jones, N. 2016, June 1. Tiny 'chipsat' spacecraft set for first flight. *Nature, 534*(7605). Retrieved from http://www.nature.com/news/tiny-chipsat-spacecraft-set-for-first-flight-1.20006

Lee, E., D'Ortenzio, M., Spremo, S., Jaroux, B., Mas, I., and Saldana, D. 2005. The *.Sat CubeSat Bus: When Three Cubes Meet. *19th Annual AIAA/USU Conference on Small Satellites.* Logan, UT: American Institute of Aeronautics and Astronautics.

Mahr, E., Tu, A., and Gupta, A. 2016. *Development of the Small Satellite Cost Model 2014.* IEEE Aerospace Conference Proceedings. Retrieved from http://ieeexplore.ieee.org/document/7500515/

Matousek, S. 2014, November 21. *Mars Cube One (MarCO): The First Planetary Cubesat Mission.* Jet Propulsion Laboratory. Pasadena, Ca. Retrieved from: marscubesatworkshop.jpl.nasa.gov/static/files/presentation/Asmar-Matousek/07-MarsCubeWorkshop-MarCO-update.pdf

NASA. 2016, November 10. *CYGNSS Press Kit.* Retrieved from NASA Web Site: www.nasa.gov/sites/default/files/atoms/files/cygnss_presskit.pdf

NASA Space Science Data Coordinated Archive. 2017. *NASA's Explorer Program Satellites.* Retrieved from: nssdc.gsfc.nasa.gov/multi/explorer.html

NASA CubeSat Launch Initiative Website. 2017, February. *About CubeSat Launch Initiative.* Retrieved from www.nasa.gov/content/about-cubesat-launch-initiative

Petro, A. 2014, July 28. *Small Spacecraft Technology Markets & Motivations.* Retrieved from www.nasa.gov/sites/default/files/files/APetro_NAC_Smallsat_Market.pdf

Pierce, D. et al. 2015, June 22. NASA perspectives on Cubesats and Highlighted Activities. Retrieved from National Academies website: sites.nationalacademies.org/cs/groups/ssbsite/documents/webpage/ssb_166651.pdf

Swank, A. 2009. COTSAT Small Spacecraft Cost Optimization for Government and Commercial Use. *AIAA Space 2009 Conference.* Pasadena, CA: American Institute of Aeronautics and Astronautics.

Index

B

C

F

G

H

I

N

O

P

T

U

V

W

Z